DIMENSION	METRIC	METR[illegible]
Power, heat transfer rate	1 W = 1 J/s 1 kW = 1000 W = 1.341 hp 1 hp‡ = 745.7 W	1 kW [illegible] 1 hp = [illegible] = [illegible] = [illegible] kW 1 boiler hp = 33,475 Btu/h 1 Btu/h = 1.055056 kJ/h 1 ton of refrigeration = 200 Btu/min
Pressure	1 Pa = 1 N/m^2 1 kPa = 10^3 Pa = 10^{-3} MPa 1 atm = 101.325 kPa = 1.01325 bars = 760 mmHg at 0°C = 1.03323 kgf/cm^2 1 mmHg = 0.1333 kPa	1 Pa = 1.4504×10^{-4} psia = 0.020886 lbf/ft^2 1 psia = 144 lbf/ft^2 = 6.894757 kPa 1 atm = 14.696 psia = 29.92 inHg at 30°F 1 inHg = 3.387 kPa
Specific heat	1 kJ/kg · °C = 1 kJ/kg · K = 1 J/g · °C	1 Btu/lbm · °F = 4.1868 kJ/kg · °C 1 Btu/lbmol · R = 4.1868 kJ/kmol · K 1 kJ/kg · °C = 0.23885 Btu/lbm · °F = 0.23885 Btu/lbm · R
Specific volume	1 m^3/kg = 1000 L/kg = 1000 cm^3/g	1 m^3/kg = 16.02 ft^3/lbm 1 ft^3/lbm = 0.062428 m^3/kg
Temperature	T(K) = T(°C) + 273.15 ΔT(K) = ΔT(°C)	T(R) = T(°F) + 459.67 = 1.8T(K) T(°F) = 1.8 T(°C) + 32 ΔT(°F) = ΔT(R) = 1.8* ΔT(K)
Thermal conductivity	1 W/m · °C = 1 W/m · K	1 W/m · °C = 0.57782 Btu/h · ft · °F
Thermal resistance	1°C/W = 1 K/W	1 K/W = 0.52750°F/h · Btu
Velocity	1 m/s = 3.60 km/h	1 m/s = 3.2808 ft/s = 2.237 mi/h 1 mi/h = 1.46667 ft/s 1 mi/h = 1.609 km/h
Viscosity, dynamic	1 kg/m · s = 1 N · s/m^2 = 1 Pa · s = 10 poise	1 kg/m · s = 2419.1 lbf/ft · h = 0.020886 lbf · s/ft^2 = 5.8016×10^{-6} lbf · h/ft^2
Viscosity, kinematic	1 m^2/s = 10^4 cm^2/s 1 stoke = 1 cm^2/s = 10^{-4} m^2/s	1 m^2/s = 10.764 ft^2/s = 3.875×10^4 ft^2/h 1 m^2/s = 10.764 ft^2/s
Volume	1 m^3 = 1000 L = 10^6 cm^3 (cc)	1 m^3 = 6.1024×10^4 in^3 = 35.315 ft^3 = 264.17 gal (U.S.) 1 U.S. gallon = 231 in^3 = 3.7854 L 1 fl ounce = 29.5735 cm^3 = 0.0295735 L 1 U.S. gallon = 128 fl ounces

Exact conversion factor between metric and English units.

Mechanical horsepower. The electrical horsepower is taken to be exactly 746 W.

Fundamentals of THERMAL-FLUID SCIENCES

Fundamentals of THERMAL-FLUID SCIENCES

THIRD EDITION

YUNUS A. ÇENGEL
Department of Mechanical Engineering
University of Nevada, Reno

ROBERT H. TURNER
Department of Mechanical Engineering
University of Nevada, Reno

JOHN M. CIMBALA
Department of Mechanical and Nuclear Engineering
The Pennsylvania State University

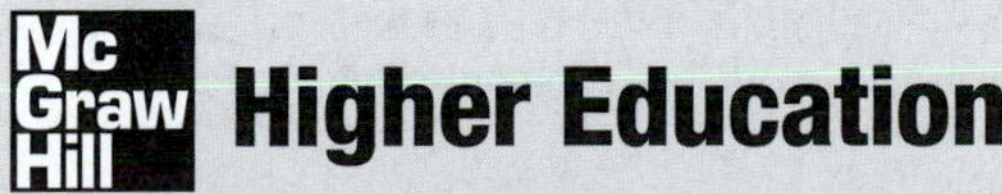

Boston Burr Ridge, IL Dubuque, IA Madison, WI New York
San Francisco St. Louis Bangkok Bogotá Caracas Kuala Lumpur
Lisbon London Madrid Mexico City Milan Montreal New Delhi
Santiago Seoul Singapore Sydney Taipei Toronto

FUNDAMENTALS OF THERMAL-FLUID SCIENCES, THIRD EDITION

Published by McGraw-Hill, a business unit of The McGraw-Hill Companies, Inc., 1221 Avenue of the Americas, New York, NY 10020.

This book is printed on acid-free paper.

1 2 3 4 5 6 7 8 9 0 DOW/DOW 0 9 8 7

ISBN 978–0–07–352925–7
MHID 0–07–352925–7

Global Publisher: *Raghothaman Srinivasan*
Executive Editor: *Michael Hackett*
Senior Sponsoring Editor: *Bill Stenquist*
Director of Development: *Kristine Tibbetts*
Developmental Editor: *Lorraine K. Buczek*
Executive Marketing Manager: *Michael Weitz*
Senior Project Manager: *Sheila M. Frank*
Senior Production Supervisor: *Sherry L. Kane*
Associate Media Producer: *Christina Nelson*
Designer: *Laurie B. Janssen*
(USE) Cover Image: *Corbis/Fire & Ice*
Senior Photo Research Coordinator: *Lori Hancock*
Compositor: *RPK Editorial Services*
Typeface: *10.5/12 Times Roman*
Printer: *R. R. Donnelley Willard, OH*

Library of Congress Cataloging-in-Publication Data

Çengel, Yunus A.
Fundamentals of thermal-fluid sciences / Yunus A. Çengel, Robert H. Turner, John M. Cimbala. — 3rd ed.
p. cm.
Includes index.
ISBN 978-0-07-352925-7 — ISBN 0-07-352925-7 (hard copy : alk. paper)
1. Thermodynamics. 2. Heat—Transmission. 3. Fluid mechanics. I. Turner, Robert H. II. Cimbala, John M. III. Title.

TJ265.C42 2008
621.4′02—dc22 2007013452

www.mhhe.com

He who will not economize will have to agonize.
—Confucius

We must think as people of action and act as people of thought.
—Henri Bergson

The world is a dangerous place to live; not because of the people who are evil, but because of the people who don't do anything about it.
—Albert Einstein

Since there is unanimity in what is good but conflict in what is better, sometimes good is better than the better.
—Said Nursi

Men die of fright and live of confidence.
—Henry David Thoreau

They may forget what you said, but they will never forget how you make them feel.
—Carol Buchner

Nothing will ever be attempted if all possible objections must first be overcome.
—Samuel Johnson

He who is devoid of the power to forgive is devoid of the power to love. There is some good in the worst of us and some evil in the best of us. When we discover this, we are less prone to hate our enemies.
—Martin Luther King, Jr.

The art of being wise is knowing what to overlook.
—William James

A leader is one who knows the way, goes the way, and shows the way.
—John C. Maxwell

Obstacles are those frightful things you see when you take your eyes off your goal.
—Henry Ford

Never worry about numbers. Help one person at a time, and always start with the person nearest you.
—Mother Teresa

The highest reward for a man's toil is not what he gets for it but what he becomes by it.
—John Ruskin

And in the end it's not the years in your life that count. It's the life in your years.
—Abraham Lincoln

About the Authors

Yunus A. Çengel is Professor Emeritus of Mechanical Engineering at the University of Nevada, Reno. He received his Ph.D. in mechanical engineering from North Carolina State University. His research areas are renewable energy, desalination, exergy analysis, heat transfer enhancement, radiation heat transfer, and energy conservation. He served as the director of the Industrial Assessment Center (IAC) at the University of Nevada, Reno, from 1996 to 2000. He has led teams of engineering students to numerous manufacturing facilities in Northern Nevada and California to industrial assessments, and has prepared energy conservation, waste minimization, and productivity enhancement reports for them.

Dr. Çengel is the author or coauthor of the widely adopted textbooks *Thermodynamics: An Engineering Approach,* 6th edition (2008), *Heat and Mass Transfer: A Practical Approach,* 3rd edition (2007), *Introduction to Thermodynamics and Heat Transfer,* 2nd edition (2008), *Fluid Mechanics: Fundamentals and Applications* (2006), *Essentials of Fluids Mechanics: Fundamentals and Applications* (2008), all published by McGraw-Hill. Some of his textbooks have been translated to Chinese, Japanese, Korean, Spanish, Turkish, Italian, Greek, Tai, and French.

Dr. Çengel is the recipient of several outstanding teacher awards, and he has received the ASEE Meriam/Wiley Distinguished Author Award for excellence in authorship in 1992 and again in 2000.

Dr. Çengel is a registered Professional Engineer in the State of Nevada, and is a member of the American Society of Mechanical Engineers (ASME) and the American Society for Engineering Education (ASEE).

Robert H. Turner is Professor Emeritus of Mechanical Engineering at the University of Nevada, Reno (UNR). He earned a B.S. and M.S. from the University of California at Berkeley, and his Ph.D. from UCLA, all in mechanical engineering. He worked in industry for 18 years, including nine years at Cal Tech's Jet Propulsion Laboratory (JPL). Dr. Turner then joined the University of Nevada in 1983. His research interests include solar and renewable energy applications, thermal sciences, and energy conservation. He established and was the first director of the Industrial Assessment Center at the University of Nevada.

For 20 years Dr. Turner has designed the solar components of many houses. In 1994–95, in a cooperative effort between UNR and Erciyes University in Kayseri, Turkey, he designed and oversaw construction of the fully instrumented Solar Research Laboratory at Erciyes University, featuring 130 square meters of site-integrated solar collectors. His interest in applications has led Dr. Turner to maintain an active consulting practice.

Dr. Turner is a registered Professional Engineer and is a member of the American Society of Mechanical Engineers (ASME) and the American Society of Heating, Refrigeration, and Air Conditioning Engineers (ASHRAE).

John M. Cimbala is Professor of Mechanical Engineering at The Pennsylvania State University, University Park. He received his B.S. in Aerospace Engineering from Penn State and his M.S. in Aeronautics from the California Institute of Technology (CalTech). He received his Ph.D. in Aeronautics from CalTech in 1984 under the supervision of Professor Anatol Roshko, to whom he will be forever grateful. His research areas include experimental and computational fluid mechanics and heat transfer, turbulence, turbulence modeling, turbomachinery, indoor air quality, and air pollution control. During the academic year 1993–94, Professor Cimbala took a sabbatical leave from the University and worked at NASA Langley Research Center, where he advanced his knowledge of computational fluid dynamics (CFD) and turbulence modeling.

Dr. Cimbala is the co-author of three other textbooks: *Fluid Mechanics: Fundamentals and Applications* (2006), *Essentials of Fluid Mechanics: Fundamentals and Applications* (2008), both published by McGraw-Hill, and *Indoor Air Quality Engineering: Environmental Health and Control of Indoor Pollutants* (2003), published by Marcel-Dekker, Inc. He has also contributed to parts of other books and is the author or co-author of dozens of journal and conference papers. More information can be found at www.mne.psu.edu/cimbala.

Professor Cimbala is the recipient of several outstanding teaching awards and views his book writing as an extension of his love of teaching. He is a member of the American Institute of Aeronautics and Astronautics (AIAA), the American Society of Mechanical Engineers (ASME), the American Society for Engineering Education (ASEE), and the American Physical Society (APS).

CONTENTS

CHAPTER FIVE
ENERGY ANALYSIS OF CLOSED SYSTEMS 159

CHAPTER SIX
MASS AND ENERGY ANALYSIS OF CONTROL VOLUMES 201

CHAPTER SEVEN
THE SECOND LAW OF THERMODYNAMICS 253

CHAPTER EIGHT
ENTROPY 297

PART 2 FLUID MECHANICS 373

CHAPTER NINE
INTRODUCTION TO FLUID MECHANICS 375

PART 3 HEAT TRANSFER 623

CHAPTER SIXTEEN
MECHANISMS OF HEAT TRANSFER 625

CHAPTER SEVENTEEN
STEADY HEAT CONDUCTION 653

CHAPTER EIGHTEEN
TRANSIENT HEAT CONDUCTION 723

CHAPTER NINETEEN
FORCED CONVECTION 777

CHAPTER TWENTY
NATURAL CONVECTION 833

CHAPTER TWENTY-ONE
RADIATION HEAT TRANSFER 875

CHAPTER TWENTY-TWO

HEAT EXCHANGERS 935

APPENDIX 1

PROPERTY TABLES AND CHARTS (SI UNITS) 987

APPENDIX 2

PROPERTY TABLES AND CHARTS (ENGLISH UNITS) 1031

PREFACE

BACKGROUND

This text is an abbreviated version of standard thermodynamics, fluid mechanics, and heat transfer texts, covering topics that the engineering students are most likely to need in their professional lives. The thermodynamics portion of this text is based on the text *Thermodynamics: An Engineering Approach* by Y. A. Çengel and M. A. Boles, the fluid mechanics portion is based on *Fluid Mechanics: Fundamentals and Applications* by Y. A. Çengel and J. M. Cimbala, and the heat transfer portion is based on *Heat and Mass Transfer: A Practical Approach* by Y. A. Çengel, all published by McGraw-Hill. Most chapters are practically independent of each other and can be covered in any order. The text is well-suited for curriculums that have a common introductory course or a two-course sequence on thermal-fluid sciences or on thermodynamics and heat transfer.

It is recognized that all topics of thermodynamics, fluid mechanics, and heat transfer cannot be covered adequately in a typical three-semester-hour course, and, therefore, sacrifices must be made from the depth if not from the breadth. Selecting the right topics and finding the proper level of depth and breadth are no small challenge for the instructors, and this text is intended to serve as the ground for such selection. Students in a combined thermal-fluids course can gain a basic understanding of energy and energy interactions, various mechanisms of heat transfer, and fundamentals of fluid flow. Such a course can also instill in students the confidence and the background to do further reading of their own and to be able to communicate effectively with specialists in thermal-fluid sciences.

OBJECTIVES

This book is intended for use as a textbook in a first course in thermal-fluid sciences for undergraduate engineering students in their junior or senior year, and as a reference book for practicing engineers. Students are assumed to have an adequate background in calculus, physics, and engineering mechanics. The objectives of this text are

- To cover the *basic principles* of thermodynamics, fluid mechanics, and heat transfer.
- To present numerous and diverse real-world *engineering examples* to give students a feel for how thermal-fluid sciences are applied in engineering practice.
- To develop an *intuitive understanding* of thermal-fluid sciences by emphasizing the physics and physical arguments.

The text contains sufficient material to give instructors flexibility and to accommodate their preferences on the right blend of thermodynamics, fluid mechanics, and heat transfer for their students. By careful selection of topics, an instructor can spend one-third, one-half, or two-thirds of the course on thermodynamics and the rest on selected topics of fluid mechanics and heat transfer.

PHILOSOPHY AND GOAL

The philosophy that contributed to the warm reception of the first edition of this book has remained unchanged. Namely, our goal is to offer an engineering textbook that

- Communicates directly to the minds of tomorrow's engineers in a *simple yet precise* manner.
- Leads students towards a clear understanding and firm grasp of the *basic principles* of thermal-fluid sciences.
- Encourages *creative thinking* and development of a *deeper understanding* and *intuitive feel* for thermal-fluid sciences.
- Is *read* by students with *interest* and *enthusiasm* rather than being used as an aid to solve problems.

Special effort has been made to appeal to readers' natural curiosity and to help students explore the various facets of the exciting subject area of thermal-fluid sciences. The enthusiastic response we received from the users of the previous editions—from small colleges to large universities all over the world—indicates that our objectives have largely been achieved. It is our philosophy that the best way to learn is by practice. Therefore, special effort is made throughout the book to reinforce material that was presented earlier.

Yesterday's engineers spent a major portion of their time substituting values into the formulas and obtaining numerical results. However, now formula manipulations and number crunching are being left to computers. Tomorrow's engineer will need to have a clear understanding and a firm grasp of the *basic principles* so that he or she can understand even the most complex problems, formulate them, and interpret the results. A conscious effort is made to emphasize these basic principles while also providing students with a look at how modern tools are used in engineering practice.

NEW IN THIS EDITION

All the popular features of the previous editions are retained while new ones are added. The main body of the text remains largely unchanged except that two new chapters are added, and two chapters are removed. The most significant changes in this edition are highlighted below.

EARLY INTRODUCTION OF THE FIRST LAW OF THERMODYNAMICS

The first law of thermodynamics is now introduced early in new Chapter 3, "Energy, Energy Transfer, and General Energy Analysis." This introductory chapter sets the framework of establishing a general understanding of various forms of energy, mechanisms of energy transfer, the concept of energy balance, thermo-economics, energy conversion, and conversion efficiency using familiar settings that involve mostly electrical and mechanical forms of energy. It also exposes students to some exciting real-world applications of thermodynamics early in the course, and helps them establish a sense of the monetary value of energy.

SEPARATE COVERAGE OF CLOSED SYSTEM AND CONTROL VOLUME ENERGY ANALYSES

The energy analysis of closed systems is now presented in a separate chapter, Chapter 5, together with the boundary work and the discussion of specific heats for both ideal gases and incompressible substances. The conservation of mass is now covered together with conservation of energy in the new Chapter 6, "Mass and Energy Analysis of Control Volumes."

A NEW CHAPTER ON FLUID KINEMATICS

The all new Chapter 11 covers topics related to fluid kinematics, such as the differences between Lagrangian and Eulerian descriptions of fluid flows, flow patterns, flow visualization, vorticity and rotationality, and the Reynolds transport theorem.

UPDATED STEAM AND REFRIGERANT-134A TABLES

The steam and refrigerant-134a tables are updated using the most current property data from EES. Tables A-4 through A-8 and A-11 through A-13, as well as their counterparts in English units have all been revised. All the examples and homework problems in the text that involve steam or refrigerant-134a are also revised to reflect the small changes in steam and refrigerant properties. An added advantage of this update is that students will get the same result when solving problems whether they use steam or refrigerant properties from EES or property tables in the Appendices.

LEARNING OBJECTIVES

Each chapter now begins with an *overview* of the material to be covered and chapter-specific *learning objectives* to introduce the material and to set goals.

CONTENT CHANGES AND REORGANIZATION

The noteworthy changes in various chapters are summarized below for those who are familiar with the previous edition.

- The first half of Chapter 1 is updated while the second half is greatly revised. A new example and some new discussions are added to the section on Dimensions and Units, and the remaining part of the chapter is reduced into a single section on Problem Solving Technique.
- Chapter 2 is also revised considerably, and the material is updated. A new section *Density and Specific Gravity* and a new subsection *The International Temperature Scale of 1990* are added. The sections *Forms of Energy* and *Energy and the Environment* are moved to the new Chap. 3.
- The new Chapter 3 mainly consists of the sections *Forms of Energy*, *Energy and the Environment*, *Energy Transfer by Heat*, *Energy Transfer by Work*, *Mechanical Forms of Energy*, *The First Law of Thermodynamics*, and *Energy Conversion Efficiencies* moved from various chapters in the previous edition.
- Chapter 4 *Properties of Pure Substances* is essentially the previous edition Chapter 3, except that the last three sections on *specific heats* are moved to new Chap. 5.
- Chapter 5 *Energy Analysis of Closed Systems* involves *Moving Boundary Work*, *Specific Heats*, and *Energy Balance for Closed Systems*.

- Chapter 6 *Mass and Energy Analysis of Control Volumes* consists of *Mass Balance for Control Volumes*, *Flow Work and the Energy of a Flowing Fluid*, and sections on *Energy Balance for Steady- and Unsteady-Flow Systems*.
- Chapter 7 *The Second Law of Thermodynamics* is identical to the previous edition Chap. 6, except the section *Energy Conversion Efficiencies* is moved to new Chap. 3.
- Chapter 8 on *Entropy* is essentially identical to Chap. 7 in the previous edition.
- The chapters titled *Power and Refrigeration Cycles* (Chapter 8) and *Gas Mixtures and Psychometrics* (Chapter 9) are deleted. The former is available for downloading from the web site as a PDF file if needed.
- The first chapter in Fluid Mechanics (Chap. 9) is completely revised, and renamed *Introduction and Properties of Fluids.*
- Chapter 10 *Fluid Statics* is slightly revised, and the section *Fluids in Rigid Body Motion* is deleted.
- A new Chapter 11 *Fluid Kinematics* is added.
- The remaining chapters on fluid mechanics (Chaps. 12 through 15) are updated with no major changes in content or organization.
- The chapters on heat conduction, forced convection, and heat exchangers are updated with only minor changes.
- In Chapter 20 *Natural Convection,* the section on combined natural and forced convection is deleted.
- The two chapters on thermal radiation are now combined into a single Chapter 21 *Radiation Heat Transfer*. The sections on *Radiation Intensity, Solar Radiation,* and *Radiation Shields* are deleted.
- In Appendices 1 and 2, the steam and refrigerant-134a tables (Tables 4 through 8 and 11 through 13) are entirely revised, but the table numbers are kept the same. Appendix 3 *Introduction to EES* is moved to the Student Resources DVD that comes packaged free with the text.
- The conversion factors on the inner cover pages and the physical constants are updated, and some nomenclature symbols are revised.

LEARNING TOOLS

EMPHASIS ON PHYSICS

A distinctive feature of this book is its emphasis on the physical aspects of subject matter in addition to mathematical representations and manipulations. The authors believe that the emphasis in undergraduate education should remain on *developing a sense of underlying physical mechanisms* and a *mastery of solving practical problems* that an engineer is likely to face in the real world. Developing an intuitive understanding should also make the course a more motivating and worthwhile experience for the students.

EFFECTIVE USE OF ASSOCIATION

An observant mind should have no difficulty understanding engineering sciences. After all, the principles of engineering sciences are based on our *everyday experiences* and *experimental observations.* A more physical, intuitive approach is used throughout this text. Frequently, *parallels are drawn* between the subject matter and students' everyday experiences so that they can relate the subject matter to what they already know.

SELF-INSTRUCTING

The material in the text is introduced at a level that an average student can follow comfortably. It speaks to students, not over students. In fact, it is *self-instructive.* Noting that the principles of science are based on experimental observations, most of the derivations in this text are largely based on physical arguments, and thus they are easy to follow and understand.

EXTENSIVE USE OF ARTWORK

Figures are important learning tools that help the students "get the picture." The text makes effective use of graphics, and it contains a great number of figures and illustrations. Figures attract attention and stimulate curiosity and interest. Some of the figures in this text are intended to serve as a means of emphasizing some key concepts that would otherwise go unnoticed; some serve as page summaries.

CHAPTER OPENERS AND SUMMARIES

Each chapter begins with an overview of the material to be covered and chapter objectives. A *summary* is included at the end of each chapter for a quick review of basic concepts and important relations.

NUMEROUS WORKED-OUT EXAMPLES

Each chapter contains several worked-out *examples* that clarify the material and illustrate the use of the basic principles. An *intuitive* and *systematic* approach is used in the solution of the example problems, with particular attention to the proper use of units.

A WEALTH OF REAL-WORLD END-OF-CHAPTER PROBLEMS

The end-of-chapter problems are grouped under specific topics in the order they are covered to make problem selection easier for both instructors and students. Within each group of problems are *Concept Questions*, indicated by "C" to check the students' level of understanding of basic concepts. The problems under *Review Problems* are more comprehensive in nature and are not directly tied to any specific section of a chapter—in some cases they require review of material learned in previous chapters. The problems under the *Design and Essay Problems* title are intended to encourage students to make engineering judgments, to conduct independent exploration of topics of interest, and to communicate their findings in a professional manner. Several economics- and safety-related problems are incorporated throughout to enhance cost and safety awareness among engineering students. Answers to selected problems are listed immediately following the problem for convenience to students.

A SYSTEMATIC SOLUTION PROCEDURE

A well-structured approach is used in problem solving while maintaining an informal conversational style. The problem is first stated and the objectives are identified, and the assumptions made are stated together with their justifications. The properties needed to solve the problem are listed separately.

Numerical values are used together with their units to emphasize that numbers without units are meaningless, and unit manipulations are as important as manipulating the numerical values with a calculator. The significance of the findings is discussed following the solutions. This approach is also used consistently in the solutions presented in the Instructor's Solutions Manual.

RELAXED SIGN CONVENTION

The use of a formal sign convention for heat and work is abandoned as it often becomes counterproductive. A physically meaningful and engaging approach is adopted for interactions instead of a mechanical approach. Subscripts "in" and "out," rather than the plus and minus signs, are used to indicate the directions of interactions.

A CHOICE OF SI ALONE OR SI / ENGLISH UNITS

In recognition of the fact that English units are still widely used in some industries, both SI and English units are used in this text, with an emphasis on SI. The material in this text can be covered using combined SI/English units or SI units alone, depending on the preference of the instructor. The property tables and charts in the appendixes are presented in both units, except the ones that involve dimensionless quantities. Problems, tables, and charts in English units are designated by "E" after the number for easy recognition, and they can be ignored easily by the SI users.

CONVERSION FACTORS

Frequently used conversion factors and physical constants are listed on the inner cover pages of the text for easy reference.

SUPPLEMENTS

The following supplements are available to the adopters of the book.

ENGINEERING EQUATION SOLVER (EES) DVD

(Limited Academic Version packaged free with every new copy of the text) Developed by Sanford Klein and William Beckman from the University of Wisconsin–Madison, this software combines equation-solving capability and engineering property data. EES can do optimization, parametric analysis, and linear and nonlinear regression, and provides publication-quality plotting capabilities. Thermodynamic and transport properties for air, water, and many other fluids are built in, and EES allows the user to enter property data or functional relationships. Some problems are solved using EES, and complete solutions together with parametric studies are included on the enclosed DVD. To obtain the full version of EES, contact your McGraw-Hill representative or visit *www.mhhe.com/ees.*

ONLINE LEARNING CENTER (www.mhhe.com/cengel)

Web support is provided for the text on our Online Learning Center. Visit this web site for general text information, errata, and author information. The site also includes resources for students including a list of helpful web links. The instructor side of the site includes the solutions manual, the text's images in PowerPoint form, and more!

ACKNOWLEDGMENTS

We would like to acknowledge with appreciation the numerous and valuable comments, suggestions, criticisms, and praise of these academic evaluators:

Thomas M. Adams
Rose-Hulman Institute of Technology

J. Iwan D. Alexander
Case Western Reserve University

Farruhk S. Alvi
Florida A&M University–Florida State University

Michael Amitay
Rensselaer Polytechnic Institute

Pradeep Kumar Bansal
University of Auckland, New Zealand

Kevin W. Cassel
Illinois Institute of Technology

John M. Cimbala
Pennsylvania State University

Subrat Das, Swinburne
University of Technology

Tahsin Engin
Sakarya University, Turkey

Richard S. Figliola
Clemson University

Mehmet Kanoğlu
Gaziantep University, Turkey

Thomas M. Kiehne
University of Texas at Austin

Joseph M. Kmec
Purdue University

William E. Lee III
University of South Florida

Frank K. Lu
University of Texas at Arlington

Richard S. Miller
Clemson University

T. Terry Ng
University of Toledo

Jim A. Nicell
McGill University, Montreal, Canada

Narender P. Reddy
University of Akron

Arthur E. Ruggles
University of Tennessee

Chiang Shih
FAMU–Florida State University

Brian E. Thompson
Rensselaer Polytechnic Institute

Their suggestions have greatly helped to improve the quality of this text. Special thanks are due to Mehmet Kanoglu of the University of Gaziantep, Turkey, for his valuable contributions and his critical review of the manuscript and for his special attention to accuracy and detail. We also would like to thank our students who provided plenty of feedback from their perspectives. Finally, we would like to express our appreciation to our wives and our children for their continued patience, understanding, and support throughout the preparation of this text.

Yunus A. Çengel
Robert H. Turner
John M. Cimbala

Nomenclature

a	Acceleration, m/s^2
A_s	Surface area, m^2
A_c	Cross-sectional area, m^2
Bi	Biot number
C	Speed of sound, m/s
C	Specific heat, kJ/kg · K
C_c, C_h	Heat capacity rate, W/°C
C_D	Drag coefficient
C_f	Fanning friction factor or friction coefficient
C_L	Lift coefficient
c_p	Constant-pressure specific heat, kJ/kg · K
c_v	Constant-volume specific heat, kJ/kg · K
COP	Coefficient of performance
COP_{HP}	Coefficient of performance of a heat pump
COP_{R}	Coefficient of performance of a refrigerator
d, D	Diameter, m
D_h	Hydraulic diameter, m
e	Specific total energy, kJ/kg
erfc	Complementary error function
E	Total energy, kJ
E_b	Blackbody emissive flux, W/m^2
EER	Energy efficiency rating
f	Darcy friction factor
f_λ	Blackbody radiation function
F	Force, N
F_D	Drag force, N
F_{ij}, $F_{i \to j}$	View factor
F_L	Lift force, N
Fo	Fourier number
g	Gravitational acceleration, m/s^2
G	Incident radiation, W/m^2
Gr	Grashof number
h	Convection heat transfer coefficient, W/m^2 · °C
h	Specific enthalpy, $u + Pv$, kJ/kg
h_c	Thermal contact conductance, W/m^2 · °C
h_{fg}	Enthalpy of vaporization, kJ/kg
h_{if}	Latent heat of fusion, kJ/kg
h_L	Head loss, m
H	Total enthalpy, $U + PV$, kJ
I	Electric current, A; Bessel function
J	Radiosity, W/m^2; Bessel function
k	Specific heat ratio, c_p/c_v
k	Thermal conductivity
k_{eff}	Effective thermal conductivity, W/m · °C
k_s	Spring constant
ke	Specific kinetic energy, $V^2/2$, kJ/kg
K_{loss}	Minor loss coefficient
KE	Total kinetic energy, $mV^2/2$, kJ
L	Length; half thickness of a plane wall
L_c	Characteristic or corrected length
L_h	Hydrodynamic entry length
L_t	Thermal entry length
m	Mass, kg
$\dot{m}$	Mass flow rate, kg/s
M	Molar mass, kg/kmol
M	Moment of force
Ma	Mach number
n	Polytropic exponent
N	Number of moles, kmol

NTU	Number of transfer units
Nu	Nusselt number
p	Perimeter or wetted perimeter, m
pe	Specific potential energy, gz, kJ/kg
P	Pressure, kPa
P_{cr}	Critical pressure, kPa
P_r	Relative pressure
P_R	Reduced pressure
P_v	Vapor pressure, kPa
PE	Total potential energy, mgz, kJ
Pr	Prandtl number
q	Heat transfer per unit mass, kJ/kg
$\dot{q}$	Heat flux, W/m^2
Q	Total heat transfer, kJ
$\dot{Q}$	Heat transfer rate, kW
Q_H	Heat transfer with high-temperature body, kJ
Q_L	Heat transfer with low-temperature body, kJ
r_{cr}	Critical radius of insulation, m
R	Gas constant, kJ/kg · K
R	Radius, m; Thermal resistance, °C/W
R_c	Thermal contact resistance, m^2 · °C/W
R_f	Fouling factor, m^2 · °C/W
R_u	Universal gas constant, kJ/kmol · K
R-value	R-value of insulation
Ra	Rayleigh number
Re	Reynolds number
s	Specific entropy, kJ/kg · K
s_{gen}	Specific entropy generation, kJ/kg · K
S	Total entropy, kJ/K; Conduction shape factor
S_{gen}	Total entropy generation, kJ/K
t	Time, s; Thickness, m
T	Temperature, °C or K
T	Torque, N · m
T_b	Bulk fluid temperature, °C
T_{cr}	Critical temperature, K
T_f	Film temperature, °C
T_H	Temperature of high-temperature body, K
T_i	Initial temperature, °C
T_∞	Free stream or surrounding medium temperature, °C
T_L	Temperature of low-temperature body, K
T_R	Reduced temperature
T_s	Surface temperature, °C or K
T_{sat}	Saturation temperature, °C
u	Specific internal energy, kJ/kg
u, v	x- and y-components of velocity, m/s
U	Total internal energy, kJ
U	Overall heat transfer coefficient, W/m^2 · °C
v	Specific volume, m^3/kg
v_{cr}	Critical specific volume, m^3/kg
v_r	Relative specific volume
V	Total volume, m^3
$\dot{V}$	Volume flow rate, m^3/s
V	Velocity, m/s
V_m	Mean velocity, m/s
V_∞	Free-stream velocity, m/s
w	Work per unit mass, kJ/kg
W	Total work, kJ
$\dot{W}$	Power, kW
W_{in}	Work input, kJ
W_{out}	Work output, kJ
x	Quality
z	Elevation, m
Z	Compressibility factor

Greek Letters

α	Absorptivity
α	Thermal diffusivity, m^2/s
α_s	Solar absorptivity
β	Volume expansivity, 1/K
δ	Velocity boundary layer thickness, m
δ_t	Thermal boundary layer thickness, m
ΔP	Pressure drop, Pa
ΔT_{lm}	Log mean temperature difference, °C
ε	Emissivity; heat exchanger or fin effectiveness
ε	Mean surface roughness, m
ζ	Vorticity, 1/s
η_{fin}	Fin efficiency
η_{th}	Thermal efficiency
θ	Dimensionless temperature
μ	Dynamic viscosity, kg/m · s or N · s/m^2
ν	Kinematic viscosity $= \mu/\rho$, m^2/s; frequency, 1/s
ρ	Density, kg/m^3
σ	Stefan–Boltzmann constant
σ_n	Normal stress, N/m^2
σ_s	Surface tension, N/m
τ	Shear stress, N/m^2
τ	Transmissivity; Fourier number
τ_w	Wall shear stress, N/m^2

Subscripts

a	Air
abs	Absolute
act	Actual
atm	Atmospheric
b	Boundary; bulk fluid
cond	Conduction
conv	Convection
cr	Critical point
cv	Control volume
e	Exit conditions
f	Saturated liquid; film
fg	Difference in property between saturated liquid and saturated vapor
g	Saturated vapor
H	High temperature as in T_H and Q_H
i	Inlet, initial, or indoor conditions
L	Low temperature as in T_L and Q_L
o	Outlet or outdoor conditions
r	Relative
rad	Radiation
s	Surface
surr	Surrounding surfaces
sat	Saturated
sys	System
v	Water vapor
1	Initial or inlet state
2	Final or exit state
∞	Property of a far field; free-flow conditions

Superscripts

· (over dot)	Quantity per unit time
– (over bar)	Quantity per unit mole
° (circle)	Standard reference state

Chapter 1

INTRODUCTION AND OVERVIEW

Many engineering systems involve the transfer, transport, and conversion of energy, and the sciences that deal with these subjects are broadly referred to as *thermal-fluid sciences*. Thermal-fluid sciences are usually studied under the subcategories of *thermodynamics, heat transfer,* and *fluid mechanics*. We start this chapter with an overview of these sciences, and give some historical background. Then we review the unit systems that will be used, and discuss dimensional homogeneity. We then present an intuitive systematic *problem solving technique* that can be used as a model in solving engineering problems, followed by a discussion of the proper place of software packages in engineering. Finally, we discuss accuracy and significant digits in engineering measurements and calculations.

Objectives

Objectives of this chapter are to:

- Be acquainted with the engineering sciences thermodynamics, heat transfer, and fluid mechanics, and understand the basic concepts of thermal-fluid sciences,
- Be comfortable with the metric SI and English units commonly used in engineering,
- Develop an intuitive systematic problem-solving technique,
- Learn the proper use of software packages in engineering, and
- Develop an understanding of accuracy and significant digits in calculations.

1–1 ▪ INTRODUCTION TO THERMAL-FLUID SCIENCES

The word *thermal* stems from the Greek word *therme,* which means *heat.* Therefore, thermal sciences can loosely be defined as the sciences that deal with heat. The recognition of different forms of energy and its transformations has forced this definition to be broadened. Today, the physical sciences that deal with energy and the transfer, transport, and conversion of energy are usually referred to as **thermal-fluid sciences** or just **thermal sciences**. Traditionally, the thermal-fluid sciences are studied under the subcategories of thermodynamics, heat transfer, and fluid mechanics. In this book we present the basic principles of these sciences, and apply them to situations that engineers are likely to encounter in their practice.

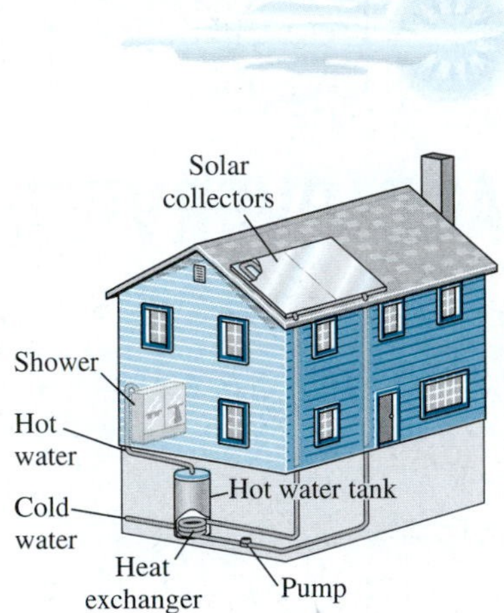

FIGURE 1–1

The design of many engineering systems, such as this solar hot water system, involves thermal-fluid sciences.

The design and analysis of most thermal systems such as power plants, automotive engines, and refrigerators involve all categories of thermal-fluid sciences as well as other sciences (Fig. 1–1). For example, designing the radiator of a car involves the determination of the amount of energy transfer from a knowledge of the properties of the coolant using *thermodynamics,* the determination of the size and shape of the inner tubes and the outer fins using *heat transfer,* and the determination of the size and type of the water pump using *fluid mechanics.* Of course the determination of the materials and the thickness of the tubes requires the use of material science as well as strength of materials. The reason for studying different sciences separately is simply to facilitate learning without being overwhelmed. Once the basic principles are mastered, they can then be synthesized by solving comprehensive real-world practical problems. But first we present an overview of thermal-fluid sciences.

Application Areas of Thermal-Fluid Sciences

All activities in nature involve some interaction between energy and matter; thus it is hard to imagine an area that does not relate to thermal-fluid sciences in some manner. Therefore, developing a good understanding of basic principles of thermal-fluid sciences has long been an essential part of engineering education.

Thermal-fluid sciences are commonly encountered in many engineering systems and other aspects of life, and one does not need to go very far to see some application areas of them. In fact, one does not need to go anywhere. The heart is constantly pumping blood to all parts of the human body, various energy conversions occur in trillions of body cells, and the body heat generated is constantly rejected to the environment. The human comfort is closely tied to the rate of this metabolic heat rejection. We try to control this heat transfer rate by adjusting our clothing to the environmental conditions. Also, any defect in the heart and the circulatory system is a major cause for alarm.

Other applications of thermal sciences are right where one lives. An ordinary house is, in some respects, an exhibition hall filled with wonders of thermal-fluid sciences. Many ordinary household utensils and appliances are designed, in whole or in part, by using the principles of thermal-fluid sciences. Some examples include the electric or gas range, the heating and air-conditioning systems, the refrigerator, the humidifier, the pressure cooker, the water heater, the shower, the iron, the plumbing and sprinkling systems,

The human body

Air conditioning systems

Airplanes

Automobile radiators

Power plants

Refrigeration systems

FIGURE 1–2

Some application areas of thermal-fluid sciences.

A/C unit, fridge, radiator: © The McGraw-Hill Companies, Inc./Jill Braaten, photographer; Plane: © Vol. 14/PhotoDisc; Humans: © Vol. 121/PhotoDisc; Power plant: © Corbis Royalty Free

and even the computer, the TV, and the DVD player. On a larger scale, thermal-fluid sciences play a major part in the design and analysis of automotive engines, rockets, jet engines, and conventional or nuclear power plants, solar collectors, the transportation of water, crude oil, and natural gas, the water distribution systems in cities, and the design of vehicles from ordinary cars to airplanes (Fig. 1–2). The energy-efficient home that you may be living in, for example, is designed on the basis of minimizing heat loss in winter and heat gain in summer. The size, location, and the power input of the fan of your computer is also selected after a thermodynamic, heat transfer, and fluid flow analysis of the computer.

1–2 ▪ THERMODYNAMICS

Thermodynamics can be defined as the science of *energy*. Although everybody has a feeling of what energy is, it is difficult to give a precise definition for it. Energy can be viewed as the ability to cause changes.

The name *thermodynamics* stems from the Greek words *therme* (heat) and *dynamis* (power), which is most descriptive of the early efforts to convert heat into power. Today the same name is broadly interpreted to include all aspects of energy and energy transformations, including power generation, refrigeration, and relationships among the properties of matter.

One of the most fundamental laws of nature is the **conservation of energy principle**. It simply states that during an interaction, energy can change from one form to another but the total amount of energy remains constant. That is, energy cannot be created or destroyed. A rock falling off a cliff, for example, picks up speed as a result of its potential energy being converted to kinetic energy (Fig. 1–3). The conservation of energy principle also forms the backbone of the diet industry: A person who has a greater energy input (food) than energy output (exercise) will gain weight (store energy in the form of fat), and a person who has a smaller energy input than output will lose weight (Fig. 1–4). The change in the energy content of a body or any other system is equal to the difference between the energy input and the energy output, and the energy balance is expressed as $E_{in} - E_{out} = \Delta E$.

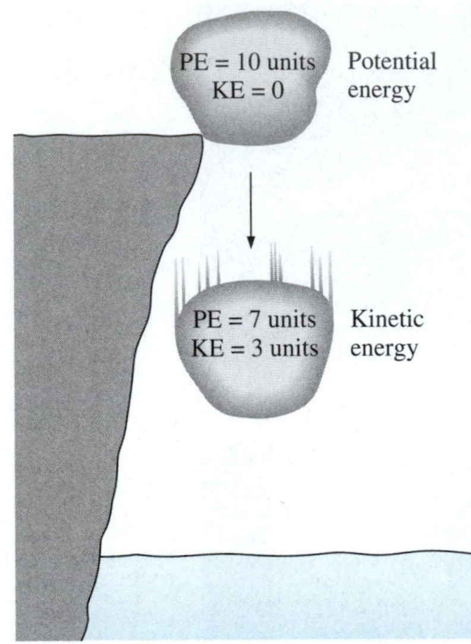

FIGURE 1–3

Energy cannot be created or destroyed; it can only change forms (the first law).

The **first law of thermodynamics** is simply an expression of the conservation of energy principle, and it asserts that *energy* is a thermodynamic property. The **second law of thermodynamics** asserts that energy has *quality* as well as *quantity,* and actual processes occur in the direction of decreasing quality of energy. For example, a cup of hot coffee left on a table eventually cools, but a cup of cool coffee in the same room never gets hot by itself (Fig. 1–5). The high-temperature energy of the coffee is degraded (transformed into a less useful form at a lower temperature) once it is transferred to the surrounding air.

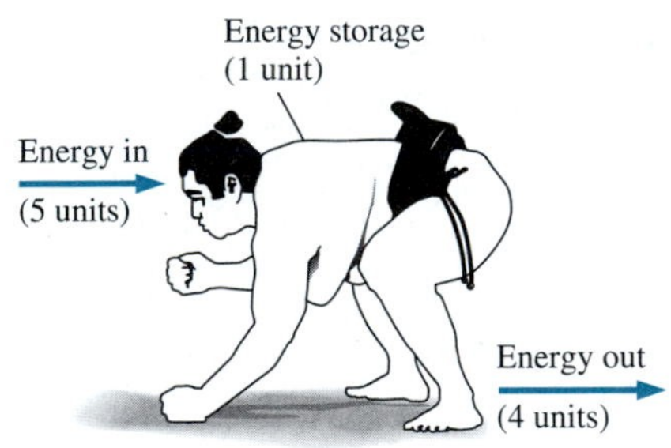

FIGURE 1–4

Conservation of energy principle for the human body.

Although the principles of thermodynamics have been in existence since the creation of the universe, thermodynamics did not emerge as a science until the construction of the first successful atmospheric steam engines in England by Thomas Savery in 1697 and Thomas Newcomen in 1712. These engines were very slow and inefficient, but they opened the way for the development of a new science.

The first and second laws of thermodynamics emerged simultaneously in the 1850s, primarily out of the works of William Rankine, Rudolph Clausius, and Lord Kelvin (formerly William Thomson). The term *thermodynamics* was first used in a publication by Lord Kelvin in 1849. The first thermodynamic textbook was written in 1859 by William Rankine, a professor at the University of Glasgow.

It is well-known that a substance consists of a large number of particles called *molecules.* The properties of the substance naturally depend on the behavior of these particles. For example, the pressure of a gas in a container is the result of momentum transfer between the molecules and the walls of the container. However, one does not need to know the behavior of the gas particles to determine the pressure in the container. It would be sufficient to attach a pressure gage to the container. This macroscopic approach to the study of thermodynamics that does not require a knowledge of the behavior of individual particles is called **classical thermodynamics**. It provides a direct and easy way to the solution of engineering problems. A more elaborate approach, based on the average behavior of large groups of individual particles, is called **statistical thermodynamics**. This microscopic approach is rather involved and is used in this text only in the supporting role.

1–3 ■ HEAT TRANSFER

We all know from experience that a cold canned drink left in a room warms up and a warm canned drink put in a refrigerator cools down. This is accomplished by the transfer of *energy* from the warm medium to the cold one. The energy transfer is always from the higher temperature medium to the lower temperature one, and the energy transfer stops when the two mediums reach the same temperature.

Energy exists in various forms. In heat transfer, we are primarily interested in heat, which is *the form of energy that can be transferred from one system to another as a result of temperature difference.* The science that deals with the determination of the *rates* of such energy transfers is **heat transfer**.

You may be wondering why we need the science of heat transfer. After all, we can determine the amount of heat transfer for any system undergoing any process using a thermodynamic analysis alone. The reason is that thermodynamics is concerned with the *amount* of heat transfer as a system undergoes a process from one equilibrium state to another, and it gives no indication about *how long* the process will take. But in engineering, we are often interested in the *rate* of heat transfer, which is the topic of the science of *heat transfer.* A thermodynamic analysis simply tells us how much heat must be transferred to realize a specified change of state to satisfy the conservation of energy principle.

In practice we are more concerned about the rate of heat transfer (heat transfer per unit time) than we are with the amount of it. For example, we can determine the amount of heat transferred from a thermos bottle as the hot coffee inside cools from 90°C to 80°C by a thermodynamic analysis alone. But a typical user or designer of a thermos is primarily interested in *how long* it will be before the hot coffee inside cools to 80°C, and a thermodynamic analysis cannot answer this question. Determining the rates of heat transfer to or from a system and thus the times of cooling or heating, as well as the variation of the temperature, is the subject of *heat transfer* (Fig. 1–6).

Thermodynamics deals with equilibrium states and changes from one equilibrium state to another. Heat transfer, on the other hand, deals with systems that lack thermal equilibrium, and thus it is a *nonequilibrium* phenomenon. Therefore, the study of heat transfer cannot be based on the principles of thermodynamics alone. However, the laws of thermodynamics lay the framework for the science of heat transfer. The *first law* requires that the rate of energy transfer into a system be equal to the rate of increase of the energy of that system. The *second law* requires that heat be transferred in the direction of decreasing temperature. This is analogous to a car parked on an inclined road; it must go downhill in the direction of decreasing elevation when its brakes are released. It is also analogous to the electric current flowing in the direction of decreasing voltage or the fluid flowing in the direction of decreasing pressure.

The basic requirement for heat transfer is the presence of a *temperature difference.* There can be no net heat transfer between two mediums that are at the same temperature. The temperature difference is the *driving force* for heat transfer; just as the *voltage difference* is the driving force for electric current, and *pressure difference* is the driving force for fluid flow. The rate

FIGURE 1–5
Heat flows in the direction of decreasing temperature.

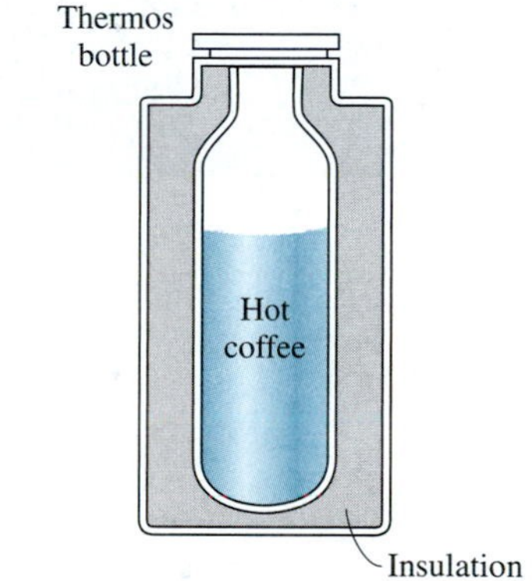

FIGURE 1–6
We are normally interested in how long it takes for the hot coffee in a thermos to cool to a certain temperature, which cannot be determined from a thermodynamic analysis alone.

of heat transfer in a certain direction depends on the magnitude of the *temperature gradient* (the temperature difference per unit length or the rate of change of temperature) in that direction. The larger the temperature gradient, the higher the rate of heat transfer.

1–4 ▪ FLUID MECHANICS

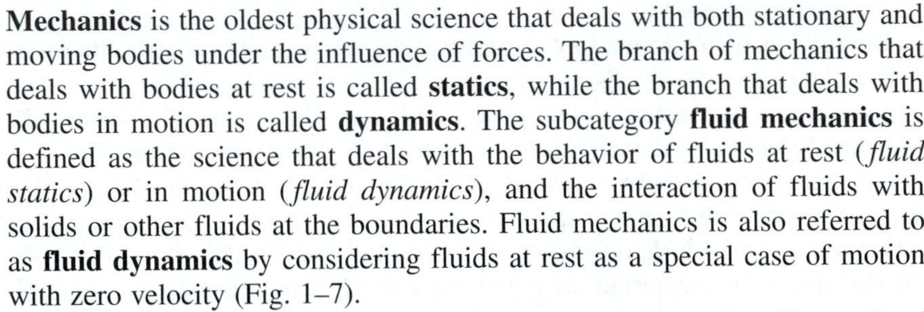

Mechanics is the oldest physical science that deals with both stationary and moving bodies under the influence of forces. The branch of mechanics that deals with bodies at rest is called **statics**, while the branch that deals with bodies in motion is called **dynamics**. The subcategory **fluid mechanics** is defined as the science that deals with the behavior of fluids at rest (*fluid statics*) or in motion (*fluid dynamics*), and the interaction of fluids with solids or other fluids at the boundaries. Fluid mechanics is also referred to as **fluid dynamics** by considering fluids at rest as a special case of motion with zero velocity (Fig. 1–7).

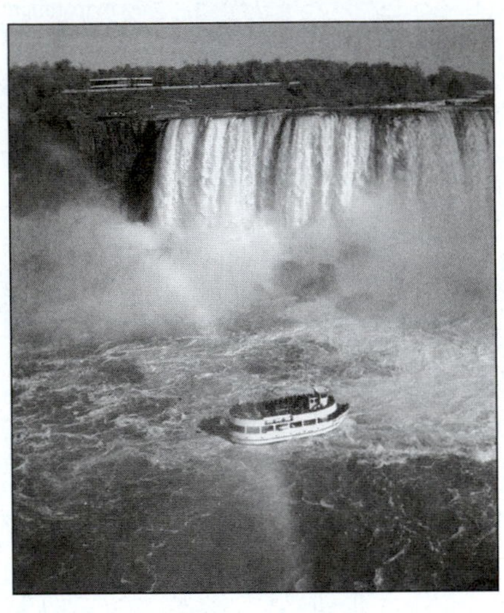

FIGURE 1–7
Fluid mechanics deals with liquids and gases in motion or at rest.
© Vol. 16/Photo Disc.

Fluid mechanics itself is also divided into several categories. The study of the motion of fluids that are practically incompressible (such as liquids, especially water, and gases at low speeds) is usually referred to as **hydrodynamics**. A subcategory of hydrodynamics is **hydraulics**, which deals with liquid flows in pipes and open channels. **Gas dynamics** deals with the flow of fluids that undergo significant density changes, such as the flow of gases through nozzles at high speeds. The category **aerodynamics** deals with the flow of gases (especially air) over bodies such as aircraft, rockets, and automobiles at high or low speeds. Some other specialized categories such as **meteorology**, **oceanography**, and **hydrology** deal with naturally occurring flows.

You will recall from physics that a substance exists in three primary phases: solid, liquid, and gas. (At very high temperatures, it also exists as plasma.) A substance in the liquid or gas phase is referred to as a **fluid**. Distinction between a solid and a fluid is made on the basis of the substance's ability to resist an applied shear (or tangential) stress that tends to change its shape. A solid can resist an applied shear stress by deforming, whereas *a fluid deforms continuously under the influence of shear stress, no matter how small.* In solids stress is proportional to *strain*, but in fluids stress is proportional to *strain rate*. When a constant shear force is applied, a solid eventually stops deforming, at some fixed strain angle, whereas a fluid never stops deforming but instead establishes a certain rate of strain.

Consider a rectangular rubber block tightly placed between two plates. As the upper plate is pulled with a force F while the lower plate is held fixed, the rubber block deforms, as shown in Fig. 1–8. The angle of deformation α (called the *shear strain* or *angular displacement*) increases in proportion to the applied force F. Assuming there is no slip between the rubber and the plates, the upper surface of the rubber is displaced by an amount equal to the displacement of the upper plate while the lower surface remains stationary. In equilibrium, the net force acting on the upper plate in the horizontal direction must be zero, and thus a force equal and opposite to F must be acting on the plate. This opposing force that develops at the plate–rubber interface due to friction is expressed as $F = \tau A$, where τ is the shear stress and A is the contact area between the upper plate and the rubber. When the force is removed, the rubber returns to its original position. This phenomenon would also be observed with other solids such as a steel block provided that the applied force does not exceed the elastic range. If this experiment were repeated with a fluid (with two large parallel plates placed in a large body of water, for example), the fluid layer in contact with the upper plate would move with the plate continuously at the velocity of the plate no matter how small the force F is. The fluid velocity decreases with depth because of friction between fluid layers, reaching zero at the lower plate.

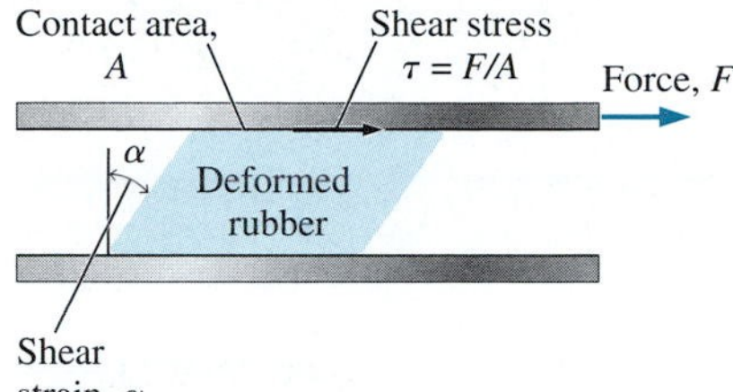

FIGURE 1–8
Deformation of a rubber block placed between two parallel plates under the influence of a shear force. The shear stress shown is that on the rubber—an equal but opposite shear stress acts on the upper plate.

You will recall from statics that **stress** is defined as force per unit area and is determined by dividing the force by the area upon which it acts. The normal component of the force acting on a surface per unit area is called the **normal stress**, and the tangential component of a force acting on a surface per unit area is called **shear stress** (Fig. 1–9). In a fluid at rest, the normal stress is called **pressure**. A fluid at rest is at a state of zero shear stress. When walls are removed or a liquid container is tilted, a shear develops as the liquid moves to re-establish a horizontal free surface.

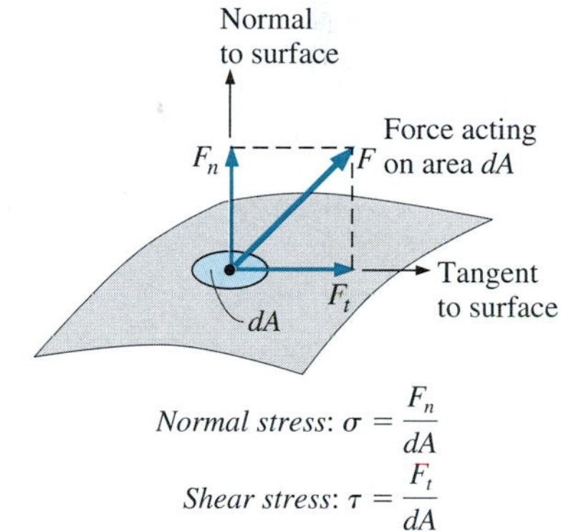

FIGURE 1–9
The normal stress and shear stress at the surface of a fluid element. For fluids at rest, the shear stress is zero and pressure is the only normal stress.

In a liquid, groups of molecules can move relative to each other, but the volume remains relatively constant because of the strong cohesive forces between the molecules. As a result, a liquid takes the shape of the container it is in, and it forms a free surface in a larger container in a gravitational field. A gas, on the other hand, expands until it encounters the walls of the container and fills the entire available space. This is because the gas molecules are widely spaced, and the cohesive forces between them are very small (Fig. 1–10).

Although solids and fluids are easily distinguished in most cases, this distinction is not so clear in some borderline cases. For example, *asphalt* appears and behaves as a solid since it resists shear stress for short periods of time. When these forces are exerted over extended periods of time, however, the asphalt deforms slowly. Some plastics, lead, and slurry mixtures exhibit similar behavior. Such borderline cases are beyond the scope of this text. The fluids we deal with in this text will be clearly recognizable as fluids.

1–5 ▪ IMPORTANCE OF DIMENSIONS AND UNITS

Any physical quantity can be characterized by **dimensions**. The magnitudes assigned to the dimensions are called **units**. Some basic dimensions such as mass m, length L, time t, and temperature T are selected as **primary**, **basic**, or **fundamental dimensions**, while others such as velocity V, energy E, and volume V are expressed in terms of the primary dimensions and are called **secondary dimensions**, or **derived dimensions**.

A number of unit systems have been developed over the years. Despite strong efforts in the scientific and engineering community to unify the world with a single unit system, two sets of units are still in common use today: the **English system**, which is also known as the *United States Customary System* (USCS), and the metric **SI** (from *Le Système International d' Unités*), which is also known as the *International System*. The SI is a simple and logical system based on a decimal relationship between the various units, and it is being used for scientific and engineering work in most of the industrialized nations, including England. The English system, however, has no systematic numerical base, and various units in this system are related to

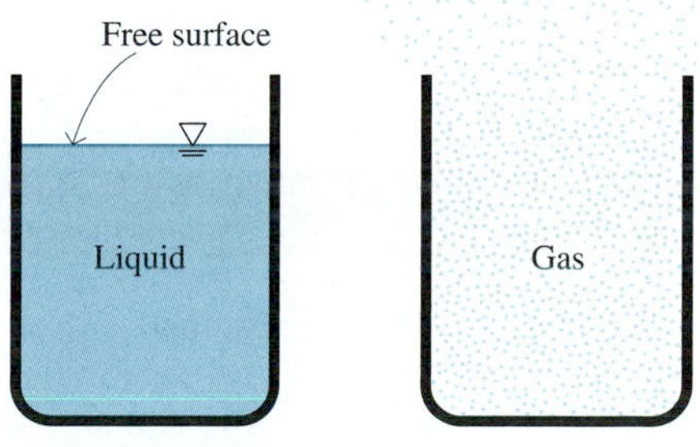

FIGURE 1–10
Unlike a liquid, a gas does not form a free surface, and it expands to fill the entire available space.

each other rather arbitrarily (12 in = 1 ft, 1 mile = 5280 ft, 4 qt = 1 gal, etc.), which makes it confusing and difficult to learn. The United States is the only industrialized country that has not yet fully converted to the metric system.

The systematic efforts to develop a universally acceptable system of units dates back to 1790 when the French National Assembly charged the French Academy of Sciences to come up with such a unit system. An early version of the metric system was soon developed in France, but it did not find universal acceptance until 1875 when *The Metric Convention Treaty* was prepared and signed by 17 nations, including the United States. In this international treaty, meter and gram were established as the metric units for length and mass, respectively, and a *General Conference of Weights and Measures* (CGPM) was established that was to meet every six years. In 1960, the CGPM produced the SI, which was based on six fundamental quantities, and their units were adopted in 1954 at the Tenth General Conference of Weights and Measures: *meter* (m) for length, *kilogram* (kg) for mass, *second* (s) for time, *ampere* (A) for electric current, *degree Kelvin* (°K) for temperature, and *candela* (cd) for luminous intensity (amount of light). In 1971, the CGPM added a seventh fundamental quantity and unit: *mole* (mol) for the amount of matter.

Based on the notational scheme introduced in 1967, the degree symbol was officially dropped from the absolute temperature unit, and all unit names were to be written without capitalization even if they were derived from proper names (Table 1–1). However, the abbreviation of a unit was to be capitalized if the unit was derived from a proper name. For example, the SI unit of force, which is named after Sir Isaac Newton (1647–1723), is *newton* (not Newton), and it is abbreviated as N. Also, the full name of a unit may be pluralized, but its abbreviation cannot. For example, the length of an object can be 5 m or 5 meters, *not* 5 ms or 5 meter. Finally, no period is to be used in unit abbreviations unless they appear at the end of a sentence. For example, the proper abbreviation of meter is m (not m.).

The recent move toward the metric system in the United States seems to have started in 1968 when Congress, in response to what was happening in the rest of the world, passed a Metric Study Act. Congress continued to promote a voluntary switch to the metric system by passing the Metric Conversion Act in 1975. A trade bill passed by Congress in 1988 set a September 1992 deadline for all federal agencies to convert to the metric system. However, the deadlines were relaxed later with no clear plans for the future.

TABLE 1–1

The seven fundamental (or primary) dimensions and their units in SI

Dimension	Unit
Length	meter (m)
Mass	kilogram (kg)
Time	second (s)
Temperature	kelvin (K)
Electric current	ampere (A)
Amount of light	candela (cd)
Amount of matter	mole (mol)

The industries that are heavily involved in international trade (such as the automotive, soft drink, and liquor industries) have been quick in converting to the metric system for economic reasons (having a single worldwide design, fewer sizes, smaller inventories, etc.). Today, nearly all the cars manufactured in the United States are metric. Most car owners probably do not realize this until they try an English socket wrench on a metric bolt. Most industries, however, resisted the change, thus slowing down the conversion process.

Presently the United States is a dual-system society, and it will stay that way until the transition to the metric system is completed. This puts an extra burden on today's engineering students, since they are expected to retain their understanding of the English system while learning, thinking, and

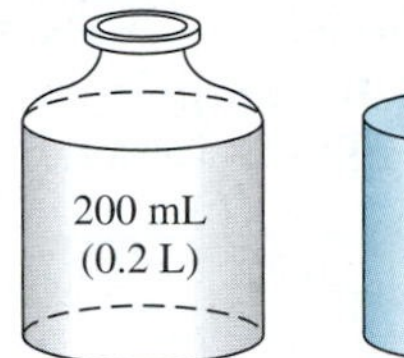

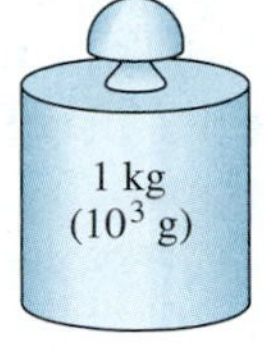

1 MΩ
(10^6 Ω)

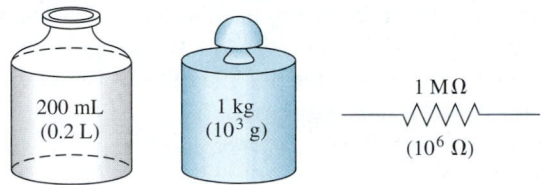

FIGURE 1–11
The SI unit prefixes are used in all branches of engineering.

working in terms of the SI. Given the position of the engineers in the transition period, both unit systems are used in this text, with particular emphasis on SI units.

As pointed out, the SI is based on a decimal relationship between units. The prefixes used to express the multiples of the various units are listed in Table 1–2. They are standard for all units, and the student is encouraged to memorize them because of their widespread use (Fig. 1–11).

TABLE 1–2

Standard prefixes in SI units

Multiple	Prefix
10^{12}	tera, T
10^{9}	giga, G
10^{6}	mega, M
10^{3}	kilo, k
10^{2}	hecto, h
10^{1}	deka, da
10^{-1}	deci, d
10^{-2}	centi, c
10^{-3}	milli, m
10^{-6}	micro, μ
10^{-9}	nano, n
10^{-12}	pico, p

Some SI and English Units

In SI, the units of mass, length, and time are the kilogram (kg), meter (m), and second (s), respectively. The respective units in the English system are the pound-mass (lbm), foot (ft), and second (s). The pound symbol *lb* is actually the abbreviation of *libra*, which was the ancient Roman unit of weight. The English retained this symbol even after the end of the Roman occupation of Britain in 410. The mass and length units in the two systems are related to each other by

$$1 \text{ lbm} = 0.45359 \text{ kg}$$
$$1 \text{ ft} = 0.3048 \text{ m}$$

In the English system, force is usually considered to be one of the primary dimensions and is assigned a nonderived unit. This is a source of confusion and error that necessitates the use of a dimensional constant (g_c) in many formulas. To avoid this nuisance, we consider force to be a secondary dimension whose unit is derived from an equation based on Newton's second law, i.e.,

$$\text{Force} = (\text{Mass})\,(\text{Acceleration})$$

or

$$F = ma \tag{1-1}$$

In SI, the force unit is the newton (N), and it is defined as the *force required to accelerate a mass of 1 kg at a rate of 1 m/s²*. In the English system, the force unit is the **pound-force** (lbf) and is defined as the *force required to accelerate a mass of 32.174 lbm (1 slug) at a rate of 1 ft/s²* (Fig. 1–12). That is,

$$1 \text{ N} = 1 \text{ kg} \cdot \text{m/s}^2$$
$$1 \text{ lbf} = 32.174 \text{ lbm} \cdot \text{ft/s}^2$$

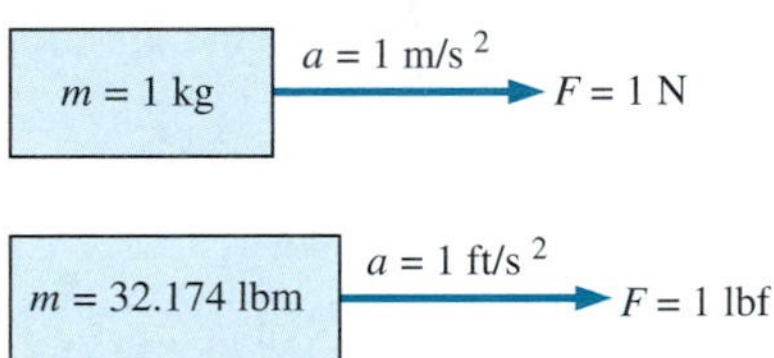

FIGURE 1–12
The definition of the force units.

A force of 1 N is roughly equivalent to the weight of a small apple (m = 102 g), whereas a force of 1 lbf is roughly equivalent to the weight of four apples (m_{total} = 454 g), as shown in Fig. 1–13. Another force unit in common use in Europe is the *kilogram-force* (kgf), which is the weight of 1 kg mass at sea level (1 kgf = 9.807 N).

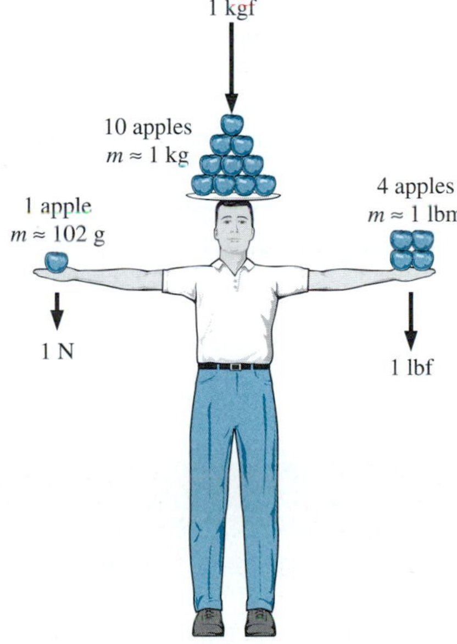

FIGURE 1–13
The relative magnitudes of the force units newton (N), kilogram-force (kgf), and pound-force (lbf).

FIGURE 1–14
A body weighing 150 lbf on earth would weigh only 25 lbf on the moon.

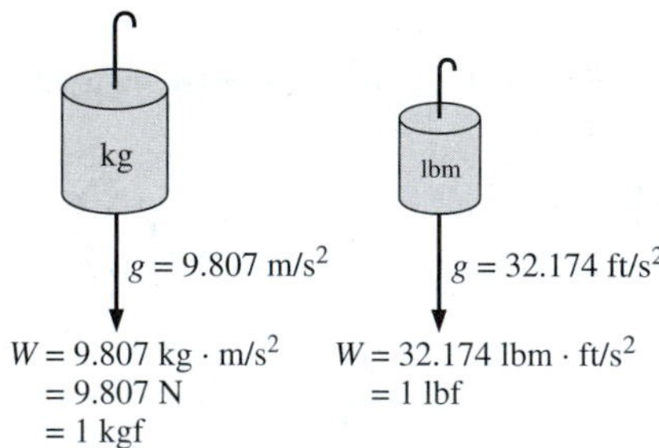

FIGURE 1–15
The weight of a unit mass at sea level.

The term **weight** is often incorrectly used to express mass, particularly by the "weight watchers." Unlike mass, weight W is a *force*. It is the gravitational force applied to a body, and its magnitude is determined from an equation based on Newton's second law,

$$W = mg \quad \text{(N)} \tag{1–2}$$

where m is the mass of the body, and g is the local gravitational acceleration (g is 9.807 m/s^2 or 32.174 ft/s^2 at sea level and 45° latitude). An ordinary bathroom scale measures the gravitational force acting on a body. The weight per unit volume of a substance is called the **specific weight** γ and is determined from $\gamma = \rho g$, where ρ is density.

The mass of a body remains the same regardless of its location in the universe. Its weight, however, changes with a change in gravitational acceleration. A body weighs less on top of a mountain since g decreases (by a small amount) with altitude. On the surface of the moon, an astronaut weighs about one-sixth of what she or he normally weighs on earth (Fig. 1–14).

At sea level a mass of 1 kg weighs 9.807 N, as illustrated in Fig. 1–15. A mass of 1 lbm, however, weighs 1 lbf, which misleads people to believe that pound-mass and pound-force can be used interchangeably as pound (lb), which is a major source of error in the English system.

It should be noted that the *gravity force* acting on a mass is due to the *attraction* between the masses, and thus it is proportional to the magnitudes of the masses and inversely proportional to the square of the distance between them. Therefore, the gravitational acceleration g at a location depends on the *local density* of the earth's crust, the *distance* to the center of the earth, and to a lesser extent, the positions of the moon and the sun. The value of g varies with location from 9.8295 m/s^2 at 4500 m below sea level to 7.3218 m/s^2 at 100,000 m above sea level. However, at altitudes up to 30,000 m, the variation of g from the sea-level value of 9.807 m/s^2 is less than 1 percent. Therefore, for most practical purposes, the gravitational acceleration can be approximated to be *constant* at 9.81 m/s^2. It is interesting to note that at locations below sea level, the value of g increases with distance below the sea level, reaches a maximum at about 4500 m, and then starts decreasing. (What do you think the value of g is at the center of the earth?)

The primary cause of confusion between mass and weight is that mass is usually measured *indirectly* by measuring the *gravity force* it exerts. This approach also assumes that the forces exerted by other effects such as air buoyancy and fluid motion are negligible. This is like measuring the distance to a star by measuring its red shift, or measuring the altitude of an airplane by measuring barometric pressure. Both of these are also indirect measurements. The correct *direct* way of measuring mass is to compare it to a known mass. This is cumbersome, however, and it is mostly used for calibration and measuring precious metals.

Work, which is a form of energy, can simply be defined as force times distance; therefore, it has the unit "newton-meter (N · m)," which is called a joule (J). That is,

$$1 \text{ J} = 1 \text{ N} \cdot \text{m} \tag{1–3}$$

A more common unit for energy in SI is the kilojoule (1 kJ = 10^3 J). In the English system, the energy unit is the **Btu** (British thermal unit), which is defined as the energy required to raise the temperature of 1 lbm of water at

68°F by 1°F. The magnitudes of the kilojoule and Btu are very nearly the same (1 Btu = 1.0551 kJ). In the metric system, the amount of energy needed to raise the temperature of 1 g of water at 14.5°C by 1°C is defined as 1 **calorie** (cal), and 1 cal = 4.1868 J. Don't confuse this calorie unit with the Calories that you eat (1 Calorie = 1000 calories).

Dimensional Homogeneity

We all know from grade school that apples and oranges do not add. But we somehow manage to do it (by mistake, of course). In engineering, all equations must be *dimensionally homogeneous*. That is, every term in an equation must have the same dimensions (Fig. 1–16). If, at some stage of an analysis, we find ourselves in a position to add two quantities that have different dimensions (or units), it is a clear indication that we have made an error at an earlier stage. So checking dimensions (or units) can serve as a valuable tool to spot errors.

FIGURE 1–16
To be dimensionally homogeneous, all the terms in an equation must have the same dimensions.

© *Reprinted with special permission of King Features Syndicate.*

EXAMPLE 1–1 Spotting Errors from Unit Inconsistencies

While solving a problem, a person ended up with the following equation at some stage:

$$E = 25 \text{ kJ} + 7 \text{ kJ/kg}$$

where E is the total energy and has the unit of kilojoules. Determine how to correct the error.

Solution During an analysis, a relation with inconsistent units is obtained. A correction is to be found, and the probable cause of the error is to be determined.

Analysis The two terms on the right-hand side do not have the same units, and therefore they cannot be added to obtain the total energy (Fig. 1–17). Multiplying the last term by mass will eliminate the kilograms in the denominator, and the whole equation will become dimensionally homogeneous; that is, every term in the equation will have the same dimensions and units.

Discussion This error was most likely caused by forgetting to multiply the last term by mass at an earlier stage.

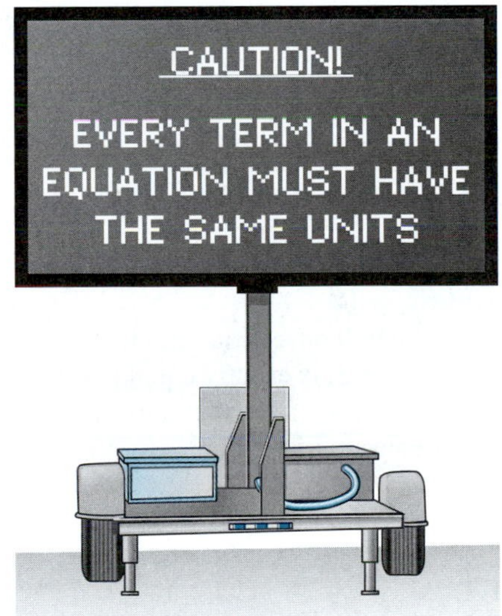

FIGURE 1–17
Always check the units in your calculations.

We all know from experience that units can give terrible headaches if they are not used carefully in solving a problem. However, with some attention and skill, units can be used to our advantage. They can be used to check formulas; sometimes they can even be used to *derive* formulas, as illustrated in the following example.

EXAMPLE 1–2 Obtaining Formulas from Unit Considerations

A tank is filled with oil whose density is $\rho = 850 \text{ kg/m}^3$. If the volume of the tank is $V = 2 \text{ m}^3$, determine the amount of mass m in the tank.

Solution The volume of an oil tank is given. The mass of oil is to be determined.

Assumptions Oil is a nearly incompressible substance and thus its density is constant.

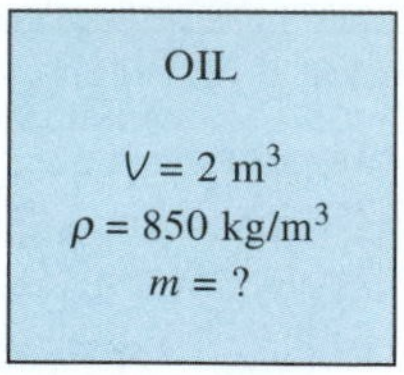

OIL

$V = 2\ \text{m}^3$

$\rho = 850\ \text{kg/m}^3$

$m = ?$

FIGURE 1–18
Schematic for Example 1–2.

Analysis A sketch of the system just described is given in Fig. 1–18. Suppose we forgot the formula that relates mass to density and volume. However, we know that mass has the unit of kilograms. That is, whatever calculations we do, we should end up with the unit of kilograms. Putting the given information into perspective, we have

$$\rho = 850\ \text{kg/m}^3 \quad \text{and} \quad V = 2\ \text{m}^3$$

It is obvious that we can eliminate m^3 and end up with kg by multiplying these two quantities. Therefore, the formula we are looking for should be

$$m = \rho V$$

Thus,

$$m = (850\ \text{kg/m}^3)(2\ \text{m}^3) = \mathbf{1700\ kg}$$

Discussion Note that this approach may not work for more complicated formulas. Nondimensional constants may also be present in the formulas, and these cannot be derived from unit considerations alone.

You should keep in mind that a formula that is not dimensionally homogeneous is definitely wrong, but a dimensionally homogeneous formula is not necessarily right.

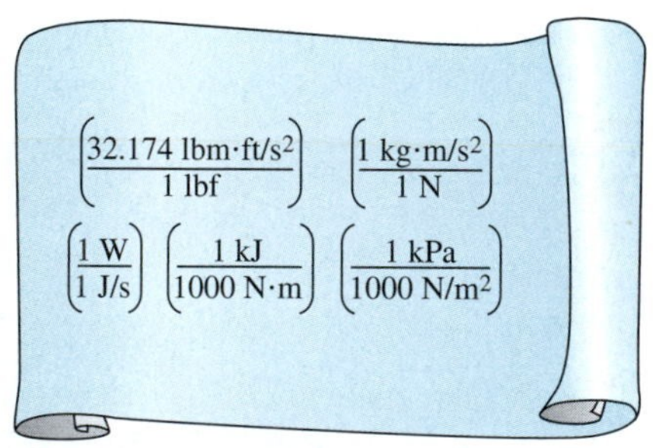

FIGURE 1–19
Every unity conversion ratio (as well as its inverse) is exactly equal to one. Shown here are a few commonly used unity conversion ratios.

Unity Conversion Ratios

Just as all nonprimary dimensions can be formed by suitable combinations of primary dimensions, *all nonprimary units (**secondary units**) can be formed by combinations of primary units.* Force units, for example, can be expressed as

$$\text{N} = \text{kg}\frac{\text{m}}{\text{s}^2} \quad \text{and} \quad \text{lbf} = 32.174\ \text{lbm}\frac{\text{ft}}{\text{s}^2}$$

They can also be expressed more conveniently as **unity conversion ratios** as

$$\frac{\text{N}}{\text{kg}\cdot\text{m/s}^2} = 1 \quad \text{and} \quad \frac{\text{lbf}}{32.174\ \text{lbm}\cdot\text{ft/s}^2} = 1$$

Unity conversion ratios are identically equal to 1 and are unitless, and thus such ratios (or their inverses) can be inserted conveniently into any calculation to properly convert units (Fig. 1–19). You are encouraged to always use unity conversion ratios such as those given here when converting units. Some textbooks insert the archaic gravitational constant g_c defined as $g_c = 32.174\ \text{lbm}\cdot\text{ft/lbf}\cdot\text{s}^2 = \text{kg}\cdot\text{m/N}\cdot\text{s}^2 = 1$ into equations in order to force units to match. This practice leads to unnecessary confusion and is strongly discouraged by the present authors. We recommend that you instead use unity conversion ratios.

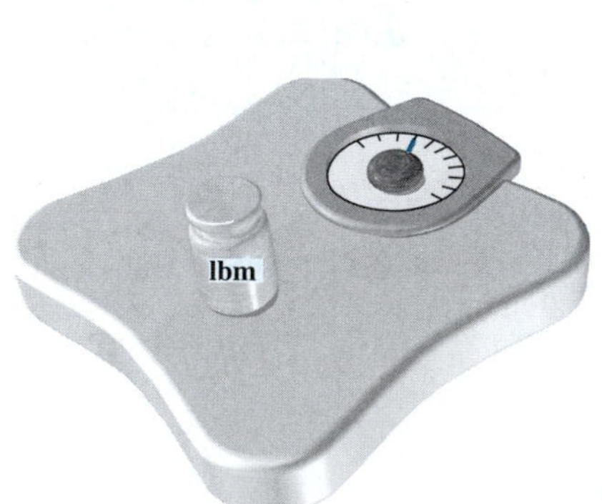

FIGURE 1–20
A mass of 1 lbm weighs 1 lbf on earth.

EXAMPLE 1–3 **The Weight of One Pound-Mass**

Using unity conversion ratios, show that 1.00 lbm weighs 1.00 lbf on earth (Fig. 1–20).

Solution A mass of 1.00 lbm is subjected to standard earth gravity. Its weight in lbf is to be determined.
Assumptions Standard sea-level conditions are assumed.
Properties The gravitational constant is $g = 32.174\ \text{ft/s}^2$.
Analysis We apply Newton's second law to calculate the weight (force) that corresponds to the known mass and acceleration. The weight of any object is equal to its mass times the local value of gravitational acceleration. Thus,

$$W = mg = (1.00\ \text{lbm})(32.174\ \text{ft/s}^2)\left(\frac{1\ \text{lbf}}{32.174\ \text{lbm}\cdot\text{ft/s}^2}\right) = \mathbf{1.00\ lbf}$$

Discussion The quantity in large parentheses in the above equation is a unity conversion ratio. Mass is the same regardless of its location. However, on some other planet with a different value of gravitational acceleration, the weight of 1 lbm would differ from that calculated here.

FIGURE 1–21
A quirk in the metric system of units.

When you buy a box of breakfast cereal, the printing may say "Net weight: One pound (454 grams)." (See Fig. 1–21.) Technically, this means that the cereal inside the box weighs 1.00 lbf on earth and has a *mass* of 453.6 g (0.4536 kg). Using Newton's second law, the actual weight of the cereal on earth is

$$W = mg = (453.6\ \text{g})(9.81\ \text{m/s}^2)\left(\frac{1\ \text{N}}{1\ \text{kg}\cdot\text{m/s}^2}\right)\left(\frac{1\ \text{kg}}{1000\ \text{g}}\right) = 4.49\ \text{N}$$

1–6 ▪ PROBLEM-SOLVING TECHNIQUE

The first step in learning any science is to grasp the fundamentals and to gain a sound knowledge of it. The next step is to master the fundamentals by testing this knowledge. This is done by solving significant real-world problems. Solving such problems, especially complicated ones, requires a systematic approach. By using a step-by-step approach, an engineer can reduce the solution of a complicated problem into the solution of a series of simple problems (Fig. 1–22). When you are solving a problem, we recommend that you use the following steps zealously as applicable. This will help you avoid some of the common pitfalls associated with problem solving.

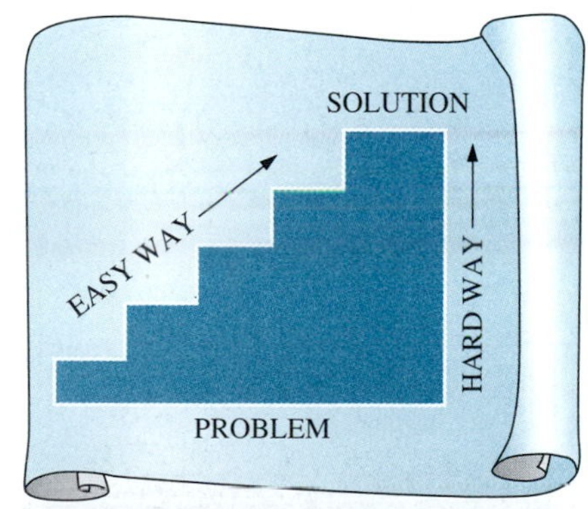

FIGURE 1–22
A step-by-step approach can greatly simplify problem solving.

Step 1: Problem Statement

In your own words, briefly state the problem, the key information given, and the quantities to be found. This is to make sure that you understand the problem and the objectives before you attempt to solve the problem.

Step 2: Schematic

Draw a realistic sketch of the physical system involved, and list the relevant information on the figure. The sketch does not have to be something elaborate, but it should resemble the actual system and show the key features. Indicate any energy and mass interactions with the surroundings. Listing the given information on the sketch helps one to see the entire problem at once.

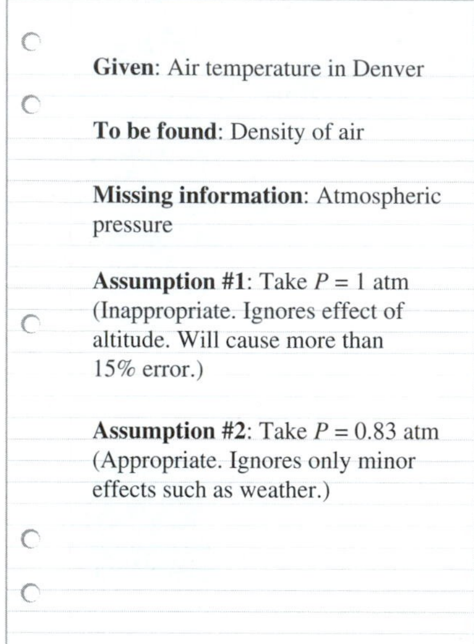

Given: Air temperature in Denver
To be found: Density of air
Missing information: Atmospheric pressure
Assumption #1: Take P = 1 atm (Inappropriate. Ignores effect of altitude. Will cause more than 15% error.)
Assumption #2: Take P = 0.83 atm (Appropriate. Ignores only minor effects such as weather.)

FIGURE 1–23

The assumptions made while solving an engineering problem must be reasonable and justifiable.

Also, check for properties that remain constant during a process (such as temperature during an isothermal process), and indicate them on the sketch.

Step 3: Assumptions and Approximations

State any appropriate assumptions and approximations made to simplify the problem to make it possible to obtain a solution. Justify the questionable assumptions. Assume reasonable values for missing quantities that are necessary. For example, in the absence of specific data for atmospheric pressure, it can be taken to be 1 atm. However, it should be noted in the analysis that the atmospheric pressure decreases with increasing elevation. For example, it drops to 0.83 atm in Denver (elevation 1610 m) (Fig. 1–23).

Step 4: Physical Laws

Apply all the relevant basic physical laws and principles (such as the conservation of mass), and reduce them to their simplest form by utilizing the assumptions made. However, the region to which a physical law is applied must be clearly identified first. For example, the increase in speed of water flowing through a nozzle is analyzed by applying conservation of mass between the inlet and outlet of the nozzle.

Step 5: Properties

Determine the unknown properties at known states necessary to solve the problem from property relations or tables. List the properties separately, and indicate their source, if applicable.

Step 6: Calculations

Substitute the known quantities into the simplified relations and perform the calculations to determine the unknowns. Pay particular attention to the units and unit cancellations, and remember that a dimensional quantity without a unit is meaningless. Also, don't give a false implication of high precision by copying all the digits from the screen of the calculator—round the results to an appropriate number of significant digits (see p. 17).

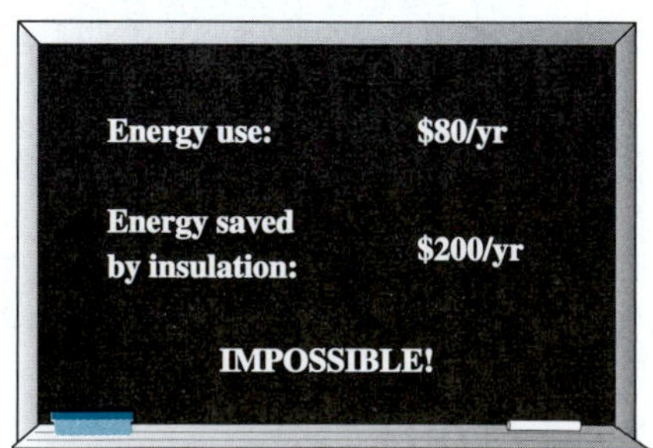

FIGURE 1–24

The results obtained from an engineering analysis must be checked for reasonableness.

Step 7: Reasoning, Verification, and Discussion

Check to make sure that the results obtained are reasonable and intuitive, and verify the validity of the questionable assumptions. Re-do the calculations that resulted in unreasonable values. For example, insulating a water heater that uses \$80 worth of natural gas a year cannot result in savings of \$200 a year (Fig. 1–24).

Also, point out the significance of the results, and discuss their implications. State the conclusions that can be drawn from the results, and any recommendations that can be made from them. Emphasize the limitations under which the results are applicable, and caution against any possible misunderstandings and using the results in situations where the underlying assumptions do not apply. For example, if you determined that wrapping a water heater with a \$20 insulation jacket will reduce the energy cost by \$30 a year, indicate that the insulation will pay for itself from the energy it saves in less than a year. However, also indicate that the analysis does not consider labor costs, and that this will be the case if you install the insulation yourself.

Keep in mind that the solutions you present to your instructors, and any engineering analysis presented to others, is a form of communication. Therefore neatness, organization, completeness, and visual appearance are of utmost importance for maximum effectiveness. Besides, neatness also serves as a great checking tool since it is very easy to spot errors and inconsistencies in neat work. Carelessness and skipping steps to save time often end up costing more time and unnecessary anxiety.

The approach described here is used in the solved example problems without explicitly stating each step, as well as in the Solutions Manual of this text. For some problems, some of the steps may not be applicable or necessary. For example, often it is not practical to list the properties separately. However, we cannot overemphasize the importance of a logical and orderly approach to problem solving. Most difficulties encountered while solving a problem are not due to a lack of knowledge; rather, they are due to a lack of organization. You are strongly encouraged to follow these steps in problem solving until you develop your own approach that works best for you.

Engineering Software Packages

You may be wondering why we are about to undertake an in-depth study of the fundamentals of another engineering science. After all, almost all such problems we are likely to encounter in practice can be solved using one of several sophisticated software packages readily available in the market today. These software packages not only give the desired numerical results, but also supply the outputs in colorful graphical form for impressive presentations. It is unthinkable to practice engineering today without using some of these packages. This tremendous computing power available to us at the touch of a button is both a blessing and a curse. It certainly enables engineers to solve problems easily and quickly, but it also opens the door for abuses and misinformation. In the hands of poorly educated people, these software packages are as dangerous as sophisticated powerful weapons in the hands of poorly trained soldiers.

Thinking that a person who can use the engineering software packages without proper training on fundamentals can practice engineering is like thinking that a person who can use a wrench can work as a car mechanic. If it were true that the engineering students do not need all these fundamental courses they are taking because practically everything can be done by computers quickly and easily, then it would also be true that the employers would no longer need high-salaried engineers since any person who knows how to use a word-processing program can also learn how to use those software packages. However, the statistics show that the need for engineers is on the rise, not on the decline, despite the availability of these powerful packages.

We should always remember that all the computing power and the engineering software packages available today are just *tools*, and tools have meaning only in the hands of masters. Having the best word-processing program does not make a person a good writer, but it certainly makes the job of a good writer much easier and makes the writer more productive (Fig. 1–25). Hand calculators did not eliminate the need to teach our children how to add

FIGURE 1–25
An excellent word-processing program does not make a person a good writer; it simply makes a good writer a more efficient writer.

or subtract, and sophisticated medical software packages did not take the place of medical school training. Neither will engineering software packages replace the traditional engineering education. They will simply cause a shift in emphasis in the courses from mathematics to physics. That is, more time will be spent in the classroom discussing the physical aspects of the problems in greater detail, and less time on the mechanics of solution procedures.

All these marvelous and powerful tools available today put an extra burden on today's engineers. They must still have a thorough understanding of the fundamentals, develop a "feel" of the physical phenomena, be able to put the data into proper perspective, and make sound engineering judgments, just like their predecessors. However, they must do it much better, and much faster, using more realistic models because of the powerful tools available today. The engineers in the past had to rely on hand calculations, slide rules, and later hand calculators and computers. Today they rely on software packages. The easy access to such power and the possibility of a simple misunderstanding or misinterpretation causing great damage make it more important today than ever to have solid training in the fundamentals of engineering. In this text we make an extra effort to put the emphasis on developing an intuitive and physical understanding of natural phenomena instead of on the mathematical details of solution procedures.

Engineering Equation Solver (EES)

EES is a program that solves systems of linear or nonlinear algebraic or differential equations numerically. It has a large library of built-in thermodynamic property functions as well as mathematical functions, and allows the user to supply additional property data. Unlike some software packages, EES does not solve engineering problems; it only solves the equations supplied by the user. Therefore, the user must understand the problem and formulate it by applying any relevant physical laws and relations. EES saves the user considerable time and effort by simply solving the resulting mathematical equations. This makes it possible to attempt significant engineering problems not suitable for hand calculations, and to conduct parametric studies quickly and conveniently. EES is a very powerful yet intuitive program that is very easy to use, as shown in Example 1–4. The use and capabilities of EES are explained in Appendix 3 on the enclosed DVD.

EXAMPLE 1–4 Solving a System of Equations with EES

The difference of two numbers is 4, and the sum of the squares of these two numbers is equal to the sum of the numbers plus 20. Determine these two numbers.

Solution Relations are given for the difference and the sum of the squares of two numbers. They are to be determined.

Analysis We start the EES program by double-clicking on its icon, open a new file, and type the following on the blank screen that appears:

```
x-y = 4

x^2+y^2 = x+y+20
```

which is an exact mathematical expression of the problem statement with x and y denoting the unknown numbers. The solution to this system of two nonlinear equations with two unknowns is obtained by a single click on the "calculator" icon on the taskbar. It gives

$$x=5 \quad \text{and} \quad y=1$$

Discussion Note that all we did is formulate the problem as we would on paper; EES took care of all the mathematical details of the solution. Also note that equations can be linear or nonlinear, and they can be entered in any order with unknowns on either side. Friendly equation solvers such as EES allow the user to concentrate on the physics of the problem without worrying about the mathematical complexities associated with the solution of the resulting system of equations.

A Remark on Significant Digits

In engineering calculations, the information given is not known to more than a certain number of significant digits, usually three digits. Consequently, the results obtained cannot possibly be accurate to more significant digits. Reporting results in more significant digits implies greater accuracy than exists, and it should be avoided.

For example, consider a 3.75-L container filled with gasoline whose density is 0.845 kg/L, and determine its mass. Probably the first thought that comes to your mind is to multiply the volume and density to obtain 3.16875 kg for the mass, which falsely implies that the mass determined is accurate to six significant digits. In reality, however, the mass cannot be more accurate than three significant digits since both the volume and the density are accurate to three significant digits only. Therefore, the result should be rounded to three significant digits, and the mass should be reported to be 3.17 kg instead of what appears in the screen of the calculator. The result 3.16875 kg would be correct only if the volume and density were given to be 3.75000 L and 0.845000 kg/L, respectively. The value 3.75 L implies that we are fairly confident that the volume is accurate within $\pm$0.01 L, and it cannot be 3.74 or 3.76 L. However, the volume can be 3.746, 3.750, 3.753, etc., since they all round to 3.75 L (Fig. 1–26). It is more appropriate to retain all the digits during *intermediate* calculations, and to do the rounding in the final step. This is also what a computer will normally do.

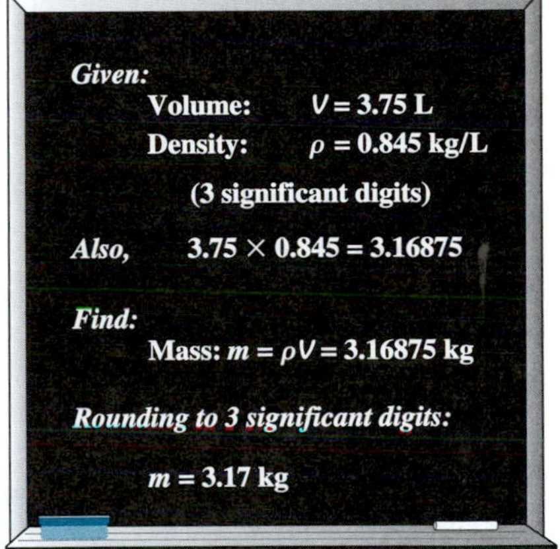

FIGURE 1–26

A result with more significant digits than that of given data falsely implies more accuracy.

When solving problems, we typically assume the given information to be accurate to at least three significant digits. Therefore, if the length of a pipe is given to be 40 m, we will assume it to be 40.0 m in order to justify using three significant digits in the final results. You should also keep in mind that all experimentally determined values are subject to measurement errors, and such errors will reflect in the results obtained. For example, if the density of a substance has an uncertainty of 2 percent, then the mass determined using this density value will also have an uncertainty of 2 percent.

You should also be aware that we sometimes knowingly introduce small errors in order to avoid the trouble of searching for more accurate data. For example, when dealing with liquid water, we just use the value of 1000 kg/m^3

for density, which is the density value of pure water at 0°C. Using this value at 75°C will result in an error of 2.5 percent since the density at this temperature is 975 kg/m^3. The minerals and impurities in the water will introduce additional error. This being the case, you should have no reservation in rounding the final results to a reasonable number of significant digits. Besides, having a few percent uncertainty in the results of engineering analysis is usually the norm, not the exception.

SUMMARY

In this chapter, some basic concepts of thermal-fluid sciences are introduced and discussed. The physical sciences that deal with energy and the transfer, transport, and conversion of energy are referred to as *thermal-fluid sciences,* and they are studied under the subcategories of thermodynamics, heat transfer, and fluid mechanics.

Thermodynamics is the science that primarily deals with energy. The *first law of thermodynamics* is simply an expression of the conservation of energy principle, and it asserts that *energy* is a thermodynamic property. The *second law of thermodynamics* asserts that energy has *quality* as well as *quantity,* and actual processes occur in the direction of decreasing quality of energy. Determining the rates of heat transfer to or from a system and thus the times of cooling or heating, as well as the variation of the temperature, is the subject of *heat transfer.* The basic requirement for heat transfer is the presence of a *temperature difference.* A substance in the liquid or gas phase is referred to as a *fluid. Fluid mechanics* is the science that deals with the behavior of fluids at rest (*fluid statics*) or in motion (*fluid dynamics*), and the interaction of fluids with solids or other fluids at the boundaries.

In engineering calculations, it is important to pay particular attention to the units of the quantities to avoid errors caused by inconsistent units, and to follow a systematic approach. It is also important to recognize that the information given is not known to more than a certain number of significant digits, and the results obtained cannot possibly be accurate to more significant digits.

When solving a problem, it is recommended that a step-by-step approach be used. Such an approach involves stating the problem, drawing a schematic, making appropriate assumptions, applying the physical laws, listing the relevant properties, making the necessary calculations, and making sure that the results are reasonable.

REFERENCES AND SUGGESTED READINGS

1. American Society for Testing and Materials. *Standards for Metric Practice.* ASTM E 380-79, January 1980.
2. Y. A. Çengel. *Heat and Mass Transfer: A Practical Approach.* 3rd ed. New York: McGraw-Hill, 2007.
3. Y. A. Çengel and M. A. Boles. *Thermodynamics: An Engineering Approach.* 6th ed. New York: McGraw-Hill, 2008.
3. Y. A. Çengel and John M. Cimbala. *Fluid Mechanics: Fundamentals and Applications.* NewYork: McGraw-Hill, 2006.

PROBLEMS*

Thermodynamics, Heat Transfer, and Fluid Mechanics

1–1C What is the difference between the classical and the statistical approaches to thermodynamics?

1–2C Why does a bicyclist pick up speed on a downhill road even when he is not pedaling? Does this violate the conservation of energy principle?

1–3C An office worker claims that a cup of cold coffee on his table warmed up to 80°C by picking up energy from the surrounding air, which is at 25°C. Is there any truth to his claim? Does this process violate any thermodynamic laws?

*Problems designated by a "C" are concept questions, and students are encouraged to answer them all. Problems designated by an "E" are in English units, and the SI users can ignore them. Problems with an icon are solved using EES, and complete solutions together with parametric studies are included on the enclosed DVD. Problems with the icon are comprehensive in nature, and are intended to be solved with a computer, preferably using the EES software that accompanies this text.

1–4C One of the most amusing things a person can experience is that in certain parts of the world a still car on neutral going uphill when its brakes are released. Such occurrences are even broadcast on TV. Can this really happen or is it a bad eyesight? How can you verify if a road is really uphill or downhill?

1–5C How does the science of heat transfer differ from the science of thermodynamics?

1–6C What is the driving force for (*a*) heat transfer, (*b*) electric current, and (*c*) fluid flow?

1–7C Why is heat transfer a nonequilibrium phenomenon?

1–8C Can there be any heat transfer between two bodies that are at the same temperature but at different pressures?

1–9C Define stress, normal stress, shear stress, and pressure.

Mass, Force, and Units

1–10C What is the difference between pound-mass and pound-force?

1–11C Explain why the light-year has the dimension of length.

1–12C What is the net force acting on a car cruising at a constant velocity of 70 km/h (*a*) on a level road and (*b*) on an uphill road?

1–13E A man weighs 180 lbf at a location where $g = 32.10 \text{ ft/s}^2$. Determine his weight on the moon, where $g = 5.47 \text{ ft/s}^2$. *Answer: 30.7 lbf*

1–14 Determine the mass and the weight of the air contained in a room whose dimensions are 6 m × 6 m × 8 m. Assume the density of the air is 1.16 kg/m^3. *Answers: 334.1 kg, 3277 N*

1–15 At 45° latitude, the gravitational acceleration as a function of elevation z above sea level is given by $g = a - bz$, where $a = 9.807 \text{ m/s}^2$ and $b = 3.32 \times 10^{-6} \text{ s}^{-2}$. Determine the height above sea level where the weight of an object will decrease by 1 percent. *Answer: 29,539 m*

1–16E If the mass of an object is 10 lbm, what is its weight, in lbf, at a location where $g = 32.0 \text{ ft/s}^2$?

1–17 The acceleration of high-speed aircraft is sometimes expressed in *g*'s (in multiples of the standard acceleration of gravity). Determine the net upward force, in N, that a 90-kg man would experience in an aircraft whose acceleration is 6 *g*'s.

1–18 A 5-kg rock is thrown upward with a force of 150 N at a location where the local gravitational acceleration is 9.79 m/s^2. Determine the acceleration of the rock, in m/s^2.

1–19 Solve Prob. 1–18 using EES (or other) software. Print out the entire solution, including the numerical results with proper units.

1–20 The value of the gravitational acceleration g decreases with elevation from 9.807 m/s^2 at sea level to 9.767 m/s^2 at an altitude of 13,000 m, where large passenger planes cruise. Determine the percent reduction in the weight of an airplane cruising at 13,000 m relative to its weight at sea level.

Modeling and Solving Engineering Problems

1–21C How do rating problems in heat transfer differ from the sizing problems?

1–22C What is the difference between the analytical and experimental approach to engineering problems? Discuss the advantages and disadvantages of each approach?

1–23C What is the importance of modeling in engineering? How are the mathematical models for engineering processes prepared?

1–24C When modeling an engineering process, how is the right choice made between a simple but crude and a complex but accurate model? Is the complex model necessarily a better choice since it is more accurate?

Solving Engineering Problems and EES

1–25C What is the value of the engineering software packages in (*a*) engineering education and (*b*) engineering practice?

1–26 Determine a positive real root of this equation using EES:

$$2x^3 - 10x^{0.5} - 3x = -3$$

1–27 Solve this system of two equations with two unknowns using EES:

$$x^3 - y^2 = 7.75$$

$$3xy + y = 3.5$$

1–28 Solve this system of three equations with three unknowns using EES:

$$2x - y + z = 5$$

$$3x^2 + 2y = z + 2$$

$$xy + 2z = 8$$

1–29 Solve this system of three equations with three unknowns using EES:

$$x^2y - z = 1$$

$$x - 3y^{0.5} + xz = -2$$

$$x + y - z = 2$$

1–30E Specific heat is defined as the amount of energy needed to increase the temperature of a unit mass of a substance by one degree. The specific heat of water at room temperature is 4.18 kJ/kg · °C in SI unit system. Using the unit conversion function capability of EES, express the specific heat of water in (*a*) kJ/kg · K, (*b*) Btu/lbm · °F, (*c*) Btu/lbm · R, and (*d*) kcal/kg · °C units. *Answers:* (*a*) 4.18, (*b*) (*c*) (*d*) 0.9984

Review Problems

1–31 A lunar exploration module weighs 4000 N at a location where $g = 9.8$ m/s^2. Determine the weight of this module in newtons when it is on the moon where $g = 1.64$ m/s^2.

1–32 The weight of bodies may change somewhat from one location to another as a result of the variation of the gravitational acceleration g with elevation. Accounting for this variation using the relation in Prob. 1–15, determine the weight of an 80-kg person at sea level ($z = 0$), in Denver ($z = 1610$ m), and on the top of Mount Everest ($z = 8848$ m).

1–33E A man goes to a traditional market to buy a steak for dinner. He finds a 12-oz steak (1 lbm = 16 oz) for $3.15. He then goes to the adjacent international market and finds a 320-g steak of identical quality for $2.80. Which steak is the better buy?

1–34E The reactive force developed by a jet engine to push an airplane forward is called thrust, and the thrust developed by the engine of a Boeing 777 is about 85,000 lbf. Express this thrust in N and kgf.

Design and Essay Problems

1–35 Write an essay on the various mass- and volume-measurement devices used throughout history. Also, explain the development of the modern units for mass and volume.

PART 1

THERMODYNAMICS

Chapter 2

INTRODUCTION AND BASIC CONCEPTS

Every science has a unique vocabulary associated with it, and thermodynamics is no exception. Precise definition of basic concepts forms a sound foundation for the development of a science and prevents possible misunderstandings. We start this chapter with a discussion of some basic concepts such as *system, state, state postulate, equilibrium, and process*. We also discuss *temperature* and *temperature scales* with particular emphasis on the International Temperature Scale of 1990. We then present *pressure,* which is the normal force exerted by a fluid per unit area and discuss *absolute* and *gage* pressures, the variation of pressure with depth, and pressure measurement devices, such as manometers and barometers. Careful study of these concepts is essential for a good understanding of the topics in the following chapters.

Objectives

The objectives of this chapter are to:

- Identify the unique vocabulary associated with thermodynamics through the precise definition of basic concepts to form a sound foundation for the development of the principles of thermodynamics.
- Explain the basic concepts of thermodynamics such as system, state, state postulate, equilibrium, process, and cycle.
- Review concepts of temperature, temperature scales, pressure, and absolute and gage pressure.

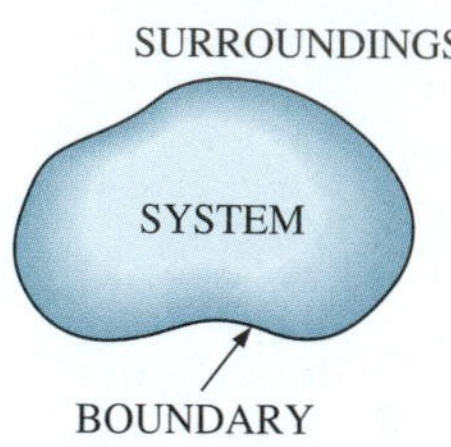

FIGURE 2–1

System, surroundings, and boundary.

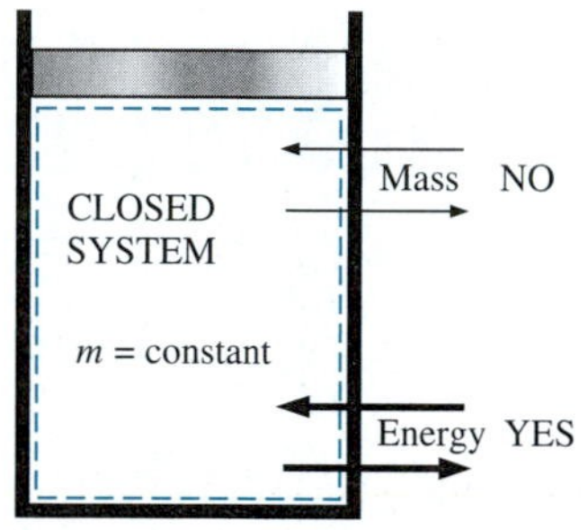

FIGURE 2–2

Mass cannot cross the boundaries of a closed system, but energy can.

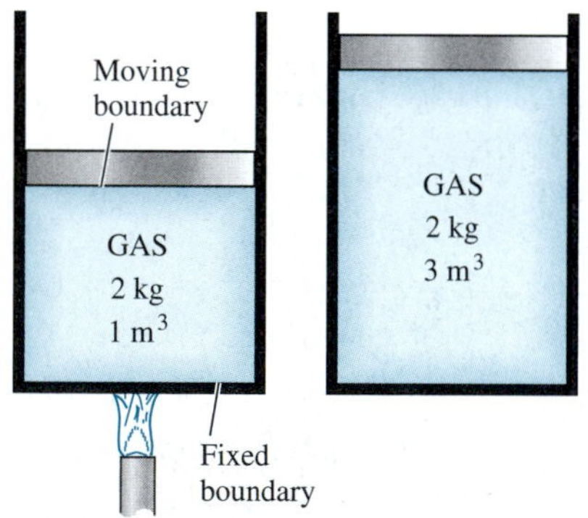

FIGURE 2–3

A closed system with a moving boundary.

2–1 ▪ SYSTEMS AND CONTROL VOLUMES

A system is defined as a *quantity of matter or a region in space chosen for study*. The mass or region outside the system is called the **surroundings**. The real or imaginary surface that separates the system from its surroundings is called the **boundary**. These terms are illustrated in Fig. 2–1. The boundary of a system can be *fixed* or *movable*. Note that the boundary is the contact surface shared by both the system and the surroundings. Mathematically speaking, the boundary has zero thickness, and thus it can neither contain any mass nor occupy any volume in space.

Systems may be considered to be *closed* or *open*, depending on whether a fixed mass or a fixed volume in space is chosen for study. A **closed system** (also known as a **control mass**) consists of a fixed amount of mass, and no mass can cross its boundary. That is, no mass can enter or leave a closed system, as shown in Fig. 2–2. But energy, in the form of heat or work, can cross the boundary; and the volume of a closed system does not have to be fixed. If, as a special case, even energy is not allowed to cross the boundary, that system is called an **isolated system**.

Consider the piston-cylinder device shown in Fig. 2–3. Let us say that we would like to find out what happens to the enclosed gas when it is heated. Since we are focusing our attention on the gas, it is our system. The inner surfaces of the piston and the cylinder form the boundary, and since no mass is crossing this boundary, it is a closed system. Notice that energy may cross the boundary, and part of the boundary (the inner surface of the piston, in this case) may move. Everything outside the gas, including the piston and the cylinder, is the surroundings.

An **open system**, or a **control volume**, as it is often called, is a properly selected region in space. It usually encloses a device that involves mass flow such as a compressor, turbine, or nozzle. Flow through these devices is best studied by selecting the region within the device as the control volume. Both mass and energy can cross the boundary of a control volume.

A large number of engineering problems involve mass flow in and out of a system and, therefore, are modeled as *control volumes*. A water heater, a car radiator, a turbine, and a compressor all involve mass flow and should be analyzed as control volumes (open systems) instead of as control masses (closed systems). In general, *any arbitrary region in space* can be selected as a control volume. There are no concrete rules for the selection of control volumes, but the proper choice certainly makes the analysis much easier. If we were to analyze the flow of air through a nozzle, for example, a good choice for the control volume would be the region within the nozzle.

The boundaries of a control volume are called a *control surface*, and they can be real or imaginary. In the case of a nozzle, the inner surface of the nozzle forms the real part of the boundary, and the entrance and exit areas form the imaginary part, since there are no physical surfaces there (Fig. 2–4*a*).

A control volume can be fixed in size and shape, as in the case of a nozzle, or it may involve a moving boundary, as shown in Fig. 2–4*b*. Most control volumes, however, have fixed boundaries and thus do not involve any moving boundaries. A control volume can also involve heat and work interactions just as a closed system, in addition to mass interaction.

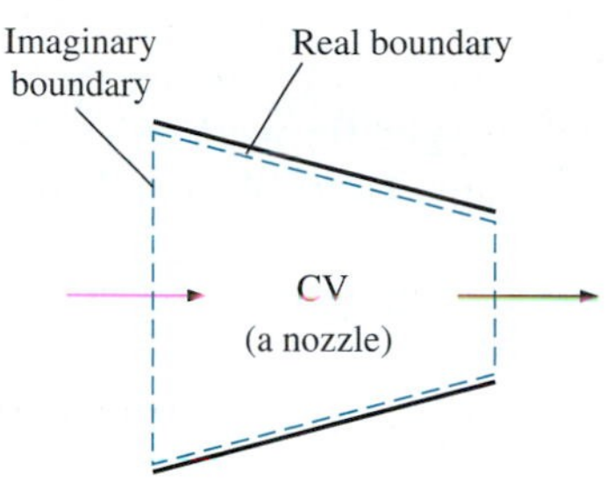

(*a*) A control volume with real and imaginary boundaries

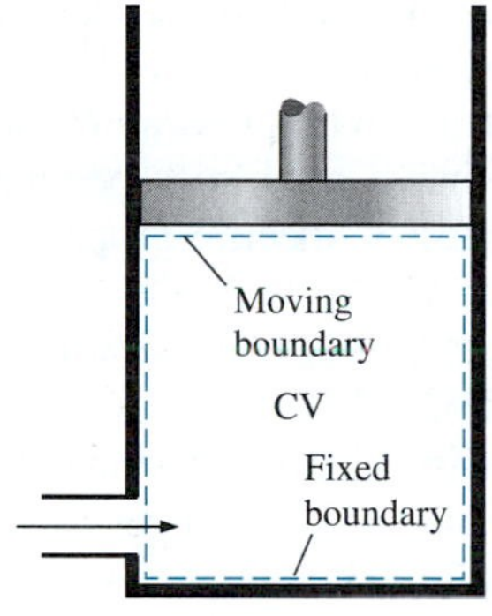

(*b*) A control volume with fixed and moving boundaries

FIGURE 2–4

A control volume can involve fixed, moving, real, and imaginary boundaries.

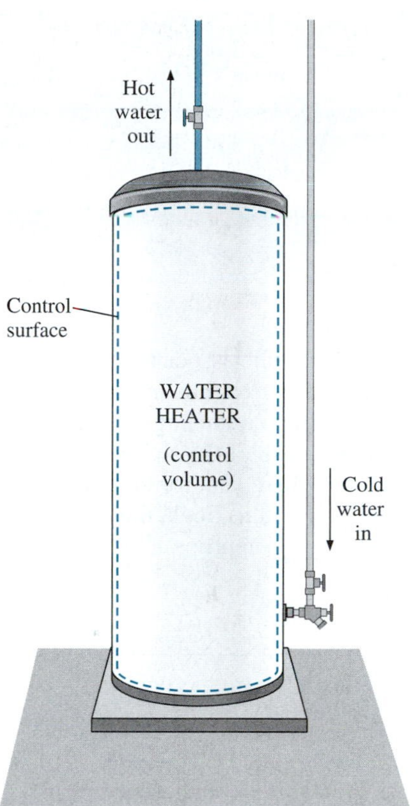

FIGURE 2–5

An open system (a control volume) with one inlet and one exit.

As an example of an open system, consider the water heater shown in Fig. 2–5. Let us say that we would like to determine how much heat we must transfer to the water in the tank in order to supply a steady stream of hot water. Since hot water will leave the tank and be replaced by cold water, it is not convenient to choose a fixed mass as our system for the analysis. Instead, we can concentrate our attention on the volume formed by the interior surfaces of the tank and consider the hot and cold water streams as mass leaving and entering the control volume. The interior surfaces of the tank form the control surface for this case, and mass is crossing the control surface at two locations.

In an engineering analysis, the system under study *must* be defined carefully. In most cases, the system investigated is quite simple and obvious, and defining the system may seem like a tedious and unnecessary task. In other cases, however, the system under study may be rather involved, and a proper choice of the system may greatly simplify the analysis.

2–2 ▪ PROPERTIES OF A SYSTEM

Any characteristic of a system is called a **property**. Some familiar properties are pressure P, temperature T, volume V, and mass m. The list can be extended to include less familiar ones such as viscosity, thermal conductivity, modulus of elasticity, thermal expansion coefficient, electric resistivity, and even velocity and elevation.

Properties are considered to be either *intensive* or *extensive*. **Intensive properties** are those that are independent of the mass of the system, such as temperature, pressure, and density. **Extensive properties** are those whose values depend on the size—or extent—of the system. Total mass, total volume V, and total momentum are some examples of extensive properties. An easy way to determine whether a property is intensive or extensive is to divide the system into two equal parts with an imaginary partition, as shown in Fig. 2–6. Each part will have the same value of intensive properties as the original system, but half the value of the extensive properties.

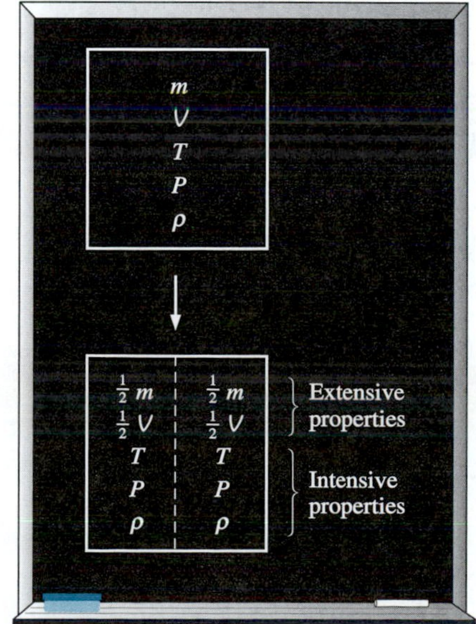

FIGURE 2–6

Criteria to differentiate intensive and extensive properties.

FIGURE 2–7
The length scale associated with most flows, such as seagulls in flight, is orders of magnitude larger than the mean free path of the air molecules. Therefore, here, and for all fluid flows considered in this book, the continuum idealization is appropriate.

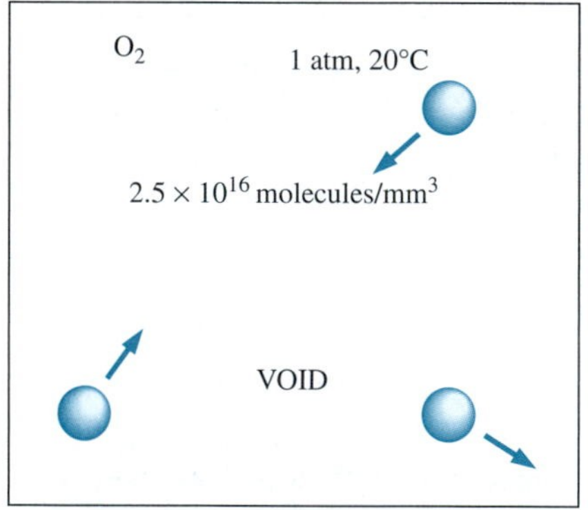

FIGURE 2–8
Despite the large gaps between molecules, a substance can be treated as a continuum because of the very large number of molecules even in an extremely small volume.

Generally, uppercase letters are used to denote extensive properties (with mass m being a major exception), and lowercase letters are used for intensive properties (with pressure P and temperature T being the obvious exceptions).

Extensive properties per unit mass are called **specific properties**. Some examples of specific properties are specific volume ($v = V/m$) and specific total energy ($e = E/m$).

Continuum

Matter is made up of atoms that are widely spaced in the gas phase. Yet it is very convenient to disregard the atomic nature of a substance and view it as continuous, homogeneous matter with no holes, that is, a **continuum**. The continuum idealization allows us to treat properties as point functions and to assume that the properties vary continually in space with no jump discontinuities. This idealization is valid as long as the size of the system we deal with is large relative to the space between the molecules (Fig. 2–7). This is the case in practically all problems, except some specialized ones. The continuum idealization is implicit in many statements we make, such as "the density of water in a glass is the same at any point."

To have a sense of the distances involved at the molecular level, consider a container filled with oxygen at atmospheric conditions. The diameter of the oxygen molecule is about 3×10^{-10} m and its mass is 5.3×10^{-26} kg. Also, the *mean free path* of oxygen at 1 atm pressure and 20°C is 6.3×10^{-8} m. That is, an oxygen molecule travels, on average, a distance of 6.3×10^{-8} m (about 200 times its diameter) before it collides with another molecule.

Also, there are about 2.5×10^{16} molecules of oxygen in the tiny volume of 1 mm^3 at 1 atm pressure and 20°C (Fig. 2–8). The continuum model is applicable as long as the characteristic length of the system (such as its diameter) is much larger than the mean free path of the molecules. At very high vacuums or very high elevations, the mean free path may become large (for example, it is about 0.1 m for atmospheric air at an elevation of 100 km). For such cases the **rarefied gas flow theory** should be used, and the impact of individual molecules should be considered. In this text we limit our consideration to substances that can be modeled as a continuum.

2–3 ▪ DENSITY AND SPECIFIC GRAVITY

Density is defined as *mass per unit volume* (Fig. 2–9). That is,

Density:
$$\rho = \frac{m}{V} \qquad (\text{kg/m}^3) \tag{2–1}$$

The reciprocal of density is the **specific volume** v, which is defined as *volume per unit mass*. That is,

$$v = \frac{V}{m} = \frac{1}{\rho} \tag{2–2}$$

For a differential volume element of mass δm and volume δV, density can be expressed as $\rho = \delta m/\delta V$.

The density of a substance, in general, depends on temperature and pressure. The density of most gases is proportional to pressure and inversely proportional to temperature. Liquids and solids, on the other hand, are essentially incompressible substances, and the variation of their density with pressure is usually negligible. At 20°C, for example, the density of water changes from 998 kg/m^3 at 1 atm to 1003 kg/m^3 at 100 atm, a change of just 0.5 percent. The density of liquids and solids depends more strongly on temperature than it does on pressure. At 1 atm, for example, the density of water changes from 998 kg/m^3 at 20°C to 975 kg/m^3 at 75°C, a change of 2.3 percent, which can still be neglected in many engineering analyses.

Sometimes the density of a substance is given relative to the density of a well-known substance. Then it is called **specific gravity**, or **relative density**, and is defined as *the ratio of the density of a substance to the density of some standard substance at a specified temperature* (usually water at 4°C, for which $\rho_{H_2O} = 1000$ kg/m^3). That is,

Specific gravity:
$$\text{SG} = \frac{\rho}{\rho_{H_2O}} \tag{2–3}$$

Note that the specific gravity of a substance is a dimensionless quantity. However, in SI units, the numerical value of the specific gravity of a substance is exactly equal to its density in g/cm^3 or kg/L (or 0.001 times the density in kg/m^3) since the density of water at 4°C is 1 g/cm^3 = 1 kg/L = 1000 kg/m^3. The specific gravity of mercury at 20°C, for example, is 13.6. Therefore, its density at 20°C is 13.6 g/cm^3 = 13.6 kg/L = 13,600 kg/m^3. The specific gravities of some substances at 20°C are given in Table 2–1. Note that substances with specific gravities less than 1 are lighter than water, and thus they would float on water (if immiscible).

The weight of a unit volume of a substance is called **specific weight**, or **weight density**, and is expressed as

Specific weight:
$$\gamma_s = \rho g \qquad (\text{N/m}^3) \tag{2–4}$$

where g is the gravitational acceleration.

The densities of liquids are essentially constant, and thus liquids can often be approximated as being incompressible substances during most processes without sacrificing much in accuracy.

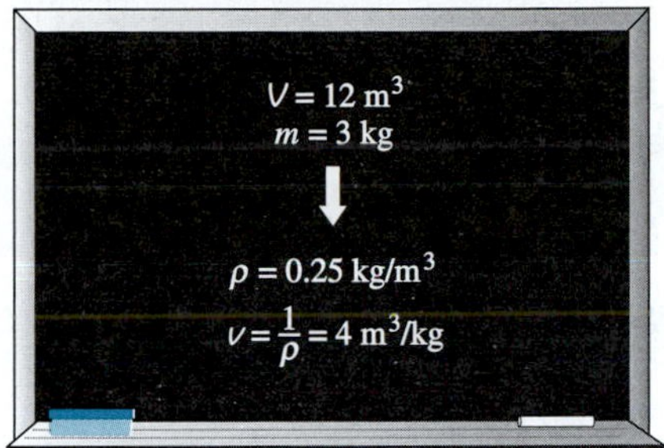

FIGURE 2–9
Density is mass per unit volume; specific volume is volume per unit mass.

TABLE 2–1
The specific gravity of some substances at 20°C and 1 atm unless stated otherwise

Substance	SG
Water	1.0
Blood (at 37°C)	1.06
Seawater	1.025
Gasoline	0.68
Ethyl alcohol	0.790
Mercury	13.6
Balsa wood	0.17
Dense oak wood	0.93
Gold	19.3
Bones	1.7–2.0
Ice (at 0° C)	0.916
Air	0.001204

2–4 ▪ STATE AND EQUILIBRIUM

Consider a system not undergoing any change. At this point, all the properties can be measured or calculated throughout the entire system, which gives us a set of properties that completely describes the condition, or the **state**, of the system. At a given state, all the properties of a system have fixed values. If the value of even one property changes, the state will change to a different one. In Fig. 2–10 a system is shown at two different states.

Thermodynamics deals with *equilibrium* states. The word **equilibrium** implies a state of balance. In an equilibrium state there are no unbalanced potentials (or driving forces) within the system. A system in equilibrium experiences no changes when it is isolated from its surroundings.

There are many types of equilibrium, and a system is not in thermodynamic equilibrium unless the conditions of all the relevant types of equilibrium are satisfied. For example, a system is in **thermal equilibrium** if the temperature is the same throughout the entire system, as shown in Fig. 2–11. That is, the system involves no temperature differential, which is the driving

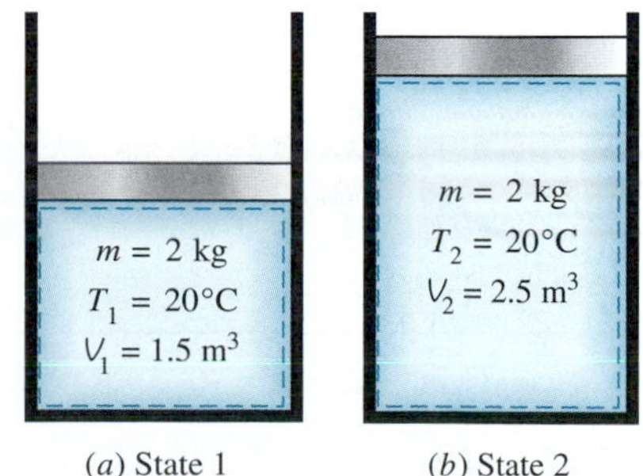

FIGURE 2–10
A system at two different states.

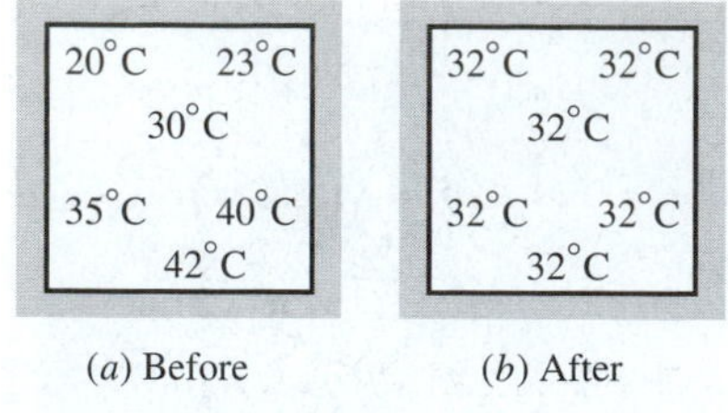

FIGURE 2–11

A closed system reaching thermal equilibrium.

force for heat flow. **Mechanical equilibrium** is related to pressure, and a system is in mechanical equilibrium if there is no change in pressure at any point of the system with time. However, the pressure may vary within the system with elevation as a result of gravitational effects. For example, the higher pressure at a bottom layer is balanced by the extra weight it must carry, and, therefore, there is no imbalance of forces. The variation of pressure as a result of gravity in most thermodynamic systems is relatively small and usually disregarded. If a system involves two phases, it is in **phase equilibrium** when the mass of each phase reaches an equilibrium level and stays there. Finally, a system is in **chemical equilibrium** if its chemical composition does not change with time, that is, no chemical reactions occur. A system is not in equilibrium unless all the relevant equilibrium criteria are satisfied.

The State Postulate

As noted earlier, the state of a system is described by its properties. But we know from experience that we do not need to specify all the properties in order to fix a state. Once a sufficient number of properties are specified, the rest of the properties assume certain values automatically. That is, specifying a certain number of properties is sufficient to fix a state. The number of properties required to fix the state of a system is given by the **state postulate**:

> The state of a simple compressible system is completely specified by two independent, intensive properties.

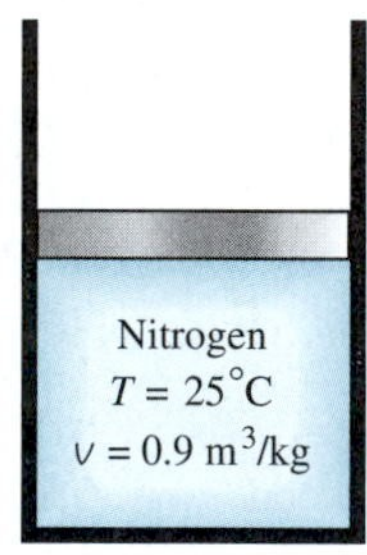

FIGURE 2–12

The state of nitrogen is fixed by two independent, intensive properties.

A system is called a **simple compressible system** in the absence of electrical, magnetic, gravitational, motion, and surface tension effects. These effects are due to external force fields and are negligible for most engineering problems. Otherwise, an additional property needs to be specified for each effect that is significant. If the gravitational effects are to be considered, for example, the elevation z needs to be specified in addition to the two properties necessary to fix the state.

The state postulate requires that the two properties specified be independent to fix the state. Two properties are **independent** if one property can be varied while the other one is held constant. Temperature and specific volume, for example, are always independent properties, and together they can fix the state of a simple compressible system (Fig. 2–12). Temperature and pressure, however, are independent properties for single-phase systems, but are dependent properties for multiphase systems. At sea level (P = 1 atm), water boils at 100°C, but on a mountaintop where the pressure is lower, water boils at a lower temperature. That is, $T = f(P)$ during a phase-change process; thus, temperature and pressure are not sufficient to fix the state of a two-phase system. Phase-change processes are discussed in detail in Chap. 4.

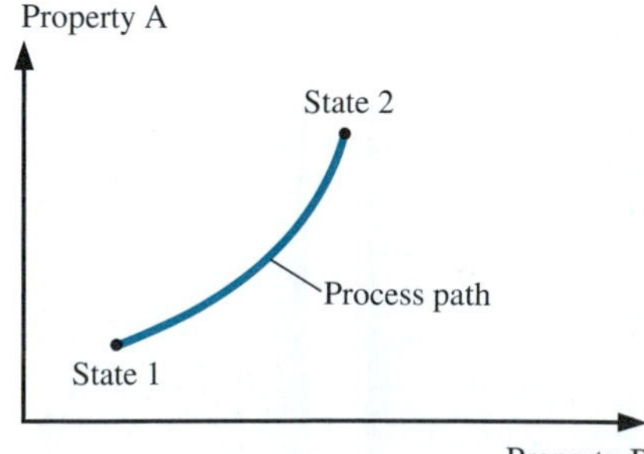

FIGURE 2–13

A process between states 1 and 2 and the process path.

2–5 ▪ PROCESSES AND CYCLES

Any change that a system undergoes from one equilibrium state to another is called a **process**, and the series of states through which a system passes during a process is called the **path** of the process (Fig. 2–13). To describe a process completely, one should specify the initial and final states of the process, as well as the path it follows, and the interactions with the surroundings.

When a process proceeds in such a manner that the system remains infinitesimally close to an equilibrium state at all times, it is called a **quasi-static**, or **quasi-equilibrium**, **process**. A quasi-equilibrium process can be viewed as a sufficiently slow process that allows the system to adjust itself internally so that properties in one part of the system do not change any faster than those at other parts.

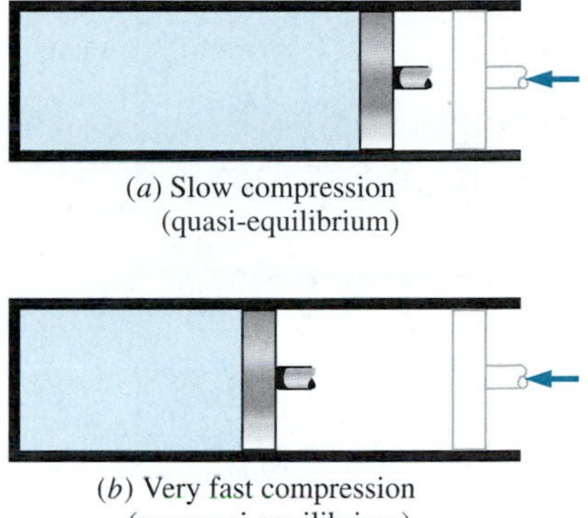

FIGURE 2–14

Quasi-equilibrium and nonquasi-equilibrium compression processes.

This is illustrated in Fig. 2–14. When a gas in a piston-cylinder device is compressed suddenly, the molecules near the face of the piston do not have enough time to escape and they will have to pile up in a small region in front of the piston, thus creating a high-pressure region there. Because of this pressure difference, the system can no longer be said to be in equilibrium, and this makes the entire process nonquasi-equilibrium. However, if the piston is moved slowly, the molecules have sufficient time to redistribute and there will not be a molecule pileup in front of the piston. As a result, the pressure inside the cylinder will always be nearly uniform and will rise at the same rate at all locations. Since equilibrium is maintained at all times, this is a quasi-equilibrium process.

It should be pointed out that a quasi-equilibrium process is an idealized process and is not a true representation of an actual process. But many actual processes closely approximate it, and they can be modeled as quasi-equilibrium with negligible error. Engineers are interested in quasiequilibrium processes for two reasons. First, they are easy to analyze; second, work-producing devices deliver the most work when they operate on quasi-equilibrium processes. Therefore, quasi-equilibrium processes serve as standards to which actual processes can be compared.

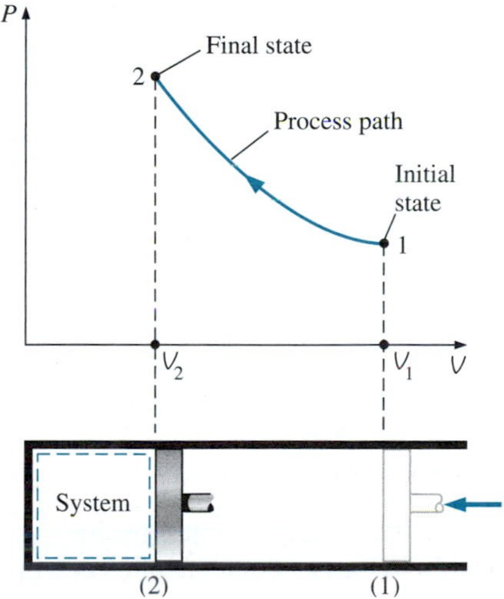

FIGURE 2–15

The *P*-*V* diagram of a compression process.

Process diagrams plotted by employing thermodynamic properties as coordinates are very useful in visualizing the processes. Some common properties that are used as coordinates are temperature *T*, pressure *P*, and volume *V* (or specific volume *v*). Figure 2–15 shows the *P*-*V* diagram of a compression process of a gas.

Note that the process path indicates a series of equilibrium states through which the system passes during a process and has significance for quasi-equilibrium processes only. For nonquasi-equilibrium processes, we are not able to characterize the entire system by a single state, and thus we cannot speak of a process path for a system as a whole. A nonquasi-equilibrium process is denoted by a dashed line between the initial and final states instead of a solid line.

The prefix *iso-* is often used to designate a process for which a particular property remains constant. An **isothermal process**, for example, is a process during which the temperature *T* remains constant; an **isobaric process** is a process during which the pressure *P* remains constant; and an **isochoric** (or **isometric**) **process** is a process during which the specific volume *v* remains constant.

A system is said to have undergone a **cycle** if it returns to its initial state at the end of the process. That is, for a cycle the initial and final states are identical.

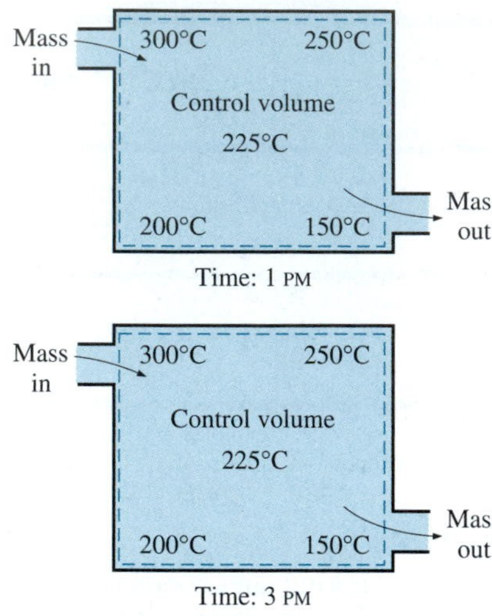

FIGURE 2–16

During a steady-flow process, fluid properties within the control volume may change with position but not with time.

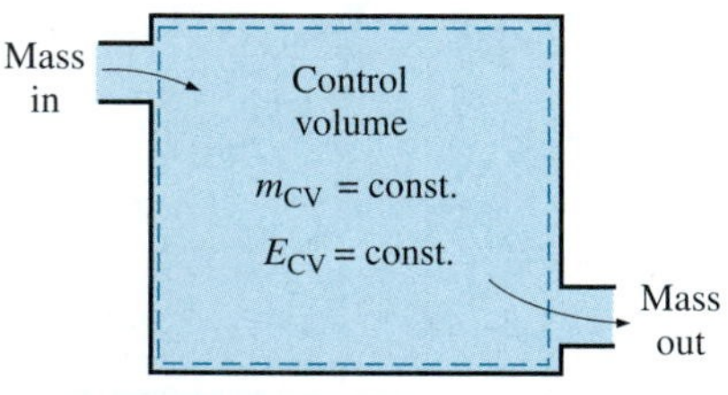

FIGURE 2–17

Under steady-flow conditions, the mass and energy contents of a control volume remain constant.

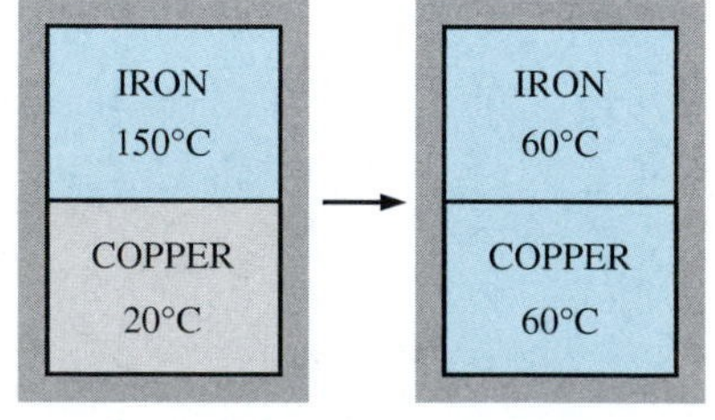

FIGURE 2–18

Two bodies reaching thermal equilibrium after being brought into contact in an isolated enclosure.

The Steady-Flow Process

The terms *steady* and *uniform* are used frequently in engineering, and thus it is important to have a clear understanding of their meanings. The term *steady* implies *no change with time*. The opposite of steady is *unsteady,* or *transient*. The term *uniform,* however, implies *no change with location* over a specified region. These meanings are consistent with their everyday use (steady girlfriend, uniform properties, etc.).

A large number of engineering devices operate for long periods of time under the same conditions, and they are classified as *steady-flow devices*. Processes involving such devices can be represented reasonably well by a somewhat idealized process, called the **steady-flow process**, which is defined as a *process during which a fluid flows through a control volume steadily* (Fig. 2–16). That is, the fluid properties can change from point to point within the control volume, but at any fixed point they remain the same during the entire process. Therefore, the volume V, the mass m, and the total energy content E of the control volume remain constant during a steady-flow process (Fig. 2–17).

Steady-flow conditions can be closely approximated by devices that are intended for continuous operation such as turbines, pumps, boilers, condensers, and heat exchangers or power plants or refrigeration systems. Some cyclic devices, such as reciprocating engines or compressors, do not satisfy any of the conditions stated above since the flow at the inlets and the exits are pulsating and not steady. However, the fluid properties vary with time in a periodic manner, and the flow through these devices can still be analyzed as a steady-flow process by using time-averaged values for the properties.

2–6 ▪ TEMPERATURE AND THE ZEROTH LAW OF THERMODYNAMICS

Although we are familiar with temperature as a measure of "hotness" or "coldness," it is not easy to give an exact definition for it. Based on our physiological sensations, we express the level of temperature qualitatively with words like *freezing cold, cold, warm, hot,* and *red-hot.* However, we cannot assign numerical values to temperatures based on our sensations alone. Furthermore, our senses may be misleading. A metal chair, for example, feels much colder than a wooden one even when both are at the same temperature.

Fortunately, several properties of materials change with temperature in a *repeatable* and *predictable* way, and this forms the basis for accurate temperature measurement. The commonly used mercury-in-glass thermometer, for example, is based on the expansion of mercury with temperature. Temperature is also measured by using several other temperature-dependent properties.

It is a common experience that a cup of hot coffee left on the table eventually cools off and a cold drink eventually warms up. That is, when a body is brought into contact with another body that is at a different temperature, heat is transferred from the body at higher temperature to the one at lower temperature until both bodies attain the same temperature (Fig. 2–18). At that point, the heat transfer stops, and the two bodies are said to have reached **thermal equilibrium**. The equality of temperature is the only requirement for thermal equilibrium.

The **zeroth law of thermodynamics** states that if two bodies are in thermal equilibrium with a third body, they are also in thermal equilibrium with each other. It may seem silly that such an obvious fact is called one of the basic laws of thermodynamics. However, it cannot be concluded from the other laws of thermodynamics, and it serves as a basis for the validity of temperature measurement. By replacing the third body with a thermometer, the zeroth law can be restated as *two bodies are in thermal equilibrium if both have the same temperature reading even if they are not in contact.*

The zeroth law was first formulated and labeled by R. H. Fowler in 1931. As the name suggests, its value as a fundamental physical principle was recognized more than half a century after the formulation of the first and the second laws of thermodynamics. It was named the zeroth law since it should have preceded the first and the second laws of thermodynamics.

Temperature Scales

Temperature scales enable us to use a common basis for temperature measurements, and several have been introduced throughout history. All temperature scales are based on some easily reproducible states such as the freezing and boiling points of water, which are also called the *ice point* and the *steam point,* respectively. A mixture of ice and water that is in equilibrium with air saturated with vapor at 1 atm pressure is said to be at the ice point, and a mixture of liquid water and water vapor (with no air) in equilibrium at 1 atm pressure is said to be at the steam point.

The temperature scales used in the SI and in the English system today are the **Celsius scale** (formerly called the *centigrade scale;* in 1948 it was renamed after the Swedish astronomer A. Celsius, 1702–1744, who devised it) and the **Fahrenheit scale** (named after the German instrument maker G. Fahrenheit, 1686–1736), respectively. On the Celsius scale, the ice and steam points were originally assigned the values of 0 and 100°C, respectively. The corresponding values on the Fahrenheit scale are 32 and 212°F. These are often referred to as *two-point scales* since temperature values are assigned at two different points.

In thermodynamics, it is very desirable to have a temperature scale that is independent of the properties of any substance or substances. Such a temperature scale is called a **thermodynamic temperature scale**, which is developed later in conjunction with the second law of thermodynamics. The thermodynamic temperature scale in the SI is the **Kelvin scale**, named after Lord Kelvin (1824–1907). The temperature unit on this scale is the **kelvin**, which is designated by K (not °K; the degree symbol was officially dropped from kelvin in 1967). The lowest temperature on the Kelvin scale is absolute zero, or 0 K. Then it follows that only one nonzero reference point needs to be assigned to establish the slope of this linear scale. Using nonconventional refrigeration techniques, scientists have approached absolute zero kelvin (they achieved 0.000000002 K in 1989).

The thermodynamic temperature scale in the English system is the **Rankine scale**, named after William Rankine (1820–1872). The temperature unit on this scale is the **rankine**, which is designated by R.

A temperature scale that turns out to be nearly identical to the Kelvin scale is the **ideal-gas temperature scale**. The temperatures on this scale are measured using a **constant-volume gas thermometer**, which is basically a rigid vessel filled with a gas, usually hydrogen or helium, at low pressure. This thermometer is based on the principle that *at low pressures, the temperature of a gas is proportional to its pressure at constant volume*. That is, the temperature of a gas of fixed volume varies *linearly* with pressure at sufficiently low pressures. Then the relationship between the temperature and the pressure of the gas in the vessel can be expressed as

$$T = a + bP \tag{2-5}$$

where the values of the constants a and b for a gas thermometer are determined experimentally. Once a and b are known, the temperature of a medium can be calculated from this relation by immersing the rigid vessel of the gas thermometer into the medium and measuring the gas pressure when thermal equilibrium is established between the medium and the gas in the vessel whose volume is held constant.

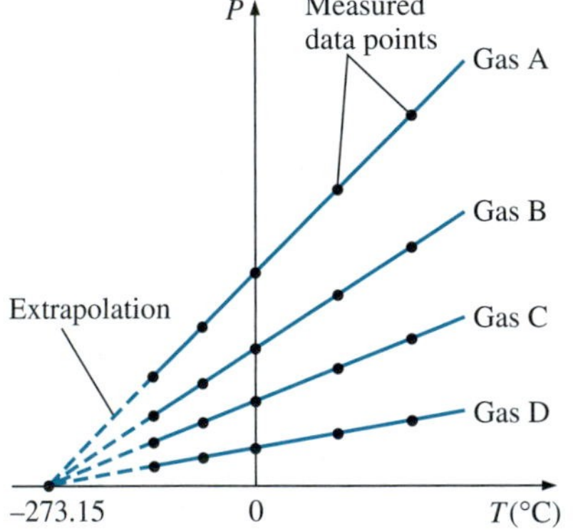

FIGURE 2–19

P versus *T* plots of the experimental data obtained from a constant-volume gas thermometer using four different gases at different (but low) pressures.

An ideal-gas temperature scale can be developed by measuring the pressures of the gas in the vessel at two reproducible points (such as the ice and the steam points) and assigning suitable values to temperatures at those two points. Considering that only one straight line passes through two fixed points on a plane, these two measurements are sufficient to determine the constants a and b in Eq. 2–5. Then the unknown temperature T of a medium corresponding to a pressure reading P can be determined from that equation by a simple calculation. The values of the constants will be different for each thermometer, depending on the type and the amount of the gas in the vessel, and the temperature values assigned at the two reference points. If the ice and steam points are assigned the values 0°C and 100°C, respectively, then the gas temperature scale will be identical to the Celsius scale. In this case the value of the constant a (which corresponds to an absolute pressure of zero) is determined to be −273.15°C regardless of the type and the amount of the gas in the vessel of the gas thermometer. That is, on a P-T diagram, all the straight lines passing through the data points in this case will intersect the temperature axis at −273.15°C when extrapolated, as shown in Fig. 2–19. This is the lowest temperature that can be obtained by a gas thermometer, and thus we can obtain an *absolute gas temperature scale* by assigning a value of zero to the constant a in Eq. 2–5. In that case Eq. 2–5 reduces to $T = bP$, and thus we need to specify the temperature at only *one* point to define an absolute gas temperature scale.

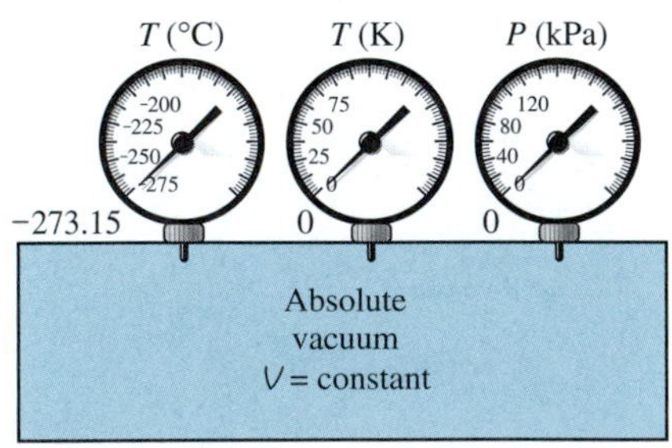

FIGURE 2–20

A constant-volume gas thermometer would read −273.15°C at absolute zero pressure.

It should be noted that the absolute gas temperature scale is not a thermodynamic temperature scale, since it cannot be used at very low temperatures (due to condensation) and at very high temperatures (due to dissociation and ionization). However, absolute gas temperature is identical to the thermodynamic temperature in the temperature range in which the gas thermometer can be used, and thus we can view the thermodynamic temperature scale at this point as an absolute gas temperature scale that utilizes an "ideal" or "imaginary" gas that always acts as a low-pressure gas regardless of the temperature. If such a gas thermometer existed, it would read zero kelvin at absolute zero pressure, which corresponds to −273.15°C on the Celsius scale (Fig. 2–20).

The Kelvin scale is related to the Celsius scale by

$$T(\text{K}) = T(°\text{C}) + 273.15 \tag{2–6}$$

The Rankine scale is related to the Fahrenheit scale by

$$T(\text{R}) = T(°\text{F}) + 459.67 \tag{2–7}$$

It is common practice to round the constant in Eq. 2–6 to 273 and that in Eq. 2–7 to 460.

The temperature scales in the two unit systems are related by

$$T(\text{R}) = 1.8T(\text{K}) \tag{2–8}$$

$$T(°\text{F}) = 1.8T(°\text{C}) + 32 \tag{2–9}$$

A comparison of various temperature scales is given in Fig. 2–21.

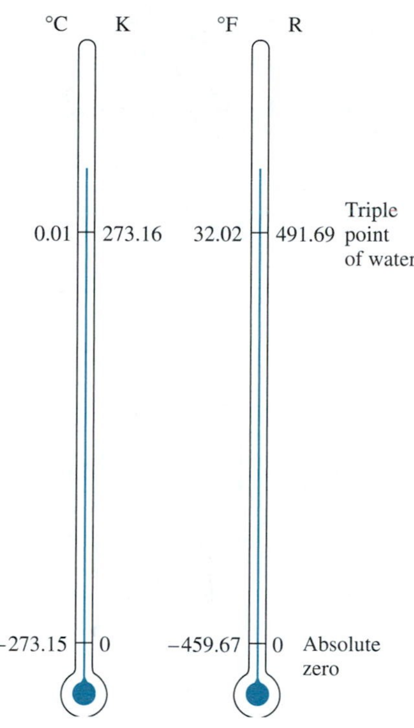

FIGURE 2–21
Comparison of temperature scales.

The reference temperature chosen in the original Kelvin scale was 273.15 K (or 0°C), which is the temperature at which water freezes (or ice melts) and water exists as a solid–liquid mixture in equilibrium under standard atmospheric pressure (the *ice point*). At the Tenth General Conference on Weights and Measures in 1954, the reference point was changed to a much more precisely reproducible point, the *triple point* of water (the state at which all three phases of water coexist in equilibrium), which is assigned the value 273.16 K. The Celsius scale was also redefined at this conference in terms of the ideal-gas temperature scale and a single fixed point, which is again the triple point of water with an assigned value of 0.01°C. The boiling temperature of water (the *steam point*) was experimentally determined to be again 100.00°C, and thus the new and old Celsius scales were in good agreement.

The International Temperature Scale of 1990 (ITS-90)

The *International Temperature Scale of 1990,* which supersedes the International Practical Temperature Scale of 1968 (IPTS-68), 1948 (ITPS-48), and 1927 (ITS-27), was adopted by the International Committee of Weights and Measures at its meeting in 1989 at the request of the Eighteenth General Conference on Weights and Measures. The ITS-90 is similar to its predecessors except that it is more refined with updated values of fixed temperatures, has an extended range, and conforms more closely to the thermodynamic temperature scale. On this scale, the unit of thermodynamic temperature T is again the kelvin (K), defined as the fraction 1/273.16 of the thermodynamic temperature of the triple point of water, which is the sole defining fixed point of both the ITS-90 and the Kelvin scale and is the most important thermometric fixed point used in the calibration of thermometers to ITS-90.

The unit of Celsius temperature is the degree Celsius (°C), which is by definition equal in magnitude to the kelvin (K). A temperature difference may be expressed in kelvins or degrees Celsius. The ice point remains the same at 0°C (273.15°C) in both ITS-90 and ITPS-68, but the steam point is 99.975°C in ITS-90 (with an uncertainly of ±0.005°C) whereas it was 100.000°C in IPTS-68. The change is due to precise measurements made by gas thermometry by paying particular attention to the effect of sorption (the impurities in a gas absorbed by the walls of the bulb at the reference temperature being desorbed at higher temperatures, causing the measured gas pressure to increase).

The ITS-90 extends upward from 0.65 K to the highest temperature practically measurable in terms of the Planck radiation law using monochromatic radiation. It is based on specifying definite temperature values on a number of fixed and easily reproducible points to serve as benchmarks and expressing the variation of temperature in a number of ranges and subranges in functional form.

In ITS-90, the temperature scale is considered in four ranges. In the range of 0.65 to 5 K, the temperature scale is defined in terms of the vapor pressure—temperature relations for ^{3}He and ^{4}He. Between 3 and 24.5561 K (the triple point of neon), it is defined by means of a properly calibrated helium gas thermometer. From 13.8033 K (the triple point of hydrogen) to 1234.93 K (the freezing point of silver), it is defined by means of platinum resistance thermometers calibrated at specified sets of defining fixed points. Above 1234.93 K, it is defined in terms of the Planck radiation law and a suitable defining fixed point such as the freezing point of gold (1337.33 K).

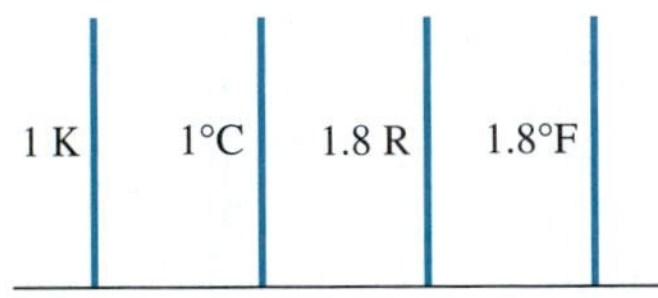

FIGURE 2–22

Comparison of magnitudes of various temperature units.

We emphasize that the magnitudes of each division of 1 K and 1°C are identical (Fig. 2–22). Therefore, when we are dealing with temperature differences ΔT, the temperature interval on both scales is the same. Raising the temperature of a substance by 10°C is the same as raising it by 10 K. That is,

$$\Delta T(\text{K}) = \Delta T(°\text{C}) \tag{2–10}$$

$$\Delta T(\text{R}) = \Delta T(°\text{F}) \tag{2–11}$$

Some thermodynamic relations involve the temperature T and often the question arises of whether it is in K or °C. If the relation involves temperature differences (such as $a = b\Delta T$), it makes no difference and either can be used. However, if the relation involves temperatures only instead of temperature differences (such as $a = bT$) then K must be used. When in doubt, it is always safe to use K because there are virtually no situations in which the use of K is incorrect, but there are many thermodynamic relations that will yield an erroneous result if °C is used.

Example 2–1 Expressing Temperature Rise in Different Units

During a heating process, the temperature of a system rises by 10°C. Express this rise in temperature in K, °F, and R.

Solution The temperature rise of a system is to be expressed in different units.

Analysis This problem deals with temperature changes, which are identical in Kelvin and Celsius scales. Then,

$$\Delta T(\text{K}) = \Delta T(°\text{C}) = \mathbf{10\ K}$$

The temperature changes in Fahrenheit and Rankine scales are also identical and are related to the changes in Celsius and Kelvin scales through Eqs. 2–8 and 2–11:

$$\Delta T(\text{R}) = 1.8\ \Delta T(\text{K}) = (1.8)(10) = \mathbf{18\ R}$$

and

$$\Delta T(°F) = \Delta T(R) = \mathbf{18°F}$$

Discussion Note that the units °C and K are interchangeable when dealing with temperature differences.

2–7 ▪ PRESSURE

Pressure is defined as *a normal force exerted by a fluid per unit area.* We speak of pressure only when we deal with a gas or a liquid. The counterpart of pressure in solids is *normal stress*. Since pressure is defined as force per unit area, it has the unit of newtons per square meter (N/m^2), which is called a **pascal** (Pa). That is,

$$1 \text{ Pa} = 1 \text{ N/m}^2$$

The pressure unit pascal is too small for pressures encountered in practice. Therefore, its multiples *kilopascal* (1 kPa = 10^3 Pa) and *megapascal* (1 MPa = 10^6 Pa) are commonly used. Three other pressure units commonly used in practice, especially in Europe, are *bar*, *standard atmosphere*, and *kilogram-force per square centimeter*:

$$1 \text{ bar} = 10^5 \text{ Pa} = 0.1 \text{ MPa} = 100 \text{ kPa}$$

$$1 \text{ atm} = 101{,}325 \text{ Pa} = 101.325 \text{ kPa} = 1.01325 \text{ bars}$$

$$\begin{aligned} 1 \text{ kgf/cm}^2 &= 9.807 \text{ N/cm}^2 = 9.807 \times 10^4 \text{ N/m}^2 = 9.807 \times 10^4 \text{ Pa} \\ &= 0.9807 \text{ bar} \\ &= 0.9679 \text{ atm} \end{aligned}$$

Note the pressure units bar, atm, and kgf/cm^2 are almost equivalent to each other. In the English system, the pressure unit is *pound-force per square inch* (lbf/in^2, or psi), and 1 atm = 14.696 psi. The pressure units kgf/cm^2 and lbf/in^2 are also denoted by kg/cm^2 and lb/in^2, respectively, and they are commonly used in tire gages. It can be shown that 1 kgf/cm^2 = 14.223 psi.

When used for solids, pressure is synonymous to *normal stress*, which is force acting perpendicular to the surface per unit area. For example, a 150-pound person with a total foot imprint area of 50 in^2 exerts a pressure of 150 lbf/50 in^2 = 3.0 psi on the floor (Fig. 2–23). If the person stands on one foot, the pressure doubles. If the person gains excessive weight, he or she is likely to encounter foot discomfort because of the increased pressure on the foot (the size of the foot does not change with weight gain). This also explains how a person can walk on fresh snow without sinking by wearing large snowshoes, and how a person cuts with little effort when using a sharp knife.

The actual pressure at a given position is called the **absolute pressure**, and it is measured relative to absolute vacuum (i.e., absolute zero pressure). Most pressure-measuring devices, however, are calibrated to read zero in the atmosphere (Fig. 2–24), and so they indicate the difference between the absolute pressure and the local atmospheric pressure. This difference is called the **gage pressure**. P_{gage} can be positive or negative, but pressures

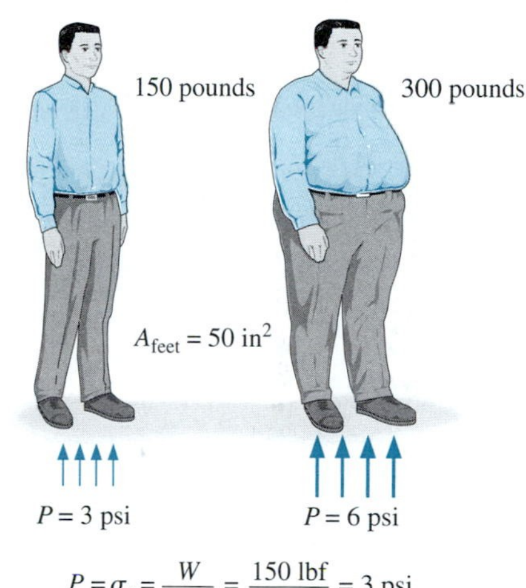

FIGURE 2–23
The normal stress (or "pressure") on the feet of a chubby person is much greater than on the feet of a slim person.

FIGURE 2–24
Some basic pressure gages.
Dresser Instruments, Dresser, Inc. Used by permission.

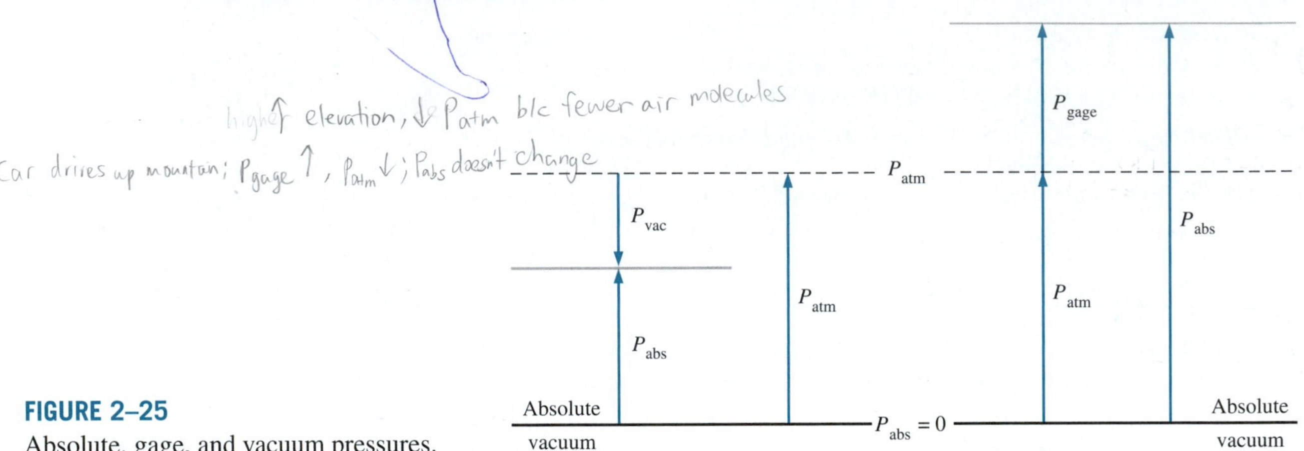

FIGURE 2–25
Absolute, gage, and vacuum pressures.

below atmospheric pressure are sometimes called **vacuum pressures** and are measured by vacuum gages that indicate the difference between the atmospheric pressure and the absolute pressure. Absolute, gage, and vacuum pressures are related to each other by

$$P_{gage} = P_{abs} - P_{atm} \quad \text{(2–12)}$$

$$P_{vac} = P_{atm} - P_{abs} \quad \text{(2–13)}$$

This is illustrated in Fig. 2–25.

Like other pressure gages, the gage used to measure the air pressure in an automobile tire reads the gage pressure. Therefore, the common reading of 32 psi (2.25 kgf/cm^2) indicates a pressure of 32 psi above the atmospheric pressure. At a location where the atmospheric pressure is 14.3 psi, for example, the absolute pressure in the tire is 32 + 14.3 = 46.3 psi.

In thermodynamic relations and tables, absolute pressure is almost always used. Throughout this text, the pressure P will denote *absolute pressure* unless specified otherwise. Often the letters "a" (for absolute pressure) and "g" (for gage pressure) are added to pressure units (such as psia and psig) to clarify what is meant.

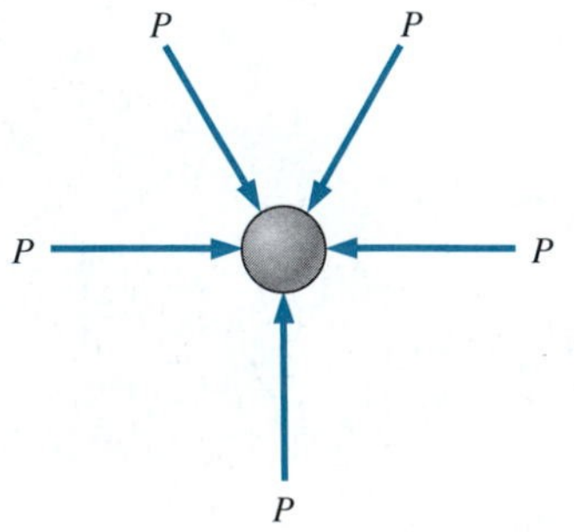

FIGURE 2–26
Pressure is a *scalar* quantity, not a vector; the pressure at a point in a fluid is the same in all directions.

EXAMPLE 2–2 Absolute Pressure of a Vacuum Chamber

A vacuum gage connected to a chamber reads 5.8 psi at a location where the atmospheric pressure is 14.5 psi. Determine the absolute pressure in the chamber.

Solution The gage pressure of a vacuum chamber is given. The absolute pressure in the chamber is to be determined.
Analysis The absolute pressure is easily determined from Eq. 2–13 to be

$$P_{abs} = P_{atm} - P_{vac} = 14.5 - 5.8 = \mathbf{8.7\ psi}$$

Discussion Note that the *local* value of the atmospheric pressure is used when determining the absolute pressure.

Pressure at a Point

Pressure is the *compressive force* per unit area, and it gives the impression of being a vector. However, pressure at any point in a fluid is the same in all directions (Fig. 2–26). That is, it has magnitude but not a specific direction, and thus it is a scalar quantity. In other words, *the pressure at a point in a fluid has the same magnitude in all directions*. It can be shown in the absence of shear forces that this result is applicable to fluids in motion as well as fluids at rest.

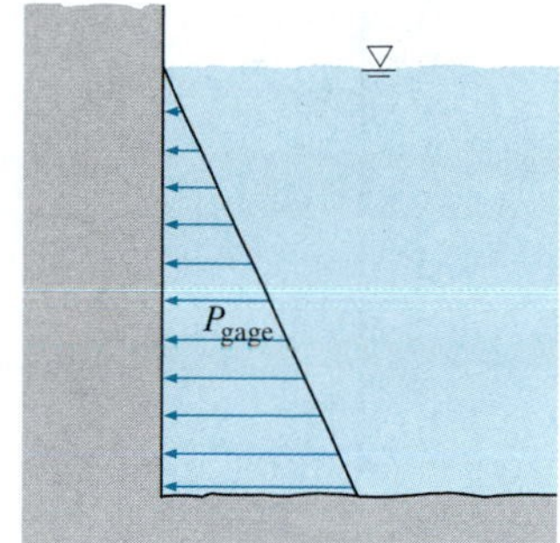

FIGURE 2–27
The pressure of a fluid at rest increases with depth (as a result of added weight).

Variation of Pressure with Depth

It will come as no surprise to you that pressure in a fluid at rest does not change in the horizontal direction. This can be shown easily by considering a thin horizontal layer of fluid and doing a force balance in any horizontal direction. However, this is not the case in the vertical direction in a gravity field. Pressure in a fluid increases with depth because more fluid rests on deeper layers, and the effect of this "extra weight" on a deeper layer is balanced by an increase in pressure (Fig. 2–27).

To obtain a relation for the variation of pressure with depth, consider a rectangular fluid element of height Δz, length Δx, and unit depth ($\Delta y = 1$ unit into the page) in equilibrium, as shown in Fig. 2–28. Assuming the density of the fluid ρ to be constant, a force balance in the vertical z-direction gives

$$\sum F_z = ma_z = 0: \qquad P_2\,\Delta x\Delta y - P_1\,\Delta x\Delta y - \rho g\,\Delta x\,\Delta z\Delta y = 0 \tag{2–14}$$

where $W = mg = \rho g\,\Delta x\,\Delta z\,\Delta y$ is the weight of the fluid element. Dividing by $\Delta x\,\Delta y$ and rearranging gives

$$\Delta P = P_2 - P_1 = \rho g\,\Delta z = \gamma_s\,\Delta z \tag{2–15}$$

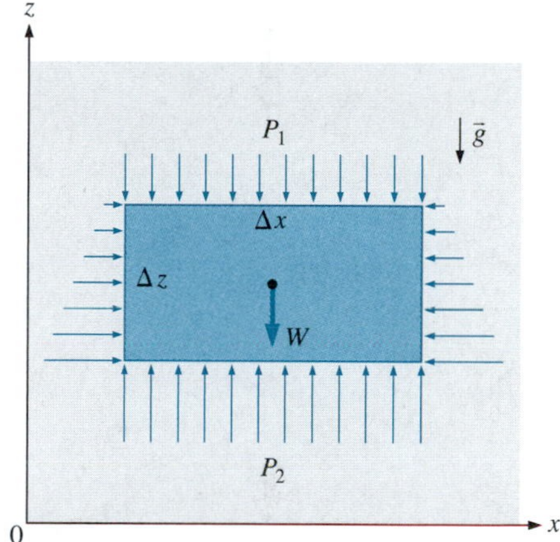

FIGURE 2–28
Free-body diagram of a rectangular fluid element in equilibrium.

where $\gamma_s = \rho g$ is the *specific weight* of the fluid. Thus, we conclude that the pressure difference between two points in a constant density fluid is proportional to the vertical distance Δz between the points and the density ρ of the fluid. In other words, *pressure in a static fluid increases linearly with depth*. This is what a diver experiences when diving deeper in a lake. For a given fluid, the vertical distance Δz is sometimes used as a measure of pressure, and it is called the *pressure head*.

We also conclude from Eq. 2–15 that for small to moderate distances, the variation of pressure with height is negligible for gases because of their low density. The pressure in a tank containing a gas, for example, can be considered to be uniform since the weight of the gas is too small to make a significant difference. Also, the pressure in a room filled with air can be assumed to be constant (Fig. 2–29).

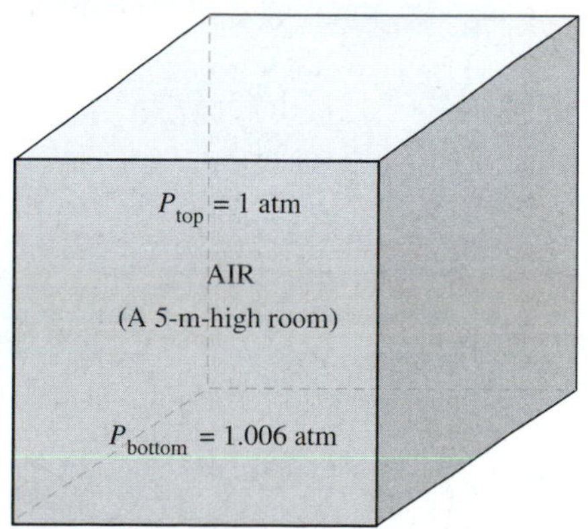

FIGURE 2–29
In a room filled with a gas, the variation of pressure with height is negligible.

If we take point 1 to be at the free surface of a liquid open to the atmosphere (Fig. 2–30), where the pressure is the atmospheric pressure P_{atm}, then the pressure at a depth h from the free surface becomes

$$P = P_{atm} + \rho gh \qquad \text{or} \qquad P_{gage} = \rho gh \tag{2–16}$$

Liquids are essentially incompressible substances, and thus the variation of density with depth is negligible. This is also the case for gases when the elevation change is not very large. The variation of density of liquids or

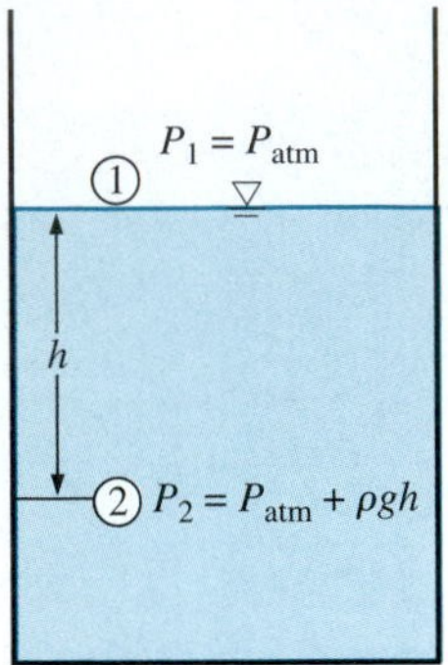

FIGURE 2–30
Pressure in a liquid at rest increases linearly with distance from the free surface.

gases with temperature can be significant, however, and may need to be considered when high accuracy is desired. Also, at great depths such as those encountered in oceans, the change in the density of a liquid can be significant because of the compression by the tremendous amount of liquid weight above.

The gravitational acceleration g varies from 9.807 m/s^2 at sea level to 9.764 m/s^2 at an elevation of 14,000 m where large passenger planes cruise. This is a change of just 0.4 percent in this extreme case. Therefore, g can be assumed to be constant with negligible error.

For fluids whose density changes significantly with elevation, a relation for the variation of pressure with elevation can be obtained by dividing Eq. 2–14 by $\Delta x\ \Delta y\ \Delta z$, and taking the limit as $\Delta z \to 0$. It gives

$$\frac{dP}{dz} = -\rho g \tag{2–17}$$

The negative sign is due to our taking the positive z direction to be upward so that dP is negative when dz is positive since pressure decreases in an upward direction. When the variation of density with elevation is known, the pressure difference between points 1 and 2 can be determined by integration to be

$$\Delta P = P_2 - P_1 = -\int_1^2 \rho g\, dz \tag{2–18}$$

For constant density and constant gravitational acceleration, this relation reduces to Eq. 2–15, as expected.

Pressure in a fluid at rest is independent of the shape or cross section of the container. It changes with the vertical distance, but remains constant in other directions. Therefore, the pressure is the same at all points on a horizontal plane in a given fluid. The Dutch mathematician Simon Stevin (1548–1620) published in 1586 the principle illustrated in Fig. 2–31. Note

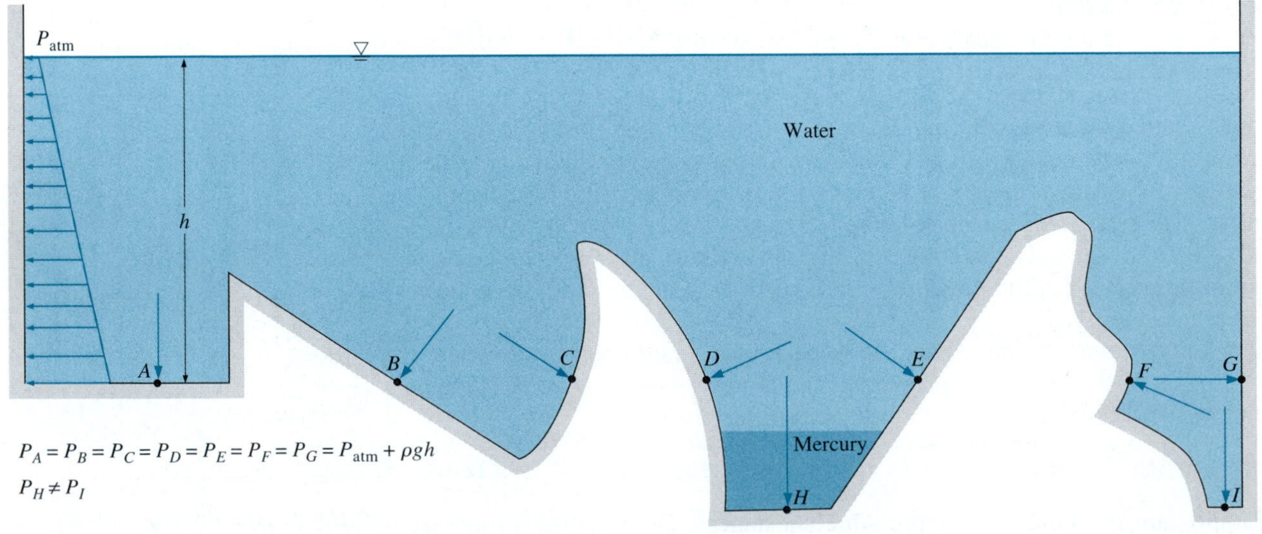

FIGURE 2–31
The pressure is the same at all points on a horizontal plane in a given fluid at rest regardless of geometry, provided that the points are interconnected by the same fluid.

that the pressures at points A, B, C, D, E, F, and G are the same since they are at the same depth, and they are interconnected by the same static fluid. However, the pressures at points H and I are not the same since these two points cannot be interconnected by the same fluid (i.e., we cannot draw a curve from point I to point H while remaining in the same fluid at all times), although they are at the same depth. (Can you tell at which point the pressure is higher?) Also, the pressure force exerted by the fluid is always normal to the surface at the specified points.

A consequence of the pressure in a fluid remaining constant in the horizontal direction is that *the pressure applied to a confined fluid increases the pressure throughout by the same amount.* This is called **Pascal's law**, after Blaise Pascal (1623–1662). Pascal also knew that the force applied by a fluid is proportional to the surface area. He realized that two hydraulic cylinders of different areas could be connected, and the larger could be used to exert a proportionally greater force than that applied to the smaller. "Pascal's machine" has been the source of many inventions that are a part of our daily lives such as hydraulic brakes and lifts. This is what enables us to lift a car easily by one arm, as shown in Fig. 2–32. Noting that $P_1 = P_2$ since both pistons are at the same level (the effect of small height differences is negligible, especially at high pressures), the ratio of output force to input force is determined to be

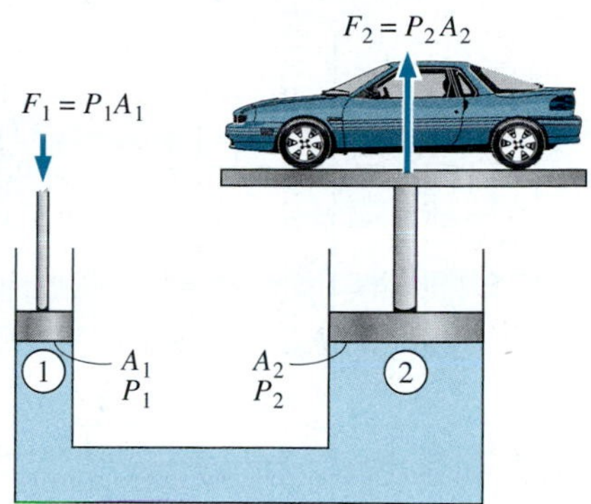

FIGURE 2–32
Lifting of a large weight by a small force by the application of Pascal's law.

$$P_1 = P_2 \quad \rightarrow \quad \frac{F_1}{A_1} = \frac{F_2}{A_2} \quad \rightarrow \quad \frac{F_2}{F_1} = \frac{A_2}{A_1} \tag{2–19}$$

The area ratio A_2/A_1 is called the *ideal mechanical advantage* of the hydraulic lift. Using a hydraulic car jack with a piston area ratio of $A_2/A_1 = 100$, for example, a person can lift a 1000-kg car by applying a force of just 10 kgf (= 90.8 N).

2–8 ▪ PRESSURE MEASUREMENT DEVICES

The Barometer

Atmospheric pressure is measured by a device called a **barometer**; thus, the atmospheric pressure is often referred to as the *barometric pressure*.

The Italian Evangelista Torricelli (1608–1647) was the first to conclusively prove that the atmospheric pressure can be measured by inverting a mercury-filled tube into a mercury container that is open to the atmosphere, as shown in Fig. 2–33. The pressure at point B is equal to the atmospheric pressure, and the pressure at C can be taken to be zero since there is only mercury vapor above point C and the pressure is very low relative to P_{atm} and can be neglected to an excellent approximation. Writing a force balance in the vertical direction gives

$$P_{atm} = \rho g h \tag{2–20}$$

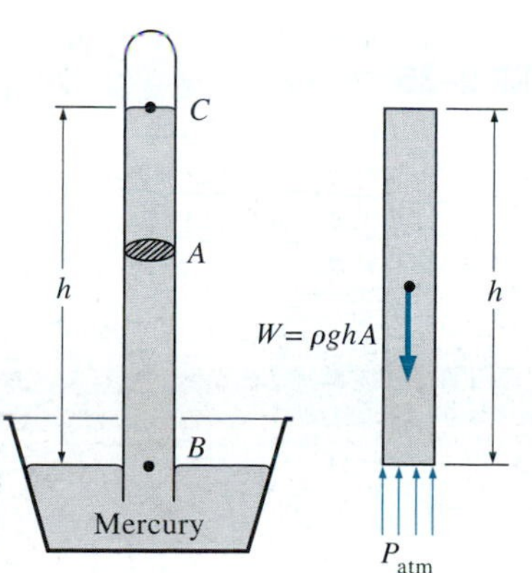

FIGURE 2–33
The basic barometer.

where ρ is the density of mercury, g is the local gravitational acceleration, and h is the height of the mercury column above the free surface. Note that the length and the cross-sectional area of the tube (unless it is so small that surface tension effects are significant) have no effect on the height of the fluid column of a barometer (Fig. 2–34).

A frequently used pressure unit is the *standard atmosphere*, which is defined as the pressure produced by a column of mercury 760 mm in height at 0°C (ρ_{Hg} = 13,595 kg/m^3) under standard gravitational acceleration

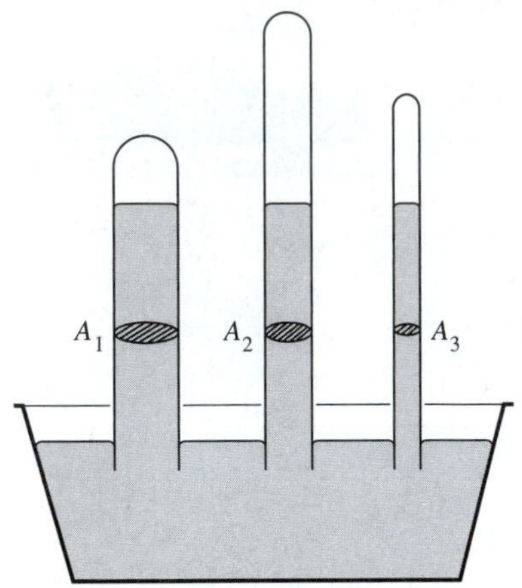

FIGURE 2–34
The length or the cross-sectional area of the tube has no effect on the height of the fluid column of a barometer, provided that the tube diameter is large enough to avoid surface tension (capillary) effects.

(g = 9.807 m/s²). If water instead of mercury were used to measure the standard atmospheric pressure, a water column of about 10.3 m would be needed. Pressure is sometimes expressed (especially by weather forecasters) in terms of the height of the mercury column. The standard atmospheric pressure, for example, is 760 mmHg (29.92 inHg) at 0°C. The unit mmHg is also called the **torr** in honor of Torricelli. Therefore, 1 atm = 760 torr and 1 torr = 133.3 Pa.

The atmospheric pressure P_{atm} changes from 101.325 kPa at sea level to 89.88, 79.50, 54.05, 26.5, and 5.53 kPa at altitudes of 1000, 2000, 5000, 10,000, and 20,000 meters, respectively. The typical atmospheric pressure in Denver (elevation = 1610 m), for example, is 83.4 kPa. Remember that the atmospheric pressure at a location is simply the weight of the air above that location per unit surface area. Therefore, it changes not only with elevation but also with weather conditions.

The decline of atmospheric pressure with elevation has far-reaching ramifications in daily life. For example, cooking takes longer at high altitudes since water boils at a lower temperature at lower atmospheric pressures. Nose bleeding is a common experience at high altitudes since the difference between the blood pressure and the atmospheric pressure is larger in this case, and the delicate walls of veins in the nose are often unable to withstand this extra stress.

For a given temperature, the density of air is lower at high altitudes, and thus a given volume contains less air and less oxygen. So it is no surprise that we tire more easily and experience breathing problems at high altitudes. To compensate for this effect, people living at higher altitudes develop more efficient lungs. Similarly, a 2.0-L car engine will act like a 1.7-L car engine at 1500 m altitude (unless it is turbocharged) because of the 15 percent drop in pressure and thus 15 percent drop in the density of air (Fig. 2–35). A fan or compressor will displace 15 percent less air at that altitude for the same volume displacement rate. Therefore, larger cooling fans may need to be selected for operation at high altitudes to ensure the specified mass flow rate. The lower pressure and thus lower density also affects lift and drag: airplanes need a longer runway at high altitudes to develop the required lift, and they climb to very high altitudes for cruising for reduced drag and thus better fuel efficiency.

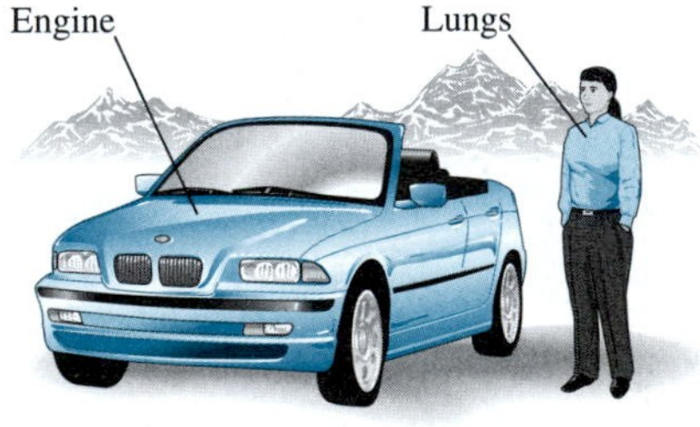

FIGURE 2–35
At high altitudes, a car engine generates less power and a person gets less oxygen because of the lower density of air.

EXAMPLE 2–3 **Measuring Atmospheric Pressure with a Barometer**

Determine the atmospheric pressure at a location where the barometric reading is 740 mm Hg and the gravitational acceleration is g = 9.81 m/s². Assume the temperature of mercury to be 10°C, at which its density is 13,570 kg/m³.

Solution The barometric reading at a location in height of mercury column is given. The atmospheric pressure is to be determined.

Assumptions The temperature of mercury is assumed to be 10°C.

Properties The density of mercury is given to be 13,570 kg/m³.

Analysis From Eq. 2–20, the atmospheric pressure is determined to be

$$P_{\text{atm}} = \rho g h$$

$$= (13{,}570\ \text{kg/m}^3)(9.81\ \text{m/s}^2)(0.74\ \text{m})\left(\frac{1\ \text{N}}{1\ \text{kg}\cdot\text{m/s}^2}\right)\left(\frac{1\ \text{kPa}}{1000\ \text{N/m}^2}\right)$$

$$= \mathbf{98.5\ kPa}$$

Discussion Note that density changes with temperature, and thus this effect should be considered in calculations.

The Manometer

We notice from Eq. 2–15 that an elevation change of Δz in a fluid at rest corresponds to $\Delta P/\rho g$, which suggests that a fluid column can be used to measure pressure differences. A device based on this principle is called a **manometer**, and it is commonly used to measure small and moderate pressure differences. A manometer consists of a glass or plastic U-tube containing one or more fluids such as mercury, water, alcohol, or oil. To keep the size of the manometer to a manageable level, heavy fluids such as mercury are used if large pressure differences are anticipated.

Consider the manometer shown in Fig. 2–36 that is used to measure the pressure in the tank. Since the gravitational effects of gases are negligible, the pressure anywhere in the tank and at position 1 has nearly the same value. Furthermore, since pressure in a fluid does not vary in the horizontal direction within a fluid, the pressure at point 2 is the same as the pressure at point 1, $P_2 = P_1$.

The differential fluid column of height h is in static equilibrium, and it is open to the atmosphere. Then the pressure at point 2 is determined directly from Eq. 2–16 to be

$$P_2 = P_{\text{atm}} + \rho g h \tag{2-21}$$

where ρ is the density of the fluid in the tube. Note that the cross-sectional area of the tube has no effect on the differential height h, and thus the pressure exerted by the fluid. However, the diameter of the tube should be large enough (more than a few millimeters) to ensure that the surface tension effect and thus the capillary rise is negligible.

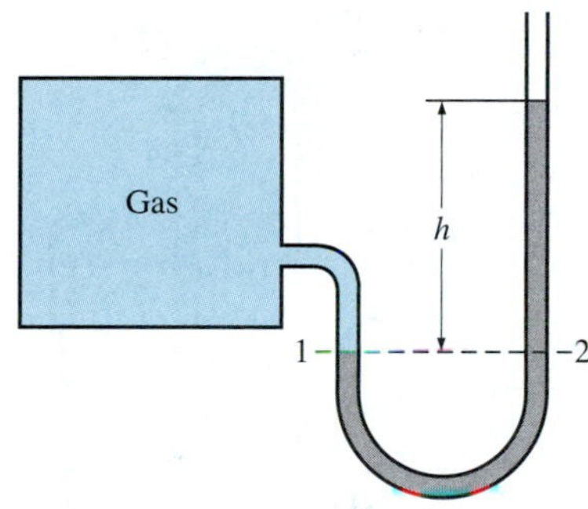

FIGURE 2–36
The basic manometer.

EXAMPLE 2–4 Measuring Pressure with a Manometer

A manometer is used to measure the pressure in a tank. The fluid used has a specific gravity of 0.85, and the manometer column height is 55 cm, as shown in Fig. 2–37. If the local atmospheric pressure is 96 kPa, determine the absolute pressure within the tank.

Solution The reading of a manometer attached to a tank and the atmospheric pressure are given. The absolute pressure in the tank is to be determined.

Assumptions The fluid in the tank is a gas whose density is much lower than the density of manometer fluid.

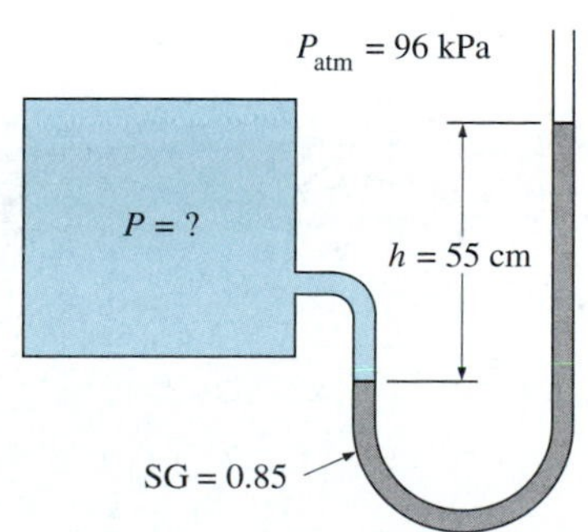

FIGURE 2–37
Schematic for Example 2–4.

Properties The specific gravity of the manometer fluid is given to be 0.85. We take the standard density of water.

Analysis The density of the fluid is obtained by multiplying its specific gravity by the density of water.

$$\rho = \text{SG}\,(\rho_{H_2O}) = (0.85)(1000\text{ kg/m}^3) = 850\text{ kg/m}^3$$

Then from Eq. 2–21,

$$\begin{aligned} P &= P_{atm} + \rho g h \\ &= 96\text{ kPa} + (850\text{ kg/m}^3)(9.81\text{ m/s}^2)(0.55\text{ m})\left(\frac{1\text{ N}}{1\text{ kg}\cdot\text{m/s}^2}\right)\left(\frac{1\text{ kPa}}{1000\text{ N/m}^2}\right) \\ &= \mathbf{100.6\ kPa} \end{aligned}$$

Discussion Note that the gage pressure in the tank is 4.6 kPa.

EXAMPLE 2–5 Measuring Pressure with a Multifluid Manometer

The water in a tank is pressurized by air, and the pressure is measured by a multifluid manometer as shown in Fig. 2–38. The tank is located on a mountain at an altitude of 1400 m where the atmospheric pressure is 85.6 kPa. Determine the air pressure in the tank if h_1 = 0.1 m, h_2 = 0.2 m, and h_3 = 0.35 m. Take the densities of water, oil, and mercury to be 1000 kg/m^3, 850 kg/m^3, and 13,600 kg/m^3, respectively.

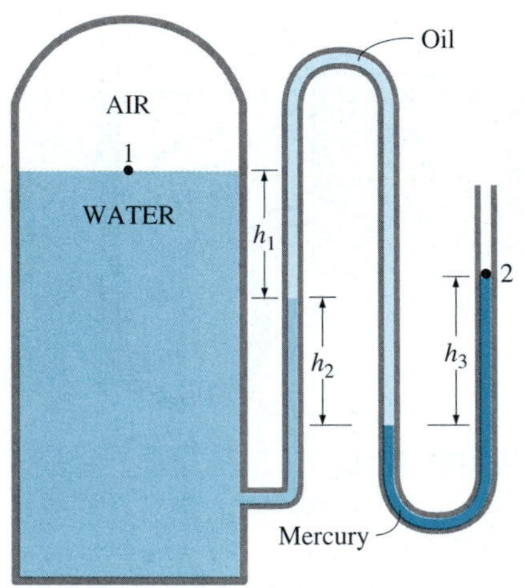

FIGURE 2–38
Schematic for Example 2–5; drawing not to scale.

Solution The pressure in a pressurized water tank is measured by a multifluid manometer. The air pressure in the tank is to be determined.

Assumption The air pressure in the tank is uniform (i.e., its variation with elevation is negligible due to its low density), and thus we can determine the pressure at the air–water interface.

Properties The densities of water, oil, and mercury are given to be 1000 kg/m^3, 850 kg/m^3, and 13,600 kg/m^3, respectively.

Analysis Starting with the pressure at point 1 at the air–water interface, moving along the tube by adding or subtracting the ρgh terms until we reach point 2, and setting the result equal to P_{atm} since the tube is open to the atmosphere gives

$$P_1 + \rho_{water} g h_1 + \rho_{oil} g h_2 - \rho_{mercury} g h_3 = P_2 = P_{atm}$$

Solving for P_1 and substituting,

$$\begin{aligned} P_1 &= P_{atm} - \rho_{water} g h_1 - \rho_{oil} g h_2 + \rho_{mercury} g h_3 \\ &= P_{atm} + g(\rho_{mercury} h_3 - \rho_{water} h_1 - \rho_{oil} h_2) \\ &= 85.6\text{ kPa} + (9.81\text{ m/s}^2)[(13{,}600\text{ kg/m}^3)(0.35\text{ m}) - (1000\text{ kg/m}^3)(0.1\text{ m}) \\ &\quad - (850\text{ kg/m}^3)(0.2\text{ m})]\left(\frac{1\text{ N}}{1\text{ kg}\cdot\text{m/s}^2}\right)\left(\frac{1\text{ kPa}}{1000\text{ N/m}^2}\right) \\ &= \mathbf{130\ kPa} \end{aligned}$$

Discussion Note that jumping horizontally from one tube to the next and realizing that pressure remains the same in the same fluid simplifies the analysis considerably. Also note that mercury is a toxic fluid, and mercury manometers and thermometers are being replaced by ones with safer fluids because of the risk of exposure to mercury vapor during an accident.

Some manometers use a slanted or inclined tube in order to increase the resolution (precision) when reading the fluid height. Such devices are called **inclined manometers**.

Many engineering problems and some manometers involve multiple immiscible fluids of different densities stacked on top of each other. Such systems can be analyzed easily by remembering that (1) the pressure change across a fluid column of height h is $\Delta P = \rho gh$, (2) pressure increases downward in a given fluid and decreases upward (i.e., $P_{bottom} > P_{top}$), and (3) two points at the same elevation in a continuous fluid at rest are at the same pressure.

The last principle, which is a result of *Pascal's law*, allows us to "jump" from one fluid column to the next in manometers without worrying about pressure change as long as we stay in the same continuous fluid and the fluid is at rest. Then the pressure at any point can be determined by starting with a point of known pressure and adding or subtracting ρgh terms as we advance toward the point of interest. For example, the pressure at the bottom of the tank in Fig. 2–39 can be determined by starting at the free surface where the pressure is P_{atm}, moving downward until we reach point 1 at the bottom, and setting the result equal to P_1. It gives

$$P_{atm} + \rho_1 gh_1 + \rho_2 gh_2 + \rho_3 gh_3 = P_1$$

In the special case of all fluids having the same density, this relation reduces to Eq. 2–21, as expected.

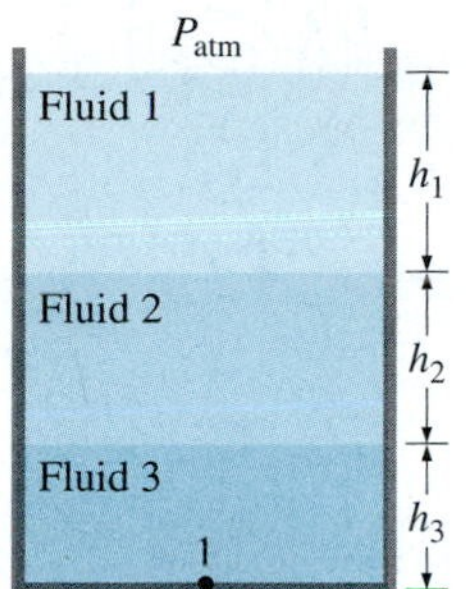

FIGURE 2–39
In stacked-up fluid layers, the pressure change across a fluid layer of density ρ and height h is ρgh.

Manometers are particularly well-suited to measure pressure drops across a horizontal flow section between two specified points due to the presence of a device such as a valve or heat exchanger or any resistance to flow. This is done by connecting the two legs of the manometer to these two points, as shown in Fig. 2–40. The working fluid can be either a gas or a liquid whose density is ρ_1. The density of the manometer fluid is ρ_2, and the differential fluid height is h. The two fluids must be immiscible, and ρ_2 must be greater than ρ_1.

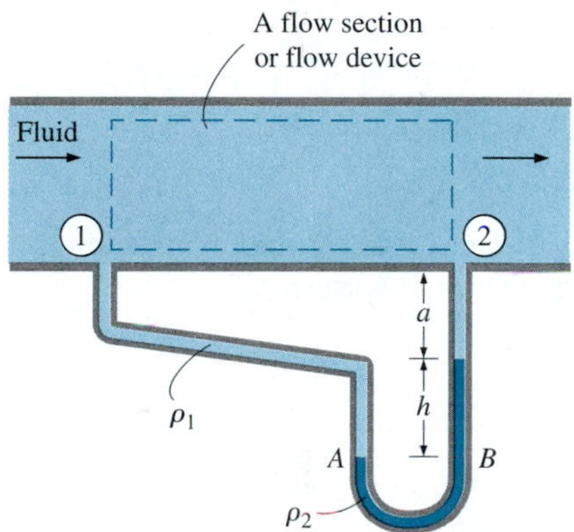

FIGURE 2–40
Measuring the pressure drop across a flow section or a flow device by a differential manometer.

A relation for the pressure difference $P_1 - P_2$ can be obtained by starting at point 1 with P_1, moving along the tube by adding or subtracting the ρgh terms until we reach point 2, and setting the result equal to P_2:

$$P_1 + \rho_1 g(a + h) - \rho_2 gh - \rho_1 ga = P_2 \qquad \textbf{(2–22)}$$

Note that we jumped from point A horizontally to point B and ignored the part underneath since the pressure at both points is the same. Simplifying,

$$P_1 - P_2 = (\rho_2 - \rho_1)gh \qquad \textbf{(2–23)}$$

Note that the distance a has been included in the analysis even though it has no effect on the result. Also, when the fluid flowing in the pipe is a gas, then $\rho_1 << \rho_2$ and the relation in Eq. 2–23 simplifies to $P_1 - P_2 \cong \rho_2 gh$.

Other Pressure Measurement Devices

Another type of commonly used mechanical pressure measurement device is the **Bourdon tube**, named after the French engineer and inventor Eugene Bourdon (1808–1884), which consists of a bent, coiled, or twisted hollow metal tube whose end is closed and connected to a dial indicator needle (Fig. 2–41). When the tube is open to the atmosphere, the tube is undeflected, and the needle on the dial at this state is calibrated to read zero

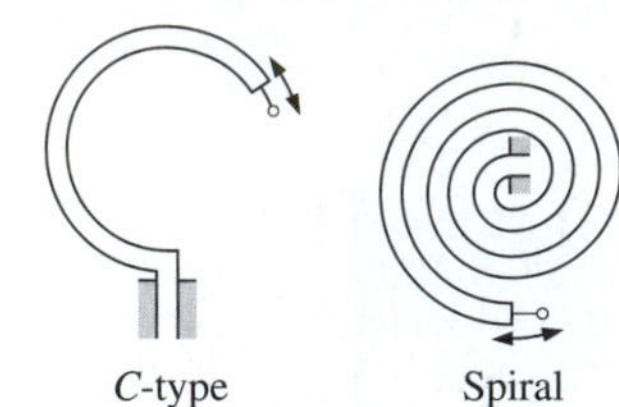

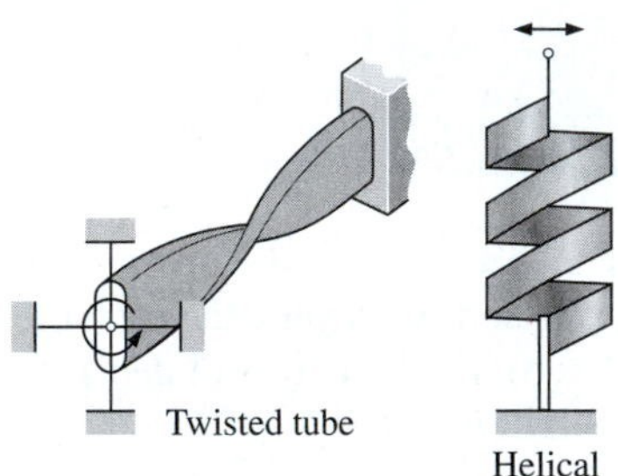

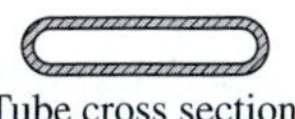

FIGURE 2–41
Various types of Bourdon tubes used to measure pressure.

EXAMPLE 2–6 Analyzing a Multifluid Manometer with EES

Reconsider the multifluid manometer discussed in Example 2–5. Determine the air pressure in the tank using EES. Also determine what the differential fluid height h_3 would be for the same air pressure if the mercury in the last column were replaced by seawater with a density of 1030 kg/m^3.

Solution The pressure in a water tank is measured by a multifluid manometer. The air pressure in the tank and the differential fluid height h_3 if mercury is replaced by seawater are to be determined using EES.

Analysis We start the EES program by double-clicking on its icon, open a new file, and type the following on the blank screen that appears (we express the atmospheric pressure in Pa for unit consistency):

```
g=9.81
Patm=85600
h1=0.1;   h2=0.2;   h3=0.35
rw=1000;   roil=850;   rm=13600
P1+rw*g*h1+roil*g*h2-rm*g*h3=Patm
```

Here P_1 is the only unknown, and it is determined by EES to be

$$P_1 = 129647 \text{ Pa} \cong \mathbf{130 \text{ kPa}}$$

which is identical to the result obtained before. The height of the fluid column h_3 when mercury is replaced by seawater is determined easily by replacing "*h*3=0.35" by "*P*1=129647" and "*r*m=13600" by "*r*m=1030," and clicking on the calculator symbol. It gives

$$h_3 = \mathbf{4.62 \text{ m}}$$

Discussion Note that we used the screen like a paper pad and wrote down the relevant information together with the applicable relations in an organized manner. EES did the rest. Equations can be written on separate lines or on the same line by separating them by semicolons, and blank or comment lines can be inserted for readability. EES makes it very easy to ask "what if" questions, and to perform parametric studies, as explained in Appendix 3 on the DVD.

EES also has the capability to check the equations for unit consistency if units are supplied together with numerical values. Units can be specified within brackets [ ] after the specified value. When this feature is utilized, the previous equations would take the following form:

```
g=9.81 [m/s^2]
Patm=85600 [Pa]
h1=0.1 [m];   h2=0.2 [m];   h3=0.35 [m]
rw=1000 [kg/m^3];   roil=850 [kg/m^3];   rm=13600 [kg/m^3]
P1+rw*g*h1+roil*g*h2-rm*g*h3=Patm
```

(gage pressure). When the fluid inside the tube is pressurized, the tube stretches and moves the needle in proportion to the pressure applied.

Electronics have made their way into every aspect of life, including pressure measurement devices. Modern pressure sensors, called **pressure transducers**, use various techniques to convert the pressure effect to an electrical effect such as a change in voltage, resistance, or capacitance. Pressure transducers are smaller and faster, and they can be more sensitive, reliable, and precise than their mechanical counterparts. They can measure pressures from less than a millionth of 1 atm to several thousands of atm.

A wide variety of pressure transducers is available to measure gage, absolute, and differential pressures in a wide range of applications. *Gage pressure transducers* use the atmospheric pressure as a reference by venting the back side of the pressure-sensing diaphragm to the atmosphere, and they give a zero signal output at atmospheric pressure regardless of altitude. *Absolute pressure transducers* are calibrated to have a zero signal output at full vacuum. *Differential pressure transducers* measure the pressure difference between two locations directly instead of using two pressure transducers and taking their difference.

Strain-gage pressure transducers work by having a diaphragm deflect between two chambers open to the pressure inputs. As the diaphragm stretches in response to a change in pressure difference across it, the strain gage stretches and a Wheatstone bridge circuit amplifies the output. A capacitance transducer works similarly, but capacitance change is measured instead of resistance change as the diaphragm stretches.

Piezoelectric transducers, also called solid-state pressure transducers, work on the principle that an electric potential is generated in a crystalline substance when it is subjected to mechanical pressure. This phenomenon, first discovered by brothers Pierre and Jacques Curie in 1880, is called the piezoelectric (or press-electric) effect. Piezoelectric pressure transducers have a much faster frequency response compared to diaphragm units and are very suitable for high-pressure applications, but they are generally not as sensitive as are diaphragm-type transducers, especially at low pressures.

EXAMPLE 2–7 Effect of Piston Weight on Pressure in a Cylinder

The piston of a vertical piston–cylinder device containing a gas has a mass of 60 kg and a cross-sectional area of 0.04 m^2, as shown in Fig. 2–42. The local atmospheric pressure is 0.97 bar, and the gravitational acceleration is 9.81 m/s^2. (*a*) Determine the pressure inside the cylinder. (*b*) If some heat is transferred to the gas and its volume is doubled, do you expect the pressure inside the cylinder to change?

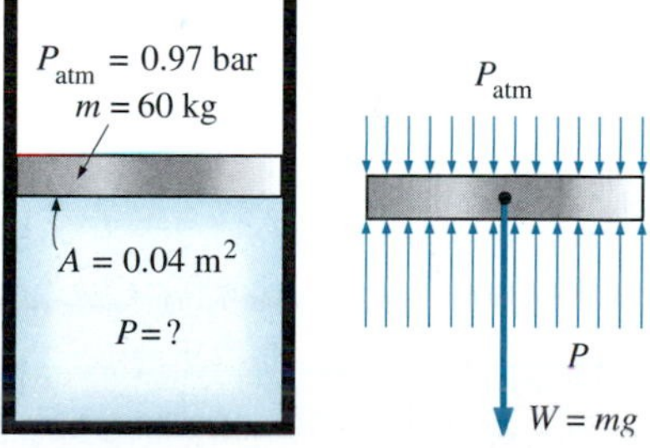

FIGURE 2–42
Schematic for Example 2–7, and the free-body diagram of the piston.

Solution A gas is contained in a vertical cylinder with a heavy piston. The pressure inside the cylinder and the effect of volume change on pressure are to be determined.

Assumptions Friction between the piston and the cylinder is negligible.

Analysis (*a*) The gas pressure in the piston–cylinder device depends on the atmospheric pressure and the weight of the piston. Drawing the free-body diagram of the piston as shown in Fig. 2–42 and balancing the vertical forces yield

$$PA = P_{atm}A + W$$

Solving for P and substituting,

$$\begin{aligned} P &= P_{atm} + \frac{mg}{A} \\ &= 0.97 \text{ bar} + \frac{(60 \text{ kg})(9.81 \text{ m/s}^2)}{0.04 \text{ m}^2}\left(\frac{1 \text{ N}}{1 \text{ kg}\cdot\text{m/s}^2}\right)\left(\frac{1 \text{ bar}}{10^5 \text{ N/m}^2}\right) \\ &= \mathbf{1.12 \text{ bars}} \end{aligned}$$

(*b*) The volume change will have no effect on the free-body diagram drawn in part (*a*), and therefore the pressure inside the cylinder will remain the same.

Discussion If the gas behaves as an ideal gas, the absolute temperature doubles when the volume is doubled at constant pressure.

EXAMPLE 2–8 Hydrostatic Pressure in a Solar Pond with Variable Density

Solar ponds are small artificial lakes of a few meters deep that are used to store solar energy. The rise of heated (and thus less dense) water to the surface is prevented by adding salt at the pond bottom. In a typical salt gradient solar pond, the density of water increases in the gradient zone, as shown in Fig. 2–43, and the density can be expressed as

$$\rho = \rho_0\sqrt{1 + \tan^2\left(\frac{\pi}{4}\frac{z}{H}\right)}$$

where ρ_0 is the density on the water surface, z is the vertical distance measured downward from the top of the gradient zone, and H is the thickness of the gradient zone. For $H = 4$ m, $\rho_0 = 1040$ kg/m^3, and a thickness of 0.8 m for the surface zone, calculate the gage pressure at the bottom of the gradient zone.

Solution The variation of density of saline water in the gradient zone of a solar pond with depth is given. The gage pressure at the bottom of the gradient zone is to be determined.

Assumptions The density in the surface zone of the pond is constant.

Properties The density of brine on the surface is given to be 1040 kg/m^3.

Analysis We label the top and the bottom of the gradient zone as 1 and 2, respectively. Noting that the density of the surface zone is constant, the gage pressure at the bottom of the surface zone (which is the top of the gradient zone) is

$$P_1 = \rho g h_1 = (1040\text{ kg/m}^3)(9.81\text{ m/s}^2)(0.8\text{ m})\left(\frac{1\text{ kN}}{1000\text{ kg}\cdot\text{m/s}^2}\right) = 8.16\text{ kPa}$$

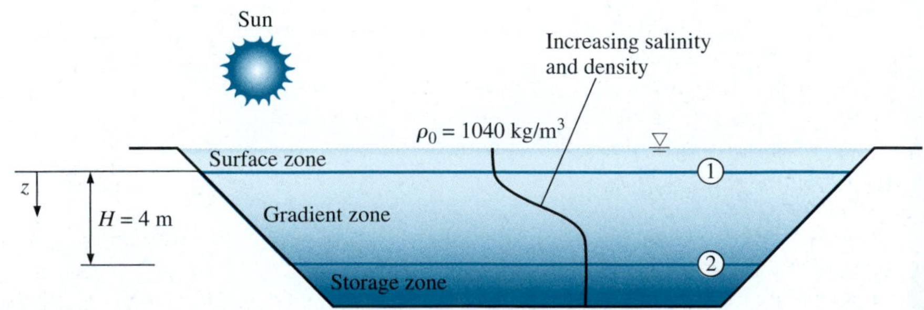

FIGURE 2–43
Schematic for Example 2–8.

since 1 kN/m² = 1 kPa. The differential change in hydrostatic pressure across a vertical distance of dz is given by

$$dP = \rho g \, dz$$

Integrating from the top of the gradient zone (point 1 where $z = 0$) to any location z in the gradient zone (no subscript) gives

$$P - P_1 = \int_0^z \rho g \, dz \quad \rightarrow \quad P = P_1 + \int_0^z \rho_0 \sqrt{1 + \tan^2\left(\frac{\pi}{4}\frac{z}{H}\right)} \, g \, dz$$

Performing the integration gives the variation of gage pressure in the gradient zone to be

$$P = P_1 + \rho_0 g \frac{4H}{\pi} \sinh^{-1}\left(\tan \frac{\pi}{4}\frac{z}{H}\right)$$

Then the pressure at the bottom of the gradient zone ($z = H = 4$ m) becomes

$$P_2 = 8.16 \text{ kPa} + (1040 \text{ kg/m}^3)(9.81 \text{ m/s}^2)\frac{4(4 \text{ m})}{\pi}$$

$$\times \sinh^{-1}\left(\tan \frac{\pi}{4}\frac{4}{4}\right)\left(\frac{1 \text{ kN}}{1000 \text{ kg} \cdot \text{m/s}^2}\right)$$

$$= \mathbf{54.0 \text{ kPa (gage)}}$$

Discussion The variation of gage pressure in the gradient zone with depth is plotted in Fig. 2–44. The dashed line indicates the hydrostatic pressure for the case of constant density at 1040 kg/m³ and is given for reference. Note that the variation of pressure with depth is not linear when density varies with depth.

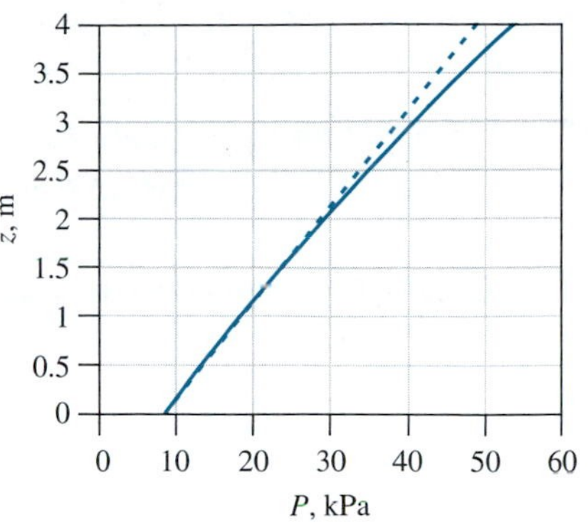

FIGURE 2–44

The variation of gage pressure with depth in the gradient zone of the solar pond.

SUMMARY

In this chapter, the basic concepts of thermodynamics are introduced and discussed. A system of fixed mass is called a *closed system,* or *control mass,* and a system that involves mass transfer across its boundaries is called an *open system,* or *control volume*. The mass-dependent properties of a system are called *extensive properties* and the others *intensive properties*. *Density* is mass per unit volume, and *specific volume* is volume per unit mass.

A system is said to be in *thermodynamic equilibrium* if it maintains thermal, mechanical, phase, and chemical equilibrium. Any change from one state to another is called a *process*. A process with identical end states is called a *cycle.* During a *quasi-static* or *quasi-equilibrium process*, the system remains practically in equilibrium at all times. The state of a simple, compressible system is completely specified by two independent, intensive properties.

The *zeroth law of thermodynamics* states that two bodies are in thermal equilibrium if both have the same temperature reading even if they are not in contact.

The temperature scales used in the SI and the English system today are the *Celsius scale* and the *Fahrenheit scale*, respectively. They are related to absolute temperature scales by

$$T(\text{K}) = T(°\text{C}) + 273.15$$

$$T(\text{R}) = T(°\text{F}) + 459.67$$

The magnitudes of each division of 1 K and 1°C are identical, and so are the magnitudes of each division of 1 R and 1°F. Therefore,

$$\Delta T(\text{K}) = \Delta T(°\text{C}) \quad \text{and} \quad \Delta T(\text{R}) = \Delta T(°\text{F})$$

The normal force exerted by a fluid per unit area is called *pressure,* and its unit is the *pascal,* 1 Pa = 1 N/m². The pressure relative to absolute vacuum is called the *absolute pressure*, and the difference between the absolute pressure and the local atmospheric pressure is called the *gage pressure.* Pressures below atmospheric pressure are called *vacuum pressures.* The absolute, gage, and vacuum pressures are related by

$$P_{gage} = P_{abs} - P_{atm} \qquad \text{(for pressures above or below } P_{atm}\text{)}$$

$$P_{vac} = P_{atm} - P_{abs} \qquad \text{(for pressures below } P_{atm}\text{)}$$

The pressure at a point in a fluid has the same magnitude in all directions. The variation of pressure with elevation is given by

$$\frac{dP}{dz} = -\rho g$$

where the positive z direction is taken to be upward. When the density of the fluid is constant, the pressure difference across a fluid layer of thickness Δz is

$$\Delta P = P_2 - P_1 = \rho g\, \Delta z$$

The absolute and gage pressures in a liquid open to the atmosphere at a depth h from the free surface are

$$P = P_{atm} + \rho gh \quad \text{or} \quad P_{gage} = \rho gh$$

Small to moderate pressure differences are measured by a *manometer*. The pressure in a stationary fluid remains constant in the horizontal direction. *Pascal's principle* states that the pressure applied to a confined fluid increases the pressure throughout by the same amount. The atmospheric pressure is measured by a *barometer* and is given by

$$P_{atm} = \rho gh$$

where h is the height of the liquid column.

REFERENCES AND SUGGESTED READINGS

1. A. Bejan. *Advanced Engineering Thermodynamics.* 3rd ed. New York: Wiley, 2006.

2. J. A. Schooley. *Thermometry*. Boca Raton, FL: CRC Press, 1986.

PROBLEMS*

Systems, Properties, State, and Processes

2–1C You have been asked to do a metabolism (energy) analysis of a person. How would you define the system for this purpose? What type of system is this?

2–2C You are trying to understand how a reciprocating air compressor (a piston-cylinder device) works. What system would you use? What type of system is this?

2–3C How would you define a system to determine the rate at which an automobile adds carbon dioxide to the atmosphere?

2–4C How would you define a system to determine the temperature rise created in a lake when a portion of its water is used to cool a nearby electrical power plant?

2–5C What is the difference between intensive and extensive properties?

2–6C The specific weight of a system is defined as the weight per unit volume (note that this definition violates the normal specific property-naming convention). Is the specific weight an extensive or intensive property?

2–7C Is the number of moles of a substance contained in a system an extensive or intensive property?

2–8C For a system to be in thermodynamic equilibrium, do the temperature and the pressure have to be the same everywhere?

2–9C What is a quasi-equilibrium process? What is its importance in engineering?

2–10C Define the isothermal, isobaric, and isochoric processes.

---

*Problems designated by a "C" are concept questions, and students are encouraged to answer them all. Problems designated by an "E" are in English units, and the SI users can ignore them. Problems with the icon are solved using EES, and complete solutions together with parametric studies are included on the enclosed DVD. Problems with the icon are comprehensive in nature and are intended to be solved with a computer, preferably using the EES software that accompanies this text.

2–11C What is the state postulate?

2–12C How would you describe the state of the air in the atmosphere? What kind of process does this air undergo from a cool morning to a warm afternoon?

2–13C What is a steady-flow process?

2–14 EES The density of atmospheric air varies with elevation, decreasing with increasing altitude. (*a*) Using the data given in the table, obtain a relation for the variation of density with elevation, and calculate the density at an elevation of 7000 m. (*b*) Calculate the mass of the atmosphere using the correlation you obtained. Assume the earth to be a perfect sphere with a radius of 6377 km, and take the thickness of the atmosphere to be 25 km.

z, km	ρ, kg/m^3
6377	1.225
6378	1.112
6379	1.007
6380	0.9093
6381	0.8194
6382	0.7364
6383	0.6601
6385	0.5258
6387	0.4135
6392	0.1948
6397	0.08891
6402	0.04008

Temperature

2–15C What is the zeroth law of thermodynamics?

2–16C What are the ordinary and absolute temperature scales in the SI and the English system?

2–17C Consider an alcohol and a mercury thermometer that read exactly 0°C at the ice point and 100°C at the steam point. The distance between the two points is divided into 100 equal parts in both thermometers. Do you think these thermometers will give exactly the same reading at a temperature of, say, 60°C? Explain.

2–18 The deep body temperature of a healthy person is 37°C. What is it in kelvins?

2–19E Consider a system whose temperature is 18°C. Express this temperature in R, K, and °F.

2–20 The temperature of a system rises by 15°C during a heating process. Express this rise in temperature in kelvins.

2–21E Steam enters a heat exchanger at 300 K. What is the temperature of this steam in °F?

2–22E The temperature of the lubricating oil in an automobile engine is measured as 150°F. What is the temperature of this oil in °C?

2–23E Heated air is at 150°C. What is the temperature of this air in °F?

2–24E Humans are most comfortable when the temperature is between 65°F and 75°F. Express these temperature limits in °C. Convert the size of this temperature range (10°F) to K, °C, and R. Is there any difference in the size of this range as measured in relative or absolute units?

Pressure, Manometer, and Barometer

2–25C What is the difference between gage pressure and absolute pressure?

2–26C A health magazine reported that physicians measured 100 adults' blood pressure using two different arm positions: parallel to the body (along the side) and perpendicular to the body (straight out). Readings in the parallel position were up to 10 percent higher than those in the perpendicular position, regardless of whether the patient was standing, sitting, or lying down. Explain the possible cause for the difference.

2–27C Someone claims that the absolute pressure in a liquid of constant density doubles when the depth is doubled. Do you agree? Explain.

2–28C A tiny steel cube is suspended in water by a string. If the lengths of the sides of the cube are very small, how would you compare the magnitudes of the pressures on the top, bottom, and side surfaces of the cube?

2–29C Express Pascal's law, and give a real-world example of it.

2–30E The maximum safe air pressure of a tire is typically written on the tire itself. The label on a tire indicates that the maximum pressure is 35 psi (gage). Express this maximum pressure in kPa.

FIGURE P2–30E

2–31 The pressure in a compressed air storage tank is 1500 kPa. What is the tank's pressure in (*a*) kN and m units; (*b*) kg, m, and s units; and (*c*) kg, km, and s units?

2–32E The absolute pressure in a compressed air tank is 200 kPa. What is this pressure in psia?

2–33E A manometer measures a pressure difference as 40 inches of water. What is this pressure difference in pounds-force per square inch, psi? *Answer:* 1.44 psi

2–34 The pressure of the helium inside a toy balloon is 1000 mm Hg. What is this pressure in kPa?

2–35 The water in a tank is pressurized by air, and the pressure is measured by a multifluid manometer as shown in Fig. P2–35. Determine the gage pressure of air in the tank if h_1 = 0.2 m, h_2 = 0.3 m, and h_3 = 0.46 m. Take the densities of water, oil, and mercury to be 1000 kg/m^3, 850 kg/m^3, and 13,600 kg/m^3, respectively.

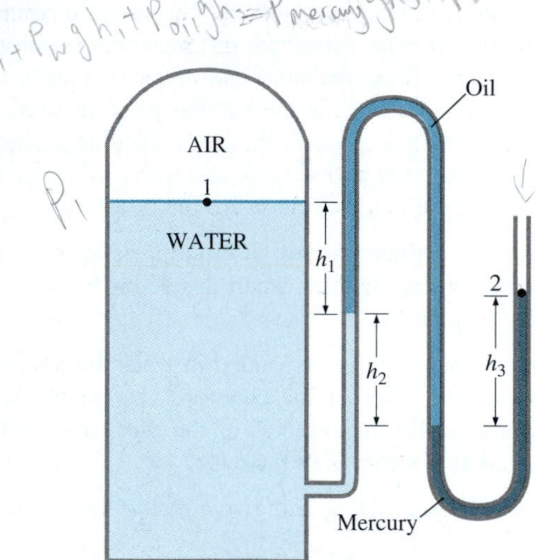

FIGURE P2–35

2–36 Determine the atmospheric pressure at a location where the barometric reading is 750 mm Hg. Take the density of mercury to be 13,600 kg/m^3.

2–37 The gage pressure in a liquid at a depth of 3 m is read to be 28 kPa. Determine the gage pressure in the same liquid at a depth of 9 m.

2–38 The absolute pressure in water at a depth of 5 m is read to be 145 kPa. Determine (*a*) the local atmospheric pressure, and (*b*) the absolute pressure at a depth of 5 m in a liquid whose specific gravity is 0.85 at the same location.

2–39E Show that 1 kgf/cm^2 = 14.223 psi.

2–40E The diameters of the pistons shown in Fig. P2–40E are D_1 = 3 in and D_2 = 2 in. Determine the pressure in chamber 3, in psia, when the other pressures are P_1 = 150 psia and P_2 = 200 psia.

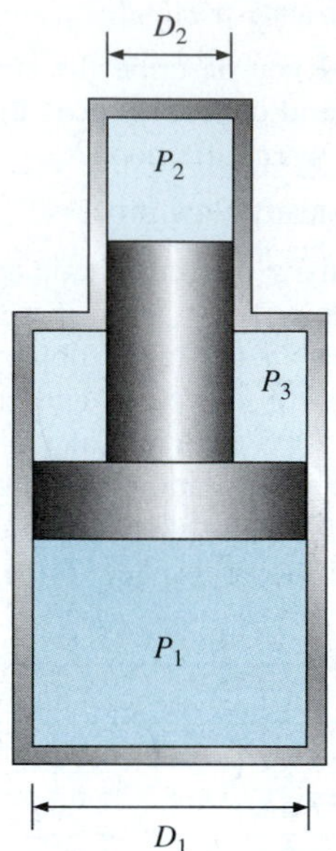

FIGURE P2–40E

2–41 In Fig. P2–40E, the piston diameters are D_1 = 10 cm and D_2 = 4 cm. If P_1 = 1000 kPa and P_3= 500 kPa, what is the pressure in chamber 2, in kPa?

2–42 The piston diameters in Fig. P2–40E are D_1 = 10 cm and D_2 = 4 cm. When the pressure in chamber 2 is 2000 kPa and the pressure in chamber 3 is 700 kPa, what is the pressure in chamber 1, in kPa? *Answer:* 908 kPa

2–43 Consider a 70-kg woman who has a total foot imprint area of 400 cm^2. She wishes to walk on the snow, but the snow cannot withstand pressures greater than 0.5 kPa. Determine the minimum size of the snowshoes needed (imprint area per shoe) to enable her to walk on the snow without sinking.

2–44 A vacuum gage connected to a tank reads 15 kPa at a location where the barometric reading is 750 mm Hg. Determine the absolute pressure in the tank. Take ρ_{Hg} = 13,590 kg/m^3. *Answer:* 85.0 kPa

2–45E A pressure gage connected to a tank reads 50 psi at a location where the barometric reading is 29.1 in Hg. Determine the absolute pressure in the tank. Take ρ_{Hg} = 848.4 lbm/ft^3. *Answer:* 64.3 psia

2–46 A pressure gage connected to a tank reads 500 kPa at a location where the atmospheric pressure is 94 kPa. Determine the absolute pressure in the tank.

2–47 The barometer of a mountain hiker reads 930 mbars at the beginning of a hiking trip and 780 mbars at the end. Neglecting the effect of altitude on local gravitational acceleration, determine the vertical distance climbed. Assume an average air density of 1.20 kg/m^3. *Answer:* 1274 m

2–48 The basic barometer can be used to measure the height of a building. If the barometric readings at the top and at the bottom of a building are 730 and 755 mm Hg, respectively, determine the height of the building. Take the densities of air and mercury to be 1.18 kg/m^3 and 13,600 kg/m^3, respectively.

FIGURE P2–48

© *Vol. 74/Corbis*

2–49 Solve Prob. 2–48 using EES (or other) software. Print out the entire solution, including the numerical results with proper units.

2–50 Determine the pressure exerted on a diver at 30 m below the free surface of the sea. Assume a barometric pressure of 101 kPa and a specific gravity of 1.03 for seawater. *Answer:* 404.0 kPa

2–51 A gas is contained in a vertical, frictionless piston–cylinder device. The piston has a mass of 4 kg and a cross-sectional area of 35 cm^2. A compressed spring above the piston exerts a force of 60 N on the piston. If the atmospheric pressure is 95 kPa, determine the pressure inside the cylinder. *Answer:* 123.4 kPa

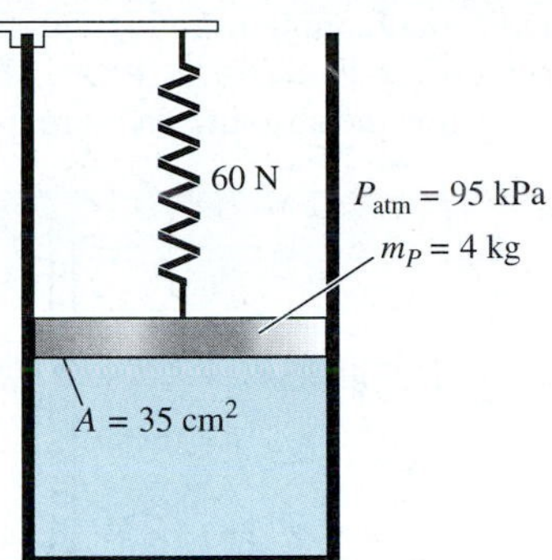

FIGURE P2–51

2–52 Reconsider Prob. 2–51. Using EES (or other) software, investigate the effect of the spring force in the range of 0 to 500 N on the pressure inside the cylinder. Plot the pressure against the spring force, and discuss the results.

2–53 Both a gage and a manometer are attached to a gas tank to measure its pressure. If the reading on the pressure gage is 80 kPa, determine the distance between the two fluid levels of the manometer if the fluid is (*a*) mercury (ρ = 13,600 kg/m^3) or (*b*) water (ρ = 1000 kg/m^3).

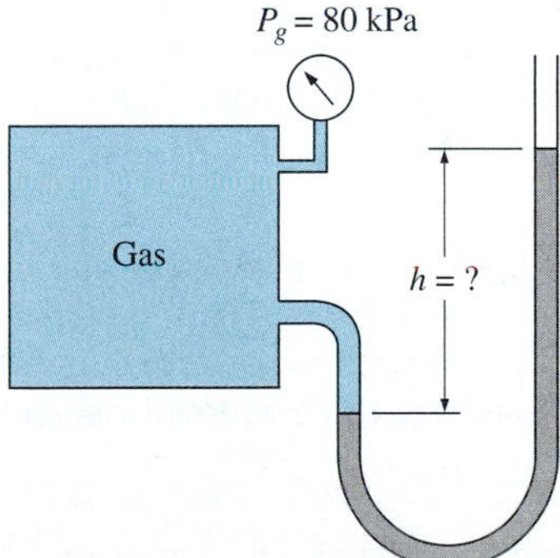

FIGURE P2–53

2–54 Reconsider Prob. 2–53. Using EES (or other) software, investigate the effect of the manometer fluid density in the range of 800 to 13,000 kg/m^3 on the differential fluid height of the manometer. Plot the differential fluid height against the density, and discuss the results.

2–55 A manometer containing oil (ρ = 850 kg/m^3) is attached to a tank filled with air. If the oil-level difference between the two columns is 60 cm and the atmospheric pressure is 98 kPa, determine the absolute pressure of the air in the tank. *Answer:* 103 kPa

2–56 A mercury manometer (ρ = 13,600 kg/m^3) is connected to an air duct to measure the pressure inside. The difference in the manometer levels is 15 mm, and the atmospheric pressure is 100 kPa. (*a*) Judging from Fig. P2–56, determine if the pressure in the duct is above or below the atmospheric pressure. (*b*) Determine the absolute pressure in the duct.

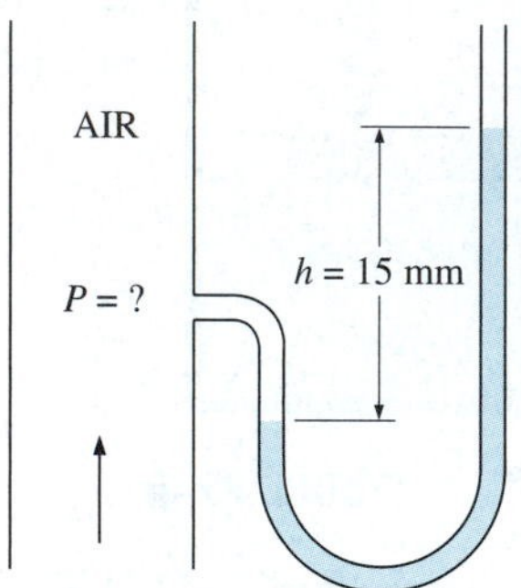

FIGURE P2–56

2–57 Repeat Prob. 2–56 for a differential mercury height of 45 mm.

2–58E Blood pressure is usually measured by wrapping a closed air-filled jacket equipped with a pressure gage around the upper arm of a person at the level of the heart. Using a mercury manometer and a stethoscope, the systolic pressure (the maximum pressure when the heart is pumping) and the diastolic pressure (the minimum pressure when the heart is resting) are measured in mm Hg. The systolic and diastolic pressures of a healthy person are about 120 mm Hg and 80 mm Hg, respectively, and are indicated as 120/80. Express both of these gage pressures in kPa, psi, and meter water column.

2–59 The maximum blood pressure in the upper arm of a healthy person is about 120 mm Hg. If a vertical tube open to the atmosphere is connected to the vein in the arm of the person, determine how high the blood will rise in the tube. Take the density of the blood to be 1050 kg/m^3.

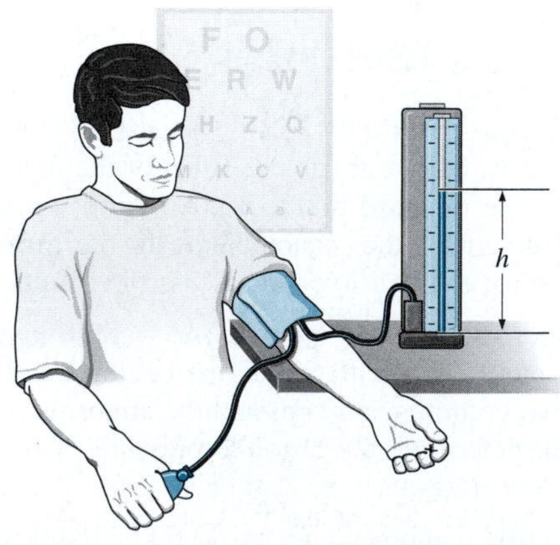

FIGURE P2–59

2–60 Consider a 1.8-m-tall man standing vertically in water and completely submerged in a pool. Determine the difference between the pressures acting at the head and at the toes of this man, in kPa.

2–61 Consider a U-tube whose arms are open to the atmosphere. Now water is poured into the U-tube from one arm, and light oil (ρ = 790 kg/m^3) from the other. One arm contains 70-cm-high water, while the other arm contains both fluids with an oil-to-water height ratio of 4. Determine the height of each fluid in that arm.

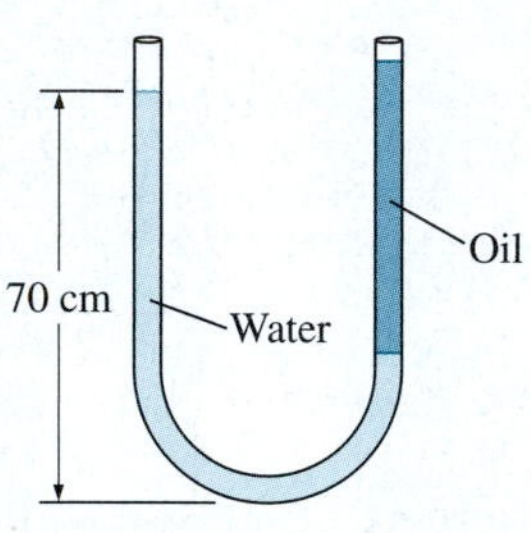

FIGURE P2–61

2–62 Freshwater and seawater flowing in parallel horizontal pipelines are connected to each other by a double U-tube manometer, as shown in Fig. P2–62. Determine the pressure difference between the two pipelines. Take the density of seawater at that location to be ρ = 1035 kg/m^3. Can the air column be ignored in the analysis?

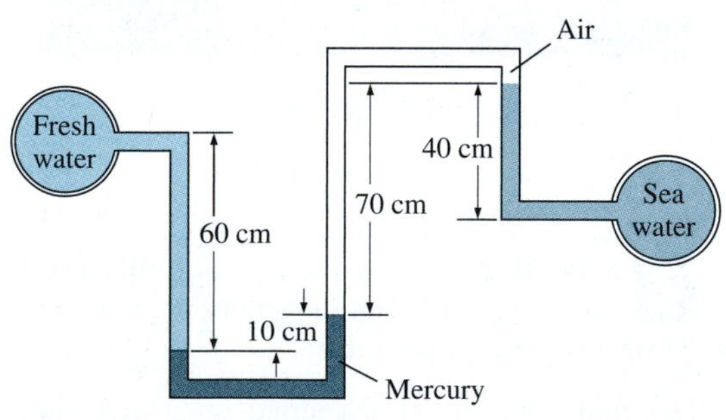

FIGURE P2–62

2–63 Repeat Prob. 2–62 by replacing the air with oil whose specific gravity is 0.72.

2–64 Calculate the absolute pressure, P_1, of the manometer shown in Fig. P2–64 in kPa. The local atmospheric pressure is 758 mm Hg.

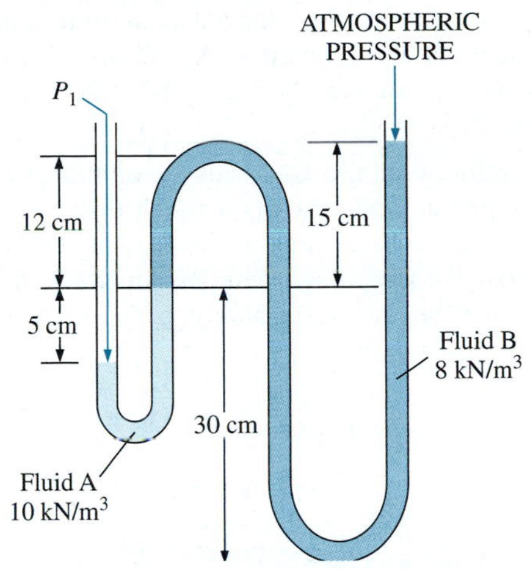

FIGURE P2–64

2–65 Consider the manometer in Fig. P2–64. If the specific weight of fluid A is 100 kN/m³, what is the absolute pressure, in kPa, indicated by the manometer when the local atmospheric pressure is 90 kPa?

2–66 Consider the manometer in Fig. P2–64. If the specific weight of fluid B is 20 kN/m³, what is the absolute pressure, in kPa, indicated by the manometer when the local atmospheric pressure is 745 mm Hg?

2–67 The gage pressure of the air in the tank shown in Fig. P2–67 is measured to be 80 kPa. Determine the differential height h of the mercury column.

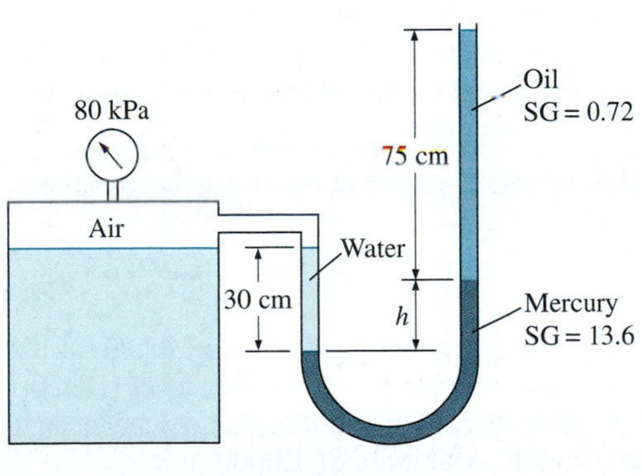

FIGURE P2–67

2–68 Repeat Prob. 2–67 for a gage pressure of 40 kPa.

2–69 The top part of a water tank is divided into two compartments, as shown in Fig. P2–69. Now a fluid with an unknown density is poured into one side, and the water level rises a certain amount on the other side to compensate for this effect. Based on the final fluid heights shown on the figure, determine the density of the fluid added. Assume the liquid does not mix with water.

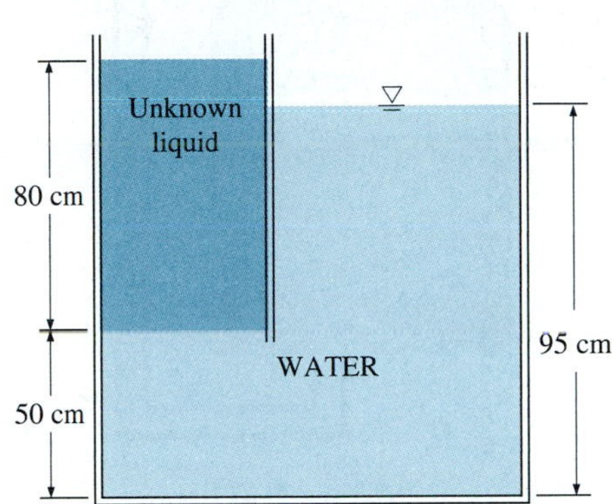

FIGURE P2–69

2–70 Consider the system shown in Fig. P2–70. If a change of 0.7 kPa in the pressure of air causes the brine–mercury interface in the right column to drop by 5 mm in the brine level in the right column while the pressure in the brine pipe remains constant, determine the ratio of A_2/A_1.

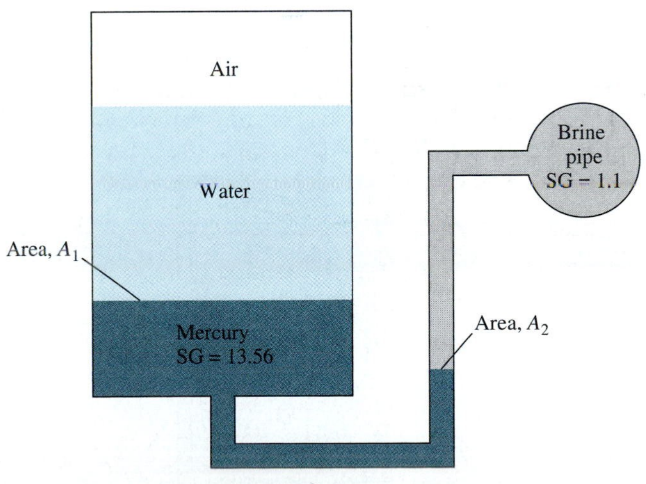

FIGURE P2–70

2–71 A multifluid container is connected to a U-tube, as shown in Fig. P2–71. For the given specific gravities and fluid column heights, determine the gage pressure at A. Also determine the height of a mercury column that would create the same pressure at A. *Answers:* 0.471 kPa, 0.353 cm

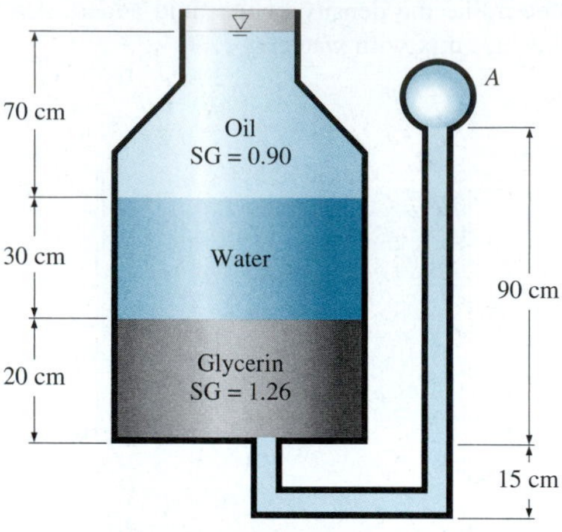

FIGURE P2–71

Review Problems

2–72 The force generated by a spring is given by $F = kx$, where k is the spring constant and x is the deflection of the spring. The spring of Fig. P2–72 has a spring constant of 8 kN/cm. The pressures are $P_1 = 5000$ kPa, $P_2 = 10{,}000$ kPa, and $P_3 = 1000$ kPa. If the piston diameters are $D_1 = 8$ cm and $D_2 = 3$ cm, how far will the spring be deflected? *Answer:* 1.72 cm

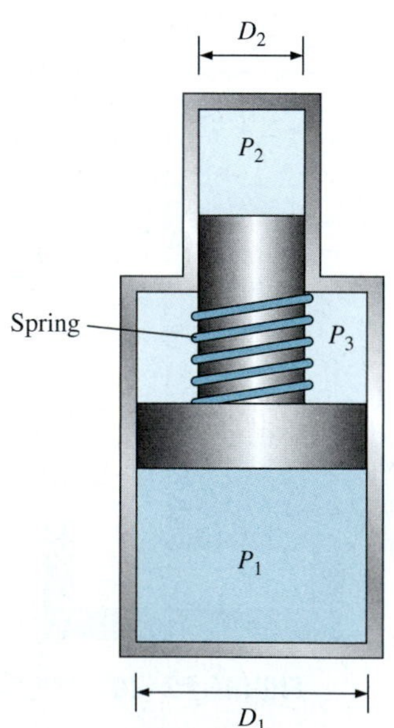

FIGURE P2–72

2–73 In Fig. P2–72, the spring has a spring constant of 12 kN/cm and has been extended 5 cm. The piston diameters are $D_1 = 7$ cm and $D_2 = 3$ cm. Determine P_1 when $P_2 = 8$ MPa and $P_3 = 0.3$ MPa.

2–74E In Fig. P2–72, the spring has a spring constant of 200 lbf/in and has been compressed 2 inches. The piston diameters are $D_1 = 5$ in and $D_2 = 2$ in. What is P_2 when $P_1 = 100$ psia and $P_3 = 20$ psia?

2–75 The pilot of an airplane reads the altitude 3000 m and the absolute pressure 58 kPa when flying over a city. Calculate the local atmospheric pressure in that city in kPa and in mm Hg. Take the densities of air and mercury to be 1.15 kg/m^3 and 13,600 kg/m^3, respectively.

FIGURE P2–75

2–76E The efficiency of a refrigerator increases by 3 percent for each °C rise in the minimum temperature in the device. What is the increase in the efficiency for each (*a*) K, (*b*) °F, and (*c*) R rise in temperature?

2–77E The boiling temperature of water decreases by about 3°C for each 1000-m rise in altitude. What is the decrease in the boiling temperature in (*a*) K, (*b*) °F, and (*c*) R for each 1000-m rise in altitude?

2–78E Hyperthermia of 5°C (i.e., 5°C rise above the normal body temperature) is considered fatal. Express this fatal level of hyperthermia in (*a*) K, (*b*) °F, and (*c*) R.

2–79E A house is losing heat at a rate of 4500 kJ/h per °C temperature difference between the indoor and the outdoor temperatures. Express the rate of heat loss from this house per (*a*) K, (*b*) °F, and (*c*) R difference between the indoor and the outdoor temperature.

2–80 The average temperature of the atmosphere in the world is approximated as a function of altitude by the relation

$$T_{atm} = 288.15 - 6.5z$$

where T_{atm} is the temperature of the atmosphere in K and z is the altitude in km with $z = 0$ at sea level. Determine the average temperature of the atmosphere outside an airplane that is cruising at an altitude of 12,000 m.

2–81 Joe Smith, an old-fashioned engineering student, believes that the boiling point of water is best suited for use as the reference point on temperature scales. Unhappy that the boiling point corresponds to some odd number in the current absolute temperature scales, he has proposed a new absolute temperature scale that he calls the Smith scale. The temperature unit on this scale is *smith*, denoted by S, and the boiling point of water on this scale is assigned to be 1000 S. From a thermodynamic point of view, discuss if it is an acceptable temperature scale. Also, determine the ice point of water on the Smith scale and obtain a relation between the Smith and Celsius scales.

2–82E It is well-known that cold air feels much colder in windy weather than what the thermometer reading indicates because of the "chilling effect" of the wind. This effect is due to the increase in the convection heat transfer coefficient with increasing air velocities. The *equivalent wind chill temperature* in °F is given by [ASHRAE, *Handbook of Fundamentals* (Atlanta, GA, 1993), p. 8.15]

$$T_{\text{equiv}} = 91.4 - (91.4 - T_{\text{ambient}}) \times (0.475 - 0.0203V + 0.304\sqrt{V})$$

where V is the wind velocity in mi/h and T_{ambient} is the ambient air temperature in °F in calm air, which is taken to be air with light winds at speeds up to 4 mi/h. The constant 91.4°F in the given equation is the mean skin temperature of a resting person in a comfortable environment. Windy air at temperature T_{ambient} and velocity V will feel as cold as the calm air at temperature T_{equiv}. Using proper conversion factors, obtain an equivalent relation in SI units where V is the wind velocity in km/h and T_{ambient} is the ambient air temperature in °C.

Answer: $T_{equiv} = 33.0 - (33.0 - T_{ambient}) \times (0.475 - 0.0126V + 0.240\sqrt{V})$

2–83E Reconsider Prob. 2–82E. Using EES (or other) software, plot the equivalent wind chill temperatures in °F as a function of wind velocity in the range of 4 to 100 mph for the ambient temperatures of 20, 40, and 60°F. Discuss the results.

2–84 An air-conditioning system requires a 20-m-long section of 15-cm diameter duct work to be laid underwater. Determine the upward force the water will exert on the duct. Take the densities of air and water to be 1.3 kg/m^3 and 1000 kg/m^3, respectively.

2–85 Balloons are often filled with helium gas because it weighs only about one-seventh of what air weighs under identical conditions. The buoyancy force, which can be expressed as $F_b = \rho_{\text{air}} g V_{\text{balloon}}$, will push the balloon upward. If the balloon has a diameter of 10 m and carries two people, 70 kg each, determine the acceleration of the balloon when it is first released. Assume the density of air is $\rho = 1.16$ kg/m^3, and neglect the weight of the ropes and the cage. *Answer:* 16.5 m/s^2

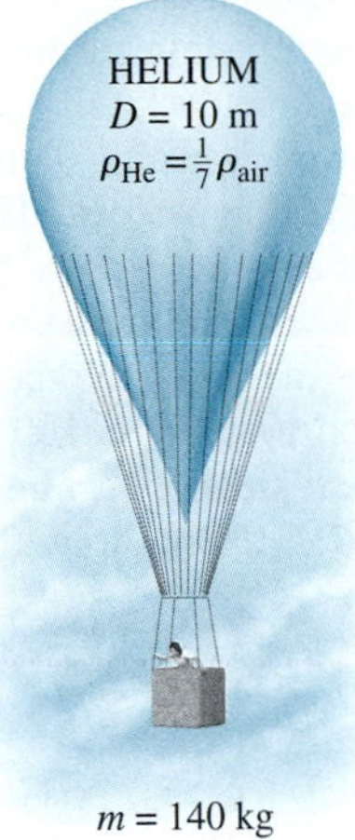

FIGURE P2–85

2–86 Reconsider Prob. 2–85. Using EES (or other) software, investigate the effect of the number of people carried in the balloon on acceleration. Plot the acceleration against the number of people, and discuss the results.

2–87 Determine the maximum amount of load, in kg, the balloon described in Prob. 2–85 can carry. *Answer:* 520.5 kg

2–88E The pressure in a steam boiler is given to be 92 kgf/cm^2. Express this pressure in psi, kPa, atm, and bars.

2–89 The lower half of a 10-m-high cylindrical container is filled with water ($\rho = 1000$ kg/m^3) and the upper half with oil that has a specific gravity of 0.85. Determine the pressure difference between the top and bottom of the cylinder. *Answer:* 90.7 kPa

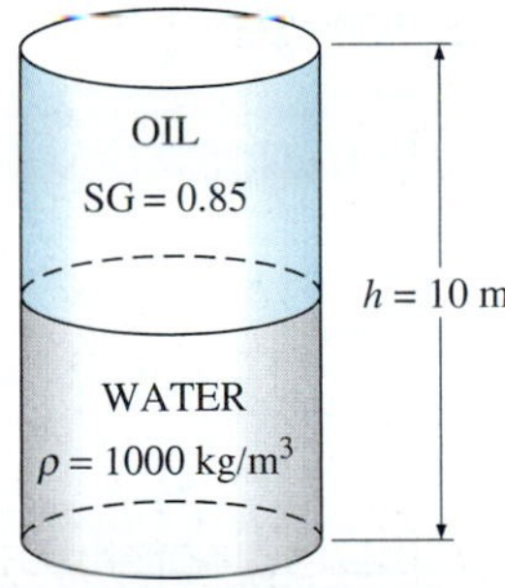

FIGURE P2–89

2–90 A vertical, frictionless piston–cylinder device contains a gas at 250 kPa absolute pressure. The atmospheric pressure outside is 100 kPa, and the piston area is 30 cm^2. Determine the mass of the piston.

2–91 A pressure cooker cooks a lot faster than an ordinary pan by maintaining a higher pressure and temperature inside.

The lid of a pressure cooker is well sealed, and steam can escape only through an opening in the middle of the lid. A separate metal piece, the petcock, sits on top of this opening and prevents steam from escaping until the pressure force overcomes the weight of the petcock. The periodic escape of the steam in this manner prevents any potentially dangerous pressure buildup and keeps the pressure inside at a constant value. Determine the mass of the petcock of a pressure cooker whose operation pressure is 100 kPa gage and has an opening cross-sectional area of 4 mm². Assume an atmospheric pressure of 101 kPa, and draw the free-body diagram of the petcock. *Answer:* 40.8 g

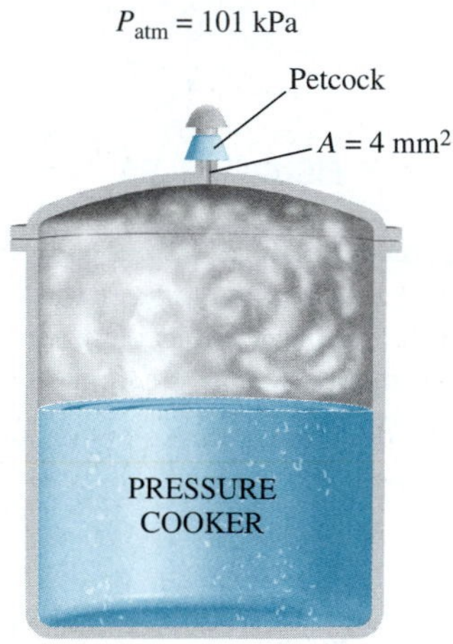

FIGURE P2–91

2–92 A glass tube is attached to a water pipe, as shown in Fig. P2–92. If the water pressure at the bottom of the tube is 115 kPa and the local atmospheric pressure is 92 kPa, determine how high the water will rise in the tube, in m. Take the density of water to be 1000 kg/m³.

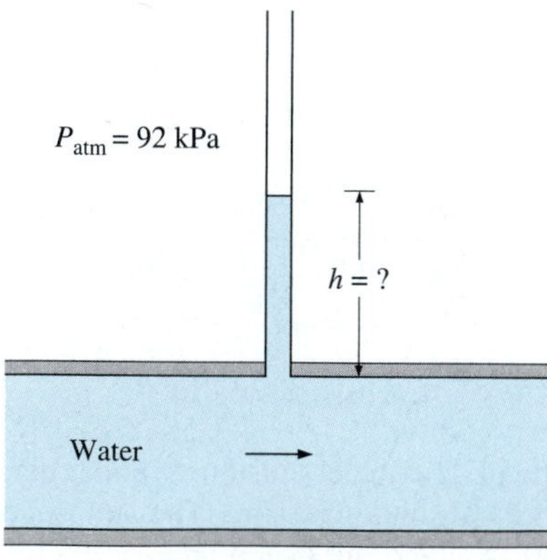

FIGURE P2–92

2–93 The average atmospheric pressure on earth is approximated as a function of altitude by the relation $P_{atm} = 101.325 (1 - 0.02256z)^{5.256}$, where P_{atm} is the atmospheric pressure in kPa and z is the altitude in km with $z = 0$ at sea level. Determine the approximate atmospheric pressures at Atlanta (z = 306 m), Denver (z = 1610 m), Mexico City (z = 2309 m), and the top of Mount Everest (z = 8848 m).

2–94 When measuring small pressure differences with a manometer, often one arm of the manometer is inclined to improve the accuracy of reading. (The pressure difference is still proportional to the *vertical* distance and not the actual length of the fluid along the tube.) The air pressure in a circular duct is to be measured using a manometer whose open arm is inclined 35° from the horizontal, as shown in Fig. P2–94. The density of the liquid in the manometer is 0.81 kg/L, and the vertical distance between the fluid levels in the two arms of the manometer is 8 cm. Determine the gage pressure of air in the duct and the length of the fluid column in the inclined arm above the fluid level in the vertical arm.

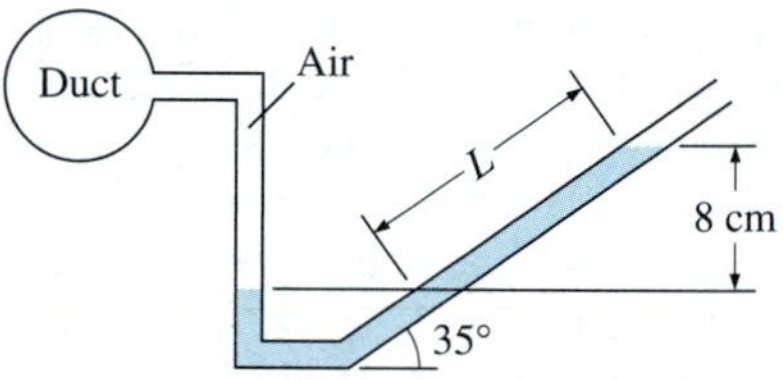

FIGURE P2–94

2–95E Consider a U-tube whose arms are open to the atmosphere. Now equal volumes of water and light oil (ρ = 49.3 lbm/ft³) are poured from different arms. A person blows from the oil side of the U-tube until the contact surface of the two fluids moves to the bottom of the U-tube, and thus the liquid levels in the two arms are the same. If the fluid height in each arm is 30 in, determine the gage pressure the person exerts on the oil by blowing.

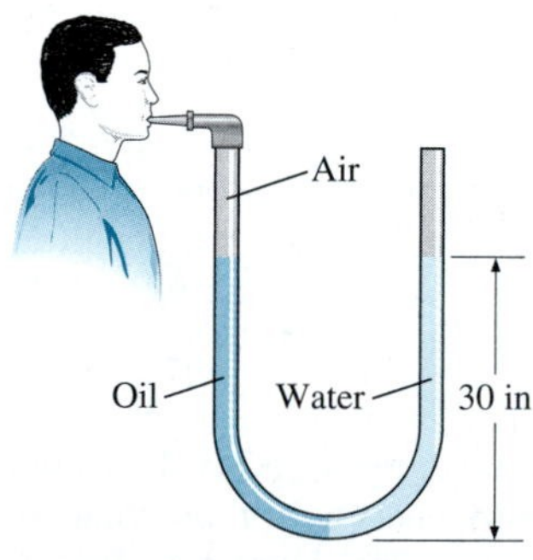

FIGURE P2–95E

2–96 Intravenous infusions are usually driven by gravity by hanging the fluid bottle at sufficient height to counteract the blood pressure in the vein and to force the fluid into the body. The higher the bottle is raised, the higher the flow rate of the fluid will be. (*a*) If it is observed that the fluid and the blood pressures balance each other when the bottle is 1.2 m above the arm level, determine the gage pressure of the blood. (*b*) If the gage pressure of the fluid at the arm level needs to be 20 kPa for sufficient flow rate, determine how high the bottle must be placed. Take the density of the fluid to be 1020 kg/m^3.

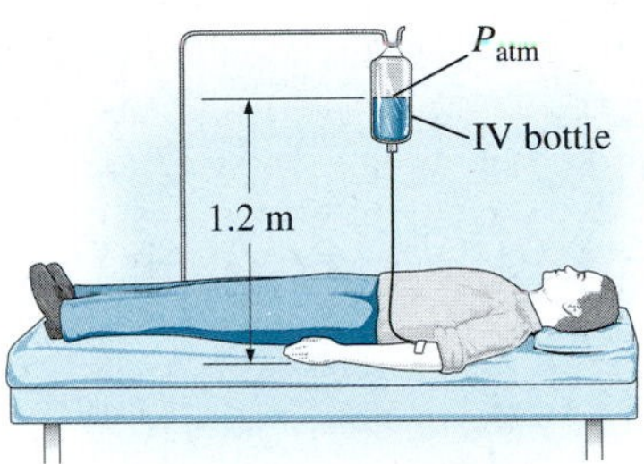

FIGURE P2–96

2–97E A water pipe is connected to a double-U manometer as shown in Fig. P2–97E at a location where the local atmospheric pressure is 14.2 psia. Determine the absolute pressure at the center of the pipe.

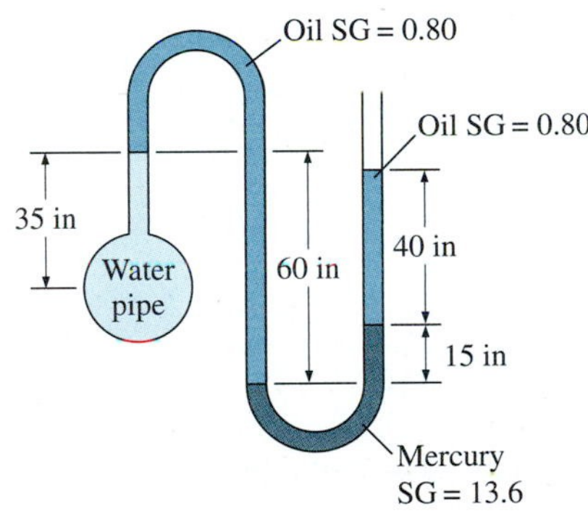

FIGURE P2–97E

2–98 It is well-known that the temperature of the atmosphere varies with altitude. In the troposphere, which extends to an altitude of 11 km, for example, the variation of temperature can be approximated by $T = T_0 - \beta z$, where T_0 is the temperature at sea level, which can be taken to be 288.15 K, and $\beta = 0.0065$ K/m. The gravitational acceleration also changes with altitude as $g(z) = g_0/(1 + z/6{,}370{,}320)^2$ where $g_0 = 9.807$ m/s^2 and z is the elevation from sea level in m. Obtain a relation for the variation of pressure in the troposphere (*a*) by ignoring and (*b*) by considering the variation of *g* with altitude.

2–99 The variation of pressure with density in a thick gas layer is given by $P = C\rho^n$, where C and n are constants. Noting that the pressure change across a differential fluid layer of thickness dz in the vertical z-direction is given as $dP = -\rho g\,dz$, obtain a relation for pressure as a function of elevation z. Take the pressure and density at $z = 0$ to be P_0 and ρ_0, respectively.

2–100 Pressure transducers are commonly used to measure pressure by generating analog signals usually in the range of 4 mA to 20 mA or 0 V-dc to 10 V-dc in response to applied pressure. The system whose schematic is shown in Fig. P2–100 can be used to calibrate pressure transducers. A rigid container is filled with pressurized air, and pressure is measured by the manometer attached. A valve is used to regulate the pressure in the container. Both the pressure and the electric signal are measured simultaneously for various settings, and the results are tabulated. For the given set of measurements, obtain the calibration curve in the form of $P = aI + b$, where a and b are constants, and calculate the pressure that corresponds to a signal of 10 mA.

Δh, mm	28.0	181.5	297.8	413.1	765.9
I, mA	4.21	5.78	6.97	8.15	11.76

Δh, mm	1027	1149	1362	1458	1536
I, mA	14.43	15.68	17.86	18.84	19.64

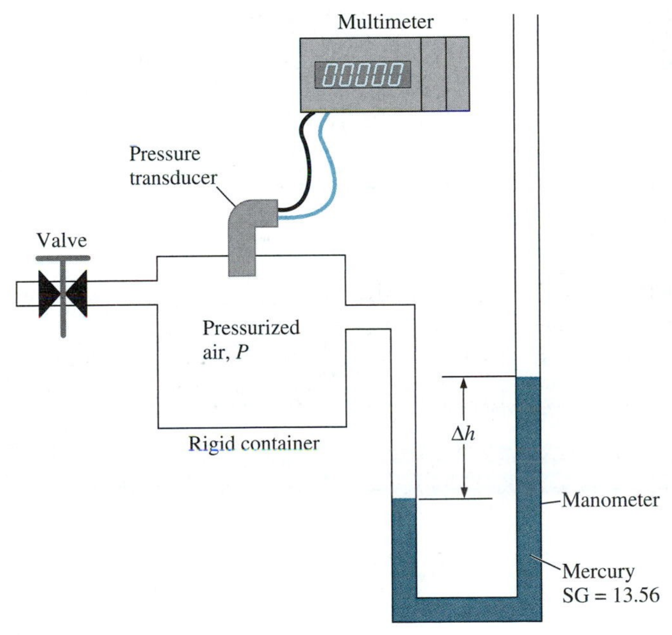

FIGURE P2–100

Design and Essay Problems

2–101 Write an essay on different temperature measurement devices. Explain the operational principle of each device, its advantages and disadvantages, its cost, and its range of applicability. Which device would you recommend for use in the following cases: taking the temperatures of patients in a doctor's office, monitoring the variations of temperature of a car engine block at several locations, and monitoring the temperatures in the furnace of a power plant?

Chapter 3

ENERGY, ENERGY TRANSFER, AND GENERAL ENERGY ANALYSIS

Whether we realize it or not, energy is an important part of most aspects of daily life. The quality of life, and even its sustenance, depends on the availability of energy. Therefore, it is important to have a good understanding of the sources of energy, the conversion of energy from one form to another, and the ramifications of these conversions.

Energy exists in numerous forms such as thermal, mechanical, electric, chemical, and nuclear. Even mass can be considered a form of energy. Energy can be transferred to or from a closed system (a fixed mass) in two distinct forms: *heat* and *work*. For control volumes, energy can also be transferred by mass flow. An energy transfer to or from a closed system is *heat* if it is caused by a temperature difference. Otherwise it is *work*, and it is caused by a force acting through a distance.

We start this chapter with a discussion of various forms of energy and energy transfer by heat. We then introduce various forms of work and discuss energy transfer by work. We continue with developing a general intuitive expression for the *first law of thermodynamics*, also known as the *conservation of energy principle*, which is one of the most fundamental principles in nature, and we then demonstrate its use. Finally, we discuss the efficiencies of some familiar energy conversion processes, and examine the impact on energy conversion on the environment. Detailed treatments of the first law of thermodynamics for closed systems and control volumes are given in Chaps. 5 and 6, respectively.

Objectives

The objectives of this chapter are to:

- Introduce the concept of energy and define its various forms.
- Discuss the nature of internal energy.
- Define the concept of heat and the terminology associated with energy transfer by heat.
- Define the concept of work, including electrical work and several forms of mechanical work.
- Introduce the first law of thermodynamics, energy balances, and mechanisms of energy transfer to or from a system.
- Determine that a fluid flowing across a control surface of a control volume carries energy across the control surface in addition to any energy transfer across the control surface that may be in the form of heat and/or work.
- Define energy conversion efficiencies.
- Discuss the implications of energy conversion on the environment.

FIGURE 3–1

A refrigerator operating with its door open in a well-sealed and well-insulated room.

3–1 ▪ INTRODUCTION

Since our high school years, we have been familiar with the conservation of energy principle, which is an expression of the first law of thermodynamics. We are told repeatedly that energy cannot be created or destroyed during a process; it can only change from one form to another. This seems simple enough, but let's test ourselves to see how well we understand and truly believe in this principle.

Consider a room whose door and windows are tightly closed, and whose walls are well-insulated so that heat loss or gain through the walls is negligible. Now let's place a refrigerator in the middle of the room with its door open, and plug it into a wall outlet (Fig. 3–1). You may even use a small fan to circulate the air in order to maintain temperature uniformity in the room. Now, what do you think will happen to the average temperature of air in the room? Will it be increasing or decreasing? Or will it remain constant?

Probably the first thought that comes to mind is that the average air temperature in the room will decrease as the warmer room air mixes with the air cooled by the refrigerator. Some may draw our attention to the heat generated by the motor of the refrigerator, and may argue that the average air temperature may rise if this heating effect is greater than the cooling effect. But they will get confused if it is stated that the motor is made of superconducting materials, and thus there is hardly any heat generation in the motor.

Heated discussion may continue with no end in sight until we remember the conservation of energy principle that we take for granted: If we take the entire room—including the air and the refrigerator—as the system, which is an adiabatic closed system since the room is well-sealed and well-insulated, the only energy interaction involved is the electrical energy crossing the system boundary and entering the room. The conservation of energy requires the energy content of the room to increase by an amount equal to the amount of the electrical energy drawn by the refrigerator, which can be measured by an ordinary electric meter. The refrigerator or its motor does not store this energy. Therefore, this energy must now be in the room air, and it will manifest itself as a rise in the air temperature. The temperature rise of air can be calculated on the basis of the conservation of energy principle using the properties of air and the amount of electrical energy consumed. What do you think would happen if we had a window air conditioning unit instead of a refrigerator placed in the middle of this room? What if we operated a fan in this room instead (Fig. 3–2)?

FIGURE 3–2

A fan running in a well-sealed and well-insulated room will raise the temperature of air in the room.

Note that energy is conserved during the process of operating the refrigerator placed in a room—the electrical energy is converted into an equivalent amount of thermal energy stored in the room air. If energy is already conserved, then what are all those speeches on energy conservation and the measures taken to conserve energy? Actually, by "energy conservation" what is meant is the conservation of the *quality* of energy, not the quantity. Electricity, which is of the highest quality of energy, for example, can always be converted to an equal amount of thermal energy (also called *heat*). But only a small fraction of thermal energy, which is the lowest quality of energy, can be converted back to electricity, as we discuss in Chap. 7. Think about the things that you can do with the electrical energy that the refrigerator has consumed, and the air in the room that is now at a higher temperature.

Now if asked to name the energy transformations associated with the operation of a refrigerator, we may still have a hard time answering because all we see is electrical energy entering the refrigerator and heat dissipated from the refrigerator to the room air. Obviously there is need to study the various forms of energy first, and this is exactly what we do next, followed by a study of the mechanisms of energy transfer.

3–2 ▪ FORMS OF ENERGY

Energy can exist in numerous forms such as thermal, mechanical, kinetic, potential, electric, magnetic, chemical, and nuclear, and their sum constitutes the **total energy** E of a system. The total energy of a system on a *unit mass* basis is denoted by e and is expressed as

$$e = \frac{E}{m} \qquad \text{(kJ/kg)} \tag{3–1}$$

Thermodynamics provides no information about the absolute value of the total energy. It deals only with the *change* of the total energy, which is what matters in engineering problems. Thus the total energy of a system can be assigned a value of zero ($E = 0$) at some convenient reference point. The change in total energy of a system is independent of the reference point selected. The decrease in the potential energy of a falling rock, for example, depends on only the elevation difference and not the reference level selected.

In thermodynamic analysis, it is often helpful to classify the various forms of energy that make up the total energy of a system into two groups: *macroscopic* and *microscopic*. The **macroscopic** forms of energy are those a system possesses as a whole with respect to some outside reference frame, such as kinetic and potential energies (Fig. 3–3). The **microscopic** forms of energy are those related to the molecular structure of a system and the degree of the molecular activity, and they are independent of outside reference frames. The sum of all the microscopic forms of energy is called the **internal energy** of a system and is denoted by U.

FIGURE 3–3

The macroscopic energy of an object changes with velocity and elevation.

The term *energy* was coined in 1807 by Thomas Young, and its use in thermodynamics was proposed in 1852 by Lord Kelvin. The term *internal energy* and its symbol U first appeared in the works of Rudolph Clausius and William Rankine in the second half of the nineteenth century, and it eventually replaced the alternative terms *inner work, internal work,* and *intrinsic energy* commonly used at the time.

The macroscopic energy of a system is related to motion and the influence of some external effects such as gravity, magnetism, electricity, and surface tension. The energy that a system possesses as a result of its motion relative to some reference frame is called **kinetic energy** (KE). When all parts of a system move with the same velocity, the kinetic energy is expressed as

$$\text{KE} = m\frac{V^2}{2} \qquad \text{(kJ)} \tag{3–2}$$

or, on a unit mass basis,

$$\mathrm{ke} = \frac{V^2}{2} \quad \text{(kJ/kg)} \tag{3-3}$$

where V denotes the magnitude of the velocity of the system relative to some fixed reference frame. The kinetic energy of a rotating solid body is given by $\frac{1}{2}I\omega^2$ where I is the moment of inertia of the body and ω is the angular velocity magnitude.

The energy that a system possesses as a result of its elevation in a gravitational field is called **potential energy** (PE) and is expressed as

$$\mathrm{PE} = mgz \quad \text{(kJ)} \tag{3-4}$$

or, on a unit mass basis,

$$\mathrm{pe} = gz \quad \text{(kJ/kg)} \tag{3-5}$$

where g is the gravitational acceleration and z is the elevation of the center of gravity of a system relative to some arbitrarily selected reference level.

The magnetic, electric, and surface tension effects are significant only in some specialized cases and are usually ignored. In the absence of such effects, the total energy of a system consists of the kinetic, potential, and internal energies and is expressed as

$$E = U + \mathrm{KE} + \mathrm{PE} = U + m\frac{V^2}{2} + mgz \quad \text{(kJ)} \tag{3-6}$$

or, on a unit mass basis,

$$e = u + \mathrm{ke} + \mathrm{pe} = u + \frac{V^2}{2} + gz \quad \text{(kJ/kg)} \tag{3-7}$$

Most closed systems remain stationary during a process and thus experience no change in their kinetic and potential energies. Closed systems whose velocity and elevation of the center of gravity remain constant during a process are frequently referred to as **stationary systems**. The change in the total energy ΔE of a stationary system is identical to the change in its internal energy ΔU. In this text, a closed system is assumed to be stationary unless stated otherwise.

Control volumes typically involve fluid flow for long periods of time, and it is convenient to express the energy flow associated with a fluid stream in the rate form. This is done by incorporating the **mass flow rate** $\dot{m}$, which is *the amount of mass flowing through a cross section per unit time*. It is related to the **volume flow rate** $\dot{V}$, which is the volume of a fluid flowing through a cross section per unit time, by

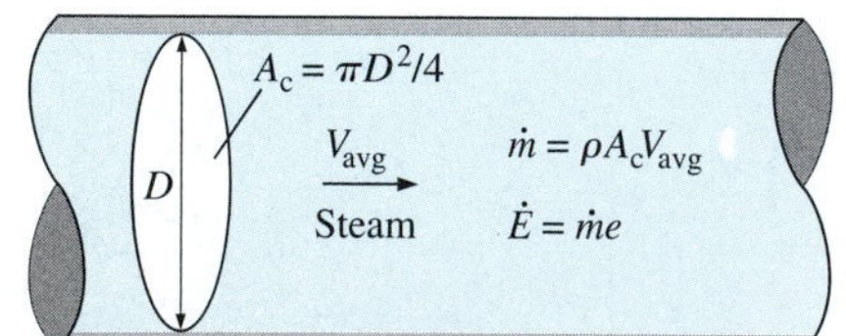

FIGURE 3–4

Mass and energy flow rates associated with the flow of steam in a pipe of inner diameter D with an average velocity of V_{avg}.

Mass flow rate: $$\dot{m} = \rho\dot{V} = \rho A_c V_{avg} \quad \text{(kg/s)} \tag{3-8}$$

which is analogous to $m = \rho V$. Here ρ is the fluid density, A_c is the cross-sectional area of flow, and V_{avg} is the average flow velocity normal to A_c. The dot over a symbol is used to indicate *time rate* throughout the book. Then the energy flow rate associated with a fluid flowing at a rate of $\dot{m}$ is (Fig. 3–4)

Energy flow rate: $$\dot{E} = \dot{m}e \quad \text{(kJ/s or kW)} \tag{3-9}$$

which is analogous to $E = me$.

Some Physical Insight to Internal Energy

Internal energy is defined earlier as the sum of all the *microscopic* forms of energy of a system. It is related to the *molecular structure* and the degree of *molecular activity* and can be viewed as the sum of the *kinetic* and *potential* energies of the molecules.

To have a better understanding of internal energy, let us examine a system at the molecular level. The molecules of a gas move through space with some velocity, and thus possess some kinetic energy. This is known as the *translational energy*. The atoms of polyatomic molecules rotate about an axis, and the energy associated with this rotation is the *rotational kinetic energy*. The atoms of a polyatomic molecule may also vibrate about their common center of mass, and the energy associated with this back-and-forth motion is the *vibrational kinetic energy*. For gases, the kinetic energy is mostly due to translational and rotational motions, with vibrational motion becoming significant at higher temperatures. The electrons in an atom rotate about the nucleus, and thus possess *rotational kinetic energy*. Electrons at outer orbits have larger kinetic energies. Electrons also spin about their axes, and the energy associated with this motion is the *spin energy*. Other particles in the nucleus of an atom also possess spin energy. The portion of the internal energy of a system associated with the kinetic energies of the molecules is called the **sensible energy** (Fig. 3–5). The average velocity and the degree of activity of the molecules are proportional to the temperature of the gas. Therefore, at higher temperatures, the molecules possess higher kinetic energies, and as a result the system has a higher internal energy.

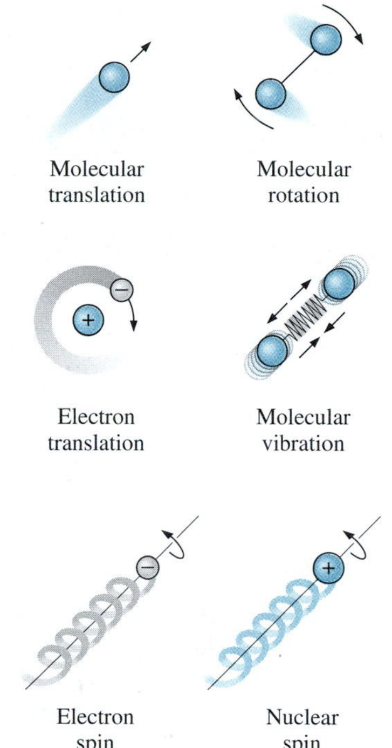

FIGURE 3–5

The various forms of microscopic energies that make up *sensible* energy.

The internal energy is also associated with various *binding forces* between the molecules of a substance, between the atoms within a molecule, and between the particles within an atom and its nucleus. The forces that bind the *molecules* to each other are, as one would expect, strongest in solids and weakest in gases. If sufficient energy is added to the molecules of a solid or liquid, the molecules overcome these molecular forces and break away, turning the substance into a gas. This is a phase-change process. Because of this added energy, a system in the gas phase is at a higher internal energy level than it is in the solid or the liquid phase. The internal energy associated with the phase of a system is called the **latent energy**. The phase-change process can occur without a change in the chemical composition of a system. Most practical problems fall into this category, and one does not need to pay any attention to the forces binding the atoms in a molecule to each other.

An atom consists of neutrons and positively charged protons bound together by very strong nuclear forces in the nucleus, and negatively charged electrons orbiting around it. The internal energy associated with the atomic bonds in a molecule is called **chemical energy**. During a chemical reaction, such as a combustion process, some chemical bonds are destroyed while others are formed. As a result, the internal energy changes. The nuclear forces are much larger than the forces that bind the electrons to the nucleus. The tremendous amount of energy associated with the strong bonds within the nucleus of the atom itself is called **nuclear energy** (Fig. 3–6). Obviously, we need not be concerned with nuclear energy in thermodynamics unless, of course, we deal with fusion or fission reactions. A chemical reaction involves changes in the structure of the electrons of the atoms, but a nuclear reaction involves changes in the core or nucleus. Therefore, an

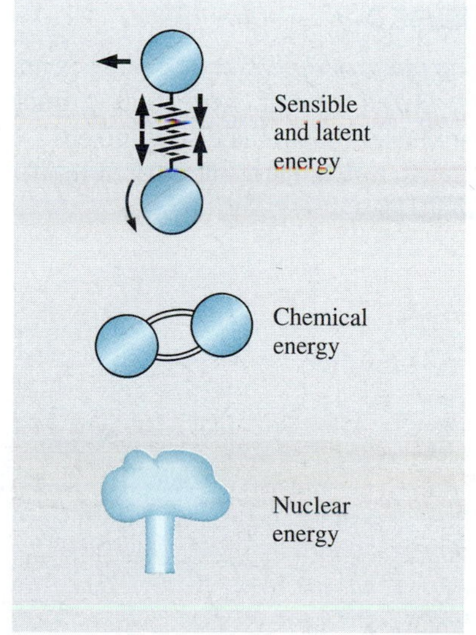

FIGURE 3–6

The internal energy of a system is the sum of all forms of the microscopic energies.

atom preserves its identity during a chemical reaction but loses it during a nuclear reaction. Atoms may also possess *electric* and *magnetic dipole-moment energies* when subjected to external electric and magnetic fields due to the twisting of the magnetic dipoles produced by the small electric currents associated with the orbiting electrons.

The forms of energy already discussed, which constitute the total energy of a system, can be *contained* or *stored* in a system, and thus can be viewed as the *static* forms of energy. The forms of energy not stored in a system can be viewed as the *dynamic* forms of energy or as *energy interactions*. The dynamic forms of energy are recognized at the system boundary as they cross it, and they represent the energy gained or lost by a system during a process. The only two forms of energy interactions associated with a closed system are **heat transfer** and **work**. An energy interaction is heat transfer if its driving force is a temperature difference. Otherwise it is work, as explained in the next section. A control volume can also exchange energy via mass transfer since any time mass is transferred into or out of a system, the energy content of the mass is also transferred with it.

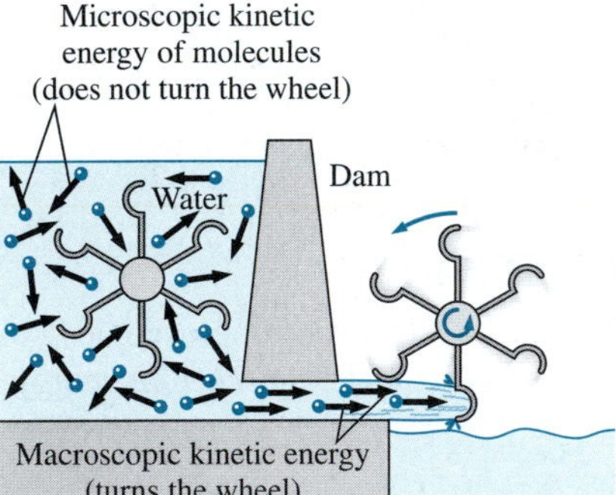

FIGURE 3–7

The *macroscopic* kinetic energy is an organized form of energy and is much more useful than the disorganized *microscopic* kinetic energies of the molecules.

In daily life, we frequently refer to the sensible and latent forms of internal energy as *heat*, and we talk about heat content of bodies. In thermodynamics, however, we usually refer to those forms of energy as **thermal energy** to prevent any confusion with *heat transfer*.

Distinction should be made between the macroscopic kinetic energy of an object as a whole and the microscopic kinetic energies of its molecules that constitute the sensible internal energy of the object (Fig. 3–7). The kinetic energy of an object is an *organized* form of energy associated with the orderly motion of all molecules in one direction in a straight path or around an axis. In contrast, the kinetic energies of the molecules are completely *random* and highly *disorganized*. As you will see in later chapters, the organized energy is much more valuable than the disorganized energy, and a major application area of thermodynamics is the conversion of disorganized energy (heat) into organized energy (work). You will also see that the organized energy can be converted to disorganized energy completely, but only a fraction of disorganized energy can be converted to organized energy by specially built devices called *heat engines* (like car engines and power plants). A similar argument can be given for the macroscopic potential energy of an object as a whole and the microscopic potential energies of the molecules.

More on Nuclear Energy

The best known fission reaction involves the split of the uranium atom (the U-235 isotope) into other elements and is commonly used to generate electricity in nuclear power plants (443 of them in 2006, generating 370,000 MW worldwide), to power nuclear submarines, aircraft carriers, and even to power spacecraft.

The percentage of electricity produced by nuclear power is 78 percent in France, 25 percent in Japan, 28 percent in Germany, and 20 percent in the United States. The first nuclear chain reaction was achieved by Enrico Fermi in 1942, and the first large-scale nuclear reactors were built in 1944 for the purpose of producing material for nuclear weapons. When a uranium-235 atom absorbs a neutron and splits during a fission process, it produces a cesium-140 atom, a rubidium-93 atom, 3 neutrons, and 3.2×10^{-11} J of energy. In practical terms, the complete fission of 1 kg of uranium-235 releases 6.73×10^{10} kJ of heat, which is more than the heat released when 3000 tons of coal are burned. Therefore, for the same amount of fuel, a nuclear fission reaction releases several million times more energy than a chemical reaction. The safe disposal of used nuclear fuel, however, remains a concern.

Nuclear energy by fusion is released when two small nuclei combine into a larger one. The huge amount of energy radiated by the sun and the other stars is thought to originate from such a fusion process that involves the combination of two hydrogen atoms into a helium atom. When two heavy hydrogen (deuterium) nuclei combine during a fusion process, they produce a helium-3 atom, a free neutron, and 5.1×10^{-13} J of energy (Fig. 3–8).

Fusion reactions are much more difficult to achieve in practice because of the strong repulsion between the positively charged nuclei, called the *Coulomb repulsion*. To overcome this repulsive force and to enable the two nuclei to fuse together, the energy level of the nuclei must be raised by heating them to about 100 million °C. But such high temperatures are found only in the stars or in exploding atomic bombs (the A-bomb). In fact, the uncontrolled fusion reaction in a hydrogen bomb (the H-bomb) is initiated by a small atomic bomb. The uncontrolled fusion reaction was achieved in the early 1950s, but all the efforts since then to achieve controlled fusion by massive lasers, powerful magnetic fields, and electric currents to generate a net production of power have failed.

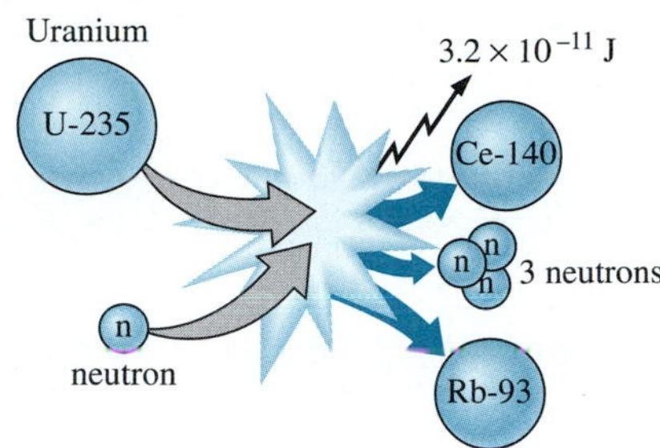

(a) Fission of uranium

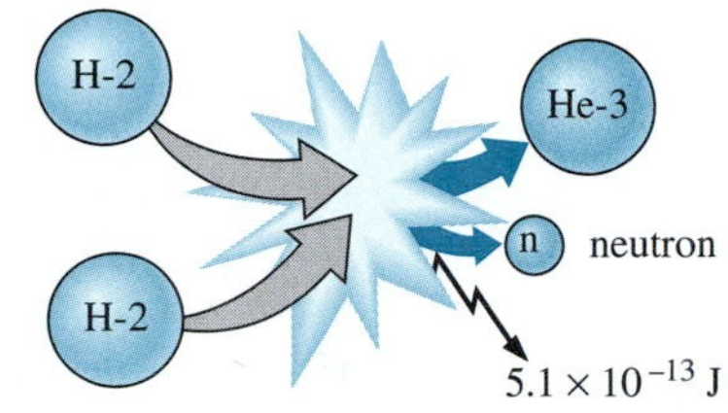

(b) Fusion of hydrogen

FIGURE 3–8

The fission of uranium and the fusion of hydrogen during nuclear reactions, and the release of nuclear energy.

EXAMPLE 3–1 A Car Powered by Nuclear Fuel

An average car consumes about 5 L of gasoline a day, and the capacity of the fuel tank of a car is about 50 L. Therefore, the car needs to be refueled once every 10 days. Also, the density of gasoline ranges from 0.68 to 0.78 kg/L, and its lower heating value is about 44,000 kJ/kg (that is, 44,000 kJ of heat is released when 1 kg of gasoline is completely burned). Suppose all the problems associated with the radioactivity and waste disposal of nuclear fuels are resolved, and a car is to be powered by U-235. If a new car comes equipped with 0.1-kg of the nuclear fuel U-235, determine if this car will ever need refueling under average driving conditions (Fig. 3–9).

Solution A car powered by nuclear energy comes equipped with nuclear fuel. It is to be determined if this car will ever need refueling.

Assumptions **1** Gasoline is an incompressible substance with an average density of 0.75 kg/L. **2** Nuclear fuel is completely converted to thermal energy.

Analysis The mass of gasoline used per day by the car is

$$\dot{m}_{\text{gasoline}} = (\rho\dot{V})_{\text{gasoline}} = (0.75\text{ kg/L})(5\text{ L/day}) = 3.75\text{ kg/day}$$

Noting that the heating value of gasoline is 44,000 kJ/kg, the energy supplied to the car per day is

$$\dot{E} = (\dot{m}_{\text{gasoline}})(\text{Heating value})$$
$$= (3.75\text{ kg/day})(44{,}000\text{ kJ/kg}) = 165{,}000\text{ kJ/day}$$

FIGURE 3–9

Schematic for Example 3–1.

The complete fission of 0.1 kg of uranium-235 releases

$$(6.73 \times 10^{10}\ \text{kJ/kg})(0.1\ \text{kg}) = 6.73 \times 10^{9}\ \text{kJ}$$

of heat, which is sufficient to meet the energy needs of the car for

$$\text{No. of days} = \frac{\text{Energy content of fuel}}{\text{Daily energy use}} = \frac{6.73 \times 10^{9}\ \text{kJ}}{165{,}000\ \text{kJ/day}} = \mathbf{40{,}790\ days}$$

which is equivalent to about 112 years. Considering that no car will last more than 100 years, this car will never need refueling. It appears that nuclear fuel of the size of a cherry is sufficient to power a car during its lifetime.

Discussion Note that this problem is not quite realistic since the necessary critical mass cannot be achieved with such a small amount of fuel. Further, all of the uranium cannot be converted in fission, again because of the critical mass problems after partial conversion.

Mechanical Energy

Many engineering systems are designed to transport a fluid from one location to another at a specified flow rate, velocity, and elevation difference, and the system may generate mechanical work in a turbine or it may consume mechanical work in a pump or fan during this process. These systems do not involve the conversion of nuclear, chemical, or thermal energy to mechanical energy. Also, they do not involve heat transfer in any significant amount, and they operate essentially at constant temperature. Such systems can be analyzed conveniently by considering the *mechanical forms of energy* only and the frictional effects that cause the mechanical energy to be lost (i.e., to be converted to thermal energy that usually cannot be used for any useful purpose).

The **mechanical energy** can be defined as *the form of energy that can be converted to mechanical work completely and directly by an ideal mechanical device such as an ideal turbine*. Kinetic and potential energies are the familiar forms of mechanical energy. Thermal energy is not mechanical energy, however, since it cannot be converted to work directly and completely (the second law of thermodynamics).

A pump transfers mechanical energy to a fluid by raising its pressure, and a turbine extracts mechanical energy from a fluid by dropping its pressure. Therefore, the pressure of a flowing fluid is also associated with its mechanical energy. In fact, the pressure unit Pa is equivalent to Pa $=$ N/m^2 $=$ N $\cdot$ m/m^3 $=$ J/m^3, which is energy per unit volume, and the product $P\mathrm{v}$ or its equivalent P/ρ has the unit J/kg, which is energy per unit mass. Note that pressure itself is not a form of energy. But a pressure force acting on a fluid through a distance produces work, called *flow work*, in the amount of P/ρ per unit mass. Flow work is expressed in terms of fluid properties, and it is convenient to view it as part of the energy of a flowing fluid and call it *flow energy*. Therefore, the mechanical energy of a flowing fluid can be expressed on a unit mass basis as

$$e_{\text{mech}} = \frac{P}{\rho} + \frac{V^2}{2} + gz \qquad \textbf{(3–10)}$$

where P/ρ is the *flow energy*, $V^2/2$ is the *kinetic energy*, and gz is the *potential energy* of the fluid, all per unit mass. It can also be expressed in rate form as

$$\dot{E}_{\text{mech}} = \dot{m}e_{\text{mech}} = \dot{m}\left(\frac{P}{\rho} + \frac{V^2}{2} + gz\right) \quad \textbf{(3–11)}$$

where $\dot{m}$ is the mass flow rate of the fluid. Then the mechanical energy change of a fluid during incompressible (ρ = constant) flow becomes

$$\Delta e_{\text{mech}} = \frac{P_2 - P_1}{\rho} + \frac{V_2^2 - V_1^2}{2} + g(z_2 - z_1) \qquad \text{(kJ/kg)} \quad \textbf{(3–12)}$$

and

$$\Delta\dot{E}_{\text{mech}} = \dot{m}\Delta e_{\text{mech}} = \dot{m}\left(\frac{P_2 - P_1}{\rho} + \frac{V_2^2 - V_1^2}{2} + g(z_2 - z_1)\right) \qquad \text{(kW)} \quad \textbf{(3–13)}$$

Therefore, the mechanical energy of a fluid does not change during flow if its pressure, density, velocity, and elevation remain constant. In the absence of any losses, the mechanical energy change represents the mechanical work supplied to the fluid (if $\Delta e_{\text{mech}} > 0$) or extracted from the fluid (if $\Delta e_{\text{mech}} < 0$).

EXAMPLE 3–2 Wind Energy

A site evaluated for a wind farm is observed to have steady winds at a speed of 8.5 m/s (Fig. 3–10). Determine the wind energy (*a*) per unit mass, (*b*) for a mass of 10 kg, and (*c*) per second for a flow rate of 1154 kg/s for air.

Solution A site with a specified wind speed is considered. Wind energy per unit mass, for a specified mass, and for a given mass flow rate of air are to be determined.

Assumptions Wind flows steadily at the specified speed.

Analysis The only harvestable form of energy of atmospheric air is the kinetic energy, which is captured by a wind turbine.

(*a*) Wind energy per unit mass of air is

$$e = \text{ke} = \frac{V^2}{2} = \frac{(8.5 \text{ m/s})^2}{2}\left(\frac{1 \text{ J/kg}}{1 \text{ m}^2/\text{s}^2}\right) = \mathbf{36.1\ J/kg}$$

(*b*) Wind energy for an air mass of 10 kg is

$$E = me = (10 \text{ kg})(36.1 \text{ J/kg}) = \mathbf{361\ J}$$

(*c*) Wind power (energy per unit time) for a mass flow rate of 1154 kg/s is

$$\dot{E} = \dot{m}e = (1154 \text{ kg/s})(36.1 \text{ J/kg})\left(\frac{1 \text{ kW}}{1000 \text{ J/s}}\right) = \mathbf{41.7\ kW}$$

Discussion It can be shown that the specified mass flow rate corresponds to a 12-m diameter flow section when the air density is 1.2 kg/m³. Therefore, a wind turbine with a wind span diameter of 12 m has a power generation potential of 41.7 kW. Real wind turbines convert only about one-third of this potential to electric power, due to inefficiencies.

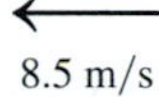

8.5 m/s

FIGURE 3–10

Potential site for a wind farm as discussed in Example 3–2.

3–3 ▪ ENERGY TRANSFER BY HEAT

Energy can cross the boundary of a closed system in two distinct forms: *heat* and *work* (Fig. 3–11). It is important to distinguish between these two forms of energy. Therefore, they will be discussed first, to form a sound basis for the development of the laws of thermodynamics.

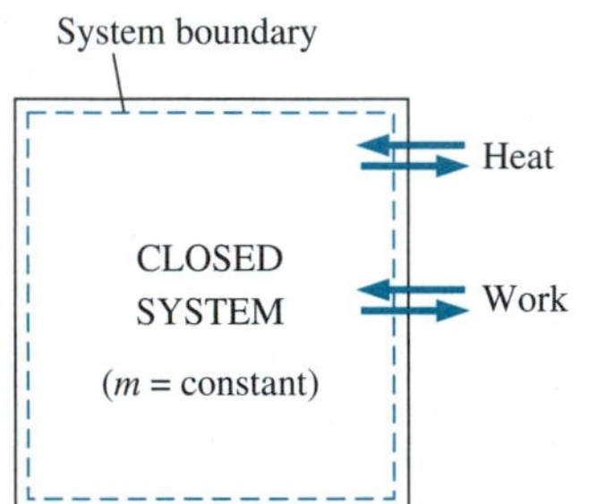

FIGURE 3–11

Energy can cross the boundaries of a closed system in the form of heat and work.

We know from experience that a can of cold soda left on a table eventually warms up and that a hot baked potato on the same table cools down. When a body is left in a medium that is at a different temperature, energy transfer takes place between the body and the surrounding medium until thermal equilibrium is established, that is, the body and the medium reach the same temperature. The direction of energy transfer is always from the higher temperature body to the lower temperature one. Once the temperature equality is established, energy transfer stops. In the processes described above, energy is said to be transferred in the form of heat.

Heat is defined as *the form of energy that is transferred between two systems (or a system and its surroundings) by virtue of a temperature difference* (Fig. 3–12). That is, an energy interaction is heat only if it takes place because of a temperature difference. Then it follows that there cannot be any heat transfer between two systems that are at the same temperature.

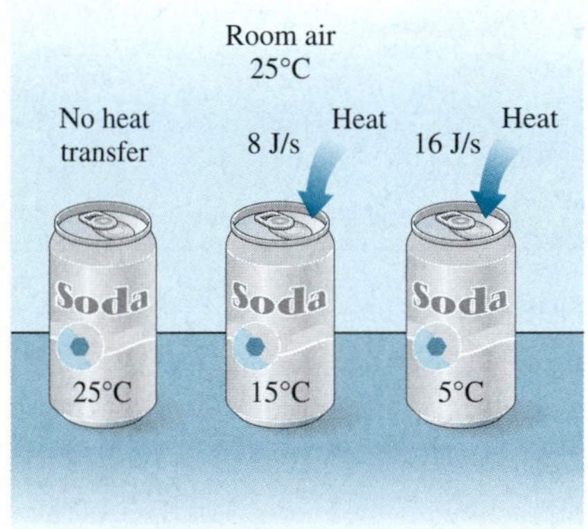

FIGURE 3–12

Temperature difference is the driving force for heat transfer. The larger the temperature difference, the higher is the rate of heat transfer.

Several phrases in common use today—such as heat flow, heat addition, heat rejection, heat absorption, heat removal, heat gain, heat loss, heat storage, heat generation, electrical heating, resistance heating, frictional heating, gas heating, heat of reaction, liberation of heat, specific heat, sensible heat, latent heat, waste heat, body heat, process heat, heat sink, and heat source—are not consistent with the strict thermodynamic meaning of the term *heat*, which limits its use to the *transfer* of thermal energy during a process. However, these phrases are deeply rooted in our vocabulary, and they are used by both ordinary people and scientists without causing any misunderstanding since they are usually interpreted properly instead of being taken literally. (Besides, no acceptable alternatives exist for some of these phrases.) For example, the phrase *body heat* is understood to mean *the thermal energy content* of a body. Likewise, *heat flow* is understood to mean *the transfer of thermal energy,* not the flow of a fluidlike substance called heat, although the latter incorrect interpretation, which is based on the caloric theory, is the origin of this phrase. Also, the transfer of heat into a system is frequently referred to as *heat addition* and the transfer of heat out of a system as *heat rejection*. Perhaps there are thermodynamic reasons for being so reluctant to replace *heat* by *thermal energy*: It takes less time and energy to say, write, and comprehend *heat* than it does *thermal energy*.

Heat is energy in transition. It is recognized only as it crosses the boundary of a system. Consider the hot baked potato one more time. The potato contains energy, but this energy is heat transfer only as it passes through the skin of the potato (the system boundary) to reach the air, as shown in Fig. 3–13. Once in the surroundings, the transferred heat becomes part of the internal energy of the surroundings. Thus, in thermodynamics, the term *heat* simply means *heat transfer*.

A process during which there is no heat transfer is called an **adiabatic process** (Fig. 3–14). The word *adiabatic* comes from the Greek word *adiabatos,* which means *not to be passed.* There are two ways a process can be adiabatic: Either the system is well insulated so that only a negligible amount of heat can pass through the boundary, or both the system and the surroundings are at the same temperature and therefore there is no driving force (temperature difference) for heat transfer. An adiabatic process should not be confused with an isothermal process. Even though there is no heat transfer during an adiabatic process, the energy content and thus the temperature of a system can still be changed by other means such as work.

As a form of energy, heat has energy units, kJ (or Btu) being the most common one. The amount of heat transferred during the process between two states (states 1 and 2) is denoted by Q_{12}, or just Q. Heat transfer *per unit mass* of a system is denoted q and is determined from

$$q = \frac{Q}{m} \qquad \text{(kJ/kg)} \tag{3-14}$$

Sometimes it is desirable to know the *rate of heat transfer* (the amount of heat transferred per unit time) instead of the total heat transferred over some time interval (Fig. 3–15). The heat transfer rate is denoted $\dot{Q}$, where the overdot stands for the time derivative, or "per unit time." The heat transfer rate $\dot{Q}$ has the unit kJ/s, which is equivalent to kW. When $\dot{Q}$ varies with time, the amount of heat transfer during a process is determined by integrating $\dot{Q}$ over the time interval of the process:

$$Q = \int_{t_1}^{t_2} \dot{Q}\, dt \qquad \text{(kJ)} \tag{3-15}$$

When $\dot{Q}$ remains constant during a process, this relation reduces to

$$Q = \dot{Q}\, \Delta t \qquad \text{(kJ)} \tag{3-16}$$

where $\Delta t = t_2 - t_1$ is the time interval during which the process takes place.

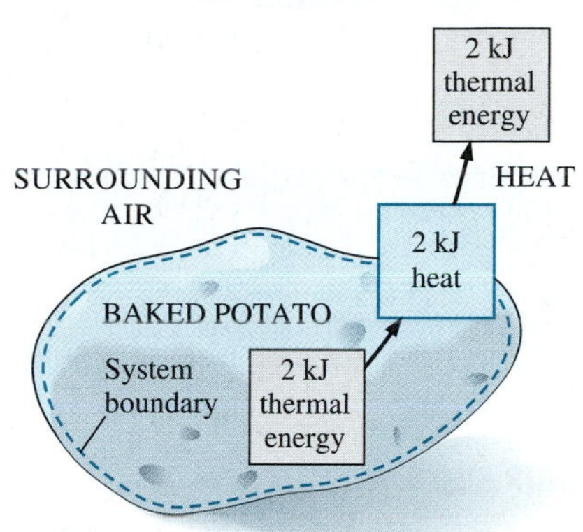

FIGURE 3–13

Energy is recognized as heat transfer only as it crosses the system boundary.

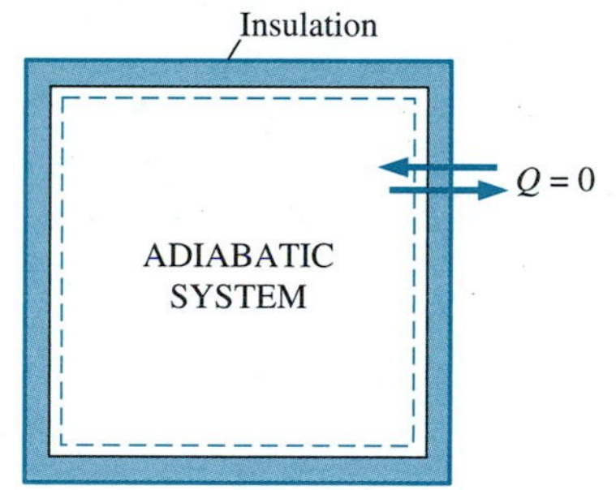

FIGURE 3–14

During an adiabatic process, a system exchanges no heat with its surroundings.

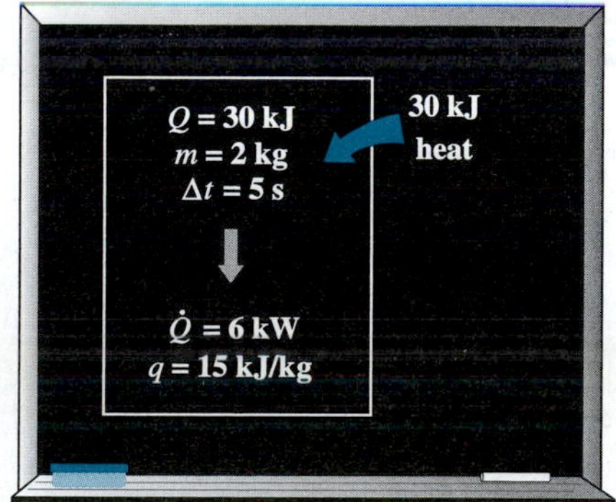

FIGURE 3–15

The relationships among q, Q, and $\dot{Q}$.

Historical Background on Heat

Heat has always been perceived to be something that produces in us a sensation of warmth, and one would think that the nature of heat is one of the first things understood by mankind. However, it was only in the middle of the nineteenth century that we had a true physical understanding of the nature of heat, thanks to the development at that time of the **kinetic theory**, which treats molecules as tiny balls that are in motion and thus possess kinetic energy. Heat is then defined as the energy associated with the random motion of atoms and molecules. Although it was suggested in the eighteenth and early nineteenth centuries that heat is the manifestation of motion at the molecular level (called the *live force*), the prevailing view of heat until the middle of the nineteenth century was based on the caloric theory proposed by the French chemist Antoine Lavoisier (1744–1794) in 1789. The caloric theory asserts that heat is a fluidlike substance called the **caloric** that is a massless, colorless, odorless, and tasteless substance that can be poured from one body into another (Fig. 3–16). When caloric was added to a body, its

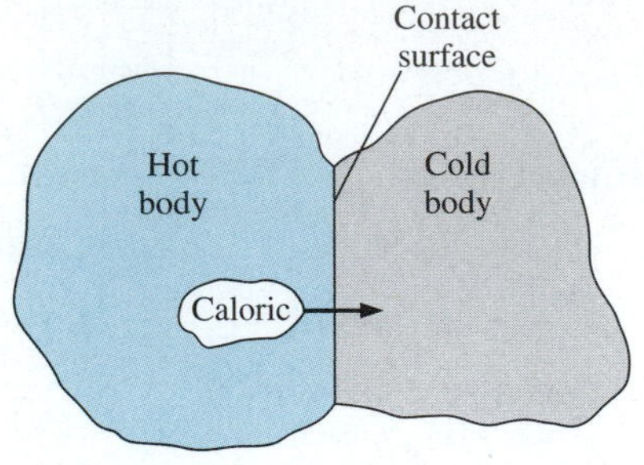

FIGURE 3–16

In the early nineteenth century, heat was thought to be an invisible fluid called the *caloric* that flowed from warmer bodies to the cooler ones.

temperature increased; and when caloric was removed from a body, its temperature decreased. When a body could not contain any more caloric, much the same way as when a glass of water could not dissolve any more salt or sugar, the body was said to be saturated with caloric. This interpretation gave rise to the terms *saturated liquid* and *saturated vapor* that are still in use today.

The caloric theory came under attack soon after its introduction. It maintained that heat is a substance that could not be created or destroyed. Yet it was known that heat can be generated indefinitely by rubbing one's hands together or rubbing two pieces of wood together. In 1798, the American Benjamin Thompson (Count Rumford) (1754–1814) showed in his papers that heat can be generated continuously through friction. The validity of the caloric theory was also challenged by several others. But it was the careful experiments of the Englishman James P. Joule (1818–1889) published in 1843 that finally convinced the skeptics that heat was not a substance after all, and thus put the caloric theory to rest. Although the caloric theory was totally abandoned in the middle of the nineteenth century, it contributed greatly to the development of thermodynamics and heat transfer.

Heat is transferred by three mechanisms: conduction, convection, and radiation. **Conduction** is the transfer of energy from the more energetic particles of a substance to the adjacent less energetic ones as a result of interaction between particles. **Convection** is the transfer of energy between a solid surface and the adjacent fluid that is in motion, and it involves the combined effects of conduction and fluid motion. **Radiation** is the transfer of energy due to the emission of electromagnetic waves (or photons).

3–4 ▪ ENERGY TRANSFER BY WORK

Work, like heat, is an energy interaction between a system and its surroundings. As mentioned earlier, energy can cross the boundary of a closed system in the form of heat or work. Therefore, *if the energy crossing the boundary of a closed system is not heat, it must be work.* Heat is easy to recognize: Its driving force is a temperature difference between the system and its surroundings. Then we can simply say that an energy interaction that is not caused by a temperature difference between a system and its surroundings is work. More specifically, *work is the energy transfer associated with a force acting through a distance.* A rising piston, a rotating shaft, and an electric wire crossing the system boundaries are all associated with work interactions.

Work is also a form of energy transferred like heat and, therefore, has energy units such as kJ. The work done during a process between states 1 and 2 is denoted by W_{12}, or simply W. The work done *per unit mass* of a system is denoted by w and is expressed as

$$w = \frac{W}{m} \qquad \text{(kJ/kg)} \tag{3–17}$$

The work done *per unit time* is called **power** and is denoted $\dot{W}$ (Fig. 3–17). The unit of power is kJ/s, or kW.

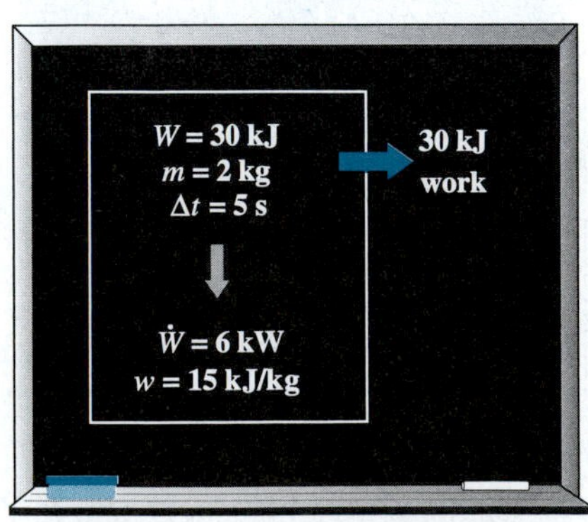

FIGURE 3–17

The relationships among w, W, and $\dot{W}$.

Heat and work are *directional quantities*, and thus the complete description of a heat or work interaction requires the specification of both the *magnitude* and *direction*. One way of doing that is to adopt a sign convention. The generally accepted **formal sign convention** for heat and work interactions is as follows: *heat transfer to a system and work done by a system are positive; heat transfer from a system and work done on a system are negative*. Another way is to use the subscripts *in* and *out* to indicate direction (Fig. 3–18). For example, a work input of 5 kJ can be expressed as W_{in} = 5 kJ, while a heat loss of 3 kJ can be expressed as Q_{out} = 3 kJ. When the direction of a heat or work interaction is not known, we can simply *assume* a direction for the interaction (using the subscript *in* or *out*) and solve for it. A positive result indicates the assumed direction is right. A negative result, on the other hand, indicates that the direction of the interaction is the opposite of the assumed direction. This is just like assuming a direction for an unknown force when solving a statics problem, and reversing the direction when a negative result is obtained for the force. We will use this *intuitive approach* in this book as it eliminates the need to adopt a formal sign convention and the need to carefully assign negative values to some interactions.

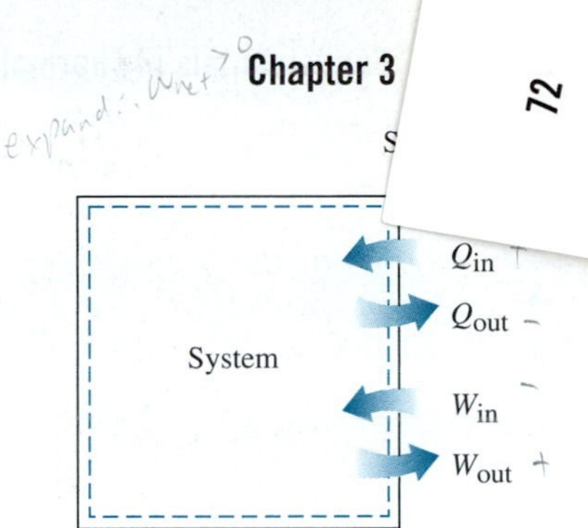

FIGURE 3–18

Specifying the directions of heat and work.

Note that a quantity that is transferred to or from a system during an interaction is not a property since the amount of such a quantity depends on more than just the state of the system. Heat and work are *energy transfer mechanisms* between a system and its surroundings, and there are many similarities between them:

1. Both are recognized at the boundaries of a system as they cross the boundaries. That is, both heat and work are *boundary* phenomena.
2. Neither can be possessed by a system. Systems possess energy, but not heat or work.
3. Both are associated with a *process,* not a state. Unlike properties, heat or work has no meaning at a state.
4. Both are *path functions* (i.e., their magnitudes depend on the path followed during a process as well as the end states).

Path functions have **inexact differentials** designated by the symbol δ. Therefore, a differential amount of heat or work is represented by δQ or δW, respectively, instead of dQ or dW. Properties, however, are **point functions** (i.e., they depend on the state only, and not on how a system reaches that state), and they have **exact differentials** designated by the symbol *d*. A small change in volume, for example, is represented by dV, and the total volume change during a process between states 1 and 2 is

$$\int_1^2 dV = V_2 - V_1 = \Delta V$$

That is, the volume change during process 1–2 is always the volume at state 2 minus the volume at state 1, regardless of the path followed (Fig. 3–19). The total work done during process 1–2, however, is

$$\int_1^2 \delta W = W_{12} \qquad (not\ \Delta W)$$

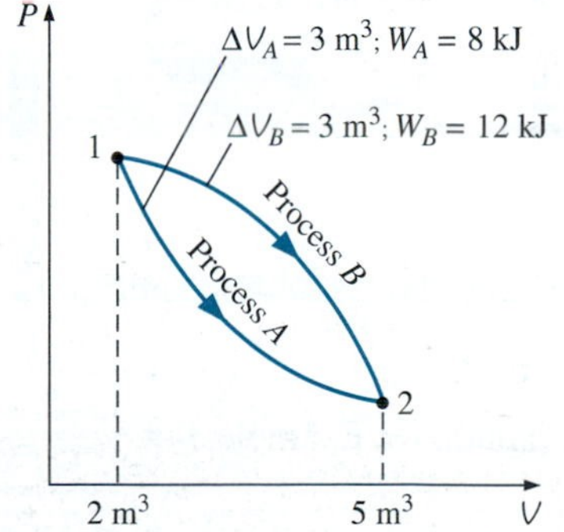

FIGURE 3–19

Properties are point functions; but heat and work are path functions (their magnitudes depend on the path followed).

That is, the total work is obtained by following the process path and adding the differential amounts of work (δW) done along the way. The integral of δW *is not* $W_2 - W_1$ (i.e., the work at state 2 minus work at state 1), which is meaningless since work is not a property and systems do not possess work at a state.

FIGURE 3–20
Schematic for Example 3–3.

EXAMPLE 3–3 Burning of a Candle in an Insulated Room

A candle is burning in a well-insulated room. Taking the room (the air plus the candle) as the system, determine (*a*) if there is any heat transfer during this burning process and (*b*) if there is any change in the internal energy of the system.

Solution A candle burning in a well-insulated room is considered. It is to be determined whether there is any heat transfer and any change in internal energy.

Analysis (*a*) The interior surfaces of the room form the system boundary, as indicated by the dashed lines in Fig. 3–20. As pointed out earlier, heat is recognized as it crosses the boundaries. Since the room is well insulated, we have an adiabatic system and no heat passes through the boundaries. Therefore, $Q = 0$ for this process.

(*b*) The internal energy involves energies that exist in various forms (sensible, latent, chemical, nuclear). During the process just described, part of the chemical energy is converted to sensible energy. Since there is no increase or decrease in the total internal energy of the system, $\Delta U = 0$ for this process.

Discussion There *is* heat transfer from the candle flame to the air, but not across the system boundary.

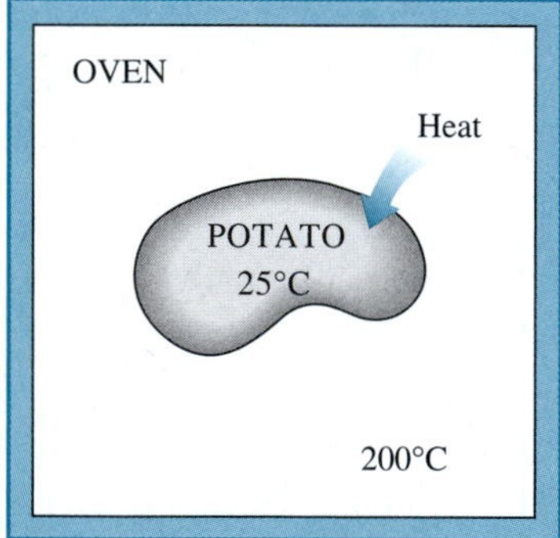

FIGURE 3–21
Schematic for Example 3–4.

EXAMPLE 3–4 Heating of a Potato in an Oven

A potato initially at room temperature (25°C) is being baked in an oven that is maintained at 200°C, as shown in Fig. 3–21. Is there any heat transfer during this baking process?

Solution A potato is being baked in an oven. It is to be determined whether there is any heat transfer during this process.

Analysis This is not a well-defined problem since the system is not specified. Let us assume that we are observing the potato, which will be our system. Then the skin of the potato can be viewed as the system boundary. Part of the energy in the oven will pass through the skin to the potato. Since the driving force for this energy transfer is a temperature difference, this is a heat transfer process.

Discussion The oven air itself must be heated somehow.

EXAMPLE 3–5 Heating of an Oven by Work Transfer

A well-insulated electric oven is being heated through its heating element. If the entire oven, including the heating element, is taken to be the system, determine whether this is a heat or work interaction.

Solution A well-insulated electric oven is being heated by its heating element. It is to be determined whether this is a heat or work interaction.

Analysis For this problem, the interior surfaces of the oven form the system boundary, as shown in Fig. 3–22. The energy content of the oven obviously increases during this process, as evidenced by a rise in temperature. This energy transfer to the oven is not caused by a temperature difference between the oven and the surrounding air. Instead, it is caused by *electrons* crossing the system boundary and thus doing work. Therefore, this is a work interaction.

Discussion There *is* heat transfer from the heating element to the oven air, but this heat does not cross the system boundary.

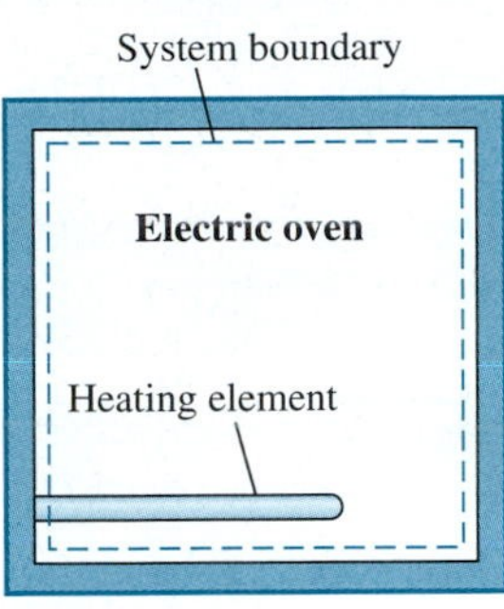

FIGURE 3–22
Schematic for Example 3–5.

EXAMPLE 3–6 Heating of an Oven by Heat Transfer

Answer the question in Example 3–5 if the system is taken as only the air in the oven without the heating element.

Solution The question in Example 3–5 is to be reconsidered by taking the system to be only the air in the oven.

Analysis This time, the system boundary includes the outer surface of the heating element and does not cut through it, as shown in Fig. 3–23. Therefore, no electrons cross the system boundary at any point. Instead, the energy generated in the interior of the heating element is transferred to the air around it as a result of the temperature difference between the heating element and the air in the oven. Therefore, this is a heat transfer process.

Discussion For both cases, the amount of energy transfer to the air is the same. These two examples show that an energy transfer can be heat or work, depending on how the system is selected.

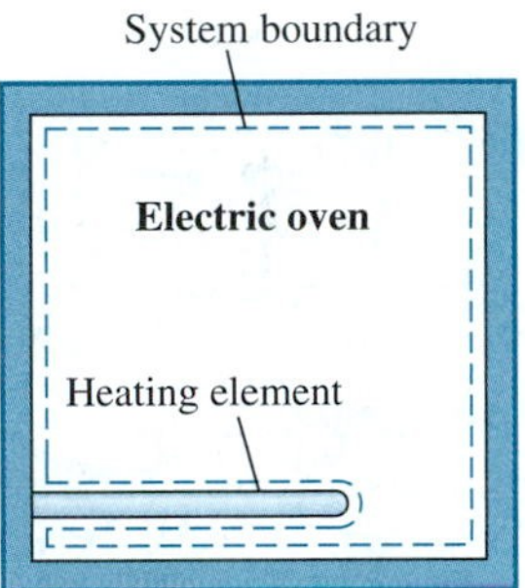

FIGURE 3–23
Schematic for Example 3–6.

Electrical Work

It was pointed out in Example 3–5 that electrons crossing the system boundary do electrical work on the system. In an electric field, electrons in a wire move under the effect of electromotive forces, doing work. When N coulombs of electrical charge move through a potential difference $\mathbf{V}$, the electrical work done is

$$W_e = \mathbf{V}N$$

which can also be expressed in the rate form as

$$\dot{W}_e = \mathbf{V}I \qquad \text{(W)} \tag{3-18}$$

where $\dot{W}_e$ is the **electrical power** and I is the number of electrical charges flowing per unit time, that is, the *current* (Fig. 3–24). In general, both $\mathbf{V}$ and I vary with time, and the electrical work done during a time interval Δt is expressed as

$$W_e = \int_1^2 \mathbf{V}I\,dt \qquad \text{(kJ)} \tag{3-19}$$

When both $\mathbf{V}$ and I remain constant during the time interval Δt, it reduces to

$$W_e = \mathbf{V}I\,\Delta t \qquad \text{(kJ)} \tag{3-20}$$

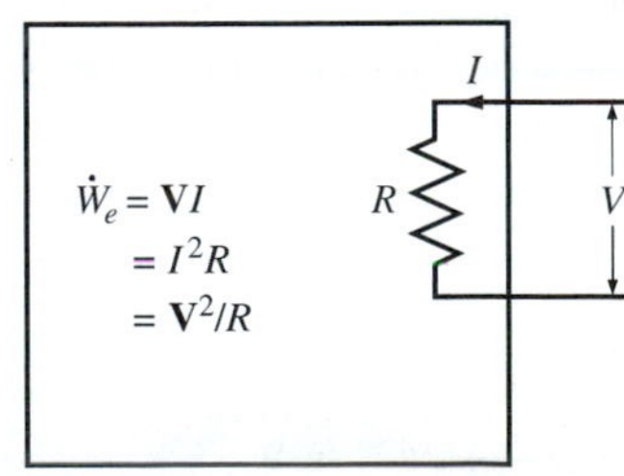

FIGURE 3–24
Electrical power in terms of resistance R, current I, and potential difference $\mathbf{V}$.

3–5 ▪ MECHANICAL FORMS OF WORK

There are several different ways of doing work, each in some way related to a force acting through a distance (Fig. 3–25). In elementary mechanics, the work done by a constant force F on a body displaced a distance s in the direction of the force is given by

$$W = Fs \qquad \text{(kJ)} \tag{3-21}$$

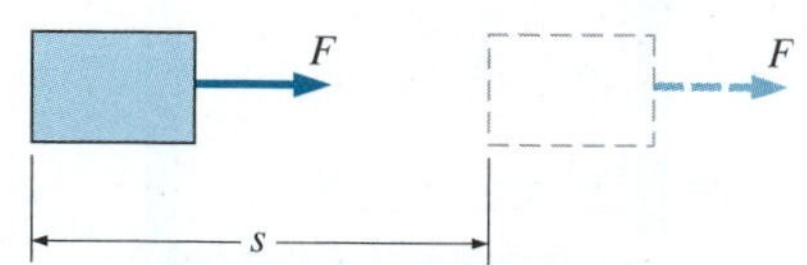

FIGURE 3–25

The work done is proportional to the force applied (F) and the distance traveled (s).

If the force F is not constant, the work done is obtained by adding (i.e., integrating) the differential amounts of work,

$$W = \int_1^2 F\,ds \qquad \text{(kJ)} \tag{3-22}$$

Obviously one needs to know how the force varies with displacement to perform this integration. Equations 3–21 and 3–22 give only the magnitude of the work. The sign is easily determined from physical considerations: The work done on a system by an external force acting in the direction of motion is negative, and work done by a system against an external force acting in the opposite direction to motion is positive.

There are two requirements for a work interaction between a system and its surroundings to exist: (1) there must be a *force* acting on the boundary, and (2) the boundary must *move.* Therefore, the presence of forces on the boundary without any displacement of the boundary does not constitute a work interaction. Likewise, the displacement of the boundary without any force to oppose or drive this motion (such as the expansion of a gas into an evacuated space) is not a work interaction since no energy is transferred.

In many thermodynamic problems, mechanical work is the only form of work involved. It is associated with the movement of the boundary of a system or with the movement of the entire system as a whole (Fig. 3–26). Some common forms of mechanical work are discussed next.

FIGURE 3–26

If there is no movement, no work is done.

Shaft Work

Energy transmission with a rotating shaft is very common in engineering practice (Fig. 3–27). Often the torque T applied to the shaft is constant, which means that the force F applied is also constant. For a specified constant torque, the work done during n revolutions is determined as follows: A force F acting through a moment arm r generates a torque T of (Fig. 3–28)

$$\mathrm{T} = Fr \quad \rightarrow \quad F = \frac{\mathrm{T}}{r} \tag{3-23}$$

This force acts through a distance s, which is related to the radius r by

$$s = (2\pi r)n \tag{3-24}$$

Then the shaft work is determined from

$$W_{\text{sh}} = Fs = \left(\frac{\mathrm{T}}{r}\right)(2\pi rn) = 2\pi n\mathrm{T} \qquad \text{(kJ)} \tag{3-25}$$

The power transmitted through the shaft is the shaft work done per unit time, which can be expressed as

$$\dot{W}_{\text{sh}} = 2\pi\dot{n}\mathrm{T} \qquad \text{(kW)} \tag{3-26}$$

where $\dot{n}$ is the number of revolutions per unit time.

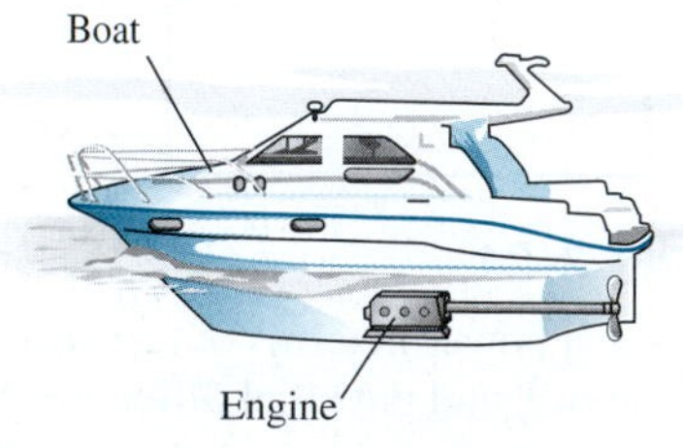

FIGURE 3–27

Energy transmission through rotating shafts is commonly encountered in practice.

EXAMPLE 3–7 Power Transmission by the Shaft of a Car

Determine the power transmitted through the shaft of a car when the torque applied is 200 N · m and the shaft rotates at a rate of 4000 revolutions per minute (rpm).

Solution The torque and the rpm for a car engine are given. The power transmitted is to be determined.

Analysis A sketch of the car is given in Fig. 3–29. The shaft power is determined directly from

$$\dot{W}_{sh} = 2\pi \dot{n} T = (2\pi)\left(4000 \frac{1}{\text{min}}\right)(200\ \text{N} \cdot \text{m})\left(\frac{1\ \text{min}}{60\ \text{s}}\right)\left(\frac{1\ \text{kJ}}{1000\ \text{N} \cdot \text{m}}\right)$$

$$= \mathbf{83.8\ kW} \quad \text{(or 112 hp)}$$

Discussion Note that the power transmitted by a shaft is proportional to the torque and the rotational speed.

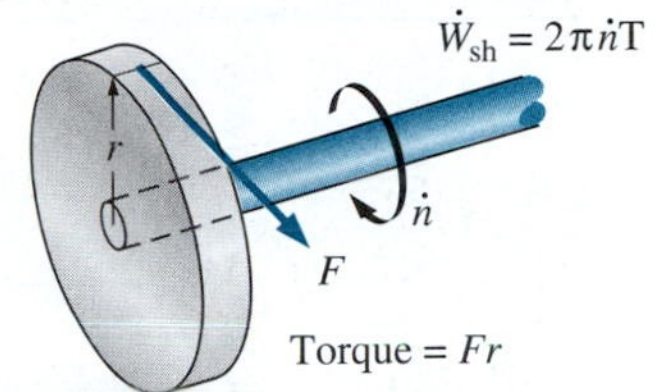

FIGURE 3–28

Shaft power is proportional to the torque applied and the number of revolutions of the shaft per unit time.

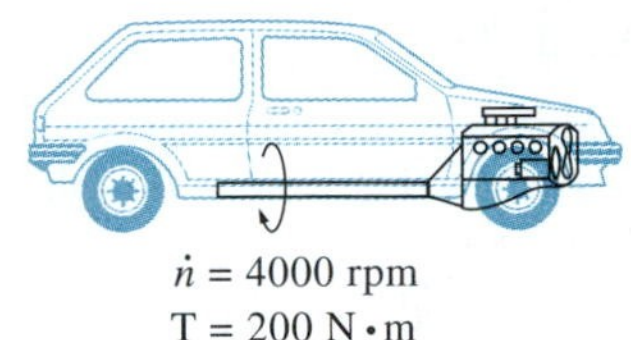

FIGURE 3–29

Schematic for Example 3–7.

Spring Work

It is common knowledge that when a force is applied on a spring, the length of the spring changes (Fig. 3–30). When the length of the spring changes by a differential amount dx under the influence of a force F, the work done is

$$\delta W_{\text{spring}} = F\,dx \tag{3–27}$$

To determine the total spring work, we need to know a functional relationship between F and x. For linear elastic springs, the displacement x is proportional to the force applied (Fig. 3–31). That is,

$$F = kx \qquad \text{(kN)} \tag{3–28}$$

where k is the spring constant and has the unit kN/m. The displacement x is measured from the undisturbed position of the spring (that is, $x = 0$ when $F = 0$). Substituting Eq. 3–28 into Eq. 3–27 and integrating yield

$$W_{\text{spring}} = \tfrac{1}{2}k(x_2^2 - x_1^2) \qquad \text{(kJ)} \tag{3–29}$$

where x_1 and x_2 are the initial and the final displacements of the spring, respectively, measured from the undisturbed position of the spring.

There are many other forms of mechanical work. Next we introduce some of them briefly.

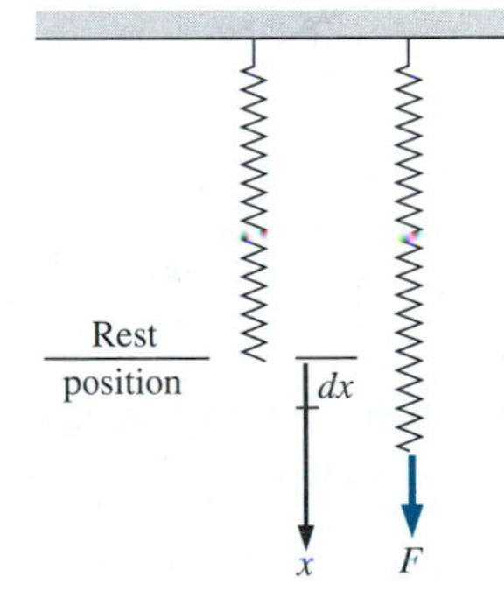

FIGURE 3–30

Elongation of a spring under the influence of a force.

Work Done on Elastic Solid Bars

Solids are often modeled as linear springs because under the action of a force they contract or elongate, as shown in Fig. 3–32, and when the force is lifted, they return to their original lengths, like a spring. This is true as long as the force is in the elastic range, that is, not large enough to cause permanent (plastic) deformations. Therefore, the equations given for a linear spring can also be used for elastic solid bars. Alternately, we can determine

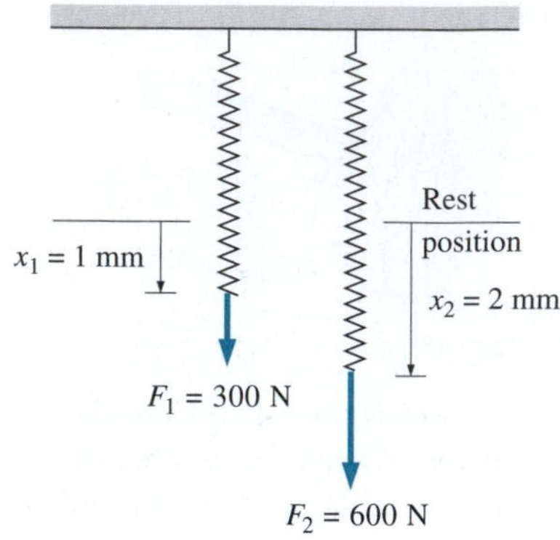

FIGURE 3–31

The displacement of a linear spring doubles when the force is doubled.

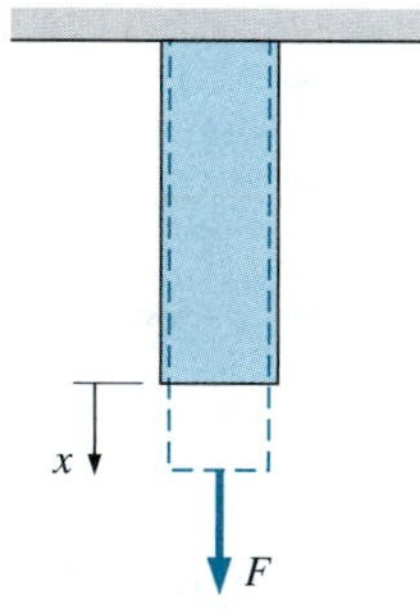

FIGURE 3–32

Solid bars behave as springs under the influence of a force.

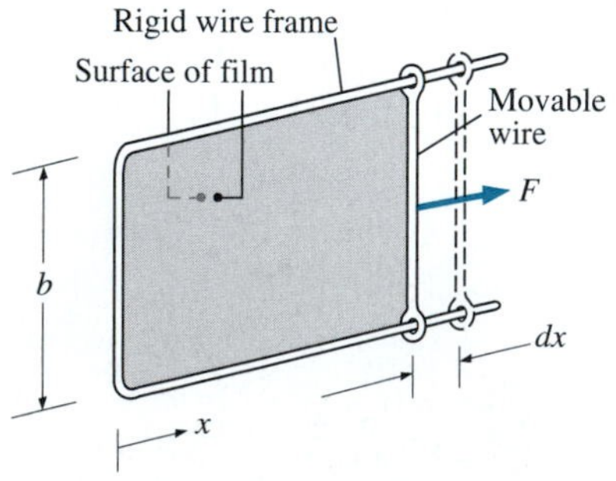

FIGURE 3–33

Stretching a liquid film with a movable wire.

the work associated with the expansion or contraction of an elastic solid bar by replacing pressure P by its counterpart in solids, *normal stress* $\sigma_n = F/A$, in the work expression:

$$W_{\text{elastic}} = \int_1^2 F\,dx = \int_1^2 \sigma_n A\,dx \qquad \text{(kJ)} \tag{3–30}$$

where A is the cross-sectional area of the bar. Note that the normal stress has pressure units.

Work Associated with the Stretching of a Liquid Film

Consider a liquid film such as soap film suspended on a wire frame (Fig. 3–33). We know from experience that it takes some force to stretch this film by the movable portion of the wire frame. This force is used to overcome the microscopic forces between molecules at the liquid–air interfaces. These microscopic forces are perpendicular to any line in the surface, and the force generated by these forces per unit length is called the **surface tension** σ_s, whose unit is N/m. Therefore, the work associated with the stretching of a film is also called *surface tension work.* It is determined from

$$W_{\text{surface}} = \int_1^2 \sigma_s\,dA \qquad \text{(kJ)} \tag{3–31}$$

where $dA = 2b\,dx$ is the change in the surface area of the film. The factor 2 is due to the fact that the film has two surfaces in contact with air. The force acting on the movable wire as a result of surface tension effects is $F = 2b\sigma_s$ where σ_s is the surface tension force per unit length.

Work Done to Raise or to Accelerate a Body

When a body is raised in a gravitational field, its potential energy increases. Likewise, when a body is accelerated, its kinetic energy increases. The conservation of energy principle requires that an equivalent amount of energy must be transferred to the body being raised or accelerated. Remember that energy can be transferred to a given mass by heat and work, and the energy transferred in this case obviously is not heat since it is not driven by a temperature difference. Therefore, it must be work. Then we conclude that (1) the work transfer needed to raise a body is equal to the change in the potential energy of the body, and (2) the work transfer needed to accelerate a body is equal to the change in the kinetic energy of the body (Fig. 3–34). Similarly, the potential or kinetic energy of a body represents the work that can be obtained from the body as it is lowered to the reference level or decelerated to zero velocity.

This discussion together with the consideration for friction and other losses form the basis for determining the required power rating of motors used to drive devices such as elevators, escalators, conveyor belts, and ski lifts. It also plays a primary role in the design of automotive and aircraft engines, and in the determination of the maximum amount of hydroelectric power that can be produced from a given water reservoir, which is simply the potential energy of the water relative to the location of the hydraulic turbine.

EXAMPLE 3–8 Power Needs of a Car to Climb a Hill

Consider a 1200-kg car cruising steadily on a level road at 90 km/h. Now the car starts climbing a hill that is sloped 30° from the horizontal (Fig. 3–35). If the velocity of the car is to remain constant during climbing, determine the additional power that must be delivered by the engine.

Solution A car is to climb a hill while maintaining a constant velocity. The additional power needed is to be determined.

Analysis The additional power required is simply the work that needs to be done per unit time to raise the elevation of the car, which is equal to the change in the potential energy of the car per unit time:

$$\dot{W}_g = mg\ \Delta z/\Delta t = mgV_{\text{vertical}}$$

$$= (1200\text{ kg})(9.81\text{ m/s}^2)(90\text{ km/h})(\sin 30°)\left(\frac{1\text{ m/s}}{3.6\text{ km/h}}\right)\left(\frac{1\text{ kJ/kg}}{1000\text{ m}^2/\text{s}^2}\right)$$

$$= 147\text{ kJ/s} = \mathbf{147\ kW} \qquad \text{(or 197 hp)}$$

Discussion Note that the car engine will have to produce almost 200 hp of additional power while climbing the hill if the car is to maintain its velocity.

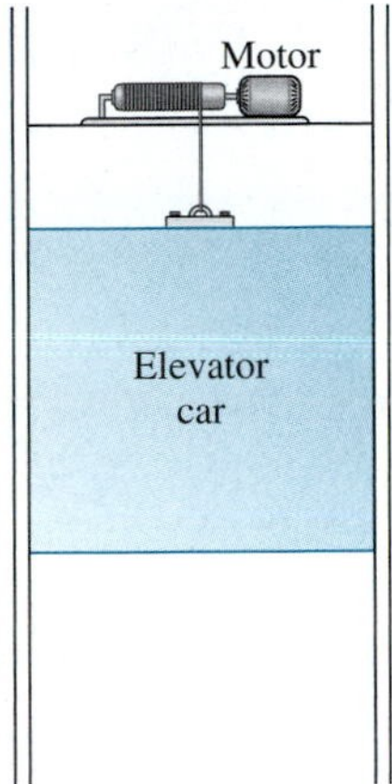

FIGURE 3–34

The energy transferred to a body while being raised is equal to the change in its potential energy.

EXAMPLE 3–9 Power Needs of a Car to Accelerate

Determine the average power required to accelerate a 900-kg car shown in Fig. 3–36 from rest to a velocity of 80 km/h in 20 s on a level road.

Solution The average power required to accelerate a car to a specified velocity is to be determined.

Analysis The work needed to accelerate a body is simply the change in the kinetic energy of the body,

$$W_a = \tfrac{1}{2}m(V_2^2 - V_1^2) = \tfrac{1}{2}(900\text{ kg})\left[\left(\frac{80{,}000\text{ m}}{3600\text{ s}}\right)^2 - 0^2\right]\left(\frac{1\text{ kJ/kg}}{1000\text{ m}^2/\text{s}^2}\right)$$

$$= 222\text{ kJ}$$

The average power is determined from

$$\dot{W}_a = \frac{W_a}{\Delta t} = \frac{222\text{ kJ}}{20\text{ s}} = \mathbf{11.1\ kW} \qquad \text{(or 14.9 hp)}$$

Discussion This is in addition to the power required to overcome friction, rolling resistance, and other irreversibilities.

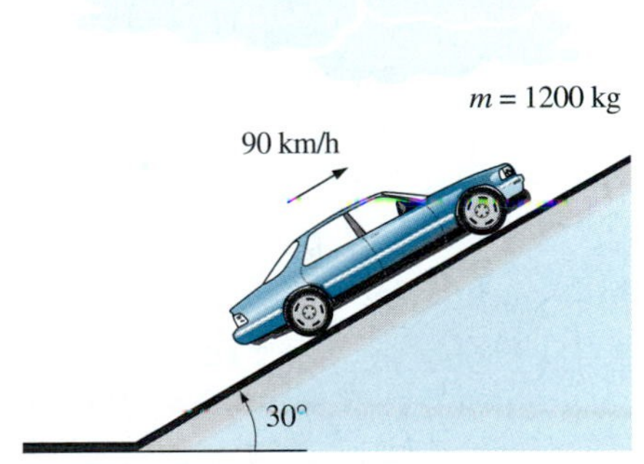

FIGURE 3–35

Schematic for Example 3–8.

Nonmechanical Forms of Work

The treatment in Section 3–5 represents a fairly comprehensive coverage of mechanical forms of work except the *moving boundary work* that is covered in Chap. 5. But some work modes encountered in practice are not mechanical in nature. However, these nonmechanical work modes can be treated in a similar manner by identifying a *generalized force F* acting in the direction

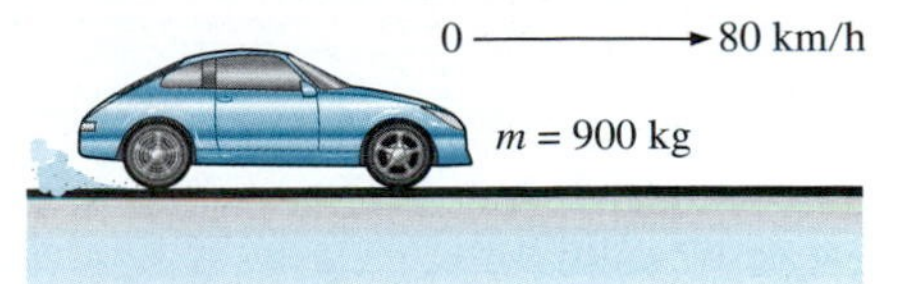

FIGURE 3–36

Schematic for Example 3–9.

of a *generalized displacement x*. Then the work associated with the differential displacement under the influence of this force is determined from $\delta W = F dx$.

Some examples of nonmechanical work modes are **electrical work**, where the generalized force is the *voltage* (the electrical potential) and the generalized displacement is the *electrical charge,* as discussed earlier; **magnetic work**, where the generalized force is the *magnetic field strength* and the generalized displacement is the total *magnetic dipole moment;* and **electrical polarization work**, where the generalized force is the *electric field strength* and the generalized displacement is the *polarization of the medium* (the sum of the electric dipole rotation moments of the molecules). Detailed consideration of these and other nonmechanical work modes can be found in specialized books on these topics.

3–6 ▪ THE FIRST LAW OF THERMODYNAMICS

So far, we have considered various forms of energy such as heat Q, work W, and total energy E individually, and no attempt is made to relate them to each other during a process. The *first law of thermodynamics,* also known as *the conservation of energy principle,* provides a sound basis for studying the relationships among the various forms of energy and energy interactions. Based on experimental observations, the first law of thermodynamics states that *energy can be neither created nor destroyed during a process; it can only change forms.* Therefore, every bit of energy should be accounted for during a process.

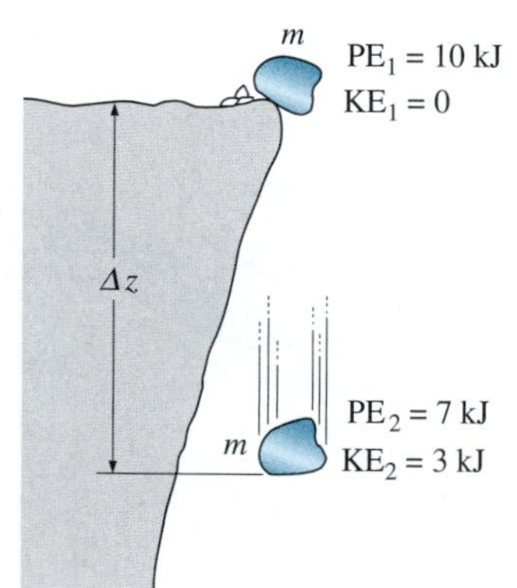

FIGURE 3–37
Energy cannot be created or destroyed; it can only change forms.

We all know that a rock at some elevation possesses some potential energy, and part of this potential energy is converted to kinetic energy as the rock falls (Fig. 3–37). Experimental data show that the decrease in potential energy ($mg\,\Delta z$) exactly equals the increase in kinetic energy $[m(V_2^2 - V_1^2)/2]$ when the air resistance is negligible, thus confirming the conservation of energy principle for mechanical energy.

Consider a system undergoing a series of *adiabatic* processes from a specified state 1 to another specified state 2. Being adiabatic, these processes obviously cannot involve any heat transfer, but they may involve several kinds of work interactions. Careful measurements during these experiments indicate the following: *For all adiabatic processes between two specified states of a closed system, the net work done is the same regardless of the nature of the closed system and the details of the process.* Considering that there are an infinite number of ways to perform work interactions under adiabatic conditions, this statement appears to be very powerful, with a potential for far-reaching implications. This statement, which is largely based on the experiments of Joule in the first half of the nineteenth century, cannot be drawn from any other known physical principle and is recognized as a fundamental principle. This principle is called the **first law of thermodynamics** or just the **first law**.

A major consequence of the first law is the existence and the definition of the property *total energy E.* Considering that the net work is the same for all adiabatic processes of a closed system between two specified states, the value of the net work must depend on the end states of the system only, and thus it must correspond to a change in a property of the system. This property is the *total energy.* Note that the first law makes no reference to the value of the total energy of a closed system at a state. It simply states that the *change* in the total energy during an adiabatic process must be equal to the net work done. Therefore, any convenient arbitrary value can be assigned to total energy at a specified state to serve as a reference point.

Implicit in the first law statement is the conservation of energy. Although the essence of the first law is the existence of the property *total energy,* the first law is often viewed as a statement of the *conservation of energy* principle. Next we develop the first law or the conservation of energy relation with the help of some familiar examples using intuitive arguments.

First, we consider some processes that involve heat transfer but no work interactions. The potato baked in the oven is a good example for this case (Fig. 3–38). As a result of heat transfer to the potato, the energy of the potato increases. If we disregard any mass transfer (moisture loss from the potato), the increase in the total energy of the potato becomes equal to the amount of heat transfer. That is, if 5 kJ of heat is transferred to the potato, the energy increase of the potato must also be 5 kJ.

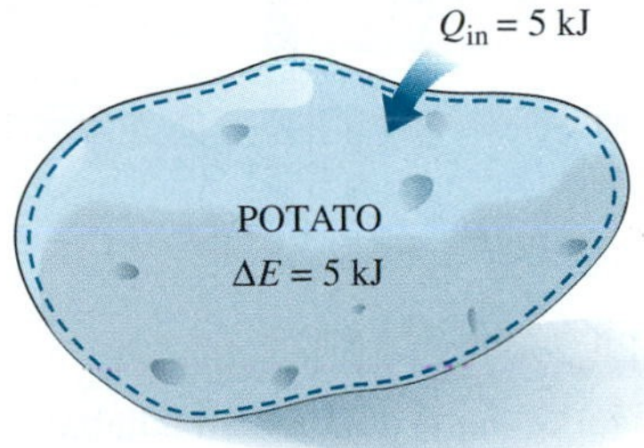

FIGURE 3–38

The increase in the energy of a potato in an oven is equal to the amount of heat transferred to it.

As another example, consider the heating of water in a pot on top of a range (Fig. 3–39). If 15 kJ of heat is transferred to the water from the heating element and 3 kJ of it is lost from the water to the surrounding air, the increase in energy of the water is equal to the net heat transfer to water, which is 12 kJ.

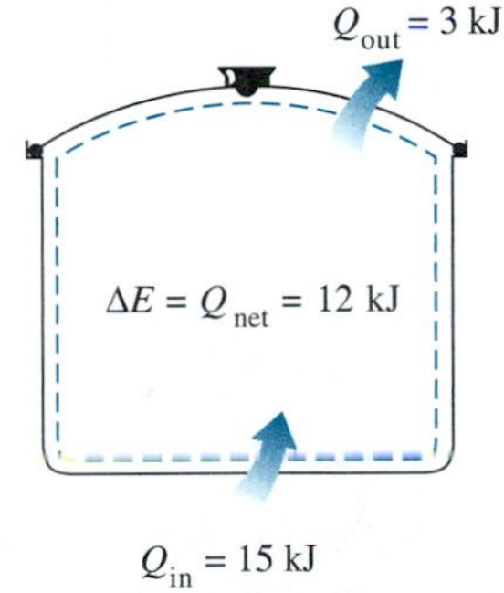

FIGURE 3–39

In the absence of any work interactions, the energy change of a system is equal to the net heat transfer.

Now consider a well-insulated (i.e., adiabatic) room heated by an electric heater as our system (Fig. 3–40). As a result of electrical work done, the energy of the system increases. Since the system is adiabatic and cannot have any heat transfer to or from the surroundings ($Q = 0$), the conservation of energy principle dictates that the electrical work done on the system must equal the increase in energy of the system.

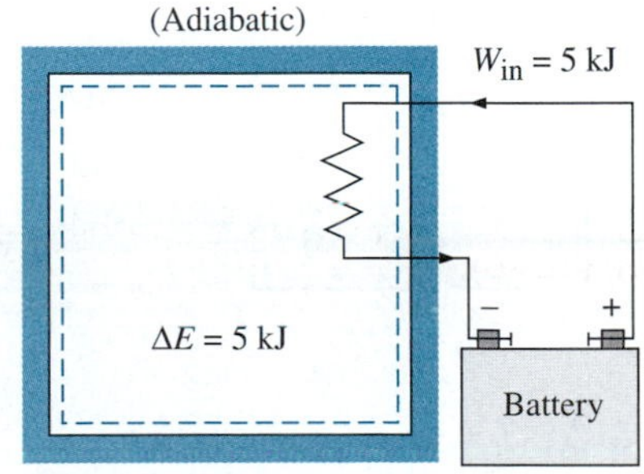

FIGURE 3–40

The work (e.g., electrical) done on an adiabatic system is equal to the increase in the energy of the system.

Next, let us replace the electric heater with a paddle wheel (Fig. 3–41). As a result of the stirring process, the energy of the system increases. Again, since there is no heat interaction between the system and its surroundings ($Q = 0$), the shaft work done on the system must show up as an increase in the energy of the system.

Many of you have probably noticed that the temperature of air rises when it is compressed (Fig. 3–42). This is because energy is transferred to the air in the form of boundary work. In the absence of any heat transfer ($Q = 0$), the entire boundary work will be stored in the air as part of its total energy. The conservation of energy principle again requires that the increase in the energy of the system be equal to the boundary work done on the system.

We can extend these discussions to systems that involve various heat and work interactions simultaneously. For example, if a system gains 12 kJ of heat during a process while 6 kJ of work is done on it, the increase in the energy of the system during that process is 18 kJ (Fig. 3–43). That is, the change in the energy of a system during a process is simply equal to the net energy transfer to (or from) the system.

Energy Balance

In the light of the preceding discussions, the conservation of energy principle can be expressed as follows: *The net change (increase or decrease) in the total energy of the system during a process is equal to the difference*

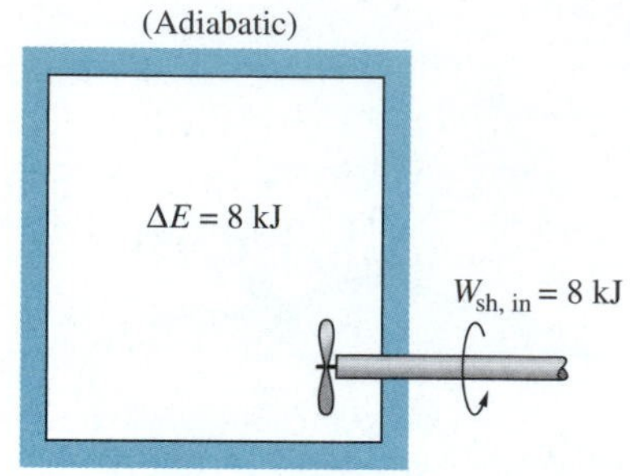

FIGURE 3–41

The work (shaft) done on an adiabatic system is equal to the increase in the energy of the system.

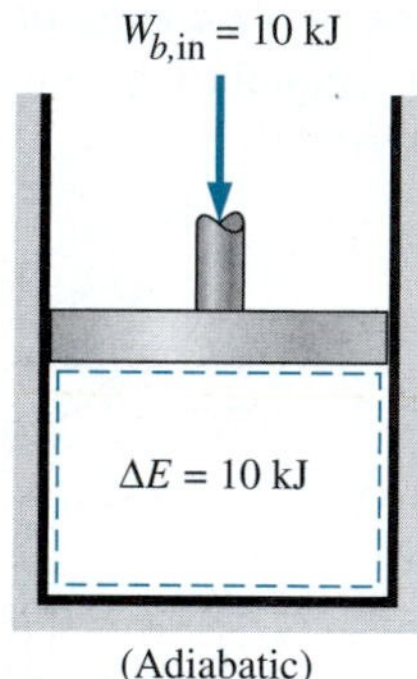

FIGURE 3–42

The work (boundary) done on an adiabatic system is equal to the increase in the energy of the system.

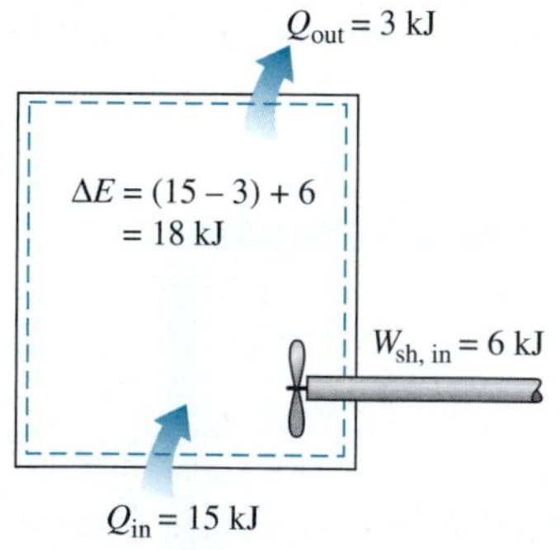

FIGURE 3–43

The energy change of a system during a process is equal to the *net* work and heat transfer between the system and its surroundings.

between the total energy entering and the total energy leaving the system during that process. That is,

$$\left(\begin{array}{c}\text{Total energy}\\ \text{entering the system}\end{array}\right) - \left(\begin{array}{c}\text{Total energy}\\ \text{leaving the system}\end{array}\right) = \left(\begin{array}{c}\text{Change in the total}\\ \text{energy of the system}\end{array}\right)$$

or

$$E_{in} - E_{out} = \Delta E_{system}$$

This relation is often referred to as the **energy balance** and is applicable to any kind of system undergoing any kind of process. The successful use of this relation to solve engineering problems depends on understanding the various forms of energy and recognizing the forms of energy transfer.

Energy Change of a System, ΔE_{system}

The determination of the energy change of a system during a process involves the evaluation of the energy of the system at the beginning and at the end of the process, and taking their difference. That is,

$$\text{Energy change} = \text{Energy at final state} - \text{Energy at initial state}$$

or

$$\Delta E_{system} = E_{final} - E_{initial} = E_2 - E_1 \tag{3–32}$$

Note that energy is a property, and the value of a property does not change unless the state of the system changes. Therefore, the energy change of a system is zero if the state of the system does not change during the process. Also, energy can exist in numerous forms such as internal (sensible, latent, chemical, and nuclear), kinetic, potential, electric, and magnetic, and their sum constitutes the *total energy E* of a system. In the absence of electric, magnetic, and surface tension effects (i.e., for simple compressible systems), the change in the total energy of a system during a process is the sum of the changes in its internal, kinetic, and potential energies and can be expressed as

$$\Delta E = \Delta U + \Delta KE + \Delta PE \tag{3–33}$$

where

$$\Delta U = m(u_2 - u_1)$$
$$\Delta KE = \tfrac{1}{2}m(V_2^2 - V_1^2)$$
$$\Delta PE = mg(z_2 - z_1)$$

When the initial and final states are specified, the values of the specific internal energies u_1 and u_2 can be determined directly from the property tables or thermodynamic property relations.

Most systems encountered in practice are stationary, that is, they do not involve any changes in their velocity or elevation during a process (Fig. 3–44). Thus, for **stationary systems**, the changes in kinetic and potential energies are zero (that is, $\Delta KE = \Delta PE = 0$), and the total energy change relation in Eq. 3–33 reduces to $\Delta E = \Delta U$ for such systems. Also, the energy of a system during a process changes even if only one form of its energy changes while the other forms of energy remain unchanged.

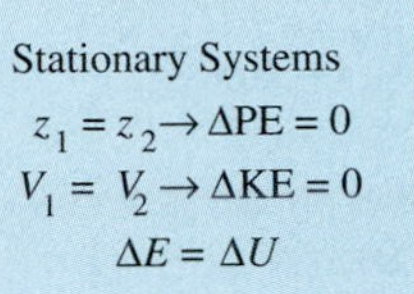

FIGURE 3–44
For stationary systems, $\Delta KE = \Delta PE = 0$; thus $\Delta E = \Delta U$.

Mechanisms of Energy Transfer, E_{in} and E_{out}

Energy can be transferred to or from a system in three forms: *heat, work,* and *mass flow.* Energy interactions are recognized at the system boundary as they cross it, and they represent the energy gained or lost by a system during a process. The only two forms of energy interactions associated with a fixed mass or closed system are *heat transfer* and *work.*

1. **Heat Transfer,** Q Heat transfer to a system (heat gain) increases the energy of the molecules and thus the internal energy of the system, and heat transfer from a system (heat loss) decreases it since the energy transferred out as heat comes from the energy of the molecules of the system.
2. **Work Transfer,** W An energy interaction that is not caused by a temperature difference between a system and its surroundings is work. A rising piston, a rotating shaft, and an electrical wire crossing the system boundaries are all associated with work interactions. Work transfer to a system (i.e., work done on a system) increases the energy of the system, and work transfer from a system (i.e., work done by the system) decreases it since the energy transferred out as work comes from the energy contained in the system. Car engines and hydraulic, steam, or gas turbines produce work while compressors, pumps, and mixers consume work.
3. **Mass Flow,** m Mass flow in and out of the system serves as an additional mechanism of energy transfer. When some mass enters a system, the energy of the system increases because mass carries energy with it. Likewise, when some mass leaves the system, the energy contained within the system decreases because the leaving mass takes out some energy with it. For example, when some hot water is taken out of a water heater and is replaced by the same amount of cold water, the energy content of the hot-water tank (the control volume) decreases as a result of this mass interaction (Fig. 3–45).

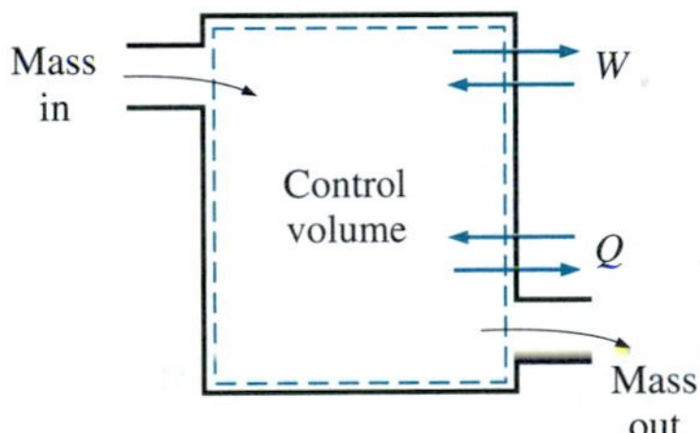

FIGURE 3–45
The energy content of a control volume can be changed by mass flow as well as heat and work interactions.

Noting that energy can be transferred in the forms of heat, work, and mass, and that the net transfer of a quantity is equal to the difference between the amounts transferred in and out, the energy balance can be written more explicitly as

$$E_{in} - E_{out} = (Q_{in} - Q_{out}) + (W_{in} - W_{out}) + (E_{mass,in} - E_{mass,out}) = \Delta E_{system} \quad \textbf{(2–34)}$$

where the subscripts "in" and "out" denote quantities that enter and leave the system, respectively. All six quantities on the right side of the equation represent "amounts," and thus they are *positive* quantities. The direction of any energy transfer is described by the subscripts "in" and "out."

The heat transfer Q is zero for adiabatic systems, the work transfer W is zero for systems that involve no work interactions, and the energy transport with mass E_{mass} is zero for systems that involve no mass flow across their boundaries (i.e., closed systems).

The energy balance for any system undergoing any kind of process can be expressed more compactly as

$$\underbrace{E_{in} - E_{out}}_{\substack{\text{Net energy transfer} \\ \text{by heat, work, and mass}}} = \underbrace{\Delta E_{system}}_{\substack{\text{Change in internal, kinetic,} \\ \text{potential, etc., energies}}} \quad \text{(kJ)} \tag{3–35}$$

or, in the **rate form**, as

$$\underbrace{\dot{E}_{in} - \dot{E}_{out}}_{\substack{\text{Rate of net energy transfer} \\ \text{by heat, work, and mass}}} = \underbrace{dE_{system}/dt}_{\substack{\text{Rate of change in internal,} \\ \text{kinetic, potential, etc., energies}}} \quad \text{(kW)} \tag{3–36}$$

For constant rates, the total quantities during a time interval Δt are related to the quantities per unit time as

$$Q = \dot{Q}\,\Delta t, \quad W = \dot{W}\,\Delta t, \quad \text{and} \quad \Delta E = (dE/dt)\,\Delta t \quad \text{(kJ)} \tag{3–37}$$

The energy balance can be expressed on a **per unit mass** basis as

$$e_{in} - e_{out} = \Delta e_{system} \quad \text{(kJ/kg)} \tag{3–38}$$

which is obtained by dividing all the quantities in Eq. 3–35 by the mass m of the system. Energy balance can also be expressed in the differential form as

$$\delta E_{in} - \delta E_{out} = dE_{system} \quad \text{or} \quad \delta e_{in} - \delta e_{out} = de_{system} \tag{3–39}$$

For a closed system undergoing a **cycle**, the initial and final states are identical, and thus $\Delta E_{system} = E_2 - E_1 = 0$. Then the energy balance for a cycle simplifies to $E_{in} - E_{out} = 0$ or $E_{in} = E_{out}$. Noting that a closed system does not involve any mass flow across its boundaries, the energy balance for a cycle can be expressed in terms of heat and work interactions as

$$W_{net,out} = Q_{net,in} \quad \text{or} \quad \dot{W}_{net,out} = \dot{Q}_{net,in} \quad \text{(for a cycle)} \tag{3–40}$$

That is, the net work output during a cycle is equal to net heat input (Fig. 3–46).

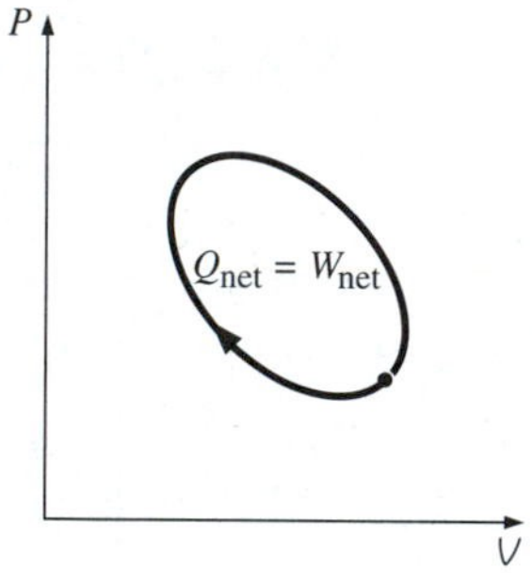

FIGURE 3–46
For a cycle $\Delta E = 0$, thus $Q = W$.

EXAMPLE 3–10 Cooling of a Hot Fluid in a Tank

A rigid tank contains a hot fluid that is cooled while being stirred by a paddle wheel. Initially, the internal energy of the fluid is 800 kJ. During the cooling process, the fluid loses 500 kJ of heat, and the paddle wheel does 100 kJ of work on the fluid. Determine the final internal energy of the fluid. Neglect the energy stored in the paddle wheel.

Solution A fluid in a rigid tank looses heat while being stirred. The final internal energy of the fluid is to be determined.

Assumptions **1** The tank is stationary and thus the kinetic and potential energy changes are zero, $\Delta KE = \Delta PE = 0$. Therefore, $\Delta E = \Delta U$ and internal energy is the only form of the system's energy that may change during this process. **2** Energy stored in the paddle wheel is negligible.

Analysis Take the contents of the tank as the *system* (Fig. 3–47). This is a *closed system* since no mass crosses the boundary during the process. We observe that the volume of a rigid tank is constant, and thus there is no moving boundary work. Also, heat is lost from the system and shaft work is done on the system. Applying the energy balance on the system gives

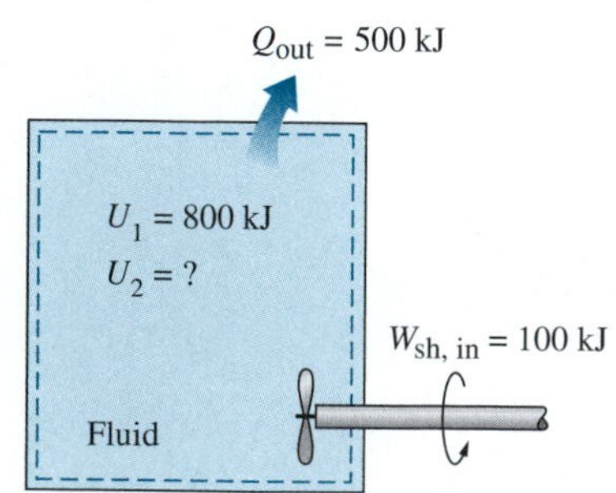

FIGURE 3–47
Schematic for Example 3–10.

$$\underbrace{E_{\text{in}} - E_{\text{out}}}_{\substack{\text{Net energy transfer}\\ \text{by heat, work, and mass}}} = \underbrace{\Delta E_{\text{system}}}_{\substack{\text{Change in internal, kinetic,}\\ \text{potential, etc., energies}}}$$

$$W_{\text{sh,in}} - Q_{\text{out}} = \Delta U = U_2 - U_1$$

$$100 \text{ kJ} - 500 \text{ kJ} = U_2 - 800 \text{ kJ}$$

$$U_2 = \mathbf{400 \ kJ}$$

Therefore, the final internal energy of the system is 400 kJ.

EXAMPLE 3–11 Acceleration of Air by a Fan

A fan that consumes 20 W of electric power when operating is claimed to discharge air from a ventilated room at a rate of 1.0 kg/s at a discharge velocity of 8 m/s (Fig. 3–48). Determine if this claim is reasonable.

FIGURE 3–48

Schematic for Example 3–11.

© Vol. 0557/PhotoDisc

Solution A fan is claimed to increase the velocity of air to a specified value while consuming electric power at a specified rate. The validity of this claim is to be investigated.

Assumptions The ventilating room is relatively calm, and air velocity in it is negligible.

Analysis First, let's examine the energy conversions involved: The motor of the fan converts part of the electrical power it consumes to mechanical (shaft) power, which is used to rotate the fan blades in air. The blades are shaped such that they impart a large fraction of the mechanical power of the shaft to air by mobilizing it. In the limiting ideal case of no losses (no conversion of electrical and mechanical energy to thermal energy) in steady operation, the electric power input is equal to the rate of increase of the kinetic energy of air. Therefore, for a control volume that encloses the fan-motor unit, the energy balance can be written as

$$\underbrace{\dot{E}_{\text{in}} - \dot{E}_{\text{out}}}_{\substack{\text{Rate of net energy transfer}\\ \text{by heat, work, and mass}}} = \underbrace{dE_{\text{system}} / dt^{\nearrow 0 \text{ (steady)}}}_{\substack{\text{Rate of change in internal, kinetic,}\\ \text{potential, etc., energies}}} = 0 \rightarrow \dot{E}_{\text{in}} = \dot{E}_{\text{out}}$$

$$\dot{W}_{\text{elect, in}} = \dot{m}_{\text{air}} \text{ke}_{\text{out}} = \dot{m}_{\text{air}} \frac{V_{\text{out}}^2}{2}$$

Solving for V_{out} and substituting gives the maximum air outlet velocity to be

$$V_{\text{out}} = \sqrt{\frac{2\dot{W}_{\text{elect,in}}}{\dot{m}_{\text{air}}}} = \sqrt{\frac{2(20 \text{ J/s})}{1.0 \text{ kg/s}} \left(\frac{1 \text{ m}^2/\text{s}^2}{1 \text{ J/kg}}\right)} = 6.3 \text{ m/s}$$

which is less than 8 m/s. Therefore, the claim is **false.**

Discussion The conservation of energy principle requires the energy to be preserved as it is converted from one form to another, and it does not allow any energy to be created or destroyed during a process. From the first law point of view, there is nothing wrong with the conversion of the entire electrical energy into kinetic energy. Therefore, the first law has no objection to air velocity reaching 6.3 m/s—but this is the upper limit. Any claim of higher velocity is in violation of the first law, and thus impossible. In reality, the air velocity will be considerably lower than 6.3 m/s because of the losses associated with the conversion of electrical energy to mechanical shaft energy, and the conversion of mechanical shaft energy to kinetic energy of air.

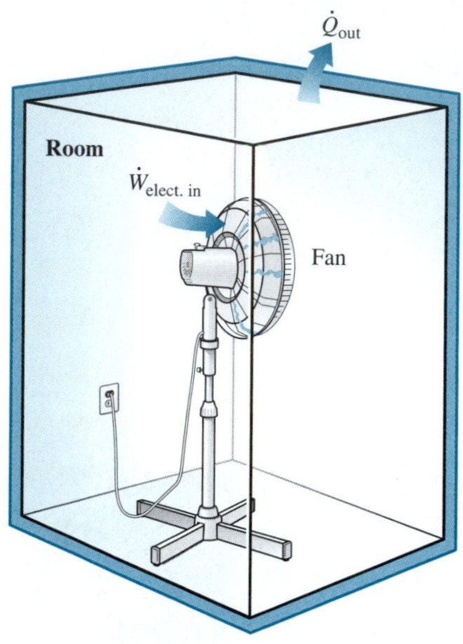

FIGURE 3–49

Schematic for Example 3–12.

EXAMPLE 3–12 Heating Effect of a Fan

A room is initially at the outdoor temperature of 25°C. Now a large fan that consumes 200 W of electricity when running is turned on (Fig. 3–49). The heat transfer rate between the room and the outdoor air is given as $\dot{Q} = UA(T_i - T_o)$ where $U = 6$ W/m$^2 \cdot$ °C is the overall heat transfer coefficient, $A = 30$ m^2 is the exposed surface area of the room, and T_i and T_o are the indoor and outdoor air temperatures, respectively. Determine the indoor air temperature when steady operating conditions are established.

Solution A large fan is turned on and kept on in a room that loses heat to the outdoors. The indoor air temperature is to be determined when steady operation is reached.

Assumptions **1** Heat transfer through the floor is negligible. **2** There are no other energy interactions involved.

Analysis The electricity consumed by the fan is energy input for the room, and thus the room gains energy at a rate of 200 W. As a result, the room air temperature tends to rise. But as the room air temperature rises, the rate of heat loss from the room increases until the rate of heat loss equals the electric power consumption. At that point, the temperature of the room air, and thus the energy content of the room, remains constant, and the conservation of energy for the room becomes

$$\underbrace{\dot{E}_{in} - \dot{E}_{out}}_{\substack{\text{Rate of net energy transfer}\\ \text{by heat, work, and mass}}} = \underbrace{dE_{system} / dt^{\nearrow 0\,(steady)}}_{\substack{\text{Rate of change in internal, kinetic,}\\ \text{potential, etc., energies}}} = 0 \quad \rightarrow \quad \dot{E}_{in} = \dot{E}_{out}$$

$$\dot{W}_{elect,in} = \dot{Q}_{out} = UA(T_i - T_o)$$

Substituting,

$$200 \text{ W} = (6 \text{ W/m}^2 \cdot °\text{C})(30 \text{ m}^2)(T_i - 25°\text{C})$$

which yields

$$T_i = \mathbf{26.1°C}$$

Therefore, the room air temperature will remain constant after it reaches 26.1°C.

Discussion Note that a 200-W fan heats a room just like a 200-W resistance heater. In the case of a fan, the motor converts part of the electric energy it draws into mechanical energy in the form of a rotating shaft while the remaining part is dissipated as heat to the room air because of the motor inefficiency (no motor converts 100 percent of the electric energy it receives to mechanical energy, although some large motors come close with a conversion efficiency of over 97 percent). Part of the mechanical energy of the shaft is converted to kinetic energy of air through the blades, which is then converted to thermal energy as air molecules slow down because of friction. At the end, the entire electric energy drawn by the fan motor is converted to thermal energy of air, which manifests itself as a rise in temperature.

FIGURE 3–50

Fluorescent lamps lighting a classroom as discussed in Example 3–13.

EXAMPLE 3–13 Annual Lighting Cost of a Classroom

The lighting needs of a classroom are met by 30 fluorescent lamps, each consuming 80 W of electricity (Fig. 3–50). The lights in the classroom are kept on for 12 hours a day and 250 days a year. For a unit electricity cost of

7 cents per kWh, determine annual energy cost of lighting for this classroom. Also, discuss the effect of lighting on the heating and air-conditioning requirements of the room.

Solution The lighting of a classroom by fluorescent lamps is considered. The annual electricity cost of lighting for this classroom is to be determined, and the lighting's effect on the heating and air-conditioning requirements is to be discussed.

Assumptions The effect of voltage fluctuations is negligible so that each fluorescent lamp consumes its rated power.

Analysis The electric power consumed by the lamps when all are on and the number of hours they are kept on per year are

$$\begin{aligned}\text{Lighting power} &= (\text{Power consumed per lamp}) \times (\text{No. of lamps})\\ &= (80\ \text{W/lamp})(30\ \text{lamps})\\ &= 2400\ \text{W} = 2.4\ \text{kW}\end{aligned}$$

$$\text{Operating hours} = (12\ \text{h/day})(250\ \text{days/year}) = 3000\ \text{h/year}$$

Then the amount and cost of electricity used per year become

$$\begin{aligned}\text{Lighting energy} &= (\text{Lighting power})(\text{Operating hours})\\ &= (2.4\ \text{kW})(3000\ \text{h/year}) = 7200\ \text{kWh/year}\end{aligned}$$

$$\begin{aligned}\text{Lighting cost} &= (\text{Lighting energy})(\text{Unit cost})\\ &= (7200\ \text{kWh/year})(\$0.07/\text{kWh}) = \mathbf{\$504/year}\end{aligned}$$

Light is absorbed by the surfaces it strikes and is converted to thermal energy. Disregarding the light that escapes through the windows, the entire 2.4 kW of electric power consumed by the lamps eventually becomes part of thermal energy of the classroom. Therefore, the lighting system in this room reduces the heating requirements by 2.4 kW, but increases the air-conditioning load by 2.4 kW.

Discussion Note that the annual lighting cost of this classroom alone is over \$500. This shows the importance of energy conservation measures. If incandescent light bulbs were used instead of fluorescent tubes, the lighting costs would be four times as much since incandescent lamps use four times as much power for the same amount of light produced. However, the heating requirements would also be reduced proportionally, and there would be no net change in overall energy cost in the winter.

EXAMPLE 3–14 Conservation of Energy for an Oscillating Steel Ball

The motion of a steel ball in a hemispherical bowl of radius *h* shown in Fig. 3–51 is to be analyzed. The ball is initially held at the highest location at point *A*, and then it is released. Obtain relations for the conservation of energy of the ball for the cases of frictionless and actual motions.

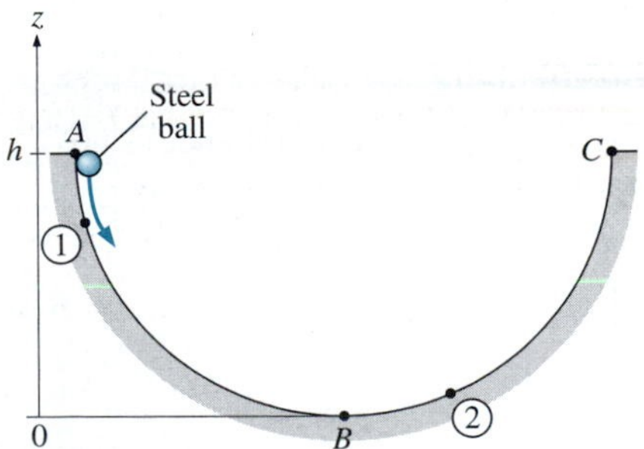

FIGURE 3–51
Schematic for Example 3–14.

Solution A steel ball is released in a bowl. Relations for the energy balance are to be obtained.

Assumptions The motion is frictionless, and thus friction between the ball, the bowl, and the air is negligible.

Analysis When the ball is released, it accelerates under the influence of gravity, reaches a maximum velocity (and minimum elevation) at point *B* at the bottom of the bowl, and moves up toward point *C* on the opposite side. In the ideal case of frictionless motion, the ball oscillates between points *A* and *C*. The actual motion involves the conversion of the kinetic and potential energies of the ball to each other, together with overcoming resistance to motion due to friction (doing frictional work). The general energy balance for any system undergoing any process is

$$\underbrace{E_{in} - E_{out}}_{\substack{\text{Net energy transfer}\\ \text{by heat, work, and mass}}} = \underbrace{\Delta E_{system}}_{\substack{\text{Change in internal, kinetic,}\\ \text{potential, etc., energies}}}$$

Then the energy balance for the ball for a process from point 1 to point 2 becomes

$$-w_{friction} = (\text{ke}_2 + \text{pe}_2) - (\text{ke}_1 + \text{pe}_1)$$

or

$$\frac{V_1^2}{2} + gz_1 = \frac{V_2^2}{2} + gz_2 + w_{friction}$$

since there is no energy transfer by heat or mass and no change in the internal energy of the ball (the heat generated by frictional heating is dissipated to the surrounding air). The frictional work term $w_{friction}$ is often expressed as e_{loss} to represent the loss (conversion) of mechanical energy into thermal energy.

For the idealized case of frictionless motion, the last relation reduces to

$$\frac{V_1^2}{2} + gz_1 = \frac{V_2^2}{2} + gz_2 \quad \text{or} \quad \frac{V^2}{2} + gz = C = \text{constant}$$

where the value of the constant is $C = gh$. That is, *when the frictional effects are negligible, the sum of the kinetic and potential energies of the ball remains constant.*

Discussion This is certainly a more intuitive and convenient form of the conservation of energy equation for this and other similar processes such as the swinging motion of the pendulum of a wall clock.

FIGURE 3–52

The definition of performance is not limited to thermodynamics only.

3–7 ▪ ENERGY CONVERSION EFFICIENCIES

Efficiency is one of the most frequently used terms in thermodynamics, and it indicates how well an energy conversion or transfer process is accomplished. Efficiency is also one of the most frequently misused terms in thermodynamics and a source of misunderstandings. This is because efficiency is often used without being properly defined first. Here we clarify this further, and define some efficiencies commonly used in practice.

Performance or efficiency, in general, can be expressed in terms of the desired output and the required input as (Fig. 3–52)

$$\text{Performance} = \frac{\text{Desired output}}{\text{Required input}} \tag{3–41}$$

If you are shopping for a water heater, a knowledgeable salesperson will tell you that the efficiency of a conventional electric water heater is about 90 percent (Fig. 3–53). You may find this confusing, since the heating elements of electric water heaters are resistance heaters, and the efficiency of all resistance heaters is 100 percent as they convert all the electrical energy they consume into thermal energy. A knowledgeable salesperson will clarify this by explaining that the heat losses from the hot-water tank to the surrounding air amount to 10 percent of the electrical energy consumed, and the **efficiency of a water heater** is defined as the ratio of the *energy delivered to the house by hot water* to the *energy supplied to the water heater.* A clever salesperson may even talk you into buying a more expensive water heater with thicker insulation that has an efficiency of 94 percent. If you are a knowledgeable consumer and have access to natural gas, you will probably purchase a gas water heater whose efficiency is only 55 percent since a gas unit costs about the same as an electric unit to purchase and install, but the annual energy cost of a gas unit is much less than that of an electric unit. Even though the efficiency is lower, natural gas is a less expensive source of energy compared to electricity.

Water heater

Type	Efficiency
Gas, conventional	55%
Gas, high-efficiency	90%
Electric, conventional	90%
Electric, high-efficiency	94%

FIGURE 3–53

Typical efficiencies of conventional and high-efficiency electric and natural gas water heaters.

© The McGraw-Hill Companies, Inc./Jill Braaten, photographer

Perhaps you are wondering how the efficiency for a gas water heater is defined, and why it is much lower than the efficiency of an electric heater. As a general rule, the efficiency of equipment that involves the combustion of a fuel is based on the **heating value of the fuel**, which is *the amount of heat released when a unit amount of fuel at room temperature is completely burned and the combustion products are cooled to the room temperature* (Fig. 3–54). Then the performance of combustion equipment can be characterized by **combustion efficiency**, defined as

$$\eta_{\text{combustion}} = \frac{Q}{\text{HV}} = \frac{\text{Amount of heat released during combustion}}{\text{Heating value of the fuel burned}} \tag{3–42}$$

A combustion efficiency of 100 percent indicates that the fuel is burned completely and the stack gases leave the combustion chamber at room temperature, and thus the amount of heat released during a combustion process is equal to the heating value of the fuel.

Most fuels contain hydrogen, which forms water when burned, and the heating value of a fuel will be different, depending on whether the water in combustion products is in the liquid or vapor form. The heating value is called the *lower heating value,* or LHV, when the water leaves as a vapor, and the *higher heating value,* or HHV, when the water in the combustion gases is completely condensed and thus the heat of vaporization is also recovered. The difference between these two heating values is equal to the product of the amount of water and the enthalpy of vaporization of water at room temperature. For example, the lower and higher heating values of gasoline are 44,000 kJ/kg and 47,300 kJ/kg, respectively. An efficiency definition should make it clear whether it is based on the higher or lower heating value of the fuel. Efficiencies of cars and jet engines are normally based on *lower heating values* since water normally leaves as a vapor in the exhaust gases, and it is not practical to try to recuperate the heat of vaporization. Efficiencies of furnaces, on the other hand, are based on *higher heating values.*

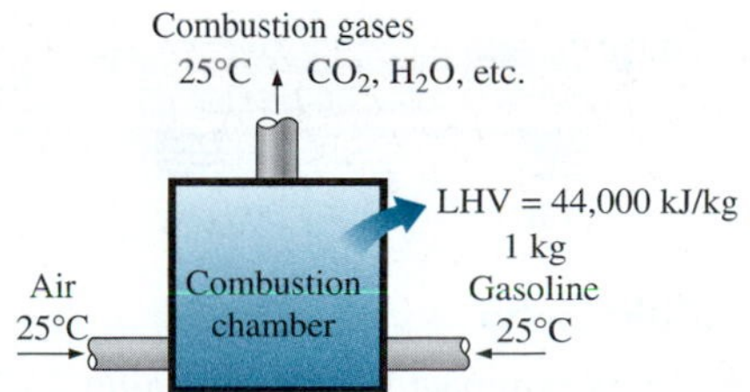

FIGURE 3–54

The definition of the heating value of gasoline.

The efficiency of space heating systems of residential and commercial buildings is usually expressed in terms of the **annual fuel utilization efficiency**, or **AFUE**, which accounts for the combustion efficiency as well as other losses such as heat losses to unheated areas and start-up and cooldown losses. The AFUE of most new heating systems is about 85 percent, although the AFUE of some old heating systems is under 60 percent. The

TABLE 3–1

The efficacy of different lighting systems

Type of lighting	Efficacy, lumens/W
Combustion	
Candle	0.2
Incandescent	
Ordinary	6–20
Halogen	16–25
Fluorescent	
Ordinary	40–60
High output	70–90
Compact	50–80
High-intensity discharge	
Mercury vapor	50–60
Metal halide	56–125
High-pressure sodium	100–150
Low-pressure sodium	up to 200

AFUE of some new high-efficiency furnaces exceeds 96 percent, but the high cost of such furnaces cannot be justified for locations with mild to moderate winters. Such high efficiencies are achieved by reclaiming most of the heat in the flue gases, condensing the water vapor, and discharging the flue gases at temperatures as low as 38°C (or 100°F) instead of about 200°C (or 400°F) for the conventional models.

For *car engines,* the work output is understood to be the power delivered by the crankshaft. But for power plants, the work output can be the mechanical power at the turbine exit, or the electrical power output of the generator.

A generator is a device that converts mechanical energy to electrical energy, and the effectiveness of a generator is characterized by the **generator efficiency**, which is the ratio of the *electrical power output* to the *mechanical power input.* The *thermal efficiency* of a power plant, which is of primary interest in thermodynamics, is usually defined as the ratio of the net shaft work output of the turbine to the heat input to the working fluid. The effects of other factors are incorporated by defining an **overall efficiency** for the power plant as the ratio of the *net electrical power output* to the *rate of fuel energy input.* That is,

$$\eta_{\text{overall}} = \eta_{\text{combustion}}\,\eta_{\text{thermal}}\,\eta_{\text{generator}} = \frac{\dot{W}_{\text{net,electric}}}{\text{HHV} \times \dot{m}_{\text{net}}} \qquad \textbf{(3–43)}$$

The overall efficiencies are about 26–30 percent for gasoline automotive engines, 34–40 percent for diesel engines, and up to 60 percent for large power plants.

We are all familiar with the conversion of electrical energy to *light* by incandescent lightbulbs, fluorescent tubes, and high-intensity discharge lamps. The efficiency for the conversion of electricity to light can be defined as the ratio of the energy converted to light to the electrical energy consumed. For example, common incandescent lightbulbs convert about 10 percent of the electrical energy they consume to light; the rest of the energy consumed is dissipated as heat, which adds to the cooling load of the air conditioner in summer. (But it reduces the heating load in the winter.) However, it is more common to express the effectiveness of this conversion process by **lighting efficacy**, which is defined as the *amount of light output in lumens per W of electricity consumed.*

The efficacy of different lighting systems is given in Table 3–1. Note that a compact fluorescent lightbulb produces about four times as much light as an incandescent lightbulb per W, and thus a 15-W fluorescent bulb can replace a 60-W incandescent lightbulb (Fig. 3–55). Also, a compact fluorescent bulb lasts about 10,000 h, which is 10 times as long as an incandescent bulb, and it plugs directly into the socket of an incandescent lamp. Therefore, despite their higher initial cost, compact fluorescents reduce the lighting costs considerably through reduced electricity consumption. Sodium-filled high-intensity discharge lamps provide the most efficient lighting, but their use is limited to outdoor use because of their yellowish light.

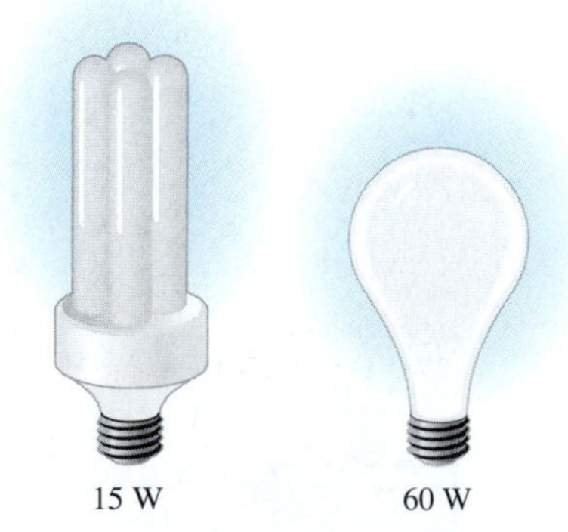

FIGURE 3–55

A 15-W compact fluorescent lamp provides as much light as a 60-W incandescent lamp.

We can also define efficiency for cooking appliances since they convert electrical or chemical energy to heat for cooking. The **efficiency of a cooking appliance** is defined as the ratio of the *useful energy transferred to the* *food to the energy consumed by the appliance* (Fig. 3–56). Electric ranges are more efficient than gas ranges, but it is much cheaper to cook with natural gas than with electricity because of the lower unit cost of natural gas (Table 3–2).

The cooking efficiency depends on user habits as well as the individual appliances. Convection and microwave ovens are inherently more efficient than conventional ovens. On average, convection ovens save about *one-third* and microwave ovens save about *two-thirds* of the energy used by conventional ovens. The cooking efficiency can be increased by using the smallest oven for baking, using a pressure cooker, using an electric slow cooker for stews and soups, using the smallest pan that will do the job, using the smaller heating element for small pans on electric ranges, using flat-bottomed pans on electric burners to assure good contact, keeping burner drip pans clean and shiny, defrosting frozen foods in the refrigerator before cooking, avoiding preheating unless it is necessary, keeping the pans covered during cooking, using timers and thermometers to avoid overcooking, using the self-cleaning feature of ovens right after cooking, and keeping inside surfaces of microwave ovens clean.

Using energy-efficient appliances and practicing energy conservation measures help our pocketbooks by reducing our utility bills. It also helps the **environment** by reducing the amount of pollutants emitted to the atmosphere during the combustion of fuel at home or at the power plants where electricity is generated. The combustion of *each therm of natural gas* (see Example 3–15) produces 6.4 kg of carbon dioxide, which causes global climate change; 4.7 g of nitrogen oxides and 0.54 g of hydrocarbons, which cause smog; 2.0 g of carbon monoxide, which is toxic; and 0.030 g of sulfur dioxide, which causes acid rain. Each therm of natural gas saved eliminates the emission of these pollutants while saving \$0.60 for the average consumer in the United States. Each kWh of electricity conserved saves 0.4 kg of coal and 1.0 kg of CO_2 and 15 g of SO_2 from a coal power plant.

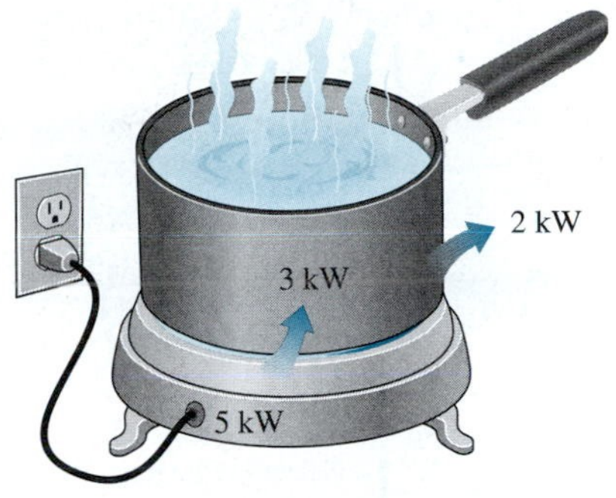

$$\text{Efficiency} = \frac{\text{Energy utilized}}{\text{Energy supplied to appliance}}$$

$$= \frac{3\text{ kWh}}{5\text{ kWh}} = 0.60$$

FIGURE 3–56

The efficiency of a cooking appliance represents the fraction of the energy supplied to the appliance that is transferred to the food.

TABLE 3–2

Energy costs of cooking a casserole with different appliances*

[From A. Wilson and J. Morril, *Consumer Guide to Home Energy Savings,* Washington, DC: American Council for an Energy-Efficient Economy, 1996, p. 192.]

Cooking appliance	Cooking temperature	Cooking time	Energy used	Cost of energy
Electric oven	350°F (177°C)	1 h	2.0 kWh	\$0.16
Convection oven (elect.)	325°F (163°C)	45 min	1.39 kWh	\$0.11
Gas oven	350°F (177°C)	1 h	0.112 therm	\$0.07
Frying pan	420°F (216°C)	1 h	0.9 kWh	\$0.07
Toaster oven	425°F (218°C)	50 min	0.95 kWh	\$0.08
Electric slow cooker	200°F (93°C)	7 h	0.7 kWh	\$0.06
Microwave oven	"High"	15 min	0.36 kWh	\$0.03

*Assumes a unit cost of \$0.08/kWh for electricity and \$0.60/therm for gas.

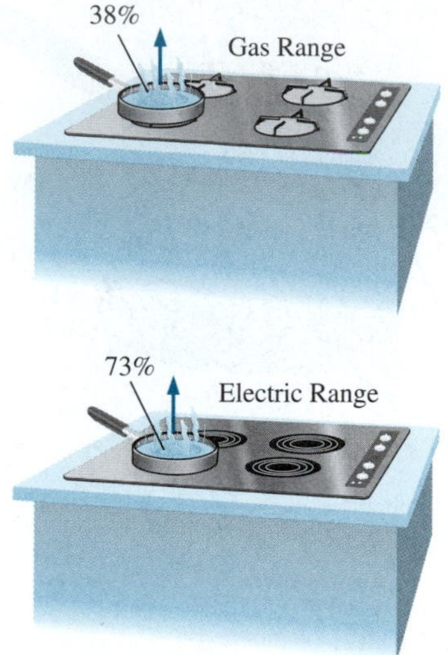

FIGURE 3–57

Schematic of the 73 percent efficient electric heating unit and 38 percent efficient gas burner discussed in Example 3–15.

EXAMPLE 3–15 Cost of Cooking with Electric and Gas Ranges

The efficiency of cooking appliances affects the internal heat gain from them since an inefficient appliance consumes a greater amount of energy for the same task, and the excess energy consumed shows up as heat in the living space. The efficiency of open burners is determined to be 73 percent for electric units and 38 percent for gas units (Fig. 3–57). Consider a 2-kW electric burner at a location where the unit costs of electricity and natural gas are \$0.09/kWh and \$0.55/therm, respectively. Determine the rate of energy consumption by the burner and the unit cost of utilized energy for both electric and gas burners.

Solution The operation of electric and gas ranges is considered. The rate of energy consumption and the unit cost of utilized energy are to be determined.

Analysis The efficiency of the electric heater is given to be 73 percent. Therefore, a burner that consumes 2 kW of electrical energy supplies

$$\dot{Q}_{\text{utilized}} = (\text{Energy input}) \times (\text{Efficiency}) = (2\text{ kW})(0.73) = \mathbf{1.46\ kW}$$

of useful energy. The unit cost of utilized energy is inversely proportional to the efficiency, and is determined from

$$\text{Cost of utilized energy} = \frac{\text{Cost of energy input}}{\text{Efficiency}} = \frac{\$0.09/\text{kWh}}{0.73} = \mathbf{\$0.123/kWh}$$

Noting that the efficiency of a gas burner is 38 percent, the energy input to a gas burner that supplies utilized energy at the same rate (1.46 kW) is

$$\dot{Q}_{\text{input, gas}} = \frac{\dot{Q}_{\text{utilized}}}{\text{Efficiency}} = \frac{1.46\text{ kW}}{0.38} = \mathbf{3.84\ kW} \qquad (= 13{,}100\text{ Btu/h})$$

since 1 kW = 3412 Btu/h. Therefore, a gas burner should have a rating of at least 13,100 Btu/h to perform as well as the electric unit.

Noting that 1 therm = 29.3 kWh, the unit cost of utilized energy in the case of a gas burner is determined to be

$$\text{Cost of utilized energy} = \frac{\text{Cost of energy input}}{\text{Efficiency}} = \frac{\$0.55/29.3\text{ kWh}}{0.38}$$
$$= \mathbf{\$0.049/kWh}$$

Discussion The cost of utilized gas is less than half of the unit cost of utilized electricity. Therefore, despite its higher efficiency, cooking with an electric burner would cost more than twice as much compared to a gas burner in this case. This explains why cost-conscious consumers always ask for gas appliances, and it is not wise to use electricity for heating purposes.

Efficiencies of Mechanical and Electrical Devices

The transfer of mechanical energy is usually accomplished by a rotating shaft, and thus mechanical work is often referred to as *shaft work*. A pump or a fan receives shaft work (usually from an electric motor) and transfers it to the fluid as mechanical energy (less frictional losses). A turbine, on the other hand, converts the mechanical energy of a fluid to shaft work. In the absence of any irreversibilities such as friction, mechanical energy can be converted entirely from one mechanical form to another, and the **mechanical efficiency** of a device or process can be defined as (Fig. 3–58)

$$\eta_{\text{mech}} = \frac{\text{Mechanical energy output}}{\text{Mechanical energy input}} = \frac{E_{\text{mech,out}}}{E_{\text{mech,in}}} = 1 - \frac{E_{\text{mech,loss}}}{E_{\text{mech,in}}} \tag{3–44}$$

A conversion efficiency of less than 100 percent indicates that conversion is less than perfect and some losses have occurred during conversion. A mechanical efficiency of 97 percent indicates that 3 percent of the mechanical energy input is converted to thermal energy as a result of frictional heating, and this manifests itself as a slight rise in the temperature of the fluid.

In fluid systems, we are usually interested in increasing the pressure, velocity, and/or elevation of a fluid. This is done by *supplying mechanical energy* to the fluid by a pump, a fan, or a compressor (we will refer to all of them as pumps). Or we are interested in the reverse process of *extracting mechanical energy* from a fluid by a turbine and producing mechanical power in the form of a rotating shaft that can drive a generator or any other rotary device. The degree of perfection of the conversion process between the mechanical work supplied or extracted and the mechanical energy of the fluid is expressed by the **pump efficiency** and **turbine efficiency**, defined as

$$\eta_{\text{pump}} = \frac{\text{Mechanical energy increase of the fluid}}{\text{Mechanical energy input}} = \frac{\Delta \dot{E}_{\text{mech,fluid}}}{\dot{W}_{\text{shaft,in}}} = \frac{\dot{W}_{\text{pump},u}}{\dot{W}_{\text{pump}}} \tag{3–45}$$

where $\Delta \dot{E}_{\text{mech,fluid}} = \dot{E}_{\text{mech,out}} - \dot{E}_{\text{mech,in}}$ is the rate of increase in the mechanical energy of the fluid, which is equivalent to the **useful pumping power** $\dot{W}_{\text{pump},u}$ supplied to the fluid, and

$$\eta_{\text{turbine}} = \frac{\text{Mechanical energy output}}{\text{Mechanical energy decrease of the fluid}} = \frac{\dot{W}_{\text{shaft,out}}}{|\Delta \dot{E}_{\text{mech,fluid}}|} = \frac{\dot{W}_{\text{turbine}}}{\dot{W}_{\text{turbine},e}} \tag{3–46}$$

where $|\Delta \dot{E}_{\text{mech,fluid}}| = \dot{E}_{\text{mech,in}} - \dot{E}_{\text{mech,out}}$ is the rate of decrease in the mechanical energy of the fluid, which is equivalent to the mechanical power extracted from the fluid by the turbine $\dot{W}_{\text{turbine},e}$, and we use the absolute value sign to avoid negative values for efficiencies. A pump or turbine efficiency of 100 percent indicates perfect conversion between the shaft work and the mechanical energy of the fluid, and this value can be approached (but never attained) as the frictional effects are minimized.

Electrical energy is commonly converted to *rotating mechanical energy* by electric motors to drive fans, compressors, robot arms, car starters, and so forth. The effectiveness of this conversion process is characterized by the *motor efficiency* η_{motor}, which is the ratio of the *mechanical energy output* of the motor to the *electrical energy input.* The full-load motor efficiencies range from about 35 percent for small motors to over 97 percent for large high-efficiency motors. The difference between the electrical energy consumed and the mechanical energy delivered is dissipated as waste heat.

The mechanical efficiency should not be confused with the **motor efficiency** and the **generator efficiency**, which are defined as

Motor:

$$\eta_{\text{motor}} = \frac{\text{Mechanical power output}}{\text{Electric power input}} = \frac{\dot{W}_{\text{shaft,out}}}{\dot{W}_{\text{elect,in}}} \tag{3–47}$$

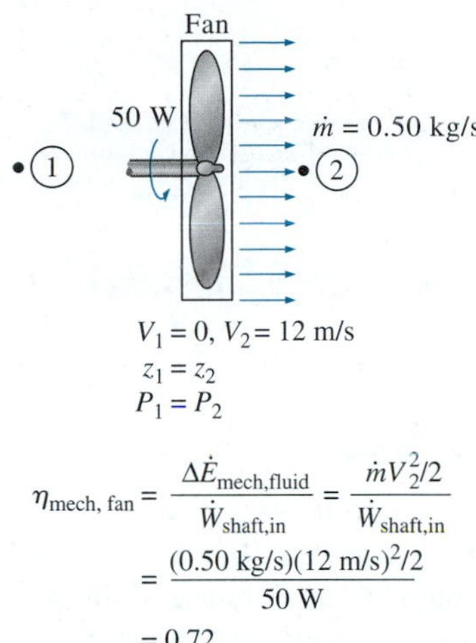

$$\eta_{\text{mech, fan}} = \frac{\Delta \dot{E}_{\text{mech,fluid}}}{\dot{W}_{\text{shaft,in}}} = \frac{\dot{m}V_2^2/2}{\dot{W}_{\text{shaft,in}}} = \frac{(0.50\ \text{kg/s})(12\ \text{m/s})^2/2}{50\ \text{W}} = 0.72$$

FIGURE 3–58

The mechanical efficiency of a fan is the ratio of the kinetic energy of air at the fan exit to the mechanical power input.

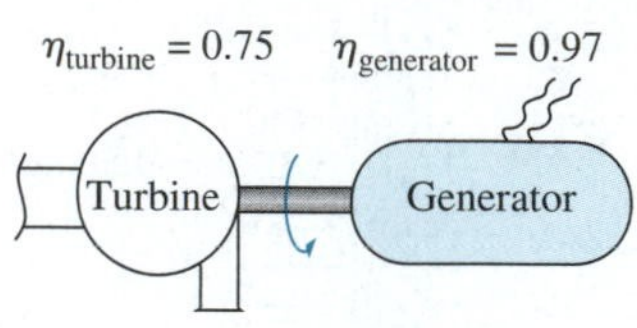

FIGURE 3–59

The overall efficiency of a turbine–generator is the product of the efficiency of the turbine and the efficiency of the generator, and represents the fraction of the mechanical energy of the fluid converted to electric energy.

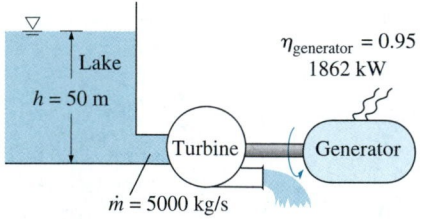

and

$$\textit{Generator:} \qquad \eta_{\text{generator}} = \frac{\text{Electric power output}}{\text{Mechanical power input}} = \frac{\dot{W}_{\text{elect,out}}}{\dot{W}_{\text{shaft,in}}} \tag{3–48}$$

A pump is usually packaged together with its motor, and a turbine with its generator. Therefore, we are usually interested in the **combined** or **overall efficiency** of pump–motor and turbine–generator combinations (Fig. 3–59), which are defined as

$$\eta_{\text{pump-motor}} = \eta_{\text{pump}}\eta_{\text{motor}} = \frac{\dot{W}_{\text{pump},u}}{\dot{W}_{\text{elect,in}}} = \frac{\Delta \dot{E}_{\text{mech,fluid}}}{\dot{W}_{\text{elect,in}}} \tag{3–49}$$

and

$$\eta_{\text{turbine-gen}} = \eta_{\text{turbine}}\eta_{\text{generator}} = \frac{\dot{W}_{\text{elect,out}}}{\dot{W}_{\text{turbine},e}} = \frac{\dot{W}_{\text{elect,out}}}{|\Delta \dot{E}_{\text{mech,fluid}}|} \tag{3–50}$$

All the efficiencies just defined range between 0 and 100 percent. The lower limit of 0 percent corresponds to the conversion of the entire mechanical or electric energy input to thermal energy, and the device in this case functions like a resistance heater. The upper limit of 100 percent corresponds to the case of perfect conversion with no friction or other irreversibilities, and thus no conversion of mechanical or electric energy to thermal energy.

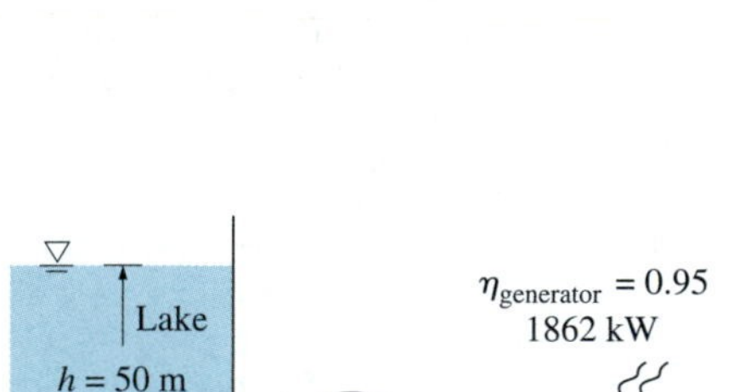

FIGURE 3–60

Schematic for Example 3–16.

EXAMPLE 3–16 Performance of a Hydraulic Turbine–Generator

The water in a large lake is to be used to generate electricity by the installation of a hydraulic turbine–generator at a location where the depth of the water is 50 m (Fig. 3–60). Water is to be supplied at a rate of 5000 kg/s. If the electric power generated is measured to be 1862 kW and the generator efficiency is 95 percent, determine (*a*) the overall efficiency of the turbine–generator, (*b*) the mechanical efficiency of the turbine, and (*c*) the shaft power supplied by the turbine to the generator.

Solution A hydraulic turbine–generator is to generate electricity from the water of a lake. The overall efficiency, the turbine efficiency, and the turbine shaft power are to be determined.

Assumptions **1** The elevation of the lake remains constant. **2** The mechanical energy of water at the turbine exit is negligible.

Properties The density of water can be taken to be $\rho = 1000$ kg/m^3.

Analysis (*a*) We take the bottom of the lake as the reference level for convenience. Then the potential energy of water at the surface becomes *gh*. Considering that the kinetic energy of water at the lake surface and the turbine exit are negligible and the pressure at both locations is the atmospheric pressure, the change in the mechanical energy of water between lake surface and the turbine exit becomes

$$e_{\text{mech,in}} - e_{\text{mech,out}} = gh - 0 = gh = (9.81 \text{ m/s}^2)(50 \text{ m})\left(\frac{1 \text{ kJ/kg}}{1000 \text{ m}^2/\text{s}^2}\right)$$
$$= 0.491 \text{ kJ/kg}$$

Then the rate at which mechanical energy is supplied to the turbine by the fluid and the overall efficiency become

$$|\Delta \dot{E}_{\text{mech,fluid}}| = \dot{m}(e_{\text{mech,in}} - e_{\text{mech,out}}) = (5000 \text{ kg/s})(0.491 \text{ kJ/kg}) = 2455 \text{ kW}$$

$$\eta_{\text{overall}} = \eta_{\text{turbine-gen}} = \frac{\dot{W}_{\text{elect,out}}}{|\Delta \dot{E}_{\text{mech,fluid}}|} = \frac{1862 \text{ kW}}{2455 \text{ kW}} = \mathbf{0.76}$$

(*b*) Knowing the overall and generator efficiencies, the mechanical efficiency of the turbine is determined from

$$\eta_{\text{turbine-gen}} = \eta_{\text{turbine}}\eta_{\text{generator}} \rightarrow \eta_{\text{turbine}} = \frac{\eta_{\text{turbine-gen}}}{\eta_{\text{generator}}} = \frac{0.76}{0.95} = \mathbf{0.80}$$

(*c*) The shaft power output is determined from the definition of mechanical efficiency,

$$\dot{W}_{\text{shaft,out}} = \eta_{\text{turbine}}|\Delta \dot{E}_{\text{mech,fluid}}| = (0.80)(2455 \text{ kW}) = \mathbf{1964 \text{ kW}}$$

Discussion Note that the lake supplies 2455 kW of mechanical energy to the turbine, which converts 1964 kW of it to shaft work that drives the generator, which generates 1862 kW of electric power. There are losses associated with each component.

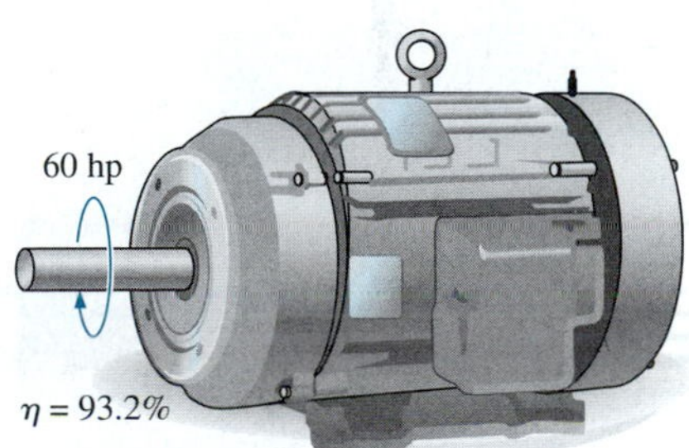

FIGURE 3–61

Schematic for Example 3–17.

EXAMPLE 3–17 Cost Savings Associated with High-Efficiency Motors

A 60-hp electric motor (a motor that delivers 60 hp of shaft power at full load) that has an efficiency of 89.0 percent is worn out and is to be replaced by a 93.2 percent efficient high-efficiency motor (Fig. 3–61). The motor operates 3500 hours a year at full load. Taking the unit cost of electricity to be $0.08/kWh, determine the amount of energy and money saved as a result of installing the high-efficiency motor instead of the standard motor. Also, determine the simple payback period if the purchase prices of the standard and high-efficiency motors are $4520 and $5160, respectively.

Solution A worn-out standard motor is to be replaced by a high-efficiency one. The amount of electrical energy and money saved as well as the simple payback period are to be determined.

Assumptions The load factor of the motor remains constant at 1 (full load) when operating.

Analysis The electric power drawn by each motor and their difference can be expressed as

$$\dot{W}_{\text{electric in,standard}} = \dot{W}_{\text{shaft}}/\eta_{\text{st}} = (\text{Rated power})(\text{Load factor})/\eta_{\text{st}}$$

$$\dot{W}_{\text{electric in,efficient}} = \dot{W}_{\text{shaft}}/\eta_{\text{eff}} = (\text{Rated power})(\text{Load factor})/\eta_{\text{eff}}$$

$$\text{Power savings} = \dot{W}_{\text{electric in,standard}} - \dot{W}_{\text{electric in,efficient}}$$

$$= (\text{Rated power})(\text{Load factor})(1/\eta_{\text{st}} - 1/\eta_{\text{eff}})$$

where η_{st} is the efficiency of the standard motor, and η_{eff} is the efficiency of the comparable high-efficiency motor. Then the annual energy and cost savings associated with the installation of the high-efficiency motor become

$$\begin{aligned}\text{Energy savings} &= (\text{Power savings})(\text{Operating hours})\\ &= (\text{Rated power})(\text{Operating hours})(\text{Load factor})(1/\eta_{st} - 1/\eta_{eff})\\ &= (60\text{ hp})(0.7457\text{ kW/hp})(3500\text{ h/year})(1)(1/0.89 - 1/0.93.2)\\ &= \mathbf{7929\ kWh/year}\end{aligned}$$

$$\begin{aligned}\text{Cost savings} &= (\text{Energy savings})(\text{Unit cost of energy})\\ &= (7929\text{ kWh/year})(\$0.08/\text{ kWh})\\ &= \mathbf{\$634/year}\end{aligned}$$

Also,

$$\text{Excess initial cost} = \text{Purchase price differential} = \$5160 - \$4520 = \$640$$

This gives a simple payback period of

$$\text{Simple payback period} = \frac{\text{Excess initial cost}}{\text{Annual cost savings}} = \frac{\$640}{\$634/\text{year}} = \mathbf{1.01\ year}$$

Discussion Note that the high-efficiency motor pays for its price differential within about one year from the electrical energy it saves. Considering that the service life of electric motors is several years, the purchase of the higher efficiency motor is definitely indicated in this case. Be careful, however. If the motor is used in the winter, the "wasted" heat is not really wasted, since it heats the building, reducing the heating load. In the summer, however, the savings are enhanced by reducing the cooling load.

FIGURE 3–62

Energy conversion processes are often accompanied by environmental pollution.

© Corbis Royalty Free

3–8 ■ ENERGY AND ENVIRONMENT

The conversion of energy from one form to another often affects the environment and the air we breathe in many ways, and thus the study of energy is not complete without considering its impact on the environment (Fig. 3–62). Fossil fuels such as coal, oil, and natural gas have been powering the industrial development and the amenities of modern life that we enjoy since the 1700s, but this has not been without any undesirable side effects. From the soil we farm and the water we drink to the air we breathe, the environment has been paying a heavy toll for it. Pollutants emitted during the combustion of fossil fuels are responsible for smog, acid rain, and global warming and climate change. The environmental pollution has reached such high levels that it became a serious threat to vegetation, wild life, and human health. Air pollution has been the cause of numerous health problems including asthma and cancer. It is estimated that over 60,000 people in the United States alone die each year due to heart and lung diseases related to air pollution.

Hundreds of elements and compounds such as benzene and formaldehyde are known to be emitted during the combustion of coal, oil, natural gas, and wood in electric power plants, engines of vehicles, furnaces, and even fireplaces. Some compounds are added to liquid fuels for various reasons (such as MTBE to raise the octane number of the fuel and also to oxygenate the fuel in winter months to reduce urban smog). The largest source of air pollution is the motor vehicles, and the pollutants released by the vehicles are usually grouped as hydrocarbons (HC), nitrogen oxides (NO_x), and carbon monoxide (CO) (Fig. 3–63). The HC emissions are a large component of volatile organic compounds (VOCs) emissions, and the two terms are generally used interchangeably for motor vehicle emissions. A significant portion of the VOC or HC emissions are caused by the evaporation of fuels during refueling or spillage during spitback or by evaporation from gas tanks with faulty caps that do not close tightly. The solvents, propellants, and household cleaning products that contain benzene, butane, or other HC products are also significant sources of HC emissions.

FIGURE 3–63

Motor vehicles are the largest source of air pollution.

The increase of environmental pollution at alarming rates and the rising awareness of its dangers made it necessary to control it by legislation and international treaties. In the United States, the Clean Air Act of 1970 (whose passage was aided by the 14-day smog alert in Washington that year) set limits on pollutants emitted by large plants and vehicles. These early standards focused on emissions of hydrocarbons, nitrogen oxides, and carbon monoxide. The new cars were required to have catalytic converters in their exhaust systems to reduce HC and CO emissions. As a side benefit, the removal of lead from gasoline to permit the use of catalytic converters led to a significant reduction in toxic lead emissions.

Emission limits for HC, NO_x, and CO from cars have been declining steadily since 1970. The Clean Air Act of 1990 made the requirements on emissions even tougher, primarily for ozone, CO, nitrogen dioxide, and particulate matter (PM). As a result, today's industrial facilities and vehicles emit a fraction of the pollutants they used to emit a few decades ago. The HC emissions of cars, for example, decreased from about 8 gpm (grams per mile) in 1970 to 0.4 gpm in 1980 and about 0.1 gpm in 1999. This is a significant reduction since many of the gaseous toxics from motor vehicles and liquid fuels are hydrocarbons.

Children are most susceptible to the damages caused by air pollutants since their organs are still developing. They are also exposed to more pollution since they are more active, and thus they breathe faster. People with heart and lung problems, especially those with asthma, are most affected by air pollutants. This becomes apparent when the air pollution levels in their neighborhoods rise to high levels.

Ozone and Smog

If you live in a metropolitan area such as Los Angeles, you are probably familiar with urban smog—the dark yellow or brown haze that builds up in a large stagnant air mass and hangs over populated areas on calm hot summer days. *Smog* is made up mostly of ground-level ozone (O_3), but it also contains numerous other chemicals, including carbon monoxide (CO), particulate matter such as soot and dust, volatile organic compounds (VOCs) such as benzene, butane, and other hydrocarbons. The harmful ground-level ozone should not be confused with the useful ozone layer high in the stratosphere that protects the earth from the sun's harmful ultraviolet rays. Ozone at ground level is a pollutant with several adverse health effects.

The primary source of both nitrogen oxides and hydrocarbons is motor vehicles. Hydrocarbons and nitrogen oxides react in the presence of sunlight on hot calm days to form ground-level ozone, which is the primary component of smog (Fig. 3–64). The smog formation usually peaks in late afternoons when the temperatures are highest and there is plenty of sunlight. Although ground-level smog and ozone form in urban areas with heavy traffic or industry, the prevailing winds can transport them several hundred miles to other cities. This shows that pollution knows no boundaries, and it is a global problem.

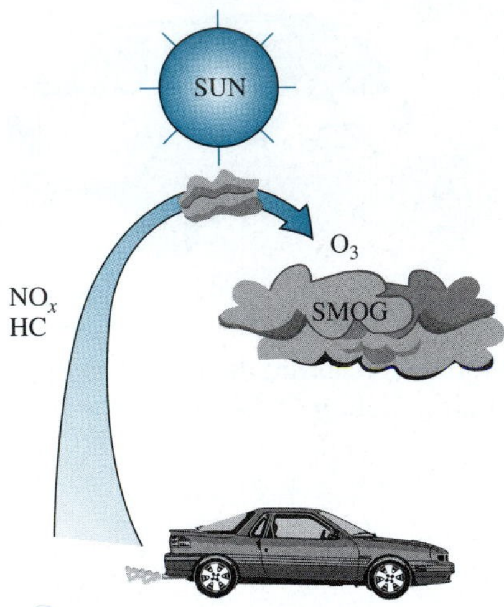

FIGURE 3–64

Ground-level ozone, which is the primary component of smog, forms when HC and NO_x react in the presence of sunlight in hot calm days.

Ozone irritates eyes and damages the air sacs in the lungs where oxygen and carbon dioxide are exchanged, causing eventual hardening of this soft and spongy tissue. It also causes shortness of breath, wheezing, fatigue, headaches, and nausea, and aggravates respiratory problems such as asthma. Every exposure to ozone does a little damage to the lungs, just like cigarette smoke, eventually reducing the individual's lung capacity. Staying indoors and minimizing physical activity during heavy smog minimizes damage. Ozone also harms vegetation by damaging leaf tissues. To improve the air quality in areas with the worst ozone problems, reformulated gasoline (RFG) that contains at least 2 percent oxygen was introduced. The use of RFG has resulted in significant reduction in the emission of ozone and other pollutants, and its use is mandatory in many smog-prone areas.

The other serious pollutant in smog is *carbon monoxide,* which is a colorless, odorless, poisonous gas. It is mostly emitted by motor vehicles, and it can build to dangerous levels in areas with heavy congested traffic. It deprives the body's organs from getting enough oxygen by binding with the red blood cells that would otherwise carry oxygen. At low levels, carbon monoxide decreases the amount of oxygen supplied to the brain and other organs and muscles, slows body reactions and reflexes, and impairs judgment. It poses a serious threat to people with heart disease because of the fragile condition of the circulatory system and to fetuses because of the oxygen needs of the developing brain. At high levels, it can be fatal, as evidenced by numerous deaths caused by cars that are warmed up in closed garages or by exhaust gases leaking into the cars.

Smog also contains suspended particulate matter such as dust and soot emitted by vehicles and industrial facilities. Such particles irritate the eyes and the lungs since they may carry compounds such as acids and metals.

Acid Rain

Fossil fuels are mixtures of various chemicals, including small amounts of sulfur. The sulfur in the fuel reacts with oxygen to form sulfur dioxide (SO_2), which is an air pollutant. The main source of SO_2 is the electric power plants that burn high-sulfur coal. The Clean Air Act of 1970 has limited the SO_2 emissions severely, which forced the plants to install SO_2 scrubbers, to switch to low-sulfur coal, or to gasify the coal and recover the sulfur. Motor vehicles also contribute to SO_2 emissions since gasoline and diesel fuel also contain small amounts of sulfur. Volcanic eruptions and hot springs also release sulfur oxides (the cause of the rotten egg smell).

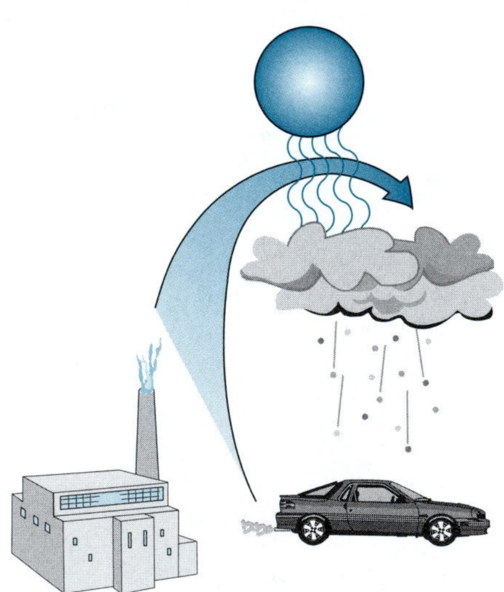

FIGURE 3–65

Sulfuric acid and nitric acid are formed when sulfur oxides and nitric oxides react with water vapor and other chemicals high in the atmosphere in the presence of sunlight.

The sulfur oxides and nitric oxides react with water vapor and other chemicals high in the atmosphere in the presence of sunlight to form sulfuric and nitric acids (Fig. 3–65). The acids formed usually dissolve in the suspended water droplets in clouds or fog. These acid-laden droplets, which can be as acidic as lemon juice, are washed from the air on to the soil by rain or snow. This is known as **acid rain**. The soil is capable of neutralizing a certain amount of acid, but the amounts produced by the power plants using inexpensive high-sulfur coal has exceeded this capability, and as a result many lakes and rivers in industrial areas such as New York, Pennsylvania, and Michigan have become too acidic for fish to grow. Forests in those areas also experience a slow death due to absorbing the acids through their leaves, needles, and roots. Even marble structures deteriorate due to acid rain. The magnitude of the problem was not recognized until the early 1970s, and serious measures have been taken since then to reduce the sulfur dioxide emissions drastically by installing scrubbers in plants and by desulfurizing coal before combustion.

The Greenhouse Effect: Global Warming and Climate Change

You have probably noticed that when you leave your car under direct sunlight on a sunny day, the interior of the car gets much warmer than the air outside, and you may have wondered why the car acts like a heat trap. This is because glass at thicknesses encountered in practice transmits over 90 percent of radiation in the visible range and is practically opaque (nontransparent) to radiation in the longer wavelength infrared regions. Therefore, glass allows the solar radiation to enter freely but blocks the infrared radiation emitted by the interior surfaces. This causes a rise in the interior temperature as a result of the thermal energy buildup in the car. This heating effect is known as the **greenhouse effect**, since it is utilized primarily in greenhouses.

The greenhouse effect is also experienced on a larger scale on earth. The surface of the earth, which warms up during the day as a result of the absorption of solar energy, cools down at night by radiating part of its energy into deep space as infrared radiation. Carbon dioxide (CO_2), water vapor, and trace amounts of some other gases such as methane and nitrogen oxides act like a blanket and keep the earth warm at night by blocking the heat radiated from the earth (Fig. 3–66). Therefore, they are called "greenhouse gases," with CO_2 being the primary component. Water vapor is usually taken out of this list since it comes down as rain or snow as part of the water cycle and human activities in producing water (such as the burning of fossil fuels) do not make much difference on its concentration in the atmosphere (which is mostly due to evaporation from rivers, lakes, oceans, etc.). CO_2 is different, however, in that people's activities do make a difference in CO_2 concentration in the atmosphere.

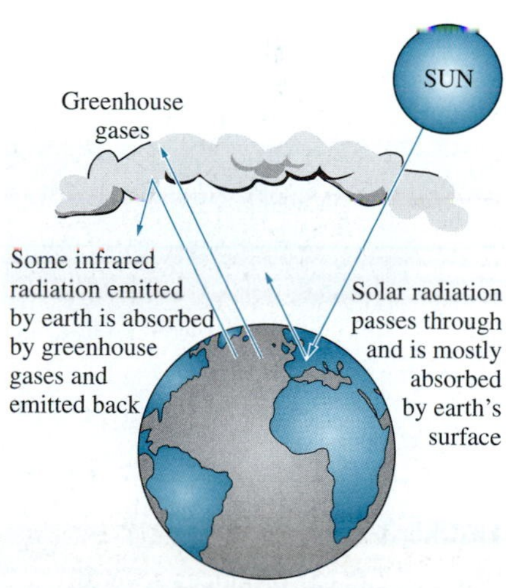
SUN
Greenhouse gases
Some infrared radiation emitted by earth is absorbed by greenhouse gases and emitted back
Solar radiation passes through and is mostly absorbed by earth's surface

FIGURE 3–66

The greenhouse effect on earth.

The greenhouse effect makes life on earth possible by keeping the earth warm (about 30°C warmer). However, excessive amounts of these gases disturb the delicate balance by trapping too much energy, which causes the average temperature of the earth to rise and the climate at some localities to change. These undesirable consequences of the greenhouse effect are referred to as **global warming** or **global climate change**.

The global climate change is due, in part, to the excessive use of fossil fuels such as coal, petroleum products, and natural gas in electric power generation, transportation, buildings, and manufacturing, and it has been a concern in recent decades. In 1995, a total of 6.5 billion tons of carbon was released to the atmosphere as CO_2. The current concentration of CO_2 in the atmosphere is about 360 ppm (or 0.36 percent). This is 20 percent higher than the level a century ago, and it is projected to increase to over 700 ppm by the year 2100. Under normal conditions, vegetation consumes CO_2 and releases O_2 during the photosynthesis process, and thus keeps the CO_2 concentration in the atmosphere in check. A mature, growing tree consumes about 12 kg of CO_2 a year and exhales enough oxygen to support a family of four. However, deforestation and the huge increase in the CO_2 production in recent decades disturbed this balance.

In a 1995 report, the world's leading climate scientists concluded that the earth has already warmed about 0.5°C during the last century, and they estimate that the earth's temperature will rise another 2°C by the year 2100. A rise of this magnitude is feared to cause severe changes in weather patterns with storms and heavy rains and flooding at some parts and drought in others, major floods due to the melting of ice at the poles, loss of wetlands and coastal areas due to rising sea levels, variations in water supply, changes in the ecosystem due to the inability of some animal and plant species to adjust to the changes, increases in epidemic diseases due to the warmer temperatures, and adverse side effects on human health and socioeconomic conditions in some areas.

The seriousness of these threats has moved the United Nations to establish a committee on climate change. A world summit in 1992 in Rio de Janeiro, Brazil, attracted world attention to the problem. The agreement prepared by the committee in 1992 to control greenhouse gas emissions was signed by 162 nations. In the 1997 meeting in Kyoto (Japan), the world's industrialized countries adopted the Kyoto protocol and committed to reduce their CO_2 and other greenhouse gas emissions by 5 percent below the 1990 levels by 2008 to 2012. This can be done by increasing conservation efforts and improving conversion efficiencies, while meeting new energy demands by the use of renewable energy (such as hydroelectric, solar, wind, and geothermal energy) rather than by fossil fuels.

FIGURE 3–67

The average car produces several times its weight in CO_2 every year (it is driven 12,000 miles a year, consumes 600 gallons of gasoline, and produces 20 lbm of CO_2 per gallon).

The United States is the largest contributor of greenhouse gases, with over 5 tons of carbon emissions per person per year. A major source of greenhouse gas emissions is transportation. Each liter of gasoline burned by a vehicle produces about 2.5 kg of CO_2 (or, each gallon of gasoline burned produces about 20 lbm of CO_2). An average car in the United States is driven about 12,000 miles a year, and it consumes about 600 gallons of gasoline. Therefore, a car emits about 12,000 lbm of CO_2 to the atmosphere a year, which is about four times the weight of a typical car (Fig. 3–67). This and other emissions can be reduced significantly by buying an energy-efficient car that burns less fuel over the same distance, and by driving sensibly. Saving fuel also saves money and the environment. For example, choosing a vehicle that gets 30 rather than 20 miles per gallon will prevent 2 tons of CO_2 from being released to the atmosphere every year while reducing the fuel cost by $500 per year (under average driving conditions of 12,000 miles a year and at a fuel cost of $2.50/gal).

It is clear from these discussions that considerable amounts of pollutants are emitted as the chemical energy in fossil fuels is converted to thermal, mechanical, or electrical energy via combustion, and thus power plants, motor vehicles, and even stoves take the blame for air pollution. In contrast, no pollution is emitted as electricity is converted to thermal, chemical, or mechanical energy, and thus electric cars are often touted as "zero emission" vehicles and their widespread use is seen by some as the ultimate solution to the air pollution problem. It should be remembered, however, that the electricity used by the electric cars is generated somewhere else mostly by burning fuel and thus emitting pollution. Therefore, each time an electric car consumes 1 kWh of electricity, it bears the responsibility for the pollutions emitted as 1 kWh of electricity (plus the conversion and transmission losses) is generated elsewhere. The electric cars can be claimed to be zero emission vehicles only when the electricity they consume is generated by emission-free renewable resources such as hydroelectric, solar, wind, and geothermal energy (Fig. 3–68). Therefore, the use of renewable energy should be encouraged worldwide, with incentives, as necessary, to make the earth a better place in which to live. The advancements in thermodynamics have contributed greatly in recent decades to improve conversion efficiencies (in some cases doubling them) and thus to reduce pollution. As individuals, we can also help by practicing energy conservation measures and by making energy efficiency a high priority in our purchases.

FIGURE 3–68

Renewable energies such as wind are called "green energy" since they emit no pollutants or greenhouse gases.

EXAMPLE 3–18 Reducing Air Pollution by Geothermal Heating

A geothermal power plant in Nevada is generating electricity using geothermal water extracted at 180°C, and reinjected back to the ground at 85°C. It is proposed to utilize the reinjected brine for heating the residential and commercial buildings in the area, and calculations show that the geothermal heating system can save 18 million therms of natural gas a year. Determine the amount of NO_x and CO_2 emissions the geothermal system will save a year. Take the average NO_x and CO_2 emissions of gas furnaces to be 0.0047 kg/therm and 6.4 kg/therm, respectively.

Solution The gas heating systems in an area are being replaced by a geothermal district heating system. The amounts of NO_x and CO_2 emissions saved per year are to be determined.

Analysis The amounts of emissions saved per year are equivalent to the amounts emitted by furnaces when 18 million therms of natural gas are burned,

$$\begin{aligned} NO_x \text{ savings} &= (NO_x \text{ emission per therm})(\text{No. of therms per year}) \\ &= (0.0047 \text{ kg/therm})(18 \times 10^6 \text{ therm/year}) \\ &= \mathbf{8.5 \times 10^4 \text{ kg/year}} \end{aligned}$$

$$\begin{aligned} CO_2 \text{ savings} &= (CO_2 \text{ emission per therm})(\text{No. of therms per year}) \\ &= (6.4 \text{ kg/therm})(18 \times 10^6 \text{ therm/year}) \\ &= \mathbf{1.2 \times 10^8 \text{ kg/year}} \end{aligned}$$

Discussion A typical car on the road generates about 8.5 kg of NO_x and 6000 kg of CO_2 a year. Therefore the environmental impact of replacing the gas heating systems in the area by the geothermal heating system is equivalent to taking 10,000 cars off the road for NO_x emission and taking 20,000 cars off the road for CO_2 emission. The proposed system should have a significant effect on reducing smog in the area.

SUMMARY

The sum of all forms of energy of a system is called *total energy,* which consists of internal, kinetic, and potential energy for simple compressible systems. *Internal energy* represents the molecular energy of a system and may exist in sensible, latent, chemical, and nuclear forms.

Mass flow rate $\dot{m}$ is defined as the amount of mass flowing through a cross section per unit time. It is related to the *volume flow rate* $\dot{V}$, which is the volume of a fluid flowing through a cross section per unit time, by

$$\dot{m} = \rho \dot{V} = \rho A_c V_{\text{avg}}$$

The energy flow rate associated with a fluid flowing at a rate of $\dot{m}$ is

$$\dot{E} = \dot{m}e$$

which is analogous to $E = me$.

The *mechanical energy* is defined as *the form of energy that can be converted to mechanical work completely and directly by a mechanical device such as an ideal turbine*. It is expressed on a unit mass basis and rate form as

$$e_{\text{mech}} = \frac{P}{\rho} + \frac{V^2}{2} + gz$$

and

$$\dot{E}_{\text{mech}} = \dot{m}e_{\text{mech}} = \dot{m}\left(\frac{P}{\rho} + \frac{V^2}{2} + gz\right)$$

where P/ρ is the *flow energy*, $V^2/2$ is the *kinetic energy*, and gz is the *potential energy* of the fluid per unit mass.

Energy can cross the boundaries of a closed system in the form of heat or work. For control volumes, energy can also be transported by mass. If the energy transfer is due to a temperature difference between a closed system and its surroundings, it is *heat;* otherwise, it is *work.*

Work is the energy transferred as a force acts on a system through a distance. Various forms of work are expressed as follows:

Electrical work: $W_e = \mathbf{V}I\,\Delta t$

Shaft work: $W_{\text{sh}} = 2\pi n\mathrm{T}$

Spring work: $W_{\text{spring}} = \frac{1}{2}k(x_2^2 - x_1^2)$

The *first law of thermodynamics* is essentially an expression of the conservation of energy principle, also called the *energy balance*. The general mass and energy balances for *any system* undergoing *any process* can be expressed as

$$\underbrace{E_{\text{in}} - E_{\text{out}}}_{\substack{\text{Net energy transfer}\\ \text{by heat, work, and mass}}} = \underbrace{\Delta E_{\text{system}}}_{\substack{\text{Change in internal, kinetic,}\\ \text{potential, etc., energies}}}$$

It can also be expressed in the *rate form* as

$$\underbrace{\dot{E}_{\text{in}} - \dot{E}_{\text{out}}}_{\substack{\text{Rate of net energy transfer}\\ \text{by heat, work, and mass}}} = \underbrace{dE_{\text{system}}/dt}_{\substack{\text{Rate of change in internal,}\\ \text{kinetic, potential, etc., energies}}}$$

The efficiencies of various devices are defined as

$$\eta_{\text{pump}} = \frac{\Delta\dot{E}_{\text{mech,fluid}}}{\dot{W}_{\text{shaft,in}}} = \frac{\dot{W}_{\text{pump},u}}{\dot{W}_{\text{pump}}}$$

$$\eta_{\text{turbine}} = \frac{\dot{W}_{\text{shaft,out}}}{|\Delta\dot{E}_{\text{mech,fluid}}|} = \frac{\dot{W}_{\text{turbine}}}{\dot{W}_{\text{turbine},e}}$$

$$\eta_{\text{motor}} = \frac{\text{Mechanical power output}}{\text{Electric power input}} = \frac{\dot{W}_{\text{shaft,out}}}{\dot{W}_{\text{elect,in}}}$$

$$\eta_{\text{generator}} = \frac{\text{Electric power output}}{\text{Mechanical power input}} = \frac{\dot{W}_{\text{elect,out}}}{\dot{W}_{\text{shaft,in}}}$$

$$\eta_{\text{pump-motor}} = \eta_{\text{pump}}\eta_{\text{motor}} = \frac{\Delta\dot{E}_{\text{mech,fluid}}}{\dot{W}_{\text{elect,in}}}$$

$$\eta_{\text{turbine-gen}} = \eta_{\text{turbine}}\eta_{\text{generator}} = \frac{\dot{W}_{\text{elect,out}}}{|\Delta\dot{E}_{\text{mech,fluid}}|}$$

The conversion of energy from one form to another is often associated with adverse effects on the environment, and environmental impact should be an important consideration in the conversion and utilization of energy.

REFERENCES AND SUGGESTED READINGS

1. ASHRAE *Handbook of Fundamentals*. SI version. Atlanta, GA: American Society of Heating, Refrigerating, and Air-Conditioning Engineers, Inc., 1993.
2. Y. A. Çengel. "An Intuitive and Unified Approach to Teaching Thermodynamics." ASME International Mechanical Engineering Congress and Exposition, Atlanta, Georgia, AES-Vol. 36, pp. 251–260, November 17–22, 1996.

PROBLEMS*

Forms of Energy

3–1C Consider the falling of a rock off a cliff into seawater, and eventually settling at the bottom of the sea. Starting with the potential energy of the rock, identify the energy transfers and transformations involved during this process.

3–2C Natural gas, which is mostly methane CH_4, is a fuel and a major energy source. Can we say the same about hydrogen gas, H_2?

3–3C What is the difference between the macroscopic and microscopic forms of energy?

3–4C What is total energy? Identify the different forms of energy that constitute the total energy.

3–5C How are heat, internal energy, and thermal energy related to each other?

3–6C What is mechanical energy? How does it differ from thermal energy? What are the forms of mechanical energy of a fluid stream?

3–7E The specific kinetic energy of a moving mass is given by ke $= V^2/2$, where V is the velocity of the mass. Determine the specific kinetic energy of a mass whose velocity is 100 ft/s, in Btu/lbm. *Answer:* 0.2 Btu/lbm

3–8 Determine the specific kinetic energy of a mass whose velocity is 30 m/s, in kJ/kg.

3–9E Calculate the total potential energy, in Btu, of an object that is 20 ft below a datum level at a location where $g = 31.7$ ft/s^2 and which has a mass of 100 lbm.

3–10 Determine the specific potential energy, in kJ/kg, of an object 50 m above a datum in a location where $g = 9.8$ m/s^2.

3–11 An object whose mass is 100 kg is located 20 m above a datum level in a location where standard gravitational acceleration exists. Determine the total potential energy, in kJ, of this object.

3–12 Consider a river flowing toward a lake at an average velocity of 3 m/s at a rate of 500 m^3/s at a location 90 m above the lake surface. Determine the total mechanical energy of the river water per unit mass and the power generation potential of the entire river at that location.

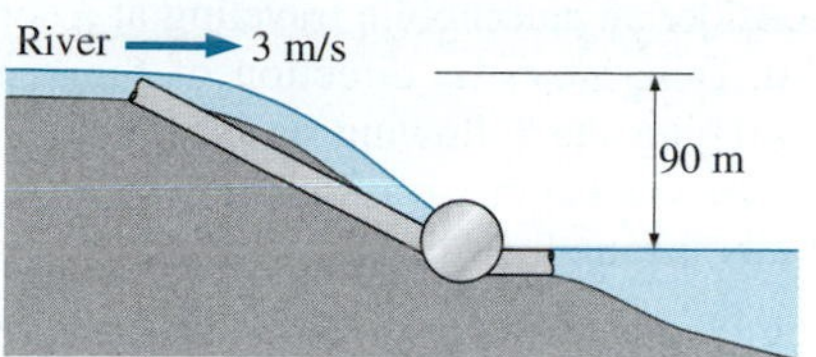

FIGURE P3–12

3–13 Electric power is to be generated by installing a hydraulic turbine–generator at a site 120 m below the free surface of a large water reservoir that can supply water at a rate of 1500 kg/s steadily. Determine the power generation potential.

3–14 At a certain location, wind is blowing steadily at 10 m/s. Determine the mechanical energy of air per unit mass and the power generation potential of a wind turbine with 60-m-diameter blades at that location. Take the air density to be 1.25 kg/m^3.

3–15 A water jet that leaves a nozzle at 60 m/s at a flow rate of 120 kg/s is to be used to generate power by striking the buckets located on the perimeter of a wheel. Determine the power generation potential of this water jet.

3–16 Two sites are being considered for wind power generation. In the first site, the wind blows steadily at 7 m/s for 3000 hours per year, whereas in the second site the wind blows at 10 m/s for 2000 hours per year. Assuming the wind velocity is negligible at other times for simplicity, determine which is a better site for wind power generation. *Hint:* Note that the mass flow rate of air is proportional to wind velocity.

3–17 A river flowing steadily at a rate of 240 m^3/s is considered for hydroelectric power generation. It is determined that a dam can be built to collect water and release it from an elevation difference of 50 m to generate power. Determine how much power can be generated from this river water after the dam is filled.

3–18 A person gets into an elevator at the lobby level of a hotel together with his 30-kg suitcase, and gets out at the 10th floor 35 m above. Determine the amount of energy consumed by the motor of the elevator that is now stored in the suitcase.

Energy Transfer by Heat and Work

3–19C In what forms can energy cross the boundaries of a closed system?

3–20C When is the energy crossing the boundaries of a closed system heat and when is it work?

3–21C What is an adiabatic process? What is an adiabatic system?

3–22C What are point and path functions? Give some examples.

---

*Problems designated by a "C" are concept questions, and students are encouraged to answer them all. Problems designated by an "E" are in English units, and the SI users can ignore them. Problems with the icon are solved using EES, and complete solutions together with parametric studies are included on the enclosed DVD. Problems with the icon are comprehensive in nature, and are intended to be solved with a computer, preferably using the EES software that accompanies this text.

3–23C What is the caloric theory? When and why was it abandoned?

3–24C Consider an automobile traveling at a constant speed along a road. Determine the direction of the heat and work interactions, taking the following as the system: (*a*) the car radiator, (*b*) the car engine, (*c*) the car wheels, (*d*) the road, and (*e*) the air surrounding the car.

3–25C The length of a spring can be changed by (*a*) applying a force to it or (*b*) changing its temperature (i.e., thermal expansion). What type of energy interaction between the system (spring) and surroundings is required to change the length of the spring in these two ways?

3–26C Consider an electric refrigerator located in a room. Determine the direction of the work and heat interactions (in or out) when the following are taken as the system: (*a*) the contents of the refrigerator, (*b*) all parts of the refrigerator including the contents, and (*c*) everything contained within the room during a winter day.

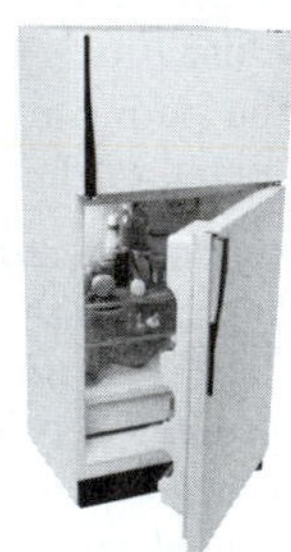

FIGURE P3–26C

© PhotoDisc/Punchstock

3–27C A personal computer is to be examined from a thermodynamic perspective. Determine the direction of the work and heat transfers (in or out) when the (*a*) keyboard, (*b*) monitor, (*c*) processing unit, and (*d*) all of these are taken as the system.

3–28 A small electrical motor produces 10 W of mechanical power. What is this power in (*a*) N, m, and s units; and (*b*) kg, m, and s units? *Answers:* (*a*) 10 N · m/s, (*b*) 10 kg · m^2/s^3

3–29E A model aircraft internal-combustion engine produces 10 W of power. How much power is this in (*a*) lbf · ft/s and (*b*) hp?

Mechanical Forms of Work

3–30C A car is accelerated from rest to 85 km/h in 10 s. Would the energy transferred to the car be different if it were accelerated to the same speed in 5 s?

3–31 Determine the energy required to accelerate an 800-kg car from rest to 100 km/h on a level road. *Answer:* 309 kJ

3–32E A construction crane lifts a prestressed concrete beam weighing 2 tons from the ground to the top of piers that are 18 ft above the ground. Determine the amount of work done considering (*a*) the beam and (*b*) the crane as the system. Express your answers in both lbf · ft and Btu.

3–33 A man whose mass is 100 kg pushes a cart whose mass, including its contents, is 100 kg up a ramp that is inclined at an angle of 20° from the horizontal. The local gravitational acceleration is 9.8 m/s^2. Determine the work, in kJ, needed to move along this ramp a distance of 100 m considering (*a*) the man and (*b*) the cart and its contents as the system.

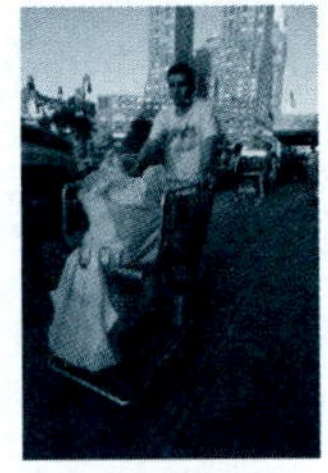

FIGURE P3–33

© The McGraw-Hill Companies, Inc./Lars A. Niki, photographer

3–34E The force F required to compress a spring a distance x is given by $F - F_0 = kx$ where k is the spring constant and F_0 is the preload. Determine the work required to compress a spring whose spring constant is $k = 200$ lbf/in a distance of one inch starting from its free length where $F_0 = 0$ lbf. Express your answer in both lbf · ft and Btu.

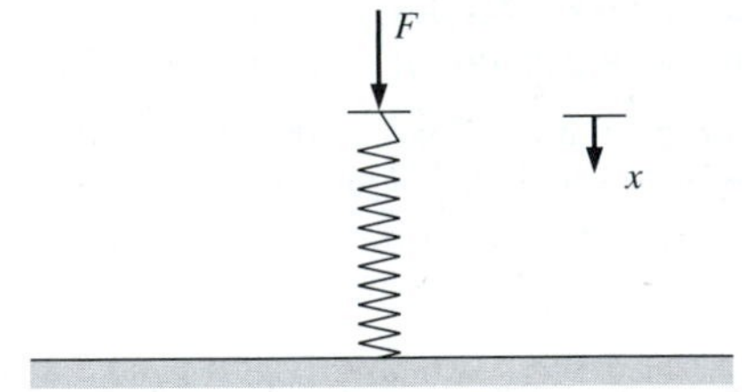

FIGURE P3–34E

3–35 As a spherical ammonia vapor bubble rises in liquid ammonia, its diameter changes from 1 cm to 3 cm. Calculate the amount of work produced by this bubble, in kJ, if the surface-tension of ammonia is 0.02 N/m. *Answer:* 5.03×10^{-8} kJ

3–36 A steel rod of 0.5 cm diameter and 10 m length is stretched 3 cm. Young's modulus for this steel is 21 kN/cm^2. How much work, in kJ, is required to stretch this rod?

3–37E A spring whose spring constant is 200 lbf/in has an initial force of 100 lbf acting on it. Determine the work, in Btu, required to compress it another 1 inch.

3–38 How much work, in kJ, can a spring whose spring constant is 3 kN/cm produce after it has been compressed 3 cm from its unloaded length?

3–39 A ski lift has a one-way length of 1 km and a vertical rise of 200 m. The chairs are spaced 20 m apart, and each chair can seat three people. The lift is operating at a steady speed of 10 km/h. Neglecting friction and air drag and assuming that the average mass of each loaded chair is 250 kg, determine the power required to operate this ski lift. Also estimate the power required to accelerate this ski lift in 5 s to its operating speed when it is first turned on.

3–40 Determine the power required for a 2000-kg car to climb a 100-m-long uphill road with a slope of 30° (from horizontal) in 10 s (*a*) at a constant velocity, (*b*) from rest to a final velocity of 30 m/s, and (*c*) from 35 m/s to a final velocity of 5 m/s. Disregard friction, air drag, and rolling resistance. *Answers:* (*a*) 98.1 kW, (*b*) 188 kW, (*c*) −21.9 kW

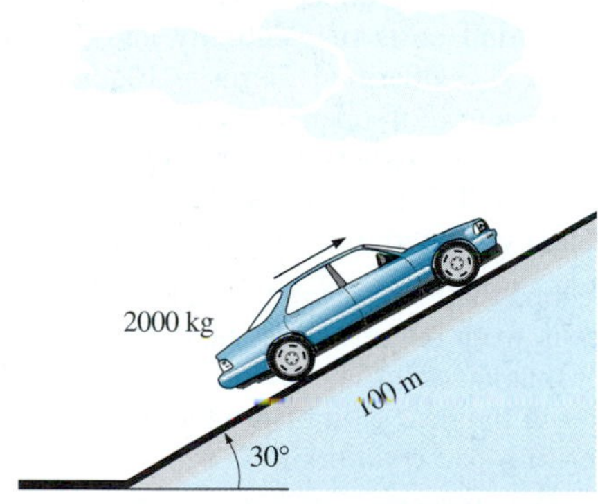

FIGURE P3–40

3–41 A damaged 1200-kg car is being towed by a truck. Neglecting the friction, air drag, and rolling resistance, determine the extra power required (*a*) for constant velocity on a level road, (*b*) for constant velocity of 50 km/h on a 30° (from horizontal) uphill road, and (*c*) to accelerate on a level road from stop to 90 km/h in 12 s. *Answers:* (*a*) 0, (*b*) 81.7 kW, (*c*) 31.3 kW

The First Law of Thermodynamics

3–42C For a cycle, is the net work necessarily zero? For what kind of systems will this be the case?

3–43C On a hot summer day, a student turns his fan on when he leaves his room in the morning. When he returns in the evening, will the room be warmer or cooler than the neighboring rooms? Why? Assume all the doors and windows are kept closed.

3–44C What are the different mechanisms for transferring energy to or from a control volume?

3–45E One way to improve the fuel efficiency of a car is to use tires that have a lower rolling resistance—tires that roll with less resistance, and highway tests at 65 mph showed that tires with the lowest rolling resistance can improve the fuel efficiency by nearly 2 mpg (miles per gallon). Consider a car that gets 25 mpg on high rolling resistance tires and is driven 15,000 miles per year. For a fuel cost of $2.20/gal, determine how much money will be saved per year by switching to low rolling resistance tires.

3–46 An adiabatic closed system is accelerated from 0 m/s to 30 m/s. Determine the specific energy change of this system, in kJ/kg.

3–47 An adiabatic closed system is raised 100 m at a location where the gravitational acceleration is 9.8 m/s^2. Determine the energy change of this system, in kJ/kg.

3–48E A water pump increases the water pressure from 10 psia to 50 psia. Determine the power input required, in hp, to pump 1.2 ft^3/s of water. Does the water temperature at the inlet have any significant effect on the required flow power? *Answer:* 12.6 hp

3–49 An automobile moving through the air causes the air velocity (measured with respect to the car) to decrease and fill a larger flow channel. An automobile has an effective flow channel area of 3 m^2. The car is traveling at 90 km/h on a day when the barometric pressure is 75 cm of mercury and the temperature is 30°C. Behind the car, the air velocity (with respect to the car) is measured to be 82 km/h, and the temperature is 30°C. Determine the power required to move this car through the air and the area of the effective flow channel behind the car.

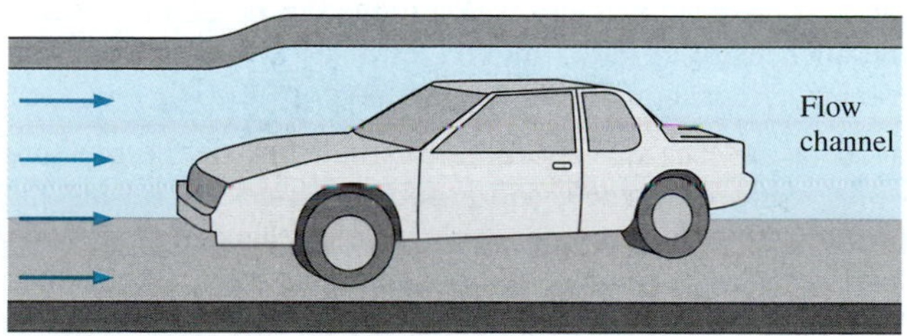

FIGURE P3–49

3–50 A classroom that normally contains 40 people is to be air-conditioned with window air-conditioning units of 5-kW cooling capacity. A person at rest may be assumed to dissipate heat at a rate of about 360 kJ/h. There are 10 lightbulbs in the room, each with a rating of 100 W. The rate of heat transfer to the classroom through the walls and the windows is estimated to be 15,000 kJ/h. If the room air is to be maintained at a constant temperature of 21°C, determine the number of window air-conditioning units required.
Answer: 2 units

3–51 The lighting needs of a storage room are being met by 6 fluorescent light fixtures, each fixture containing four lamps rated at 60 W each. All the lamps are on during operating hours of the facility, which are 6 AM to 6 PM 365 days a year. The storage room is actually used for an average of 3 h a day. If the price of electricity is \$0.08/kWh, determine the amount of energy and money that will be saved as a result of installing motion sensors. Also, determine the simple payback period if the purchase price of the sensor is \$32 and it takes 1 hour to install it at a cost of \$40.

3–52 A university campus has 200 classrooms and 400 faculty offices. The classrooms are equipped with 12 fluorescent tubes, each consuming 110 W, including the electricity used by the ballasts. The faculty offices, on average, have half as many tubes. The campus is open 240 days a year. The classrooms and faculty offices are not occupied an average of 4 h a day, but the lights are kept on. If the unit cost of electricity is \$0.082/kWh, determine how much the campus will save a year if the lights in the classrooms and faculty offices are turned off during unoccupied periods.

3–53 Consider a room that is initially at the outdoor temperature of 20°C. The room contains a 100-W lightbulb, a 110-W TV set, a 200-W refrigerator, and a 1000-W iron. Assuming no heat transfer through the walls, determine the rate of increase of the energy content of the room when all of these electric devices are on.

3–54 A fan is to accelerate quiescent air to a velocity of 10 m/s at a rate of 4 m^3/s. Determine the minimum power that must be supplied to the fan. Take the density of air to be 1.18 kg/m^3. *Answer: 236 W*

3–55E Consider a fan located in a 3 ft × 3 ft square duct. Velocities at various points at the outlet are measured, and the average flow velocity is determined to be 22 ft/s. Taking the air density to 0.075 lbm/ft^3, estimate the minimum electric power consumption of the fan motor.

3–56 EES The driving force for fluid flow is the pressure difference, and a pump operates by raising the pressure of a fluid (by converting the mechanical shaft work to flow energy). A gasoline pump is measured to consume 5.2 kW of electric power when operating. If the pressure differential between the outlet and inlet of the pump is measured to be 5 kPa and the changes in velocity and elevation are negligible, determine the maximum possible volume flow rate of gasoline.

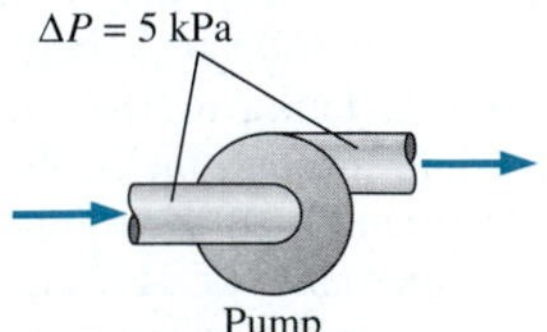

FIGURE P3–56

3–57 An escalator in a shopping center is designed to move 30 people, 75 kg each, at a constant speed of 0.8 m/s at 45° slope. Determine the minimum power input needed to drive this escalator. What would your answer be if the escalator velocity were to be doubled?

Energy Conversion Efficiencies

3–58C What is mechanical efficiency? What does a mechanical efficiency of 100 percent mean for a hydraulic turbine?

3–59C How is the combined pump–motor efficiency of a pump and motor system defined? Can the combined pump–motor efficiency be greater than either the pump or the motor efficiency?

3–60C Define turbine efficiency, generator efficiency, and combined turbine–generator efficiency.

3–61C Can the combined turbine-generator efficiency be greater than either the turbine efficiency or the generator efficiency? Explain.

3–62 Consider a 3-kW hooded electric open burner in an area where the unit costs of electricity and natural gas are \$0.07/kWh and \$1.20/therm (1 therm = 105,500 kJ), respectively. The efficiency of open burners can be taken to be 73 percent for electric burners and 38 percent for gas burners. Determine the rate of energy consumption and the unit cost of utilized energy for both electric and gas burners.

3–63 A 75-hp (shaft output) motor that has an efficiency of 91.0 percent is worn out and is replaced by a high-efficiency 75-hp motor that has an efficiency of 95.4 percent. Determine the reduction in the heat gain of the room due to higher efficiency under full-load conditions.

3–64 A 90-hp (shaft output) electric car is powered by an electric motor mounted in the engine compartment. If the motor has an average efficiency of 91 percent, determine the rate of heat supply by the motor to the engine compartment at full load.

3–65 A 75-hp (shaft output) motor that has an efficiency of 91.0 percent is worn out and is to be replaced by a high-efficiency motor that has an efficiency of 95.4 percent. The motor operates 4368 hours a year at a load factor of 0.75. Taking the cost of electricity to be \$0.08/kWh, determine the amount of energy and money saved as a result of installing the high-efficiency motor instead of the standard motor. Also, determine the simple payback period if the purchase prices of the standard and high-efficiency motors are \$5449 and \$5520, respectively.

3–66E The steam requirements of a manufacturing facility are being met by a boiler whose rated heat input is 3.6×10^6 Btu/h. The combustion efficiency of the boiler is measured to be 0.7 by a hand-held flue gas analyzer. After tuning up the boiler, the combustion efficiency rises to 0.8. The boiler operates 1500 hours a year intermittently. Taking the unit cost of energy to be \$4.35/10^6 Btu, determine the annual energy and cost savings as a result of tuning up the boiler.

3–67E Reconsider Prob. 3–66E. Using EES (or other) software, study the effects of the unit cost of energy and combustion efficiency on the annual energy used and the cost savings. Let the efficiency vary from 0.6 to 0.9, and the unit cost to vary from \$4 to \$6 per million Btu. Plot the annual energy used and the cost savings against the efficiency for unit costs of \$4, \$5, and \$6 per million Btu, and discuss the results.

3–68 An exercise room has eight weight-lifting machines that have no motors and four treadmills each equipped with a 2.5-hp (shaft output) motor. The motors operate at an average load factor of 0.7, at which their efficiency is 0.77. During peak evening hours, all 12 pieces of exercising equipment are used continuously, and there are also two people doing light exercises while waiting in line for one piece of the equipment. Assuming the average rate of heat dissipation from people in an exercise room is 525 W, determine the rate of heat gain of the exercise room from people and the equipment at peak load conditions.

3–69 Consider a classroom for 55 students and one instructor, each generating heat at a rate of 100 W. Lighting is provided by 18 fluorescent lightbulbs, 40 W each, and the ballasts consume an additional 10 percent. Determine the rate of internal heat generation in this classroom when it is fully occupied.

3–70 A room is cooled by circulating chilled water through a heat exchanger located in a room. The air is circulated through the heat exchanger by a 0.25-hp (shaft output) fan. Typical efficiency of small electric motors driving 0.25-hp equipment is 54 percent. Determine the rate of heat supply by the fan–motor assembly to the room.

3–71 Electric power is to be generated by installing a hydraulic turbine–generator at a site 70 m below the free surface of a large water reservoir that can supply water at a rate of 1500 kg/s steadily. If the mechanical power output of the turbine is 800 kW and the electric power generation is 750 kW, determine the turbine efficiency and the combined turbine–generator efficiency of this plant. Neglect losses in the pipes.

3–72 At a certain location, wind is blowing steadily at 12 m/s. Determine the mechanical energy of air per unit mass and the power generation potential of a wind turbine with 50-m-diameter blades at that location. Also determine the actual electric power generation assuming an overall efficiency of 30 percent. Take the air density to be 1.25 kg/m^3.

3–73 Reconsider Prob. 3–72. Using EES (or other) software, investigate the effect of wind velocity and the blade span diameter on wind power generation. Let the velocity vary from 5 to 20 m/s in increments of 5 m/s, and the diameter vary from 20 to 80 m in increments of 20 m. Tabulate the results, and discuss their significance.

3–74 A wind turbine is rotating at 15 rpm under steady winds flowing through the turbine at a rate of 42,000 kg/s. The tip velocity of the turbine blade is measured to be 250 km/h. If 180 kW power is produced by the turbine, determine (*a*) the average velocity of the air and (*b*) the conversion efficiency of the turbine. Take the density of air to be 1.31 kg/m^3.

3–75 Water is pumped from a lake to a storage tank 20 m above at a rate of 70 L/s while consuming 20.4 kW of electric power. Disregarding any frictional losses in the pipes and any changes in kinetic energy, determine (*a*) the overall efficiency of the pump–motor unit and (*b*) the pressure difference between the inlet and the exit of the pump.

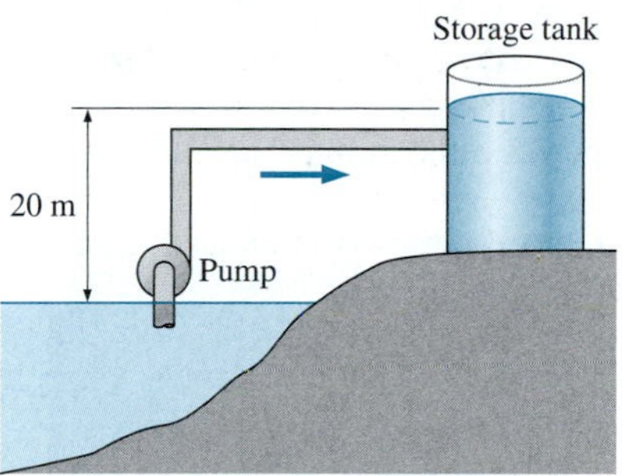

FIGURE P3–75

3–76 Large wind turbines with blade span diameters of over 100 m are available for electric power generation. Consider a wind turbine with a blade span diameter of 100 m installed at a site subjected to steady winds at 8 m/s. Taking the overall efficiency of the wind turbine to be 32 percent and the air density to be 1.25 kg/m^3, determine the electric power generated by this wind turbine. Also, assuming steady winds of 8 m/s during a 24-hour period, determine the amount of electric energy and the revenue generated per day for a unit price of \$0.06/kWh for electricity.

3–77E A water pump delivers 3 hp of shaft power when operating. If the pressure differential between the outlet and the inlet of the pump is measured to be 1.2 psi when the flow rate is 8 ft^3/s and the changes in velocity and elevation are negligible, determine the mechanical efficiency of this pump.

3–78 Water is pumped from a lower reservoir to a higher reservoir by a pump that provides 20 kW of shaft power. The free surface of the upper reservoir is 45 m higher than that of the lower reservoir. If the flow rate of water is measured to be 0.03 m^3/s, determine mechanical power that is converted to thermal energy during this process due to frictional effects.

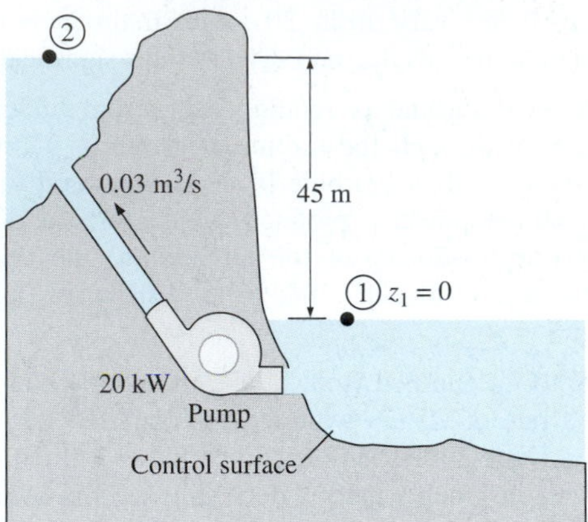

FIGURE P3–78

3–79 The water behind Hoover Dam in Nevada is 206 m higher than the Colorado River below it. At what rate must water pass through the hydraulic turbines of this dam to produce 100 MW of power if the turbines are 100 percent efficient?

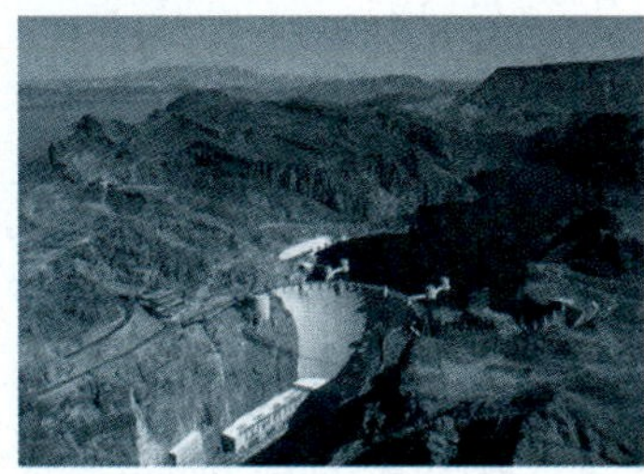

FIGURE P3–79

Photo by Lynn Betts, USDA Natural Resources Conservation Service

3–80 An oil pump is drawing 35 kW of electric power while pumping oil with ρ = 860 kg/m^3 at a rate of 0.1 m^3/s. The inlet and outlet diameters of the pipe are 8 cm and 12 cm, respectively. If the pressure rise of oil in the pump is measured to be 400 kPa and the motor efficiency is 90 percent, determine the mechanical efficiency of the pump.

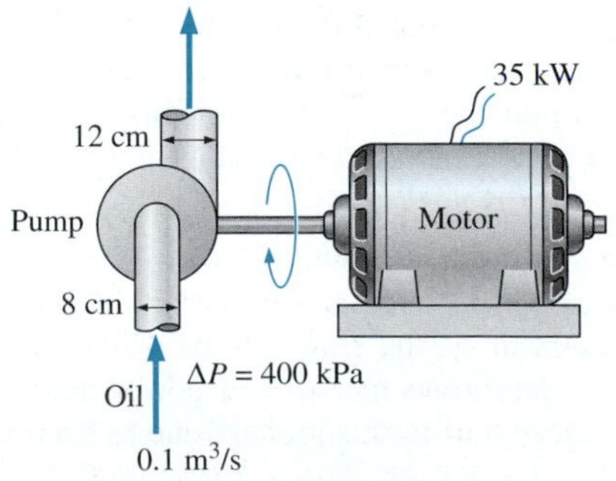

FIGURE P3–80

3–81E A 73-percent efficient pump with a power input of 12 hp is pumping water from a lake to a nearby pool at a rate of 1.2 ft^3/s through a constant-diameter pipe. The free surface of the pool is 35 ft above that of the lake. Determine the mechanical power used to overcome frictional effects in piping. *Answer:* 4.0 hp

Energy and Environment

3–82C How does energy conversion affect the environment? What are the primary chemicals that pollute the air? What is the primary source of these pollutants?

3–83C What is smog? What does it consist of? How does ground-level ozone form? What are the adverse effects of ozone on human health?

3–84C What is acid rain? Why is it called a "rain"? How do the acids form in the atmosphere? What are the adverse effects of acid rain on the environment?

3–85C What is the greenhouse effect? How does the excess CO_2 gas in the atmosphere cause the greenhouse effect? What are the potential long-term consequences of greenhouse effect? How can we combat this problem?

3–86C Why is carbon monoxide a dangerous air pollutant? How does it affect human health at low and at high levels?

3–87E A Ford Taurus driven 15,000 miles a year will use about 715 gallons of gasoline compared to a Ford Explorer that would use 940 gallons. About 19.7 lbm of CO_2, which causes global warming, is released to the atmosphere when a gallon of gasoline is burned. Determine the extra amount of CO_2 production a man is responsible for during a 5-year period if he trades his Taurus for an Explorer.

3–88 When a hydrocarbon fuel is burned, almost all of the carbon in the fuel burns completely to form CO_2 (carbon dioxide), which is the principal gas causing the greenhouse effect and thus global climate change. On average, 0.59 kg of CO_2 is produced for each kWh of electricity generated from a power plant that burns natural gas. A typical new household refrigerator uses about 700 kWh of electricity per year. Determine the amount of CO_2 production that is due to the refrigerators in a city with 200,000 households.

3–89 Repeat Prob. 3–88 assuming the electricity is produced by a power plant that burns coal. The average production of CO_2 in this case is 1.1 kg per kWh.

3–90E Consider a household that uses 11,000 kWh of electricity per year and 1500 gallons of fuel oil during a heating season. The average amount of CO_2 produced is 26.4 lbm/gallon of fuel oil and 1.54 lbm/kWh of electricity. If this household reduces its oil and electricity usage by 15 percent as a result of implementing some energy conservation measures, determine the reduction in the amount of CO_2 emissions by that household per year.

3–91 A typical car driven 20,000 km a year emits to the atmosphere about 11 kg per year of NO_x (nitrogen oxides), which cause smog in major population areas. Natural gas burned in the furnace emits about 4.3 g of NO_x per therm (1 therm = 105,500 kJ), and the electric power plants emit about 7.1 g of NO_x per kWh of electricity produced. Consider a household that has two cars and consumes 9000 kWh of electricity and 1200 therms of natural gas. Determine the amount of NO_x emission to the atmosphere per year for which this household is responsible.

FIGURE P3–91

Review Problems

3–92 Consider a homeowner who is replacing his 25-year-old natural gas furnace that has an efficiency of 55 percent. The homeowner is considering a conventional furnace that has an efficiency of 82 percent and costs $1600 and a high-efficiency furnace that has an efficiency of 95 percent and costs $2700. The homeowner would like to buy the high-efficiency furnace if the savings from the natural gas pay for the additional cost in less than 8 years. If the homeowner presently pays $1200 a year for heating, determine if he should buy the conventional or high-efficiency model.

3–93 Wind energy has been used since 4000 BC to power sailboats, grind grain, pump water for farms, and, more recently, generate electricity. In the United States alone, more than 6 million small windmills, most of them under 5 hp, have been used since the 1850s to pump water. Small windmills have been used to generate electricity since 1900, but the development of modern wind turbines occurred only recently in response to the energy crises in the early 1970s. The cost of wind power has dropped an order of magnitude from about $0.50/kWh in the early 1980s to about $0.05/kWh in the mid-1990s, which is about the price of electricity generated at coal-fired power plants. Areas with an average wind speed of 6 m/s (or 14 mph) are potential sites for economical wind power generation. Commercial wind turbines generate from 100 kW to 3.2 MW of electric power each at peak design conditions. The blade span (or rotor) diameter of the 3.2 MW wind turbine built by Boeing Engineering is 320 ft (97.5 m). The rotation speed of rotors of wind turbines is usually under 40 rpm (under 20 rpm for large turbines). Altamont Pass in California is the world's largest wind farm with 15,000 modern wind turbines. This farm and two others in California produced 2.8 billion kWh of electricity in 1991, which is enough power to meet the electricity needs of San Francisco.

In 2003, 8133 MW of new wind energy generating capacity were installed worldwide, bringing the world's total wind energy capacity to 39,294 MW. The United States, Germany, Denmark, and Spain account for over 75 percent of current wind energy generating capacity worldwide. Denmark uses wind turbines to supply 10 percent of its national electricity.

Many wind turbines currently in operation have just two blades. This is because at tip speeds of 100 to 200 mph, the efficiency of the two-bladed turbine approaches the theoretical maximum, and the increase in the efficiency by adding a third or fourth blade is so little that they do not justify the added cost and weight.

Consider a wind turbine with an 80-m-diameter rotor that is rotating at 20 rpm under steady winds at an average velocity of 30 km/h. Assuming the turbine has an efficiency of 35 percent (i.e., it converts 35 percent of the kinetic energy of the wind to electricity), determine (*a*) the power produced, in kW; (*b*) the tip speed of the blade, in km/h; and (*c*) the revenue generated by the wind turbine per year if the electric power produced is sold to the utility at $0.06/kWh. Take the density of air to be 1.20 kg/m^3.

FIGURE P3–93

© Vol. 57/PhotoDisc

3–94 Repeat Prob. 3–93 for an average wind velocity of 25 km/h.

3–95E The energy contents, unit costs, and typical conversion efficiencies of various energy sources for use in water heaters are given as follows: 1025 Btu/ft^3, $0.012/ft^3, and 55 percent for natural gas; 138,700 Btu/gal, $1.15/gal, and 55 percent for heating oil; and 1 kWh/kWh, $0.084/kWh, and 90 percent for electric heaters, respectively. Determine the lowest-cost energy source for water heaters.

3–96 A homeowner is considering these heating systems for heating his house: Electric resistance heating with \$0.09/kWh and 1 kWh = 3600 kJ, gas heating with \$1.24/therm and 1 therm = 105,500 kJ, and oil heating with \$1.25/gal and 1 gal of oil = 138,500 kJ. Assuming efficiencies of 100 percent for the electric furnace and 87 percent for the gas and oil furnaces, determine the heating system with the lowest energy cost.

3–97 A typical household pays about \$1200 a year on energy bills, and the U.S. Department of Energy estimates that 46 percent of this energy is used for heating and cooling, 15 percent for heating water, 15 percent for refrigerating and freezing, and the remaining 24 percent for lighting, cooking, and running other appliances. The heating and cooling costs of a poorly insulated house can be reduced by up to 30 percent by adding adequate insulation. If the cost of insulation is \$200, determine how long it will take for the insulation to pay for itself from the energy it saves.

3–98 The U.S. Department of Energy estimates that up to 10 percent of the energy use of a house can be saved by caulking and weatherstripping doors and windows to reduce air leaks at a cost of about \$50 for materials for an average home with 12 windows and 2 doors. Caulking and weatherstripping every gas-heated home properly would save enough energy to heat about 4 million homes. The savings can be increased by installing storm windows. Determine how long it will take for the caulking and weatherstripping to pay for itself from the energy they save for a house whose annual energy use is \$1100.

3–99 The force F required to compress a spring a distance x is given by $F - F_0 = kx$ where k is the spring constant and F_0 is the preload. Determine the work, in kJ, required to compress a spring a distance of 1 cm when its spring constant is 300 N/cm and the spring is initially compressed by a force of 100 N.

3–100 The force required to expand the gas in a gas spring a distance x is given by

$$F = \frac{\text{Constant}}{x^k}$$

where the constant is determined by the geometry of this device and k is determined by the gas used in the device. Such a gas spring is arranged to have a constant of 1000 N · m$^{1.3}$ and $k = 1.3$. Determine the work, in kJ, required to compress this spring from 0.1 m to 0.3 m. *Answer:* 1.87 kJ

3–101E A man weighing 180 lbf pushes a block weighing 100 lbf along a horizontal plane. The dynamic coefficient of friction between the block and plane is 0.2. Assuming that the block is moving at constant speed, calculate the work required to move the block a distance of 100 ft considering (*a*) the man and (*b*) the block as the system. Express your answers in both lbf · ft and Btu.

3–102E Water is pumped from a 200-ft-deep well into a 100-ft-high storage tank. Determine the power, in kW, that would be required to pump 200 gallons per minute.

3–103 A grist mill of the 1800s employed a water wheel that was 10 m high; 400 liters per minute of water flowed on to the wheel near the top. How much power, in kW, could this water wheel have produced? *Answer:* 0.654 kW

3–104 Windmills slow the air and cause it to fill a larger channel as it passes through the blades. Consider a circular windmill with a 7-m-diameter rotor in a 10 m/s wind on a day when the atmospheric pressure is 100 kPa and the temperature is 20°C. The wind speed behind the windmill is measured at 9 m/s. Determine the diameter of the wind channel downstream from the rotor and the power produced by this windmill, presuming that the air is incompressible.

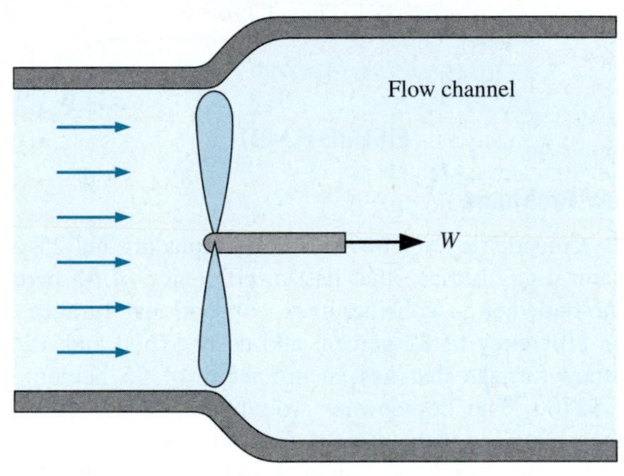

FIGURE P3–104

3–105 In a hydroelectric power plant, 100 m^3/s of water flows from an elevation of 120 m to a turbine, where electric power is generated. The overall efficiency of the turbine–generator is 80 percent. Disregarding frictional losses in piping, estimate the electric power output of this plant. *Answer:* 94.2 MW

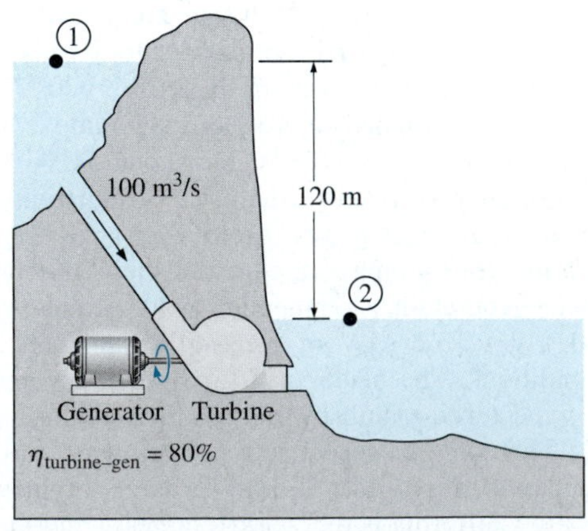

FIGURE P3–105

3–106 The demand for electric power is usually much higher during the day than it is at night, and utility companies often sell power at night at much lower prices to encourage consumers to use the available power generation capacity and to avoid building new expensive power plants that will be used only a short time during peak periods. Utilities are also willing to purchase power produced during the day from private parties at a high price.

Suppose a utility company is selling electric power for \$0.03/kWh at night and is willing to pay \$0.08/kWh for power produced during the day. To take advantage of this opportunity, an entrepreneur is considering building a large reservoir 40 m above the lake level, pumping water from the lake to the reservoir at night using cheap power, and letting the water flow from the reservoir back to the lake during the day, producing power as the pump–motor operates as a turbine–generator during reverse flow. Preliminary analysis shows that a water flow rate of 2 m^3/s can be used in either direction. The combined pump–motor and turbine–generator efficiencies are expected to be 75 percent each. Disregarding the frictional losses in piping and assuming the system operates for 10 h each in the pump and turbine modes during a typical day, determine the potential revenue this pump–turbine system can generate per year.

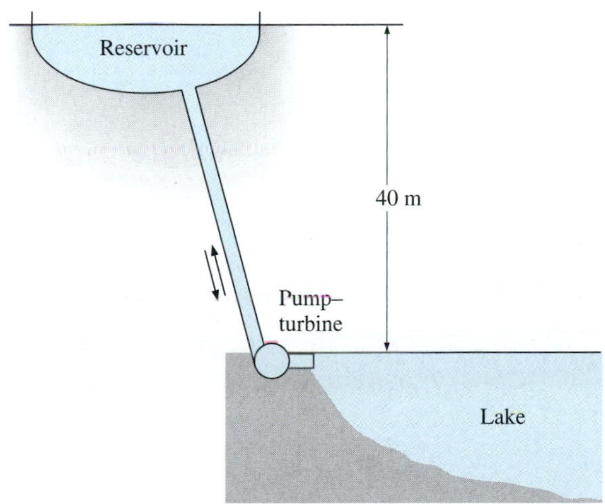

FIGURE P3–106

3–107 A diesel engine with an engine volume of 4.0 L and an engine speed of 2500 rpm operates on an air–fuel ratio of 18 kg air/kg fuel. The engine uses light diesel fuel that contains 750 ppm (parts per million) of sulfur by mass. All of this sulfur is exhausted to the environment where the sulfur is converted to sulfurous acid (H_2SO_3). If the rate of the air entering the engine is 336 kg/h, determine the mass flow rate of sulfur in the exhaust. Also, determine the mass flow rate of sulfurous acid added to the environment if for each kmol of sulfur in the exhaust, one kmol sulfurous acid will be added to the environment.

3–108 Leaded gasoline contains lead that ends up in the engine exhaust. Lead is a very toxic engine emission. The use of leaded gasoline in the United States has been unlawful for most vehicles since the 1980s. However, leaded gasoline is still used in some parts of the world. Consider a city with 10,000 cars using leaded gasoline. The gasoline contains 0.15 g/L of lead and 35 percent of lead is exhausted to the environment. Assuming that an average car travels 15,000 km per year with a gasoline consumption of 10 L/100 km, determine the amount of lead put into the atmosphere per year in that city. *Answer:* 788 kg

Design and Essay Problems

3–109 An average vehicle puts out nearly 20 lbm of carbon dioxide into the atmosphere for every gallon of gasoline it burns, and thus one thing we can do to reduce global warming is to buy a vehicle with higher fuel economy. A U.S. government publication states that a vehicle that gets 25 rather than 20 miles per gallon will prevent 10 tons of carbon dioxide from being released over the lifetime of the vehicle. Making reasonable assumptions, evaluate if this is a reasonable claim or a gross exaggeration.

3–110 Solar energy reaching the earth is about 1350 W/m^2 outside the earth's atmosphere, and 950 W/m^2 on earth's surface normal to the sun on a clear day. Someone is marketing 2 m × 3 m photovoltaic cell panels with the claim that a single panel can meet the electricity needs of a house. How do you evaluate this claim? Photovoltaic cells have a conversion efficiency of about 15 percent.

3–111 Find out the prices of heating oil, natural gas, and electricity in your area, and determine the cost of each per kWh of energy supplied to the house as heat. Go through your utility bills and determine how much money you spent for heating last January. Also determine how much your January heating bill would be for each of the heating systems if you had the latest and most efficient system installed.

3–112 Prepare a report on the heating systems available in your area for residential buildings. Discuss the advantages and disadvantages of each system and compare their initial and operating costs. What are the important factors in the selection of a heating system? Give some guidelines. Identify the conditions under which each heating system would be the best choice in your area.

3–113 The performance of a device is defined as the ratio of the desired output to the required input, and this definition can be extended to nontechnical fields. For example, your performance in this course can be viewed as the grade you earn relative to the effort you put in. If you have been investing a lot of time in this course and your grades do not reflect it, you are performing poorly. In that case, perhaps you should try to find out the underlying cause and how to correct the problem. Give three other definitions of performance from nontechnical fields and discuss them.

3–114 Some engineers have suggested that air compressed into tanks can be used to propel personal transportation vehicles. Current compressed-air tank technology permits us to compress and safely hold air at up to 4000 psia. Tanks made of composite materials require about 10 lbm of construction materials for each 1 ft^3 of stored gas. Approximately 0.01 hp is required per pound of vehicle weight to move a vehicle at a speed of 30 miles per hour. What is the maximum range that this vehicle can have? Account for the weight of the tanks only and assume perfect conversion of the energy in the compressed air.

3–115 Pressure changes across atmospheric weather fronts are typically a few centimeters of mercury, while the temperature changes are typically 2-20°C. Develop a plot of front pressure change versus front temperature change that will cause a maximum wind velocity of 10 m/s or more.

Chapter 4

PROPERTIES OF PURE SUBSTANCES

We start this chapter with the introduction of the concept of a *pure substance* and a discussion of the physics of phase-change processes. We then illustrate the various property diagrams and *P-v-T* surfaces of pure substances. After demonstrating the use of the property tables, the hypothetical substance *ideal gas* and the *ideal-gas equation of state* are discussed. The *compressibility factor,* which accounts for the deviation of real gases from ideal-gas behavior, is introduced, and some of the best-known equations of state such as the van der Waals, Beattie-Bridgeman, and Benedict-Webb-Rubin equations are presented.

Objectives

The objectives of this chapter are to:

- Introduce the concept of a pure substance.
- Discuss the physics of phase-change processes.
- Illustrate the *P-v*, *T-v*, and *P-T* property diagrams and *P-v-T* surfaces of pure substances.
- Demonstrate the procedures for determining thermodynamic properties of pure substances from tables of property data.
- Describe the hypothetical substance "ideal gas" and the ideal-gas equation of state.
- Apply the ideal-gas equation of state in the solution of typical problems.
- Introduce the compressibility factor, which accounts for the deviation of real gases from ideal-gas behavior.
- Present some of the most widely used equations of state.

4–1 ▪ PURE SUBSTANCE

A substance that has a fixed chemical composition throughout is called a **pure substance**. Water, nitrogen, helium, and carbon dioxide, for example, are all pure substances.

A pure substance does not have to be of a single chemical element or compound, however. A mixture of various chemical elements or compounds also qualifies as a pure substance as long as the mixture is homogeneous. Air, for example, is a mixture of several gases, but it is often considered to be a pure substance because it has a uniform chemical composition (Fig. 4–1). However, a mixture of oil and water is not a pure substance. Since oil is not soluble in water, it collects on top of the water, forming two chemically dissimilar regions.

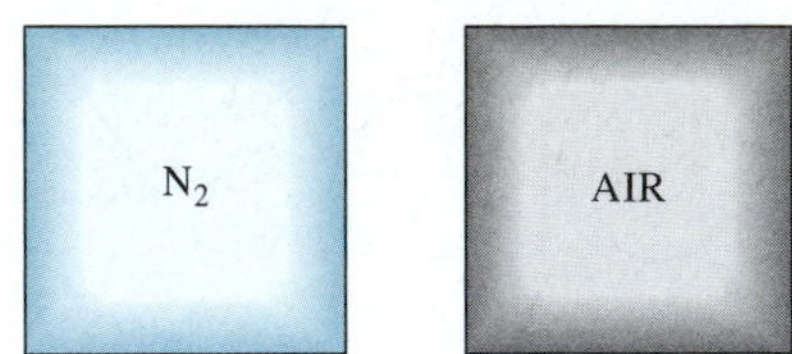

FIGURE 4–1

Nitrogen and gaseous air are pure substances.

A mixture of two or more phases of a pure substance is still a pure substance as long as the chemical composition of all phases is the same (Fig. 4–2). A mixture of ice and liquid water, for example, is a pure substance because both phases have the same chemical composition. A mixture of liquid air and gaseous air, however, is not a pure substance since the composition of liquid air is different from the composition of gaseous air, and thus the mixture is no longer chemically homogeneous. This is due to different components in air condensing at different temperatures at a specified pressure.

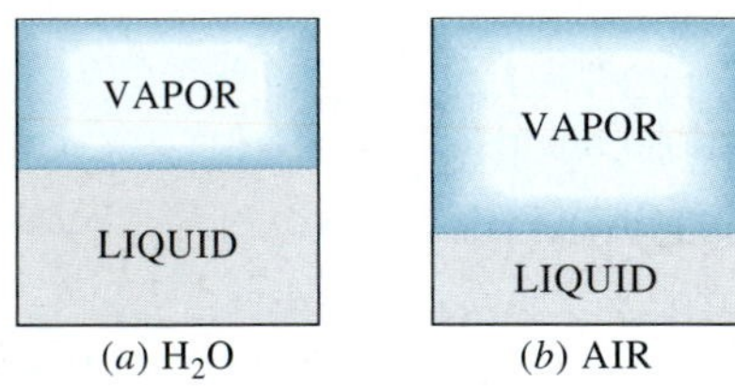

FIGURE 4–2

A mixture of liquid and gaseous water is a pure substance, but a mixture of liquid and gaseous air is not.

4–2 ▪ PHASES OF A PURE SUBSTANCE

We all know from experience that substances exist in different phases. At room temperature and pressure, copper is a solid, mercury is a liquid, and nitrogen is a gas. Under different conditions, each may appear in a different phase. Even though there are three principal phases—solid, liquid, and gas—a substance may have several phases within a principal phase, each with a different molecular structure. Carbon, for example, may exist as graphite or diamond in the solid phase. Helium has two liquid phases; iron has three solid phases. Ice may exist at seven different phases at high pressures. A phase is identified as having a distinct molecular arrangement that is homogeneous throughout and separated from the others by easily identifiable boundary surfaces. The two phases of H_2O in iced water (solid and liquid) represent a good example of this.

When studying phases or phase changes in thermodynamics, one does not need to be concerned with the molecular structure and behavior of different phases. However, it is very helpful to have some understanding of the molecular phenomena involved in each phase, and a brief discussion of phase transformations follows.

Intermolecular bonds are strongest in solids and weakest in gases. One reason is that molecules in solids are closely packed together, whereas in gases they are separated by relatively large distances.

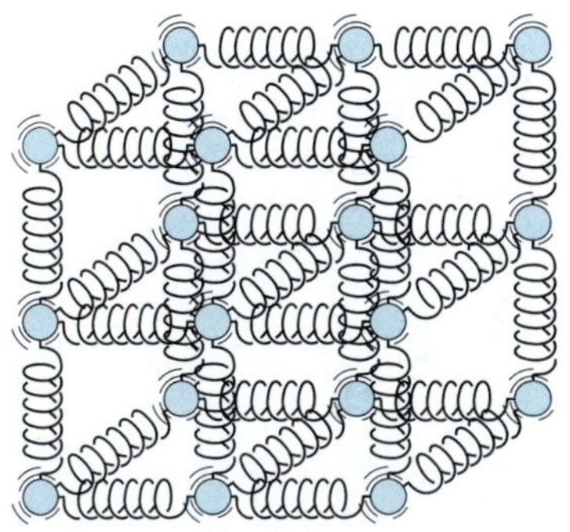

FIGURE 4–3

The molecules in a solid are kept at their positions by the large springlike intermolecular forces.

The molecules in a **solid** are arranged in a three-dimensional pattern (lattice) that is repeated throughout (Fig. 4–3). Because of the small distances between molecules in a solid, the attractive forces of molecules on each other are large and keep the molecules at fixed positions (Fig. 4–4). Note that the attractive forces between molecules turn to repulsive forces as the distance between the molecules approaches zero, thus preventing the molecules from piling up on top of each other. Even though the molecules in a solid cannot move relative to each other, they continually oscillate about their equilibrium positions. The velocity of the molecules during these oscillations depends on the temperature. At sufficiently high temperatures, the velocity (and thus the momentum) of the molecules may reach a point where the intermolecular forces are partially overcome and groups of molecules break away (Fig. 4–5). This is the beginning of the melting process.

The molecular spacing in the **liquid** phase is not much different from that of the solid phase, except the molecules are no longer at fixed positions relative to each other and they can rotate and translate freely. In a liquid, the intermolecular forces are weaker relative to solids, but still relatively strong compared with gases. The distances between molecules generally experience a slight increase as a solid turns liquid, with water being a notable exception.

In the **gas** phase, the molecules are far apart from each other, and a molecular order is nonexistent. Gas molecules move about at random, continually colliding with each other and the walls of the container they are in. Particularly at low densities, the intermolecular forces are very small, and collisions are the only mode of interaction between the molecules. Molecules in the gas phase are at a considerably higher energy level than they are in the liquid or solid phases. Therefore, the gas must release a large amount of its energy before it can condense or freeze.

FIGURE 4–4

In a solid, the attractive and repulsive forces between the molecules tend to maintain them at relatively constant distances from each other.

4–3 ▪ PHASE-CHANGE PROCESSES OF PURE SUBSTANCES

There are many practical situations where two phases of a pure substance coexist in equilibrium. Water exists as a mixture of liquid and vapor in the boiler and the condenser of a steam power plant. The refrigerant turns from liquid to vapor in the freezer of a refrigerator. Even though many home owners consider the freezing of water in underground pipes as the most

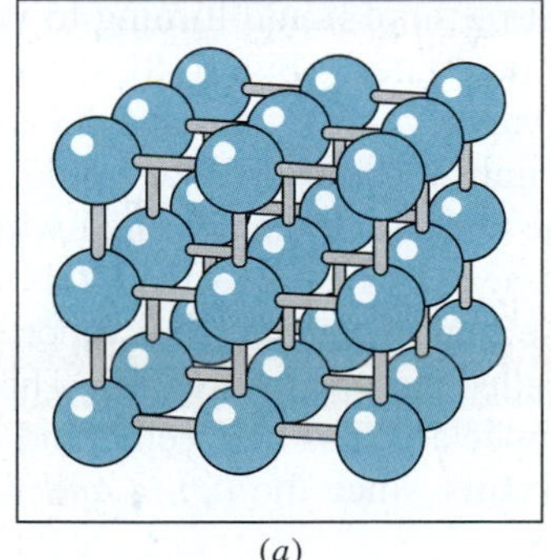

(*a*)

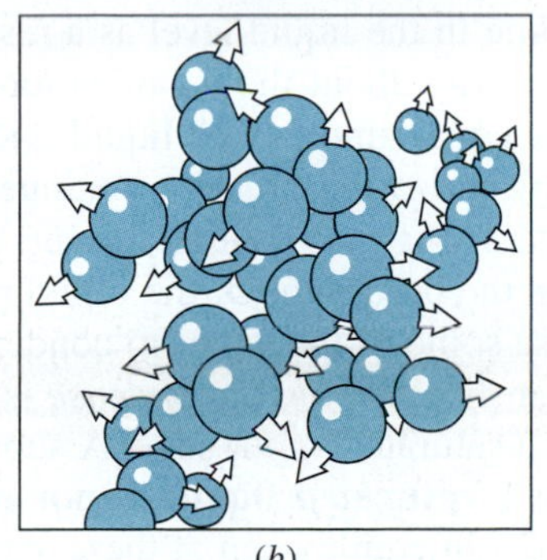

(*b*)

(*c*)

FIGURE 4–5

The arrangement of atoms in different phases: (*a*) molecules are at relatively fixed positions in a solid, (*b*) groups of molecules move about each other in the liquid phase, and (*c*) molecules move about at random in the gas phase.

important phase-change process, attention in this section is focused on the liquid and vapor phases and their mixture. As a familiar substance, water is used to demonstrate the basic principles involved. Remember, however, that all pure substances exhibit the same general behavior.

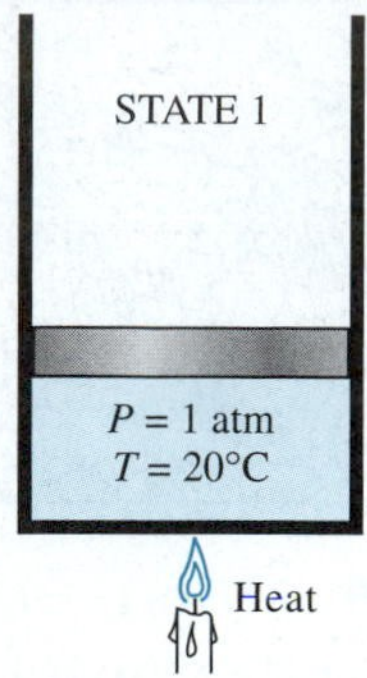

FIGURE 4–6

At 1 atm and 20°C, water exists in the liquid phase (*compressed liquid*).

Compressed Liquid and Saturated Liquid

Consider an ideal frictionless piston–cylinder device containing liquid water at 20°C and 1 atm pressure (state 1, Fig. 4–6). Under these conditions, water exists in the liquid phase, and it is called a **compressed liquid**, or a **subcooled liquid**, meaning that it is *not about to vaporize.* Heat is now transferred to the water until its temperature rises to, say, 40°C. As the temperature rises, the liquid water expands slightly, and so its specific volume increases. To accommodate this expansion, the piston moves up slightly. The pressure in the cylinder remains constant at 1 atm during this process since it depends on the outside barometric pressure and the weight of the piston, both of which are constant. Water is still a compressed liquid at this state since it has not started to vaporize.

As more heat is transferred, the temperature keeps rising until it reaches 100°C (state 2, Fig. 4–7). At this point water is still a liquid, but any further heat addition would cause some of the liquid to vaporize. That is, a phase-change process from liquid to vapor is about to take place. A liquid that is *about to vaporize* is called a **saturated liquid**. Therefore, state 2 is a saturated liquid state.

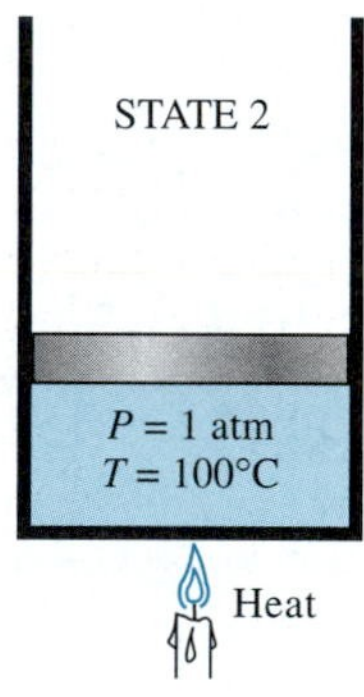

FIGURE 4–7

At 1 atm pressure and 100°C, water exists as a liquid that is ready to vaporize (*saturated liquid*).

Saturated Vapor and Superheated Vapor

Once boiling starts, the temperature stops rising until the liquid is completely vaporized. That is, the temperature remains constant during the entire phase-change process if the pressure is held constant. This can easily be verified by placing a thermometer into boiling pure water on top of a stove. At sea level (P = 1 atm), the thermometer will always read 100°C if the pan is uncovered or covered with a light lid. During a boiling process, the only change we observe is a large increase in the volume and a steady decline in the liquid level as a result of more liquid turning to vapor.

Midway about the vaporization line (state 3, Fig. 4–8), the cylinder contains equal amounts of liquid and vapor. As we continue transferring heat, the vaporization process continues until the last drop of liquid is vaporized (state 4, Fig. 4–9). At this point, the entire cylinder is filled with vapor that is on the borderline of the liquid phase. Any heat loss from this vapor would cause some of the vapor to condense (phase change from vapor to liquid). A vapor that is *about to condense* is called a **saturated vapor**. Therefore, state 4 is a saturated vapor state. A substance at states between 2 and 4 is referred to as a **saturated liquid–vapor mixture** since the *liquid and vapor phases coexist* in equilibrium at these states.

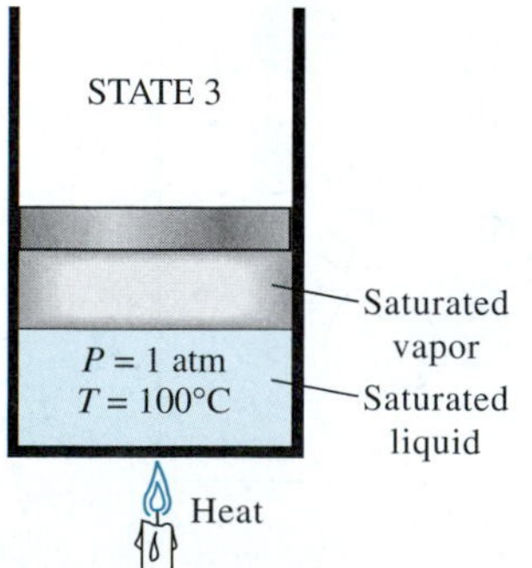

FIGURE 4–8

As more heat is transferred, part of the saturated liquid vaporizes (*saturated liquid–vapor mixture*).

Once the phase-change process is completed, we are back to a single-phase region again (this time vapor), and further transfer of heat results in an increase in both the temperature and the specific volume (Fig. 4–10). At state 5, the temperature of the vapor is, let us say, 300°C; and if we transfer some heat from the vapor, the temperature may drop somewhat but no condensation takes place as long as the temperature remains above 100°C (for

P = 1 atm). A vapor that is *not about to condense* (i.e., not a saturated vapor) is called a **superheated vapor**. Therefore, water at state 5 is a superheated vapor. This constant-pressure phase-change process is illustrated on a T-v diagram in Fig. 4–11.

If the entire process described here is reversed by cooling the water while maintaining the pressure at the same value, the water will go back to state 1, retracing the same path, and in so doing, the amount of heat released will exactly match the amount of heat added during the heating process.

In our daily life, water implies liquid water and steam implies water vapor. In thermodynamics, however, both water and steam usually mean only one thing: H_2O.

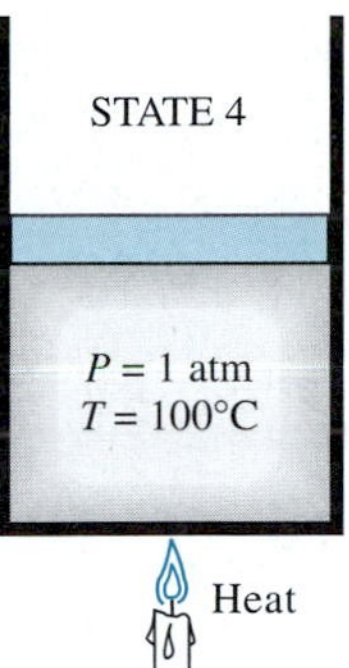

FIGURE 4–9

At 1 atm pressure, the temperature remains constant at 100°C until the last drop of liquid is vaporized (*saturated vapor*).

Saturation Temperature and Saturation Pressure

It probably came as no surprise to you that water started to boil at 100°C. Strictly speaking, the statement "water boils at 100°C" is incorrect. The correct statement is "water boils at 100°C at 1 atm pressure." The only reason water started boiling at 100°C was because we held the pressure constant at 1 atm (101.325 kPa). If the pressure inside the cylinder were raised to 500 kPa by adding weights on top of the piston, water would start boiling at 151.8°C. That is, *the temperature at which water starts boiling depends on the pressure; therefore, if the pressure is fixed, so is the boiling temperature.*

At a given pressure, the temperature at which a pure substance changes phase is called the **saturation temperature** T_{sat}. Likewise, at a given temperature, the pressure at which a pure substance changes phase is called the **saturation pressure** P_{sat}. At a pressure of 101.325 kPa, T_{sat} is 99.97°C. Conversely, at a temperature of 99.97°C, P_{sat} is 101.325 kPa. (At 100.00°C, P_{sat} is 101.42 kPa in the ITS-90 discussed in Chap. 2.)

Saturation tables that list the saturation pressure against the temperature (or the saturation temperature against the pressure) are available for practically all substances. A partial listing of such a table is given in

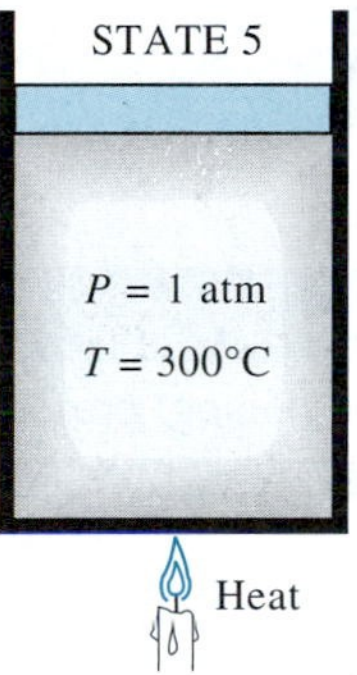

FIGURE 4–10

As more heat is transferred, the temperature of the vapor starts to rise (*superheated vapor*).

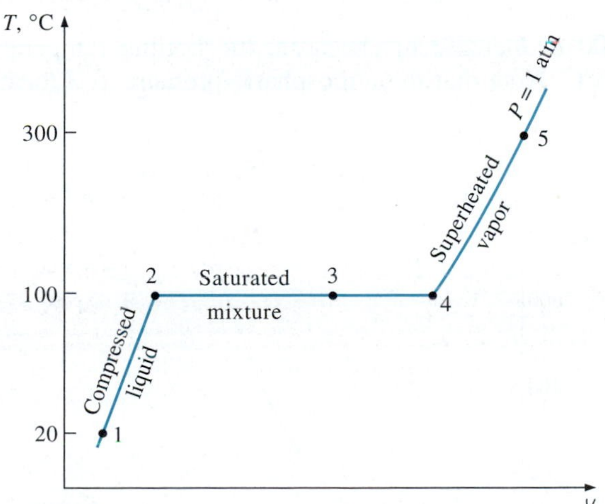

FIGURE 4–11

T-v diagram for the heating process of water at constant pressure.

Table 4–1 for water. This table indicates that the pressure of water changing phase (boiling or condensing) at 25°C must be 3.17 kPa, and the pressure of water must be maintained at 3976 kPa (about 40 atm) to have it boil at 250°C. Also, water can be frozen by dropping its pressure below 0.61 kPa, as discussed in more detail in Section 4.4.

It takes a large amount of energy to melt a solid or vaporize a liquid. The amount of energy absorbed or released during a phase-change process is called the **latent heat**. More specifically, the amount of energy absorbed during melting is called the **latent heat of fusion** and is equivalent to the amount of energy released during freezing. Similarly, the amount of energy absorbed during vaporization is called the **latent heat of vaporization** and is equivalent to the energy released during condensation. The magnitudes of the latent heats depend on the temperature or pressure at which the phase change occurs. At 1 atm pressure, the latent heat of fusion of water is 333.7 kJ/kg and the latent heat of vaporization is 2256.5 kJ/kg.

During a phase-change process, pressure and temperature are obviously dependent properties, and there is a definite relation between them, that is, $T_{sat} = f(P_{sat})$. A plot of T_{sat} versus P_{sat}, such as the one given for water in Fig. 4–12, is called a **liquid–vapor saturation curve**. A curve of this kind is characteristic of all pure substances.

It is clear from Fig. 4–12 that T_{sat} increases with P_{sat}. Thus, a substance at higher pressure boils at a higher temperature. In the kitchen, higher boiling temperatures mean shorter cooking times and energy savings. A beef stew, for example, may take 1 to 2 h to cook in a regular pan that operates at 1 atm pressure, but only 20 min in a pressure cooker operating at 3 atm absolute pressure (corresponding boiling temperature: 134°C).

The atmospheric pressure, and thus the boiling temperature of water, decreases with elevation. Therefore, it takes longer to cook at higher altitudes than it does at sea level (unless a pressure cooker is used). For example, the standard atmospheric pressure at an elevation of 2000 m is 79.50 kPa, which corresponds to a boiling temperature of 93.3°C as opposed to 100°C at sea level (zero elevation). The variation of the boiling temperature of water with altitude at standard atmospheric conditions is given in Table 4–2. For each 1000 m increase in elevation, the boiling temperature drops by a little over 3°C. Note that the atmospheric pressure at a location, and thus

TABLE 4–1

Saturation (boiling) pressure of water at various temperatures

Temperature, T, °C	Saturation pressure, P_{sat}, kPa
−10	0.26
−5	0.40
0	0.61
5	0.87
10	1.23
15	1.71
20	2.34
25	3.17
30	4.25
40	7.39
50	12.35
100	101.4
150	476.2
200	1555
250	3976
300	8588

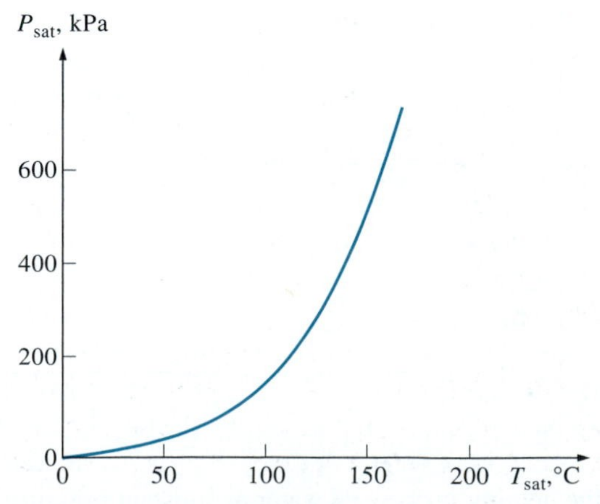

FIGURE 4–12

The liquid–vapor saturation curve of a pure substance (numerical values are for water).

the boiling temperature, changes slightly with the weather conditions. But the corresponding change in the boiling temperature is no more than about 1°C.

TABLE 4–2

Variation of the standard atmospheric pressure and the boiling (saturation) temperature of water with altitude

Elevation, m	Atmospheric pressure, kPa	Boiling temperature, °C
0	101.33	100.0
1,000	89.55	96.5
2,000	79.50	93.3
5,000	54.05	83.3
10,000	26.50	66.3
20,000	5.53	34.7

Some Consequences of T_{sat} and P_{sat} Dependence

We mentioned earlier that a substance at a specified pressure boils at the saturation temperature corresponding to that pressure. This phenomenon allows us to control the boiling temperature of a substance by simply controlling the pressure, and it has numerous applications in practice. Below we give some examples. The natural drive to achieve phase equilibrium by allowing some liquid to evaporate is at work behind the scenes.

Consider a sealed can of *liquid refrigerant-134a* in a room at 25°C. If the can has been in the room long enough, the temperature of the refrigerant in the can is also 25°C. Now, if the lid is opened slowly and some refrigerant is allowed to escape, the pressure in the can starts dropping until it reaches the atmospheric pressure. If you are holding the can, you will notice its temperature dropping rapidly, and even ice forming outside the can if the air is humid. A thermometer inserted in the can would register −26°C when the pressure drops to 1 atm, which is the saturation temperature of refrigerant-134a at that pressure. The temperature of the liquid refrigerant remains at −26°C until the last drop of it vaporizes.

Another aspect of this interesting physical phenomenon is that a liquid cannot vaporize unless it absorbs energy in the amount of the latent heat of vaporization, which is 217 kJ/kg for refrigerant-134a at 1 atm. Therefore, the rate of vaporization of the refrigerant depends on the rate of heat transfer to the can: the larger the rate of heat transfer, the higher the rate of vaporization. The rate of heat transfer to the can and thus the rate of vaporization of the refrigerant can be minimized by insulating the can heavily. In the limiting case of no heat transfer, the refrigerant would remain in the can as a liquid at −26°C indefinitely.

The boiling temperature of *nitrogen* at atmospheric pressure is −196°C (see Table A–3*a*). This means the temperature of liquid nitrogen exposed to the atmosphere must be −196°C since some nitrogen will be evaporating. The temperature of liquid nitrogen remains constant at −196°C until it is depleted. For this reason, nitrogen is commonly used in low-temperature scientific studies (such as superconductivity) and cryogenic applications to maintain a test chamber at a constant temperature of −196°C. This is done by placing the test chamber into a liquid nitrogen bath that is open to the atmosphere. Any heat transfer from the environment to the test section is absorbed by the nitrogen, which evaporates isothermally and keeps the test chamber temperature constant at −196°C (Fig. 4–13). The entire test section must be insulated heavily to minimize heat transfer and thus liquid nitrogen consumption. Liquid nitrogen is also used for medical purposes to burn off unsightly spots (e.g., warts) on the skin. This is done by soaking a cotton swab in liquid nitrogen and wetting the target area with it. As the nitrogen evaporates, it freezes the affected skin by rapidly absorbing heat from it.

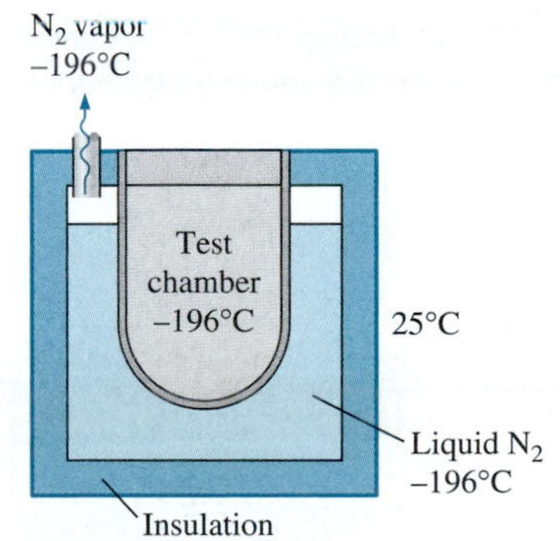

FIGURE 4–13

The temperature of liquid nitrogen exposed to the atmosphere remains constant at −196°C, and thus it maintains the test chamber at −196°C.

A practical way of cooling leafy vegetables is **vacuum cooling**, which is based on *reducing the pressure* of the sealed cooling chamber to the saturation pressure at the desired low temperature, and evaporating some water

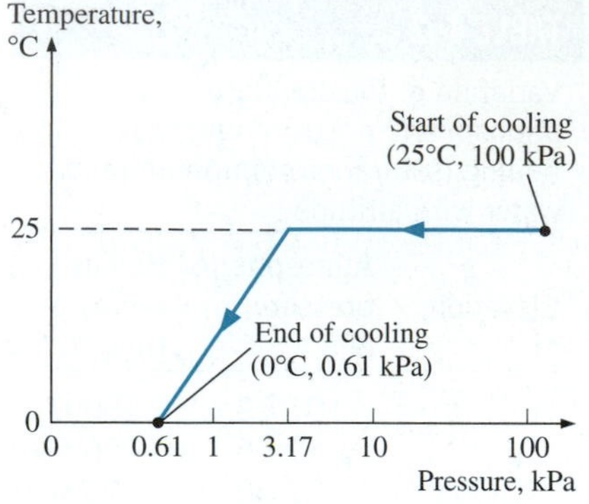

FIGURE 4–14

The variation of the temperature of fruits and vegetables with pressure during vacuum cooling from 25°C to 0°C.

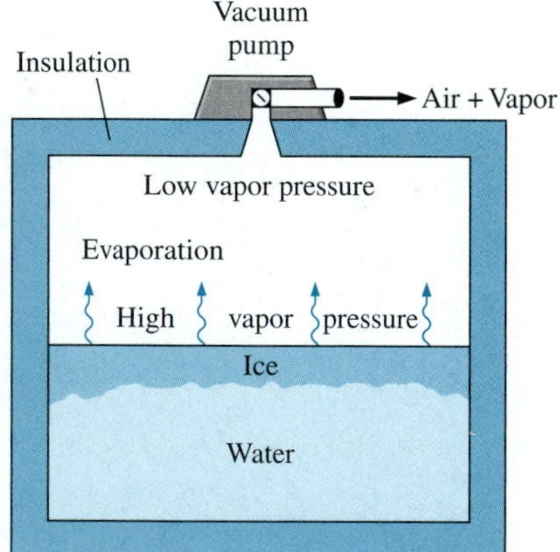

FIGURE 4–15

In 1775, ice was made by evacuating the air space in a water tank.

from the products to be cooled. The heat of vaporization during evaporation is absorbed from the products, which lowers the products' temperature. The saturation pressure of water at 0°C is 0.61 kPa, and the products can be cooled to 0°C by lowering the pressure to this level. The cooling rate can be increased by lowering the pressure below 0.61 kPa, but this is not desirable because of the danger of freezing and the added cost.

In vacuum cooling, there are two distinct stages. In the first stage, the products at ambient temperature, say at 25°C, are loaded into the chamber, and the operation begins. The temperature in the chamber remains constant until the *saturation pressure* is reached, which is 3.17 kPa at 25°C. In the second stage that follows, saturation conditions are maintained inside at progressively *lower pressures* and the corresponding *lower temperatures* until the desired temperature is reached (Fig. 4–14).

Vacuum cooling is usually more expensive than the conventional refrigerated cooling, and its use is limited to applications that result in much faster cooling. Products with large surface area per unit mass and a high tendency to release moisture such as lettuce and spinach are well-suited for vacuum cooling. Products with low surface area to mass ratio are not suitable, especially those that have relatively impervious peels such as tomatoes and cucumbers. Some products such as mushrooms and green peas can be vacuum cooled successfully by wetting them first.

The vacuum cooling just described becomes **vacuum freezing** if the vapor pressure in the vacuum chamber is dropped below 0.61 kPa, the saturation pressure of water at 0°C. The idea of making ice by using a vacuum pump is nothing new. Dr. William Cullen actually made ice in Scotland in 1775 by evacuating the air in a water tank (Fig. 4–15).

Package icing is commonly used in small-scale cooling applications to remove heat and keep the products cool during transit by taking advantage of the large latent heat of fusion of water, but its use is limited to products that are not harmed by contact with ice. Also, ice provides *moisture* as well as *refrigeration.*

4–4 ▪ PROPERTY DIAGRAMS FOR PHASE-CHANGE PROCESSES

The variations of properties during phase-change processes are best studied and understood with the help of property diagrams. Next, we develop and discuss the *T*-*v*, *P*-*v*, and *P*-*T* diagrams for pure substances.

1 The *T*-*v* Diagram

The phase-change process of water at 1 atm pressure was described in detail in the last section and plotted on a *T*-*v* diagram in Fig. 4–11. Now we repeat this process at different pressures to develop the *T*-*v* diagram.

Let us add weights on top of the piston until the pressure inside the cylinder reaches 1 MPa. At this pressure, water has a somewhat smaller specific volume than it does at 1 atm pressure. As heat is transferred to the water at this new pressure, the process follows a path that looks very much like the process path at 1 atm pressure, as shown in Fig. 4–16, but there are some noticeable differences. First, water starts boiling at a much higher tempera-

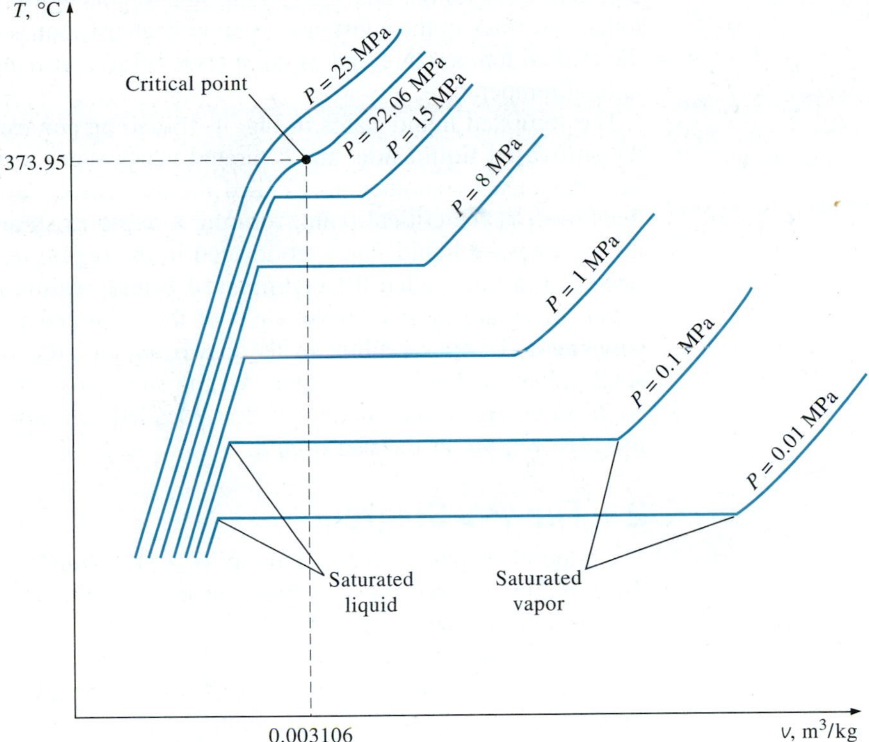

FIGURE 4–16
T-v diagram of constant-pressure phase-change processes of a pure substance at various pressures (numerical values are for water).

ture (179.9°C) at this pressure. Second, the specific volume of the saturated liquid is larger and the specific volume of the saturated vapor is smaller than the corresponding values at 1 atm pressure. That is, thc horizontal line that connects the saturated liquid and saturated vapor states is much shorter.

As the pressure is increased further, this saturation line continues to shrink, as shown in Fig. 4–16, and it becomes a point when the pressure reaches 22.06 MPa for the case of water. This point is called the **critical point**, and it is defined as *the point at which the saturated liquid and saturated vapor states are identical.*

The temperature, pressure, and specific volume of a substance at the critical point are called, respectively, the *critical temperature* T_{cr}, *critical pressure* P_{cr}, and *critical specific volume* v_{cr}. The critical-point properties of water are $P_{cr} = 22.06$ MPa, $T_{cr} = 373.95$°C, and $v_{cr} = 0.003106$ m^3/kg. For helium, they are 0.23 MPa, −267.85°C, and 0.01444 m^3/kg. The critical properties for various substances are given in Table A–1 in the appendix.

At pressures above the critical pressure, there is not a distinct phase-change process (Fig. 4–17). Instead, the specific volume of the substance continually increases, and at all times there is only one phase present. Eventually, it resembles a vapor, but we can never tell when the change has occurred. Above the critical state, there is no line that separates the com-

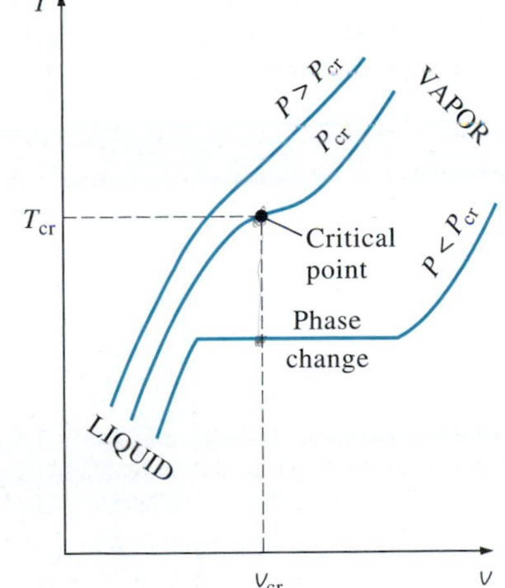

FIGURE 4–17
At supercritical pressures ($P > P_{cr}$), there is no distinct phase-change (boiling) process.

pressed liquid region and the superheated vapor region. However, it is customary to refer to the substance as superheated vapor at temperatures above the critical temperature and as compressed liquid at temperatures below the critical temperature.

The saturated liquid states in Fig. 4–16 can be connected by a line called the **saturated liquid line**, and saturated vapor states in the same figure can be connected by another line, called the **saturated vapor line**. These two lines meet at the critical point, forming a dome as shown in Fig. 4–18. All the compressed liquid states are located in the region to the left of the saturated liquid line, called the **compressed liquid region**. All the superheated vapor states are located to the right of the saturated vapor line, called the **superheated vapor region**. In these two regions, the substance exists in a single phase, a liquid or a vapor. All the states that involve both phases in equilibrium are located under the dome, called the **saturated liquid–vapor mixture region**, or the **wet region**.

2 The *P*-v Diagram

The general shape of the *P*-v diagram of a pure substance is very much like the *T*-v diagram, but the T = constant lines on this diagram have a downward trend, as shown in Fig. 4–19.

Consider again a piston–cylinder device that contains liquid water at 1 MPa and 150°C. Water at this state exists as a compressed liquid. Now the weights on top of the piston are removed one by one so that the pressure inside the cylinder decreases gradually (Fig. 4–20). The water is allowed to exchange heat with the surroundings so its temperature remains constant. As the pressure decreases, the volume of the water increases slightly. When the

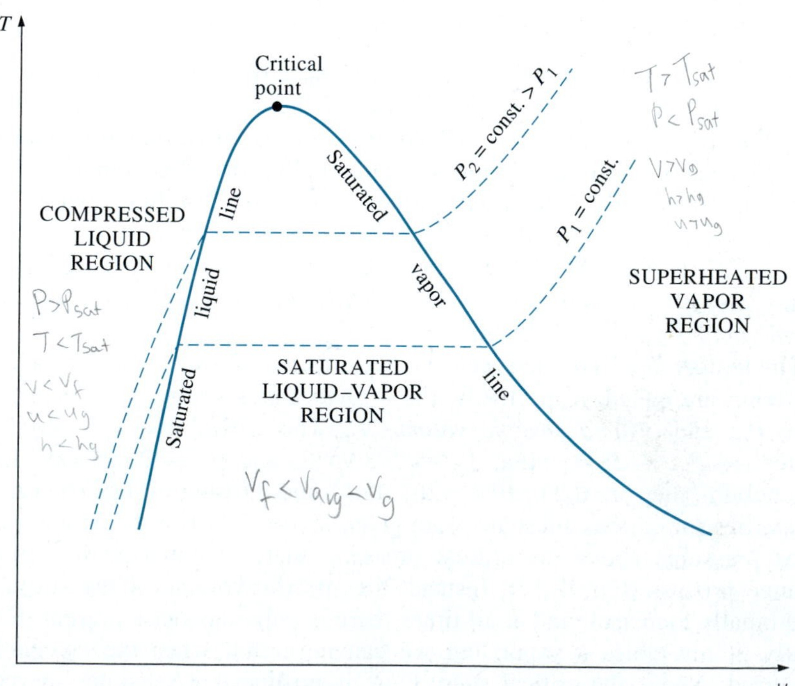

FIGURE 4–18

T-v diagram of a pure substance.

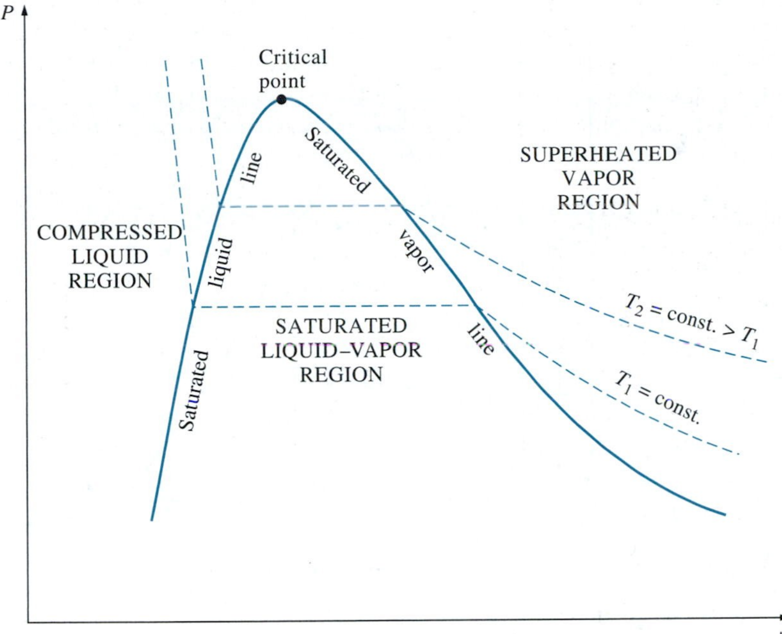

FIGURE 4–19

P-*v* diagram of a pure substance.

pressure reaches the saturation-pressure value at the specified temperature (0.4762 MPa), the water starts to boil. During this vaporization process, heat is being added, but both the temperature and the pressure remain constant, and the specific volume increases. Once the last drop of liquid is vaporized, further reduction in pressure results in a further increase in specific volume. Notice that during the phase-change process, we did not remove any weights. Doing so would cause the pressure and therefore the temperature to drop [since $T_{sat} = f(P_{sat})$], and the process would no longer be isothermal.

When the process is repeated for other temperatures, similar paths are obtained for the phase-change processes. Connecting the saturated liquid and the saturated vapor states by a curve, we obtain the *P*-*v* diagram of a pure substance, as shown in Fig. 4–19.

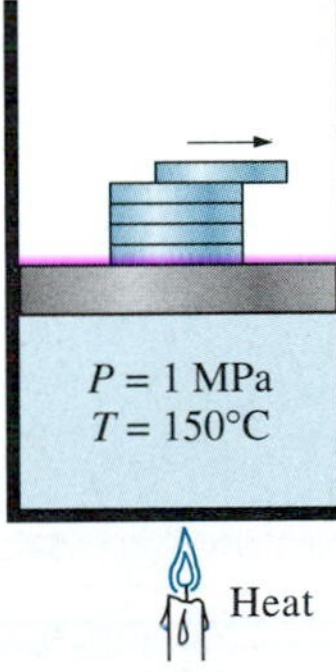

FIGURE 4–20

The pressure in a piston–cylinder device can be reduced by reducing the weight of the piston.

Extending the Diagrams to Include the Solid Phase

The two equilibrium diagrams developed so far represent the equilibrium states involving the liquid and the vapor phases only. However, these diagrams can easily be extended to include the solid phase as well as the solid–liquid and the solid–vapor saturation regions. The basic principles discussed in conjunction with the liquid–vapor phase-change process apply equally to the solid–liquid and solid–vapor phase-change processes. Most substances contract during a solidification (i.e., freezing) process. Others, like water, expand as they freeze. The *P*-*v* diagrams for both groups of substances are given in Figs. 4–21 and 4–22. These two diagrams differ only in

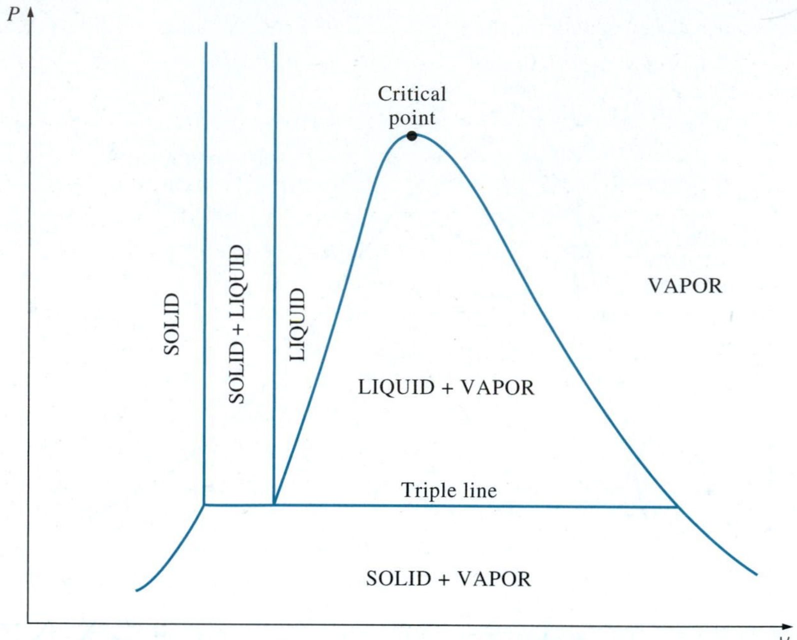

FIGURE 4–21

P-*v* diagram of a substance that contracts on freezing.

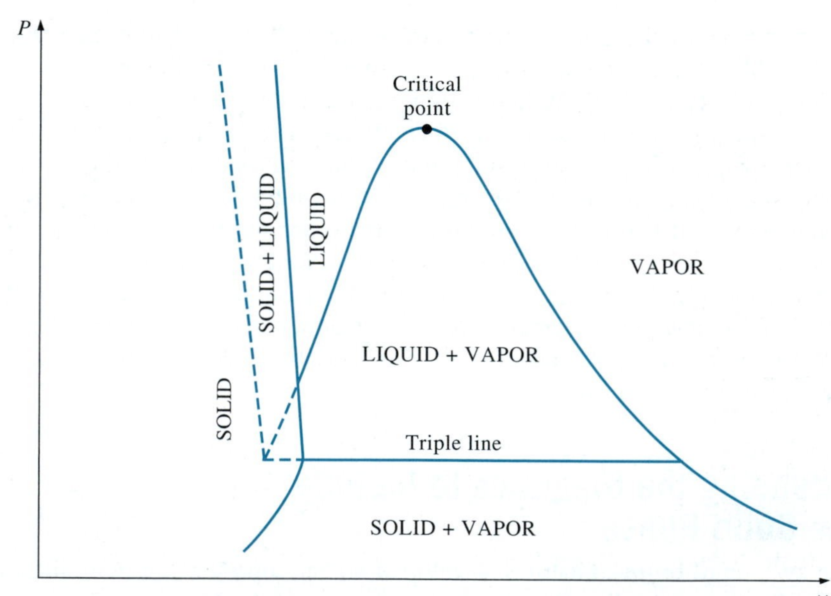

FIGURE 4–22

P-*v* diagram of a substance that expands on freezing (such as water).

the solid–liquid saturation region. The *T*-*v* diagrams look very much like the *P*-*v* diagrams, especially for substances that contract on freezing.

The fact that water expands upon freezing has vital consequences in nature. If water contracted on freezing as most other substances do, the ice formed would be heavier than the liquid water, and it would settle to the bottom of rivers, lakes, and oceans instead of floating at the top. The sun's rays would never reach these ice layers, and the bottoms of many rivers, lakes, and oceans would be covered with ice at times, seriously disrupting marine life.

We are all familiar with two phases being in equilibrium, but under some conditions all three phases of a pure substance coexist in equilibrium (Fig. 4–23). On *P*-*v* or *T*-*v* diagrams, these triple-phase states form a line called the **triple line**. The states on the triple line of a substance have the same pressure and temperature but different specific volumes. The triple line appears as a point on the *P*-*T* diagrams and, therefore, is often called the **triple point**. The triple-point temperatures and pressures of various substances are given in Table 4–3. For water, the triple-point temperature and pressure are 0.01°C and 0.6117 kPa, respectively. That is, all three phases of water coexist in equilibrium only if the temperature and pressure have precisely these values. No substance can exist in the liquid phase in stable equilibrium at pressures below the triple-point pressure. The same can be said for temperature for substances that contract on freezing. However,

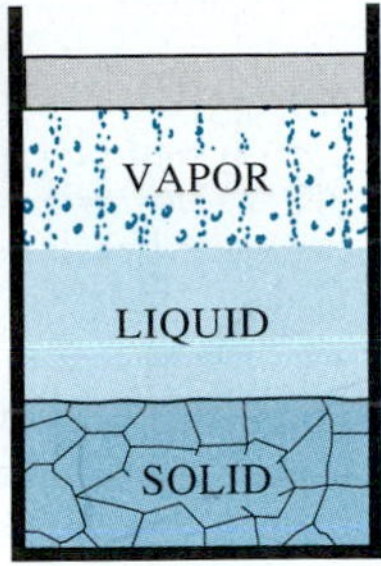

FIGURE 4–23

At triple-point pressure and temperature, a substance exists in three phases in equilibrium.

TABLE 4–3

Triple-point temperatures and pressures of various substances

Substance	Formula	T_{tp}, K	P_{tp}, kPa
Acetylene	C_2H_2	192.4	120
Ammonia	NH_3	195.40	6.076
Argon	A	83.81	68.9
Carbon (graphite)	C	3900	10,100
Carbon dioxide	CO_2	216.55	517
Carbon monoxide	CO	68.10	15.37
Deuterium	D_2	18.63	17.1
Ethane	C_2H_6	89.89	8×10^{-4}
Ethylene	C_2H_4	104.0	0.12
Helium 4 (λ point)	He	2.19	5.1
Hydrogen	H_2	13.84	7.04
Hydrogen chloride	HCl	158.96	13.9
Mercury	Hg	234.2	1.65×10^{-7}
Methane	CH_4	90.68	11.7
Neon	Ne	24.57	43.2
Nitric oxide	NO	109.50	21.92
Nitrogen	N_2	63.18	12.6
Nitrous oxide	N_2O	182.34	87.85
Oxygen	O_2	54.36	0.152
Palladium	Pd	1825	3.5×10^{-3}
Platinum	Pt	2045	2.0×10^{-4}
Sulfur dioxide	SO_2	197.69	1.67
Titanium	Ti	1941	5.3×10^{-3}
Uranium hexafluoride	UF_6	337.17	151.7
Water	H_2O	273.16	0.61
Xenon	Xe	161.3	81.5
Zinc	Zn	692.65	0.065

Source: Data from National Bureau of Standards (U.S.) Circ., 500 (1952).

FIGURE 4–24

At low pressures (below the triple-point value), solids evaporate without melting first (*sublimation*).

substances at high pressures can exist in the liquid phase at temperatures below the triple-point temperature. For example, water cannot exist in liquid form in equilibrium at atmospheric pressure at temperatures below 0°C, but it can exist as a liquid at −20°C at 200 MPa pressure. Also, ice exists at seven different solid phases at pressures above 100 MPa.

There are two ways a substance can pass from the solid to vapor phase: either it melts first into a liquid and subsequently evaporates, or it evaporates directly without melting first. The latter occurs at pressures below the triple-point value, since a pure substance cannot exist in the liquid phase at those pressures (Fig. 4–24). Passing from the solid phase directly into the vapor phase is called **sublimation**. For substances that have a triple-point pressure above the atmospheric pressure such as solid CO_2 (dry ice), sublimation is the only way to change from the solid to vapor phase at atmospheric conditions.

3 The *P-T* Diagram

Figure 4–25 shows the *P-T* diagram of a pure substance. This diagram is often called the **phase diagram** since all three phases are separated from each other by three lines. The sublimation line separates the solid and vapor regions, the vaporization line separates the liquid and vapor regions, and the melting (or fusion) line separates the solid and liquid regions. These three lines meet at the triple point, where all three phases coexist in equilibrium. The vaporization line ends at the critical point because no distinction can be made between liquid and vapor phases above the critical point. Substances that expand and contract on freezing differ only in the melting line on the *P-T* diagram.

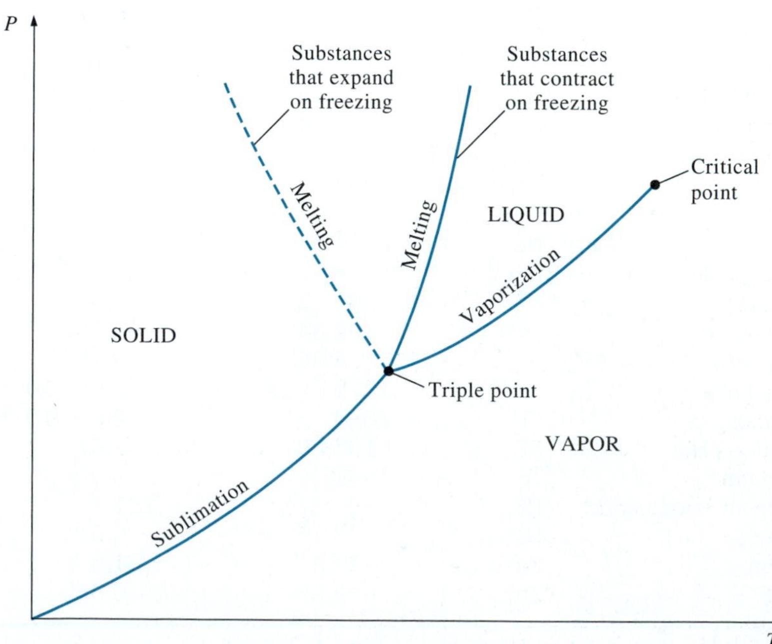

FIGURE 4–25

P-T diagram of pure substances.

The *P-v-T* Surface

The state of a simple compressible substance is fixed by any two independent, intensive properties. Once the two appropriate properties are fixed, all the other properties become dependent properties. Remembering that any equation with two independent variables in the form $z = z(x, y)$ represents a surface in space, we can represent the *P*-v-*T* behavior of a substance as a surface in space, as shown in Figs. 4–26 and 4–27. Here *T* and v may be

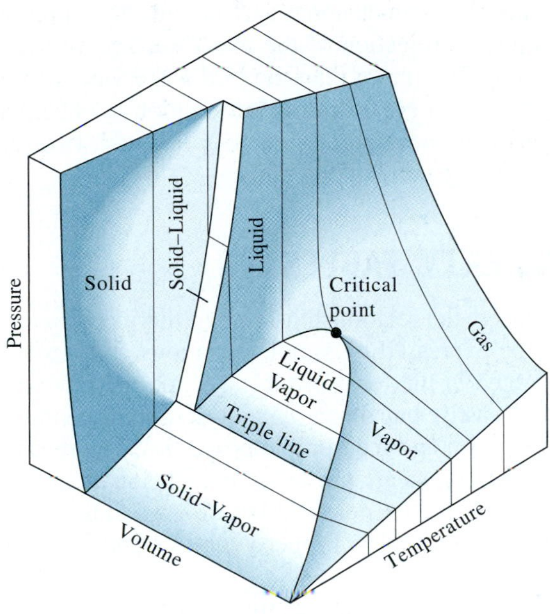

FIGURE 4–26
P-v-*T* surface of a substance that *contracts* on freezing.

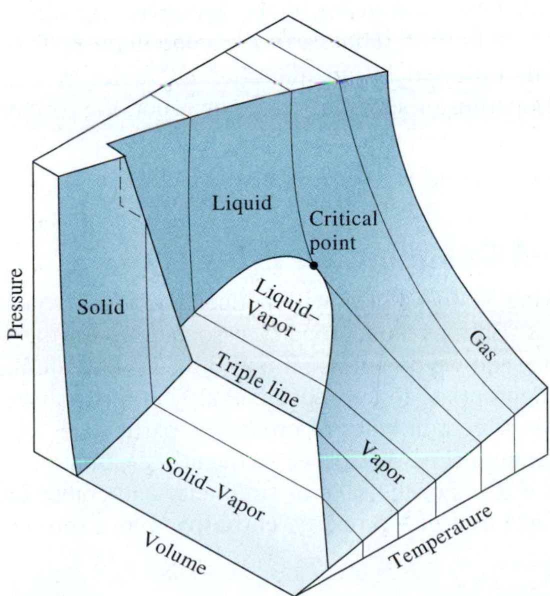

FIGURE 4–27
P-v-*T* surface of a substance that *expands* on freezing (like water).

viewed as the independent variables (the base) and P as the dependent variable (the height).

All the points on the surface represent equilibrium states. All states along the path of a quasi-equilibrium process lie on the P-v-T surface since such a process must pass through equilibrium states. The single-phase regions appear as curved surfaces on the P-v-T surface, and the two-phase regions as surfaces perpendicular to the P-T plane. This is expected since the projections of two-phase regions on the P-T plane are lines.

All the two-dimensional diagrams we have discussed so far are merely projections of this three-dimensional surface onto the appropriate planes. A P-v diagram is just a projection of the P-v-T surface on the P-v plane, and a T-v diagram is nothing more than the bird's-eye view of this surface. The P-v-T surfaces present a great deal of information at once, but in a thermodynamic analysis it is more convenient to work with two-dimensional diagrams, such as the P-v and T-v diagrams.

4–5 ▪ PROPERTY TABLES

For most substances, the relationships among thermodynamic properties are too complex to be expressed by simple equations. Therefore, properties are frequently presented in the form of tables. Some thermodynamic properties can be measured easily, but others cannot and are calculated by using the relations between them and measurable properties. The results of these measurements and calculations are presented in tables in a convenient format. In the following discussion, the steam tables are used to demonstrate the use of thermodynamic property tables. Property tables of other substances are used in the same manner.

For each substance, the thermodynamic properties are listed in more than one table. In fact, a separate table is prepared for each region of interest such as the superheated vapor, compressed liquid, and saturated (mixture) regions. Property tables are given in the appendix in both SI and English units. The tables in English units carry the same number as the corresponding tables in SI, followed by an identifier E. Tables A–6 and A–6E, for example, list properties of superheated water vapor, the former in SI and the latter in English units. Before we get into the discussion of property tables, we define a new property called *enthalpy*.

Enthalpy—A Combination Property

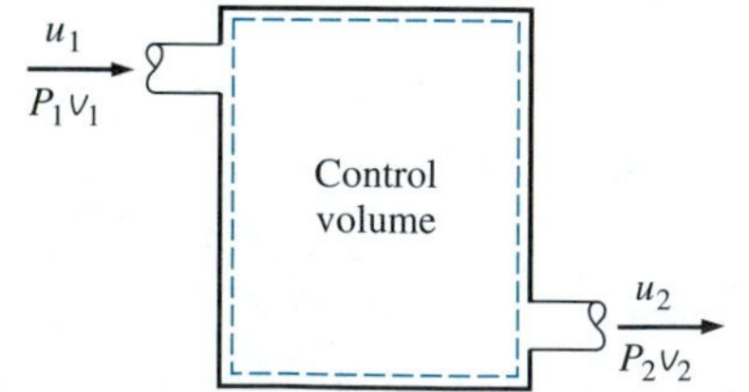

FIGURE 4–28

The combination $u + Pv$ is frequently encountered in the analysis of control volumes.

A person looking at the tables will notice two new properties: enthalpy h and entropy s. Entropy is a property associated with the second law of thermodynamics, and we do not use it until it is properly defined in Chap. 8. However, it is appropriate to introduce enthalpy at this point.

In the analysis of certain types of processes, particularly in power generation and refrigeration (Fig. 4–28), we frequently encounter the combination of properties $u + Pv$. For the sake of simplicity and convenience, this combination is defined as a new property, **enthalpy**, and given the symbol h:

$$h = u + Pv \qquad \text{(kJ/kg)} \tag{4–1}$$

or,

$$H = U + PV \qquad \text{(kJ)} \tag{4–2}$$

Both the total enthalpy H and specific enthalpy h are simply referred to as enthalpy since the context clarifies which one is meant. Notice that the equations given above are dimensionally homogeneous. That is, the unit of the pressure–volume product may differ from the unit of the internal energy by only a factor (Fig. 4–29). For example, it can be easily shown that 1 kPa · m^3 = 1 kJ. In some tables encountered in practice, the internal energy u is frequently not listed, but it can always be determined from $u = h - Pv$.

The widespread use of the property enthalpy is due to Professor Richard Mollier, who recognized the importance of the group $u + Pv$ in the analysis of steam turbines and in the representation of the properties of steam in tabular and graphical form (as in the famous Mollier chart). Mollier referred to the group $u + Pv$ as *heat content* and *total heat*. These terms were not quite consistent with the modern thermodynamic terminology and were replaced in the 1930s by the term *enthalpy* (from the Greek word *enthalpien*, which means *to heat*).

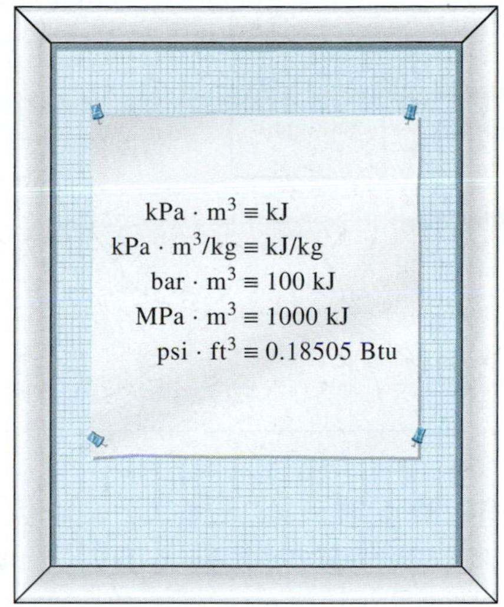

FIGURE 4–29
The product *pressure* × *volume* has energy units, and the product *pressure* × *specific volume* has energy per unit mass units.

1a Saturated Liquid and Saturated Vapor States

The properties of saturated liquid and saturated vapor for water are listed in Tables A–4 and A–5. Both tables give the same information. The only difference is that in Table A–4 properties are listed under temperature and in Table A–5 under pressure. Therefore, it is more convenient to use Table A–4 when *temperature* is given and Table A–5 when *pressure* is given. The use of Table A–4 is illustrated in Fig. 4–30.

The subscript f is used to denote properties of a saturated liquid, and the subscript g to denote the properties of saturated vapor. These symbols are commonly used in thermodynamics and originated from the German words for liquid and vapor. Another subscript commonly used is fg, which denotes the difference between the saturated vapor and saturated liquid values of the same property. For example,

$$v_f = \text{specific volume of saturated liquid}$$

$$v_g = \text{specific volume of saturated vapor}$$

$$v_{fg} = \text{difference between } v_g \text{ and } v_f \text{ (that is, } v_{fg} = v_g - v_f)$$

The quantity h_{fg} is called the **enthalpy of vaporization** (or latent heat of vaporization). It represents the amount of energy needed to vaporize a unit mass of saturated liquid at a given temperature or pressure. It decreases as the temperature or pressure increases and becomes zero at the critical point.

Temp. °C T	Sat. press. kPa P_{sat}	Specific volume m^3/kg: Sat. liquid v_f	Sat. vapor v_g
85	57.868	0.001032	2.8261
90	70.183	0.001036	2.3593
95	84.609	0.001040	1.9808

Temperature
Corresponding saturation pressure
Specific volume of saturated liquid
Specific volume of saturated vapor

FIGURE 4–30
A partial list of Table A–4.

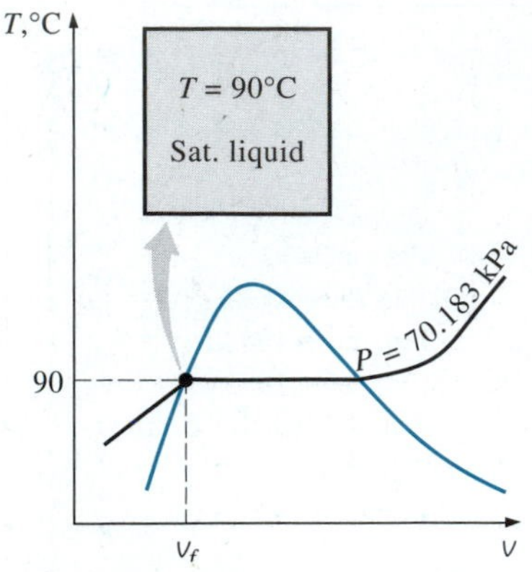

FIGURE 4–31

Schematic and T-v diagram for Example 4–1.

EXAMPLE 4–1 Pressure of Saturated Liquid in a Tank

A rigid tank contains 50 kg of saturated liquid water at 90°C. Determine the pressure in the tank and the volume of the tank.

Solution A rigid tank contains saturated liquid water. The pressure and volume of the tank are to be determined.

Analysis The state of the saturated liquid water is shown on a T-v diagram in Fig. 4–31. Since saturation conditions exist in the tank, the pressure must be the saturation pressure at 90°C:

$$P = P_{\text{sat @ 90°C}} = \mathbf{70.183\ kPa} \qquad \text{(Table A–4)}$$

The specific volume of the saturated liquid at 90°C is

$$v = v_{f\text{ @ 90°C}} = 0.001036\ \text{m}^3/\text{kg} \qquad \text{(Table A–4)}$$

Then the total volume of the tank becomes

$$V = mv = (50\ \text{kg})(0.001036\ \text{m}^3/\text{kg}) = \mathbf{0.0518\ m^3}$$

Discussion Only one property (T) is needed since we are told that the state is a saturated liquid.

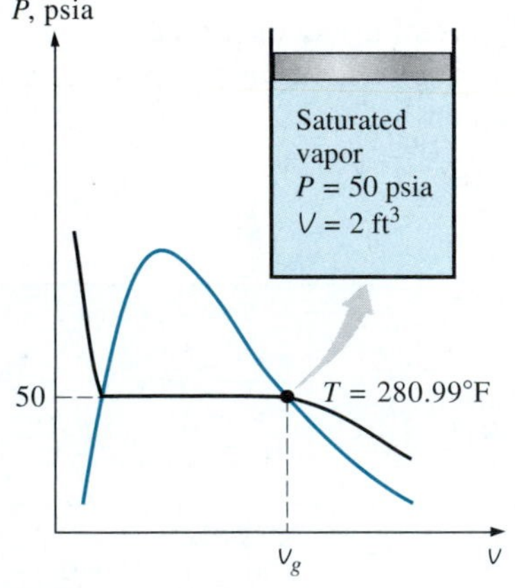

FIGURE 4–32

Schematic and P-v diagram for Example 4–2.

EXAMPLE 4–2 Temperature of Saturated Vapor in a Cylinder

A piston–cylinder device contains 2 ft^3 of saturated water vapor at 50-psia pressure. Determine the temperature and the mass of the vapor inside the cylinder.

Solution A cylinder contains saturated water vapor. The temperature and the mass of vapor are to be determined.

Analysis The state of the saturated water vapor is shown on a P-v diagram in Fig. 4–32. Since the cylinder contains saturated vapor at 50 psia, the temperature inside must be the saturation temperature at this pressure:

$$T = T_{\text{sat @ 50 psia}} = \mathbf{280.99°F} \qquad \text{(Table A–5E)}$$

The specific volume of the saturated vapor at 50 psia is

$$v = v_{g\text{ @ 50 psia}} = 8.5175\ \text{ft}^3/\text{lbm} \qquad \text{(Table A–5E)}$$

Then the mass of water vapor inside the cylinder becomes

$$m = \frac{V}{v} = \frac{2\ \text{ft}^3}{8.5175\ \text{ft}^3/\text{lbm}} = \mathbf{0.235\ lbm}$$

Discussion Along the saturated vapor line, temperature is a unique function of pressure.

EXAMPLE 4–3 Volume and Energy Change during Evaporation

A mass of 200 g of saturated liquid water is completely vaporized at a constant pressure of 100 kPa. Determine (*a*) the volume change and (*b*) the amount of energy transferred to the water.

Solution Saturated liquid water is vaporized at constant pressure. The volume change and the energy transferred are to be determined.

Analysis (*a*) The process described is illustrated on a *P*-*v* diagram in Fig. 4–33. The volume change per unit mass during a vaporization process is v_{fg}, which is the difference between v_g and v_f. Reading these values from Table A–5 at 100 kPa and substituting yield

$$v_{fg} = v_g - v_f = 1.6941 - 0.001043 = 1.6931 \text{ m}^3/\text{kg}$$

Thus,

$$\Delta V = m v_{fg} = (0.2 \text{ kg})(1.6931 \text{ m}^3/\text{kg}) = \mathbf{0.3386 \text{ m}^3}$$

(*b*) The amount of energy needed to vaporize a unit mass of a substance at a given pressure is the enthalpy of vaporization at that pressure, which is h_{fg} = 2257.5 kJ/kg for water at 100 kPa. Thus, the amount of energy transferred is

$$m h_{fg} = (0.2 \text{ kg})(2257.5 \text{ kJ/kg}) = \mathbf{451.5 \text{ kJ}}$$

Discussion Note that we have considered the first four decimal digits of v_{fg} and disregarded the rest. This is because v_g has significant numbers to the first four decimal places only, and we do not know the numbers in the other decimal places. Copying all the digits from the calculator would mean that we are assuming v_g = 1.694100, which is not necessarily the case. It could very well be that v_g = 1.694138 since this number, too, would truncate to 1.6941. All the digits in our result (1.6931) are significant. But if we did not truncate the result, we would obtain v_{fg} = 1.693057, which falsely implies that our result is accurate to the sixth decimal place.

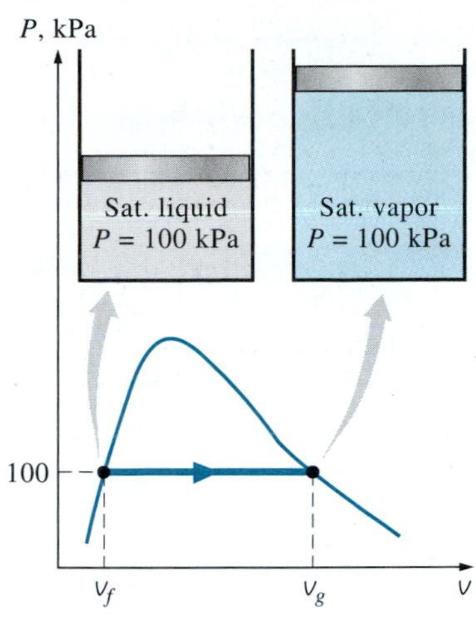

FIGURE 4–33

Schematic and *P*-*v* diagram for Example 4–3.

1b Saturated Liquid–Vapor Mixture

During a vaporization process, a substance exists as part liquid and part vapor. That is, it is a mixture of saturated liquid and saturated vapor (Fig. 4–34). To analyze this mixture properly, we need to know the proportions of the liquid and vapor phases in the mixture. This is done by defining a new property called the **quality** x as the ratio of the mass of vapor to the total mass of the mixture:

$$x = \frac{m_{\text{vapor}}}{m_{\text{total}}} \qquad \textbf{(4–3)}$$

where

$$m_{\text{total}} = m_{\text{liquid}} + m_{\text{vapor}} = m_f + m_g$$

Quality has significance for *saturated mixtures* only. It has no meaning in the compressed liquid or superheated vapor regions. Its value is between 0 and 1. The quality of a system that consists of *saturated liquid* is 0 (or 0 percent), and the quality of a system consisting of *saturated vapor* is 1 (or 100 percent). In saturated mixtures, quality can serve as one of the two independent intensive properties needed to describe a state. Note that *the properties of the saturated liquid are the same whether it exists alone or in a mixture with saturated vapor*. During the vaporization process, only the amount of saturated liquid changes, not its properties. The same can be said about a saturated vapor.

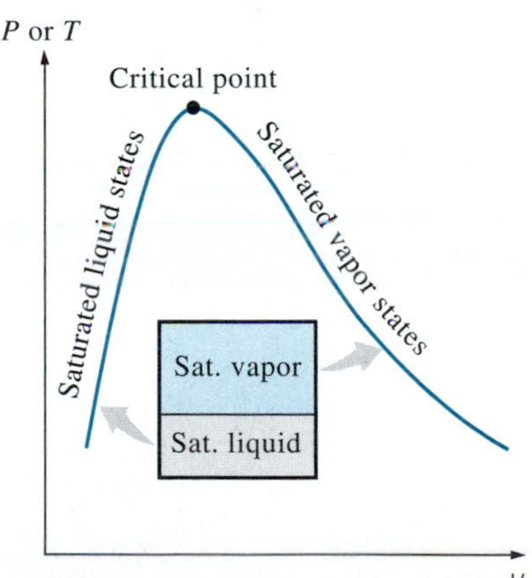

FIGURE 4–34

The relative amounts of liquid and vapor phases in a saturated mixture are specified by the *quality x*.

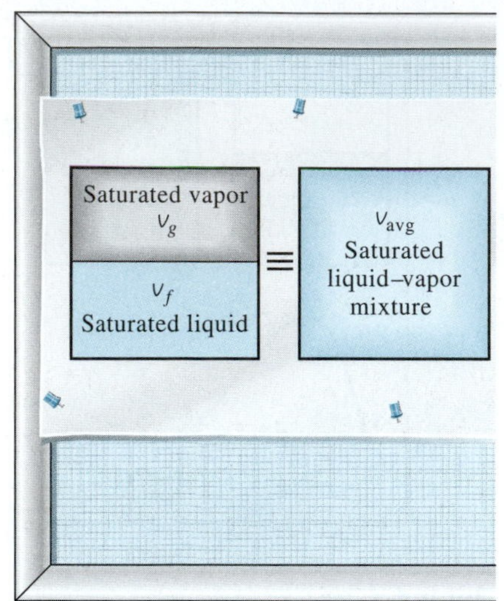

FIGURE 4–35

A two-phase system can be treated as a homogeneous mixture for convenience.

A saturated mixture can be treated as a combination of two subsystems: the saturated liquid and the saturated vapor. However, the amount of mass for each phase is usually not known. Therefore, it is often more convenient to imagine that the two phases are well mixed, forming a homogeneous mixture (Fig. 4–35). Then the properties of this "mixture" are simply the average properties of the saturated liquid–vapor mixture under consideration. Here is how it is done.

Consider a tank that contains a saturated liquid–vapor mixture. The volume occupied by saturated liquid is V_f, and the volume occupied by saturated vapor is V_g. The total volume V is the sum of the two:

$$V = V_f + V_g$$

$$V = mv \longrightarrow m_t v_{avg} = m_f v_f + m_g v_g$$

$$m_f = m_t - m_g \longrightarrow m_t v_{avg} = (m_t - m_g)v_f + m_g v_g$$

Dividing by m_t yields

$$v_{avg} = (1 - x)v_f + xv_g$$

since $x = m_g/m_t$. This relation can also be expressed as

$$v_{avg} = v_f + xv_{fg} \qquad (\text{m}^3/\text{kg}) \tag{4–4}$$

where $v_{fg} = v_g - v_f$. Solving for quality, we obtain

$$x = \frac{v_{avg} - v_f}{v_{fg}} \tag{4–5}$$

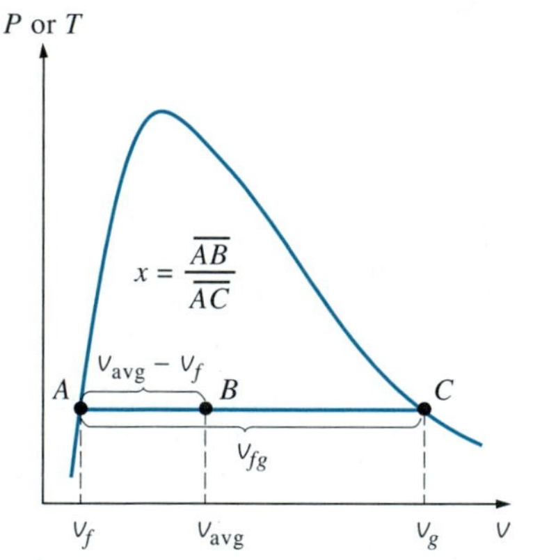

FIGURE 4–36

Quality is related to the horizontal distances on P-v and T-v diagrams.

Based on this equation, quality is related to the horizontal distances on a P-v or T-v diagram (Fig. 4–36). At a given temperature or pressure, the numerator of Eq. 4–5 is the distance between the actual state and the saturated liquid state, and the denominator is the length of the entire horizontal line that connects the saturated liquid and saturated vapor states. A state of 50 percent quality lies in the middle of this horizontal line.

The analysis given above can be repeated for internal energy and enthalpy with the following results:

$$u_{avg} = u_f + xu_{fg} \qquad (\text{kJ/kg}) \tag{4–6}$$

$$h_{avg} = h_f + xh_{fg} \qquad (\text{kJ/kg}) \tag{4–7}$$

All the results are of the same format, and they can be summarized in a single equation as

$$y_{avg} = y_f + xy_{fg}$$

where y is v, u, or h. The subscript "avg" (for "average") is usually dropped for simplicity. The values of the average properties of the mixtures are always *between* the values of the saturated liquid and the saturated vapor properties (Fig. 4–37). That is,

$$y_f \leq y_{avg} \leq y_g$$

Finally, all the saturated-mixture states are located under the saturation curve, and to analyze saturated mixtures, all we need are saturated liquid and saturated vapor data (Tables A–4 and A–5 in the case of water).

EXAMPLE 4–4 Pressure and Volume of a Saturated Mixture

A rigid tank contains 10 kg of water at 90°C. If 8 kg of the water is in the liquid form and the rest is in the vapor form, determine (*a*) the pressure in the tank and (*b*) the volume of the tank.

Solution A rigid tank contains a saturated mixture of water. The pressure and the volume of the tank are to be determined.

Analysis (*a*) The state of the saturated liquid–vapor mixture is shown in Fig. 4–38. Since the two phases coexist in equilibrium, we have a saturated mixture, and the pressure must be the saturation pressure at the given temperature:

$$P = P_{\text{sat @ 90°C}} = \mathbf{70.183\ kPa} \qquad \text{(Table A–4)}$$

(*b*) At 90°C, we have $v_f = 0.001036$ m³/kg and $v_g = 2.3593$ m³/kg (Table A-4). One way of finding the volume of the tank is to determine the volume occupied by each phase and then add them:

$$V = V_f + V_g = m_f v_f + m_g v_g$$

$$= (8\text{ kg})(0.001036\text{ m}^3/\text{kg}) + (2\text{ kg})(2.3593\text{ m}^3/\text{kg})$$

$$= \mathbf{4.73\ m^3}$$

Another way is to first determine the quality x, then the average specific volume v, and finally the total volume:

$$x = \frac{m_g}{m_t} = \frac{2\text{ kg}}{10\text{ kg}} = 0.2$$

$$v = v_f + x v_{fg}$$

$$= 0.001036\text{ m}^3/\text{kg} + (0.2)[(2.3593 - 0.001036)\text{ m}^3/\text{kg}]$$

$$= 0.473\text{ m}^3/\text{kg}$$

and

$$V = mv = (10\text{ kg})(0.473\text{ m}^3/\text{kg}) = 4.73\text{ m}^3$$

Discussion The first method appears to be easier in this case since the masses of each phase are given. In most cases, however, the masses of each phase are not available, and the second method becomes more convenient. Also note that we have dropped the "avg" subscript for convenience.

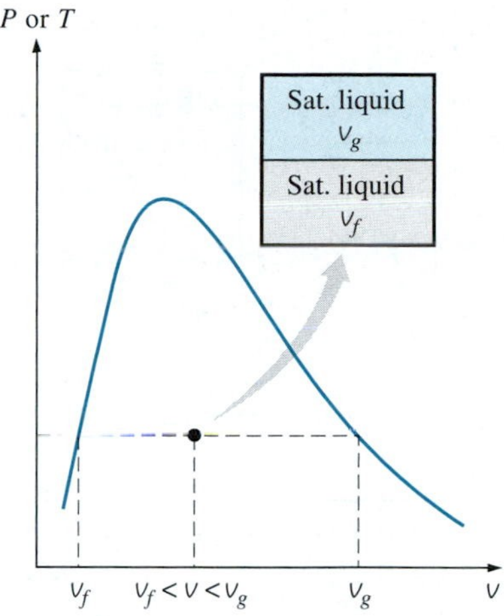

FIGURE 4–37

The v value of a saturated liquid–vapor mixture lies between the v_f and v_g values at the specified T or P.

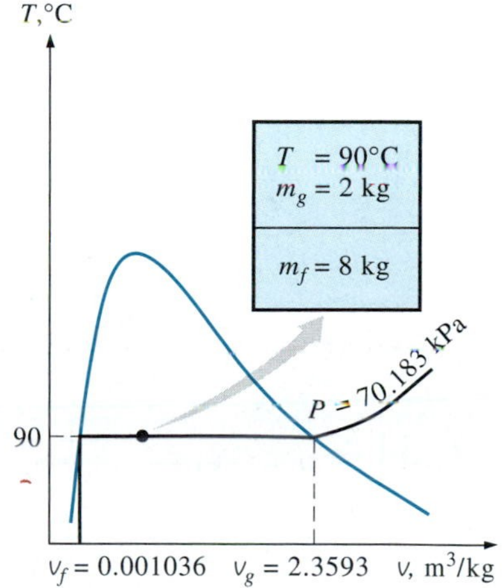

FIGURE 4–38

Schematic and T-v diagram for Example 4–4.

EXAMPLE 4–5 Properties of Saturated Liquid–Vapor Mixture

An 80-L vessel contains 4 kg of refrigerant-134a at a pressure of 160 kPa. Determine (*a*) the temperature, (*b*) the quality, (*c*) the enthalpy of the refrigerant, and (*d*) the volume occupied by the vapor phase.

Solution A vessel is filled with refrigerant-134a. Some properties of the refrigerant are to be determined.

Analysis (*a*) The state of the saturated liquid–vapor mixture is shown in Fig. 4–39. At this point we do not know whether the refrigerant is in the compressed liquid, superheated vapor, or saturated mixture region. This can

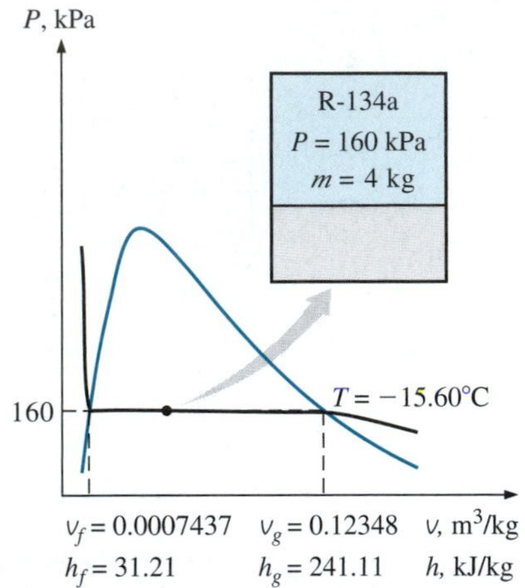

FIGURE 4–39

Schematic and P-v diagram for Example 4–5.

be determined by comparing a suitable property to the saturated liquid and saturated vapor values. From the information given, we determine the specific volume:

$$v = \frac{V}{m} = \frac{0.080\ \text{m}^3}{4\ \text{kg}} = 0.02\ \text{m}^3/\text{kg}$$

At 160 kPa, we read

$$v_f = 0.0007437\ \text{m}^3/\text{kg}$$
$$v_g = 0.12348\ \text{m}^3/\text{kg} \qquad \text{(Table A–12)}$$

Obviously, $v_f < v < v_g$, and the refrigerant is in the saturated mixture region. Thus, the temperature must be the saturation temperature at the specified pressure:

$$T = T_{\text{sat @ 160 kPa}} = \mathbf{-15.60°C}$$

(*b*) The quality is determined from

$$x = \frac{v - v_f}{v_{fg}} = \frac{0.02 - 0.0007437}{0.12348 - 0.0007437} = \mathbf{0.157}$$

(*c*) At 160 kPa, we also read from Table A–12 that h_f = 31.21 kJ/kg and h_{fg} = 209.90 kJ/kg. Then,

$$\begin{aligned} h &= h_f + xh_{fg} \\ &= 31.21\ \text{kJ/kg} + (0.157)(209.90\ \text{kJ/kg}) \\ &= \mathbf{64.2\ kJ/kg} \end{aligned}$$

(*d*) The mass of the vapor is

$$m_g = xm_t = (0.157)(4\ \text{kg}) = 0.628\ \text{kg}$$

and the volume occupied by the vapor phase is

$$V_g = m_g v_g = (0.628\ \text{kg})(0.12348\ \text{m}^3/\text{kg}) = \mathbf{0.0775\ m^3}\ \text{(or 77.5 L)}$$

Discussion The rest of the volume (2.5 L) is occupied by the liquid.

T,°C	v m³/kg	u kJ/kg	h kJ/kg
	P = 0.1 MPa (99.61°C)		
Sat.	1.6941	2505.6	2675.0
100	1.6959	2506.2	2675.8
150	1.9367	2582.9	2776.6
⋮	⋮	⋮	⋮
1300	7.2605	4687.2	5413.3
	P = 0.5 MPa (151.83°C)		
Sat.	0.37483	2560.7	2748.1
200	0.42503	2643.3	2855.8
250	0.47443	2723.8	2961.0

FIGURE 4–40

A partial listing of Table A–6.

Property tables are also available for saturated solid–vapor mixtures. Properties of saturated ice–water vapor mixtures, for example, are listed in Table A–8. Saturated solid–vapor mixtures can be handled just as saturated liquid–vapor mixtures.

2 Superheated Vapor

In the region to the right of the saturated vapor line and at temperatures above the critical point temperature, a substance exists as superheated vapor. Since the superheated region is a single-phase region (vapor phase only), temperature and pressure are no longer dependent properties and they can conveniently be used as the two independent properties in the tables. The format of the superheated vapor tables is illustrated in Fig. 4–40.

In these tables, the properties are listed against temperature for selected pressures starting with the saturated vapor data. The saturation temperature is given in parentheses following the pressure value.

Compared to saturated vapor, superheated vapor is characterized by

Lower pressures ($P < P_{sat}$ at a given T)
Higher tempreatures ($T > T_{sat}$ at a given P)
Higher specific volumes ($v > v_g$ at a given P or T)
Higher internal energies ($u > u_g$ at a given P or T)
Higher enthalpies ($h > h_g$ at a given P or T)

EXAMPLE 4–6 Internal Energy of Superheated Vapor

Determine the internal energy of water at 20 psia and 400°F.

Solution The internal energy of water at a specified state is to be determined.

Analysis At 20 psia, the saturation temperature is 227.92°F. Since $T > T_{sat}$, the water is in the superheated vapor region. Then the internal energy at the given temperature and pressure is determined from the superheated vapor table (Table A–6E) to be

$$u = \mathbf{1145.1\ Btu/lbm}$$

Discussion In problems like this, *interpolation* is usually required, since the tables do not list all probable combinations of temperature and pressure.

EXAMPLE 4–7 Temperature of Superheated Vapor

Determine the temperature of water at a state of P = 0.5 MPa and h = 2890 kJ/kg.

Solution The temperature of water at a specified state is to be determined.

Analysis At 0.5 MPa, the enthalpy of saturated water vapor is h_g = 2748.1 kJ/kg. Since $h > h_g$, as shown in Fig. 4–41, we again have superheated vapor. Under 0.5 MPa in Table A–6 we read

T, °C	*h*, kJ/kg
200	2855.8
250	2961.0

Obviously, the temperature is between 200 and 250°C. By linear interpolation it is determined to be

$$T = \mathbf{216.3°C}$$

Discussion It is also obvious from Fig. 4–41 that $T > T_{sat}$ at the given pressure.

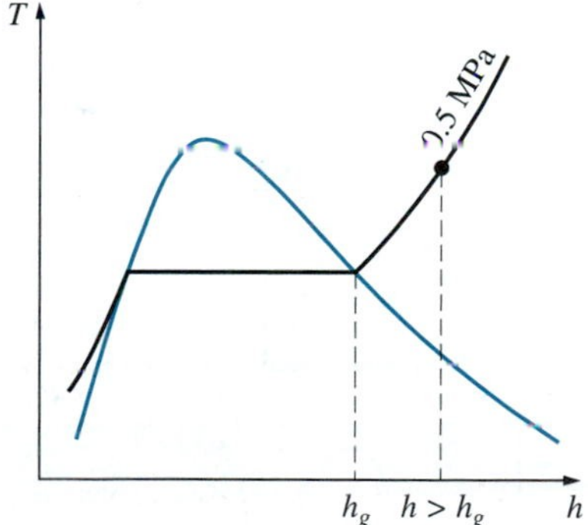

FIGURE 4–41

At a specified P, superheated vapor exists at a higher h than the saturated vapor (Example 4–7).

3 Compressed Liquid

Compressed liquid tables are not as commonly available, and Table A–7 is the only compressed liquid table in this text. The format of Table A–7 is very much like the format of the superheated vapor tables. One reason for the lack of compressed liquid data is the relative independence of compressed liquid properties from pressure. Variation of properties of compressed liquid with pressure is very mild. Increasing the pressure 100 times often causes properties to change less than 1 percent.

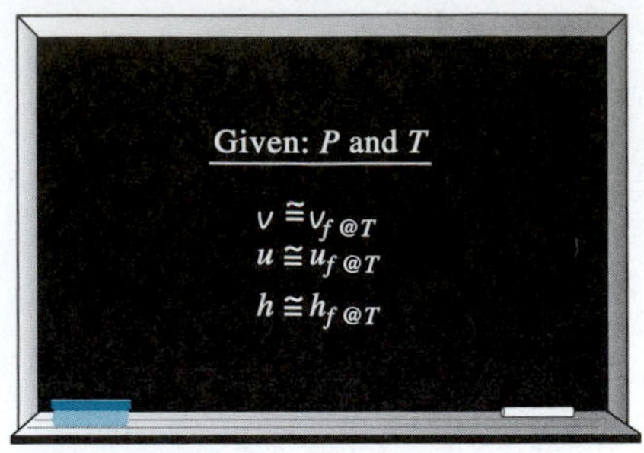

FIGURE 4–42

A compressed liquid may be approximated as a saturated liquid at the given temperature.

In general, a compressed liquid (also called a subcooled liquid) is characterized by

Higher pressures ($P > P_{sat}$ at a given T)
Lower tempreatures ($T < T_{sat}$ at a given P)
Lower specific volumes ($v < v_f$ at a given P or T)
Lower internal energies ($u < u_f$ at a given P or T)
Lower enthalpies ($h < h_f$ at a given P or T)

But unlike superheated vapor, the compressed liquid properties are not much different from the corresponding saturated liquid values.

In the absence of compressed liquid data, a general approximation is *to treat compressed liquid as saturated liquid at the given temperature* (Fig. 4–42). This is because the compressed liquid properties depend on temperature much more strongly than they do on pressure. Thus,

$$y \cong y_{f @ T} \tag{4–8}$$

for compressed liquids, where y is v, u, or h. Of these three properties, the property whose value is most sensitive to variations in the pressure is the enthalpy h. Although the above approximation results in negligible error in v and u, the error in h may reach undesirable levels. However, the error in h at low to moderate pressures and temperatures can be reduced significantly by evaluating it from

$$h \cong h_{f @ T} + v_{f @ T}(P - P_{sat @ T}) \tag{4–9}$$

instead of taking it to be just h_f. Note, however, that the approximation in Eq. 4–9 does not yield any significant improvement at moderate to high temperatures and pressures, and it may even backfire and result in greater error due to overcorrection at very high temperatures and pressures (*see* Kostic, Ref. 4).

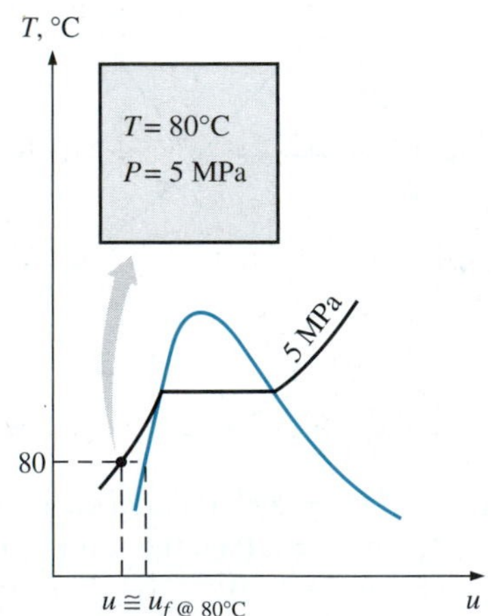

FIGURE 4–43

Schematic and *T-u* diagram for Example 4–8.

EXAMPLE 4–8 Approximating Compressed Liquid as Saturated Liquid

Determine the internal energy of compressed liquid water at 80°C and 5 MPa, using (*a*) data from the compressed liquid table and (*b*) saturated liquid data. What is the error involved in the second case?

Solution The exact and approximate values of the internal energy of liquid water are to be determined.

Analysis At 80°C, the saturation pressure of water is 47.416 kPa, and since 5 MPa $> P_{sat}$, we obviously have compressed liquid, as shown in Fig. 4–43.

(*a*) From the compressed liquid table (Table A–7)

$$\left.\begin{array}{l} P = 5 \text{ MPa} \\ T = 80^\circ\text{C} \end{array}\right\} \quad u = \mathbf{333.82 \ kJ/kg}$$

(*b*) From the saturation table (Table A–4), we read

$$u \cong u_{f @ 80^\circ C} = \mathbf{334.97 \ kJ/kg}$$

The error involved is

$$\frac{334.97 - 333.82}{333.82} \times 100 = \mathbf{0.34\%}$$

Discussion Here, P is more than 100 times greater than P_{sat}, yet the error in u is less than 1 percent.

Reference State and Reference Values

The values of u, h, and s cannot be measured directly, and they are calculated from measurable properties using the relations between thermodynamic properties. However, those relations give the *changes* in properties, not the values of properties at specified states. Therefore, we need to choose a convenient *reference state* and assign a value of *zero* for a convenient property or properties at that state. For water, the state of saturated liquid at 0.01°C is taken as the reference state, and the internal energy and entropy are assigned zero values at that state. For refrigerant-134a, the state of saturated liquid at −40°C is taken as the reference state, and the enthalpy and entropy are assigned zero values at that state. Note that some properties may have negative values as a result of the reference state chosen.

It should be mentioned that sometimes different tables list different values for some properties at the same state as a result of using a different reference state. However, in thermodynamics we are concerned with the *changes* in properties, and the reference state chosen is of no consequence in calculations as long as we use values from a single consistent set of tables or charts.

EXAMPLE 4–9 The Use of Steam Tables to Determine Properties

Determine the missing properties and the phase descriptions in the following table for water:

	T, °C	P, kPa	u, kJ/kg	x	Phase description
(*a*)		200		0.6	
(*b*)	125		1600		
(*c*)		1000	2950		
(*d*)	75	500			
(*e*)		850		0.0	

Solution Properties and phase descriptions of water are to be determined at various states.

Analysis (*a*) The quality is given to be $x = 0.6$, which implies that 60 percent of the mass is in the vapor phase and the remaining 40 percent is in the liquid phase. Therefore, we have saturated liquid–vapor mixture at a pressure of 200 kPa. Then the temperature must be the saturation temperature at the given pressure:

$$T = T_{\text{sat @ 200 kPa}} = \mathbf{120.21°C} \qquad \text{(Table A–5)}$$

At 200 kPa, we also read from Table A–5 that u_f = 504.50 kJ/kg and u_{fg} = 2024.6 kJ/kg. Then the average internal energy of the mixture is

$$\begin{aligned} u &= u_f + xu_{fg} \\ &= 504.50\ \text{kJ/kg} + (0.6)(2024.6\ \text{kJ/kg}) \\ &= \mathbf{1719.26\ kJ/kg} \end{aligned}$$

(*b*) This time the temperature and the internal energy are given, but we do not know which table to use to determine the missing properties because we have no clue as to whether we have a saturated mixture, compressed liquid, or superheated vapor. To determine the region we are in, we first go to the saturation table (Table A–4) and determine the u_f and u_g values at the given temperature. At 125°C, we read u_f = 524.83 kJ/kg and u_g = 2534.3 kJ/kg. Next we compare the given u value to these u_f and u_g values, keeping in mind that

if	$u < u_f$	we have a *compressed liquid*
if	$u_f \leq u \leq u_g$	we have a *saturated mixture*
if	$u > u_g$	we have a *superheated vapor*

In our case the given u value is 1600, which falls between the u_f and u_g values at 125°C. Therefore, we have a saturated liquid–vapor mixture. Then the pressure must be the saturation pressure at the given temperature:

$$P = P_{\text{sat @ 125°C}} = \mathbf{232.23\ kPa} \qquad \text{(Table A–4)}$$

The quality is determined from

$$x = \frac{u - u_f}{u_{fg}} = \frac{1600 - 524.83}{2009.5} = \mathbf{0.535}$$

The criteria above for determining whether we have a compressed liquid, saturated mixture, or superheated vapor can also be used when enthalpy h or specific volume v is given instead of internal energy u, or when pressure is given instead of temperature.

(*c*) This is similar to case (*b*), except pressure is given instead of temperature. Following the argument given above, we read the u_f and u_g values at the specified pressure. At 1 MPa, we have u_f = 761.39 kJ/kg and u_g = 2582.8 kJ/kg. The specified u value is 2950 kJ/kg, which is greater than the u_g value at 1 MPa. Therefore, we have a superheated vapor, and the temperature at this state is determined from the superheated vapor table by interpolation to be

$$T = \mathbf{395.2°C} \qquad \text{(Table A–6)}$$

We would leave the quality column blank or write "NA" (not applicable) in this case since quality has no meaning for a superheated vapor.

(*d*) In this case the temperature and pressure are given, but again we cannot tell which table to use to determine the missing properties because we do not know whether we have a saturated mixture, compressed liquid, or superheated vapor. To determine the region we are in, we go to the saturation table (Table A–5) and determine the saturation temperature value at the given pressure. At 500 kPa, we have T_{sat} = 151.83°C. We then compare the given T value to this T_{sat} value, keeping in mind that

if	$T < T_{\text{sat @ given } P}$	we have a *compressed liquid*
if	$T = T_{\text{sat @ given } P}$	we have a *saturated mixture*
if	$T > T_{\text{sat @ given } P}$	we have a *superheated vapor*

In our case, the given T value is 75°C, which is less than the T_{sat} value at the specified pressure. Therefore, we have a compressed liquid (Fig. 4–44), and normally we would determine the internal energy value from the compressed liquid table. But in this case the given pressure is much lower than the lowest pressure value in the compressed liquid table (which is 5 MPa), and therefore we are justified to treat the compressed liquid as saturated liquid at the given temperature (*not* pressure):

$$u \cong u_{f\text{ @ 75°C}} = \mathbf{313.99\ kJ/kg} \qquad \text{(Table A–4)}$$

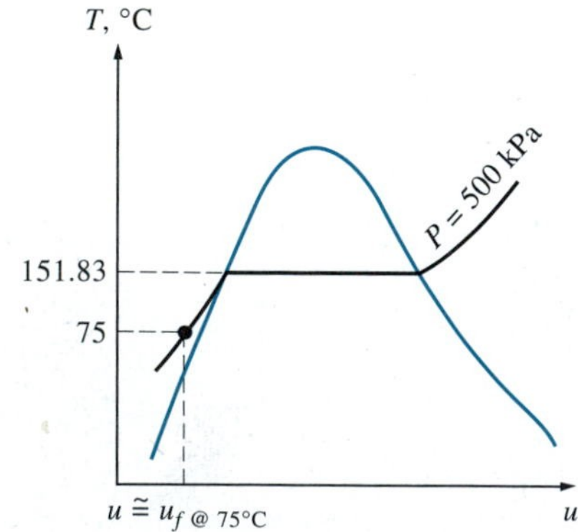

FIGURE 4–44

At a given P and T, a pure substance exists as a compressed liquid if $T < T_{\text{sat @ }P}$.

We would leave the quality column blank or write "NA" in this case since quality has no meaning in the compressed liquid region.

(*e*) The quality is given to be $x = 0$, and thus we have saturated liquid at the specified pressure of 850 kPa. Then the temperature must be the saturation temperature at the given pressure, and the internal energy must have the saturated liquid value:

$$T = T_{\text{sat @ 850 kPa}} = \mathbf{172.94°C}$$

$$u = u_{f\text{ @ 850 kPa}} = \mathbf{731.00\ kJ/kg} \qquad \text{(Table A–5)}$$

Discussion You would do well to practice "filling in the blanks" for other cases.

4–6 ▪ THE IDEAL-GAS EQUATION OF STATE

Property tables provide very accurate information about the properties, but they are bulky and vulnerable to typographical errors. A more practical and desirable approach would be to have some simple relations among the properties that are sufficiently general and accurate.

Any equation that relates the pressure, temperature, and specific volume of a substance is called an **equation of state**. Property relations that involve other properties of a substance at equilibrium states are also referred to as equations of state. There are several equations of state, some simple and others very complex. The simplest and best-known equation of state for substances in the gas phase is the ideal-gas equation of state. This equation predicts the *P*-*v*-*T* behavior of a gas quite accurately within some properly selected region.

Gas and *vapor* are often used as synonymous words. The vapor phase of a substance is customarily called a *gas* when it is above the critical temperature. *Vapor* usually implies a gas that is not far from a state of condensation.

In 1662, Robert Boyle, an Englishman, observed during his experiments with a vacuum chamber that the pressure of gases is inversely proportional to their volume. In 1802, Frenchmen J. Charles and J. Gay-Lussac experimentally determined that at low pressures the volume of a gas is proportional to its temperature. That is,

$$P = R\left(\frac{T}{v}\right)$$

or

$$Pv = RT \qquad \text{(4–10)}$$

where the constant of proportionality R is called the **gas constant**. Equation 4–10 is called the **ideal-gas equation of state**, or simply the **ideal-gas relation**, and a gas that obeys this relation is called an **ideal gas** (formerly, it had been called a perfect gas). In this equation, P is the absolute pressure, T is the absolute temperature, and v is the specific volume.

The gas constant R is different for each gas (Fig. 4–45) and is determined from

$$R = \frac{R_u}{M} \qquad (\text{kJ/kg}\cdot\text{K or kPa}\cdot\text{m}^3/\text{kg}\cdot\text{K})$$

where R_u is the **universal gas constant** and M is the molar mass (also

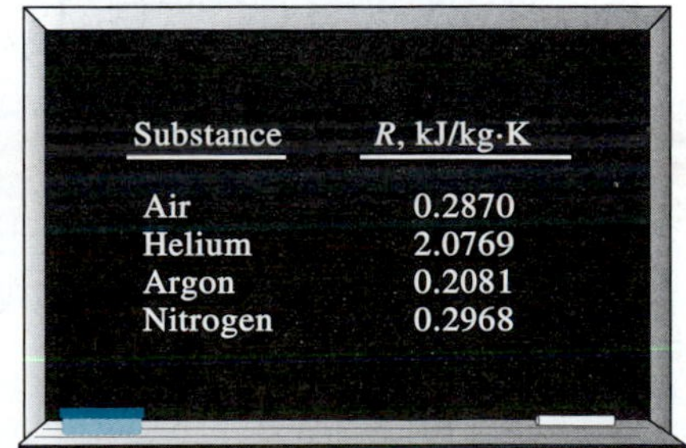

Substance	R, kJ/kg·K
Air	0.2870
Helium	2.0769
Argon	0.2081
Nitrogen	0.2968

FIGURE 4–45
Different substances have different gas constants.

called *molecular weight*) of the gas. The constant R_u is the same for all substances, and its value is

$$R_u = \begin{cases} 8.31447 \text{ kJ/kmol}\cdot\text{K} \\ 8.31447 \text{ kPa}\cdot\text{m}^3\text{/kmol}\cdot\text{K} \\ 0.0831447 \text{ bar}\cdot\text{m}^3\text{/kmol}\cdot\text{K} \\ 1.98588 \text{ Btu/lbmol}\cdot\text{R} \\ 10.7316 \text{ psia}\cdot\text{ft}^3\text{/lbmol}\cdot\text{R} \\ 1545.37 \text{ ft}\cdot\text{lbf/lbmol}\cdot\text{R} \end{cases} \tag{4–11}$$

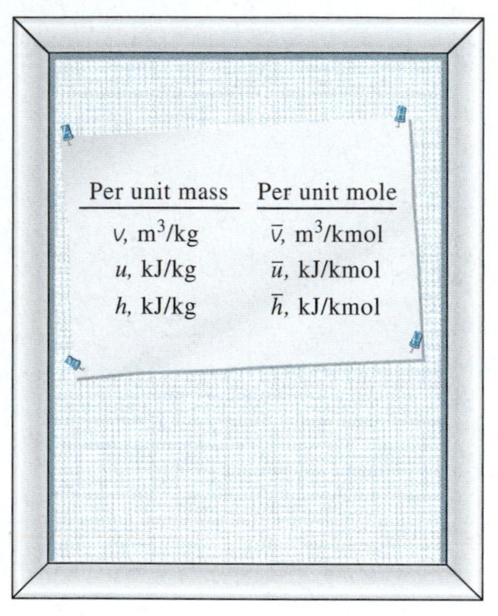

Per unit mass
Per unit mole
v, m³/kg
v̄, m³/kmol
u, kJ/kg
ū, kJ/kmol
h, kJ/kg
h̄, kJ/kmol

FIGURE 4–46

Properties per unit mole (or kmol) are denoted with a bar on the top.

The **molar mass** M can simply be defined as *the mass of one mole* (also called a *gram-mole*, abbreviated gmol) *of a substance in grams*, or *the mass of one kmol* (also called a *kilogram-mole*, abbreviated kgmol) *in kilograms*. In English units, it is the mass of 1 lbmol in lbm. Notice that the molar mass of a substance has the same numerical value in both unit systems because of the way it is defined. When we say the molar mass of nitrogen is 28, it simply means the mass of 1 kmol of nitrogen is 28 kg, or the mass of 1 lbmol of nitrogen is 28 lbm. That is, M = 28 kg/kmol = 28 lbm/lbmol. The mass of a system is equal to the product of its molar mass M and the mole number (number of moles) N:

$$m = MN \qquad \text{(kg)} \tag{4–12}$$

The values of R and M for several substances are given in Table A–1.

The ideal-gas equation of state can be written in several different forms:

$$V = mv \longrightarrow PV = mRT \tag{4–13}$$

$$mR = (MN)R = NR_u \longrightarrow PV = NR_uT \tag{4–14}$$

$$V = N\bar{v} \longrightarrow P\bar{v} = R_uT \tag{4–15}$$

where $\bar{v}$ is the molar specific volume, that is, the volume per unit mole (in m³/kmol or ft³/lbmol). A bar above a property denotes values on a *unit-mole basis* throughout this text (Fig. 4–46).

By writing Eq. 4–13 twice for a fixed mass and simplifying, the properties of an ideal gas at two different states are related to each other by

$$\frac{P_1V_1}{T_1} = \frac{P_2V_2}{T_2} \tag{4–16}$$

FIGURE 4–47

The ideal-gas relation often is not applicable to real gases; thus, care should be exercised when using it.

An ideal gas is an *imaginary* substance that obeys the relation $Pv = RT$ (Fig. 4–47). It has been experimentally observed that the ideal-gas relation given closely approximates the P-v-T behavior of real gases at low densities. At low pressures and high temperatures, the density of a gas decreases, and the gas behaves as an ideal gas under these conditions. What constitutes low pressure and high temperature is explained later.

In the range of practical interest, many familiar gases such as air, nitrogen, oxygen, hydrogen, helium, argon, neon, krypton, and even heavier gases such as carbon dioxide can be treated as ideal gases with negligible error (often less than 1 percent). Dense gases such as water vapor in steam power plants and refrigerant vapor in refrigerators, however, should not be treated as ideal gases. Instead, the property tables should be used for these substances.

EXAMPLE 4–10 Mass of Air in a Room

Determine the mass of the air in a vacant room whose dimensions are 4 m × 5 m × 6 m at 100 kPa and 25°C.

Solution The mass of air in a room is to be determined.

Analysis A sketch of the room is given in Fig. 4–48. Air at specified conditions can be treated as an ideal gas. From Table A–1, the gas constant of air is $R = 0.287 \text{ kPa} \cdot \text{m}^3/\text{kg} \cdot \text{K}$, and the absolute temperature is $T = 25°\text{C} + 273 = 298 \text{ K}$. The volume of the room is

$$V = (4 \text{ m})(5 \text{ m})(6 \text{ m}) = 120 \text{ m}^3$$

The mass of air in the room is determined from the ideal-gas relation to be

$$m = \frac{PV}{RT} = \frac{(100 \text{ kPa})(120 \text{ m}^3)}{(0.287 \text{ kPa} \cdot \text{m}^3/\text{kg} \cdot \text{K})(298 \text{ K})} = \mathbf{140.3 \text{ kg}}$$

Discussion This is about the same mass as two average men!

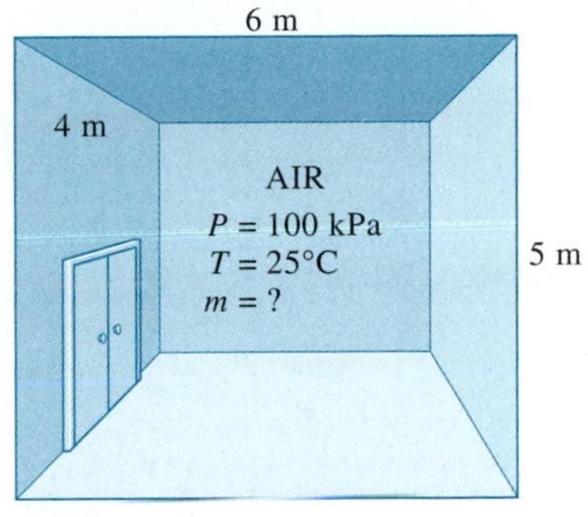

FIGURE 4–48

Schematic for Example 4–10.

Is Water Vapor an Ideal Gas?

This question cannot be answered with a simple yes or no. The error involved in treating water vapor as an ideal gas is calculated and plotted in Fig. 4–49. It is clear from this figure that at pressures below 10 kPa, water vapor can be treated as an ideal gas, regardless of its temperature, with negligible error (less than 0.1 percent). At higher pressures, however, the ideal-gas assumption yields unacceptable errors, particularly in the vicinity of the critical point and the saturated vapor line (over 100 percent error). Therefore, in air-conditioning applications, the water vapor in the air can be treated as an ideal gas with essentially no error since the pressure of the water vapor is very low. In steam power plant applications, however, the pressures involved are usually very high; therefore, ideal-gas relations should not be used.

4–7 ▪ COMPRESSIBILITY FACTOR—A MEASURE OF DEVIATION FROM IDEAL-GAS BEHAVIOR

The ideal-gas equation is very simple and thus very convenient to use. However, as illustrated in Fig. 4–49, gases deviate from ideal-gas behavior significantly at states near the saturation region and the critical point. This deviation from ideal-gas behavior at a given temperature and pressure can accurately be accounted for by the introduction of a correction factor called the **compressibility factor** Z defined as

$$Z = \frac{Pv}{RT} \tag{4–17}$$

or

$$Pv = ZRT \tag{4–18}$$

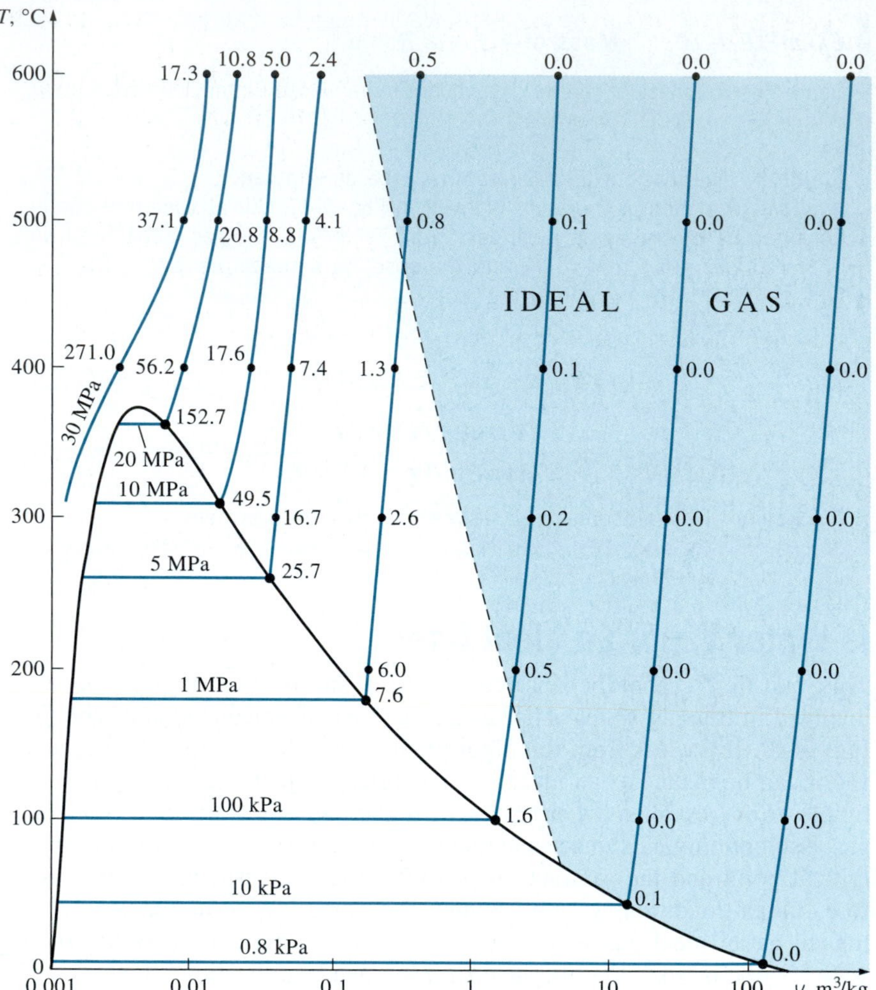

FIGURE 4–49

Percentage of error to one decimal place ($[|v_{\text{table}} - v_{\text{ideal}}|/v_{\text{table}}] \times 100$) involved in assuming steam to be an ideal gas, and the region where steam can be treated as an ideal gas with less than 1 percent error.

It can also be expressed as

$$Z = \frac{v_{\text{actual}}}{v_{\text{ideal}}} \tag{4–19}$$

where $v_{\text{ideal}} = RT/P$. Obviously, $Z = 1$ for ideal gases. For real gases Z can be greater than or less than unity (Fig. 4–50). The farther away Z is from unity, the more the gas deviates from ideal-gas behavior.

We have said that gases follow the ideal-gas equation closely at low pressures and high temperatures. But what exactly constitutes low pressure or high temperature? Is -100°C a low temperature? It definitely is for most substances but not for air. Air (or nitrogen) can be treated as an ideal gas at this temperature and atmospheric pressure with an error under 1 percent. This is because nitrogen is well over its critical temperature (-147°C) and away from the saturation region. At this temperature and pressure, however, most substances would exist in the solid phase. Therefore, the pressure or temperature of a substance is high or low relative to its critical temperature or pressure.

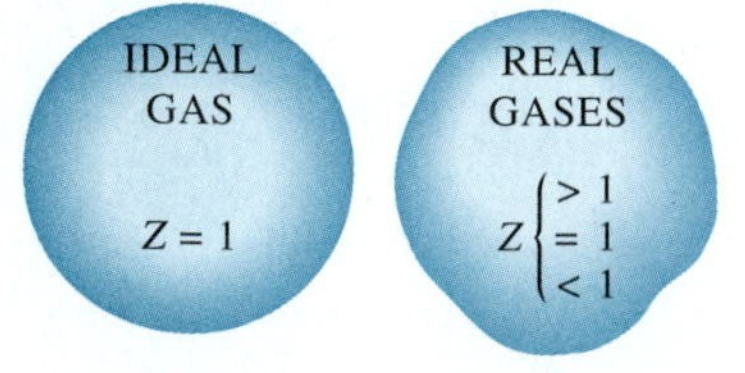

FIGURE 4–50

The compressibility factor is unity for ideal gases.

Gases behave differently at a given temperature and pressure, but they behave very much the same at temperatures and pressures normalized with respect to their critical temperatures and pressures. The normalization is done (always using *absolute* pressure and temperature) as

$$P_R = \frac{P}{P_{cr}} \quad \text{and} \quad T_R = \frac{T}{T_{cr}} \tag{4-20}$$

Here P_R is called the **reduced pressure** and T_R the **reduced temperature**. The Z factor for all gases is approximately the same at the same reduced pressure and temperature. This is called the **principle of corresponding states**. In Fig. 4–51, the experimentally determined Z values are plotted against P_R and T_R for several gases. The gases seem to obey the principle of corresponding states reasonably well. By curve-fitting all the data, we

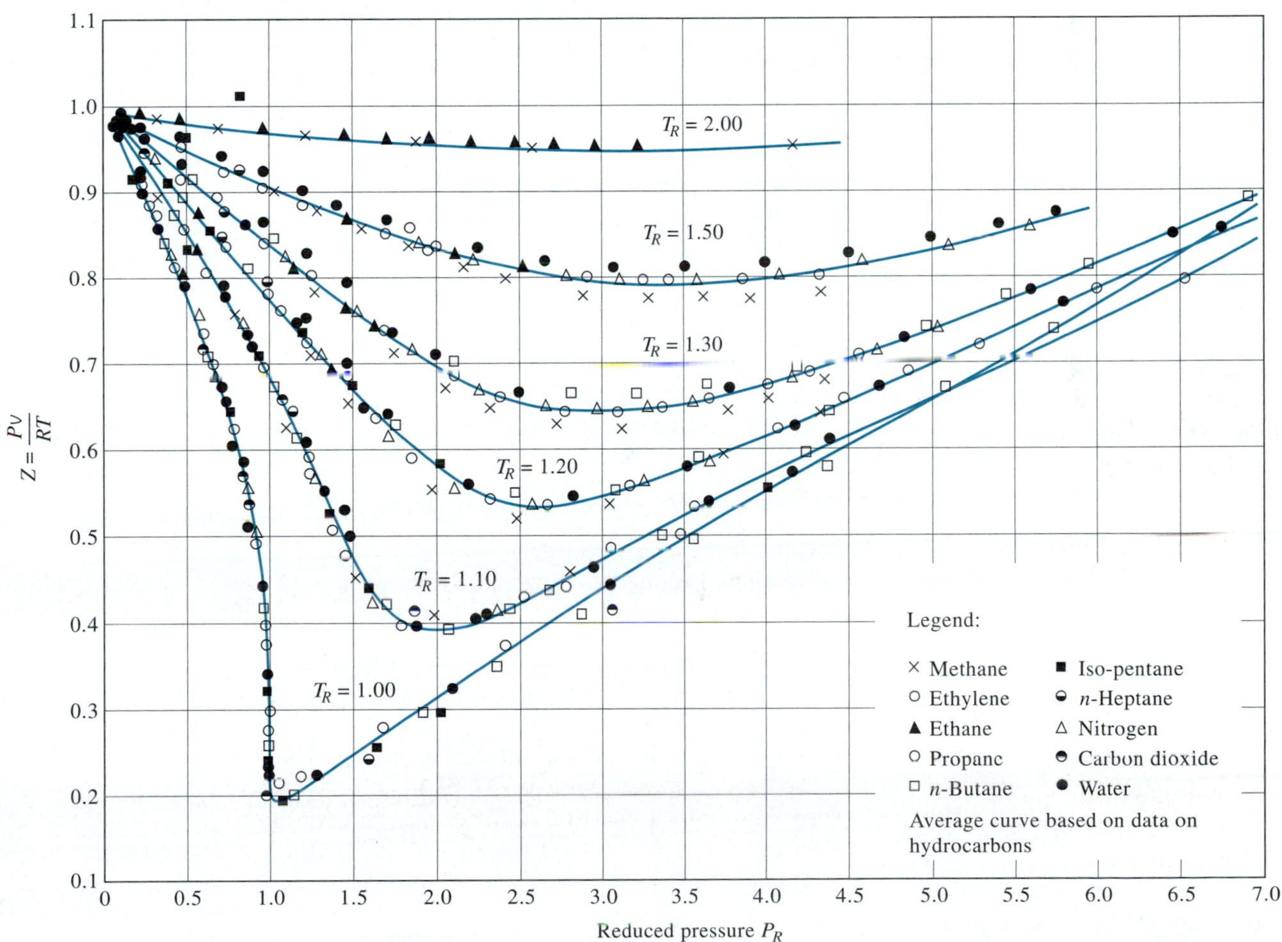

FIGURE 4–51

Comparison of Z factors for various gases.

Source: *Gour-Jen Su, "Modified Law of Corresponding States,"* Ind. Eng. Chem. *(international ed.) 38 (1946), p. 803.*

obtain the **generalized compressibility chart** that can be used for all gases (Fig. A–28).

The following observations can be made from the generalized compressibility chart:

1. At very low pressures ($P_R \ll 1$), a gas behaves as an ideal gas regardless of temperature (Fig. 4–52),
2. At high temperatures ($T_R > 2$), ideal-gas behavior can be approximated with good accuracy regardless of pressure (except when $P_R \gg 1$).
3. The deviation of a gas from ideal-gas behavior is greatest in the vicinity of the critical point (Fig. 4–53).

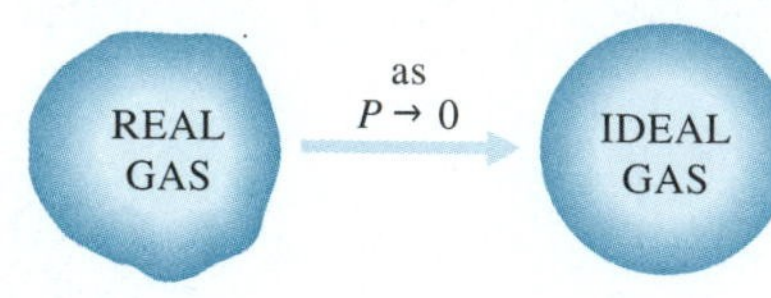

FIGURE 4–52

At very low pressures, all gases approach ideal-gas behavior (regardless of their temperature).

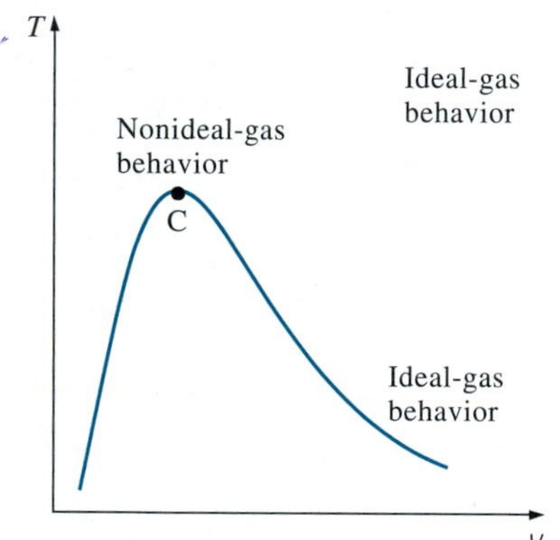

FIGURE 4–53

Gases deviate from the ideal-gas behavior the most in the neighborhood of the critical point.

EXAMPLE 4–11 The Use of Generalized Charts

Determine the specific volume of refrigerant-134a at 1 MPa and 50°C, using (*a*) the ideal-gas equation of state and (*b*) the generalized compressibility chart. Compare the values obtained to the actual value of 0.021796 m³/kg and determine the error involved in each case.

Solution The specific volume of refrigerant-134a is to be determined assuming ideal- and nonideal-gas behavior.

Analysis The gas constant, the critical pressure, and the critical temperature of refrigerant-134a are determined from Table A–1 to be

$$R = 0.0815\ \text{kPa}\cdot\text{m}^3/\text{kg}\cdot\text{K}$$

$$P_{\text{cr}} = 4.059\ \text{MPa}$$

$$T_{\text{cr}} = 374.2\ \text{K}$$

(*a*) The specific volume of refrigerant-134a under the ideal-gas assumption is

$$v = \frac{RT}{P} = \frac{(0.0815\ \text{kPa}\cdot\text{m}_3/\text{kg}\cdot\text{K})(323\ \text{K})}{1000\ \text{kPa}} = \mathbf{0.026325\ m^3/kg}$$

Therefore, treating the refrigerant-134a vapor as an ideal gas would result in an error of (0.026325 − 0.021796)/0.021796 = **0.208**, or 20.8 percent in this case.

(*b*) To determine the correction factor *Z* from the compressibility chart, we first need to calculate the reduced pressure and temperature:

$$\left.\begin{aligned} P_R &= \frac{P}{P_{\text{cr}}} = \frac{1\ \text{MPa}}{4.059\ \text{MPa}} = 0.246 \\ T_R &= \frac{T}{T_{\text{cr}}} = \frac{323\ \text{K}}{374.2\ \text{K}} = 0.863 \end{aligned}\right\} \quad Z = 0.84$$

Thus

$$v = Zv_{\text{ideal}} = (0.84)(0.026325\ \text{m}^3/\text{kg}) = \mathbf{0.022113\ m^3/kg}$$

Discussion The error in this result is less than **2 percent.** Therefore, in the absence of tabulated data, the generalized compressibility chart can be used with confidence.

When P and v, or T and v, are given instead of P and T, the generalized compressibility chart can still be used to determine the third property, but it would involve tedious trial and error. Therefore, it is necessary to define one more reduced property called the **pseudo-reduced specific volume** v_R as

$$v_R = \frac{v_{\text{actual}}}{RT_{\text{cr}}/P_{\text{cr}}} \quad \text{(4–21)}$$

Note that v_R is defined differently from P_R and T_R. It is related to T_{cr} and P_{cr} instead of v_{cr}. Lines of constant v_R are also added to the compressibility charts, and this enables one to determine T or P without having to resort to time-consuming iterations (Fig. 4–54).

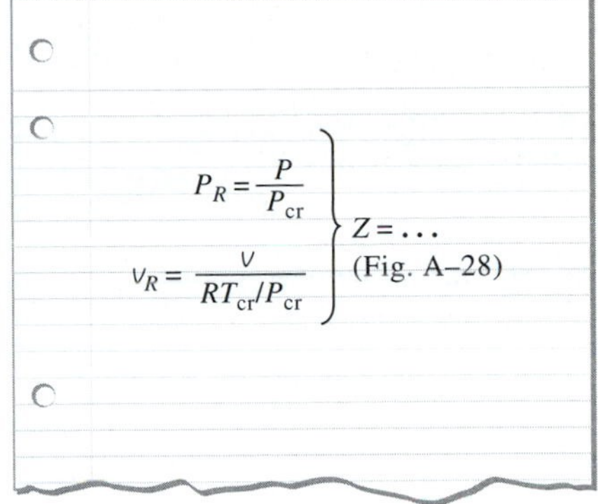

FIGURE 4–54

The compressibility factor can also be determined from a knowledge of P_R and v_R.

EXAMPLE 4–12 Using Generalized Charts to Determine Pressure

Determine the pressure of water vapor at 600°F and 0.51431 ft³/lbm, using (*a*) the steam tables, (*b*) the ideal-gas equation, and (*c*) the generalized compressibility chart.

Solution The pressure of water vapor is to be determined in three different ways.

Analysis A sketch of the system is given in Fig. 4–55. The gas constant, the critical pressure, and the critical temperature of steam are determined from Table A–1E to be

$$R = 0.5956 \text{ psia} \cdot \text{ft}^3/\text{lbm} \cdot \text{R}$$

$$P_{\text{cr}} = 3200 \text{ psia}$$

$$T_{\text{cr}} = 1164.8 \text{ R}$$

(*a*) The pressure at the specified state is determined from Table A–6E to be

$$\left.\begin{aligned} v &= 0.51431 \text{ ft}^3/\text{lbm} \\ T &= 600°\text{F} \end{aligned}\right\} \quad P = \mathbf{1000 \text{ psia}}$$

This is the experimentally determined value, and thus it is the most accurate.

(*b*) The pressure of steam under the ideal-gas assumption is determined from the ideal-gas relation to be

$$P = \frac{RT}{v} = \frac{(0.5956 \text{ psia} \cdot \text{ft}^3/\text{lbm} \cdot \text{R})(1060 \text{ R})}{0.51431 \text{ ft}^3/\text{lbm}} = \mathbf{1228 \text{ psia}}$$

Therefore, treating the steam as an ideal gas would result in an error of (1228 − 1000)/1000 = 0.228, or 22.8 percent in this case.

(*c*) To determine the correction factor Z from the compressibility chart (Fig. A–28), we first need to calculate the pseudo-reduced specific volume and the reduced temperature:

$$\left.\begin{aligned} v_R &= \frac{v_{\text{actual}}}{RT_{\text{cr}}/P_{\text{cr}}} = \frac{(0.51431 \text{ ft}^3/\text{lbm})(3200 \text{ psia})}{(0.5956 \text{ psia} \cdot \text{ft}^3/\text{lbm} \cdot \text{R})(1164.8 \text{ R})} = 2.372 \\ T_R &= \frac{T}{T_{\text{cr}}} = \frac{1060 \text{ R}}{1164.8 \text{ R}} = 0.91 \end{aligned}\right\} \quad P_R = 0.33$$

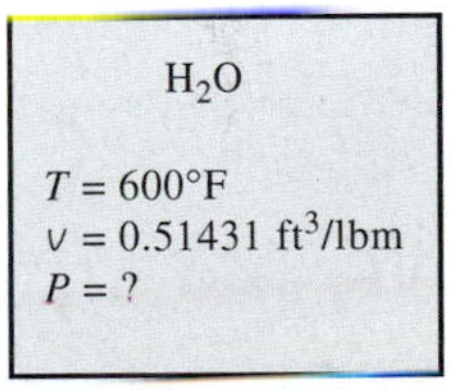

FIGURE 4–55

Schematic for Example 4–12.

H_2O
$T = 600°\text{F}$
$v = 0.51431 \text{ ft}^3/\text{lbm}$
$P = ?$

Thus,

$$P = P_R P_{cr} = (0.33)(3200 \text{ psia}) = \mathbf{1056 \text{ psia}}$$

Discussion Using the compressibility chart reduced the error from 22.8 to 5.6 percent, which is acceptable for most engineering purposes (Fig. 4–56). A bigger chart, of course, would give better resolution and reduce the reading errors. Notice that we did not have to determine Z in this problem since we could read P_R directly from the chart.

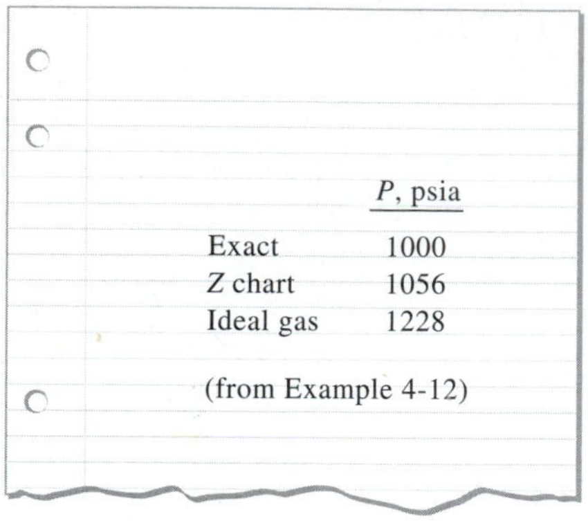

	P, psia
Exact	1000
Z chart	1056
Ideal gas	1228

(from Example 4-12)

FIGURE 4–56
Results obtained by using the compressibility chart are usually within a few percent of actual values.

4–8 ▪ OTHER EQUATIONS OF STATE

The ideal-gas equation of state is very simple, but its range of applicability is limited. It is desirable to have equations of state that represent the P-v-T behavior of substances accurately over a larger region with no limitations. Such equations are naturally more complicated. Several equations have been proposed for this purpose (Fig. 4–57), but we shall discuss only three: the *van der Waals* equation because it is one of the earliest, the *Beattie-Bridgeman* equation of state because it is one of the best known and is reasonably accurate, and the *Benedict-Webb-Rubin* equation because it is one of the more recent and is very accurate.

Van der Waals Equation of State

The van der Waals equation of state was proposed in 1873, and it has two constants that are determined from the behavior of a substance at the critical point. It is given by

$$\left(P + \frac{a}{v^2}\right)(v - b) = RT \qquad \text{(4–22)}$$

Van der Waals intended to improve the ideal-gas equation of state by including two of the effects not considered in the ideal-gas model: the *intermolecular attraction forces* and the *volume occupied by the molecules themselves*. The term a/v^2 accounts for the intermolecular forces, and b accounts for the volume occupied by the gas molecules. In a room at atmospheric pressure and temperature, the volume actually occupied by molecules is only about one-thousandth of the volume of the room. As the pressure increases, the volume occupied by the molecules becomes an increasingly significant part of the total volume. Van der Waals proposed to correct this by replacing v in the ideal-gas relation with the quantity $v - b$, where b represents the volume occupied by the gas molecules per unit mass.

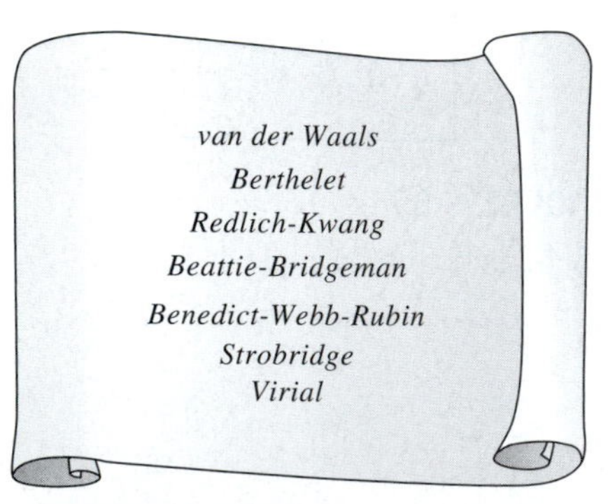

FIGURE 4–57
Several equations of state have been proposed throughout history.

The determination of the two constants appearing in this equation is based on the observation that the critical isotherm on a P-v diagram has a horizontal inflection point at the critical point (Fig. 4–58). Thus, the first and the second derivatives of P with respect to v at the critical point must be zero. That is,

$$\left(\frac{\partial P}{\partial v}\right)_{T=T_{cr}=\text{const}} = 0 \quad \text{and} \quad \left(\frac{\partial^2 P}{\partial v^2}\right)_{T=T_{cr}=\text{const}} = 0$$

By performing the differentiations and eliminating v_{cr}, the constants a and b are determined to be

$$a = \frac{27R^2T_{cr}^2}{64P_{cr}} \quad \text{and} \quad b = \frac{RT_{cr}}{8P_{cr}} \tag{4–23}$$

The constants a and b can be determined for any substance from the critical-point data alone (Table A–1).

The accuracy of the van der Waals equation of state is often inadequate, but it can be improved by using values of a and b that are based on the actual behavior of the gas over a wider range instead of a single point. Despite its limitations, the van der Waals equation of state has a historical value in that it was one of the first attempts to model the behavior of real gases. The van der Waals equation of state can also be expressed on a unit-mole basis by replacing the v in Eq. 4–22 by $\overline{v}$ and the R in Eqs. 4–22 and 4–23 by R_u.

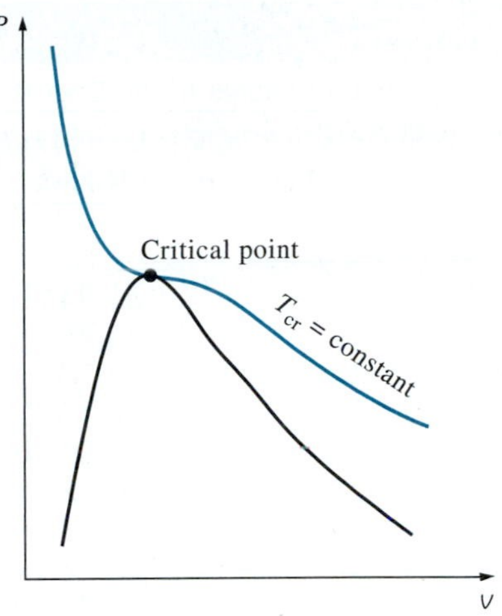

FIGURE 4–58

The critical isotherm of a pure substance has an inflection point at the critical state.

Beattie-Bridgeman Equation of State

The Beattie-Bridgeman equation, proposed in 1928, is an equation of state based on five experimentally determined constants. It is expressed as

$$P = \frac{R_uT}{\overline{v}^2}\left(1 - \frac{c}{\overline{v}T^3}\right)(\overline{v} + B) - \frac{A}{\overline{v}^2} \tag{4–24}$$

where

$$A = A_0\left(1 - \frac{a}{\overline{v}}\right) \quad \text{and} \quad B = B_0\left(1 - \frac{b}{\overline{v}}\right) \tag{4–25}$$

The constants appearing in the above equation are given in Table 4–4 for various substances. The Beattie-Bridgeman equation is known to be reasonably accurate for densities up to about $0.8\rho_{cr}$, where ρ_{cr} is the density of the substance at the critical point.

Benedict-Webb-Rubin Equation of State

Benedict, Webb, and Rubin extended the Beattie-Bridgeman equation in 1940 by raising the number of constants to eight. It is expressed as

$$P = \frac{R_uT}{\overline{v}} + \left(B_0R_uT - A_0 - \frac{C_0}{T^2}\right)\frac{1}{\overline{v}^2} + \frac{bR_uT - a}{\overline{v}^3} + \frac{a\alpha}{\overline{v}^6} + \frac{c}{\overline{v}^3T^2}\left(1 + \frac{\gamma}{\overline{v}^2}\right)e^{-\gamma/\overline{v}^2} \tag{4–26}$$

The values of the constants appearing in this equation are given in Table 4–4. This equation can handle substances at densities up to about $2.5\rho_{cr}$. In 1962, Strobridge further extended this equation by raising the number of constants to 16 (Fig. 4–59).

Virial Equation of State

The equation of state of a substance can also be expressed in a series form as

$$P = \frac{RT}{v} + \frac{a(T)}{v^2} + \frac{b(T)}{v^3} + \frac{c(T)}{v^4} + \frac{d(T)}{v^5} + \dots \tag{4–27}$$

TABLE 4–4

Constants that appear in the Beattie-Bridgeman and the Benedict-Webb-Rubin equations of state

(*a*) When P is in kPa, $\bar{v}$ is in m^3/kmol, T is in K, and $R_u = 8.314$ kPa · m^3/kmol · K, the five constants in the Beattie-Bridgeman equation are as follows:

Gas	A_0	a	B_0	b	c
Air	131.8441	0.01931	0.04611	−0.001101	4.34×10^4
Argon, Ar	130.7802	0.02328	0.03931	0.0	5.99×10^4
Carbon dioxide, CO_2	507.2836	0.07132	0.10476	0.07235	6.60×10^5
Helium, He	2.1886	0.05984	0.01400	0.0	40
Hydrogen, H_2	20.0117	−0.00506	0.02096	−0.04359	504
Nitrogen, N_2	136.2315	0.02617	0.05046	−0.00691	4.20×10^4
Oxygen, O_2	151.0857	0.02562	0.04624	0.004208	4.80×10^4

Source: Gordon J. Van Wylen and Richard E. Sonntag, *Fundamentals of Classical Thermodynamics*, English/SI Version, 3rd ed. (New York: John Wiley & Sons, 1986), p. 46, table 4.3.

(*b*) When P is in kPa, $\bar{v}$ is in m^3/kmol, T is in K, and $R_u = 8.314$ kPa · m^3/kmol · K, the eight constants in the Benedict-Webb-Rubin equation are as follows:

Gas	a	A_0	b	B_0	c	C_0	α	γ
n-Butane, C_4H_{10}	190.68	1021.6	0.039998	0.12436	3.205×10^7	1.006×10^8	1.101×10^{-3}	0.0340
Carbon dioxide, CO_2	13.86	277.30	0.007210	0.04991	1.511×10^6	1.404×10^7	8.470×10^{-5}	0.00539
Carbon monoxide, CO	3.71	135.87	0.002632	0.05454	1.054×10^5	8.673×10^5	1.350×10^{-4}	0.0060
Methane, CH_4	5.00	187.91	0.003380	0.04260	2.578×10^5	2.286×10^6	1.244×10^{-4}	0.0060
Nitrogen, N_2	2.54	106.73	0.002328	0.04074	7.379×10^4	8.164×10^5	1.272×10^{-4}	0.0053

Source: Kenneth Wark, *Thermodynamics*, 4th ed. (New York: McGraw-Hill, 1983), p. 815, table A-21M. Originally published in H. W. Cooper and J. C. Goldfrank, *Hydrocarbon Processing* 46, no. 12 (1967), p. xxx.

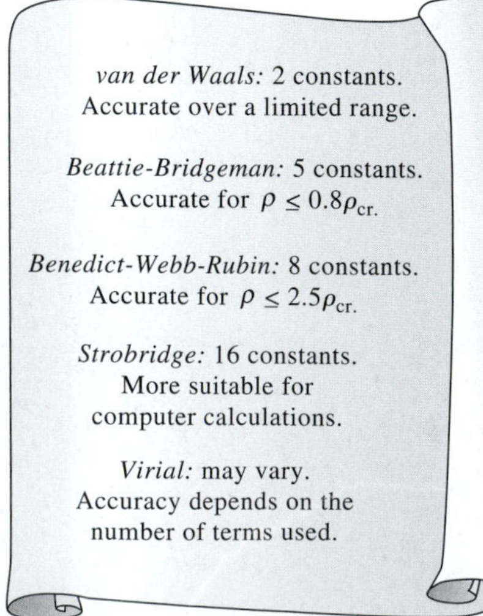

van der Waals: 2 constants.
Accurate over a limited range.
Beattie-Bridgeman: 5 constants.
Accurate for $\rho \leq 0.8\rho_{cr}$.
Benedict-Webb-Rubin: 8 constants.
Accurate for $\rho \leq 2.5\rho_{cr}$.
Strobridge: 16 constants.
More suitable for
computer calculations.
Virial: may vary.
Accuracy depends on the
number of terms used.

FIGURE 4–59

Complex equations of state represent the *P-v-T* behavior of gases more accurately over a wider range.

This and similar equations are called the *virial equations of state*, and the coefficients $a(T)$, $b(T)$, $c(T)$, and so on, that are functions of temperature alone are called *virial coefficients*. These coefficients can be determined experimentally or theoretically from statistical mechanics. Obviously, as the pressure approaches zero, all the virial coefficients vanish and the equation reduces to the ideal-gas equation of state. The *P-v-T* behavior of a substance can be represented accurately with the virial equation of state over a wider range by including a sufficient number of terms. The equations of state discussed here are applicable to the gas phase of the substances only, and thus should not be used for liquids or liquid–vapor mixtures.

Complex equations represent the *P-v-T* behavior of substances reasonably well and are very suitable for digital computer applications. For hand calculations, however, it is suggested that the reader use the property tables or the simpler equations of state for convenience. This is particularly true for specific-volume calculations since all the earlier equations are implicit in v and require a trial-and-error approach. The accuracy of the van der Waals,

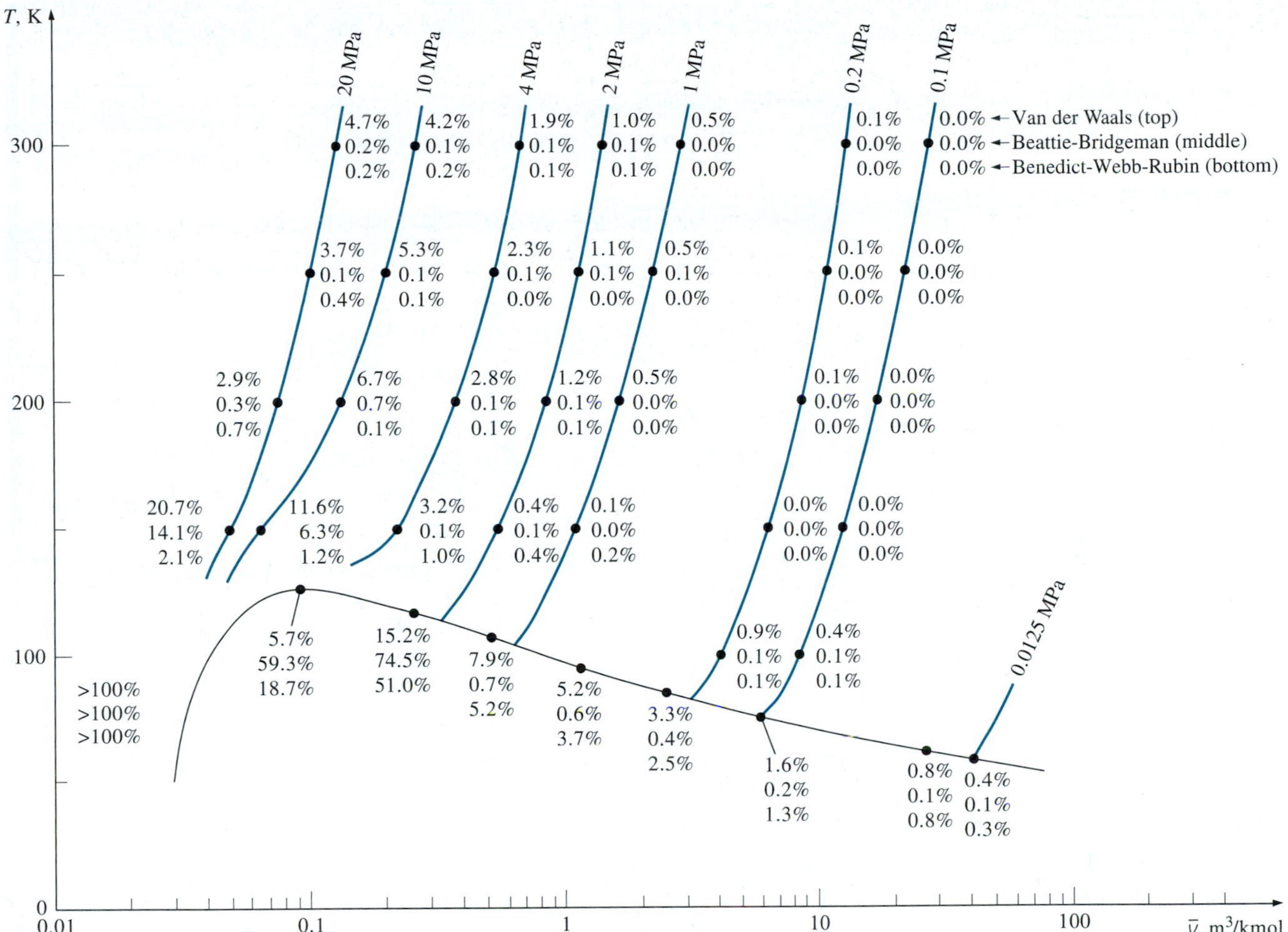

FIGURE 4–60

Percentage of error involved in various equations of state for nitrogen (% error $= [(|v_{\text{table}} - v_{\text{equation}}|)/v_{\text{table}}] \times 100$).

Beattie-Bridgeman, and Benedict-Webb-Rubin equations of state is illustrated in Fig. 4–60. It is apparent from this figure that the Benedict-Webb-Rubin equation of state is usually the most accurate.

EXAMPLE 4–13 Different Methods of Evaluating Gas Pressure

Predict the pressure of nitrogen gas at $T = 175$ K and $v = 0.00375$ m^3/kg on the basis of (*a*) the ideal-gas equation of state, (*b*) the van der Waals equation of state, (*c*) the Beattie-Bridgeman equation of state, and (*d*) the Benedict-Webb-Rubin equation of state. Compare the values obtained to the experimentally determined value of 10,000 kPa.

Solution The pressure of nitrogen gas is to be determined using four different equations of state.

Properties The gas constant of nitrogen gas is 0.2968 kPa · m³/kg · K (Table A–1).

Analysis (*a*) Using the ideal-gas equation of state, the pressure is found to be

$$P = \frac{RT}{v} = \frac{(0.2968 \text{ kPa} \cdot \text{m}^3/\text{kg} \cdot \text{K})(175 \text{ K})}{0.00375 \text{ m}^3/\text{kg}} = \mathbf{13{,}851 \text{ kPa}}$$

which is in error by 38.5 percent.

(*b*) The van der Waals constants for nitrogen are determined from Eq. 4–23 to be

$$a = 0.175 \text{ m}^6 \cdot \text{kPa/kg}^2$$

$$b = 0.00138 \text{ m}^3/\text{kg}$$

From Eq. 4–22,

$$P = \frac{RT}{v - b} - \frac{a}{v^2} = \mathbf{9471 \text{ kPa}}$$

which is in error by 5.3 percent.

(*c*) The constants in the Beattie-Bridgeman equation are determined from Table 4–4 to be

$$A = 102.29$$

$$B = 0.05378$$

$$c = 4.2 \times 10^4$$

Also, $\bar{v} = Mv = (28.013 \text{ kg/mol})(0.00375 \text{ m}^3/\text{kg}) = 0.10505 \text{ m}^3/\text{kmol}$. Substituting these values into Eq. 4–24, we obtain

$$P = \frac{R_u T}{\bar{v}^2}\left(1 - \frac{c}{\bar{v}T^3}\right)(\bar{v} + B) - \frac{A}{\bar{v}^2} = \mathbf{10{,}110 \text{ kPa}}$$

which is in error by 1.1 percent.

(*d*) The constants in the Benedict-Webb-Rubin equation are determined from Table 4–4 to be

$$a = 2.54 \qquad A_0 = 106.73$$

$$b = 0.002328 \qquad B_0 = 0.04074$$

$$c = 7.379 \times 10^4 \qquad C_0 = 8.164 \times 10^5$$

$$\alpha = 1.272 \times 10^{-4} \qquad \gamma = 0.0053$$

Substituting these values into Eq. 4–26 gives

$$P = \frac{R_u T}{\bar{v}} + \left(B_0 R_u T - A_0 - \frac{C_0}{T^2}\right)\frac{1}{\bar{v}^2} + \frac{bR_u T - a}{\bar{v}^3}$$

$$+ \frac{a\alpha}{\bar{v}^6} + \frac{c}{\bar{v}^3 T^2}\left(1 + \frac{\gamma}{\bar{v}^2}\right)e^{-\gamma/\bar{v}^2}$$

$$= \mathbf{10{,}009 \text{ kPa}}$$

which is in error by only 0.09 percent. Thus, the accuracy of the Benedict-Webb-Rubin equation of state is rather impressive in this case.

Discussion Many modern computer codes (such as EES) have built-in property functions and eliminate the need for this kind of manual calculation.

SUMMARY

A substance that has a fixed chemical composition throughout is called a *pure substance.* A pure substance exists in different phases depending on its energy level. In the liquid phase, a substance that is not about to vaporize is called a *compressed* or *subcooled liquid.* In the gas phase, a substance that is not about to condense is called a *superheated vapor.* During a phase-change process, the temperature and pressure of a pure substance are dependent properties. At a given pressure, a substance changes phase at a fixed temperature, called the *saturation temperature.* Likewise, at a given temperature, the pressure at which a substance changes phase is called the *saturation pressure.* During a boiling process, both the liquid and the vapor phases coexist in equilibrium, and under this condition the liquid is called *saturated liquid* and the vapor *saturated vapor.*

In a saturated liquid–vapor mixture, the mass fraction of vapor is called the *quality* and is expressed as

$$x = \frac{m_{\text{vapor}}}{m_{\text{total}}}$$

Quality may have values between 0 (saturated liquid) and 1 (saturated vapor). It has no meaning in the compressed liquid or superheated vapor regions. In the saturated mixture region, the average value of any intensive property y is determined from

$$y = y_f + xy_{fg}$$

where f stands for saturated liquid and g for saturated vapor.

In the absence of compressed liquid data, a general approximation is to treat a compressed liquid as a saturated liquid at the given *temperature* (not pressure),

$$y \cong y_{f@T}$$

where y stands for v, u, or h.

The state beyond which there is no distinct vaporization process is called the *critical point.* At supercritical pressures, a substance gradually and uniformly expands from the liquid to vapor phase. All three phases of a substance coexist in equilibrium at states along the *triple line* characterized by triple-line temperature and pressure. The compressed liquid has lower v, u, and h values than the saturated liquid at the same T or P. Likewise, superheated vapor has higher v, u, and h values than the saturated vapor at the same T or P.

Any relation among the pressure, temperature, and specific volume of a substance is called an *equation of state.* The simplest and best-known equation of state is the *ideal-gas equation of state,* given as

$$Pv = RT$$

where R is the gas constant. Caution should be exercised in using this relation since an ideal gas is a fictitious substance. Real gases exhibit ideal-gas behavior at relatively low pressures and high temperatures.

The deviation from ideal-gas behavior can be properly accounted for by using the *compressibility factor Z,* defined as

$$Z = \frac{Pv}{RT} \quad \text{or} \quad Z = \frac{v_{\text{actual}}}{v_{\text{ideal}}}$$

The Z factor is approximately the same for all gases at the same *reduced temperature* and *reduced pressure,* which are defined as

$$T_R = \frac{T}{T_{\text{cr}}} \quad \text{and} \quad P_R = \frac{P}{P_{\text{cr}}}$$

where P_{cr} and T_{cr} are the critical pressure and temperature, respectively. This is known as the *principle of corresponding states.* When either P or T is unknown, it can be determined from the compressibility chart with the help of the *pseudo-reduced specific volume,* defined as

$$v_R = \frac{v_{\text{actual}}}{RT_{\text{cr}}/P_{\text{cr}}}$$

The *P-v-T* behavior of substances can be represented more accurately by more complex equations of state. Three of the best known are

van der Waals: $$\left(P + \frac{a}{v^2}\right)(v - b) = RT$$

where

$$a = \frac{27R^2T_{\text{cr}}^2}{64P_{\text{cr}}} \quad \text{and} \quad b = \frac{RT_{\text{cr}}}{8P_{\text{cr}}}$$

Beattie-Bridgeman: $$P = \frac{R_uT}{\bar{v}^2}\left(1 - \frac{c}{\bar{v}T^3}\right)(\bar{v} + B) - \frac{A}{\bar{v}^2}$$

where

$$A = A_0\left(1 - \frac{a}{\bar{v}}\right) \quad \text{and} \quad B = B_0\left(1 - \frac{b}{\bar{v}}\right)$$

Benedict-Webb-Rubin:

$$P = \frac{R_uT}{\bar{v}} + \left(B_0R_uT - A_0 - \frac{C_0}{T^2}\right)\frac{1}{\bar{v}^2} + \frac{bR_uT - a}{\bar{v}^3} + \frac{a\alpha}{\bar{v}^6} + \frac{c}{\bar{v}^3T^2}\left(1 + \frac{\gamma}{\bar{v}^2}\right)e^{-\gamma/\bar{v}^2}$$

where R_u is the universal gas constant and $\bar{v}$ is the molar specific volume.

REFERENCES AND SUGGESTED READINGS

1. ASHRAE *Handbook of Fundamentals*. SI version. Atlanta, GA: American Society of Heating, Refrigerating, and Air-Conditioning Engineers, Inc., 1993.
2. ASHRAE *Handbook of Refrigeration*. SI version. Atlanta, GA: American Society of Heating, Refrigerating, and Air-Conditioning Engineers, Inc., 1994.
3. A. Bejan. *Advanced Engineering Thermodynamics*. 3rd ed. New York: Wiley, 2006.
4. "M. Kostic. *Analysis of Enthalpy Approximation for Compressed Liquid Water*, Journal of Heat Transfer, Vol. 128, p. 421–426, May 2006."

PROBLEMS*

Pure Substances, Phase-Change Processes, Property Diagrams

4–1C A propane tank is filled with a mixture of liquid and vapor propane. Can the contents of this tank be considered a pure substance? Explain.

4–2C What is the difference between saturated liquid and compressed liquid?

4–3C What is the difference between saturated vapor and superheated vapor?

4–4C Is there any difference between the intensive properties of saturated vapor at a given temperature and the vapor of a saturated mixture at the same temperature?

4–5C If the pressure of a substance is increased during a boiling process, will the temperature also increase or will it remain constant? Why?

4–6C Why are the temperature and pressure dependent properties in the saturated mixture region?

4–7C What is the difference between the critical point and the triple point?

4–8C Is it possible to have water vapor at $-10°C$?

4–9C A househusband is cooking beef stew for his family in a pan that is (*a*) uncovered, (*b*) covered with a light lid, and (*c*) covered with a heavy lid. For which case will the cooking time be the shortest? Why?

4–10C How does the boiling process at supercritical pressures differ from the boiling process at subcritical pressures?

*Problems designated by a "C" are concept questions, and students are encouraged to answer them all. Problems designated by an "E" are in English units, and the SI users can ignore them. Problems with the icon are solved using EES, and complete solutions together with parametric studies are included on the enclosed DVD. Problems with the icon are comprehensive in nature, and are intended to be solved with a computer, preferably using the EES software that accompanies this text.

Property Tables

4–11C A perfectly fitting pot and its lid often stick after cooking, and it becomes very difficult to open the lid when the pot cools down. Explain why this happens and what you would do to open the lid.

4–12C It is well known that warm air in a cooler environment rises. Now consider a warm mixture of air and gasoline on top of an open gasoline can. Do you think this gas mixture will rise in a cooler environment?

4–13C In 1775, Dr. William Cullen made ice in Scotland by evacuating the air in a water tank. Explain how that device works, and discuss how the process can be made more efficient.

4–14C Does the amount of heat absorbed as 1 kg of saturated liquid water boils at 100°C have to be equal to the amount of heat released as 1 kg of saturated water vapor condenses at 100°C?

4–15C Does the reference point selected for the properties of a substance have any effect on thermodynamic analysis? Why?

4–16C What is the physical significance of h_{fg}? Can it be obtained from a knowledge of h_f and h_g? How?

4–17C Is it true that it takes more energy to vaporize 1 kg of saturated liquid water at 100°C than it would at 120°C?

4–18C What is quality? Does it have any meaning in the superheated vapor region?

4–19C Which process requires more energy: completely vaporizing 1 kg of saturated liquid water at 1 atm pressure or completely vaporizing 1 kg of saturated liquid water at 8 atm pressure?

4–20C Does h_{fg} change with pressure? How?

4–21C Can quality be expressed as the ratio of the volume occupied by the vapor phase to the total volume? Explain.

4–22C In the absence of compressed liquid tables, how is the specific volume of a compressed liquid at a given P and T determined?

4–23 Complete this table for H_2O:

T, °C	P, kPa	v, m³/kg	Phase description
50		4.16	
	200		Saturated vapor
250	400		
110	600		

4–24 Reconsider Prob. 4–23. Using EES (or other) software, determine the missing properties of water. Repeat the solution for refrigerant-134a, refrigerant-22, and ammonia.

4–25E Complete this table for H_2O:

T, °F	P, psia	u, Btu/lbm	Phase description
300		782	
	40		Saturated liquid
500	120		
400	400		

4–26E Reconsider Prob. 4–25E. Using EES (or other) software, determine the missing properties of water. Repeat the solution for refrigerant-134a, refrigerant-22, and ammonia.

4–27 Complete this table for H_2O:

T, °C	P, kPa	h, kJ/kg	x	Phase description
	200		0.7	
140		1800		
	950		0.0	
80	500			
	800	3162.2		

4–28 Complete this table for refrigerant-134a:

T, °C	P, kPa	v, m³/kg	Phase description
−8	320		
30		0.015	
	180		Saturated vapor
80	600		

4–29 Complete this table for refrigerant-134a:

T, °C	P, kPa	u, kJ/kg	Phase description
20		95	
−12			Saturated liquid
	400	300	
8	600		

4–30E Complete this table for refrigerant-134a:

T, °F	P, psia	h, Btu/lbm	x	Phase description
	80	78		
15			0.6	
10	70			
	180	129.46		
110			1.0	

4–31 A piston–cylinder device contains 0.85 kg of refrigerant-134a at −10°C. The piston that is free to move has a mass of 12 kg and a diameter of 25 cm. The local atmospheric pressure is 88 kPa. Now, heat is transferred to refrigerant-134a until the temperature is 15°C. Determine (*a*) the final pressure, (*b*) the change in the volume of the cylinder, and (*c*) the change in the enthalpy of the refrigerant-134a.

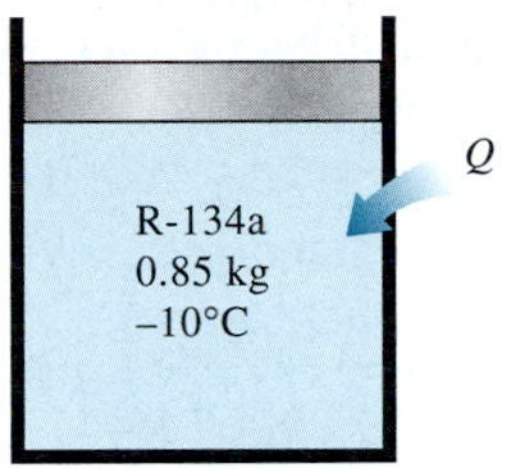

FIGURE P4–31

4–32E One pound-mass of water fills a 2.29-ft³ rigid container at an initial pressure of 250 psia. The container is then cooled to 100°F. Determine the initial temperature and final pressure of the water.

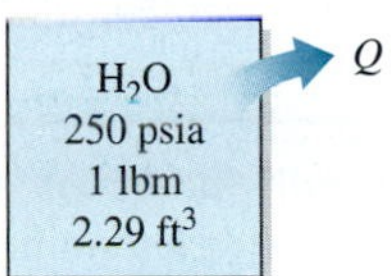

FIGURE P4–32E

4–33 One kilogram of R-134a fills a 0.14-m³ weighted piston-cylinder device at a temperature of –26.4°C. The container is now heated until the temperature is 100°C. Determine the final volume of the R-134a. *Answer:* 0.3014 m³

4–34 One kilogram of water vapor at 200 kPa fills the 1.1989-m³ left chamber of a partitioned system shown in Fig. P4–34. The right chamber has twice the volume of the left and is initially evacuated. Determine the pressure of the water after the partition has been removed and enough heat has been transferred so that the temperature of the water is 3°C.

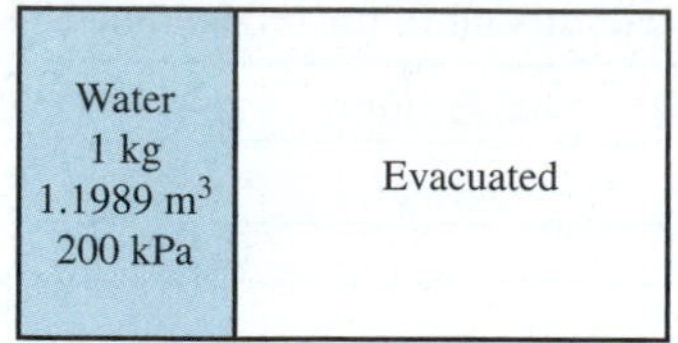

FIGURE P4–34

4–35E One pound-mass of water fills a 2.3615-ft^3 weighted piston-cylinder device at a temperature of 400°F. The piston-cylinder device is now cooled until its temperature is 100°F. Determine the final pressure of water, in psia, and the volume, in ft^3. *Answers:* 200 psia, 0.01613 ft^3

4–36 Ten kilograms of R-134a fill a 1.595-m^3 weighted piston-cylinder device at a temperature of –26.4°C. The container is now heated until the temperature is 100°C. Determine the final volume of the R-134a.

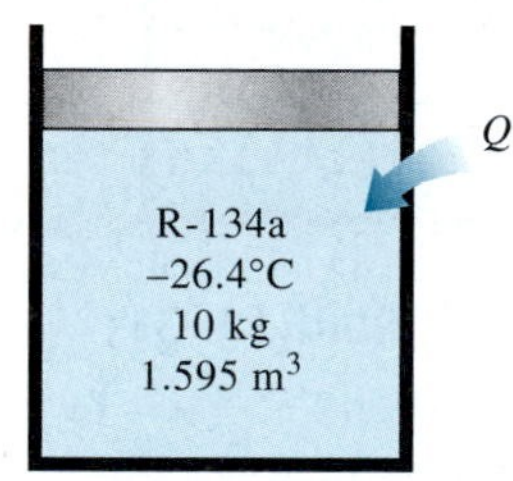

FIGURE P4–36

4–37 What is the specific internal energy of water at 50 kPa and 200°C?

4–38 What is the specific volume of water at 5 MPa and 100°C? What would it be if the incompressible-liquid approximation is used? Determine the accuracy of this approximation.

4–39E One pound-mass of water fills a container whose volume is 2 ft^3. The pressure in the container is 100 psia. Calculate the total internal energy and enthalpy in the container. *Answers:* 661 Btu, 698 Btu

4–40 Three kilograms of water in a container have a pressure of 100 kPa and temperature of 360°C. What is the volume of this container?

4–41 10-kg of R-134a at 300 kPa fills a rigid container whose volume is 14 L. Determine the temperature and total enthalpy in the container. The container is now heated until the pressure is 600 kPa. Determine the temperature and total enthalpy when the heating is completed.

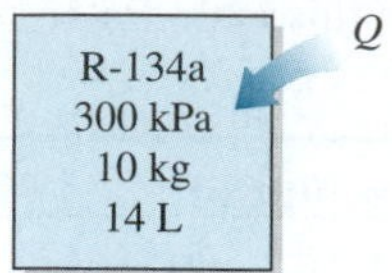

FIGURE P4–41

4–42 100-kg of R-134a at 200 kPa are contained in a piston-cylinder device whose volume is 12.322 m^3. The piston is now moved until the volume is one-half its original size. This is done such that the pressure of the R-134a does not change. Determine the final temperature and the change in the total internal energy of the R-134a.

4–43 The spring-loaded piston-cylinder device shown in Fig. P4–43 is filled with 0.5 kg of water vapor that is initially at 4 MPa and 400°C. Initially, the spring exerts no force against the piston. The spring constant in the spring force relation $F = kx$ is $k = 0.9$ kN/cm and the piston diameter is $D = 20$ cm. The water now undergoes a process until its volume is one-half of the original volume. Calculate the final temperature and the specific enthalpy of the water. *Answers:* 220°C, 1721 kJ/kg

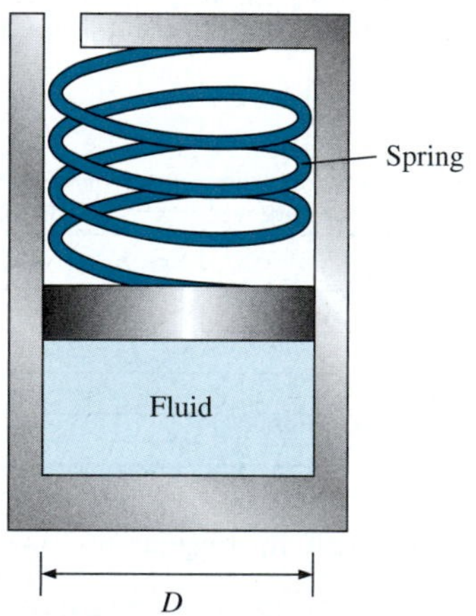

FIGURE P4–43

4–44E The atmospheric pressure at a location is usually specified at standard conditions, but it changes with the weather conditions. As the weather forecasters frequently state, the atmospheric pressure drops during stormy weather and it rises during clear and sunny days. If the pressure difference between the two extreme conditions is given to be 0.3 in of mercury, determine how much the boiling temperatures of water will vary as the weather changes from one extreme to the other.

4–45 A person cooks a meal in a 30-cm-diameter pot that is covered with a well-fitting lid and lets the food cool to the room temperature of 20°C. The total mass of the food and the pot is 8 kg. Now the person tries to open the pan by lifting the lid up. Assuming no air has leaked into the pan during cooling, determine if the lid will open or the pan will move up together with the lid.

4–46 Water is boiled at 1 atm pressure in a 25-cm-internal-diameter stainless steel pan on an electric range. If it is observed that the water level in the pan drops by 10 cm in 45 min, determine the rate of heat transfer to the pan.

4–47 Repeat Prob. 4–46 for a location at 2000-m elevation where the standard atmospheric pressure is 79.5 kPa.

4–48 Saturated steam coming off the turbine of a steam power plant at 30°C condenses on the outside of a 3-cm-outer-diameter, 35-m-long tube at a rate of 45 kg/h. Determine the rate of heat transfer from the steam to the cooling water flowing through the pipe.

4–49 Water in a 5-cm-deep pan is observed to boil at 98°C. At what temperature will the water in a 40-cm-deep pan boil? Assume both pans are full of water.

4–50 Water is being heated in a vertical piston–cylinder device. The piston has a mass of 20 kg and a cross-sectional area of 100 cm^2. If the local atmospheric pressure is 100 kPa, determine the temperature at which the water starts boiling.

4–51 A rigid tank with a volume of 2.5 m^3 contains 15 kg of saturated liquid–vapor mixture of water at 75°C. Now the water is slowly heated. Determine the temperature at which the liquid in the tank is completely vaporized. Also, show the process on a T-v diagram with respect to saturation lines. *Answer:* 187.0°C

4–52 A rigid vessel contains 2 kg of refrigerant-134a at 800 kPa and 120°C. Determine the volume of the vessel and the total internal energy. *Answers:* 0.0753 m^3, 655.7 kJ

4–53 A 0.5-m^3 vessel contains 10 kg of refrigerant-134a at −20°C. Determine (*a*) the pressure, (*b*) the total internal energy, and (*c*) the volume occupied by the liquid phase. *Answers:* (*a*) 132.82 kPa, (*b*) 904.2 kJ, (*c*) 0.00489 m^3

4–54 A piston–cylinder device contains 0.1 m^3 of liquid water and 0.9 m^3 of water vapor in equilibrium at 800 kPa. Heat is transferred at constant pressure until the temperature reaches 350°C.

(*a*) What is the initial temperature of the water?
(*b*) Determine the total mass of the water.
(*c*) Calculate the final volume.
(*d*) Show the process on a P-v diagram with respect to saturation lines.

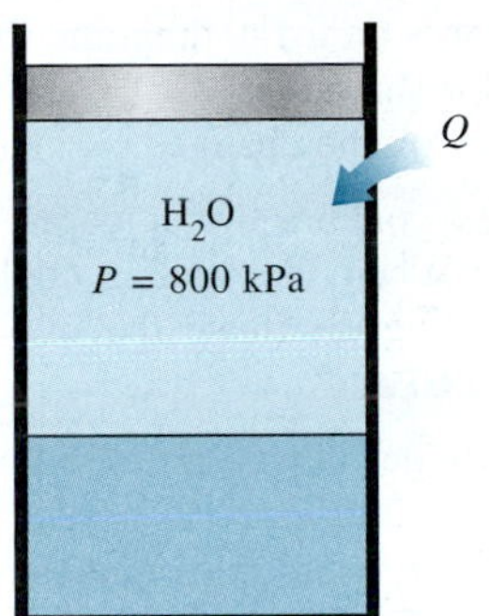

FIGURE P4–54

4–55 Reconsider Prob. 4–54. Using EES (or other) software, investigate the effect of pressure on the total mass of water in the tank. Let the pressure vary from 0.1 MPa to 1 MPa. Plot the total mass of water against pressure, and discuss the results. Also, show the process in Prob. 4–54 on a P-v diagram using the property plot feature of EES.

4–56E Superheated water vapor at 180 psia and 500°F is allowed to cool at constant volume until the temperature drops to 250°F. At the final state, determine (*a*) the pressure, (*b*) the quality, and (*c*) the enthalpy. Also, show the process on a T-v diagram with respect to saturation lines. *Answers:* (*a*) 29.84 psia, (*b*) 0.219, (*c*) 426.0 Btu/lbm

4–57E Reconsider Prob. 4–56E. Using EES (or other) software, investigate the effect of initial pressure on the quality of water at the final state. Let the pressure vary from 100 psi to 300 psi. Plot the quality against initial pressure, and discuss the results. Also, show the process in Prob. 4–56E on a T-v diagram using the property plot feature of EES.

4–58 A 0.3-m^3 rigid vessel initially contains saturated liquid–vapor mixture of water at 150°C. The water is now heated until it reaches the critical state. Determine the mass of the liquid water and the volume occupied by the liquid at the initial state. *Answers:* 96.10 kg, 0.105 m^3

4–59 Determine the specific volume, internal energy, and enthalpy of compressed liquid water at 100°C and 15 MPa using the saturated liquid approximation. Compare these values to the ones obtained from the compressed liquid tables.

4–60 Reconsider Prob. 4–59. Using EES (or other) software, determine the indicated properties of compressed liquid, and compare them to those obtained using the saturated liquid approximation.

4–61 A piston–cylinder device contains 0.8 kg of steam at 300°C and 1 MPa. Steam is cooled at constant pressure until one-half of the mass condenses.

(*a*) Show the process on a *T*-*v* diagram.
(*b*) Find the final temperature.
(*c*) Determine the volume change.

4–62 A rigid tank contains water vapor at 250°C and an unknown pressure. When the tank is cooled to 150°C, the vapor starts condensing. Estimate the initial pressure in the tank. *Answer:* 0.60 MPa

4–63 A piston-cylinder device initially contains 1.4-kg saturated liquid water at 200°C. Now heat is transferred to the water until the volume quadruples and the cylinder contains saturated vapor only. Determine (*a*) the volume of the tank, (*b*) the final temperature and pressure, and (*c*) the internal energy change of the water.

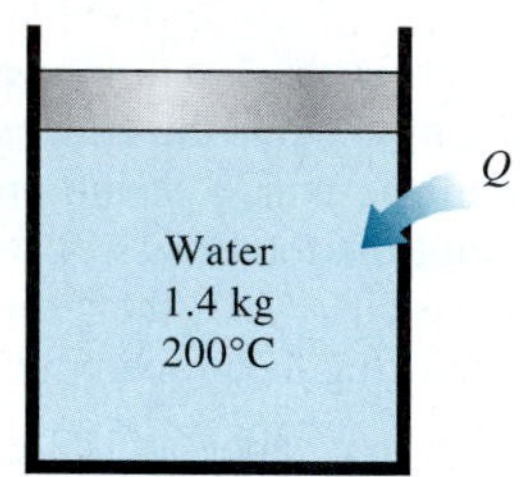

FIGURE P4–63

4–64 A piston–cylinder device initially contains steam at 3.5 MPa, superheated by 5°C. Now, steam loses heat to the surroundings and the piston moves down hitting a set of stops at which point the cylinder contains saturated liquid water. The cooling continues until the cylinder contains water at 200°C. Determine (*a*) the initial temperature, (*b*) the enthalpy change per unit mass of the steam by the time the piston first hits the stops, and (*c*) the final pressure and the quality (if mixture).

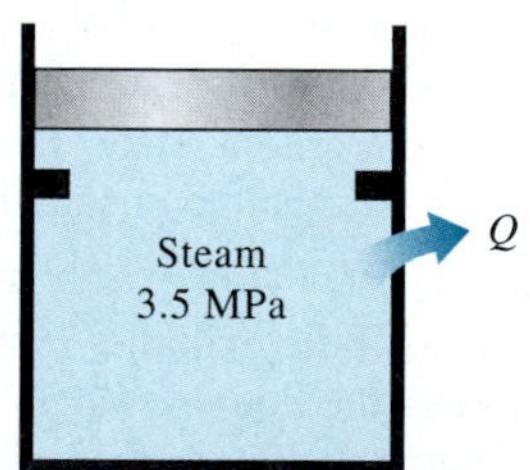

FIGURE P4–64

4–65E How much error would one expect in determining the specific enthalpy by applying the incompressible-liquid approximation to water at 1500 psia and 400°F?

4–66 How much error would result in calculating the specific volume and enthalpy of water at 10 MPa and 100°C by using the incompressible-liquid approximation?

4–67 What is the specific volume of R-134a at 20°C and 700 kPa? What is the internal energy at that state?

4–68 100 grams of R-134a initially fill a weighted piston-cylinder device at 60 kPa and −20°C. The device is then heated until the temperature is 100°C. Determine the change in the device's volume as a result of the heating. *Answer:* 0.0168 m^3

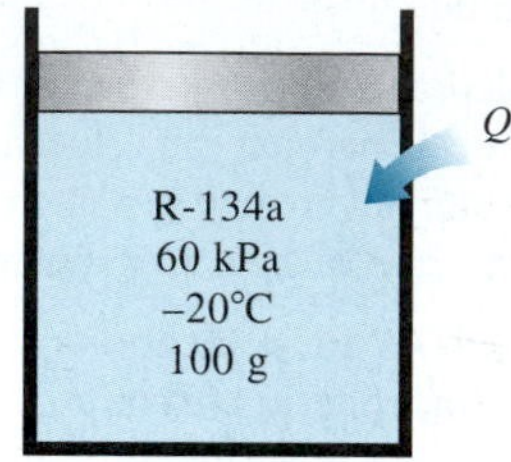

FIGURE P4–68

4–69 [EES] Pressure-enthalpy state diagrams are useful for studying refrigeration and like systems. Using EES (or other) software and actual property data, plot a pressure-enthalpy diagram for R-134a that includes the saturation lines. Also, sketch isothermal and constant-entropy processes on this diagram.

Ideal Gas

4–70C Propane and methane are commonly used for heating in winter, and the leakage of these fuels, even for short periods, poses a fire danger for homes. Which gas leakage do you think poses a greater risk for fire? Explain.

4–71C Under what conditions is the ideal-gas assumption suitable for real gases?

4–72C What is the difference between R and R_u? How are these two related?

4–73C What is the difference between mass and molar mass? How are these two related?

4–74E What is the specific volume of oxygen at 25 psia and 80°F?

4–75 A 100-L container is filled with 1 kg of air at a temperature of 27°C. What is the pressure in the container?

4–76E A mass of 1-lbm of argon is maintained at 200 psia and 100°F in a tank. What is the volume of the tank?

4–77 A spherical balloon with a diameter of 6 m is filled with helium at 20°C and 200 kPa. Determine the mole number and the mass of the helium in the balloon. *Answers:* 9.28 kmol, 37.15 kg

4–78 [EES] Reconsider Prob. 4–77. Using EES (or other) software, investigate the effect of the balloon diameter on the mass of helium contained in the balloon for the pressures of (*a*) 100 kPa and (*b*) 200 kPa. Let the diameter vary from 5 m to 15 m. Plot the mass of helium against the diameter for both cases.

4–79 The pressure in an automobile tire depends on the temperature of the air in the tire. When the air temperature is 25°C, the pressure gage reads 210 kPa. If the volume of the tire is 0.025 m^3, determine the pressure rise in the tire when the air temperature in the tire rises to 50°C. Also, determine the amount of air that must be bled off to restore pressure to its original value at this temperature. Assume the atmospheric pressure is 100 kPa.

FIGURE P4–79

4–80 A 1-m^3 tank containing air at 25°C and 500 kPa is connected through a valve to another tank containing 5 kg of air at 35°C and 200 kPa. Now the valve is opened, and the entire system is allowed to reach thermal equilibrium with the surroundings, which are at 20°C. Determine the volume of the second tank and the final equilibrium pressure of air. *Answers:* 2.21 m^3, 284.1 kPa

4–81E In an informative article in a magazine it is stated that tires lose roughly 1 psi of pressure for every 10°F drop in outside temperature. Investigate if this is a valid statement.

4–82 A mass of 10-g of oxygen fill a weighted piston-cylinder device at 20 kPa and 100°C. The device is now cooled until the temperature is 0°C. Determine the change of the volume of the device during this cooling.

4–83 A mass of 0.1 kg of helium fills a 0.2 m^3 rigid vessel at 350 kPa. The vessel is heated until the pressure is 700 kPa. Calculate the temperature change of helium (in °C and K) as a result of this heating. *Answers:* 337°C, 337 K

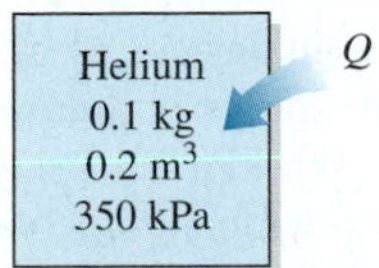

FIGURE P4–83

4–84 Argon in the amount of 0.2 kg fills a 0.05-m^3 piston-cylinder device at 400 kPa. The piston is now moved by changing the weights until the volume is twice its original size. During this process, argon's temperature is maintained constant. Determine the final pressure in the device.

Compressibility Factor

4–85C What is the physical significance of the compressibility factor Z?

4–86C What is the principle of corresponding states?

4–87C How are the reduced pressure and reduced temperature defined?

4–88 Determine the specific volume of superheated water vapor at 10 MPa and 400°C, using (*a*) the ideal-gas equation, (*b*) the generalized compressibility chart, and (*c*) the steam tables. Also determine the error involved in the first two cases. *Answers:* (*a*) 0.03106 m^3/kg, 17.6 percent; (*b*) 0.02609 m^3/kg, 1.2 percent; (*c*) 0.02644 m^3/kg

4–89 EES Reconsider Prob. 4–88. Solve the problem using the generalized compressibility factor feature of the EES software. Again using EES, compare the specific volume of water for the three cases at 10 MPa over the temperature range of 325 to 600°C in 25°C intervals. Plot the percent error involved in the ideal-gas approximation against temperature, and discuss the results.

4–90 Determine the specific volume of refrigerant-134a vapor at 0.9 MPa and 70°C based on (*a*) the ideal-gas equation, (*b*) the generalized compressibility chart, and (*c*) data from tables. Also, determine the error involved in the first two cases.

4–91 Determine the specific volume of nitrogen gas at 10 MPa and 150 K based on (*a*) the ideal-gas equation and (*b*) the generalized compressibility chart. Compare these results with the experimental value of 0.002388 m^3/kg, and determine the error involved in each case. *Answers:* (*a*) 0.004452 m^3/kg, 86.4 percent; (*b*) 0.002404 m^3/kg, 0.7 percent

4–92 Determine the specific volume of superheated water vapor at 3.5 MPa and 450°C based on (*a*) the ideal-gas equation, (*b*) the generalized compressibility chart, and (*c*) the steam tables. Determine the error involved in the first two cases.

4–93E Ethane in a rigid vessel is to be heated from 50 psia and 100°F until its temperature is 600°F. What is the final pressure of the ethane as predicted by the compressibility chart?

4–94 Ethylene is heated at constant pressure from 5 MPa and 20°C to 200°C. Using the compressibility chart, determine the change in the ethylene's specific volume as a result of this heating. *Answer:* 0.0172 m^3/kg

4–95 Saturated water vapor at 350°C is heated at constant pressure until its volume has doubled. Determine the final temperature using the ideal gas equation of state, the compressibility charts, and the steam tables.

4–96E Saturated water vapor at 400°F is heated at constant pressure until its volume has doubled. Determine the final temperature using the ideal gas equation of state, the compressibility charts, and the steam tables.

4–97 Methane at 8 MPa and 300 K is heated at constant pressure until its volume has increased by 50 percent. Determine the final temperature using the ideal gas equation of state and the compressibility factor. Which of these two results is more accurate?

4–98 What is the percentage of error involved in treating carbon dioxide at 3 MPa and 10°C as an ideal gas? *Answer:* 25 percent

4–99 Carbon dioxide gas enters a pipe at 3 MPa and 500 K at a rate of 2 kg/s. CO_2 is cooled at constant pressure as it flows in the pipe and the temperature of CO_2 drops to 450 K at the exit. Determine the volume flow rate and the density of carbon dioxide at the inlet and the volume flow rate at the exit of the pipe using (*a*) the ideal-gas equation and (*b*) the generalized compressibility chart. Also, determine (*c*) the error involved in the first case.

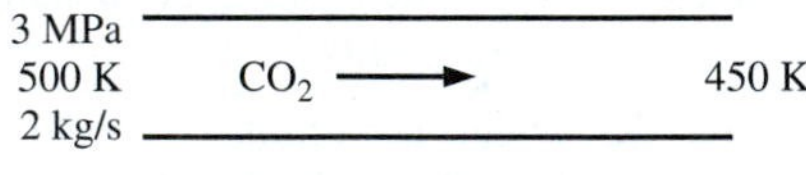

FIGURE P4–99

Other Equations of State

4–100C What is the physical significance of the two constants that appear in the van der Waals equation of state? On what basis are they determined?

4–101E 1-lbm of carbon dioxide is heated in a constant pressure apparatus. Initially, the carbon dioxide is at 1000 psia and 200°F and it is heated until its temperature becomes 800°F. Determine the final volume of the carbon dioxide treating it as (*a*) an ideal gas and (*b*) a Benedict-Webb-Rubin gas.

4–102 Methane is heated in a rigid container from 100 kPa and 20°C to 400°C. Determine the final pressure of the methane treating it as (*a*) an ideal gas and (*b*) a Benedict-Webb-Rubin gas.

4–103E Carbon monoxide is heated in a rigid container from 14.7 psia and 70°F to 800°F. Determine the final pressure of the carbon monoxide treating it as (*a*) an ideal gas and (*b*) a Benedict-Webb-Rubin gas.

4–104 1-kg of carbon dioxide is compressed from 1 MPa and 200°C to 3 MPa in a piston-cylinder device arranged to execute a polytropic process for which $PV^{1.2}$ = constant. Determine the final temperature treating the carbon dioxide as (*a*) an ideal gas and (*b*) a van der Waals gas.

4–105E Refrigerant-134a at 100 psia has a specific volume of 0.54022 ft³/lbm. Determine the temperature of the refrigerant based on (*a*) the ideal-gas equation, (*b*) the van der Waals equation, and (*c*) the refrigerant tables.

4–106 Nitrogen at 150 K has a specific volume of 0.041884 m³/kg. Determine the pressure of the nitrogen, using (*a*) the ideal-gas equation and (*b*) the Beattie-Bridgeman equation. Compare your results to the experimental value of 1000 kPa. *Answers:* (*a*) 1063 kPa, (*b*) 1000.4 kPa

4–107 Reconsider Prob. 4–106. Using EES (or other) software, compare the pressure results of the ideal-gas and Beattie-Bridgeman equations with nitrogen data supplied by EES. Plot temperature versus specific volume for a pressure of 1000 kPa with respect to the saturated liquid and saturated vapor lines of nitrogen over the range of 110 K $< T <$ 150 K.

Review Problems

4–108 The combustion in a gasoline engine may be approximated by a constant volume heat addition process. There exists the air–fuel mixture in the cylinder before the combustion and the combustion gases after it, and both may be approximated as air, an ideal gas. In a gasoline engine, the cylinder conditions are 1.8 MPa and 450°C before the combustion and 1300°C after it. Determine the pressure at the end of the combustion process. *Answer:* 3916 kPa

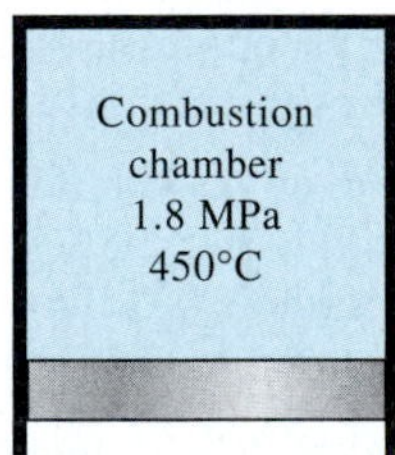

FIGURE P4–108

4–109 A rigid tank contains an ideal gas at 300 kPa and 600 K. Now half of the gas is withdrawn from the tank and the gas is found at 100 kPa at the end of the process. Determine (*a*) the final temperature of the gas and (*b*) the final pressure if no mass was withdrawn from the tank and the same final temperature was reached at the end of the process.

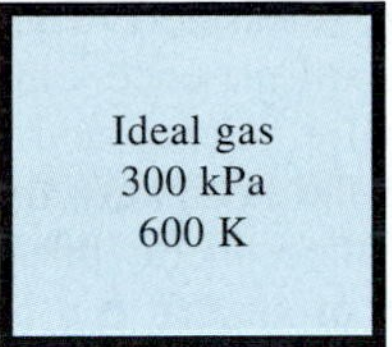

FIGURE P4–109

4–110 Carbon-dioxide gas at 3 MPa and 500 K flows steadily in a pipe at a rate of 0.4 kmol/s. Determine (*a*) the volume and mass flow rates and the density of carbon dioxide at this state. If CO_2 is cooled at constant pressure as it flows in the pipe so that the temperature of CO_2 drops to 450 K at the exit of the pipe, determine (*b*) the volume flow rate at the exit of the pipe.

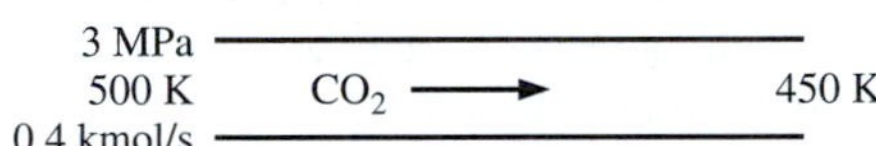

FIGURE P4–110

4–111 Combustion in a diesel engine may be modeled as a constant-pressure heat addition process with air in the cylinder before and after combustion. Consider a diesel engine with cylinder conditions of 950 K and 75 cm^3 before combustion, and 150 cm^3 after it. The engine operates with an air–fuel ratio of 22 kg air/kg fuel (the mass of the air divided by the mass of the fuel). Determine the temperature after the combustion process.

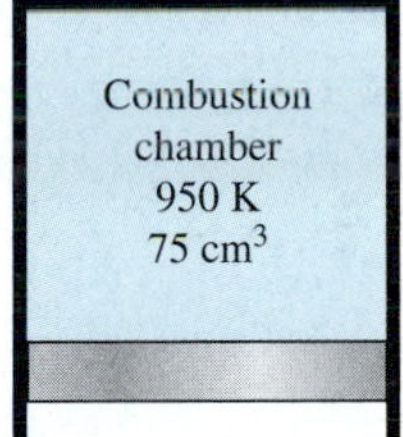

FIGURE P4–111

4–112 One kilogram of R-134a fills a 0.1450 m^3 rigid container at an initial temperature of −40°C. The container is then heated until the pressure is 200 kPa. Determine the initial pressure and final temperature. *Answers:* 51.25 kPa, 90°C

4–113E One pound-mass of water fills a 2.649 ft^3 weighted piston-cylinder device at a temperature of 400°F. The piston-cylinder device is now cooled until its temperature is 100°F. Determine the final pressure and volume of the water.

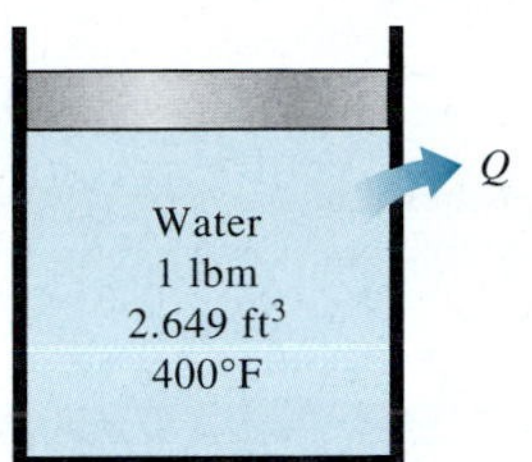

FIGURE P4–113E

4–114 The piston diameters in Fig. P4–114 are D_1 = 10 cm and D_2 = 4 cm. Chamber 1 contains 1 kg of helium, chamber 2 is filled with condensing water vapor, and chamber 3 is evacuated. The entire assembly is placed in an environment whose temperature is 200°C. Determine the volume of chamber 1 when thermodynamic equilibrium has been established. *Answer:* 3.95 m^3

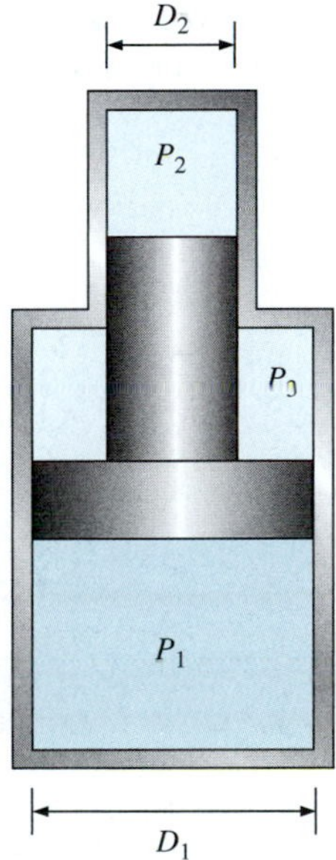

FIGURE P4–114

4–115 Liquid propane is commonly used as a fuel for heating homes, powering vehicles such as forklifts, and filling portable picnic tanks. Consider a propane tank that initially contains 5 L of liquid propane at the environment temperature of 20°C. If a hole develops in the connecting tube of a propane tank and the propane starts to leak out, determine the temperature of propane when the pressure in the tank drops to 1 atm. Also, determine the total amount of heat transfer from the environment to the tank to vaporize the entire propane in the tank.

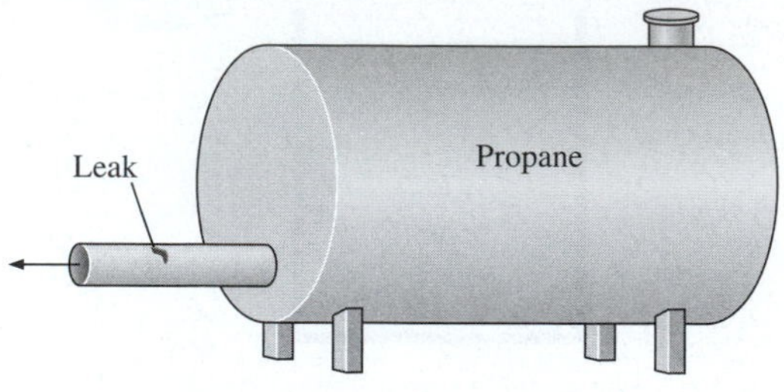

FIGURE P4–115

4–116 Repeat Prob. 4–115 for isobutane.

4–117 A tank contains helium at 100°C and 10 kPa gage. The helium is heated in a process by heat transfer from the surroundings such that the helium reaches a final equilibrium state at 300°C. Determine the final gage pressure of the helium. Assume atmospheric pressure is 100 kPa.

Design and Essay Problems

4–118 In an article on tire maintenance, it is stated that tires lose air over time, and pressure losses as high as 90 kPa (13 psi) per year are measured. The article recommends checking tire pressure at least once a month to avoid low tire pressure that hurts fuel efficiency and causes uneven thread wear on tires. Taking the beginning tire pressure to be 220 kPa (gage) and the atmospheric pressure to be 100 kPa, determine the fraction of air that can be lost from a tire per year.

4–119 It is well known that water freezes at 0°C at atmospheric pressure. The mixture of liquid water and ice at 0°C is said to be at stable equilibrium since it cannot undergo any changes when it is isolated from its surroundings. However, when water is free of impurities and the inner surfaces of the container are smooth, the temperature of water can be lowered to −2°C or even lower without any formation of ice at atmospheric pressure. But at that state even a small disturbance can initiate the formation of ice abruptly, and the water temperature stabilizes at 0°C following this sudden change. The water at −2°C is said to be in a *metastable state.* Write an essay on metastable states and discuss how they differ from stable equilibrium states.

Chapter 5

ENERGY ANALYSIS OF CLOSED SYSTEMS

In Chap. 3, we considered various forms of energy and energy transfer, and we developed a general relation for the conservation of energy principle or energy balance. Then in Chap. 4, we learned how to determine the thermodynamics properties of substances. In this chapter, we apply the energy balance relation to systems that do not involve any mass flow across their boundaries; that is, closed systems.

We start this chapter with a discussion of the *moving boundary work* or *P dV work* commonly encountered in reciprocating devices such as automotive engines and compressors. We continue by applying the *general energy balance* relation, which is simply expressed as $E_{in} - E_{out} = \Delta E_{system}$, to systems that involve a pure substance. Then we define *specific heats*, obtain relations for the internal energy and enthalpy of *ideal gases* in terms of specific heats and temperature changes, and perform energy balances on various systems that involve ideal gases. We repeat this for systems that involve solids and liquids, which are approximated as *incompressible substances.*

const v.

$Q_{in} = \Delta U$

$C_v = \frac{Q_{in}}{m \Delta T} = \frac{\Delta U}{m \Delta T}$

const. P

$Q_{in} = \Delta U + W_{out}$

$Q_{in} = (U_2 + PV_2) - (U_1 + PV_1)$

$C_p = \frac{Q_{in}}{m \Delta T} = \frac{\Delta H}{m \Delta T}$

Objectives

The objectives of this chapter are to:

- Examine the moving boundary work or *P dV* work commonly encountered in reciprocating devices such as automotive engines and compressors.
- Identify the first law of thermodynamics as simply a statement of the conservation of energy principle for closed (fixed mass) systems.
- Develop the general energy balance applied to closed systems.
- Define the specific heat at constant volume and the specific heat at constant pressure.
- Relate the specific heats to the calculation of the changes in internal energy and enthalpy of ideal gases.
- Describe incompressible substances and determine the changes in their internal energy and enthalpy.
- Solve energy balance problems for closed (fixed mass) systems that involve heat and work interactions for general pure substances, ideal gases, and incompressible substances.

5–1 ▪ MOVING BOUNDARY WORK

One form of mechanical work frequently encountered in practice is associated with the expansion or compression of a gas in a piston–cylinder device. During this process, part of the boundary (the inner face of the piston) moves back and forth. Therefore, the expansion and compression work is often called **moving boundary work**, or simply **boundary work** (Fig. 5–1). Some call it the $P\,dV$ work for reasons explained later. Moving boundary work is the primary form of work involved in *automobile engines*. During their expansion, the combustion gases force the piston to move, which in turn forces the crankshaft to rotate.

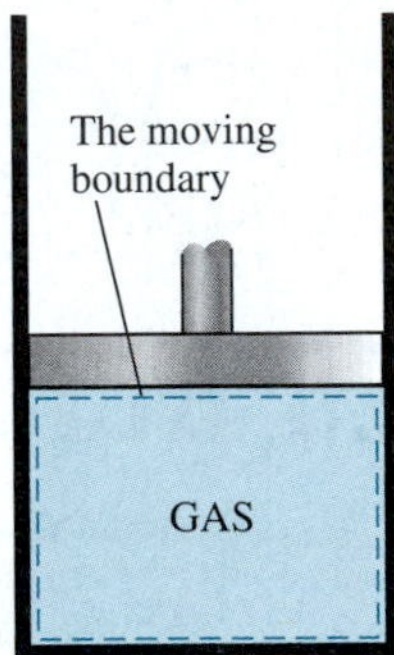

FIGURE 5–1

The work associated with a moving boundary is called *boundary work.*

The moving boundary work associated with real engines or compressors cannot be determined exactly from a thermodynamic analysis alone because the piston usually moves at very high speeds, making it difficult for the gas inside to maintain equilibrium. Then the states through which the system passes during the process cannot be specified, and no process path can be drawn. Work, being a path function, cannot be determined analytically without a knowledge of the path. Therefore, the boundary work in real engines or compressors is determined by direct measurements.

In this section, we analyze the moving boundary work for a *quasi-equilibrium process*, a process during which the system remains nearly in equilibrium at all times. A quasi-equilibrium process, also called a *quasi-static process*, is closely approximated by real engines when the piston moves at low velocities. Under identical conditions, the work output of the engines is found to be a maximum, and the work input to the compressors to be a minimum when quasi-equilibrium processes are used in place of nonquasi-equilibrium processes. Below, the work associated with a moving boundary is evaluated for a quasi-equilibrium process.

Consider the gas enclosed in the piston–cylinder device shown in Fig. 5–2. The initial pressure of the gas is P, the total volume is V, and the cross-sectional area of the piston is A. If the piston is allowed to move a distance ds in a quasi-equilibrium manner, the differential work done during this process is

$$\delta W_b = F\,ds = PA\,ds = P\,dV \tag{5–1}$$

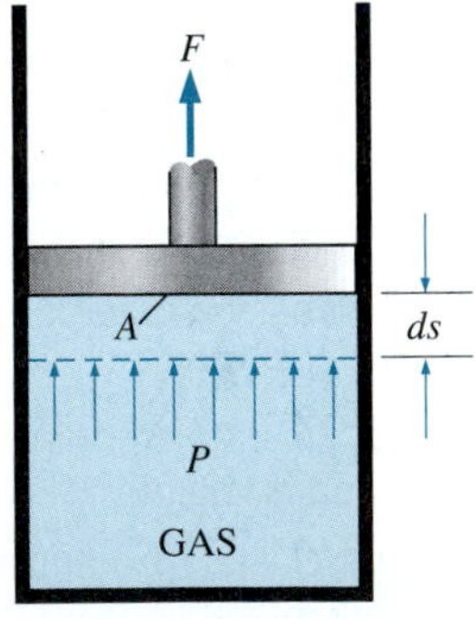

FIGURE 5–2

A gas does a differential amount of work δW_b as it forces the piston to move by a differential amount ds.

That is, the boundary work in the differential form is equal to the product of the absolute pressure P and the differential change in the volume dV of the system. This expression also explains why the moving boundary work is sometimes called the $P\,dV$ work.

Note in Eq. 5–1 that P is the absolute pressure, which is always positive. However, the volume change dV is positive during an expansion process (volume increasing) and negative during a compression process (volume decreasing). Thus, the boundary work is positive during an expansion process and negative during a compression process. Therefore, Eq. 5–1 can be viewed as an expression for boundary work output, $W_{b,\text{out}}$. A negative result indicates boundary work input (compression).

The total boundary work done during the entire process as the piston moves is obtained by adding all the differential works from the initial state to the final state:

$$W_b = \int_1^2 P\,dV \qquad \text{(kJ)} \tag{5–2}$$

This integral can be evaluated only if we know the functional relationship between P and V during the process. That is, $P = f(V)$ should be available. Note that $P = f(V)$ is simply the equation of the process path on a P-V diagram.

The quasi-equilibrium expansion process described is shown on a P-V diagram in Fig. 5–3. On this diagram, the differential area dA is equal to $P\,dV$, which is the differential work. The total area A under the process curve 1–2 is obtained by integration:

$$\text{Area} = A = \int_1^2 dA = \int_1^2 P\,dV \tag{5-3}$$

A comparison of this equation with Eq. 5–2 reveals that *the area under the process curve on a P-V diagram is equal, in magnitude, to the work done during a quasi-equilibrium expansion or compression process of a closed system.* (On the P-v diagram, it represents the boundary work done per unit mass.)

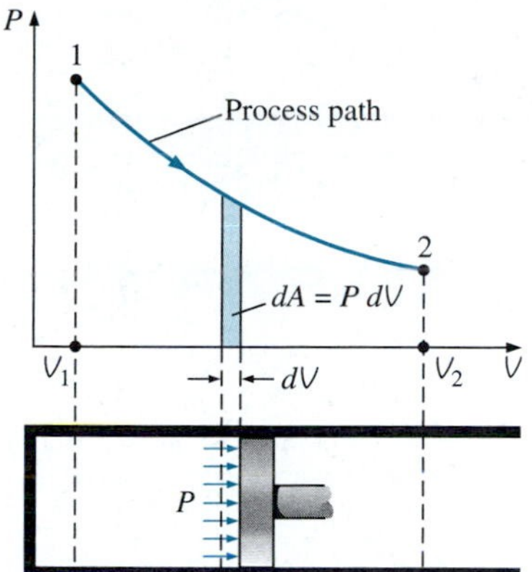

FIGURE 5–3

The area under the process curve on a P-V diagram represents the boundary work.

A gas can follow several different paths as it expands from state 1 to state 2. In general, each path has a different area underneath it, and since this area represents the magnitude of the work, the work done is different for each process (Fig. 5–4). This is expected, since work is a path function (i.e., it depends on the path followed as well as the end states). If work were not a path function, no cyclic devices (car engines, power plants) could operate as work-producing devices. The work produced by these devices during one part of the cycle would have to be consumed during another part, and there would be no net work output. The cycle shown in Fig. 5–5 produces a net work output because the work done by the system during the expansion process (area under path A) is greater than the work done on the system during the compression part of the cycle (area under path B), and the difference between these two is the net work done during the cycle (the colored area).

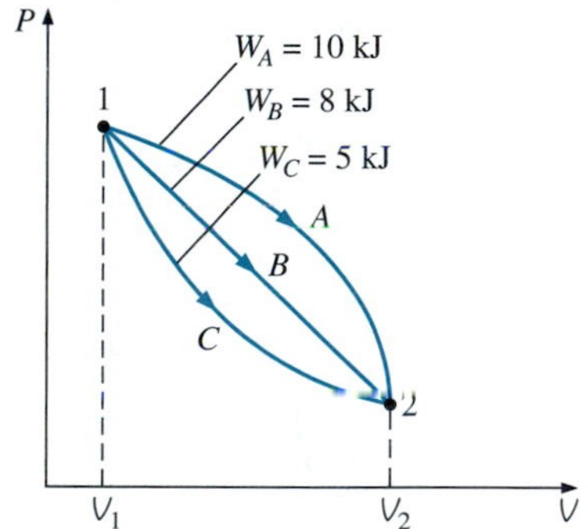

FIGURE 5–4

The boundary work done during a process depends on the path followed as well as the end states.

If the relationship between P and V during an expansion or a compression process is given in terms of experimental data instead of in a functional form, obviously we cannot perform the integration analytically. But we can always plot the P-V diagram of the process, using these data points, and calculate the area underneath graphically to determine the work done.

Strictly speaking, the pressure P in Eq. 5–2 is the pressure at the inner surface of the piston. It becomes equal to the pressure of the gas in the cylinder only if the process is quasi-equilibrium and thus the entire gas in the cylinder is at the same pressure at any given time. Equation 5–2 can also be used for nonquasi-equilibrium processes provided that the pressure *at the inner face of the piston* is used for P. (Besides, we cannot speak of the pressure of a *system* during a nonquasi-equilibrium process since properties are defined for equilibrium states only.) Therefore, we generalize the boundary work relation by expressing it as

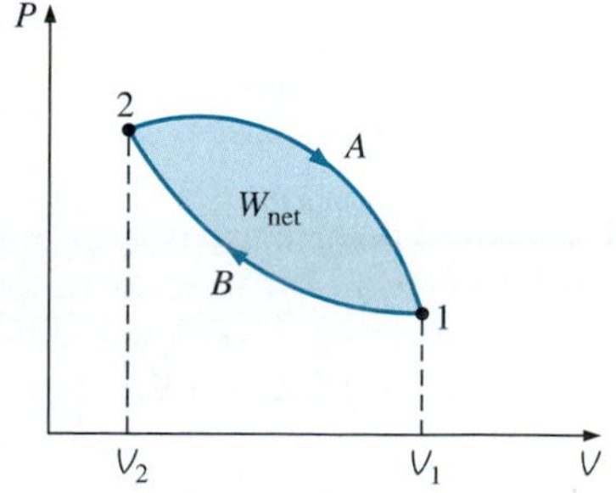

FIGURE 5–5

The net work done during a cycle is the difference between the work done by the system and the work done on the system.

$$W_b = \int_1^2 P_i\,dV \tag{5-4}$$

where P_i is the pressure at the inner face of the piston.

Note that work is a mechanism for energy interaction between a system and its surroundings, and W_b represents the amount of energy transferred from the system during an expansion process (or to the system during a compression process). Therefore, it has to appear somewhere else and we must be able to account for it since energy is conserved. In a car engine, for example, the boundary work done by the expanding hot gases is used to overcome friction between the piston and the cylinder, to push atmospheric air out of the way, and to rotate the crankshaft. Therefore,

$$W_b = W_{\text{friction}} + W_{\text{atm}} + W_{\text{crank}} = \int_1^2 (F_{\text{friction}} + P_{\text{atm}}A + F_{\text{crank}})\,dx \quad \text{(5–5)}$$

Of course the work used to overcome friction appears as frictional heat and the energy transmitted through the crankshaft is transmitted to other components (such as the wheels) to perform certain functions. But note that the energy transferred by the system as work must equal the energy received by the crankshaft, the atmosphere, and the energy used to overcome friction.

The use of the boundary work relation is not limited to the quasi-equilibrium processes of gases only. It can also be used for solids and liquids.

EXAMPLE 5–1 Boundary Work for a Constant-Volume Process

A rigid tank contains air at 500 kPa and 150°C. As a result of heat transfer to the surroundings, the temperature and pressure inside the tank drop to 65°C and 400 kPa, respectively. Determine the boundary work done during this process.

Solution Air in a rigid tank is cooled, and both the pressure and temperature drop. The boundary work done is to be determined.

Analysis A sketch of the system and the *P-V* diagram of the process are shown in Fig. 5–6. The boundary work can be determined from Eq. 5–2 to be

$$W_b = \int_1^2 P\,\cancelto{0}{dV} = \mathbf{0}$$

Discussion This is expected since a rigid tank has a constant volume and $dV = 0$ in this equation. Therefore, there is no boundary work done during this process. That is, the boundary work done during a constant-volume process is always zero. This is also evident from the *P-V* diagram of the process (the area under the process curve is zero).

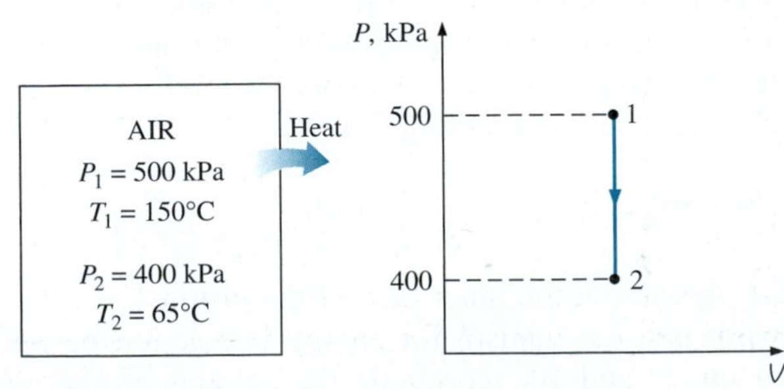

FIGURE 5–6
Schematic and *P-V* diagram for Example 5–1.

EXAMPLE 5–2 Boundary Work for a Constant-Pressure Process

A frictionless piston–cylinder device contains 10 lbm of steam at 60 psia and 320°F. Heat is now transferred to the steam until the temperature reaches 400°F. If the piston is not attached to a shaft and its mass is constant, determine the work done by the steam during this process.

Solution Steam in a piston cylinder device is heated and the temperature rises at constant pressure. The boundary work done is to be determined.

Analysis A sketch of the system and the *P-v* diagram of the process are shown in Fig. 5–7.

Assumption The expansion process is quasi-equilibrium.

Analysis Even though it is not explicitly stated, the pressure of the steam within the cylinder remains constant during this process since both the atmospheric pressure and the weight of the piston remain constant. Therefore, this is a constant-pressure process, and, from Eq. 5–2

$$W_b = \int_1^2 P\,dV = P_0 \int_1^2 dV = P_0(V_2 - V_1) \tag{5-6}$$

or

$$W_b = mP_0(v_2 - v_1)$$

since $V = mv$. From the superheated vapor table (Table A–6E), the specific volumes are determined to be $v_1 = 7.4863$ ft³/lbm at state 1 (60 psia, 320°F) and $v_2 = 8.3548$ ft³/lbm at state 2 (60 psia, 400°F). Substituting these values yields

$$W_b = (10\text{ lbm})(60\text{ psia})[(8.3548 - 7.4863)\text{ ft}^3/\text{lbm}]\left(\frac{1\text{ Btu}}{5.404\text{ psia}\cdot\text{ft}^3}\right)$$

$$= \mathbf{96.4\ Btu}$$

Discussion The positive sign indicates that the work is done by the system. That is, the steam uses 96.4 Btu of its energy to do this work. The magnitude of this work could also be determined by calculating the area under the process curve on the *P-V* diagram, which is simply $P_0\,\Delta V$ for this case.

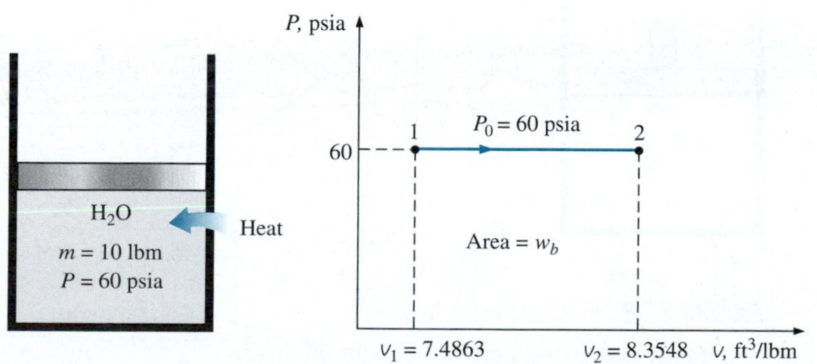

FIGURE 5–7
Schematic and *P-v* diagram for Example 5–2.

EXAMPLE 5–3 Isothermal Compression of an Ideal Gas

A piston–cylinder device initially contains 0.4 m^3 of air at 100 kPa and 80°C. The air is now compressed to 0.1 m^3 in such a way that the temperature inside the cylinder remains constant. Determine the work done during this process.

Solution Air in a piston–cylinder device is compressed isothermally. The boundary work done is to be determined.

Analysis A sketch of the system and the *P-V* diagram of the process are shown in Fig. 5–8.

Assumptions **1** The compression process is quasi-equilibrium. **2** At specified conditions, air can be considered to be an ideal gas since it is at a high temperature and low pressure relative to its critical-point values.

Analysis For an ideal gas at constant temperature T_0,

$$PV = mRT_0 = C \quad \text{or} \quad P = \frac{C}{V}$$

where C is a constant. Substituting this into Eq. 5–2, we have

$$W_b = \int_1^2 P\,dV = \int_1^2 \frac{C}{V}\,dV = C\int_1^2 \frac{dV}{V} = C\ln\frac{V_2}{V_1} = P_1V_1\ln\frac{V_2}{V_1} \tag{5–7}$$

In Eq. 5–7, P_1V_1 can be replaced by P_2V_2 or mRT_0. Also, V_2/V_1 can be replaced by P_1/P_2 for this case since $P_1V_1 = P_2V_2$.

Substituting the numerical values into Eq. 5–7 yields

$$W_b = (100\text{ kPa})(0.4\text{ m}^3)\left(\ln\frac{0.1}{0.4}\right)\left(\frac{1\text{ kJ}}{1\text{ kPa}\cdot\text{m}^3}\right)$$

$$= \mathbf{-55.5\ kJ}$$

Discussion The negative sign indicates that this work is done on the system (a work input), which is always the case for compression processes.

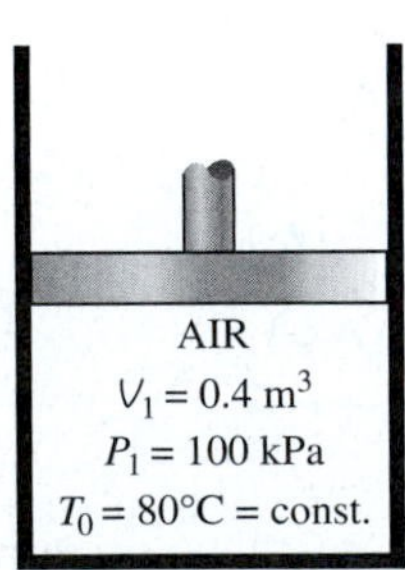

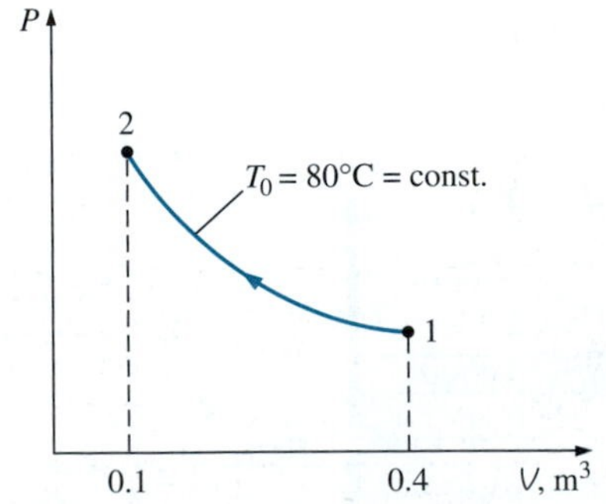

FIGURE 5–8

Schematic and *P-V* diagram for Example 5–3.

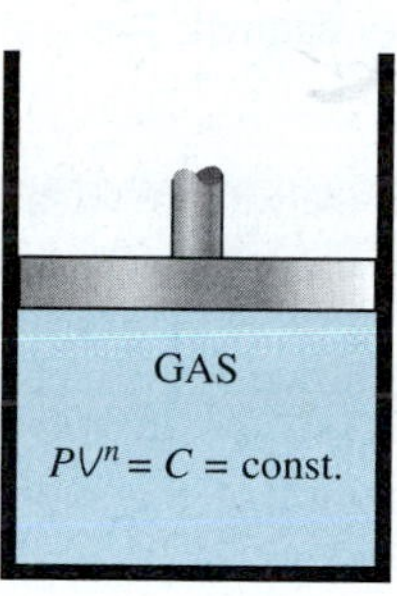

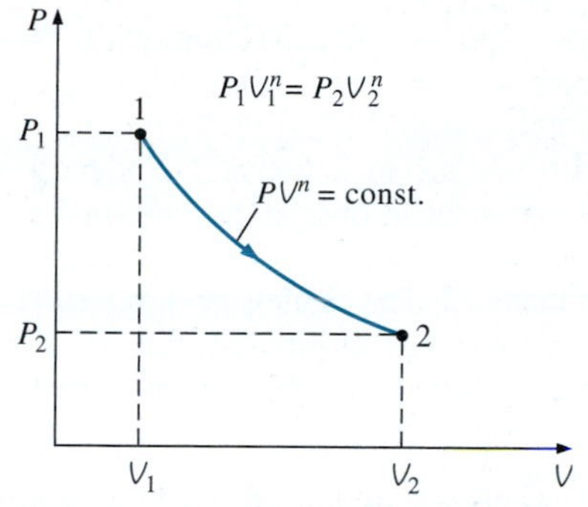

FIGURE 5–9

Schematic and P-V diagram for a polytropic process.

Polytropic Process

During actual expansion and compression processes of gases, pressure and volume are often related by $PV^n = C$, where n and C are constants. A process of this kind is called a **polytropic process** (Fig. 5–9). Below we develop a general expression for the work done during a polytropic process. The pressure for a polytropic process can be expressed as

$$P = CV^{-n} \tag{5-8}$$

Substituting this relation into Eq. 5–2, we obtain

$$W_b = \int_1^2 P\,dV = \int_1^2 CV^{-n}\,dV = C\frac{V_2^{-n+1} - V_1^{-n+1}}{-n+1} = \frac{P_2V_2 - P_1V_1}{1-n} \tag{5-9}$$

since $C = P_1V_1^n = P_2V_2^n$. For an ideal gas ($PV = mRT$), this equation can also be written as

$$W_b = \frac{mR(T_2 - T_1)}{1-n} \qquad n \neq 1 \qquad \text{(kJ)} \tag{5-10}$$

For the special case of $n = 1$ the boundary work becomes

$$W_b = \int_1^2 P\,dV = \int_1^2 CV^{-1}\,dV = PV\ln\left(\frac{V_2}{V_1}\right)$$

For an ideal gas this result is equivalent to the isothermal process discussed in the previous example.

EXAMPLE 5–4 Expansion of a Gas against a Spring

A piston–cylinder device contains 0.05 m^3 of a gas initially at 200 kPa. At this state, a linear spring that has a spring constant of 150 kN/m is touching the piston but exerting no force on it. Now heat is transferred to the gas, causing the piston to rise and to compress the spring until the volume inside the cylinder doubles. If the cross-sectional area of the piston is 0.25 m^2, determine (*a*) the final pressure inside the cylinder, (*b*) the total work done by the gas, and (*c*) the fraction of this work done against the spring to compress it.

Solution A gas in a piston–cylinder device equipped with a linear spring expands as a result of heating. The final gas pressure, the total work done, and the fraction of the work done to compress the spring are to be determined.

Assumptions **1** The expansion process is quasi-equilibrium. **2** The spring is linear in the range of interest.

Analysis A sketch of the system and the *P-V* diagram of the process are shown in Fig. 5–10.

(*a*) The enclosed volume at the final state is

$$V_2 = 2V_1 = (2)(0.05 \text{ m}^3) = 0.1 \text{ m}^3$$

Then the displacement of the piston (and of the spring) becomes

$$x = \frac{\Delta V}{A} = \frac{(0.1 - 0.05) \text{ m}^3}{0.25 \text{ m}^2} = 0.2 \text{ m}$$

The force applied by the linear spring at the final state is

$$F = kx = (150 \text{ kN/m})(0.2 \text{ m}) = 30 \text{ kN}$$

The additional pressure applied by the spring on the gas at this state is

$$P = \frac{F}{A} = \frac{30 \text{ kN}}{0.25 \text{ m}^2} = 120 \text{ kPa}$$

Without the spring, the pressure of the gas would remain constant at 200 kPa while the piston is rising. But under the effect of the spring, the pressure rises linearly from 200 kPa to

$$200 + 120 = \mathbf{320 \text{ kPa}}$$

at the final state.

(*b*) An easy way of finding the work done is to plot the process on a *P-V* diagram and find the area under the process curve. From Fig. 5–10 the area under the process curve (a trapezoid) is determined to be

$$W = \text{area} = \frac{(200 + 320) \text{ kPa}}{2}[(0.1 - 0.05) \text{ m}^3]\left(\frac{1 \text{ kJ}}{1 \text{ kPa}\cdot\text{m}^3}\right) = \mathbf{13 \text{ kJ}}$$

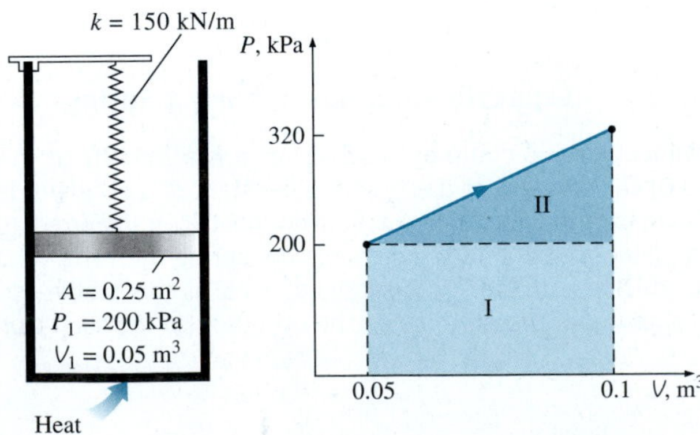

FIGURE 5–10

Schematic and *P-V* diagram for Example 5–4.

Note that the work is done by the system.

(*c*) The work represented by the rectangular area (region I) is done against the piston and the atmosphere, and the work represented by the triangular area (region II) is done against the spring. Thus,

$$W_{spring} = \tfrac{1}{2}[(320 - 200)\text{ kPa}](0.05\text{ m}^3)\left(\frac{1\text{ kJ}}{1\text{ kPa}\cdot\text{m}^3}\right) = \mathbf{3.00\text{ kJ}}$$

Discussion This result could also be obtained from

$$W_{spring} = \tfrac{1}{2}k(x_2^2 - x_1^2) = \tfrac{1}{2}(150\text{ kN/m})[(0.2\text{ m})^2 - 0^2]\left(\frac{1\text{ kJ}}{1\text{ kN}\cdot\text{m}}\right) = 3.00\text{ kJ}$$

5–2 ▪ ENERGY BALANCE FOR CLOSED SYSTEMS

Energy balance for any system undergoing any kind of process was expressed as (see Chap. 3)

$$\underbrace{E_{in} - E_{out}}_{\substack{\text{Net energy transfer}\\ \text{by heat, work, and mass}}} = \underbrace{\Delta E_{system}}_{\substack{\text{Change in internal, kinetic,}\\ \text{potential, etc., energies}}} \quad \text{(kJ)} \tag{5–11}$$

or, in the **rate form**, as

$$\underbrace{\dot{E}_{in} - \dot{E}_{out}}_{\substack{\text{Rate of net energy transfer}\\ \text{by heat, work, and mass}}} = \underbrace{dE_{system}/dt}_{\substack{\text{Rate of change in internal,}\\ \text{kinetic, potential, etc., energies}}} \quad \text{(kW)} \tag{5–12}$$

For constant rates, the total quantities during a time interval Δt are related to the quantities per unit time as

$$Q = \dot{Q}\,\Delta t, \quad W = \dot{W}\,\Delta t, \quad \text{and} \quad \Delta E = (dE/dt)\Delta t \quad \text{(kJ)} \tag{5–13}$$

The energy balance can be expressed on a **per unit mass** basis as

$$e_{in} - e_{out} = \Delta e_{system} \quad \text{(kJ/kg)} \tag{5–14}$$

which is obtained by dividing all the quantities in Eq. 5–11 by the mass m of the system. Energy balance can also be expressed in the differential form as

$$\delta E_{in} - \delta E_{out} = dE_{system} \quad \text{or} \quad \delta e_{in} - \delta e_{out} = de_{system} \tag{5–15}$$

For a closed system undergoing a **cycle**, the initial and final states are identical, and thus $\Delta E_{system} = E_2 - E_1 = 0$. Then the energy balance for a cycle simplifies to $E_{in} - E_{out} = 0$ or $E_{in} = E_{out}$. Noting that a closed system does not involve any mass flow across its boundaries, the energy balance for a cycle can be expressed in terms of heat and work interactions as

$$W_{net,out} = Q_{net,in} \quad \text{or} \quad \dot{W}_{net,out} = \dot{Q}_{net,in} \quad \text{(for a cycle)} \tag{5–16}$$

That is, the net work output during a cycle is equal to net heat input (Fig. 5–11).

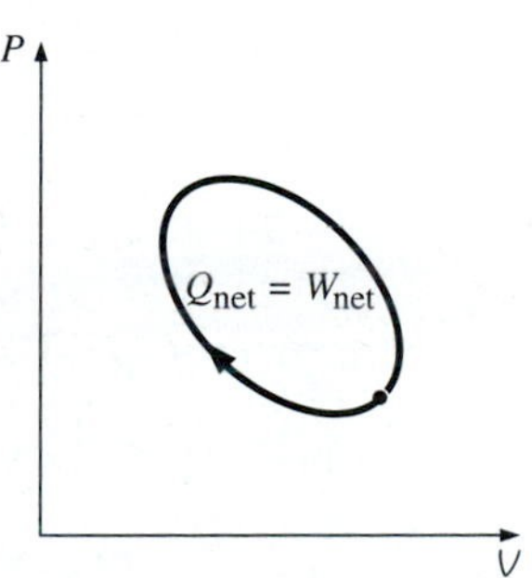

FIGURE 5–11
For a cycle $\Delta E = 0$, thus $Q = W$.

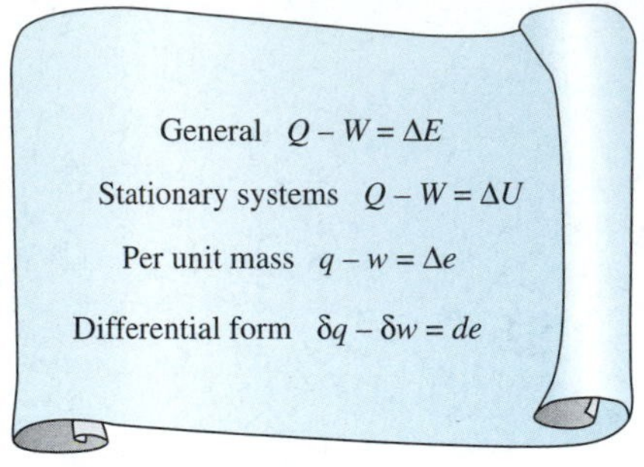

General $Q - W = \Delta E$
Stationary systems $Q - W = \Delta U$
Per unit mass $q - w = \Delta e$
Differential form $\delta q - \delta w = de$

FIGURE 5–12

Various forms of the first-law relation for closed systems.

The energy balance (or the first-law) relations already given are intuitive in nature and are easy to use when the magnitudes and directions of heat and work transfers are known. However, when performing a general analytical study or solving a problem that involves an unknown heat or work interaction, we need to assume a direction for the heat or work interactions. In such cases, it is common practice to use the classical thermodynamics sign convention and to assume heat to be transferred *into the system* (heat input) in the amount of Q and work to be done *by the system* (work output) in the amount of W, and then to solve the problem. The energy balance relation in that case for a closed system becomes

$$Q_{\text{net,in}} - W_{\text{net,out}} = \Delta E_{\text{system}} \quad \text{or} \quad Q - W = \Delta E \tag{5–17}$$

where $Q = Q_{\text{net,in}} = Q_{\text{in}} - Q_{\text{out}}$ is the *net heat input* and $W = W_{\text{net,out}} = W_{\text{out}} - W_{\text{in}}$ is the *net work output.* Obtaining a negative quantity for Q or W simply means that the assumed direction for that quantity is wrong and should be reversed. Various forms of this "traditional" first-law relation for closed systems are given in Fig. 5–12.

The first law cannot be proven mathematically, but no process in nature is known to have violated the first law, and this should be taken as sufficient evidence of its validity. Note that if it were possible to prove the first law on the basis of other physical principles, the first law then would be a consequence of those principles instead of being a fundamental physical law itself.

As energy quantities, heat and work are not that different, and you probably wonder why we keep distinguishing them. After all, the change in the energy content of a system is equal to the amount of energy that crosses the system boundaries, and it makes no difference whether the energy crosses the boundary as heat or work. It seems as if the first-law relations would be much simpler if we had just one quantity that we could call *energy interaction* to represent both heat and work. Well, from the first-law point of view, heat and work are not different at all. From the second-law point of view, however, heat and work are very different, as is discussed in later chapters.

EXAMPLE 5–5 Electric Heating of a Gas at Constant Pressure

A piston–cylinder device contains 25 g of saturated water vapor that is maintained at a constant pressure of 300 kPa. A resistance heater within the cylinder is turned on and passes a current of 0.2 A for 5 min from a 120-V source. At the same time, a heat loss of 3.7 kJ occurs. (*a*) Show that for a closed system the boundary work W_b and the change in internal energy ΔU in the first-law relation can be combined into one term, ΔH, for a constant-pressure process. (*b*) Determine the final temperature of the steam.

Solution Saturated water vapor in a piston–cylinder device expands at constant pressure as a result of heating. It is to be shown that $\Delta U + W_b = \Delta H$, and the final temperature is to be determined.

Assumptions **1** The tank is stationary and thus the kinetic and potential energy changes are zero, $\Delta KE = \Delta PE = 0$. Therefore, $\Delta E = \Delta U$ and internal energy is the only form of energy of the system that may change during this process. **2** Electrical wires constitute a very small part of the system, and thus the energy change of the wires can be neglected.

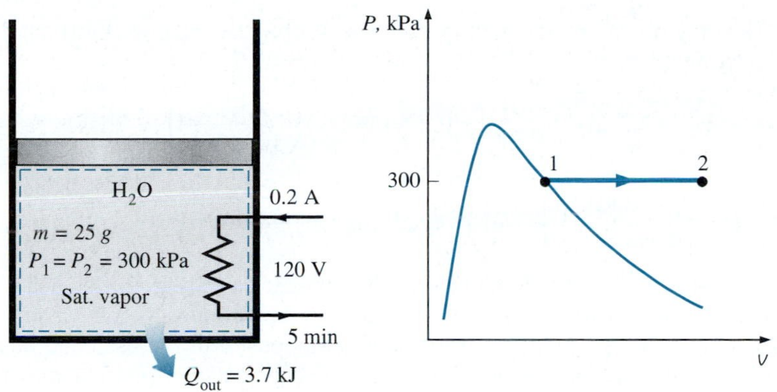

FIGURE 5–13

Schematic and P-v diagram for Example 5–5.

Analysis We take the contents of the cylinder, including the resistance wires, as the *system* (Fig. 5–13). This is a *closed system* since no mass crosses the system boundary during the process. We observe that a piston–cylinder device typically involves a moving boundary and thus boundary work W_b. The pressure remains constant during the process and thus $P_2 = P_1$. Also, heat is lost from the system and electrical work W_e is done on the system.

(*a*) This part of the solution involves a general analysis for a closed system undergoing a quasi-equilibrium constant-pressure process, and thus we consider a general closed system. We take the direction of heat transfer Q to be into the system and the work W to be done by the system. We also express the work as the sum of boundary and other forms of work (such as electrical and shaft). Then the energy balance can be expressed as

$$\underbrace{E_{in} - E_{out}}_{\substack{\text{Net energy transfer}\\ \text{by heat, work, and mass}}} = \underbrace{\Delta E_{system}}_{\substack{\text{Change in internal, kinetic,}\\ \text{potential, etc., energies}}}$$

$$Q - W = \Delta U + \Delta KE^{\nearrow 0} + \Delta PE^{\nearrow 0}$$

$$Q - W_{other} - W_b = U_2 - U_1$$

For a constant-pressure process, the boundary work is given as $W_b = P_0(V_2 - V_1)$. Substituting this into the preceding relation gives

$$Q - W_{other} - P_0(V_2 - V_1) = U_2 - U_1$$

However,

$$P_0 = P_2 = P_1 \quad \rightarrow \quad Q - W_{other} = (U_2 + P_2V_2) - (U_1 + P_1V_1)$$

Also $H = U + PV$, and thus

$$Q - W_{other} = H_2 - H_1 \qquad \text{(kJ)} \qquad \textbf{(5–18)}$$

which is the desired relation (Fig. 5–14). This equation is very convenient to use in the analysis of closed systems undergoing a constant-pressure quasi-equilibrium process since the boundary work is automatically taken care of by the enthalpy terms, and one no longer needs to determine it separately.

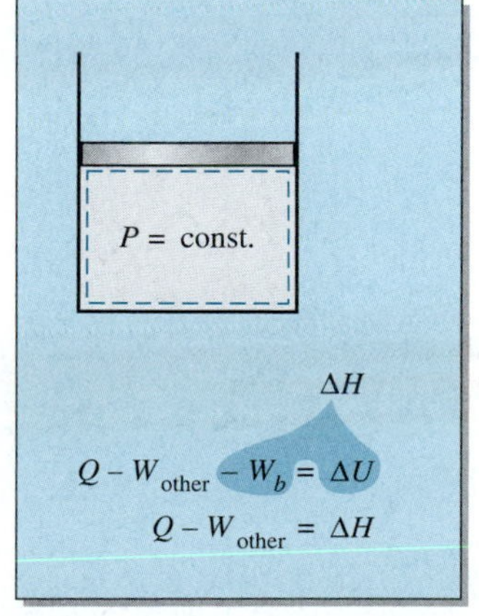

FIGURE 5–14

For a closed system undergoing a quasi-equilibrium, P = constant process, $\Delta U + W_b = \Delta H$.

(*b*) The only other form of work in this case is the electrical work, which can be determined from

$$W_e = \mathbf{V}I\ \Delta t = (120\ \text{V})(0.2\ \text{A})(300\ \text{s})\left(\frac{1\ \text{kJ/s}}{1000\ \text{VA}}\right) = 7.2\ \text{kJ}$$

$$\textit{State 1:}\quad \left.\begin{array}{l} P_1 = 300\ \text{kPa} \\ \text{sat. vapor}\end{array}\right\}\quad h_1 = h_{g\ @\ 300\ \text{kPa}} = 2724.9\ \text{kJ/kg} \qquad \text{(Table A–5)}$$

The enthalpy at the final state can be determined directly from Eq. 5–18 by expressing heat transfer from the system and work done on the system as negative quantities (since their directions are opposite to the assumed directions). Alternately, we can use the general energy balance relation with the simplification that the boundary work is considered automatically by replacing ΔU by ΔH for a constant-pressure expansion or compression process:

$$\underbrace{E_{\text{in}} - E_{\text{out}}}_{\substack{\text{Net energy transfer}\\ \text{by heat, work, and mass}}} = \underbrace{\Delta E_{\text{system}}}_{\substack{\text{Change in internal, kinetic,}\\ \text{potential, etc., energies}}}$$

$$W_{e,\text{in}} - Q_{\text{out}} - W_b = \Delta U$$

$$W_{e,\text{in}} - Q_{\text{out}} = \Delta H = m(h_2 - h_1) \qquad (\text{since } P = \text{constant})$$

$$7.2\ \text{kJ} - 3.7\ \text{kJ} = (0.025\ \text{kg})(h_2 - 2724.9)\ \text{kJ/kg}$$

$$h_2 = 2864.9\ \text{kJ/kg}$$

Now the final state is completely specified since we know both the pressure and the enthalpy. The temperature at this state is

$$\textit{State 2:}\quad \left.\begin{array}{l} P_2 = 300\ \text{kPa} \\ h_2 = 2864.9\ \text{kJ/kg}\end{array}\right\}\quad T_2 = \mathbf{200°C} \qquad \text{(Table A–6)}$$

Therefore, the steam will be at 200°C at the end of this process.

Discussion Strictly speaking, the potential energy change of the steam is not zero for this process since the center of gravity of the steam rose somewhat. Assuming an elevation change of 1 m (which is rather unlikely), the change in the potential energy of the steam would be 0.0002 kJ, which is very small compared to the other terms in the first-law relation. Therefore, in problems of this kind, the potential energy term for gases is always neglected.

EXAMPLE 5–6 Unrestrained Expansion of Water

A rigid tank is divided into two equal parts by a partition. Initially, one side of the tank contains 5 kg of water at 200 kPa and 25°C, and the other side is evacuated. The partition is then removed, and the water expands into the entire tank. The water is allowed to exchange heat with its surroundings until the temperature in the tank returns to the initial value of 25°C. Determine (*a*) the volume of the tank, (*b*) the final pressure, and (*c*) the heat transfer for this process.

Solution One half of a rigid tank is filled with liquid water while the other side is evacuated. The partition between the two parts is removed and water is allowed to expand and fill the entire tank while the temperature is maintained constant. The volume of tank, the final pressure, and the heat transfer are to be to determined.

Assumptions **1** The system is stationary and thus the kinetic and potential energy changes are zero, $\Delta KE = \Delta PE = 0$ and $\Delta E = \Delta U$. **2** The direction of heat transfer is to the system (heat gain, Q_{in}). A negative result for Q_{in} indicates the assumed direction is wrong and thus it is a heat loss. **3** The volume of the rigid tank is constant, and thus there is no energy transfer as boundary work. **4** There is no electrical, shaft, or any other kind of work involved.

Analysis We take the contents of the tank, including the evacuated space, as the *system* (Fig. 5–15). This is a *closed system* since no mass crosses the system boundary during the process. We observe that the water fills the entire tank when the partition is removed (possibly as a liquid–vapor mixture).

(*a*) Initially the water in the tank exists as a compressed liquid since its pressure (200 kPa) is greater than the saturation pressure at 25°C (3.1698 kPa). Approximating the compressed liquid as a saturated liquid at the given temperature, we find

$$v_1 \cong v_{f @ 25°C} = 0.001003 \text{ m}^3/\text{kg} \cong 0.001 \text{ m}^3/\text{kg} \qquad \text{(Table A–4)}$$

Then the initial volume of the water is

$$V_1 = mv_1 = (5 \text{ kg})(0.001 \text{ m}^3/\text{kg}) = 0.005 \text{ m}^3$$

The total volume of the tank is twice this amount:

$$V_{tank} = (2)(0.005 \text{ m}^3) = \mathbf{0.01 \text{ m}^3}$$

(*b*) At the final state, the specific volume of the water is

$$v_2 = \frac{V_2}{m} = \frac{0.01 \text{ m}^3}{5 \text{ kg}} = 0.002 \text{ m}^3/\text{kg}$$

which is twice the initial value of the specific volume. This result is expected since the volume doubles while the amount of mass remains constant.

$$\text{At 25°C:} \quad v_f = 0.001003 \text{ m}^3/\text{kg} \quad \text{and} \quad v_g = 43.340 \text{ m}^3/\text{kg} \quad \text{(Table A–4)}$$

Since $v_f < v_2 < v_g$, the water is a saturated liquid–vapor mixture at the final state, and thus the pressure is the saturation pressure at 25°C:

$$P_2 = P_{sat @ 25°C} = \mathbf{3.1698 \text{ kPa}} \qquad \text{(Table A–4)}$$

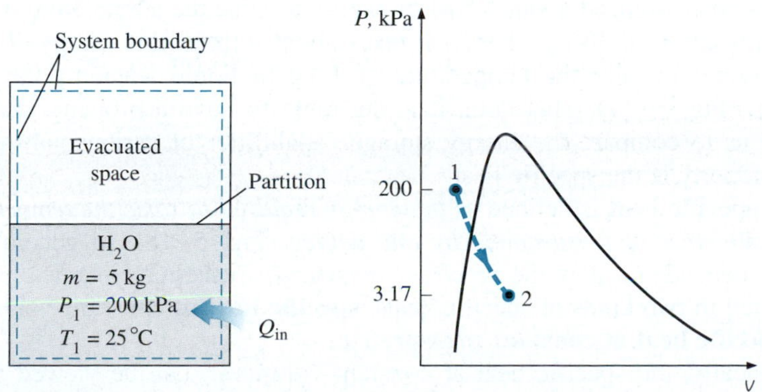

FIGURE 5–15

Schematic and P-v diagram for Example 5–6.

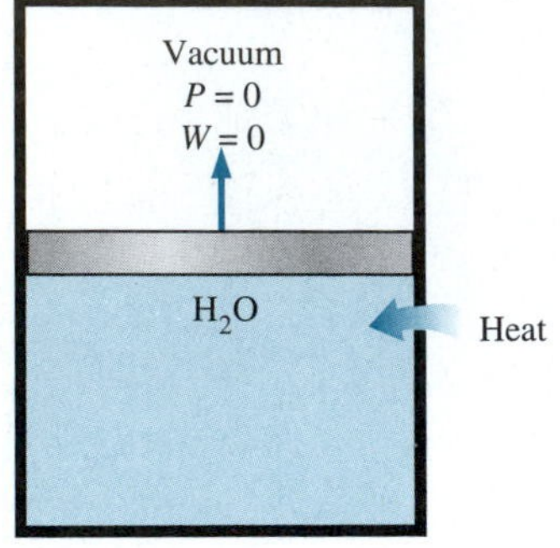

FIGURE 5–16

Expansion against a vacuum involves no work and thus no energy transfer.

(*c*) Under stated assumptions and observations, the energy balance on the system can be expressed as

$$\underbrace{E_{in} - E_{out}}_{\substack{\text{Net energy transfer} \\ \text{by heat, work, and mass}}} = \underbrace{\Delta E_{system}}_{\substack{\text{Change in internal, kinetic,} \\ \text{potential, etc., energies}}}$$

$$Q_{in} = \Delta U = m(u_2 - u_1)$$

Notice that even though the water is expanding during this process, the system chosen involves fixed boundaries only (the dashed lines) and therefore the moving boundary work is zero (Fig. 5–16). Then $W = 0$ since the system does not involve any other forms of work. (Can you reach the same conclusion by choosing the water as our system?) Initially,

$$u_1 \cong u_{f @ 25°C} = 104.83 \text{ kJ/kg}$$

The quality at the final state is determined from the specific volume information:

$$x_2 = \frac{v_2 - v_f}{v_{fg}} = \frac{0.002 - 0.001}{43.34 - 0.001} = 2.3 \times 10^{-5}$$

Then

$$\begin{aligned} u_2 &= u_f + x_2 u_{fg} \\ &= 104.83 \text{ kJ/kg} + (2.3 \times 10^{-5})(2304.3 \text{ kJ/kg}) \\ &= 104.88 \text{ kJ/kg} \end{aligned}$$

Substituting yields

$$Q_{in} = (5 \text{ kg})[(104.88 - 104.83) \text{ kJkg}] = \mathbf{0.25 \text{ kJ}}$$

Discussion The positive sign indicates that the assumed direction is correct, and heat is transferred to the water.

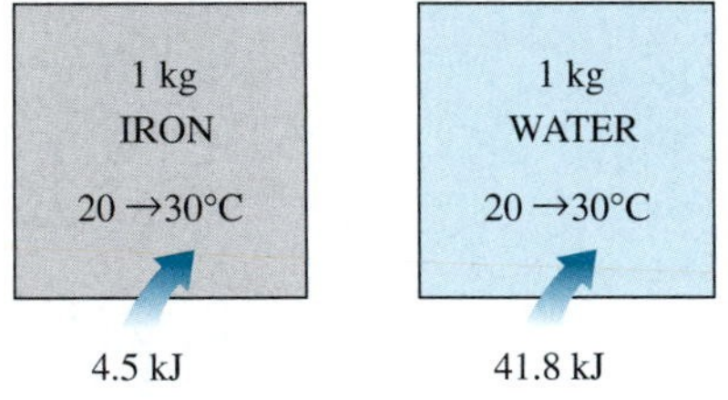

FIGURE 5–17

It takes different amounts of energy to raise the temperature of different substances by the same amount.

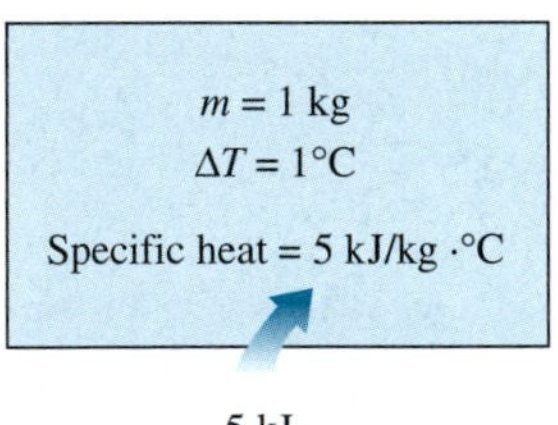

FIGURE 5–18

Specific heat is the energy required to raise the temperature of a unit mass of a substance by one degree in a specified way.

5–3 ▪ SPECIFIC HEATS

We know from experience that it takes different amounts of energy to raise the temperature of identical masses of different substances by one degree. For example, we need about 4.5 kJ of energy to raise the temperature of 1 kg of iron from 20 to 30°C, whereas it takes about 9 times this energy (41.8 kJ to be exact) to raise the temperature of 1 kg of liquid water by the same amount (Fig. 5–17). Therefore, it is desirable to have a property that will enable us to compare the energy storage capabilities of various substances. This property is the specific heat.

The **specific heat** is defined as *the energy required to raise the temperature of a unit mass of a substance by one degree* (Fig. 5–18). In general, this energy depends on how the process is executed. In thermodynamics, we are interested in two kinds of specific heats: **specific heat at constant volume** c_v and **specific heat at constant pressure** c_p.

Physically, the specific heat at constant volume c_v can be viewed as *the energy required to raise the temperature of the unit mass of a substance by one degree as the volume is maintained constant*. The energy required to do the same as the pressure is maintained constant is the specific heat at constant pressure c_p. This is illustrated in Fig. 5–19. The specific heat at constant pressure c_p is always greater than c_v because at constant pressure the system is allowed to expand and the energy for this expansion work must also be supplied to the system.

Now we attempt to express the specific heats in terms of other thermodynamic properties. First, consider a fixed mass in a stationary closed system undergoing a constant-volume process (and thus no expansion or compression work is involved). The conservation of energy principle $e_{in} - e_{out} = \Delta e_{system}$ for this process can be expressed in the differential form as

$$\delta e_{in} - \delta e_{out} = du$$

The left-hand side of this equation represents the net amount of energy transferred to the system. From the definition of c_v, this energy must be equal to $c_v\, dT$, where dT is the differential change in temperature. Thus,

$$c_v\, dT = du \qquad \text{at constant volume}$$

or

$$c_v = \left(\frac{\partial u}{\partial T}\right)_v \tag{5–19}$$

Similarly, an expression for the specific heat at constant pressure c_p can be obtained by considering a constant-pressure expansion or compression process. It yields

$$c_p = \left(\frac{\partial h}{\partial T}\right)_p \tag{5–20}$$

Equations 5–19 and 5–20 are the defining equations for c_v and c_p, and their interpretation is given in Fig. 5–20.

Note that c_v and c_p are expressed in terms of other properties; thus, they must be properties themselves. Like any other property, the specific heats of a substance depend on the state that, in general, is specified by two independent, intensive properties. That is, the energy required to raise the temperature of a substance by one degree is different at different temperatures and pressures (Fig. 5–21). But this difference is usually not very large.

A few observations can be made from Eqs. 5–19 and 5–20. First, these equations are *property relations* and as such *are independent of the type of processes*. They are valid for *any* substance undergoing *any* process. The only relevance c_v has to a constant-volume process is that c_v happens to be the energy transferred to a system during a constant-volume process per unit mass per unit degree rise in temperature. This is how the values of c_v are determined. This is also how the name *specific heat at constant volume* originated. Likewise, the energy transferred to a system per unit mass per unit temperature rise during a constant-pressure process happens to be equal to c_p. This is how the values of c_p can be determined and also explains the origin of the name *specific heat at constant pressure*.

Another observation that can be made from Eqs. 5–19 and 5–20 is that c_v is related to the changes in *internal energy* and c_p to the changes in *enthalpy*. In fact, it would be more proper to define c_v as *the change in the internal energy of a substance per unit change in temperature at constant*

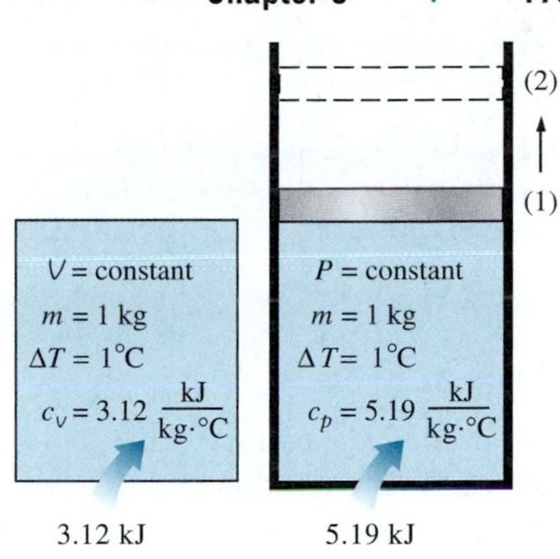

FIGURE 5–19
Constant-volume and constant-pressure specific heats c_v and c_p (values given are for helium gas).

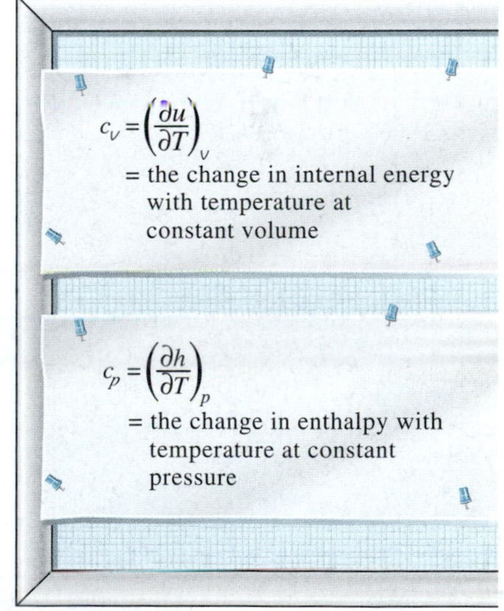

FIGURE 5–20
Formal definitions of c_v and c_p.

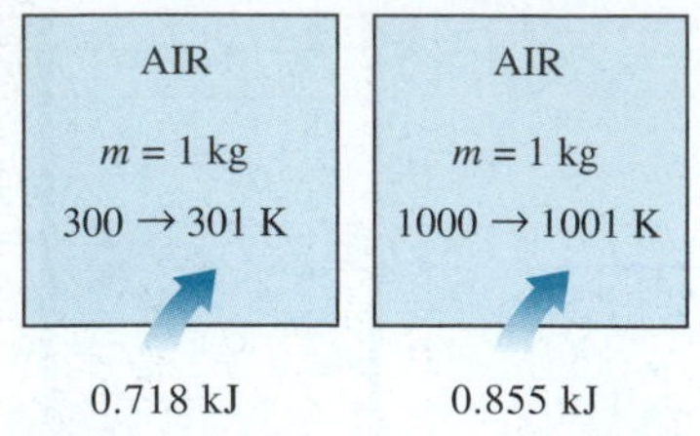

FIGURE 5–21

The specific heat of a substance changes with temperature.

volume. Likewise, c_p can be defined as *the change in the enthalpy of a substance per unit change in temperature at constant pressure*. In other words, c_v is a measure of the variation of internal energy of a substance with temperature, and c_p is a measure of the variation of enthalpy of a substance with temperature.

Both the internal energy and enthalpy of a substance can be changed by the transfer of *energy* in any form, with heat being only one of them. Therefore, the term *specific energy* is probably more appropriate than the term *specific heat*, which implies that energy is transferred (and stored) in the form of heat.

A common unit for specific heats is kJ/kg · °C or kJ/kg · K. Notice that these two units are *identical* since $\Delta T(°C) = \Delta T(K)$, and 1°C change in temperature is equivalent to a change of 1 K. The specific heats are sometimes given on a *molar basis*. They are then denoted by $\bar{c}_v$ and $\bar{c}_p$ and have the unit kJ/kmol · °C or kJ/kmol · K.

5–4 ▪ INTERNAL ENERGY, ENTHALPY, AND SPECIFIC HEATS OF IDEAL GASES

We defined an ideal gas as a gas whose temperature, pressure, and specific volume are related by

$$Pv = RT$$

It has been demonstrated mathematically and experimentally (Joule, 1843) that for an ideal gas the internal energy is a function of the temperature only. That is,

$$u = u(T) \tag{5-21}$$

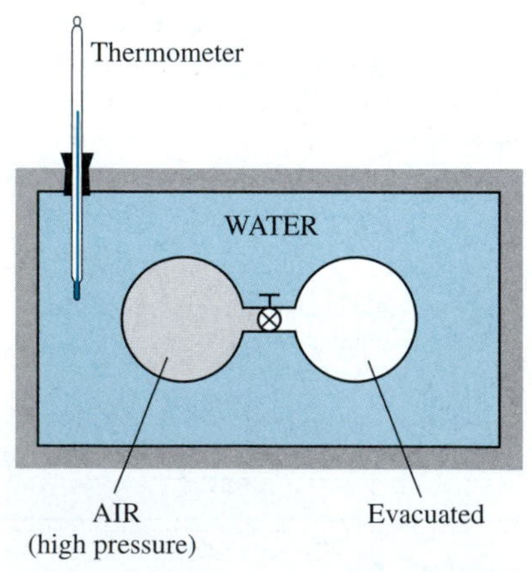

FIGURE 5–22

Schematic of the experimental apparatus used by Joule.

In his classical experiment, Joule submerged two tanks connected with a pipe and a valve in a water bath, as shown in Fig. 5–22. Initially, one tank contained air at a high pressure and the other tank was evacuated. When thermal equilibrium was attained, he opened the valve to let air pass from one tank to the other until the pressures equalized. Joule observed no change in the temperature of the water bath and assumed that no heat was transferred to or from the air. Since there was also no work done, he concluded that the internal energy of the air did not change even though the volume and the pressure changed. Therefore, he reasoned, the internal energy is a function of temperature only and not a function of pressure or specific volume. (Joule later showed that for gases that deviate significantly from ideal-gas behavior, the internal energy is not a function of temperature alone.)

Using the definition of enthalpy and the equation of state of an ideal gas, we have

$$\left.\begin{aligned} h &= u + Pv \\ Pv &= RT \end{aligned}\right\} \quad h = u + RT$$

Since R is constant and $u = u(T)$, it follows that the enthalpy of an ideal gas is also a function of temperature only:

$$h = h(T) \tag{5-22}$$

Since u and h depend only on temperature for an ideal gas, the specific heats c_v and c_p also depend, at most, on temperature only. Therefore, at a given temperature, u, h, c_v, and c_p of an ideal gas have fixed values regardless of the volume or pressure (Fig. 5–23). Thus, for ideal gases, the partial derivatives in Eqs. 5–19 and 5–20 can be replaced by ordinary derivatives. Then the differential changes in the internal energy and enthalpy of an ideal gas can be expressed as

$$du = c_v(T)\,dT \tag{5–23}$$

and

$$dh = c_p(T)\,dT \tag{5–24}$$

The change in internal energy or enthalpy for an ideal gas during a process from state 1 to state 2 is determined by integrating these equations:

$$\Delta u = u_2 - u_1 = \int_1^2 c_v(T)\,dT \qquad \text{(kJ/kg)} \tag{5–25}$$

and

$$\Delta h = h_2 - h_1 = \int_1^2 c_p(T)\,dT \qquad \text{(kJ/kg)} \tag{5–26}$$

To carry out these integrations, we need to have relations for c_v and c_p as functions of temperature.

At low pressures, all real gases approach ideal-gas behavior, and therefore their specific heats depend on temperature only. The specific heats of real gases at low pressures are called *ideal-gas specific heats*, or *zero-pressure specific heats*, and are often denoted c_{p0} and c_{v0}. Accurate analytical expressions for ideal-gas specific heats, based on direct measurements or calculations from statistical behavior of molecules, are available and are given as third-degree polynomials in the appendix (Table A–2*c*) for several gases. A plot of $\bar{c}_{p0}(T)$ data for some common gases is given in Fig. 5–24.

The use of ideal-gas specific heat data is limited to low pressures, but these data can also be used at moderately high pressures with reasonable accuracy as long as the gas does not deviate from ideal-gas behavior significantly.

The integrations in Eqs. 5–25 and 5–26 are straightforward but rather time-consuming and thus impractical. To avoid these laborious calculations, u and h data for a number of gases have been tabulated over small temperature intervals. These tables are obtained by choosing an arbitrary reference point and performing the integrations in Eqs. 5–25 and 5–26 by treating state 1 as the reference state. In the ideal-gas tables given in the appendix, zero kelvin is chosen as the reference state, and both the enthalpy and the internal energy are assigned zero values at that state (Fig. 5–25). The choice of the reference state has no effect on Δu or Δh calculations. The u and h data are given in kJ/kg for air (Table A–21) and usually in kJ/kmol for other gases. The unit kJ/kmol is very convenient in the thermodynamic analysis of chemical reactions.

Some observations can be made from Fig. 5–24. First, the specific heats of gases with complex molecules (molecules with two or more atoms) are higher and increase with temperature. Also, the variation of specific heats

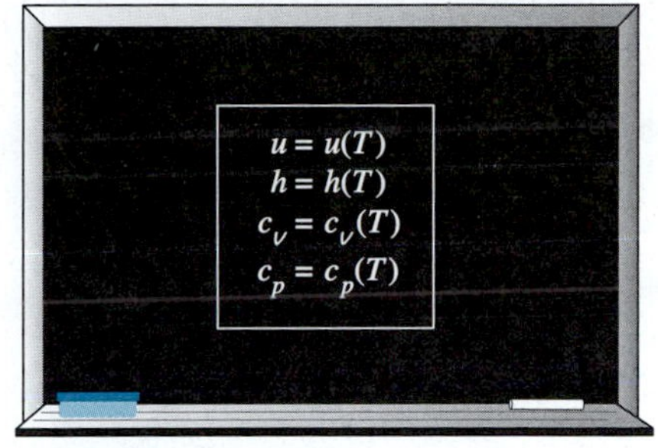

FIGURE 5–23

For ideal gases, u, h, c_v, and c_p vary with temperature only.

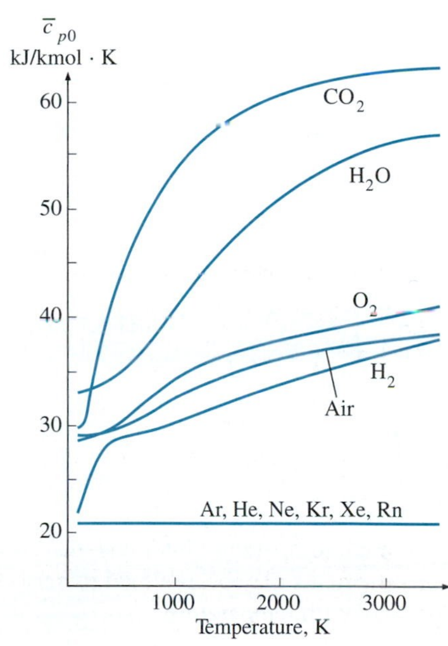

FIGURE 5–24

Ideal-gas constant-pressure specific heats for some gases (see Table A–2*c* for c_p equations).

AIR

T, K	u, kJ/kg	h, kJ/kg
0	0	0
.	.	.
.	.	.
300	214.07	300.19
310	221.25	310.24
.	.	.
.	.	.

FIGURE 5–25

In the preparation of ideal-gas tables, 0 K is chosen as the reference temperature.

with temperature is smooth and may be approximated as linear over small temperature intervals (a few hundred degrees or less). Therefore the specific heat functions in Eqs. 5–25 and 5–26 can be replaced by the constant average specific heat values. Then the integrations in these equations can be performed, yielding

$$u_2 - u_1 = c_{v,\text{avg}}(T_2 - T_1) \qquad \text{(kJ/kg)} \tag{5-27}$$

and

$$h_2 - h_1 = c_{p,\text{avg}}(T_2 - T_1) \qquad \text{(kJ/kg)} \tag{5-28}$$

The specific heat values for some common gases are listed as a function of temperature in Table A–2*b*. The average specific heats $c_{p,\text{avg}}$ and $c_{v,\text{avg}}$ are evaluated from this table at the average temperature $(T_1 + T_2)/2$, as shown in Fig. 5–26. If the final temperature T_2 is not known, the specific heats may be evaluated at T_1 or at the anticipated average temperature. Then T_2 can be determined by using these specific heat values. The value of T_2 can be refined, if necessary, by evaluating the specific heats at the new average temperature.

Another way of determining the average specific heats is to evaluate them at T_1 and T_2 and then take their average. Usually both methods give reasonably good results, and one is not necessarily better than the other.

Another observation that can be made from Fig. 5–24 is that the ideal-gas specific heats of *monatomic gases* such as argon, neon, and helium remain constant over the entire temperature range. Thus, Δu and Δh of monatomic gases can easily be evaluated from Eqs. 5–27 and 5–28.

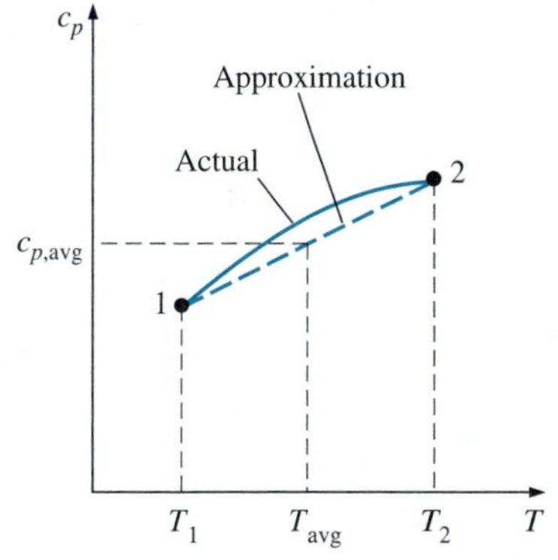

FIGURE 5–26

For small temperature intervals, the specific heats may be assumed to vary linearly with temperature.

Note that the Δu and Δh relations given previously are not restricted to any kind of process. They are valid for all processes. The presence of the constant-volume specific heat c_v in an equation should not lead one to believe that this equation is valid for a constant-volume process only. On the contrary, the relation $\Delta u = c_{v,\text{avg}}\,\Delta T$ is valid for *any* ideal gas undergoing *any* process (Fig. 5–27). A similar argument can be given for c_p and Δh.

To summarize, there are three ways to determine the internal energy and enthalpy changes of ideal gases (Fig. 5–28):

1. By using the tabulated u and h data. This is the easiest and most accurate way when tables are readily available.
2. By using the c_v or c_p relations as a function of temperature and performing the integrations. This is very inconvenient for hand calculations but quite desirable for computerized calculations. The results obtained are very accurate.
3. By using average specific heats. This is very simple and certainly very convenient when property tables are not available. The results obtained are reasonably accurate if the temperature interval is not very large.

Specific Heat Relations of Ideal Gases

A special relationship between c_p and c_v for ideal gases can be obtained by differentiating the relation $h = u + RT$, which yields

$$dh = du + R\,dT$$

Replacing dh by $c_p\,dT$ and du by $c_v\,dT$ and dividing the resulting expression by dT, we obtain

$$c_p = c_v + R \qquad (\text{kJ/kg}\cdot\text{K}) \tag{5–29}$$

This is an important relationship for ideal gases since it enables us to determine c_v from a knowledge of c_p and the gas constant R.

When the specific heats are given on a molar basis, R in the above equation should be replaced by the universal gas constant R_u (Fig. 5–29).

$$\bar{c}_p = \bar{c}_v + R_u \qquad (\text{kJ/kmol}\cdot\text{K}) \tag{5–30}$$

At this point, we introduce another ideal-gas property called the **specific heat ratio** k, defined as

$$k = \frac{c_p}{c_v} \tag{5–31}$$

The specific ratio also varies with temperature, but this variation is very mild. For monatomic gases, its value is essentially constant at 1.667. Many diatomic gases, including air, have a specific heat ratio of about 1.4 at room temperature.

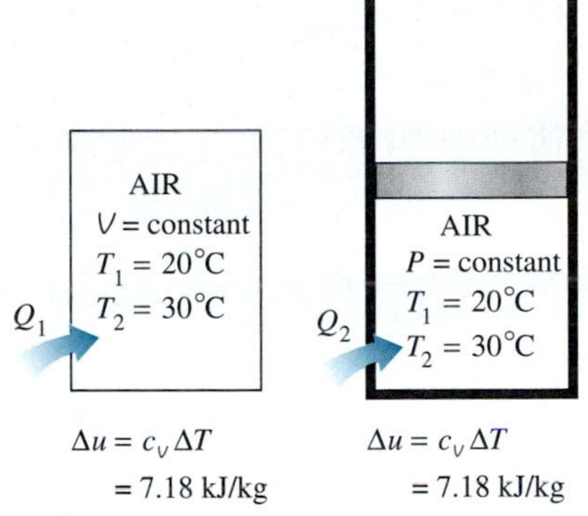

FIGURE 5–27
The relation $\Delta u = c_v\,\Delta T$ is valid for *any* kind of process, constant-volume or not.

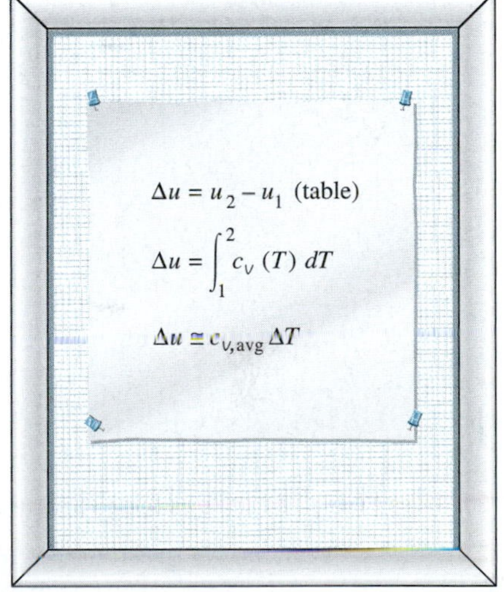

FIGURE 5–28
Three ways of calculating Δu.

EXAMPLE 5–7 Evaluation of the Δu of an Ideal Gas

Air at 300 K and 200 kPa is heated at constant pressure to 600 K. Determine the change in internal energy of air per unit mass, using (*a*) data from the air table (Table A–21), (*b*) the functional form of the specific heat (Table A–2*c*), and (*c*) the average specific heat value (Table A–2*b*).

Solution The internal energy change of air is to be determined in three different ways.

Assumptions At specified conditions, air can be considered to be an ideal gas since it is at a high temperature and low pressure relative to its critical-point values.

Analysis The internal energy change Δu of ideal gases depends on the initial and final temperatures only, and not on the type of process. Thus, the following solution is valid for any kind of process.

(*a*) One way of determining the change in internal energy of air is to read the u values at T_1 and T_2 from Table A–21 and take the difference:

$$u_1 = u_{@\,300\,\text{K}} = 214.07\ \text{kJ/kg}$$

$$u_2 = u_{@\,600\,\text{K}} = 434.78\ \text{kJ/kg}$$

Thus, the exact answer is

$$\Delta u = u_2 - u_1 = (434.78 - 214.07)\ \text{kJ/kg} = \mathbf{220.71\ kJ/kg}$$

(*b*) The $\bar{c}_p(T)$ of air is given in Table A–2*c* in the form of a third-degree polynomial expressed as

$$\bar{c}_p(T) = a + bT + cT^2 + dT^3$$

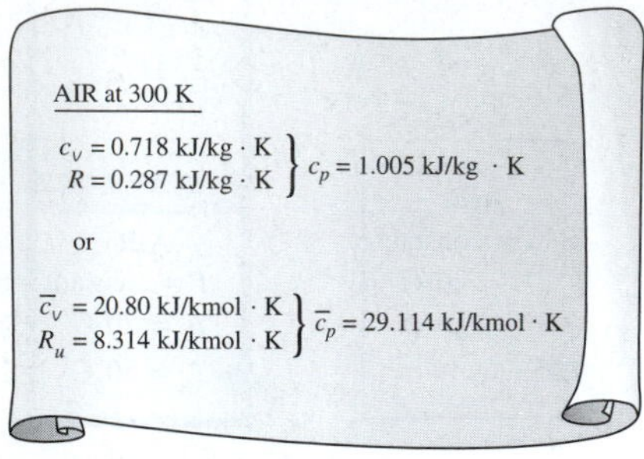

FIGURE 5–29

The c_p of an ideal gas can be determined from a knowledge of c_v and R.

where $a = 28.11$, $b = 0.1967 \times 10^{-2}$, $c = 0.4802 \times 10^{-5}$, and $d = -1.966 \times 10^{-9}$. From Eq. 5–30,

$$\bar{c}_v(T) = \bar{c}_p - R_u = (a - R_u) + bT + cT^2 + dT^3$$

From Eq. 5–25,

$$\Delta\bar{u} = \int_1^2 \bar{c}_v(T) \cdot dT = \int_{T_1}^{T_2} [(a - R_u) + bT + cT^2 + dT^3]\, dT$$

Performing the integration and substituting the values, we obtain

$$\Delta\bar{u} = 6447 \text{ kJ/kmol}$$

The change in the internal energy on a unit-mass basis is determined by dividing this value by the molar mass of air (Table A–1):

$$\Delta u = \frac{\Delta\bar{u}}{M} = \frac{6447 \text{ kJ/kmol}}{28.97 \text{ kg/kmol}} = \mathbf{222.5 \text{ kJ/kg}}$$

which differs from the tabulated value by 0.8 percent.

(*c*) The average value of the constant-volume specific heat $c_{v,\text{avg}}$ is determined from Table A–2*b* at the average temperature of $(T_1 + T_2)/2 = 450$ K to be

$$c_{v,\text{avg}} = c_{v\,@\,450\text{ K}} = 0.733 \text{ kJ/kg} \cdot \text{K}$$

Thus,

$$\Delta u = c_{v,\text{avg}}(T_2 - T_1) = (0.733 \text{ kJ/kg} \cdot \text{K})[(600 - 300)\text{K}]$$
$$= 220 \text{ kJ/kg}$$

Discussion This answer differs from the tabulated value (220.71 kJ/kg) by only 0.4 percent. This close agreement is not surprising since the assumption that c_v varies linearly with temperature is a reasonable one at temperature intervals of only a few hundred degrees. If we had used the c_v value at $T_1 = 300$ K instead of at T_{avg}, the result would be 215.4 kJ/kg, which is in error by about 2 percent. Errors of this magnitude are acceptable for most engineering purposes.

EXAMPLE 5–8 Heating of a Gas in a Tank by Stirring

An insulated rigid tank initially contains 1.5 lbm of helium at 80°F and 50 psia. A paddle wheel with a power rating of 0.02 hp is operated within the tank for 30 min. Determine (*a*) the final temperature and (*b*) the final pressure of the helium gas.

Solution Helium gas in an insulated rigid tank is stirred by a paddle wheel. The final temperature and pressure of helium are to be determined.

Assumptions **1** Helium is an ideal gas since it is at a very high temperature relative to its critical-point value of −451°F. **2** Constant specific heats can be used for helium. **3** The system is stationary and thus the kinetic and potential energy changes are zero, $\Delta KE = \Delta PE = 0$ and $\Delta E = \Delta U$. **4** The volume of the tank is constant, and thus there is no boundary work. **5** The system is adiabatic and thus there is no heat transfer.

Analysis We take the contents of the tank as the *system* (Fig. 5–30). This is a *closed system* since no mass crosses the system boundary during the process. We observe that there is shaft work done on the system.

(*a*) The amount of paddle-wheel work done on the system is

$$W_{sh} = \dot{W}_{sh}\,\Delta t = (0.02\text{ hp})(0.5\text{ h})\left(\frac{2545\text{ Btu/h}}{1\text{ hp}}\right) = 25.45\text{ Btu}$$

Under the stated assumptions and observations, the energy balance on the system can be expressed as

$$\underbrace{E_{in} - E_{out}}_{\substack{\text{Net energy transfer}\\ \text{by heat, work, and mass}}} = \underbrace{\Delta E_{system}}_{\substack{\text{Change in internal, kinetic,}\\ \text{potential, etc., energies}}}$$

$$W_{sh,in} = \Delta U = m(u_2 - u_1) = mc_{v,avg}(T_2 - T_1)$$

As we pointed out earlier, the ideal-gas specific heats of monatomic gases (helium being one of them) are constant. The c_v value of helium is determined from Table A–2E*a* to be c_v = 0.753 Btu/lbm · °F. Substituting this and other known quantities into the above equation, we obtain

$$25.45\text{ Btu} = (1.5\text{ lbm})(0.753\text{ Btu/lbm}\cdot{}^\circ\text{F})(T_2 - 80^\circ\text{F})$$

$$T_2 = \mathbf{102.5^\circ F}$$

(*b*) The final pressure is determined from the ideal-gas relation

$$\frac{P_1V_1}{T_1} = \frac{P_2V_2}{T_2}$$

where V_1 and V_2 are identical and cancel out. Then the final pressure becomes

$$\frac{50\text{ psia}}{(80 + 460)\text{ R}} = \frac{P_2}{(102.5 + 460)\text{R}}$$

$$P_2 = \mathbf{52.1\ psia}$$

Discussion Note that the pressure in the ideal-gas relation is always the absolute pressure.

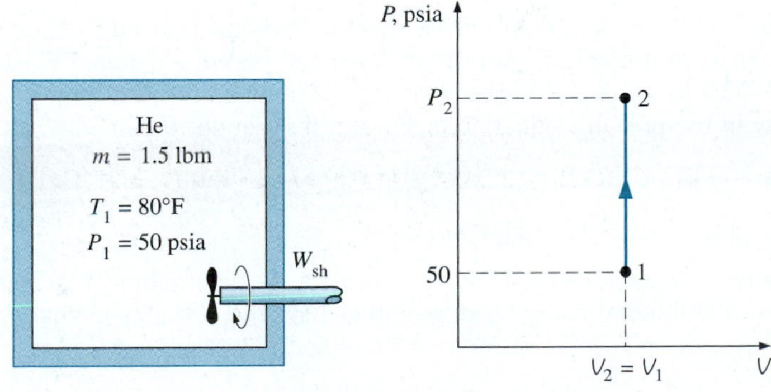

FIGURE 5–30
Schematic and *P-V* diagram for Example 5–8.

EXAMPLE 5–9 Heating of a Gas by a Resistance Heater

A piston–cylinder device initially contains 0.5 m^3 of nitrogen gas at 400 kPa and 27°C. An electric heater within the device is turned on and is allowed to pass a current of 2 A for 5 min from a 120-V source. Nitrogen expands at constant pressure, and a heat loss of 2800 J occurs during the process. Determine the final temperature of nitrogen.

Solution Nitrogen gas in a piston–cylinder device is heated by an electric resistance heater. Nitrogen expands at constant pressure while some heat is lost. The final temperature of nitrogen is to be determined.

Assumptions **1** Nitrogen is an ideal gas since it is at a high temperature and low pressure relative to its critical-point values of −147°C, and 3.39 MPa. **2** The system is stationary and thus the kinetic and potential energy changes are zero, $\Delta \text{KE} = \Delta \text{PE} = 0$ and $\Delta E = \Delta U$. **3** The pressure remains constant during the process and thus $P_2 = P_1$. **4** Nitrogen has constant specific heats at room temperature.

Analysis We take the contents of the cylinder as the *system* (Fig. 5–31). This is a *closed system* since no mass crosses the system boundary during the process. We observe that a piston–cylinder device typically involves a moving boundary and thus boundary work, W_b. Also, heat is lost from the system and electrical work W_e is done on the system.

First, let us determine the electrical work done on the nitrogen:

$$W_e = \mathbf{V}I\,\Delta t = (120\text{ V})(2\text{ A})(5 \times 60\text{ s})\left(\frac{1\text{ kJ/s}}{1000\text{ VA}}\right) = 72\text{ kJ}$$

The mass of nitrogen is determined from the ideal-gas relation:

$$m = \frac{P_1 V_1}{RT_1} = \frac{(400\text{ kPa})(0.5\text{ m}^3)}{(0.297\text{ kPa}\cdot\text{m}^3/\text{kg}\cdot\text{K})(300\text{ K})} = 2.245\text{ kg}$$

Under the stated assumptions and observations, the energy balance on the system can be expressed as

$$\underbrace{E_{\text{in}} - E_{\text{out}}}_{\substack{\text{Net energy transfer}\\ \text{by heat, work, and mass}}} = \underbrace{\Delta E_{\text{system}}}_{\substack{\text{Change in internal, kinetic,}\\ \text{potential, etc., energies}}}$$

$$W_{e,\text{in}} - Q_{\text{out}} - W_{b,\text{out}} = \Delta U$$

$$W_{e,\text{in}} - Q_{\text{out}} = \Delta H = m(h_2 - h_1) = mc_p(T_2 - T_1)$$

since $\Delta U + W_b \equiv \Delta H$ for a closed system undergoing a quasi-equilibrium expansion or compression process at constant pressure. From Table A–2a, $c_p = 1.039$ kJ/kg · K for nitrogen at room temperature. The only unknown quantity in the previous equation is T_2, and it is found to be

$$72\text{ kJ} - 2.8\text{ kJ} = (2.245\text{ kg})(1.039\text{ kJ/kg}\cdot\text{K})(T_2 - 27°\text{C})$$

$$T_2 = \mathbf{56.7°C}$$

Discussion We could also solve this problem by determining the boundary work and the internal energy change rather than the enthalpy change.

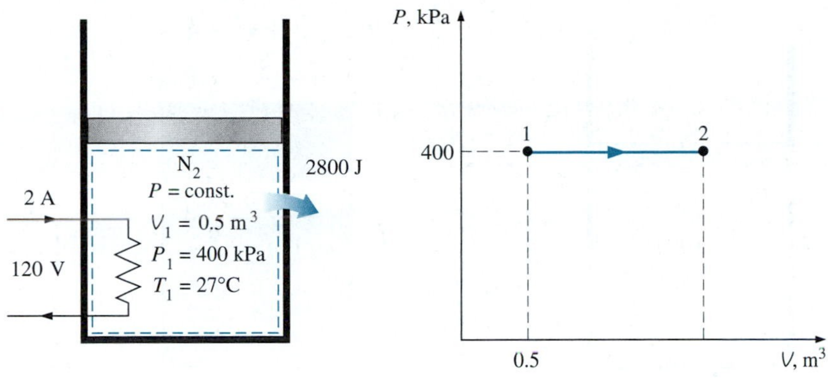

FIGURE 5–31

Schematic and P-V diagram for Example 5–9.

EXAMPLE 5–10 Heating of a Gas at Constant Pressure

A piston–cylinder device initially contains air at 150 kPa and 27°C. At this state, the piston is resting on a pair of stops, as shown in Fig. 5–32, and the enclosed volume is 400 L. The mass of the piston is such that a 350-kPa pressure is required to move it. The air is now heated until its volume has doubled. Determine (*a*) the final temperature, (*b*) the work done by the air, and (*c*) the total heat transferred to the air.

Solution Air in a piston cylinder device with a set of stops is heated until its volume is doubled. The final temperature, work done, and the total heat transfer are to be determined.

Assumptions **1** Air is an ideal gas since it is at a high temperature and low pressure relative to its critical-point values. **2** The system is stationary and thus the kinetic and potential energy changes are zero, $\Delta KE = \Delta PE = 0$ and $\Delta E = \Delta U$. **3** The volume remains constant until the piston starts moving, and the pressure remains constant afterwards. **4** There are no electrical, shaft, or other forms of work involved.

Analysis We take the contents of the cylinder as the *system* (Fig. 5–32). This is a *closed system* since no mass crosses the system boundary during the process. We observe that a piston-cylinder device typically involves a moving boundary and thus boundary work, W_b. Also, the boundary work is done by the system, and heat is transferred to the system.

(*a*) The final temperature can be determined easily by using the ideal-gas relation between states 1 and 3 in the following form:

$$\frac{P_1V_1}{T_1} = \frac{P_3V_3}{T_3} \longrightarrow \frac{(150\text{ kPa})(V_1)}{300\text{ K}} = \frac{(350\text{ kPa})(2V_1)}{T_3}$$

$$T_3 = \mathbf{1400\ K}$$

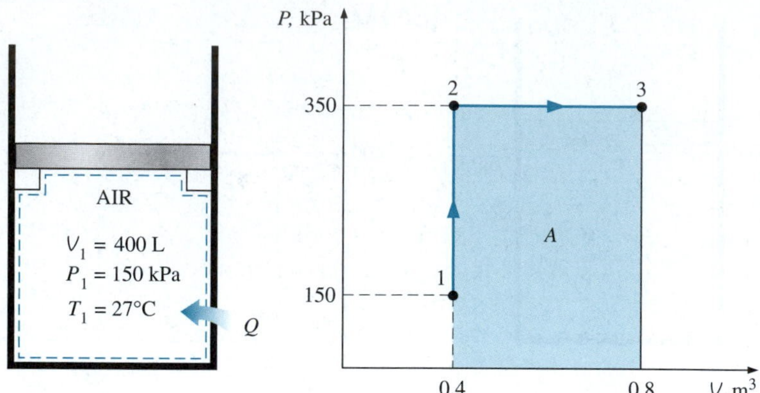

FIGURE 5–32

Schematic and P-V diagram for Example 5–10.

(*b*) The work done could be determined by integration, but for this case it is much easier to find it from the area under the process curve on a *P*-*V* diagram, shown in Fig. 5–32:

$$A = (V_2 - V_1)P_2 = (0.4\ \text{m}^3)(350\ \text{kPa}) = 140\ \text{m}^3\cdot\text{kPa}$$

Therefore,

$$W_{13} = \mathbf{140\ kJ}$$

The work is done by the system (to raise the piston and to push the atmospheric air out of the way), and thus it is work output.

(*c*) Under the stated assumptions and observations, the energy balance on the system between the initial and final states (process 1–3) can be expressed as

$$\underbrace{E_{\text{in}} - E_{\text{out}}}_{\substack{\text{Net energy transfer}\\ \text{by heat, work, and mass}}} = \underbrace{\Delta E_{\text{system}}}_{\substack{\text{Change in internal, kinetic,}\\ \text{potential, etc., energies}}}$$

$$Q_{\text{in}} - W_{b,\text{out}} = \Delta U = m(u_3 - u_1)$$

The mass of the system can be determined from the ideal-gas relation:

$$m = \frac{P_1V_1}{RT_1} = \frac{(150\ \text{kPa})(0.4\ \text{m}^3)}{(0.287\ \text{kPa}\cdot\text{m}^3/\text{kg}\cdot\text{K})(300\ \text{K})} = 0.697\ \text{kg}$$

The internal energies are determined from the air table (Table A–21) to be

$$u_1 = u_{@\ 300\ \text{K}} = 214.07\ \text{kJ/kg}$$

$$u_3 = u_{@\ 1400\ \text{K}} = 1113.52\ \text{kJ/kg}$$

Thus,

$$Q_{\text{in}} - 140\ \text{kJ} = (0.697\ \text{kg})[(1113.52 - 214.07)\ \text{kJ/kg}]$$

$$Q_{\text{in}} = \mathbf{767\ kJ}$$

Discussion The positive sign verifies that heat is transferred to the system.

5–5 ■ INTERNAL ENERGY, ENTHALPY, AND SPECIFIC HEATS OF SOLIDS AND LIQUIDS

A substance whose specific volume (or density) is constant is called an **incompressible substance**. The specific volumes of solids and liquids essentially remain constant during a process (Fig. 5–33). Therefore, liquids and solids can be approximated as incompressible substances without sacrificing much in accuracy. The constant-volume assumption should be taken to imply that the energy associated with the volume change is negligible compared with other forms of energy. Otherwise, this assumption would be ridiculous for studying the thermal stresses in solids (caused by volume change with temperature) or analyzing liquid-in-glass thermometers.

It can be mathematically shown that the constant-volume and constant-pressure specific heats are identical for incompressible substances (Fig. 5–34). Therefore, for solids and liquids, the subscripts on c_p and c_v can be dropped, and both specific heats can be represented by a single symbol c. That is,

$$c_p = c_v = c \qquad \textbf{(5–32)}$$

This result could also be deduced from the physical definitions of constant-volume and constant-pressure specific heats. Specific heat values for several common liquids and solids are given in Table A–3.

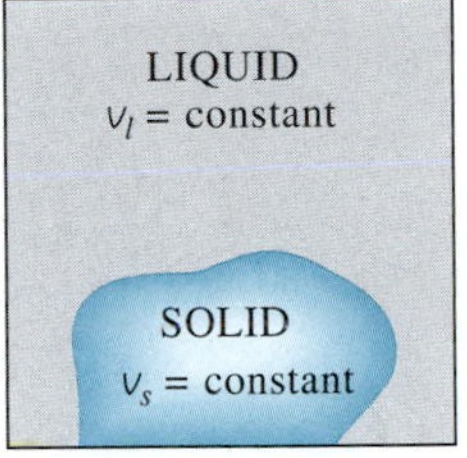

FIGURE 5–33

The specific volumes of incompressible substances remain constant during a process.

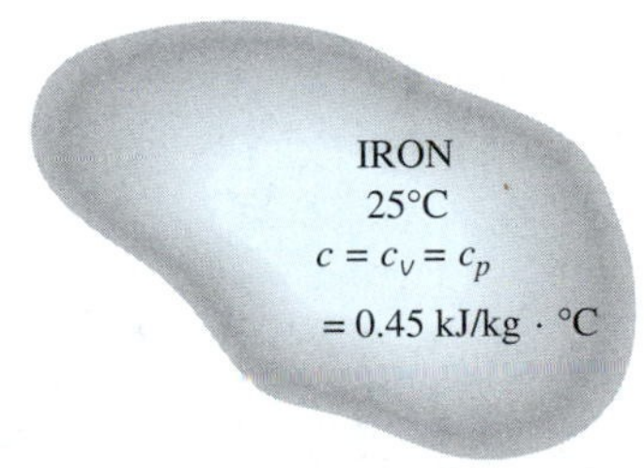

FIGURE 5–34

The c_v and c_p values of incompressible substances are identical and are denoted by c.

Internal Energy Changes

Like those of ideal gases, the specific heats of incompressible substances depend on temperature only. Thus, the partial differentials in the defining equation of c_v can be replaced by ordinary differentials, which yield

$$du = c_v\,dT = c(T)\,dT \qquad \textbf{(5–33)}$$

The change in internal energy between states 1 and 2 is then obtained by integration:

$$\Delta u = u_2 - u_1 = \int_1^2 c(T)\,dT \qquad \text{(kJ/kg)} \qquad \textbf{(5–34)}$$

The variation of specific heat c with temperature should be known before this integration can be carried out. For small temperature intervals, a c value at the average temperature can be used and treated as a constant, yielding

$$\Delta u \cong c_{\text{avg}}(T_2 - T_1) \qquad \text{(kJ/kg)} \qquad \textbf{(5–35)}$$

Enthalpy Changes

Using the definition of enthalpy $h = u + Pv$ and noting that v = constant, the differential form of the enthalpy change of incompressible substances can be determined by differentiation to be

$$dh = du + v\,dP + P\,\overset{\nearrow 0}{dv} = du + v\,dP \qquad \textbf{(5–36)}$$

Integrating,

$$\Delta h = \Delta u + v\,\Delta P \cong c_{\text{avg}}\,\Delta T + v\,\Delta P \qquad \text{(kJ/kg)} \qquad \textbf{(5–37)}$$

For *solids*, the term $v\,\Delta P$ is insignificant and thus $\Delta h = \Delta u \cong c_{avg}\Delta T$. For *liquids*, two special cases are commonly encountered:

1. *Constant-pressure processes,* as in heaters ($\Delta P = 0$): $\Delta h = \Delta u \cong c_{avg}\Delta T$
2. *Constant-temperature processes,* as in pumps ($\Delta T = 0$): $\Delta h = v\,\Delta P$

For a process between states 1 and 2, the last relation can be expressed as $h_2 - h_1 = v(P_2 - P_1)$. By taking state 2 to be the compressed liquid state at a given T and P and state 1 to be the saturated liquid state at the same temperature, the enthalpy of the compressed liquid can be expressed as

$$h_{@P,T} \cong h_{f@T} + v_{f@T}(P - P_{sat@T}) \tag{5-38}$$

as discussed in Chap. 4. This is an improvement over the assumption that the enthalpy of the compressed liquid could be taken as h_f at the given temperature (that is, $h_{@P,T} \cong h_{f@T}$). However, the contribution of the last term is often very small, and is neglected. (Note that at high temperature and pressures, Eq. 5–38 may overcorrect the enthalpy and result in a larger error than the approximation $h \cong h_{f@T}$.)

EXAMPLE 5–11 Enthalpy of a Compressed Liquid

Determine the enthalpy of liquid water at 100°C and 15 MPa (*a*) by using compressed liquid tables, (*b*) by approximating it as a saturated liquid, and (*c*) by using the correction given by Eq. 5–38.

Solution The enthalpy of liquid water is to be determined exactly and approximately.

Analysis At 100°C, the saturation pressure of water is 101.42 kPa, and since $P > P_{sat}$, the water exists as a compressed liquid at the specified state.

(*a*) From compressed liquid tables, we read

$$\left.\begin{aligned} P &= 15\text{ MPa} \\ T &= 100^\circ\text{C} \end{aligned}\right\} \quad h = \mathbf{430.39\ kJ/kg} \qquad \text{(Table A–7)}$$

This is the exact value.

(*b*) Approximating the compressed liquid as a saturated liquid at 100°C, as is commonly done, we obtain

$$h \cong h_{f@100^\circ C} = \mathbf{419.17\ kJ/kg}$$

This value is in error by about 2.6 percent.

(*c*) From Eq. 5–38,

$$h_{@P,T} \cong h_{f@T} + v_{f@T}(P - P_{sat@T})$$

$$= (419.17\text{ kJ/kg}) + (0.001\text{ m}^3\text{ kg})[(15{,}000 - 101.42)\text{ kPa}]\left(\frac{1\text{ kJ}}{1\text{ kPa}\cdot\text{m}^3}\right)$$

$$= \mathbf{434.07\ kJ/kg}$$

Discussion Note that the correction term reduced the error from 2.6 to about 1 percent in this case. However, this improvement in accuracy is often not worth the extra effort involved.

EXAMPLE 5–12 Cooling of an Iron Block by Water

A 50-kg iron block at 80°C is dropped into an insulated tank that contains 0.5 m^3 of liquid water at 25°C. Determine the temperature when thermal equilibrium is reached.

Solution An iron block is dropped into water in an insulated tank. The final temperature when thermal equilibrium is reached is to be determined.

Assumptions **1** Both water and the iron block are incompressible substances. **2** Constant specific heats at room temperature can be used for water and the iron. **3** The system is stationary and thus the kinetic and potential energy changes are zero, $\Delta KE = \Delta PE = 0$ and $\Delta E = \Delta U$. **4** There are no electrical, shaft, or other forms of work involved. **5** The system is well-insulated and thus there is no heat transfer.

Analysis We take the entire contents of the tank as the *system* (Fig. 5–35). This is a *closed system* since no mass crosses the system boundary during the process. We observe that the volume of a rigid tank is constant, and thus there is no boundary work. The energy balance on the system can be expressed as

$$\underbrace{E_{in} - E_{out}}_{\substack{\text{Net energy transfer}\\ \text{by heat, work, and mass}}} = \underbrace{\Delta E_{system}}_{\substack{\text{Change in internal, kinetic,}\\ \text{potential, etc., energies}}}$$

$$0 = \Delta U$$

The total internal energy U is an extensive property, and therefore it can be expressed as the sum of the internal energies of the parts of the system. Then the total internal energy change of the system becomes

$$\Delta U_{sys} = \Delta U_{iron} + \Delta U_{water} = 0$$

$$[mc(T_2 - T_1)]_{iron} + [mc(T_2 - T_1)]_{water} = 0$$

The specific volume of liquid water at or about room temperature can be taken to be 0.001 m^3/kg. Then the mass of the water is

$$m_{water} = \frac{V}{v} = \frac{0.5 \text{ m}^3}{0.001 \text{ m}_3\text{/kg}} = 500 \text{ kg}$$

The specific heats of iron and liquid water are determined from Table A–3 to be $c_{iron} = 0.45$ kJ/kg · °C and $c_{water} = 4.18$ kJ/kg · °C. Substituting these values into the energy equation, we obtain

$$(50 \text{ kg})(0.45 \text{ kJ/kg} \cdot °\text{C})(T_2 - 80°\text{C}) + (500 \text{ kg})(4.18 \text{ kJ/kg} \cdot °\text{C})(T_2 - 25°\text{C}) = 0$$

$$T_2 = \mathbf{25.6°C}$$

Therefore, when thermal equilibrium is established, both the water and iron are at 25.6°C.

Discussion The small rise in water temperature is due to its large mass and large specific heat.

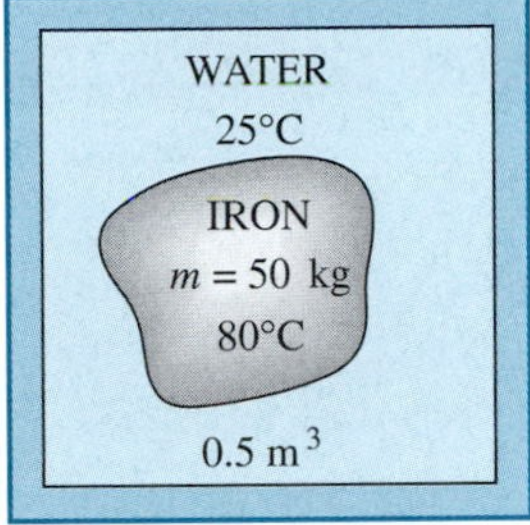

FIGURE 5–35
Schematic for Example 5–12.

EXAMPLE 5–13 Temperature Rise due to Slapping

If you ever slapped someone or got slapped yourself, you probably remember the burning sensation. Imagine you had the unfortunate occasion of being slapped by an angry person, which caused the temperature of the affected area of your face to rise by 1.8°C (ouch!). Assuming the slapping hand has a mass of 1.2 kg and about 0.150 kg of the tissue on the face and the hand is affected by the incident, estimate the velocity of the hand just before impact. Take the specific heat of the tissue to be 3.8 kJ/kg · °C.

Solution The face of a person is slapped. For the specified temperature rise of the affected part, the impact velocity of the hand is to be determined.

Assumptions **1** The hand is brought to a complete stop after the impact. **2** The face takes the blow without significant movement. **3** No heat is transferred from the affected area to the surroundings, and thus the process is adiabatic. **4** No work is done on or by the system. **5** The potential energy change is zero, $\Delta PE = 0$ and $\Delta E = \Delta U + \Delta KE$.

Analysis We analyze this incident in a professional manner without involving any emotions. First, we identify the system, draw a sketch of it, and state our observations about the specifics of the problem. We take the hand and the affected portion of the face as the *system* (Fig. 5–36). This is a *closed system* since it involves a fixed amount of mass (no mass transfer). We observe that the kinetic energy of the hand decreases during the process, as evidenced by a decrease in velocity from initial value to zero, while the internal energy of the affected area increases, as evidenced by an increase in the temperature. There seems to be no significant energy transfer between the system and its surroundings during this process.

FIGURE 5–36

Schematic for Example 5–13.

Under the stated assumptions and observations, the energy balance on the system can be expressed as

$$\underbrace{E_{in} - E_{out}}_{\substack{\text{Net energy transfer} \\ \text{by heat, work, and mass}}} = \underbrace{\Delta E_{system}}_{\substack{\text{Change in internal, kinetic,} \\ \text{potential, etc., energies}}}$$

$$0 = \Delta U_{\text{affected tissue}} + \Delta KE_{\text{hand}}$$

$$0 = (mc\Delta T)_{\text{affected tissue}} + [m(0 - V^2)/2]_{\text{hand}}$$

That is, the decrease in the kinetic energy of the hand must be equal to the increase in the internal energy of the affected area. Solving for the velocity and substituting the given quantities, the impact velocity of the hand is determined to be

$$V_{\text{hand}} = \sqrt{\frac{2(mc\Delta T)_{\text{affected tissue}}}{m_{\text{hand}}}}$$

$$= \sqrt{\frac{2(0.15\text{ kg})(3.8\text{ kJ/kg}\cdot{}^\circ\text{C})(1.8^\circ\text{C})}{1.2\text{ kg}}\left(\frac{1000\text{ m}^2/\text{s}^2}{1\text{ kJ/kg}}\right)}$$

$$= \mathbf{41.4\ m/s}\text{ (or 149 km/h)}$$

Discussion Reconstruction of events such as this by making appropriate assumptions are commonly used in forensic engineering.

SUMMARY

Work is the energy transferred as a force acts on a system through a distance. The most common form of mechanical work is the *boundary work,* which is the work associated with the expansion and compression of substances. On a *P*-*V* diagram, the area under the process curve represents the boundary work for a quasi-equilibrium process. Various forms of boundary work are expressed as follows:

(1) General $$W_b = \int_1^2 P\,dV$$

(2) Isobaric process

$$W_b = P_0(V_2 - V_1) \quad (P_1 = P_2 = P_0 = \text{constant})$$

(3) Polytropic process

$$W_b = \frac{P_2V_2 - P_1V_1}{1-n} \quad (n \neq 1) \quad (PV^n = \text{constant})$$

(4) Isothernal process of an ideal gas

$$W_b = P_1V_1 \ln \frac{V_2}{V_1} = mRT_0 \ln \frac{V_2}{V_1} \quad (PV = mRT_0 = \text{constant})$$

The first law of thermodynamics is essentially an expression of the conservation of energy principle, also called the energy balance. The general energy balances for *any system* undergoing *any process* can be expressed as

$$\underbrace{E_{in} - E_{out}}_{\substack{\text{Net energy transfer} \\ \text{by heat, work, and mass}}} = \underbrace{\Delta E_{system}}_{\substack{\text{Change in internal, kinetic,} \\ \text{potential, etc., energies}}}$$

It can also be expressed in the *rate form* as

$$\underbrace{\dot{E}_{in} - \dot{E}_{out}}_{\substack{\text{Rate of net energy transfer} \\ \text{by heat, work, and mass}}} = \underbrace{dE_{system}/dt}_{\substack{\text{Rate of change in internal,} \\ \text{kinetic, potential, etc., energies}}}$$

Taking heat transfer *to* the system and work done *by* the system to be positive quantities, the energy balance for a closed system can also be expressed as

$$Q - W = \Delta U + \Delta KE + \Delta PE$$

where

$$W = W_{other} + W_b$$

$$\Delta U = m(u_2 - u_1)$$

$$\Delta KE = \tfrac{1}{2}m(V_2^2 - V_1^2)$$

$$\Delta PE = mg(z_2 - z_1)$$

For a *constant-pressure process,* $W_b + \Delta U = \Delta H$. Thus,

$$Q - W_{other} = \Delta H + \Delta KE + \Delta PE$$

The amount of energy needed to raise the temperature of a unit mass of a substance by one degree is called the *specific heat at constant volume* c_v for a constant-volume process and the *specific heat at constant pressure* c_p for a constant-pressure process. They are defined as

$$c_v = \left(\frac{\partial u}{\partial T}\right)_v \quad \text{and} \quad c_p = \left(\frac{\partial h}{\partial T}\right)_p$$

For ideal gases u, h, c_v, and c_p are functions of temperature alone. The Δu and Δh of ideal gases are expressed as

$$\Delta u = u_2 - u_1 = \int_1^2 c_v(T)\,dT \cong c_{v,avg}(T_2 - T_1)$$

$$\Delta h = h_2 - h_1 = \int_1^2 c_p(T)\,dT \cong c_{p,avg}(T_2 - T_1)$$

For ideal gases, c_v and c_p are related by

$$c_p = c_v + R$$

where R is the gas constant. The *specific heat ratio* k is defined as

$$k = \frac{c_p}{c_v}$$

For *incompressible substances* (liquids and solids), both the constant-pressure and constant-volume specific heats are identical and denoted by c:

$$c_p = c_v = c$$

The Δu and Δh of imcompressible substances are given by

$$\Delta u = \int_1^2 c(T)\,dT \cong c_{avg}(T_2 - T_1)$$

$$\Delta h = \Delta u + v\Delta P$$

REFERENCES AND SUGGESTED READINGS

1. ASHRAE *Handbook of Fundamentals.* SI version. Atlanta, GA: American Society of Heating, Refrigerating, and Air-Conditioning Engineers, Inc., 1993.
2. ASHRAE *Handbook of Refrigeration.* SI version. Atlanta, GA: American Society of Heating, Refrigerating, and Air-Conditioning Engineers, Inc., 1994.

PROBLEMS*

Moving Boundary Work

5–1C On a *P*-*v* diagram, what does the area under the process curve represent?

5–2C Is the boundary work associated with constant-volume systems always zero?

5–3C An ideal gas at a given state expands to a fixed final volume first at constant pressure and then at constant temperature. For which case is the work done greater?

5–4C Show that 1 kPa · m^3 = 1 kJ.

5–5 The volume of 1 kg of helium in a piston-cylinder device is initially 5 m^3. Now helium is compressed to 3 m^3 while its pressure is maintained constant at 200 kPa. Determine the initial and final temperatures of helium as well as the work required to compress it, in kJ.

5–6 Calculate the total work, in kJ, for process 1-3 shown in Fig. P5–6 when the system consists of 2 kg of nitrogen.

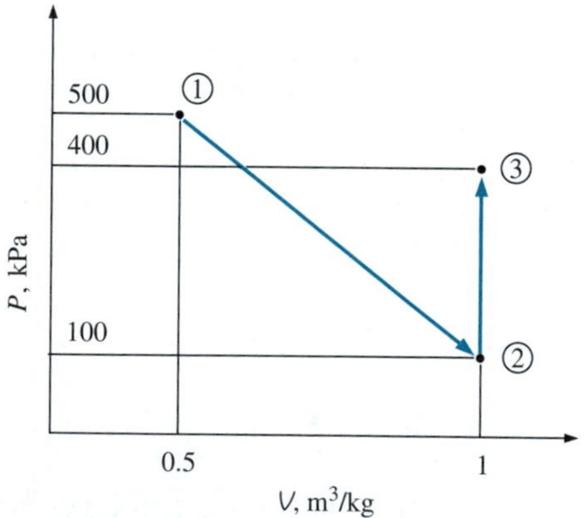

FIGURE P5–6

*Problems designated by a "C" are concept questions, and students are encouraged to answer them all. Problems designated by an "E" are in English units, and the SI users can ignore them. Problems with the icon are solved using EES, and complete solutions together with parametric studies are included on the enclosed DVD. Problems with the icon are comprehensive in nature, and are intended to be solved with a computer, preferably using the EES software that accompanies this text.

5–7E Calculate the total work, in Btu, produced by the process of Fig. P5–7E.

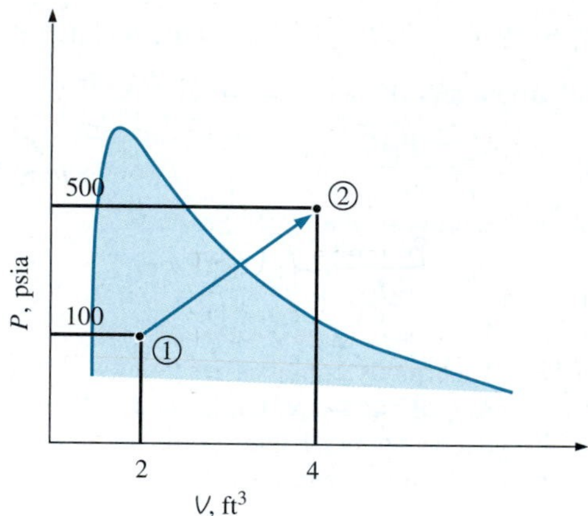

FIGURE P5–7E

5–8 A piston–cylinder device initially contains 0.07 m^3 of nitrogen gas at 130 kPa and 120°C. The nitrogen is now expanded polytropically to a state of 100 kPa and 100°C. Determine the boundary work done during this process.

5–9 A piston–cylinder device with a set of stops initially contains 0.3 kg of steam at 1.0 MPa and 400°C. The location of the stops corresponds to 60 percent of the initial volume. Now the steam is cooled. Determine the compression work if the final state is (*a*) 1.0 MPa and 250°C and (*b*) 500 kPa. (*c*) Also determine the temperature at the final state in part (*b*).

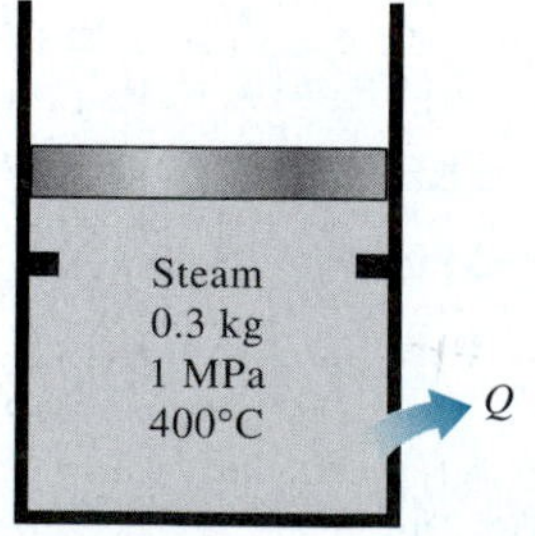

FIGURE P5–9

5–10 A piston–cylinder device initially contains 0.07 m^3 of nitrogen gas at 130 kPa and 120°C. The nitrogen is now expanded to a pressure of 100 kPa polytropically with a polytropic exponent whose value is equal to the specific heat ratio (called *isentropic expansion*). Determine the final temperature and the boundary work done during this process.

5–11 A mass of 5 kg of saturated water vapor at 300 kPa is heated at constant pressure until the temperature reaches 200°C. Calculate the work done by the steam during this process. *Answer:* 165.9 kJ

5–12 A frictionless piston–cylinder device initially contains 200 L of saturated liquid refrigerant-134a. The piston is free to move, and its mass is such that it maintains a pressure of 900 kPa on the refrigerant. The refrigerant is now heated until its temperature rises to 70°C. Calculate the work done during this process. *Answer:* 5571 kJ

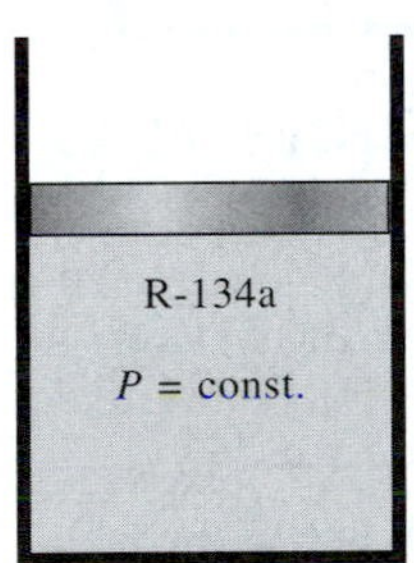

FIGURE P5–12

5–13 Reconsider Prob. 5–12. Using EES (or other) software, investigate the effect of pressure on the work done. Let the pressure vary from 400 to 1200 kPa. Plot the work done versus the pressure, and discuss the results. Explain why the plot is not linear. Also plot the process described in Prob. 5–12 on the *P-v* diagram.

5–14E A frictionless piston–cylinder device contains 16 lbm of superheated water vapor at 40 psia and 600°F. Steam is now cooled at constant pressure until 70 percent of it, by mass, condenses. Determine the work done during this process.

5–15 A mass of 2.4 kg of air at 150 kPa and 12°C is contained in a gas-tight, frictionless piston–cylinder device. The air is now compressed to a final pressure of 600 kPa. During the process, heat is transferred from the air such that the temperature inside the cylinder remains constant. Calculate the work input during this process. *Answer:* 272 kJ

5–16E During an expansion process, the pressure of a gas changes from 15 to 100 psia according to the relation $P = aV + b$, where $a = 5$ psia/ft^3 and b is a constant. If the initial volume of the gas is 7 ft^3, calculate the work done during the process. *Answer:* 181 Btu

5–17 During some actual expansion and compression processes in piston–cylinder devices, the gases have been observed to satisfy the relationship $PV^n = C$, where n and C are constants. Calculate the work done when a gas expands from 150 kPa and 0.03 m^3 to a final volume of 0.2 m^3 for the case of $n = 1.3$.

5–18 Reconsider Prob. 5–17. Using the EES (or other) software, plot the process described in the problem on a *P-V* diagram, and investigate the effect of the polytropic exponent n on the boundary work. Let the polytropic exponent vary from 1.1 to 1.6. Plot the boundary work versus the polytropic exponent, and discuss the results.

5–19 A frictionless piston–cylinder device contains 2 kg of nitrogen at 100 kPa and 300 K. Nitrogen is now compressed slowly according to the relation $PV^{1.4}$ = constant until it reaches a final temperature of 360 K. Calculate the work input during this process. *Answer:* 89 kJ

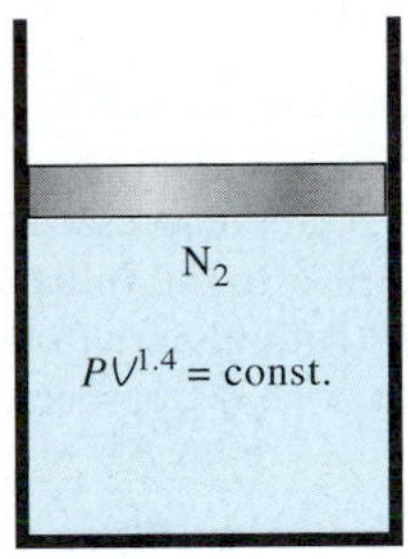

FIGURE P5–19

5–20 The equation of state of a gas is given as $\bar{v}(P + 10/\bar{v}^2) = R_uT$, where the units of $\bar{v}$ and P are m^3/kmol and kPa, respectively. Now 0.5 kmol of this gas is expanded in a quasi-equilibrium manner from 2 to 4 m^3 at a constant temperature of 300 K. Determine (*a*) the unit of the quantity 10 in the equation and (*b*) the work done during this isothermal expansion process.

5–21 Reconsider Prob. 5–20. Using the integration feature of the EES software, calculate the work done, and compare your result with the "hand-calculated" result obtained in Prob. 5–20. Plot the process described in the problem on a *P-v* diagram.

5–22 Carbon dioxide contained in a piston–cylinder device is compressed from 0.3 to 0.1 m^3. During the process, the pressure and volume are related by $P = aV^{-2}$, where $a = 8$ kPa · m^6. Calculate the work done on the carbon dioxide during this process. *Answer:* 53.3 kJ

5–23 Determine the boundary work done by a gas during an expansion process if the pressure and volume values at various states are measured to be 300 kPa, 1 L; 290 kPa, 1.1 L; 270 kPa, 1.2 L; 250 kPa, 1.4 L; 220 kPa, 1.7 L; and 200 kPa, 2 L.

5–24 A piston–cylinder device initially contains 0.25 kg of nitrogen gas at 130 kPa and 120°C. The nitrogen is now expanded isothermally to a pressure of 100 kPa. Determine the boundary work done during this process. *Answer:* 7.65 kJ

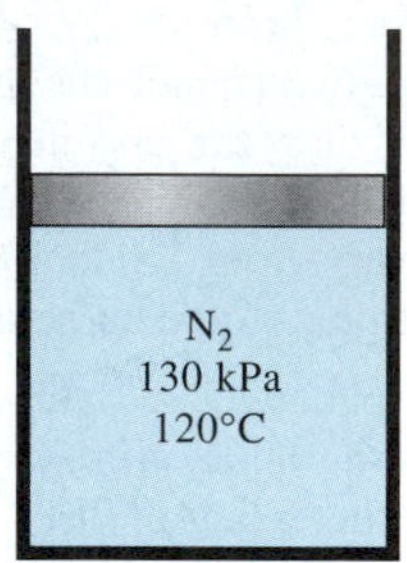

FIGURE P5–24

5–25 A piston–cylinder device contains 0.15 kg of air initially at 2 MPa and 350°C. The air is first expanded isothermally to 500 kPa, then compressed polytropically with a polytropic exponent of 1.2 to the initial pressure, and finally compressed at the constant pressure to the initial state. Determine the boundary work for each process and the net work of the cycle.

5–26 1-kg of water that is initially at 90°C with a quality of 10 percent occupies a spring-loaded piston–cylinder device, such as that in Fig. P5–26. This device is now heated until the pressure rises to 800 kPa and the temperature is 250°C. Determine the total work produced during this process, in kJ. *Answer:* 24.5 kJ

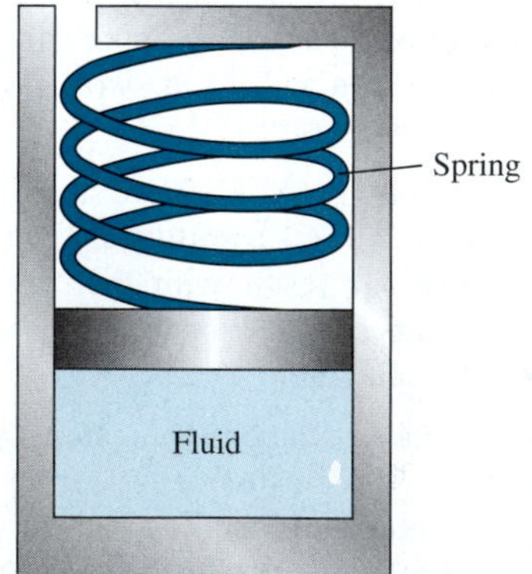

FIGURE P5–26

5–27 0.5-kg water that is initially at 1 MPa and 10 percent quality occupies a spring-loaded piston–cylinder device. This device is now cooled until the water is a saturated liquid at 100°C. Calculate the total work produced during this process, in kJ.

5–28 Argon is compressed in a polytropic process with $n = 1.2$ from 120 kPa and 30°C to 1200 kPa in a piston-cylinder device. Determine the final temperature of the argon.

Closed System Energy Analysis

5–29 Saturated water vapor at 200°C is isothermally condensed to a saturated liquid in a piston–cylinder device. Calculate the heat transfer and the work done during this process, in kJ/kg. *Answers:* 1940 kJ/kg, 196 kJ/kg

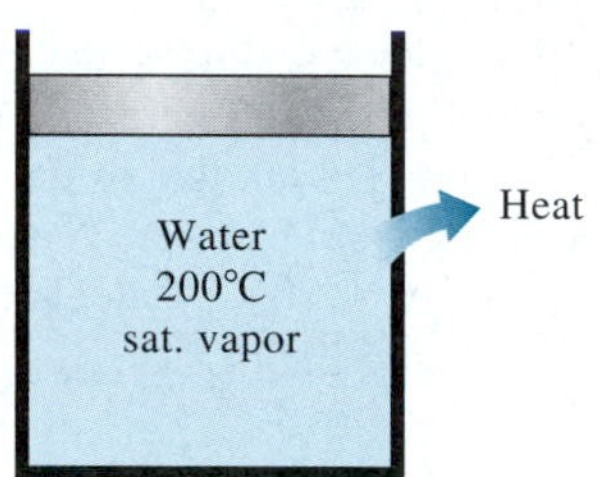

FIGURE P5–29

5–30E A closed system undergoes a process in which there is no internal energy change. During this process, the system produces 1.6×10^6 ft · lbf of work. Calculate the heat transfer for this process, in Btu.

5–31 Complete each line of the table below on the basis of the conservation of energy principle for a closed system.

Q_{in} kJ	W_{out} kJ	E_1 kJ	E_2 kJ	m kg	$e_2 - e_1$ kJ/kg
280	—	1020	860	3	—
−350	130	550	—	5	—
—	260	300	—	2	−150
300	—	750	500	1	—
—	−200	—	300	2	−100

5–32 A substance is contained in a well-insulated rigid container that is equipped with a stirring device, as shown in Fig. P5–32. Determine the change in the internal energy of this substance when 15 kJ of work is applied to the stirring device.

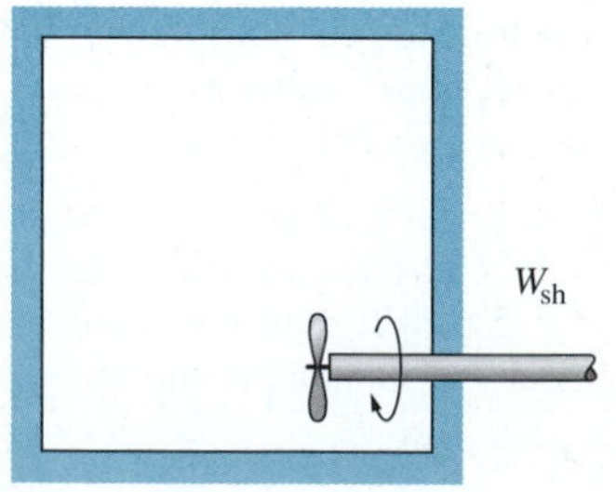

FIGURE P5–32

5–33 A rigid container equipped with a stirring device contains 1.5 kg of motor oil. Determine the rate of specific energy increase when heat is transferred to the oil at a rate of 1 W, and 1.5 W of power is applied to the stirring device.

5–34E A rigid 1-ft^3 vessel contains R-134a originally at -20°F and 27.7 percent quality. The refrigerant is then heated until its temperature is 100°F. Calculate the heat transfer required to do this. *Answer:* 84.7 Btu

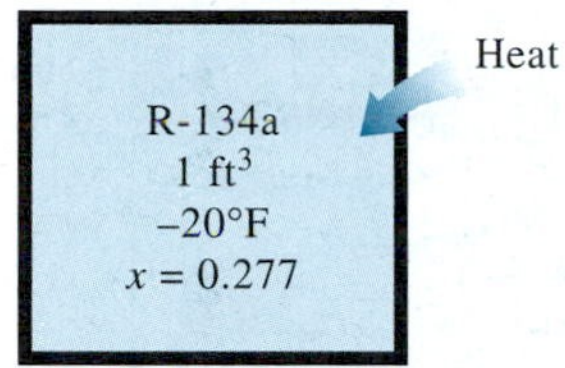

FIGURE P5–34E

5–35 A well-insulated rigid tank contains 5 kg of a saturated liquid–vapor mixture of water at 100 kPa. Initially, three-quarters of the mass is in the liquid phase. An electric resistor placed in the tank is connected to a 110-V source, and a current of 8 A flows through the resistor when the switch is turned on. Determine how long it will take to vaporize all the liquid in the tank. Also, show the process on a *T-v* diagram with respect to saturation lines.

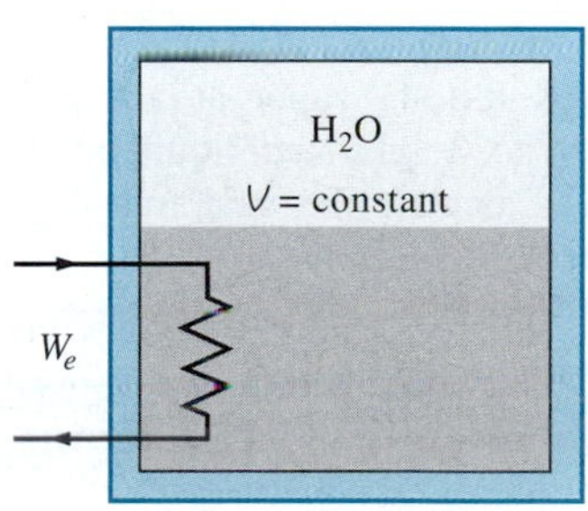

FIGURE P5–35

5–36 Reconsider Prob. 5–35. Using EES (or other) software, investigate the effect of the initial mass of water on the length of time required to completely vaporize the liquid. Let the initial mass vary from 1 to 10 kg. Plot the vaporization time against the initial mass, and discuss the results.

5–37 A piston–cylinder device contains 5 kg of refrigerant-134a at 800 kPa and 70°C. The refrigerant is now cooled at constant pressure until it exists as a liquid at 15°C. Determine the amount of heat loss and show the process on a *T-v* diagram with respect to saturation lines. *Answer:* 1173 kJ

5–38E A piston–cylinder device contains 0.5 lbm of water initially at 120 psia and 2 ft^3. Now 200 Btu of heat is transferred to the water while its pressure is held constant. Determine the final temperature of the water. Also, show the process on a *T-v* diagram with respect to saturation lines.

5–39 An insulated piston–cylinder device contains 5 L of saturated liquid water at a constant pressure of 175 kPa. Water is stirred by a paddle wheel while a current of 8 A flows for 45 min through a resistor placed in the water. If one-half of the liquid is evaporated during this constant-pressure process and the paddle-wheel work amounts to 400 kJ, determine the voltage of the source. Also, show the process on a *P-v* diagram with respect to saturation lines. *Answer:* 224 V

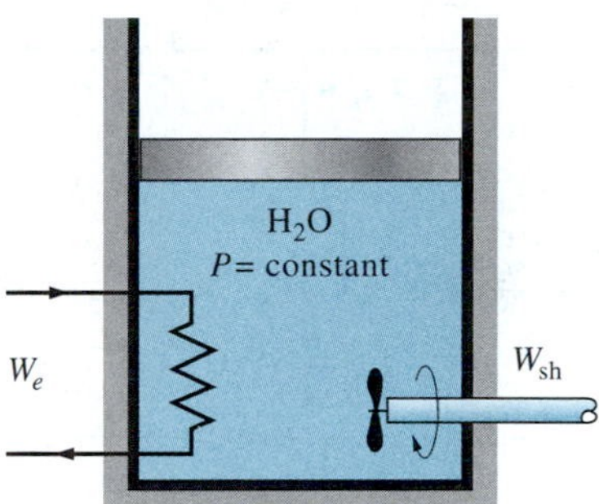

FIGURE P5–39

5–40 A piston–cylinder device initially contains steam at 200 kPa, 200°C, and 0.5 m^3. At this state, a linear spring ($F \propto x$) is touching the piston but exerts no force on it. Heat is now slowly transferred to the steam, causing the pressure and the volume to rise to 500 kPa and 0.6 m^3, respectively. Show the process on a *P-v* diagram with respect to saturation lines and determine (*a*) the final temperature, (*b*) the work done by the steam, and (*c*) the total heat transferred. *Answers:* (*a*) 1132°C, (*b*) 35 kJ, (*c*) 808 kJ

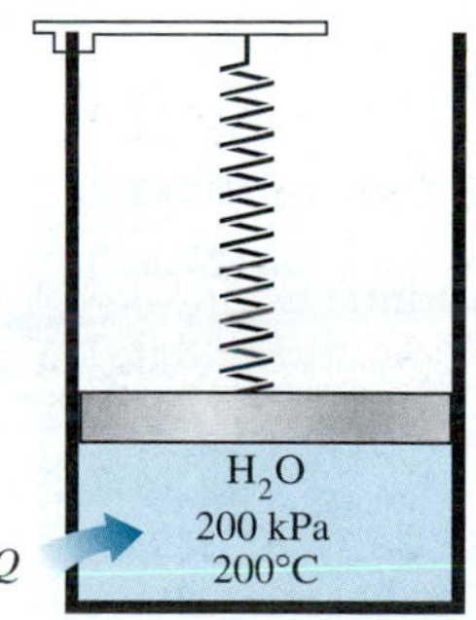

FIGURE P5–40

5–41 Reconsider Prob. 5–40. Using EES (or other) software, investigate the effect of the initial temperature of steam on the final temperature, the work done, and the total heat transfer. Let the initial temperature vary from 150 to 250°C. Plot the final results against the initial temperature, and discuss the results.

5–42 Two tanks (Tank A and Tank B) are separated by a partition. Initially Tank A contains 2-kg steam at 1 MPa and 300°C while Tank B contains 3-kg saturated liquid–vapor mixture at 150°C with a vapor mass fraction of 50 percent. Now the partition is removed and the two sides are allowed to mix until the mechanical and thermal equilibrium are established. If the pressure at the final state is 300 kPa, determine (*a*) the temperature and quality of the steam (if mixture) at the final state and (*b*) the amount of heat lost from the tanks.

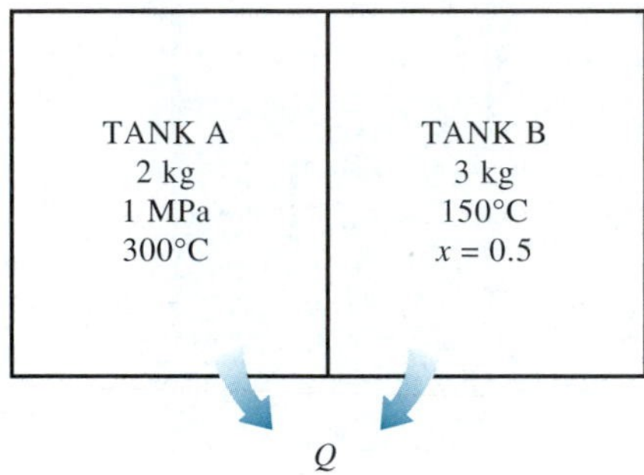

FIGURE P5–42

5–43 A 30-L electrical radiator containing heating oil is placed in a 50-m^3 room. Both the room and the oil in the radiator are initially at 10°C. The radiator with a rating of 1.8 kW is now turned on. At the same time, heat is lost from the room at an average rate of 0.35 kJ/s. After some time, the average temperature is measured to be 20°C for the air in the room, and 50°C for the oil in the radiator. Taking the density and the specific heat of the oil to be 950 kg/m^3 and 2.2 kJ/kg · °C, respectively, determine how long the heater is kept on. Assume the room is well-sealed so that there are no air leaks.

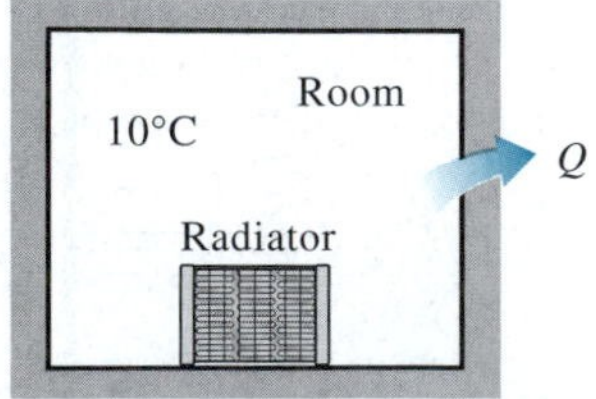

FIGURE P5–43

5–44 2-kg of saturated liquid water at 150°C is heated at constant pressure in a piston–cylinder device until it is saturated vapor. Determine the heat transfer required for this process.

5–45 Steam at 75 kPa and 13 percent quality is contained in a spring-loaded piston–cylinder device, as shown in Fig. P5–45, with an initial volume of 2 m^3. Steam is now heated until its volume is 5 m^3 and its pressure is 300 kPa. Determine the heat transferred to and the work produced by the steam during this process.

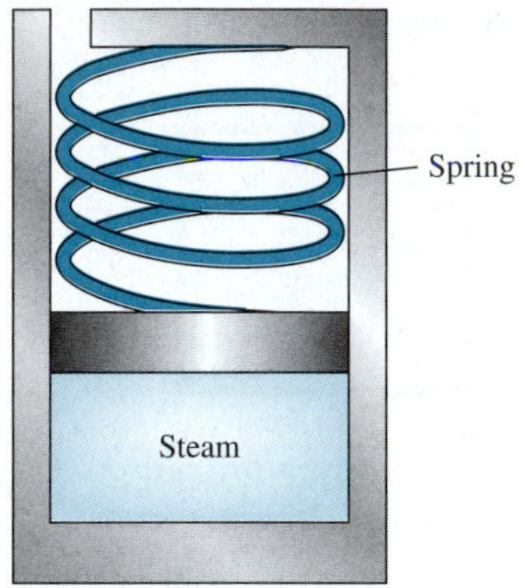

FIGURE P5–45

5–46 Refrigerant-134a at 600 kPa and 150°C is contained in a spring-loaded piston–cylinder device with an initial volume of 0.3 m^3. The refrigerant is now cooled until its temperature is −30°C and its volume is 0.1 m^3. Determine the heat transferred to and the work produced by the refrigerant during this process. *Answers:* 1849 kJ heat out, 68.4 kJ work in

5–47E Saturated R-134a vapor at 100°F is condensed at constant pressure to a saturated liquid in a closed piston–cylinder system. Calculate the heat transfer and work done during this process, in Btu/lbm.

5–48 2-kg of saturated liquid R-134a with an initial temperature of −10°C is contained in a well-insulated, weighted piston–cylinder device. This device contains an electrical resistor, as shown in Fig. P5–48, to which 10 volts are applied causing a current of 2 amperes to flow through the resistor.

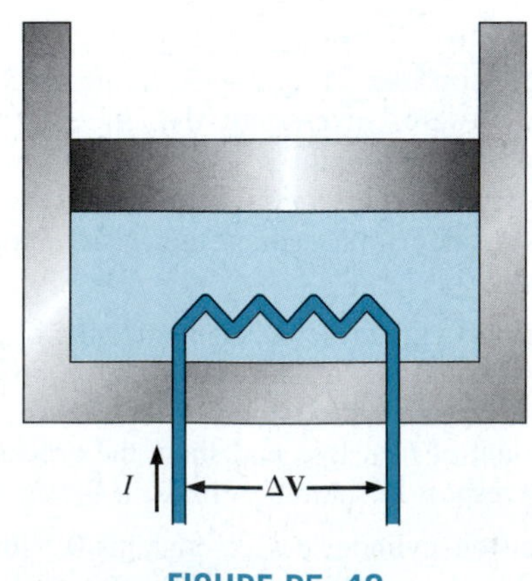

FIGURE P5–48

Determine the time required for the refrigerant to be converted to a saturated vapor, and the final temperature.

Specific Heats, Δu, and Δh of Ideal Gases

5–49C Is the relation $\Delta u = mc_{v,\text{avg}}\Delta T$ restricted to constant-volume processes only, or can it be used for any kind of process of an ideal gas?

5–50C Is the relation $\Delta h = mc_{p,\text{avg}}\Delta T$ restricted to constant-pressure processes only, or can it be used for any kind of process of an ideal gas?

5–51C Show that for an ideal gas $\bar{c}_p = \bar{c}_v + R_u$.

5–52C Is the energy required to heat air from 295 to 305 K the same as the energy required to heat it from 345 to 355 K? Assume the pressure remains constant in both cases.

5–53C In the relation $\Delta u = mc_v \Delta T$, what is the correct unit of c_v — kJ/kg · °C or kJ/kg · K?

5–54C A fixed mass of an ideal gas is heated from 50 to 80°C at a constant pressure of (*a*) 1 atm and (*b*) 3 atm. For which case do you think the energy required will be greater? Why?

5–55C A fixed mass of an ideal gas is heated from 50 to 80°C at a constant volume of (*a*) 1 m^3 and (*b*) 3 m^3. For which case do you think the energy required will be greater? Why?

5–56C When undergoing a given temperature change, which of the two gases—air or oxygen—experiences the largest change in its (*a*) enthalpy, *h* and (*b*) internal energy, *u*?

5–57 A mass of 10 g of nitrogen is contained in the spring-loaded piston–cylinder device shown in Fig. P5–57. The spring constant is 1 kN/m, and the piston diameter is 10 cm. When the spring exerts no force against the piston, the nitrogen is at 120 kPa and 27°C. The device is now heated until its volume is 10 percent greater than the original volume. Determine the change in the specific internal energy and enthalpy of the nitrogen. *Answers:* 46.8 kJ/kg, 65.5 kJ/kg

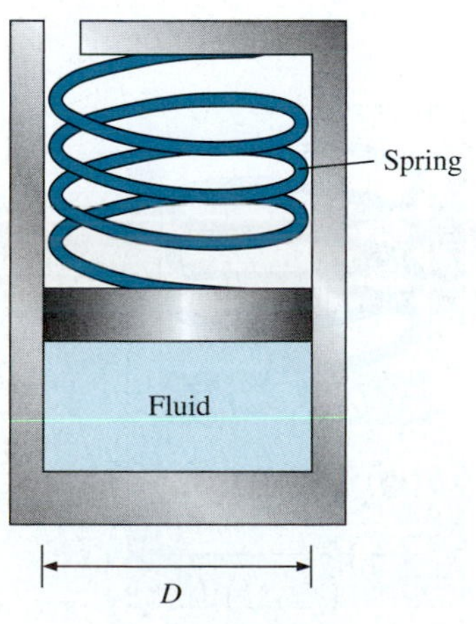

FIGURE P5–57

5–58E What is the change in the internal energy, in Btu/lbm, of air as its temperature changes from 100 to 200°F? Is there any difference if the temperature were to change from 0 to 100°F?

5–59 The temperature of 2 kg of neon is increased from 20°C to 180°C. Calculate the change in the total internal energy of the neon, in kJ. Would the internal energy change be any different if the neon were replaced with argon?

5–60 Calculate the change in the enthalpy of argon, in kJ/kg, when it is cooled from 400 to 100°C. If neon had undergone this same change of temperature, would its enthalpy change have been any different?

5–61 Neon is compressed from 100 kPa and 20°C to 500 kPa in an isothermal compressor. Determine the change in the specific volume and specific enthalpy of neon caused by this compression.

5–62E Determine the enthalpy change Δh of oxygen, in Btu/lbm, as it is heated from 800 to 1500 R, using (*a*) the empirical specific heat equation as a function of temperature (Table A–2E*c*), (*b*) the c_p value at the average temperature (Table A–2E*b*), and (*c*) the c_p value at room temperature (Table A–2E*a*).

Answers: (*a*) 170.1 Btu/lbm, (*b*) 178.5 Btu/lbm, (*c*) 153.3 Btu/lbm

5–63 Determine the internal energy change Δu of hydrogen, in kJ/kg, as it is heated from 200 to 800 K, using (*a*) the empirical specific heat equation as a function of temperature (Table A–2*c*), (*b*) the c_v value at the average temperature (Table A–2*b*), and (*c*) the c_v value at room temperature (Table A–2*a*).

Closed-System Energy Analysis: Ideal Gases

5–64C Is it possible to compress an ideal gas isothermally in an adiabatic piston–cylinder device? Explain.

5–65E A rigid tank contains 20 lbm of air at 50 psia and 80°F. The air is now heated until its pressure doubles. Determine (*a*) the volume of the tank and (*b*) the amount of heat transfer. *Answers:* (*a*) 80 ft^3, (*b*) 1898 Btu

5–66 A 4-m × 5-m × 6-m room is to be heated by a baseboard resistance heater. It is desired that the resistance heater be able to raise the air temperature in the room from 7 to 23°C within 15 min. Assuming no heat losses from the room and an atmospheric pressure of 100 kPa, determine the required power of the resistance heater. Assume constant specific heats at room temperature. *Answer:* 1.91 kW

5–67 A student living in a 4-m × 6-m × 6-m dormitory room turns on her 150-W fan before she leaves the room on a summer day, hoping that the room will be cooler when she comes back in the evening. Assuming all the doors and windows are tightly closed and disregarding any heat transfer through the walls and the windows, determine the temperature in the room when she comes back 10 h later. Use specific heat values at room temperature, and assume the

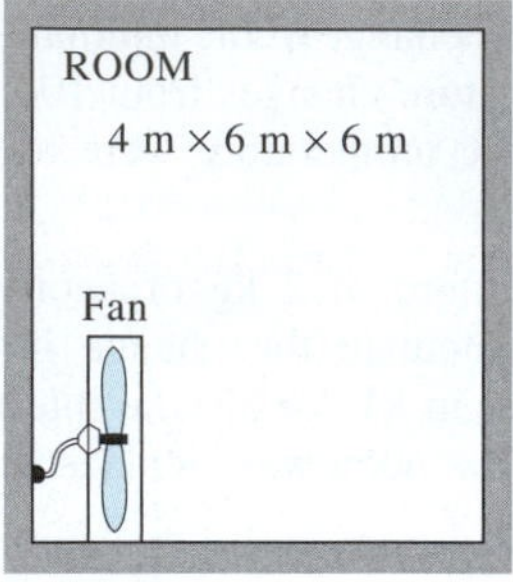

FIGURE P5–67

room to be at 100 kPa and 15°C in the morning when she leaves. *Answer:* 58.2°C

5–68E A 3-ft³ adiabatic rigid container is divided into two equal volumes by a thin membrane, as shown in Fig. P5–68E. Initially, one of these chambers is filled with air at 100 psia and 100°F while the other chamber is evacuated. Determine the internal energy change of the air when the membrane is ruptured. Also determine the final air pressure in the container.

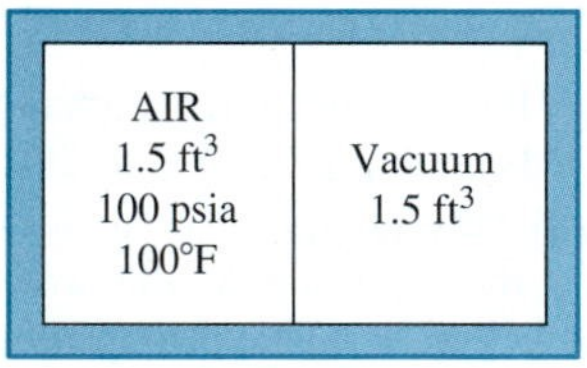

FIGURE P5–68E

AIR 1.5 ft³ 100 psia 100°F | Vacuum 1.5 ft³

5–69 Nitrogen in a rigid vessel is cooled by rejecting 100 kJ/kg of heat. Determine the internal energy change of the nitrogen, in kJ/kg.

5–70E Nitrogen at 100 psia and 300°F in a rigid container is cooled until its pressure is 50 psia. Determine the work done and the heat transferred during this process, in Btu/lbm. *Answers:* 0 Btu/lbm, 67.3 Btu/lbm

5–71 1-kg of oxygen is heated from 25 to 300°C. Determine the amount of heat transfer required when this is done during a (*a*) constant-volume process and (*b*) isobaric process.

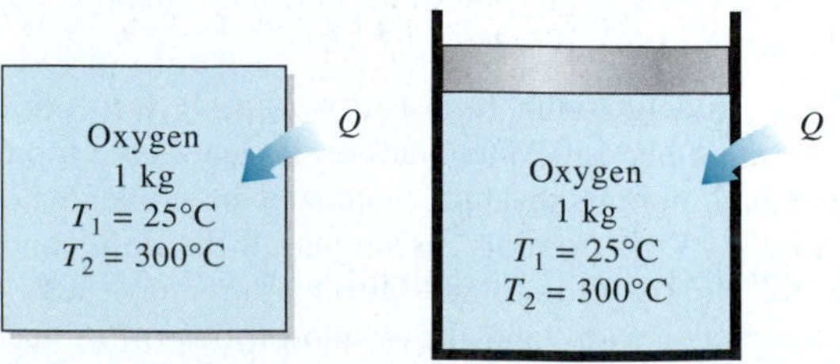

FIGURE P5–71

5–72 A closed system containing 2-kg of air undergoes an isothermal process from 600 kPa and 200°C to 80 kPa. Determine the initial volume of this system, the work done, and the heat transfer during this process. *Answers:* 0.453 m³, 547 kJ, 547 kJ

5–73 A piston–cylinder device containing argon gas as the system undergoes an isothermal process from 200 kPa and 100°C to 50 kPa. During the process, 1500 kJ of heat is transferred to the system. Determine the mass of this system and the amount of work produced.

5–74 Argon is compressed in a polytropic process with n = 1.2 from 120 kPa and 30°C to 1200 kPa in a piston–cylinder device. Determine the work produced and heat transferred during this compression process, in kJ/kg.

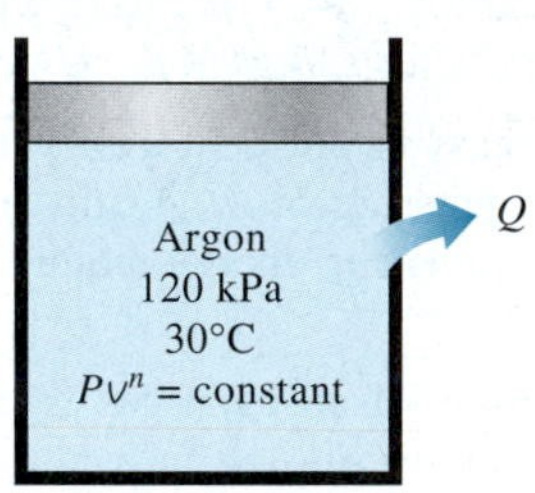

FIGURE P5–74

5–75E A piston–cylinder device containing carbon-dioxide gas undergoes an isobaric process from 15 psia and 80°F to 200°F. Determine the work and the heat transfer associated with this process, in Btu/lbm. *Answers:* 5.42 Btu/lbm, 24.4 Btu/lbm

5–76 A spring-loaded piston–cylinder device contains 5 kg of helium as the system, as shown in Fig. P5–76. This system is heated from 100 kPa and 20°C to 800 kPa and 160°C. Determine the heat transferred to and work produced by this system.

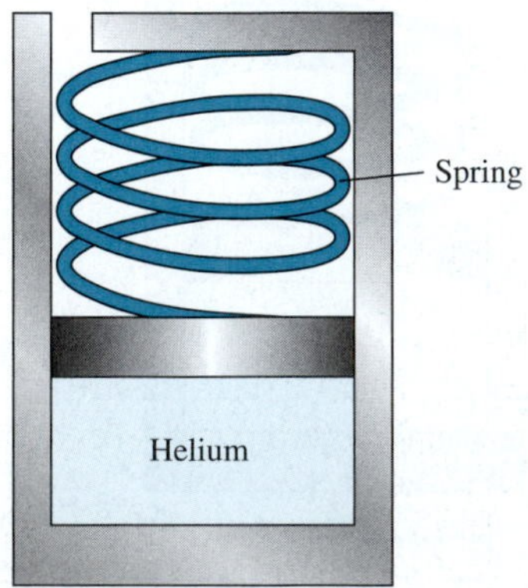

FIGURE P5–76

5–77 Air is contained in a variable-load piston device equipped with a paddle wheel. Initially, air is at 500 kPa and 27°C. The paddle wheel is now turned by an external electric motor until 50 kJ/kg of work has been transferred to air. During this process, heat is transferred to maintain a constant air temperature while allowing the gas volume to triple. Calculate the required amount of heat transfer, in kJ/kg. *Answer:* 122 kJ/kg

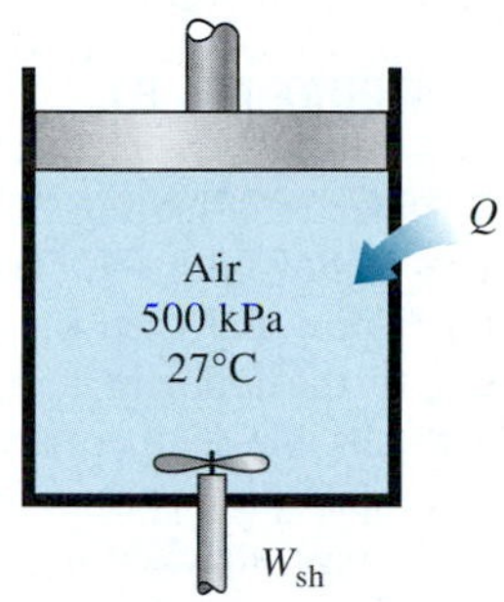

FIGURE P5–77

5–78 A piston–cylinder device whose piston is resting on top of a set of stops initially contains 0.5 kg of helium gas at 100 kPa and 25°C. The mass of the piston is such that 500 kPa of pressure is required to raise it. How much heat must be transferred to the helium before the piston starts rising? *Answer:* 1857 kJ

5–79 A mass of 15 kg of air in a piston–cylinder device is heated from 25 to 77°C by passing current through a resistance heater inside the cylinder. The pressure inside the cylinder is held constant at 300 kPa during the process, and a heat loss of 60 kJ occurs. Determine the electric energy supplied, in kWh. *Answer:* 0.235 kWh

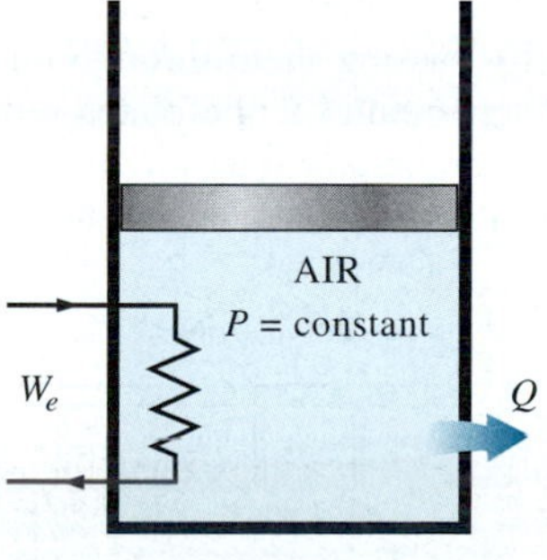

FIGURE P5–79

5–80 An insulated piston–cylinder device initially contains 0.3 m^3 of carbon dioxide at 200 kPa and 27°C. An electric switch is turned on, and a 110-V source supplies current to a resistance heater inside the cylinder for a period of 10 min. The pressure is held constant during the process, while the volume is doubled. Determine the current that passes through the resistance heater.

5–81 A piston–cylinder device contains 0.8 kg of nitrogen initially at 100 kPa and 27°C. The nitrogen is now compressed slowly in a polytropic process during which $PV^{1.3}$ = constant until the volume is reduced by one-half. Determine the work done and the heat transfer for this process.

5–82 Reconsider Prob. 5–81. Using EES (or other) software, plot the process described in the problem on a *P-V* diagram, and investigate the effect of the polytropic exponent *n* on the boundary work and heat transfer. Let the polytropic exponent vary from 1.1 to 1.6. Plot the boundary work and the heat transfer versus the polytropic exponent, and discuss the results.

5–83 A room is heated by a baseboard resistance heater. When the heat losses from the room on a winter day amount to 6500 kJ/h, the air temperature in the room remains constant even though the heater operates continuously. Determine the power rating of the heater, in kW.

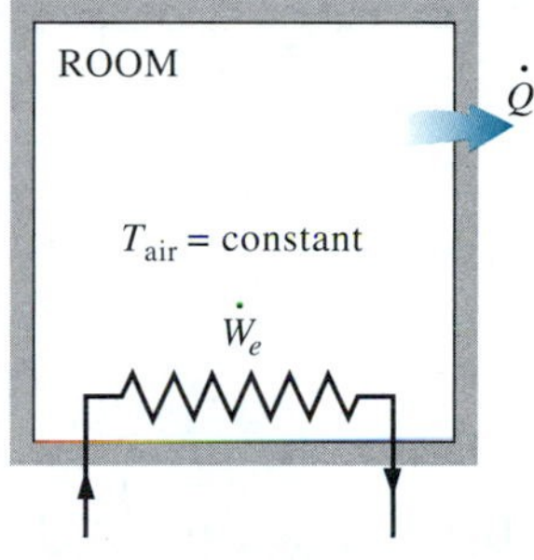

FIGURE P5–83

5–84 A piston–cylinder device, whose piston is resting on a set of stops, initially contains 3 kg of air at 200 kPa and 27°C. The mass of the piston is such that a pressure of 400 kPa is required to move it. Heat is now transferred to the air until its volume doubles. Determine the work done by the air and the total heat transferred to the air during this process. Also show the process on a *P-v* diagram. *Answers:* 516 kJ, 2674 kJ

5–85 A piston–cylinder device, with a set of stops on the top, initially contains 3 kg of air at 200 kPa and 27°C. Heat is now transferred to the air, and the piston rises until it hits the stops, at which point the volume is twice the initial volume. More heat is transferred until the pressure inside the cylinder also doubles. Determine the work done and the amount of heat transfer for this process. Also, show the process on a *P-v* diagram.

Closed-System Energy Analysis: Solids and Liquids

5–86 A 1-kg block of iron is heated from 20°C to 80°C. What is the change in the iron's total internal energy and enthalpy?

5–87E The state of liquid water is changed from 50 psia and 50°F to 2000 psia and 100°F. Determine the change in the internal energy and enthalpy of water on the basis of the (*a*) compressed liquid tables, (*b*) incompressible substance approximation and property tables, and (*c*) specific-heat model.

5–88E During a picnic on a hot summer day, all the cold drinks disappeared quickly, and the only available drinks were those at the ambient temperature of 75°F. In an effort to cool a 12-fluid-oz drink in a can, a person grabs the can and starts shaking it in the iced water of the chest at 32°F. Using the properties of water for the drink, determine the mass of ice that will melt by the time the canned drink cools to 45°F.

5–89 Consider a 1000-W iron whose base plate is made of 0.5-cm-thick aluminum alloy 2024-T6 (ρ = 2770 kg/m^3 and c_p = 875 J/kg · °C). The base plate has a surface area of 0.03 m^2. Initially, the iron is in thermal equilibrium with the ambient air at 22°C. Assuming 85 percent of the heat generated in the resistance wires is transferred to the plate, determine the minimum time needed for the plate temperature to reach 140°C.

FIGURE P5–89

© *Vol. 58/PhotoDisc*

5–90 Stainless steel ball bearings (ρ = 8085 kg/m^3 and c_p = 0.480 kJ/kg · °C) having a diameter of 1.2 cm are to be quenched in water at a rate of 800 per minute. The balls leave the oven at a uniform temperature of 900°C and are exposed to air at 25°C for a while before they are dropped into the water. If the temperature of the balls drops to 850°C prior to quenching, determine the rate of heat transfer from the balls to the air.

5–91 Carbon steel balls (ρ = 7833 kg/m^3 and c_p = 0.465 kJ/kg · °C) 8 mm in diameter are annealed by heating them first to 900°C in a furnace, and then allowing them to cool slowly to 100°C in ambient air at 35°C. If 2500 balls are to be annealed per hour, determine the total rate of heat transfer from the balls to the ambient air. *Answer:* 542 W

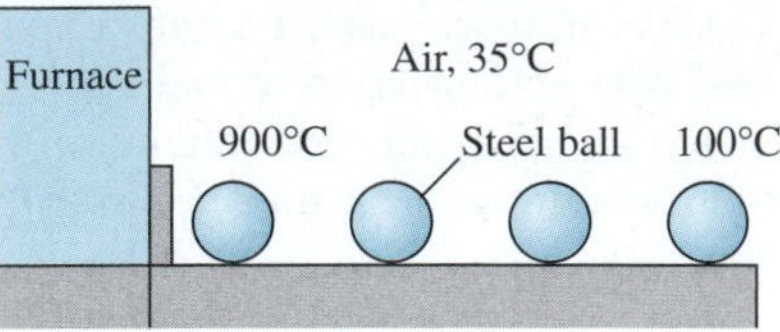

FIGURE P5–91

5–92 An electronic device dissipating 30 W has a mass of 20 g and a specific heat of 850 J/kg · °C. The device is lightly used, and it is on for 5 min and then off for several hours, during which it cools to the ambient temperature of 25°C. Determine the highest possible temperature of the device at the end of the 5-min operating period. What would your answer be if the device were attached to a 0.2-kg aluminum heat sink? Assume the device and the heat sink to be nearly isothermal.

5–93 EES Reconsider Prob. 5–92. Using EES (or other) software, investigate the effect of the mass of the heat sink on the maximum device temperature. Let the mass of heat sink vary from 0 to 1 kg. Plot the maximum temperature against the mass of heat sink, and discuss the results.

5–94 An ordinary egg can be approximated as a 5.5-cm-diameter sphere. The egg is initially at a uniform temperature of 8°C and is dropped into boiling water at 97°C. Taking the properties of the egg to be ρ = 1020 kg/m^3 and c_p = 3.32 kJ/kg · °C, determine how much heat is transferred to the egg by the time the average temperature of the egg rises to 80°C.

5–95E In a production facility, 1.2-in-thick 2-ft × 2-ft square brass plates (ρ = 532.5 lbm/ft^3 and c_p = 0.091 Btu/lbm · °F) that are initially at a uniform temperature of 75°F are heated by passing them through an oven at 1300°F at a rate of 300 per minute. If the plates remain in the oven until their average temperature rises to 1000°F, determine the rate of heat transfer to the plates in the furnace.

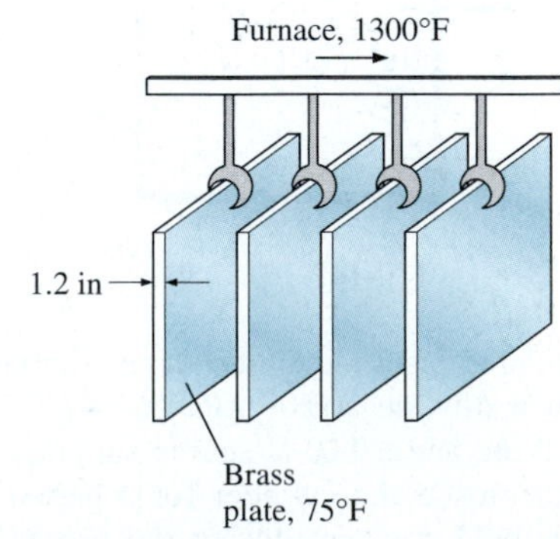

FIGURE P5–95E

5–96 Long cylindrical steel rods (ρ = 7833 kg/m^3 and c_p = 0.465 kJ/kg · °C) of 10-cm diameter are heat-treated by drawing them at a velocity of 3 m/min through an oven maintained at 900°C. If the rods enter the oven at 30°C and leave at a mean temperature of 700°C, determine the rate of heat transfer to the rods in the oven.

Review Problems

5–97 Which of the two gases—neon or air—requires the least amount of work when compressed in a closed system from P_1 to P_2 using a polytropic process with n = 1.5?

5–98 10-kg of nitrogen is heated from 20°C to 250°C. Determine the total amount of heat required when this is done in (*a*) a constant-volume process and (*b*) an isobaric process.

5–99E Air in the amount of 1 lbm is contained in a well-insulated, rigid vessel equipped with a stirring paddle wheel. The initial state of this air is 30 psia and 40°F. How much work, in Btu, must be transferred to the air with the paddle wheel to raise the air pressure to 50 psia? Also, what is the final temperature of air?

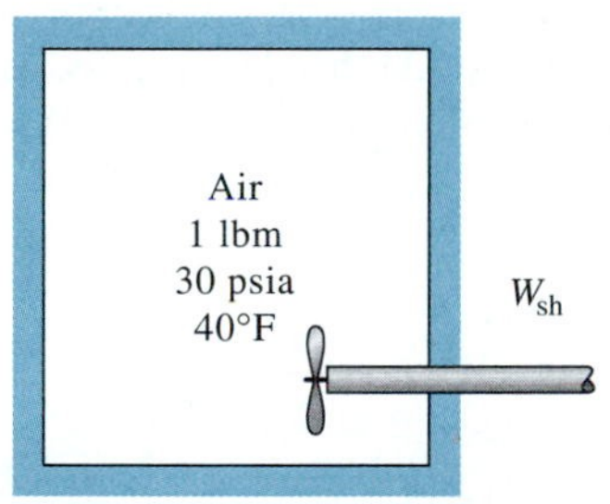

FIGURE P5–99E

5–100 Does the concept of a specific heat property (either at constant volume or constant pressure) have a meaning for substances that are undergoing a phase change? Why or why not?

5–101 A rigid tank contains a gas mixture with a specific heat of c_v = 0.748 kJ/kg · K. The mixture is cooled from 200 kPa and 200°C until its pressure is 100 kPa. Determine the heat transfer during this process, in kJ/kg.

5–102 A well-insulated rigid vessel contains 3 kg of saturated liquid water at 40°C. The vessel also contains an electrical resistor that draws 10 amperes when 50 volts are applied. Determine the final temperature in the vessel after the resistor has been operating for 30 minutes. *Answer:* 119°C

5–103 Derive a general expression for the work produced by an ideal gas as it undergoes a polytropic process in a closed system from initial state 1 to final state 2. Your result should be in terms of the initial pressure and temperature and the final pressure as well as the gas constant R and the polytropic exponent n.

5–104 A frictionless piston–cylinder device initially contains air at 200 kPa and 0.2 m^3. At this state, a linear spring ($F \propto x$) is touching the piston but exerts no force on it. The air is now heated to a final state of 0.5 m^3 and 800 kPa. Determine (*a*) the total work done by the air and (*b*) the work done against the spring. Also, show the process on a P-V diagram. *Answers:* (*a*) 150 kJ, (*b*) 90 kJ

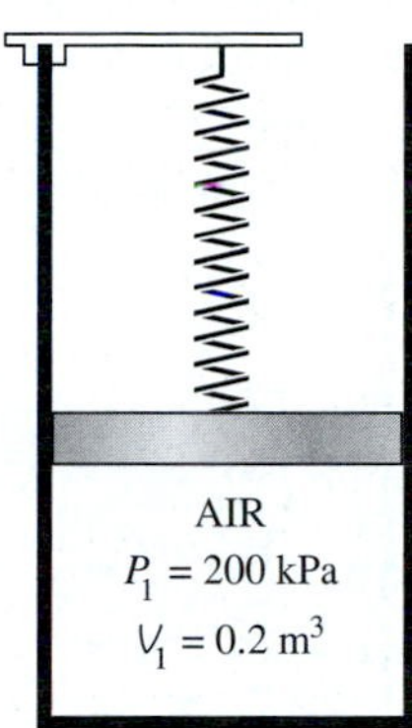

FIGURE P5–104

5–105 A mass of 5 kg of saturated liquid–vapor mixture of water is contained in a piston–cylinder device at 125 kPa. Initially, 2 kg of the water is in the liquid phase and the rest is in the vapor phase. Heat is now transferred to the water, and the piston, which is resting on a set of stops, starts moving when the pressure inside reaches 300 kPa. Heat transfer continues until the total volume increases by 20 percent. Determine (*a*) the initial and final temperatures, (*b*) the mass of liquid water when the piston first starts moving, and (*c*) the work done during this process. Also, show the process on a P-V diagram.

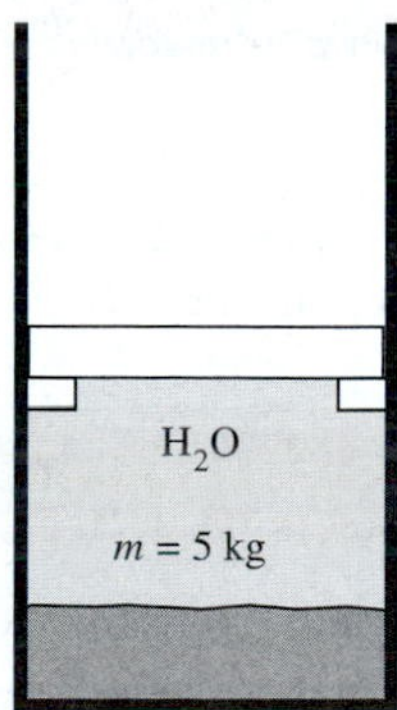

FIGURE P5–105

5–106E A spherical balloon contains 10 lbm of air at 30 psia and 800 R. The balloon material is such that the pressure inside is always proportional to the square of the diameter. Determine the work done when the volume of the balloon doubles as a result of heat transfer. *Answer:* 715 Btu

5–107E Reconsider Prob. 5–106E. Using the integration feature of the EES software, determine the work done. Compare the result with your "hand-calculated" result.

5–108 A mass of 12 kg of saturated refrigerant-134a vapor is contained in a piston–cylinder device at 240 kPa. Now 300 kJ of heat is transferred to the refrigerant at constant pressure while a 110-V source supplies current to a resistor within the cylinder for 6 min. Determine the current supplied if the final temperature is 70°C. Also, show the process on a *T*-v diagram with respect to the saturation lines. *Answer:* 12.8 A

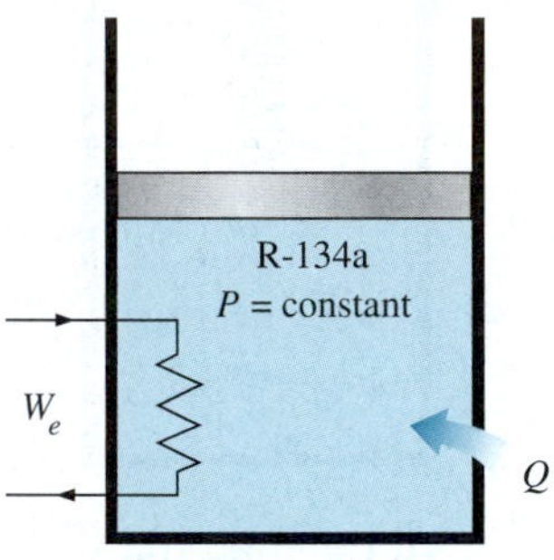

FIGURE P5–108

5–109 A mass of 0.2 kg of saturated refrigerant-134a is contained in a piston–cylinder device at 200 kPa. Initially, 75 percent of the mass is in the liquid phase. Now heat is transferred to the refrigerant at constant pressure until the cylinder contains vapor only. Show the process on a *P*-v diagram with respect to saturation lines. Determine (*a*) the volume occupied by the refrigerant initially, (*b*) the work done, and (*c*) the total heat transfer.

5–110 A piston–cylinder device contains helium gas initially at 150 kPa, 20°C, and 0.5 m³. The helium is now compressed in a polytropic process (PV^n = constant) to 400 kPa and 140°C. Determine the heat loss or gain during this process. *Answer:* 11.2 kJ loss

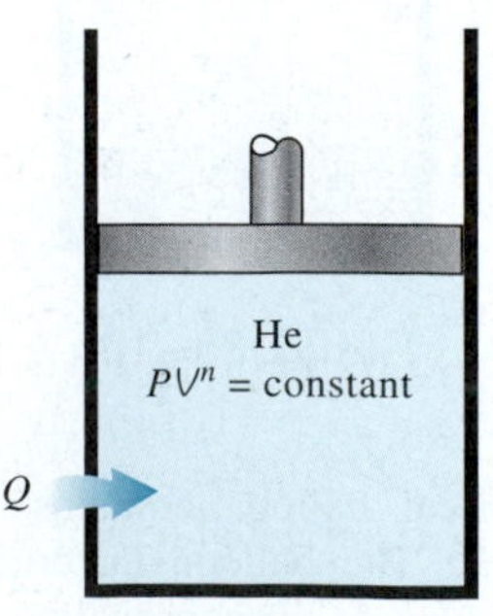

FIGURE P5–110

5–111 Nitrogen gas is expanded in a polytropic process with n = 1.35 from 2 MPa and 1200 K to 120 kPa in a piston–cylinder device. How much work is produced and heat is transferred during this expansion process, in kJ/kg?

5–112 Which of the two gases—neon or air—produces the greatest amount of work when expanded from P_1 to P_2 in a closed system polytropic process with n = 1.2?

5–113 A frictionless piston–cylinder device and a rigid tank initially contain 12 kg of an ideal gas each at the same temperature, pressure, and volume. It is desired to raise the temperatures of both systems by 15°C. Determine the amount of extra heat that must be supplied to the gas in the cylinder which is maintained at constant pressure to achieve this result. Assume the molar mass of the gas is 25.

5–114 A passive solar house that is losing heat to the outdoors at an average rate of 50,000 kJ/h is maintained at 22°C at all times during a winter night for 10 h. The house is to be heated by 50 glass containers each containing 20 L of water that is heated to 80°C during the day by absorbing solar energy. A thermostat-controlled 15-kW back-up electric resistance heater turns on whenever necessary to keep the house at 22°C. (*a*) How long did the electric heating system run that night? (*b*) How long would the electric heater run that night if the house incorporated no solar heating? *Answers:* (*a*) 4.77 h, (*b*) 9.26 h

FIGURE P5–114

5–115 An 1800-W electric resistance heating element is immersed in 40 kg of water initially at 20°C. Determine how long it will take for this heater to raise the water temperature to 80°C.

5–116 One ton (1000 kg) of liquid water at 80°C is brought into a well-insulated and well-sealed 4-m × 5-m × 6-m room initially at 22°C and 100 kPa. Assuming constant specific heats for both air and water at room temperature, determine the final equilibrium temperature in the room. *Answer:* 78.6°C

5–117 A 4-m × 5-m × 6-m room is to be heated by one ton (1000 kg) of liquid water contained in a tank that is placed in the room. The room is losing heat to the outside at an average rate of 8000 kJ/h. The room is initially at 20°C and 100 kPa and is maintained at an average temperature of 20°C at all times. If the hot water is to meet the heating requirements of this room for a 24-h period, determine the minimum temperature of the water when it is first brought into the room. Assume constant specific heats for both air and water at room temperature.

5–118 The energy content of a certain food is to be determined in a bomb calorimeter that contains 3 kg of water by burning a 2-g sample of it in the presence of 100 g of air in the reaction chamber. If the water temperature rises by 3.2°C when equilibrium is established, determine the energy content of the food, in kJ/kg, by neglecting the thermal energy stored in the reaction chamber and the energy supplied by the mixer. What is a rough estimate of the error involved in neglecting the thermal energy stored in the reaction chamber? *Answer:* 20,060 kJ/kg

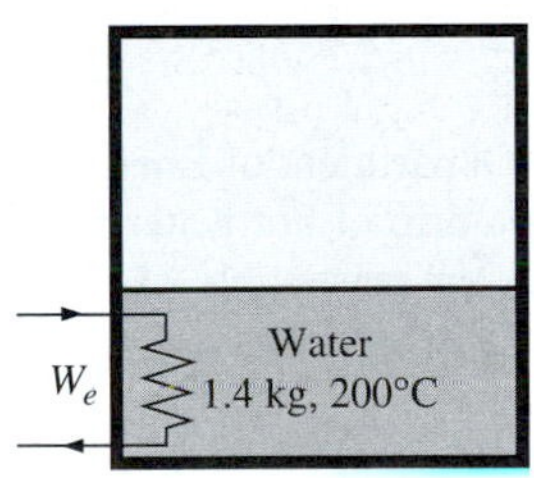

FIGURE P5–118

5–119 A 68-kg man whose average body temperature is 39°C drinks 1 L of cold water at 3°C in an effort to cool down. Taking the average specific heat of the human body to be 3.6 kJ/kg · °C, determine the drop in the average body temperature of this person under the influence of this cold water.

5–120 A 0.2-L glass of water at 20°C is to be cooled with ice to 5°C. Determine how much ice needs to be added to the water, in grams, if the ice is at (*a*) 0°C and (*b*) −8°C. Also determine how much water would be needed if the cooling is to be done with cold water at 0°C. The melting temperature and the heat of fusion of ice at atmospheric pressure are 0°C and 333.7 kJ/kg, respectively, and the density of water is 1 kg/L.

5–121 EES Reconsider Prob. 5–120. Using EES (or other) software, investigate the effect of the initial temperature of the ice on the final mass required. Let the ice temperature vary from –20 to 0°C. Plot the mass of ice against the initial temperature of ice, and discuss the results.

5–122 One kilogram of carbon dioxide is compressed from 1 MPa and 200°C to 3 MPa in a piston–cylinder device arranged to execute a polytropic process $PV^{1.5}$ = constant. Determine the final temperature treating the carbon dioxide as (*a*) an ideal gas and (*b*) a van der Waals gas. *Answers:* (*a*) 682.1 K, (*b*) 680.9 K

FIGURE P5–122

5–123 Two adiabatic chambers, 2 m^3 each, are interconnected by a valve, as shown in Fig. P5–123, with one chamber containing oxygen at 1000 kPa and 127°C and the other chamber evacuated. The valve is now opened until the oxygen fills both chambers and both tanks have the same pressure. Determine the total internal energy change and the final pressure in the tanks.

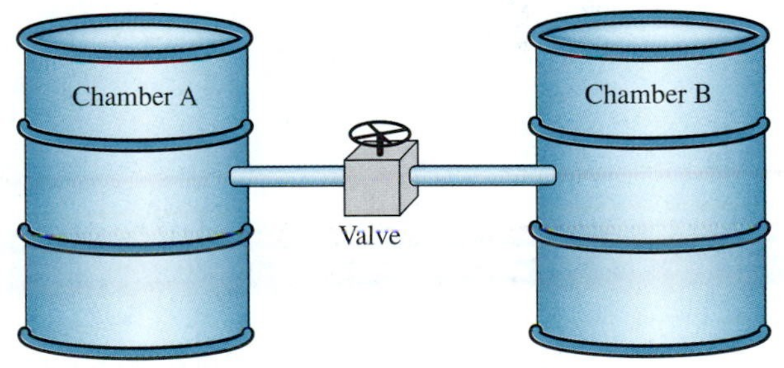

FIGURE P5–123

Design and Essay Problems

5–124 Compressed gases and phase-changing liquids are used to store energy in rigid containers. What are the advantages and disadvantages of each substance as a means of storing energy?

5–125 Someone has suggested that the device shown in Fig. P5–125 be used to move the maximum force F against the spring, which has a spring constant of k. This is accomplished by changing the temperature of the liquid–vapor mixture in the container. You are to design such a device to close sun-blocking window shutters that require a maximum force of 0.5 lbf. The piston must move 6 inches to close these shutters completely. You elect to use R-134a as the working fluid and arrange the liquid–vapor mixture container such that the temperature changes from 70°F when shaded from the sun to 100°F when exposed to the full sun. Select the sizes of the various components in this system to do this task. Also select the necessary spring constant and the amount of R-134a to be used.

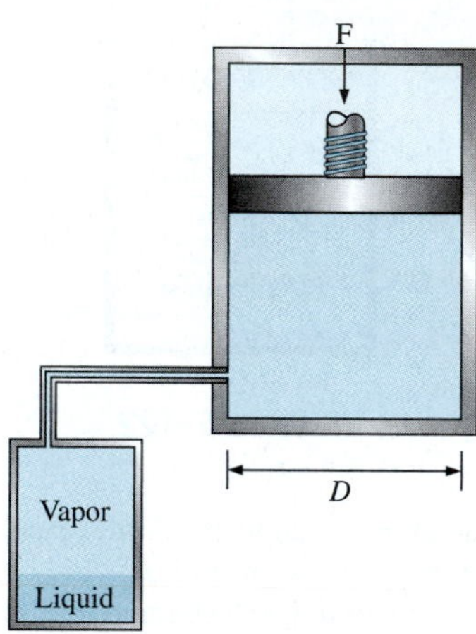

FIGURE P5–125

5–126 Design an experiment complete with instrumentation to determine the specific heats of a liquid using a resistance heater. Discuss how the experiment will be conducted, what measurements need to be taken, and how the specific heats will be determined. What are the sources of error in your system? How can you minimize the experimental error? How would you modify this system to determine the specific heat of a solid?

5–127 You are asked to design a heating system for a swimming pool that is 2 m deep, 25 m long, and 25 m wide. Your client desires that the heating system be large enough to raise the water temperature from 20 to 30°C in 3 h. The rate of heat loss from the water to the air at the outdoor design conditions is determined to be 960 W/m^2, and the heater must also be able to maintain the pool at 30°C at those conditions. Heat losses to the ground are expected to be small and can be disregarded. The heater considered is a natural gas furnace whose efficiency is 80 percent. What heater size (in kW input) would you recommend to your client?

5–128 It is claimed that fruits and vegetables are cooled by 6°C for each percentage point of weight loss as moisture during vacuum cooling. Using calculations, demonstrate if this claim is reasonable.

5–129 A 1982 U.S. Department of Energy article (FS #204) states that a leak of one drip of hot water per second can cost $1.00 per month. Making reasonable assumptions about the drop size and the unit cost of energy, determine if this claim is reasonable.

Chapter 6

MASS AND ENERGY ANALYSIS OF CONTROL VOLUMES

In Chap. 5, we applied the general energy balance relation expressed as $E_{in} - E_{out} = \Delta E_{system}$ to closed systems. In this chapter, we extend the energy analysis to systems that involve mass flow across their boundaries i.e., control volumes, with particular emphasis to steady-flow systems.

We start this chapter with the development of the general *conservation of mass* relation for control volumes, and we continue with a discussion of flow work and the energy of fluid streams. We then apply the energy balance to systems that involve *steady-flow processes* and analyze the common steady-flow devices such as nozzles, diffusers, compressors, turbines, throttling devices, mixing chambers, and heat exchangers. Finally, we apply the energy balance to general *unsteady-flow processes* such as the charging and discharging of vessels.

Objectives

The objectives of this chapter are to:

- Develop the conservation of mass principle.
- Apply the conservation of mass principle to various systems including steady- and unsteady-flow control volumes.
- Apply the first law of thermodynamics as the statement of the conservation of energy principle to control volumes.
- Identify the energy carried by a fluid stream crossing a control surface as the sum of internal energy, flow work, kinetic energy, and potential energy of the fluid and relate the combination of the internal energy and the flow work to the property enthalpy.
- Solve energy balance problems for common steady-flow devices such as nozzles, compressors, turbines, throttling valves, mixers, heaters, and heat exchangers.
- Apply the energy balance to general unsteady-flow processes with particular emphasis on the uniform-flow process as the model for commonly encountered charging and discharging processes.

6–1 ▪ CONSERVATION OF MASS

Conservation of mass is one of the most fundamental principles in nature. We are all familiar with this principle, and it is not difficult to understand. As the saying goes, You cannot have your cake and eat it too! A person does not have to be a scientist to figure out how much vinegar-and-oil dressing is obtained by mixing 100 g of oil with 25 g of vinegar. Even chemical equations are balanced on the basis of the conservation of mass principle. When 16 kg of oxygen reacts with 2 kg of hydrogen, 18 kg of water is formed (Fig. 6–1). In an electrolysis process, the water separates back to 2 kg of hydrogen and 16 kg of oxygen.

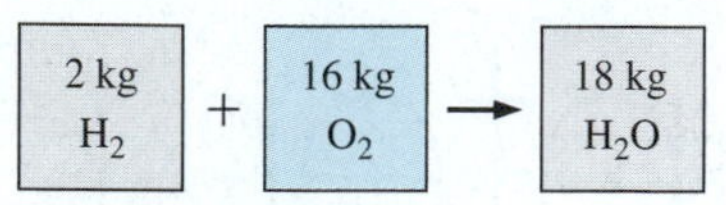

FIGURE 6–1

Mass is conserved even during chemical reactions.

Mass, like energy, is a conserved property, and it cannot be created or destroyed during a process. However, mass m and energy E can be converted to each other according to the well-known formula proposed by Albert Einstein (1879–1955):

$$E = mc^2 \qquad \text{(6–1)}$$

where c is the speed of light in a vacuum, which is $c = 2.9979 \times 10^8$ m/s. This equation suggests that the mass of a system changes when its energy changes. However, for all energy interactions encountered in practice, with the exception of nuclear reactions, the change in mass is extremely small and cannot be detected by even the most sensitive devices. For example, when 1 kg of water is formed from oxygen and hydrogen, the amount of energy released is 15,879 kJ, which corresponds to a mass of 1.76×10^{-10} kg. A mass of this magnitude is beyond the accuracy required by practically all engineering calculations and thus can be disregarded.

For *closed systems*, the conservation of mass principle is implicitly used by requiring that the mass of the system remain constant during a process. For *control volumes*, however, mass can cross the boundaries, and so we must keep track of the amount of mass entering and leaving the control volume.

Mass and Volume Flow Rates

The amount of mass flowing through a cross section per unit time is called the **mass flow rate** and is denoted by $\dot{m}$. The dot over a symbol is used to indicate *time rate of change*, as explained in Chap. 3.

A fluid usually flows into or out of a control volume through pipes or ducts. The differential mass flow rate of fluid flowing across a small area element dA_c on a flow cross section is proportional to dA_c itself, the fluid density ρ, and the component of the flow velocity normal to dA_c, which we denote as V_n, and is expressed as (Fig. 6–2)

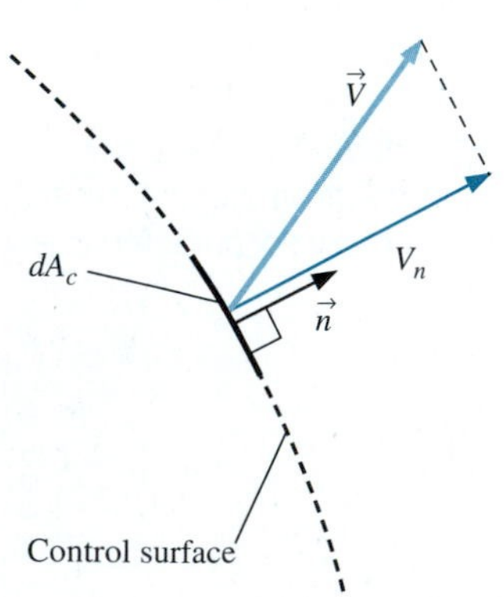

FIGURE 6–2

The normal velocity V_n for a surface is the component of velocity perpendicular to the surface.

$$\delta\dot{m} = \rho V_n \, dA_c \qquad \text{(6–2)}$$

Note that both δ and d are used to indicate differential quantities, but δ is typically used for quantities (such as heat, work, and mass transfer) that are *path functions* and have *inexact differentials*, while d is used for quantities (such as properties) that are *point functions* and have *exact differentials*. For flow through an annulus of inner radius r_1 and outer radius r_2, for example, $\int_1^2 dA_c = A_{c2} - A_{c1} = \pi(r_2^2 - r_1^2)$ but $\int_1^2 \delta\dot{m} = \dot{m}_{\text{total}}$ (total mass flow rate through the annulus), not $\dot{m}_2 - \dot{m}_1$. For specified values of r_1 and r_2, the value of the integral of dA_c is fixed (thus the names point function and exact differential), but this is not the case for the integral of $\delta\dot{m}$ (thus the names path function and inexact differential).

The mass flow rate through the entire cross-sectional area of a pipe or duct is obtained by integration:

$$\dot{m} = \int_{A_c} \delta\dot{m} = \int_{A_c} \rho V_n \, dA_c \qquad \text{(kg/s)} \tag{6–3}$$

While Eq. 6–3 is always valid (in fact it is *exact*), it is not always practical for engineering analyses because of the integral. We would like instead to express mass flow rate in terms of average values over a cross section of the pipe. In a general compressible flow, both ρ and V_n vary across the pipe. In many practical applications, however, the density is essentially uniform over the pipe cross section, and we can take ρ outside the integral of Eq. 6–3. Velocity, however, is *never* uniform over a cross section of a pipe because of the fluid sticking to the surface and thus having zero velocity at the wall (the no-slip condition). Rather, the velocity varies from zero at the walls to some maximum value at or near the centerline of the pipe. We define the **average velocity** V_{avg} as the average value of V_n across the entire cross section (Fig. 6–3),

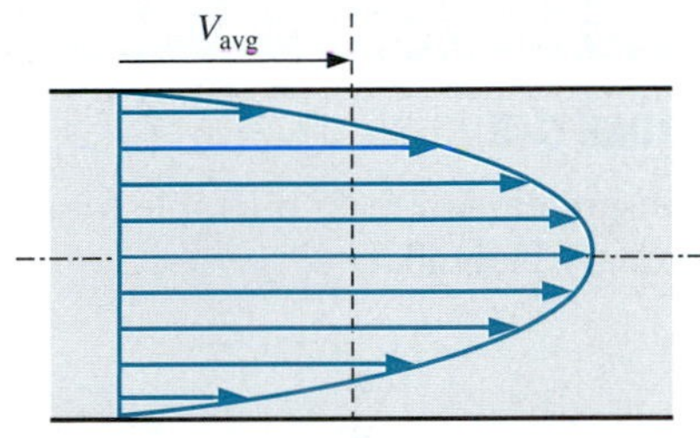

FIGURE 6–3
The average velocity V_{avg} is defined as the average speed through a cross section.

Average velocity:

$$V_{avg} = \frac{1}{A_c} \int_{A_c} V_n \, dA_c \tag{6–4}$$

where A_c is the area of the cross section normal to the flow direction. Note that if the velocity were V_{avg} all through the cross section, the mass flow rate would be identical to that obtained by integrating the actual velocity profile. Thus for incompressible flow or even for compressible flow where ρ is uniform across A_c, Eq. 6–3 becomes

$$\dot{m} = \rho V_{avg} A_c \qquad \text{(kg/s)} \tag{6–5}$$

For compressible flow, we can think of ρ as the bulk average density over the cross section, and then Eq. 6–5 can still be used as a reasonable approximation.

For simplicity, we drop the subscript on the average velocity. Unless otherwise stated, V denotes the average velocity in the flow direction. Also, A_c denotes the cross-sectional area normal to the flow direction.

The volume of the fluid flowing through a cross section per unit time is called the **volume flow rate** $\dot{V}$ (Fig. 6–4) and is given by

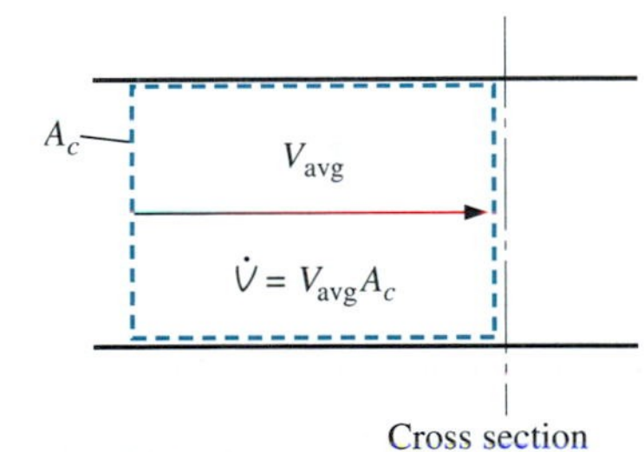

FIGURE 6–4
The volume flow rate is the volume of fluid flowing through a cross section per unit time.

$$\dot{V} = \int_{A_c} V_n \, dA_c = V_{avg} A_c = VA_c \qquad \text{(m}^3\text{/s)} \tag{6–6}$$

An early form of Eq. 6–6 was published in 1628 by the Italian monk Benedetto Castelli (circa 1577–1644). Note that most fluid mechanics textbooks use Q instead of $\dot{V}$ for volume flow rate. We use $\dot{V}$ to avoid confusion with heat transfer.

The mass and volume flow rates are related by

$$\dot{m} = \rho\dot{V} = \frac{\dot{V}}{v} \tag{6–7}$$

where v is the specific volume. This relation is analogous to $m = \rho V = V/v$, which is the relation between the mass and the volume of a fluid in a container.

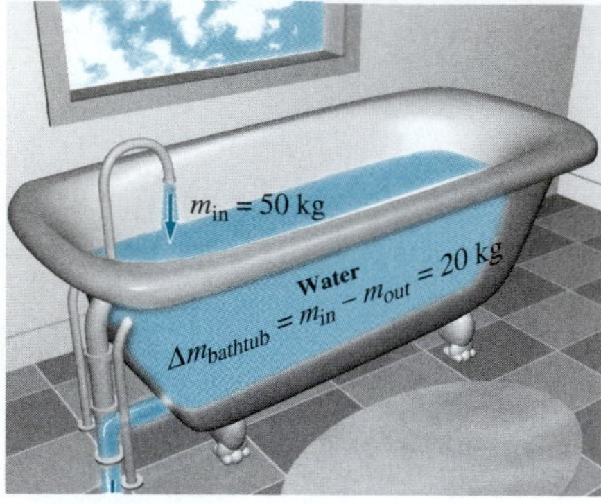

FIGURE 6–5

Conservation of mass principle for an ordinary bathtub.

Conservation of Mass Principle

The **conservation of mass principle** for a control volume can be expressed as: *The net mass transfer to or from a control volume during a time interval Δt is equal to the net change (increase or decrease) in the total mass within the control volume during Δt.* That is,

$$\left(\begin{array}{c}\text{Total mass entering}\\ \text{the CV during } \Delta t\end{array}\right) - \left(\begin{array}{c}\text{Total mass leaving}\\ \text{the CV during } \Delta t\end{array}\right) = \left(\begin{array}{c}\text{Net change in mass}\\ \text{within the CV during } \Delta t\end{array}\right)$$

or

$$m_{in} - m_{out} = \Delta m_{CV} \qquad \text{(kg)} \tag{6–8}$$

where $\Delta m_{CV} = m_{final} - m_{initial}$ is the change in the mass of the control volume during the process (Fig. 6–5). It can also be expressed in *rate form* as

$$\dot{m}_{in} - \dot{m}_{out} = dm_{CV}/dt \qquad \text{(kg/s)} \tag{6–9}$$

where $\dot{m}_{in}$ and $\dot{m}_{out}$ are the total rates of mass flow into and out of the control volume, and dm_{CV}/dt is the time rate of change of mass within the control volume boundaries. Equations 6–8 and 6–9 are often referred to as the **mass balance** and are applicable to any control volume undergoing any kind of process.

Consider a control volume of arbitrary shape, as shown in Fig. 6–6. The mass of a differential volume dV within the control volume is $dm = \rho\, dV$. The total mass within the control volume at any instant in time t is determined by integration to be

Total mass within the CV:
$$m_{CV} = \int_{CV} \rho\, dV \tag{6–10}$$

Then the time rate of change of the amount of mass within the control volume can be expressed as

Rate of change of mass within the CV:
$$\frac{dm_{CV}}{dt} = \frac{d}{dt}\int_{CV} \rho\, dV \tag{6–11}$$

For the special case of no mass crossing the control surface (i.e., the control volume resembles a closed system), the conservation of mass principle reduces to that of a system that can be expressed as $dm_{CV}/dt = 0$. This relation is valid whether the control volume is fixed, moving, or deforming.

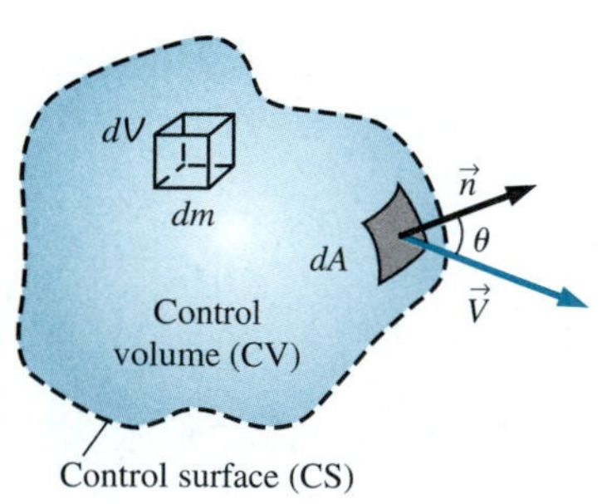

FIGURE 6–6

The differential control volume dV and the differential control surface dA used in the derivation of the conservation of mass relation.

Now consider mass flow into or out of the control volume through a differential area dA on the control surface of a fixed control volume. Let $\vec{n}$ be the outward unit vector of dA normal to dA and $\vec{V}$ be the flow velocity at dA relative to a fixed coordinate system, as shown in Fig. 6–6. In general, the velocity may cross dA at an angle θ off the normal of dA, and the mass flow rate is proportional to the normal component of velocity $\vec{V}_n = \vec{V}\cos\theta$ ranging from a maximum outflow of $\vec{V}$ for $\theta = 0$ (flow is normal to dA) to a minimum of zero for $\theta = 90°$ (flow is tangent to dA) to a maximum *inflow* of $\vec{V}$ for $\theta = 180°$ (flow is normal to dA but in the opposite direction). Making use of the concept of dot product of two vectors, the magnitude of the normal component of velocity can be expressed as

Normal component of velocity:
$$V_n = V\cos\theta = \vec{V}\cdot\vec{n} \tag{6–12}$$

The mass flow rate through dA is proportional to the fluid density ρ, normal velocity V_n, and the flow area dA, and can be expressed as

Differential mass flow rate:
$$\delta\dot{m} = \rho V_n\, dA = \rho(V\cos\theta)\, dA = \rho(\vec{V}\cdot\vec{n})\, dA \tag{6–13}$$

The net flow rate into or out of the control volume through the entire control surface is obtained by integrating $\delta\dot{m}$ over the entire control surface,

Net mass flow rate:
$$\dot{m}_{\text{net}} = \int_{CS} \delta\dot{m} = \int_{CS} \rho V_n \, dA = \int_{CS} \rho(\vec{V} \cdot \vec{n}) \, dA \quad \textbf{(6–14)}$$

Note that $\vec{V} \cdot \vec{n} = V \cos\theta$ is positive for $\theta < 90°$ (outflow) and negative for $\theta > 90°$ (inflow). Therefore, the direction of flow is automatically accounted for, and the surface integral in Eq. 6–14 directly gives the *net* mass flow rate. A positive value for $\dot{m}_{\text{net}}$ indicates net outflow, and a negative value indicates a net inflow of mass.

Rearranging Eq. 6–9 as $dm_{CV}/dt + \dot{m}_{\text{out}} - \dot{m}_{\text{in}} = 0$, the conservation of mass relation for a fixed control volume can then be expressed as

General conservation of mass:
$$\frac{d}{dt}\int_{CV} \rho \, dV + \int_{CS} \rho(\vec{V} \cdot \vec{n}) \, dA = 0 \quad \textbf{(6–15)}$$

It states that *the time rate of change of mass within the control volume plus the net mass flow rate through the control surface is equal to zero.*

Splitting the surface integral in Eq. 6–15 into two parts—one for the outgoing flow streams (positive) and one for the incoming streams (negative)—the general conservation of mass relation can also be expressed as

$$\frac{d}{dt}\int_{CV} \rho \, dV + \sum_{\text{out}} \int_A \rho V_n \, dA - \sum_{\text{in}} \int_A \rho V_n \, dA = 0 \quad \textbf{(6–16)}$$

where A represents the area for an inlet or outlet, and the summation signs are used to emphasize that *all* the inlets and outlets are to be considered. Using the definition of mass flow rate, Eq. 6–16 can also be expressed as

$$\frac{d}{dt}\int_{CV} \rho \, dV = \sum_{\text{in}} \dot{m} - \sum_{\text{out}} \dot{m} \quad \text{or} \quad \frac{dm_{CV}}{dt} = \sum_{\text{in}} \dot{m} - \sum_{\text{out}} \dot{m} \quad \textbf{(6–17)}$$

Equations 6–15 and 6–16 are also valid for moving or deforming control volumes provided that the *absolute velocity* $\vec{V}$ is replaced by the *relative velocity* $\vec{V}_r$, which is the fluid velocity relative to the control surface.

Mass Balance for Steady-Flow Processes

During a steady-flow process, the total amount of mass contained within a control volume does not change with time (m_{CV} = constant). Then the conservation of mass principle requires that the total amount of mass entering a control volume equal the total amount of mass leaving it. For a garden hose nozzle in steady operation, for example, the amount of water entering the nozzle per unit time is equal to the amount of water leaving it per unit time.

When dealing with steady-flow processes, we are not interested in the amount of mass that flows in or out of a device over time; instead, we are interested in the amount of mass flowing per unit time, that is, *the mass flow rate* $\dot{m}$. *The conservation of mass principle* for a general steady-flow system with multiple inlets and outlets can be expressed in rate form as (Fig. 6–7)

Steady flow:
$$\sum_{\text{in}} \dot{m} = \sum_{\text{out}} \dot{m} \quad \text{(kg/s)} \quad \textbf{(6–18)}$$

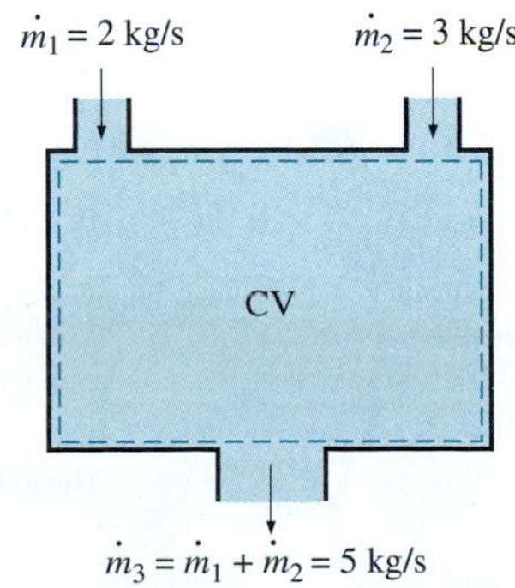

FIGURE 6–7

Conservation of mass principle for a two-inlet–one-outlet steady-flow system.

It states that *the total rate of mass entering a control volume is equal to the total rate of mass leaving it.*

Many engineering devices such as nozzles, diffusers, turbines, compressors, and pumps involve a single stream (only one inlet and one outlet). For these cases, we denote the inlet state by the subscript 1 and the outlet state by the subscript 2, and drop the summation signs. Then Eq. 6–18 reduces, for *single-stream steady-flow systems,* to

Steady flow (single stream): $$\dot{m}_1 = \dot{m}_2 \quad \rightarrow \quad \rho_1 V_1 A_1 = \rho_2 V_2 A_2 \tag{6–19}$$

Special Case: Incompressible Flow

The conservation of mass relations can be simplified even further when the fluid is incompressible, which is usually the case for liquids. Canceling the density from both sides of the general steady-flow relation gives

Steady, incompressible flow: $$\sum_{\text{in}} \dot{V} = \sum_{\text{out}} \dot{V} \qquad (\text{m}^3/\text{s}) \tag{6–20}$$

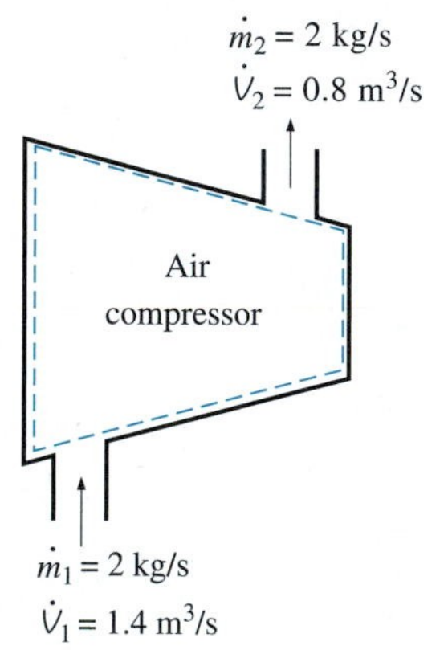

FIGURE 6–8

During a steady-flow process, volume flow rates are not necessarily conserved although mass flow rates are.

For single-stream steady-flow systems it becomes

Steady, incompressible flow (single stream): $$\dot{V}_1 = \dot{V}_2 \rightarrow V_1 A_1 = V_2 A_2 \tag{6–21}$$

It should always be kept in mind that there is no such thing as a "conservation of volume" principle. Therefore, the volume flow rates into and out of a steady-flow device may be different. The volume flow rate at the outlet of an air compressor is much less than that at the inlet even though the mass flow rate of air through the compressor is constant (Fig. 6–8). This is due to the higher density of air at the compressor exit. For steady flow of liquids, however, the volume flow rates, as well as the mass flow rates, remain constant since liquids are essentially incompressible (constant-density) substances. Water flow through the nozzle of a garden hose is an example of the latter case.

The conservation of mass principle is based on experimental observations and requires every bit of mass to be accounted for during a process. If you can balance your checkbook (by keeping track of deposits and withdrawals, or by simply observing the "conservation of money" principle), you should have no difficulty applying the conservation of mass principle to engineering systems.

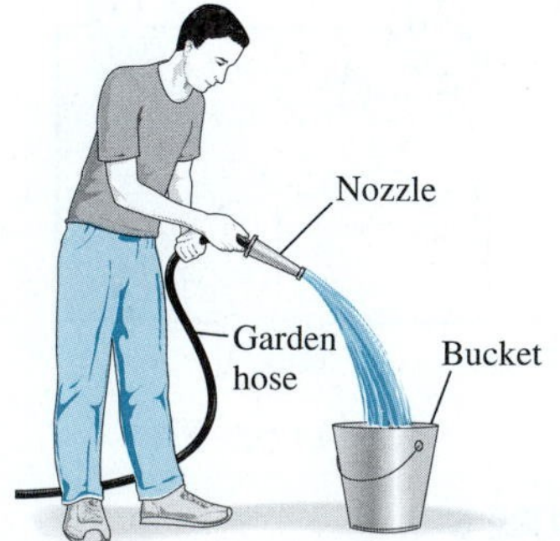

FIGURE 6–9

Schematic for Example 6–1.

EXAMPLE 6–1 Water Flow through a Garden Hose Nozzle

A garden hose attached with a nozzle is used to fill a 10-gal bucket. The inner diameter of the hose is 2 cm, and it reduces to 0.8 cm at the nozzle exit (Fig. 6–9). If it takes 50 s to fill the bucket with water, determine (*a*) the volume and mass flow rates of water through the hose, and (*b*) the average velocity of water at the nozzle exit.

Solution A garden hose is used to fill a water bucket. The volume and mass flow rates of water and the exit velocity are to be determined.

Assumptions **1** Water is an incompressible substance. **2** Flow through the hose is steady. **3** There is no waste of water by splashing.

Properties We take the density of water to be 1000 kg/m³ = 1 kg/L.

Analysis (*a*) Noting that 10 gal of water are discharged in 50 s, the volume and mass flow rates of water are

$$\dot{V} = \frac{V}{\Delta t} = \frac{10\text{ gal}}{50\text{ s}}\left(\frac{3.7854\text{ L}}{1\text{ gal}}\right) = \mathbf{0.757\ L/s}$$

$$\dot{m} = \rho\dot{V} = (1\text{ kg/L})(0.757\text{ L/s}) = \mathbf{0.757\ kg/s}$$

(*b*) The cross-sectional area of the nozzle exit is

$$A_e = \pi r_e^2 = \pi(0.4\text{ cm})^2 = 0.5027\text{ cm}^2 = 0.5027 \times 10^{-4}\text{ m}^2$$

The volume flow rate through the hose and the nozzle is constant. Then the average velocity of water at the nozzle exit becomes

$$V_e = \frac{\dot{V}}{A_e} = \frac{0.757\text{ L/s}}{0.5027 \times 10^{-4}\text{ m}^2}\left(\frac{1\text{ m}^3}{1000\text{ L}}\right) = \mathbf{15.1\ m/s}$$

Discussion It can be shown that the average velocity in the hose is 2.4 m/s. Therefore, the nozzle increases the water velocity by over six times.

EXAMPLE 6–2 Discharge of Water from a Tank

A 4-ft-high, 3-ft-diameter cylindrical water tank whose top is open to the atmosphere is initially filled with water. Now the discharge plug near the bottom of the tank is pulled out, and a water jet whose diameter is 0.5 in streams out (Fig. 6–10). The average velocity of the jet is given by $V = \sqrt{2gh}$, where h is the height of water in the tank measured from the center of the hole (a variable) and g is the gravitational acceleration. Determine how long it will take for the water level in the tank to drop to 2 ft from the bottom.

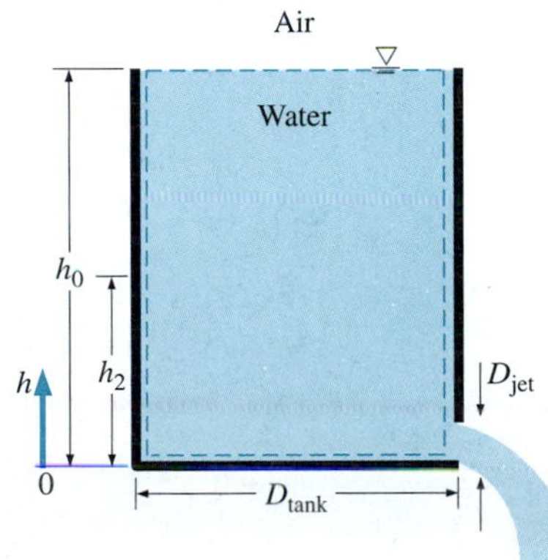

FIGURE 6–10
Schematic for Example 6–2.

Solution The plug near the bottom of a water tank is pulled out. The time it takes for half of the water in the tank to empty is to be determined.

Assumptions **1** Water is an incompressible substance. **2** The distance between the bottom of the tank and the center of the hole is negligible compared to the total water height. **3** The gravitational acceleration is 32.2 ft/s².

Analysis We take the volume occupied by water as the control volume. The size of the control volume decreases in this case as the water level drops, and thus this is a variable control volume. (We could also treat this as a fixed control volume that consists of the interior volume of the tank by disregarding the air that replaces the space vacated by the water.) This is obviously an unsteady-flow problem since the properties (such as the amount of mass) within the control volume change with time.

The conservation of mass relation for a control volume undergoing any process is given in the rate form as

$$\dot{m}_{in} - \dot{m}_{out} = \frac{dm_{CV}}{dt} \tag{1}$$

During this process no mass enters the control volume ($\dot{m}_{in} = 0$), and the mass flow rate of discharged water can be expressed as

$$\dot{m}_{out} = (\rho V A)_{out} = \rho\sqrt{2gh}A_{jet} \tag{2}$$

where $A_{jet} = \pi D_{jet}^2/4$ is the cross-sectional area of the jet, which is constant. Noting that the density of water is constant, the mass of water in the tank at any time is

$$m_{CV} = \rho V = \rho A_{tank} h \quad \textbf{(3)}$$

where $A_{tank} = \pi D_{tank}^2/4$ is the base area of the cylindrical tank. Substituting Eqs. 2 and 3 into the mass balance relation (Eq. 1) gives

$$-\rho\sqrt{2gh}A_{jet} = \frac{d(\rho A_{tank}h)}{dt} \rightarrow -\rho\sqrt{2gh}(\pi D_{jet}^2/4) = \frac{\rho(\pi D_{tank}^2/4)\,dh}{dt}$$

Canceling the densities and other common terms and separating the variables give

$$dt = -\frac{D_{tank}^2}{D_{jet}^2}\frac{dh}{\sqrt{2gh}}$$

Integrating from $t = 0$ at which $h = h_0$ to $t = t$ at which $h = h_2$ gives

$$\int_0^t dt = -\frac{D_{tank}^2}{D_{jet}^2\sqrt{2g}}\int_{h_0}^{h_2}\frac{dh}{\sqrt{h}} \rightarrow t = \frac{\sqrt{h_0} - \sqrt{h_2}}{\sqrt{g/2}}\left(\frac{D_{tank}}{D_{jet}}\right)^2$$

Substituting, the time of discharge is

$$t = \frac{\sqrt{4\text{ ft}} - \sqrt{2\text{ ft}}}{\sqrt{32.2/2\text{ ft/s}^2}}\left(\frac{3 \times 12\text{ in}}{0.5\text{ in}}\right)^2 = 757\text{ s} = \mathbf{12.6\ min}$$

Therefore, half of the tank is emptied in 12.6 min after the discharge hole is unplugged.

Discussion Using the same relation with $h_2 = 0$ gives $t = 43.1$ min for the discharge of the entire amount of water in the tank. Therefore, emptying the bottom half of the tank takes much longer than emptying the top half. This is due to the decrease in the average discharge velocity of water with decreasing h.

6–2 ▪ FLOW WORK AND THE ENERGY OF A FLOWING FLUID

Unlike closed systems, control volumes involve mass flow across their boundaries, and some work is required to push the mass into or out of the control volume. This work is known as the **flow work**, or **flow energy**, and is necessary for maintaining a continuous flow through a control volume.

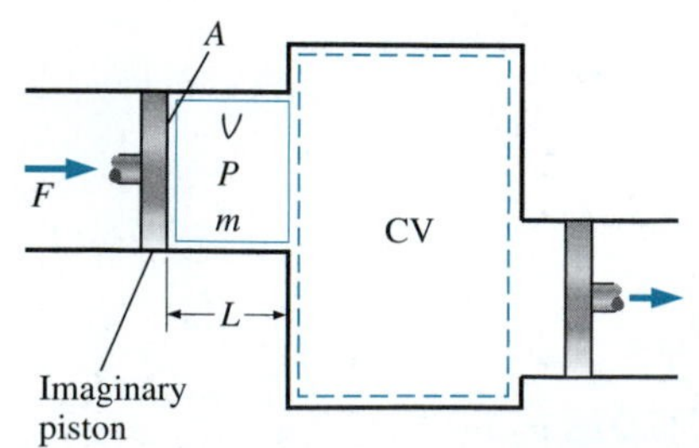

FIGURE 6–11
Schematic for flow work.

To obtain a relation for flow work, consider a fluid element of volume V as shown in Fig. 6–11. The fluid immediately upstream forces this fluid element to enter the control volume; thus, it can be regarded as an imaginary piston. The fluid element can be chosen to be sufficiently small so that it has uniform properties throughout.

If the fluid pressure is P and the cross-sectional area of the fluid element is A (Fig. 6–12), the force applied on the fluid element by the imaginary piston is

$$F = PA \quad \textbf{(6–22)}$$

To push the entire fluid element into the control volume, this force must act through a distance L. Thus, the work done in pushing the fluid element across the boundary (i.e., the flow work) is

$$W_{\text{flow}} = FL = PAL = PV \qquad \text{(kJ)} \tag{6–23}$$

The flow work per unit mass is obtained by dividing both sides of this equation by the mass of the fluid element:

$$w_{\text{flow}} = Pv \qquad \text{(kJ/kg)} \tag{6–24}$$

The flow work relation is the same whether the fluid is pushed into or out of the control volume (Fig. 6–13).

It is interesting that unlike other work quantities, flow work is expressed in terms of properties. In fact, it is the product of two properties of the fluid. For that reason, some people view it as a *combination property* (like enthalpy) and refer to it as *flow energy*, *convected energy*, or *transport energy* instead of flow work. Others, however, argue rightfully that the product Pv represents energy for flowing fluids only and does not represent any form of energy for nonflow (closed) systems. Therefore, it should be treated as work. This controversy is not likely to end, but it is comforting to know that both arguments yield the same result for the energy balance equation. In the discussions that follow, we consider the flow energy to be part of the energy of a flowing fluid, since this greatly simplifies the energy analysis of control volumes.

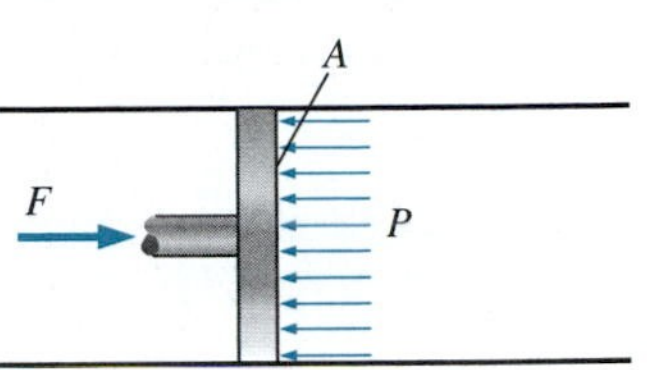

FIGURE 6–12
In the absence of acceleration, the force applied on a fluid by a piston is equal to the force applied on the piston by the fluid.

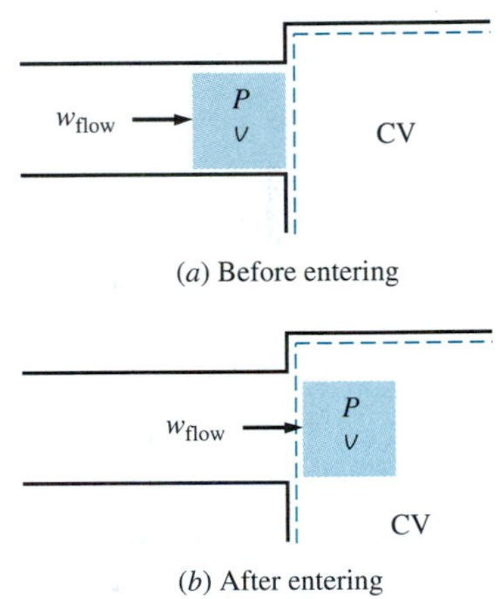

FIGURE 6–13
Flow work is the energy needed to push a fluid into or out of a control volume, and it is equal to Pv.

Total Energy of a Flowing Fluid

As we discussed in Chap. 3, the total energy of a simple compressible system consists of three parts: internal, kinetic, and potential energies (Fig. 6–14). On a unit-mass basis, it is expressed as

$$e = u + \text{ke} + \text{pe} = u + \frac{V^2}{2} + gz \qquad \text{(kJ/kg)} \tag{6–25}$$

where V is the velocity and z is the elevation of the system relative to some external reference point.

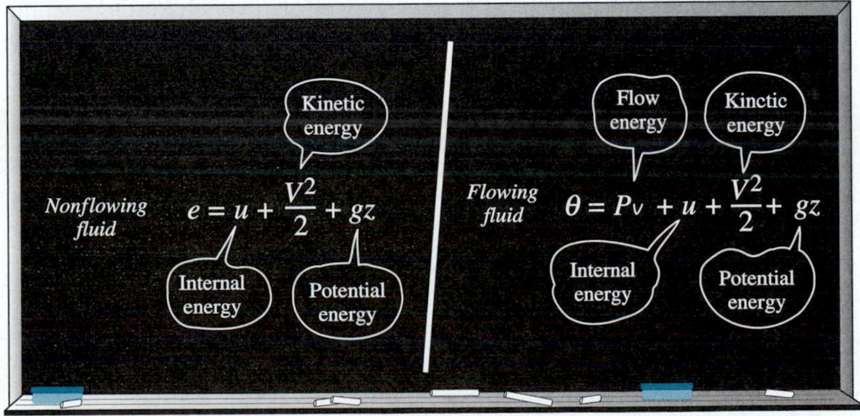

FIGURE 6–14
The total energy consists of three parts for a nonflowing fluid and four parts for a flowing fluid.

The fluid entering or leaving a control volume possesses an additional form of energy—the *flow energy* $P\text{v}$, as already discussed. Then the total energy of a **flowing fluid** on a unit-mass basis (denoted by θ) becomes

$$\theta = P\text{v} + e = P\text{v} + (u + \text{ke} + \text{pe}) \qquad \text{(6–26)}$$

But the combination $P\text{v} + u$ has been previously defined as the enthalpy h. So the relation in Eq. 6–26 reduces to

$$\theta = h + \text{ke} + \text{pe} = h + \frac{V^2}{2} + gz \qquad \text{(kJ/kg)} \qquad \text{(6–27)}$$

By using the enthalpy instead of the internal energy to represent the energy of a flowing fluid, one does not need to be concerned about the flow work. The energy associated with pushing the fluid into or out of the control volume is automatically taken care of by enthalpy. In fact, this is the main reason for defining the property enthalpy. From now on, the energy of a fluid stream flowing into or out of a control volume is represented by Eq. 6–27, and no reference will be made to flow work or flow energy.

Energy Transport by Mass

Noting that θ is total energy per unit mass, the total energy of a flowing fluid of mass m is simply $m\theta$, provided that the properties of the mass m are uniform. Also, when a fluid stream with uniform properties is flowing at a mass flow rate of $\dot{m}$, the rate of energy flow with that stream is $\dot{m}\theta$ (Fig. 6–15). That is,

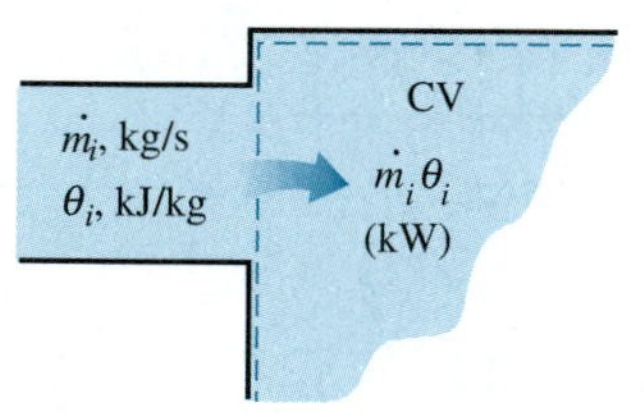

FIGURE 6–15

The product $\dot{m}_i\theta_i$ is the energy transported into control volume by mass per unit time.

Amount of energy transport:
$$E_{\text{mass}} = m\theta = m\left(h + \frac{V^2}{2} + gz\right) \qquad \text{(kJ)} \qquad \text{(6–28)}$$

Rate of energy transport:
$$\dot{E}_{\text{mass}} = \dot{m}\theta = \dot{m}\left(h + \frac{V^2}{2} + gz\right) \qquad \text{(kW)} \qquad \text{(6–29)}$$

When the kinetic and potential energies of a fluid stream are negligible, as is often the case, these relations simplify to $E_{\text{mass}} = mh$ and $\dot{E}_{\text{mass}} = \dot{m}h$.

In general, the total energy transported by mass into or out of the control volume is not easy to determine since the properties of the mass at each inlet or exit may be changing with time as well as over the cross section. Thus, the only way to determine the energy transport through an opening as a result of mass flow is to consider sufficiently small differential masses δm that have uniform properties and to add their total energies during flow.

Again noting that θ is total energy per unit mass, the total energy of a flowing fluid of mass δm is $\theta\ \delta m$. Then the total energy transported by mass through an inlet or exit ($m_i\theta_i$ and $m_e\theta_e$) is obtained by integration. At an inlet, for example, it becomes

$$E_{\text{in,mass}} = \int_{m_i} \theta_i\ \delta m_i = \int_{m_i} \left(h_i + \frac{V_i^2}{2} + gz_i\right) \delta m_i \qquad \text{(6–30)}$$

Most flows encountered in practice can be approximated as being steady and one-dimensional (with h_i, V_i, and z_i being *average* properties.)

EXAMPLE 6–3 Energy Transport by Mass

Steam is leaving a 4-L pressure cooker whose operating pressure is 150 kPa (Fig. 6–16). It is observed that the amount of liquid in the cooker has decreased by 0.6 L in 40 min after the steady operating conditions are established, and the cross-sectional area of the exit opening is 8 mm^2. Determine (*a*) the mass flow rate of the steam and the exit velocity, (*b*) the total and flow energies of the steam per unit mass, and (*c*) the rate at which energy leaves the cooker by steam.

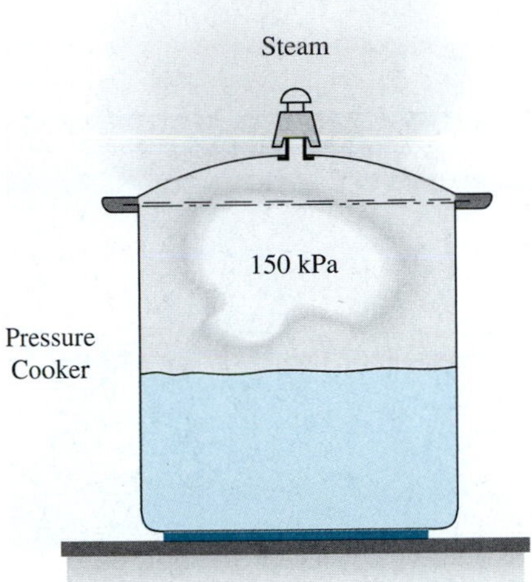

FIGURE 6–16
Schematic for Example 6–3.

Solution Steam leaves a pressure cooker at a specified pressure. The velocity, flow rate, the total and flow energies, and the rate of energy transfer by mass are to be determined.

Assumptions **1** The flow is steady, and the initial start-up period is disregarded. **2** The kinetic and potential energies are negligible, and thus they are not considered. **3** Saturation conditions exist within the cooker at all times so that steam leaves the cooker as a saturated vapor at the cooker pressure.

Properties The properties of saturated liquid water and water vapor at 150 kPa are v_f = 0.001053 m^3/kg, v_g = 1.1594 m^3/kg, u_g = 2519.2 kJ/kg, and h_g = 2693.1 kJ/kg (Table A–5).

Analysis (*a*) Saturation conditions exist in a pressure cooker at all times after the steady operating conditions are established. Therefore, the liquid has the properties of saturated liquid and the exiting steam has the properties of saturated vapor at the operating pressure. The amount of liquid that has evaporated, the mass flow rate of the exiting steam, and the exit velocity are

$$m = \frac{\Delta V_{\text{liquid}}}{v_f} = \frac{0.6 \text{ L}}{0.001053 \text{ m}^3/\text{kg}}\left(\frac{1 \text{ m}^3}{1000 \text{ L}}\right) = 0.570 \text{ kg}$$

$$\dot{m} = \frac{m}{\Delta t} = \frac{0.570 \text{ kg}}{40 \text{ min}} = 0.0142 \text{ kg/min} = \mathbf{2.37 \times 10^{-4} \text{ kg/s}}$$

$$V = \frac{\dot{m}}{\rho_g A_c} = \frac{\dot{m} v_g}{A_c} = \frac{(2.37 \times 10^{-4} \text{ kg/s})(1.1594 \text{ m}^3/\text{kg})}{8 \times 10^{-6} \text{ m}^2} = \mathbf{34.3 \text{ m/s}}$$

(*b*) Noting that $h = u + Pv$ and that the kinetic and potential energies are disregarded, the flow and total energies of the exiting steam are

$$e_{\text{flow}} = Pv = h - u = 2693.1 - 2519.2 = \mathbf{173.9 \text{ kJ/kg}}$$

$$\theta = h + \text{ke} + \text{pe} \cong h = \mathbf{2693.1 \text{ kJ/kg}}$$

Note that the kinetic energy in this case is ke $= V^2/2 = (34.3 \text{ m/s})^2/2 = 588 \text{ m}^2/\text{s}^2 = 0.588$ kJ/kg, which is small compared to enthalpy.

(*c*) The rate at which energy is leaving the cooker by mass is simply the product of the mass flow rate and the total energy of the exiting steam per unit mass,

$$\dot{E}_{\text{mass}} = \dot{m}\theta = (2.37 \times 10^{-4} \text{ kg/s})(2693.1 \text{ kJ/kg}) = 0.638 \text{ kJ/s} = \mathbf{0.638 \text{ kW}}$$

Discussion The numerical value of the energy leaving the cooker with steam alone does not mean much since this value depends on the reference point selected for enthalpy (it could even be negative). The significant quantity is the difference between the enthalpies of the exiting vapor and the liquid inside (which is h_{fg}) since it relates directly to the amount of energy supplied to the cooker.

6–3 ■ ENERGY ANALYSIS OF STEADY-FLOW SYSTEMS

A large number of engineering devices such as turbines, compressors, and nozzles operate for long periods of time under the same conditions once the transient start-up period is completed and steady operation is established, and they are classified as *steady-flow devices* (Fig. 6–17). Processes involving such devices can be represented reasonably well by a somewhat idealized process, called the **steady-flow process**, which was defined in Chap. 2 as *a process during which a fluid flows through a control volume steadily*. That is, the fluid properties can change from point to point within the control volume, but at any point, they remain constant during the entire process. (Remember, *steady* means *no change with time*.)

FIGURE 6–17

Many engineering systems such as power plants operate under steady conditions.

© Vol. 57/PhotoDisc

During a steady-flow process, no intensive or extensive properties *within the control volume* change with time. Thus, the volume V, the mass m, and the total energy content E of the control volume remain constant (Fig. 6–18). As a result, the boundary work is zero for steady-flow systems (since V_{CV} = constant), and the total mass or energy entering the control volume must be equal to the total mass or energy leaving it (since m_{CV} = constant and E_{CV} = constant). These observations greatly simplify the analysis.

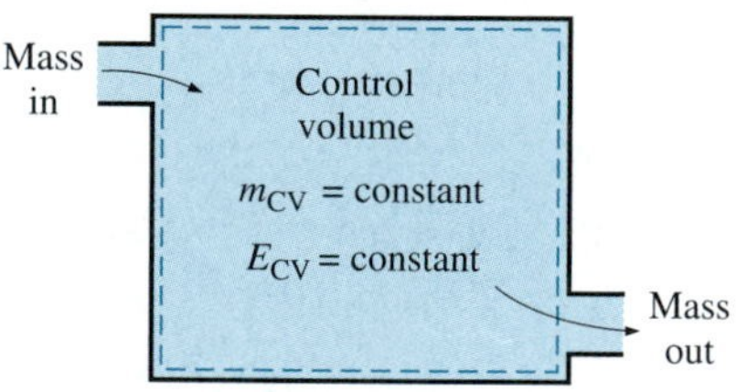

FIGURE 6–18

Under steady-flow conditions, the mass and energy contents of a control volume remain constant.

The fluid properties at an inlet or exit remain constant during a steady-flow process. The properties may, however, be different at different inlets and exits. They may even vary over the cross section of an inlet or an exit. However, all properties, including the velocity and elevation, must remain constant with time at a fixed point at an inlet or exit. It follows that the mass flow rate of the fluid at an opening must remain constant during a steady-flow process (Fig. 6–19). As an added simplification, the fluid properties at an opening are usually approximated as uniform (at some average value) over the cross section. Thus, the fluid properties at an inlet or exit may be specified by the average single values. Also, the *heat* and *work* interactions between a steady-flow system and its surroundings do not change with time. Thus, the power delivered by a system and the rate of heat transfer to or from a system remain constant during a steady-flow process.

The *mass balance* for a general steady-flow system was given in Sec. 6–1 as

$$\sum_{in} \dot{m} = \sum_{out} \dot{m} \qquad \text{(kg/s)} \tag{6–31}$$

The mass balance for a single-stream (one-inlet and one-outlet) steady-flow system was given as

$$\dot{m}_1 = \dot{m}_2 \quad \rightarrow \quad \rho_1 V_1 A_1 = \rho_2 V_2 A_2 \tag{6–32}$$

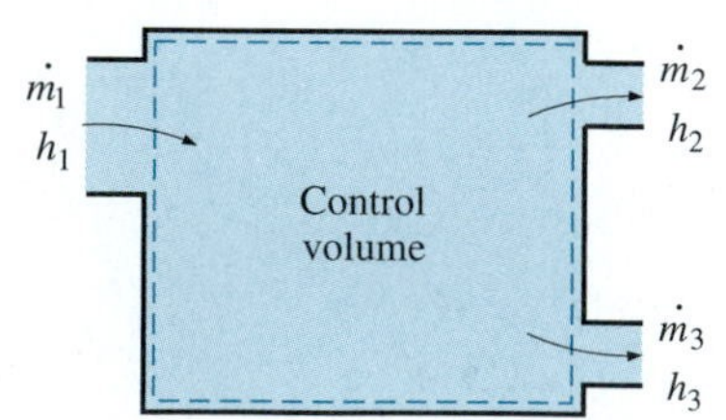

FIGURE 6–19

Under steady-flow conditions, the fluid properties at an inlet or exit remain constant (do not change with time).

where the subscripts 1 and 2 denote the inlet and the exit states, respectively, ρ is density, V is the average flow velocity in the flow direction, and A is the cross-sectional area normal to the flow direction.

During a steady-flow process, the total energy content of a control volume remains constant (E_{CV} = constant), and thus the change in the total energy of the control volume is zero ($\Delta E_{CV} = 0$). Therefore, the amount of energy entering a control volume in all forms (by heat, work, and mass) must be equal to the amount of energy leaving it. Then the rate form of the general energy balance reduces for a steady-flow process to

$$\underbrace{\dot{E}_{in} - \dot{E}_{out}}_{\substack{\text{Rate of net energy transfer} \\ \text{by heat, work, and mass}}} = \underbrace{dE_{system}/dt^{\nearrow 0 \text{ (steady)}}}_{\substack{\text{Rate of change in internal, kinetic,} \\ \text{potential, etc., energies}}} = 0 \quad \text{(6–33)}$$

or

$$\textit{Energy balance:} \quad \underbrace{\dot{E}_{in}}_{\substack{\text{Rate of net energy transfer in} \\ \text{by heat, work, and mass}}} = \underbrace{\dot{E}_{out}}_{\substack{\text{Rate of net energy transfer out} \\ \text{by heat, work, and mass}}} \quad \text{(kW)} \quad \text{(6–34)}$$

Noting that energy can be transferred by heat, work, and mass only, the energy balance in Eq. 6–34 for a general steady-flow system can also be written more explicitly as

$$\dot{Q}_{in} + \dot{W}_{in} + \sum_{in} \dot{m}\theta = \dot{Q}_{out} + \dot{W}_{out} + \sum_{out} \dot{m}\theta \quad \text{(6–35)}$$

or

$$\dot{Q}_{in} + \dot{W}_{in} + \underbrace{\sum_{in} \dot{m}\left(h + \frac{V^2}{2} + gz\right)}_{\text{for each inlet}} = \dot{Q}_{out} + \dot{W}_{out} + \underbrace{\sum_{out} \dot{m}\left(h + \frac{V^2}{2} + gz\right)}_{\text{for each exit}} \quad \text{(6–36)}$$

since the energy of a flowing fluid per unit mass is $\theta = h + \text{ke} + \text{pe} = h + V^2/2 + gz$. The energy balance relation for steady-flow systems first appeared in 1859 in a German thermodynamics book written by Gustav Zeuner.

Consider, for example, an ordinary electric hot-water heater under steady operation, as shown in Fig. 6–20. A cold-water stream with a mass flow rate $\dot{m}$ is continuously flowing into the water heater, and a hot-water stream of the same mass flow rate is continuously flowing out of it. The water heater (the control volume) is losing heat to the surrounding air at a rate of $\dot{Q}_{out}$, and the electric heating element is supplying electrical work (heating) to the water at a rate of $\dot{W}_{in}$. On the basis of the conservation of energy principle, we can say that the water stream experiences an increase in its total energy as it flows through the water heater that is equal to the electric energy supplied to the water minus the heat losses.

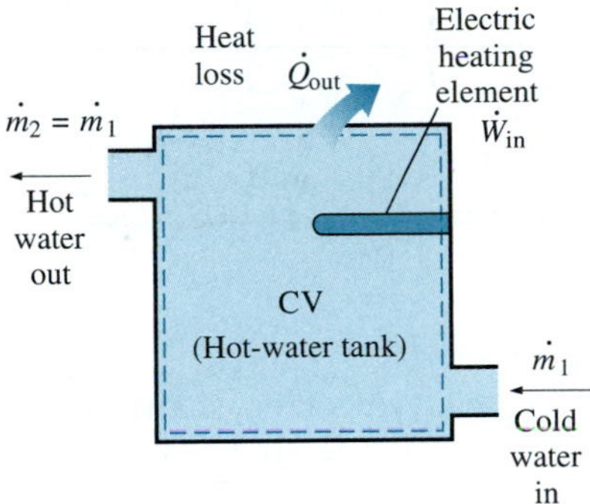

FIGURE 6–20

A water heater in steady operation.

The energy balance relation just given is intuitive in nature and is easy to use when the magnitudes and directions of heat and work transfers are known. When performing a general analytical study or solving a problem that involves an unknown heat or work interaction, however, we need to assume a direction for the heat or work interactions. In such cases, it is common practice to assume heat to be transferred *into the system* (heat input) at a rate of $\dot{Q}$, and work produced *by the system* (work output) at a rate of $\dot{W}$, and then solve the problem. The first-law or energy balance relation in that case for a general steady-flow system becomes

$$\dot{Q} - \dot{W} = \underbrace{\sum_{out} \dot{m}\left(h + \frac{V^2}{2} + gz\right)}_{\text{for each exit}} - \underbrace{\sum_{in} \dot{m}\left(h + \frac{V^2}{2} + gz\right)}_{\text{for each inlet}} \quad \text{(6–37)}$$

Obtaining a negative quantity for $\dot{Q}$ or $\dot{W}$ simply means that the assumed direction is wrong and should be reversed. For single-stream devices, the steady-flow energy balance equation becomes

$$\dot{Q} - \dot{W} = \dot{m}\left[h_2 - h_1 + \frac{V_2^2 - V_1^2}{2} + g(z_2 - z_1)\right] \quad \text{(6–38)}$$

Dividing Eq. 6–38 by $\dot{m}$ gives the energy balance on a unit-mass basis as

$$q - w = h_2 - h_1 + \frac{V_2^2 - V_1^2}{2} + g(z_2 - z_1) \qquad \textbf{(6–39)}$$

where $q = \dot{Q}/\dot{m}$ and $w = \dot{W}/\dot{m}$ are the heat transfer and work done per unit mass of the working fluid, respectively. When the fluid experiences negligible changes in its kinetic and potential energies (that is, $\Delta \text{ke} \cong 0$, $\Delta \text{pe} \cong 0$), the energy balance equation is reduced further to

$$q - w = h_2 - h_1 \qquad \textbf{(6–40)}$$

The various terms appearing in the above equations are as follows:

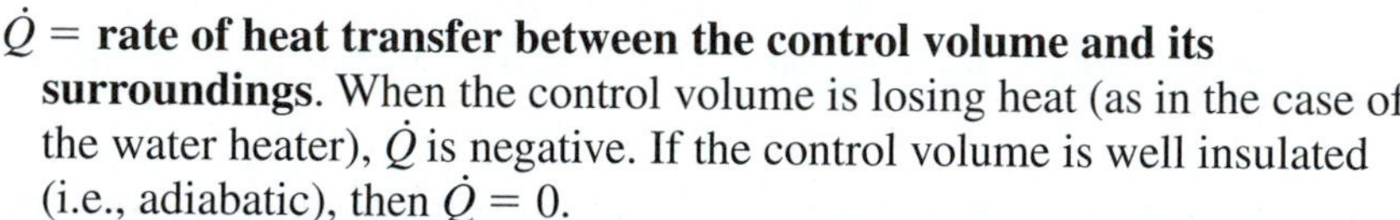

$\dot{Q}$ = **rate of heat transfer between the control volume and its surroundings**. When the control volume is losing heat (as in the case of the water heater), $\dot{Q}$ is negative. If the control volume is well insulated (i.e., adiabatic), then $\dot{Q} = 0$.

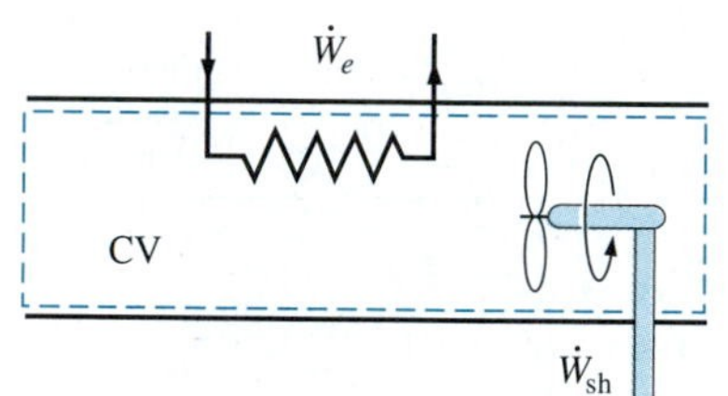

FIGURE 6–21

Under steady operation, shaft work and electrical work are the only forms of work a simple compressible system may involve.

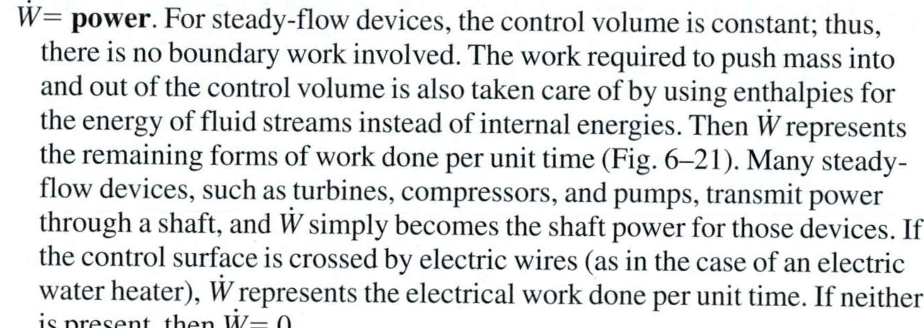

$\dot{W}$ = **power**. For steady-flow devices, the control volume is constant; thus, there is no boundary work involved. The work required to push mass into and out of the control volume is also taken care of by using enthalpies for the energy of fluid streams instead of internal energies. Then $\dot{W}$ represents the remaining forms of work done per unit time (Fig. 6–21). Many steady-flow devices, such as turbines, compressors, and pumps, transmit power through a shaft, and $\dot{W}$ simply becomes the shaft power for those devices. If the control surface is crossed by electric wires (as in the case of an electric water heater), $\dot{W}$ represents the electrical work done per unit time. If neither is present, then $\dot{W} = 0$.

$\Delta h = h_2 - h_1$. The enthalpy change of a fluid can easily be determined by reading the enthalpy values at the exit and inlet states from the tables. For ideal gases, it can be approximated by $\Delta h = c_{p,\text{avg}}(T_2 - T_1)$. Note that (kg/s)(kJ/kg) ≡ kW.

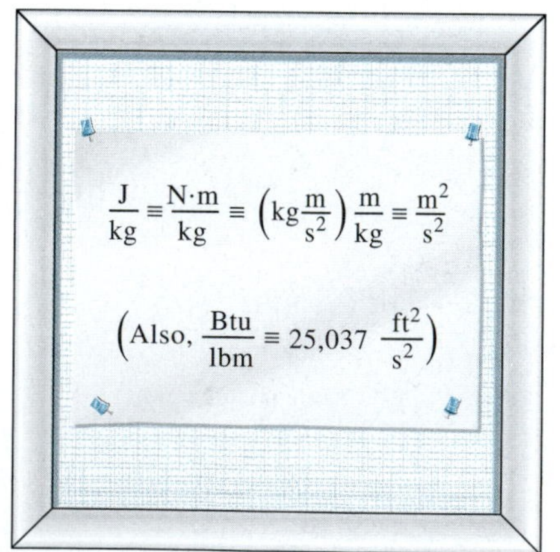

FIGURE 6–22

The units m^2/s^2 and J/kg are equivalent.

$\Delta \text{ke} = (V_2^2 - V_1^2)/2$. The unit of kinetic energy is m^2/s^2, which is equivalent to J/kg (Fig. 6–22). The enthalpy is usually given in kJ/kg. To add these two quantities, the kinetic energy should be expressed in kJ/kg. This is easily accomplished by dividing it by 1000. A velocity of 45 m/s corresponds to a kinetic energy of only 1 kJ/kg, which is a very small value compared with the enthalpy values encountered in practice. Thus, the kinetic energy term at low velocities can usually be neglected. When a fluid stream enters and leaves a steady-flow device at about the same velocity ($V_1 \cong V_2$), the change in the kinetic energy is close to zero regardless of the velocity. Caution should be exercised at high velocities, however, since small changes in velocities may cause significant changes in kinetic energy (Fig. 6–23).

$\Delta \text{pe} = g(z_2 - z_1)$. A similar argument can be given for the potential energy term. A potential energy change of 1 kJ/kg corresponds to an elevation difference of 102 m. The elevation difference between the inlet and exit of most industrial devices such as turbines and compressors is well below this value, and the potential energy term is nearly always neglected for these devices. The only time the potential energy term is significant is when a process involves pumping a fluid to high elevations and we are interested in the required pumping power.

6–4 ▪ SOME STEADY-FLOW ENGINEERING DEVICES

Many engineering devices operate essentially under the same conditions for long periods of time. The components of a steam power plant (turbines, compressors, heat exchangers, and pumps), for example, operate nonstop for months before the system is shut down for maintenance (Fig. 6–24). Therefore, these devices can be conveniently analyzed as steady-flow devices.

In this section, some common steady-flow devices are described, and the thermodynamic aspects of the flow through them are analyzed. The conservation of mass and the conservation of energy principles for these devices are illustrated with examples.

V_1 m/s	V_2 m/s	Δke kJ/kg
0	45	1
50	67	1
100	110	1
200	205	1
500	502	1

FIGURE 6–23

At very high velocities, even small changes in velocities can cause significant changes in the kinetic energy of the fluid.

1 Nozzles and Diffusers

Nozzles and diffusers are commonly utilized in jet engines, rockets, spacecraft, and even garden hoses. A **nozzle** is a device that *increases the velocity of a fluid* at the expense of pressure. A **diffuser** is a device that *increases the pressure of a fluid* by slowing it down. That is, nozzles and diffusers perform opposite tasks. The cross-sectional area of a nozzle decreases in the flow direction for subsonic flows and increases for supersonic flows. The reverse is true for diffusers.

The rate of heat transfer between the fluid flowing through a nozzle or a diffuser and the surroundings is usually very small ($\dot{Q} \approx 0$) since the fluid has high velocities, and thus it does not spend enough time in the device for any significant heat transfer to take place. Nozzles and diffusers typically involve no work ($\dot{W} = 0$) and any change in potential energy is negligible ($\Delta \text{pe} \cong 0$). But nozzles and diffusers usually involve very high velocities, and as a fluid passes through a nozzle or diffuser, it experiences large changes in its velocity (Fig. 6–25). Therefore, the kinetic energy changes must be accounted for in analyzing the flow through these devices ($\Delta \text{ke} \neq 0$).

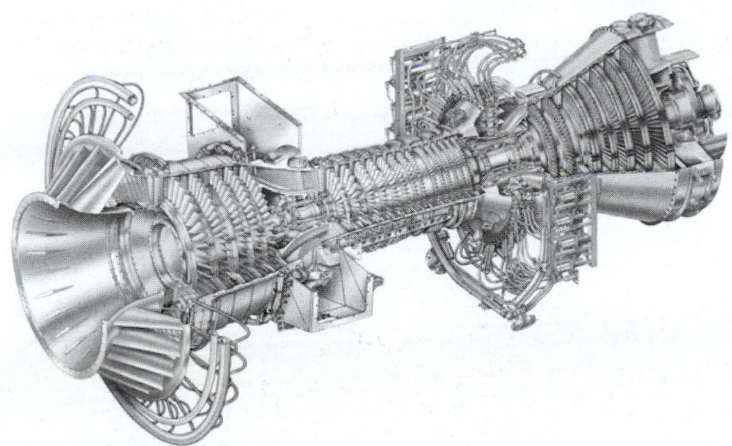

FIGURE 6–24

A modern land-based gas turbine used for electric power production. This is a General Electric LM5000 turbine. It has a length of 6.2 m, it weighs 12.5 tons, and produces 55.2 MW at 3600 rpm with steam injection.

Courtesy of GE Power Systems

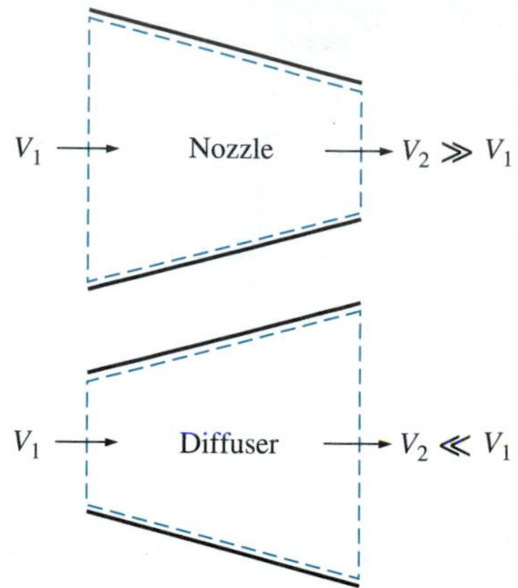

FIGURE 6–25

Nozzles and diffusers are shaped so that they cause large changes in fluid velocities and thus kinetic energies.

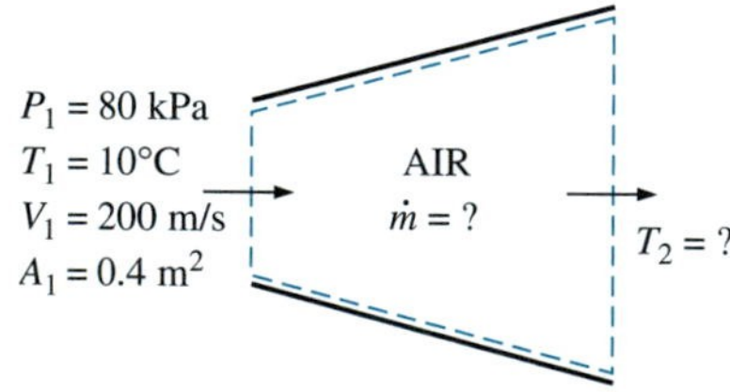

FIGURE 6–26

Schematic for Example 6–4.

EXAMPLE 6–4 Deceleration of Air in a Diffuser

Air at 10°C and 80 kPa enters the diffuser of a jet engine steadily with a velocity of 200 m/s. The inlet area of the diffuser is 0.4 m^2. The air leaves the diffuser with a velocity that is very small compared with the inlet velocity. Determine (*a*) the mass flow rate of the air and (*b*) the temperature of the air leaving the diffuser.

Solution Air enters the diffuser of a jet engine steadily at a specified velocity. The mass flow rate of air and the temperature at the diffuser exit are to be determined.

Assumptions **1** This is a steady-flow process since there is no change with time at any point and thus $\Delta m_{CV} = 0$ and $\Delta E_{CV} = 0$. **2** Air is an ideal gas since it is at a high temperature and low pressure relative to its critical-point values. **3** The potential energy change is zero, $\Delta pe = 0$. **4** Heat transfer is negligible. **5** Kinetic energy at the diffuser exit is negligible. **6** There are no work interactions.

Analysis We take the *diffuser* as the system (Fig. 6–26). This is a *control volume* since mass crosses the system boundary during the process. We observe that there is only one inlet and one exit and thus $\dot{m}_1 = \dot{m}_2 = \dot{m}$.

(*a*) To determine the mass flow rate, we need to find the specific volume of the air first. This is determined from the ideal-gas relation at the inlet conditions:

$$v_1 = \frac{RT_1}{P_1} = \frac{(0.287\ \text{kPa}\cdot\text{m}^3/\text{kg}\cdot\text{K})(283\ \text{K})}{80\ \text{kPa}} = 1.015\ \text{m}^3/\text{kg}$$

Then,

$$\dot{m} = \frac{1}{v_1}V_1A_1 = \frac{1}{1.015\ \text{m}^3/\text{kg}}(200\ \text{m/s})(0.4\ \text{m}^2) = \mathbf{78.8\ kg/s}$$

Since the flow is steady, the mass flow rate through the entire diffuser remains constant at this value.

(*b*) Under stated assumptions and observations, the energy balance for this steady-flow system can be expressed in the rate form as

$$\underbrace{\dot{E}_{\text{in}} - \dot{E}_{\text{out}}}_{\substack{\text{Rate of net energy transfer}\\ \text{by heat, work, and mass}}} = \underbrace{dE_{\text{system}}/dt^{\nearrow 0\ \text{(steady)}}}_{\substack{\text{Rate of change in internal, kinetic,}\\ \text{potential, etc., energies}}} = 0$$

$$\dot{E}_{\text{in}} = \dot{E}_{\text{out}}$$

$$\dot{m}\left(h_1 + \frac{V_1^2}{2}\right) = \dot{m}\left(h_2 + \frac{V_2^2}{2}\right) \quad (\text{since } \dot{Q} \cong 0,\ \dot{W} = 0,\ \text{and } \Delta\text{pe} \cong 0)$$

$$h_2 = h_1 - \frac{V_2^2 - V_1^2}{2}$$

The exit velocity of a diffuser is usually small compared with the inlet velocity ($V_2 \ll V_1$); thus, the kinetic energy at the exit can be neglected. The enthalpy of air at the diffuser inlet is determined from the air table (Table A–21) to be

$$h_1 = h_{@\ 283\ \text{K}} = 283.14\ \text{kJ/kg}$$

Substituting, we get

$$h_2 = 283.14 \text{ kJ/kg} - \frac{0 - (200 \text{ m/s})^2}{2}\left(\frac{1 \text{ kJ/kg}}{1000 \text{ m}^2/\text{s}^2}\right)$$

$$= 303.14 \text{ kJ/kg}$$

From Table A–21, the temperature corresponding to this enthalpy value is

$$T_2 = \mathbf{303 \text{ K}}$$

Discussion This result shows that the temperature of the air increases by about 20°C as it is slowed down in the diffuser. The temperature rise of the air is mainly due to the conversion of kinetic energy to internal energy.

EXAMPLE 6–5 Acceleration of Steam in a Nozzle

Steam at 250 psia and 700°F steadily enters a nozzle whose inlet area is 0.2 ft². The mass flow rate of steam through the nozzle is 10 lbm/s. Steam leaves the nozzle at 200 psia with a velocity of 900 ft/s. Heat losses from the nozzle per unit mass of the steam are estimated to be 1.2 Btu/lbm. Determine (*a*) the inlet velocity and (*b*) the exit temperature of the steam.

Solution Steam enters a nozzle steadily at a specified flow rate and velocity. The inlet velocity of steam and the exit temperature are to be determined.

Assumptions **1** This is a steady-flow process since there is no change with time at any point and thus $\Delta m_{CV} = 0$ and $\Delta E_{CV} = 0$. **2** There are no work interactions. **3** The potential energy change is zero, $\Delta pe = 0$.

Analysis We take the *nozzle* as the system (Fig. 6–26A). This is a *control volume* since mass crosses the system boundary during the process. We observe that there is only one inlet and one exit and thus $\dot{m}_1 = \dot{m}_2 = \dot{m}$.

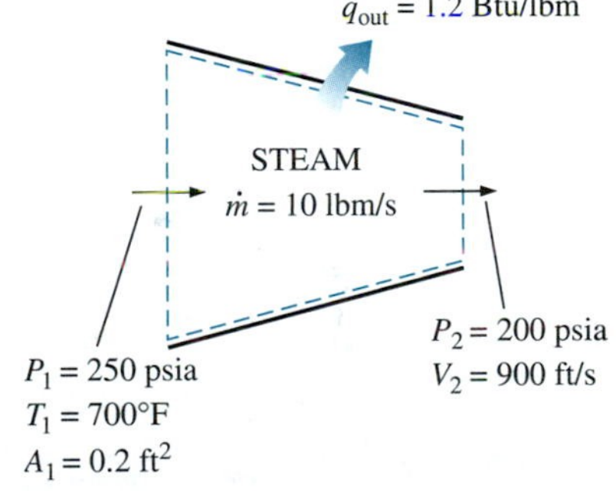

FIGURE 6–26A

Schematic for Example 6–5.

(*a*) The specific volume and enthalpy of steam at the nozzle inlet are

$$\left.\begin{array}{l} P_1 = 250 \text{ psia} \\ T_1 = 700°\text{F} \end{array}\right\} \begin{array}{l} v_1 = 2.6883 \text{ ft}^3/\text{lbm} \\ h_1 = 1371.4 \text{ Btu/lbm} \end{array} \quad \text{(Table A–6E)}$$

Then,

$$\dot{m} = \frac{1}{v_1} V_1 A_1$$

$$10 \text{ lbm/s} = \frac{1}{2.6883 \text{ ft}^3/\text{lbm}} (V_1)(0.2 \text{ ft}^2)$$

$$V_1 = \mathbf{134.4 \text{ ft/s}}$$

(*b*) Under stated assumptions and observations, the energy balance for this steady-flow system can be expressed in the rate form as

$$\underbrace{\dot{E}_{in} - \dot{E}_{out}}_{\substack{\text{Rate of net energy transfer} \\ \text{by heat, work, and mass}}} = \underbrace{dE_{system}/dt}_{\substack{\text{Rate of change in internal, kinetic,} \\ \text{potential, etc., energies}}} \overset{\nearrow 0 \text{ (steady)}}{} = 0$$

$$\dot{E}_{in} = \dot{E}_{out}$$

$$\dot{m}\left(h_1 + \frac{V_1^2}{2}\right) = \dot{Q}_{out} + \dot{m}\left(h_2 + \frac{V_2^2}{2}\right) \qquad (\text{since } \dot{W} = 0, \text{ and } \Delta pe \cong 0)$$

Dividing by the mass flow rate $\dot{m}$ and substituting, h_2 is determined to be

$$h_2 = h_1 - q_{out} - \frac{V_2^2 - V_1^2}{2}$$

$$= (1371.4 - 1.2)\ \text{Btu/lbm} - \frac{(900\ \text{ft/s})^2 - (134.4\ \text{ft/s})^2}{2}\left(\frac{1\ \text{Btu/lbm}}{25{,}037\ \text{ft}^2/\text{s}^2}\right)$$

$$= 1354.4\ \text{Btu/lbm}$$

Then,

$$\left.\begin{aligned} P_2 &= 200\ \text{psia} \\ h_2 &= 1354.4\ \text{Btu/lbm} \end{aligned}\right\}\quad T_2 = \mathbf{662.0°F} \qquad \text{(Table A–6E)}$$

Discussion Note that the temperature of steam drops by 38.0°F as it flows through the nozzle. This drop in temperature is mainly due to the conversion of internal energy to kinetic energy. (The heat loss is too small to cause any significant effect in this case.)

2 Turbines and Compressors

In steam, gas, or hydroelectric power plants, the device that drives the electric generator is the turbine. As the fluid passes through the turbine, work is done against the blades, which are attached to the shaft. As a result, the shaft rotates, and the turbine produces work.

Compressors, as well as pumps and fans, are devices used to increase the pressure of a fluid. Work is supplied to these devices from an external source through a rotating shaft. Therefore, compressors involve work inputs. Even though these three devices function similarly, they do differ in the tasks they perform. A *fan* increases the pressure of a gas slightly and is mainly used to mobilize a gas. A *compressor* is capable of compressing the gas to very high pressures. *Pumps* work very much like compressors except that they handle liquids instead of gases.

Note that turbines produce power output whereas compressors, pumps, and fans require power input. Heat transfer from turbines is usually negligible ($\dot{Q} \approx 0$) since they are typically well insulated. Heat transfer is also negligible for compressors unless there is intentional cooling. Potential energy changes are negligible for all of these devices ($\Delta pe \cong 0$). The velocities involved in these devices, with the exception of turbines and fans, are usually too low to cause any significant change in the kinetic energy ($\Delta ke \cong 0$). The fluid velocities encountered in most turbines are very high, and the fluid experiences a significant change in its kinetic energy. However, this change is usually very small relative to the change in enthalpy, and thus it is often disregarded.

EXAMPLE 6–6 Compressing Air by a Compressor

Air at 100 kPa and 280 K is compressed steadily to 600 kPa and 400 K. The mass flow rate of the air is 0.02 kg/s, and a heat loss of 16 kJ/kg occurs during the process. Assuming the changes in kinetic and potential energies are negligible, determine the necessary power input to the compressor.

Solution Air is compressed steadily by a compressor to a specified temperature and pressure. The power input to the compressor is to be determined.

Assumptions **1** This is a steady-flow process since there is no change with time at any point and thus $\Delta m_{CV} = 0$ and $\Delta E_{CV} = 0$. **2** Air is an ideal gas since it is at a high temperature and low pressure relative to its critical-point values. **3** The kinetic and potential energy changes are zero, $\Delta ke = \Delta pe = 0$.

Analysis We take the *compressor* as the system (Fig. 6–27). This is a *control volume* since mass crosses the system boundary during the process. We observe that there is only one inlet and one exit and thus $\dot{m}_1 = \dot{m}_2 = \dot{m}$. Also, heat is lost from the system and work is supplied to the system.

Under stated assumptions and observations, the energy balance for this steady-flow system can be expressed in the rate form as

$$\underbrace{\dot{E}_{in} - \dot{E}_{out}}_{\substack{\text{Rate of net energy transfer}\\ \text{by heat, work, and mass}}} = \underbrace{dE_{system}/dt}_{\substack{\text{Rate of change in internal, kinetic,}\\ \text{potential, etc., energies}}}^{\nearrow 0 \text{ (steady)}} = 0$$

$$\dot{E}_{in} = \dot{E}_{out}$$

$$\dot{W}_{in} + \dot{m}h_1 = \dot{Q}_{out} + \dot{m}h_2 \qquad (\text{since } \Delta ke = \Delta pe \cong 0)$$

$$\dot{W}_{in} = \dot{m}q_{out} + \dot{m}(h_2 - h_1)$$

The enthalpy of an ideal gas depends on temperature only, and the enthalpies of the air at the specified temperatures are determined from the air table (Table A–21) to be

$$h_1 = h_{@\ 280\ K} = 280.13 \text{ kJ/kg}$$

$$h_2 = h_{@\ 400\ K} = 400.98 \text{ kJ/kg}$$

Substituting, the power input to the compressor is determined to be

$$\dot{W}_{in} = (0.02 \text{ kg/s})(16 \text{ kJ/kg}) + (0.02 \text{ kg/s})(400.98 - 280.13) \text{ kJ/kg}$$

$$= \mathbf{2.74\ kW}$$

Discussion Note that the mechanical energy input to the compressor manifests itself as a rise in enthalpy of air and heat loss from the compressor.

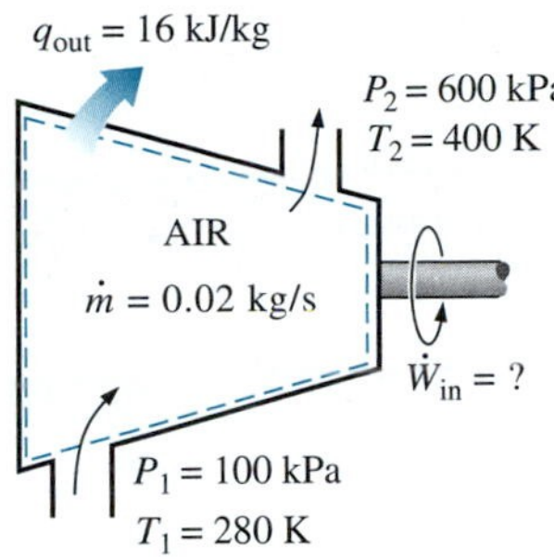

FIGURE 6–27

Schematic for Example 6–6.

EXAMPLE 6–7 Power Generation by a Steam Turbine

The power output of an adiabatic steam turbine is 5 MW, and the inlet and the exit conditions of the steam are as indicated in Fig. 6–28.

(*a*) Compare the magnitudes of Δh, Δke, and Δpe.

(*b*) Determine the work done per unit mass of the steam flowing through the turbine.

(*c*) Calculate the mass flow rate of the steam.

Solution The inlet and exit conditions of a steam turbine and its power output are given. The changes in kinetic energy, potential energy, and enthalpy of steam, as well as the work done per unit mass and the mass flow rate of steam are to be determined.

Assumptions **1** This is a steady-flow process since there is no change with time at any point and thus $\Delta m_{CV} = 0$ and $\Delta E_{CV} = 0$. **2** The system is adiabatic and thus there is no heat transfer.

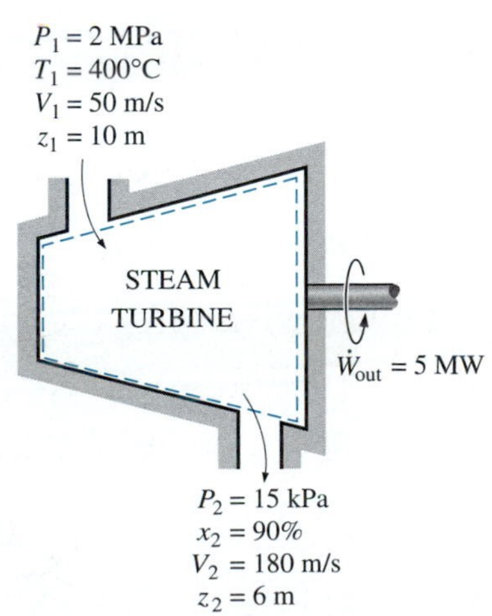

FIGURE 6–28

Schematic for Example 6–7.

Analysis We take the *turbine* as the system. This is a *control volume* since mass crosses the system boundary during the process. We observe that there is only one inlet and one exit and thus $\dot{m}_1 = \dot{m}_2 = \dot{m}$. Also, work is done by the system. The inlet and exit velocities and elevations are given, and thus the kinetic and potential energies are to be considered.

(*a*) At the inlet, steam is in a superheated vapor state, and its enthalpy is

$$\left.\begin{array}{l} P_1 = 2 \text{ MPa} \\ T_1 = 400^\circ\text{C} \end{array}\right\} \quad h_1 = 3248.4 \text{ kJ/kg} \qquad \text{(Table A–6)}$$

At the turbine exit, we obviously have a saturated liquid–vapor mixture at 15-kPa pressure. The enthalpy at this state is

$$h_2 = h_f + x_2 h_{fg} = [225.94 + (0.9)(2372.3)] \text{ kJ/kg} = 2361.01 \text{ kJ/kg}$$

Then

$$\Delta h = h_2 - h_1 = (2361.01 - 3248.4) \text{ kJ/kg} = \mathbf{-887.39 \text{ kJ/kg}}$$

$$\Delta\text{ke} = \frac{V_2^2 - V_1^2}{2} = \frac{(180 \text{ m/s})^2 - (50 \text{ m/s})^2}{2}\left(\frac{1 \text{ kJ/kg}}{1000 \text{ m}^2/\text{s}^2}\right) = \mathbf{14.95 \text{ kJ/kg}}$$

$$\Delta\text{pe} = g(z_2 - z_1) = (9.81 \text{ m/s}^2)[(6 - 10) \text{ m}]\left(\frac{1 \text{ kJ/kg}}{1000 \text{ m}^2/\text{s}^2}\right) = \mathbf{-0.04 \text{ kJ/kg}}$$

(*b*) The energy balance for this steady-flow system can be expressed in the rate form as

$$\underbrace{\dot{E}_{\text{in}} - \dot{E}_{\text{out}}}_{\substack{\text{Rate of net energy transfer} \\ \text{by heat, work, and mass}}} = \underbrace{dE_{\text{system}}/dt^{\nearrow 0 \text{ (steady)}}}_{\substack{\text{Rate of change in internal, kinetic,} \\ \text{potential, etc., energies}}} = 0$$

$$\dot{E}_{\text{in}} = \dot{E}_{\text{out}}$$

$$\dot{m}\left(h_1 + \frac{V_1^2}{2} + gz_1\right) = \dot{W}_{\text{out}} + \dot{m}\left(h_2 + \frac{V_2^2}{2} + gz_2\right) \qquad (\text{since } \dot{Q} = 0)$$

Dividing by the mass flow rate $\dot{m}$ and substituting, the work done by the turbine per unit mass of the steam is determined to be

$$w_{\text{out}} = -\left[(h_2 - h_1) + \frac{V_2^2 - V_1^2}{2} + g(z_2 - z_1)\right] = -(\Delta h + \Delta\text{ke} + \Delta\text{pe})$$

$$= -[-887.39 + 14.95 - 0.04] \text{ kJ/kg} = \mathbf{872.48 \text{ kJ/kg}}$$

(*c*) The required mass flow rate for a 5-MW power output is

$$\dot{m} = \frac{\dot{W}_{\text{out}}}{w_{\text{out}}} = \frac{5000 \text{ kJ/s}}{872.48 \text{ kJ/kg}} = \mathbf{5.73 \text{ kg/s}}$$

Discussion Two observations can be made from these results. First, the change in potential energy is insignificant in comparison to the changes in enthalpy and kinetic energy. This is typical for most engineering devices. Second, as a result of low pressure and thus high specific volume, the steam velocity at the turbine exit can be very high. Yet the change in kinetic energy is a small fraction of the change in enthalpy (less than 2 percent in our case) and is therefore often neglected.

3 Throttling Valves

Throttling valves are *any kind of flow-restricting devices* that cause a significant pressure drop in the fluid. Some familiar examples are ordinary adjustable valves, capillary tubes, and porous plugs (Fig. 6–29). Unlike turbines, they produce a pressure drop without involving any work. The pressure drop in the fluid is often accompanied by a *large drop in temperature*, and for that reason throttling devices are commonly used in refrigeration and air-conditioning applications. The magnitude of the temperature drop (or, sometimes, the temperature rise) during a throttling process is governed by a property called the *Joule-Thomson coefficient.*

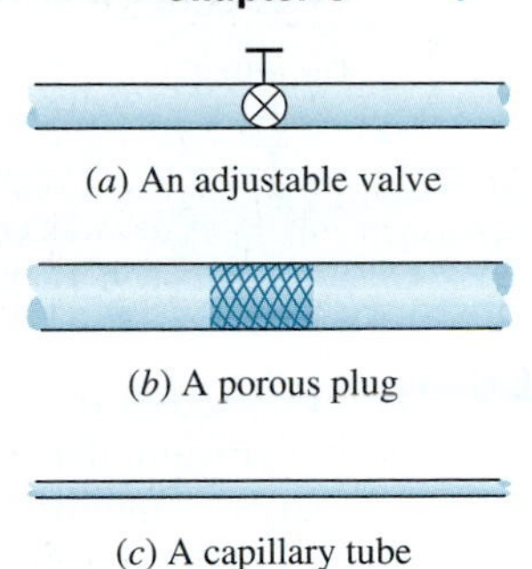

(*a*) An adjustable valve

(*b*) A porous plug

(*c*) A capillary tube

FIGURE 6–29

Throttling valves are devices that cause large pressure drops in the fluid.

Throttling valves are usually small devices, and the flow through them may be assumed to be adiabatic ($q \cong 0$) since there is neither sufficient time nor large enough area for any effective heat transfer to take place. Also, there is no work done ($w = 0$), and the change in potential energy, if any, is very small ($\Delta pe \cong 0$). Even though the exit velocity is often considerably higher than the inlet velocity, in many cases, the increase in kinetic energy is insignificant ($\Delta ke \cong 0$). Then the conservation of energy equation for this single-stream steady-flow device reduces to

$$h_2 \cong h_1 \qquad \text{(kJ/kg)} \tag{6–41}$$

That is, enthalpy values at the inlet and exit of a throttling valve are the same. For this reason, a throttling valve is sometimes called an *isenthalpic device*. Note, however, that for throttling devices with large exposed surface areas such as capillary tubes, heat transfer may be significant.

To gain some insight into how throttling affects fluid properties, let us express Eq. 6–41 as follows:

$$u_1 + P_1 v_1 = u_2 + P_2 v_2$$

or

$$\text{Internal energy} + \text{Flow energy} = \text{Constant}$$

Thus the final outcome of a throttling process depends on which of the two quantities increases during the process. If the flow energy increases during the process ($P_2 v_2 > P_1 v_1$), it can do so at the expense of the internal energy. As a result, internal energy decreases, which is usually accompanied by a drop in temperature. If the product Pv decreases, the internal energy and the temperature of a fluid will increase during a throttling process. In the case of an ideal gas, $h = h(T)$, and thus the temperature has to remain constant during a throttling process (Fig. 6–30).

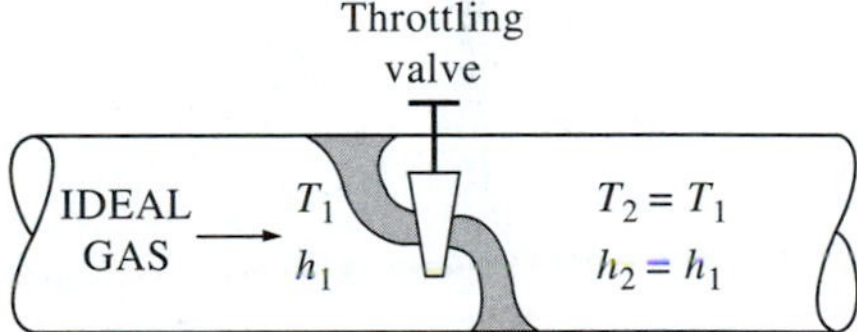

FIGURE 6–30

The temperature of an ideal gas does not change during a throttling (h = constant) process since $h = h(T)$.

EXAMPLE 6–8 Expansion of Refrigerant-134a in a Refrigerator

Refrigerant-134a enters the capillary tube of a refrigerator as saturated liquid at 0.8 MPa and is throttled to a pressure of 0.12 MPa. Determine the quality of the refrigerant at the final state and the temperature drop during this process.

Solution Refrigerant-134a that enters a capillary tube as saturated liquid is throttled to a specified pressure. The exit quality of the refrigerant and the temperature drop are to be determined.

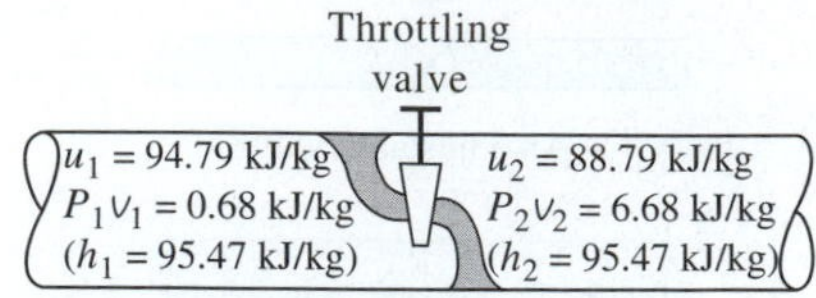

FIGURE 6–31

During a throttling process, the enthalpy (flow energy + internal energy) of a fluid remains constant. But internal and flow energies may be converted to each other.

Assumptions **1** Heat transfer from the tube is negligible. **2** Kinetic energy change of the refrigerant is negligible.

Analysis A capillary tube is a simple flow-restricting device that is commonly used in refrigeration applications to cause a large pressure drop in the refrigerant. Flow through a capillary tube is a throttling process; thus, the enthalpy of the refrigerant remains constant (Fig. 6–31).

$$\text{At inlet:}\quad \left.\begin{array}{l} P_1 = 0.8\text{ MPa} \\ \text{sat. liquid} \end{array}\right\} \quad \begin{array}{l} T_1 = T_{\text{sat @ 0.8 MPa}} = 31.31^\circ\text{C} \\ h_1 = h_{f\text{ @ 0.8 MPa}} = 95.47\text{ kJ/kg} \end{array} \qquad \text{(Table A–12)}$$

$$\text{At exit:}\quad \begin{array}{l} P_2 = 0.12\text{ MPa} \\ (h_2 = h_1) \end{array} \quad \longrightarrow \quad \begin{array}{l} h_f = 22.49\text{ kJ/kg} \qquad T_{\text{sat}} = -22.32^\circ\text{C} \\ h_g = 236.97\text{ kJ/kg} \end{array}$$

Obviously $h_f < h_2 < h_g$; thus, the refrigerant exists as a saturated mixture at the exit state. The quality at this state is

$$x_2 = \frac{h_2 - h_f}{h_{fg}} = \frac{95.47 - 22.49}{236.97 - 22.49} = \mathbf{0.340}$$

Since the exit state is a saturated mixture at 0.12 MPa, the exit temperature must be the saturation temperature at this pressure, which is −22.32°C. Then the temperature change for this process becomes

$$\Delta T = T_2 - T_1 = (-22.32 - 31.31)^\circ\text{C} = \mathbf{-53.63^\circ C}$$

Discussion Note that the temperature of the refrigerant drops by 53.63°C during this throttling process. Also note that 34.0 percent of the refrigerant vaporizes during this throttling process, and the energy needed to vaporize this refrigerant is absorbed from the refrigerant itself.

4a Mixing Chambers

In engineering applications, mixing two streams of fluids is not a rare occurrence. The section where the mixing process takes place is commonly referred to as a **mixing chamber**. The mixing chamber does not have to be a distinct "chamber." An ordinary T-elbow or a Y-elbow in a shower, for example, serves as the mixing chamber for the cold- and hot-water streams (Fig. 6–32).

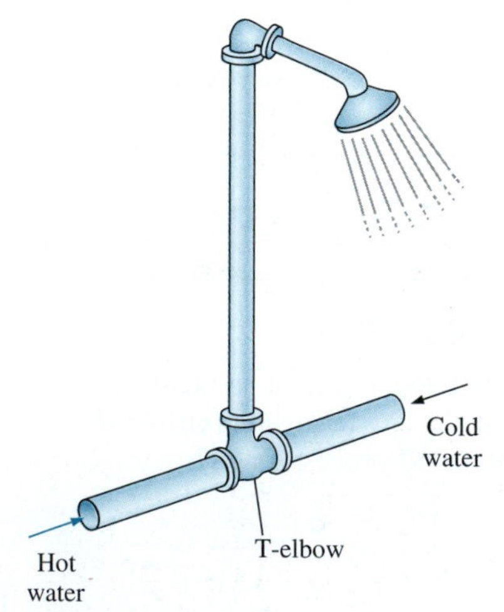

FIGURE 6–32

The T-elbow of an ordinary shower serves as the mixing chamber for the hot- and the cold-water streams.

The conservation of mass principle for a mixing chamber requires that the sum of the incoming mass flow rates equal the mass flow rate of the outgoing mixture.

Mixing chambers are usually well insulated ($q \cong 0$) and usually do not involve any kind of work ($w = 0$). Also, the kinetic and potential energies of the fluid streams are usually negligible (ke $\cong$ 0, pe $\cong$ 0). Then all there is left in the energy equation is the total energies of the incoming streams and the outgoing mixture. The conservation of energy principle requires that these two equal each other. Therefore, the conservation of energy equation becomes analogous to the conservation of mass equation for this case.

EXAMPLE 6–9 Mixing of Hot and Cold Waters in a Shower

Consider an ordinary shower where hot water at 140°F is mixed with cold water at 50°F. If it is desired that a steady stream of warm water at 110°F be supplied, determine the ratio of the mass flow rates of the hot to cold water. Assume the heat losses from the mixing chamber to be negligible and the mixing to take place at a pressure of 20 psia.

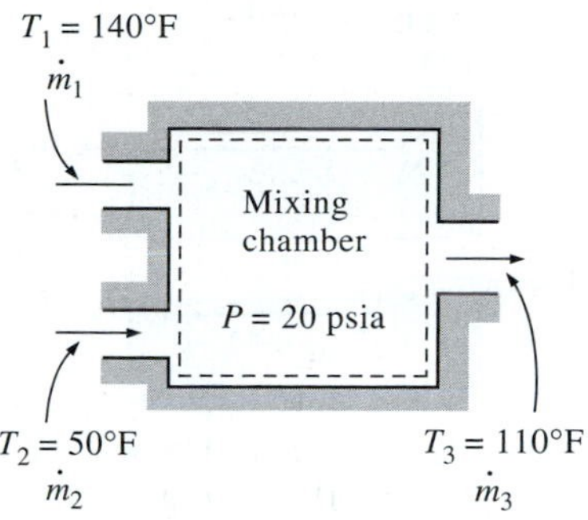

FIGURE 6–33
Schematic for Example 6–9.

Solution In a shower, cold water is mixed with hot water at a specified temperature. For a specified mixture temperature, the ratio of the mass flow rates of the hot to cold water is to be determined.

Assumptions **1** This is a steady-flow process since there is no change with time at any point and thus $\Delta m_{CV} = 0$ and $\Delta E_{CV} = 0$. **2** The kinetic and potential energies are negligible, ke ≅ pe ≅ 0. **3** Heat losses from the system are negligible and thus $\dot{Q} \cong 0$. **4** There is no work interaction involved.

Analysis We take the *mixing chamber* as the system (Fig. 6–33). This is a *control volume* since mass crosses the system boundary during the process. We observe that there are two inlets and one exit.

Under the stated assumptions and observations, the mass and energy balances for this steady-flow system can be expressed in the rate form as follows:

Mass balance:
$$\dot{m}_{in} - \dot{m}_{out} = dm_{system}/dt \overset{\nearrow 0 \text{ (steady)}}{} = 0$$
$$\dot{m}_{in} = \dot{m}_{out} \rightarrow \dot{m}_1 + \dot{m}_2 = \dot{m}_3$$

Energy balance:
$$\underbrace{\dot{E}_{in} - \dot{E}_{out}}_{\substack{\text{Rate of net energy transfer}\\ \text{by heat, work, and mass}}} = \underbrace{dE_{system}/dt \overset{\nearrow 0 \text{ (steady)}}{}}_{\substack{\text{Rate of change in internal, kinetic,}\\ \text{potential, etc., energies}}} = 0$$
$$\dot{E}_{in} = \dot{E}_{out}$$
$$\dot{m}_1 h_1 + \dot{m}_2 h_2 = \dot{m}_3 h_3 \quad (\text{since } \dot{Q} \cong 0,\ \dot{W} = 0,\ \text{ke} \cong \text{pe} \cong 0)$$

Combining the mass and energy balances,
$$\dot{m}_1 h_1 + \dot{m}_2 h_2 = (\dot{m}_1 + \dot{m}_2) h_3$$

Dividing this equation by $\dot{m}_2$ yields
$$y h_1 + h_2 = (y + 1) h_3$$

where $y = \dot{m}_1/\dot{m}_2$ is the desired mass flow rate ratio.

The saturation temperature of water at 20 psia is 227.92°F. Since the temperatures of all three streams are below this value ($T < T_{sat}$), the water in all three streams exists as a compressed liquid (Fig. 6–34). A compressed liquid can be approximated as a saturated liquid at the given temperature. Thus,

$$h_1 \cong h_{f @ 140°F} = 107.99 \text{ Btu/lbm}$$
$$h_2 \cong h_{f @ 50°F} = 18.07 \text{ Btu/lbm}$$
$$h_3 \cong h_{f @ 110°F} = 78.02 \text{ Btu/lbm}$$

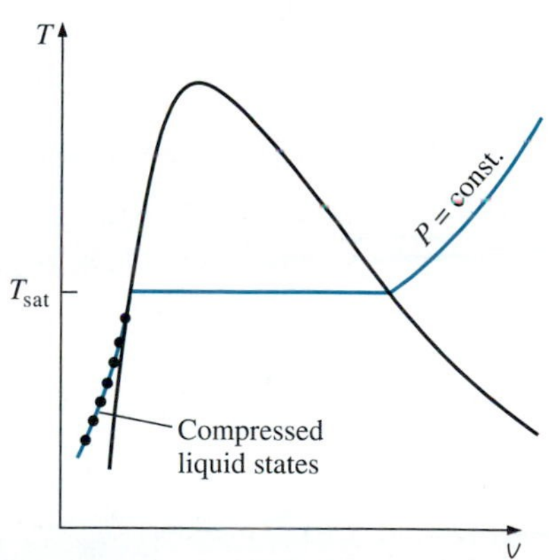

FIGURE 6–34
A substance exists as a compressed liquid at temperatures below the saturation temperatures at the given pressure.

Solving for y and substituting yields
$$y = \frac{h_3 - h_2}{h_1 - h_3} = \frac{78.02 - 18.07}{107.99 - 78.02} = \mathbf{2.0}$$

Discussion Note that the mass flow rate of the hot water must be twice the mass flow rate of the cold water for the mixture to leave at 110°F.

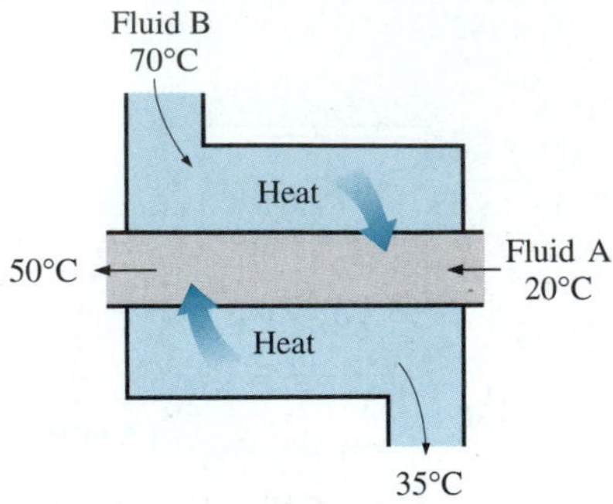

FIGURE 6–35

A heat exchanger can be as simple as two concentric pipes.

4b Heat Exchangers

As the name implies, **heat exchangers** are devices where two moving fluid streams exchange heat without mixing. Heat exchangers are widely used in various industries, and they come in various designs.

The simplest form of a heat exchanger is a *double-tube* (also called *tube-and-shell*) *heat exchanger*, shown in Fig. 6–35. It is composed of two concentric pipes of different diameters. One fluid flows in the inner pipe, and the other in the annular space between the two pipes. Heat is transferred from the hot fluid to the cold one through the wall separating them. Sometimes the inner tube makes a couple of turns inside the shell to increase the heat transfer area, and thus the rate of heat transfer. The mixing chambers discussed earlier are sometimes classified as *direct-contact* heat exchangers.

The conservation of mass principle for a heat exchanger in steady operation requires that the sum of the inbound mass flow rates equal the sum of the outbound mass flow rates. This principle can also be expressed as follows: *Under steady operation, the mass flow rate of each fluid stream flowing through a heat exchanger remains constant.*

Heat exchangers typically involve no work interactions ($w = 0$) and negligible kinetic and potential energy changes ($\Delta ke \cong 0$, $\Delta pe \cong 0$) for each fluid stream. The heat transfer rate associated with heat exchangers depends on how the control volume is selected. Heat exchangers are intended for heat transfer between two fluids *within* the device, and the outer shell is usually well insulated to prevent any heat loss to the surrounding medium.

When the entire heat exchanger is selected as the control volume, $\dot{Q}$ becomes zero, since the boundary for this case lies just beneath the insulation and little or no heat crosses the boundary (Fig. 6–36). If, however, only one of the fluids is selected as the control volume, then heat will cross this boundary as it flows from one fluid to the other and $\dot{Q}$ will not be zero. In fact, $\dot{Q}$ in this case will be the rate of heat transfer between the two fluids.

EXAMPLE 6–10 Cooling of Refrigerant-134a by Water

Refrigerant-134a is to be cooled by water in a condenser. The refrigerant enters the condenser with a mass flow rate of 6 kg/min at 1 MPa and 70°C and leaves at 35°C. The cooling water enters at 300 kPa and 15°C and leaves

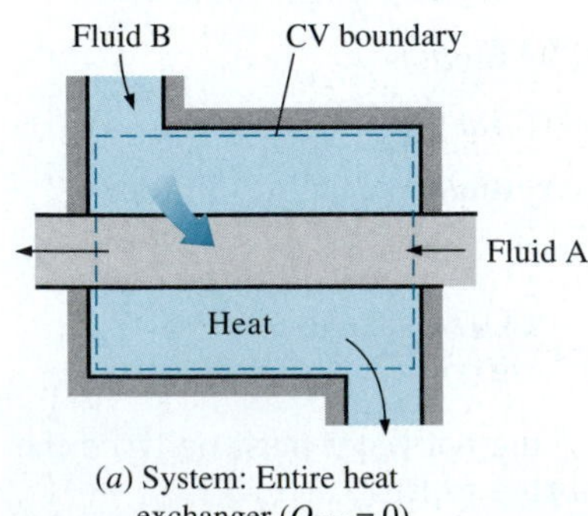

(*a*) System: Entire heat exchanger ($Q_{CV} = 0$)

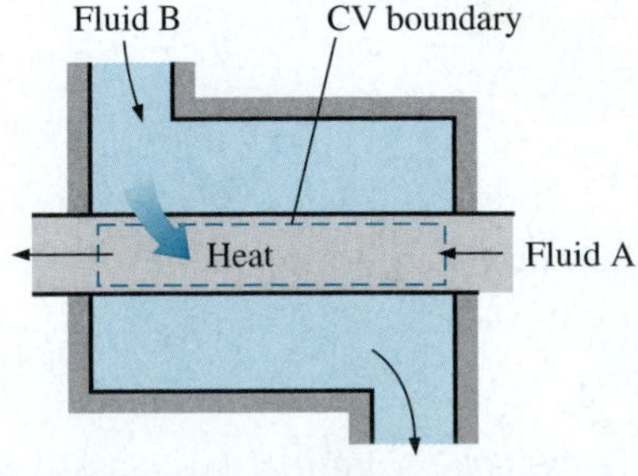

(*b*) System: Fluid A ($Q_{CV} \neq 0$)

FIGURE 6–36

The heat transfer associated with a heat exchanger may be zero or nonzero depending on how the control volume is selected.

at 25°C. Neglecting any pressure drops, determine (*a*) the mass flow rate of the cooling water required and (*b*) the heat transfer rate from the refrigerant to water.

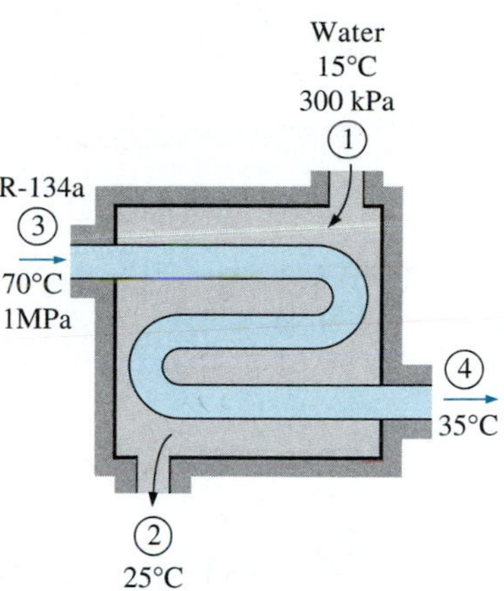

FIGURE 6–37

Schematic for Example 6–10.

Solution Refrigerant-134a is cooled by water in a condenser. The mass flow rate of the cooling water and the rate of heat transfer from the refrigerant to the water are to be determined.

Assumptions **1** This is a steady-flow process since there is no change with time at any point and thus $\Delta m_{CV} = 0$ and $\Delta E_{CV} = 0$. **2** The kinetic and potential energies are negligible, ke ≅ pe ≅ 0. **3** Heat losses from the system are negligible and thus $\dot{Q} \cong 0$. **4** There is no work interaction.

Analysis We take the *entire heat exchanger* as the system (Fig. 6–37). This is a *control volume* since mass crosses the system boundary during the process. In general, there are several possibilities for selecting the control volume for multiple-stream steady-flow devices, and the proper choice depends on the situation at hand. We observe that there are two fluid streams (and thus two inlets and two exits) but no mixing.

(*a*) Under the stated assumptions and observations, the mass and energy balances for this steady-flow system can be expressed in the rate form as follows:

Mass balance: $$\dot{m}_{in} = \dot{m}_{out}$$

for each fluid stream since there is no mixing. Thus,

$$\dot{m}_1 = \dot{m}_2 = \dot{m}_w$$

$$\dot{m}_3 = \dot{m}_4 = \dot{m}_R$$

Energy balance: $$\underbrace{\dot{E}_{in} - \dot{E}_{out}}_{\substack{\text{Rate of net energy transfer}\\ \text{by heat, work, and mass}}} = \underbrace{dE_{system}/dt^{\nearrow 0\ (\text{steady})}}_{\substack{\text{Rate of change in internal, kinetic,}\\ \text{potential, etc., energies}}} = 0$$

$$\dot{E}_{in} = \dot{E}_{out}$$

$$\dot{m}_1 h_1 + \dot{m}_3 h_3 = \dot{m}_2 h_2 + \dot{m}_4 h_4 \qquad (\text{since } \dot{Q} \cong 0,\ \dot{W} = 0,\ \text{ke} \cong \text{pe} \cong 0)$$

Combining the mass and energy balances and rearranging give

$$\dot{m}_w(h_1 - h_2) = \dot{m}_R(h_4 - h_3)$$

Now we need to determine the enthalpies at all four states. Water exists as a compressed liquid at both the inlet and the exit since the temperatures at both locations are below the saturation temperature of water at 300 kPa (133.52°C). Approximating the compressed liquid as a saturated liquid at the given temperatures, we have

$$\left.\begin{aligned} h_1 &\cong h_{f@\,15^\circ C} = 62.982\ \text{kJ/kg} \\ h_2 &\cong h_{f@\,25^\circ C} = 104.83\ \text{kJ/kg} \end{aligned}\right. \qquad (\text{Table A–4})$$

The refrigerant enters the condenser as a superheated vapor and leaves as a compressed liquid at 35°C. From refrigerant-134a tables,

$$\left.\begin{aligned} P_3 &= 1\ \text{MPa} \\ T_3 &= 70^\circ C \end{aligned}\right\} \quad h_3 = 303.85\ \text{kJ/kg} \qquad (\text{Table A–13})$$

$$\left.\begin{aligned} P_4 &= 1\ \text{MPa} \\ T_4 &= 35^\circ C \end{aligned}\right\} \quad h_4 \cong h_{f@\,35^\circ C} = 100.87\ \text{kJ/kg} \qquad (\text{Table A–11})$$

Substituting, we find

$$\dot{m}_w(62.982 - 104.83)\text{ kJ/kg} = (6\text{ kg/min})[(100.87 - 303.85)\text{ kJ/kg}]$$

$$\dot{m}_w = \mathbf{29.1\ kg/min}$$

(*b*) To determine the heat transfer from the refrigerant to the water, we have to choose a control volume whose boundary lies on the path of heat transfer. We can choose the volume occupied by either fluid as our control volume. For no particular reason, we choose the volume occupied by the water. All the assumptions stated earlier apply, except that the heat transfer is no longer zero. Then assuming heat to be transferred to water, the energy balance for this single-stream steady-flow system reduces to

$$\underbrace{\dot{E}_{in} - \dot{E}_{out}}_{\text{Rate of net energy transfer by heat, work, and mass}} = \underbrace{dE_{system}/dt}_{\text{Rate of change in internal, kinetic, potential, etc., energies}}^{\nearrow 0 \text{ (steady)}} = 0$$

$$\dot{E}_{in} = \dot{E}_{out}$$

$$\dot{Q}_{w,\text{ in}} + \dot{m}_w h_1 = \dot{m}_w h_2$$

Rearranging and substituting,

$$\dot{Q}_{w,\text{ in}} = \dot{m}_w(h_2 - h_1) = (29.1\text{ kg/min})[(104.83 - 62.982)\text{ kJ/kg}]$$

$$= \mathbf{1218\ kJ/min}$$

Discussion Had we chosen the volume occupied by the refrigerant as the control volume (Fig. 6–38), we would have obtained the same result for $\dot{Q}_{R,out}$ since the heat gained by the water is equal to the heat lost by the refrigerant.

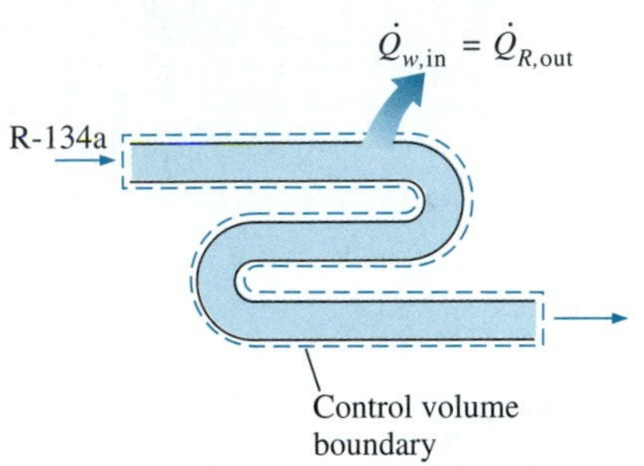

FIGURE 6–38

In a heat exchanger, the heat transfer depends on the choice of the control volume.

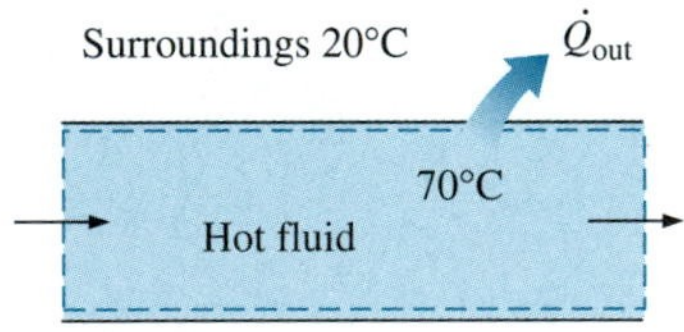

FIGURE 6–39

Heat losses from a hot fluid flowing through an uninsulated pipe or duct to the cooler environment may be very significant.

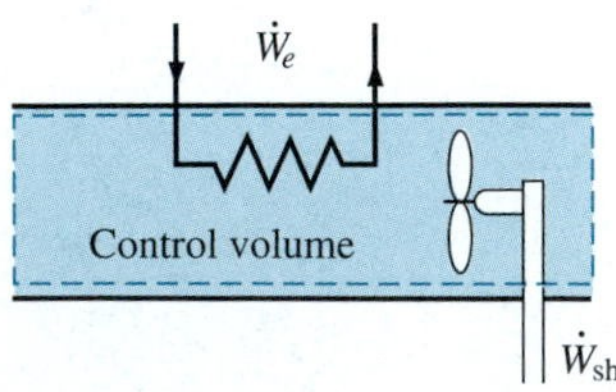

FIGURE 6–40

Pipe or duct flow may involve more than one form of work at the same time.

5 Pipe and Duct Flow

The transport of liquids or gases in pipes and ducts is of great importance in many engineering applications. Flow through a pipe or a duct usually satisfies the steady-flow conditions and thus can be analyzed as a steady-flow process. This, of course, excludes the transient start-up and shut-down periods. The control volume can be selected to coincide with the interior surfaces of the portion of the pipe or the duct that we are interested in analyzing.

Under normal operating conditions, the amount of heat gained or lost by the fluid may be very significant, particularly if the pipe or duct is long (Fig. 6–39). Sometimes heat transfer is desirable and is the sole purpose of the flow. Water flow through the pipes in the furnace of a power plant, the flow of refrigerant in a freezer, and the flow in heat exchangers are some examples of this case. At other times, heat transfer is undesirable, and the pipes or ducts are insulated to prevent any heat loss or gain, particularly when the temperature difference between the flowing fluid and the surroundings is large. Heat transfer in this case is negligible.

If the control volume involves a heating section (electric wires), a fan, or a pump (shaft), the work interactions should be considered (Fig. 6–40). Of these, fan work is usually small and often neglected in energy analysis.

The velocities involved in pipe and duct flow are relatively low, and the kinetic energy changes are usually insignificant. This is particularly true when the pipe or duct diameter is constant and the heating effects are negligible. Kinetic energy changes may be significant, however, for gas flow in ducts with variable cross-sectional areas especially when the compressibility effects are significant. The potential energy term may also be significant when the fluid undergoes a considerable elevation change as it flows in a pipe or duct.

EXAMPLE 6–11 Electric Heating of Air in a House

The electric heating systems used in many houses consist of a simple duct with resistance heaters. Air is heated as it flows over resistance wires. Consider a 15-kW electric heating system. Air enters the heating section at 100 kPa and 17°C with a volume flow rate of 150 m^3/min. If heat is lost from the air in the duct to the surroundings at a rate of 200 W, determine the exit temperature of air.

Solution The electric heating system of a house is considered. For specified electric power consumption and air flow rate, the air exit temperature is to be determined.

Assumptions **1** This is a steady-flow process since there is no change with time at any point and thus $\Delta m_{CV} = 0$ and $\Delta E_{CV} = 0$. **2** Air is an ideal gas since it is at a high temperature and low pressure relative to its critical-point values. **3** The kinetic and potential energy changes are negligible, $\Delta ke \cong \Delta pe \cong 0$. **4** Constant specific heats at room temperature can be used for air.

Analysis We take the *heating section portion of the duct* as the system (Fig. 6–41). This is a *control volume* since mass crosses the system boundary during the process. We observe that there is only one inlet and one exit and thus $\dot{m}_1 = \dot{m}_2 = \dot{m}$. Also, heat is lost from the system and electrical work is supplied to the system.

At temperatures encountered in heating and air-conditioning applications, Δh can be replaced by $c_p\,\Delta T$ where $c_p = 1.005$ kJ/kg · °C—the value at room temperature—with negligible error (Fig. 6–42). Then the energy balance for this steady-flow system can be expressed in the rate form as

$$\underbrace{\dot{E}_{in} - \dot{E}_{out}}_{\substack{\text{Rate of net energy transfer}\\ \text{by heat, work, and mass}}} = \underbrace{dE_{system}/dt}_{\substack{\text{Rate of change in internal, kinetic,}\\ \text{potential, etc., energies}}} \overset{\nearrow 0 \text{ (steady)}}{} = 0$$

$$\dot{E}_{in} = \dot{E}_{out}$$

$$\dot{W}_{e,in} + \dot{m}h_1 = \dot{Q}_{out} + \dot{m}h_2 \qquad (\text{since } \Delta ke \cong \Delta pe \cong 0)$$

$$\dot{W}_{e,in} - \dot{Q}_{out} = \dot{m}c_p(T_2 - T_1)$$

From the ideal-gas relation, the specific volume of air at the inlet of the duct is

$$v_1 = \frac{RT_1}{P_1} = \frac{(0.287 \text{ kPa} \cdot \text{m}^3/\text{kg} \cdot \text{K})(290 \text{ K})}{100 \text{ kPa}} = 0.832 \text{ m}^3/\text{kg}$$

The mass flow rate of the air through the duct is determined from

$$\dot{m} = \frac{\dot{V}_1}{v_1} = \frac{150 \text{ m}^3/\text{min}}{0.832 \text{ m}^3/\text{kg}}\left(\frac{1 \text{ min}}{60 \text{ s}}\right) = 3.0 \text{ kg/s}$$

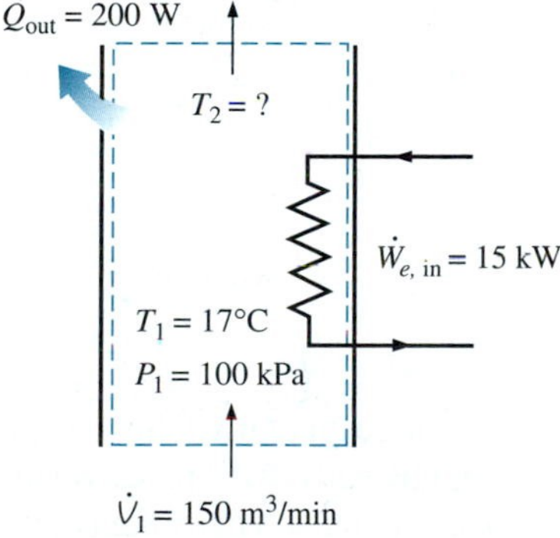

FIGURE 6–41
Schematic for Example 6–11.

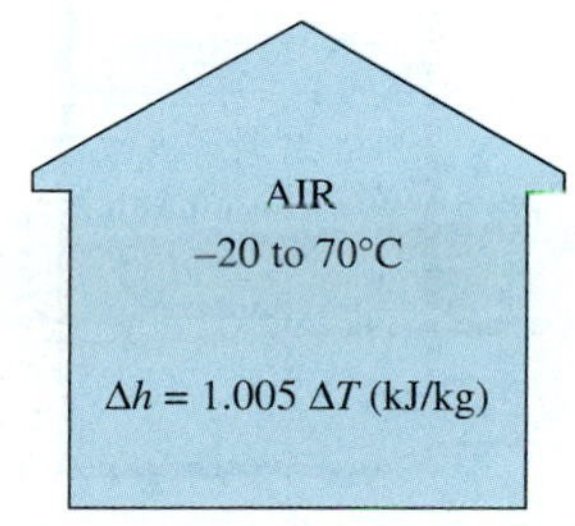

FIGURE 6–42
The error involved in $\Delta h = c_p\,\Delta T$, where $c_p = 1.005$ kJ/kg · °C, is less than 0.5 percent for air in the temperature range −20 to 70°C.

Substituting the known quantities, the exit temperature of the air is determined to be

$$(15\text{ kJ/s}) - (0.2\text{ kJ/s}) = (3\text{ kg/s})(1.005\text{ kJ/kg}\cdot{}^\circ\text{C})(T_2 - 17)^\circ\text{C}$$

$$T_2 = \mathbf{21.9^\circ C}$$

Discussion Note that heat loss from the duct reduces the exit temperature of the air.

6–5 ▪ ENERGY ANALYSIS OF UNSTEADY-FLOW PROCESSES

During a steady-flow process, no changes occur within the control volume with time; thus, one does not need to be concerned about what is going on within the boundaries. Not having to worry about any changes within the control volume with time greatly simplifies the analysis.

Many processes of interest, however, involve *changes* within the control volume with time. Such processes are called *unsteady-flow,* or *transient-flow,* processes. The steady-flow relations developed earlier are obviously not applicable to these processes. When an unsteady-flow process is analyzed, it is important to keep track of the mass and energy contents of the control volume as well as the energy interactions across the boundary.

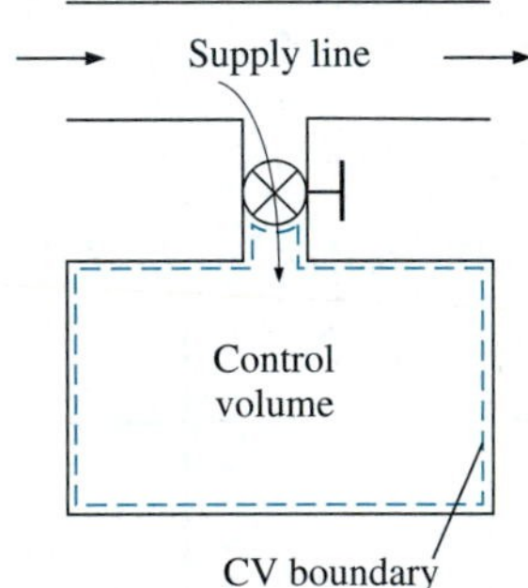

FIGURE 6–43

Charging of a rigid tank from a supply line is an unsteady-flow process since it involves changes within the control volume.

Some familiar unsteady-flow processes are the charging of rigid vessels from supply lines (Fig. 6–43), discharging a fluid from a pressurized vessel, driving a gas turbine with pressurized air stored in a large container, inflating tires or balloons, and even cooking with an ordinary pressure cooker.

Unlike steady-flow processes, unsteady-flow processes start and end over some finite time period instead of continuing indefinitely. Therefore in this section, we deal with changes that occur over some time interval Δt instead of with the rate of changes (changes per unit time). An unsteady-flow system, in some respects, is similar to a closed system, except that the mass within the system boundaries does not remain constant during a process.

Another difference between steady- and unsteady-flow systems is that steady-flow systems are fixed in space, size, and shape. Unsteady-flow systems, however, are not (Fig. 6–44). They are usually stationary; that is, they are fixed in space, but they may involve moving boundaries and thus boundary work.

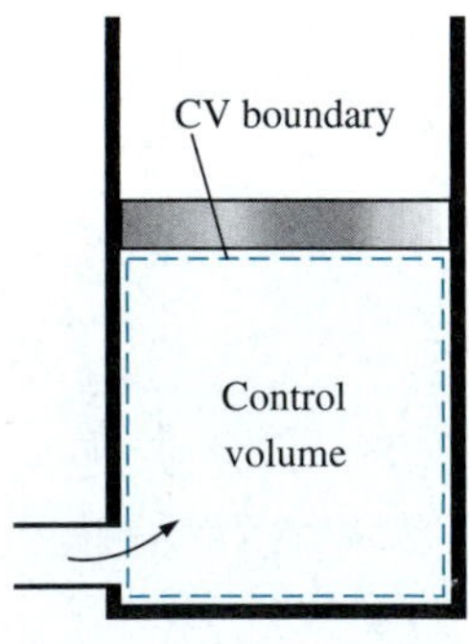

FIGURE 6–44

The shape and size of a control volume may change during an unsteady-flow process.

The *mass balance* for any system undergoing any process can be expressed as (*see* Sec. 6–1)

$$m_{\text{in}} - m_{\text{out}} = \Delta m_{\text{system}} \qquad (\text{kg}) \tag{6–42}$$

where $\Delta m_{\text{system}} = m_{\text{final}} - m_{\text{initial}}$ is the change in the mass of the system. For control volumes, it can also be expressed more explicitly as

$$m_i - m_e = (m_2 - m_1)_{\text{CV}} \tag{6–43}$$

where i = inlet, e = exit, 1 = initial state, and 2 = final state of the control volume. Often one or more terms in the equation above are zero. For example, $m_i = 0$ if no mass enters the control volume during the process, $m_e = 0$ if no mass leaves, and $m_1 = 0$ if the control volume is initially evacuated.

The energy content of a control volume changes with time during an unsteady-flow process. The magnitude of change depends on the amount of energy transfer across the system boundaries as heat and work as well as on the amount of energy transported into and out of the control volume by mass during the process. When analyzing an unsteady-flow process, we must keep track of the energy content of the control volume as well as the energies of the incoming and outgoing flow streams.

The general energy balance was given earlier as

Energy balance:

$$\underbrace{E_{\text{in}} - E_{\text{out}}}_{\substack{\text{Net energy transfer} \\ \text{by heat, work, and mass}}} = \underbrace{\Delta E_{\text{system}}}_{\substack{\text{Change in internal, kinetic,} \\ \text{potential, etc., energies}}} \quad \text{(kJ)} \tag{6-44}$$

The general unsteady-flow process, in general, is difficult to analyze because the properties of the mass at the inlets and exits may change during a process. Most unsteady-flow processes, however, can be represented reasonably well by the **uniform-flow process**, which involves the following idealization: *The fluid flow at any inlet or exit is uniform and steady, and thus the fluid properties do not change with time or position over the cross section of an inlet or exit. If they do, they are averaged and treated as constants for the entire process.*

Note that unlike the steady-flow systems, the state of an unsteady-flow system may change with time, and that the state of the mass leaving the control volume at any instant is the same as the state of the mass in the control volume at that instant. The initial and final properties of the control volume can be determined from the knowledge of the initial and final states, which are completely specified by two independent intensive properties for simple compressible systems.

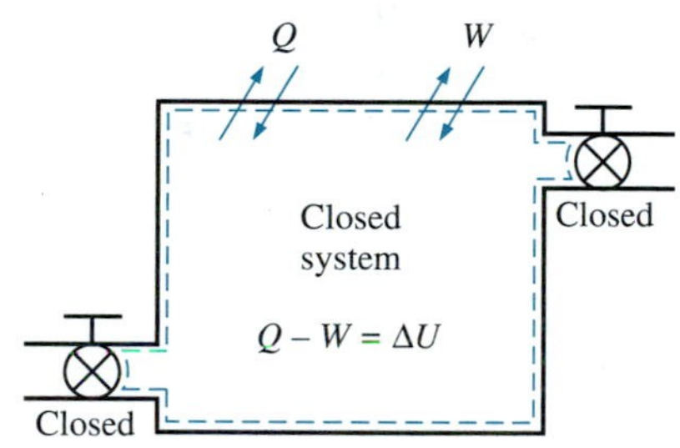

FIGURE 6–45

The energy equation of a uniform-flow system reduces to that of a closed system when all the inlets and exits are closed.

Then the energy balance for a uniform-flow system can be expressed explicitly as

$$\left(Q_{\text{in}} + W_{\text{in}} + \sum_{\text{in}} m\theta\right) - \left(Q_{\text{out}} + W_{\text{out}} + \sum_{\text{out}} m\theta\right) = (m_2e_2 - m_1e_1)_{\text{system}} \tag{6-45}$$

where $\theta = h + \text{ke} + \text{pe}$ is the energy of a fluid stream at any inlet or exit per unit mass, and $e = u + \text{ke} + \text{pe}$ is the energy of the nonflowing fluid within the control volume per unit mass. When the kinetic and potential energy changes associated with the control volume and fluid streams are negligible, as is usually the case, the energy balance above simplifies to

$$Q - W = \sum_{\text{out}} mh - \sum_{\text{in}} mh + (m_2u_2 - m_1u_1)_{\text{system}} \tag{6-46}$$

where $Q = Q_{\text{net,in}} = Q_{\text{in}} - Q_{\text{out}}$ is the net heat input and $W = W_{\text{net,out}} = W_{\text{out}} - W_{\text{in}}$ is the net work output. Note that if no mass enters or leaves the control volume during a process ($m_i = m_e = 0$, and $m_1 = m_2 = m$), this equation reduces to the energy balance relation for closed systems (Fig. 6–45). Also note that an unsteady-flow system may involve boundary work as well as electrical and shaft work (Fig. 6–46).

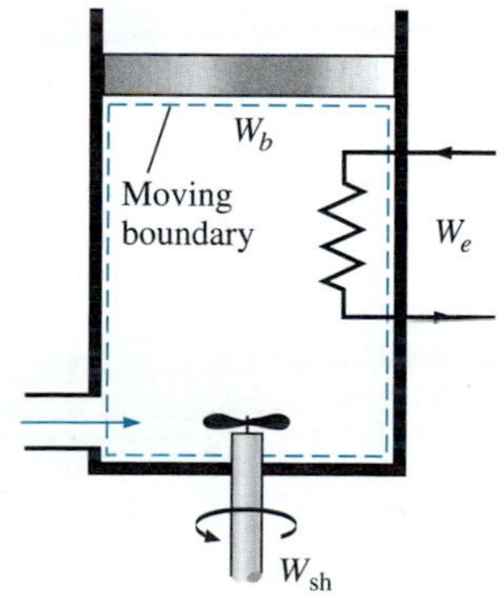

FIGURE 6–46

A uniform-flow system may involve electrical, shaft, and boundary work all at once.

Although both the steady-flow and uniform-flow processes are somewhat idealized, many actual processes can be approximated reasonably well by one of these with satisfactory results. The degree of satisfaction depends on the desired accuracy and the degree of validity of the assumptions made.

EXAMPLE 6–12 Charging of a Rigid Tank by Steam

A rigid, insulated tank that is initially evacuated is connected through a valve to a supply line that carries steam at 1 MPa and 300°C. Now the valve is opened, and steam is allowed to flow slowly into the tank until the pressure reaches 1 MPa, at which point the valve is closed. Determine the final temperature of the steam in the tank.

Solution A valve connecting an initially evacuated tank to a steam line is opened, and steam flows in until the pressure inside rises to the line level. The final temperature in the tank is to be determined.

Assumptions **1** This process can be analyzed as a *uniform-flow process* since the properties of the steam entering the control volume remain constant during the entire process. **2** The kinetic and potential energies of the streams are negligible, ke ≅ pe ≅ 0. **3** The tank is stationary and thus its kinetic and potential energy changes are zero; that is, $\Delta KE = \Delta PE = 0$ and $\Delta E_{system} = \Delta U_{system}$. **4** There are no boundary, electrical, or shaft work interactions involved. **5** The tank is well insulated and thus there is no heat transfer.

Analysis We take the *tank* as the system (Fig. 6–47(a)). This is a *control volume* since mass crosses the system boundary during the process. We observe that this is an unsteady-flow process since changes occur within the control volume. The control volume is initially evacuated and thus $m_1 = 0$ and $m_1u_1 = 0$. Also, there is one inlet and no exits for mass flow.

Noting that microscopic energies of flowing and nonflowing fluids are represented by enthalpy h and internal energy u, respectively, the mass and energy balances for this uniform-flow system can be expressed as

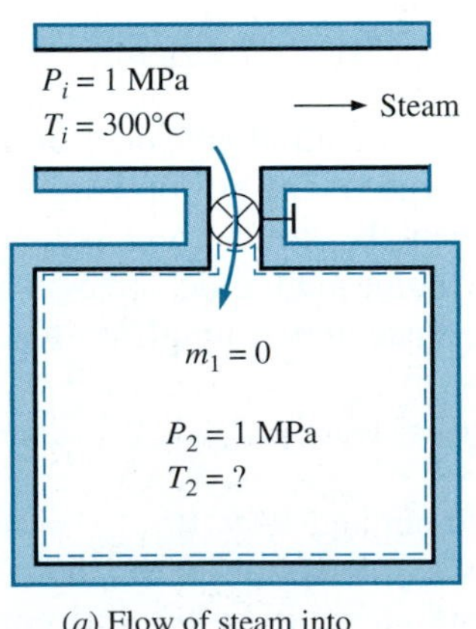

(*a*) Flow of steam into an evacuated tank

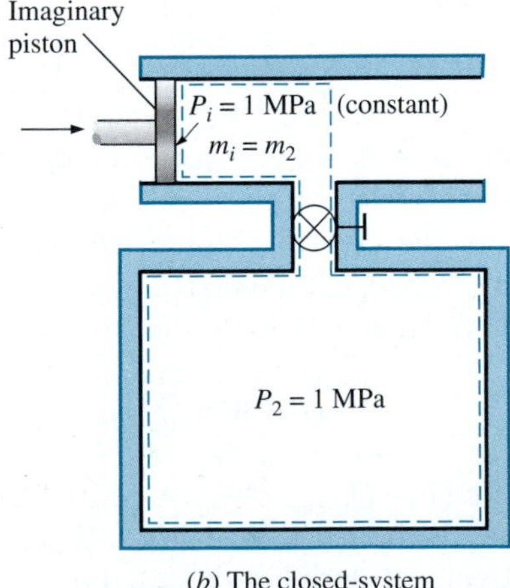

(*b*) The closed-system equivalence

FIGURE 6–47
Schematic for Example 6–12.

Mass balance: $m_{in} - m_{out} = \Delta m_{system} \quad \rightarrow \quad m_i = m_2 - m_1^{\nearrow 0} = m_2$

Energy balance:
$$\underbrace{E_{in} - E_{out}}_{\substack{\text{Net energy transfer} \\ \text{by heat, work, and mass}}} = \underbrace{\Delta E_{system}}_{\substack{\text{Change in internal, kinetic,} \\ \text{potential, etc., energies}}}$$

$$m_i h_i = m_2 u_2 \qquad (\text{since } W = Q = 0, \text{ ke} \cong \text{pe} \cong 0, m_1 = 0)$$

Combining the mass and energy balances gives

$$u_2 = h_i$$

That is, the final internal energy of the steam in the tank is equal to the enthalpy of the steam entering the tank. The enthalpy of the steam at the inlet state is

$$\left.\begin{array}{l} P_i = 1 \text{ MPa} \\ T_i = 300°\text{C} \end{array}\right\} \quad h_i = 3051.6 \text{ kJ/kg} \qquad (\text{Table A–6})$$

which is equal to u_2. Since we now know two properties at the final state, it is fixed and the temperature at this state is determined from the same table to be

$$\left.\begin{array}{l} P_2 = 1 \text{ MPa} \\ u_2 = 3051.6 \text{ kJ/kg} \end{array}\right\} \quad T_2 = \mathbf{456.1°C}$$

Discussion Note that the temperature of the steam in the tank has increased by 156.1°C. This result may be surprising at first, and you may be wondering where the energy to raise the temperature of the steam came from. The answer lies in the enthalpy term $h = u + Pv$. Part of the energy represented by enthalpy is the flow energy Pv, and this flow energy is converted to sensible internal energy once the flow ceases to exist in the control volume, and it shows up as an increase in temperature (Fig. 6–48).

Alternative solution This problem can also be solved by considering the region within the tank and the mass that is destined to enter the tank as a closed system, as shown in Fig. 6–47*b*. Since no mass crosses the boundaries, viewing this as a closed system is appropriate.

During the process, the steam upstream (the imaginary piston) will push the enclosed steam in the supply line into the tank at a constant pressure of 1 MPa. Then the boundary work done during this process is

$$W_{b,in} = -\int_1^2 P_i\, dV = -P_i(V_2 - V_1) = -P_i[V_{tank} - (V_{tank} + V_i)] = P_i V_i$$

where V_i is the volume occupied by the steam before it enters the tank and P_i is the pressure at the moving boundary (the imaginary piston face). The energy balance for the closed system gives

$$\underbrace{E_{in} - E_{out}}_{\substack{\text{Net energy transfer} \\ \text{by heat, work, and mass}}} = \underbrace{\Delta E_{system}}_{\substack{\text{Change in internal, kinetic,} \\ \text{potential, etc., energies}}}$$

$$W_{b,in} = \Delta U$$

$$m_i P_i V_i = m_2 u_2 - m_i u_i$$

$$u_2 = u_i + P_i V_i = h_i$$

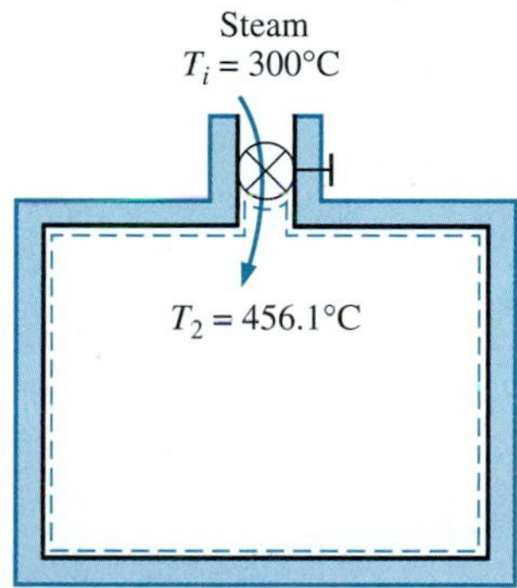

FIGURE 6–48

The temperature of steam rises from 300 to 456.1°C as it enters a tank as a result of flow energy being converted to internal energy.

since the initial state of the system is simply the line conditions of the steam. This result is identical to the one obtained with the uniform-flow analysis. Once again, the temperature rise is caused by the so-called flow energy or flow work, which is the energy required to move the fluid during flow.

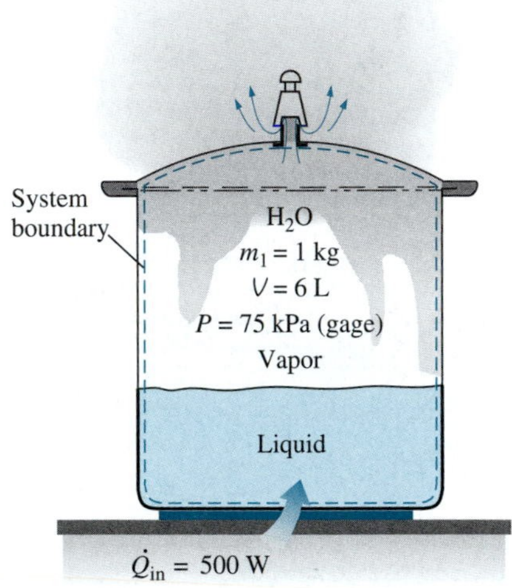

FIGURE 6–49

Schematic for Example 6–13.

EXAMPLE 6–13 Cooking with a Pressure Cooker

A pressure cooker is a pot that cooks food much faster than ordinary pots by maintaining a higher pressure and temperature during cooking. The pressure inside the pot is controlled by a pressure regulator (the petcock) that keeps the pressure at a constant level by periodically allowing some steam to escape, thus preventing any excess pressure buildup.

Pressure cookers, in general, maintain a gage pressure of 2 atm (or 3 atm absolute) inside. Therefore, pressure cookers cook at a temperature of about 133°C (or 271°F) instead of 100°C (or 212°F), cutting the cooking time by as much as 70 percent while minimizing the loss of nutrients. The newer pressure cookers use a spring valve with several pressure settings rather than a weight on the cover.

A certain pressure cooker has a volume of 6 L and an operating pressure of 75 kPa gage. Initially, it contains 1 kg of water. Heat is supplied to the pressure cooker at a rate of 500 W for 30 min after the operating pressure is reached. Assuming an atmospheric pressure of 100 kPa, determine (*a*) the temperature at which cooking takes place and (*b*) the amount of water left in the pressure cooker at the end of the process.

Solution Heat is transferred to a pressure cooker at a specified rate for a specified time period. The cooking temperature and the water remaining in the cooker are to be determined.

Assumptions **1** This process can be analyzed as a *uniform-flow process* since the properties of the steam leaving the control volume remain constant during the entire cooking process. **2** The kinetic and potential energies of the streams are negligible, ke ≅ pe ≅ 0. **3** The pressure cooker is stationary and thus its kinetic and potential energy changes are zero; that is, $\Delta KE = \Delta PE = 0$ and $\Delta E_{system} = \Delta U_{system}$. **4** The pressure (and thus temperature) in the pressure cooker remains constant. **5** Steam leaves as a saturated vapor at the cooker pressure. **6** There are no boundary, electrical, or shaft work interactions involved. **7** Heat is transferred to the cooker at a constant rate.

Analysis We take the *pressure cooker* as the system (Fig. 6–49). This is a *control volume* since mass crosses the system boundary during the process. We observe that this is an unsteady-flow process since changes occur within the control volume. Also, there is one exit and no inlets for mass flow.

(*a*) The absolute pressure within the cooker is

$$P_{abs} = P_{gage} + P_{atm} = 75 + 100 = 175 \text{ kPa}$$

FIGURE 6–50

As long as there is liquid in a pressure cooker, the saturation conditions exist and the temperature remains constant at the saturation temperature.

Since saturation conditions exist in the cooker at all times (Fig. 6–50), the cooking temperature must be the saturation temperature corresponding to this pressure. From Table A–5, it is

$$T = T_{sat\ @\ 175\ kPa} = \mathbf{116.04°C}$$

which is about 16°C higher than the ordinary cooking temperature.

(*b*) Noting that the microscopic energies of flowing and nonflowing fluids are represented by enthalpy h and internal energy u, respectively, the mass and energy balances for this uniform-flow system can be expressed as

Mass balance:

$$m_{\text{in}} - m_{\text{out}} = \Delta m_{\text{system}} \rightarrow -m_e = (m_2 - m_1)_{\text{CV}} \quad \text{or} \quad m_e = (m_1 - m_2)_{\text{CV}}$$

Energy balance:

$$\underbrace{E_{\text{in}} - E_{\text{out}}}_{\substack{\text{Net energy transfer} \\ \text{by heat, work, and mass}}} = \underbrace{\Delta E_{\text{system}}}_{\substack{\text{Change in internal, kinetic,} \\ \text{potential, etc., energies}}}$$

$$Q_{\text{in}} - m_e h_e = (m_2 u_2 - m_1 u_1)_{\text{CV}} \qquad (\text{since } W = 0,\ \text{ke} \cong \text{pe} \cong 0)$$

Combining the mass and energy balances gives

$$Q_{\text{in}} = (m_1 - m_2)h_e + (m_2 u_2 - m_1 u_1)_{\text{CV}}$$

The amount of heat transfer during this process is found from

$$Q_{\text{in}} = \dot{Q}_{\text{in}}\,\Delta t = (0.5 \text{ kJ/s})(30 \times 60 \text{ s}) = 900 \text{ kJ}$$

Steam leaves the pressure cooker as saturated vapor at 175 kPa at all times (Fig. 6–51). Thus,

$$h_e = h_{g\ @\ 175\ \text{kPa}} = 2700.2 \text{ kJ/kg}$$

The initial internal energy is found after the quality is determined:

$$v_1 = \frac{V}{m_1} = \frac{0.006 \text{ m}^3}{1 \text{ kg}} = 0.006 \text{ m}^3/\text{kg}$$

$$x_1 = \frac{v_1 - v_f}{v_{fg}} = \frac{0.006 - 0.001}{1.004 - 0.001} = 0.00499$$

Thus,

$$u_1 = u_f + x_1 u_{fg} = 486.82 + (0.00499)(2037.7) \text{ kJ/kg} = 497 \text{ kJ/kg}$$

and

$$U_1 = m_1 u_1 = (1 \text{ kg})(497 \text{ kJ/kg}) = 497 \text{ kJ}$$

The mass of the system at the final state is $m_2 = V/v_2$. Substituting this into the energy equation yields

$$Q_{\text{in}} = \left(m_1 - \frac{V}{v_2}\right)h_e + \left(\frac{V}{v_2}u_2 - m_1 u_1\right)$$

There are two unknowns in this equation, u_2 and v_2. Thus we need to relate them to a single unknown before we can determine these unknowns. Assuming there is still some liquid water left in the cooker at the final state (i.e., saturation conditions exist), v_2 and u_2 can be expressed as

$$v_2 = v_f + x_2 v_{fg} = 0.001 + x_2(1.004 - 0.001) \text{ m}^3/\text{kg}$$

$$u_2 = u_f + x_2 u_{fg} = 486.82 + x_2(2037.7) \text{ kJ/kg}$$

Recall that during a boiling process at constant pressure, the properties of each phase remain constant (only the amounts change). When these expressions are substituted into the above energy equation, x_2 becomes the only unknown, and it is determined to be

$$x_2 = 0.009$$

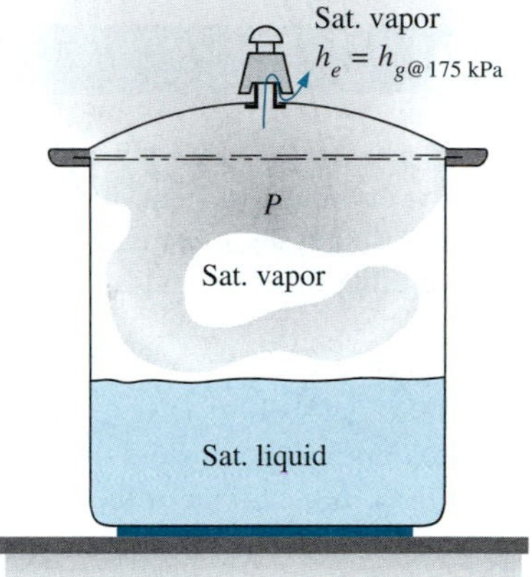

FIGURE 6–51

In a pressure cooker, the enthalpy of the exiting steam is $h_{g\ @\ 175\ \text{kPa}}$ (enthalpy of the saturated vapor at the given pressure).

Thus,

$$v_2 = 0.001 + (0.009)(1.004 - 0.001)\ \text{m}^3/\text{kg} = 0.010\ \text{m}^3/\text{kg}$$

and

$$m_2 = \frac{V}{v_2} = \frac{0.006\ \text{m}^3}{0.01\ \text{m}^3/\text{kg}} = \mathbf{0.6\ kg}$$

Therefore, after 30 min there is 0.6 kg water (liquid + vapor) left in the pressure cooker.

Discussion Note that almost half of the water in the pressure cooker has evaporated during cooking.

SUMMARY

The *conservation of mass principle* states that the net mass transfer to or from a system during a process is equal to the net change (increase or decrease) in the total mass of the system during that process, and is expressed as

$$m_{\text{in}} - m_{\text{out}} = \Delta m_{\text{system}} \quad \text{and} \quad \dot{m}_{\text{in}} - \dot{m}_{\text{out}} = dm_{\text{system}}/dt$$

where $\Delta m_{\text{system}} = m_{\text{final}} - m_{\text{initial}}$ is the change in the mass of the system during the process, $\dot{m}_{\text{in}}$ and $\dot{m}_{\text{out}}$ are the total rates of mass flow into and out of the system, and dm_{system}/dt is the rate of change of mass within the system boundaries. The relations above are also referred to as the *mass balance* and are applicable to any system undergoing any kind of process.

The amount of mass flowing through a cross section per unit time is called the *mass flow rate*, and is expressed as

$$\dot{m} = \rho V A$$

where ρ = density of fluid, V = average fluid velocity normal to A, and A = cross-sectional area normal to flow direction. The volume of the fluid flowing through a cross section per unit time is called the *volume flow rate* and is expressed as

$$\dot{V} = VA = \dot{m}/\rho$$

The work required to push a unit mass of fluid into or out of a control volume is called *flow work* or *flow energy*, and is expressed as $w_{\text{flow}} = Pv$. In the analysis of control volumes, it is convenient to combine the flow energy and internal energy into *enthalpy*. Then the total energy of a flowing fluid is expressed as

$$\theta = h + \text{ke} + \text{pe} = h + \frac{V^2}{2} + gz$$

The total energy transported by a flowing fluid of mass m with uniform properties is $m\theta$. The rate of energy transport by a fluid with a mass flow rate of $\dot{m}$ is $\dot{m}\theta$. When the kinetic and potential energies of a fluid stream are negligible, the amount and rate of energy transport become $E_{\text{mass}} = mh$ and $\dot{E}_{\text{mass}} = \dot{m}h$, respectively.

The *first law of thermodynamics* is essentially an expression of the conservation of energy principle, also called the *energy balance*. The general mass and energy balances for *any system* undergoing *any process* can be expressed as

$$\underbrace{E_{\text{in}} - E_{\text{out}}}_{\substack{\text{Net energy transfer}\\ \text{by heat, work, and mass}}} = \underbrace{\Delta E_{\text{system}}}_{\substack{\text{Changes in internal, kinetic,}\\ \text{potential, etc., energies}}}$$

It can also be expressed in the *rate form* as

$$\underbrace{\dot{E}_{\text{in}} - \dot{E}_{\text{out}}}_{\substack{\text{Rate of net energy transfer}\\ \text{by heat, work, and mass}}} = \underbrace{dE_{\text{system}}/dt}_{\substack{\text{Rate of change in internal, kinetic,}\\ \text{potential, etc., energies}}}$$

Thermodynamic processes involving control volumes can be considered in two groups: steady-flow processes and unsteady-flow processes. During a *steady-flow process*, the fluid flows through the control volume steadily, experiencing no change with time at a fixed position. The mass and energy content of the control volume remain constant during a steady-flow process. Taking heat transfer *to* the system and work done *by* the system to be positive quantities, the conservation of mass and energy equations for steady-flow processes are expressed as

$$\sum_{\text{in}} \dot{m} = \sum_{\text{out}} \dot{m}$$

$$\dot{Q} - \dot{W} = \underbrace{\sum_{\text{out}} \dot{m}\left(h + \frac{V^2}{2} + gz\right)}_{\text{for each exit}} - \underbrace{\sum_{\text{in}} \dot{m}\left(h + \frac{V^2}{2} + gz\right)}_{\text{for each inlet}}$$

These are the most general forms of the equations for steady-flow processes. For single-stream (one-inlet–one-exit) systems such as nozzles, diffusers, turbines, compressors, and pumps, they simplify to

$$\dot{m}_1 = \dot{m}_2 \ \rightarrow\ \frac{1}{v_1} V_1 A_1 = \frac{1}{v_2} V_2 A_2$$

$$\dot{Q} - \dot{W} = \dot{m}\left[h_2 - h_1 + \frac{V_2^2 - V_1^2}{2} + g(z_2 - z_1)\right]$$

In these relations, subscripts 1 and 2 denote the inlet and exit states, respectively.

Most unsteady-flow processes can be modeled as a *uniform-flow process*, which requires that the fluid flow at any inlet or exit is uniform and steady, and thus the fluid properties do not change with time or position over the cross section of an inlet or exit. If they do, they are averaged and treated as constants for the entire process. When kinetic and potential energy changes associated with the control volume and the fluid streams are negligible, the mass and energy balance relations for a uniform-flow system are expressed as

$$m_{in} - m_{out} = \Delta m_{system}$$

$$Q - W = \sum_{out} mh - \sum_{in} mh + (m_2u_2 - m_1u_1)_{system}$$

where $Q = Q_{net,in} = Q_{in} - Q_{out}$ is the net heat input and $W = W_{net,out} = W_{out} - W_{in}$ is the net work output.

When solving thermodynamic problems, it is recommended that the general form of the energy balance $E_{in} - E_{out} = \Delta E_{system}$ be used for all problems, and simplify it for the particular problem instead of using the specific relations given here for different processes.

REFERENCES AND SUGGESTED READINGS

1. ASHRAE Handbook of Fundamentals. SI version. Atlanta, GA: American Society of Heating, Refrigerating, and Air-Conditioning Engineers, Inc., 1993.

2. ASHRAE Handbook of Refrigeration. SI version. Atlanta, GA: American Society of Heating, Refrigerating, and Air-Conditioning Engineers, Inc., 1994.

3. Y. A. Çengel and J. M. Cimbala, *Fluid Mechanics: Fundamentals and Applications.* New York: McGraw-Hill, 2006.

PROBLEMS*

Conservation of Mass

6–1C Name four physical quantities that are conserved and two quantities that are not conserved during a process.

6–2C Define mass and volume flow rates. How are they related to each other?

6–3C Does the amount of mass entering a control volume have to be equal to the amount of mass leaving during an unsteady-flow process?

6–4C When is the flow through a control volume steady?

6–5C Consider a device with one inlet and one outlet. If the volume flow rates at the inlet and at the outlet are the same, is the flow through this device necessarily steady? Why?

6–6E A garden hose attached with a nozzle is used to fill a 20-gal bucket. The inner diameter of the hose is 1 in and it reduces to 0.5 in at the nozzle exit. If the average velocity in the hose is 8 ft/s, determine (*a*) the volume and mass flow rates of water through the hose, (*b*) how long it will take to fill the bucket with water, and (*c*) the average velocity of water at the nozzle exit.

6–7 Air enters a nozzle steadily at 2.21 kg/m³ and 40 m/s and leaves at 0.762 kg/m³ and 180 m/s. If the inlet area of the nozzle is 90 cm², determine (*a*) the mass flow rate through the nozzle, and (*b*) the exit area of the nozzle. *Answers:* (*a*) 0.796 kg/s, (*b*) 58 cm²

6–8E A steam pipe is to transport 200 lbm/s of steam at 200 psia and 600°F. Calculate the minimum diameter this pipe can have so that the steam velocity does not exceed 59 ft/s. *Answer:* 3.63 ft

6–9 A pump increases the water pressure from 70 kPa at the inlet to 700 kPa at the outlet. Water enters this pump at 15°C through a 1-cm-diameter opening and exits through a 1.5-cm-diameter opening. Determine the velocity of the water at the inlet and outlet when the mass flow rate through the pump is 0.5 kg/s. Will these velocities change significantly if the inlet temperature is raised to 40°C?

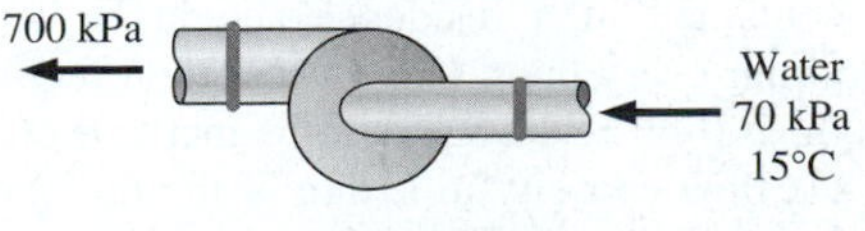

FIGURE P6–9

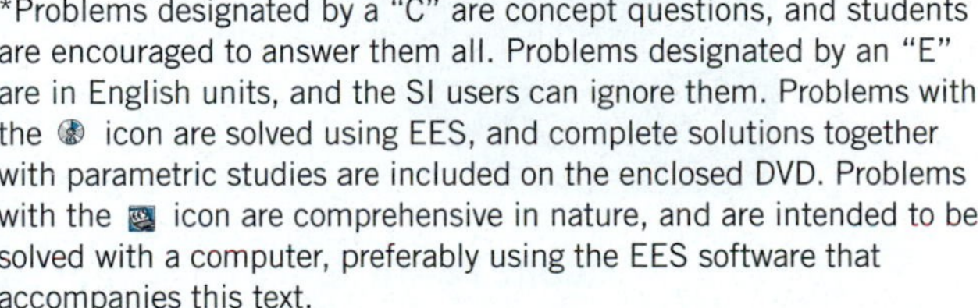
*Problems designated by a "C" are concept questions, and students are encouraged to answer them all. Problems designated by an "E" are in English units, and the SI users can ignore them. Problems with the icon are solved using EES, and complete solutions together with parametric studies are included on the enclosed DVD. Problems with the icon are comprehensive in nature, and are intended to be solved with a computer, preferably using the EES software that accompanies this text.

6–10 A hair dryer is basically a duct of constant diameter in which a few layers of electric resistors are placed. A small fan pulls the air in and forces it through the resistors where it is heated. If the density of air is 1.20 kg/m^3 at the inlet and 1.05 kg/m^3 at the exit, determine the percent increase in the velocity of air as it flows through the dryer.

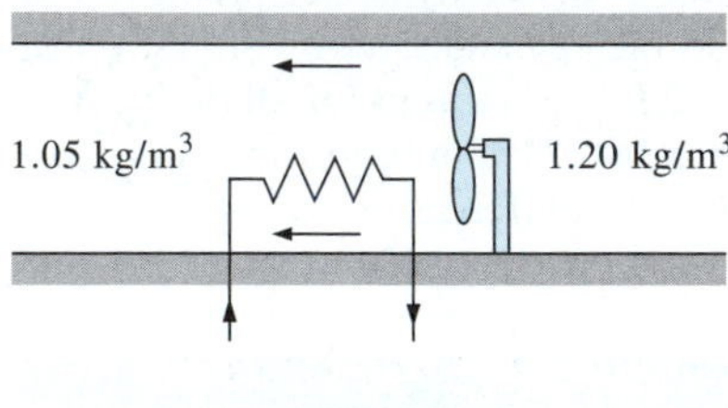

FIGURE P6–10

6–11 A 1-m^3 rigid tank initially contains air whose density is 1.18 kg/m^3. The tank is connected to a high-pressure supply line through a valve. The valve is opened, and air is allowed to enter the tank until the density in the tank rises to 7.20 kg/m^3. Determine the mass of air that has entered the tank. *Answer:* 6.02 kg

6–12 A smoking lounge is to accommodate 15 heavy smokers. The minimum fresh air requirement for smoking lounges is specified to be 30 L/s per person (ASHRAE, Standard 62, 1989). Determine the minimum required flow rate of fresh air that needs to be supplied to the lounge, and the diameter of the duct if the air velocity is not to exceed 8 m/s.

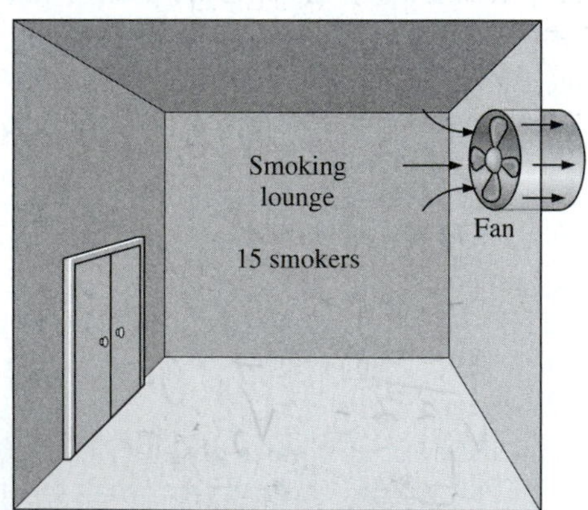

FIGURE P6–12

6–13 The minimum fresh air requirement of a residential building is specified to be 0.35 air change per hour (ASHRAE, Standard 62, 1989). That is, 35 percent of the entire air contained in a residence should be replaced by fresh outdoor air every hour. If the ventilation requirement of a 2.7-m-high, 200-m^2 residence is to be met entirely by a fan, determine the flow capacity in L/min of the fan that needs to be installed. Also determine the diameter of the duct if the air velocity is not to exceed 6 m/s.

6–14 A cyclone separator like that in Fig. P6–14 is used to remove fine solid particles, such as fly ash, that are suspended in a gas stream. In the flue-gas system of an electrical power plant, the weight fraction of fly ash in the exhaust gases is approximately 0.001. Determine the mass flow rates at the two outlets (flue gas and fly ash) when 10 kg/s of flue gas and ash mixture enters this unit. Also determine the amount of fly ash collected per year.

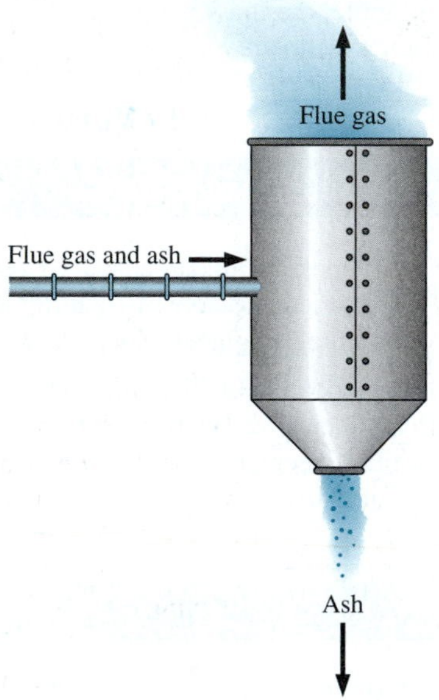

FIGURE P6–14

6–15 Air enters the 1-m^2 inlet of an aircraft engine at 100 kPa and 20°C with a velocity of 180 m/s. Determine the volume flow rate, in m^3/s, at the engine's inlet and the mass flow rate, in kg/s, at the engine's exit.

6–16 A spherical hot-air balloon is initially filled with air at 120 kPa and 35°C with an initial diameter of 3 m. Air enters this balloon at 120 kPa and 35°C with a velocity of 2 m/s through a 1-m diameter opening. How many minutes will it take to inflate this balloon to a 15-m diameter when the pressure and temperature of the air in the balloon remain the same as the air entering the balloon? *Answer:* 18.6 min

FIGURE P6–16

6–17 Water enters the constant 130-mm inside-diameter tubes of a boiler at 7 MPa and 65°C, and leaves the tubes at 6 MPa and 450°C with a velocity of 80 m/s. Calculate the velocity of the water at the tube inlet and the inlet volume flow rate.

6–18 Refrigerant-134a enters a 28-cm diameter pipe steadily at 200 kPa and 20°C with a velocity of 5 m/s. The refrigerant gains heat as it flows and leaves the pipe at 180 kPa and 40°C. Determine (*a*) the volume flow rate of the refrigerant at the inlet, (*b*) the mass flow rate of the refrigerant, and (*c*) the velocity and volume flow rate at the exit.

Flow Work and Energy Transfer by Mass

6–19C What are the different mechanisms for transferring energy to or from a control volume?

6–20C What is flow energy? Do fluids at rest possess any flow energy?

6–21C How do the energies of a flowing fluid and a fluid at rest compare? Name the specific forms of energy associated with each case.

6–22E Steam is leaving a pressure cooker whose operating pressure is 30 psia. It is observed that the amount of liquid in the cooker has decreased by 0.4 gal in 45 minutes after the steady operating conditions are established, and the cross-sectional area of the exit opening is 0.15 in^2. Determine (*a*) the mass flow rate of the steam and the exit velocity, (*b*) the total and flow energies of the steam per unit mass, and (*c*) the rate at which energy is leaving the cooker by steam.

6–23 Air flows steadily in a pipe at 300 kPa, 77°C, and 25 m/s at a rate of 18 kg/min. Determine (*a*) the diameter of the pipe, (*b*) the rate of flow energy, (*c*) the rate of energy transport by mass, and (*d*) the error involved in part (*c*) if the kinetic energy is neglected.

6–24E A water pump increases the water pressure from 10 psia to 50 psia. Determine the flow work, in Btu/lbm, required by the pump.

6–25 An air compressor compresses 10 L of air at 120 kPa and 20°C to 1000 kPa and 300°C. Determine the flow work, in kJ/kg, required by the compressor. *Answer:* 80.4 kJ/kg

Steady-Flow Energy Balance: Nozzles and Diffusers

6–26C How is a steady-flow system characterized?

6–27C Can a steady-flow system involve boundary work?

6–28C A diffuser is an adiabatic device that decreases the kinetic energy of the fluid by slowing it down. What happens to this *lost* kinetic energy?

6–29C The kinetic energy of a fluid increases as it is accelerated in an adiabatic nozzle. Where does this energy come from?

6–30C Is heat transfer to or from the fluid desirable as it flows through a nozzle? How will heat transfer affect the fluid velocity at the nozzle exit?

6–31 Air enters an adiabatic nozzle steadily at 300 kPa, 200°C, and 30 m/s and leaves at 100 kPa and 180 m/s. The inlet area of the nozzle is 80 cm^2. Determine (*a*) the mass flow rate through the nozzle, (*b*) the exit temperature of the air, and (*c*) the exit area of the nozzle. *Answers:* (*a*) 0.5304 kg/s, (*b*) 184.6°C, (*c*) 38.7 cm^2

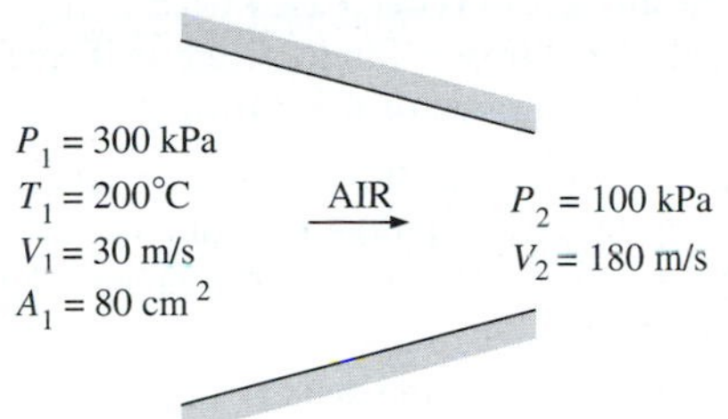

FIGURE P6–31

6–32 Reconsider Prob. 6–31. Using EES (or other) software, investigate the effect of the inlet area on the mass flow rate, exit temperature, and the exit area. Let the inlet area vary from 50 cm^2 to 150 cm^2. Plot the final results against the inlet area, and discuss the results.

6–33E The stators in a gas turbine are designed to increase the kinetic energy of the gas passing through them adiabatically. Air enters a set of these nozzles at 300 psia and 700°F with a velocity of 80 ft/s and exits at 250 psia and 645°F. Calculate the velocity at the exit of the nozzles.

6–34 The diffuser in a jet engine is designed to decrease the kinetic energy of the air entering the engine compressor without any work or heat interactions. Calculate the velocity at the exit of a diffuser when air at 100 kPa and 20°C enters it with a velocity of 500 m/s and the exit state is 200 kPa and 90°C.

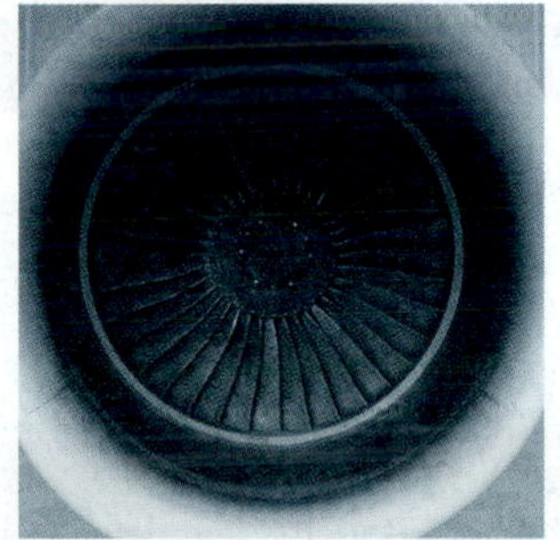

FIGURE P6–34

6–35 Steam at 5 MPa and 400°C enters a nozzle steadily with a velocity of 80 m/s, and it leaves at 2 MPa and 300°C. The inlet area of the nozzle is 50 cm^2, and heat is being lost at a rate of 120 kJ/s. Determine (*a*) the mass flow rate of the steam, (*b*) the exit velocity of the steam, and (*c*) the exit area of the nozzle.

6–36 Steam at 3 MPa and 400°C enters an adiabatic nozzle steadily with a velocity of 40 m/s and leaves at 2.5 MPa and 300 m/s. Determine (*a*) the exit temperature and (*b*) the ratio of the inlet to exit area A_1/A_2.

6–37 Air at 80 kPa and 127°C enters an adiabatic diffuser steadily at a rate of 6000 kg/h and leaves at 100 kPa. The velocity of the air stream is decreased from 230 to 30 m/s as it passes through the diffuser. Find (*a*) the exit temperature of the air and (*b*) the exit area of the diffuser.

6–38E Air at 13 psia and 20°F enters an adiabatic diffuser steadily with a velocity of 600 ft/s and leaves with a low velocity at a pressure of 14.5 psia. The exit area of the diffuser is 5 times the inlet area. Determine (*a*) the exit temperature and (*b*) the exit velocity of the air.

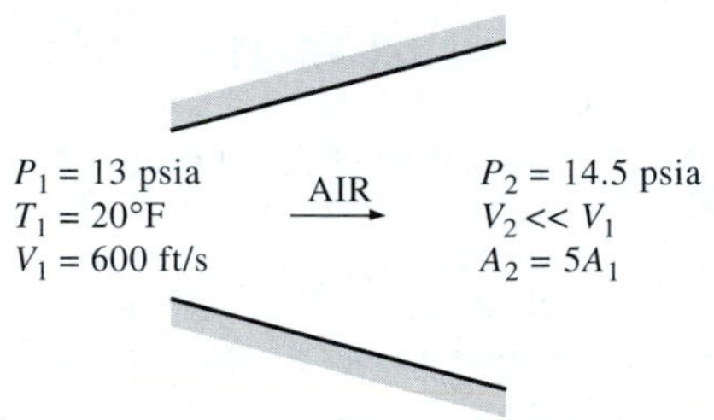

FIGURE P6–38E

6–39 Carbon dioxide enters an adiabatic nozzle steadily at 1 MPa and 500°C with a mass flow rate of 6000 kg/h and leaves at 100 kPa and 450 m/s. The inlet area of the nozzle is 40 cm². Determine (*a*) the inlet velocity and (*b*) the exit temperature. *Answers:* (*a*) 60.8 m/s, (*b*) 685.8 K

6–40 Refrigerant-134a at 700 kPa and 120°C enters an adiabatic nozzle steadily with a velocity of 20 m/s and leaves at 400 kPa and 30°C. Determine (*a*) the exit velocity and (*b*) the ratio of the inlet to exit area A_1/A_2.

6–41 Nitrogen gas at 60 kPa and 7°C enters an adiabatic diffuser steadily with a velocity of 200 m/s and leaves at 85 kPa and 22°C. Determine (*a*) the exit velocity of the nitrogen and (*b*) the ratio of the inlet to exit area A_1/A_2.

6–42 Reconsider Prob. 6–41. Using EES (or other) software, investigate the effect of the inlet velocity on the exit velocity and the ratio of the inlet-to-exit area. Let the inlet velocity vary from 180 to 260 m/s. Plot the final results against the inlet velocity, and discuss the results.

6–43 Refrigerant-134a enters a diffuser steadily as saturated vapor at 800 kPa with a velocity of 120 m/s, and it leaves at 900 kPa and 40°C. The refrigerant is gaining heat at a rate of 2 kJ/s as it passes through the diffuser. If the exit area is 80 percent greater than the inlet area, determine (*a*) the exit velocity and (*b*) the mass flow rate of the refrigerant. *Answers:* (*a*) 60.8 m/s, (*b*) 1.308 kg/s

6–44 Steam enters a nozzle at 400°C and 800 kPa with a velocity of 10 m/s, and leaves at 300°C and 200 kPa while losing heat at a rate of 25 kW. For an inlet area of 800 cm², determine the velocity and the volume flow rate of the steam at the nozzle exit. *Answers:* 606 m/s, 2.74 m³/s

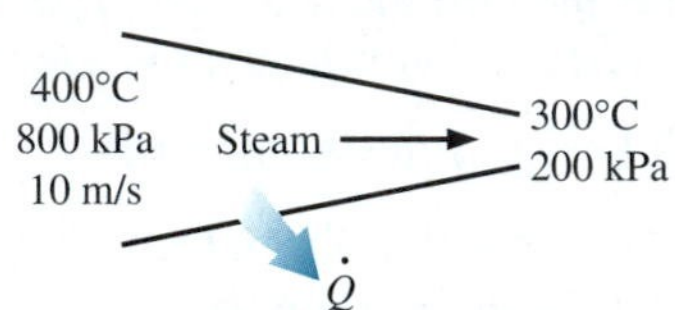

FIGURE P6–44

Turbines and Compressors

6–45C Consider an adiabatic turbine operating steadily. Does the work output of the turbine have to be equal to the decrease in the energy of the steam flowing through it?

6–46C Consider an air compressor operating steadily. How would you compare the volume flow rates of the air at the compressor inlet and exit?

6–47C Will the temperature of air rise as it is compressed by an adiabatic compressor? Why?

6–48C Somebody proposes the following system to cool a house in the summer: Compress the regular outdoor air, let it cool back to the outdoor temperature, pass it through a turbine, and discharge the cold air leaving the turbine into the house. From a thermodynamic point of view, is the proposed system sound?

6–49 Air is expanded from 1000 kPa and 600°C at the inlet of a steady-flow turbine to 100 kPa and 200°C at the outlet. The inlet area and velocity are 0.1 m² and 30 m/s, respectively, and the outlet velocity is 10 m/s. Determine the mass flow rate and outlet area.

6–50E Air enters a gas turbine at 150 psia and 700°F and leaves at 15 psia and 100°F. Determine the inlet and outlet volume flow rates when the mass flow rate through this turbine is 5 lbm/s.

6–51 An adiabatic air compressor compresses 10 L/S of air at 120 kPa and 20°C to 1000 kPa and 300°C. Determine (*a*) the work required by the compressor, in kJ/kg, and (*b*) the power required to drive the air compressor, in kW.

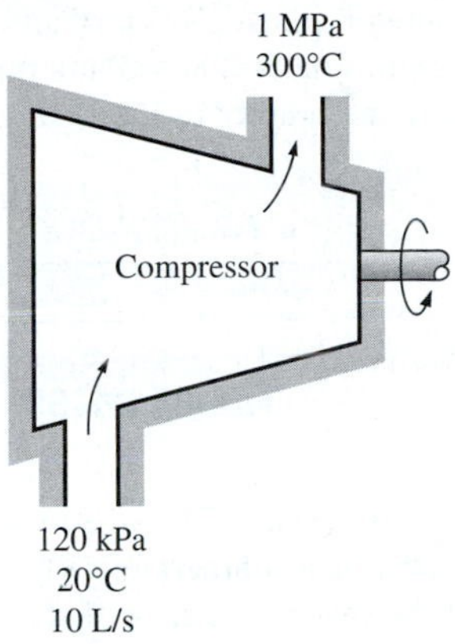

FIGURE P6–51

6–52 Steam flows steadily through an adiabatic turbine. The inlet conditions of the steam are 10 MPa, 450°C, and 80 m/s, and the exit conditions are 10 kPa, 92 percent quality, and 50 m/s. The mass flow rate of the steam is 12 kg/s. Determine (*a*) the change in kinetic energy, (*b*) the power output, and (*c*) the turbine inlet area. *Answers:* (*a*) −1.95 kJ/kg, (*b*) 10.2 MW, (*c*) 0.00447 m^2

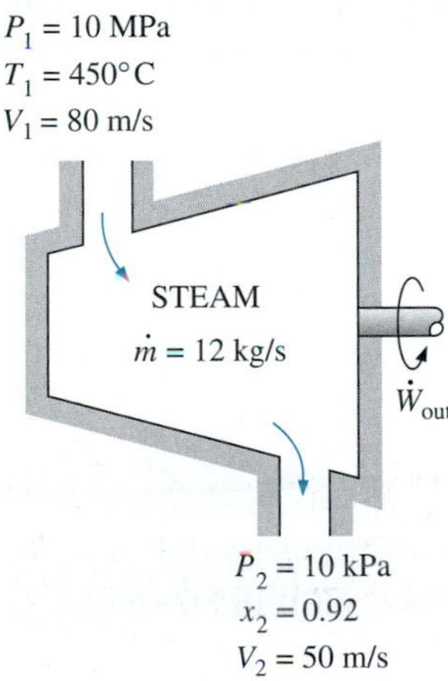

FIGURE P6–52

6–53 Reconsider Prob. 6–52. Using EES (or other) software, investigate the effect of the turbine exit pressure on the power output of the turbine. Let the exit pressure vary from 10 to 200 kPa. Plot the power output against the exit pressure, and discuss the results.

6–54 Steam enters an adiabatic turbine at 10 MPa and 500°C and leaves at 10 kPa with a quality of 90 percent. Neglecting the changes in kinetic and potential energies, determine the mass flow rate required for a power output of 5 MW. *Answer:* 4.852 kg/s

6–55E Steam flows steadily through a turbine at a rate of 45,000 lbm/h, entering at 1000 psia and 900°F and leaving at 5 psia as saturated vapor. If the power generated by the turbine is 4 MW, determine the rate of heat loss from the steam.

6–56 Steam enters an adiabatic turbine at 8 MPa and 500°C at a rate of 3 kg/s and leaves at 20 kPa. If the power output of the turbine is 2.5 MW, determine the temperature of the steam at the turbine exit. Neglect kinetic energy changes. *Answer:* 60.1°C

6–57 Argon gas enters an adiabatic turbine steadily at 900 kPa and 450°C with a velocity of 80 m/s and leaves at 150 kPa with a velocity of 150 m/s. The inlet area of the turbine is 60 cm^2. If the power output of the turbine is 250 kW, determine the exit temperature of the argon.

6–58 Helium is to be compressed from 120 kPa and 310 K to 700 kPa and 430 K. A heat loss of 20 kJ/kg occurs during the compression process. Neglecting kinetic energy changes, determine the power input required for a mass flow rate of 90 kg/min.

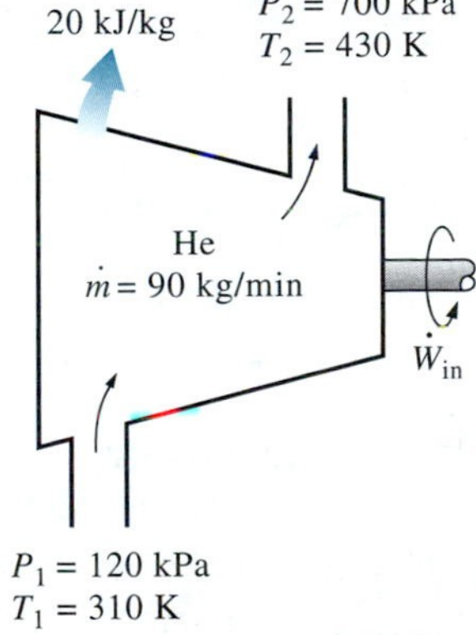

FIGURE P6–58

6–59 Carbon dioxide enters an adiabatic compressor at 100 kPa and 300 K at a rate of 0.5 kg/s and leaves at 600 kPa and 450 K. Neglecting kinetic energy changes, determine (*a*) the volume flow rate of the carbon dioxide at the compressor inlet and (*b*) the power input to the compressor. *Answers:* (*a*) 0.28 m^3/s, (*b*) 68.8 kW

6–60 An adiabatic gas turbine expands air at 1000 kPa and 500°C to 100 kPa and 150°C. Air enters the turbine through a 0.2-m^2 opening with an average velocity of 40 m/s, and exhausts through a 1-m^2 opening. Determine (*a*) the mass flow rate of air through the turbine and (*b*) the power produced by the turbine. *Answers:* (*a*) 36.1 kg/s, (*b*) 13.3 MW

6–61 Air is compressed by an adiabatic compressor from 100 kPa and 20°C to 1.8 MPa and 400°C. Air enters the compressor through a 0.15-m^2 opening with a velocity of 30 m/s. It exits through a 0.08-m^2 opening. Calculate the mass flow rate of air and the required power input.

6–62E Air is expanded by an adiabatic gas turbine from 500 psia and 800°F to 60 psia and 250°F. If the volume flow rate at the exit is 50 ft^3/s; the inlet area is 0.6 ft^2; and the outlet area is 1.2 ft^2; determine the power produced by this turbine.

6–63 A portion of the steam passing through a steam turbine is sometimes removed for the purposes of feedwater heating as shown in Fig. P6–63. Consider an adiabatic steam turbine with 12.5 MPa and 550°C steam entering at a rate of 20 kg/s. Steam is bled from this turbine at 1000 kPa and 200°C with a mass flow rate of 1 kg/s. The remaining steam leaves the turbine at 100 kPa and 100°C. Determine the power produced by this turbine. *Answer:* 15,860 kW

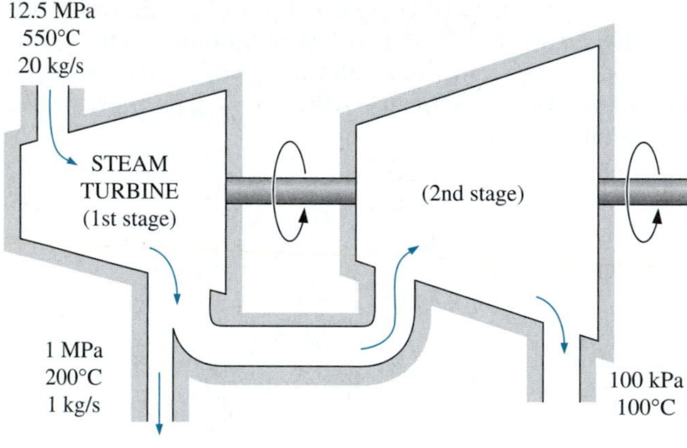

FIGURE P6–63

Throttling Valves

6–64C During a throttling process, the temperature of a fluid drops from 30 to −20°C. Can this process occur adiabatically?

6–65C Would you expect the temperature of air to drop as it undergoes a steady-flow throttling process? Explain.

6–66C Would you expect the temperature of a liquid to change as it is throttled? Explain.

6–67C Someone claims, based on temperature measurements, that the temperature of a fluid rises during a throttling process in a well-insulated valve with negligible friction. How do you evaluate this claim? Does this process violate any thermodynamic laws?

6–68 An adiabatic capillary tube is used in some refrigeration systems to drop the pressure of the refrigerant from the condenser level to the evaporator level. The R-134a enters the capillary tube as a saturated liquid at 50°C, and leaves at −12°C. Determine the quality of the refrigerant at the inlet of the evaporator.

6–69 Saturated liquid-vapor mixture of water, called wet steam, in a steam line at 2000 kPa is throttled to 100 kPa and 120°C. What is the quality in the steam line? *Answer:* 0.957

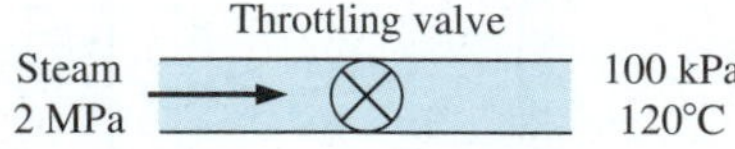

FIGURE P6–69

6–70 Refrigerant-134a at 800 kPa and 25°C is throttled to a temperature of −20°C. Determine the pressure and the internal energy of the refrigerant at the final state. *Answers:* 133 kPa, 80.7 kJ/kg

6–71 A well-insulated valve is used to throttle steam from 8 MPa and 500°C to 6 MPa. Determine the final temperature of the steam. *Answer:* 490.1°C

6–72 Reconsider Prob. 6–71. Using EES (or other) software, investigate the effect of the exit pressure of steam on the exit temperature after throttling. Let the exit pressure vary from 6 to 1 MPa. Plot the exit temperature of steam against the exit pressure, and discuss the results.

6–73E Air at 200 psia and 90°F is throttled to the atmospheric pressure of 14.7 psia. Determine the final temperature of the air.

6–74 Carbon dioxide gas enters a throttling valve at 5 MPa and 100°C and leaves at 100 kPa. Determine the temperature change during this process if CO_2 is assumed to be (*a*) an ideal gas and (*b*) a real gas. Real gas properties of CO_2 may be obtained from EES.

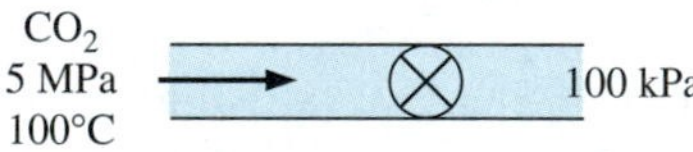

FIGURE P6–74

Mixing Chambers and Heat Exchangers

6–75C When two fluid streams are mixed in a mixing chamber, can the mixture temperature be lower than the temperature of both streams? Explain.

6–76C Consider a steady-flow mixing process. Under what conditions will the energy transported into the control volume by the incoming streams be equal to the energy transported out of it by the outgoing stream?

6–77C Consider a steady-flow heat exchanger involving two different fluid streams. Under what conditions will the amount of heat lost by one fluid be equal to the amount of heat gained by the other?

6–78 A hot-water stream at 80°C enters a mixing chamber with a mass flow rate of 0.5 kg/s where it is mixed with a stream of cold water at 20°C. If it is desired that the mixture leave the chamber at 42°C, determine the mass flow rate of the cold-water stream. Assume all the streams are at a pressure of 250 kPa. *Answer:* 0.865 kg/s

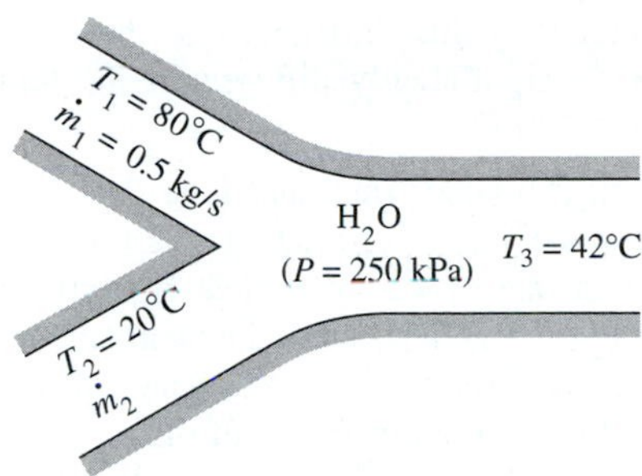

FIGURE P6–78

6–79E Water at 50°F and 50 psia is heated in a chamber by mixing it with saturated water vapor at 50 psia. If both streams enter the mixing chamber at the same mass flow rate, determine the temperature and the quality of the exiting stream. *Answers:* 281°F, 0.374

6–80 A stream of refrigerant-134a at 1 MPa and 12°C is mixed with another stream at 1 MPa and 60°C. If the mass flow rate of the cold stream is twice that of the hot one, determine the temperature and the quality of the exit stream.

6–81 Reconsider Prob. 6–80. Using EES (or other) software, investigate the effect of the mass flow rate of the cold stream of R-134a on the temperature and the quality of the exit stream. Let the ratio of the mass flow rate of the cold stream to that of the hot stream vary from 1 to 4. Plot the mixture temperature and quality against the cold-to-hot mass flow rate ratio, and discuss the results.

6–82 Refrigerant-134a at 700 kPa, 70°C, and 8 kg/min is cooled by water in a condenser until it exists as a saturated liquid at the same pressure. The cooling water enters the condenser at 300 kPa and 15°C and leaves at 25°C at the same pressure. Determine the mass flow rate of the cooling water required to cool the refrigerant. *Answer:* 42.0 kg/min

6–83E In a steam heating system, air is heated by being passed over some tubes through which steam flows steadily. Steam enters the heat exchanger at 30 psia and 400°F at a rate of 15 lbm/min and leaves at 25 psia and 212°F. Air enters at 14.7 psia and 80°F and leaves at 130°F. Determine the volume flow rate of air at the inlet.

6–84 Steam enters the condenser of a steam power plant at 20 kPa and a quality of 95 percent with a mass flow rate of 20,000 kg/h. It is to be cooled by water from a nearby river by circulating the water through the tubes within the condenser. To prevent thermal pollution, the river water is not allowed to experience a temperature rise above 10°C. If the steam is to leave the condenser as saturated liquid at 20 kPa, determine the mass flow rate of the cooling water required. *Answer:* 297.7 kg/s

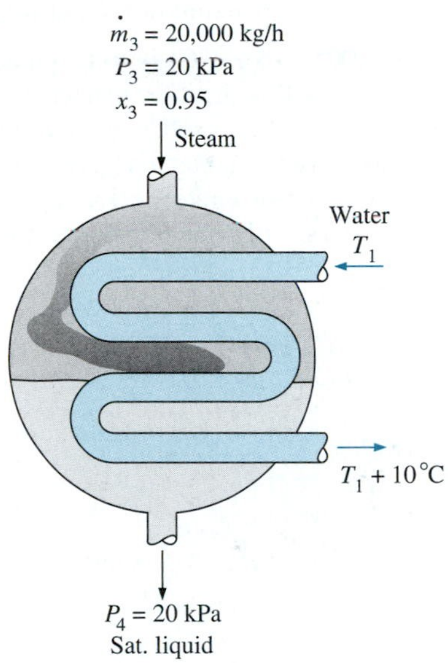

FIGURE P6–84

6–85 A heat exchanger is to cool ethylene glycol (c_p = 2.56 kJ/kg · °C) flowing at a rate of 2 kg/s from 80°C to 40°C by water (c_p = 4.18 kJ/kg · °C) that enters at 20°C and leaves at 55°C. Determine (*a*) the rate of heat transfer and (*b*) the mass flow rate of water.

6–86 Reconsider Prob. 6–85. Using EES (or other) software, investigate the effect of the inlet temperature of cooling water on the mass flow rate of water. Let the inlet temperature vary from 10 to 40°C, and assume the exit temperature to remain constant. Plot the mass flow rate of water against the inlet temperature, and discuss the results.

6–87 A thin-walled double-pipe counter-flow heat exchanger is used to cool oil (c_p = 2.20 kJ/kg · °C) from 150 to 40°C at a rate of 2 kg/s by water (c_p = 4.18 kJ/kg · °C) that enters at 22°C at a rate of 1.5 kg/s. Determine the rate of heat transfer in the heat exchanger and the exit temperature of water.

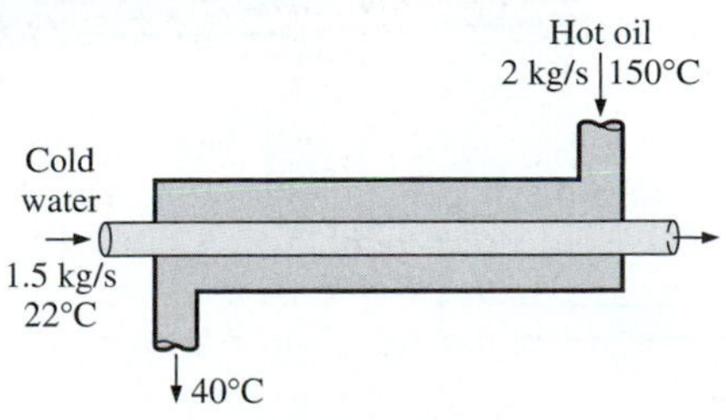

FIGURE P6–87

6–88 Cold water (c_p = 4.18 kJ/kg · °C) leading to a shower enters a thin-walled double-pipe counter-flow heat exchanger at 15°C at a rate of 0.60 kg/s and is heated to 45°C by hot water (c_p = 4.19 kJ/kg · °C) that enters at 100°C at a rate of 3 kg/s. Determine the rate of heat transfer in the heat exchanger and the exit temperature of the hot water.

6–89 Air (c_p = 1.005 kJ/kg · °C) is to be preheated by hot exhaust gases in a cross-flow heat exchanger before it enters the furnace. Air enters the heat exchanger at 95 kPa and 20°C at a rate of 0.8 m³/s. The combustion gases (c_p = 1.10 kJ/kg · °C) enter at 180°C at a rate of 1.1 kg/s and leave at 95°C. Determine the rate of heat transfer to the air and its outlet temperature.

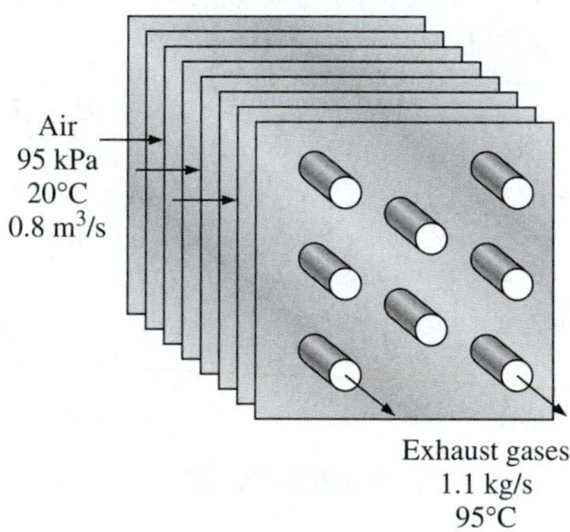

FIGURE P6–89

6–90 An adiabatic open feedwater heater in an electrical power plant mixes 0.2 kg/s of steam at 100 kPa and 160°C with 10 kg/s of feedwater at 100 kPa and 50°C to produce feedwater at 100 kPa and 60°C at the outlet. Determine the outlet mass flow rate and the outlet velocity when the outlet pipe diameter is 0.03 m.

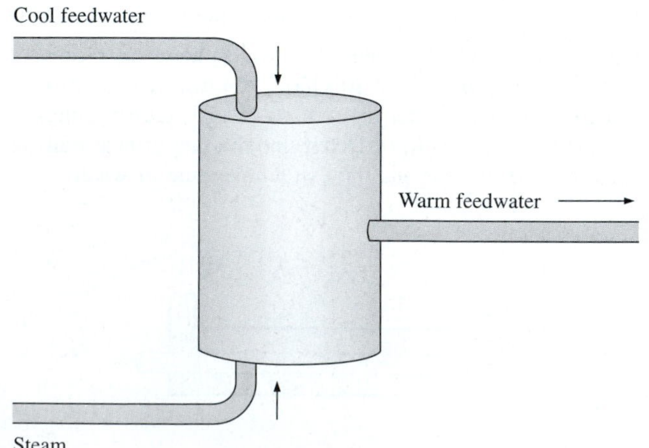

FIGURE P6–90

6–91E An open feedwater heater such as that shown in Fig. P6–90 heats the feedwater by mixing it with hot steam. Consider an electrical power plant with an open feedwater heater that mixes 0.1 lbm/s of steam at 10 psia and 200°F with 2.0 lbm/s of feedwater at 10 psia and 100°F to produce 10 psia and 120°F feedwater at the outlet. The diameter of the outlet pipe is 0.5 ft. Determine the mass flow rate and feedwater velocity at the outlet. Would the outlet flow rate and velocity be significantly different if the temperature at the outlet were 180°F?

6–92 A well-insulated shell-and-tube heat exchanger is used to heat water (c_p = 4.18 kJ/kg · °C) in the tubes from 20 to 70°C at a rate of 4.5 kg/s. Heat is supplied by hot oil (c_p = 2.30 kJ/kg · °C) that enters the shell side at 170°C at a rate of 10 kg/s. Determine the rate of heat transfer in the heat exchanger and the exit temperature of oil.

6–93E Steam is to be condensed on the shell side of a heat exchanger at 85°F. Cooling water enters the tubes at 60°F at a rate of 138 lbm/s and leaves at 73°F. Assuming the heat exchanger to be well-insulated, determine the rate of heat transfer in the heat exchanger and the rate of condensation of the steam.

6–94 An air-conditioning system involves the mixing of cold air and warm outdoor air before the mixture is routed to the conditioned room in steady operation. Cold air enters the mixing chamber at 5°C and 105 kPa at a rate of 1.25 m³/s while warm air enters at 34°C and 105 kPa. The air leaves the room at 24°C. The ratio of the mass flow rates of the hot to cold air streams is 1.6. Using variable specific heats, determine (*a*) the mixture temperature at the inlet of the room and (*b*) the rate of heat gain of the room.

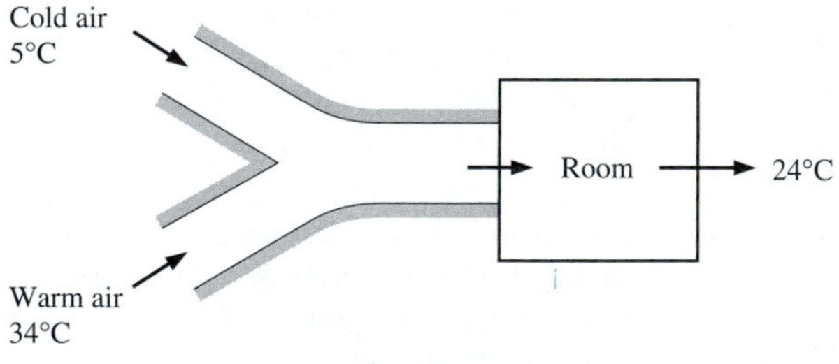

FIGURE P6–94

6–95 Hot exhaust gases of an internal combustion engine are to be used to produce saturated water vapor at 2 MPa pressure. The exhaust gases enter the heat exchanger at 400°C at a rate of 32 kg/min while water enters at 15°C. The heat exchanger is not well insulated, and it is estimated that 10 percent of heat given up by the exhaust gases is lost to the surroundings. If the mass flow rate of the exhaust gases is

15 times that of the water, determine (*a*) the temperature of the exhaust gases at the heat exchanger exit and (*b*) the rate of heat transfer to the water. Use the constant specific heat properties of air for the exhaust gases.

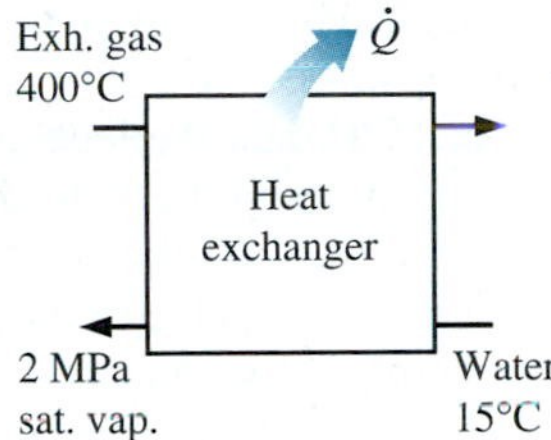

FIGURE P6–95

6–96 Two streams of water are mixed in an insulated container to form a third stream leaving the container. The first stream has a flow rate of 30 kg/s and a temperature of 90°C. The flow rate of the second stream is 200 kg/s, and its temperature is 50°C. What is the temperature of the third stream?

6–97 A chilled-water heat-exchange unit is designed to cool 5 m^3/s of air at 100 kPa and 30°C to 100 kPa and 18°C by using water at 8°C. Determine the maximum water outlet temperature when the mass flow rate of the water is 2 kg/s. *Answer:* 16.3°C

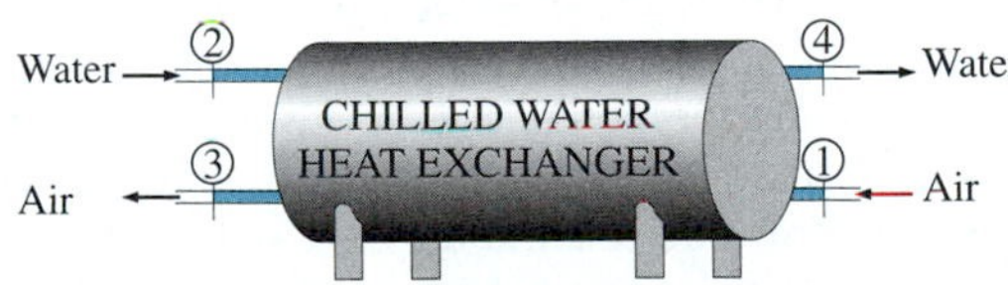

FIGURE P6–97

Pipe and Duct Flow

6–98 Saturated liquid water is heated in a steady-flow steam boiler at a constant pressure of 5 MPa at a rate of 10 kg/s to an outlet temperature of 350°C. Determine the rate of heat transfer in the boiler.

6–99 A 110-volt electrical heater is used to warm 0.3 m^3/s of air at 100 kPa and 15°C to 100 kPa and 30°C. How much current in amperes must be supplied to this heater?

6–100E The fan on a personal computer draws 0.5 ft^3/s of air at 14.7 psia and 70°F through the box containing the CPU and other components. Air leaves at 14.7 psia and 80°F. Calculate the electrical power, in kW, dissipated by the PC components. *Answer:* 0.0948 kW

FIGURE P6–100E

6–101 A desktop computer is to be cooled by a fan. The electronic components of the computer consume 60 W of power under full-load conditions. The computer is to operate in environments at temperatures up to 45°C and at elevations up to 3400 m where the average atmospheric pressure is 66.63 kPa. The exit temperature of air is not to exceed 60°C to meet the reliability requirements. Also, the average velocity of air is not to exceed 110 m/min at the exit of the computer case where the fan is installed to keep the noise level down. Determine the flow rate of the fan that needs to be installed and the diameter of the casing of the fan.

6–102 Repeat Prob. 6–101 for a computer that consumes 100 W of power.

6–103E Water enters the tubes of a cold plate at 95°F with an average velocity of 60 ft/min and leaves at 105°F. The diameter of the tubes is 0.25 in. Assuming 15 percent of the heat generated is dissipated from the components to the surroundings by convection and radiation, and the remaining 85 percent is removed by the cooling water, determine the amount of heat generated by the electronic devices mounted on the cold plate. *Answer:* 263 W

6–104 The components of an electronic system dissipating 180 W are located in a 1.4-m-long horizontal duct whose cross section is 20 cm × 20 cm. The components in the duct are cooled by forced air that enters the duct at 30°C and 1 atm at a rate of 0.6 m^3/min and leaves at 40°C. Determine the rate of heat transfer from the outer surfaces of the duct to the ambient. *Answer:* 63 W

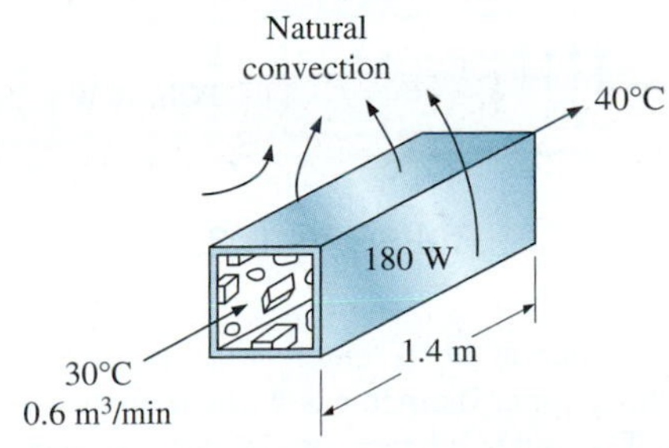

FIGURE P6–104

6–105 Repeat Prob. 6–104 for a circular horizontal duct of diameter 10 cm.

6–106E The hot-water needs of a household are to be met by heating water at 55°F to 180°F by a parabolic solar collector at a rate of 4 lbm/s. Water flows through a 1.25-in-diameter thin aluminum tube whose outer surface is black-anodized in order to maximize its solar absorption ability. The centerline of the tube coincides with the focal line of the collector, and a glass sleeve is placed outside the tube to minimize the heat losses. If solar energy is transferred to water at a net rate of 400 Btu/h per ft length of the tube, determine the required length of the parabolic collector to meet the hot-water requirements of this house.

6–107 Consider a hollow-core printed circuit board 12 cm high and 18 cm long, dissipating a total of 20 W. The width of the air gap in the middle of the PCB is 0.25 cm. If the cooling air enters the 12-cm-wide core at 32°C and 1 atm at a rate of 0.8 L/s, determine the average temperature at which the air leaves the hollow core. *Answer:* 53.4°C

6–108 A computer cooled by a fan contains eight PCBs, each dissipating 10 W power. The height of the PCBs is 12 cm and the length is 18 cm. The cooling air is supplied by a 25-W fan mounted at the inlet. If the temperature rise of air as it flows through the case of the computer is not to exceed 10°C, determine (*a*) the flow rate of the air that the fan needs to deliver and (*b*) the fraction of the temperature rise of air that is due to the heat generated by the fan and its motor. *Answers:* (*a*) 0.0104 kg/s, (*b*) 24 percent

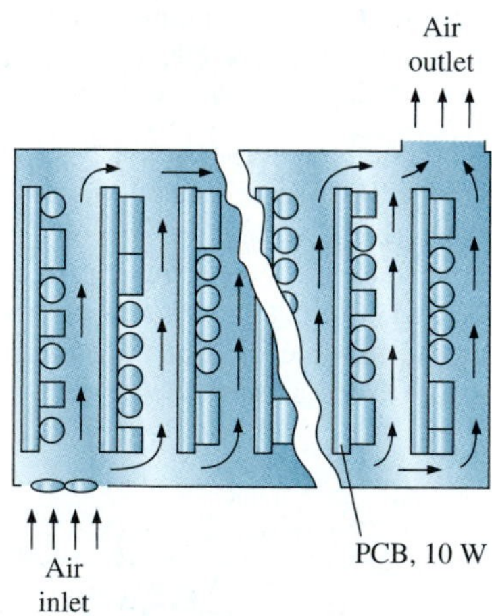

FIGURE P6–108

6–109 Hot water at 90°C enters a 15-m section of a cast iron pipe whose inner diameter is 4 cm at an average velocity of 0.8 m/s. The outer surface of the pipe is exposed to the cold air at 10°C in a basement. If water leaves the basement at 88°C, determine the rate of heat loss from the water.

6–110 Reconsider Prob. 6–109. Using EES (or other) software, investigate the effect of the inner pipe diameter on the rate of heat loss. Let the pipe diameter vary from 1.5 to 7.5 cm. Plot the rate of heat loss against the diameter, and discuss the results.

6–111 A 5-m × 6-m × 8-m room is to be heated by an electric resistance heater placed in a short duct in the room. Initially, the room is at 15°C, and the local atmospheric pressure is 98 kPa. The room is losing heat steadily to the outside at a rate of 200 kJ/min. A 200-W fan circulates the air steadily through the duct and the electric heater at an average mass flow rate of 50 kg/min. The duct can be assumed to be adiabatic, and there is no air leaking in or out of the room. If it takes 15 min for the room air to reach an average temperature of 25°C, find (*a*) the power rating of the electric heater and (*b*) the temperature rise that the air experiences each time it passes through the heater.

6–112 A house has an electric heating system that consists of a 300-W fan and an electric resistance heating element placed in a duct. Air flows steadily through the duct at a rate of 0.6 kg/s and experiences a temperature rise of 7°C. The rate of heat loss from the air in the duct is estimated to be 300 W. Determine the power rating of the electric resistance heating element. *Answer:* 4.22 kW

6–113 A hair dryer is basically a duct in which a few layers of electric resistors are placed. A small fan pulls the air in and forces it through the resistors where it is heated. Air enters a 1200-W hair dryer at 100 kPa and 22°C and leaves at 47°C. The cross-sectional area of the hair dryer at the exit is 60 cm^2. Neglecting the power consumed by the fan and the heat losses through the walls of the hair dryer, determine (*a*) the volume flow rate of air at the inlet and (*b*) the velocity of the air at the exit. *Answers:* (*a*) 0.0404 m^3/s, (*b*) 7.31 m/s

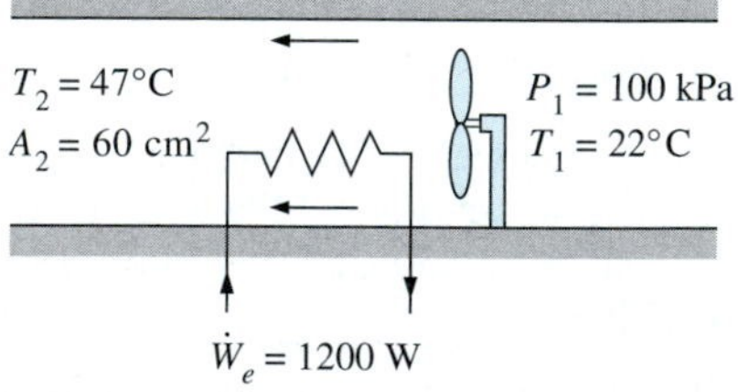

FIGURE P6–113

6–114 Reconsider Prob. 6–113. Using EES (or other) software, investigate the effect of the exit cross-sectional area of the hair dryer on the exit velocity. Let the exit area vary from 25 to 75 cm^2. Plot the exit velocity against the exit cross-sectional area, and discuss the results. Include the effect of the flow kinetic energy in the analysis.

6–115 The ducts of an air heating system pass through an unheated area. As a result of heat losses, the temperature of the air in the duct drops by 4°C. If the mass flow rate of air is 120 kg/min, determine the rate of heat loss from the air to the cold environment.

6–116 Steam enters a long, horizontal pipe with an inlet diameter of $D_1 = 12$ cm at 1 MPa and 300°C with a velocity of 2 m/s. Farther downstream, the conditions are 800 kPa and 250°C, and the diameter is $D_2 = 10$ cm. Determine (*a*) the mass flow rate of the steam and (*b*) the rate of heat transfer. *Answers:* (*a*) 0.0877 kg/s, (*b*) 8.87 kJ/s

6–117 Steam enters an insulated pipe at 200 kPa and 200°C and leaves at 150 kPa and 150°C. The inlet-to-outlet diameter ratio for the pipe is $D_1/D_2 = 1.80$. Determine the inlet and exit velocities of the steam.

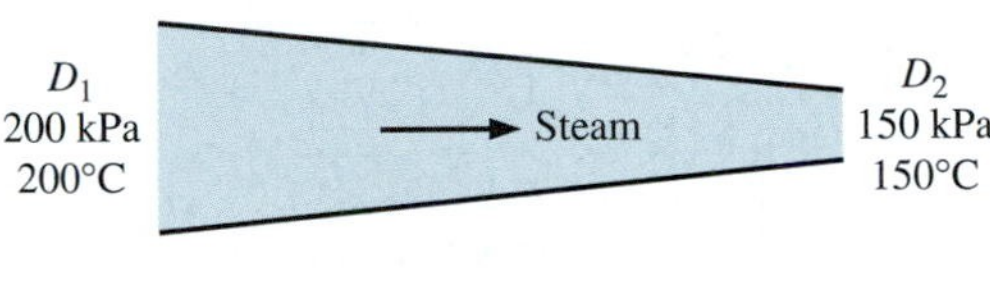

FIGURE P6–117

6–118E Water enters a boiler at 500 psia as a saturated liquid and leaves at 600°F at the same pressure. Calculate the heat transfer per unit mass of water.

6–119 Refrigerant-134a enters the condenser of a refrigerator at 1200 kPa and 80°C, and leaves as a saturated liquid at the same pressure. Determine the heat transfer from the refrigerant per unit mass.

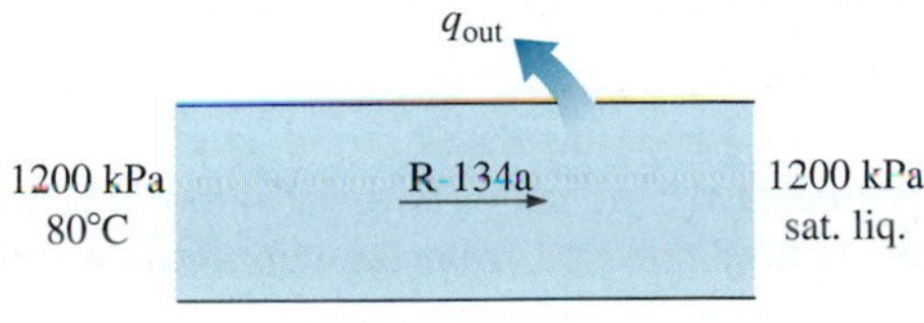

FIGURE P6–119

6–120 Saturated liquid water is heated at constant pressure in a steady-flow device until it is a saturated vapor. Calculate the heat transfer, in kJ/kg, when the vaporization is done at a pressure of 800 kPa.

Charging and Discharging Processes

6–121 A rigid, insulated tank that is initially evacuated is connected through a valve to a supply line that carries helium at 200 kPa and 120°C. Now the valve is opened, and helium is allowed to flow into the tank until the pressure reaches 200 kPa, at which point the valve is closed. Determine the flow work of the helium in the supply line and the final temperature of the helium in the tank. *Answers:* 816 kJ/kg, 655 K

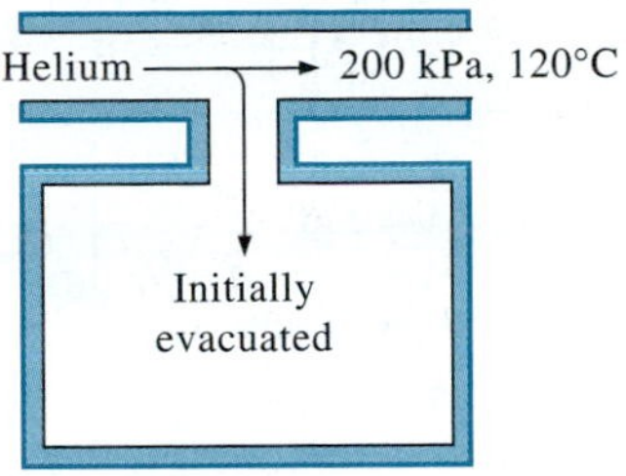

FIGURE P6–121

6–122 Consider an 8-L evacuated rigid bottle that is surrounded by the atmosphere at 100 kPa and 17°C. A valve at the neck of the bottle is now opened and the atmospheric air is allowed to flow into the bottle. The air trapped in the bottle eventually reaches thermal equilibrium with the atmosphere as a result of heat transfer through the wall of the bottle. The valve remains open during the process so that the trapped air also reaches mechanical equilibrium with the atmosphere. Determine the net heat transfer through the wall of the bottle during this filling process. *Answer:* $Q_{out} = 0.8$ kJ

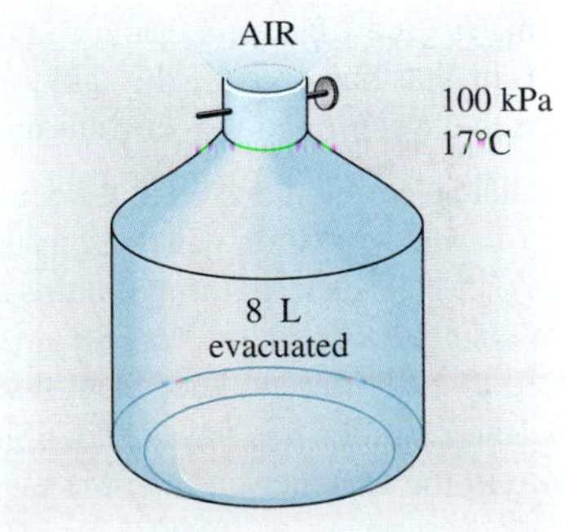

FIGURE P6–122

6–123 An insulated rigid tank is initially evacuated. A valve is opened, and atmospheric air at 95 kPa and 17°C enters the tank until the pressure in the tank reaches 95 kPa, at which point the valve is closed. Determine the final temperature of the air in the tank. Assume constant specific heats. *Answer:* 406 K

6–124 A 2-m³ rigid tank initially contains air at 100 kPa and 22°C. The tank is connected to a supply line through a valve. Air is flowing in the supply line at 600 kPa and 22°C. The valve is opened, and air is allowed to enter the tank until the pressure in the tank reaches the line pressure, at which point the valve is closed. A thermometer placed in the tank indicates

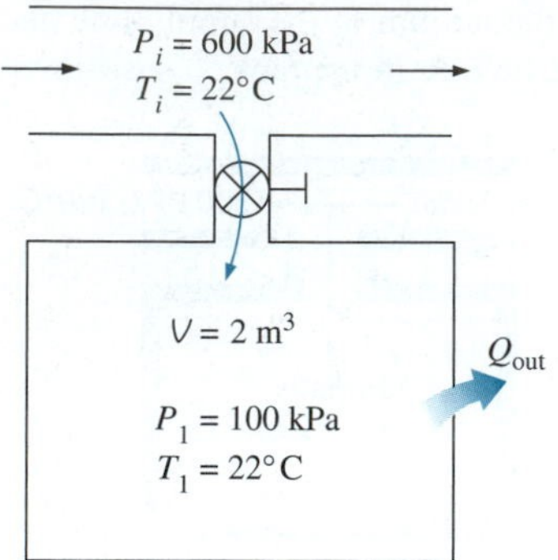

FIGURE P6–124

that the air temperature at the final state is 77°C. Determine (*a*) the mass of air that has entered the tank and (*b*) the amount of heat transfer. *Answers:* (*a*) 9.58 kg, (*b*) 339 kJ

6–125 A 0.2-m^3 rigid tank initially contains refrigerant-134a at 8°C. At this state, 70 percent of the mass is in the vapor phase, and the rest is in the liquid phase. The tank is connected by a valve to a supply line where refrigerant at 1 MPa and 100°C flows steadily. Now the valve is opened slightly, and the refrigerant is allowed to enter the tank. When the pressure in the tank reaches 800 kPa, the entire refrigerant in the tank exists in the vapor phase only. At this point the valve is closed. Determine (*a*) the final temperature in the tank, (*b*) the mass of refrigerant that has entered the tank, and (*c*) the heat transfer between the system and the surroundings.

6–126 An insulated, vertical piston–cylinder device initially contains 10 kg of water, 6 kg of which is in the vapor phase. The mass of the piston is such that it maintains a constant pressure of 200 kPa inside the cylinder. Now steam at 0.5 MPa and 350°C is allowed to enter the cylinder from a supply line until all the liquid in the cylinder has vaporized. Determine (*a*) the final temperature in the cylinder and (*b*) the mass of the steam that has entered. *Answers:* (*a*) 120.2°C, (*b*) 19.07 kg

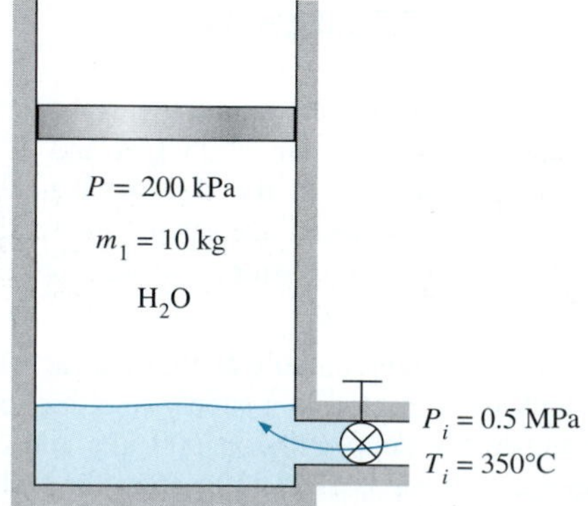

FIGURE P6–126

6–127E A scuba diver's 2-ft^3 air tank is to be filled with air from a compressed air line at 120 psia and 100°F. Initially, the air in this tank is at 20 psia and 70°F. Presuming that the tank is well insulated, determine the temperature and mass in the tank when it is filled to 120 psia.

6–128 An air-conditioning system is to be filled from a rigid container that initially contains 5 kg of liquid R-134a at 24°C. The valve connecting this container to the air-conditioning system is now opened until the mass in the container is 0.25 kg, at which time the valve is closed. During this time, only liquid R-134a flows from the container. Presuming that the process is isothermal while the valve is open, determine the final quality of the R-134a in the container and the total heat transfer. *Answers:* 0.506, 22.6 kJ

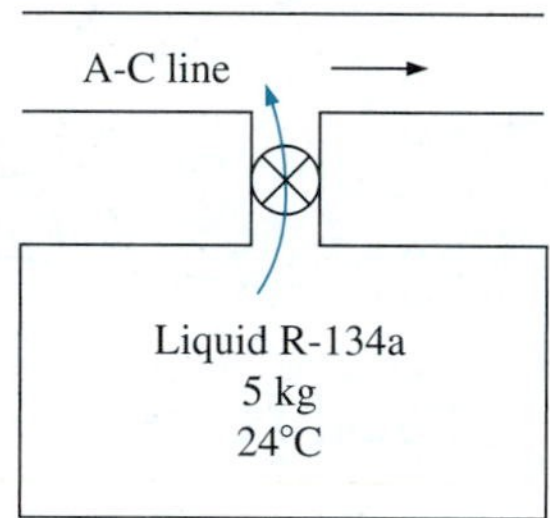

FIGURE P6–128

6–129E Oxygen is supplied to a medical facility from ten 3-ft^3 compressed oxygen tanks. Initially, these tanks are at 2000 psia and 80°F. The oxygen is removed from these tanks slowly enough that the temperature in the tanks remains at 80°F. After two weeks, the pressure in the tanks is 100 psia. Determine the mass of oxygen used and the total heat transfer to the tanks.

6–130 The air in an insulated, rigid compressed-air tank whose volume is 0.5 m^3 is initially at 4000 kPa and 20°C. Enough air is now released from the tank to reduce the pressure to 2000 kPa. Following this release, what is the temperature of the remaining air in the tank?

FIGURE P6–130

6–131E The weighted piston of the device shown in Fig. P6–131E maintains the pressure of the piston-cylinder contents at 200 psia. Initially, this system contains no mass. The valve is now opened, and steam from the line flows into the cylinder until the volume is 10 ft^3. This process is adiabatic, and the steam in the line remains at 300 psia and 450°F. Determine the final temperature (and quality if appropriate) of the steam in the cylinder and the total work produced as the device is filled.

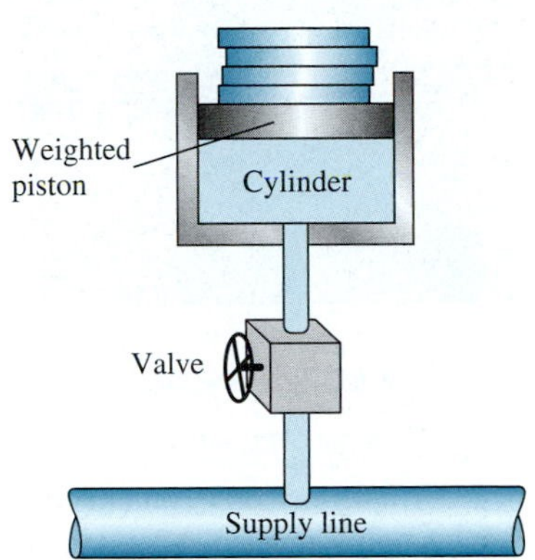

FIGURE P6–131E

6–132E Repeat Prob. 6–131E when the supply line carries oxygen at 300 psia and 450°F. *Answers:* 450°F, 370 Btu

6–133 A 0.12-m^3 rigid tank initially contains refrigerant-134a at 1 MPa and 100 percent quality. The tank is connected by a valve to a supply line that carries refrigerant-134a at 1.2 MPa and 36°C. Now the valve is opened, and the refrigerant is allowed to enter the tank. The valve is closed when it is observed that the tank contains saturated liquid at 1.2 MPa. Determine (*a*) the mass of the refrigerant that has entered the tank and (*b*) the amount of heat transfer. *Answers:* (*a*) 128.4 kg, (*b*) 1057 kJ

6–134 A 0.3-m^3 rigid tank is filled with saturated liquid water at 200°C. A valve at the bottom of the tank is opened, and liquid is withdrawn from the tank. Heat is transferred to the water such that the temperature in the tank remains constant. Determine the amount of heat that must be transferred by the time one-half of the total mass has been withdrawn.

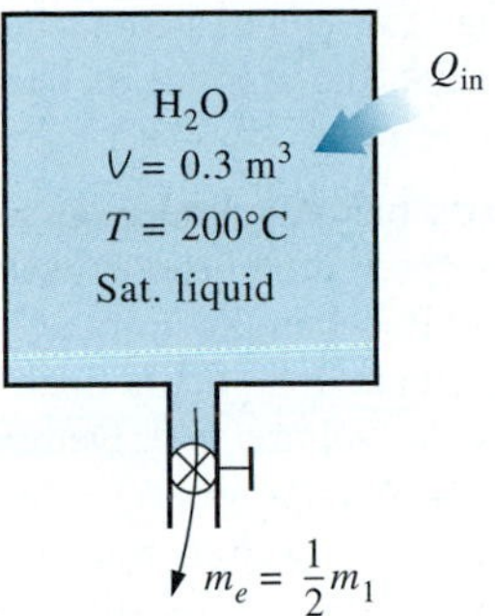

FIGURE P6–134

6–135 A 0.12-m^3 rigid tank contains saturated refrigerant-134a at 800 kPa. Initially, 25 percent of the volume is occupied by liquid and the rest by vapor. A valve at the bottom of the tank is now opened, and liquid is withdrawn from the tank. Heat is transferred to the refrigerant such that the pressure inside the tank remains constant. The valve is closed when no liquid is left in the tank and vapor starts to come out. Determine the total heat transfer for this process. *Answer:* 201.2 kJ

6–136E A 4-ft^3 rigid tank contains saturated refrigerant-134a at 100 psia. Initially, 20 percent of the volume is occupied by liquid and the rest by vapor. A valve at the top of the tank is now opened, and vapor is allowed to escape slowly from the tank. Heat is transferred to the refrigerant such that the pressure inside the tank remains constant. The valve is closed when the last drop of liquid in the tank is vaporized. Determine the total heat transfer for this process.

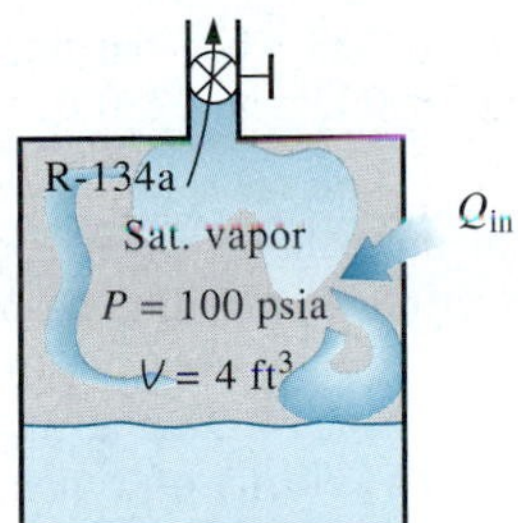

FIGURE P6–136E

6–137 A 0.2-m^3 rigid tank equipped with a pressure regulator contains steam at 2 MPa and 300°C. The steam in the tank is now heated. The regulator keeps the steam pressure constant by letting out some steam, but the temperature inside rises. Determine the amount of heat transferred when the steam temperature reaches 500°C.

6–138 An adiabatic piston-cylinder device equipped with a spring maintains the pressure inside at 300 kPa when the volume is 0, and 3000 kPa when the volume is 5 m^3. The device is connected to a steam line maintained at 1500 kPa and 200°C, and initially the volume is 0. Determine the final temperature (and quality if appropriate) when valve is opened and steam in the line is allowed to enter the cylinder until the pressure inside matches that in the line. Also determine the total work produced during this adiabatic filling process.

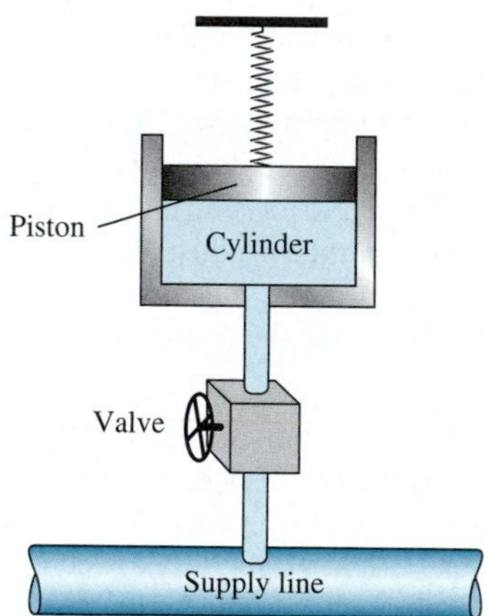

FIGURE P6–138

6–139 Repeat Prob. 6–138 when the line is filled with air that is maintained at 2000 kPa and 250°C.

6–140 The air-release flap on a hot-air balloon is used to release hot air from the balloon when appropriate. On one hot-air balloon, the air release opening has an area of 0.5 m^2, and the filling opening has an area of 1 m^2. During a two minute adiabatic flight maneuver, hot air enters the balloon at 100 kPa and 35°C with a velocity of 2 m/s; the air in the balloon remains at 100 kPa and 35°C; and air leaves the balloon through the air-release flap at velocity 1 m/s. At the start of this maneuver, the volume of the balloon is 75 m^3. Determine the final volume of the balloon and work produced by the air inside the balloon as it expands the balloon skin.

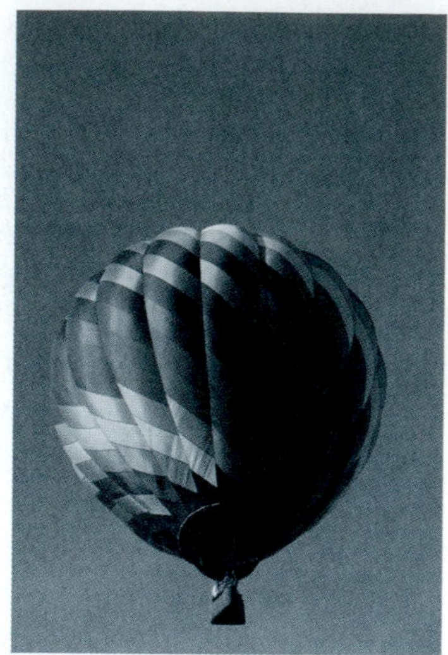

FIGURE P6–140

6–141 An insulated 0.08-m^3 tank contains helium at 2 MPa and 80°C. A valve is now opened, allowing some helium to escape. The valve is closed when one-half of the initial mass has escaped. Determine the final temperature and pressure in the tank. *Answers:* 225 K, 637 kPa

6–142E An insulated 60-ft^3 rigid tank contains air at 75 psia and 120°F. A valve connected to the tank is now opened, and air is allowed to escape until the pressure inside drops to 30 psia. The air temperature during this process is maintained constant by an electric resistance heater placed in the tank. Determine the electrical work done during this process.

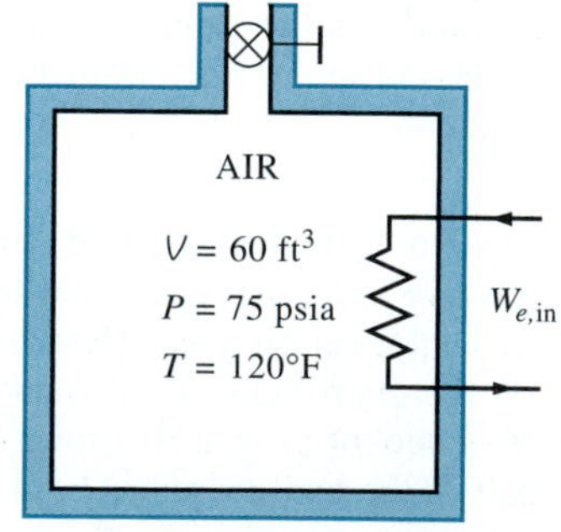

FIGURE P6–142E

6–143 A vertical piston–cylinder device initially contains 0.2 m^3 of air at 20°C. The mass of the piston is such that it maintains a constant pressure of 300 kPa inside. Now a valve connected to the cylinder is opened, and air is allowed to escape until the volume inside the cylinder is decreased by one-half. Heat transfer takes place during the process so that the temperature of the air in the cylinder remains constant.

Determine (*a*) the amount of air that has left the cylinder and (*b*) the amount of heat transfer. *Answers:* (*a*) 0.357 kg, (*b*) 0

6–144 A balloon initially contains 65 m^3 of helium gas at atmospheric conditions of 100 kPa and 22°C. The balloon is connected by a valve to a large reservoir that supplies helium gas at 150 kPa and 25°C. Now the valve is opened, and helium is allowed to enter the balloon until pressure equilibrium with the helium at the supply line is reached. The material of the balloon is such that its volume increases linearly with pressure. If no heat transfer takes place during this process, determine the final temperature in the balloon. *Answer:* 334 K

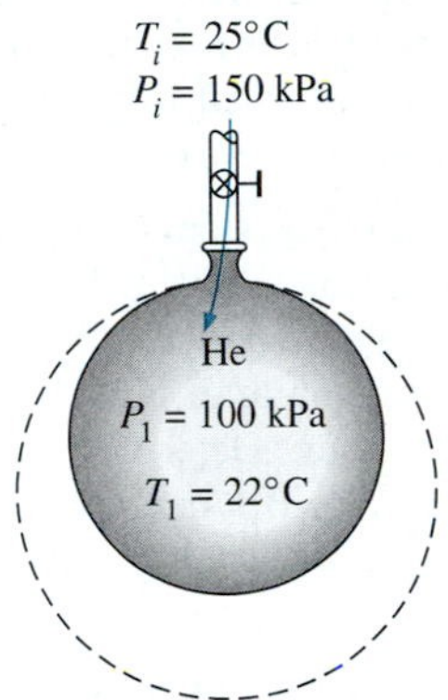

FIGURE P6–144

6–145 An insulated vertical piston–cylinder device initially contains 0.8 m^3 of refrigerant-134a at 1.2 MPa and 120°C. A linear spring at this point applies full force to the piston. A valve connected to the cylinder is now opened, and refrigerant is allowed to escape. The spring unwinds as the piston moves down, and the pressure and volume drop to 0.6 MPa and 0.5 m^3 at the end of the process. Determine (*a*) the amount of refrigerant that has escaped and (*b*) the final temperature of the refrigerant.

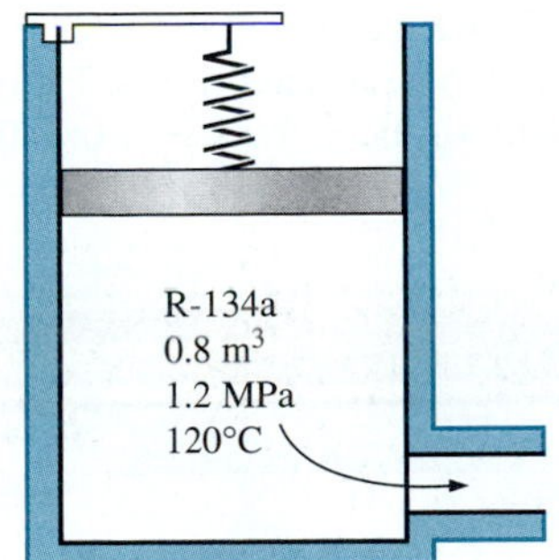

FIGURE P6–145

6–146 A piston–cylinder device initially contains 0.6 kg of steam with a volume of 0.1 m^3. The mass of the piston is such that it maintains a constant pressure of 800 kPa. The cylinder is connected through a valve to a supply line that carries steam at 5 MPa and 500°C. Now the valve is opened and steam is allowed to flow slowly into the cylinder until the volume of the cylinder doubles and the temperature in the cylinder reaches 250°C, at which point the valve is closed. Determine (*a*) the mass of steam that has entered and (*b*) the amount of heat transfer.

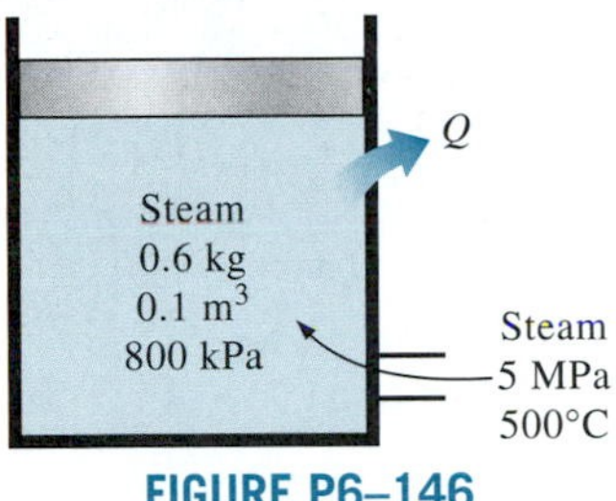

FIGURE P6–146

Review Problems

6–147 A D_0 = 10-m-diameter tank is initially filled with water 2 m above the center of a D = 10-cm-diameter valve near the bottom. The tank surface is open to the atmosphere, and the tank drains through a L = 100-m-long pipe connected to the valve. The friction factor of the pipe is given to be f = 0.015, and the discharge velocity is expressed as $V = \sqrt{\dfrac{2gz}{1.5 + fL/D}}$ where z is the water height above the center of the valve. Determine (*a*) the initial discharge velocity from the tank and (*b*) the time required to empty the tank. The tank can be considered to be empty when the water level drops to the center of the valve.

6–148 Underground water is being pumped into a pool whose cross section is 3 m × 4 m while water is discharged through a 5-cm-diameter orifice at a constant average velocity of 5 m/s. If the water level in the pool rises at a rate of 1.5 cm/min, determine the rate at which water is supplied to the pool, in m^3/s.

6–149 The velocity of a liquid flowing in a circular pipe of radius R varies from zero at the wall to a maximum at the pipe center. The velocity distribution in the pipe can be represented as $V(r)$, where r is the radial distance from the pipe center. Based on the definition of mass flow rate $\dot{m}$, obtain a relation for the average velocity in terms of $V(r)$, R, and r.

6–150 Air at 4.18 kg/m^3 enters a nozzle that has an inlet-to-exit area ratio of 2:1 with a velocity of 120 m/s and leaves with a velocity of 380 m/s. Determine the density of air at the exit. *Answer:* 2.64 kg/m^3

6–151 An air compressor compresses 10 L/s of air at 120 kPa and 20°C to 1000 kPa and 300°C while consuming 4.5 kW of power. How much of this power is being used to increase the pressure of the air versus the power needed to move the fluid through the compressor? *Answers:* 3.35 kW, 1.15 kW

6–152 A steam turbine operates with 1.6 MPa and 350°C steam at its inlet and saturated vapor at 30°C at its exit. The mass flow rate of the steam is 16 kg/s, and the turbine produces 9000 kW of power. Determine the rate at which heat is lost through the casing of this turbine.

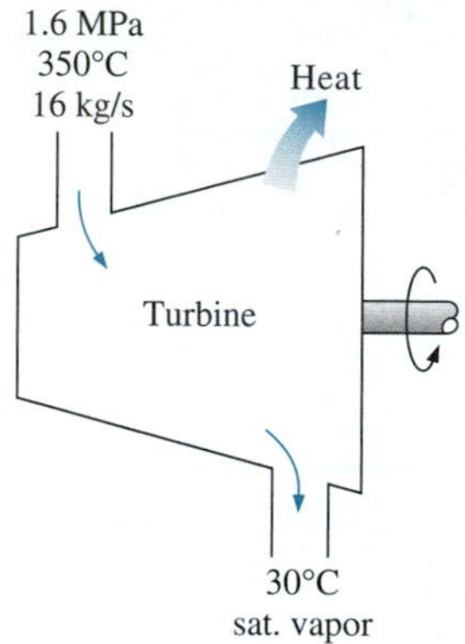

FIGURE P6–152

6–153E Nitrogen gas flows through a long, constant-diameter adiabatic pipe. It enters at 100 psia and 120°F and leaves at 50 psia and 70°F. Calculate the velocity of the nitrogen at the pipe's inlet and outlet.

6–154 A 110-V electric hot-water heater warms 0.1 L/s of water from 15 to 20°C. Calculate the current in amperes that must be supplied to this heater. *Answer:* 19.0 A

6–155 Steam enters a long, insulated pipe at 1400 kPa, 350°C, and 10 m/s, and exits at 1000 kPa. The diameter of the pipe is 0.15 m at the inlet, and 0.1 m at the exit. Calculate the mass flow rate of the steam and its speed at the pipe outlet.

6–156 The air flow in a compressed air line is divided into two equal streams by a T-fitting in the line. The compressed air enters this 2.5-cm diameter fitting at 1.6 MPa and 40°C with a velocity of 50 m/s. Each outlet has the same diameter as the inlet, and the air at these outlets has a pressure of 1.4 MPa and a temperature of 36°C. Determine the velocity of the air at the outlets and the rate of change of flow energy (flow power) across the T-fitting.

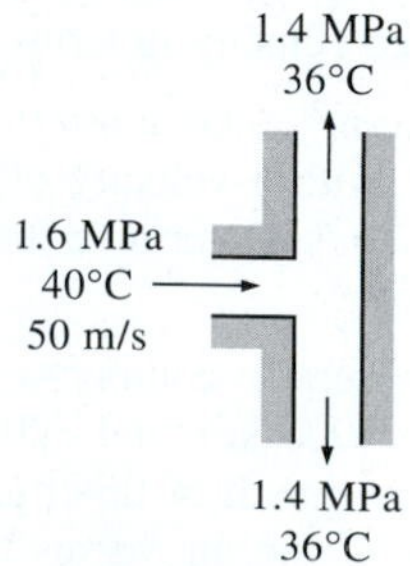

FIGURE P6–156

6–157 Air enters a pipe at 50°C and 200 kPa and leaves at 40°C and 150 kPa. It is estimated that heat is lost from the pipe in the amount of 3.3 kJ per kg of air flowing in the pipe. The diameter ratio for the pipe is $D_1/D_2 = 1.8$. Using constant specific heats for air, determine the inlet and exit velocities of the air. *Answers:* 28.6 m/s, 120 m/s

6–158 Cold water enters a steam generator at 20°C and leaves as saturated vapor at 150°C. Determine the fraction of heat used in the steam generator to preheat the liquid water from 20°C to the saturation temperature of 150°C.

6–159 Cold water enters a steam generator at 20°C and leaves as saturated vapor at the boiler pressure. At what pressure will the amount of heat needed to preheat the water to saturation temperature be equal to the heat needed to vaporize the liquid at the boiler pressure?

6–160E A refrigeration system is being designed to cool eggs (ρ = 67.4 lbm/ft³ and c_p = 0.80 Btu/lbm · °F) with an average mass of 0.14 lbm from an initial temperature of 90°F to a final average temperature of 50°F by air at 34°F at a rate of 10,000 eggs per hour. Determine (*a*) the rate of heat removal from the eggs, in Btu/h and (*b*) the required volume flow rate of air, in ft³/h, if the temperature rise of air is not to exceed 10°F.

6–161 A glass bottle washing facility uses a well-agitated hot-water bath at 55°C that is placed on the ground. The bottles enter at a rate of 800 per minute at an ambient temperature of 20°C and leave at the water temperature. Each bottle has a mass of 150 g and removes 0.2 g of water as it leaves the bath wet. Make-up water is supplied at 15°C. Disregarding any heat losses from the outer surfaces of the bath, determine the rate at which (*a*) water and (*b*) heat must be supplied to maintain steady operation.

6–162 Repeat Prob. 6–161 for a water bath temperature of 50°C.

6–163 Long aluminum wires of diameter 3 mm (ρ = 2702 kg/m³ and c_p = 0.896 kJ/kg · °C) are extruded at a temperature of 350°C and are cooled to 50°C in atmospheric air at 30°C. If the wire is extruded at a velocity of 10 m/min, determine the rate of heat transfer from the wire to the extrusion room.

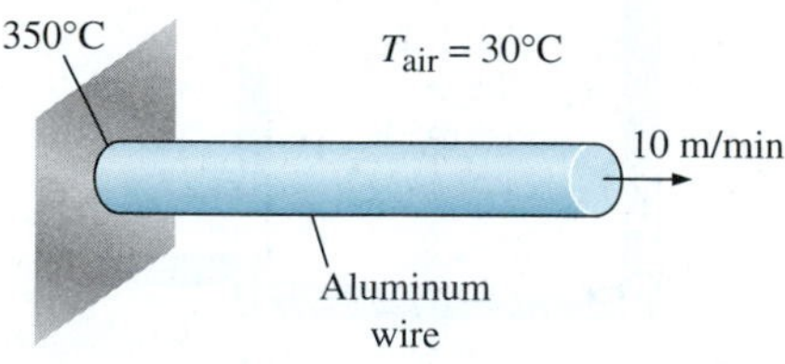

FIGURE P6–163

6–164 Repeat Prob. 6–163 for a copper wire (ρ = 8950 kg/m³ and c_p = 0.383 kJ/kg · °C).

6–165 Steam at 40°C condenses on the outside of a 5-m-long, 3-cm-diameter thin horizontal copper tube by cooling water that enters the tube at 25°C at an average velocity of 2 m/s and leaves at 35°C. Determine the rate of condensation of steam. *Answer:* 0.0245 kg/s

FIGURE P6–165

6–166 A fan is powered by a 0.5-hp motor and delivers air at a rate of 85 m^3/min. Determine the highest value for the average velocity of air mobilized by the fan. Take the density of air to be 1.18 kg/m^3.

6–167 A liquid R-134a bottle has an internal volume of 0.001 m^3. Initially it contains 0.4 kg of R-134a (saturated mixture) at 26°C. A valve is opened and R-134a vapor only (no liquid) is allowed to escape slowly such that temperature remains constant until the mass of R-134a remaining is 0.1 kg. Find the heat transfer necessary with the surroundings to maintain the temperature and pressure of the R-134a constant.

6–168 Steam enters a turbine steadily at 10 MPa and 550°C with a velocity of 60 m/s and leaves at 25 kPa with a quality of 95 percent. A heat loss of 30 kJ/kg occurs during the process. The inlet area of the turbine is 150 cm^2, and the exit area is 1400 cm^2. Determine (*a*) the mass flow rate of the steam, (*b*) the exit velocity, and (*c*) the power output.

6–169 Reconsider Prob. 6–168. Using EES (or other) software, investigate the effects of turbine exit area and turbine exit pressure on the exit velocity and power output of the turbine. Let the exit pressure vary from 10 to 50 kPa (with the same quality), and the exit area to vary from 1000 to 3000 cm^2. Plot the exit velocity and the power outlet against the exit pressure for the exit areas of 1000, 2000, and 3000 cm^2, and discuss the results.

6–170E Refrigerant-134a enters an adiabatic compressor at 15 psia and 20°F with a volume flow rate of 10 ft^3/s and leaves at a pressure of 100 psia. The power input to the compressor is 45 hp. Find (*a*) the mass flow rate of the refrigerant and (*b*) the exit temperature.

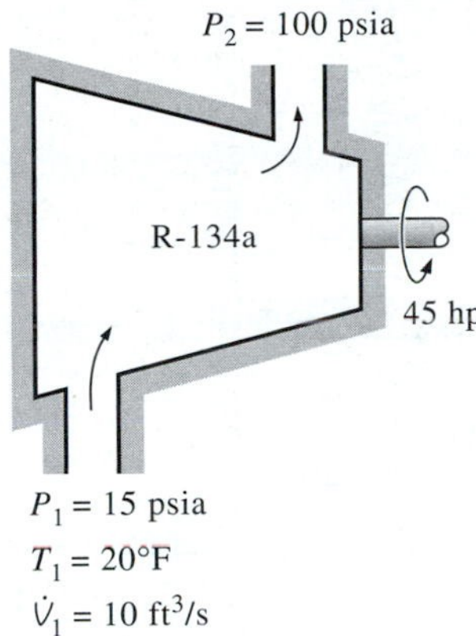

FIGURE P6–170

6–171 In large gas-turbine power plants, air is preheated by the exhaust gases in a heat exchanger called the *regenerator* before it enters the combustion chamber. Air enters the regenerator at 1 MPa and 550 K at a mass flow rate of 800 kg/min. Heat is transferred to the air at a rate of 3200 kJ/s. Exhaust gases enter the regenerator at 140 kPa and 800 K and leave at 130 kPa and 600 K. Treating the exhaust gases as air, determine (*a*) the exit temperature of the air and (*b*) the mass flow rate of exhaust gases. *Answers:* (*a*) 775 K, (*b*) 14.9 kg/s

6–172 It is proposed to have a water heater that consists of an insulated pipe of 5-cm diameter and an electric resistor inside. Cold water at 20°C enters the heating section steadily at a rate of 30 L/min. If water is to be heated to 55°C, determine (*a*) the power rating of the resistance heater and (*b*) the average velocity of the water in the pipe.

6–173 An insulated vertical piston–cylinder device initially contains 0.2 m^3 of air at 200 kPa and 22°C. At this state, a linear spring touches the piston but exerts no force on it. The cylinder is connected by a valve to a line that supplies air at 800 kPa and 22°C. The valve is opened, and air from the high-pressure line is allowed to enter the cylinder. The valve is turned off when the pressure inside the cylinder reaches 600 kPa. If the enclosed volume inside the cylinder doubles during this process, determine (*a*) the mass of air that entered the cylinder, and (*b*) the final temperature of the air inside the cylinder.

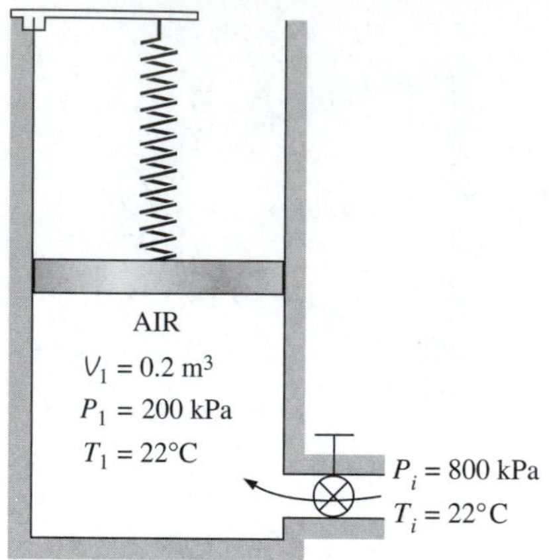

FIGURE P6–173

Design and Essay Problems

6–174E Pneumatic nail drivers used in construction require 0.02 ft^3 of air at 100 psia and 1 Btu of energy to drive a single nail. You have been assigned the task of designing a compressed-air storage tank with enough capacity to drive 500 nails. The pressure in this tank cannot exceed 500 psia, and the temperature cannot exceed that normally found at a construction site. What is the maximum pressure to be used in the tank and what is the tank's volume?

6–175 You have been given the responsibility of picking a steam turbine for an electrical-generation station that is to produce 300 MW of electrical power that will sell for $0.05 per kilowatt-hour. The boiler will produce steam at 700 psia and 700°F, and the condenser is planned to operate at 80°F. The cost of generating and condensing the steam is $0.01 per kilowatt-hour of electricity produced. You have narrowed your selection to the three turbines in the table below. Your criterion for selection is to pay for the equipment as quickly as possible. Which turbine should you choose?

Turbine	Capacity (MW)	η	Cost ($Million)	Operating Cost ($/kWh)
A	50	0.9	5	0.01
B	100	0.92	11	0.01
C	100	0.93	10.5	0.015

6–176E You are to design a small, directional control rocket to operate in space by providing as many as 100 bursts of 5 seconds each with a mass flow rate of 0.5 lbm/s at a velocity of 400 ft/s. Storage tanks that will contain up to 3000 psia are available, and the tanks will be located in an environment whose temperature is 40°F. Your design criterion is to minimize the volume of the storage tank. Should you use a compressed-air or an R-134a system?

6–177 An air cannon uses compressed air to propel a projectile from rest to a final velocity. Consider an air cannon that is to accelerate a 10-gram projectile to a speed of 300 m/s using compressed air, whose temperature cannot exceed 20°C. The volume of the storage tank is not to exceed 0.1 m^3. Select the storage volume size and maximum storage pressure that requires the minimum amount of energy to fill the tank.

6–178 To maintain altitude, the temperature of the air inside a hot-air balloon must remain within a 1°C band, while the volume cannot vary by more than 1 percent. At a 300-m altitude, the air in a 1000 m^3 hot-air balloon needs to maintain a 35°C average temperature. This balloon loses heat at a rate of 3 kW through the fabric. When the burner is activated, it adds 30 kg/s of air at 200°C and 100 kPa to the balloon. When the flap that allows air to escape is opened, air leaves the balloon at a rate of 20 kg/s. Design the burner and exhaust-flap control cycles (on time and off time) necessary to maintain the balloon at a 300-m altitude.

Chapter 7

THE SECOND LAW OF THERMODYNAMICS

To this point, we have focused our attention on the first law of thermodynamics, which requires that energy be conserved during a process. In this chapter, we introduce the second law of thermodynamics, which asserts that processes occur in a certain direction and that energy has quality as well as quantity. A process cannot take place unless it satisfies both the first and second laws of thermodynamics. In this chapter, the thermal energy reservoirs, reversible and irreversible processes, heat engines, refrigerators, and heat pumps are introduced first. Various statements of the second law are followed by a discussion of perpetual-motion machines and the thermodynamic temperature scale. The Carnot cycle is introduced next, and the Carnot principles are discussed. Finally, the idealized Carnot heat engines, refrigerators, and heat pumps are examined.

Objectives

The objectives of this chapter are to:

- Introduce the second law of thermodynamics.
- Identify valid processes as those that satisfy both the first and second laws of thermodynamics.
- Discuss thermal energy reservoirs, reversible and irreversible processes, heat engines, refrigerators, and heat pumps.
- Describe the Kelvin–Planck and Clausius statements of the second law of thermodynamics.
- Discuss the concepts of perpetual-motion machines.
- Apply the second law of thermodynamics to cycles and cyclic devices.
- Apply the second law to develop the absolute thermodynamic temperature scale.
- Describe the Carnot cycle.
- Examine the Carnot principles, idealized Carnot heat engines, refrigerators, and heat pumps.
- Determine the expressions for the thermal efficiencies and coefficients of performance for reversible heat engines, heat pumps, and refrigerators.

7–1 ▪ INTRODUCTION TO THE SECOND LAW

In Chaps. 5 and 6, we applied the *first law of thermodynamics*, or the *conservation of energy principle*, to processes involving closed and open systems. As pointed out repeatedly in those chapters, energy is a conserved property, and no process is known to have taken place in violation of the first law of thermodynamics. Therefore, it is reasonable to conclude that a process must satisfy the first law to occur. However, as explained here, satisfying the first law alone does not ensure that the process will actually take place.

FIGURE 7–1

A cup of hot coffee does not get hotter in a cooler room.

It is common experience that a cup of hot coffee left in a cooler room eventually cools off (Fig. 7–1). This process satisfies the first law of thermodynamics since the amount of energy lost by the coffee is equal to the amount gained by the surrounding air. Now let us consider the reverse process—the hot coffee getting even hotter in a cooler room as a result of heat transfer from the room air. We all know that this process never takes place. Yet, doing so would not violate the first law as long as the amount of energy lost by the air is equal to the amount gained by the coffee.

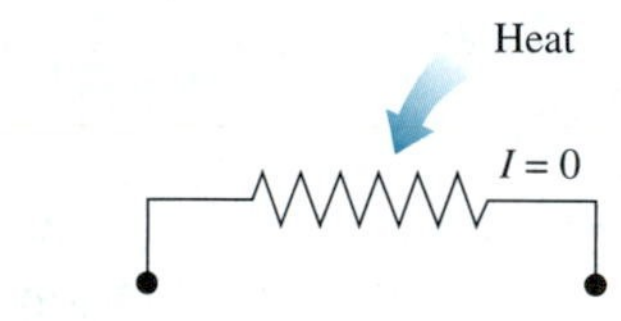

FIGURE 7–2

Transferring heat to a wire will not generate electricity.

As another familiar example, consider the heating of a room by the passage of electric current through a resistor (Fig. 7–2). Again, the first law dictates that the amount of electric energy supplied to the resistance wires be equal to the amount of energy transferred to the room air as heat. Now let us attempt to reverse this process. It will come as no surprise that transferring some heat to the wires does not cause an equivalent amount of electric energy to be generated in the wires.

Finally, consider a paddle-wheel mechanism that is operated by the fall of a mass (Fig. 7–3). The paddle wheel rotates as the mass falls and stirs a fluid within an insulated container. As a result, the potential energy of the mass decreases, and the internal energy of the fluid increases in accordance with the conservation of energy principle. However, the reverse process, raising the mass by transferring heat from the fluid to the paddle wheel, does not occur in nature, although doing so would not violate the first law of thermodynamics.

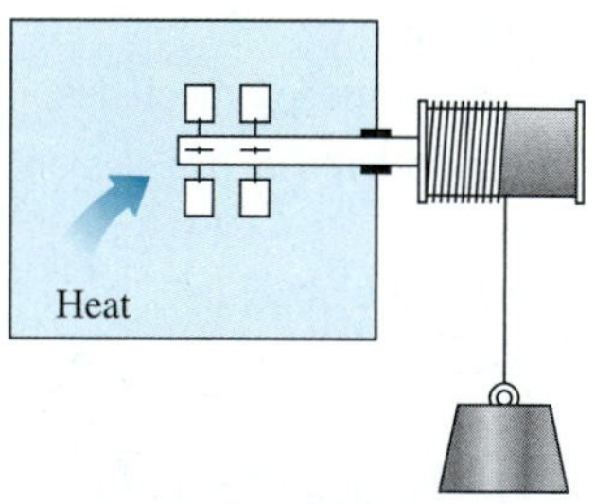

FIGURE 7–3

Transferring heat to a paddle wheel will not cause it to rotate.

It is clear from these arguments that processes proceed in a *certain direction* and not in the reverse direction (Fig. 7–4). The first law places no restriction on the direction of a process, but satisfying the first law does not ensure that the process can actually occur. This inadequacy of the first law to identify whether a process can take place is remedied by introducing another general principle, the *second law of thermodynamics*. We show later in this chapter that the reverse processes discussed above violate the second law of thermodynamics. This violation is easily detected with the help of a property, called *entropy*, defined in Chap. 8. A process cannot occur unless it satisfies both the first and the second laws of thermodynamics (Fig. 7–5).

FIGURE 7–4

Processes occur in a certain direction, and not in the reverse direction.

There are numerous valid statements of the second law of thermodynamics. Two such statements are presented and discussed later in this chapter in relation to some engineering devices that operate on cycles.

The use of the second law of thermodynamics is not limited to identifying the direction of processes, however. The second law also asserts that energy has *quality* as well as quantity. The first law is concerned with the quantity of energy and the transformations of energy from one form to another with no regard to its quality. Preserving the quality of energy is a major concern to engineers, and the second law provides the necessary means to determine the quality as well as the degree of degradation of energy during a process. As discussed later in this chapter, more of high-temperature energy can be converted to work, and thus it has a higher quality than the same amount of energy at a lower temperature.

The second law of thermodynamics is also used in determining the *theoretical limits* for the performance of commonly used engineering systems, such as heat engines and refrigerators, as well as predicting the *degree of completion* of chemical reactions.

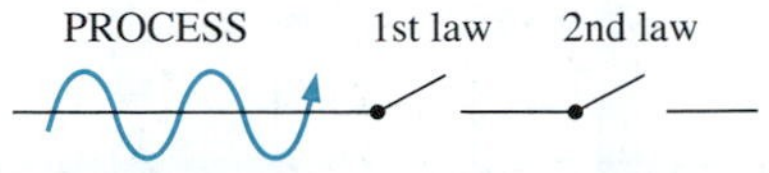

PROCESS 1st law 2nd law

FIGURE 7–5

A process must satisfy both the first and second laws of thermodynamics to proceed.

7–2 ▪ THERMAL ENERGY RESERVOIRS

In the development of the second law of thermodynamics, it is very convenient to have a hypothetical body with a relatively large *thermal energy capacity* (mass × specific heat) that can supply or absorb finite amounts of heat without undergoing any change in temperature. Such a body is called a **thermal energy reservoir**, or just a reservoir. In practice, large bodies of water such as oceans, lakes, and rivers as well as the atmospheric air can be modeled accurately as thermal energy reservoirs because of their large thermal energy storage capabilities or thermal masses (Fig. 7–6). The *atmosphere*, for example, does not warm up as a result of heat losses from residential buildings in winter. Likewise, megajoules of waste energy dumped in large rivers by power plants do not cause any significant change in water temperature.

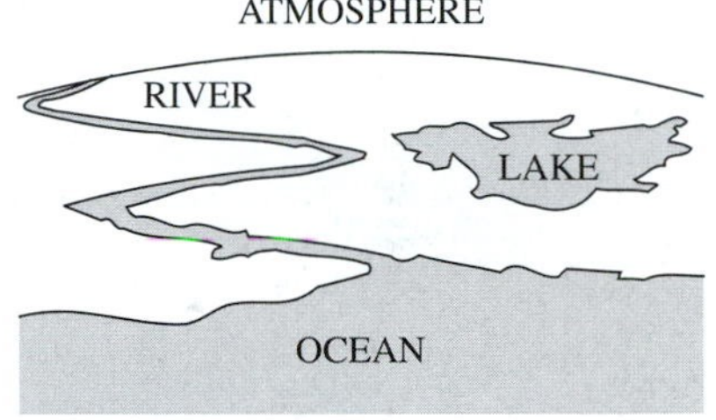

FIGURE 7–6

Bodies with relatively large thermal masses can be modeled as thermal energy reservoirs.

A *two-phase system* can be modeled as a reservoir also since it can absorb and release large quantities of heat while remaining at constant temperature. Another familiar example of a thermal energy reservoir is the *industrial furnace*. The temperatures of most furnaces are carefully controlled, and they are capable of supplying large quantities of thermal energy as heat in an essentially isothermal manner. Therefore, they can be modeled as reservoirs.

A body does not actually have to be very large to be considered a reservoir. Any physical body whose thermal energy capacity is large relative to the amount of energy it supplies or absorbs can be modeled as one. The air in a room, for example, can be treated as a reservoir in the analysis of the heat dissipation from a TV set in the room, since the amount of heat transfer from the TV set to the room air is not large enough to have a noticeable effect on the room air temperature.

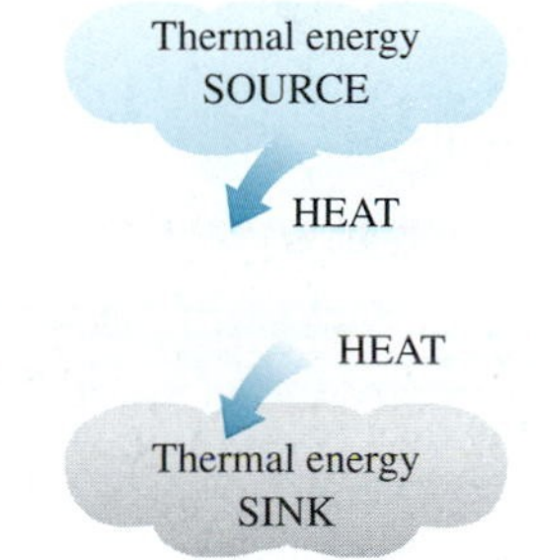

FIGURE 7–7

A source supplies energy in the form of heat, and a sink absorbs it.

A reservoir that supplies energy in the form of heat is called a **source**, and one that absorbs energy in the form of heat is called a **sink** (Fig. 7–7). Thermal energy reservoirs are often referred to as **heat reservoirs** since they supply or absorb energy in the form of heat.

Heat transfer from industrial sources to the environment is of major concern to environmentalists as well as to engineers. Irresponsible management of waste energy can significantly increase the temperature of portions of the environment, causing what is called *thermal pollution*. If it is not carefully controlled, thermal pollution can seriously disrupt marine life in lakes and rivers. However, by careful design and management, the waste energy dumped into large bodies of water can be used to improve the quality of marine life by keeping the local temperature increases within safe and desirable levels.

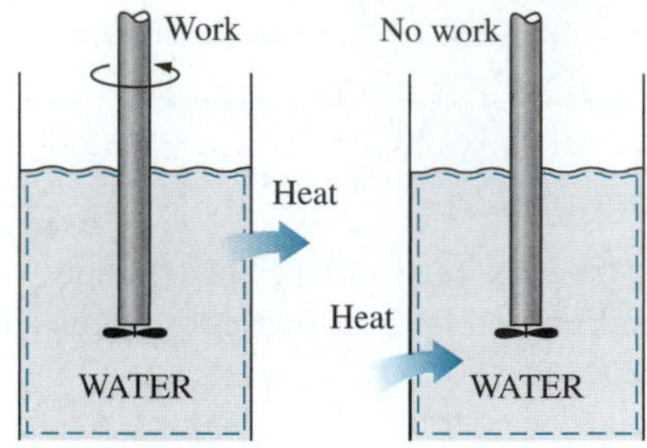

FIGURE 7–8

Work can always be converted to heat directly and completely, but the reverse is not true.

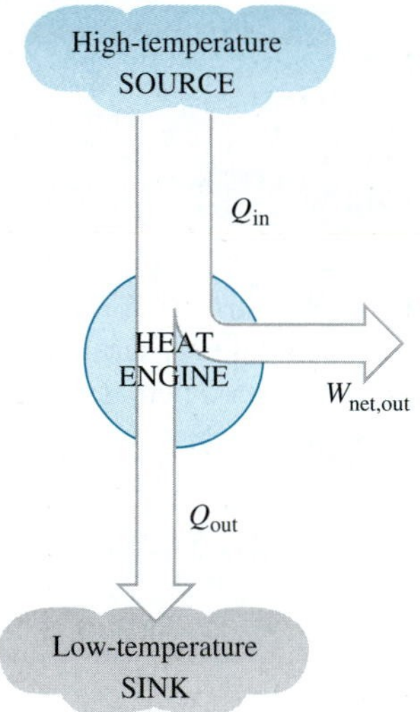

FIGURE 7–9

Part of the heat received by a heat engine is converted to work, while the rest is rejected to a sink.

7–3 ▪ HEAT ENGINES

As pointed out earlier, work can easily be converted to other forms of energy, but converting other forms of energy to work is not that easy. The mechanical work done by the shaft shown in Fig. 7–8, for example, is first converted to the internal energy of the water. This energy may then leave the water as heat. We know from experience that any attempt to reverse this process will fail. That is, transferring heat to the water does not cause the shaft to rotate. From this and other observations, we conclude that work can be converted to heat directly and completely, but converting heat to work requires the use of some special devices. These devices are called **heat engines**.

Heat engines differ considerably from one another, but all can be characterized by the following (Fig. 7–9):

1. They receive heat from a high-temperature source (solar energy, oil furnace, nuclear reactor, etc.).
2. They convert part of this heat to work (usually in the form of a rotating shaft).
3. They reject the remaining waste heat to a low-temperature sink (the atmosphere, rivers, etc.).
4. They operate on a cycle.

Heat engines and other cyclic devices usually involve a fluid to and from which heat is transferred while undergoing a cycle. This fluid is called the **working fluid**.

The term *heat engine* is often used in a broader sense to include work-producing devices that do not operate in a thermodynamic cycle. Engines that involve internal combustion such as gas turbines and car engines fall into this category. These devices operate in a mechanical cycle but not in a thermodynamic cycle since the working fluid (the combustion gases) does not undergo a complete cycle. Instead of being cooled to the initial temperature, the exhaust gases are purged and replaced by fresh air-and-fuel mixture at the end of the cycle.

The work-producing device that best fits into the definition of a heat engine is the *steam power plant*, which is an external-combustion engine. That is, combustion takes place outside the engine, and the thermal energy released during this process is transferred to the steam as heat. The schematic of a basic steam power plant is shown in Fig. 7–10. This is a rather simplified diagram, and the discussion of actual steam power plants is given in later chapters. The various quantities shown on this figure are as follows:

Q_{in} = amount of heat supplied to steam in boiler from a high-temperature source (furnace)

Q_{out} = amount of heat rejected from steam in condenser to a low-temperature sink (the atmosphere, a river, etc.)

W_{out} = amount of work delivered by steam as it expands in turbine

W_{in} = amount of work required to compress water to boiler pressure

Notice that the directions of the heat and work interactions are indicated by the subscripts *in* and *out*. Therefore, all four of the described quantities are always *positive*.

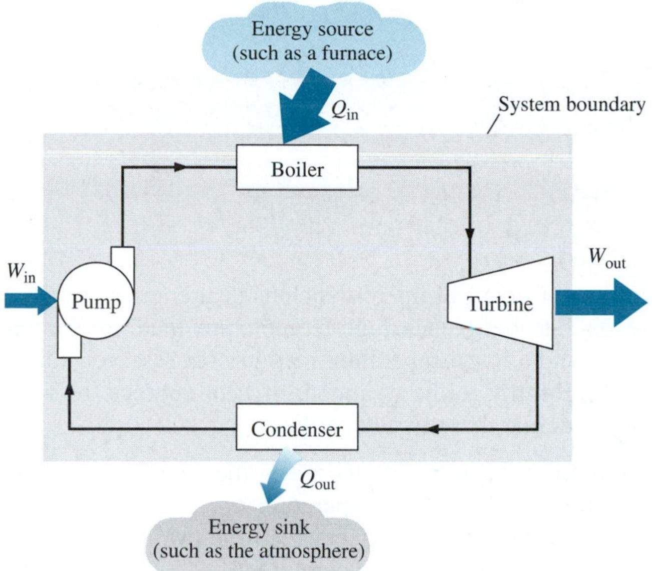

FIGURE 7–10

Schematic of a steam power plant.

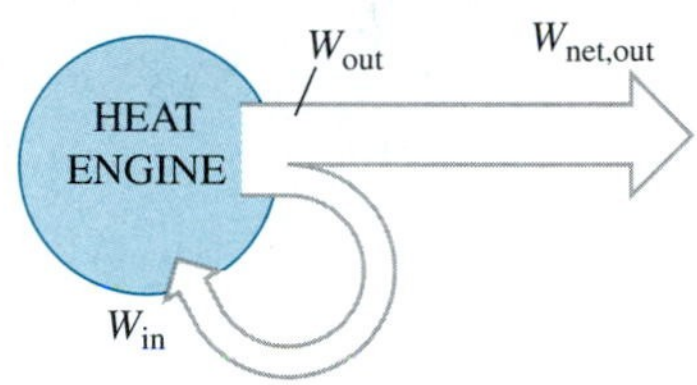

FIGURE 7–11

A portion of the work output of a heat engine is consumed internally to maintain continuous operation.

The net work output of this power plant is simply the difference between the total work output of the plant and the total work input (Fig. 7–11):

$$W_{net,out} = W_{out} - W_{in} \qquad (kJ) \tag{7–1}$$

The net work can also be determined from the heat transfer data alone. The four components of the steam power plant involve mass flow in and out, and therefore they should be treated as open systems. These components, together with the connecting pipes, however, always contain the same fluid (not counting the steam that may leak out, of course). No mass enters or leaves this combination system, which is indicated by the shaded area on Fig. 7–10; thus, it can be analyzed as a closed system. Recall that for a closed system undergoing a cycle, the change in internal energy ΔU is zero, and therefore the net work output of the system is also equal to the net heat transfer to the system:

$$W_{net,out} = Q_{in} - Q_{out} \qquad (kJ) \tag{7–2}$$

Thermal Efficiency

In Eq. 7–2, Q_{out} represents the magnitude of the energy wasted in order to complete the cycle. But Q_{out} is never zero; thus, the net work output of a heat engine is always less than the amount of heat input. That is, only part of the heat transferred to the heat engine is converted to work. The fraction of the heat input that is converted to net work output is a measure of the performance of a heat engine and is called the **thermal efficiency** η_{th} (Fig. 7–12).

For heat engines, the desired output is the net work output, and the required input is the amount of heat supplied to the working fluid. Then the thermal efficiency of a heat engine can be expressed as

$$\text{Thermal efficiency} = \frac{\text{Net work output}}{\text{Total heat input}} \tag{7–3}$$

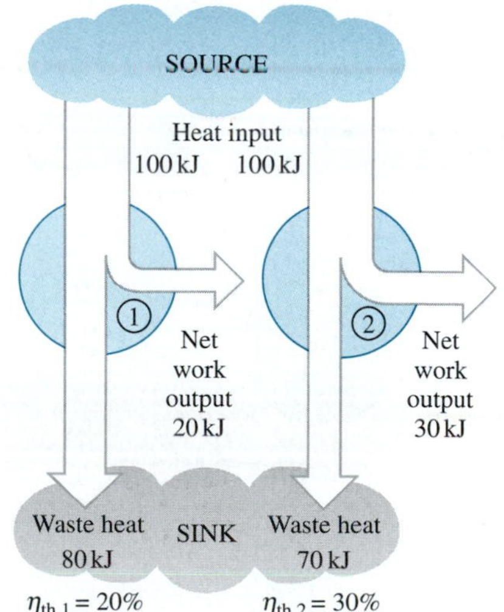

FIGURE 7–12

Some heat engines perform better than others (convert more of the heat they receive to work).

or

$$\eta_{th} = \frac{W_{net,out}}{Q_{in}} \tag{7-4}$$

It can also be expressed as

$$\eta_{th} = 1 - \frac{Q_{out}}{Q_{in}} \tag{7-5}$$

since $W_{net,out} = Q_{in} - Q_{out}$.

Cyclic devices of practical interest such as heat engines, refrigerators, and heat pumps operate between a high-temperature medium (or reservoir) at temperature T_H and a low-temperature medium (or reservoir) at temperature T_L. To bring uniformity to the treatment of heat engines, refrigerators, and heat pumps, we define these two quantities:

Q_H = magnitude of heat transfer between the cyclic device and the high-temperature medium at temperature T_H

Q_L = magnitude of heat transfer between the cyclic device and the low-temperature medium at temperature T_L

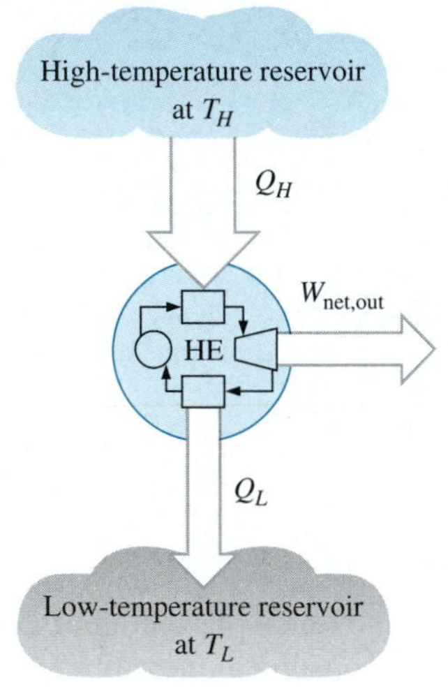

FIGURE 7–13

Schematic of a heat engine.

Notice that both Q_L and Q_H are defined as *magnitudes* and therefore are positive quantities. The direction of Q_H and Q_L is easily determined by inspection. Then the net work output and thermal efficiency relations for any heat engine (shown in Fig. 7–13) can also be expressed as

$$W_{net,out} = Q_H - Q_L$$

and

$$\eta_{th} = \frac{W_{net,out}}{Q_H}$$

or

$$\eta_{th} = 1 - \frac{Q_L}{Q_H} \tag{7-6}$$

The thermal efficiency of a heat engine is always less than unity since both Q_L and Q_H are defined as positive quantities.

Thermal efficiency is a measure of how efficiently a heat engine converts the heat that it receives to work. Heat engines are built for the purpose of converting heat to work, and engineers are constantly trying to improve the efficiencies of these devices since increased efficiency means less fuel consumption and thus lower fuel bills and less pollution.

The thermal efficiencies of work-producing devices are relatively low. Ordinary spark-ignition automobile engines have a thermal efficiency of about 25 percent. That is, an automobile engine converts about 25 percent of the chemical energy of the gasoline to mechanical work. This number is as high as 40 percent for diesel engines and large gas-turbine plants and as high as 60 percent for large combined gas-steam power plants. Thus, even with the most efficient heat engines available today, almost one-half of the energy supplied ends up in the rivers, lakes, or the atmosphere as waste or useless energy (Fig. 7–14).

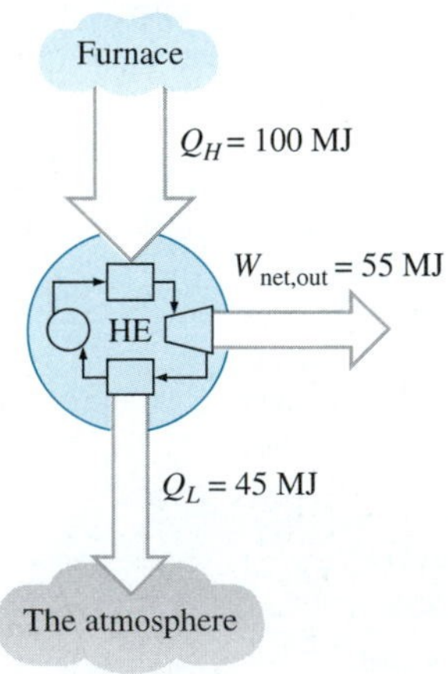

FIGURE 7–14

Even the most efficient heat engines reject almost one-half of the energy they receive as waste heat.

Can We Save Q_{out}?

In a steam power plant, the condenser is the device where large quantities of waste heat is rejected to rivers, lakes, or the atmosphere. Then one may ask, can we not just take the condenser out of the plant and save all that waste energy? The answer to this question is, unfortunately, a firm *no* for the simple reason that without a heat rejection process in a condenser, the cycle cannot be completed. (Cyclic devices such as steam power plants cannot run continuously unless the cycle is completed.) This is demonstrated next with the help of a simple heat engine.

Consider the simple heat engine shown in Fig. 7–15 that is used to lift weights. It consists of a piston–cylinder device with two sets of stops. The working fluid is the gas contained within the cylinder. Initially, the gas temperature is 30°C. The piston, which is loaded with the weights, is resting on top of the lower stops. Now 100 kJ of heat is transferred to the gas in the cylinder from a source at 100°C, causing it to expand and to raise the loaded piston until the piston reaches the upper stops, as shown in the figure. At this point, the load is removed, and the gas temperature is observed to be 90°C.

The work done on the load during this expansion process is equal to the increase in its potential energy, say 15 kJ. Even under ideal conditions (weightless piston, no friction, no heat losses, and quasi-equilibrium expansion), the amount of heat supplied to the gas is greater than the work done since part of the heat supplied is used to raise the temperature of the gas.

Now let us try to answer this question: *Is it possible to transfer the 85 kJ of excess heat at 90°C back to the reservoir at 100°C for later use?* If it is, then we will have a heat engine that can have a thermal efficiency of 100 percent under ideal conditions. The answer to this question is again *no*, for the very simple reason that heat is always transferred from a high-temperature medium to a low-temperature one, and never the other way around. Therefore, we cannot cool this gas from 90 to 30°C by transferring heat to a reservoir at 100°C. Instead, we have to bring the system into contact with a low-temperature reservoir, say at 20°C, so that the gas can return to its initial state by rejecting its 85 kJ of excess energy as heat to this reservoir. This energy cannot be recycled, and it is properly called *waste energy.*

We conclude from this discussion that every heat engine must *waste* some energy by transferring it to a low-temperature reservoir in order to complete

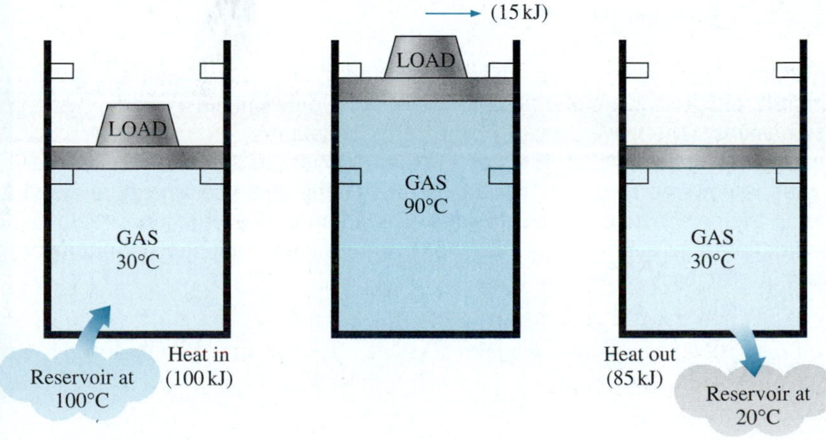

FIGURE 7–15
A heat-engine cycle cannot be completed without rejecting some heat to a low-temperature sink.

the cycle, even under idealized conditions. The requirement that a heat engine exchange heat with at least two reservoirs for continuous operation forms the basis for the Kelvin–Planck expression of the second law of thermodynamics discussed later in this section.

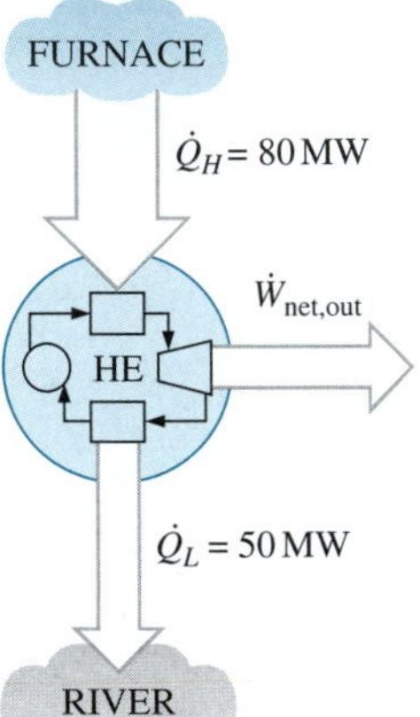

FIGURE 7–16
Schematic for Example 7–1.

EXAMPLE 7–1 Net Power Production of a Heat Engine

Heat is transferred to a heat engine from a furnace at a rate of 80 MW. If the rate of waste heat rejection to a nearby river is 50 MW, determine the net power output and the thermal efficiency for this heat engine.

Solution The rates of heat transfer to and from a heat engine are given. The net power output and the thermal efficiency are to be determined.
Assumptions Heat losses through the pipes and other components are negligible.
Analysis A schematic of the heat engine is given in Fig. 7–16. The furnace serves as the high-temperature reservoir for this heat engine and the river as the low-temperature reservoir. The given quantities can be expressed as

$$\dot{Q}_H = 80 \text{ MW} \quad \text{and} \quad \dot{Q}_L = 50 \text{ MW}$$

The net power output of this heat engine is

$$\dot{W}_{\text{net,out}} = \dot{Q}_H - \dot{Q}_L = (80 - 50) \text{ MW} = \mathbf{30 \text{ MW}}$$

Then the thermal efficiency is easily determined to be

$$\eta_{\text{th}} = \frac{\dot{W}_{\text{net,out}}}{\dot{Q}_H} = \frac{30 \text{ MW}}{80 \text{ MW}} = \mathbf{0.375} \text{ (or 37.5\%)}$$

Discussion Note that the heat engine converts 37.5 percent of the heat it receives to work.

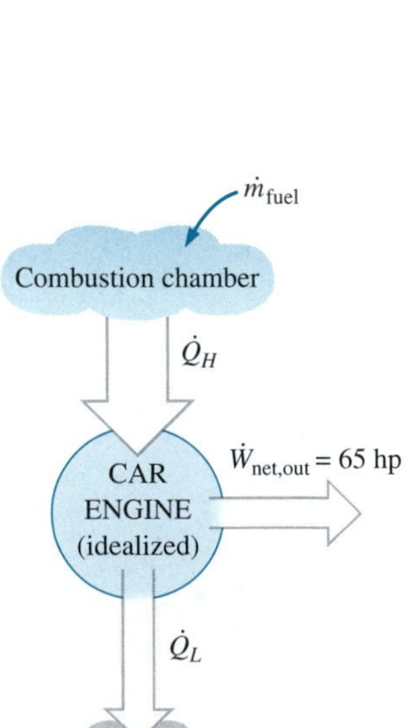

FIGURE 7–17
Schematic for Example 7–2.

EXAMPLE 7–2 Fuel Consumption Rate of a Car

A car engine with a power output of 65 hp has a thermal efficiency of 24 percent. Determine the fuel consumption rate of this car if the fuel has a heating value of 19,000 Btu/lbm (that is, 19,000 Btu of energy is released for each lbm of fuel burned).

Solution The power output and the efficiency of a car engine are given. The rate of fuel consumption of the car is to be determined.
Assumptions The power output of the car is constant.
Analysis A schematic of the car engine is given in Fig. 7–17. The car engine is powered by converting 24 percent of the chemical energy released during the combustion process to work. The amount of energy input required to produce a power output of 65 hp is determined from the definition of thermal efficiency to be

$$\dot{Q}_H = \frac{\dot{W}_{\text{net,out}}}{\eta_{\text{th}}} = \frac{65 \text{ hp}}{0.24}\left(\frac{2545 \text{ Btu/h}}{1 \text{ hp}}\right) = 689{,}270 \text{ Btu/h}$$

To supply energy at this rate, the engine must burn fuel at a rate of

$$\dot{m} = \frac{689{,}270 \text{ Btu/h}}{19{,}000 \text{ Btu/lbm}} = \mathbf{36.3 \text{ lbm/h}}$$

since 19,000 Btu of thermal energy is released for each lbm of fuel burned.

Discussion Note that if the thermal efficiency of the car could be doubled, the rate of fuel consumption would be reduced by half.

The Second Law of Thermodynamics: Kelvin–Planck Statement

We have demonstrated earlier with reference to the heat engine shown in Fig. 7–15 that, even under ideal conditions, a heat engine must reject some heat to a low-temperature reservoir in order to complete the cycle. That is, no heat engine can convert all the heat it receives to useful work. This limitation on the thermal efficiency of heat engines forms the basis for the Kelvin–Planck statement of the second law of thermodynamics, which is expressed as follows:

> It is impossible for any device that operates on a cycle to receive heat from a single reservoir and produce a net amount of work.

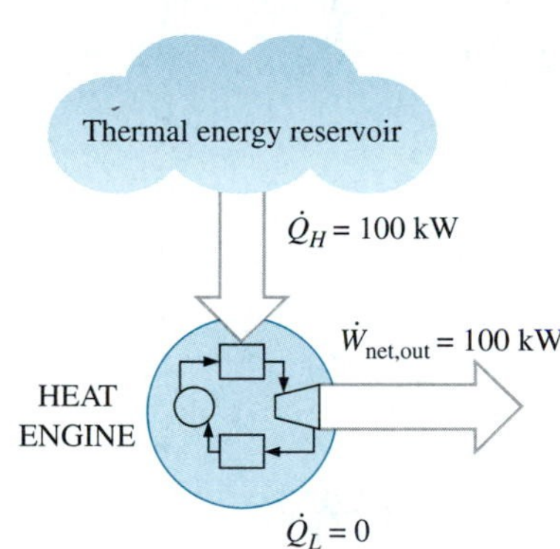

FIGURE 7–18

A heat engine that violates the Kelvin–Planck statement of the second law.

That is, a heat engine must exchange heat with a low-temperature sink as well as a high-temperature source to keep operating. The Kelvin–Planck statement can also be expressed as *no heat engine can have a thermal efficiency of 100 percent* (Fig. 7–18), or as *for a power plant to operate, the working fluid must exchange heat with the environment as well as the furnace.*

Note that the impossibility of having a 100 percent efficient heat engine is not due to friction or other dissipative effects. It is a limitation that applies to both the idealized and the actual heat engines. Later in this chapter, we develop a relation for the maximum thermal efficiency of a heat engine. We also demonstrate that this maximum value depends on the reservoir temperatures only.

7–4 ▪ REFRIGERATORS AND HEAT PUMPS

We all know from experience that heat is transferred in the direction of decreasing temperature, that is, from high-temperature mediums to low-temperature ones. This heat transfer process occurs in nature without requiring any devices. The reverse process, however, cannot occur by itself. The transfer of heat from a low-temperature medium to a high-temperature one requires special devices called **refrigerators**.

Refrigerators, like heat engines, are cyclic devices. The working fluid used in the refrigeration cycle is called a **refrigerant**. The most frequently used refrigeration cycle is the *vapor-compression refrigeration cycle*, which involves four main components: a compressor, a condenser, an expansion valve, and an evaporator, as shown in Fig. 7–19.

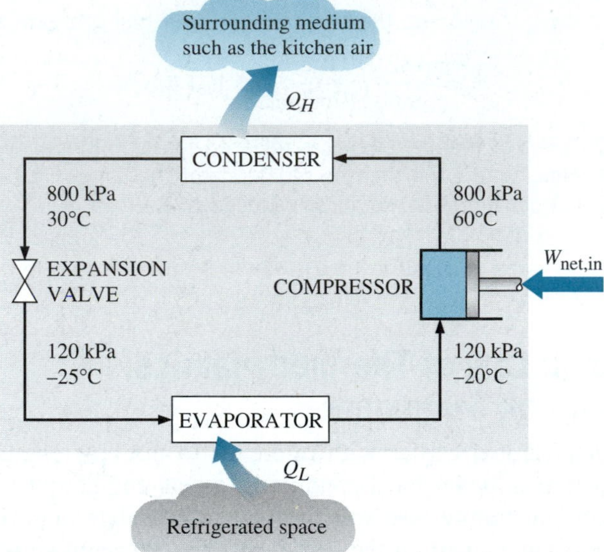

FIGURE 7–19

Basic components of a refrigeration system and typical operating conditions.

The refrigerant enters the compressor as a vapor and is compressed to the condenser pressure. It leaves the compressor at a relatively high temperature and cools down and condenses as it flows through the coils of the condenser by rejecting heat to the surrounding medium. It then enters a capillary tube where its pressure and temperature drop drastically due to the throttling effect. The low-temperature refrigerant then enters the evaporator, where it evaporates by absorbing heat from the refrigerated space. The cycle is completed as the refrigerant leaves the evaporator and reenters the compressor.

In a household refrigerator, the freezer compartment where heat is absorbed by the refrigerant serves as the evaporator, and the coils usually behind the refrigerator where heat is dissipated to the kitchen air serve as the condenser.

A refrigerator is shown schematically in Fig. 7–20. Here Q_L is the magnitude of the heat removed from the refrigerated space at temperature T_L, Q_H is the magnitude of the heat rejected to the warm environment at temperature T_H, and $W_{net,in}$ is the net work input to the refrigerator. As discussed before, Q_L and Q_H represent magnitudes and thus are positive quantities.

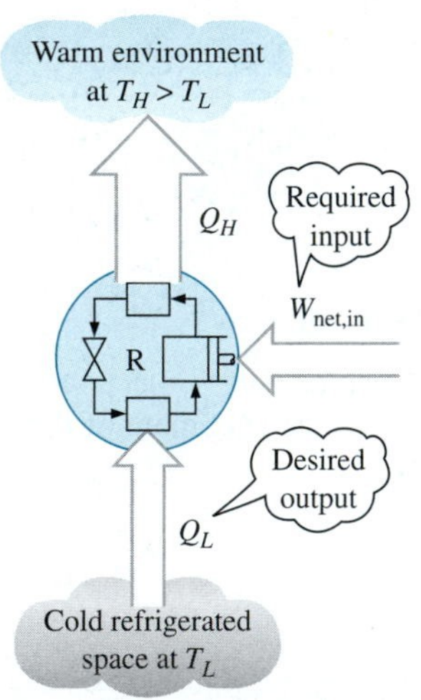

FIGURE 7–20

The objective of a refrigerator is to remove Q_L from the cooled space.

Coefficient of Performance

The *efficiency* of a refrigerator is expressed in terms of the **coefficient of performance** (COP), denoted by COP_R. The objective of a refrigerator is to remove heat (Q_L) from the refrigerated space. To accomplish this objective, it requires a work input of $W_{net,in}$. Then the COP of a refrigerator can be expressed as

$$COP_R = \frac{\text{Desired output}}{\text{Required input}} = \frac{Q_L}{W_{net,in}} \quad (7\text{–}7)$$

This relation can also be expressed in rate form by replacing Q_L by $\dot{Q}_L$ and $W_{net,in}$ by $\dot{W}_{net,in}$.

The conservation of energy principle for a cyclic device requires that

$$W_{net,in} = Q_H - Q_L \quad \text{(kJ)} \quad (7\text{–}8)$$

Then the COP relation becomes

$$\mathrm{COP_R} = \frac{Q_L}{Q_H - Q_L} = \frac{1}{Q_H/Q_L - 1} \tag{7-9}$$

Notice that the value of $\mathrm{COP_R}$ can be *greater than unity*. That is, the amount of heat removed from the refrigerated space can be greater than the amount of work input. This is in contrast to the thermal efficiency, which can never be greater than 1. In fact, one reason for expressing the efficiency of a refrigerator by another term—the coefficient of performance—is the desire to avoid the oddity of having efficiencies greater than unity.

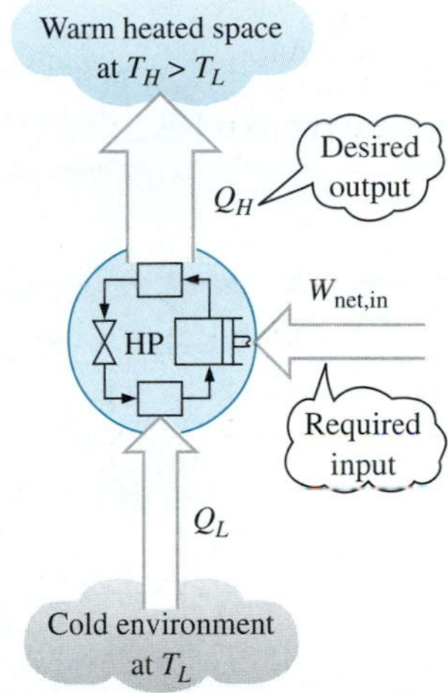

FIGURE 7–21

The objective of a heat pump is to supply heat Q_H into the warmer space.

Heat Pumps

Another device that transfers heat from a low-temperature medium to a high-temperature one is the **heat pump**, shown schematically in Fig. 7–21. Refrigerators and heat pumps operate on the same cycle but differ in their objectives. The objective of a refrigerator is to maintain the refrigerated space at a low temperature by removing heat from it. Discharging this heat to a higher-temperature medium is merely a necessary part of the operation, not the purpose. The objective of a heat pump, however, is to maintain a heated space at a high temperature. This is accomplished by absorbing heat from a low-temperature source, such as well water or cold outside air in winter, and supplying this heat to the high-temperature medium such as a house (Fig. 7–22).

An ordinary refrigerator that is placed in the window of a house with its door open to the cold outside air in winter will function as a heat pump since it will try to cool the outside by absorbing heat from it and rejecting this heat into the house through the coils behind it (Fig. 7–23).

The measure of performance of a heat pump is also expressed in terms of the **coefficient of performance** $\mathrm{COP_{HP}}$, defined as

$$\mathrm{COP_{HP}} = \frac{\text{Desired output}}{\text{Required input}} = \frac{Q_H}{W_{\mathrm{net,in}}} \tag{7-10}$$

which can also be expressed as

$$\mathrm{COP_{HP}} = \frac{Q_H}{Q_H - Q_L} = \frac{1}{1 - Q_L/Q_H} \tag{7-11}$$

A comparison of Eqs. 7–7 and 7–10 reveals that

$$\mathrm{COP_{HP}} = \mathrm{COP_R} + 1 \tag{7-12}$$

for fixed values of Q_L and Q_H. This relation implies that the coefficient of performance of a heat pump is always greater than unity since $\mathrm{COP_R}$ is a positive quantity. That is, a heat pump will function, at worst, as a resistance heater, supplying as much energy to the house as it consumes. In reality, however, part of Q_H is lost to the outside air through piping and other devices, and $\mathrm{COP_{HP}}$ may drop below unity when the outside air temperature is too low. When this happens, the system usually switches to a resistance heating mode. Most heat pumps in operation today have a seasonally averaged COP of 2 to 3.

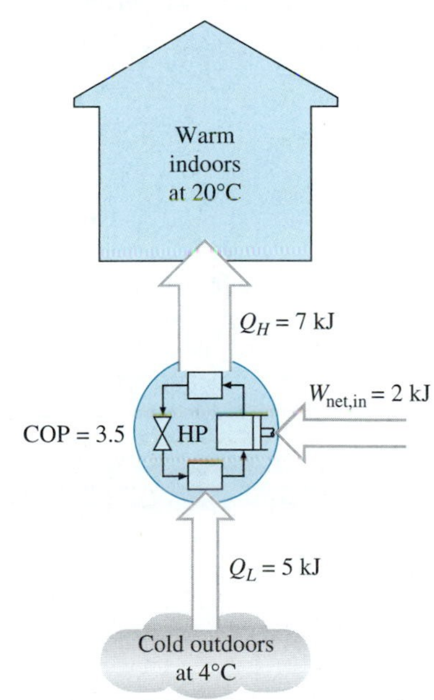

FIGURE 7–22

The work supplied to a heat pump is used to extract energy from the cold outdoors and carry it into the warm indoors.

FIGURE 7–23

When installed backward, an air conditioner functions as a heat pump.

© Reprinted with special permission of King Features Syndicate.

Most existing heat pumps use the cold outside air as the heat source in winter, and they are referred to as *air-source heat pumps.* The COP of such heat pumps is about 3.0 at design conditions. Air-source heat pumps are not appropriate for cold climates since their efficiency drops considerably when temperatures are below the freezing point. In such cases, geothermal (also called ground-source) heat pumps that use the ground as the heat source can be used. Geothermal heat pumps require the burial of pipes in the ground 1 to 2 m deep. Such heat pumps are more expensive to install, but they are also more efficient (up to 45 percent more efficient than air-source heat pumps). The COP of ground-source heat pumps is about 4.0.

Air conditioners are basically refrigerators whose refrigerated space is a room or a building instead of the food compartment. A window air-conditioning unit cools a room by absorbing heat from the room air and discharging it to the outside. The same air-conditioning unit can be used as a heat pump in winter by installing it backwards as shown in Fig. 7–23. In this mode, the unit absorbs heat from the cold outside and delivers it to the room. Air-conditioning systems that are equipped with proper controls and a reversing valve operate as air conditioners in summer and as heat pumps in winter.

The performance of refrigerators and air conditioners in the United States is often expressed in terms of the **energy efficiency rating** (EER), which is the amount of heat removed from the cooled space in Btu's for 1 Wh (watt-hour) of electricity consumed. Considering that 1 kWh = 3412 Btu and thus 1 Wh = 3.412 Btu, a unit that removes 1 kWh of heat from the cooled space for each kWh of electricity it consumes (COP = 1) will have an EER of 3.412. Therefore, the relation between EER and COP is

$$\text{EER} = 3.412\ \text{COP}_R$$

Most air conditioners have an EER between 8 and 12 (a COP of 2.3 to 3.5). A high-efficiency heat pump manufactured by the Trane Company using a reciprocating variable-speed compressor is reported to have a COP of 3.3 in the heating mode and an EER of 16.9 (COP of 5.0) in the air-conditioning mode. Variable-speed compressors and fans allow the unit to operate at maximum efficiency for varying heating/cooling needs and weather conditions as determined by a microprocessor. In the air-conditioning mode, for example, they operate at higher speeds on hot days and at lower speeds on cooler days, enhancing both efficiency and comfort.

The EER or COP of a refrigerator decreases with decreasing refrigeration temperature. Therefore, it is not economical to refrigerate to a lower temperature than needed. The COPs of refrigerators are in the range of 2.7–3.0 for cutting and preparation rooms; 2.3–2.6 for meat, deli, dairy, and produce; 1.2–1.5 for frozen foods; and 1.0–1.2 for ice cream units. Note that the COP of freezers is about half of the COP of meat refrigerators, and thus it costs twice as much to cool the meat products with refrigerated air that is cold enough to cool frozen foods. It is good energy conservation practice to use separate refrigeration systems to meet different refrigeration needs.

EXAMPLE 7–3 Heat Rejection by a Refrigerator

The food compartment of a refrigerator, shown in Fig. 7–24, is maintained at 4°C by removing heat from it at a rate of 360 kJ/min. If the required power input to the refrigerator is 2 kW, determine (*a*) the coefficient of performance of the refrigerator and (*b*) the rate of heat rejection to the room that houses the refrigerator.

Solution The power consumption of a refrigerator is given. The COP and the rate of heat rejection are to be determined.
Assumptions Steady operating conditions exist.
Analysis (*a*) The coefficient of performance of the refrigerator is

$$\text{COP}_\text{R} = \frac{\dot{Q}_L}{\dot{W}_{\text{net,in}}} = \frac{360 \text{ kJ/min}}{2 \text{ kW}}\left(\frac{1 \text{ kW}}{60 \text{ kJ/min}}\right) = \mathbf{3}$$

That is, 3 kJ of heat is removed from the refrigerated space for each kJ of work supplied.
(*b*) The rate at which heat is rejected to the room that houses the refrigerator is determined from the conservation of energy relation for cyclic devices,

$$\dot{Q}_H = \dot{Q}_L + \dot{W}_{\text{net,in}} = 360 \text{ kJ/min} + (2 \text{ kW})\left(\frac{60 \text{ kJ/min}}{1 \text{ kW}}\right) = \mathbf{480 \text{ kJ/min}}$$

Discussion Notice that both the energy removed from the refrigerated space as heat and the energy supplied to the refrigerator as electrical work eventually show up in the room air and become part of the internal energy of the air. This demonstrates that energy can change from one form to another, can move from one place to another, but is never destroyed during a process.

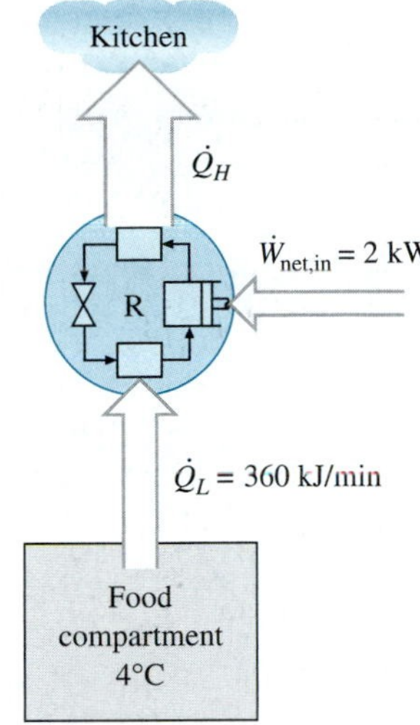

FIGURE 7–24
Schematic for Example 7–3.

EXAMPLE 7–4 Heating a House by a Heat Pump

A heat pump is used to meet the heating requirements of a house and maintain it at 20°C. On a day when the outdoor air temperature drops to −2°C, the house is estimated to lose heat at a rate of 80,000 kJ/h. If the heat pump under these conditions has a COP of 2.5, determine (*a*) the power consumed by the heat pump and (*b*) the rate at which heat is absorbed from the cold outdoor air.

Solution The COP of a heat pump is given. The power consumption and the rate of heat absorption are to be determined.
Assumptions Steady operating conditions exist.
Analysis (*a*) The power consumed by this heat pump, shown in Fig. 7–25, is determined from the definition of the coefficient of performance to be

$$\dot{W}_{\text{net,in}} = \frac{\dot{Q}_H}{\text{COP}_\text{HP}} = \frac{80{,}000 \text{ kJ/h}}{2.5} = \mathbf{32{,}000 \text{ kJ/h}} \text{ (or 8.9 kW)}$$

(*b*) The house is losing heat at a rate of 80,000 kJ/h. If the house is to be maintained at a constant temperature of 20°C, the heat pump must deliver

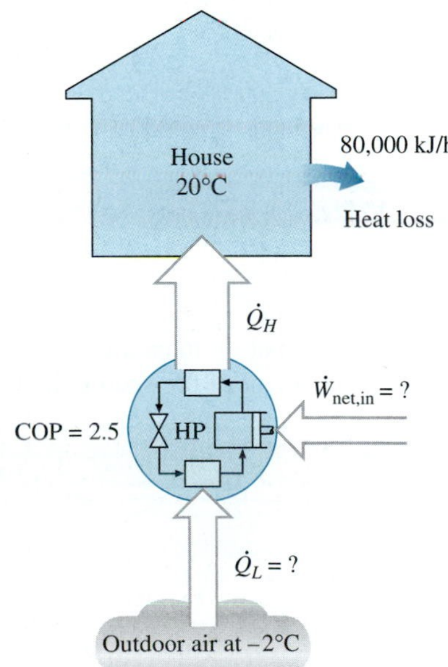

FIGURE 7–25
Schematic for Example 7–4.

heat to the house at the same rate, that is, at a rate of 80,000 kJ/h. Then the rate of heat transfer from the outdoor becomes

$$\dot{Q}_L = \dot{Q}_H - \dot{W}_{\text{net,in}} = (80{,}000 - 32{,}000)\ \text{kJ/h} = \mathbf{48{,}000\ kJ/h}$$

Discussion Note that 48,000 of the 80,000 kJ/h heat delivered to the house is actually extracted from the cold outdoor air. Therefore, we are paying only for the 32,000-kJ/h energy that is supplied as electrical work to the heat pump. If we were to use an electric resistance heater instead, we would have to supply the entire 80,000 kJ/h to the resistance heater as electric energy. This would mean a heating bill that is 2.5 times higher. This explains the popularity of heat pumps as heating systems and why they are preferred to simple electric resistance heaters despite their considerably higher initial cost.

The Second Law of Thermodynamics: Clausius Statement

There are two classical statements of the second law—the Kelvin–Planck statement, which is related to heat engines and discussed in the preceding section, and the Clausius statement, which is related to refrigerators or heat pumps. The Clausius statement is expressed as follows:

> It is impossible to construct a device that operates in a cycle and produces no effect other than the transfer of heat from a lower-temperature body to a higher-temperature body.

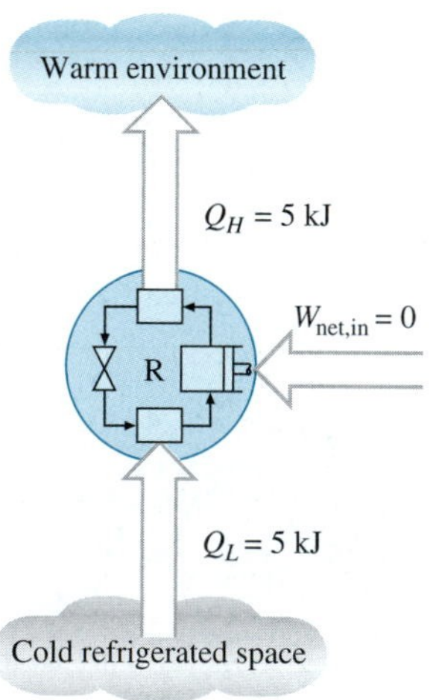

FIGURE 7–26

A refrigerator that violates the Clausius statement of the second law.

It is common knowledge that heat does not, of its own volition, transfer from a cold medium to a warmer one. The Clausius statement does not imply that a cyclic device that transfers heat from a cold medium to a warmer one is impossible to construct. In fact, this is precisely what a common household refrigerator does. It simply states that a refrigerator cannot operate unless its compressor is driven by an external power source, such as an electric motor (Fig. 7–26). This way, the net effect on the surroundings involves the consumption of some energy in the form of work, in addition to the transfer of heat from a colder body to a warmer one. That is, it leaves a trace in the surroundings. Therefore, a household refrigerator is in complete compliance with the Clausius statement of the second law.

Both the Kelvin–Planck and the Clausius statements of the second law are negative statements, and a negative statement cannot be proved. Like any other physical law, the second law of thermodynamics is based on experimental observations. To date, no experiment has been conducted that contradicts the second law, and this should be taken as sufficient proof of its validity.

Equivalence of the Two Statements

The Kelvin–Planck and the Clausius statements are equivalent in their consequences, and either statement can be used as the expression of the second law of thermodynamics. Any device that violates the Kelvin–Planck statement also violates the Clausius statement, and vice versa. This can be demonstrated as follows.

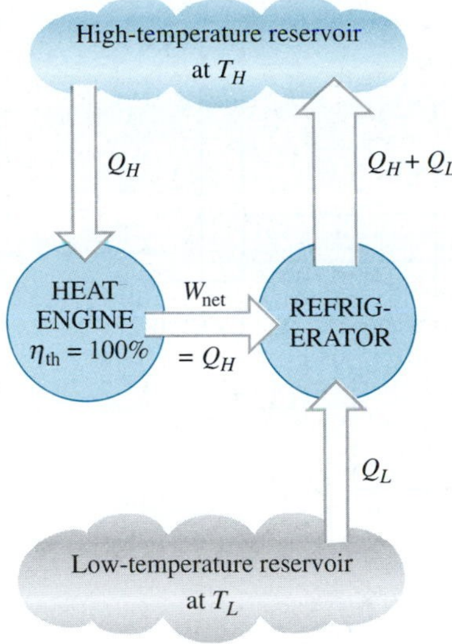

(*a*) A refrigerator that is powered by a 100 percent efficient heat engine

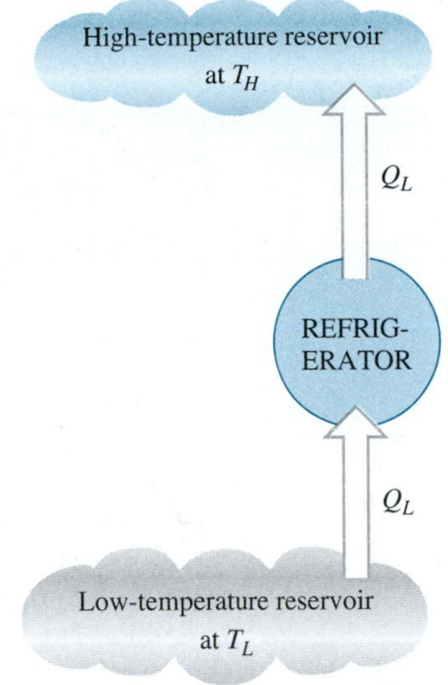

(*b*) The equivalent refrigerator

FIGURE 7–27
Proof that the violation of the Kelvin–Planck statement leads to the violation of the Clausius statement.

Consider the heat-engine-refrigerator combination shown in Fig. 7–27*a*, operating between the same two reservoirs. The heat engine is assumed to have, in violation of the Kelvin–Planck statement, a thermal efficiency of 100 percent, and therefore it converts all the heat Q_H it receives to work W. This work is now supplied to a refrigerator that removes heat in the amount of Q_L from the low-temperature reservoir and rejects heat in the amount of $Q_L + Q_H$ to the high-temperature reservoir. During this process, the high-temperature reservoir receives a net amount of heat Q_L (the difference between $Q_L + Q_H$ and Q_H). Thus, the combination of these two devices can be viewed as a refrigerator, as shown in Fig. 7–27*b*, that transfers heat in an amount of Q_L from a cooler body to a warmer one without requiring any input from outside. This is clearly a violation of the Clausius statement. Therefore, a violation of the Kelvin–Planck statement results in the violation of the Clausius statement.

It can also be shown in a similar manner that a violation of the Clausius statement leads to the violation of the Kelvin–Planck statement. Therefore, the Clausius and the Kelvin–Planck statements are two equivalent expressions of the second law of thermodynamics.

7–5 ▪ PERPETUAL-MOTION MACHINES

We have repeatedly stated that a process cannot take place unless it satisfies both the first and second laws of thermodynamics. Any device that violates either law is called a **perpetual-motion machine**, and despite numerous attempts, no perpetual-motion machine is known to have worked. But this has not stopped inventors from trying to create new ones.

A device that violates the first law of thermodynamics (by *creating* energy) is called a **perpetual-motion machine of the first kind** (PMM1), and a device that violates the second law of thermodynamics is called a **perpetual-motion machine of the second kind** (PMM2).

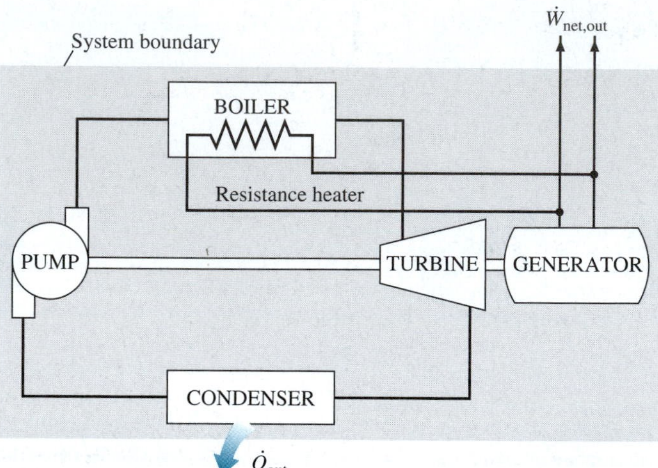

FIGURE 7–28

A perpetual-motion machine that violates the first law of thermodynamics (PMM1).

Consider the steam power plant shown in Fig. 7–28. It is proposed to heat the steam by resistance heaters placed inside the boiler, instead of by the energy supplied from fossil or nuclear fuels. Part of the electricity generated by the plant is to be used to power the resistors as well as the pump. The rest of the electric energy is to be supplied to the electric network as the net work output. The inventor claims that once the system is started, this power plant will produce electricity indefinitely without requiring any energy input from the outside.

Well, here is an invention that could solve the world's energy problem—if it works, of course. A careful examination of this invention reveals that the system enclosed by the shaded area is continuously supplying energy to the outside at a rate of $\dot{Q}_{out} + \dot{W}_{net,out}$ without receiving any energy. That is, this system is creating energy at a rate of $\dot{Q}_{out} + \dot{W}_{net,out}$, which is clearly a violation of the first law. Therefore, this wonderful device is nothing more than a PMM1 and does not warrant any further consideration.

Now let us consider another novel idea by the same inventor. Convinced that energy cannot be created, the inventor suggests the following modification that will greatly improve the thermal efficiency of that power plant without violating the first law. Aware that more than one-half of the heat transferred to the steam in the furnace is discarded in the condenser to the environment, the inventor suggests getting rid of this wasteful component and sending the steam to the pump as soon as it leaves the turbine, as shown in Fig. 7–29. This way, all the heat transferred to the steam in the boiler will be converted to work, and thus the power plant will have a theoretical efficiency of 100 percent. The inventor realizes that some heat losses and friction between the moving components are unavoidable and that these effects will hurt the efficiency somewhat, but still expects the efficiency to be no less than 80 percent (as opposed to 40 percent in most actual power plants) for a carefully designed system.

Well, the possibility of doubling the efficiency would certainly be very tempting to plant managers and, if not properly trained, they would probably give this idea a chance, since intuitively they see nothing wrong with it. A student of thermodynamics, however, will immediately label this

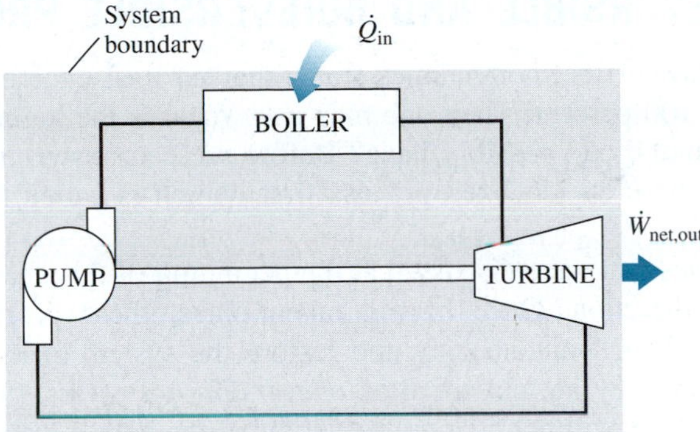

FIGURE 7–29

A perpetual-motion machine that violates the second law of thermodynamics (PMM2).

device as a PMM2, since it works on a cycle and does a net amount of work while exchanging heat with a single reservoir (the furnace) only. It satisfies the first law but violates the second law, and therefore it will not work.

Countless perpetual-motion machines have been proposed throughout history, and many more are being proposed. Some proposers have even gone so far as to patent their inventions, only to find out that what they actually have in their hands is a worthless piece of paper.

Some perpetual-motion machine inventors were very successful in fund-raising. For example, a Philadelphia carpenter named J. W. Kelly collected millions of dollars between 1874 and 1898 from investors in his *hydropneumatic-pulsating-vacu-engine*, which supposedly could push a railroad train 3000 miles on 1 L of water. Of course, it never did. After his death in 1898, the investigators discovered that the demonstration machine was powered by a hidden motor. Recently a group of investors was set to invest $2.5 million into a mysterious *energy augmentor*, which multiplied whatever power it took in, but their lawyer wanted an expert opinion first. Confronted by the scientists, the "inventor" fled the scene without even attempting to run his demo machine.

Tired of applications for perpetual-motion machines, the U.S. Patent Office decreed in 1918 that it would no longer consider any perpetual-motion machine applications. However, several such patent applications were still filed, and some made it through the patent office undetected. Some applicants whose patent applications were denied sought legal action. For example, in 1982 the U.S. Patent Office dismissed as just another perpetual-motion machine a huge device that involves several hundred kilograms of rotating magnets and kilometers of copper wire that is supposed to be generating more electricity than it is consuming from a battery pack. However, the inventor challenged the decision, and in 1985 the National Bureau of Standards finally tested the machine just to certify that it is battery-operated. However, it did not convince the inventor that his machine will not work.

The proposers of perpetual-motion machines generally have innovative minds, but they usually lack formal engineering training, which is very unfortunate. No one is immune from being deceived by an innovative perpetual-motion machine. As the saying goes, however, if something sounds too good to be true, it probably is.

7–6 ▪ REVERSIBLE AND IRREVERSIBLE PROCESSES

The second law of thermodynamics states that no heat engine can have an efficiency of 100 percent. Then one may ask, What is the highest efficiency that a heat engine can possibly have? Before we can answer this question, we need to define an idealized process first, which is called the *reversible process*.

The processes that were discussed at the beginning of this chapter occurred in a certain direction. Once having taken place, these processes cannot reverse themselves spontaneously and restore the system to its initial state. For this reason, they are classified as *irreversible processes*. Once a cup of hot coffee cools, it will not heat up by retrieving the heat it lost from the surroundings. If it could, the surroundings, as well as the system (coffee), would be restored to their original condition, and this would be a reversible process.

(*a*) Frictionless pendulum

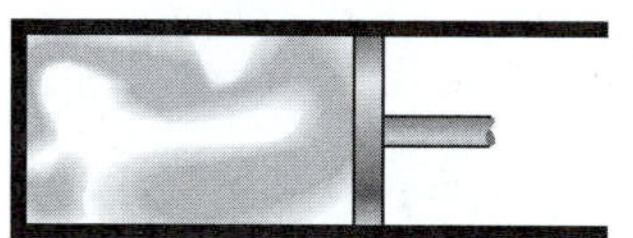

(*b*) Quasi-equilibrium expansion and compression of a gas

FIGURE 7–30

Two familiar reversible processes.

A **reversible process** is defined as a *process that can be reversed without leaving any trace on the surroundings* (Fig. 7–30). That is, both the system *and* the surroundings are returned to their initial states at the end of the reverse process. This is possible only if the net heat *and* net work exchange between the system and the surroundings is zero for the combined (original and reverse) process. Processes that are not reversible are called **irreversible processes**.

It should be pointed out that a system can be restored to its initial state following a process, regardless of whether the process is reversible or irreversible. But for reversible processes, this restoration is made without leaving any net change on the surroundings, whereas for irreversible processes, the surroundings usually do some work on the system and therefore does not return to their original state.

Reversible processes actually do not occur in nature. They are merely *idealizations* of actual processes. Reversible processes can be approximated by actual devices, but they can never be achieved. That is, all the processes occurring in nature are irreversible. You may be wondering, then, *why* we are bothering with such fictitious processes. There are two reasons. First, they are easy to analyze, since a system passes through a series of equilibrium states during a reversible process; second, they serve as idealized models to which actual processes can be compared.

In daily life, the concepts of Mr. Right and Ms. Right are also idealizations, just like the concept of a reversible (perfect) process. People who insist on finding Mr. or Ms. Right to settle down are bound to remain Mr. or Ms. Single for the rest of their lives. The possibility of finding the perfect prospective mate is no higher than the possibility of finding a perfect (reversible) process. Likewise, a person who insists on perfection in friends is bound to have no friends.

Engineers are interested in reversible processes because work-producing devices such as car engines and gas or steam turbines *deliver the most work*, and work-consuming devices such as compressors, fans, and pumps *consume the least work* when reversible processes are used instead of irreversible ones (Fig. 7–31).

Reversible processes can be viewed as *theoretical limits* for the corresponding irreversible ones. Some processes are more irreversible than others. We may never be able to have a reversible process, but we can certainly

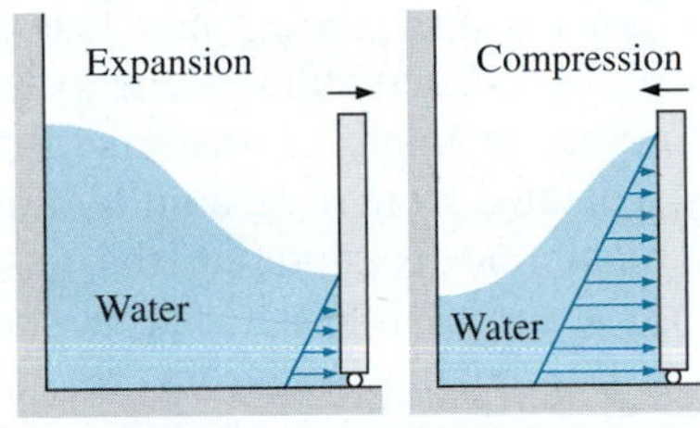

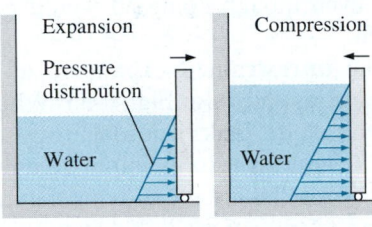

(a) Slow (reversible) process

(b) Fast (irreversible) process

FIGURE 7–31

Reversible processes deliver the most and consume the least work.

approach it. The more closely we approximate a reversible process, the more work delivered by a work-producing device or the less work required by a work-consuming device.

The concept of reversible processes leads to the definition of the **second-law efficiency** for actual processes, which is the degree of approximation to the corresponding reversible processes. This enables us to compare the performance of different devices that are designed to do the same task on the basis of their efficiencies. The better the design, the lower the irreversibilities and the higher the second-law efficiency.

Irreversibilities

The factors that cause a process to be irreversible are called **irreversibilities**. They include friction, unrestrained expansion, mixing of two fluids, heat transfer across a finite temperature difference, electric resistance, inelastic deformation of solids, and chemical reactions. The presence of any of these effects renders a process irreversible. A reversible process involves none of these. Some of the frequently encountered irreversibilities are discussed briefly below.

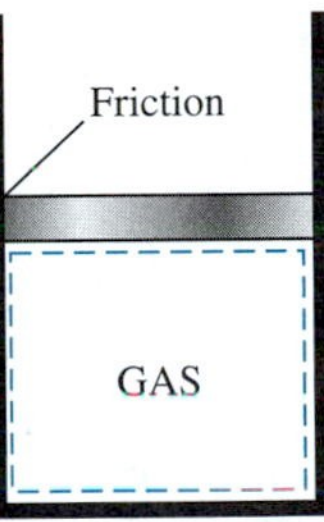

FIGURE 7–32

Friction renders a process irreversible.

Friction is a familiar form of irreversibility associated with bodies in motion. When two bodies in contact are forced to move relative to each other (a piston in a cylinder, for example, as shown in Fig. 7–32), a friction force that opposes the motion develops at the interface of these two bodies, and some work is needed to overcome this friction force. The energy supplied as work is eventually converted to heat during the process and is transferred to the bodies in contact, as evidenced by a temperature rise at the interface. When the direction of the motion is reversed, the bodies are restored to their original position, but the interface does not cool, and heat is not converted back to work. Instead, more of the work is converted to heat while overcoming the friction forces that also oppose the reverse motion. Since the system (the moving bodies) and the surroundings cannot be returned to their original states, this process is irreversible. Therefore, any process that involves friction is irreversible. The larger the friction forces involved, the more irreversible the process is.

Friction does not always involve two solid bodies in contact. It is also encountered between a fluid and solid and even between the layers of a fluid moving at different velocities. A considerable fraction of the power produced by a car engine is used to overcome the friction (the drag force) between the air and the external surfaces of the car, and it eventually becomes part of the internal energy of the air. It is not possible to reverse this process and recover that lost power, even though doing so would not violate the conservation of energy principle.

Another example of irreversibility is the **unrestrained expansion of a gas** separated from a vacuum by a membrane, as shown in Fig. 7–33. When the membrane is ruptured, the gas fills the entire tank. The only way to restore the system to its original state is to compress it to its initial volume, while transferring heat from the gas until it reaches its initial temperature. From the conservation of energy considerations, it can easily be shown that the amount of heat transferred from the gas equals the amount of work done on the gas by the surroundings. The restoration of the surroundings involves conversion of this heat completely to work, which would violate the second law. Therefore, unrestrained expansion of a gas is an irreversible process.

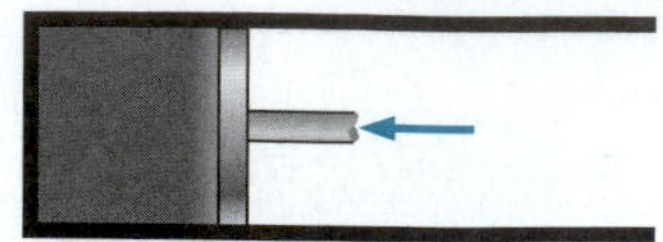

(*a*) Fast compression

(*b*) Fast expansion

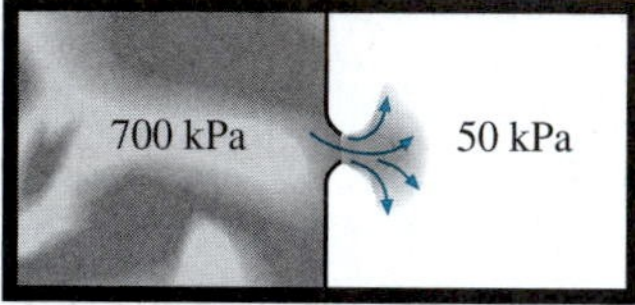

(*c*) Unrestrained expansion

FIGURE 7–33

Irreversible compression and expansion processes.

A third form of irreversibility familiar to us all is **heat transfer** through a finite temperature difference. Consider a can of cold soda left in a warm room (Fig. 7–34). Heat is transferred from the warmer room air to the cooler soda. The only way this process can be reversed and the soda restored to its original temperature is to provide refrigeration, which requires some work input. At the end of the reverse process, the soda will be restored to its initial state, but the surroundings will not be. The internal energy of the surroundings will increase by an amount equal in magnitude to the work supplied to the refrigerator. The restoration of the surroundings to the initial state can be done only by converting this excess internal energy completely to work, which is impossible to do without violating the second law. Since only the system, not both the system and the surroundings, can be restored to its initial condition, heat transfer through a finite temperature difference is an irreversible process.

Heat transfer can occur only when there is a temperature difference between a system and its surroundings. Therefore, it is physically impossible to have a reversible heat transfer process. But a heat transfer process becomes less and less irreversible as the temperature difference between the two bodies approaches zero. Then heat transfer through a differential temperature difference dT can be considered to be reversible. As dT approaches zero, the process can be reversed in direction (at least theoretically) without requiring any refrigeration. Notice that reversible heat transfer is a conceptual process and cannot be duplicated in the real world.

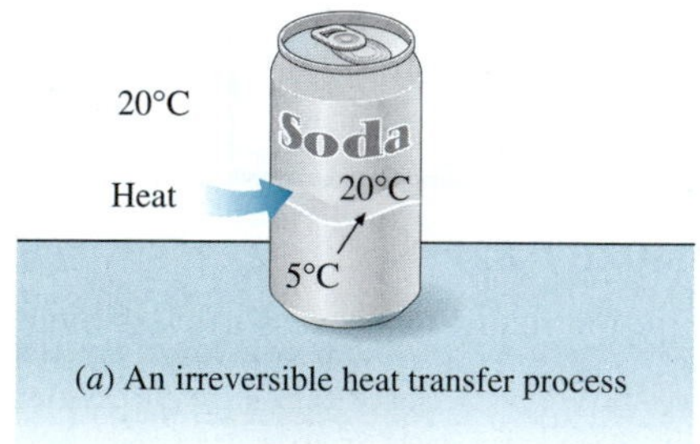

(*a*) An irreversible heat transfer process

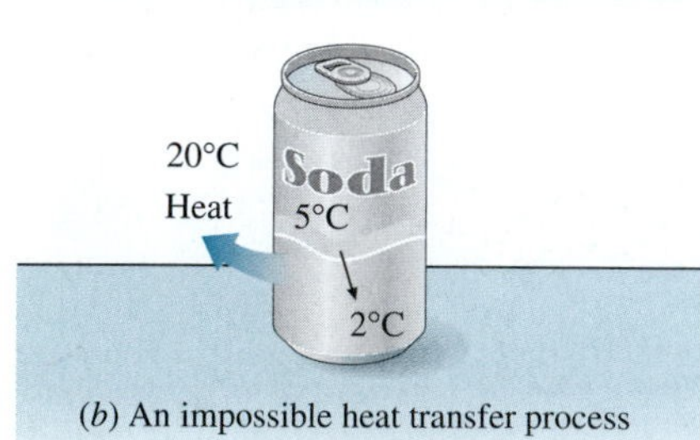

(*b*) An impossible heat transfer process

FIGURE 7–34

(*a*) Heat transfer through a temperature difference is irreversible, and (*b*) the reverse process is impossible.

The smaller the temperature difference between two bodies, the smaller the heat transfer rate will be. Any significant heat transfer through a small temperature difference requires a very large surface area and a very long time. Therefore, even though approaching reversible heat transfer is desirable from a thermodynamic point of view, it is impractical and not economically feasible.

Internally and Externally Reversible Processes

A typical process involves interactions between a system and its surroundings, and a reversible process involves no irreversibilities associated with either of them.

A process is called **internally reversible** if no irreversibilities occur within the boundaries of the system during the process. During an internally reversible process, a system proceeds through a series of equilibrium states, and when the process is reversed, the system passes through exactly the same equilibrium states while returning to its initial state. That is, the paths of the forward and reverse processes coincide for an internally reversible process. The quasi-equilibrium process is an example of an internally reversible process.

A process is called **externally reversible** if no irreversibilities occur outside the system boundaries during the process. Heat transfer between a reservoir and a system is an externally reversible process if the outer surface of the system is at the temperature of the reservoir.

A process is called **totally reversible**, or simply **reversible**, if it involves no irreversibilities within the system or its surroundings (Fig. 7–35). A totally reversible process involves no heat transfer through a finite temperature difference, no nonquasi-equilibrium changes, and no friction or other dissipative effects.

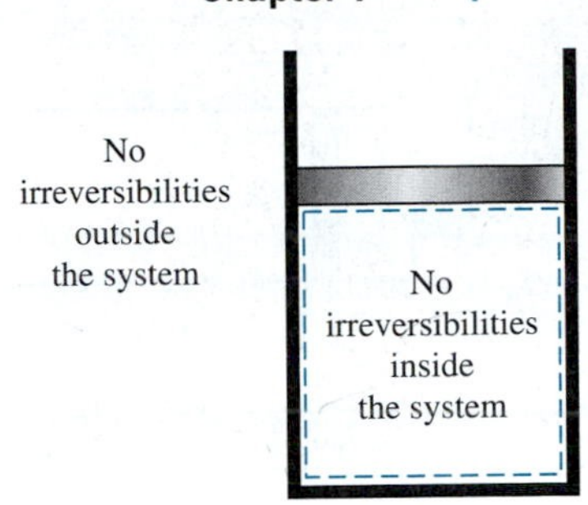

FIGURE 7–35

A reversible process involves no internal and external irreversibilities.

As an example, consider the transfer of heat to two identical systems that are undergoing a constant-pressure (thus constant-temperature) phase-change process, as shown in Fig. 7–36. Both processes are internally reversible, since both take place isothermally and both pass through exactly the same equilibrium states. The first process shown is externally reversible also, since heat transfer for this process takes place through an infinitesimal temperature difference dT. The second process, however, is externally irreversible, since it involves heat transfer through a finite temperature difference ΔT.

7–7 ▪ THE CARNOT CYCLE

We mentioned earlier that heat engines are cyclic devices and that the working fluid of a heat engine returns to its initial state at the end of each cycle. Work is done by the working fluid during one part of the cycle and on the working fluid during another part. The difference between these two is the net work delivered by the heat engine. The efficiency of a heat-engine cycle greatly depends on how the individual processes that make up the cycle are executed. The net work, thus the cycle efficiency, can be maximized by using processes that require the least amount of work and deliver the most,

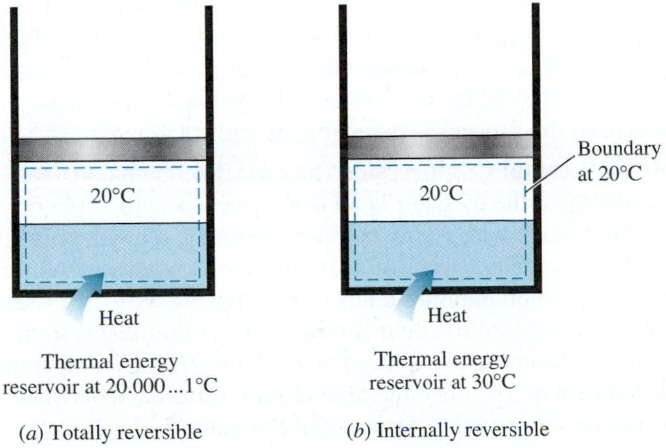

FIGURE 7–36

Totally and interally reversible heat transfer processes.

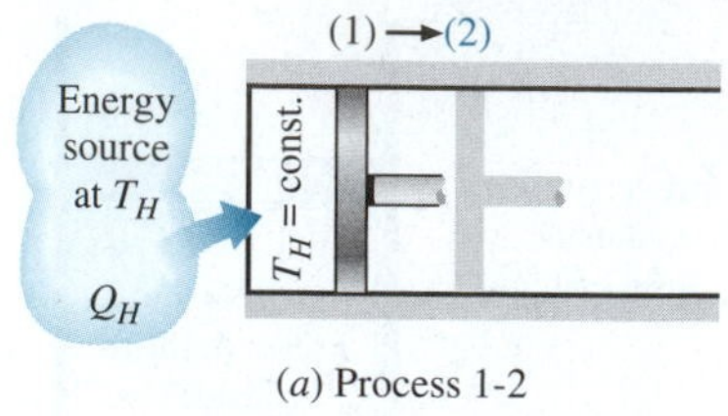

(*a*) Process 1-2

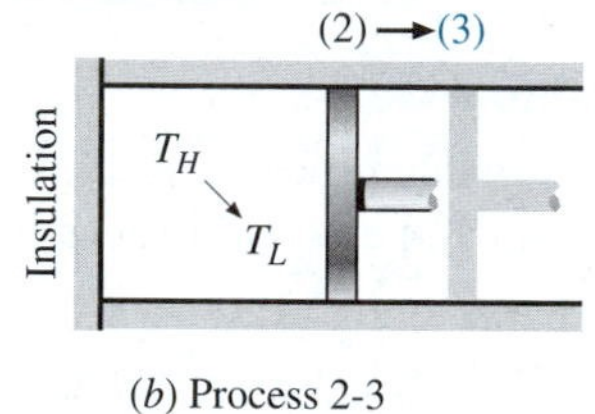

(*b*) Process 2-3

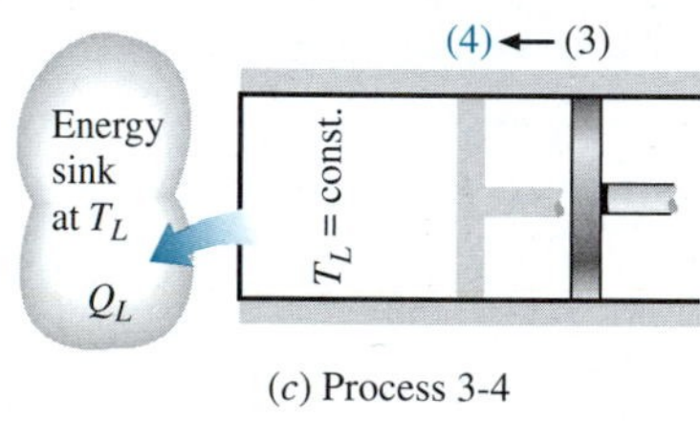

(*c*) Process 3-4

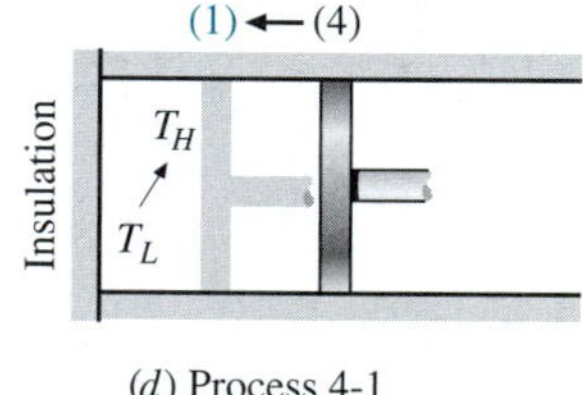

(*d*) Process 4-1

FIGURE 7–37

Execution of the Carnot cycle in a closed system.

that is, by using *reversible processes*. Therefore, it is no surprise that the most efficient cycles are reversible cycles, that is, cycles that consist entirely of reversible processes.

Reversible cycles cannot be achieved in practice because the irreversibilities associated with each process cannot be eliminated. However, reversible cycles provide upper limits on the performance of real cycles. Heat engines and refrigerators that work on reversible cycles serve as models to which actual heat engines and refrigerators can be compared. Reversible cycles also serve as starting points in the development of actual cycles and are modified as needed to meet certain requirements.

Probably the best known reversible cycle is the **Carnot cycle**, first proposed in 1824 by French engineer Sadi Carnot. The theoretical heat engine that operates on the Carnot cycle is called the **Carnot heat engine**. The Carnot cycle is composed of four reversible processes—two isothermal and two adiabatic—and it can be executed either in a closed or a steady-flow system.

Consider a closed system that consists of a gas contained in an adiabatic piston–cylinder device, as shown in Fig. 7–37. The insulation of the cylinder head is such that it may be removed to bring the cylinder into contact with reservoirs to provide heat transfer. The four reversible processes that make up the Carnot cycle are as follows:

Reversible Isothermal Expansion (process 1-2, T_H = constant). Initially (state 1), the temperature of the gas is T_H and the cylinder head is in close contact with a source at temperature T_H. The gas is allowed to expand slowly, doing work on the surroundings. As the gas expands, the temperature of the gas tends to decrease. But as soon as the temperature drops by an infinitesimal amount dT, some heat is transferred from the reservoir into the gas, raising the gas temperature to T_H. Thus, the gas temperature is kept constant at T_H. Since the temperature difference between the gas and the reservoir never exceeds a differential amount dT, this is a reversible heat transfer process. It continues until the piston reaches position 2. The amount of total heat transferred to the gas during this process is Q_H.

Reversible Adiabatic Expansion (process 2-3, temperature drops from T_H to T_L). At state 2, the reservoir that was in contact with the cylinder head is removed and replaced by insulation so that the system becomes adiabatic. The gas continues to expand slowly, doing work on the surroundings until its temperature drops from T_H to T_L (state 3). The piston is assumed to be frictionless and the process to be quasi-equilibrium, so the process is reversible as well as adiabatic.

Reversible Isothermal Compression (process 3-4, T_L = constant). At state 3, the insulation at the cylinder head is removed, and the cylinder is brought into contact with a sink at temperature T_L. Now the piston is pushed inward by an external force, doing work on the gas. As the gas is compressed, its temperature tends to rise. But as soon as it rises by an infinitesimal amount dT, heat is transferred from the gas to the sink, causing the gas temperature to drop to T_L. Thus, the gas temperature remains constant at T_L. Since the temperature difference between the gas and the sink never exceeds a differential amount dT, this is a reversible heat transfer process. It continues until the piston reaches state 4. The amount of heat rejected from the gas during this process is Q_L.

Reversible Adiabatic Compression (process 4-1, temperature rises from T_L to T_H). State 4 is such that when the low-temperature reservoir is removed, the insulation is put back on the cylinder head, and the gas is compressed in a reversible manner, the gas returns to its initial state (state 1). The temperature rises from T_L to T_H during this reversible adiabatic compression process, which completes the cycle.

The *P-V* diagram of this cycle is shown in Fig. 7–38. Remembering that on a *P-V* diagram the area under the process curve represents the boundary work for quasi-equilibrium (internally reversible) processes, we see that the area under curve 1-2-3 is the work done by the gas during the expansion part of the cycle, and the area under curve 3-4-1 is the work done on the gas during the compression part of the cycle. The area enclosed by the path of the cycle (area 1-2-3-4-1) is the difference between these two and represents the net work done during the cycle.

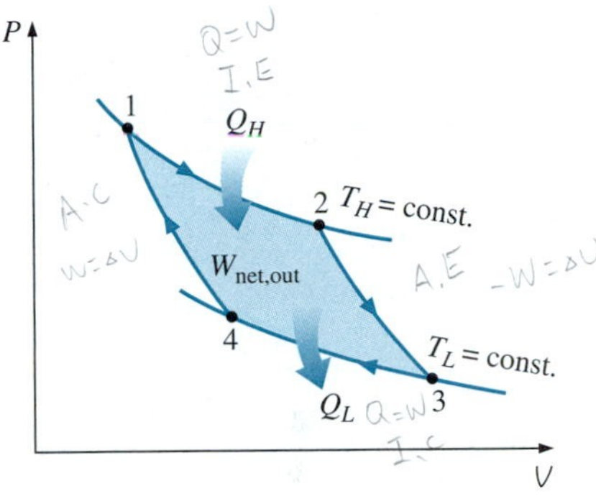

FIGURE 7–38

P-V diagram of the Carnot cycle.

Notice that if we acted stingily and compressed the gas at state 3 adiabatically instead of isothermally in an effort *to save* Q_L, we would end up back at state 2, retracing the process path 3-2. By doing so we would save Q_L, but we would not be able to obtain any net work output from this engine. This illustrates once more the necessity of a heat engine exchanging heat with at least two reservoirs at different temperatures to operate in a cycle and produce a net amount of work.

The Carnot cycle can also be executed in a steady-flow system. It is discussed in later chapters in conjunction with other power cycles.

Being a reversible cycle, the Carnot cycle is the most efficient cycle operating between two specified temperature limits. Even though the Carnot cycle cannot be achieved in reality, the efficiency of actual cycles can be improved by attempting to approximate the Carnot cycle more closely.

The Reversed Carnot Cycle

The Carnot heat-engine cycle just described is a totally reversible cycle. Therefore, all the processes that comprise it can be *reversed*, in which case it becomes the **Carnot refrigeration cycle**. This time, the cycle remains exactly the same, except that the directions of any heat and work interactions are reversed: Heat in the amount of Q_L is absorbed from the low-temperature reservoir, heat in the amount of Q_H is rejected to a high-temperature reservoir, and a work input of $W_{net,in}$ is required to accomplish all this.

The *P-V* diagram of the reversed Carnot cycle is the same as the one given for the Carnot cycle, except that the directions of the processes are reversed, as shown in Fig. 7–39.

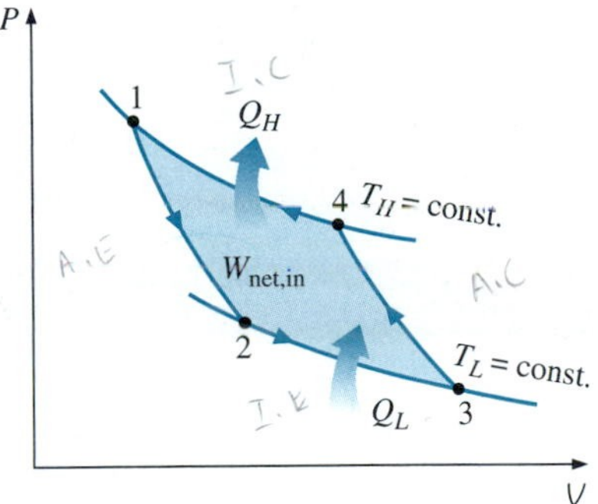

FIGURE 7–39

P-V diagram of the reversed Carnot cycle.

7–8 ▪ THE CARNOT PRINCIPLES

The second law of thermodynamics puts limits on the operation of cyclic devices as expressed by the Kelvin–Planck and Clausius statements. A heat engine cannot operate by exchanging heat with a single reservoir, and a refrigerator cannot operate without a net energy input from an external source.

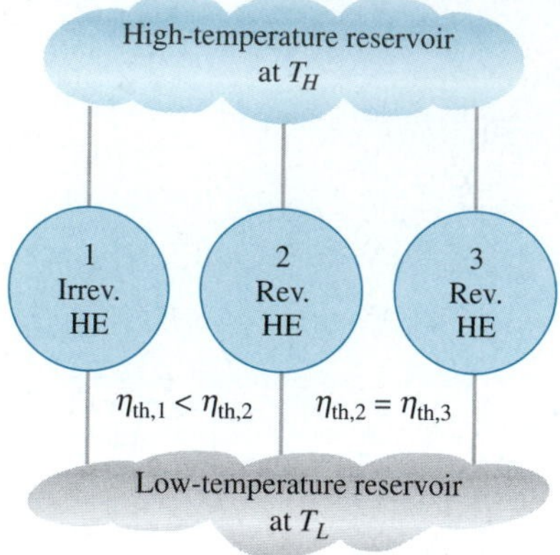

FIGURE 7–40
The Carnot principles.

We can draw valuable conclusions from these statements. Two conclusions pertain to the thermal efficiency of reversible and irreversible (i.e., actual) heat engines, and they are known as the **Carnot principles** (Fig. 7–40), expressed as follows:

1. The efficiency of an irreversible heat engine is always less than the efficiency of a reversible one operating between the same two reservoirs.
2. The efficiencies of all reversible heat engines operating between the same two reservoirs are the same.

These two statements can be proved by demonstrating that the violation of either statement results in the violation of the second law of thermodynamics.

To prove the first statement, consider two heat engines operating between the same reservoirs, as shown in Fig. 7–41. One engine is reversible and the other is irreversible. Now each engine is supplied with the same amount of heat Q_H. The amount of work produced by the reversible heat engine is W_{rev}, and the amount produced by the irreversible one is W_{irrev}.

In violation of the first Carnot principle, we assume that the irreversible heat engine is more efficient than the reversible one (that is, $\eta_{th,irrev} > \eta_{th,rev}$) and thus delivers more work than the reversible one. Now let the reversible heat engine be reversed and operate as a refrigerator. This refrigerator will receive a work input of W_{rev} and reject heat to the high-temperature reservoir. Since the refrigerator is rejecting heat in the amount of Q_H to the high-temperature reservoir and the irreversible heat engine is receiving the same amount of heat from this reservoir, the net heat exchange for this reservoir is zero. Thus, it could be eliminated by having the refrigerator discharge Q_H directly into the irreversible heat engine.

Now considering the refrigerator and the irreversible engine together, we have an engine that produces a net work in the amount of $W_{irrev} - W_{rev}$

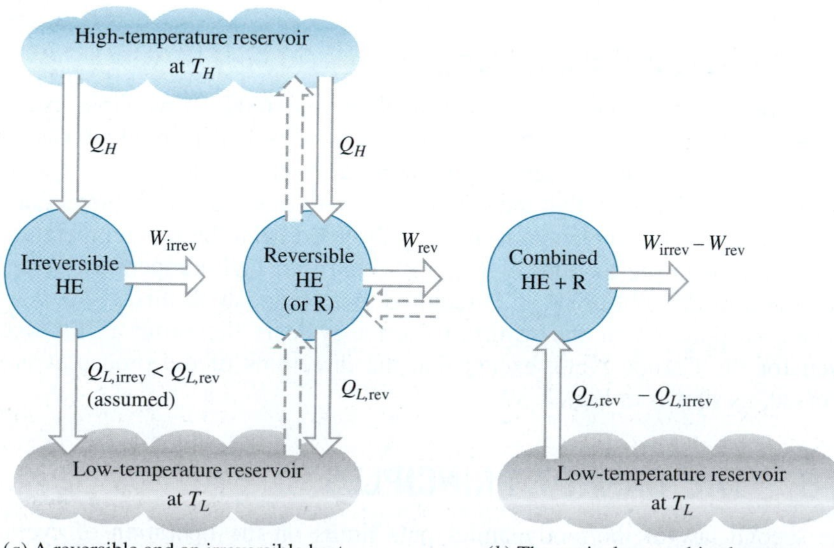

(*a*) A reversible and an irreversible heat engine operating between the same two reservoirs (the reversible heat engine is then reversed to run as a refrigerator)

(*b*) The equivalent combined system

FIGURE 7–41
Proof of the first Carnot principle.

while exchanging heat with a single reservoir—a violation of the Kelvin–Planck statement of the second law. Therefore, our initial assumption that $\eta_{\text{th,irrev}} > \eta_{\text{th,rev}}$ is incorrect. Then we conclude that no heat engine can be more efficient than a reversible heat engine operating between the same reservoirs.

The second Carnot principle can also be proved in a similar manner. This time, let us replace the irreversible engine by another reversible engine that is more efficient and thus delivers more work than the first reversible engine. By following through the same reasoning, we end up having an engine that produces a net amount of work while exchanging heat with a single reservoir, which is a violation of the second law. Therefore, we conclude that no reversible heat engine can be more efficient than a reversible one operating between the same two reservoirs, regardless of how the cycle is completed or the kind of working fluid used.

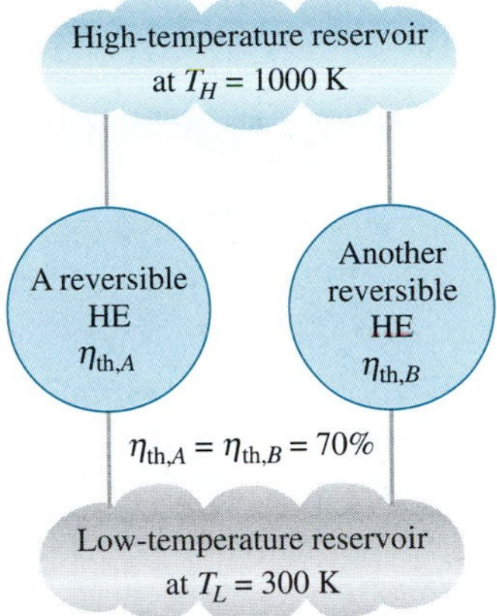

FIGURE 7–42

All reversible heat engines operating between the same two reservoirs have the same efficiency (the second Carnot principle).

7–9 ▪ THE THERMODYNAMIC TEMPERATURE SCALE

A temperature scale that is independent of the properties of the substances that are used to measure temperature is called a **thermodynamic temperature scale**. Such a temperature scale offers great conveniences in thermodynamic calculations, and its derivation is given below using some reversible heat engines.

The second Carnot principle discussed in Section 7–8 states that all reversible heat engines have the same thermal efficiency when operating between the same two reservoirs (Fig. 7–42). That is, the efficiency of a reversible engine is independent of the working fluid employed and its properties, the way the cycle is executed, or the type of reversible engine used. Since energy reservoirs are characterized by their temperatures, the thermal efficiency of reversible heat engines is a function of the reservoir temperatures only. That is,

$$\eta_{\text{th,rev}} = g(T_H, T_L)$$

or

$$\frac{Q_H}{Q_L} = f(T_H, T_L) \tag{7–13}$$

since $\eta_{\text{th}} = 1 - Q_L/Q_H$. In these relations T_H and T_L are the temperatures of the high- and low-temperature reservoirs, respectively.

The functional form of $f(T_H, T_L)$ can be developed with the help of the three reversible heat engines shown in Fig. 7–43. Engines A and C are supplied with the same amount of heat Q_1 from the high-temperature reservoir at T_1. Engine C rejects Q_3 to the low-temperature reservoir at T_3. Engine B receives the heat Q_2 rejected by engine A at temperature T_2 and rejects heat in the amount of Q_3 to a reservoir at T_3.

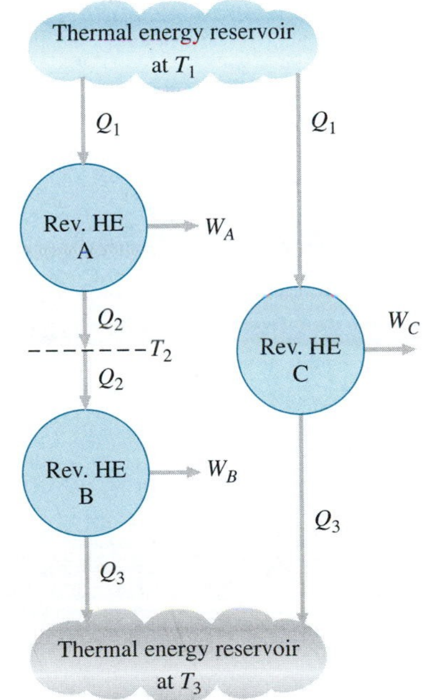

FIGURE 7–43

The arrangement of heat engines used to develop the thermodynamic temperature scale.

The amounts of heat rejected by engines B and C must be the same since engines A and B can be combined into one reversible engine operating between the same reservoirs as engine C and thus the combined engine will have the same efficiency as engine C. Since the heat input to engine C is the same as the heat input to the combined engines A and B, both systems must reject the same amount of heat.

Applying Eq. 7–13 to all three engines separately, we obtain

$$\frac{Q_1}{Q_2} = f(T_1, T_2), \quad \frac{Q_2}{Q_3} = f(T_2, T_3), \quad \text{and} \quad \frac{Q_1}{Q_3} = f(T_1, T_3)$$

Now consider the identity

$$\frac{Q_1}{Q_3} = \frac{Q_1}{Q_2}\frac{Q_2}{Q_3}$$

which corresponds to

$$f(T_1, T_3) = f(T_1, T_2) \cdot f(T_2, T_3)$$

A careful examination of this equation reveals that the left-hand side is a function of T_1 and T_3, and therefore the right-hand side must also be a function of T_1 and T_3 only, and not T_2. That is, the value of the product on the right-hand side of this equation is independent of the value of T_2. This condition will be satisfied only if the function f has the following form:

$$f(T_1, T_2) = \frac{\phi(T_1)}{\phi(T_2)} \quad \text{and} \quad f(T_2, T_3) = \frac{\phi(T_2)}{\phi(T_3)}$$

so that $\phi(T_2)$ will cancel from the product of $f(T_1, T_2)$ and $f(T_2, T_3)$, yielding

$$\frac{Q_1}{Q_3} = f(T_1, T_3) = \frac{\phi(T_1)}{\phi(T_3)} \tag{7–14}$$

This relation is much more specific than Eq. 7–13 for the functional form of Q_1/Q_3 in terms of T_1 and T_3.

For a reversible heat engine operating between two reservoirs at temperatures T_H and T_L, Eq. 7–14 can be written as

$$\frac{Q_H}{Q_L} = \frac{\phi(T_H)}{\phi(T_L)} \tag{7–15}$$

This is the only requirement that the second law places on the ratio of heat transfers to and from the reversible heat engines. Several functions $\phi(T)$ satisfy this equation, and the choice is completely arbitrary. Lord Kelvin first proposed taking $\phi(T) = T$ to define a thermodynamic temperature scale as (Fig. 7–44)

$$\left(\frac{Q_H}{Q_L}\right)_{\text{rev}} = \frac{T_H}{T_L} \tag{7–16}$$

This temperature scale is called the **Kelvin scale**, and the temperatures on this scale are called **absolute temperatures**. On the Kelvin scale, the temperature ratios depend on the ratios of heat transfer between a reversible heat engine and the reservoirs and are independent of the physical properties of any substance. On this scale, temperatures vary between zero and infinity.

The thermodynamic temperature scale is not completely defined by Eq. 7–16 since it gives us only a ratio of absolute temperatures. We also need to know the magnitude of a kelvin. At the International Conference on

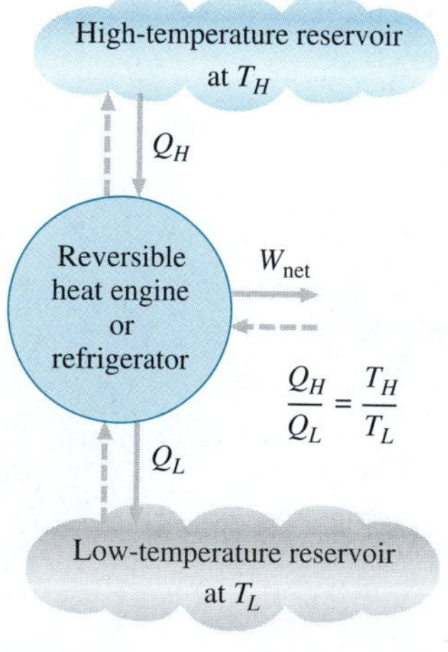

FIGURE 7–44

For reversible cycles, the heat transfer ratio Q_H/Q_L can be replaced by the absolute temperature ratio T_H/T_L.

Weights and Measures held in 1954, the triple point of water (the state at which all three phases of water exist in equilibrium) was assigned the value 273.16 K (Fig. 7–45). The *magnitude of a kelvin* is defined as 1/273.16 of the temperature interval between absolute zero and the triple-point temperature of water. The magnitudes of temperature units on the Kelvin and Celsius scales are identical (1 K ≡ 1°C). The temperatures on these two scales differ by a constant 273.15:

$$T(°\text{C}) = T(\text{K}) - 273.15 \tag{7–17}$$

Even though the thermodynamic temperature scale is defined with the help of the reversible heat engines, it is not possible, nor is it practical, to actually operate such an engine to determine numerical values on the absolute temperature scale. Absolute temperatures can be measured accurately by other means, such as the constant-volume ideal-gas thermometer together with extrapolation techniques as discussed in Chap. 2. The validity of Eq. 7–16 can be demonstrated from physical considerations for a reversible cycle using an ideal gas as the working fluid.

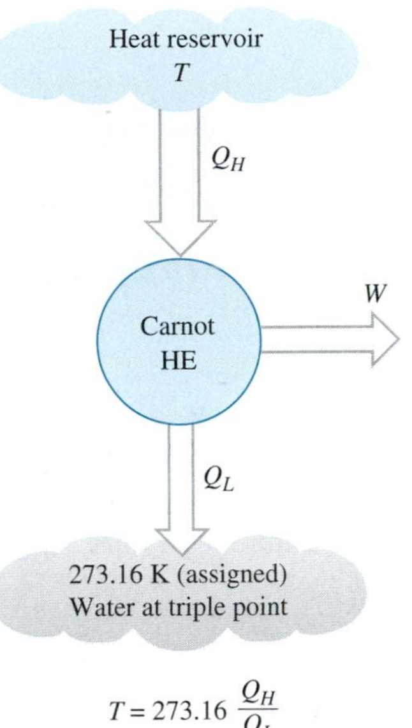

FIGURE 7–45
A conceptual experimental setup to determine thermodynamic temperatures on the Kelvin scale by measuring heat transfers Q_H and Q_L.

7–10 ▪ THE CARNOT HEAT ENGINE

The hypothetical heat engine that operates on the reversible Carnot cycle is called the **Carnot heat engine**. The thermal efficiency of any heat engine, reversible or irreversible, is given by Eq. 7–6 as

$$\eta_{\text{th}} = 1 - \frac{Q_L}{Q_H}$$

where Q_H is heat transferred to the heat engine from a high-temperature reservoir at T_H, and Q_L is heat rejected to a low-temperature reservoir at T_L. For reversible heat engines, the heat transfer ratio in the above relation can be replaced by the ratio of the absolute temperatures of the two reservoirs, as given by Eq. 7–16. Then the efficiency of a Carnot engine, or any reversible heat engine, becomes

$$\eta_{\text{th,rev}} = 1 - \frac{T_L}{T_H} \tag{7–18}$$

This relation is often referred to as the **Carnot efficiency**, since the Carnot heat engine is the best known reversible engine. *This is the highest efficiency a heat engine operating between the two thermal energy reservoirs at temperatures T_L and T_H can have* (Fig. 7–46). All irreversible (i.e., actual) heat engines operating between these temperature limits (T_L and T_H) have lower efficiencies. An actual heat engine cannot reach this maximum theoretical efficiency value because it is impossible to completely eliminate all the irreversibilities associated with the actual cycle.

Note that T_L and T_H in Eq. 7–18 are *absolute temperatures.* Using °C or °F for temperatures in this relation gives results grossly in error.

The thermal efficiencies of actual and reversible heat engines operating between the same temperature limits compare as follows (Fig. 7–47):

$$\eta_{\text{th}} \begin{cases} < \eta_{\text{th, rev}} & \text{irreversible heat engine} \\ = \eta_{\text{th, rev}} & \text{reversible heat engine} \\ > \eta_{\text{th, rev}} & \text{impossible heat engine} \end{cases} \tag{7–19}$$

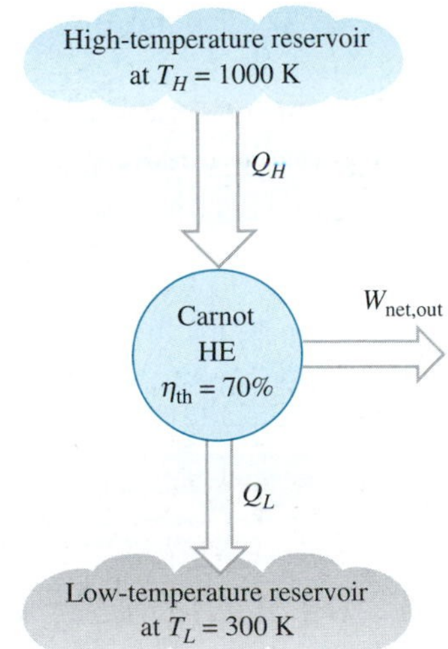

FIGURE 7–46
The Carnot heat engine is the most efficient of all heat engines operating between the same high- and low-temperature reservoirs.

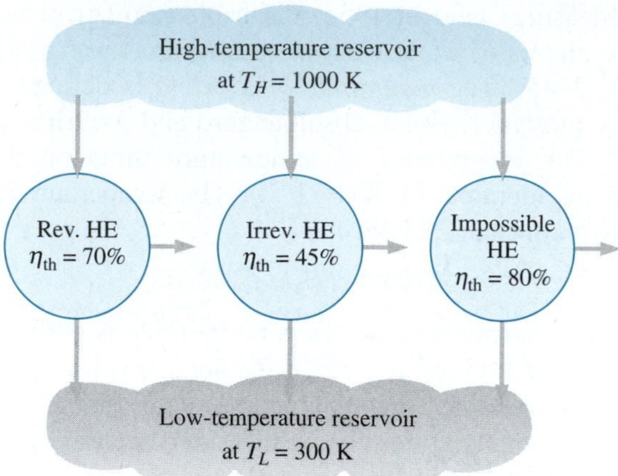

FIGURE 7–47

No heat engine can have a higher efficiency than a reversible heat engine operating between the same high- and low-temperature reservoirs.

Most work-producing devices (heat engines) in operation today have efficiencies under 40 percent, which appear low relative to 100 percent. However, when the performance of actual heat engines is assessed, the efficiencies should not be compared to 100 percent; instead, they should be compared to the efficiency of a reversible heat engine operating between the same temperature limits—because this is the true theoretical upper limit for the efficiency, not 100 percent.

The maximum efficiency of a steam power plant operating between $T_H = 1000$ K and $T_L = 300$ K is 70 percent, as determined from Eq. 7–18. Compared with this value, an actual efficiency of 40 percent does not seem so bad, even though there is still plenty of room for improvement.

It is obvious from Eq. 7–18 that the efficiency of a Carnot heat engine increases as T_H is increased, or as T_L is decreased. This is to be expected since as T_L decreases, so does the amount of heat rejected, and as T_L approaches zero, the Carnot efficiency approaches unity. This is also true for actual heat engines. *The thermal efficiency of actual heat engines can be maximized by supplying heat to the engine at the highest possible temperature* (limited by material strength) *and rejecting heat from the engine at the lowest possible temperature* (limited by the temperature of the cooling medium such as rivers, lakes, or the atmosphere).

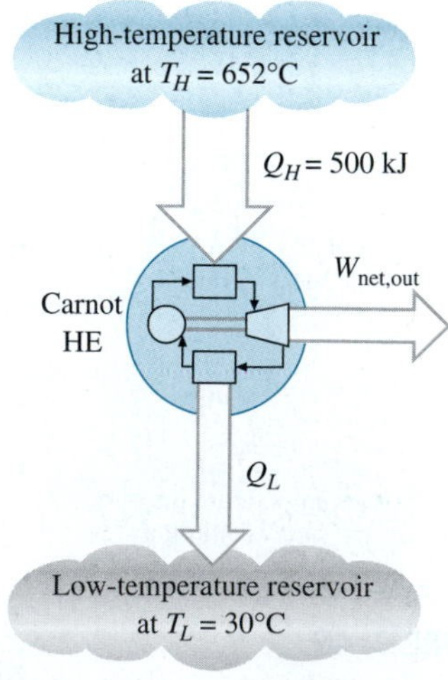

FIGURE 7–48

Schematic for Example 7–5.

EXAMPLE 7–5 Analysis of a Carnot Heat Engine

A Carnot heat engine, shown in Fig. 7–48, receives 500 kJ of heat per cycle from a high-temperature source at 652°C and rejects heat to a low-temperature sink at 30°C. Determine (*a*) the thermal efficiency of this Carnot engine and (*b*) the amount of heat rejected to the sink per cycle.

Solution The heat supplied to a Carnot heat engine is given. The thermal efficiency and the heat rejected are to be determined.

Analysis (*a*) The Carnot heat engine is a reversible heat engine, and so its efficiency can be determined from Eq. 7–18 to be

$$\eta_{th,C} = \eta_{th,rev} = 1 - \frac{T_L}{T_H} = 1 - \frac{(30 + 273)\text{ K}}{(652 + 273)\text{ K}} = \mathbf{0.672}$$

That is, this Carnot heat engine converts 67.2 percent of the heat it receives to work.

(*b*) The amount of heat rejected Q_L by this reversible heat engine is easily determined from Eq. 7–16 to be

$$Q_{L,\text{rev}} = \frac{T_L}{T_H} Q_{H,\text{rev}} = \frac{(30 + 273)\text{ K}}{(652 + 273)\text{ K}}(500\text{ kJ}) = \mathbf{164\ kJ}$$

Discussion Note that this Carnot heat engine rejects to a low-temperature sink 164 kJ of the 500 kJ of heat it receives during each cycle.

The Quality of Energy

The Carnot heat engine in Example 7–5 receives heat from a source at 925 K and converts 67.2 percent of it to work while rejecting the rest (32.8 percent) to a sink at 303 K. Now let us examine how the thermal efficiency varies with the source temperature when the sink temperature is held constant.

The thermal efficiency of a Carnot heat engine that rejects heat to a sink at 303 K is evaluated at various source temperatures using Eq. 7–18 and is listed in Fig. 7–49. Clearly the thermal efficiency decreases as the source temperature is lowered. When heat is supplied to the heat engine at 500 instead of 925 K, for example, the thermal efficiency drops from 67.2 to 39.4 percent. That is, the fraction of heat that can be converted to work drops to 39.4 percent when the temperature of the source drops to 500 K. When the source temperature is 350 K, this fraction becomes a mere 13.4 percent.

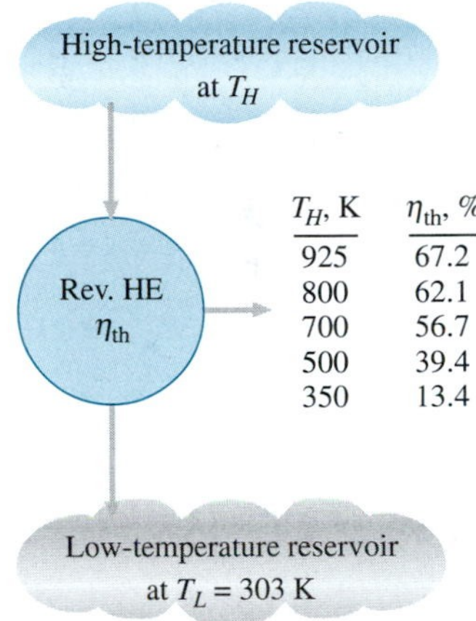

FIGURE 7–49
The fraction of heat that can be converted to work as a function of source temperature (for T_L = 303 K).

These efficiency values show that energy has **quality** as well as quantity. It is clear from the thermal efficiency values in Fig. 7–49 that *more of the high-temperature thermal energy can be converted to work. Therefore, the higher the temperature, the higher the quality of the energy* (Fig. 7–50).

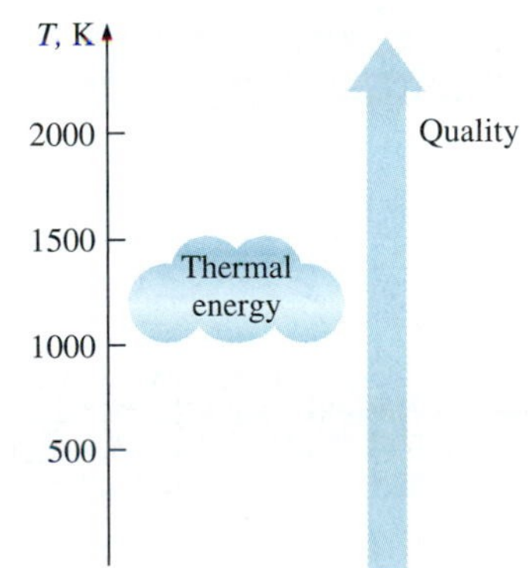

FIGURE 7–50
The higher the temperature of the thermal energy, the higher its quality.

Large quantities of solar energy, for example, can be stored in large bodies of water called *solar ponds* at about 350 K. This stored energy can then be supplied to a heat engine to produce work (electricity). However, the efficiency of solar pond power plants is very low (under 5 percent) because of the low quality of the energy stored in the source, and the construction and maintenance costs are relatively high. Therefore, they are not competitive even though the energy supply of such plants is free. The temperature (and thus the quality) of the solar energy stored could be raised by utilizing concentrating collectors, but the equipment cost in that case becomes very high.

Work is a more valuable form of energy than heat since 100 percent of work can be converted to heat, but only a fraction of heat can be converted to work. When heat is transferred from a high-temperature body to a lower-temperature one, it is degraded since less of it now can be converted to work. For example, if 100 kJ of heat is transferred from a body at 1000 K to a body at 300 K, at the end we will have 100 kJ of thermal energy stored at 300 K, which has no practical value. But if this conversion is made through a heat engine, up to 1 − 300/1000 = 70 percent of it could be converted to work, which is a more valuable form of energy. Thus 70 kJ of work potential is wasted as a result of this heat transfer, and energy is degraded.

Quantity versus Quality in Daily Life

At times of energy crisis, we are bombarded with speeches and articles on how to "conserve" energy. Yet we all know that the *quantity* of energy is already conserved. What is not conserved is the *quality* of energy, or the work potential of energy. Wasting energy is synonymous to converting it to a less useful form. One unit of high-quality energy can be more valuable than three units of lower-quality energy. For example, a finite amount of thermal energy at high temperature is more attractive to power plant engineers than a vast amount of thermal energy at low temperature, such as the energy stored in the upper layers of the oceans at tropical climates.

As part of our culture, we seem to be fascinated by quantity, and little attention is given to quality. However, quantity alone cannot give the whole picture, and we need to consider quality as well. That is, we need to look at something from both the first- and second-law points of view when evaluating something, even in nontechnical areas. Below we present some ordinary events and show their relevance to the second law of thermodynamics.

Consider two students Andy and Wendy. Andy has 10 friends who never miss his parties and are always around during fun times. However, they seem to be busy when Andy needs their help. Wendy, on the other hand, has five friends. They are never too busy for her, and she can count on them at times of need. Let us now try to answer the question, *Who has more friends?* From the first-law point of view, which considers quantity only, it is obvious that Andy has more friends. However, from the second-law point of view, which considers quality as well, there is no doubt that Wendy is the one with more friends.

Another example with which most people will identify is the multibillion-dollar diet industry, which is primarily based on the first law of thermodynamics. However, considering that 90 percent of the people who lose weight gain it back quickly, with interest, suggests that the first law alone does not give the whole picture. This is also confirmed by studies that show that calories that come from fat are more likely to be stored as fat than the calories that come from carbohydrates and protein. A Stanford study found that body weight was related to fat calories consumed and not calories per se. A Harvard study found no correlation between calories eaten and degree of obesity. A major Cornell University survey involving 6500 people in nearly all provinces of China found that the Chinese eat more—gram for gram, calorie for calorie—than Americans do, but they weigh less, with less body fat. Studies indicate that the metabolism rates and hormone levels change noticeably in the mid-30s. Some researchers concluded that prolonged dieting teaches a body to survive on fewer calories, making it more *fuel efficient*. This probably explains why the dieters gain more weight than they lost once they go back to their normal eating levels.

People who seem to be eating whatever they want, whenever they want, without gaining weight are living proof that the calorie-counting technique (the first law) leaves many questions on dieting unanswered. Obviously, more research focused on the second-law effects of dieting is needed before we can fully understand the weight-gain and weight-loss process.

It is tempting to judge things on the basis of their *quantity* instead of their *quality* since assessing quality is much more difficult than assessing quantity. However, assessments made on the basis of quantity only (the first law) may be grossly inadequate and misleading.

7–11 ▪ THE CARNOT REFRIGERATOR AND HEAT PUMP

A refrigerator or a heat pump that operates on the reversed Carnot cycle is called a **Carnot refrigerator**, or a **Carnot heat pump**. The coefficient of performance of any refrigerator or heat pump, reversible or irreversible, is given by Eqs. 7–9 and 7–11 as

$$\text{COP}_\text{R} = \frac{1}{Q_H/Q_L - 1} \quad \text{and} \quad \text{COP}_\text{HP} = \frac{1}{1 - Q_L/Q_H}$$

where Q_L is the amount of heat absorbed from the low-temperature medium and Q_H is the amount of heat rejected to the high-temperature medium. The COPs of all reversible refrigerators or heat pumps can be determined by replacing the heat transfer ratios in the above relations by the ratios of the absolute temperatures of the high- and low-temperature reservoirs, as expressed by Eq. 7–16. Then the COP relations for reversible refrigerators and heat pumps become

$$\text{COP}_\text{R,rev} = \frac{1}{T_H/T_L - 1} \qquad \text{(7–20)}$$

and

$$\text{COP}_\text{HP,rev} = \frac{1}{1 - T_L/T_H} \qquad \text{(7–21)}$$

These are the highest coefficients of performance that a refrigerator or a heat pump operating between the temperature limits of T_L and T_H can have. All actual refrigerators or heat pumps operating between these temperature limits (T_L and T_H) have lower coefficients of performance (Fig. 7–51).

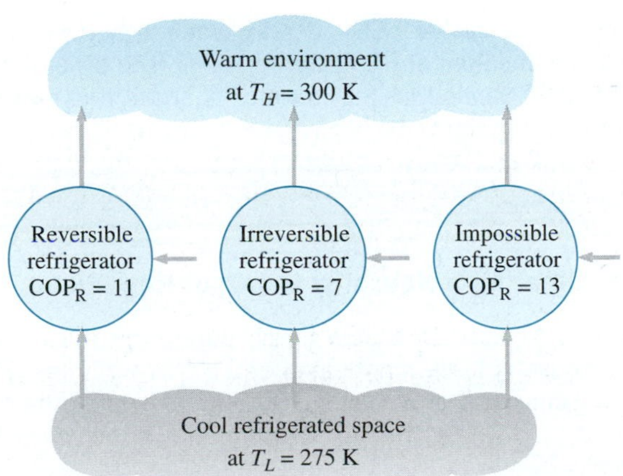

FIGURE 7–51

No refrigerator can have a higher COP than a reversible refrigerator operating between the same temperature limits.

The coefficients of performance of actual and reversible refrigerators operating between the same temperature limits can be compared as follows:

$$\text{COP}_R \begin{cases} < \text{COP}_{R,\text{rev}} & \text{irreversible refrigerator} \\ = \text{COP}_{R,\text{rev}} & \text{reversible refrigerator} \\ > \text{COP}_{R,\text{rev}} & \text{impossible refrigerator} \end{cases} \tag{7-22}$$

A similar relation can be obtained for heat pumps by replacing all COP_R's in Eq. 7–22 by COP_{HP}.

The COP of a reversible refrigerator or heat pump is the maximum theoretical value for the specified temperature limits. Actual refrigerators or heat pumps may approach these values as their designs are improved, but they can never reach them.

As a final note, the COPs of both the refrigerators and the heat pumps decrease as T_L decreases. That is, it requires more work to absorb heat from lower-temperature media. As the temperature of the refrigerated space approaches zero, the amount of work required to produce a finite amount of refrigeration approaches infinity and COP_R approaches zero.

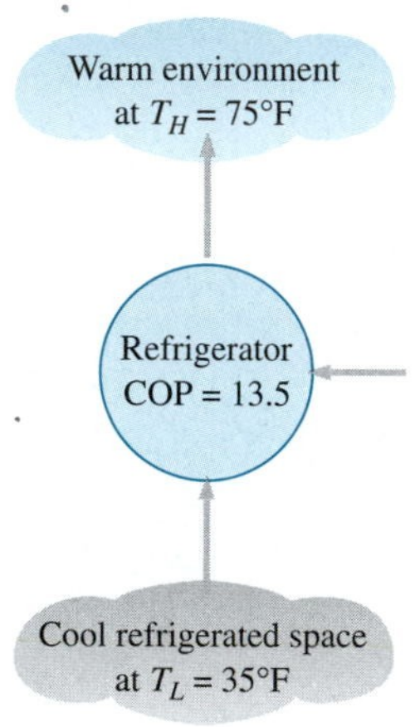

FIGURE 7–52
Schematic for Example 7–6.

EXAMPLE 7–6 A Questionable Claim for a Refrigerator

An inventor claims to have developed a refrigerator that maintains the refrigerated space at 35°F while operating in a room where the temperature is 75°F and that has a COP of 13.5. Is this claim reasonable?

Solution An extraordinary claim made for the performance of a refrigerator is to be evaluated.

Assumptions Steady operating conditions exist.

Analysis The performance of this refrigerator (shown in Fig. 7–52) can be evaluated by comparing it with a reversible refrigerator operating between the same temperature limits:

$$\text{COP}_{R,\max} = \text{COP}_{R,\text{rev}} = \frac{1}{T_H/T_L - 1}$$

$$= \frac{1}{(75 + 460\text{ R})/(35 + 460\text{ R}) - 1} = 12.4$$

Discussion This is the highest COP a refrigerator can have when absorbing heat from a cool medium at 35°F and rejecting it to a warmer medium at 75°F. Since the COP claimed by the inventor is above this maximum value, **the claim is *false*.**

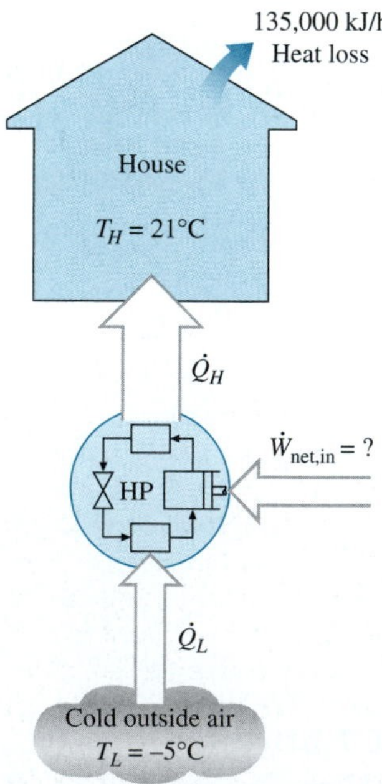

FIGURE 7–53
Schematic for Example 7–7.

EXAMPLE 7–7 Heating a House by a Carnot Heat Pump

A heat pump is to be used to heat a house during the winter, as shown in Fig. 7–53. The house is to be maintained at 21°C at all times. The house is estimated to be losing heat at a rate of 135,000 kJ/h when the outside temperature drops to −5°C. Determine the minimum power required to drive this heat pump.

Solution A heat pump maintains a house at a constant temperature. The required minimum power input to the heat pump is to be determined.

Assumptions Steady operating conditions exist.

Analysis The heat pump must supply heat to the house at a rate of $Q_H = 135{,}000$ kJ/h $= 37.5$ kW. The power requirements are minimum when a reversible heat pump is used to do the job. The COP of a reversible heat pump operating between the house and the outside air is

$$\text{COP}_{\text{HP,rev}} = \frac{1}{1 - T_L/T_H} = \frac{1}{1 - (-5 + 273\text{ K})/(21 + 273\text{ K})} = 11.3$$

Then the required power input to this reversible heat pump becomes

$$\dot{W}_{\text{net,in}} = \frac{\dot{Q}_H}{\text{COP}_{\text{HP}}} = \frac{37.5\text{ kW}}{11.3} = \mathbf{3.32\ kW}$$

Discussion This reversible heat pump can meet the heating requirements of this house by consuming electric power at a rate of 3.32 kW only. If this house were to be heated by electric resistance heaters instead, the power consumption would jump up 11.3 times to 37.5 kW. This is because in resistance heaters the electric energy is converted to heat at a one-to-one ratio. With a heat pump, however, energy is absorbed from the outside and carried to the inside using a refrigeration cycle that consumes only 3.32 kW. Notice that the heat pump does not create energy. It merely transports it from one medium (the cold outdoors) to another (the warm indoors).

SUMMARY

The *second law of thermodynamics* states that processes occur in a certain direction, not in any direction. A process does not occur unless it satisfies both the first and the second laws of thermodynamics. Bodies that can absorb or reject finite amounts of heat isothermally are called *thermal energy reservoirs* or *heat reservoirs*.

Work can be converted to heat directly, but heat can be converted to work only by some devices called *heat engines*. The *thermal efficiency* of a heat engine is defined as

$$\eta_{\text{th}} = \frac{W_{\text{net,out}}}{Q_H} = 1 - \frac{Q_L}{Q_H}$$

where $W_{\text{net,out}}$ is the net work output of the heat engine, Q_H is the amount of heat supplied to the engine, and Q_L is the amount of heat rejected by the engine.

Refrigerators and heat pumps are devices that absorb heat from low-temperature media and reject it to higher-temperature ones. The performance of a refrigerator or a heat pump is expressed in terms of the *coefficient of performance*, which is defined as

$$\text{COP}_{\text{R}} = \frac{Q_L}{W_{\text{net,in}}} = \frac{1}{Q_H/Q_L - 1}$$

$$\text{COP}_{\text{HP}} = \frac{Q_H}{W_{\text{net,in}}} = \frac{1}{1 - Q_L/Q_H}$$

The *Kelvin–Planck statement* of the second law of thermodynamics states that no heat engine can produce a net amount of work while exchanging heat with a single reservoir only. The *Clausius statement* of the second law states that no device can transfer heat from a cooler body to a warmer one without leaving an effect on the surroundings.

Any device that violates the first or the second law of thermodynamics is called a *perpetual-motion machine*.

A process is said to be *reversible* if both the system and the surroundings can be restored to their original conditions. Any other process is *irreversible*. The effects such as friction, non-quasi-equilibrium expansion or compression, and heat transfer through a finite temperature difference render a process irreversible and are called *irreversibilities*.

The *Carnot cycle* is a reversible cycle that is composed of four reversible processes, two isothermal and two adiabatic. The *Carnot principles* state that the thermal efficiencies of all reversible heat engines operating between the same two reservoirs are the same, and that no heat engine is more efficient than a reversible one operating between the same two reservoirs. These statements form the basis for establishing a *thermodynamic temperature scale* related to the heat transfers between a reversible device and the high- and low-temperature reservoirs by

$$\left(\frac{Q_H}{Q_L}\right)_{\text{rev}} = \frac{T_H}{T_L}$$

Therefore, the Q_H/Q_L ratio can be replaced by T_H/T_L for reversible devices, where T_H and T_L are the absolute temperatures of the high- and low-temperature reservoirs, respectively.

A heat engine that operates on the reversible Carnot cycle is called a *Carnot heat engine*. The thermal efficiency of a Carnot heat engine, as well as all other reversible heat engines, is given by

$$\eta_{\text{th,rev}} = 1 - \frac{T_L}{T_H}$$

This is the maximum efficiency a heat engine operating between two reservoirs at temperatures T_H and T_L can have.

The COPs of reversible refrigerators and heat pumps are given in a similar manner as

$$\text{COP}_{\text{R,rev}} = \frac{1}{T_H/T_L - 1}$$

and

$$\text{COP}_{\text{HP,rev}} = \frac{1}{1 - T_L/T_H}$$

Again, these are the highest COPs a refrigerator or a heat pump operating between the temperature limits of T_H and T_L can have.

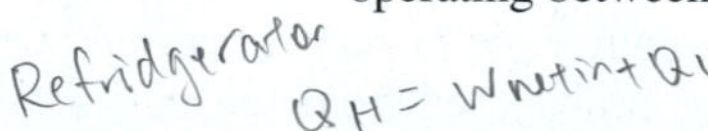

REFERENCES AND SUGGESTED READINGS

1. *ASHRAE Handbook of Refrigeration,* SI version. Atlanta, GA: American Society of Heating, Refrigerating, and Air-Conditioning Engineers, Inc. 1994.
2. D. Stewart. "Wheels Go Round and Round, but Always Run Down." November 1986, *Smithsonian*, pp. 193–208.

PROBLEMS*

Second Law of Thermodynamics and Thermal Energy Reservoirs

7–1C A mechanic claims to have developed a car engine that runs on water instead of gasoline. What is your response to this claim?

7–2C Describe an imaginary process that satisfies the first law but violates the second law of thermodynamics.

7–3C Describe an imaginary process that satisfies the second law but violates the first law of thermodynamics.

7–4C Describe an imaginary process that violates both the first and the second laws of thermodynamics.

7–5C An experimentalist claims to have raised the temperature of a small amount of water to 150°C by transferring heat from high-pressure steam at 120°C. Is this a reasonable claim? Why? Assume no refrigerator or heat pump is used in the process.

7–6C What is a thermal energy reservoir? Give some examples.

7–7C Consider the process of baking potatoes in a conventional oven. Can the hot air in the oven be treated as a thermal energy reservoir? Explain.

7–8C Consider the energy generated by a TV set. What is a suitable choice for a thermal energy reservoir?

Heat Engines and Thermal Efficiency

7–9C Is it possible for a heat engine to operate without rejecting any waste heat to a low-temperature reservoir? Explain.

7–10C What are the characteristics of all heat engines?

7–11C Consider a pan of water being heated (*a*) by placing it on an electric range and (*b*) by placing a heating element in

* Problems designated by a "C" are concept questions, and students are encouraged to answer them all. Problems designated by an "E" are in English units, and the SI users can ignore them. Problems with the icon are solved using EES, and complete solutions together with parametric studies are included on the enclosed DVD. Problems with the icon are comprehensive in nature, and are intended to be solved with a computer, preferably using the EES software that accompanies this text.

the water. Which method is a more efficient way of heating water? Explain.

7–12C Baseboard heaters are basically electric resistance heaters and are frequently used in space heating. A home owner claims that her 5-year-old baseboard heaters have a conversion efficiency of 100 percent. Is this claim in violation of any thermodynamic laws? Explain.

7–13C What is the Kelvin–Planck expression of the second law of thermodynamics?

7–14C Does a heat engine that has a thermal efficiency of 100 percent necessarily violate (*a*) the first law and (*b*) the second law of thermodynamics? Explain.

7–15C In the absence of any friction and other irreversibilities, can a heat engine have an efficiency of 100 percent? Explain.

7–16C Are the efficiencies of all the work-producing devices, including the hydroelectric power plants, limited by the Kelvin–Planck statement of the second law? Explain.

7–17 A 600-MW steam power plant, which is cooled by a nearby river, has a thermal efficiency of 40 percent. Determine the rate of heat transfer to the river water. Will the actual heat transfer rate be higher or lower than this value? Why?

7–18 A heat engine has a total heat input of 1.3 kJ and a thermal efficiency of 35 percent. How much work will it produce?

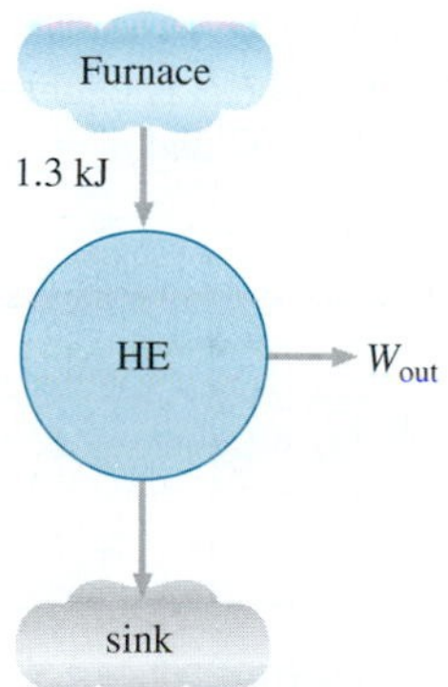

FIGURE P7–18

7–19E A heat engine that propels a ship produces 500 Btu/lbm of work while rejecting 300 Btu/lbm of heat. What is its thermal efficiency?

7–20 A heat engine that pumps water out of an underground mine accepts 500 kJ of heat and produces 200 kJ of work. How much heat does it reject, in kJ?

7–21 A heat engine with a thermal efficiency of 40 percent rejects 1000 kJ/kg of heat. How much heat does it receive? *Answer:* 1667 kJ/kg

7–22 A steam power plant with a power output of 150 MW consumes coal at a rate of 60 tons/h. If the heating value of the coal is 30,000 kJ/kg, determine the overall efficiency of this plant. *Answer:* 30.0 percent

7–23 An automobile engine consumes fuel at a rate of 28 L/h and delivers 60 kW of power to the wheels. If the fuel has a heating value of 44,000 kJ/kg and a density of 0.8 g/cm^3, determine the efficiency of this engine. *Answer:* 21.9 percent

7–24E Solar energy stored in large bodies of water, called solar ponds, is being used to generate electricity. If such a solar power plant has an efficiency of 4 percent and a net power output of 350 kW, determine the average value of the required solar energy collection rate, in Btu/h.

7–25 In 2001, the United States produced 51 percent of its electricity in the amount of 1.878×10^{12} kWh from coal-fired power plants. Taking the average thermal efficiency to be 34 percent, determine the amount of thermal energy rejected by the coal-fired power plants in the United States that year.

7–26 The Department of Energy projects that between the years 1995 and 2010, the United States will need to build new power plants to generate an additional 150,000 MW of electricity to meet the increasing demand for electric power. One possibility is to build coal-fired power plants, which cost \$1300 per kW to construct and have an efficiency of 34 percent. Another possibility is to use the clean-burning Integrated Gasification Combined Cycle (IGCC) plants where the coal is subjected to heat and pressure to gasify it while removing sulfur and particulate matter from it. The gaseous coal is then burned in a gas turbine, and part of the waste heat from the exhaust gases is recovered to generate steam for the steam turbine. Currently the construction of IGCC plants costs about \$1500 per kW, but their efficiency is about 45 percent. The average heating value of the coal is about 28,000,000 kJ per ton (that is, 28,000,000 kJ of heat is released when 1 ton of coal is burned). If the IGCC plant is to recover its cost difference from fuel savings in five years, determine what the price of coal should be in \$ per ton.

7–27 EES Reconsider Prob. 7–26. Using EES (or other) software, investigate the price of coal for varying simple payback periods, plant construction costs, and operating efficiency.

7–28 Repeat Prob. 7–26 for a simple payback period of three years instead of five years.

7–29 A coal-burning steam power plant produces a net power of 300 MW with an overall thermal efficiency of 32 percent. The actual gravimetric air–fuel ratio in the furnace is calculated to be 12 kg air/kg fuel. The heating value of the coal is 28,000 kJ/kg. Determine (*a*) the amount of coal consumed during a 24-hour period and (*b*) the rate of air flowing through the furnace. *Answers:* (*a*) 2.89×10^6 kg, (*b*) 402 kg/s

Refrigerators and Heat Pumps

7–30C What is the difference between a refrigerator and a heat pump?

7–31C What is the difference between a refrigerator and an air conditioner?

7–32C In a refrigerator, heat is transferred from a lower-temperature medium (the refrigerated space) to a higher-temperature one (the kitchen air). Is this a violation of the second law of thermodynamics? Explain.

7–33C A heat pump is a device that absorbs energy from the cold outdoor air and transfers it to the warmer indoors. Is this a violation of the second law of thermodynamics? Explain.

7–34C Define the coefficient of performance of a refrigerator in words. Can it be greater than unity?

7–35C Define the coefficient of performance of a heat pump in words. Can it be greater than unity?

7–36C A heat pump that is used to heat a house has a COP of 2.5. That is, the heat pump delivers 2.5 kWh of energy to the house for each 1 kWh of electricity it consumes. Is this a violation of the first law of thermodynamics? Explain.

7–37C A refrigerator has a COP of 1.5. That is, the refrigerator removes 1.5 kWh of energy from the refrigerated space for each 1 kWh of electricity it consumes. Is this a violation of the first law of thermodynamics? Explain.

7–38C What is the Clausius expression of the second law of thermodynamics?

7–39C Show that the Kelvin–Planck and the Clausius expressions of the second law are equivalent.

7–40 A household refrigerator with a COP of 1.2 removes heat from the refrigerated space at a rate of 60 kJ/min. Determine (*a*) the electric power consumed by the refrigerator and (*b*) the rate of heat transfer to the kitchen air. *Answers:* (*a*) 0.83 kW, (*b*) 110 kJ/min

7–41E A commercial heat pump removes 10,000 Btu/h from the source, rejects 15,090 Btu/h to the sink, and requires 2 hp of power. What is this heat pump's coefficient of performance?

7–42 The coefficient of performance of a residential heat pump is 1.6. Calculate the heating effect, in kJ/s, this heat pump will produce when it consumes 2 kW of electrical power.

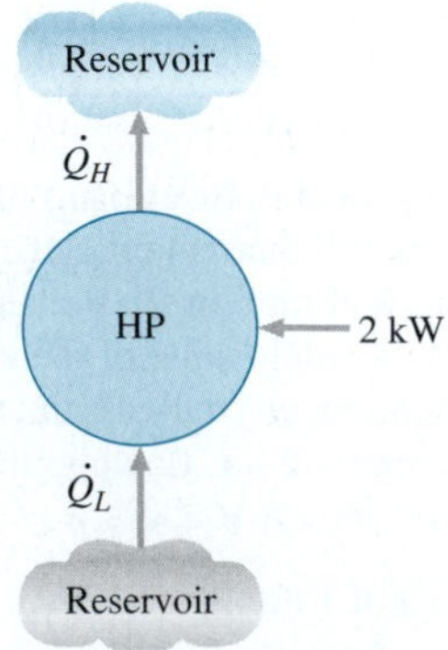

FIGURE P7–42

7–43 A refrigerator used for cooling food in a grocery store is to produce a 10,000 kJ/h cooling effect, and it has a coefficient of performance of 1.35. How many kilowatts of power will this refrigerator require to operate? *Answer:* 2.06 kW

7–44 A food freezer is to produce a 5 kW cooling effect, and its COP is 1.3. How many kW of power will this refrigerator require for operation?

7–45 A heat pump has a COP of 1.7. Determine the heat transferred to and from this heat pump when 50 kJ of work is supplied.

7–46E Water enters an ice machine at 55°F and leaves as ice at 25°F. If the COP of the ice machine is 2.4 during this operation, determine the required power input for an ice production rate of 28 lbm/h. (169 Btu of energy needs to be removed from each lbm of water at 55°F to turn it into ice at 25°F.)

7–47 A household refrigerator that has a power input of 450 W and a COP of 2.5 is to cool five large watermelons, 10 kg each, to 8°C. If the watermelons are initially at 20°C, determine how long it will take for the refrigerator to cool them. The watermelons can be treated as water whose specific heat is 4.2 kJ/kg · °C. Is your answer realistic or optimistic? Explain. *Answer:* 2240 s

7–48 When a man returns to his well-sealed house on a summer day, he finds that the house is at 32°C. He turns on the air conditioner, which cools the entire house to 20°C in 15 min. If the COP of the air-conditioning system is 2.5, determine the power drawn by the air conditioner. Assume the entire mass within the house is equivalent to 800 kg of air for which c_v = 0.72 kJ/kg · °C and c_p = 1.0 kJ/kg · °C.

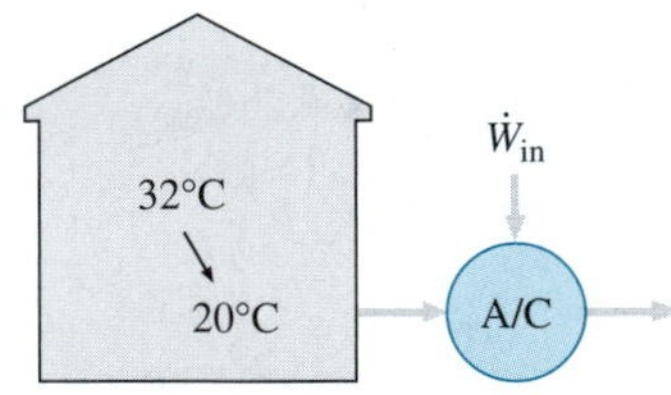

FIGURE P7–48

7–49 Reconsider Prob. 7–48. Using EES (or other) software, determine the power input required by the air conditioner to cool the house as a function for air-conditioner EER ratings in the range 9 to 16. Discuss your results and include representative costs of air-conditioning units in the EER rating range.

7–50 A house that was heated by electric resistance heaters consumed 1200 kWh of electric energy in a winter month. If this house were heated instead by a heat pump that has an average COP of 2.4, determine how much money the home owner would have saved that month. Assume a price of 8.5¢/kWh for electricity.

7–51E A heat pump with a COP of 1.4 is to produce a 100,000 Btu/h heating effect. How much power does this device require, in hp?

7–52 A food refrigerator is to provide a 15,000 kJ/h cooling effect while rejecting 22,000 kJ/h of heat. Calculate the COP of this refrigerator. *Answer:* 2.14

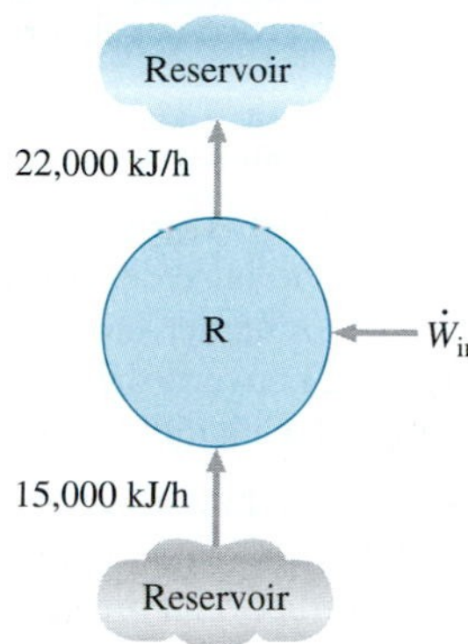

FIGURE P7–52

7–53 An automotive air conditioner produces a 1-kW cooling effect while consuming 0.75 kW of power. What is the rate at which heat is rejected from this air conditioner?

7–54 A heat pump is used to maintain a house at a constant temperature of 23°C. The house is losing heat to the outside air through the walls and the windows at a rate of 60,000 kJ/h while the energy generated within the house from people, lights, and appliances amounts to 4000 kJ/h. For a COP of 2.5, determine the required power input to the heat pump. *Answer:* 6.22 kW

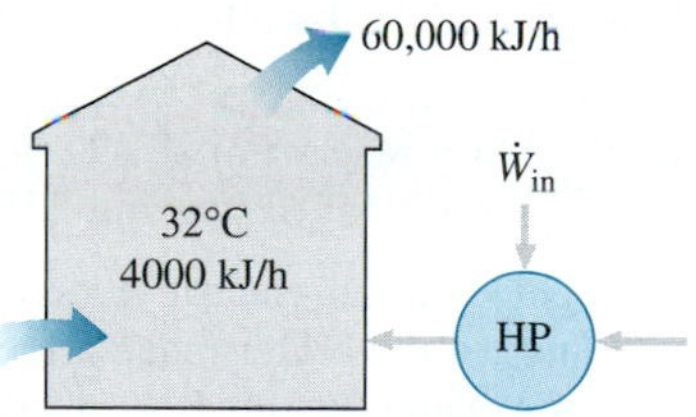

FIGURE P7–54

7–55E Consider an office room that is being cooled adequately by a 12,000 Btu/h window air conditioner. Now it is decided to convert this room into a computer room by installing several computers, terminals, and printers with a total rated power of 3.5 kW. The facility has several 4000 Btu/h air conditioners in storage that can be installed to meet the additional cooling requirements. Assuming a usage factor of 0.4 (i.e., only 40 percent of the rated power will be consumed at any given time) and additional occupancy of four people, each generating heat at a rate of 100 W, determine how many of these air conditioners need to be installed to the room.

7–56 Consider a building whose annual air-conditioning load is estimated to be 120,000 kWh in an area where the unit cost of electricity is \$0.10/kWh. Two air conditioners are considered for the building. Air conditioner A has a seasonal average COP of 3.2 and costs \$5500 to purchase and install. Air conditioner B has a seasonal average COP of 5.0 and costs \$7000 to purchase and install. All else being equal, determine which air conditioner is a better buy.

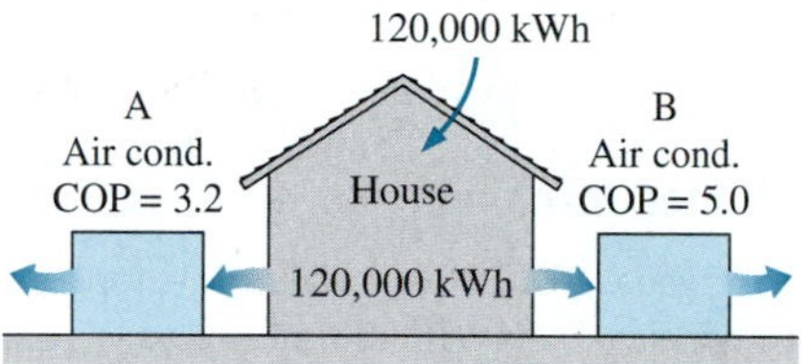

FIGURE P7–56

7–57 Refrigerant-134a enters the condenser of a residential heat pump at 800 kPa and 35°C at a rate of 0.018 kg/s and leaves at 800 kPa as a saturated liquid. If the compressor consumes 1.2 kW of power, determine (*a*) the COP of the heat pump and (*b*) the rate of heat absorption from the outside air.

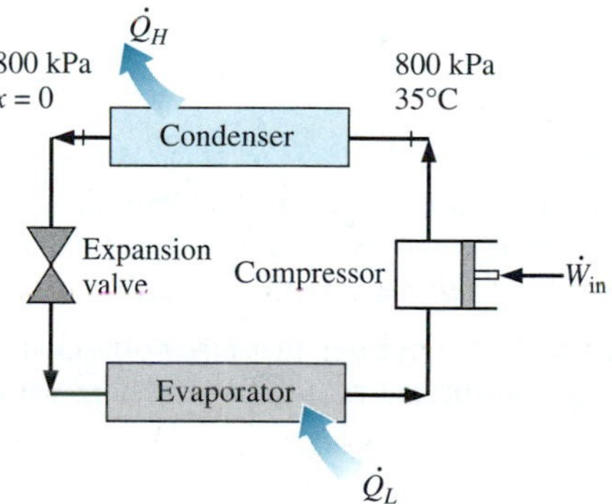

FIGURE P7–57

7–58 Refrigerant-134a enters the evaporator coils placed at the back of the freezer section of a household refrigerator at 120 kPa with a quality of 20 percent and leaves at 120 kPa and −20°C. If the compressor consumes 450 W of power and the COP the refrigerator is 1.2, determine (*a*) the mass flow rate of the refrigerant and (*b*) the rate of heat rejected to the kitchen air. *Answers:* (*a*) 0.00311 kg/s, (*b*) 990 W

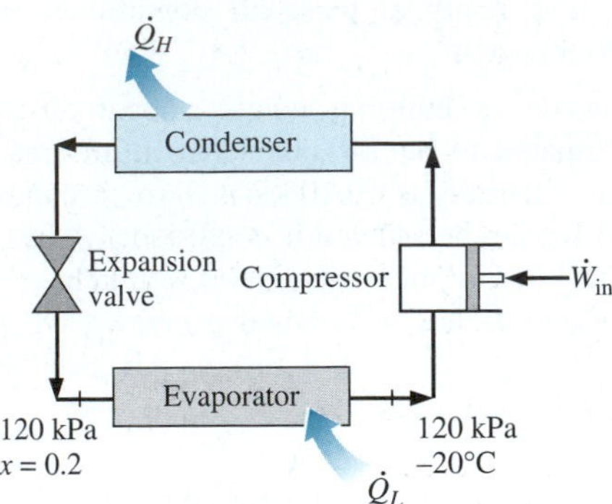

FIGURE P7–58

Perpetual-Motion Machines

7–59C An inventor claims to have developed a resistance heater that supplies 1.2 kWh of energy to a room for each kWh of electricity it consumes. Is this a reasonable claim, or has the inventor developed a perpetual-motion machine? Explain.

7–60C It is common knowledge that the temperature of air rises as it is compressed. An inventor thought about using this high-temperature air to heat buildings. He used a compressor driven by an electric motor. The inventor claims that the compressed hot-air system is 25 percent more efficient than a resistance heating system that provides an equivalent amount of heating. Is this claim valid, or is this just another perpetual-motion machine? Explain.

Reversible and Irreversible Processes

7–61C A block slides down an inclined plane with friction and no restraining force. Is this process reversible or irreversible? Justify your answer.

7–62C Show that processes that use work for mixing are irreversible by considering an adiabatic system whose contents are stirred by turning a paddle wheel inside the system (e.g., stirring a cake mix with an electric mixer).

7–63C Show that processes involving rapid chemical reactions are irreversible by considering the combustion of a natural gas (e.g., methane) and air mixture in a rigid container.

7–64C A cold canned drink is left in a warmer room where its temperature rises as a result of heat transfer. Is this a reversible process? Explain.

7–65C Why are engineers interested in reversible processes even though they can never be achieved?

7–66C Why does a nonquasi-equilibrium compression process require a larger work input than the corresponding quasi-equilibrium one?

7–67C Why does a nonquasi-equilibrium expansion process deliver less work than the corresponding quasi-equilibrium one?

7–68C How do you distinguish between internal and external irreversibilities?

7–69C Is a reversible expansion or compression process necessarily quasi-equilibrium? Is a quasi-equilibrium expansion or compression process necessarily reversible? Explain.

The Carnot Cycle and Carnot Principles

7–70C What are the four processes that make up the Carnot cycle?

7–71C What are the two statements known as the Carnot principles?

7–72C Somebody claims to have developed a new reversible heat-engine cycle that has a higher theoretical efficiency than the Carnot cycle operating between the same temperature limits. How do you evaluate this claim?

7–73C Somebody claims to have developed a new reversible heat-engine cycle that has the same theoretical efficiency as the Carnot cycle operating between the same temperature limits. Is this a reasonable claim?

7–74C Is it possible to develop (*a*) an actual and (*b*) a reversible heat-engine cycle that is more efficient than a Carnot cycle operating between the same temperature limits? Explain.

Carnot Heat Engines

7–75C Is there any way to increase the efficiency of a Carnot heat engine other than by increasing T_H or decreasing T_L?

7–76C Consider two actual power plants operating with solar energy. Energy is supplied to one plant from a solar pond at 80°C and to the other from concentrating collectors that raise the water temperature to 600°C. Which of these power plants will have a higher efficiency? Explain.

7–77 From a work-production perspective, which is more valuable: (*a*) thermal energy reservoirs at 675 K and 325 K or (*b*) thermal energy reservoirs at 625 K and 275 K?

7–78 You are an engineer in an electric-generation station. You know that the flames in the boiler reach a temperature of 1200 K and that cooling water at 300 K is available from a nearby river. What is the maximum efficiency your plant will ever achieve?

7–79 As the engineer of the previous problem, you also know that the metallurgical temperature limit for the blades in the turbine is 1000 K before they will incur excessive creep. Now what is the maximum efficiency for this plant?

7–80 A heat engine operates between a source at 550°C and a sink at 25°C. If heat is supplied to the heat engine at a steady rate of 1200 kJ/min, determine the maximum power output of this heat engine.

7–81 Reconsider Prob. 7–80. Using EES (or other) software, study the effects of the temperatures of the heat source and the heat sink on the power produced and the cycle thermal efficiency. Let the source temperature vary from 300 to 1000°C, and the sink temperature to vary from 0 to 50°C. Plot the power produced and the cycle efficiency against the source temperature for sink temperatures of 0°C, 25°C, and 50°C, and discuss the results.

7–82E A heat engine is operating on a Carnot cycle and has a thermal efficiency of 75 percent. The waste heat from this engine is rejected to a nearby lake at 60°F at a rate of 800 Btu/min. Determine (*a*) the power output of the engine and (*b*) the temperature of the source. *Answers:* (*a*) 56.6 hp, (*b*) 2080 R

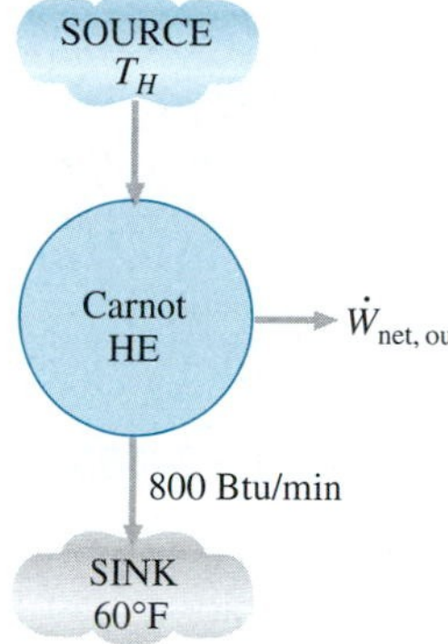

FIGURE P7–82E

7–83E A completely reversible heat engine operates with a source at 1500 R and a sink at 500 R. At what rate must heat be supplied to this engine, in Btu/h, for it to produce 5 hp of power? *Answer:* 19,100 Btu/h

7–84E An inventor claims to have devised a cyclical engine for use in space vehicles that operates with a nuclear-fuel-generated energy source whose temperature is 1000 R and a sink at 550 R that radiates waste heat to deep space. He also claims that this engine produces 5 hp while rejecting heat at a rate of 15,000 Btu/h. Is this claim valid?

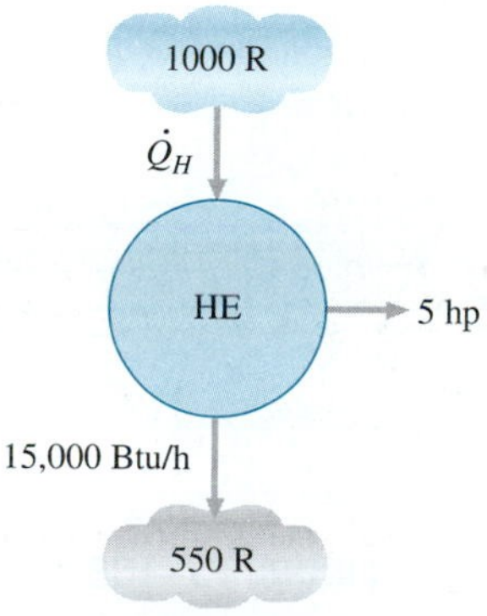

FIGURE P7–84E

7–85 It is claimed that the efficiency of a completely reversible heat engine can be doubled by doubling the temperature of the energy source. Justify the validity of this claim.

7–86 An inventor claims to have developed a heat engine that receives 700 kJ of heat from a source at 500 K and produces 300 kJ of net work while rejecting the waste heat to a sink at 290 K. Is this a reasonable claim? Why?

7–87E An experimentalist claims that, based on his measurements, a heat engine receives 300 Btu of heat from a source of 900 R, converts 160 Btu of it to work, and rejects the rest as waste heat to a sink at 540 R. Are these measurements reasonable? Why?

7–88 A geothermal power plant uses geothermal water extracted at 160°C at a rate of 440 kg/s as the heat source and produces 22 MW of net power. If the environment temperature is 25°C, determine (*a*) the actual thermal efficiency, (*b*) the maximum possible thermal efficiency, and (*c*) the actual rate of heat rejection from this power plant.

Carnot Refrigerators and Heat Pumps

7–89C How can we increase the COP of a Carnot refrigerator?

7–90C What is the highest COP that a refrigerator operating between temperature levels T_L and T_H can have?

7–91C In an effort to conserve energy in a heat-engine cycle, somebody suggests incorporating a refrigerator that will absorb some of the waste energy Q_L and transfer it to the energy source of the heat engine. Is this a smart idea? Explain.

7–92C It is well established that the thermal efficiency of a heat engine increases as the temperature T_L at which heat is rejected from the heat engine decreases. In an effort to increase the efficiency of a power plant, somebody suggests refrigerating the cooling water before it enters the condenser, where heat rejection takes place. Would you be in favor of this idea? Why?

7–93C It is well known that the thermal efficiency of heat engines increases as the temperature of the energy source increases. In an attempt to improve the efficiency of a power plant, somebody suggests transferring heat from the available energy source to a higher-temperature medium by a heat pump before energy is supplied to the power plant. What do you think of this suggestion? Explain.

7–94 Determine the minimum work per unit of heat transfer from the source reservoir that is required to drive a heat pump with thermal energy reservoirs at 460 K and 535 K.

7–95E A thermodynamicist claims to have developed a heat engine with 50 percent thermal efficiency when operating with thermal energy reservoirs at 1260 R and 510 R. Is this claim valid?

7–96 Derive an expression for the COP of a completely reversible refrigerator in terms of the thermal energy reservoir temperatures, T_L and T_H.

7–97 A completely reversible refrigerator is driven by a 10-kW compressor and operates with thermal energy reservoirs at 250 K and 300 K. Calculate the rate of cooling provided by this refrigerator. *Answer:* 50 kW

7–98 A refrigerator is to remove heat from the cooled space at a rate of 300 kJ/min to maintain its temperature at −8°C. If the air surrounding the refrigerator is at 25°C, determine the minimum power input required for this refrigerator. *Answer:* 0.623 kW

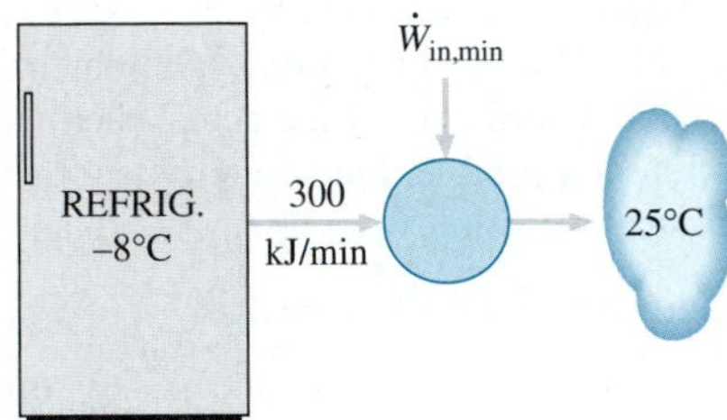

FIGURE P7–98

7–99 A Carnot refrigerator operates in a room in which the temperature is 25°C. The refrigerator consumes 500 W of power when operating and has a COP of 4.5. Determine (*a*) the rate of heat removal from the refrigerated space and (*b*) the temperature of the refrigerated space. *Answers:* (*a*) 135 kJ/min, (*b*) −29.2°C

7–100 An inventor claims to have developed a refrigeration system that removes heat from the closed region at −12°C and transfers it to the surrounding air at 25°C while maintaining a COP of 6.5. Is this claim reasonable? Why?

7–101E An air-conditioning system is used to maintain a house at 75°F when the temperature outside is 95°F. The house is gaining heat through the walls and the windows at a rate of 800 Btu/min, and the heat generation rate within the house from people, lights, and appliances amounts to 100 Btu/min. Determine the minimum power input required for this air-conditioning system. *Answer:* 0.79 hp

7–102 A heat pump is used to heat a house and maintain it at 24°C. On a winter day when the outdoor air temperature is −5°C, the house is estimated to lose heat at a rate of 80,000 kJ/h. Determine the minimum power required to operate this heat pump.

7–103 A heat pump is used to maintain a house at 22°C by extracting heat from the outside air on a day when the outside air temperature is 2°C. The house is estimated to lose heat at a rate of 110,000 kJ/h, and the heat pump consumes 5 kW of electric power when running. Is this heat pump powerful enough to do the job?

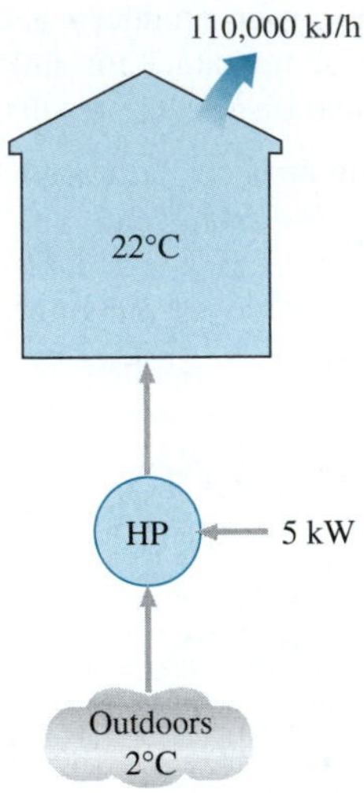

FIGURE P7–103

7–104E A completely reversible refrigerator operates between thermal energy reservoirs at 450 R and 540 R. How many kilowatts of power are required for this device to produce a 15,000-Btu/h cooling effect?

7–105 Calculate and plot the COP of a completely reversible refrigerator as a function of the temperature of the sink up to 500 K with the temperature of the source fixed at 250 K.

7–106 A completely reversible heat pump has a COP of 1.6 and a sink temperature of 300 K. Calculate (*a*) the temperature of the source and (*b*) the rate of heat transfer to the sink when 1.5 kW of power is supplied to this heat pump.

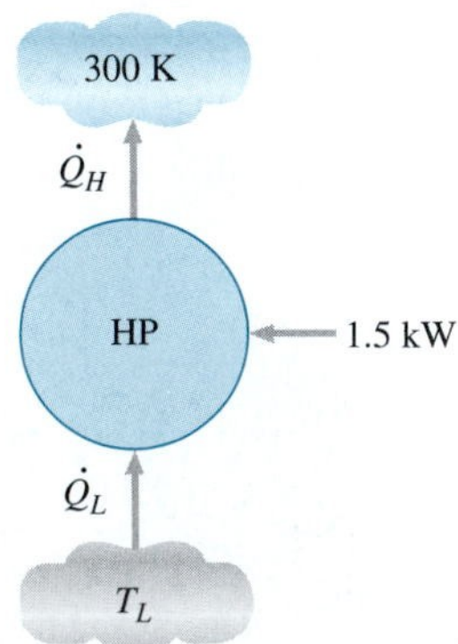

FIGURE P7–106

7–107 The structure of a house is such that it loses heat at a rate of 5400 kJ/h per °C difference between the indoors and outdoors. A heat pump that requires a power input of 6 kW is used to maintain this house at 21°C. Determine the lowest outdoor temperature for which the heat pump can meet the heating requirements of this house. *Answer:* −13.3°C

7–108 The performance of a heat pump degrades (i.e., its COP decreases) as the temperature of the heat source decreases. This makes using heat pumps at locations with severe weather conditions unattractive. Consider a house that is heated and maintained at 20°C by a heat pump during the winter. What is the maximum COP for this heat pump if heat is extracted from the outdoor air at (*a*) 10°C, (*b*) −5°C, and (*c*) −30°C?

7–109E A heat pump is to be used for heating a house in winter. The house is to be maintained at 78°F at all times. When the temperature outdoors drops to 25°F, the heat losses from the house are estimated to be 55,000 Btu/h. Determine the minimum power required to run this heat pump if heat is extracted from (*a*) the outdoor air at 25°F and (*b*) the well water at 50°F.

7–110 A Carnot heat pump is to be used to heat a house and maintain it at 20°C in winter. On a day when the average outdoor temperature remains at about 2°C, the house is estimated to lose heat at a rate of 82,000 kJ/h. If the heat pump consumes 8 kW of power while operating, determine (*a*) how long the heat pump ran on that day; (*b*) the total heating costs, assuming an average price of 8.5¢/kWh for electricity; and (*c*) the heating cost for the same day if resistance heating is used instead of a heat pump. *Answers:* (*a*) 4.19 h, (*b*) $2.85, (*c*) $46.47

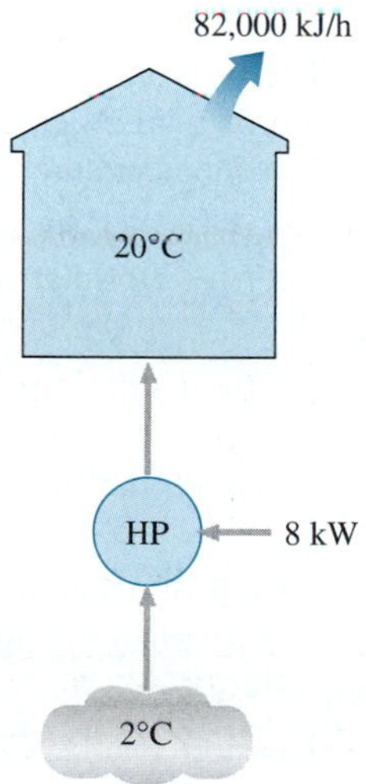

FIGURE P7–110

7–111 A Carnot heat engine receives heat from a reservoir at 900°C at a rate of 800 kJ/min and rejects the waste heat to the ambient air at 27°C. The entire work output of the heat engine is used to drive a refrigerator that removes heat from the refrigerated space at −5°C and transfers it to the same ambient air at 27°C. Determine (*a*) the maximum rate of heat removal from the refrigerated space and (*b*) the total rate of heat rejection to the ambient air. *Answers:* (*a*) 4982 kJ/min, (*b*) 5782 kJ/min

7–112E A Carnot heat engine receives heat from a reservoir at 1700°F at a rate of 700 Btu/min and rejects the waste heat to the ambient air at 80°F. The entire work output of the heat engine is used to drive a refrigerator that removes heat from the refrigerated space at 20°F and transfers it to the same ambient air at 80°F. Determine (*a*) the maximum rate of heat removal from the refrigerated space and (*b*) the total rate of heat rejection to the ambient air. *Answers:* (*a*) 4200 Btu/min, (*b*) 4900 Btu/min

7–113 An air-conditioner with refrigerant-134a as the working fluid is used to keep a room at 26°C by rejecting the waste heat to the outdoor air at 34°C. The room gains heat through the walls and the windows at a rate of 250 kJ/min while the heat generated by the computer, TV, and lights amounts to 900 W. The refrigerant enters the compressor at 500 kPa as a saturated vapor at a rate of 100 L/min and leaves at 1200 kPa and 50°C. Determine (*a*) the actual COP, (*b*) the maximum COP, and (*c*) the minimum volume flow rate of the refrigerant at the compressor inlet for the same compressor inlet and exit conditions. *Answers:* (*a*) 6.59, (*b*) 37.4, (*c*) 17.6 L/min

Review Problems

7–114 Consider a Carnot heat-engine cycle executed in a steady-flow system using steam as the working fluid. The cycle has a thermal efficiency of 30 percent, and steam changes from saturated liquid to saturated vapor at 275°C during the heat addition process. If the mass flow rate of the steam is 3 kg/s, determine the net power output of this engine, in kW.

7–115 A heat pump with a COP of 2.4 is used to heat a house. When running, the heat pump consumes 8 kW of electric power. If the house is losing heat to the outside at an average rate of 40,000 kJ/h and the temperature of the house is 3°C when the heat pump is turned on, determine how long it will take for the temperature in the house to rise to 22°C. Assume the house is well sealed (i.e., no air leaks) and take the entire mass within the house (air, furniture, etc.) to be equivalent to 2000 kg of air.

7–116 A manufacturer of ice cream freezers claims that its product has a coefficient of performance of 1.3 while freezing ice cream at 250 K when the surrounding environment is at 300 K. Is this claim valid?

7–117 A heat pump designer claims to have an air-source heat pump whose coefficient of performance is 1.8 when heating a building whose interior temperature is 300 K and when the atmospheric air surrounding the building is at 260 K. Is this claim valid?

7–118E A heat pump creates a heating effect of 100,000 Btu/h for a space maintained at 530 R while using 3 kW of electrical power. What is the minimum temperature of the source that satisfies the second law of thermodynamics? *Answer:* 476 R

7–119 A thermodynamicist claims to have developed a heat pump with a COP of 1.7 when operating with thermal energy reservoirs at 273 K and 293 K. Is this claim valid?

7–120 Consider a Carnot heat-engine cycle executed in a closed system using 0.01 kg of refrigerant-134a as the working fluid. The cycle has a thermal efficiency of 15 percent, and the refrigerant-134a changes from saturated liquid to saturated vapor at 50°C during the heat addition process. Determine the net work output of this engine per cycle.

7–121 A heat pump with a COP of 2.8 is used to heat an air-tight house. When running, the heat pump consumes 5 kW of power. If the temperature in the house is 7°C when the heat pump is turned on, how long will it take for the heat pump to raise the temperature of the house to 22°C? Is this answer realistic or optimistic? Explain. Assume the entire mass within the house (air, furniture, etc.) is equivalent to 1500 kg of air. *Answer:* 19.2 min

7–122 A promising method of power generation involves collecting and storing solar energy in large artificial lakes a few meters deep, called solar ponds. Solar energy is absorbed by all parts of the pond, and the water temperature rises everywhere. The top part of the pond, however, loses to the atmosphere much of the heat it absorbs, and as a result, its temperature drops. This cool water serves as insulation for the bottom part of the pond and helps trap the energy there. Usually, salt is planted at the bottom of the pond to prevent the rise of this hot water to the top. A power plant that uses an organic fluid, such as alcohol, as the working fluid can be operated between the top and the bottom portions of the pond. If the water temperature is 35°C near the surface and 80°C near the bottom of the pond, determine the maximum thermal efficiency that this power plant can have. Is it realistic to use 35 and 80°C for temperatures in the calculations? Explain. *Answer:* 12.7 percent

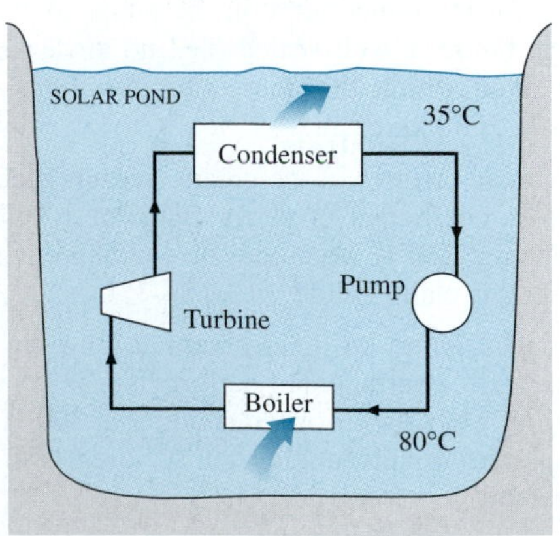

FIGURE P7–122

7–123 Consider a Carnot heat-engine cycle executed in a closed system using 0.0103 kg of steam as the working fluid. It is known that the maximum absolute temperature in the cycle is twice the minimum absolute temperature, and the net work output of the cycle is 25 kJ. If the steam changes from saturated vapor to saturated liquid during heat rejection, determine the temperature of the steam during the heat rejection process.

7–124 Reconsider Prob. 7–123. Using EES (or other) software, investigate the effect of the net work output on the required temperature of the steam during the heat rejection process. Let the work output vary from 15 to 25 kJ.

7–125 Consider a Carnot refrigeration cycle executed in a closed system in the saturated liquid–vapor mixture region using 0.96 kg of refrigerant-134a as the working fluid. It is known that the maximum absolute temperature in the cycle is 1.2 times the minimum absolute temperature, and the net work input to the cycle is 22 kJ. If the refrigerant changes from saturated vapor to saturated liquid during the heat rejection process, determine the minimum pressure in the cycle.

7–126 Reconsider Prob. 7–125. Using EES (or other) software, investigate the effect of the net work input on the minimum pressure. Let the work input vary from 10 to 30 kJ. Plot the minimum pressure in the refrigeration cycle as a function of net work input, and discuss the results.

7–127 Consider two Carnot heat engines operating in series. The first engine receives heat from the reservoir at 1800 K and rejects the waste heat to another reservoir at temperature *T*. The second engine receives this energy rejected by the first one, converts some of it to work, and rejects the rest to a reservoir at 300 K. If the thermal efficiencies of both engines are the same, determine the temperature *T*. *Answer:* 735 K

7–128 Prove that a refrigerator's COP cannot exceed that of a completely reversible refrigerator that shares the same thermal energy reservoirs.

7–129E Calculate and plot the thermal efficiency of a completely reversible heat engine as a function of the source temperature up to 2000 R with the sink temperature fixed at 500 R.

7–130 Prove that the COP of all completely reversible refrigerators must be the same when the reservoir temperatures are the same.

7–131 Derive an expression for the COP of a completely reversible heat pump in terms of the thermal energy reservoir temperatures, T_L and T_H.

7–132 A Carnot heat engine receives heat at 750 K and rejects the waste heat to the environment at 300 K. The entire work output of the heat engine is used to drive a Carnot refrigerator that removes heat from the cooled space at −15°C at a rate of 400 kJ/min and rejects it to the same environment at 300 K. Determine (*a*) the rate of heat supplied to the heat engine and (*b*) the total rate of heat rejection to the environment.

7–133 Reconsider Prob. 7–132. Using EES (or other) software, investigate the effects of the heat engine source temperature, the environment temperature, and the cooled space temperature on the required heat supply to the heat engine and the total rate of heat rejection to the environment. Let the source temperature vary from 500 to 1000 K, the environment temperature vary from 275 to 325 K, and the cooled space temperature vary from −20 to 0°C. Plot the required heat supply against the source temperature for the cooled space temperature of −15°C and environment temperatures of 275, 300, and 325 K, and discuss the results.

7–134 A heat engine operates between two reservoirs at 800 and 20°C. One-half of the work output of the heat engine is used to drive a Carnot heat pump that removes heat from the cold surroundings at 2°C and transfers it to a house maintained at 22°C. If the house is losing heat at a rate of 62,000 kJ/h, determine the minimum rate of heat supply to the heat engine required to keep the house at 22°C.

7–135 Consider a Carnot refrigeration cycle executed in a closed system in the saturated liquid–vapor mixture region using 0.8 kg of refrigerant-134a as the working fluid. The maximum and the minimum temperatures in the cycle are 20°C and −8°C, respectively. It is known that the refrigerant is saturated liquid at the end of the heat rejection process, and the net work input to the cycle is 15 kJ. Determine the fraction of the mass of the refrigerant that vaporizes during the heat addition process, and the pressure at the end of the heat rejection process.

7–136 Consider a Carnot heat-pump cycle executed in a steady-flow system in the saturated liquid–vapor mixture region using refrigerant-134a flowing at a rate of 0.264 kg/s as the working fluid. It is known that the maximum absolute temperature in the cycle is 1.25 times the minimum absolute temperature, and the net power input to the cycle is 7 kW. If the refrigerant changes from saturated vapor to saturated liquid during the heat rejection process, determine the ratio of the maximum to minimum pressures in the cycle.

7–137 A Carnot heat engine is operating between a source at T_H and a sink at T_L. If it is desired to double the thermal efficiency of this engine, what should the new source temperature be? Assume the sink temperature is held constant.

7–138E The "Energy Guide" label on a washing machine indicates that the washer will use $33 worth of hot water if the water is heated by a gas water heater at a natural gas rate of $1.21/therm. If the water is heated from 60 to 130°F, determine how many gallons of hot water an average family uses per week. Disregard the electricity consumed by the washer, and take the efficiency of the gas water heater to be 58 percent.

7–139 A typical electric water heater has an efficiency of 90 percent and costs $390 a year to operate at a unit cost of electricity of $0.08/kWh. A typical heat pump–powered water heater has a COP of 2.2 but costs about $800 more to install. Determine how many years it will take for the heat pump water heater to pay for its cost differential from the energy it saves.

Water heater

© *The McGraw-Hill Companies, Inc. Jill Braaten, photographer*

FIGURE P7–139

7–140 Reconsider Prob. 7–139. Using EES (or other) software, investigate the effect of the heat pump COP on the yearly operation costs and the number of years required to break even. Let the COP vary from 2 to 5. Plot the payback period against the COP and discuss the results.

7–141 A homeowner is trying to decide between a high-efficiency natural gas furnace with an efficiency of 97 percent and a ground-source heat pump with a COP of 3.5. The unit costs of electricity and natural gas are $0.092/kWh and $1.42/therm (1 therm = 105,500 kJ). Determine which system will have a lower energy cost.

Design and Essay Problems

7–142 Show that the work produced by a reversible process exceeds that produced by an equivalent irreversible process by considering a weight moving down a plane both with and without friction.

7–143 The sun supplies electromagnetic energy to the earth. It appears to have an effective temperature of approximately 5800 K. On a clear summer day in North America, the energy incident on a surface facing the sun is approximately 0.95 kW/m^2. The electromagnetic solar energy can be converted into thermal energy by being absorbed on a darkened surface. How might you characterize the work potential of the sun's energy when it is to be used to produce work?

7–144 In the search to reduce thermal pollution and take advantage of renewable energy sources, some people have proposed that we take advantage of such sources as discharges from electrical power plants, geothermal energy, and ocean thermal energy. Although many of these sources contain an enormous amount of energy, the amount of work they are capable of producing is limited. How might you use the work potential to assign an "energy quality" to these proposed sources? Test your proposed "energy quality" measure by applying it to the ocean thermal source, where the temperature 30 m below the surface is perhaps 5ºC lower than at the surface. Apply it also to the geothermal water source, where the temperature 2 to 3 km below the surface is perhaps 150ºC hotter than at the surface.

7–145 The maximum work that can be produced by a heat engine may be increased by lowering the temperature of its sink. Hence, it seems logical to use refrigerators to cool the sink and reduce its temperature. What is the fallacy in this logic?

7–146 Using a timer (or watch) and a thermometer, conduct the following experiment to determine the rate of heat gain of your refrigerator. First make sure that the door of the refrigerator is not opened for at least a few hours so that steady operating conditions are established. Start the timer when the refrigerator stops running and measure the time Δt_1 it stays off before it kicks in. Then measure the time Δt_2 it stays on. Noting that the heat removed during Δt_2 is equal to the heat gain of the refrigerator during $\Delta t_1 + \Delta t_2$ and using the power consumed by the refrigerator when it is running, determine the average rate of heat gain for your refrigerator, in W. Take the COP (coefficient of performance) of your refrigerator to be 1.3 if it is not available.

7–147 Design a hydrocooling unit that can cool fruits and vegetables from 30 to 5°C at a rate of 20,000 kg/h under the following conditions:

The unit will be of flood type, which will cool the products as they are conveyed into the channel filled with water. The products will be dropped into the channel filled with water at one end and be picked up at the other end. The channel can be as wide as 3 m and as high as 90 cm. The water is to be circulated and cooled by the evaporator section of a refrigeration system. The refrigerant temperature inside the coils is to be −2°C, and the water temperature is not to drop below 1°C and not to exceed 6°C.

Assuming reasonable values for the average product density, specific heat, and porosity (the fraction of air volume in a box), recommend reasonable values for (*a*) the water velocity through the channel and (*b*) the refrigeration capacity of the refrigeration system.

Clausius inequality

$\oint \left(\frac{\delta Q}{T}\right)_C = 0$ reversible $= \Delta S$

" < 0 irreversible $< \Delta S$

" > 0 impossible

$\Delta S = S_2 - S_1 = \int \frac{\delta Q}{T} + S_{gen}$

ΔS same for reversible and irreversible process

Irreversibilities:
- friction
- unrestrained expansion
- mixing 2 fluids
- heat transfer across a finite T difference
- electric resistance
- inelastic deformation of solids
- chemical reaction

$S = 0$ at $T = 0K$ absolute zero

Mechanism	Possibility of S transfer
Heat	Yes (open and closed)
Work	No — can be generated
Mass	Yes (open)

$\Delta S = S_{in} - S_{out} + S_{gen}$

S_{gen} is path dependent

3 ways to increase entropy:
1) heat transfers
2) irreversibilities
3) mass flow

$\Delta S = 0$ when:
1) cycle
2) irreversible process + losing heat
3) closed reversible adiabatic process
4) open, steady flow, reversible adiabatic w/ a single stream.

No process is truly reversible

Chapter 8

ENTROPY

In Chap. 7, we introduced the second law of thermodynamics and applied it to cycles and cyclic devices. In this chapter, we apply the second law to processes. The first law of thermodynamics deals with the property *energy* and the conservation of it. The second law leads to the definition of a new property called *entropy*. Entropy is a somewhat abstract property, and it is difficult to give a physical description of it without considering the microscopic state of the system. Entropy is best understood and appreciated by studying its uses in commonly encountered engineering processes, and this is what we intend to do.

This chapter starts with a discussion of the Clausius inequality, which forms the basis for the definition of entropy, and continues with the increase of entropy principle. Unlike energy, entropy is a nonconserved property, and there is no such thing as *conservation of entropy*. Next, the entropy changes that take place during processes for pure substances, incompressible substances, and ideal gases are discussed, and a special class of idealized processes, called *isentropic processes*, is examined. Then, the reversible steady-flow work and the isentropic efficiencies of various engineering devices such as turbines and compressors are considered. Finally, entropy balance is introduced and applied to various systems.

Objectives

The objectives of this chapter are to:

- Apply the second law of thermodynamics to processes.
- Define a new property called *entropy* to quantify the second-law effects.
- Establish the *increase of entropy principle*.
- Calculate the entropy changes that take place during processes for pure substances, incompressible substances, and ideal gases.
- Examine a special class of idealized processes, called *isentropic processes*, and develop the property relations for these processes.
- Derive the reversible steady-flow work relations.
- Develop the isentropic efficiencies for various steady-flow devices.
- Introduce and apply the entropy balance to various systems.

8–1 ▪ ENTROPY

The second law of thermodynamics often leads to expressions that involve inequalities. An irreversible (i.e., actual) heat engine, for example, is less efficient than a reversible one operating between the same two thermal energy reservoirs. Likewise, an irreversible refrigerator or a heat pump has a lower coefficient of performance (COP) than a reversible one operating between the same temperature limits. Another important inequality that has major consequences in thermodynamics is the **Clausius inequality**. It was first stated by the German physicist R. J. E. Clausius (1822–1888), one of the founders of thermodynamics, and is expressed as

$$\oint \frac{\delta Q}{T} \leq 0$$

That is, *the cyclic integral of $\delta Q/T$ is always less than or equal to zero*. This inequality is valid for all cycles, reversible or irreversible. The symbol $\oint$ (integral symbol with a circle in the middle) is used to indicate that the integration is to be performed over the entire cycle. Any heat transfer to or from a system can be considered to consist of differential amounts of heat transfer. Then the cyclic integral of $\delta Q/T$ can be viewed as the sum of all these differential amounts of heat transfer divided by the temperature at the boundary.

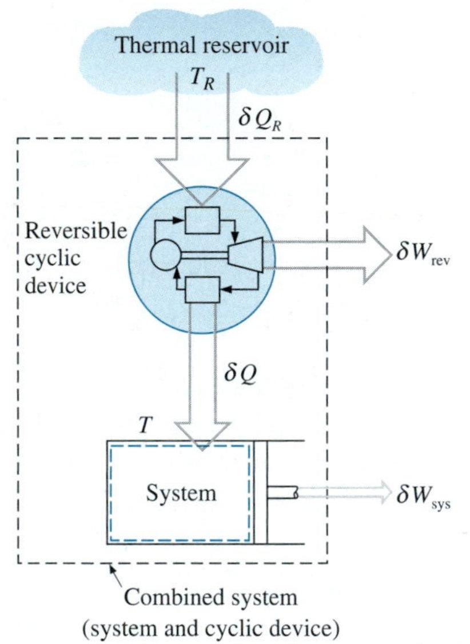

FIGURE 8–1

The system considered in the development of the Clausius inequality.

To demonstrate the validity of the Clausius inequality, consider a system connected to a thermal energy reservoir at a constant thermodynamic (i.e., absolute) temperature of T_R through a *reversible* cyclic device (Fig. 8–1). The cyclic device receives heat δQ_R from the reservoir and supplies heat δQ to the system whose temperature at that part of the boundary is T (a variable) while producing work δW_{rev}. The system produces work δW_{sys} as a result of this heat transfer. Applying the energy balance to the combined system identified by dashed lines yields

$$\delta W_C = \delta Q_R - dE_C$$

where δW_C is the total work of the combined system ($\delta W_{rev} + \delta W_{sys}$) and dE_C is the change in the total energy of the combined system. Considering that the cyclic device is a *reversible* one, we have

$$\frac{\delta Q_R}{T_R} = \frac{\delta Q}{T}$$

where the sign of δQ is determined with respect to the system (positive if *to* the system and negative if *from* the system) and the sign of δQ_R is determined with respect to the reversible cyclic device. Eliminating δQ_R from the two relations above yields

$$\delta W_C = T_R \frac{\delta Q}{T} - dE_C$$

We now let the system undergo a cycle while the cyclic device undergoes an integral number of cycles. Then the preceding relation becomes

$$W_C = T_R \oint \frac{\delta Q}{T}$$

since the cyclic integral of energy (the net change in the energy, which is a property, during a cycle) is zero. Here W_C is the cyclic integral of δW_C, and it represents the net work for the combined cycle.

It appears that the combined system is exchanging heat with a single thermal energy reservoir while involving (producing or consuming) work W_C during a cycle. On the basis of the Kelvin–Planck statement of the second law, which states that *no system can produce a net amount of work while operating in a cycle and exchanging heat with a single thermal energy reservoir*, we reason that W_C cannot be a work output, and thus it cannot be a positive quantity. Considering that T_R is the thermodynamic temperature and thus a positive quantity, we must have

$$\oint \frac{\delta Q}{T} \leq 0 \tag{8–1}$$

which is the *Clausius inequality*. This inequality is valid for all thermodynamic cycles, reversible or irreversible, including the refrigeration cycles.

If no irreversibilities occur within the system as well as the reversible cyclic device, then the cycle undergone by the combined system is internally reversible. As such, it can be reversed. In the reversed cycle case, all the quantities have the same magnitude but the opposite sign. Therefore, the work W_C, which could not be a positive quantity in the regular case, cannot be a negative quantity in the reversed case. Then it follows that $W_{C,\text{int rev}} = 0$ since it cannot be a positive or negative quantity, and therefore

$$\oint \left(\frac{\delta Q}{T}\right)_{\text{int rev}} = 0 \tag{8–2}$$

for internally reversible cycles. Thus, we conclude that *the equality in the Clausius inequality holds for totally or just internally reversible cycles and the inequality for the irreversible ones*.

To develop a relation for the definition of entropy, let us examine Eq. 8–2 more closely. Here we have a quantity whose cyclic integral is zero. Let us think for a moment what kind of quantities can have this characteristic. We know that the cyclic integral of *work* is not zero. (It is a good thing that it is not. Otherwise, heat engines that work on a cycle such as steam power plants would produce zero net work.) Neither is the cyclic integral of heat.

Now consider the volume occupied by a gas in a piston–cylinder device undergoing a cycle, as shown in Fig. 8–2. When the piston returns to its initial position at the end of a cycle, the volume of the gas also returns to its initial value. Thus the net change in volume during a cycle is zero. This is also expressed as

$$\oint dV = 0 \tag{8–3}$$

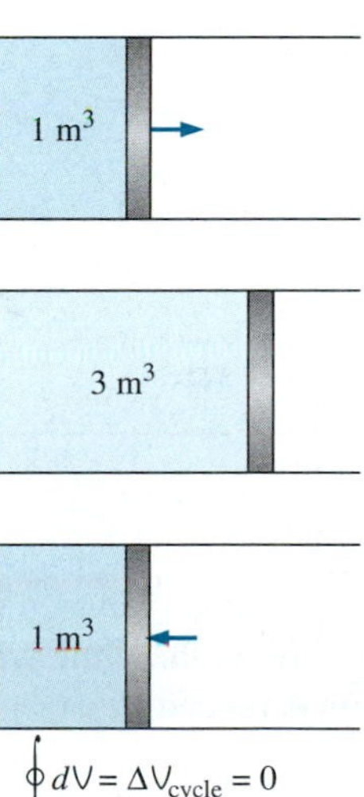

FIGURE 8–2
The net change in volume (a property) during a cycle is always zero.

That is, the cyclic integral of volume (or any other property) is zero. Conversely, a quantity whose cyclic integral is zero depends on the *state* only and not the process path, and thus it is a property. Therefore, the quantity $(\delta Q/T)_{\text{int rev}}$ must represent a property in the differential form.

Clausius realized in 1865 that he had discovered a new thermodynamic property, and he chose to name this property **entropy**. It is designated S and is defined as

$$dS = \left(\frac{\delta Q}{T}\right)_{\text{int rev}} \qquad \text{(kJ/K)} \tag{8–4}$$

Entropy is an extensive property of a system and sometimes is referred to as *total entropy*. Entropy per unit mass, designated s, is an intensive property and has the unit kJ/kg · K. The term *entropy* is generally used to refer to both total entropy and entropy per unit mass since the context usually clarifies which one is meant.

The entropy change of a system during a process can be determined by integrating Eq. 8–4 between the initial and the final states:

$$\Delta S = S_2 - S_1 = \int_1^2 \left(\frac{\delta Q}{T}\right)_{\text{int rev}} \quad \text{(kJ/K)} \tag{8–5}$$

Notice that we have actually defined the *change* in entropy instead of entropy itself, just as we defined the change in energy instead of the energy itself when we developed the first-law relation. Absolute values of entropy are determined on the basis of the third law of thermodynamics, which is discussed later in this chapter. Engineers are usually concerned with the *changes* in entropy. Therefore, the entropy of a substance can be assigned a zero value at some arbitrarily selected reference state, and the entropy values at other states can be determined from Eq. 8–5 by choosing state 1 to be the reference state ($S = 0$) and state 2 to be the state at which entropy is to be determined.

To perform the integration in Eq. 8–5, one needs to know the relation between Q and T during a process. This relation is often not available, and the integral in Eq. 8–5 can be performed for a few cases only. For the majority of cases we have to rely on tabulated data for entropy.

Note that entropy is a property, and like all other properties, it has fixed values at fixed states. Therefore, the entropy change ΔS between two specified states is the same no matter what path, reversible or irreversible, is followed during a process (Fig. 8–3).

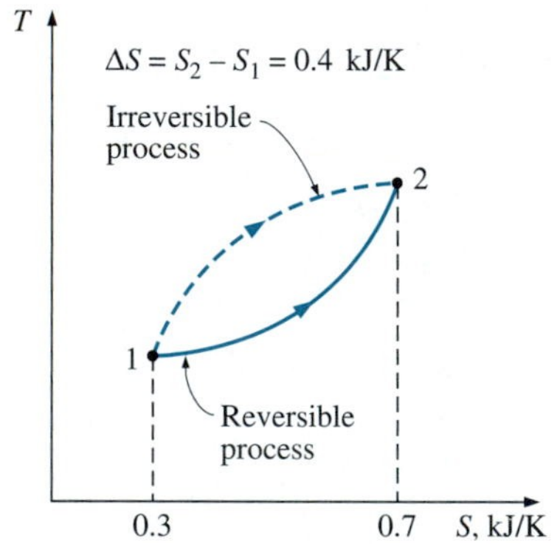

FIGURE 8–3

The entropy change between two specified states is the same whether the process is reversible or irreversible.

Also note that the integral of $\delta Q/T$ gives us the value of entropy change *only if* the integration is carried out along an *internally reversible* path between the two states. The integral of $\delta Q/T$ along an irreversible path is not a property, and in general, different values will be obtained when the integration is carried out along different irreversible paths. Therefore, even for irreversible processes, the entropy change should be determined by carrying out this integration along some convenient *imaginary* internally reversible path between the specified states.

A Special Case: Internally Reversible Isothermal Heat Transfer Processes

Recall that isothermal heat transfer processes are internally reversible. Therefore, the entropy change of a system during an internally reversible isothermal heat transfer process can be determined by performing the integration in Eq. 8–5:

$$\Delta S = \int_1^2 \left(\frac{\delta Q}{T}\right)_{\text{int rev}} = \int_1^2 \left(\frac{\delta Q}{T_0}\right)_{\text{int rev}} = \frac{1}{T_0}\int_1^2 (\delta Q)_{\text{int rev}}$$

which reduces to

$$\Delta S = \frac{Q}{T_0} \quad \text{(kJ/K)} \tag{8–6}$$

where T_0 is the constant temperature of the system and Q is the heat transfer for the internally reversible process. Equation 8–6 is particularly useful for determining the entropy changes of thermal energy reservoirs that can absorb or supply heat indefinitely at a constant temperature.

Notice that the entropy change of a system during an internally reversible isothermal process can be positive or negative, depending on the direction of heat transfer. Heat transfer to a system increases the entropy of a system, whereas heat transfer from a system decreases it. In fact, losing heat is the only way the entropy of a system can be decreased.

EXAMPLE 8–1 Entropy Change during an Isothermal Process

A piston–cylinder device contains a liquid–vapor mixture of water at 300 K. During a constant-pressure process, 750 kJ of heat is transferred to the water. As a result, part of the liquid in the cylinder vaporizes. Determine the entropy change of the water during this process.

Solution Heat is transferred to a liquid–vapor mixture of water in a piston–cylinder device at constant pressure. The entropy change of water is to be determined.

Assumptions No irreversibilities occur within the system boundaries during the process.

Analysis We take the *entire water* (liquid + vapor) in the cylinder as the system (Fig. 8–4). This is a *closed system* since no mass crosses the system boundary during the process. We note that the temperature of the system remains constant at 300 K during this process since the temperature of a pure substance remains constant at the saturation value during a phase-change process at constant pressure.

The system undergoes an internally reversible, isothermal process, and thus its entropy change can be determined directly from Eq. 8–6 to be

$$\Delta S_{\text{sys,isothermal}} = \frac{Q}{T_{\text{sys}}} = \frac{750 \text{ kJ}}{300 \text{ K}} = \mathbf{2.5\ kJ/K}$$

Discussion Note that the entropy change of the system is positive, as expected, since heat transfer is *to* the system.

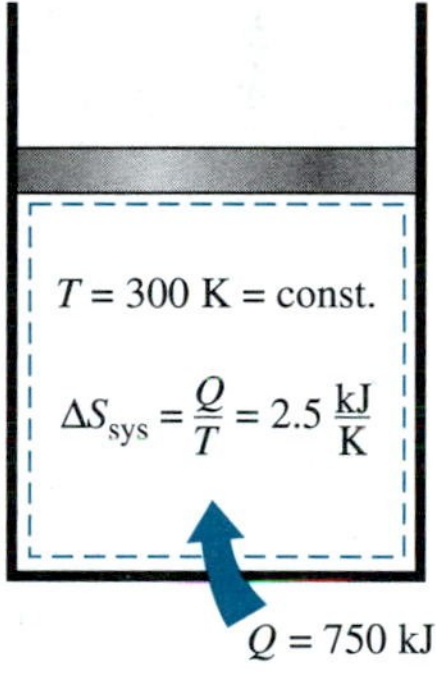

FIGURE 8–4

Schematic for Example 8–1.

8–2 ▪ THE INCREASE OF ENTROPY PRINCIPLE

Consider a cycle that is made up of two processes: process 1-2, which is arbitrary (reversible or irreversible), and process 2-1, which is internally reversible, as shown in Figure 8–5. From the Clausius inequality,

$$\oint \frac{\delta Q}{T} \leq 0$$

or

$$\int_1^2 \frac{\delta Q}{T} + \int_2^1 \left(\frac{\delta Q}{T}\right)_{\text{int rev}} \leq 0$$

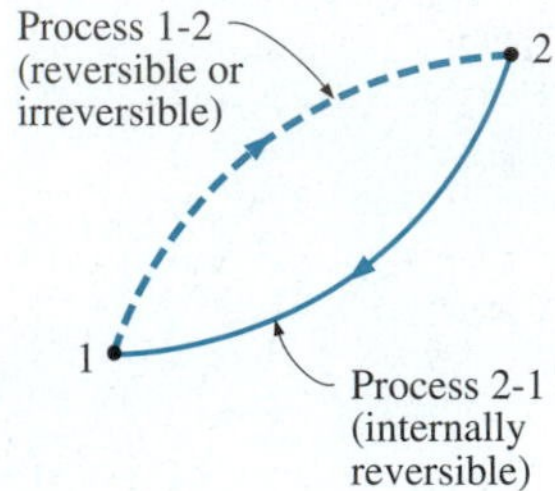

FIGURE 8–5

A cycle composed of a reversible and an irreversible process.

The second integral in the previous relation is recognized as the entropy change $S_1 - S_2$. Therefore,

$$\int_1^2 \frac{\delta Q}{T} + S_1 - S_2 \leq 0$$

which can be rearranged as

$$S_2 - S_1 \geq \int_1^2 \frac{\delta Q}{T} \tag{8–7}$$

It can also be expressed in differential form as

$$dS \geq \frac{\delta Q}{T} \tag{8–8}$$

where the equality holds for an internally reversible process and the inequality for an irreversible process. We may conclude from these equations that the entropy change of a closed system during an irreversible process is greater than the integral of $\delta Q/T$ evaluated for that process. In the limiting case of a reversible process, these two quantities become equal. We again emphasize that T in these relations is the *thermodynamic temperature* at the *boundary* where the differential heat δQ is transferred between the system and the surroundings.

The quantity $\Delta S = S_2 - S_1$ represents the *entropy change* of the system. For a reversible process, it becomes equal to $\int_1^2 \delta Q/T$, which represents the *entropy transfer* with heat.

The inequality sign in the preceding relations is a constant reminder that the entropy change of a closed system during an irreversible process is always greater than the entropy transfer. That is, some entropy is *generated* or *created* during an irreversible process, and this generation is due entirely to the presence of irreversibilities. The entropy generated during a process is called **entropy generation** and is denoted by S_{gen}. Noting that the difference between the entropy change of a closed system and the entropy transfer is equal to entropy generation, Eq. 8–7 can be rewritten as an equality as

$$\Delta S_{sys} = S_2 - S_1 = \int_1^2 \frac{\delta Q}{T} + S_{gen} \tag{8–9}$$

Note that the entropy generation S_{gen} is always a *positive* quantity or zero. Its value depends on the process, and thus it is *not* a property of the system. Also, in the absence of any entropy transfer, the entropy change of a system is equal to the entropy generation.

Equation 8–7 has far-reaching implications in thermodynamics. For an isolated system (or simply an adiabatic closed system), the heat transfer is zero, and Eq. 8–7 reduces to

$$\Delta S_{isolated} \geq 0 \tag{8–10}$$

This equation can be expressed as *the entropy of an isolated system during a process always increases or, in the limiting case of a reversible process, remains constant.* In other words, it *never* decreases. This is known as the **increase of entropy principle**. Note that in the absence of any heat transfer, entropy change is due to irreversibilities only, and their effect is always to increase entropy.

Entropy is an extensive property, and thus the total entropy of a system is equal to the sum of the entropies of the parts of the system. An isolated system may consist of any number of subsystems (Fig. 8–6). A system and its surroundings, for example, constitute an isolated system since both can be enclosed by a sufficiently large arbitrary boundary across which there is no heat, work, or mass transfer (Fig. 8–7). Therefore, a system and its surroundings can be viewed as the two subsystems of an isolated system, and the entropy change of this isolated system during a process is the sum of the entropy changes of the system and its surroundings, which is equal to the entropy generation since an isolated system involves no entropy transfer. That is,

$$S_{gen} = \Delta S_{total} = \Delta S_{sys} + \Delta S_{surr} \geq 0 \tag{8–11}$$

where the equality holds for reversible processes and the inequality for irreversible ones. Note that ΔS_{surr} refers to the change in the entropy of the surroundings as a result of the occurrence of the process under consideration.

Since no actual process is truly reversible, we can conclude that some entropy is generated during a process, and therefore the entropy of the universe, which can be considered to be an isolated system, is continuously increasing. The more irreversible a process, the larger the entropy generated during that process. No entropy is generated during reversible processes ($S_{gen} = 0$).

Entropy increase of the universe is a major concern not only to engineers but also to philosophers, theologians, economists, and environmentalists since entropy is viewed as a measure of the disorder (or "mixed-up-ness") in the universe.

The increase of entropy principle does not imply that the entropy of a system cannot decrease. The entropy change of a system *can* be negative during a process (Fig. 8–8), but entropy generation cannot. The increase of entropy principle can be summarized as follows:

$$S_{gen} \begin{cases} > 0 & \text{Irreversible process} \\ = 0 & \text{Reversible process} \\ < 0 & \text{Impossible process} \end{cases}$$

This relation serves as a criterion in determining whether a process is reversible, irreversible, or impossible.

Things in nature have a tendency to change until they attain a state of equilibrium. The increase of entropy principle dictates that the entropy of an isolated system increases until the entropy of the system reaches a *maximum* value. At that point, the system is said to have reached an equilibrium state since the increase of entropy principle prohibits the system from undergoing any change of state that results in a decrease in entropy.

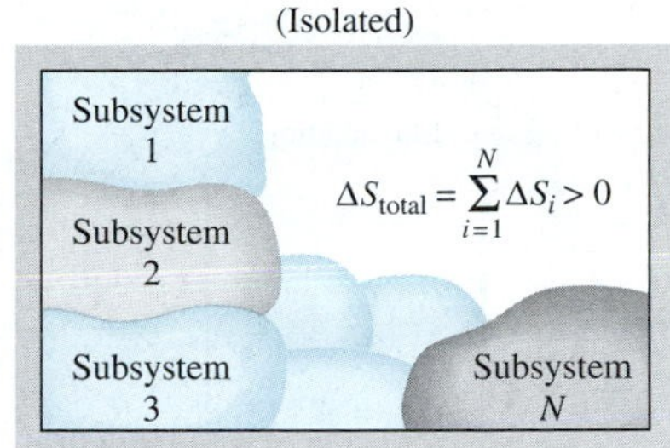

FIGURE 8–6

The entropy change of an isolated system is the sum of the entropy changes of its components, and is never less than zero.

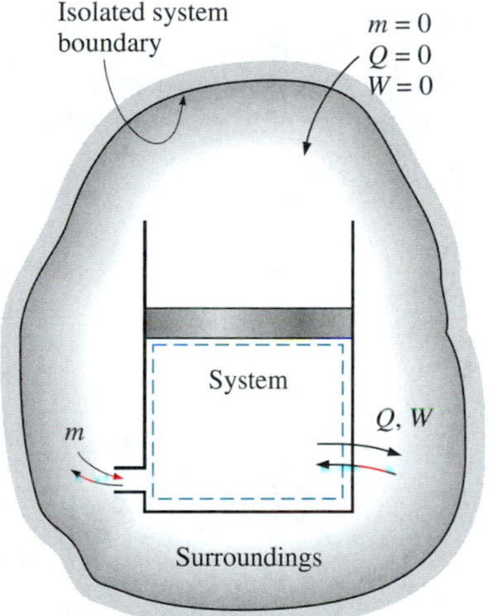

FIGURE 8–7

A system and its surroundings form an isolated system.

Some Remarks about Entropy

In light of the preceding discussions, we draw the following conclusions:

1. Processes can occur in a *certain* direction only, not in *any* direction. A process must proceed in the direction that complies with the increase of entropy principle, that is, $S_{gen} \geq 0$. A process that violates this principle is impossible. This principle often forces chemical reactions to come to a halt before reaching completion.

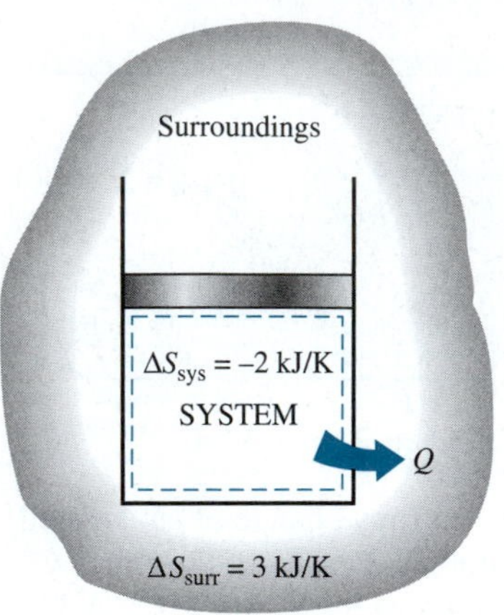

FIGURE 8–8

The entropy change of a system can be negative, but the entropy generation cannot.

2. Entropy is a *nonconserved property*, and there is *no* such thing as the *conservation of entropy principle*. Entropy is conserved during the idealized reversible processes only and increases during *all* actual processes.
3. The performance of engineering systems is degraded by the presence of irreversibilities, and *entropy generation* is a measure of the magnitudes of the irreversibilities present during that process. The greater the extent of irreversibilities, the greater the entropy generation. Therefore, entropy generation can be used as a quantitative measure of irreversibilities associated with a process. It is also used to establish criteria for the performance of engineering devices. This point is illustrated further in Example 8–2.

EXAMPLE 8–2 Entropy Generation during Heat Transfer Processes

A heat source at 800 K loses 2000 kJ of heat to a sink at (*a*) 500 K and (*b*) 750 K. Determine which heat transfer process is more irreversible.

Solution Heat is transferred from a heat source to two heat sinks at different temperatures. The heat transfer process that is more irreversible is to be determined.

Analysis A sketch of the reservoirs is shown in Fig. 8–9. Both cases involve heat transfer through a finite temperature difference, and therefore both are irreversible. The magnitude of the irreversibility associated with each process can be determined by calculating the total entropy change for each case. The total entropy change for a heat transfer process involving two reservoirs (a source and a sink) is the sum of the entropy changes of each reservoir since the two reservoirs form an adiabatic system.

Or do they? The problem statement gives the impression that the two reservoirs are in direct contact during the heat transfer process. But this cannot be the case since the temperature at a point can have only one value, and thus it cannot be 800 K on one side of the point of contact and 500 K on the other side. In other words, the temperature function cannot have a jump discontinuity. Therefore, it is reasonable to assume that the two reservoirs are separated by a partition through which the temperature drops from 800 K on one side to 500 K (or 750 K) on the other. Therefore, the entropy change of the partition should also be considered when evaluating the total entropy change for this process. However, considering that entropy is a property and the values of properties depend on the state of a system, we can argue that the entropy change of the partition is zero since the partition appears to have undergone a *steady* process and thus experienced no change in its properties at any point. We base this argument on the fact that the temperature on both sides of the partition and thus throughout remains constant during this process. Therefore, we are justified to assume that $\Delta S_{\text{partition}} = 0$ since the entropy (as well as the energy) content of the partition remains constant during this process.

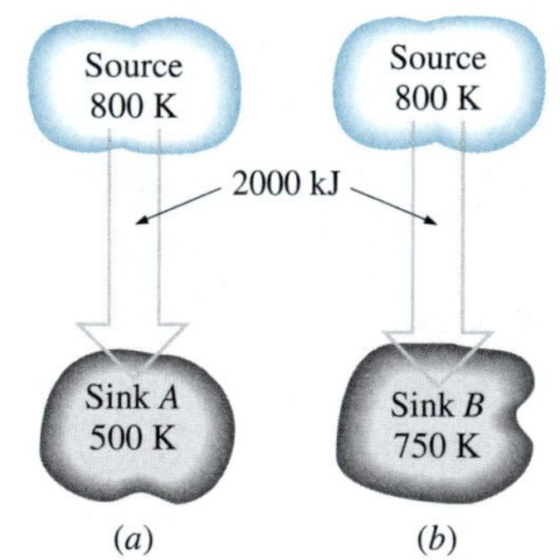

FIGURE 8–9

Schematic for Example 8–2.

The entropy change for each reservoir can be determined from Eq. 8–6 since each reservoir undergoes an internally reversible, isothermal process.
(*a*) For the heat transfer process to a sink at 500 K:

$$\Delta S_{\text{source}} = \frac{Q_{\text{source}}}{T_{\text{source}}} = \frac{-2000 \text{ kJ}}{800 \text{ K}} = -2.5 \text{ kJ/K}$$

$$\Delta S_{\text{sink}} = \frac{Q_{\text{sink}}}{T_{\text{sink}}} = \frac{2000 \text{ kJ}}{500 \text{ K}} = +4.0 \text{ kJ/K}$$

and

$$S_{\text{gen}} = \Delta S_{\text{total}} = \Delta S_{\text{source}} + \Delta S_{\text{sink}} = (-2.5 + 4.0) \text{ kJ/K} = \mathbf{1.5 \text{ kJ/K}}$$

Therefore, 1.5 kJ/K of entropy is generated during this process. Noting that both reservoirs have undergone internally reversible processes, the entire entropy generation took place in the partition.
(*b*) Repeating the calculations in part (*a*) for a sink temperature of 750 K, we obtain

$$\Delta S_{\text{source}} = -2.5 \text{ kJ/k}$$

$$\Delta S_{\text{sink}} = +2.7 \text{ kJ/K}$$

and

$$S_{\text{gen}} = \Delta S_{\text{total}} = (-2.5 + 2.7) \text{ kJ/K} = \mathbf{0.2 \text{ kJ/K}}$$

The total entropy change for the process in part (*b*) is smaller, and therefore it is less irreversible. This is expected since the process in (*b*) involves a smaller temperature difference and thus a smaller irreversibility.
Discussion The irreversibilities associated with both processes could be eliminated by operating a Carnot heat engine between the source and the sink. For this case it can be shown that $\Delta S_{\text{total}} = 0$.

8–3 ▪ ENTROPY CHANGE OF PURE SUBSTANCES

Entropy is a property, and thus the value of entropy of a system is fixed once the state of the system is fixed. Specifying two intensive independent properties fixes the state of a simple compressible system, and thus the value of entropy, as well as the values of other properties at that state. Starting with its defining relation, the entropy change of a substance can be expressed in terms of other properties (see Sec. 8–7). But in general, these relations are too complicated and are not practical to use for hand calculations. Therefore, using a suitable reference state, the entropies of substances are evaluated from measurable property data following rather involved computations, and the results are tabulated in the same manner as the other properties such as v, u, and h (Fig. 8–10).

The entropy values in the property tables are given relative to an arbitrary reference state. In steam tables the entropy of saturated liquid s_f at 0.01°C is assigned the value of zero. For refrigerant-134a, the zero value is assigned to saturated liquid at −40°C. The entropy values become negative at temperatures below the reference value.

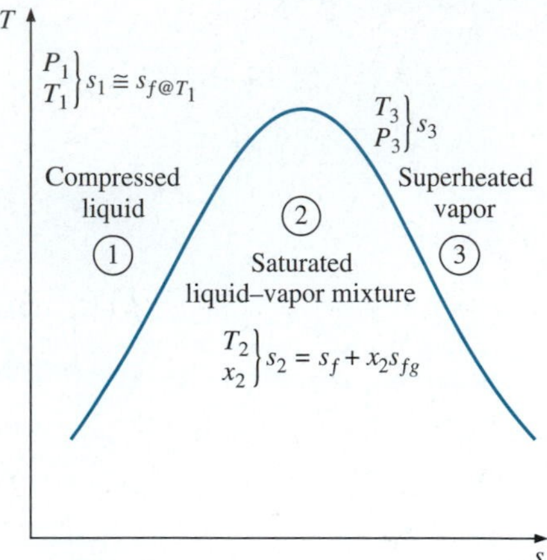

FIGURE 8–10

The entropy of a pure substance is determined from the tables (like other properties).

The value of entropy at a specified state is determined just like any other property. In the compressed liquid and superheated vapor regions, it can be obtained directly from the tables at the specified state. In the saturated mixture region, it is determined from

$$s = s_f + xs_{fg} \qquad \text{(kJ/kg·K)}$$

where x is the quality and s_f and s_{fg} values are listed in the saturation tables. In the absence of compressed liquid data, the entropy of the compressed liquid can be approximated by the entropy of the saturated liquid at the given temperature:

$$s_{@\,T,P} \cong s_{f\,@\,T} \qquad \text{(kJ/kg·K)}$$

The entropy change of a specified mass m (a closed system) during a process is simply

$$\Delta S = m\Delta s = m(s_2 - s_1) \qquad \text{(kJ/K)} \tag{8–12}$$

which is the difference between the entropy values at the final and initial states.

When studying the second-law aspects of processes, entropy is commonly used as a coordinate on diagrams such as the *T-s* and *h-s* diagrams. The general characteristics of the *T-s* diagram of pure substances are shown in Fig. 8–11 using data for water. Notice from this diagram that the constant-volume lines are steeper than the constant-pressure lines and the constant-pressure lines are parallel to the constant-temperature lines in the saturated liquid–vapor mixture region. Also, the constant-pressure lines almost coincide with the saturated liquid line in the compressed liquid region.

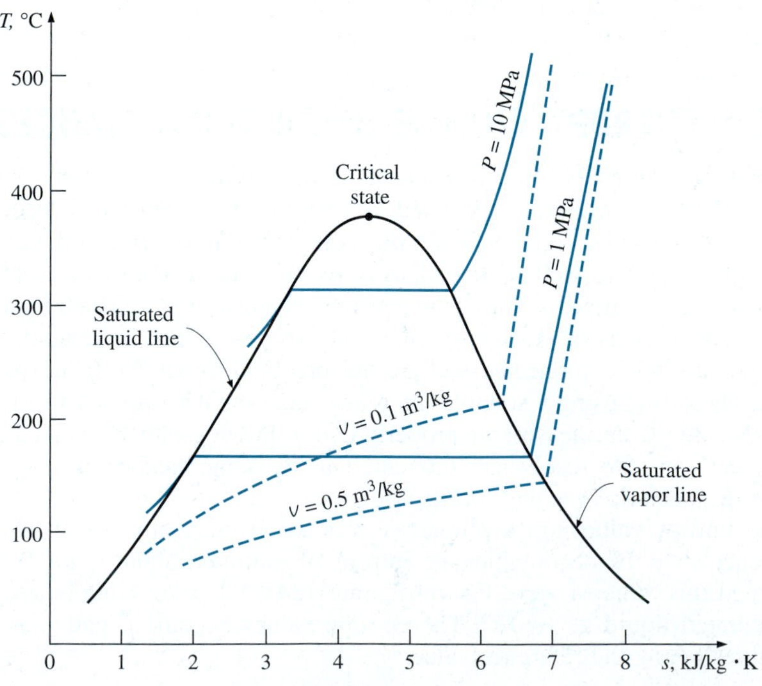

FIGURE 8–11

Schematic of the *T-s* diagram for water.

EXAMPLE 8–3 Entropy Change of a Substance in a Tank

A rigid tank contains 5 kg of refrigerant-134a initially at 20°C and 140 kPa. The refrigerant is now cooled while being stirred until its pressure drops to 100 kPa. Determine the entropy change of the refrigerant during this process.

Solution The refrigerant in a rigid tank is cooled while being stirred. The entropy change of the refrigerant is to be determined.

Assumptions The volume of the tank is constant and thus $v_2 = v_1$.

Analysis We take the refrigerant in the tank as the *system* (Fig. 8–12). This is a *closed system* since no mass crosses the system boundary during the process. We note that the change in entropy of a substance during a process is simply the difference between the entropy values at the final and initial states. The initial state of the refrigerant is completely specified.

Recognizing that the specific volume remains constant during this process, the properties of the refrigerant at both states are

$$\textit{State 1:}\quad \left.\begin{array}{l} P_1 = 140 \text{ kPa} \\ T_1 = 20°\text{C} \end{array}\right\} \quad \begin{array}{l} s_1 = 1.0624 \text{ kJ/kg} \cdot \text{K} \\ v_1 = 0.16544 \text{ m}^3\text{/kg} \end{array}$$

$$\textit{State 2:}\quad \left.\begin{array}{l} P_2 = 100 \text{ kPa} \\ (v_2 = v_1) \end{array}\right\} \quad \begin{array}{l} v_f = 0.0007259 \text{ m}^3\text{/kg} \\ v_g = 0.19254 \text{ m}^3\text{/kg} \end{array}$$

The refrigerant is a saturated liquid–vapor mixture at the final state since $v_f < v_2 < v_g$ at 100 kPa pressure. Therefore, we need to determine the quality first:

$$x_2 = \frac{v_2 - v_f}{v_{fg}} = \frac{0.16544 - 0.0007259}{0.19254 - 0.0007259} = 0.859$$

Thus,

$$s_2 = s_f + x_2 s_{fg} = 0.07188 + (0.859)(0.87995) = 0.8278 \text{ kJ/kg} \cdot \text{K}$$

Then the entropy change of the refrigerant during this process is

$$\begin{aligned} \Delta S = m(s_2 - s_1) &= (5 \text{ kg})(0.8278 - 1.0624) \text{ kJ/kg} \cdot \text{K} \\ &= \mathbf{-1.173 \text{ kJ/K}} \end{aligned}$$

Discussion The negative sign indicates that the entropy of the system is decreasing during this process. This is not a violation of the second law, however, since it is the *entropy generation* S_{gen} that cannot be negative.

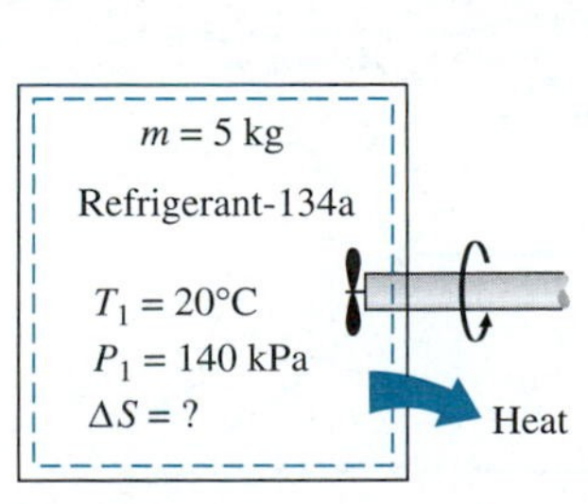

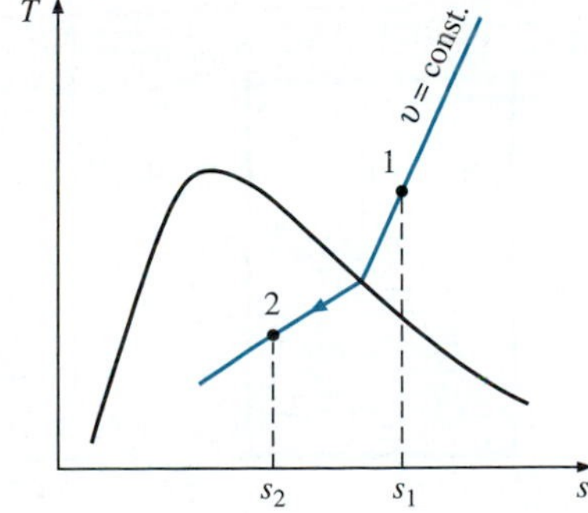

FIGURE 8–12

Schematic and *T-s* diagram for Example 8–3.

EXAMPLE 8–4 Entropy Change during a Constant-Pressure Process

A piston–cylinder device initially contains 3 lbm of liquid water at 20 psia and 70°F. The water is now heated at constant pressure by the addition of 3450 Btu of heat. Determine the entropy change of the water during this process.

Solution Liquid water in a piston–cylinder device is heated at constant pressure. The entropy change of water is to be determined.

Assumptions **1** The tank is stationary and thus the kinetic and potential energy changes are zero, $\Delta KE = \Delta PE = 0$. **2** The process is quasi-equilibrium. **3** The pressure remains constant during the process and thus $P_2 = P_1$.

Analysis We take the water in the cylinder as the *system* (Fig. 8–13). This is a *closed system* since no mass crosses the system boundary during the process. We note that a piston–cylinder device typically involves a moving boundary and thus boundary work W_b. Also, heat is transferred to the system.

Water exists as a compressed liquid at the initial state since its pressure is greater than the saturation pressure of 0.3632 psia at 70°F. By approximating the compressed liquid as a saturated liquid at the given temperature, the properties at the initial state are

State 1:
$$\left.\begin{aligned} P_1 &= 20\text{ psia} \\ T_1 &= 70°\text{F} \end{aligned}\right\} \quad \begin{aligned} s_1 &\cong s_{f\,@\,70°\text{F}} = 0.07459\text{ Btu/lbm}\cdot\text{R} \\ h_1 &\cong h_{f\,@\,70°\text{F}} = 38.08\text{ Btu/lbm} \end{aligned}$$

At the final state, the pressure is still 20 psia, but we need one more property to fix the state. This property is determined from the energy balance,

$$\underbrace{E_{\text{in}} - E_{\text{out}}}_{\substack{\text{Net energy transfer} \\ \text{by heat, work, and mass}}} = \underbrace{\Delta E_{\text{system}}}_{\substack{\text{Change in internal, kinetic,} \\ \text{potential, etc., energies}}}$$

$$Q_{\text{in}} - W_b = \Delta U$$

$$Q_{\text{in}} = \Delta H = m(h_2 - h_1)$$

$$3450\text{ Btu} = (3\text{ lbm})(h_2 - 38.08\text{ Btu/lbm})$$

$$h_2 = 1188.1\text{ Btu/lbm}$$

since $\Delta U + W_b = \Delta H$ for a constant-pressure quasi-equilibrium process. Then,

State 2:
$$\left.\begin{aligned} P_2 &= 20\text{ psia} \\ h_2 &= 1188.1\text{ Btu/lbm} \end{aligned}\right\} \quad \begin{aligned} &s_2 = 1.7761\text{ Btu/lbm}\cdot\text{R} \\ &(\text{Table A-6E, interpolation}) \end{aligned}$$

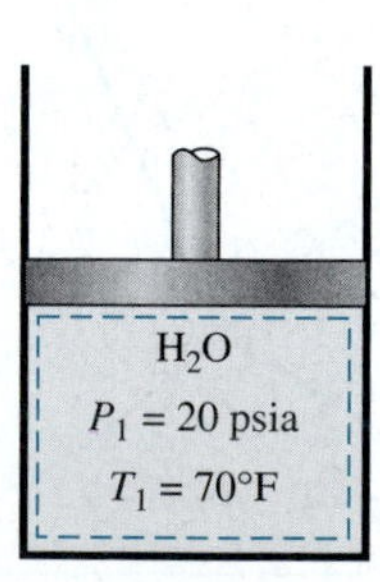

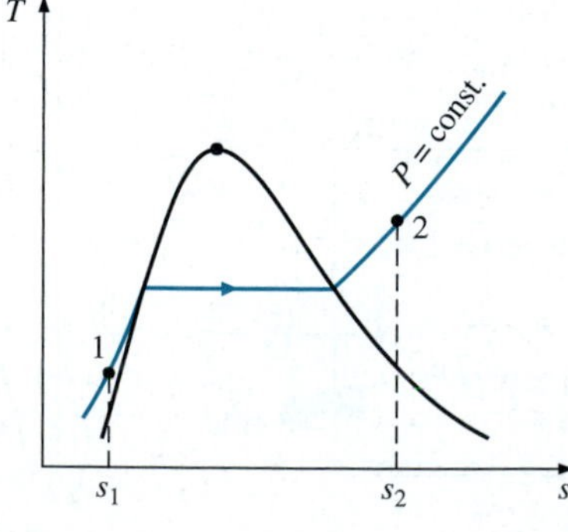

FIGURE 8–13
Schematic and *T-s* diagram for Example 8–4.

Therefore, the entropy change of water during this process is

$$\Delta S = m(s_2 - s_1) = (3\text{ lbm})(1.7761 - 0.07459)\text{ Btu/lbm}\cdot\text{R}$$
$$= \mathbf{5.105\ Btu/R}$$

8–4 ▪ ISENTROPIC PROCESSES

We mentioned earlier that the entropy of a fixed mass can be changed by (1) heat transfer and (2) irreversibilities. Then it follows that the entropy of a fixed mass does not change during a process that is *internally reversible* and *adiabatic* (Fig. 8–14). A process during which the entropy remains constant is called an **isentropic process**. It is characterized by

Isentropic process: $$\Delta s = 0 \quad \text{or} \quad s_2 = s_1 \qquad (\text{kJ/kg}\cdot\text{K}) \tag{8-13}$$

That is, a substance will have the same entropy value at the end of the process as it does at the beginning if the process is carried out in an isentropic manner.

Many engineering systems or devices such as pumps, turbines, nozzles, and diffusers are essentially adiabatic in their operation, and they perform best when the irreversibilities, such as the friction associated with the process, are minimized. Therefore, an isentropic process can serve as an appropriate model for actual processes. Also, isentropic processes enable us to define efficiencies for processes to compare the actual performance of these devices to the performance under idealized conditions.

It should be recognized that a *reversible adiabatic* process is necessarily isentropic ($s_2 = s_1$), but an *isentropic* process is not necessarily a reversible adiabatic process. (The entropy increase of a substance during a process as a result of irreversibilities may be offset by a decrease in entropy as a result of heat losses, for example.) However, the term *isentropic process* is customarily used in thermodynamics to imply an *internally reversible, adiabatic process*.

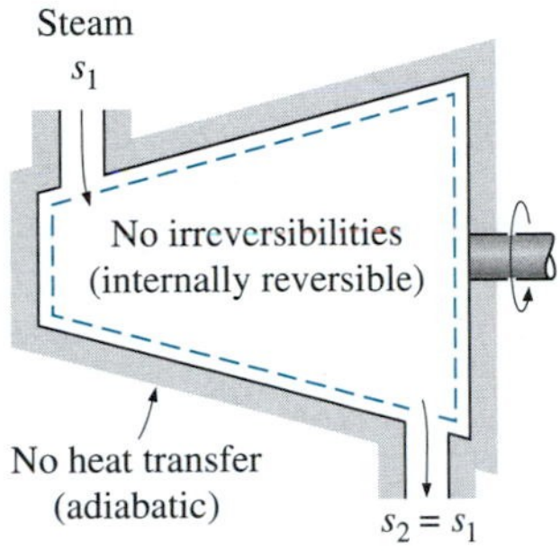

FIGURE 8–14
During an internally reversible, adiabatic (isentropic) process, the entropy remains constant.

EXAMPLE 8–5 Isentropic Expansion of Steam in a Turbine

Steam enters an adiabatic turbine at 5 MPa and 450°C and leaves at a pressure of 1.4 MPa. Determine the work output of the turbine per unit mass of steam if the process is reversible.

Solution Steam is expanded in an adiabatic turbine to a specified pressure in a reversible manner. The work output of the turbine is to be determined.

Assumptions **1** This is a steady-flow process since there is no change with time at any point and thus $\Delta m_{CV} = 0$, $\Delta E_{CV} = 0$, and $\Delta S_{CV} = 0$. **2** The process is reversible. **3** Kinetic and potential energies are negligible. **4** The turbine is adiabatic and thus there is no heat transfer.

Analysis We take the *turbine* as the system (Fig. 8–15). This is a *control volume* since mass crosses the system boundary during the process. We note that there is only one inlet and one exit, and thus $\dot{m}_1 = \dot{m}_2 = \dot{m}$.

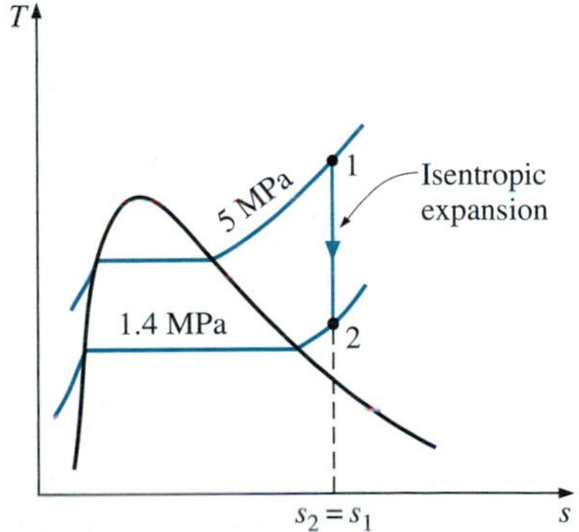

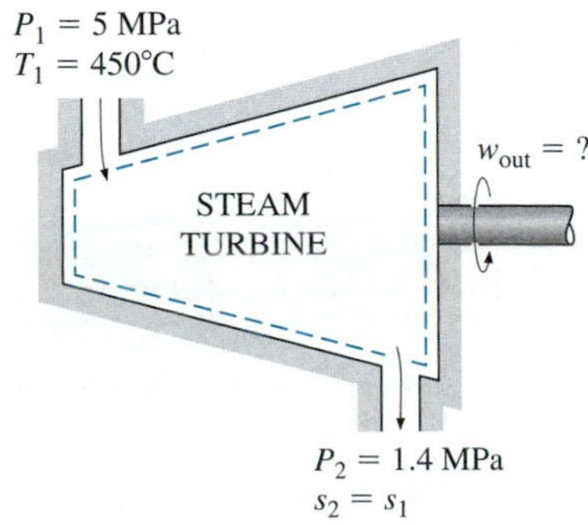

FIGURE 8–15
Schematic and *T-s* diagram for Example 8–5.

The power output of the turbine is determined from the rate form of the energy balance,

$$\underbrace{\dot{E}_{in} - \dot{E}_{out}}_{\substack{\text{Rate of net energy transfer} \\ \text{by heat, work, and mass}}} = \underbrace{dE_{system}/dt^{\nearrow 0 \text{ (steady)}}}_{\substack{\text{Rate of change in internal, kinetic,} \\ \text{potential, etc., energies}}} = 0$$

$$\dot{E}_{in} = \dot{E}_{out}$$

$$\dot{m}h_1 = \dot{W}_{out} + \dot{m}h_2 \qquad (\text{since } \dot{Q} = 0, \text{ke} \cong \text{pe} \cong 0)$$

$$\dot{W}_{out} = \dot{m}(h_1 - h_2)$$

The inlet state is completely specified since two properties are given. But only one property (pressure) is given at the final state, and we need one more property to fix it. The second property comes from the observation that the process is reversible and adiabatic, and thus isentropic. Therefore, $s_2 = s_1$, and

State 1: $\left.\begin{aligned} P_1 &= 5 \text{ MPa} \\ T_1 &= 450°\text{C} \end{aligned}\right\} \begin{aligned} h_1 &= 3317.2 \text{ kJ/kg} \\ s_1 &= 6.8210 \text{ kJ/kg} \cdot \text{K} \end{aligned}$

State 2: $\left.\begin{aligned} P_2 &= 1.4 \text{ MPa} \\ s_2 &= s_1 \end{aligned}\right\} h_2 = 2967.4 \text{ kJ/kg}$

Then the work output of the turbine per unit mass of the steam becomes

$$w_{out} = h_1 - h_2 = 3317.2 - 2967.4 = \mathbf{349.8 \text{ kJ/kg}}$$

8–5 ▪ PROPERTY DIAGRAMS INVOLVING ENTROPY

Property diagrams serve as great visual aids in the thermodynamic analysis of processes. We have used *P*-*v* and *T*-*v* diagrams extensively in previous chapters in conjunction with the first law of thermodynamics. In the second-law analysis, it is very helpful to plot the processes on diagrams for which one of the coordinates is entropy. The two diagrams commonly used in the second-law analysis are the *temperature-entropy* and the *enthalpy-entropy* diagrams.

Consider the defining equation of entropy (Eq. 8–4). It can be rearranged as

$$\delta Q_{int\ rev} = T\,dS \qquad \text{(kJ)} \tag{8–14}$$

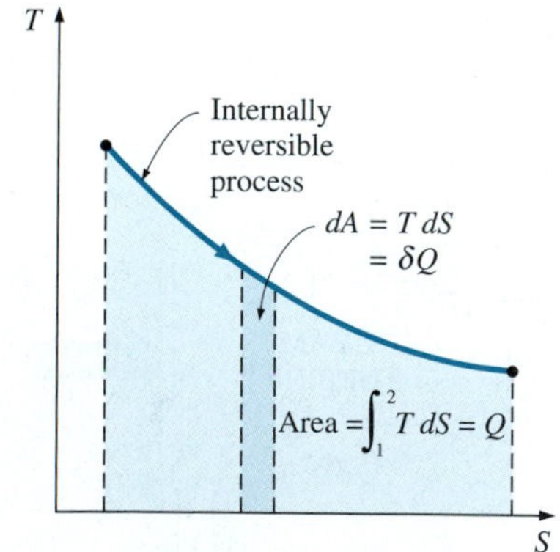

FIGURE 8–16

On a *T-S* diagram, the area under the process curve represents the heat transfer for internally reversible processes.

As shown in Fig. 8–16, $\delta Q_{rev\ int}$ corresponds to a differential area on a *T-S* diagram. The total heat transfer during an internally reversible process is determined by integration to be

$$Q_{int\ rev} = \int_1^2 T\,dS \qquad \text{(kJ)} \tag{8–15}$$

which corresponds to the area under the process curve on a *T-S* diagram. Therefore, we conclude that *the area under the process curve on a T-S diagram represents heat transfer during an internally reversible process.* This is somewhat analogous to reversible boundary work being represented by the area under the process curve on a P-V diagram. Note that the area under the process curve represents heat transfer for processes that are internally (or totally) reversible. The area has no meaning for irreversible processes.

Equations 8–14 and 8–15 can also be expressed on a unit-mass basis as

$$\delta q_{\text{int rev}} = T\,ds \qquad (\text{kJ/kg}) \tag{8–16}$$

and

$$q_{\text{int rev}} = \int_1^2 T\,ds \qquad (\text{kJ/kg}) \tag{8–17}$$

To perform the integrations in Eqs. 8–15 and 8–17, one needs to know the relationship between T and s during a process. One special case for which these integrations can be performed easily is the *internally reversible isothermal process*. It yields

$$Q_{\text{int rev}} = T_0\,\Delta S \qquad (\text{kJ}) \tag{8–18}$$

or

$$q_{\text{int rev}} = T_0\,\Delta s \qquad (\text{kJ/kg}) \tag{8–19}$$

where T_0 is the constant temperature and ΔS is the entropy change of the system during the process.

An isentropic process on a T-s diagram is easily recognized as a *vertical-line segment*. This is expected since an isentropic process involves no heat transfer, and therefore the area under the process path must be zero (Fig. 8–17). The T-s diagrams serve as valuable tools for visualizing the second-law aspects of processes and cycles, and thus they are frequently used in thermodynamics. The T-s diagram of water is given in the appendix in Fig. A–9.

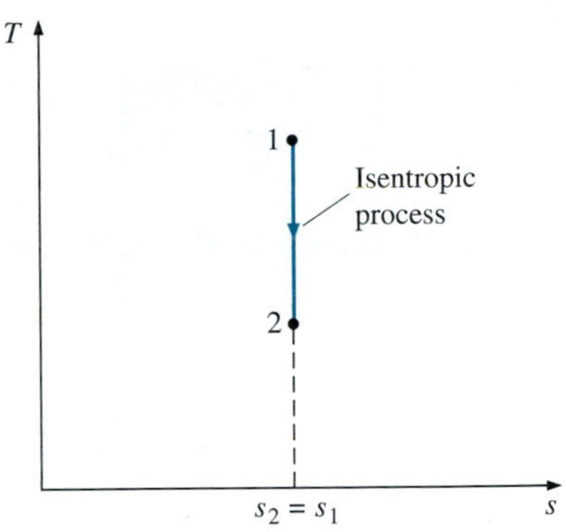

FIGURE 8–17

The isentropic process appears as a *vertical* line segment on a T-s diagram.

Another diagram commonly used in engineering is the enthalpy-entropy diagram, which is quite valuable in the analysis of steady-flow devices such as turbines, compressors, and nozzles. The coordinates of an h-s diagram represent two properties of major interest: enthalpy, which is a primary property in the first-law analysis of the steady-flow devices, and entropy, which is the property that accounts for irreversibilities during adiabatic processes. In analyzing the steady flow of steam through an adiabatic turbine, for example, the vertical distance between the inlet and the exit states Δh is a measure of the work output of the turbine, and the horizontal distance Δs is a measure of the irreversibilities associated with the process (Fig. 8–18).

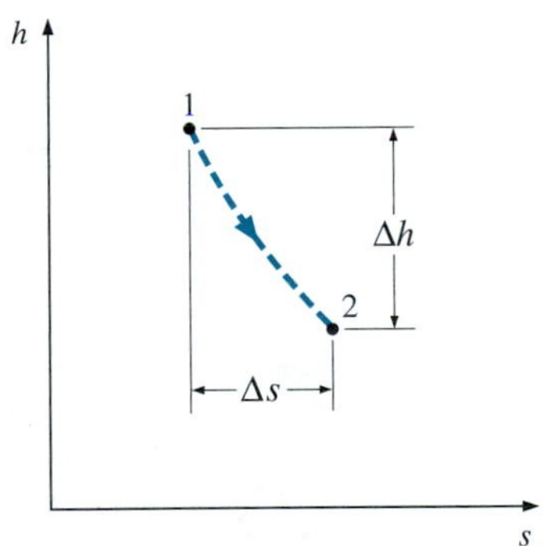

FIGURE 8–18

For adiabatic steady-flow devices, the vertical distance Δh on an h-s diagram is a measure of work, and the horizontal distance Δs is a measure of irreversibilities.

The h-s diagram is also called a **Mollier diagram** after the German scientist R. Mollier (1863–1935). An h-s diagram is given in the appendix for steam in Fig. A–10.

EXAMPLE 8–6 The *T-S* Diagram of the Carnot Cycle

Show the Carnot cycle on a T-S diagram and indicate the areas that represent the heat supplied Q_H, heat rejected Q_L, and the net work output $W_{\text{net,out}}$ on this diagram.

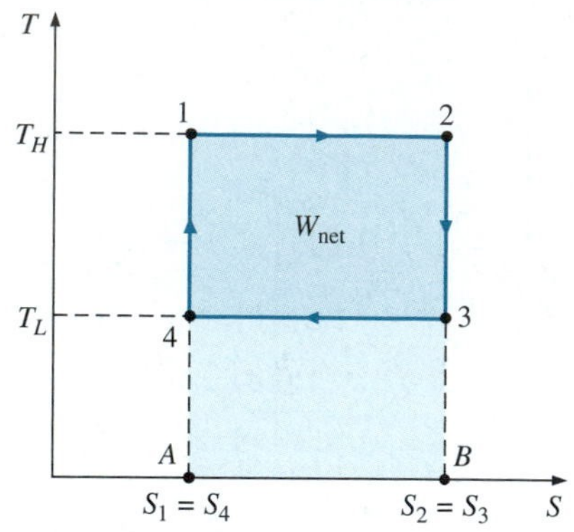

FIGURE 8–19

The *T-S* diagram of a Carnot cycle (Example 8–6).

Solution The Carnot cycle is to be shown on a *T-S* diagram, and the areas that represent Q_H, Q_L, and $W_{net,out}$ are to be indicated.

Analysis Recall that the Carnot cycle is made up of two reversible isothermal (T = constant) processes and two isentropic (s = constant) processes. These four processes form a rectangle on a *T-S* diagram, as shown in Fig. 8–19.

On a *T-S* diagram, the area under the process curve represents the heat transfer for that process. Thus the area *A*12*B* represents Q_H, the area *A*43*B* represents Q_L, and the difference between these two (the area in color) represents the net work since

$$W_{net,out} = Q_H - Q_L$$

Therefore, the area enclosed by the path of a cycle (area 1234) on a *T-S* diagram represents the net work. Recall that the area enclosed by the path of a cycle also represents the net work on a *P-V* diagram.

8–6 ▪ WHAT IS ENTROPY?

It is clear from the previous discussion that entropy is a useful property and serves as a valuable tool in the second-law analysis of engineering devices. But this does not mean that we know and understand entropy well. Because we do not. In fact, we cannot even give an adequate answer to the question, What is entropy? Not being able to describe entropy fully, however, does not take anything away from its usefulness. We could not define *energy* either, but it did not interfere with our understanding of energy transformations and the conservation of energy principle. Granted, entropy is not a household word like energy. But with continued use, our understanding of entropy will deepen, and our appreciation of it will grow. The next discussion should shed some light on the physical meaning of entropy by considering the microscopic nature of matter.

Entropy can be viewed as a measure of *molecular disorder*, or *molecular randomness*. As a system becomes more disordered, the positions of the molecules become less predictable and the entropy increases. Thus, it is not surprising that the entropy of a substance is lowest in the solid phase and highest in the gas phase (Fig. 8–20). In the solid phase, the molecules of a substance continually oscillate about their equilibrium positions, but they cannot move relative to each other, and their position at any instant can be predicted with good certainty. In the gas phase, however, the molecules move about at random, collide with each other, and change direction, making it extremely difficult to predict accurately the microscopic state of a system at any instant. Associated with this molecular chaos is a high value of entropy.

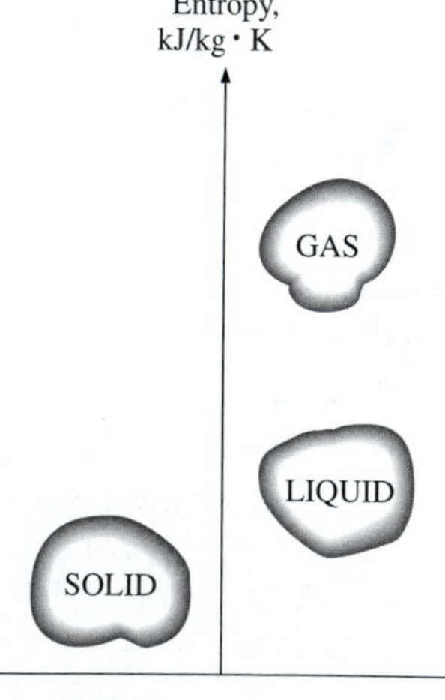

FIGURE 8–20

The level of molecular disorder (entropy) of a substance increases as it melts or evaporates.

When viewed microscopically (from a statistical thermodynamics point of view), an isolated system that appears to be at a state of equilibrium may exhibit a high level of activity because of the continual motion of the molecules. To each state of macroscopic equilibrium there corresponds a large number of possible microscopic states or molecular configurations. The entropy of a system is related to the total number of possible microscopic states of that system, called *thermodynamic probability p*, by the **Boltzmann relation**, expressed as

$$S = k \ln p \tag{8–20}$$

where $k = 1.3806 \times 10^{-23}$ J/K is the **Boltzmann constant**. Therefore, from a microscopic point of view, the entropy of a system increases whenever the molecular randomness or uncertainty (i.e., molecular probability) of a system increases. Thus, entropy is a measure of molecular disorder, and the molecular disorder of an isolated system increases anytime it undergoes a process.

As mentioned earlier, the molecules of a substance in solid phase continually oscillate, creating an uncertainty about their position. These oscillations, however, fade as the temperature is decreased, and the molecules supposedly become motionless at absolute zero. This represents a state of ultimate molecular order (and minimum energy). Therefore, *the entropy of a pure crystalline substance at absolute zero temperature is zero* since there is no uncertainty about the state of the molecules at that instant (Fig. 8–21). This statement is known as the **third law of thermodynamics**. The third law of thermodynamics provides an absolute reference point for the determination of entropy. The entropy determined relative to this point is called **absolute entropy**, and it is extremely useful in the thermodynamic analysis of chemical reactions. Notice that the entropy of a substance that is not pure crystalline (such as a solid solution) is not zero at absolute zero temperature. This is because more than one molecular configuration exists for such substances, which introduces some uncertainty about the microscopic state of the substance.

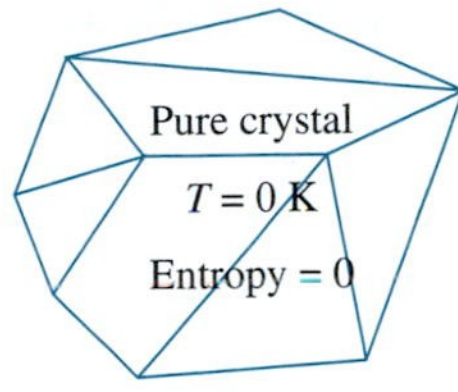

FIGURE 8–21
A pure crystalline substance at absolute zero temperature is in perfect order, and its entropy is zero (the third law of thermodynamics).

Molecules in the gas phase possess a considerable amount of kinetic energy. However, we know that no matter how large their kinetic energies are, the gas molecules do not rotate a paddle wheel inserted into the container and produce work. This is because the gas molecules, and the energy they possess, are disorganized. Probably the number of molecules trying to rotate the wheel in one direction at any instant is equal to the number of molecules that are trying to rotate it in the opposite direction, causing the wheel to remain motionless. Therefore, we cannot extract any useful work directly from disorganized energy (Fig. 8–22).

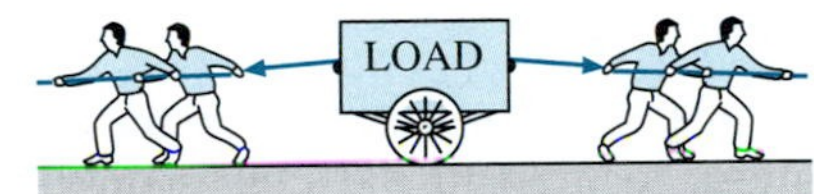

FIGURE 8–22
Disorganized energy does not create much useful effect, no matter how large it is.

Now consider a rotating shaft shown in Fig. 8–23. This time the energy of the molecules is completely organized since the molecules of the shaft are rotating in the same direction together. This organized energy can readily be used to perform useful tasks such as raising a weight or generating electricity. Being an organized form of energy, work is free of disorder or randomness and thus free of entropy. *There is no entropy transfer associated with energy transfer as work.* Therefore, in the absence of any friction, the process of raising a weight by a rotating shaft (or a flywheel) does not produce any entropy. Any process that does not produce a net entropy is reversible, and thus the process just described can be reversed by lowering the weight. Therefore, energy is not degraded during this process, and no potential to do work is lost.

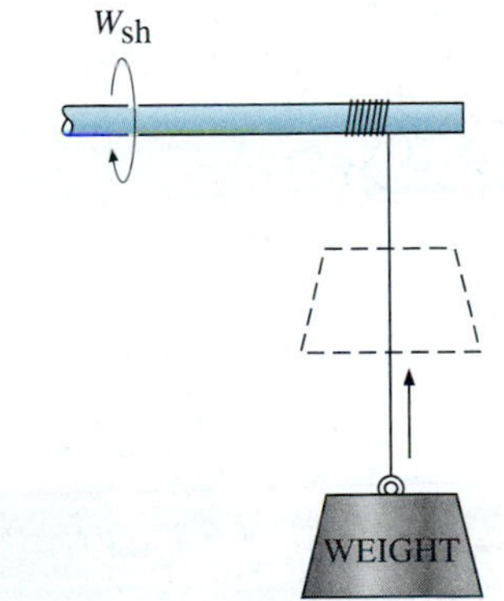

FIGURE 8–23
In the absence of friction, raising a weight by a rotating shaft does not create any disorder (entropy), and thus energy is not degraded during this process.

Instead of raising a weight, let us operate the paddle wheel in a container filled with a gas, as shown in Fig. 8–24. The paddle-wheel work in this case

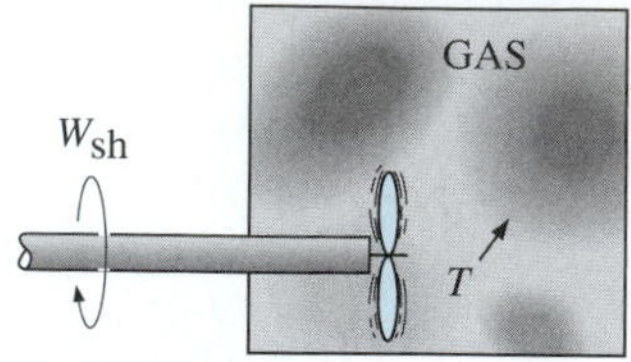

FIGURE 8–24

The paddle-wheel work done on a gas increases the level of disorder (entropy) of the gas, and thus energy is degraded during this process.

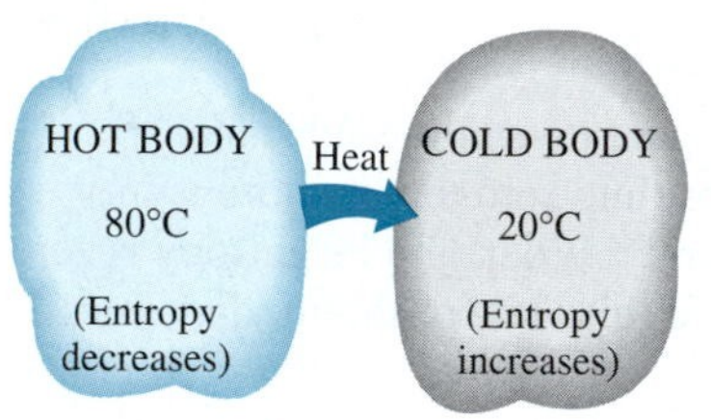

FIGURE 8–25

During a heat transfer process, the net entropy increases. (The increase in the entropy of the cold body more than offsets the decrease in the entropy of the hot body.)

FIGURE 8–26

The use of entropy (disorganization, uncertainty) is not limited to thermodynamics.

is converted to the internal energy of the gas, as evidenced by a rise in gas temperature, creating a higher level of molecular disorder in the container. This process is quite different from raising a weight since the organized paddle-wheel energy is now converted to a highly disorganized form of energy, which cannot be converted back to the paddle wheel as the rotational kinetic energy. Only a portion of this energy can be converted to work by partially reorganizing it through the use of a heat engine. Therefore, energy is degraded during this process, the ability to do work is reduced, molecular disorder is produced, and associated with all this is an increase in entropy.

The *quantity* of energy is always preserved during an actual process (the first law), but the *quality* is bound to decrease (the second law). This decrease in quality is always accompanied by an increase in entropy. As an example, consider the transfer of 10 kJ of energy as heat from a hot medium to a cold one. At the end of the process, we still have the 10 kJ of energy, but at a lower temperature and thus at a lower quality.

Heat is, in essence, a form of *disorganized energy*, and some disorganization (entropy) flows with heat (Fig. 8–25). As a result, the entropy and the level of molecular disorder or randomness of the hot body decreases with the entropy and the level of molecular disorder of the cold body increases. The second law requires that the increase in entropy of the cold body be greater than the decrease in entropy of the hot body, and thus the net entropy of the combined system (the cold body and the hot body) increases. That is, the combined system is at a state of greater disorder at the final state. Thus we can conclude that processes can occur only in the direction of increased overall entropy or molecular disorder. That is, the entire universe is getting more and more chaotic every day.

Entropy and Entropy Generation in Daily Life

The concept of entropy can also be applied to other areas. Entropy can be viewed as a measure of disorder or disorganization in a system. Likewise, entropy generation can be viewed as a measure of disorder or disorganization generated during a process. The concept of entropy is not used in daily life nearly as extensively as the concept of energy, even though entropy is readily applicable to various aspects of daily life. The extension of the entropy concept to nontechnical fields is not a novel idea. It has been the topic of several articles, and even some books. Next we present several ordinary events and show their relevance to the concept of entropy and entropy generation.

Efficient people lead low-entropy (highly organized) lives. They have a place for everything (minimum uncertainty), and it takes minimum energy for them to locate something. Inefficient people, on the other hand, are disorganized and lead high-entropy lives. It takes them minutes (if not hours) to find something they need, and they are likely to create a bigger disorder as they are searching since they will probably conduct the search in a disorganized manner (Fig. 8–26). People leading high-entropy lifestyles are always on the run, and never seem to catch up.

You probably noticed (with frustration) that some people seem to *learn* fast and remember well what they learn. We can call this type of learning organized or low-entropy learning. These people make a conscientious effort to file the new information properly by relating it to their existing knowledge base and creating a solid information network in their minds. On the other hand, people who throw the information into their minds as they study, with no effort to secure it, may *think* they are learning. They are bound to discover otherwise when they need to locate the information, for example, during a test. It is not easy to retrieve information from a database that is, in a sense, in the gas phase. Students who have blackouts during tests should reexamine their study habits.

A *library* with a good shelving and indexing system can be viewed as a low-entropy library because of the high level of organization. Likewise, a library with a poor shelving and indexing system can be viewed as a high-entropy library because of the high level of disorganization. A library with no indexing system is like no library, since a book is of no value if it cannot be found.

Consider two identical buildings, each containing one million books. In the first building, the books are *piled* on top of each other, whereas in the second building they are *highly organized, shelved, and indexed* for easy reference. There is no doubt about which building a student will prefer to go to for checking out a certain book. Yet, some may argue from the first-law point of view that these two buildings are equivalent since the mass and knowledge content of the two buildings are identical, despite the high level of disorganization (entropy) in the first building. This example illustrates that any realistic comparisons should involve the second-law point of view.

Two *textbooks* that seem to be identical because both cover basically the same topics and present the same information may actually be *very* different depending on *how* they cover the topics. After all, two seemingly identical cars are not so identical if one goes only half as many miles as the other one on the same amount of fuel. Likewise, two seemingly identical books are not so identical if it takes twice as long to learn a topic from one of them as it does from the other. Thus, comparisons made on the basis of the first law only may be highly misleading.

Having a disorganized (high-entropy) *army* is like having no army at all. It is no coincidence that the command centers of any armed forces are among the primary targets during a war. One army that consists of 10 divisions is 10 times more powerful than 10 armies each consisting of a single division. Likewise, one country that consists of 10 states is more powerful than 10 countries, each consisting of a single state. The *United States* would not be such a powerful country if there were 50 independent countries in its place instead of a single country with 50 states. The European Union has the potential to be a new economic and political superpower. The old cliché "divide and conquer" can be rephrased as "increase the entropy and conquer."

FIGURE 8–27

As in mechanical systems, friction in the workplace is bound to generate entropy and reduce performance.

We know that mechanical friction is always accompanied by entropy generation, and thus reduced performance. We can generalize this to daily life: *friction in the workplace* with fellow workers is bound to generate entropy, and thus adversely affect performance (Fig. 8–27). It results in reduced productivity.

We also know that *unrestrained expansion* (or explosion) and uncontrolled electron exchange (chemical reactions) generate entropy and are highly irreversible. Likewise, unrestrained opening of the mouth to scatter angry words is highly irreversible since this generates entropy, and it can cause considerable damage. A person who gets up in anger is bound to sit down at a loss. Hopefully, someday we will be able to come up with some procedures to quantify entropy generated during nontechnical activities, and maybe even pinpoint its primary sources and magnitude.

8–7 ▪ THE *T ds* RELATIONS

Recall that the quantity $(\delta Q/T)_{\text{int rev}}$ corresponds to a differential change in the property *entropy*. The entropy change for a process, then, can be evaluated by integrating $\delta Q/T$ along some imaginary internally reversible path between the actual end states. For isothermal internally reversible processes, this integration is straightforward. But when the temperature varies during the process, we have to have a relation between δQ and T to perform this integration. Finding such relations is what we intend to do in this section.

The differential form of the conservation of energy equation for a closed stationary system (a fixed mass) containing a simple compressible substance can be expressed for an internally reversible process as

$$\delta Q_{\text{int rev}} - \delta W_{\text{int rev,out}} = dU \tag{8–21}$$

But

$$\delta Q_{\text{int rev}} = T\,dS$$
$$\delta W_{\text{int rev,out}} = P\,dV$$

Thus,

$$T\,dS = dU + P\,dV \qquad \text{(kJ)} \tag{8–22}$$

or

$$T\,ds = du + P\,dv \qquad \text{(kJ/kg)} \tag{8–23}$$

This equation is known as the first *T ds*, or *Gibbs*, *equation*. Notice that the only type of work interaction a simple compressible system may involve as it undergoes an internally reversible process is the boundary work.

The second *T ds* equation is obtained by eliminating *du* from Eq. 8–23 by using the definition of enthalpy ($h = u + Pv$):

$$\left.\begin{array}{lcl} h = u + Pv & \longrightarrow & dh = du + P\,dv + v\,dP \\ \text{(Eq. 8–23)} & \longrightarrow & T\,ds = du + P\,dv \end{array}\right\} T\,ds = dh - v\,dP \tag{8–24}$$

Equations 8–23 and 8–24 are extremely valuable since they relate entropy changes of a system to the changes in other properties. Unlike Eq. 8–4, they are property relations and therefore are independent of the type of the processes.

These *T ds* relations are developed with an internally reversible process in mind since the entropy change between two states must be evaluated along a reversible path. However, the results obtained are valid for both reversible and irreversible processes since entropy is a property and the change in a property between two states is independent of the type of process the system undergoes. Equations 8–23 and 8–24 are relations between the properties of a unit mass of a simple compressible system as it undergoes a change of state, and they are applicable whether the change occurs in a closed or an open system (Fig. 8–28).

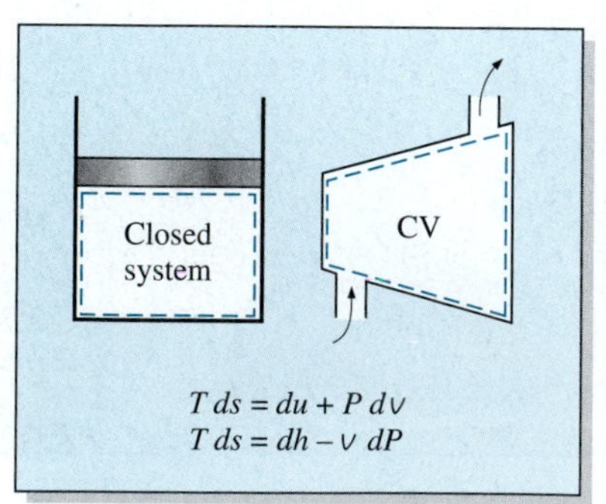

FIGURE 8–28
The *T ds* relations are valid for both reversible and irreversible processes and for both closed and open systems.

Explicit relations for differential changes in entropy are obtained by solving for ds in Eqs. 8–23 and 8–24:

$$ds = \frac{du}{T} + \frac{P\,dv}{T} \tag{8–25}$$

and

$$ds = \frac{dh}{T} - \frac{v\,dP}{T} \tag{8–26}$$

The entropy change during a process can be determined by integrating either of these equations between the initial and the final states. To perform these integrations, however, we must know the relationship between du or dh and the temperature (such as $du = c_v\,dT$ and $dh = c_p\,dT$ for ideal gases) as well as the equation of state for the substance (such as the ideal-gas equation of state $Pv = RT$). For substances for which such relations exist, the integration of Eq. 8–25 or 8–26 is straightforward. For other substances, we have to rely on tabulated data.

The $T\,ds$ relations for nonsimple systems, that is, systems that involve more than one mode of quasi-equilibrium work, can be obtained in a similar manner by including all the relevant quasi-equilibrium work modes.

8–8 ▪ ENTROPY CHANGE OF LIQUIDS AND SOLIDS

Recall that liquids and solids can be approximated as *incompressible substances* since their specific volumes remain nearly constant during a process. Thus, $dv \cong 0$ for liquids and solids, and Eq. 8–25 for this case reduces to

$$ds = \frac{du}{T} = \frac{c\,dT}{T} \tag{8–27}$$

since $c_p = c_v = c$ and $du = c\,dT$ for incompressible substances. Then the entropy change during a process is determined by integration to be

Liquids, solids:
$$s_2 - s_1 = \int_1^2 c(T)\frac{dT}{T} \cong c_{avg} \ln\frac{T_2}{T_1} \quad \text{(kJ/kg·K)} \tag{8–28}$$

where c_{avg} is the *average* specific heat of the substance over the given temperature interval. Note that the entropy change of a truly incompressible substance depends on temperature only and is independent of pressure.

Equation 8–28 can be used to determine the entropy changes of solids and liquids with reasonable accuracy. However, for liquids that expand considerably with temperature, it may be necessary to consider the effects of volume change in calculations. This is especially the case when the temperature change is large.

A relation for isentropic processes of liquids and solids is obtained by setting the entropy change relation above equal to zero. It gives

Isentropic:
$$s_2 - s_1 = c_{avg} \ln\frac{T_2}{T_1} = 0 \quad \rightarrow \quad T_2 = T_1 \tag{8–29}$$

That is, the temperature of a truly incompressible substance remains constant during an isentropic process. Therefore, the isentropic process of an incompressible substance is also isothermal. This behavior is closely approximated by liquids and solids.

EXAMPLE 8–7 Effect of Density of a Liquid on Entropy

Liquid methane is commonly used in various cryogenic applications. The critical temperature of methane is 191 K (or −82°C), and thus methane must be maintained below 191 K to keep it in liquid phase. The properties of liquid methane at various temperatures and pressures are given in Table 8–1. Determine the entropy change of liquid methane as it undergoes a process from 110 K and 1 MPa to 120 K and 5 MPa (*a*) using tabulated properties and (*b*) approximating liquid methane as an incompressible substance. What is the error involved in the latter case?

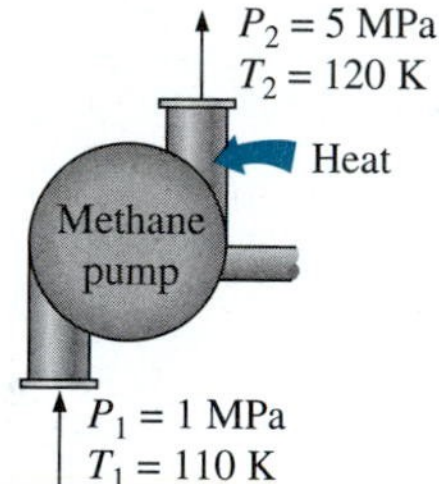

FIGURE 8–29

Schematic for Example 8–7.

Solution Liquid methane undergoes a process between two specified states. The entropy change of methane is to be determined by using actual data and by assuming methane to be incompressible.

Analysis (*a*) We consider a unit mass of liquid methane (Fig. 8–29). The properties of the methane at the initial and final states are

State 1:
$$\left.\begin{aligned} P_1 &= 1 \text{ MPa} \\ T_1 &= 110 \text{ K} \end{aligned}\right\} \quad \begin{aligned} s_1 &= 4.875 \text{ kJ/kg}\cdot\text{K} \\ c_{p1} &= 3.471 \text{ kJ/kg}\cdot\text{K} \end{aligned}$$

State 2:
$$\left.\begin{aligned} P_2 &= 5 \text{ MPa} \\ T_2 &= 120 \text{ K} \end{aligned}\right\} \quad \begin{aligned} s_2 &= 5.145 \text{ kJ/kg}\cdot\text{K} \\ c_{p2} &= 3.486 \text{ kJ/kg}\cdot\text{K} \end{aligned}$$

Therefore,

$$\Delta s = s_2 - s_1 = 5.145 - 4.875 = \mathbf{0.270\ kJ/kg\cdot K}$$

(*b*) Approximating liquid methane as an incompressible substance, its entropy change is determined to be

$$\Delta s = c_{\text{avg}} \ln \frac{T_2}{T_1} = (3.4785 \text{ kJ/kg}\cdot\text{K}) \ln \frac{120 \text{ K}}{110 \text{ K}} = \mathbf{0.303\ kJ/kg\cdot K}$$

since

$$c_{\text{avg}} = \frac{c_{p1} + c_{p2}}{2} = \frac{3.471 + 3.486}{2} = 3.4785 \text{ kJ/kg}\cdot\text{K}$$

TABLE 8–1

Properties of liquid methane

Temp., T, K	Pressure, P, MPa	Density, ρ, kg/m^3	Enthalpy, h, kJ/kg	Entropy, s, kJ/kg · K	Specific heat, c_p, kJ/kg · K
110	0.5	425.3	208.3	4.878	3.476
	1.0	425.8	209.0	4.875	3.471
	2.0	426.6	210.5	4.867	3.460
	5.0	429.1	215.0	4.844	3.432
120	0.5	410.4	243.4	5.185	3.551
	1.0	411.0	244.1	5.180	3.543
	2.0	412.0	245.4	5.171	3.528
	5.0	415.2	249.6	5.145	3.486

Therefore, the error involved in approximating liquid methane as an incompressible substance is

$$\text{Error} = \frac{|\Delta s_{\text{actual}} - \Delta s_{\text{ideal}}|}{\Delta s_{\text{actual}}} = \frac{|0.270 - 0.303|}{0.270} = \mathbf{0.122\ (or\ 12.2\%)}$$

Discussion This result is not surprising since the density of liquid methane changes during this process from 425.8 to 415.2 kg/m^3 (about 3 percent), which makes us question the validity of the incompressible substance assumption. Still, this assumption enables us to obtain reasonably accurate results with less effort, which proves to be very convenient in the absence of compressed liquid data.

EXAMPLE 8–8 Economics of Replacing a Valve by a Turbine

A cryogenic manufacturing facility handles liquid methane at 115 K and 5 MPa at a rate of 0.280 m^3/s . A process requires dropping the pressure of liquid methane to 1 MPa, which is done by throttling the liquid methane by passing it through a flow resistance such as a valve. A recently hired engineer proposes to replace the throttling valve by a turbine in order to produce power while dropping the pressure to 1 MPa. Using data from Table 8–1, determine the maximum amount of power that can be produced by such a turbine. Also, determine how much this turbine will save the facility from electricity usage costs per year if the turbine operates continuously (8760 h/yr) and the facility pays $0.075/kWh for electricity.

FIGURE 8–30

A 1.0-MW liquified natural gas (LNG) turbine with 95-cm turbine runner diameter being installed in a cryogenic test facility.

Courtesy of Ebara International Corporation, Cryodynamics Division, Sparks, Nevada.

Solution Liquid methane is expanded in a turbine to a specified pressure at a specified rate. The maximum power that this turbine can produce and the amount of money it can save per year are to be determined.

Assumptions **1** This is a steady-flow process since there is no change with time at any point and thus $\Delta m_{CV} = 0$, $\Delta E_{CV} = 0$, and $\Delta S_{CV} = 0$. **2** The turbine is adiabatic and thus there is no heat transfer. **3** The process is reversible. **4** Kinetic and potential energies are negligible.

Analysis We take the *turbine* as the system (Fig. 8–30). This is a *control volume* since mass crosses the system boundary during the process. We note that there is only one inlet and one exit and thus $\dot{m}_1 = \dot{m}_2 = \dot{m}$.

The assumptions above are reasonable since a turbine is normally well insulated and it must involve no irreversibilities for best performance and thus *maximum* power production. Therefore, the process through the turbine must be *reversible adiabatic* or *isentropic*. Then, $s_2 = s_1$ and

$$\textit{State 1:} \quad \left.\begin{array}{l} P_1 = 5\text{ MPa} \\ T_1 = 115\text{ K} \end{array}\right\} \begin{array}{l} h_1 = 232.3\text{ kJ/kg} \\ s_1 = 4.9945\text{ kJ/kg}\cdot\text{K} \\ \rho_1 = 422.15\text{ kg/s} \end{array}$$

$$\textit{State 2:} \quad \left.\begin{array}{l} P_2 = 1\text{ MPa} \\ s_2 = s_1 \end{array}\right\} \quad h_2 = 222.8\text{ kJ/kg}$$

Also, the mass flow rate of liquid methane is

$$\dot{m} = \rho_1 \dot{V}_1 = (422.15\text{ kg/m}^3)(0.280\text{ m}^3/\text{s}) = 118.2\text{ kg/s}$$

Then the power output of the turbine is determined from the rate form of the energy balance to be

$$\underbrace{\dot{E}_{in} - \dot{E}_{out}}_{\substack{\text{Rate of net energy transfer}\\ \text{by heat, work, and mass}}} = \underbrace{dE_{system}/dt^{\nearrow 0 \text{ (steady)}}}_{\substack{\text{Rate of change in internal,}\\ \text{kinetic, potential, etc., energies}}} = 0$$

$$\dot{E}_{in} = \dot{E}_{out}$$

$$\dot{m}h_1 = \dot{W}_{out} + \dot{m}h_2 \qquad (\text{since } \dot{Q} = 0, \text{ke} \cong \text{pe} \cong 0)$$

$$\dot{W}_{out} = \dot{m}(h_1 - h_2)$$

$$= (118.2 \text{ kg/s})(232.3 - 222.8) \text{ kJ/kg}$$

$$= \mathbf{1123 \text{ kW}}$$

For continuous operation ($365 \times 24 = 8760$ h), the amount of power produced per year is

$$\text{Annual power production} = \dot{W}_{out} \times \Delta t = (1123 \text{ kW})(8760 \text{ h/yr})$$

$$= 0.9837 \times 10^7 \text{ kWh/yr}$$

At \$0.075/kWh, the amount of money this turbine can save the facility is

$$\text{Annual power savings} = (\text{Annual power production})(\text{Unit cost of power})$$

$$= (0.9837 \times 10^7 \text{ kWh/yr})(\$0.075/\text{kWh})$$

$$= \mathbf{\$737{,}800/yr}$$

That is, this turbine can save the facility \$737,800 a year by simply taking advantage of the potential that is currently being wasted by a throttling valve, and the engineer who made this observation should be rewarded.

Discussion This example shows the importance of the property entropy since it enabled us to quantify the work potential that is being wasted. In practice, the turbine will not be isentropic, and thus the power produced will be less. The analysis above gave us the upper limit. An actual turbine-generator assembly can utilize about 80 percent of the potential and produce more than 900 kW of power while saving the facility more than \$600,000 a year.

It can also be shown that the temperature of methane drops to 113.9 K (a drop of 1.1 K) during the isentropic expansion process in the turbine instead of remaining constant at 115 K as would be the case if methane were assumed to be an incompressible substance. The temperature of methane would rise to 116.6 K (a rise of 1.6 K) during the throttling process.

8–9 ▪ THE ENTROPY CHANGE OF IDEAL GASES

An expression for the entropy change of an ideal gas can be obtained from Eq. 8–25 or 8–26 by employing the property relations for ideal gases (Fig. 8–31). By substituting $du = c_v\, dT$ and $P = RT/v$ into Eq. 8–25, the differential entropy change of an ideal gas becomes

$$ds = c_v \frac{dT}{T} + R\frac{dv}{v} \qquad \textbf{(8–30)}$$

The entropy change for a process is obtained by integrating this relation between the end states:

$$s_2 - s_1 = \int_1^2 c_v(T)\frac{dT}{T} + R\ln\frac{v_2}{v_1} \tag{8–31}$$

A second relation for the entropy change of an ideal gas is obtained in a similar manner by substituting $dh = c_p\,dT$ and $v = RT/P$ into Eq. 8–26 and integrating. The result is

$$s_2 - s_1 = \int_1^2 c_p(T)\frac{dT}{T} - R\ln\frac{P_2}{P_1} \tag{8–32}$$

FIGURE 8–31

A broadcast from channel IG.

© *Vol. 1/PhotoDisc*

The specific heats of ideal gases, with the exception of monatomic gases, depend on temperature, and the integrals in Eqs. 8–31 and 8–32 cannot be performed unless the dependence of c_v and c_p on temperature is known. Even when the $c_v(T)$ and $c_p(T)$ functions are available, performing long integrations every time entropy change is calculated is not practical. Then two reasonable choices are left: either perform these integrations by simply assuming constant specific heats or evaluate those integrals once and tabulate the results. Both approaches are presented next.

Constant Specific Heats (Approximate Analysis)

Assuming constant specific heats for ideal gases is a common approximation, and we used this assumption before on several occasions. It usually simplifies the analysis greatly, and the price we pay for this convenience is some loss in accuracy. The magnitude of the error introduced by this assumption depends on the situation at hand. For example, for monatomic ideal gases such as helium, the specific heats are independent of temperature, and therefore the constant-specific-heat assumption introduces no error. For ideal gases whose specific heats vary almost linearly in the temperature range of interest, the possible error is minimized by using specific heat values evaluated at the average temperature (Fig. 8–32). The results obtained in this way usually are sufficiently accurate if the temperature range is not greater than a few hundred degrees.

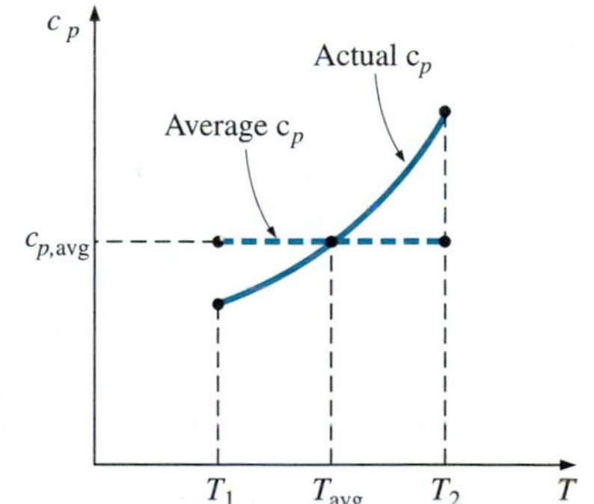

FIGURE 8–32

Under the constant-specific-heat assumption, the specific heat is assumed to be constant at some average value.

The entropy-change relations for ideal gases under the constant-specific-heat assumption are easily obtained by replacing $c_v(T)$ and $c_p(T)$ in Eqs. 8–31 and 8–32 by $c_{v,\text{avg}}$ and $c_{p,\text{avg}}$, respectively, and performing the integrations. We obtain

$$s_2 - s_1 = c_{v,\text{avg}}\ln\frac{T_2}{T_1} + R\ln\frac{v_2}{v_1} \qquad \text{(kJ/kg·K)} \tag{8–33}$$

and

$$s_2 - s_1 = c_{p,\text{avg}}\ln\frac{T_2}{T_1} - R\ln\frac{P_2}{P_1} \qquad \text{(kJ/kg·K)} \tag{8–34}$$

Entropy changes can also be expressed on a unit-mole basis by multiplying these relations by molar mass:

$$\bar{s}_2 - \bar{s}_1 = \bar{c}_{v,\text{avg}}\ln\frac{T_2}{T_1} + R_u\ln\frac{v_2}{v_1} \qquad \text{(kJ/kmol·K)} \tag{8–35}$$

and

$$\bar{s}_2 - \bar{s}_1 = \bar{c}_{p,\text{avg}} \ln \frac{T_2}{T_1} - R_u \ln \frac{P_2}{P_1} \qquad (\text{kJ/kmol} \cdot \text{K}) \tag{8–36}$$

Variable Specific Heats (Exact Analysis)

When the temperature change during a process is large and the specific heats of the ideal gas vary nonlinearly within the temperature range, the assumption of constant specific heats may lead to considerable errors in entropy-change calculations. For those cases, the variation of specific heats with temperature should be properly accounted for by utilizing accurate relations for the specific heats as a function of temperature. The entropy change during a process is then determined by substituting these $c_v(T)$ or $c_p(T)$ relations into Eq. 8–31 or 8–32 and performing the integrations.

Instead of performing these laborious integrals each time we have a new process, it is convenient to perform these integrals once and tabulate the results. For this purpose, we choose absolute zero as the reference temperature and define a function $s°$ as

$$s° = \int_0^T c_p(T) \frac{dT}{T} \tag{8–37}$$

Obviously, $s°$ is a function of temperature alone, and its value is zero at absolute zero temperature. The values of $s°$ are calculated at various temperatures, and the results are tabulated in the appendix as a function of temperature for air. Given this definition, the integral in Eq. 8–32 becomes

$$\int_1^2 c_p(T) \frac{dT}{T} = s_2° - s_1° \tag{8–38}$$

where $s_2°$ is the value of $s°$ at T_2 and $s_1°$ is the value at T_1. Thus,

$$s_2 - s_1 = s_2° - s_1° - R \ln \frac{P_2}{P_1} \qquad (\text{kJ/kg} \cdot \text{K}) \tag{8–39}$$

It can also be expressed on a unit-mole basis as

$$\bar{s}_2 - \bar{s}_1 = \bar{s}_2° - \bar{s}_1° - R_u \ln \frac{P_2}{P_1} \qquad (\text{kJ/kmol} \cdot \text{K}) \tag{8–40}$$

Note that unlike internal energy and enthalpy, the entropy of an ideal gas varies with specific volume or pressure as well as the temperature. Therefore, entropy cannot be tabulated as a function of temperature alone. The $s°$ values in the tables account for the temperature dependence of entropy (Fig. 8–33). The variation of entropy with pressure is accounted for by the last term in Eq. 8–39. Another relation for entropy change can be developed based on Eq. 8–31, but this would require the definition of another function and tabulation of its values, which is not practical.

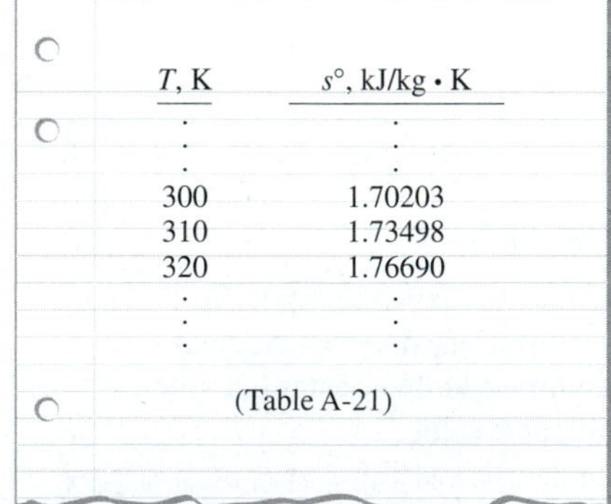

T, K	$s°$, kJ/kg · K
.	.
.	.
300	1.70203
310	1.73498
320	1.76690
.	.
.	.

(Table A-21)

FIGURE 8–33
The entropy of an ideal gas depends on both T and P. The function $s°$ represents only the temperature-dependent part of entropy.

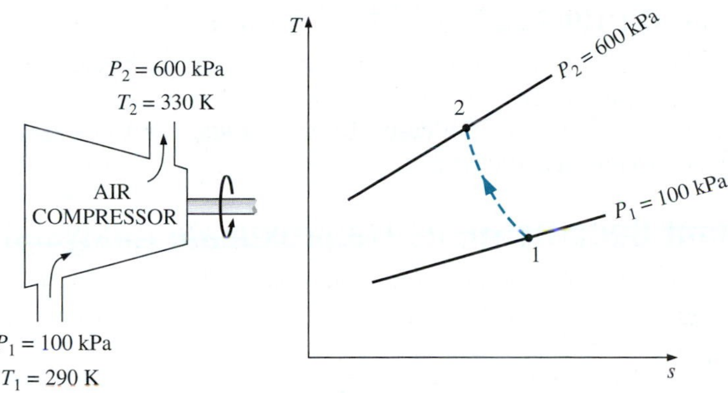

FIGURE 8–34

Schematic and *T-s* diagram for Example 8–9.

EXAMPLE 8–9 Entropy Change of an Ideal Gas

Air is compressed from an initial state of 100 kPa and 17°C to a final state of 600 kPa and 57°C. Determine the entropy change of air during this compression process by using (*a*) property values from the air table and (*b*) average specific heats.

Solution Air is compressed between two specified states. The entropy change of air is to be determined by using tabulated property values and also by using average specific heats.

Assumptions Air is an ideal gas since it is at a high temperature and low pressure relative to its critical-point values. Therefore, entropy change relations developed under the ideal-gas assumption are applicable.

Analysis A sketch of the system and the *T-s* diagram for the process are given in Fig. 8–34. We note that both the initial and the final states of air are completely specified.

(*a*) The properties of air are given in the air table (Table A–21). Reading $s°$ values at given temperatures and substituting, we find

$$s_2 - s_1 = s_2^\circ - s_1^\circ - R \ln \frac{P_2}{P_1}$$

$$= [(1.79783 - 1.66802)\ \text{kJ/kg}\cdot\text{K}] - (0.287\ \text{kJ/kg}\cdot\text{K}) \ln \frac{600\ \text{kPa}}{100\ \text{kPa}}$$

$$= \mathbf{-0.3844\ kJ/kg\cdot K}$$

(*b*) The entropy change of air during this process can also be determined approximately from Eq. 8–34 by using a c_p value at the average temperature of 37°C (Table A–2b) and treating it as a constant:

$$s_2 - s_1 = c_{p,\text{avg}} \ln \frac{T_2}{T_1} - R \ln \frac{P_2}{P_1}$$

$$= (1.006\ \text{kJ/kg}\cdot\text{K}) \ln \frac{330\ \text{K}}{290\ \text{K}} - (0.287\ \text{kJ/kg}\cdot\text{K}) \ln \frac{600\ \text{kPa}}{100\ \text{kPa}}$$

$$= \mathbf{-0.3842\ kJ/kg\cdot K}$$

Discussion The two results above are almost identical since the change in temperature during this process is relatively small (Fig. 8–35). When the temperature change is large, however, they may differ significantly. For those cases, Eq. 8–39 should be used instead of Eq. 8–34 since it accounts for the variation of specific heats with temperature.

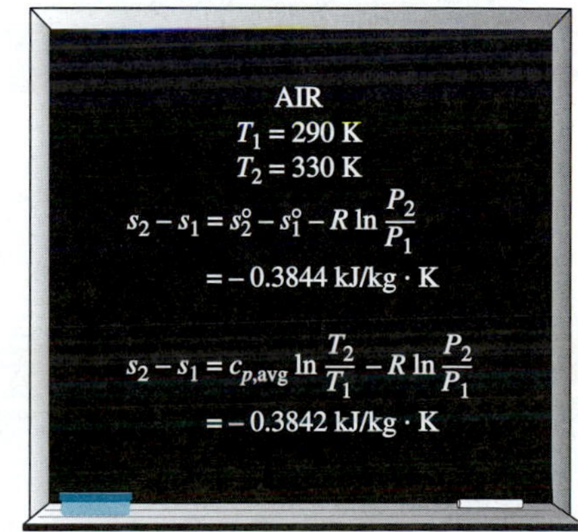

AIR
$T_1 = 290$ K
$T_2 = 330$ K
$s_2 - s_1 = s_2^\circ - s_1^\circ - R \ln \frac{P_2}{P_1}$
$= -0.3844$ kJ/kg · K
$s_2 - s_1 = c_{p,\text{avg}} \ln \frac{T_2}{T_1} - R \ln \frac{P_2}{P_1}$
$= -0.3842$ kJ/kg · K

FIGURE 8–35

For small temperature differences, the exact and approximate relations for entropy changes of ideal gases give almost identical results.

Isentropic Processes of Ideal Gases

Several relations for the isentropic processes of ideal gases can be obtained by setting the entropy-change relations developed previously equal to zero. Again, this is done first for the case of constant specific heats and then for the case of variable specific heats.

Constant Specific Heats (Approximate Analysis)

When the constant-specific-heat assumption is valid, the isentropic relations for ideal gases are obtained by setting Eqs. 8–33 and 8–34 equal to zero. From Eq. 8–33,

$$\ln \frac{T_2}{T_1} = -\frac{R}{c_v} \ln \frac{v_2}{v_1}$$

which can be rearranged as

$$\ln \frac{T_2}{T_1} = \ln \left(\frac{v_1}{v_2}\right)^{R/c_v} \quad \textbf{(8–41)}$$

or

$$\left(\frac{T_2}{T_1}\right)_{s=\text{const.}} = \left(\frac{v_1}{v_2}\right)^{k-1} \quad \text{(ideal gas)} \quad \textbf{(8–42)}$$

since $R = c_p - c_v$, $k = c_p/c_v$, and thus $R/c_v = k - 1$.

Equation 8–42 is the *first isentropic relation* for ideal gases under the constant-specific-heat assumption. The *second isentropic relation* is obtained in a similar manner from Eq. 8–34 with the following result:

$$\left(\frac{T_2}{T_1}\right)_{s=\text{const.}} = \left(\frac{P_2}{P_1}\right)^{(k-1)/k} \quad \text{(ideal gas)} \quad \textbf{(8–43)}$$

The *third isentropic relation* is obtained by substituting Eq. 8–43 into Eq. 8–42 and simplifying:

$$\left(\frac{P_2}{P_1}\right)_{s=\text{const.}} = \left(\frac{v_1}{v_2}\right)^{k} \quad \text{(ideal gas)} \quad \textbf{(8–44)}$$

Equations 8–42 through 8–44 can also be expressed in a compact form as

$$Tv^{k-1} = \text{constant} \quad \textbf{(8–45)}$$

$$TP^{(1-k)/k} = \text{constant} \quad \text{(ideal gas)} \quad \textbf{(8–46)}$$

$$Pv^{k} = \text{constant} \quad \textbf{(8–47)}$$

The specific heat ratio k, in general, varies with temperature, and thus an average k value for the given temperature range should be used.

Note that the ideal-gas isentropic relations above, as the name implies, are strictly valid for isentropic processes only when the constant-specific-heat assumption is appropriate (Fig. 8–36).

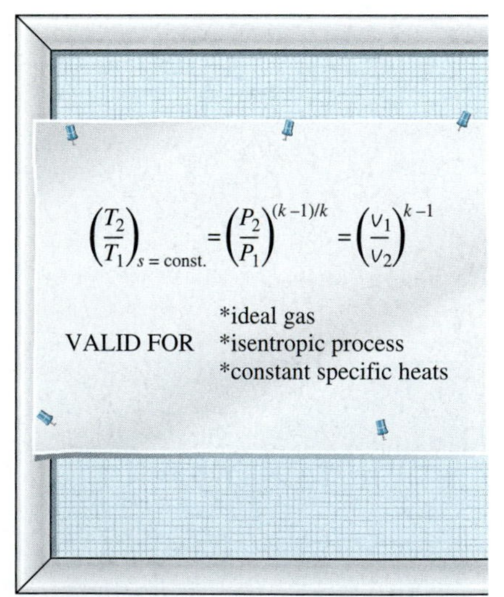

FIGURE 8–36
The isentropic relations of ideal gases are valid for the isentropic processes of ideal gases only.

Variable Specific Heats (Exact Analysis)

When the constant-specific-heat assumption is not appropriate, the isentropic relations developed previously yields results that are not quite accurate. For such cases, we should use an isentropic relation obtained from Eq. 8–39 that accounts for the variation of specific heats with temperature. Setting this equation equal to zero gives

$$0 = s_2^\circ - s_1^\circ - R\ln\frac{P_2}{P_1}$$

or

$$s_2^\circ = s_1^\circ + R\ln\frac{P_2}{P_1} \tag{8–48}$$

where s_2° is the s° value at the end of the isentropic process.

Relative Pressure and Relative Specific Volume

Equation 8–48 provides an accurate way of evaluating property changes of ideal gases during isentropic processes since it accounts for the variation of specific heats with temperature. However, it involves tedious iterations when the volume ratio is given instead of the pressure ratio. This is quite an inconvenience in optimization studies, which usually require numerous repetitive calculations. To remedy this deficiency, we define two new dimensionless quantities associated with isentropic processes.

The definition of the first is based on Eq. 8–48, which can be rearranged as

$$\frac{P_2}{P_1} = \exp\frac{s_2^\circ - s_1^\circ}{R}$$

or

$$\frac{P_2}{P_1} = \frac{\exp(s_2^\circ/R)}{\exp(s_1^\circ/R)}$$

The quantity $\exp(s^\circ/R)$ is defined as the **relative pressure** P_r. With this definition, the last relation becomes

$$\left(\frac{P_2}{P_1}\right)_{s=\text{const.}} = \frac{P_{r2}}{P_{r1}} \tag{8–49}$$

Note that the relative pressure P_r is a *dimensionless* quantity that is a function of temperature only since s° depends on temperature alone. Therefore, values of P_r can be tabulated against temperature. This is done for air in Table A–21. The use of P_r data is illustrated in Fig. 8–37.

Sometimes specific volume ratios are given instead of pressure ratios. This is particularly the case when automotive engines are analyzed. In such cases, one needs to work with volume ratios. Therefore, we define another quantity related to specific volume ratios for isentropic processes. This is done by utilizing the ideal-gas relation and Eq. 8–49:

$$\frac{P_1 v_1}{T_1} = \frac{P_2 v_2}{T_2} \rightarrow \frac{v_2}{v_1} = \frac{T_2}{T_1}\frac{P_1}{P_2} = \frac{T_2}{T_1}\frac{P_{r1}}{P_{r2}} = \frac{T_2/P_{r2}}{T_1/P_{r1}}$$

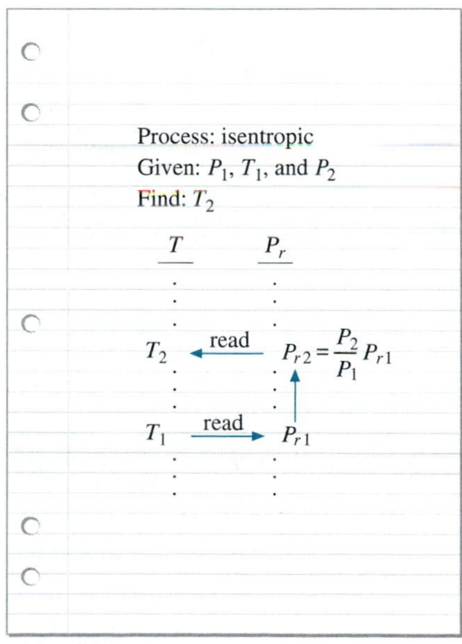

FIGURE 8–37

The use of P_r data for calculating the final temperature during an isentropic process.

The quantity T/P_r is a function of temperature only and is defined as **relative specific volume** v_r. Thus,

$$\left(\frac{v_2}{v_1}\right)_{s=\text{const.}} = \frac{v_{r2}}{v_{r1}} \quad \text{(8–50)}$$

Equations 8–49 and 8–50 are strictly valid for isentropic processes of ideal gases only. They account for the variation of specific heats with temperature and therefore give more accurate results than Eqs. 8–42 through 8–47. The values of P_r and v_r are listed for air in Table A–21.

EXAMPLE 8–10 Isentropic Compression of Air in a Car Engine

Air is compressed in a car engine from 22°C and 95 kPa in a reversible and adiabatic manner. If the compression ratio V_1/V_2 of this engine is 8, determine the final temperature of the air.

Solution Air is compressed in a car engine isentropically. For a given compression ratio, the final air temperature is to be determined.

Assumptions At specified conditions, air can be treated as an ideal gas. Therefore, the isentropic relations for ideal gases are applicable.

Analysis A sketch of the system and the *T-s* diagram for the process are given in Fig. 8–38.

This process is easily recognized as being isentropic since it is both reversible and adiabatic. The final temperature for this isentropic process can be determined from Eq. 8–50 with the help of relative specific volume data (Table A–21), as illustrated in Fig. 8–39.

For closed systems: $\dfrac{v_2}{v_1} = \dfrac{V_2}{V_1}$

At $T_1 = 295$ K: $v_{r1} = 647.9$

From Eq. 8–50: $v_{r2} = v_{r1}\left(\dfrac{v_2}{v_1}\right) = (647.9)\left(\dfrac{1}{8}\right) = 80.99 \rightarrow T_2 = \mathbf{662.7\ K}$

Therefore, the temperature of air will increase by 367.7°C during this process.

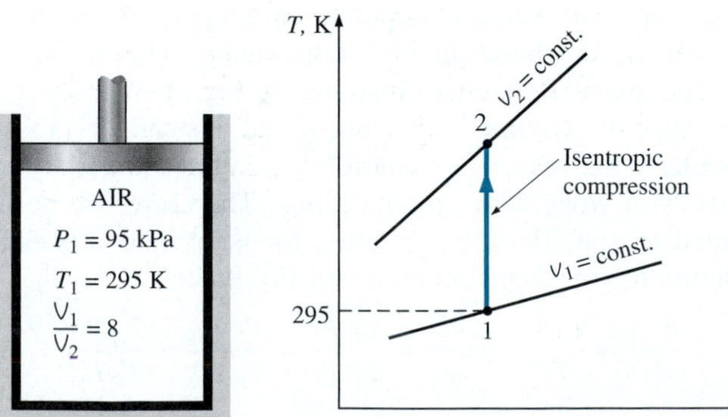

FIGURE 8–38

Schematic and *T-s* diagram for Example 8–10.

Alternative Solution The final temperature could also be determined from Eq. 8–42 by assuming constant specific heats for air:

$$\left(\frac{T_2}{T_1}\right)_{s=\text{const.}} = \left(\frac{v_1}{v_2}\right)^{k-1}$$

The specific heat ratio k also varies with temperature, and we need to use the value of k corresponding to the average temperature. However, the final temperature is not given, and so we cannot determine the average temperature in advance. For such cases, calculations can be started with a k value at the initial or the anticipated average temperature. This value could be refined later, if necessary, and the calculations can be repeated. We know that the temperature of the air will rise considerably during this adiabatic compression process, so we *guess* the average temperature to be about 450 K. The k value at this anticipated average temperature is determined from Table A–2*b* to be 1.391. Then the final temperature of air becomes

$$T_2 = (295\text{ K})(8)^{1.391-1} = 665.2\text{ K}$$

This gives an average temperature value of 480.1 K, which is sufficiently close to the assumed value of 450 K. Therefore, it is not necessary to repeat the calculations by using the k value at this average temperature.

The result obtained by assuming constant specific heats for this case is in error by about 0.4 percent, which is rather small. This is not surprising since the temperature change of air is relatively small (only a few hundred degrees) and the specific heats of air vary almost linearly with temperature in this temperature range.

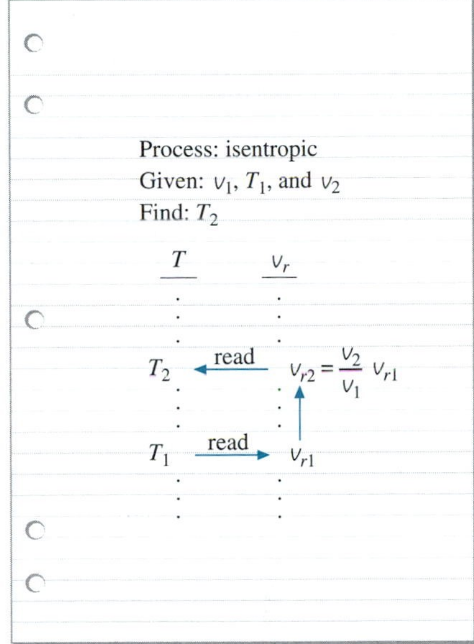

FIGURE 8–39

The use of v_r data for calculating the final temperature during an isentropic process (Example 8–10).

EXAMPLE 8–11 Isentropic Compression of an Ideal Gas

Helium gas is compressed by an adiabatic compressor from an initial state of 14 psia and 50°F to a final temperature of 320°F in a reversible manner. Determine the exit pressure of helium.

Solution Helium is compressed from a given state to a specified pressure isentropically. The exit pressure of helium is to be determined.

Assumptions At specified conditions, helium can be treated as an ideal gas. Therefore, the isentropic relations developed earlier for ideal gases are applicable.

Analysis A sketch of the system and the *T-s* diagram for the process are given in Fig. 8–40.

The specific heat ratio k of helium is 1.667 and is independent of temperature in the region where it behaves as an ideal gas. Thus the final pressure of helium can be determined from Eq. 8–43:

$$P_2 = P_1\left(\frac{T_2}{T_1}\right)^{k/(k-1)} = (14\text{ psia})\left(\frac{780\text{ R}}{510\text{ R}}\right)^{1.667/0.667} = \mathbf{40.5\text{ psia}}$$

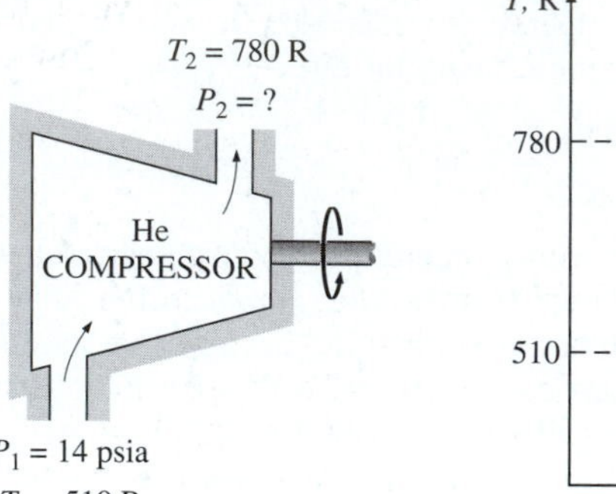

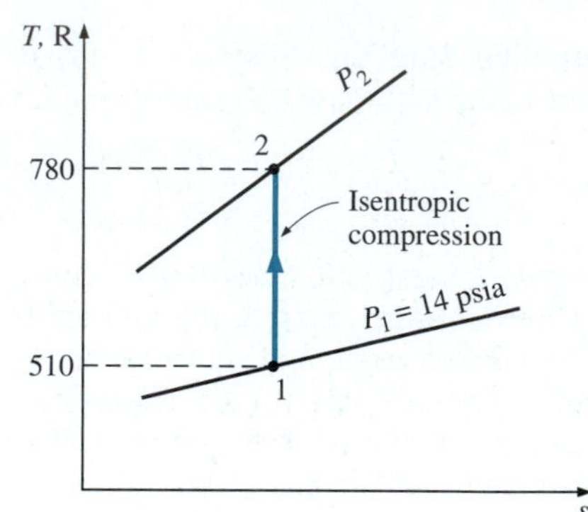

FIGURE 8–40

Schematic and *T-s* diagram for Example 8–11.

8–10 ▪ REVERSIBLE STEADY-FLOW WORK

The work done during a process depends on the path followed as well as on the properties at the end states. Recall that reversible (quasi-equilibrium) moving boundary work associated with closed systems is expressed in terms of the fluid properties as

$$W_b = \int_1^2 P\, dV$$

We mentioned that the quasi-equilibrium work interactions lead to the maximum work output for work-producing devices and the minimum work input for work-consuming devices.

It would also be very insightful to express the work associated with steady-flow devices in terms of fluid properties.

Taking the positive direction of work to be from the system (work output), the energy balance for a steady-flow device undergoing an internally reversible process can be expressed in differential form as

$$\delta q_{\text{rev}} - \delta w_{\text{rev}} = dh + d\text{ke} + d\text{pe}$$

But

$$\left.\begin{aligned} \delta q_{\text{rev}} &= T\, ds \qquad &\text{(Eq. 8–16)} \\ T\, ds &= dh - v\, dP \qquad &\text{(Eq. 8–24)} \end{aligned}\right\} \quad \delta q_{\text{rev}} = dh - v\, dP$$

Substituting this into the relation above and canceling *dh* yield

$$-\delta w_{\text{rev}} = v\, dP + d\text{ke} + d\text{pe}$$

Integrating, we find

$$w_{\text{rev}} = -\int_1^2 v\, dP - \Delta\text{ke} - \Delta\text{pe} \qquad \text{(kJ/kg)} \tag{8–51}$$

When the changes in kinetic and potential energies are negligible, this equation reduces to

$$w_{\text{rev}} = -\int_1^2 v\, dP \qquad \text{(kJ/kg)} \tag{8–52}$$

Equations 8–51 and 8–52 are relations for the *reversible work output* associated with an internally reversible process in a steady-flow device. They will give a negative result when work is done on the system. To avoid the negative sign, Eq. 8–51 can be written for work input to steady-flow devices such as compressors and pumps as

$$w_{\text{rev,in}} = \int_1^2 v\, dP + \Delta\text{ke} + \Delta\text{pe} \tag{8–53}$$

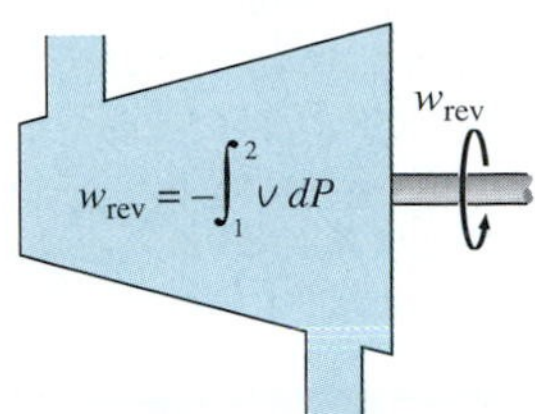

(*a*) Steady-flow system

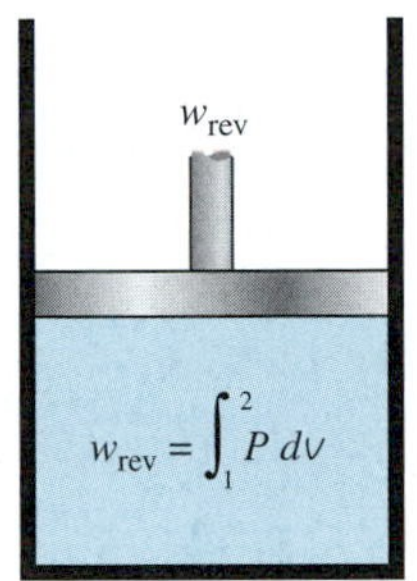

(*b*) Closed system

FIGURE 8–41

Reversible work relations for steady-flow and closed systems.

The resemblance between the $v\, dP$ in these relations and $P\, dv$ is striking. They should not be confused with each other, however, since $P\, dv$ is associated with reversible boundary work in closed systems (Fig. 8–41).

Obviously, one needs to know v as a function of P for the given process to perform the integration. When the working fluid is *incompressible*, the specific volume v remains constant during the process and can be taken out of the integration. Then Eq. 8–51 simplifies to

$$w_{\text{rev}} = -v(P_2 - P_1) - \Delta\text{ke} - \Delta\text{pe} \qquad \text{(kJ/kg)} \tag{8–54}$$

For the steady flow of a liquid through a device that involves no work interactions (such as a nozzle or a pipe section), the work term is zero, and the equation above can be expressed as

$$v(P_2 - P_1) + \frac{V_2^2 - V_1^2}{2} + g(z_2 - z_1) = 0 \tag{8–55}$$

which is known as the **Bernoulli equation** in fluid mechanics. It is developed for an internally reversible process and thus is applicable to incompressible fluids that involve no irreversibilities such as friction or shock waves. This equation can be modified, however, to incorporate these effects.

Equation 8–52 has far-reaching implications in engineering regarding devices that produce or consume work steadily such as turbines, compressors, and pumps. It is obvious from this equation that the reversible steady-flow work is closely associated with the specific volume of the fluid flowing through the device. *The larger the specific volume, the larger the reversible work produced or consumed by the steady flow device* (Fig. 8–42). This conclusion is equally valid for actual steady-flow devices. Therefore, every effort should be made to keep the specific volume of a fluid as small as possible during a compression process to minimize the work input and as large as possible during an expansion process to maximize the work output.

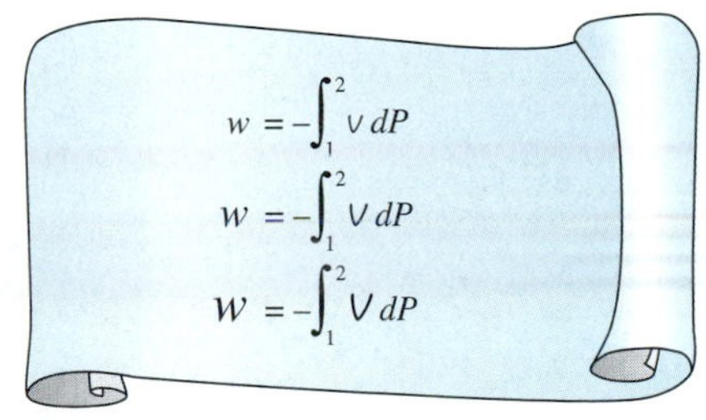

FIGURE 8–42

The larger the specific volume, the greater the work produced (or consumed) by a steady-flow device.

In steam or gas power plants, the pressure rise in the pump or compressor is equal to the pressure drop in the turbine if we disregard the pressure losses in various other components. In steam power plants, the pump handles liquid, which has a very small specific volume, and the turbine handles vapor, whose specific volume is many times larger. Therefore, the work output of the turbine is much larger than the work input to the pump. This is one of the reasons for the wide-spread use of steam power plants in electric power generation.

If we were to compress the steam exiting the turbine back to the turbine inlet pressure before cooling it first in the condenser in order to "save" the heat rejected, we would have to supply all the work produced by the turbine back to the compressor. In reality, the required work input would be even greater than the work output of the turbine because of the irreversibilities present in both processes.

In gas power plants, the working fluid (typically air) is compressed in the gas phase, and a considerable portion of the work output of the turbine is consumed by the compressor. As a result, a gas power plant delivers less net work per unit mass of the working fluid.

EXAMPLE 8–12 Compressing a Substance in the Liquid versus Gas Phases

Determine the compressor work input required to compress steam isentropically from 100 kPa to 1 MPa, assuming that the steam exists as (*a*) saturated liquid and (*b*) saturated vapor at the inlet state.

Solution Steam is to be compressed from a given pressure to a specified pressure isentropically. The work input is to be determined for the cases of steam being a saturated liquid and saturated vapor at the inlet.

Assumptions **1** Steady operating conditions exist. **2** Kinetic and potential energy changes are negligible. **3** The process is given to be isentropic.

Analysis We take first the turbine and then the pump as the *system.* Both are *control volumes* since mass crosses the boundary. Sketches of the pump and the turbine together with the *T-s* diagram are given in Fig. 8–43.

(*a*) In this case, steam is a saturated liquid initially, and its specific volume is

$$v_1 = v_{f\,@\,100\text{ kPa}} = 0.001043\ \text{m}^3/\text{kg} \qquad \text{(Table A–5)}$$

which remains essentially constant during the process. Thus,

$$\begin{aligned} w_{\text{rev,in}} &= \int_1^2 v\,dP \cong v_1(P_2 - P_1) \\ &= (0.001043\ \text{m}^3/\text{kg})[(1000 - 100)\ \text{kPa}]\left(\frac{1\ \text{kJ}}{1\ \text{kPa}\cdot\text{m}^3}\right) \\ &= \mathbf{0.94\ kJ/kg} \end{aligned}$$

(*b*) This time, steam is a saturated vapor initially and remains a vapor during the entire compression process. Since the specific volume of a gas changes considerably during a compression process, we need to know how v varies with P to perform the integration in Eq. 8–53. This relation, in general, is not readily available. But for an isentropic process, it is easily obtained from the

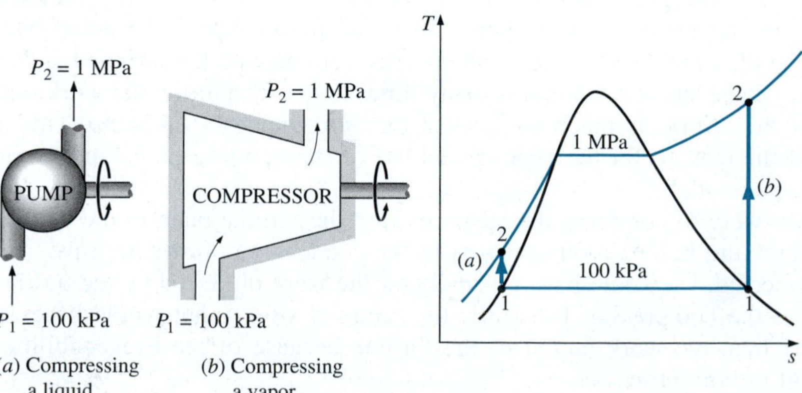

FIGURE 8–43
Schematic and *T-s* diagram for Example 8–12.

second $T\,ds$ relation by setting $ds = 0$:

$$\left.\begin{aligned} T\,ds &= dh - v\,dP \quad \text{(Eq. 8-24)} \\ ds &= 0 \quad \text{(isentropic process)} \end{aligned}\right\} \quad v\,dP = dh$$

Thus,

$$w_{\text{rev,in}} = \int_1^2 v\,dP = \int_1^2 dh = h_2 - h_1$$

This result could also be obtained from the energy balance relation for an isentropic steady-flow process. Next we determine the enthalpies:

State 1:
$$\left.\begin{aligned} P_1 &= 100 \text{ kPa} \\ &\text{(sat. vapor)} \end{aligned}\right\} \quad \begin{aligned} h_1 &= 2675.0 \text{ kJ/kg} \\ s_1 &= 7.3589 \text{ kJ/kg} \cdot \text{K} \end{aligned} \qquad \text{(Table A–5)}$$

State 2:
$$\left.\begin{aligned} P_2 &= 1 \text{ MPa} \\ s_2 &= s_1 \end{aligned}\right\} \quad h_2 = 3194.5 \text{ kJ/kg} \qquad \text{(Table A–6)}$$

Thus,

$$w_{\text{rev,in}} = (3194.5 - 2675.0) \text{ kJ/kg} = \mathbf{519.5 \text{ kJ/kg}}$$

Discussion Note that compressing steam in the vapor form would require over 500 times more work than compressing it in the liquid form between the same pressure limits.

Proof that Steady-Flow Devices Deliver the Most and Consume the Least Work when the Process Is Reversible

We have shown in Chap. 7 that cyclic devices (heat engines, refrigerators, and heat pumps) deliver the most work and consume the least when reversible processes are used. Now we demonstrate that this is also the case for individual devices such as turbines and compressors in steady operation.

Consider two steady-flow devices, one reversible and the other irreversible, operating between the same inlet and exit states. Again taking heat transfer to the system and work done by the system to be positive quantities, the energy balance for each of these devices can be expressed in the differential form as

Actual: $$\delta q_{\text{act}} - \delta w_{\text{act}} = dh + d\text{ke} + d\text{pe}$$

Reversible: $$\delta q_{\text{rev}} - \delta w_{\text{rev}} = dh + d\text{ke} + d\text{pe}$$

The right-hand sides of these two equations are identical since both devices are operating between the same end states. Thus,

$$\delta q_{\text{act}} - \delta w_{\text{act}} = \delta q_{\text{rev}} - \delta w_{\text{rev}}$$

or

$$\delta w_{\text{rev}} - \delta w_{\text{act}} = \delta q_{\text{rev}} - \delta q_{\text{act}}$$

However,

$$\delta q_{\text{rev}} = T\,ds$$

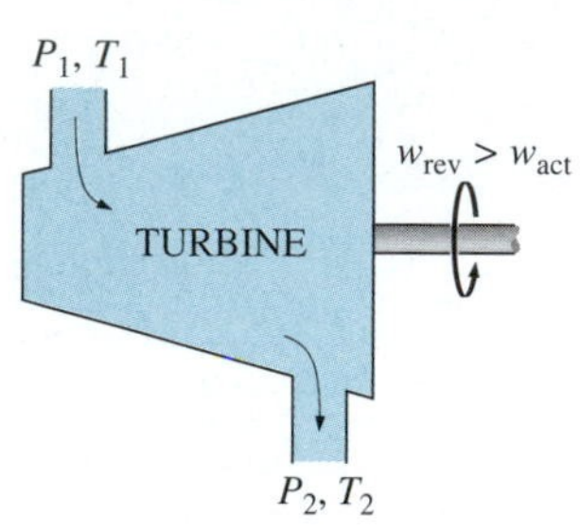

FIGURE 8–44

A reversible turbine delivers more work than an irreversible one if both operate between the same end states.

Substituting this relation into the preceding equation and dividing each term by T, we obtain

$$\frac{\delta w_{rev} - \delta w_{act}}{T} = ds - \frac{\delta q_{act}}{T} \geq 0$$

since

$$ds \geq \frac{\delta q_{act}}{T}$$

Also, T is the absolute temperature, which is always positive. Thus,

$$\partial w_{rev} \geq \partial w_{act}$$

or

$$w_{rev} \geq w_{act}$$

Therefore, work-producing devices such as turbines (w is positive) deliver more work, and work-consuming devices such as pumps and compressors (w is negative) require less work when they operate reversibly (Fig. 8–44).

8–11 ▪ MINIMIZING THE COMPRESSOR WORK

We have just shown that the work input to a compressor is minimized when the compression process is executed in an internally reversible manner. When the changes in kinetic and potential energies are negligible, the compressor work is given by (Eq. 8–53)

$$w_{rev,in} = \int_1^2 v\, dP \qquad \textbf{(8–56)}$$

Obviously one way of minimizing the compressor work is to approximate an internally reversible process as much as possible by minimizing the irreversibilities such as friction, turbulence, and nonquasi-equilibrium compression. The extent to which this can be accomplished is limited by economic considerations. A second (and more practical) way of reducing the compressor work is to keep the specific volume of the gas as small as possible during the compression process. This is done by maintaining the temperature of the gas as low as possible during compression since the specific volume of a gas is proportional to temperature. Therefore, reducing the work input to a compressor requires that the gas be cooled as it is compressed.

To have a better understanding of the effect of cooling during the compression process, we compare the work input requirements for three kinds of processes: *an isentropic process* (involves no cooling), *a polytropic process* (involves some cooling), and *an isothermal process* (involves maximum cooling). Assuming all three processes are executed between the same pressure levels (P_1 and P_2) in an internally reversible manner and the gas behaves as an ideal gas ($Pv = RT$) with constant specific heats, we see that the compression work is determined by performing the integration in Eq. 8–56 for each case, with the following results:

Isentropic ($P\mathrm{v}^k$ = constant):

$$w_{\text{comp,in}} = \frac{kR(T_2 - T_1)}{k - 1} = \frac{kRT_1}{k - 1}\left[\left(\frac{P_2}{P_1}\right)^{(k-1)/k} - 1\right] \tag{8–57a}$$

Polytropic ($P\mathrm{v}^n$ = constant):

$$w_{\text{comp,in}} = \frac{nR(T_2 - T_1)}{n - 1} = \frac{nRT_1}{n - 1}\left[\left(\frac{P_2}{P_1}\right)^{(n-1)/n} - 1\right] \tag{8–57b}$$

Isothermal ($P\mathrm{v}$ = constant):

$$w_{\text{comp,in}} = RT \ln \frac{P_2}{P_1} \tag{8–57c}$$

The three processes are plotted on a P-v diagram in Fig. 8–45 for the same inlet state and exit pressure. On a P-v diagram, the area to the left of the process curve is the integral of $\mathrm{v}\,dP$. Thus it is a measure of the steady-flow compression work. It is interesting to observe from this diagram that of the three internally reversible cases considered, the adiabatic compression ($P\mathrm{v}^k$ = constant) requires the maximum work and the isothermal compression (T = constant or $P\mathrm{v}$ = constant) requires the minimum. The work input requirement for the polytropic case ($P\mathrm{v}^n$ = constant) is between these two and decreases as the polytropic exponent n is decreased, by increasing the heat rejection during the compression process. If sufficient heat is removed, the value of n approaches unity and the process becomes isothermal. One common way of cooling the gas during compression is to use cooling jackets around the casing of the compressors.

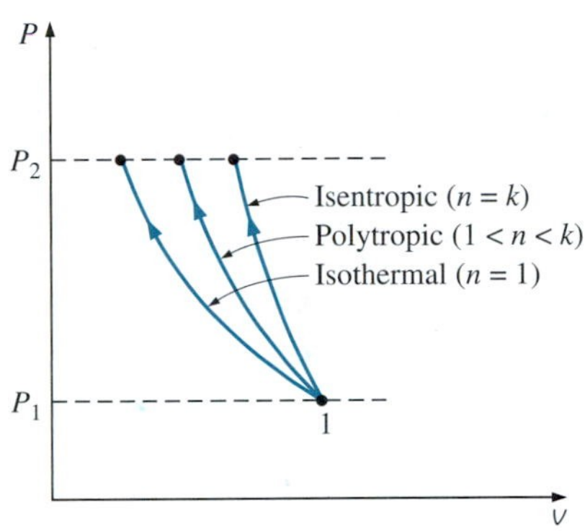

FIGURE 8–45

P-v diagrams of isentropic, polytropic, and isothermal compression processes between the same pressure limits.

Multistage Compression with Intercooling

It is clear from these arguments that cooling a gas as it is compressed is desirable since this reduces the required work input to the compressor. However, often it is not possible to have adequate cooling through the casing of the compressor, and it becomes necessary to use other techniques to achieve effective cooling. One such technique is **multistage compression with intercooling**, where the gas is compressed in stages and cooled between each stage by passing it through a heat exchanger called an *intercooler*. Ideally, the cooling process takes place at constant pressure, and the gas is cooled to the initial temperature T_1 at each intercooler. Multistage compression with intercooling is especially attractive when a gas is to be compressed to very high pressures.

The effect of intercooling on compressor work is graphically illustrated on P-v and T-s diagrams in Fig. 8–46 for a two-stage compressor. The gas is compressed in the first stage from P_1 to an intermediate pressure P_x, cooled at constant pressure to the initial temperature T_1, and compressed in the second stage to the final pressure P_2. The compression processes, in general, can be modeled as polytropic ($P\mathrm{v}^n$ = constant) where the value of n varies between k and 1. The colored area on the P-v diagram represents the work saved as a result of two-stage compression with intercooling. The process paths for single-stage isothermal and polytropic processes are also shown for comparison.

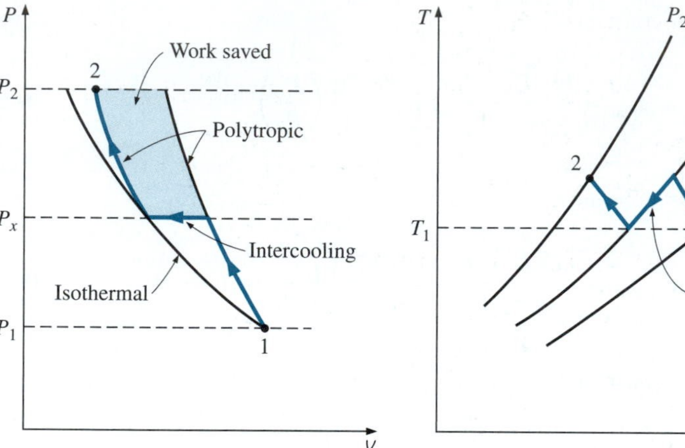

FIGURE 8–46

P-*v* and *T*-*s* diagrams for a two-stage steady-flow compression process.

The size of the colored area (the saved work input) varies with the value of the intermediate pressure P_x, and it is of practical interest to determine the conditions under which this area is maximized. The total work input for a two-stage compressor is the sum of the work inputs for each stage of compression, as determined from Eq. 8–57*b*:

$$w_{\text{comp,in}} = w_{\text{comp I,in}} + w_{\text{comp II,in}} \tag{8–58}$$

$$= \frac{nRT_1}{n-1}\left[\left(\frac{P_x}{P_1}\right)^{(n-1)/n} - 1\right] + \frac{nRT_1}{n-1}\left[\left(\frac{P_2}{P_x}\right)^{(n-1)/n} - 1\right]$$

The only variable in this equation is P_x. The P_x value that minimizes the total work is determined by differentiating this expression with respect to P_x and setting the resulting expression equal to zero. It yields

$$P_x = (P_1P_2)^{1/2} \quad \text{or} \quad \frac{P_x}{P_1} = \frac{P_2}{P_x} \tag{8–59}$$

That is, *to minimize compression work during two-stage compression, the pressure ratio across each stage of the compressor must be the same*. When this condition is satisfied, the compression work at each stage becomes identical, that is, $w_{\text{comp I,in}} = w_{\text{comp II,in}}$.

EXAMPLE 8–13 Work Input for Various Compression Processes

Air is compressed steadily by a reversible compressor from an inlet state of 100 kPa and 300 K to an exit pressure of 900 kPa. Determine the compressor work per unit mass for (*a*) isentropic compression with $k = 1.4$, (*b*) polytropic compression with $n = 1.3$, (*c*) isothermal compression, and (*d*) ideal two-stage compression with intercooling with a polytropic exponent of 1.3.

Solution Air is compressed reversibly from a specified state to a specified pressure. The compressor work is to be determined for the cases of isentropic, polytropic, isothermal, and two-stage compression.

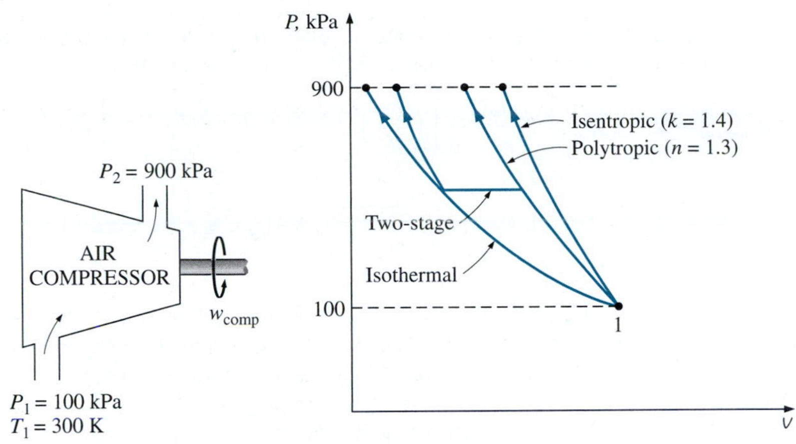

FIGURE 8–47

Schematic and *P-v* diagram for Example 8–13.

Assumptions **1** Steady operating conditions exist. **2** At specified conditions, air can be treated as an ideal gas. **3** Kinetic and potential energy changes are negligible.

Analysis We take the compressor to be the system. This is a control volume since mass crosses the boundary. A sketch of the system and the *T-s* diagram for the process are given in Fig. 8–47.

The steady-flow compression work for all these four cases is determined by using the relations developed earlier in this section:

(*a*) Isentropic compression with $k = 1.4$:

$$w_{\text{comp,in}} = \frac{kRT_1}{k-1}\left[\left(\frac{P_2}{P_1}\right)^{(k-1)/k} - 1\right]$$

$$= \frac{(1.4)(0.287\ \text{kJ/kg}\cdot\text{K})(300\ \text{K})}{1.4-1}\left[\left(\frac{900\ \text{kPa}}{100\ \text{kPa}}\right)^{(1.4-1)/1.4} - 1\right]$$

$$= \mathbf{263.2\ kJ/kg}$$

(*b*) Polytropic compression with $n = 1.3$:

$$w_{\text{comp,in}} = \frac{nRT_1}{n-1}\left[\left(\frac{P_2}{P_1}\right)^{(n-1)/n} - 1\right]$$

$$= \frac{(1.3)(0.287\ \text{kJ/kg}\cdot\text{K})(300\ \text{K})}{1.3-1}\left[\left(\frac{900\ \text{kPa}}{100\ \text{kPa}}\right)^{(1.3-1)/1.3} - 1\right]$$

$$= \mathbf{246.4\ kJ/kg}$$

(*c*) Isothermal compression:

$$w_{\text{comp,in}} = RT\ln\frac{P_2}{P_1} = (0.287\ \text{kJ/kg}\cdot\text{K})(300\ \text{K})\ln\frac{900\ \text{kPa}}{100\ \text{kPa}}$$

$$= \mathbf{189.2\ kJ/kg}$$

(*d*) Ideal two-stage compression with intercooling ($n = 1.3$): In this case, the pressure ratio across each stage is the same, and its value is

$$P_x = (P_1P_2)^{1/2} = [(100\ \text{kPa})(900\ \text{kPa})]^{1/2} = 300\ \text{kPa}$$

The compressor work across each stage is also the same. Thus the total compressor work is twice the compression work for a single stage:

$$w_{\text{comp,in}} = 2w_{\text{comp I,in}} = 2\,\frac{nRT_1}{n-1}\left[\left(\frac{P_x}{P_1}\right)^{(n-1)/n} - 1\right]$$

$$= \frac{2(1.3)(0.287\ \text{kJ/kg}\cdot\text{K})(300\ \text{K})}{1.3-1}\left[\left(\frac{300\ \text{kPa}}{100\ \text{kPa}}\right)^{(1.3-1)/1.3} - 1\right]$$

$$= \mathbf{215.3\ kJ/kg}$$

Discussion Of all four cases considered, the isothermal compression requires the minimum work and the isentropic compression the maximum. The compressor work is decreased when two stages of polytropic compression are utilized instead of just one. As the number of compressor stages is increased, the compressor work approaches the value obtained for the isothermal case.

8–12 ■ ISENTROPIC EFFICIENCIES OF STEADY-FLOW DEVICES

We mentioned repeatedly that irreversibilities inherently accompany all actual processes and that their effect is always to downgrade the performance of devices. In engineering analysis, it would be very desirable to have some parameters that would enable us to quantify the degree of degradation of energy in these devices. In the last chapter we did this for cyclic devices, such as heat engines and refrigerators, by comparing the actual cycles to the idealized ones, such as the Carnot cycle. A cycle that was composed entirely of reversible processes served as the *model cycle* to which the actual cycles could be compared. This idealized model cycle enabled us to determine the theoretical limits of performance for cyclic devices under specified conditions and to examine how the performance of actual devices suffered as a result of irreversibilities.

Now we extend the analysis to discrete engineering devices working under steady-flow conditions, such as turbines, compressors, and nozzles, and we examine the degree of degradation of energy in these devices as a result of irreversibilities. However, first we need to define an ideal process that serves as a model for the actual processes.

Although some heat transfer between these devices and the surrounding medium is unavoidable, many steady-flow devices are intended to operate under adiabatic conditions. Therefore, the model process for these devices should be an adiabatic one. Furthermore, an ideal process should involve no irreversibilities since the effect of irreversibilities is always to downgrade the performance of engineering devices. Thus, the ideal process that can serve as a suitable model for adiabatic steady-flow devices is the *isentropic* process (Fig. 8–48).

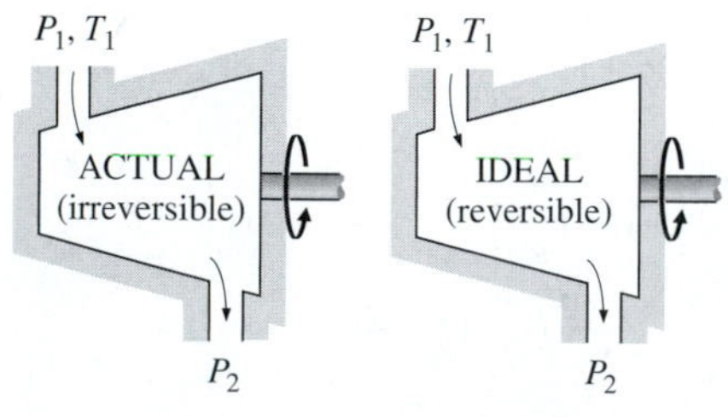

FIGURE 8–48

The isentropic process involves no irreversibilities and serves as the ideal process for adiabatic devices.

The more closely the actual process approximates the idealized isentropic process, the better the device performs. Thus, it would be desirable to have a parameter that expresses quantitatively how efficiently an actual device approximates an idealized one. This parameter is the **isentropic** or **adiabatic efficiency**, which is a measure of the deviation of actual processes from the corresponding idealized ones.

Isentropic efficiencies are defined differently for different devices since each device is set up to perform different tasks. Next we define the isentropic efficiencies of turbines, compressors, and nozzles by comparing the actual performance of these devices to their performance under isentropic conditions for the same inlet state and exit pressure.

Isentropic Efficiency of Turbines

For a turbine under steady operation, the inlet state of the working fluid and the exhaust pressure are fixed. Therefore, the ideal process for an adiabatic turbine is an isentropic process between the inlet state and the exhaust pressure. The desired output of a turbine is the work produced, and the **isentropic efficiency of a turbine** is defined as *the ratio of the actual work output of the turbine to the work output that would be achieved if the process between the inlet state and the exit pressure were isentropic:*

$$\eta_T = \frac{\text{Actual turbine work}}{\text{Isentropic turbine work}} = \frac{w_a}{w_s} \quad \text{(8–60)}$$

Usually the changes in kinetic and potential energies associated with a fluid stream flowing through a turbine are small relative to the change in enthalpy and can be neglected. Then the work output of an adiabatic turbine simply becomes the change in enthalpy, and Eq. 8–60 becomes

$$\eta_T \cong \frac{h_1 - h_{2a}}{h_1 - h_{2s}} \quad \text{(8–61)}$$

where h_{2a} and h_{2s} are the enthalpy values at the exit state for actual and isentropic processes, respectively (Fig. 8–49).

The value of η_T greatly depends on the design of the individual components that make up the turbine. Well-designed, large turbines have isentropic efficiencies above 90 percent. For small turbines, however, it may drop even below 70 percent. The value of the isentropic efficiency of a turbine is determined by measuring the actual work output of the turbine and by calculating the isentropic work output for the measured inlet conditions and the exit pressure. This value can then be used conveniently in the design of power plants.

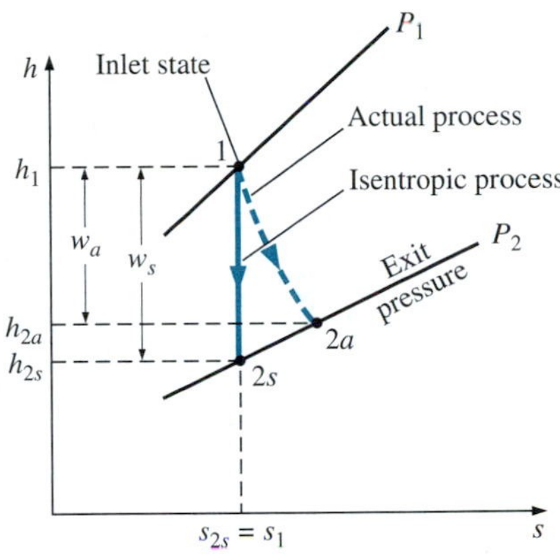

FIGURE 8–49

The *h-s* diagram for the actual and isentropic processes of an adiabatic turbine.

EXAMPLE 8–14 Isentropic Efficiency of a Steam Turbine

Steam enters an adiabatic turbine steadily at 3 MPa and 400°C and leaves at 50 kPa and 100°C. If the power output of the turbine is 2 MW, determine (*a*) the isentropic efficiency of the turbine and (*b*) the mass flow rate of the steam flowing through the turbine.

Solution Steam flows steadily in a turbine between inlet and exit states. For a specified power output, the isentropic efficiency and the mass flow rate are to be determined.

Assumptions **1** Steady operating conditions exist. **2** The changes in kinetic and potential energies are negligible.

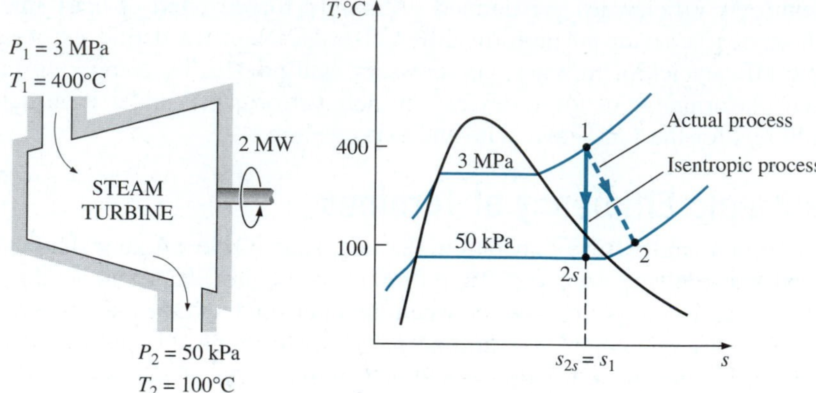

FIGURE 8–50
Schematic and *T-s* diagram for Example 8–14.

Analysis A sketch of the system and the *T-s* diagram of the process are given in Fig. 8–50.

(*a*) The enthalpies at various states are

$$\text{State 1:} \quad \left.\begin{aligned} P_1 &= 3\text{ MPA} \\ T_1 &= 400°\text{C} \end{aligned}\right\} \quad \begin{aligned} h_1 &= 3231.7\text{ kJ/kg} \\ s_1 &= 6.9235\text{ kJ/kg}\cdot\text{K} \end{aligned} \quad \text{(Table A–6)}$$

$$\text{State 2a:} \quad \left.\begin{aligned} P_{2a} &= 50\text{ kPa} \\ T_{2a} &= 100°\text{C} \end{aligned}\right\} \quad h_{2a} = 2682.4\text{ kJ/kg} \quad \text{(Table A–6)}$$

The exit enthalpy of the steam for the isentropic process h_{2s} is determined from the requirement that the entropy of the steam remain constant ($s_{2s} = s_1$):

$$\text{State 2s:} \quad \begin{aligned} P_{2s} &= 50\text{ kPa} \\ (s_{2s} &= s_1) \end{aligned} \quad \rightarrow \quad \begin{aligned} s_f &= 1.0912\text{ kJ/kg}\cdot\text{K} \\ s_g &= 7.5931\text{ kJ/kg}\cdot\text{K} \end{aligned} \quad \text{(Table A–5)}$$

Obviously, at the end of the isentropic process steam exists as a saturated mixture since $s_f < s_{2s} < s_g$. Thus we need to find the quality at state 2*s* first:

$$x_{2s} = \frac{s_{2s} - s_f}{s_{fg}} = \frac{6.9235 - 1.0912}{6.5019} = 0.897$$

and

$$h_{2s} = h_f + x_{2s}h_{fg} = 340.54 + 0.897(2304.7) = 2407.9\text{ kJ/kg}$$

By substituting these enthalpy values into Eq. 8–61, the isentropic efficiency of this turbine is determined to be

$$\eta_T \cong \frac{h_1 - h_{2a}}{h_1 - h_{2s}} = \frac{3231.7 - 2682.4}{3231.7 - 2407.9} = \mathbf{0.667,\ or\ 66.7\%}$$

(*b*) The mass flow rate of steam through this turbine is determined from the energy balance for steady-flow systems:

$$\dot{E}_{\text{in}} = \dot{E}_{\text{out}}$$

$$\dot{m}h_1 = \dot{W}_{a,\text{out}} + \dot{m}h_{2a}$$

$$\dot{W}_{a,\text{out}} = \dot{m}(h_1 - h_{2a})$$

$$2\text{ MW}\left(\frac{1000\text{ kJ/s}}{1\text{ MW}}\right) = \dot{m}(3231.7 - 2682.4)\text{ kJ/kg}$$

$$\dot{m} = \mathbf{3.64\ kg/s}$$

Isentropic Efficiencies of Compressors and Pumps

The **isentropic efficiency of a compressor** is defined as *the ratio of the work input required to raise the pressure of a gas to a specified value in an isentropic manner to the actual work input:*

$$\eta_C = \frac{\text{Isentropic compressor work}}{\text{Actual compressor work}} = \frac{w_s}{w_a} \tag{8–62}$$

Notice that the isentropic compressor efficiency is defined with the *isentropic work input in the numerator* instead of in the denominator. This is because w_s is a smaller quantity than w_a, and this definition prevents η_C from becoming greater than 100 percent, which would falsely imply that the actual compressors performed better than the isentropic ones. Also notice that the inlet conditions and the exit pressure of the gas are the same for both the actual and the isentropic compressor.

When the changes in kinetic and potential energies of the gas being compressed are negligible, the work input to an adiabatic compressor becomes equal to the change in enthalpy, and Eq. 8–62 for this case becomes

$$\eta_C \cong \frac{h_{2s} - h_1}{h_{2a} - h_1} \tag{8–63}$$

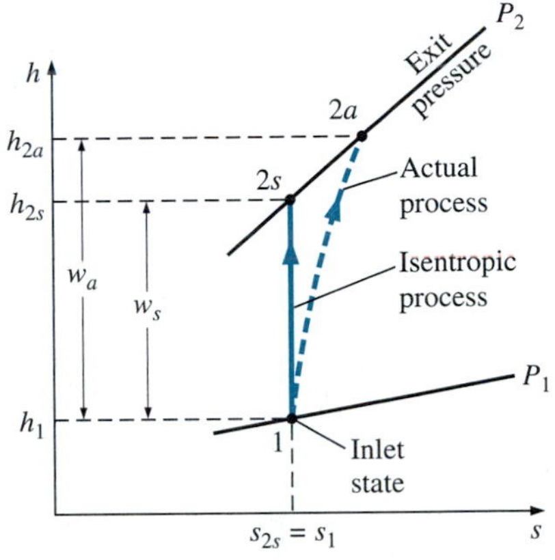

FIGURE 8–51

The *h-s* diagram of the actual and isentropic processes of an adiabatic compressor.

where h_{2a} and h_{2s} are the enthalpy values at the exit state for actual and isentropic compression processes, respectively, as illustrated in Fig. 8–51. Again, the value of η_C greatly depends on the design of the compressor. Well-designed compressors have isentropic efficiencies that range from 80 to 90 percent.

When the changes in potential and kinetic energies of a liquid are negligible, the isentropic efficiency of a pump is defined similarly as

$$\eta_P = \frac{w_s}{w_a} = \frac{v(P_2 - P_1)}{h_{2a} - h_1} \tag{8–64}$$

When no attempt is made to cool the gas as it is compressed, the actual compression process is nearly adiabatic and the reversible adiabatic (i.e., isentropic) process serves well as the ideal process. However, sometimes *compressors are cooled intentionally* by utilizing fins or a water jacket placed around the casing to reduce the work input requirements (Fig. 8–52). In this case, the isentropic process is not suitable as the model process since the device is no longer adiabatic and the isentropic compressor efficiency defined above is meaningless. A realistic model process for compressors that are intentionally cooled during the compression process is the *reversible isothermal process.* Then we can conveniently define an **isothermal efficiency** for such cases by comparing the actual process to a reversible isothermal one:

$$\eta_C = \frac{w_t}{w_a} \tag{8–65}$$

where w_t and w_a are the required work inputs to the compressor for the reversible isothermal and actual cases, respectively.

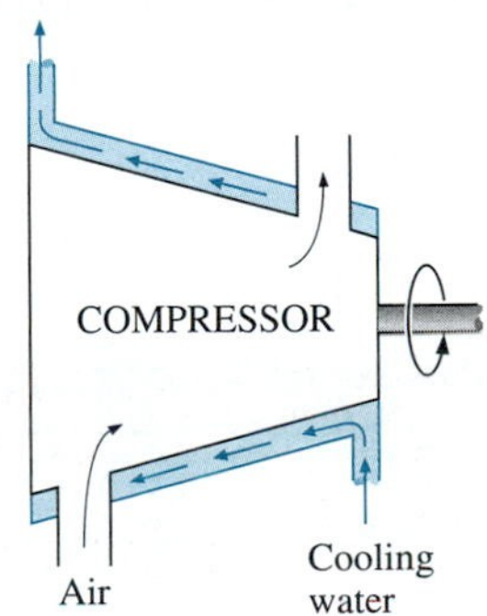

FIGURE 8–52

Compressors are sometimes intentionally cooled to minimize the work input.

EXAMPLE 8–15 Effect of Efficiency on Compressor Power Input

Air is compressed by an adiabatic compressor from 100 kPa and 12°C to a pressure of 800 kPa at a steady rate of 0.2 kg/s. If the isentropic efficiency of the compressor is 80 percent, determine (*a*) the exit temperature of air and (*b*) the required power input to the compressor.

Solution Air is compressed to a specified pressure at a specified rate. For a given isentropic efficiency, the exit temperature and the power input are to be determined.

Assumptions **1** Steady operating conditions exist. **2** Air is an ideal gas. **3** The changes in kinetic and potential energies are negligible.

Analysis A sketch of the system and the *T-s* diagram of the process are given in Fig. 8–53.

(*a*) We know only one property (pressure) at the exit state, and we need to know one more to fix the state and thus determine the exit temperature. The property that can be determined with minimal effort in this case is h_{2a} since the isentropic efficiency of the compressor is given. At the compressor inlet,

$$T_1 = 285 \text{ K} \quad \rightarrow \quad h_1 = 285.14 \text{ kJ/kg} \qquad \text{(Table A–21)}$$

$$(P_{r1} = 1.1584)$$

The enthalpy of the air at the end of the isentropic compression process is determined by using one of the isentropic relations of ideal gases,

$$P_{r2} = P_{r1}\left(\frac{P_2}{P_1}\right) = 1.1584\left(\frac{800 \text{ kPa}}{100 \text{ kPa}}\right) = 9.2672$$

and

$$P_{r2} = 9.2672 \quad \rightarrow \quad h_{2s} = 517.05 \text{ kJ/kg}$$

Substituting the known quantities into the isentropic efficiency relation, we have

$$\eta_C \cong \frac{h_{2s} - h_1}{h_{2a} - h_1} \quad \rightarrow \quad 0.80 = \frac{(517.05 - 285.14) \text{ kJ/kg}}{(h_{2a} - 285.14) \text{ kJ/kg}}$$

Thus,

$$h_{2a} = 575.03 \text{ kJ/kg} \quad \rightarrow \quad T_{2a} = \mathbf{569.5 \text{ K}}$$

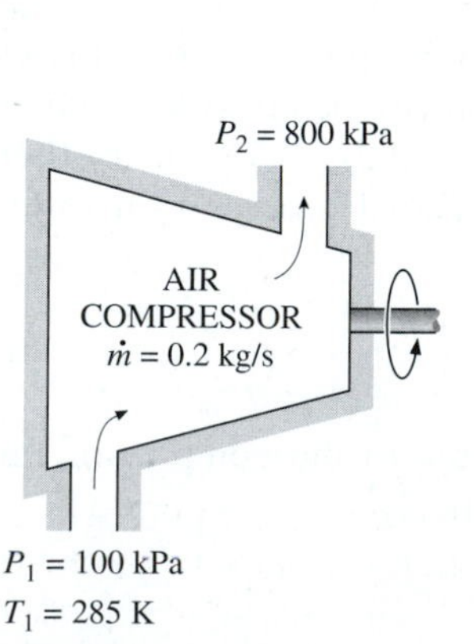

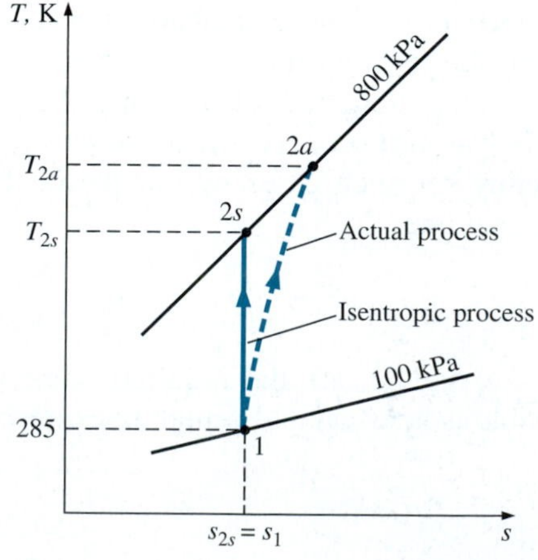

FIGURE 8–53

Schematic and *T-s* diagram for Example 8–15.

(*b*) The required power input to the compressor is determined from the energy balance for steady-flow devices,

$$\dot{E}_{in} = \dot{E}_{out}$$

$$\dot{m}h_1 + \dot{W}_{a,in} = \dot{m}h_{2a}$$

$$\dot{W}_{a,in} = \dot{m}(h_{2a} - h_1)$$

$$= (0.2\ \text{kg/s})[(575.03 - 285.14)\ \text{kJ/kg}]$$

$$= \mathbf{58.0\ kW}$$

Discussion Notice that in determining the power input to the compressor, we used h_{2a} instead of h_{2s} since h_{2a} is the actual enthalpy of the air as it exits the compressor. The quantity h_{2s} is a hypothetical enthalpy value that the air would have if the process were isentropic.

Isentropic Efficiency of Nozzles

Nozzles are essentially adiabatic devices and are used to accelerate a fluid. Therefore, the isentropic process serves as a suitable model for nozzles. The **isentropic efficiency of a nozzle** is defined as *the ratio of the actual kinetic energy of the fluid at the nozzle exit to the kinetic energy value at the exit of an isentropic nozzle for the same inlet state and exit pressure*. That is,

$$\eta_N = \frac{\text{Actual KE at nozzle exit}}{\text{Isentropic KE at nozzle exit}} = \frac{V_{2a}^2}{V_{2s}^2} \qquad \text{(8–66)}$$

Note that the exit pressure is the same for both the actual and isentropic processes, but the exit state is different.

Nozzles involve no work interactions, and the fluid experiences little or no change in its potential energy as it flows through the device. If, in addition, the inlet velocity of the fluid is small relative to the exit velocity, the energy balance for this steady-flow device reduces to

$$h_1 = h_{2a} + \frac{V_{2a}^2}{2}$$

Then the isentropic efficiency of the nozzle can be expressed in terms of enthalpies as

$$\eta_N \cong \frac{h_1 - h_{2a}}{h_1 - h_{2s}} \qquad \text{(8–67)}$$

where h_{2a} and h_{2s} are the enthalpy values at the nozzle exit for the actual and isentropic processes, respectively (Fig. 8–54). Isentropic efficiencies of nozzles are typically above 90 percent, and nozzle efficiencies above 95 percent are not uncommon.

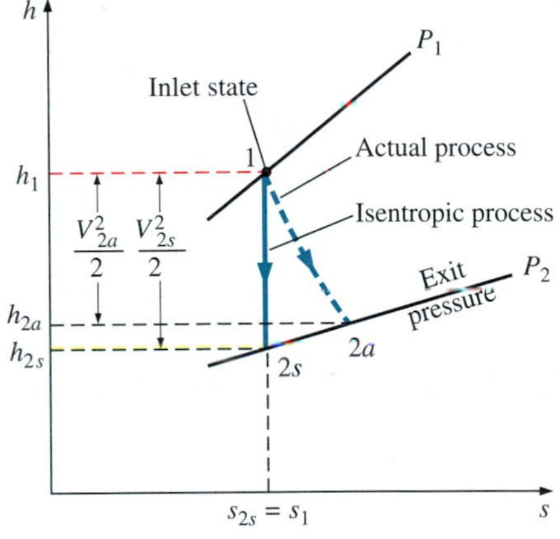

FIGURE 8–54

The *h-s* diagram of the actual and isentropic processes of an adiabatic nozzle.

EXAMPLE 8–16 **Effect of Efficiency on Nozzle Exit Velocity**

Air at 200 kPa and 950 K enters an adiabatic nozzle at low velocity and is discharged at a pressure of 80 kPa. If the isentropic efficiency of the nozzle is 92 percent, determine (*a*) the maximum possible exit velocity, (*b*) the exit temperature, and (*c*) the actual exit velocity of the air. Assume constant specific heats for air.

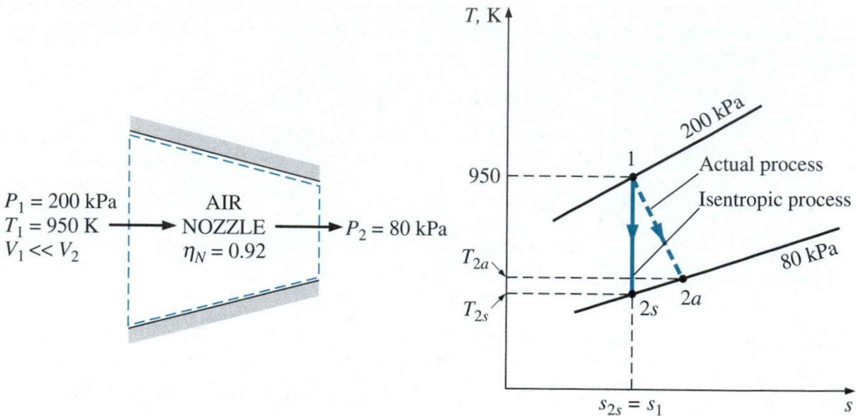

FIGURE 8–55

Schematic and *T-s* diagram for Example 8–16.

Solution The acceleration of air in a nozzle is considered. For specified exit pressure and isentropic efficiency, the maximum and actual exit velocities and the exit temperature are to be determined.

Assumptions **1** Steady operating conditions exist. **2** Air is an ideal gas. **3** The inlet kinetic energy is negligible.

Analysis A sketch of the system and the *T-s* diagram of the process are given in Fig. 8–55.

The temperature of air will drop during this acceleration process because some of its internal energy is converted to kinetic energy. This problem can be solved accurately by using property data from the air table. But we will assume constant specific heats (thus sacrifice some accuracy) to demonstrate their use. Let us guess the average temperature of the air to be about 800 K. Then the average values of c_p and k at this anticipated average temperature are determined from Table A–2*b* to be $c_p = 1.099$ kJ/kg · K and $k = 1.354$.

(*a*) The exit velocity of the air will be a maximum when the process in the nozzle involves no irreversibilities. The exit velocity in this case is determined from the steady-flow energy equation. However, first we need to determine the exit temperature. For the isentropic process of an ideal gas we have:

$$\frac{T_{2s}}{T_1} = \left(\frac{P_{2s}}{P_1}\right)^{(k-1)/k}$$

or

$$T_{2s} = T_1\left(\frac{P_{2s}}{P_1}\right)^{(k-1)/k} = (950\text{ K})\left(\frac{80\text{ kPa}}{200\text{ kPa}}\right)^{0.354/1.354} = 748\text{ K}$$

This gives an average temperature of 849 K, which is somewhat higher than the assumed average temperature (800 K). This result could be refined by reevaluating the k value at 749 K and repeating the calculations, but it is not warranted since the two average temperatures are sufficiently close (doing so would change the temperature by only 1.5 K, which is not significant).

Now we can determine the isentropic exit velocity of the air from the energy balance for this isentropic steady-flow process:

$$e_{\text{in}} = e_{\text{out}}$$

$$h_1 + \frac{V_1^2}{2} = h_{2s} + \frac{V_{2s}^2}{2}$$

or

$$V_{2s} = \sqrt{2(h_1 - h_{2s})} = \sqrt{2c_{p,\text{avg}}(T_1 - T_{2s})}$$

$$= \sqrt{2(1.099\ \text{kJ/kg}\cdot\text{K})[(950 - 748)\ \text{K}]\left(\frac{1000\ \text{m}^2/\text{s}^2}{1\ \text{kJ/kg}}\right)}$$

$$= \mathbf{666\ m/s}$$

(*b*) The actual exit temperature of the air is higher than the isentropic exit temperature evaluated above and is determined from

$$\eta_N \cong \frac{h_1 - h_{2a}}{h_1 - h_{2s}} = \frac{c_{p,\text{avg}}(T_1 - T_{2a})}{c_{p,\text{avg}}(T_1 - T_{2s})}$$

or

$$0.92 = \frac{950 - T_{2a}}{950 - 748} \quad \rightarrow \quad T_{2a} = \mathbf{764\ K}$$

That is, the temperature is 16 K higher at the exit of the actual nozzle as a result of irreversibilities such as friction. It represents a loss since this rise in temperature comes at the expense of kinetic energy (Fig. 8–56).

(*c*) The actual exit velocity of air can be determined from the definition of isentropic efficiency of a nozzle,

$$\eta_N = \frac{V_{2a}^2}{V_{2s}^2} \quad \rightarrow \quad V_{2a} = \sqrt{\eta_N V_{2s}^2} = \sqrt{0.92(666\ \text{m/s})^2} = \mathbf{639\ m/s}$$

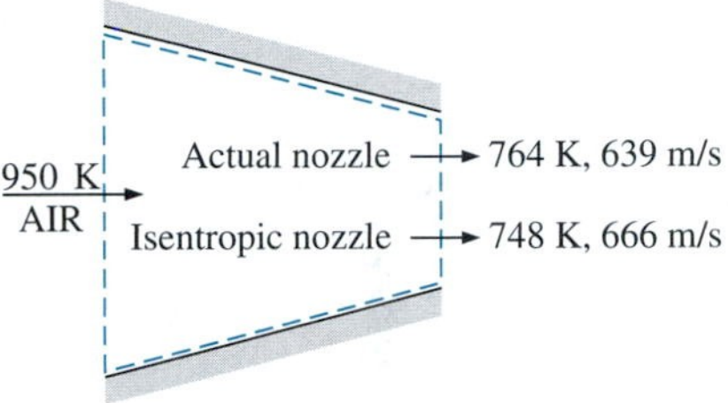

FIGURE 8–56

A substance leaves actual nozzles at a higher temperature (thus a lower velocity) as a result of friction.

8–13 ▪ ENTROPY BALANCE

The property *entropy* is a measure of molecular disorder or randomness of a system, and the second law of thermodynamics states that entropy can be created but it cannot be destroyed. Therefore, the entropy change of a system during a process is greater than the entropy transfer by an amount equal to the entropy generated during the process within the system, and the *increase of entropy principle* for any system is expressed as (Fig. 8–57)

$$\begin{pmatrix}\text{Total}\\ \text{entropy}\\ \text{entering}\end{pmatrix} - \begin{pmatrix}\text{Total}\\ \text{entropy}\\ \text{leaving}\end{pmatrix} + \begin{pmatrix}\text{Total}\\ \text{entropy}\\ \text{generated}\end{pmatrix} = \begin{pmatrix}\text{Change in the}\\ \text{total entropy}\\ \text{of the system}\end{pmatrix}$$

or

$$S_{\text{in}} - S_{\text{out}} + S_{\text{gen}} = \Delta S_{\text{system}} \tag{8–68}$$

which is a verbal statement of Eq. 8–9. This relation is often referred to as the **entropy balance** and is applicable to any system undergoing any process. The entropy balance relation above can be stated as: *the entropy change of a system during a process is equal to the net entropy transfer through the system boundary and the entropy generated within the system.* Next we discuss the various terms in that relation.

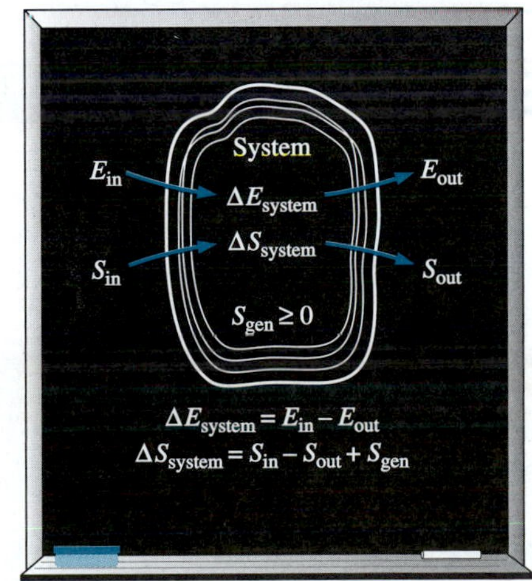

FIGURE 8–57

Energy and entropy balances for a system.

Entropy Change of a System, ΔS_{system}

Despite the reputation of entropy as being vague and abstract and the intimidation associated with it, entropy balance is actually easier to deal with than energy balance since, unlike energy, entropy does not exist in various forms. Therefore, the determination of entropy change of a system during a process involves evaluating entropy of the system at the beginning and at the end of the process and taking their difference. That is,

$$\text{Entropy change} = \text{Entropy at final state} - \text{Entropy at initial state}$$

or

$$\Delta S_{\text{system}} = S_{\text{final}} - S_{\text{initial}} = S_2 - S_1 \qquad \textbf{(8–69)}$$

Note that entropy is a property, and the value of a property does not change unless the state of the system changes. Therefore, the entropy change of a system is zero if the state of the system does not change during the process. For example, the entropy change of steady-flow devices such as nozzles, compressors, turbines, pumps, and heat exchangers is zero during steady operation.

When the properties of the system are not uniform, the entropy of the system can be determined by integration from

$$S_{\text{system}} = \int s\,\delta m = \int_V s\rho\, dV \qquad \textbf{(8–70)}$$

where V is the volume of the system and ρ is density.

Mechanisms of Entropy Transfer, S_{in} and S_{out}

Entropy can be transferred to or from a system by two mechanisms: *heat transfer* and *mass flow* (in contrast, energy is transferred by work also). Entropy transfer is recognized at the system boundary as it crosses the boundary, and it represents the entropy gained or lost by a system during a process. The only form of entropy interaction associated with a fixed mass or closed system is *heat transfer*, and thus the entropy transfer for an adiabatic closed system is zero.

1 Heat Transfer

Heat is, in essence, a form of disorganized energy, and some disorganization (entropy) will flow with heat. Heat transfer to a system increases the entropy of that system and thus the level of molecular disorder or randomness, and heat transfer from a system decreases it. In fact, heat rejection is the only way the entropy of a fixed mass can be decreased. The ratio of the heat transfer Q at a location to the absolute temperature T at that location is called the *entropy flow* or *entropy transfer* and is expressed as (Fig. 8–58)

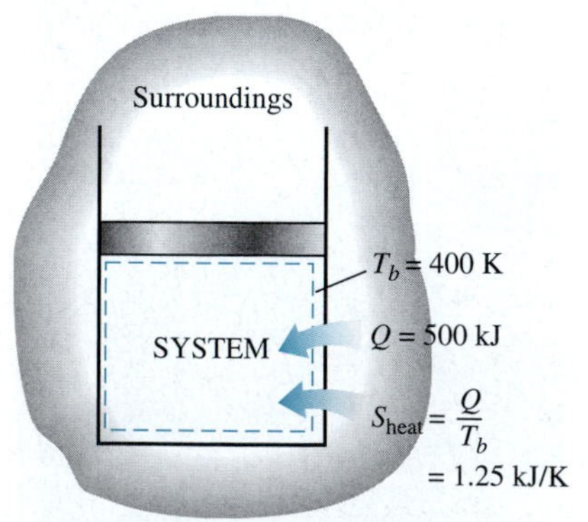

FIGURE 8–58

Heat transfer is always accompanied by entropy transfer in the amount of Q/T, where T is the boundary temperature.

Entropy transfer by heat transfer:

$$S_{\text{heat}} = \frac{Q}{T} \quad (T = \text{constant}) \qquad \textbf{(8–71)}$$

The quantity Q/T represents the entropy transfer accompanied by heat transfer, and the direction of entropy transfer is the same as the direction of heat transfer since thermodynamic temperature T is always a positive quantity.

When the temperature T is not constant, the entropy transfer during a process 1-2 can be determined by integration (or by summation if appropriate) as

$$S_{\text{heat}} = \int_1^2 \frac{\delta Q}{T} \cong \sum \frac{Q_k}{T_k} \quad \textbf{(8–72)}$$

where Q_k is the heat transfer through the boundary at temperature T_k at location k.

When two systems are in contact, the entropy transfer from the warmer system is equal to the entropy transfer into the cooler one at the point of contact. That is, no entropy can be created or destroyed at the boundary since the boundary has no thickness and occupies no volume.

Note that **work** is entropy-free, and no entropy is transferred by work. Energy is transferred by both heat and work, whereas entropy is transferred only by heat. That is,

Entropy transfer by work: $$S_{\text{work}} = 0 \quad \textbf{(8–73)}$$

The first law of thermodynamics makes no distinction between heat transfer and work; it considers them as *equals*. The distinction between heat transfer and work is brought out by the second law: *an energy interaction that is accompanied by entropy transfer is heat transfer, and an energy interaction that is not accompanied by entropy transfer is work.* That is, no entropy is exchanged during a work interaction between a system and its surroundings. Thus, only *energy* is exchanged during work interaction whereas both *energy* and *entropy* are exchanged during heat transfer (Fig. 8–59).

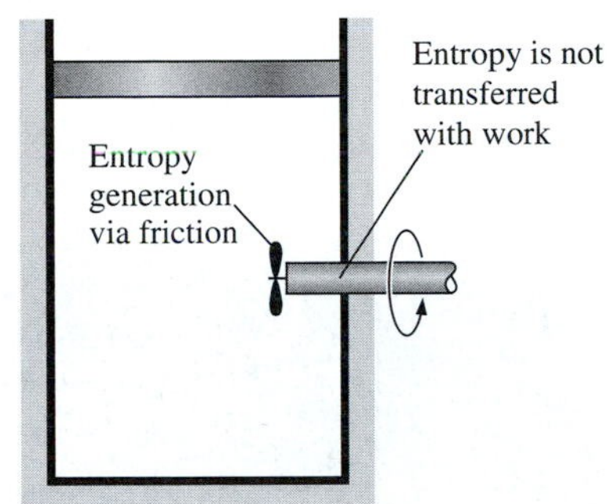

FIGURE 8–59

No entropy accompanies work as it crosses the system boundary. But entropy may be generated within the system as work is dissipated into a less useful form of energy.

2 Mass Flow

Mass contains entropy as well as energy, and the entropy and energy contents of a system are proportional to the mass. (When the mass of a system is doubled, so are the entropy and energy contents of the system.) Both entropy and energy are carried into or out of a system by streams of matter, and the rates of entropy and energy transport into or out of a system are proportional to the mass flow rate. Closed systems do not involve any mass flow and thus any entropy transfer by mass. When a mass in the amount of m enters or leaves a system, entropy in the amount of ms, where s is the specific entropy (entropy per unit mass entering or leaving), accompanies it (Fig. 8–60). That is,

Entropy transfer by mass flow: $$S_{\text{mass}} = ms \quad \textbf{(8–74)}$$

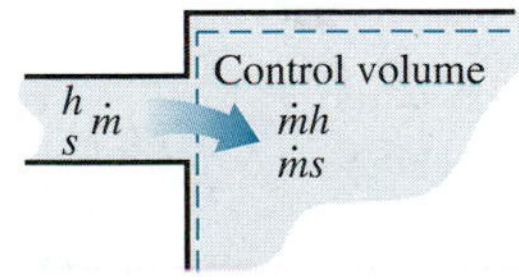

FIGURE 8–60

Mass contains entropy as well as energy, and thus mass flow into or out of system is always accompanied by energy and entropy transfer.

Therefore, the entropy of a system increases by ms when mass in the amount of m enters and decreases by the same amount when the same amount of mass at the same state leaves the system. When the properties of the mass change during the process, the entropy transfer by mass flow can be determined by integration from

$$\dot{S}_{\text{mass}} = \int_{A_c} s\rho V_n \, dA_c \quad \text{and} \quad S_{\text{mass}} = \int s\,\delta m = \int_{\Delta t} \dot{S}_{\text{mass}}\, dt \quad \textbf{(8–75)}$$

where A_c is the cross-sectional area of the flow and V_n is the local velocity normal to dA_c.

Entropy Generation, S_{gen}

Irreversibilities such as friction, mixing, chemical reactions, heat transfer through a finite temperature difference, unrestrained expansion, nonquasi-equilibrium compression, or expansion always cause the entropy of a system to increase, and entropy generation is a measure of the entropy created by such effects during a process.

For a *reversible process* (a process that involves no irreversibilities), the entropy generation is zero and thus the *entropy change* of a system is equal to the *entropy transfer*. Therefore, the entropy balance relation in the reversible case becomes analogous to the energy balance relation, which states that *energy change* of a system during a process is equal to the *energy transfer* during that process. However, note that the energy change of a system equals the energy transfer for *any* process, but the entropy change of a system equals the entropy transfer only for a *reversible* process.

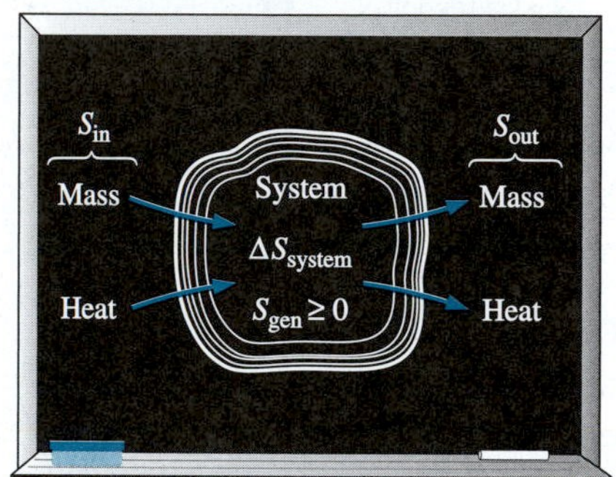

FIGURE 8–61
Mechanisms of entropy transfer for a general system.

The entropy transfer by heat Q/T is zero for adiabatic systems, and the entropy transfer by mass ms is zero for systems that involve no mass flow across their boundary (i.e., closed systems).

Entropy balance for *any system* undergoing *any process* can be expressed more explicitly as

$$\underbrace{S_{in} - S_{out}}_{\substack{\text{Net entropy transfer}\\ \text{by heat and mass}}} + \underbrace{S_{gen}}_{\substack{\text{Entropy}\\ \text{generation}}} = \underbrace{\Delta S_{system}}_{\substack{\text{Change}\\ \text{in entropy}}} \qquad (\text{kJ/K}) \tag{8–76}$$

or, in the **rate form**, as

$$\underbrace{\dot{S}_{in} - \dot{S}_{out}}_{\substack{\text{Rate of net entropy}\\ \text{transfer by heat}\\ \text{and mass}}} + \underbrace{\dot{S}_{gen}}_{\substack{\text{Rate of entropy}\\ \text{generation}}} = \underbrace{dS_{system}/dt}_{\substack{\text{Rate of change}\\ \text{in entropy}}} \qquad (\text{kW/K}) \tag{8–77}$$

where the rates of entropy transfer by heat transferred at a rate of $\dot{Q}$ and mass flowing at a rate of $\dot{m}$ are $\dot{S}_{heat} = \dot{Q}/T$ and $\dot{S}_{mass} = \dot{m}s$. The entropy balance can also be expressed on a **unit-mass basis** as

$$(s_{in} - s_{out}) + s_{gen} = \Delta s_{system} \qquad (\text{kJ/kg} \cdot \text{K}) \tag{8–78}$$

where all the quantities are expressed per unit mass of the system. Note that for a *reversible process*, the entropy generation term S_{gen} drops out from all of the relations above.

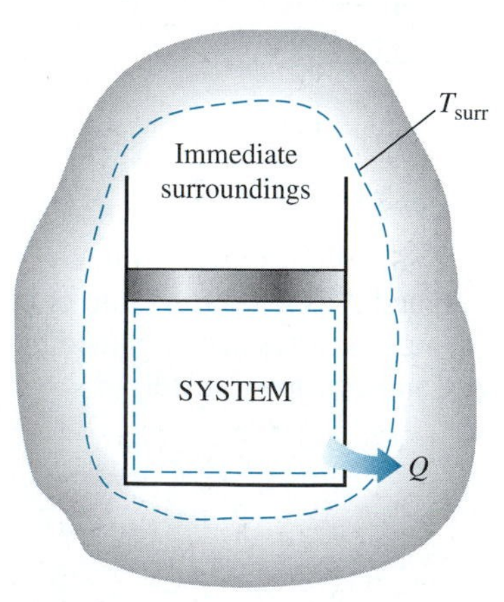

FIGURE 8–62
Entropy generation outside system boundaries can be accounted for by writing an entropy balance on an extended system that includes the system and its immediate surroundings.

The term S_{gen} represents the entropy generation *within the system boundary* only (Fig. 8–61), and not the entropy generation that may occur outside the system boundary during the process as a result of external irreversibilities. Therefore, a process for which $S_{gen} = 0$ is *internally reversible*, but not necessarily *totally* reversible. The *total* entropy generated during a process can be determined by applying the entropy balance to an *extended system* that includes the system itself and its immediate surroundings where external irreversibilities might be occurring (Fig. 8–62). Also, the entropy change in this case is equal to the sum of the entropy change of the system and the entropy change of the immediate surroundings. Note that under steady conditions, the state and thus the entropy of the immediate surroundings (let us call it the "buffer zone") at any point does not change during the process, and the entropy change of the buffer zone is zero. The entropy change of the buffer zone, if any, is usually small relative to the entropy change of the system, and thus it is usually disregarded.

When evaluating the entropy transfer between an extended system and the surroundings, the boundary temperature of the extended system is simply taken to be the *environment temperature*.

Closed Systems

A closed system involves *no mass flow* across its boundaries, and its entropy change is simply the difference between the initial and final entropies of the system. The *entropy change* of a closed system is due to the *entropy transfer* accompanying heat transfer and the *entropy generation* within the system boundaries. Taking the positive direction of heat transfer to be *to* the system, the general entropy balance relation (Eq. 8–76) can be expressed for a closed system as

Closed system:
$$\sum \frac{Q_k}{T_k} + S_{\text{gen}} = \Delta S_{\text{system}} = S_2 - S_1 \qquad (\text{kJ/K}) \qquad \textbf{(8–79)}$$

The entropy balance relation above can be stated as:

> The entropy change of a closed system during a process is equal to the sum of the net entropy transferred through the system boundary by heat transfer and the entropy generated within the system boundaries.

For an *adiabatic process* ($Q = 0$), the entropy transfer term in the above relation drops out and the entropy change of the closed system becomes equal to the entropy generation within the system boundaries. That is,

Adiabatic closed system:
$$S_{\text{gen}} = \Delta S_{\text{adiabatic system}} \qquad \textbf{(8–80)}$$

Noting that any closed system and its surroundings can be treated as an adiabatic system and the total entropy change of a system is equal to the sum of the entropy changes of its parts, the entropy balance for a closed system and its surroundings can be written as

System + Surroundings:
$$S_{\text{gen}} = \sum \Delta S = \Delta S_{\text{system}} + \Delta S_{\text{surroundings}} \qquad \textbf{(8–81)}$$

where $\Delta S_{\text{system}} = m(s_2 - s_1)$ and the entropy change of the surroundings can be determined from $\Delta S_{\text{surr}} = Q_{\text{surr}}/T_{\text{surr}}$ if its temperature is constant. At initial stages of studying entropy and entropy transfer, it is more instructive to start with the general form of the entropy balance (Eq. 8–76) and to simplify it for the problem under consideration. The specific relations above are convenient to use after a certain degree of intuitive understanding of the material is achieved.

Control Volumes

The entropy balance relations for control volumes differ from those for closed systems in that they involve one more mechanism of entropy exchange: *mass flow across the boundaries*. As mentioned earlier, mass possesses entropy as well as energy, and the amounts of these two extensive properties are proportional to the amount of mass (Fig. 8–63).

Taking the positive direction of heat transfer to be *to* the system, the general entropy balance relations (Eqs. 8–76 and 8–77) can be expressed for control volumes as

$$\sum \frac{Q_k}{T_k} + \sum m_i s_i - \sum m_e s_e + S_{\text{gen}} = (S_2 - S_1)_{\text{CV}} \qquad (\text{kJ/K}) \qquad \textbf{(8–82)}$$

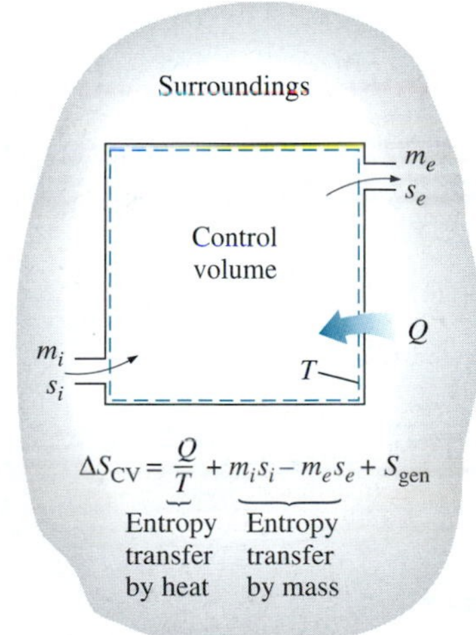

FIGURE 8–63
The entropy of a control volume changes as a result of mass flow as well as heat transfer.

or, in the rate form, as

$$\sum \frac{\dot{Q}_k}{T_k} + \sum \dot{m}_i s_i - \sum \dot{m}_e s_e + \dot{S}_{gen} = dS_{CV}/dt \qquad (kW/K) \qquad (8\text{–}83)$$

This entropy balance relation can be stated as:

> The rate of entropy change within the control volume during a process is equal to the sum of the rate of entropy transfer through the control volume boundary by heat transfer, the net rate of entropy transfer into the control volume by mass flow, and the rate of entropy generation within the boundaries of the control volume as a result of irreversibilities.

$s_e \geq s_i$

s_i

FIGURE 8–64

The entropy of a substance always increases (or remains constant in the case of a reversible process) as it flows through a single-stream, adiabatic, steady-flow device.

Most control volumes encountered in practice such as turbines, compressors, nozzles, diffusers, heat exchangers, pipes, and ducts operate steadily, and thus they experience no change in their entropy. Therefore, the entropy balance relation for a general **steady-flow process** can be obtained from Eq. 8–83 by setting $dS_{CV}/dt = 0$ and rearranging to give

Steady-flow:
$$\dot{S}_{gen} = \sum \dot{m}_e s_e - \sum \dot{m}_i s_i - \sum \frac{\dot{Q}_k}{T_k} \qquad (8\text{–}84)$$

For *single-stream* (one inlet and one exit) steady-flow devices, the entropy balance relation simplifies to

Steady-flow, single-stream:
$$\dot{S}_{gen} = \dot{m}(s_e - s_i) - \sum \frac{\dot{Q}_k}{T_k} \qquad (8\text{–}85)$$

For the case of an *adiabatic* single-stream device, the entropy balance relation further simplifies to

Steady-flow, single-stream, adiabatic:
$$\dot{S}_{gen} = \dot{m}(s_e - s_i) \qquad (8\text{–}86)$$

which indicates that the specific entropy of the fluid must increase as it flows through an adiabatic device since $\dot{S}_{gen} \geq 0$ (Fig. 8–64). If the flow through the device is *reversible* and *adiabatic*, then the entropy remains constant, $s_e = s_i$, regardless of the changes in other properties.

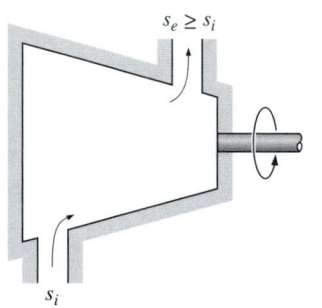

EXAMPLE 8–17 Entropy Generation in a Wall

Consider steady heat transfer through a 5-m × 7-m brick wall of a house of thickness 30 cm. On a day when the temperature of the outdoors is 0°C, the house is maintained at 27°C. The temperatures of the inner and outer surfaces of the brick wall are measured to be 20°C and 5°C, respectively, and the rate of heat transfer through the wall is 1035 W. Determine the rate of entropy generation in the wall, and the rate of total entropy generation associated with this heat transfer process.

Solution Steady heat transfer through a wall is considered. For specified heat transfer rate, wall temperatures, and environment temperatures, the entropy generation rate within the wall and the total entropy generation rate are to be determined.

Assumptions **1** The process is steady, and thus the rate of heat transfer through the wall is constant. **2** Heat transfer through the wall is one-dimensional.

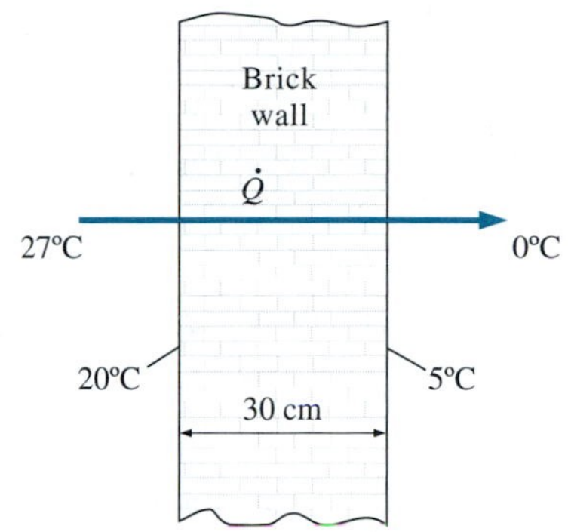

FIGURE 8–65
Schematic for Example 8–17.

Analysis We first take the *wall* as the system (Fig. 8–65). This is a *closed system* since no mass crosses the system boundary during the process. We note that the entropy change of the wall is zero during this process since the state and thus the entropy of the wall do not change anywhere in the wall. Heat and entropy are entering from one side of the wall and leaving from the other side.

The rate form of the entropy balance for the wall simplifies to

$$\underbrace{\dot{S}_{in} - \dot{S}_{out}}_{\substack{\text{Rate of net entropy}\\\text{transfer by heat}\\\text{and mass}}} + \underbrace{\dot{S}_{gen}}_{\substack{\text{Rate of entropy}\\\text{generation}}} = \underbrace{dS_{system}/dt}_{\substack{\text{Rate of change}\\\text{in entropy}}} \overset{\nearrow 0 \text{ (steady)}}{}$$

$$\left(\frac{\dot{Q}}{T}\right)_{in} - \left(\frac{\dot{Q}}{T}\right)_{out} + \dot{S}_{gen} = 0$$

$$\frac{1035 \text{ W}}{293 \text{ K}} - \frac{1035 \text{ W}}{278 \text{ K}} + \dot{S}_{gen} = 0$$

Therefore, the rate of entropy generation in the wall is

$$\dot{S}_{gen,wall} = \mathbf{0.191 \ W/K}$$

Note that entropy transfer by heat at any location is *Q*/*T* at that location, and the direction of entropy transfer is the same as the direction of heat transfer.

To determine the rate of total entropy generation during this heat transfer process, we extend the system to include the regions on both sides of the wall that experience a temperature change. Then one side of the system boundary becomes room temperature while the other side becomes the temperature of the outdoors. The entropy balance for this *extended system* (system + immediate surroundings) is the same as that given above, except the two boundary temperatures are now 300 and 273 K instead of 293 and 278 K, respectively. Then the rate of total entropy generation becomes

$$\frac{1035 \text{ W}}{300 \text{ K}} - \frac{1035 \text{ W}}{273 \text{ K}} + \dot{S}_{gen,total} = 0 \quad \rightarrow \quad \dot{S}_{gen,total} = \mathbf{0.341 \ W/K}$$

Discussion Note that the entropy change of this extended system is also zero since the state of air does not change at any point during the process. The differences between the two entropy generations is 0.150 W/K, and it represents the entropy generated in the air layers on both sides of the wall. The entropy generation in this case is entirely due to irreversible heat transfer through a finite temperature difference.

EXAMPLE 8–18 Entropy Generation during a Throttling Process

Steam at 7 MPa and 450°C is throttled in a valve to a pressure of 3 MPa during a steady-flow process. Determine the entropy generated during this process and check if the increase of entropy principle is satisfied.

Solution Steam is throttled to a specified pressure. The entropy generated during this process is to be determined, and the validity of the increase of entropy principle is to be verified.

Assumptions **1** This is a steady-flow process since there is no change with time at any point and thus $\Delta m_{CV} = 0$, $\Delta E_{CV} = 0$, and $\Delta S_{CV} = 0$. **2** Heat transfer to or from the valve is negligible. **3** The kinetic and potential energy changes are negligible, $\Delta ke = \Delta pe = 0$.

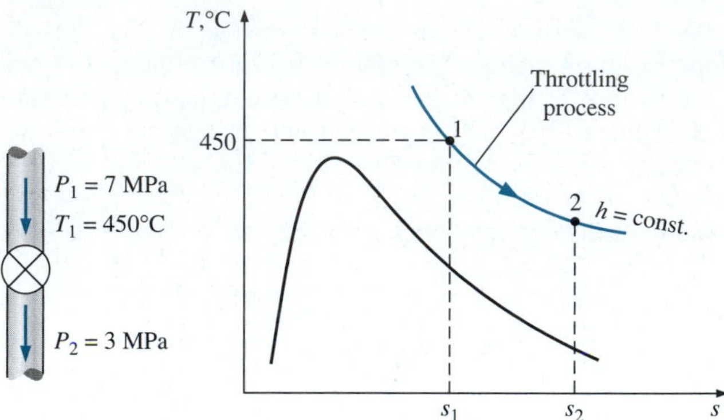

FIGURE 8–66

Schematic and T-s diagram for Example 8–18.

Analysis We take the throttling valve as the *system* (Fig. 8–66). This is a *control volume* since mass crosses the system boundary during the process. We note that there is only one inlet and one exit and thus $\dot{m}_1 = \dot{m}_2 = \dot{m}$. Also, the enthalpy of a fluid remains nearly constant during a throttling process and thus $h_2 \cong h_1$.

The entropy of the steam at the inlet and the exit states is determined from the steam tables to be

$$\text{State 1:} \quad \left.\begin{aligned} P_1 &= 7 \text{ MPa} \\ T_1 &= 450°\text{C} \end{aligned}\right\} \quad \begin{aligned} h_1 &= 3288.3 \text{ kJ/kg} \\ s_1 &= 6.6353 \text{ kJ/kg}\cdot\text{K} \end{aligned}$$

$$\text{State 2:} \quad \left.\begin{aligned} P_2 &= 3 \text{ MPa} \\ h_2 &= h_1 \end{aligned}\right\} \quad s_2 = 7.0046 \text{ kJ/kg}\cdot\text{K}$$

Then the entropy generation per unit mass of the steam is determined from the entropy balance applied to the throttling valve,

$$\underbrace{\dot{S}_{\text{in}} - \dot{S}_{\text{out}}}_{\substack{\text{Rate of net entropy} \\ \text{transfer by heat} \\ \text{and mass}}} + \underbrace{\dot{S}_{\text{gen}}}_{\substack{\text{Rate of entropy} \\ \text{generation}}} = \underbrace{dS_{\text{system}}/dt}_{\substack{\text{Rate of change} \\ \text{in entropy}}} \nearrow^{0 \text{ (steady)}}$$

$$\dot{m}s_1 - \dot{m}s_2 + \dot{S}_{\text{gen}} = 0$$

$$\dot{S}_{\text{gen}} = \dot{m}(s_2 - s_1)$$

Dividing by mass flow rate and substituting gives

$$s_{\text{gen}} = s_2 - s_1 = 7.0046 - 6.6353 = \mathbf{0.3693\ kJ/kg\cdot K}$$

This is the amount of entropy generated per unit mass of steam as it is throttled from the inlet state to the final pressure, and it is caused by unrestrained expansion. The increase of entropy principle is obviously satisfied during this process since the entropy generation is positive.

EXAMPLE 8–19 Entropy Generated when a Hot Block Is Dropped in a Lake

A 50-kg block of iron casting at 500 K is thrown into a large lake that is at a temperature of 285 K. The iron block eventually reaches thermal equilibrium with the lake water. Assuming an average specific heat of 0.45 kJ/kg · K for the iron, determine (*a*) the entropy change of the iron block, (*b*) the entropy change of the lake water, and (*c*) the entropy generated during this process.

Solution A hot iron block is thrown into a lake, and cools to the lake temperature. The entropy changes of the iron and of the lake as well as the entropy generated during this process are to be determined.

Assumptions **1** Both the water and the iron block are incompressible substances. **2** Constant specific heats can be used for the water and the iron. **3** The kinetic and potential energy changes of the iron are negligible, $\Delta KE = \Delta PE = 0$ and thus $\Delta E = \Delta U$.

Properties The specific heat of the iron is 0.45 kJ/kg · K (Table A–3).

Analysis We take the *iron casting* as the system (Fig. 8–67). This is a *closed system* since no mass crosses the system boundary during the process.

To determine the entropy change for the iron block and for the lake, first we need to know the final equilibrium temperature. Given that the thermal energy capacity of the lake is very large relative to that of the iron block, the lake will absorb all the heat rejected by the iron block without experiencing any change in its temperature. Therefore, the iron block will cool to 285 K during this process while the lake temperature remains constant at 285 K.

(*a*) The entropy change of the iron block can be determined from

$$\Delta S_{\text{iron}} = m(s_2 - s_1) = mc_{\text{avg}} \ln \frac{T_2}{T_1}$$

$$= (50 \text{ kg})(0.45 \text{ kJ/kg} \cdot \text{K}) \ln \frac{285 \text{ K}}{500 \text{ K}}$$

$$= \mathbf{-12.65 \text{ kJ/K}}$$

(*b*) The temperature of the lake water remains constant during this process at 285 K. Also, the amount of heat transfer from the iron block to the lake is determined from an energy balance on the iron block to be

$$\underbrace{E_{\text{in}} - E_{\text{out}}}_{\substack{\text{Net energy transfer}\\ \text{by heat, work, and mass}}} = \underbrace{\Delta E_{\text{system}}}_{\substack{\text{Change in internal, kinetic,}\\ \text{potential, etc., energies}}}$$

$$-Q_{\text{out}} = \Delta U = mc_{\text{avg}}(T_2 - T_1)$$

or

$$Q_{\text{out}} = mc_{\text{avg}}(T_1 - T_2) = (50 \text{ kg})(0.45 \text{ kJ/kg} \cdot \text{K})(500 - 285) \text{ K} = 4838 \text{ kJ}$$

Then the entropy change of the lake becomes

$$\Delta S_{\text{lake}} = \frac{Q_{\text{lake}}}{T_{\text{lake}}} = \frac{+4838 \text{ kJ}}{285 \text{ K}} = \mathbf{16.97 \text{ kJ/K}}$$

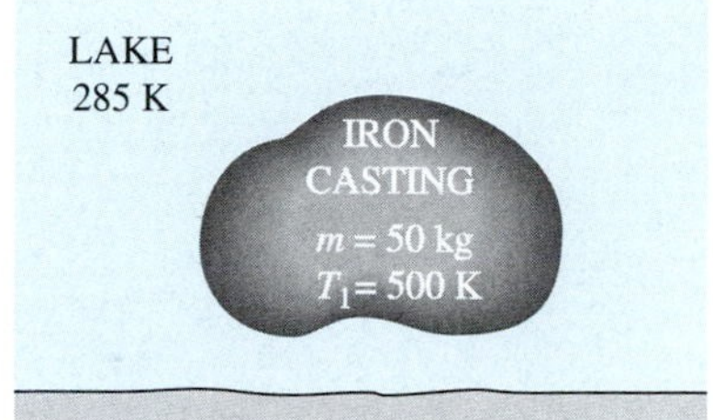

FIGURE 8–67
Schematic for Example 8–19.

(*c*) The entropy generated during this process can be determined by applying an entropy balance on an *extended system* that includes the iron block and its immediate surroundings so that the boundary temperature of the extended system is at 285 K at all times:

$$\underbrace{S_{in} - S_{out}}_{\substack{\text{Net entropy transfer}\\\text{by heat and mass}}} + \underbrace{S_{gen}}_{\substack{\text{Entropy}\\\text{generation}}} = \underbrace{\Delta S_{system}}_{\substack{\text{Change}\\\text{in entropy}}}$$

$$-\frac{Q_{out}}{T_b} + S_{gen} = \Delta S_{system}$$

or

$$S_{gen} = \frac{Q_{out}}{T_b} + \Delta S_{system} = \frac{4838 \text{ kJ}}{285 \text{ K}} - (12.65 \text{ kJ/K}) = \mathbf{4.32 \text{ kJ/K}}$$

Discussion The entropy generated can also be determined by taking the iron block and the entire lake as the system, which is an isolated system, and applying an entropy balance. An isolated system involves no heat or entropy transfer, and thus the entropy generation in this case becomes equal to the total entropy change,

$$S_{gen} = \Delta S_{total} = \Delta S_{system} + \Delta S_{lake} = -12.65 + 16.97 = 4.32 \text{ kJ/K}$$

which is the same result obtained above.

EXAMPLE 8–20 Entropy Generation in a Mixing Chamber

Water at 20 psia and 50°F enters a mixing chamber at a rate of 300 lbm/min where it is mixed steadily with steam entering at 20 psia and 240°F. The mixture leaves the chamber at 20 psia and 130°F, and heat is lost to the surrounding air at 70°F at a rate of 180 Btu/min. Neglecting the changes in kinetic and potential energies, determine the rate of entropy generation during this process.

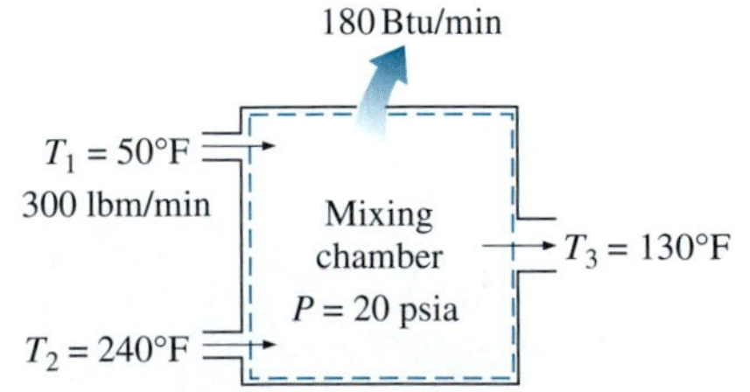

FIGURE 8–68
Schematic for Example 8–20.

Solution Water and steam are mixed in a chamber that is losing heat at a specified rate. The rate of entropy generation during this process is to be determined.

Assumptions **1** This is a steady-flow process since there is no change with time at any point and thus $\Delta m_{CV} = 0$, $\Delta E_{CV} = 0$, and $\Delta S_{CV} = 0$. **2** There are no work interactions involved. **3** The kinetic and potential energies are negligible, ke ≅ pe ≅ 0.

Analysis We take the *mixing chamber* as the system (Fig. 8–68). This is a *control volume* since mass crosses the system boundary during the process. We note that there are two inlets and one exit.

Under the stated assumptions and observations, the mass and energy balances for this steady-flow system can be expressed in the rate form as follows:

Mass balance: $\dot{m}_{in} - \dot{m}_{out} = dm_{system}/dt \overset{\nearrow 0 \text{ (steady)}}{=} 0 \rightarrow \dot{m}_1 + \dot{m}_2 = \dot{m}_3$

Energy balance:

$$\underbrace{\dot{E}_{in} - \dot{E}_{out}}_{\substack{\text{Rate of net energy transfer} \\ \text{by heat, work, and mass}}} = \underbrace{dE_{system}/dt \overset{\nearrow 0 \text{ (steady)}}{=} 0}_{\substack{\text{Rate of change in internal, kinetic,} \\ \text{potential, etc., energies}}}$$

$$\dot{E}_{in} = \dot{E}_{out}$$

$$\dot{m}_1 h_1 + \dot{m}_2 h_2 = \dot{m}_3 h_3 + \dot{Q}_{out} \qquad (\text{since } \dot{W} = 0, \text{ke} \cong \text{pe} \cong 0)$$

Combining the mass and energy balances gives

$$\dot{Q}_{out} = \dot{m}_1 h_1 + \dot{m}_2 h_2 - (\dot{m}_1 + \dot{m}_2) h_3$$

The desired properties at the specified states are determined from the steam tables to be

State 1: $\left.\begin{array}{l} P_1 = 20 \text{ psia} \\ T_1 = 50°\text{F} \end{array}\right\} \begin{array}{l} h_1 \cong h_{f @ 50°\text{F}} = 18.07 \text{ Btu/lbm} \\ s_1 \cong s_{f @ 50°\text{F}} = 0.03609 \text{ Btu/lbm} \cdot \text{R} \end{array}$

State 2: $\left.\begin{array}{l} P_2 = 20 \text{ psia} \\ T_2 = 240°\text{F} \end{array}\right\} \begin{array}{l} h_2 \cong 1162.3 \text{ Btu/lbm} \\ s_2 \cong 1.7406 \text{ Btu/lbm} \cdot \text{R} \end{array}$

State 3: $\left.\begin{array}{l} P_3 = 20 \text{ psia} \\ T_3 = 130°\text{F} \end{array}\right\} \begin{array}{l} h_3 \cong h_{f @ 130°\text{F}} = 97.99 \text{ Btu/lbm} \\ s_3 \cong s_{f @ 130°\text{F}} = 0.18174 \text{ Btu/lbm} \cdot \text{R} \end{array}$

Substituting,

$$180 \text{ Btu/min} = [300 \times 18.07 + \dot{m}_2 \times 1162.3 - (300 + \dot{m}_2) \times 97.99] \text{ Btu/min}$$

which gives

$$\dot{m}_2 = 22.7 \text{ lbm/min}$$

The rate of entropy generation during this process can be determined by applying the rate form of the entropy balance on an *extended system* that includes the mixing chamber and its immediate surroundings so that the boundary temperature of the extended system is 70°F = 530 R:

$$\underbrace{\dot{S}_{in} - \dot{S}_{out}}_{\substack{\text{Rate of net entropy} \\ \text{transfer by heat} \\ \text{and mass}}} + \underbrace{\dot{S}_{gen}}_{\substack{\text{Rate of entropy} \\ \text{generation}}} = \underbrace{dS_{system}/dt}_{\substack{\text{Rate of change} \\ \text{in entropy}}}$$

$$\dot{m}_1 s_1 + \dot{m}_2 s_2 - \dot{m}_3 s_3 - \frac{\dot{Q}_{out}}{T_b} + \dot{S}_{gen} = 0$$

Substituting, the rate of entropy generation is determined to be

$$\dot{S}_{gen} = \dot{m}_3 s_3 - \dot{m}_1 s_1 - \dot{m}_2 s_2 + \frac{\dot{Q}_{out}}{T_b}$$

$$= (322.7 \times 0.18174 - 300 \times 0.03609 - 22.7 \times 1.7406)\ \text{Btu/min} \cdot \text{R} + \frac{180\ \text{Btu/min}}{530\ \text{R}}$$

$$= \mathbf{8.65\ Btu/min \cdot R}$$

Discussion Note that entropy is generated during this process at a rate of 8.65 Btu/min · R. This entropy generation is caused by the mixing of two fluid streams (an irreversible process) and the heat transfer between the mixing chamber and the surroundings through a finite temperature difference (another irreversible process).

SUMMARY

The second law of thermodynamics leads to the definition of a new property called *entropy*, which is a quantitative measure of microscopic disorder for a system. Any quantity whose cyclic integral is zero is a property, and entropy is defined as

$$dS = \left(\frac{dQ}{T}\right)_{\text{int rev}}$$

For the special case of an internally reversible, isothermal process, it gives

$$\Delta S = \frac{Q}{T_0}$$

The inequality part of the Clausius inequality combined with the definition of entropy yields an inequality known as the *increase of entropy principle*, expressed as

$$S_{gen} \geq 0$$

where S_{gen} is the *entropy generated* during the process. Entropy change is caused by heat transfer, mass flow, and irreversibilities. Heat transfer to a system increases the entropy, and heat transfer from a system decreases it. The effect of irreversibilities is always to increase the entropy.

The *entropy-change* and *isentropic relations* for a process can be summarized as follows:

1. *Pure substances:*

Any process: $\Delta s = s_2 - s_1$

Isentropic process: $s_2 = s_1$

2. *Incompressible substances:*

Any process: $s_2 - s_1 = c_{avg} \ln \frac{T_2}{T_1}$

Isentropic process: $T_2 = T_1$

3. *Ideal gases:*

a. Constant specific heats (approximate treatment):

Any process:

$$s_2 - s_1 = c_{v,avg} \ln \frac{T_2}{T_1} + R \ln \frac{v_2}{v_1}$$

$$s_2 - s_1 = c_{p,avg} \ln \frac{T_2}{T_1} - R \ln \frac{P_2}{P_1}$$

Isentropic process:

$$\left(\frac{T_2}{T_1}\right)_{s=\text{const.}} = \left(\frac{v_1}{v_2}\right)^{k-1}$$

$$\left(\frac{T_2}{T_1}\right)_{s=\text{const.}} = \left(\frac{P_2}{P_1}\right)^{(k-1)/k}$$

$$\left(\frac{P_2}{P_1}\right)_{s=\text{const.}} = \left(\frac{v_1}{v_2}\right)^{k}$$

b. Variable specific heats (exact treatment):

Any process:

$$s_2 - s_1 = s_2^\circ - s_1^\circ - R \ln \frac{P_2}{P_1}$$

Isentropic process:

$$s_2^\circ = s_1^\circ + R \ln \frac{P_2}{P_1}$$

$$\left(\frac{P_2}{P_1}\right)_{s=\text{const.}} = \frac{P_{r2}}{P_{r1}}$$

$$\left(\frac{v_2}{v_1}\right)_{s=\text{const.}} = \frac{v_{r2}}{v_{r1}}$$

where P_r is the *relative pressure* and v_r is the *relative specific volume*. The function $s°$ depends on temperature only.

The *steady-flow work* for a reversible process can be expressed in terms of the fluid properties as

$$w_{\text{rev}} = -\int_1^2 v\,dP - \Delta\text{ke} - \Delta\text{pe}$$

For incompressible substances (v = constant) it simplifies to

$$w_{\text{rev}} = -v(P_2 - P_1) - \Delta\text{ke} - \Delta\text{pe}$$

The work done during a steady-flow process is proportional to the specific volume. Therefore, v should be kept as small as possible during a compression process to minimize the work input and as large as possible during an expansion process to maximize the work output.

The reversible work inputs to a compressor compressing an ideal gas from T_1, P_1 to P_2 in an isentropic (Pv^k = constant), polytropic (Pv^n = constant), or isothermal (Pv = constant) manner, are determined by integration for each case with the following results:

Isentropic: $$w_{\text{comp,in}} = \frac{kR(T_2 - T_1)}{k-1} = \frac{kRT_1}{k-1}\left[\left(\frac{P_2}{P_1}\right)^{(k-1)/k} - 1\right]$$

Polytropic: $$w_{\text{comp,in}} = \frac{nR(T_2 - T_1)}{n-1} = \frac{nRT_1}{n-1}\left[\left(\frac{P_2}{P_1}\right)^{(n-1)/n} - 1\right]$$

Isothermal: $$w_{\text{comp,in}} = RT\ln\frac{P_2}{P_1}$$

The work input to a compressor can be reduced by using multistage compression with intercooling. For maximum savings from the work input, the pressure ratio across each stage of the compressor must be the same.

Most steady-flow devices operate under adiabatic conditions, and the ideal process for these devices is the isentropic process. The parameter that describes how efficiently a device approximates a corresponding isentropic device is called *isentropic* or *adiabatic efficiency*. It is expressed for turbines, compressors, and nozzles as follows:

$$\eta_T = \frac{\text{Actual turbine work}}{\text{Isentropic turbine work}} = \frac{w_a}{w_s} \cong \frac{h_1 - h_{2a}}{h_1 - h_{2s}}$$

$$\eta_C = \frac{\text{Isentropic compressor work}}{\text{Actual compressor work}} = \frac{w_s}{w_a} \cong \frac{h_{2s} - h_1}{h_{2a} - h_1}$$

$$\eta_N = \frac{\text{Actual KE at nozzle exit}}{\text{Isentropic KE at nozzle exit}} = \frac{V_{2a}^2}{V_{2s}^2} \cong \frac{h_1 - h_{2a}}{h_1 - h_{2s}}$$

In the relations above, h_{2a} and h_{2s} are the enthalpy values at the exit state for actual and isentropic *processes*, respectively.

The entropy balance for any system undergoing any process can be expressed in the general form as

$$\underbrace{S_{\text{in}} - S_{\text{out}}}_{\substack{\text{Net entropy transfer}\\ \text{by heat and mass}}} + \underbrace{S_{\text{gen}}}_{\substack{\text{Entropy}\\ \text{generation}}} = \underbrace{\Delta S_{\text{system}}}_{\substack{\text{Change}\\ \text{in entropy}}}$$

or, in the *rate form,* as

$$\underbrace{\dot{S}_{\text{in}} - \dot{S}_{\text{out}}}_{\substack{\text{Rate of net entropy}\\ \text{transfer by}\\ \text{heat and mass}}} + \underbrace{\dot{S}_{\text{gen}}}_{\substack{\text{Rate of entropy}\\ \text{generation}}} = \underbrace{dS_{\text{system}}/dt}_{\substack{\text{Rate of change}\\ \text{in entropy}}}$$

For a general *steady-flow process* it simplifies to

$$\dot{S}_{\text{gen}} = \sum \dot{m}_e s_e - \sum \dot{m}_i s_i - \sum \frac{\dot{Q}_k}{T_k}$$

REFERENCES AND SUGGESTED READINGS

1. A. Bejan. *Advanced Engineering Thermodynamics*. 3rd ed. New York: Wiley Interscience, 2006.
2. A. Bejan. *Entropy Generation through Heat and Fluid Flow*. New York: Wiley Interscience, 1982.
3. Y. A. Çengel and H. Kimmel. "Optimization of Expansion in Natural Gas Liquefaction Processes." *LNG Journal*, U.K., May–June, 1998.
4. Y. Çerci, Y. A. Çengel, and R. H. Turner, "Reducing the Cost of Compressed Air in Industrial Facilities." *International Mechanical Engineering Congress and Exposition*, San Francisco, California, November 12–17, 1995.
5. W. F. E. Feller. *Air Compressors: Their Installation, Operation, and Maintenance*. New York: McGraw-Hill, 1944.
6. D. W. Nutter, A. J. Britton, and W. M. Heffington. "Conserve Energy to Cut Operating Costs." *Chemical Engineering*, September 1993, pp. 128–137.
7. J. Rifkin. *Entropy*. New York: The Viking Press, 1980.

PROBLEMS*

Entropy and the Increase of Entropy Principle

8–1C Does a cycle for which $\oint \delta Q > 0$ violate the Clausius inequality? Why?

8–2C Does the cyclic integral of work have to be zero (i.e., does a system have to produce as much work as it consumes to complete a cycle)? Explain.

8–3C A system undergoes a process between two fixed states first in a reversible manner and then in an irreversible manner. For which case is the entropy change greater? Why?

8–4C Is the value of the integral $\int_1^2 \delta Q/T$ the same for all processes between states 1 and 2? Explain.

8–5C Is the value of the integral $\int_1^2 \delta Q/T$ the same for all reversible processes between states 1 and 2? Why?

8–6C To determine the entropy change for an irreversible process between states 1 and 2, should the integral $\int_1^2 \delta Q/T$ be performed along the actual process path or an imaginary reversible path? Explain.

8–7C Is an isothermal process necessarily internally reversible? Explain your answer with an example.

8–8C How do the values of the integral $\int_1^2 \delta Q/T$ compare for a reversible and irreversible process between the same end states?

8–9C The entropy of a hot baked potato decreases as it cools. Is this a violation of the increase of entropy principle? Explain.

8–10C Is it possible to create entropy? Is it possible to destroy it?

8–11C When a system is adiabatic, what can be said about the entropy change of the substance in the system?

8–12C Work is entropy free, and sometimes the claim is made that work will not change the entropy of a fluid passing through an adiabatic steady-flow system with a single inlet and outlet. Is this a valid claim?

8–13C A piston–cylinder device contains helium gas. During a reversible, isothermal process, the entropy of the helium will (*never, sometimes, always*) increase.

8–14C A piston–cylinder device contains nitrogen gas. During a reversible, adiabatic process, the entropy of the nitrogen will (*never, sometimes, always*) increase.

8–15C A piston–cylinder device contains superheated steam. During an actual adiabatic process, the entropy of the steam will (*never, sometimes, always*) increase.

8–16C The entropy of steam will (*increase, decrease, remain the same*) as it flows through an actual adiabatic turbine.

8–17C The entropy of the working fluid of the ideal Carnot cycle (*increases, decreases, remains the same*) during the isothermal heat addition process.

8–18C The entropy of the working fluid of the ideal Carnot cycle (*increases, decreases, remains the same*) during the isothermal heat rejection process.

8–19C During a heat transfer process, the entropy of a system (*always, sometimes, never*) increases.

8–20C Is it possible for the entropy change of a closed system to be zero during an irreversible process? Explain.

8–21C What three different mechanisms can cause the entropy of a control volume to change?

8–22C Steam is accelerated as it flows through an actual adiabatic nozzle. The entropy of the steam at the nozzle exit will be (*greater than, equal to, less than*) the entropy at the nozzle inlet.

8–23 A rigid tank contains an ideal gas at 40°C that is being stirred by a paddle wheel. The paddle wheel does 200 kJ of work on the ideal gas. It is observed that the temperature of the ideal gas remains constant during this process as a result of heat transfer between the system and the surroundings at 30°C. Determine the entropy change of the ideal gas.

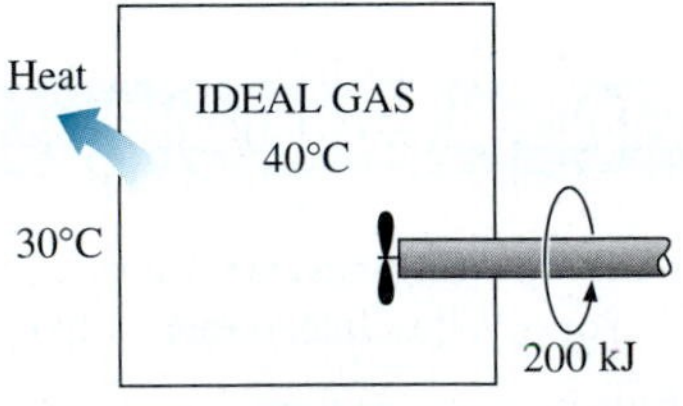

FIGURE P8–23

8–24 Air is compressed by a 12-kW compressor from P_1 to P_2. The air temperature is maintained constant at 25°C during this process as a result of heat transfer to the surrounding medium at 10°C. Determine the rate of entropy change of the air. State the assumptions made in solving this problem. *Answer:* −0.0403 kW/K

*Problems designated by a "C" are concept questions, and students are encouraged to answer them all. Problems designated by an "E" are in English units, and the SI users can ignore them. Problems with the icon are solved using EES, and complete solutions together with parametric studies are included on the enclosed DVD. Problems with the icon are comprehensive in nature, and are intended to be solved with a computer, preferably using the EES software that accompanies this text.

8–25 Heat in the amount of 100 kJ is transferred directly from a hot reservoir at 1200 K to a cold reservoir at 600 K. Calculate the entropy change of the two reservoirs and determine if the increase of entropy principle is satisfied.

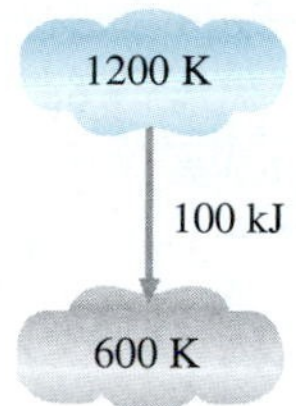

FIGURE P8–25

8–26 In the previous problem, assume that the heat is transferred from the cold reservoir to the hot reservoir contrary to the Clausius statement of the second law. Prove that this violates the increase of entropy principle—as it must according to Clausius.

8–27 Heat is transferred at a rate of 2 kW from a hot reservoir at 800 K to a cold reservoir at 300 K. Calculate the rate at which the entropy of the two reservoirs change and determine if the second law is satisfied. *Answer:* 0.00417 kW/K

8–28E A completely reversible air conditioner provides 36,000 Btu/h of cooling for a space maintained at 70°F while rejecting heat to the environmental air at 110°F. Calculate the rate at which the entropies of the two reservoirs change and verify that this air conditioner satisfies the increase of entropy principle.

8–29 A completely reversible heat pump produces heat at a rate of 100 kW to warm a house maintained at 21°C. The exterior air, which is at 10°C, serves as the source. Calculate the rate of entropy change of the two reservoirs and determine if this heat pump satisfies the second law according to the increase of entropy principle.

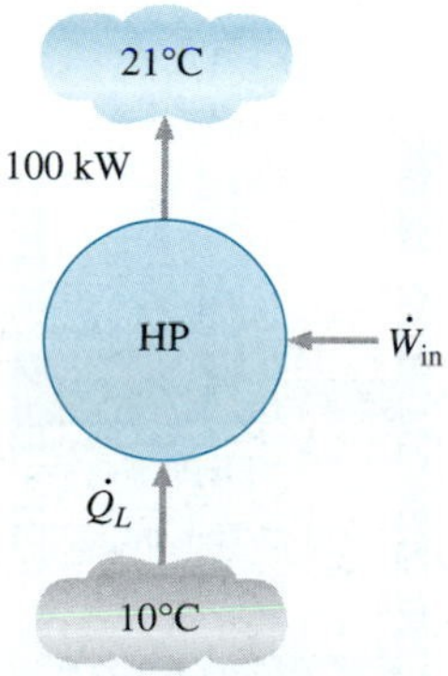

FIGURE P8–29

8–30E During the isothermal heat rejection process of a Carnot cycle, the working fluid experiences an entropy change of −0.7 Btu/R. If the temperature of the heat sink is 95°F, determine (*a*) the amount of heat transfer, (*b*) the entropy change of the sink, and (*c*) the total entropy change for this process. *Answers:* (*a*) 388.5 Btu, (*b*) 0.7 Btu/R, (*c*) 0

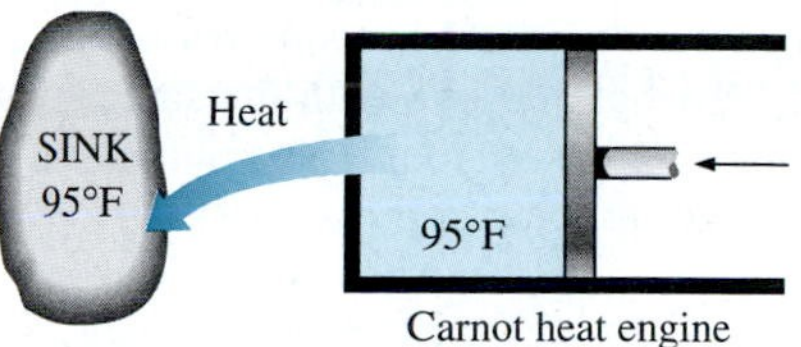

FIGURE P8–30E

8–31 Refrigerant-134a enters the coils of the evaporator of a refrigeration system as a saturated liquid–vapor mixture at a pressure of 160 kPa. The refrigerant absorbs 180 kJ of heat from the cooled space, which is maintained at −5°C, and leaves as saturated vapor at the same pressure. Determine (*a*) the entropy change of the refrigerant, (*b*) the entropy change of the cooled space, and (*c*) the total entropy change for this process.

Entropy Changes of Pure Substances

8–32C Is a process that is internally reversible and adiabatic necessarily isentropic? Explain.

8–33 A rigid vessel is filled with a fluid from a source whose properties remain constant. How does the entropy of the surroundings change if the vessel is filled such that the specific entropy of the vessel contents remains constant?

8–34 A rigid vessel filled with a fluid is allowed to leak some fluid out through an opening. During this process, the specific entropy of the remaining fluid remains constant. How does the entropy of the environment change during this process?

8–35E R-134a vapor enters to a turbine at 250 psia and 175°F. The temperature of R-134a is reduced to 20°F in this turbine while its specific entropy remains constant. Determine the change in the enthalpy of R-134a as it passes through the turbine.

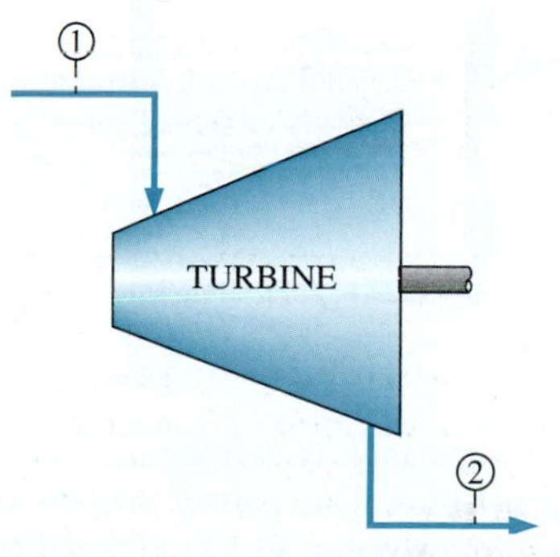

FIGURE P8–35E

8–36E 2-lbm of water at 300 psia fill a weighted piston-cylinder device whose volume is 2.5 ft^3. The water is then heated at constant pressure until the temperature reaches 500°F. Determine the resulting change in the water's total entropy. *Answer:* 0.474 Btu/R

8–37 Water vapor enters a compressor at 35 kPa and 160°C, and leaves at 300 kPa with the same specific entropy as at the inlet. What is the temperature and the specific enthalpy of water at the compressor exit?

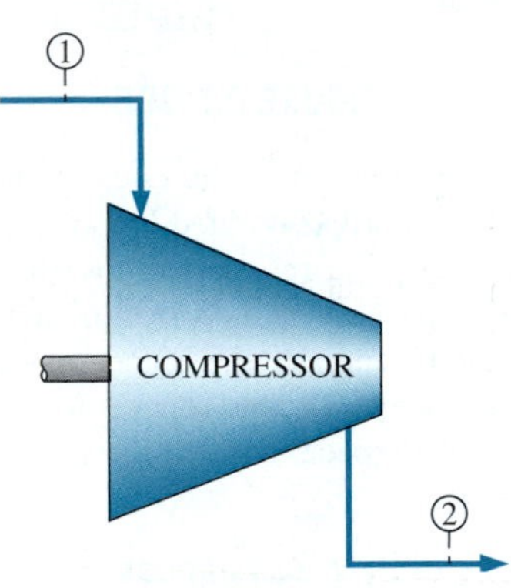

FIGURE P8–37

8–38E 1-lbm of R-134a is expanded isentropically in a closed system from 100 psia and 100°F to 10 psia. Determine the total heat transfer and work production for this process.

8–39 A well-insulated rigid tank contains 2 kg of a saturated liquid–vapor mixture of water at 100 kPa. Initially, three-quarters of the mass is in the liquid phase. An electric resistance heater placed in the tank is now turned on and kept on until all the liquid in the tank is vaporized. Determine the entropy change of the steam during this process. *Answer:* 8.10 kJ/K

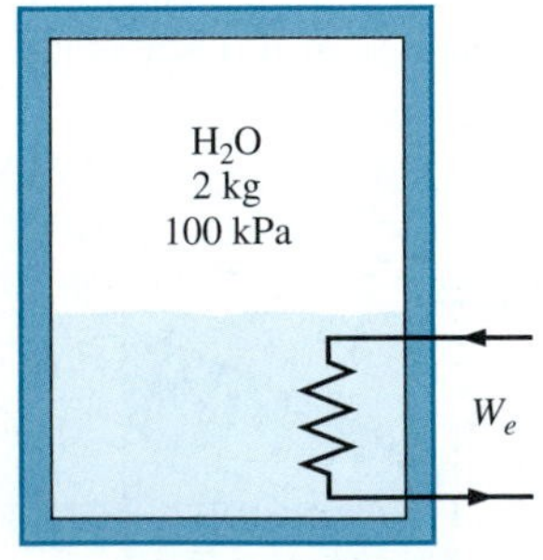

FIGURE P8–39

8–40 A rigid tank is divided into two equal parts by a partition. One part of the tank contains 1.5 kg of compressed liquid water at 300 kPa and 60°C while the other part is evacuated. The partition is now removed, and the water expands to fill the entire tank. Determine the entropy change of water during this process, if the final pressure in the tank is 15 kPa. *Answer:* −0.114 kJ/K

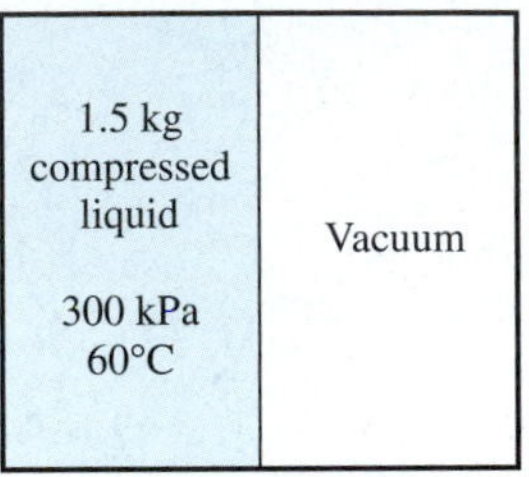

FIGURE P8–40

8–41 Reconsider Prob. 8–40. Using EES (or other) software, evaluate and plot the entropy generated as a function of surrounding temperature, and determine the values of the surrounding temperatures that are valid for this problem. Let the surrounding temperature vary from 0 to 100°C. Discuss your results.

8–42E A piston–cylinder device contains 2 lbm of refrigerant-134a at 120 psia and 100°F. The refrigerant is now cooled at constant pressure until it exists as a liquid at 50°F. Determine the entropy change of the refrigerant during this process.

8–43 An insulated piston–cylinder device contains 5 L of saturated liquid water at a constant pressure of 150 kPa. An electric resistance heater inside the cylinder is now turned on, and 2200 kJ of energy is transferred to the steam. Determine the entropy change of the water during this process. *Answer:* 5.72 kJ/K

8–44 An insulated piston–cylinder device contains 0.05 m^3 of saturated refrigerant-134a vapor at 0.8-MPa pressure. The refrigerant is now allowed to expand in a reversible manner until the pressure drops to 0.4 MPa. Determine (*a*) the final temperature in the cylinder and (*b*) the work done by the refrigerant.

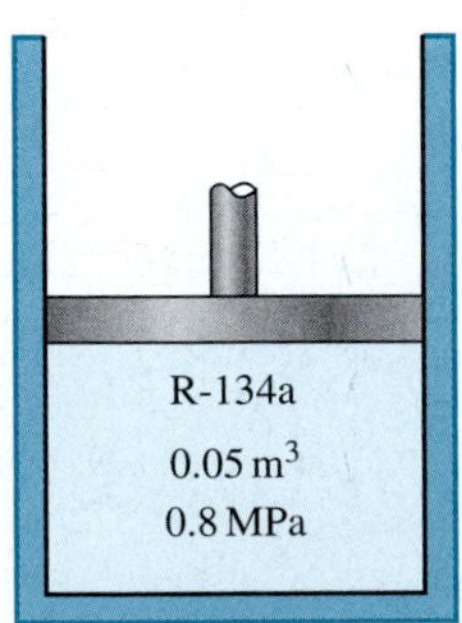

FIGURE P8–44

8–45 Reconsider Prob. 8–44. Using EES (or other) software, evaluate and plot the work done by the refrigerant as a function of final pressure as it varies from 0.8 to 0.4 MPa. Compare the work done for this process to one for which the temperature is constant over the same pressure range. Discuss your results.

8–46 Refrigerant-134a enters an adiabatic compressor as saturated vapor at 160 kPa at a rate of 2 m^3/min and is compressed to a pressure of 900 kPa. Determine the minimum power that must be supplied to the compressor.

8–47 A heavily insulated piston–cylinder device contains 0.05 m^3 of steam at 300 kPa and 150°C. Steam is now compressed in a reversible manner to a pressure of 1 MPa. Determine the work done on the steam during this process.

8–48 Reconsider Prob. 8–47. Using EES (or other) software, evaluate and plot the work done on the steam as a function of final pressure as the pressure varies from 300 kPa to 1 MPa.

8–49 A piston–cylinder device contains 1.2 kg of saturated water vapor at 200°C. Heat is now transferred to steam, and steam expands reversibly and isothermally to a final pressure of 800 kPa. Determine the heat transferred and the work done during this process.

8–50 Reconsider Prob. 8–49. Using EES (or other) software, evaluate and plot the heat transferred to the steam and the work done as a function of final pressure as the pressure varies from the initial value to the final value of 800 kPa.

8–51 Water at 10°C and 81.4 percent quality is compressed isentropically in a closed system to 3 MPa. How much work does this process require, in kJ/kg?

8–52 Refrigerant-134a at 240 kPa and 20°C undergoes an isothermal process in a closed system until its quality is 20 percent. On per unit mass basis, determine how much work and heat transfer are required. *Answers:* 37.0 kJ/kg, 172 kJ/kg

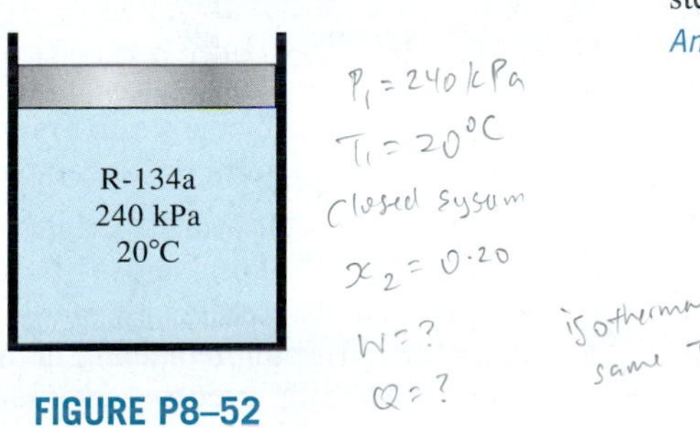

FIGURE P8–52

8–53 Determine the total heat transfer for the reversible process 1-3 shown in Fig. P8–53.

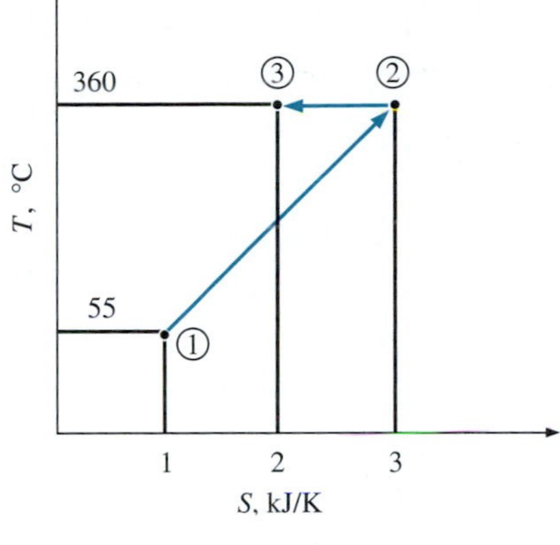

FIGURE P8–53

8–54 Determine the total heat transfer for the reversible process 1-2 shown in Fig. P8–54.

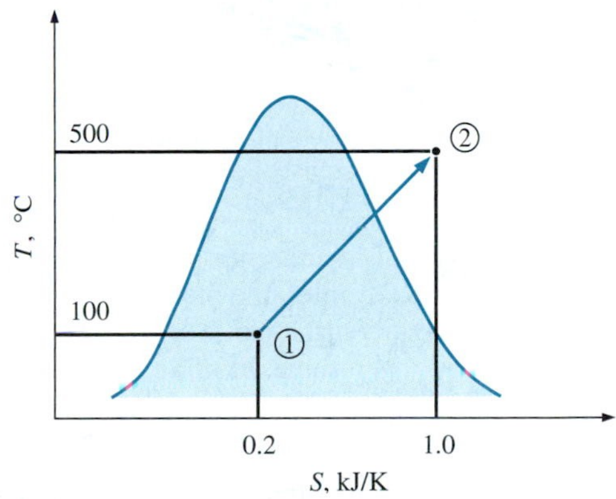

FIGURE P8–54

8–55 Calculate the heat transfer, in kJ/kg, for the reversible steady-flow process 1-3 shown on a *T-s* diagram in Fig. P8–55. *Answer:* 341 kJ/kg

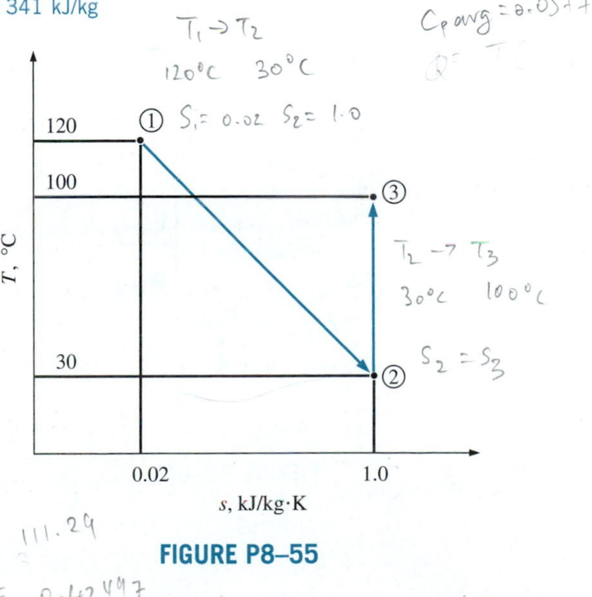

FIGURE P8–55

8–56E Calculate the heat transfer, in Btu/lbm, for the reversible steady-flow process 1-2 shown on a *T-s* diagram in Fig. P8–56E.

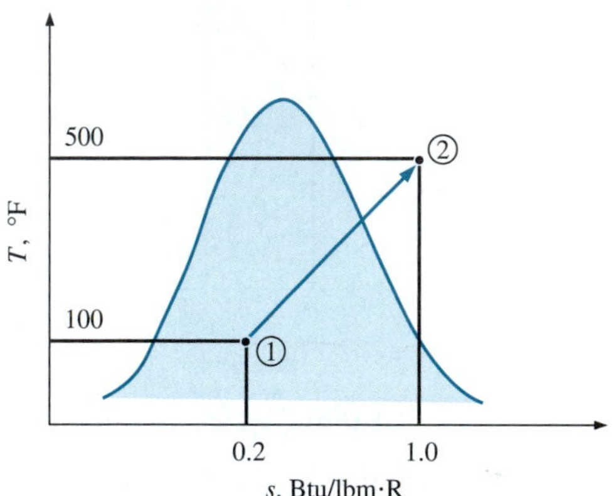

FIGURE P8–56E

8–57 Using the relation $ds = (\delta Q/T)_{\text{int rev}}$ for the definition of entropy, calculate the change in the specific entropy of R-134a as it is heated at a constant pressure of 200 kPa from a saturated liquid to a saturated vapor. Use the R-134a tables to verify your answer.

8–58 Steam is expanded in an isentropic turbine with a single inlet and outlet. At the inlet, the steam is at 2 MPa and 360°C. The steam pressure at the outlet is 100 kPa. Calculate the work produced by this turbine, in kJ/kg.

8–59E The compressor in a refrigerator compresses saturated R-134a vapor at 0°F to 200 psia. Calculate the work required by this compressor, in Btu/lbm, when the compression process is isentropic.

8–60 An isentropic steam turbine processes 5 kg/s of steam at 4 MPa, which is exhausted at 50 kPa and 100°C. Five percent of this flow is diverted for feedwater heating at 700 kPa. Determine the power produced by this turbine, in kW. *Answer:* 6328 kW

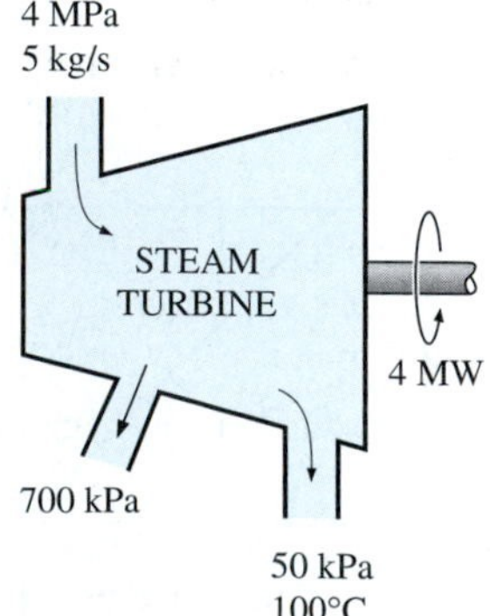

FIGURE P8–60

8–61 A rigid, 100-L steam cooker is arranged with a pressure relief valve set to release vapor and maintain the pressure once the pressure inside the cooker reaches 200 kPa. Initially, this cooker is filled with water at 200 kPa with a quality of 10 percent. Heat is now added until the quality inside the cooker is 50 percent. Determine the minimum entropy change of the thermal energy reservoir supplying this heat.

8–62 In the preceding problem, the water is stirred at the same time that it is being heated. Determine the minimum entropy change of the heat-supplying source if 100 kJ of work is done on the water as it is being heated.

8–63 A piston–cylinder device contains 5 kg of steam at 100°C with a quality of 50 percent. This steam undergoes two processes as follows:

1-2 Heat is transferred to the steam in a reversible manner while the temperature is held constant until the steam exists as a saturated vapor.

2-3 The steam expands in an adiabatic, reversible process until the pressure is 15 kPa.

(*a*) Sketch these processes with respect to the saturation lines on a single *T-s* diagram.

(*b*) Determine the heat transferred to the steam in process 1-2, in kJ.

(*c*) Determine the work done by the steam in process 2-3, in kJ.

8–64 Steam at 6000 kPa and 500°C enters a steady-flow turbine. The steam expands in the turbine while doing work until the pressure is 1000 kPa. When the pressure is 1000 kPa, 10 percent of the steam is removed from the turbine for other uses. The remaining 90 percent of the steam continues to expand through the turbine while doing work and leaves the turbine at 10 kPa. The entire expansion process by the steam through the turbine is reversible and adiabatic.

(*a*) Sketch the process on a *T-s* diagram with respect to the saturation lines. Be sure to label the data states and the lines of constant pressure.

(*b*) If the turbine has an isentropic efficiency of 85 percent, what is the work done by the steam as it flows through the turbine per unit mass of steam flowing into the turbine, in kJ/kg?

8–65E A 1.2-ft^3 well-insulated rigid can initially contains refrigerant-134a at 140 psia and 70°F. Now a crack develops in the can, and the refrigerant starts to leak out slowly, Assuming the refrigerant remaining in the can has undergone a reversible, adiabatic process, determine the final mass in the can when the pressure drops to 20 psia.

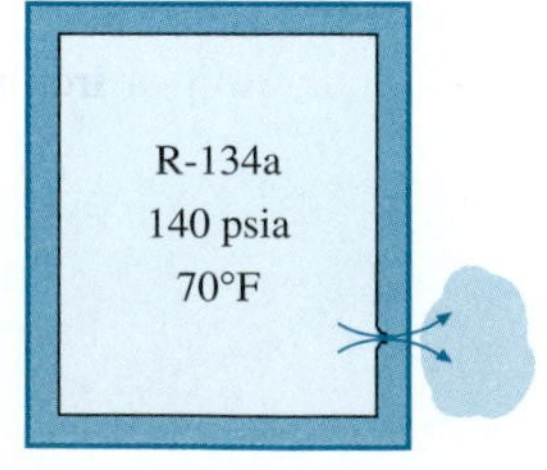

FIGURE P8–65E

Entropy Change of Incompressible Substances

8–66C Consider two solid blocks, one hot and the other cold, brought into contact in an adiabatic container. After a while, thermal equilibrium is established in the container as a result of heat transfer. The first law requires that the amount of energy lost by the hot solid be equal to the amount of energy gained by the cold one. Does the second law require that the decrease in entropy of the hot solid be equal to the increase in entropy of the cold one?

8–67 A 50-kg copper block initially at 80°C is dropped into an insulated tank that contains 120 L of water at 25°C. Determine the final equilibrium temperature and the total entropy change for this process.

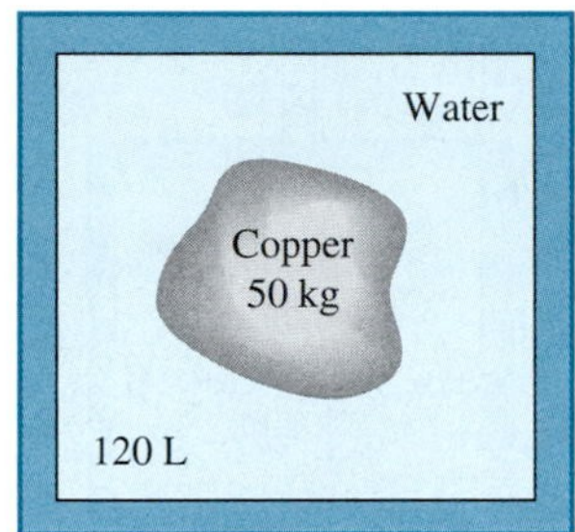

FIGURE P8–67

8–68 Ten grams of computer chips with a specific heat of 0.3 kJ/kg·K are initially at 20ºC. These chips are cooled by placement in 5 grams of saturated liquid R-134a at −40ºC. Presuming that the pressure remains constant while the chips are being cooled, determine the entropy change of (*a*) the chips, (*b*) the R-134a, and (*c*) the entire system. Is this process possible? Why?

8–69 A 25-kg iron block initially at 350°C is quenched in an insulated tank that contains 100 kg of water at 18°C. Assuming the water that vaporizes during the process condenses back in the tank, determine the total entropy change during this process.

8–70 A 20-kg aluminum block initially at 200°C is brought into contact with a 20-kg block of iron at 100°C in an insulated enclosure. Determine the final equilibrium temperature and the total entropy change for this process. *Answers:* 168.4°C, 0.169 kJ/K

8–71 Reconsider Prob. 8–70. Using EES (or other) software, study the effect of the mass of the iron block on the final equilibrium temperature and the total entropy change for the process. Let the mass of the iron vary from 1 to 10 kg. Plot the equilibrium temperature and the total entropy change as a function of iron mass, and discuss the results.

8–72 A 50-kg iron block and a 20-kg copper block, both initially at 80°C, are dropped into a large lake at 15°C. Thermal equilibrium is established after a while as a result of heat transfer between the blocks and the lake water. Determine the total entropy change for this process.

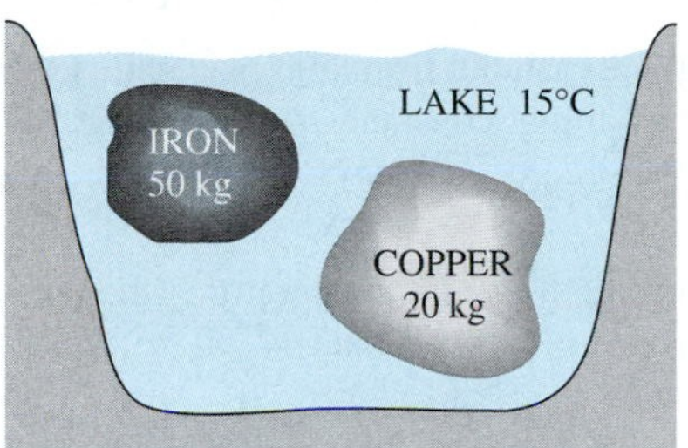

FIGURE P8–72

8–73 An adiabatic pump is to be used to compress saturated liquid water at 10 kPa to a pressure to 15 MPa in a reversible manner. Determine the work input using (*a*) entropy data from the compressed liquid table, (*b*) inlet specific volume and pressure values, (*c*) average specific volume and pressure values. Also, determine the errors involved in parts (*b*) and (*c*).

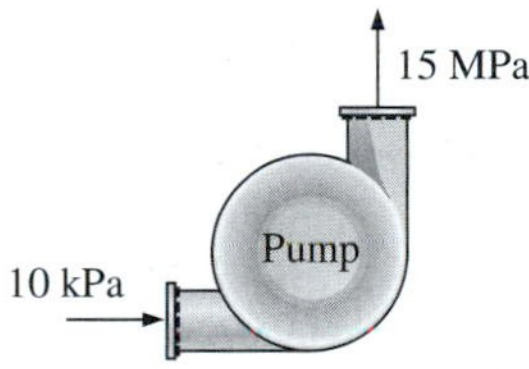

FIGURE P8–73

Entropy Change of Ideal Gases

8–74C Prove that the two relations for entropy change of ideal gases under the constant-specific-heat assumption (Eqs. 8–33 and 8–34) are equivalent.

8–75C Starting with the second *T ds* relation (Eq. 8–26), obtain Eq. 8–34 for the entropy change of ideal gases under the constant-specific-heat assumption.

8–76C Some properties of ideal gases such as internal energy and enthalpy vary with temperature only [that is, $u = u(T)$ and $h = h(T)$]. Is this also the case for entropy?

8–77C Starting with Eq. 8–34, obtain Eq. 8–43.

8–78C What are P_r and v_r called? Is their use limited to isentropic processes? Explain.

8–79C Can the entropy of an ideal gas change during an isothermal process?

8–80C An ideal gas undergoes a process between two specified temperatures, first at constant pressure and then at constant volume. For which case will the ideal gas experience a larger entropy change? Explain.

8–81 What is the difference between entropies of oxygen at 150 kPa and 39°C and oxygen at 150 kPa and 337°C per unit mass basis.

8–82 Which of the two gases—helium or nitrogen—experiences the greatest entropy change as its state is changed from 2000 kPa and 427°C to 200 kPa and 27°C?

8–83E Air is expanded from 200 psia and 500°F to 100 psia and 50°F. Assuming constant specific heats, determine the change in the specific entropy of air. *Answer:* −0.106 Btu/lbm · R

8–84 Determine the final temperature when air is expanded isentropically from 1000 kPa and 477°C to 100 kPa in a piston-cylinder device.

8–85E Air is expanded isentropically from 100 psia and 500°F to 20 psia in a closed system. Determine its final temperature.

8–86 Which of the two gases—helium or nitrogen—has the highest final temperature as it is compressed isentropically from 100 kPa and 25°C to 1 MPa in a closed system?

8–87 Which of the two gases—neon or air—has the lowest final temperature as it is expanded isentropically from 1000 kPa and 500°C to 100 kPa in a piston–cylinder device?

8–88 An insulated piston–cylinder device initially contains 300 L of air at 120 kPa and 17°C. Air is now heated for 15 min by a 200-W resistance heater placed inside the cylinder. The pressure of air is maintained constant during this process. Determine the entropy change of air, assuming (*a*) constant specific heats and (*b*) variable specific heats.

8–89 A piston–cylinder device contains 1.2 kg of nitrogen gas at 120 kPa and 27°C. The gas is now compressed slowly in a polytropic process during which $PV^{1.3}$ = constant. The process ends when the volume is reduced by one-half. Determine the entropy change of nitrogen during this process. *Answer:* −0.0617 kJ/K

8–90 Reconsider Prob. 8–89. Using EES (or other) software, investigate the effect of varying the polytropic exponent from 1 to 1.4 on the entropy change of the nitrogen. Show the processes on a common *P*-*v* diagram.

8–91E A mass of 15 lbm of helium undergoes a process from an initial state of 50 ft³/lbm and 80°F to a final state of 10 ft³/lbm and 200°F. Determine the entropy change of helium during this process, assuming (*a*) the process is reversible and (*b*) the process is irreversible.

8–92 An insulated rigid tank is divided into two equal parts by a partition. Initially, one part contains 5 kmol of an ideal gas at 250 kPa and 40°C, and the other side is evacuated. The partition is now removed, and the gas fills the entire tank. Determine the total entropy change during this process. *Answer:* 28.81 kJ/K

8–93 Air is compressed in a piston–cylinder device from 100 kPa and 17°C to 800 kPa in a reversible, adiabatic process. Determine the final temperature and the work done during this process, assuming (*a*) constant specific heats and (*b*) variable specific heats for air. *Answers:* (*a*) 525.3 K, 171.1 kJ/kg, (*b*) 522.4 K, 169.3 kJ/kg

8–94 Reconsider Prob. 8–93. Using EES (or other) software, evaluate and plot the work done and final temperature during the compression process as functions of the final pressure for the two cases as the final pressure varies from 100 to 800 kPa.

8–95 An insulated rigid tank contains 4 kg of argon gas at 450 kPa and 30°C. A valve is now opened, and argon is allowed to escape until the pressure inside drops to 200 kPa. Assuming the argon remaining inside the tank has undergone a reversible, adiabatic process, determine the final mass in the tank. *Answer:* 2.46 kg

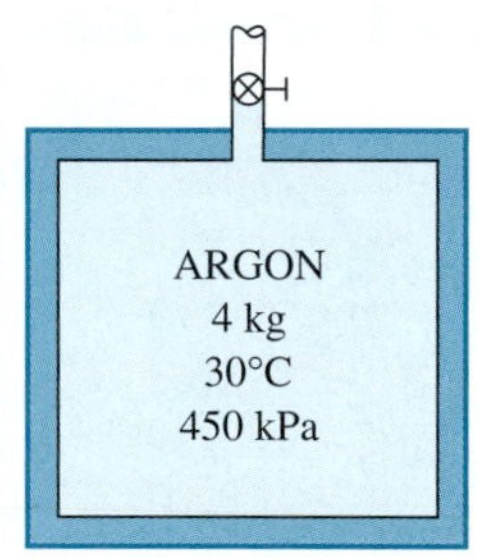

FIGURE P8–95

8–96 Reconsider Prob. 8–95. Using EES (or other) software, investigate the effect of the final pressure on the final mass in the tank as the pressure varies from 450 to 150 kPa, and plot the results.

8–97E Air enters an adiabatic nozzle at 60 psia, 540°F, and 200 ft/s and exits at 12 psia. Assuming air to be an ideal gas with variable specific heats and disregarding any irreversibilities, determine the exit velocity of the air.

8–98E Air enters an isentropic turbine at 150 psia and 900°F through a 0.5-ft² inlet section with a velocity of 500 ft/s. It leaves at 15 psia with a velocity of 100 ft/s. Calculate the air temperature at the turbine exit and the power produced, in hp, by this turbine.

8–99 Nitrogen at 120 kPa and 30°C is compressed to 600 kPa in an adiabatic compressor. Calculate the minimum work needed for this process, in kJ/kg. *Answer:* 184 kJ/kg

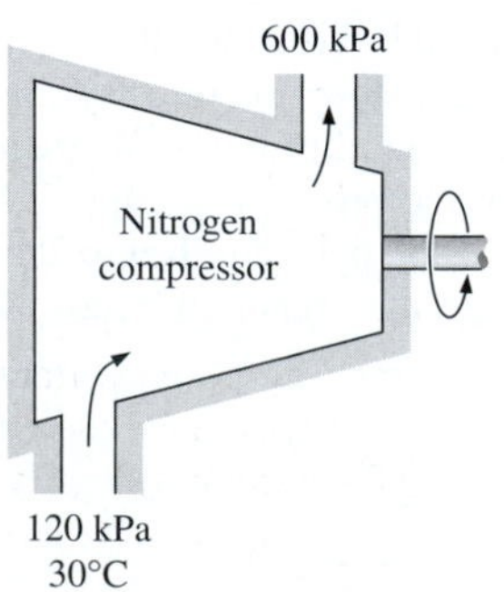

FIGURE P8–99

8–100 Oxygen at 300 kPa and 90°C flowing at an average velocity of 3 m/s is expanded in an adiabatic nozzle. What is the maximum velocity of the oxygen at the outlet of this nozzle when the outlet pressure is 120 kPa? *Answer:* 390 m/s

8–101 Air is expanded in an adiabatic nozzle during a polytropic process with $n = 1.3$. It enters the nozzle at 700 kPa and 100°C with a velocity of 30 m/s and exits at a pressure of 200 kPa. Calculate the air temperature and velocity at the nozzle exit.

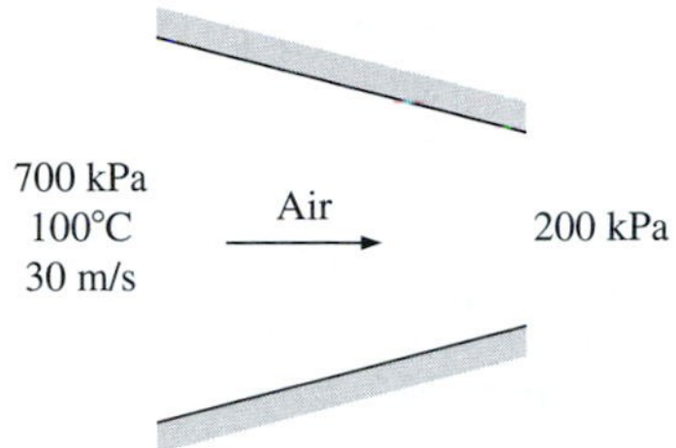

FIGURE P8–101

8–102 Repeat Prob. 8–101 for the polytropic exponent of $n = 1.2$.

8–103E The well-insulated container shown in Fig. P8–103E is initially evacuated. The supply line contains air that is maintained at 200 psia and 100°F. The valve is opened until the pressure in the container is the same as the pressure in the supply line. Determine the minimum temperature in the container when the valve is closed.

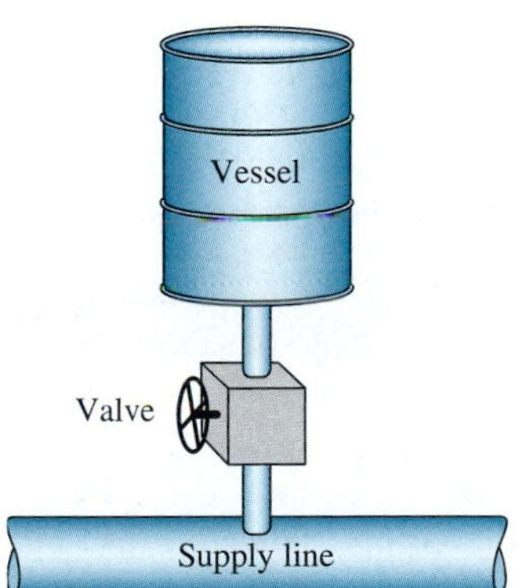

FIGURE P8–103E

8–104 A container filled with 45 kg of liquid water at 95°C is placed in a 90-m^3 room that is initially at 12°C. Thermal equilibrium is established after a while as a result of heat transfer between the water and the air in the room. Using constant specific heats, determine (*a*) the final equilibrium temperature, (*b*) the amount of heat transfer between the water and the air in the room, and (*c*) the entropy generation. Assume the room is well sealed and heavily insulated.

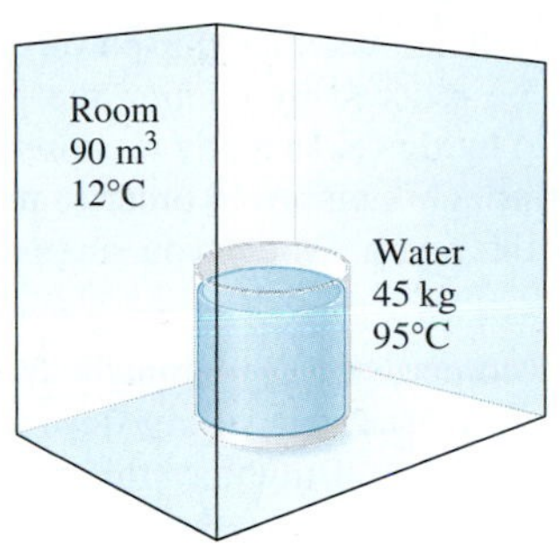

FIGURE P8–104

8–105 Air at 800 kPa and 400°C enters a steady-flow nozzle with a low velocity and leaves at 100 kPa. If the air undergoes an adiabatic expansion process through the nozzle, what is the maximum velocity of the air at the nozzle exit, in m/s?

8–106 An ideal gas at 100 kPa and 27°C enters a steady-flow compressor. The gas is compressed to 400 kPa, and 10 percent of the mass that entered the compressor is removed for some other use. The remaining 90 percent of the inlet gas is compressed to 600 kPa before leaving the compressor. The entire compression process is assumed to be reversible and adiabatic. The power supplied to the compressor is measured to be 32 kW. If the ideal gas has constant specific heats such that $c_v = 0.8$ kJ/kg · K and $c_p = 1.1$ kJ/kg · K, (*a*) sketch the compression process on a *T-s* diagram, (*b*) determine the temperature of the gas at the two compressor exits, in K, and (*c*) determine the mass flow rate of the gas into the compressor, in kg/s.

8–107 A constant-volume tank contains 5 kg of air at 100 kPa and 327°C. The air is cooled to the surroundings temperature of 27°C. Assume constant specific heats at 300 K. (*a*) Determine the entropy change of the air in the tank during the process, in kJ/K, (*b*) determine the net entropy change of the universe due to this process, in kJ/K, and (*c*) sketch the processes for the air in the tank and the surroundings on a single *T-s* diagram. Be sure to label the initial and final states for both processes.

Reversible Steady-Flow Work

8–108C In large compressors, the gas is frequently cooled while being compressed to reduce the power consumed by the compressor. Explain how cooling the gas during a compression process reduces the power consumption.

8–109C The turbines in steam power plants operate essentially under adiabatic conditions. A plant engineer suggests to end this practice. She proposes to run cooling water through the outer surface of the casing to cool the steam as it flows through the turbine. This way, she reasons, the entropy of the steam will decrease, the performance of the turbine will improve, and as a result the work output of the turbine will increase. How would you evaluate this proposal?

8–110C It is well known that the power consumed by a compressor can be reduced by cooling the gas during compression. Inspired by this, somebody proposes to cool the liquid as it flows through a pump, in order to reduce the power consumption of the pump. Would you support this proposal? Explain.

8–111E Air is compressed isothermally from 16 psia and 75°F to 120 psia in a reversible steady-flow device. Calculate the work required, in Btu/lbm, for this compression. *Answer:* 73.9 Btu/lbm

8–112 Saturated water vapor at 150°C is compressed in a reversible steady-flow device to 1000 kPa while its specific volume remains constant. Determine the work required, in kJ/kg.

8–113 Calculate the work produced, in kJ/kg, for the reversible steady-flow process 1-3 shown in Fig. P8–113.

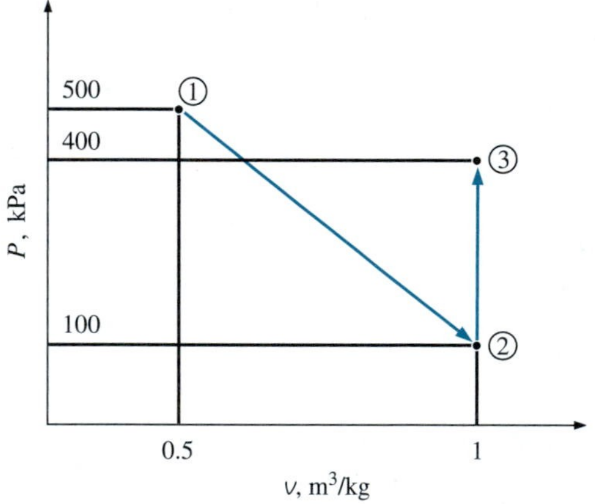

FIGURE P8–113

8–114E Calculate the work produced, in Btu/lbm, for the reversible steady-flow process 1-2 shown in Fig. P8-114.

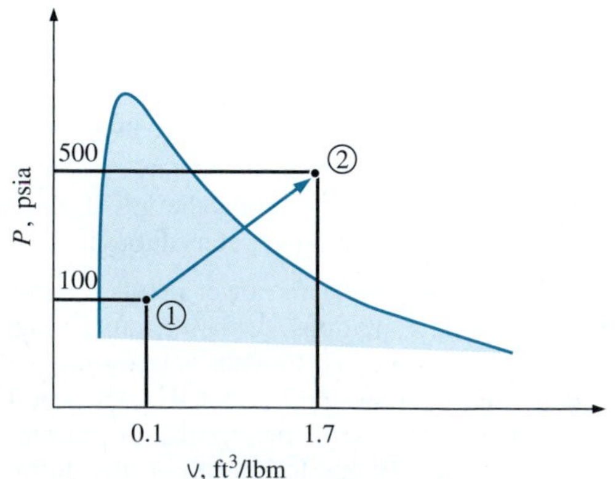

FIGURE P8–114E

8–115 Liquid water enters a 25-kW pump at 100-kPa pressure at a rate of 5 kg/s. Determine the highest pressure the liquid water can have at the exit of the pump. Neglect the kinetic and potential energy changes of water, and take the specific volume of water to be 0.001 m^3/kg. *Answer:* 5100 kPa

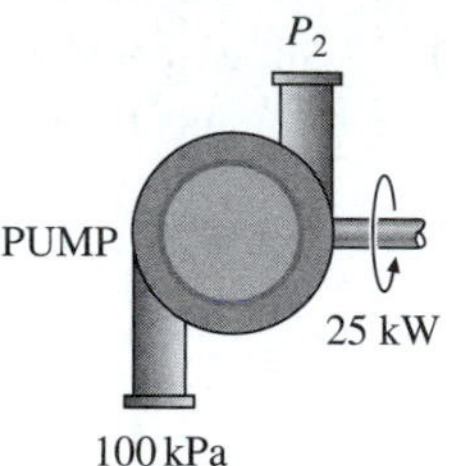

FIGURE P8–115

8–116 Consider a steam power plant that operates between the pressure limits of 10 MPa and 20 kPa. Steam enters the pump as saturated liquid and leaves the turbine as saturated vapor. Determine the ratio of the work delivered by the turbine to the work consumed by the pump. Assume the entire cycle to be reversible and the heat losses from the pump and the turbine to be negligible.

8–117 Reconsider Prob. 8–116. Using EES (or other) software, investigate the effect of the quality of the steam at the turbine exit on the net work output. Vary the quality from 0.5 to 1.0, and plot the net work output as a function of this quality.

8–118 Liquid water at 120 kPa enters a 7-kW pump where its pressure is raised to 5 MPa. If the elevation difference between the exit and the inlet levels is 10 m, determine the highest mass flow rate of liquid water this pump can handle. Neglect the kinetic energy change of water, and take the specific volume of water to be 0.001 m^3/kg.

8–119E Helium gas is compressed from 14 psia and 70°F to 120 psia at a rate of 5 ft^3/s. Determine the power input to the compressor, assuming the compression process to be (*a*) isentropic, (*b*) polytropic with $n = 1.2$, (*c*) isothermal, and (*d*) ideal two-stage polytropic with $n = 1.2$.

8–120E Reconsider Prob. 8–119E. Using EES (or other) software, evaluate and plot the work of compression and entropy change of the helium as functions of the polytropic exponent as it varies from 1 to 1.667. Discuss your results.

8–121 The compression stages in the axial compressor of the industrial gas turbine are close coupled, making intercooling very impractical. To cool the air in such compressors and to reduce the compression power, it is proposed to spray water mist with drop size on the order of 5 microns into the air stream as it is compressed and to cool the air continuously as the water evaporates. Although the collision of water droplets with turbine blades is a concern, experience with steam turbines indicates that they can cope with water droplet concentrations of up to 14 percent. Assuming air is compressed isentropically at a rate of 2 kg/s from 300 K and 100 kPa to 1200 kPa and the water is injected at a temperature of 20°C at a rate of 0.2 kg/s, determine the reduction in the exit temperature of the compressed air and the compressor power saved. Assume the water vaporizes completely before leaving the compressor, and assume an average mass flow rate of 2.1 kg/s throughout the compressor.

8–122 Reconsider Prob. 8–121. The water-injected compressor is used in a gas turbine power plant. It is claimed that the power output of a gas turbine will increase because of the increase in the mass flow rate of the gas (air + water vapor) through the turbine. Do you agree?

Isentropic Efficiencies of Steady-Flow Devices

8–123C Describe the ideal process for an (*a*) adiabatic turbine, (*b*) adiabatic compressor, and (*c*) adiabatic nozzle, and define the isentropic efficiency for each device.

8–124C Is the isentropic process a suitable model for compressors that are cooled intentionally? Explain.

8–125C On a *T-s* diagram, does the actual exit state (state 2) of an adiabatic turbine have to be on the right-hand side of the isentropic exit state (state 2*s*)? Why?

8–126 100-kg of saturated steam at 100 kPa is to be adiabatically compressed in a closed system to 1000 kPa. How much work is required if the isentropic compression efficiency is 90 percent? *Answer:* 44,200 kJ

8–127E 10-lbm of R-134a is expanded without any heat transfer in a closed system from 120 psia and 100°F to 20 psia. If the isentropic expansion efficiency is 95 percent, what is the final volume of this steam?

8–128 Steam at 3 MPa and 400°C is expanded to 30 kPa in an adiabatic turbine with an isentropic efficiency of 92 percent. Determine the power produced by this turbine, in kW, when the mass flow rate is 2 kg/s.

8–129 Repeat Prob. 8–128 for a turbine efficiency of 90 percent.

8–130 An adiabatic steady-flow device compresses argon at 200 kPa and 27°C to 2 MPa. If the argon leaves this compressor at 550°C, what is the isentropic efficiency of the compressor?

8–131 The adiabatic compressor of a refrigeration system compresses saturated R-134a vapor at 0°C to 600 kPa and 50°C. What is the isentropic efficiency of this compressor?

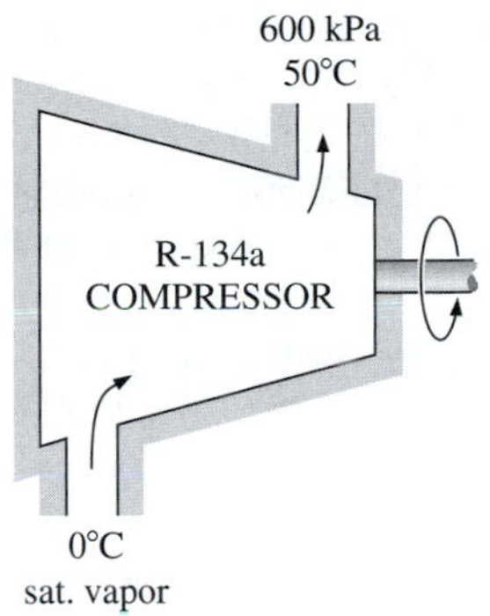

FIGURE P8–131

8–132 Steam enters an adiabatic turbine at 7 MPa, 600°C, and 80 m/s and leaves at 50 kPa, 150°C, and 140 m/s. If the power output of the turbine is 6 MW, determine (*a*) the mass flow rate of the steam flowing through the turbine and (*b*) the isentropic efficiency of the turbine. *Answers:* (*a*) 6.95 kg/s, (*b*) 73.4 percent

8–133 Argon gas enters an adiabatic turbine at 800°C and 1.5 MPa at a rate of 80 kg/min and exhausts at 200 kPa. If the power output of the turbine is 370 kW, determine the isentropic efficiency of the turbine.

8–134E Combustion gases enter an adiabatic gas turbine at 1540°F and 120 psia and leave at 60 psia with a low velocity. Treating the combustion gases as air and assuming an isentropic efficiency of 82 percent, determine the work output of the turbine. *Answer:* 71.7 Btu/lbm

8–135 Refrigerant-134a enters an adiabatic compressor as saturated vapor at 120 kPa at a rate of 0.3 m^3/min and exits at 1-MPa pressure. If the isentropic efficiency of the compressor is 80 percent, determine (*a*) the temperature of the refrigerant at the exit of the compressor and (*b*) the power input, in kW. Also, show the process on a *T-s* diagram with respect to saturation lines.

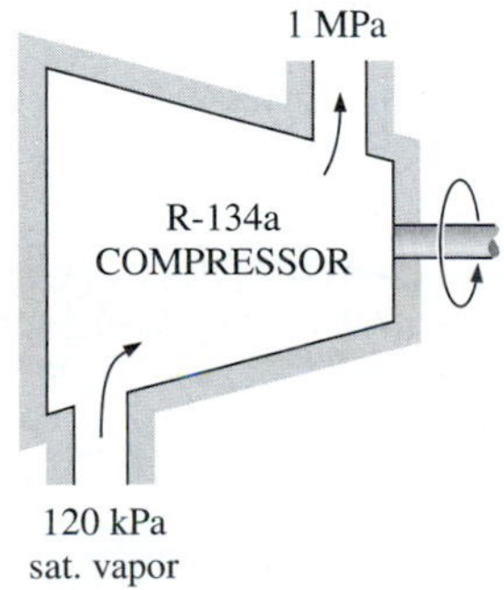

FIGURE P8–135

8–136 Reconsider Prob. 8–135. Using EES (or other) software, redo the problem by including the effects of the kinetic energy of the flow by assuming an inlet-to-exit area ratio of 1.5 for the compressor when the compressor exit pipe inside diameter is 2 cm.

8–137 Air enters an adiabatic compressor at 100 kPa and 17°C at a rate of 2.4 m^3/s, and it exits at 257°C. The compressor has an isentropic efficiency of 84 percent. Neglecting the changes in kinetic and potential energies, determine (*a*) the exit pressure of air and (*b*) the power required to drive the compressor.

8–138 Air is compressed by an adiabatic compressor from 95 kPa and 27°C to 600 kPa and 277°C. Assuming variable specific heats and neglecting the changes in kinetic and potential energies, determine (*a*) the isentropic efficiency of the compressor and (*b*) the exit temperature of air if the process were reversible. *Answers:* (*a*) 81.9 percent, (*b*) 505.5 K

8–139E Argon gas enters an adiabatic compressor at 20 psia and 90°F with a velocity of 60 ft/s, and it exits at 200 psia and 240 ft/s. If the isentropic efficiency of the compressor is 80 percent, determine (*a*) the exit temperature of the argon and (*b*) the work input to the compressor.

8–140E Air enters an adiabatic nozzle at 60 psia and 1020°F with low velocity and exits at 800 ft/s. If the isentropic efficiency of the nozzle is 90 percent, determine the exit temperature and pressure of the air.

8–141E EES Reconsider Prob. 8–140E. Using EES (or other) software, study the effect of varying the nozzle isentropic efficiency from 0.8 to 1.0 on both the exit temperature and pressure of the air, and plot the results.

8–142 The exhaust nozzle of a jet engine expands air at 300 kPa and 180°C adiabatically to 100 kPa. Determine the air velocity at the exit when the inlet velocity is low and the nozzle isentropic efficiency is 96 percent.

8–143E An adiabatic diffuser at the inlet of a jet engine increases the pressure of the air that enters the diffuser at 13 psia and 30°F to 20 psia. What will the air velocity at the diffuser exit be if the diffuser isentropic efficiency is 82 percent and the diffuser inlet velocity is 1000 ft/s? *Answer:* 606 ft/s

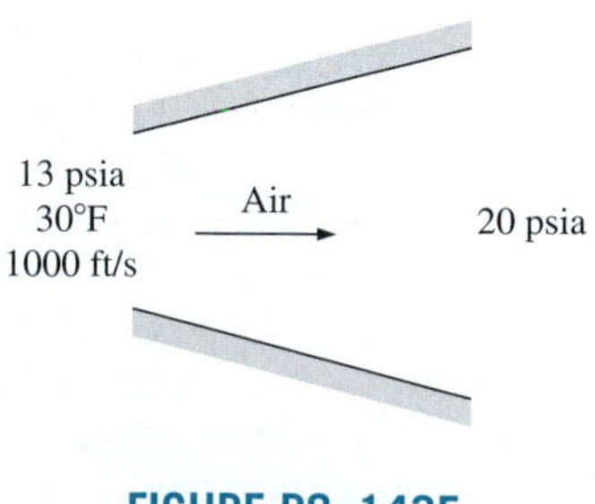

FIGURE P8–143E

Entropy Balance

8–144 Refrigerant-134a is throttled from 900 kPa and 35°C to 200 kPa. Heat is lost from the refrigerant in the amount of 0.8 kJ/kg to the surroundings at 25°C. Determine (*a*) the exit temperature of the refrigerant and (*b*) the entropy generation during this process.

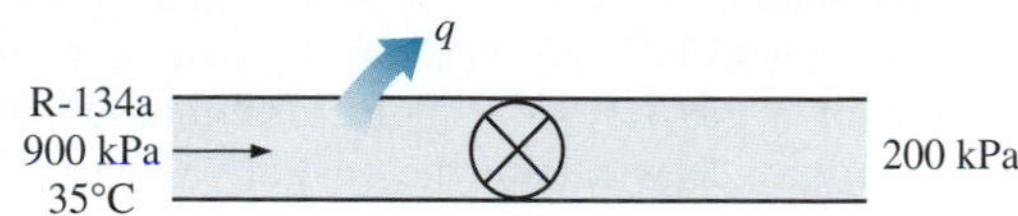

FIGURE P8–144

8–145 A rigid tank contains 7.5 kg of saturated water mixture at 400 kPa. A valve at the bottom of the tank is now opened, and liquid is withdrawn from the tank. Heat is transferred to the steam such that the pressure inside the tank remains constant. The valve is closed when no liquid is left in the tank. If it is estimated that a total of 5 kJ of heat is transferred to the tank, determine (*a*) the quality of steam in the tank at the initial state, (*b*) the amount of mass that has escaped, and (*c*) the entropy generation during this process if heat is supplied to the tank from a source at 500°C.

8–146 Consider a family of four, with each person taking a 5-min shower every morning. The average flow rate through the shower head is 12 L/min. City water at 15°C is heated to 55°C in an electric water heater and tempered to 42°C by cold water at the T-elbow of the shower before being routed to the shower head. Determine the amount of entropy generated by this family per year as a result of taking daily showers.

8–147 Steam is to be condensed in the condenser of a steam power plant at a temperature of 60°C with cooling water from a nearby lake, which enters the tubes of the condenser at 18°C at a rate of 75 kg/s and leaves at 27°C. Assuming the condenser to be perfectly insulated, determine (*a*) the rate of condensation of the steam and (*b*) the rate of entropy generation in the condenser. *Answers:* (*a*) 1.20 kg/s, (*b*) 1.06 kW/K

8–148 Cold water (c_p = 4.18 kJ/kg · °C) leading to a shower enters a well-insulated, thin-walled, double-pipe, counter-flow heat exchanger at 15°C at a rate of 0.25 kg/s and is heated to 45°C by hot water (c_p = 4.19 kJ/kg · °C) that enters at 100°C at a rate of 3 kg/s. Determine (*a*) the rate of heat transfer and (*b*) the rate of entropy generation in the heat exchanger.

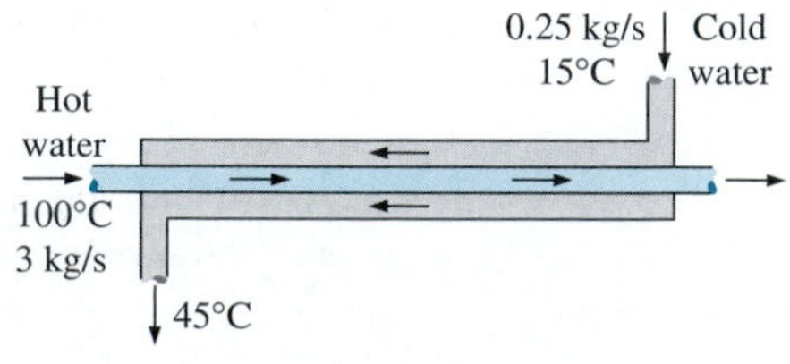

FIGURE P8–148

8–149 Air (c_p = 1.005 kJ/kg · °C) is to be preheated by hot exhaust gases in a cross-flow heat exchanger before it enters the furnace. Air enters the heat exchanger at 95 kPa and 20°C at a rate of 1.6 m^3/s. The combustion gases (c_p = 1.10 kJ/kg · °C) enter at 180°C at a rate of 2.2 kg/s and leave at 95°C. Determine (*a*) the rate of heat transfer to the air, (*b*) the outlet temperature of the air, and (*c*) the rate of entropy generation.

8–150 A well-insulated, shell-and-tube heat exchanger is used to heat water (c_p = 4.18 kJ/kg · °C) in the tubes from 20 to 70°C at a rate of 4.5 kg/s. Heat is supplied by hot oil (c_p = 2.30 kJ/kg · °C) that enters the shell side at 170°C at a rate of 10 kg/s. Disregarding any heat loss from the heat exchanger, determine (*a*) the exit temperature of the oil and (*b*) the rate of entropy generation in the heat exchanger.

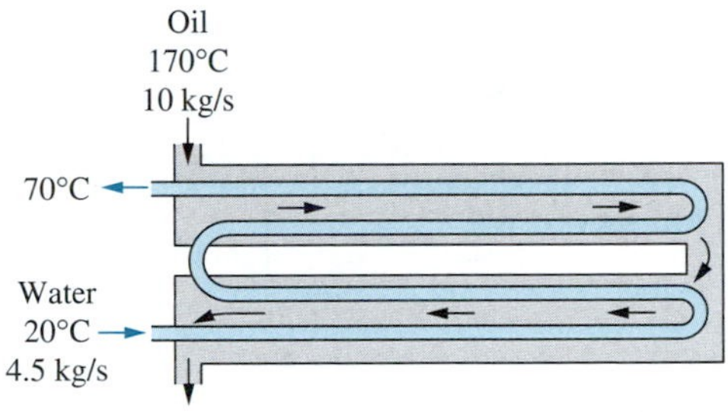

FIGURE P8–150

8–151E Refrigerant-134a is expanded adiabatically from 100 psia and 100°F to a saturated vapor at 10 psia. Determine the entropy generation for this process, in Btu/lbm·R.

8–152 In an ice-making plant, water at 0°C is frozen at atmospheric pressure by evaporating saturated R-134a liquid at −10°C. The refrigerant leaves this evaporator as a saturated vapor, and the plant is sized to produce ice at 0°C at a rate of 4000 kg/h. Determine the rate of entropy generation in this plant. *Answer:* 0.0505 kW/K

FIGURE P8–152

8–153 Air in a large building is kept warm by heating it with steam in a heat exchanger. Saturated water vapor enters this unit at 35°C at a rate of 10,000 kg/h and leaves as a liquid at 32°C. Air at 1-atm pressure enters the unit at 20°C and leaves at 30°C. Determine the rate of entropy generation associated with this process.

8–154 Oxygen enters an insulated 12-cm-diameter pipe with a velocity of 70 m/s. At the pipe entrance, the oxygen is at 240 kPa and 20°C; and, at the exit, it is at 200 kPa and 18°C. Calculate the rate at which entropy is generated in the pipe.

8–155 Nitrogen is compressed by an adiabatic compressor from 100 kPa and 17°C to 600 kPa and 227°C. Calculate the entropy generation for this process, in kJ/kg·K.

8–156E Steam is to be condensed on the shell side of a heat exchanger at 120°F. Cooling water enters the tubes at 60°F at a rate of 92 lbm/s and leaves at 73°F. Assuming the heat exchanger to be well-insulated, determine (*a*) the rate of heat transfer in the heat exchanger and (*b*) the rate of entropy generation in the heat exchanger.

8–157 In a dairy plant, milk at 4°C is pasteurized continuously at 72°C at a rate of 12 L/s for 24 hours a day and 365 days a year. The milk is heated to the pasteurizing temperature by hot water heated in a natural-gas-fired boiler that has an efficiency of 82 percent. The pasteurized milk is then cooled by cold water at 18°C before it is finally refrigerated back to 4°C. To save energy and money, the plant installs a regenerator that has an effectiveness of 82 percent. If the cost of natural gas is \$1.04/therm (1 therm = 105,500 kJ), determine how much energy and money the regenerator will save this company per year and the annual reduction in entropy generation.

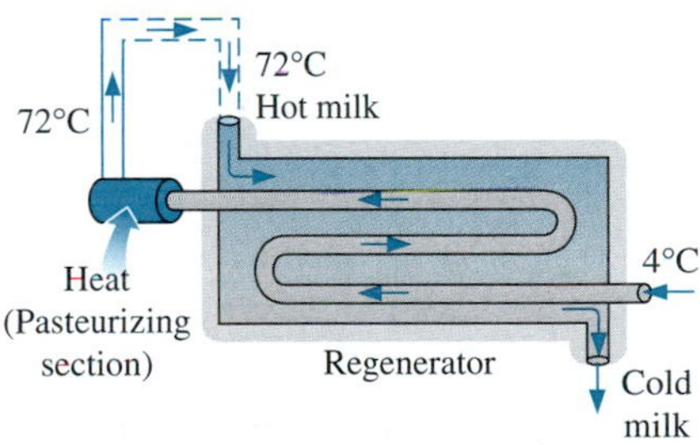

FIGURE P8–157

8–158 Stainless-steel ball bearings (ρ = 8085 kg/m^3 and c_p = 0.480 kJ/kg · °C) having a diameter of 1.2 cm are to be quenched in water at a rate of 1400 per minute. The balls leave the oven at a uniform temperature of 900°C and are exposed to air at 30°C for a while before they are dropped into the water. If the temperature of the balls drops to 850°C prior to quenching, determine (*a*) the rate of heat transfer from the balls to the air and (*b*) the rate of entropy generation due to heat loss from the balls to the air.

8–159 An ordinary egg can be approximated as a 5.5-cm-diameter sphere. The egg is initially at a uniform temperature of 8°C and is dropped into boiling water at 97°C. Taking the properties of the egg to be ρ = 1020 kg/m^3 and c_p = 3.32 kJ/kg · °C, determine (*a*) how much heat is transferred to the egg by the time the average temperature of the egg rises to 70°C and (*b*) the amount of entropy generation associated with this heat transfer process.

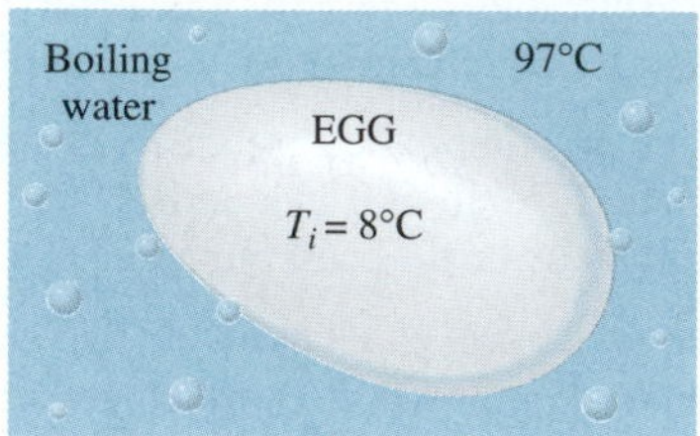

FIGURE P8–159

8–160 Long cylindrical steel rods (ρ = 7833 kg/m^3 and c_p = 0.465 kJ/kg · °C) of 10-cm diameter are heat treated by drawing them at a velocity of 3 m/min through a 7-m-long oven maintained at 900°C. If the rods enter the oven at 30°C and leave at 700°C, determine (*a*) the rate of heat transfer to the rods in the oven and (*b*) the rate of entropy generation associated with this heat transfer process.

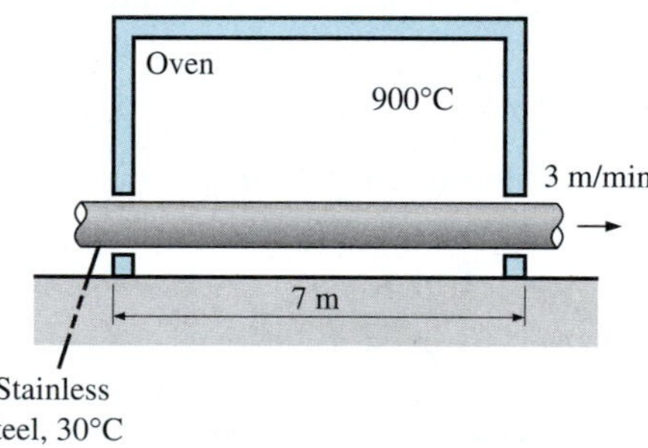

FIGURE P8–160

8–161 The inner and outer surfaces of a 5-m × 7-m brick wall of thickness 20 cm are maintained at temperatures of 20°C and 5°C, respectively. If the rate of heat transfer through the wall is 1890 W, determine the rate of entropy generation within the wall.

8–162 For heat transfer purposes, a standing man can be modeled as a 30-cm-diameter, 170-cm-long vertical cylinder with both the top and bottom surfaces insulated and with the side surface at an average temperature of 34°C. If the rate of heat loss from this man to the environment at 20°C is 336 W, determine the rate of entropy transfer from the body of this person accompanying heat transfer, in W/K.

8–163 A 1000-W iron is left on the ironing board with its base exposed to the air at 20°C. If the surface temperature is 400°C, determine the rate of entropy generation during this process in steady operation. How much of this entropy generation occurs within the iron?

8–164E A frictionless piston–cylinder device contains saturated liquid water at 25-psia pressure. Now 400 Btu of heat is transferred to water from a source at 900°F, and part of the liquid vaporizes at constant pressure. Determine the total entropy generated during this process, in Btu/R.

8–165E Steam enters a diffuser at 20 psia and 240°F with a velocity of 900 ft/s and exits as saturated vapor at 240°F and 100 ft/s. The exit area of the diffuser is 1 ft^2. Determine (*a*) the mass flow rate of the steam and (*b*) the rate of entropy generation during this process. Assume an ambient temperature of 77°F.

8–166 Steam expands in a turbine steadily at a rate of 25,000 kg/h, entering at 6 MPa and 450°C and leaving at 20 kPa as saturated vapor. If the power generated by the turbine is 4 MW, determine the rate of entropy generation for this process. Assume the surrounding medium is at 25°C. *Answer:* 11.0 kW/K

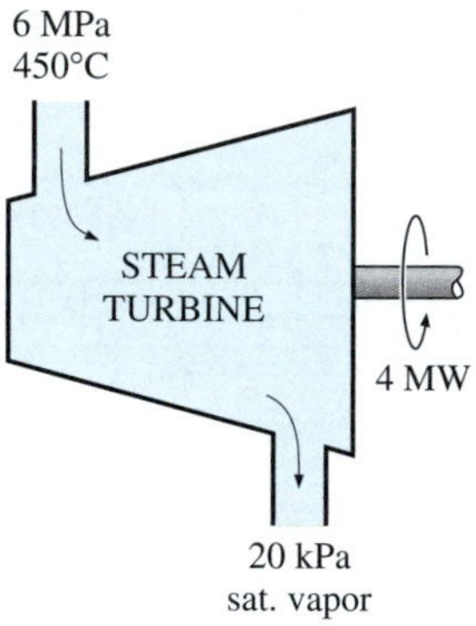

6 MPa
450°C
STEAM TURBINE
4 MW
20 kPa
sat. vapor

FIGURE P8–166

8–167 A hot-water stream at 70°C enters an adiabatic mixing chamber with a mass flow rate of 3.6 kg/s, where it is mixed with a stream of cold water at 20°C. If the mixture leaves the chamber at 42°C, determine (*a*) the mass flow rate of the cold water and (*b*) the rate of entropy generation during this adiabatic mixing process. Assume all the streams are at a pressure of 200 kPa.

8–168 Liquid water at 200 kPa and 20°C is heated in a chamber by mixing it with superheated steam at 200 kPa and 150°C. Liquid water enters the mixing chamber at a rate of 2.5 kg/s, and the chamber is estimated to lose heat to the surrounding air at 25°C at a rate of 1200 kJ/min. If the mixture leaves the mixing chamber at 200 kPa and 60°C, determine (*a*) the mass flow rate of the superheated steam and (*b*) the rate of entropy generation during this mixing process. *Answers:* (*a*) 0.166 kg/s, (*b*) 0.333 kW/K

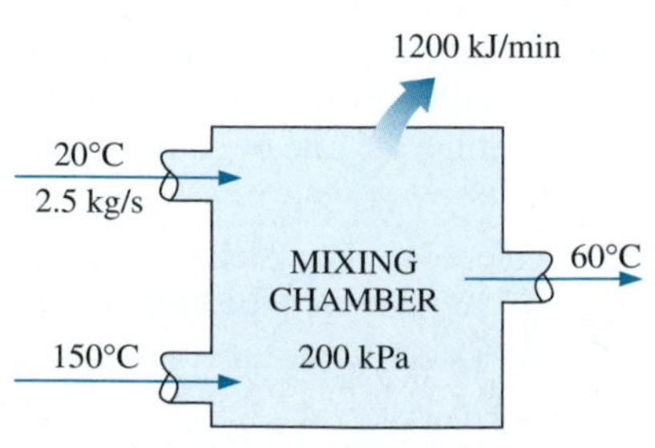

FIGURE P8–168

8–169 A 0.3-m^3 rigid tank is filled with saturated liquid water at 150°C. A valve at the bottom of the tank is now opened, and one-half of the total mass is withdrawn from the tank in the liquid form. Heat is transferred to water from a source at 200°C so that the temperature in the tank remains constant. Determine (*a*) the amount of heat transfer and (*b*) the total entropy generation for this process.

8–170E An iron block of unknown mass at 185°F is dropped into an insulated tank that contains 0.8 ft^3 of water at 70°F. At the same time, a paddle wheel driven by a 200-W motor is activated to stir the water. Thermal equilibrium is established after 10 min with a final temperature of 75°F. Determine (*a*) the mass of the iron block and (*b*) the entropy generated during this process.

8–171E Air enters a compressor at ambient conditions of 15 psia and 60°F with a low velocity and exits at 150 psia, 620°F, and 350 ft/s. The compressor is cooled by the ambient air at 60°F at a rate of 1500 Btu/min. The power input to the compressor is 400 hp. Determine (*a*) the mass flow rate of air and (*b*) the rate of entropy generation.

8–172 Steam enters an adiabatic nozzle at 4 MPa and 450°C with a velocity of 70 m/s and exits at 3 MPa and 320 m/s. If the nozzle has an inlet area of 7 cm^2, determine (*a*) the exit temperature and (*b*) the rate of entropy generation for this process. *Answers:* (*a*) 422.3°C, (*b*) 0.0361 kW/K

Review Problems

8–173E A heat engine whose thermal efficiency is 40 percent uses a hot reservoir at 1300 R and a cold reservoir at 500 R. Calculate the entropy change of the two reservoirs when 1 Btu of heat is transferred from the hot reservoir to the engine. Does this engine satisfy the increase of entropy principle? If the thermal efficiency of the heat engine is increased to 70 percent, will the increase of entropy principle still be satisfied?

8–174 A refrigerator with a coefficient of performance of 4 transfers heat from a cold region at –20ºC to a hot region at 30ºC. Calculate the total entropy change of the regions when 1 kJ of heat is transferred from the cold region. Is the second law satisfied? Will this refrigerator still satisfy the second law if its coefficient of performance is 6?

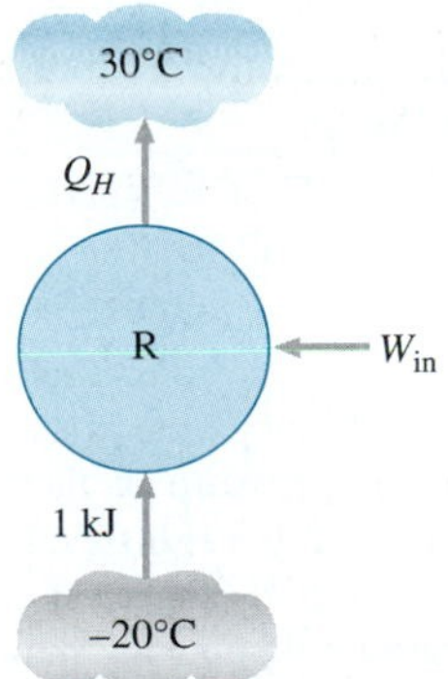

FIGURE P8–174

8–175 A proposed heat pump design creates a heating effect of 25 kW while using 5 kW of electrical power. The thermal energy reservoirs are at 300 K and 260 K. Is this possible according to the increase of entropy principle?

8–176 What is the minimum internal energy that steam can achieve as it is expanded adiabatically in a closed system from 1500 kPa and 320ºC to 100 kPa?

8–177E Is it possible to expand water at 30 psia and 70 percent quality to 10 psia in a closed system undergoing an isothermal, reversible process while exchanging heat with an energy reservoir at 300ºF?

8–178E Can 2 lbm of air in a closed system at 16 psia and 100ºF be compressed adiabatically to 100 psia and a volume of 4 ft^3?

8–179 What is the maximum volume that 3 kg of oxygen at 950 kPa and 373ºC can be adiabatically expanded to in a piston–cylinder device if the final pressure is to be 100 kPa? *Answer:* 2.66 m^3

8–180E A 100-lbm block of a solid material whose specific heat is 0.5 Btu/lbm·R is at 70ºF. It is heated with 10 lbm of saturated water vapor that has a constant pressure of 14.7 psia. Determine the final temperature of the block and water, and the entropy change of (*a*) the block, (*b*) the water, and (*c*) the entire system. Is this process possible? Why?

8–181 One kilogram of air is in a piston–cylinder apparatus that can only exchange heat with a reservoir at 300 K. Initially this air is at 100 kPa and 27ºC. Someone claims that the air can be compressed to 250 kPa and 27ºC. Determine if this claim is valid by performing a second-law analysis of the claimed process.

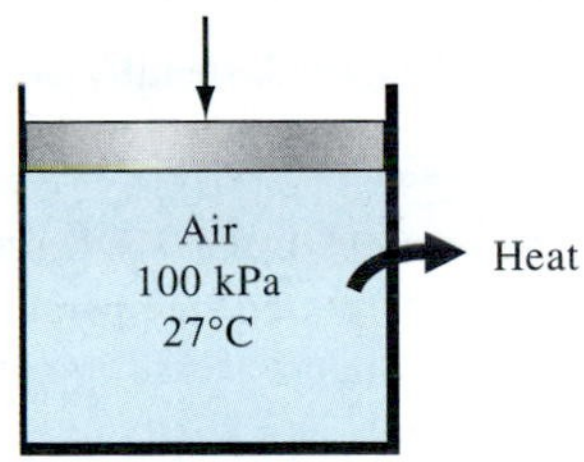

FIGURE P8–181

8–182 A rigid tank contains 1.5 kg of water at 120°C and 500 kPa. Now 22 kJ of shaft work is done on the system and the final temperature in the tank is 95°C. If the entropy change of water is zero and the surroundings are at 15°C, determine (*a*) the final pressure in the tank, (*b*) the amount of heat transfer between the tank and the surroundings, and (*c*) the entropy generation during this process. *Answers:* (*a*) 84.6 kPa, (*b*) 38.5 kJ, (*c*) 0.134 kJ/K

8–183 A horizontal cylinder is separated into two compartments by an adiabatic, frictionless piston. One side contains 0.2 m^3 of nitrogen and the other side contains 0.1 kg of helium, both initially at 20°C and 95 kPa. The sides of the cylinder and the helium end are insulated. Now heat is added to the nitrogen side from a reservoir at 500°C until the pressure of the helium rises to 120 kPa. Determine (*a*) the final temperature of the helium, (*b*) the final volume of the nitrogen, (*c*) the heat transferred to the nitrogen, and (*d*) the entropy generation during this process.

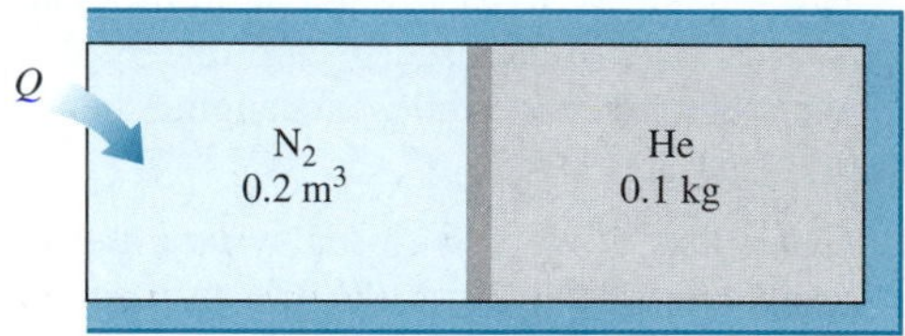

FIGURE P8–183

8–184 A 0.8-m³ rigid tank contains carbon dioxide (CO_2) gas at 250 K and 100 kPa. A 500-W electric resistance heater placed in the tank is now turned on and kept on for 40 min after which the pressure of CO_2 is measured to be 175 kPa. Assuming the surroundings to be at 300 K and using constant specific heats, determine (*a*) the final temperature of CO_2, (*b*) the net amount of heat transfer from the tank, and (*c*) the entropy generation during this process.

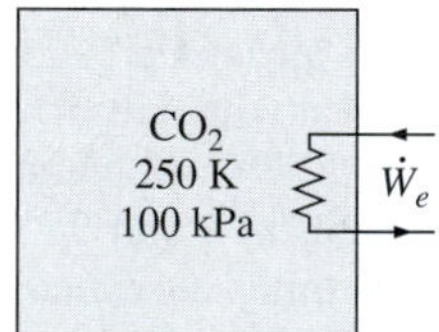

FIGURE P8–184

8–185 Helium gas is throttled steadily from 500 kPa and 70°C. Heat is lost from the helium in the amount of 2.5 kJ/kg to the surroundings at 25°C and 100 kPa. If the entropy of the helium increases by 0.25 kJ/kg · K in the valve, determine (*a*) the exit pressure and temperature and (*b*) the entropy generation during this process. *Answers:* (*a*) 442 kPa, 69.5°C, (*b*) 0.258 kJ/kg · K

8–186 Refrigerant-134a enters a compressor as a saturated vapor at 200 kPa at a rate of 0.03 m³/s and leaves at 700 kPa. The power input to the compressor is 10 kW. If the surroundings at 20°C experience an entropy increase of 0.008 kW/K, determine (*a*) the rate of heat loss from the compressor, (*b*) the exit temperature of the refrigerant, and (*c*) the rate of entropy generation.

8–187 Air at 500 kPa and 400 K enters an adiabatic nozzle at a velocity of 30 m/s and leaves at 300 kPa and 350 K. Using variable specific heats, determine (*a*) the isentropic efficiency, (*b*) the exit velocity, and (*c*) the entropy generation.

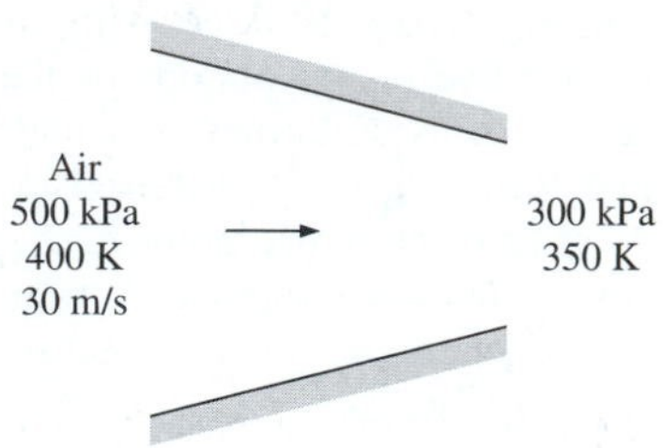

FIGURE P8–187

8–188 Show that the difference between the reversible steady-flow work and reversible moving boundary work is equal to the flow energy.

8–189 An insulated tank containing 0.4 m³ of saturated water vapor at 500 kPa is connected to an initially evacuated, insulated piston–cylinder device. The mass of the piston is such that a pressure of 150 kPa is required to raise it. Now the valve is opened slightly, and part of the steam flows to the cylinder, raising the piston. This process continues until the pressure in the tank drops to 150 kPa. Assuming the steam that remains in the tank to have undergone a reversible adiabatic process, determine the final temperature (*a*) in the rigid tank and (*b*) in the cylinder.

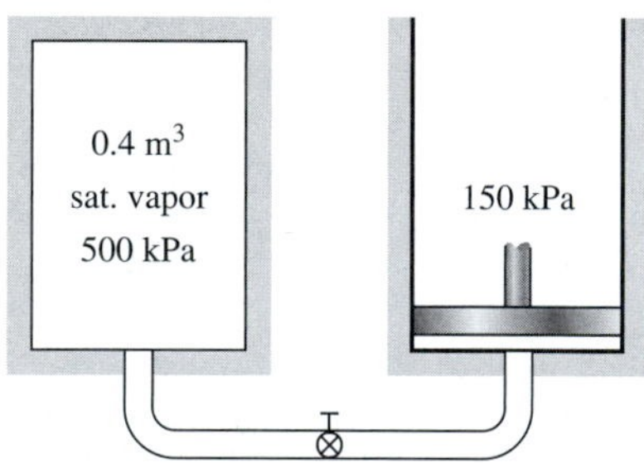

FIGURE P8–189

8–190 Carbon dioxide is compressed in a reversible, isothermal process from 100 kPa and 20°C to 400 kPa using a steady-flow device with one inlet and one outlet. Determine the work required and the heat transfer, both in kJ/kg, for this compression.

8–191 Reconsider Prob. 8–190. Determine the change in the work and heat transfer when the compression process is isentropic rather than isothermal.

8–192 The compressor of a refrigerator compresses saturated R-134a vapor at -10°C to 800 kPa. How much work, in kJ/kg, does this process require when the process is isentropic?

8–193 An adiabatic capillary tube is used in some refrigeration systems to drop the pressure of the refrigerant from the condenser level to the evaporator level. R-134a enters the capillary tube as a saturated liquid at 50°C, and leaves at –12°C. Determine the rate of entropy generation in the capillary tube for a mass flow rate of 0.2 kg/s. *Answer:* 0.0077 kW/K

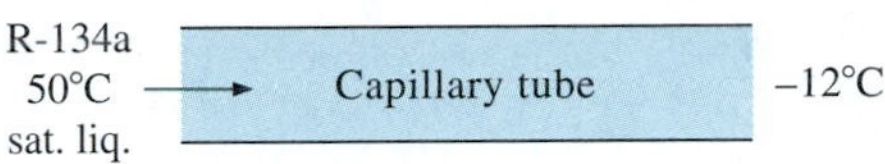

FIGURE P8–193

8–194 A steam turbine is equipped to bleed 6 percent of the inlet steam for feedwater heating. It is operated with 4 MPa and 350°C steam at the inlet, a bleed pressure of 800 kPa, and an exhaust pressure of 30 kPa. Calculate the work produced by this turbine when the isentropic efficiency between the inlet and bleed point is 97 percent and the isentropic efficiency between the bleed point and exhaust is 95 percent. What is the overall isentropic efficiency of the turbine? *Hint:* Treat this turbine as two separate turbines, with one operating between the inlet and bleed conditions and the other operating between the bleed and exhaust conditions.

8–195 Air is compressed steadily by a compressor from 100 kPa and 17°C to 700 kPa at a rate of 5 kg/min. Determine the minimum power input required if the process is (*a*) adiabatic and (*b*) isothermal. Assume air to be an ideal gas with variable specific heats, and neglect the changes in kinetic and potential energies. *Answers:* (*a*) 18.0 kW, (*b*) 13.5 kW

8–196 Refrigerant-134a at 140 kPa and −10°C is compressed by an adiabatic 0.7-kW compressor to an exit state of 700 kPa and 50°C. Neglecting the changes in kinetic and potential energies, determine (*a*) the isentropic efficiency of the compressor, (*b*) the volume flow rate of the refrigerant at the compressor inlet, in L/min, and (*c*) the maximum volume flow rate at the inlet conditions that this adiabatic 0.7-kW compressor can handle without violating the second law.

8–197E Helium gas enters a nozzle whose isentropic efficiency is 94 percent with a low velocity, and it exits at 14 psia, 180°F, and 1000 ft/s. Determine the pressure and temperature at the nozzle inlet.

Design and Essay Problems

8–198 It is well-known that the temperature of a gas rises while it is compressed as a result of the energy input in the form of compression work. At high compression ratios, the air temperature may rise above the autoignition temperature of some hydrocarbons, including some lubricating oil. Therefore, the presence of some lubricating oil vapor in high-pressure air raises the possibility of an explosion, creating a fire hazard. The concentration of the oil within the compressor is usually too low to create a real danger. However, the oil that collects on the inner walls of exhaust piping of the compressor may cause an explosion. Such explosions have largely been eliminated by using the proper lubricating oils, carefully designing the equipment, intercooling between compressor stages, and keeping the system clean.

A compressor is to be designed for an industrial application in Los Angeles. If the compressor exit temperature is not to exceed 250°C for safety consideration, determine the maximum allowable compression ratio that is safe for all possible weather conditions for that area.

8–199 Obtain the following information about a power plant that is closest to your town: the net power output; the type and amount of fuel; the power consumed by the pumps, fans, and other auxiliary equipment; stack gas losses; temperatures at several locations; and the rate of heat rejection at the condenser. Using these and other relevant data, determine the rate of entropy generation in that power plant.

8–200 Enthalpy-entropy state diagrams (also known as *Mollier diagrams*) are more useful than temperature-entropy state diagrams for analyzing some systems. Using EES (or other) software and actual property data, plot an enthalpy-entropy diagram for water that includes the saturation lines. Also, sketch isothermal, isobaric, and constant volume processes on this diagram.

8–201 You are given the task of compressing a fixed quantity of gas from P_1, T_1 to P_2, T_1. You wish to do this using a polytropic process coupled with a constant pressure P_2 heat-transfer process. Determine the polytropic process that minimizes the entropy generation.

possibility of S transfer

Heat — Yes (closed & open)
- Heat transfer to a system ↗ entropy
- " " from " ↘ "

Work — No (closed & open)
- No entropy is transferred to a system by work
- Entropy only generated within a system as work is dissipated as heat

Mass — Yes (open systems only)
- Mass contains entropy & energy
- Entropy carried into or out of system by mass flow

Maxwell's demon:

- Initially same T for both chambers
- Opens frictionless door for fast molecules
- Heat flows without temp diff (from colder to hot)
- Trapdoor jiggles uncontrollably, opens randomly (Brownian motion)
- Evolution

$Tds = dh - vdP \rightarrow \Delta s = c_v \ln \frac{T_2}{T_1} + R \ln \frac{V_2}{V_1}$

$Tds = du + PdV \rightarrow \Delta s = c_p \ln \frac{T_2}{T_1} - R \ln \frac{P_2}{P_1}$

cond:

$$\dot{Q} = -kA \frac{(T_2 - T_1)}{L} \qquad \alpha = \frac{k}{\rho c_p}$$

conv.:

$$\dot{Q}_{conv} = hA_s (T_s - T_\infty)$$

$$\dot{Q} = \frac{T_H - T_C}{\Sigma R}$$

PART

2

FLUID MECHANICS

Chapter 9

INTRODUCTION TO FLUID MECHANICS

In the second part of the text we present the fundamentals of fluid mechanics. In this introductory chapter, we introduce the basic concepts commonly used in the analysis of fluid flow to avoid any misunderstandings. We start with a discussion of the no-slip condition, which is responsible for the development of boundary layers adjacent to the solid surfaces, and the no-temperature-jump condition. We continue with a discussion of the numerous ways of classification of fluid flow, such as *viscous versus inviscid regions of flow, internal versus external flow, compressible versus incompressible flow, laminar versus turbulent flow, natural versus forced flow,* and *steady versus unsteady flow,* followed by a brief history of the development of fluid mechanics. Then we discuss the properties *vapor pressure, speed of sound*, and *viscosity*, which play a dominant role in most aspects of fluid flow. Finally, we present the property *surface tension*, and determine the *capillary rise* from static equilibrium conditions.

Objectives

The objectives of this chapter are to:

- Understand the basic concepts of Fluid Mechanics
- Recognize the various types of fluid flow problems encountered in practice.
- Learn the history of Fluid Mechanics
- Understand the vapor pressure and its role in the occurrence of cavitation.
- Have a working knowledge of viscosity and the consequences of the frictional effects it causes in fluid flow.
- Calculate the capillary rises and drops due to the surface tension effects.

9–1 ▪ THE NO-SLIP CONDITION

Fluid flow is often confined by solid surfaces, and it is important to understand how the presence of solid surfaces affects fluid flow. We know that water in a river cannot flow through large rocks, and must go around them. That is, the water velocity normal to the rock surface must be zero, and water approaching the surface normally comes to a complete stop at the surface. What is not as obvious is that water approaching the rock at any angle also comes to a complete stop at the rock surface, and thus the tangential velocity of water at the surface is also zero.

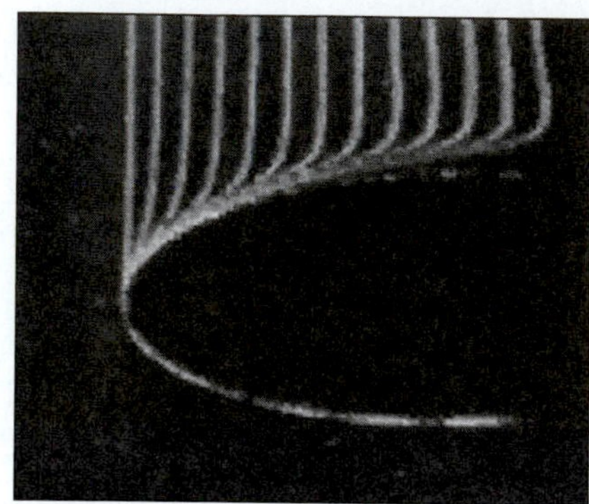

FIGURE 9–1

The development of a velocity profile due to the no-slip condition as a fluid flows over a blunt nose.

"Hunter Rouse: Laminar and Turbulent Flow Film." Copyright IIHR-Hydroscience & Engineering, The University of Iowa. Used by permission.

Consider the flow of a fluid in a stationary pipe or over a solid surface that is nonporous (i.e., impermeable to the fluid). All experimental observations indicate that a fluid in motion comes to a complete stop at the surface and assumes a zero velocity relative to the surface. That is, a fluid in direct contact with a solid "sticks" to the surface, and there is no slip. This is known as the **no-slip condition**. The fluid property responsible for the no-slip condition and the development of the boundary layer is *viscosity* and is discussed in Section 9–6.

The photograph in Fig. 9–1 clearly shows the evolution of a velocity gradient as a result of the fluid sticking to the surface of a blunt nose. The layer that sticks to the surface slows the adjacent fluid layer because of viscous forces between the fluid layers, which slows the next layer, and so on. A consequence of the no-slip condition is that all velocity profiles must have zero values with respect to the surface at the points of contact between a fluid and a solid surface (Fig. 9–2). Therefore, the no-slip condition is responsible for the development of the velocity profile. The flow region adjacent to the wall in which the viscous effects (and thus the velocity gradients) are significant is called the **boundary layer**. Another consequence of the no-slip condition is the *surface drag*, or *skin friction drag*, which is the force a fluid exerts on a surface in the flow direction.

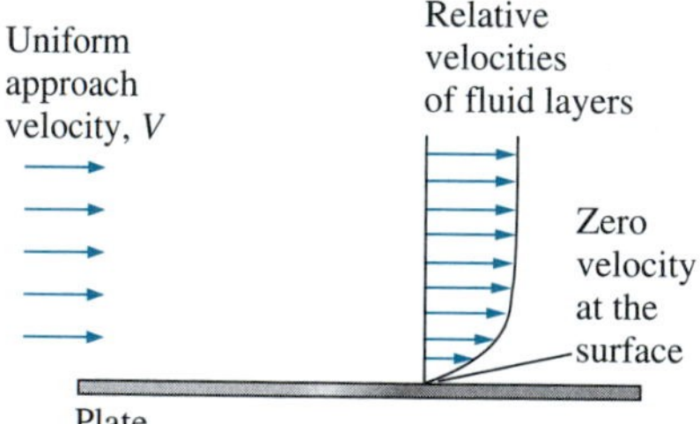

FIGURE 9–2

A fluid flowing over a stationary surface comes to a complete stop at the surface because of the no-slip condition.

When a fluid is forced to flow over a curved surface, such as the back side of a cylinder, the boundary layer may no longer remain attached to the surface and separates from the surface—a process called **flow separation** (Fig. 9–3). We emphasize that the no-slip condition applies *everywhere* along the surface, even downstream of the separation point.

A phenomenon similar to the no-slip condition occurs in heat transfer. When two bodies at different temperatures are brought into contact, heat transfer occurs such that both bodies assume the same temperature at the points of contact. Therefore, a fluid and a solid surface have the same

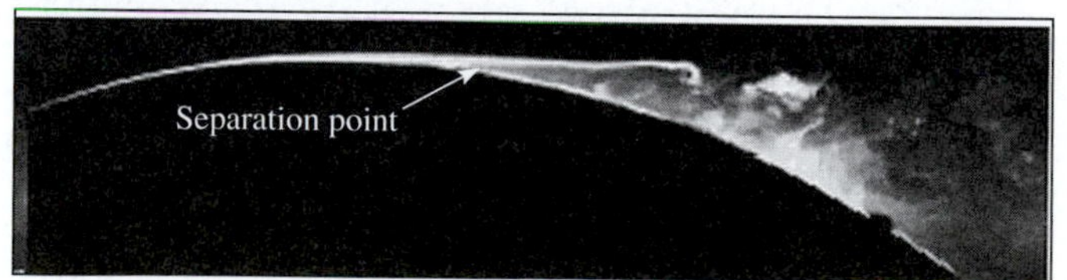

FIGURE 9–3

Flow separation during flow over a curved surface.

From G. M. Homsy et al, "Multi-Media Fluid Mechanics," Cambridge Univ. Press (2001). ISBN 0-521-78748-3. Reprinted by permission.

temperature at the points of contact. This is known as **no-temperature-jump condition**.

9–2 ▪ CLASSIFICATION OF FLUID FLOWS

Earlier we defined *fluid mechanics* as the science that deals with the behavior of fluids at rest or in motion, and the interaction of fluids with solids or other fluids at the boundaries. There is a wide variety of fluid flow problems encountered in practice, and it is usually convenient to classify them on the basis of some common characteristics to make it feasible to study them in groups. There are many ways to classify fluid flow problems, and here we present some general categories.

Viscous versus Inviscid Regions of Flow

When two fluid layers move relative to each other, a friction force develops between them and the slower layer tries to slow down the faster layer. This internal resistance to flow is quantified by the fluid property *viscosity*, which is a measure of internal stickiness of the fluid. Viscosity is caused by cohesive forces between the molecules in liquids and by molecular collisions in gases. There is no fluid with zero viscosity, and thus all fluid flows involve viscous effects to some degree. Flows in which the frictional effects are significant are called **viscous flows**. In many flows of practical interest, there are *regions* (typically regions not close to solid surfaces) where viscous forces are negligibly small compared to inertial or pressure forces. Neglecting the viscous terms in such **inviscid flow regions** greatly simplifies the analysis without much loss in accuracy.

The development of viscous and inviscid regions of flow as a result of inserting a flat plate parallel into a fluid stream of uniform velocity is shown in Fig. 9–4. The fluid sticks to the plate on both sides because of the no-slip condition, and the thin boundary layer in which the viscous effects are significant near the plate surface is the *viscous flow region*. The region of flow on both sides away from the plate and unaffected by the presence of the plate is the *inviscid flow region*.

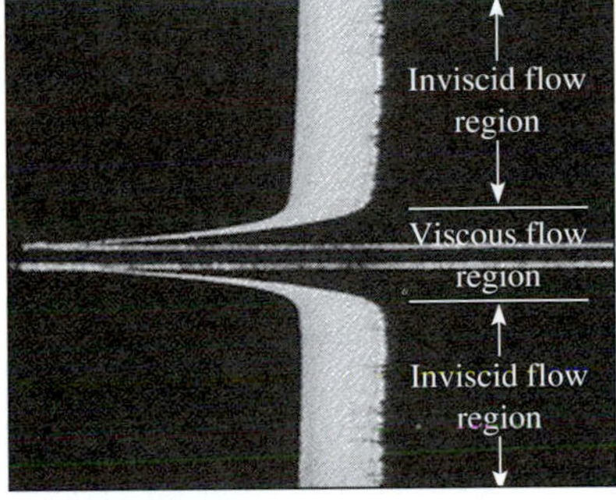

FIGURE 9–4
The flow of an originally uniform fluid stream over a flat plate, and the regions of viscous flow (next to the plate on both sides) and inviscid flow (away from the plate).
Fundamentals of Boundary Layers, National Committee from Fluid Mechanics Films, © Education Development Center.

Internal versus External Flow

A fluid flow is classified as being internal or external, depending on whether the fluid flows in a confined space or over a surface. The flow of an unbounded fluid over a surface such as a plate, a wire, or a pipe is **external flow**. The flow in a pipe or duct is **internal flow** if the fluid is completely bounded by solid surfaces. Water flow in a pipe, for example, is internal flow, and airflow over a ball is external flow (Fig. 9–5). The flow of liquids in a duct is called *open-channel flow* if the duct is only partially filled with the liquid and there is a free surface. The flows of water in rivers and irrigation ditches are examples of such flows.

Internal flows are dominated by the influence of viscosity throughout the flow field. In external flows the viscous effects are limited to boundary layers near solid surfaces and to wake regions downstream of bodies.

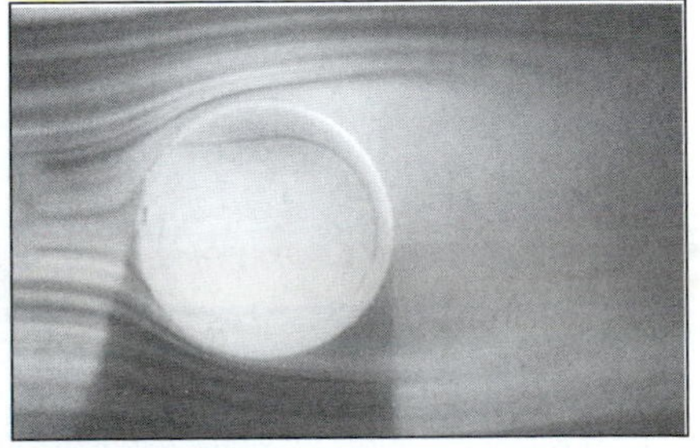

FIGURE 9–5
External flow over a tennis ball, and the turbulent wake region behind.
Courtesy NASA and Cislunar Aerospace, Inc.

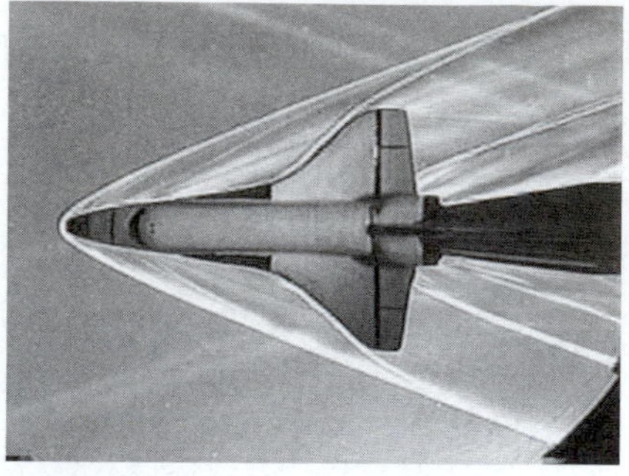

FIGURE 9–6

Schlieren image of a small model of the space shuttle orbiter being tested at Mach 3 in the supersonic wind tunnel of the Penn State Gas Dynamics Lab. Several *oblique shocks* are seen in the air surrounding the spacecraft.

Photo by G. S. Settles, Penn State University. Used by permission.

Compressible versus Incompressible Flow

A flow is classified as being *compressible* or *incompressible*, depending on the level of variation of density during flow. Incompressibility is an approximation in which a flow is said to be **incompressible** if the density remains nearly constant throughout. Therefore, the volume of every portion of fluid remains unchanged over the course of its motion when the flow is incompressible.

The densities of liquids are essentially constant, and thus the flow of liquids is typically incompressible. Therefore, liquids are usually referred to as *incompressible substances*. A pressure of 210 atmospheres (atm), for example, causes the density of liquid water at 1 atm to change by just 1 percent. Gases, on the other hand, are highly compressible. A pressure change of just 0.01 atm, for example, causes a change of 1 percent in the density of atmospheric air.

When analyzing rockets, spacecraft, and other systems that involve high-speed gas flows (Fig. 9–6), the flow speed is often expressed in terms of the dimensionless **Mach number** defined as

$$\text{Ma} = \frac{V}{c} = \frac{\text{Speed of flow}}{\text{Speed of sound}}$$

where c is the **speed of sound** whose value is 346 m/s in air at room temperature at sea level. A flow is called **sonic** when Ma = 1, **subsonic** when Ma < 1, **supersonic** when Ma > 1, and **hypersonic** when Ma >> 1. The Mach number is *dimensionless*—the dimensions of the numerator and those of the denominator cancel each other out.

Liquid flows are incompressible to a high level of accuracy, but the level of variation in density in gas flows and the consequent level of approximation made when modeling gas flows as incompressible depends on the Mach number. Gas flows can often be approximated as incompressible if the density changes are under about 5 percent, which is usually the case when Ma < 0.3. Therefore, the compressibility effects of air at room temperature can be neglected at speeds under about 100 m/s.

Small density changes of liquids corresponding to large pressure changes can still have important consequences. The irritating "water hammer" in a water pipe, for example, is caused by the vibrations of the pipe generated by the reflection of pressure waves following the sudden closing of the valves.

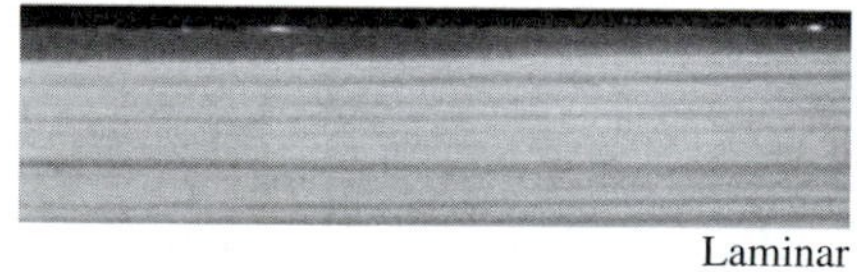

Laminar

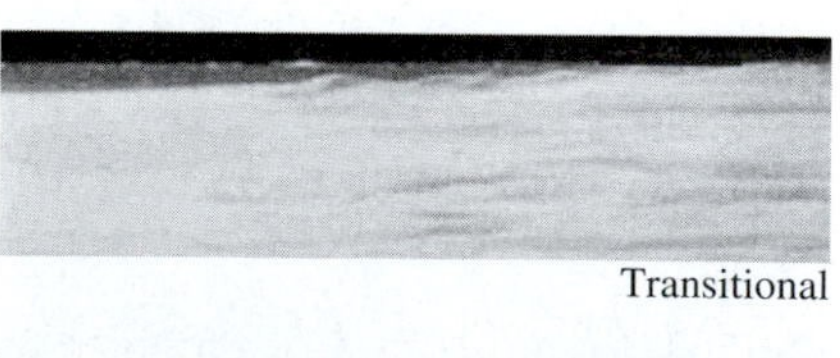

Transitional

Turbulent

FIGURE 9–7

Laminar, transitional, and turbulent flows.

Courtesy ONERA, photograph by Werlé.

Laminar versus Turbulent Flow

Some flows are smooth and orderly while others are rather chaotic. The highly ordered fluid motion characterized by smooth layers of fluid is called **laminar**. The word *laminar* comes from the movement of adjacent fluid particles together in "laminae." The flow of high-viscosity fluids such as oils at low velocities is typically laminar. The highly disordered fluid motion that typically occurs at high velocities and is characterized by velocity fluctuations is called **turbulent** (Fig. 9–7). The flow of low-viscosity fluids such as air at high velocities is typically turbulent. A flow that alternates between being laminar and turbulent is called **transitional**. The experiments conducted by Osborne Reynolds in the 1880s resulted in the establishment of the dimensionless **Reynolds number**, **Re**, as the key parameter for the determination of the flow regime in pipes.

Natural (or Unforced) versus Forced Flow

A fluid flow is said to be natural or forced, depending on how the fluid motion is initiated. In **forced flow**, a fluid is forced to flow over a surface or in a pipe by external means such as a pump or a fan. In **natural flows**, fluid motion is due to natural means such as the buoyancy effect, which manifests itself as the rise of warmer (and thus lighter) fluid and the fall of cooler (and thus denser) fluid (Fig. 9–8). In solar hot-water systems, for example, the thermosiphoning effect is commonly used to replace pumps by placing the water tank sufficiently above the solar collectors.

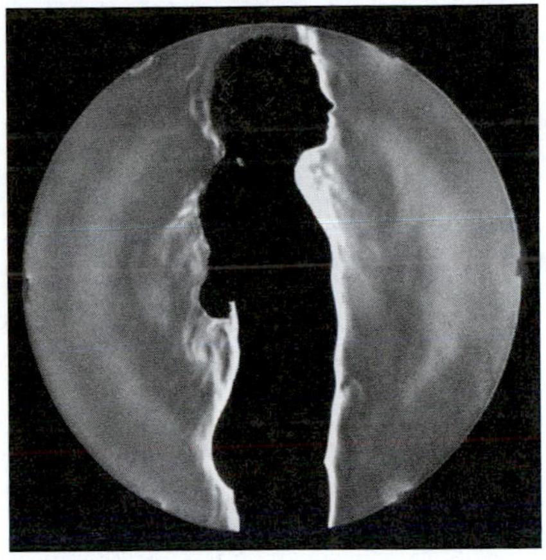

FIGURE 9–8

In this schlieren image of a girl in a swimming suit, the rise of lighter, warmer air adjacent to her body indicates that humans and warm-blooded animals are surrounded by thermal plumes of rising warm air.

G. S. Settles, Gas Dynamics Lab, Penn State University. Used by permission.

Steady versus Unsteady Flow

The terms *steady* and *uniform* are used frequently in engineering, and thus it is important to have a clear understanding of their meanings. The term **steady** implies *no change at a point with time*. The opposite of steady is **unsteady**. The term **uniform** implies *no change with location* over a specified region. These meanings are consistent with their everyday use (steady girlfriend, uniform distribution, etc.).

The terms *unsteady* and *transient* are often used interchangeably, but these terms are not synonyms. In fluid mechanics, *unsteady* is the most general term that applies to any flow that is not steady, but **transient** is typically used for developing flows. When a rocket engine is fired up, for example, there are transient effects (the pressure builds up inside the rocket engine, the flow accelerates, etc.) until the engine settles down and operates steadily. The term **periodic** refers to the kind of unsteady flow in which the flow oscillates about a steady mean.

Many devices such as turbines, compressors, boilers, condensers, and heat exchangers operate for long periods of time under the same conditions, and they are classified as *steady-flow devices*. (Note that the flow field near the rotating blades of a turbomachine is of course unsteady, but we consider the overall flow field rather than the details at some localities when we classify devices.) During steady flow, the fluid properties can change from point to point within a device, but at any fixed point they remain constant. Therefore, the volume, the mass, and the total energy content of a steady-flow device or flow section remain constant in steady operation.

Steady-flow conditions can be closely approximated by devices that are intended for continuous operation such as turbines, pumps, boilers, condensers, and heat exchangers of power plants or refrigeration systems. Some cyclic devices, such as reciprocating engines or compressors, do not satisfy the steady-flow conditions since the flow at the inlets and the exits is pulsating and not steady. However, the fluid properties vary with time in a periodic manner, and the flow through these devices can still be analyzed as a steady-flow process by using time-averaged values for the properties.

Some fascinating visualizations of fluid flow are provided in the book *An Album of Fluid Motion* by Milton Van Dyke (1982). A nice illustration of an unsteady-flow field is shown in Fig. 9–9, taken from Van Dyke's book. Figure 9–9*a* is an instantaneous snapshot from a high-speed motion picture; it reveals large, alternating, swirling, turbulent eddies that are shed into the

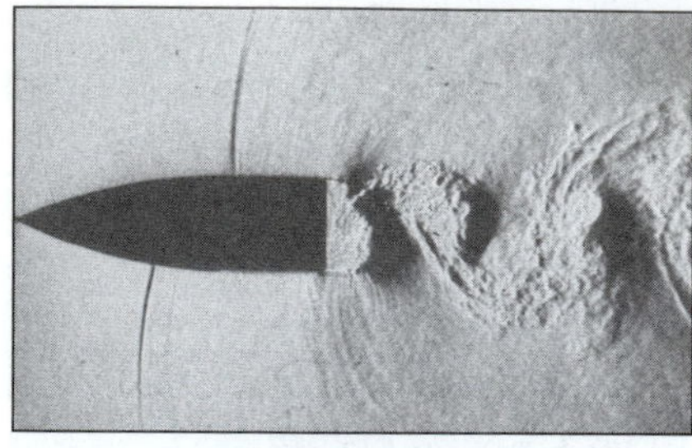

(*a*)

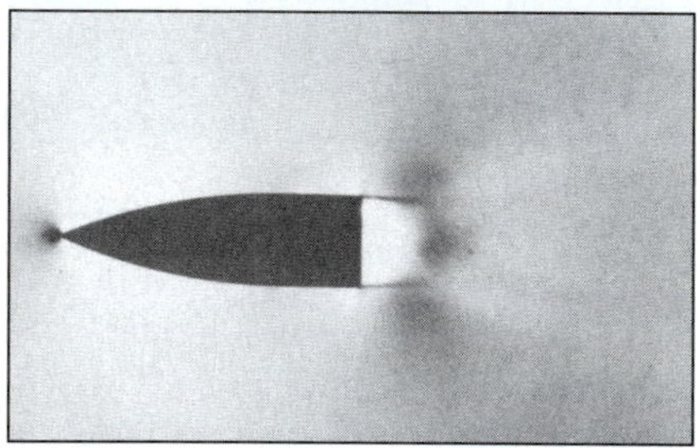

(*b*)

FIGURE 9–9

Oscillating wake of a blunt-based airfoil at Mach number 0.6. Photo (*a*) is an instantaneous image, while photo (*b*) is a long-exposure (time-averaged) image.

(*a*) *Dyment, A., Flodrops, J. P. & Gryson, P. 1982 in* Flow Visualization II, *W. Merzkirch, ed., 331–336. Washington:* Hemisphere. *Used by permission of Arthur Dyment.*

(*b*) *Dyment, A. & Gryson, P. 1978 in* Inst. Mèc. Fluides Lille, *No. 78-5. Used by permission of Arthur Dyment.*

periodically oscillating wake from the blunt base of the object. The eddies produce shock waves that move upstream alternately over the top and bottom surfaces of the airfoil in an unsteady fashion. Figure 9–9*b* shows the *same* flow field, but the film is exposed for a longer time so that the image is time averaged over 12 cycles. The resulting time-averaged flow field appears "steady" since the details of the unsteady oscillations have been lost in the long exposure.

One of the most important jobs of an engineer is to determine whether it is sufficient to study only the time-averaged "steady" flow features of a problem, or whether a more detailed study of the unsteady features is required. If the engineer were interested only in the overall properties of the flow field (such as the time-averaged drag coefficient, the mean velocity, and mean pressure fields), a time-averaged description like that of Fig. 9–9*b*, a time-averaged experimental measurement, or an analytical or numerical calculation of the time-averaged flow field would be sufficient. However, if the engineer were interested in details about the unsteady-flow field, such as flow-induced vibrations, unsteady pressure fluctuations, or the sound waves emitted from the turbulent eddies or the shock waves, a time-averaged description of the flow field would be insufficient.

Most of the analytical and computational examples provided in this textbook deal with steady or time-averaged flows, although we occasionally point out some relevant unsteady-flow features as well when appropriate.

One-, Two-, and Three-Dimensional Flows

A flow field is best characterized by its velocity distribution, and thus a flow is said to be one-, two-, or three-dimensional if the flow velocity varies in one, two, or three primary dimensions, respectively. A typical fluid flow involves a three-dimensional geometry, and the velocity may vary in all three dimensions, rendering the flow three-dimensional [$\vec{V}$ (x, y, z) in rectangular or $\vec{V}$ (r, θ, z) in cylindrical coordinates]. However, the variation of velocity in certain directions can be small relative to the variation in other directions and can be ignored with negligible error. In such cases, the flow can be modeled conveniently as being one- or two-dimensional, which is easier to analyze.

Consider steady flow of a fluid entering from a large tank into a circular pipe. The fluid velocity everywhere on the pipe surface is zero because of the no-slip condition, and the flow is two-dimensional in the entrance region of the pipe since the velocity changes in both the r- and z-directions, but not in the θ-direction. The velocity profile develops fully and remains unchanged after some distance from the inlet (about 10 pipe diameters in turbulent flow, and less in laminar pipe flow, as in Fig. 9–10), and the flow in this region is said to be *fully developed*. The fully developed flow in a circular pipe is *one-dimensional* since the velocity varies in the radial r-direction but not in the angular θ- or axial z-directions, as shown in Fig. 9–10. That is, the velocity profile is the same at any axial z-location, and it is symmetric about the axis of the pipe.

Note that the dimensionality of the flow also depends on the choice of coordinate system and its orientation. The pipe flow discussed, for example, is one-dimensional in cylindrical coordinates, but two-dimensional in Cartesian coordinates—illustrating the importance of choosing the most

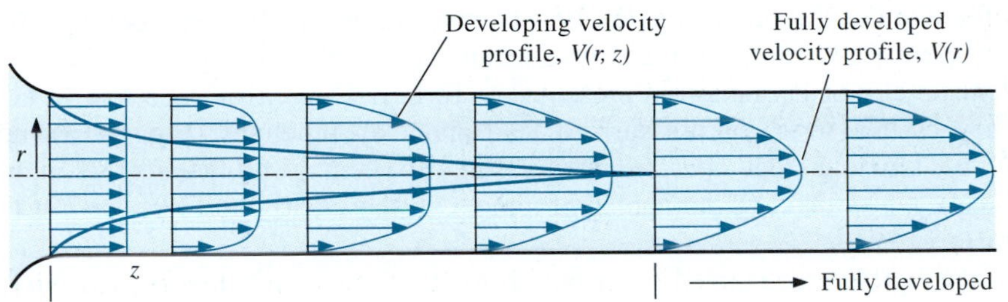

FIGURE 9–10

The development of the velocity profile in a circular pipe. $V = V(r, z)$ and thus the flow is two-dimensional in the entrance region, and becomes one-dimensional downstream when the velocity profile fully develops and remains unchanged in the flow direction, $V = V(r)$.

appropriate coordinate system. Also note that even in this simple flow, the velocity cannot be uniform across the cross section of the pipe because of the no-slip condition. However, at a well-rounded entrance to the pipe, the velocity profile may be approximated as being nearly uniform across the pipe, since the velocity is nearly constant at all radii except very close to the pipe wall.

A flow may be approximated as *two-dimensional* when the aspect ratio is large and the flow does not change appreciably along the longer dimension. For example, the flow of air over a car antenna can be considered two-dimensional except near its ends since the antenna's length is much greater than its diameter, and the airflow hitting the antenna is fairly uniform (Fig. 9–11).

FIGURE 9–11

Flow over a car antenna is approximately two-dimensional except near the top and bottom of the antenna.

EXAMPLE 9–1 Axisymmetric Flow over a Bullet

Consider a bullet piercing through calm air. Determine if the time-averaged airflow over the bullet during its flight is one-, two-, or three-dimensional (Fig. 9–12).

Solution It is to be determined whether airflow over a bullet is one-, two-, or three-dimensional.

Assumptions There are no significant winds and the bullet is not spinning.

Analysis The bullet possesses an axis of symmetry and is therefore an axisymmetric body. The airflow upstream of the bullet is parallel to this axis, and we expect the time-averaged airflow to be rotationally symmetric about the axis—such flows are said to be axisymmetric. The velocity in this case varies with axial distance z and radial distance r, but not with angle θ. Therefore, the time-averaged airflow over the bullet is **two-dimensional**.

Discussion While the time-averaged airflow is axisymmetric, the *instantaneous* airflow is not, as illustrated in Fig. 9–9.

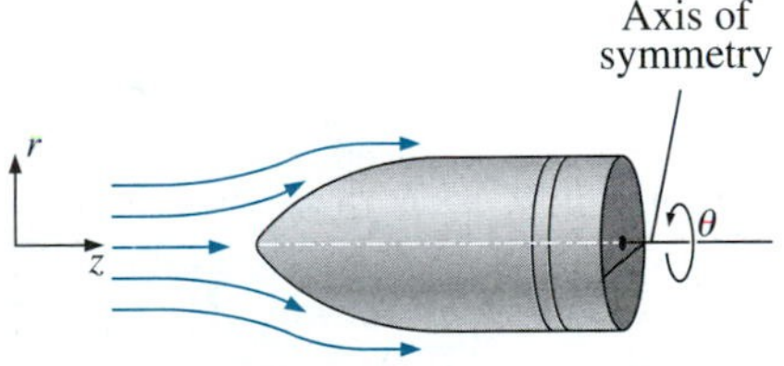

FIGURE 9–12

Axisymmetric flow over a bullet.

9–3 ▪ A BRIEF HISTORY OF FLUID MECHANICS[1]

One of the first engineering problems humankind faced as cities were developed was the supply of water for domestic use and irrigation of crops. Our urban lifestyles can be retained only with abundant water, and it is clear from archeology that every successful civilization of prehistory invested in the construction and maintenance of water systems. The Roman aqueducts, some

[1] This section is contributed by Professor Glenn Brown of Oklahoma State University.

FIGURE 9–13

Segment of Pergamon pipeline. Each clay pipe section was 13 to 18 cm in diameter.

Courtesy Gunther Garbrecht. Used by permission.

FIGURE 9–14

A mine hoist powered by a reversible water wheel.

G. Agricola, De Re Metalica, *Basel, 1556.*

of which are still in use, are the best known examples. However, perhaps the most impressive engineering from a technical viewpoint was done at the Hellenistic city of Pergamon in present-day Turkey. There, from 283 to 133 BC, they built a series of pressurized lead and clay pipelines (Fig. 9–13), up to 45 km long that operated at pressures exceeding 1.7 MPa (180 m of head). Unfortunately, the names of almost all these early builders are lost to history.

The earliest recognized contribution to fluid mechanics theory was made by the Greek mathematician Archimedes (285–212 BC). He formulated and applied the buoyancy principle in history's first nondestructive test to determine the gold content of the crown of King Hiero I. The Romans built great aqueducts and educated many conquered people on the benefits of clean water, but overall had a poor understanding of fluids theory. (Perhaps they shouldn't have killed Archimedes when they sacked Syracuse.)

During the Middle Ages, the application of fluid machinery slowly but steadily expanded. Elegant piston pumps were developed for dewatering mines, and the watermill and windmill were perfected to grind grain, forge metal, and for other tasks. For the first time in recorded human history, significant work was being done without the power of a muscle supplied by a person or animal, and these inventions are generally credited with enabling the later industrial revolution. Again the creators of most of the progress are unknown, but the devices themselves were well documented by several technical writers such as Georgius Agricola (Fig. 9–14).

The Renaissance brought continued development of fluid systems and machines, but more importantly, the scientific method was perfected and adopted throughout Europe. Simon Stevin (1548–1617), Galileo Galilei (1564–1642), Edme Mariotte (1620–1684), and Evangelista Torricelli (1608–1647) were among the first to apply the method to fluids as they investigated hydrostatic pressure distributions and vacuums. That work was integrated and refined by the brilliant mathematician and philosopher, Blaise Pascal (1623– 1662). The Italian monk, Benedetto Castelli (1577–1644) was the first person to publish a statement of the continuity principle for fluids. Besides formulating his equations of motion for solids, Sir Isaac Newton (1643–1727) applied his laws to fluids and explored fluid inertia and resistance, free jets, and viscosity. That effort was built upon by Daniel Bernoulli (1700–1782) and his associate Leonard Euler (1707–1783). Together, their work defined the energy and momentum equations. Bernoulli's 1738 classic treatise *Hydrodynamica* may be considered the first fluid mechanics text. Finally, Jean d'Alembert (1717–1789) developed the idea of velocity and acceleration components, a differential expression of continuity, and his "paradox" of zero resistance to steady uniform motion.

The development of fluid mechanics theory up through the end of the eighteenth century had little impact on engineering since fluid properties and parameters were poorly quantified, and most theories were abstractions that could not be quantified for design purposes. That was to change with the development of the French school of engineering led by Riche de Prony (1755–1839). Prony (still known for his brake to measure power) and his associates in Paris at the École Polytechnique and the École des Ponts et Chaussées were the first to integrate calculus and scientific theory into the engineering curriculum, which became the model for the rest of the world. (So now you know whom to blame for your painful freshman year.) Antonie Chezy (1718–1798), Louis Navier (1785–1836), Gaspard Coriolis (1792–1843), Henry Darcy (1803–1858), and many other contributors to fluid engineering and theory were students and/or instructors at the schools.

By the mid nineteenth century, fundamental advances were coming on several fronts. The physician Jean Poiseuille (1799–1869) had accurately measured flow in capillary tubes for multiple fluids, while in Germany Gotthilf Hagen (1797–1884) had differentiated between laminar and turbulent flow in pipes. In England, Lord Osborne Reynolds (1842–1912) continued that work and developed the dimensionless number that bears his name. Similarly, in parallel to the early work of Navier, George Stokes (1819–1903) completed the general equations of fluid motion with friction that take their names. William Froude (189–1879) almost single-handedly developed the procedures for, and proved the value of physical model testing. American expertise had become equal to the Europeans as demonstrated by James Francis's (1815–1892) and Lester Pelton's (1829–1908) pioneering work in turbines and Clemens Herschel's (1842–1930) invention of the Venturi meter.

In addition to Reynolds and Stokes, many notable contributions were made to fluid theory in the late nineteenth century by Irish and English scientists, including William Thomson, Lord Kelvin (1824–1907), William Strutt, Lord Rayleigh (1842–1919), and Sir Horace Lamb (1849–1934). These individuals investigated a large number of problems, including dimensional analysis, irrotational flow, vortex motion, cavitation, and waves. In a broader sense, their work also explored the links between fluid mechanics, thermodynamics, and heat transfer.

FIGURE 9–15
The Wright brothers take flight at Kitty Hawk.
National Air and Space Museum/ Smithsonian Institution.

The dawn of the twentieth century brought two monumental developments. First, in 1903, the self-taught Wright brothers (Wilbur, 1867–1912; Orville, 1871–1948) invented the airplane through application of theory and determined experimentation. Their primitive invention was complete and contained all the major aspects of modern craft (Fig. 9–15). The Navier–Stokes equations were of little use up to this time because they were too difficult to solve. In a pioneering paper in 1904, the German Ludwig Prandtl (1875–1953) showed that fluid flows can be divided into a layer near the walls, the *boundary layer*, where the friction effects are significant and an outer layer where such effects are negligible and the simplified Euler and Bernoulli equations are applicable. His students, Theodor von Kármán (1881–1963), Paul Blasius (1883–1970), Johann Nikuradse (1894–1979), and others, built on that theory in both hydraulic and aerodynamic applications. (During World War II, both sides benefited from the theory as Prandtl remained in Germany while his best student, the Hungarian-born Theodor von Kármán, worked in America.)

The mid twentieth century could be considered a golden age of fluid mechanics applications. Existing theories were adequate for the tasks at hand, and fluid properties and parameters were well defined. These supported a huge expansion of the aeronautical, chemical, industrial, and water resources sectors; each of which pushed fluid mechanics in new directions. Fluid mechanics research and work in the late twentieth century were dominated by the development of the digital computer in America. The ability to solve large complex problems, such as global climate modeling or the optimization of a turbine blade, has provided a benefit to our society that the eighteenth-century developers of fluid mechanics could never have imagined (Fig. 9–16). The principles presented in the following chapters have been applied to flows ranging from a moment at the microscopic scale to 50 years of simulation for an entire river basin. It is truly mind boggling.

Where will fluid mechanics go in the twenty-first century? Frankly, even a limited extrapolation beyond the present would be sheer folly. However, if history tells us anything, it is that engineers will be applying what they know to benefit society, researching what they don't know, and having a great time in the process.

FIGURE 9–16

The Oklahoma Wind Power Center near Woodward consists of 68 turbines, 1.5 MW each.

Courtesy Steve Stadler, Oklahoma Wind Power Initiative. Used by permission.

9–4 ▪ VAPOR PRESSURE AND CAVITATION

It is well-established that temperature and pressure are dependent properties for pure substances during phase-change processes, and there is one-to-one correspondence between temperatures and pressures. At a given pressure, the temperature at which a pure substance changes phase is called the **saturation temperature** T_{sat}. Likewise, at a given temperature, the pressure at which a pure substance changes phase is called the **saturation pressure** P_{sat}. At a pressure of 101.325 kPa, T_{sat} is 99.97°C. Conversely, at a temperature of 99.97°C, P_{sat} is 101.325 kPa. (At 100.00°C, P_{sat} is 101.42 kPa in the ITS-90 discussed in Chap. 2.)

The **vapor pressure** P_v of a pure substance is defined as *the pressure exerted by its vapor in phase equilibrium with its liquid at a given temperature.* P_v is a property of the pure substance, and turns out to be identical to the saturation pressure P_{sat} of the liquid ($P_v = P_{sat}$). We must be careful not to confuse vapor pressure with *partial pressure*. **Partial pressure** is defined as *the pressure of a gas or vapor in a mixture with other gases*. For example, atmospheric air is a mixture of dry air and water vapor, and atmospheric pressure is the sum of the partial pressure of dry air and the partial pressure of water vapor. The partial pressure of water vapor constitutes a small fraction (usually under 3 percent) of the atmospheric pressure since air is mostly nitrogen and oxygen. The partial pressure of a vapor must be less than or equal to the vapor pressure if there is no liquid present. However, when both vapor and liquid are present and the system is in phase equilibrium, the partial pressure of the vapor must equal the vapor pressure, and the system is said to be *saturated*. The rate of evaporation from open water bodies such as lakes is controlled by the difference between the vapor pressure and the partial pressure. For example, the vapor pressure of water at 20°C is 2.34 kPa. Therefore, a bucket of water at 20°C left in a room with dry air at 1 atm will continue evaporating until one of two things happens: the water evaporates away (there is not enough water to establish phase equilibrium in the room), or the evaporation stops when the partial pressure of the water vapor in the room rises to 2.34 kPa at which point phase equilibrium is established.

For phase-change processes between the liquid and vapor phases of a pure substance, the saturation pressure and the vapor pressure are equivalent since the vapor is pure. Note that the pressure value would be the same whether it is measured in the vapor or liquid phase (provided that it is measured at a location close to the liquid–vapor interface to avoid the hydrostatic effects). Vapor pressure increases with temperature. Thus, a substance at higher pressure boils at higher temperature. For example, water boils at

134°C in a pressure cooker operating at 3 atm absolute pressure, but it boils at 93°C in an ordinary pan at a 2000-m elevation, where the atmospheric pressure is 0.8 atm. The saturation (or vapor) pressures are given in Appendices 1 and 2 for various substances. An abridged table for water is given in Table 9–1 for easy reference.

The reason for our interest in vapor pressure is the possibility of the liquid pressure in liquid-flow systems dropping below the vapor pressure at some locations, and the resulting unplanned vaporization. For example, water at 10°C may vaporize and form bubbles at locations (such as the tip regions of impellers or suction sides of pumps) where the pressure drops below 1.23 kPa. The vapor bubbles (called **cavitation bubbles** since they form "cavities" in the liquid) collapse as they are swept away from the low-pressure regions, generating highly destructive, extremely high-pressure waves. This phenomenon, which is a common cause for drop in performance and even the erosion of impeller blades, is called **cavitation**, and it is an important consideration in the design of hydraulic turbines and pumps (Fig. 9–17).

Cavitation must be avoided (or at least minimized) in most flow systems since it reduces performance, generates annoying vibrations and noise, and causes damage to equipment. [We note that some flow systems use cavitation to their *advantage*, e.g., high speed "supercavitating" torpedoes.] The pressure spikes resulting from the large number of bubbles collapsing near a solid surface over a long period of time may cause erosion, surface pitting, fatigue failure, and the eventual destruction of the components or machinery. The presence of cavitation in a flow system can be sensed by its characteristic tumbling sound.

TABLE 9–1

Saturation (or vapor) pressure of water at various temperatures

Temperature T, °C	Saturation Pressure P_{sat}, kPa
−10	0.26
−5	0.40
0	0.61
5	0.87
10	1.23
15	1.71
20	2.34
25	3.17
30	4.25
40	7.39
50	12.35
100	101.4
150	476.2
200	1555
250	3976
300	8588

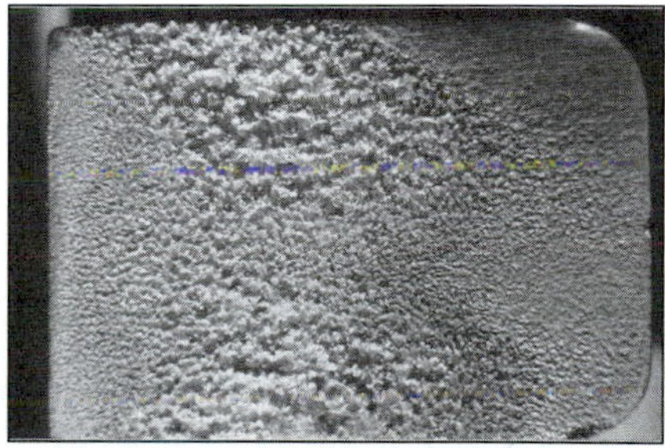

FIGURE 9–17

Cavitation damage on a 16-mm by 23-mm aluminum sample tested at 60 m/s for 2.5 h. The sample was located at the cavity collapse region downstream of a cavity generator specifically designed to produce high damage potential.

Photograph by David Stinebring, ARL/Pennsylvania State University. Used by permission.

EXAMPLE 9–2 Minimum Pressure to Avoid Cavitation

In a water distribution system, the temperature of water is observed to be as high as 30°C. Determine the minimum pressure allowed in the system to avoid cavitation.

Solution The minimum pressure in a water distribution system to avoid cavitation is to be determined.

Properties The vapor pressure of water at 30°C is 4.25 kPa (Table A-4).

Analysis To avoid cavitation, the pressure anywhere in the flow should not be allowed to drop below the vapor (or saturation) pressure at the given temperature. That is,

$$P_{min} = P_{sat@30°C} = \mathbf{4.25\ kPa}$$

Therefore, the pressure should be maintained above 4.25 kPa everywhere in the flow.

Discussion Note that the vapor pressure increases with increasing temperature, and thus the risk of cavitation is greater at higher fluid temperatures.

9–5 ▪ COMPRESSIBILITY AND SPEED OF SOUND

An important parameter in the study of compressible flow is the **speed of sound** (or the **sonic speed**), which is the speed at which an infinitesimally small pressure wave travels through a medium. The pressure wave may be caused by a small disturbance, which creates a slight rise in local pressure.

To obtain a relation for the speed of sound in a medium, consider a duct that is filled with a fluid at rest, as shown in Fig. 9–18. A piston fitted in the duct is now moved to the right with a constant incremental velocity dV, creating a sonic wave. The wave front moves to the right through the fluid at the speed of sound c and separates the moving fluid adjacent to the piston from the fluid still at rest. The fluid to the left of the wave front experiences an incremental change in its thermodynamic properties, while the fluid on the right of the wave front maintains its original thermodynamic properties, as shown in Fig. 9–18.

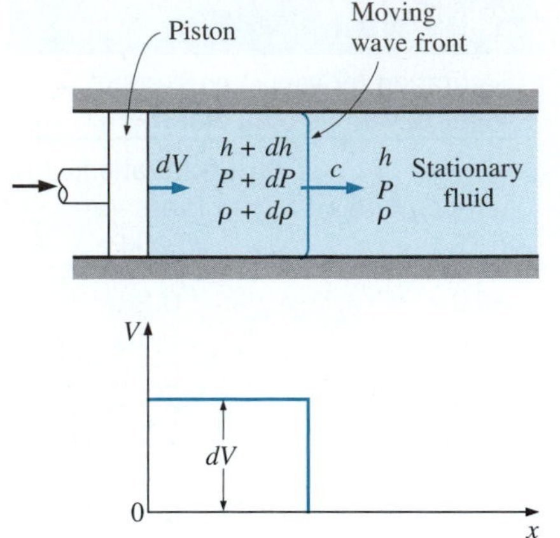

FIGURE 9–18

Propagation of a small pressure wave along a duct.

To simplify the analysis, consider a control volume that encloses the wave front and moves with it, as shown in Fig. 9–19. To an observer traveling with the wave front, the fluid to the right appears to be moving toward the wave front with a speed of c and the fluid to the left to be moving away from the wave front with a speed of $c - dV$. Of course, the observer sees the control volume that encloses the wave front (and herself or himself) as stationary, and the observer is witnessing a steady-flow process. The mass balance for this single-stream, steady-flow process is expressed as

$$\dot{m}_{\text{right}} = \dot{m}_{\text{left}}$$

or

$$\rho A c = (\rho + d\rho)A(c - dV)$$

By canceling the cross-sectional (or flow) area A and neglecting the higher-order terms, this equation reduces to

$$c\, d\rho - \rho\, dV = 0 \qquad \textbf{(a)}$$

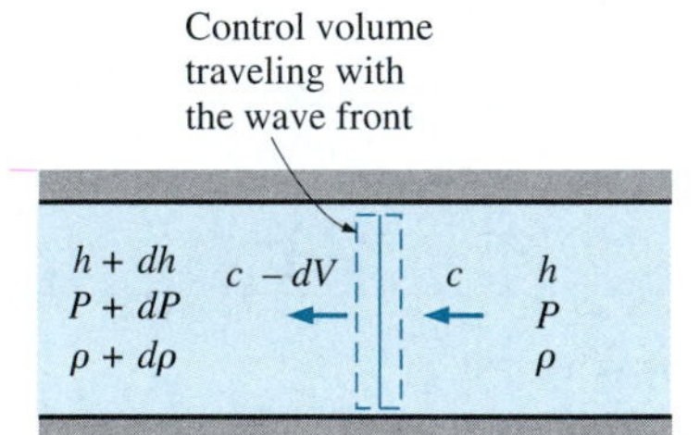

FIGURE 9–19

Control volume moving with the small pressure wave along a duct.

No heat or work crosses the boundaries of the control volume during this steady-flow process, and the potential energy change can be neglected. Then the steady-flow energy balance $e_{\text{in}} = e_{\text{out}}$ becomes

$$h + \frac{c^2}{2} = h + dh + \frac{(c - dV)^2}{2}$$

which yields

$$dh - c\, dV = 0 \qquad \textbf{(b)}$$

where we have neglected the second-order term dV^2. The amplitude of the ordinary sonic wave is very small and does not cause any appreciable change in the pressure and temperature of the fluid. Therefore, the propagation of a sonic wave is not only adiabatic but also very nearly isentropic. Then the thermodynamic relation $T\, ds = dh - dP/\rho$ reduces to

$$T\, \cancel{ds}^{\,0} = dh - \frac{dP}{\rho}$$

or

$$dh = \frac{dP}{\rho} \qquad \textbf{(c)}$$

Combining Eqs. a, b, and c yields the desired expression for the speed of sound as

$$c^2 = \frac{dP}{d\rho} \qquad \text{at } s = \text{constant}$$

or

$$c^2 = \left(\frac{\partial P}{\partial \rho}\right)_s \tag{9–1}$$

It is left as an exercise for the reader to show, by using thermodynamic property relations (see Çengel and Boles, 2008) that Eq. 9–1 can also be written as

$$c^2 = k\left(\frac{\partial P}{\partial \rho}\right)_T \tag{9–2}$$

where k is the specific heat ratio of the fluid $k = c_p/c_v$. Note that the speed of sound in a fluid is a function of the thermodynamic properties of that fluid.

When the fluid is an ideal gas ($P = \rho RT$), the differentiation in Eq. 9–2 can easily be performed to yield

$$c^2 = k\left(\frac{\partial P}{\partial \rho}\right)_T = k\left[\frac{\partial(\rho RT)}{\partial \rho}\right]_T = kRT$$

or

$$c = \sqrt{kRT} \tag{9–3}$$

Noting that the gas constant R has a fixed value for a specified ideal gas and the specific heat ratio k of an ideal gas is, at most, a function of temperature, we see that the speed of sound in a specified ideal gas is a function of temperature alone (Fig. 9–20).

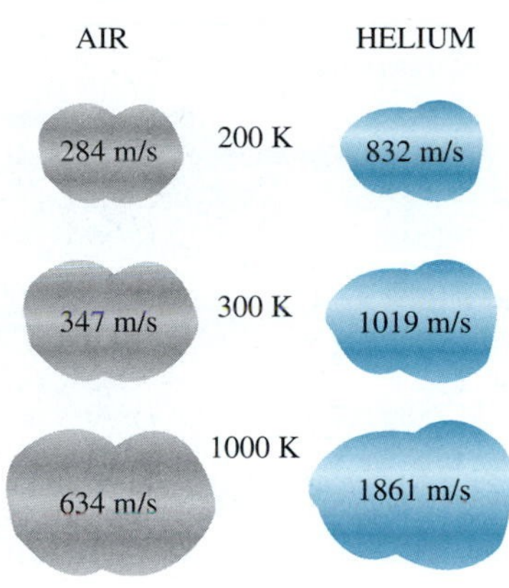

FIGURE 9–20

The speed of sound changes with temperature and varies with the fluid.

EXAMPLE 9–3 Mach Number of Air Entering a Diffuser

Air enters a diffuser shown in Fig. 9–21 with a velocity of 200 m/s. Determine (*a*) the speed of sound and (*b*) the Mach number at the diffuser inlet when the air temperature is 30°C.

Solution Air enters a diffuser with a high velocity. The speed of sound and the Mach number are to be determined at the diffuser inlet.

Assumption Air at specified conditions behaves as an ideal gas.

Properties The gas constant of air is $R = 0.287$ kJ/kg · K, and its specific heat ratio at 30°C is 1.4 (Table A-2*a*).

Analysis We note that the speed of sound in a gas varies with temperature, which is given to be 30°C.

(*a*) The speed of sound in air at 30°C is determined from Eq. 9–3 to be

$$c = \sqrt{kRT} = \sqrt{(1.4)(0.287 \text{ kJ/kg} \cdot \text{K})(303 \text{ K})\left(\frac{1000 \text{ m}^2/\text{s}^2}{1 \text{ kJ/kg}}\right)} = \mathbf{349 \text{ m/s}}$$

(*b*) Then the Mach number becomes

$$\text{Ma} = \frac{V}{c} = \frac{200 \text{ m/s}}{349 \text{ m/s}} = \mathbf{0.573}$$

Discussion The flow at the diffuser inlet is subsonic since Ma < 1.

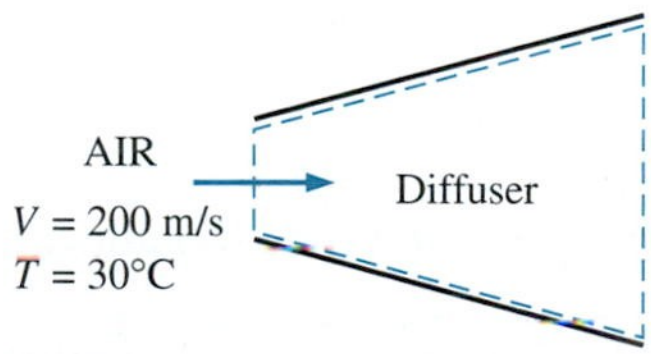

FIGURE 9–21

Schematic for Example 9–3.

9–6 ▪ VISCOSITY

When two solid bodies in contact move relative to each other, a friction force develops at the contact surface in the direction opposite to motion. To move a table on the floor, for example, we have to apply a force to the table in the horizontal direction large enough to overcome the friction force. The magnitude of the force needed to move the table depends on the *friction coefficient* between the table legs and the floor.

FIGURE 9–22

A fluid moving relative to a body exerts a drag force on the body, partly because of friction caused by viscosity.

The situation is similar when a fluid moves relative to a solid or when two fluids move relative to each other. We move with relative ease in air, but not so in water. Moving in oil would be even more difficult, as can be observed by the slower downward motion of a glass ball dropped in a tube filled with oil. It appears that there is a property that represents the internal resistance of a fluid to motion or the "fluidity," and that property is the **viscosity**. The force a flowing fluid exerts on a body in the flow direction is called the **drag force**, and the magnitude of this force depends, in part, on viscosity (Fig. 9–22).

To obtain a relation for viscosity, consider a fluid layer between two very large parallel plates (or equivalently, two parallel plates immersed in a large body of a fluid) separated by a distance ℓ (Fig. 9–23). Now a constant parallel force F is applied to the upper plate while the lower plate is held fixed. After the initial transients, it is observed that the upper plate moves continuously under the influence of this force at a constant velocity V. The fluid in contact with the upper plate sticks to the plate surface and moves with it at the same velocity, and the shear stress τ acting on this fluid layer is

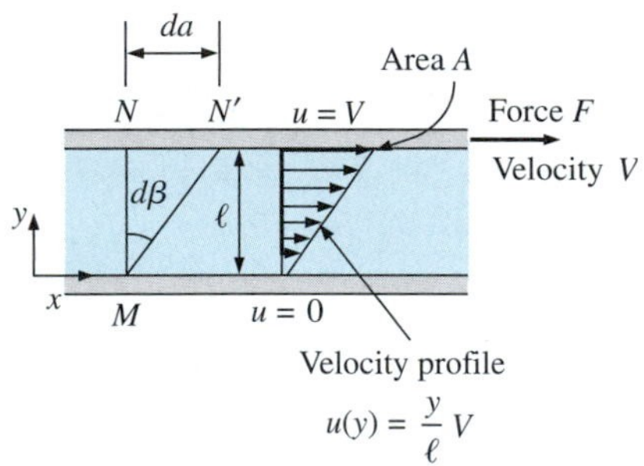

FIGURE 9–23

The behavior of a fluid in laminar flow between two parallel plates when the upper plate moves with a constant velocity.

$$\tau = \frac{F}{A} \tag{9–4}$$

where A is the contact area between the plate and the fluid. Note that the fluid layer deforms continuously under the influence of shear stress.

The fluid in contact with the lower plate assumes the velocity of that plate, which is zero (because of the no-slip condition—see Section 9–1). In steady laminar flow, the fluid velocity between the plates varies linearly between 0 and V, and thus the *velocity profile* and the *velocity gradient* are

$$u(y) = \frac{y}{\ell}V \quad \text{and} \quad \frac{du}{dy} = \frac{V}{\ell} \tag{9–5}$$

where y is the vertical distance from the lower plate.

During a differential time interval dt, the sides of fluid particles along a vertical line MN rotate through a differential angle $d\beta$ while the upper plate moves a differential distance $da = V\,dt$. The angular displacement or deformation (or shear strain) can be expressed as

$$d\beta \approx \tan(d\beta) = \frac{da}{\ell} = \frac{V\,dt}{\ell} = \frac{du}{dy}dt \tag{9–6}$$

Rearranging, the rate of deformation under the influence of shear stress τ becomes

$$\frac{d\beta}{dt} = \frac{du}{dy} \tag{9–7}$$

Thus we conclude that the rate of deformation of a fluid element is equivalent to the velocity gradient du/dy. Further, it can be verified experimentally that for most fluids the rate of deformation (and thus the velocity gradient) is directly proportional to the shear stress τ,

$$\tau \propto \frac{d(d\beta)}{dt} \quad \text{or} \quad \tau \propto \frac{du}{dy} \tag{9-8}$$

Fluids for which the rate of deformation is proportional to the shear stress are called **Newtonian fluids** after Sir Isaac Newton, who expressed it first in 1687. Most common fluids such as water, air, gasoline, and oils are Newtonian fluids. Blood and liquid plastics are examples of non-Newtonian fluids.

In one-dimensional shear flow of Newtonian fluids, shear stress can be expressed by the linear relationship

Shear stress:
$$\tau = \mu \frac{du}{dy} \quad (\text{N/m}^2) \tag{9-9}$$

where the constant of proportionality μ is called the **coefficient of viscosity** or the **dynamic** (or **absolute**) **viscosity** of the fluid, whose unit is kg/m · s, or equivalently, N · s/m^2 (or Pa · s where Pa is the pressure unit pascal). A common viscosity unit is **poise**, which is equivalent to 0.1 Pa · s (*centipoise* is one-hundredth of a poise). The viscosity of water at 20°C is 1 centipoise, and thus the unit centipoise serves as a useful reference. A plot of shear stress versus the rate of deformation (velocity gradient) for a Newtonian fluid is a straight line whose slope is the viscosity of the fluid, as shown in Fig. 9–24. Note that viscosity is independent of the rate of deformation for Newtonian fluids. Since the rate of deformation is proportional to the strain rate, Fig. 9–24 reveals that viscosity is actually a coefficient in a stress-strain rate relationship.

The **shear force** acting on a Newtonian fluid layer (or, by Newton's third law, the force acting on the plate) is

Shear force:
$$F = \tau A = \mu A \frac{du}{dy} \quad (\text{N}) \tag{9-10}$$

where again A is the contact area between the plate and the fluid. Then the force F required to move the upper plate in Fig. 9–23 at a constant velocity of V while the lower plate remains stationary is

$$F = \mu A \frac{V}{\ell} \quad (\text{N}) \tag{9-11}$$

This relation can alternately be used to calculate μ when the force F is measured. Therefore, the experimental setup just described can be used to measure the viscosity of fluids. Note that under identical conditions, the force F would be very different for different fluids.

For non-Newtonian fluids, the relationship between shear stress and rate of deformation is not linear, as shown in Fig. 9–25. The slope of the curve on the τ versus du/dy chart is referred to as the *apparent viscosity* of the fluid. Fluids for which the apparent viscosity increases with the rate of deformation (such as solutions with suspended starch or sand) are referred to as *dilatant* or *shear thickening fluids*, and those that exhibit the opposite

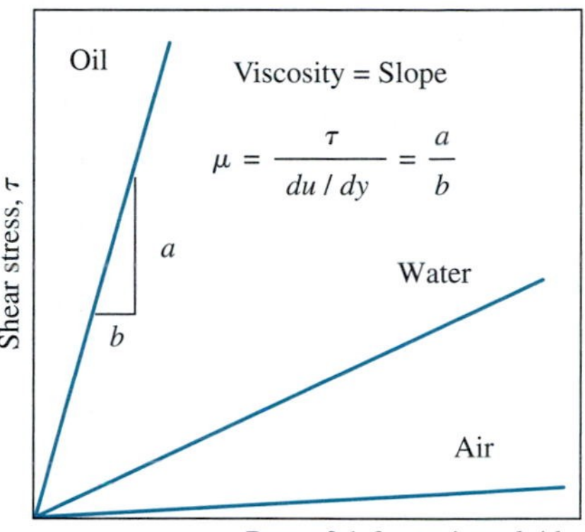

FIGURE 9–24
The rate of deformation (velocity gradient) of a Newtonian fluid is proportional to shear stress, and the constant of proportionality is the viscosity.

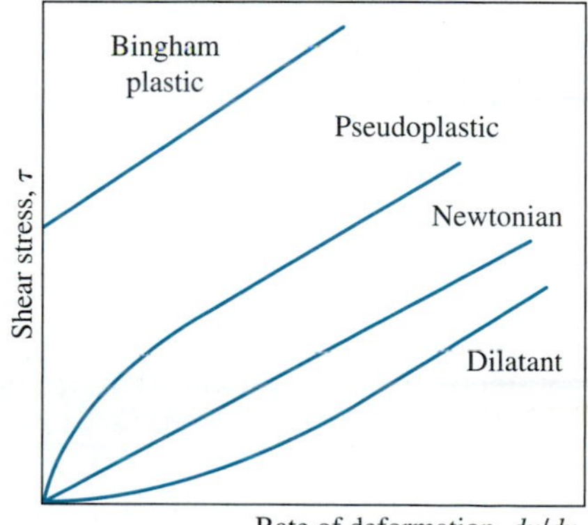

FIGURE 9–25
Variation of shear stress with the rate of deformation for Newtonian and non-Newtonian fluids (the slope of a curve at a point is the apparent viscosity of the fluid at that point).

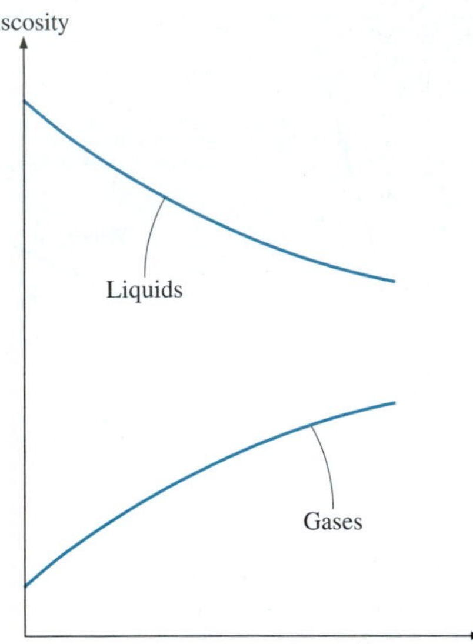

FIGURE 9–26

The viscosity of liquids decreases and the viscosity of gases increases with temperature.

TABLE 9–2

Dynamic viscosities of some fluids at 1 atm and 20°C (unless otherwise stated)

Fluid	Dynamic Viscosity μ, kg/m · s
Glycerin:	
−20°C	134.0
0°C	10.5
20°C	1.52
40°C	0.31
Engine oil:	
SAE 10W	0.10
SAE 10W30	0.17
SAE 30	0.29
SAE 50	0.86
Mercury	0.0015
Ethyl alcohol	0.0012
Water:	
0°C	0.0018
20°C	0.0010
100°C (liquid)	0.00028
100°C (vapor)	0.000012
Blood, 37°C	0.00040
Gasoline	0.00029
Ammonia	0.00015
Air	0.000018
Hydrogen, 0°C	0.0000088

behavior (the fluid becoming less viscous as it is sheared harder, such as some paints, polymer solutions, and fluids with suspended particles) are referred to as *pseudoplastic* or *shear thinning fluids*. Some materials such as toothpaste can resist a finite shear stress and thus behave as a solid, but deform continuously when the shear stress exceeds the yield stress and behave as a fluid. Such materials are referred to as Bingham plastics after E. C. Bingham, who did pioneering work on fluid viscosity for the U.S. National Bureau of Standards in the early twentieth century.

In fluid mechanics and heat transfer, the ratio of dynamic viscosity to density appears frequently. For convenience, this ratio is given the name **kinematic viscosity** ν and is expressed as $\nu = \mu/\rho$. Two common units of kinematic viscosity are m^2/s and **stoke** (1 stoke $= 1\ cm^2/s = 0.0001\ m^2/s$).

In general, the viscosity of a fluid depends on both temperature and pressure, although the dependence on pressure is rather weak. For *liquids*, both the dynamic and kinematic viscosities are practically independent of pressure, and any small variation with pressure is usually disregarded, except at extremely high pressures. For *gases*, this is also the case for dynamic viscosity (at low to moderate pressures), but not for kinematic viscosity since the density of a gas is proportional to its pressure.

The viscosity of a fluid is a measure of its "resistance to deformation." Viscosity is due to the internal frictional force that develops between different layers of fluids as they are forced to move relative to each other. Viscosity is caused by the cohesive forces between the molecules in liquids and by the molecular collisions in gases, and it varies greatly with temperature. The viscosity of liquids decreases with temperature, whereas the viscosity of gases increases with temperature (Fig. 9–26). This is because in a liquid the molecules possess more energy at higher temperatures, and they can oppose the large cohesive intermolecular forces more strongly. As a result, the energized liquid molecules can move more freely.

In a gas, on the other hand, the intermolecular forces are negligible, and the gas molecules at high temperatures move randomly at higher velocities. This results in more molecular collisions per unit volume per unit time and therefore in greater resistance to flow.

The viscosities of some fluids at room temperature are listed in Table 9–2. Note that the viscosities of different fluids differ by several orders of magnitude. Also note that it is more difficult to move an object in a higher-viscosity fluid such as engine oil than it is in a lower-viscosity fluid such as water. Liquids, in general, are much more viscous than gases. The viscosity of a fluid is directly related to the pumping power needed to transport a fluid in a pipe or to move a body (such as a car in air or a submarine in the sea) through a fluid.

Consider a fluid layer of thickness ℓ within a small gap between two concentric cylinders, such as the thin layer of oil in a journal bearing. The gap between the cylinders can be modeled as two parallel flat plates separated by the fluid. Noting that torque is $\mathrm{T} = FR$ (force times the moment arm, which is the radius R of the inner cylinder in this case), the tangential velocity is $V = \omega R$ (angular velocity times the radius), and taking the wetted surface area of the inner cylinder to be $A = 2\pi RL$ by disregarding the shear stress acting on the two ends of the inner cylinder, torque can be expressed as

$$\mathrm{T} = FR = \mu \frac{2\pi R^3 \omega L}{\ell} = \mu \frac{4\pi^2 R^3 \dot{n} L}{\ell} \tag{9–12}$$

where L is the length of the cylinder and $\dot{n}$ is the number of revolutions per unit time, which is usually expressed in rpm (revolutions per minute). Note that the angular distance traveled during one rotation is 2π rad, and thus the relation between the angular velocity in rad/min and the rpm is $\omega = 2\pi\dot{n}$. Equation 9–12 can be used to calculate the viscosity of a fluid by measuring torque at a specified angular velocity. Therefore, two concentric cylinders can be used as a *viscometer*, a device that measures viscosity.

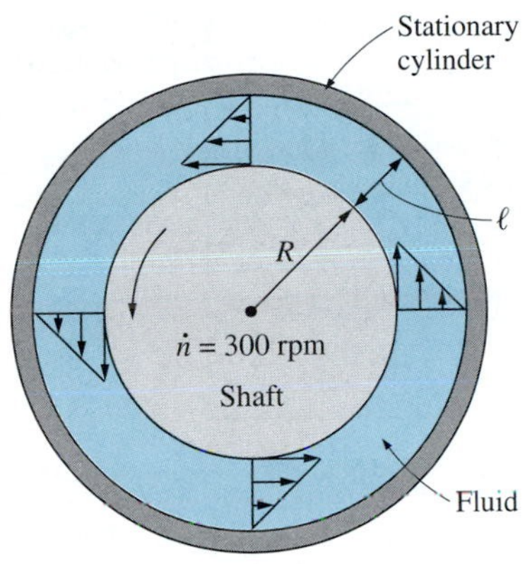

FIGURE 9–27

Schematic for Example 9–4 (not to scale).

EXAMPLE 9–4 Determining the Viscosity of a Fluid

The viscosity of a fluid is to be measured by a viscometer constructed of two 40-cm-long concentric cylinders (Fig. 9–27). The outer diameter of the inner cylinder is 12 cm, and the gap between the two cylinders is 0.15 cm. The inner cylinder is rotated at 300 rpm, and the torque is measured to be 1.8 N · m. Determine the viscosity of the fluid.

Solution The torque and the rpm of a double cylinder viscometer are given. The viscosity of the fluid is to be determined.

Assumptions **1** The inner cylinder is completely submerged in the fluid. **2** The viscous effects on the two ends of the inner cylinder are negligible.

Analysis The velocity profile is linear only when the curvature effects are negligible, and the profile can be approximated as being linear in this case since $\ell/R = 0.025 \ll 1$. Solving Eq. 9–12 for viscosity and substituting the given values, the viscosity of the fluid is determined to be

$$\mu = \frac{\mathbf{T}\ell}{4\pi^2 R^3 \dot{n} L} = \frac{(1.8\ \text{N}\cdot\text{m})(0.0015\ \text{m})}{4\pi^2(0.06\ \text{m})^3(300/60\ 1/\text{s})(0.4\ \text{m})} = \mathbf{0.158\ N\cdot s/m^2}$$

Discussion Viscosity is a strong function of temperature, and a viscosity value without a corresponding temperature is of little usefulness. Therefore, the temperature of the fluid should have also been measured during this experiment, and reported with this calculation.

(*a*)

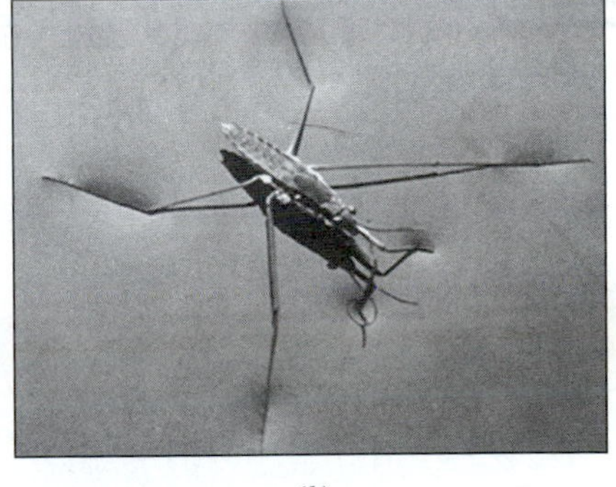

(*b*)

FIGURE 9–28

Some consequences of surface tension.

(a) © Pegasus/Visuals Unlimited.
(b) © Dennis Drenner/Visuals Unlimited.

9–7 ▪ SURFACE TENSION AND CAPILLARY EFFECT

It is often observed that a drop of blood forms a hump on a horizontal glass; a drop of mercury forms a near-perfect sphere and can be rolled just like a steel ball over a smooth surface; water droplets from rain or dew hang from branches or leaves of trees; a liquid fuel injected into an engine forms a mist of spherical droplets; water dripping from a leaky faucet falls as nearly spherical droplets; a soap bubble released into the air forms a nearly spherical shape; and water beads up into small drops on flower petals (Fig. 9–28*a*).

In these and other observances, liquid droplets behave like small balloons filled with the liquid, and the surface of the liquid acts like a stretched elastic membrane under tension. The pulling force that causes this tension acts parallel to the surface and is due to the attractive forces between the molecules of the liquid. The magnitude of this force per unit length is called **surface tension** (or *coefficient of surface tension*) σ_s and is usually expressed in the unit N/m (or lbf/ft in English units). This effect is also called *surface energy* [per unit area] and is expressed in the equivalent unit of N · m/m² or

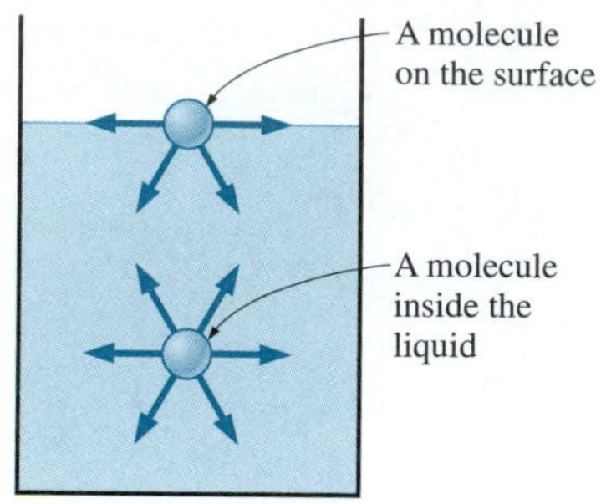

FIGURE 9–29

Attractive forces acting on a liquid molecule at the surface and deep inside the liquid.

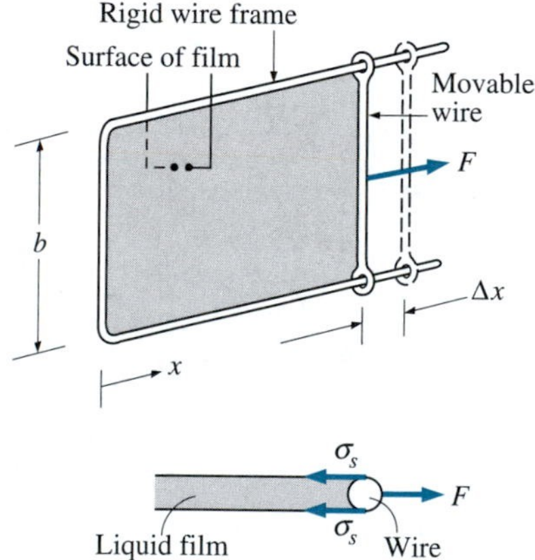

FIGURE 9–30

Stretching a liquid film with a U-shaped wire, and the forces acting on the movable wire of length b.

J/m^2. In this case, σ_s represents the stretching work that needs to be done to increase the surface area of the liquid by a unit amount.

To visualize how surface tension arises, we present a microscopic view in Fig. 9–29 by considering two liquid molecules, one at the surface and one deep within the liquid body. The attractive forces applied on the interior molecule by the surrounding molecules balance each other because of symmetry. But the attractive forces acting on the surface molecule are not symmetric, and the attractive forces applied by the gas molecules above are usually very small. Therefore, there is a net attractive force acting on the molecule at the surface of the liquid, which tends to pull the molecules on the surface toward the interior of the liquid. This force is balanced by the repulsive forces from the molecules below the surface that are trying to be compressed. The result is that the liquid minimizes its surface area. This is the reason for the tendency of liquid droplets to attain a spherical shape, which has the minimum surface area for a given volume.

You also may have observed, with amusement, that some insects can land on water or even walk on water (Fig. 9–28*b*) and that small steel needles can float on water. These phenomena are made possible by surface tension which balances the weights of these objects.

To understand the surface tension effect better, consider a liquid film (such as the film of a soap bubble) suspended on a U-shaped wire frame with a movable side (Fig. 9–30). Normally, the liquid film tends to pull the movable wire inward in order to minimize its surface area. A force F needs to be applied on the movable wire in the opposite direction to balance this pulling effect. Both sides of the film are surfaces exposed to air, and thus the length along which the tension acts in this case is $2b$. Then a force balance on the movable wire gives $F = 2b\sigma_s$, and thus the surface tension can be expressed as

$$\sigma_s = \frac{F}{2b} \tag{9-13}$$

Note that for $b = 0.5$ m, the measured force F (in N) is simply the surface tension in N/m. An apparatus of this kind with sufficient precision can be used to measure the surface tension of various liquids.

In the U-shaped wire frame apparatus, the movable wire is pulled to stretch the film and increase its surface area. When the movable wire is pulled a distance Δx, the surface area increases by $\Delta A = 2b\,\Delta x$, and the work done W during this stretching process is

$$W = \text{Force} \times \text{Distance} = F\,\Delta x = 2b\sigma_s\,\Delta x = \sigma_s\,\Delta A$$

where we have assumed that the force remains constant over the small distance. This result can also be interpreted as *the surface energy of the film is increased by an amount $\sigma_s\,\Delta A$ during this stretching process*, which is consistent with the alternative interpretation of σ_s as surface energy per unit area. This is similar to a rubber band having more potential (elastic) energy after it is stretched further. In the case of liquid film, the work is used to move liquid molecules from the interior parts to the surface against the attraction forces of other molecules. Therefore, surface tension also can be defined as *the work done per unit increase in the surface area of the liquid.*

The surface tension varies greatly from substance to substance, and with temperature for a given substance, as shown in Table 9–3. At 20°C, for example, the surface tension is 0.073 N/m for water and 0.440 N/m for mercury surrounded by atmospheric air. The surface tension of mercury is large enough that mercury droplets form nearly spherical balls that can be rolled like a solid ball on a smooth surface. The surface tension of a liquid, in general, decreases with temperature and becomes zero at the critical point (and thus there is no distinct liquid–vapor interface at temperatures above the critical point). The effect of pressure on surface tension is usually negligible.

The surface tension of a substance can be changed considerably by *impurities*. Therefore, certain chemicals, called *surfactants*, can be added to a liquid to decrease its surface tension. For example, soaps and detergents lower the surface tension of water and enable it to penetrate the small openings between fibers for more effective washing. But this also means that devices whose operation depends on surface tension (such as heat pipes) can be destroyed by the presence of impurities due to poor workmanship.

We speak of surface tension for liquids only at liquid–liquid or liquid–gas interfaces. Therefore, it is imperative that the adjacent liquid or gas be specified when specifying surface tension. Surface tension determines the size of the liquid droplets that form, and so a droplet that keeps growing by the addition of more mass will break down when the surface tension can no longer hold it together. This is like a balloon that bursts while being inflated when the pressure inside rises above the strength of the balloon material.

A curved interface indicates a pressure difference (or "pressure jump") across the interface with pressure being higher on the concave side. Consider, for example, a droplet of liquid in air, an air bubble in water, or a soap bubble in air. The excess pressure ΔP above the atmospheric pressure can be determined by considering the free-body diagram of half the droplet or soap bubble (Fig. 9–31). Noting that surface tension acts along the circumference and the pressure acts on the area, horizontal force balances for the droplet or air bubble and the soap bubble give

Droplet or air bubble:
$$(2\pi R)\sigma_s = (\pi R^2)\Delta P_{\text{droplet}} \rightarrow \Delta P_{\text{droplet}} = P_i - P_o = \frac{2\sigma_s}{R} \qquad \text{(9–14)}$$

Soap bubble:
$$2(2\pi R)\sigma_s = (\pi R^2)\Delta P_{\text{bubble}} \rightarrow \Delta P_{\text{bubble}} = P_i - P_o = \frac{4\sigma_s}{R} \qquad \text{(9–15)}$$

where P_i and P_o are the pressures inside and outside the droplet or soap bubble, respectively. When the droplet or soap bubble is in the atmosphere, P_o is simply atmospheric pressure. The extra factor of 2 in the force balance for the soap bubble is due to the existence of a soap film with *two* surfaces (inner and outer surfaces) and thus two circumferences in the cross section.

The excess pressure in a droplet of liquid in a gas (or air bubble of gas in a liquid) also can be determined by considering a differential increase in the radius of the droplet due to the addition of a differential amount of mass and interpreting the surface tension as the increase in the surface energy per unit area. Then the increase in the surface energy of the droplet during this differential expansion process becomes

$$\delta W_{\text{surface}} = \sigma_s\, dA = \sigma_s\, d(4\pi R^2) = 8\pi R\sigma_s\, dR$$

TABLE 9–3

Surface tension of some fluids in air at 1 atm and 20°C (unless otherwise stated)

Fluid	Surface Tension σ_s, N/m*
†Water:	
0°C	0.076
20°C	0.073
100°C	0.059
300°C	0.014
Glycerin	0.063
SAE 30 oil	0.035
Mercury	0.440
Ethyl alcohol	0.023
Blood, 37°C	0.058
Gasoline	0.022
Ammonia	0.021
Soap solution	0.025
Kerosene	0.028

* Multiply by 0.06852 to convert to lbf/ft.
† See Appendix for more precise data for water.

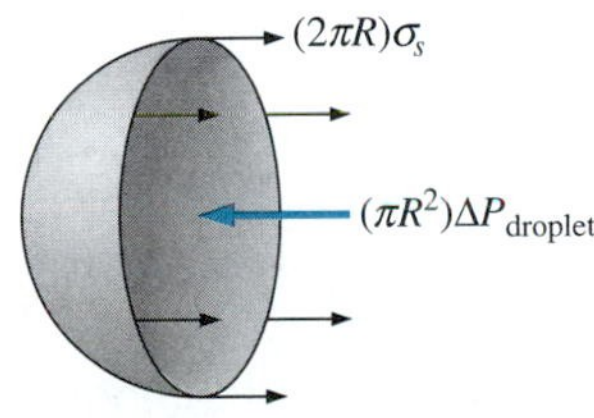

(*a*) Half a droplet or air bubble

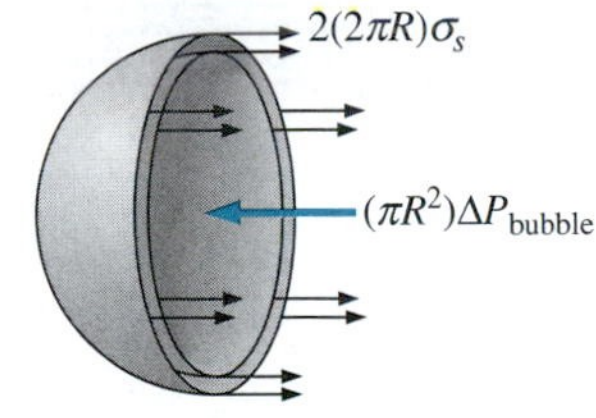

(*b*) Half a soap bubble

FIGURE 9–31

The free-body diagram of half a droplet or air bubble and half a soap bubble.

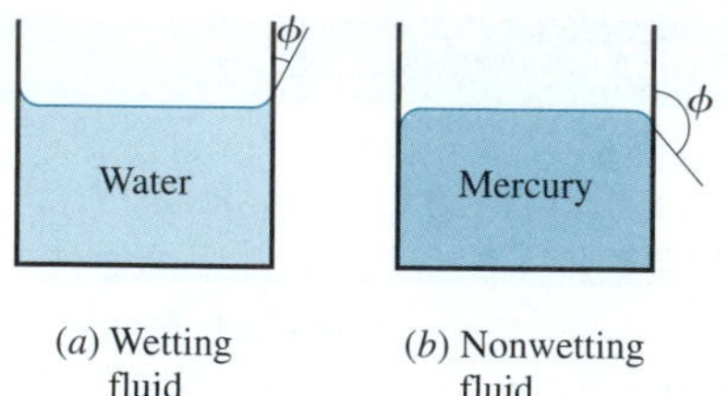

FIGURE 9–32

The contact angle for wetting and nonwetting fluids.

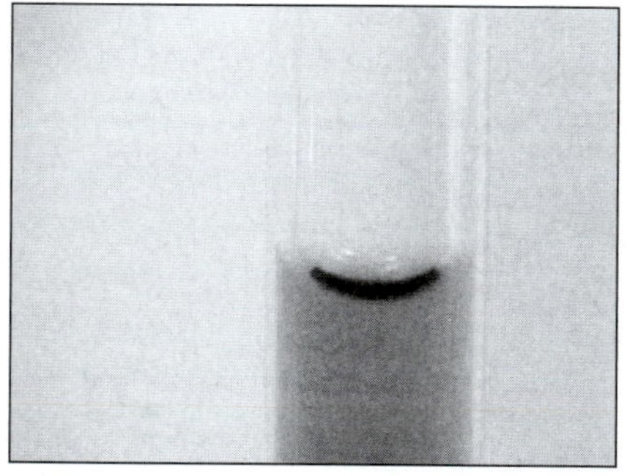

FIGURE 9–33

The meniscus of colored water in a 4-mm-inner-diameter glass tube. Note that the edge of the meniscus meets the wall of the capillary tube at a very small contact angle.

Photo by Gabrielle Trembley, Pennsylvania State University. Used by permission.

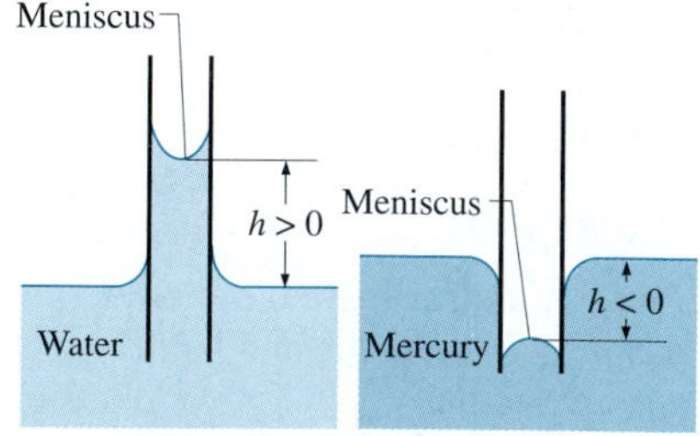

FIGURE 9–34

The capillary rise of water and the capillary fall of mercury in a small-diameter glass tube.

The expansion work done during this differential process is determined by multiplying the force by distance to obtain

$$\delta W_{\text{expansion}} = \text{Force} \times \text{Distance} = F\,dR = (\Delta PA)\,dR = 4\pi R^2\,\Delta P\,dR$$

Equating the two expressions above gives $\Delta P_{\text{droplet}} = 2\sigma_s/R$, which is the same relation obtained before and given in Eq. 9–14. Note that the excess pressure in a droplet is inversely proportional to the radius.

Capillary Effect

Another interesting consequence of surface tension is the **capillary effect**, which is the rise or fall of a liquid in a small-diameter tube inserted into the liquid. Such narrow tubes or confined flow channels are called **capillaries**. The rise of kerosene through a cotton wick inserted into the reservoir of a kerosene lamp is due to this effect. The capillary effect is also partially responsible for the rise of water to the top of tall trees. The curved free surface of a liquid in a capillary tube is called the **meniscus**.

It is commonly observed that water in a glass container curves up slightly at the edges where it touches the glass surface; but the opposite occurs for mercury: it curves down at the edges (Fig. 9–32). This effect is usually expressed by saying that water *wets* the glass (by sticking to it) while mercury does not. The strength of the capillary effect is quantified by the **contact** (or *wetting*) **angle** ϕ, defined as *the angle that the tangent to the liquid surface makes with the solid surface at the point of contact*. The surface tension force acts along this tangent line toward the solid surface. A liquid is said to wet the surface when $\phi < 90°$ and not to wet the surface when $\phi > 90°$. In atmospheric air, the contact angle of water (and most other organic liquids) with glass is nearly zero, $\phi \approx 0°$ (Fig. 9–33). Therefore, the surface tension force acts upward on water in a glass tube along the circumference, tending to pull the water up. As a result, water rises in the tube until the weight of the liquid in the tube above the liquid level of the reservoir balances the surface tension force. The contact angle is 130° for mercury–glass and 26° for kerosene–glass in air. Note that the contact angle, in general, is different in different environments (such as another gas or liquid in place of air).

The phenomenon of the capillary effect can be explained microscopically by considering *cohesive forces* (the forces between like molecules, such as water and water) and *adhesive forces* (the forces between unlike molecules, such as water and glass). The liquid molecules at the solid–liquid interface are subjected to both cohesive forces by other liquid molecules and adhesive forces by the molecules of the solid. The relative magnitudes of these forces determine whether a liquid wets a solid surface or not. Obviously, the water molecules are more strongly attracted to the glass molecules than they are to other water molecules, and thus water tends to rise along the glass surface. The opposite occurs for mercury, which causes the liquid surface near the glass wall to be suppressed (Fig. 9–34).

The magnitude of the capillary rise in a circular tube can be determined from a force balance on the cylindrical liquid column of height h in the tube (Fig. 9–35). The bottom of the liquid column is at the same level as the free surface of the reservoir, and thus the pressure there must be atmospheric pressure. This balances the atmospheric pressure acting at the top surface, and thus these two effects cancel each other. The weight of the liquid column is approximately

$$W = mg = \rho \mathcal{V} g = \rho g(\pi R^2 h)$$

Equating the vertical component of the surface tension force to the weight gives

$$W = F_{\text{surface}} \rightarrow \rho g(\pi R^2 h) = 2\pi R \sigma_s \cos \phi$$

Solving for h gives the capillary rise to be

Capillary rise: $$h = \frac{2\sigma_s}{\rho g R} \cos \phi \qquad (R = \text{constant}) \tag{9-16}$$

This relation is also valid for nonwetting liquids (such as mercury in glass) and gives the capillary drop. In this case $\phi > 90°$ and thus $\cos \phi < 0$, which makes h negative. Therefore, a negative value of capillary rise corresponds to a capillary drop (Fig. 9–34).

Note that the capillary rise is inversely proportional to the radius of the tube. Therefore, the thinner the tube is, the greater the rise (or fall) of the liquid in the tube. In practice, the capillary effect (for water) is usually negligible in tubes whose diameter is greater than 1 cm. When pressure measurements are made using manometers and barometers, it is important to use sufficiently large tubes to minimize the capillary effect. The capillary rise is also inversely proportional to the density of the liquid, as expected. Therefore, in general, lighter liquids experience greater capillary rises. Finally, it should be kept in mind that Eq. 9–16 is derived for constant-diameter tubes and should not be used for tubes of variable cross section.

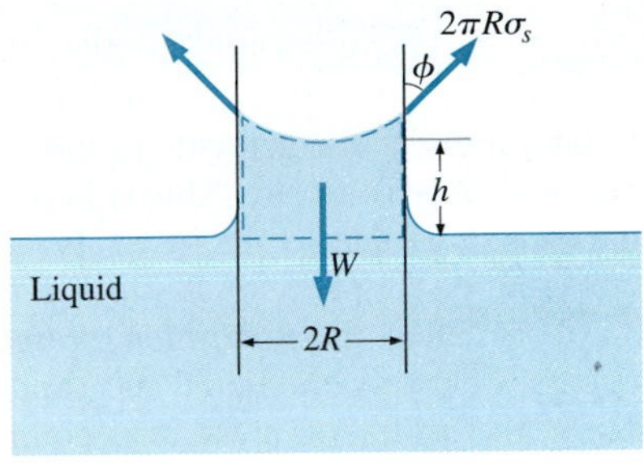

FIGURE 9–35

The forces acting on a liquid column that has risen in a tube due to the capillary effect.

EXAMPLE 9–5 The Capillary Rise of Water in a Tube

A 0.6-mm-diameter glass tube is inserted into water at 20°C in a cup. Determine the capillary rise of water in the tube (Fig. 9–36).

Solution The rise of water in a slender tube as a result of the capillary effect is to be determined.

Assumptions **1** There are no impurities in the water and no contamination on the surfaces of the glass tube. **2** The experiment is conducted in atmospheric air.

Properties The surface tension of water at 20°C is 0.073 N/m (Table 9–3). The contact angle of water with glass is 0° (from preceding text). We take the density of liquid water to be 1000 kg/m^3.

Analysis The capillary rise is determined directly from Eq. 9–16 by substituting the given values, yielding

$$h = \frac{2\sigma_s}{\rho g R} \cos \phi = \frac{2(0.073 \text{ N/m})}{(1000 \text{ kg/m}^3)(9.81 \text{ m/s}^2)(0.3 \times 10^{-3} \text{ m})} (\cos 0°) \left(\frac{1 \text{ kg} \cdot \text{m/s}^2}{1 \text{ N}}\right)$$

$$= 0.050 \text{ m} = \mathbf{5.0 \text{ cm}}$$

Therefore, water rises in the tube 5 cm above the liquid level in the cup.

Discussion Note that if the tube diameter were 1 cm, the capillary rise would be 0.3 mm, which is hardly noticeable to the eye. Actually, the capillary rise in a large-diameter tube occurs only at the rim. The center does not rise at all. Therefore, the capillary effect can be ignored for large-diameter tubes.

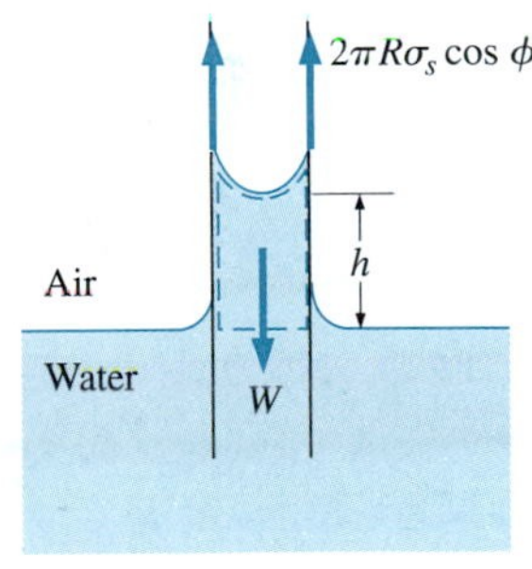

FIGURE 9–36

Schematic for Example 9–5.

SUMMARY

A fluid in direct contact with a solid surface sticks to the surface and there is no slip. This is known as the *no-slip condition*, and it is due to the viscosity of the fluid. At a given temperature, the pressure at which a pure substance changes phase is called the *saturation pressure*. For phase-change processes between the liquid and vapor phases of a pure substance, the saturation pressure is commonly called the *vapor pressure* P_v.

At a given temperature, the pressure at which a pure substance changes phase is called the *saturation pressure*. For phase-change processes between the liquid and vapor phases of a pure substance, the saturation pressure is commonly called the *vapor pressure* P_v. Vapor bubbles that form in the low-pressure regions in a liquid (a phenomenon called *cavitation*) collapse as they are swept away from the low-pressure regions, generating highly destructive, extremely high-pressure waves.

The velocity at which an infinitesimally small pressure wave travels through a medium is the *speed of sound*. For an ideal gas it is expressed as

$$c = \sqrt{\left(\frac{\partial P}{\partial \rho}\right)_s} = \sqrt{kRT}$$

The *viscosity* of a fluid is a measure of its resistance to deformation. The tangential force per unit area is called *shear stress* and is expressed for simple shear flow between plates (one-dimensional flow) as

$$\tau = \mu \frac{du}{dy}$$

where μ is the coefficient of viscosity or the *dynamic* (or *absolute*) *viscosity* of the fluid, u is the velocity component in the flow direction, and y is the direction normal to the flow direction. Fluids that obey this linear relationship are called *Newtonian fluids*. The ratio of dynamic viscosity to density is called the *kinematic viscosity* ν.

The pulling effect on the liquid molecules at an interface caused by the attractive forces of molecules per unit length is called *surface tension* σ_s. The excess pressure ΔP inside a spherical droplet or soap bubble, respectively, is given by

$$\Delta P_{\text{droplet}} = P_i - P_o = \frac{2\sigma_s}{R} \quad \text{and} \quad \Delta P_{\text{soap bubble}} = P_i - P_o = \frac{4\sigma_s}{R}$$

where P_i and P_o are the pressures inside and outside the droplet or soap bubble. The rise or fall of a liquid in a small-diameter tube inserted into the liquid due to surface tension is called the *capillary effect*. The capillary rise or drop is given by

$$h = \frac{2\sigma_s}{\rho g R} \cos \phi$$

where ϕ is the *contact angle*. The capillary rise is inversely proportional to the radius of the tube; for water, it is negligible for tubes whose diameter is larger than about 1 cm.

REFERENCES AND SUGGESTED READING

1. E. C. Bingham. "An Investigation of the Laws of Plastic Flow," *U.S. Bureau of Standards Bulletin*, 13, pp. 309–353, 1916.
2. Y. A. Çengel and M. A. Boles. *Thermodynamics: An Engineering Approach*, 6th ed. New York: McGraw-Hill, 2008.
3. D. C. Giancoli. *Physics*, 3rd ed. Upper Saddle River, NJ: Prentice Hall, 1991.
4. Y. S. Touloukian, S. C. Saxena, and P. Hestermans. *Thermophysical Properties of Matter, The TPRC Data Series*, Vol. 11, *Viscosity*. New York: Plenum, 1975.
5. L. Trefethen. "Surface Tension in Fluid Mechanics." In *Illustrated Experiments in Fluid Mechanics*. Cambridge, MA: MIT Press, 1972.
6. *The U.S. Standard Atmosphere*. Washington, DC: U.S. Government Printing Office, 1976.
7. M. Van Dyke. *An Album of Fluid Motion*. Stanford, CA: Parabolic Press, 1982.
8. C. L. Yaws, X. Lin, and L. Bu. "Calculate Viscosities for 355 Compounds. An Equation Can Be Used to Calculate Liquid Viscosity as a Function of Temperature," *Chemical Engineering*, 101, no. 4, pp. 119–1128, April 1994.
9. C. L. Yaws. *Handbook of Viscosity*. 3 Vols. Houston, TX: Gulf Publishing, 1994.
10. G. M. Homsy, H. Aref, K. S. Breuer, S. Hochgreb, J. R. Koseff, B. R. Munson, K. G. Powell, C. R. Robertson, and S. T. Thoroddsen, *Multi-Media Fluid Mechanics* (CD). Cambridge: Cambridge University Press, 2000.

PROBLEMS*

No Slip Condition and Classification of Fluid Flows

9–1C Define internal, external, and open-channel flows.

9–2C Define incompressible flow and incompressible fluid. Must the flow of a compressible fluid necessarily be treated as compressible?

9–3C What is the no-slip condition? What causes it?

9–4C What is forced flow? How does it differ from natural flow? Is flow caused by wind forced or natural flow?

Vapor Pressure and Cavitation

9–5C What is vapor pressure? How is it related to saturation pressure?

9–6C Does water boil at higher temperatures at higher pressures? Explain.

9–7C If the pressure of a substance is increased during a boiling process, will the temperature also increase or will it remain constant? Why?

9–8C What is cavitation? What causes it?

9–9 In a piping system, the water temperature remains under 40°C. Determine the minimum pressure allowed in the system to avoid cavitation.

9–10 The analysis of a propeller that operates in water at 20°C shows that the pressure at the tips of the propeller drops to 2 kPa at high speeds. Determine if there is a danger of cavitation for this propeller.

9–11E The analysis of a propeller that operates in water at 70°F shows that the pressure at the tips of the propeller drops to 0.1 psia at high speeds. Determine if there is a danger of cavitation for this propeller.

9–12 A pump is used to transport water to a higher reservoir. If the water temperature is 25°C, determine the lowest pressure that can exist in the pump without cavitation.

*Problems designated by a "C" are concept questions, and students are encouraged to answer them all. Problems designated by an "E" are in English units, and the SI users can ignore them. Problems with an icon are solved using EES, and complete solutions together with parametric studies are included on the enclosed DVD. Problems with the icon are comprehensive in nature, and are intended to be solved with a computer, preferably using the EES software that accompanies this text.

Speed of Sound

9–13C What is sound? How is it generated? How does it travel? Can sound waves travel in a vacuum?

9–14C Is it realistic to assume that the propagation of sound waves is an isentropic process? Explain.

9–15C In which medium does a sound wave travel faster: in cool air or in warm air?

9–16C In which medium will sound travel fastest for a given temperature: air, helium, or argon?

9–17 Determine the speed of sound in air at (*a*) 300 K and (*b*) 1000 K. Also determine the Mach number of an aircraft moving in air at a velocity of 240 m/s for both cases.

9–18 Assuming ideal gas behavior, determine the speed of sound in refrigerant-134a at 0.1 MPa and 60°C.

9–19 The isentropic process for an ideal gas is expressed as Pv^k = constant. Using this process equation and the definition of the speed of sound (Eq. 9–1), obtain the expression for the speed of sound for an ideal gas (Eq. 9–3).

9–20 Air expands isentropically from 1.5 MPa and 60°C to 0.4 MPa. Calculate the ratio of the initial to final speed of sound. *Answer:* 1.21

9–21 Repeat Prob. 9–20 for helium gas.

9–22E Air expands isentropically from 170 psia and 200°F to 60 psia. Calculate the ratio of the initial to final speed of sound. *Answer:* 1.16

Viscosity

9–23C What is viscosity? What is the cause of it in liquids and in gases? Do liquids or gases have higher dynamic viscosities?

9–24C What is a Newtonian fluid? Is water a Newtonian fluid?

9–25C Consider two identical small glass balls dropped into two identical containers, one filled with water and the other with oil. Which ball will reach the bottom of the container first? Why?

9–26C How does the dynamic viscosity of (*a*) liquids and (*b*) gases vary with temperature?

9–27C How does the kinematic viscosity of (*a*) liquids and (*b*) gases vary with temperature?

9–28 A 50-cm × 30-cm × 20-cm block weighing 150 N is to be moved at a constant velocity of 0.8 m/s on an inclined surface with a friction coefficient of 0.27. (*a*) Determine the force F that needs to be applied in the horizontal direction. (*b*) If a 0.4-mm-thick oil film with a dynamic viscosity of

0.012 Pa · s is applied between the block and inclined surface, determine the percent reduction in the required force.

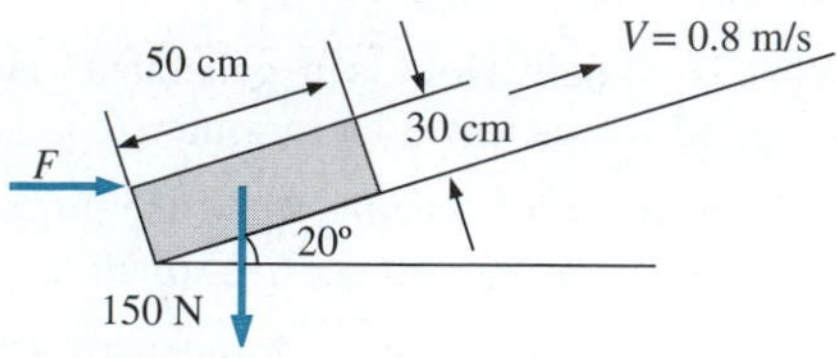

FIGURE P9–28

9–29 Consider the flow of a fluid with viscosity μ through a circular pipe. The velocity profile in the pipe is given as $u(r) = u_{max}(1 - r^n/R^n)$, where u_{max} is the maximum flow velocity, which occurs at the centerline; r is the radial distance from the centerline; and $u(r)$ is the flow velocity at any position r. Develop a relation for the drag force exerted on the pipe wall by the fluid in the flow direction per unit length of the pipe.

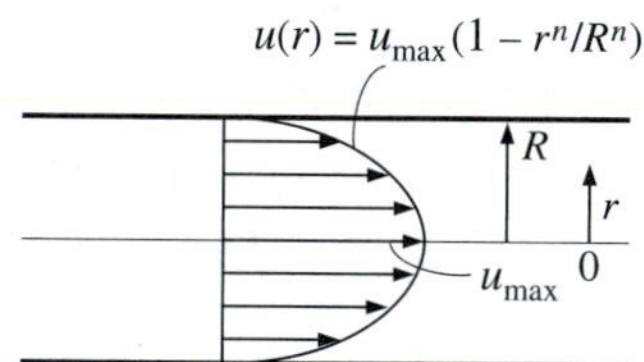

FIGURE P9–29

9–30 A thin 20-cm × 20-cm flat plate is pulled at 1 m/s horizontally through a 3.6-mm-thick oil layer sandwiched between two plates, one stationary and the other moving at a constant velocity of 0.3 m/s, as shown in Fig. P9–30. The dynamic viscosity of oil is 0.027 Pa · s. Assuming the velocity in each oil layer to vary linearly, (*a*) plot the velocity profile and find the location where the oil velocity is zero and (*b*) determine the force that needs to be applied on the plate to maintain this motion.

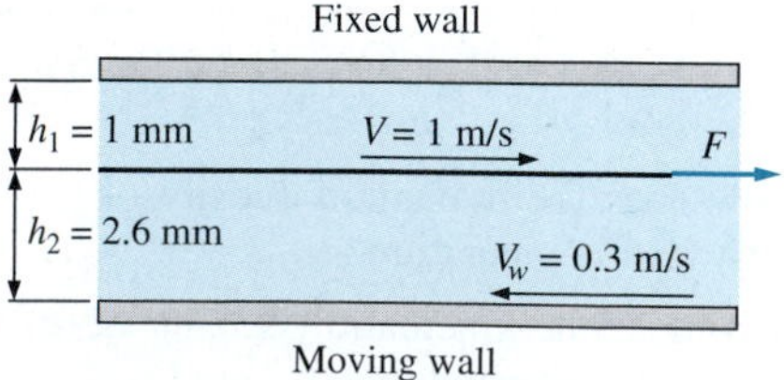

FIGURE P9–30

9–31 A frustum-shaped body is rotating at a constant angular speed of 200 rad/s in a container filled with SAE 10W oil at 20°C (μ = 0.1 Pa · s), as shown in Fig. P9–31. If the thickness of the oil film on all sides is 1.2 mm, determine the power required to maintain this motion. Also determine the reduction in the required power input when the oil temperature rises to 80°C (μ = 0.0078 Pa · s).

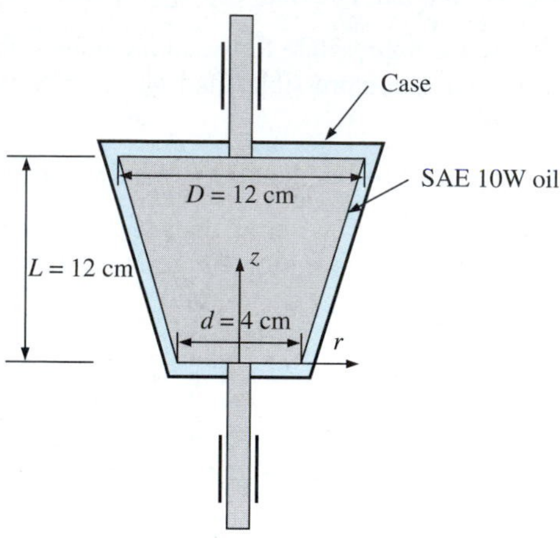

FIGURE P9–31

9–32 The clutch system shown in Fig. P9–32 is used to transmit torque through a 3-mm-thick oil film with μ = 0.38 N · s/m^2 between two identical 30-cm-diameter disks. When the driving shaft rotates at a speed of 1450 rpm, the driven shaft is observed to rotate at 1398 rpm. Assuming a linear velocity profile for the oil film, determine the transmitted torque.

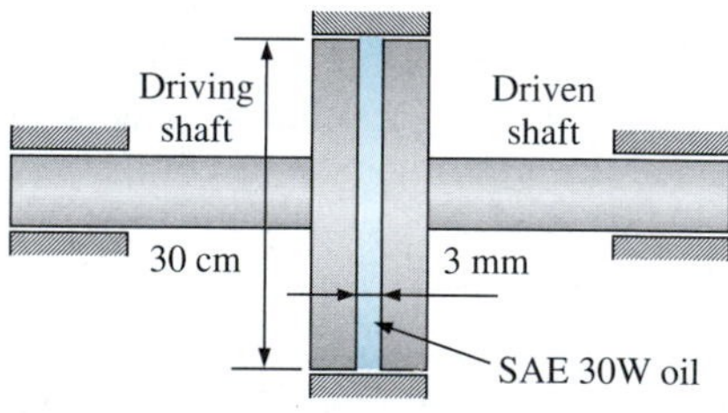

FIGURE P9–32

9–33 Reconsider Prob. 9–32. Using EES (or other) software, investigate the effect of oil film thickness on the torque transmitted. Let the film thickness vary from 0.1 mm to 10 mm. Plot your results, and state your conclusions.

9–34 The viscosity of some fluids changes when a strong electric field is applied on them. This phenomenon is known as the electrorheological (ER) effect, and fluids that exhibit such behavior are known as ER fluids. The Bingham plastic model for shear stress, which is expressed as $\tau = \tau_y + \mu(du/dy)$ is widely used to describe ER fluid behavior because of its simplicity. One of the most promising applications of ER fluids is the ER clutch. A typical multidisk ER clutch consists of several equally spaced steel disks of inner radius R_1 and outer radius R_2, N of them attached to the input shaft. The gap h between the parallel disks is filled with a viscous fluid. (*a*) Find a relationship for the torque generated by the clutch when the output shaft is stationary and (*b*) calculate the torque for an ER clutch with $N = 11$ for $R_1 = 50$ mm, $R_2 = 200$ mm, and $\dot{n} = 2400$ rpm if the fluid is SAE 10 with $\mu = 0.1$ Pa · s, $\tau_y = 2.5$ kPa, and $h = 1.2$ mm. *Answer:* (*b*) 2060 N · m

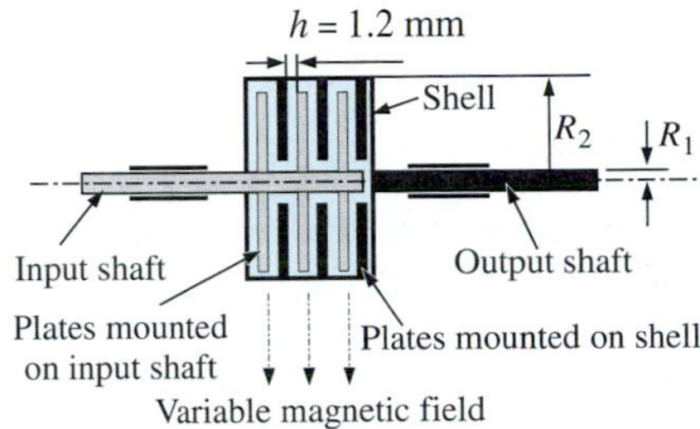

FIGURE P9–34

9–35 The viscosity of some fluids, called magnetorheological (MR) fluids, changes when a magnetic field is applied. Such fluids involve micron-sized magnetizable particles suspended in an appropriate carrier liquid, and are suitable for use in controllable hydraulic clutches. See Fig. P9–34. The MR fluids can have much higher viscosities than the ER fluids, and they often exhibit shear-thinning behavior in which the viscosity of the fluid decreases as the applied shear force increases. This behavior is also known as pseudoplastic behavior, and can be successfully represented by Herschel–Bulkley constitutive model expressed as $\tau = \tau_y + K(du/dy)^m$. Here τ is the shear stress applied, τ_y is the yield stress, K is the consistency index, and m is the power index. For a Herschel–Bulkley fluid with $\tau_y = 900$ Pa, $K = 58$ Pa · s^m, and $m = 0.82$, (*a*) find a relationship for the torque transmitted by an MR clutch for N plates attached to the input shaft when the input shaft is rotating at an angular speed of ω while the output shaft is stationary and (*b*) calculate the torque transmitted by such a clutch with $N = 11$ plates for $R_1 = 50$ mm, $R_2 = 200$ mm, $\dot{n} = 2400$ rpm, and $h = 1.2$ mm.

9–36 The viscosity of a fluid is to be measured by a viscometer constructed of two 75-cm-long concentric cylinders. The outer diameter of the inner cylinder is 15 cm, and the gap between the two cylinders is 0.12 cm. The inner cylinder is rotated at 200 rpm, and the torque is measured to be 0.8 N · m. Determine the viscosity of the fluid.

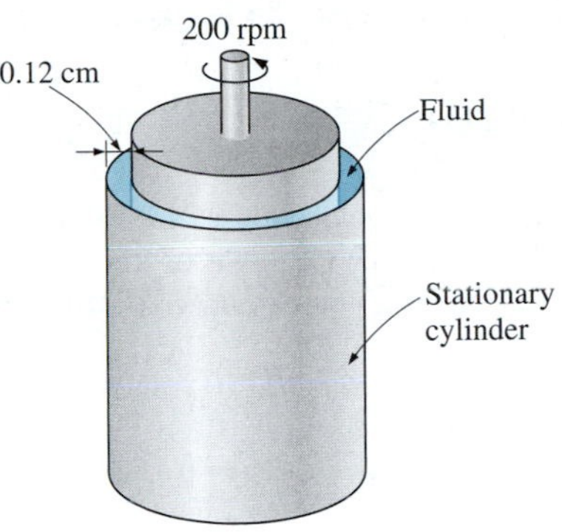

FIGURE P9–36

9–37E The viscosity of a fluid is to be measured by a viscometer constructed of two 3-ft-long concentric cylinders. The inner diameter of the outer cylinder is 6 in, and the gap between the two cylinders is 0.05 in. The outer cylinder is rotated at 250 rpm, and the torque is measured to be 1.2 lbf · ft. Determine the viscosity of the fluid. *Answer:* 0.000648 lbf · s/ft^2

9–38 In regions far from the entrance, fluid flow through a circular pipe is one-dimensional, and the velocity profile for laminar flow is given by $u(r) = u_{max}(1 - r^2/R^2)$, where R is the radius of the pipe, r is the radial distance from the center of the pipe, and u_{max} is the maximum flow velocity, which occurs at the center. Obtain (*a*) a relation for the drag force applied by the fluid on a section of the pipe of length L and (*b*) the value of the drag force for water flow at 20°C with $R = 0.08$ m, $L = 15$ m, $u_{max} = 3$ m/s, and $\mu = 0.0010$ kg/m · s.

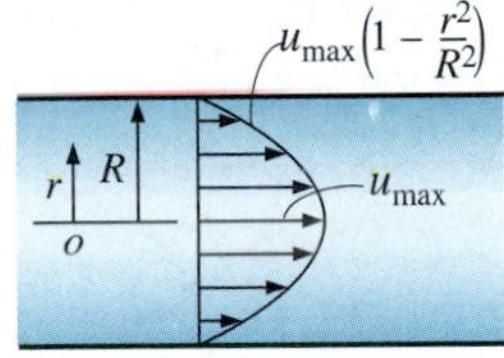

FIGURE P9–38

9–39 Repeat Prob. 9–38 for $u_{max} = 5$ m/s. *Answer:* (*b*) 0.942 N

Surface Tension and Capillary Effect

9–40C What is surface tension? What is it caused by? Why is the surface tension also called surface energy?

9–41C Consider a soap bubble. Is the pressure inside the bubble higher or lower than the pressure outside?

9–42C What is the capillary effect? What is it caused by? How is it affected by the contact angle?

9–43C A small-diameter tube is inserted into a liquid whose contact angle is 110°. Will the level of liquid in the tube be higher or lower than the level of the rest of the liquid? Explain.

9–44C Is the capillary rise greater in small- or large-diameter tubes?

9–45E A 0.03-in-diameter glass tube is inserted into kerosene at 68°F. The contact angle of kerosene with a glass surface is 26°. Determine the capillary rise of kerosene in the tube. *Answer:* 0.65 in

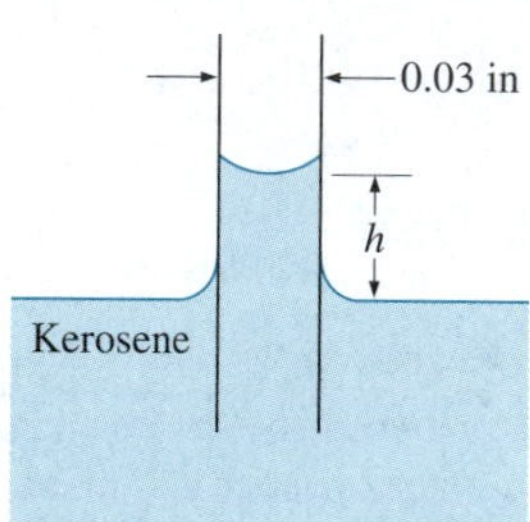

FIGURE P9–45E

9–46 A 1.9-mm-diameter tube is inserted into an unknown liquid whose density is 960 kg/m³, and it is observed that the liquid rises 5 mm in the tube, making a contact angle of 15°. Determine the surface tension of the liquid.

9–47 Determine the gage pressure inside a soap bubble of diameter (*a*) 0.2 cm and (*b*) 5 cm at 20°C.

9–48 Nutrients dissolved in water are carried to upper parts of plants by tiny tubes partly because of the capillary effect. Determine how high the water solution will rise in a tree in a 0.005-mm-diameter tube as a result of the capillary effect. Treat the solution as water at 20°C with a contact angle of 15°. *Answer:* 5.75 m

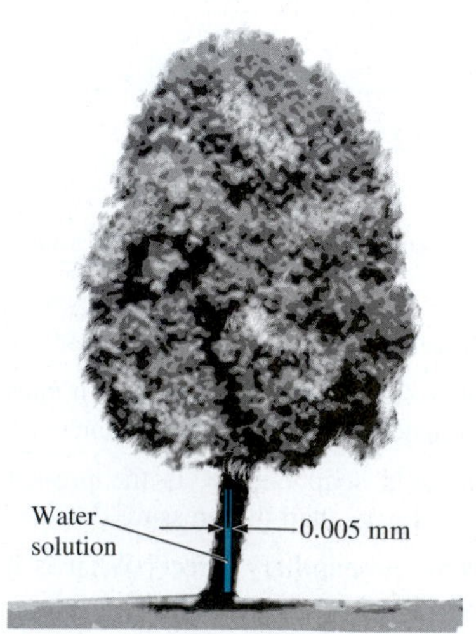

FIGURE P9–48

9–49 The surface tension of a liquid is to be measured using a liquid film suspended on a U-shaped wire frame with an 8-cm-long movable side. If the force needed to move the wire is 0.012 N, determine the surface tension of this liquid in air.

9–50 Contrary to what you might expect, a solid steel ball can float on water due to the surface tension effect. Determine the maximum diameter of a steel ball that would float on water at 20°C. What would your answer be for an aluminum ball? Take the densities of steel and aluminum balls to be 7800 kg/m³ and 2700 kg/m³, respectively.

Review Problems

9–51E The pressure on the suction side of pumps is typically low, and the surfaces on that side of the pump are susceptible to cavitation, especially at high fluid temperatures. If the minimum pressure on the suction side of a water pump is 0.95 psia absolute, determine the maximum water temperature to avoid the danger of cavitation.

9–52 A closed tank is partially filled with water at 60°C. If the air above the water is completely evacuated, determine the absolute pressure in the evacuated space. Assume the temperature to remain constant.

9–53 The variation of the dynamic viscosity of water with absolute temperature is given as

T, K	μ, Pa · s
273.15	1.787×10^{-3}
278.15	1.519×10^{-3}
283.15	1.307×10^{-3}
293.15	1.002×10^{-3}
303.15	7.975×10^{-4}
313.15	6.529×10^{-4}
333.15	4.665×10^{-4}
353.15	3.547×10^{-4}
373.15	2.828×10^{-4}

Using tabulated data, develop a relation for viscosity in the form of $\mu = \mu(T) = A + BT + CT^2 + DT^3 + ET^4$. Using the relation developed, predict the dynamic viscosity of water at 50°C at which the reported value is 5.468×10^{-4} Pa · s. Compare your result with the results of Andrade's equation, which is given in the form of $\mu = D \cdot e^{B/T}$, where D and B are constants whose values are to be determined using the viscosity data given.

9–54 Consider laminar flow of a Newtonian fluid of viscosity μ between two parallel plates. The flow is one-dimensional, and the velocity profile is given as $u(y) = 4u_{max}[y/h - (y/h)^2]$, where y is the vertical coordinate from the bottom surface, h is the distance between the two plates, and u_{max} is the maximum flow velocity that occurs at midplane. Develop a relation for the drag force exerted on both plates by the fluid in the flow direction per unit area of the plates.

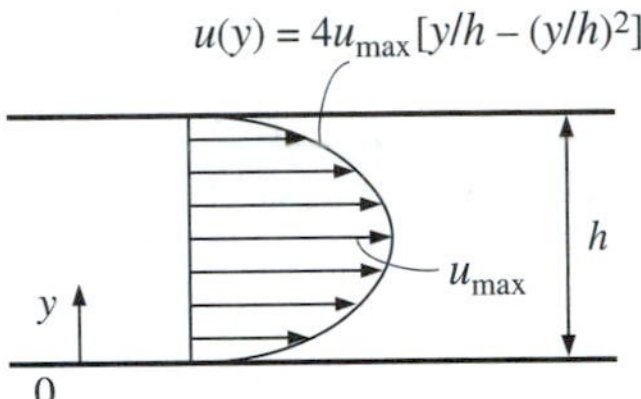

FIGURE P9–54

9–55 Some non-Newtonian fluids behave as a Bingham plastic for which shear stress can be expressed as $\tau = \tau_y + \mu(du/dr)$. For laminar flow of a Bingham plastic in a horizontal pipe of radius R, the velocity profile is given as $u(r) = (\Delta P/4\mu L)(r^2 - R^2) + (\tau_y/\mu)(r - R)$, where $\Delta P/L$ is the constant pressure drop along the pipe per unit length, μ is the dynamic viscosity, r is the radial distance from the centerline, and τ_y is the yield stress of Bingham plastic. Determine (*a*) the shear stress at the pipe wall and (*b*) the drag force acting on a pipe section of length L.

9–56 In some damping systems, a circular disk immersed in oil is used as a damper, as shown in Fig. P9–56. Show that the damping torque is proportional to angular speed in accordance with the relation $T_{damping} = C\omega$ where $C = 0.5\pi\mu(1/a + 1/b)R^4$. Assume linear velocity profiles on both sides of the disk and neglect the tip effects.

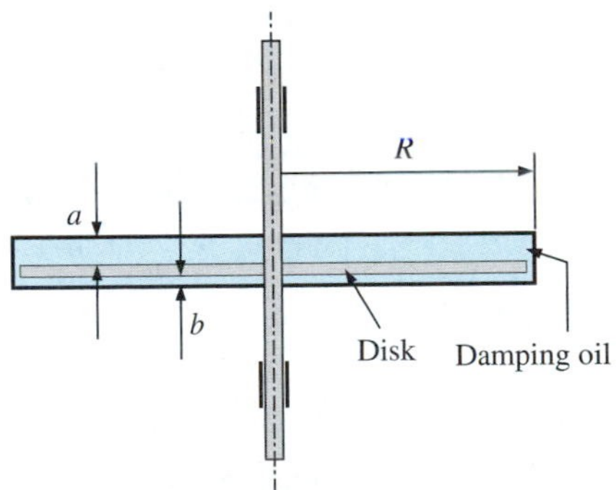

FIGURE P9–56

9–57E A 0.9-in-diameter glass tube is inserted into mercury, which makes a contact angle of 140° with glass. Determine the capillary drop of mercury in the tube at 68°F. *Answer:* 0.0175 in

9–58 Derive a relation for the capillary rise of a liquid between two large parallel plates a distance t apart inserted into the liquid vertically. Take the contact angle to be ϕ.

9–59 Consider a 30-cm-long journal bearing that is lubricated with oil whose viscosity is 0.1 kg/m · s at 20°C at the beginning of operation and 0.008 kg/m · s at the anticipated steady operating temperature of 80°C. The diameter of the shaft is 8 cm, and the average gap between the shaft and the journal is 0.08 cm. Determine the torque needed to overcome the bearing friction initially and during steady operation when the shaft is rotated at 500 rpm.

Design and Essay Problems

9–60 Design an experiment to measure the viscosity of liquids using a vertical funnel with a cylindrical reservoir of height h and a narrow flow section of diameter D and length L. Making appropriate assumptions, obtain a relation for viscosity in terms of easily measurable quantities such as density and volume flow rate.

9–61 Write an essay on the rise of the fluid to the top of the trees by capillary and other effects.

9–62 Write an essay on the oils used in car engines in different seasons and their viscosities.

Chapter 10

FLUID STATICS

This chapter deals with forces applied by fluids at rest. The fluid property responsible for those forces is *pressure*, which is a normal force exerted by a fluid per unit area. We start this chapter with a discussion of the *hydrostatic forces* applied on submerged bodies with plane or curved surfaces. We then consider the *buoyant force* applied by fluids on submerged or floating bodies, and discuss the *stability* of such bodies. This chapter makes extensive use of force balances for bodies in static equilibrium, and it will be helpful if the relevant topics from statics are first reviewed.

Objectives

The objectives of this chapter are to:

- Calculate the forces exerted by a fluid at rest on plane or curved submerged surfaces
- Analyze the stability of floating and submerged bodies.

10–1 ▪ INTRODUCTION

Fluid statics deals with problems associated with fluids at rest. The fluid can be either gaseous or liquid. Fluid statics is generally referred to as *hydrostatics* when the fluid is a liquid and as *aerostatics* when the fluid is a gas. In fluid statics, there is no relative motion between adjacent fluid layers, and thus there are no shear (tangential) stresses in the fluid trying to deform it. The only stress we deal with in fluid statics is the *normal stress*, which is the pressure, and the variation of pressure is due only to the weight of the fluid. Therefore, the topic of fluid statics has significance only in gravity fields, and the force relations developed naturally involve the gravitational acceleration g. The force exerted on a surface by a fluid at rest is normal to the surface at the point of contact since there is no relative motion between the fluid and the solid surface, and thus there are no shear forces acting parallel to the surface.

Fluid statics is used to determine the forces acting on floating or submerged bodies and the forces developed by devices like hydraulic presses and car jacks. The design of many engineering systems such as water dams and liquid storage tanks requires the determination of the forces acting on the surfaces using fluid statics. The complete description of the resultant hydrostatic force acting on a submerged surface requires the determination of the magnitude, the direction, and the line of action of the force. In Sections 10–2 and 10–3, we consider the forces acting on both plane and curved surfaces of submerged bodies due to pressure.

10–2 ▪ HYDROSTATIC FORCES ON SUBMERGED PLANE SURFACES

FIGURE 10–1
Hoover Dam.
Courtesy United States Department of the Interior, Bureau of Reclamation-Lower Colorado Region.

A plate, such as a gate valve in a dam, the wall of a liquid storage tank, or the hull of a ship at rest, is subjected to fluid pressure distributed over its surface when exposed to a liquid (Fig. 10–1). On a *plane* surface, the hydrostatic forces form a system of parallel forces, and we often need to determine the *magnitude* of the force and its *point of application*, which is called the **center of pressure**. In most cases, the other side of the plate is open to the atmosphere (such as the dry side of a gate), and thus atmospheric pressure acts on both sides of the plate, yielding a zero resultant. In such cases, it is convenient to subtract atmospheric pressure and work with the gage pressure only (Fig. 10–2). For example, $P_{gage} = \rho gh$ at the bottom of the lake.

Consider the top surface of a flat plate of arbitrary shape completely submerged in a liquid, as shown in Fig. 10–3 together with its normal view. The plane of this surface (normal to the page) intersects the horizontal free surface with an angle θ, and we take the line of intersection to be the x-axis (out of the page). The absolute pressure above the liquid is P_0, which is the local atmospheric pressure P_{atm} if the liquid is open to the atmosphere (but P_0 may be different than P_{atm} if the space above the liquid is evacuated or pressurized). Then the absolute pressure at any point on the plate is

$$P = P_0 + \rho gh = P_0 + \rho gy \sin \theta \qquad \textbf{(10–1)}$$

where h is the vertical distance of the point from the free surface and y is the distance of the point from the x-axis (from point O in Fig. 10–3). The

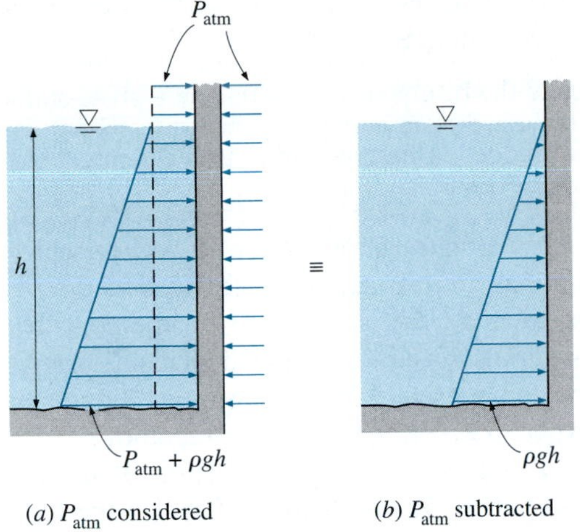

FIGURE 10–2
When analyzing hydrostatic forces on submerged surfaces, the atmospheric pressure can be subtracted for simplicity when it acts on both sides of the structure.

resultant hydrostatic force F_R acting on the surface is determined by integrating the force $P\,dA$ acting on a differential area dA over the entire surface area,

$$F_R = \int_A P\,dA = \int_A (P_0 + \rho g y \sin\theta)\,dA = P_0 A + \rho g \sin\theta \int_A y\,dA \qquad \textbf{(10–2)}$$

But the *first moment of area* $\int_A y\,dA$ is related to the *y*-coordinate of the centroid (or center) of the surface by

$$y_C = \frac{1}{A}\int_A y\,dA \qquad \textbf{(10–3)}$$

Substituting,

$$F_R = (P_0 + \rho g y_C \sin\theta)A = (P_0 + \rho g h_C)A = P_C A = P_{avg} A \qquad \textbf{(10–4)}$$

where $P_C = P_0 + \rho g h_C$ is the pressure at the centroid of the surface, which is equivalent to the *average* pressure P_{avg} on the surface, and $h_C = y_C \sin\theta$

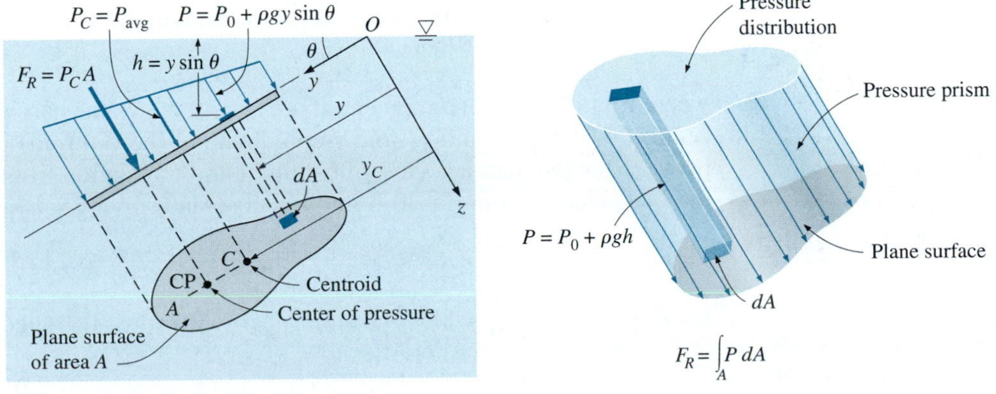

FIGURE 10–3
Hydrostatic force on an inclined plane surface completely submerged in a liquid.

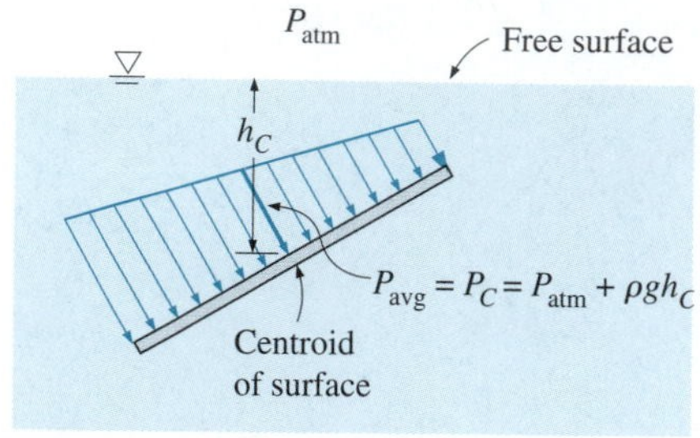

FIGURE 10–4
The pressure at the centroid of a surface is equivalent to the *average* pressure on the surface.

is the *vertical distance* of the centroid from the free surface of the liquid (Fig. 10–4). Thus we conclude that:

> The magnitude of the resultant force acting on a plane surface of a completely submerged plate in a homogeneous (constant density) fluid is equal to the product of the pressure P_C at the centroid of the surface and the area A of the surface (Fig. 10–5).

The pressure P_0 is usually atmospheric pressure, which cancels out in most force calculations since it acts on both sides of the plate. When this is not the case, a practical way of accounting for the contribution of P_0 to the resultant force is simply to add an equivalent depth $h_{equiv} = P_0/\rho g$ to h_C; that is, to assume the presence of an additional liquid layer of thickness h_{equiv} on top of the liquid with absolute vacuum above.

Next we need to determine the line of action of the resultant force F_R. Two parallel force systems are equivalent if they have the same magnitude and the same moment about any point. The line of action of the resultant hydrostatic force, in general, does not pass through the centroid of the surface—it lies underneath where the pressure is higher. The point of intersection of the line of action of the resultant force and the surface is the **center of pressure**. The vertical location of the line of action is determined by equating the moment of the resultant force to the moment of the distributed pressure force about the x-axis. It gives

$$y_P F_R = \int_A yP\, dA = \int_A y(P_0 + \rho g y \sin\theta)\, dA = P_0 \int_A y\, dA + \rho g \sin\theta \int_A y^2\, dA$$

or

$$y_P F_R = P_0 y_C A + \rho g \sin\theta\, I_{xx,\,O} \qquad \textbf{(10–5)}$$

where y_P is the distance of the center of pressure from the x-axis (point O in Fig. 10–5) and $I_{xx,\,O} = \int_A y^2\, dA$ is the *second moment of area* (also called the *area moment of inertia*) about the x-axis. The second moments of area are widely available for common shapes in engineering handbooks, but they are usually given about the axes passing through the centroid of the area. Fortunately, the second moments of area about two parallel axes are related to each other by the *parallel axis theorem,* which in this case is expressed as

$$I_{xx,\,O} = I_{xx,\,C} + y_C^2 A \qquad \textbf{(10–6)}$$

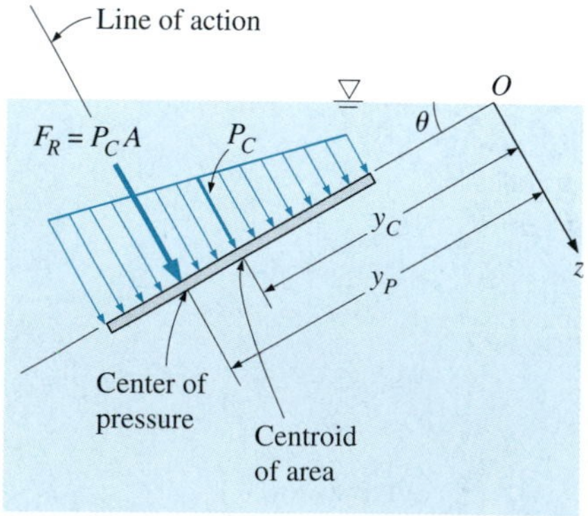

where $I_{xx,\,C}$ is the second moment of area about the x-axis passing through the centroid of the area and y_C (the y-coordinate of the centroid) is the distance between the two parallel axes. Substituting the F_R relation from Eq. 10–4 and the $I_{xx,\,O}$ relation from Eq. 10–5 into Eq. 10–6 and solving for y_P gives

$$y_P = y_C + \frac{I_{xx,\,C}}{[y_C + P_0/(\rho g \sin\theta)]A} \qquad \textbf{(10–7)}$$

For the usual case in which the atmospheric pressure is applied to both sides of the plate, Eq. 10–7 simplifies to

$$y_P = y_C + \frac{I_{xx,\,C}}{y_C A} \qquad \textbf{(10–8)}$$

Knowing y_P, the vertical distance of the center of pressure from the free surface is determined from $h_P = y_P \sin\theta$.

Line of action
$F_R = P_C A$
P_C
O
θ
y_C
y_P
z
Center of pressure
Centroid of area

FIGURE 10–5
The resultant force acting on a plane surface is equal to the product of the pressure at the centroid of the surface and the surface area, and its line of action passes through the center of pressure.

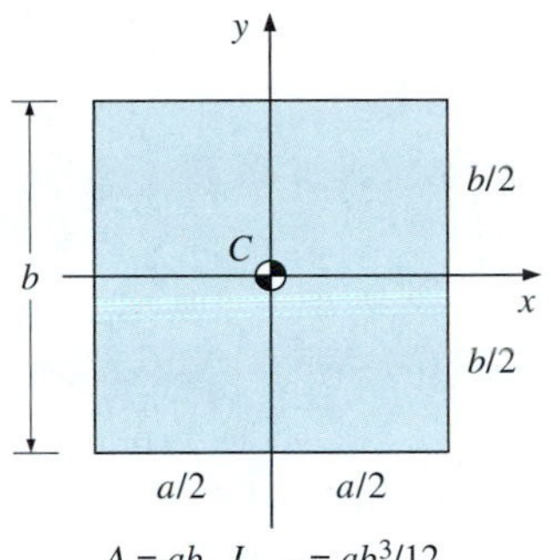

$A = ab,\ I_{xx,\,C} = ab^3/12$

(*a*) Rectangle

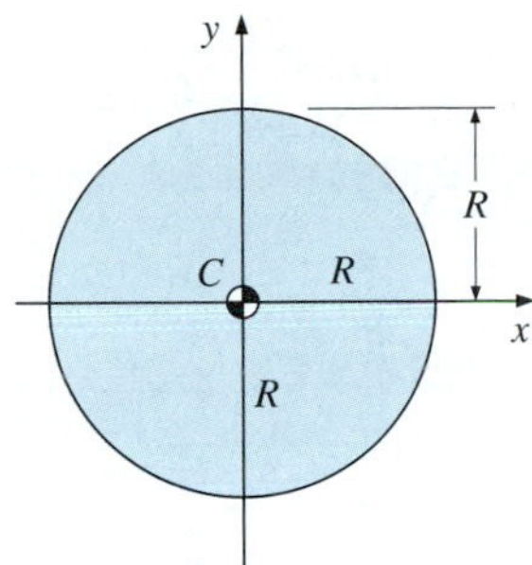

$A = \pi R^2,\ I_{xx,\,C} = \pi R^4/4$

(*b*) Circle

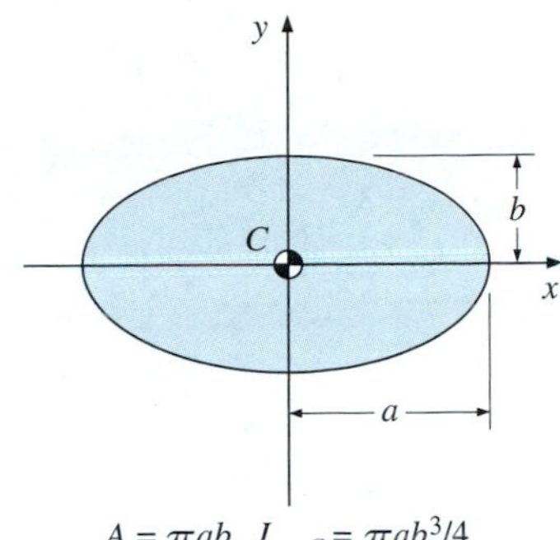

$A = \pi ab,\ I_{xx,\,C} = \pi ab^3/4$

(*c*) Ellipse

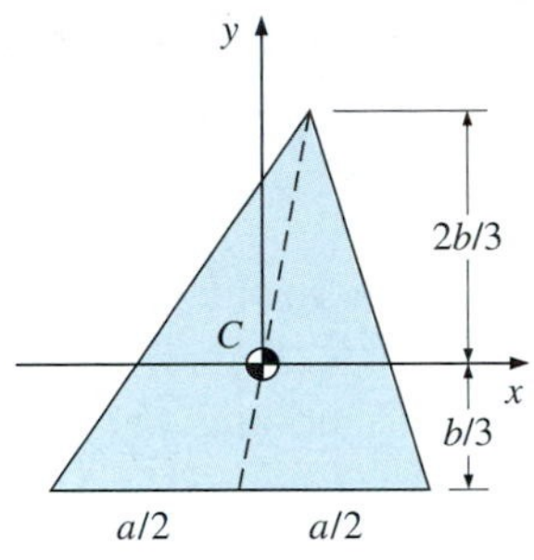

$A = ab/2,\ I_{xx,\,C} = ab^3/36$

(*d*) Triangle

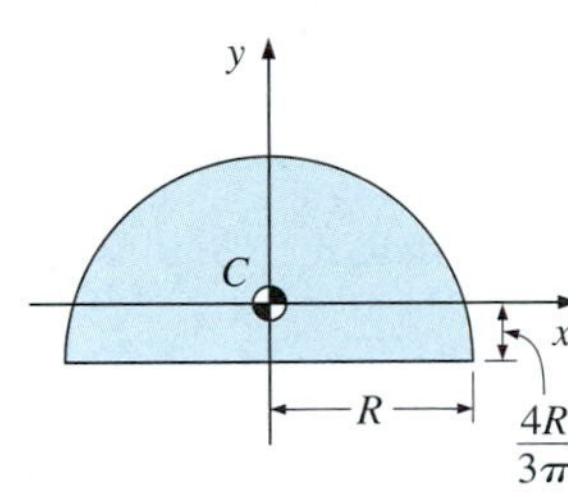

$A = \pi R^2/2,\ I_{xx,\,C} = 0.109757R^4$

(*e*) Semicircle

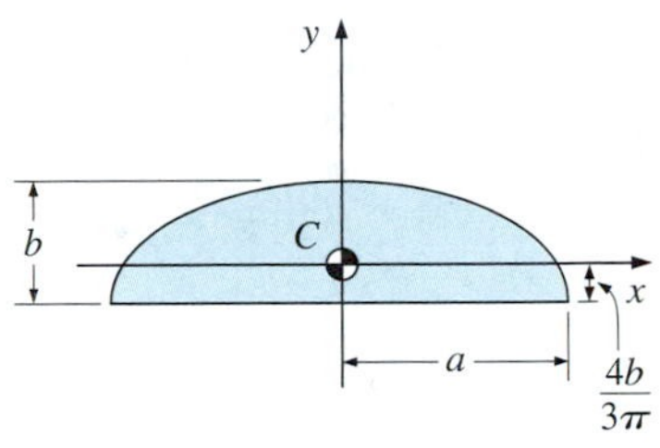

$A = \pi ab/2,\ I_{xx,\,C} = 0.109757ab^3$

(*f*) Semiellipse

FIGURE 10–6

The centroid and the centroidal moments of inertia for some common geometries.

The $I_{xx,\,C}$ values for some common areas are given in Fig. 10–6. For areas that possess symmetry about the *y*-axis, the center of pressure lies on the *y*-axis directly below the centroid. The location of the center of pressure in such cases is simply the point on the surface of the vertical plane of symmetry at a distance h_P from the free surface.

Special Case: Submerged Rectangular Plate

Consider a completely submerged rectangular flat plate of height b and width a tilted at an angle θ from the horizontal and whose top edge is horizontal and is at a distance s from the free surface along the plane of the plate, as shown in Fig. 10–7*a*. The resultant hydrostatic force on the upper surface is equal to the average pressure, which is the pressure at the midpoint of the surface, times the surface area *A*. That is,

Tilted rectangular plate: $$F_R = P_C A = [P_0 + \rho g(s + b/2)\sin\theta]ab \quad \textbf{(10–9)}$$

The force acts at a vertical distance of $h_P = y_P \sin\theta$ from the free surface directly beneath the centroid of the plate where, from Eq. 10–8,

$$y_P = s + \frac{b}{2} + \frac{ab^3/12}{[s + b/2 + P_0/(\rho g \sin\theta)]ab}$$

$$= s + \frac{b}{2} + \frac{b^2}{12[s + b/2 + P_0/(\rho g \sin\theta)]} \quad \textbf{(10–10)}$$

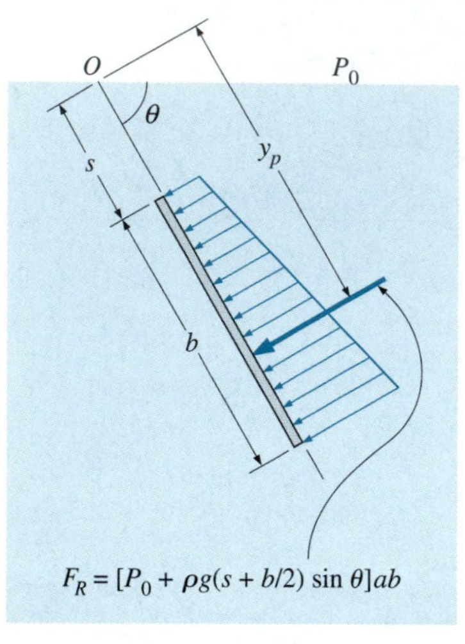

(*a*) Tilted plate

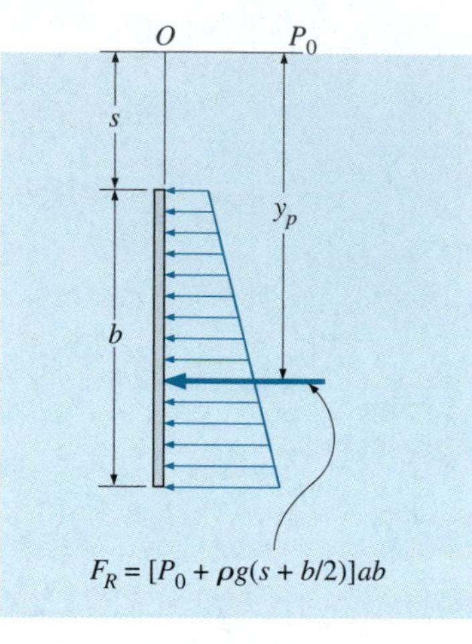

(*b*) Vertical plate

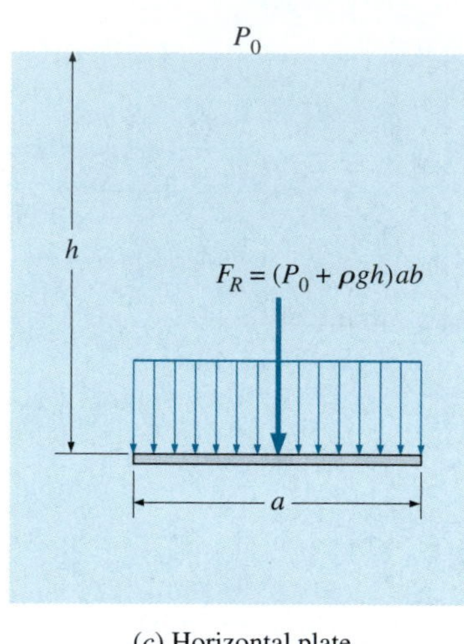

(*c*) Horizontal plate

FIGURE 10–7
Hydrostatic force acting on the top surface of a submerged rectangular plate for tilted, vertical, and horizontal cases.

When the upper edge of the plate is at the free surface and thus $s = 0$, Eq. 10–9 reduces to

Tilted rectangular plate ($s = 0$): $$F_R = [P_0 + \rho g(b \sin\theta)/2]ab \qquad \textbf{(10–11)}$$

For a completely submerged *vertical* plate ($\theta = 90°$) whose top edge is horizontal, the hydrostatic force can be obtained by setting $\sin\theta = 1$ (Fig. 10–7*b*)

Vertical rectangular plate: $$F_R = [P_0 + \rho g(s + b/2)]ab \qquad \textbf{(10–12)}$$

Vertical rectangular plate ($s = 0$): $$F_R = (P_0 + \rho gb/2)ab \qquad \textbf{(10–13)}$$

When the effect of P_0 cancels since it acts on both sides of the plate, the hydrostatic force on a vertical rectangular surface of height b whose top edge is horizontal and at the free surface is $F_R = \rho gab^2/2$ acting at a distance of $2b/3$ from the free surface directly beneath the centroid of the plate.

The pressure distribution on a submerged *horizontal* surface is uniform, and its magnitude is $P = P_0 + \rho gh$, where h is the distance of the surface from the free surface. Therefore, the hydrostatic force acting on a horizontal rectangular surface is

Horizontal rectangular plate: $$F_R = (P_0 + \rho gh)ab \qquad \textbf{(10–14)}$$

and it acts through the midpoint of the plate (Fig. 10–7*c*).

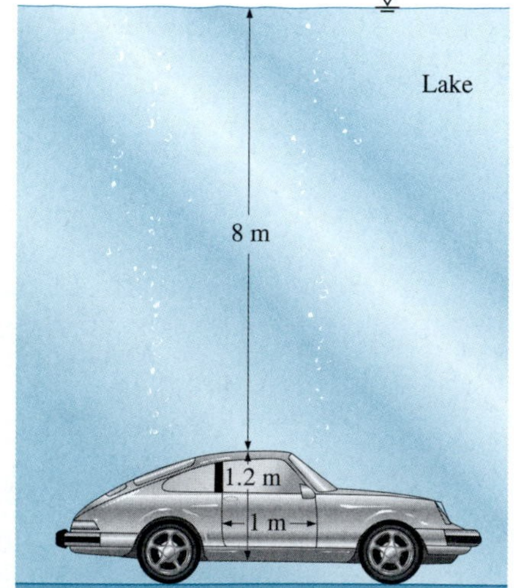

FIGURE 10–8
Schematic for Example 10–1.

EXAMPLE 10–1 Hydrostatic Force Acting on the Door of a Submerged Car

A heavy car plunges into a lake during an accident and lands at the bottom of the lake on its wheels (Fig. 10–8). The door is 1.2 m high and 1 m wide, and the top edge of the door is 8 m below the free surface of the water. Determine the hydrostatic force on the door and the location of the pressure center, and discuss if the driver can open the door.

Solution A car is submerged in water. The hydrostatic force on the door is to be determined, and the likelihood of the driver opening the door is to be assessed.

Assumptions **1** The bottom surface of the lake is horizontal. **2** The passenger cabin is well-sealed so that no water leaks inside. **3** The door can be approximated as a vertical rectangular plate. **4** The pressure in the passenger cabin remains at atmospheric value since there is no water leaking in, and thus no compression of the air inside. Therefore, atmospheric pressure cancels out in the calculations since it acts on both sides of the door. **5** The weight of the car is larger than the buoyant force acting on it.

Properties We take the density of lake water to be 1000 kg/m³ throughout.

Analysis The average (gage) pressure on the door is the pressure value at the centroid (midpoint) of the door and is determined to be

$$P_{avg} = P_C = \rho g h_C = \rho g(s + b/2)$$

$$= (1000 \text{ kg/m}^3)(9.81 \text{ m/s}^2)(8 + 1.2/2 \text{ m})\left(\frac{1 \text{ kN}}{1000 \text{ kg} \cdot \text{m/s}^2}\right)$$

$$= \mathbf{84.4 \text{ kN/m}^2}$$

Then the resultant hydrostatic force on the door becomes

$$F_R = P_{avg}A = (84.4 \text{ kN/m}^2)(1 \text{ m} \times 1.2 \text{ m}) = \mathbf{101.3 \text{ kN}}$$

The pressure center is directly under the midpoint of the door, and its distance from the surface of the lake is determined from Eq. 10–10 with the P_0 term removed, which yields

$$y_P = s + \frac{b}{2} + \frac{b^2}{12(s + b/2)} = 8 + \frac{1.2}{2} + \frac{1.2^2}{12(8 + 1.2/2)} = \mathbf{8.61 \text{ m}}$$

Discussion A strong person can lift 100 kg, whose weight is 981 N or about 1 kN. Also, the person can apply the force at a point farthest from the hinges (1 m farther) for maximum effect and generate a moment of 1 kN · m. The resultant hydrostatic force acts under the midpoint of the door, and thus a distance of 0.5 m from the hinges. This generates a moment of 50.6 kN · m, which is about 50 times the moment the driver can possibly generate. Therefore, it is impossible for the driver to open the door of the car. The driver's best bet is to wait for some water to leak in, or let some water in (by rolling the window down a little, for example) and to keep his or her head close to the ceiling. The driver should be able to open the door shortly before the car is filled with water since at that point the pressures on both sides of the door are nearly the same and opening the door in water is almost as easy as opening it in air.

10–3 ▪ HYDROSTATIC FORCES ON SUBMERGED CURVED SURFACES

For a submerged curved surface, the determination of the resultant hydrostatic force is more involved since it typically requires integration of the pressure forces that change direction along the curved surface.

The easiest way to determine the resultant hydrostatic force F_R acting on a two-dimensional curved surface is to determine the horizontal and vertical components F_H and F_V separately. This is done by considering the free-body diagram of the liquid block enclosed by the curved surface and the two

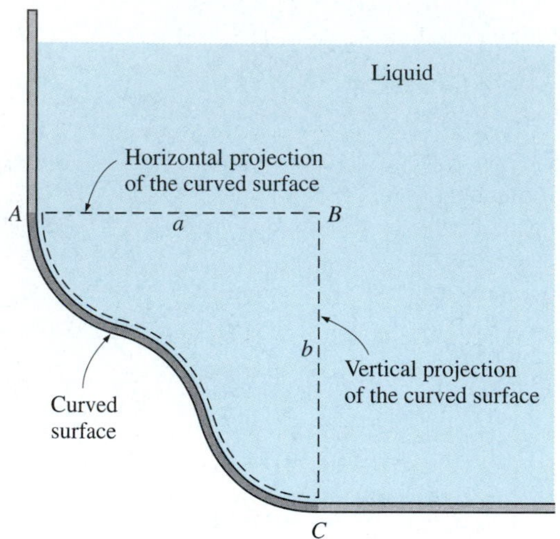

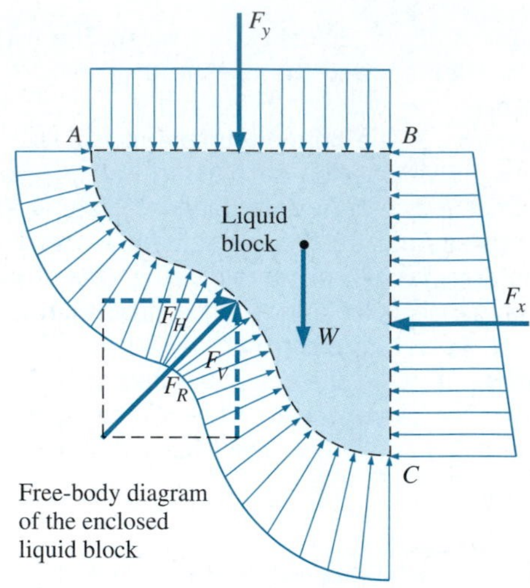

FIGURE 10–9
Determination of the hydrostatic force acting on a submerged curved surface.

plane surfaces (one horizontal and one vertical) passing through the two ends of the curved surface, as shown in Fig. 10–9. Note that the vertical surface of the liquid block considered is simply the projection of the curved surface on a *vertical plane*, and the horizontal surface is the projection of the curved surface on a *horizontal plane.* The resultant force acting on the curved solid surface is then equal and opposite to the force acting on the curved liquid surface (Newton's third law).

The force acting on the imaginary horizontal or vertical plane surface and its line of action can be determined as discussed in Section 10–2. The weight of the enclosed liquid block of volume V is simply $W = \rho g V$, and it acts downward through the centroid of this volume. Noting that the liquid block is in static equilibrium, the force balances in the horizontal and vertical directions give

Horizontal force component on curved surface: $$F_H = F_x \qquad \textbf{(10–15)}$$

Vertical force component on curved surface: $$F_V = F_y + W \qquad \textbf{(10–16)}$$

where the summation $F_y + W$ is a vector addition (i.e., add magnitudes if both act in the same direction and subtract if they act in opposite directions). Thus, we conclude that

1. The horizontal component of the hydrostatic force acting on a curved surface is equal (in both magnitude and the line of action) to the hydrostatic force acting on the vertical projection of the curved surface.
2. The vertical component of the hydrostatic force acting on a curved surface is equal to the hydrostatic force acting on the horizontal projection of the curved surface, plus (minus, if acting in the opposite direction) the weight of the enclosed liquid block.

The magnitude of the resultant hydrostatic force acting on the curved surface is $F_R = \sqrt{F_H^2 + F_V^2}$, and the tangent of the angle it makes with the horizontal is $\tan \alpha = F_V/F_H$. The exact location of the line of action of the resultant force (e.g., its distance from one of the end points of the curved surface) can be determined by taking a moment about an appropriate point. These discussions are valid for all curved surfaces regardless of whether they are above or below the liquid. Note that in the case of a *curved surface above a liquid*, the weight of the liquid is *subtracted* from the vertical component of the hydrostatic force since they act in opposite directions (Fig. 10–10).

When the curved surface is a *circular arc* (full circle or any part of it), the resultant hydrostatic force acting on the surface always passes through the center of the circle. This is because the pressure forces are normal to the surface, and all lines normal to the surface of a circle pass through the center of the circle. Thus, the pressure forces form a concurrent force system at the center, which can be reduced to a single equivalent force at that point (Fig. 10–11).

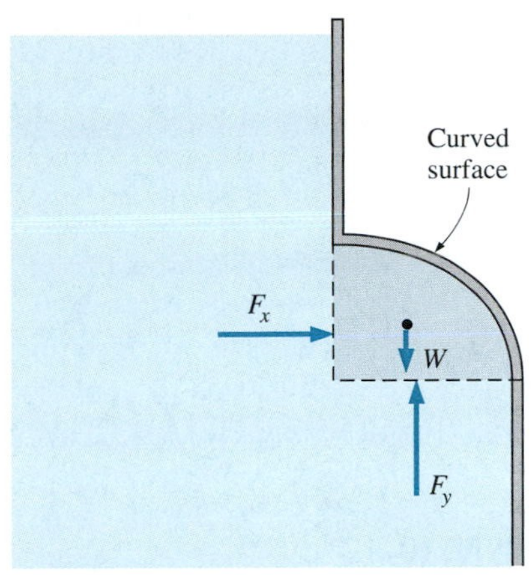

FIGURE 10–10
When a curved surface is above the liquid, the weight of the liquid and the vertical component of the hydrostatic force act in the opposite directions.

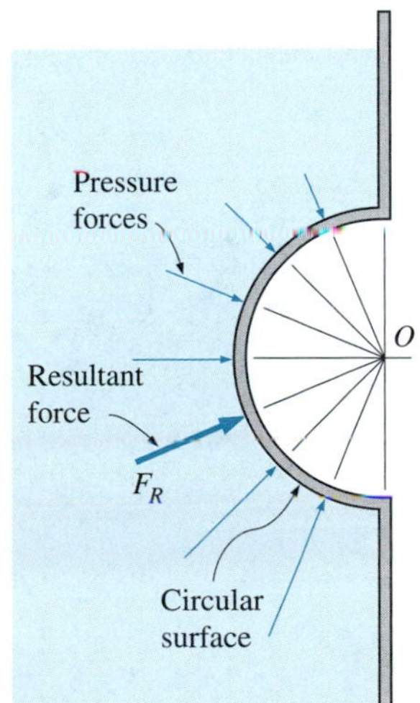

FIGURE 10–11
The hydrostatic force acting on a circular surface always passes through the center of the circle since the pressure forces are normal to the surface and they all pass through the center.

EXAMPLE 10–2 A Gravity-Controlled Cylindrical Gate

A long solid cylinder of radius 0.8 m hinged at point *A* is used as an automatic gate, as shown in Fig. 10–12. When the water level reaches 5 m, the gate opens by turning about the hinge at point *A*. Determine (*a*) the hydrostatic force acting on the cylinder and its line of action when the gate opens and (*b*) the weight of the cylinder per m length of the cylinder.

Solution The height of a water reservoir is controlled by a cylindrical gate hinged to the reservoir. The hydrostatic force on the cylinder and the weight of the cylinder per m length are to be determined.

Assumptions **1** Friction at the hinge is negligible. **2** Atmospheric pressure acts on both sides of the gate, and thus it cancels out.

Properties We take the density of water to be 1000 kg/m³ throughout.

Analysis (*a*) We consider the free-body diagram of the liquid block enclosed by the circular surface of the cylinder and its vertical and horizontal projections. The hydrostatic forces acting on the vertical and horizontal plane surfaces as well as the weight of the liquid block are determined as

Horizontal force on vertical surface:

$$F_H = F_x = P_{avg}A = \rho g h_C A = \rho g(s + R/2)A$$

$$= (1000\ \text{kg/m}^3)(9.81\ \text{m/s}^2)(4.2 + 0.8/2\ \text{m})(0.8\ \text{m} \times 1\ \text{m})\left(\frac{1\ \text{kN}}{1000\ \text{kg} \cdot \text{m/s}^2}\right)$$

$$= \mathbf{36.1\ kN}$$

Vertical force on horizontal surface (upward):

$$F_y = P_{avg}A = \rho g h_C A = \rho g h_{bottom} A$$

$$= (1000\ \text{kg/m}^3)(9.81\ \text{m/s}^2)(5\ \text{m})(0.8\ \text{m} \times 1\ \text{m})\left(\frac{1\ \text{kN}}{1000\ \text{kg} \cdot \text{m/s}^2}\right)$$

$$= 39.2\ \text{kN}$$

Weight of fluid block per m length (downward):

$$W = mg = \rho g V = \rho g(R^2 - \pi R^2/4)(1\ \text{m})$$

$$= (1000\ \text{kg/m}^3)(9.81\ \text{m/s}^2)(0.8\ \text{m})^2(1 - \pi/4)(1\ \text{m})\left(\frac{1\ \text{kN}}{1000\ \text{kg} \cdot \text{m/s}^2}\right)$$

$$= 1.3\ \text{kN}$$

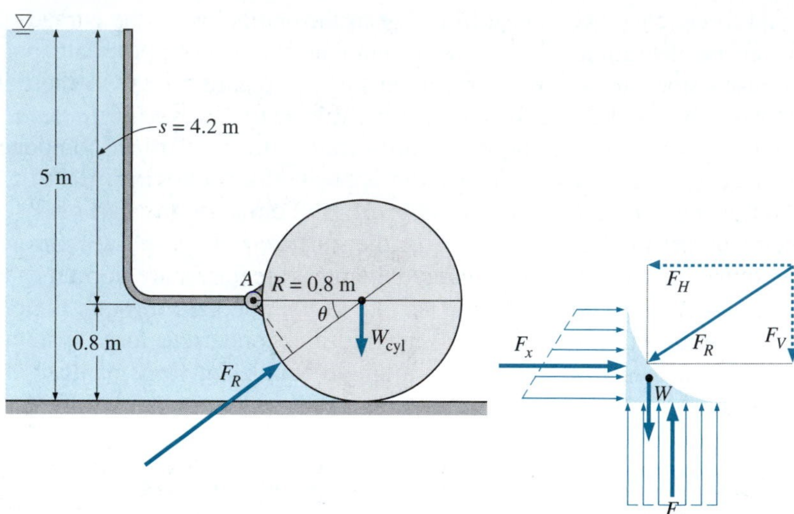

FIGURE 10–12
Schematic for Example 10–2 and the free-body diagram of the liquid underneath the cylinder.

Therefore, the net upward vertical force is

$$F_V = F_y - W = 39.2 - 1.3 = 37.9 \text{ kN}$$

Then the magnitude and direction of the hydrostatic force acting on the cylindrical surface become

$$F_R = \sqrt{F_H^2 + F_V^2} = \sqrt{36.1^2 + 37.9^2} = \mathbf{52.3 \text{ kN}}$$

$$\tan\theta = F_V/F_H = 37.9/36.1 = 1.05 \rightarrow \theta = 46.4°$$

Therefore, the magnitude of the hydrostatic force acting on the cylinder is 52.3 kN per m length of the cylinder, and its line of action passes through the center of the cylinder making an angle 46.4° with the horizontal.

(*b*) When the water level is 5 m high, the gate is about to open and thus the reaction force at the bottom of the cylinder is zero. Then the forces other than those at the hinge acting on the cylinder are its weight, acting through the center, and the hydrostatic force exerted by water. Taking a moment about point *A* at the location of the hinge and equating it to zero gives

$$F_R R \sin\theta - W_{cyl} R = 0 \rightarrow W_{cyl} = F_R \sin\theta = (52.3 \text{ kN}) \sin 46.4° = \mathbf{37.9 \text{ kN}}$$

Discussion The weight of the cylinder per m length is determined to be 37.9 kN. It can be shown that this corresponds to a mass of 3863 kg per m length and to a density of 1921 kg/m³ for the material of the cylinder.

10–4 ▪ BUOYANCY AND STABILITY

It is a common experience that an object feels lighter and weighs less in a liquid than it does in air. This can be demonstrated easily by weighing a heavy object in water by a waterproof spring scale. Also, objects made of wood or other light materials float on water. These and other observations suggest that a fluid exerts an upward force on a body immersed in it. This force that tends to lift the body is called the **buoyant force** and is denoted by F_B.

The buoyant force is caused by the increase of pressure with depth in a fluid. Consider, for example, a flat plate of thickness h submerged in a liquid of density ρ_f parallel to the free surface, as shown in Fig. 10–13. The area of the top (and also bottom) surface of the plate is A, and its distance to the free surface is s. The pressures at the top and bottom surfaces of the plate are $\rho_f gs$ and $\rho_f g(s + h)$, respectively. Then the hydrostatic force $F_{\text{top}} = \rho_f gsA$ acts downward on the top surface, and the larger force $F_{\text{bottom}} = \rho_f g(s + h)A$ acts upward on the bottom surface of the plate. The difference between these two forces is a net upward force, which is the *buoyant force*,

$$F_B = F_{\text{bottom}} - F_{\text{top}} = \rho_f g(s + h)A - \rho_f gsA = \rho_f ghA = \rho_f gV \quad \text{(10–17)}$$

where $V = hA$ is the volume of the plate. But the relation $\rho_f gV$ is simply the weight of the liquid whose volume is equal to the volume of the plate. Thus, we conclude that *the buoyant force acting on the plate is equal to the weight of the liquid displaced by the plate.* For a fluid with constant density, the buoyant force is independent of the distance of the body from the free surface. It is also independent of the density of the solid body.

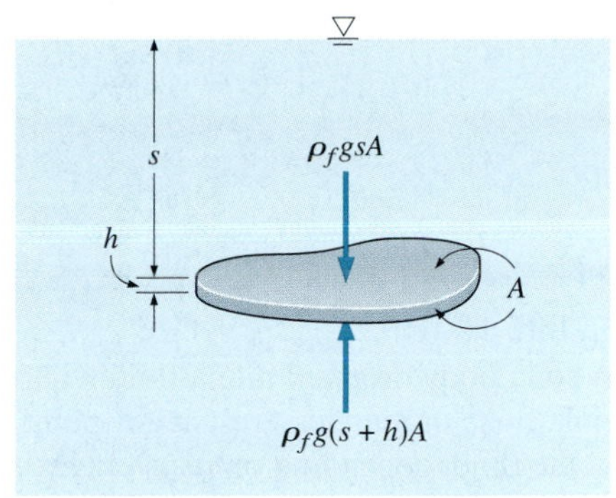

FIGURE 10–13
A flat plate of uniform thickness h submerged in a liquid parallel to the free surface.

The relation in Eq. 10–17 is developed for a simple geometry, but it is valid for any body regardless of its shape. This can be shown mathematically by a force balance, or simply by this argument: Consider an arbitrarily shaped solid body submerged in a fluid at rest and compare it to a body of fluid of the same shape indicated by dotted lines at the same vertical location (Fig. 10–14). The buoyant forces acting on these two bodies are the same since the pressure distributions, which depend only on elevation, are the same at the boundaries of both. The imaginary fluid body is in static equilibrium, and thus the net force and net moment acting on it are zero. Therefore, the upward buoyant force must be equal to the weight of the imaginary fluid body whose volume is equal to the volume of the solid body. Further, the weight and the buoyant force must have the same line of action to have a zero moment. This is known as **Archimedes' principle**, after the Greek mathematician Archimedes (287–212 BC), and is expressed as

> The buoyant force acting on a body immersed in a fluid is equal to the weight of the fluid displaced by the body, and it acts upward through the centroid of the displaced volume.

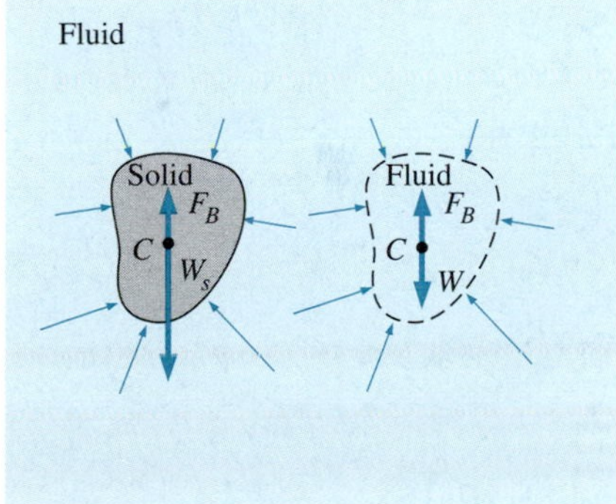

FIGURE 10–14
The buoyant forces acting on a solid body submerged in a fluid and on a fluid body of the same shape at the same depth are identical. The buoyant force F_B acts upward through the centroid C of the displaced volume and is equal in magnitude to the weight W of the displaced fluid, but is opposite in direction. For a solid of uniform density, its weight W_s also acts through the centroid, but its magnitude is not necessarily equal to that of the fluid it displaces. (Here $W_s > W$ and thus $W_s > F_B$; this solid body would sink.)

For *floating* bodies, the weight of the entire body must be equal to the buoyant force, which is the weight of the fluid whose volume is equal to the volume of the submerged portion of the floating body. That is,

$$F_B = W \rightarrow \rho_f gV_{\text{sub}} = \rho_{\text{avg, body}} gV_{\text{total}} \rightarrow \frac{V_{\text{sub}}}{V_{\text{total}}} = \frac{\rho_{\text{avg, body}}}{\rho_f} \quad \text{(10–18)}$$

Therefore, the submerged volume fraction of a floating body is equal to the ratio of the average density of the body to the density of the fluid. Note that when the density ratio is equal to or greater than one, the floating body becomes completely submerged.

It follows from these discussions that a body immersed in a fluid (1) remains at rest at any location in the fluid where its average density is equal to the density of the fluid, (2) sinks to the bottom when its average density is

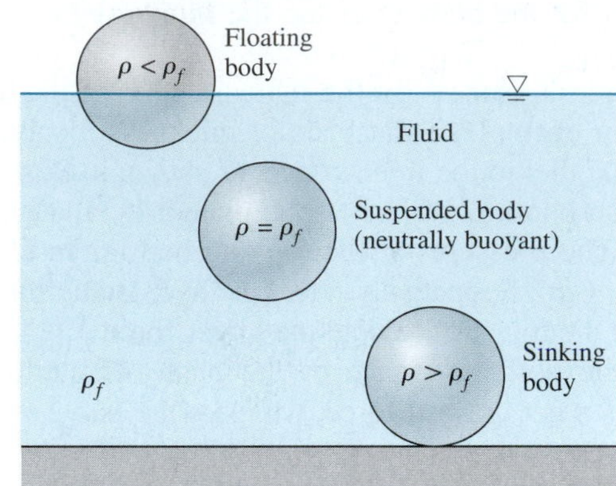

FIGURE 10–15
A solid body dropped into a fluid will sink, float, or remain at rest at any point in the fluid, depending on its average density relative to the density of the fluid.

greater than the density of the fluid, and (3) rises to the surface of the fluid and floats when the average density of the body is less than the density of the fluid (Fig. 10–15).

The buoyant force is proportional to the density of the fluid, and thus we might think that the buoyant force exerted by gases such as air is negligible. This is certainly the case in general, but there are significant exceptions. For example, the volume of a person is about 0.1 m^3, and taking the density of air to be 1.2 kg/m^3, the buoyant force exerted by air on the person is

$$F_B = \rho_f g V = (1.2 \text{ kg/m}^3)(9.81 \text{ m/s}^2)(0.1 \text{ m}^3) \cong 1.2 \text{ N}$$

The weight of an 80-kg person is $80 \times 9.81 = 788$ N. Therefore, ignoring the buoyancy in this case results in an error in weight of just 0.15 percent, which is negligible. But the buoyancy effects in gases dominate some important natural phenomena such as the rise of warm air in a cooler environment and thus the onset of natural convection currents, the rise of hot-air or helium balloons, and air movements in the atmosphere. A helium balloon, for example, rises as a result of the buoyancy effect until it reaches an altitude where the density of air (which decreases with altitude) equals the density of helium in the balloon—assuming the balloon does not burst by then, and ignoring the weight of the balloon's skin. Hot air balloons (Fig. 10–16) work by similar principles.

FIGURE 10–16
The altitude of a hot air balloon is controlled by the temperature difference between the air inside and outside the balloon, since warm air is less dense than cold air. When the balloon is neither rising nor falling, the upward buoyant force exactly balances the downward weight.
© Vol. 1/PhotoDisc/Getty Images

Archimedes' principle is also used in geology by considering the continents to be floating on a sea of magma.

EXAMPLE 10–3 Measuring Specific Gravity by a Hydrometer

If you have a seawater aquarium, you have probably used a small cylindrical glass tube with some lead-weight at its bottom to measure the salinity of the water by simply watching how deep the tube sinks. Such a device that floats in a vertical position and is used to measure the specific gravity of a liquid is called a *hydrometer* (Fig. 10–17). The top part of the hydrometer extends above the liquid surface, and the divisions on it allow one to read the specific gravity directly. The hydrometer is calibrated such that in pure water it reads exactly 1.0 at the air–water interface. (*a*) Obtain a relation for the specific gravity of a liquid as a function of distance Δz from the mark corresponding to pure water and (*b*) determine the mass of lead that must be poured into a 1-cm-diameter, 20-cm-long hydrometer if it is to float halfway (the 10-cm mark) in pure water.

Solution The specific gravity of a liquid is to be measured by a hydrometer. A relation between specific gravity and the vertical distance from the reference level is to be obtained, and the amount of lead that needs to be added into the tube for a certain hydrometer is to be determined.

Assumptions 1 The weight of the glass tube is negligible relative to the weight of the lead added. 2 The curvature of the tube bottom is disregarded.

Properties We take the density of pure water to be 1000 kg/m³.

Analysis (*a*) Noting that the hydrometer is in static equilibrium, the buoyant force F_B exerted by the liquid must always be equal to the weight W of the hydrometer. In pure water (subscript w), we let the vertical distance between the bottom of the hydrometer and the free surface of water be z_0. Setting $F_B = F_{B,w} = W$ in this case gives

$$W_{\text{hydro}} = F_{B,w} = \rho_w g V_{\text{sub}} = \rho_w g A z_0 \tag{1}$$

where A is the cross-sectional area of the tube, and ρ_w is the density of pure water.

In a fluid lighter than water ($\rho_f < \rho_w$), the hydrometer will sink deeper, and the liquid level will be a distance of Δz above z_0. Again setting $F_B = W$ gives

$$W_{\text{hydro}} = F_{B,f} = \rho_f g V_{\text{sub}} = \rho_f g A(z_0 + \Delta z) \tag{2}$$

This relation is also valid for fluids heavier than water by taking Δz to be a negative quantity. Setting Eqs. (1) and (2) here equal to each other since the weight of the hydrometer is constant and rearranging gives

$$\rho_w g A z_0 = \rho_f g A(z_0 + \Delta z) \quad \rightarrow \quad \text{SG}_f = \frac{\rho_f}{\rho_w} = \frac{z_0}{z_0 + \Delta z}$$

which is the relation between the specific gravity of the fluid and Δz. Note that z_0 is constant for a given hydrometer and Δz is negative for fluids heavier than pure water.

(*b*) Disregarding the weight of the glass tube, the amount of lead that needs to be added to the tube is determined from the requirement that the weight of the lead be equal to the buoyant force. When the hydrometer is floating with half of it submerged in water, the buoyant force acting on it is

$$F_B = \rho_w g V_{\text{sub}}$$

Equating F_B to the weight of lead gives

$$W = mg = \rho_w g V_{\text{sub}}$$

Solving for m and substituting, the mass of lead is determined to be

$$m = \rho_w V_{\text{sub}} = \rho_w(\pi R^2 h_{\text{sub}}) = (1000 \text{ kg/m}^3)[\pi(0.005 \text{ m})^2(0.1 \text{ m})] = \mathbf{0.00785 \text{ kg}}$$

Discussion Note that if the hydrometer were required to sink only 5 cm in water, the required mass of lead would be one-half of this amount. Also, the assumption that the weight of the glass tube is negligible is questionable since the mass of lead is only 7.85 g.

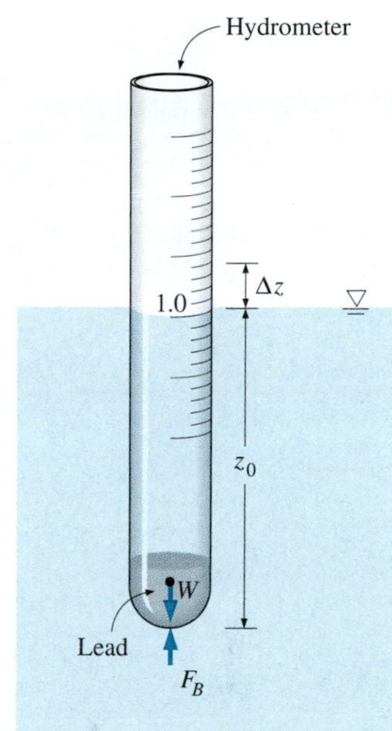

FIGURE 10–17
Schematic for Example 10–3.

EXAMPLE 10–4 Weight Loss of an Object in Seawater

A crane is used to lower weights into the sea (density = 1025 kg/m³) for an underwater construction project (Fig. 10–18). Determine the tension in the rope of the crane due to a rectangular 0.4-m × 0.4-m × 3-m concrete block (density = 2300 kg/m³) when it is (*a*) suspended in the air and (*b*) completely immersed in water.

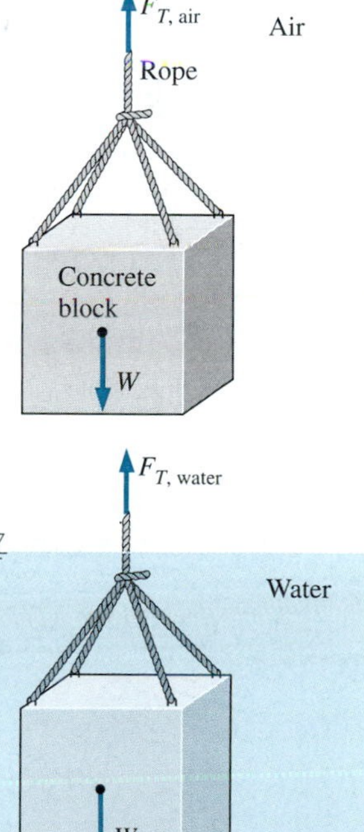

FIGURE 10–18
Schematic for Example 10–4.

FIGURE 10–19
For floating bodies such as ships, stability is an important consideration for safety.
© Corbis/vol. 96.

Solution A concrete block is lowered into the sea. The tension in the rope is to be determined before and after the block is in water.

Assumptions **1** The buoyant force in air is negligible. **2** The weight of the ropes is negligible.

Properties The densities are given to be 1025 kg/m³ for seawater and 2300 kg/m³ for concrete.

Analysis (*a*) Consider a free-body diagram of the concrete block. The forces acting on the concrete block in air are its weight and the upward pull action (tension) by the rope. These two forces must balance each other, and thus the tension in the rope must be equal to the weight of the block:

$$V = (0.4\text{ m})(0.4\text{ m})(3\text{ m}) = 0.48\text{ m}^3$$

$$F_{T,\text{ air}} = W = \rho_{\text{concrete}}\, gV$$

$$= (2300\text{ kg/m}^3)(9.81\text{ m/s}^2)(0.48\text{ m}^3)\left(\frac{1\text{ kN}}{1000\text{ kg}\cdot\text{m/s}^2}\right) = \mathbf{10.8\text{ kN}}$$

(*b*) When the block is immersed in water, there is the additional force of buoyancy acting upward. The force balance in this case gives

$$F_B = \rho_f gV = (1025\text{ kg/m}^3)(9.81\text{ m/s}^2)(0.48\text{ m}^3)\left(\frac{1\text{ kN}}{1000\text{ kg}\cdot\text{m/s}^2}\right) = \mathbf{4.8\text{ kN}}$$

$$F_{T,\text{ water}} = W - F_B = 10.8 - 4.8 = \mathbf{6.0\text{ kN}}$$

Discussion Note that the weight of the concrete block, and thus the tension of the rope, decreases by (10.8 − 6.0)/10.8 = 55 percent in water.

Stability of Immersed and Floating Bodies

An important application of the buoyancy concept is the assessment of the stability of immersed and floating bodies with no external attachments. This topic is of great importance in the design of ships and submarines (Fig. 10–19). Here we provide some general qualitative discussions on vertical and rotational stability.

We use the "ball on the floor" analogy to explain the fundamental concepts of stability and instability. Shown in Fig. 10–20 are three balls at rest on the floor. Case (*a*) is **stable** since any small disturbance (someone moves the ball to the right or left) generates a restoring force (due to gravity) that returns it to its initial position. Case (*b*) is **neutrally stable** because if someone moves the ball to the right or left, it would stay put at its new location. It has no tendency to move back to its original location, nor does it continue to move away. Case (*c*) is a situation in which the ball may be at rest at the moment, but any disturbance, even an infinitesimal one, causes the ball to roll off the hill—it does not return to its original position; rather it *diverges* from it. This situation is **unstable**. What about a case where the ball is on an *inclined* floor? It is not appropriate to discuss stability for this case since the ball is not in a state of equilibrium. In other words, it cannot be at rest and would roll down the hill even without any disturbance.

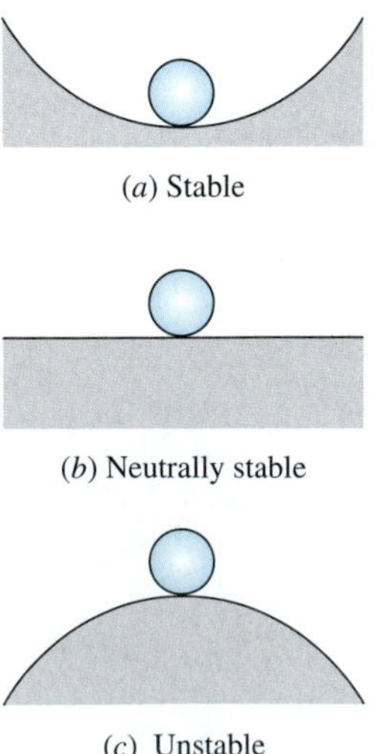

FIGURE 10–20
Stability is easily understood by analyzing a ball on the floor.

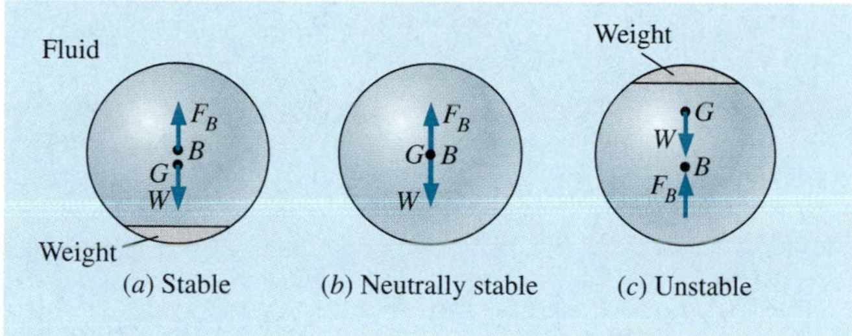

FIGURE 10–21
An immersed neutrally buoyant body is (*a*) stable if the center of gravity *G* is directly below the center of buoyancy *B* of the body, (*b*) neutrally stable if *G* and *B* are coincident, and (*c*) unstable if *G* is directly above *B*.

For an immersed or floating body in static equilibrium, the weight and the buoyant force acting on the body balance each other, and such bodies are inherently stable in the *vertical direction.* If an immersed neutrally buoyant body is raised or lowered to a different depth in an incompressible fluid, the body will remain in equilibrium at that location. If a floating body is raised or lowered somewhat by a vertical force, the body will return to its original position as soon as the external effect is removed. Therefore, a floating body possesses vertical stability, while an immersed neutrally buoyant body is neutrally stable since it does not return to its original position after a disturbance.

The *rotational stability* of an *immersed body* depends on the relative locations of the *center of gravity G* of the body and the *center of buoyancy B,* which is the centroid of the displaced volume. An immersed body is *stable* if the body is bottom-heavy and thus point *G* is directly below point *B* (Fig. 10–21a). A rotational disturbance of the body in such cases produces a *restoring moment* to return the body to its original stable position. Thus, a stable design for a submarine calls for the engines and the cabins for the crew to be located at the lower half in order to shift the weight to the bottom as much as possible. Hot-air or helium balloons (which can be viewed as being immersed in air) are also stable since the cage that carries the load is at the bottom. An immersed body whose center of gravity *G* is directly above point *B* (Fig. 10–21c) is *unstable*, and any disturbance will cause this body to turn upside down. A body for which *G* and *B* coincide (Fig. 10–21b) is *neutrally stable.* This is the case for bodies whose density is constant throughout. For such bodies, there is no tendency to overturn or right themselves.

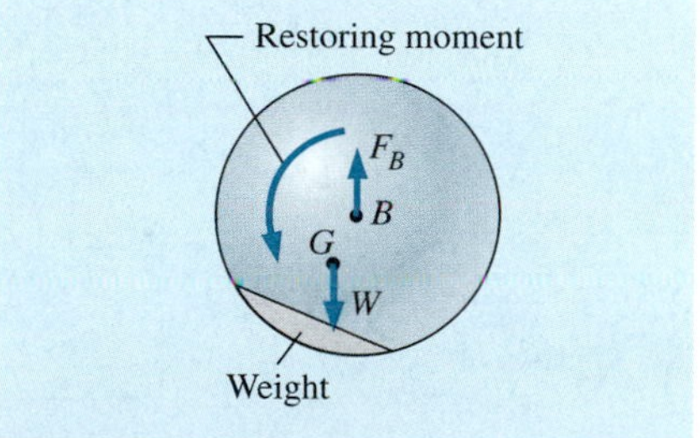

FIGURE 10–22
When the center of gravity *G* of an immersed neutrally buoyant body is not vertically aligned with the center of buoyancy *B* of the body, it is not in an equilibrium state and would rotate to its stable state, even without any disturbance.

What about a case where the center of gravity is not vertically aligned with the center of buoyancy (Fig. 10–22)? It is not appropriate to discuss stability for this case since the body is not in a state of equilibrium. In other words, it cannot be at rest and would rotate toward its stable state even without any disturbance. The restoring moment in the case shown in Fig. 10–22 is counterclockwise and causes the body to rotate counterclockwise so as to align point *G* vertically with point *B*. Note that there may be some oscillation, but eventually the body settles down at its stable equilibrium state [case (*a*) of Fig. 10–21]. The stability of the body of Fig. 10–22 is analogous to that of the ball on an inclined floor. Can you predict what would happen if the weight in the body of Fig. 10–22 were on the opposite side of the body?

The rotational stability criteria are similar for *floating bodies.* Again, if the floating body is bottom-heavy and thus the center of gravity *G* is directly below the center of buoyancy *B*, the body is always stable. But unlike

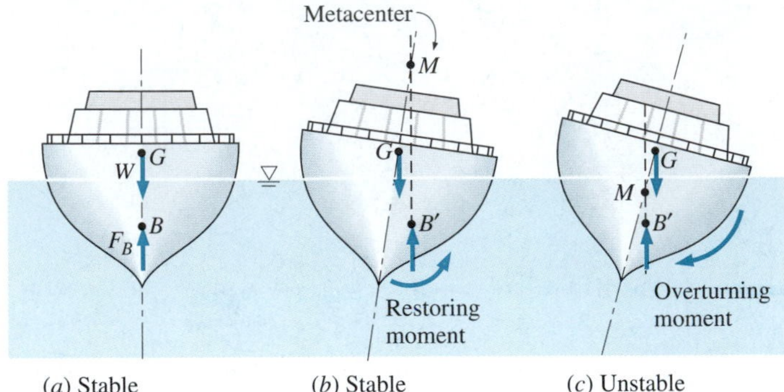

FIGURE 10–23
A floating body is *stable* if the body is bottom-heavy and thus the center of gravity G is below the centroid B of the body, or if the metacenter M is above point G. However, the body is *unstable* if point M is below point G.

immersed bodies, a floating body may still be stable when G is directly above B (Fig. 10–23). This is because the centroid of the displaced volume shifts to the side to a point B' during a rotational disturbance while the center of gravity G of the body remains unchanged. If point B' is sufficiently far, these two forces create a restoring moment and return the body to the original position. A measure of stability for floating bodies is the **metacentric height** GM, which is the distance between the center of gravity G and the metacenter M—the intersection point of the lines of action of the buoyant force through the body before and after rotation. The metacenter may be considered to be a fixed point for most hull shapes for small rolling angles up to about 20°. A floating body is stable if point M is above point G, and thus GM is positive, and unstable if point M is below point G, and thus GM is negative. In the latter case, the weight and the buoyant force acting on the tilted body generate an overturning moment instead of a restoring moment, causing the body to capsize. The length of the metacentric height GM above G is a measure of the stability: the larger it is, the more stable is the floating body.

SUMMARY

Fluid statics deals with problems associated with fluids at rest, and it is called *hydrostatics* when the fluid is a liquid. The magnitude of the resultant force acting on a plane surface of a completely submerged plate in a homogeneous fluid is equal to the product of the pressure P_C at the centroid of the surface and the area A of the surface and is expressed as

$$F_R = (P_0 + \rho g h_C)A = P_C A = P_{\text{avg}} A$$

where $h_C = y_C \sin\theta$ is the *vertical distance* of the centroid from the free surface of the liquid. The pressure P_0 at the surface of the liquid is usually atmospheric pressure, which cancels out in most cases since it acts on both sides of the plate. The point of intersection of the line of action of the resultant force and the surface is the *center of pressure.* The vertical location of the line of action of the resultant force is given by

$$y_P = y_C + \frac{I_{xx,\,C}}{[y_C + P_0/(\rho g \sin\theta)]A}$$

where $I_{xx,\,C}$ is the second moment of area about the x-axis passing through the centroid of the area.

A fluid of density P_f exerts an upward force on a body immersed in it. This force is called the *buoyant force* and is expressed as

$$F_B = \rho_f g V$$

where V is the volume of the body. This is known as *Archimedes' principle* and is expressed as: the buoyant force acting on a body immersed in a fluid is equal to the weight of the fluid displaced by the body; it acts upward through the centroid of the displaced volume. In a fluid with constant density, the buoyant force is independent of the distance of the body from the free surface. For *floating* bodies, the submerged volume fraction of the body is equal to the ratio of the average density of the body to the density of the fluid.

Pressure is a fundamental property, and it is hard to imagine a significant fluid flow problem that does not involve pressure. Therefore, you will see this property in all chapters in the rest of this book. The consideration of hydrostatic forces acting on plane or curved surfaces, however, is mostly limited to this chapter.

REFERENCES AND SUGGESTED READING

1. F. P. Beer, E. R. Johnston, Jr., E. R. Eisenberg, and G. H. Staab. *Vector Mechanics for Engineers, Statics*, 7th ed. New York: McGraw-Hill, 2004.

2. D. C. Giancoli. *Physics*, 3rd ed. Upper Saddle River, NJ: Prentice Hall, 1991.

PROBLEMS*

Fluid Statics: Hydrostatic Forces on Plane and Curved Surfaces

10–1C Define the resultant hydrostatic force acting on a submerged surface, and the center of pressure.

10–2C Someone claims that she can determine the magnitude of the hydrostatic force acting on a plane surface submerged in water regardless of its shape and orientation if she knew the vertical distance of the centroid of the surface from the free surface and the area of the surface. Is this a valid claim? Explain.

10–3C A submerged horizontal flat plate is suspended in water by a string attached at the centroid of its upper surface. Now the plate is rotated 45° about an axis that passes through its centroid. Discuss the change on the hydrostatic force acting on the top surface of this plate as a result of this rotation. Assume the plate remains submerged at all times.

10–4C You may have noticed that dams are much thicker at the bottom. Explain why dams are built that way.

10–5C Consider a submerged curved surface. Explain how you would determine the horizontal component of the hydrostatic force acting on this surface.

10–6C Consider a submerged curved surface. Explain how you would determine the vertical component of the hydrostatic force acting on this surface.

10–7C Consider a circular surface subjected to hydrostatic forces by a constant density liquid. If the magnitudes of the horizontal and vertical components of the resultant hydrostatic force are determined, explain how you would find the line of action of this force.

10–8 Consider a heavy car submerged in water in a lake with a flat bottom. The driver's side door of the car is 1.1 m high and 0.9 m wide, and the top edge of the door is 8 m below the water surface. Determine the net force acting on the door (normal to its surface) and the location of the pressure center if (*a*) the car is well-sealed and it contains air at atmospheric pressure and (*b*) the car is filled with water.

10–9E A long, solid cylinder of radius 2 ft hinged at point *A* is used as an automatic gate, as shown in Fig. P10–9E. When the water level reaches 15 ft, the cylindrical gate opens by turning about the hinge at point *A*. Determine (*a*) the hydrostatic force acting on the cylinder and its line of action when the gate opens and (*b*) the weight of the cylinder per ft length of the cylinder.

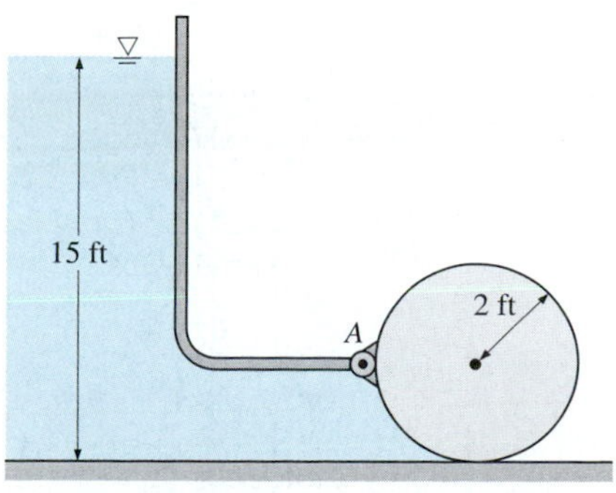

FIGURE P10–9E

* Problems designated by a "C" are concept questions, and students are encouraged to answer them all. Problems designated by an "E" are in English units, and the SI users can ignore them. Problems with the icon are solved using EES, and complete solutions together with parametric studies are included on the enclosed DVD. Problems with the icon are comprehensive in nature and are intended to be solved with a computer, preferably using the EES software that accompanies this text.

10–10 Consider a 4-m-long, 4-m-wide, and 1.5-m-high aboveground swimming pool that is filled with water to the rim. (*a*) Determine the hydrostatic force on each wall and the distance of the line of action of this force from the ground. (*b*) If the height of the walls of the pool is doubled and the pool is filled, will the hydrostatic force on each wall double or quadruple? Why? *Answer:* (*a*) 44.1 kN

10–11E Consider a 200-ft-high, 1200-ft-wide dam filled to capacity. Determine (*a*) the hydrostatic force on the dam and (*b*) the force per unit area of the dam near the top and near the bottom.

10–12 A room in the lower level of a cruise ship has a 30-cm-diameter circular window. If the midpoint of the window is 5 m below the water surface, determine the hydrostatic force acting on the window, and the pressure center. Take the specific gravity of seawater to be 1.025. *Answers:* 3554 N, 5.001 m

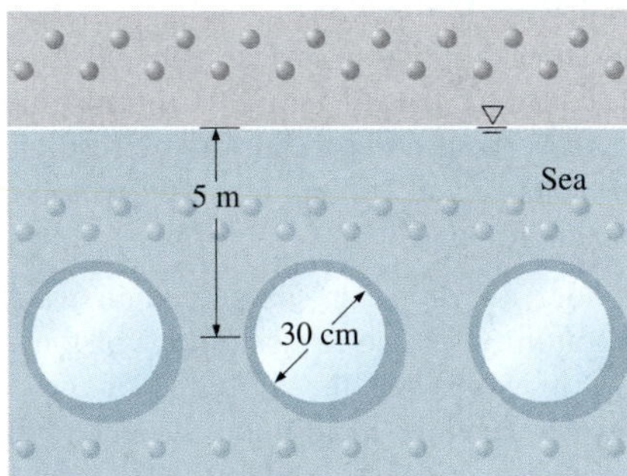

FIGURE P10–12

10–13 The water side of the wall of a 100-m-long dam is a quarter circle with a radius of 10 m. Determine the hydrostatic force on the dam and its line of action when the dam is filled to the rim.

10–14 A 5-m-high, 5-m-wide rectangular plate blocks the end of a 4-m-deep freshwater channel, as shown in Fig. P10–14. The plate is hinged about a horizontal axis along its upper edge through a point *A* and is restrained from opening by a fixed ridge at point *B*. Determine the force exerted on the plate by the ridge.

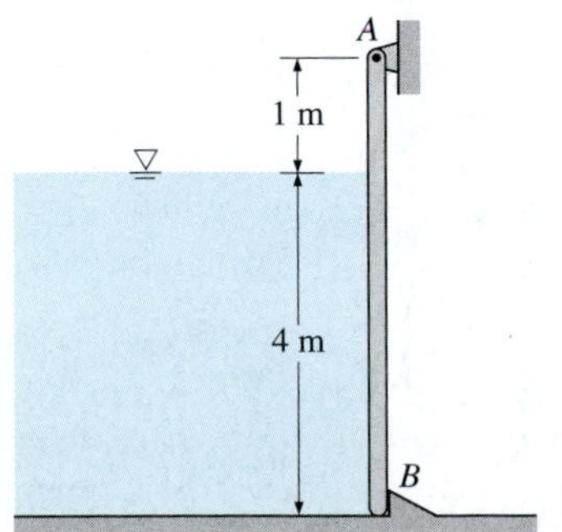

FIGURE P10–14

10–15 Reconsider Prob. 10–14. Using EES (or other) software, investigate the effect of water depth on the force exerted on the plate by the ridge. Let the water depth vary from 0 m to 5 m in increments of 0.5 m. Tabulate and plot your results.

10–16E The flow of water from a reservoir is controlled by a 5-ft-wide L-shaped gate hinged at point *A*, as shown in Fig. P10–16E. If it is desired that the gate open when the water height is 12 ft, determine the mass of the required weight *W*. *Answer:* 30,900 lbm

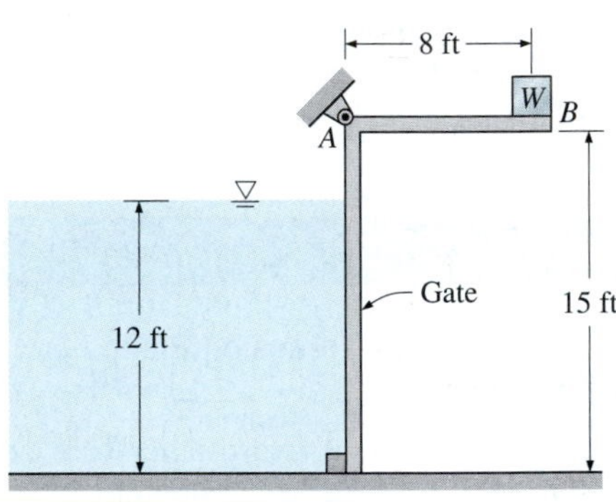

FIGURE P10–16E

10–17E Repeat Prob. 10–16E for a water height of 8 ft.

10–18 A water trough of semicircular cross section of radius 0.5 m consists of two symmetric parts hinged to each other at the bottom, as shown in Fig. P10–18. The two parts are held together by a cable and turnbuckle placed every 3 m along the length of the trough. Calculate the tension in each cable when the trough is filled to the rim.

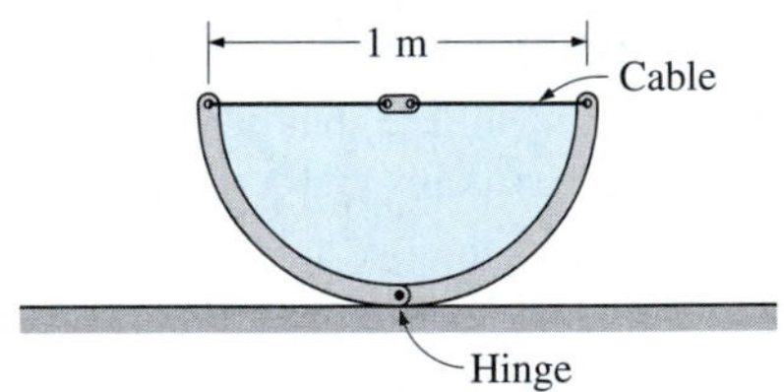

FIGURE P10–18

10–19 The two sides of a V-shaped water trough are hinged to each other at the bottom where they meet, as shown in Fig. P10–19, making an angle of 45° with the ground from both sides. Each side is 0.75 m wide, and the two parts are held together by a cable and turnbuckle placed every 6 m along the length of the trough. Calculate the tension in each cable when the trough is filled to the rim. *Answer:* 5510 N

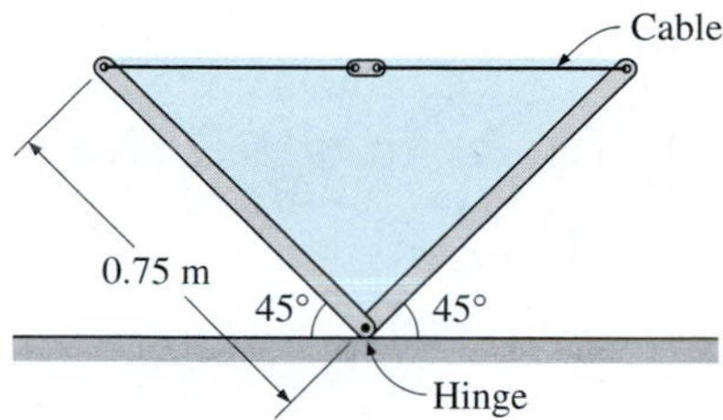

FIGURE P10–19

10–20 Repeat Prob. 10–19 for the case of a partially filled trough with a water height of 0.4 m directly above the hinge.

10–21 A retaining wall against a mud slide is to be constructed by placing 0.8-m-high and 0.2-m-wide rectangular concrete blocks (ρ = 2700 kg/m^3) side by side, as shown in Fig. P10–21. The friction coefficient between the ground and the concrete blocks is f = 0.3, and the density of the mud is about 1800 kg/m^3. There is concern that the concrete blocks may slide or tip over the lower left edge as the mud level rises. Determine the mud height at which (*a*) the blocks will overcome friction and start sliding and (*b*) the blocks will tip over.

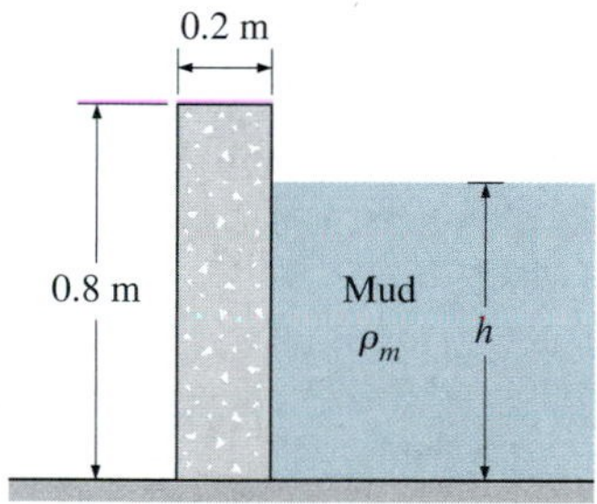

FIGURE P10–21

10–22 Repeat Prob. 10–21 for 0.4-m-wide concrete blocks.

10–23 A 4-m-long quarter-circular gate of radius 3 m and of negligible weight is hinged about its upper edge *A*, as shown in Fig. P10–23. The gate controls the flow of water over the ledge at *B*, where the gate is pressed by a spring. Determine the minimum spring force required to keep the gate closed when the water level rises to *A* at the upper edge of the gate.

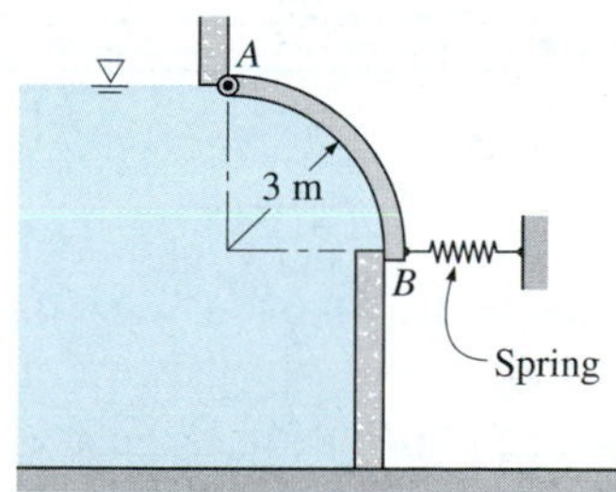

FIGURE P10–23

10–24 Repeat Prob. 10–23 for a radius of 4 m for the gate. *Answer:* 314 kN

Buoyancy

10–25C What is buoyant force? What causes it? What is the magnitude of the buoyant force acting on a submerged body whose volume is V? What are the direction and the line of action of the buoyant force?

10–26C Consider two identical spherical balls submerged in water at different depths. Will the buoyant forces acting on these two balls be the same or different? Explain.

10–27C Consider two 5-cm-diameter spherical balls—one made of aluminum, the other of iron—submerged in water. Will the buoyant forces acting on these two balls be the same or different? Explain.

10–28C Consider a 3-kg copper cube and a 3-kg copper ball submerged in a liquid. Will the buoyant forces acting on these two bodies be the same or different? Explain.

10–29C Discuss the stability of (*a*) a submerged and (*b*) a floating body whose center of gravity is above the center of buoyancy.

10–30 The density of a liquid is to be determined by an old 1-cm-diameter cylindrical hydrometer whose division marks are completely wiped out. The hydrometer is first dropped in water, and the water level is marked. The hydrometer is then dropped into the other liquid, and it is observed that the mark for water has risen 0.5 cm above the liquid–air interface. If the height of the water mark is 10 cm, determine the density of the liquid.

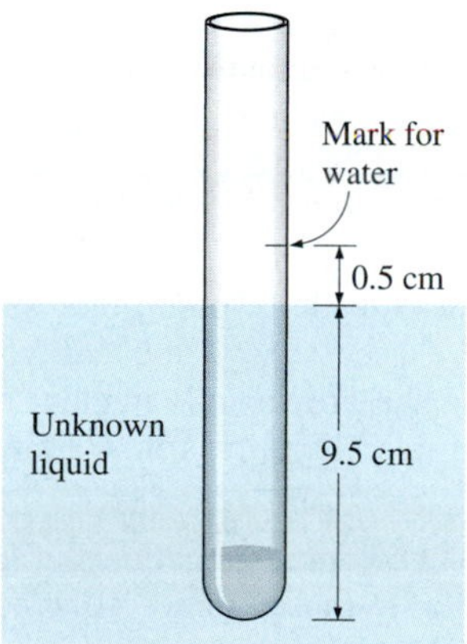

FIGURE P10–30

10–31E A crane is used to lower weights into a lake for an underwater construction project. Determine the tension in the rope of the crane due to a 3-ft-diameter spherical steel block (density = 494 lbm/ft^3) when it is (*a*) suspended in the air and (*b*) completely immersed in water.

10–32 The volume and the average density of an irregularly shaped body are to be determined by using a spring scale. The body weighs 7200 N in air and 4790 N in water. Determine the volume and the density of the body. State your assumptions.

10–33 Consider a large cubic ice block floating in seawater. The specific gravities of ice and seawater are 0.92 and 1.025, respectively. If a 10-cm-high portion of the ice block extends above the surface of the water, determine the height of the ice block below the surface. *Answer:* 87.6 cm

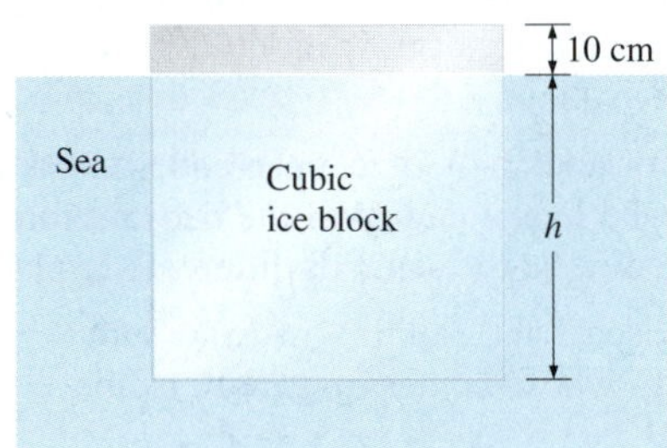

FIGURE P10–33

10–34 A 170-kg granite rock ($\rho = 2700$ kg/m^3) is dropped into a lake. A man dives in and tries to lift the rock. Determine how much force the man needs to apply to lift it from the bottom of the lake. Do you think he can do it?

10–35 It is said that Archimedes discovered his principle during a bath while thinking about how he could determine if King Hiero's crown was actually made of pure gold. While in the bathtub, he conceived the idea that he could determine the average density of an irregularly shaped object by weighing it in air and also in water. If the crown weighed 3.20 kgf (= 31.4 N) in air and 2.95 kgf (= 28.9 N) in water, determine if the crown is made of pure gold. The density of gold is 19,300 kg/m^3. Discuss how you can solve this problem without weighing the crown in water but by using an ordinary bucket with no calibration for volume. You may weigh anything in air.

10–36 One of the common procedures in fitness programs is to determine the fat-to-muscle ratio of the body. This is based on the principle that the muscle tissue is denser than the fat tissue, and, thus, the higher the average density of the body, the higher is the fraction of muscle tissue. The average density of the body can be determined by weighing the person in air and also while submerged in water in a tank. Treating all tissues and bones (other than fat) as muscle with an equivalent density of ρ_{muscle}, obtain a relation for the volume fraction of body fat x_{fat}. *Answer:* $x_{\text{fat}} = (\rho_{\text{muscle}} - \rho_{\text{avg}})/(\rho_{\text{muscle}} - \rho_{\text{fat}})$.

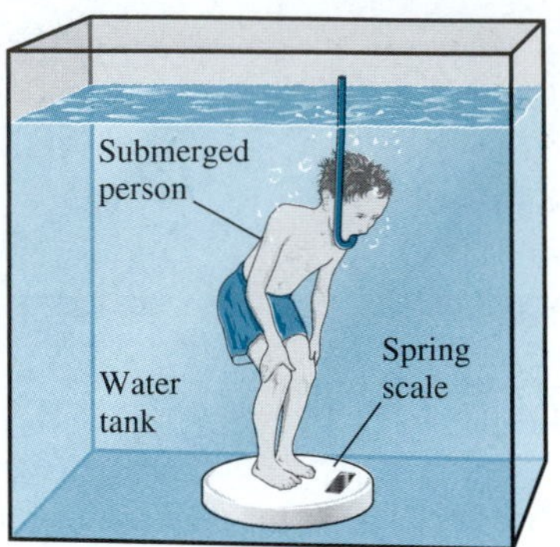

FIGURE P10–36

10–37 The hull of a boat has a volume of 150 m^3, and the total mass of the boat when empty is 8560 kg. Determine how much load this boat can carry without sinking (*a*) in a lake and (*b*) in seawater with a specific gravity of 1.03.

Review Problems

10–38 The density of a floating body can be determined by tying weights to the body until both the body and the weights are completely submerged, and then weighing them separately in air. Consider a wood log that weighs 1540 N in air. If it takes 34 kg of lead ($\rho = 11{,}300$ kg/m^3) to completely sink the log and the lead in water, determine the average density of the log. *Answer:* 835 kg/m^3

10–39 The 200-kg, 5-m-wide rectangular gate shown in Fig. P10–39 is hinged at *B* and leans against the floor at *A* making an angle of 45° with the horizontal. The gate is to be opened from its lower edge by applying a normal force at its center. Determine the minimum force *F* required to open the water gate. *Answer:* 522 kN

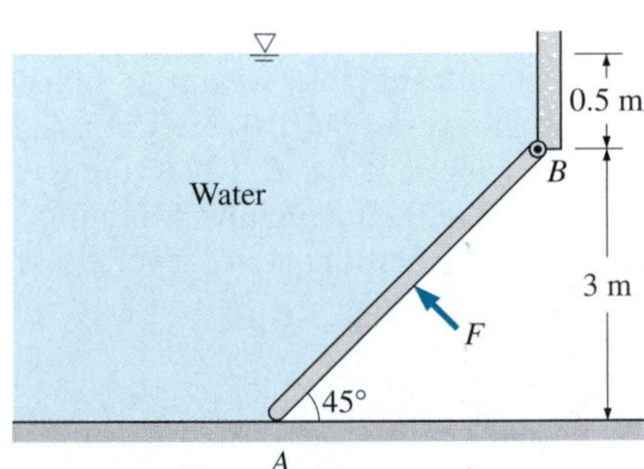

FIGURE P10–39

10–40 Repeat Prob. 10–39 for a water height of 1.2 m above the hinge at *B*.

10–41 A 3-m-high, 6-m-wide rectangular gate is hinged at the top edge at *A* and is restrained by a fixed ridge at *B*.

Determine the hydrostatic force exerted on the gate by the 5-m-high water and the location of the pressure center.

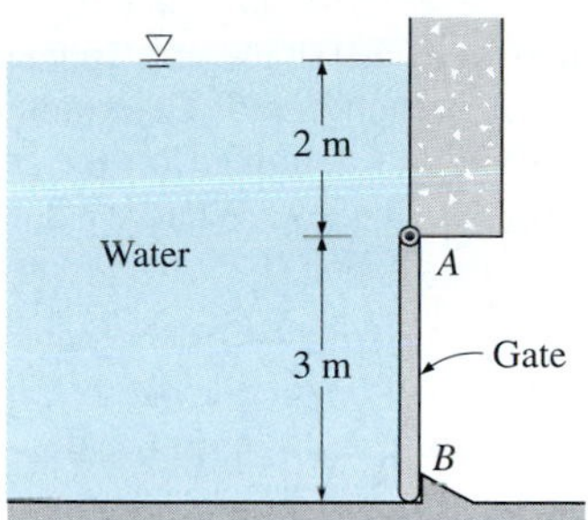

FIGURE P10–41

10–42 Repeat Prob. 10–41 for a total water height of 2 m.

10–43E A semicircular 30-ft-diameter tunnel is to be built under a 150-ft-deep, 800-ft-long lake, as shown in Fig. P10–43E. Determine the total hydrostatic force acting on the roof of the tunnel.

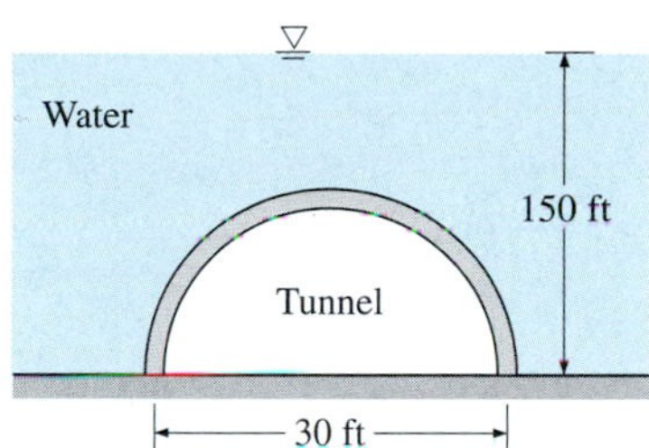

FIGURE P10–43E

10–44 A 50-ton, 6-m-diameter hemispherical dome on a level surface is filled with water, as shown in Fig. P10–44. Someone claims that he can lift this dome by making use of Pascal's law by attaching a long tube to the top and filling it with water. Determine the required height of water in the tube to lift the dome. Disregard the weight of the tube and the water in it. *Answer: 0.77 m*

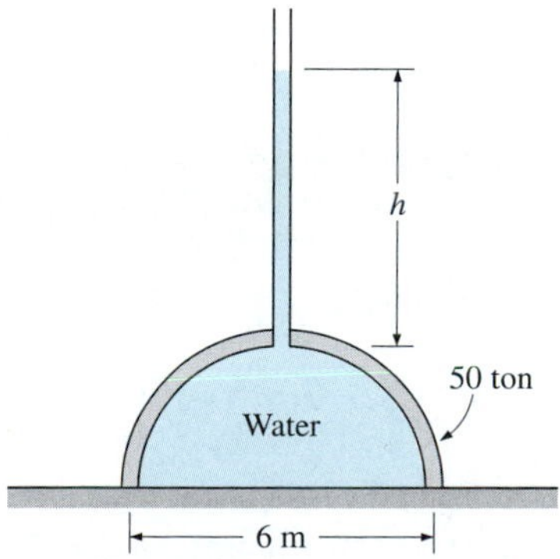

FIGURE P10–44

10–45 The water in a 25-m-deep reservoir is kept inside by a 150-m-wide wall whose cross section is an equilateral triangle, as shown in Fig. P10–45. Determine (*a*) the total force (hydrostatic + atmospheric) acting on the inner surface of the wall and its line of action and (*b*) the magnitude of the horizontal component of this force. Take P_{atm} = 100 kPa.

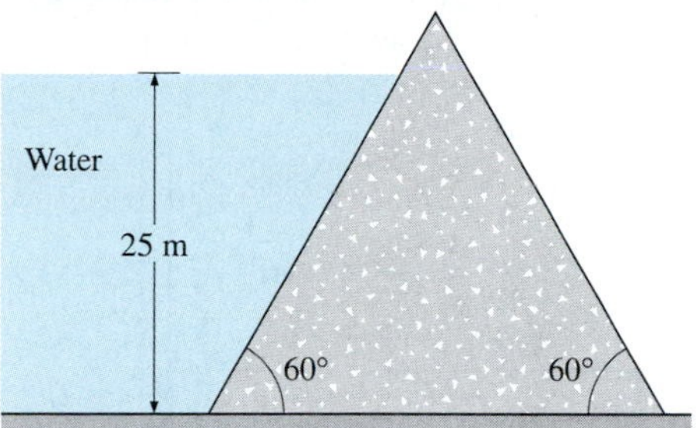

FIGURE P10–45

10–46 An elastic air balloon having a diameter of 30 cm is attached to the base of a container partially filled with water at +4°C, as shown in Fig. P10–46. If the pressure of air above water is gradually increased from 100 kPa to 1.6 MPa, will the force on the cable change? If so, what is the percent change in the force? Assume the pressure on the free surface and the diameter of the balloon are related by $P = CD^n$, where C is a constant and $n = -2$. The weight of the balloon and the air in it is negligible. *Answer: 98.4 percent*

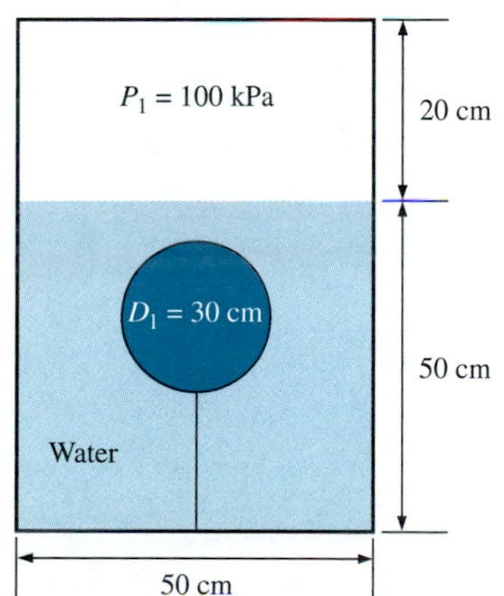

FIGURE P10–46

10–47 Reconsider Prob. 10–46. Using EES (or other) software, investigate the effect of air pressure above water on the cable force. Let this pressure vary from 0.1 MPa to 10 MPa. Plot the cable force versus the air pressure.

10–48 The average density of icebergs is about 917 kg/m³. (*a*) Determine the percentage of the total volume of an iceberg submerged in seawater of density 1042 kg/m^3. (*b*) Although icebergs are mostly submerged, they are observed to turn over. Explain how this can happen. (Hint: Consider the temperatures of icebergs and seawater.)

10–49 A cylindrical container whose weight is 79 N is inverted and pressed into the water, as shown in Fig. P10–49. Determine the differential height *h* of the manometer and the force *F* needed to hold the container at the position shown.

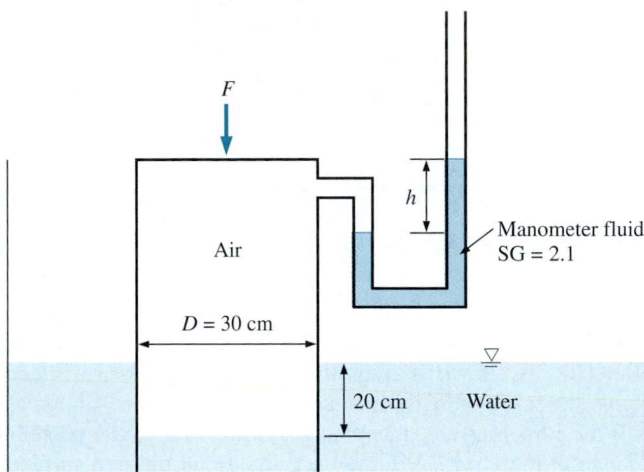

FIGURE P10–49

Design and Essay Problems

10–50 Shoes are to be designed to enable people of up to 80 kg to walk on freshwater or seawater. The shoes are to be made of blown plastic in the shape of a sphere, a (American) football, or a loaf of French bread. Determine the equivalent diameter of each shoe and comment on the proposed shapes from the stability point of view. What is your assessment of the marketability of these shoes?

10–51 The volume of a rock is to be determined without using any volume measurement devices. Explain how you would do this with a waterproof spring scale.

Chapter 11

FLUID KINEMATICS

Fluid kinematics deals with describing the motion of fluids without necessarily considering the forces and moments that *cause* the motion. In this chapter, we introduce several kinematic concepts related to flowing fluids. We discuss the *material derivative* and its role in transforming the conservation equations from the *Lagrangian description of fluid flow* (following a *fluid particle*) to the *Eulerian description of fluid flow* (pertaining to a *flow field*). We then discuss various ways to visualize flow fields—*streamlines, streaklines, pathlines, timelines*, and various surface flow visualization methods. The concepts of *vorticity, rotationality, and irrotationality* in fluid flows are then discussed. Finally, we discuss the *Reynolds transport theorem* (*RTT*), emphasizing its role in transforming the equations of motion from those following a *system* to those pertaining to fluid flow into and out of a *control volume*.

Objectives

The objectives of this chapter are to:

- Understand the role of the material derivative in transforming between Lagrangian and Eulerian descriptions
- Distinguish between various types of flow visualizations
- Distinguish between rotational and irrotational regions of flow based on the flow property vorticity
- Understand the usefulness of the Reynolds transport theorem

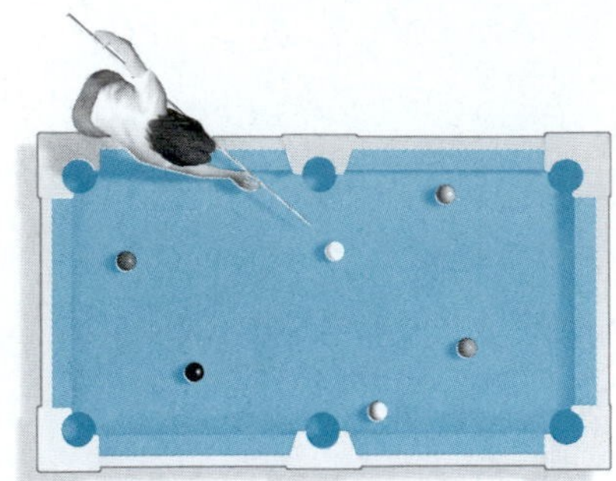

FIGURE 11–1
With a small number of objects, such as billiard balls on a pool table, individual objects can be tracked.

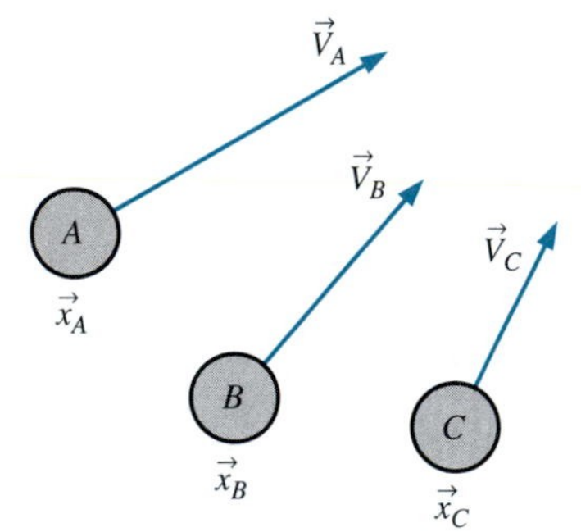

FIGURE 11–2
In the Lagrangian description, one must keep track of the position and velocity of individual particles.

11–1 ▪ LAGRANGIAN AND EULERIAN DESCRIPTIONS

The subject called **kinematics** concerns the study of *motion*. In fluid dynamics, *fluid kinematics* is the study of how fluids flow and how to describe fluid motion. From a fundamental point of view, there are two distinct ways to describe motion. The first and most familiar method is the one you learned in high school physics—to follow the path of individual objects. For example, we have all seen physics experiments in which a ball on a pool table or a puck on an air hockey table collides with another ball or puck or with the wall (Fig.11–1). Newton's laws are used to describe the motion of such objects, and we can accurately predict where they go and how momentum and kinetic energy are exchanged from one object to another. The kinematics of such experiments involves keeping track of the **position vector** of each object, $\vec{x}_A$, $\vec{x}_B$, . . . , and the **velocity vector** of each object, $\vec{V}_A$, $\vec{V}_B$, . . . , as functions of time (Fig. 11–2). When this method is applied to a flowing fluid, we call it the **Lagrangian description** of fluid motion after the Italian mathematician Joseph Louis Lagrange (1736–1813). Lagrangian analysis is analogous to the (closed) **system analysis** of thermodynamics; namely, we follow a mass of fixed identity. The Lagrangian description requires us to track the position and velocity of each individual fluid parcel, which we refer to as a **fluid particle**, and take to be a parcel of fixed identity.

As you can imagine, this method of describing motion is much more difficult for fluids than for billiard balls! First of all we cannot easily define and identify particles of fluid as they move around. Secondly, a fluid is a **continuum** (from a macroscopic point of view), so interactions between fluid particles are not as easy to describe as are interactions between distinct objects like billiard balls or air hockey pucks. Furthermore, the fluid parcels continually *deform* as they move in the flow.

From a *microscopic* point of view, a fluid is composed of *billions* of molecules that are continuously banging into one another, somewhat like billiard balls; but the task of following even a subset of these molecules is quite difficult, even for our fastest and largest computers. Nevertheless, there are many practical applications of the Lagrangian description, such as the tracking of passive scalars in a flow to model contaminant transport, rarefied gas dynamics calculations concerning reentry of a spaceship into the earth's atmosphere, and flow visualization methods based on particle tracking (as discussed in Section 11–2).

A more common method of describing fluid flow is the **Eulerian description** of fluid motion, named after the Swiss mathematician Leonhard Euler (1707–1783). In the Eulerian description of fluid flow, a finite volume called a **flow domain** or **control volume** is defined, through which fluid flows in and out. Instead of tracking individual fluid particles, we define **field variables**, functions of space and time, within the control volume. The field variable at a particular location at a particular time is the value of the variable for whichever fluid particle happens to occupy that location at that time. For example, the **pressure field** is a **scalar field variable**; for general unsteady three-dimensional fluid flow in Cartesian coordinates,

Pressure field: $$P = P(x, y, z, t) \qquad \textbf{(11–1)}$$

We define the **velocity field** as a **vector field variable** in similar fashion,

Velocity field: $$\vec{V} = \vec{V}(x, y, z, t) \quad \textbf{(11–2)}$$

Likewise, the **acceleration field** is also a vector field variable,

Acceleration field: $$\vec{a} = \vec{a}(x, y, z, t) \quad \textbf{(11–3)}$$

Collectively, these (and other) field variables define the **flow field**. The velocity field of Eq. 11–2 can be expanded in Cartesian coordinates (x, y, z), $(\vec{i}, \vec{j}, \vec{k})$ as

$$\vec{V} = (u, v, w) = u(x, y, z, t)\vec{i} + v(x, y, z, t)\vec{j} + w(x, y, z, t)\vec{k} \quad \textbf{(11–4)}$$

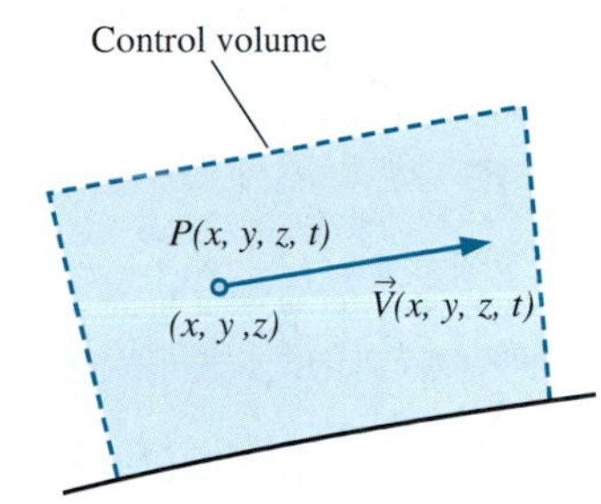

FIGURE 11–3
In the Eulerian description, one defines field variables, such as the pressure field and the velocity field, at any location and instant in time.

A similar expansion can be performed for the acceleration field of Eq. 11–3. In the Eulerian description, all such field variables are defined at any location (x, y, z) in the control volume and at any instant in time t (Fig. 11–3). In the Eulerian description we don't really care what happens to individual fluid particles; rather we are concerned with the pressure, velocity, acceleration, etc., of whichever fluid particle happens to be at the location of interest at the time of interest.

The difference between these two descriptions is made clearer by imagining a person standing beside a river, measuring its properties. In the Lagrangian approach, he throws in a probe that moves downstream with the water. In the Eulerian approach, he anchors the probe at a fixed location in the water.

While there are many occasions in which the Lagrangian description is useful, the Eulerian description is often more convenient for fluid mechanics applications. Furthermore, experimental measurements are generally more suited to the Eulerian description. In a wind tunnel, for example, velocity or pressure probes are usually placed at a fixed location in the flow, measuring $\vec{V}(x, y, z, t)$ or $P(x, y, z, t)$. However, whereas the equations of motion in the Lagrangian description following individual fluid particles are well known (e.g., Newton's second law), the equations of motion of fluid flow are not so readily apparent in the Eulerian description and must be carefully derived. We do this for control volume (integral) analysis via the Reynolds transport theorem at the end of this chapter.

EXAMPLE 11–1 **A Steady Two-Dimensional Velocity Field**

A steady, incompressible, two-dimensional velocity field is given by

$$\rightarrow V = (u, v) = (0.5 + 0.8x)\vec{i} + (1.5 - 0.8y)\vec{j} \quad \textbf{(1)}$$

where the x- and y-coordinates are in meters and the magnitude of velocity is in m/s. A **stagnation point** is defined as *a point in the flow field where the velocity is identically zero.* (*a*) Determine if there are any stagnation points in this flow field and, if so, where? (*b*) Sketch velocity vectors at several locations in the domain between $x = -2$ m to 2 m and $y = 0$ m to 5 m; qualitatively describe the flow field.

Solution For the given velocity field, the location(s) of stagnation point(s) are to be determined. Several velocity vectors are to be sketched and the velocity field is to be described.

Assumptions **1** The flow is steady and incompressible. **2** The flow is two-dimensional, implying no z-component of velocity and no variation of u or v with z.

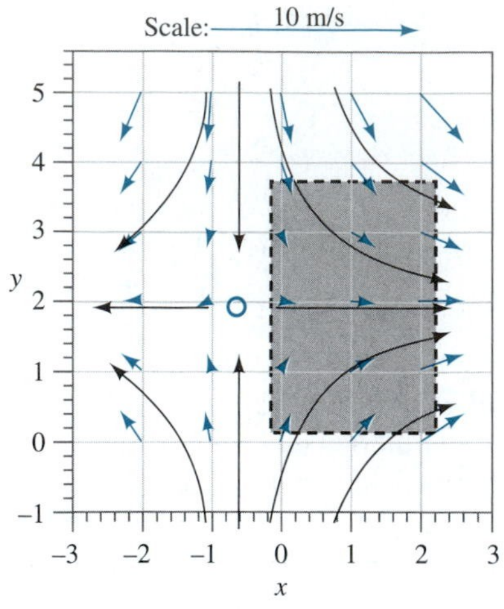

FIGURE 11–4
Velocity vectors for the velocity field of Example 11–1. The scale is shown by the top arrow, and the solid black curves represent the approximate shapes of some streamlines, based on the calculated velocity vectors. The stagnation point is indicated by the blue circle. The shaded region represents a portion of the flow field that can approximate flow into an inlet (Fig. 11–5).

Analysis (*a*) Since $\vec{V}$ is a vector, *all* its components must equal zero in order for $\vec{V}$ itself to be zero. Using Eq. 11–4 and setting Eq. 1 equal to zero,

Stagnation point:
$$u = 0.5 + 0.8x = 0 \quad \rightarrow \quad x = -0.625 \text{ m}$$
$$v = 1.5 - 0.8y = 0 \quad \rightarrow \quad y = 1.875 \text{ m}$$

Yes. There is one stagnation point located at $x = -0.625$ **m,** $y = 1.875$ **m.**

(*b*) The *x*- and *y*-components of velocity are calculated from Eq. 1 for several (*x, y*) locations in the specified range. For example, at the point (x = 2 m, y = 3 m), u = 2.10 m/s and v = −0.900 m/s. The magnitude of velocity (the *speed*) at that point is 2.28 m/s. At this and at an array of other locations, the velocity vector is constructed from its two components, the results of which are shown in Fig. 11–4. The flow can be described as stagnation point flow in which flow enters from the top and bottom and spreads out to the right and left about a horizontal line of symmetry at y = 1.875 m. The stagnation point of part (*a*) is indicated by the blue circle in Fig. 11–4.

If we look only at the shaded portion of Fig. 11–4, this flow field models a converging, accelerating flow from the left to the right. Such a flow might be encountered, for example, near the submerged bell mouth inlet of a hydroelectric dam (Fig. 11–5). The useful portion of the given velocity field may be thought of as a first-order approximation of the shaded portion of the physical flow field of Fig. 11–5.

Discussion It can be verified that this flow field is physically valid because it satisfies the differential equation for conservation of mass (see Çengel & Cimbala, 2006).

Acceleration Field

As you should recall from your study of thermodynamics, the fundamental conservation laws (such as conservation of mass and the first law of thermodynamics) are expressed for a *system* of fixed identity (also called a *closed system*). In cases where analysis of a *control volume* (also called an *open system*) is more convenient than system analysis, it is necessary to rewrite these fundamental laws into forms applicable to the control volume. The same principle applies here. In fact, there is a direct analogy between systems versus control volumes in thermodynamics and Lagrangian versus Eulerian descriptions in fluid dynamics. The equations of motion for fluid flow (such as Newton's second law) are written for a fluid particle, which we also call a **material particle**. If we were to follow a particular fluid particle as it moves around in the flow, we would be employing the Lagrangian description, and the equations of motion would be directly applicable. For example, we would define the particle's location in space in terms of a **material position vector** ($x_{\text{particle}}(t)$, $y_{\text{particle}}(t)$, $z_{\text{particle}}(t)$). However, some mathematical manipulation is then necessary to convert the equations of motion into forms applicable to the Eulerian description.

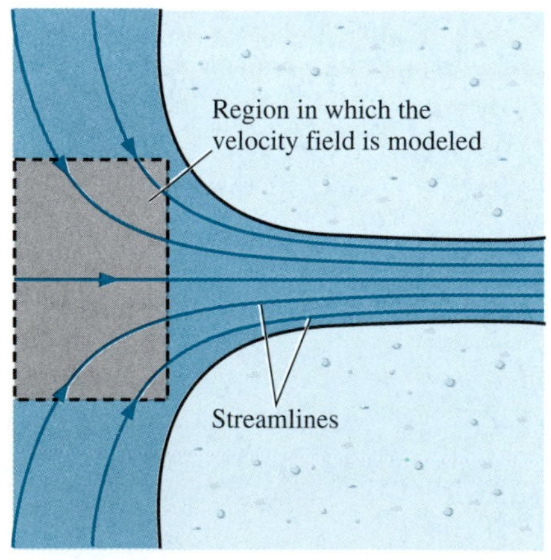

FIGURE 11–5
Flow field near the bell mouth inlet of a hydroelectric dam; a portion of the velocity field of Example 11–1 may be used as a first-order approximation of this physical flow field. The shaded region corresponds to that of Fig. 11–4.

Consider, for example, Newton's second law applied to our fluid particle,

Newton's second law:
$$\vec{F}_{\text{particle}} = m_{\text{particle}}\vec{a}_{\text{particle}} \tag{11–5}$$

where $\vec{F}_{\text{particle}}$ is the net force acting on the fluid particle, m_{particle} is its mass, and $\vec{a}_{\text{particle}}$ is its acceleration (Fig. 11–6). By definition, the acceleration of the fluid particle is the time derivative of the particle's velocity,

Acceleration of a fluid particle:
$$\vec{a}_{\text{particle}} = \frac{d\vec{V}_{\text{particle}}}{dt} \tag{11–6}$$

However, at any instant in time t, the velocity of the particle is the same as the local value of the velocity *field* at the location ($x_{\text{particle}}(t)$, $y_{\text{particle}}(t)$, $z_{\text{particle}}(t)$) of the particle, since the fluid particle moves with the fluid by definition. In other words, $\vec{V}_{\text{particle}}(t) \equiv \vec{V}(x_{\text{particle}}(t), y_{\text{particle}}(t), z_{\text{particle}}(t), t)$. To take the time derivative in Eq. 11–6, we must therefore use the *chain rule*, since the dependent variable ($\vec{V}$) is a function of *four* independent variables (x_{particle}, y_{particle}, z_{particle}, and t),

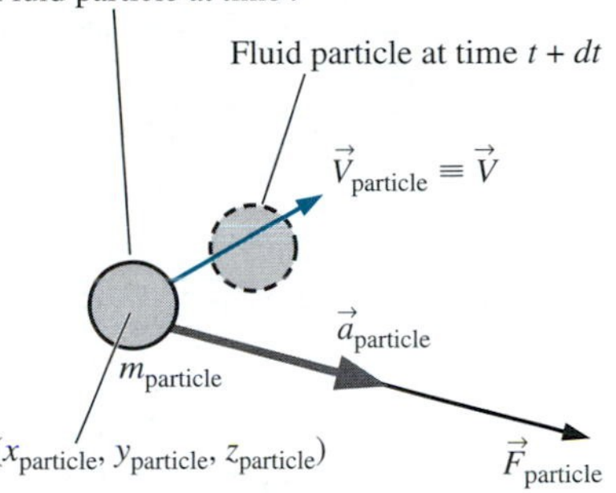

FIGURE 11–6
Newton's second law applied to a fluid particle; the acceleration vector (gray arrow) is in the same direction as the force vector (black arrow), but the velocity vector (blue arrow) may act in a different direction.

$$\vec{a}_{\text{particle}} = \frac{d\vec{V}_{\text{particle}}}{dt} = \frac{d\vec{V}}{dt} = \frac{d\vec{V}(x_{\text{particle}}, y_{\text{particle}}, z_{\text{particle}}, t)}{dt}$$
$$= \frac{\partial \vec{V}}{\partial t}\frac{dt}{dt} + \frac{\partial \vec{V}}{\partial x_{\text{particle}}}\frac{dx_{\text{particle}}}{dt} + \frac{\partial \vec{V}}{\partial y_{\text{particle}}}\frac{dy_{\text{particle}}}{dt} + \frac{\partial \vec{V}}{\partial z_{\text{particle}}}\frac{dz_{\text{particle}}}{dt} \quad \text{(11–7)}$$

In Eq. 11–7, ∂ is the **partial derivative operator** and d is the **total derivative operator**. Consider the second term on the right-hand side of Eq. 11–7. Since the acceleration is defined as that *following a fluid particle* (Lagrangian description), the rate of change of the particle's x-position with respect to time is $dx_{\text{particle}}/dt = u$ (Fig. 11–7), where u is the x-component of the velocity vector defined by Eq. 11–4. Similarly, $dy_{\text{particle}}/dt = v$ and $dz_{\text{particle}}/dt = w$. Furthermore, at any instant in time under consideration, the material position vector (x_{particle}, y_{particle}, z_{particle}) of the fluid particle in the Lagrangian frame is equal to the position vector (x, y, z) in the Eulerian frame. Equation 11–7 thus becomes

$$\vec{a}_{\text{particle}}(x, y, z, t) = \frac{d\vec{V}}{dt} = \frac{\partial \vec{V}}{\partial t} + u\frac{\partial \vec{V}}{\partial x} + v\frac{\partial \vec{V}}{\partial y} + w\frac{\partial \vec{V}}{\partial z} \quad \text{(11–8)}$$

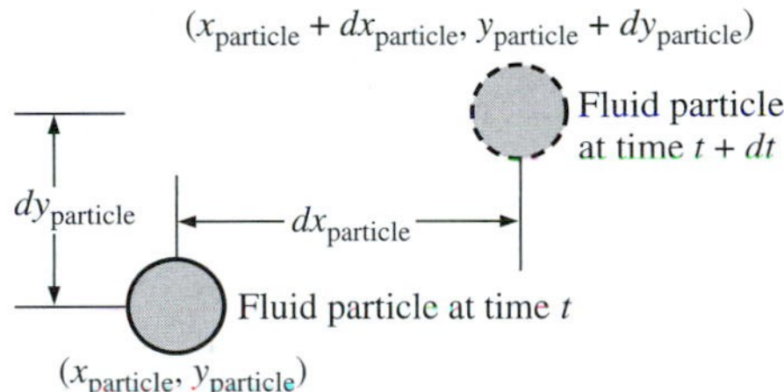

FIGURE 11–7
When following a fluid particle, the x-component of velocity, u, is defined as dx_{particle}/dt. Similarly, $v = dy_{\text{particle}}/dt$ and $w = dz_{\text{particle}}/dt$. Movement is shown here only in two dimensions for simplicity.

where we have also used the (obvious) fact that $dt/dt = 1$. Finally, at any instant in time t, the acceleration field of Eq. 11–3 must equal the acceleration of the fluid particle that happens to occupy the location (x, y, z) at that time t. Why? Because the fluid particle is by definition accelerating with the fluid flow. Hence, *we may replace* $\vec{a}_{\text{particle}}$ *with* $\vec{a}(x, y, z, t)$ *in Eqs. 11–7 and 11–8 to transform from the Lagrangian to the Eulerian frame of reference.* In vector form, Eq. 11–8 is written as

Acceleration of a fluid particle expressed as a field variable:

$$\vec{a}(x, y, z, t) = \frac{d\vec{V}}{dt} = \frac{\partial \vec{V}}{\partial t} + (\vec{V} \cdot \vec{\nabla})\vec{V} \quad \text{(11–9)}$$

where $\vec{\nabla}$ is the **gradient operator** or **del operator**, a vector operator that is defined in Cartesian coordinates as

Gradient or del operator:
$$\vec{\nabla} = \left(\frac{\partial}{\partial x}, \frac{\partial}{\partial y}, \frac{\partial}{\partial z}\right) = \vec{i}\frac{\partial}{\partial x} + \vec{j}\frac{\partial}{\partial y} + \vec{k}\frac{\partial}{\partial z} \quad \text{(11–10)}$$

In Cartesian coordinates then, the components of the acceleration vector are

Cartesian coordinates:
$$a_x = \frac{\partial u}{\partial t} + u\frac{\partial u}{\partial x} + v\frac{\partial u}{\partial y} + w\frac{\partial u}{\partial z}$$
$$a_y = \frac{\partial v}{\partial t} + u\frac{\partial v}{\partial x} + v\frac{\partial v}{\partial y} + w\frac{\partial v}{\partial z} \quad \text{(11–11)}$$
$$a_z = \frac{\partial w}{\partial t} + u\frac{\partial w}{\partial x} + v\frac{\partial w}{\partial y} + w\frac{\partial w}{\partial z}$$

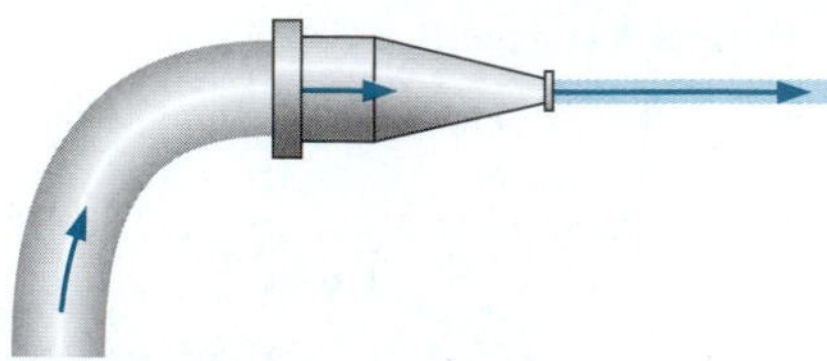

FIGURE 11–8
Flow of water through the nozzle of a garden hose illustrates that fluid particles may accelerate, even in a steady flow. In this example, the exit speed of the water is much higher than the water speed in the hose, implying that fluid particles have accelerated even though the flow is steady.

The first term on the right-hand side of Eq. 11–9, $\partial\vec{V}/\partial t$, is called the **local acceleration** and is nonzero only for unsteady flows. The second term, $(\vec{V}\cdot\vec{\nabla})\vec{V}$, is called the **advective acceleration** (sometimes the **convective acceleration**); *this term can be nonzero even for steady flows*. It accounts for the effect of the fluid particle moving (advecting or convecting) to a new location in the flow, where the velocity field is different. For example, consider steady flow of water through a garden hose nozzle (Fig. 11–8). We define *steady* in the Eulerian frame of reference to be when properties at any point in the flow field do not change with respect to time. Since the velocity at the exit of the nozzle is larger than that at the nozzle entrance, fluid particles clearly accelerate, even though the flow is steady. The acceleration is nonzero because of the advective acceleration terms in Eq. 11–9. Note that while the flow is steady from the point of view of a fixed observer in the Eulerian reference frame, it is *not* steady from the Lagrangian reference frame moving with a fluid particle that enters the nozzle and accelerates as it passes through the nozzle.

Material Derivative

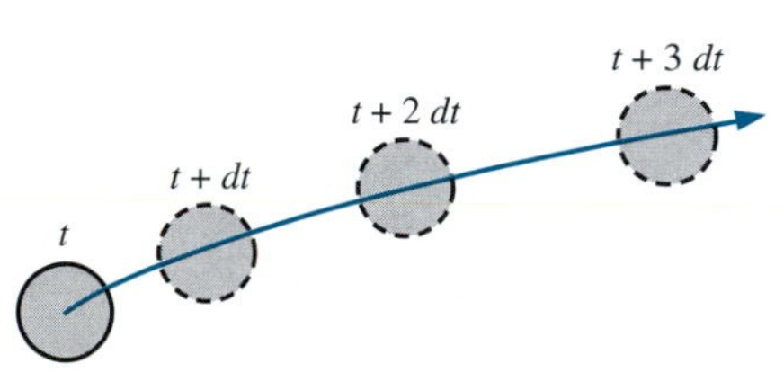

FIGURE 11–9
The material derivative D/Dt is defined by following a fluid particle as it moves throughout the flow field. In this illustration, the fluid particle is accelerating to the right as it moves up and to the right.

The total derivative operator d/dt in Eq. 11–9 is given a special name, the **material derivative**; it is assigned a special notation, D/Dt, in order to emphasize that it is formed by *following a fluid particle as it moves through the flow field* (Fig. 11–9). Other names for the material derivative include **total, particle, Lagrangian, Eulerian,** and **substantial derivative**.

Material derivative:
$$\frac{D}{Dt}=\frac{d}{dt}=\frac{\partial}{\partial t}+(\vec{V}\cdot\vec{\nabla}) \tag{11-12}$$

When we apply the material derivative of Eq. 11–12 to the velocity field, the result is the acceleration field as expressed by Eq. 11–9, which is thus sometimes called the **material acceleration**.

Material acceleration:
$$\vec{a}(x, y, z, t)=\frac{D\vec{V}}{Dt}=\frac{d\vec{V}}{dt}=\frac{\partial\vec{V}}{\partial t}+(\vec{V}\cdot\vec{\nabla})\vec{V} \tag{11-13}$$

Equation 11–12 can also be applied to other fluid properties besides velocity, both scalars and vectors. For example, the material derivative of pressure is written as

Material derivative of pressure:
$$\frac{DP}{Dt}=\frac{dP}{dt}=\frac{\partial P}{\partial t}+(\vec{V}\cdot\vec{\nabla})P \tag{11-14}$$

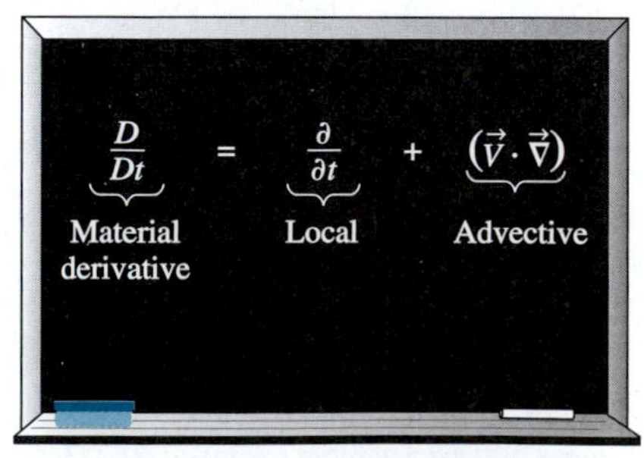

FIGURE 11–10
The material derivative D/Dt is composed of a *local* or *unsteady* part and a *convective* or *advective* part.

Equation 11–14 represents the time rate of change of pressure following a fluid particle as it moves through the flow and contains both local (unsteady) and advective components (Fig. 11–10).

EXAMPLE 11–2 Material Acceleration of a Steady Velocity Field

Consider the steady, incompressible, two-dimensional velocity field of Example 11–1. (*a*) Calculate the material acceleration at the point (x = 2 m, y = 3 m). (*b*) Sketch the material acceleration vectors at the same array of x- and y-values as in Example 11–1.

Solution For the given velocity field, the material acceleration vector is to be calculated at a particular point and plotted at an array of locations in the flow field.

Assumptions **1** The flow is steady and incompressible. **2** The flow is two-dimensional, implying no *z*-component of velocity and no variation of *u* or *v* with *z*.

Analysis (*a*) Using the velocity field of Eq. 1 of Example 11–1 and the equation for material acceleration components in Cartesian coordinates (Eq. 11–11), we write expressions for the two nonzero components of the acceleration vector:

$$a_x = \frac{\partial u}{\partial t} + \underbrace{u\frac{\partial u}{\partial x}} + \underbrace{v\frac{\partial u}{\partial y}} + w\frac{\partial u}{\partial z}$$

$$= 0 + (0.5 + 0.8x)(0.8) + (1.5 - 0.8y)(0) + 0 = (0.4 + 0.64x)\text{ m/s}^2$$

and

$$a_y = \frac{\partial v}{\partial t} + \underbrace{u\frac{\partial v}{\partial x}} + \underbrace{v\frac{\partial v}{\partial y}} + w\frac{\partial v}{\partial z}$$

$$= 0 + (0.5 + 0.8x)(0) + (1.5 - 0.8y)(-0.8) + 0 = (-1.2 + 0.64y)\text{ m/s}^2$$

At the point (x = 2 m, y = 3 m), **a_x = 1.68 m/s² and a_y = 0.720 m/s².**

(*b*) The equations in part (*a*) are applied to an array of *x*- and *y*-values in the flow domain within the given limits, and the acceleration vectors are plotted in Fig. 11–11.

Discussion The acceleration field is nonzero, even though the flow is *steady*. Above the stagnation point (above y = 1.875 m), the acceleration vectors plotted in Fig. 11–11 point upward, increasing in magnitude away from the stagnation point. To the right of the stagnation point (to the right of x = −0.625 m), the acceleration vectors point to the right, again increasing in magnitude away from the stagnation point. This agrees qualitatively with the velocity vectors of Fig. 11–4 and the streamlines sketched in Fig. 11–11; namely, in the upper-right portion of the flow field, fluid particles are accelerated in the upper-right direction and therefore veer in the counterclockwise direction due to **centripetal acceleration** toward the upper right. The flow below y = 1.875 m is a mirror image of the flow above this symmetry line, and flow to the left of x = −0.625 m is a mirror image of the flow to the right of this symmetry line.

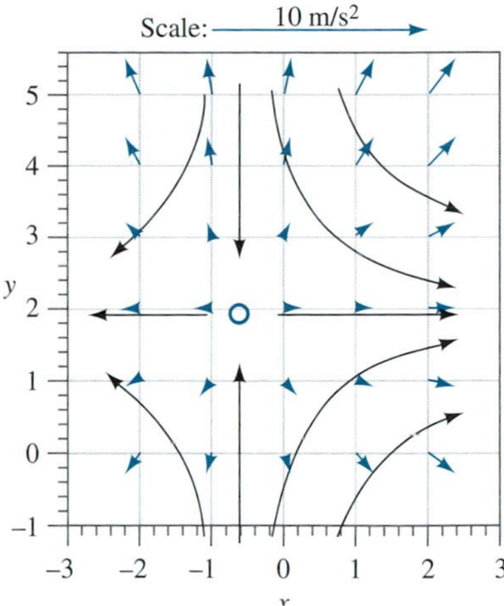

FIGURE 11–11

Acceleration vectors for the velocity field of Examples 11–1 and 11–2. The scale is shown by the top arrow, and the solid black curves represent the approximate shapes of some streamlines, based on the calculated velocity vectors (see Fig. 11–4). The stagnation point is indicated by the color circle.

11–2 ▪ FLOW PATTERNS AND FLOW VISUALIZATION

While quantitative study of fluid dynamics requires advanced mathematics, much can be learned from **flow visualization**—the visual examination of flow field features. Flow visualization is useful not only in physical experiments (Fig. 11–12), but in *numerical* solutions as well [**computational fluid dynamics (CFD)**]. In fact, the very first thing an engineer using CFD does after obtaining a numerical solution is simulate some form of flow visualization, so that he or she can see the "whole picture" rather than merely a list of numbers and quantitative data. Why? Because the human mind is designed to rapidly process an incredible amount of visual information; as they say, a

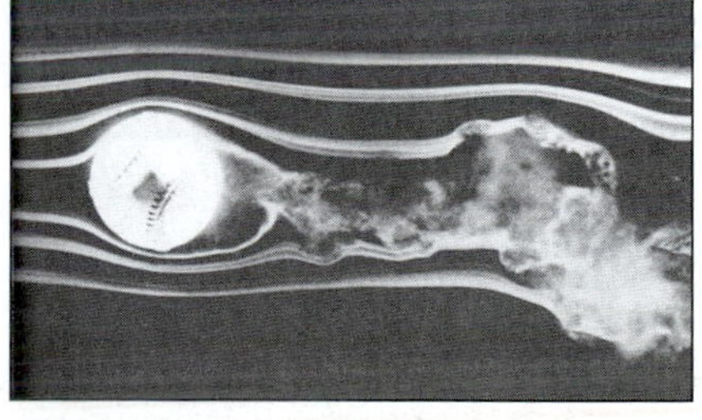

FIGURE 11–12

Spinning baseball. The late F. N. M. Brown devoted many years to developing and using smoke visualization in wind tunnels at the University of Notre Dame. Here the flow speed is about 77 ft/s and the ball is rotated at 630 rpm.

Photograph courtesy of T. J. Mueller.

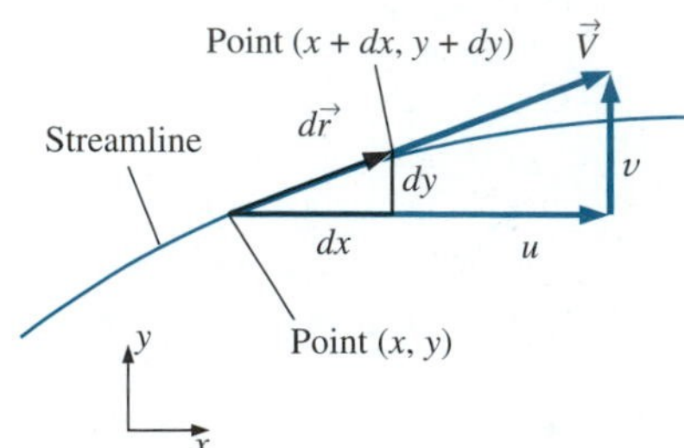

FIGURE 11–13

For two-dimensional flow in the xy-plane, arc length $d\vec{r} = (dx, dy)$ along a *streamline* is everywhere tangent to the local instantaneous velocity vector $\vec{V} = (u, v)$.

picture is worth a thousand words. There are many types of flow patterns that can be visualized, both physically (experimentally) and/or computationally.

Streamlines

A **streamline** is a curve that is everywhere tangent to the instantaneous local velocity vector.

Streamlines are useful as indicators of the instantaneous direction of fluid motion throughout the flow field. For example, regions of recirculating flow and separation of a fluid off of a solid wall are easily identified by the streamline pattern. Streamlines cannot be directly observed experimentally except in steady flow fields, in which they are coincident with pathlines and streaklines, to be discussed next. Mathematically, however, we can write a simple expression for a streamline based on its definition.

Consider an infinitesimal arc length $d\vec{r} = dx\vec{i} + dy\vec{j} + dz\vec{k}$ along a streamline; $d\vec{r}$ must be parallel to the local velocity vector $\vec{V} = u\vec{i} + v\vec{j} + w\vec{k}$ by definition of the streamline. By simple geometric arguments using similar triangles, we know that the components of $d\vec{r}$ must be proportional to those of $\vec{V}$ (Fig. 11–13). Hence,

Equation for a streamline:

$$\frac{dr}{V} = \frac{dx}{u} = \frac{dy}{v} = \frac{dz}{w} \tag{11–15}$$

where dr is the magnitude of $d\vec{r}$ and V is the speed, the magnitude of $\vec{V}$. Equation 11–15 is illustrated in two dimensions for simplicity in Fig. 11–13. For a known velocity field, we integrate Eq. 11–15 to obtain equations for the streamlines. In two dimensions, (x, y), (u, v), the following differential equation is obtained:

Streamline in the xy-plane:

$$\left(\frac{dy}{dx}\right)_{\text{along a streamline}} = \frac{v}{u} \tag{11–16}$$

In some simple cases, Eq. 11–16 may be solvable analytically; in the general case, it must be solved numerically. In either case, an arbitrary constant of integration appears. Each chosen value of the constant represents a different streamline. The *family* of curves that satisfy Eq. 11–16 therefore represents streamlines of the flow field.

EXAMPLE 11–3 **Streamlines in the xy-Plane—An Analytical Solution**

For the steady, incompressible, two-dimensional velocity field of Example 11–1, plot several streamlines in the right half of the flow ($x > 0$) and compare to the velocity vectors plotted in Fig. 11–4.

Solution An analytical expression for streamlines is to be generated and plotted in the right half of the flow.

Assumptions **1** The flow is steady and incompressible. **2** The flow is two-dimensional, implying no z-component of velocity and no variation of u or v with z.

Analysis Equation 11–16 is applicable here; thus, along a streamline,

$$\frac{dy}{dx} = \frac{v}{u} = \frac{1.5 - 0.8y}{0.5 + 0.8x}$$

We solve this differential equation by separation of variables:

$$\frac{dy}{1.5 - 0.8y} = \frac{dx}{0.5 + 0.8x} \quad \rightarrow \quad \int \frac{dy}{1.5 - 0.8y} = \int \frac{dx}{0.5 + 0.8x}$$

After some algebra, we solve for y as a function of x along a streamline,

$$y = \frac{C}{0.8(0.5 + 0.8x)} + 1.875$$

where C is a constant of integration that can be set to various values in order to plot the streamlines. Several streamlines of the given flow field are shown in Fig. 11–14.

Discussion The velocity vectors of Fig. 11–4 are superimposed on the streamlines of Fig. 11–14; the agreement is excellent in the sense that the velocity vectors point everywhere tangent to the streamlines. Note that speed cannot be determined directly from the streamlines alone.

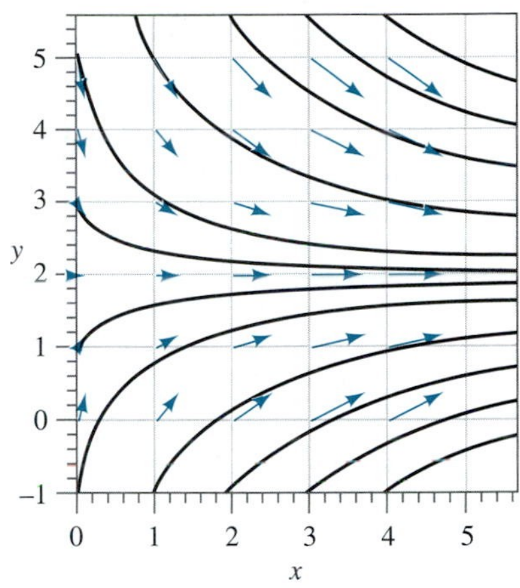

FIGURE 11–14

Streamlines (solid black curves) for the velocity field of Example 11–3; velocity vectors of Fig. 11–4 (color arrows) are superimposed for comparison.

Pathlines

A **pathline** is the actual path traveled by an individual fluid particle over some time period.

Pathlines are the easiest of the flow patterns to understand. A pathline is a Lagrangian concept in that we simply follow the path of an individual fluid particle as it moves around in the flow field (Fig. 11–15). Thus, a pathline is the same as the fluid particle's material position vector ($x_{particle}(t)$, $y_{particle}(t)$, $z_{particle}(t)$), discussed in Section 11–1, traced out over some finite time interval. In a physical experiment, you can imagine a tracer fluid particle that is marked somehow—either by color or brightness—such that it is easily distinguishable from surrounding fluid particles. Now imagine a camera with the shutter open for a certain time period, $t_{start} < t < t_{end}$, in which the particle's path is recorded; the resulting curve is called a pathline. An intriguing example is shown in Fig. 11–16 for the case of waves moving along the surface of water in a tank. Neutrally buoyant white **tracer particles** are suspended in the water, and a time-exposure photograph is taken for one complete wave period. The result is pathlines that are elliptical in shape, showing that fluid particles bob up and down and forward and backward, but return to their

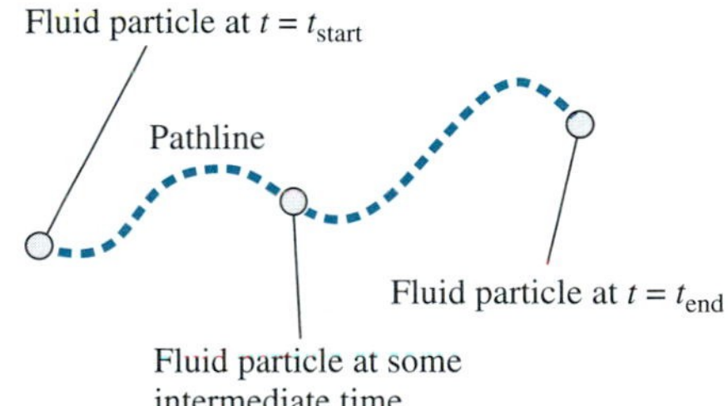

FIGURE 11–15

A *pathline* is formed by following the actual path of a fluid particle.

FIGURE 11–16

Pathlines produced by white tracer particles suspended in water and captured by time-exposure photography; as waves pass horizontally, each particle moves in an elliptical path during one wave period.

Wallet, A. & Ruellan, F. 1950, La Houille Blanche *5:483–489. Used by permission.*

original position upon completion of one wave period; there is no net forward motion. You may have experienced something similar while bobbing up and down on ocean waves.

Pathlines can also be calculated numerically for a known velocity field. Specifically, the location of the tracer particle is integrated over time from some starting location $\vec{x}_{\text{start}}$ and starting time t_{start} to some later time t.

Tracer particle location at time t:

$$\vec{x} = \vec{x}_{\text{start}} + \int_{t_{\text{start}}}^{t} \vec{V}\, dt \qquad \textbf{(11–17)}$$

When Eq. 11–17 is calculated for t between t_{start} and t_{end}, a plot of $\vec{x}(t)$ is the pathline of the fluid particle during that time interval, as illustrated in Fig. 11–15. For some simple flow fields, Eq. 11–17 can be integrated analytically. For more complex flows, we must perform a numerical integration.

If the velocity field is steady, individual fluid particles will follow streamlines. Thus, *for steady flow, pathlines are identical to streamlines.*

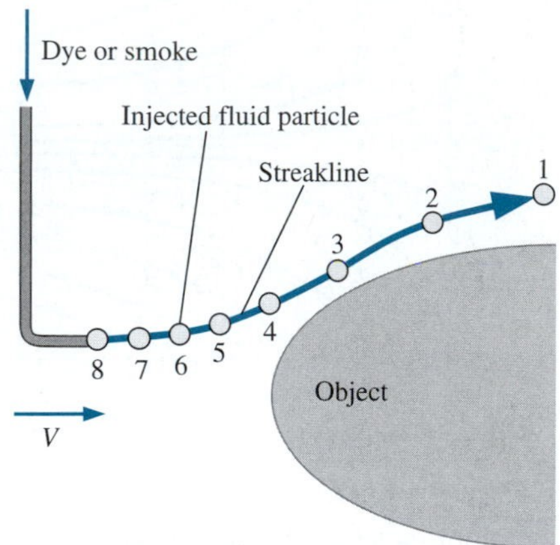

FIGURE 11–17

A *streakline* is formed by continuous introduction of dye or smoke from a point in the flow. Labeled tracer particles (1 through 8) were introduced sequentially.

Streaklines

A **streakline** is the locus of fluid particles that have passed sequentially through a prescribed point in the flow.

Streaklines are the most common flow pattern generated in a physical experiment. If you insert a small tube into a flow and introduce a continuous stream of tracer fluid (dye in a water flow or smoke in an airflow), the observed pattern is a streakline. Figure 11–17 shows a tracer being injected into a free-stream flow containing an object, such as a wing. The circles represent individual injected tracer fluid particles, released at a uniform time interval. As the particles are forced out of the way by the object, they accelerate around the shoulder of the object, as indicated by the increased distance between individual tracer particles in that region. The streakline is formed by connecting all the circles into a smooth curve. In physical experiments in a wind or water tunnel, the smoke or dye is injected *continuously*, not as individual particles, and the resulting flow pattern is by definition a streakline. In Fig. 11–17, tracer particle 1 was released at an earlier time than tracer particle 2, and so on. The location of an individual tracer particle is determined by the surrounding velocity field from the moment of its injection into the flow until the present time. If the flow is unsteady, the surrounding velocity field changes, and we cannot expect the resulting streakline to resemble a streamline or pathline at any given instant in time. However, *if the flow is steady, streamlines, pathlines, and streaklines are identical* (Fig. 11–18).

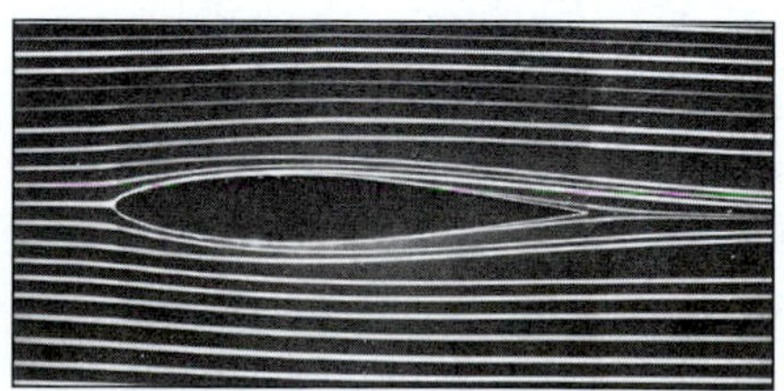

FIGURE 11–18

Streaklines produced by colored fluid introduced upstream; since the flow is steady, these streaklines are the same as streamlines and pathlines.

Courtesy ONERA. Photograph by Werlé.

Streaklines are often confused with streamlines or pathlines. While the three flow patterns are identical in steady flow, they can be quite different in unsteady flow. The main difference is that a streamline represents an *instantaneous* flow pattern at a given instant in time, while a streakline and a pathline are flow patterns that have some *age* and thus a *time history* associated with them. A streakline is an instantaneous snapshot of a *time-integrated* flow pattern. A pathline, on the other hand, is the *time-exposed* flow path of an individual particle over some time period.

For a known velocity field, a streakline can be generated numerically. We need to follow the paths of a continuous stream of tracer particles from the time of their injection into the flow until the present time, using Eq. 11–17. Mathematically, the location of a tracer particle is integrated over time from the time of its injection t_{inject} to the present time t_{present}. Equation 11–17 becomes

Integrated tracer particle location:

$$\vec{x} = \vec{x}_{\text{injection}} + \int_{t_{\text{inject}}}^{t_{\text{present}}} \vec{V}\,dt \qquad \textbf{(11–18)}$$

In a complex unsteady flow, the time integration must be performed numerically as the velocity field changes with time. When the locus of tracer particle locations at $t = t_{\text{present}}$ is connected by a smooth curve, the result is the desired streakline.

EXAMPLE 11–4 Comparison of Flow Patterns in an Unsteady Flow

An *unsteady*, incompressible, two-dimensional velocity field is given by

$$\vec{V} = (u, v) = (0.5 + 0.8x)\vec{i} + (1.5 + 2.5\sin(\omega t) - 0.8y)\vec{j} \qquad (1)$$

where the angular frequency ω is equal to 2π rad/s (a physical frequency of 1 Hz). This velocity field is identical to that of Eq. 1 of Example 11–1 except for the additional periodic term in the v-component of velocity. In fact, since the period of oscillation is 1 s, when time t is any integral multiple of $\frac{1}{2}$ s ($t = 0, \frac{1}{2}, 1, \frac{3}{2}, 2, \ldots$ s), the sine term in Eq. 1 is zero and the velocity field is instantaneously identical to that of Example 11–1. Physically, we imagine flow into a large bell mouth inlet that is oscillating up and down at a frequency of 1 Hz. Consider two complete cycles of flow from $t = 0$ s to $t = 2$ s. Compare instantaneous streamlines at $t = 2$ s to pathlines and streaklines generated during the time period from $t = 0$ s to $t = 2$ s.

Solution Streamlines, pathlines, and streaklines are to be generated and compared for the given unsteady velocity field.

Assumptions **1** The flow is incompressible. **2** The flow is two-dimensional, implying no z-component of velocity and no variation of u or v with z.

Analysis The instantaneous streamlines at $t = 2$ s are identical to those of Fig. 11–14, and several of them are replotted in Fig. 11–19. To simulate pathlines, we use the Runge–Kutta numerical integration technique to march in time from $t = 0$ s to $t = 2$ s, tracing the path of fluid particles released at three locations: ($x = 0.5$ m, $y = 0.5$ m), ($x = 0.5$ m, $y = 2.5$ m), and ($x = 0.5$ m, $y = 4.5$ m). These pathlines are shown in Fig. 11–19, along with the streamlines. Finally, streaklines are simulated by following the paths of *many* fluid tracer particles released at the given three locations at times between $t = 0$ s and $t = 2$ s, and connecting the locus of their positions at $t = 2$ s. These streaklines are also plotted in Fig. 11–19.

Discussion Since the flow is unsteady, the streamlines, pathlines, and streaklines are *not* coincident. In fact, they differ significantly from each other. Note that the streaklines and pathlines are wavy due to the undulating v-component of velocity. Two complete periods of oscillation have occurred between $t = 0$ s and $t = 2$ s, as verified by a careful look at the pathlines and streaklines. The streamlines have no such waviness since they have no time history; they represent an instantaneous snapshot of the velocity field at $t = 2$ s.

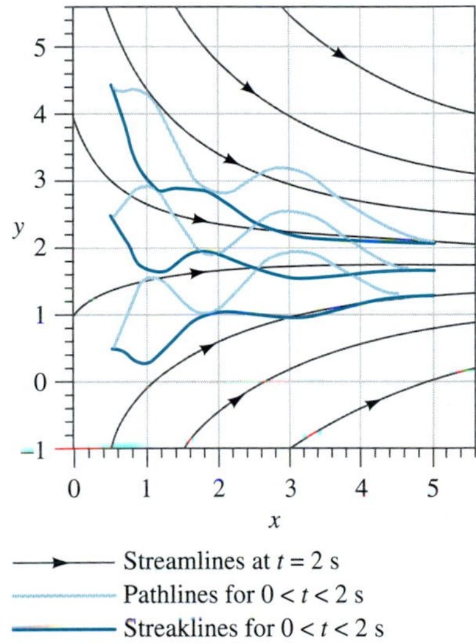

FIGURE 11–19

Streamlines, pathlines, and streaklines for the oscillating velocity field of Example 11–4. The streaklines and pathlines are wavy because of their integrated time history, but the streamlines are not wavy since they represent an instantaneous snapshot of the velocity field.

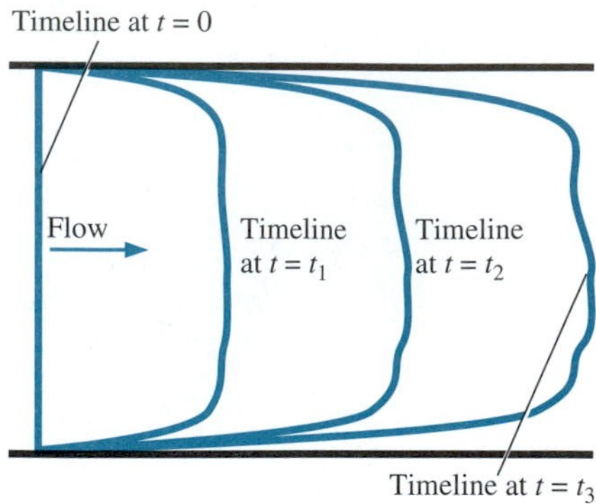

FIGURE 11–20

Timelines are formed by marking a line of fluid particles, and then watching that line move (and deform) through the flow field; timelines are shown at $t = 0$, t_1, t_2, and t_3.

Timelines

A **timeline** is a set of adjacent fluid particles that were marked at the same (earlier) instant in time.

Timelines are particularly useful in situations where the uniformity of a flow (or lack thereof) is to be examined. Figure 11–20 illustrates timelines in a channel flow between two parallel walls. Because of friction at the walls, the fluid velocity there is zero (the no-slip condition), and the top and bottom of the timeline are anchored at their starting locations. In regions of the flow away from the walls, the marked fluid particles move at the local fluid velocity, deforming the timeline. In the example of Fig. 11–20, the speed near the center of the channel is fairly uniform, but small deviations tend to amplify with time as the timeline stretches. Timelines can be generated experimentally in a water channel through use of a **hydrogen bubble wire**. When a short burst of electric current is sent through the cathode wire, electrolysis of the water occurs and tiny hydrogen gas bubbles form at the wire. Since the bubbles are so small, their buoyancy is nearly negligible, and the bubbles follow the water flow nicely (Fig. 11–21).

Surface Flow Visualization Techniques

Finally, we briefly mention some flow visualization techniques that are useful along solid surfaces. The direction of fluid flow immediately above a solid surface can be visualized with **tufts**—short, flexible strings glued to the surface at one end that point in the flow direction. Tufts are especially useful for locating regions of flow separation, where the flow direction suddenly reverses.

A technique called **surface oil visualization** can be used for the same purpose—oil placed on the surface forms streaks called **friction lines** that indicate the direction of flow. If it rains lightly when your car is dirty (especially in the winter when salt is on the roads), you may have noticed streaks along the hood and sides of the car, or even on the windshield. This is similar to what is observed with surface oil visualization.

Lastly, there are pressure-sensitive and temperature-sensitive paints that enable researchers to observe the pressure or temperature distribution along solid surfaces.

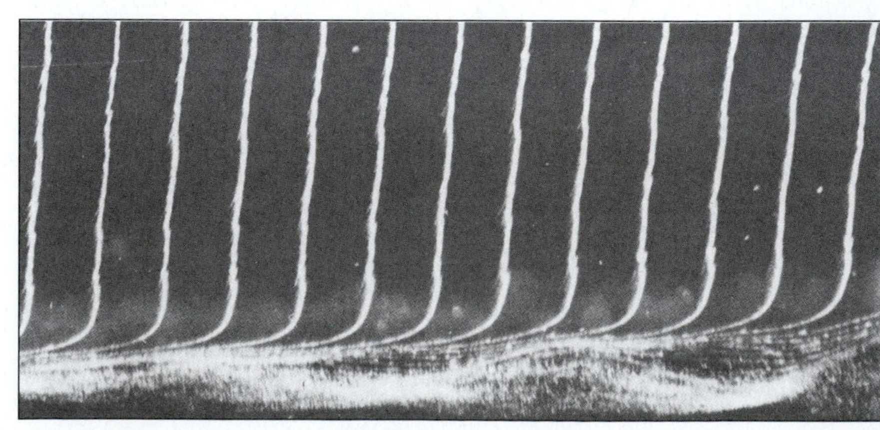

FIGURE 11–21

Timelines produced by a hydrogen bubble wire are used to visualize the boundary layer velocity profile shape. Flow is from left to right, and the hydrogen bubble wire is located to the left of the field of view. Bubbles near the wall reveal a flow instability that leads to turbulence.

Bippes, H. 1972 Sitzungsber, Heidelb. Akad. Wiss. Math. Naturwiss. *Kl., no. 3, 103–180; NASA TM-75243, 1978.*

11–3 ▪ VORTICITY AND ROTATIONALITY

Another kinematic property of great importance to the analysis of fluid flows is the **vorticity vector**, defined mathematically as the curl of the velocity vector $\vec{V}$,

Vorticity vector:
$$\vec{\zeta} = \vec{\nabla} \times \vec{V} = \text{curl}(\vec{V}) \tag{11–19}$$

Physically, you can tell the direction of the vorticity vector by using the right-hand rule for cross product (Fig. 11–22). The symbol ζ used for vorticity is the Greek letter *zeta*. You should note that this symbol for vorticity is *not* universal among fluid mechanics textbooks; some authors use the Greek letter *omega* (ω) while still others use uppercase *omega* (Ω). In this book, $\vec{\omega}$ is used to denote the rate of rotation vector (angular velocity vector) of a fluid element. It turns out that the rate of rotation vector is equal to half of the vorticity vector,

Rate of rotation vector:
$$\vec{\omega} = \frac{1}{2}\vec{\nabla} \times \vec{V} = \frac{1}{2}\text{curl}(\vec{V}) = \frac{\vec{\zeta}}{2} \tag{11–20}$$

Thus, *vorticity is a measure of rotation of a fluid particle*. Specifically,

> **Vorticity** is equal to twice the angular velocity of a fluid particle (Fig. 11–23).

If the vorticity at a point in a flow field is nonzero, the fluid particle that happens to occupy that point in space is rotating; the flow in that region is called **rotational**. Likewise, if the vorticity in a region of the flow is zero (or negligibly small), fluid particles there are not rotating; the flow in that region is called **irrotational**. Physically, fluid particles in a rotational region of flow rotate end over end as they move along in the flow. For example, fluid particles within the viscous boundary layer near a solid wall are rotational (and thus have nonzero vorticity), while fluid particles outside the boundary layer are irrotational (and their vorticity is zero). Both of these cases are illustrated in Fig. 11–24.

Rotation of fluid elements is associated with wakes, boundary layers, flow through turbomachinery (fans, turbines, compressors, etc.), and flow with heat transfer. The vorticity of a fluid element cannot change except through the action of viscosity, nonuniform heating (temperature gradients), or other nonuniform phenomena. Thus if a flow originates in an irrotational region, it remains irrotational until some nonuniform process alters it. For example, air entering an inlet from quiescent (still) surroundings is irrotational and

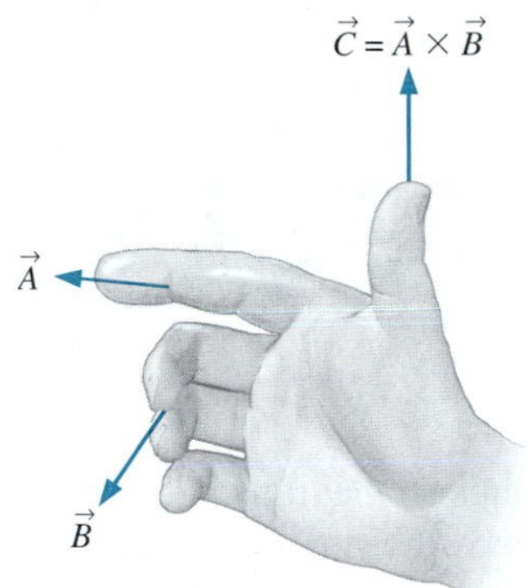

FIGURE 11–22
The direction of a vector cross product is determined by the right-hand rule.

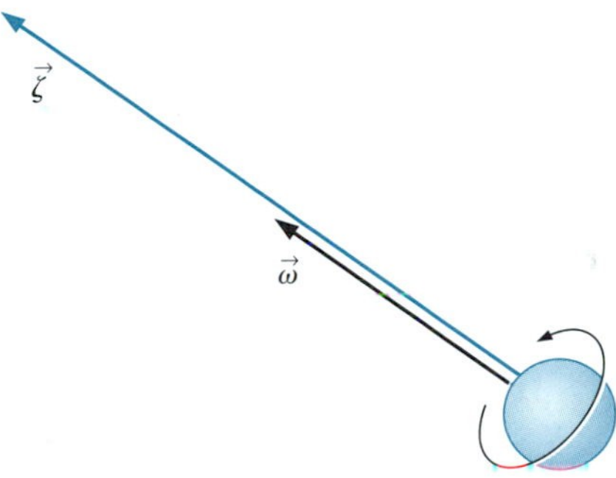

FIGURE 11–23
The *vorticity vector* is equal to twice the angular velocity vector of a rotating fluid particle.

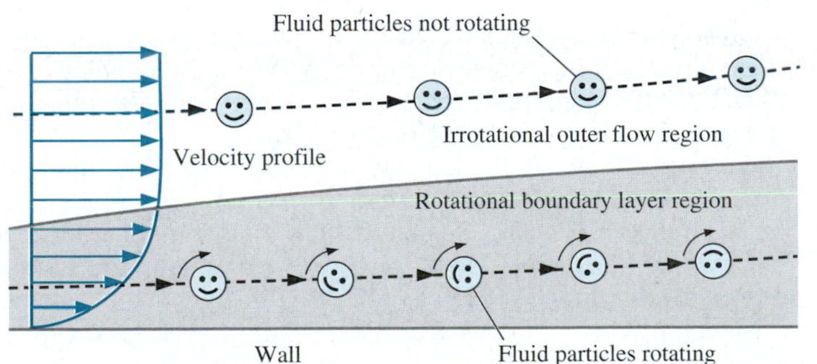

FIGURE 11–24
The difference between rotational and irrotational flow: fluid elements in a rotational region of the flow rotate, but those in an irrotational region of the flow do not.

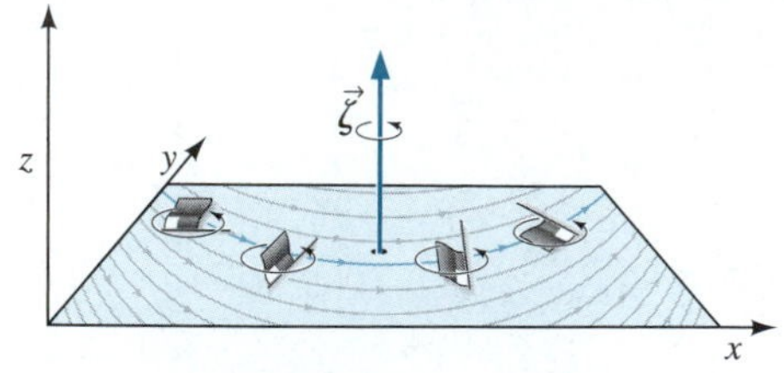

FIGURE 11–25

For a two-dimensional flow in the xy-plane, the vorticity vector always points in the z- or $-z$-direction. In this illustration, the flag-shaped fluid particle rotates in the counterclockwise direction as it moves in the xy-plane; its vorticity points in the positive z-direction as shown.

remains so unless it encounters an object in its path or is subjected to nonuniform heating. If a region of flow can be approximated as irrotational, the equations of motion are greatly simplified (see Çengel & Cimbala, 2006).

In Cartesian coordinates, $(\vec{i}, \vec{j}, \vec{k})$, (x, y, z), and (u, v, w), Eq. 11–19 is expanded as follows:

Vorticity vector in Cartesian coordinates:

$$\vec{\zeta} = \left(\frac{\partial w}{\partial y} - \frac{\partial v}{\partial z}\right)\vec{i} + \left(\frac{\partial u}{\partial z} - \frac{\partial w}{\partial x}\right)\vec{j} + \left(\frac{\partial v}{\partial x} - \frac{\partial u}{\partial y}\right)\vec{k} \tag{11–21}$$

If the flow is two-dimensional in the xy-plane, the z-component of velocity (w) is zero and neither u nor v varies with z. Thus the first two components of Eq. 11–21 are identically zero and the vorticity reduces to

Two-dimensional flow in Cartesian coordinates:

$$\vec{\zeta} = \left(\frac{\partial v}{\partial x} - \frac{\partial u}{\partial y}\right)\vec{k} \tag{11–22}$$

Note that if a flow is two-dimensional in the xy-plane, the vorticity vector must point in either the z- or $-z$-direction (Fig. 11–25).

EXAMPLE 11–5 Determination of Rotationality in a Two-Dimensional Flow

Consider the following steady, incompressible, two-dimensional velocity field:

$$\vec{V} = (u, v) = x^2\vec{i} + (-2xy - 1)\vec{j} \tag{1}$$

Is this flow rotational or irrotational? Sketch some streamlines in the first quadrant and discuss.

Solution We are to determine whether a flow with a given velocity field is rotational or irrotational, and we are to draw some streamlines in the first quadrant.

Analysis Since the flow is two-dimensional, Eq. 11–22 is valid. Thus,

Vorticity:

$$\vec{\zeta} = \left(\frac{\partial v}{\partial x} - \frac{\partial u}{\partial y}\right)\vec{k} = (-2y - 0)\vec{k} = -2y\vec{k} \tag{2}$$

Since the vorticity is nonzero, this flow is **rotational**. In Fig. 11–26 we plot several streamlines of the flow in the first quadrant; we see that fluid moves downward and to the right. The translation and deformation of a fluid parcel is also shown: at $\Delta t = 0$, the fluid parcel is square, at $\Delta t = 0.25$ s, it has moved and deformed, and at $\Delta t = 0.50$ s, the parcel has moved farther and is further deformed. In particular, the right-most portion of the fluid parcel moves faster to the right and faster downward compared to the left-most portion, stretching the parcel in the x-direction and squashing it in the vertical direction. It is clear that there is also a net *clockwise* rotation of the fluid parcel, which agrees with the result of Eq. 2.

Discussion From Eq. 11–20, individual fluid particles rotate at an angular velocity equal to $\vec{\omega} = -y\vec{k}$, half of the vorticity vector. Since $\vec{\omega}$ is not constant, this flow is *not* solid-body rotation. Rather, $\vec{\omega}$ is a linear function of y. Further analysis reveals that this flow field is incompressible; the shaded areas representing the fluid parcel in Fig. 11–26 remain constant at all three instants in time.

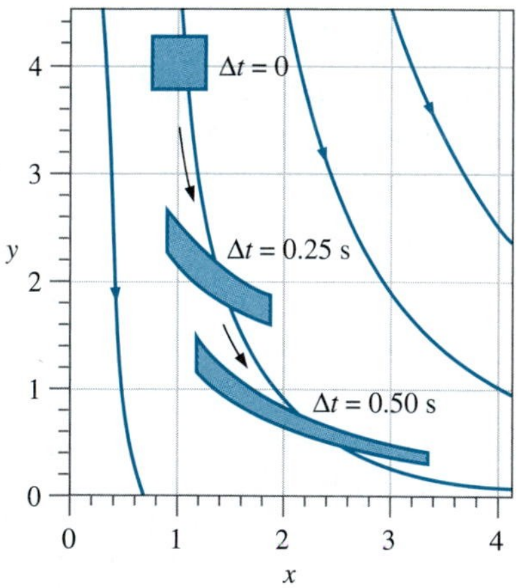

FIGURE 11–26

Deformation of an initially square fluid parcel subjected to the velocity field of Example 11–5 for a time period of 0.25 s and 0.50 s. Several streamlines are also plotted in the first quadrant. It is clear that this flow is *rotational*.

In cylindrical coordinates, $(\vec{e}_r, \vec{e}_\theta, \vec{e}_z)$, (r, θ, z), and (u_r, u_θ, u_z), Eq. 11–19 is expanded as

Vorticity vector in cylindrical coordinates:

$$\vec{\zeta} = \left(\frac{1}{r}\frac{\partial u_z}{\partial \theta} - \frac{\partial u_\theta}{\partial z}\right)\vec{e}_r + \left(\frac{\partial u_r}{\partial z} - \frac{\partial u_z}{\partial r}\right)\vec{e}_\theta + \frac{1}{r}\left(\frac{\partial (r u_\theta)}{\partial r} - \frac{\partial u_r}{\partial \theta}\right)\vec{e}_z \tag{11–23}$$

For two-dimensional flow in the $r\theta$-plane, Eq. 11–23 reduces to

Two-dimensional flow in cylindrical coordinates:

$$\vec{\zeta} = \frac{1}{r}\left(\frac{\partial (r u_\theta)}{\partial r} - \frac{\partial u_r}{\partial \theta}\right)\vec{k} \tag{11–24}$$

where $\vec{k}$ is used as the unit vector in the z-direction in place of $\vec{e}_z$. Note that if a flow is two-dimensional in the $r\theta$-plane, the vorticity vector must point in either the z- or $-z$-direction (Fig. 11–27).

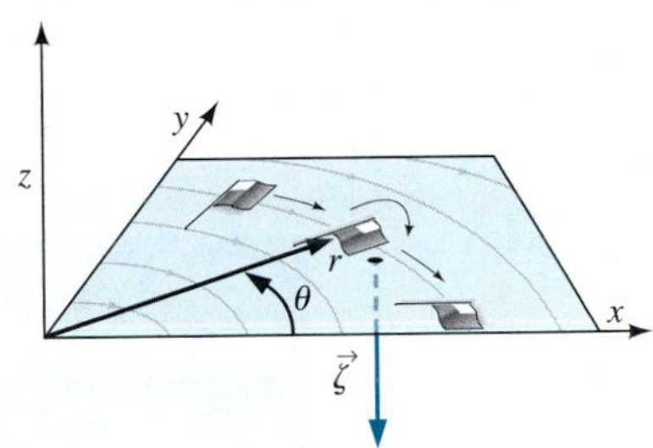

FIGURE 11–27

For a two-dimensional flow in the $r\theta$-plane, the vorticity vector always points in the z (or $-z$) direction. In this illustration, the flag-shaped fluid particle rotates in the clockwise direction as it moves in the $r\theta$-plane; its vorticity points in the $-z$-direction as shown.

Comparison of Two Circular Flows

Not all flows with circular streamlines are rotational. To illustrate this point, we consider two incompressible, steady, two-dimensional flows, both of which have circular streamlines in the $r\theta$-plane:

Flow A—solid-body rotation: $$u_r = 0 \quad \text{and} \quad u_\theta = \omega r \tag{11–25}$$

Flow B—line vortex: $$u_r = 0 \quad \text{and} \quad u_\theta = \frac{K}{r} \tag{11–26}$$

where ω and K are constants. (Alert readers will note that u_θ in Eq. 11–26 is infinite at $r = 0$, which is of course physically impossible; we ignore the region close to the origin to avoid this problem.) Since the radial component of velocity is zero in both cases, the streamlines are circles about the origin. The velocity profiles for the two flows, along with their streamlines, are sketched in Fig. 11–28. We now calculate and compare the vorticity field for each of these flows, using Eq. 11 24.

Flow A—solid-body rotation: $$\vec{\zeta} = \frac{1}{r}\left(\frac{\partial(\omega r^2)}{\partial r} - 0\right)\vec{k} = 2\omega\vec{k} \tag{11–27}$$

Flow B—line vortex: $$\vec{\zeta} = \frac{1}{r}\left(\frac{\partial(K)}{\partial r} - 0\right)\vec{k} = 0 \tag{11–28}$$

Not surprisingly, the vorticity for solid-body rotation is nonzero. In fact, it is a constant of magnitude twice the angular velocity and pointing in the same direction. (This agrees with Eq. 11–20.) *Flow A is rotational.* Physically, this means that individual fluid particles rotate as they revolve around the origin (Fig. 11–28*a*). By contrast, the vorticity of the line vortex is identically zero everywhere (except right at the origin, which is a mathematical singularity). *Flow B is irrotational.* Physically, fluid particles do *not* rotate as they revolve in circles about the origin (Fig. 11–28*b*).

A simple analogy can be made between flow A and a merry-go-round or roundabout, and flow B and a Ferris wheel (Fig. 11–29). As children revolve around a roundabout, they also rotate at the same angular velocity

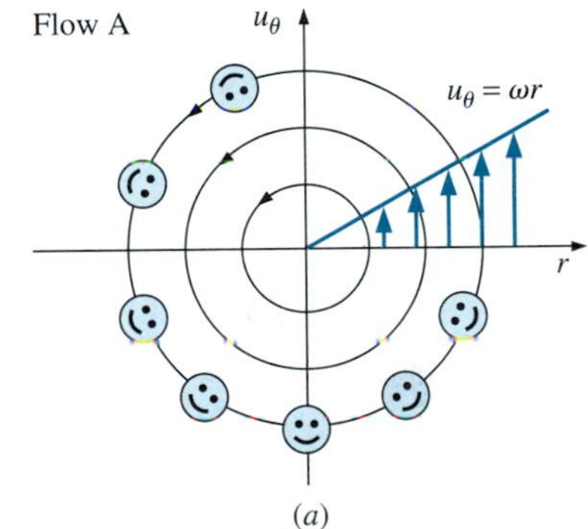

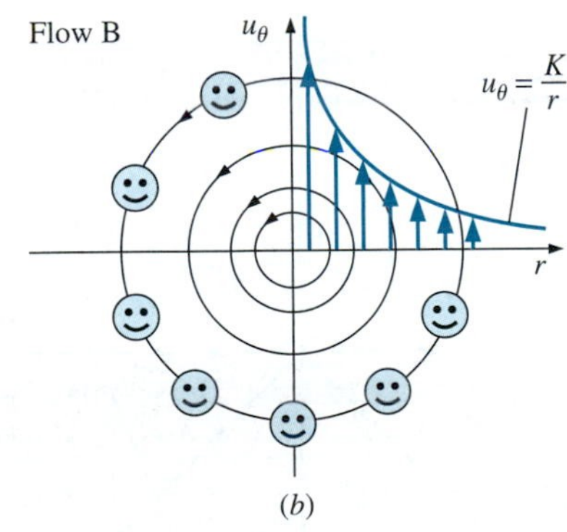

FIGURE 11–28

Streamlines and velocity profiles for (*a*) flow A, solid-body rotation and (*b*) flow B, a line vortex. Flow A is rotational, but flow B is irrotational everywhere except at the origin.

(a)

(b)

FIGURE 11–29
A simple analogy: (*a*) *rotational* circular flow is analogous to a roundabout, while (*b*) *irrotational* circular flow is analogous to a Ferris wheel.
© Robb Gregg

as that of the ride itself. This is analogous to a rotational flow. In contrast, children on a Ferris wheel always remain oriented in an upright position as they trace out their circular path. This is analogous to an irrotational flow.

EXAMPLE 11–6 Determination of Rotationality of a Line Sink

A simple two-dimensional velocity field called a **line sink** is often used to simulate fluid being sucked into a line along the z-axis. Suppose the volume flow rate per unit length along the z-axis, $\dot{V}/L$, is known, where $\dot{V}$ is a negative quantity. In two dimensions in the $r\theta$-plane,

$$\textit{Line sink:}\qquad u_r = \frac{\dot{V}}{2\pi L}\frac{1}{r} \qquad \text{and} \qquad u_\theta = 0 \tag{1}$$

Draw several streamlines of the flow and calculate the vorticity. Is this flow rotational or irrotational?

Solution Streamlines of the given flow field are to be sketched and the rotationality of the flow is to be determined.

Analysis Since there is only radial flow and no tangential flow, we know immediately that all streamlines must be rays into the origin. Several streamlines are sketched in Fig. 11–30. The vorticity is calculated from Eq. 11–33:

$$\vec{\zeta} = \frac{1}{r}\left(\frac{\partial(ru_\theta)}{\partial r} - \frac{\partial}{\partial \theta}u_r\right)\vec{k} = \frac{1}{r}\left(0 - \frac{\partial}{\partial \theta}\left(\frac{\dot{V}}{2\pi L}\frac{1}{r}\right)\right)\vec{k} = 0 \tag{2}$$

Since the vorticity vector is everywhere zero, this flow field is **irrotational**.

Discussion Many practical flow fields involving suction, such as flow into inlets and hoods, can be approximated quite accurately by assuming irrotational flow (Heinsohn and Cimbala, 2003).

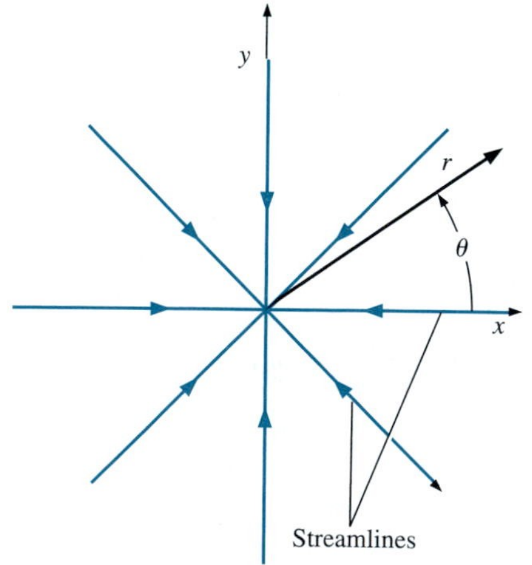

FIGURE 11–30
Streamlines in the $r\theta$-plane for the case of a line sink.

11–4 ■ THE REYNOLDS TRANSPORT THEOREM

In thermodynamics and solid mechanics we often work with a *system* (also called a *closed system*), defined as a *quantity of matter of fixed identity*. In fluid dynamics, it is more common to work with a *control volume* (also called an *open system*), defined as a *region in space chosen for study*. The size and shape of a system may change during a process, but no mass crosses its boundaries. A control volume, on the other hand, allows mass to flow in or out across its boundaries, which are called the **control surface**. A control volume may also move and deform during a process, but many real-world applications involve fixed, nondeformable control volumes.

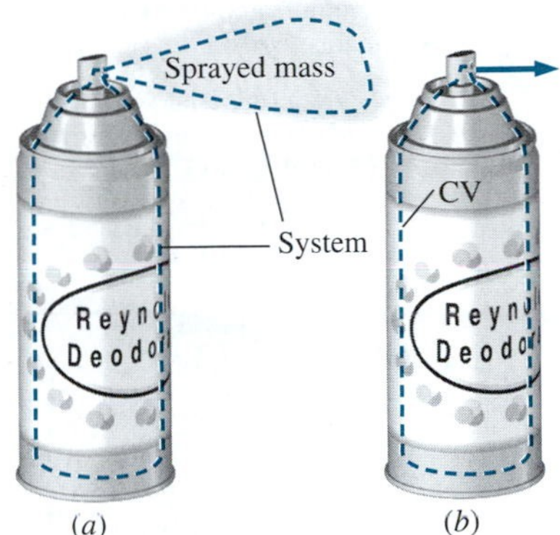

FIGURE 11–31

Two methods of analyzing the spraying of deodorant from a spray can: (*a*) We follow the fluid as it moves and deforms. This is the *system approach*—no mass crosses the boundary, and the total mass of the system remains fixed. (*b*) We consider a fixed interior volume of the can. This is the *control volume approach*—mass crosses the boundary.

Figure 11–31 illustrates both a system and a control volume for the case of deodorant being sprayed from a spray can. When analyzing the spraying process, a natural choice for our analysis is either the moving, deforming fluid (a system) or the volume bounded by the inner surfaces of the can (a control volume). These two choices are identical before the deodorant is sprayed. When some contents of the can are discharged, the system approach considers the discharged mass as part of the system and tracks it (a difficult job indeed); thus the mass of the system remains constant. Conceptually, this is equivalent to attaching a flat balloon to the nozzle of the can and letting the spray inflate the balloon. The inner surface of the balloon now becomes part of the boundary of the system. The control volume approach, however, is not concerned at all with the deodorant that has escaped the can (other than its properties at the exit), and thus the mass of the control volume decreases during this process while its volume remains constant. Therefore, the system approach treats the spraying process as an expansion of the system's volume, whereas the control volume approach considers it as a fluid discharge through the control surface of the fixed control volume.

Most principles of fluid mechanics are adopted from solid mechanics, where the physical laws dealing with the time rates of change of extensive properties are expressed for systems. In fluid mechanics, it is usually more convenient to work with control volumes, and thus there is a need to relate the changes in a control volume to the changes in a system. The relationship between the time rates of change of an extensive property for a system and for a control volume is expressed by the **Reynolds transport theorem (RTT)**, which provides the link between the system and control volume approaches (Fig. 11–32). RTT is named after the English engineer, Osborne Reynolds (1842–1912), who did much to advance its application in fluid mechanics.

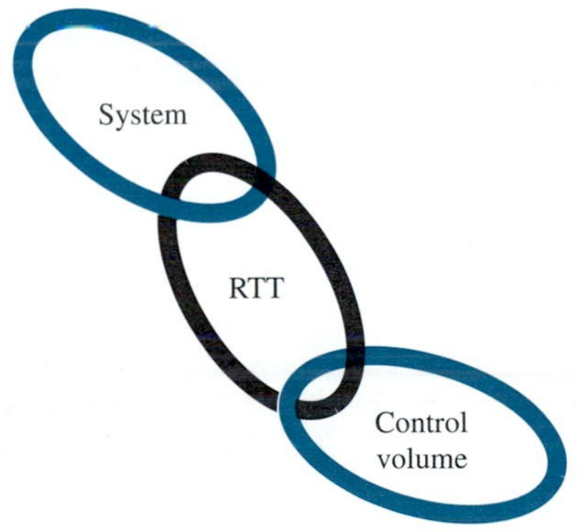

FIGURE 11–32

The *Reynolds transport theorem* (RTT) provides a link between the system approach and the control volume approach.

The general form of the Reynolds transport theorem can be derived by considering a system with an arbitrary shape and arbitrary interactions, but the derivation is rather involved. To help you grasp the fundamental meaning of the theorem, we derive it first in a straightforward manner using a simple geometry and then generalize the results.

Consider flow from left to right through a diverging (expanding) portion of a flow field as sketched in Fig. 11–33. The upper and lower bounds of the fluid under consideration are *streamlines* of the flow, and we assume uniform flow through any cross section between these two streamlines. We choose the control volume to be fixed between sections (1) and (2) of the flow field. Both (1) and (2) are normal to the direction of flow. At some initial time t, the

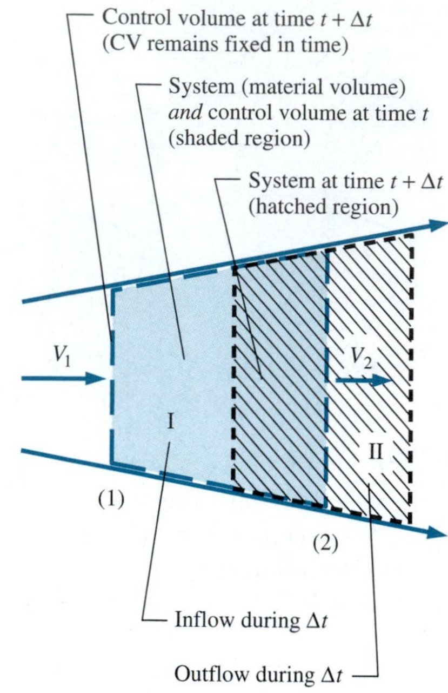

FIGURE 11–33

A moving *system* (hatched region) and a fixed *control volume* (shaded region) in a diverging portion of a flow field at times t and $t + \Delta t$. The upper and lower bounds are streamlines of the flow.

system coincides with the control volume, and thus the system and control volume are identical (the shaded region in Fig. 11–33). During time interval Δt, the system moves in the flow direction at uniform speeds V_1 at section (1) and V_2 at section (2). The system at this later time is indicated by the hatched region. The region uncovered by the system during this motion is designated as section I (part of the CV), and the new region covered by the system is designated as section II (not part of the CV). Therefore, at time $t + \Delta t$, the system consists of the same fluid, but it occupies the region CV − I + II. The control volume is fixed in space, and thus it remains as the shaded region marked CV at all times.

Let B represent any **extensive property** (such as mass, energy, or momentum), and let $b = B/m$ represent the corresponding **intensive property**. Noting that extensive properties are additive, the extensive property B of the system at times t and $t + \Delta t$ can be expressed as

$$B_{\text{sys}, t} = B_{\text{CV}, t} \qquad \text{(the system and CV concide at time } t\text{)}$$

$$B_{\text{sys}, t+\Delta t} = B_{\text{CV}, t+\Delta t} - B_{\text{I}, t+\Delta t} + B_{\text{II}, t+\Delta t}$$

Subtracting the first equation from the second one and dividing by Δt gives

$$\frac{B_{\text{sys}, t+\Delta t} - B_{\text{sys}, t}}{\Delta t} = \frac{B_{\text{CV}, t+\Delta t} - B_{\text{CV}, t}}{\Delta t} - \frac{B_{\text{I}, t+\Delta t}}{\Delta t} + \frac{B_{\text{II}, t+\Delta t}}{\Delta t}$$

Taking the limit as $\Delta t \to 0$, and using the definition of derivative, we get

$$\frac{dB_{\text{sys}}}{dt} = \frac{dB_{\text{CV}}}{dt} - \dot{B}_{\text{in}} + \dot{B}_{\text{out}} \qquad \textbf{(11–29)}$$

or

$$\frac{dB_{\text{sys}}}{dt} = \frac{dB_{\text{CV}}}{dt} - b_1\rho_1 V_1 A_1 + b_2\rho_2 V_2 A_2$$

since

$$B_{\text{I}, t+\Delta t} = b_1 m_{\text{I}, t+\Delta t} = b_1\rho_1 V_{\text{I}, t+\Delta t} = b_1\rho_1 V_1\,\Delta t\,A_1$$

$$B_{\text{II}, t+\Delta t} = b_2 m_{\text{II}, t+\Delta t} = b_2\rho_2 V_{\text{II}, t+\Delta t} = b_2\rho_2 V_2\,\Delta t\,A_2$$

and

$$\dot{B}_{\text{in}} = \dot{B}_{\text{I}} = \lim_{\Delta t \to 0} \frac{B_{\text{I}, t+\Delta t}}{\Delta t} = \lim_{\Delta t \to 0} \frac{b_1\rho_1 V_1\,\Delta t\,A_1}{\Delta t} = b_1\rho_1 V_1\,A_1$$

$$\dot{B}_{\text{out}} = \dot{B}_{\text{II}} = \lim_{\Delta t \to 0} \frac{B_{\text{II}, t+\Delta t}}{\Delta t} = \lim_{\Delta t \to 0} \frac{b_2\rho_2 V_2\,\Delta t\,A_2}{\Delta t} = b_2\rho_2 V_2\,A_2$$

where A_1 and A_2 are the cross-sectional areas at locations 1 and 2. Equation 11–29 states that *the time rate of change of the property B of the system is equal to the time rate of change of B of the control volume plus the net flux of B out of the control volume by mass crossing the control surface*. This is the desired relation since it relates the change of a property of a system to the change of that property for a control volume. Note that Eq. 11–29 applies at any instant in time, where it is assumed that the system and the control volume occupy the same space at that particular instant in time.

The influx $\dot{B}_{\text{in}}$ and outflux $\dot{B}_{\text{out}}$ of the property B in this case are easy to determine since there is only one inlet and one outlet, and the velocities are normal to the surfaces at sections (1) and (2). In general, however, we may have several inlet and outlet ports, and the velocity may not be normal to the control surface at the point of entry. Also, the velocity may not be uniform. To generalize the process, we consider a differential surface area dA on the control surface and denote its **unit outer normal** by $\vec{n}$. The flow rate of property b through dA is $\rho b\vec{V} \cdot \vec{n}\, dA$ since the dot product $\vec{V} \cdot \vec{n}$ gives the normal component of the velocity. Then the net rate of outflow through the entire control surface is determined by integration to be (Fig. 11–34)

$$\dot{B}_{\text{net}} = \dot{B}_{\text{out}} - \dot{B}_{\text{in}} = \int_{\text{CS}} \rho b\vec{V} \cdot \vec{n}\, dA \quad \text{(inflow if negative)} \tag{11–30}$$

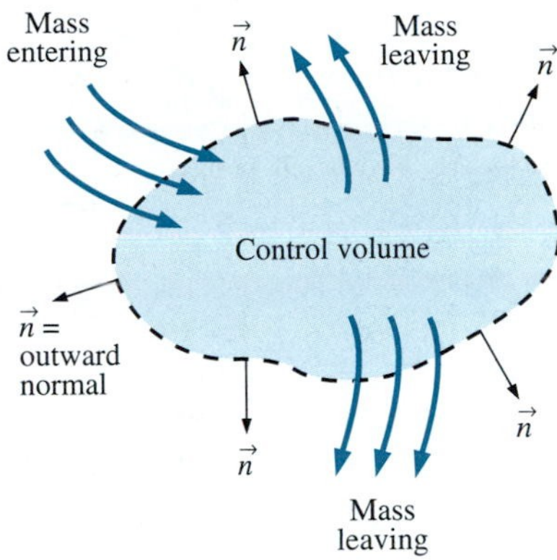

FIGURE 11–34
The integral of $b\rho\vec{V} \cdot \vec{n}\, dA$ over the control surface gives the net amount of the property B flowing out of the control volume (into the control volume if it is negative) per unit time.

An important aspect of this relation is that it automatically subtracts the inflow from the outflow, as explained next. The dot product of the velocity vector at a point on the control surface and the outer normal at that point is $\vec{V} \cdot \vec{n} = |\vec{V}||\vec{n}| \cos\theta = |\vec{V}| \cos\theta$, where θ is the angle between the velocity vector and the outer normal, as shown in Fig. 11–35. For $\theta < 90°$, we have $\cos\theta > 0$ and thus $\vec{V} \cdot \vec{n} > 0$ for outflow of mass from the control volume, and for $\theta > 90°$, we have $\cos\theta < 0$ and thus $\vec{V} \cdot \vec{n} < 0$ for inflow of mass into the control volume. Therefore, the differential quantity $\rho b\vec{V} \cdot \vec{n} dA$ is positive for mass flowing out of the control volume, and negative for mass flowing into the control volume, and its integral over the entire control surface gives the rate of net outflow of the property B by mass.

The properties within the control volume may vary with position, in general. In such a case, the total amount of property B within the control volume must be determined by integration:

$$B_{\text{CV}} = \int_{\text{CV}} \rho b\, dV \tag{11–31}$$

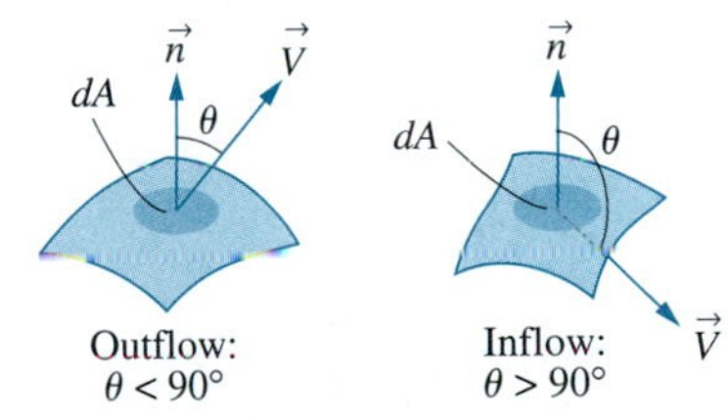

$\vec{V} \cdot \vec{n} = |\vec{V}||\vec{n}| \cos\theta = V\cos\theta$
If $\theta < 90°$, then $\cos\theta > 0$ (outflow).
If $\theta > 90°$, then $\cos\theta < 0$ (inflow).
If $\theta = 90°$, then $\cos\theta = 0$ (no flow).

FIGURE 11–35
Outflow and inflow of mass across the differential area of a control surface.

The term dB_{CV}/dt in Eq. 11–29 is thus equal to $\dfrac{d}{dt}\displaystyle\int_{\text{CV}} \rho b\, dV$, and represents the time rate of change of the property B content of the control volume. A positive value for dB_{CV}/dt indicates an increase in the B content, and a negative value indicates a decrease. Substituting Eqs. 11–30 and 11–31 into Eq. 11–29 yields the Reynolds transport theorem, also known as the *system-to-control-volume transformation* for a fixed control volume:

RTT, fixed CV:

$$\frac{dB_{\text{sys}}}{dt} = \frac{d}{dt}\int_{\text{CV}} \rho b\, dV + \int_{\text{CS}} \rho b\vec{V} \cdot \vec{n}\, dA \tag{11–32}$$

Since the control volume is not moving or deforming with time, the time derivative on the right-hand side can be moved inside the integral, since the domain of integration does not change with time. (In other words, it is irrelevant whether we differentiate or integrate first.) But the time derivative in that case must be expressed as a *partial* derivative ($\partial/\partial t$) since density and the quantity b may depend on the position within the control volume. Thus, an alternate form of the Reynolds transport theorem for a fixed control volume is

Alternate RTT, fixed CV:

$$\frac{dB_{\text{sys}}}{dt} = \int_{\text{CV}} \frac{\partial}{\partial t}(\rho b)\, dV + \int_{\text{CS}} \rho b\vec{V} \cdot \vec{n}\, dA \tag{11–33}$$

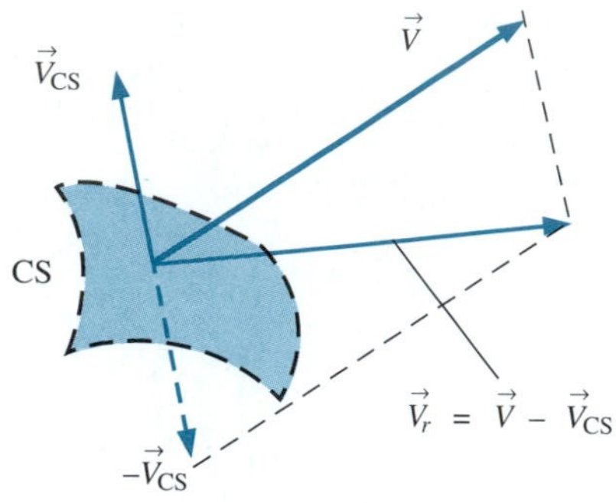

FIGURE 11–36

Relative velocity crossing a control surface is found by vector addition of the absolute velocity of the fluid and the negative of the local velocity of the control surface.

Equations 11–32 and 33 were derived for a *fixed* control volume. However, many practical systems such as turbine and propeller blades involve nonfixed control volumes. It turns out that Eq. 11–33 is also valid for the most general case of a moving and/or deforming control volume, provided that velocity vector $\vec{V}$ is an *absolute* velocity (as viewed from a fixed reference frame). Fortunately, Eq. 11–32 is also valid for moving and/or deforming control volumes provided that the absolute fluid velocity $\vec{V}$ in the last term is replaced by the **relative velocity** $\vec{V}_r$,

Relative velocity:
$$\vec{V}_r = \vec{V} - \vec{V}_{CS} \tag{11-34}$$

where $\vec{V}_{CS}$ is the local velocity of the control surface (Fig. 11–36). The most general form of the Reynolds transport theorem is thus

RTT, nonfixed CV:
$$\frac{dB_{sys}}{dt} = \frac{d}{dt}\int_{CV} \rho b \, dV + \int_{CS} \rho b \vec{V}_r \cdot \vec{n} \, dA \tag{11-35}$$

Note that for a control volume that moves and/or deforms with time, the time derivative must be applied *after* integration, as in Eq. 11–35. As a simple example of a moving control volume, consider a toy car moving at a constant absolute velocity $\vec{V}_{car}$ = 10 km/h to the right. A high-speed jet of water (absolute velocity = $\vec{V}_{jet}$ = 25 km/h to the right) strikes the back of the car and propels it (Fig. 11–37). If we draw a control volume around the car, the relative velocity is $\vec{V}_r = 25 - 10 = 15$ km/h to the right. This represents the velocity at which an observer moving with the control volume (moving with the car) would observe the fluid crossing the control surface. In other words, $\vec{V}_r$ is the fluid velocity expressed relative to a coordinate system moving *with* the control volume.

During steady flow, the amount of the property B within the control volume remains constant in time, and thus the time derivative in Eq. 11–35 becomes zero. Then the Reynolds transport theorem reduces to

RTT, steady flow:
$$\frac{dB_{sys}}{dt} = \int_{CS} \rho b \vec{V}_r \cdot \vec{n} \, dA \tag{11-36}$$

Note that unlike the control volume, the property B content of the system may still change with time during a steady process. But in this case the change must be equal to the net property transported by mass across the control surface (an advective rather than an unsteady effect).

In most practical engineering applications of the RTT, fluid crosses the boundary of the control volume at a finite number of well-defined inlets and outlets (Fig. 11–38). In such cases, it is convenient to cut the control surface directly across each inlet and outlet and replace the surface integral in Eq. 11–35 with approximate algebraic expressions at each inlet and outlet based on the *average* values of fluid properties crossing the boundary. We define ρ_{avg}, b_{avg}, and $V_{r,\,avg}$ as the average values of ρ, b, and V_r, respectively, across an inlet or outlet of cross-sectional area A [e.g., $b_{avg} = (1/A)\int_A b \, dA$]. The surface integrals in the RTT (Eq. 11–35), when applied over an inlet or outlet of cross-sectional area A, are then *approximated* by pulling property b out of the surface integral and replacing it with its average. This yields

Absolute reference frame:

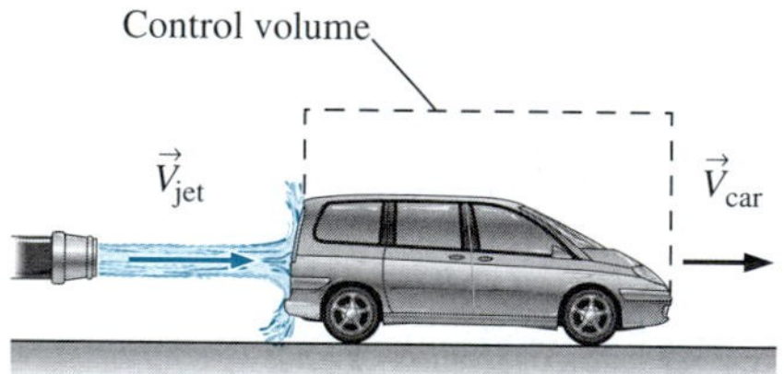

Relative reference frame:

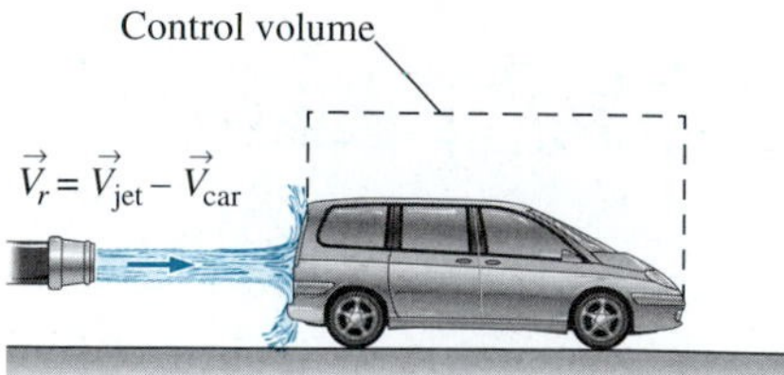

FIGURE 11–37

Reynolds transport theorem applied to a control volume moving at constant velocity.

$$\int_A \rho b \vec{V}_r \cdot \vec{n}\, dA \cong b_{\text{avg}} \int_A \rho \vec{V}_r \cdot \vec{n}\, dA = b_{\text{avg}} \dot{m}_r$$

where $\dot{m}_r$ is the mass flow rate through the inlet or outlet relative to the (moving) control surface. The approximation in this equation is exact when property b is uniform over cross-sectional area A. Equation 11–35 thus becomes

$$\frac{dB_{\text{sys}}}{dt} = \frac{d}{dt}\int_{\text{CV}} \rho b\, dV + \underbrace{\sum_{\text{out}} \dot{m}_r b_{\text{avg}}}_{\text{for each outlet}} - \underbrace{\sum_{\text{in}} \dot{m}_r b_{\text{avg}}}_{\text{for each inlet}} \qquad \text{(11–37)}$$

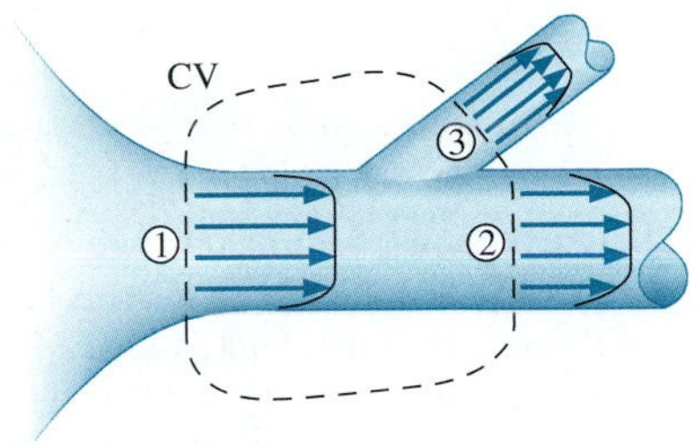

FIGURE 11–38

An example control volume in which there is one well-defined inlet (1) and two well-defined outlets (2 and 3). In such cases, the control surface integral in the RTT can be more conveniently written in terms of the average values of fluid properties crossing each inlet and outlet.

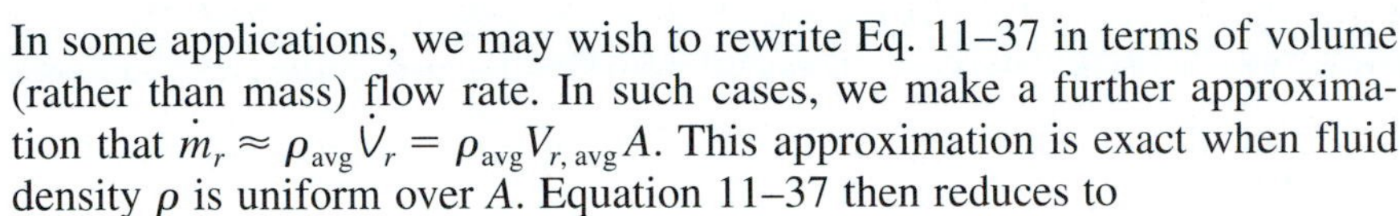

In some applications, we may wish to rewrite Eq. 11–37 in terms of volume (rather than mass) flow rate. In such cases, we make a further approximation that $\dot{m}_r \approx \rho_{\text{avg}} \dot{V}_r = \rho_{\text{avg}} V_{r,\text{avg}} A$. This approximation is exact when fluid density ρ is uniform over A. Equation 11–37 then reduces to

Approximate RTT for well-defined inlets and outlets:

$$\frac{dB_{\text{sys}}}{dt} = \frac{d}{dt}\int_{\text{CV}} \rho b\, dV + \underbrace{\sum_{\text{out}} \rho_{\text{avg}} b_{\text{avg}} V_{r,\text{avg}} A}_{\text{for each outlet}} - \underbrace{\sum_{\text{in}} \rho_{\text{avg}} b_{\text{avg}} V_{r,\text{avg}} A}_{\text{for each inlet}} \qquad \text{(11–38)}$$

Note that these approximations simplify the analysis greatly but may not always be accurate, especially in cases where the velocity distribution across the inlet or outlet is not very uniform (e.g., pipe flows; Fig. 11–38). In particular, the control surface integral of Eq. 11–35 becomes *nonlinear* when property b contains a velocity term (e.g., when applying RTT to the linear momentum equation, $b = \vec{V}$), and the approximation of Eq. 11–38 leads to errors.

Equations 11–37 and 11–38 apply to fixed *or* moving control volumes, but as discussed previously, the *relative velocity* must be used for the case of a nonfixed control volume. In Eq. 11–37 for example, the mass flow rate $\dot{m}_r$ is relative to the (moving) control surface, hence the r subscript.

SUMMARY

Fluid kinematics is concerned with describing fluid motion, without necessarily analyzing the forces responsible for such motion. There are two fundamental descriptions of fluid motion—*Lagrangian* and *Eulerian*. In a Lagrangian description, we follow individual fluid particles or collections of fluid particles, while in the Eulerian description, we define a *control volume* through which fluid flows in and out. We transform equations of motion from Lagrangian to Eulerian through use of the *material derivative* for infinitesimal fluid particles and through use of the *Reynolds transport theorem (RTT)* for systems of finite volume. For some extensive property B or its corresponding intensive property b,

Material derivative: $$\frac{Db}{Dt} = \frac{\partial b}{\partial t} + (\vec{V} \cdot \vec{\nabla})b$$

General RTT, nonfixed CV:

$$\frac{dB_{\text{sys}}}{dt} = \frac{d}{dt}\int_{\text{CV}} \rho b \, dV + \int_{\text{CS}} \rho b \vec{V}_r \cdot \vec{n} \, dA$$

In both equations, the total change of the property following a fluid particle or following a system is composed of two parts: a *local* (unsteady) part and an *advective* (movement) part.

There are various ways to visualize and analyze flow fields—*streamlines*, *streaklines*, *pathlines*, *timelines*, and *surface imaging*. We define each of these and provide examples in this chapter. In general unsteady flow, streamlines, streaklines, and pathlines differ, but *in steady flow, streamlines, streaklines, and pathlines are coincident*.

Vorticity is a property of fluid flows that indicates the *rotationality* of fluid particles.

Vorticity vector: $\vec{\zeta} = \vec{\nabla} \times \vec{V} = \text{curl}(\vec{V}) = 2\vec{\omega}$

A region of flow is *irrotational* if the vorticity is zero in that region.

REFERENCES AND SUGGESTED READING

1. R. J. Heinsohn and J. M. Cimbala. *Indoor Air Quality Engineering*. New York: Marcel-Dekker, 2003.
2. P. K. Kundu. *Fluid Mechanics*. San Diego, CA: Academic Press, 1990.
3. W. Merzkirch. *Flow Visualization*, 2nd ed. Orlando, FL: Academic Press, 1987.
4. M. Van Dyke. *An Album of Fluid Motion*. Stanford, CA: The Parabolic Press, 1982.
5. F. M. White. *Viscous Fluid Flow*, 2nd ed. New York: McGraw-Hill, 1991.
6. Çengel, Y. A. and Cimbala, J. M. *Fluid Mechanics—Fundamentals and Applications.* New York, McGraw-Hill, 2006.

PROBLEMS*

Introductory Problems

11–1C What does the word *kinematics* mean? Explain what the study of *fluid kinematics* involves.

11–2 Consider steady flow of water through an axisymmetric garden hose nozzle (Fig. P11–2). Along the centerline of the nozzle, the water speed increases from u_{entrance} to u_{exit} as sketched. Measurements reveal that the centerline water speed increases parabolically through the nozzle. Write an equation for centerline speed $u(x)$, based on the parameters given here, from $x = 0$ to $x = L$.

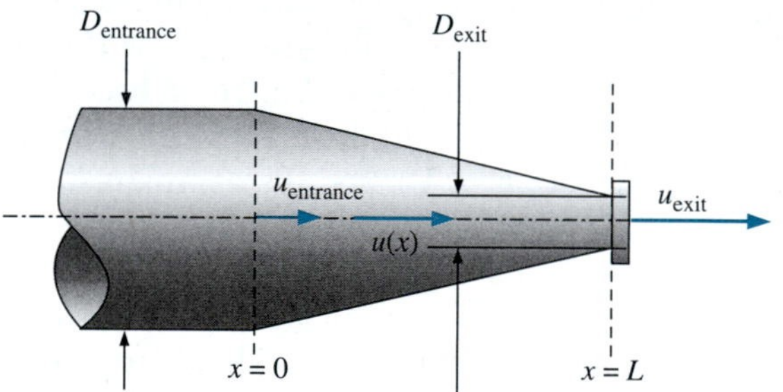

FIGURE P11–2

11–3 Consider the following steady, two-dimensional velocity field:

$$\vec{V} = (u, v) = (0.5 + 1.2x)\vec{i} + (-2.0 - 1.2y)\vec{j}$$

Is there a stagnation point in this flow field? If so, where is it? *Answer*: $x = -0.417$, $y = -1.67$

11–4 Consider the following steady, two-dimensional velocity field:

$$\vec{V} = (u, v) = (a^2 - (b - cx)^2)\vec{i} + (-2cby + 2c^2xy)\vec{j}$$

Is there a stagnation point in this flow field? If so, where is it?

---

* Problems designated by a "C" are concept questions, and students are encouraged to answer them all. Problems designated by an "E" are in English units, and the SI users can ignore them. Problems with the EES icon are solved using EES, and complete solutions together with parametric studies are included on the enclosed DVD. Problems with the computer icon are comprehensive in nature and are intended to be solved with a computer, preferably using the EES software that accompanies this text.

Lagrangian and Eulerian Descriptions

11–5C What is the *Lagrangian description* of fluid motion?

11–6C Is the Lagrangian method of fluid flow analysis more similar to study of a system or a control volume? Explain.

11–7C What is the *Eulerian description* of fluid motion? How does it differ from the Lagrangian description?

11–8C A stationary probe is placed in a fluid flow and measures pressure and temperature as functions of time at one location in the flow (Fig. P11–8C). Is this a Lagrangian or an Eulerian measurement? Explain.

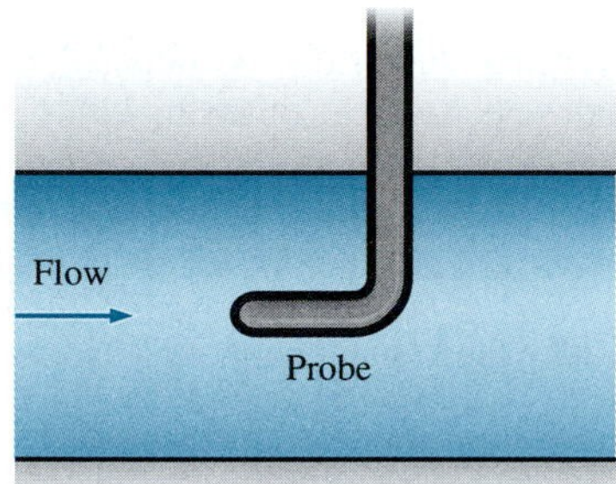

FIGURE P11–8C

11–9C A tiny neutrally buoyant electronic pressure probe is released into the inlet pipe of a water pump and transmits 2000 pressure readings per second as it passes through the pump. Is this a Lagrangian or an Eulerian measurement? Explain.

11–10C A weather balloon is launched into the atmosphere by meteorologists. When the balloon reaches an altitude where it is neutrally buoyant, it transmits information about weather conditions to monitoring stations on the ground (Fig. P11–10C). Is this a Lagrangian or an Eulerian measurement? Explain.

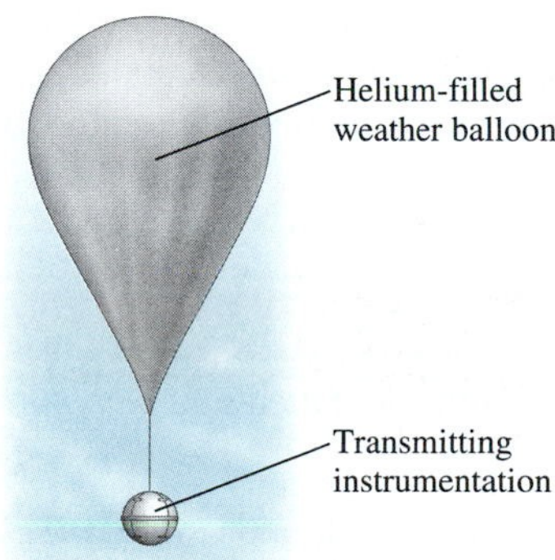

FIGURE P11–10C

11–11C A Pitot-static probe can often be seen protruding from the underside of an airplane (Fig. P11–11C). As the airplane flies, the probe measures relative wind speed. Is this a Lagrangian or an Eulerian measurement? Explain.

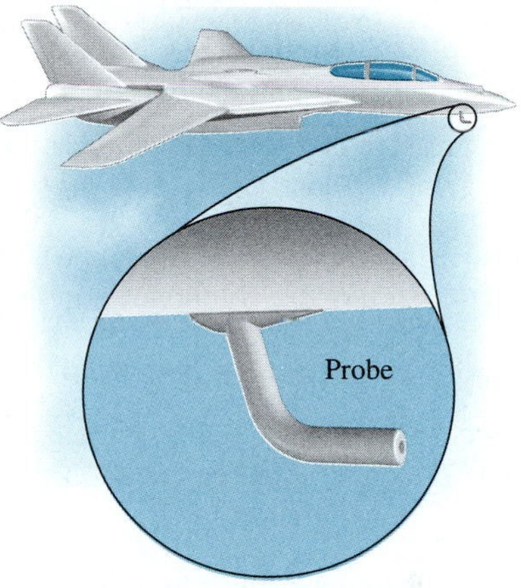

FIGURE P11–11C

11–12C Is the Eulerian method of fluid flow analysis more similar to study of a system or a control volume? Explain.

11–13C Define a *steady flow field* in the Eulerian reference frame. In such a steady flow, is it possible for a fluid particle to experience a nonzero acceleration?

11–14C List at least three other names for the material derivative, and write a brief explanation about why each name is appropriate.

11–15 Consider steady, incompressible, two-dimensional flow through a converging duct (Fig. P11–15). A simple approximate velocity field for this flow is

$$\vec{V} = (u, v) = (U_0 + bx)\vec{i} - by\vec{j}$$

where U_0 is the horizontal speed at $x = 0$. Note that this equation ignores viscous effects along the walls but is a reasonable approximation throughout the majority of the flow field. Calculate the material acceleration for fluid particles passing through this duct. Give your answer in two ways: (1) as acceleration components a_x and a_y and (2) as acceleration vector $\vec{a}$.

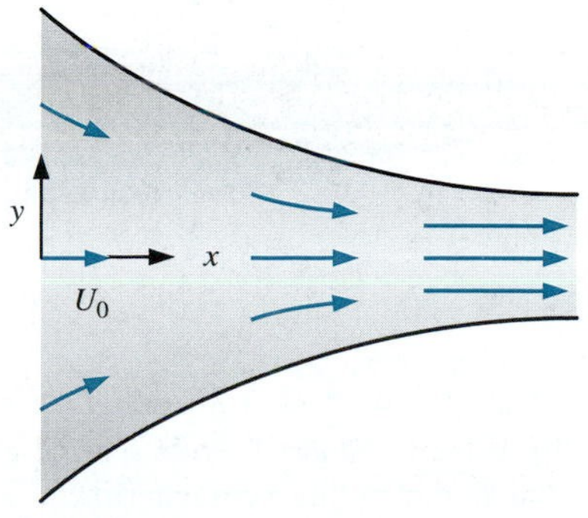

FIGURE P11–15

11–16 Converging duct flow is modeled by the steady, two-dimensional velocity field of Prob. 11–15. The pressure field is given by

$$P = P_0 - \frac{\rho}{2}\left[2U_0 bx + b^2(x^2 + y^2)\right]$$

where P_0 is the pressure at $x = 0$. Generate an expression for the rate of change of pressure *following a fluid particle.*

11–17 A steady, incompressible, two-dimensional velocity field is given by the following components in the xy-plane:

$$u = 1.1 + 2.8x + 0.65y \qquad v = 0.98 - 2.1x - 2.8y$$

Calculate the acceleration field (find expressions for acceleration components a_x and a_y), and calculate the acceleration at the point $(x, y) = (-2, 3)$. *Answers:* $a_x = -9.233$, $a_y = 14.37$

11–18 A steady, incompressible, two-dimensional velocity field is given by the following components in the xy-plane:

$$u = 0.20 + 1.3x + 0.85y \qquad v = -0.50 + 0.95x - 1.3y$$

Calculate the acceleration field (find expressions for acceleration components a_x and a_y) and calculate the acceleration at the point $(x, y) = (1, 2)$.

11–19 For the velocity field of Prob. 11–2, calculate the fluid acceleration along the nozzle centerline as a function of x and the given parameters.

11–20 Consider steady flow of air through the diffuser portion of a wind tunnel (Fig. P11–20). Along the centerline of the diffuser, the air speed decreases from $u_{entrance}$ to u_{exit} as sketched. Measurements reveal that the centerline air speed decreases parabolically through the diffuser. Write an equation for centerline speed $u(x)$, based on the parameters given here, from $x = 0$ to $x = L$.

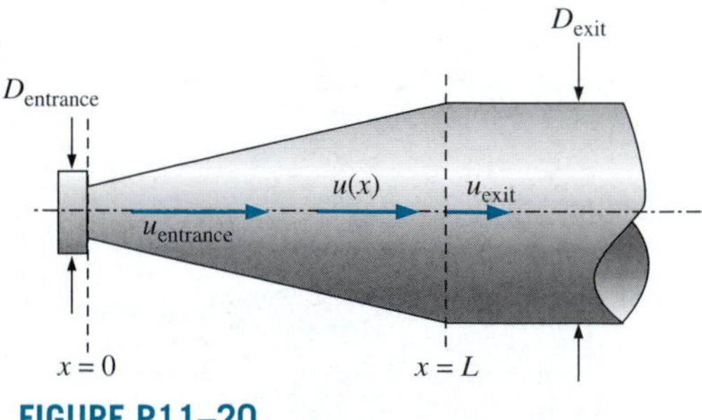

FIGURE P11–20

11–21 For the velocity field of Prob. 11–20, calculate the fluid acceleration along the diffuser centerline as a function of x and the given parameters. For $L = 2.0$ m, $u_{entrance} = 30.0$ m/s, and $u_{exit} = 5.0$ m/s, calculate the acceleration at $x = 0$ and $x = 1.0$ m. *Answers:* 0, -297 m/s^2

Flow Patterns and Flow Visualization

11–22C What is the definition of a *streamline*? What do streamlines indicate?

11–23 Converging duct flow (Fig. P11–15) is modeled by the steady, two-dimensional velocity field of Prob. 11–15. Generate an analytical expression for the flow streamlines. *Answer:* $y = C/(U_0 + bx)$

11–24E Converging duct flow is modeled by the steady, two-dimensional velocity field of Prob. 11–15. For the case in which $U_0 = 5.0$ ft/s and $b = 4.6\ \text{s}^{-1}$, plot several streamlines from $x = 0$ ft to 5 ft and $y = -3$ ft to 3 ft. Be sure to show the *direction* of the streamlines.

11–25C Consider the visualization of flow over a 12° cone in Fig. P11–25C. Are we seeing streamlines, streaklines, pathlines, or timelines? Explain.

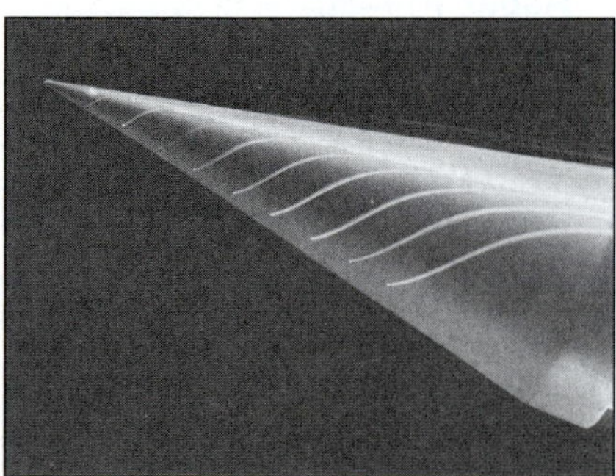

FIGURE P11–25C

Visualization of flow over a 12° cone at a 16° angle of attack at a Reynolds number of 15,000. The visualization is produced by colored fluid injected into water from ports in the body.

Courtesy ONERA. Photograph by Werlé.

11–26C What is the definition of a *pathline*? What do pathlines indicate?

11–27C What is the definition of a *streakline*? How do streaklines differ from streamlines?

11–28C Consider the visualization of flow over a 15° delta wing in Fig. P11–28C. Are we seeing streamlines, streaklines, pathlines, or timelines? Explain.

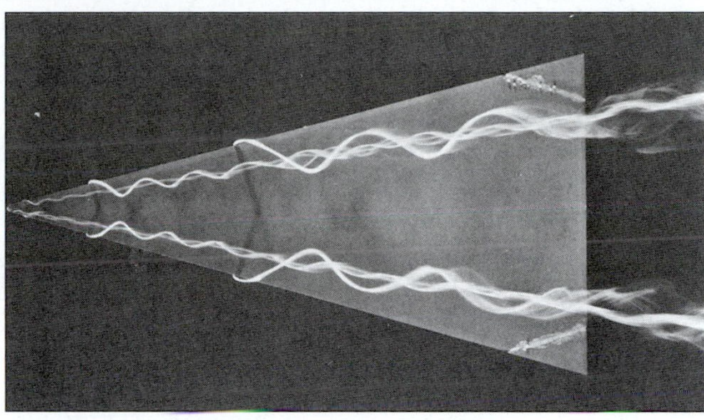

FIGURE P11–28C
Visualization of flow over a 15° delta wing at a 20° angle of attack at a Reynolds number of 20,000. The visualization is produced by colored fluid injected into water from ports on the underside of the wing.
Courtesy ONERA. Photograph by Werlé.

11–29C Consider the visualization of ground vortex flow in Fig. P11–29C. Are we seeing streamlines, streaklines, pathlines, or timelines? Explain.

FIGURE P11–29C
Visualization of ground vortex flow. A high-speed round air jet impinges on the ground in the presence of a free-stream flow of air from left to right. (The ground is at the bottom of the picture.) The portion of the jet that travels upstream forms a recirculating flow known as a **ground vortex.** The visualization is produced by a smoke wire mounted vertically to the left of the field of view.
Photo by John M. Cimbala.

11–30C Consider the visualization of flow over a sphere in Fig. P11–30C. Are we seeing streamlines, streaklines, pathlines, or timelines? Explain.

FIGURE P11–30C
Visualization of flow over a sphere at a Reynolds number of 15,000. The visualization is produced by a time exposure of air bubbles in water.
Courtesy ONERA. Photograph by Werlé.

11–31C What is the definition of a *timeline*? How can timelines be produced in a water channel? Name an application where timelines are more useful than streaklines.

11–32 Consider the following steady, incompressible, two-dimensional velocity field:

$$\vec{V} = (u, v) = (0.5 + 1.2x)\vec{i} + (-2.0 - 1.2y)\vec{j}$$

Generate an analytical expression for the flow streamlines and draw several streamlines in the upper-right quadrant from $x = 0$ to 5 and $y = 0$ to 6.

11–33 Consider the steady, incompressible, two-dimensional velocity field of Prob. 11–32. Generate a velocity vector plot in the upper-right quadrant from $x = 0$ to 5 and $y = 0$ to 6.

11–34 Consider the steady, incompressible, two-dimensional velocity field of Prob. 11–32. Generate a vector plot of the acceleration field in the upper-right quadrant from $x = 0$ to 5 and $y = 0$ to 6.

11–35 A steady, incompressible, two-dimensional velocity field is given by

$$\vec{V} = (u, v) = (1 + 2.5x + y)\vec{i} + (-0.5 - 1.5x - 2.5y)\vec{j}$$

where the x- and y-coordinates are in m and the magnitude of velocity is in m/s.

(*a*) Determine if there are any stagnation points in this flow field, and if so, where they are.

(*b*) Sketch velocity vectors at several locations in the upper-right quadrant for $x = 0$ m to 4 m and $y = 0$ m to 4 m; qualitatively describe the flow field.

11–36 Consider the steady, incompressible, two-dimensional velocity field of Prob. 11–35.

(*a*) Calculate the material acceleration at the point ($x = 2$ m, $y = 3$ m). *Answers:* $a_x = 11.5$ m/s^2, $a_y = 14.0$ m/s^2

(*b*) Sketch the material acceleration vectors at the same array of x- and y-values as in Prob. 11–35.

Vorticity and Rotationality

11–37C Explain the relationship between vorticity and rotationality.

11–38 Converging duct flow (Fig. P11–15) is modeled by the steady, two-dimensional velocity field of Prob. 11–15. Is this flow field rotational or irrotational? Show all your work. *Answer*: irrotational

11–39 A cylindrical tank of water rotates in solid-body rotation, counterclockwise about its vertical axis (Fig. P11–39) at angular speed $\dot{n}$ = 360 rpm. Calculate the vorticity of fluid particles in the tank. *Answer*: $75.4\,\vec{k}$ rad/s

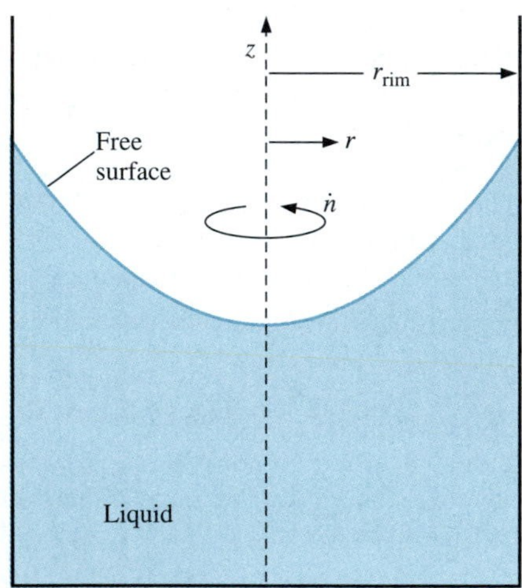

FIGURE P11–39

11–40 A cylindrical tank of water rotates about its vertical axis (Fig. P11–39). A PIV system is used to measure the vorticity field of the flow. The measured value of vorticity in the z-direction is −55.4 rad/s and is constant to within ±0.5 percent everywhere that it is measured. Calculate the angular speed of rotation of the tank in rpm. Is the tank rotating clockwise or counterclockwise about the vertical axis?

11–41 A cylindrical tank of radius r_{rim} = 0.35 m rotates about its vertical axis (Fig. P11–39). The tank is partially filled with oil. The speed of the rim is 2.6 m/s in the counterclockwise direction (looking from the top), and the tank has been spinning long enough to be in solid-body rotation. For any fluid particle in the tank, calculate the magnitude of the component of vorticity in the vertical z-direction. *Answer*: 15.0 rad/s

11–42 Consider the following steady, three-dimensional velocity field:

$$\vec{V} = (u, v, w)$$
$$= (3.0 + 2.0x - y)\vec{i} + (2.0x - 2.0y)\vec{j} + (0.5xy)\vec{k}$$

Calculate the vorticity vector as a function of space (x, y, z).

11–43 Consider fully developed **Couette flow**—flow between two infinite parallel plates separated by distance h, with the top plate moving and the bottom plate stationary as illustrated in Fig. P11–43. The flow is steady, incompressible, and two-dimensional in the xy-plane. The velocity field is given by

$$\vec{V} = (u, v) = V\frac{y}{h}\vec{i} + 0\vec{j}$$

Is this flow rotational or irrotational? If it is rotational, calculate the vorticity component in the z-direction. Do fluid particles in this flow rotate clockwise or counterclockwise? *Answers*: yes, $-V/h$, clockwise

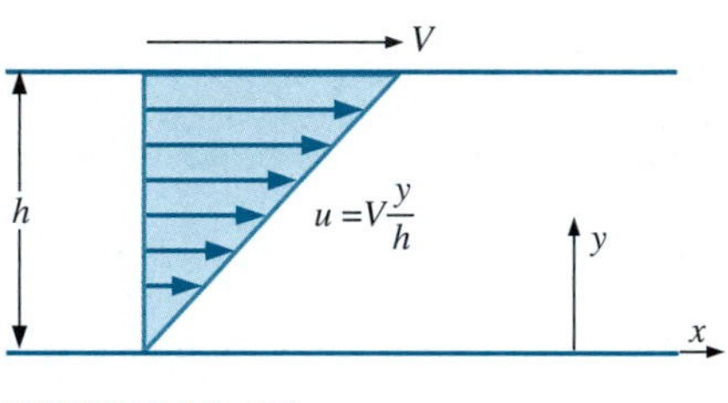

FIGURE P11–43

Reynolds Transport Theorem

11–44C True or false: For each statement, choose whether the statement is true or false and discuss your answer briefly.

(*a*) The Reynolds transport theorem is useful for transforming conservation equations from their naturally occurring control volume forms to their system forms.

(*b*) The Reynolds transport theorem is applicable only to nondeforming control volumes.

(*c*) The Reynolds transport theorem can be applied to both steady and unsteady flow fields.

(*d*) The Reynolds transport theorem can be applied to both scalar and vector quantities.

11–45 Consider the general form of the Reynolds transport theorem (RTT) given by

$$\frac{dB_{sys}}{dt} = \frac{d}{dt}\int_{CV} \rho b\, dV + \int_{CS} \rho b \vec{V}_r \cdot \vec{n}\, dA$$

where $\vec{V}_r$ is the velocity of the fluid relative to the control surface. Let B_{sys} be the mass m of a system of fluid particles. We know that for a system, $dm/dt = 0$ since no mass can enter or leave the system by definition. Use the given equation to derive the equation of conservation of mass for a control volume.

11–46 Consider the general form of the Reynolds transport theorem (RTT) given by Prob. 11–45. Let B_{sys} be the linear momentum $m\vec{V}$ of a system of fluid particles. We know that for a system, Newton's second law is

$$\sum \vec{F} = m\vec{a} = m\frac{d\vec{V}}{dt} = \frac{d}{dt}(m\vec{V})_{\text{sys}}$$

Use the equation of Prob. 11–45 and this equation to derive the equation of conservation of linear momentum for a control volume.

11–47 Consider the general form of the Reynolds transport theorem (RTT) given in Prob. 11–45. Let B_{sys} be the angular momentum $\vec{H} = \vec{r} \times m\vec{V}$ of a system of fluid particles, where $\vec{r}$ is the moment arm. We know that for a system, conservation of angular momentum can be expressed as

$$\sum \vec{M} = \frac{d}{dt}\vec{H}_{\text{sys}}$$

where $\Sigma \vec{M}$ is the net moment applied to the system. Use the equation given in Prob. 11–45 and this equation to derive the equation of conservation of angular momentum for a control volume.

Review Problems

11–48 Consider fully developed two-dimensional **Poiseuille flow**—flow between two infinite parallel plates separated by distance h, with both the top plate and bottom plate stationary, and a forced pressure gradient dP/dx driving the flow as illustrated in Fig. P11–48. (dP/dx is constant and negative.)

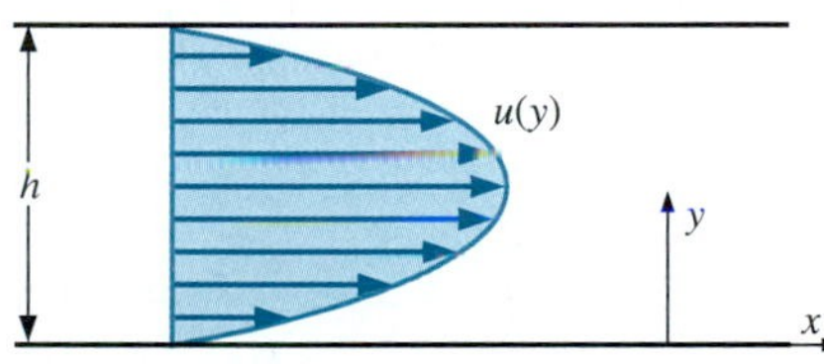

FIGURE P11–48

The flow is steady, incompressible, and two-dimensional in the xy-plane. The velocity components are given by

$$u = \frac{1}{2\mu}\frac{dP}{dx}(y^2 - hy) \qquad v = 0$$

where μ is the fluid's viscosity. Is this flow rotational or irrotational? If it is rotational, calculate the vorticity component in the z-direction. Do fluid particles in this flow rotate clockwise or counterclockwise?

11–49 Consider the two-dimensional Poiseuille flow of Prob. 11–48. The fluid between the plates is water at 40°C. Let the gap height h = 1.6 mm and the pressure gradient dP/dx = −230 N/m³. Calculate and plot seven *pathlines* from t = 0 to t = 10 s. The fluid particles are released at x = 0 and at y = 0.2, 0.4, 0.6, 0.8, 1.0, 1.2, and 1.4 mm.

11–50 Consider the two-dimensional Poiseuille flow of Prob. 11–48. The fluid between the plates is water at 40°C. Let the gap height h = 1.6 mm and the pressure gradient dP/dx = −230 N/m³. Calculate and plot seven *streaklines* generated from a dye rake that introduces dye streaks at x = 0 and at y = 0.2, 0.4, 0.6, 0.8, 1.0, 1.2, and 1.4 mm (Fig. P11–50). The dye is introduced from t = 0 to t = 10 s, and the streaklines are to be plotted at t = 10 s.

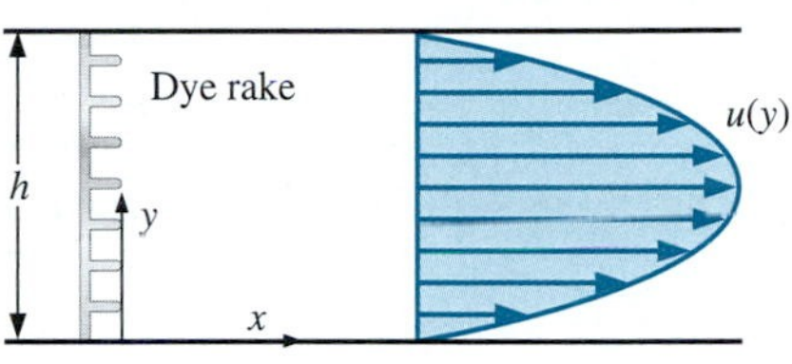

FIGURE P11–50

11–51 Repeat Prob. 11–50 except that the dye is introduced from t = 0 to t = 10 s, and the streaklines are to be plotted at t = 12 s instead of 10 s.

11–52 Consider the two-dimensional Poiseuille flow of Prob. 11–48. The fluid between the plates is water at 40°C. Let the gap height h = 1.6 mm and the pressure gradient dP/dx = −230 N/m³. Imagine a hydrogen bubble wire stretched vertically through the channel at x = 0 (Fig. P11–52). The wire is pulsed on and off such that bubbles are produced periodically to create *timelines*. Five distinct timelines are generated at t = 0, 2.5, 5.0, 7.5, and 10.0 s. Calculate and plot what these five timelines look like at time t = 12.5 s.

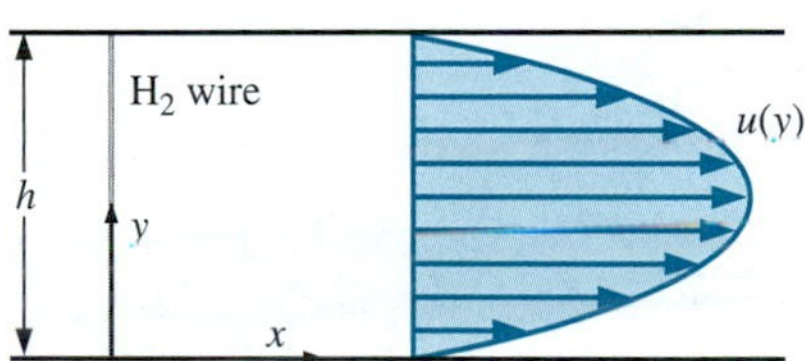

FIGURE P11–52

11–53 Consider fully developed axisymmetric Poiseuille flow—flow in a round pipe of radius R (diameter D = 2R), with a forced pressure gradient dP/dx driving the flow as illustrated in Fig. P11–53. (dP/dx is constant and negative.) The flow is steady, incompressible, and axisymmetric about the x-axis. The velocity components are given by

$$u = \frac{1}{4\mu}\frac{dP}{dx}(r^2 - R^2) \qquad u_r = 0 \qquad u_\theta = 0$$

where μ is the fluid's viscosity. Is this flow rotational or irrotational? If it is rotational, calculate the vorticity component in the circumferential (θ) direction and discuss the sign of the rotation.

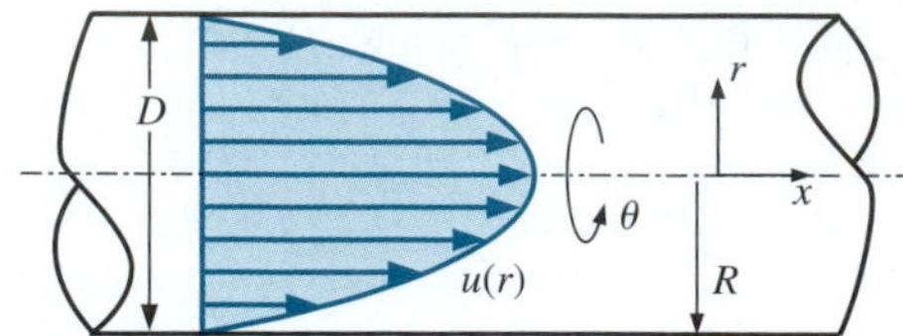

FIGURE P11–53

11–54 We approximate the flow of air into a vacuum cleaner attachment by the following velocity components in the centerplane (the xy-plane):

$$u = \frac{-\dot{V}x}{\pi L}\frac{x^2 + y^2 + b^2}{x^4 + 2x^2y^2 + 2x^2b^2 + y^4 - 2y^2b^2 + b^4}$$

and

$$v = \frac{-\dot{V}y}{\pi L}\frac{x^2 + y^2 - b^2}{x^4 + 2x^2y^2 + 2x^2b^2 + y^4 - 2y^2b^2 + b^4}$$

where b is the distance of the attachment above the floor, L is the length of the attachment, and $\dot{V}$ is the volume flow rate of air being sucked up into the hose (Fig. P11–54). Determine the location of any stagnation point(s) in this flow field. *Answer:* at the origin

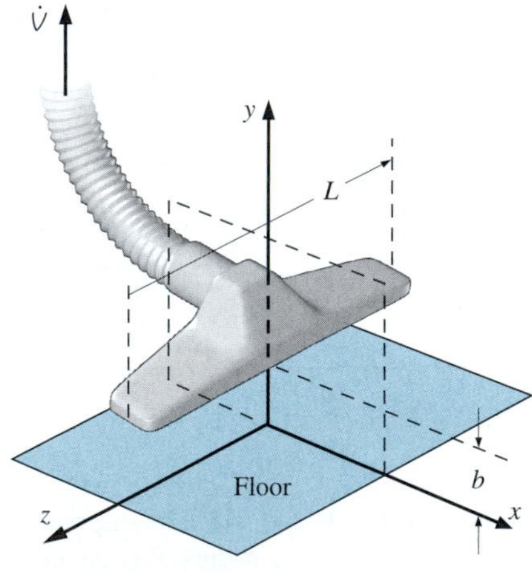

FIGURE P11–54

11–55 Consider the vacuum cleaner of Prob. 11–54. For the case where $b = 2.0$ cm, $L = 35$ cm, and $\dot{V} = 0.1098$ m^3/s, create a velocity vector plot in the upper half of the xy-plane from $x = -3$ cm to 3 cm and from $y = 0$ cm to 2.5 cm. Draw as many vectors as you need to get a good feel of the flow field. *Note*: The velocity is infinite at the point $(x, y) = (0, 2.0$ cm$)$, so do not attempt to draw a velocity vector at that point.

11–56 Consider the approximate velocity field given for the vacuum cleaner of Prob. 11–54. Calculate the flow speed along the floor. Dust particles on the floor are most likely to be sucked up by the vacuum cleaner at the location of maximum speed. Where is that location? Do you think the vacuum cleaner will do a good job at sucking up dust directly below the inlet (at the origin)? Why or why not?

11–57 Consider a steady, two-dimensional flow field in the xy-plane whose x-component of velocity is given by

$$u = a + b(x - c)^2$$

where a, b, and c are constants with appropriate dimensions. Of what form does the y-component of velocity need to be in order for the flow field to be incompressible? In other words, generate an expression for v as a function of x, y, and the constants of the given equation such that the flow is incompressible. *Answer:* $-2b(x - c)y + f(x)$

11–58 There are numerous occasions in which a fairly uniform free-stream flow encounters a long circular cylinder aligned normal to the flow (Fig. P11–58). Examples include air flowing around a car antenna, wind blowing against a flag pole or telephone pole, wind hitting electrical wires, and ocean currents impinging on the submerged round beams that support oil platforms. In all these cases, the flow at the rear of the cylinder is separated and unsteady, and usually turbulent. However, the flow in the front half of the cylinder is much more steady and predictable. In fact, except for a very thin boundary layer near the cylinder surface, the flow field may be approximated by the following steady, two-dimensional velocity components in the xy- or $r\theta$-plane:

$$u_r = V\cos\theta\left(1 - \frac{a^2}{r^2}\right) \qquad u_\theta = -V\sin\theta\left(1 + \frac{a^2}{r^2}\right)$$

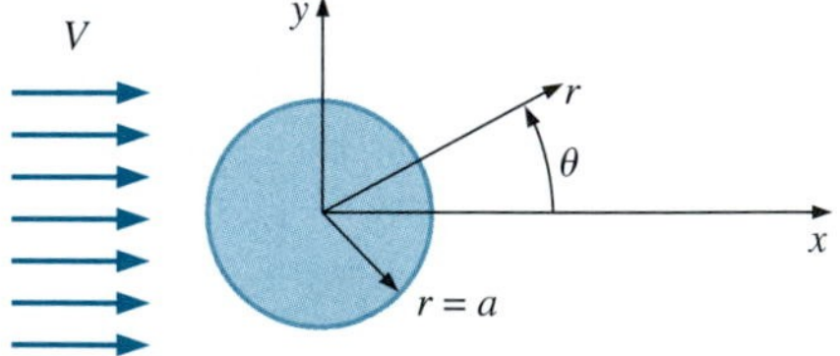

FIGURE P11–58

Is this flow field rotational or irrotational? Explain.

11–59 Consider the flow field of Prob. 11–58 (flow over a circular cylinder). Consider only the front half of the flow ($x < 0$). There is one stagnation point in the front half of the flow field. Where is it? Give your answer in both cylindrical (r, θ) coordinates and Cartesian (x, y) coordinates.

11–60 EES Consider the upstream half ($x < 0$) of the flow field of Prob. 11–58 (flow over a circular cylinder). We introduce a parameter called the **stream function** ψ, which is *constant along streamlines* in two-dimensional flows such as the one being considered here (Fig. P11–60). The velocity field of Prob. 11–58 corresponds to a stream function given by

$$\psi = V \sin\theta\left(r - \frac{a^2}{r}\right)$$

(*a*) Setting ψ to a constant, generate an equation for a streamline. (Hint: Use the quadratic rule to solve for r as a function of θ.)

(*b*) For the particular case in which $V = 1.00$ m/s and cylinder radius $a = 10.0$ cm, plot several streamlines in the upstream half of the flow ($90° < \theta < 270°$). For consistency, plot in the range -0.4 m $< x < 0$ m, -0.2 m $< y < 0.2$ m, with stream function values evenly spaced between -0.16 m^2/s and 0.16 m^2/s.

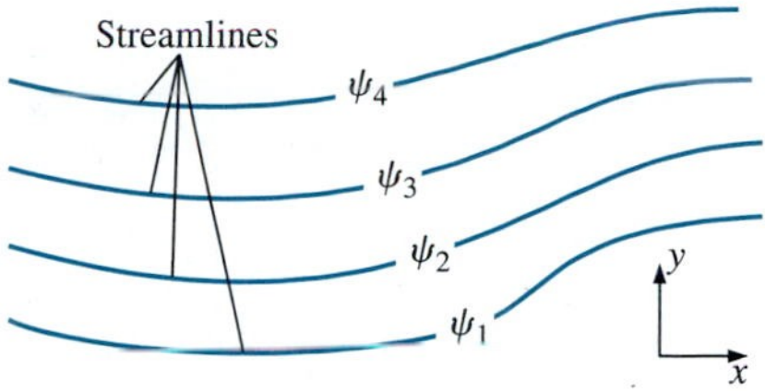

FIGURE P11–60

Chapter 12

BERNOULLI AND ENERGY EQUATIONS

This chapter deals with two equations commonly used in fluid mechanics: Bernoulli and energy equations. The *Bernoulli equation* is concerned with the conservation of kinetic, potential, and flow energies of a fluid stream and their conversion to each other in regions of flow where net viscous forces are negligible and where other restrictive conditions apply. The *energy equation* is a statement of the conservation of energy principle. In fluid mechanics, it is found convenient to separate *mechanical energy* from *thermal energy* and to consider the conversion of mechanical energy to thermal energy as a result of frictional effects as *mechanical energy loss*. Then the energy equation becomes the *mechanical energy balance*.

We start this chapter with a discussion of various forms of mechanical energy and the efficiency of mechanical work devices such as pumps and turbines. Then we derive the Bernoulli equation by applying Newton's second law to a fluid element along a streamline and demonstrate its use in a variety of applications. We continue with the development of the energy equation in a form suitable for use in fluid mechanics and introduce the concept of *head loss*. Finally, we apply the energy equation to various engineering systems.

Objectives

The objectives of this chapter are to:

- Recognize various forms of mechanical energy, and work with energy conversion efficiencies
- Understand the use and limitations of the Bernoulli equation, and apply it to solve a variety of fluid flow problems
- Work with the energy equation expressed in terms of heads, and use it to determine turbine power output and pumping power requirements

12–1 ▪ MECHANICAL ENERGY AND EFFICIENCY

Many fluid systems are designed to transport a fluid from one location to another at a specified flow rate, velocity, and elevation difference, and the system may generate mechanical work in a turbine or it may consume mechanical work in a pump or fan during this process. These systems do not involve the conversion of nuclear, chemical, or thermal energy to mechanical energy. Also, they do not involve heat transfer in any significant amount, and they operate essentially at constant temperature. Such systems can be analyzed conveniently by considering only the *mechanical forms of energy* and the frictional effects that cause the mechanical energy to be lost (i.e., to be converted to thermal energy that usually cannot be used for any useful purpose).

The **mechanical energy** is defined as *the form of energy that can be converted to mechanical work completely and directly by an ideal mechanical device such as an ideal turbine*. Kinetic and potential energies are the familiar forms of mechanical energy. Thermal energy is not mechanical energy, however, since it cannot be converted to work directly and completely (the second law of thermodynamics).

A pump transfers mechanical energy to a fluid by raising its pressure, and a turbine extracts mechanical energy from a fluid by dropping its pressure. Therefore, the pressure of a flowing fluid is also associated with its mechanical energy. In fact, the pressure unit Pa is equivalent to Pa = N/m^2 = N · m/m^3 = J/m^3, which is energy per unit volume, and the product $P\text{v}$ or its equivalent P/ρ has the unit J/kg, which is energy per unit mass. Note that pressure itself is not a form of energy. But a pressure force acting on a fluid through a distance produces work, called *flow work*, in the amount of P/ρ per unit mass. Flow work is expressed in terms of fluid properties, and it is convenient to view it as part of the energy of a flowing fluid and call it *flow energy*. Therefore, the mechanical energy of a flowing fluid can be expressed on a unit-mass basis as

$$e_{\text{mech}} = \frac{P}{\rho} + \frac{V^2}{2} + gz$$

where P/ρ is the *flow energy*, $V^2/2$ is the *kinetic energy*, and gz is the *potential energy* of the fluid, all per unit mass. Then the mechanical energy change of a fluid during incompressible flow becomes

$$\Delta e_{\text{mech}} = \frac{P_2 - P_1}{\rho} + \frac{V_2^2 - V_1^2}{2} + g(z_2 - z_1) \qquad \text{(kJ/kg)} \tag{12–1}$$

Therefore, the mechanical energy of a fluid does not change during flow if its pressure, density, velocity, and elevation remain constant. In the absence of any irreversible losses, the mechanical energy change represents the mechanical work supplied to the fluid (if $\Delta e_{\text{mech}} > 0$) or extracted from the fluid (if $\Delta e_{\text{mech}} < 0$). The maximum (ideal) power generated by a turbine, for example, is $\dot{W}_{\text{max}} = \dot{m}\Delta e_{\text{mech}}$, as shown in Fig. 12–1.

Consider a container of height h filled with water, as shown in Fig. 12–2, with the reference level selected at the bottom surface. The gage pressure and the potential energy per unit mass are, respectively, $P_A = 0$ and $\text{pe}_A = gh$

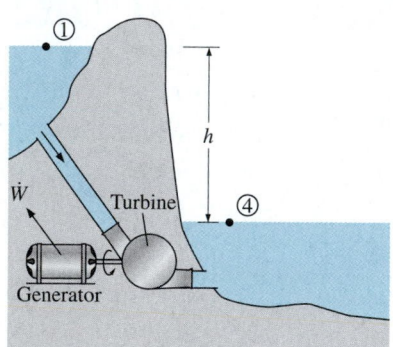

$$\dot{W}_{\text{max}} = \dot{m}\Delta e_{\text{mech}} = \dot{m}g(z_1 - z_4) = \dot{m}gh$$

since $P_1 \approx P_4 = P_{\text{atm}}$ and $V_1 = V_4 \approx 0$

(*a*)

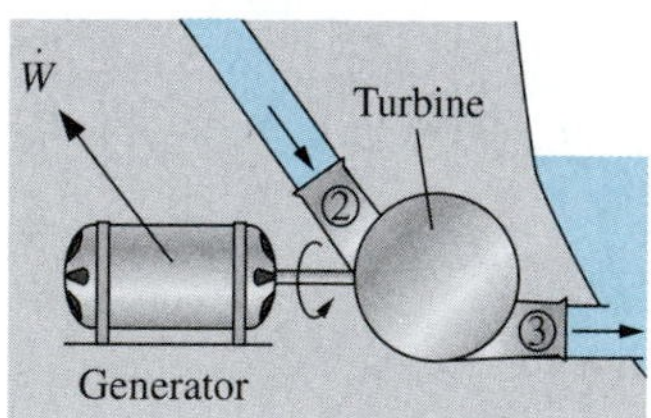

$$\dot{W}_{\text{max}} = \dot{m}\Delta e_{\text{mech}} = \dot{m}\frac{(P_2 - P_3)}{\rho} = \dot{m}\frac{\Delta P}{\rho}$$

since $V_2 \approx V_3$ and $z_2 \approx z_3$

(*b*)

FIGURE 12–1

Mechanical energy is illustrated by an ideal hydraulic turbine coupled with an ideal generator. In the absence of irreversible losses, the maximum produced power is proportional to (*a*) the change in water surface elevation from the upstream to the downstream reservoir or (*b*) (close-up view) the drop in water pressure from just upstream to just downstream of the turbine.

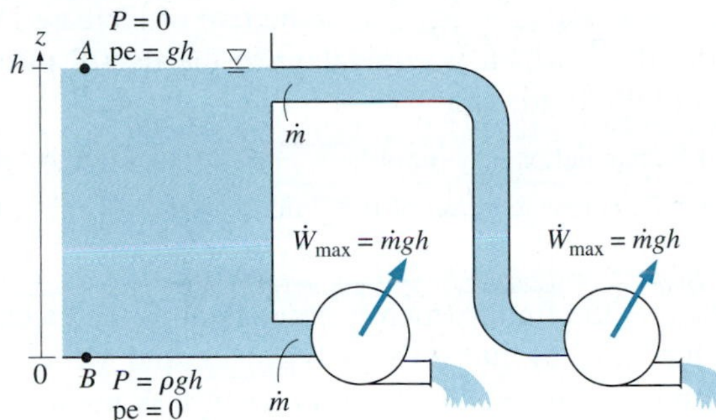

FIGURE 12–2

The mechanical energy of water at the bottom of a container is equal to the mechanical energy at any depth including the free surface of the container.

at point A at the free surface, and $P_B = \rho gh$ and $\text{pe}_B = 0$ at point B at the bottom of the container. An ideal hydraulic turbine would produce the same work per unit mass $w_{\text{turbine}} = gh$ whether it receives water (or any other fluid with constant density) from the top or from the bottom of the container. Note that we are also assuming ideal flow (no irreversible losses) through the pipe leading from the tank to the turbine and negligible kinetic energy at the turbine outlet. Therefore, the total mechanical energy of water at the bottom is equivalent to that at the top.

The transfer of mechanical energy is usually accomplished by a rotating shaft, and thus mechanical work is often referred to as *shaft work*. A pump or a fan receives shaft work (usually from an electric motor) and transfers it to the fluid as mechanical energy (less frictional losses). A turbine, on the other hand, converts the mechanical energy of a fluid to shaft work. In the absence of any irreversibilities such as friction, mechanical energy can be converted entirely from one mechanical form to another, and the **mechanical efficiency** of a device or process can be defined as (Fig. 12–3)

$$\eta_{\text{mech}} = \frac{\text{Mechanical energy output}}{\text{Mechanical energy input}} = \frac{E_{\text{mech, out}}}{E_{\text{mech, in}}} = 1 - \frac{E_{\text{mech, loss}}}{E_{\text{mech, in}}} \quad \text{(12–2)}$$

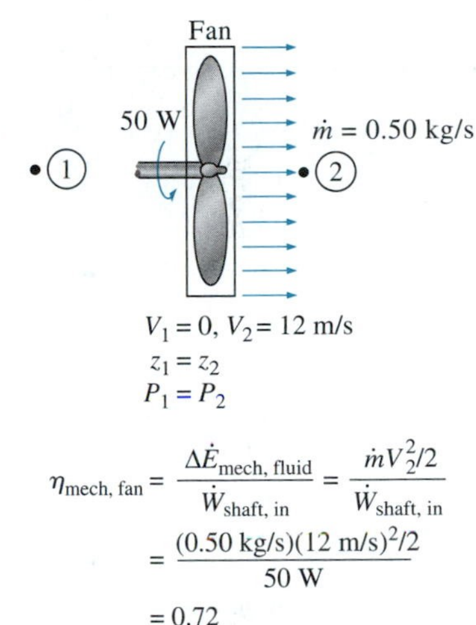

FIGURE 12–3

The mechanical efficiency of a fan is the ratio of the kinetic energy of air at the fan exit to the mechanical power input.

A conversion efficiency of less than 100 percent indicates that conversion is less than perfect and some losses have occurred during conversion. A mechanical efficiency of 97 percent indicates that 3 percent of the mechanical energy input is converted to thermal energy as a result of frictional heating, and this manifests itself as a slight rise in the temperature of the fluid.

In fluid systems, we are usually interested in increasing the pressure, velocity, and/or elevation of a fluid. This is done by *supplying mechanical energy* to the fluid by a pump, a fan, or a compressor (we refer to all of them as pumps). Or we are interested in the reverse process of *extracting mechanical energy* from a fluid by a turbine and producing mechanical power in the form of a rotating shaft that can drive a generator or any other rotary device. The degree of perfection of the conversion process between the mechanical work supplied or extracted and the mechanical energy of the fluid is expressed by the **pump efficiency** and **turbine efficiency**, defined as

$$\eta_{\text{pump}} = \frac{\text{Mechanical energy increase of the fluid}}{\text{Mechanical energy input}} = \frac{\Delta \dot{E}_{\text{mech, fluid}}}{\dot{W}_{\text{shaft, in}}} = \frac{\dot{W}_{\text{pump}, u}}{\dot{W}_{\text{pump}}} \quad \text{(12–3)}$$

where $\Delta\dot{E}_{\text{mech, fluid}} = \dot{E}_{\text{mech, out}} - \dot{E}_{\text{mech, in}}$ is the rate of increase in the mechanical energy of the fluid, which is equivalent to the **useful pumping power** $\dot{W}_{\text{pump, }u}$ supplied to the fluid, and

$$\eta_{\text{turbine}} = \frac{\text{Mechanical energy output}}{\text{Mechanical energy decrease of the fluid}} = \frac{\dot{W}_{\text{shaft, out}}}{|\Delta\dot{E}_{\text{mech, fluid}}|} = \frac{\dot{W}_{\text{turbine}}}{\dot{W}_{\text{turbine, }e}} \quad \text{(12–4)}$$

where $|\Delta\dot{E}_{\text{mech, fluid}}| = \dot{E}_{\text{mech, in}} - \dot{E}_{\text{mech, out}}$ is the rate of decrease in the mechanical energy of the fluid, which is equivalent to the mechanical power extracted from the fluid by the turbine $\dot{W}_{\text{turbine, }e}$, and we use the absolute value sign to avoid negative values for efficiencies. A pump or turbine efficiency of 100 percent indicates perfect conversion between the shaft work and the mechanical energy of the fluid, and this value can be approached (but never attained) as the frictional effects are minimized.

The mechanical efficiency should not be confused with the **motor efficiency** and the **generator efficiency**, which are defined as

Motor:

$$\eta_{\text{motor}} = \frac{\text{Mechanical power output}}{\text{Electric power input}} = \frac{\dot{W}_{\text{shaft, out}}}{\dot{W}_{\text{elect, in}}} \quad \text{(12–5)}$$

and

Generator:

$$\eta_{\text{generator}} = \frac{\text{Electric power output}}{\text{Mechanical power input}} = \frac{\dot{W}_{\text{elect, out}}}{\dot{W}_{\text{shaft, in}}} \quad \text{(12–6)}$$

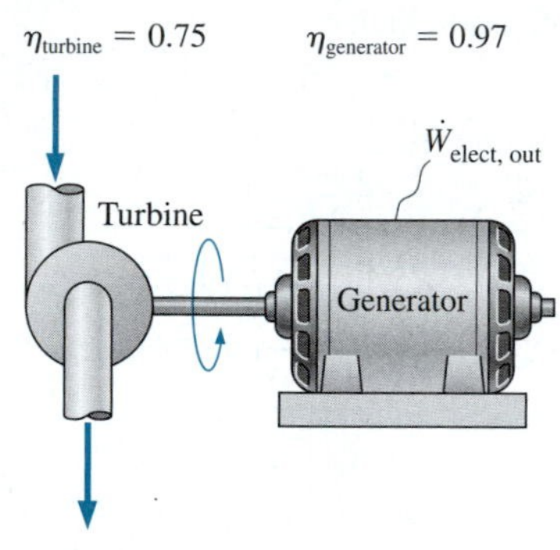

FIGURE 12–4

The overall efficiency of a turbine–generator is the product of the efficiency of the turbine and the efficiency of the generator, and represents the fraction of the mechanical energy of the fluid converted to electric energy.

A pump is usually packaged together with its motor, and a turbine with its generator. Therefore, we are usually interested in the **combined** or **overall efficiency** of pump–motor and turbine–generator combinations (Fig. 12–4), which are defined as

$$\eta_{\text{pump-motor}} = \eta_{\text{pump}}\,\eta_{\text{motor}} = \frac{\dot{W}_{\text{pump, }u}}{\dot{W}_{\text{elect, in}}} = \frac{\Delta\dot{E}_{\text{mech, fluid}}}{\dot{W}_{\text{elect, in}}} \quad \text{(12–7)}$$

and

$$\eta_{\text{turbine-gen}} = \eta_{\text{turbine}}\,\eta_{\text{generator}} = \frac{\dot{W}_{\text{elect, out}}}{\dot{W}_{\text{turbine, }e}} = \frac{\dot{W}_{\text{elect, out}}}{|\Delta\dot{E}_{\text{mech, fluid}}|} \quad \text{(12–8)}$$

All the efficiencies just defined range between 0 and 100 percent. The lower limit of 0 percent corresponds to the conversion of the entire mechanical or electric energy input to thermal energy, and the device in this case functions like a resistance heater. The upper limit of 100 percent corresponds to the case of perfect conversion with no friction or other irreversibilities, and thus no conversion of mechanical or electric energy to thermal energy.

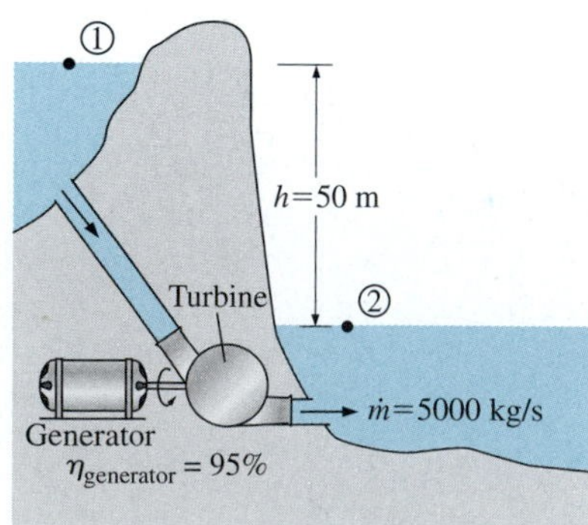

FIGURE 12–5

Schematic for Example 12–1.

EXAMPLE 12–1 Performance of a Hydraulic Turbine–Generator

The water in a large lake is to be used to generate electricity by the installation of a hydraulic turbine–generator. The elevation difference between the free surfaces upstream and downstream of the dam is 50 m (Fig. 12–5). Water is to be supplied at a rate of 5000 kg/s. If the electric power generated is measured to be 1862 kW and the generator efficiency is 95 percent, determine (*a*) the overall efficiency of the turbine–generator, (*b*) the mechanical efficiency of the turbine, and (*c*) the shaft power supplied by the turbine to the generator.

Solution A hydraulic turbine–generator is to generate electricity from the water of a lake. The overall efficiency, the turbine efficiency, and the shaft power are to be determined.

Assumptions 1 The elevation of the lake and that of the discharge site remain constant.

Properties The density of water is taken to be $\rho = 1000\ \text{kg/m}^3$.

Analysis (*a*) We perform our analysis from inlet (1) at the free surface of the lake to outlet (2) at the free surface of the downstream discharge site. At both free surfaces the pressure is atmospheric and the velocity is negligibly small. The change in the water's mechanical energy per unit mass is then

$$e_{\text{mech, in}} - e_{\text{mech, out}} = \underbrace{\frac{P_{\text{in}} - P_{\text{out}}}{\rho}}_{0} + \underbrace{\frac{V_{\text{in}}^2 - V_{\text{out}}^2}{2}}_{0} + g(z_{\text{in}} - z_{\text{out}})$$

$$= gh$$

$$= (9.81\ \text{m/s}^2)(50\ \text{m})\left(\frac{1\ \text{kJ/kg}}{1000\ \text{m}^2/\text{s}^2}\right) = 0.491\ \text{kJ/kg}$$

Then the rate at which mechanical energy is supplied to the turbine by the fluid and the overall efficiency become

$$|\Delta \dot{E}_{\text{mech, fluid}}| = \dot{m}(e_{\text{mech, in}} - e_{\text{mech, out}}) = (5000\ \text{kg/s})(0.491\ \text{kJ/kg}) = 2455\ \text{kW}$$

$$\eta_{\text{overall}} = \eta_{\text{turbine–gen}} = \frac{\dot{W}_{\text{elect, out}}}{|\Delta \dot{E}_{\text{mech, fluid}}|} = \frac{1862\ \text{kW}}{2455\ \text{kW}} = \mathbf{0.76}$$

(*b*) Knowing the overall and generator efficiencies, the mechanical efficiency of the turbine is determined from

$$\eta_{\text{turbine–gen}} = \eta_{\text{turbine}}\,\eta_{\text{generator}} \rightarrow \eta_{\text{turbine}} = \frac{\eta_{\text{turbine–gen}}}{\eta_{\text{generator}}} = \frac{0.76}{0.95} = \mathbf{0.80}$$

(*c*) The shaft power output is determined from the definition of mechanical efficiency,

$$\dot{W}_{\text{shaft, out}} = \eta_{\text{turbine}}|\Delta \dot{E}_{\text{mech, fluid}}| = (0.80)(2455\ \text{kW}) = \mathbf{1964\ kW}$$

Discussion Note that the lake supplies 2455 kW of mechanical energy to the turbine, which converts 1964 kW of it to shaft work that drives the generator, which generates 1862 kW of electric power. There are irreversible losses through each component.

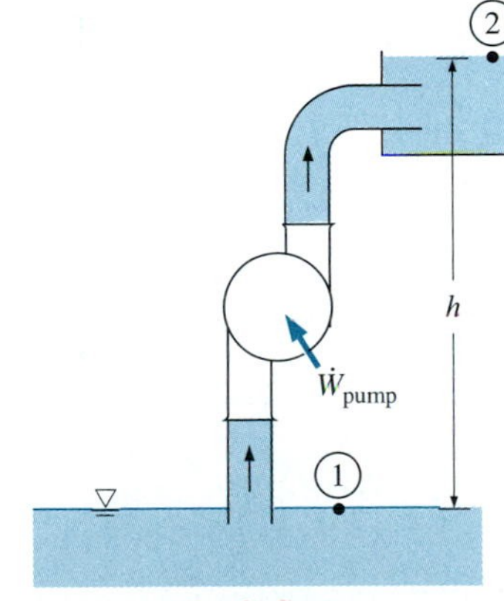

Steady flow
$V_1 - V_2 \approx 0$
$z_2 = z_1 + h$
$P_1 = P_2 = P_{\text{atm}}$

$\dot{E}_{\text{mech, in}} = \dot{E}_{\text{mech, out}} + \dot{E}_{\text{mech, loss}}$
$\dot{W}_{\text{pump}} + \dot{m}gz_1 = \dot{m}gz_2 + \dot{E}_{\text{mech, loss}}$
$\dot{W}_{\text{pump}} = \dot{m}gh + \dot{E}_{\text{mech, loss}}$

FIGURE 12–6

Most fluid flow problems involve mechanical forms of energy only, and such problems are conveniently solved by using a *mechanical energy* balance.

Most processes encountered in practice involve only certain forms of energy, and in such cases it is more convenient to work with the simplified versions of the energy balance. For systems that involve only *mechanical forms of energy* and its transfer as *shaft work*, the conservation of energy principle can be expressed conveniently as

$$E_{\text{mech, in}} - E_{\text{mech, out}} = \Delta E_{\text{mech, system}} + E_{\text{mech, loss}} \quad \textbf{(12–9)}$$

where $E_{\text{mech, loss}}$ represents the conversion of mechanical energy to thermal energy due to irreversibilities such as friction. For a system in steady operation, the mechanical energy balance becomes $\dot{E}_{\text{mech, in}} = \dot{E}_{\text{mech, out}} + \dot{E}_{\text{mech, loss}}$ (Fig. 12–6).

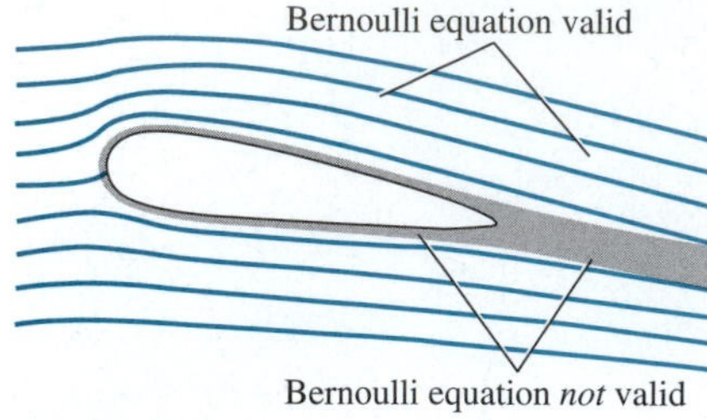

FIGURE 12–7

The *Bernoulli equation* is an *approximate* equation that is valid only in *inviscid regions of flow* where net viscous forces are negligibly small compared to inertial, gravitational, or pressure forces. Such regions occur outside of *boundary layers* and *wakes*.

12–2 ▪ THE BERNOULLI EQUATION

The **Bernoulli equation** is *an approximate relation between pressure, velocity, and elevation*, and is valid in *regions of steady, incompressible flow where net frictional forces are negligible* (Fig. 12–7). Despite its simplicity, it has proven to be a very powerful tool in fluid mechanics. In this section, we derive the Bernoulli equation by applying the *conservation of linear momentum principle*, and we demonstrate both its usefulness and its limitations.

The key approximation in the derivation of the Bernoulli equation is that *viscous effects are negligibly small compared to inertial, gravitational, and pressure effects*. Since all fluids have viscosity (there is no such thing as an "inviscid fluid"), this approximation cannot be valid for an entire flow field of practical interest. In other words, we cannot apply the Bernoulli equation *everywhere* in a flow, no matter how small the fluid's viscosity. However, it turns out that the approximation *is* reasonable in certain *regions* of many practical flows. We refer to such regions as *inviscid regions of flow*, and we stress that they are *not* regions where the fluid itself is inviscid or frictionless, but rather they are regions where net viscous or frictional forces are negligibly small compared to other forces acting on fluid particles.

Care must be exercised when applying the Bernoulli equation since it is an approximation that applies only to inviscid regions of flow. In general, frictional effects are always important very close to solid walls (*boundary layers*) and directly downstream of bodies (*wakes*). Thus, the Bernoulli approximation is typically useful in flow regions outside of boundary layers and wakes, where the fluid motion is governed by the combined effects of pressure and gravity forces.

The motion of a particle and the path it follows are described by the *velocity vector* as a function of time and space coordinates and the initial position of the particle. When the flow is *steady* (no change with time at a specified location), all particles that pass through the same point follow the same path (which is the *streamline*), and the velocity vectors remain tangent to the path at every point.

Derivation of the Bernoulli Equation

Consider the motion of a fluid particle in a flow field in steady flow. Applying Newton's second law (which is referred to as the *linear momentum equation* in fluid mechanics) in the s-direction on a particle moving along a streamline gives

$$\sum F_s = ma_s \tag{12–10}$$

In regions of flow where net frictional forces are negligible, there is no pump or turbine, and no heat transfer along the streamline, the significant forces acting in the s-direction are the pressure (acting on both sides) and the component of the weight of the particle in the s-direction (Fig. 12–8). Therefore, Eq. 12–10 becomes

$$P\,dA - (P + dP)\,dA - W \sin\theta = mV\frac{dV}{ds} \tag{12–11}$$

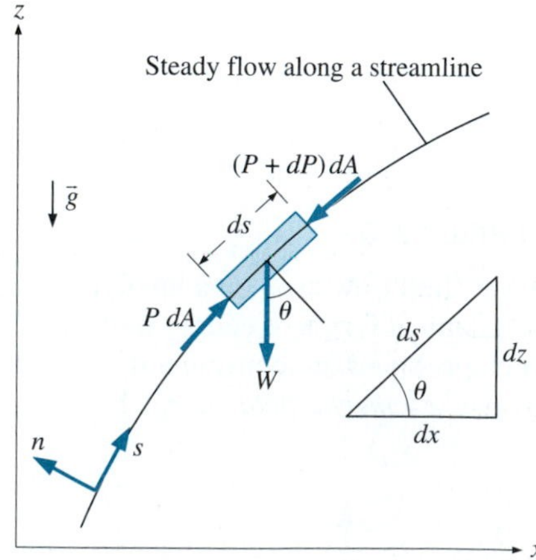

FIGURE 12–8

The forces acting on a fluid particle along a streamline.

where θ is the angle between the normal of the streamline and the vertical z-axis at that point, $m = \rho V = \rho\, dA\, ds$ is the mass, $W = mg = \rho g\, dA\, ds$

is the weight of the fluid particle, and $\sin\theta = dz/ds$. Substituting,

$$-dP\,dA - \rho g\,dA\,ds\,\frac{dz}{ds} = \rho\,dA\,ds\,V\,\frac{dV}{ds} \tag{12–12}$$

Canceling dA from each term and simplifying,

$$-dP - \rho g\,dz = \rho V\,dV \tag{12–13}$$

Noting that $V\,dV = \frac{1}{2}\,d(V^2)$ and dividing each term by ρ gives

$$\frac{dP}{\rho} + \tfrac{1}{2}\,d(V^2) + g\,dz = 0 \tag{12–14}$$

The last two terms are exact differentials. In the case of incompressible flow, the first term also becomes an exact differential, and integration gives

Steady, incompressible flow: $$\frac{P}{\rho} + \frac{V^2}{2} + gz = \text{constant (along a streamline)} \tag{12–15}$$

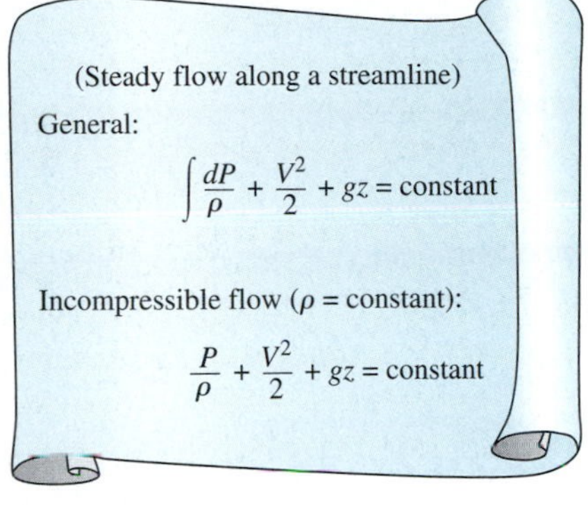

FIGURE 12–9

The Bernoulli equation is derived assuming incompressible flow, and thus it should not be used for flows with significant compressibility effects.

This is the famous **Bernoulli equation** (Fig. 12–9), which is commonly used in fluid mechanics for steady, incompressible flow along a streamline in inviscid regions of flow. The Bernoulli equation was first stated in words by the Swiss mathematician Daniel Bernoulli (1700–1782) in a text written in 1738 when he was working in St. Petersburg, Russia. It was later derived in equation form by his associate Leonhard Euler in 1755.

The value of the constant in Eq. 12–15 can be evaluated at any point on the streamline where the pressure, density, velocity, and elevation are known. The Bernoulli equation can also be written between any two points on the same streamline as

Steady, incompressible flow: $$\frac{P_1}{\rho} + \frac{V_1^2}{2} + gz_1 = \frac{P_2}{\rho} + \frac{V_2^2}{2} + gz_2 \tag{12–16}$$

We recognize $V^2/2$ as *kinetic energy, gz* as *potential energy*, and P/ρ as *flow energy*, all per unit mass. Therefore, the Bernoulli equation can be viewed as an expression of *mechanical energy balance* and can be stated as follows (Fig. 12–10):

> The sum of the kinetic, potential, and flow energies of a fluid particle is constant along a streamline during steady flow when compressibility and frictional effects are negligible.

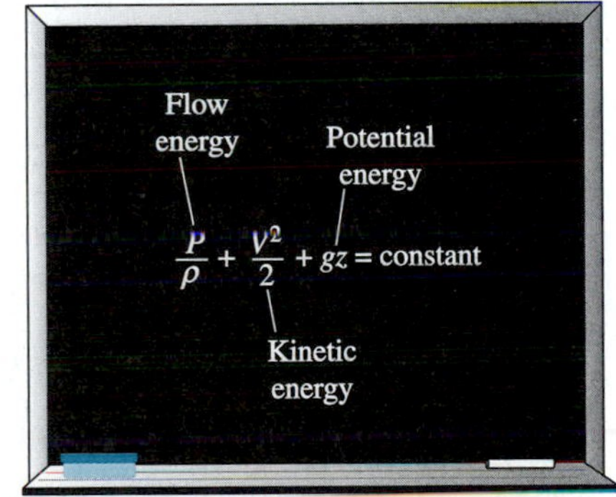

FIGURE 12–10

The Bernoulli equation states that the sum of the kinetic, potential, and flow energies of a fluid particle is constant along a streamline during steady flow.

The kinetic, potential, and flow energies are the mechanical forms of energy, as discussed in Section 12–1, and the Bernoulli equation can be viewed as the "conservation of mechanical energy principle." This is equivalent to the general conservation of energy principle for systems that do not involve any conversion of mechanical energy and thermal energy to each other, and thus the mechanical energy and thermal energy are conserved separately. The Bernoulli equation states that during steady, incompressible flow with negligible friction, the various forms of mechanical energy are converted to each other, but their sum remains constant. In other words, there is no dissipation of mechanical energy during such flows since there is no friction that converts mechanical energy to sensible thermal (internal) energy.

Recall that energy is transferred to a system as work when a force is applied to a system through a distance. In the light of Newton's second law of motion, the Bernoulli equation can also be viewed as: *The work done by the pressure and gravity forces on the fluid particle is equal to the increase in the kinetic energy of the particle.*

The Bernoulli equation is obtained from the conservation of momentum for a fluid particle moving along a streamline. It can also be obtained from the *first law of thermodynamics* applied to a steady-flow system, as shown in Section 12–4.

Despite the highly restrictive approximations used in its derivation, the Bernoulli equation is commonly used in practice since a variety of practical fluid flow problems can be analyzed to reasonable accuracy with it. This is because many flows of practical engineering interest are steady (or at least steady in the mean), compressibility effects are relatively small, and net frictional forces are negligible in regions of interest in the flow.

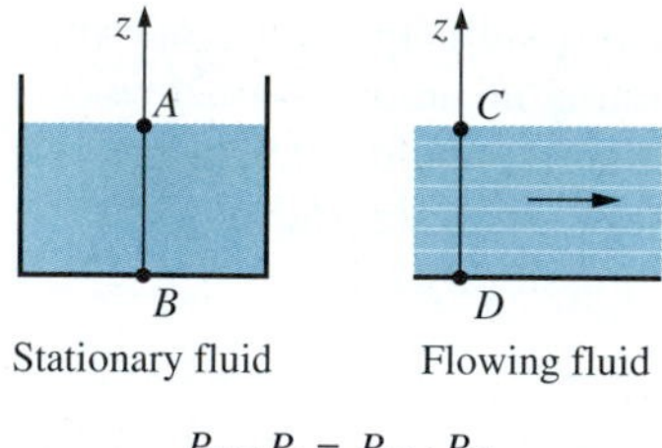

FIGURE 12–11

The variation of pressure with elevation in steady, incompressible flow along a straight line is the same as that in the stationary fluid (but this is not the case for a curved flow section).

Force Balance across Streamlines

It is left as an exercise to show that a force balance in the direction n normal to the streamline yields the following relation applicable *across* the streamlines for steady, incompressible flow:

$$\frac{P}{\rho} + \int \frac{V^2}{R}\,dn + gz = \text{constant} \qquad \text{(across streamlines)} \qquad \textbf{(12–17)}$$

For flow along a straight line, $R \to \infty$ and Eq. 12–17 reduces to $P/\rho + gz =$ constant or $P = -\rho gz +$ constant, which is an expression for the variation of hydrostatic pressure with vertical distance for a stationary fluid body. Therefore, the variation of pressure with elevation in steady, incompressible flow along a straight line is the same as that in the stationary fluid (Fig. 12–11).

Static, Dynamic, and Stagnation Pressures

The Bernoulli equation states that the sum of the flow, kinetic, and potential energies of a fluid particle along a streamline is constant. Therefore, the kinetic and potential energies of the fluid can be converted to flow energy (and vice versa) during flow, causing the pressure to change. This phenomenon can be made more visible by multiplying the Bernoulli equation by the density ρ,

$$P + \rho\frac{V^2}{2} + \rho gz = \text{constant (along a streamline)} \qquad \textbf{(12–18)}$$

Each term in this equation has pressure units, and thus each term represents some kind of pressure:

- P is the **static pressure** (it does not incorporate any dynamic effects); it represents the actual thermodynamic pressure of the fluid. This is the same as the pressure used in thermodynamics and property tables.
- $\rho V^2/2$ is the **dynamic pressure**; it represents the pressure rise when the fluid in motion is brought to a stop isentropically.
- ρgz is the **hydrostatic pressure** term, which is not pressure in a real sense since its value depends on the reference level selected; it accounts for the elevation effects, i.e., of fluid weight on pressure. (Be careful of sign—unlike hydrostatic pressure ρgh which *increases* with fluid depth, the hydrostatic pressure term ρgz *decreases* with fluid depth.)

The sum of the static, dynamic, and hydrostatic pressures is called the **total pressure**. Therefore, the Bernoulli equation states that *the total pressure along a streamline is constant.*

The sum of the static and dynamic pressures is called the **stagnation pressure**, and it is expressed as

$$P_{stag} = P + \rho\frac{V^2}{2} \qquad \text{(kPa)} \tag{12–19}$$

The stagnation pressure represents the pressure at a point where the fluid is brought to a complete stop isentropically. The static, dynamic, and stagnation pressures are shown in Fig. 12–12. When static and stagnation pressures are measured at a specified location, the fluid velocity at that location can be calculated from

$$V = \sqrt{\frac{2(P_{stag} - P)}{\rho}} \tag{12–20}$$

Equation 12–20 is useful in the measurement of flow velocity when a combination of a static pressure tap and a Pitot tube is used, as illustrated in Fig. 12–12. A **static pressure tap** is simply a small hole drilled into a wall such that the plane of the hole is parallel to the flow direction. It measures the static pressure. A **Pitot tube** is a small tube with its open end aligned *into* the flow so as to sense the full impact pressure of the flowing fluid. It measures the stagnation pressure. In situations in which the static and stagnation pressure of a flowing *liquid* are greater than atmospheric pressure, a vertical transparent tube called a **piezometer tube** (or simply a **piezometer**) can be attached to the pressure tap and to the Pitot tube, as sketched in Fig. 12–12. The liquid rises in the piezometer tube to a column height (*head*) that is proportional to the pressure being measured. If the pressures to be measured are below atmospheric, or if measuring pressures in *gases*, piezometer tubes do not work. However, the static pressure tap and Pitot tube can still be used, but they must be connected to some other kind of pressure measurement device such as a U-tube manometer or a pressure transducer. Sometimes it is convenient to integrate static pressure holes on a Pitot probe. The result is a **Pitot-static probe**, as shown in Fig. 12–13. A Pitot-static probe connected to a pressure transducer or a manometer measures the dynamic pressure (and thus fluid velocity).

When a stationary body is immersed in a flowing stream, the fluid is brought to a stop at the nose of the body (the **stagnation point**). The flow streamline that extends from far upstream to the stagnation point is called the **stagnation streamline** (Fig. 12–14). For a two-dimensional flow in the *xy*-plane, the stagnation point is actually a *line* parallel the *z*-axis, and the stagnation streamline is actually a *surface* that separates fluid that flows *over* the body from fluid that flows *under* the body. In an incompressible flow, the fluid decelerates nearly isentropically from its free-stream velocity to zero at the stagnation point, and the pressure at the stagnation point is thus the stagnation pressure.

Limitations on the Use of the Bernoulli Equation

The Bernoulli equation (Eq. 12–15) is one of the most frequently used and *misused* equations in fluid mechanics. Its versatility, simplicity, and ease of

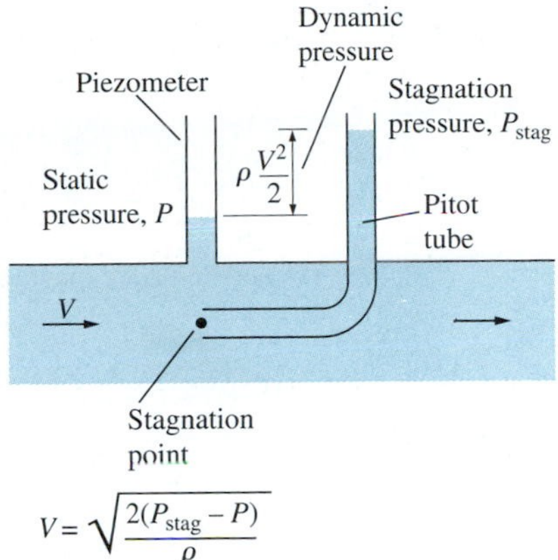

FIGURE 12–12
The static, dynamic, and stagnation pressures.

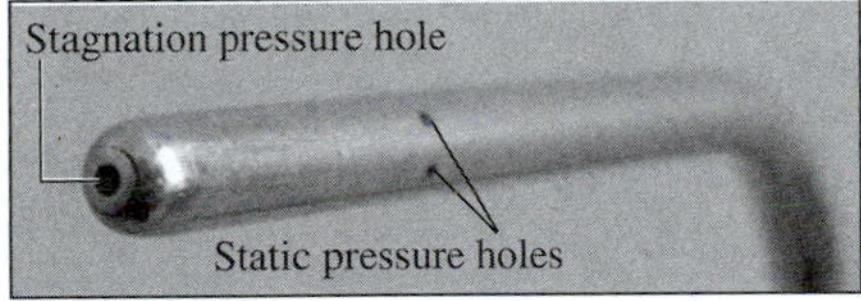

FIGURE 12–13
Close-up of a Pitot-static probe, showing the stagnation pressure hole and two of the five static circumferential pressure holes.
Photo by Po-Ya Abel Chuang. Used by permission.

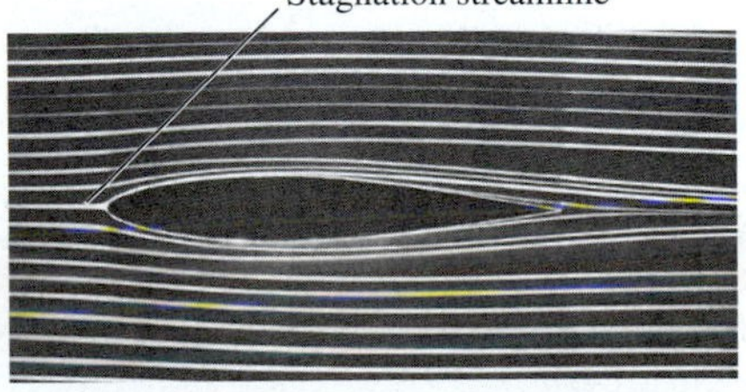

FIGURE 12–14
Streaklines produced by colored fluid introduced upstream of an airfoil; since the flow is steady, the streaklines are the same as streamlines and pathlines. The stagnation streamline is marked.
Courtesy ONERA. Photograph by Werlé.

use make it a very valuable tool for use in analysis, but the same attributes also make it very tempting to misuse. Therefore, it is important to understand the restrictions on its applicability and observe the limitations on its use, as explained here:

1. **Steady flow** The first limitation on the Bernoulli equation is that it is applicable to *steady flow*. Therefore, it should not be used during transient start-up and shut-down periods, or during periods of change in the flow conditions.
2. **Frictionless flow** Every flow involves some friction, no matter how small, and *frictional effects* may or may not be negligible. The situation is complicated even more by the amount of error that can be tolerated. In general, frictional effects are negligible for short flow sections with large cross sections, especially at low flow velocities. Frictional effects are usually significant in long and narrow flow passages, in the wake region downstream of an object, and in *diverging flow sections* such as diffusers because of the increased possibility of the fluid separating from the walls in such geometries. Frictional effects are also significant near solid surfaces, and thus the Bernoulli equation is usually applicable along a streamline in the core region of the flow, but not along a streamline close to the surface (Fig. 12–15).

   A component that disturbs the streamlined structure of flow and thus causes considerable mixing and backflow such as a sharp entrance of a tube or a partially closed valve in a flow section can make the Bernoulli equation inapplicable.
3. **No shaft work** The Bernoulli equation was derived from a force balance on a particle moving along a streamline. Therefore, the Bernoulli equation is not applicable in a flow section that involves a pump, turbine, fan, or any other machine or impeller since such devices destroy the streamlines and carry out energy interactions with the fluid particles. When the flow section considered involves any of these

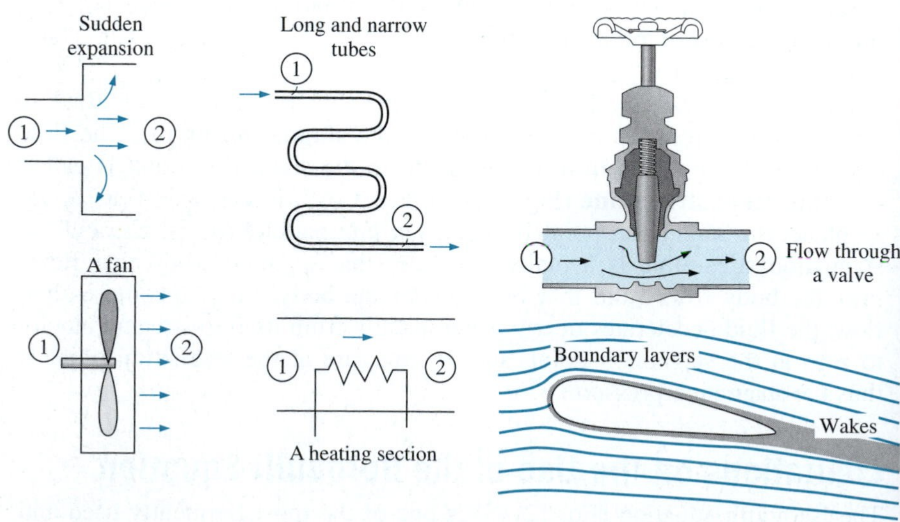

FIGURE 12–15

Frictional effects and components that disturb the streamlined structure of flow in a flow section make the Bernoulli equation invalid. It should *not* be used in any of the flows shown here.

devices, the energy equation should be used instead to account for the shaft work input or output. However, the Bernoulli equation can still be applied to a flow section prior to or past a machine (assuming, of course, that the other restrictions on its use are satisfied). In such cases, the Bernoulli constant changes from upstream to downstream of the device.

4. **Incompressible flow** One of the assumptions used in the derivation of the Bernoulli equation is that ρ = constant and thus the flow is incompressible. This condition is satisfied by liquids and also by gases at Mach numbers less than about 0.3 since compressibility effects and thus density variations of gases are negligible at such relatively low velocities.
5. **No heat transfer** The density of a gas is inversely proportional to temperature, and thus the Bernoulli equation should not be used for flow sections that involve significant temperature change such as heating or cooling sections.
6. **Flow along a streamline** Strictly speaking, the Bernoulli equation $P/\rho + V^2/2 + gz = C$ is applicable along a streamline, and the value of the constant C is generally different for different streamlines. When a region of the flow is *irrotational* and there is negligibly small *vorticity* in the flow field, the value of the constant C remains the same for all streamlines, and the Bernoulli equation becomes applicable *across* streamlines as well (Fig. 12–16). Therefore, we do not need to be concerned about the streamlines when the flow is irrotational, and we can apply the Bernoulli equation between any two points in the irrotational region of the flow.

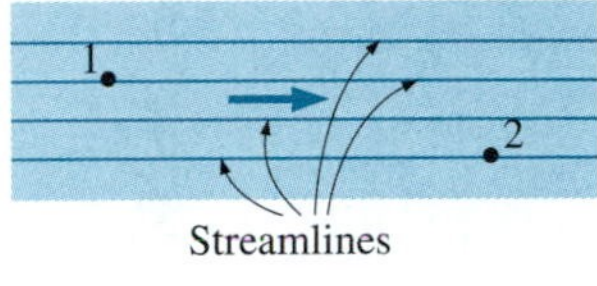

$$\frac{P_1}{\rho} + \frac{V_1^2}{2} + gz_1 = \frac{P_2}{\rho} + \frac{V_2^2}{2} + gz_2$$

FIGURE 12–16
When the flow is irrotational, the Bernoulli equation becomes applicable between any two points along the flow (not just on the same streamline).

We derived the Bernoulli equation by considering two-dimensional flow in the *xz*-plane for simplicity, but the equation is valid for general three-dimensional flow as well, as long as it is applied along the same streamline. We should always keep in mind the assumptions used in the derivation of the Bernoulli equation and make sure that they are not violated.

Hydraulic Grade Line (HGL) and Energy Grade Line (EGL)

It is often convenient to represent the level of mechanical energy graphically using *heights* to facilitate visualization of the various terms of the Bernoulli equation. This is done by dividing each term of the Bernoulli equation by g to give

$$\frac{P}{\rho g} + \frac{V^2}{2g} + z = H = \text{constant} \qquad \text{(along a streamline)} \qquad \textbf{(12–21)}$$

Each term in this equation has the dimension of length and represents some kind of "head" of a flowing fluid as follows:

- $P/\rho g$ is the **pressure head**; it represents the height of a fluid column that produces the static pressure P.

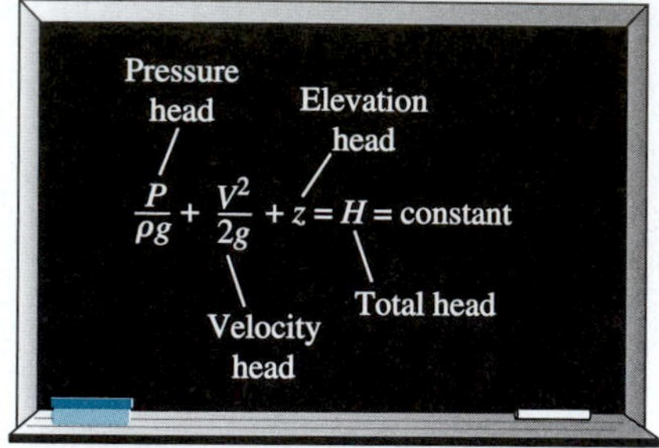

FIGURE 12–17

An alternative form of the Bernoulli equation is expressed in terms of heads as: *The sum of the pressure, velocity, and elevation heads is constant along a streamline.*

- $V^2/2g$ is the **velocity head**; it represents the elevation needed for a fluid to reach the velocity V during frictionless free fall.
- z is the **elevation head**; it represents the potential energy of the fluid.

Also, H is the **total head** for the flow. Therefore, the Bernoulli equation can be expressed in terms of heads as: *The sum of the pressure, velocity, and elevation heads along a streamline is constant during steady flow when the compressibility and frictional effects are negligible* (Fig. 12–17).

If a piezometer (which measures static pressure) is tapped into a pipe, as shown in Fig. 12–18, the liquid would rise to a height of $P/\rho g$ above the pipe center. The *hydraulic grade line* (HGL) is obtained by doing this at several locations along the pipe and drawing a curve through the liquid levels in the piezometers. The vertical distance above the pipe center is a measure of pressure within the pipe. Similarly, if a Pitot tube (measures static + dynamic pressure) is tapped into a pipe, the liquid would rise to a height of $P/\rho g + V^2/2g$ above the pipe center, or a distance of $V^2/2g$ above the HGL. The *energy grade line* (EGL) is obtained by doing this at several locations along the pipe and drawing a curve through the liquid levels in the Pitot tubes.

Noting that the fluid also has elevation head z (unless the reference level is taken to be the centerline of the pipe), the HGL and EGL are defined as follows: The line that represents the sum of the static pressure and the elevation heads, $P/\rho g + z$, is called the **hydraulic grade line**. The line that represents the total head of the fluid, $P/\rho g + V^2/2g + z$, is called the **energy grade line**. The difference between the heights of EGL and HGL is equal to the dynamic head, $V^2/2g$. We note the following about the HGL and EGL:

- For *stationary bodies* such as reservoirs or lakes, the EGL and HGL coincide with the free surface of the liquid. The elevation of the free surface z in such cases represents both the EGL and the HGL since the velocity is zero and the static (gage) pressure is zero.
- The EGL is always a distance $V^2/2g$ above the HGL. These two curves approach each other as the velocity decreases, and they diverge as the velocity increases. The height of the HGL decreases as the velocity increases, and vice versa.
- In an *idealized Bernoulli-type flow*, EGL is horizontal and its height remains constant. This would also be the case for HGL when the flow velocity is constant (Fig. 12–19).

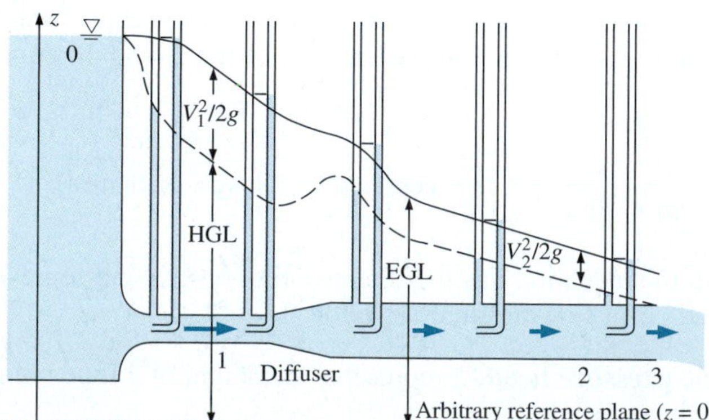

FIGURE 12–18

The *hydraulic grade line* (HGL) and the *energy grade line* (EGL) for free discharge from a reservoir through a horizontal pipe with a diffuser.

- For *open-channel flow*, the HGL coincides with the free surface of the liquid, and the EGL is a distance $V^2/2g$ above the free surface.
- At a *pipe exit*, the pressure head is zero (atmospheric pressure) and thus the HGL coincides with the pipe outlet (location 3 on Fig. 12–18).
- The *mechanical energy loss* due to frictional effects (conversion to thermal energy) causes the EGL and HGL to slope downward in the direction of flow. The slope is a measure of the head loss in the pipe. A component, such as a valve, that generates significant frictional effects causes a sudden drop in both EGL and HGL at that location.
- A *steep jump* occurs in EGL and HGL whenever mechanical energy is added to the fluid (by a pump, for example). Likewise, a *steep drop* occurs in EGL and HGL whenever mechanical energy is removed from the fluid (by a turbine, for example), as shown in Fig. 12–20.
- The (gage) pressure of a fluid is zero at locations where the HGL *intersects* the fluid. The pressure in a flow section that lies above the HGL is negative, and the pressure in a section that lies below the HGL is positive (Fig. 12–21). Therefore, an accurate drawing of a piping system and the HGL can be used to determine the regions where the pressure in the pipe is negative (below the atmospheric pressure).

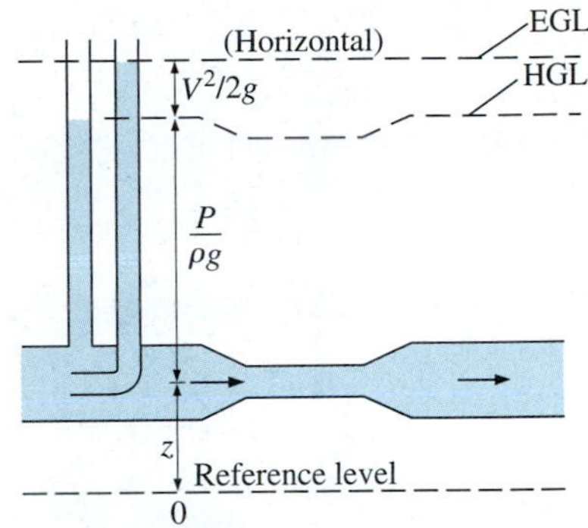

FIGURE 12–19

In an idealized Bernoulli-type flow, EGL is horizontal and its height remains constant. But this is not the case for HGL when the flow velocity varies along the flow.

The last remark enables us to avoid situations in which the pressure drops below the vapor pressure of the liquid (which may cause *cavitation*). Proper consideration is necessary in the placement of a liquid pump to ensure that the suction side pressure does not fall too low, especially at elevated temperatures where vapor pressure is higher than it is at low temperatures.

Now we examine Fig. 12–18 more closely. At point 0 (at the liquid surface), EGL and HGL are even with the liquid surface since there is no flow there. HGL decreases rapidly as the liquid accelerates into the pipe; however, EGL decreases very slowly through the well-rounded pipe inlet. EGL declines continually along the flow direction due to friction and other irreversible losses in the flow. EGL cannot increase in the flow direction unless energy is supplied to the fluid. HGL can rise or fall in the flow direction, but can never exceed EGL. HGL rises in the diffuser section as the velocity decreases, and the static pressure recovers somewhat; the total pressure does *not* recover, however, and EGL decreases through the diffuser. The difference between EGL and HGL is $V_1^2/2g$ at point 1, and $V_2^2/2g$ at point 2. Since $V_1 > V_2$, the difference between the two grade lines is larger at point 1 than at point 2. The downward slope of both grade lines is larger for the smaller diameter section of pipe since the frictional head loss is greater. Finally, HGL decays to the liquid surface at the outlet since the pressure there is atmospheric. However, EGL is still higher than HGL by the amount $V_2^2/2g$ since $V_3 = V_2$ at the outlet.

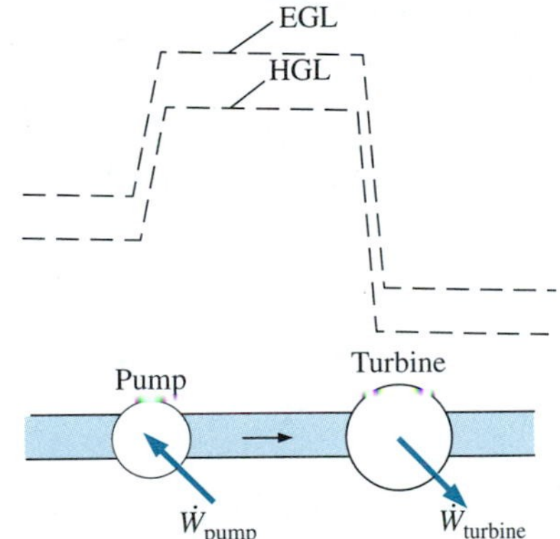

FIGURE 12–20

A *steep jump* occurs in EGL and HGL whenever mechanical energy is added to the fluid by a pump, and a *steep drop* occurs whenever mechanical energy is removed from the fluid by a turbine.

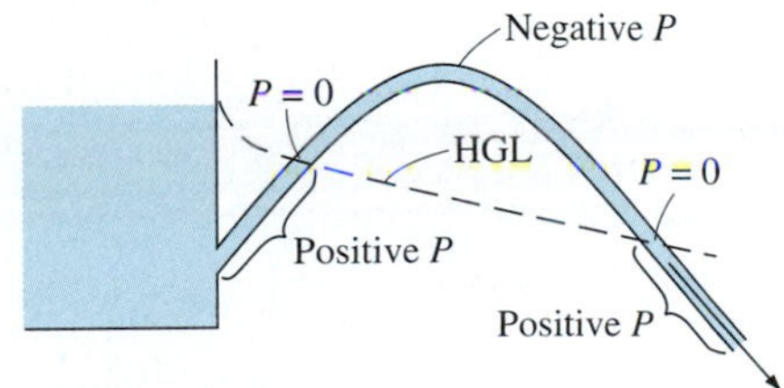

FIGURE 12–21

The (gage) pressure of a fluid is zero at locations where the HGL *intersects* the fluid, and the pressure is negative (vacuum) in a flow section that lies above the HGL.

Applications of the Bernouli Equation

So far, we have discussed the fundamental aspects of the Bernoulli equation. Now we demonstrate its use in a wide range of applications through examples.

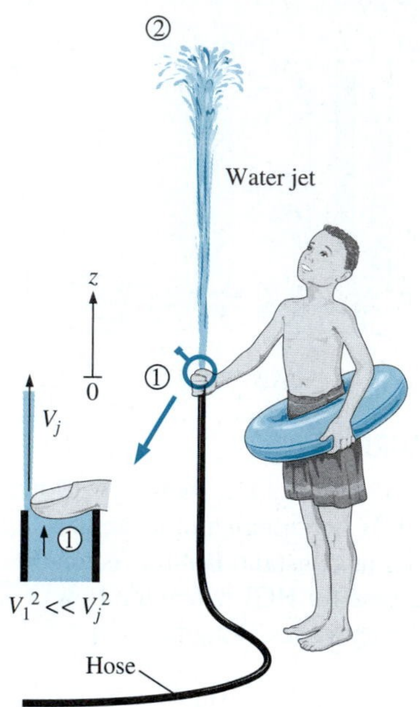

FIGURE 12–22

Schematic for Example 12–2. Inset shows a magnified view of the hose outlet region.

EXAMPLE 12–2 Spraying Water into the Air

Water is flowing from a garden hose (Fig. 12–22). A child places his thumb to cover most of the hose outlet, causing a thin jet of high-speed water to emerge. The pressure in the hose just upstream of his thumb is 400 kPa. If the hose is held upward, what is the maximum height that the jet could achieve?

Solution Water from a hose attached to the water main is sprayed into the air. The maximum height the water jet can rise is to be determined.

Assumptions **1** The flow exiting into the air is steady, incompressible, and irrotational (so that the Bernoulli equation is applicable). **2** The surface tension effects are negligible. **3** The friction between the water and air is negligible. **4** The irreversibilities that occur at the outlet of the hose due to abrupt contraction are not taken into account.

Properties We take the density of water to be 1000 kg/m^3.

Analysis This problem involves the conversion of flow, kinetic, and potential energies to each other without involving any pumps, turbines, and wasteful components with large frictional losses, and thus it is suitable for the use of the Bernoulli equation. The water height will be maximum under the stated assumptions. The velocity inside the hose is relatively low ($V_1^2 \ll V_2^2$, and thus $V_1 \cong 0$) and we take elevation just below the hose outlet as the reference level ($z_1 = 0$). At the top of the water trajectory $V_2 = 0$, and atmospheric pressure pertains. Then the Bernoulli equation simplifies to

$$\frac{P_1}{\rho g} + \frac{V_1^2}{2g}\nearrow^{\approx 0} + z_1\nearrow^{0} = \frac{P_2}{\rho g} + \frac{V_2^2}{2g}\nearrow^{0} + z_2 \quad \rightarrow \quad \frac{P_1}{\rho g} = \frac{P_{atm}}{\rho g} + z_2$$

Solving for z_2 and substituting,

$$z_2 = \frac{P_1 - P_{atm}}{\rho g} = \frac{P_{1,\,gage}}{\rho g} = \frac{400 \text{ kPa}}{(1000 \text{ kg/m}^3)(9.81 \text{ m/s}^2)}\left(\frac{1000 \text{ N/m}^2}{1 \text{ kPa}}\right)\left(\frac{1 \text{ kg}\cdot\text{m/s}^2}{1 \text{ N}}\right)$$

$$= \mathbf{40.8 \text{ m}}$$

Therefore, the water jet can rise as high as 40.8 m into the sky in this case.

Discussion The result obtained by the Bernoulli equation represents the upper limit and should be interpreted accordingly. It tells us that the water cannot possibly rise more than 40.8 m, and, in all likelihood, the rise will be much less than 40.8 m due to irreversible losses that we neglected.

EXAMPLE 12–3 Water Discharge from a Large Tank

A large tank open to the atmosphere is filled with water to a height of 5 m from the outlet tap (Fig. 12–23). A tap near the bottom of the tank is now opened, and water flows out from the smooth and rounded outlet. Determine the maximum water velocity at the outlet.

Solution A tap near the bottom of a tank is opened. The maximum exit velocity of water from the tank is to be determined.

Assumptions **1** The flow is incompressible and irrotational (except very close to the walls). **2** The water drains slowly enough that the flow can be approximated as steady (actually quasi-steady when the tank begins to drain).

Analysis This problem involves the conversion of flow, kinetic, and potential energies to each other without involving any pumps, turbines, and wasteful components with large frictional losses, and thus it is suitable for the use of the Bernoulli equation. We take point 1 to be at the free surface of water so that $P_1 = P_{atm}$ (open to the atmosphere), $V_1^2 \ll V_2^2$ and thus $V_1 \cong 0$ (the tank is very large relative to the outlet), and $z_1 = 5$ m and $z_2 = 0$ (we take the reference level at the center of the outlet). Also, $P_2 = P_{atm}$ (water discharges into the atmosphere). Then the Bernoulli equation simplifies to

$$\frac{P_1}{\rho g} + \frac{V_1^2}{2g}\overset{\approx 0}{\nearrow} + z_1 = \frac{P_2}{\rho g} + \frac{V_2^2}{2g} + z_2\overset{0}{\nearrow} \quad \rightarrow \quad z_1 = \frac{V_2^2}{2g}$$

Solving for V_2 and substituting,

$$V_2 = \sqrt{2gz_1} = \sqrt{2(9.81 \text{ m/s}^2)(5 \text{ m})} = \mathbf{9.9 \text{ m/s}}$$

The relation $V = \sqrt{2gz}$ is called the **Toricelli equation**.

Therefore, the water leaves the tank with an initial maximum velocity of 9.9 m/s. This is the same velocity that would manifest if a solid were dropped a distance of 5 m in the absence of air friction drag. (What would the velocity be if the tap were at the bottom of the tank instead of on the side?)

Discussion If the orifice were sharp-edged instead of rounded, then the flow would be disturbed, and the average exit velocity would be less than 9.9 m/s. Care must be exercised when attempting to apply the Bernoulli equation to situations where abrupt expansions or contractions occur since the friction and flow disturbance in such cases may not be negligible. From conservation of mass, $(V_1/V_2)^2 = (D_2/D_1)^4$. So, for example, if $D_2/D_1 = 0.1$, then $(V_1/V_2)^2 = 0.0001$, and our approximation that $V_1^2 \ll V_2^2$ is justified.

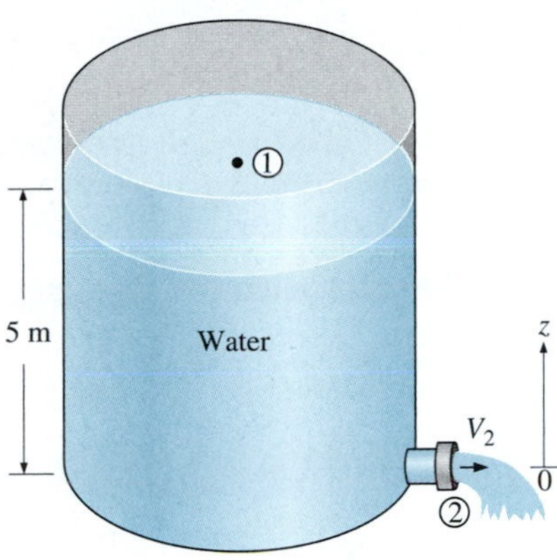

FIGURE 12–23

Schematic for Example 12–3.

EXAMPLE 12–4 Velocity Measurement by a Pitot Tube

A piezometer and a Pitot tube are tapped into a horizontal water pipe, as shown in Fig. 12–24, to measure static and stagnation (static + dynamic) pressures. For the indicated water column heights, determine the velocity at the center of the pipe.

Solution The static and stagnation pressures in a horizontal pipe are measured. The velocity at the center of the pipe is to be determined.

Assumptions **1** The flow is steady and incompressible. **2** Points 1 and 2 are close enough together that the irreversible energy loss between these two points is negligible, and thus we can use the Bernoulli equation.

Analysis We take points 1 and 2 along the centerline of the pipe, with point 1 directly under the piezometer and point 2 at the tip of the Pitot tube. This is a steady flow with straight and parallel streamlines, and the gage pressures at points 1 and 2 can be expressed as

$$P_1 = \rho g(h_1 + h_2)$$

$$P_2 = \rho g(h_1 + h_2 + h_3)$$

Noting that point 2 is a stagnation point and thus $V_2 = 0$ and $z_1 = z_2$, the application of the Bernoulli equation between points 1 and 2 gives

$$\frac{P_1}{\rho g} + \frac{V_1^2}{2g} + \not{z}_1 = \frac{P_2}{\rho g} + \frac{V_2^2}{2g}\overset{0}{\nearrow} + \not{z}_2 \quad \rightarrow \quad \frac{V_1^2}{2g} = \frac{P_2 - P_1}{\rho g}$$

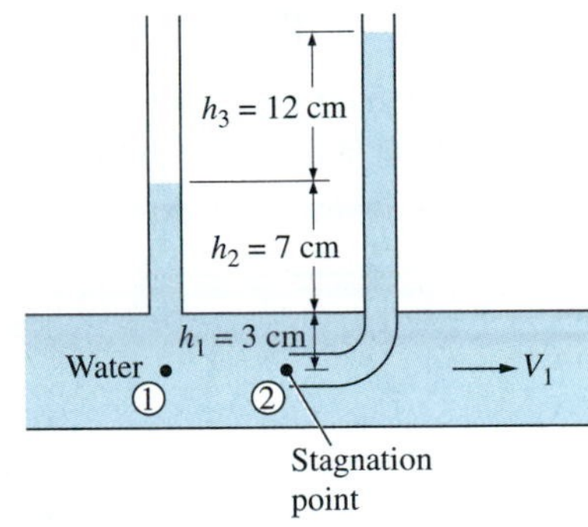

FIGURE 12–24

Schematic for Example 12–4.

Substituting the P_1 and P_2 expressions gives

$$\frac{V_1^2}{2g} = \frac{P_2 - P_1}{\rho g} = \frac{\rho g(h_1 + h_2 + h_3) - \rho g(h_1 + h_2)}{\rho g} = h_3$$

Solving for V_1 and substituting,

$$V_1 = \sqrt{2gh_3} = \sqrt{2(9.81 \text{ m/s}^2)(0.12 \text{ m})} = \mathbf{1.53 \text{ m/s}}$$

Discussion Note that to determine the flow velocity, all we need is to measure the height of the excess fluid column in the Pitot tube compared to that in the piezometer tube.

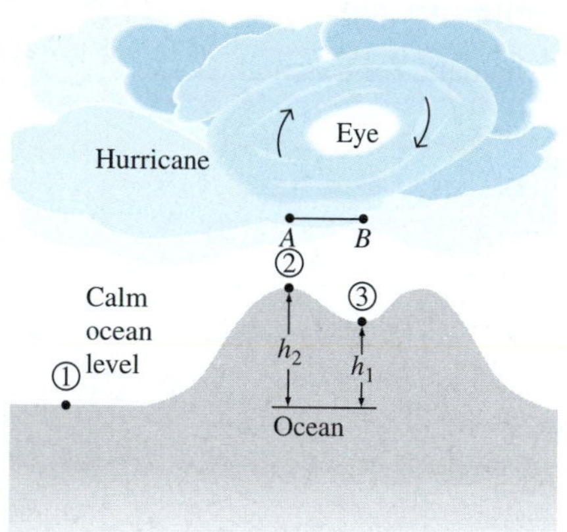

FIGURE 12–25

Schematic for Example 12–5. The vertical scale is greatly exaggerated.

EXAMPLE 12–5 The Rise of the Ocean Due to a Hurricane

A hurricane is a tropical storm formed over the ocean by low atmospheric pressures. As a hurricane approaches land, inordinate ocean swells (very high tides) accompany the hurricane. A Class-5 hurricane features winds in excess of 155 mph, although the wind velocity at the center "eye" is very low.

Figure 12–25 depicts a hurricane hovering over the ocean swell below. The atmospheric pressure 200 mi from the eye is 30.0 in Hg (at point 1, generally normal for the ocean) and the winds are calm. The hurricane atmospheric pressure at the eye of the storm is 22.0 in Hg. Estimate the ocean swell at (*a*) the eye of the hurricane at point 3 and (*b*) point 2, where the wind velocity is 155 mph. Take the density of seawater and mercury to be 64 lbm/ft^3 and 848 lbm/ft^3, respectively, and the density of air at normal sea-level temperature and pressure to be 0.076 lbm/ft^3.

Solution A hurricane is moving over the ocean. The amount of ocean swell at the eye and at active regions of the hurricane are to be determined.

Assumptions **1** The airflow within the hurricane is steady, incompressible, and irrotational (so that the Bernoulli equation is applicable). (This is certainly a very questionable assumption for a highly turbulent flow, but it is justified in the discussion.) **2** The effect of water sucked into the air is negligible.

Properties The densities of air at normal conditions, seawater, and mercury are given to be 0.076 lbm/ft^3, 64 lbm/ft^3, and 848 lbm/ft^3, respectively.

Analysis (*a*) Reduced atmospheric pressure over the water causes the water to rise. Thus, decreased pressure at point 2 relative to point 1 causes the ocean water to rise at point 2. The same is true at point 3, where the storm air velocity is negligible. The pressure difference given in terms of the mercury column height is expressed in terms of the seawater column height by

$$\Delta P = (\rho g h)_{\text{Hg}} = (\rho g h)_{\text{sw}} \rightarrow h_{\text{sw}} = \frac{\rho_{\text{Hg}}}{\rho_{\text{sw}}} h_{\text{Hg}}$$

Then the pressure difference between points 1 and 3 in terms of the seawater column height becomes

$$h_1 = \frac{\rho_{\text{Hg}}}{\rho_{\text{sw}}} h_{\text{Hg}} = \left(\frac{848 \text{ lbm/ft}^3}{64 \text{ lbm/ft}^3}\right)[(30 - 22) \text{ in Hg}]\left(\frac{1 \text{ ft}}{12 \text{ in}}\right) = \mathbf{8.83 \text{ ft}}$$

which is equivalent to the storm surge at the *eye of the hurricane* since the wind velocity there is negligible and there are no dynamic effects.

(*b*) To determine the additional rise of ocean water at point 2 due to the high winds at that point, we write the Bernoulli equation between points *A* and *B*, which are on top of points 2 and 3, respectively. Noting that $V_B \cong 0$ (the eye region of the hurricane is relatively calm) and $z_A = z_B$ (both points are on the same horizontal line), the Bernoulli equation simplifies to

$$\frac{P_A}{\rho g} + \frac{V_A^2}{2g} + \cancel{z_A} = \frac{P_B}{\rho g} + \frac{V_B^2}{2g}\nearrow^{0} + \cancel{z_B} \quad \rightarrow \quad \frac{P_B - P_A}{\rho g} = \frac{V_A^2}{2g}$$

Substituting,

$$\frac{P_B - P_A}{\rho g} = \frac{V_A^2}{2g} = \frac{(155 \text{ mph})^2}{2(32.2 \text{ ft/s}^2)}\left(\frac{1.4667 \text{ ft/s}}{1 \text{ mph}}\right)^2 = 803 \text{ ft}$$

where ρ is the density of air in the hurricane. Noting that the density of an ideal gas at constant temperature is proportional to absolute pressure and the density of air at the normal atmospheric pressure of 14.7 psia $\cong$ 30 in Hg is 0.076 lbm/ft^3, the density of air in the hurricane is

$$\rho_{\text{air}} = \frac{P_{\text{air}}}{P_{\text{atm air}}}\rho_{\text{atm air}} = \left(\frac{22 \text{ in Hg}}{30 \text{ in Hg}}\right)(0.076 \text{ lbm/ft}^3) = 0.056 \text{ lbm/ft}^3$$

Using the relation developed above in part (*a*), the seawater column height equivalent to 803 ft of air column height is determined to be

$$h_{\text{dynamic}} = \frac{\rho_{\text{air}}}{\rho_{\text{sw}}}h_{\text{air}} = \left(\frac{0.056 \text{ lbm/ft}^3}{64 \text{ lbm/ft}^3}\right)(803 \text{ ft}) = 0.70 \text{ ft}$$

Therefore, the pressure at point 2 is 0.70 ft seawater column lower than the pressure at point 3 due to the high wind velocities, causing the ocean to rise an additional 0.70 ft. Then the total storm surge at point 2 becomes

$$h_2 = h_1 + h_{\text{dynamic}} = 8.83 + 0.70 = \mathbf{9.53 \text{ ft}}$$

Discussion This problem involves highly turbulent flow and the intense breakdown of the streamlines, and thus the applicability of the Bernoulli equation in part (*b*) is questionable. Furthermore, the flow in the eye of the storm is *not* irrotational, and the Bernoulli equation constant changes across streamlines. The Bernoulli analysis can be thought of as the limiting, ideal case, and shows that the rise of seawater due to high-velocity winds cannot be more than 0.70 ft.

The wind power of hurricanes is not the only cause of damage to coastal areas. Ocean flooding and erosion from excessive tides is just as serious, as are high waves generated by the storm turbulence and energy.

12–3 ▪ GENERAL ENERGY EQUATION

One of the most fundamental laws in nature is the **first law of thermodynamics**, also known as the **conservation of energy principle**, which provides a sound basis for studying the relationships among the various forms of energy and energy interactions. It states that *energy can be neither created nor destroyed during a process; it can only change forms*. Therefore, every bit of energy must be accounted for during a process. The conservation of energy principle for any system can be expressed simply as $E_{\text{in}} - E_{\text{out}} = \Delta E$.

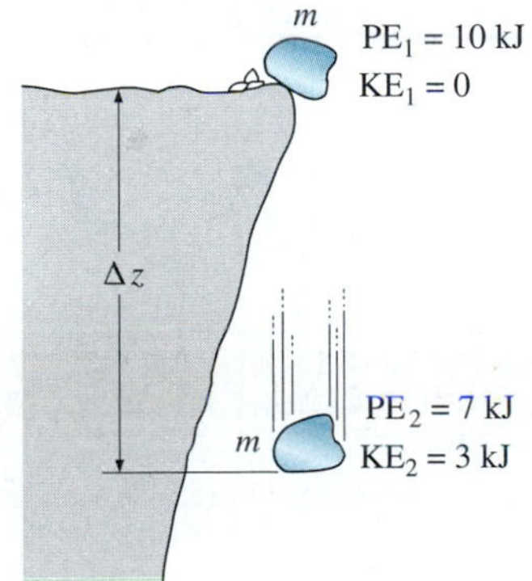

FIGURE 12–26

Energy cannot be created or destroyed during a process; it can only change forms.

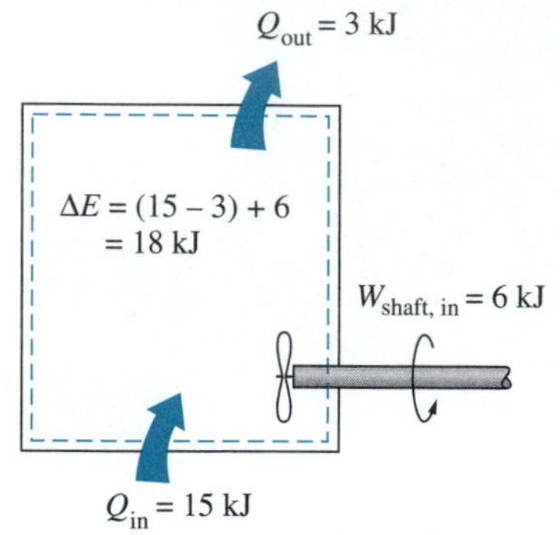

FIGURE 12–27

The energy change of a system during a process is equal to the *net* work and heat transfer between the system and its surroundings.

The transfer of any quantity (such as mass, momentum, and *energy*) is recognized *at the boundary* as the quantity *crosses the boundary*. A quantity is said to *enter* a system (or control volume) if it crosses the boundary from the outside to the inside, and to *exit* the system if it moves in the reverse direction. A quantity that moves from one location to another within a system is not considered as a transferred quantity in an analysis since it does not enter or exit the system. Therefore, it is important to specify the system and thus clearly identify its boundaries before an engineering analysis is performed.

The energy content of a fixed quantity of mass (a closed system) can be changed by two mechanisms: *heat transfer Q* and *work W*. Then the conservation of energy for a fixed quantity of mass can be expressed in rate form as (Fig. 12–27)

$$\dot{Q}_{\text{net in}} + \dot{W}_{\text{net in}} = \frac{dE_{\text{sys}}}{dt} \quad \text{or} \quad \dot{Q}_{\text{net in}} + \dot{W}_{\text{net in}} = \frac{d}{dt}\int_{\text{sys}} \rho e\, dV \tag{12-22}$$

where $\dot{Q}_{\text{net in}} = \dot{Q}_{\text{in}} - \dot{Q}_{\text{out}}$ is the net rate of heat transfer to the system (negative, if from the system), $\dot{W}_{\text{net in}} = \dot{W}_{\text{in}} - \dot{W}_{\text{out}}$ is the net power input to the system in all forms (negative, if power output) and dE_{sys}/dt is the rate of change of the total energy content of the system. For simple compressible systems, total energy consists of internal, kinetic, and potential energies, and it is expressed on a unit-mass basis as (see Chap. 3)

$$e = u + \text{ke} + \text{pe} = u + \frac{V^2}{2} + gz \tag{12-23}$$

Note that total energy is a property, and its value does not change unless the state of the system changes.

Energy Transfer by Heat, *Q*

In daily life, we frequently refer to the sensible and latent forms of internal energy as *heat*, and talk about the heat content of bodies. Scientifically the more correct name for these forms of energy is *thermal energy*. For single-phase substances, a change in the thermal energy of a given mass results in a change in temperature, and thus temperature is a good representative of thermal energy. Thermal energy tends to move naturally in the direction of decreasing temperature. The transfer of energy from one system to another as a result of a temperature difference is called **heat transfer.** The warming up of a canned drink in a warmer room, for example, is due to heat transfer (Fig. 12–28). The time rate of heat transfer is called **heat transfer rate** and is denoted by $\dot{Q}$.

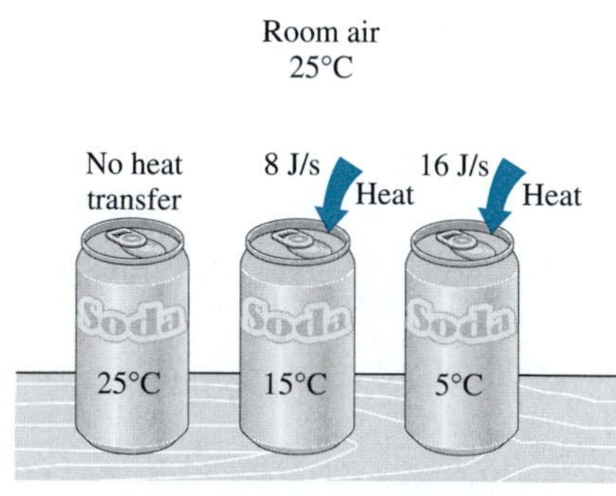

FIGURE 12–28

Temperature difference is the driving force for heat transfer. The larger the temperature difference, the higher is the rate of heat transfer.

The direction of heat transfer is always from the higher-temperature body to the lower-temperature one. Once temperature equality is established, heat transfer stops. There cannot be any net heat transfer between two systems (or a system and its surroundings) that are at the same temperature.

A process during which there is no heat transfer is called an **adiabatic process**. There are two ways a process can be adiabatic: Either the system is well insulated so that only a negligible amount of heat can pass through the system boundary, or both the system and the surroundings are at the same temperature and therefore there is no driving force (temperature difference) for net heat transfer. An adiabatic process should not be confused with an isothermal process. Even though there is no heat transfer during an adiabatic process, the energy content and thus the temperature of a system can still be changed by other means such as work.

Energy Transfer by Work, *W*

An energy interaction is **work** if it is associated with a force acting through a distance. A rising piston, a rotating shaft, and an electric wire crossing the system boundary are all associated with work interactions. The time rate of doing work is called **power** and is denoted by $\dot{W}$. Car engines and hydraulic, steam, and gas turbines produce work; compressors, pumps, fans, and mixers consume work.

Work-consuming devices transfer energy to the fluid, and thus increase the energy of the fluid. A fan in a room, for example, mobilizes the air and increases its kinetic energy. The electric energy a fan consumes is first converted to mechanical energy by its motor that forces the shaft of the blades to rotate. This mechanical energy is then transferred to the air, as evidenced by the increase in air velocity. This energy transfer to air has nothing to do with a temperature difference, so it cannot be heat transfer. Therefore, it must be work. Air discharged by the fan eventually comes to a stop and thus loses its mechanical energy as a result of friction between air particles of different velocities. But this is not a "loss" in the real sense; it is simply the conversion of mechanical energy to an equivalent amount of thermal energy (which is of limited value, and thus the term *loss*) in accordance with the conservation of energy principle. If a fan runs a long time in a sealed room, we can sense the buildup of this thermal energy by a rise in air temperature.

A system may involve numerous forms of work, and the total work can be expressed as

$$W_{total} = W_{shaft} + W_{pressure} + W_{viscous} + W_{other} \qquad \textbf{(12–24)}$$

where W_{shaft} is the work transmitted by a rotating shaft, $W_{pressure}$ is the work done by the pressure forces on the control surface, $W_{viscous}$ is the work done by the normal and shear components of viscous forces on the control surface, and W_{other} is the work done by other forces such as electric, magnetic, and surface tension, which are insignificant for simple compressible systems and are not considered in this text. We do not consider $W_{viscous}$ either since moving walls, such as fan blades or turbine runners, are usually *inside* the control volume, and are not part of the control surface. But it should be kept in mind that the work done by shear forces as the blades shear through the fluid may need to be considered in a refined analysis of turbomachinery.

Shaft Work

Many flow systems involve a machine such as a pump, a turbine, a fan, or a compressor whose shaft protrudes through the control surface, and the work transfer associated with all such devices is simply referred to as *shaft work*

W_{shaft}. The power transmitted via a rotating shaft is proportional to the shaft torque T_{shaft} and is expressed as

$$\dot{W}_{\text{shaft}} = \omega T_{\text{shaft}} = 2\pi \dot{n} T_{\text{shaft}} \tag{12-25}$$

where ω is the angular speed of the shaft in rad/s and $\dot{n}$ is defined as the number of revolutions of the shaft per unit time, often expressed in rev/min or rpm.

Work Done by Pressure Forces

Consider a gas being compressed in the piston-cylinder device shown in Fig. 12–29*a*. When the piston moves down a differential distance ds under the influence of the pressure force PA, where A is the cross-sectional area of the piston, the boundary work done *on* the system is $\delta W_{\text{boundary}} = PA\, ds$. Dividing both sides of this relation by the differential time interval dt gives the time rate of boundary work (i.e., *power*),

$$\delta \dot{W}_{\text{pressure}} = \delta \dot{W}_{\text{boundary}} = PAV_{\text{piston}}$$

where $V_{\text{piston}} = ds/dt$ is the piston velocity, which is the velocity of the moving boundary at the piston face.

Now consider a material chunk of fluid (a system) of arbitrary shape, which moves with the flow and is free to deform under the influence of pressure, as shown in Fig. 12–29*b*. Pressure always acts inward and normal to the surface, and the pressure force acting on a differential area dA is $P\, dA$. Again noting that work is force times distance and distance traveled per unit time is velocity, the time rate at which work is done by pressure forces on this differential part of the system is

$$\delta \dot{W}_{\text{pressure}} = -P\, dA\, V_n = -P\, dA(\vec{V} \cdot \vec{n}) \tag{12-26}$$

since the normal component of velocity through the differential area dA is $V_n = V \cos\theta = \vec{V} \cdot \vec{n}$. Note that $\vec{n}$ is the outer normal of dA, and thus the quantity $\vec{V} \cdot \vec{n}$ is positive for expansion and negative for compression. The negative sign in Eq. 12–26 ensures that work done by pressure forces is positive when it is done *on* the system, and negative when it is done *by* the system, which agrees with our sign convention. The total rate of work done by pressure forces is obtained by integrating $\delta \dot{W}_{\text{pressure}}$ over the entire surface A,

$$\dot{W}_{\text{pressure, net in}} = -\int_A P(\vec{V} \cdot \vec{n})\, dA = -\int_A \frac{P}{\rho} \rho(\vec{V} \cdot \vec{n})\, dA \tag{12-27}$$

In light of these discussions, the net power transfer is expressed as

$$\dot{W}_{\text{net in}} = \dot{W}_{\text{shaft, net in}} + \dot{W}_{\text{pressure, net in}} = \dot{W}_{\text{shaft, net in}} - \int_A P(\vec{V} \cdot \vec{n})\, dA \tag{12-28}$$

Then the rate form of the conservation of energy relation for a closed system becomes

$$\dot{Q}_{\text{net in}} + \dot{W}_{\text{shaft, net in}} + \dot{W}_{\text{pressure, net in}} = \frac{dE_{\text{sys}}}{dt} \tag{12-29}$$

To obtain a relation for the conservation of energy for a *control volume,* we apply the Reynolds transport theorem by replacing B with total energy

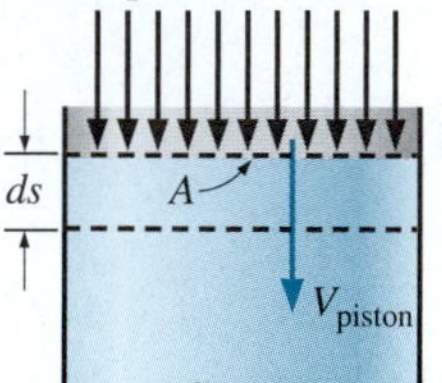

(*a*)

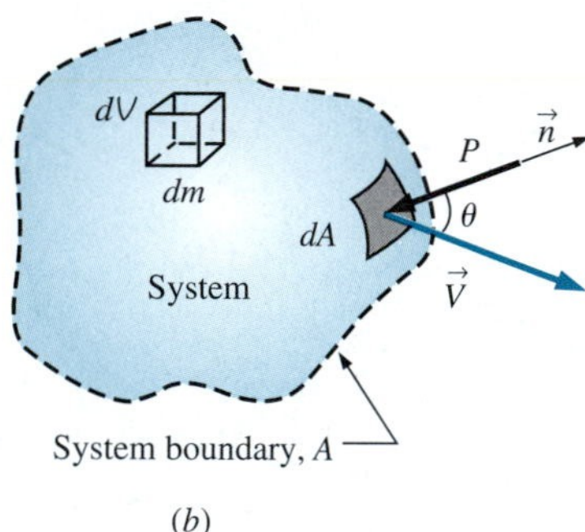

(*b*)

FIGURE 12–29

The pressure force acting on (*a*) the moving boundary of a system in a piston-cylinder device, and (*b*) the differential surface area of a system of arbitrary shape.

E, and b with total energy per unit mass e, which is $e = u + \text{ke} + \text{pe} = u + V^2/2 + gz$ (Fig. 12–30). This yields

$$\frac{dE_{sys}}{dt} = \frac{d}{dt}\int_{CV} e\rho \, dV + \int_{CS} e\rho(\vec{V}_r \cdot \vec{n})A \tag{12–30}$$

Substituting the left-hand side of Eq. 12–29 into Eq. 12–30, the general form of the energy equation that applies to fixed, moving, or deforming control volumes becomes

$$\dot{Q}_{\text{net in}} + \dot{W}_{\text{shaft, net in}} + \dot{W}_{\text{pressure, net in}} = \frac{d}{dt}\int_{CV} e\rho \, dV + \int_{CS} e\rho(\vec{V}_r \cdot \vec{n})\, dA \tag{12–31}$$

which can be stated as

$$\begin{pmatrix}\text{The net rate of energy}\\ \text{transfer into a CV by}\\ \text{heat and work transfer}\end{pmatrix} = \begin{pmatrix}\text{The time rate of}\\ \text{change of the energy}\\ \text{content of the CV}\end{pmatrix} + \begin{pmatrix}\text{The net flow rate of}\\ \text{energy out of the control}\\ \text{surface by mass flow}\end{pmatrix}$$

Here $\vec{V}_r = \vec{V} - \vec{V}_{CS}$ is the fluid velocity relative to the control surface, and the product $\rho(\vec{V}_r \cdot \vec{n})\, dA$ represents the mass flow rate through area element dA into or out of the control volume. Again noting that $\vec{n}$ is the outer normal of dA, the quantity $\vec{V}_r \cdot \vec{n}$ and thus mass flow is positive for outflow and negative for inflow.

Substituting the surface integral for the rate of pressure work from Eq. 12–27 into Eq. 12–31 and combining it with the surface integral on the right give

$$\dot{Q}_{\text{net in}} + \dot{W}_{\text{shaft, net in}} = \frac{d}{dt}\int_{CV} e\rho \, dV + \int_{CS}\left(\frac{P}{\rho} + e\right)\rho(\vec{V}_r \cdot \vec{n})\, dA \tag{12–32}$$

This is a very convenient form for the energy equation since pressure work is now combined with the energy of the fluid crossing the control surface and we no longer have to deal with pressure work.

The term $P/\rho = Pv = w_{\text{flow}}$ is the **flow work**, which is the work per unit mass associated with pushing a fluid into or out of a control volume. Note that the fluid velocity at a solid surface is equal to the velocity of the solid surface because of the no-slip condition. As a result, the pressure work along the portions of the control surface that coincide with nonmoving solid surfaces is zero. Therefore, pressure work for fixed control volumes can exist only along the imaginary part of the control surface where the fluid enters and leaves the control volume, i.e., inlets and outlets.

For a fixed control volume (no motion or deformation of control volume), $\vec{V}_r = \vec{V}$ and the energy equation Eq. 12–32 becomes

Fixed CV: $$\dot{Q}_{\text{net in}} + \dot{W}_{\text{shaft, net in}} = \frac{d}{dt}\int_{CV} e\rho \, dV + \int_{CS}\left(\frac{P}{\rho} + e\right)\rho(\vec{V} \cdot \vec{n})\, dA \tag{12–33}$$

This equation is not in a convenient form for solving practical engineering problems because of the integrals, and thus it is desirable to rewrite it in terms of average velocities and mass flow rates through inlets and outlets. If $P/\rho + e$ is nearly uniform across an inlet or outlet, we can simply take it

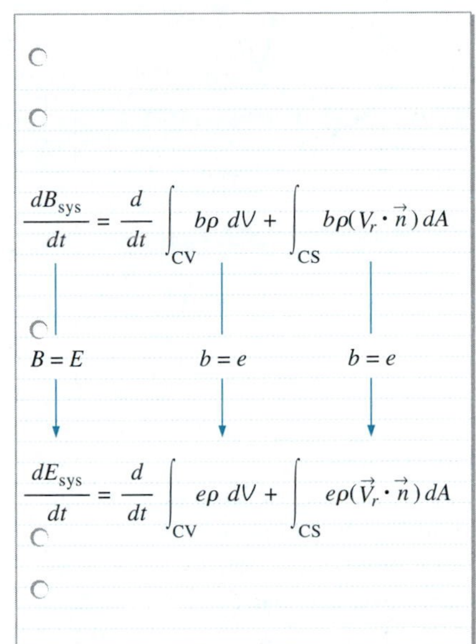

FIGURE 12–30
The conservation of energy equation is obtained by replacing B in the Reynolds transport theorem by energy E and b by e.

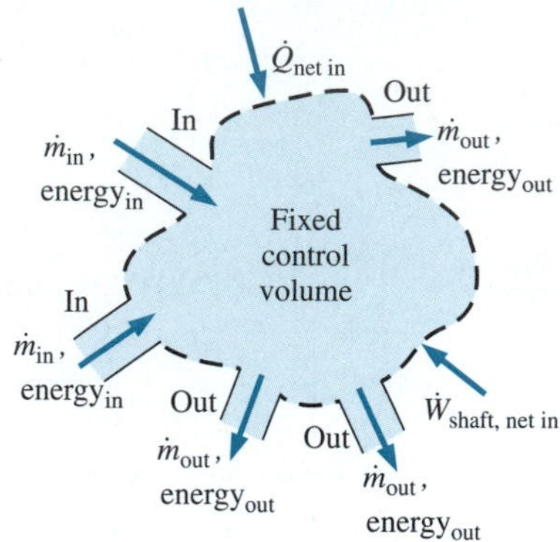

FIGURE 12–31
In a typical engineering problem, the control volume may contain many inlets and outlets; energy flows in at each inlet, and energy flows out at each outlet. Energy also enters the control volume through net heat transfer and net shaft work.

outside the integral. Noting that $\dot{m} = \int_{A_c} \rho(\vec{V} \cdot \vec{n})\, dA_c$ is the mass flow rate across an inlet or outlet, the rate of inflow or outflow of energy through the inlet or outlet can be approximated as $\dot{m}(P/\rho + e)$. Then the energy equation becomes (Fig. 12–31)

$$\dot{Q}_{\text{net in}} + \dot{W}_{\text{shaft, net in}} = \frac{d}{dt}\int_{\text{CV}} e\rho\, dV + \sum_{\text{out}} \dot{m}\left(\frac{P}{\rho} + e\right) - \sum_{\text{in}} \dot{m}\left(\frac{P}{\rho} + e\right) \quad \textbf{(12–34)}$$

where $e = u + V^2/2 + gz$ (Eq. 12–23) is the total energy per unit mass for both the control volume and flow streams. Then,

$$\dot{Q}_{\text{net in}} + \dot{W}_{\text{shaft, net in}} = \frac{d}{dt}\int_{\text{CV}} e\rho\, d \mp \sum_{\text{out}} \dot{m}\left(\frac{P}{\rho} + u + \frac{V^2}{2} + gz\right) - \sum_{\text{in}} \dot{m}\left(\frac{P}{\rho} + u + \frac{V^2}{2} + gz\right) \quad \textbf{(12–35)}$$

or

$$\dot{Q}_{\text{net in}} + \dot{W}_{\text{shaft, net in}} = \frac{d}{dt}\int_{\text{CV}} e\rho\, dV + \sum_{\text{out}} \dot{m}\left(h + \frac{V^2}{2} + gz\right) - \sum_{\text{in}} \dot{m}\left(h + \frac{V^2}{2} + gz\right) \quad \textbf{(12–36)}$$

where we used the definition of enthalpy $h = u + Pv = u + P/\rho$. The last two equations are fairly general expressions of conservation of energy, but their use is still limited to fixed control volumes, uniform flow at inlets and outlets, and negligible work due to viscous forces and other effects. Also, the subscript "net in" stands for "net input," and thus any heat or work transfer is positive if *to* the system and negative if *from* the system.

12–4 ▪ ENERGY ANALYSIS OF STEADY FLOWS

For steady flows, the time rate of change of the energy content of the control volume is zero, and Eq. 12–36 simplifies to

$$\dot{Q}_{\text{net in}} + \dot{W}_{\text{shaft, net in}} = \sum_{\text{out}} \dot{m}\left(h + \frac{V^2}{2} + gz\right) - \sum_{\text{in}} \dot{m}\left(h + \frac{V^2}{2} + gz\right) \quad \textbf{(12–37)}$$

It states that *the net rate of energy transfer to a control volume by heat transfer and work during steady flow is equal to the difference between the rates of outgoing and incoming energy flows with mass.*

Many practical problems involve just one inlet and one outlet (Fig. 12–32). The mass flow rate for such **single-stream devices** remains constant, and Eq. 12–37 reduces to

$$\dot{Q}_{\text{net in}} + \dot{W}_{\text{shaft, net in}} = \dot{m}\left(h_2 - h_1 + \frac{V_2^2 - V_1^2}{2} + g(z_2 - z_1)\right) \quad \textbf{(12–38)}$$

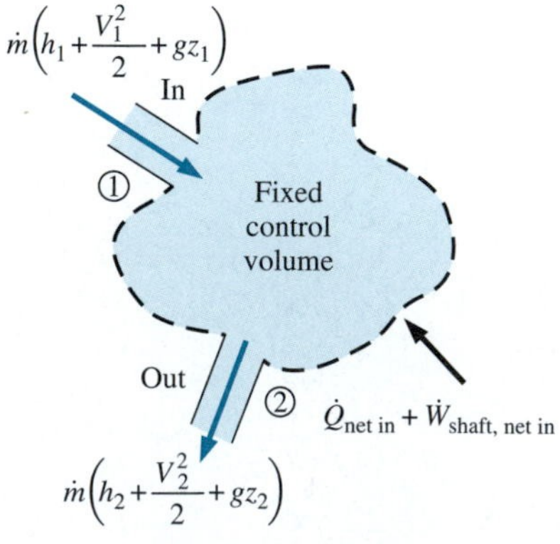

FIGURE 12–32
A control volume with only one inlet and one outlet and energy interactions.

where subscripts 1 and 2 refer to the inlet and outlet, respectively. The steady-flow energy equation on a unit-mass basis is obtained by dividing Eq. 12–38 by the mass flow rate $\dot{m}$,

$$q_{\text{net in}} + w_{\text{shaft, net in}} = h_2 - h_1 + \frac{V_2^2 - V_1^2}{2} + g(z_2 - z_1) \quad \textbf{(12–39)}$$

where $q_{\text{net in}} = \dot{Q}_{\text{net in}}/\dot{m}$ is the net heat transfer to the fluid per unit mass and $w_{\text{shaft, net in}} = \dot{W}_{\text{shaft, net in}}/\dot{m}$ is the net shaft work input to the fluid per unit mass. Using the definition of enthalpy $h = u + P/\rho$ and rearranging, the steady-flow energy equation can also be expressed as

$$w_{\text{shaft, net in}} + \frac{P_1}{\rho_1} + \frac{V_1^2}{2} + gz_1 = \frac{P_2}{\rho_2} + \frac{V_2^2}{2} + gz_2 + (u_2 - u_1 - q_{\text{net in}}) \quad \textbf{(12–40)}$$

where u is the *internal energy*, P/ρ is the *flow energy*, $V^2/2$ is the *kinetic energy*, and gz is the *potential energy* of the fluid, all per unit mass. These relations are valid for both compressible and incompressible flows.

The left side of Eq. 12–40 represents the mechanical energy input, while the first three terms on the right side represent the mechanical energy output. If the flow is ideal with no irreversibilities such as friction, the total mechanical energy must be conserved, and the term in parentheses $(u_2 - u_1 - q_{\text{net in}})$ must equal zero. That is,

Ideal flow (no mechanical energy loss):
$$q_{\text{net in}} = u_2 - u_1 \quad \textbf{(12–41)}$$

Any increase in $u_2 - u_1$ above $q_{\text{net in}}$ is due to the irreversible conversion of mechanical energy to thermal energy, and thus $u_2 - u_1 - q_{\text{net in}}$ represents the mechanical energy loss (Fig. 12–33). That is,

Mechanical energy loss:
$$e_{\text{mech, loss}} = u_2 - u_1 - q_{\text{net in}} \quad \textbf{(12–42)}$$

For single-phase fluids (a gas or a liquid), we have $u_2 - u_1 = c_v(T_2 - T_1)$ where c_v is the constant-volume specific heat.

The steady-flow energy equation on a unit-mass basis can be written conveniently as a **mechanical energy** balance as

$$e_{\text{mech, in}} = e_{\text{mech, out}} + e_{\text{mech, loss}} \quad \textbf{(12–43)}$$

or

$$w_{\text{shaft, net in}} + \frac{P_1}{\rho_1} + \frac{V_1^2}{2} + gz_1 = \frac{P_2}{\rho_2} + \frac{V_2^2}{2} + gz_2 + e_{\text{mech, loss}} \quad \textbf{(12–44)}$$

Noting that $w_{\text{shaft, net in}} = w_{\text{pump}} - w_{\text{turbine}}$, the mechanical energy balance can be written more explicitly as

$$\frac{P_1}{\rho_1} + \frac{V_1^2}{2} + gz_1 + w_{\text{pump}} = \frac{P_2}{\rho_2} + \frac{V_2^2}{2} + gz_2 + w_{\text{turbine}} + e_{\text{mech, loss}} \quad \textbf{(12–45)}$$

where w_{pump} is the mechanical work input (due to the presence of a pump, fan, compressor, etc.) and w_{turbine} is the mechanical work output (due to a turbine). When the flow is incompressible, either absolute or gage pressure can be used for P since P_{atm}/ρ would appear on both sides and would cancel out.

Multiplying Eq. 12–45 by the mass flow rate $\dot{m}$ gives

$$\dot{m}\left(\frac{P_1}{\rho_1} + \frac{V_1^2}{2} + gz_1\right) + \dot{W}_{\text{pump}} = \dot{m}\left(\frac{P_2}{\rho_2} + \frac{V_2^2}{2} + gz_2\right) + \dot{W}_{\text{turbine}} + \dot{E}_{\text{mech, loss}} \quad \textbf{(12–46)}$$

where $\dot{W}_{\text{pump}}$ is the shaft power input through the pump's shaft, $\dot{W}_{\text{turbine}}$ is the shaft power output through the turbine's shaft, and $\dot{E}_{\text{mech, loss}}$ is the *total*

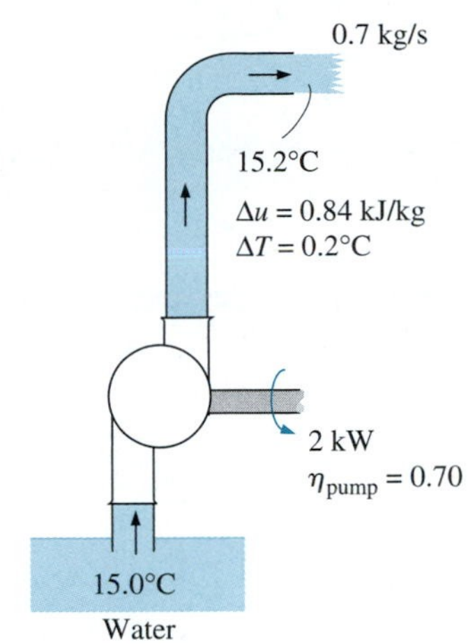

FIGURE 12–33
The lost mechanical energy in a fluid flow system results in an increase in the internal energy of the fluid and thus in a rise of fluid temperature.

mechanical power loss, which consists of pump and turbine losses as well as the frictional losses in the piping network. That is,

$$\dot{E}_{\text{mech, loss}} = \dot{E}_{\text{mech loss, pump}} + \dot{E}_{\text{mech loss, turbine}} + \dot{E}_{\text{mech loss, piping}}$$

By convention, irreversible pump and turbine losses are treated separately from irreversible losses due to other components of the piping system. Thus the energy equation can be expressed in its most common form in terms of *heads* by dividing each term in Eq. 12–46 by $\dot{m}g$. The result is

$$\frac{P_1}{\rho_1 g} + \frac{V_1^2}{2g} + z_1 + h_{\text{pump, }u} = \frac{P_2}{\rho_2 g} + \frac{V_2^2}{2g} + z_2 + h_{\text{turbine, }e} + h_L \qquad \textbf{(12–47)}$$

where

- $h_{\text{pump, }u} = \dfrac{w_{\text{pump, }u}}{g} = \dfrac{\dot{W}_{\text{pump, }u}}{\dot{m}g} = \dfrac{\eta_{\text{pump}}\dot{W}_{\text{pump}}}{\dot{m}g}$ is the *useful head delivered to the fluid by the pump*. Because of irreversible losses in the pump, $h_{\text{pump, }u}$ is *less* than $\dot{W}_{\text{pump}}/\dot{m}g$ by the factor η_{pump}.
- $h_{\text{turbine, }e} = \dfrac{w_{\text{turbine, }e}}{g} = \dfrac{\dot{W}_{\text{turbine, }e}}{\dot{m}g} = \dfrac{\dot{W}_{\text{turbine}}}{\eta_{\text{turbine}}\dot{m}g}$ is the *extracted head removed from the fluid by the turbine*. Because of irreversible losses in the turbine, $h_{\text{turbine, }e}$ is greater than $\dot{W}_{\text{turbine}}/\dot{m}g$ by the factor η_{turbine}.
- $h_L = \dfrac{e_{\text{mech loss, piping}}}{g} = \dfrac{\dot{E}_{\text{mech loss, piping}}}{\dot{m}g}$ is the irreversible *head loss* between 1 and 2 due to all components of the piping system other than the pump or turbine.

Note that the head loss h_L represents the frictional losses associated with fluid flow in piping, and it does not include the losses that occur within the pump or turbine due to the inefficiencies of these devices—these losses are taken into account by η_{pump} and η_{turbine}. Equation 12–47 is illustrated schematically in Fig. 12–34.

The *pump head is* zero if the piping system does not involve a pump, a fan, or a compressor, and the *turbine head* is zero if the system does not involve a turbine.

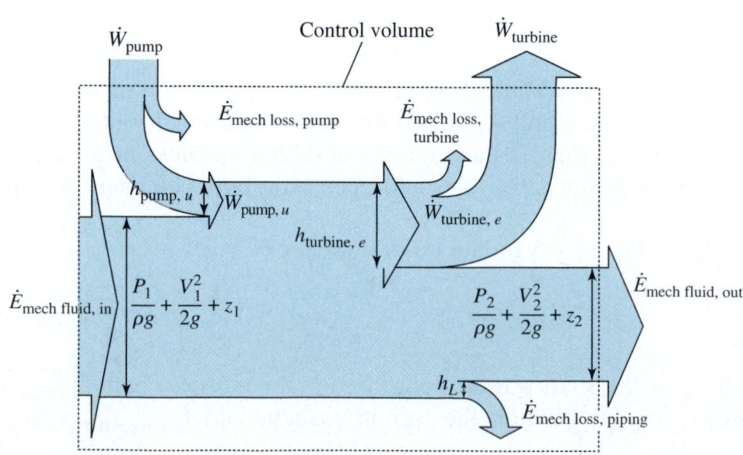

FIGURE 12–34
Mechanical energy flow chart for a fluid flow system that involves a pump and a turbine. Vertical dimensions show each energy term expressed as an equivalent column height of fluid, i.e., *head*, corresponding to each term of Eq. 12–47.

Special Case: Incompressible Flow with No Mechanical Work Devices and Negligible Friction

When piping losses are negligible, there is negligible dissipation of mechanical energy into thermal energy, and thus $h_L = e_{\text{mech loss, piping}}/g \cong 0$. Also, $h_{\text{pump, }u} = h_{\text{turbine, }e} = 0$ when there are no mechanical work devices such as fans, pumps, or turbines. Then Eq. 12–47 reduces to

$$\frac{P_1}{\rho g} + \frac{V_1^2}{2g} + z_1 = \frac{P_2}{\rho g} + \frac{V_2^2}{2g} + z_2 \quad \text{or} \quad \frac{P}{\rho g} + \frac{V^2}{2g} + z = \text{constant} \qquad \textbf{(12–48)}$$

which is the **Bernoulli equation** derived earlier using Newton's second law of motion. Thus, the Bernoulli equation can be thought of as a *degenerate* form of the energy equation.

Kinetic Energy Correction Factor, α

The average flow velocity V_{avg} was defined such that the relation $\rho V_{\text{avg}}A$ gives the actual mass flow rate. Therefore, there is no such thing as a correction factor for mass flow rate. However, as Gaspard Coriolis (1792–1843) showed, the kinetic energy of a fluid stream obtained from $V^2/2$ is not the same as the actual kinetic energy of the fluid stream since the square of a sum is not equal to the sum of the squares of its components (Fig. 12–35). This error can be corrected by replacing the kinetic energy terms $V^2/2$ in the energy equation by $\alpha V_{\text{avg}}^2/2$, where α is the **kinetic energy correction factor**. By using equations for the variation of velocity with the radial distance, it can be shown that the correction factor is 2.0 for fully developed laminar pipe flow, and it ranges between 1.04 and 1.11 for fully developed turbulent flow in a round pipe.

The kinetic energy correction factors are often ignored (i.e., α is set equal to 1) in an elementary analysis since (1) most flows encountered in practice are turbulent, for which the correction factor is near unity, and (2) the kinetic energy terms are often small relative to the other terms in the energy equation, and multiplying them by a factor less than 2.0 does not make much difference. When the velocity and thus the kinetic energy are high, the flow turns turbulent, and a unity correction factor is more appropriate. However, you should keep in mind that you may encounter some situations for which these factors *are* significant, especially when the flow is laminar. Therefore, we recommend that you always include the kinetic energy correction factor when analyzing fluid flow problems. When the kinetic energy correction factors are included, the energy equations for *steady incompressible flow* (Eqs. 12–46 and 12–47) become

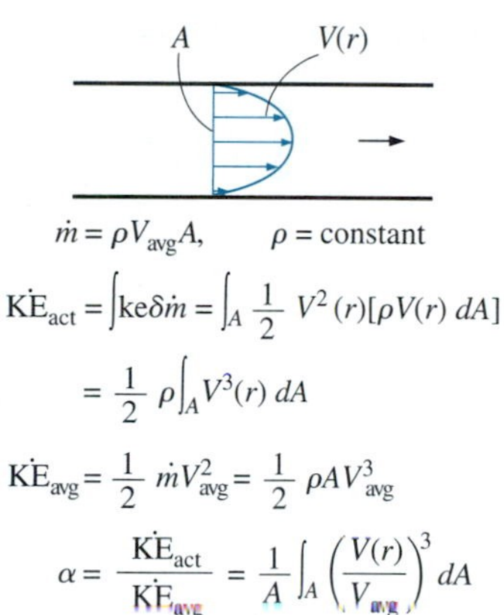

FIGURE 12–35
The determination of the *kinetic energy correction factor* using the actual velocity distribution $V(r)$ and the average velocity V_{avg} at a cross section.

$$\dot{m}\left(\frac{P_1}{\rho} + \alpha_1\frac{V_1^2}{2} + gz_1\right) + \dot{W}_{\text{pump}} = \dot{m}\left(\frac{P_2}{\rho} + \alpha_2\frac{V_2^2}{2} + gz_2\right) + \dot{W}_{\text{turbine}} + \dot{E}_{\text{mech, loss}} \qquad \textbf{(12–49)}$$

$$\frac{P_1}{\rho g} + \alpha_1\frac{V_1^2}{2g} + z_1 + h_{\text{pump, }u} = \frac{P_2}{\rho g} + \alpha_2\frac{V_2^2}{2g} + z_2 + h_{\text{turbine, }e} + h_L \qquad \textbf{(12–50)}$$

If the flow at an inlet or outlet is fully developed turbulent pipe flow, we recommend using $\alpha = 1.05$ as a reasonable estimate of the correction factor. This leads to a more conservative estimate of head loss, and it does not take much additional effort to include α in the equations.

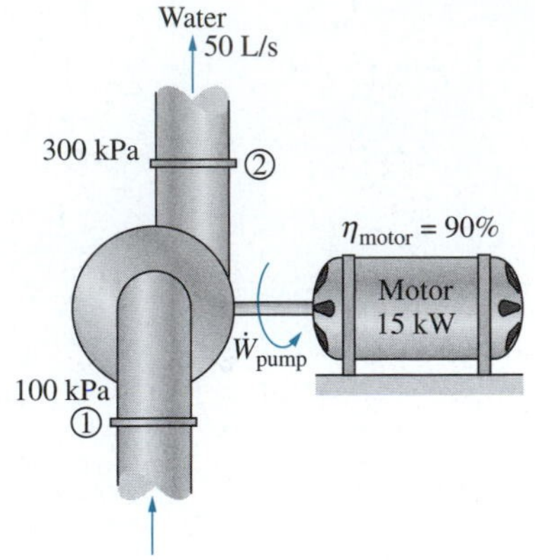

FIGURE 12–36
Schematic for Example 12–6.

EXAMPLE 12–6 Pumping Power and Frictional Heating in a Pump

The pump of a water distribution system is powered by a 15-kW electric motor whose efficiency is 90 percent (Fig. 12–36). The water flow rate through the pump is 50 L/s. The diameters of the inlet and outlet pipes are the same, and the elevation difference across the pump is negligible. If the absolute pressures at the inlet and outlet of the pump are measured to be 100 kPa and 300 kPa, respectively, determine (*a*) the mechanical efficiency of the pump and (*b*) the temperature rise of water as it flows through the pump due to the mechanical inefficiency.

Solution The pressures across a pump are measured. The mechanical efficiency of the pump and the temperature rise of water are to be determined.

Assumptions **1** The flow is steady and incompressible. **2** The pump is driven by an external motor so that the heat generated by the motor is dissipated to the atmosphere. **3** The elevation difference between the inlet and outlet of the pump is negligible, $z_1 \cong z_2$. **4** The inlet and outlet diameters are the same and thus the average inlet and outlet velocities are equal, $V_1 = V_2$. **5** The kinetic energy correction factors are equal, $\alpha_1 = \alpha_2$.

Properties We take the density of water to be 1 kg/L = 1000 kg/m^3 and its specific heat to be 4.18 kJ/kg · °C.

Analysis (*a*) The mass flow rate of water through the pump is

$$\dot{m} = \rho\dot{V} = (1 \text{ kg/L})(50 \text{ L/s}) = 50 \text{ kg/s}$$

The motor draws 15 kW of power and is 90 percent efficient. Thus the mechanical (shaft) power it delivers to the pump is

$$\dot{W}_{\text{pump, shaft}} = \eta_{\text{motor}}\dot{W}_{\text{electric}} = (0.90)(15 \text{ kW}) = 13.5 \text{ kW}$$

To determine the mechanical efficiency of the pump, we need to know the increase in the mechanical energy of the fluid as it flows through the pump, which is

$$\Delta\dot{E}_{\text{mech, fluid}} = \dot{E}_{\text{mech, out}} - \dot{E}_{\text{mech, in}} = \dot{m}\left(\frac{P_2}{\rho} + \alpha_2\frac{V_2^2}{2} + gz_2\right) - \dot{m}\left(\frac{P_1}{\rho} + \alpha_1\frac{V_1^2}{2} + gz_1\right)$$

Simplifying it for this case and substituting the given values,

$$\Delta\dot{E}_{\text{mech, fluid}} = \dot{m}\left(\frac{P_2 - P_1}{\rho}\right) = (50 \text{ kg/s})\left(\frac{(300 - 100) \text{ kPa}}{1000 \text{ kg/m}^3}\right)\left(\frac{1 \text{ kJ}}{1 \text{ kPa} \cdot \text{m}^3}\right) = 10.0 \text{ kW}$$

Then the mechanical efficiency of the pump becomes

$$\eta_{\text{pump}} = \frac{\dot{W}_{\text{pump, }u}}{\dot{W}_{\text{pump, shaft}}} = \frac{\Delta\dot{E}_{\text{mech, fluid}}}{\dot{W}_{\text{pump, shaft}}} = \frac{10.0 \text{ kW}}{13.5 \text{ kW}} = \mathbf{0.741} \quad \text{or} \quad \mathbf{74.1\%}$$

(*b*) Of the 13.5-kW mechanical power supplied by the pump, only 10.0 kW is imparted to the fluid as mechanical energy. The remaining 3.5 kW is converted to thermal energy due to frictional effects, and this "lost" mechanical energy manifests itself as a heating effect in the fluid,

$$\dot{E}_{\text{mech, loss}} = \dot{W}_{\text{pump, shaft}} - \Delta\dot{E}_{\text{mech, fluid}} = 13.5 - 10.0 = 3.5 \text{ kW}$$

The temperature rise of water due to this mechanical inefficiency is determined from the thermal energy balance, $\dot{E}_{\text{mech, loss}} = \dot{m}(u_2 - u_1) = \dot{m}c\Delta T$. Solving for ΔT,

$$\Delta T = \frac{\dot{E}_{\text{mech, loss}}}{\dot{m}c} = \frac{3.5 \text{ kW}}{(50 \text{ kg/s})(4.18 \text{ kJ/kg} \cdot {}^\circ\text{C})} = \mathbf{0.017^\circ C}$$

Therefore, the water experiences a temperature rise of 0.017°C due to mechanical inefficiency, which is very small, as it flows through the pump.

Discussion In an actual application, the temperature rise of water will probably be less since part of the heat generated will be transferred to the casing of the pump and from the casing to the surrounding air. If the entire pump motor were submerged in water, then the 3.5 kW dissipated to the air due to motor inefficiency would also be transferred to the surrounding water as heat. This would cause the water temperature to rise more.

EXAMPLE 12–7 Hydroelectric Power Generation from a Dam

In a hydroelectric power plant, 100 m³/s of water flows from an elevation of 120 m to a turbine, where electric power is generated (Fig. 12–37). The total irreversible head loss in the piping system from point 1 to point 2 (excluding the turbine unit) is determined to be 35 m. If the overall efficiency of the turbine–generator is 80 percent, estimate the electric power output.

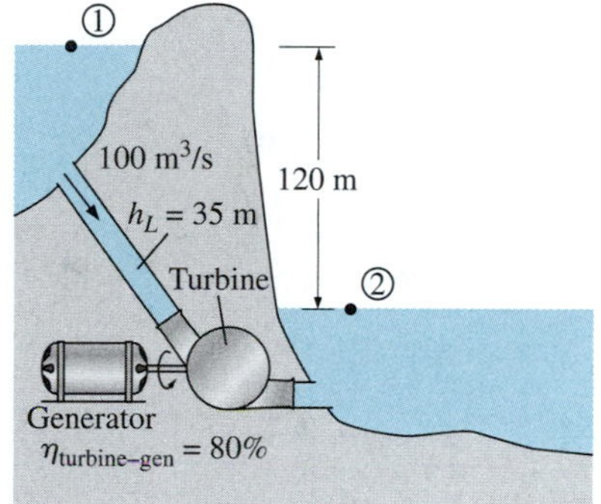

FIGURE 12–37
Schematic for Example 12–7.

Solution The available head, flow rate, head loss, and efficiency of a hydroelectric turbine are given. The electric power output is to be determined.

Assumptions **1** The flow is steady and incompressible. **2** Water levels at the reservoir and the discharge site remain constant.

Properties We take the density of water to be 1000 kg/m³.

Analysis The mass flow rate of water through the turbine is

$$\dot{m} = \rho \dot{V} = (1000 \text{ kg/m}^3)(100 \text{ m}^3\text{/s}) = 10^5 \text{ kg/s}$$

We take point 2 as the reference level, and thus $z_2 = 0$. Also, both points 1 and 2 are open to the atmosphere ($P_1 = P_2 = P_{\text{atm}}$) and the flow velocities are negligible at both points ($V_1 = V_2 = 0$). Then the energy equation for steady, incompressible flow reduces to

$$\frac{P_1}{\rho g} + \alpha_1 \frac{V_1^2}{2g} + z_1 + h_{\text{pump, }u} = \frac{P_2}{\rho g} + \alpha_2 \frac{V_2^2}{2g} + z_2 + h_{\text{turbine, }e} + h_L$$

or

$$h_{\text{turbine, }e} = z_1 - h_L$$

Substituting, the extracted turbine head and the corresponding turbine power are

$$h_{\text{turbine, }e} = z_1 - h_L = 120 - 35 = 85 \text{ m}$$

$$\dot{W}_{\text{turbine, }e} = \dot{m} g h_{\text{turbine, }e} = (10^5 \text{ kg/s})(9.81 \text{ m/s}^2)(85 \text{ m})\left(\frac{1 \text{ kJ/kg}}{1000 \text{ m}^2\text{/s}^2}\right) = 83{,}400 \text{ kW}$$

Therefore, a perfect turbine–generator would generate 83,400 kW of electricity from this resource. The electric power generated by the actual unit is

$$\dot{W}_{\text{electric}} = \eta_{\text{turbine–gen}} \dot{W}_{\text{turbine, }e} = (0.80)(83.4 \text{ MW}) = \mathbf{66.7 \text{ MW}}$$

Discussion Note that the power generation would increase by almost 1 MW for each percentage point improvement in the efficiency of the turbine–generator unit.

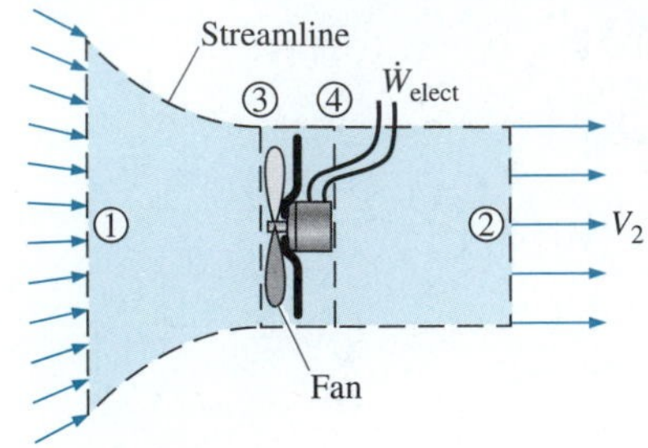

FIGURE 12–38
Schematic for Example 12–8.

EXAMPLE 12–8 Fan Selection for Air Cooling of a Computer

A fan is to be selected to cool a computer case whose dimensions are 12 cm × 40 cm × 40 cm (Fig. 12–38). Half of the volume in the case is expected to be filled with components and the other half to be air space. A 5-cm-diameter hole is available at the back of the case for the installation of the fan that is to replace the air in the void spaces of the case once every second. Small low-power fan–motor combined units are available in the market and their efficiency is estimated to be 30 percent. Determine (*a*) the wattage of the fan–motor unit to be purchased and (*b*) the pressure difference across the fan. Take the air density to be 1.20 kg/m^3.

Solution A fan is to cool a computer case by completely replacing the air inside once every second. The power of the fan and the pressure difference across it are to be determined.

Assumptions **1** The flow is steady and incompressible. **2** Losses other than those due to the inefficiency of the fan–motor unit are negligible. **3** The flow at the outlet is fairly uniform except near the center (due to the wake of the fan motor), and the kinetic energy correction factor at the outlet is 1.10.

Properties The density of air is given to be 1.20 kg/m^3.

Analysis (*a*) Noting that half of the volume of the case is occupied by the components, the air volume in the computer case is

$$V = \text{(Void fraction)(Total case volume)}$$
$$= 0.5(12\text{ cm} \times 40\text{ cm} \times 40\text{ cm}) = 9600\text{ cm}^3$$

Therefore, the volume and mass flow rates of air through the case are

$$\dot{V} = \frac{V}{\Delta t} = \frac{9600\text{ cm}^3}{1\text{ s}} = 9600\text{ cm}^3\text{/s} = 9.6 \times 10^{-3}\text{ m}^3\text{/s}$$

$$\dot{m} = \rho\dot{V} = (1.20\text{ kg/m}^3)(9.6 \times 10^{-3}\text{ m}^3\text{/s}) = 0.0115\text{ kg/s}$$

The cross-sectional area of the opening in the case and the average air velocity through the outlet are

$$A = \frac{\pi D^2}{4} = \frac{\pi(0.05\text{ m})^2}{4} = 1.96 \times 10^{-3}\text{ m}^2$$

$$V = \frac{\dot{V}}{A} = \frac{9.6 \times 10^{-3}\text{ m}^3\text{/s}}{1.96 \times 10^{-3}\text{ m}^2} = 4.90\text{ m/s}$$

We draw the control volume around the fan such that both the inlet and the outlet are at atmospheric pressure ($P_1 = P_2 = P_{atm}$), as shown in Fig. 12–38, and the inlet section 1 is large and far from the fan so that the flow velocity at the inlet section is negligible ($V_1 \cong 0$). Noting that $z_1 = z_2$ and frictional losses in flow are disregarded, the mechanical losses consist of fan losses only and the energy equation (Eq. 12–49) simplifies to

$$\dot{m}\left(\cancel{\frac{P_1}{\rho}} + \alpha_1\frac{V_1^2}{2}\stackrel{\nearrow 0}{} + \cancel{gz_1}\right) + \dot{W}_{fan} = \dot{m}\left(\cancel{\frac{P_2}{\rho}} + \alpha_2\frac{V_2^2}{2} + \cancel{gz_2}\right) + \dot{W}_{turbine}\stackrel{\nearrow 0}{} + \dot{E}_{mech\ loss,\ fan}$$

Solving for $\dot{W}_{fan} - \dot{E}_{mech\ loss,\ fan} = \dot{W}_{fan,\ u}$ and substituting,

$$\dot{W}_{fan,\ u} = \dot{m}\alpha_2\frac{V_2^2}{2} = (0.0115\text{ kg/s})(1.10)\frac{(4.90\text{ m/s})^2}{2}\left(\frac{1\text{ N}}{1\text{ kg}\cdot\text{m/s}^2}\right) = 0.152\text{ W}$$

Then the required electric power input to the fan is determined to be

$$\dot{W}_{\text{elect}} = \frac{\dot{W}_{\text{fan, }u}}{\eta_{\text{fan–motor}}} = \frac{0.152\ \text{W}}{0.3} = \mathbf{0.506\ W}$$

Therefore, a fan–motor rated at about a half watt is adequate for this job.
(*b*) To determine the pressure difference across the fan unit, we take points 3 and 4 to be on the two sides of the fan on a horizontal line. This time again $z_3 = z_4$ and $V_3 = V_4$ since the fan is a narrow cross section, and the energy equation reduces to

$$\dot{m}\frac{P_3}{\rho} + \dot{W}_{\text{fan}} = \dot{m}\frac{P_4}{\rho} + \dot{E}_{\text{mech loss, fan}} \quad \rightarrow \quad \dot{W}_{\text{fan, }u} = \dot{m}\frac{P_4 - P_3}{\rho}$$

Solving for $P_4 - P_3$ and substituting,

$$P_4 - P_3 = \frac{\rho \dot{W}_{\text{fan, }u}}{\dot{m}} = \frac{(1.2\ \text{kg/m}^3)(0.152\ \text{W})}{0.0115\ \text{kg/s}}\left(\frac{1\ \text{Pa}\cdot\text{m}^3}{1\ \text{Ws}}\right) = \mathbf{15.8\ Pa}$$

Therefore, the pressure rise across the fan is 15.8 Pa.

Discussion The efficiency of the fan–motor unit is given to be 30 percent, which means 30 percent of the electric power $\dot{W}_{\text{electric}}$ consumed by the unit is converted to useful mechanical energy while the rest (70 percent) is "lost" and converted to thermal energy. Also, a more powerful fan is required in an actual system to overcome frictional losses inside the computer case. Note that if we had ignored the kinetic energy correction factor at the outlet, the required electrical power and pressure rise would have been 10 percent lower in this case (0.460 W and 14.4 Pa, respectively).

SUMMARY

This chapter deals with the Bernoulli and energy equations and their applications.

Mechanical energy is the form of energy associated with the velocity, elevation, and pressure of the fluid, and it can be converted to mechanical work completely and directly by an ideal mechanical device. The efficiencies of various devices are defined as

$$\eta_{\text{pump}} = \frac{\Delta\dot{E}_{\text{mech, fluid}}}{\dot{W}_{\text{shaft, in}}} = \frac{\dot{W}_{\text{pump, }u}}{\dot{W}_{\text{pump}}}$$

$$\eta_{\text{turbine}} = \frac{\dot{W}_{\text{shaft, out}}}{|\Delta\dot{E}_{\text{mech, fluid}}|} = \frac{\dot{W}_{\text{turbine}}}{\dot{W}_{\text{turbine, }e}}$$

$$\eta_{\text{motor}} = \frac{\text{Mechanical power output}}{\text{Electric power input}} = \frac{\dot{W}_{\text{shaft, out}}}{\dot{W}_{\text{elect, in}}}$$

$$\eta_{\text{generator}} = \frac{\text{Electric power output}}{\text{Mechanical power input}} = \frac{\dot{W}_{\text{elect, out}}}{\dot{W}_{\text{shaft, in}}}$$

$$\eta_{\text{pump–motor}} = \eta_{\text{pump}}\eta_{\text{motor}} = \frac{\Delta\dot{E}_{\text{mech, fluid}}}{\dot{W}_{\text{elect, in}}} = \frac{\dot{W}_{\text{pump, }u}}{\dot{W}_{\text{elect, in}}}$$

$$\eta_{\text{turbine–gen}} = \eta_{\text{turbine}}\eta_{\text{generator}} = \frac{\dot{W}_{\text{elect, out}}}{|\Delta\dot{E}_{\text{mech, fluid}}|} = \frac{\dot{W}_{\text{elect, out}}}{\dot{W}_{\text{turbine, }e}}$$

The *Bernoulli equation* is a relation between pressure, velocity, and elevation in steady, incompressible flow, and is expressed along a streamline and in regions where net viscous forces are negligible as

$$\frac{P}{\rho} + \frac{V^2}{2} + gz = \text{constant}$$

The Bernoulli equation can also be considered as an expression of mechanical energy balance, stated as: *The sum of the kinetic, potential, and flow energies of a fluid particle is constant along a streamline during steady flow when the compressibility and frictional effects are negligible.* Multiplying the Bernoulli equation by density gives

$$P + \rho\frac{V^2}{2} + \rho gz = \text{constant}$$

where P is the *static pressure*, which represents the actual pressure of the fluid; $\rho V^2/2$ is the *dynamic pressure*, which represents the pressure rise when the fluid in motion is brought to a stop; and ρgz is the *hydrostatic pressure*, which accounts for the effects of fluid weight on pressure. The sum of the static, dynamic, and hydrostatic pressures is called the *total pressure*. The Bernoulli equation states that *the total pressure along a streamline is constant*. The sum of the static and dynamic pressures is called the *stagnation pressure*, which represents the pressure at a point where the fluid is brought to a complete stop in a frictionless manner. The Bernoulli equation can also be represented in terms of "heads" by dividing each term by g,

$$\frac{P}{\rho g} + \frac{V^2}{2g} + z = H = \text{constant}$$

where $P/\rho g$ is the *pressure head*, which represents the height of a fluid column that produces the static pressure P; $V^2/2g$ is the *velocity head*, which represents the elevation needed for a fluid to reach the velocity V during frictionless free fall; and z is the *elevation head*, which represents the potential energy of the fluid. Also, H is the *total head* for the flow. The curve that represents the sum of the static pressure and the elevation heads, $P/\rho g + z$, is called the *hydraulic grade line* (HGL), and the curve that represents the total head of the fluid, $P/\rho g + V^2/2g + z$, is called the *energy grade line* (EGL).

The *energy equation* for steady, incompressible flow is expressed as

$$\frac{P_1}{\rho g} + \alpha_1 \frac{V_1^2}{2g} + z_1 + h_{\text{pump, }u} = \frac{P_2}{\rho g} + \alpha_2 \frac{V_2^2}{2g} + z_2 + h_{\text{turbine, }e} + h_L$$

where

$$h_{\text{pump, }u} = \frac{w_{\text{pump, }u}}{g} = \frac{\dot{W}_{\text{pump, }u}}{\dot{m}g} = \frac{\eta_{\text{pump}}\dot{W}_{\text{pump}}}{\dot{m}g}$$

$$h_{\text{turbine, }e} = \frac{w_{\text{turbine, }e}}{g} = \frac{\dot{W}_{\text{turbine, }e}}{\dot{m}g} = \frac{\dot{W}_{\text{turbine}}}{\eta_{\text{turbine}}\dot{m}g}$$

$$h_L = \frac{e_{\text{mech loss, piping}}}{g} = \frac{\dot{E}_{\text{mech loss, piping}}}{\dot{m}g}$$

$$e_{\text{mech, loss}} = u_2 - u_1 - q_{\text{net in}}$$

The Bernoulli and energy equations are two of the most fundamental relations in fluid mechanics.

REFERENCES AND SUGGESTED READING

1. R. C. Dorf, ed. in chief. *The Engineering Handbook*. Boca Raton, FL: CRC Press, 1995.

2. R. L. Panton. *Incompressible Flow*, 2nd ed. New York: Wiley, 1996.

3. M. Van Dyke. *An Album of Fluid Motion*. Stanford, CA: The Parabolic Press, 1982.

PROBLEMS*

Mechanical Energy and Efficiency

12–1C What is mechanical energy? How does it differ from thermal energy? What are the forms of mechanical energy of a fluid stream?

12–2C What is mechanical efficiency? What does a mechanical efficiency of 100 percent mean for a hydraulic turbine?

12–3C How is the combined pump–motor efficiency of a pump and motor system defined? Can the combined pump–motor efficiency be greater than either the pump or the motor efficiency?

12–4C Define turbine efficiency, generator efficiency, and combined turbine–generator efficiency.

12–5 Consider a river flowing toward a lake at an average velocity of 3 m/s at a rate of 500 m^3/s at a location 90 m above the lake surface. Determine the total mechanical energy of the river water per unit mass and the power generation potential of the entire river at that location. *Answer:* 444 MW

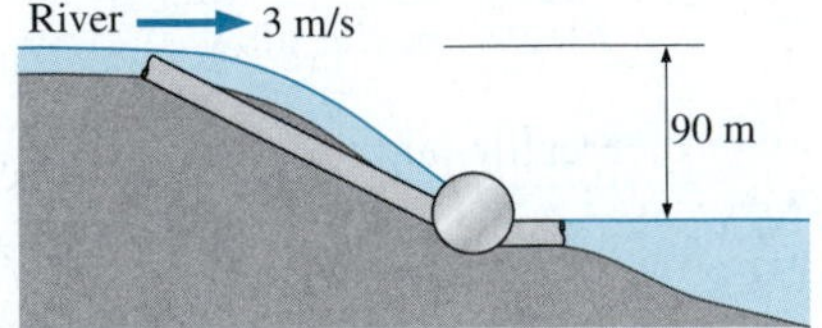

FIGURE P12–5

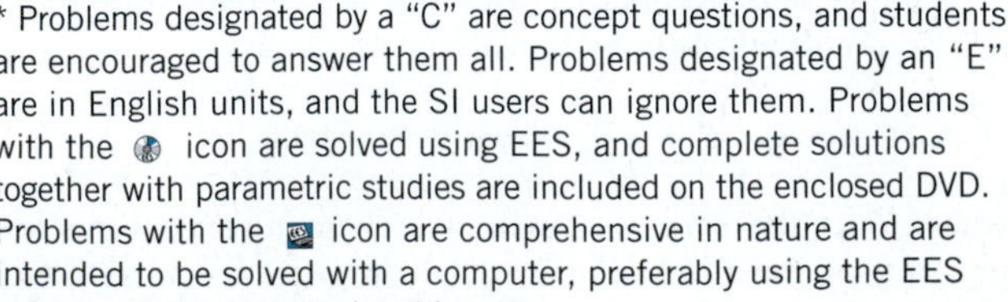
* Problems designated by a "C" are concept questions, and students are encouraged to answer them all. Problems designated by an "E" are in English units, and the SI users can ignore them. Problems with the icon are solved using EES, and complete solutions together with parametric studies are included on the enclosed DVD. Problems with the icon are comprehensive in nature and are intended to be solved with a computer, preferably using the EES software that accompanies this text.

12–6 Electric power is to be generated by installing a hydraulic turbine–generator at a site 70 m below the free surface of a large water reservoir that can supply water at a rate of 1500 kg/s steadily. If the mechanical power output of the turbine is 800 kW and the electric power generation is 750 kW, determine the turbine efficiency and the combined turbine–generator efficiency of this plant. Neglect losses in the pipes.

12–7 At a certain location, wind is blowing steadily at 12 m/s. Determine the mechanical energy of air per unit mass and the power generation potential of a wind turbine with 50-m-diameter blades at that location. Also determine the actual electric power generation assuming an overall efficiency of 30 percent. Take the air density to be 1.25 kg/m^3.

12–8 Reconsider Prob. 12–7. Using EES (or other) software, investigate the effect of wind velocity and the blade span diameter on wind power generation. Let the velocity vary from 5 to 20 m/s in increments of 5 m/s, and the diameter to vary from 20 to 80 m in increments of 20 m. Tabulate the results, and discuss their significance.

12–9E A differential thermocouple with sensors at the inlet and exit of a pump indicates that the temperature of water rises 0.072°F as it flows through the pump at a rate of 1.5 ft^3/s. If the shaft power input to the pump is 27 hp, determine the mechanical efficiency of the pump. *Answer*: 64.7 percent

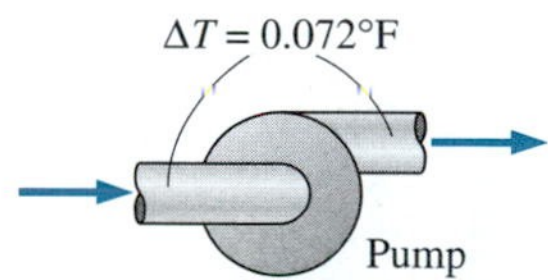

FIGURE P12–9E

12–10 Water is pumped from a lake to a storage tank 20 m above at a rate of 70 L/s while consuming 20.4 kW of electric power. Disregarding any frictional losses in the pipes and any changes in kinetic energy, determine (*a*) the overall efficiency of the pump–motor unit and (*b*) the pressure difference between the inlet and the exit of the pump.

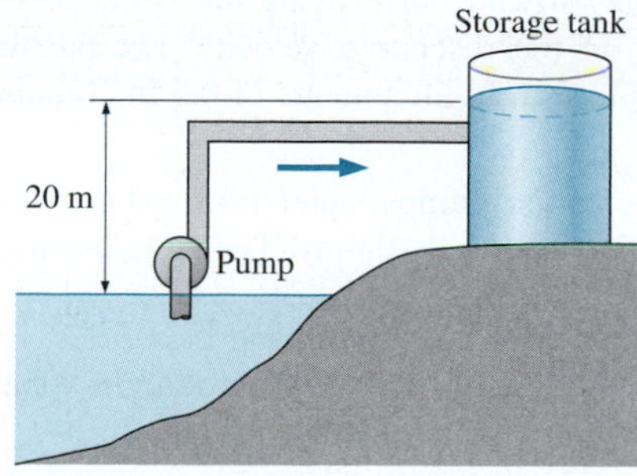

FIGURE P12–10

Bernoulli Equation

12–11C What is streamwise acceleration? How does it differ from normal acceleration? Can a fluid particle accelerate in steady flow?

12–12C Express the Bernoulli equation in three different ways using (*a*) energies, (*b*) pressures, and (*c*) heads.

12–13C What are the three major assumptions used in the derivation of the Bernoulli equation?

12–14C Define static, dynamic, and hydrostatic pressure. Under what conditions is their sum constant for a flow stream?

12–15C What is stagnation pressure? Explain how it can be measured.

12–16C Define pressure head, velocity head, and elevation head for a fluid stream and express them for a fluid stream whose pressure is P, velocity is V, and elevation is z.

12–17C What is the hydraulic grade line? How does it differ from the energy grade line? Under what conditions do both lines coincide with the free surface of a liquid?

12–18C How is the location of the hydraulic grade line determined for open-channel flow? How is it determined at the outlet of a pipe discharging to the atmosphere?

12–19C The water level of a tank on a building roof is 20 m above the ground. A hose leads from the tank bottom to the ground. The end of the hose has a nozzle, which is pointed straight up. What is the maximum height to which the water could rise? What factors would reduce this height?

12–20C In a certain application, a siphon must go over a high wall. Can water or oil with a specific gravity of 0.8 go over a higher wall? Why?

12–21C Explain how and why a siphon works. Someone proposes siphoning cold water over a 7-m-high wall. Is this feasible? Explain.

12–22C A student siphons water over a 8.5-m-high wall at sea level. She then climbs to the summit of Mount Shasta (elevation 4390 m, P_{atm} = 58.5 kPa) and attempts the same experiment. Comment on her prospects for success.

12–23C A glass manometer with oil as the working fluid is connected to an air duct as shown in Fig. P12–23C. Will the oil levels in the manometer be as in Fig. P12–23C*a* or *b*? Explain. What would your response be if the flow direction is reversed?

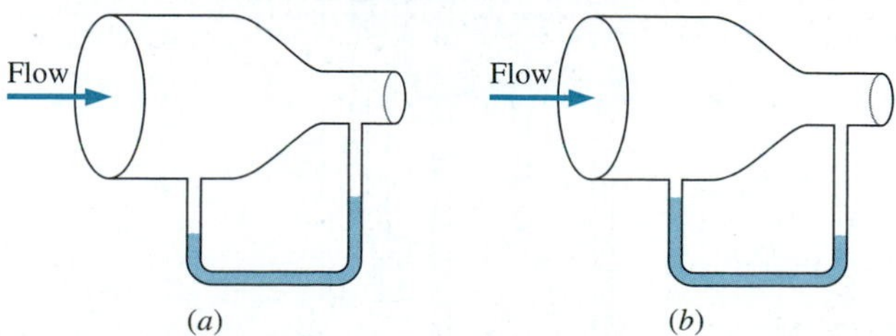

FIGURE P12–23C

12–24C The velocity of a fluid flowing in a pipe is to be measured by two different Pitot-type mercury manometers shown in Fig. P12–24C. Would you expect both manometers to predict the same velocity for flowing water? If not, which would be more accurate? Explain. What would your response be if air were flowing in the pipe instead of water?

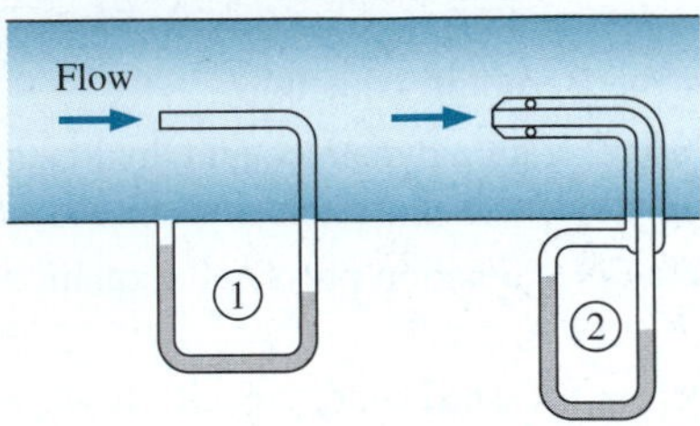

FIGURE P12–24C

12–25 In cold climates, water pipes may freeze and burst if proper precautions are not taken. In such an occurrence, the exposed part of a pipe on the ground ruptures, and water shoots up to 34 m. Estimate the gage pressure of water in the pipe. State your assumptions and discuss if the actual pressure is more or less than the value you predicted.

12–26 A Pitot-static probe is used to measure the velocity of an aircraft flying at 3000 m. If the differential pressure reading is 3 kPa, determine the velocity of the aircraft.

12–27 While traveling on a dirt road, the bottom of a car hits a sharp rock and a small hole develops at the bottom of its gas tank. If the height of the gasoline in the tank is 30 cm, determine the initial velocity of the gasoline at the hole. Discuss how the velocity will change with time and how the flow will be affected if the lid of the tank is closed tightly. *Answer:* 2.43 m/s

12–28E The drinking water needs of an office are met by large water bottles. One end of a 0.25-in-diameter plastic hose is inserted into the bottle placed on a high stand, while the other end with an on/off valve is maintained 2 ft below the bottom of the bottle. If the water level in the bottle is 1.5 ft when it is full, determine how long it will take at the minimum to fill an 8-oz glass (= 0.00835 ft^3) (*a*) when the bottle is first opened and (*b*) when the bottle is almost empty. Neglect frictional losses.

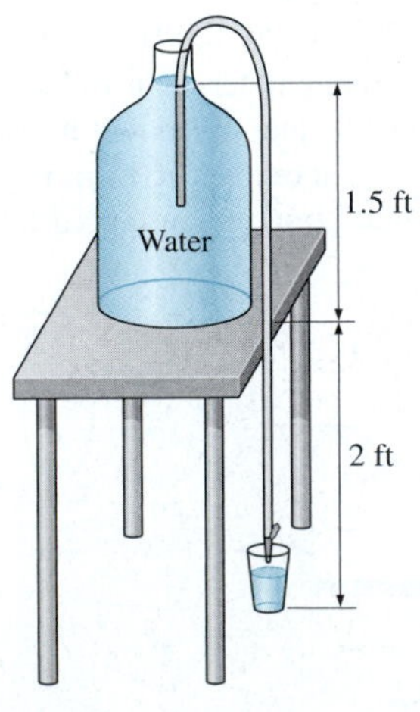

FIGURE P12–28E

12–29 A piezometer and a Pitot tube are tapped into a 3-cm-diameter horizontal water pipe, and the height of the water columns are measured to be 20 cm in the piezometer and 35 cm in the Pitot tube (both measured from the top surface of the pipe). Determine the velocity at the center of the pipe.

12–30 The diameter of a cylindrical water tank is D_o and its height is H. The tank is filled with water, which is open to the atmosphere. An orifice of diameter D with a smooth entrance (i.e., no losses) is open at the bottom. Develop a relation for the time required for the tank (*a*) to empty halfway and (*b*) to empty completely.

12–31 A pressurized tank of water has a 10-cm-diameter orifice at the bottom, where water discharges to the atmosphere. The water level is 3 m above the outlet. The tank air pressure above the water level is 300 kPa (absolute) while the atmospheric pressure is 100 kPa. Neglecting frictional effects, determine the initial discharge rate of water from the tank. *Answer:* 0.168 m^3/s

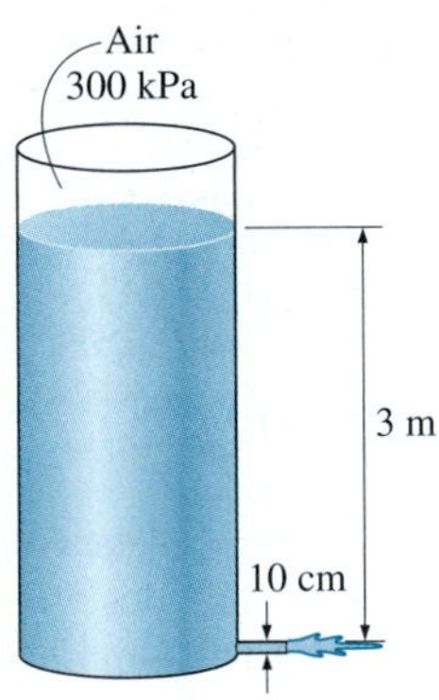

FIGURE P12–31

12–32 Reconsider Prob. 12–31. Using EES (or other) software, investigate the effect of water height in the tank on the discharge velocity. Let the water height vary from 0 to 5 m in increments of 0.5 m. Tabulate and plot the results.

12–33E A siphon pumps water from a large reservoir to a lower tank that is initially empty. The tank also has a rounded orifice 15 ft below the reservoir surface where the water leaves the tank. Both the siphon and the orifice diameters are 2 in. Ignoring frictional losses, determine to what height the water will rise in the tank at equilibrium.

12–34 Water enters a tank of diameter D_T steadily at a mass flow rate of $\dot{m}_{in}$. An orifice at the bottom with diameter D_o

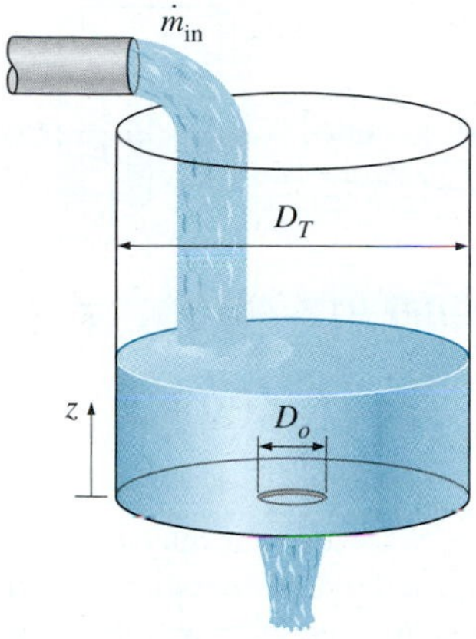

FIGURE P12–34

allows water to escape. The orifice has a rounded entrance, so the frictional losses are negligible. If the tank is initially empty, (*a*) determine the maximum height that the water will reach in the tank and (*b*) obtain a relation for water height z as a function of time.

12–35E Water flows through a horizontal pipe at a rate of 1 gal/s. The pipe consists of two sections of diameters 4 in and 2 in with a smooth reducing section. The pressure difference between the two pipe sections is measured by a mercury manometer. Neglecting frictional effects, determine the differential height of mercury between the two pipe sections. *Answer:* 0.52 in

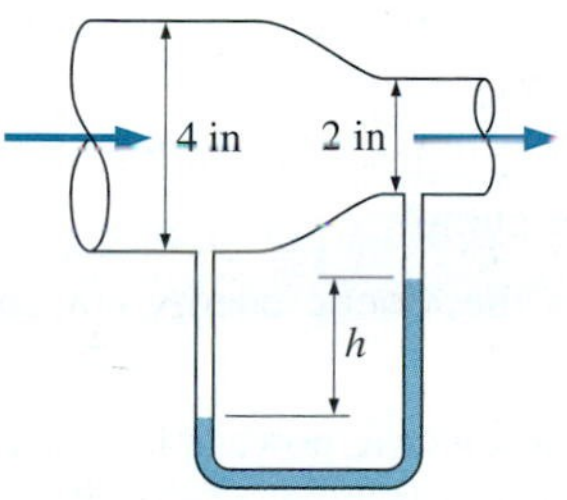

FIGURE P12–35E

12–36 An airplane is flying at an altitude of 12,000 m. Determine the gage pressure at the stagnation point on the nose of the plane if the speed of the plane is 200 km/h. How would you solve this problem if the speed were 1050 km/h? Explain.

12–37 The air velocity in the duct of a heating system is to be measured by a Pitot-static probe inserted into the duct parallel to flow. If the differential height between the water columns connected to the two outlets of the probe is 2.4 cm, determine (*a*) the flow velocity and (*b*) the pressure rise at the tip of the probe. The air temperature and pressure in the duct are 45°C and 98 kPa, respectively.

12–38 The water in a 10-m-diameter, 2-m-high above-ground swimming pool is to be emptied by unplugging a 3-cm-diameter, 25-m-long horizontal pipe attached to the bottom of the pool. Determine the maximum discharge rate of water through the pipe. Also, explain why the actual flow rate will be less.

12–39 Reconsider Prob. 12–38. Determine how long it will take to empty the swimming pool completely. *Answer:* 19.7 h

12–40 Reconsider Prob. 12–39. Using EES (or other) software, investigate the effect of the discharge pipe diameter on the time required to empty the pool completely. Let the diameter vary from 1 to 10 cm in increments of 1 cm. Tabulate and plot the results.

12–41 Air at 110 kPa and 50°C flows upward through a 6-cm-diameter inclined duct at a rate of 45 L/s. The duct diameter is then reduced to 4 cm through a reducer. The pressure change across the reducer is measured by a water manometer. The elevation difference between the two points on the pipe where the two arms of the manometer are attached is 0.20 m. Determine the differential height between the fluid levels of the two arms of the manometer.

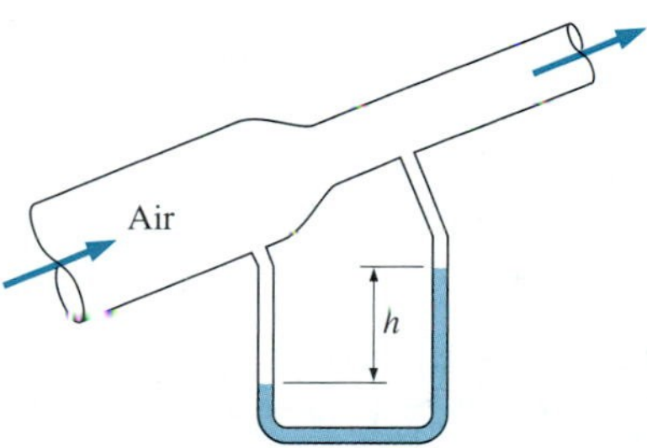

FIGURE P12–41

12–42E Air is flowing through a venturi meter whose diameter is 2.6 in at the entrance part (location 1) and 1.8 in at the throat (location 2). The gage pressure is measured to be 12.2 psia at the entrance and 11.8 psia at the throat. Neglecting frictional effects, show that the volume flow rate can be expressed as

$$\dot{V} = A_2\sqrt{\frac{2(P_1 - P_2)}{\rho(1 - A_2^2/A_1^2)}}$$

and determine the flow rate of air. Take the air density to be 0.075 lbm/ft^3.

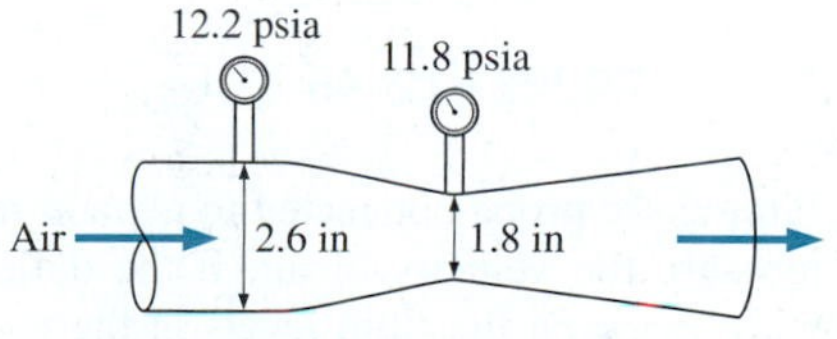

FIGURE P12–42E

12–43 The water pressure in the mains of a city at a particular location is 400 kPa gage. Determine if this main can serve water to neighborhoods that are 50 m above this location.

12–44 A handheld bicycle pump can be used as an atomizer to generate a fine mist of paint or pesticide by forcing air at a high velocity through a small hole and placing a short tube between the liquid reservoir and the high-speed air jet whose low pressure drives the liquid up through the tube. In such an atomizer, the hole diameter is 0.3 cm, the vertical distance between the liquid level in the tube and the hole is 10 cm, and the bore (diameter) and the stroke of the air pump are 5 cm and 20 cm, respectively. If the atmospheric conditions are 20°C and 95 kPa, determine the minimum speed that the piston must be moved in the cylinder during pumping to initiate the atomizing effect. The liquid reservoir is open to the atmosphere.

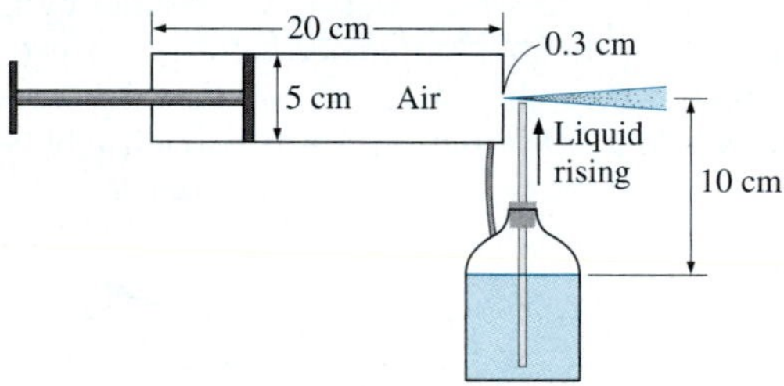

FIGURE P12–44

12–45 The water level in a tank is 20 m above the ground. A hose is connected to the bottom of the tank, and the nozzle at the end of the hose is pointed straight up. The tank cover is airtight, and the air pressure above the water surface is 2 atm gage. The system is at sea level. Determine the maximum height to which the water stream could rise. *Answer:* 40.7 m

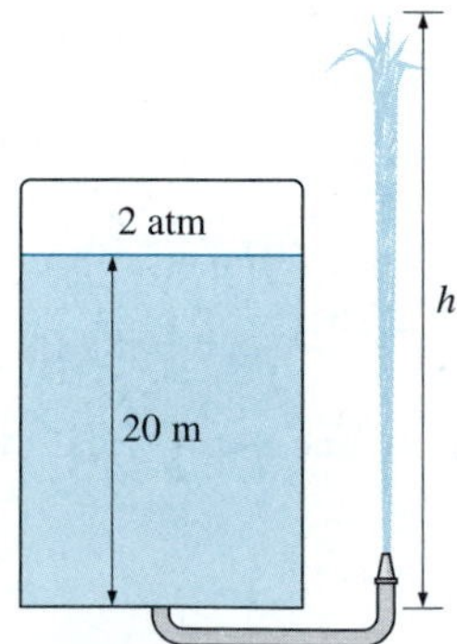

FIGURE P12–45

12–46 A Pitot-static probe connected to a water manometer is used to measure the velocity of air. If the deflection (the vertical distance between the fluid levels in the two arms) is 7.3 cm, determine the air velocity. Take the density of air to be 1.25 kg/m^3.

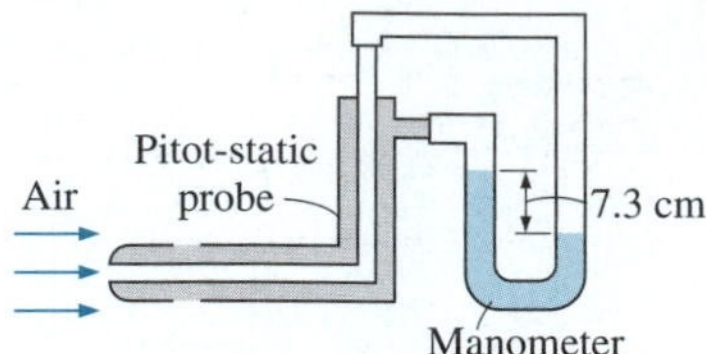

FIGURE P12–46

12–47E The air velocity in a duct is measured by a Pitot-static probe connected to a differential pressure gage. If the air is at 13.4 psia absolute and 70°F and the reading of the differential pressure gage is 0.15 psi, determine the air velocity. *Answer:* 143 ft/s

12–48 In a hydroelectric power plant, water enters the turbine nozzles at 700 kPa absolute with a low velocity. If the nozzle outlets are exposed to atmospheric pressure of 100 kPa, determine the maximum velocity to which water can be accelerated by the nozzles before striking the turbine blades.

Energy Equation

12–49C Consider the steady adiabatic flow of an incompressible fluid. Can the temperature of the fluid decrease during flow? Explain.

12–50C Consider the steady adiabatic flow of an incompressible fluid. If the temperature of the fluid remains constant during flow, is it accurate to say that the frictional effects are negligible?

12–51C What is irreversible head loss? How is it related to the mechanical energy loss?

12–52C What is useful pump head? How is it related to the power input to the pump?

12–53C What is the kinetic energy correction factor? Is it significant?

12–54E In a hydroelectric power plant, water flows from an elevation of 240 ft to a turbine, where electric power is generated. For an overall turbine–generator efficiency of 83 percent, determine the minimum flow rate required to generate 100 kW of electricity. *Answer:* 370 lbm/s

12–55E Reconsider Prob. 12–54E. Determine the flow rate of water if the irreversible head loss of the piping system between the free surfaces of the source and the sink is 36 ft.

12–56 EES A fan is to be selected to ventilate a bathroom whose dimensions are 2 m × 3 m × 3 m. The air velocity is not to exceed 8 m/s to minimize vibration and noise. The combined efficiency of the fan–motor unit to be used can be taken to be 50 percent. If the fan is to replace the entire volume of air in 10 min, determine (*a*) the wattage of the fan–motor unit to be purchased, (*b*) the diameter of the

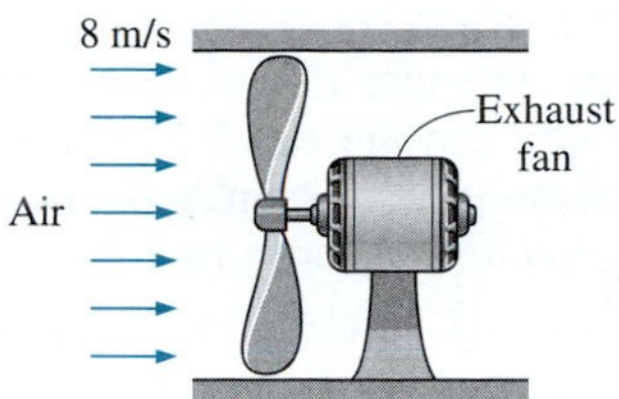

FIGURE P12–56

fan casing, and (*c*) the pressure difference across the fan. Take the air density to be 1.25 kg/m^3 and disregard the effect of the kinetic energy correction factors.

12–57 Water is being pumped from a large lake to a reservoir 25 m above at a rate of 25 L/s by a 10-kW (shaft) pump. If the irreversible head loss of the piping system is 7 m, determine the mechanical efficiency of the pump. *Answer:* 78.5 percent

12–58 Reconsider Prob. 12–57. Using EES (or other) software, investigate the effect of irreversible head loss on the mechanical efficiency of the pump. Let the head loss vary from 0 to 15 m in increments of 1 m. Plot the results, and discuss them.

12–59 A 7-hp (shaft) pump is used to raise water to a 15-m higher elevation. If the mechanical efficiency of the pump is 82 percent, determine the maximum volume flow rate of water.

12–60 Water flows at a rate of 0.035 m^3/s in a horizontal pipe whose diameter is reduced from 15 cm to 8 cm by a reducer. If the pressure at the centerline is measured to be 470 kPa and 440 kPa before and after the reducer, respectively, determine the irreversible head loss in the reducer. Take the kinetic energy correction factors to be 1.05. *Answer:* 0.68 m

12–61 The water level in a tank is 20 m above the ground. A hose is connected to the bottom of the tank, and the nozzle at the end of the hose is pointed straight up. The tank is at sea level, and the water surface is open to the atmosphere. In the line leading from the tank to the nozzle is a pump, which increases the pressure of water. If the water jet rises to a height of 27 m from the ground, determine the minimum pressure rise supplied by the pump to the water line.

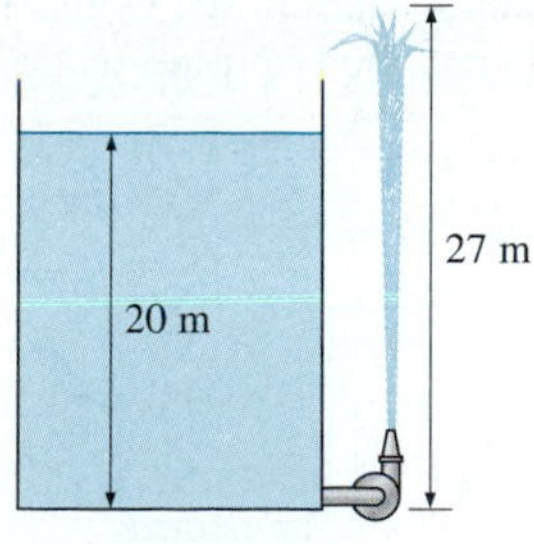

FIGURE P12–61

12–62 A hydraulic turbine has 85 m of head available at a flow rate of 0.25 m^3/s, and its overall turbine–generator efficiency is 78 percent. Determine the electric power output of this turbine.

12–63 The demand for electric power is usually much higher during the day than it is at night, and utility companies often sell power at night at much lower prices to encourage consumers to use the available power generation capacity and to avoid building new expensive power plants that will be used only a short time during peak periods. Utilities are also willing to purchase power produced during the day from private parties at a high price.

Suppose a utility company is selling electric power for \$0.03/kWh at night and is willing to pay \$0.08/kWh for power produced during the day. To take advantage of this opportunity, an entrepreneur is considering building a large reservoir 40 m above the lake level, pumping water from the lake to the reservoir at night using cheap power, and letting the water flow from the reservoir back to the lake during the day, producing power as the pump–motor operates as a turbine–generator during reverse flow. Preliminary analysis shows that a water flow rate of 2 m^3/s can be used in either direction, and the irreversible head loss of the piping system is 4 m. The combined pump–motor and turbine–generator efficiencies are expected to be 75 percent each. Assuming the system operates for 10 h each in the pump and turbine modes during a typical day, determine the potential revenue this pump–turbine system can generate per year.

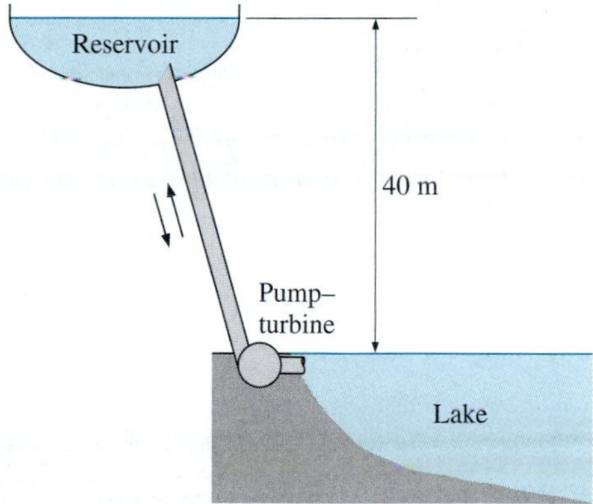

FIGURE P12–63

12–64 Water flows at a rate of 20 L/s through a horizontal pipe whose diameter is constant at 3 cm as shown in Fig. P12–64. The pressure drop across a valve in the pipe is measured to be 2 kPa. Determine the irreversible head loss of the valve, and the useful pumping power needed to overcome the resulting pressure drop. *Answers:* 0.204 m, 40 W

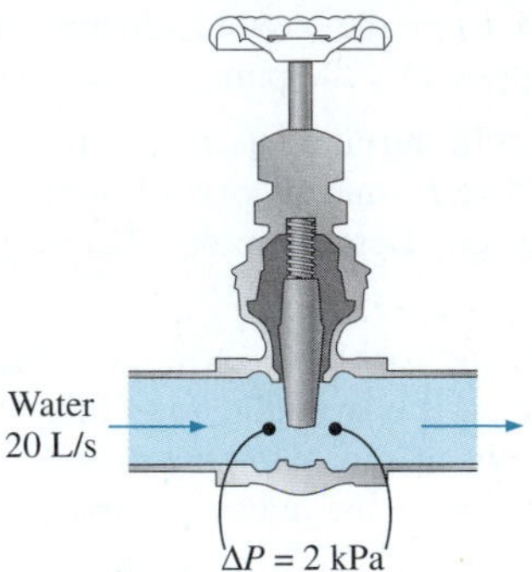

FIGURE P12–64

12–65 Water enters a hydraulic turbine through a 30-cm-diameter pipe at a rate of 0.6 m^3/s and exits through a 25-cm-diameter pipe. The pressure drop in the turbine is measured by a mercury manometer to be 1.2 m. For a combined turbine–generator efficiency of 83 percent, determine the net electric power output. Disregard the effect of the kinetic energy correction factors.

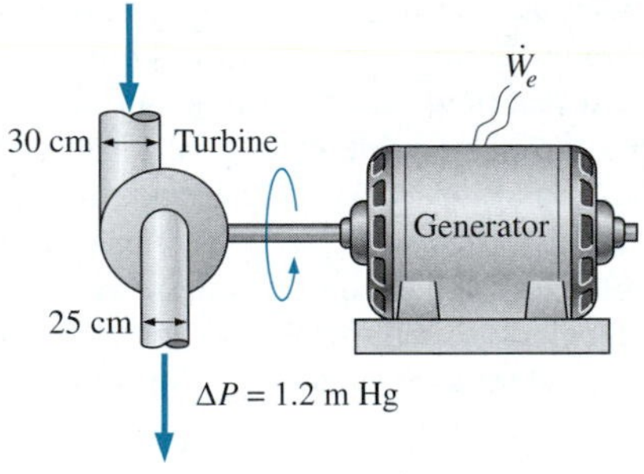

FIGURE P12–65

12–66 The velocity profile for turbulent flow in a circular pipe is usually approximated as $u(r) = u_{max}(1 - r/R)^{1/n}$, where $n = 7$. Determine the kinetic energy correction factor for this flow. *Answer:* 1.06

12–67 An oil pump is drawing 35 kW of electric power while pumping oil with $\rho = 860$ kg/m^3 at a rate of 0.1 m^3/s. The inlet and outlet diameters of the pipe are 8 cm and 12 cm, respectively. If the pressure rise of oil in the pump is measured to be 400 kPa and the motor efficiency is 90 percent, determine the mechanical efficiency of the pump. Take the kinetic energy correction factor to be 1.05.

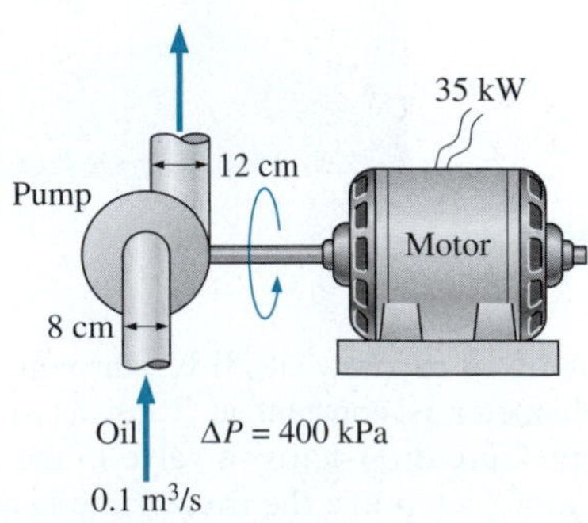

FIGURE P12–67

12–68E A 73-percent efficient 12-hp pump is pumping water from a lake to a nearby pool at a rate of 1.2 ft^3/s through a constant-diameter pipe. The free surface of the pool is 35 ft above that of the lake. Determine the irreversible head loss of the piping system, in ft, and the mechanical power used to overcome it.

12–69 A fireboat is to fight fires at coastal areas by drawing seawater with a density of 1030 kg/m^3 through a 20-cm-diameter pipe at a rate of 0.1 m^3/s and discharging it through a hose nozzle with an exit diameter of 5 cm. The total irreversible head loss of the system is 3 m, and the position of the nozzle is 4 m above sea level. For a pump efficiency of 70 percent, determine the required shaft power input to the pump and the water discharge velocity. *Answers:* 201 kW, 50.9 m/s

FIGURE P12–69

Review Problems

12–70 A pressurized 2-m-diameter tank of water has a 10-cm-diameter orifice at the bottom, where water discharges to the atmosphere. The water level initially is 3 m above the outlet. The tank air pressure above the water level is maintained at 450 kPa absolute and the atmospheric pressure is 100 kPa. Neglecting frictional effects, determine (*a*) how long it will take for half of the water in the tank to be discharged and (*b*) the water level in the tank after 10 s.

12–71 Air flows through a pipe at a rate of 200 L/s. The pipe consists of two sections of diameters 20 cm and 10 cm with a smooth reducing section that connects them. The pressure difference between the two pipe sections is measured by

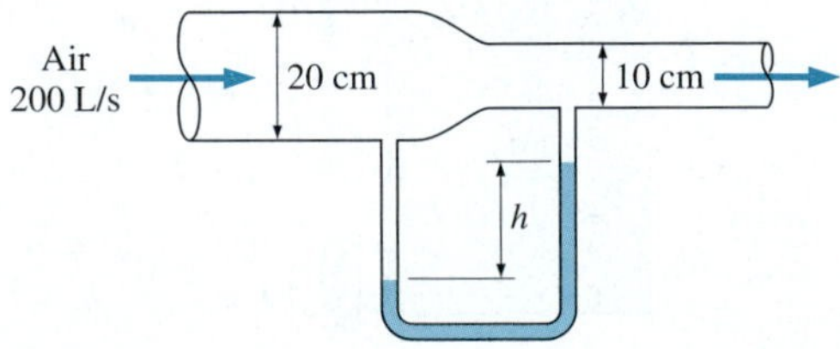

FIGURE P12–71

a water manometer. Neglecting frictional effects, determine the differential height of water between the two pipe sections. Take the air density to be 1.20 kg/m^3. *Answer:* 3.7 cm

12–72 A wind tunnel draws atmospheric air at 20°C and 101.3 kPa by a large fan located near the exit of the tunnel. If the air velocity in the tunnel is 80 m/s, determine the pressure in the tunnel.

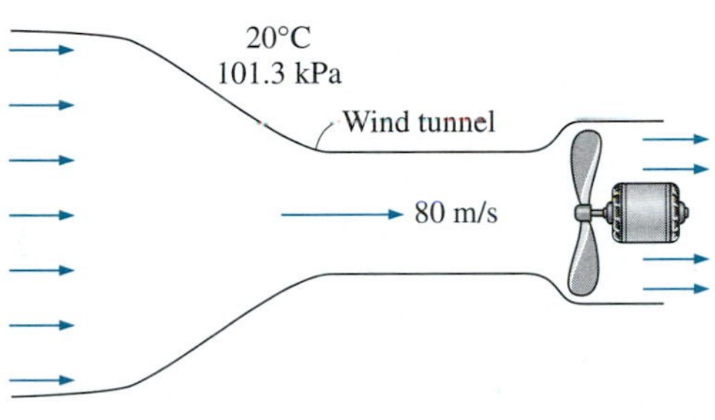

FIGURE P12–72

12–73 Water flows at a rate of 0.025 m^3/s in a horizontal pipe whose diameter increases from 6 to 11 cm by an enlargement section. If the head loss across the enlargement section is 0.45 m and the kinetic energy correction factor at both the inlet and the outlet is 1.05, determine the pressure change.

Design and Essay Problems

12–74 Computer-aided designs, the use of better materials, and better manufacturing techniques have resulted in a tremendous increase in the efficiency of pumps, turbines, and electric motors. Contact several pump, turbine, and motor manufacturers and obtain information about the efficiency of their products. In general, how does efficiency vary with rated power of these devices?

12–75 Using a handheld bicycle pump to generate an air jet, a soda can as the water reservoir, and a straw as the tube, design and build an atomizer. Study the effects of various parameters such as the tube length, the diameter of the exit hole, and the pumping speed on performance.

12–76 Using a flexible drinking straw and a ruler, explain how you would measure the water flow velocity in a river.

12–77 The power generated by a wind turbine is proportional to the cube of the wind velocity. Inspired by the acceleration of a fluid in a nozzle, someone proposes to install a reducer casing to capture the wind energy from a larger area and accelerate it before the wind strikes the turbine blades, as shown in Fig. P12–77. Evaluate if the proposed modification should be given a consideration in the design of new wind turbines.

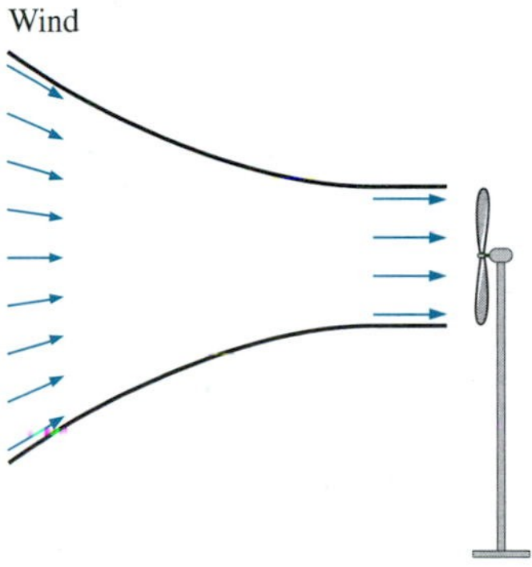

FIGURE P12–77

Chapter 13

MOMENTUM ANALYSIS OF FLOW SYSTEMS

When dealing with engineering problems, it is desirable to obtain fast and accurate solutions at minimal cost. Most engineering problems, including those associated with fluid flow, can be analyzed using one of three basic approaches: differential, experimental, and control volume. In *differential approaches*, the problem is formulated accurately using differential quantities, but the solution of the resulting differential equations is difficult, usually requiring the use of numerical methods with extensive computer codes. *Experimental approaches* complemented with dimensional analysis are highly accurate, but they are typically time-consuming and expensive. The *finite control volume approach* described in this chapter is remarkably fast and simple and usually gives answers that are sufficiently accurate for most engineering purposes. Therefore, despite the approximations involved, the basic finite control volume analysis performed with paper and pencil has always been an indispensable tool for engineers.

In Chap. 12, the control volume energy analysis of fluid flow systems was presented. In this chapter, we present the finite control volume momentum analysis of fluid flow problems. First we give an overview of Newton's laws and the conservation relations for linear and angular momentum. Then using the Reynolds transport theorem, we develop the linear momentum and angular momentum equations for control volumes and use them to determine the forces and torques associated with fluid flow.

Objectives

The objectives of this chapter are to:

- Identify the various kinds of forces and moments acting on a control volume
- Use control volume analysis to determine the forces associated with fluid flow
- Use control volume analysis to determine the moments caused by fluid flow and the torque transmitted

13–1 ▪ NEWTON'S LAWS

Newton's laws are relations between motions of bodies and the forces acting on them. Newton's first law states that *a body at rest remains at rest, and a body in motion remains in motion at the same velocity in a straight path when the net force acting on it is zero.* Therefore, a body tends to preserve its state of inertia. Newton's second law states that *the acceleration of a body is proportional to the net force acting on it and is inversely proportional to its mass.* Newton's third law states that *when a body exerts a force on a second body, the second body exerts an equal and opposite force on the first.* Therefore, the direction of an exposed reaction force depends on the body taken as the system.

For a rigid body of mass m, Newton's second law is expressed as

Newton's second law:
$$\vec{F} = m\vec{a} = m\frac{d\vec{V}}{dt} = \frac{d(m\vec{V})}{dt} \tag{13–1}$$

where $\vec{F}$ is the net force acting on the body and $\vec{a}$ is the acceleration of the body under the influence of $\vec{F}$.

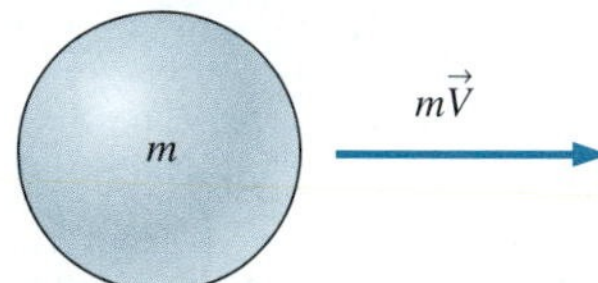

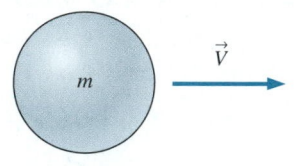

FIGURE 13–1
Linear momentum is the product of mass and velocity, and its direction is the direction of velocity.

The product of the mass and the velocity of a body is called the *linear momentum* or just the *momentum* of the body. The momentum of a rigid body of mass m moving with velocity $\vec{V}$ is $m\vec{V}$ (Fig. 13–1). Then Newton's second law expressed in Eq. 13–1 can also be stated as *the rate of change of the momentum of a body is equal to the net force acting on the body* (Fig. 13–2). This statement is more in line with Newton's original statement of the second law, and it is more appropriate for use in fluid mechanics when studying the forces generated as a result of velocity changes of fluid streams. Therefore, in fluid mechanics, Newton's second law is usually referred to as the *linear momentum equation.*

The momentum of a system remains constant only when the net force acting on it is zero, and thus the momentum of such systems is conserved. This is known as the *conservation of momentum principle*. This principle has proven to be a very useful tool when analyzing collisions such as those between balls; between balls and rackets, bats, or clubs; and between atoms or subatomic particles; and explosions such as those that occur in rockets, missiles, and guns. In fluid mechanics, however, the net force acting on a system is typically *not* zero, and we prefer to work with the linear momentum equation rather than the conservation of momentum principle.

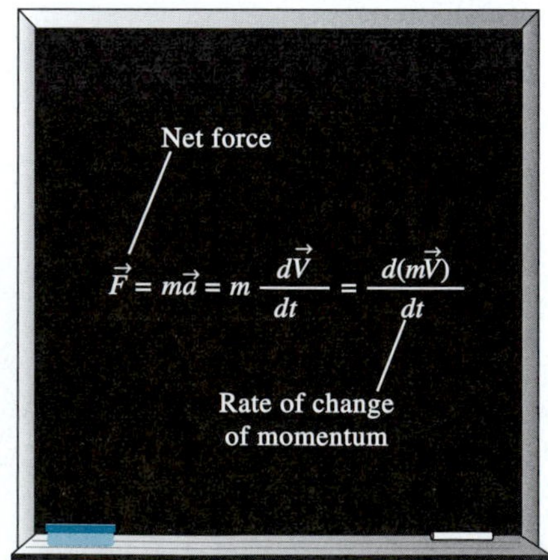

FIGURE 13–2
Newton's second law is also expressed as *the rate of change of the momentum of a body is equal to the net force acting on it.*

Note that force, acceleration, velocity, and momentum are vector quantities, and as such they have direction as well as magnitude. Also, momentum is a constant multiple of velocity, and thus the direction of momentum is the direction of velocity (Fig. 13–1). Any vector equation can be written in scalar form for a specified direction using magnitudes, e.g., $F_x = ma_x = d(mV_x)/dt$ in the x-direction.

The counterpart of Newton's second law for rotating rigid bodies is expressed as $\vec{M} = I\vec{\alpha}$, where $\vec{M}$ is the net moment or torque applied on the body, I is the moment of inertia of the body about the axis of rotation, and $\vec{\alpha}$ is the angular acceleration. It can also be expressed in terms of the rate of change of angular momentum $d\vec{H}/dt$ as

Angular momentum equation:
$$\vec{M} = I\vec{\alpha} = I\frac{d\vec{\omega}}{dt} = \frac{d(I\vec{\omega})}{dt} = \frac{d\vec{H}}{dt} \tag{13–2}$$

where $\vec{\omega}$ is the angular velocity. For a rigid body rotating about a fixed x-axis, the angular momentum equation can be written in scalar form as

Angular momentum about x-axis:

$$M_x = I_x \frac{d\omega_x}{dt} = \frac{dH_x}{dt} \quad \textbf{(13–3)}$$

The angular momentum equation can be stated as *the rate of change of the angular momentum of a body is equal to the net torque acting on it* (Fig. 13–3).

The total angular momentum of a rotating body remains constant when the net torque acting on it is zero, and thus the angular momentum of such systems is conserved. This is known as the *conservation of angular momentum principle* and is expressed as $I\omega$ = constant. Many interesting phenomena such as ice skaters spinning faster when they bring their arms close to their bodies and divers rotating faster when they curl after the jump can be explained easily with the help of the conservation of angular momentum principle (in both cases, the moment of inertia I is decreased and thus the angular velocity ω is increased as the outer parts of the body are brought closer to the axis of rotation).

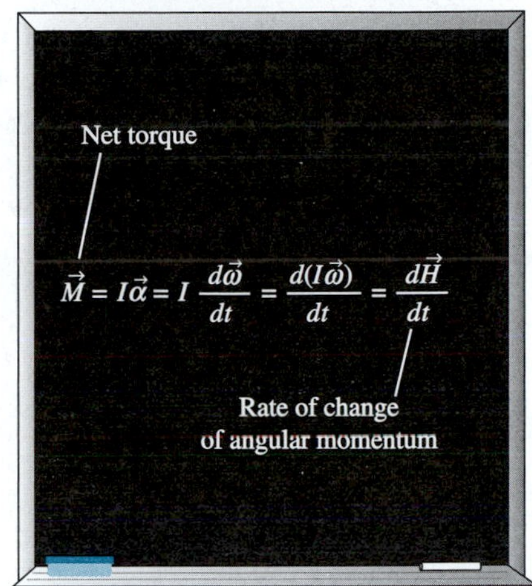

FIGURE 13–3

The rate of change of the angular momentum of a body is equal to the net torque acting on it.

13–2 ▪ CHOOSING A CONTROL VOLUME

We now briefly discuss how to *wisely* select a control volume. A control volume can be selected as any arbitrary region in space through which fluid flows, and its bounding control surface can be fixed, moving, and even deforming during flow. The application of a basic conservation law is simply a systematic procedure for bookkeeping or accounting of the quantity under consideration, and thus it is extremely important that the boundaries of the control volume are well defined during an analysis. Also, the flow rate of any quantity into or out of a control volume depends on the flow velocity *relative to the control surface*, and thus it is essential to know if the control volume remains at rest during flow or if it moves.

Many flow systems involve stationary hardware firmly fixed to a stationary surface, and such systems are best analyzed using *fixed* control volumes. When determining the reaction force acting on a tripod holding the nozzle of a hose, for example, a natural choice for the control volume is one that passes perpendicularly through the nozzle exit flow and through the bottom of the tripod legs (Fig. 13–4*a*). This is a fixed control volume, and the water velocity relative to a fixed point on the ground is the same as the water velocity relative to the nozzle exit plane.

When analyzing flow systems that are moving or deforming, it is usually more convenient to allow the control volume to *move* or *deform*. When determining the thrust developed by the jet engine of an airplane cruising at constant velocity, for example, a wise choice of control volume is one that encloses the airplane and cuts through the nozzle exit plane (Fig. 13–4*b*). The control volume in this case moves with velocity $\vec{V}_{CV}$, which is identical to the cruising velocity of the airplane relative to a fixed point on earth. When determining the flow rate of exhaust gases leaving the nozzle, the proper velocity to use is the velocity of the exhaust gases relative to the nozzle exit plane, that is, the *relative velocity* $\vec{V}_r$. Since the entire control volume moves at velocity

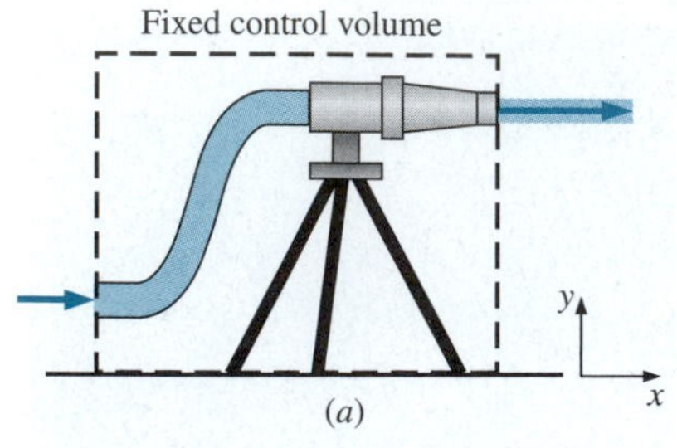

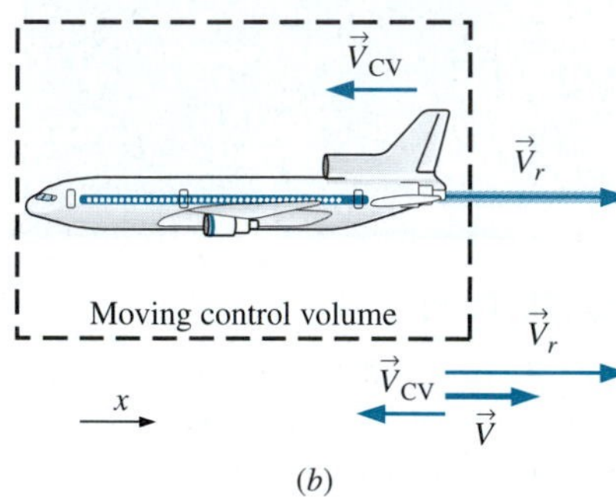

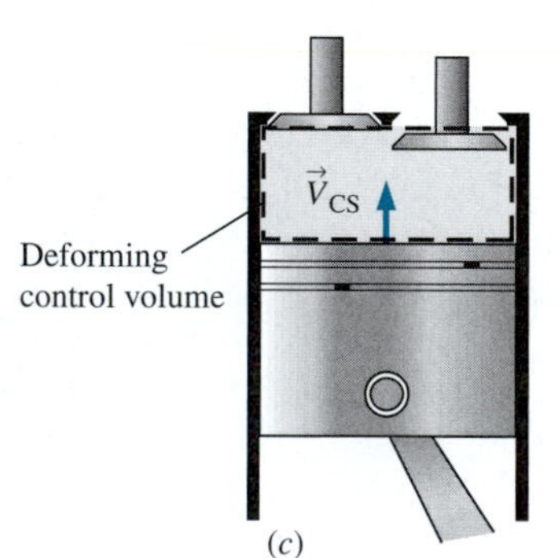

FIGURE 13–4

Examples of (*a*) fixed, (*b*) moving, and (*c*) deforming control volumes.

$\vec{V}_{CV}$, the relative velocity becomes $\vec{V}_r = \vec{V} - \vec{V}_{CV}$, where $\vec{V}$ is the *absolute velocity* of the exhaust gases, i.e., the velocity relative to a fixed point on earth. Note that $\vec{V}_r$ is the fluid velocity expressed relative to a coordinate system moving *with* the control volume. Also, this is a vector equation, and velocities in opposite directions have opposite signs. For example, if the airplane is cruising at 500 km/h to the left, and the velocity of the exhaust gases is 800 km/h to the right relative to the ground, the velocity of the exhaust gases relative to the nozzle exit is

$$\vec{V}_r = \vec{V} - \vec{V}_{CV} = 800\vec{i} - (-500\vec{i}) = 1300\vec{i} \text{ km/h}$$

That is, the exhaust gases leave the nozzle at 1300 km/h to the right relative to the nozzle exit (in the direction opposite to that of the airplane); this is the velocity that should be used when evaluating the outflow of exhaust gases through the control surface (Fig. 13–4*b*). Note that the exhaust gases would appear motionless to an observer on the ground if the relative velocity were equal in magnitude to the airplane velocity.

When analyzing the purging of exhaust gases from a reciprocating internal combustion engine, a wise choice for the control volume is one that comprises the space between the top of the piston and the cylinder head (Fig. 13–4*c*). This is a *deforming* control volume, since part of the control surface moves relative to other parts. The relative velocity for an inlet or outlet on the deforming part of a control surface (there are no such inlets or outlets in Fig. 13–4*c*) is then given by $\vec{V}_r = \vec{V} - \vec{V}_{CS}$ where $\vec{V}$ is the absolute fluid velocity and $\vec{V}_{CS}$ is the control surface velocity, both relative to a fixed point outside the control volume. Note that $\vec{V}_{CS} = \vec{V}_{CV}$ for moving but nondeforming control volumes, and $\vec{V}_{CS} = \vec{V}_{CV} = 0$ for fixed ones.

13–3 ▪ FORCES ACTING ON A CONTROL VOLUME

The forces acting on a control volume consist of **body forces** that act throughout the entire body of the control volume (such as gravity, electric, and magnetic forces) and **surface forces** that act on the control surface (such as pressure and viscous forces and reaction forces at points of contact). Only external forces are considered in the analysis. Internal forces (such as the pressure force between a fluid and the inner surfaces of the flow section) are not considered in a control volume analysis unless they are exposed by passing the control surface through that area.

In control volume analysis, the sum of all forces acting on the control volume at a particular instant in time is represented by $\Sigma \vec{F}$ and is expressed as

Total force acting on control volume:
$$\sum \vec{F} = \sum \vec{F}_{\text{body}} + \sum \vec{F}_{\text{surface}} \tag{13–4}$$

Body forces act on each volumetric portion of the control volume. The body force acting on a differential element of fluid of volume dV within the control volume is shown in Fig. 13–5, and we must perform a volume integral to account for the net body force on the entire control volume. *Surface forces* act on each portion of the control surface. A differential surface element of area dA and unit outward normal $\vec{n}$ on the control surface is shown in Fig. 13–5, along with the surface force acting on it. We must perform an area integral to obtain the net surface force acting on the entire control surface. As sketched, the surface force may act in a direction independent of that of the outward normal vector.

The most common body force is that of **gravity**, which exerts a downward force on every differential element of the control volume. While other body forces, such as electric and magnetic forces, may be important in some analyses, we consider only gravitational forces here.

The differential body force $d\vec{F}_{\text{body}} = d\vec{F}_{\text{gravity}}$ acting on the small fluid element shown in Fig. 13–6 is simply its weight,

Gravitational force acting on a fluid element:
$$d\vec{F}_{\text{gravity}} = \rho \vec{g}\, dV \quad (13\text{–}5)$$

where ρ is the average density of the element and $\vec{g}$ is the gravitational vector. In Cartesian coordinates we adopt the convention that $\vec{g}$ acts in the negative z-direction, as in Fig. 13–6, so that

Gravitational vector in Cartesian coordinates:
$$\vec{g} = -g\vec{k} \quad (13\text{–}6)$$

Note that the coordinate axes in Fig. 13–6 are oriented so that the gravity vector acts *downward* in the $-z$-direction. On earth at sea level, the gravitational constant g is equal to 9.807 m/s^2. Since gravity is the only body force being considered, integration of Eq. 13–5 yields

Total body force acting on control volume:
$$\sum \vec{F}_{\text{body}} = \int_{\text{CV}} \rho \vec{g}\, dV = m_{\text{CV}}\vec{g} \quad (13\text{–}7)$$

Surface forces are not as simple to analyze since they consist of both *normal* and *tangential* components. **Normal stresses** are composed of pressure (which always acts inwardly normal) and viscous stresses. **Shear stresses** are composed entirely of viscous stresses.

A careful selection of the control volume enables us to write the total force acting on the control volume, $\Sigma \vec{F}$, as the sum of more readily available quantities like weight, pressure, and reaction forces. We recommend the following for control volume analysis:

Total force:
$$\underbrace{\sum \vec{F}}_{\text{total force}} = \underbrace{\sum \vec{F}_{\text{gravity}}}_{\text{body force}} + \underbrace{\sum \vec{F}_{\text{pressure}} + \sum \vec{F}_{\text{viscous}} + \sum \vec{F}_{\text{other}}}_{\text{surface forces}} \quad (13\text{–}8)$$

The first term on the right-hand side of Eq. 13–8 is the body force *weight*, since gravity is the only body force we are considering. The other three terms combine to form the net surface force; they are pressure forces, viscous forces, and "other" forces acting on the control surface. $\Sigma \vec{F}_{\text{other}}$ is composed of reaction forces required to turn the flow; forces at bolts, cables, struts, or walls through which the control surface cuts; etc.

All these surface forces arise as the control volume is isolated from its surroundings for analysis, and the effect of any detached object is accounted for by a force at that location. This is similar to drawing a free-body diagram in your statics and dynamics classes. We should choose the control volume such that forces that are not of interest remain internal, and thus they do not complicate the analysis. A well-chosen control volume exposes only the forces that are to be determined (such as reaction forces) and a minimum number of other forces.

A common simplication in the application of Newton's laws of motion is to subtract the *atmospheric pressure* and work with gage pressures. This is

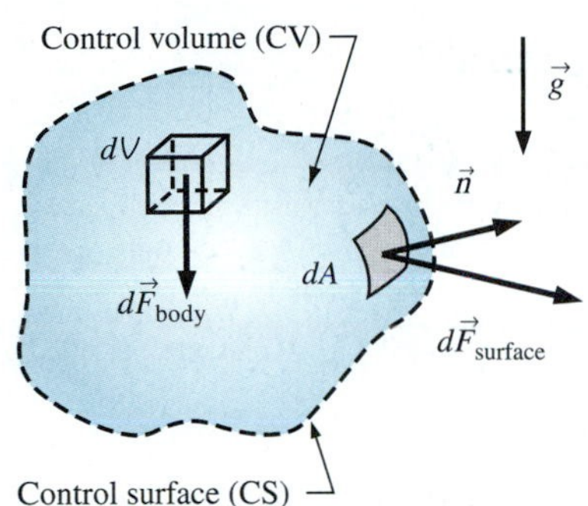

FIGURE 13–5

The total force acting on a control volume is composed of body forces and surface forces; body force is shown on a differential volume element, and surface force is shown on a differential surface element.

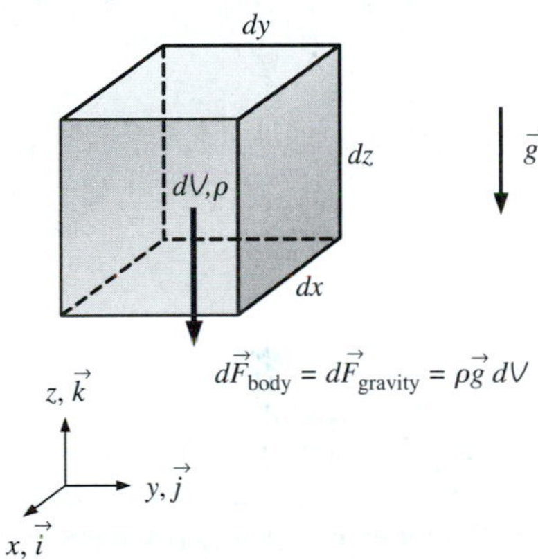

FIGURE 13–6

The gravitational force acting on a differential volume element of fluid is equal to its weight; the axes have been rotated so that the gravity vector acts *downward* in the negative z-direction.

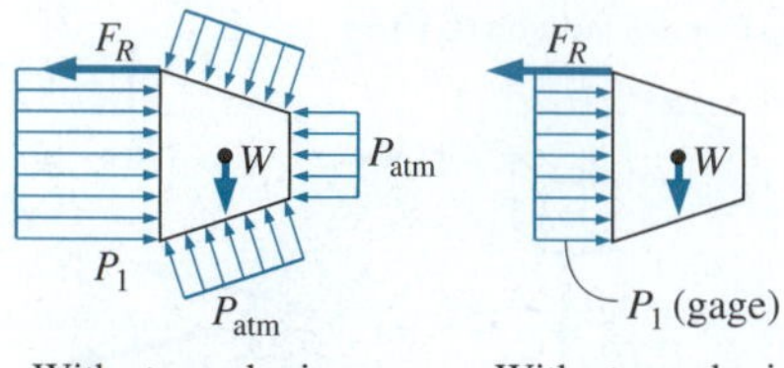

FIGURE 13–7

Atmospheric pressure acts in all directions, and thus it can be ignored when performing force balances since its effect cancels out in every direction.

because atmospheric pressure acts in all directions, and its effect cancels out in every direction (Fig. 13–7). This means we can also ignore the pressure forces at outlet sections where the fluid is discharged to the atmosphere since the discharge pressure in such cases is very near atmospheric pressure at subsonic velocities.

As an example of how to wisely choose a control volume, consider control volume analysis of water flowing steadily through a faucet with a partially closed gate valve spigot (Fig. 13–8). It is desired to calculate the net force on the flange to ensure that the flange bolts are strong enough. There are many possible choices for the control volume. Some engineers restrict their control volumes to the fluid itself, as indicated by CV A (the colored control volume). With this control volume, there are pressure forces that vary along the control surface, there are viscous forces along the pipe wall and at locations inside the valve, and there is a body force, namely, the weight of the water in the control volume. Fortunately, to calculate the net force on the flange, we do *not* need to integrate the pressure and viscous stresses all along the control surface. Instead, we can lump the unknown pressure and viscous forces together into one reaction force, representing the net force of the walls on the water. This force, plus the weight of the faucet and the water, is equal to the net force on the flange. (We must be very careful with our signs, of course.)

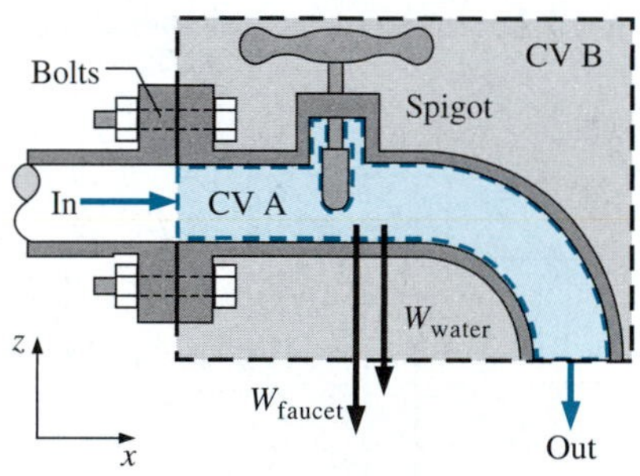

FIGURE 13–8

Cross section through a faucet assembly, illustrating the importance of choosing a control volume wisely; CV B is much easier to work with than CV A.

When choosing a control volume, you are not limited to the fluid alone. Often it is more convenient to slice the control surface *through* solid objects such as walls, struts, or bolts as illustrated by CV B (the gray control volume) in Fig. 13–8. A control volume may even surround an entire object, like the one shown here. Control volume B is a wise choice because we are not concerned with any details of the flow or even the geometry inside the control volume. For the case of CV B, we assign a net reaction force acting at the portions of the control surface that slice through the flange bolts. Then, the only other things we need to know are the gage pressure of the water at the flange (the inlet to the control volume) and the weights of the water and the faucet assembly. The pressure everywhere else along the control surface is atmospheric (zero gage pressure) and cancels out.

13–4 ▪ THE LINEAR MOMENTUM EQUATION

Newton's second law for a system of mass m subjected to a net force $\vec{F}$ is expressed from Eq. 14–1 as

$$\sum \vec{F} = m\vec{a} = m\frac{d\vec{V}}{dt} = \frac{d}{dt}(m\vec{V}) \tag{13–9}$$

where $m\vec{V}$ is the **linear momentum** of the system. Noting that both the density and velocity may change from point to point within the system, Newton's second law can be expressed more generally as

$$\sum \vec{F} = \frac{d}{dt}\int_{\text{sys}} \rho\vec{V}\,dV \tag{13–10}$$

where $\rho\vec{V}\,dV$ is the momentum of a differential element dV, which has mass $\delta m = \rho dV$.

Therefore, Newton's second law can be stated as *the sum of all external forces acting on a system is equal to the time rate of change of linear momentum of the system*. This statement is valid for a coordinate system that is at rest or moves with a constant velocity, called an *inertial coordinate system* or *inertial reference frame*. Accelerating systems such as aircraft during takeoff are best analyzed using noninertial (or accelerating) coordinate systems fixed to the aircraft. Note that Eq. 11–4 is a vector relation, and thus the quantities $\vec{F}$ and $\vec{V}$ have direction as well as magnitude.

Equation 13–10 is for a given mass of a solid or fluid and is of limited use in fluid mechanics since most flow systems are analyzed using control volumes. The *Reynolds transport theorem* developed in Section 11–4 provides the necessary tools to shift from the system formulation to the control volume formulation. Setting $b = \vec{V}$ and thus $B = m\vec{V}$, the Reynolds transport theorem is expressed for linear momentum as (Fig. 13–9)

$$\frac{d(m\vec{V})_{sys}}{dt} = \frac{d}{dt}\int_{CV} \rho\vec{V}\, dV + \int_{CS} \rho\vec{V}\,(\vec{V}_r \cdot \vec{n})dA \qquad \text{(13–11)}$$

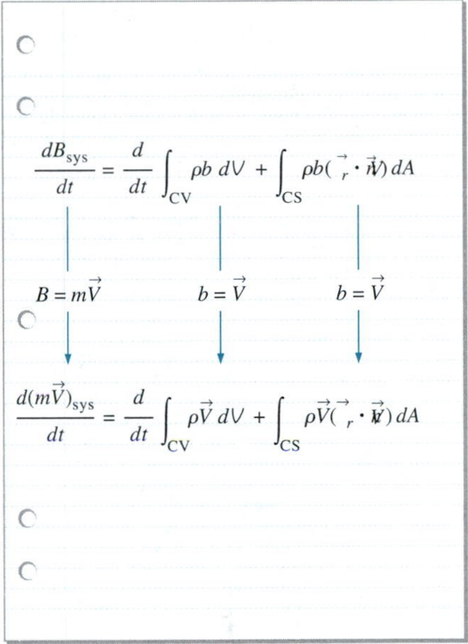

FIGURE 13–9
The linear momentum equation is obtained by replacing B in the Reynolds transport theorem by the momentum $m\vec{V}$, and b by the momentum per unit mass $\vec{V}$.

The left-hand side of this equation is, from Eq. 13–9, equal to $\Sigma\vec{F}$. Substituting, the general form of the linear momentum equation that applies to fixed, moving, or deforming control volumes is

General:
$$\sum \vec{F} = \frac{d}{dt}\int_{CV} \rho\vec{V}\, dV + \int_{CS} \rho\vec{V}(\vec{V}_r \cdot \vec{n})dA \qquad \text{(13–12)}$$

which is stated in words as

$$\begin{pmatrix}\text{The sum of all}\\ \text{external forces}\\ \text{acting on a CV}\end{pmatrix} = \begin{pmatrix}\text{The time rate of change}\\ \text{of the linear momentum}\\ \text{of the contents of the CV}\end{pmatrix} + \begin{pmatrix}\text{The net flow rate of}\\ \text{linear momentum out of the}\\ \text{control surface by mass flow}\end{pmatrix}$$

Here $\vec{V}_r = \vec{V} - \vec{V}_{CS}$ is the fluid velocity relative to the control surface (for use in mass flow rate calculations at all locations where the fluid crosses the control surface), and $\vec{V}$ is the fluid velocity as viewed from an inertial reference frame. The product $\rho(\vec{V}_r \cdot \vec{n})\, dA$ represents the mass flow rate through area element dA into or out of the control volume.

For a fixed control volume (no motion or deformation of control volume), $\vec{V}_r = \vec{V}$ and the linear momentum equation becomes

Fixed CV:
$$\sum \vec{F} = \frac{d}{dt}\int_{CV} \rho\vec{V}\, dV + \int_{CS} \rho\vec{V}(\vec{V} \cdot \vec{n})dA \qquad \text{(13–13)}$$

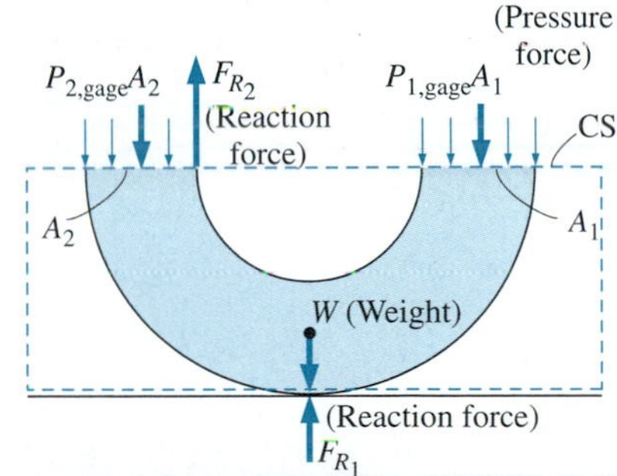

FIGURE 13–10
In most flow systems, the sum of forces $\Sigma\vec{F}$ consists of weights, pressure forces, and reaction forces. Gage pressures are used here since atmospheric pressure cancels out on all sides of the control surface.

Note that the momentum equation is a *vector equation*, and thus each term should be treated as a vector. Also, the components of this equation can be resolved along orthogonal coordinates (such as *x*, *y*, and *z* in the Cartesian coordinate system) for convenience. The sum of forces $\Sigma\vec{F}$ in most cases consists of weights, pressure forces, and reaction forces (Fig. 13–10). The momentum equation is commonly used to calculate the forces (usually on support systems or connectors) induced by the flow.

Special Cases

During *steady flow*, the amount of momentum within the control volume remains constant, and thus the time rate of change of linear momentum of the contents of the control volume (the second term of Eq. 13–10) is zero. Thus,

Steady flow:
$$\sum \vec{F} = \int_{CS} \rho \vec{V}\,(\vec{V}_r \cdot \vec{n})\,dA \tag{13–14}$$

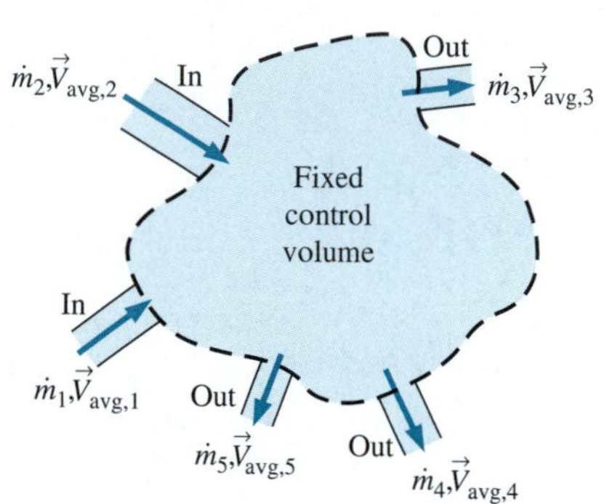

FIGURE 13–11

In a typical engineering problem, the control volume may contain many inlets and outlets; at each inlet or outlet we define the mass flow rate $\dot{m}$ and the average velocity $\vec{V}_{avg}$.

Most momentum problems considered in this text are steady.

For a case in which a non-deforming control volume moves at constant velocity (an inertial reference frame), the *first* $\vec{V}$ in Eq. 13–18 may *also* be taken relative to the moving control surface. While Eq. 13–13 is exact for fixed control volumes, it is not always convenient when solving practical engineering problems because of the integrals. Instead, as we did for conservation of mass, we would like to rewrite Eq. 13–13 in terms of average velocities and mass flow rates through inlets and outlets. In other words, our desire is to rewrite the equation in *algebraic* rather than *integral* form. In many practical applications, fluid crosses the boundaries of the control volume at one or more inlets and one or more outlets, and carries with it some momentum into or out of the control volume. For simplicity, we always draw our control surface such that it slices normal to the inflow or outflow velocity at each such inlet or outlet (Fig. 13–11).

The mass flow rate $\dot{m}$ into or out of the control volume across an inlet or outlet at which ρ is nearly constant is

Mass flow rate across an inlet or outlet:
$$\dot{m} = \int_{A_c} \rho(\vec{V} \cdot \vec{n})\,dA_c = \rho V_{avg} A_c \tag{13–15}$$

Comparing Eq. 13–15 to Eq. 13–13, we notice an extra velocity in the control surface integral of Eq. 13–13. If $\vec{V}$ were uniform ($\vec{V} = \vec{V}_{avg}$) across the inlet or outlet, we could simply take it outside the integral. Then we could write the rate of inflow or outflow of momentum through the inlet or outlet in simple algebraic form,

Momentum flow rate across a uniform inlet or outlet:

$$\int_{A_c} \rho \vec{V}(\vec{V} \cdot \vec{n})\,dA_c = \rho V_{avg} A_c \vec{V}_{avg} = \dot{m}\vec{V}_{avg} \tag{13–16}$$

The uniform flow approximation is reasonable at some inlets and outlets, e.g., the well-rounded entrance to a pipe, the flow at the entrance to a wind tunnel test section, and a slice through a water jet moving at nearly uniform speed through air (Fig. 13–12). At each such inlet or outlet, Eq. 13–16 can be applied directly.

Momentum-Flux Correction Factor, β

Unfortunately, the velocity across most inlets and outlets of practical engineering interest is *not* uniform. Nevertheless, it turns out that we can still convert the control surface integral of Eq. 13–13 into algebraic form, but a

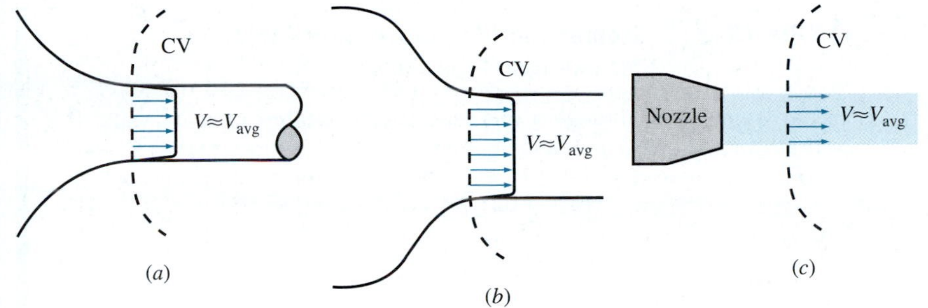

FIGURE 13–12
Examples of inlets or outlets in which the uniform flow approximation is reasonable: (*a*) the well-rounded entrance to a pipe, (*b*) the entrance to a wind tunnel test section, and (*c*) a slice through a free water jet in air.

dimensionless correction factor β, called the **momentum-flux correction factor**, is required, as first shown by the French scientist Joseph Boussinesq (1842–1929). The algebraic form of Eq. 13–13 for a fixed control volume is then written as

$$\sum \vec{F} = \frac{d}{dt}\int_{CV} \rho \vec{V}\, dV + \sum_{out} \beta \dot{m} \vec{V}_{avg} - \sum_{in} \beta \dot{m} \vec{V}_{avg} \tag{13–17}$$

where a unique value of momentum-flux correction factor is applied to each inlet and outlet in the control surface. Note that $\beta = 1$ *for the case of uniform flow* over an inlet or outlet, as in Fig. 13–12. For the general case, we define β such that the integral form of the momentum flux into or out of the control surface at an inlet or outlet of cross-sectional area A_c can be expressed in terms of mass flow rate $\dot{m}$ through the inlet or outlet and average velocity $\vec{V}_{avg}$ through the inlet or outlet,

Momentum flux across an inlet or outlet:
$$\int_{A_c} \rho \vec{V}(\vec{V} \cdot \vec{n})\, dA_c = \beta \dot{m} \vec{V}_{avg} \tag{13–18}$$

For the case in which density is uniform over the inlet or outlet and $\vec{V}$ is in the same direction as $\vec{V}_{avg}$ over the inlet or outlet, we solve Eq. 13–18 for β,

$$\beta = \frac{\int_{A_c} \rho V(\vec{V} \cdot \vec{n})\, dA_c}{\dot{m} V_{avg}} = \frac{\int_{A_c} \rho V(\vec{V} \cdot \vec{n})\, dA_c}{\rho V_{avg} A_c V_{avg}} \tag{13–19}$$

where we have substituted $\rho V_{avg} A_c$ for $\dot{m}$ in the denominator. The densities cancel and since V_{avg} is constant, it can be brought inside the integral. Furthermore, if the control surface slices normal to the inlet or outlet area, we have $(\vec{V} \cdot \vec{n})\, dA_c = V\, dA_c$. Then, Eq. 13–19 simplifies to

Momentum-flux correction factor:
$$\beta = \frac{1}{A_c}\int_{A_c} \left(\frac{V}{V_{avg}}\right)^2 dA_c \tag{13–20}$$

It may be shown that β is always greater than or equal to unity.

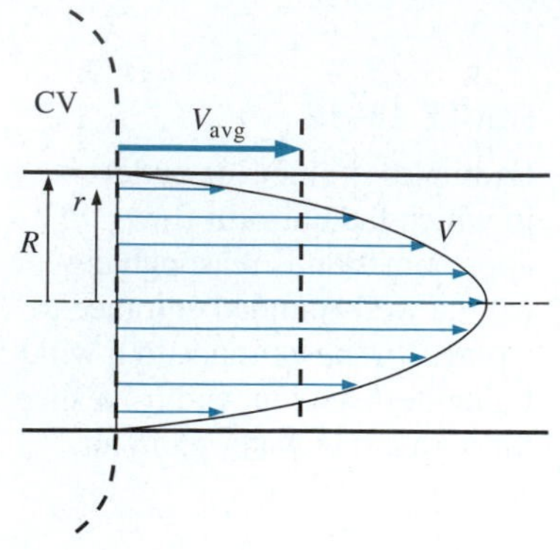

FIGURE 13–13

Velocity profile over a cross section of a pipe in which the flow is fully developed and laminar.

EXAMPLE 13–1 Momentum-Flux Correction Factor for Laminar Pipe Flow

Consider laminar flow through a very long straight section of round pipe. It is shown in Chap. 14 that the velocity profile through a cross-sectional area of the pipe is parabolic (Fig. 13–13), with the axial velocity component given by

$$V = 2V_{avg}\left(1 - \frac{r^2}{R^2}\right) \tag{1}$$

where R is the radius of the inner wall of the pipe and V_{avg} is the average velocity. Calculate the momentum-flux correction factor through a cross section of the pipe for the case in which the pipe flow represents an outlet of the control volume, as sketched in Fig. 13–13.

Solution For a given velocity distribution we are to calculate the momentum-flux correction factor.

Assumptions **1** The flow is incompressible and steady. **2** The control volume slices through the pipe normal to the pipe axis, as sketched in Fig. 13–13.

Analysis We substitute the given velocity profile for V in Eq. 13–20 and integrate, noting that $dA_c = 2\pi r\, dr$,

$$\beta = \frac{1}{A_c}\int_{A_c}\left(\frac{V}{V_{avg}}\right)^2 dA_c = \frac{4}{\pi R^2}\int_0^R \left(1 - \frac{r^2}{R^2}\right)^2 2\pi r\, dr \tag{2}$$

Defining a new integration variable $y = 1 - r^2/R^2$ and thus $dy = -2r\, dr/R^2$ (also, $y = 1$ at $r = 0$, and $y = 0$ at $r = R$) and performing the integration, the momentum-flux correction factor for fully developed laminar flow becomes

Laminar flow:
$$\beta = -4\int_1^0 y^2\, dy = -4\left[\frac{y^3}{3}\right]_1^0 = \frac{4}{3} \tag{3}$$

Discussion We have calculated β for an outlet, but the same result would have been obtained if we had considered the cross section of the pipe as an *inlet* to the control volume.

From Example 13–1 we see that β is not very close to unity for fully developed laminar pipe flow, and ignoring β could potentially lead to significant error. If we were to perform the same kind of integration as in Example 13–1 but for fully developed *turbulent* rather than laminar pipe flow, we would find that β ranges from about 1.01 to 1.04. Since these values are so close to unity, many practicing engineers completely disregard the momentum-flux correction factor. While the neglect of β in turbulent flow calculations may have an insignificant effect on the final results, it is wise to keep it in our equations. Doing so not only improves the accuracy of our calculations, but reminds us to include the momentum-flux correction factor when solving laminar flow control volume problems.

Steady Flow

If the flow is also *steady*, the time derivative term in Eq. 13–17 vanishes and we are left with

Steady linear momentum equation:
$$\sum \vec{F} = \sum_{\text{out}} \beta \dot{m} \vec{V} - \sum_{\text{in}} \beta \dot{m} \vec{V} \qquad \textbf{(13–21)}$$

where we have dropped the subscript "avg" from average velocity. Equation 13–21 states that *the net force acting on the control volume during steady flow is equal to the difference between the rates of outgoing and incoming momentum flows*. This statement is illustrated in Fig. 13–14. It can also be expressed for any direction, since Eq. 13–21 is a vector equation.

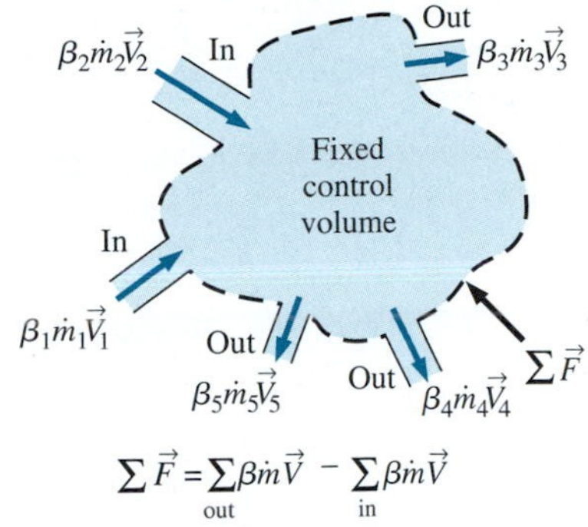

FIGURE 13–14
The net force acting on the control volume during steady flow is equal to the difference between the outgoing and the incoming momentum fluxes.

Steady Flow with One Inlet and One Outlet

Many practical engineering problems involve just one inlet and one outlet (Fig. 13–15). The mass flow rate for such **single-stream systems** remains constant, and Eq. 13–21 reduces to

One inlet and one outlet:
$$\sum \vec{F} = \dot{m}\,(\beta_2 \vec{V}_2 - \beta_1 \vec{V}_1) \qquad \textbf{(13–22)}$$

where we have adopted the usual convention that subscript 1 implies the inlet and subscript 2 the outlet, and $\vec{V}_1$ and $\vec{V}_2$ denote the *average* velocities across the inlet and outlet, respectively.

We emphasize again that all the preceding relations are *vector* equations, and thus all the additions and subtractions are *vector* additions and subtractions. Recall that subtracting a vector is equivalent to adding it after reversing its direction (Fig. 13–16). When writing the momentum equation for a specified coordinate direction (such as the x-axis), we use the projections of the vectors on that axis. For example, Eq. 13–22 is written along the x-coordinate as

Along x-coordinate:
$$\sum F_x = \dot{m}(\beta_2 V_{2,x} - \beta_1 V_{1,x}) \qquad \textbf{(13–23)}$$

where $\sum F_x$ is the vector sum of the x-components of the forces, and $V_{2,x}$ and $V_{1,x}$ are the x-components of the outlet and inlet average velocities of the fluid stream, respectively. The force or velocity components in the positive x-direction are positive quantities, and those in the negative x-direction are negative quantities. Also, it is good practice to take the direction of unknown forces in the positive directions (unless the problem is very straightforward). A negative value obtained for an unknown force indicates that the assumed direction is wrong and should be reversed.

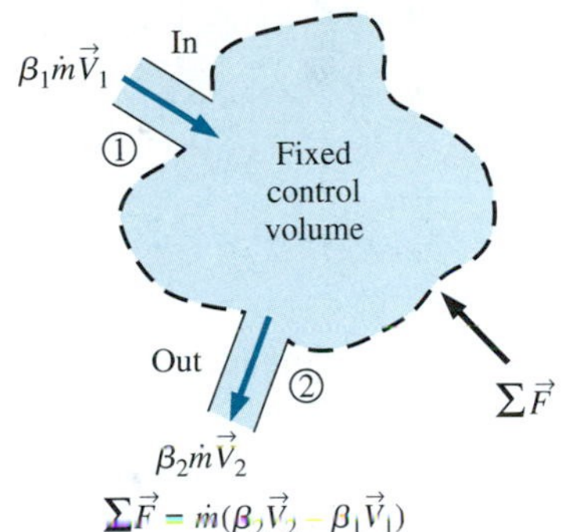

FIGURE 13–15
A control volume with only one inlet and one outlet.

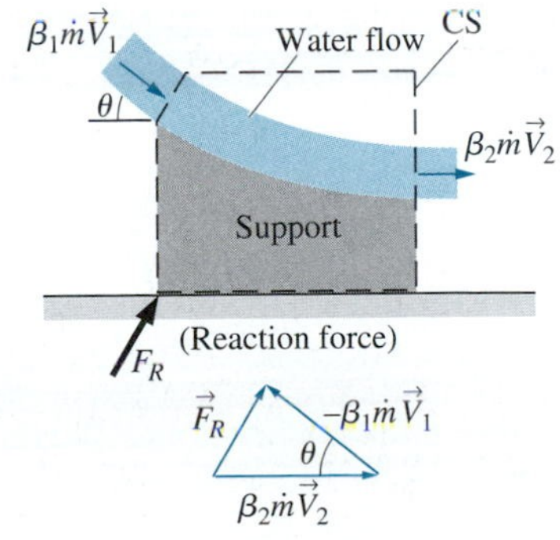

FIGURE 13–16
The determination by vector addition of the reaction force on the support caused by a change of direction of water.

Flow with No External Forces

An interesting situation arises when there are no external forces such as weight, pressure, and reaction forces acting on the body in the direction of motion—a common situation for space vehicles and satellites. For a control volume with multiple inlets and outlets, Eq. 13–17 reduces in this case to

No external forces:
$$0 = \frac{d(m\vec{V})_{\text{CV}}}{dt} + \sum_{\text{out}} \beta \dot{m} \vec{V} - \sum_{\text{in}} \beta \dot{m} \vec{V} \qquad (13\text{–}24)$$

This is an expression of the conservation of momentum principle, which can be stated as *in the absence of external forces, the rate of change of the momentum of a control volume is equal to the difference between the rates of incoming and outgoing momentum flow rates.*

When the mass m of the control volume remains nearly constant, the first term of the Eq. 13–24 simply becomes mass times acceleration since

$$\frac{d(m\vec{V})_{CV}}{dt} = m_{CV}\frac{d\vec{V}_{CV}}{dt} = (m\vec{a})_{CV} = m_{CV}\vec{a}$$

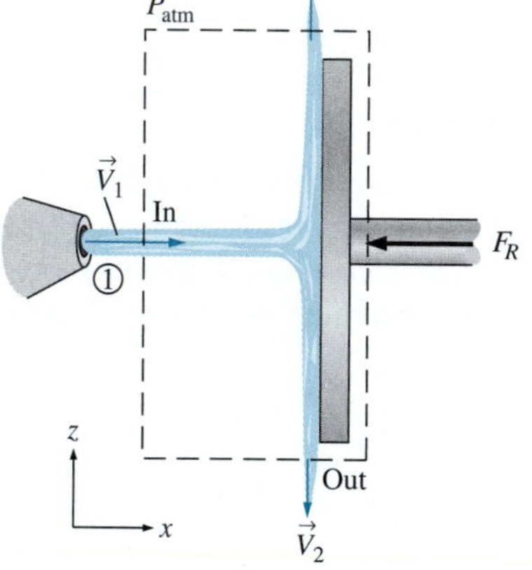

FIGURE 13–17
Schematic for Example 13–2.

Therefore, the control volume in this case can be treated as a solid body (a fixed-mass system), with a net thrusting force (or just **thrust**) of

Thrust: $$\vec{F}_{thrust} = m_{CV}\vec{a} = \sum_{in} \beta \dot{m}\vec{V} - \sum_{out} \beta \dot{m}\vec{V} \qquad (13\text{–}25)$$

acting on the body. This approach can be used to determine the linear acceleration of space vehicles when a rocket is fired (Fig. 13–17).

Recall that thrust is a mechanical force typically generated through the reaction of an accelerating fluid. In the jet engine of an aircraft, for example, hot exhaust gases are accelerated by the action of expansion and outflow of gases through the back of the engine, and a thrusting force is produced by reaction in the opposite direction. The generation of thrust is based on Newton's third law of motion, which states that for every action at a point there is an equal and opposite reaction. In the case of a jet engine, if the engine exerts a force on exhaust gases, then the exhaust gases exert an equal force on the engine in the opposite direction. That is, the pushing force exerted on the departing gases by the engine is equal to the thrusting force the departing gases exert on the remaining mass of the aircraft in the opposite direction, $\vec{F}_{thrust} = -\vec{F}_{push}$. On the free body diagram of an aircraft, the effect of outgoing exhaust gases is accounted for by the insertion of a force in the opposite direction of motion of exhaust gases.

EXAMPLE 13–2 The Force to Hold a Deflector Elbow in Place

A reducing elbow is used to deflect water flow at a rate of 14 kg/s in a horizontal pipe upward 30° while accelerating it (Fig. 13–18). The elbow discharges water into the atmosphere. The cross-sectional area of the elbow is 113 cm^2 at the inlet and 7 cm^2 at the outlet. The elevation difference between the centers of the outlet and the inlet is 30 cm. The weight of the elbow and the water in it is considered to be negligible. Determine (*a*) the gage pressure at the center of the inlet of the elbow and (*b*) the anchoring force needed to hold the elbow in place.

Solution A reducing elbow deflects water upward and discharges it to the atmosphere. The pressure at the inlet of the elbow and the force needed to hold the elbow in place are to be determined.

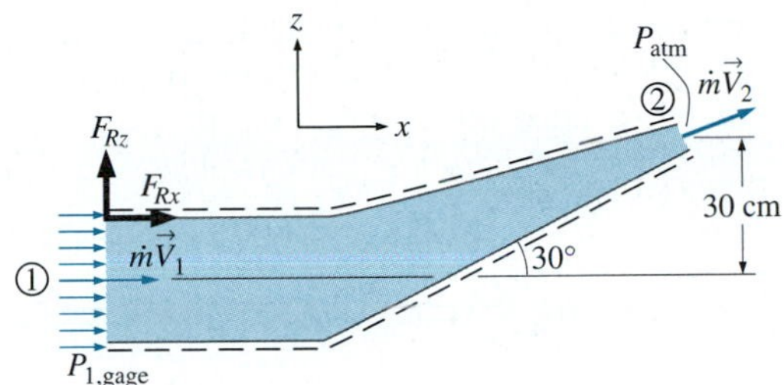

FIGURE 13–18
Schematic for Example 13–2.

Assumptions **1** The flow is steady, and the frictional effects are negligible. **2** The weight of the elbow and the water in it is negligible. **3** The water is discharged to the atmosphere, and thus the gage pressure at the outlet is zero. **4** The flow is turbulent and fully developed at both the inlet and outlet of the control volume, and we take the momentum-flux correction factor to be $\beta = 1.03$ (as a conservative estimate).

Properties We take the density of water to be 1000 kg/m³.

Analysis (*a*) We take the elbow as the control volume and designate the inlet by 1 and the outlet by 2. We also take the *x*- and *z*-coordinates as shown. The continuity equation for this one-inlet, one-outlet, steady-flow system is $\dot{m}_1 = \dot{m}_2 = \dot{m} = 14$ kg/s. Noting that $\dot{m} = \rho AV$, the inlet and outlet average velocities of water are

$$V_1 = \frac{\dot{m}}{\rho A_1} = \frac{14 \text{ kg/s}}{(1000 \text{ kg/m}^3)(0.0113 \text{ m}^2)} = 1.24 \text{ m/s}$$

$$V_2 = \frac{\dot{m}}{\rho A_2} = \frac{14 \text{ kg/s}}{(1000 \text{ kg/m}^3)(7 \times 10^{-4} \text{ m}^2)} = 20.0 \text{ m/s}$$

We use the Bernoulli equation (Chap. 12) as a first approximation to calculate the pressure. In Chap. 14 we will learn how to account for frictional losses along the walls. Taking the center of the inlet cross section as the reference level ($z_1 = 0$) and noting that $P_2 = P_{atm}$, the Bernoulli equation for a streamline going through the center of the elbow is expressed as

$$\frac{P_1}{\rho g} + \frac{V_1^2}{2g} + z_1 = \frac{P_2}{\rho g} + \frac{V_2^2}{2g} + z_2$$

$$P_1 - P_2 = \rho g\left(\frac{V_2^2 - V_1^2}{2g} + z_2 - z_1\right)$$

$$P_1 - P_{atm} = (1000 \text{ kg/m}^3)(9.81 \text{ m/s}^2)$$

$$\times\left(\frac{(20 \text{ m/s})^2 - (1.24 \text{ m/s})^2}{2(9.81 \text{ m/s}^2)} + 0.3 - 0\right)\left(\frac{1 \text{ kN}}{1000 \text{ kg}\cdot\text{m/s}^2}\right)$$

$$P_{1,\text{ gage}} = 202.2 \text{ kN/m}^2 = \mathbf{202.2 \text{ kPa}} \quad \text{(gage)}$$

(*b*) The momentum equation for steady one-dimensional flow is

$$\sum \vec{F} = \sum_{\text{out}} \beta \dot{m}\vec{V} - \sum_{\text{in}} \beta \dot{m}\vec{V}$$

We let the x- and z-components of the anchoring force of the elbow be F_{Rx} and F_{Rz}, and assume them to be in the positive direction. We also use gage pressure since the atmospheric pressure acts on the entire control surface. Then the momentum equations along the x- and z-axes become

$$F_{Rx} + P_{1,\text{ gage}}A_1 = \beta \dot{m} V_2 \cos\theta - \beta \dot{m} V_1$$

$$F_{Rz} = \beta \dot{m} V_2 \sin\theta$$

Solving for F_{Rx} and F_{Rz}, and substituting the given values,

$$F_{Rx} = \beta \dot{m}(V_2 \cos\theta - V_1) - P_{1,\text{ gage}}A_1$$

$$= 1.03(14 \text{ kg/s})[(20 \cos 30° - 1.24) \text{ m/s}]\left(\frac{1 \text{ N}}{1 \text{ kg} \cdot \text{m/s}^2}\right)$$

$$- (202{,}200 \text{ N/m}^2)(0.0113 \text{ m}^2)$$

$$= 232 - 2285 = \mathbf{-2053 \text{ N}}$$

$$F_{Rz} = \beta \dot{m} V_2 \sin\theta = (1.03)(14 \text{ kg/s})(20 \sin 30° \text{ m/s})\left(\frac{1 \text{ N}}{1 \text{ kg} \cdot \text{m/s}^2}\right) = \mathbf{144 \text{ N}}$$

The negative result for F_{Rx} indicates that the assumed direction is wrong, and it should be reversed. Therefore, F_{Rx} acts in the negative x-direction.

Discussion There is a nonzero pressure distribution along the inside walls of the elbow, but since the control volume is outside the elbow, these pressures do not appear in our analysis. The weight of the elbow and the water in it could be added to the vertical force for better accuracy. The actual value of $P_{1,\text{ gage}}$ will be higher than that calculated here because of frictional and other irreversible losses in the elbow.

EXAMPLE 13–3 The Force to Hold a Reversing Elbow in Place

The deflector elbow in Example 13–2 is replaced by a reversing elbow such that the fluid makes a 180° U-turn before it is discharged, as shown in Fig. 13–19. The elevation difference between the centers of the inlet and the exit sections is still 0.3 m. Determine the anchoring force needed to hold the elbow in place.

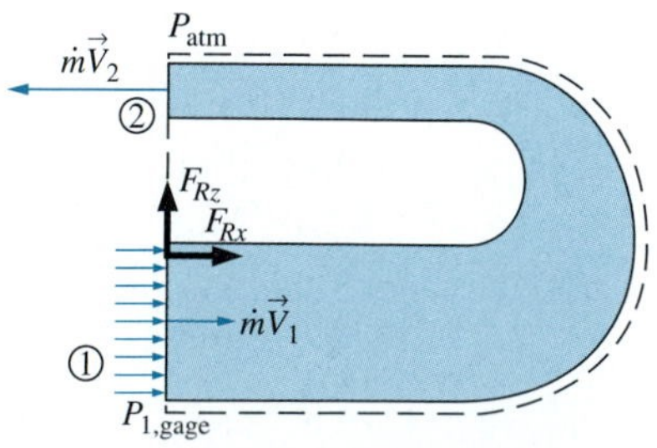

FIGURE 13–19
Schematic for Example 13–3.

Solution The inlet and the outlet velocities and the pressure at the inlet of the elbow remain the same, but the vertical component of the anchoring force at the connection of the elbow to the pipe is zero in this case (F_{Rz} = 0) since there is no other force or momentum flux in the vertical direction (we are neglecting the weight of the elbow and the water). The horizontal component of the anchoring force is determined from the momentum equation written in the x-direction. Noting that the outlet velocity is negative since it is in the negative x-direction, we have

$$F_{Rx} + P_{1,\text{ gage}}A_1 = \beta_2 \dot{m}(-V_2) - \beta_1 \dot{m} V_1 = -\beta \dot{m}(V_2 + V_1)$$

Solving for F_{Rx} and substituting the known values,

$$F_{Rx} = -\beta\dot{m}(V_2 + V_1) - P_{1,\,\text{gage}}A_1$$
$$= -(1.03)(14\ \text{kg/s})[(20 + 1.24)\ \text{m/s}]\left(\frac{1\ \text{N}}{1\ \text{kg}\cdot\text{m/s}^2}\right) - (202{,}200\ \text{N/m}^2)(0.0113\ \text{m}^2)$$
$$= -306 - 2285 = \mathbf{-2591\ N}$$

Therefore, the horizontal force on the flange is 2591 N acting in the negative *x*-direction (the elbow is trying to separate from the pipe). This force is equivalent to the weight of about 260 kg mass, and thus the connectors (such as bolts) used must be strong enough to withstand this force.

Discussion The reaction force in the *x*-direction is larger than that of Example 13–2 since the walls turn the water over a much greater angle. If the reversing elbow is replaced by a straight nozzle (like one used by firefighters) such that water is discharged in the positive *x*-direction, the momentum equation in the *x*-direction becomes

$$F_{Rx} + P_{1,\,\text{gage}}A_1 = \beta\dot{m}V_2 - \beta\dot{m}V_1 \quad \rightarrow \quad F_{Rx} = \beta\dot{m}(V_2 - V_1) - P_{1,\,\text{gage}}A_1$$

since both V_1 and V_2 are in the positive *x*-direction. This shows the importance of using the correct sign (positive if in the positive direction and negative if in the opposite direction) for velocities and forces.

EXAMPLE 13–4 Water Jet Striking a Stationary Plate

Water is accelerated by a nozzle to an average speed of 20 m/s, and strikes a stationary vertical plate at a rate of 10 kg/s with a normal velocity of 20 m/s (Fig. 13–20). After the strike, the water stream splatters off in all directions in the plane of the plate. Determine the force needed to prevent the plate from moving horizontally due to the water stream.

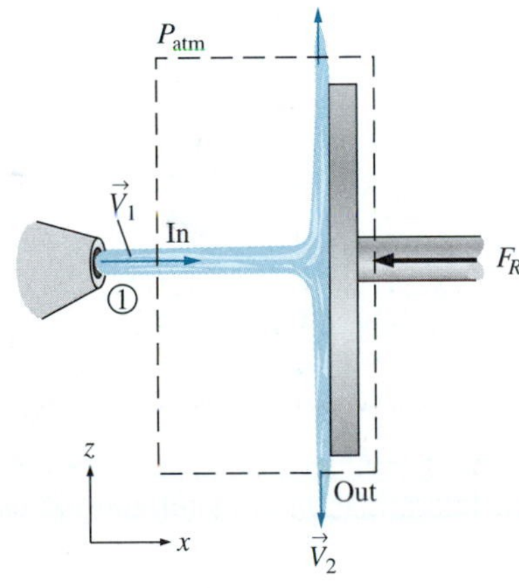

FIGURE 13–20
Schematic for Example 13–4.

Solution A water jet strikes a vertical stationary plate normally. The force needed to hold the plate in place is to be determined.

Assumptions **1** The flow of water at the nozzle outlet is steady. **2** The water splatters in directions normal to the approach direction of the water jet. **3** The water jet is exposed to the atmosphere, and thus the pressure of the water jet and the splattered water leaving the control volume is atmospheric pressure, which is disregarded since it acts on the entire system. **4** The vertical forces and momentum fluxes are not considered since they have no effect on the horizontal reaction force. **5** The effect of the momentum-flux correction factor is negligible, and thus $\beta \cong 1$ at the inlet.

Analysis We draw the control volume for this problem such that it contains the entire plate and cuts through the water jet and the support bar normally. The momentum equation for steady one-dimensional flow is given as

$$\sum \vec{F} = \sum_{\text{out}} \beta\dot{m}\vec{V} - \sum_{\text{in}} \beta\dot{m}\vec{V}$$

Writing it for this problem along the *x*-direction (without forgetting the negative sign for forces and velocities in the negative *x*-direction) and noting that $V_{1,x} = V_1$ and $V_{2,x} = 0$ gives

$$-F_R = 0 - \beta\dot{m}V_1$$

Substituting the given values,

$$F_R = \beta \dot{m} V_1 = (1)(10 \text{ kg/s})(20 \text{ m/s})\left(\frac{1 \text{ N}}{1 \text{ kg} \cdot \text{m/s}^2}\right) = \mathbf{200\ N}$$

Therefore, the support must apply a 200-N horizontal force (equivalent to the weight of about a 20-kg mass) in the negative *x*-direction (the opposite direction of the water jet) to hold the plate in place.

Discussion The plate absorbs the full brunt of the momentum of the water jet since the *x*-direction momentum at the outlet of the control volume is zero. If the control volume were drawn instead along the interface between the water and the plate, there would be additional (unknown) pressure forces in the analysis. By cutting the control volume through the support, we avoid having to deal with this additional complexity. This is an example of a "wise" choice of control volume.

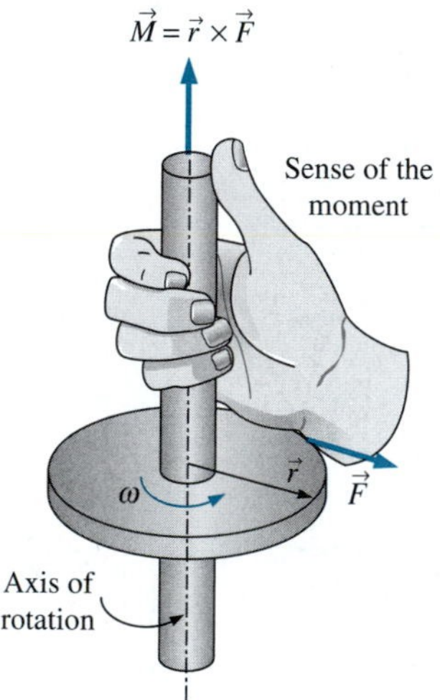

FIGURE 13–21
The determination of the direction of the moment by the right-hand rule.

EXAMPLE 13–5 Repositioning of a Satellite

An orbiting satellite has a mass of m_{sat} = 5000 kg and is traveling at a constant speed of V_o. To alter its orbit, an attached rocket discharges m_f = 100 kg of gases from the reaction of solid fuel at a velocity V_f = 3000 m/s relative to the satellite in a direction opposite to V_o (Fig. 13–21). The fuel discharge rate is constant for 2 s. Determine (*a*) the acceleration of the satellite during this 2-s period, (*b*) the change of velocity of the satellite during this time period, and (*c*) the thrust exerted on the satellite.

Solution The rocket of a satellite is fired in the opposite direction to motion. The acceleration, the velocity change, and the thrust are to be determined.

Assumptions **1** The flow of combustion gases is steady and one-dimensional during the firing period. **2** There are no external forces acting on the satellite, and the effect of the pressure force at the nozzle exit is negligible. **3** The mass of discharged fuel is negligible relative to the mass of the satellite, and thus the satellite may be treated as a solid body with a constant mass. **4** The nozzle is well-designed such that the effect of the momentum-flux correction factor is negligible, and thus $\beta \cong 1$.

Analysis (*a*) We choose a reference frame in which the control volume moves with the satellite. The the velocities of fluid streams become simply their velocities relative to the moving body. We take the direction of motion of the satellite as the positive direction along the *x*-axis. There are no external forces acting on the satellite and its mass is nearly constant. Therefore, the satellite can be treated as a solid body with constant mass, and the momentum equation in this case is simply Eq. 13–24.

$$0 = \frac{d(m\vec{V})_{CV}}{dt} + \sum_{out} \beta \dot{m} \vec{V} - \sum_{in} \beta \dot{m} \vec{V} \quad \rightarrow \quad m_{sat}\frac{d\vec{V}_{sat}}{dt} = -\dot{m}_f \vec{V}_f$$

Noting that the motion is on a straight line and the discharged gases move in the negative *x*-direction, we can write the momentum equation using magnitudes as

$$m_{sat}\frac{dV_{sat}}{dt} = \dot{m}_f V_f \quad \rightarrow \quad \frac{dV_{sat}}{dt} = \frac{\dot{m}_f}{m_{sat}} V_f = \frac{m_f/\Delta t}{m_{sat}} V_f$$

Substituting, the acceleration of the satellite during the first 2 s is determined to be

$$a_{sat} = \frac{dV_{sat}}{dt} = \frac{m_f/\Delta t}{m_{sat}}V_f = \frac{(100\ kg)/(2\ s)}{5000\ kg}(3000\ m/s) = \mathbf{30\ m/s^2}$$

(*b*) Knowing acceleration, which is constant, the velocity change of the satellite during the first 2 s is determined from the definition of acceleration $a_{sat} = dV_{sat}/dt$ to be

$$dV_{sat} = a_{sat}dt \quad \rightarrow \quad \Delta V_{sat} = a_{sat}\Delta t = (30\ m/s^2)(2\ s) = \mathbf{60\ m/s}$$

(*c*) The thrust exerted on the satellite is, from Eq. 13–31,

$$F_{sat} = 0 - \dot{m}_f(-V_f) = -(100/2\ kg/s)(-3000\ m/s)\left(\frac{1\ kN}{1000\ kg\cdot m/s^2}\right) = \mathbf{150\ kN}$$

Discussion Note that if this satellite were attached somewhere, it would exert a force of 150 kN (equivalent to the weight of 15 tons of mass) to its support. This can be verified by taking the satellite as the system and applying the momentum equation.

13–5 ▪ REVIEW OF ROTATIONAL MOTION AND ANGULAR MOMENTUM

The motion of a rigid body can be considered to be the combination of the translational motion of its center of mass and rotational motion about its center of mass. The translational motion can be analyzed using the linear momentum equation, Eq. 13–12. Now we discuss the rotational motion—a motion during which all points in the body move in circles about the axis of rotation. Rotational motion is described with angular quantities such as the angular distance θ, angular velocity ω, and angular acceleration α.

The amount of rotation of a point in a body is expressed in terms of the angle θ swept by a line of length r that connects the point to the axis of rotation and is perpendicular to the axis. The angle θ is expressed in radians (rad), which is the arc length corresponding to θ on a circle of unit radius. Noting that the circumference of a circle of radius r is $2\pi r$, the angular distance traveled by any point in a rigid body during a complete rotation is 2π rad. The physical distance traveled by a point along its circular path is $l = \theta r$, where r is the normal distance of the point from the axis of rotation and θ is the angular distance in rad. Note that 1 rad corresponds to $360/(2\pi) \cong 57.3°$.

Angular velocity ω is the angular distance traveled per unit time, and angular acceleration α is the rate of change of angular velocity. They are expressed as (Fig. 13–22),

$$\omega = \frac{d\theta}{dt} = \frac{d(l/r)}{dt} = \frac{1}{r}\frac{dl}{dt} = \frac{V}{r} \quad \text{and} \quad \alpha = \frac{d\omega}{dt} = \frac{d^2\theta}{dt^2} = \frac{1}{r}\frac{dV}{dt} = \frac{a_t}{r} \tag{13–26}$$

or

$$V = r\omega \quad \text{and} \quad a_t = r\alpha \tag{13–27}$$

where V is the linear velocity and a_t is the linear acceleration in the tangential direction for a point located at a distance r from the axis of rotation.

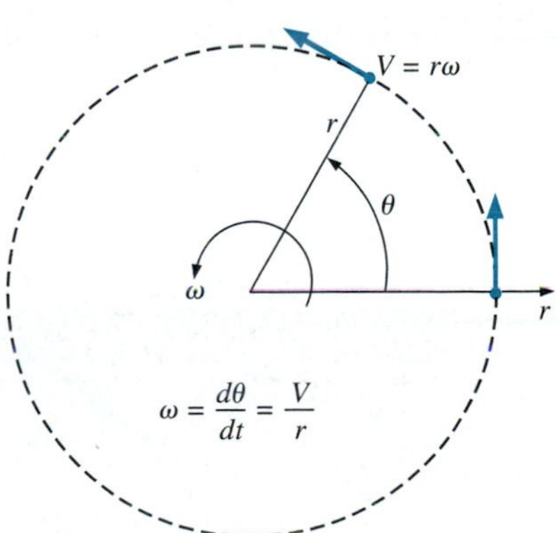

FIGURE 13–22
The relations between angular distance θ, angular velocity ω, and linear velocity V.

Note that ω and α are the same for all points of a rotating rigid body, but V and a_t are not (they are proportional to r).

Newton's second law requires that there must be a force acting in the tangential direction to cause angular acceleration. The strength of the rotating effect, called the *moment* or *torque*, is proportional to the magnitude of the force and its distance from the axis of rotation. The perpendicular distance from the axis of rotation to the line of action of the force is called the *moment arm*, and the torque M acting on a point mass m at a normal distance r from the axis of rotation is expressed as

$$M = rF_t = rma_t = mr^2\alpha \tag{13–28}$$

Mass, m ⟷ Moment of inertia, I
Linear acceleration, $\vec{a}$ ⟷ Angular acceleration, $\vec{\alpha}$
Linear velocity, $\vec{V}$ ⟷ Angular velocity, $\vec{\omega}$
Linear momentum ⟷ Angular momentum
$m\vec{V}$ ⟷ $I\vec{\omega}$

Force, $\vec{F}$ ⟷ Torque, $\vec{M}$
$\vec{F} = m\vec{a}$ ⟷ $\vec{M} = I\vec{\alpha}$
Moment of force, $\vec{M}$ ⟷ Moment of momentum, $\vec{H}$
$\vec{M} = \vec{r} \times \vec{F}$ ⟷ $\vec{H} = \vec{r} \times m\vec{V}$

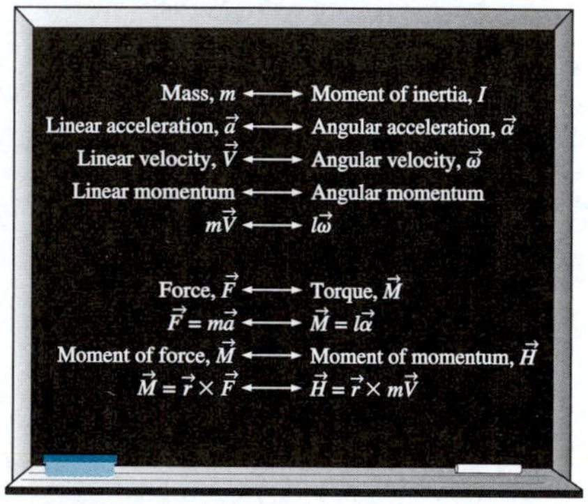

FIGURE 13–23

Analogy between corresponding linear and angular quantities.

The total torque acting on a rotating rigid body about an axis can be determined by integrating the torques acting on differential masses δm over the entire body to give

Torque:
$$M = \int_{\text{mass}} r^2\alpha\, \delta m = \left[\int_{\text{mass}} r^2\, \delta m\right]\alpha = I\alpha \tag{13–29}$$

where I is the *moment of inertia* of the body about the axis of rotation, which is a measure of the inertia of a body against rotation. The relation $M = I\alpha$ is the counterpart of Newton's second law, with torque replacing force, moment of inertia replacing mass, and angular acceleration replacing linear acceleration (Fig. 13–23). Note that unlike mass, the rotational inertia of a body also depends on the distribution of the mass of the body with respect to the axis of rotation. Therefore, a body whose mass is closely packed about its axis of rotation has a small resistance against angular acceleration, while a body whose mass is concentrated at its periphery has a large resistance against angular acceleration. A flywheel is a good example of the latter.

The linear momentum of a body of mass m having a velocity V is mV, and the direction of linear momentum is identical to the direction of velocity. Noting that the moment of a force is equal to the product of the force and the normal distance, the moment of momentum, called the **angular momentum**, of a point mass m about an axis can be expressed as $H = rmV = r^2m\omega$, where r is the normal distance from the axis of rotation to the line of action of the momentum vector (Fig. 13–24). Then the total angular momentum of a rotating rigid body can be determined by integration to be

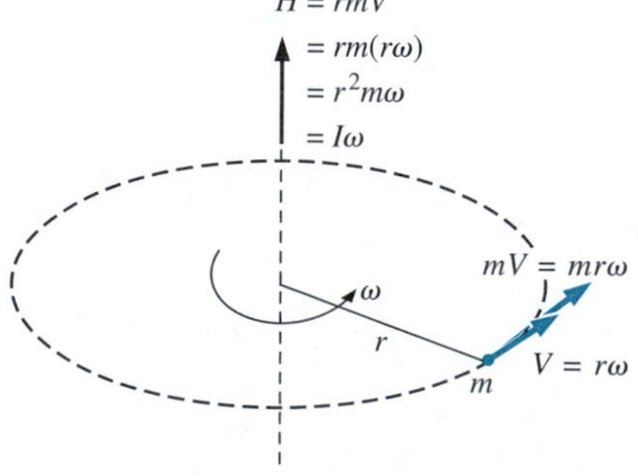

FIGURE 13–24

Angular momentum of point mass m rotating at angular velocity ω at distance r from the axis of rotation.

Angular momentum:
$$H = \int_{\text{mass}} r^2\omega\, \delta m = \left[\int_{\text{mass}} r^2\, \delta m\right]\omega = I\omega \tag{13–30}$$

where again I is the *moment of inertia* of the body about the axis of rotation. It can also be expressed in vector form as

$$\vec{H} = I\vec{\omega} \tag{13–31}$$

Note that the angular velocity $\vec{\omega}$ is the same at every point of a rigid body.

Newton's second law $\vec{F} = m\vec{a}$ was expressed in terms of the rate of change of linear momentum in Eq. 13–1 as $\vec{F} = d(m\vec{V})/dt$. Likewise, the counterpart of Newton's second law for rotating bodies $\vec{M} = I\vec{\alpha}$ is expressed in Eq. 13–2 in terms of the rate of change of angular momentum as

Angular momentum equation:
$$\vec{M} = I\vec{\alpha} = I\frac{d\vec{\omega}}{dt} = \frac{d(I\vec{\omega})}{dt} = \frac{d\vec{H}}{dt} \tag{13–32}$$

where $\vec{M}$ is the net torque applied on the body about the axis of rotation.

The angular velocity of rotating machinery is typically expressed in rpm (number of revolutions per minute) and denoted by $\dot{n}$. Noting that velocity is distance traveled per unit time and the angular distance traveled during each revolution is 2π, the angular velocity of rotating machinery is $\omega = 2\pi\dot{n}$ rad/min or

Angular velocity versus rpm: $$\omega = \frac{2\pi\dot{n}}{60} \quad \text{(rad/s)} \tag{13–33}$$

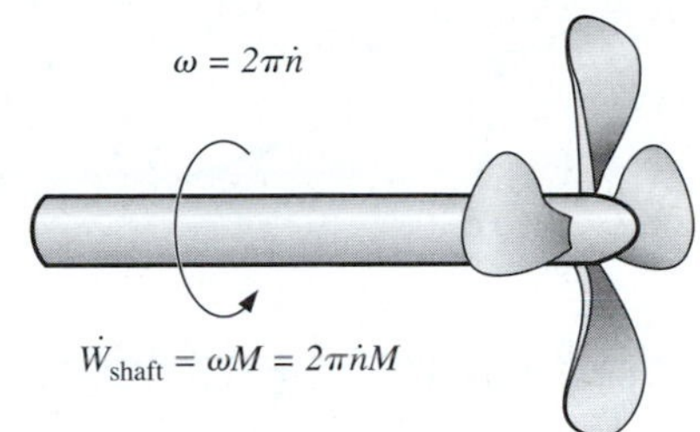

FIGURE 13–25
The relations between angular velocity, rpm, and the power transmitted through a shaft.

Consider a constant force F acting in the tangential direction on the outer surface of a shaft of radius r rotating at an rpm of $\dot{n}$. Noting that work W is force times distance, and power $\dot{W}$ is work done per unit time and thus force times velocity, we have $\dot{W}_{shaft} = FV = Fr\omega = M\omega$. Therefore, the power transmitted by a shaft rotating at an rpm of $\dot{n}$ under the influence of an applied torque M is (Fig. 13–25)

Shaft power: $$\dot{W}_{shaft} = \omega M = 2\pi\dot{n}M \quad \text{(W)} \tag{13–34}$$

The kinetic energy of a body of mass m during translational motion is $KE = \frac{1}{2}mV^2$. Noting that $V = r\omega$, the rotational kinetic energy of a body of mass m at a distance r from the axis of rotation is $KE = \frac{1}{2}mr^2\omega^2$. The total rotational kinetic energy of a rotating rigid body about an axis can be determined by integrating the rotational kinetic energies of differential masses dm over the entire body to give

Rotational kinetic energy: $$KE_r = \tfrac{1}{2}I\omega^2 \tag{13–35}$$

where again I is the moment of inertia of the body and ω is the angular velocity.

During rotational motion, the direction of velocity changes even when its magnitude remains constant. Velocity is a vector quantity, and thus a change in direction constitutes a change in velocity with time, and thus acceleration. This is called **centripetal acceleration**. Its magnitude is

$$a_r = \frac{V^2}{r} = r\omega^2$$

Centripetal acceleration is directed toward the axis of rotation (opposite direction of radial acceleration), and thus the radial acceleration is negative. Noting that acceleration is a constant multiple of force, centripetal acceleration is the result of a force acting on an element of the body toward the axis of rotation, known as the **centripetal force**, whose magnitude is $F_r = mV^2/r$. Tangential and radial accelerations are perpendicular to each other (since the radial and tangential directions are perpendicular), and the total linear acceleration is determined by their vector sum, $\vec{a} = \vec{a}_t + \vec{a}_r$. For a body rotating at constant angular velocity, the only acceleration is the centripetal acceleration. The centripetal force does not produce torque since its line of action intersects the axis of rotation.

13–6 ▪ THE ANGULAR MOMENTUM EQUATION

The linear momentum equation discussed in Section 13–4 is useful in determining the relationship between the linear momentum of flow streams and the resultant forces. Many engineering problems involve the moment of the

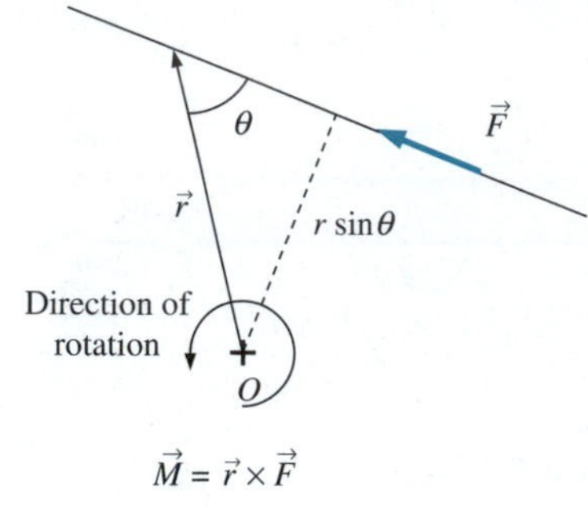

FIGURE 13–26

The moment of a force $\vec{F}$ about a point O is the vector product of the position vector $\vec{r}$ and $\vec{F}$.

linear momentum of flow streams, and the rotational effects caused by them. Such problems are best analyzed by the *angular momentum equation*, also called the *moment of momentum equation*. An important class of fluid devices, called *turbomachines*, which include centrifugal pumps, turbines, and fans, is analyzed by the angular momentum equation.

The *moment of a force* $\vec{F}$ about a point O is the vector (or cross) product (Fig. 13–26)

Moment of a force:
$$\vec{M} = \vec{r} \times \vec{F} \tag{13-36}$$

where $\vec{r}$ is the position vector from point O to any point on the line of action of $\vec{F}$. The vector product of two vectors is a vector whose line of action is normal to the plane that contains the crossed vectors ($\vec{r}$ and $\vec{F}$ in this case) and whose magnitude is

Magnitude of the moment of a force:
$$M = Fr \sin\theta \tag{13-37}$$

where θ is the angle between the lines of action of the vectors $\vec{r}$ and $\vec{F}$. Therefore, the magnitude of the moment about point O is equal to the magnitude of the force multiplied by the normal distance of the line of action of the force from the point O. The sense of the moment vector $\vec{M}$ is determined by the right-hand rule: when the fingers of the right hand are curled in the direction that the force tends to cause rotation, the thumb points the direction of the moment vector (Fig. 13–27). Note that a force whose line of action passes through point O produces zero moment about point O.

The vector product of $\vec{r}$ and the momentum vector $m\vec{V}$ gives the *moment of momentum*, also called the *angular momentum*, about a point O as

Moment of momentum:
$$\vec{H} = \vec{r} \times m\vec{V} \tag{13-38}$$

Therefore, $\vec{r} \times \vec{V}$ represents the angular momentum per unit mass, and the angular momentum of a differential mass $\delta m = \rho\, dV$ is $d\vec{H} = (\vec{r} \times \vec{V})\rho\, dV$. Then the angular momentum of a system is determined by integration to be

Moment of momentum (system):
$$\vec{H}_{sys} = \int_{sys} (\vec{r} \times \vec{V})\rho\, dV \tag{13-39}$$

The rate of change of the moment of momentum is

Rate of change of moment of momentum:
$$\frac{d\vec{H}_{sys}}{dt} = \frac{d}{dt}\int_{sys} (\vec{r} \times \vec{V})\rho\, dV \tag{13-40}$$

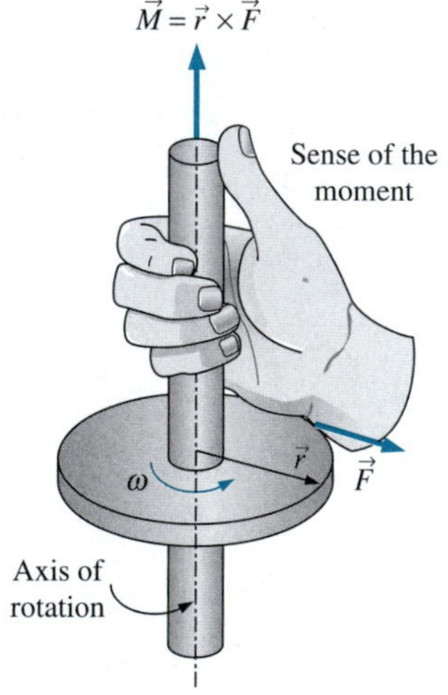

FIGURE 13–27

The determination of the direction of the moment by the right-hand rule.

The angular momentum equation for a system was expressed in Eq. 13–2 as

$$\sum \vec{M} = \frac{d\vec{H}_{sys}}{dt} \tag{13-41}$$

where $\sum \vec{M} = \sum(\vec{r} \times \vec{F})$ is the net torque applied on the system, which is the vector sum of the moments of all forces acting on the system, and $d\vec{H}_{sys}/dt$ is the rate of change of the angular momentum of the system. Equation 13–41 is stated as *the rate of change of angular momentum of a system is equal to the net torque acting on the system.* This equation is valid for a fixed quantity of mass and an inertial reference frame, i.e., a reference frame that is fixed or moves with a constant velocity in a straight path.

The general control volume formulation of the angular momentum equation is obtained by setting $b = \vec{r} \times \vec{V}$ and thus $B = \vec{H}$ in the general Reynolds transport theorem. It gives (Fig. 13–28)

$$\frac{d\vec{H}_{sys}}{dt} = \frac{d}{dt}\int_{CV} (\vec{r} \times \vec{V})\rho\, dV + \int_{CS} (\vec{r} \times \vec{V})\rho(\vec{V}_r \cdot \vec{n})\, dA \tag{13–42}$$

The left-hand side of this equation is, from Eq. 13–41, equal to $\sum \vec{M}$. Substituting, the angular momentum equation for a general control volume (stationary or moving, fixed shape or distorting) is

General:
$$\sum \vec{M} = \frac{d}{dt}\int_{CV} (\vec{r} \times \vec{V})\rho\, dV + \int_{CS} (\vec{r} \times \vec{V})\rho(\vec{V}_r \cdot \vec{n})\, dA \tag{13–43}$$

which can be stated as

$$\begin{pmatrix}\text{The sum of all}\\ \text{external moments}\\ \text{acting on a CV}\end{pmatrix} = \begin{pmatrix}\text{The time rate of change}\\ \text{of the angular momentum}\\ \text{of the contents of the CV}\end{pmatrix} + \begin{pmatrix}\text{The net flow rate of}\\ \text{angular momentum}\\ \text{out of the control}\\ \text{surface by mass flow}\end{pmatrix}$$

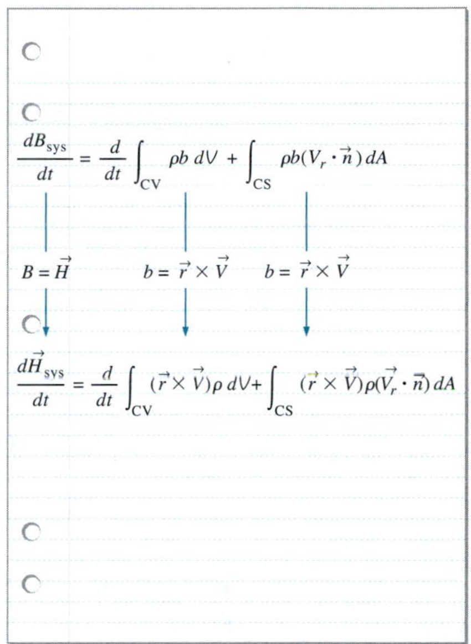

FIGURE 13–28
The angular momentum equation is obtained by replacing B in the Reynolds transport theorem by the angular momentum $\vec{H}$, and b by the angular momentum per unit mass $\vec{r} \times \vec{V}$.

Again, $\vec{V}_r = \vec{V} - \vec{V}_{CS}$ is the fluid velocity relative to the control surface (for use in mass flow rate calculations at all locations where the fluid crosses the control surface), and $\vec{V}$ is the fluid velocity as viewed from a fixed reference frame. The product $\rho(\vec{V}_r \cdot \vec{n})dA$ represents the mass flow rate through dA into or out of the control volume, depending on the sign.

For a fixed control volume (no motion or deformation of control volume), $\vec{V}_r = \vec{V}$ and the angular momentum equation becomes

Fixed CV:
$$\sum \vec{M} = \frac{d}{dt}\int_{CV} (\vec{r} \times \vec{V})\rho\, dV + \int_{CS} (\vec{r} \times \vec{V})\rho(\vec{V} \cdot \vec{n})\, dA \tag{13–44}$$

Also, note that the forces acting on the control volume consist of *body forces* that act throughout the entire body of the control volume such as gravity, and *surface forces* that act on the control surface such as the pressure and reaction forces at points of contact. The net torque consists of the moments of these forces as well as the torques applied on the control volume.

Special Cases

During *steady flow*, the amount of angular momentum within the control volume remains constant, and thus the time rate of change of angular momentum of the contents of the control volume is zero. Then,

Steady flow:
$$\sum \vec{M} = \int_{CS} (\vec{r} \times \vec{V})\rho(\vec{V}_r \cdot \vec{n})\, dA \tag{13–45}$$

In many practical applications, the fluid crosses the boundaries of the control volume at a certain number of inlets and outlets, and it is convenient to replace the area integral by an algebraic expression written in terms of the average properties over the cross-sectional areas where the fluid enters or leaves the control volume. In such cases, the angular momentum flow rate can be expressed as the difference in the angular momentum of outgoing and incoming streams. Furthermore, in many cases the moment arm $\vec{r}$ is either constant along the inlet or outlet (as in radial flow turbomachines) or is large compared to the diameter of the inlet or outlet pipe (as in rotating lawn sprinklers). In such cases, the *average* value of $\vec{r}$ is used throughout the cross-sectional area of the inlet or outlet. Then, an approximate form of the angular momentum equation in terms of average properties at inlets and outlets becomes

$$\sum \vec{M} = \frac{d}{dt}\int_{CV} (\vec{r} \times \vec{V})\rho\, dV + \sum_{out} \vec{r} \times \dot{m}\vec{V} - \sum_{in} \vec{r} \times \dot{m}\vec{V} \quad \textbf{(13–46)}$$

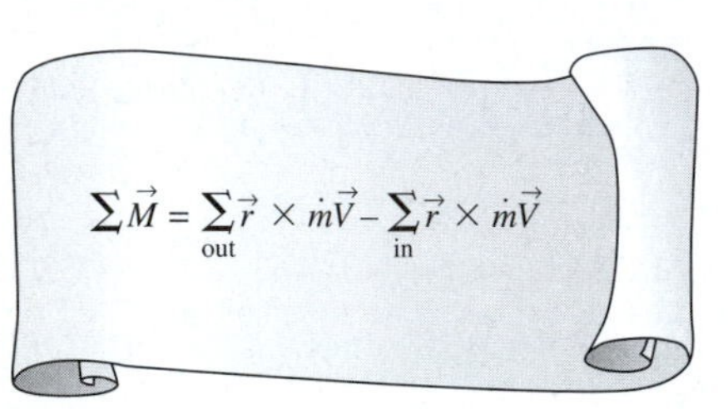

FIGURE 13–29
The net torque acting on a control volume during steady flow is equal to the difference between the outgoing and incoming angular momentum flows.

You may be wondering why we don't introduce a correction factor into Eq. 13–46, like we did for conservation of energy (Chap. 12) and for conservation of linear momentum (Section 13–4). The reason is that the cross product of $\vec{r}$ and $\dot{m}\vec{V}$ is dependent on problem geometry, so such a correction factor would vary from problem to problem. Thus, whereas we can readily calculate a kinetic energy flux correction factor and a momentum flux correction factor for fully developed pipe flow that can be applied to various problems, we cannot do so for angular momentum. Fortunately, in many problems of practical engineering interest, the error associated with using average values of radius and velocity is small, and the approximation of Eq. 13–46 is reasonable.

If the flow is *steady*, Eq. 13–46 further reduces to (Fig. 13–29)

Steady flow:
$$\sum \vec{M} = \sum_{out} \vec{r} \times \dot{m}\vec{V} - \sum_{in} \vec{r} \times \dot{m}\vec{V} \quad \textbf{(13–47)}$$

It states that *the net torque acting on the control volume during steady flow is equal to the difference between the outgoing and incoming angular momentum flow rates.* This statement can also be expressed for any specified direction.

In many problems, all the significant forces and momentum flows are in the same plane, and thus all give rise to moments in the same plane and about the same axis. For such cases, Eq. 13–47 can be expressed in scalar form as

$$\sum M = \sum_{out} r\dot{m}V - \sum_{in} r\dot{m}V \quad \textbf{(13–48)}$$

where r represents the average normal distance between the point about which moments are taken and the line of action of the force or velocity, provided that the sign convention for the moments is observed. That is, all moments in the counterclockwise direction are positive, and all moments in the clockwise direction are negative.

Radial-Flow Devices

Many rotary-flow devices such as centrifugal pumps and fans involve flow in the radial direction normal to the axis of rotation and are called *radial-flow devices*. In a centrifugal pump, for example, the fluid enters the device in the axial direction through the eye of the impeller, turns outward as it flows through the passages between the blades of the impeller, collects in the scroll, and is discharged in the tangential direction, as shown in Fig. 13–30. Axial-flow devices are easily analyzed using the linear momentum equation.

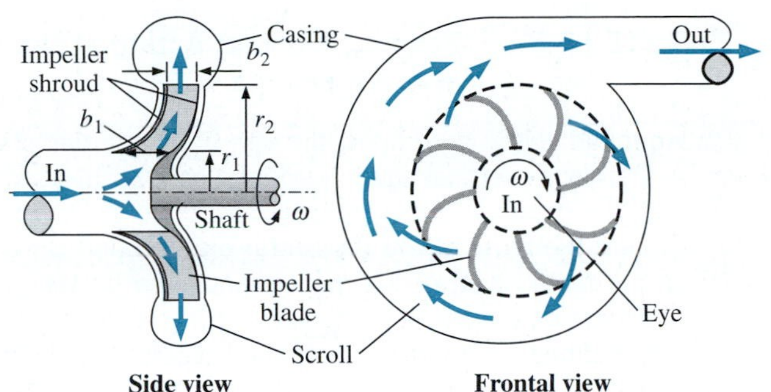

FIGURE 13–30

Side and frontal views of a typical centrifugal pump.

But radial-flow devices involve large changes in angular momentum of the fluid and are best analyzed with the help of the angular momentum equation.

To analyze the centrifugal pump, we choose the annular region that encloses the impeller section as the control volume, as shown in Fig. 13–31. Note that the average flow velocity, in general, has normal and tangential components at both the inlet and the outlet of the impeller section. Also, when the shaft rotates at angular velocity ω, the impeller blades have tangential velocity ωr_1 at the inlet and ωr_2 at the outlet. For steady incompressible flow, the conservation of mass equation is written as

$$\dot{V}_1 = \dot{V}_2 = \dot{V} \quad \rightarrow \quad (2\pi r_1 b_1)V_{1,n} = (2\pi r_2 b_2)V_{2,n} \tag{13–49}$$

where b_1 and b_2 are the flow widths at the inlet where $r = r_1$ and the outlet where $r = r_2$, respectively. (Note that the actual circumferential cross-sectional area is somewhat less than $2\pi rb$ since the blade thickness is not zero.) Then the average normal components $V_{1,n}$ and $V_{2,n}$ of absolute velocity can be expressed in terms of the volumetric flow rate $\dot{V}$ as

$$V_{1,n} = \frac{\dot{V}}{2\pi r_1 b_1} \quad \text{and} \quad V_{2,n} = \frac{\dot{V}}{2\pi r_2 b_2} \tag{13–50}$$

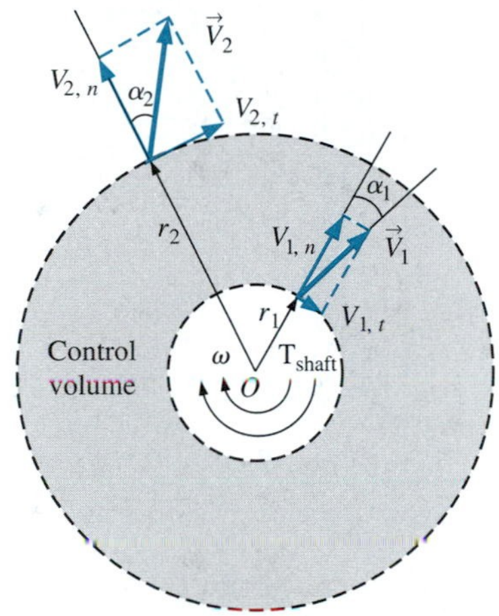

FIGURE 13–31

An annular control volume that encloses the impeller section of a centrifugal pump.

The normal velocity components $V_{1,n}$ and $V_{2,n}$ as well as pressure acting on the inner and outer circumferential areas pass through the shaft center, and thus they do not contribute to torque about the origin. Then only the tangential velocity components contribute to torque, and the application of the angular momentum equation $\sum M = \sum_{\text{out}} r\dot{m}V - \sum_{\text{in}} r\dot{m}V$ to the control volume gives

$$T_{shaft} = \dot{m}(r_2 V_{2,t} - r_1 V_{1,t}) \tag{13–51}$$

which is known as **Euler's turbine formula**. When angles α_1 and α_2 between the direction of absolute flow velocities and the radial direction are known, Eq. 13–51 becomes

$$T_{shaft} = \dot{m}(r_2 V_2 \sin \alpha_2 - r_1 V_1 \sin \alpha_1) \tag{13–52}$$

In the idealized case of the tangential fluid velocity being equal to the blade angular velocity both at the inlet and the exit, we have $V_{1,t} = \omega r_1$ and $V_{2,t} = \omega r_2$, and the torque becomes

$$T_{shaft,\ ideal} = \dot{m}\omega(r_2^2 - r_1^2) \tag{13–53}$$

where $\omega = 2\pi\dot{n}$ is the angular velocity of the blades. When the torque is known, the shaft power can be determined from $\dot{W}_{shaft} = \omega T_{shaft} = 2\pi\dot{n}T_{shaft}$.

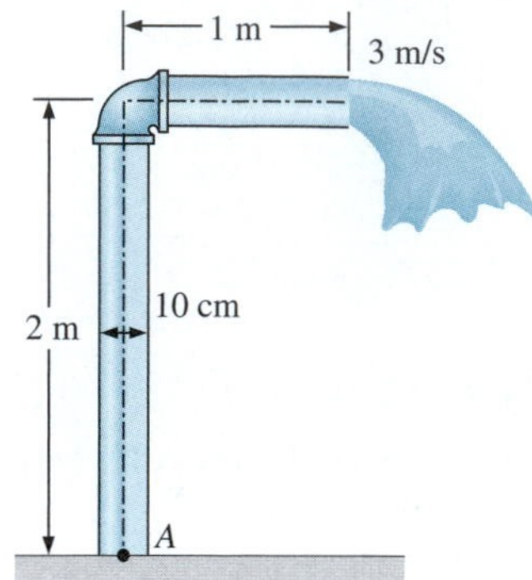

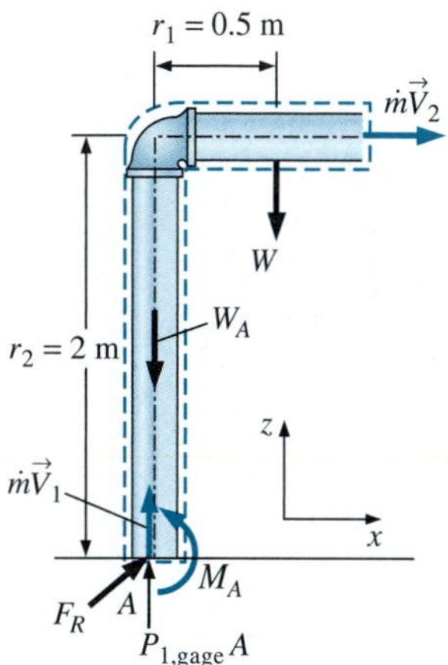

FIGURE 13–32

Schematic for Example 13–6 and the free-body diagram.

EXAMPLE 13–6 Bending Moment Acting at the Base of a Water Pipe

Underground water is pumped through a 10-cm-diameter pipe that consists of a 2-m-long vertical and 1-m-long horizontal section, as shown in Fig. 13–32. Water discharges to atmospheric air at an average velocity of 3 m/s, and the mass of the horizontal pipe section when filled with water is 12 kg per meter length. The pipe is anchored on the ground by a concrete base. Determine the bending moment acting at the base of the pipe (point *A*) and the required length of the horizontal section that would make the moment at point *A* zero.

Solution Water is pumped through a piping section. The moment acting at the base and the required length of the horizontal section to make this moment zero is to be determined.

Assumptions **1** The flow is steady. **2** The water is discharged to the atmosphere, and thus the gage pressure at the outlet is zero. **3** The pipe diameter is small compared to the moment arm, and thus we use average values of radius and velocity at the outlet.

Properties We take the density of water to be 1000 kg/m³.

Analysis We take the entire L-shaped pipe as the control volume, and designate the inlet by 1 and the outlet by 2. We also take the *x*- and *z*-coordinates as shown. The control volume and the reference frame are fixed.

The conservation of mass equation for this one-inlet, one-outlet, steady-flow system is $\dot{m}_1 = \dot{m}_2 = \dot{m}$, and $V_1 = V_2 = V$ since A_c = constant. The mass flow rate and the weight of the horizontal section of the pipe are

$$\dot{m} = \rho A_c V = (1000\ \text{kg/m}^3)[\pi(0.10\ \text{m})^2/4](3\ \text{m/s}) = 23.56\ \text{kg/s}$$

$$W = mg = (12\ \text{kg/m})(1\ \text{m})(9.81\ \text{m/s}^2)\left(\frac{1\ \text{N}}{1\ \text{kg}\cdot\text{m/s}^2}\right) = 117.7\ \text{N}$$

To determine the moment acting on the pipe at point *A*, we need to take the moment of all forces and momentum flows about that point. This is a steady-flow problem, and all forces and momentum flows are in the same plane. Therefore, the angular momentum equation in this case is expressed as

$$\sum M = \sum_{\text{out}} r\dot{m}V - \sum_{\text{in}} r\dot{m}V$$

where *r* is the average moment arm, *V* is the average speed, all moments in the counterclockwise direction are positive, and all moments in the clockwise direction are negative.

The free-body diagram of the L-shaped pipe is given in Fig. 13–32. Noting that the moments of all forces and momentum flows passing through point *A* are zero, the only force that yields a moment about point *A* is the weight *W* of the horizontal pipe section, and the only momentum flow that yields a moment is the outlet stream (both are negative since both moments are in the clockwise direction). Then the angular momentum equation about point *A* becomes

$$M_A - r_1 W = -r_2\dot{m}V_2$$

Solving for M_A and substituting give

$$\begin{aligned} M_A &= r_1 W - r_2\dot{m}V_2 \\ &= (0.5\ \text{m})(118\ \text{N}) - (2\ \text{m})(23.56\ \text{kg/s})(3\ \text{m/s})\left(\frac{1\ \text{N}}{1\ \text{kg}\cdot\text{m/s}^2}\right) \\ &= \mathbf{-82.5\ N\cdot m} \end{aligned}$$

The negative sign indicates that the assumed direction for M_A is wrong and should be reversed. Therefore, a moment of 82.5 N · m acts at the stem of the pipe in the clockwise direction. That is, the concrete base must apply a 82.5 N · m moment on the pipe stem in the clockwise direction to counteract the excess moment caused by the exit stream.

The weight of the horizontal pipe is $w = W/L = 117.7$ N per m length. Therefore, the weight for a length of L m is Lw with a moment arm of $r_1 = L/2$. Setting $M_A = 0$ and substituting, the length L of the horizontal pipe that would cause the moment at the pipe stem to vanish is determined to be

$$0 = r_1 W - r_2 \dot{m} V_2 \quad \rightarrow \quad 0 = (L/2)Lw - r_2 \dot{m} V_2$$

or

$$L = \sqrt{\frac{2r_2 \dot{m} V_2}{w}} = \sqrt{\frac{2(2 \text{ m})(23.56 \text{ kg/s})(3 \text{ m/s})}{117.7 \text{ N/m}}\left(\frac{\text{N}}{\text{kg} \cdot \text{m/s}^2}\right)} = \mathbf{1.55\ m}$$

Discussion Note that the pipe weight and the momentum of the exit stream cause opposing moments at point A. This example shows the importance of accounting for the moments of momentums of flow streams when performing a dynamic analysis and evaluating the stresses in pipe materials at critical cross sections.

EXAMPLE 13–7 Power Generation from a Sprinkler System

A large lawn sprinkler with four identical arms is to be converted into a turbine to generate electric power by attaching a generator to its rotating head, as shown in Fig. 13–33. Water enters the sprinkler from the base along the axis of rotation at a rate of 20 L/s and leaves the nozzles in the tangential direction. The sprinkler rotates at a rate of 300 rpm in a horizontal plane. The diameter of each jet is 1 cm, and the normal distance between the axis of rotation and the center of each nozzle is 0.6 m. Estimate the electric power produced.

Solution A four-armed sprinkler is used to generate electric power. For a specified flow rate and rotational speed, the power produced is to be determined.

Assumptions **1** The flow is cyclically steady (i.e., steady from a frame of reference rotating with the sprinkler head). **2** The water is discharged to the atmosphere, and thus the gage pressure at the nozzle exit is zero. **3** Generator losses and air drag of rotating components are neglected. **4** The nozzle diameter is small compared to the moment arm, and thus we use average values of radius and velocity at the outlet.

Properties We take the density of water to be 1000 kg/m³ = 1 kg/L.

Analysis We take the disk that encloses the sprinkler arms as the control volume, which is a stationary control volume.

The conservation of mass equation for this steady-flow system is $\dot{m}_1 = \dot{m}_2 = \dot{m}_{\text{total}}$. Noting that the four nozzles are identical, we have $\dot{m}_{\text{nozzle}} = \dot{m}_{\text{total}}/4$ or $\dot{V}_{\text{nozzle}} = \dot{V}_{\text{total}}/4$ since the density of water is constant. The average jet exit velocity relative to the nozzle is

$$V_{\text{jet},r} = \frac{\dot{V}_{\text{nozzle}}}{A_{\text{jet}}} = \frac{5 \text{ L/s}}{[\pi(0.01 \text{ m})^2/4]}\left(\frac{1 \text{ m}^3}{1000 \text{ L}}\right) = 63.66 \text{ m/s}$$

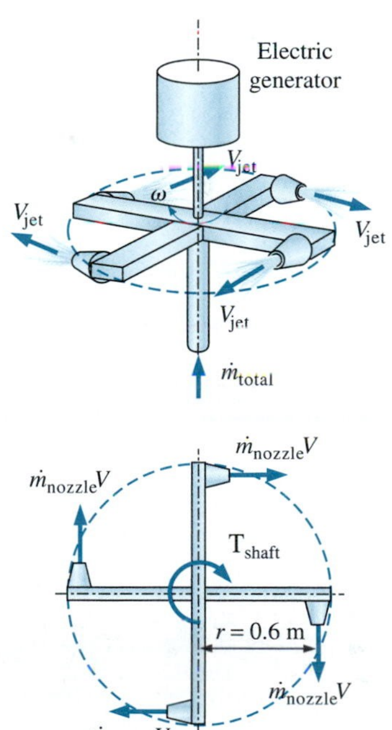

FIGURE 13–33
Schematic for Example 13–7 and the free-body diagram.

The angular and tangential velocities of the nozzles are

$$\omega = 2\pi\dot{n} = 2\pi(300 \text{ rev/min})\left(\frac{1 \text{ min}}{60 \text{ s}}\right) = 31.42 \text{ rad/s}$$

$$V_{\text{nozzle}} = r\omega = (0.6 \text{ m})(31.42 \text{ rad/s}) = 18.85 \text{ m/s}$$

Note that water in the nozzle is also moving at a velocity of 18.85 m/s in the opposite direction when it is discharged. The absolute velocity of the water jet (velocity relative to a fixed location on earth) is the vector sum of its relative velocity (jet velocity relative to the nozzle) and the absolute nozzle velocity,

$$\vec{V}_{\text{jet}} = \vec{V}_{\text{jet}, r} + \vec{V}_{\text{nozzle}}$$

All of these three velocities are in the tangential direction, and taking the direction of jet flow as positive, the vector equation above is written in scalar from using magnitudes as

$$V_{\text{jet}} = V_{\text{jet}, r} - V_{\text{nozzle}} = 63.66 - 18.85 = 44.81 \text{ m/s}$$

Noting that this is a cyclically steady-flow problem, and all forces and momentum flows are in the same plane, the angular momentum equation is approximated as $\sum M = \sum_{\text{out}} r\dot{m}V - \sum_{\text{in}} r\dot{m}V$, where r is the moment arm, all moments in the counterclockwise direction are positive, and all moments in the clockwise direction are negative.

The free-body diagram of the disk that contains the sprinkler arms is given in Fig. 13–33. Note that the moments of all forces and momentum flows passing through the axis of rotation are zero. The momentum flows via the water jets leaving the nozzles yield a moment in the clockwise direction and the effect of the generator on the control volume is a moment also in the clockwise direction (thus both are negative). Then the angular momentum equation about the axis of rotation becomes

$$-\text{T}_{\text{shaft}} = -4r\dot{m}_{\text{nozzle}}V_{\text{jet}} \qquad \text{or} \qquad \text{T}_{\text{shaft}} = r\dot{m}_{\text{total}}V_{\text{jet}}$$

Substituting, the torque transmitted through the shaft is determined to be

$$\text{T}_{\text{shaft}} = r\dot{m}_{\text{total}}V_{\text{jet}} = (0.6 \text{ m})(20 \text{ kg/s})(44.81 \text{ m/s})\left(\frac{1 \text{ N}}{1 \text{ kg} \cdot \text{m/s}^2}\right) = 537.7 \text{ N} \cdot \text{m}$$

since $\dot{m}_{\text{total}} = \rho\dot{V}_{\text{total}} = (1 \text{ kg/L})(20 \text{ L/s}) = 20 \text{ kg/s}$.

Then the power generated becomes

$$\dot{W} = 2\pi\dot{n}\text{T}_{\text{shaft}} = \omega\text{T}_{\text{shaft}} = (31.42 \text{ rad/s})(537.7 \text{ N} \cdot \text{m})\left(\frac{1 \text{ kW}}{1000 \text{ N} \cdot \text{m/s}}\right) = \mathbf{16.9 \text{ kW}}$$

Therefore, this sprinkler-type turbine has the potential to produce 16.9 kW of power.

Discussion: To put the result obtained in perspective, we consider two limiting cases. In the first limiting case, the sprinkler is stuck and thus the angular velocity is zero. The torque developed will be maximum in this case since $V_{\text{nozzle}} = 0$ and thus $V_{\text{jet}} = V_{\text{jet}, r} = 63.66$ m/s, giving $\text{T}_{\text{shaft, max}} = 764$ N · m. But the power generated will be zero since the generator shaft does not rotate.

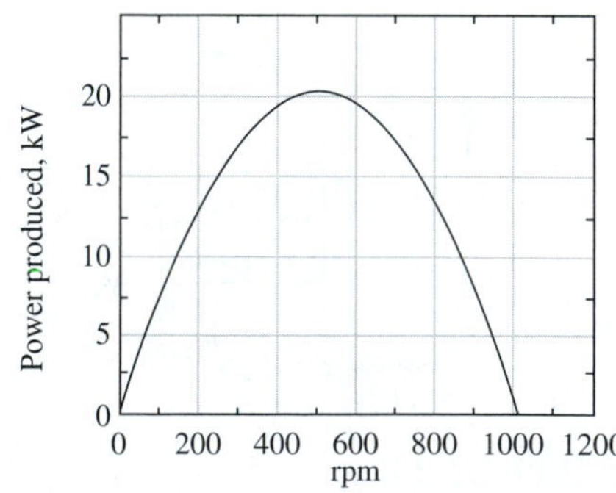

FIGURE 13–34
The variation of power produced with angular speed.

In the second limiting case, the sprinkler shaft is disconnected from the generator (and thus both the useful torque and power generation are zero) and the shaft accelerates until it reaches an equilibrium velocity. Setting $T_{shaft} = 0$ in the angular momentum equation gives the absolute water jet velocity (jet velocity relative to an observer on earth) to be zero, $V_{jet} = 0$. Therefore, the relative velocity $V_{jet,\,r}$ and absolute velocity V_{nozzle} are equal but in opposite direction (as if the nozzle is sliding backward over a stationary jet-water) so that the jet-water absolute tangential velocity (and thus torque) is zero and the water mass is "dropping down as a downward fountain" under gravity with zero angular momentum (around the axis of rotation). The angular speed of the sprinkler in this case is

$$\dot{n} = \frac{\omega}{2\pi} = \frac{V_{nozzle}}{2\pi r} = \frac{63.66\text{ m/s}}{2\pi(0.6\text{ m})}\left(\frac{60\text{ s}}{1\text{ min}}\right) = 1013\text{ rpm}$$

Of course the $T_{shaft} = 0$ case is only possible for an ideal frictionless nozzle (i.e., 100% nozzle efficiency, as a no-load ideal turbine). Otherwise there will be a resisting torque due to friction of water, shaft, and surrounding air.

The variation of power produced with angular speed is plotted in Fig. 13–34. Note that the power produced increases with increasing rpm, reaches a maximum (at about 500 rpm in this case), and then decreases. The actual power produced will be less than this due to generator inefficiency, and other irreversible losses such as the fluid friction within the nozzle, shaft friction, and aerodynamic drag.

SUMMARY

This chapter deals mainly with the conservation of momentum for finite control volumes. The forces acting on the control volume consist of *body forces* that act throughout the entire body of the control volume (such as gravity, electric, and magnetic forces) and *surface forces* that act on the control surface (such as the pressure forces and reaction forces at points of contact). The sum of all forces acting on the control volume at a particular instant in time is represented by $\sum \vec{F}$ and is expressed as

$$\underbrace{\sum \vec{F}}_{\text{total force}} = \underbrace{\sum \vec{F}_{\text{gravity}}}_{\text{body force}} + \underbrace{\sum \vec{F}_{\text{pressure}} + \sum \vec{F}_{\text{viscous}} + \sum \vec{F}_{\text{other}}}_{\text{surface forces}}$$

Newton's second law can be stated as *the sum of all external forces acting on a system is equal to the time rate of change of linear momentum of the system.* Setting $b = \vec{V}$ and thus $B = m\vec{V}$ in the Reynolds transport theorem and utilizing Newton's second law gives the *linear momentum equation* for a control volume as

$$\sum \vec{F} = \frac{d}{dt}\int_{CV} \rho\vec{V}\,dV + \int_{CS} \rho\vec{V}(\vec{V}_r \cdot \vec{n})\,dA$$

which reduces to the following special cases:

Steady flow: $$\sum \vec{F} = \int_{CS} \rho\vec{V}(\vec{V}_r \cdot \vec{n})\,dA$$

Unsteady flow (*algebraic form*):

$$\sum \vec{F} = \frac{d}{dt}\int_{CV} \rho\vec{V}\,dV + \sum_{out} \beta\dot{m}\vec{V} - \sum_{in} \beta\dot{m}\vec{V}$$

Steady flow (*algebraic form*): $$\sum \vec{F} = \sum_{out} \beta\dot{m}\vec{V} - \sum_{in} \beta\dot{m}\vec{V}$$

where β is the momentum-flux correction factor. A control volume whose mass m remains constant can be treated as a solid body (a fixed-mass system), with a net thrusting force (or just **thrust**) of $\vec{F}_{thrust} = m_{CV}\vec{a} = \sum_{in} \beta\dot{m}\vec{V} - \sum_{out} \beta\dot{m}\vec{V}$ acting on it.

Newton's second law can also be stated as *the rate of change of angular momentum of a system is equal to the net torque acting on the system.* Setting $b = \vec{r} \times \vec{V}$ and thus $B = \vec{H}$ in the general Reynolds transport theorem gives the *angular momentum equation* as

$$\sum \vec{M} = \frac{d}{dt}\int_{CV} (\vec{r} \times \vec{V})\rho \, dV + \int_{CS} (\vec{r} \times \vec{V})\rho(\vec{V}_r \cdot \vec{n})\, dA$$

which reduces to the following special cases:

Steady flow:
$$\sum \vec{M} = \int_{CS} (\vec{r} \times \vec{V})\rho(\vec{V}_r \cdot \vec{n})\, dA$$

Unsteady flow (algebraic form):

$$\sum \vec{M} = \frac{d}{dt}\int_{CV} (\vec{r} \times \vec{V})\rho \, dV + \sum_{out} \vec{r} \times \dot{m}\vec{V} - \sum_{in} \vec{r} \times \dot{m}\vec{V}$$

Steady and uniform flow:

$$\sum \vec{M} = \sum_{out} \vec{r} \times \dot{m}\vec{V} - \sum_{in} \vec{r} \times \dot{m}\vec{V}$$

Scalar form for one direction:

$$\sum M = \sum_{out} r\dot{m}V - \sum_{in} r\dot{m}V$$

The linear and angular momentum equations are of fundamental importance in the analysis of turbomachinery.

REFERENCES AND SUGGESTED READING

1. P. K. Kundu. *Fluid Mechanics*. San Diego, CA: Academic Press, 1990.

2. T. Wright. *Fluid Machinery Performance, Analysis, and Design*. Boca Raton, FL: CRC Press, 1999.

PROBLEMS*

Newton's Laws and Conservation of Momentum

13–1C Express Newton's first, second, and third laws.

13–2C Is momentum a vector? If so, in what direction does it point?

13–3C Express the conservation of momentum principle. What can you say about the momentum of a body if the net force acting on it is zero?

13–4C Express Newton's second law of motion for rotating bodies. What can you say about the angular velocity and angular momentum of a rotating nonrigid body of constant mass if the net torque acting on it is zero?

13–5C Consider two rigid bodies having the same mass and angular speed. Do you think these two bodies must have the same angular momentum? Explain.

Linear Momentum Equation

13–6C Explain the importance of the Reynolds transport theorem in fluid mechanics, and describe how the linear momentum equation is obtained from it.

13–7C Describe body forces and surface forces, and explain how the net force acting on control volume is determined. Is fluid weight a body force or surface force? How about pressure?

13–8C How do surface forces arise in the momentum analysis of a control volume? How can we minimize the number of surface forces exposed during analysis?

13–9C What is the importance of the momentum-flux correction factor in the momentum analysis of flow systems? For which type of flow is it significant and must it be considered in analysis: laminar flow, turbulent flow, or jet flow?

13–10C Write the momentum equation for steady one-dimensional flow for the case of no external forces and explain the physical significance of its terms.

13–11C In the application of the momentum equation, explain why we can usually disregard the atmospheric pressure and work with gage pressures only.

13–12C Two firefighters are fighting a fire with identical water hoses and nozzles, except that one is holding the hose straight so that the water leaves the nozzle in the same direction it comes, while the other holds it backward so that the water makes a U-turn before being discharged. Which firefighter will experience a greater reaction force?

13–13C A rocket in space (no friction or resistance to motion) can expel gases relative to itself at some high velocity V. Is V the upper limit to the rocket's ultimate velocity?

13–14C Describe in terms of momentum and airflow why a helicopter hovers.

* Problems designated by a "C" are concept questions, and students are encouraged to answer them all. Problems designated by an "E" are in English units, and the SI users can ignore them. Problems with the EES icon are solved using EES, and complete solutions together with parametric studies are included on the enclosed DVD. Problems with the computer icon are comprehensive in nature and are intended to be solved with a computer, preferably using the EES software that accompanies this text.

FIGURE P13–14C

13–15C Does it take more, equal, or less power for a helicopter to hover at the top of a high mountain than it does at sea level? Explain.

13–16C In a given location, would a helicopter require more energy in summer or winter to achieve a specified performance? Explain.

13–17C A horizontal water jet from a nozzle of constant exit cross section impinges normally on a stationary vertical flat plate. A certain force F is required to hold the plate against the water stream. If the water velocity is doubled, will the necessary holding force also be doubled? Explain.

13–18C A constant-velocity horizontal water jet from a stationary nozzle impinges normally on a vertical flat plate that is held in a nearly frictionless track. As the water jet hits the plate, it begins to move due to the water force. Will the acceleration of the plate remain constant or change? Explain.

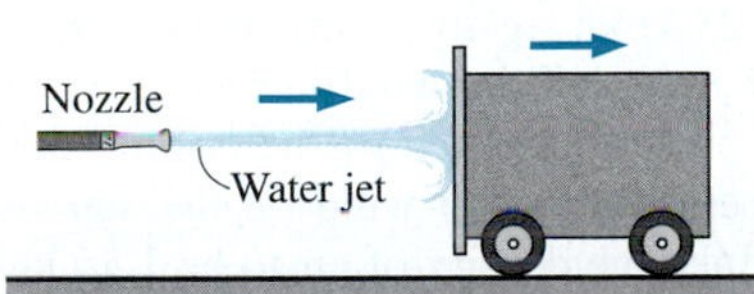

FIGURE P13–18C

13–19C A horizontal water jet of constant velocity V from a stationary nozzle impinges normally on a vertical flat plate that is held in a nearly frictionless track. As the water jet hits the plate, it begins to move due to the water force. What is the highest velocity the plate can attain? Explain.

13–20 Show that the force exerted by a liquid jet on a stationary nozzle as it leaves with a velocity V is proportional to V^2 or, alternatively, to $\dot{m}^2$. Assume the jet stream is perpendicular to the incoming liquid flow line.

13–21 A horizontal water jet of constant velocity V impinges normally on a vertical flat plate and splashes off the sides in the vertical plane. The plate is moving toward the oncoming water jet with velocity $\frac{1}{2}V$. If a force F is required to maintain the plate stationary, how much force is required to move the plate toward the water jet?

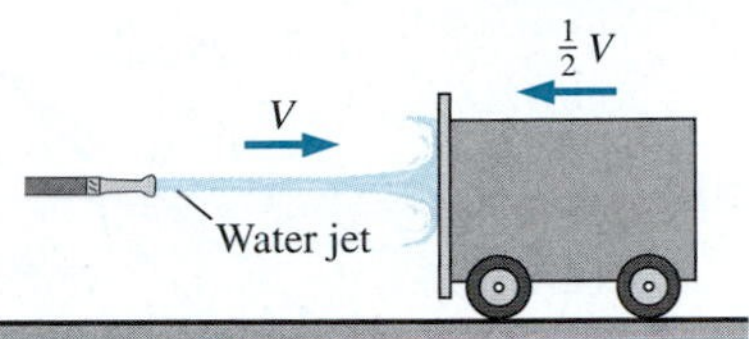

FIGURE P13–21

13–22 A 90° elbow is used to direct water flow at a rate of 25 kg/s in a horizontal pipe upward. The diameter of the entire elbow is 10 cm. The elbow discharges water into the atmosphere, and thus the pressure at the exit is the local atmospheric pressure. The elevation difference between the centers of the exit and the inlet of the elbow is 35 cm. The weight of the elbow and the water in it is considered to be negligible. Determine (*a*) the gage pressure at the center of the inlet of the elbow and (*b*) the anchoring force needed to hold the elbow in place. Take the momentum-flux correction factor to be 1.03.

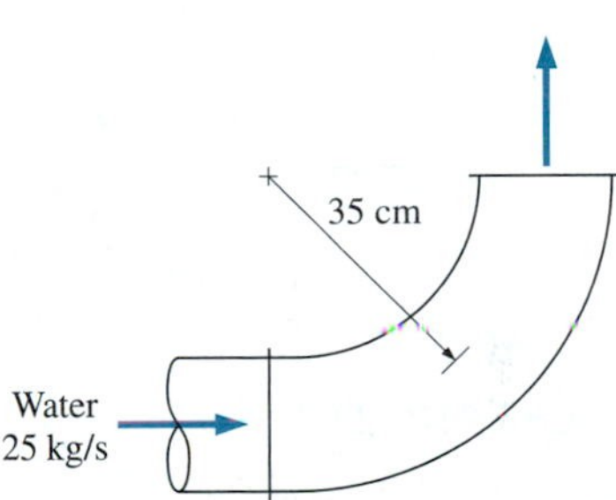

FIGURE P13–22

13–23 Repeat Prob. 13–22 for the case of another (identical) elbow being attached to the existing elbow so that the fluid makes a U-turn. *Answers:* (*a*) 6.87 kPa, (*b*) 218 N

13–24E A horizontal water jet impinges against a vertical flat plate at 30 ft/s and splashes off the sides in the vertical plane. If a horizontal force of 350 lbf is required to hold the plate against the water stream, determine the volume flow rate of the water.

13–25 A reducing elbow is used to deflect water flow at a rate of 30 kg/s in a horizontal pipe upward by an angle $\theta = 45°$ from the flow direction while accelerating it. The elbow discharges water into the atmosphere. The cross-sectional area of the elbow is 150 cm^2 at the inlet and 25 cm^2 at the exit. The elevation difference between the centers of the exit and the inlet is 40 cm. The mass of the elbow and the water in it is 50 kg. Determine the anchoring force needed to hold the elbow in place. Take the momentum-flux correction factor to be 1.03.

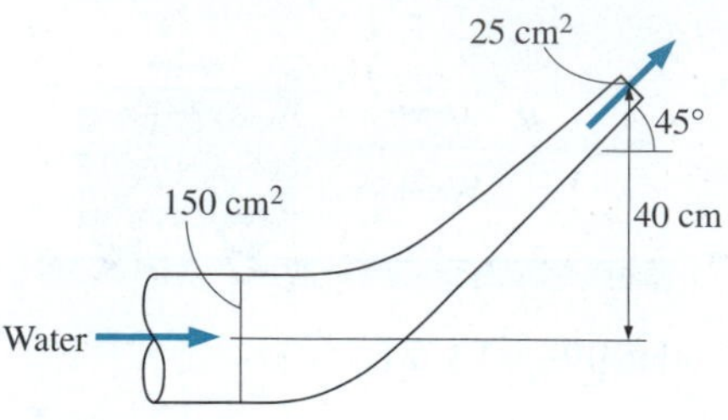

FIGURE P13–25

13–26 Repeat Prob. 13–25 for the case of $\theta = 110°$.

13–27 Water accelerated by a nozzle to 15 m/s strikes the vertical back surface of a cart moving horizontally at a constant velocity of 5 m/s in the flow direction. The mass flow rate of water through the stationary nozzle is 25 kg/s. After the strike, the water stream splatters off in all directions in the plane of the back surface. (*a*) Determine the force that needs to be applied by the brakes of the cart to prevent it from accelerating. (*b*) If this force were used to generate power instead of wasting it on the brakes, determine the maximum amount of power that can be generated. *Answers:* (*a*) −167 N, (*b*) 833 W

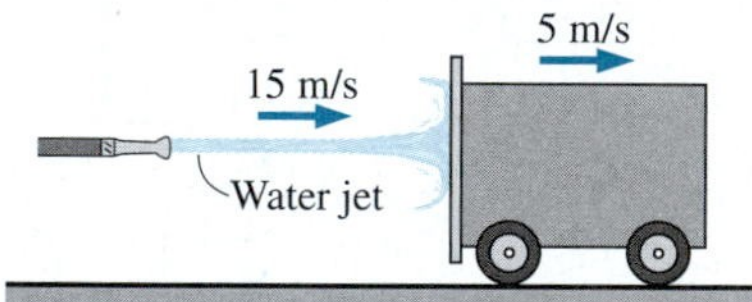

FIGURE P13–27

13–28 Reconsider Prob. 13–27. If the mass of the cart is 300 kg and the brakes fail, determine the acceleration of the cart when the water first strikes it. Assume the mass of water that wets the back surface is negligible.

13–29E A 100-ft^3/s water jet is moving in the positive *x*-direction at 20 ft/s. The stream hits a stationary splitter, such that half of the flow is diverted upward at 45° and the other half is directed downward, and both streams have a final speed of 20 ft/s. Disregarding gravitational effects, determine the *x*- and *z*-components of the force required to hold the splitter in place against the water force.

13–30E Reconsider Prob. 13–29E. Using EES (or other) software, investigate the effect of the splitter angle on the force exerted on the splitter in the incoming flow direction. Let the half splitter angle vary from 0° to 180° in increments of 10°. Tabulate and plot your results, and draw some conclusions.

13–31 A horizontal 5-cm-diameter water jet with a velocity of 18 m/s impinges normally upon a vertical plate of mass 1000 kg. The plate is held in a nearly frictionless track and is initially stationary. When the jet strikes the plate, the plate begins to move in the direction of the jet. The water always splatters in the plane of the retreating plate. Determine (*a*) the acceleration of the plate when the jet first strikes it (time = 0), (*b*) the time it will take for the plate to reach a velocity of 9 m/s, and (*c*) the plate velocity 20 s after the jet first strikes the plate. Assume the velocity of the jet relative to the plate remains constant.

13–32 Water flowing in a horizontal 30-cm-diameter pipe at 5 m/s and 300 kPa gage enters a 90° bend reducing section, which connects to a 15-cm-diameter vertical pipe. The inlet of the bend is 50 cm above the exit. Neglecting any frictional and gravitational effects, determine the net resultant force exerted on the reducer by the water. Take the momentum-flux correction factor to be 1.04.

13–33 Commercially available large wind turbines have blade span diameters as large as 100 m and generate over 3 MW of electric power at peak design conditions. Consider a wind turbine with a 90-m blade span subjected to 25-km/h steady winds. If the combined turbine–generator efficiency of the wind turbine is 32 percent, determine (*a*) the power generated by the turbine and (*b*) the horizontal force exerted by the wind on the supporting mast of the turbine. Take the density of air to be 1.25 kg/m^3, and disregard frictional effects.

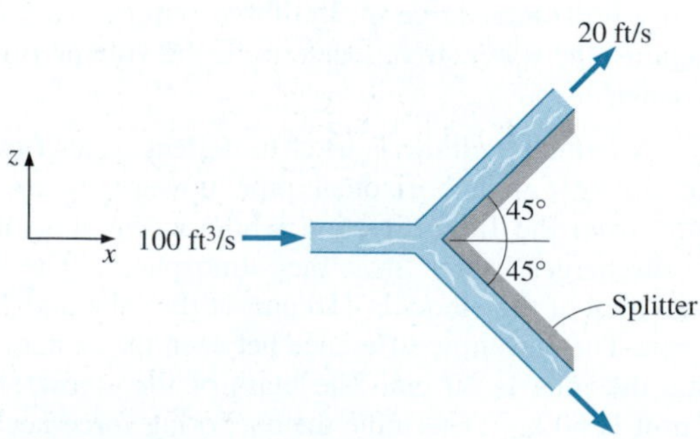

FIGURE P13–29E

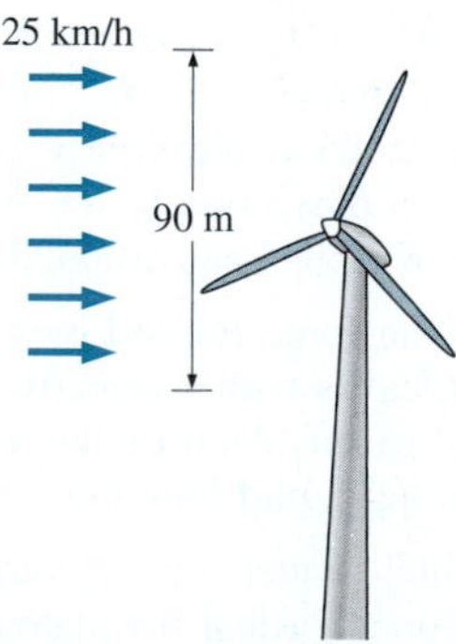

FIGURE P13–33

13–34E A 3-in-diameter horizontal water jet having a velocity of 140 ft/s strikes a curved plate, which deflects the water 180° at the same speed. Ignoring the frictional effects, determine the force required to hold the plate against the water stream.

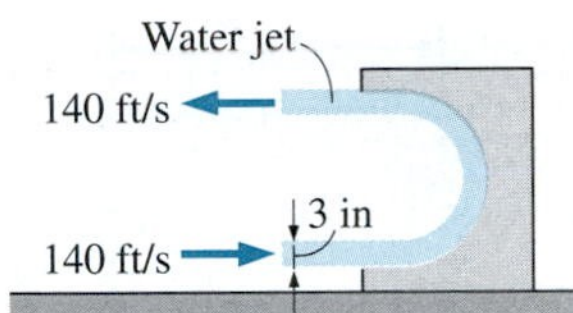

FIGURE P13–34E

13–35E A 3-in-diameter horizontal jet of water, with velocity 140 ft/s, strikes a bent plate, which deflects the water by 135° from its original direction. How much force is required to hold the plate against the water stream and what is its direction? Disregard frictional and gravitational effects.

13–36 Firefighters are holding a nozzle at the end of a hose while trying to extinguish a fire. If the nozzle exit diameter is 6 cm and the water flow rate is 5 m³/min, determine (*a*) the average water exit velocity and (*b*) the horizontal resistance force required of the firefighters to hold the nozzle. *Answers:* (*a*) 29.5 m/s, (*b*) 2457 N

FIGURE P13–36

13–37 A 5-cm-diameter horizontal jet of water with a velocity of 30 m/s relative to the ground strikes a flat plate that is moving in the same direction as the jet at a velocity of 10 m/s. The water splatters in all directions in the plane of the plate. How much force does the water stream exert on the plate?

13–38 Reconsider Prob. 13–37. Using EES (or other) software, investigate the effect of the plate velocity on the force exerted on the plate. Let the plate velocity vary from 0 to 30 m/s, in increments of 3 m/s. Tabulate and plot your results.

13–39E A fan with 24-in-diameter blades moves 2000 cfm (cubic feet per minute) of air at 70°F at sea level. Determine (*a*) the force required to hold the fan and (*b*) the minimum power input required for the fan. Choose the control volume sufficiently large to contain the fan, and the gage pressure and the air velocity on the inlet side to be zero. Assume air approaches the fan through a large area with negligible velocity and air exits the fan with a uniform velocity at atmospheric pressure through an imaginary cylinder whose diameter is the fan blade diameter. *Answers:* (*a*) 0.82 lbf, (*b*) 5.91 W

13–40 An unloaded helicopter of mass 10,000 kg hovers at sea level while it is being loaded. In the unloaded hover mode, the blades rotate at 400 rpm. The horizontal blades above the helicopter cause a 15-m-diameter air mass to move downward at an average velocity proportional to the overhead blade rotational velocity (rpm). A load of 15,000 kg is loaded onto the helicopter, and the helicopter slowly rises. Determine (*a*) the volumetric airflow rate downdraft that the helicopter generates during unloaded hover and the required power input and (*b*) the rpm of the helicopter blades to hover with the 15,000-kg load and the required power input. Take the density of atmospheric air to be 1.18 kg/m³. Assume air approaches the blades from the top through a large area with negligible velocity and air is forced by the blades to move down with a uniform velocity through an imaginary cylinder whose base is the blade span area.

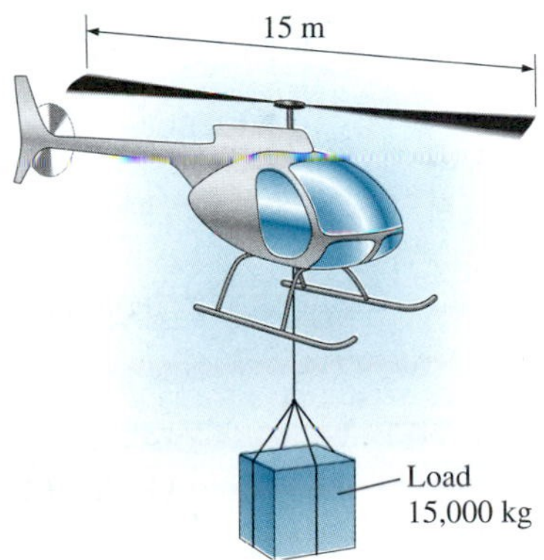

FIGURE P13–40

13–41 Reconsider the helicopter in Prob. 13–40, except that it is hovering on top of a 3000-m-high mountain where the air density is 0.79 kg/m³. Noting that the unloaded helicopter blades must rotate at 400 rpm to hover at sea level, determine the blade rotational velocity to hover at the higher altitude. Also determine the percent increase in the required power input to hover at 3000-m altitude relative to that at sea level. *Answers:* 489 rpm, 22 percent

13–42 A sluice gate, which controls flow rate in a channel by simply raising or lowering a vertical plate, is commonly used in irrigation systems. A force is exerted on the gate due to the difference between the water heights y_1 and y_2 and the flow velocities V_1 and V_2 upstream and downstream from the gate, respectively. Disregarding the wall shear forces at the channel surfaces, develop relations for V_1, V_2, and the force acting on a sluice gate of width w during steady and uniform flow.

Answer: $F_R = \dot{m}(V_1 - V_2) + \frac{w}{2}\rho g\,(y_1^2 - y_2^2)$

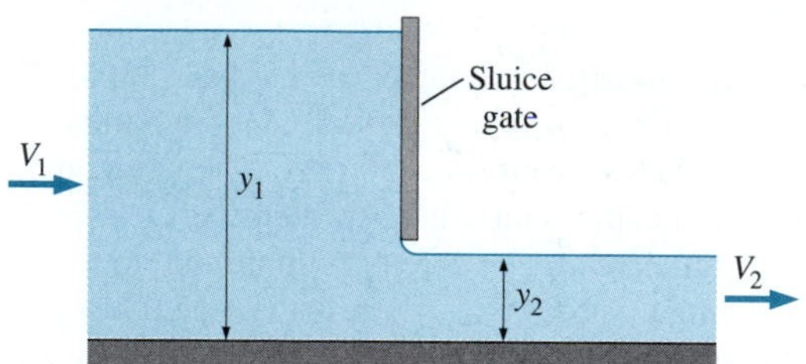

FIGURE P13–42

13–43 Water enters a centrifugal pump axially at atmospheric pressure at a rate of 0.12 m^3/s and at a velocity of 7 m/s, and leaves in the normal direction along the pump casing, as shown in Fig. P13–43. Determine the force acting on the shaft (which is also the force acting on the bearing of the shaft) in the axial direction.

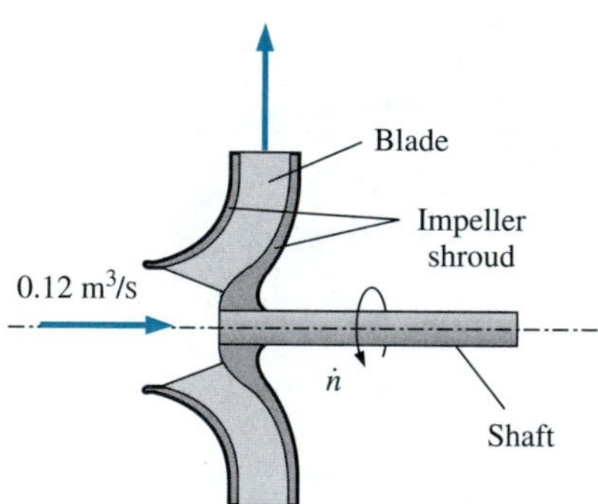

FIGURE P13–43

Angular Momentum Equation

13–44C How is the angular momentum equation obtained from Reynolds transport equations?

13–45C Express the unsteady angular momentum equation in vector form for a control volume that has a constant moment of inertia I, no external moments applied, one outgoing uniform flow stream of velocity $\vec{V}$, and mass flow rate $\dot{m}$.

13–46C Express the angular momentum equation in scalar form about a specified axis of rotation for a fixed control volume for steady and uniform flow.

13–47 Water is flowing through a 12-cm-diameter pipe that consists of a 3-m-long vertical and 2-m-long horizontal section with a 90° elbow at the exit to force the water to be discharged downward, as shown in Fig. P13–47, in the vertical direction. Water discharges to atmospheric air at a velocity of 4 m/s, and the mass of the pipe section when filled with water is 15 kg per meter length. Determine the moment acting at the intersection of the vertical and horizontal sections of the pipe (point A). What would your answer be if the flow were discharged upward instead of downward?

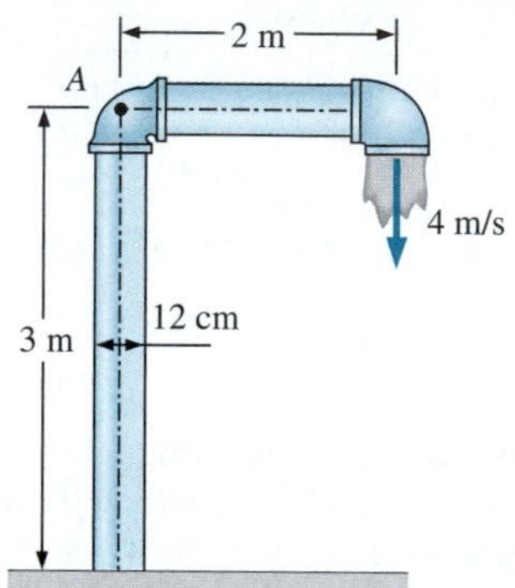

FIGURE P13–47

13–48E A large lawn sprinkler with two identical arms is used to generate electric power by attaching a generator to its rotating head. Water enters the sprinkler from the base along the axis of rotation at a rate of 8 gal/s and leaves the nozzles in the tangential direction. The sprinkler rotates at a rate of 250 rpm in a horizontal plane. The diameter of each jet is 0.5 in, and the normal distance between the axis of rotation and the center of each nozzle is 2 ft. Determine the maximum possible electric power produced.

13–49E Reconsider the lawn sprinkler in Prob. 13–48E. If the rotating head is somehow stuck, determine the moment acting on the head.

13–50 A lawn sprinkler with three identical arms is used to water a garden by rotating in a horizontal plane by the impulse caused by water flow. Water enters the sprinkler along the axis of rotation at a rate of 40 L/s and leaves the 1.2-cm-diameter nozzles in the tangential direction. The bearing applies a retarding torque of T_0 = 50 N · m due to friction at the anticipated operating speeds. For a normal distance of 40 cm between the axis of rotation and the center of the nozzles, determine the angular velocity of the sprinkler shaft.

13–51 Pelton wheel turbines are commonly used in hydroelectric power plants to generate electric power. In these turbines, a high-speed jet at a velocity of V_j impinges on buckets, forcing the wheel to rotate. The buckets reverse the direction of the jet, and the jet leaves the bucket making an angle β with the direction of the jet, as shown in Fig. P13–51. Show that the power produced by a Pelton wheel of radius r rotating steadily at an angular velocity of ω is $\dot{W}_{\text{shaft}} = \rho\omega r\dot{V}(V_j - \omega r)(1 - \cos\beta)$, where ρ is the density and $\dot{V}$ is the volume flow rate of the fluid. Obtain the numerical value for ρ = 1000 kg/m^3, r = 2 m, $\dot{V}$ = 10 m^3/s, $\dot{n}$ = 150 rpm, β = 160°, and V_j = 50 m/s.

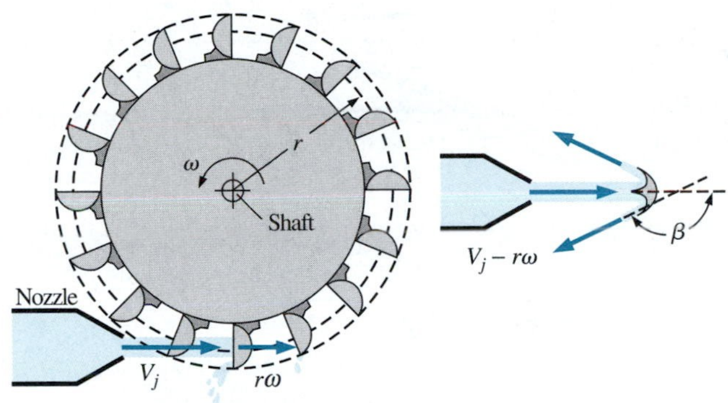

FIGURE P13–51

13–52 Reconsider Prob. 13–51. The turbine will have the maximum efficiency when $\beta = 180°$, but this is not practical. Investigate the effect of β on the power generation by allowing it to vary from 0° to 180°. Do you think we are wasting a large fraction of power by using buckets with a β of 160°?

13–53 The impeller of a centrifugal blower has a radius of 15 cm and a blade width of 6.1 cm at the inlet, and a radius of 30 cm and a blade width of 3.4 cm at the outlet. The blower delivers atmospheric air at 20°C and 95 kPa. Disregarding any losses and assuming the tangential components of air velocity at the inlet and the outlet to be equal to the impeller velocity at respective locations, determine the volumetric flow rate of air when the rotational speed of the shaft is 800 rpm and the power consumption of the blower is 120 W. Also determine the normal components of velocity at the inlet and outlet of the impeller.

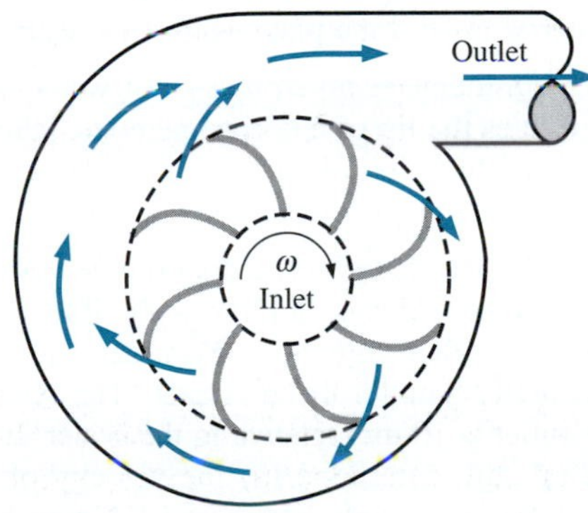

FIGURE P13–53

13–54 Consider a centrifugal blower that has a radius of 20 cm and a blade width of 8.2 cm at the impeller inlet, and a radius of 45 cm and a blade width of 5.6 cm at the outlet. The blower delivers air at a rate of 0.70 m^3/s at a rotational speed of 700 rpm. Assuming the air to enter the impeller in radial direction and to exit at an angle of 50° from the radial direction, determine the minimum power consumption of the blower. Take the density of air to be 1.25 kg/m^3.

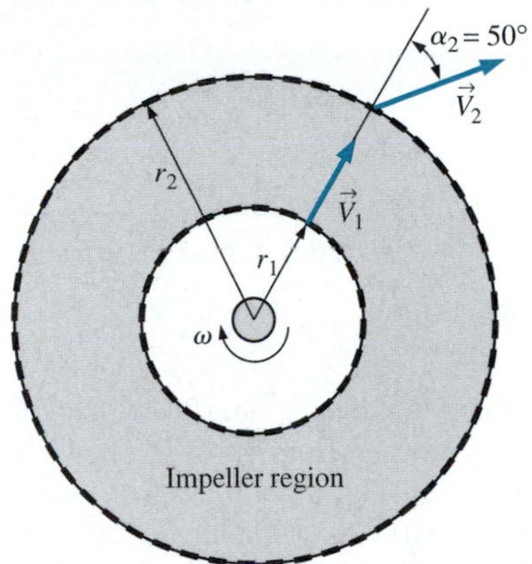

FIGURE P13–54

13–55 Reconsider Prob. 13–54. For the specified flow rate, investigate the effect of discharge angle α_2 on the minimum power input requirements. Assume the air to enter the impeller in the radial direction ($\alpha_1 = 0°$), and vary α_2 from 0° to 85° in increments of 5°. Plot the variation of power input versus α_2, and discuss your results.

13–56E Water enters the impeller of a centrifugal pump radially at a rate of 80 cfm (cubic feet per minute) when the shaft is rotating at 500 rpm. The tangential component of absolute velocity of water at the exit of the 2-ft outer diameter impeller is 180 ft/s. Determine the torque applied to the impeller.

13–57 The impeller of a centrifugal pump has inner and outer diameters of 13 and 30 cm, respectively, and a flow rate of 0.15 m^3/s at a rotational speed of 1200 rpm. The blade width of the impeller is 8 cm at the inlet and 3.5 cm at the outlet. If water enters the impeller in the radial direction and exits at an angle of 60° from the radial direction, determine the minimum power requirement for the pump.

Review Problems

13–58 Water is flowing into and discharging from a pipe U-section as shown in Fig. P13–58. At flange (1), the total absolute pressure is 200 kPa, and 30 kg/s flows into the pipe. At flange (2), the total pressure is 150 kPa. At location (3), 8 kg/s of water discharges to the atmosphere, which is at 100 kPa. Determine the total x- and z-forces at the two flanges connecting the pipe. Discuss the significance of gravity force for this problem. Take the momentum-flux correction factor to be 1.03.

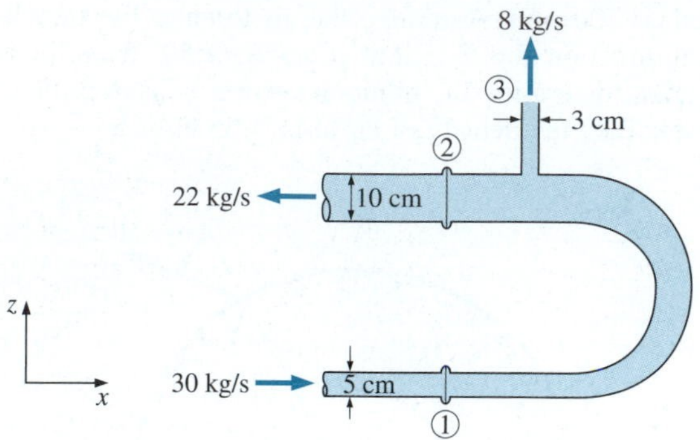

FIGURE P13–58

13–59 A tripod holding a nozzle, which directs a 5-cm-diameter stream of water from a hose, is shown in Fig. P13–59. The nozzle mass is 10 kg when filled with water. The tripod is rated to provide 1800 N of holding force. A firefighter was standing 60 cm behind the nozzle and was hit by the nozzle when the tripod suddenly failed and released the nozzle. You have been hired as an accident reconstructionist and, after testing the tripod, have determined that as water flow rate increased, it did collapse at 1800 N. In your final report you must state the water velocity and the flow rate consistent with the failure and the nozzle velocity when it hit the firefighter. *Answers*: 30.3 m/s, 0.0595 m^3/s, 14.7 m/s

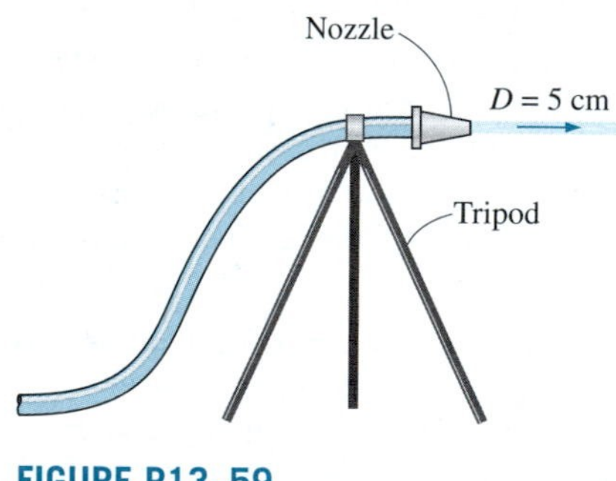

FIGURE P13–59

13–60 Consider an airplane with a jet engine attached to the tail section that expels combustion gases at a rate of 18 kg/s with a velocity of $V = 250$ m/s relative to the plane. During landing, a thrust reverser (which serves as a brake for the aircraft and facilitates landing on a short runway) is lowered in the path of the exhaust jet, which deflects the exhaust from rearward to 160°. Determine (*a*) the thrust (forward force) that the engine produces prior to the insertion of the thrust reverser and (*b*) the braking force produced after the thrust reverser is deployed.

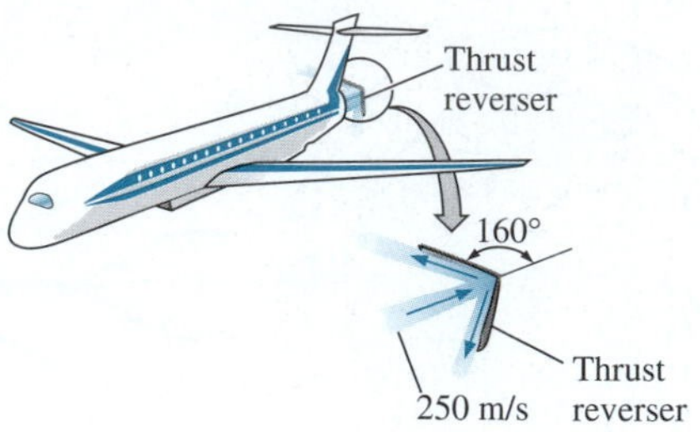

FIGURE P13–60

13–61 Reconsider Prob. 13–60. Using EES (or other) software, investigate the effect of thrust reverser angle on the braking force exerted on the airplane. Let the reverser angle vary from 0° (no reversing) to 180° (full reversing) in increments of 10°. Tabulate and plot your results and draw conclusions.

13–62E A spacecraft cruising in space at a constant velocity of 1500 ft/s has a mass of 18,000 lbm. To slow down the spacecraft, a solid fuel rocket is fired, and the combustion gases leave the rocket at a constant rate of 150 lbm/s at a velocity of 5000 ft/s in the same direction as the spacecraft for a period of 5 s. Assuming the mass of the spacecraft remains constant, determine (*a*) the deceleration of the spacecraft during this 5-s period, (*b*) the change of velocity of the spacecraft during this time period, and (*c*) the thrust exerted on the spacecraft.

13–63 A 5-cm-diameter horizontal water jet having a velocity of 30 m/s strikes a vertical stationary flat plate. The water splatters in all directions in the plane of the plate. How much force is required to hold the plate against the water stream?

13–64 A 5-cm-diameter horizontal jet of water, with velocity 30 m/s, strikes the tip of a horizontal cone, which deflects the water by 45° from its original direction. How much force is required to hold the cone against the water stream?

13–65 A 60-kg ice skater is standing on ice with ice skates (negligible friction). She is holding a flexible hose (essentially weightless) that directs a 2-cm-diameter stream of water horizontally parallel to her skates. The water velocity at the hose outlet is 10 m/s relative to the skater. If she is initially standing still, determine (*a*) the velocity of the skater and the distance she travels in 5 s and (*b*) how long it will take to move 5 m and the velocity at that moment. *Answers*: (*a*) 2.62 m/s, 6.54 m, (*b*) 4.4 s, 2.3 m/s

FIGURE P13–65

13–66 Indiana Jones needs to ascend a 10-m-high building. There is a large hose filled with pressurized water hanging down from the building top. He builds a square platform and mounts four 5-cm-diameter nozzles pointing down at each corner. By connecting hose branches, a water jet with 15-m/s velocity can be produced from each nozzle. Jones, the platform, and the nozzles have a combined mass of 150 kg. Determine (*a*) the minimum water jet velocity needed to raise the system, (*b*) how long it takes for the system to rise 10 m when the water jet velocity is 15 m/s and the velocity of the platform at that moment, and (*c*) how much higher will the momentum raise Jones if he shuts off the water at the moment the platform reaches 10 m above the ground. How much time does he have to jump from the platform to the roof? *Answers:* (*a*) 13.7 m/s, (*b*) 3.2 s, (*c*) 2.1 m, 1.3 s

FIGURE P13–66

13–67E An engineering student considers using a fan as a levitation demonstration. She plans to face the box-enclosed fan so the air blast is directed face down through a 3-ft-diameter blade span area. The system weighs 5 lbf, and the student will secure the system from rotating. By increasing the power to the fan, she plans to increase the blade rpm and air exit velocity until the exhaust provides sufficient upward force to cause the box fan to hover in the air. Determine (*a*) the air exit velocity to produce 5 lbf, (*b*) the volumetric flow rate needed, and (*c*) the minimum mechanical power that must be supplied to the airstream. Take the air density to be 0.078 lbm/ft^3.

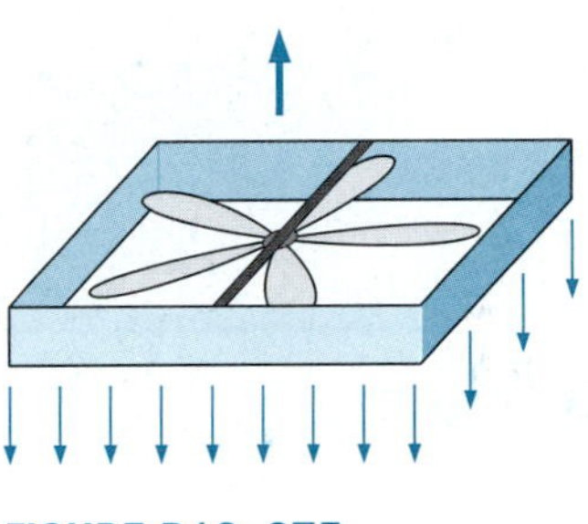

FIGURE P13–67E

13–68 A soldier jumps from a plane and opens his parachute when his velocity reaches the terminal velocity V_T. The parachute slows him down to his landing velocity of V_F. After the parachute is deployed, the air resistance is proportional to the velocity squared (i.e., $F = kV^2$). The soldier, his parachute, and his gear have a total mass of m. Show that $k = mg/V_F^2$ and develop a relation for the soldier's velocity after he opens the parachute at time $t = 0$.

Answer: $V = V_F \dfrac{V_T + V_F + (V_T - V_F)e^{-2gt/V_F}}{V_T + V_F - (V_T - V_F)e^{-2gt/V_F}}$

FIGURE P13–68

13–69 A horizontal water jet with a flow rate of $\dot{V}$ and cross-sectional area of A drives a covered cart of mass m_c along a level and nearly frictionless path. The jet enters a hole at the rear of the cart and all water that enters the cart is retained, increasing the system mass. The relative velocity between the jet of constant velocity V_J and the cart of variable velocity V is $V_J - V$. If the cart is initially empty and stationary when the jet action is initiated, develop a relation (integral form is acceptable) for cart velocity versus time.

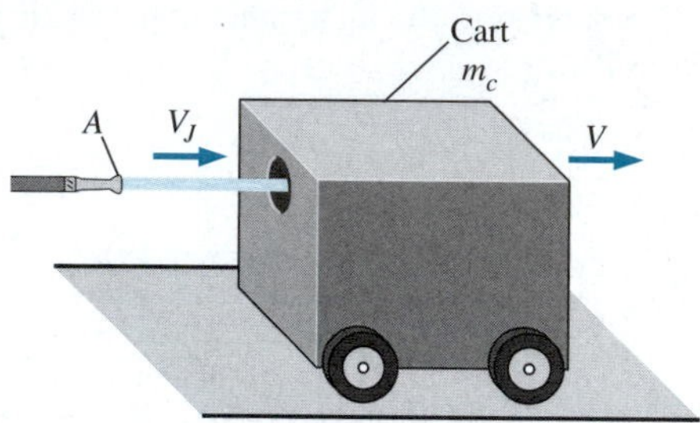

FIGURE P13–69

13–70 Nearly frictionless vertical guide rails maintain a plate of mass m_p in a horizontal position, such that it can slide freely in the vertical direction. A nozzle directs a water stream of area A against the plate underside. The water jet splatters in the plate plane, applying an upward force against the plate. The water flow rate $\dot{m}$ (kg/s) can be controlled. Assume that distances are short, so the velocity of the rising jet can be considered constant with height. (*a*) Determine the minimum mass flow rate $\dot{m}_{min}$ necessary to just levitate the plate and obtain a relation for the steady-state velocity of the upward moving plate for $\dot{m} > \dot{m}_{min}$. (*b*) At time $t = 0$, the plate is at rest, and the water jet with $\dot{m} > \dot{m}_{min}$ is suddenly turned on. Apply a force balance to the plate and obtain the integral that relates velocity to time (do not solve).

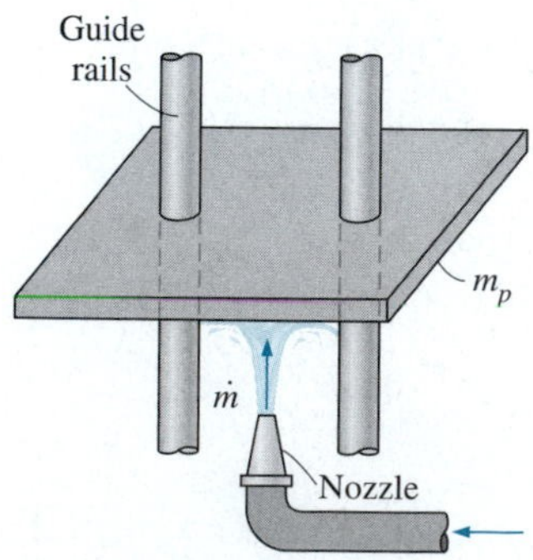

FIGURE P13–70

13–71 Water enters a mixed flow pump axially at a rate of 0.2 m³/s and at a velocity of 5 m/s, and is discharged to the atmosphere at an angle of 60° from the horizontal, as shown in Fig. P13–71. If the discharge flow area is half the inlet area, determine the force acting on the shaft in the axial direction.

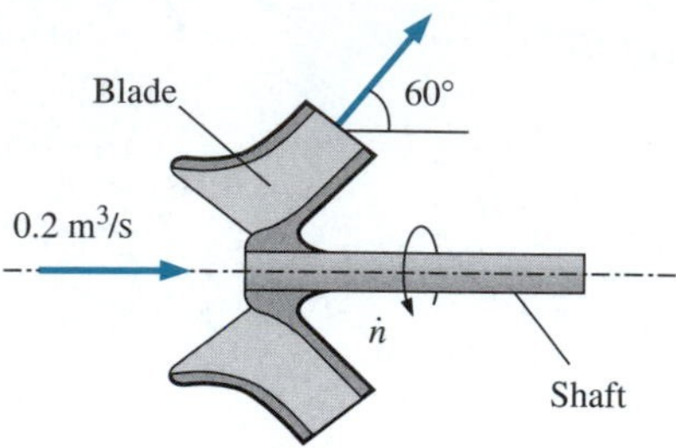

FIGURE P13–71

13–72 Water accelerated by a nozzle enters the impeller of a turbine through its outer edge of diameter D with a velocity of V making an angle α with the radial direction at a mass flow rate of $\dot{m}$. Water leaves the impeller in the radial direction. If the angular speed of the turbine shaft is $\dot{n}$, show that the maximum power that can be generated by this radial turbine is $\dot{W}_{shaft} = \pi\dot{n}\dot{m}DV \sin \alpha$.

13–73 Water enters a two-armed lawn sprinkler along the vertical axis at a rate of 60 L/s, and leaves the sprinkler nozzles as 2-cm diameter jets at an angle of θ from the tangential direction, as shown in Fig. P13–73. The length of each sprinkler arm is 0.45 m. Disregarding any frictional effects, determine the rate of rotation $\dot{n}$ of the sprinkler in rev/min for (*a*) $\theta = 0°$, (*b*) $\theta = 30°$, and (*c*) $\theta = 60°$.

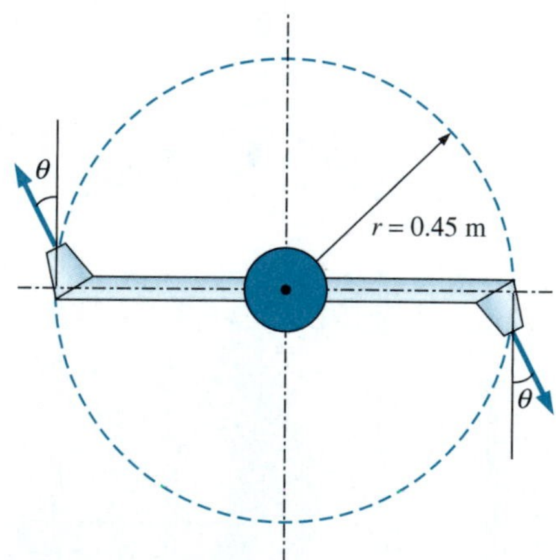

FIGURE P13–73

13–74 Reconsider Prob. 13–73. For the specified flow rate, investigate the effect of discharge angle θ on the rate of rotation $\dot{n}$ by varying θ from 0° to 90° in increments of 10°. Plot the rate of rotation versus θ, and discuss your results.

13–75 A stationary water tank of diameter D is mounted on wheels and is placed on a nearly frictionless level surface. A smooth hole of diameter D_o near the bottom of the tank allows water to jet horizontally and rearward and the water jet force propels the system forward. The water in the tank is much heavier than the tank-and-wheel assembly, so only the mass of water remaining in the tank needs to be considered in this problem. Considering the decrease in the mass of water with time, develop relations for (*a*) the acceleration, (*b*) the velocity, and (*c*) the distance traveled by the system as a function of time.

Design and Essay Problem

13–76 Visit a fire station and obtain information about flow rates through hoses and discharge diameters. Using this information, calculate the impulse force to which the firefighters are subjected.

Chapter 14

INTERNAL FLOW

Fluid flow is classified as *external* or *internal*, depending on whether the fluid is forced to flow over a surface or in a conduit. Internal and external flows exhibit very different characteristics. In this chapter we consider *internal flow* where the conduit is completely filled with the fluid, and flow is driven primarily by a pressure difference. This should not be confused with *open-channel flow* where the conduit is partially filled by the fluid and thus the flow is partially bounded by solid surfaces, as in an irrigation ditch, and flow is driven by gravity alone.

We start this chapter with a general physical description of internal flow through pipes and ducts including the *entrance region* and the *fully developed* region. We continue with a discussion of the dimensionless *Reynolds number* and its physical significance. We then introduce the *pressure drop* correlations associated with pipe flow for both laminar and turbulent flows. Finally we discuss minor losses and determine the pressure drop and pumping power requirements for real-world piping systems.

Objectives

The objectives of this chapter are to:

- Have a deeper understanding of laminar and turbulent flow in pipes and the analysis of fully developed flow
- Calculate the major and minor losses associated with pipe flow in piping networks and determine the pumping power requirements

14–1 ▪ INTRODUCTION

Liquid or gas flow through *pipes* or *ducts* is commonly used in heating and cooling applications and fluid distribution networks. The fluid in such applications is usually forced to flow by a fan or pump through a flow section. We pay particular attention to *friction*, which is directly related to the *pressure drop* and *head loss* during flow through pipes and ducts. The pressure drop is then used to determine the pumping power requirement. A typical piping system involves pipes of different diameters connected to each other by various fittings or elbows to route the fluid, valves to control the flow rate, and pumps to pressurize the fluid.

The terms *pipe*, *duct*, and *conduit* are usually used interchangeably for flow sections. In general, flow sections of circular cross section are referred to as *pipes* (especially when the fluid is a liquid), and flow sections of noncircular cross section as *ducts* (especially when the fluid is a gas). Small-diameter pipes are usually referred to as *tubes*. Given this uncertainty, we will use more descriptive phrases (such as *a circular pipe* or *a rectangular duct*) whenever necessary to avoid any misunderstandings.

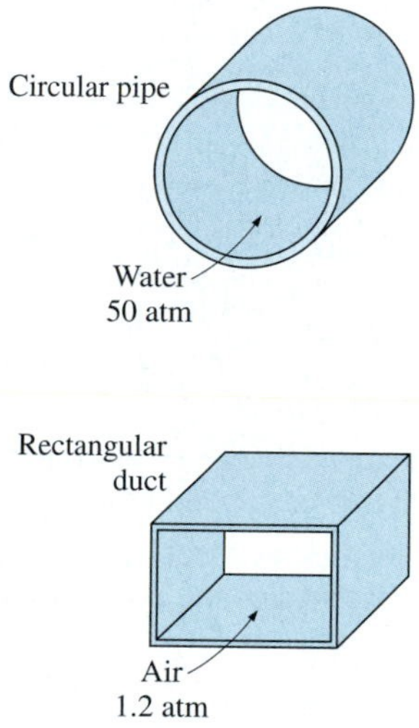

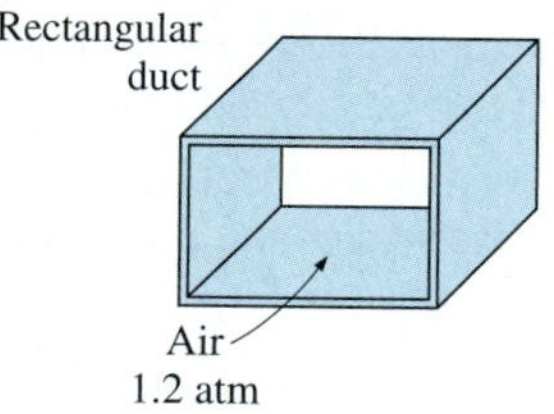

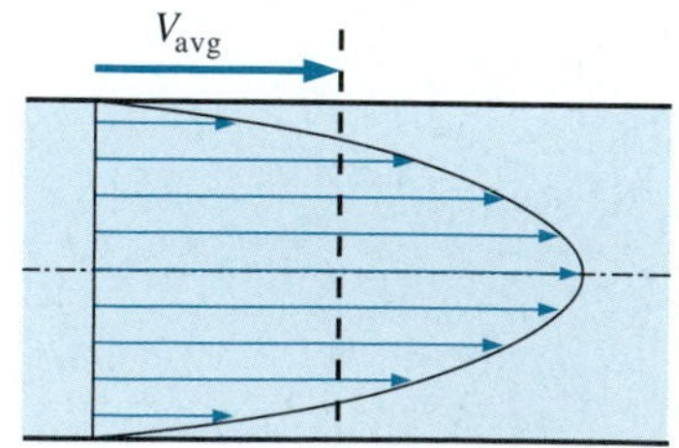

FIGURE 14–1

Circular pipes can withstand large pressure differences between the inside and the outside without undergoing any significant distortion, but noncircular pipes cannot.

You have probably noticed that most fluids, especially liquids, are transported in *circular pipes*. This is because pipes with a circular cross section can withstand large pressure differences between the inside and the outside without undergoing significant distortion. *Noncircular pipes* are usually used in applications such as the heating and cooling systems of buildings where the pressure difference is relatively small, the manufacturing and installation costs are lower, and the available space is limited for ductwork (Fig. 14–1).

Although the theory of fluid flow is reasonably well understood, theoretical solutions are obtained only for a few simple cases such as fully developed laminar flow in a circular pipe. Therefore, we must rely on experimental results and empirical relations for most fluid flow problems rather than closed-form analytical solutions. Noting that the experimental results are obtained under carefully controlled laboratory conditions and that no two systems are exactly alike, we must not be so naive as to view the results obtained as "exact." An error of 10 percent (or more) in friction factors calculated using the relations in this chapter is the "norm" rather than the "exception."

FIGURE 14–2

Average velocity V_{avg} is defined as the average speed through a cross section. For fully developed laminar pipe flow, V_{avg} is half of the maximum velocity.

The fluid velocity in a pipe changes from *zero* at the wall because of the no-slip condition to a maximum at the pipe center. In fluid flow, it is convenient to work with an *average* velocity V_{avg}, which remains constant in incompressible flow when the cross-sectional area of the pipe is constant (Fig. 14–2). The average velocity in heating and cooling applications may change somewhat because of changes in density with temperature. But, in practice, we evaluate the fluid properties at some average temperature and treat them as constants. The convenience of working with constant properties usually more than justifies the slight loss in accuracy.

Also, the friction between the fluid particles in a pipe does cause a slight rise in fluid temperature as a result of the mechanical energy being converted to sensible thermal energy. But this temperature rise due to *frictional heating* is usually too small to warrant any consideration in calculations and thus is disregarded. For example, in the absence of any heat transfer, no noticeable difference can be detected between the inlet and outlet temperatures of water flowing in a pipe. The primary consequence of friction in fluid flow is pressure drop, and thus any significant temperature change in the fluid is due to heat transfer.

The value of the average velocity V_{avg} at some streamwise cross-section is determined from the requirement that the *conservation of mass* principle be satisfied (Fig. 14–2). That is,

$$\dot{m} = \rho V_{avg} A_c = \int_{A_c} \rho u(r)\, dA_c \tag{14–1}$$

where $\dot{m}$ is the mass flow rate, ρ is the density, A_c is the cross-sectional area, and $u(r)$ is the velocity profile. Then the average velocity for incompressible flow in a circular pipe of radius R is expressed as

$$V_{avg} = \frac{\int_{A_c} \rho u(r)\, dA_c}{\rho A_c} = \frac{\int_0^R \rho u(r) 2\pi r\, dr}{\rho \pi R^2} = \frac{2}{R^2} \int_0^R u(r) r\, dr \tag{14–2}$$

Therefore, when we know the flow rate or the velocity profile, the average velocity can be determined easily.

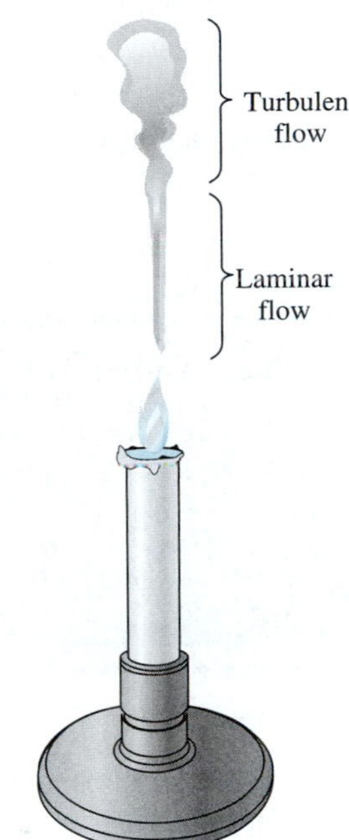

FIGURE 14–3

Laminar and turbulent flow regimes of candle smoke.

14–2 ▪ LAMINAR AND TURBULENT FLOWS

If you have been around smokers, you probably noticed that the cigarette smoke rises in a smooth plume for the first few centimeters and then starts fluctuating randomly in all directions as it continues its rise. Other plumes behave similarly (Fig. 14–3). Likewise, a careful inspection of flow in a pipe reveals that the fluid flow is streamlined at low velocities but turns chaotic as the velocity is increased above a critical value, as shown in Fig. 14–4. The flow regime in the first case is said to be **laminar,** characterized by *smooth streamlines* and *highly ordered motion*, and **turbulent** in the second case, where it is characterized by *velocity fluctuations* and *highly disordered motion*. The **transition** from laminar to turbulent flow does not occur suddenly; rather, it occurs over some region in which the flow fluctuates between laminar and turbulent flows before it becomes fully turbulent. Most flows encountered in practice are turbulent. Laminar flow is encountered when highly viscous fluids such as oils flow in small pipes or narrow passages.

We can verify the existence of these laminar, transitional, and turbulent flow regimes by injecting some dye streaks into the flow in a glass pipe, as the British engineer Osborne Reynolds (1842–1912) did over a century ago. We observe that the dye streak forms a *straight and smooth line* at low velocities when the flow is laminar (we may see some blurring because of molecular diffusion), has *bursts of fluctuations* in the transitional regime, and *zigzags rapidly and randomly* when the flow becomes fully turbulent. These zigzags and the dispersion of the dye are indicative of the fluctuations in the main flow and the rapid mixing of fluid particles from adjacent layers.

The *intense mixing* of the fluid in turbulent flow as a result of rapid fluctuations enhances momentum transfer between fluid particles, which increases the friction force on the pipe wall and thus the required pumping power. The friction factor reaches a maximum when the flow becomes fully turbulent.

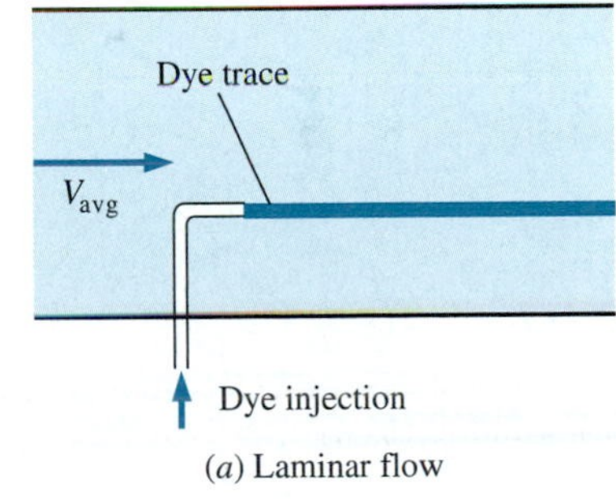

(*a*) Laminar flow

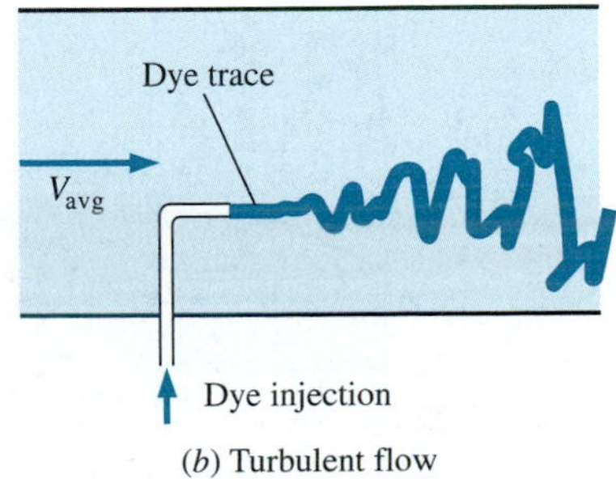

(*b*) Turbulent flow

FIGURE 14–4

The behavior of colored fluid injected into the flow in laminar and turbulent flows in a pipe.

Reynolds Number

The transition from laminar to turbulent flow depends on the *geometry*, *surface roughness*, *flow velocity*, *surface temperature*, and *type of fluid*, among other things. After exhaustive experiments in the 1880s, Osborne Reynolds discovered that the flow regime depends mainly on the ratio of *inertial forces* to *viscous forces* in the fluid. This ratio is called the **Reynolds number** and is expressed for internal flow in a circular pipe as (Fig. 14–5)

$$\text{Re} = \frac{\text{Inertial forces}}{\text{Viscous forces}} = \frac{V_{\text{avg}}D}{\nu} = \frac{\rho V_{\text{avg}}D}{\mu} \tag{14–3}$$

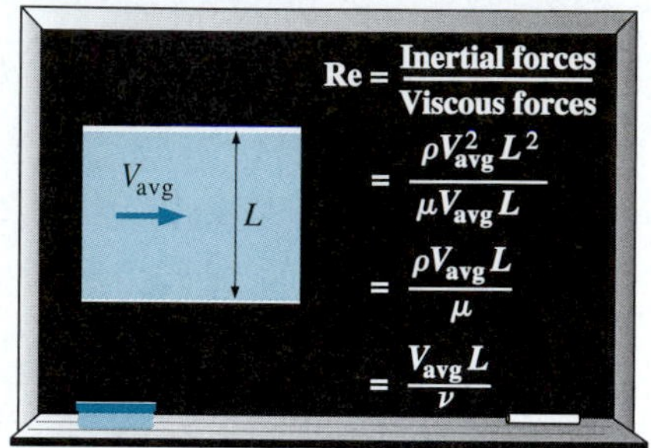

FIGURE 14–5
The Reynolds number can be viewed as the ratio of inertial forces to viscous forces acting on a fluid element.

where V_{avg} = average flow velocity (m/s), D = characteristic length of the geometry (diameter in this case, in m), and $\nu = \mu/\rho$ = kinematic viscosity of the fluid (m²/s). Note that the Reynolds number is a *dimensionless* quantity. Also, kinematic viscosity has the unit m²/s, and can be viewed as *viscous diffusivity* or *diffusivity for momentum*.

At large Reynolds numbers, the inertial forces, which are proportional to the fluid density and the square of the fluid velocity, are large relative to the viscous forces, and thus the viscous forces cannot prevent the random and rapid fluctuations of the fluid. At *small* or *moderate* Reynolds numbers, however, the viscous forces are large enough to suppress these fluctuations and to keep the fluid "in line." Thus the flow is *turbulent* in the first case and *laminar* in the second.

The Reynolds number at which the flow becomes turbulent is called the **critical Reynolds number**, Re_{cr}. The value of the critical Reynolds number is different for different geometries and flow conditions. For internal flow in a circular pipe, the generally accepted value of the critical Reynolds number is $\text{Re}_{\text{cr}} = 2300$.

For flow through noncircular pipes, the Reynolds number is based on the **hydraulic diameter** D_h defined as (Fig. 14–6)

Hydraulic diameter:

$$D_h = \frac{4A_c}{p} \tag{14–4}$$

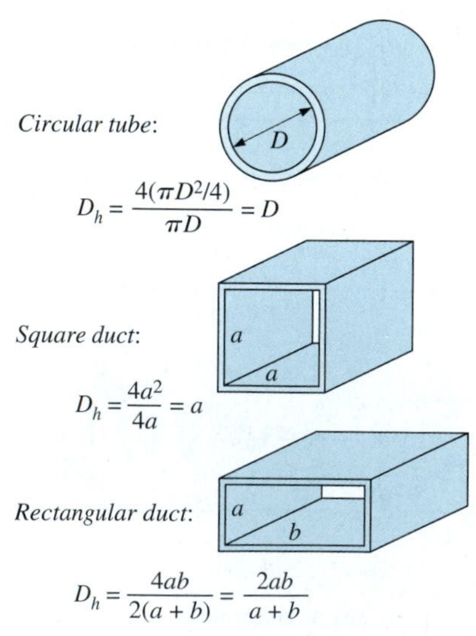

FIGURE 14–6
The hydraulic diameter $D_h = 4A_c/p$ is defined such that it reduces to ordinary diameter for circular tubes.

where A_c is the cross-sectional area of the pipe and p is its wetted perimeter. The hydraulic diameter is defined such that it reduces to ordinary diameter D for circular pipes,

Circular pipes:

$$D_h = \frac{4A_c}{p} = \frac{4(\pi D^2/4)}{\pi D} = D$$

It certainly is desirable to have precise values of Reynolds numbers for laminar, transitional, and turbulent flows, but this is not the case in practice. It turns out that the transition from laminar to turbulent flow also depends on the degree of disturbance of the flow by *surface roughness*, *pipe vibrations*, and *fluctuations in the flow*. Under most practical conditions, the flow in a circular pipe is laminar for $\text{Re} \lesssim 2300$, turbulent for $\text{Re} \gtrsim 10{,}000$, and transitional in between. That is,

$$\text{Re} \lesssim 2300 \qquad \text{laminar flow}$$
$$2300 \lesssim \text{Re} \lesssim 10{,}000 \qquad \text{transitional flow}$$
$$\text{Re} \gtrsim 10{,}000 \qquad \text{turbulent flow}$$

In transitional flow, the flow switches between laminar and turbulent in a disorderly fashion (Fig. 14–7). It should be kept in mind that laminar flow can be maintained at much higher Reynolds numbers in very smooth pipes by avoiding flow disturbances and pipe vibrations. In such carefully controlled experiments, laminar flow has been maintained at Reynolds numbers of up to 100,000.

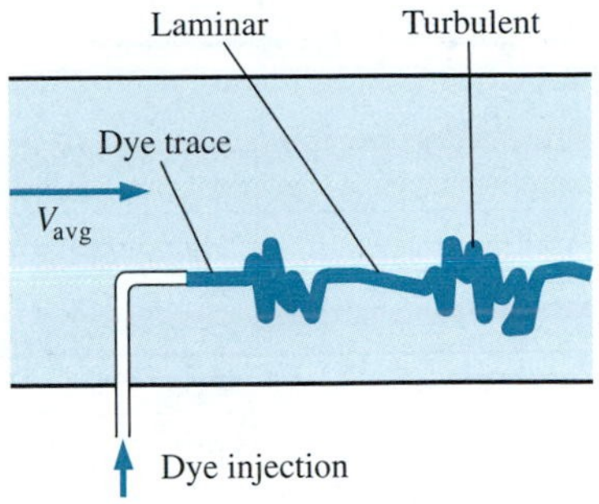

FIGURE 14–7
In the transitional flow region of $2300 \le \text{Re} \le 10{,}000$, the flow switches between laminar and turbulent seemingly randomly.

14–3 ▪ THE ENTRANCE REGION

Consider a fluid entering a circular pipe at a uniform velocity. Because of the no-slip condition, the fluid particles in the layer in contact with the wall of the pipe come to a complete stop. This layer also causes the fluid particles in the adjacent layers to slow down gradually as a result of friction. To make up for this velocity reduction, the velocity of the fluid at the midsection of the pipe has to increase to keep the mass flow rate through the pipe constant. As a result, a velocity gradient develops along the pipe.

The region of the flow in which the effects of the viscous shearing forces caused by fluid viscosity are felt is called the **velocity boundary layer** or just the **boundary layer**. The hypothetical boundary surface divides the flow in a pipe into two regions: the **boundary layer region**, in which the viscous effects and the velocity changes are significant, and the **irrotational (core) flow region**, in which the frictional effects are negligible and the velocity remains essentially constant in the radial direction.

The thickness of this boundary layer increases in the flow direction until the boundary layer reaches the pipe center and thus fills the entire pipe, as shown in Fig. 14–8. The region from the pipe inlet to the point at which the boundary layer merges at the centerline is called the **hydrodynamic entrance region**, and the length of this region is called the **hydrodynamic entry length** L_h. Flow in the entrance region is called *hydrodynamically developing flow* since this is the region where the velocity profile develops. The region beyond the entrance region in which the velocity profile is fully developed and remains unchanged is called the **hydrodynamically fully developed region**. The flow is said to be **fully developed** when the normalized temperature profile remains unchanged as well. Hydrodynamically fully developed flow is equivalent to fully developed flow when the fluid in the pipe is not heated or cooled since the fluid temperature in this case

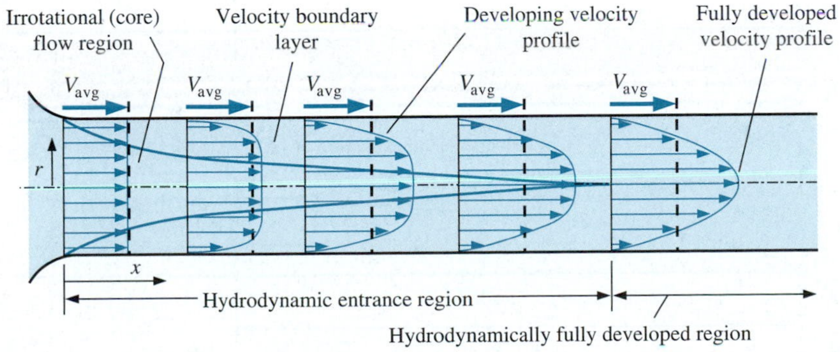

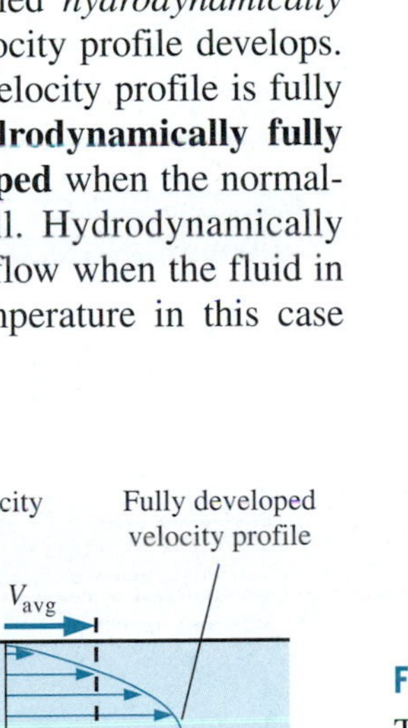

FIGURE 14–8
The development of the velocity boundary layer in a pipe. (The developed average velocity profile is parabolic in laminar flow, as shown, but somewhat flatter or fuller in turbulent flow.)

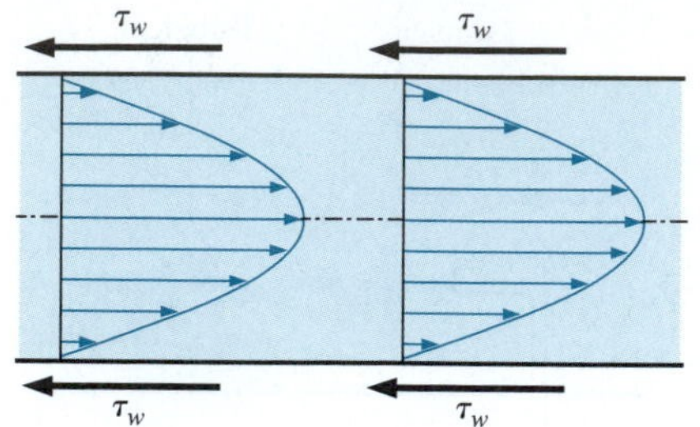

FIGURE 14–9

In the fully developed flow region of a pipe, the velocity profile does not change downstream, and thus the wall shear stress remains constant as well.

remains essentially constant throughout. The velocity profile in the fully developed region is *parabolic* in laminar flow and somewhat *flatter* (or *fuller*) in turbulent flow due to eddy motion and more vigorous mixing in the radial direction. The time-averaged velocity profile remains unchanged when the flow is fully developed, and thus

$$\textit{Hydrodynamically fully developed:} \qquad \frac{\partial u(r, x)}{\partial x} = 0 \quad \rightarrow \quad u = u(r) \tag{14–5}$$

The shear stress at the pipe wall τ_w is related to the slope of the velocity profile at the surface. Noting that the velocity profile remains unchanged in the hydrodynamically fully developed region, the wall shear stress also remains constant in that region (Fig. 14–9).

Consider fluid flow in the hydrodynamic entrance region of a pipe. The wall shear stress is the *highest* at the pipe inlet where the thickness of the boundary layer is smallest, and decreases gradually to the fully developed value, as shown in Fig. 14–10. Therefore, the pressure drop is *higher* in the entrance regions of a pipe, and the effect of the entrance region is always to *increase* the average friction factor for the entire pipe. This increase may be significant for short pipes but is negligible for long ones.

Entry Lengths

The hydrodynamic entry length is usually taken to be the distance from the pipe entrance to where the wall shear stress (and thus the friction factor) reaches within about 2 percent of the fully developed value. In *laminar flow*, the nondimensional hydrodynamic entry length is given approximately as [see Kays and Crawford (1993) and Shah and Bhatti (1987)]

$$\frac{L_{h,\,\text{laminar}}}{D} \cong 0.05\text{Re} \tag{14–6}$$

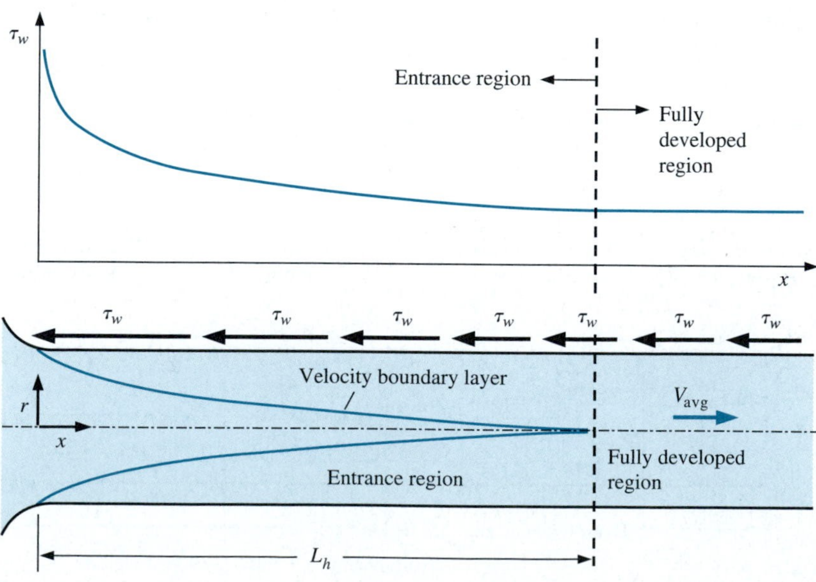

FIGURE 14–10

The variation of wall shear stress in the flow direction for flow in a pipe from the entrance region into the fully developed region.

For Re = 20, the hydrodynamic entry length is about the size of the diameter, but increases linearly with velocity. In the limiting laminar case of Re = 2300, the hydrodynamic entry length is 115*D*.

In *turbulent flow*, the intense mixing during random fluctuations usually overshadows the effects of molecular diffusion. The nondimensional hydrodynamic entry length for turbulent flow is approximated as [see Bhatti and Shah (1987) and Zhi-qing (1982)]

$$\frac{L_{h,\,\text{turbulent}}}{D} = 1.359\text{Re}^{1/4} \tag{14–7}$$

The entry length is much shorter in turbulent flow, as expected, and its dependence on the Reynolds number is weaker. In many pipe flows of practical engineering interest, the entrance effects become insignificant beyond a pipe length of 10 diameters, and the nondimensional hydrodynamic entry length is approximated as

$$\frac{L_{h,\,\text{turbulent}}}{D} \approx 10 \tag{14–8}$$

Correlations for calculating the frictional head losses in entrance regions are available in the literature. However, the pipes used in practice are usually several times the length of the entrance region, and thus the flow through the pipes is often assumed to be fully developed for the entire length of the pipe. This simplistic approach gives *reasonable* results for long pipes but sometimes poor results for short ones since it underpredicts the wall shear stress and thus the friction factor.

14–4 ▪ LAMINAR FLOW IN PIPES

We mentioned in Section 14–2 that flow in pipes is laminar for Re $\lesssim$ 2300, and that the flow is fully developed if the pipe is sufficiently long (relative to the entry length) so that the entrance effects are negligible. In this section we consider steady, laminar, incompressible flow of a fluid with constant properties in the fully developed region of a straight circular pipe. We obtain the momentum equation by applying a momentum balance to a differential volume element, and obtain the velocity profile by solving it. Then we use it to obtain a relation for the friction factor. An important aspect of the analysis here is that it is one of the few available for viscous flow.

In fully developed laminar flow, each fluid particle moves at a constant axial velocity along a streamline and the velocity profile $u(r)$ remains unchanged in the flow direction. There is no motion in the radial direction, and thus the velocity component in the direction normal to the pipe axis is everywhere zero. There is no acceleration since the flow is steady and fully developed.

Now consider a ring-shaped differential volume element of radius r, thickness dr, and length dx oriented coaxially with the pipe, as shown in Fig. 14–11. The volume element involves only pressure and viscous effects and thus the pressure and shear forces must balance each other. The pressure force acting on a submerged plane surface is the product of the pressure at the centroid of the surface and the surface area. A force balance on the volume element in the flow direction gives

$$(2\pi r\,dr\,P)_x - (2\pi r\,dr\,P)_{x+dx} + (2\pi r\,dx\,\tau)_r - (2\pi r\,dx\,\tau)_{r+dr} = 0 \tag{14–9}$$

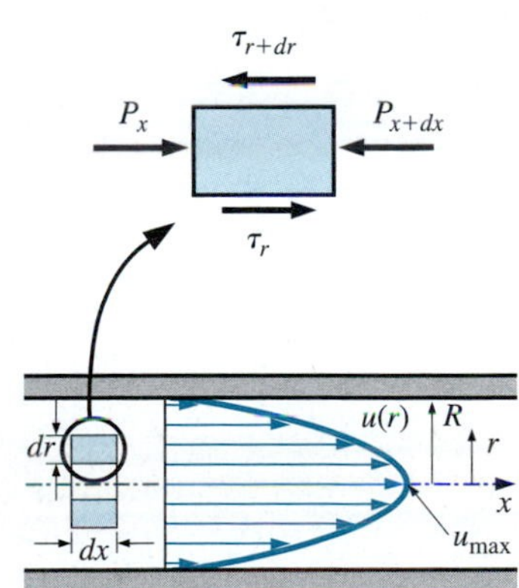

FIGURE 14–11
Free-body diagram of a ring-shaped differential fluid element of radius r, thickness dr, and length dx oriented coaxially with a horizontal pipe in fully developed laminar flow.

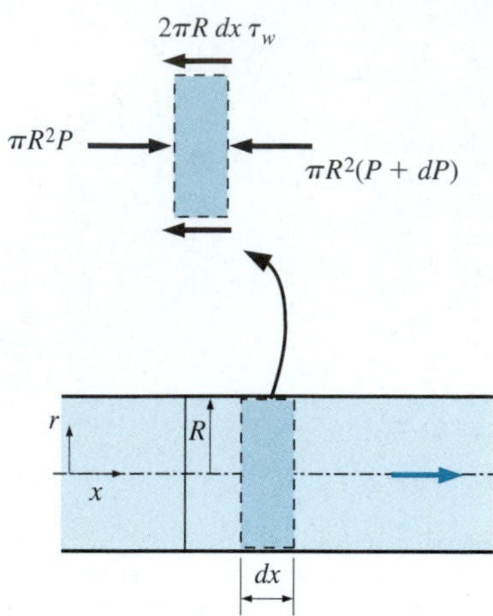

Force balance:

$$\pi R^2 P - \pi R^2 (P + dP) - 2\pi R\,dx\,\tau_w = 0$$

Simplifying:

$$\frac{dP}{dx} = -\frac{2\tau_w}{R}$$

FIGURE 14–12

Free-body diagram of a fluid disk element of radius R and length dx in fully developed laminar flow in a horizontal pipe.

which indicates that in fully developed flow in a horizontal pipe, the viscous and pressure forces balance each other. Dividing by $2\pi dr dx$ and rearranging,

$$r\frac{P_{x+dx} - P_x}{dx} + \frac{(r\tau)_{r+dr} - (r\tau)_r}{dr} = 0 \tag{14–10}$$

Taking the limit as $dr, dx \to 0$ gives

$$r\frac{dP}{dx} + \frac{d(r\tau)}{dr} = 0 \tag{14–11}$$

Substituting $\tau = -\mu(du/dr)$ and taking $\mu =$ constant gives the desired equation,

$$\frac{\mu}{r}\frac{d}{dr}\left(r\frac{du}{dr}\right) = \frac{dP}{dx} \tag{14–12}$$

The quantity du/dr is negative in pipe flow, and the negative sign is included to obtain positive values for τ. (Or, $du/dr = -du/dy$ if we set $y = R - r$.) The left side of Eq. 14–12 is a function of r, and the right side is a function of x. The equality must hold for any value of r and x, and an equality of the form $f(r) = g(x)$ can be satisfied only if both $f(r)$ and $g(x)$ are equal to the same constant. Thus we conclude that $dP/dx =$ constant. This can be verified by writing a force balance on a volume element of radius R and thickness dx (a slice of the pipe), which gives (Fig. 14–12)

$$\frac{dP}{dx} = -\frac{2\tau_w}{R} \tag{14–13}$$

Here τ_w is constant since the viscosity and the velocity profile are constants in the fully developed region. Therefore, $dP/dx =$ constant.

Equation 14–12 is solved by rearranging and integrating it twice to give

$$u(r) = \frac{r^2}{4\mu}\left(\frac{dP}{dx}\right) + C_1 \ln r + C_2 \tag{14–14}$$

The velocity profile $u(r)$ is obtained by applying the boundary conditions $\partial u/\partial r = 0$ at $r = 0$ (because of symmetry about the centerline) and $u = 0$ at $r = R$ (the no-slip condition at the pipe wall). We get

$$u(r) = -\frac{R^2}{4\mu}\left(\frac{dP}{dx}\right)\left(1 - \frac{r^2}{R^2}\right) \tag{14–15}$$

Therefore, the velocity profile in fully developed laminar flow in a pipe is *parabolic* with a maximum at the centerline and a minimum (zero) at the pipe wall. Also, the axial velocity u is positive for any r, and thus the axial pressure gradient dP/dx must be negative (i.e., pressure must decrease in the flow direction because of viscous effects).

The average velocity is determined from its definition by substituting Eq. 14–15 into Eq. 14–2, and performing the integration, yielding

$$V_{avg} = \frac{2}{R^2}\int_0^R u(r) r\,dr = \frac{-2}{R^2}\int_0^R \frac{R^2}{4\mu}\left(\frac{dP}{dx}\right)\left(1 - \frac{r^2}{R^2}\right) r\,dr = -\frac{R^2}{8\mu}\left(\frac{dP}{dx}\right) \tag{14–16}$$

Combining the last two equations, the velocity profile is rewritten as

$$u(r) = 2V_{avg}\left(1 - \frac{r^2}{R^2}\right) \tag{14–17}$$

This is a convenient form for the velocity profile since V_{avg} can be determined easily from the flow rate information.

The maximum velocity occurs at the centerline and is determined from Eq. 14–17 by substituting $r = 0$,

$$u_{max} = 2V_{avg} \quad (14\text{–}18)$$

Therefore, *the average velocity in fully developed laminar pipe flow is one-half of the maximum velocity.*

Pressure Drop and Head Loss

A quantity of interest in the analysis of pipe flow is the *pressure drop* ΔP since it is directly related to the power requirements of the fan or pump to maintain flow. We note that dP/dx = constant, and integrating from $x = x_1$ where the pressure is P_1 to $x = x_1 + L$ where the pressure is P_2 gives

$$\frac{dP}{dx} = \frac{P_2 - P_1}{L} \quad (14\text{–}19)$$

Substituting Eq. 14–19 into the V_{avg} expression in Eq. 14–16, the pressure drop is expressed as

Laminar flow:
$$\Delta P = P_1 - P_2 = \frac{8\mu L V_{avg}}{R^2} = \frac{32\mu L V_{avg}}{D^2} \quad (14\text{–}20)$$

The symbol Δ is typically used to indicate the difference between the final and initial values, like $\Delta y = y_2 - y_1$. But in fluid flow, ΔP is used to designate pressure drop, and thus it is $P_1 - P_2$. A pressure drop due to viscous effects represents an irreversible pressure loss, and it is called **pressure loss** ΔP_L to emphasize that it is a *loss* (just like the head loss h_L, which as we shall see is proportional to ΔP_L).

Note from Eq. 14–20 that the pressure drop is proportional to the viscosity μ of the fluid, and ΔP would be zero if there were no friction. Therefore, the drop of pressure from P_1 to P_2 in this case is due entirely to viscous effects, and Eq. 14–20 represents the pressure loss ΔP_L when a fluid of viscosity μ flows through a pipe of constant diameter D and length L at average velocity V_{avg}.

In practice, it is convenient to express the pressure loss for all types of fully developed internal flows (laminar or turbulent flows, circular or noncircular pipes, smooth or rough surfaces, horizontal or inclined pipes) as (Fig. 14–13)

Pressure loss:
$$\Delta P_L = f\frac{L}{D}\frac{\rho V_{avg}^2}{2} \quad (14\text{–}21)$$

where $\rho V_{avg}^2/2$ is the *dynamic pressure* and f is the **Darcy friction factor**,

$$f = \frac{8\tau_w}{\rho V_{avg}^2} \quad (14\text{–}22)$$

It is also called the **Darcy–Weisbach friction factor**, named after the Frenchman Henry Darcy (1803–1858) and the German Julius Weisbach (1806–1871), the two engineers who provided the greatest contribution in its development. It should not be confused with the *friction coefficient* C_f

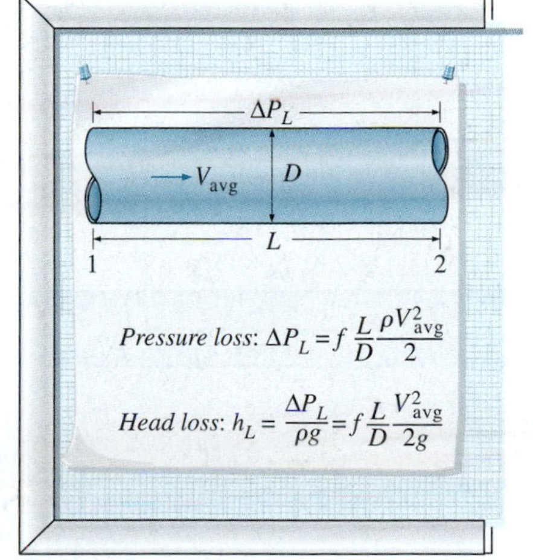

FIGURE 14–13
The relation for pressure loss (and head loss) is one of the most general relations in fluid mechanics, and it is valid for laminar or turbulent flows, circular or noncircular pipes, and pipes with smooth or rough surfaces.

[also called the *Fanning friction factor*, named after the American engineer John Fanning (1837–1911)], which is defined as $C_f = 2\tau_w/(\rho V_{avg}^2) = f/4$.

Setting Eqs. 14–20 and 14–21 equal to each other and solving for f gives the friction factor for fully developed laminar flow in a circular pipe,

Circular pipe, laminar:
$$f = \frac{64\mu}{\rho D V_{avg}} = \frac{64}{\text{Re}} \tag{14–23}$$

This equation shows that *in laminar flow, the friction factor is a function of the Reynolds number only and is independent of the roughness of the pipe surface*.

In the analysis of piping systems, pressure losses are commonly expressed in terms of the *equivalent fluid column height*, called the **head loss** h_L. Noting from fluid statics that $\Delta P = \rho g h$ and thus a pressure difference of ΔP corresponds to a fluid height of $h = \Delta P/\rho g$, the *pipe head loss* is obtained by dividing ΔP_L by ρg to give

Head loss:
$$h_L = \frac{\Delta P_L}{\rho g} = f\frac{L}{D}\frac{V_{avg}^2}{2g} \tag{14–24}$$

The head loss h_L represents *the additional height that the fluid needs to be raised by a pump in order to overcome the frictional losses in the pipe*. The head loss is caused by viscosity, and it is directly related to the wall shear stress. Equations 14–21, 14–22 and 14–24 are valid for both laminar and turbulent flows in both circular and noncircular pipes, but Eq. 14–23 is valid only for fully developed laminar flow in circular pipes.

Once the pressure loss (or head loss) is known, the required pumping power *to overcome the pressure loss* is determined from

$$\dot{W}_{\text{pump}, L} = \dot{V}\Delta P_L = \dot{V}\rho g h_L = \dot{m} g h_L \tag{14–25}$$

where $\dot{V}$ is the volume flow rate and $\dot{m}$ is the mass flow rate.

The average velocity for laminar flow in a horizontal pipe is, from Eq. 14–20,

Horizontal pipe:
$$V_{avg} = \frac{(P_1 - P_2)R^2}{8\mu L} = \frac{(P_1 - P_2)D^2}{32\mu L} = \frac{\Delta P\, D^2}{32\mu L} \tag{14–26}$$

Then the volume flow rate for laminar flow through a horizontal pipe of diameter D and length L becomes

$$\dot{V} = V_{avg}A_c = \frac{(P_1 - P_2)R^2}{8\mu L}\pi R^2 = \frac{(P_1 - P_2)\pi D^4}{128\mu L} = \frac{\Delta P\,\pi D^4}{128\mu L} \tag{14–27}$$

This equation is known as **Poiseuille's law,** and this flow is called *Hagen–Poiseuille flow* in honor of the works of G. Hagen (1797–1884) and J. Poiseuille (1799–1869) on the subject. Note from Eq. 14–27 that *for a specified flow rate, the pressure drop and thus the required pumping power is proportional to the length of the pipe and the viscosity of the fluid, but it is inversely proportional to the fourth power of the radius (or diameter) of the pipe*. Therefore, the pumping power requirement for a piping system can be reduced by a factor of 16 by doubling the pipe diameter (Fig. 14–14). Of course the benefits of the reduction in the energy costs must be weighed against the increased cost of construction due to using a larger-diameter pipe.

The pressure drop ΔP equals the pressure loss ΔP_L in the case of a horizontal pipe, but this is not the case for inclined pipes or pipes with variable

$\dot{W}_{pump} = 16$ hp

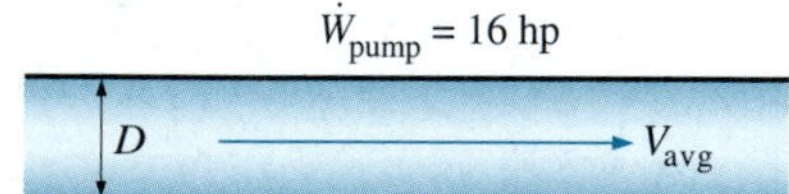

$\dot{W}_{pump} = 1$ hp

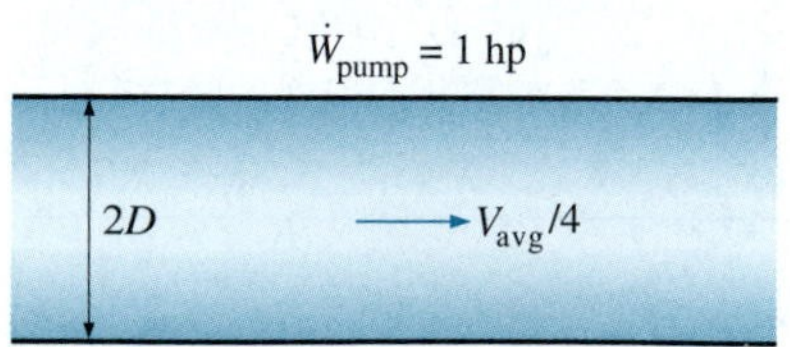

FIGURE 14–14
The pumping power requirement for a laminar flow piping system can be reduced by a factor of 16 by doubling the pipe diameter.

cross-sectional area. This can be demonstrated by writing the energy equation for steady, incompressible one-dimensional flow in terms of heads as (see Chap. 12)

$$\frac{P_1}{\rho g} + \alpha_1 \frac{V_1^2}{2g} + z_1 + h_{\text{pump}, u} = \frac{P_2}{\rho g} + \alpha_2 \frac{V_2^2}{2g} + z_2 + h_{\text{turbine}, e} + h_L \qquad \text{(14–28)}$$

where $h_{\text{pump}, u}$ is the useful pump head delivered to the fluid, $h_{\text{turbine}, e}$ is the turbine head extracted from the fluid, h_L is the irreversible head loss between sections 1 and 2, V_1 and V_2 are the average velocities at sections 1 and 2, respectively, and α_1 and α_2 are the *kinetic energy correction factors* at sections 1 and 2 (it can be shown that $\alpha = 2$ for fully developed laminar flow and about 1.05 for fully developed turbulent flow). Equation 14–28 can be rearranged as

$$P_1 - P_2 = \rho(\alpha_2 V_2^2 - \alpha_1 V_1^2)/2 + \rho g[(z_2 - z_1) + h_{\text{turbine}, e} - h_{\text{pump}, u} + h_L] \qquad \text{(14–29)}$$

Therefore, the pressure drop $\Delta P = P_1 - P_2$ and pressure loss $\Delta P_L = \rho g h_L$ for a given flow section are equivalent if (1) the flow section is horizontal so that there are no hydrostatic or gravity effects ($z_1 = z_2$), (2) the flow section does not involve any work devices such as a pump or a turbine since they change the fluid pressure ($h_{\text{pump}, u} = h_{\text{turbine}, e} = 0$), (3) the cross-sectional area of the flow section is constant and thus the average flow velocity is constant ($V_1 = V_2$), and (4) the velocity profiles at sections 1 and 2 are the same shape ($\alpha_1 = \alpha_2$).

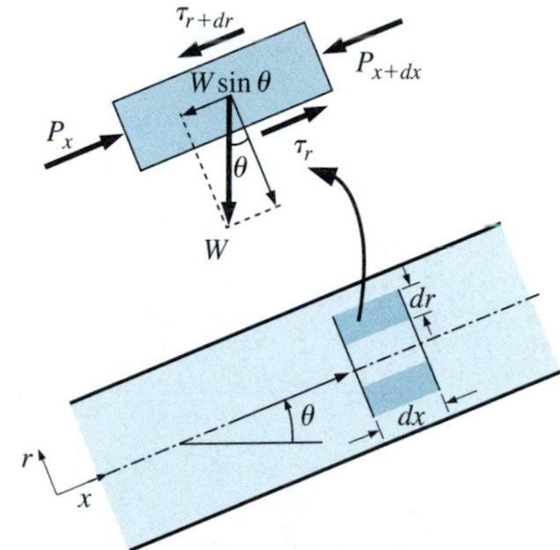

FIGURE 14–15
Free-body diagram of a ring-shaped differential fluid element of radius r, thickness dr, and length dx oriented coaxially with an inclined pipe in fully developed laminar flow.

Inclined Pipes

Relations for inclined pipes can be obtained in a similar manner from a force balance in the direction of flow. The only additional force in this case is the component of the fluid weight in the flow direction, whose magnitude is

$$W_x = W \sin \theta = \rho g V_{\text{element}} \sin \theta = \rho g(2\pi r\, dr\, dx) \sin \theta \qquad \text{(14–30)}$$

where θ is the angle between the horizontal and the flow direction (Fig. 14–15). The force balance in Eq. 14–9 now becomes

$$(2\pi r\, dr\, P)_x - (2\pi r\, dr\, P)_{x+dx} + (2\pi r\, dx\, \tau)_r$$
$$- (2\pi r\, dx\, \tau)_{r+dr} - \rho g(2\pi r\, dr\, dx) \sin \theta = 0 \qquad \text{(14–31)}$$

which results in the differential equation

$$\frac{\mu}{r} \frac{d}{dr}\left(r \frac{du}{dr}\right) = \frac{dP}{dx} + \rho g \sin \theta \qquad \text{(14–32)}$$

Following the same solution procedure as previously, the velocity profile is

$$u(r) = -\frac{R^2}{4\mu}\left(\frac{dP}{dx} + \rho g \sin \theta\right)\left(1 - \frac{r^2}{R^2}\right) \qquad \text{(14–33)}$$

From Eq. 14–33, the *average velocity* and the *volume flow rate* relations for laminar flow through inclined pipes are, respectively,

$$V_{\text{avg}} = \frac{(\Delta P - \rho g L \sin \theta) D^2}{32 \mu L} \quad \text{and} \quad \dot{V} = \frac{(\Delta P - \rho g L \sin \theta)\pi D^4}{128 \mu L} \qquad \text{(14–34)}$$

which are identical to the corresponding relations for horizontal pipes, except that ΔP is replaced by $\Delta P - \rho g L \sin \theta$. Therefore, the results already obtained for horizontal pipes can also be used for inclined pipes provided that ΔP is replaced by $\Delta P - \rho g L \sin \theta$ (Fig. 14–16). Note that $\theta > 0$ and thus $\sin \theta > 0$ for uphill flow, and $\theta < 0$ and thus $\sin \theta < 0$ for downhill flow.

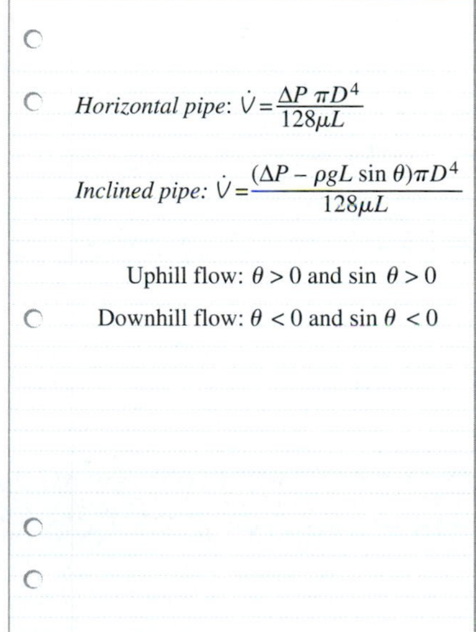

Horizontal pipe: $\dot{V} = \frac{\Delta P\, \pi D^4}{128 \mu L}$
Inclined pipe: $\dot{V} = \frac{(\Delta P - \rho g L \sin \theta)\pi D^4}{128 \mu L}$
Uphill flow: $\theta > 0$ and $\sin \theta > 0$
Downhill flow: $\theta < 0$ and $\sin \theta < 0$

FIGURE 14–16
The relations developed for fully developed laminar flow through horizontal pipes can also be used for inclined pipes by replacing ΔP with $\Delta P - \rho g L \sin \theta$.

In inclined pipes, the combined effect of pressure difference and gravity drives the flow. Gravity helps downhill flow but opposes uphill flow. Therefore, much greater pressure differences need to be applied to maintain a specified flow rate in uphill flow although this becomes important only for liquids, because the density of gases is generally low. In the special case of *no flow* ($\dot{V} = 0$), we have $\Delta P = \rho g L \sin\theta$, which is what we would obtain from fluid statics (Chap. 10).

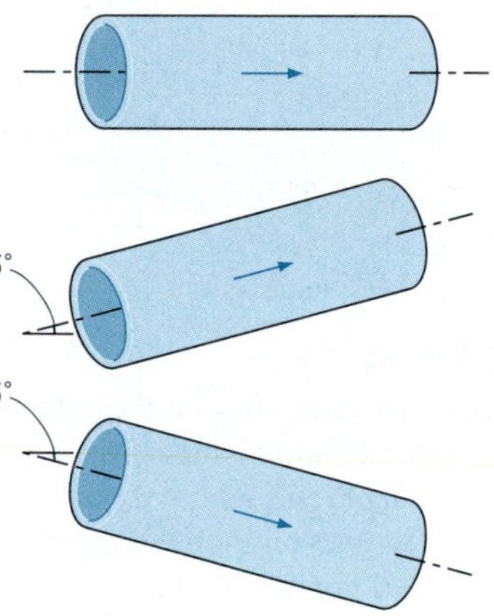

EXAMPLE 14–1 Flow Rates in Horizontal and Inclined Pipes

Oil at 20°C (ρ = 888 kg/m^3 and μ = 0.800 kg/m · s) is flowing steadily through a 5-cm-diameter 40-m-long pipe (Fig. 14–17). The pressure at the pipe inlet and outlet are measured to be 745 and 97 kPa, respectively. Determine the flow rate of oil through the pipe assuming the pipe is (*a*) horizontal, (*b*) inclined 15° upward, (*c*) inclined 15° downward. Also verify that the flow through the pipe is laminar.

Horizontal

+15°

−15°

FIGURE 14–17
Schematic for Example 14–1.

Solution The pressure readings at the inlet and outlet of a pipe are given. The flow rates are to be determined for three different orientations, and the flow is to be shown to be laminar.

Assumptions **1** The flow is steady, laminar, and incompressible. **2** The entrance effects are negligible, and thus the flow is fully developed. **3** The pipe involves no components such as bends, valves, and connectors. **4** The piping section involves no work devices such as a pump or a turbine.

Properties The density and dynamic viscosity of oil are given to be ρ = 888 kg/m^3 and μ = 0.800 kg/m · s, respectively.

Analysis The pressure drop across the pipe and the pipe cross-sectional area are

$$\Delta P = P_1 - P_2 = 745 - 97 = 648 \text{ kPa}$$

$$A_c = \pi D^2/4 = \pi(0.05 \text{ m})^2/4 = 0.001963 \text{ m}^2$$

(*a*) The flow rate for all three cases is determined from Eq. 14–34,

$$\dot{V} = \frac{(\Delta P - \rho g L \sin\theta)\pi D^4}{128\mu L}$$

where θ is the angle the pipe makes with the horizontal. For the horizontal case, θ = 0 and thus sin θ = 0. Therefore,

$$\dot{V}_{\text{horiz}} = \frac{\Delta P\,\pi D^4}{128\mu L} = \frac{(648 \text{ kPa})\pi(0.05 \text{ m})^4}{128(0.800 \text{ kg/m}\cdot\text{s})(40 \text{ m})}\left(\frac{1000 \text{ N/m}^2}{1 \text{ kPa}}\right)\left(\frac{1 \text{ kg}\cdot\text{m/s}^2}{1 \text{ N}}\right)$$

$$= \mathbf{0.00311 \text{ m}^3/\text{s}}$$

(*b*) For uphill flow with an inclination of 15°, we have θ = +15°, and

$$\dot{V}_{\text{uphill}} = \frac{(\Delta P - \rho g L \sin\theta)\pi D^4}{128\mu L}$$

$$= \frac{[648{,}000 \text{ Pa} - (888 \text{ kg/m}^3)(9.81 \text{ m/s}^2)(40 \text{ m})\sin 15°]\pi(0.05 \text{ m})^4}{128(0.800 \text{ kg/m}\cdot\text{s})(40 \text{ m})}\left(\frac{1 \text{ kg}\cdot\text{m/s}^2}{1 \text{ Pa}\cdot\text{m}^2}\right)$$

$$= \mathbf{0.00267 \text{ m}^3/\text{s}}$$

(*c*) For downhill flow with an inclination of 15°, we have $\theta = -15°$, and

$$\dot{V}_{\text{downhill}} = \frac{(\Delta P - \rho g L \sin\theta)\pi D^4}{128\mu L}$$

$$= \frac{[648{,}000 \text{ Pa} - (888 \text{ kg/m}^3)(9.81 \text{ m/s}^2)(40 \text{ m})\sin(-15°)]\pi(0.05 \text{ m})^4}{128(0.800 \text{ kg/m}\cdot\text{s})(40 \text{ m})}\left(\frac{1 \text{ kg}\cdot\text{m/s}^2}{1 \text{ Pa}\cdot\text{m}^2}\right)$$

$$= \mathbf{0.00354\ m^3/s}$$

The flow rate is the highest for the downhill flow case, as expected. The average fluid velocity and the Reynolds number in this case are

$$V_{\text{avg}} = \frac{\dot{V}}{A_c} = \frac{0.00354 \text{ m}^3/\text{s}}{0.001963 \text{ m}^2} = 1.80 \text{ m/s}$$

$$\text{Re} = \frac{\rho V_{\text{avg}} D}{\mu} = \frac{(888 \text{ kg/m}^3)(1.80 \text{ m/s})(0.05 \text{ m})}{0.800 \text{ kg/m}\cdot\text{s}} = 100$$

which is much less than 2300. Therefore, the flow is *laminar* for all three cases and the analysis is valid. Note that until we calculated V_{avg}, we could not calculate Re to verify that the flow is laminar.

Discussion When solving pipe flow problems, you should *always* calculate the Reynolds number to determine the flow regime—whether the flow is laminar or turbulent. The flow is driven by the combined effect of pressure difference and gravity. As can be seen from the flow rates we calculated, gravity opposes uphill flow, but enhances downhill flow. Gravity has no effect on the flow rate in the horizontal case. Downhill flow can occur even in the absence of an applied pressure difference. For the case of $P_1 = P_2 = 97$ kPa (i.e., no applied pressure difference), the pressure throughout the entire pipe would remain constant at 97 Pa, and the fluid would flow through the pipe at a rate of 0.00043 m³/s under the influence of gravity. The flow rate increases as the tilt angle of the pipe from the horizontal is increased in the negative direction and would reach its maximum value when the pipe is vertical.

EXAMPLE 14–2 Pressure Drop and Head Loss in a Pipe

Water at 40°F ($\rho = 62.42$ lbm/ft³ and $\mu = 1.038 \times 10^{-3}$ lbm/ft · s) is flowing through a 0.12-in- (= 0.010 ft) diameter 30-ft-long horizontal pipe steadily at an average velocity of 3.0 ft/s (Fig. 14–18). Determine (*a*) the head loss, (*b*) the pressure drop, and (*c*) the pumping power requirement to overcome this pressure drop.

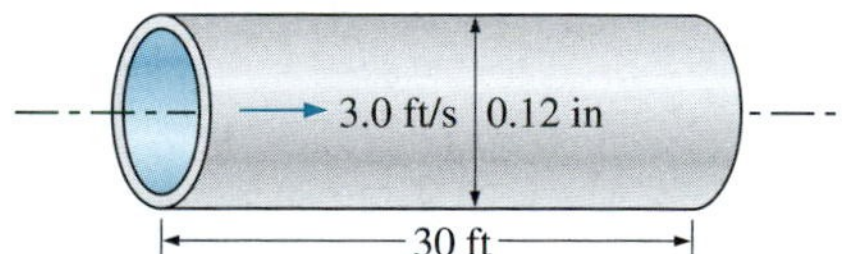

FIGURE 14–18
Schematic for Example 14–2.

Solution The average flow velocity in a pipe is given. The head loss, the pressure drop, and the pumping power are to be determined.

Assumptions **1** The flow is steady and incompressible. **2** The entrance effects are negligible, and thus the flow is fully developed. **3** The pipe involves no components such as bends, valves, and connectors.

Properties The density and dynamic viscosity of water are given to be $\rho = 62.42$ lbm/ft³ and $\mu = 1.038 \times 10^{-3}$ lbm/ft · s, respectively.

Analysis (*a*) First we need to determine the flow regime. The Reynolds number is

$$\text{Re} = \frac{\rho V_{\text{avg}} D}{\mu} = \frac{(62.42 \text{ lbm/ft}^3)(3 \text{ ft/s})(0.01 \text{ ft})}{1.038 \times 10^{-3} \text{ lbm/ft}\cdot\text{s}} = 1803$$

which is less than 2300. Therefore, the flow is laminar. Then the friction factor and the head loss become

$$f = \frac{64}{\text{Re}} = \frac{64}{1803} = 0.0355$$

$$h_L = f\frac{L}{D}\frac{V_{avg}^2}{2g} = 0.0355\frac{30 \text{ ft}}{0.01 \text{ ft}}\frac{(3 \text{ ft/s})^2}{2(32.2 \text{ ft/s}^2)} = \mathbf{14.9 \text{ ft}}$$

(*b*) Noting that the pipe is horizontal and its diameter is constant, the pressure drop in the pipe is due entirely to the frictional losses and is equivalent to the pressure loss,

$$\Delta P = \Delta P_L = f\frac{L}{D}\frac{\rho V_{avg}^2}{2} = 0.0355\frac{30 \text{ ft}}{0.01 \text{ ft}}\frac{(62.42 \text{ lbm/ft}^3)(3 \text{ ft/s})^2}{2}\left(\frac{1 \text{ lbf}}{32.2 \text{ lbm}\cdot\text{ft/s}^2}\right)$$

$$= \mathbf{929 \text{ lbf/ft}^2 = 6.45 \text{ psi}}$$

(*c*) The volume flow rate and the pumping power requirements are

$$\dot{V} = V_{avg}A_c = V_{avg}(\pi D^2/4) = (3 \text{ ft/s})[\pi(0.01 \text{ ft})^2/4] = 0.000236 \text{ ft}^3\text{/s}$$

$$\dot{W}_{pump} = \dot{V}\,\Delta P = (0.000236 \text{ ft}^3\text{/s})(929 \text{ lbf/ft}^2)\left(\frac{1 \text{ W}}{0.737 \text{ lbf}\cdot\text{ft/s}}\right) = \mathbf{0.30 \text{ W}}$$

Therefore, power input in the amount of 0.30 W is needed to overcome the frictional losses in the flow due to viscosity.

Discussion The pressure rise provided by a pump is often listed by a pump manufacturer in units of head (Chap. 11). Thus, the pump in this flow needs to provide 14.9 ft of water head in order to overcome the irreversible head loss.

14–5 ▪ TURBULENT FLOW IN PIPES

Most flows encountered in engineering practice are turbulent, and thus it is important to understand how turbulence affects wall shear stress. However, turbulent flow is a complex mechanism dominated by fluctuations, and despite tremendous amounts of work done in this area by researchers, turbulent flow is still not fully understood. Therefore, we must rely on experiments and the empirical or semi-empirical correlations developed for various situations.

Turbulent flow is characterized by disorderly and rapid fluctuations of swirling regions of fluid, called **eddies**, throughout the flow. These fluctuations provide an additional mechanism for momentum and energy transfer. In laminar flow, fluid particles flow in an orderly manner along pathlines, and momentum and energy are transferred across streamlines by molecular diffusion. In turbulent flow, the swirling eddies transport mass, momentum, and energy to other regions of flow much more rapidly than molecular diffusion, greatly enhancing mass, momentum, and heat transfer. As a result, turbulent flow is associated with much higher values of friction, heat transfer, and mass transfer coefficients (Fig. 14–19).

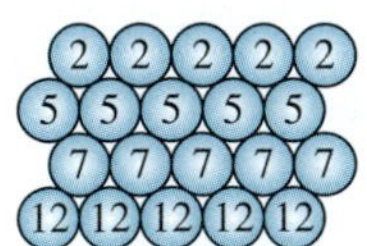

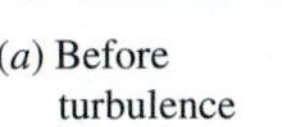

(*a*) Before turbulence

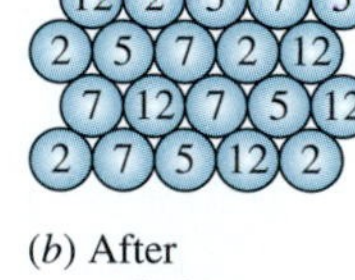

(*b*) After turbulence

FIGURE 14–19
The intense mixing in turbulent flow brings fluid particles at different momentums into close contact and thus enhances momentum transfer.

Turbulent Velocity Profile

Unlike laminar flow, the expressions for the velocity profile in a turbulent flow are based on both analysis and measurements, and thus they are semi-empirical in nature with constants determined from experimental data.

Consider fully developed turbulent flow in a pipe, and let u denote the time-averaged velocity in the axial direction.

Typical velocity profiles for fully developed laminar and turbulent pipe flows are given in Fig. 14–20. Note that the velocity profile is parabolic in laminar flow but is much fuller in turbulent flow, with a sharp drop near the pipe wall. Turbulent flow along a wall can be considered to consist of four regions, characterized by the distance from the wall (Fig. 14–20). The very thin layer next to the wall where viscous effects are dominant is the **viscous** (or **laminar** or **linear** or **wall**) sublayer. The velocity profile in this layer is very nearly *linear*, and the flow is streamlined. Next to the viscous sublayer is the **buffer layer**, in which turbulent effects are becoming significant, but the flow is still dominated by viscous effects. Above the buffer layer is the **overlap** (or **transition**) **layer**, also called the **inertial sublayer**, in which the turbulent effects are much more significant, but still not dominant. Above that is the **outer** (or **turbulent**) **layer** in the remaining part of the flow in which turbulent effects dominate over molecular diffusion (viscous) effects.

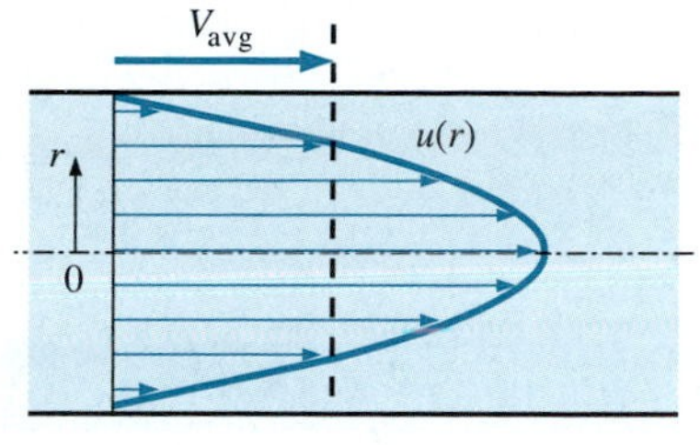

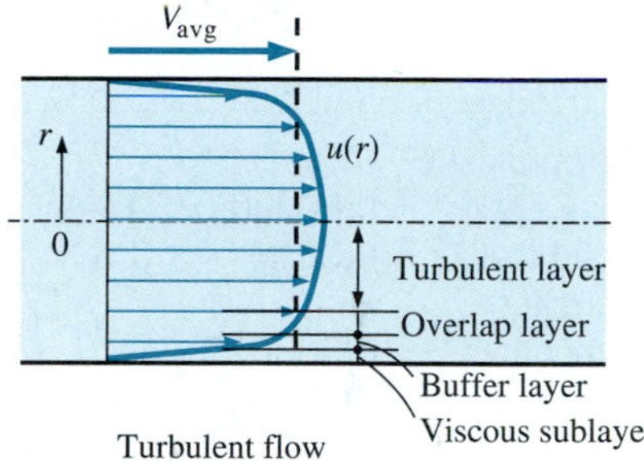

FIGURE 14–20
The velocity profile in fully developed pipe flow is parabolic in laminar flow, but much fuller in turbulent flow.

Flow characteristics are quite different in different regions, and thus it is difficult to come up with an analytic relation for the velocity profile for the entire flow as we did for laminar flow. The best approach in the turbulent case turns out to be to identify the key variables and functional forms using dimensional analysis, and then to use experimental data to determine the numerical values of any constants. Several equations have been generated in this manner to approximate the velocity profile shape in fully developed turbulent pipe flow. Discussion of these equations is beyond the scope of the present text.

The Moody Chart

The friction factor in fully developed turbulent pipe flow depends on the Reynolds number and the **relative roughness** ε/D, which is the ratio of the mean height of roughness of the pipe to the pipe diameter. The functional form of this dependence cannot be obtained from a theoretical analysis, and all available results are obtained from painstaking experiments using artificially roughened surfaces (usually by gluing sand grains of a known size on the inner surfaces of the pipes). Most such experiments were conducted by Prandtl's student J. Nikuradse in 1933, followed by the works of others. The friction factor was calculated from the measurements of the flow rate and the pressure drop.

The experimental results obtained are presented in tabular, graphical, and functional forms obtained by curve-fitting experimental data. In 1939, Cyril F. Colebrook (1910–1997) combined the available data for transition and turbulent flow in smooth as well as rough pipes into the following implicit relation known as the **Colebrook equation**:

$$\frac{1}{\sqrt{f}} = -2.0 \log\left(\frac{\varepsilon/D}{3.7} + \frac{2.51}{\text{Re}\sqrt{f}}\right) \qquad \text{(turbulent flow)} \qquad \textbf{(14–35)}$$

We note that the logarithm in Eq. 14–35 is a base 10 rather than a natural logarithm. In 1942, the American engineer Hunter Rouse (1906–1996) verified Colebrook's equation and produced a graphical plot of f as a function of Re and the product $\text{Re}\sqrt{f}$. He also presented the laminar flow relation and a table of commercial pipe roughness. Two years later, Lewis F. Moody (1880–1953) redrew Rouse's diagram into the form commonly used today.

The now famous **Moody chart** is given in the appendix as Fig. A–27. It presents the Darcy friction factor for pipe flow as a function of the Reynolds number and ε/D over a wide range. It is probably one of the most widely accepted and used charts in engineering. Although it is developed for circular pipes, it can also be used for noncircular pipes by replacing the diameter with the hydraulic diameter.

Commercially available pipes differ from those used in the experiments in that the roughness of pipes in the market is not uniform and it is difficult to give a precise description of it. Equivalent roughness values for some commercial pipes are given in Table 14–1 as well as on the Moody chart. It should be kept in mind that these values are for new pipes, and the relative roughness of pipes may increase with use as a result of corrosion, scale buildup, and precipitation. As a result, the friction factor may increase by a factor of 5 to 10. Actual operating conditions must be considered in the design of piping systems. Also, the Moody chart and its equivalent Colebrook equation involve several uncertainties (the roughness size, experimental error, curve fitting of data, etc.), and thus the results obtained should not be treated as "exact." They are usually considered to be accurate to ±15 percent over the entire range in the figure.

TABLE 14–1

Equivalent roughness values for new commercial pipes*

Material	Roughness, ε ft	Roughness, ε mm
Glass, plastic	0 (smooth)	
Concrete	0.003–0.03	0.9–9
Wood stave	0.0016	0.5
Rubber, smoothed	0.000033	0.01
Copper or brass tubing	0.000005	0.0015
Cast iron	0.00085	0.26
Galvanized iron	0.0005	0.15
Wrought iron	0.00015	0.046
Stainless steel	0.000007	0.002
Commercial steel	0.00015	0.045

* The uncertainty in these values can be as much as ±60 percent.

The Colebrook equation is implicit in f, and thus the determination of the friction factor requires iteration. An approximate explicit relation for f was given by S. E. Haaland in 1983 as

$$\frac{1}{\sqrt{f}} \cong -1.8 \log\left[\frac{6.9}{\text{Re}} + \left(\frac{\varepsilon/D}{3.7}\right)^{1.11}\right] \quad \text{(14–36)}$$

The results obtained from this relation are within 2 percent of those obtained from the Colebrook equation. If more accurate results are desired, Eq. 14–36 can be used as a good *first guess* in a Newton iteration when using a programmable calculator or a spreadsheet to solve for f with Eq. 14–35.

We make the following observations from the Moody chart:

- For laminar flow, the friction factor decreases with increasing Reynolds number, and it is independent of surface roughness.
- The friction factor is a minimum for a smooth pipe (but still not zero because of the no-slip condition) and increases with roughness (Fig. 14–21). The Colebrook equation in this case ($\varepsilon = 0$) reduces to the **Prandtl equation** expressed as $1/\sqrt{f} = 2.0 \log(\text{Re}\sqrt{f}) - 0.8$.
- The transition region from the laminar to turbulent regime ($2300 < \text{Re} < 10{,}000$) is indicated by the shaded area in the Moody chart (Fig. A–12). The flow in this region may be laminar or turbulent, depending on flow disturbances, or it may alternate between laminar and turbulent, and thus the friction factor may also alternate between the values for laminar and turbulent flow. The data in this range are the least reliable. At small relative roughnesses, the friction factor increases in the transition region and approaches the value for smooth pipes.
- At very large Reynolds numbers (to the right of the dashed line on the Moody chart) the friction factor curves corresponding to specified relative roughness curves are nearly horizontal, and thus the friction factors are independent of the Reynolds number. The flow in that region is called

Relative Roughness, ε/D	Friction Factor, f
0.0*	0.0119
0.00001	0.0119
0.0001	0.0134
0.0005	0.0172
0.001	0.0199
0.005	0.0305
0.01	0.0380
0.05	0.0716

* Smooth surface. All values are for Re = 10^6 and are calculated from the Colebrook equation.

FIGURE 14–21

The friction factor is minimum for a smooth pipe and increases with roughness, as seen here for Re = 10^6.

fully rough turbulent flow or just *fully rough flow* because the thickness of the viscous sublayer decreases with increasing Reynolds number, and it becomes so thin that it is negligibly small compared to the surface roughness height. The viscous effects in this case are produced in the main flow primarily by the protruding roughness elements, and the contribution of the viscous sublayer is negligible. The Colebrook equation in the *fully rough* zone (Re $\rightarrow \infty$) reduces to the **von Kármán equation** expressed as $1/\sqrt{f} = -2.0 \log[(\varepsilon/D)/3.7]$, which is explicit in f. Some authors call this zone *completely* (or *fully*) *turbulent flow*, but this is misleading since the flow to the left of the dashed blue line in Fig. A–27 is also fully turbulent.

In calculations, we should make sure that we use the actual internal diameter of the pipe, which may be different than the nominal diameter. For example, the internal diameter of a steel pipe whose nominal diameter is 1 in is 1.049 in (Table 14–2).

TABLE 14–2

Standard sizes for Schedule 40 steel pipes

Nominal Size, in	Actual Inside Diameter, in
$\frac{1}{8}$	0.269
$\frac{1}{4}$	0.364
$\frac{3}{8}$	0.493
$\frac{1}{2}$	0.622
$\frac{3}{4}$	0.824
1	1.049
$1\frac{1}{2}$	1.610
2	2.067
$2\frac{1}{2}$	2.469
3	3.068
5	5.047
10	10.02

Types of Fluid Flow Problems

In the design and analysis of piping systems that involve the use of the Moody chart (or the Colebrook equation), we usually encounter three types of problems (the fluid and the roughness of the pipe are assumed to be specified in all cases) (Fig. 14–22):

1. Determining the **pressure drop** (or head loss) when the pipe length and diameter are given for a specified flow rate (or velocity)
2. Determining the **flow rate** when the pipe length and diameter are given for a specified pressure drop (or head loss)
3. Determining the **pipe diameter** when the pipe length and flow rate are given for a specified pressure drop (or head loss)

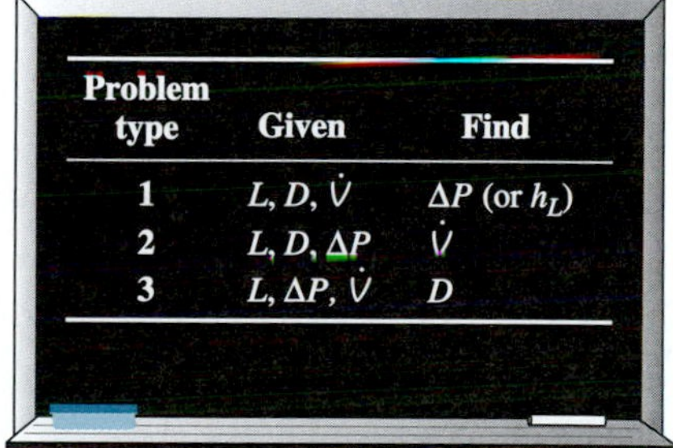

Problem type	Given	Find
1	$L, D, \dot{V}$	ΔP (or h_L)
2	$L, D, \Delta P$	$\dot{V}$
3	$L, \Delta P, \dot{V}$	D

FIGURE 14–22
The three types of problems encountered in pipe flow.

Problems of the *first type* are straightforward and can be solved directly by using the Moody chart. Problems of the *second type* and *third type* are commonly encountered in engineering design (in the selection of pipe diameter, for example, that minimizes the sum of the construction and pumping costs), but the use of the Moody chart with such problems requires an iterative approach—an equation solver is recommended if available.

In problems of the *second type*, the diameter is given but the flow rate is unknown. A good guess for the friction factor in that case is obtained from the completely turbulent flow region for the given roughness. This is true for large Reynolds numbers, which is often the case in practice. Once the flow rate is obtained, the friction factor is corrected using the Moody chart or the Colebrook equation, and the process is repeated until the solution converges. (Typically only a few iterations are required for convergence to three or four digits of precision.)

In problems of the *third type*, the diameter is not known and thus the Reynolds number and the relative roughness cannot be calculated. Therefore, we start calculations by assuming a pipe diameter. The pressure drop calculated for the assumed diameter is then compared to the specified pressure drop, and calculations are repeated with another pipe diameter in an iterative fashion until convergence.

To avoid tedious iterations in head loss, flow rate, and diameter calculations, Swamee and Jain proposed the following explicit relations in 1976 that are accurate to within 2 percent of the Moody chart:

$$h_L = 1.07\frac{\dot{V}^2L}{gD^5}\left\{\ln\left[\frac{\varepsilon}{3.7D} + 4.62\left(\frac{\nu D}{\dot{V}}\right)^{0.9}\right]\right\}^{-2} \qquad \begin{matrix}10^{-6} < \varepsilon/D < 10^{-2}\\ 3000 < \text{Re} < 3 \times 10^8\end{matrix} \tag{14–37}$$

$$\dot{V} = -0.965\left(\frac{gD^5h_L}{L}\right)^{0.5}\ln\left[\frac{\varepsilon}{3.7D} + \left(\frac{3.17\nu^2L}{gD^3h_L}\right)^{0.5}\right] \qquad \text{Re} > 2000 \tag{14–38}$$

$$D = 0.66\left[\varepsilon^{1.25}\left(\frac{L\dot{V}^2}{gh_L}\right)^{4.75} + \nu\dot{V}^{9.4}\left(\frac{L}{gh_L}\right)^{5.2}\right]^{0.04} \qquad \begin{matrix}10^{-6} < \varepsilon/D < 10^{-2}\\ 5000 < \text{Re} < 3 \times 10^8\end{matrix} \tag{14–39}$$

Note that all quantities are dimensional and the units simplify to the desired unit (for example, to m or ft in the last relation) when consistent units are used. Noting that the Moody chart is accurate to within 15 percent of experimental data, we should have no reservation in using these approximate relations in the design of piping systems.

EXAMPLE 14–3 Determining the Head Loss in a Water Pipe

Water at 60°F (ρ = 62.36 lbm/ft^3 and $\mu = 7.536 \times 10^{-4}$ lbm/ft · s) is flowing steadily in a 2-in-diameter horizontal pipe made of stainless steel at a rate of 0.2 ft^3/s (Fig. 14–23). Determine the pressure drop, the head loss, and the required pumping power input for flow over a 200-ft-long section of the pipe.

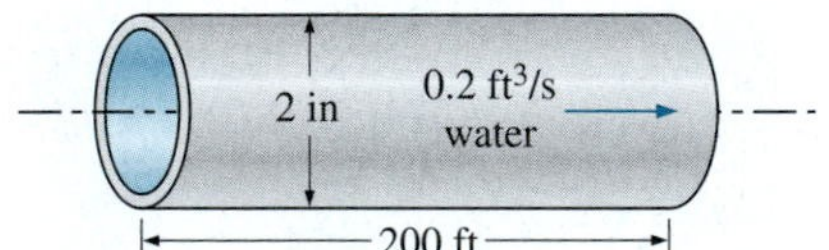

FIGURE 14–23
Schematic for Example 14–3.

Solution The flow rate through a specified water pipe is given. The pressure drop, the head loss, and the pumping power requirements are to be determined.

Assumptions **1** The flow is steady and incompressible. **2** The entrance effects are negligible, and thus the flow is fully developed. **3** The pipe involves no components such as bends, valves, and connectors. **4** The piping section involves no work devices such as a pump or a turbine.

Properties The density and dynamic viscosity of water are given to be ρ = 62.36 lbm/ft^3 and $\mu = 7.536 \times 10^{-4}$ lbm/ft · s, respectively.

Analysis We recognize this as a problem of the first type, since flow rate, pipe length, and pipe diameter are known. First we calculate the average velocity and the Reynolds number to determine the flow regime:

$$V = \frac{\dot{V}}{A_c} = \frac{\dot{V}}{\pi D^2/4} = \frac{0.2 \text{ ft}^3/\text{s}}{\pi(2/12 \text{ ft})^2/4} = 9.17 \text{ ft/s}$$

$$\text{Re} = \frac{\rho V D}{\mu} = \frac{(62.36 \text{ lbm/ft}^3)(9.17 \text{ ft/s})(2/12 \text{ ft})}{7.536 \times 10^{-4} \text{ lbm/ft} \cdot \text{s}} = 126{,}400$$

Since Re is greater than 10,000, the flow is turbulent. The relative roughness of the pipe is calculated using Table 14–1

$$\varepsilon/D = \frac{0.000007 \text{ ft}}{2/12 \text{ ft}} = 0.000042$$

The friction factor corresponding to this relative roughness and Reynolds number can simply be determined from the Moody chart. To avoid any reading error, we determine f from the Colebrook equation:

$$\frac{1}{\sqrt{f}} = -2.0 \log\left(\frac{\varepsilon/D}{3.7} + \frac{2.51}{\text{Re}\sqrt{f}}\right) \rightarrow \frac{1}{\sqrt{f}} = -2.0 \log\left(\frac{0.000042}{3.7} + \frac{2.51}{126{,}400\sqrt{f}}\right)$$

Using an equation solver or an iterative scheme, the friction factor is determined to be f = 0.0174. Then the pressure drop (which is equivalent to pressure loss in this case), head loss, and the required power input become

$$\Delta P = \Delta P_L = f\frac{L}{D}\frac{\rho V^2}{2} = 0.0174\,\frac{200\text{ ft}}{2/12\text{ ft}}\,\frac{(62.36\text{ lbm/ft}^3)(9.17\text{ ft/s})^2}{2}\left(\frac{1\text{ lbf}}{32.2\text{ lbm}\cdot\text{ft/s}^2}\right)$$

$$= \mathbf{1700\ lbf/ft^2} = \mathbf{11.8\ psi}$$

$$h_L = \frac{\Delta P_L}{\rho g} = f\frac{L}{D}\frac{V^2}{2g} = 0.0174\,\frac{200\text{ ft}}{2/12\text{ ft}}\,\frac{(9.17\text{ ft/s})^2}{2(32.2\text{ ft/s}^2)} = \mathbf{27.3\ ft}$$

$$\dot{W}_{\text{pump}} = \dot{V}\,\Delta P = (0.2\text{ ft}^3\text{/s})(1700\text{ lbf/ft}^2)\left(\frac{1\text{ W}}{0.737\text{ lbf}\cdot\text{ft/s}}\right) = \mathbf{461\ W}$$

Therefore, power input in the amount of 461 W is needed to overcome the frictional losses in the pipe.

Discussion It is common practice to write our final answers to three significant digits, even though we know that the results are accurate to at most two significant digits because of inherent inaccuracies in the Colebrook equation, as discussed previously. The friction factor could also be determined easily from the explicit Haaland relation (Eq. 14–36). It would give f = 0.0172, which is sufficiently close to 0.0174. Also, the friction factor corresponding to ε = 0 in this case is 0.0171, which indicates that this stainless-steel pipe can be approximated as smooth with very little error.

EXAMPLE 14–4 Determining the Diameter of an Air Duct

Heated air at 1 atm and 35°C is to be transported in a 150-m-long circular plastic duct at a rate of 0.35 m³/s (Fig. 14–24). If the head loss in the pipe is not to exceed 20 m, determine the minimum diameter of the duct.

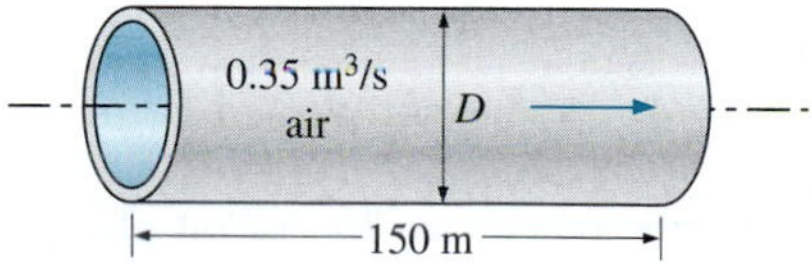

FIGURE 14–24
Schematic for Example 14–4.

Solution The flow rate and the head loss in an air duct are given. The diameter of the duct is to be determined.

Assumptions **1** The flow is steady and incompressible. **2** The entrance effects are negligible, and thus the flow is fully developed. **3** The duct involves no components such as bends, valves, and connectors. **4** Air is an ideal gas. **5** The duct is smooth since it is made of plastic. **6** The flow is turbulent (to be verified).

Properties The density, dynamic viscosity, and kinematic viscosity of air at 35°C are ρ = 1.145 kg/m³, μ = 1.895 × 10⁻⁵ kg/m · s, and ν = 1.655 × 10⁻⁵ m²/s (Table A–22).

Analysis This is a problem of the third type since it involves the determination of diameter for specified flow rate and head loss. We can solve this problem by three different approaches: (1) an iterative approach by assuming a pipe diameter, calculating the head loss, comparing the result to the specified head loss, and repeating calculations until the calculated head loss matches the specified value; (2) writing all the relevant equations (leaving the diameter as an unknown) and solving them simultaneously using an equation solver; and (3) using the third Swamee–Jain formula. We will demonstrate the use of the last two approaches.

The average velocity, the Reynolds number, the friction factor, and the head loss relations are expressed as (D is in m, V is in m/s, and Re and f are dimensionless)

$$V = \frac{\dot{V}}{A_c} = \frac{\dot{V}}{\pi D^2/4} = \frac{0.35 \text{ m}^3/\text{s}}{\pi D^2/4}$$

$$\text{Re} = \frac{VD}{\nu} = \frac{VD}{1.655 \times 10^{-5} \text{ m}^2/\text{s}}$$

$$\frac{1}{\sqrt{f}} = -2.0 \log\left(\frac{\varepsilon/D}{3.7} + \frac{2.51}{\text{Re}\sqrt{f}}\right) = -2.0 \log\left(\frac{2.51}{\text{Re}\sqrt{f}}\right)$$

$$h_L = f\frac{L}{D}\frac{V^2}{2g} \quad \rightarrow \quad 20 \text{ m} = f\frac{150 \text{ m}}{D}\frac{V^2}{2(9.81 \text{ m/s}^2)}$$

The roughness is approximately zero for a plastic pipe (Table 14–1). Therefore, this is a set of four equations and four unknowns, and solving them with an equation solver such as EES gives

$$D = \mathbf{0.267 \text{ m}}, \quad f = 0.0180, \quad V = 6.24 \text{ m/s}, \quad \text{and} \quad \text{Re} = 100{,}800$$

Therefore, the diameter of the duct should be more than 26.7 cm if the head loss is not to exceed 20 m. Note that Re > 10,000, and thus the turbulent flow assumption is verified.

The diameter can also be determined directly from the third Swamee–Jain formula to be

$$D = 0.66\left[\varepsilon^{1.25}\left(\frac{L\dot{V}^2}{gh_L}\right)^{4.75} + \nu\dot{V}^{9.4}\left(\frac{L}{gh_L}\right)^{5.2}\right]^{0.04}$$

$$= 0.66\left[0 + (1.655 \times 10^{-5} \text{ m}^2/\text{s})(0.35 \text{ m}^3/\text{s})^{9.4}\left(\frac{150 \text{ m}}{(9.81 \text{ m/s}^2)(20 \text{ m})}\right)^{5.2}\right]^{0.04}$$

$$= \mathbf{0.271 \text{ m}}$$

Discussion Note that the difference between the two results is less than 2 percent. Therefore, the simple Swamee–Jain relation can be used with confidence. Finally, the first (iterative) approach requires an initial guess for D. If we use the Swamee–Jain result as our initial guess, the diameter converges to $D = 0.267$ m in short order.

EXAMPLE 14–5 Determining the Flow Rate of Air in a Duct

Reconsider Example 14–4. Now the duct length is doubled while its diameter is maintained constant. If the total head loss is to remain constant, determine the drop in the flow rate through the duct.

Solution The diameter and the head loss in an air duct are given. The drop in the flow rate is to be determined.

Analysis This is a problem of the second type since it involves the determination of the flow rate for a specified pipe diameter and head loss. The solution involves an iterative approach since the flow rate (and thus the flow velocity) is not known.

The average velocity, Reynolds number, friction factor, and the head loss relations are expressed as (D is in m, V is in m/s, and Re and f are dimensionless)

$$V = \frac{\dot{V}}{A_c} = \frac{\dot{V}}{\pi D^2/4} \quad \rightarrow \quad V = \frac{\dot{V}}{\pi(0.267 \text{ m})^2/4}$$

$$\text{Re} = \frac{VD}{\nu} \quad \rightarrow \quad \text{Re} = \frac{V(0.267 \text{ m})}{1.655 \times 10^{-5} \text{ m}^2/\text{s}}$$

$$\frac{1}{\sqrt{f}} = -2.0 \log\left(\frac{\varepsilon/D}{3.7} + \frac{2.51}{\text{Re}\sqrt{f}}\right) \quad \rightarrow \quad \frac{1}{\sqrt{f}} = -2.0 \log\left(\frac{2.51}{\text{Re}\sqrt{f}}\right)$$

$$h_L = f\frac{L}{D}\frac{V^2}{2g} \quad \rightarrow \quad 20 \text{ m} = f\frac{300 \text{ m}}{0.267 \text{ m}}\frac{V^2}{2(9.81 \text{ m/s}^2)}$$

This is a set of four equations in four unknowns and solving them with an equation solver such as EES gives

$$\dot{V} = 0.24 \text{ m}^3/\text{s}, \quad f = 0.0195, \quad V = 4.23 \text{ m/s}, \quad \text{and} \quad \text{Re} = 68{,}300$$

Then the drop in the flow rate becomes

$$\dot{V}_{\text{drop}} = \dot{V}_{\text{old}} - \dot{V}_{\text{new}} = 0.35 - 0.24 = \mathbf{0.11 \text{ m}^3/\text{s}} \quad \text{(a drop of 31 percent)}$$

Therefore, for a specified head loss (or available head or fan pumping power), the flow rate drops by about 31 percent from 0.35 to 0.24 m^3/s when the duct length doubles.

Alternative Solution If a computer is not available (as in an exam situation), another option is to set up a *manual iteration loop*. We have found that the best convergence is usually realized by first guessing the friction factor f, and then solving for the velocity V. The equation for V as a function of f is

Average velocity through the pipe: $$V = \sqrt{\frac{2gh_L}{fL/D}}$$

Once V is calculated, the Reynolds number can be calculated, from which a *corrected* friction factor is obtained from the Moody chart or the Colebrook equation. We repeat the calculations with the corrected value of f until convergence. We guess $f = 0.04$ for illustration:

Iteration	f (guess)	V, m/s	Re	Corrected f
1	0.04	2.955	4.724×10^4	0.0212
2	0.0212	4.059	6.489×10^4	0.01973
3	0.01973	4.207	6.727×10^4	0.01957
4	0.01957	4.224	6.754×10^4	0.01956
5	0.01956	4.225	6.756×10^4	0.01956

Notice that the iteration has converged to three digits in only three iterations and to four digits in only four iterations. The final results are nearly identical to those obtained with EES, yet do not require a computer.

Discussion The new flow rate can also be determined directly from the second Swamee–Jain formula to be

$$\begin{aligned} \dot{V} &= -0.965\left(\frac{gD^5h_L}{L}\right)^{0.5}\left[\ln\frac{\varepsilon}{3.7D} + \left(\frac{3.17\nu^2L}{gD^3h_L}\right)^{0.5}\right] \\ &= -0.965\left(\frac{(9.81\text{ m/s}^2)(0.267\text{ m})^5(20\text{ m})}{300\text{ m}}\right)^{0.5} \\ &\quad \times \ln\left[0 + \left(\frac{3.17(1.655\times10^{-5}\text{ m}^2\text{/s})^2(300\text{ m})}{(9.81\text{ m/s}^2)(0.267\text{ m})^3(20\text{ m})}\right)^{0.5}\right] \\ &= 0.24\text{ m}^3\text{/s} \end{aligned}$$

Note that the result from the Swamee–Jain relation is the same (to two significant digits) as that obtained with the Colebrook equation using EES or using our manual iteration technique. Therefore, the simple Swamee–Jain relation can be used with confidence.

14–6 ▪ MINOR LOSSES

The fluid in a typical piping system passes through various fittings, valves, bends, elbows, tees, inlets, exits, enlargements, and contractions in addition to the pipes. These components interrupt the smooth flow of the fluid and cause additional losses because of the flow separation and mixing they induce. In a typical system with long pipes, these losses are minor compared to the total head loss in the pipes (the **major losses**) and are called **minor losses**. Although this is generally true, in some cases the minor losses may be greater than the major losses. This is the case, for example, in systems with several turns and valves in a short distance. The head loss introduced by a completely open valve, for example, may be negligible. But a partially closed valve may cause the largest head loss in the system, as evidenced by the drop in the flow rate. Flow through valves and fittings is very complex, and a theoretical analysis is generally not plausible. Therefore, minor losses are determined experimentally, usually by the manufacturers of the components.

Pipe section with valve:

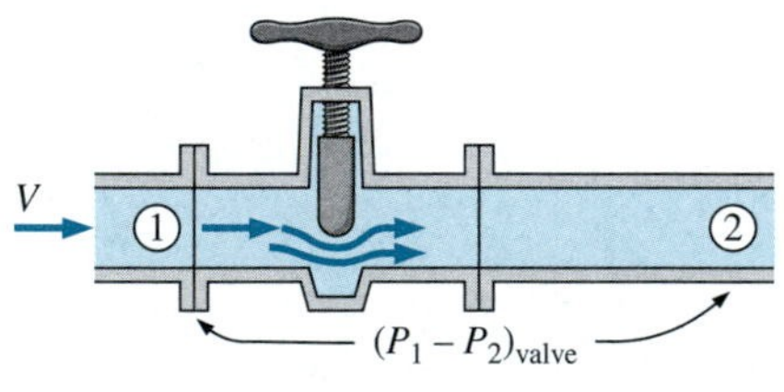

Pipe section without valve:

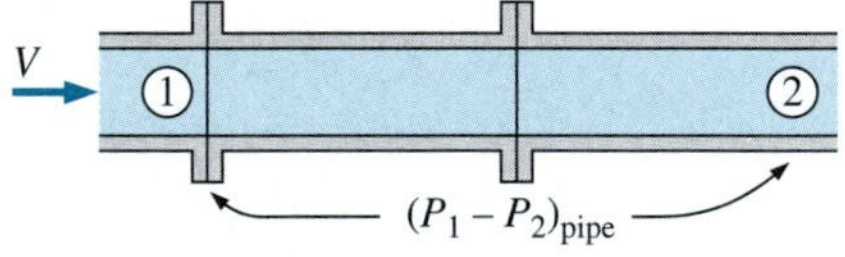

$\Delta P_L = (P_1 - P_2)_{valve} - (P_1 - P_2)_{pipe}$

FIGURE 14–25
For a constant-diameter section of a pipe with a minor loss component, the loss coefficient of the component (such as the gate valve shown) is determined by measuring the additional pressure loss it causes and dividing it by the dynamic pressure in the pipe.

Minor losses are usually expressed in terms of the **loss coefficient** K_L (also called the **resistance coefficient**), defined as (Fig. 14–25)

Loss coefficient:
$$K_L = \frac{h_L}{V^2/(2g)} \tag{14–40}$$

where h_L is the *additional* irreversible head loss in the piping system caused by insertion of the component, and is defined as $h_L = \Delta P_L/\rho g$. For example, imagine replacing the valve in Fig. 14–25 with a section of constant diameter pipe from location 1 to location 2. ΔP_L is defined as the pressure drop from 1 to 2 for the case *with* the valve, $(P_1 - P_2)_{valve}$, *minus* the pressure drop that would occur in the imaginary straight pipe section from 1 to 2 *without* the valve, $(P_1 - P_2)_{pipe}$ at the same flow rate. While the majority of the irreversible head loss occurs locally near the valve, some of it occurs downstream of the valve due to induced swirling turbulent eddies that are produced in the valve and continue downstream. These eddies "waste" mechanical energy because they are ultimately dissipated into heat while the flow in the downstream section of pipe eventually returns to fully developed conditions. When measuring minor losses in some minor loss components, such as *elbows*, for example, location 2 must be considerably far downstream (tens of pipe diameters) in order to fully account for the additional irreversible losses due to these decaying eddies.

When the pipe diameter downstream of the component *changes*, determination of the minor loss is even more complicated. In all cases, however, it is based on the *additional* irreversible loss of mechanical energy that would otherwise not exist if the minor loss component were not there. For simplicity, you may think of the minor loss as occurring *locally* across the minor loss component, but keep in mind that the component influences the flow for several pipe diameters downstream. By the way, this is the reason why most flow meter manufacturers recommend installing their flow meter at least 10 to 20 pipe diameters downstream of any elbows or valves—this allows the swirling turbulent eddies generated by the elbow or valve to largely disappear and the velocity profile to become fully developed before entering the flow meter. (Most flow meters are calibrated with a fully developed velocity profile at the flow meter inlet, and yield the best accuracy when such conditions also exist in the actual application.)

When the inlet diameter equals outlet diameter, the loss coefficient of a component can also be determined by measuring the pressure loss across the component and dividing it by the dynamic pressure, $K_L = \Delta P_L/(\frac{1}{2}\rho V^2)$. When the loss coefficient for a component is available, the head loss for that component is determined from

Minor loss:
$$h_L = K_L \frac{V^2}{2g} \tag{14–41}$$

The loss coefficient, in general, depends on the geometry of the component and the Reynolds number, just like the friction factor. However, it is usually assumed to be independent of the Reynolds number. This is a reasonable approximation since most flows in practice have large Reynolds numbers and the loss coefficients (including the friction factor) tend to be independent of the Reynolds number at large Reynolds numbers.

Minor losses are also expressed in terms of the **equivalent length** L_{equiv}, defined as (Fig. 14–26)

Equivalent length:
$$h_L = K_L \frac{V^2}{2g} = f \frac{L_{\text{equiv}}}{D} \frac{V^2}{2g} \quad \rightarrow \quad L_{\text{equiv}} = \frac{D}{f} K_L \tag{14–42}$$

where f is the friction factor and D is the diameter of the pipe that contains the component. The head loss caused by the component is equivalent to the head loss caused by a section of the pipe whose length is L_{equiv}. Therefore, the contribution of a component to the head loss can be accounted for by simply adding L_{equiv} to the total pipe length.

Both approaches are used in practice, but the use of loss coefficients is more common. Therefore, we will also use that approach in this book. Once all the loss coefficients are available, the total head loss in a piping system is determined from

Total head loss (general):
$$h_{L,\text{ total}} = h_{L,\text{ major}} + h_{L,\text{ minor}}$$
$$= \sum_i f_i \frac{L_i}{D_i} \frac{V_i^2}{2g} + \sum_j K_{L,j} \frac{V_j^2}{2g} \tag{14–43}$$

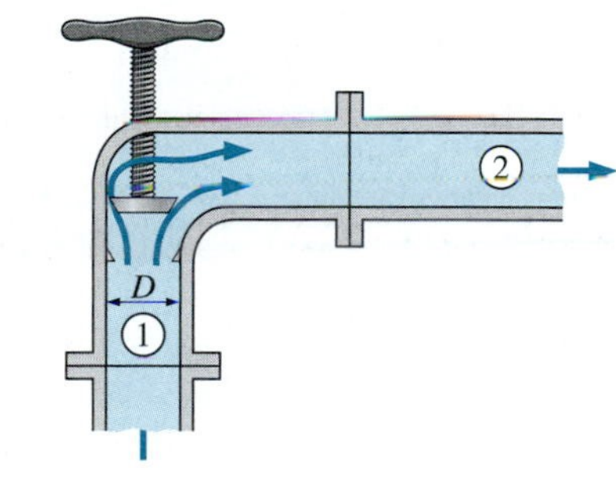

$\Delta P = P_1 - P_2 = P_3 - P_4$

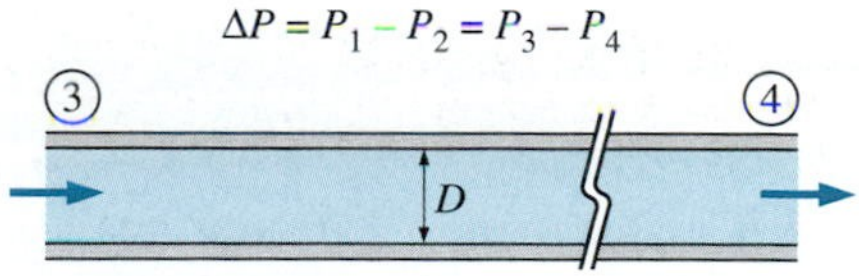

FIGURE 14–26
The head loss caused by a component (such as the angle valve shown) is equivalent to the head loss caused by a section of the pipe whose length is the equivalent length.

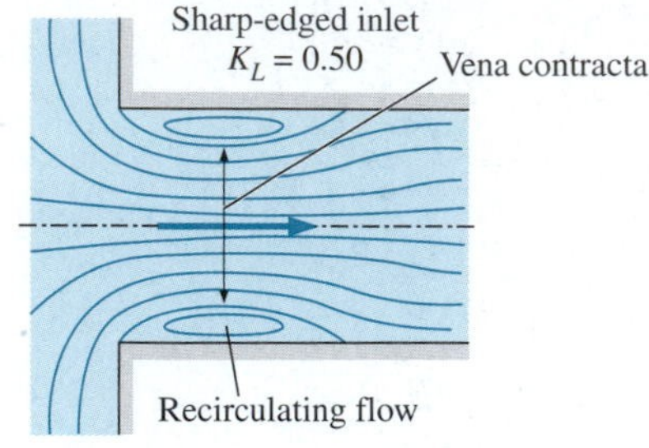

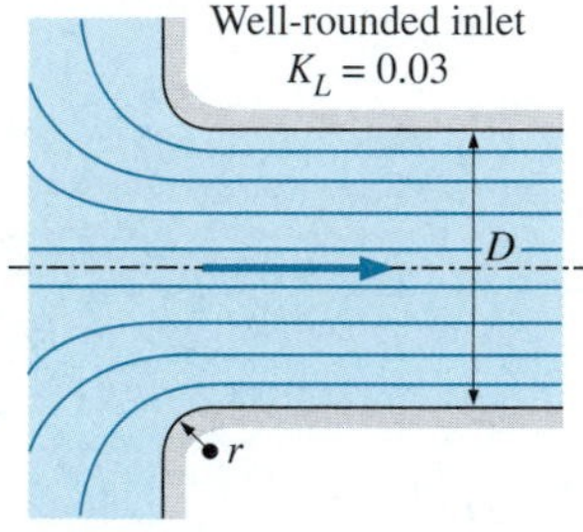

FIGURE 14–27
The head loss at the inlet of a pipe is almost negligible for well-rounded inlets ($K_L = 0.03$ for $r/D > 0.2$) but increases to about 0.50 for sharp-edged inlets.

where i represents each pipe section with constant diameter and j represents each component that causes a minor loss. If the entire piping system being analyzed has a constant diameter, Eq. 14–43 reduces to

Total head loss (D = constant):
$$h_{L,\,\text{total}} = \left(f\frac{L}{D} + \sum K_L\right)\frac{V^2}{2g} \tag{14–44}$$

where V is the average flow velocity through the entire system (note that V = constant since D = constant).

Representative loss coefficients K_L are given in Table 14–3 for inlets, exits, bends, sudden and gradual area changes, and valves. There is considerable uncertainty in these values since the loss coefficients, in general, vary with the pipe diameter, the surface roughness, the Reynolds number, and the details of the design. The loss coefficients of two seemingly identical valves by two different manufacturers, for example, can differ by a factor of 2 or more. Therefore, the particular manufacturer's data should be consulted in the final design of piping systems rather than relying on the representative values in handbooks.

The head loss at the inlet of a pipe is a strong function of geometry. It is almost negligible for well-rounded inlets ($K_L = 0.03$ for $r/D > 0.2$), but increases to about 0.50 for sharp-edged inlets (Fig. 14–27). That is, a sharp-edged inlet causes half of the velocity head to be lost as the fluid enters the pipe. This is because the fluid cannot make sharp 90° turns easily, especially at high velocities. As a result, the flow separates at the corners, and the flow is constricted into the **vena contracta** region formed in the midsection of the pipe (Fig. 14–27). Therefore, a sharp-edged inlet acts like a flow constriction. The velocity increases in the vena contracta region (and the pressure decreases) because of the reduced effective flow area and then decreases as the flow fills the entire cross section of the pipe. There would be negligible loss if the pressure were increased in accordance with Bernoulli's equation (the velocity head would simply be converted into pressure head). However, this deceleration process is far from ideal and the viscous dissipation caused by intense mixing and the turbulent eddies converts part of the kinetic energy into frictional heating, as evidenced by a slight rise in fluid temperature. The end result is a drop in velocity without much pressure recovery, and the inlet loss is a measure of this irreversible pressure drop.

Even slight rounding of the edges can result in significant reduction of K_L, as shown in Fig. 14–28. The loss coefficient rises sharply (to about K_L = 0.8) when the pipe protrudes into the reservoir since some fluid near the edge in this case is forced to make a 180° turn (we call this a *reentrant* inlet).

The loss coefficient for a submerged pipe exit is often listed in handbooks as $K_L = 1$. More precisely, however, K_L is equal to the kinetic energy correction factor α at the exit of the pipe. Although α is indeed close to 1 for fully developed *turbulent* pipe flow, it is equal to 2 for fully developed *laminar* pipe flow. To avoid possible errors when analyzing laminar pipe flow, then, it is best to always set $K_L = \alpha$ at a submerged pipe exit. At any such exit, whether laminar or turbulent, the fluid leaving the pipe loses *all* of its kinetic energy as it mixes with the reservoir fluid and eventually comes to rest through the irreversible action of viscosity. This is true regardless of the shape of the exit (Table 14–3 and Fig. 14–29). Therefore, there is no need to round the pipe exits.

TABLE 14–3

Loss coefficients K_L of various pipe components for turbulent flow (for use in the relation $h_L = K_L V^2/(2g)$, where V is the average velocity in the pipe that contains the component)*

Pipe Inlet

Reentrant: $K_L = 0.80$
($t \ll D$ and $l \approx 0.1D$)

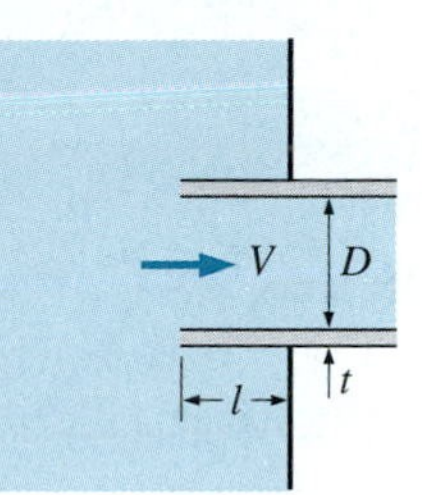

Sharp-edged: $K_L = 0.50$

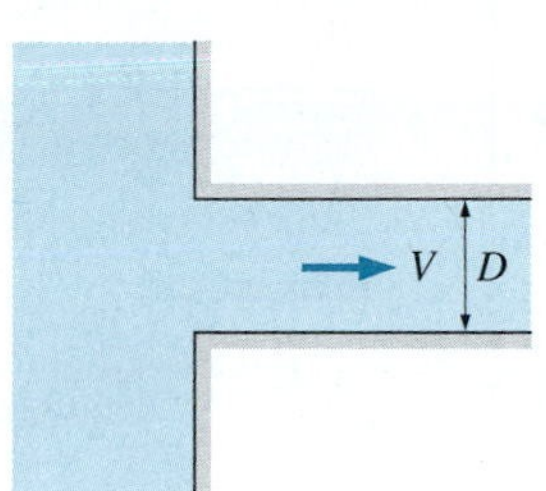

Well-rounded ($r/D > 0.2$): $K_L = 0.03$
Slightly rounded ($r/D = 0.1$): $K_L = 0.12$
(see Fig. 14–27)

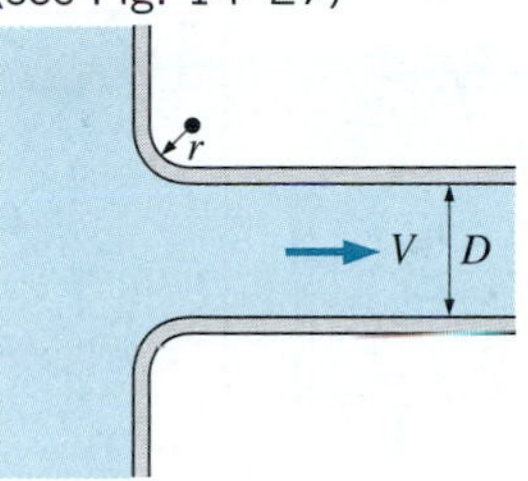

Pipe Exit

Reentrant: $K_L = \alpha$

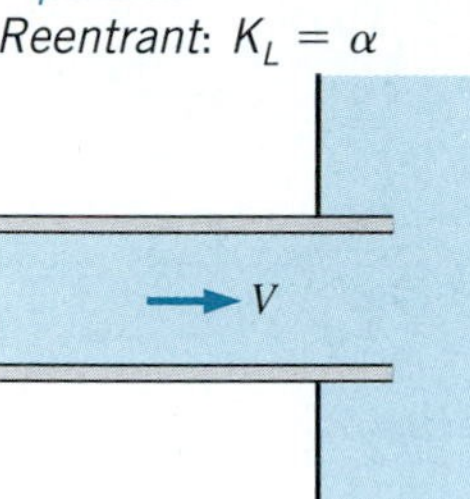

Sharp-edged: $K_L = \alpha$

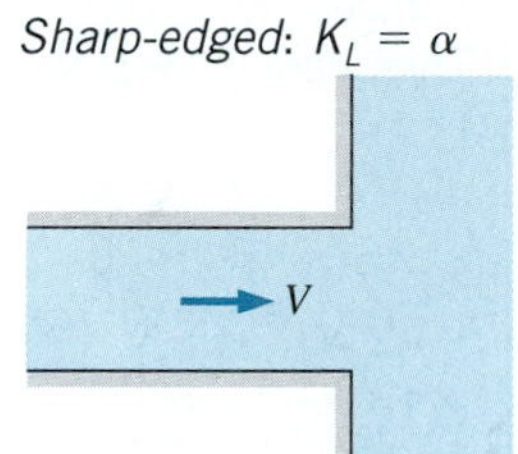

Rounded: $K_L = \alpha$

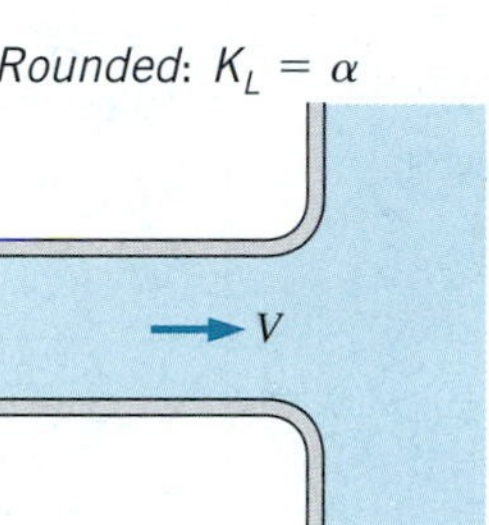

Note: The kinetic energy correction factor is $\alpha = 2$ for fully developed laminar flow, and $\alpha \approx 1.05$ for fully developed turbulent flow.

Sudden Expansion and Contraction (based on the velocity in the smaller-diameter pipe)

Sudden expansion: $K_L = \alpha\left(1 - \dfrac{d^2}{D^2}\right)^2$

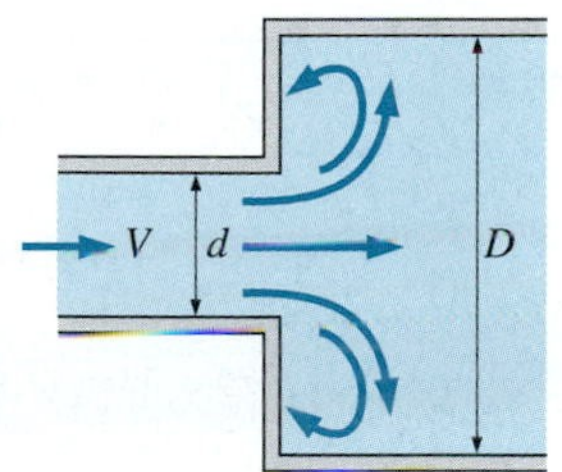

Sudden contraction: See chart.

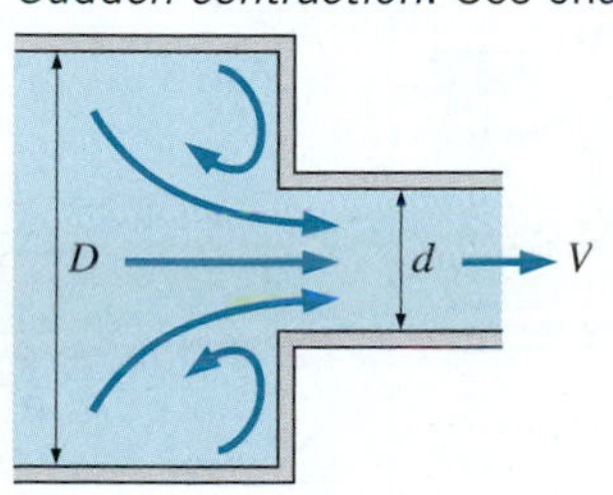

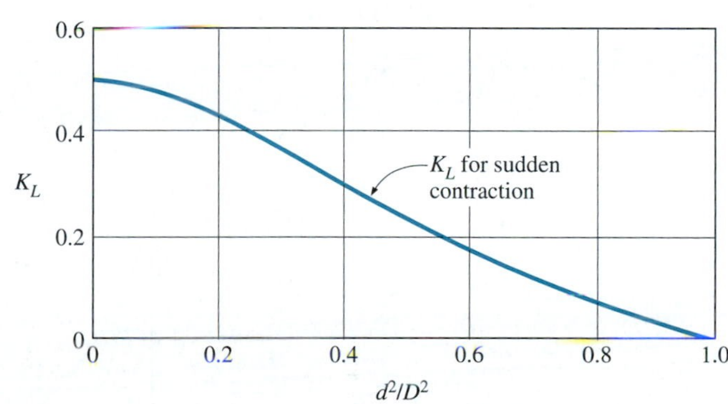

Gradual Expansion and Contraction (based on the velocity in the smaller-diameter pipe)

Expansion (for $\theta = 20°$):
$K_L = 0.30$ for $d/D = 0.2$
$K_L = 0.25$ for $d/D = 0.4$
$K_L = 0.15$ for $d/D = 0.6$
$K_L = 0.10$ for $d/D = 0.8$

Contraction:
$K_L = 0.02$ for $\theta = 30°$
$K_L = 0.04$ for $\theta = 45°$
$K_L = 0.07$ for $\theta = 60°$

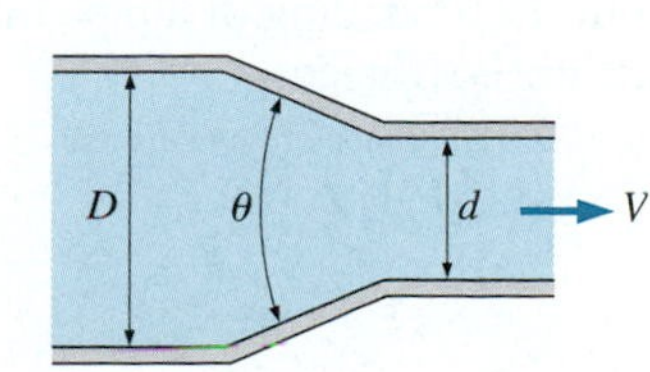

TABLE 14–3 (CONCLUDED)

Bends and Branches

90° smooth bend: Flanged: $K_L = 0.3$ Threaded: $K_L = 0.9$	*90° miter bend* (without vanes): $K_L = 1.1$	*90° miter bend* (with vanes): $K_L = 0.2$	*45° threaded elbow*: $K_L = 0.4$
V	V	V	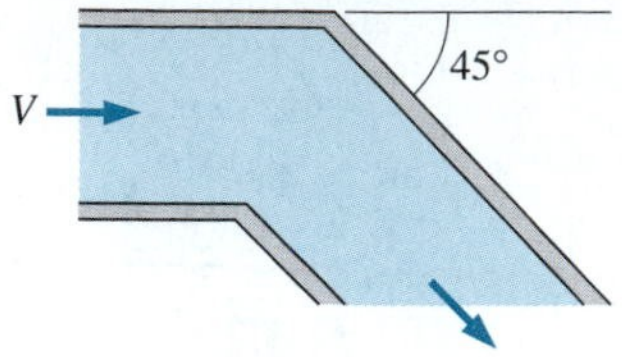 V, 45°
180° return bend: Flanged: $K_L = 0.2$ Threaded: $K_L = 1.5$	*Tee* (branch flow): Flanged: $K_L = 1.0$ Threaded: $K_L = 2.0$	*Tee* (line flow): Flanged: $K_L = 0.2$ Threaded: $K_L = 0.9$	*Threaded union*: $K_L = 0.08$
V	V	V	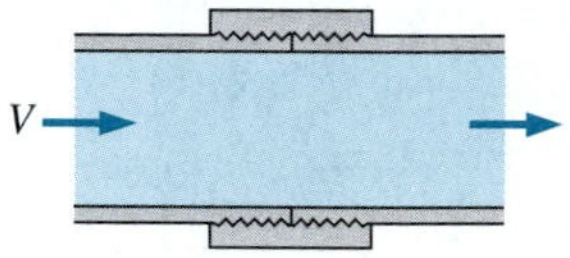 V

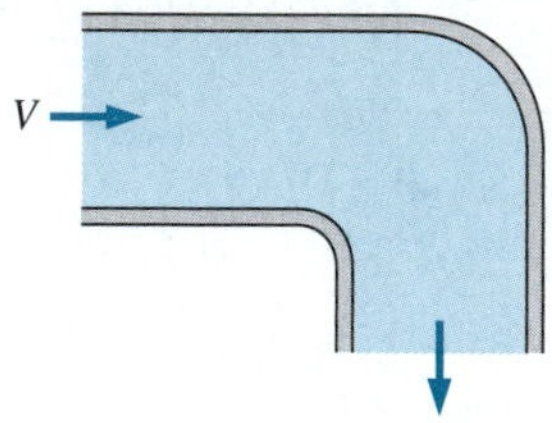

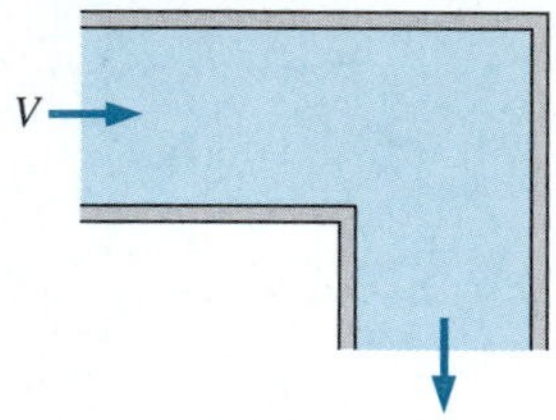

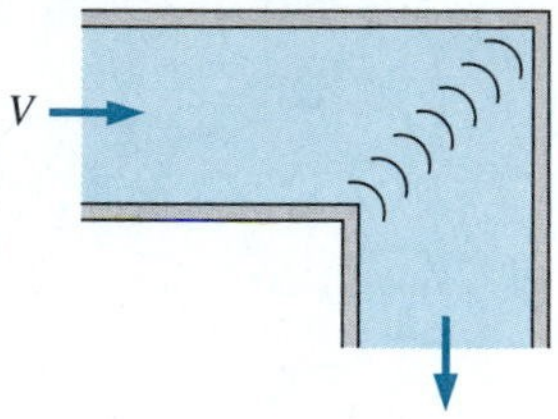

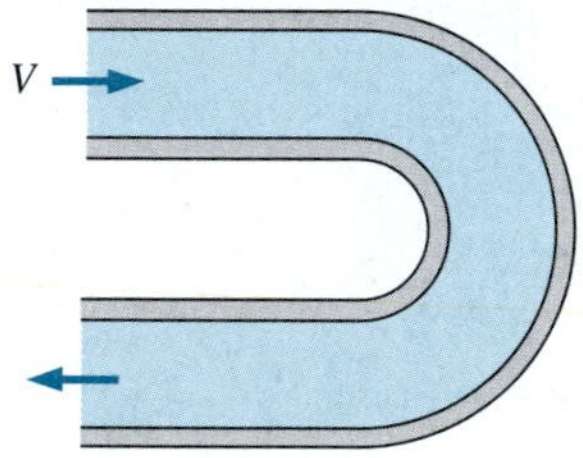

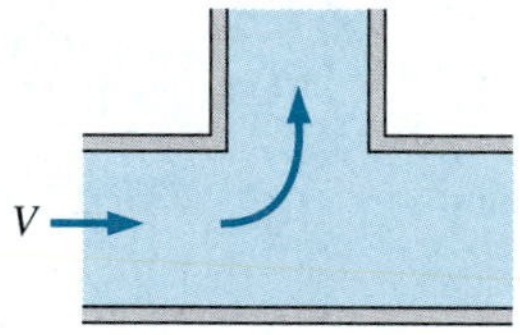

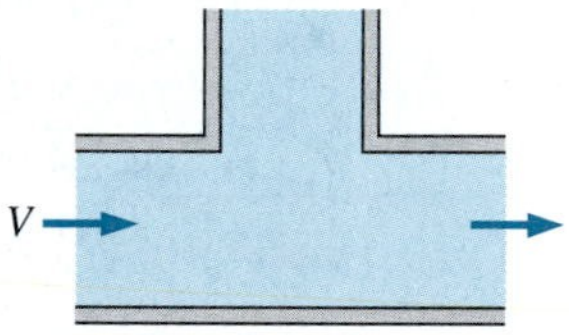

Valves

Globe valve, fully open: $K_L = 10$	*Gate valve*, fully open: $K_L = 0.2$
Angle valve, fully open: $K_L = 5$	$\frac{1}{4}$ closed: $K_L = 0.3$
Ball valve, fully open: $K_L = 0.05$	$\frac{1}{2}$ closed: $K_L = 2.1$
Swing check valve: $K_L = 2$	$\frac{3}{4}$ closed: $K_L = 17$

* These are representative values for loss coefficients. Actual values strongly depend on the design and manufacture of the components and may differ from the given values considerably (especially for valves). Actual manufacturer's data should be used in the final design.

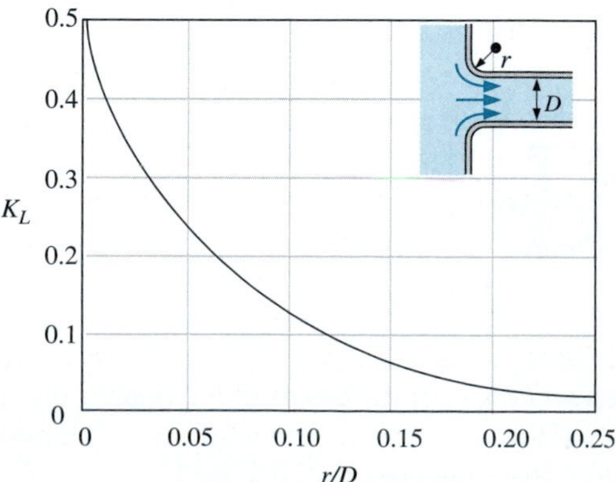

FIGURE 14–28

The effect of rounding of a pipe inlet on the loss coefficient.

From ASHRAE Handbook of Fundamentals.

Piping systems often involve *sudden* or *gradual* expansion or contraction sections to accommodate changes in flow rates or properties such as density and velocity. The losses are usually much greater in the case of *sudden* expansion and contraction (or wide-angle expansion) because of flow separation. By combining the equations of mass, momentum, and energy balance, the loss coefficient for the case of a **sudden expansion** is approximated as

$$K_L = \alpha\left(1 - \frac{A_{\text{small}}}{A_{\text{large}}}\right)^2 \quad \text{(sudden expansion)} \tag{14–45}$$

where A_{small} and A_{large} are the cross-sectional areas of the small and large pipes, respectively. Note that $K_L = 0$ when there is no area change ($A_{\text{small}} = A_{\text{large}}$) and $K_L = \alpha$ when a pipe discharges into a reservoir ($A_{\text{large}} >> A_{\text{small}}$). No such relation exists for a sudden contraction, and the K_L values in that case should be read from the chart in Table 14–3. The losses due to expansion and contraction can be reduced significantly by installing conical gradual area changers (nozzles and diffusers) between the small and large pipes. The K_L values for representative cases of gradual expansion and contraction are given in Table 14–3. Note that in head loss calculations, the velocity in the *small pipe* is to be used as the reference velocity in Eq. 14–41. Losses during expansion are usually much higher than the losses during contraction because of flow separation.

Piping systems also involve changes in direction without a change in diameter, and such flow sections are called *bends* or *elbows*. The losses in these devices are due to flow separation (just like a car being thrown off the road when it enters a turn too fast) on the inner side and the swirling secondary flows that result. The losses during changes of direction can be minimized by making the turn "easy" on the fluid by using circular arcs (like 90° elbows) instead of sharp turns (like miter bends) (Fig. 14–30). But the use of sharp turns (and thus suffering a penalty in loss coefficient) may be necessary when the turning space is limited. In such cases, the losses can be minimized by utilizing properly placed guide vanes to help the flow turn in an orderly manner without being thrown off the course. The loss coefficients for some elbows and miter bends as well as tees are given in Table 14–3. These coefficients do not include the frictional losses along the pipe bend. Such losses should be calculated as in straight pipes (using the length of the centerline as the pipe length) and added to other losses.

Valves are commonly used in piping systems to control the flow rates by simply altering the head loss until the desired flow rate is achieved. For valves it is desirable to have a very low loss coefficient when they are fully open so that they cause minimal head loss during full-load operation. Several different valve designs, each with its own advantages and disadvantages, are in common use today. The *gate valve* slides up and down like a gate, the *globe valve* closes a hole placed in the valve, the *angle valve* is a globe valve with a 90° turn, and the *check valve* allows the fluid to flow only in one direction like a diode in an electric circuit. Table 14–3 lists the representative loss coefficients of the popular designs. Note that the loss coefficient increases drastically as a valve is closed. Also, the deviation in the loss coefficients for different manufacturers is greatest for valves because of their complex geometries.

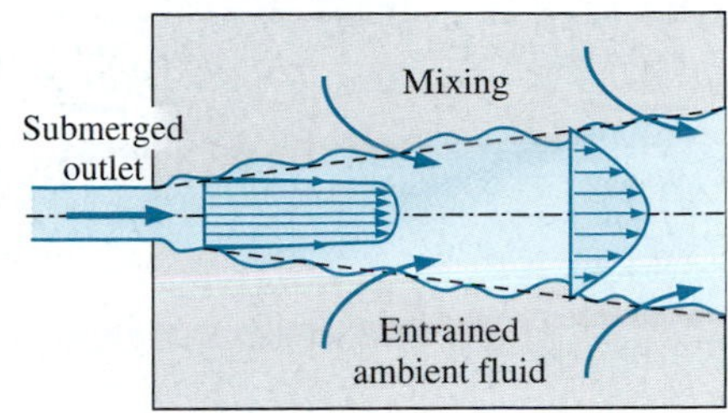

FIGURE 14–29
All the kinetic energy of the flow is "lost" (turned into thermal energy) through friction as the jet decelerates and mixes with ambient fluid downstream of a submerged outlet.

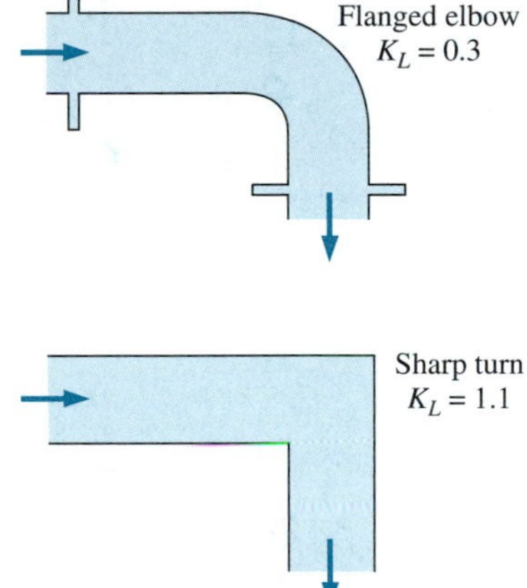

FIGURE 14–30
The losses during changes of direction can be minimized by making the turn "easy" on the fluid by using circular arcs instead of sharp turns.

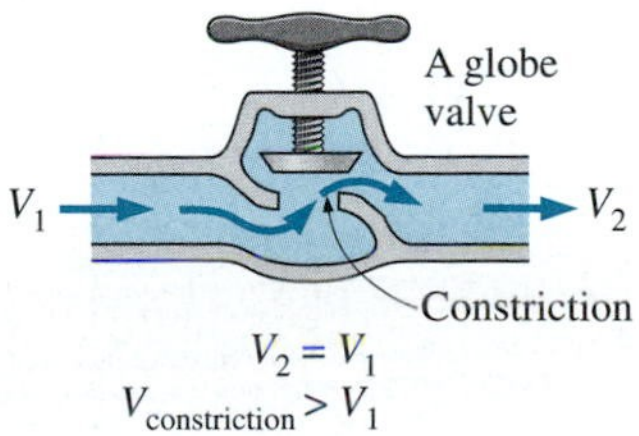

FIGURE 14–31
The large head loss in a partially closed valve is due to irreversible deceleration, flow separation, and mixing of high-velocity fluid coming from the narrow valve passage.

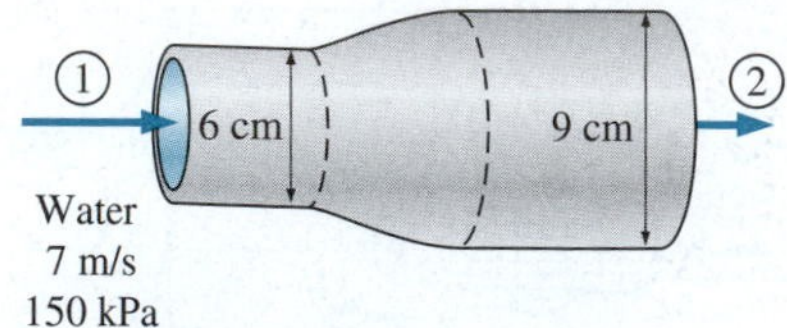

FIGURE 14–32
Schematic for Example 14–6.

EXAMPLE 14–6 Head Loss and Pressure Rise during Gradual Expansion

A 6-cm-diameter horizontal water pipe expands gradually to a 9-cm-diameter pipe (Fig. 14–32). The walls of the expansion section are angled 10° from the axis. The average velocity and pressure of water before the expansion section are 7 m/s and 150 kPa, respectively. Determine the head loss in the expansion section and the pressure in the larger-diameter pipe.

Solution A horizontal water pipe expands gradually into a larger-diameter pipe. The head loss and pressure after the expansion are to be determined.

Assumptions **1** The flow is steady and incompressible. **2** The flow at sections 1 and 2 is fully developed and turbulent with $\alpha_1 = \alpha_2 \cong 1.06$.

Properties We take the density of water to be $\rho = 1000$ kg/m^3. The loss coefficient for a gradual expansion of total included angle $\theta = 20°$ and diameter ratio $d/D = 6/9$ is $K_L = 0.133$ (by interpolation using Table 14–3).

Analysis Noting that the density of water remains constant, the downstream velocity of water is determined from conservation of mass to be

$$\dot{m}_1 = \dot{m}_2 \quad \rightarrow \quad \rho V_1 A_1 = \rho V_2 A_2 \quad \rightarrow \quad V_2 = \frac{A_1}{A_2} V_1 = \frac{D_1^2}{D_2^2} V_1$$

$$V_2 = \frac{(0.06 \text{ m})^2}{(0.09 \text{ m})^2}(7 \text{ m/s}) = 3.11 \text{ m/s}$$

Then the irreversible head loss in the expansion section becomes

$$h_L = K_L \frac{V_1^2}{2g} = (0.133)\frac{(7 \text{ m/s})^2}{2(9.81 \text{ m/s}^2)} = \mathbf{0.333 \text{ m}}$$

Noting that $z_1 = z_2$ and there are no pumps or turbines involved, the energy equation for the expansion section is expressed in terms of heads as

$$\frac{P_1}{\rho g} + \alpha_1 \frac{V_1^2}{2g} + \cancel{z_1} + \cancelto{0}{h_{\text{pump, }u}} = \frac{P_2}{\rho g} + \alpha_2 \frac{V_2^2}{2g} + \cancel{z_2} + \cancelto{0}{h_{\text{turbine, }e}} + h_L$$

$$\rightarrow \quad \frac{P_1}{\rho g} + \alpha_1 \frac{V_1^2}{2g} = \frac{P_2}{\rho g} + \alpha_2 \frac{V_2^2}{2g} + h_L$$

Solving for P_2 and substituting,

$$P_2 = P_1 + \rho\left\{\frac{\alpha_1 V_1^2 - \alpha_2 V_2^2}{2} - g h_L\right\} = (150 \text{ kPa}) + (1000 \text{ kg/m}^3)$$

$$\times \left\{\frac{1.06(7 \text{ m/s})^2 - 1.06(3.11 \text{ m/s})^2}{2} - (9.81 \text{ m/s}^2)(0.333 \text{ m})\right\}$$

$$\times \left(\frac{1 \text{ kN}}{1000 \text{ kg} \cdot \text{m/s}^2}\right)\left(\frac{1 \text{ kPa}}{1 \text{ kN/m}^2}\right)$$

$$= \mathbf{168 \text{ kPa}}$$

Therefore, despite the head (and pressure) loss, the pressure *increases* from 150 to 168 kPa after the expansion. This is due to the conversion of dynamic pressure to static pressure when the average flow velocity is decreased in the larger pipe.

Discussion It is common knowledge that higher pressure upstream is necessary to cause flow, and it may come as a surprise to you that the downstream pressure has *increased* after the expansion, despite the loss. This is because the flow is driven by the sum of the three heads that comprise the total head (namely, the pressure head, velocity head, and elevation head). During flow expansion, the higher velocity head upstream is converted to pressure head downstream, and this increase outweighs the nonrecoverable head loss. Also, you may be tempted to solve this problem using the Bernoulli equation. Such a solution would ignore the head (and the associated pressure) loss and result in an incorrect higher pressure for the fluid downstream.

14–7 ▪ PIPING NETWORKS AND PUMP SELECTION

Most piping systems encountered in practice such as the water distribution systems in cities or commercial or residential establishments involve numerous parallel and series connections as well as several sources (supply of fluid into the system) and loads (discharges of fluid from the system) (Fig. 14–33). A piping project may involve the design of a new system or the expansion of an existing system. The engineering objective in such projects is to design a piping system that will reliably deliver the specified flow rates at specified pressures at minimum total (initial plus operating and maintenance) cost. Once the layout of the system is prepared, the determination of the pipe diameters and the pressures throughout the system, while remaining within the budget constraints, typically requires solving the system repeatedly until the optimal solution is reached. Computer modeling and analysis of such systems make this tedious task a simple chore.

FIGURE 14–33
A piping network in an industrial facility.
Courtesy UMDE Engineering, Contracting, and Trading. Used by permission.

Piping systems typically involve several pipes connected to each other in series and/or in parallel, as shown in Figs. 14–34 and 14–35. When the pipes are connected **in series**, the flow rate through the entire system remains constant regardless of the diameters of the individual pipes in the system. This is a natural consequence of the conservation of mass principle for steady incompressible flow. The total head loss in this case is equal to the sum of the head losses in individual pipes in the system, including the minor losses. The expansion or contraction losses at connections are considered to belong to the smaller-diameter pipe since the expansion and contraction loss coefficients are defined on the basis of the average velocity in the smaller-diameter pipe.

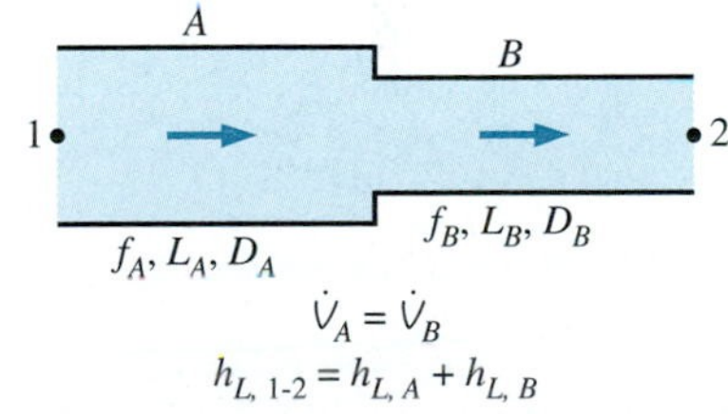

FIGURE 14–34
For pipes *in series*, the flow rate is the same in each pipe, and the total head loss is the sum of the head losses in individual pipes.

For a pipe that branches out into two (or more) **parallel pipes** and then rejoins at a junction downstream, the total flow rate is the sum of the flow

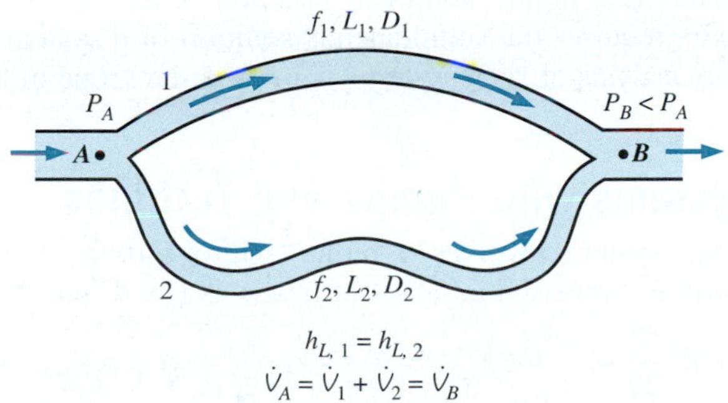

FIGURE 14–35
For pipes *in parallel*, the head loss is the same in each pipe, and the total flow rate is the sum of the flow rates in individual pipes.

rates in the individual pipes. The pressure drop (or head loss) in each individual pipe connected in parallel must be the same since $\Delta P = P_A - P_B$ and the junction pressures P_A and P_B are the same for all the individual pipes. For a system of two parallel pipes 1 and 2 between junctions A and B with negligible minor losses, this can be expressed as

$$h_{L,1} = h_{L,2} \quad \rightarrow \quad f_1 \frac{L_1}{D_1}\frac{V_1^2}{2g} = f_2 \frac{L_2}{D_2}\frac{V_2^2}{2g}$$

Then the ratio of the average velocities and the flow rates in the two parallel pipes become

$$\frac{V_1}{V_2} = \left(\frac{f_2}{f_1}\frac{L_2}{L_1}\frac{D_1}{D_2}\right)^{1/2} \quad \text{and} \quad \frac{\dot{V}_1}{\dot{V}_2} = \frac{A_{c,1}V_1}{A_{c,2}V_2} = \frac{D_1^2}{D_2^2}\left(\frac{f_2}{f_1}\frac{L_2}{L_1}\frac{D_1}{D_2}\right)^{1/2}$$

Therefore, the relative flow rates in parallel pipes are established from the requirement that the head loss in each pipe be the same. This result can be extended to any number of pipes connected in parallel. The result is also valid for pipes for which the minor losses are significant if the equivalent lengths for components that contribute to minor losses are added to the pipe length. Note that the flow rate in one of the parallel branches is proportional to its diameter to the power 5/2 and is inversely proportional to the square root of its length and friction factor.

The analysis of piping networks, no matter how complex they are, is based on two simple principles:

1. *Conservation of mass throughout the system must be satisfied.* This is done by requiring the total flow into a junction to be equal to the total flow out of the junction for all junctions in the system. Also, the flow rate must remain constant in pipes connected in series regardless of the changes in diameters.
2. *Pressure drop (and thus head loss) between two junctions must be the same for all paths between the two junctions.* This is because pressure is a point function and it cannot have two values at a specified point. In practice this rule is used by requiring that the algebraic sum of head losses in a loop (for all loops) be equal to zero. (A head loss is taken to be positive for flow in the clockwise direction and negative for flow in the counterclockwise direction.)

Therefore, the analysis of piping networks is very similar to the analysis of electric circuits (Kirchhoff's laws), with flow rate corresponding to electric current and pressure corresponding to electric potential. However, the situation is much more complex here since, unlike the electrical resistance, the "flow resistance" is a highly nonlinear function. Therefore, the analysis of piping networks requires the simultaneous solution of a system of nonlinear equations. The analysis of such systems is beyond the scope of this introductory text.

Piping Systems with Pumps and Turbines

When a piping system involves a pump and/or turbine, the steady-flow energy equation is expressed in terms of heads as (see Chap. 12)

$$\frac{P_1}{\rho g} + \alpha_1 \frac{V_1^2}{2g} + z_1 + h_{\text{pump},\,u} = \frac{P_2}{\rho g} + \alpha_2 \frac{V_2^2}{2g} + z_2 + h_{\text{turbine},\,e} + h_L \quad \textbf{(14–46)}$$

where $h_{\text{pump}, u} = w_{\text{pump}, u}/g$ is the useful pump head delivered to the fluid, $h_{\text{turbine}, e} = w_{\text{turbine}, e}/g$ is the turbine head extracted from the fluid, α is the kinetic energy correction factor whose value is about 1.05 for most (turbulent) flows encountered in practice, and h_L is the total head loss in piping (including the minor losses if they are significant) between points 1 and 2. The pump head is zero if the piping system does not involve a pump or a fan, the turbine head is zero if the system does not involve a turbine, and both are zero if the system does not involve any mechanical work-producing or work-consuming devices.

Many practical piping systems involve a pump to move a fluid from one reservoir to another. Taking points 1 and 2 to be at the *free surfaces* of the reservoirs (Fig. 14–36), the energy equation in this case reduces for the useful pump head required to

$$h_{\text{pump}, u} = (z_2 - z_1) + h_L \qquad \textbf{(14–47)}$$

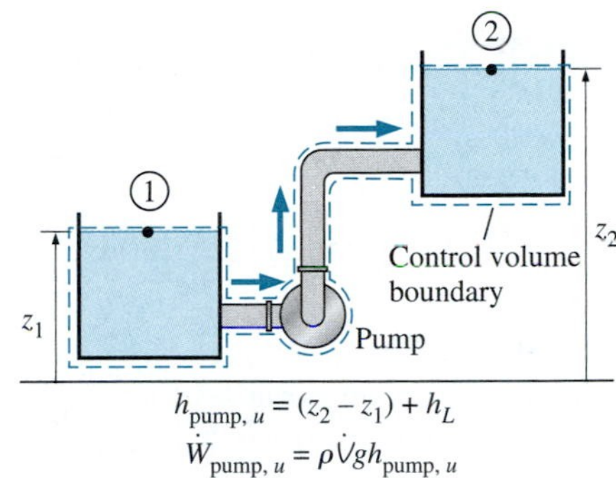

FIGURE 14–36
When a pump moves a fluid from one reservoir to another, the useful pump head requirement is equal to the elevation difference between the two reservoirs plus the head loss.

since the velocities at free surfaces are negligible for large reservoirs and the pressures are at atmospheric pressure. Therefore, the useful pump head is equal to the elevation difference between the two reservoirs plus the head loss. If the head loss is negligible compared to $z_2 - z_1$, the useful pump head is simply equal to the elevation difference between the two reservoirs. In the case of $z_1 > z_2$ (the first reservoir being at a higher elevation than the second one) with no pump, the flow is driven by gravity at a flow rate that causes a head loss equal to the elevation difference. A similar argument can be given for the turbine head for a hydroelectric power plant by replacing $h_{\text{pump}, u}$ in Eq. 14–47 by $-h_{\text{turbine}, e}$.

Once the useful pump head is known, the *mechanical power that needs to be delivered by the pump to the fluid* and the *electric power consumed by the motor of the pump* for a specified flow rate are determined from

$$\dot{W}_{\text{pump, shaft}} = \frac{\rho \dot{V} g h_{\text{pump}, u}}{\eta_{\text{pump}}} \quad \text{and} \quad \dot{W}_{\text{elect}} = \frac{\rho \dot{V} g h_{\text{pump}, u}}{\eta_{\text{pump–motor}}} \qquad \textbf{(14–48)}$$

where $\eta_{\text{pump motor}}$ is the *efficiency of the pump–motor combination*, which is the product of the pump and the motor efficiencies (Fig. 14–37). The pump–motor efficiency is defined as the ratio of the net mechanical energy delivered to the fluid by the pump to the electric energy consumed by the motor of the pump, and it typically ranges between 50 and 85 percent.

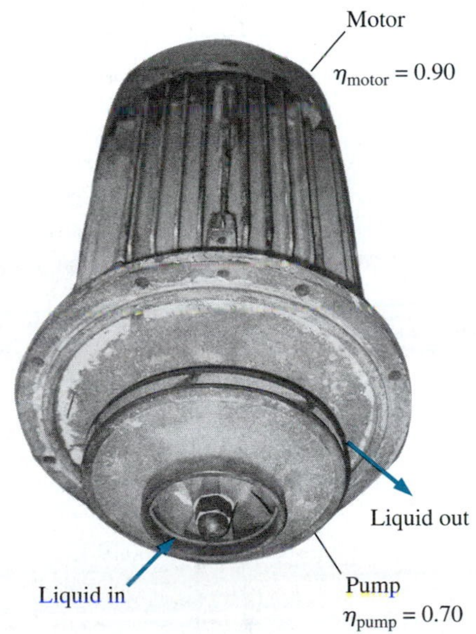

FIGURE 14–37
The efficiency of the pump–motor combination is the product of the pump and the motor efficiencies.
Courtesy Yunus Çengel

The head loss of a piping system increases (usually quadratically) with the flow rate. A plot of required useful pump head $h_{\text{pump}, u}$ as a function of flow rate is called the **system** (or **demand**) **curve**. The head produced by a pump is not a constant either. Both the pump head and the pump efficiency vary with the flow rate, and pump manufacturers supply this variation in tabular or graphical form, as shown in Fig. 14–38. These experimentally determined $h_{\text{pump}, u}$ and $\eta_{\text{pump}, u}$ versus $\dot{V}$ curves are called **characteristic** (or **supply** or **performance**) **curves**. Note that the flow rate of a pump increases as the required head decreases. The intersection point of the pump head curve with the vertical axis typically represents the *maximum head* (called the **shutoff head**) the pump can provide, while the intersection point with the horizontal axis indicates the *maximum flow rate* (called the **free delivery**) that the pump can supply.

The *efficiency* of a pump is highest for a certain range of head and flow rate combination. Therefore, a pump that can supply the required head and

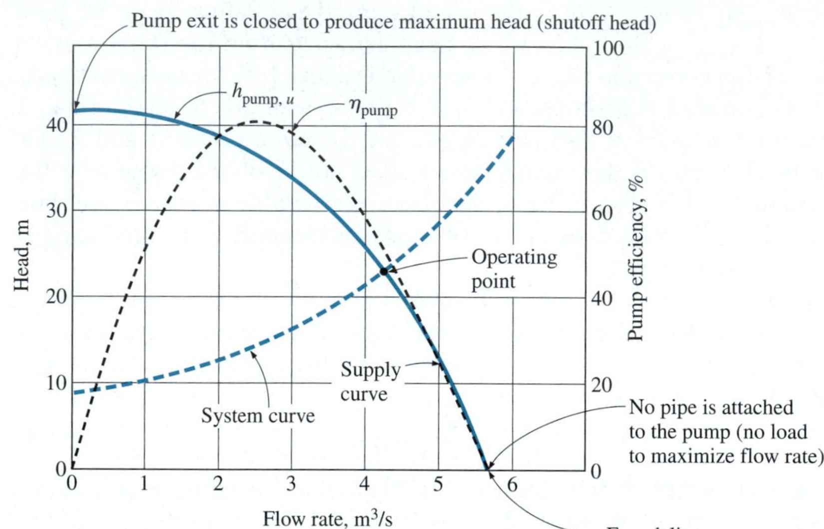

FIGURE 14–38
Characteristic pump curves for centrifugal pumps, the system curve for a piping system, and the operating point.

flow rate is not necessarily a good choice for a piping system unless the efficiency of the pump at those conditions is sufficiently high. The pump installed in a piping system will operate at the point where the *system curve* and the *characteristic curve* intersect. This point of intersection is called the **operating point**, as shown in Fig. 14–38. The useful head produced by the pump at this point matches the head requirements of the system at that flow rate. Also, the efficiency of the pump during operation is the value corresponding to that flow rate.

EXAMPLE 14–7 Pumping Water through Two Parallel Pipes

Water at 20°C is to be pumped from a reservoir (z_A = 5 m) to another reservoir at a higher elevation (z_B = 13 m) through two 36-m-long pipes connected in parallel, as shown in Fig. 14–39. The pipes are made of commercial steel, and the diameters of the two pipes are 4 and 8 cm. Water is to be pumped by a 70 percent efficient motor–pump combination that draws 8 kW of electric power during operation. The minor losses and the head loss in pipes that connect the parallel pipes to the two reservoirs are considered to be negligible.

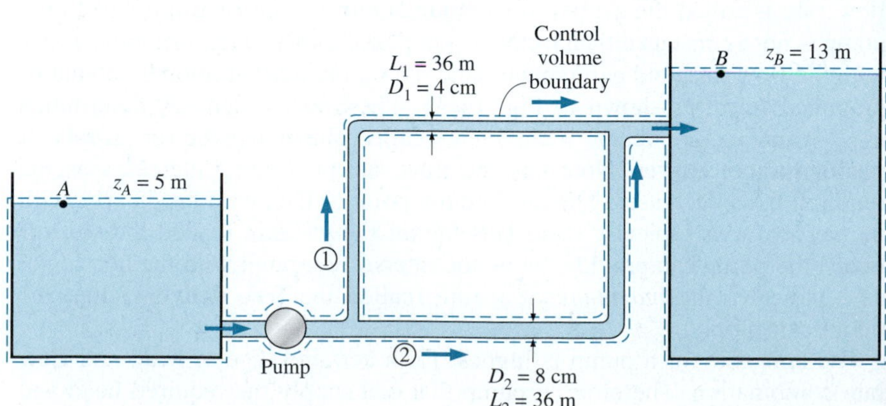

FIGURE 14–39
The piping system discussed in Example 14–7.

Determine the total flow rate between the reservoirs and the flow rate through each of the parallel pipes.

Solution The pumping power input to a piping system with two parallel pipes is given. The flow rates are to be determined.

Assumptions **1** The flow is steady (since the reservoirs are large) and incompressible. **2** The entrance effects are negligible, and thus the flow is fully developed. **3** The elevations of the reservoirs remain constant. **4** The minor losses and the head loss in pipes other than the parallel pipes are said to be negligible. **5** Flows through both pipes are turbulent (to be verified).

Properties The density and dynamic viscosity of water at 20°C are $\rho = 998$ kg/m^3 and $\mu = 1.002 \times 10^{-3}$ kg/m · s (Table A–15). The roughness of commercial steel pipe is $\varepsilon = 0.000045$ m (Table 14–1).

Analysis This problem cannot be solved directly since the velocities (or flow rates) in the pipes are not known. Therefore, we would normally use a trial-and-error approach here. However, equation solvers such as EES are widely available, and thus we simply set up the equations to be solved by an equation solver. The useful head supplied by the pump to the fluid is determined from

$$\dot{W}_{\text{elect}} = \frac{\rho \dot{V} g h_{\text{pump},\,u}}{\eta_{\text{pump}-\text{motor}}} \quad \rightarrow \quad 8000 \text{ W} = \frac{(998 \text{ kg/m}^3)\dot{V}(9.81 \text{ m/s}^2)h_{\text{pump},\,u}}{0.70} \qquad \textbf{(1)}$$

We choose points A and B at the free surfaces of the two reservoirs. Noting that the fluid at both points is open to the atmosphere (and thus $P_A = P_B = P_{\text{atm}}$) and that the fluid velocities at both points are nearly zero ($V_A \cong V_B \cong 0$) since the reservoirs are large, the energy equation for a control volume between these two points simplifies to

$$\cancel{\frac{P_A}{\rho g}} + \alpha_A \frac{V_A^2}{2g}\nearrow^{0} + z_A + h_{\text{pump},\,u} = \cancel{\frac{P_B}{\rho g}} + \alpha_B \frac{V_B^2}{2g}\nearrow^{0} + z_B + h_L$$

$$h_{\text{pump},\,u} = (z_B - z_A) + h_L$$

or

$$h_{\text{pump},\,u} = (13 - 5) + h_L \qquad \textbf{(2)}$$

where

$$h_L = h_{L,1} = h_{L,2} \qquad \textbf{(3)(4)}$$

We designate the 4-cm-diameter pipe by 1 and the 8-cm-diameter pipe by 2. Equations for the average velocity, the Reynolds number, the friction factor, and the head loss in each pipe are

$$V_1 = \frac{\dot{V}_1}{A_{c,1}} = \frac{\dot{V}_1}{\pi D_1^2/4} \quad \rightarrow \quad V_1 = \frac{\dot{V}_1}{\pi (0.04 \text{ m})^2/4} \qquad \textbf{(5)}$$

$$V_2 = \frac{\dot{V}_2}{A_{c,2}} = \frac{\dot{V}_2}{\pi D_2^2/4} \quad \rightarrow \quad V_2 = \frac{\dot{V}_2}{\pi (0.08 \text{ m})^2/4} \qquad \textbf{(6)}$$

$$\text{Re}_1 = \frac{\rho V_1 D_1}{\mu} \quad \rightarrow \quad \text{Re}_1 = \frac{(998 \text{ kg/m}^3)V_1(0.04 \text{ m})}{1.002 \times 10^{-3} \text{ kg/m} \cdot \text{s}} \qquad \textbf{(7)}$$

$$\text{Re}_2 = \frac{\rho V_2 D_2}{\mu} \quad \rightarrow \quad \text{Re}_2 = \frac{(998 \text{ kg/m}^3)V_2(0.08 \text{ m})}{1.002 \times 10^{-3} \text{ kg/m} \cdot \text{s}} \qquad \textbf{(8)}$$

$$\frac{1}{\sqrt{f_1}} = -2.0 \log\left(\frac{\varepsilon/D_1}{3.7} + \frac{2.51}{\text{Re}_1\sqrt{f_1}}\right)$$

$$\frac{1}{\sqrt{f_1}} = -2.0 \log\left(\frac{0.000045}{3.7 \times 0.04} + \frac{2.51}{\text{Re}_1\sqrt{f_1}}\right) \quad \text{(9)}$$

$$\frac{1}{\sqrt{f_2}} = -2.0 \log\left(\frac{\varepsilon/D_2}{3.7} + \frac{2.51}{\text{Re}_2\sqrt{f_2}}\right)$$

$$\frac{1}{\sqrt{f_2}} = -2.0 \log\left(\frac{0.000045}{3.7 \times 0.08} + \frac{2.51}{\text{Re}_2\sqrt{f_2}}\right) \quad \text{(10)}$$

$$h_{L,1} = f_1 \frac{L_1}{D_1}\frac{V_1^2}{2g} \quad \rightarrow \quad h_{L,1} = f_1 \frac{36 \text{ m}}{0.04 \text{ m}} \frac{V_1^2}{2(9.81 \text{ m/s}^2)} \quad \text{(11)}$$

$$h_{L,2} = f_2 \frac{L_2}{D_2}\frac{V_2^2}{2g} \quad \rightarrow \quad h_{L,2} = f_2 \frac{36 \text{ m}}{0.08 \text{ m}} \frac{V_2^2}{2(9.81 \text{ m/s}^2)} \quad \text{(12)}$$

$$\dot{V} = \dot{V}_1 + \dot{V}_2 \quad \text{(13)}$$

This is a system of 13 equations in 13 unknowns, and their simultaneous solution by an equation solver gives

$$\dot{V} = \mathbf{0.0300\ m^3/s}, \quad \dot{V}_1 = \mathbf{0.00415\ m^3/s}, \quad \dot{V}_2 = \mathbf{0.0259\ m^3/s}$$

$$V_1 = 3.30 \text{ m/s}, \quad V_2 = 5.15 \text{ m/s}, \quad h_L = h_{L,1} = h_{L,2} = 11.1 \text{ m}, \quad h_{\text{pump}} = 19.1 \text{ m}$$

$$\text{Re}_1 = 131{,}600, \quad \text{Re}_2 = 410{,}000, \quad f_1 = 0.0221, \quad f_2 = 0.0182$$

Note that Re $>$ 10,000 for both pipes, and thus the assumption of turbulent flow is verified.

Discussion The two parallel pipes are identical, except the diameter of the first pipe is half the diameter of the second one. But only 14 percent of the water flows through the first pipe. This shows the strong dependence of the flow rate on diameter. Also, it can be shown that if the free surfaces of the two reservoirs were at the same elevation (and thus $z_A = z_B$), the flow rate would increase by 20 percent from 0.0300 to 0.0361 m^3/s. Alternately, if the reservoirs were as given but the irreversible head losses were negligible, the flow rate would become 0.0715 m^3/s (an increase of 138 percent).

EXAMPLE 14–8 Gravity-Driven Water Flow in a Pipe

Water at 10°C flows from a large reservoir to a smaller one through a 5-cm-diameter cast iron piping system, as shown in Fig. 14–40. Determine the elevation z_1 for a flow rate of 6 L/s.

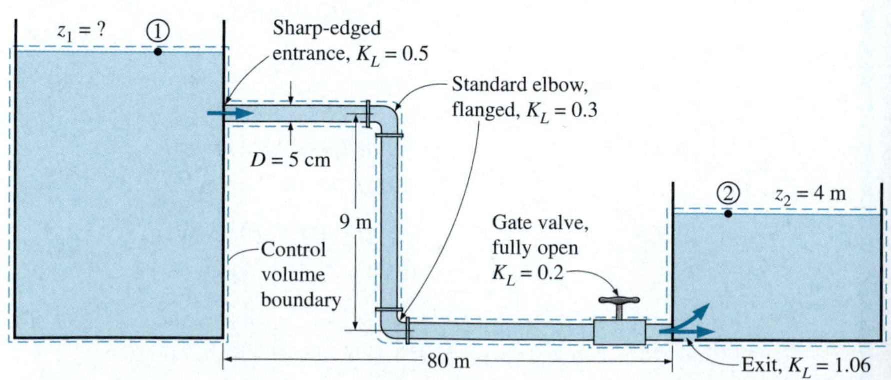

FIGURE 14–40
The piping system discussed in Example 14–8.

Solution The flow rate through a piping system connecting two reservoirs is given. The elevation of the source is to be determined.

Assumptions 1 The flow is steady and incompressible. 2 The elevations of the reservoirs remain constant. 3 There are no pumps or turbines in the line.

Properties The density and dynamic viscosity of water at 10°C are ρ = 999.7 kg/m^3 and μ = 1.307 × 10^{-3} kg/m · s (Table A–15). The roughness of cast iron pipe is ε = 0.00026 m (Table 14–1).

Analysis The piping system involves 89 m of piping, a sharp-edged entrance (K_L = 0.5), two standard flanged elbows (K_L = 0.3 each), a fully open gate valve (K_L = 0.2), and a submerged exit (K_L = 1.06). We choose points 1 and 2 at the free surfaces of the two reservoirs. Noting that the fluid at both points is open to the atmosphere (and thus $P_1 = P_2 = P_{atm}$) and that the fluid velocities at both points are nearly zero ($V_1 \cong V_2 \cong 0$), the energy equation for a control volume between these two points simplifies to

$$\frac{P_1}{\rho g} + \alpha_1 \frac{V_1^2}{2g}^{\nearrow 0} + z_1 = \frac{P_2}{\rho g} + \alpha_2 \frac{V_2^2}{2g}^{\nearrow 0} + z_2 + h_L \quad \rightarrow \quad z_1 = z_2 + h_L$$

where

$$h_L = h_{L,\,\text{total}} = h_{L,\,\text{major}} + h_{L,\,\text{minor}} = \left(f\frac{L}{D} + \sum K_L\right)\frac{V^2}{2g}$$

since the diameter of the piping system is constant. The average velocity in the pipe and the Reynolds number are

$$V = \frac{\dot{V}}{A_c} = \frac{\dot{V}}{\pi D^2/4} = \frac{0.006 \text{ m}^3/\text{s}}{\pi(0.05 \text{ m})^2/4} = 3.06 \text{ m/s}$$

$$\text{Re} = \frac{\rho V D}{\mu} = \frac{(999.7 \text{ kg/m}^3)(3.06 \text{ m/s})(0.05 \text{ m})}{1.307 \times 10^{-3} \text{ kg/m} \cdot \text{s}} = 117{,}000$$

The flow is turbulent since Re > 10,000. Noting that ε/D = 0.00026/0.05 = 0.0052, the friction factor is determined from the Colebrook equation (or the Moody chart),

$$\frac{1}{\sqrt{f}} = -2.0 \log\left(\frac{\varepsilon/D}{3.7} + \frac{2.51}{\text{Re}\sqrt{f}}\right) \quad \rightarrow \quad \frac{1}{\sqrt{f}} = -2.0 \log\left(\frac{0.0052}{3.7} + \frac{2.51}{117{,}000\sqrt{f}}\right)$$

It gives f = 0.0315. The sum of the loss coefficients is

$$\sum K_L = K_{L,\,\text{entrance}} + 2K_{L,\,\text{elbow}} + K_{L,\,\text{valve}} + K_{L,\,\text{exit}}$$
$$= 0.5 + 2 \times 0.3 + 0.2 + 1.06 = 2.36$$

Then the total head loss and the elevation of the source become

$$h_L = \left(f\frac{L}{D} + \sum K_L\right)\frac{V^2}{2g} = \left(0.0315\,\frac{89 \text{ m}}{0.05 \text{ m}} + 2.36\right)\frac{(3.06 \text{ m/s})^2}{2(9.81 \text{ m/s}^2)} = 27.9 \text{ m}$$

$$z_1 = z_2 + h_L = 4 + 27.9 = \mathbf{31.9 \text{ m}}$$

Therefore, the free surface of the first reservoir must be 31.9 m above the ground level to ensure water flow between the two reservoirs at the specified rate.

Discussion Note that fL/D = 56.1 in this case, which is about 24 times the total minor loss coefficient. Therefore, ignoring the sources of minor losses in this case would result in about 4 percent error.

It can be shown that the total head loss would be 35.9 m (instead of 27.9 m) if the valve were three-fourths closed, and it would drop to 24.8 m if the pipe between the two reservoirs were straight at the ground level (thus eliminating the elbows and the vertical section of the pipe). The head loss could be reduced further (from 24.8 to 24.6 m) by rounding the entrance. The head loss can be reduced significantly (from 27.9 to 16.0 m) by replacing the cast iron pipes by smooth pipes such as those made of plastic.

EXAMPLE 14–9 Effect of Flushing on Flow Rate from a Shower

The bathroom plumbing of a building consists of 1.5-cm-diameter copper pipes with threaded connectors, as shown in Fig. 14–41. (*a*) If the gage pressure at the inlet of the system is 200 kPa during a shower and the toilet reservoir is full (no flow in that branch), determine the flow rate of water through the shower head. (*b*) Determine the effect of flushing of the toilet on the flow rate through the shower head. Take the loss coefficients of the shower head and the reservoir to be 12 and 14, respectively.

Solution The cold-water plumbing system of a bathroom is given. The flow rate through the shower and the effect of flushing the toilet on the flow rate are to be determined.

Assumptions **1** The flow is steady and incompressible. **2** The flow is turbulent and fully developed. **3** The reservoir is open to the atmosphere. **4** The velocity heads are negligible.

Properties The properties of water at 20°C are $\rho = 998$ kg/m^3, $\mu = 1.002 \times 10^{-3}$ kg/m · s, and $\nu = \mu/\rho = 1.004 \times 10^{-6}$ m^2/s (Table A–15). The roughness of copper pipes is $\varepsilon = 1.5 \times 10^{-6}$ m (Table 14–1).

Analysis This is a problem of the second type since it involves the determination of the flow rate for a specified pipe diameter and pressure drop. The solution involves an iterative approach since the flow rate (and thus the flow velocity) is not known.

(*a*) The piping system of the shower alone involves 11 m of piping, a tee with line flow ($K_L = 0.9$), two standard elbows ($K_L = 0.9$ each), a fully open globe valve ($K_L = 10$), and a shower head ($K_L = 12$). Therefore, $\Sigma K_L = 0.9 + 2 \times 0.9 + 10 + 12 = 24.7$. Noting that the shower head is open to the atmosphere, and the velocity heads are negligible, the energy equation for a control volume between points 1 and 2 simplifies to

$$\frac{P_1}{\rho g} + \alpha_1 \frac{V_1^2}{2g} + z_1 + h_{\text{pump}, u} = \frac{P_2}{\rho g} + \alpha_2 \frac{V_2^2}{2g} + z_2 + h_{\text{turbine}, e} + h_L$$

$$\frac{P_{1,\text{ gage}}}{\rho g} = (z_2 - z_1) + h_L$$

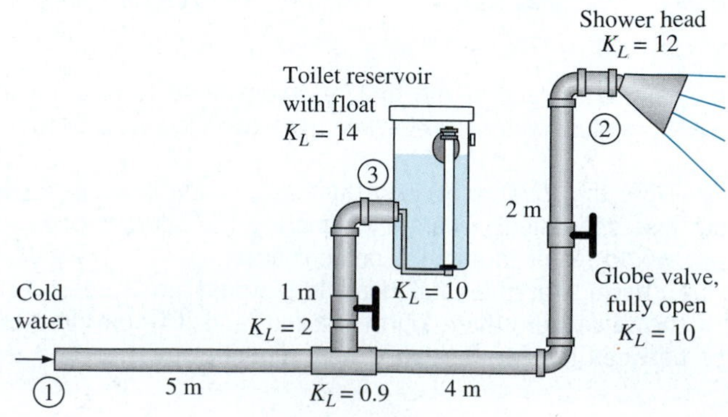

FIGURE 14–41
Schematic for Example 14–9.

Therefore, the head loss is

$$h_L = \frac{200{,}000 \text{ N/m}^2}{(998 \text{ kg/m}^3)(9.81 \text{ m/s}^2)} - 2 \text{ m} = 18.4 \text{ m}$$

Also,

$$h_L = \left(f\frac{L}{D} + \sum K_L\right)\frac{V^2}{2g} \quad \rightarrow \quad 18.4 = \left(f\frac{11 \text{ m}}{0.015 \text{ m}} + 24.7\right)\frac{V^2}{2(9.81 \text{ m/s}^2)}$$

since the diameter of the piping system is constant. Equations for the average velocity in the pipe, the Reynolds number, and the friction factor are

$$V = \frac{\dot{V}}{A_c} = \frac{\dot{V}}{\pi D^2/4} \quad \rightarrow \quad V = \frac{\dot{V}}{\pi(0.015 \text{ m})^2/4}$$

$$\text{Re} = \frac{VD}{\nu} \quad \rightarrow \quad \text{Re} = \frac{V(0.015 \text{ m})}{1.004 \times 10^{-6} \text{ m}^2/\text{s}}$$

$$\frac{1}{\sqrt{f}} = -2.0 \log\left(\frac{\varepsilon/D}{3.7} + \frac{2.51}{\text{Re}\sqrt{f}}\right)$$

$$\rightarrow \quad \frac{1}{\sqrt{f}} = -2.0 \log\left(\frac{1.5 \times 10^{-6} \text{ m}}{3.7(0.015 \text{ m})} + \frac{2.51}{\text{Re}\sqrt{f}}\right)$$

This is a set of four equations with four unknowns, and solving them with an equation solver such as EES gives

$$\dot{V} = 0.00053 \text{ m}^3/\text{s}, \quad f = 0.0218, \quad V = 2.98 \text{ m/s}, \quad \text{and} \quad \text{Re} = 44{,}550$$

Therefore, the flow rate of water through the shower head is **0.53 L/s**.

(*b*) When the toilet is flushed, the float moves and opens the valve. The discharged water starts to refill the reservoir, resulting in parallel flow after the tee connection. The head loss and minor loss coefficients for the shower branch were determined in (*a*) to be $h_{L,2} = 18.4$ m and $\Sigma K_{L,2} = 24.7$, respectively. The corresponding quantities for the reservoir branch can be determined similarly to be

$$h_{L,3} = \frac{200{,}000 \text{ N/m}^2}{(998 \text{ kg/m}^3)(9.81 \text{ m/s}^2)} - 1 \text{ m} = 19.4 \text{ m}$$

$$\Sigma K_{L,3} = 2 + 10 + 0.9 + 14 = 26.9$$

The relevant equations in this case are

$$\dot{V}_1 = \dot{V}_2 + \dot{V}_3$$

$$h_{L,2} = f_1\frac{5 \text{ m}}{0.015 \text{ m}}\frac{V_1^2}{2(9.81 \text{ m/s}^2)} + \left(f_2\frac{6 \text{ m}}{0.015 \text{ m}} + 24.7\right)\frac{V_2^2}{2(9.81 \text{ m/s}^2)} = 18.4$$

$$h_{L,3} = f_1\frac{5 \text{ m}}{0.015 \text{ m}}\frac{V_1^2}{2(9.81 \text{ m/s}^2)} + \left(f_3\frac{1 \text{ m}}{0.015 \text{ m}} + 26.9\right)\frac{V_3^2}{2(9.81 \text{ m/s}^2)} = 19.4$$

FIGURE 14–42
Flow rate of cold water through a shower may be affected significantly by the flushing of a nearby toilet.

$$V_1 = \frac{\dot{V}_1}{\pi(0.015\text{ m})^2/4}, \quad V_2 = \frac{\dot{V}_2}{\pi(0.015\text{ m})^2/4}, \quad V_3 = \frac{\dot{V}_3}{\pi(0.015\text{ m})^2/4}$$

$$\text{Re}_1 = \frac{V_1(0.015\text{ m})}{1.004 \times 10^{-6}\text{m}^2/\text{s}}, \quad \text{Re}_2 = \frac{V_2(0.015\text{ m})}{1.004 \times 10^{-6}\text{m}^2/\text{s}}, \quad \text{Re}_3 = \frac{V_3(0.015\text{ m})}{1.004 \times 10^{-6}\text{m}^2/\text{s}}$$

$$\frac{1}{\sqrt{f_1}} = -2.0\log\left(\frac{1.5 \times 10^{-6}\text{ m}}{3.7(0.015\text{ m})} + \frac{2.51}{\text{Re}_1\sqrt{f_1}}\right)$$

$$\frac{1}{\sqrt{f_2}} = -2.0\log\left(\frac{1.5 \times 10^{-6}\text{ m}}{3.7(0.015\text{ m})} + \frac{2.51}{\text{Re}_2\sqrt{f_2}}\right)$$

$$\frac{1}{\sqrt{f_3}} = -2.0\log\left(\frac{1.5 \times 10^{-6}\text{ m}}{3.7(0.015\text{ m})} + \frac{2.51}{\text{Re}_3\sqrt{f_3}}\right)$$

Solving these 12 equations in 12 unknowns simultaneously using an equation solver, the flow rates are determined to be

$$\dot{V}_1 = 0.00090\text{ m}^3/\text{s}, \quad \dot{V}_2 = 0.00042\text{ m}^3/\text{s}, \quad \text{and} \quad \dot{V}_3 = 0.00048\text{ m}^3/\text{s}$$

SUMMARY

In *internal flow*, a pipe is completely filled with a fluid. *Laminar flow* is characterized by smooth streamlines and highly ordered motion, and *turbulent flow* is characterized by velocity fluctuations and highly disordered motion. The *Reynolds number* is defined as

$$\text{Re} = \frac{\text{Inertial forces}}{\text{Viscous forces}} = \frac{V_{\text{avg}}D}{\nu} = \frac{\rho V_{\text{avg}}D}{\mu}$$

Under most practical conditions, the flow in a pipe is laminar at Re < 2300, turbulent at Re > 10,000, and transitional in between.

The region of the flow in which the effects of the viscous shearing forces are felt is called the *velocity boundary layer*. The region from the pipe inlet to the point at which the boundary layer merges at the centerline is called the *hydrodynamic entrance region*, and the length of this region is called the *hydrodynamic entry length* L_h. It is given by

$$\frac{L_{h,\text{ laminar}}}{D} \cong 0.05\text{ Re} \quad \text{and} \quad \frac{L_{h,\text{ turbulent}}}{D} \cong 10$$

The friction coefficient in the fully developed flow region remains constant. The *maximum* and *average* velocities in fully developed laminar flow in a circular pipe are

$$u_{\text{max}} = 2V_{\text{avg}} \quad \text{and} \quad V_{\text{avg}} = \frac{\Delta P D^2}{32\mu L}$$

The *volume flow rate* and the *pressure drop* for laminar flow in a horizontal pipe are

$$\dot{V} = V_{\text{avg}}A_c = \frac{\Delta P \pi D^4}{128\mu L} \quad \text{and} \quad \Delta P = \frac{32\mu L V_{\text{avg}}}{D^2}$$

These results for horizontal pipes can also be used for inclined pipes provided that ΔP is replaced by $\Delta P - \rho g L \sin\theta$,

$$V_{\text{avg}} = \frac{(\Delta P - \rho g L \sin\theta)D^2}{32\mu L} \quad \text{and}$$

$$\dot{V} = \frac{(\Delta P - \rho g L \sin\theta)\pi D^4}{128\mu L}$$

The *pressure loss* and *head loss* for all types of internal flows (laminar or turbulent, in circular or noncircular pipes, smooth or rough surfaces) are expressed as

$$\Delta P_L = f\frac{L}{D}\frac{\rho V^2}{2} \quad \text{and} \quad h_L = \frac{\Delta P_L}{\rho g} = f\frac{L}{D}\frac{V^2}{2g}$$

where $\rho V^2/2$ is the *dynamic pressure* and the dimensionless quantity f is the *friction factor*. For fully developed laminar flow in a circular pipe, the friction factor is $f = 64/\text{Re}$.

For noncircular pipes, the diameter in the previous relations is replaced by the *hydraulic diameter* defined as $D_h = 4A_c/p$, where A_c is the cross-sectional area of the pipe and p is its wetted perimeter.

In fully developed turbulent flow, the friction factor depends on the Reynolds number and the *relative roughness* ε/D. The friction factor in turbulent flow is given by the *Colebrook equation*, expressed as

$$\frac{1}{\sqrt{f}} = -2.0 \log\left(\frac{\varepsilon/D}{3.7} + \frac{2.51}{\text{Re}\sqrt{f}}\right)$$

The plot of this formula is known as the Moody chart. The design and analysis of piping systems involve the determination of the head loss, flow rate, or the pipe diameter. Tedious iterations in these calculations can be avoided by the approximate Swamee–Jain formulas expressed as

$$h_L = 1.07\frac{\dot{V}^2 L}{gD^5}\left\{\ln\left[\frac{\varepsilon}{3.7D} + 4.62\left(\frac{\nu D}{\dot{V}}\right)^{0.9}\right]\right\}^{-2}$$

$$10^{-6} < \varepsilon/D < 10^{-2}$$
$$3000 < \text{Re} < 3 \times 10^8$$

$$\dot{V} = -0.965\left(\frac{gD^5 h_L}{L}\right)^{0.5} \ln\left[\frac{\varepsilon}{3.7D} + \left(\frac{3.17\nu^2 L}{gD^3 h_L}\right)^{0.5}\right]$$

$$\text{Re} > 2000$$

$$D = 0.66\left[\varepsilon^{1.25}\left(\frac{L\dot{V}^2}{gh_L}\right)^{4.75} + \nu\dot{V}^{9.4}\left(\frac{L}{gh_L}\right)^{5.2}\right]^{0.04}$$

$$10^{-6} < \varepsilon/D < 10^{-2}$$
$$5000 < \text{Re} < 3 \times 10^8$$

The losses that occur in piping components such as fittings, valves, bends, elbows, tees, inlets, exits, expansions, and contractions are called *minor losses*. The minor losses are usually expressed in terms of the *loss coefficient* K_L. The head loss for a component is determined from

$$h_L = K_L \frac{V^2}{2g}$$

When all the loss coefficients are available, the total head loss in a piping system is determined from

$$h_{L,\text{ total}} = h_{L,\text{ major}} + h_{L,\text{ minor}} = \sum_i f_i \frac{L_i}{D_i}\frac{V_i^2}{2g} + \sum_j K_{L,j}\frac{V_j^2}{2g}$$

If the entire piping system has a constant diameter, the total head loss reduces to

$$h_{L,\text{ total}} = \left(f\frac{L}{D} + \sum K_L\right)\frac{V^2}{2g}$$

The analysis of a piping system is based on two simple principles: (1) The conservation of mass throughout the system must be satisfied and (2) the pressure drop between two points must be the same for all paths between the two points. When the pipes are connected *in series*, the flow rate through the entire system remains constant regardless of the diameters of the individual pipes. For a pipe that branches out into two (or more) *parallel pipes* and then rejoins at a junction downstream, the total flow rate is the sum of the flow rates in the individual pipes but the head loss in each branch is the same.

When a piping system involves a pump and/or turbine, the steady-flow energy equation is expressed as

$$\frac{P_1}{\rho g} + \alpha_1\frac{V_1^2}{2g} + z_1 + h_{\text{pump},\,u} = \frac{P_2}{\rho g} + \alpha_2\frac{V_2^2}{2g} + z_2 + h_{\text{turbine},\,e} + h_L$$

When the useful pump head $h_{\text{pump},\,u}$ is known, the mechanical power that needs to be supplied by the pump to the fluid and the electric power consumed by the motor of the pump for a specified flow rate are determined from

$$\dot{W}_{\text{pump, shaft}} = \frac{\rho\dot{V}gh_{\text{pump},\,u}}{\eta_{\text{pump}}} \quad \text{and} \quad \dot{W}_{\text{elect}} = \frac{\rho\dot{V}gh_{\text{pump},\,u}}{\eta_{\text{pump–motor}}}$$

where $\eta_{\text{pump–motor}}$ is the *efficiency of the pump–motor combination*, which is the product of the pump and the motor efficiencies.

The plot of the head loss versus the flow rate is called the *system curve*. The head produced by a pump is not a constant, and the curves of $h_{\text{pump},\,u}$ and η_{pump} versus $\dot{V}$ are called the *characteristic curves*. A pump installed in a piping system operates at the *operating point*, which is the point of intersection of the system curve and the characteristic curve.

REFERENCES AND SUGGESTED READING

1. M. S. Bhatti and R. K. Shah. "Turbulent and Transition Flow Convective Heat Transfer in Ducts." In *Handbook of Single-Phase Convective Heat Transfer*, ed. S. Kakaç, R. K. Shah, and W. Aung. New York: Wiley Interscience, 1987.
2. C. F. Colebrook. "Turbulent Flow in Pipes, with Particular Reference to the Transition between the Smooth and Rough Pipe Laws," *Journal of the Institute of Civil Engineers London*. 11 (1939), pp. 133–156.
3. S. E. Haaland. "Simple and Explicit Formulas for the Friction Factor in Turbulent Pipe Flow," *Journal of Fluids Engineering*, March 1983, pp. 89–90.
4. I. E. Idelchik. *Handbook of Hydraulic Resistance*, 3rd ed. Boca Raton, FL: CRC Press, 1993.

5. W. M. Kays and M. E. Crawford. *Convective Heat and Mass Transfer*, 3rd ed. New York: McGraw-Hill, 1993.
6. R. W. Miller. *Flow Measurement Engineering Handbook*, 3rd ed. New York: McGraw-Hill, 1997.
7. L. F. Moody. "Friction Factors for Pipe Flows," *Transactions of the ASME* 66 (1944), pp. 671–684.
8. O. Reynolds. "On the Experimental Investigation of the Circumstances Which Determine Whether the Motion of Water Shall Be Direct or Sinuous, and the Law of Resistance in Parallel Channels." *Philosophical Transactions of the Royal Society of London*, 174 (1883), pp. 935–982.
9. H. Schlichting. *Boundary Layer Theory*, 7th ed. New York: McGraw-Hill, 1979.
10. R. K. Shah and M. S. Bhatti. "Laminar Convective Heat Transfer in Ducts." In *Handbook of Single-Phase Convective Heat Transfer*, ed. S. Kakaç, R. K. Shah, and W. Aung. New York: Wiley Interscience, 1987.
11. P. L. Skousen. *Valve Handbook*. New York: McGraw-Hill, 1998.
12. A. J. Wheeler and A. R. Ganji. *Introduction to Engineering Experimentation*. Englewood Cliffs, NJ: Prentice-Hall, 1996.
13. W. Zhi-qing. "Study on Correction Coefficients of Laminar and Turbulent Entrance Region Effects in Round Pipes," *Applied Mathematical Mechanics*, 3 (1982), p. 433.

PROBLEMS*

Laminar and Turbulent Flow

14–1C Why are liquids usually transported in circular pipes?

14–2C What is the physical significance of the Reynolds number? How is it defined for (*a*) flow in a circular pipe of inner diameter D and (*b*) flow in a rectangular duct of cross section $a \times b$?

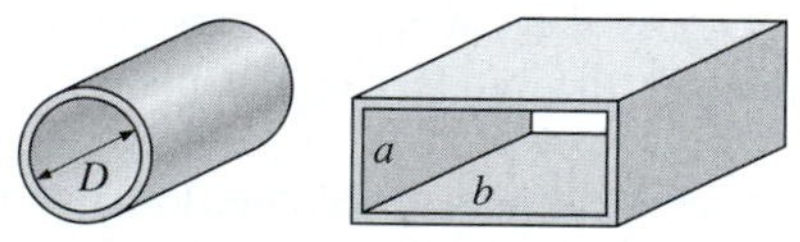

FIGURE P14–2C

14–3C Consider a person walking first in air and then in water at the same speed. For which motion will the Reynolds number be higher?

14–4C Show that the Reynolds number for flow in a circular pipe of diameter D can be expressed as $\text{Re} = 4\dot{m}/(\pi D\mu)$.

14–5C Which fluid at room temperature requires a larger pump to flow at a specified velocity in a given pipe: water or engine oil? Why?

14–6C What is the generally accepted value of the Reynolds number above which the flow in smooth pipes is turbulent?

14–7C Consider the flow of air and water in pipes of the same diameter, at the same temperature, and at the same mean velocity. Which flow is more likely to be turbulent? Why?

14–8C What is hydraulic diameter? How is it defined? What is it equal to for a circular pipe of diameter D?

14–9C How is the hydrodynamic entry length defined for flow in a pipe? Is the entry length longer in laminar or turbulent flow?

14–10C Consider laminar flow in a circular pipe. Will the wall shear stress τ_w be higher near the inlet of the pipe or near the exit? Why? What would your response be if the flow were turbulent?

14–11C How does surface roughness affect the pressure drop in a pipe if the flow is turbulent? What would your response be if the flow were laminar?

Fully Developed Flow in Pipes

14–12C How does the wall shear stress τ_w vary along the flow direction in the fully developed region in (*a*) laminar flow and (*b*) turbulent flow?

14–13C What fluid property is responsible for the development of the velocity boundary layer? For what kinds of fluids will there be no velocity boundary layer in a pipe?

14–14C In the fully developed region of flow in a circular pipe, will the velocity profile change in the flow direction?

14–15C How is the friction factor for flow in a pipe related to the pressure loss? How is the pressure loss related to the pumping power requirement for a given mass flow rate?

14–16C Someone claims that the shear stress at the center of a circular pipe during fully developed laminar flow is zero. Do you agree with this claim? Explain.

* Problems designated by a "C" are concept questions, and students are encouraged to answer them all. Problems designated by an "E" are in English units, and the SI users can ignore them. Problems with the icon are solved using EES, and complete solutions together with parametric studies are included on the enclosed DVD. Problems with the icon are comprehensive in nature and are intended to be solved with a computer, preferably using the EES software that accompanies this text.

14–17C Someone claims that in fully developed turbulent flow in a pipe, the shear stress is a maximum at the pipe wall. Do you agree with this claim? Explain.

14–18C Consider fully developed flow in a circular pipe with negligible entrance effects. If the length of the pipe is doubled, the head loss will (*a*) double, (*b*) more than double, (*c*) less than double, (*d*) reduce by half, or (*e*) remain constant.

14–19C Someone claims that the volume flow rate in a circular pipe with laminar flow can be determined by measuring the velocity at the centerline in the fully developed region, multiplying it by the cross-sectional area, and dividing the result by 2. Do you agree? Explain.

14–20C Someone claims that the average velocity in a circular pipe in fully developed laminar flow can be determined by simply measuring the velocity at $R/2$ (midway between the wall surface and the centerline). Do you agree? Explain.

14–21C Consider fully developed laminar flow in a circular pipe. If the diameter of the pipe is reduced by half while the flow rate and the pipe length are held constant, the head loss will (*a*) double, (*b*) triple, (*c*) quadruple, (*d*) increase by a factor of 8, or (*e*) increase by a factor of 16.

14–22C What is the physical mechanism that causes the friction factor to be higher in turbulent flow?

14–23C Discuss whether fully developed pipe flow is one-, two-, or three-dimensional.

14–24C The head loss for a certain circular pipe is given by $h_L = 0.0826fL(\dot{V}^2/D^5)$, where f is the friction factor (dimensionless), L is the pipe length, $\dot{V}$ is the volumetric flow rate, and D is the pipe diameter. Determine if the 0.0826 is a dimensional or dimensionless constant. Is this equation dimensionally homogeneous as it stands?

14–25C Consider fully developed laminar flow in a circular pipe. If the viscosity of the fluid is reduced by half by heating while the flow rate is held constant, how will the head loss change?

14–26C How is head loss related to pressure loss? For a given fluid, explain how you would convert head loss to pressure loss.

14–27C Consider laminar flow of air in a circular pipe with perfectly smooth surfaces. Do you think the friction factor for this flow will be zero? Explain.

14–28C Explain why the friction factor is independent of the Reynolds number at very large Reynolds numbers.

14–29E Oil at 80°F (ρ = 56.8 lbm/ft^3 and μ = 0.0278 lbm/ft · s) is flowing steadily in a 0.5-in-diameter, 120-ft-long pipe. During the flow, the pressure at the pipe inlet and exit is measured to be 120 psi and 14 psi, respectively. Determine the flow rate of oil through the pipe assuming the pipe is (*a*) horizontal, (*b*) inclined 20° upward, and (*c*) inclined 20° downward.

14–30 Oil with a density of 850 kg/m^3 and kinematic viscosity of 0.00062 m^2/s is being discharged by a 5-mm-diameter, 40-m-long horizontal pipe from a storage tank open to the atmosphere. The height of the liquid level above the center of the pipe is 3 m. Disregarding the minor losses, determine the flow rate of oil through the pipe.

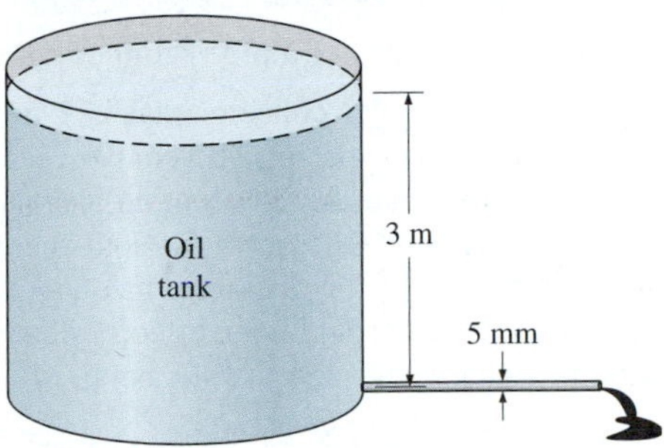

FIGURE P14–30

14–31 Water at 10°C (ρ = 999.7 kg/m^3 and μ = 1.307 × 10^{-3} kg/m · s) is flowing steadily in a 0.20-cm-diameter, 15-m-long pipe at an average velocity of 1.2 m/s. Determine (*a*) the pressure drop, (*b*) the head loss, and (*c*) the pumping power requirement to overcome this pressure drop. *Answers:* (*a*) 188 kPa, (*b*) 19.2 m, (*c*) 0.71 W

14–32 Water at 15°C (ρ = 999.1 kg/m^3 and μ = 1.138 × 10^{-3} kg/m · s) is flowing steadily in a 30-m-long and 4-cm-diameter horizontal pipe made of stainless steel at a rate of 8 L/s. Determine (*a*) the pressure drop, (*b*) the head loss, and (*c*) the pumping power requirement to overcome this pressure drop.

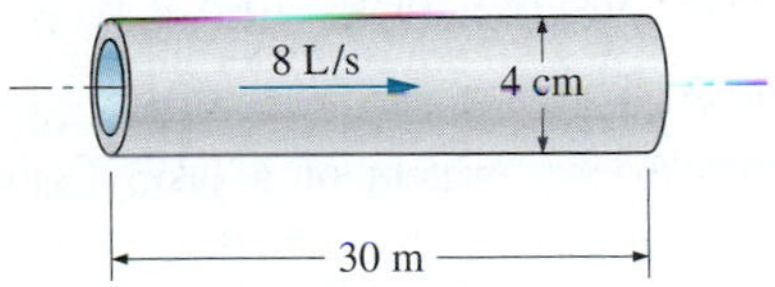

FIGURE P14–32

14–33E Heated air at 1 atm and 100°F is to be transported in a 400-ft-long circular plastic duct at a rate of 12 ft^3/s. If the head loss in the pipe is not to exceed 50 ft, determine the minimum diameter of the duct.

14–34 In fully developed laminar flow in a circular pipe, the velocity at $R/2$ (midway between the wall surface and the centerline) is measured to be 6 m/s. Determine the velocity at the center of the pipe. *Answer:* 8 m/s

14–35 The velocity profile in fully developed laminar flow in a circular pipe of inner radius R = 2 cm, in m/s, is given by $u(r) = 4(1 - r^2/R^2)$. Determine the average and maximum velocities in the pipe and the volume flow rate.

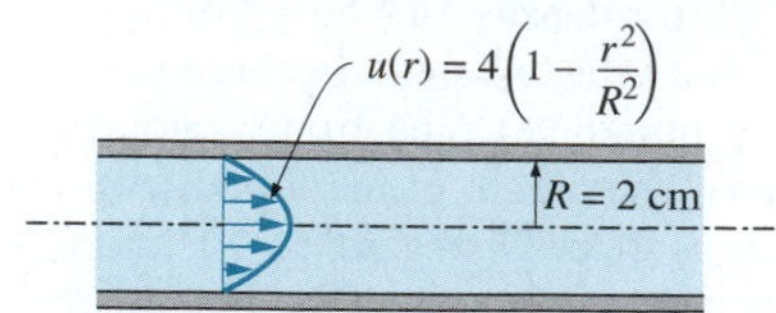

FIGURE P14–35

14–36 Repeat Prob. 14–35 for a pipe of inner radius 7 cm.

14–37 Consider an air solar collector that is 1 m wide and 5 m long and has a constant spacing of 3 cm between the glass cover and the collector plate. Air flows at an average temperature of 45°C at a rate of 0.15 m^3/s through the 1-m-wide edge of the collector along the 5-m-long passageway. Disregarding the entrance and roughness effects, determine the pressure drop in the collector and the 90° bend. *Answer:* 32.3 Pa

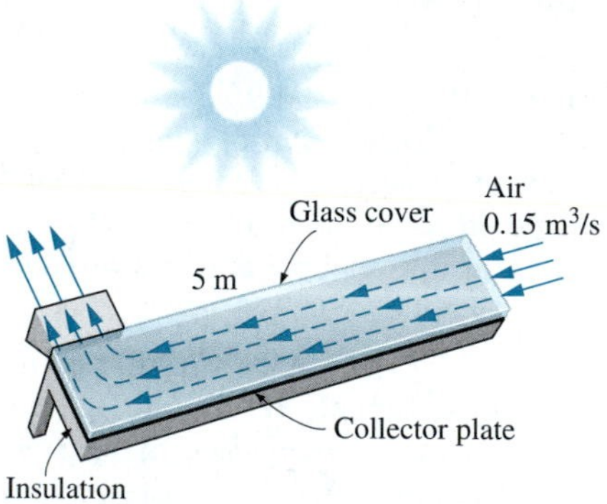

FIGURE P14–37

14–38 Consider the flow of oil with ρ = 894 kg/m^3 and μ = 2.33 kg/m · s in a 40-cm-diameter pipeline at an average velocity of 0.5 m/s. A 300-m-long section of the pipeline passes through the icy waters of a lake. Disregarding the entrance effects, determine the pumping power required to overcome the pressure losses and to maintain the flow of oil in the pipe.

14–39 Consider laminar flow of a fluid through a square channel with smooth surfaces (f = 56.92/Re). Now the average velocity of the fluid is doubled. Determine the change in the head loss of the fluid. Assume the flow regime remains unchanged.

14–40 Repeat Prob. 14–39 for turbulent flow in smooth pipes for which the friction factor is given as $f = 0.184\text{Re}^{-0.2}$. What would your answer be for fully turbulent flow in a rough pipe?

14–41 Air enters a 7-m-long section of a rectangular duct of cross section 15 cm × 20 cm made of commercial steel at 1 atm and 35°C at an average velocity of 7 m/s. Disregarding the entrance effects, determine the fan power needed to overcome the pressure losses in this section of the duct. *Answer:* 4.9 W

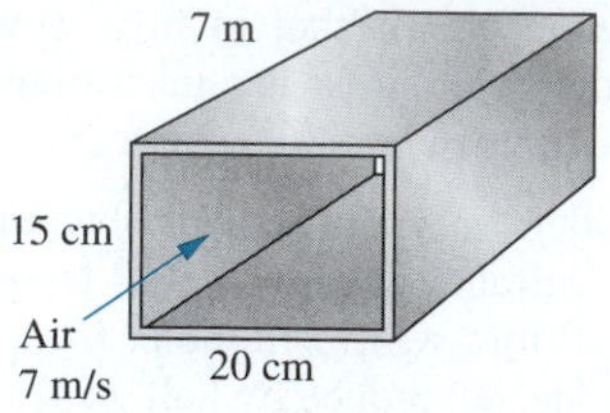

FIGURE P14–41

14–42E Water at 60°F passes through 0.75-in-internal-diameter copper tubes at a rate of 1.2 lbm/s. Determine the pumping power per ft of pipe length required to maintain this flow at the specified rate.

14–43 Oil with ρ = 876 kg/m^3 and μ = 0.24 kg/m · s is flowing through a 1.5-cm-diameter pipe that discharges into the atmosphere at 88 kPa. The absolute pressure 15 m before the exit is measured to be 135 kPa. Determine the flow rate of oil through the pipe if the pipe is (*a*) horizontal, (*b*) inclined 8° upward from the horizontal, and (*c*) inclined 8° downward from the horizontal.

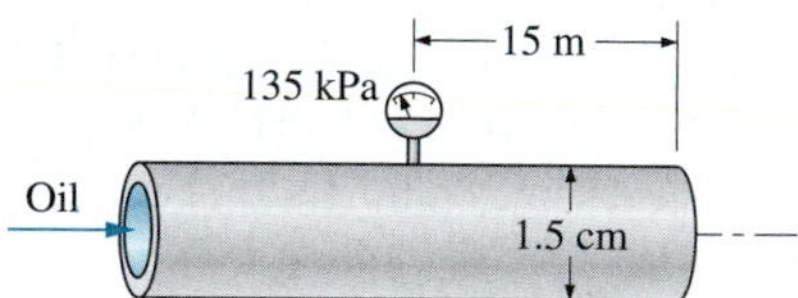

FIGURE P14–43

14–44 Glycerin at 40°C with ρ = 1252 kg/m^3 and μ = 0.27 kg/m · s is flowing through a 2-cm-diameter, 25-m-long pipe that discharges into the atmosphere at 100 kPa. The flow rate through the pipe is 0.035 L/s. (*a*) Determine the absolute pressure 25 m before the pipe exit. (*b*) At what angle θ must the pipe be inclined downward from the horizontal for the pressure in the entire pipe to be atmospheric pressure and the flow rate to be maintained the same?

14–45 In an air heating system, heated air at 40°C and 105 kPa absolute is distributed through a 0.2 m × 0.3 m rectangular duct made of commercial steel at a rate of 0.5 m^3/s. Determine the pressure drop and head loss through a 40-m-long section of the duct. *Answers:* 128 Pa, 10.8 m

14–46 Glycerin at 40°C with ρ = 1252 kg/m^3 and μ = 0.27 kg/m · s is flowing through a 5-cm-diameter horizontal smooth pipe with an average velocity of 3.5 m/s. Determine the pressure drop per 10 m of the pipe.

14–47 EES Reconsider Prob. 14–46. Using EES (or other) software, investigate the effect of the pipe diameter on the pressure drop for the same constant flow rate. Let the pipe diameter vary from 1 to 10 cm in increments of 1 cm. Tabulate and plot the results, and draw conclusions.

14–48E Air at 1 atm and 60°F is flowing through a 1 ft × 1 ft square duct made of commercial steel at a rate of

1200 cfm. Determine the pressure drop and head loss per ft of the duct.

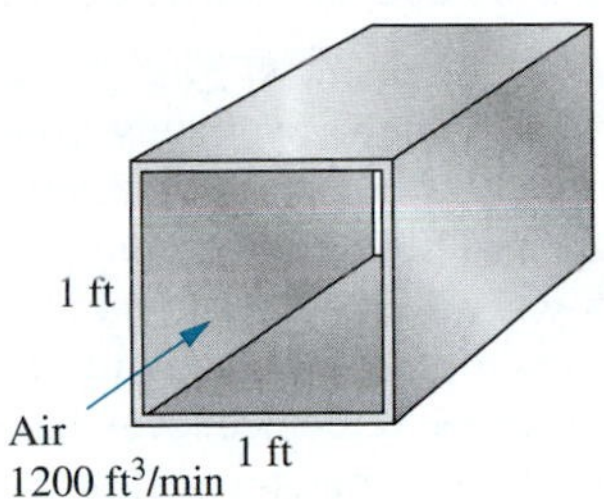

FIGURE P14–48E

14–49 Liquid ammonia at −20°C is flowing through a 30-m-long section of a 5-mm-diameter copper tube at a rate of 0.15 kg/s. Determine the pressure drop, the head loss, and the pumping power required to overcome the frictional losses in the tube. *Answers:* 4792 kPa, 734 m, 1.08 kW

14–50 Shell-and-tube heat exchangers with hundreds of tubes housed in a shell are commonly used in practice for heat transfer between two fluids. Such a heat exchanger used in an active solar hot-water system transfers heat from a water-antifreeze solution flowing through the shell and the solar collector to fresh water flowing through the tubes at an average temperature of 60°C at a rate of 15 L/s. The heat exchanger contains 80 brass tubes 1 cm in inner diameter and 1.5 m in length. Disregarding inlet, exit, and header losses, determine the pressure drop across a single tube and the pumping power required by the tube-side fluid of the heat exchanger.

After operating a long time, 1-mm-thick scale builds up on the inner surfaces with an equivalent roughness of 0.4 mm. For the same pumping power input, determine the percent reduction in the flow rate of water through the tubes.

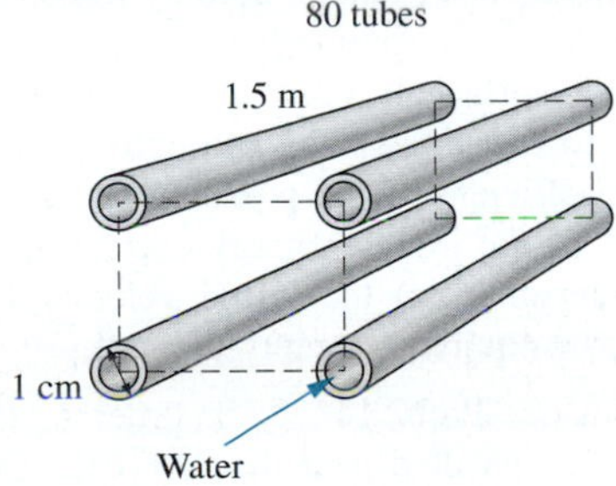

FIGURE P14–50

Minor Losses

14–51C What is minor loss in pipe flow? How is the minor loss coefficient K_L defined?

14–52C Define equivalent length for minor loss in pipe flow. How is it related to the minor loss coefficient?

14–53C The effect of rounding of a pipe inlet on the loss coefficient is (*a*) negligible, (*b*) somewhat significant, or (*c*) very significant.

14–54C The effect of rounding of a pipe exit on the loss coefficient is (*a*) negligible, (*b*) somewhat significant, or (*c*) very significant.

14–55C Which has a greater minor loss coefficient during pipe flow: gradual expansion or gradual contraction? Why?

14–56C A piping system involves sharp turns, and thus large minor head losses. One way of reducing the head loss is to replace the sharp turns by circular elbows. What is another way?

14–57C During a retrofitting project of a fluid flow system to reduce the pumping power, it is proposed to install vanes into the miter elbows or to replace the sharp turns in 90° miter elbows by smooth curved bends. Which approach will result in a greater reduction in pumping power requirements?

14–58 Water is to be withdrawn from a 3-m-high water reservoir by drilling a 1.5-cm-diameter hole at the bottom surface. Disregarding the effect of the kinetic energy correction factor, determine the flow rate of water through the hole if (*a*) the entrance of the hole is well-rounded and (*b*) the entrance is sharp-edged.

14–59 Consider flow from a water reservoir through a circular hole of diameter D at the side wall at a vertical distance H from the free surface. The flow rate through an actual hole with a sharp-edged entrance (K_L = 0.5) will be considerably less than the flow rate calculated assuming "frictionless" flow and thus zero loss for the hole. Disregarding the effect of the kinetic energy correction factor, obtain a relation for the "equivalent diameter" of the sharp-edged hole for use in frictionless flow relations.

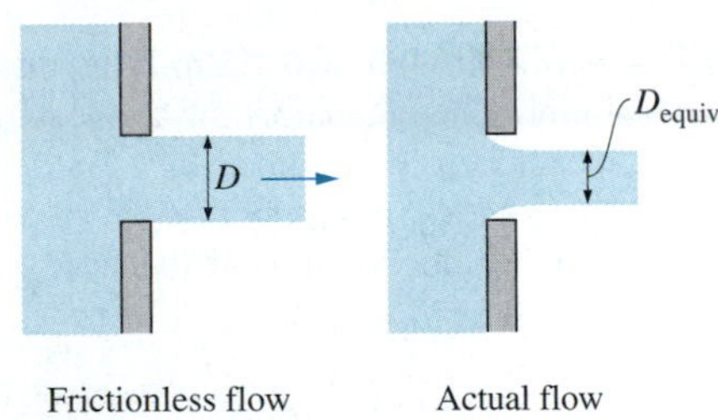

FIGURE P14–59

14–60 Repeat Prob. 14–59 for a slightly rounded entrance (K_L = 0.12).

14–61 A horizontal pipe has an abrupt expansion from D_1 = 8 cm to D_2 = 16 cm. The water velocity in the smaller section is 10 m/s and the flow is turbulent. The pressure in the smaller section is P_1 = 300 kPa. Taking the kinetic energy correction factor to be 1.06 at both the inlet and the outlet, determine the downstream pressure P_2, and estimate the error that would have occurred if Bernoulli's equation had been used. *Answers:* 322 kPa, 25 kPa

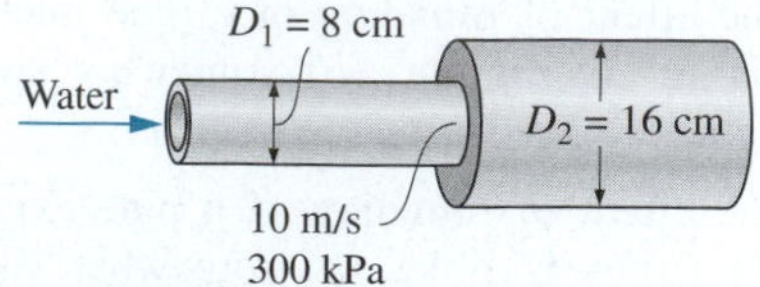

FIGURE P14–61

Piping Systems and Pump Selection

14–62C A piping system involves two pipes of different diameters (but of identical length, material, and roughness) connected in series. How would you compare the (*a*) flow rates and (*b*) pressure drops in these two pipes?

14–63C A piping system involves two pipes of different diameters (but of identical length, material, and roughness) connected in parallel. How would you compare the (*a*) flow rates and (*b*) pressure drops in these two pipes?

14–64C A piping system involves two pipes of identical diameters but of different lengths connected in parallel. How would you compare the pressure drops in these two pipes?

14–65C Water is pumped from a large lower reservoir to a higher reservoir. Someone claims that if the head loss is negligible, the required pump head is equal to the elevation difference between the free surfaces of the two reservoirs. Do you agree?

14–66C A piping system equipped with a pump is operating steadily. Explain how the operating point (the flow rate and the head loss) is established.

14–67C For a piping system, define the system curve, the characteristic curve, and the operating point on a head versus flow rate chart.

14–68 Water at 20°C is to be pumped from a reservoir (z_A = 2 m) to another reservoir at a higher elevation (z_B = 9 m) through two 25-m-long plastic pipes connected in parallel. The diameters of the two pipes are 3 cm and 5 cm. Water is to be pumped by a 68 percent efficient motor–pump unit that draws 7 kW of electric power during operation. The minor losses and the head loss in the pipes that connect the parallel pipes to the two reservoirs are considered to be negligible. Determine the total flow rate between the reservoirs and the flow rates through each of the parallel pipes.

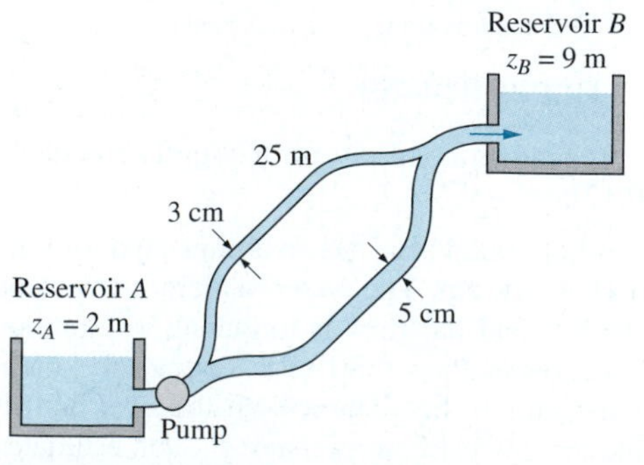

FIGURE P14–68

14–69E Water at 70°F flows by gravity from a large reservoir at a high elevation to a smaller one through a 120-ft-long, 2-in-diameter cast iron piping system that includes four standard flanged elbows, a well-rounded entrance, a sharp-edged exit, and a fully open gate valve. Taking the free surface of the lower reservoir as the reference level, determine the elevation z_1 of the higher reservoir for a flow rate of 10 ft^3/min. *Answer:* 23.1 ft

14–70 A 3-m-diameter tank is initially filled with water 2 m above the center of a sharp-edged 10-cm-diameter orifice. The tank water surface is open to the atmosphere, and the orifice drains to the atmosphere. Neglecting the effect of the kinetic energy correction factor, calculate (*a*) the initial velocity from the tank and (*b*) the time required to empty the tank. Does the loss coefficient of the orifice cause a significant increase in the draining time of the tank?

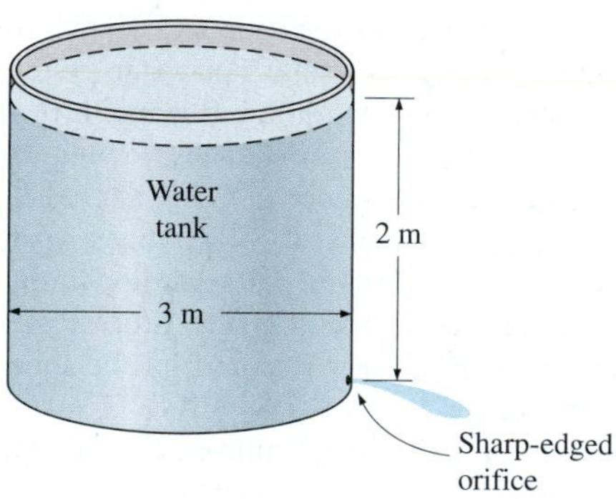

FIGURE P14–70

14–71 A 3-m-diameter tank is initially filled with water 2 m above the center of a sharp-edged 10-cm-diameter orifice. The tank water surface is open to the atmosphere, and the orifice drains to the atmosphere through a 100-m-long pipe. The friction coefficient of the pipe can be taken to be 0.015 and the effect of the kinetic energy correction factor can be neglected. Determine (*a*) the initial velocity from the tank and (*b*) the time required to empty the tank.

14–72 Reconsider Prob. 14–71. In order to drain the tank faster, a pump is installed near the tank exit. Determine how much pump power input is necessary to establish an average water velocity of 4 m/s when the tank is full at z = 2 m. Also, assuming the discharge velocity to remain constant, estimate the time required to drain the tank.

Someone suggests that it makes no difference whether the pump is located at the beginning or at the end of the pipe, and that the performance will be the same in either case, but another person argues that placing the pump near the end of the pipe may cause cavitation. The water temperature is 30°C, so the water vapor pressure is P_v = 4.246 kPa = 0.43 m-H_2O, and the system is located at sea level. Investigate if there is the possibility of cavitation and if we should be concerned about the location of the pump.

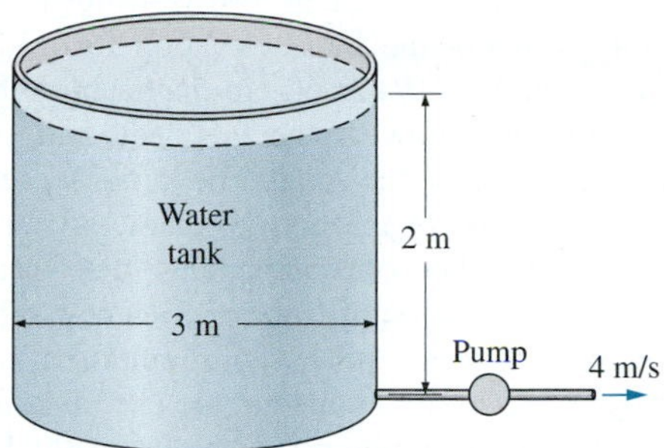

FIGURE P14–72

14–73 Oil at 20°C is flowing through a vertical glass funnel that consists of a 15-cm-high cylindrical reservoir and a 1-cm-diameter, 25-cm-high pipe. The funnel is always maintained full by the addition of oil from a tank. Assuming the entrance effects to be negligible, determine the flow rate of oil through the funnel and calculate the "funnel effectiveness," which can be defined as the ratio of the actual flow rate through the funnel to the maximum flow rate for the "frictionless" case. *Answers:* 4.09×10^{-6} m^3/s, 1.86 percent

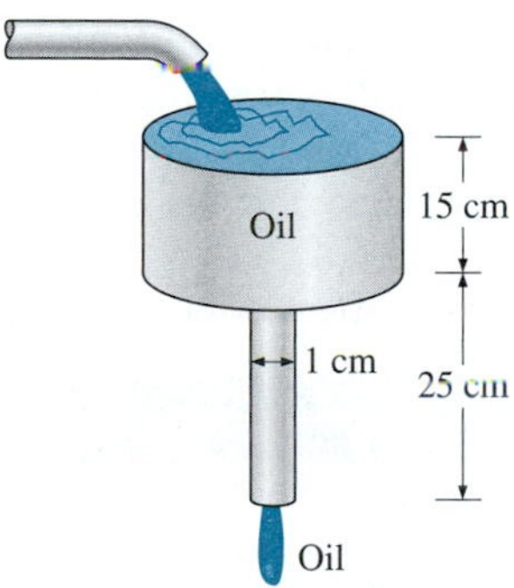

FIGURE P14–73

14–74 Repeat Prob. 14–73 assuming (*a*) the diameter of the pipe is doubled and (*b*) the length of the pipe is doubled.

14–75 Water at 15°C is drained from a large reservoir using two horizontal plastic pipes connected in series. The first pipe is 20 m long and has a 10-cm diameter, while the second pipe is 35 m long and has a 4-cm diameter. The water level in the reservoir is 18 m above the centerline of the pipe. The pipe entrance is sharp-edged, and the contraction between the two pipes is sudden. Neglecting the effect of the kinetic energy correction factor, determine the discharge rate of water from the reservoir.

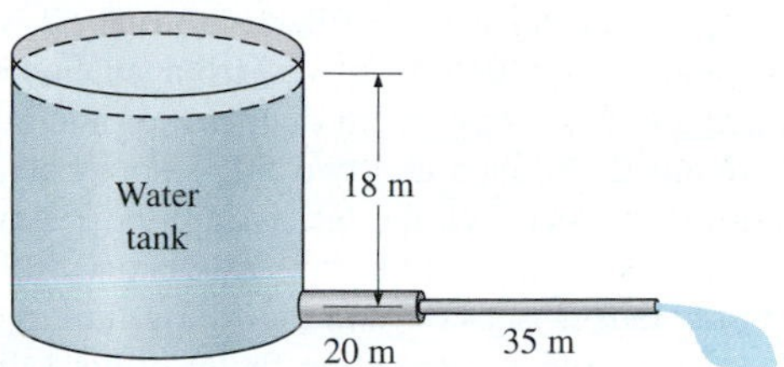

FIGURE P14–75

14–76E A farmer is to pump water at 70°F from a river to a water storage tank nearby using a 125-ft-long, 5-in-diameter plastic pipe with three flanged 90° smooth bends. The water velocity near the river surface is 6 ft/s, and the pipe inlet is placed in the river normal to the flow direction of water to take advantage of the dynamic pressure. The elevation difference between the river and the free surface of the tank is 12 ft. For a flow rate of 1.5 ft^3/s and an overall pump efficiency of 70 percent, determine the required electric power input to the pump.

14–77E EES Reconsider Prob. 14–76E. Using EES (or other) software, investigate the effect of the pipe diameter on the required electric power input to the pump. Let the pipe diameter vary from 1 to 10 in, in increments of 1 in. Tabulate and plot the results, and draw conclusions.

14–78 A water tank filled with solar-heated water at 40°C is to be used for showers in a field using gravity-driven flow. The system includes 20 m of 1.5-cm-diameter galvanized iron piping with four miter bends (90°) without vanes and a wide-open globe valve. If water is to flow at a rate of 0.7 L/s through the shower head, determine how high the water level in the tank must be from the exit level of the shower. Disregard the losses at the entrance and at the shower head, and neglect the effect of the kinetic energy correction factor.

14–79 Two water reservoirs *A* and *B* are connected to each other through a 40-m-long, 2-cm-diameter cast iron pipe with a sharp-edged entrance. The pipe also involves a swing check valve and a fully open gate valve. The water level in both reservoirs is the same, but reservoir *A* is pressurized by compressed air while reservoir *B* is open to the atmosphere at 88 kPa. If the initial flow rate through the pipe is 1.2 L/s, determine the absolute air pressure on top of reservoir *A*. Take the water temperature to be 10°C. *Answer:* 733 kPa

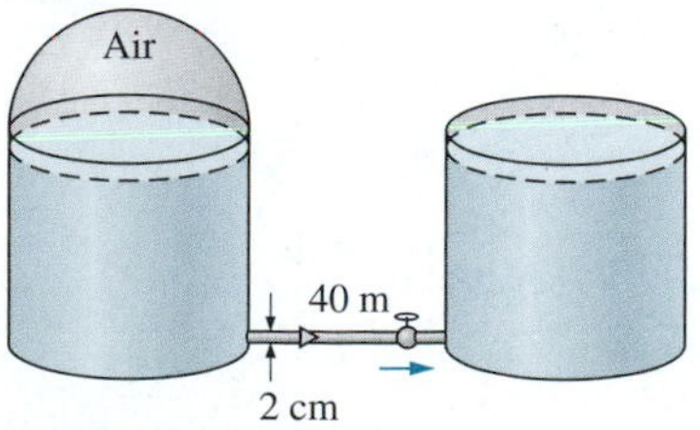

FIGURE P14–79

14–80 A vented tanker is to be filled with fuel oil with $\rho = 920$ kg/m³ and $\mu = 0.045$ kg/m · s from an underground reservoir using a 20-m-long, 5-cm-diameter plastic hose with a slightly rounded entrance and two 90° smooth bends. The elevation difference between the oil level in the reservoir and the top of the tanker where the hose is discharged is 5 m. The capacity of the tanker is 18 m³ and the filling time is 30 min. Taking the kinetic energy correction factor at hose discharge to be 1.05 and assuming an overall pump efficiency of 82 percent, determine the required power input to the pump.

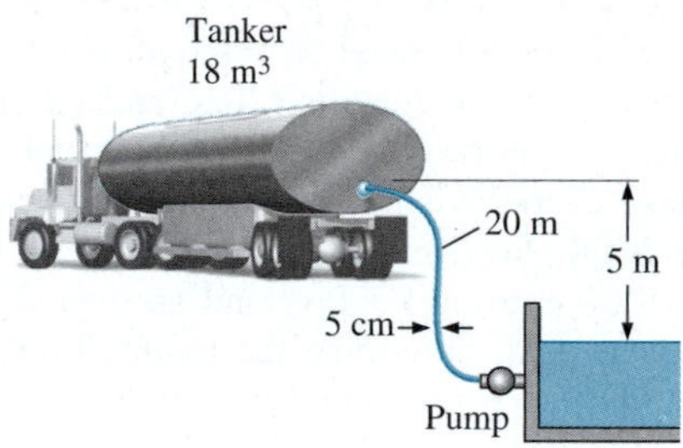

FIGURE P14–80

14–81 Two pipes of identical length and material are connected in parallel. The diameter of pipe *A* is twice the diameter of pipe *B*. Assuming the friction factor to be the same in both cases and disregarding minor losses, determine the ratio of the flow rates in the two pipes.

14–82 A certain part of cast iron piping of a water distribution system involves a parallel section. Both parallel pipes have a diameter of 30 cm, and the flow is fully turbulent. One of the branches (pipe *A*) is 1000 m long while the other branch (pipe *B*) is 3000 m long. If the flow rate through pipe *A* is 0.4 m³/s, determine the flow rate through pipe *B*. Disregard minor losses and assume the water temperature to be 15°C. Show that the flow is fully turbulent, and thus the friction factor is independent of Reynolds number. *Answer:* 0.231 m³/s

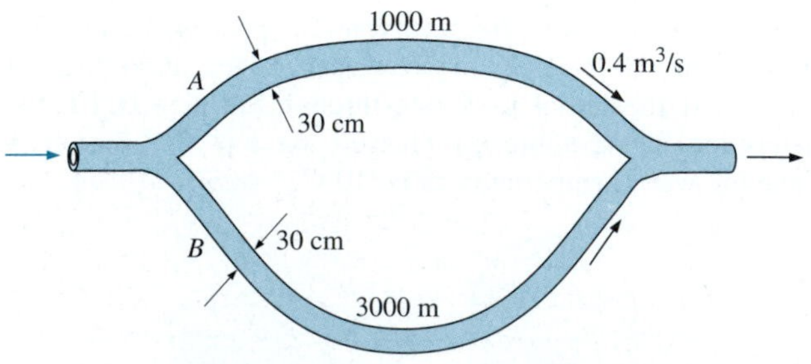

FIGURE P14–82

14–83 Repeat Prob. 14–82 assuming pipe *A* has a halfway-closed gate valve ($K_L = 2.1$) while pipe *B* has a fully open globe valve ($K_L = 10$), and the other minor losses are negligible. Assume the flow to be fully turbulent.

14–84 A geothermal district heating system involves the transport of geothermal water at 110°C from a geothermal well to a city at about the same elevation for a distance of 12 km at a rate of 1.5 m³/s in 60-cm-diameter stainless-steel pipes. The fluid pressures at the wellhead and the arrival point in the city are to be the same. The minor losses are negligible because of the large length-to-diameter ratio and the relatively small number of components that cause minor losses. (*a*) Assuming the pump–motor efficiency to be 74 percent, determine the electric power consumption of the system for pumping. Would you recommend the use of a single large pump or several smaller pumps of the same total pumping power scattered along the pipeline? Explain. (*b*) Determine the daily cost of power consumption of the system if the unit cost of electricity is $0.06/kWh. (*c*) The temperature of geothermal water is estimated to drop 0.5°C during this long flow. Determine if the frictional heating during flow can make up for this drop in temperature.

14–85 Repeat Prob. 14–84 for cast iron pipes of the same diameter.

14–86E A clothes dryer discharges air at 1 atm and 120°F at a rate of 1.2 ft³/s when its 5-in-diameter, well-rounded vent with negligible loss is not connected to any duct. Determine the flow rate when the vent is connected to a 15-ft-long, 5-in-diameter duct made of galvanized iron, with three 90° flanged smooth bends. Take the friction factor of the duct to be 0.019, and assume the fan power input to remain constant.

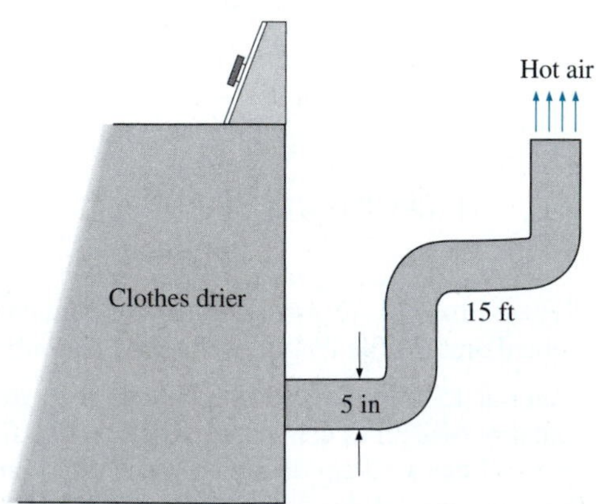

FIGURE P14–86E

14–87 In large buildings, hot water in a water tank is circulated through a loop so that the user doesn't have to wait for all the water in long piping to drain before hot water starts coming out. A certain recirculating loop involves 40-m-long, 1.2-cm-diameter cast iron pipes with six 90° threaded smooth bends and two fully open gate valves. If the average flow velocity through the loop is 2.5 m/s, determine the required power input for the recirculating pump. Take the average water temperature to be 60°C and the efficiency of the pump to be 70 percent. *Answer:* 0.217 kW

14–88 Reconsider Prob. 14–87. Using EES (or other) software, investigate the effect of the average flow velocity on the power input to the recirculating pump. Let the velocity vary from 0 to 3 m/s in increments of 0.3 m/s. Tabulate and plot the results.

14–89 Repeat Prob. 14–87 for plastic pipes.

Review Problems

14–90 The compressed air requirements of a manufacturing facility are met by a 150-hp compressor that draws in air from the outside through an 8-m-long, 20-cm-diameter duct made of thin galvanized iron sheets. The compressor takes in air at a rate of 0.27 m³/s at the outdoor conditions of 15°C and 95 kPa. Disregarding any minor losses, determine the useful power used by the compressor to overcome the frictional losses in this duct. *Answer:* 9.66 W

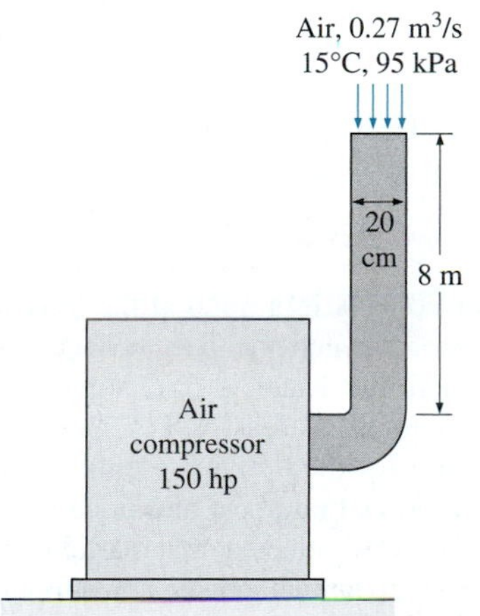

FIGURE P14–90

14–91 A house built on a riverside is to be cooled in summer by utilizing the cool water of the river. A 15-m-long section of a circular stainless-steel duct of 20-cm diameter passes through the water. Air flows through the underwater section of the duct at 3 m/s at an average temperature of 15°C. For an overall fan efficiency of 62 percent, determine the fan power needed to overcome the flow resistance in this section of the duct.

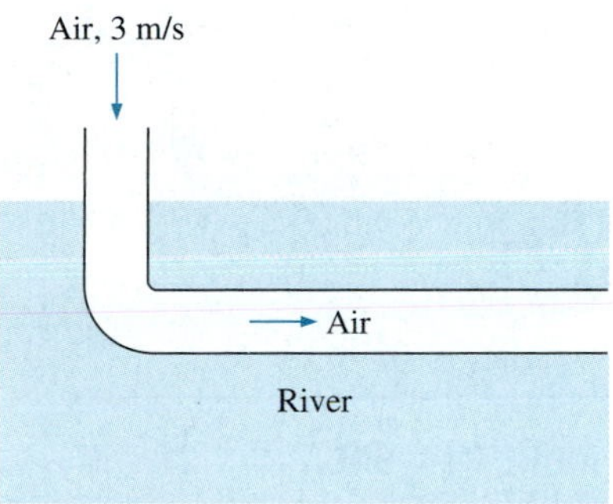

FIGURE P14–91

14–92 Water is to be withdrawn from a 5-m-high water reservoir by drilling a well-rounded 3-cm-diameter hole with negligible loss at the bottom surface and attaching a horizontal 90° bend of negligible length. Taking the kinetic energy correction factor to be 1.05, determine the flow rate of water through the bend if (*a*) the bend is a flanged smooth bend and (*b*) the bend is a miter bend without vanes. *Answers:* (*a*) 0.00603 m³/s, (*b*) 0.00478 m³/s

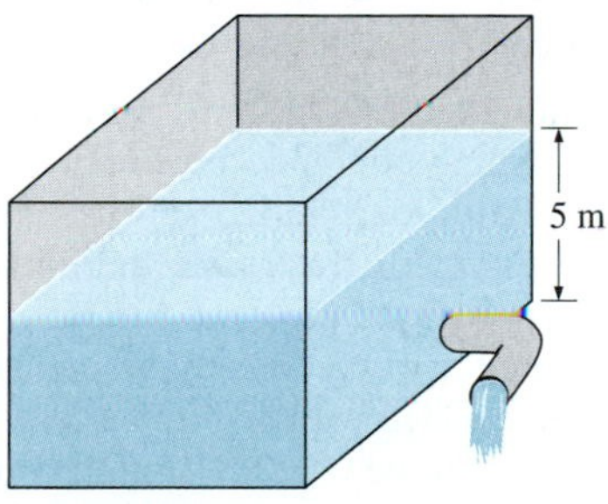

FIGURE P14–92

14–93 Two pipes of identical diameter and material are connected in parallel. The length of pipe *A* is twice the length of pipe *B*. Assuming the flow is fully turbulent in both pipes and thus the friction factor is independent of the Reynolds number and disregarding minor losses, determine the ratio of the flow rates in the two pipes. *Answer:* 0.707

14–94E A water fountain is to be installed at a remote location by attaching a cast iron pipe directly to a water main through which water is flowing at 70°F and 60 psig. The entrance to the pipe is sharp-edged, and the 50-ft-long piping system involves three 90° miter bends without vanes, a fully open gate valve, and an angle valve with a loss coefficient of 5 when fully open. If the system is to provide water at a rate of 20 gal/min and the elevation difference between the pipe and the fountain is negligible, determine the minimum diameter of the piping system. *Answer:* 0.76 in

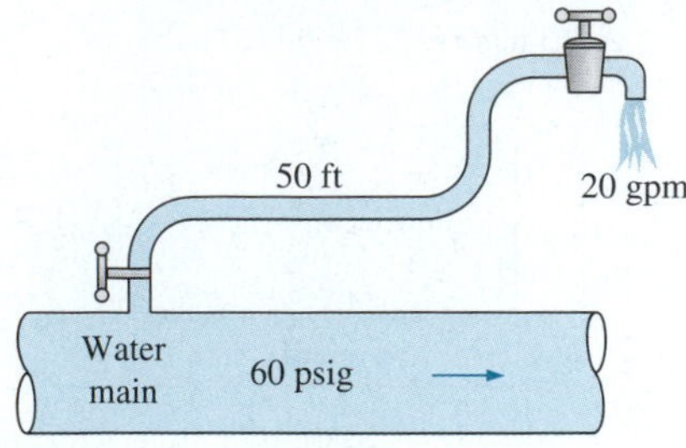

FIGURE P14–94E

14–95E Repeat Prob. 14–94E for plastic pipes.

14–96 In a hydroelectric power plant, water at 20°C is supplied to the turbine at a rate of 0.8 m^3/s through a 200-m-long, 0.35-m-diameter cast iron pipe. The elevation difference between the free surface of the reservoir and the turbine discharge is 70 m, and the combined turbine–generator efficiency is 84 percent. Disregarding the minor losses because of the large length-to-diameter ratio, determine the electric power output of this plant.

14–97 In Prob. 14–96, the pipe diameter is tripled in order to reduce the pipe losses. Determine the percent increase in the net power output as a result of this modification.

14–98 The water at 20°C in a 10-m-diameter, 2-m-high aboveground swimming pool is to be emptied by unplugging a 3-cm-diameter, 25-m-long horizontal plastic pipe attached to the bottom of the pool. Determine the initial rate of discharge of water through the pipe and the time it will take to empty the swimming pool completely assuming the entrance to the pipe is well-rounded with negligible loss. Take the friction factor of the pipe to be 0.022. Using the initial discharge velocity, check if this is a reasonable value for the friction factor. *Answers:* 1.01 L/s, 86.7 h

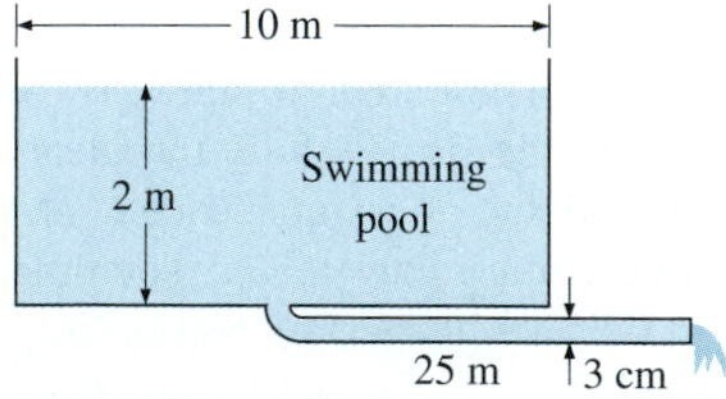

FIGURE P14–98

14–99 EES Reconsider Prob. 14–98. Using EES (or other) software, investigate the effect of the discharge pipe diameter on the time required to empty the pool completely. Let the diameter vary from 1 to 10 cm, in increments of 1 cm. Tabulate and plot the results.

14–100 Repeat Prob. 14–98 for a sharp-edged entrance to the pipe with $K_L = 0.5$. Is this "minor loss" truly "minor" or not?

14–101 A highly viscous liquid discharges from a large container through a small-diameter tube in laminar flow. Disregarding entrance effects and velocity heads, obtain a relation for the variation of fluid depth in the tank with time.

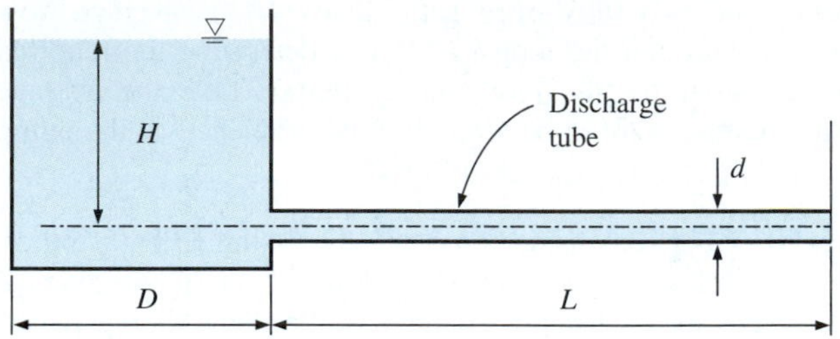

FIGURE P14–101

14–102 A student is to determine the kinematic viscosity of an oil using the system shown in Prob. 14–101. The initial fluid height in the tank is $H = 40$ cm, the tube diameter is $d = 6$ mm, the tube length is $L = 0.65$ m, and the tank diameter is $D = 0.63$ m. The student observes that it takes 2842 s for the fluid level in the tank to drop to 36 cm. Find the fluid viscosity.

Design and Essay Problems

14–103 Design an experiment to measure the viscosity of liquids using a vertical funnel with a cylindrical reservoir of height h and a narrow flow section of diameter D and length L. Making appropriate assumptions, obtain a relation for viscosity in terms of easily measurable quantities such as density and volume flow rate. Is there a need for the use of a correction factor?

14–104 A pump is to be selected for a waterfall in a garden. The water collects in a pond at the bottom, and the elevation difference between the free surface of the pond and the location where the water is discharged is 3 m. The flow rate of water is to be at least 8 L/s. Select an appropriate motor– pump unit for this job and identify three manufacturers with product model numbers and prices. Make a selection and explain why you selected that particular product. Also estimate the cost of annual power consumption of this unit assuming continuous operation.

14–105 During a camping trip you notice that water is discharged from a high reservoir to a stream in the valley through a 30-cm-diameter plastic pipe. The elevation difference between the free surface of the reservoir and the stream is 70 m. You conceive the idea of generating power from this water. Design a power plant that will produce the most power from this resource. Also, investigate the effect of power generation on the discharge rate of water. What discharge rate will maximize the power production?

Chapter 15

EXTERNAL FLOW: DRAG AND LIFT

In this chapter we consider *external flow*—flow over bodies that are immersed in a fluid, with emphasis on the resulting lift and drag forces. In external flow, the viscous effects are confined to a portion of the flow field such as the boundary layers and wakes, which are surrounded by an outer flow region that involves small velocity and temperature gradients.

When a fluid moves over a solid body, it exerts pressure forces normal to the surface and shear forces parallel to the surface of the body. We are usually interested in the *resultant* of the pressure and shear forces acting on the body rather than the details of the distributions of these forces along the entire surface of the body. The component of the resultant pressure and shear forces that acts in the flow direction is called the *drag force* (or just *drag*), and the component that acts normal to the flow direction is called the *lift force* (or just *lift*).

We start this chapter with a discussion of drag and lift, and explore the concepts of pressure drag, friction drag, and flow separation. We continue with the drag coefficients of various two- and three-dimensional geometries encountered in practice and determine the drag force using experimentally determined drag coefficients. We then examine the development of the velocity boundary layer during parallel flow over a flat surface, and develop relations for the skin friction and drag coefficients for flow over flat plates, cylinders, and spheres. Finally, we discuss the lift developed by airfoils and the factors that affect the lift characteristics of bodies.

Objectives

The objectives of this chapter are to:

- Develop an intuitive understanding of the various physical phenomena associated with external flow such as drag, friction and pressure drag, drag reduction, and lift
- Calculate the drag force associated with flow over common geometries
- Understand the effects of flow regime on the drag coefficients associated with flow over cylinders and spheres
- Understand the fundamentals of flow over airfoils, and calculate the drag and lift forces acting on airfoils

15–1 ▪ INTRODUCTION

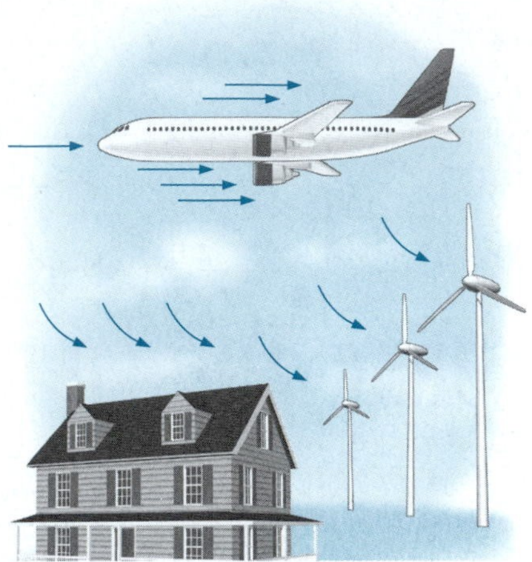

FIGURE 15–1
Flow over bodies is commonly encountered in practice.

Fluid flow over solid bodies frequently occurs in practice, and it is responsible for numerous physical phenomena such as the *drag force* acting on automobiles, power lines, trees, and underwater pipelines; the *lift* developed by airplane wings; *upward draft* of rain, snow, hail, and dust particles in high winds; the transportation of red blood cells by blood flow; the entrainment and disbursement of liquid droplets by sprays; the vibration and noise generated by bodies moving in a fluid; and the power generated by wind turbines (Fig. 15–1). Therefore, developing a good understanding of external flow is important in the design of many engineering systems such as aircraft, automobiles, buildings, ships, submarines, and all kinds of turbines. Late-model cars, for example, have been designed with particular emphasis on aerodynamics. This has resulted in significant reductions in fuel consumption and noise, and considerable improvement in handling.

Sometimes a fluid moves over a stationary body (such as the wind blowing over a building), and other times a body moves through a quiescent fluid (such as a car moving through air). These two seemingly different processes are equivalent to each other; what matters is the relative motion between the fluid and the body. Such motions are conveniently analyzed by fixing the coordinate system on the body and are referred to as **flow over bodies** or **external flow**. The aerodynamic aspects of different airplane wing designs, for example, are studied conveniently in a lab by placing the wings in a wind tunnel and blowing air over them by large fans. Also, a flow can be classified as being steady or unsteady, depending on the reference frame selected. Flow around an airplane, for example, is always unsteady with respect to the ground, but it may be considered to be steady with respect to a frame of reference moving with the airplane at cruise conditions.

The flow fields and geometries for most external flow problems are too complicated to be solved analytically, and thus we have to rely on correlations based on experimental data. The availability of high-speed computers has made it possible to conduct a series of "numerical experiments" quickly by solving the governing equations numerically, and to resort to the expensive and time-consuming testing and experimentation only in the final stages of design. Such testing is done in wind tunnels. H. F. Phillips (1845–1912) built the first wind tunnel in 1894 and measured lift and drag. In this chapter we mostly rely on relations developed experimentally.

The velocity of the fluid approaching a body is called the **free-stream velocity** and is denoted by V. It is also denoted by u_∞ or U_∞ when the flow is aligned with the x-axis since u is used to denote the x-component of velocity. The fluid velocity ranges from zero at the body surface (the no-slip condition) to the free-stream value away from the surface, and the subscript "infinity" serves as a reminder that this is the value at a distance where the presence of the body is not felt. The free-stream velocity may vary with location and time (e.g., the wind blowing past a building). But in the design and analysis, the free-stream velocity is usually assumed to be *uniform* and *steady* for convenience, and this is what we do in this chapter.

The shape of a body has a profound influence on the flow over the body and the velocity field. The flow over a body is said to be **two-dimensional** when the body is very long and of constant cross section and the flow is normal to the body. The wind blowing over a long pipe perpendicular to its axis is an example of two-dimensional flow. Note that the velocity component in the axial direction is zero in this case, and thus the velocity field is two-dimensional.

The two-dimensional idealization is appropriate when the body is sufficiently long so that the end effects are negligible and the approach flow is uniform. Another simplification occurs when the body possesses rotational symmetry about an axis in the flow direction. The flow in this case is also two-dimensional and is said to be **axisymmetric**. The front portion of a bullet piercing through air is an example of axisymmetric flow. The velocity in this case varies with the axial distance x and the radial distance r. Flow over a body that cannot be modeled as two-dimensional or axisymmetric, such as flow over a car, is **three-dimensional** (Fig. 15–2).

Flow over bodies can also be classified as **incompressible flows** (e.g., flows over automobiles, submarines, and buildings) and **compressible flows** (e.g., flows over high-speed aircraft, rockets, and missiles). Compressibility effects are negligible at low velocities (flows with Ma $\lesssim$ 0.3), and such flows can be treated as incompressible with little loss of accuracy.

Bodies subjected to fluid flow are classified as being streamlined or blunt, depending on their overall shape. A body is said to be **streamlined** if a conscious effort is made to align its shape with the anticipated streamlines in the flow. Streamlined bodies such as race cars and airplanes appear to be contoured and sleek. Otherwise, a body (such as a building) tends to block the flow and is said to be **bluff** or **blunt**. Usually it is much easier to force a streamlined body through a fluid, and thus streamlining has been of great importance in the design of vehicles and airplanes (Fig. 15–3).

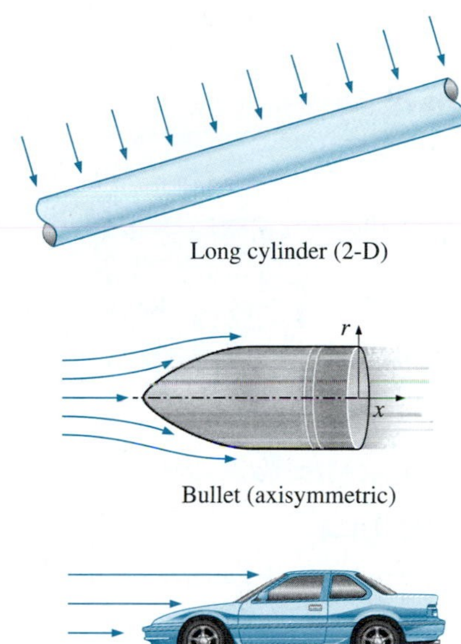

FIGURE 15–2
Two-dimensional, axisymmetric, and three-dimensional flows.

15–2 ▪ DRAG AND LIFT

It is a common experience that a body meets some resistance when it is forced to move through a fluid, especially a liquid. As you may have noticed, it is very difficult to walk in water because of the much greater resistance it offers to motion compared to air. Also, you may have seen high winds knocking down trees, power lines, and even trailers and felt the strong "push" the wind exerts on your body (Fig. 15–4). You experience the same feeling when you extend your arm out of the window of a moving car. A fluid may exert forces and moments on a body in and about various directions. The force a flowing fluid exerts on a body in the flow direction is called **drag**. The drag force can be measured directly by simply attaching the body subjected to fluid flow to a calibrated spring and measuring the displacement in the flow direction (just like measuring weight with a spring scale). More sophisticated drag-measuring devices, called drag balances, use flexible beams fitted with strain gages to measure the drag electronically.

Drag is usually an undesirable effect, like friction, and we do our best to minimize it. Reduction of drag is closely associated with the reduction of fuel consumption in automobiles, submarines, and aircraft; improved safety and durability of structures subjected to high winds; and reduction of noise and vibration. But in some cases drag produces a very beneficial effect and we try to maximize it. Friction, for example, is a "life saver" in the brakes

FIGURE 15–3
It is much easier to force a streamlined body than a blunt body through a fluid.

of automobiles. Likewise, it is the drag that makes it possible for people to parachute, for pollens to fly to distant locations, and for us all to enjoy the waves of the oceans and the relaxing movements of the leaves of trees.

A stationary fluid exerts only normal pressure forces on the surface of a body immersed in it. A moving fluid, however, also exerts tangential shear forces on the surface because of the no-slip condition caused by viscous effects. Both of these forces, in general, have components in the direction of flow, and thus the drag force is due to the combined effects of pressure and wall shear forces in the flow direction. The components of the pressure and wall shear forces in the direction *normal* to the flow tend to move the body in that direction, and their sum is called **lift**.

FIGURE 15–4

High winds knock down trees, power lines, and even people as a result of the drag force.

For two-dimensional flows, the resultant of the pressure and shear forces can be split into two components: one in the direction of flow, which is the drag force, and another in the direction normal to flow, which is the lift, as shown in Fig. 15–5. For three-dimensional flows, there is also a side force component in the direction normal to the page that tends to move the body in that direction.

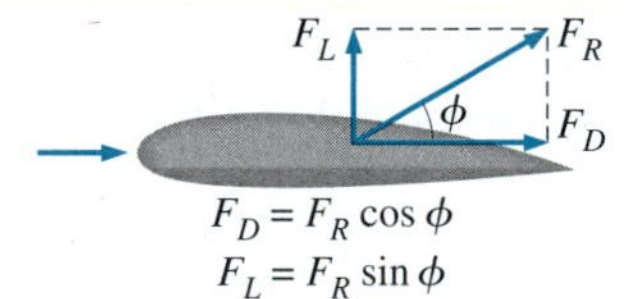

FIGURE 15–5

The resultant of the pressure and viscous forces acting on a two-dimensional body is decomposed into lift and drag forces.

The fluid forces also may generate moments and cause the body to rotate. The moment about the flow direction is called the *rolling moment*, the moment about the lift direction is called the *yawing moment*, and the moment about the side force direction is called the *pitching moment*. For bodies that possess symmetry about the lift–drag plane such as cars, airplanes, and ships, the side force, the yawing moment, and the rolling moment are zero when the wind and wave forces are aligned with the body. What remain for such bodies are the drag and lift forces and the pitching moment. For axisymmetric bodies aligned with the flow, such as a bullet, the only time-averaged force exerted by the fluid on the body is the drag force.

Both skin friction (wall shear) and pressure, in general, contribute to the drag and the lift. In the special case of a thin *flat plate* aligned parallel to the flow direction, the drag force depends on the wall shear only and is independent of pressure. When the flat plate is placed normal to the flow direction, however, the drag force depends on the pressure only and is independent of wall shear since the shear stress in this case acts in the direction normal to flow (Fig. 15–6). If the flat plate is tilted at an angle relative to the flow direction, then the drag force depends on both the pressure and the shear stress.

The wings of airplanes are shaped and positioned specifically to generate lift with minimal drag. This is done by maintaining an angle of attack during cruising, as shown in Fig. 15–7. Both lift and drag are strong functions of the angle of attack, as we discuss later in this chapter. The pressure difference between the top and bottom surfaces of the wing generates an upward force that tends to lift the wing and thus the airplane to which it is connected. For slender bodies such as wings, the shear force acts nearly parallel to the flow direction, and thus its contribution to the lift is small. The drag force for such slender bodies is mostly due to shear forces (the skin friction).

The drag and lift forces depend on the density ρ of the fluid, the upstream velocity V, and the size, shape, and orientation of the body, among other things, and it is not practical to list these forces for a variety of situations. Instead, it is more convenient to work with appropriate dimensionless numbers that represent the drag and lift characteristics of the body. These numbers are the **drag coefficient** C_D, and the **lift coefficient** C_L, and they are defined as

Drag coefficient:
$$C_D = \frac{F_D}{\frac{1}{2}\rho V^2 A} \tag{15–1}$$

Lift coefficient:
$$C_L = \frac{F_L}{\frac{1}{2}\rho V^2 A} \tag{15–2}$$

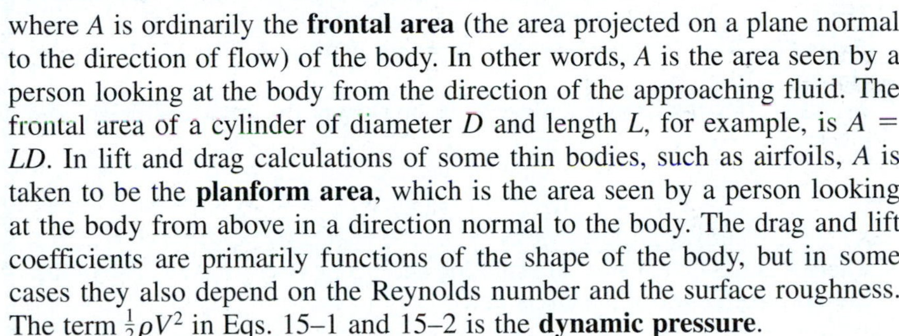

where A is ordinarily the **frontal area** (the area projected on a plane normal to the direction of flow) of the body. In other words, A is the area seen by a person looking at the body from the direction of the approaching fluid. The frontal area of a cylinder of diameter D and length L, for example, is $A = LD$. In lift and drag calculations of some thin bodies, such as airfoils, A is taken to be the **planform area**, which is the area seen by a person looking at the body from above in a direction normal to the body. The drag and lift coefficients are primarily functions of the shape of the body, but in some cases they also depend on the Reynolds number and the surface roughness. The term $\frac{1}{2}\rho V^2$ in Eqs. 15–1 and 15–2 is the **dynamic pressure**.

The local drag and lift coefficients vary along the surface as a result of the changes in the velocity boundary layer in the flow direction. We are usually interested in the drag and lift forces for the *entire* surface, which can be determined using the *average* drag and lift coefficients. Therefore, we present correlations for both local (identified with the subscript x) and average drag and lift coefficients. When relations for local drag and lift coefficients for a surface of length L are available, the *average* drag and lift coefficients for the entire surface can be determined by integration from

$$C_D = \frac{1}{L}\int_0^L C_{D,x}\,dx \tag{15–3}$$

and

$$C_L = \frac{1}{L}\int_0^L C_{L,x}\,dx \tag{15–4}$$

The forces acting on a falling body are usually the drag force, the buoyant force, and the weight of the body. When a body is dropped into the atmosphere or a lake, it first accelerates under the influence of its weight. The motion of the body is resisted by the drag force, which acts in the direction opposite to motion. As the velocity of the body increases, so does the drag force. This continues until all the forces balance each other and the net force acting on the body (and thus its acceleration) is zero. Then the velocity of the body remains constant during the rest of its fall if the properties of the fluid in the path of the body remain essentially constant. This is the maximum velocity a falling body can attain and is called the **terminal velocity** (Fig. 15–8).

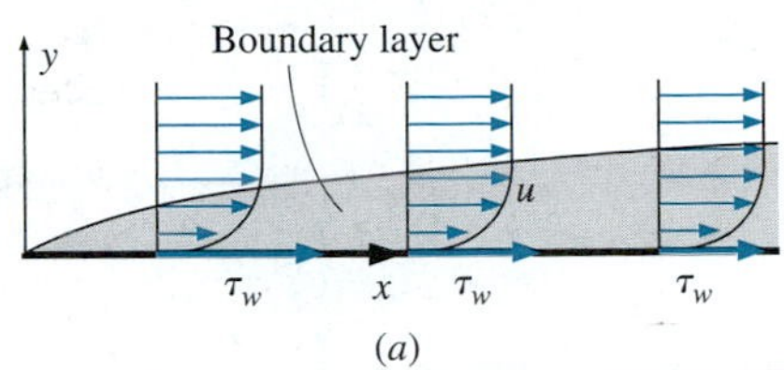

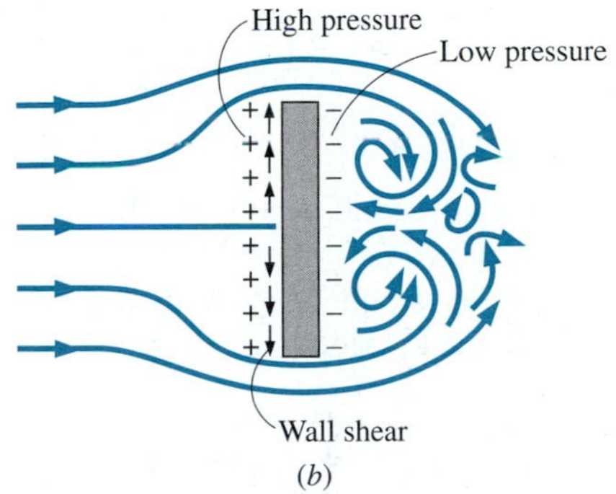

FIGURE 15–6

(*a*) Drag force acting on a flat plate parallel to the flow depends on wall shear only. (*b*) Drag force acting on a flat plate normal to the flow depends on the pressure only and is independent of the wall shear, which acts normal to the free-stream flow.

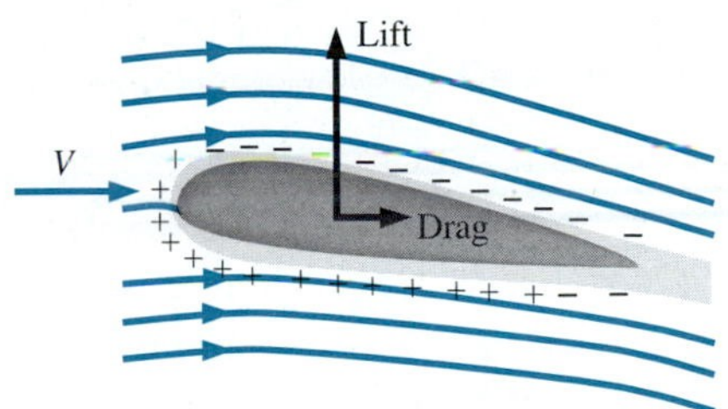

FIGURE 15–7

Airplane wings are shaped and positioned to generate sufficient lift during flight while keeping drag at a minimum. Pressures above and below atmospheric pressure are indicated by plus and minus signs, respectively.

EXAMPLE 15–1 **Measuring the Drag Coefficient of a Car**

The drag coefficient of a car at the design conditions of 1 atm, 70°F, and 60 mi/h is to be determined experimentally in a large wind tunnel in a full-scale test (Fig. 15–9). The frontal area of the car is 22.26 ft^2. If the force

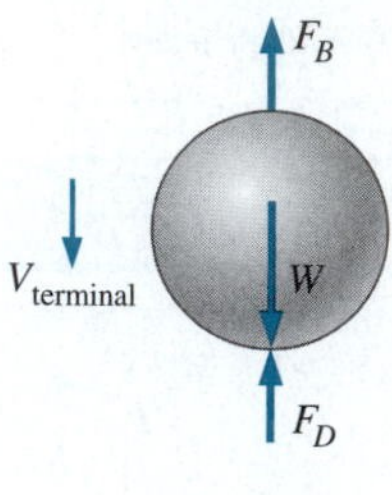

FIGURE 15–8

During a free fall, a body reaches its *terminal velocity* when the drag force equals the weight of the body minus the buoyant force.

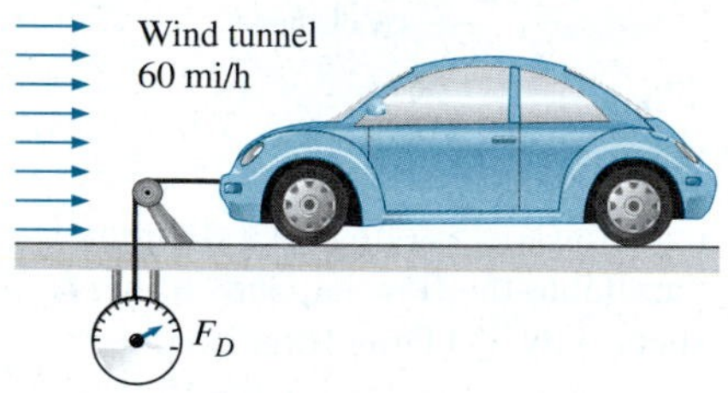

FIGURE 15–9

Schematic for Example 15–1.

acting on the car in the flow direction is measured to be 68 lbf, determine the drag coefficient of this car.

Solution The drag force acting on a car is measured in a wind tunnel. The drag coefficient of the car at test conditions is to be determined.

Assumptions **1** The flow of air is steady and incompressible. **2** The cross section of the tunnel is large enough to simulate free flow over the car. **3** The bottom of the tunnel is also moving at the speed of air to approximate actual driving conditions or this effect is negligible.

Properties The density of air at 1 atm and 70°F is $\rho = 0.07489$ lbm/ft^3 (Table A–22E).

Analysis The drag force acting on a body and the drag coefficient are given by

$$F_D = C_D A \frac{\rho V^2}{2} \quad \text{and} \quad C_D = \frac{2F_D}{\rho A V^2}$$

where A is the frontal area. Substituting and noting that 1 mi/h = 1.467 ft/s, the drag coefficient of the car is determined to be

$$C_D = \frac{2 \times (68 \text{ lbf})}{(0.07489 \text{ lbm/ft}^3)(22.26 \text{ ft}^2)(60 \times 1.467 \text{ ft/s})^2}\left(\frac{32.2 \text{ lbm} \cdot \text{ft/s}^2}{1 \text{ lbf}}\right) = \mathbf{0.34}$$

Discussion Note that the drag coefficient depends on the design conditions, and its value may be different at different conditions such as the Reynolds number. Therefore, the published drag coefficients of different vehicles can be compared meaningfully only if they are determined under similar conditions. This shows the importance of developing standard testing procedures in industry.

15–3 ▪ FRICTION AND PRESSURE DRAG

As mentioned in Section 15–2, the drag force is the net force exerted by a fluid on a body in the direction of flow due to the combined effects of wall shear and pressure forces. It is often instructive to separate the two effects, and study them separately.

The part of drag that is due directly to wall shear stress τ_w is called the **skin friction drag** (or just *friction drag* $F_{D,\text{ friction}}$) since it is caused by frictional effects, and the part that is due directly to pressure P is called the **pressure drag** (also called the *form drag* because of its strong dependence on the form or shape of the body). The friction and pressure drag coefficients are defined as

$$C_{D,\text{ friction}} = \frac{F_{D,\text{ friction}}}{\frac{1}{2}\rho V^2 A} \quad \text{and} \quad C_{D,\text{ pressure}} = \frac{F_{D,\text{ pressure}}}{\frac{1}{2}\rho V^2 A} \tag{15–5}$$

When the friction and pressure drag coefficients or forces are available, the total drag coefficient or drag force is determined by simply adding them,

$$C_D = C_{D,\text{ friction}} + C_{D,\text{ pressure}} \quad \text{and} \quad F_D = F_{D,\text{ friction}} + F_{D,\text{ pressure}} \tag{15–6}$$

The *friction drag* is the component of the wall shear force in the direction of flow, and thus it depends on the orientation of the body as well as the magnitude of the wall shear stress τ_w. The friction drag is *zero* for a flat surface normal to flow, and *maximum* for a flat surface parallel to flow since the friction drag in this case equals the total shear force on the surface. Therefore, for parallel flow over a flat surface, the drag coefficient is equal to the *friction drag coefficient*, or simply the *friction coefficient*. Friction drag is a strong function of viscosity, and increases with increasing viscosity.

The Reynolds number is inversely proportional to the viscosity of the fluid. Therefore, the contribution of friction drag to total drag for blunt bodies is less at higher Reynolds numbers and may be negligible at very high Reynolds numbers. The drag in such cases is mostly due to pressure drag. At low Reynolds numbers, most drag is due to friction drag. This is especially the case for highly streamlined bodies such as airfoils. The friction drag is also proportional to the surface area. Therefore, bodies with a larger surface area experience a larger friction drag. Large commercial airplanes, for example, reduce their total surface area and thus their drag by retracting their wing extensions when they reach cruising altitudes to save fuel. The friction drag coefficient is independent of *surface roughness* in laminar flow, but is a strong function of surface roughness in turbulent flow due to surface roughness elements protruding further into the boundary layer. The *friction drag coefficient* is analogous to the *friction factor* in pipe flow discussed in Chap. 14, and its value depends on the flow regime.

The pressure drag is proportional to the frontal area and to the *difference* between the pressures acting on the front and back of the immersed body. Therefore, the pressure drag is usually dominant for blunt bodies, small for streamlined bodies such as airfoils, and zero for thin flat plates parallel to the flow (Fig. 15–10). The pressure drag becomes most significant when the velocity of the fluid is too high for the fluid to be able to follow the curvature of the body, and thus the fluid *separates* from the body at some point and creates a very low pressure region in the back. The pressure drag in this case is due to the large pressure difference between the front and back sides of the body.

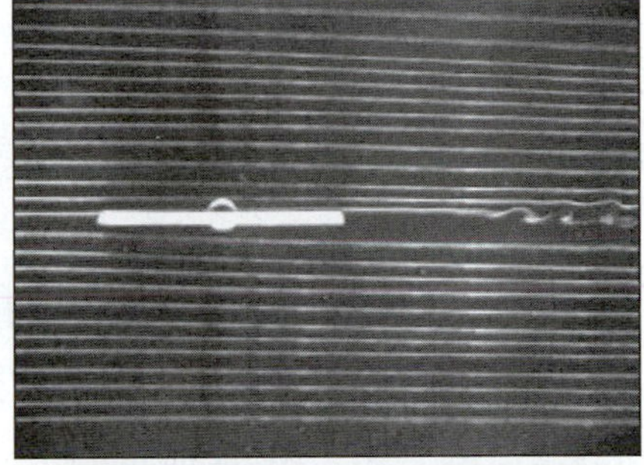

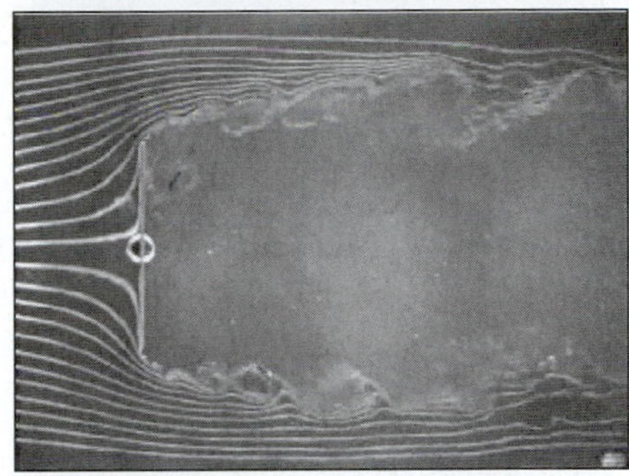

FIGURE 15–10

Drag is due entirely to *friction drag* for a flat plate parallel to the flow; it is due entirely to pressure drag for a flat plate normal to the flow; and it is due to *both* (but mostly *pressure drag*) for a cylinder normal to the flow. The total drag coefficient C_D is lowest for a parallel flat plate, highest for a vertical flat plate, and in between (but close to that of a vertical flat plate) for a cylinder.

From G. M. Homsy, et al. (2000).

Reducing Drag by Streamlining

The first thought that comes to mind to reduce drag is to streamline a body in order to reduce flow separation and thus to reduce pressure drag. Even car salespeople are quick to point out the low drag coefficients of their cars, owing to streamlining. But streamlining has opposite effects on pressure and friction drags. It decreases pressure drag by delaying boundary layer separation and thus reducing the pressure difference between the front and back of the body and increases the friction drag by increasing the surface area. The end result depends on which effect dominates. Therefore, any optimization study to reduce the drag of a body must consider both effects and must attempt to minimize the *sum* of the two, as shown in Fig. 15–11. The minimum total drag occurs at $D/L = 0.25$ for the case shown in Fig. 15–11. For the case of a circular cylinder with the same thickness as the streamlined

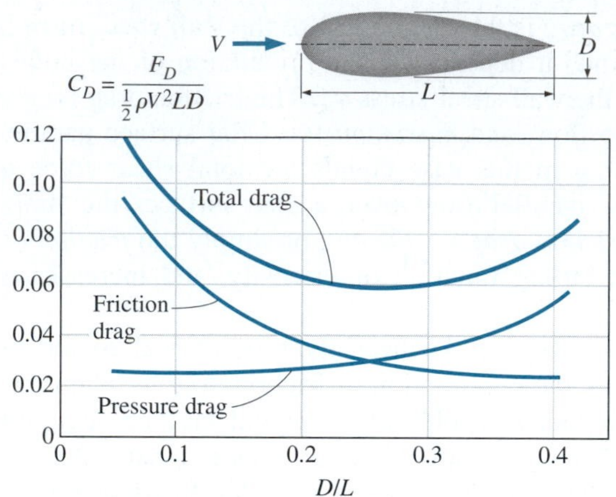

FIGURE 15–11

The variation of friction, pressure, and total drag coefficients of a streamlined strut with thickness-to-chord length ratio for Re $= 4 \times 10^4$. Note that C_D for airfoils and other thin bodies is based on *planform* area rather than frontal area.

From Abbott and von Doenhoff (1959).

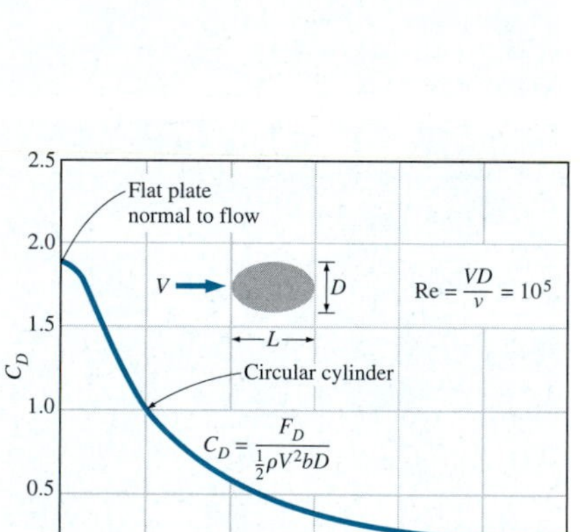

FIGURE 15–12

The variation of the drag coefficient of a long elliptical cylinder with aspect ratio. Here C_D is based on the frontal area bD where b is the width of the body.

From Blevins (1984).

shape of Fig. 15–11, the drag coefficient would be about five times as much. Therefore, it is possible to reduce the drag of a cylindrical component to one-fifth by the use of proper fairings.

The effect of streamlining on the drag coefficient is described best by considering long elliptical cylinders with different aspect (or length-to-thickness) ratios L/D, where L is the length in the flow direction and D is the thickness, as shown in Fig. 15–12. Note that the drag coefficient decreases drastically as the ellipse becomes slimmer. For the special case of $L/D = 1$ (a circular cylinder), the drag coefficient is $C_D \cong 1$ at this Reynolds number. As the aspect ratio is decreased and the cylinder resembles a flat plate, the drag coefficient increases to 1.9, the value for a flat plate normal to flow. Note that the curve becomes nearly flat for aspect ratios greater than about 4. Therefore, for a given diameter D, elliptical shapes with an aspect ratio of about $L/D \cong 4$ usually offer a good compromise between the total drag coefficient and length L. The reduction in the drag coefficient at high aspect ratios is primarily due to the boundary layer staying attached to the surface longer and the resulting pressure recovery. The friction drag on an elliptical cylinder with an aspect ratio of 4 is negligible (less than 2 percent of total drag at this Reynolds number).

As the aspect ratio of an elliptical cylinder is increased by flattening it (i.e., decreasing D while holding L constant), the drag coefficient starts increasing and tends to infinity as $L/D \rightarrow \infty$ (i.e., as the ellipse resembles a flat plate parallel to flow). This is due to the frontal area, which appears in the denominator in the definition of C_D, approaching zero. It does not mean that the drag force increases drastically (actually, the drag force decreases) as the body becomes flat. This shows that the frontal area is inappropriate for use in the drag force relations for slim bodies such as thin airfoils and flat plates. In such cases, the drag coefficient is defined on the basis of the *planform area*, which is simply the surface area (of the top or bottom surface) for a flat plate parallel to flow. This is quite appropriate since for slim bodies the drag is almost entirely due to friction drag, which is proportional to the surface area.

Streamlining has the added benefit of *reducing vibration and noise*. Streamlining should be considered only for blunt bodies that are subjected to high-velocity fluid flow (and thus high Reynolds numbers) for which flow separation is a real possibility. It is not necessary for bodies that typically involve low Reynolds number flows (e.g., creeping flows in which $\text{Re} < 1$) since the drag in those cases is almost entirely due to friction drag, and streamlining will only increase the surface area and thus the total drag. Therefore, careless streamlining may actually increase drag instead of decreasing it.

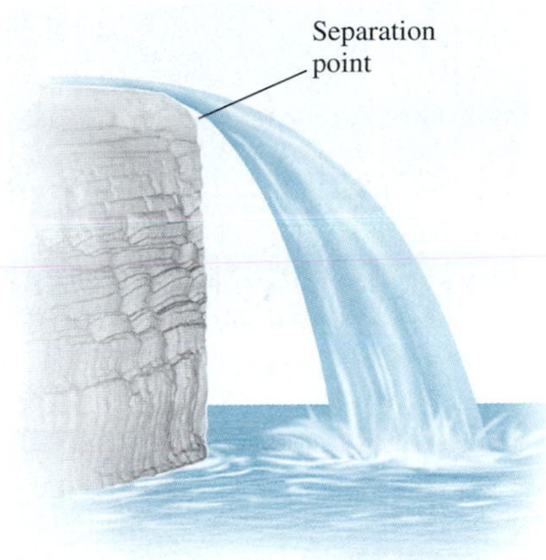

FIGURE 15–13

Flow separation in a waterfall.

Flow Separation

When driving on country roads, it is a common safety measure to slow down at sharp turns in order to avoid being thrown off the road. Many drivers have learned the hard way that a car will refuse to comply when forced to turn curves at excessive speeds. We can view this phenomenon as "the separation of cars" from roads. This phenomenon is also observed when fast vehicles jump off hills. At low velocities, the wheels of the vehicle always remain in contact with the road surface. But at high velocities, the vehicle is too fast to follow the curvature of the road and takes off at the hill, losing contact with the road.

A fluid acts much the same way when forced to flow over a curved surface at high velocities. A fluid follows the front portion of the curved surface with no problem, but it has difficulty remaining attached to the surface on the back side. At sufficiently high velocities, the fluid stream detaches itself from the surface of the body. This is called **flow separation** (Fig. 15–13). Flow can separate from a surface even if it is fully submerged in a liquid or immersed in a gas (Fig. 15–14). The location of the separation point depends on several factors such as the Reynolds number, the surface roughness, and the level of fluctuations in the free stream, and it is usually difficult to predict exactly where separation will occur unless there are sharp corners or abrupt changes in the shape of the solid surface.

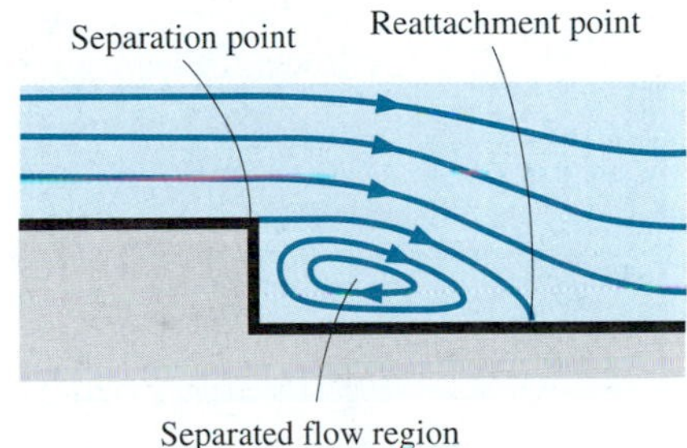

FIGURE 15–14

Flow separation over a backward-facing step along a wall.

When a fluid separates from a body, it forms a separated region between the body and the fluid stream. This low-pressure region behind the body where recirculating and backflows occur is called the **separated region**. The larger the separated region, the larger the pressure drag. The effects of flow separation are felt far downstream in the form of reduced velocity (relative to the upstream velocity). The region of flow trailing the body where the effects of the body on velocity are felt is called the **wake** (Fig. 15–15). The separated region comes to an end when the two flow streams reattach. Therefore, the separated region is an enclosed volume, whereas the wake keeps growing behind the body until the fluid in the wake region regains its velocity and the velocity profile becomes nearly flat again. Viscous and rotational effects are the most significant in the boundary layer, the separated region, and the wake.

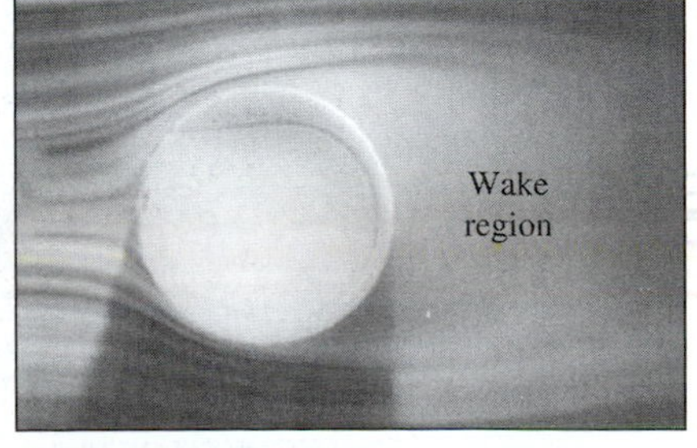

FIGURE 15–15

Flow separation and the wake region during flow over a tennis ball.

Courtesy NASA and Cislunar Aerospace, Inc.

The occurrence of separation is not limited to blunt bodies. Complete separation over the entire back surface may also occur on a streamlined body such as an airplane wing at a sufficiently large **angle of attack** (larger than

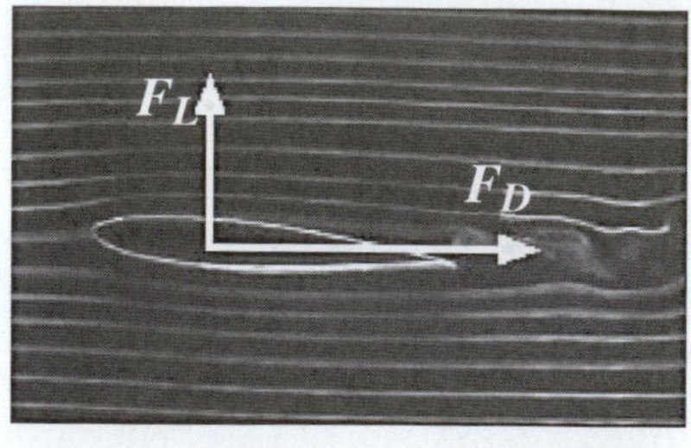

(*a*) 5°

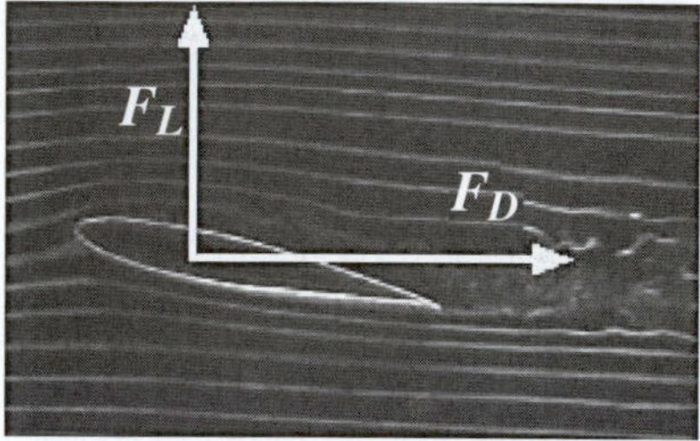

(*b*) 15°

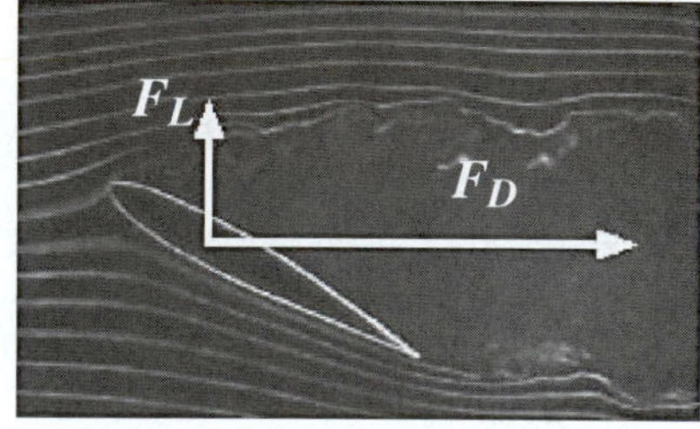

(*c*) 30°

FIGURE 15–16

At large angles of attack (usually larger than 15°), flow may separate completely from the top surface of an airfoil, reducing lift drastically and causing the airfoil to stall.

From G. M. Homsy, et al. (2000).

about 15° for most airfoils), which is the angle the incoming fluid stream makes with the **chord** (the line that connects the nose and the end) of the wing. Flow separation on the top surface of a wing reduces lift drastically and may cause the airplane to **stall**. Stalling has been blamed for many airplane accidents and loss of efficiencies in turbomachinery (Fig. 15–16).

Note that drag and lift are strongly dependent on the shape of the body, and any effect that causes the shape to change has a profound effect on the drag and lift. For example, snow accumulation and ice formation on airplane wings may change the shape of the wings sufficiently to cause significant loss of lift. This phenomenon has caused many airplanes to lose altitude and crash and many others to abort takeoff. Therefore, it has become a routine safety measure to check for ice or snow buildup on critical components of airplanes before takeoff in bad weather. This is especially important for airplanes that have waited a long time on the runway before takeoff because of heavy traffic.

An important consequence of flow separation is the formation and shedding of circulating fluid structures, called **vortices**, in the wake region. The periodic generation of these vortices downstream is referred to as **vortex shedding**. This phenomenon usually occurs during normal flow over long cylinders or spheres for Re $\gtrsim$ 90. The vibrations generated by vortices near the body may cause the body to resonate to dangerous levels if the frequency of the vortices is close to the natural frequency of the body—a situation that must be avoided in the design of equipment that is subjected to high-velocity fluid flow such as the wings of airplanes and suspended bridges subjected to steady high winds.

15–4 ▪ DRAG COEFFICIENTS OF COMMON GEOMETRIES

The concept of drag has important consequences in daily life, and the drag behavior of various natural and human-made bodies is characterized by their drag coefficients measured under typical operating conditions. Although drag is caused by two different effects (friction and pressure), it is usually difficult to determine them separately. Besides, in most cases, we are interested in the *total* drag rather than the individual drag components, and thus usually the *total* drag coefficient is reported. The determination of drag coefficients has been the topic of numerous studies (mostly experimental), and there is a huge amount of drag coefficient data in the literature for just about any geometry of practical interest.

The drag coefficient, in general, depends on the *Reynolds number*, especially for Reynolds numbers below about 10^4. At higher Reynolds numbers, the drag coefficients for many geometries remain essentially constant (Fig. 15–17). This is due to the flow at high Reynolds numbers becoming fully turbulent. However, this is not the case for rounded bodies such as circular cylinders and spheres, as we discuss later in this section. The reported drag coefficients are usually applicable only to flows at high Reynolds numbers.

The drag coefficient exhibits different behavior in the low (creeping), moderate (laminar), and high (turbulent) regions of the Reynolds number. The inertia effects are negligible in low Reynolds number flows (Re < 1), called *creeping flows*, and the fluid wraps around the body smoothly. The drag coefficient in this case is inversely proportional to the Reynolds number, and for a sphere it is determined to be

Sphere: $$C_D = \frac{24}{\text{Re}} \qquad (\text{Re} \lesssim 1) \tag{15-7}$$

Then the drag force acting on a spherical object at low Reynolds numbers becomes

$$F_D = C_D A \frac{\rho V^2}{2} = \frac{24}{\text{Re}} A \frac{\rho V^2}{2} = \frac{24}{\rho V D/\mu} \frac{\pi D^2}{4} \frac{\rho V^2}{2} = 3\pi\mu V D \tag{15-8}$$

which is known as **Stokes law**, after British mathematician and physicist G. G. Stokes (1819–1903). This relation shows that at very low Reynolds numbers, the drag force acting on spherical objects is proportional to the diameter, the velocity, and the viscosity of the fluid. This relation is often applicable to dust particles in the air and suspended solid particles in water.

The drag coefficients for low Reynolds number flows past some other geometries are given in Fig. 15–18. Note that at low Reynolds numbers, the shape of the body does not have a major influence on the drag coefficient.

The drag coefficients for various two- and three-dimensional bodies are given in Tables 15–1 and 15–2 for large Reynolds numbers. We make several observations from these tables about the drag coefficient at high Reynolds numbers. First of all, the *orientation* of the body relative to the direction of flow has a major influence on the drag coefficient. For example, the drag coefficient for flow over a hemisphere is 0.4 when the spherical

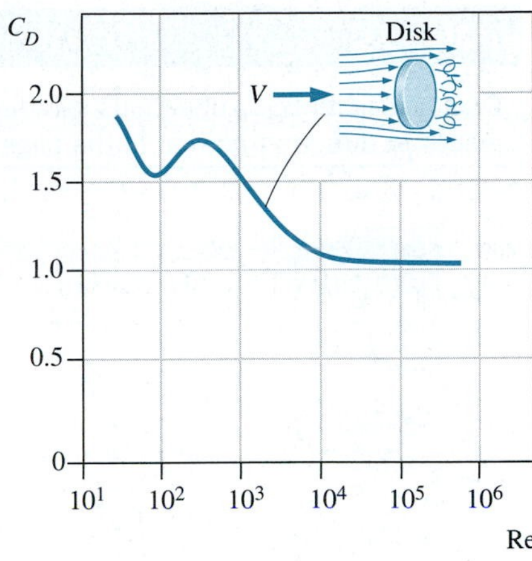

FIGURE 15–17

The drag coefficient for many (but not all) geometries remains essentially constant at Reynolds numbers above about 10^4.

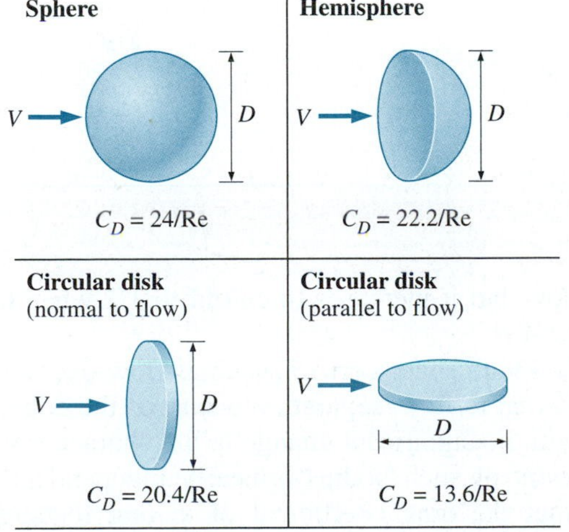

FIGURE 15–18

Drag coefficients C_D for creeping flow at low Reynolds number (Re ≤ 1 where Re $= VD/\nu$ and $A = \pi D^2/4$).

TABLE 15–1

Drag coefficients C_D of various two-dimensional bodies for Re > 10^4 based on the frontal area $A = bD$, where b is the length in direction normal to the page (for use in the drag force relation $F_D = C_D A \rho V^2/2$ where V is the upstream velocity)

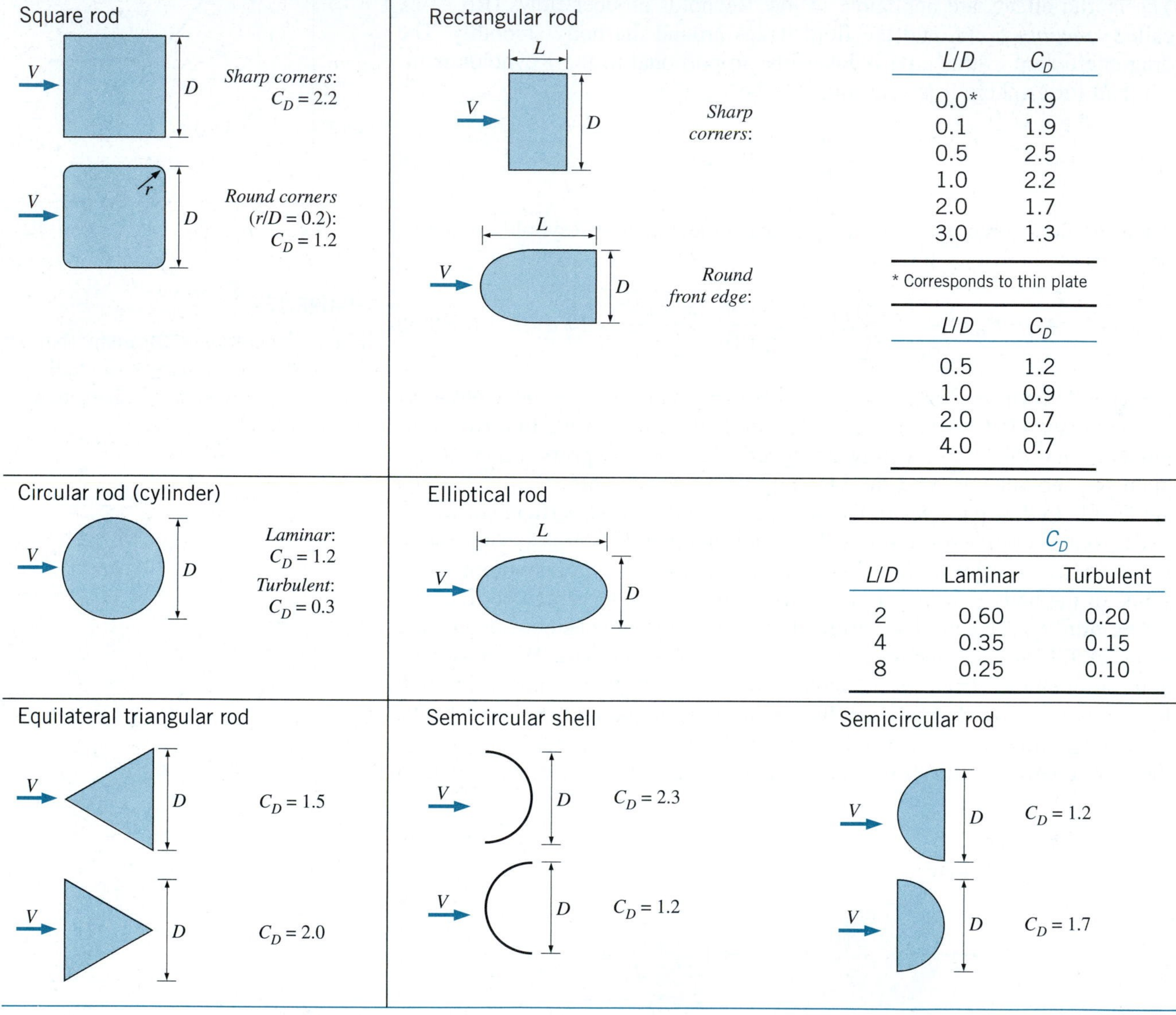

Rectangular rod, sharp corners:

L/D	C_D
0.0*	1.9
0.1	1.9
0.5	2.5
1.0	2.2
2.0	1.7
3.0	1.3

* Corresponds to thin plate

Rectangular rod, round front edge:

L/D	C_D
0.5	1.2
1.0	0.9
2.0	0.7
4.0	0.7

Elliptical rod:

L/D	C_D Laminar	C_D Turbulent
2	0.60	0.20
4	0.35	0.15
8	0.25	0.10

side faces the flow, but it increases threefold to 1.2 when the flat side faces the flow (Fig. 15–19).

For blunt bodies with sharp corners, such as flow over a rectangular block or a flat plate normal to flow, separation occurs at the edges of the front and back surfaces, with no significant change in the character of flow. Therefore, the drag coefficient of such bodies is nearly independent of the Reynolds number. Note that the drag coefficient of a long rectangular rod can be reduced almost by half from 2.2 to 1.2 by rounding the corners.

TABLE 15–2

Representative drag coefficients C_D for various three-dimensional bodies for Re > 10^4 based on the frontal area (for use in the drag force relation $F_D = C_D A \rho V^2/2$ where V is the upstream velocity)

Cube, $A = D^2$

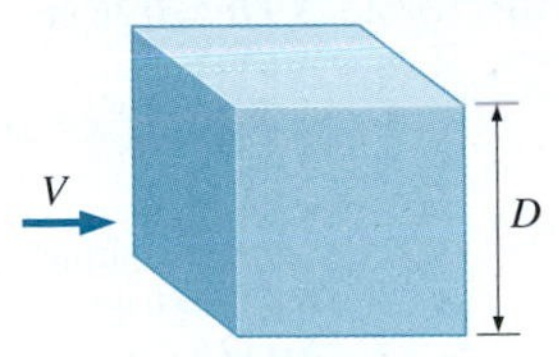

Thin circular disk, $A = \pi D^2/4$

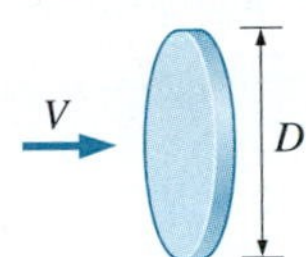

Cone (for $\theta = 30°$), $A = \pi D^2/4$

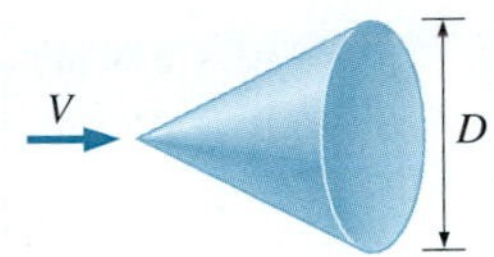

Sphere, $A = \pi D^2/4$

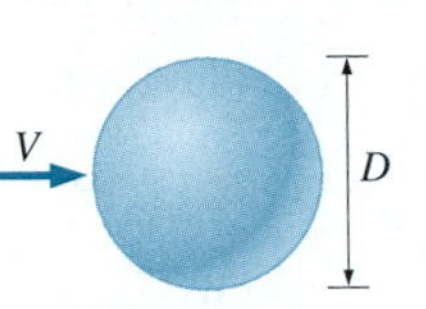

Laminar: Re $\lesssim 2 \times 10^5$, $C_D = 0.5$

Turbulent: Re $\gtrsim 2 \times 10^6$, $C_D = 0.2$

Ellipsoid, $A = \pi D^2/4$

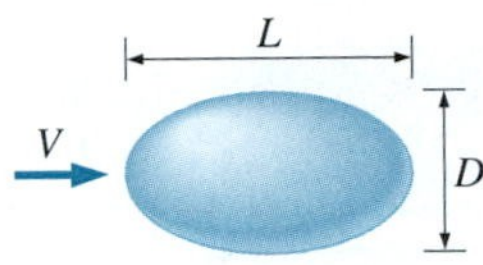

L/D	C_D Laminar Re $\lesssim 2 \times 10^5$	C_D Turbulent Re $\gtrsim 2 \times 10^6$
0.75	0.5	0.2
1	0.5	0.2
2	0.3	0.1
4	0.3	0.1
8	0.2	0.1

Hemisphere, $A = \pi D^2/4$

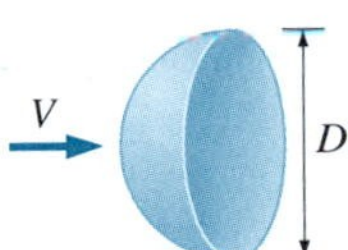

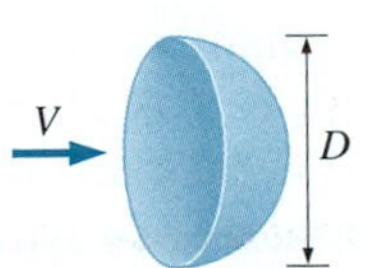

Short cylinder, vertical, $A = LD$

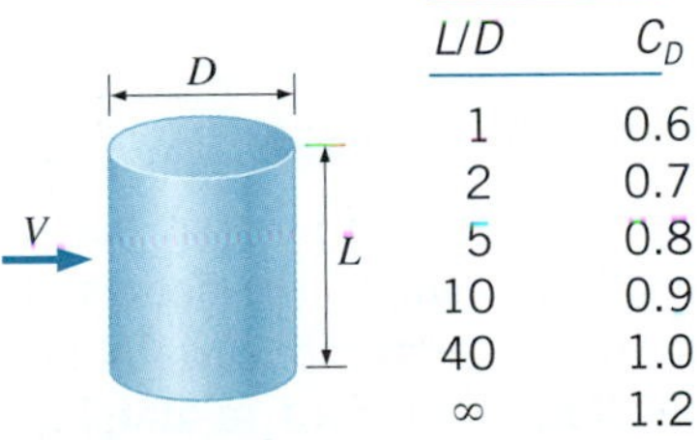

L/D	C_D
1	0.6
2	0.7
5	0.8
10	0.9
40	1.0
∞	1.2

Values are for laminar flow (Re $\lesssim 2 \times 10^5$)

Short cylinder, horizontal, $A = \pi D^2/4$

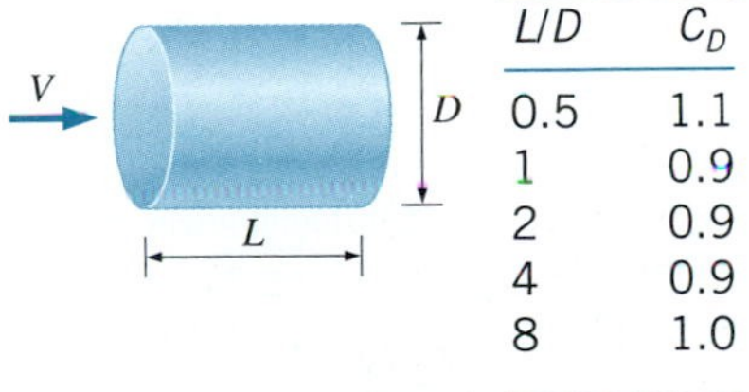

L/D	C_D
0.5	1.1
1	0.9
2	0.9
4	0.9
8	1.0

Streamlined body, $A = \pi D^2/4$

Parachute, $A = \pi D^2/4$

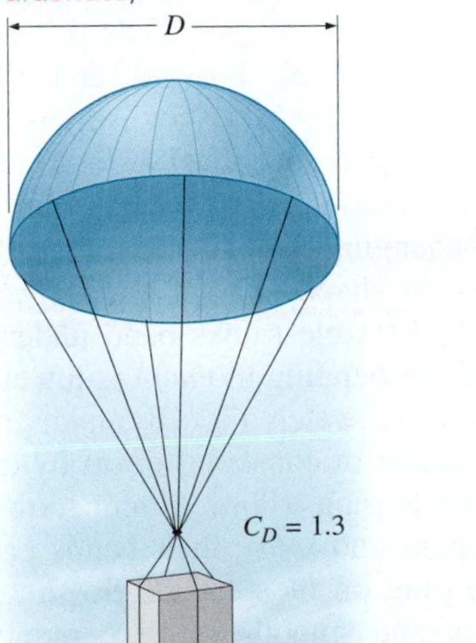

Tree, A = frontal area

V, m/s	C_D
10	0.4–1.2
20	0.3–1.0
30	0.2–0.7

TABLE 15–2 (Continued)

Person (average)	Bikes	
Standing: C_DA = 9 ft^2 = 0.84 m^2 Sitting: C_DA = 6 ft^2 = 0.56 m^2	Upright: A = 5.5 ft^2 = 0.51 m^2, C_D = 1.1 Racing: A = 3.9 ft^2 = 0.36 m^2, C_D = 0.9	Drafting: C_D = 0.5, C_D = 0.9; A = 3.9 ft^2 = 0.36 m^2, C_D = 0.50 With fairing: A = 5.0 ft^2 = 0.46 m^2, C_D = 0.12
Semitrailer, A = frontal area	**Automotive, A = frontal area**	**High-rise buildings, A = frontal area**
Without fairing: C_D = 0.96 With fairing: C_D = 0.76	Minivan: C_D = 0.4 Passenger car: C_D = 0.3	$C_D \approx$ 1.0 to 1.4

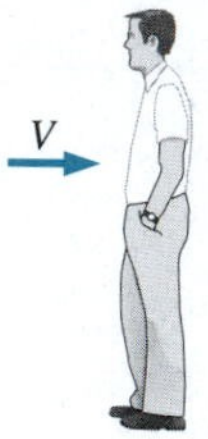

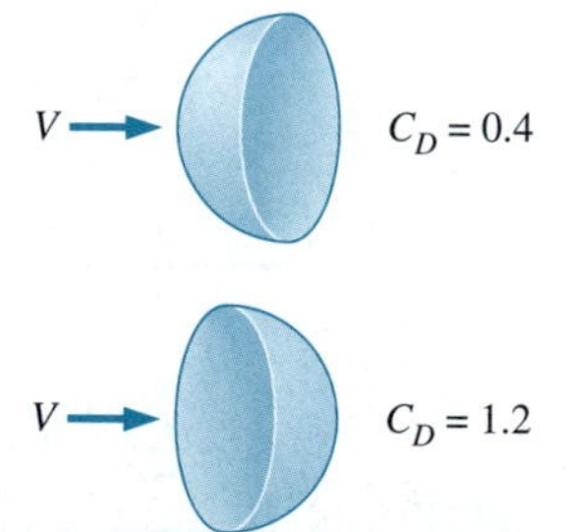

FIGURE 15–19

The drag coefficient of a body may change drastically by changing the body's orientation (and thus shape) relative to the direction of flow.

Biological Systems and Drag

The concept of drag also has important consequences for biological systems. For example, the bodies of *fish*, especially the ones that swim fast for long distances (such as dolphins), are highly streamlined to minimize drag (the drag coefficient of dolphins based on the wetted skin area is about 0.0035, comparable to the value for a flat plate in turbulent flow). So it is no surprise that we build submarines that mimic large fish. Tropical fish with fascinating beauty and elegance, on the other hand, swim short distances only. Obviously grace, not high speed and drag, was the primary consideration in their design. Birds teach us a lesson on drag reduction by extending their beak forward and folding their feet backward during flight (Fig. 15–20). Airplanes, which look somewhat like big birds, retract their wheels after takeoff in order to reduce drag and thus fuel consumption.

The flexible structure of plants enables them to reduce drag at high winds by changing their shapes. Large flat leaves, for example, curl into a low-drag conical shape at high wind speeds, while tree branches cluster to reduce drag. Flexible trunks bend under the influence of the wind to reduce drag, and the bending moment is lowered by reducing frontal area.

If you watch the Olympic games, you have probably observed many instances of conscious effort by the competitors to reduce drag. Some examples: During 100-m running, the runners hold their fingers together and straight and move their hands parallel to the direction of motion to reduce the drag on their hands. Swimmers with long hair cover their head with a tight and smooth cover to reduce head drag. They also wear well-fitting one-piece swimming suits. Horse and bicycle riders lean forward as much as they can to reduce drag (by reducing both the drag coefficient and frontal area). Speed skiers do the same thing.

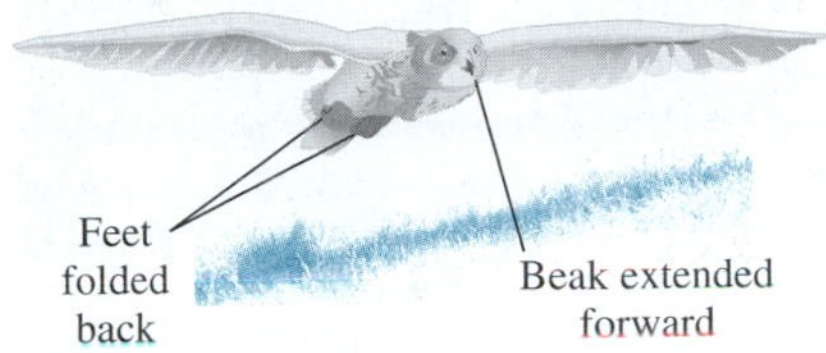

FIGURE 15–20

Birds teach us a lesson on drag reduction by extending their beak forward and folding their feet backward during flight.

Drag Coefficients of Vehicles

The term *drag coefficient* is commonly used in various areas of daily life. Car manufacturers try to attract consumers by pointing out the *low drag coefficients* of their cars (Fig. 15–21). The drag coefficients of vehicles range from about 1.0 for large semitrailers to 0.4 for minivans and 0.3 for passenger cars. In general, the more blunt the vehicle, the higher the drag coefficient. Installing a fairing reduces the drag coefficient of tractor-trailer rigs by about 20 percent by making the frontal surface more streamlined. As a rule of thumb, the percentage of fuel savings due to reduced drag is about half the percentage of drag reduction at highway speeds.

When the effect of the road on air motion is disregarded, the ideal shape of a *vehicle* is the basic *teardrop*, with a drag coefficient of about 0.1 for the turbulent flow case. But this shape needs to be modified to accommodate several necessary external components such as wheels, mirrors, axles, and door handles. Also, the vehicle must be high enough for comfort and there must be a minimum clearance from the road. Further, a vehicle cannot be too long to fit in garages and parking spaces. Controlling the material and manufacturing costs requires minimizing or eliminating any "dead" volume that cannot be utilized. The result is a shape that resembles more a box than a teardrop, and this was the shape of early cars with a drag coefficient of about 0.8 in the 1920s. This wasn't a problem in those days since the velocities were low and drag was not a major design consideration.

FIGURE 15–21

This sleek-looking Toyota Prius has a drag coefficient of 0.26—one of the lowest for a passenger car.

Courtesy Toyota.

The average drag coefficients of cars dropped to about 0.70 in the 1940s, to 0.55 in the 1970s, to 0.45 in the 1980s, and to 0.30 in the 1990s as a result of improved manufacturing techniques for metal forming and paying more attention to the shape of the car and streamlining (Fig. 15–22). The drag coefficient for well-built racing cars is about 0.2, but this is achieved after making the comfort of drivers a secondary consideration. Noting that the theoretical lower limit of C_D is about 0.1 and the value for racing cars is 0.2, it appears that there is only a little room for further improvement in the drag coefficient of passenger cars from the current value of about 0.3. For trucks and buses, the drag coefficient can be reduced further by optimizing the front and rear contours (by rounding, for example) to the extent it is practical while keeping the overall length of the vehicle the same.

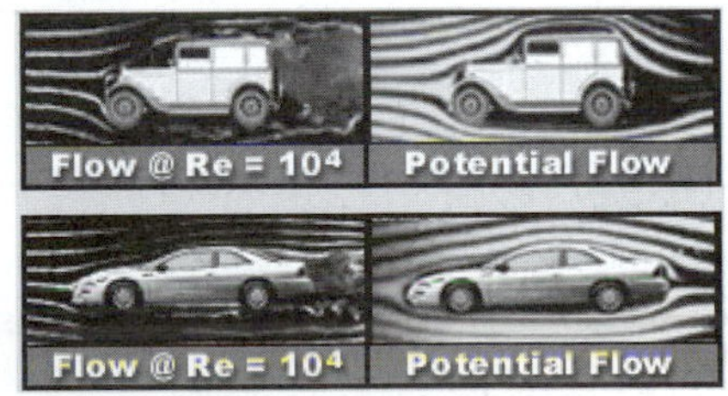

FIGURE 15–22

Streamlines around an aerodynamically designed modern car closely resemble the streamlines around the car in the ideal potential flow (assumes negligible friction), except near the rear end, resulting in a low drag coefficient.

From G. M. Homsy, et al. (2000).

When traveling as a group, a sneaky way of reducing drag is **drafting**, a phenomenon well known by bicycle riders and car racers. It involves approaching a moving body from behind and being *drafted* into the low-pressure region in the rear of the body. The drag coefficient of a racing bicyclist, for example, can be reduced from 0.9 to 0.5 (Table 15–2) by drafting, as also shown in Fig. 15–23.

We also can help reduce the overall drag of a vehicle and thus fuel consumption by being more conscientious drivers. For example, drag force is proportional to the square of velocity. Therefore, driving over the speed limit on the highways not only increases the chances of getting speeding tickets or getting into an accident, but it also increases the amount of fuel consumption per mile. Therefore, driving at moderate speeds is safe and economical. Also, anything that extends from the car, even an arm, increases the drag coefficient. Driving with the windows rolled down also increases the drag and fuel

FIGURE 15–23
The drag coefficients of bodies following other moving bodies closely can be reduced considerably due to drafting (i.e., falling into the low pressure region created by the body in front).
© Karl Weatherly/Getty Images

consumption. At highway speeds, a driver can often save fuel in hot weather by running the air conditioner instead of driving with the windows rolled down. For many low-drag automobiles, the turbulence and additional drag generated by open windows consume more fuel than does the air conditioner.

Superposition

The shapes of many bodies encountered in practice are not simple. But such bodies can be treated conveniently in drag force calculations by considering them to be composed of two or more simple bodies. A satellite dish mounted on a roof with a cylindrical bar, for example, can be considered to be a combination of a hemispherical body and a cylinder. Then the drag coefficient of the body can be determined approximately by using **superposition**. Such a simplistic approach does not account for the effects of components on each other, and thus the results obtained should be interpreted accordingly.

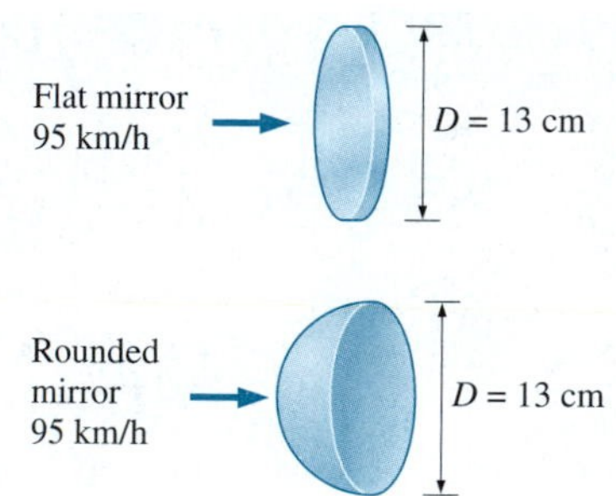

FIGURE 15–24
Schematic for Example 15–2.

EXAMPLE 15–2 Effect of Mirror Design on the Fuel Consumption of a Car

As part of the continuing efforts to reduce the drag coefficient and thus to improve the fuel efficiency of cars, the design of side rearview mirrors has changed drastically from a simple circular plate to a streamlined shape. Determine the amount of fuel and money saved per year as a result of replacing a 13-cm-diameter flat mirror by one with a hemispherical back (Fig. 15–24). Assume the car is driven 24,000 km a year at an average speed of 95 km/h. Take the density and price of gasoline to be 0.8 kg/L and $0.80/L, respectively; the heating value of gasoline to be 44,000 kJ/kg; and the overall efficiency of the engine to be 30 percent.

Solution The flat mirror of a car is replaced by one with a hemispherical back. The amount of fuel and money saved per year as a result are to be determined.

Assumptions **1** The car is driven 24,000 km a year at an average speed of 95 km/h. **2** The effect of the car body on the flow around the mirror is negligible (no interference).

Properties The densities of air and gasoline are taken to be 1.20 kg/m^3 and 800 kg/m^3, respectively. The heating value of gasoline is given to be 44,000 kJ/kg. The drag coefficients C_D are 1.1 for a circular disk and 0.4 for a hemispherical body.

Analysis The drag force acting on a body is determined from

$$F_D = C_D A \frac{\rho V^2}{2}$$

where A is the frontal area of the body, which is $A = \pi D^2/4$ for both the flat and rounded mirrors. The drag force acting on the flat mirror is

$$F_D = 1.1 \frac{\pi(0.13\text{ m})^2}{4} \frac{(1.20\text{ kg/m}^3)(95\text{ km/h})^2}{2} \left(\frac{1\text{ m/s}}{3.6\text{ km/h}}\right)^2 \left(\frac{1\text{ N}}{1\text{ kg}\cdot\text{m/s}^2}\right) = 6.10\text{ N}$$

Noting that work is force times distance, the amount of work done to overcome this drag force and the required energy input for a distance of 24,000 km are

$$W_{\text{drag}} = F_D \times L = (6.10\text{ N})(24{,}000\text{ km/year}) = 146{,}400\text{ kJ/year}$$

$$E_{\text{in}} = \frac{W_{\text{drag}}}{\eta_{\text{car}}} = \frac{146{,}400\text{ kJ/year}}{0.3} = 488{,}000\text{ kJ/year}$$

Then the amount and costs of the fuel that supplies this much energy are

$$\text{Amount of fuel} = \frac{m_{\text{fuel}}}{\rho_{\text{fuel}}} = \frac{E_{\text{in}}/\text{HV}}{\rho_{\text{fuel}}} = \frac{(488{,}000\ \text{kJ/year})/(44{,}000\ \text{kJ/kg})}{0.8\ \text{kg/L}}$$

$$= 13.9\ \text{L/year}$$

$$\text{Cost} = (\text{Amount of fuel})(\text{Unit cost}) = (13.9\ \text{L/year})(\$0.80/\text{L}) = \$11.10/\text{year}$$

That is, the car uses 13.9 L of gasoline at a cost of $11.10 per year (to 3 significant digits) to overcome the drag generated by a flat mirror extending out from the side of a car.

The drag force and the work done to overcome it are directly proportional to the drag coefficient. Then the percent reduction in the fuel consumption due to replacing the mirror is equal to the percent reduction in the drag coefficient:

$$\text{Reduction ratio} = \frac{C_{D,\ \text{flat}} - C_{D,\ \text{hemisp}}}{C_{D,\ \text{flat}}} = \frac{1.1 - 0.4}{1.1} = 0.636$$

$$\text{Fuel reduction} = (\text{Reduction ratio})(\text{Amount of fuel})$$

$$= 0.636(13.9\ \text{L/year}) = \mathbf{8.82\ L/year}$$

$$\text{Cost reduction} = (\text{Reduction ratio})(\text{Cost}) = 0.636(\$11.10/\text{year}) = \mathbf{\$7.06/year}$$

Since a typical car has two side rearview mirrors, the driver saves more than $14 per year in gasoline by replacing the flat mirrors with hemispherical ones.

Discussion Note from this example that significant reductions in drag and fuel consumption can be achieved by streamlining the shape of various components and the entire car. So it is no surprise that the sharp corners are replaced in late model cars by rounded contours. This also explains why large airplanes retract their wheels after takeoff and small airplanes use contoured fairings around their wheels.

Example 15–2 is indicative of the tremendous amount of effort put into redesigning various parts of cars such as the window moldings, the door handles, the windshield, and the front and rear ends in order to reduce aerodynamic drag. For a car moving on a level road at constant speed, the power developed by the engine is used to overcome rolling resistance, friction between moving components, aerodynamic drag, and driving the auxiliary equipment. The aerodynamic drag is negligible at low speeds, but becomes significant at speeds above about 30 mi/h. Reduction of the frontal area of the cars (to the dislike of tall drivers) has also contributed greatly to the reduction of drag and fuel consumption.

15–5 ▪ PARALLEL FLOW OVER FLAT PLATES

Consider the flow of a fluid over a *flat plate*, as shown in Fig. 15–25. Surfaces that are slightly contoured such as turbine blades can also be approximated as flat plates with reasonable accuracy. The x-coordinate is measured along the plate surface from the *leading edge* of the plate in the direction of the flow, and y is measured from the surface in the normal direction. The fluid approaches the plate in the x-direction with a uniform velocity V, which is equivalent to the velocity over the plate away from the surface.

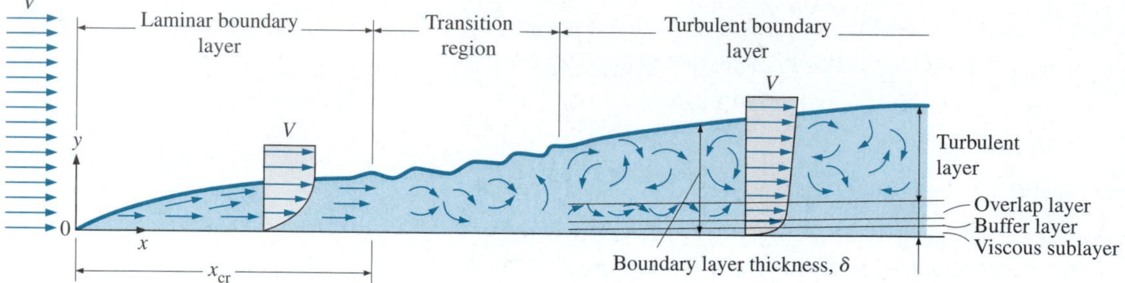

FIGURE 15–25

The development of the boundary layer for flow over a flat plate, and the different flow regimes. Not to scale.

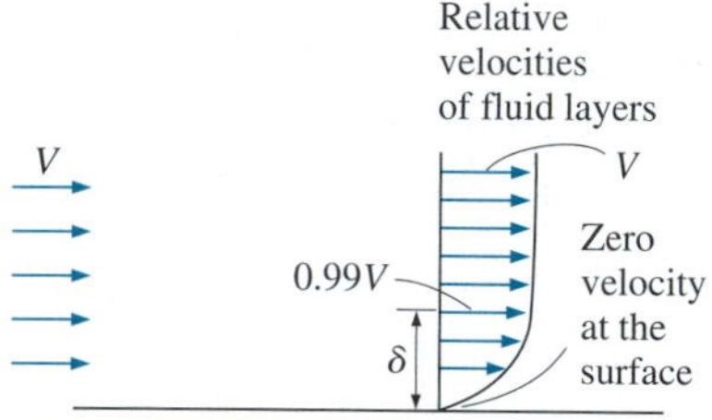

FIGURE 15–26

The development of a boundary layer on a surface is due to the no-slip condition and friction.

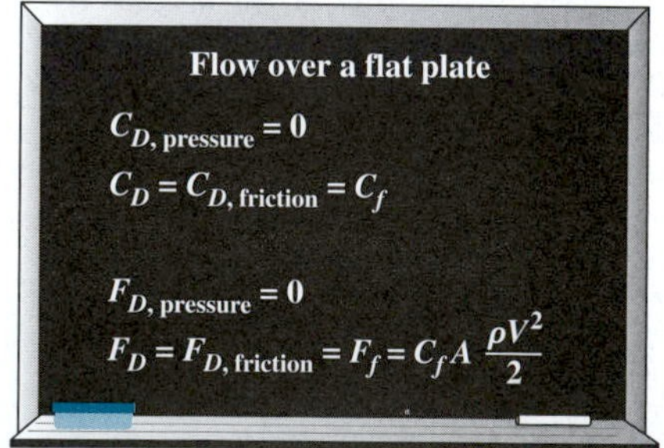

FIGURE 15–27

For parallel flow over a flat plate, the pressure drag is zero, and thus the drag coefficient is equal to the friction coefficient and the drag force is equal to the friction force.

For the sake of discussion, we consider the fluid to consist of adjacent layers piled on top of each other. The velocity of the particles in the first fluid layer adjacent to the plate becomes zero because of the no-slip condition. This motionless layer slows down the particles of the neighboring fluid layer as a result of friction between the particles of these two adjoining fluid layers at different velocities. This fluid layer then slows down the molecules of the next layer, and so on. Thus, the presence of the plate is felt up to some normal distance δ from the plate beyond which the free-stream velocity remains virtually unchanged. As a result, the x-component of the fluid velocity, u, varies from 0 at $y = 0$ to nearly V (typically $0.99\,V$) at $y = \delta$ (Fig. 15–26).

The region of the flow above the plate bounded by δ in which the effects of the viscous shearing forces caused by fluid viscosity are felt is called the **velocity boundary layer**. The *boundary layer thickness* δ is typically defined as the distance y from the surface at which $u = 0.99V$.

The hypothetical curve of $u = 0.99V$ divides the flow over a plate into two regions: the **boundary layer region**, in which the viscous effects and the velocity changes are significant, and the **irrotational flow region**, in which the frictional effects are negligible and the velocity remains essentially constant.

For parallel flow over a flat plate, the pressure drag is zero, and thus the drag coefficient is equal to the *friction drag coefficient*, or simply the *friction coefficient* (Fig. 15–27). That is,

Flat plate:
$$C_D = C_{D,\text{ friction}} = C_f \tag{15–9}$$

Once the average friction coefficient C_f is available, the drag (or friction) force over the surface is determined from

Friction force on a flat plate:
$$F_D = F_f = \tfrac{1}{2}C_f A \rho V^2 \tag{15–10}$$

where A is the surface area of the plate exposed to fluid flow. When both sides of a thin plate are subjected to flow, A becomes the total area of the top and bottom surfaces. Note that both the average friction coefficient C_f and the local friction coefficient $C_{f,\,x}$, in general, vary with location along the surface.

Typical average velocity profiles in laminar and turbulent flow are sketched in Fig. 15–25. Note that the velocity profile in turbulent flow is much fuller than that in laminar flow, with a sharp drop near the surface. The turbulent boundary layer can be considered to consist of four regions, characterized by the distance from the wall. The very thin layer next to the wall where viscous effects are dominant is the **viscous sublayer**. The velocity profile in this layer is very nearly *linear*, and the flow is nearly parallel. Next to the viscous sublayer is the **buffer layer**, in which turbulent effects are becoming significant, but the flow is still dominated by viscous effects. Above the buffer layer is the **overlap layer**, in which the turbulent effects are much more significant, but still not dominant. Above that is the **turbulent layer** in which turbulent effects dominate over viscous effects. Note that the turbulent boundary layer profile on a flat plate closely resembles the boundary layer profile in fully developed turbulent pipe flow (see Chap. 14).

The transition from laminar to turbulent flow depends on the *surface geometry*, *surface roughness*, *upstream velocity*, *surface temperature*, and the *type of fluid*, among other things, and is best characterized by the Reynolds number. The Reynolds number at a distance x from the leading edge of a flat plate is expressed as

$$\text{Re}_x = \frac{\rho V x}{\mu} = \frac{Vx}{\nu} \tag{15–11}$$

where V is the upstream velocity and x is the characteristic length of the geometry, which, for a flat plate, is the length of the plate in the flow direction. Note that unlike pipe flow, the Reynolds number varies for a flat plate along the flow, reaching $\text{Re}_L = VL/\nu$ at the end of the plate. For any point on a flat plate, the characteristic length is the distance x of the point from the leading edge in the flow direction.

For flow over a smooth flat plate, transition from laminar to turbulent begins at about $\text{Re} \cong 1 \times 10^5$, but does not become fully turbulent before the Reynolds number reaches much higher values, typically around 3×10^6. In engineering analysis, a generally accepted value for the critical Reynolds number is

$$\text{Re}_{x,\,\text{cr}} = \frac{\rho V x_{\text{cr}}}{\mu} = 5 \times 10^5$$

The actual value of the engineering critical Reynolds number for a flat plate may vary somewhat from about 10^5 to 3×10^6 depending on the surface roughness, the turbulence level, and the variation of pressure along the surface.

Friction Coefficient

The friction coefficient for laminar flow over a flat plate can be determined theoretically by solving the conservation of mass and momentum equations numerically. For turbulent flow, however, it must be determined experimentally and expressed by empirical correlations.

The local friction coefficient *varies* along the surface of the flat plate as a result of the changes in the velocity boundary layer in the flow direction. We are usually interested in the drag force on the *entire* surface, which can be determined using the *average* friction coefficient. But sometimes we are also interested in the drag force at a certain location, and in such cases, we need to know the *local* value of the friction coefficient. With this in mind, we present correlations for both local (identified with the subscript x) and average friction coefficients over a flat plate for *laminar*, *turbulent*, and *combined laminar and turbulent* flow conditions. Once the local values are available, the *average* friction coefficient for the entire plate is determined by integration,

$$C_f = \frac{1}{L}\int_0^L C_{f,\,x}\, dx \tag{15–12}$$

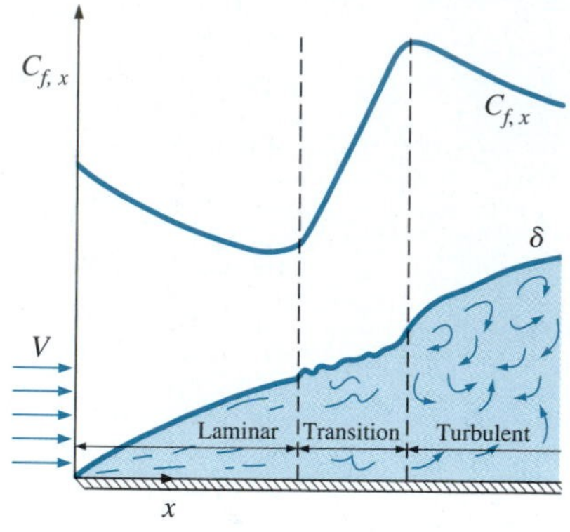

FIGURE 15–28

The variation of the local friction coefficient for flow over a flat plate. Note that the vertical scale of the boundary layer is greatly exaggerated in this sketch.

Based on analysis, the boundary layer thickness and the local friction coefficient at location x for laminar flow over a flat plate are

$$\textit{Laminar:}\quad \delta = \frac{4.91x}{\text{Re}_x^{1/2}} \quad \text{and} \quad C_{f,x} = \frac{0.664}{\text{Re}_x^{1/2}}, \qquad \text{Re}_x < 5 \times 10^5 \tag{15–13}$$

The corresponding relations for turbulent flow are

$$\textit{Turbulent:}\ \delta = \frac{0.38x}{\text{Re}_x^{1/5}} \quad \text{and} \quad C_{f,x} = \frac{0.059}{\text{Re}_x^{1/5}}, \qquad 5 \times 10^5 \le \text{Re}_x \le 10^7 \tag{15–14}$$

where x is the distance from the leading edge of the plate and $\text{Re}_x = Vx/\nu$ is the Reynolds number at location x. Note that $C_{f,x}$ is proportional to $1/\text{Re}_x^{1/2}$ and thus to $x^{-1/2}$ for laminar flow and it is proportional to $x^{-1/5}$ for turbulent flow. In either case, $C_{f,x}$ is infinite at the leading edge ($x = 0$), and therefore Eqs. 15–13 and 15–14 are not valid close to the leading edge. The variation of the boundary layer thickness δ and the friction coefficient $C_{f,x}$ along a flat plate is sketched in Fig. 15–28. The local friction coefficients are higher in turbulent flow than they are in laminar flow because of the intense mixing that occurs in the turbulent boundary layer. Note that $C_{f,x}$ reaches its highest values when the flow becomes fully turbulent, and then decreases by a factor of $x^{-1/5}$ in the flow direction, as shown in the figure.

The *average* friction coefficient over the entire plate is determined by substituting Eqs. 15–13 and 15–14 into Eq. 15–12 and performing the integrations (Fig. 15–29). We get

$$\textit{Laminar:}\qquad C_f = \frac{1.33}{\text{Re}_L^{1/2}} \qquad \text{Re}_L < 5 \times 10^5 \tag{15–15}$$

$$\textit{Turbulent:}\qquad C_f = \frac{0.074}{\text{Re}_L^{1/5}} \qquad 5 \times 10^5 \le \text{Re}_L \le 10^7 \tag{15–16}$$

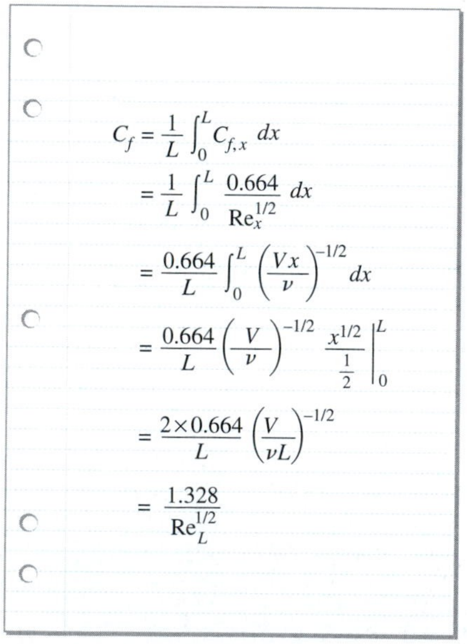

$$\begin{aligned} C_f &= \frac{1}{L}\int_0^L C_{f,x}\,dx \\ &= \frac{1}{L}\int_0^L \frac{0.664}{\text{Re}_x^{1/2}}\,dx \\ &= \frac{0.664}{L}\int_0^L \left(\frac{Vx}{\nu}\right)^{-1/2} dx \\ &= \frac{0.664}{L}\left(\frac{V}{\nu}\right)^{-1/2} \frac{x^{1/2}}{\frac{1}{2}}\Bigg|_0^L \\ &= \frac{2\times 0.664}{L}\left(\frac{V}{\nu L}\right)^{-1/2} \\ &= \frac{1.328}{\text{Re}_L^{1/2}} \end{aligned}$$

FIGURE 15–29

The average friction coefficient over a surface is determined by integrating the local friction coefficient over the entire surface. The values shown here are for a laminar flat plate boundary layer.

The first of these relations gives the average friction coefficient for the entire plate when the flow is *laminar* over the *entire* plate. The second relation gives the average friction coefficient for the entire plate only when the flow is *turbulent* over the *entire* plate, or when the laminar flow region of the plate is negligibly small relative to the turbulent flow region (that is, $x_{cr} \ll L$ where the length of the plate x_{cr} over which the flow is laminar is determined from $\text{Re}_{cr} = 5 \times 10^5 = Vx_{cr}/\nu$).

In some cases, a flat plate is sufficiently long for the flow to become turbulent, but not long enough to disregard the laminar flow region. In such cases, the *average* friction coefficient over the entire plate is determined by performing the integration in Eq. 15–12 over two parts: the laminar region $0 \le x \le x_{cr}$ and the turbulent region $x_{cr} < x \le L$ as

$$C_f = \frac{1}{L}\left(\int_0^{x_{cr}} C_{f,x,\text{ laminar}}\,dx + \int_{x_{cr}}^{L} C_{f,x,\text{ turbulent}}\,dx\right) \tag{15–17}$$

Note that we included the transition region with the turbulent region. Again taking the critical Reynolds number to be $\text{Re}_{cr} = 5 \times 10^5$ and performing these integrations after substituting the indicated expressions, the *average* friction coefficient over the *entire* plate is determined to be

$$C_f = \frac{0.074}{\text{Re}_L^{1/5}} - \frac{1742}{\text{Re}_L} \qquad 5 \times 10^5 \le \text{Re}_L \le 10^7 \tag{15–18}$$

The constants in this relation would be different for different critical Reynolds numbers. Also, the surfaces are assumed to be *smooth*, and the free stream to be of very low turbulence intensity. For laminar flow, the friction coefficient depends only on the Reynolds number, and the surface roughness has no effect. For turbulent flow, however, surface roughness causes the friction coefficient to increase severalfold, to the point that in the fully rough turbulent regime the friction coefficient is a function of surface roughness alone and is independent of the Reynolds number (Fig. 15–30). This is analogous to flow in pipes. A curve fit of experimental data for the average friction coefficient in this regime is given by Schlichting (1979) as

Fully rough turbulent regime:

$$C_f = \left(1.89 - 1.62 \log \frac{\varepsilon}{L}\right)^{-2.5} \quad \text{(15–19)}$$

where ε is the surface roughness and L is the length of the plate in the flow direction. In the absence of a better one, this relation can be used for turbulent flow on rough surfaces for $\text{Re} > 10^6$, especially when $\varepsilon/L > 10^{-4}$.

Friction coefficients C_f for parallel flow over smooth and rough flat plates are plotted in Fig. 15–31 for both laminar and turbulent flows. Note that C_f increases severalfold with roughness in turbulent flow. Also note that C_f is independent of the Reynolds number in the fully rough region. This chart is the flat-plate analog of the Moody chart for pipe flows.

Relative Roughness, ε/L	Friction Coefficient, C_f
0.0*	0.0029
1×10^{-5}	0.0032
1×10^{-4}	0.0049
1×10^{-3}	0.0084

* Smooth surface for $\text{Re} = 10^7$. Others calculated from Eq. 15–19 for fully rough flow.

FIGURE 15–30

For turbulent flow, surface roughness may cause the friction coefficient to increase severalfold.

EXAMPLE 15–3 Flow of Hot Oil over a Flat Plate

Engine oil at 40°C flows over a 5-m-long flat plate with a free-stream velocity of 2 m/s (Fig. 15–32). Determine the drag force acting on the plate per unit width.

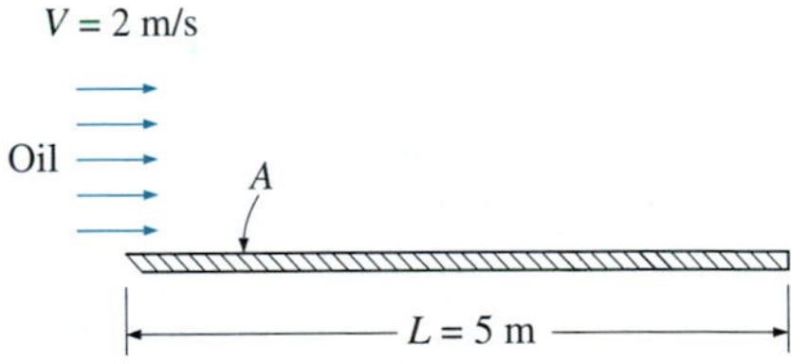

FIGURE 15–32

Schematic for Example 15–3.

Solution Engine oil flows over a flat plate. The drag force per unit width of the plate is to be determined.

Assumptions **1** The flow is steady and incompressible. **2** The critical Reynolds number is $\text{Re}_{cr} = 5 \times 10^5$.

Properties The density and kinematic viscosity of engine oil at 40°C are $\rho = 876$ kg/m³ and $\nu = 2.485 \times 10^{-4}$ m²/s (Table A–19).

Analysis Noting that $L = 5$ m, the Reynolds number at the end of the plate is

$$\text{Re}_L = \frac{VL}{\nu} = \frac{(2 \text{ m/s})(5 \text{ m})}{2.485 \times 10^{-4} \text{ m}^2/\text{s}} = 4.024 \times 10^4$$

which is less than the critical Reynolds number. Thus we have *laminar flow* over the entire plate, and the average friction coefficient is (Fig. 15–29)

$$C_f = 1.328\text{Re}_L^{-0.5} = 1.328 \times (4.024 \times 10^4)^{-0.5} = 0.00662$$

Noting that the pressure drag is zero and thus $C_D = C_f$ for parallel flow over a flat plate, the drag force acting on the plate per unit width becomes

$$F_D = C_f A \frac{\rho V^2}{2} = 0.00662(5 \times 1 \text{ m}^2) \frac{(876 \text{ kg/m}^3)(2 \text{ m/s})^2}{2} \left(\frac{1 \text{ N}}{1 \text{ kg} \cdot \text{m/s}^2}\right) = \mathbf{58.0\ N}$$

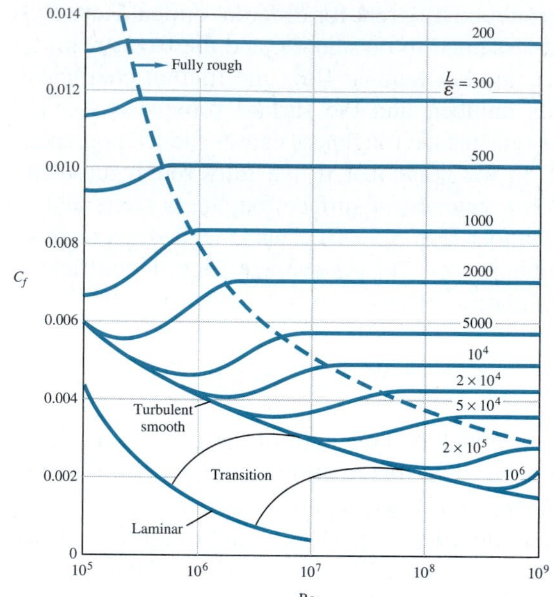

FIGURE 15–31

Friction coefficient for parallel flow over smooth and rough flat plates.

From White (2003).

The total drag force acting on the entire plate can be determined by multiplying the value just obtained by the width of the plate.

Discussion The force per unit width corresponds to the weight of a mass of about 6 kg. Therefore, a person who applies an equal and opposite force to the plate to keep it from moving will feel like he or she is using as much force as is necessary to hold a 6-kg mass from dropping.

15–6 ▪ FLOW OVER CYLINDERS AND SPHERES

Flow over cylinders and spheres is frequently encountered in practice. For example, the tubes in a shell-and-tube heat exchanger involve both *internal flow* through the tubes and *external flow* over the tubes, and both flows must be considered in the analysis of the heat exchanger. Also, many sports such as soccer, tennis, and golf involve flow over spherical balls.

The characteristic length for a circular cylinder or sphere is taken to be the *external diameter D*. Thus, the Reynolds number is defined as $Re = VD/\nu$ where V is the uniform velocity of the fluid as it approaches the cylinder or sphere. The critical Reynolds number for flow across a circular cylinder or sphere is about $Re_{cr} \cong 2 \times 10^5$. That is, the boundary layer remains laminar for $Re \lesssim 2 \times 10^5$, is "transitional" for $2 \times 10^5 \lesssim Re \lesssim 2 \times 10^6$, and becomes fully turbulent for $Re \gtrsim 2 \times 10^6$.

Cross-flow over a cylinder exhibits complex flow patterns, as shown in Fig. 15–33. The fluid approaching the cylinder branches out and encircles the cylinder, forming a boundary layer that wraps around the cylinder. The fluid particles on the midplane strike the cylinder at the stagnation point, bringing the fluid to a complete stop and thus raising the pressure at that point. The pressure decreases in the flow direction while the fluid velocity increases.

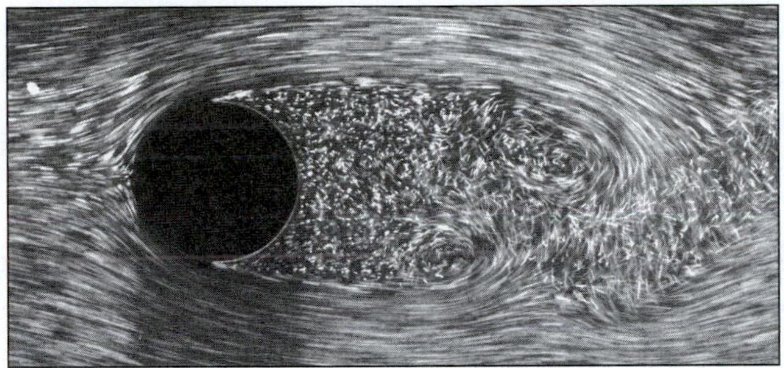

FIGURE 15–33

Laminar boundary layer separation with a turbulent wake; flow over a circular cylinder at Re = 2000.

Courtesy ONERA, photograph by Werlé.

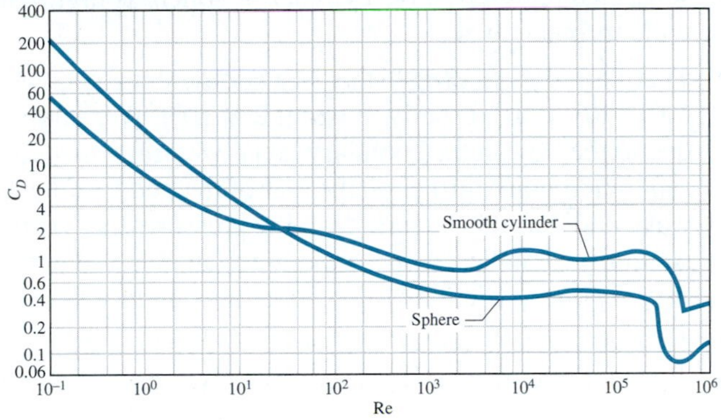

FIGURE 15–34

Average drag coefficient for cross-flow over a smooth circular cylinder and a smooth sphere.

From H. Schlichting, Boundary Layer Theory 7e. Copyright © 1979 The McGraw-Hill Companies, Inc. Used by permission.

At very low upstream velocities (Re $\lesssim$ 1), the fluid completely wraps around the cylinder and the two arms of the fluid meet on the rear side of the cylinder in an orderly manner. Thus, the fluid follows the curvature of the cylinder. At higher velocities, the fluid still hugs the cylinder on the frontal side, but it is too fast to remain attached to the surface as it approaches the top (or bottom) of the cylinder. As a result, the boundary layer detaches (separates) from the surface, forming a separation region behind the cylinder. Flow in the wake region is characterized by periodic vortex formation and pressures much lower than the stagnation point pressure.

The nature of the flow across a cylinder or sphere strongly affects the total drag coefficient C_D. Both the *friction drag* and the *pressure drag* can be significant. The high pressure in the vicinity of the stagnation point and the low pressure on the opposite side in the wake produce a net force on the body in the direction of flow. The drag force is primarily due to friction drag at low Reynolds numbers (Re < 10) and to pressure drag at high Reynolds numbers (Re > 5000). Both effects are significant at intermediate Reynolds numbers.

The average drag coefficients C_D for cross-flow over a smooth single circular cylinder and a sphere are given in Fig. 15–34. The curves exhibit different behaviors in different ranges of Reynolds numbers:

- For Re $\lesssim$ 1, we have creeping flow, and the drag coefficient decreases with increasing Reynolds number. For a sphere, it is C_D = 24/Re. There is no flow separation in this regime.
- At about Re = 10, separation starts occurring on the rear of the body with vortex shedding starting at about Re $\cong$ 90. The region of separation

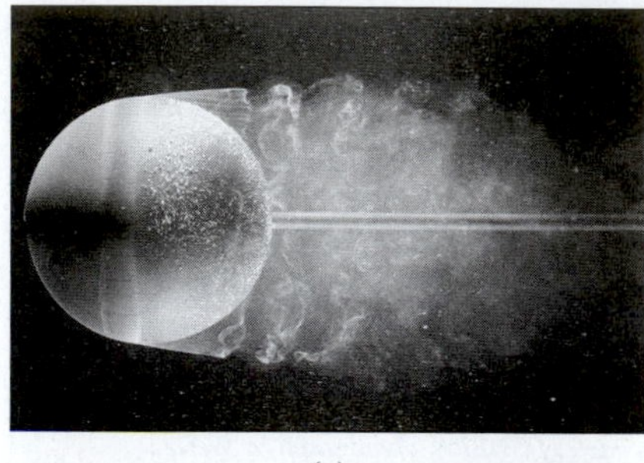

(a)

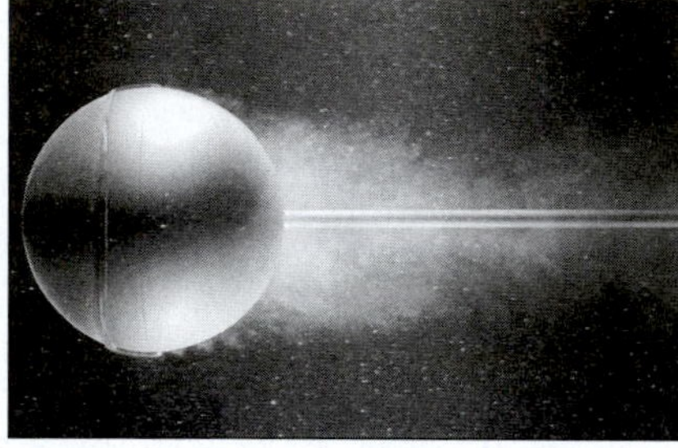

(b)

FIGURE 15–35

Flow visualization of flow over (*a*) a smooth sphere at Re = 15,000, and (*b*) a sphere at Re = 30,000 with a trip wire. The delay of boundary layer separation is clearly seen by comparing the two photographs.

Courtesy ONERA, photograph by Werlé.

increases with increasing Reynolds number up to about $Re = 10^3$. At this point, the drag is mostly (about 95 percent) due to pressure drag. The drag coefficient continues to decrease with increasing Reynolds number in this range of $10 < Re < 10^3$. (A decrease in the drag coefficient does not necessarily indicate a decrease in drag. The drag force is proportional to the square of the velocity, and the increase in velocity at higher Reynolds numbers usually more than offsets the decrease in the drag coefficient.)

- In the moderate range of $10^3 < Re < 10^5$, the drag coefficient remains relatively constant. This behavior is characteristic of bluff bodies. The flow in the boundary layer is laminar in this range, but the flow in the separated region past the cylinder or sphere is highly turbulent with a wide turbulent wake.
- There is a sudden drop in the drag coefficient somewhere in the range of $10^5 < Re < 10^6$ (usually, at about 2×10^5). This large reduction in C_D is due to the flow in the boundary layer becoming *turbulent*, which moves the separation point further on the rear of the body, reducing the size of the wake and thus the magnitude of the pressure drag. This is in contrast to streamlined bodies, which experience an increase in the drag coefficient (mostly due to friction drag) when the boundary layer becomes turbulent.
- There is a "transitional" regime for $2 \times 10^5 \lesssim Re \lesssim 2 \times 10^6$, in which C_D dips to a minimum value and then slowly rises to its final turbulent value.

Flow separation occurs at about $\theta \cong 80°$ (measured from the front stagnation point of a cylinder) when the boundary layer is *laminar* and at about $\theta \cong 140°$ when it is *turbulent* (Fig. 15–35). The delay of separation in turbulent flow is caused by the rapid fluctuations of the fluid in the transverse direction, which enables the turbulent boundary layer to travel farther along the surface before separation occurs, resulting in a narrower wake and a smaller pressure drag. Keep in mind that turbulent flow has a fuller velocity profile as compared to the laminar case, and thus it requires a stronger adverse pressure gradient to overcome the additional momentum close to the wall. In the range of Reynolds numbers where the flow changes from laminar to turbulent, even the drag force F_D decreases as the velocity (and thus the Reynolds number) increases. This results in a sudden decrease in drag of a flying body (sometimes called the *drag crisis*) and instabilities in flight.

Effect of Surface Roughness

We mentioned earlier that *surface roughness*, in general, increases the drag coefficient in turbulent flow. This is especially the case for streamlined bodies. For blunt bodies such as a circular cylinder or sphere, however, an increase in the surface roughness may actually *decrease* the drag coefficient, as shown in Fig. 15–36 for a sphere. This is done by tripping the boundary layer into turbulence at a lower Reynolds number, and thus causing the fluid to close in behind the body, narrowing the wake and reducing pressure drag considerably. This results in a much smaller drag coefficient and thus drag force for a rough-surfaced cylinder or sphere in a certain range of Reynolds number compared to a smooth one of identical size at the same velocity. At $Re = 2 \times 10^5$, for example, $C_D \cong 0.1$ for a rough sphere with $\varepsilon/D = 0.0015$, whereas $C_D \cong 0.5$ for a smooth one. Therefore, the drag coefficient in this case is reduced by a factor of 5 by simply roughening the surface.

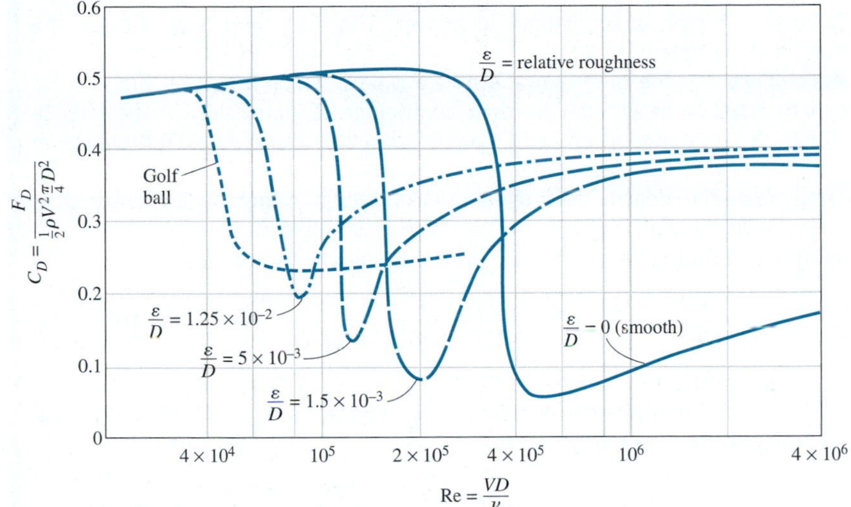

FIGURE 15–36

The effect of surface roughness on the drag coefficient of a sphere.

From Blevins (1984).

	C_D	
Re	Smooth Surface	Rough Surface, $\varepsilon/D = 0.0015$
2×10^5	0.5	0.1
10^6	0.1	0.4

FIGURE 15–37

Surface roughness may increase or decrease the drag coefficient of a spherical object, depending on the value of the Reynolds number.

Note, however, that at Re $= 10^6$, $C_D \cong 0.4$ for a very rough sphere while $C_D \cong 0.1$ for the smooth one. Obviously, roughening the sphere in this case increases the drag by a factor of 4 (Fig. 15–37).

The preceding discussion shows that roughening the surface can be used to great advantage in reducing drag, but it can also backfire on us if we are not careful—specifically, if we do not operate in the right range of the Reynolds number. With this consideration, golf balls are intentionally roughened to induce *turbulence* at a lower Reynolds number to take advantage of the sharp *drop* in the drag coefficient at the onset of turbulence in the boundary layer (the typical velocity range of golf balls is 15 to 150 m/s, and the Reynolds number is less than 4×10^5). The critical Reynolds number of dimpled golf balls is about 4×10^4. The occurrence of turbulent flow at this Reynolds number reduces the drag coefficient of a golf ball by about half, as shown in Fig. 15–36. For a given hit, this means a longer distance for the ball. Experienced golfers also give the ball a spin during the hit, which helps the rough ball develop a lift and thus travel higher and farther. A similar argument can be given for a tennis ball. For a table tennis ball, however, the speeds are slower and the ball is smaller—it never reaches speeds in the turbulent range. Therefore, the surfaces of table tennis balls are made smooth.

Once the drag coefficient is available, the drag force acting on a body in cross-flow can be determined from Eq. 15–1 where A is the *frontal area* ($A = LD$ for a cylinder of length L and $A = \pi D^2/4$ for a sphere). It should be kept in mind that free-stream turbulence and disturbances by other bodies in the flow (such as flow over tube bundles) may affect the drag coefficient significantly.

EXAMPLE 15–4 Drag Force Acting on a Pipe in a River

A 2.2-cm-outer-diameter pipe is to span across a river at a 30-m-wide section while being completely immersed in water (Fig. 15–38). The average flow velocity of water is 4 m/s and the water temperature is 15°C. Determine the drag force exerted on the pipe by the river.

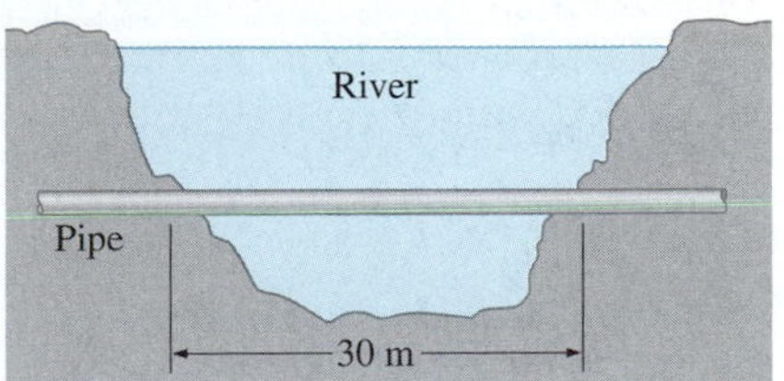

FIGURE 15–38

Schematic for Example 15–4.

Solution A pipe is submerged in a river. The drag force that acts on the pipe is to be determined.

Assumptions **1** The outer surface of the pipe is smooth so that Fig. 15–34 can be used to determine the drag coefficient. **2** Water flow in the river is steady. **3** The direction of water flow is normal to the pipe. **4** Turbulence in the river flow is not considered.

Properties The density and dynamic viscosity of water at 15°C are ρ = 999.1 kg/m^3 and μ = 1.138 × 10^{-3} kg/m · s (Table A–15).

Analysis Noting that D = 0.022 m, the Reynolds number is

$$\text{Re} = \frac{VD}{\nu} = \frac{\rho VD}{\mu} = \frac{(999.1\ \text{kg/m}^3)(4\ \text{m/s})(0.022\ \text{m})}{1.138 \times 10^{-3}\ \text{kg/m} \cdot \text{s}} = 7.73 \times 10^4$$

The drag coefficient corresponding to this value is, from Fig. 15–34, C_D = 1.0. Also, the frontal area for flow past a cylinder is $A = LD$. Then the drag force acting on the pipe becomes

$$F_D = C_D A \frac{\rho V^2}{2} = 1.0(30 \times 0.022\ \text{m}^2)\frac{(999.1\ \text{kg/m}^3)(4\ \text{m/s})^2}{2}\left(\frac{1\ \text{N}}{1\ \text{kg} \cdot \text{m/s}^2}\right)$$

$$= 5275\ \text{N} \cong \mathbf{5300\ N}$$

Discussion Note that this force is equivalent to the weight of a mass over 500 kg. Therefore, the drag force the river exerts on the pipe is equivalent to hanging a total of over 500 kg in mass on the pipe supported at its ends 30 m apart. The necessary precautions should be taken if the pipe cannot support this force. If the river were to flow at a faster speed or if turbulent fluctuations in the river were more significant, the drag force would be even larger. *Unsteady* forces on the pipe might then be significant.

15–7 ▪ LIFT

Lift was defined earlier as the component of the net force (due to viscous and pressure forces) that is perpendicular to the flow direction, and the lift coefficient was expressed as

$$C_L = \frac{F_L}{\frac{1}{2}\rho V^2 A} \qquad \textbf{(15–20)}$$

where A in this case is normally the *planform area*, which is the area that would be seen by a person looking at the body from above in a direction

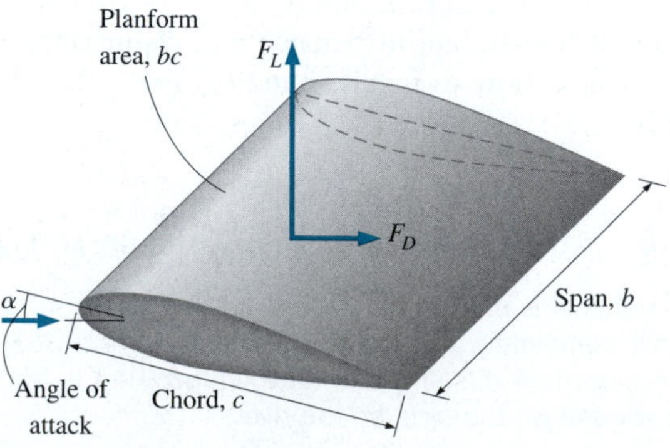

FIGURE 15–39
Definition of various terms associated with an airfoil.

normal to the body, and V is the upstream velocity of the fluid (or, equivalently, the velocity of a flying body in a quiescent fluid). For an airfoil of width (or span) b and chord length c (the length between the leading and trailing edges), the planform area is $A = bc$. The distance between the two ends of a wing or airfoil is called the **wingspan** or just the **span**. For an aircraft, the wingspan is taken to be the total distance between the tips of the two wings, which includes the width of the fuselage between the wings (Fig. 15–39). The average lift per unit planform area F_L/A is called the **wing loading**, which is simply the ratio of the weight of the aircraft to the planform area of the wings (since lift equals the weight during flying at constant altitude).

Airplane flight is based on lift, and thus developing a better understanding of lift as well as improving the lift characteristics of bodies have been the focus of numerous studies. Our emphasis in this section is on devices such as *airfoils* that are specifically designed to generate lift while keeping the drag at a minimum. But it should be kept in mind that some devices such as the *spoilers* and *inverted airfoils* on racing cars are designed for the opposite purpose of avoiding lift or even generating negative lift to improve traction and control (some early race cars actually "took off" at high speeds as a result of the lift produced, which alerted the engineers to come up with ways to reduce lift in their design).

For devices that are intended to generate lift such as airfoils, the contribution of *viscous effects* to lift is usually negligible since the bodies are streamlined, and wall shear is parallel to the surfaces of such devices and thus nearly normal to the direction of lift (Fig. 15–40). Therefore, lift in practice can be taken to be due entirely to the pressure distribution on the surfaces of the body, and thus the shape of the body has the primary influence on lift. Then the primary consideration in the design of airfoils is minimizing the average pressure at the upper surface while maximizing it at the lower surface. The Bernoulli equation can be used as a guide in identifying the high- and low-pressure regions: *Pressure is low at locations where the flow velocity is high, and pressure is high at locations where the flow velocity is low.* Also, lift at moderate angles of attack is practically independent of the surface roughness since roughness affects the wall shear, not the pressure. The contribution of shear to lift is usually significant only for very small (lightweight) bodies that fly at low velocities (and thus low Reynolds numbers).

Noting that the contribution of viscous effects to lift is negligible, we should be able to determine the lift acting on an airfoil by simply integrating the pressure distribution around the airfoil. The pressure changes in the flow direction along the surface, but it remains essentially constant through the boundary layer in a direction normal to the surface. Therefore, it seems reasonable to ignore the very thin boundary layer on the airfoil and calculate the pressure distribution around the airfoil from the relatively simple potential flow theory (zero vorticity, irrotational flow) for which net viscous forces are zero for flow past an airfoil.

The flow fields obtained from such calculations are sketched in Fig. 15–41 for both symmetrical and nonsymmetrical airfoils by ignoring the thin boundary layer. At zero angle of attack, the lift produced by the symmetrical airfoil is zero, as expected because of symmetry, and the stagnation points are at the leading and trailing edges. For the nonsymmetrical airfoil,

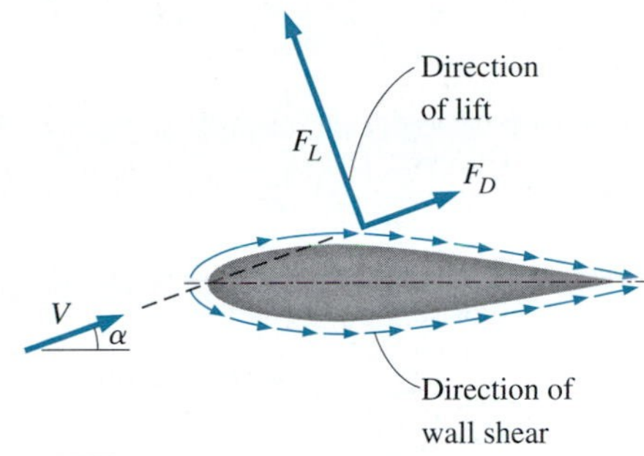

FIGURE 15–40

For airfoils, the contribution of viscous effects to lift is usually negligible since wall shear is parallel to the surfaces and thus nearly normal to the direction of lift.

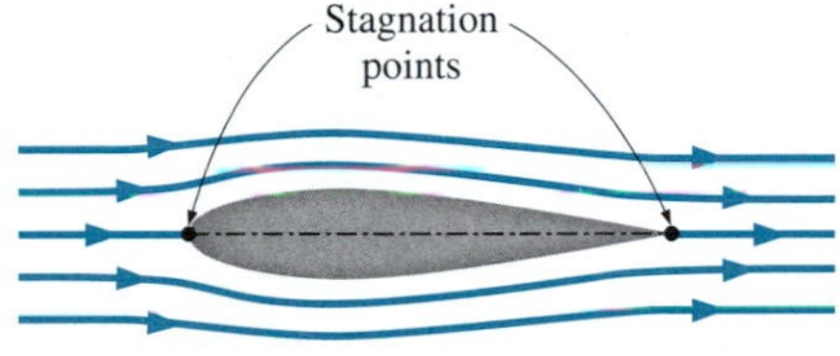

(*a*) Irrotational flow past a symmetrical airfoil (zero lift)

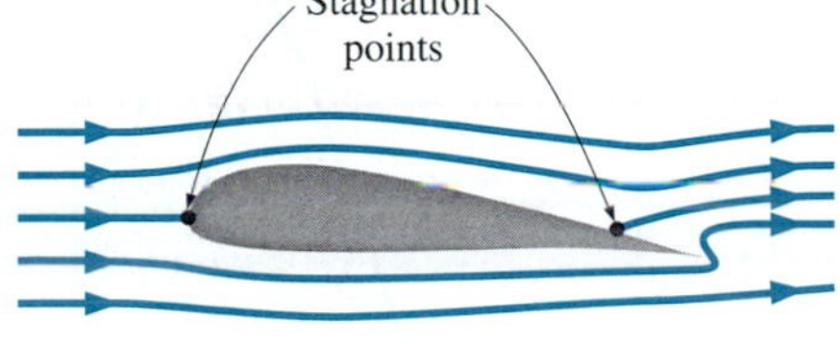

(*b*) Irrotational flow past a nonsymmetrical airfoil (zero lift)

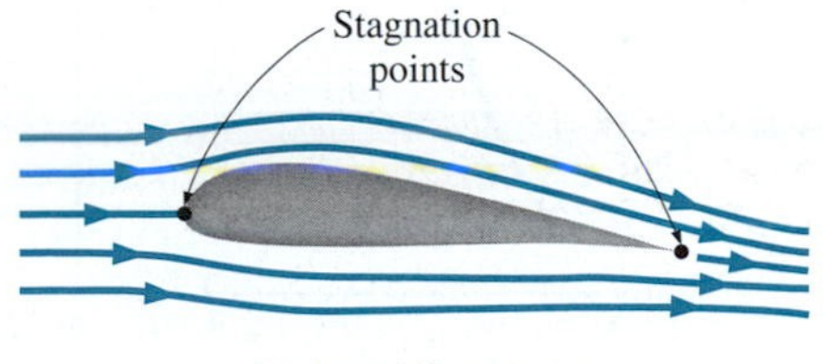

(*c*) Actual flow past a nonsymmetrical airfoil (positive lift)

FIGURE 15–41

Irrotational and actual flow past symmetrical and nonsymmetrical two-dimensional airfoils.

which is at a small angle of attack, the front stagnation point has moved down below the leading edge, and the rear stagnation point has moved up to the upper surface close to the trailing edge. To our surprise, the lift produced is calculated again to be zero—a clear contradiction of experimental observations and measurements. Obviously, the theory needs to be modified to bring it in line with the observed phenomenon.

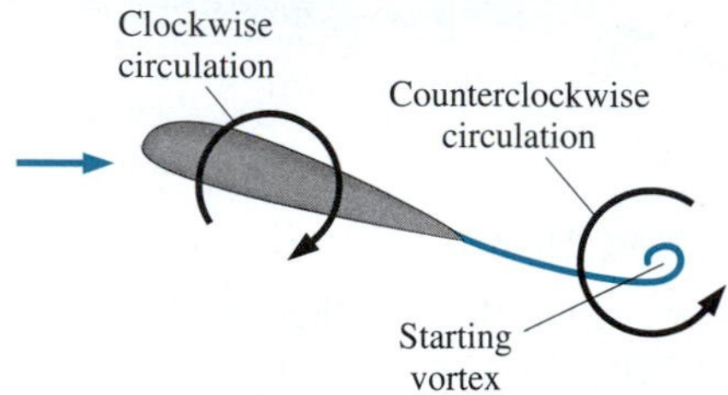

FIGURE 15–42

Shortly after a sudden increase in angle of attack, a counterclockwise starting vortex is shed from the airfoil, while clockwise circulation appears around the airfoil, causing lift to be generated.

The source of inconsistency is the rear stagnation point being at the upper surface instead of the trailing edge. This requires the lower side fluid to make a nearly U-turn and flow around the (sharp) trailing edge toward the stagnation point while remaining attached to the surface, which is a physical impossibility since the observed phenomenon is the separation of flow at sharp turns (imagine a car attempting to make this turn at high speed). Therefore, the lower side fluid separates smoothly off the trailing edge, and the upper side fluid responds by pushing the rear stagnation point downstream. In fact, the stagnation point at the upper surface moves all the way to the trailing edge. This way the two flow streams from the top and the bottom sides of the airfoil meet at the trailing edge, yielding a smooth flow downstream parallel to the trailing edge. Lift is generated because the flow velocity at the top surface is higher, and thus the pressure on that surface is lower due to the Bernoulli effect.

The potential flow theory and the observed phenomenon can be reconciled as follows: Flow starts out as predicted by theory, with no lift, but the lower fluid stream separates at the trailing edge when the velocity reaches a certain value. This forces the separated upper fluid stream to close in at the trailing edge, initiating clockwise circulation around the airfoil. This clockwise circulation increases the velocity of the upper stream while decreasing that of the lower stream, causing lift. A **starting vortex** of opposite sign (counterclockwise circulation) is then shed downstream (Fig. 15–42), and smooth streamlined flow is established over the airfoil. When the potential flow theory is modified by the addition of an appropriate amount of circulation to move the stagnation point down to the trailing edge, excellent agreement is obtained between theory and experiment for both the flow field and the lift.

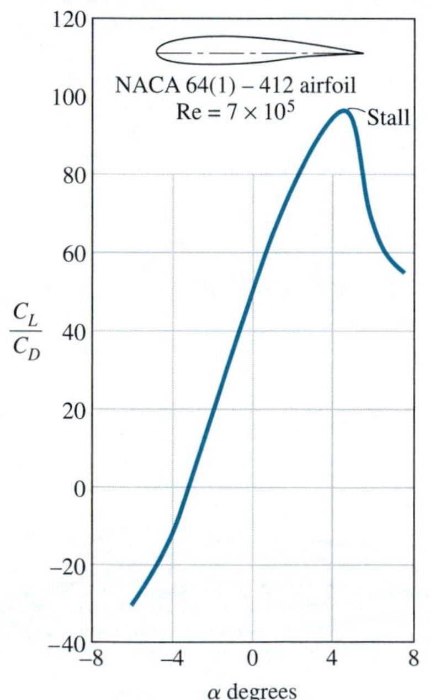

FIGURE 15–43

The variation of the lift-to-drag ratio with angle of attack for a two-dimensional airfoil.

From Abbott, von Doenhoff, and Stivers (1945).

It is desirable for airfoils to generate the most lift while producing the least drag. Therefore, a measure of performance for airfoils is the **lift-to-drag ratio**, which is equivalent to the ratio of the lift-to-drag coefficients C_L/C_D. This information is provided by either plotting C_L versus C_D for different values of the angle of attack (a lift–drag polar) or by plotting the ratio C_L/C_D versus the angle of attack. The latter is done for a particular airfoil design in Fig. 15–43. Note that the C_L/C_D ratio increases with the angle of attack until the airfoil stalls, and the value of the lift-to-drag ratio can be of the order of 100 for a two-dimensional airfoil.

One obvious way to change the lift and drag characteristics of an airfoil is to change the angle of attack. On an airplane, for example, the entire plane is pitched up to increase lift, since the wings are fixed relative to the fuselage. Another approach is to change the shape of the airfoil by the use of movable *leading edge* and *trailing edge flaps*, as is commonly done in modern large aircraft (Fig. 15–44). The flaps are used to alter the shape of the wings during takeoff and landing to maximize lift and to enable the aircraft to land or take off at low speeds. The increase in drag during this takeoff and landing is not much of a concern because of the relatively short time

(*a*) Flaps extended (takeoff)

(*b*) Flaps retracted (cruising)

FIGURE 15–44

The lift and drag characteristics of an airfoil during takeoff and landing can be changed by changing the shape of the airfoil by the use of movable flaps.

Photos by Yunus Çengel.

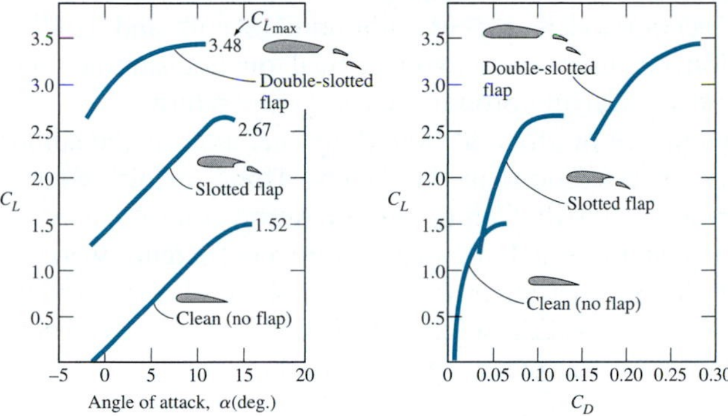

FIGURE 15–45

Effect of flaps on the lift and drag coefficients of an airfoil.

From Abbott and von Doenhoff, for NACA 23012 (1959).

periods involved. Once at cruising altitude, the flaps are retracted, and the wing is returned to its "normal" shape with minimal drag coefficient and adequate lift coefficient to minimize fuel consumption while cruising at a constant altitude. Note that even a small lift coefficient can generate a large lift force during normal operation because of the large cruising velocities of aircraft and the proportionality of lift to the square of flow velocity.

The effects of flaps on the lift and drag coefficients are shown in Fig. 15–45 for an airfoil. Note that the maximum lift coefficient increases from about 1.5 for the airfoil with no flaps to 3.5 for the double-slotted flap case. But also note that the maximum drag coefficient increases from about 0.06 for the airfoil with no flaps to about 0.3 for the double-slotted flap case. This is a five-fold increase in the drag coefficient, and the engines must work much harder to provide the necessary thrust to overcome this drag. The angle of attack of the flaps can be increased to maximize the lift coefficient. Also, the flaps extend the chord length, and thus enlarge the wing area *A*. The Boeing 727 uses a triple-slotted flap at the trailing edge and a slot at the leading edge.

The minimum flight velocity can be determined from the requirement that the total weight *W* of the aircraft be equal to lift and $C_L = C_{L,\max}$. That is,

$$W = F_L = \tfrac{1}{2} C_{L,\max} \rho V_{\min}^2 A \qquad \rightarrow \qquad V_{\min} = \sqrt{\frac{2W}{\rho C_{L,\max} A}} \qquad \textbf{(15–21)}$$

For a given weight, the landing or takeoff speed can be minimized by maximizing the product of the lift coefficient and the wing area, $C_{L,\max}A$. One way of doing that is to use flaps, as already discussed. Another way is to control the boundary layer, which can be accomplished simply by leaving flow sections (slots) between the flaps, as shown in Fig. 15–46. Slots are used to prevent the separation of the boundary layer from the upper surface of the

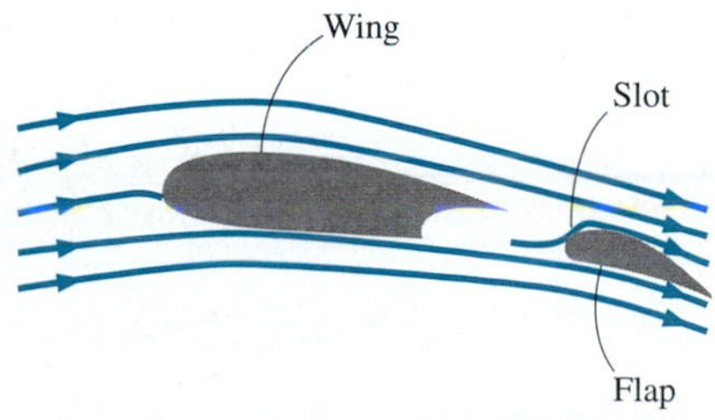

FIGURE 15–46

A flapped airfoil with a slot to prevent the separation of the boundary layer from the upper surface and to increase the lift coefficient.

wings and the flaps. This is done by allowing air to move from the high-pressure region under the wing into the low-pressure region at the top surface. Note that the lift coefficient reaches its maximum value $C_L = C_{L,\max}$, and thus the flight velocity reaches its minimum, at stall conditions, which is a region of unstable operation and must be avoided. The Federal Aviation Administration (FAA) does not allow operation below 1.2 times the stall speed for safety.

Another thing we notice from this equation is that the minimum velocity for takeoff or landing is inversely proportional to the square root of density. Noting that air density decreases with altitude (by about 15 percent at 1500 m), longer runways are required at airports at higher altitudes such as Denver to accommodate higher minimum takeoff and landing velocities. The situation becomes even more critical on hot summer days since the density of air is inversely proportional to temperature.

The development of efficient (low-drag) airfoils was the subject of intense experimental investigations in the 1930s. These airfoils were standardized by the National Advisory Committee for Aeronautics (NACA, which is now NASA), and extensive lists of data on lift coefficients were reported. The variation of the lift coefficient C_L with the angle of attack for two 2-D (infinite span) airfoils (NACA 0012 and NACA 2412) is given in Fig. 15–47. We make the following observations from this figure:

- The lift coefficient increases almost linearly with the angle of attack α, reaches a maximum at about $\alpha = 16°$, and then starts to decrease sharply. This decrease of lift with further increase in the angle of attack is called *stall*, and it is caused by flow separation and the formation of a wide wake region over the top surface of the airfoil. Stall is highly undesirable since it also increases drag.
- At zero angle of attack ($\alpha = 0°$), the lift coefficient is zero for symmetrical airfoils but nonzero for nonsymmetrical ones with greater curvature at the top surface. Therefore, planes with symmetrical wing

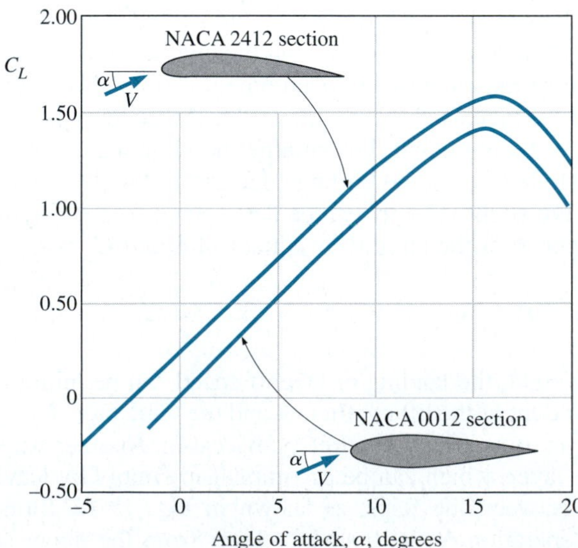

FIGURE 15–47

The variation of the lift coefficient with the angle of attack for a symmetrical and a nonsymmetrical airfoil.

From Abbott (1932).

sections must fly with their wings at higher angles of attack in order to produce the same lift.

- The lift coefficient can be increased severalfold by adjusting the angle of attack (from 0.25 at $\alpha = 0°$ for the nonsymmetrical airfoil to 1.25 at $\alpha = 10°$).
- The drag coefficient also increases with the angle of attack, often exponentially (Fig. 15–48). Therefore, large angles of attack should be used sparingly for short periods of time for fuel efficiency.

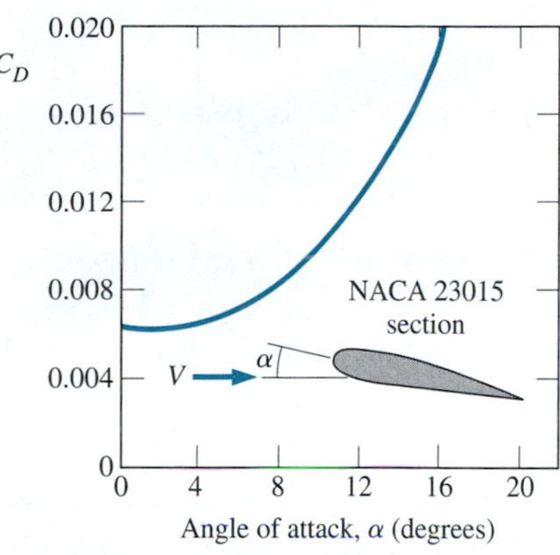

FIGURE 15–48

The variation of the drag coefficient of an airfoil with the angle of attack.

From Abbott and von Doenhoff (1959).

Finite-Span Wings and Induced Drag

For airplane wings and other airfoils of finite span, the end effects at the tips become important because of the fluid leakage between the lower and upper surfaces. The pressure difference between the lower surface (high-pressure region) and the upper surface (low-pressure region) drives the fluid at the tips upward while the fluid is swept toward the back because of the relative motion between the fluid and the wing. This results in a swirling motion that spirals along the flow, called the **tip vortex**, at the tips of both wings. Vortices are also formed along the airfoil between the tips of the wings. These distributed vortices collect toward the edges after being shed from the trailing edges of the wings and combine with the tip vortices to form two streaks of powerful **trailing vortices** along the tips of the wings (Figs. 15–49 and 15–50). Trailing vortices generated by large aircraft continue to exist for a long time for long distances (over 10 km) before they gradually disappear due to viscous dissipation. Such vortices and the accompanying downdraft are strong enough to cause a small aircraft to lose control and flip over if it flies through the wake of a larger aircraft. Therefore, following a large aircraft closely (within 10 km) poses a real danger for smaller aircraft. This issue is the controlling factor that governs the spacing of aircraft at takeoff, which limits the flight capacity at airports. In nature, this effect is used to advantage by birds that migrate in V-formation by utilizing the updraft generated by the bird in front. It has been determined that the birds in a typical flock can fly to their destination in V-formation with one-third less energy. Military jets also occasionally fly in V-formation for the same reason.

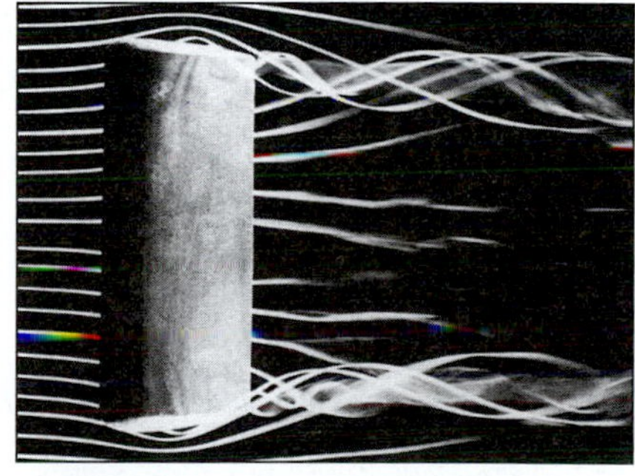

FIGURE 15–49

Trailing vortices from a rectangular wing with vortex cores leaving the trailing edge at the tips (top view).

Courtesy of The Parabolic Press, Stanford, California. Used with permission.

FIGURE 15–50

A crop duster flies through smoky air to illustrate the tip vortices produced at the tips of the wing.

NASA Langley Research Center.

(*a*) A bearded vulture with its wing feathers fanned out during flight.

(*b*) Winglets are used on this sailplane to reduce induced drag.

FIGURE 15–51

Induced drag is reduced by (*a*) wing tip feathers on bird wings and (*b*) endplates or other disruptions on airplane wings.

(a) © Vol. 44/PhotoDisc. (b) Courtesy Schempp-Hirth. Used by permission.

Tip vortices that interact with the free stream impose forces on the wing tips in all directions, including the flow direction. The component of the force in the flow direction adds to drag and is called **induced drag**. The total drag of a wing is then the sum of the induced drag (3-D effects) and the drag of the airfoil section (2-D effects).

The ratio of the square of the average span of an airfoil to the planform area is called the **aspect ratio**. For an airfoil with a rectangular planform of chord c and span b, it is expressed as

$$\text{AR} = \frac{b^2}{A} = \frac{b^2}{bc} = \frac{b}{c} \tag{15–22}$$

Therefore, the aspect ratio is a measure of how narrow an airfoil is in the flow direction. The lift coefficient of wings, in general, increases while the drag coefficient decreases with increasing aspect ratio. This is because a long narrow wing (large aspect ratio) has a shorter tip length and thus smaller tip losses and smaller induced drag than a short and wide wing of the same planform area. Therefore, bodies with large aspect ratios fly more efficiently, but they are less maneuverable because of their larger moment of inertia (owing to the greater distance from the center). Bodies with smaller aspect ratios maneuver better since the wings are closer to the central part. So it is no surprise that *fighter planes* (and fighter birds like falcons) have short and wide wings while *large commercial planes* (and soaring birds like albatrosses) have long and narrow wings.

The end effects can be minimized by attaching **endplates** or **winglets** at the tips of the wings perpendicular to the top surface. The endplates function by blocking some of the leakage around the wing tips, which results in a considerable reduction in the strength of the tip vortices and the induced drag. Wing tip feathers on birds fan out for the same purpose (Fig. 15–51).

EXAMPLE 15–5 Lift and Drag of a Commercial Airplane

A commercial airplane has a total mass of 70,000 kg and a wing planform area of 150 m^2 (Fig. 15–52). The plane has a cruising speed of 558 km/h and a cruising altitude of 12,000 m, where the air density is 0.312 kg/m^3. The plane has double-slotted flaps for use during takeoff and landing, but it cruises with all flaps retracted. Assuming the lift and the drag characteristics of the wings can be approximated by NACA 23012 (Fig. 15–45), determine (*a*) the minimum safe speed for takeoff and landing with and without extending the flaps, (*b*) the angle of attack to cruise steadily at the cruising altitude, and (*c*) the power that needs to be supplied to provide enough thrust to overcome wing drag.

Solution The cruising conditions of a passenger plane and its wing characteristics are given. The minimum safe landing and takeoff speeds, the angle of attack during cruising, and the power required are to be determined.

Assumptions **1** The drag and lift produced by parts of the plane other than the wings, such as the fuselage drag, are not considered. **2** The wings are assumed to be two-dimensional airfoil sections, and the tip effects of the wings are not considered. **3** The lift and the drag characteristics of the wings can be approximated by NACA 23012 so that Fig. 15–45 is applicable. **4** The average density of air on the ground is 1.20 kg/m^3.

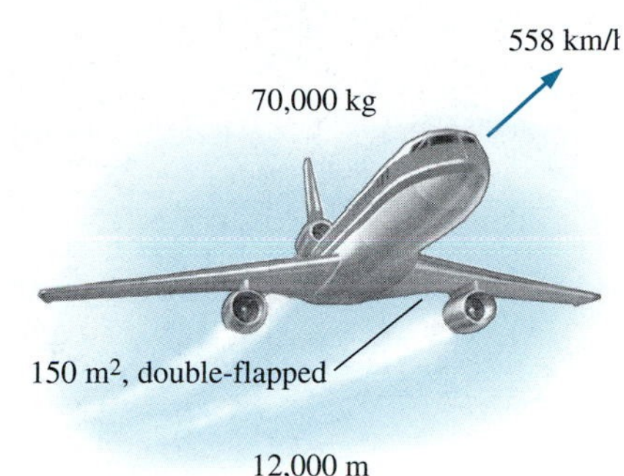

FIGURE 15–52
Schematic for Example 15–5.

Properties The density of air is 1.20 kg/m^3 on the ground and 0.312 kg/m^3 at cruising altitude. The maximum lift coefficient $C_{L,\,max}$ of the wing is 3.48 and 1.52 with and without flaps, respectively (Fig. 15–45).

Analysis (*a*) The weight and cruising speed of the airplane are

$$W = mg = (70{,}000 \text{ kg})(9.81 \text{ m/s}^2)\left(\frac{1 \text{ N}}{1 \text{ kg}\cdot\text{m/s}^2}\right) = 686{,}700 \text{ N}$$

$$V = (558 \text{ km/h})\left(\frac{1 \text{ m/s}}{3.6 \text{ km/h}}\right) = 155 \text{ m/s}$$

The minimum velocities corresponding to the stall conditions without and with flaps, respectively, are obtained from Eq. 15–21,

$$V_{\min 1} = \sqrt{\frac{2W}{\rho C_{L,\,\max 1} A}} = \sqrt{\frac{2(686{,}700 \text{ N})}{(1.2 \text{ kg/m}^3)(1.52)(150 \text{ m}^2)}\left(\frac{1 \text{ kg}\cdot\text{m/s}^2}{1 \text{ N}}\right)} = 70.9 \text{ m/s}$$

$$V_{\min 2} = \sqrt{\frac{2W}{\rho C_{L,\,\max 2} A}} = \sqrt{\frac{2(686{,}700 \text{ N})}{(1.2 \text{ kg/m}^3)(3.48)(150 \text{ m}^2)}\left(\frac{1 \text{ kg}\cdot\text{m/s}^2}{1 \text{ N}}\right)} = 46.8 \text{ m/s}$$

Then the "safe" minimum velocities to avoid the stall region are obtained by multiplying the values above by 1.2:

Without flaps: $V_{\min 1,\,\text{safe}} = 1.2V_{\min 1} = 1.2(70.9 \text{ m/s}) = 85.1 \text{ m/s} = \mathbf{306\ km/h}$

With flaps: $V_{\min 2,\,\text{safe}} = 1.2V_{\min 2} = 1.2(46.8 \text{ m/s}) = 56.2 \text{ m/s} = \mathbf{202\ km/h}$

since 1 m/s = 3.6 km/h. Note that the use of flaps allows the plane to take off and land at considerably lower velocities, and thus on a shorter runway.

(*b*) When an aircraft is cruising steadily at a constant altitude, the lift must be equal to the weight of the aircraft, $F_L = W$. Then the lift coefficient is determined to be

$$C_L = \frac{F_L}{\frac{1}{2}\rho V^2 A} = \frac{686{,}700 \text{ N}}{\frac{1}{2}(0.312 \text{ kg/m}^3)(155 \text{ m/s})^2(150 \text{ m}^2)}\left(\frac{1 \text{ kg}\cdot\text{m/s}^2}{1 \text{ N}}\right) = 1.22$$

For the case with no flaps, the angle of attack corresponding to this value of C_L is determined from Fig. 15–45 to be $\alpha \cong \mathbf{10°}$.

(*c*) When the aircraft is cruising steadily at a constant altitude, the net force acting on the aircraft is zero, and thus thrust provided by the engines must be equal to the drag force. The drag coefficient corresponding to the cruising lift coefficient of 1.22 is determined from Fig. 15–45 to be $C_D \cong 0.03$ for the case with no flaps. Then the drag force acting on the wings becomes

$$F_D = C_D A\frac{\rho V^2}{2} = (0.03)(150 \text{ m}^2)\frac{(0.312 \text{ kg/m}^3)(155 \text{ m/s})^2}{2}\left(\frac{1 \text{ kN}}{1000 \text{ kg}\cdot\text{m/s}^2}\right)$$
$$= 16.9 \text{ kN}$$

Noting that power is force times velocity (distance per unit time), the power required to overcome this drag is equal to the thrust times the cruising velocity:

$$\text{Power} = \text{Thrust} \times \text{Velocity} = F_D V = (16.9 \text{ kN})(155 \text{ m/s})\left(\frac{1 \text{ kW}}{1 \text{ kN}\cdot\text{m/s}}\right)$$
$$= \mathbf{2620\ kW}$$

Therefore, the engines must supply 2620 kW of power to overcome the drag on the wings during cruising. For a propulsion efficiency of 30 percent (i.e., 30 percent of the energy of the fuel is utilized to propel the aircraft), the plane requires energy input at a rate of 8733 kJ/s.

Discussion The power determined is the power to overcome the drag that acts on the wings only and does not include the drag that acts on the remaining parts of the aircraft (the fuselage, the tail, etc.). Therefore, the total power required during cruising will be much greater. Also, it does not consider induced drag, which can be dominant during takeoff when the angle of attack is high (Fig. 15–45 is for a 2-D airfoil, and does not include 3-D effects).

No discussion on lift and drag would be complete without mentioning the contributions of Wilbur (1867–1912) and Orville (1871–1948) Wright. The Wright Brothers are truly the most impressive engineering team of all time. Self-taught, they were well informed of the contemporary theory and practice in aeronautics. They both corresponded with other leaders in the field and published in technical journals. While they cannot be credited with developing the concepts of lift and drag, they used them to achieve the first powered, manned, heavier-than-air, controlled flight (Fig. 15–53). They succeeded, while so many before them failed, because they evaluated and designed parts separately. Before the Wrights, experimenters were building and testing whole airplanes. While intuitively appealing, the approach did not allow the determination of how to make the craft better. When a flight lasts only a moment, you can only guess at the weakness in the design. Thus, a new craft did not necessarily perform any better than its predecessor. Testing was simply one belly flop followed by another. The Wrights changed all that. They studied each part using scale and full-size models in wind tunnels and in the field. Well before the first powered flyer was assembled, they knew the area required for their best wing shape to support a plane carrying a man and the engine horsepower required to provide adequate thrust with their improved impeller. The Wright Brothers not only showed the world how to fly, they showed engineers how to use the equations presented here to design even better aircraft.

FIGURE 15–53

The Wright Brothers take flight at Kitty Hawk.

National Air and Space Museum/Smithsonian Institution.

SUMMARY

In this chapter, we study flow of fluids over immersed bodies with emphasis on the resulting lift and drag forces. A fluid may exert forces and moments on a body in and about various directions. The force a flowing fluid exerts on a body in the flow direction is called *drag* while that in the direction normal to the flow is called *lift*. The part of drag that is due directly to wall shear stress τ_w is called the *skin friction drag* since it is caused by frictional effects, and the part that is due directly to pressure P is called the *pressure drag* or *form drag* because of its strong dependence on the form or shape of the body.

The *drag coefficient* C_D and the *lift coefficient* C_L are dimensionless numbers that represent the drag and the lift characteristics of a body and are defined as

$$C_D = \frac{F_D}{\frac{1}{2}\rho V^2 A} \quad \text{and} \quad C_L = \frac{F_L}{\frac{1}{2}\rho V^2 A}$$

where A is usually the *frontal area* (the area projected on a plane normal to the direction of flow) of the body. For plates and airfoils, A is taken to be the *planform area*, which is the area that would be seen by a person looking at the body from directly above. The drag coefficient, in general, depends on the *Reynolds number*, especially for Reynolds numbers below 10^4. At higher Reynolds numbers, the drag coefficients for non-bluff geometries remain essentially constant.

A body is said to be *streamlined* if a conscious effort is made to align its shape with the anticipated streamlines in the flow in order to reduce drag. Otherwise, a body (such as a building) tends to block the flow and is said to be *blunt*. At sufficiently high velocities, the fluid stream detaches itself from the surface of the body. This is called *flow separation*. When a fluid stream separates from the body, it forms a *separated region* between the body and the fluid stream. Separation may also occur on a streamlined body such as an airplane wing at a sufficiently large *angle of attack*, which is the angle the incoming fluid stream makes with the *chord* (the line that connects the nose and the end) of the body. Flow separation on the top surface of a wing reduces lift drastically and may cause the airplane to *stall*.

The region of flow above a surface in which the effects of the viscous shearing forces caused by fluid viscosity are felt is called the *velocity boundary layer* or just the *boundary layer*. The *thickness* of the boundary layer, δ, is defined as the distance from the surface at which the velocity is $0.99V$. The hypothetical curve of velocity $0.99V$ divides the flow over a plate into two regions: the *boundary layer region*, in which the viscous effects and the velocity changes are significant, and the *irrotational outer flow region*, in which the frictional effects are negligible and the velocity remains essentially constant.

For external flow, the Reynolds number is expressed as

$$\text{Re}_L = \frac{\rho V L}{\mu} = \frac{VL}{\nu}$$

where V is the upstream velocity and L is the characteristic length of the geometry, which is the length of the plate in the flow direction for a flat plate and the diameter D for a cylinder or sphere. The *average* friction coefficients over an entire flat plate are

Laminar flow: $$C_f = \frac{1.33}{\text{Re}_L^{1/2}} \qquad \text{Re}_L < 5 \times 10^5$$

Turbulent flow: $$C_f = \frac{0.074}{\text{Re}_L^{1/5}} \qquad 5 \times 10^5 \le \text{Re}_L \le 10^7$$

If the flow is approximated as laminar up to the engineering critical number of $\text{Re}_{cr} = 5 \times 10^5$, and then turbulent beyond, the average friction coefficient over the entire flat plate becomes

$$C_f = \frac{0.074}{\text{Re}_L^{1/5}} - \frac{1742}{\text{Re}_L} \qquad 5 \times 10^5 \le \text{Re}_L \le 10^7$$

A curve fit of experimental data for the average friction coefficient in the fully rough turbulent regime is

Rough surface: $$C_f = \left(1.89 - 1.62 \log \frac{\varepsilon}{L}\right)^{-2.5}$$

where ε is the surface roughness and L is the length of the plate in the flow direction. In the absence of a better one, this relation can be used for turbulent flow on rough surfaces for $\text{Re} > 10^6$, especially when $\varepsilon/L > 10^{-4}$.

Surface roughness, in general, increases the drag coefficient in turbulent flow. For bluff bodies such as a circular cylinder or sphere, however, an increase in the surface roughness may *decrease* the drag coefficient. This is done by tripping the flow into turbulence at a lower Reynolds number, and thus causing the fluid to close in behind the body, narrowing the wake and reducing pressure drag considerably.

It is desirable for airfoils to generate the most lift while producing the least drag. Therefore, a measure of performance for airfoils is the *lift-to-drag ratio*, C_L/C_D.

The minimum safe flight velocity of an aircraft is determined from

$$V_{\min} = \sqrt{\frac{2W}{\rho C_{L,\max} A}}$$

For a given weight, the landing or takeoff speed can be minimized by maximizing the product of the lift coefficient and the wing area, $C_{L,\max}A$.

For airplane wings and other airfoils of finite span, the pressure difference between the lower and the upper surfaces drives the fluid at the tips upward. This results in a swirling eddy, called the *tip vortex*. Tip vortices that interact with the free stream impose forces on the wing tips in all directions, including the flow direction. The component of the force in the flow direction adds to drag and is called *induced drag*. The total drag of a wing is then the sum of the induced drag (3-D effects) and the drag of the airfoil section (2-D effects).

REFERENCES AND SUGGESTED READING

1. I. H. Abbott. "The Drag of Two Streamline Bodies as Affected by Protuberances and Appendages," *NACA Report* 451, 1932.
2. I. H. Abbott and A. E. von Doenhoff. *Theory of Wing Sections, Including a Summary of Airfoil Data*. New York: Dover, 1959.
3. I. H. Abbott, A. E. von Doenhoff, and L. S. Stivers. "Summary of Airfoil Data," *NACA Report* 824, Langley Field, VA, 1945.
4. J. D. Anderson. *Fundamentals of Aerodynamics*, 2nd ed. New York: McGraw-Hill, 1991.

5. R. D. Blevins. *Applied Fluid Dynamics Handbook*. New York: Van Nostrand Reinhold, 1984.

6. S. W. Churchill and M. Bernstein. "A Correlating Equation for Forced Convection from Gases and Liquids to a Circular Cylinder in Cross Flow," *Journal of Heat Transfer* 99, pp. 300–306, 1977.

7. S. Goldstein. *Modern Developments in Fluid Dynamics*. London: Oxford Press, 1938.

8. J. Happel. *Low Reynolds Number Hydrocarbons*. Englewood Cliffs, NJ: Prentice Hall, 1965.

9. S. F. Hoerner. *Fluid-Dynamic Drag*. [Published by the author.] Library of Congress No. 64, 1966.

10. G. M. Homsy, H. Aref, K. S. Breuer, S. Hochgreb, J. R. Koseff, B. R. Munson, K. G. Powell, C. R. Robertson, S. T. Thoroddsen. *Multi-Media Fluid Mechanics* (CD). Cambridge University Press, 2000.

11. W. H. Hucho. *Aerodynamics of Road Vehicles*. London: Butterworth-Heinemann, 1987.

12. H. Schlichting. *Boundary Layer Theory*, 7th ed. New York: McGraw-Hill, 1979.

13. M. Van Dyke. *An Album of Fluid Motion*. Stanford, CA: The Parabolic Press, 1982.

14. J. Vogel. *Life in Moving Fluids*, 2nd ed. Boston: Willard Grand Press, 1994.

15. F. M. White. *Fluid Mechanics*, 5th ed. New York: McGraw-Hill, 2003.

PROBLEMS*

Drag, Lift, and Drag Coefficients

15–1C Explain when an external flow is two-dimensional, three-dimensional, and axisymmetric. What type of flow is the flow of air over a car?

15–2C What is the difference between the upstream velocity and the free-stream velocity? For what types of flow are these two velocities equal to each other?

15–3C What is the difference between streamlined and blunt bodies? Is a tennis ball a streamlined or blunt body?

15–4C Name some applications in which a large drag is desired.

15–5C What is drag? What causes it? Why do we usually try to minimize it?

15–6C What is lift? What causes it? Does wall shear contribute to the lift?

15–7C During flow over a given body, the drag force, the upstream velocity, and the fluid density are measured. Explain how you would determine the drag coefficient. What area would you use in the calculations?

15–8C During flow over a given slender body such as a wing, the lift force, the upstream velocity, and the fluid density are measured. Explain how you would determine the lift coefficient. What area would you use in the calculations?

15–9C Define the frontal area of a body subjected to external flow. When is it appropriate to use the frontal area in drag and lift calculations?

15–10C Define the planform area of a body subjected to external flow. When is it appropriate to use the planform area in drag and lift calculations?

15–11C What is terminal velocity? How is it determined?

15–12C What is the difference between skin friction drag and pressure drag? Which is usually more significant for slender bodies such as airfoils?

15–13C What is the effect of surface roughness on the friction drag coefficient in laminar and turbulent flows?

15–14C In general, how does the drag coefficient vary with the Reynolds number at (*a*) low and moderate Reynolds numbers and (*b*) at high Reynolds numbers ($Re > 10^4$)?

15–15C Fairings are attached to the front and back of a cylindrical body to make it look more streamlined. What is the effect of this modification on the (*a*) friction drag, (*b*) pressure drag, and (*c*) total drag? Assume the Reynolds number is high enough so that the flow is turbulent for both cases.

* Problems designated by a "C" are concept questions, and students are encouraged to answer them all. Problems designated by an "E" are in English units, and the SI users can ignore them. Problems with the icon are solved using EES, and complete solutions together with parametric studies are included on the enclosed DVD. Problems with the icon are comprehensive in nature and are intended to be solved with a computer, preferably using the EES software that accompanies this text.

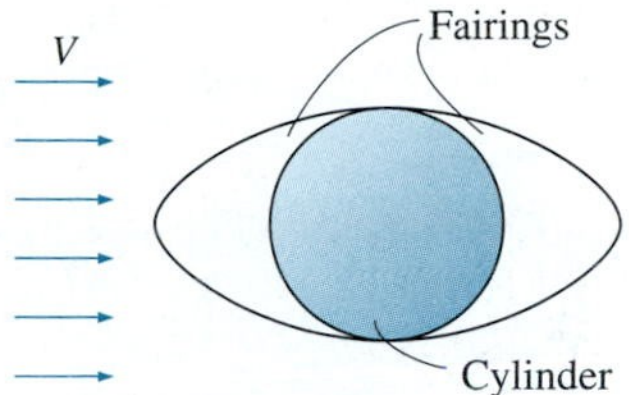

FIGURE P15–15C

15–16C What is the effect of streamlining on (*a*) friction drag and (*b*) pressure drag? Does the total drag acting on a body necessarily decrease as a result of streamlining? Explain.

15–17C What is flow separation? What causes it? What is the effect of flow separation on the drag coefficient?

15–18C What is drafting? How does it affect the drag coefficient of the drafted body?

15–19C Which car is more likely to be more fuel-efficient: the one with sharp corners or the one that is contoured to resemble an ellipse? Why?

15–20C Which bicyclist is more likely to go faster: the one who keeps his head and his body in the most upright position or the one who leans down and brings his body closer to his knees? Why?

15–21 The drag coefficient of a car at the design conditions of 1 atm, 25°C, and 90 km/h is to be determined experimentally in a large wind tunnel in a full-scale test. The height and width of the car are 1.40 m and 1.65 m, respectively. If the horizontal force acting on the car is measured to be 300 N, determine the total drag coefficient of this car. *Answer:* 0.35

15–22 A car is moving at a constant velocity of 80 km/h. Determine the upstream velocity to be used in fluid flow analysis if (*a*) the air is calm, (*b*) wind is blowing against the direction of motion of the car at 30 km/h, and (*c*) wind is blowing in the same direction of motion of the car at 50 km/h.

15–23 The resultant of the pressure and wall shear forces acting on a body is measured to be 700 N, making 35° with the direction of flow. Determine the drag and the lift forces acting on the body.

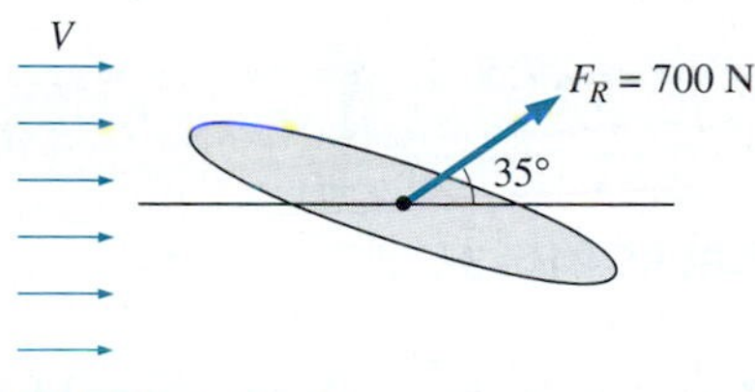

FIGURE P15–23

15–24 During a high Reynolds number experiment, the total drag force acting on a spherical body of diameter D = 12 cm subjected to airflow at 1 atm and 5°C is measured to be 5.2 N. The pressure drag acting on the body is calculated by integrating the pressure distribution (measured by the use of pressure sensors throughout the surface) to be 4.9 N. Determine the friction drag coefficient of the sphere. *Answer:* 0.0115

15–25E To reduce the drag coefficient and thus to improve the fuel efficiency, the frontal area of a car is to be reduced. Determine the amount of fuel and money saved per year as a result of reducing the frontal area from 18 to 15 ft^2. Assume the car is driven 12,000 mi a year at an average speed of 55 mi/h. Take the density and price of gasoline to be 50 lbm/ft^3 and $2.20/gal, respectively; the density of air to be 0.075 lbm/ft^3, the heating value of gasoline to be 20,000 Btu/lbm; and the overall efficiency of the engine to be 32 percent.

15–26E Reconsider Prob. 15–25E. Using EES (or other) software, investigate the effect of frontal area on the annual fuel consumption of the car. Let the frontal area vary from 10 to 30 ft^2 in increments of 2 ft^2. Tabulate and plot the results.

15–27 A circular sign has a diameter of 50 cm and is subjected to normal winds up to 150 km/h at 10°C and 100 kPa. Determine the drag force acting on the sign. Also determine the bending moment at the bottom of its pole whose height from the ground to the bottom of the sign is 1.5 m. Disregard the drag on the pole.

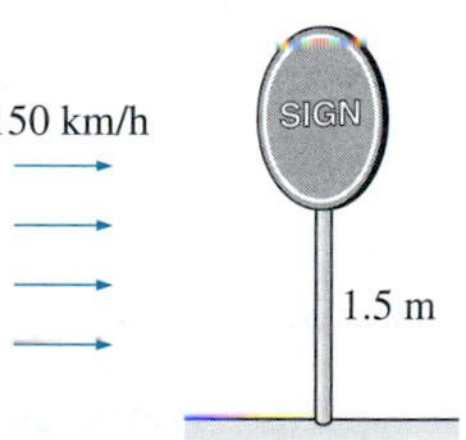

FIGURE P15–27

15–28E Wind loading is a primary consideration in the design of the supporting mechanisms of billboards, as evidenced by many billboards being knocked down during high winds. Determine the wind force acting on an 8-ft-high, 20-ft-wide billboard due to 90-mi/h winds in the normal direction when the atmospheric conditions are 14.3 psia and 40°F. *Answer:* 6684 lbf

15–29 Advertisement signs are commonly carried by taxicabs for additional income, but they also increase the fuel cost. Consider a sign that consists of a 0.30-m-high, 0.9-m-wide, and 0.9-m-long rectangular block mounted on top of a taxicab such that the sign has a frontal area of 0.3 m by 0.9 m from all four sides. Determine the increase in the annual fuel cost of this taxicab due to this sign. Assume the taxicab is driven 60,000 km a year at an average speed of 50 km/h and the overall efficiency of the engine is 28 percent. Take the density, unit price, and heating value of gasoline to be 0.75 kg/L, \$0.50/L, and 42,000 kJ/kg, respectively, and the density of air to be 1.25 kg/m^3.

FIGURE P15–29

15–30 It is proposed to meet the water needs of a recreational vehicle (RV) by installing a 2-m-long, 0.5-m-diameter cylindrical tank on top of the vehicle. Determine the additional power requirement of the RV at a speed of 95 km/h when the tank is installed such that its circular surfaces face (*a*) the front and back and (*b*) the sides of the RV. Assume atmospheric conditions are 87 kPa and 20°C. *Answers:* (*a*) 1.67 kW, (*b*) 7.55 kW

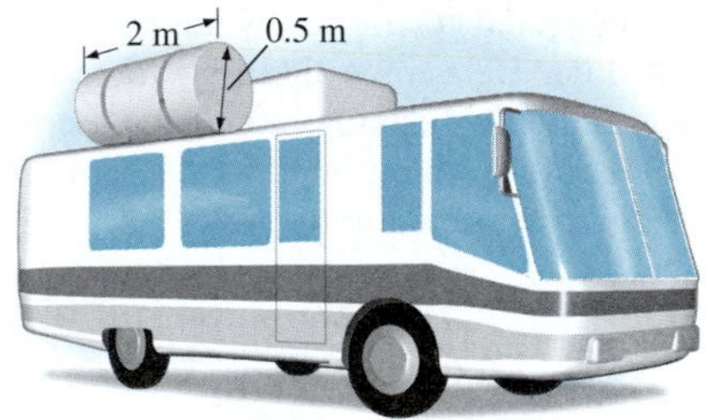

FIGURE P15–30

15–31E At highway speeds, about half of the power generated by the car's engine is used to overcome aerodynamic drag, and thus the fuel consumption is nearly proportional to the drag force on a level road. Determine the percentage increase in fuel consumption of a car per unit time when a person who normally drives at 55 mi/h now starts driving at 75 mi/h.

15–32 A 4-mm-diameter plastic sphere whose density is 1150 kg/m^3 is dropped into water at 20°C. Determine the terminal velocity of the sphere in water.

15–33 During major windstorms, high vehicles such as RVs and semis may be thrown off the road and boxcars off their tracks, especially when they are empty and in open areas. Consider a 5000-kg semi that is 8 m long, 2 m high, and 2 m wide. The distance between the bottom of the truck and the road is 0.75 m. Now the truck is exposed to winds from its side surface. Determine the wind velocity that will tip the truck over to its side. Take the air density to be 1.1 kg/m^3 and assume the weight to be uniformly distributed.

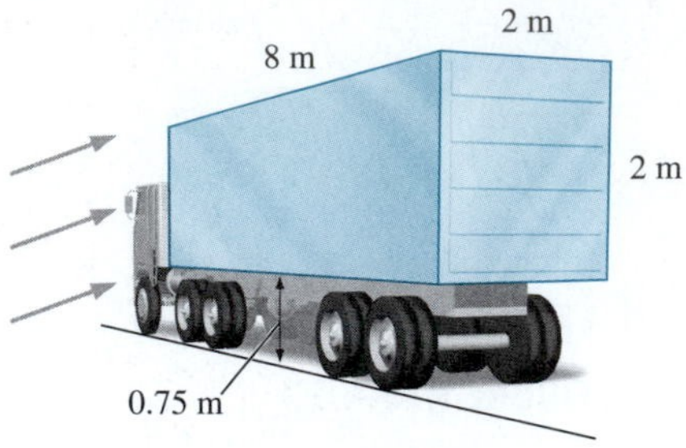

FIGURE P15–33

15–34 An 80-kg bicyclist is riding her 15-kg bicycle downhill on a road with a slope of 12° without pedaling or braking. The bicyclist has a frontal area of 0.45 m^2 and a drag coefficient of 1.1 in the upright position, and a frontal area of 0.4 m^2 and a drag coefficient of 0.9 in the racing position. Disregarding the rolling resistance and friction at the bearings, determine the terminal velocity of the bicyclist for both positions. Take the air density to be 1.25 kg/m^3. *Answers:* 90 km/h, 106 km/h

15–35 A wind turbine with two or four hollow hemispherical cups connected to a pivot is commonly used to measure wind speed. Consider a wind turbine with two 8-cm-diameter cups with a center-to-center distance of 25 cm, as shown in Fig. P15–35. The pivot is stuck as a result of some malfunction, and the cups stop rotating. For a wind speed of 15 m/s and air density of 1.25 kg/m^3, determine the maximum torque this turbine applies on the pivot.

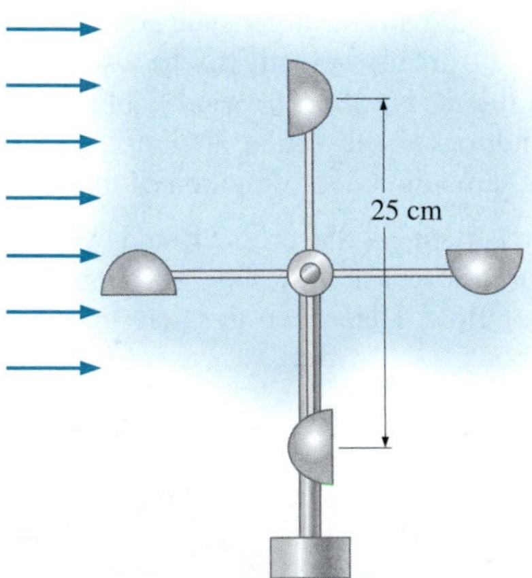

FIGURE P15–35

15–36 Reconsider Prob. 15–35. Using EES (or other) software, investigate the effect of wind speed on the torque applied on the pivot. Let the wind speed vary from 0 to 50 m/s in increments of 5 m/s. Tabulate and plot the results.

15–37E A 5-ft-diameter spherical tank completely submerged in freshwater is being towed by a ship at 12 ft/s. Assuming turbulent flow, determine the required towing power.

15–38 During steady motion of a vehicle on a level road, the power delivered to the wheels is used to overcome aerodynamic drag and rolling resistance (the product of the rolling resistance coefficient and the weight of the vehicle), assuming the friction at the bearings of the wheels is negligible. Consider a car that has a total mass of 950 kg, a drag coefficient of 0.32, a frontal area of 1.8 m^2, and a rolling resistance coefficient of 0.04. The maximum power the engine can deliver to the wheels is 80 kW. Determine (*a*) the speed at which the rolling resistance is equal to the aerodynamic drag force and (*b*) the maximum speed of this car. Take the air density to be 1.20 kg/m^3.

15–39 EES Reconsider Prob. 15–38. Using EES (or other) software, investigate the effect of car speed on the required power to overcome (*a*) rolling resistance, (*b*) the aerodynamic drag, and (*c*) their combined effect. Let the car speed vary from 0 to 150 km/h in increments of 15 km/h. Tabulate and plot the results.

15–40 A submarine can be treated as an ellipsoid with a diameter of 5 m and a length of 25 m. Determine the power required for this submarine to cruise horizontally and steadily at 40 km/h in seawater whose density is 1025 kg/m^3. Also determine the power required to tow this submarine in air whose density is 1.30 kg/m^3. Assume the flow is turbulent in both cases.

FIGURE P15–40

15–41 An 0.80-m-diameter, 1.2-m-high garbage can is found in the morning tipped over due to high winds during the night. Assuming the average density of the garbage inside to be 150 kg/m^3 and taking the air density to be 1.25 kg/m^3, estimate the wind velocity during the night when the can was tipped over. Take the drag coefficient of the can to be 0.7. *Answer*: 135 km/h

15–42E The drag coefficient of a vehicle increases when its windows are rolled down or its sunroof is opened. A sports car has a frontal area of 18 ft^2 and a drag coefficient of 0.32 when the windows and sunroof are closed. The drag coefficient increases to 0.41 when the sunroof is open. Determine the additional power consumption of the car when the sunroof is opened at (*a*) 35 mi/h and (*b*) 70 mi/h. Take the density of air to be 0.075 lbm/ft^3.

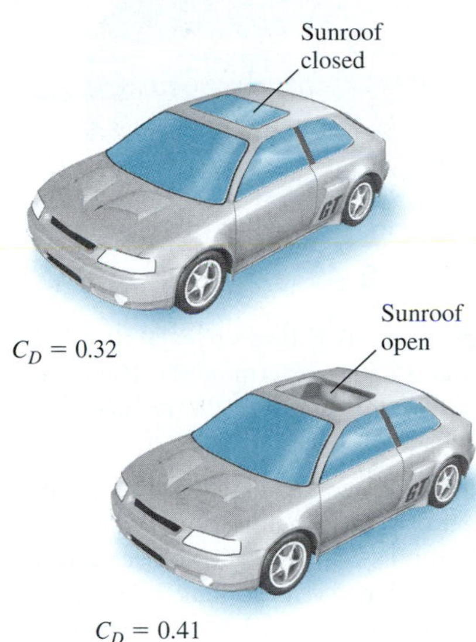

FIGURE P15–42E

Flow over Flat Plates

15–43C What fluid property is responsible for the development of the velocity boundary layer? What is the effect of the velocity on the thickness of the boundary layer?

15–44C What does the friction coefficient represent in flow over a flat plate? How is it related to the drag force acting on the plate?

15–45C Consider laminar flow over a flat plate. How does the local friction coefficient change with position?

15–46C How is the average friction coefficient determined in flow over a flat plate?

15–47E Light oil at 75°F flows over a 15-ft-long flat plate with a free-stream velocity of 6 ft/s. Determine the total drag force per unit width of the plate.

15–48 The local atmospheric pressure in Denver, Colorado (elevation 1610 m) is 83.4 kPa. Air at this pressure and at 25°C flows with a velocity of 6 m/s over a 2.5-m × 8-m flat plate. Determine the drag force acting on the top surface of the plate if the air flows parallel to the (*a*) 8-m-long side and (*b*) the 2.5-m-long side.

15–49 During a winter day, wind at 55 km/h, 5°C, and 1 atm is blowing parallel to a 4-m-high and 10-m-long wall of a house. Assuming the wall surfaces to be smooth, determine the friction drag acting on the wall. What would your answer be if the wind velocity has doubled? How realistic is it to treat the flow over side wall surfaces as flow over a flat plate? *Answers*: 16 N, 58 N

FIGURE P15–49

15–50E Air at 70°F flows over a 10-ft-long flat plate at 25 ft/s. Determine the local friction coefficient at intervals of 1 ft and plot the results against the distance from the leading edge.

15–51 The forming section of a plastics plant puts out a continuous sheet of plastic that is 1.2 m wide and 2 mm thick at a rate of 15 m/min. The sheet is subjected to airflow at a velocity of 3 m/s on both sides along its surfaces normal to the direction of motion of the sheet. The width of the air cooling section is such that a fixed point on the plastic sheet passes through that section in 2 s. Using properties of air at 1 atm and 60°C, determine the drag force the air exerts on the plastic sheet in the direction of airflow.

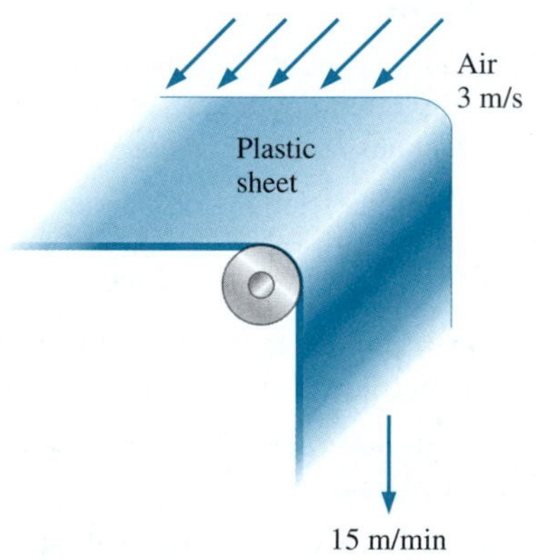

FIGURE P15–51

15–52 The top surface of the passenger car of a train moving at a velocity of 70 km/h is 3.2 m wide and 8 m long. If the outdoor air is at 1 atm and 25°C, determine the drag force acting on the top surface of the car.

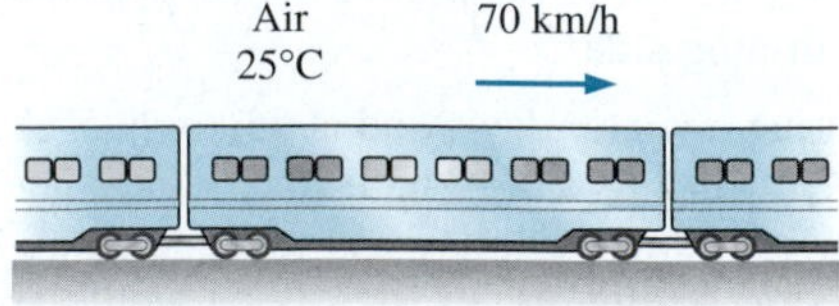

FIGURE P15–52

15–53 The weight of a thin flat plate 50 cm × 50 cm in size is balanced by a counterweight that has a mass of 2 kg, as shown in Fig. P15–53. Now a fan is turned on, and air at 1 atm and 25°C flows downward over both surfaces of the plate with a free-stream velocity of 10 m/s. Determine the mass of the counterweight that needs to be added in order to balance the plate in this case.

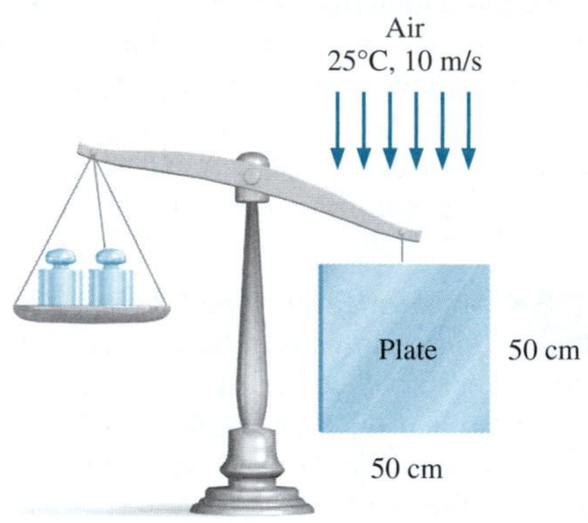

FIGURE P15–53

15–54 Consider laminar flow of a fluid over a flat plate. Now the free-stream velocity of the fluid is doubled. Determine the change in the drag force on the plate. Assume the flow to remain laminar. *Answer*: A 2.83-fold increase

15–55E Consider a refrigeration truck traveling at 65 mi/h at a location where the air temperature is at 1 atm and 80°F. The refrigerated compartment of the truck can be considered to be a 9-ft-wide, 8-ft-high, and 20-ft-long rectangular box. Assuming the airflow over the entire outer surface to be turbulent and attached (no flow separation), determine the drag force acting on the top and side surfaces and the power required to overcome this drag.

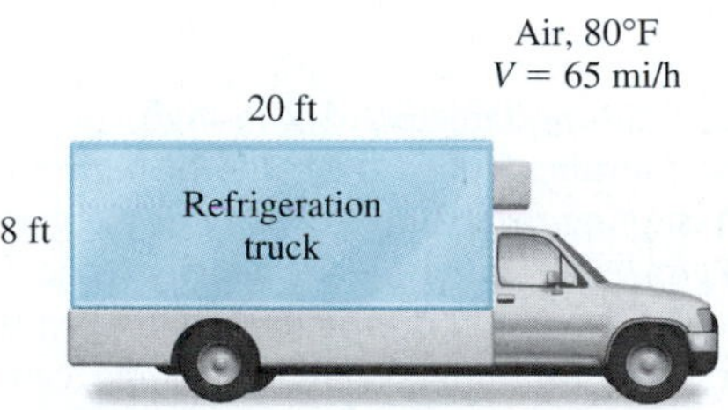

FIGURE P15–55E

15–56E Reconsider Prob. 15–55E. Using EES (or other) software, investigate the effect of truck speed on the total drag force acting on the top and side surfaces, and the power required to overcome it. Let the truck speed vary from 0 to 100 mi/h in increments of 10 mi/h. Tabulate and plot the results.

15–57 Air at 25°C and 1 atm is flowing over a long flat plate with a velocity of 8 m/s. Determine the distance from the leading edge of the plate where the flow becomes turbulent, and the thickness of the boundary layer at that location.

15–58 Repeat Prob. 15–57 for water.

Flow across Cylinders and Spheres

15–59C In flow over cylinders, why does the drag coefficient suddenly drop when the flow becomes turbulent? Isn't turbulence supposed to increase the drag coefficient instead of decreasing it?

15–60C In flow over blunt bodies such as a cylinder, how does the pressure drag differ from the friction drag?

15–61C Why is flow separation in flow over cylinders delayed in turbulent flow?

15–62E A 1.2-in-outer-diameter pipe is to span across a river at a 105-ft-wide section while being completely immersed in water. The average flow velocity of water is 10 ft/s, and the water temperature is 70°F. Determine the drag force exerted on the pipe by the river. *Answer:* 1120 lbf

15–63 A long 8-cm-diameter steam pipe passes through some area open to the winds. Determine the drag force acting on the pipe per unit of its length when the air is at 1 atm and 5°C and the wind is blowing across the pipe at a velocity of 50 km/h.

15–64E A person extends his uncovered arms into the windy air outside at 1 atm and 60°F and 20 mi/h in order to feel nature closely. Treating the arm as a 2-ft-long and 3-in-diameter cylinder, determine the drag force on both arms. *Answer:* 1.02 lbf

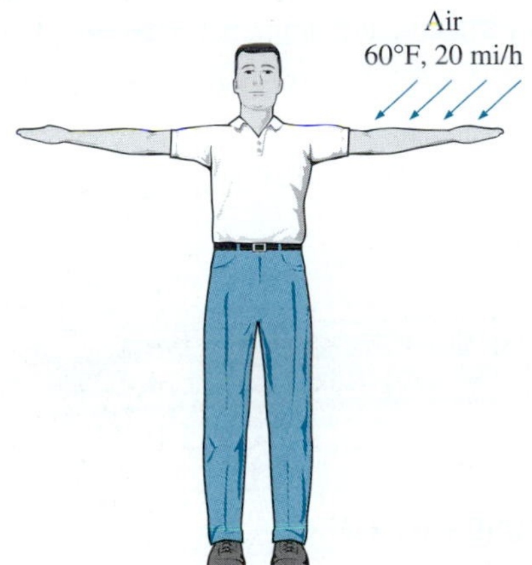

FIGURE P15–64E

15–65 A 6-mm-diameter electrical transmission line is exposed to windy air. Determine the drag force exerted on a 120-m-long section of the wire during a windy day when the air is at 1 atm and 15°C and the wind is blowing across the transmission line at 40 km/h.

15–66 Consider 0.8-cm-diameter hail that is falling freely in atmospheric air at 1 atm and 5°C. Determine the terminal velocity of the hail. Take the density of hail to be 910 kg/m^3.

15–67 A 0.1-mm-diameter dust particle whose density is 2.1 g/cm^3 is observed to be suspended in the air at 1 atm and 25°C at a fixed point. Estimate the updraft velocity of air motion at that location. Assume Stokes law to be applicable. Is this a valid assumption? *Answer:* 0.62 m/s

15–68 Dust particles of diameter 0.05 mm and density 1.8 g/cm^3 are unsettled during high winds and rise to a height of 350 m by the time things calm down. Estimate how long it will take for the dust particles to fall back to the ground in still air at 1 atm and 15°C, and their velocity. Disregard the initial transient period during which the dust particles accelerate to their terminal velocity, and assume Stokes law to be applicable.

15–69 A 2-m-long, 0.2-m-diameter cylindrical pine log (density = 513 kg/m^3) is suspended by a crane in the horizontal position. The log is subjected to normal winds of 40 km/h at 5°C and 88 kPa. Disregarding the weight of the cable and its drag, determine the angle θ the cable will make with the horizontal and the tension on the cable.

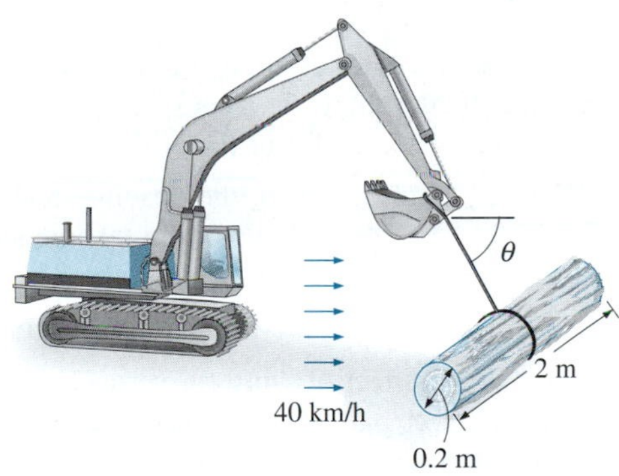

FIGURE P15–69

15–70 One of the popular demonstrations in science museums involves the suspension of a ping-pong ball by an upward air jet. Children are amused by the ball always coming back to the center when it is pushed by a finger to the side of the jet. Explain this phenomenon using the Bernoulli equation. Also determine the velocity of air if the ball has a mass of 2.6 g and a diameter of 3.8 cm. Assume air is at 1 atm and 25°C.

FIGURE P15–70

Lift

15–71C Why is the contribution of viscous effects to lift usually negligible for airfoils?

15–72C Air is flowing past a symmetrical airfoil at zero angle of attack. Will the (*a*) lift and (*b*) drag acting on the airfoil be zero or nonzero?

15–73C Air is flowing past a nonsymmetrical airfoil at zero angle of attack. Will the (*a*) lift and (*b*) drag acting on the airfoil be zero or nonzero?

15–74C Air is flowing past a symmetrical airfoil at an angle of attack of 5°. Will the (*a*) lift and (*b*) drag acting on the airfoil be zero or nonzero?

15–75C What is stall? What causes an airfoil to stall? Why are commercial aircraft not allowed to fly at conditions near stall?

15–76C Both the lift and the drag of an airfoil increase with an increase in the angle of attack. In general, which increases at a higher rate, the lift or the drag?

15–77C Why are flaps used at the leading and trailing edges of the wings of large aircraft during takeoff and landing? Can an aircraft take off or land without them?

15–78C How do flaps affect the lift and the drag of wings?

15–79C What is the effect of wing tip vortices (the air circulation from the lower part of the wings to the upper part) on the drag and the lift?

15–80C What is induced drag on wings? Can induced drag be minimized by using long and narrow wings or short and wide wings?

15–81C Explain why **endplates** or **winglets** are added to some airplane wings.

15–82E A commercial airplane has a total mass of 150,000 lbm and a wing platform area of 1800 ft^2. The plane has a cruising speed of 550 mi/h and a cruising altitude of 38,000 ft where the air density is 0.0208 lbm/ft^3. The plane has double-slotted flaps for use during takeoff and landing, but it cruises with all flaps retracted. Assuming the lift and drag characteristics of the wings can be approximated by NACA 23012, determine (*a*) the minimum safe speed for takeoff and landing with and without extending the flaps, (*b*) the angle of attack to cruise steadily at the cruising altitude, and (*c*) the power that needs to be supplied to provide enough thrust to overcome drag. Take the air density on the ground to be 0.075 lbm/ft^3.

15–83 Consider an aircraft, which takes off at 190 km/h when it is fully loaded. If the weight of the aircraft is increased by 20 percent as a result of overloading, determine the speed at which the overloaded aircraft will take off. *Answer:* 208 km/h

15–84E An airplane is consuming fuel at a rate of 5 gal/min when cruising at a constant altitude of 10,000 ft at constant speed. Assuming the drag coefficient and the engine efficiency to remain the same, determine the rate of fuel consumption at an altitude of 30,000 ft at the same speed.

15–85 A jumbo jet airplane has a mass of about 400,000 kg when fully loaded with over 400 passengers and takes off at a speed of 250 km/h. Determine the takeoff speed when the airplane has 100 empty seats. Assume each passenger with luggage is 140 kg and the wing and flap settings are maintained the same. *Answer:* 246 km/h

15–86 Reconsider Prob. 15–85. Using EES (or other) software, investigate the effect of passenger count on the takeoff speed of the aircraft. Let the number of passengers vary from 0 to 500 in increments of 50. Tabulate and plot the results.

15–87 A small aircraft has a wing area of 30 m^2, a lift coefficient of 0.45 at takeoff settings, and a total mass of 2800 kg. Determine (*a*) the takeoff speed of this aircraft at sea level at standard atmospheric conditions, (*b*) the wing loading, and (*c*) the required power to maintain a constant cruising speed of 300 km/h for a cruising drag coefficient of 0.035.

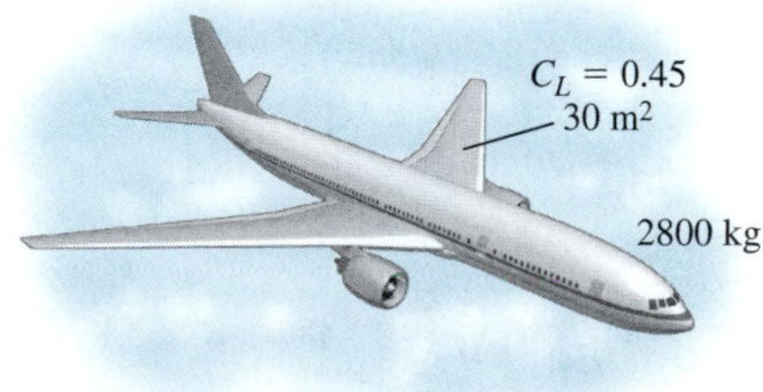

FIGURE P15–87

15–88 A small airplane has a total mass of 1800 kg and a wing area of 42 m^2. Determine the lift and drag coefficients of this airplane while cruising at an altitude of 4000 m at a constant speed of 280 km/h and generating 190 kW of power.

15–89 The NACA 64(1)–412 airfoil has a lift-to-drag ratio of 50 at 0° angle of attack, as shown in Fig. 15–43. At what angle of attack will this ratio increase to 80?

15–90 Consider a light plane that has a total weight of 15,000 N and a wing area of 46 m^2 and whose wings resemble the NACA 23012 airfoil with no flaps. Using data from Fig. 15–45, determine the takeoff speed at an angle of attack of 5° at sea level. Also determine the stall speed. *Answers:* 94 km/h, 67.4 km/h

15–91 An airplane has a mass of 50,000 kg, a wing area of 300 m^2, a maximum lift coefficient of 3.2, and a cruising drag coefficient of 0.03 at an altitude of 12,000 m. Determine (*a*) the takeoff speed at sea level, assuming it is 20 percent over the stall speed, and (*b*) the thrust that the engines must deliver for a cruising speed of 700 km/h.

Review Problems

15–92 Calculate the thickness of the boundary layer during flow over a 2.5-m-long flat plate at intervals of 25 cm and plot the boundary layer over the plate for the flow of (*a*) air, (*b*) water, and (*c*) engine oil at 1 atm and 20°C at an upstream velocity of 3 m/s.

15–93 A 1-m-external-diameter spherical tank is located outdoors at 1 atm and 25°C and is subjected to winds at 35 km/h. Determine the drag force exerted on it by the wind. *Answer:* 3.5 N

15–94 A 2-m-high, 4-m-wide rectangular advertisement panel is attached to a 4-m-wide, 0.15-m-high rectangular concrete block (density = 2300 kg/m^3) by two 5-cm-diameter, 4-m-high (exposed part) poles, as shown in Fig. P15–94. If the sign is to withstand 150 km/h winds from any direction, determine (*a*) the maximum drag force on the panel, (*b*) the drag force acting on the poles, and (*c*) the minimum length *L* of the concrete block for the panel to resist the winds. Take the density of air to be 1.30 kg/m^3.

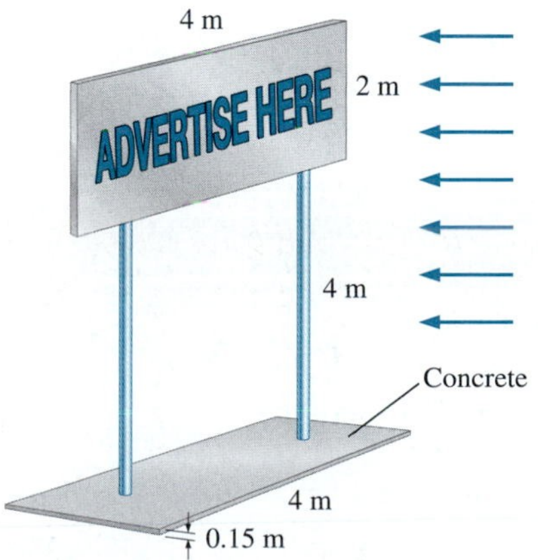

FIGURE P15–94

15–95 A plastic boat whose bottom surface can be approximated as a 1.5-m-wide, 2-m-long flat surface is to move through water at 15°C at speeds up to 30 km/h. Determine the friction drag exerted on the boat by water and the power needed to overcome it.

FIGURE P15–95

15–96 Reconsider Prob. 15–95. Using EES (or other) software, investigate the effect of boat speed on the drag force acting on the bottom surface of the boat, and the power needed to overcome it. Let the boat speed vary from 0 to 100 km/h in increments of 10 km/h. Tabulate and plot the results.

15–97 A paratrooper and his 8-m-diameter parachute weigh 950 N. Taking the average air density to be 1.2 kg/m^3, determine the terminal velocity of the paratrooper. *Answer:* 4.9 m/s

FIGURE P15–97

15–98 A 17,000-kg tractor-trailer rig has a frontal area of 9.2 m^2, a drag coefficient of 0.96, a rolling resistance coefficient of 0.05 (multiplying the weight of a vehicle by the rolling resistance coefficient gives the rolling resistance), a bearing friction resistance of 350 N, and a maximum speed of 110 km/h on a level road during steady cruising in calm weather with an air density of 1.25 kg/m^3. Now a fairing is installed to the front of the rig to suppress separation and to streamline the flow to the top surface, and the drag coefficient is reduced to 0.76. Determine the maximum speed of the rig with the fairing. *Answer*: 133 km/h

15–99 Stokes law can be used to determine the viscosity of a fluid by dropping a spherical object in it and measuring the terminal velocity of the object in that fluid. This can be done by plotting the distance traveled against time and observing when the curve becomes linear. During such an experiment a 3-mm-diameter glass ball (ρ = 2500 kg/m^3) is dropped into a fluid whose density is 875 kg/m^3, and the terminal velocity is measured to be 0.12 m/s. Disregarding the wall effects, determine the viscosity of the fluid.

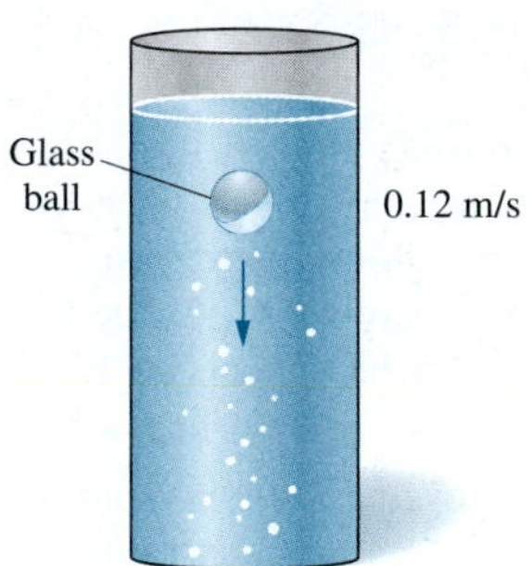

FIGURE P15–99

15–100 During an experiment, three aluminum balls (ρ_s = 2600 kg/m^3) having diameters 2, 4, and 10 mm, respectively, are dropped into a tank filled with glycerin at 22°C (ρ_f = 1274 kg/m^3 and μ = 1 kg/m · s). The terminal settling velocities of the balls are measured to be 3.2, 12.8, and 60.4 mm/s, respectively. Compare these values with the velocities predicted by Stokes law for drag force $F_D = 3\pi\mu DV$, which is valid for very low Reynolds numbers (Re $\ll$ 1). Determine the error involved for each case and assess the accuracy of Stokes law.

15–101 Repeat Prob. 15–100 by considering the more general form of Stokes law expressed as $F_D = 3\pi\mu DV + (9\pi/16)\rho_s V^2 D^2$.

Design and Essay Problems

15–102 Write a report on the history of the reduction of the drag coefficients of cars and obtain the drag coefficient data for some recent car models from the catalogs of car manufacturers.

15–103 Large commercial airplanes cruise at high altitudes (up to about 40,000 ft) to save fuel. Discuss how flying at high altitudes reduces drag and saves fuel. Also discuss why small planes fly at relatively low altitudes.

15–104 Many drivers turn off their air conditioners and roll down the car windows in hopes of saving fuel. But it is claimed that this apparent "free cooling" actually increases the fuel consumption of the car. Investigate this matter and write a report on which practice will save gasoline under what conditions.

PART

3

HEAT TRANSFER

Chapter 16

MECHANISMS OF HEAT TRANSFER

The science of thermodynamics deals with the *amount* of heat transfer as a system undergoes a process from one equilibrium state to another, and makes no reference to *how long* the process will take. But in engineering, we are often interested in the *rate* of heat transfer, which is the topic of the science of *heat transfer.*

We start this chapter with an overview of the three basic mechanisms of heat transfer, which are conduction, convection, and radiation, and discuss thermal conductivity. *Conduction* is the transfer of energy from the more energetic particles of a substance to the adjacent, less energetic ones as a result of interactions between the particles. *Convection* is the mode of heat transfer between a surface and the adjacent liquid or gas that is in motion, and it involves the combined effects of conduction and fluid motion. *Radiation* is the energy emitted by matter in the form of electromagnetic waves (or photons) as a result of the changes in the electronic configurations of the atoms or molecules. We close this chapter with a discussion of simultaneous heat transfer.

Objectives

The objectives of this chapter are to:

- Understand the basic mechanisms of heat transfer, which are conduction, convection, and radiation, and Fourier's law of heat conduction, Newton's law of cooling, and the Stefan–Boltzmann law of radiation,
- Identify the mechanisms of heat transfer that occur simultaneously in practice,
- Develop an awareness of the cost associated with heat losses, and
- Solve various heat transfer problems encountered in practice.

16–1 ■ INTRODUCTION

In Chapter 3, we defined *heat* as the form of energy that can be transferred from one system to another as a result of temperature difference. A thermodynamic analysis is concerned with the *amount* of heat transfer as a system undergoes a process from one equilibrium state to another. The science that deals with the determination of the *rates* of such energy transfers is the *heat transfer*. The transfer of energy as heat is always from the higher-temperature medium to the lower-temperature one, and heat transfer stops when the two mediums reach the same temperature.

Heat can be transferred in three different modes: *conduction, convection,* and *radiation.* All modes of heat transfer require the existence of a temperature difference, and all modes are from the high-temperature medium to a lower-temperature one. Below we give a brief description of each mode. A detailed study of these modes is given in later chapters of this text.

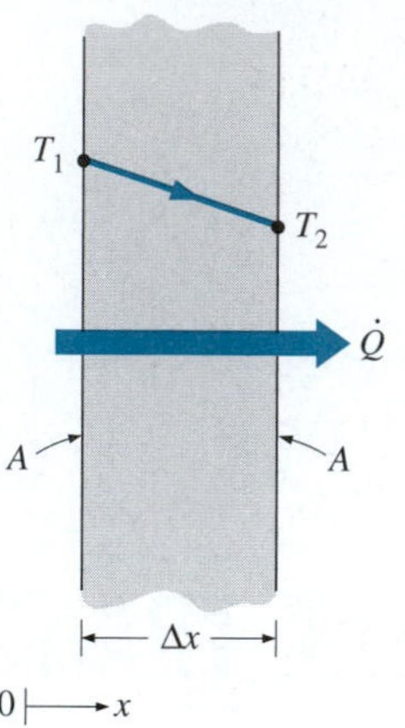

FIGURE 16–1

Heat conduction through a large plane wall of thickness Δx and area A.

16–2 ■ CONDUCTION

Conduction is the transfer of energy from the more energetic particles of a substance to the adjacent less energetic ones as a result of interactions between the particles. Conduction can take place in solids, liquids, or gases. In gases and liquids, conduction is due to the *collisions* and *diffusion* of the molecules during their random motion. In solids, it is due to the combination of *vibrations* of the molecules in a lattice and the energy transport by *free electrons.* A cold canned drink in a warm room, for example, eventually warms up to the room temperature as a result of heat transfer from the room to the drink through the aluminum can by conduction.

The *rate* of heat conduction through a medium depends on the *geometry* of the medium, its *thickness,* and the *material* of the medium, as well as the *temperature difference* across the medium. We know that wrapping a hot water tank with glass wool (an insulating material) reduces the rate of heat loss from the tank. The thicker the insulation, the smaller the heat loss. We also know that a hot water tank loses heat at a higher rate when the temperature of the room housing the tank is lowered. Further, the larger the tank, the larger the surface area and thus the rate of heat loss.

Consider steady heat conduction through a large plane wall of thickness $\Delta x = L$ and area A, as shown in Fig. 16–1. The temperature difference across the wall is $\Delta T = T_2 - T_1$. Experiments have shown that the rate of heat transfer $\dot{Q}$ through the wall is *doubled* when the temperature difference ΔT across the wall or the area A normal to the direction of heat transfer is doubled, but is *halved* when the wall thickness L is doubled. Thus we conclude that *the rate of heat conduction through a plane layer is proportional to the temperature difference across the layer and the heat transfer area, but is inversely proportional to the thickness of the layer.* That is,

$$\text{Rate of heat conduction} \propto \frac{(\text{Area})(\text{Temperature difference})}{\text{Thickness}}$$

or,

$$\dot{Q}_{\text{cond}} = kA\,\frac{T_1 - T_2}{\Delta x} = -kA\,\frac{\Delta T}{\Delta x} \qquad \text{(W)} \qquad \textbf{(16–1)}$$

where the constant of proportionality k is the **thermal conductivity** of the material, which is a *measure of the ability of a material to conduct heat*

(Fig. 16–2). In the limiting case of $\Delta x \to 0$, the equation above reduces to the differential form

$$\dot{Q}_{\text{cond}} = -kA\frac{dT}{dx} \qquad \text{(W)} \tag{16–2}$$

which is called **Fourier's law of heat conduction** after J. Fourier, who expressed it first in his heat transfer text in 1822. Here dT/dx is the **temperature gradient**, which is the slope of the temperature curve on a *T-x* diagram (the rate of change of T with x), at location x. The relation above indicates that the rate of heat conduction in a given direction is proportional to the temperature gradient in that direction. Heat is conducted in the direction of decreasing temperature, and the temperature gradient becomes negative when temperature decreases with increasing x. The *negative sign* in Eq. 16–2 ensures that heat transfer in the positive x direction is a positive quantity.

The heat transfer area A is always *normal* to the direction of heat transfer. For heat loss through a 5-m-long, 3-m-high, and 25-cm-thick wall, for example, the heat transfer area is $A = 15\ \text{m}^2$. Note that the thickness of the wall has no effect on A (Fig. 16–3).

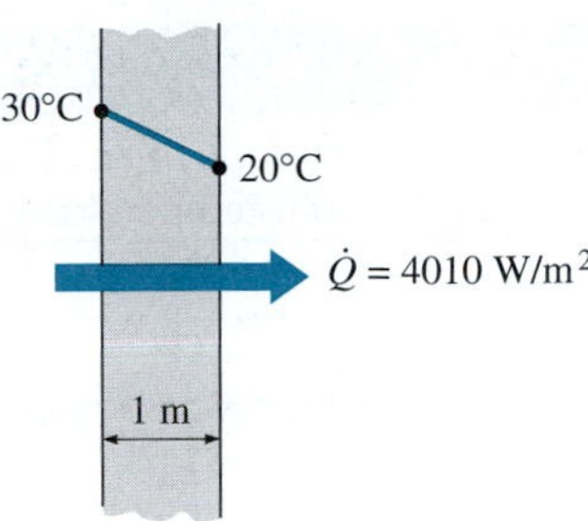

(*a*) Copper (k = 401 W/m·°C)

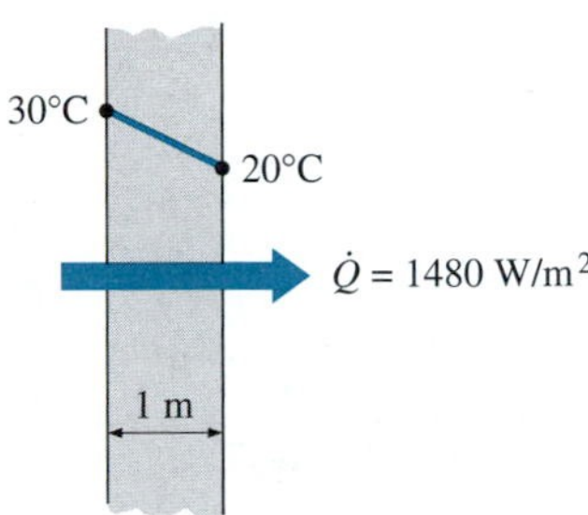

(*b*) Silicon (k = 148 W/m·°C)

FIGURE 16–2

The rate of heat conduction through a solid is directly proportional to its thermal conductivity.

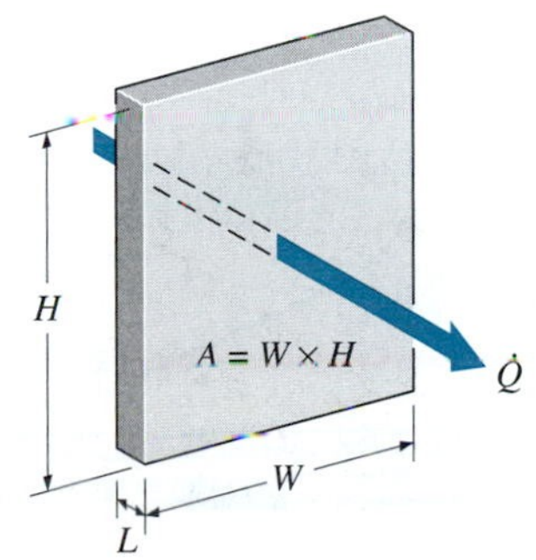

FIGURE 16–3

In heat conduction analysis, A represents the area *normal* to the direction of heat transfer.

EXAMPLE 16–1 The Cost of Heat Loss through a Roof

The roof of an electrically heated home is 6 m long, 8 m wide, and 0.25 m thick, and is made of a flat layer of concrete whose thermal conductivity is $k = 0.8$ W/m · °C (Fig. 16–4). The temperatures of the inner and the outer surfaces of the roof one night are measured to be 15°C and 4°C, respectively, for a period of 10 hours. Determine (*a*) the rate of heat loss through the roof that night and (*b*) the cost of that heat loss to the home owner if the cost of electricity is \$0.08/kWh.

Solution The inner and outer surfaces of the flat concrete roof of an electrically heated home are maintained at specified temperatures during a night. The heat loss through the roof and its cost that night are to be determined.

Assumptions **1** Steady operating conditions exist during the entire night since the surface temperatures of the roof remain constant at the specified values. **2** Constant properties can be used for the roof.

Properties The thermal conductivity of the roof is given to be $k = 0.8$ W/m · °C.

Analysis (*a*) Noting that heat transfer through the roof is by conduction and the area of the roof is $A = 6\ \text{m} \times 8\ \text{m} = 48\ \text{m}^2$, the steady rate of heat transfer through the roof is

$$\dot{Q} = kA\frac{T_1 - T_2}{L} = (0.8\ \text{W/m}\cdot{}^\circ\text{C})(48\ \text{m}^2)\frac{(15-4)^\circ\text{C}}{0.25\ \text{m}} = \mathbf{1690\ W = 1.69\ kW}$$

(*b*) The amount of heat lost through the roof during a 10-hour period and its cost is

$$Q = \dot{Q}\,\Delta t = (1.69\ \text{kW})(10\ \text{h}) = 16.9\ \text{kWh}$$

$$\text{Cost} = (\text{Amount of energy})(\text{Unit cost of energy})$$

$$= (16.9\ \text{kWh})(\$0.08/\text{kWh}) = \mathbf{\$1.35}$$

Discussion The cost to the home owner of the heat loss through the roof that night was \$1.35. The total heating bill of the house will be much larger since the heat losses through the walls are not considered in these calculations.

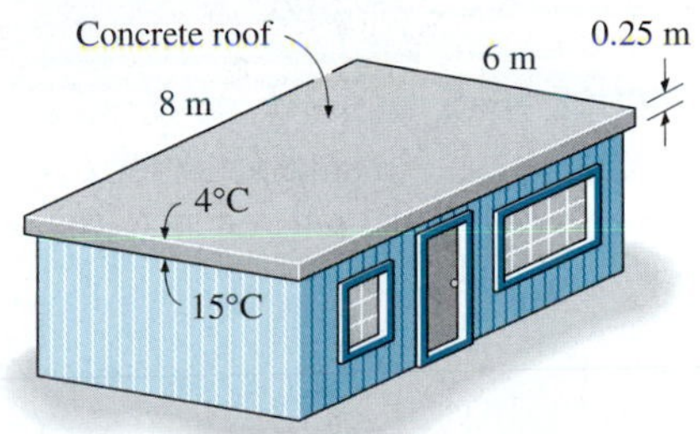

FIGURE 16–4

Schematic for Example 16–1.

TABLE 16–1

The thermal conductivities of some materials at room temperature

Material	k, W/m · °C*
Diamond	2300
Silver	429
Copper	401
Gold	317
Aluminum	237
Iron	80.2
Mercury (l)	8.54
Glass	0.78
Brick	0.72
Water (l)	0.607
Human skin	0.37
Wood (oak)	0.17
Helium (g)	0.152
Soft rubber	0.13
Glass fiber	0.043
Air (g)	0.026
Urethane, rigid foam	0.026

*Multiply by 0.5778 to convert to Btu/h · ft · °F.

Thermal Conductivity

We have seen that different materials store heat differently, and we have defined the property specific heat c_p as a measure of a material's ability to store thermal energy. For example, $c_p = 4.18$ kJ/kg · °C for water and $c_p = 0.45$ kJ/kg · °C for iron at room temperature, which indicates that water can store almost 10 times the energy that iron can per unit mass. Likewise, the thermal conductivity k is a measure of a material's ability to conduct heat. For example, $k = 0.607$ W/m · °C for water and $k = 80.2$ W/m · °C for iron at room temperature, which indicates that iron conducts heat more than 100 times faster than water can. Thus we say that water is a poor heat conductor relative to iron, although water is an excellent medium to store thermal energy.

Equation 16–1 for the rate of conduction heat transfer under steady conditions can also be viewed as the defining equation for thermal conductivity. Thus the **thermal conductivity** of a material can be defined as *the rate of heat transfer through a unit thickness of the material per unit area per unit temperature difference.* The thermal conductivity of a material is a measure of the ability of the material to conduct heat. A high value for thermal conductivity indicates that the material is a good heat conductor, and a low value indicates that the material is a poor heat conductor or *insulator.* The thermal conductivities of some common materials at room temperature are given in Table 16–1. The thermal conductivity of pure copper at room temperature is $k = 401$ W/m · °C, which indicates that a 1-m-thick copper wall will conduct heat at a rate of 401 W per m^2 area per °C temperature difference across the wall. Note that materials such as copper and silver that are good electric conductors are also good heat conductors, and have high values of thermal conductivity. Materials such as rubber, wood, and Styrofoam are poor conductors of heat and have low conductivity values.

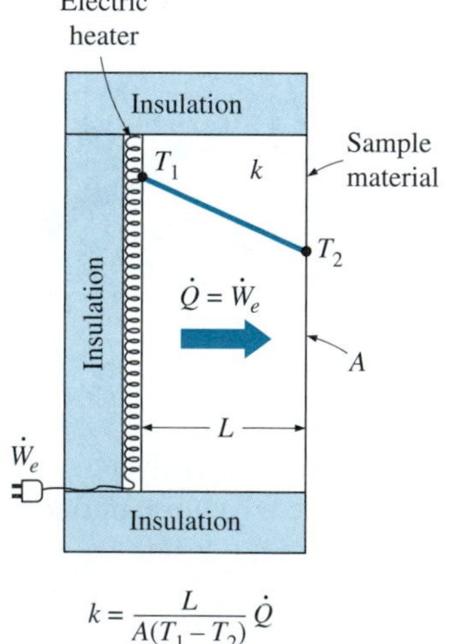

FIGURE 16–5

A simple experimental setup to determine the thermal conductivity of a material.

A layer of material of known thickness and area can be heated from one side by an electric resistance heater of known output. If the outer surfaces of the heater are well insulated, all the heat generated by the resistance heater will be transferred through the material whose conductivity is to be determined. Then measuring the two surface temperatures of the material when steady heat transfer is reached and substituting them into Eq. 16–1 together with other known quantities give the thermal conductivity (Fig. 16–5).

The thermal conductivities of materials vary over a wide range, as shown in Fig. 16–6. The thermal conductivities of gases such as air vary by a factor of 10^4 from those of pure metals such as copper. Note that pure crystals and metals have the highest thermal conductivities, and gases and insulating materials the lowest.

Temperature is a measure of the kinetic energies of the particles such as the molecules or atoms of a substance. In a liquid or gas, the kinetic energy of the molecules is due to their random translational motion as well as their vibrational and rotational motions. When two molecules possessing different kinetic energies collide, part of the kinetic energy of the more energetic (higher-temperature) molecule is transferred to the less energetic (lower-temperature) molecule, much the same as when two elastic balls of the same mass at different velocities collide, part of the kinetic energy of the faster ball is transferred to the slower one. The higher the temperature, the faster the molecules move and the higher the number of such collisions, and the better the heat transfer.

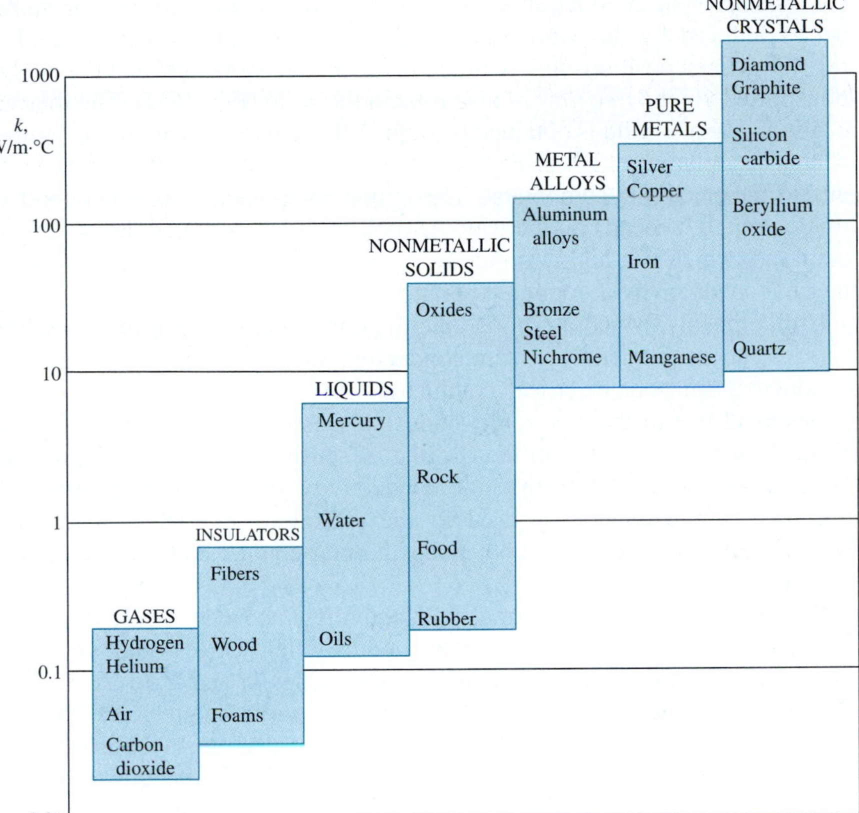

FIGURE 16–6

The range of thermal conductivity of various materials at room temperature.

The *kinetic theory* of gases predicts and the experiments confirm that the thermal conductivity of gases is proportional to the *square root of the thermodynamic temperature T*, and inversely proportional to the *square root of the molar mass M*. Therefore, the thermal conductivity of a gas increases with increasing temperature and decreasing molar mass. So it is not surprising that the thermal conductivity of helium ($M = 4$) is much higher than those of air ($M = 29$) and argon ($M = 40$).

The thermal conductivities of *gases* at 1 atm pressure are listed in Table A–23. However, they can also be used at pressures other than 1 atm, since the thermal conductivity of gases is *independent of pressure* in a wide range of pressures encountered in practice.

The mechanism of heat conduction in a *liquid* is complicated by the fact that the molecules are more closely spaced, and they exert a stronger intermolecular force field. The thermal conductivities of liquids usually lie between those of solids and gases. The thermal conductivity of a substance is normally highest in the solid phase and lowest in the gas phase. Unlike gases, the thermal conductivities of most liquids decrease with increasing temperature, with water being a notable exception. Like gases, the conductivity of liquids decreases with increasing molar mass. Liquid metals such as mercury and sodium have high thermal conductivities and are very suitable for use in applications where a high heat transfer rate to a liquid is desired, as in nuclear power plants.

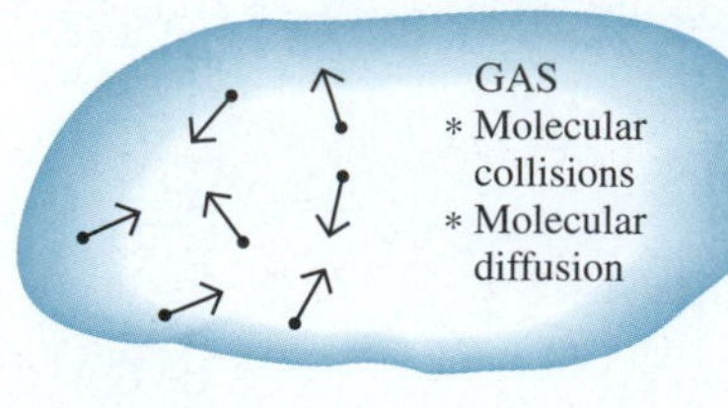

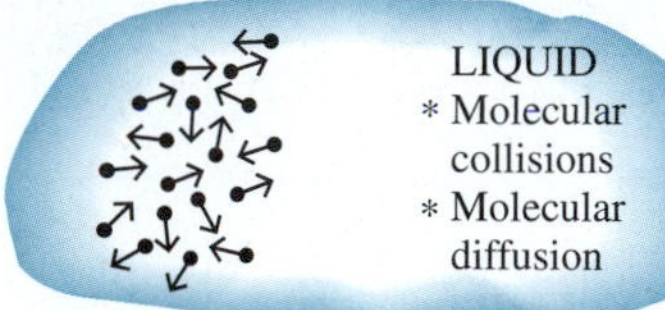

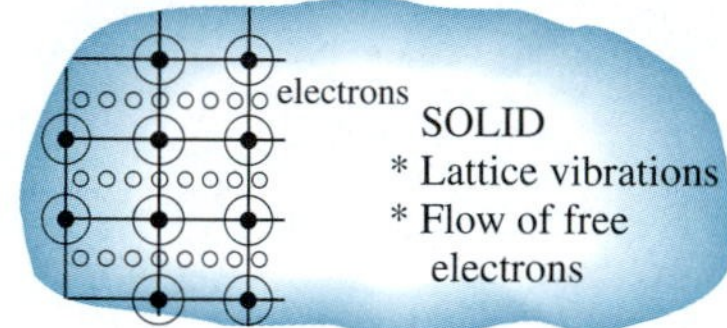

FIGURE 16–7

The mechanisms of heat conduction in different phases of a substance.

In *solids,* heat conduction is due to two effects: the *lattice vibrational waves* induced by the vibrational motions of the molecules positioned at relatively fixed positions in a periodic manner called a lattice, and the energy transported via the *free flow of electrons* in the solid (Fig. 16–7). The thermal conductivity of a solid is obtained by adding the lattice and electronic components. The relatively high thermal conductivities of pure metals are primarily due to the electronic component. The lattice component of thermal conductivity strongly depends on the way the molecules are arranged. For example, diamond, which is a highly ordered crystalline solid, has the highest known thermal conductivity at room temperature.

Unlike metals, which are good electrical and heat conductors, *crystalline solids* such as diamond and semiconductors such as silicon are good heat conductors but poor electrical conductors. As a result, such materials find widespread use in the electronics industry. Despite their higher price, diamond heat sinks are used in the cooling of sensitive electronic components because of the excellent thermal conductivity of diamond. Silicon oils and gaskets are commonly used in the packaging of electronic components because they provide both good thermal contact and good electrical insulation.

Pure metals have high thermal conductivities, and one would think that *metal alloys* should also have high conductivities. One would expect an alloy made of two metals of thermal conductivities k_1 and k_2 to have a conductivity k between k_1 and k_2. But this turns out not to be the case. The thermal conductivity of an alloy of two metals is usually much lower than that of either metal, as shown in Table 16–2. Even small amounts in a pure metal of "foreign" molecules that are good conductors themselves seriously disrupt the transfer of heat in that metal. For example, the thermal conductivity of steel containing just 1 percent of chrome is 62 W/m · °C, while the thermal conductivities of iron and chromium are 83 and 95 W/m · °C, respectively.

TABLE 16–2

The thermal conductivity of an alloy is usually much lower than the thermal conductivity of either metal of which it is composed

Pure metal or alloy	k, W/m · °C, at 300 K
Copper	401
Nickel	91
Constantan (55% Cu, 45% Ni)	23
Copper	401
Aluminum	237
Commercial bronze (90% Cu, 10% Al)	52

The thermal conductivities of materials vary with temperature (Table 16–3). The variation of thermal conductivity over certain temperature ranges is negligible for some materials, but significant for others, as shown in Fig. 16–8. The thermal conductivities of certain solids exhibit dramatic increases at temperatures near absolute zero, when these solids become *superconductors.* For example, the conductivity of copper reaches a maximum value of about 20,000 W/m · °C at 20 K, which is about 50 times the conductivity at room temperature. The thermal conductivities and other thermal properties of various materials are given in the Appendix.

The temperature dependence of thermal conductivity causes considerable complexity in conduction analysis. Therefore, it is common practice to evaluate the thermal conductivity k at the *average temperature* and treat it as a *constant* in calculations.

In heat transfer analysis, a material is normally assumed to be *isotropic*; that is, to have uniform properties in all directions. This assumption is realistic for most materials, except those that exhibit different structural characteristics in different directions, such as laminated composite materials and wood. The thermal conductivity of wood across the grain, for example, is different than that parallel to the grain.

TABLE 16–3

Thermal conductivities of materials vary with temperature

T, K	k, W/m · °C Copper	Aluminum
100	482	302
200	413	237
300	401	237
400	393	240
600	379	231
800	366	218

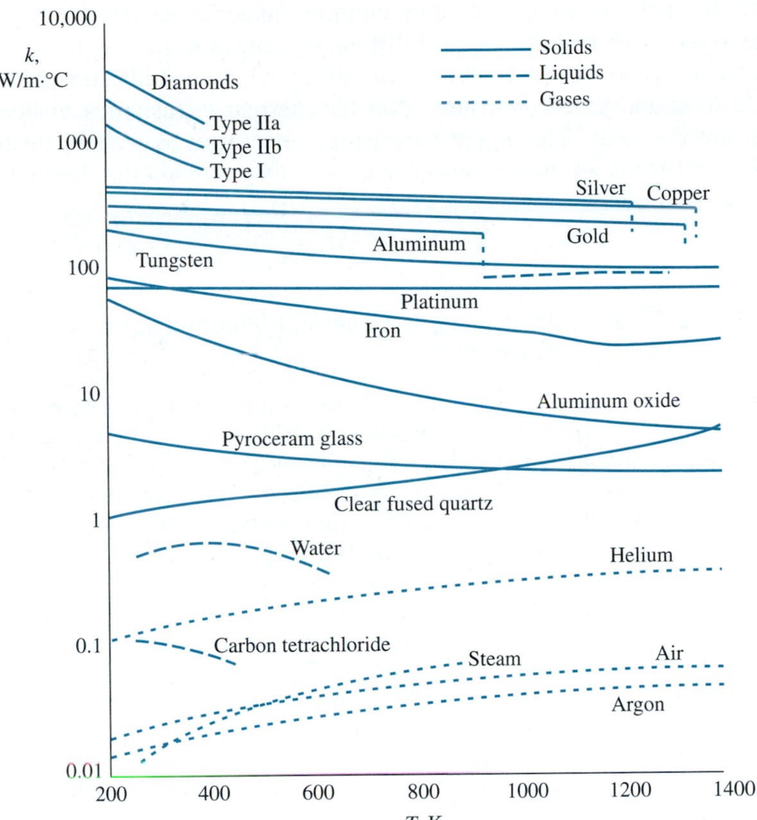

FIGURE 16–8

The variation of the thermal conductivity of various solids, liquids, and gases with temperature.

Thermal Diffusivity

The product ρc_p, which is frequently encountered in heat transfer analysis, is called the **heat capacity** of a material. Both the specific heat c_p and the heat capacity ρc_p represent the heat storage capability of a material. But c_p expresses it *per unit mass* whereas ρc_p expresses it *per unit volume*, as can be noticed from their units J/kg · °C and J/m^3 · °C, respectively.

Another material property that appears in the transient heat conduction analysis is the **thermal diffusivity**, which represents how fast heat diffuses through a material and is defined as

$$\alpha = \frac{\text{Heat conducted}}{\text{Heat stored}} = \frac{k}{\rho c_p} \qquad (\text{m}^2/\text{s}) \tag{16–3}$$

Note that the thermal conductivity k represents how well a material conducts heat, and the heat capacity ρc_p represents how much energy a material stores per unit volume. Therefore, the thermal diffusivity of a material can be viewed as the ratio of the *heat conducted* through the material to the *heat stored* per unit volume. A material that has a high thermal conductivity or a low heat capacity will obviously have a large thermal diffusivity. The larger the thermal diffusivity, the faster the propagation of heat into the medium. A small value of thermal diffusivity means that heat is mostly absorbed by the material and a small amount of heat is conducted further.

TABLE 16–4

The thermal diffusivities of some materials at room temperature

Material	α, m^2/s*
Silver	149×10^{-6}
Gold	127×10^{-6}
Copper	113×10^{-6}
Aluminum	97.5×10^{-6}
Iron	22.8×10^{-6}
Mercury (l)	4.7×10^{-6}
Marble	1.2×10^{-6}
Ice	1.2×10^{-6}
Concrete	0.75×10^{-6}
Brick	0.52×10^{-6}
Heavy soil (dry)	0.52×10^{-6}
Glass	0.34×10^{-6}
Glass wool	0.23×10^{-6}
Water (l)	0.14×10^{-6}
Beef	0.14×10^{-6}
Wood (oak)	0.13×10^{-6}

*Multiply by 10.76 to convert to ft^2/s.

The thermal diffusivities of some common materials at 20°C are given in Table 16–4. Note that the thermal diffusivity ranges from $\alpha = 0.14 \times 10^{-6}$ m^2/s for water to 149×10^{-6} m^2/s for silver, which is a difference of more than a thousand times. Also note that the thermal diffusivities of beef and water are the same. This is not surprising, since meat as well as fresh vegetables and fruits are mostly water, and thus they possess the thermal properties of water.

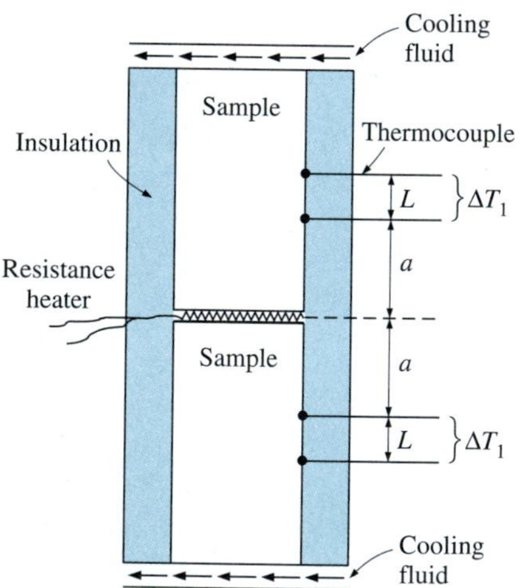

FIGURE 16–9

Apparatus to measure the thermal conductivity of a material using two identical samples and a thin resistance heater (Example 16–2).

EXAMPLE 16–2 Measuring the Thermal Conductivity of a Material

A common way of measuring the thermal conductivity of a material is to sandwich an electric thermofoil heater between two identical samples of the material, as shown in Fig. 16–9. The thickness of the resistance heater, including its cover, which is made of thin silicon rubber, is usually less than 0.5 mm. A circulating fluid such as tap water keeps the exposed ends of the samples at constant temperature. The lateral surfaces of the samples are well insulated to ensure that heat transfer through the samples is one-dimensional. Two thermocouples are embedded into each sample some distance L apart, and a differential thermometer reads the temperature drop ΔT across this distance along each sample. When steady operating conditions are reached, the total rate of heat transfer through both samples becomes equal to the electric power drawn by the heater.

In a certain experiment, cylindrical samples of diameter 5 cm and length 10 cm are used. The two thermocouples in each sample are placed 3 cm apart. After initial transients, the electric heater is observed to draw 0.4 A at 110 V, and both differential thermometers read a temperature difference of 15°C. Determine the thermal conductivity of the sample.

Solution The thermal conductivity of a material is to be determined by ensuring one-dimensional heat conduction, and by measuring temperatures when steady operating conditions are reached.

Assumptions **1** Steady operating conditions exist since the temperature readings do not change with time. **2** Heat losses through the lateral surfaces of the apparatus are negligible since those surfaces are well insulated, and thus the entire heat generated by the heater is conducted through the samples. **3** The apparatus possesses thermal symmetry.

Analysis The electrical power consumed by the resistance heater and converted to heat is

$$\dot{W}_e = \mathbf{V}I = (110 \text{ V})(0.4 \text{ A}) = 44 \text{ W}$$

The rate of heat flow through each sample is

$$\dot{Q} = \tfrac{1}{2}\dot{W}_e = \tfrac{1}{2} \times (44 \text{ W}) = 22 \text{ W}$$

since only half of the heat generated flows through each sample because of symmetry. Reading the same temperature difference across the same distance in each sample also confirms that the apparatus possesses thermal symmetry. The heat transfer area is the area normal to the direction of heat transfer, which is the cross-sectional area of the cylinder in this case:

$$A = \tfrac{1}{4}\pi D^2 = \tfrac{1}{4}\pi(0.05 \text{ m})^2 = 0.001963 \text{ m}^2$$

Noting that the temperature drops by 15°C within 3 cm in the direction of heat flow, the thermal conductivity of the sample is determined to be

$$\dot{Q} = kA\frac{\Delta T}{L} \quad \rightarrow \quad k = \frac{\dot{Q}\,L}{A\,\Delta T} = \frac{(22\text{ W})(0.03\text{ m})}{(0.001963\text{ m}^2)(15°\text{C})} = \mathbf{22.4\ W/m\cdot °C}$$

Discussion Perhaps you are wondering if we really need to use two samples in the apparatus, since the measurements on the second sample do not give any additional information. It seems like we can replace the second sample by insulation. Indeed, we do not need the second sample; however, it enables us to verify the temperature measurements on the first sample and provides thermal symmetry, which reduces experimental error.

EXAMPLE 16–3 Conversion between SI and English Units

An engineer who is working on the heat transfer analysis of a brick building in English units needs the thermal conductivity of brick. But the only value he can find from his handbooks is 0.72 W/m · °C, which is in SI units. To make matters worse, the engineer does not have a direct conversion factor between the two unit systems for thermal conductivity. Can you help him out?

Solution The situation this engineer is facing is not unique, and most engineers often find themselves in a similar position. A person must be very careful during unit conversion not to fall into some common pitfalls and to avoid some costly mistakes. Although unit conversion is a simple process, it requires utmost care and careful reasoning.

The conversion factors for W and m are straightforward and are given in conversion tables to be

$$1\text{ W} = 3.41214\text{ Btu/h}$$
$$1\text{ m} = 3.2808\text{ ft}$$

But the conversion of °C into °F is not so simple, and it can be a source of error if one is not careful. Perhaps the first thought that comes to mind is to replace °C by (°F − 32)/1.8 since T(°C) = [T(°F) − 32]/1.8. But this will be wrong since the °C in the unit W/m · °C represents *per °C change in temperature.* Noting that 1°C change in temperature corresponds to 1.8°F, the proper conversion factor to be used is

$$1°\text{C} = 1.8°\text{F}$$

Substituting, we get

$$1\text{ W/m}\cdot°\text{C} = \frac{3.41214\text{ Btu/h}}{(3.2808\text{ ft})(1.8°\text{F})} = 0.5778\text{ Btu/h}\cdot\text{ft}\cdot°\text{F}$$

which is the desired conversion factor. Therefore, the thermal conductivity of the brick in English units is

$$\begin{aligned} k_{\text{brick}} &= 0.72\text{ W/m}\cdot°\text{C} \\ &= 0.72 \times (0.5778\text{ Btu/h}\cdot\text{ft}\cdot°\text{F}) \\ &= \mathbf{0.42\ Btu/h\cdot ft\cdot °F} \end{aligned}$$

Discussion Note that the thermal conductivity value of a material in English units is about half that in SI units (Fig. 16–10). Also note that we rounded the result to two significant digits (the same number in the original value) since expressing the result in more significant digits (such as 0.4160 instead of 0.42) would falsely imply a more accurate value than the original one.

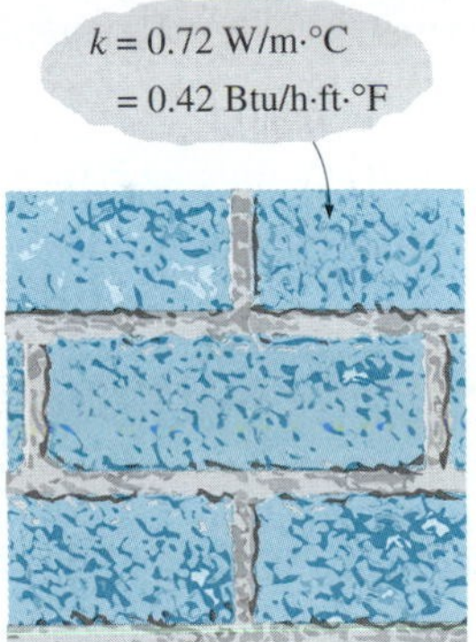

FIGURE 16–10
The thermal conductivity value in English units is obtained by multiplying the value in SI units by 0.5778.

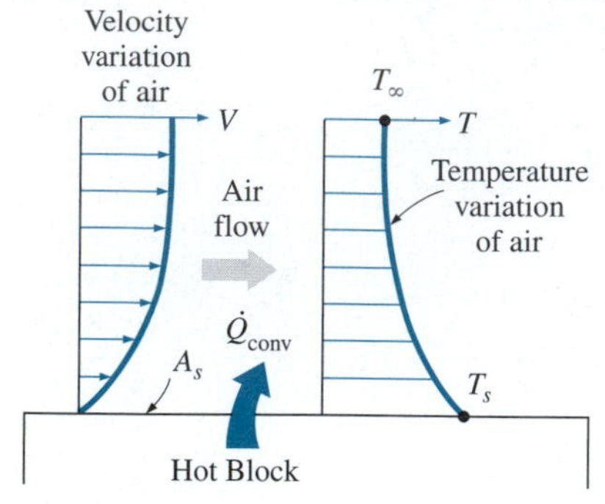

FIGURE 16–11

Heat transfer from a hot surface to air by convection.

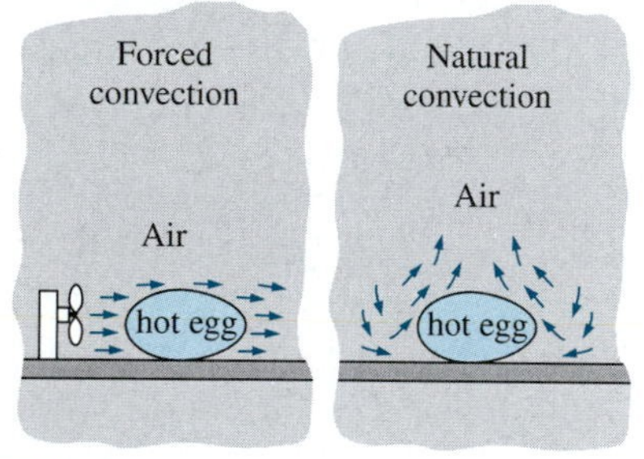

FIGURE 16–12

The cooling of a boiled egg by forced and natural convection.

TABLE 16–5

Typical values of convection heat transfer coefficient

Type of convection	h, W/m² · °C*
Free convection of gases	2–25
Free convection of liquids	10–1000
Forced convection of gases	25–250
Forced convection of liquids	50–20,000
Boiling and condensation	2500–100,000

*Multiply by 0.176 to convert to Btu/h · ft² · °F.

16–3 ▪ CONVECTION

Convection is the mode of energy transfer between a solid surface and the adjacent liquid or gas that is in motion, and it involves the combined effects of *conduction* and *fluid motion.* The faster the fluid motion, the greater the convection heat transfer. In the absence of any bulk fluid motion, heat transfer between a solid surface and the adjacent fluid is by pure conduction. The presence of bulk motion of the fluid enhances the heat transfer between the solid surface and the fluid, but it also complicates the determination of heat transfer rates.

Consider the cooling of a hot block by blowing cool air over its top surface (Fig. 16–11). Heat is first transferred to the air layer adjacent to the block by conduction. This heat is then carried away from the surface by convection, that is, by the combined effects of conduction within the air that is due to random motion of air molecules and the bulk or macroscopic motion of the air that removes the heated air near the surface and replaces it by the cooler air.

Convection is called **forced convection** if the fluid is forced to flow over the surface by external means such as a fan, pump, or the wind. In contrast, convection is called **natural** (or **free**) **convection** if the fluid motion is caused by buoyancy forces that are induced by density differences due to the variation of temperature in the fluid (Fig. 16–12). For example, in the absence of a fan, heat transfer from the surface of the hot block in Fig. 16–11 is by natural convection since any motion in the air in this case is due to the rise of the warmer (and thus lighter) air near the surface and the fall of the cooler (and thus heavier) air to fill its place. Heat transfer between the block and the surrounding air is by conduction if the temperature difference between the air and the block is not large enough to overcome the resistance of air to movement and thus to initiate natural convection currents.

Heat transfer processes that involve *change of phase* of a fluid are also considered to be convection because of the fluid motion induced during the process, such as the rise of the vapor bubbles during boiling or the fall of the liquid droplets during condensation.

Despite the complexity of convection, the rate of *convection heat transfer* is observed to be proportional to the temperature difference, and is conveniently expressed by **Newton's law of cooling** as

$$\dot{Q}_{conv} = hA_s\,(T_s - T_\infty) \qquad \text{(W)} \qquad \textbf{(16–4)}$$

where h is the *convection heat transfer coefficient* in W/m² · °C or Btu/h · ft² · °F, A_s is the surface area through which convection heat transfer takes place, T_s is the surface temperature, and T_∞ is the temperature of the fluid sufficiently far from the surface. Note that at the surface, the fluid temperature equals the surface temperature of the solid.

The convection heat transfer coefficient h is not a property of the fluid. It is an experimentally determined parameter whose value depends on all the variables influencing convection such as the surface geometry, the nature of fluid motion, the properties of the fluid, and the bulk fluid velocity. Typical values of h are given in Table 16–5.

Some people do not consider convection to be a fundamental mechanism of heat transfer since it is essentially heat conduction in the presence of fluid motion. But we still need to give this combined phenomenon a name, unless we are willing to keep referring to it as "conduction with fluid motion." Thus, it is practical to recognize convection as a separate heat transfer mechanism despite the valid arguments to the contrary.

EXAMPLE 16–4 Measuring Convection Heat Transfer Coefficient

A 2-m-long, 0.3-cm-diameter electrical wire extends across a room at 15°C, as shown in Fig. 16–13. Heat is generated in the wire as a result of resistance heating, and the surface temperature of the wire is measured to be 152°C in steady operation. Also, the voltage drop and electric current through the wire are measured to be 60 V and 1.5 A, respectively. Disregarding any heat transfer by radiation, determine the convection heat transfer coefficient for heat transfer between the outer surface of the wire and the air in the room.

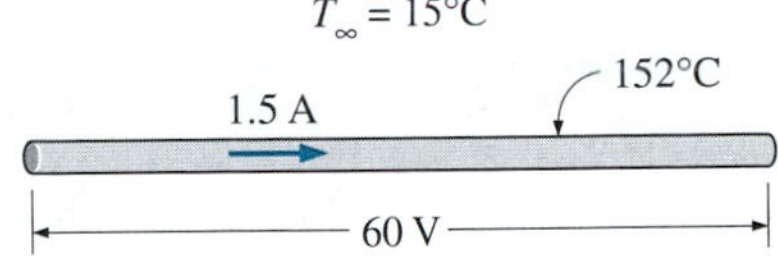

FIGURE 16–13

Schematic for Example 16–4.

Solution The convection heat transfer coefficient for heat transfer from an electrically heated wire to air is to be determined by measuring temperatures when steady operating conditions are reached and the electric power consumed.

Assumptions **1** Steady operating conditions exist since the temperature readings do not change with time. **2** Radiation heat transfer is negligible.

Analysis When steady operating conditions are reached, the rate of heat loss from the wire equals the rate of heat generation in the wire as a result of resistance heating. That is,

$$\dot{Q} = \dot{E}_{\text{generated}} = VI = (60 \text{ V})(1.5 \text{ A}) = 90 \text{ W}$$

The surface area of the wire is

$$A_s = \pi DL = \pi(0.003 \text{ m})(2 \text{ m}) = 0.01885 \text{ m}^2$$

Newton's law of cooling for convection heat transfer is expressed as

$$\dot{Q}_{\text{conv}} = hA_s\,(T_s - T_\infty)$$

Disregarding any heat transfer by radiation and thus assuming all the heat loss from the wire to occur by convection, the convection heat transfer coefficient is determined to be

$$h = \frac{\dot{Q}_{\text{conv}}}{A_s(T_s - T_\infty)} = \frac{90 \text{ W}}{(0.01885 \text{ m}^2)(152 - 15)°\text{C}} = \mathbf{34.9 \ W/m^2 \cdot °C}$$

Discussion Note that the simple setup described above can be used to determine the average heat transfer coefficients from a variety of surfaces in air. Also, heat transfer by radiation can be eliminated by keeping the surrounding surfaces at the temperature of the wire.

16–4 ▪ RADIATION

Radiation is the energy emitted by matter in the form of *electromagnetic waves* (or *photons*) as a result of the changes in the electronic configurations of the atoms or molecules. Unlike conduction and convection, the transfer of heat by radiation does not require the presence of an *intervening medium.* In fact, heat transfer by radiation is fastest (at the speed of light)

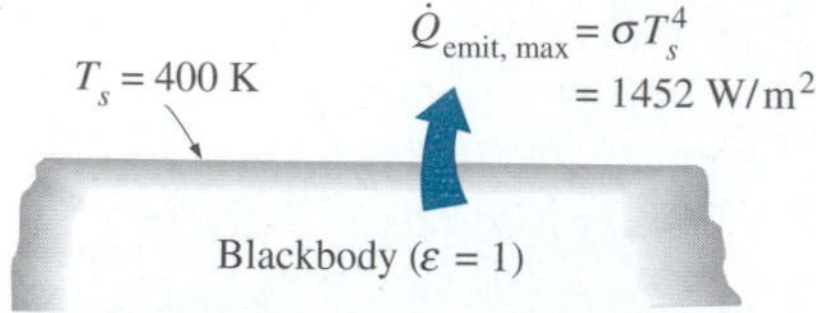

FIGURE 16–14

Blackbody radiation represents the *maximum amount of radiation that can be emitted from a surface at a specified temperature.*

TABLE 16–6

Emissivities of some materials at 300 K

Material	Emissivity
Aluminum foil	0.07
Anodized aluminum	0.82
Polished copper	0.03
Polished gold	0.03
Polished silver	0.02
Polished stainless steel	0.17
Black paint	0.98
White paint	0.90
White paper	0.92–0.97
Asphalt pavement	0.85–0.93
Red brick	0.93–0.96
Human skin	0.95
Wood	0.82–0.92
Soil	0.93–0.96
Water	0.96
Vegetation	0.92–0.96

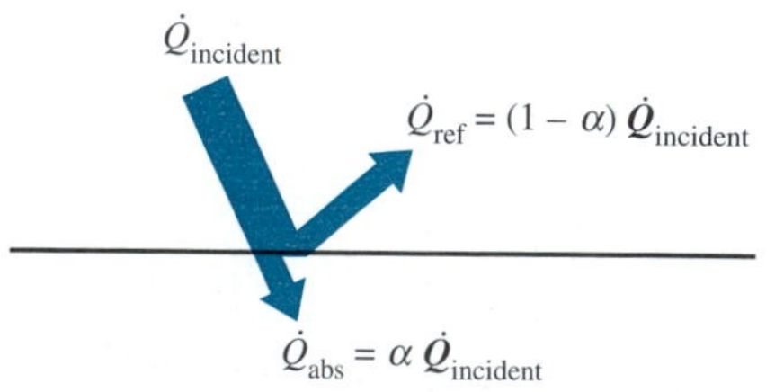

FIGURE 16–15

The absorption of radiation incident on an opaque surface of absorptivity α.

and it suffers no attenuation in a vacuum. This is how the energy of the sun reaches the earth.

In heat transfer studies we are interested in *thermal radiation,* which is the form of radiation emitted by bodies because of their temperature. It differs from other forms of electromagnetic radiation such as x-rays, gamma rays, microwaves, radio waves, and television waves that are not related to temperature. All bodies at a temperature above absolute zero emit thermal radiation.

Radiation is a *volumetric phenomenon,* and all solids, liquids, and gases emit, absorb, or transmit radiation to varying degrees. However, radiation is usually considered to be a *surface phenomenon* for solids that are opaque to thermal radiation such as metals, wood, and rocks since the radiation emitted by the interior regions of such material can never reach the surface, and the radiation incident on such bodies is usually absorbed within a few microns from the surface.

The maximum rate of radiation that can be emitted from a surface at a thermodynamic temperature T_s (in K or R) is given by the **Stefan–Boltzmann law** as

$$\dot{Q}_{emit, max} = \sigma A_s T_s^4 \qquad \text{(W)} \tag{16–5}$$

where $\sigma = 5.670 \times 10^{-8}$ W/m$^2 \cdot$ K^4 or 0.1714×10^{-8} Btu/h $\cdot$ ft$^2 \cdot$ R^4 is the *Stefan–Boltzmann constant.* The idealized surface that emits radiation at this maximum rate is called a **blackbody**, and the radiation emitted by a blackbody is called **blackbody radiation** (Fig. 16–14). The radiation emitted by all real surfaces is less than the radiation emitted by a blackbody at the same temperature, and is expressed as

$$\dot{Q}_{emit} = \varepsilon \sigma A_s T_s^4 \qquad \text{(W)} \tag{16–6}$$

where ε is the **emissivity** of the surface. The property emissivity, whose value is in the range $0 \leq \varepsilon \leq 1$, is a measure of how closely a surface approximates a blackbody for which $\varepsilon = 1$. The emissivities of some surfaces are given in Table 16–6.

Another important radiation property of a surface is its **absorptivity** α, which is the fraction of the radiation energy incident on a surface that is absorbed by the surface. Like emissivity, its value is in the range $0 \leq \alpha \leq 1$. A blackbody absorbs the entire radiation incident on it. That is, a blackbody is a perfect absorber ($\alpha = 1$) as it is a perfect emitter.

In general, both ε and α of a surface depend on the temperature and the wavelength of the radiation. **Kirchhoff's law** of radiation states that the emissivity and the absorptivity of a surface at a given temperature and wavelength are equal. In many practical applications, the surface temperature and the temperature of the source of incident radiation are of the same order of magnitude, and the average absorptivity of a surface is taken to be equal to its average emissivity. The rate at which a surface absorbs radiation is determined from (Fig. 16–15)

$$\dot{Q}_{absorbed} = \alpha \dot{Q}_{incident} \qquad \text{(W)} \tag{16–7}$$

where $\dot{Q}_{incident}$ is the rate at which radiation is incident on the surface and α is the absorptivity of the surface. For opaque (nontransparent) surfaces, the portion of incident radiation not absorbed by the surface is reflected back.

The difference between the rates of radiation emitted by the surface and the radiation absorbed is the *net* radiation heat transfer. If the rate of radiation absorption is greater than the rate of radiation emission, the surface is said to be *gaining* energy by radiation. Otherwise, the surface is said to be *losing* energy by radiation. In general, the determination of the net rate of heat transfer by radiation between two surfaces is a complicated matter since it depends on the properties of the surfaces, their orientation relative to each other, and the interaction of the medium between the surfaces with radiation.

When a surface of emissivity ε and surface area A_s at a *thermodynamic temperature* T_s is *completely enclosed* by a much larger (or black) surface at thermodynamic temperature T_{surr} separated by a gas (such as air) that does not intervene with radiation, the net rate of radiation heat transfer between these two surfaces is given by (Fig. 16–16)

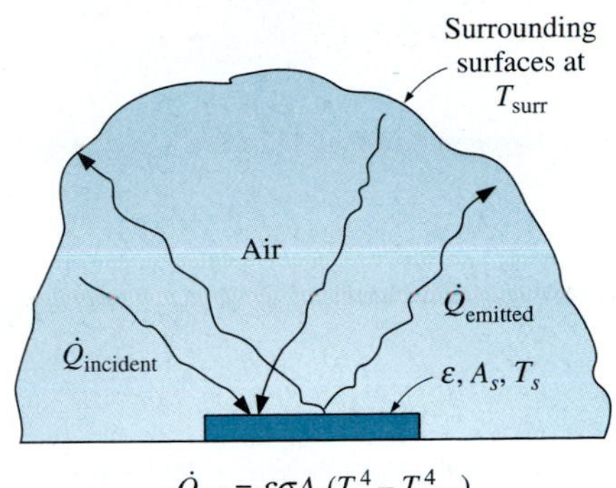

FIGURE 16–16
Radiation heat transfer between a surface and the surfaces surrounding it.

$$\dot{Q}_{rad} = \varepsilon\sigma A_s\,(T_s^4 - T_{surr}^4) \qquad \text{(W)} \qquad \text{(16–8)}$$

In this special case, the emissivity and the surface area of the surrounding surface do not have any effect on the net radiation heat transfer.

Radiation heat transfer to or from a surface surrounded by a gas such as air occurs *parallel* to conduction (or convection, if there is bulk gas motion) between the surface and the gas. Thus the total heat transfer is determined by *adding* the contributions of both heat transfer mechanisms. For simplicity and convenience, this is often done by defining a **combined heat transfer coefficient** $h_{combined}$ that includes the effects of both convection and radiation. Then the *total* heat transfer rate to or from a surface by convection and radiation is expressed as

$$\dot{Q}_{total} = h_{combined}\,A_s\,(T_s - T_\infty) \qquad \text{(W)} \qquad \text{(16–9)}$$

Note that the combined heat transfer coefficient is essentially a convection heat transfer coefficient modified to include the effects of radiation.

Radiation is usually significant relative to conduction or natural convection, but negligible relative to forced convection. Thus radiation in forced convection applications is usually disregarded, especially when the surfaces involved have low emissivities and low to moderate temperatures.

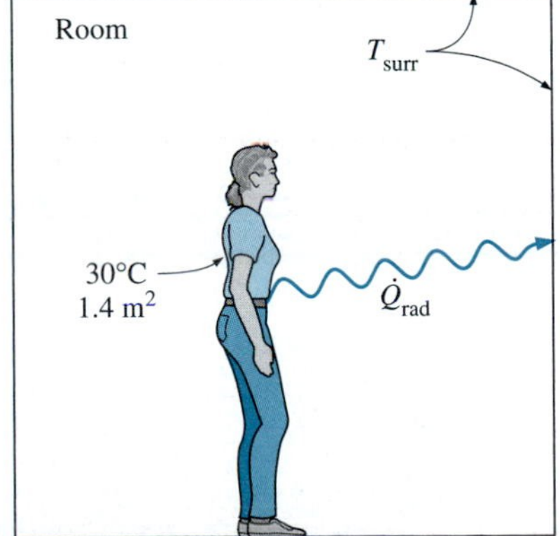

FIGURE 16–17
Schematic for Example 16–5.

EXAMPLE 16–5 Radiation Effect on Thermal Comfort

It is a common experience to feel "chilly" in winter and "warm" in summer in our homes even when the thermostat setting is kept the same. This is due to the so called "radiation effect" resulting from radiation heat exchange between our bodies and the surrounding surfaces of the walls and the ceiling.

Consider a person standing in a room maintained at 22°C at all times. The inner surfaces of the walls, floors, and the ceiling of the house are observed to be at an average temperature of 10°C in winter and 25°C in summer. Determine the rate of radiation heat transfer between this person and the surrounding surfaces if the exposed surface area and the average outer surface temperature of the person are 1.4 m^2 and 30°C, respectively (Fig. 16–17).

Solution The rates of radiation heat transfer between a person and the surrounding surfaces at specified temperatures are to be determined in summer and winter.

Assumptions **1** Steady operating conditions exist. **2** Heat transfer by convection is not considered. **3** The person is completely surrounded by the interior surfaces of the room. **4** The surrounding surfaces are at a uniform temperature.

Properties The emissivity of a person is $\varepsilon = 0.95$ (Table 16–6).

Analysis The net rates of radiation heat transfer from the body to the surrounding walls, ceiling, and floor in winter and summer are

$$\begin{aligned}\dot{Q}_{\text{rad, winter}} &= \varepsilon\sigma A_s (T_s^4 - T_{\text{surr, winter}}^4)\\ &= (0.95)(5.67 \times 10^{-8}\ \text{W/m}^2\cdot\text{K}^4)(1.4\ \text{m}^2)\\ &\quad \times [(30 + 273)^4 - (10 + 273)^4]\ \text{K}^4\\ &= \mathbf{152\ W}\end{aligned}$$

and

$$\begin{aligned}\dot{Q}_{\text{rad, summer}} &= \varepsilon\sigma A_s (T_s^4 - T_{\text{surr, summer}}^4)\\ &= (0.95)(5.67 \times 10^{-8}\ \text{W/m}^2\cdot\text{K}^4)(1.4\ \text{m}^2)\\ &\quad \times [(30 + 273)^4 - (25 + 273)^4]\ \text{K}^4\\ &= \mathbf{40.9\ W}\end{aligned}$$

Discussion Note that we must use *thermodynamic (i.e., absolute) temperatures* in radiation calculations. Also note that the rate of heat loss from the person by radiation is almost four times as large in winter than it is in summer, which explains the "chill" we feel in winter even if the thermostat setting is kept the same.

16–5 ▪ SIMULTANEOUS HEAT TRANSFER MECHANISMS

We mentioned that there are three mechanisms of heat transfer, but not all three can exist simultaneously in a medium. For example, heat transfer is only by conduction in *opaque solids,* but by conduction and radiation in *semitransparent solids.* Thus, a solid may involve conduction and radiation but not convection. However, a solid may involve heat transfer by convection and/or radiation on its surfaces exposed to a fluid or other surfaces. For example, the outer surfaces of a cold piece of rock will warm up in a warmer environment as a result of heat gain by convection (from the air) and radiation (from the sun or the warmer surrounding surfaces). But the inner parts of the rock will warm up as this heat is transferred to the inner region of the rock by conduction.

Heat transfer is by conduction and possibly by radiation in a *still fluid* (no bulk fluid motion) and by convection and radiation in a *flowing fluid.* In the absence of radiation, heat transfer through a fluid is either by conduction or convection, depending on the presence of any bulk fluid motion. Convection can be viewed as combined conduction and fluid motion, and conduction in a fluid can be viewed as a special case of convection in the absence of any fluid motion (Fig. 16–18).

Thus, when we deal with heat transfer through a *fluid,* we have either *conduction* or *convection,* but not both. Also, gases are practically transparent to radiation, except that some gases are known to absorb radiation strongly at certain wavelengths. Ozone, for example, strongly absorbs ultraviolet radiation. But in most cases, a gas between two solid surfaces does not interfere with radiation and acts effectively as a vacuum. Liquids, on the other hand, are usually strong absorbers of radiation.

Finally, heat transfer through a *vacuum* is by radiation only since conduction or convection requires the presence of a material medium.

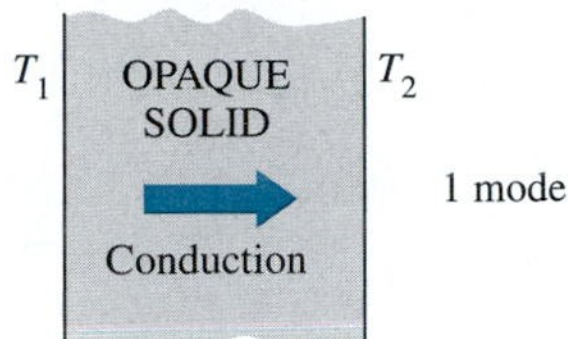

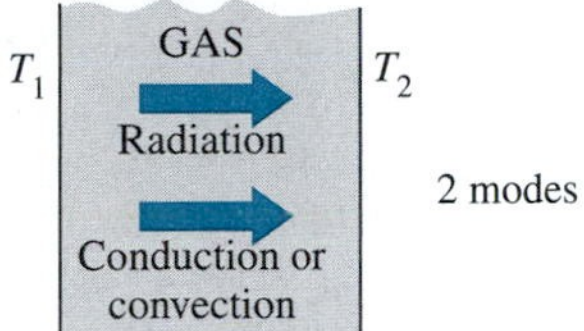

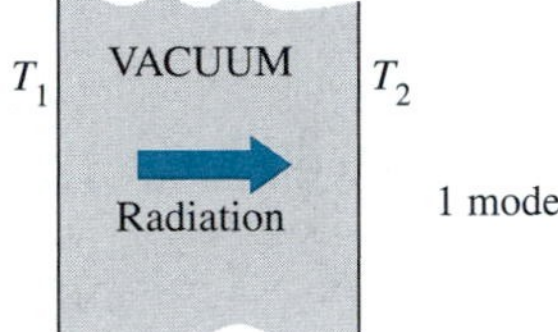

FIGURE 16–18

Although there are three mechanisms of heat transfer, a medium may involve only two of them simultaneously.

EXAMPLE 16–6 Heat Loss from a Person

Consider a person standing in a breezy room at 20°C. Determine the total rate of heat transfer from this person if the exposed surface area and the average outer surface temperature of the person are 1.6 m^2 and 29°C, respectively, and the convection heat transfer coefficient is 6 W/m^2 · K (Fig. 16–19).

Solution The total rate of heat transfer from a person by both convection and radiation to the surrounding air and surfaces at specified temperatures is to be determined.

Assumptions **1** Steady operating conditions exist. **2** The person is completely surrounded by the interior surfaces of the room. **3** The surrounding surfaces are at the same temperature as the air in the room. **4** Heat conduction to the floor through the feet is negligible.

Properties The emissivity of a person is $\varepsilon = 0.95$ (Table 16–6).

Analysis The heat transfer between the person and the air in the room is by convection (instead of conduction) since it is conceivable that the air in the vicinity of the skin or clothing warms up and rises as a result of heat transfer from the body, initiating natural convection currents. It appears that the experimentally determined value for the rate of convection heat transfer in this case is 6 W per unit surface area (m^2) per unit temperature difference (in K or °C) between the person and the air away from the person. Thus, the rate of convection heat transfer from the person to the air in the room is

$$\begin{aligned}\dot{Q}_{\text{conv}} &= hA_s(T_s - T_\infty)\\ &= (6\ \text{W/m}^2\cdot{}^\circ\text{C})(1.6\ \text{m}^2)(29 - 20)^\circ\text{C}\\ &= 86.4\ \text{W}\end{aligned}$$

The person also loses heat by radiation to the surrounding wall surfaces. We take the temperature of the surfaces of the walls, ceiling, and floor to be equal to the air temperature in this case for simplicity, but we recognize that this does not need to be the case. These surfaces may be at a higher or lower temperature than the average temperature of the room air, depending on the outdoor conditions and the structure of the walls. Considering that air does not intervene with radiation and the person is completely enclosed by the surrounding surfaces, the net rate of radiation heat transfer from the person to the surrounding walls, ceiling, and floor is

$$\begin{aligned}\dot{Q}_{\text{rad}} &= \varepsilon\sigma A_s(T_s^4 - T_{\text{surr}}^4)\\ &= (0.95)(5.67 \times 10^{-8}\ \text{W/m}^2\cdot\text{K}^4)(1.6\ \text{m}^2)\\ &\quad \times [(29 + 273)^4 - (20 + 273)^4]\ \text{K}^4\\ &= 81.7\ \text{W}\end{aligned}$$

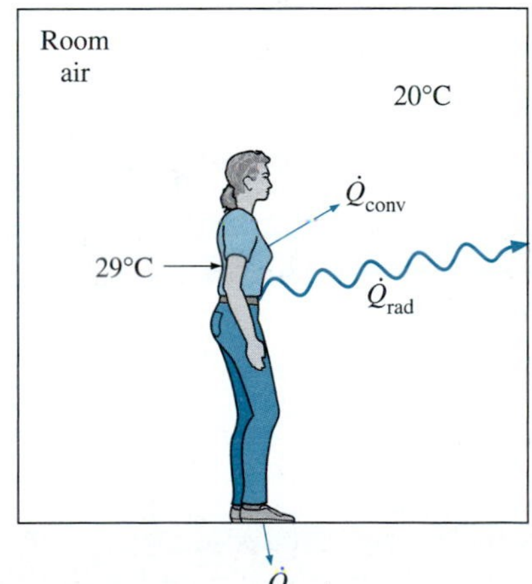

FIGURE 16–19

Heat transfer from the person described in Example 16–6.

Note that we must use *thermodynamic* temperatures in radiation calculations. Also note that we used the emissivity value for the skin and clothing at room temperature since the emissivity is not expected to change significantly at a slightly higher temperature.

Then the rate of total heat transfer from the body is determined by adding these two quantities:

$$\dot{Q}_{\text{total}} = \dot{Q}_{\text{conv}} + \dot{Q}_{\text{rad}} = (86.4 + 81.7)\text{ W} \cong \mathbf{168\ W}$$

Discussion The heat transfer would be much higher if the person were not dressed since the exposed surface temperature would be higher. Thus, an important function of the clothes is to serve as a barrier against heat transfer.

In these calculations, heat transfer through the feet to the floor by conduction, which is usually very small, is neglected. Heat transfer from the skin by perspiration, which is the dominant mode of heat transfer in hot environments, is not considered here.

Also, the units $W/m^2 \cdot °C$ and $W/m^2 \cdot K$ for heat transfer coefficient are equivalent, and can be interchanged.

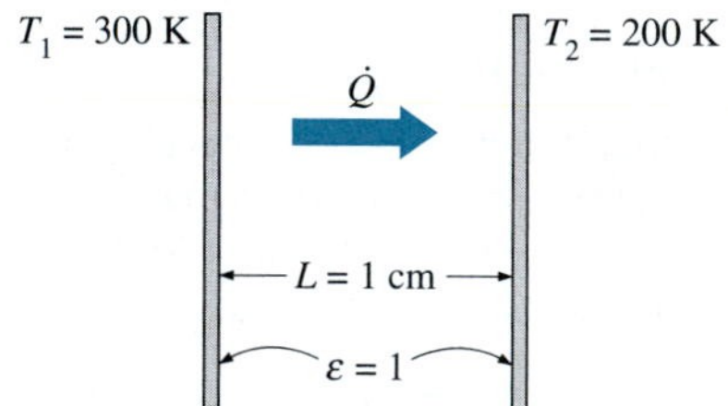

FIGURE 16–20

Schematic for Example 16–7.

EXAMPLE 16–7 Heat Transfer between Two Isothermal Plates

Consider steady heat transfer between two large parallel plates at constant temperatures of T_1 = 300 K and T_2 = 200 K that are L = 1 cm apart, as shown in Fig. 16–20. Assuming the surfaces to be black (emissivity $\varepsilon = 1$), determine the rate of heat transfer between the plates per unit surface area assuming the gap between the plates is (*a*) filled with atmospheric air, (*b*) evacuated, (*c*) filled with urethane insulation, and (*d*) filled with superinsulation that has an apparent thermal conductivity of 0.00002 W/m · K.

Solution The total rate of heat transfer between two large parallel plates at specified temperatures is to be determined for four different cases.

Assumptions **1** Steady operating conditions exist. **2** There are no natural convection currents in the air between the plates. **3** The surfaces are black and thus $\varepsilon = 1$.

Properties The thermal conductivity at the average temperature of 250 K is k = 0.0219 W/m · K for air (Table A–22), 0.026 W/m · K for urethane insulation, and 0.00002 W/m · K for the superinsulation.

Analysis (*a*) The rates of conduction and radiation heat transfer between the plates through the air layer are

$$\dot{Q}_{\text{cond}} = kA\frac{T_1 - T_2}{L} = (0.0219\text{ W/m}\cdot\text{K})(1\text{ m}^2)\frac{(300 - 200)\text{K}}{0.01\text{ m}} = 219\text{ W}$$

and

$$\dot{Q}_{\text{rad}} = \varepsilon\sigma A(T_1^4 - T_2^4)$$

$$= (1)(5.67 \times 10^{-8}\text{ W/m}^2\cdot\text{K}^4)(1\text{ m}^2)[(300\text{ K})^4 - (200\text{ K})^4] = 369\text{ W}$$

Therefore,

$$\dot{Q}_{\text{total}} = \dot{Q}_{\text{cond}} + \dot{Q}_{\text{rad}} = 219 + 369 = \mathbf{588\ W}$$

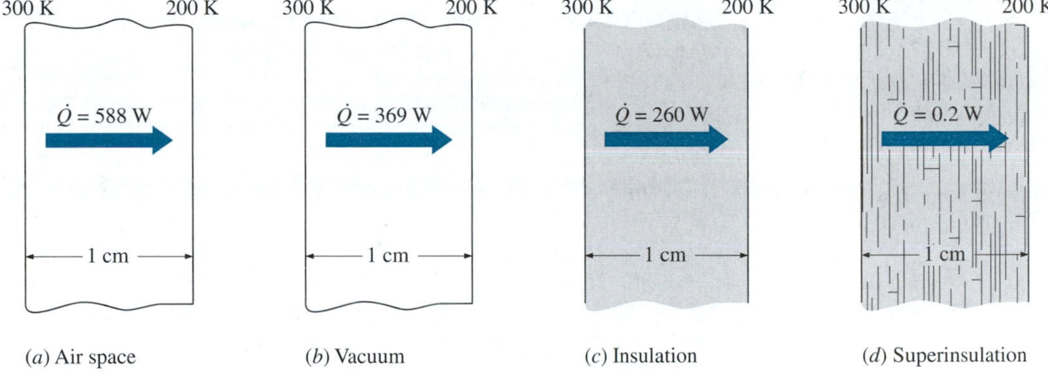

FIGURE 16–21

Different ways of reducing heat transfer between two isothermal plates, and their effectiveness.

The heat transfer rate in reality will be higher because of the natural convection currents that are likely to occur in the air space between the plates.

(*b*) When the air space between the plates is evacuated, there will be no conduction or convection, and the only heat transfer between the plates will be by radiation. Therefore,

$$\dot{Q}_{\text{total}} = \dot{Q}_{\text{rad}} = \mathbf{369\ W}$$

(*c*) An opaque solid material placed between two plates blocks direct radiation heat transfer between the plates. Also, the thermal conductivity of an insulating material accounts for the radiation heat transfer that may be occurring through the voids in the insulating material. The rate of heat transfer through the urethane insulation is

$$\dot{Q}_{\text{total}} = \dot{Q}_{\text{cond}} = kA\frac{T_1 - T_2}{L} = (0.026\ \text{W/m}\cdot\text{K})(1\ \text{m}^2)\frac{(300 - 200)\text{K}}{0.01\ \text{m}} = \mathbf{260\ W}$$

Note that heat transfer through the urethane material is less than the heat transfer through the air determined in (*a*), although the thermal conductivity of the insulation is higher than that of air. This is because the insulation blocks the radiation whereas air transmits it.

(*d*) The layers of the superinsulation prevent any direct radiation heat transfer between the plates. However, radiation heat transfer between the sheets of superinsulation does occur, and the apparent thermal conductivity of the superinsulation accounts for this effect. Therefore,

$$\dot{Q}_{\text{total}} = kA\frac{T_1 - T_2}{L} = (0.00002\ \text{W/m}\cdot\text{K})(1\ \text{m}^2)\frac{(300 - 200)\text{K}}{0.01\ \text{m}} = \mathbf{0.2\ W}$$

which is $\frac{1}{1845}$ of the heat transfer through the vacuum. The results of this example are summarized in Fig. 16–21 to put them into perspective.

Discussion This example demonstrates the effectiveness of superinsulations and explains why they are the insulation of choice in critical applications despite their high cost.

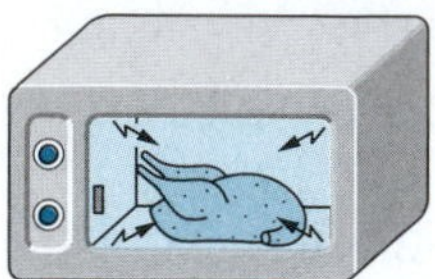

FIGURE 16–22

A chicken being cooked in a microwave oven (Example 16–8).

EXAMPLE 16–8 Heat Transfer in Conventional and Microwave Ovens

The fast and efficient cooking of microwave ovens made them one of the essential appliances in modern kitchens (Fig. 16–22). Discuss the heat transfer mechanisms associated with the cooking of a chicken in microwave and conventional ovens, and explain why cooking in a microwave oven is more efficient.

Solution Food is cooked in a microwave oven by absorbing the electromagnetic radiation energy generated by the microwave tube, called the magnetron. The radiation emitted by the magnetron is not thermal radiation, since its emission is not due to the temperature of the magnetron; rather, it is due to the conversion of electrical energy into electromagnetic radiation at a specified wavelength. The wavelength of the microwave radiation is such that it is *reflected* by metal surfaces; *transmitted* by the cookware made of glass, ceramic, or plastic; and *absorbed* and converted to internal energy by food (especially the water, sugar, and fat) molecules.

In a microwave oven, the *radiation* that strikes the chicken is absorbed by the skin of the chicken and the outer parts. As a result, the temperature of the chicken at and near the skin rises. Heat is then *conducted* toward the inner parts of the chicken from its outer parts. Of course, some of the heat absorbed by the outer surface of the chicken is lost to the air in the oven by *convection.*

In a conventional oven, the air in the oven is first heated to the desired temperature by the electric or gas heating element. This preheating may take several minutes. The heat is then transferred from the air to the skin of the chicken by *natural convection* in older ovens or by *forced convection* in the newer convection ovens that utilize a fan. The air motion in convection ovens increases the convection heat transfer coefficient and thus decreases the cooking time. Heat is then *conducted* toward the inner parts of the chicken from its outer parts as in microwave ovens.

Microwave ovens replace the slow convection heat transfer process in conventional ovens by the instantaneous radiation heat transfer. As a result, microwave ovens transfer energy to the food at full capacity the moment they are turned on, and thus they cook faster while consuming less energy.

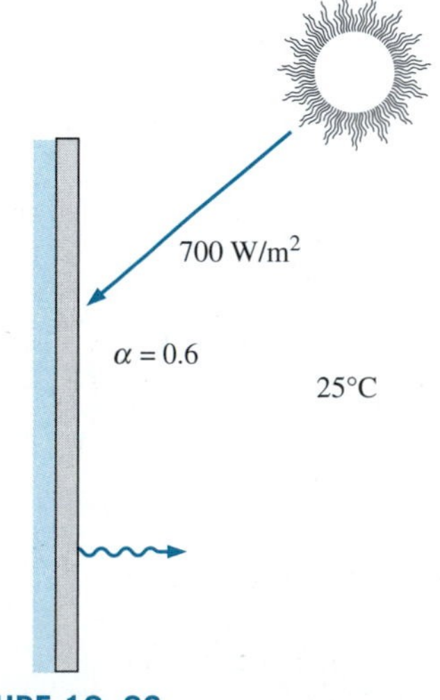

FIGURE 16–23

Schematic for Example 16–9.

EXAMPLE 16–9 Heating of a Plate by Solar Energy

A thin metal plate is insulated on the back and exposed to solar radiation at the front surface (Fig. 16–23). The exposed surface of the plate has an absorptivity of 0.6 for solar radiation. If solar radiation is incident on the plate at a rate of 700 W/m^2 and the surrounding air temperature is 25°C, determine the surface temperature of the plate when the heat loss by convection and radiation equals the solar energy absorbed by the plate. Assume the combined convection and radiation heat transfer coefficient to be 50 W/m^2 · °C.

Solution The back side of the thin metal plate is insulated and the front side is exposed to solar radiation. The surface temperature of the plate is to be determined when it stabilizes.

Assumptions **1** Steady operating conditions exist. **2** Heat transfer through the insulated side of the plate is negligible. **3** The heat transfer coefficient remains constant.

Properties The solar absorptivity of the plate is given to be $\alpha = 0.6$.

Analysis The absorptivity of the plate is 0.6, and thus 60 percent of the solar radiation incident on the plate is absorbed continuously. As a result, the temperature of the plate rises, and the temperature difference between the plate and the surroundings increases. This increasing temperature difference causes the rate of heat loss from the plate to the surroundings to increase. At some point, the rate of heat loss from the plate equals the rate of solar energy absorbed, and the temperature of the plate no longer changes. The temperature of the plate when steady operation is established is determined from

$$\dot{E}_{\text{gained}} = \dot{E}_{\text{lost}} \quad \text{or} \quad \alpha A_s \, \dot{q}_{\text{incident, solar}} = h_{\text{combined}} A_s (T_s - T_\infty)$$

Solving for T_s and substituting, the plate surface temperature is determined to be

$$T_s = T_\infty + \alpha \frac{\dot{q}_{\text{incident, solar}}}{h_{\text{combined}}} = 25°\text{C} + \frac{0.6 \times (700 \text{ W/m}^2)}{50 \text{ W/m}^2 \cdot °\text{C}} = \mathbf{33.4°C}$$

Discussion Note that the heat losses prevent the plate temperature from rising above 33.4°C. Also, the combined heat transfer coefficient accounts for the effects of both convection and radiation, and thus it is very convenient to use in heat transfer calculations when its value is known with reasonable accuracy.

SUMMARY

Heat can be transferred in three different modes: conduction, convection, and radiation. *Conduction* is the transfer of heat from the more energetic particles of a substance to the adjacent less energetic ones as a result of interactions between the particles, and is expressed by *Fourier's law of heat conduction* as

$$\dot{Q}_{\text{cond}} = -kA\frac{dT}{dx}$$

where k is the *thermal conductivity* of the material, A is the *area* normal to the direction of heat transfer, and dT/dx is the *temperature gradient.* The magnitude of the rate of heat conduction across a plane layer of thickness L is given by

$$\dot{Q}_{\text{cond}} = kA\frac{\Delta T}{L}$$

where ΔT is the temperature difference across the layer.

Convection is the mode of heat transfer between a solid surface and the adjacent liquid or gas that is in motion, and involves the combined effects of conduction and fluid motion. The rate of convection heat transfer is expressed by *Newton's law of cooling* as

$$\dot{Q}_{\text{convection}} = hA_s (T_s - T_\infty)$$

where h is the *convection heat transfer coefficient* in W/m$^2 \cdot$ K or Btu/h $\cdot$ ft$^2 \cdot$ R, A_s is the *surface area* through which convection heat transfer takes place, T_s is the *surface temperature,* and T_∞ is the *temperature of the fluid* sufficiently far from the surface.

Radiation is the energy emitted by matter in the form of electromagnetic waves (or photons) as a result of the changes in the electronic configurations of the atoms or molecules. The maximum rate of radiation that can be emitted from a surface at a thermodynamic temperature T_s is given by the *Stefan–Boltzmann law* as $\dot{Q}_{\text{emit, max}} = \sigma A_s T_s^4$ where $\sigma = 5.67 \times 10^{-8}$ W/m$^2 \cdot$ K^4 or 0.1714×10^{-8} Btu/h $\cdot$ ft$^2 \cdot$ R^4 is the *Stefan–Boltzmann constant.*

When a surface of emissivity ε and surface area A_s at a temperature T_s is completely enclosed by a much larger (or black) surface at a temperature T_{surr} separated by a gas (such as air) that does not intervene with radiation, the net rate of radiation heat transfer between these two surfaces is given by

$$\dot{Q}_{\text{rad}} = \varepsilon\sigma A_s (T_s^4 - T_{\text{surr}}^4)$$

In this case, the emissivity and the surface area of the surrounding surface do not have any effect on the net radiation heat transfer.

The rate at which a surface absorbs radiation is determined from $\dot{Q}_{absorbed} = \alpha \dot{Q}_{incident}$ where $\dot{Q}_{incident}$ is the rate at which radiation is incident on the surface and α is the absorptivity of the surface.

REFERENCES AND SUGGESTED READING

1. Y. A. Çengel. *Heat and Mass Transfer: A Practical Approach.* 3rd ed. New York: McGraw-Hill, 2007.

2. Y. A. Çengel and M. A. Boles. *Thermodynamics: An Engineering Approach.* 6th ed. New York: McGraw-Hill, 2008.

3. Robert J. Ribando. *Heat Transfer Tools.* New York: McGraw-Hill, 2002.

PROBLEMS*

Heat Transfer Mechanisms

16–1C Consider two houses that are identical, except that the walls are built using bricks in one house, and wood in the other. If the walls of the brick house are twice as thick, which house do you think will be more energy efficient?

16–2C Define thermal conductivity and explain its significance in heat transfer.

16–3C What are the mechanisms of heat transfer? How are they distinguished from each other?

16–4C What is the physical mechanism of heat conduction in a solid, a liquid, and a gas?

16–5C Consider heat transfer through a windowless wall of a house on a winter day. Discuss the parameters that affect the rate of heat conduction through the wall.

16–6C Write down the expressions for the physical laws that govern each mode of heat transfer, and identify the variables involved in each relation.

16–7C How does heat conduction differ from convection?

16–8C Does any of the energy of the sun reach the earth by conduction or convection?

16–9C How does forced convection differ from natural convection?

16–10C Define emissivity and absorptivity. What is Kirchhoff's law of radiation?

16–11C What is a blackbody? How do real bodies differ from blackbodies?

16–12C Judging from its unit W/m · K, can we define thermal conductivity of a material as the rate of heat transfer through the material per unit thickness per unit temperature difference? Explain.

16–13C Consider heat loss through the two walls of a house on a winter night. The walls are identical, except that one of them has a tightly fit glass window. Through which wall will the house lose more heat? Explain.

16–14C Which is a better heat conductor, diamond or silver?

16–15C Consider two walls of a house that are identical except that one is made of 10-cm-thick wood, while the other is made of 25-cm-thick brick. Through which wall will the house lose more heat in winter?

16–16C How do the thermal conductivity of gases and liquids vary with temperature?

16–17C Why is the thermal conductivity of superinsulation orders of magnitude lower than the thermal conductivity of ordinary insulation?

16–18C Why do we characterize the heat conduction ability of insulators in terms of their apparent thermal conductivity instead of the ordinary thermal conductivity?

16–19C Consider an alloy of two metals whose thermal conductivities are k_1 and k_2. Will the thermal conductivity of the alloy be less than k_1, greater than k_2, or between k_1 and k_2?

16–20 The inner and outer surfaces of a 4-m × 7-m brick wall of thickness 30 cm and thermal conductivity 0.69 W/m · K are maintained at temperatures of 20°C and 5°C, respectively. Determine the rate of heat transfer through the wall, in W.

* Problems designated by a "C" are concept questions, and students are encouraged to answer them all. Problems designated by an "E" are in English units, and the SI users can ignore them. Problems with the icon are solved using EES, and complete solutions together with parametric studies are included on the enclosed DVD. Problems with the icon are comprehensive in nature and are intended to be solved with a computer, preferably using the EES software that accompanies this text.

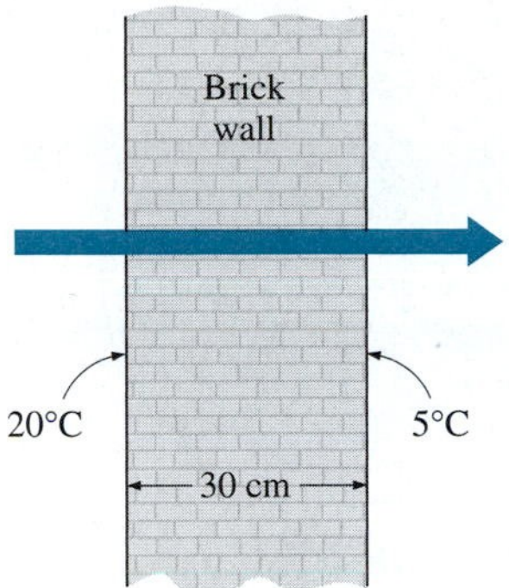

FIGURE P16–20

16–21 The inner and outer surfaces of a 0.5-cm thick 2-m × 2-m window glass in winter are 10°C and 3°C, respectively. If the thermal conductivity of the glass is 0.78 W/m · K, determine the amount of heat loss through the glass over a period of 5 h. What would your answer be if the glass were 1 cm thick? *Answers:* 78.6 MJ, 39.3 MJ

16–22 Reconsider Prob. 16–21. Using EES (or other) software, plot the amount of heat loss through the glass as a function of the window glass thickness in the range of 0.1 cm to 1.0 cm. Discuss the results.

16–23 An aluminum pan whose thermal conductivity is 237 W/m · °C has a flat bottom with diameter 15 cm and thickness 0.4 cm. Heat is transferred steadily to boiling water in the pan through its bottom at a rate of 800 W. If the inner surface of the bottom of the pan is at 105°C, determine the temperature of the outer surface of the bottom of the pan.

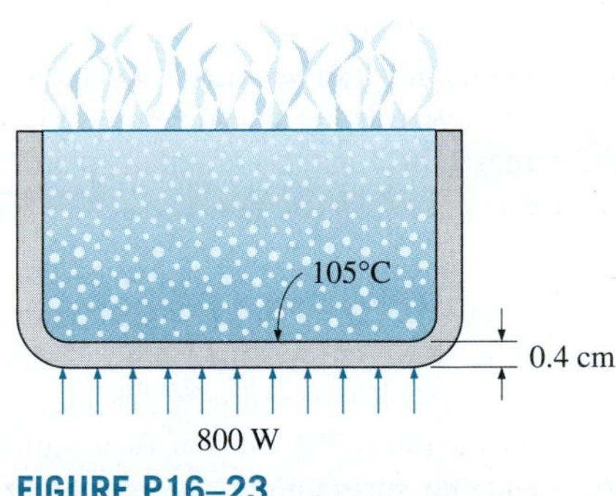

FIGURE P16–23

16–24E The north wall of an electrically heated home is 20 ft long, 10 ft high, and 1 ft thick, and is made of brick whose thermal conductivity is $k = 0.42$ Btu/h · ft · °F. On a certain winter night, the temperatures of the inner and the outer surfaces of the wall are measured to be at about 62°F and 25°F, respectively, for a period of 8 h. Determine (*a*) the rate of heat loss through the wall that night and (*b*) the cost of that heat loss to the home owner if the cost of electricity is \$0.07/kWh.

16–25 In a certain experiment, cylindrical samples of diameter 4 cm and length 7 cm are used (see Fig. 16–9). The two thermocouples in each sample are placed 3 cm apart. After initial transients, the electric heater is observed to draw 0.6 A at 110 V, and both differential thermometers read a temperature difference of 10°C. Determine the thermal conductivity of the sample. *Answer:* 78.8 W/m · °C

16–26 One way of measuring the thermal conductivity of a material is to sandwich an electric thermofoil heater between two identical rectangular samples of the material and to heavily insulate the four outer edges, as shown in the figure. Thermocouples attached to the inner and outer surfaces of the samples record the temperatures.

During an experiment, two 0.5 cm thick samples 10 cm × 10 cm in size are used. When steady operation is reached, the heater is observed to draw 25 W of electric power, and the temperature of each sample is observed to drop from 82°C at the inner surface to 74°C at the outer surface. Determine the thermal conductivity of the material at the average temperature.

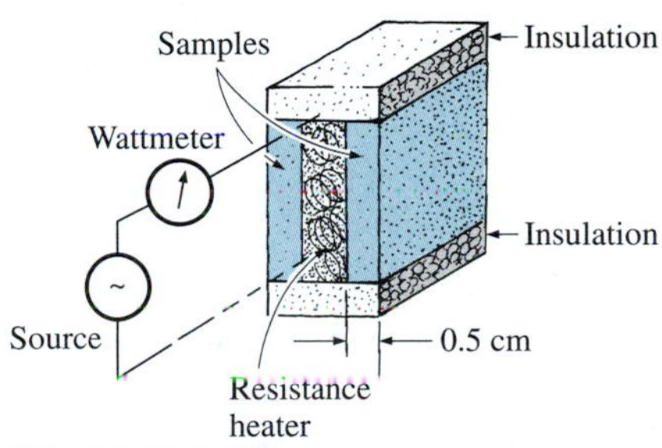

FIGURE P16–26

16–27 Repeat Prob. 16–26 for an electric power consumption of 20 W.

16–28 A heat flux meter attached to the inner surface of a 3-cm-thick refrigerator door indicates a heat flux of 25 W/m^2 through the door. Also, the temperatures of the inner and the outer surfaces of the door are measured to be 7°C and 15°C, respectively. Determine the average thermal conductivity of the refrigerator door. *Answer:* 0.0938 W/m · °C

16–29 Consider a person standing in a room maintained at 20°C at all times. The inner surfaces of the walls, floors, and ceiling of the house are observed to be at an average temperature of 12°C in winter and 23°C in summer. Determine the rates of radiation heat transfer between this person and the surrounding surfaces in both summer and winter if the exposed surface area, emissivity, and the average outer surface temperature of the person are 1.6 m^2, 0.95, and 32°C, respectively.

16–30 Reconsider Prob. 16–29. Using EES (or other) software, plot the rate of radiation heat transfer in winter as a function of the temperature of the inner surface of the room in the range of 8°C to 18°C. Discuss the results.

16–31 For heat transfer purposes, a standing man can be modeled as a 30-cm-diameter, 170-cm-long vertical cylinder with both the top and bottom surfaces insulated and with the side surface at an average temperature of 34°C. For a convection heat transfer coefficient of 20 W/m² · °C, determine the rate of heat loss from this man by convection in an environment at 18°C. *Answer:* 513 W

16–32 Hot air at 80°C is blown over a 2-m × 4-m flat surface at 30°C. If the average convection heat transfer coefficient is 55 W/m² · °C, determine the rate of heat transfer from the air to the plate, in kW. *Answer:* 22 kW

16–33 Reconsider Prob. 16–32. Using EES (or other) software, plot the rate of heat transfer as a function of the heat transfer coefficient in the range of 20 W/m² · °C to 100 W/m² · °C. Discuss the results.

16–34 The heat generated in the circuitry on the surface of a silicon chip (k = 130 W/m · °C) is conducted to the ceramic substrate to which it is attached. The chip is 6 mm × 6 mm in size and 0.5 mm thick and dissipates 3 W of power. Disregarding any heat transfer through the 0.5 mm high side surfaces, determine the temperature difference between the front and back surfaces of the chip in steady operation.

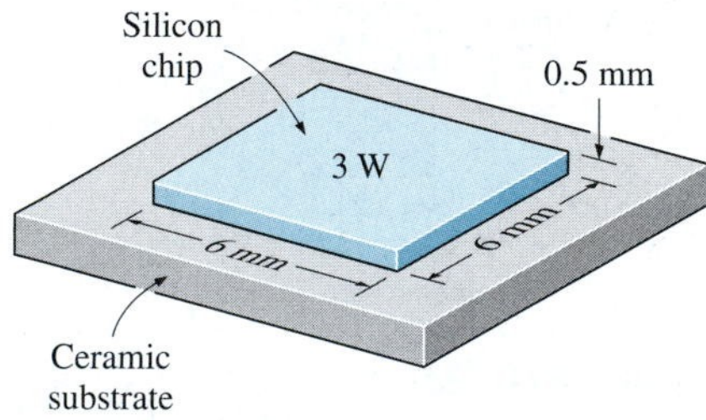

FIGURE P16–34

16–35 A 40-cm-long, 800-W electric resistance heating element with diameter 0.5 cm and surface temperature 120°C is immersed in 75 kg of water initially at 20°C. Determine how long it will take for this heater to raise the water temperature to 80°C. Also, determine the convection heat transfer coefficients at the beginning and at the end of the heating process.

16–36 A 5-cm-external-diameter, 10-m-long hot-water pipe at 80°C is losing heat to the surrounding air at 5°C by natural convection with a heat transfer coefficient of 25 W/m² · °C. Determine the rate of heat loss from the pipe by natural convection. *Answer:* 2945 W

16–37 A hollow spherical iron container with outer diameter 20 cm and thickness 0.4 cm is filled with iced water at 0°C. If the outer surface temperature is 5°C, determine the approximate rate of heat loss from the sphere, in kW, and the rate at which ice melts in the container. The heat of fusion of water is 333.7 kJ/kg.

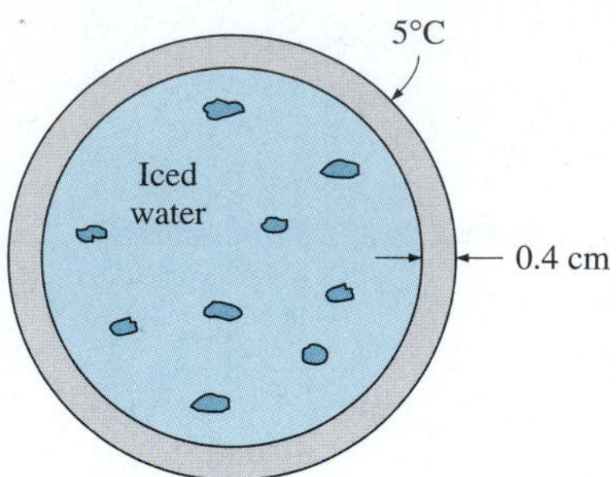

FIGURE P16–37

16–38 Reconsider Prob. 16–37. Using EES (or other) software, plot the rate at which ice melts as a function of the container thickness in the range of 0.2 cm to 2.0 cm. Discuss the results.

16–39E The inner and outer glasses of a 4-ft × 4-ft double-pane window are at 60°F and 48°F, respectively. If the 0.25-in. space between the two glasses is filled with still air, determine the rate of heat transfer through the window. *Answer:* 131 Btu/h

16–40 Two surfaces of a 2-cm-thick plate are maintained at 0°C and 80°C, respectively. If it is determined that heat is transferred through the plate at a rate of 500 W/m², determine its thermal conductivity.

16–41 Four power transistors, each dissipating 15 W, are mounted on a thin vertical aluminum plate 22 cm × 22 cm in size. The heat generated by the transistors is to be dissipated by both surfaces of the plate to the surrounding air at 25°C, which is blown over the plate by a fan. The entire plate can be assumed to be nearly isothermal, and the exposed surface area of the transistor can be taken to be equal to its base area. If the average convection heat transfer coefficient is 25 W/m² · °C, determine the temperature of the aluminum plate. Disregard any radiation effects.

16–42 An ice chest whose outer dimensions are 30 cm × 40 cm × 40 cm is made of 3-cm-thick Styrofoam (k = 0.033 W/m · °C). Initially, the chest is filled with 28 kg of ice at 0°C, and the inner surface temperature of the ice chest can be taken to be 0°C at all times. The heat of fusion of ice at 0°C is 333.7 kJ/kg, and the surrounding ambient air is at 25°C. Disregarding any heat transfer from the 40-cm × 40-cm base of the ice chest, determine how long it will take for the ice in the chest to melt completely if the outer surfaces of the ice chest are at 8°C. *Answer:* 22.9 days

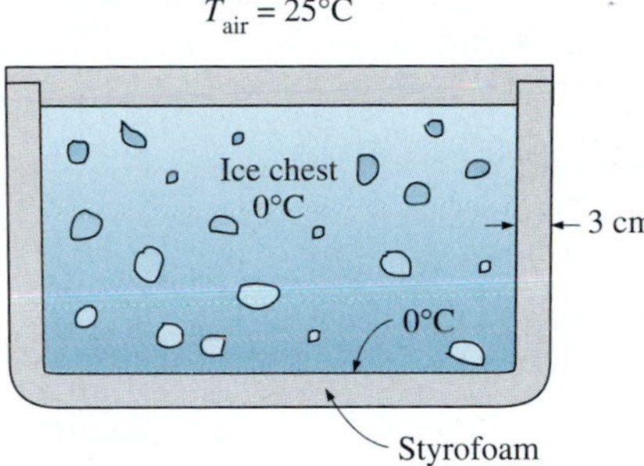

FIGURE P16–42

16–43 A transistor with a height of 0.4 cm and a diameter of 0.6 cm is mounted on a circuit board. The transistor is cooled by air flowing over it with an average heat transfer coefficient of 30 W/m^2 · °C. If the air temperature is 55°C and the transistor case temperature is not to exceed 70°C, determine the amount of power this transistor can dissipate safely. Disregard any heat transfer from the transistor base.

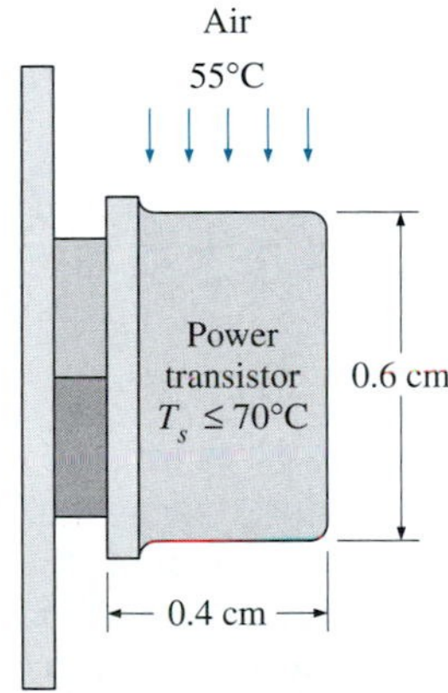

FIGURE P16–43

16–44 Reconsider Prob. 16–43. Using EES (or other) software, plot the amount of power the transistor can dissipate safely as a function of the maximum case temperature in the range of 60°C to 90°C. Discuss the results.

16–45E A 200-ft-long section of a steam pipe whose outer diameter is 4 in passes through an open space at 50°F. The average temperature of the outer surface of the pipe is measured to be 280°F, and the average heat transfer coefficient on that surface is determined to be 6 Btu/h · ft^2 · °F. Determine (*a*) the rate of heat loss from the steam pipe and (*b*) the annual cost of this energy loss if steam is generated in a natural gas furnace having an efficiency of 86 percent, and the price of natural gas is \$1.10/therm (1 therm = 100,000 Btu). *Answers:* (*a*) 289,000 Btu/h, (*b*) \$32,380/yr

16–46 The boiling temperature of nitrogen at atmospheric pressure at sea level (1 atm) is −196°C. Therefore, nitrogen is commonly used in low temperature scientific studies since the temperature of liquid nitrogen in a tank open to the atmosphere remains constant at −196°C until the liquid nitrogen in the tank is depleted. Any heat transfer to the tank results in the evaporation of some liquid nitrogen, which has a heat of vaporization of 198 kJ/kg and a density of 810 kg/m^3 at 1 atm.

Consider a 4-m-diameter spherical tank initially filled with liquid nitrogen at 1 atm and −196°C. The tank is exposed to 20°C ambient air with a heat transfer coefficient of 25 W/m^2 · °C. The temperature of the thin-shelled spherical tank is observed to be almost the same as the temperature of the nitrogen inside. Disregarding any radiation heat exchange, determine the rate of evaporation of the liquid nitrogen in the tank as a result of the heat transfer from the ambient air.

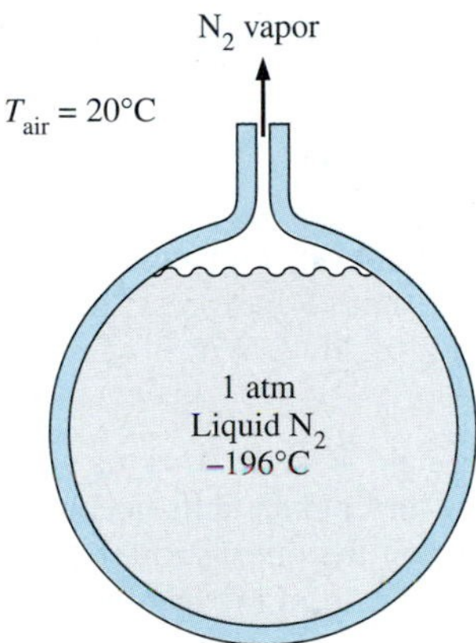

FIGURE P16–46

16–47 Repeat Prob. 16–46 for liquid oxygen, which has a boiling temperature of −183°C, a heat of vaporization of 213 kJ/kg, and a density of 1140 kg/m^3 at 1 atm pressure.

16–48 Reconsider Prob. 16–46. Using EES (or other) software, plot the rate of evaporation of liquid nitrogen as a function of the ambient air temperature in the range of 0°C to 35°C. Discuss the results.

16–49 Consider a person whose exposed surface area is 1.7 m^2, emissivity is 0.5, and surface temperature is 32°C. Determine the rate of heat loss from that person by radiation in a large room having walls at a temperature of (*a*) 300 K and (*b*) 280 K. *Answers:* (*a*) 26.7 W, (*b*) 121 W

16–50 A 0.3-cm-thick, 12-cm-high, and 18-cm-long circuit board houses 80 closely spaced logic chips on one side, each dissipating 0.06 W. The board is impregnated with copper fillings and has an effective thermal conductivity of 16 W/m · °C. All the heat generated in the chips is conducted across the circuit board and is dissipated from the back side of the board to the ambient air. Determine the temperature difference between the two sides of the circuit board. *Answer:* 0.042°C

16–51 Consider a sealed 20-cm-high electronic box whose base dimensions are 40 cm × 40 cm placed in a vacuum chamber. The emissivity of the outer surface of the box is 0.95. If the electronic components in the box dissipate a total of 100 W of power and the outer surface temperature of the box is not to exceed 55°C, determine the temperature at which the surrounding surfaces must be kept if this box is to be cooled by radiation alone. Assume the heat transfer from the bottom surface of the box to the stand to be negligible.

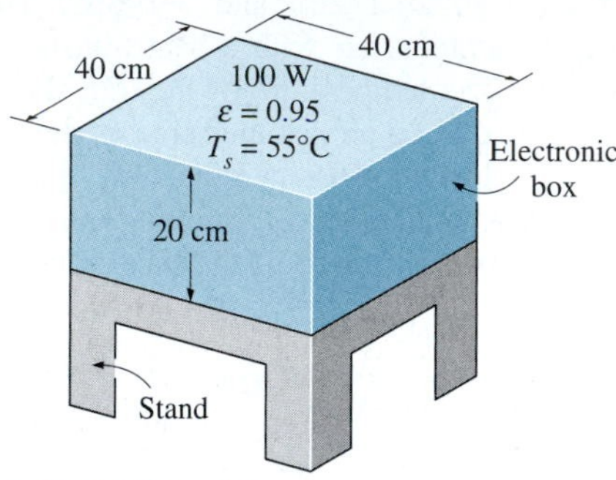

FIGURE P16–51

16–52E Using the conversion factors between W and Btu/h, m and ft, and K and R, express the Stefan–Boltzmann constant $\sigma = 5.67 \times 10^{-8}$ W/m² · K⁴ in the English unit Btu/h · ft² · R⁴.

16–53E An engineer who is working on the heat transfer analysis of a house in English units needs the convection heat transfer coefficient on the outer surface of the house. But the only value he can find from his handbooks is 14 W/m² · °C, which is in SI units. The engineer does not have a direct conversion factor between the two unit systems for the convection heat transfer coefficient. Using the conversion factors between W and Btu/h, m and ft, and °C and °F, express the given convection heat transfer coefficient in Btu/h · ft² · °F. *Answer:* 2.47 Btu/h · ft² · °F

16–54 A 2.5-cm-diameter and 8-cm-long cylindrical sample of a material is used to determine its thermal conductivity experimentally. In the thermal conductivity apparatus, the sample is placed in a well-insulated cylindrical cavity to ensure one-dimensional heat transfer in the axial direction, and a heat flux generated by a resistance heater whose electricity consumption is measured is applied on one of its faces (say, the left face). A total of 9 thermocouples are imbedded into the sample, 1 cm apart, to measure the temperatures along the sample and on its faces. When the power consumption was fixed at 83.45 W, it is observed that the thermocouple readings are stabilized at the following values:

Distance from left face, cm	Temperature, °C
0	89.38
1	83.25
2	78.28
3	74.10
4	68.25
5	63.73
6	49.65
7	44.40
8	40.00

Plot the variation of temperature along the sample, and calculate the thermal conductivity of the sample material. Based on these temperature readings, do you think steady operating conditions are established? Are there any temperature readings that do not appear right and should be discarded? Also, discuss when and how the temperature profile in a plane wall will deviate from a straight line.

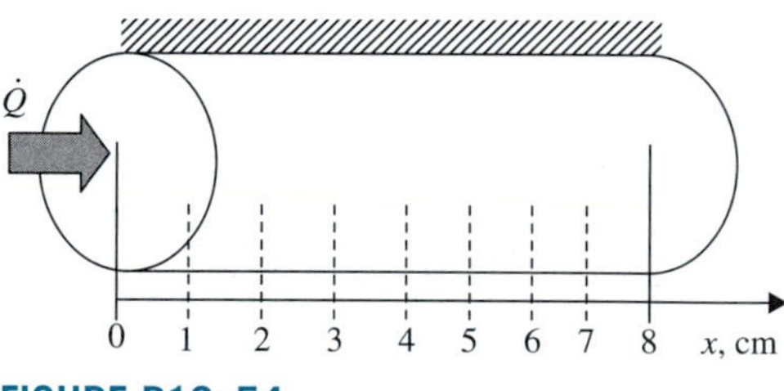

FIGURE P16–54

16–55 Water at 0°C releases 333.7 kJ/kg of heat as it freezes to ice (ρ = 920 kg/m³) at 0°C. An aircraft flying under icing conditions maintains a heat transfer coefficient of 150 W/m² · °C between the air and wing surfaces. What temperature must the wings be maintained at to prevent ice from forming on them during icing conditions at a rate of 1 mm/min or less?

Simultaneous Heat Transfer Mechanisms

16–56C Can all three modes of heat transfer occur simultaneously (in parallel) in a medium?

16–57C Can a medium involve (*a*) conduction and convection, (*b*) conduction and radiation, or (*c*) convection and radiation simultaneously? Give examples for the "yes" answers.

16–58C The deep human body temperature of a healthy person remains constant at 37°C while the temperature and the humidity of the environment change with time. Discuss the heat transfer mechanisms between the human body and the environment both in summer and winter, and explain how a person can keep cooler in summer and warmer in winter.

16–59C We often turn the fan on in summer to help us cool. Explain how a fan makes us feel cooler in the summer. Also explain why some people use ceiling fans also in winter.

16–60 Consider a person standing in a room at 23°C. Determine the total rate of heat transfer from this person if the exposed surface area and the skin temperature of the person are 1.7 m^2 and 32°C, respectively, and the convection heat transfer coefficient is 5 $W/m^2 \cdot °C$. Take the emissivity of the skin and the clothes to be 0.9, and assume the temperature of the inner surfaces of the room to be the same as the air temperature. *Answer:* 161 W

16–61 Consider steady heat transfer between two large parallel plates at constant temperatures of T_1 = 290 K and T_2 = 150 K that are L = 2 cm apart. Assuming the surfaces to be black (emissivity ε = 1), determine the rate of heat transfer between the plates per unit surface area assuming the gap between the plates is (*a*) filled with atmospheric air, (*b*) evacuated, (*c*) filled with fiberglass insulation, and (*d*) filled with superinsulation having an apparent thermal conductivity of 0.00015 W/m · °C.

16–62 The inner and outer surfaces of a 25-cm-thick wall in summer are at 27°C and 44°C, respectively. The outer surface of the wall exchanges heat by radiation with surrounding surfaces at 40°C, and convection with ambient air also at 40°C with a convection heat transfer coefficient of 8 $W/m^2 \cdot °C$. Solar radiation is incident on the surface at a rate of 150 W/m^2. If both the emissivity and the solar absorptivity of the outer surface are 0.8, determine the effective thermal conductivity of the wall.

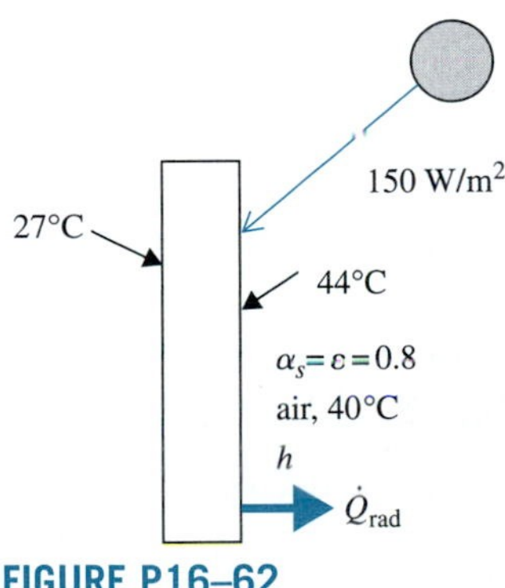

FIGURE P16–62

16–63 A 1.4-m-long, 0.2-cm-diameter electrical wire extends across a room that is maintained at 20ºC. Heat is generated in the wire as a result of resistance heating, and the surface temperature of the wire is measured to be 240ºC in steady operation. Also, the voltage drop and electric current through the wire are measured to be 110 V and 3 A, respectively. Disregarding any heat transfer by radiation, determine the convection heat transfer coefficient for heat transfer between the outer surface of the wire and the air in the room. *Answer:* 170.5 $W/m^2 \cdot °C$

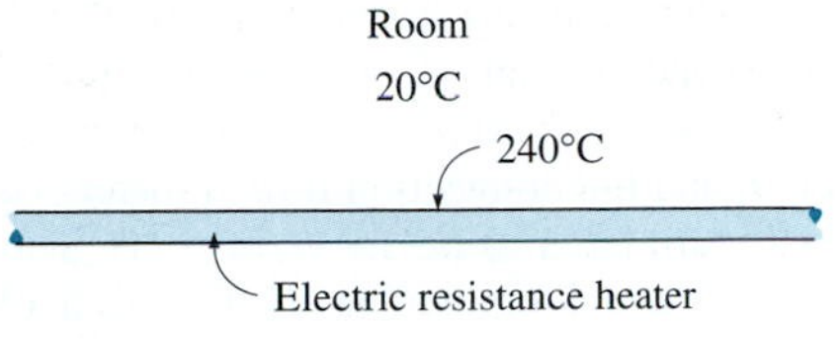

FIGURE P16–63

16–64 Reconsider Prob. 16–63. Using EES (or other) software, plot the convection heat transfer coefficient as a function of the wire surface temperature in the range of 100°C to 300°C. Discuss the results.

16–65E A 2-in-diameter spherical ball whose surface is maintained at a temperature of 170°F is suspended in the middle of a room at 70°F. If the convection heat transfer coefficient is 15 $Btu/h \cdot ft^2 \cdot °F$ and the emissivity of the surface is 0.8, determine the total rate of heat transfer from the ball.

16–66 A 1000-W iron is left on the iron board with its base exposed to the air at 20°C. The convection heat transfer coefficient between the base surface and the surrounding air is 35 $W/m^2 \cdot °C$. If the base has an emissivity of 0.6 and a surface area of 0.02 m^2, determine the temperature of the base of the iron. *Answer:* 674°C

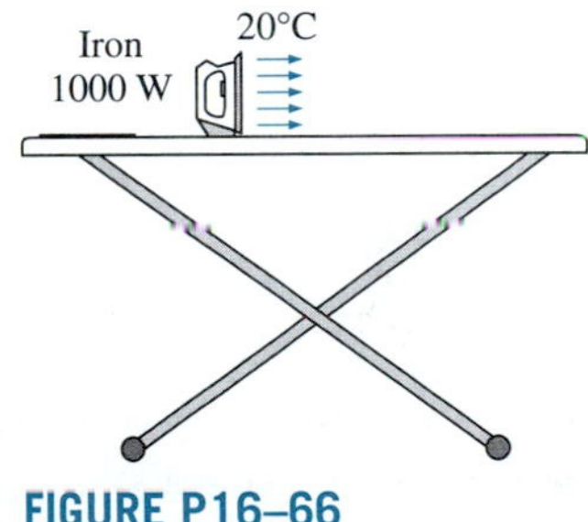

FIGURE P16–66

16–67 The outer surface of a spacecraft in space has an emissivity of 0.8 and a solar absorptivity of 0.3. If solar radiation is incident on the spacecraft at a rate of 950 W/m^2, determine the surface temperature of the spacecraft when the radiation emitted equals the solar energy absorbed.

16–68 A 3-m-internal-diameter spherical tank made of 1-cm-thick stainless steel is used to store iced water at 0°C. The tank is located outdoors at 25°C. Assuming the entire steel tank to be at 0°C and thus the thermal resistance of the tank to be negligible, determine (*a*) the rate of heat transfer to the iced water in the tank and (*b*) the amount of ice at 0°C that melts during a 24-hour period. The heat of fusion of water at atmospheric pressure is h_{if} = 333.7 kJ/kg. The emissivity of the outer surface of the tank is 0.75, and the convection heat transfer coefficient on the outer surface can be taken to be 30 $W/m^2 \cdot °C$. Assume the average surrounding surface temperature for radiation exchange to be 15°C. *Answers:* (*a*) 23.1 kW, (*b*) 5980 kg

16–69 The roof of a house consists of a 15-cm-thick concrete slab (k = 2 W/m · °C) that is 15 m wide and 20 m long. The emissivity of the outer surface of the roof is 0.9, and the convection heat transfer coefficient on that surface is estimated to be 15 W/m^2 · °C. The inner surface of the roof is maintained at 15°C. On a clear winter night, the ambient air is reported to be at 10°C while the night sky temperature for radiation heat transfer is 255 K. Considering both radiation and convection heat transfer, determine the outer surface temperature and the rate of heat transfer through the roof.

If the house is heated by a furnace burning natural gas with an efficiency of 85 percent, and the unit cost of natural gas is \$0.60/therm (1 therm = 105,500 kJ of energy content), determine the money lost through the roof that night during a 14-hour period.

16–70E Consider a flat-plate solar collector placed horizontally on the flat roof of a house. The collector is 5 ft wide and 15 ft long, and the average temperature of the exposed surface of the collector is 100°F. The emissivity of the exposed surface of the collector is 0.9. Determine the rate of heat loss from the collector by convection and radiation during a calm day when the ambient air temperature is 70°F and the effective sky temperature for radiation exchange is 50°F. Take the convection heat transfer coefficient on the exposed surface to be 2.5 Btu/h · ft^2 · °F.

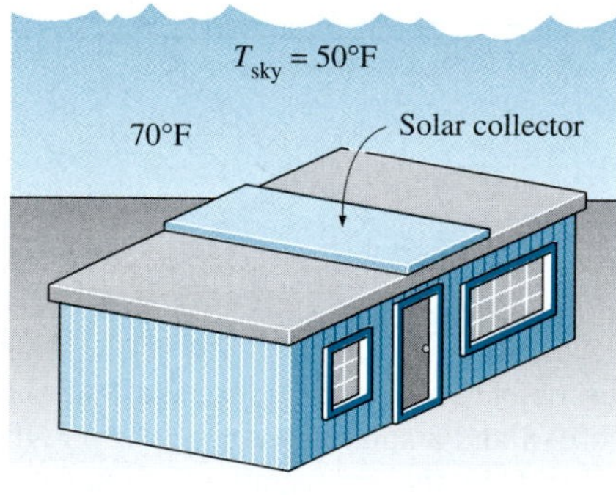

FIGURE P16–70E

Review Problems

16–71 It is well known that wind makes the cold air feel much colder as a result of the *windchill* effect that is due to the increase in the convection heat transfer coefficient with increasing air velocity. The windchill effect is usually expressed in terms of the *windchill factor,* which is the difference between the actual air temperature and the equivalent calm-air temperature. For example, a windchill factor of 20°C for an actual air temperature of 5°C means that the windy air at 5°C feels as cold as the still air at −15°C. In other words, a person will lose as much heat to air at 5°C with a windchill factor of 20°C as he or she would in calm air at −15°C.

For heat transfer purposes, a standing man can be modeled as a 30-cm-diameter, 170-cm-long vertical cylinder with both the top and bottom surfaces insulated and with the side surface at an average temperature of 34°C. For a convection heat transfer coefficient of 15 W/m^2 · °C, determine the rate of heat loss from this man by convection in still air at 20°C. What would your answer be if the convection heat transfer coefficient is increased to 50 W/m^2 · °C as a result of winds? What is the windchill factor in this case? *Answers:* 336 W, 1120 W, 32.7°C

16–72 A thin metal plate is insulated on the back and exposed to solar radiation on the front surface. The exposed surface of the plate has an absorptivity of 0.7 for solar radiation. If solar radiation is incident on the plate at a rate of 550 W/m^2 and the surrounding air temperature is 10°C, determine the surface temperature of the plate when the heat loss by convection equals the solar energy absorbed by the plate. Take the convection heat transfer coefficient to be 25 W/m^2 · °C, and disregard any heat loss by radiation.

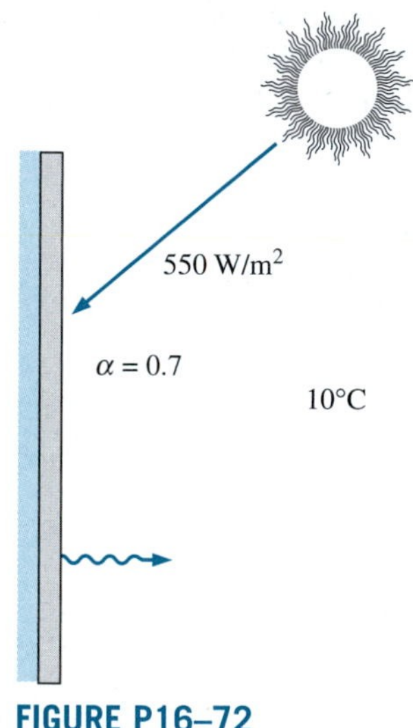

FIGURE P16–72

16–73 A 4-m × 5-m × 6-m room is to be heated by one ton (1000 kg) of liquid water contained in a tank placed in the room. The room is losing heat to the outside at an average rate of 10,000 kJ/h. The room is initially at 20°C and 100 kPa, and is maintained at an average temperature of 20°C at all times. If the hot water is to meet the heating requirements of this room for a 24-h period, determine the minimum temperature of the water when it is first brought into the room. Assume constant specific heats for both air and water at room temperature. *Answer:* 77.4°C

16–74 Consider a 3-m × 3-m × 3-m cubical furnace whose top and side surfaces closely approximate black surfaces at a temperature of 1200 K. The base surface has an emissivity of ε = 0.7, and is maintained at 800 K. Determine the net rate of radiation heat transfer to the base surface from the top and side surfaces. *Answer:* 594 kW

16–75 Consider a refrigerator whose dimensions are 1.8 m × 1.2 m × 0.8 m and whose walls are 3 cm thick. The refrigerator consumes 600 W of power when operating and has a COP of 1.5. It is observed that the motor of the refrigerator remains on for 5 min and then is off for 15 min periodically. If the average temperatures at the inner and outer surfaces of the refrigerator are 6°C and 17°C, respectively, determine the average thermal conductivity of the refrigerator walls. Also, determine the annual cost of operating this refrigerator if the unit cost of electricity is \$0.08/kWh.

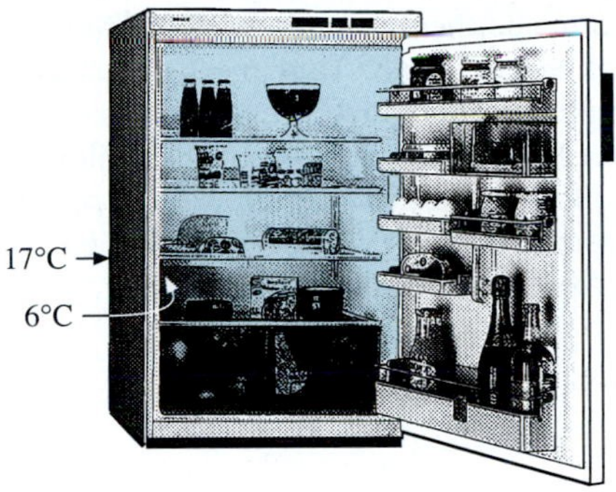

FIGURE P16–75

16–76 Engine valves (c_p = 440 J/kg · °C and ρ = 7840 kg/m^3) are to be heated from 40°C to 800°C in 5 min in the heat treatment section of a valve manufacturing facility. The valves have a cylindrical stem with a diameter of 8 mm and a length of 10 cm. The valve head and the stem may be assumed to be of equal surface area, with a total mass of 0.0788 kg. For a single valve, determine (*a*) the amount of heat transfer, (*b*) the average rate of heat transfer, (*c*) the average heat flux, and (*d*) the number of valves that can be heat treated per day if the heating section can hold 25 valves and it is used 10 h per day.

16–77 Consider a flat-plate solar collector placed on the roof of a house. The temperatures at the inner and outer surfaces of the glass cover are measured to be 28°C and 25°C, respectively. The glass cover has a surface area of 2.5 m^2, a thickness of 0.6 cm, and a thermal conductivity of 0.7 W/m · °C. Heat is lost from the outer surface of the cover by convection and radiation with a convection heat transfer coefficient of 10 W/m^2 · °C and an ambient temperature of 15°C. Determine the fraction of heat lost from the glass cover by radiation.

16–78 The rate of heat loss through a unit surface area of a window per unit temperature difference between the indoors and the outdoors is called the *U*-factor. The value of the *U*-factor ranges from about 1.25 W/m^2 · °C (or 0.22 Btu/h · ft^2 · °F) for low-*e* coated, argon-filled, quadruple-pane windows to 6.25 W/m^2 · °C (or 1.1 Btu/h · ft^2 · °F) for a single-pane window with aluminum frames. Determine the range for the rate of heat loss through a 1.2-m × 1.8-m window of a house that is maintained at 20°C when the outdoor air temperature is −8°C.

16–79 Reconsider Prob. 16–78. Using EES (or other) software, plot the rate of heat loss through the window as a function of the *U*-factor. Discuss the results.

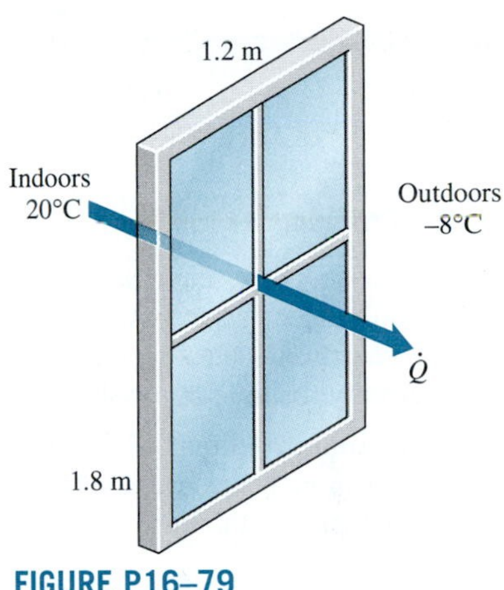

FIGURE P16–79

16–80 Consider a house in Atlanta, Georgia, that is maintained at 22°C and has a total of 20 m^2 of window area. The windows are double-door type with wood frames and metal spacers and have a *U*-factor of 2.5 W/m^2 · °C (see Prob. 16–78 for the definition of *U*-factor). The winter average temperature of Atlanta is 11.3°C. Determine the average rate of heat loss through the windows in winter.

16–81 A 50-cm-long, 2-mm-diameter electric resistance wire submerged in water is used to determine the boiling heat transfer coefficient in water at 1 atm experimentally. The wire temperature is measured to be 130°C when a wattmeter indicates the electric power consumed to be 4.1 kW. Using Newton's law of cooling, determine the boiling heat transfer coefficient.

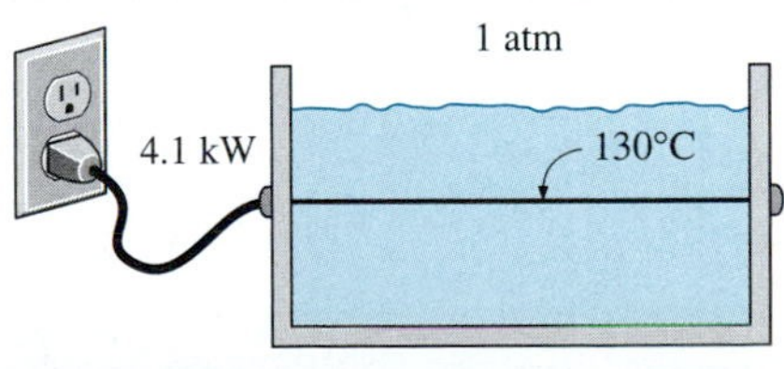

FIGURE P16–81

16–82 An electric heater with the total surface area of 0.25 m^2 and emissivity 0.75 is in a room where the air has a temperature of 20°C and the walls are at 10°C. When the heater consumes 500 W of electric power, its surface has a steady temperature of 120°C. Determine the temperature of the heater surface when it consumes 700 W. Solve the problem (*a*) assuming negligible radiation and (*b*) taking radiation into consideration. Based on your results, comment on the assumption made in part (*a*)

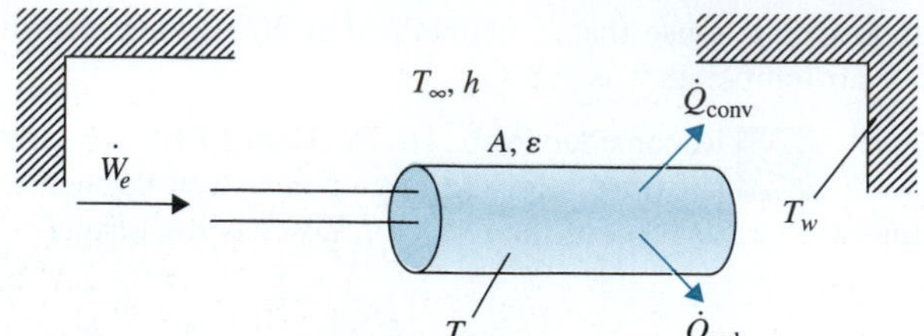

FIGURE P16–82

Design and Essay Problems

16–83 Write an essay on how microwave ovens work, and explain how they cook much faster than conventional ovens. Discuss whether conventional electric or microwave ovens consume more electricity for the same task.

16–84 Using information from the utility bills for the coldest month last year, estimate the average rate of heat loss from your house for that month. In your analysis, consider the contribution of the internal heat sources such as people, lights, and appliances. Identify the primary sources of heat loss from your house and propose ways of improving the energy efficiency of your house.

16–85 Conduct this experiment to determine the combined heat transfer coefficient between an incandescent lightbulb and the surrounding air and surfaces using a 60-W lightbulb. You will need a thermometer, which can be purchased in a hardware store, and a metal glue. You will also need a piece of string and a ruler to calculate the surface area of the lightbulb. First, measure the air temperature in the room, and then glue the tip of the thermocouple wire of the thermometer to the glass of the lightbulb. Turn the light on and wait until the temperature reading stabilizes. The temperature reading will give the surface temperature of the lightbulb. Assuming 10 percent of the rated power of the bulb is converted to light and is transmitted by the glass, calculate the heat transfer coefficient from Newton's law of cooling.

Chapter 17

STEADY HEAT CONDUCTION

In heat transfer analysis, we are often interested in the rate of heat transfer through a medium under steady conditions and surface temperatures. Such problems can be solved easily by the introduction of the *thermal resistance concept* in an analogous manner to electrical circuit problems. In this case, the thermal resistance corresponds to electrical resistance, temperature difference corresponds to voltage, and the heat transfer rate corresponds to electric current.

We start this chapter with *one-dimensional steady heat conduction* in a plane wall, a cylinder, and a sphere, and develop relations for *thermal resistances* in these geometries. We also develop thermal resistance relations for convection and radiation conditions at the boundaries. We apply this concept to heat conduction problems in *multilayer* plane walls, cylinders, and spheres and generalize it to systems that involve heat transfer in two or three dimensions. We also discuss the *thermal contact resistance* and the *overall heat transfer coefficient* and develop relations for the critical radius of insulation for a cylinder and a sphere. Finally, we discuss steady heat transfer from *finned surfaces* and some complex geometrics commonly encountered in practice through the use of *conduction shape factors.*

Objectives

The objectives of this chapter are to:

- Understand the concept of thermal resistance and its limitations, and develop thermal resistance networks for practical heat conduction problems,
- Solve steady conduction problems that involve multilayer rectangular, cylindrical, or spherical geometries,
- Develop an intuitive understanding of thermal contact resistance, and circumstances under which it may be significant,
- Identify applications in which insulation may actually increase heat transfer,
- Analyze finned surfaces, and assess how efficiently and effectively fins enhance heat transfer, and
- Solve multidimensional practical heat conduction problems using conduction shape factors.

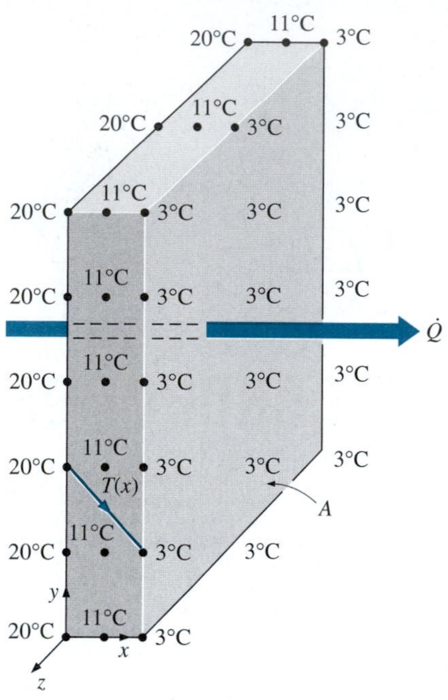

FIGURE 17–1

Heat transfer through a wall is one-dimensional when the temperature of the wall varies in one direction only.

17–1 ▪ STEADY HEAT CONDUCTION IN PLANE WALLS

Consider steady heat conduction through the walls of a house during a winter day. We know that heat is continuously lost to the outdoors through the wall. We intuitively feel that heat transfer through the wall is in the *normal direction* to the wall surface, and no significant heat transfer takes place in the wall in other directions (Fig. 17–1).

Recall that heat transfer in a certain direction is driven by the *temperature gradient* in that direction. There is no heat transfer in a direction in which there is no change in temperature. Temperature measurements at several locations on the inner or outer wall surface will confirm that a wall surface is nearly *isothermal.* That is, the temperatures at the top and bottom of a wall surface as well as at the right and left ends are almost the same. Therefore, there is no heat transfer through the wall from the top to the bottom, or from left to right, but there is considerable temperature difference between the inner and the outer surfaces of the wall, and thus significant heat transfer in the direction from the inner surface to the outer one.

The small thickness of the wall causes the temperature gradient in that direction to be large. Further, if the air temperatures in and outside the house remain constant, then heat transfer through the wall of a house can be modeled as *steady* and *one-dimensional.* The temperature of the wall in this case depends on one direction only (say the x-direction) and can be expressed as $T(x)$.

Noting that heat transfer is the only energy interaction involved in this case and there is no heat generation, the *energy balance* for the wall can be expressed as

$$\begin{pmatrix}\text{Rate of}\\ \text{heat transfer}\\ \text{into the wall}\end{pmatrix} - \begin{pmatrix}\text{Rate of}\\ \text{heat transfer}\\ \text{out of the wall}\end{pmatrix} = \begin{pmatrix}\text{Rate of change}\\ \text{of the energy}\\ \text{of the wall}\end{pmatrix}$$

or

$$\dot{Q}_{\text{in}} - \dot{Q}_{\text{out}} = \frac{dE_{\text{wall}}}{dt} \qquad \textbf{(17–1)}$$

But $dE_{\text{wall}}/dt = 0$ for *steady* operation, since there is no change in the temperature of the wall with time at any point. Therefore, the rate of heat transfer into the wall must be equal to the rate of heat transfer out of it. In other words, *the rate of heat transfer through the wall must be constant,* $\dot{Q}_{\text{cond, wall}} =$ constant.

Consider a plane wall of thickness L and average thermal conductivity k. The two surfaces of the wall are maintained at constant temperatures of T_1 and T_2. For one-dimensional steady heat conduction through the wall, we have $T(x)$. Then Fourier's law of heat conduction for the wall can be expressed as

$$\dot{Q}_{\text{cond, wall}} = -kA\frac{dT}{dx} \qquad \text{(W)} \qquad \textbf{(17–2)}$$

where the rate of conduction heat transfer $\dot{Q}_{\text{cond, wall}}$ and the wall area A are constant. Thus $dT/dx =$ constant, which means that *the temperature through the wall varies linearly with x.* That is, the temperature distribution in the wall under steady conditions is a *straight line* (Fig. 17–2).

Separating the variables in the preceding equation and integrating from $x = 0$, where $T(0) = T_1$, to $x = L$, where $T(L) = T_2$, we get

$$\int_{x=0}^{L} \dot{Q}_{\text{cond, wall}}\, dx = -\int_{T=T_1}^{T_2} kA\, dT$$

Performing the integrations and rearranging gives

$$\dot{Q}_{\text{cond, wall}} = kA\frac{T_1 - T_2}{L} \qquad \text{(W)} \tag{17-3}$$

which is identical to Eq. 1–21. Again, *the rate of heat conduction through a plane wall is proportional to the average thermal conductivity, the wall area, and the temperature difference, but is inversely proportional to the wall thickness.* Also, once the rate of heat conduction is available, the temperature $T(x)$ at any location x can be determined by replacing T_2 in Eq. 17–3 by T, and L by x.

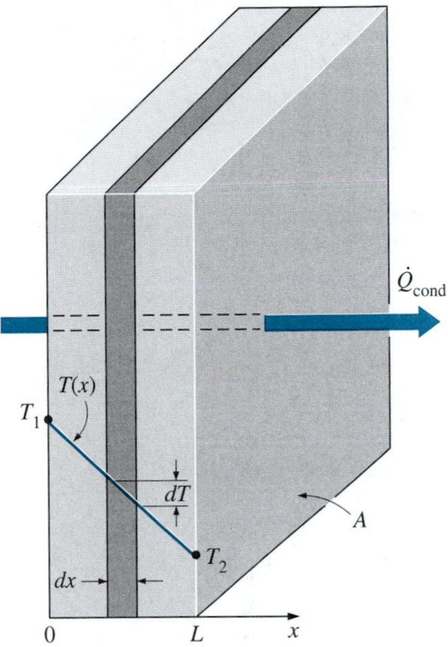

FIGURE 17–2
Under steady conditions, the temperature distribution in a plane wall is a straight line.

Thermal Resistance Concept

Equation 17–3 for heat conduction through a plane wall can be rearranged as

$$\dot{Q}_{\text{cond, wall}} = \frac{T_1 - T_2}{R_{\text{wall}}} \qquad \text{(W)} \tag{17-4}$$

where

$$R_{\text{wall}} = \frac{L}{kA} \qquad (°\text{C/W}) \tag{17-5}$$

is the *thermal resistance* of the wall against heat conduction or simply the **conduction resistance** of the wall. Note that the thermal resistance of a medium depends on the *geometry* and the *thermal properties* of the medium.

This equation for heat transfer is analogous to the relation for *electric current flow I*, expressed as

$$I = \frac{\mathbf{V}_1 - \mathbf{V}_2}{R_e} \tag{17-6}$$

where $R_e = L/\sigma_e A$ is the *electric resistance* and $\mathbf{V}_1 - \mathbf{V}_2$ is the *voltage difference* across the resistance (σ_e is the electrical conductivity). Thus, the *rate of heat transfer* through a layer corresponds to the *electric current*, the *thermal resistance* corresponds to *electrical resistance,* and the *temperature difference* corresponds to *voltage difference* across the layer (Fig. 17–3).

Consider convection heat transfer from a solid surface of area A_s and temperature T_s to a fluid whose temperature sufficiently far from the surface is T_∞, with a convection heat transfer coefficient h. Newton's law of cooling for convection heat transfer rate $\dot{Q}_{\text{conv}} = hA_s(T_s - T_\infty)$ can be rearranged as

$$\dot{Q}_{\text{conv}} = \frac{T_s - T_\infty}{R_{\text{conv}}} \qquad \text{(W)} \tag{17-7}$$

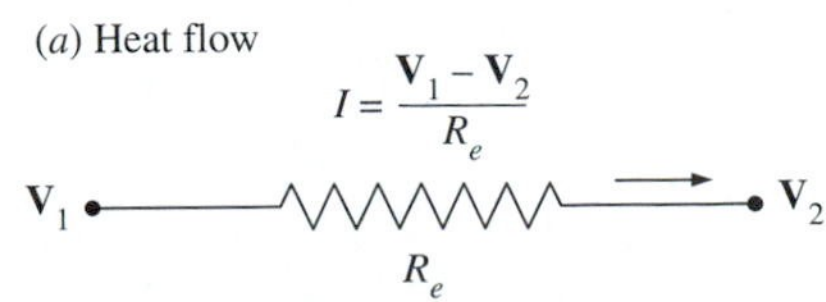

FIGURE 17–3
Analogy between thermal and electrical resistance concepts.

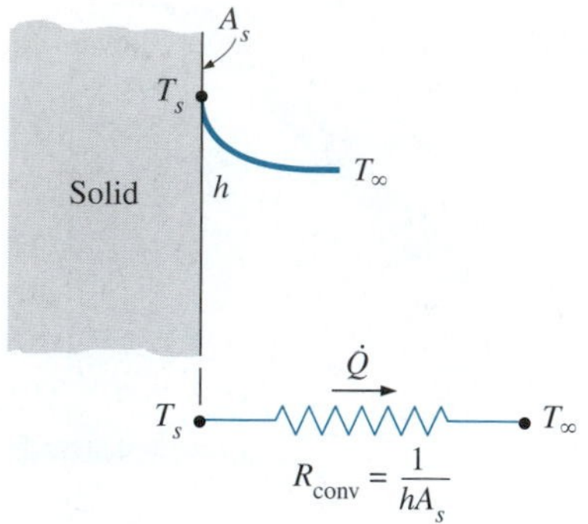

FIGURE 17–4
Schematic for convection resistance at a surface.

where

$$R_{conv} = \frac{1}{hA_s} \qquad (°C/W) \tag{17–8}$$

is the *thermal resistance* of the surface against heat convection, or simply the **convection resistance** of the surface (Fig. 17–4). Note that when the convection heat transfer coefficient is very large ($h \to \infty$), the convection resistance becomes *zero* and $T_s \approx T_\infty$. That is, the surface offers *no resistance to convection*, and thus it does not slow down the heat transfer process. This situation is approached in practice at surfaces where boiling and condensation occur. Also note that the surface does not have to be a plane surface. Equation 17–8 for convection resistance is valid for surfaces of any shape, provided that the assumption of h = constant and uniform is reasonable.

When the wall is surrounded by a gas, the *radiation effects*, which we have ignored so far, can be significant and may need to be considered. The rate of radiation heat transfer between a surface of emissivity ε and area A_s at temperature T_s and the surrounding surfaces at some average temperature T_{surr} can be expressed as

$$\dot{Q}_{rad} = \varepsilon\sigma A_s(T_s^4 - T_{surr}^4) = h_{rad} A_s(T_s - T_{surr}) = \frac{T_s - T_{surr}}{R_{rad}} \qquad (W) \tag{17–9}$$

where

$$R_{rad} = \frac{1}{h_{rad} A_s} \qquad (K/W) \tag{17–10}$$

is the *thermal resistance* of a surface against radiation, or the **radiation resistance**, and

$$h_{rad} = \frac{\dot{Q}_{rad}}{A_s(T_s - T_{surr})} = \varepsilon\sigma(T_s^2 + T_{surr}^2)(T_s + T_{surr}) \qquad (W/m^2 \cdot K) \tag{17–11}$$

is the **radiation heat transfer coefficient**. Note that both T_s and T_{surr} *must* be in K in the evaluation of h_{rad}. The definition of the radiation heat transfer coefficient enables us to express radiation conveniently in an analogous manner to convection in terms of a temperature difference. But h_{rad} depends strongly on temperature while h_{conv} usually does not.

A surface exposed to the surrounding air involves convection and radiation simultaneously, and the total heat transfer at the surface is determined by adding (or subtracting, if in the opposite direction) the radiation and convection components. The convection and radiation resistances are parallel to each other, as shown in Fig. 17–5, and may cause some complication in the thermal resistance network. When $T_{surr} \approx T_\infty$, the radiation effect can properly be accounted for by replacing h in the convection resistance relation by

$$h_{combined} = h_{conv} + h_{rad} \qquad (W/m^2 \cdot K) \tag{17–12}$$

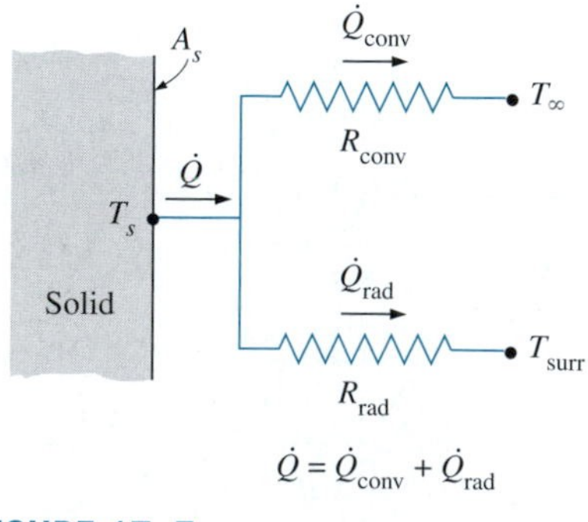

FIGURE 17–5
Schematic for convection and radiation resistances at a surface.

where $h_{combined}$ is the **combined heat transfer coefficient**. This way all complications associated with radiation are avoided.

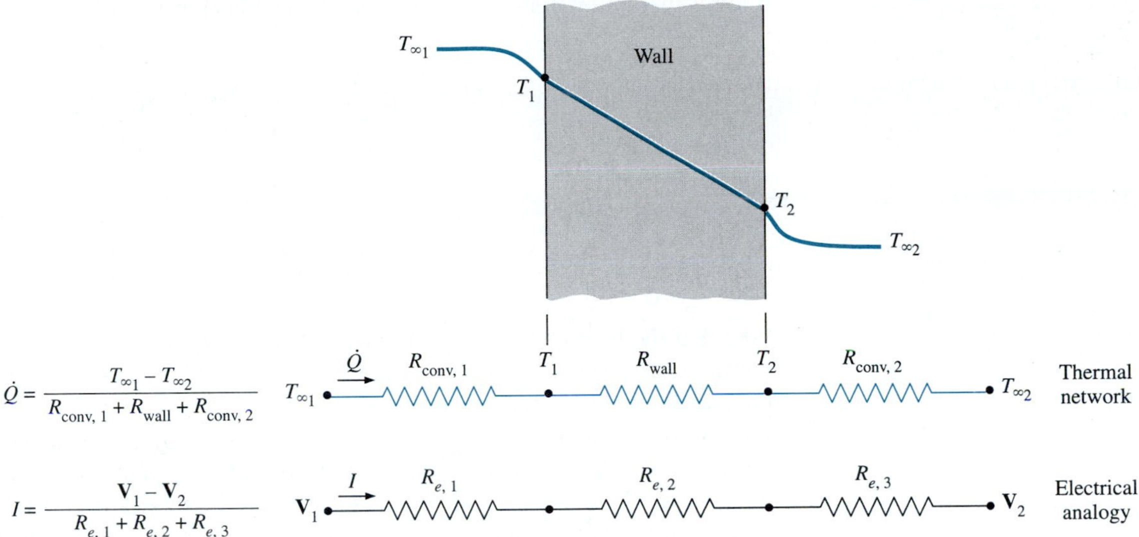

FIGURE 17–6
The thermal resistance network for heat transfer through a plane wall subjected to convection on both sides, and the electrical analogy.

Thermal Resistance Network

Now consider steady one-dimensional heat transfer through a plane wall of thickness L, area A, and thermal conductivity k that is exposed to convection on both sides to fluids at temperatures $T_{\infty 1}$ and $T_{\infty 2}$ with heat transfer coefficients h_1 and h_2, respectively, as shown in Fig. 17–6. Assuming $T_{\infty 2} < T_{\infty 1}$, the variation of temperature will be as shown in the figure. Note that the temperature varies linearly in the wall, and asymptotically approaches $T_{\infty 1}$ and $T_{\infty 2}$ in the fluids as we move away from the wall.

Under steady conditions we have

$$\begin{pmatrix}\text{Rate of}\\ \textit{heat convection}\\ \text{into the wall}\end{pmatrix} = \begin{pmatrix}\text{Rate of}\\ \textit{heat conduction}\\ \text{through the wall}\end{pmatrix} = \begin{pmatrix}\text{Rate of}\\ \textit{heat convection}\\ \text{from the wall}\end{pmatrix}$$

or

$$\dot{Q} = h_1 A(T_{\infty 1} - T_1) = kA\frac{T_1 - T_2}{L} = h_2 A(T_2 - T_{\infty 2}) \qquad \textbf{(17–13)}$$

which can be rearranged as

$$\dot{Q} = \frac{T_{\infty 1} - T_1}{1/h_1 A} = \frac{T_1 - T_2}{L/kA} = \frac{T_2 - T_{\infty 2}}{1/h_2 A}$$

$$= \frac{T_{\infty 1} - T_1}{R_{\text{conv, 1}}} = \frac{T_1 - T_2}{R_{\text{wall}}} = \frac{T_2 - T_{\infty 2}}{R_{\text{conv, 2}}} \qquad \textbf{(17–14)}$$

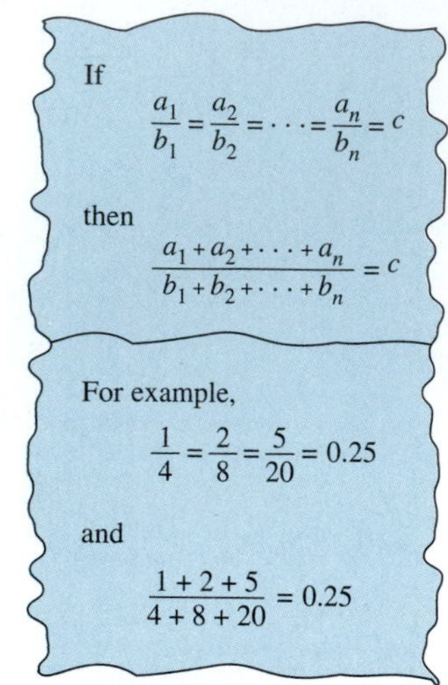

FIGURE 17–7
A useful mathematical identity.

Adding the numerators and denominators yields (Fig. 17–7)

$$\dot{Q} = \frac{T_{\infty 1} - T_{\infty 2}}{R_{\text{total}}} \qquad \text{(W)} \tag{17–15}$$

where

$$R_{\text{total}} = R_{\text{conv, 1}} + R_{\text{wall}} + R_{\text{conv, 2}} = \frac{1}{h_1 A} + \frac{L}{kA} + \frac{1}{h_2 A} \qquad (^\circ\text{C/W}) \tag{17–16}$$

Note that the heat transfer area A is constant for a plane wall, and the rate of heat transfer through a wall separating two mediums is equal to the temperature difference divided by the total thermal resistance between the mediums. Also note that the thermal resistances are in *series,* and the equivalent thermal resistance is determined by simply *adding* the individual resistances, just like the electrical resistances connected in series. Thus, the electrical analogy still applies. We summarize this as *the rate of steady heat transfer between two surfaces is equal to the temperature difference divided by the total thermal resistance between those two surfaces.*

Another observation that can be made from Eq. 17–15 is that the ratio of the temperature drop to the thermal resistance across any layer is constant, and thus the temperature drop across any layer is proportional to the thermal resistance of the layer. The larger the resistance, the larger the temperature drop. In fact, the equation $\dot{Q} = \Delta T/R$ can be rearranged as

$$\Delta T = \dot{Q}R \qquad (^\circ\text{C}) \tag{17–17}$$

which indicates that the *temperature drop* across any layer is equal to the *rate of heat transfer* times the *thermal resistance* across that layer (Fig. 17–8). You may recall that this is also true for voltage drop across an electrical resistance when the electric current is constant.

It is sometimes convenient to express heat transfer through a medium in an analogous manner to Newton's law of cooling as

$$\dot{Q} = UA\,\Delta T \qquad \text{(W)} \tag{17–18}$$

where U is the **overall heat transfer coefficient**. A comparison of Eqs. 17–15 and 17–18 reveals that

$$UA = \frac{1}{R_{\text{total}}} \qquad (^\circ\text{C/K}) \tag{17–19}$$

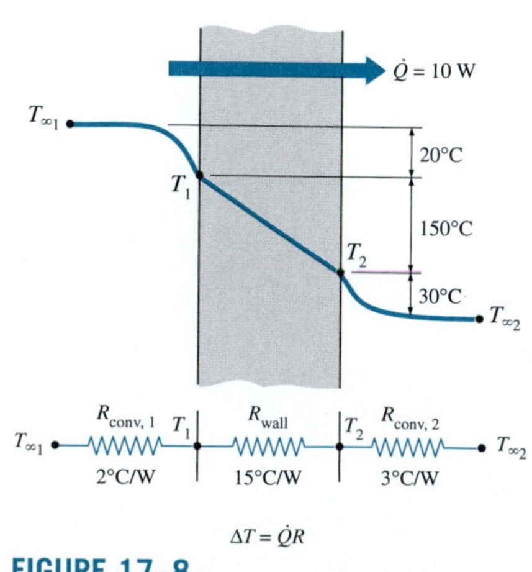

FIGURE 17–8
The temperature drop across a layer is proportional to its thermal resistance.

Therefore, for a unit area, the overall heat transfer coefficient is equal to the inverse of the total thermal resistance.

Note that we do not need to know the surface temperatures of the wall in order to evaluate the rate of steady heat transfer through it. All we need to know is the convection heat transfer coefficients and the fluid temperatures on both sides of the wall. The *surface temperature* of the wall can be determined as described above using the thermal resistance concept, but by taking the surface at which the temperature is to be determined as one of the terminal surfaces. For example, once $\dot{Q}$ is evaluated, the surface temperature T_1 can be determined from

$$\dot{Q} = \frac{T_{\infty 1} - T_1}{R_{\text{conv, 1}}} = \frac{T_{\infty 1} - T_1}{1/h_1 A} \tag{17–20}$$

Multilayer Plane Walls

In practice we often encounter plane walls that consist of several layers of different materials. The thermal resistance concept can still be used to determine the rate of steady heat transfer through such *composite* walls. As you may have already guessed, this is done by simply noting that the conduction resistance of each wall is L/kA connected in series, and using the electrical analogy. That is, by dividing the *temperature difference* between two surfaces at known temperatures by the *total thermal resistance* between them.

Consider a plane wall that consists of two layers (such as a brick wall with a layer of insulation). The rate of steady heat transfer through this two-layer composite wall can be expressed as (Fig. 17–9)

$$\dot{Q} = \frac{T_{\infty 1} - T_{\infty 2}}{R_{\text{total}}} \tag{17–21}$$

where R_{total} is the *total thermal resistance*, expressed as

$$\begin{aligned} R_{\text{total}} &= R_{\text{conv, 1}} + R_{\text{wall, 1}} + R_{\text{wall, 2}} + R_{\text{conv, 2}} \\ &= \frac{1}{h_1 A} + \frac{L_1}{k_1 A} + \frac{L_2}{k_2 A} + \frac{1}{h_2 A} \end{aligned} \tag{17–22}$$

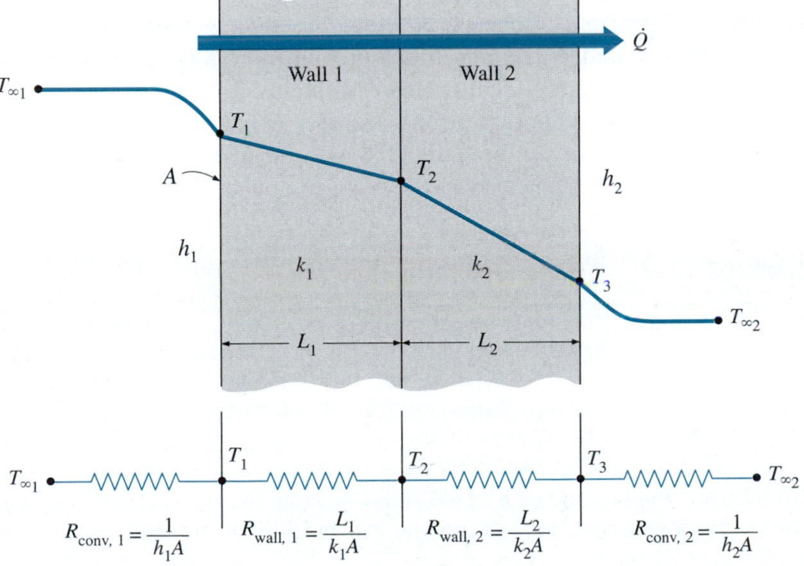

FIGURE 17–9
The thermal resistance network for heat transfer through a two-layer plane wall subjected to convection on both sides.

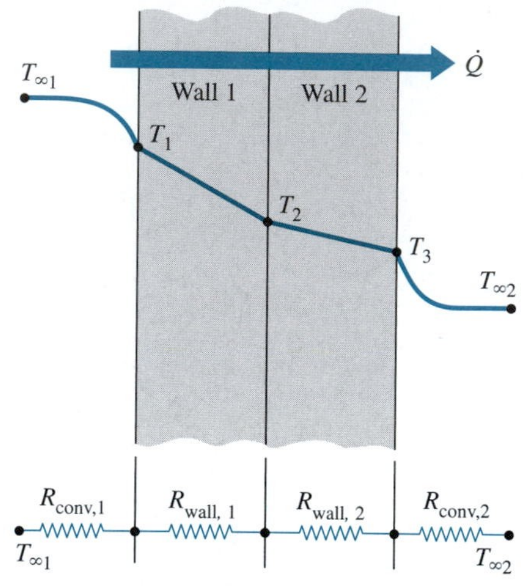

To find T_1: $\dot{Q} = \dfrac{T_{\infty 1} - T_1}{R_{\text{conv},1}}$

To find T_2: $\dot{Q} = \dfrac{T_{\infty 1} - T_2}{R_{\text{conv},1} + R_{\text{wall, 1}}}$

To find T_3: $\dot{Q} = \dfrac{T_3 - T_{\infty 2}}{R_{\text{conv},2}}$

FIGURE 17–10
The evaluation of the surface and interface temperatures when $T_{\infty 1}$ and $T_{\infty 2}$ are given and $\dot{Q}$ is calculated.

The subscripts 1 and 2 in the R_{wall} relations above indicate the first and the second layers, respectively. We could also obtain this result by following the approach already used for the single-layer case by noting that the rate of steady heat transfer $\dot{Q}$ through a multilayer medium is constant, and thus it must be the same through each layer. Note from the thermal resistance network that the resistances are *in series*, and thus the *total thermal resistance* is simply the *arithmetic sum* of the individual thermal resistances in the path of heat transfer.

This result for the *two-layer* case is analogous to the *single-layer* case, except that an *additional resistance* is added for the *additional layer.* This result can be extended to plane walls that consist of *three* or *more layers* by adding an *additional resistance* for each *additional layer.*

Once $\dot{Q}$ is *known,* an unknown surface temperature T_j at any surface or interface j can be determined from

$$\dot{Q} = \frac{T_i - T_j}{R_{\text{total}, i-j}} \quad \textbf{(17–23)}$$

where T_i is a *known* temperature at location i and $R_{\text{total}, i-j}$ is the total thermal resistance between locations i and j. For example, when the fluid temperatures $T_{\infty 1}$ and $T_{\infty 2}$ for the two-layer case shown in Fig. 17–9 are available and $\dot{Q}$ is calculated from Eq. 17–21, the interface temperature T_2 between the two walls can be determined from (Fig. 17–10)

$$\dot{Q} = \frac{T_{\infty 1} - T_2}{R_{\text{conv}, 1} + R_{\text{wall}, 1}} = \frac{T_{\infty 1} - T_2}{\dfrac{1}{h_1 A} + \dfrac{L_1}{k_1 A}} \quad \textbf{(17–24)}$$

The temperature drop across a layer is easily determined from Eq. 17–17 by multiplying $\dot{Q}$ by the thermal resistance of that layer.

The thermal resistance concept is widely used in practice because it is intuitively easy to understand and it has proven to be a powerful tool in the solution of a wide range of heat transfer problems. But its use is limited to systems through which the rate of heat transfer $\dot{Q}$ remains *constant*; that is, to systems involving *steady* heat transfer with *no heat generation* (such as resistance heating or chemical reactions) within the medium.

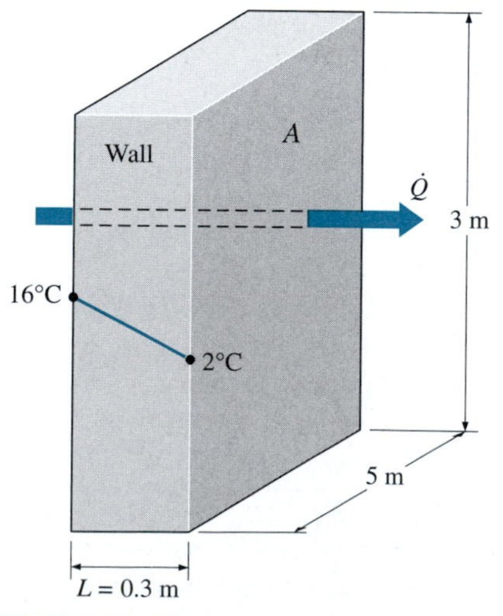

FIGURE 17–11
Schematic for Example 17–1.

EXAMPLE 17–1 Heat Loss through a Wall

Consider a 3-m-high, 5-m-wide, and 0.3-m-thick wall whose thermal conductivity is $k = 0.9$ W/m · K (Fig. 17–11). On a certain day, the temperatures of the inner and the outer surfaces of the wall are measured to be 16°C and 2°C, respectively. Determine the rate of heat loss through the wall on that day.

SOLUTION The two surfaces of a wall are maintained at specified temperatures. The rate of heat loss through the wall is to be determined.

Assumptions **1** Heat transfer through the wall is steady since the surface temperatures remain constant at the specified values. **2** Heat transfer is one-dimensional since any significant temperature gradients exist in the direction from the indoors to the outdoors. **3** Thermal conductivity is constant.

Properties The thermal conductivity is given to be $k = 0.9$ W/m · K.

Analysis Noting that heat transfer through the wall is by conduction and the area of the wall is $A = 3 \text{ m} \times 5 \text{ m} = 15 \text{ m}^2$, the steady rate of heat transfer through the wall can be determined from Eq. 17–3 to be

$$\dot{Q} = kA\frac{T_1 - T_2}{L} = (0.9 \text{ W/m} \cdot {}^\circ\text{C})(15 \text{ m}^2)\frac{(16 - 2)^\circ\text{C}}{0.3 \text{ m}} = \mathbf{630 \text{ W}}$$

We could also determine the steady rate of heat transfer through the wall by making use of the thermal resistance concept from

$$\dot{Q} = \frac{\Delta T_{\text{wall}}}{R_{\text{wall}}}$$

where

$$R_{\text{wall}} = \frac{L}{kA} = \frac{0.3 \text{ m}}{(0.9 \text{ W/m} \cdot {}^\circ\text{C})(15 \text{ m}^2)} = 0.02222^\circ\text{C/W}$$

Substituting, we get

$$\dot{Q} = \frac{(16 - 2)^\circ\text{C}}{0.02222^\circ\text{C/W}} = 630 \text{ W}$$

Discussion This is the same result obtained earlier. Note that heat conduction through a plane wall with specified surface temperatures can be determined directly and easily without utilizing the thermal resistance concept. However, the thermal resistance concept serves as a valuable tool in more complex heat transfer problems, as you will see in the following examples. Also, the units W/m · °C and W/m · K for thermal conductivity are equivalent, and thus interchangeable. This is also the case for °C and K for temperature differences.

EXAMPLE 17–2 Heat Loss through a Single-Pane Window

Consider a 0.8-m-high and 1.5-m-wide glass window with a thickness of 8 mm and a thermal conductivity of $k = 0.78$ W/m · K. Determine the steady rate of heat transfer through this glass window and the temperature of its inner surface for a day during which the room is maintained at 20°C while the temperature of the outdoors is −10°C. Take the heat transfer coefficients on the inner and outer surfaces of the window to be $h_1 = 10$ W/m^2 · °C and $h_2 = 40$ W/m^2 · °C, which includes the effects of radiation.

SOLUTION Heat loss through a window glass is considered. The rate of heat transfer through the window and the inner surface temperature are to be determined.

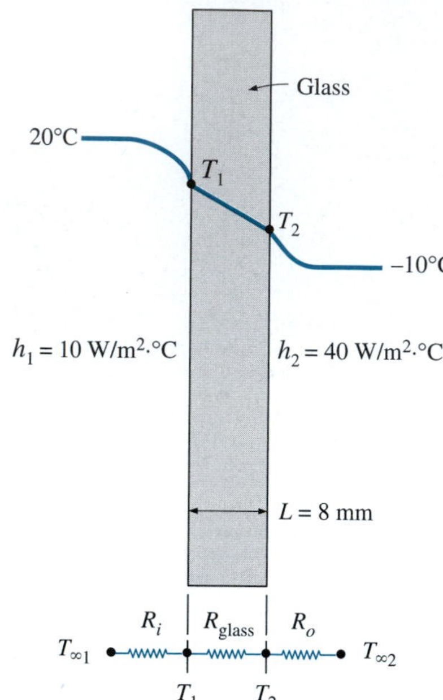

FIGURE 17–12
Schematic for Example 17–2.

Assumptions **1** Heat transfer through the window is steady since the surface temperatures remain constant at the specified values. **2** Heat transfer through the wall is one-dimensional since any significant temperature gradients exist in the direction from the indoors to the outdoors. **3** Thermal conductivity is constant.

Properties The thermal conductivity is given to be k = 0.78 W/m · K.

Analysis This problem involves conduction through the glass window and convection at its surfaces, and can best be handled by making use of the thermal resistance concept and drawing the thermal resistance network, as shown in Fig. 17–12. Noting that the area of the window is A = 0.8 m × 1.5 m = 1.2 m², the individual resistances are evaluated from their definitions to be

$$R_i = R_{\text{conv, 1}} = \frac{1}{h_1 A} = \frac{1}{(10 \text{ W/m}^2 \cdot {}^\circ\text{C})(1.2 \text{ m}^2)} = 0.08333^\circ\text{C/W}$$

$$R_{\text{glass}} = \frac{L}{kA} = \frac{0.008 \text{ m}}{(0.78 \text{ W/m} \cdot {}^\circ\text{C})(1.2 \text{ m}^2)} = 0.00855^\circ\text{C/W}$$

$$R_o = R_{\text{conv, 2}} = \frac{1}{h_2 A} = \frac{1}{(40 \text{ W/m}^2 \cdot {}^\circ\text{C})(1.2 \text{ m}^2)} = 0.02083^\circ\text{C/W}$$

Noting that all three resistances are in series, the total resistance is

$$\begin{aligned} R_{\text{total}} &= R_{\text{conv, 1}} + R_{\text{glass}} + R_{\text{conv, 2}} = 0.08333 + 0.00855 + 0.02083 \\ &= 0.1127^\circ\text{C/W} \end{aligned}$$

Then the steady rate of heat transfer through the window becomes

$$\dot{Q} = \frac{T_{\infty 1} - T_{\infty 2}}{R_{\text{total}}} = \frac{[20 - (-10)]^\circ\text{C}}{0.1127^\circ\text{C/W}} = \mathbf{266\ W}$$

Knowing the rate of heat transfer, the inner surface temperature of the window glass can be determined from

$$\dot{Q} = \frac{T_{\infty 1} - T_1}{R_{\text{conv, 1}}} \longrightarrow \begin{aligned} T_1 &= T_{\infty 1} - \dot{Q}R_{\text{conv, 1}} \\ &= 20^\circ\text{C} - (266 \text{ W})(0.08333^\circ\text{C/W}) \\ &= \mathbf{-2.2^\circ C} \end{aligned}$$

Discussion Note that the inner surface temperature of the window glass is −2.2°C even though the temperature of the air in the room is maintained at 20°C. Such low surface temperatures are highly undesirable since they cause the formation of fog or even frost on the inner surfaces of the glass when the humidity in the room is high.

EXAMPLE 17–3 Heat Loss through Double-Pane Windows

Consider a 0.8-m-high and 1.5-m-wide double-pane window consisting of two 4-mm-thick layers of glass (k = 0.78 W/m · K) separated by a 10-mm-wide stagnant air space (k = 0.026 W/m · K). Determine the steady rate of heat transfer through this double-pane window and the temperature of its inner surface for a day during which the room is maintained at 20°C while the temperature of the outdoors is −10°C. Take the convection heat transfer coefficients on the inner and outer surfaces of the window to be h_1 = 10 W/m² · °C and h_2 = 40 W/m² · °C, which includes the effects of radiation.

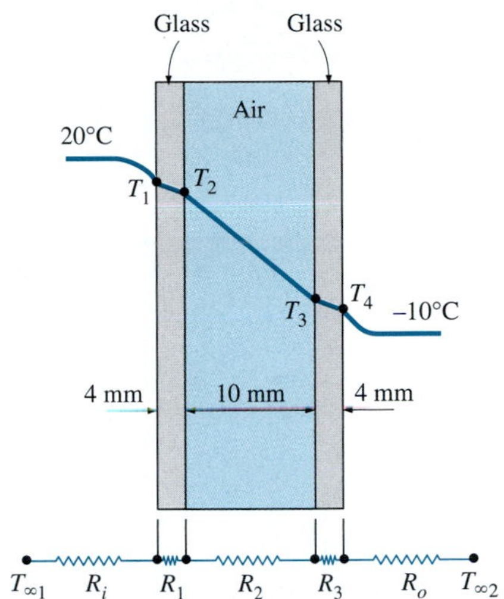

FIGURE 17–13
Schematic for Example 17–3.

Solution A double-pane window is considered. The rate of heat transfer through the window and the inner surface temperature are to be determined.

Analysis This example problem is identical to the previous one except that the single 8-mm-thick window glass is replaced by two 4-mm-thick glasses that enclose a 10-mm-wide stagnant air space. Therefore, the thermal resistance network of this problem involves two additional conduction resistances corresponding to the two additional layers, as shown in Fig. 17–13. Noting that the area of the window is again A = 0.8 m × 1.5 m = 1.2 m², the individual resistances are evaluated from their definitions to be

$$R_i = R_{\text{conv, 1}} = \frac{1}{h_1 A} = \frac{1}{(10\ \text{W/m}^2 \cdot {}^\circ\text{C})(1.2\ \text{m}^2)} = 0.08333^\circ\text{C/W}$$

$$R_1 = R_3 = R_{\text{glass}} = \frac{L_1}{k_1 A} = \frac{0.004\ \text{m}}{(0.78\ \text{W/m} \cdot {}^\circ\text{C})(1.2\ \text{m}^2)} = 0.00427^\circ\text{C/W}$$

$$R_2 = R_{\text{air}} = \frac{L_2}{k_2 A} = \frac{0.01\ \text{m}}{(0.026\ \text{W/m} \cdot {}^\circ\text{C})(1.2\ \text{m}^2)} = 0.3205^\circ\text{C/W}$$

$$R_o = R_{\text{conv, 2}} = \frac{1}{h_2 A} = \frac{1}{(40\ \text{W/m}^2 \cdot {}^\circ\text{C})(1.2\ \text{m}^2)} = 0.02083^\circ\text{C/W}$$

Noting that all three resistances are in series, the total resistance is

$$\begin{aligned} R_{\text{total}} &= R_{\text{conv, 1}} + R_{\text{glass, 1}} + R_{\text{air}} + R_{\text{glass, 2}} + R_{\text{conv, 2}} \\ &= 0.08333 + 0.00427 + 0.3205 + 0.00427 + 0.02083 \\ &= 0.4332^\circ\text{C/W} \end{aligned}$$

Then the steady rate of heat transfer through the window becomes

$$\dot{Q} = \frac{T_{\infty 1} - T_{\infty 2}}{R_{\text{total}}} = \frac{[20 - (-10)]^\circ\text{C}}{0.4332^\circ\text{C/W}} = \mathbf{69.2\ W}$$

which is about one-fourth of the result obtained in the previous example. This explains the popularity of the double- and even triple-pane windows in cold climates. The drastic reduction in the heat transfer rate in this case is due to the large thermal resistance of the air layer between the glasses.

The inner surface temperature of the window in this case will be

$$T_1 = T_{\infty 1} - \dot{Q}R_{\text{conv, 1}} = 20^\circ\text{C} - (69.2\ \text{W})(0.08333^\circ\text{C/W}) = \mathbf{14.2^\circ C}$$

which is considerably higher than the −2.2°C obtained in the previous example. Therefore, a double-pane window will rarely get fogged. A double-pane window will also reduce the heat gain in summer, and thus reduce the air-conditioning costs.

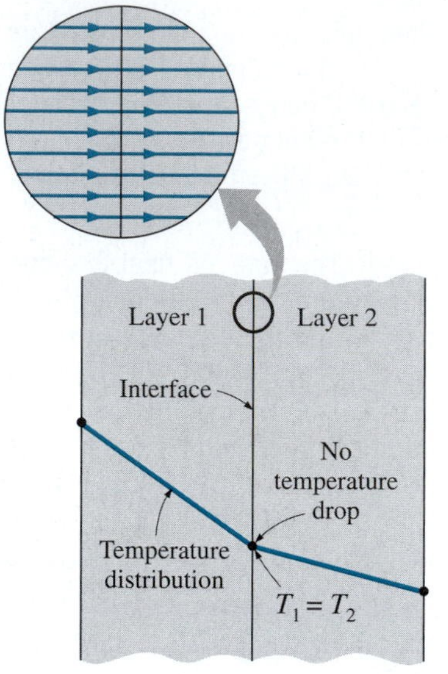

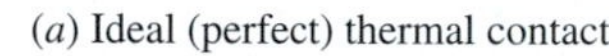
(*a*) Ideal (perfect) thermal contact

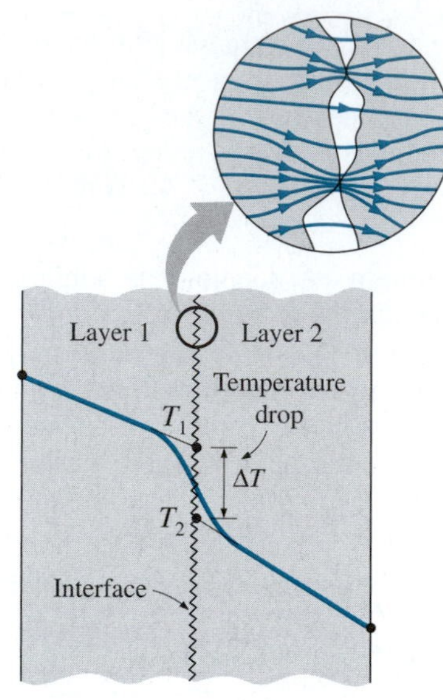

(*b*) Actual (imperfect) thermal contact

FIGURE 17–14
Temperature distribution and heat flow lines along two solid plates pressed against each other for the case of perfect and imperfect contact.

17–2 ▪ THERMAL CONTACT RESISTANCE

In the analysis of heat conduction through multilayer solids, we assumed "perfect contact" at the interface of two layers, and thus no temperature drop at the interface. This would be the case when the surfaces are perfectly smooth and they produce a perfect contact at each point. In reality, however, even flat surfaces that appear smooth to the eye turn out to be rather rough when examined under a microscope, as shown in Fig. 17–14, with numerous peaks and valleys. That is, a surface is *microscopically rough* no matter how smooth it appears to be.

When two such surfaces are pressed against each other, the peaks form good material contact but the valleys form voids filled with air. As a result, an interface contains numerous *air gaps* of varying sizes that act as *insulation* because of the low thermal conductivity of air. Thus, an interface offers some resistance to heat transfer, and this resistance per unit interface area is called the **thermal contact resistance**, R_c. The value of R_c is determined experimentally using a setup like the one shown in Fig. 17–15, and as expected, there is considerable scatter of data because of the difficulty in characterizing the surfaces.

FIGURE 17–15
A typical experimental setup for the determination of thermal contact resistance
From Song et al. (1993).

Consider heat transfer through two metal rods of cross-sectional area A that are pressed against each other. Heat transfer through the interface of these two rods is the sum of the heat transfers through the *solid contact spots* and the *gaps* in the noncontact areas and can be expressed as

$$\dot{Q} = \dot{Q}_{\text{contact}} + \dot{Q}_{\text{gap}} \tag{17–25}$$

It can also be expressed in an analogous manner to Newton's law of cooling as

$$\dot{Q} = h_c A \, \Delta T_{\text{interface}} \tag{17–26}$$

where A is the apparent interface area (which is the same as the cross-sectional area of the rods) and $\Delta T_{\text{interface}}$ is the effective temperature difference at the interface. The quantity h_c, which corresponds to the convection heat transfer coefficient, is called the **thermal contact conductance** and is expressed as

$$h_c = \frac{\dot{Q}/A}{\Delta T_{\text{interface}}} \qquad (\text{W/m}^2 \cdot {}^\circ\text{C}) \qquad \textbf{(17–27)}$$

It is related to thermal contact resistance by

$$R_c = \frac{1}{h_c} = \frac{\Delta T_{\text{interface}}}{\dot{Q}/A} \qquad (\text{m}^2 \cdot {}^\circ\text{C/W}) \qquad \textbf{(17–28)}$$

That is, thermal contact resistance is the inverse of thermal contact conductance. Usually, thermal contact conductance is reported in the literature, but the concept of thermal contact resistance serves as a better vehicle for explaining the effect of interface on heat transfer. Note that R_c represents thermal contact resistance *per unit area.* The thermal resistance for the entire interface is obtained by dividing R_c by the apparent interface area A.

The thermal contact resistance can be determined from Eq. 17–28 by measuring the temperature drop at the interface and dividing it by the heat flux under steady conditions. The value of thermal contact resistance depends on the *surface roughness* and the *material properties* as well as the *temperature* and *pressure* at the interface and the *type of fluid* trapped at the interface. The situation becomes more complex when plates are fastened by bolts, screws, or rivets since the interface pressure in this case is nonuniform. The thermal contact resistance in that case also depends on the plate thickness, the bolt radius, and the size of the contact zone. Thermal contact resistance is observed to *decrease* with *decreasing surface roughness* and *increasing interface pressure,* as expected. Most experimentally determined values of the thermal contact resistance fall between 0.000005 and 0.0005 $\text{m}^2 \cdot {}^\circ\text{C/W}$ (the corresponding range of thermal contact conductance is 2000 to 200,000 $\text{W/m}^2 \cdot {}^\circ\text{C}$).

When we analyze heat transfer in a medium consisting of two or more layers, the first thing we need to know is whether the thermal contact resistance is *significant* or not. We can answer this question by comparing the magnitudes of the thermal resistances of the layers with typical values of thermal contact resistance. For example, the thermal resistance of a 1-cm-thick layer of an insulating material per unit surface area is

$$R_{c,\,\text{insulation}} = \frac{L}{k} = \frac{0.01\ \text{m}}{0.04\ \text{W/m} \cdot {}^\circ\text{C}} = 0.25\ \text{m}^2 \cdot {}^\circ\text{C/W}$$

whereas for a 1-cm-thick layer of copper, it is

$$R_{c,\,\text{copper}} = \frac{L}{k} = \frac{0.01\ \text{m}}{386\ \text{W/m} \cdot {}^\circ\text{C}} = 0.000026\ \text{m}^2 \cdot {}^\circ\text{C/W}$$

Comparing the values above with typical values of thermal contact resistance, we conclude that thermal contact resistance is significant and can even dominate the heat transfer for good heat conductors such as metals, but

TABLE 17–1

Thermal contact conductance for aluminum plates with different fluids at the interface for a surface roughness of 10 μm and interface pressure of 1 atm (from Fried, 1969).

Fluid at the interface	Contact conductance, h_c, W/m^2 · K
Air	3640
Helium	9520
Hydrogen	13,900
Silicone oil	19,000
Glycerin	37,700

can be disregarded for poor heat conductors such as insulations. This is not surprising since insulating materials consist mostly of air space just like the interface itself.

The thermal contact resistance can be minimized by applying a thermally conducting liquid called a *thermal grease* such as silicon oil on the surfaces before they are pressed against each other. This is commonly done when attaching electronic components such as power transistors to heat sinks. The thermal contact resistance can also be reduced by replacing the air at the interface by a *better conducting gas* such as helium or hydrogen, as shown in Table 17–1.

Another way to minimize the contact resistance is to insert a *soft metallic foil* such as tin, silver, copper, nickel, or aluminum between the two surfaces. Experimental studies show that the thermal contact resistance can be reduced by a factor of up to 7 by a metallic foil at the interface. For maximum effectiveness, the foils must be very thin. The effect of metallic coatings on thermal contact conductance is shown in Fig. 17–16 for various metal surfaces.

There is considerable uncertainty in the contact conductance data reported in the literature, and care should be exercised when using them. In Table 17–2 some experimental results are given for the contact conductance between similar and dissimilar metal surfaces for use in preliminary design calculations. Note that the *thermal contact conductance* is *highest* (and thus the contact resistance is lowest) for *soft metals* with *smooth surfaces* at *high pressure.*

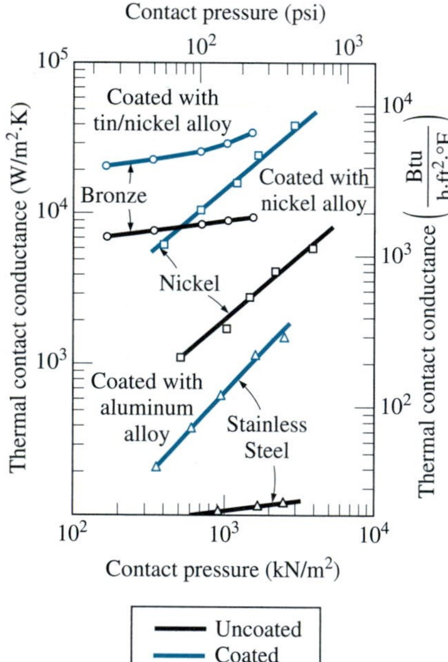

FIGURE 17–16
Effect of metallic coatings on thermal contact conductance
From Peterson (1987).

EXAMPLE 17–4 Equivalent Thickness for Contact Resistance

The thermal contact conductance at the interface of two 1-cm-thick aluminum plates is measured to be 11,000 W/m^2 · K. Determine the thickness of the aluminum plate whose thermal resistance is equal to the thermal resistance of the interface between the plates (Fig. 17–17).

Solution The thickness of the aluminum plate whose thermal resistance is equal to the thermal contact resistance is to be determined.

Properties The thermal conductivity of aluminum at room temperature is $k = 237$ W/m · K (Table A–24).

Analysis Noting that thermal contact resistance is the inverse of thermal contact conductance, the thermal contact resistance is

$$R_c = \frac{1}{h_c} = \frac{1}{11{,}000 \text{ W/m}^2 \cdot \text{K}} = 0.909 \times 10^{-4} \text{ m}^2 \cdot \text{K/W}$$

For a unit surface area, the thermal resistance of a flat plate is defined as

$$R = \frac{L}{k}$$

where L is the thickness of the plate and k is the thermal conductivity. Setting $R = R_c$, the equivalent thickness is determined from the relation above to be

$$L = kR_c = (237 \text{ W/m} \cdot \text{K})(0.909 \times 10^{-4} \text{ m}^2 \cdot \text{K/W}) = 0.0215 \text{ m} = \mathbf{2.15 \text{ cm}}$$

TABLE 17–2

Thermal contact conductance of some metal surfaces in air (from various sources)

Material	Surface condition	Roughness, μm	Temperature, °C	Pressure, MPa	h_c,* W/m^2 · °C
Identical Metal Pairs					
416 Stainless steel	Ground	2.54	90–200	0.17–2.5	3800
304 Stainless steel	Ground	1.14	20	4–7	1900
Aluminum	Ground	2.54	150	1.2–2.5	11,400
Copper	Ground	1.27	20	1.2–20	143,000
Copper	Milled	3.81	20	1–5	55,500
Copper (vacuum)	Milled	0.25	30	0.17–7	11,400
Dissimilar Metal Pairs					
Stainless steel–				10	2900
Aluminum		20–30	20	20	3600
Stainless steel–				10	16,400
Aluminum		1.0–2.0	20	20	20,800
Steel Ct-30–				10	50,000
Aluminum	Ground	1.4–2.0	20	15–35	59,000
Steel Ct-30–				10	4800
Aluminum	Milled	4.5–7.2	20	30	8300
				5	42,000
Aluminum-Copper	Ground	1.17–1.4	20	15	56,000
				10	12,000
Aluminum-Copper	Milled	4.4–4.5	20	20–35	22,000

*Divide the given values by 5.678 to convert to Btu/h · ft^2 · °F.

Discussion Note that the interface between the two plates offers as much resistance to heat transfer as a 2.15-cm-thick aluminum plate. It is interesting that the thermal contact resistance in this case is greater than the sum of the thermal resistances of both plates.

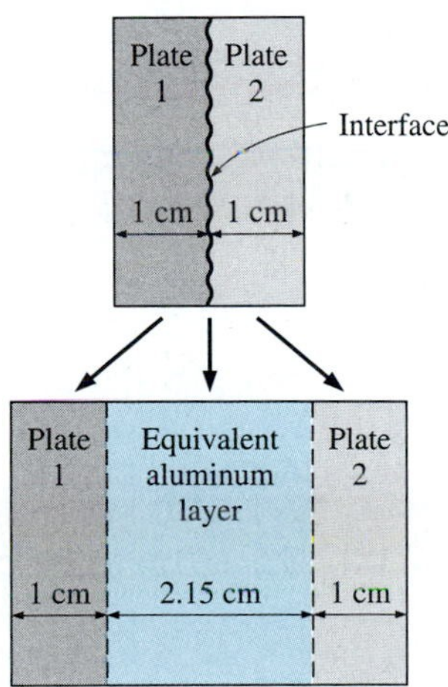

FIGURE 17–17
Schematic for Example 17–4.

EXAMPLE 17–5 Contact Resistance of Transistors

Four identical power transistors with aluminum casing are attached on one side of a 1-cm-thick 20-cm × 20-cm square copper plate (k = 386 W/m · °C) by screws that exert an average pressure of 6 MPa (Fig. 17–18). The base area of each transistor is 8 cm^2, and each transistor is placed at the center of a 10-cm × 10-cm quarter section of the plate. The interface roughness is estimated to be about 1.5 μm. All transistors are covered by a thick Plexiglas layer, which is a poor conductor of heat, and thus all the heat generated at the junction of the transistor must be dissipated to the ambient at 20°C through the back surface of the copper plate. The combined convection/radiation heat transfer coefficient at the back surface can be taken to be 25 W/m^2 · °C. If the case temperature of the transistor is not to exceed 70°C, determine the

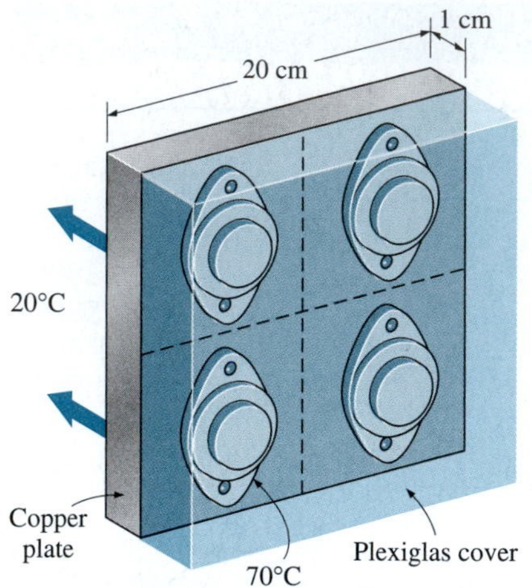

FIGURE 17–18
Schematic for Example 17–5.

maximum power each transistor can dissipate safely, and the temperature jump at the case-plate interface.

Solution Four identical power transistors are attached on a copper plate. For a maximum case temperature of 70°C, the maximum power dissipation and the temperature jump at the interface are to be determined.

Assumptions **1** Steady operating conditions exist. **2** Heat transfer can be approximated as being one-dimensional, although it is recognized that heat conduction in some parts of the plate will be two-dimensional since the plate area is much larger than the base area of the transistor. But the large thermal conductivity of copper will minimize this effect. **3** All the heat generated at the junction is dissipated through the back surface of the plate since the transistors are covered by a thick Plexiglas layer. **4** Thermal conductivities are constant.

Properties The thermal conductivity of copper is given to be k = 386 W/m · °C. The contact conductance is obtained from Table 17–2 to be h_c = 42,000 W/m^2 · °C, which corresponds to copper-aluminum interface for the case of 1.17–1.4 μm roughness and 5 MPa pressure, which is sufficiently close to what we have.

Analysis The contact area between the case and the plate is given to be 8 cm^2, and the plate area for each transistor is 100 cm^2. The thermal resistance network of this problem consists of three resistances in series (interface, plate, and convection), which are determined to be

$$R_{\text{interface}} = \frac{1}{h_c A_c} = \frac{1}{(42{,}000\ \text{W/m}^2 \cdot {}^\circ\text{C})(8 \times 10^{-4}\ \text{m}^2)} = 0.030^\circ\text{C/W}$$

$$R_{\text{plate}} = \frac{L}{kA} = \frac{0.01\ \text{m}}{(386\ \text{W/m} \cdot {}^\circ\text{C})(0.01\ \text{m}^2)} = 0.0026^\circ\text{C/W}$$

$$R_{\text{conv}} = \frac{1}{h_o A} = \frac{1}{(25\ \text{W/m}^2 \cdot {}^\circ\text{C})(0.01\ \text{m}^2)} = 4.0^\circ\text{C/W}$$

The total thermal resistance is then

$$R_{\text{total}} = R_{\text{interface}} + R_{\text{plate}} + R_{\text{ambient}} = 0.030 + 0.0026 + 4.0 = 4.0326^\circ\text{C/W}$$

Note that the thermal resistance of a copper plate is very small and can be ignored altogether. Then the rate of heat transfer is determined to be

$$\dot{Q} = \frac{\Delta T}{R_{\text{total}}} = \frac{(70 - 20)^\circ\text{C}}{4.0326^\circ\text{C/W}} = \mathbf{12.4\ W}$$

Therefore, the power transistor should not be operated at power levels greater than 12.4 W if the case temperature is not to exceed 70°C.

The temperature jump at the interface is determined from

$$\Delta T_{\text{interface}} = \dot{Q}R_{\text{interface}} = (12.4\ \text{W})(0.030^\circ\text{C/W}) = \mathbf{0.37^\circ C}$$

which is not very large. Therefore, even if we eliminate the thermal contact resistance at the interface completely, we lower the operating temperature of the transistor in this case by less than 0.4°C.

17–3 ▪ GENERALIZED THERMAL RESISTANCE NETWORKS

The *thermal resistance* concept or the *electrical analogy* can also be used to solve steady heat transfer problems that involve parallel layers or combined series-parallel arrangements. Although such problems are often two- or even three-dimensional, approximate solutions can be obtained by assuming one-dimensional heat transfer and using the thermal resistance network.

Consider the composite wall shown in Fig. 17–19, which consists of two parallel layers. The thermal resistance network, which consists of two parallel resistances, can be represented as shown in the figure. Noting that the total heat transfer is the sum of the heat transfers through each layer, we have

$$\dot{Q} = \dot{Q}_1 + \dot{Q}_2 = \frac{T_1 - T_2}{R_1} + \frac{T_1 - T_2}{R_2} = (T_1 - T_2)\left(\frac{1}{R_1} + \frac{1}{R_2}\right) \quad \text{(17–29)}$$

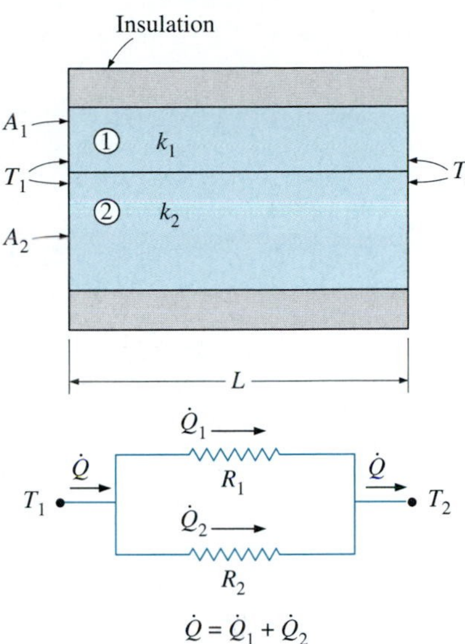

FIGURE 17–19
Thermal resistance network for two parallel layers.

Utilizing electrical analogy, we get

$$\dot{Q} = \frac{T_1 - T_2}{R_{\text{total}}} \quad \text{(17–30)}$$

where

$$\frac{1}{R_{\text{total}}} = \frac{1}{R_1} + \frac{1}{R_2} \longrightarrow R_{\text{total}} = \frac{R_1 R_2}{R_1 + R_2} \quad \text{(17–31)}$$

since the resistances are in parallel.

Now consider the combined series-parallel arrangement shown in Fig. 17–20. The total rate of heat transfer through this composite system can again be expressed as

$$\dot{Q} = \frac{T_1 - T_\infty}{R_{\text{total}}} \quad \text{(17–32)}$$

where

$$R_{\text{total}} = R_{12} + R_3 + R_{\text{conv}} = \frac{R_1 R_2}{R_1 + R_2} + R_3 + R_{\text{conv}} \quad \text{(17–33)}$$

and

$$R_1 = \frac{L_1}{k_1 A_1} \qquad R_2 = \frac{L_2}{k_2 A_2} \qquad R_3 = \frac{L_3}{k_3 A_3} \qquad R_{\text{conv}} = \frac{1}{hA_3} \quad \text{(17–34)}$$

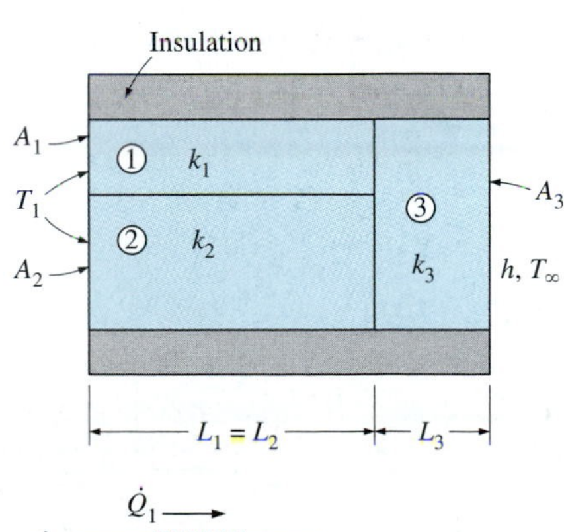

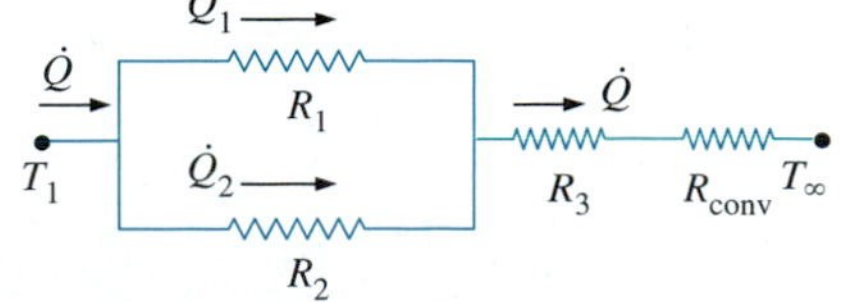

FIGURE 17–20
Thermal resistance network for combined series-parallel arrangement.

Once the individual thermal resistances are evaluated, the total resistance and the total rate of heat transfer can easily be determined from the relations above.

The result obtained is somewhat approximate, since the surfaces of the third layer are probably not isothermal, and heat transfer between the first two layers is likely to occur.

Two assumptions commonly used in solving complex multidimensional heat transfer problems by treating them as one-dimensional (say, in the

x-direction) using the thermal resistance network are (1) any plane wall normal to the x-axis is *isothermal* (i.e., to assume the temperature to vary in the x-direction only) and (2) any plane parallel to the x-axis is *adiabatic* (i.e., to assume heat transfer to occur in the x-direction only). These two assumptions result in different resistance networks, and thus different (but usually close) values for the total thermal resistance and thus heat transfer. The actual result lies between these two values. In geometries in which heat transfer occurs predominantly in one direction, either approach gives satisfactory results.

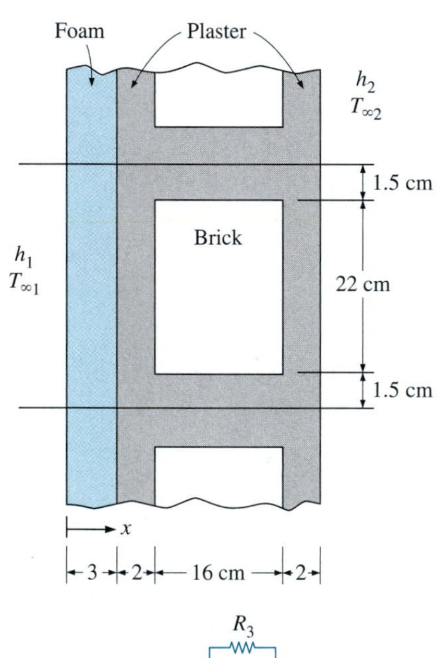

FIGURE 17–21
Schematic for Example 17–6.

EXAMPLE 17–6 Heat Loss through a Composite Wall

A 3-m-high and 5-m-wide wall consists of long 16-cm × 22-cm cross section horizontal bricks (k = 0.72 W/m · °C) separated by 3-cm-thick plaster layers (k = 0.22 W/m · °C). There are also 2-cm-thick plaster layers on each side of the brick and a 3-cm-thick rigid foam (k = 0.026 W/m · °C) on the inner side of the wall, as shown in Fig. 17–21. The indoor and the outdoor temperatures are 20°C and −10°C, respectively, and the convection heat transfer coefficients on the inner and the outer sides are h_1 = 10 W/m² · °C and h_2 = 25 W/m² · °C, respectively. Assuming one-dimensional heat transfer and disregarding radiation, determine the rate of heat transfer through the wall.

Solution The composition of a composite wall is given. The rate of heat transfer through the wall is to be determined.

Assumptions **1** Heat transfer is steady since there is no indication of change with time. **2** Heat transfer can be approximated as being one-dimensional since it is predominantly in the x-direction. **3** Thermal conductivities are constant. **4** Heat transfer by radiation is negligible.

Properties The thermal conductivities are given to be k = 0.72 W/m · °C for bricks, k = 0.22 W/m · °C for plaster layers, and k = 0.026 W/m · °C for the rigid foam.

Analysis There is a pattern in the construction of this wall that repeats itself every 25-cm distance in the vertical direction. There is no variation in the horizontal direction. Therefore, we consider a 1-m-deep and 0.25-m-high portion of the wall, since it is representative of the entire wall.

Assuming any cross section of the wall normal to the x-direction to be *isothermal,* the thermal resistance network for the representative section of the wall becomes as shown in Fig. 17–21. The individual resistances are evaluated as:

$$R_i = R_{\text{conv, 1}} = \frac{1}{h_1 A} = \frac{1}{(10\ \text{W/m}^2 \cdot {}^\circ\text{C})(0.25 \times 1\ \text{m}^2)} = 0.40^\circ\text{C/W}$$

$$R_1 = R_{\text{foam}} = \frac{L}{kA} = \frac{0.03\ \text{m}}{(0.026\ \text{W/m} \cdot {}^\circ\text{C})(0.25 \times 1\ \text{m}^2)} = 4.62^\circ\text{C/W}$$

$$R_2 = R_6 = R_{\text{plaster, side}} = \frac{L}{kA} = \frac{0.02\ \text{m}}{(0.22\ \text{W/m} \cdot {}^\circ\text{C})(0.25 \times 1\ \text{m}^2)}$$
$$= 0.36^\circ\text{C/W}$$

$$R_3 = R_5 = R_{\text{plaster, center}} = \frac{L}{kA} = \frac{0.16\ \text{m}}{(0.22\ \text{W/m} \cdot {}^\circ\text{C})(0.015 \times 1\ \text{m}^2)}$$
$$= 48.48^\circ\text{C/W}$$

$$R_4 = R_{brick} = \frac{L}{kA} = \frac{0.16 \text{ m}}{(0.72 \text{ W/m} \cdot °\text{C})(0.22 \times 1 \text{ m}^2)} = 1.01°\text{C/W}$$

$$R_o = R_{conv,\,2} = \frac{1}{h_2 A} = \frac{1}{(25 \text{ W/m}^2 \cdot °\text{C})(0.25 \times 1 \text{ m}^2)} = 0.16°\text{C/W}$$

The three resistances R_3, R_4, and R_5 in the middle are parallel, and their equivalent resistance is determined from

$$\frac{1}{R_{mid}} = \frac{1}{R_3} + \frac{1}{R_4} + \frac{1}{R_5} = \frac{1}{48.48} + \frac{1}{1.01} + \frac{1}{48.48} = 1.03 \text{ W/°C}$$

which gives

$$R_{mid} = 0.97°\text{C/W}$$

Now all the resistances are in series, and the total resistance is

$$\begin{aligned} R_{total} &= R_i + R_1 + R_2 + R_{mid} + R_6 + R_o \\ &= 0.40 + 4.62 + 0.36 + 0.97 + 0.36 + 0.16 \\ &= 6.87°\text{C/W} \end{aligned}$$

Then the steady rate of heat transfer through the wall becomes

$$\dot{Q} = \frac{T_{\infty 1} - T_{\infty 2}}{R_{total}} = \frac{[20 - (-10)]°\text{C}}{6.87°\text{C/W}} = 4.37 \text{ W} \qquad \text{(per 0.25 m}^2 \text{ surface area)}$$

or $4.37/0.25 = 17.5$ W per m^2 area. The total area of the wall is $A = 3$ m $\times$ 5 m $=$ 15 m^2. Then the rate of heat transfer through the entire wall becomes

$$\dot{Q}_{total} = (17.5 \text{ W/m}^2)(15 \text{ m}^2) = \mathbf{263 \text{ W}}$$

Of course, this result is approximate, since we assumed the temperature within the wall to vary in one direction only and ignored any temperature change (and thus heat transfer) in the other two directions.

Discussion In the above solution, we assumed the temperature at any cross section of the wall normal to the x-direction to be *isothermal.* We could also solve this problem by going to the other extreme and assuming the surfaces parallel to the x-direction to be *adiabatic.* The thermal resistance network in this case will be as shown in Fig. 17–22. By following the approach outlined above, the total thermal resistance in this case is determined to be $R_{total} =$ 6.97°C/W, which is very close to the value 6.85°C/W obtained before. Thus either approach gives roughly the same result in this case. This example demonstrates that either approach can be used in practice to obtain satisfactory results.

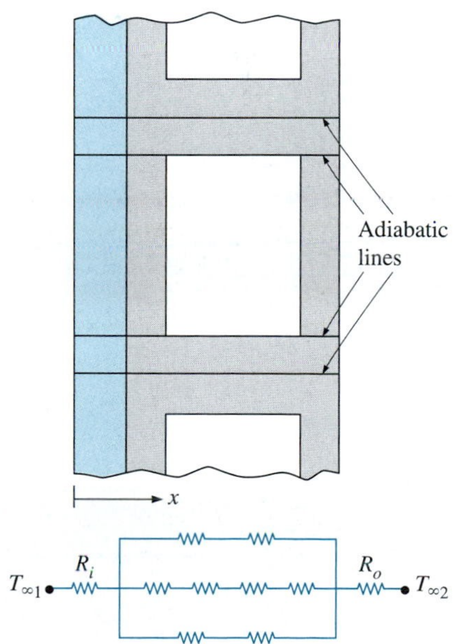

FIGURE 17–22
Alternative thermal resistance network for Example 17–6 for the case of surfaces parallel to the primary direction of heat transfer being adiabatic.

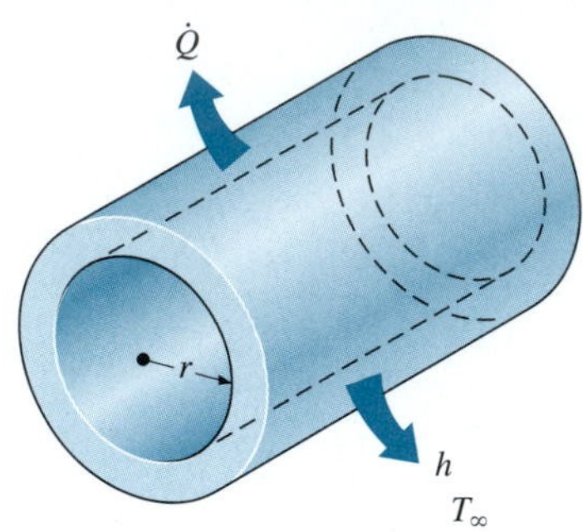

FIGURE 17–23
Heat is lost from a hot-water pipe to the air outside in the radial direction, and thus heat transfer from a long pipe is one-dimensional.

17–4 ▪ HEAT CONDUCTION IN CYLINDERS AND SPHERES

Consider steady heat conduction through a hot-water pipe. Heat is continuously lost to the outdoors through the wall of the pipe, and we intuitively feel that heat transfer through the pipe is in the normal direction to the pipe surface and no significant heat transfer takes place in the pipe in other directions (Fig. 17–23). The wall of the pipe, whose thickness is rather small, separates two fluids at different temperatures, and thus the temperature gradient in the radial direction is relatively large. Further, if the fluid temperatures inside and outside the pipe remain constant, then heat transfer through the pipe is *steady.* Thus heat transfer through the pipe can be modeled as *steady* and *one-dimensional.* The temperature of the pipe in this case depends on one direction only (the radial r-direction) and can be expressed as $T = T(r)$. The temperature is independent of the azimuthal angle or the axial distance. This situation is approximated in practice in long cylindrical pipes and spherical containers.

In *steady* operation, there is no change in the temperature of the pipe with time at any point. Therefore, the rate of heat transfer into the pipe must be equal to the rate of heat transfer out of it. In other words, heat transfer through the pipe must be constant, $\dot{Q}_{\text{cond, cyl}} = \text{constant}$.

Consider a long cylindrical layer (such as a circular pipe) of inner radius r_1, outer radius r_2, length L, and average thermal conductivity k (Fig. 17–24). The two surfaces of the cylindrical layer are maintained at constant temperatures T_1 and T_2. There is no heat generation in the layer and the thermal conductivity is constant. For one-dimensional heat conduction through the cylindrical layer, we have $T(r)$. Then Fourier's law of heat conduction for heat transfer through the cylindrical layer can be expressed as

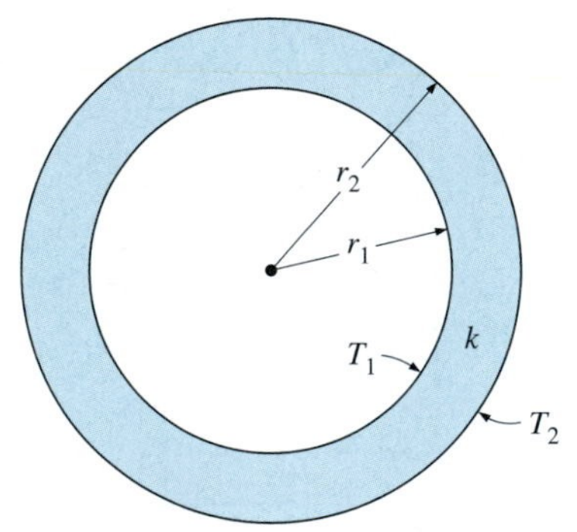

FIGURE 17–24
A long cylindrical pipe (or spherical shell) with specified inner and outer surface temperatures T_1 and T_2.

$$\dot{Q}_{\text{cond, cyl}} = -kA\frac{dT}{dr} \qquad \text{(W)} \tag{17–35}$$

where $A = 2\pi rL$ is the heat transfer area at location r. Note that A depends on r, and thus it *varies* in the direction of heat transfer. Separating the variables in the above equation and integrating from $r = r_1$, where $T(r_1) = T_1$, to $r = r_2$, where $T(r_2) = T_2$, gives

$$\int_{r=r_1}^{r_2} \frac{\dot{Q}_{\text{cond, cyl}}}{A}\,dr = -\int_{T=T_1}^{T_2} k\,dT \tag{17–36}$$

Substituting $A = 2\pi rL$ and performing the integrations give

$$\dot{Q}_{\text{cond, cyl}} = 2\pi Lk\frac{T_1 - T_2}{\ln(r_2/r_1)} \qquad \text{(W)} \tag{17–37}$$

since $\dot{Q}_{\text{cond, cyl}} = \text{constant}$. This equation can be rearranged as

$$\dot{Q}_{\text{cond, cyl}} = \frac{T_1 - T_2}{R_{\text{cyl}}} \qquad \text{(W)} \tag{17–38}$$

where

$$R_{\text{cyl}} = \frac{\ln(r_2/r_1)}{2\pi Lk} = \frac{\ln(\text{Outer radius/Inner radius})}{2\pi \times \text{Length} \times \text{Thermal conductivity}} \tag{17–39}$$

is the *thermal resistance* of the cylindrical layer against heat conduction, or simply the *conduction resistance* of the cylinder layer.

We can repeat the analysis for a *spherical layer* by taking $A = 4\pi r^2$ and performing the integrations in Eq. 17–36. The result can be expressed as

$$\dot{Q}_{\text{cond, sph}} = \frac{T_1 - T_2}{R_{\text{sph}}} \tag{17–40}$$

where

$$R_{\text{sph}} = \frac{r_2 - r_1}{4\pi r_1 r_2 k} = \frac{\text{Outer radius} - \text{Inner radius}}{4\pi(\text{Outer radius})(\text{Inner radius})(\text{Thermal conductivity})} \tag{17–41}$$

is the *thermal resistance* of the spherical layer against heat conduction, or simply the *conduction resistance* of the spherical layer.

Now consider steady one-dimensional heat transfer through a cylindrical or spherical layer that is exposed to convection on both sides to fluids at temperatures $T_{\infty 1}$ and $T_{\infty 2}$ with heat transfer coefficients h_1 and h_2, respectively, as shown in Fig. 17–25. The thermal resistance network in this case consists of one conduction and two convection resistances in series, just like the one for the plane wall, and the rate of heat transfer under steady conditions can be expressed as

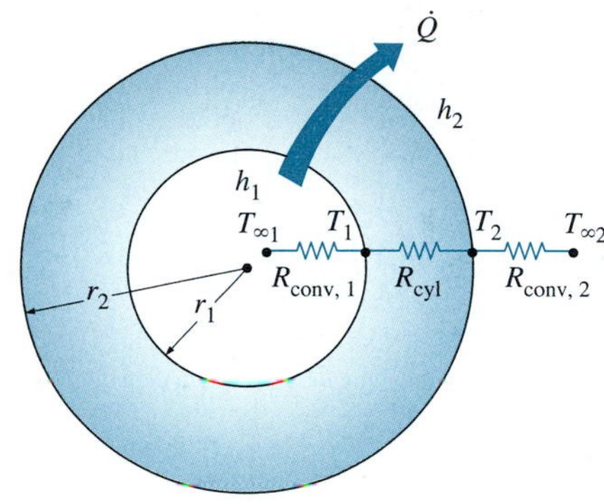

FIGURE 17–25
The thermal resistance network for a cylindrical (or spherical) shell subjected to convection from both the inner and the outer sides.

$$\dot{Q} = \frac{T_{\infty 1} - T_{\infty 2}}{R_{\text{total}}} \tag{17–42}$$

where

$$\begin{aligned} R_{\text{total}} &= R_{\text{conv, 1}} + R_{\text{cyl}} + R_{\text{conv, 2}} \\ &= \frac{1}{(2\pi r_1 L)h_1} + \frac{\ln(r_2/r_1)}{2\pi Lk} + \frac{1}{(2\pi r_2 L)h_2} \end{aligned} \tag{17–43}$$

for a *cylindrical* layer, and

$$\begin{aligned} R_{\text{total}} &= R_{\text{conv, 1}} + R_{\text{sph}} + R_{\text{conv, 2}} \\ &= \frac{1}{(4\pi r_1^2)h_1} + \frac{r_2 - r_1}{4\pi r_1 r_2 k} + \frac{1}{(4\pi r_2^2)h_2} \end{aligned} \tag{17–44}$$

for a spherical layer. Note that A in the convection resistance relation $R_{\text{conv}} = 1/hA$ is the *surface area at which convection occurs.* It is equal to $A = 2\pi rL$ for a cylindrical surface and $A = 4\pi r^2$ for a spherical surface of radius r. Also note that the thermal resistances are in series, and thus the total thermal resistance is determined by simply adding the individual resistances, just like the electrical resistances connected in series.

Multilayered Cylinders and Spheres

Steady heat transfer through multilayered cylindrical or spherical shells can be handled just like multilayered plane walls discussed earlier by simply adding an *additional resistance* in series for each *additional layer.* For example, the steady heat transfer rate through the three-layered composite cylinder of length L shown in Fig. 17–26 with convection on both sides can be expressed as

$$\dot{Q} = \frac{T_{\infty 1} - T_{\infty 2}}{R_{\text{total}}} \tag{17–45}$$

where R_{total} is the *total thermal resistance*, expressed as

$$\begin{aligned} R_{\text{total}} &= R_{\text{conv, 1}} + R_{\text{cyl, 1}} + R_{\text{cyl, 2}} + R_{\text{cyl, 3}} + R_{\text{conv, 2}} \\ &= \frac{1}{h_1 A_1} + \frac{\ln(r_2/r_1)}{2\pi L k_1} + \frac{\ln(r_3/r_2)}{2\pi L k_2} + \frac{\ln(r_4/r_3)}{2\pi L k_3} + \frac{1}{h_2 A_4} \end{aligned} \tag{17–46}$$

where $A_1 = 2\pi r_1 L$ and $A_4 = 2\pi r_4 L$. Equation 17–46 can also be used for a three-layered spherical shell by replacing the thermal resistances of cylindrical layers by the corresponding spherical ones. Again, note from the thermal resistance network that the resistances are in series, and thus the total thermal resistance is simply the *arithmetic sum* of the individual thermal resistances in the path of heat flow.

Once $\dot{Q}$ is known, we can determine any intermediate temperature T_j by applying the relation $\dot{Q} = (T_i - T_j)/R_{\text{total}, i-j}$ across any layer or layers such that T_i is a *known* temperature at location i and $R_{\text{total}, i-j}$ is the total thermal resistance between locations i and j (Fig. 17–27). For example, once $\dot{Q}$ has been calculated, the interface temperature T_2 between the first and second cylindrical layers can be determined from

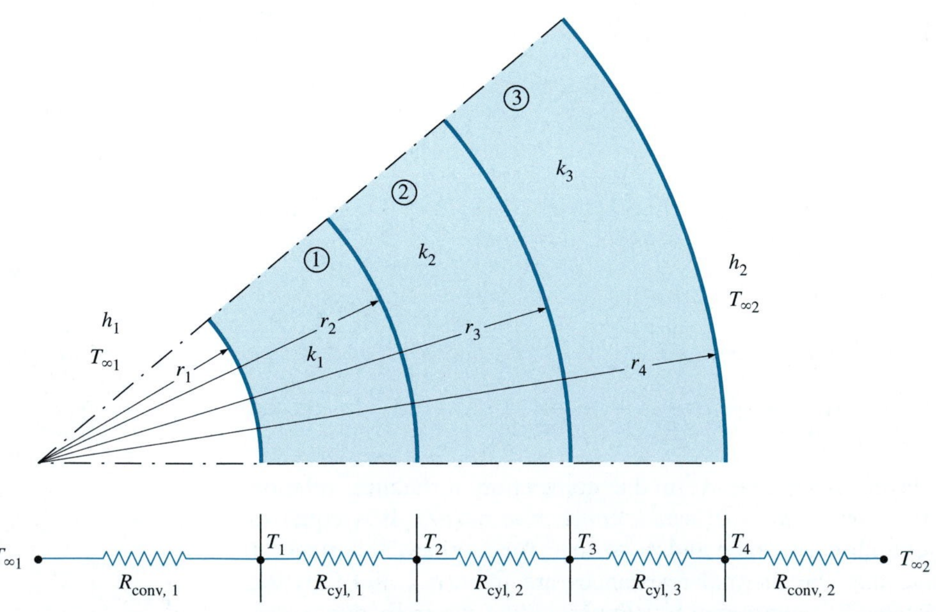

FIGURE 17–26

The thermal resistance network for heat transfer through a three-layered composite cylinder subjected to convection on both sides.

$$\dot{Q} = \frac{T_{\infty 1} - T_2}{R_{\text{conv, 1}} + R_{\text{cyl, 1}}} = \frac{T_{\infty 1} - T_2}{\dfrac{1}{h_1(2\pi r_1 L)} + \dfrac{\ln(r_2/r_1)}{2\pi L k_1}} \tag{17–47}$$

We could also calculate T_2 from

$$\dot{Q} = \frac{T_2 - T_{\infty 2}}{R_2 + R_3 + R_{\text{conv, 2}}} = \frac{T_2 - T_{\infty 2}}{\dfrac{\ln(r_3/r_2)}{2\pi L k_2} + \dfrac{\ln(r_4/r_3)}{2\pi L k_3} + \dfrac{1}{h_o(2\pi r_4 L)}} \tag{17–48}$$

Although both relations give the same result, we prefer the first one since it involves fewer terms and thus less work.

The thermal resistance concept can also be used for *other geometries*, provided that the proper conduction resistances and the proper surface areas in convection resistances are used.

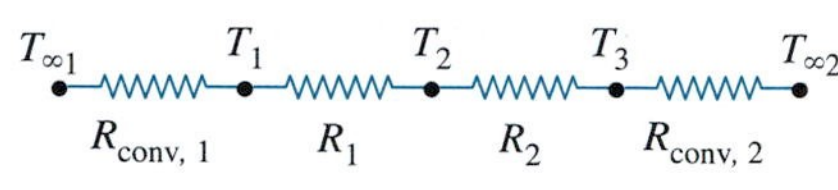

$$\dot{Q} = \frac{T_{\infty 1} - T_1}{R_{\text{conv,1}}}$$
$$= \frac{T_{\infty 1} - T_2}{R_{\text{conv,1}} + R_1}$$
$$= \frac{T_1 - T_3}{R_1 + R_2}$$
$$= \frac{T_2 - T_3}{R_2}$$
$$= \frac{T_2 - T_{\infty 2}}{R_2 + R_{\text{conv,2}}}$$
$$= \cdots$$

FIGURE 17–27
The ratio $\Delta T/R$ across any layer is equal to $\dot{Q}$, which remains constant in one-dimensional steady conduction.

EXAMPLE 17–7 Heat Transfer to a Spherical Container

A 3-m internal diameter spherical tank made of 2-cm-thick stainless steel (k = 15 W/m · °C) is used to store iced water at $T_{\infty 1}$ = 0°C. The tank is located in a room whose temperature is $T_{\infty 2}$ = 22°C. The walls of the room are also at 22°C. The outer surface of the tank is black and heat transfer between the outer surface of the tank and the surroundings is by natural convection and radiation. The convection heat transfer coefficients at the inner and the outer surfaces of the tank are h_1 = 80 W/m² · °C and h_2 = 10 W/m² · °C, respectively. Determine (*a*) the rate of heat transfer to the iced water in the tank and (*b*) the amount of ice at 0°C that melts during a 24-h period.

Solution A spherical container filled with iced water is subjected to convection and radiation heat transfer at its outer surface. The rate of heat transfer and the amount of ice that melts per day are to be determined.

Assumptions **1** Heat transfer is steady since the specified thermal conditions at the boundaries do not change with time. **2** Heat transfer is one-dimensional since there is thermal symmetry about the midpoint. **3** Thermal conductivity is constant.

Properties The thermal conductivity of steel is given to be k = 15 W/m · °C. The heat of fusion of water at atmospheric pressure is h_{if} = 333.7 kJ/kg. The outer surface of the tank is black and thus its emissivity is ε = 1.

Analysis (*a*) The thermal resistance network for this problem is given in Fig. 17–28. Noting that the inner diameter of the tank is D_1 = 3 m and the outer diameter is D_2 = 3.04 m, the inner and the outer surface areas of the tank are

$$A_1 = \pi D_1^2 = \pi(3 \text{ m})^2 = 28.3 \text{ m}^2$$
$$A_2 = \pi D_2^2 = \pi(3.04 \text{ m})^2 = 29.0 \text{ m}^2$$

Also, the radiation heat transfer coefficient is given by

$$h_{\text{rad}} = \varepsilon\sigma(T_2^2 + T_{\infty 2}^2)(T_2 + T_{\infty 2})$$

But we do not know the outer surface temperature T_2 of the tank, and thus we cannot calculate h_{rad}. Therefore, we need to assume a T_2 value now and

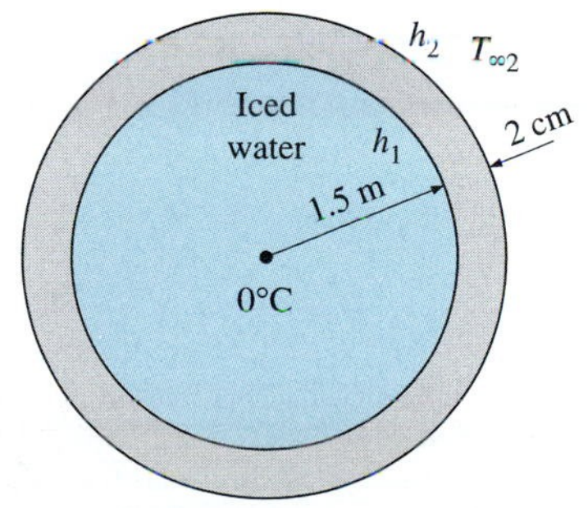

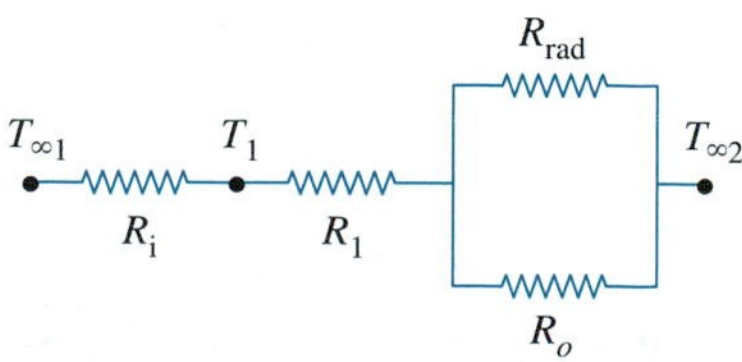

FIGURE 17–28
Schematic for Example 17–7.

check the accuracy of this assumption later. We will repeat the calculations if necessary using a revised value for T_2.

We note that T_2 must be between 0°C and 22°C, but it must be closer to 0°C, since the heat transfer coefficient inside the tank is much larger. Taking T_2 = 5°C = 278 K, the radiation heat transfer coefficient is determined to be

$$h_{rad} = (1)(5.67 \times 10^{-8}\ W/m^2 \cdot K^4)[(295\ K)^2 + (278\ K)^2][(295 + 278)\ K]$$
$$= 5.34\ W/m^2 \cdot K = 5.34\ W/m^2 \cdot °C$$

Then the individual thermal resistances become

$$R_i = R_{conv,\,1} = \frac{1}{h_1 A_1} = \frac{1}{(80\ W/m^2 \cdot °C)(28.3\ m^2)} = 0.000442°C/W$$

$$R_1 = R_{sphere} = \frac{r_2 - r_1}{4\pi k r_1 r_2} = \frac{(1.52 - 1.50)\ m}{4\pi\ (15\ W/m \cdot °C)(1.52\ m)(1.50\ m)}$$
$$= 0.000047°C/W$$

$$R_o = R_{conv,\,2} = \frac{1}{h_2 A_2} = \frac{1}{(10\ W/m^2 \cdot °C)(29.0\ m^2)} = 0.00345°C/W$$

$$R_{rad} = \frac{1}{h_{rad} A_2} = \frac{1}{(5.34\ W/m^2 \cdot °C)(29.0\ m^2)} = 0.00646°C/W$$

The two parallel resistances R_o and R_{rad} can be replaced by an equivalent resistance R_{equiv} determined from

$$\frac{1}{R_{equiv}} = \frac{1}{R_o} + \frac{1}{R_{rad}} = \frac{1}{0.00345} + \frac{1}{0.00646} = 444.7\ W/°C$$

which gives

$$R_{equiv} = 0.00225°C/W$$

Now all the resistances are in series, and the total resistance is

$$R_{total} = R_i + R_1 + R_{equiv} = 0.000442 + 0.000047 + 0.00225 = 0.00274°C/W$$

Then the steady rate of heat transfer to the iced water becomes

$$\dot{Q} = \frac{T_{\infty 2} - T_{\infty 1}}{R_{total}} = \frac{(22 - 0)°C}{0.00274°C/W} = \mathbf{8029\ W} \qquad (\text{or } \dot{Q} = 8.029\ kJ/s)$$

To check the validity of our original assumption, we now determine the outer surface temperature from

$$\dot{Q} = \frac{T_{\infty 2} - T_2}{R_{equiv}} \longrightarrow T_2 = T_{\infty 2} - \dot{Q}R_{equiv}$$
$$= 22°C - (8029\ W)(0.00225°C/W) = 4°C$$

which is sufficiently close to the 5°C assumed in the determination of the radiation heat transfer coefficient. Therefore, there is no need to repeat the calculations using 4°C for T_2.

(*b*) The total amount of heat transfer during a 24-h period is

$$Q = \dot{Q}\,\Delta t = (8.029 \text{ kJ/s})(24 \times 3600 \text{ s}) = 693{,}700 \text{ kJ}$$

Noting that it takes 333.7 kJ of energy to melt 1 kg of ice at 0°C, the amount of ice that will melt during a 24-h period is

$$m_{\text{ice}} = \frac{Q}{h_{if}} = \frac{693{,}700 \text{ kJ}}{333.7 \text{ kJ/kg}} = \mathbf{2079 \text{ kg}}$$

Therefore, about 2 metric tons of ice will melt in the tank every day.

Discussion An easier way to deal with combined convection and radiation at a surface when the surrounding medium and surfaces are at the same temperature is to add the radiation and convection heat transfer coefficients and to treat the result as the convection heat transfer coefficient. That is, to take $h = 10 + 5.34 = 15.34 \text{ W/m}^2 \cdot \text{°C}$ in this case. This way, we can ignore radiation since its contribution is accounted for in the convection heat transfer coefficient. The convection resistance of the outer surface in this case would be

$$R_{\text{combined}} = \frac{1}{h_{\text{combined}} A_2} = \frac{1}{(15.34 \text{ W/m}^2 \cdot \text{°C})(29.0 \text{ m}^2)} = 0.00225\text{°C/W}$$

which is identical to the value obtained for equivalent resistance for the parallel convection and the radiation resistances.

EXAMPLE 17–8 Heat Loss through an Insulated Steam Pipe

Steam at $T_{\infty 1} = 320\text{°C}$ flows in a cast iron pipe ($k = 80 \text{ W/m} \cdot \text{°C}$) whose inner and outer diameters are $D_1 = 5$ cm and $D_2 = 5.5$ cm, respectively. The pipe is covered with 3-cm-thick glass wool insulation with $k = 0.05 \text{ W/m} \cdot \text{°C}$. Heat is lost to the surroundings at $T_{\infty 2} = 5\text{°C}$ by natural convection and radiation, with a combined heat transfer coefficient of $h_2 = 18 \text{ W/m}^2 \cdot \text{°C}$. Taking the heat transfer coefficient inside the pipe to be $h_1 = 60 \text{ W/m}^2 \cdot \text{°C}$, determine the rate of heat loss from the steam per unit length of the pipe. Also determine the temperature drops across the pipe shell and the insulation.

Solution A steam pipe covered with glass wool insulation is subjected to convection on its surfaces. The rate of heat transfer per unit length and the temperature drops across the pipe and the insulation are to be determined.

Assumptions **1** Heat transfer is steady since there is no indication of any change with time. **2** Heat transfer is one-dimensional since there is thermal symmetry about the centerline and no variation in the axial direction. **3** Thermal conductivities are constant. **4** The thermal contact resistance at the interface is negligible.

Properties The thermal conductivities are given to be $k = 80 \text{ W/m} \cdot \text{°C}$ for cast iron and $k = 0.05 \text{ W/m} \cdot \text{°C}$ for glass wool insulation.

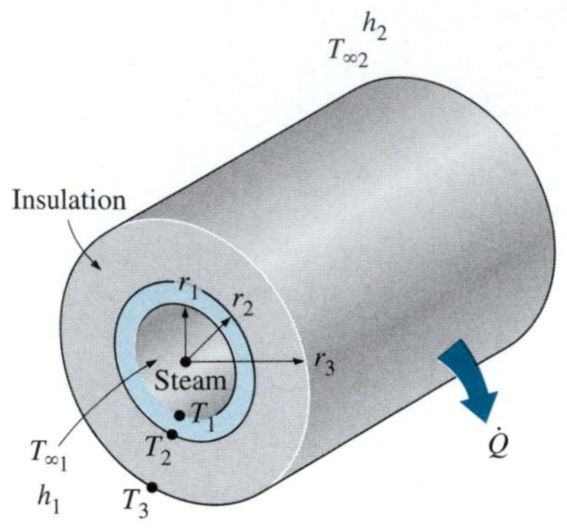

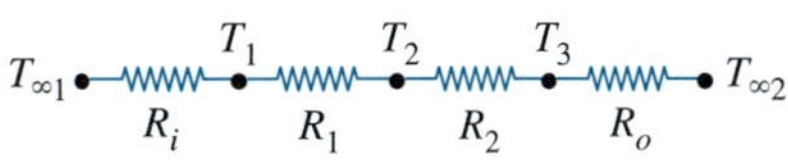

FIGURE 17–29
Schematic for Example 17–8.

Analysis The thermal resistance network for this problem involves four resistances in series and is given in Fig. 17–29. Taking $L = 1$ m, the areas of the surfaces exposed to convection are determined to be

$$A_1 = 2\pi r_1 L = 2\pi(0.025 \text{ m})(1 \text{ m}) = 0.157 \text{ m}^2$$
$$A_3 = 2\pi r_3 L = 2\pi(0.0575 \text{ m})(1 \text{ m}) = 0.361 \text{ m}^2$$

Then the individual thermal resistances become

$$R_i = R_{\text{conv, 1}} = \frac{1}{h_1 A_1} = \frac{1}{(60 \text{ W/m}^2 \cdot {}^\circ\text{C})(0.157 \text{ m}^2)} = 0.106^\circ\text{C/W}$$

$$R_1 = R_{\text{pipe}} = \frac{\ln(r_2/r_1)}{2\pi k_1 L} = \frac{\ln(2.75/2.5)}{2\pi(80 \text{ W/m} \cdot {}^\circ\text{C})(1 \text{ m})} = 0.0002^\circ\text{C/W}$$

$$R_2 = R_{\text{insulation}} = \frac{\ln(r_3/r_2)}{2\pi k_2 L} = \frac{\ln(5.75/2.75)}{2\pi(0.05 \text{ W/m} \cdot {}^\circ\text{C})(1 \text{ m})} = 2.35^\circ\text{C/W}$$

$$R_o = R_{\text{conv, 2}} = \frac{1}{h_2 A_3} = \frac{1}{(18 \text{ W/m}^2 \cdot {}^\circ\text{C})(0.361 \text{ m}^2)} = 0.154^\circ\text{C/W}$$

Noting that all resistances are in series, the total resistance is determined to be

$$R_{\text{total}} = R_i + R_1 + R_2 + R_o = 0.106 + 0.0002 + 2.35 + 0.154 = 2.61^\circ\text{C/W}$$

Then the steady rate of heat loss from the steam becomes

$$\dot{Q} = \frac{T_{\infty 1} - T_{\infty 2}}{R_{\text{total}}} = \frac{(320 - 5)^\circ\text{C}}{2.61^\circ\text{C/W}} = \mathbf{121\ W} \qquad \text{(per m pipe length)}$$

The heat loss for a given pipe length can be determined by multiplying the above quantity by the pipe length L.

The temperature drops across the pipe and the insulation are determined from Eq. 17–17 to be

$$\Delta T_{\text{pipe}} = \dot{Q}R_{\text{pipe}} = (121 \text{ W})(0.0002^\circ\text{C/W}) = \mathbf{0.02^\circ C}$$
$$\Delta T_{\text{insulation}} = \dot{Q}R_{\text{insulation}} = (121 \text{ W})(2.35^\circ\text{C/W}) = \mathbf{284^\circ C}$$

That is, the temperatures between the inner and the outer surfaces of the pipe differ by 0.02°C, whereas the temperatures between the inner and the outer surfaces of the insulation differ by 284°C.

Discussion Note that the thermal resistance of the pipe is too small relative to the other resistances and can be neglected without causing any significant error. Also note that the temperature drop across the pipe is practically zero, and thus the pipe can be assumed to be isothermal. The resistance to heat flow in insulated pipes is primarily due to insulation.

17–5 ▪ CRITICAL RADIUS OF INSULATION

We know that adding more insulation to a wall or to the attic always decreases heat transfer. The thicker the insulation, the lower the heat transfer rate. This is expected, since the heat transfer area A is constant, and adding insulation always increases the thermal resistance of the wall without increasing the convection resistance.

Adding insulation to a cylindrical pipe or a spherical shell, however, is a different matter. The additional insulation increases the conduction resistance of the insulation layer but decreases the convection resistance of the surface because of the increase in the outer surface area for convection. The heat transfer from the pipe may increase or decrease, depending on which effect dominates.

Consider a cylindrical pipe of outer radius r_1 whose outer surface temperature T_1 is maintained constant (Fig. 17–30). The pipe is now insulated with a material whose thermal conductivity is k and outer radius is r_2. Heat is lost from the pipe to the surrounding medium at temperature T_∞, with a convection heat transfer coefficient h. The rate of heat transfer from the insulated pipe to the surrounding air can be expressed as (Fig. 17–31)

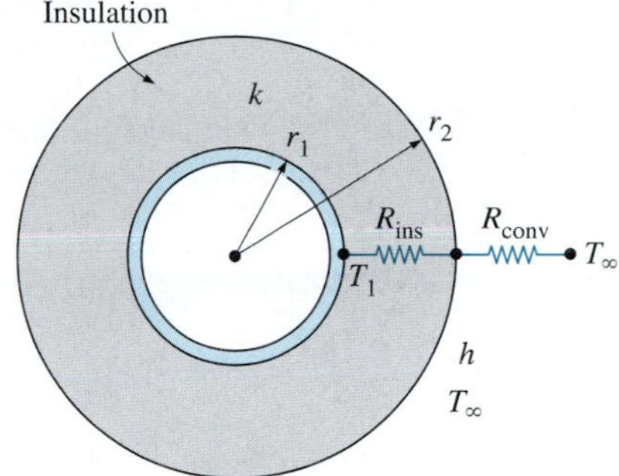

FIGURE 17–30
An insulated cylindrical pipe exposed to convection from the outer surface and the thermal resistance network associated with it.

$$\dot{Q} = \frac{T_1 - T_\infty}{R_{ins} + R_{conv}} = \frac{T_1 - T_\infty}{\dfrac{\ln(r_2/r_1)}{2\pi Lk} + \dfrac{1}{h(2\pi r_2 L)}} \qquad (17\text{–}49)$$

The variation of $\dot{Q}$ with the outer radius of the insulation r_2 is plotted in Fig. 17–31. The value of r_2 at which $\dot{Q}$ reaches a maximum is determined from the requirement that $d\dot{Q}/dr_2 = 0$ (zero slope). Performing the differentiation and solving for r_2 yields the **critical radius of insulation** for a cylindrical body to be

$$r_{cr,\,cylinder} = \frac{k}{h} \qquad \text{(m)} \qquad (17\text{–}50)$$

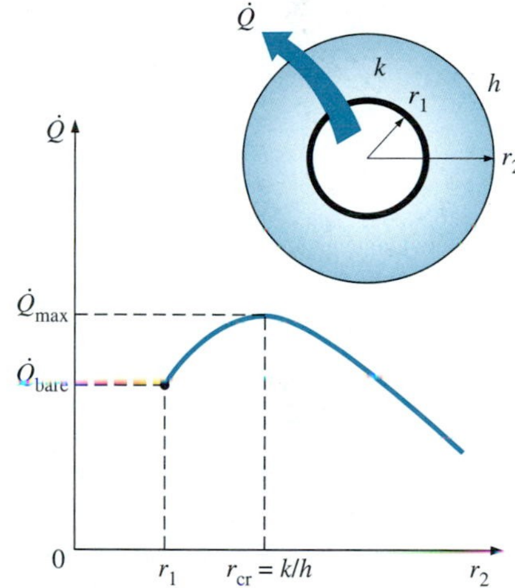

FIGURE 17–31
The variation of heat transfer rate with the outer radius of the insulation r_2 when $r_1 < r_{cr}$.

Note that the critical radius of insulation depends on the thermal conductivity of the insulation k and the external convection heat transfer coefficient h. The rate of heat transfer from the cylinder increases with the addition of insulation for $r_2 < r_{cr}$, reaches a maximum when $r_2 = r_{cr}$, and starts to decrease for $r_2 > r_{cr}$. Thus, insulating the pipe may actually increase the rate of heat transfer from the pipe instead of decreasing it when $r_2 < r_{cr}$.

The important question to answer at this point is whether we need to be concerned about the critical radius of insulation when insulating hot-water pipes or even hot-water tanks. Should we always check and make sure that the outer radius of insulation sufficiently exceeds the critical radius before we install any insulation? Probably not, as explained here.

The value of the critical radius r_{cr} is the largest when k is large and h is small. Noting that the lowest value of h encountered in practice is about 5 W/m^2 · °C for the case of natural convection of gases, and that the thermal conductivity of common insulating materials is about 0.05 W/m^2 · °C, the largest value of the critical radius we are likely to encounter is

$$r_{cr,\,max} = \frac{k_{max,\,insulation}}{h_{min}} \approx \frac{0.05 \text{ W/m} \cdot {}^\circ\text{C}}{5 \text{ W/m}^2 \cdot {}^\circ\text{C}} = 0.01 \text{ m} = 1 \text{ cm}$$

This value would be even smaller when the radiation effects are considered. The critical radius would be much less in forced convection, often less than 1 mm, because of much larger h values associated with forced convection. Therefore, we can insulate hot-water or steam pipes freely without worrying about the possibility of increasing the heat transfer by insulating the pipes.

The radius of electric wires may be smaller than the critical radius. Therefore, the plastic electrical insulation may actually *enhance* the heat transfer from electric wires and thus keep their steady operating temperatures at lower and thus safer levels.

The discussions above can be repeated for a sphere, and it can be shown in a similar manner that the critical radius of insulation for a spherical shell is

$$r_{\text{cr, sphere}} = \frac{2k}{h} \qquad \text{(17–51)}$$

where k is the thermal conductivity of the insulation and h is the convection heat transfer coefficient on the outer surface.

EXAMPLE 17–9 Heat Loss from an Insulated Electric Wire

A 3-mm-diameter and 5-m-long electric wire is tightly wrapped with a 2-mm-thick plastic cover whose thermal conductivity is $k = 0.15$ W/m · °C. Electrical measurements indicate that a current of 10 A passes through the wire and there is a voltage drop of 8 V along the wire. If the insulated wire is exposed to a medium at $T_\infty = 30$°C with a heat transfer coefficient of $h = 12$ W/m^2 · °C, determine the temperature at the interface of the wire and the plastic cover in steady operation. Also determine whether doubling the thickness of the plastic cover will increase or decrease this interface temperature.

SOLUTION An electric wire is tightly wrapped with a plastic cover. The interface temperature and the effect of doubling the thickness of the plastic cover on the interface temperature are to be determined.

Assumptions **1** Heat transfer is steady since there is no indication of any change with time. **2** Heat transfer is one-dimensional since there is thermal symmetry about the centerline and no variation in the axial direction. **3** Thermal conductivities are constant. **4** The thermal contact resistance at the interface is negligible. **5** Heat transfer coefficient incorporates the radiation effects, if any.

Properties The thermal conductivity of plastic is given to be $k = 0.15$ W/m · °C.

Analysis Heat is generated in the wire and its temperature rises as a result of resistance heating. We assume heat is generated uniformly throughout the wire and is transferred to the surrounding medium in the radial direction. In steady operation, the rate of heat transfer becomes equal to the heat generated within the wire, which is determined to be

$$\dot{Q} = \dot{W}_e = \mathbf{V}I = (8\text{ V})(10\text{ A}) = 80\text{ W}$$

The thermal resistance network for this problem involves a conduction resistance for the plastic cover and a convection resistance for the outer surface in series, as shown in Fig. 17–32. The values of these two resistances are

$$A_2 = (2\pi r_2)L = 2\pi(0.0035\text{ m})(5\text{ m}) = 0.110\text{ m}^2$$

$$R_{\text{conv}} = \frac{1}{hA_2} = \frac{1}{(12\text{ W/m}^2\cdot\text{°C})(0.110\text{ m}^2)} = 0.76\text{°C/W}$$

$$R_{\text{plastic}} = \frac{\ln(r_2/r_1)}{2\pi kL} = \frac{\ln(3.5/1.5)}{2\pi(0.15\text{ W/m}\cdot\text{°C})(5\text{ m})} = 0.18\text{°C/W}$$

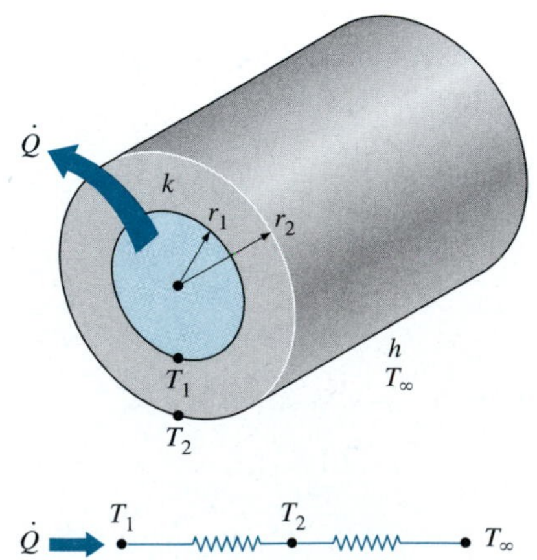

FIGURE 17–32
Schematic for Example 17–9.

and therefore

$$R_{\text{total}} = R_{\text{plastic}} + R_{\text{conv}} = 0.76 + 0.18 = 0.94°\text{C/W}$$

Then the interface temperature can be determined from

$$\dot{Q} = \frac{T_1 - T_\infty}{R_{\text{total}}} \longrightarrow \quad T_1 = T_\infty + \dot{Q}R_{\text{total}}$$
$$= 30°\text{C} + (80\text{ W})(0.94°\text{C/W}) = \mathbf{105°C}$$

Note that we did not involve the electrical wire directly in the thermal resistance network, since the wire involves heat generation.

To answer the second part of the question, we need to know the critical radius of insulation of the plastic cover. It is determined from Eq. 17–50 to be

$$r_{\text{cr}} = \frac{k}{h} = \frac{0.15\text{ W/m}\cdot°\text{C}}{12\text{ W/m}^2\cdot°\text{C}} = 0.0125\text{ m} = 12.5\text{ mm}$$

which is larger than the radius of the plastic cover. Therefore, increasing the thickness of the plastic cover will *enhance* heat transfer until the outer radius of the cover reaches 12.5 mm. As a result, the rate of heat transfer $\dot{Q}$ will *increase* when the interface temperature T_1 is held constant, or T_1 will *decrease* when $\dot{Q}$ is held constant, which is the case here.

Discussion It can be shown by repeating the calculations above for a 4-mm-thick plastic cover that the interface temperature drops to 90.6°C when the thickness of the plastic cover is doubled. It can also be shown in a similar manner that the interface reaches a minimum temperature of 83°C when the outer radius of the plastic cover equals the critical radius.

17–6 ■ HEAT TRANSFER FROM FINNED SURFACES

The rate of heat transfer from a surface at a temperature T_s to the surrounding medium at T_∞ is given by Newton's law of cooling as

$$\dot{Q}_{\text{conv}} = hA_s(T_s - T_\infty)$$

where A_s is the heat transfer surface area and h is the convection heat transfer coefficient. When the temperatures T_s and T_∞ are fixed by design considerations, as is often the case, there are *two ways* to increase the rate of heat transfer: to increase the *convection heat transfer coefficient h* or to increase the *surface area A_s*. Increasing h may require the installation of a pump or fan, or replacing the existing one with a larger one, but this approach may or may not be practical. Besides, it may not be adequate. The alternative is to increase the surface area by attaching to the surface *extended surfaces* called *fins* made of highly conductive materials such as aluminum. Finned surfaces are manufactured by extruding, welding, or wrapping a thin metal sheet on a surface. Fins enhance heat transfer from a surface by exposing a larger surface area to convection and radiation.

Finned surfaces are commonly used in practice to enhance heat transfer, and they often increase the rate of heat transfer from a surface severalfold.

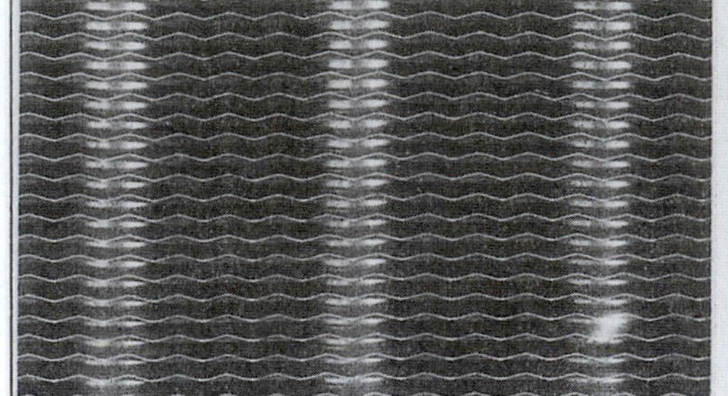

FIGURE 17–33
The thin plate fins of a car radiator greatly increase the rate of heat transfer to the air.
© *Yunus Çengel, photo by James Kleiser.*

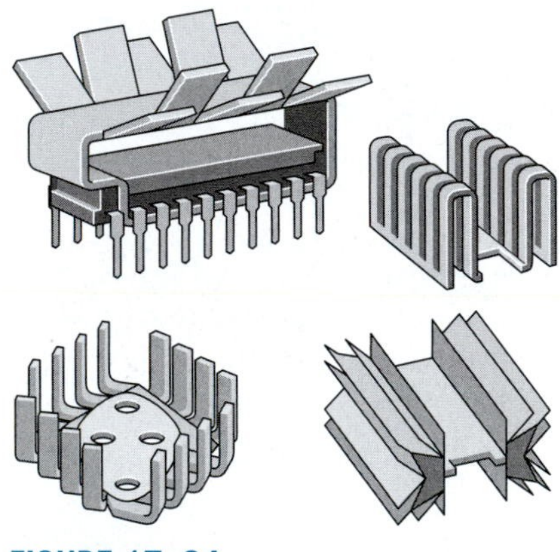

FIGURE 17–34
Some innovative fin designs.

The car radiator shown in Fig. 17–33 is an example of a finned surface. The closely packed thin metal sheets attached to the hot-water tubes increase the surface area for convection and thus the rate of convection heat transfer from the tubes to the air many times. There are a variety of innovative fin designs available in the market, and they seem to be limited only by imagination (Fig. 17–34).

In the analysis of fins, we consider *steady* operation with *no heat generation* in the fin, and we assume the thermal conductivity k of the material to remain constant. We also assume the convection heat transfer coefficient h to be *constant* and *uniform* over the entire surface of the fin for convenience in the analysis. We recognize that the convection heat transfer coefficient h, in general, varies along the fin as well as its circumference, and its value at a point is a strong function of the *fluid motion* at that point. The value of h is usually much lower at the *fin base* than it is at the *fin tip* because the fluid is surrounded by solid surfaces near the base, which seriously disrupt its motion to the point of "suffocating" it, while the fluid near the fin tip has little contact with a solid surface and thus encounters little resistance to flow. Therefore, adding too many fins on a surface may actually decrease the overall heat transfer when the decrease in h offsets any gain resulting from the increase in the surface area.

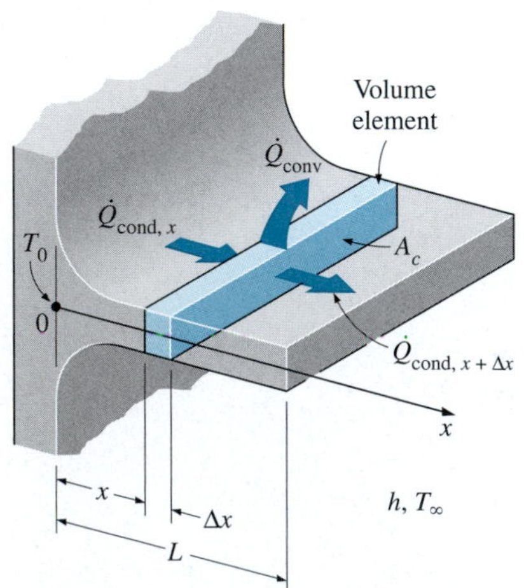

FIGURE 17–35
Volume element of a fin at location x having a length of Δx, cross-sectional area of A_c, and perimeter of p.

Fin Equation

Consider a volume element of a fin at location x having a length of Δx, cross-sectional area of A_c, and a perimeter of p, as shown in Fig. 17–35. Under steady conditions, the energy balance on this volume element can be expressed as

$$\begin{pmatrix}\text{Rate of } \textit{heat} \\ \textit{conduction} \text{ into} \\ \text{the element at } x\end{pmatrix} = \begin{pmatrix}\text{Rate of } \textit{heat} \\ \textit{conduction} \text{ from the} \\ \text{element at } x + \Delta x\end{pmatrix} + \begin{pmatrix}\text{Rate of } \textit{heat} \\ \textit{convection} \text{ from} \\ \text{the element}\end{pmatrix}$$

or

$$\dot{Q}_{cond,\,x} = \dot{Q}_{cond,\,x+\Delta x} + \dot{Q}_{conv}$$

where

$$\dot{Q}_{\text{conv}} = h(p\,\Delta x)(T - T_\infty)$$

Substituting and dividing by Δx, we obtain

$$\frac{\dot{Q}_{\text{cond},\,x+\Delta x} - \dot{Q}_{\text{cond},\,x}}{\Delta x} + hp(T - T_\infty) = 0 \tag{17–52}$$

Taking the limit as $\Delta x \to 0$ gives

$$\frac{d\dot{Q}_{\text{cond}}}{dx} + hp(T - T_\infty) = 0 \tag{17–53}$$

From Fourier's law of heat conduction we have

$$\dot{Q}_{\text{cond}} = -kA_c\frac{dT}{dx} \tag{17–54}$$

where A_c is the cross-sectional area of the fin at location x. Substitution of this relation into Eq. 17–53 gives the differential equation governing heat transfer in fins,

$$\frac{d}{dx}\left(kA_c\frac{dT}{dx}\right) - hp(T - T_\infty) = 0 \tag{17–55}$$

In general, the cross-sectional area A_c and the perimeter p of a fin vary with x, which makes this differential equation difficult to solve. In the special case of *constant cross section* and *constant thermal conductivity*, the differential equation 17–55 reduces to

$$\frac{d^2\theta}{dx^2} - m^2\theta = 0 \tag{17–56}$$

where

$$m^2 = \frac{hp}{kA_c} \tag{17–57}$$

and $\theta = T - T_\infty$ is the *temperature excess.* At the fin base we have $\theta_b = T_b - T_\infty$.

Equation 17–56 is a linear, homogeneous, second-order differential equation with constant coefficients. A fundamental theory of differential equations states that such an equation has two linearly independent solution functions, and its general solution is the linear combination of those two solution functions. A careful examination of the differential equation reveals that subtracting a constant multiple of the solution function θ from its second derivative yields zero. Thus we conclude that the function θ and its second derivative must be *constant multiples* of each other. The only functions whose derivatives are constant multiples of the functions themselves are the *exponential functions* (or a linear combination of exponential functions such as sine and cosine hyperbolic functions). Therefore, the solution functions of the differential equation above are the exponential functions e^{-mx} or e^{mx} or constant multiples of them. This can be verified by direct substitution. For example, the second derivative of e^{-mx} is m^2e^{-mx}, and its substitution into Eq. 17–56 yields zero. Therefore, the general solution of the differential equation Eq. 17–56 is

$$\theta(x) = C_1 e^{mx} + C_2 e^{-mx} \tag{17–58}$$

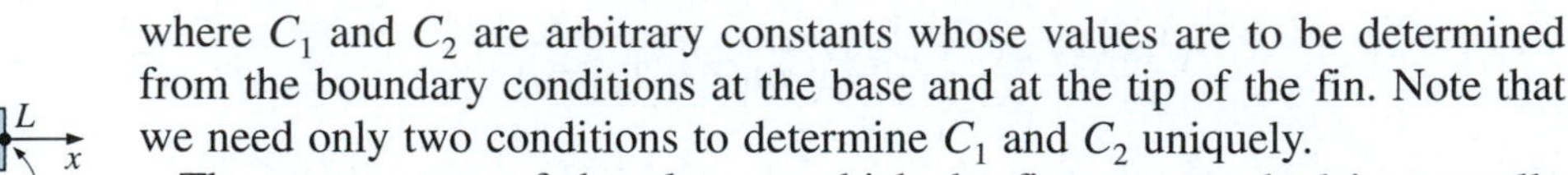

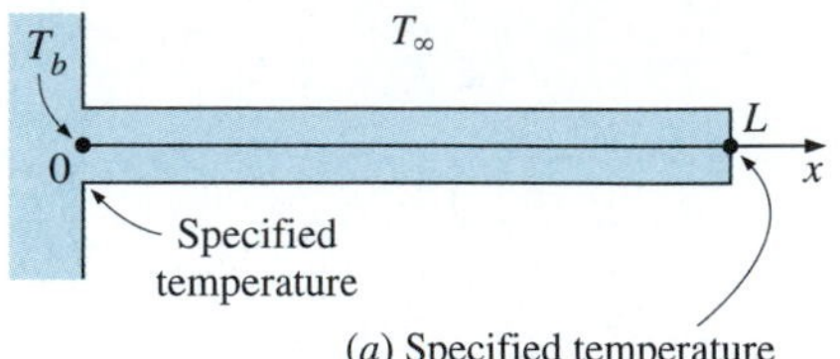

(*a*) Specified temperature
(*b*) Negligible heat loss
(*c*) Convection
(*d*) Convection and radiation

FIGURE 17–36
Boundary conditions at the fin base and the fin tip.

where C_1 and C_2 are arbitrary constants whose values are to be determined from the boundary conditions at the base and at the tip of the fin. Note that we need only two conditions to determine C_1 and C_2 uniquely.

The temperature of the plate to which the fins are attached is normally known in advance. Therefore, at the fin base we have a *specified temperature* boundary condition, expressed as

Boundary condition at fin base: $$\theta(0) = \theta_b = T_b - T_\infty \tag{17–59}$$

At the fin tip we have several possibilities, including specified temperature, negligible heat loss (idealized as an adiabatic tip), convection, and combined convection and radiation (Fig. 17–36). Next, we consider each case separately.

1 Infinitely Long Fin ($T_{\text{fin tip}} = T_\infty$)

For a sufficiently long fin of *uniform* cross section (A_c = constant), the temperature of the fin at the fin tip approaches the environment temperature T_∞ and thus θ approaches zero. That is,

Boundary condition at fin tip: $$\theta(L) = T(L) - T_\infty = 0 \quad \text{as} \quad L \to \infty$$

This condition is satisfied by the function e^{-mx}, but not by the other prospective solution function e^{mx} since it tends to infinity as x gets larger. Therefore, the general solution in this case will consist of a constant multiple of e^{-mx}. The value of the constant multiple is determined from the requirement that at the fin base where $x = 0$ the value of θ is θ_b. Noting that $e^{-mx} = e^0 = 1$, the proper value of the constant is θ_b, and the solution function we are looking for is $\theta(x) = \theta_b e^{-mx}$. This function satisfies the differential equation as well as the requirements that the solution reduce to θ_b at the fin base and approach zero at the fin tip for large x. Noting that $\theta = T - T_\infty$ and $m = \sqrt{hp/kA_c}$, the variation of temperature along the fin in this case can be expressed as

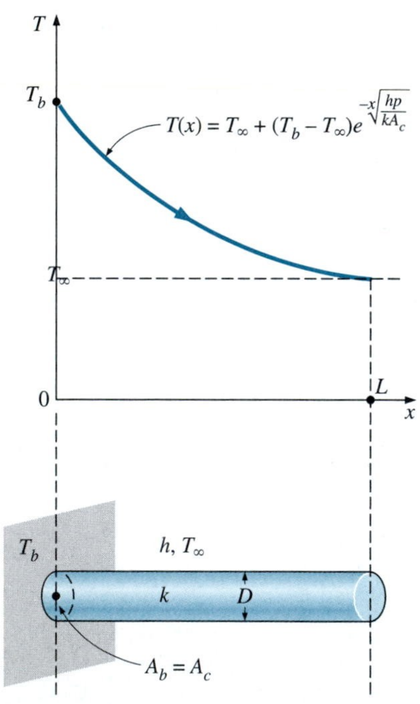

FIGURE 17–37
A long circular fin of uniform cross section and the variation of temperature along it.

Very long fin: $$\frac{T(x) - T_\infty}{T_b - T_\infty} = e^{-mx} = e^{-x\sqrt{hp/kA_c}} \tag{17–60}$$

Note that the temperature along the fin in this case decreases *exponentially* from T_b to T_∞, as shown in Fig. 17–37. The steady rate of *heat transfer* from the entire fin can be determined from Fourier's law of heat conduction

Very long fin: $$\dot{Q}_{\text{long fin}} = -kA_c \left.\frac{dT}{dx}\right|_{x=0} = \sqrt{hpkA_c}\,(T_b - T_\infty) \tag{17–61}$$

where p is the perimeter, A_c is the cross-sectional area of the fin, and x is the distance from the fin base. Alternatively, the rate of heat transfer from the fin could also be determined by considering heat transfer from a differential volume element of the fin and integrating it over the entire surface of the fin:

$$\dot{Q}_{\text{fin}} = \int_{A_{\text{fin}}} h[T(x) - T_\infty]\, dA_{\text{fin}} = \int_{A_{\text{fin}}} h\theta(x)\, dA_{\text{fin}} \tag{17–62}$$

The two approaches described are equivalent and give the same result since, under steady conditions, the heat transfer from the exposed surfaces of the fin is equal to the heat transfer to the fin at the base (Fig. 17–38).

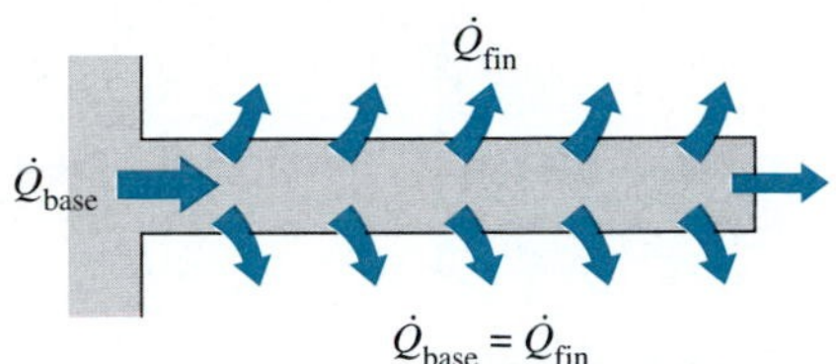

FIGURE 17–38
Under steady conditions, heat transfer from the exposed surfaces of the fin is equal to heat conduction to the fin at the base.

2 Negligible Heat Loss from the Fin Tip (Adiabatic fin tip, $\dot{Q}_{\text{fin tip}} = 0$)

Fins are not likely to be so long that their temperature approaches the surrounding temperature at the tip. A more realistic situation is for heat transfer from the fin tip to be negligible since the heat transfer from the fin is proportional to its surface area, and the surface area of the fin tip is usually a negligible fraction of the total fin area. Then the fin tip can be assumed to be adiabatic, and the condition at the fin tip can be expressed as

Boundary condition at fin tip: $$\left.\frac{d\theta}{dx}\right|_{x=L} = 0 \quad \textbf{(17–63)}$$

The condition at the fin base remains the same as expressed in Eq. 17–59. The application of these two conditions on the general solution (Eq. 17–58) yields, after some manipulations, this relation for the temperature distribution:

Adiabatic fin tip: $$\frac{T(x) - T_\infty}{T_b - T_\infty} = \frac{\cosh m(L - x)}{\cosh mL} \quad \textbf{(17–64)}$$

The rate of heat transfer from the fin can be determined again from Fourier's law of heat conduction:

Adiabatic fin tip: $$\dot{Q}_{\text{adiabatic tip}} = -kA_c \left.\frac{dT}{dx}\right|_{x=0} = \sqrt{hpkA_c}\,(T_b - T_\infty) \tanh mL \quad \textbf{(17–65)}$$

Note that the heat transfer relations for the very long fin and the fin with negligible heat loss at the tip differ by the factor tanh mL, which approaches 1 as L becomes very large.

3 Convection (or Combined Convection and Radiation) from Fin Tip

The fin tips, in practice, are exposed to the surroundings, and thus the proper boundary condition for the fin tip is convection that also includes the effects of radiation. The fin equation can still be solved in this case using the convection at the fin tip as the second boundary condition, but the analysis becomes more involved, and it results in rather lengthy expressions for the temperature distribution and the heat transfer. Yet, in general, the fin tip area is a small fraction of the total fin surface area, and thus the complexities involved can hardly justify the improvement in accuracy.

A practical way of accounting for the heat loss from the fin tip is to replace the *fin length L* in the relation for the *insulated tip* case by a **corrected length** defined as (Fig. 17–39)

Corrected fin length: $$L_c = L + \frac{A_c}{p} \quad \textbf{(17–66)}$$

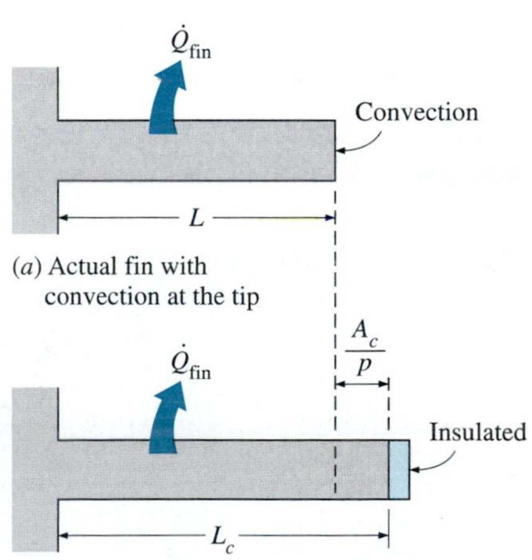

FIGURE 17–39
Corrected fin length L_c is defined such that heat transfer from a fin of length L_c with insulated tip is equal to heat transfer from the actual fin of length L with convection at the fin tip.

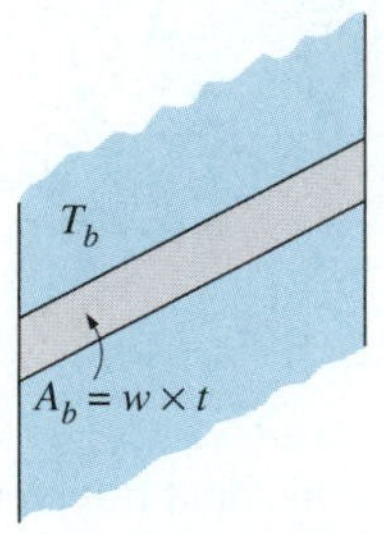

(*a*) Surface without fins

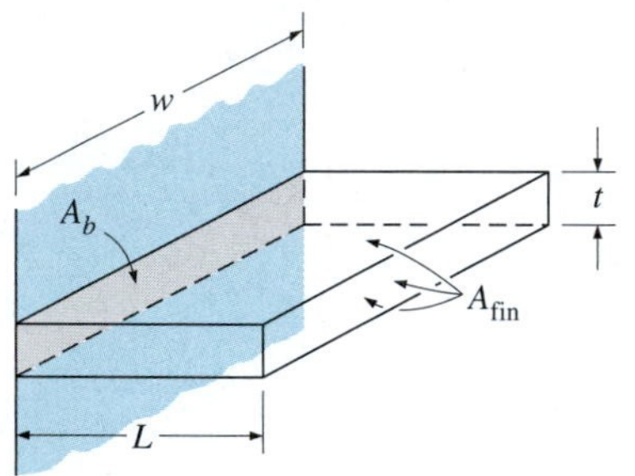

(*b*) Surface with a fin

$A_{fin} = 2 \times w \times L + w \times t$
$\cong 2 \times w \times L$

FIGURE 17–40
Fins enhance heat transfer from a surface by enhancing surface area.

where A_c is the cross-sectional area and p is the perimeter of the fin at the tip. Multiplying the relation above by the perimeter gives $A_{corrected} = A_{fin\ (lateral)} + A_{tip}$, which indicates that the fin area determined using the corrected length is equivalent to the sum of the lateral fin area plus the fin tip area.

The corrected length approximation gives very good results when the variation of temperature near the fin tip is small (which is the case when $mL \geq 1$) and the heat transfer coefficient at the fin tip is about the same as that at the lateral surface of the fin. Therefore, *fins subjected to convection at their tips can be treated as fins with insulated tips by replacing the actual fin length by the corrected length in Eqs. 17–64 and 17–65.*

Using the proper relations for A_c and p, the corrected lengths for rectangular and cylindrical fins are easily determined to be

$$L_{c,\ rectangular\ fin} = L + \frac{t}{2} \quad \text{and} \quad L_{c,\ cylindrical\ fin} = L + \frac{D}{4}$$

where t is the thickness of the rectangular fins and D is the diameter of the cylindrical fins.

Fin Efficiency

Consider the surface of a *plane wall* at temperature T_b exposed to a medium at temperature T_∞. Heat is lost from the surface to the surrounding medium by convection with a heat transfer coefficient of h. Disregarding radiation or accounting for its contribution in the convection coefficient h, heat transfer from a surface area A_s is expressed as $\dot{Q} = hA_s(T_s - T_\infty)$.

Now let us consider a fin of constant cross-sectional area $A_c = A_b$ and length L that is attached to the surface with a perfect contact (Fig. 17–40). This time heat is transfered from the surface to the fin *by conduction* and from the fin to the surrounding medium *by convection* with the same heat transfer coefficient h. The temperature of the fin is T_b at the fin base and gradually decreases toward the fin tip. Convection from the fin surface causes the temperature at any cross section to drop somewhat from the midsection toward the outer surfaces. However, the cross-sectional area of the fins is usually very small, and thus the temperature at any cross section can be considered to be uniform. Also, the fin tip can be assumed for convenience and simplicity to be adiabatic by using the corrected length for the fin instead of the actual length.

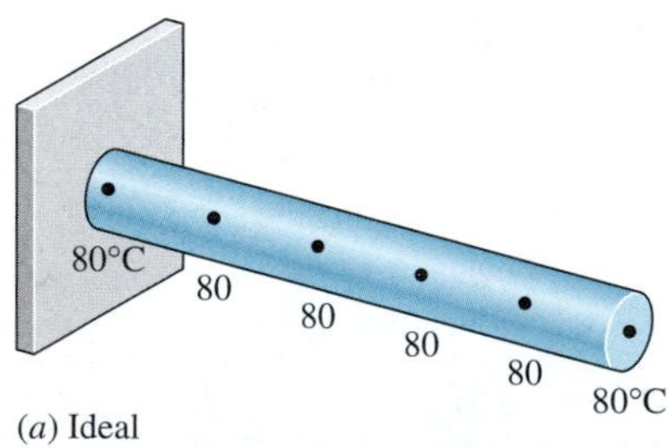

(*a*) Ideal

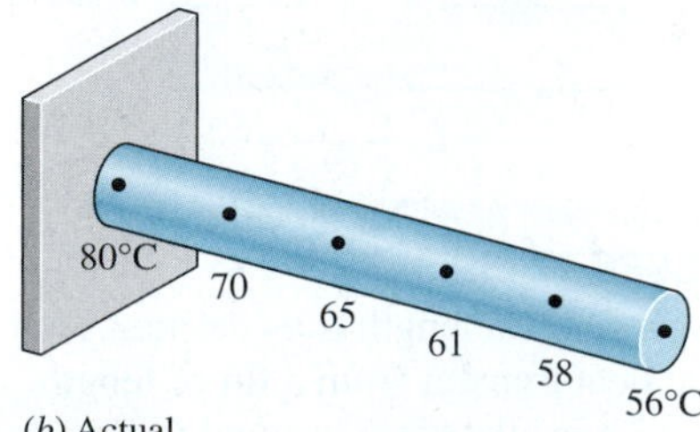

(*b*) Actual

FIGURE 17–41
Ideal and actual temperature distribution along a fin.

In the limiting case of *zero thermal resistance* or *infinite thermal conductivity* ($k \to \infty$), the temperature of the fin is uniform at the base value of T_b. The heat transfer from the fin is *maximum* in this case and can be expressed as

$$\dot{Q}_{fin,\ max} = hA_{fin}(T_b - T_\infty) \tag{17–67}$$

In reality, however, the temperature of the fin drops along the fin, and thus the heat transfer from the fin is less because of the decreasing temperature difference $T(x) - T_\infty$ toward the fin tip, as shown in Fig. 17–41. To account for the effect of this decrease in temperature on heat transfer, we define a **fin efficiency** as

$$\eta_{fin} = \frac{\dot{Q}_{fin}}{\dot{Q}_{fin,\ max}} = \frac{\text{Actual heat transfer rate from the fin}}{\text{Ideal heat transfer rate from the fin if the entire fin were at base temperature}} \tag{17–68}$$

TABLE 17–3

Efficiency and surface areas of common fin configurations

Straight rectangular fins

$m = \sqrt{2h/kt}$
$L_c = L + t/2$
$A_{\text{fin}} = 2wL_c$

$$\eta_{\text{fin}} = \frac{\tanh mL_c}{mL_c}$$

Straight triangular fins

$m = \sqrt{2h/kt}$

$A_{\text{fin}} = 2w\sqrt{L^2 + (t/2)^2}$

$$\eta_{\text{fin}} = \frac{1}{mL}\frac{I_1(2mL)}{I_0(2mL)}$$

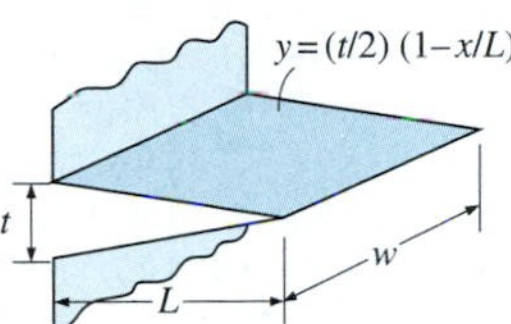

Straight parabolic fins

$m = \sqrt{2h/kt}$
$A_{\text{fin}} = wL[C_1 + (L/t)\ln(t/L + C_1)]$
$C_1 = \sqrt{1 + (t/L)^2}$

$$\eta_{\text{fin}} = \frac{2}{1 + \sqrt{(2mL)^2 + 1}}$$

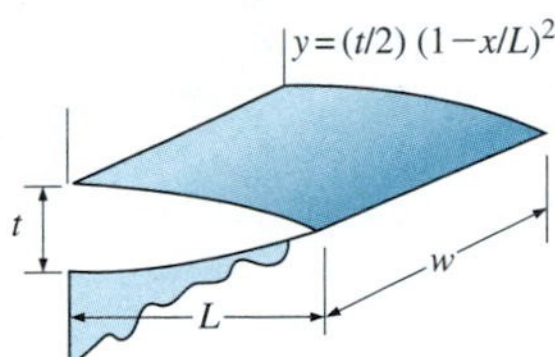

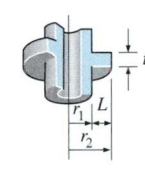

Circular fins of rectangular profile

$m = \sqrt{2h/kt}$
$r_{2c} = r_2 + t/2$
$A_{fin} = 2\pi(r_{2c}^2 - r_1^2)$

$$\eta_{\text{fin}} = C_2\frac{K_1(mr_1)I_1(mr_{2c}) - I_1(mr_1)K_1(mr_{2c})}{I_0(mr_1)K_1(mr_{2c}) + K_0(mr_1)I_1(mr_{2c})}$$

$$C_2 = \frac{2r_1/m}{r_{2c}^2 - r_1^2}$$

Pin fins of rectangular profile

$m = \sqrt{4h/kD}$
$L_c = L + D/4$
$A_{\text{fin}} = \pi DL_c$

$$\eta_{\text{fin}} = \frac{\tanh mL_c}{mL_c}$$

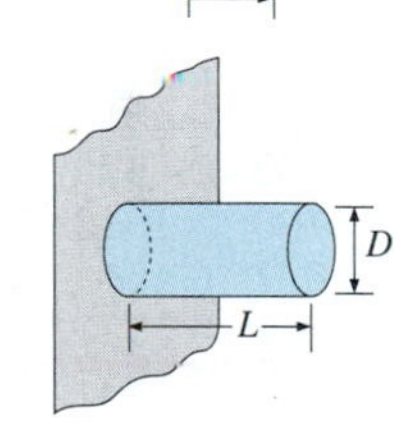

Pin fins of triangular profile

$m = \sqrt{4h/kD}$

$A_{\text{fin}} = \frac{\pi D}{2}\sqrt{L^2 + (D/2)^2}$

$$\eta_{\text{fin}} = \frac{2}{mL}\frac{I_2(2mL)}{I_1(2mL)}$$

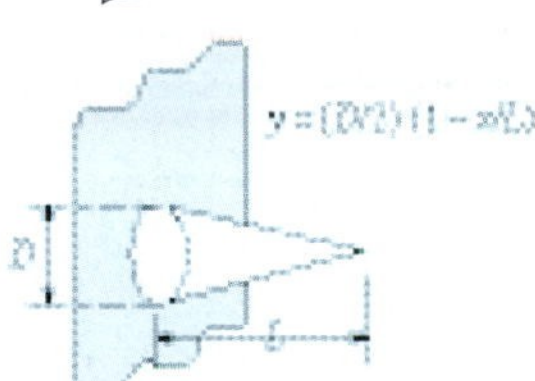

Pin fins of parabolic profile

$m = \sqrt{4h/kD}$

$A_{\text{fin}} = \frac{\pi L^3}{8D}[C_3C_4 - \frac{L}{2D}\ln(2DC_4/L + C_3)]$

$C_3 = 1 + 2(D/L)^2$
$C_4 = \sqrt{1 + (D/L)^2}$

$$\eta_{\text{fin}} = \frac{2}{1 + \sqrt{(2mL/3)^2 + 1}}$$

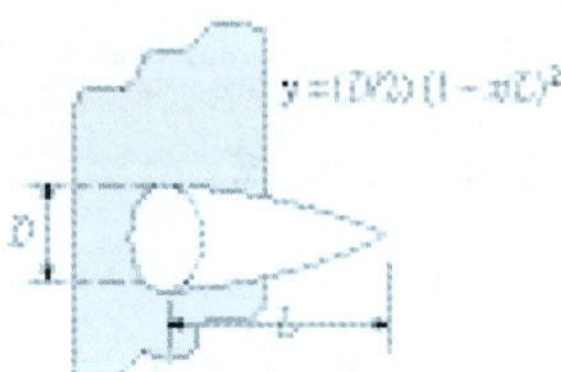

Pin fins of parabolic profile (blunt tip)

$m = \sqrt{4h/kD}$

$A_{\text{fin}} = \frac{\pi D^4}{96L^2}\left\{[16(L/D)^2 + 1]^{3/2} - 1\right\}$

$$\eta_{\text{fin}} = \frac{3}{2mL}\frac{I_1(4mL/3)}{I_0(4mL/3)}$$

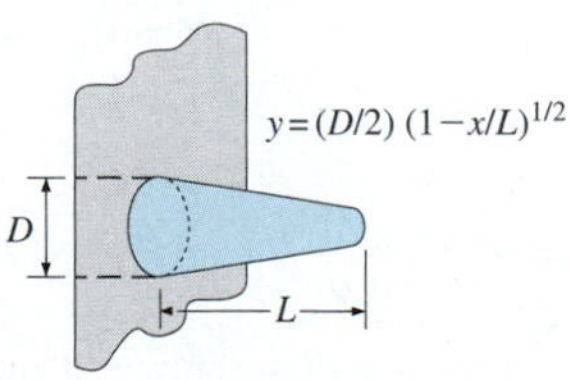

TABLE 17–4

Modified Bessel functions of the first and second kinds*

x	$e^{-x}I_0(x)$	$e^{-x}I_1(x)$	$e^{x}K_0(x)$	$e^{x}K_1(x)$
0.0	1.0000	0.0000	—	—
0.2	0.8269	0.0823	2.1408	5.8334
0.4	0.6974	0.1368	1.6627	3.2587
0.6	0.5993	0.1722	1.4167	2.3739
0.8	0.5241	0.1945	1.2582	1.9179
1.0	0.4658	0.2079	1.1445	1.6362
1.2	0.4198	0.2153	1.0575	1.4429
1.4	0.3831	0.2185	0.9881	1.3011
1.6	0.3533	0.2190	0.9309	1.1919
1.8	0.3289	0.2177	0.8828	1.1048
2.0	0.3085	0.2153	0.8416	1.0335
2.2	0.2913	0.2121	0.8057	0.9738
2.4	0.2766	0.2085	0.7740	0.9229
2.6	0.2639	0.2047	0.7459	0.8790
2.8	0.2528	0.2007	0.7206	0.8405
3.0	0.2430	0.1968	0.6978	0.8066
3.2	0.2343	0.1930	0.6770	0.7763
3.4	0.2264	0.1892	0.6580	0.7491
3.6	0.2193	0.1856	0.6405	0.7245
3.8	0.2129	0.1821	0.6243	0.7021
4.0	0.2070	0.1788	0.6093	0.6816
4.2	0.2016	0.1755	0.5953	0.6627
4.4	0.1966	0.1725	0.5823	0.6454
4.6	0.1919	0.1695	0.5701	0.6292
4.8	0.1876	0.1667	0.5586	0.6143
5.0	0.1835	0.1640	0.5478	0.6003
5.2	0.1797	0.1614	0.5376	0.5872
5.4	0.1762	0.1589	0.5280	0.5749
5.6	0.1728	0.1565	0.5188	0.5634
5.8	0.1697	0.1542	0.5101	0.5525
6.0	0.1667	0.1521	0.5019	0.5422
6.5	0.1598	0.1469	0.4828	0.5187
7.0	0.1537	0.1423	0.4658	0.4981
7.5	0.1483	0.1380	0.4505	0.4797
8.0	0.1434	0.1341	0.4366	0.4631
8.5	0.1390	0.1305	0.4239	0.4482
9.0	0.1350	0.1272	0.4123	0.4346
9.5	0.1313	0.1241	0.4016	0.4222
10.0	0.1278	0.1213	0.3916	0.4108

*Evaluated from EES using the mathematical functions Bessel_I(x) and Bessel_K(x)

or

$$\dot{Q}_{\text{fin}} = \eta_{\text{fin}}\,\dot{Q}_{\text{fin, max}} = \eta_{\text{fin}}\,hA_{\text{fin}}\,(T_b - T_\infty) \quad \text{(17–69)}$$

where A_{fin} is the total surface area of the fin. This relation enables us to determine the heat transfer from a fin when its efficiency is known. For the cases of constant cross section of *very long fins* and *fins with adiabatic tips*, the fin efficiency can be expressed as

$$\eta_{\text{long fin}} = \frac{\dot{Q}_{\text{fin}}}{\dot{Q}_{\text{fin, max}}} = \frac{\sqrt{hpkA_c}\,(T_b - T_\infty)}{hA_{\text{fin}}\,(T_b - T_\infty)} = \frac{1}{L}\sqrt{\frac{kA_c}{hp}} = \frac{1}{mL} \quad \text{(17–70)}$$

and

$$\eta_{\text{adiabatic tip}} = \frac{\dot{Q}_{\text{fin}}}{\dot{Q}_{\text{fin, max}}} = \frac{\sqrt{hpkA_c}\,(T_b - T_\infty)\tanh aL}{hA_{\text{fin}}\,(T_b - T_\infty)} = \frac{\tanh mL}{mL} \quad \text{(17–71)}$$

since $A_{\text{fin}} = pL$ for fins with constant cross section. Equation 17–71 can also be used for fins subjected to convection provided that the fin length L is replaced by the corrected length L_c.

Fin efficiency relations are developed for fins of various profiles, listed in Table 17–3. The mathematical functions I and K that appear in some of these relations are the *modified Bessel functions*, and their values are given in Table 17–4. Efficiencies are plotted in Fig. 17–42 for fins on a *plain surface* and in Fig. 17–43 for *circular fins* of constant thickness. For most fins of constant thickness encountered in practice, the fin thickness t is too small relative to the fin length L, and thus the fin tip area is negligible.

Note that fins with triangular and parabolic profiles contain less material and are more efficient than the ones with rectangular profiles, and thus are more suitable for applications requiring minimum weight such as space applications.

An important consideration in the design of finned surfaces is the selection of the proper *fin length L*. Normally the *longer* the fin, the *larger* the heat transfer area and thus the *higher* the rate of heat transfer from the fin. But also the larger the fin, the bigger the mass, the higher the price, and the larger the fluid friction. Therefore, increasing the length of the fin beyond a certain value cannot be justified unless the added benefits outweigh the added cost. Also, the fin efficiency decreases with increasing fin length because of the decrease in fin temperature with length. Fin lengths that cause the fin efficiency to drop below 60 percent usually cannot be justified economically and should be avoided. The efficiency of most fins used in practice is above 90 percent.

Fin Effectiveness

Fins are used to *enhance* heat transfer, and the use of fins on a surface cannot be recommended unless the enhancement in heat transfer justifies the added cost and complexity associated with the fins. In fact, there is no assurance that adding fins on a surface will *enhance* heat transfer. The performance of the fins is judged on the basis of the enhancement in heat trans-

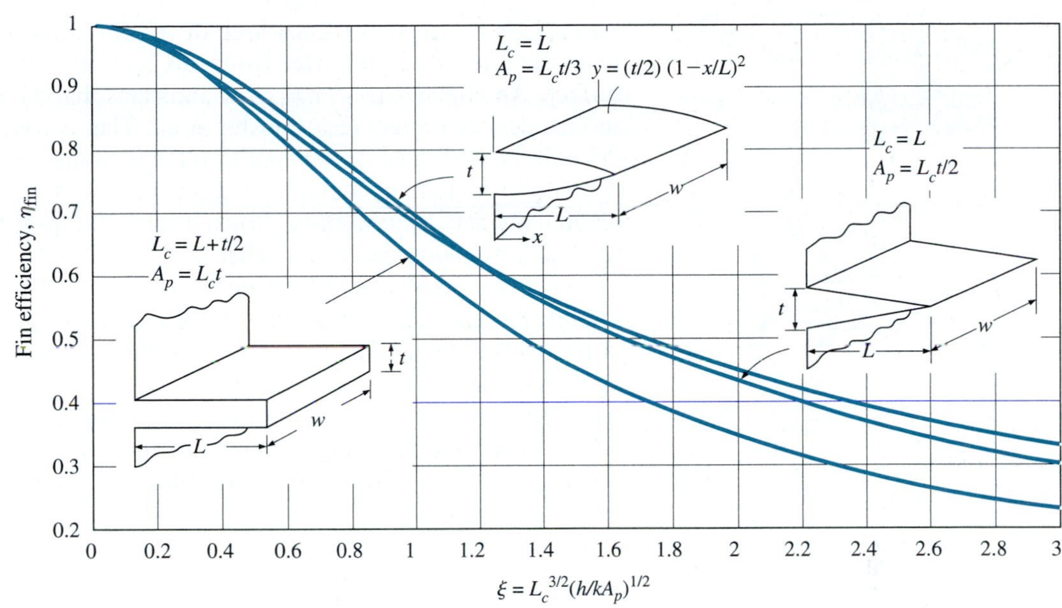

FIGURE 17–42
Efficiency of straight fins of rectangular, triangular, and parabolic profiles.

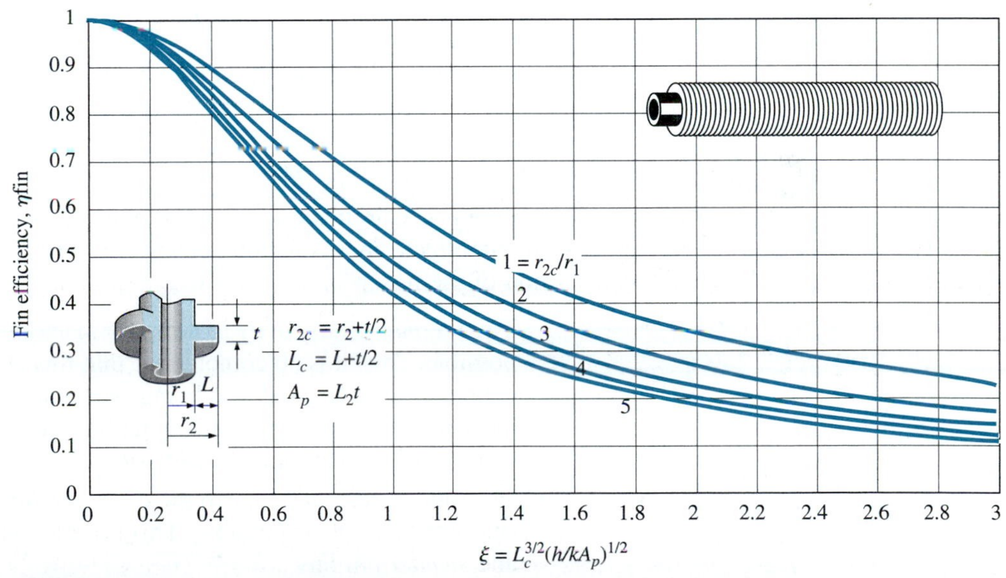

FIGURE 17–43
Efficiency of annular fins of constant thickness t.

fer relative to the no-fin case. The performance of fins is expressed in terms of the *fin effectiveness* ε_{fin} defined as (Fig. 17–44)

$$\varepsilon_{fin} = \frac{\dot{Q}_{fin}}{\dot{Q}_{no\ fin}} = \frac{\dot{Q}_{fin}}{hA_b(T_b - T_\infty)} = \frac{\text{Heat transfer rate from the fin of } \textit{base area } A_b}{\text{Heat transfer rate from the surface of } \textit{area } A_b} \quad (17\text{–}72)$$

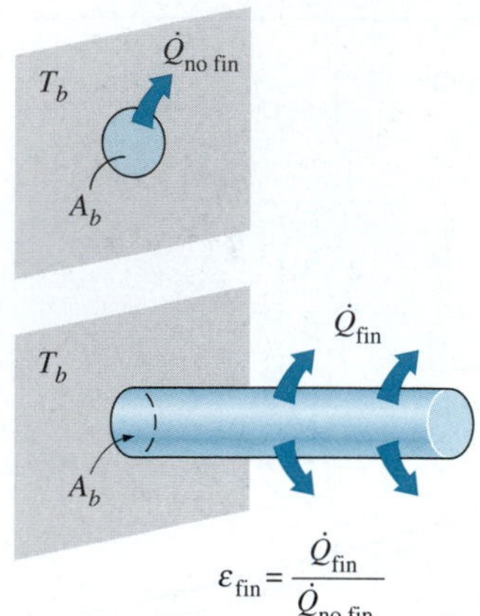

FIGURE 17–44
The effectiveness of a fin.

Here, A_b is the cross-sectional area of the fin at the base and $\dot{Q}_{\text{no fin}}$ represents the rate of heat transfer from this area if no fins are attached to the surface. An effectiveness of $\varepsilon_{\text{fin}} = 1$ indicates that the addition of fins to the surface does not affect heat transfer at all. That is, heat conducted to the fin through the base area A_b is equal to the heat transferred from the same area A_b to the surrounding medium. An effectiveness of $\varepsilon_{\text{fin}} < 1$ indicates that the fin actually acts as *insulation*, slowing down the heat transfer from the surface. This situation can occur when fins made of low thermal conductivity materials are used. An effectiveness of $\varepsilon_{\text{fin}} > 1$ indicates that fins are *enhancing* heat transfer from the surface, as they should. However, the use of fins cannot be justified unless ε_{fin} is sufficiently larger than 1. Finned surfaces are designed on the basis of *maximizing* effectiveness for a specified cost or *minimizing* cost for a desired effectiveness.

Note that both the fin efficiency and fin effectiveness are related to the performance of the fin, but they are different quantities. However, they are related to each other by

$$\varepsilon_{\text{fin}} = \frac{\dot{Q}_{\text{fin}}}{\dot{Q}_{\text{no fin}}} = \frac{\dot{Q}_{\text{fin}}}{hA_b\,(T_b - T_\infty)} = \frac{\eta_{\text{fin}}\,hA_{\text{fin}}\,(T_b - T_\infty)}{hA_b\,(T_b - T_\infty)} = \frac{A_{\text{fin}}}{A_b}\,\eta_{\text{fin}} \qquad \text{(17–73)}$$

Therefore, the fin effectiveness can be determined easily when the fin efficiency is known, or vice versa.

The rate of heat transfer from a sufficiently *long* fin of *uniform* cross section under steady conditions is given by Eq. 17–61. Substituting this relation into Eq. 17–72, the effectiveness of such a long fin is determined to be

$$\varepsilon_{\text{long fin}} = \frac{\dot{Q}_{\text{fin}}}{\dot{Q}_{\text{no fin}}} = \frac{\sqrt{hpkA_c}\,(T_b - T_\infty)}{hA_b\,(T_b - T_\infty)} = \sqrt{\frac{kp}{hA_c}} \qquad \text{(17–74)}$$

since $A_c = A_b$ in this case. We can draw several important conclusions from the fin effectiveness relation above for consideration in the design and selection of the fins:

- The *thermal conductivity k* of the fin material should be as high as possible. Thus it is no coincidence that fins are made from metals, with copper, aluminum, and iron being the most common ones. Perhaps the most widely used fins are made of aluminum because of its low cost and weight and its resistance to corrosion.
- The ratio of the *perimeter* to the *cross-sectional area* of the fin p/A_c should be as high as possible. This criterion is satisfied by *thin* plate fins and *slender* pin fins.
- The use of fins is *most effective* in applications involving a *low convection heat transfer coefficient.* Thus, the use of fins is more easily justified when the medium is a *gas* instead of a liquid and the heat transfer is by *natural convection* instead of by forced convection. Therefore, it is no coincidence that in liquid-to-gas heat exchangers such as the car radiator, fins are placed on the *gas* side.

When determining the rate of heat transfer from a finned surface, we must consider the *unfinned portion* of the surface as well as the *fins.* Therefore, the rate of heat transfer for a surface containing *n* fins can be expressed as

$$\begin{aligned}\dot{Q}_{\text{total, fin}} &= \dot{Q}_{\text{unfin}} + \dot{Q}_{\text{fin}}\\ &= hA_{\text{unfin}}(T_b - T_\infty) + \eta_{\text{fin}} hA_{\text{fin}}(T_b - T_\infty)\\ &= h(A_{\text{unfin}} + \eta_{\text{fin}} A_{\text{fin}})(T_b - T_\infty) \qquad \text{(17–75)}\end{aligned}$$

We can also define an **overall effectiveness** for a finned surface as the ratio of the total heat transfer from the finned surface to the heat transfer from the same surface if there were no fins,

$$\varepsilon_{\text{fin, overall}} = \frac{\dot{Q}_{\text{total, fin}}}{\dot{Q}_{\text{total, no fin}}} = \frac{h(A_{\text{unfin}} + \eta_{\text{fin}} A_{\text{fin}})(T_b - T_\infty)}{hA_{\text{no fin}}(T_b - T_\infty)} \qquad \text{(17–76)}$$

where $A_{\text{no fin}}$ is the area of the surface when there are no fins, A_{fin} is the total surface area of all the fins on the surface, and A_{unfin} is the area of the unfinned portion of the surface (Fig. 17–45). Note that the overall fin effectiveness depends on the fin density (number of fins per unit length) as well as the effectiveness of the individual fins. The overall effectiveness is a better measure of the performance of a finned surface than the effectiveness of the individual fins.

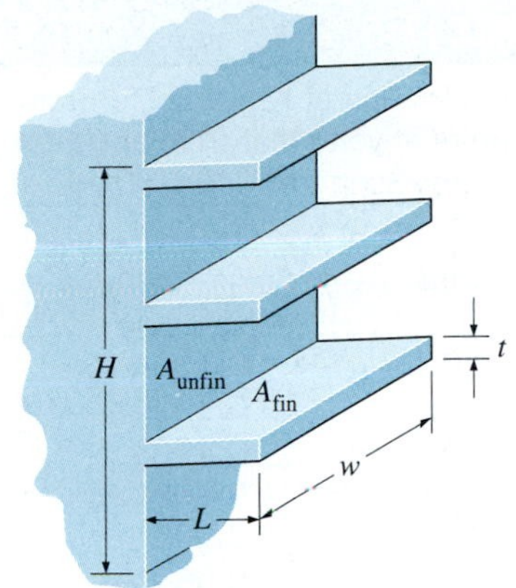

$A_{\text{no fin}} = w \times H$
$A_{\text{unfin}} = w \times H - 3 \times (t \times w)$
$A_{\text{fin}} = 2 \times L \times w + t \times w$
$\cong 2 \times L \times w$ (one fin)

FIGURE 17–45
Various surface areas associated with a rectangular surface with three fins.

Proper Length of a Fin

An important step in the design of a fin is the determination of the appropriate length of the fin once the fin material and the fin cross section are specified. You may be tempted to think that the longer the fin, the larger the surface area and thus the higher the rate of heat transfer. Therefore, for maximum heat transfer, the fin should be infinitely long. However, the temperature drops along the fin exponentially and reaches the environment temperature at some length. The part of the fin beyond this length does not contribute to heat transfer since it is at the temperature of the environment, as shown in Fig. 17–46. Therefore, designing such an "extra long" fin is out of the question since it results in material waste, excessive weight, and increased size and thus increased cost with no benefit in return (in fact, such a long fin will hurt performance since it will suppress fluid motion and thus reduce the convection heat transfer coefficient). Fins that are so long that the temperature approaches the environment temperature cannot be recommended either since the little increase in heat transfer at the tip region cannot justify the disproportionate increase in the weight and cost.

To get a sense of the proper length of a fin, we compare heat transfer from a fin of finite length to heat transfer from an infinitely long fin under the same conditions. The ratio of these two heat transfers is

Heat transfer ratio:
$$\frac{\dot{Q}_{\text{fin}}}{\dot{Q}_{\text{long fin}}} = \frac{\sqrt{hpkA_c}\,(T_b - T_\infty)\tanh mL}{\sqrt{hpkA_c}\,(T_b - T_\infty)} = \tanh mL \qquad \text{(17–77)}$$

Using a hand calculator, the values of tanh mL are evaluated for some values of mL and the results are given in Table 17–5. We observe from the table that heat transfer from a fin increases with mL almost linearly at first, but the curve reaches a plateau later and reaches a value for the infinitely long fin at about $mL = 5$. Therefore, a fin whose length is $L = \frac{1}{5}m$ can be considered to be an infinitely long fin. We also observe that reducing the fin length by half in that case (from $mL = 5$ to $mL = 2.5$) causes a drop of just 1 percent in heat

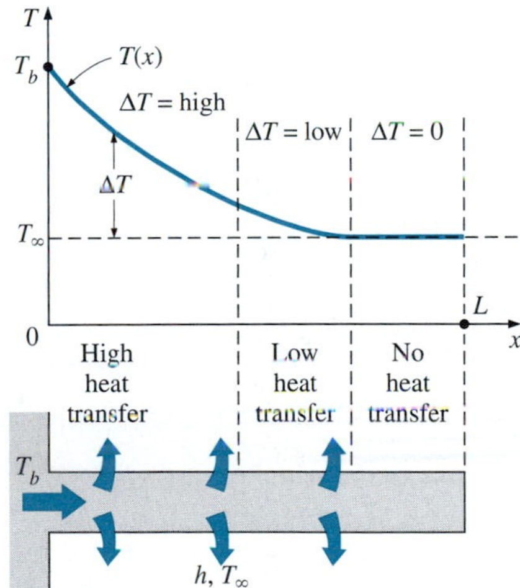

FIGURE 17–46
Because of the gradual temperature drop along the fin, the region near the fin tip makes little or no contribution to heat transfer.

transfer. We certainly would not hesitate sacrificing 1 percent in heat transfer performance in return for 50 percent reduction in the size and possibly the cost of the fin. In practice, a fin length that corresponds to about $mL = 1$ will transfer 76.2 percent of the heat that can be transferred by an infinitely long fin, and thus it should offer a good compromise between heat transfer performance and the fin size.

TABLE 17–5

The variation of heat transfer from a fin relative to that from an infinitely long fin

mL	$\dfrac{\dot{Q}_{\text{fin}}}{\dot{Q}_{\text{long fin}}} = \tanh mL$
0.1	0.100
0.2	0.197
0.5	0.462
1.0	0.762
1.5	0.905
2.0	0.964
2.5	0.987
3.0	0.995
4.0	0.999
5.0	1.000

TABLE 17–6

Combined natural convection and radiation thermal resistance of various heat sinks used in the cooling of electronic devices between the heat sink and the surroundings. All fins are made of aluminum 6063T-5, are black anodized, and are 76 mm (3 in) long.

HS 5030

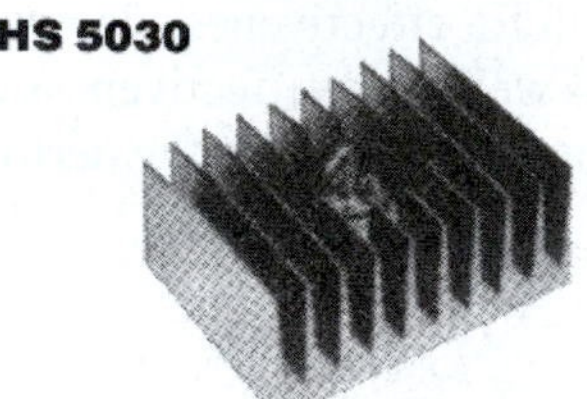

$R = 0.9$°C/W (vertical)
$R = 1.2$°C/W (horizontal)

Dimensions: 76 mm × 105 mm × 44 mm
Surface area: 677 cm^2

HS 6065

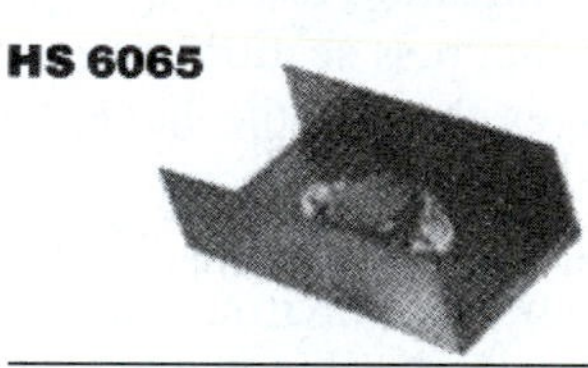

$R = 5$°C/W

Dimensions: 76 mm × 38 mm × 24 mm
Surface area: 387 cm^2

HS 6071

$R = 1.4$°C/W (vertical)
$R = 1.8$°C/W (horizontal)

Dimensions: 76 mm × 92 mm × 26 mm
Surface area: 968 cm^2

HS 6105

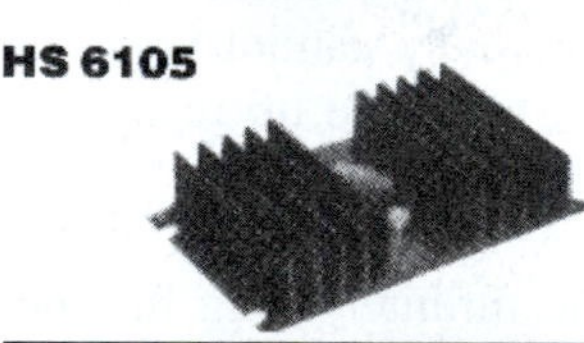

$R = 1.8$°C/W (vertical)
$R = 2.1$°C/W (horizontal)

Dimensions: 76 mm × 127 mm × 91 mm
Surface area: 677 cm^2

HS 6115

$R = 1.1$°C/W (vertical)
$R = 1.3$°C/W (horizontal)

Dimensions: 76 mm × 102 mm × 25 mm
Surface area: 929 cm^2

HS 7030

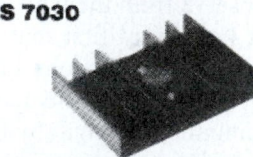

$R = 2.9$°C/W (vertical)
$R = 3.1$°C/W (horizontal)

Dimensions: 76 mm × 97 mm × 19 mm
Surface area: 290 cm^2

A common approximation used in the analysis of fins is to assume the fin temperature to vary in one direction only (along the fin length) and the temperature variation along other directions is negligible. Perhaps you are wondering if this one-dimensional approximation is a reasonable one. This is certainly the case for fins made of thin metal sheets such as the fins on a car radiator, but we wouldn't be so sure for fins made of thick materials. Studies have shown that the error involved in one-dimensional fin analysis is negligible (less than about 1 percent) when

$$\frac{h\delta}{k} < 0.2$$

where δ is the characteristic thickness of the fin, which is taken to be the plate thickness t for rectangular fins and the diameter D for cylindrical ones.

Specially designed finned surfaces called *heat sinks,* which are commonly used in the cooling of electronic equipment, involve one-of-a-kind complex geometries, as shown in Table 17–6. The heat transfer performance of heat sinks is usually expressed in terms of their *thermal resistances* R in °C/W, which is defined as

$$\dot{Q}_{\text{fin}} = \frac{T_b - T_\infty}{R} = hA_{\text{fin}}\,\eta_{\text{fin}}\,(T_b - T_\infty) \tag{17–78}$$

A small value of thermal resistance indicates a small temperature drop across the heat sink, and thus a high fin efficiency.

EXAMPLE 17–10 Maximum Power Dissipation of a Transistor

Power transistors that are commonly used in electronic devices consume large amounts of electric power. The failure rate of electronic components increases almost exponentially with operating temperature. As a rule of thumb, the failure rate of electronic components is halved for each 10°C reduction in the junction operating temperature. Therefore, the operating temperature of electronic components is kept below a safe level to minimize the risk of failure.

The sensitive electronic circuitry of a power transistor at the junction is protected by its case, which is a rigid metal enclosure. Heat transfer characteristics of a power transistor are usually specified by the manufacturer in terms of the case-to-ambient thermal resistance, which accounts for both the natural convection and radiation heat transfers.

The case-to-ambient thermal resistance of a power transistor that has a maximum power rating of 10 W is given to be 20°C/W. If the case temperature of the transistor is not to exceed 85°C, determine the power at which this transistor can be operated safely in an environment at 25°C.

Solution The maximum power rating of a transistor whose case temperature is not to exceed 85°C is to be determined.

Assumptions **1** Steady operating conditions exist. **2** The transistor case is isothermal at 85°C.

Properties The case-to-ambient thermal resistance is given to be 20°C/W.

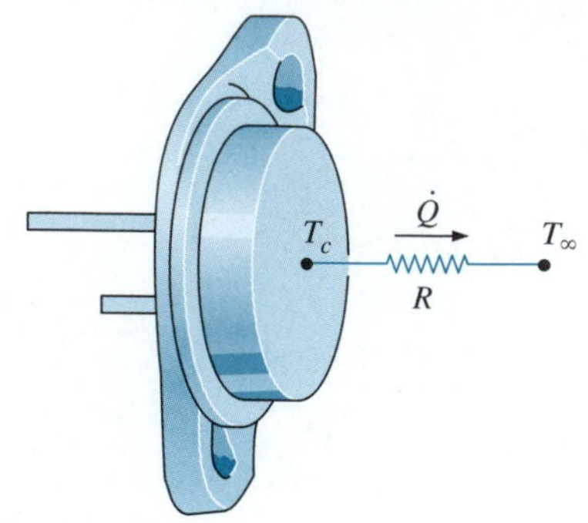

FIGURE 17–47
Schematic for Example 17–10.

Analysis The power transistor and the thermal resistance network associated with it are shown in Fig. 17–47. We notice from the thermal resistance network that there is a single resistance of 20°C/W between the case at T_c = 85°C and the ambient at T_∞ = 25°C, and thus the rate of heat transfer is

$$\dot{Q} = \left(\frac{\Delta T}{R}\right)_{\text{case-ambient}} = \frac{T_c - T_\infty}{R_{\text{case-ambient}}} = \frac{(85 - 25)°\text{C}}{20°\text{C/W}} = \mathbf{3\ W}$$

Therefore, this power transistor should not be operated at power levels above 3 W if its case temperature is not to exceed 85°C.

Discussion This transistor can be used at higher power levels by attaching it to a heat sink (which lowers the thermal resistance by increasing the heat transfer surface area, as discussed in the next example) or by using a fan (which lowers the thermal resistance by increasing the convection heat transfer coefficient).

EXAMPLE 17–11 Selecting a Heat Sink for a Transistor

A 60-W power transistor is to be cooled by attaching it to one of the commercially available heat sinks shown in Table 17–6. Select a heat sink that will allow the case temperature of the transistor not to exceed 90°C in the ambient air at 30°C.

Solution A commercially available heat sink from Table 17–6 is to be selected to keep the case temperature of a transistor below 90°C.

Assumptions **1** Steady operating conditions exist. **2** The transistor case is isothermal at 90°C. **3** The contact resistance between the transistor and the heat sink is negligible.

Analysis The rate of heat transfer from a 60-W transistor at full power is $\dot{Q}$ = 60 W. The thermal resistance between the transistor attached to the heat sink and the ambient air for the specified temperature difference is determined to be

$$\dot{Q} = \frac{\Delta T}{R} \longrightarrow R = \frac{\Delta T}{\dot{Q}} = \frac{(90 - 30)°\text{C}}{60\ \text{W}} = 1.0°\text{C/W}$$

Therefore, the thermal resistance of the heat sink should be below 1.0°C/W. An examination of Table 17–6 reveals that the HS 5030, whose thermal resistance is 0.9°C/W in the vertical position, is the only heat sink that will meet this requirement.

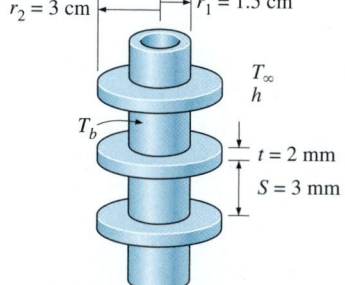

r_2 = 3 cm
r_1 = 1.5 cm
T_∞
h
T_b
t = 2 mm
S = 3 mm

FIGURE 17–48
Schematic for Example 17–12.

EXAMPLE 17–12 Effect of Fins on Heat Transfer from Steam Pipes

Steam in a heating system flows through tubes whose outer diameter is D_1 = 3 cm and whose walls are maintained at a temperature of 120°C. Circular aluminum alloy fins (k = 180 W/m · °C) of outer diameter D_2 = 6 cm and constant thickness t = 2 mm are attached to the tube, as shown in Fig. 17–48. The space between the fins is 3 mm, and thus there are 200 fins per meter length of the tube. Heat is transferred to the surrounding air at T_∞ = 25°C, with a combined heat transfer coefficient of $h = 60\ \text{W/m}^2 \cdot {}^\circ\text{C}$. Determine the increase in heat transfer from the tube per meter of its length as a result of adding fins.

Solution Circular aluminum alloy fins are to be attached to the tubes of a heating system. The increase in heat transfer from the tubes per unit length as a result of adding fins is to be determined.

Assumptions **1** Steady operating conditions exist. **2** The heat transfer coefficient is uniform over the entire fin surfaces. **3** Thermal conductivity is constant. **4** Heat transfer by radiation is negligible.

Properties The thermal conductivity of the fins is given to be $k = 180\ \text{W/m} \cdot {}^\circ\text{C}$.

Analysis In the case of no fins, heat transfer from the tube per meter of its length is determined from Newton's law of cooling to be

$$A_{\text{no fin}} = \pi D_1 L = \pi(0.03\ \text{m})(1\ \text{m}) = 0.0942\ \text{m}^2$$

$$\begin{aligned}\dot{Q}_{\text{no fin}} &= hA_{\text{no fin}}(T_b - T_\infty)\\ &= (60\ \text{W/m}^2 \cdot {}^\circ\text{C})(0.0942\ \text{m}^2)(120 - 25)^\circ\text{C}\\ &= 537\ \text{W}\end{aligned}$$

The efficiency of the circular fins attached to a circular tube is plotted in Fig. 17–43. Noting that $L = \frac{1}{2}(D_2 - D_1) = \frac{1}{2}(0.06 - 0.03) = 0.015$ m in this case, we have

$$r_{2c} = r_2 + t/2 = 0.03 + 0.002/2 = 0.031\ \text{m}$$

$$L_c = L + t/2 = 0.015 + 0.002/2 = 0.016\ \text{m}$$

$$A_p = L_c t = (0.016\ \text{m})(0.002\ \text{m}) = 3.20 \times 10^{-5}\ \text{m}^2$$

$$\frac{r_{2c}}{r_1} = \frac{0.031\ \text{m}}{0.015\ \text{m}} = 2.07$$

$$L_c^{3/2}\sqrt{\frac{h}{kA_p}} = (0.016\ \text{m})^{3/2}\sqrt{\frac{60\ \text{W/m}^2 \cdot {}^\circ\text{C}}{(180\ \text{W/m} \cdot {}^\circ\text{C})(3.20 \times 10^{-5}\ \text{m}^2)}} = 0.207$$

$$\eta_{\text{fin}} = 0.96$$

$$\begin{aligned}A_{\text{fin}} &= 2\pi(r_{2c}^2 - r_1^2) = 2\pi[(0.031\ \text{m})^2 - (0.015\ \text{m})^2]\\ &= 0.004624\ \text{m}^2\end{aligned}$$

$$\begin{aligned}\dot{Q}_{\text{fin}} &= \eta_{\text{fin}}\dot{Q}_{\text{fin, max}} = \eta_{\text{fin}} hA_{\text{fin}}(T_b - T_\infty)\\ &= 0.96(60\ \text{W/m}^2 \cdot {}^\circ\text{C})(0.004624\ \text{m}^2)(120 - 25)^\circ\text{C}\\ &= 25.3\ \text{W}\end{aligned}$$

Heat transfer from the unfinned portion of the tube is

$$A_{\text{unfin}} = \pi D_1 S = \pi(0.03\ \text{m})(0.003\ \text{m}) = 0.000283\ \text{m}^2$$

$$\begin{aligned}\dot{Q}_{\text{unfin}} &= hA_{\text{unfin}}(T_b - T_\infty)\\ &= (60\ \text{W/m}^2 \cdot {}^\circ\text{C})(0.000283\ \text{m}^2)(120 - 25)^\circ\text{C}\\ &= 1.6\ \text{W}\end{aligned}$$

Noting that there are 200 fins and thus 200 interfin spacings per meter length of the tube, the total heat transfer from the finned tube becomes

$$\dot{Q}_{\text{total, fin}} = n(\dot{Q}_{\text{fin}} + \dot{Q}_{\text{unfin}}) = 200(25.3 + 1.6)\text{ W} = 5380\text{ W}$$

Therefore, the increase in heat transfer from the tube per meter of its length as a result of the addition of fins is

$$\dot{Q}_{\text{increase}} = \dot{Q}_{\text{total, fin}} - \dot{Q}_{\text{no fin}} = 5380 - 537 = \mathbf{4843\text{ W}} \qquad \text{(per m tube length)}$$

Discussion The overall effectiveness of the finned tube is

$$\varepsilon_{\text{fin, overall}} = \frac{\dot{Q}_{\text{total, fin}}}{\dot{Q}_{\text{total, no fin}}} = \frac{5380\text{ W}}{537\text{ W}} = 10.0$$

That is, the rate of heat transfer from the steam tube increases by a factor of 10 as a result of adding fins. This explains the widespread use of finned surfaces.

17–7 ▪ HEAT TRANSFER IN COMMON CONFIGURATIONS

So far, we have considered heat transfer in *simple* geometries such as large plane walls, long cylinders, and spheres. This is because heat transfer in such geometries can be approximated as *one-dimensional*, and simple analytical solutions can be obtained easily. But many problems encountered in practice are two- or three-dimensional and involve rather complicated geometries for which no simple solutions are available.

An important class of heat transfer problems for which simple solutions are obtained encompasses those involving two surfaces maintained at *constant* temperatures T_1 and T_2. The steady rate of heat transfer between these two surfaces is expressed as

$$Q = Sk(T_1 - T_2) \tag{17–79}$$

where S is the **conduction shape factor**, which has the dimension of *length*, and k is the thermal conductivity of the medium between the surfaces. The conduction shape factor depends on the *geometry* of the system only.

Conduction shape factors have been determined for a number of configurations encountered in practice and are given in Table 17–7 for some common cases. More comprehensive tables are available in the literature. Once the value of the shape factor is known for a specific geometry, the total steady heat transfer rate can be determined from the equation above using the specified two constant temperatures of the two surfaces and the thermal conductivity of the medium between them. Note that conduction shape factors are applicable only when heat transfer between the two surfaces is by *conduction*. Therefore, they cannot be used when the medium between the surfaces is a liquid or gas, which involves natural or forced convection currents.

A comparison of Eqs. 17–4 and 17–79 reveals that the conduction shape factor S is related to the thermal resistance R by $R = 1/kS$ or $S = 1/kR$. Thus, these two quantities are the inverse of each other when the thermal conductivity of the medium is unity. The use of the conduction shape factors is illustrated with Examples 17–13 and 17–14.

TABLE 17–7

Conduction shape factors S for several configurations for use in $\dot{Q} = kS(T_1 - T_2)$ to determine the steady rate of heat transfer through a medium of thermal conductivity k between the surfaces at temperatures T_1 and T_2

(1) Isothermal cylinder of length L buried in a semi-infinite medium ($L >> D$ and $z > 1.5D$)

$$S = \frac{2\pi L}{\ln(4z/D)}$$

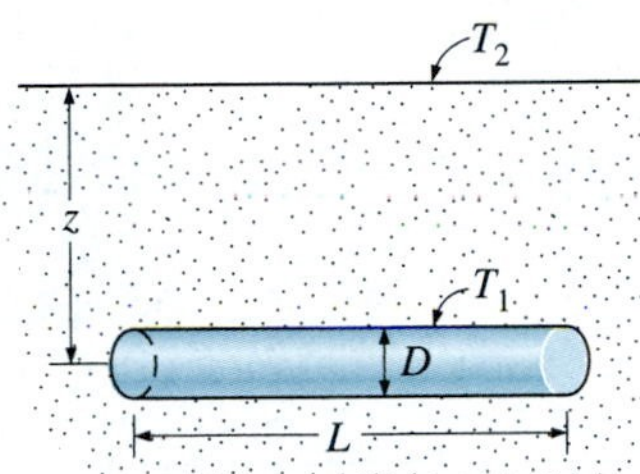

(2) Vertical isothermal cylinder of length L buried in a semi-infinite medium ($L >> D$)

$$S = \frac{2\pi L}{\ln(4L/D)}$$

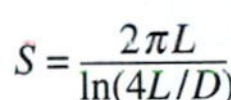

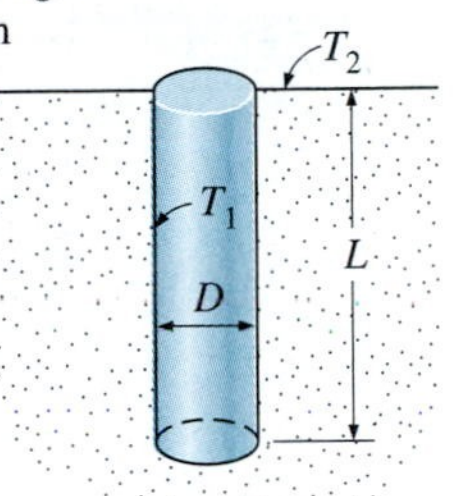

(3) Two parallel isothermal cylinders placed in an infinite medium ($L >> D_1, D_2, z$)

$$S = \frac{2\pi L}{\cosh^{-1}\left(\frac{4z^2 - D_1^2 - D_2^2}{2D_1D_2}\right)}$$

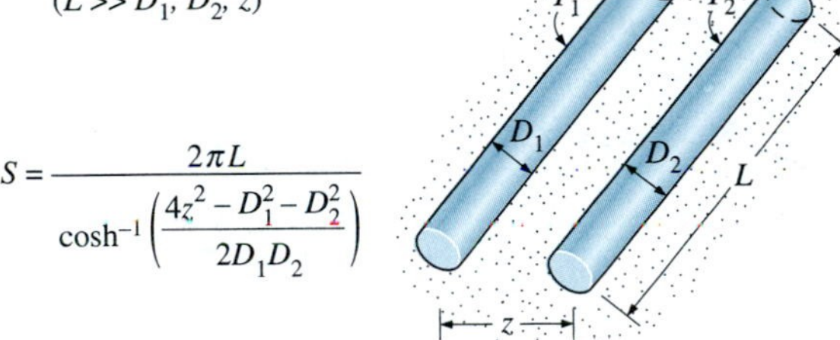

(4) A row of equally spaced parallel isothermal cylinders buried in a semi-infinite medium ($L >> D$, z, and $w > 1.5D$)

$$S = \frac{2\pi L}{\ln\left(\frac{2w}{\pi D}\sinh\frac{2\pi z}{w}\right)}$$

(per cylinder)

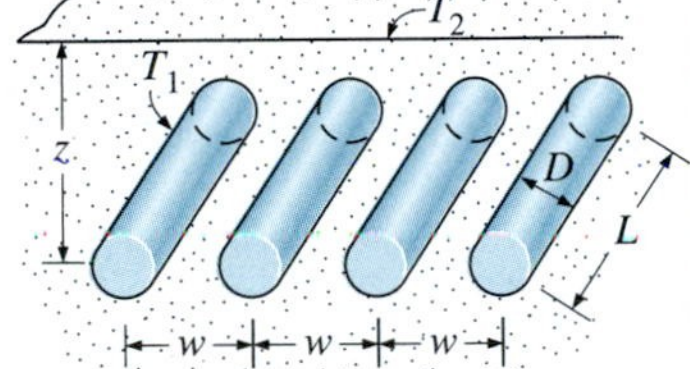

(5) Circular isothermal cylinder of length L in the midplane of an infinite wall ($z > 0.5D$)

$$S = \frac{2\pi L}{\ln(8z/\pi D)}$$

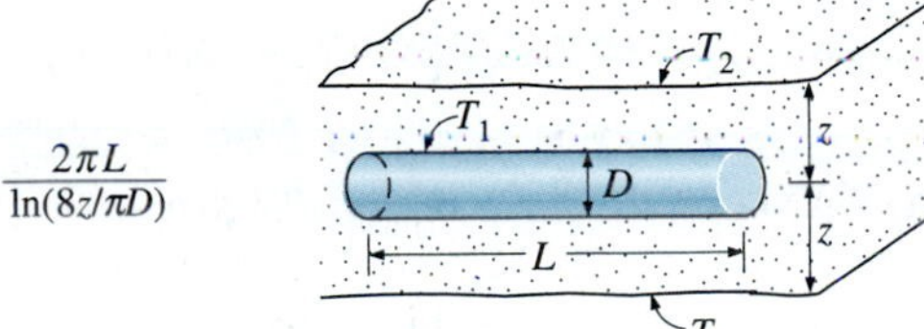

(6) Circular isothermal cylinder of length L at the center of a square solid bar of the same length

$$S = \frac{2\pi L}{\ln(1.08w/D)}$$

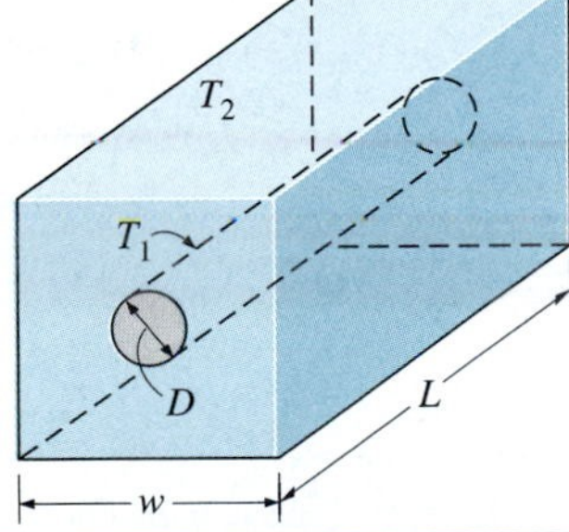

(7) Eccentric circular isothermal cylinder of length L in a cylinder of the same length ($L > D_2$)

$$S = \frac{2\pi L}{\cosh^{-1}\left(\frac{D_1^2 + D_2^2 - 4z^2}{2D_1D_2}\right)}$$

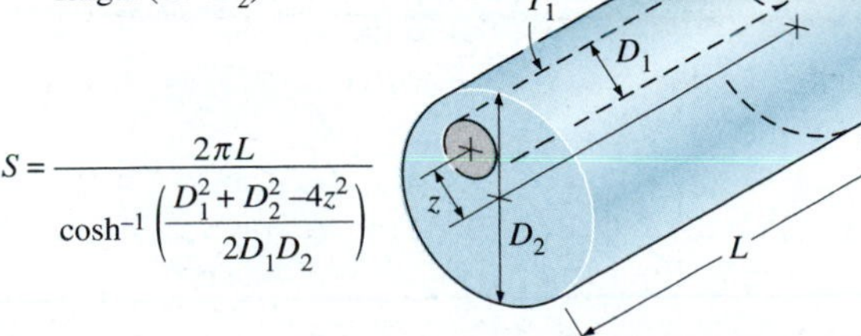

(8) Large plane wall

$$S = \frac{A}{L}$$

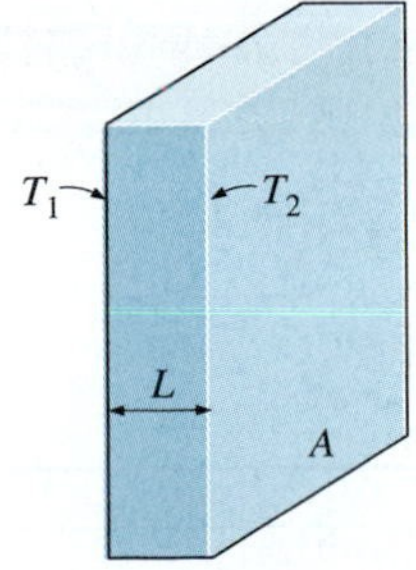

(continued)

TABLE 17–7 (*Continued*)

(9) A long cylindrical layer

$$S = \frac{2\pi L}{\ln (D_2/D_1)}$$

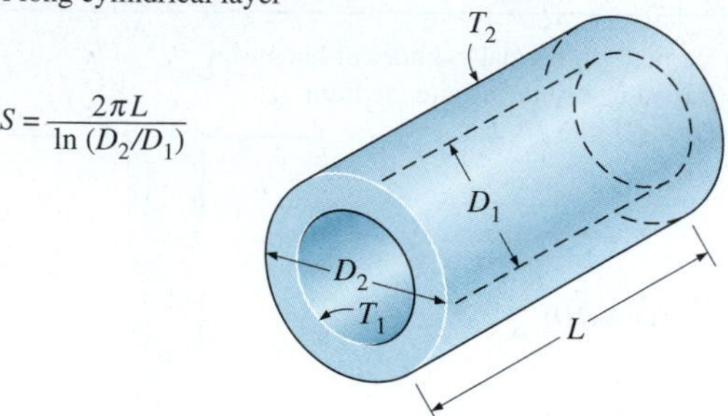

(10) A square flow passage

(*a*) For $a/b > 1.4$,

$$S = \frac{2\pi L}{0.93 \ln (0.948a/b)}$$

(*b*) For $a/b < 1.41$,

$$S = \frac{2\pi L}{0.785 \ln (a/b)}$$

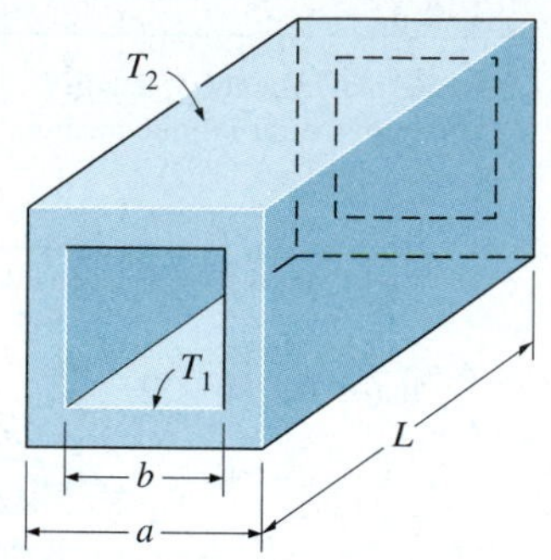

(11) A spherical layer

$$S = \frac{2\pi D_1 D_2}{D_2 - D_1}$$

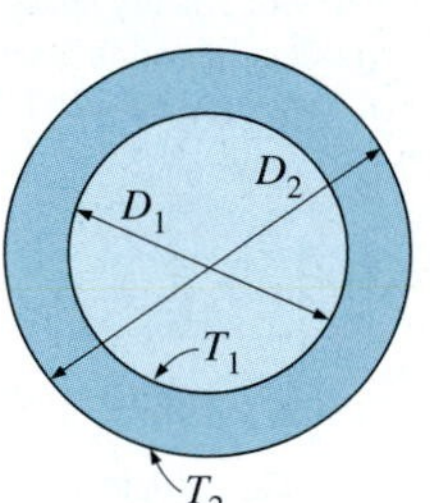

(12) Disk buried parallel to the surface in a semi-infinite medium ($z >> D$)

$$S = 4\,D$$

($S = 2D$ when $z = 0$)

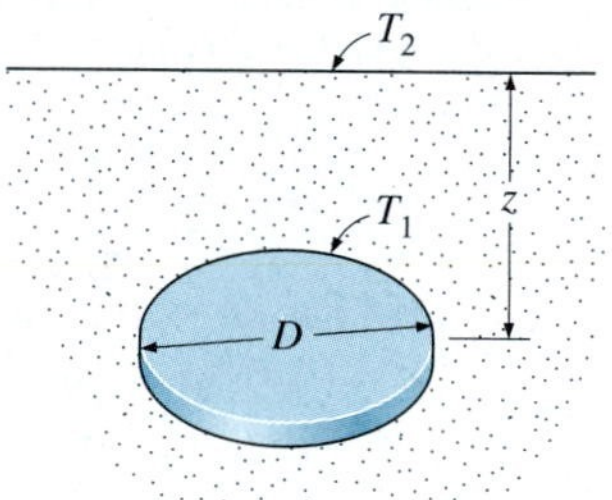

(13) The edge of two adjoining walls of equal thickness

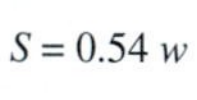

$$S = 0.54\,w$$

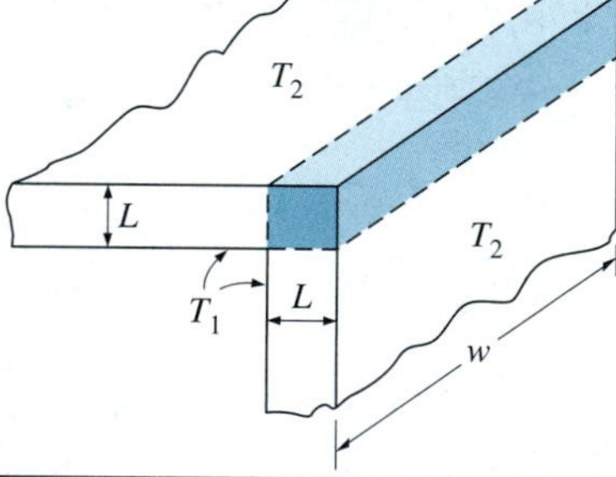

(14) Corner of three walls of equal thickness

$$S = 0.15L$$

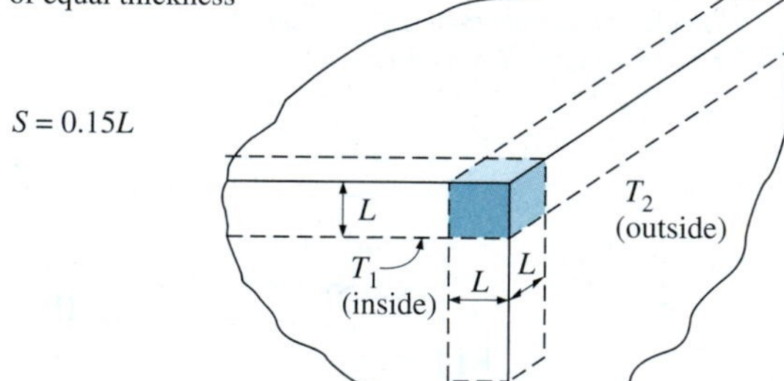

(15) Isothermal sphere buried in a semi-infinite medium

$$S = \frac{2\pi D}{1 - 0.25D/z}$$

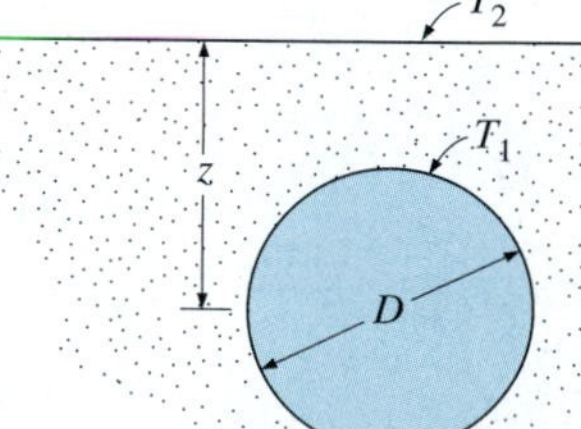

(16) Isothermal sphere buried in a semi-infinite medium at T_2 whose surface is insulated

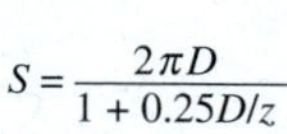

$$S = \frac{2\pi D}{1 + 0.25D/z}$$

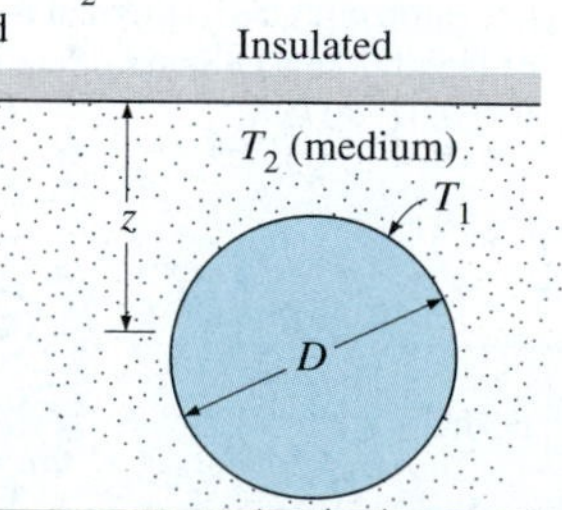

EXAMPLE 17–13 Heat Loss from Buried Steam Pipes

A 30-m-long, 10-cm-diameter hot-water pipe of a district heating system is buried in the soil 50 cm below the ground surface, as shown in Fig. 17–49. The outer surface temperature of the pipe is 80°C. Taking the surface temperature of the earth to be 10°C and the thermal conductivity of the soil at that location to be 0.9 W/m · °C, determine the rate of heat loss from the pipe.

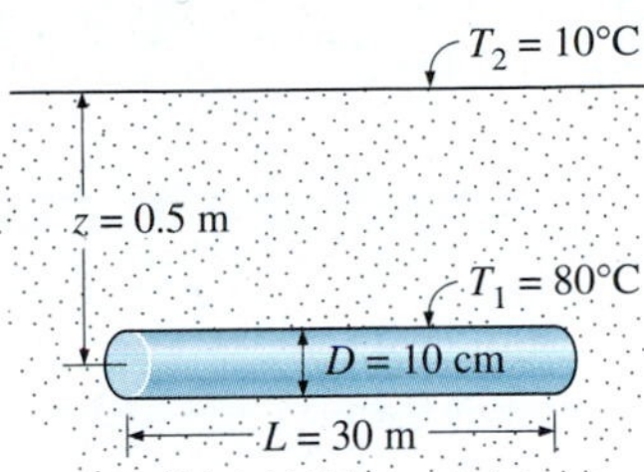

FIGURE 17–49
Schematic for Example 17–13.

Solution The hot-water pipe of a district heating system is buried in the soil. The rate of heat loss from the pipe is to be determined.

Assumptions **1** Steady operating conditions exist. **2** Heat transfer is two-dimensional (no change in the axial direction). **3** Thermal conductivity of the soil is constant.

Properties The thermal conductivity of the soil is given to be k = 0.9 W/m · °C.

Analysis The shape factor for this configuration is given in Table 17–7 to be

$$S = \frac{2\pi L}{\ln(4z/D)}$$

since $z > 1.5D$, where z is the distance of the pipe from the ground surface, and D is the diameter of the pipe. Substituting,

$$S = \frac{2\pi \times (30 \text{ m})}{\ln(4 \times 0.5/0.1)} = 62.9 \text{ m}$$

Then the steady rate of heat transfer from the pipe becomes

$$\dot{Q} = Sk(T_1 - T_2) = (62.9 \text{ m})(0.9 \text{ W/m} \cdot \text{°C})(80 - 10)\text{°C} = \mathbf{3963 \text{ W}}$$

Discussion Note that this heat is conducted from the pipe surface to the surface of the earth through the soil and then transferred to the atmosphere by convection and radiation.

EXAMPLE 17–14 Heat Transfer between Hot- and Cold-Water Pipes

A 5-m-long section of hot- and cold-water pipes run parallel to each other in a thick concrete layer, as shown in Fig. 17–50. The diameters of both pipes are 5 cm, and the distance between the centerline of the pipes is 30 cm. The surface temperatures of the hot and cold pipes are 70°C and 15°C, respectively. Taking the thermal conductivity of the concrete to be k = 0.75 W/m · °C, determine the rate of heat transfer between the pipes.

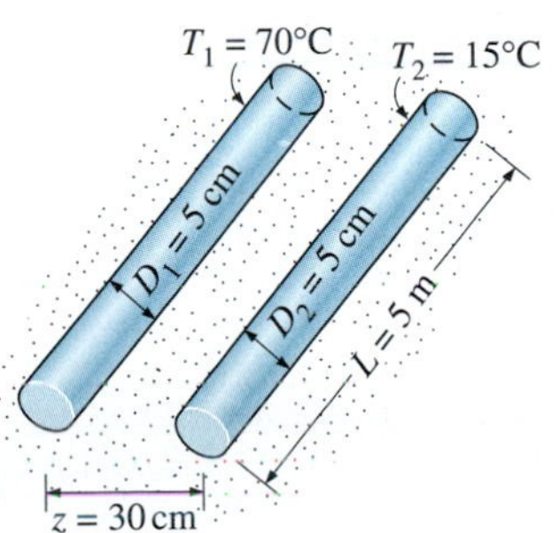

FIGURE 17–50
Schematic for Example 17–14.

Solution Hot- and cold-water pipes run parallel to each other in a thick concrete layer. The rate of heat transfer between the pipes is to be determined.

Assumptions **1** Steady operating conditions exist. **2** Heat transfer is two-dimensional (no change in the axial direction). **3** Thermal conductivity of the concrete is constant.

Properties The thermal conductivity of concrete is given to be k = 0.75 W/m · °C.

Analysis The shape factor for this configuration is given in Table 17–7 to be

$$S = \frac{2\pi L}{\cosh^{-1}\left(\dfrac{4z^2 - D_1^2 - D_2^2}{2D_1D_2}\right)}$$

where z is the distance between the centerlines of the pipes and L is their length. Substituting,

$$S = \frac{2\pi \times (5\text{ m})}{\cosh^{-1}\left(\dfrac{4 \times 0.3^2 - 0.05^2 - 0.05^2}{2 \times 0.05 \times 0.05}\right)} = 6.34\text{ m}$$

Then the steady rate of heat transfer between the pipes becomes

$$\dot{Q} = Sk(T_1 - T_2) = (6.34\text{ m})(0.75\text{ W/m}\cdot{}^\circ\text{C})(70 - 15^\circ)\text{C} = \mathbf{262\ W}$$

Discussion We can reduce this heat loss by placing the hot- and cold-water pipes further away from each other.

It is well known that insulation reduces heat transfer and saves energy and money. Decisions on the right amount of insulation are based on a heat transfer analysis, followed by an economic analysis to determine the "monetary value" of energy loss. This is illustrated with Example 17–15.

EXAMPLE 17–15 Cost of Heat Loss through Walls in Winter

Consider an electrically heated house whose walls are 9 ft high and have an R-value of insulation of 13 (i.e., a thickness-to-thermal conductivity ratio of $L/k = 13\ \text{h}\cdot\text{ft}^2\cdot{}^\circ\text{F/Btu}$). Two of the walls of the house are 40 ft long and the others are 30 ft long. The house is maintained at 75°F at all times, while the temperature of the outdoors varies. Determine the amount of heat lost through the walls of the house on a certain day during which the average temperature of the outdoors is 45°F. Also, determine the cost of this heat loss to the home owner if the unit cost of electricity is \$0.075/kWh. For combined convection and radiation heat transfer coefficients, use the ASHRAE (American Society of Heating, Refrigeration, and Air Conditioning Engineers) recommended values of $h_i = 1.46\ \text{Btu/h}\cdot\text{ft}^2\cdot{}^\circ\text{F}$ for the inner surface of the walls and $h_o = 4.0\ \text{Btu/h}\cdot\text{ft}^2\cdot{}^\circ\text{F}$ for the outer surface of the walls under 15 mph wind conditions in winter.

Solution An electrically heated house with R-13 insulation is considered. The amount of heat lost through the walls and its cost are to be determined.

Assumptions **1** The indoor and outdoor air temperatures have remained at the given values for the entire day so that heat transfer through the walls is steady. **2** Heat transfer through the walls is one-dimensional since any significant temperature gradients in this case exists in the direction from the indoors to the outdoors. **3** The radiation effects are accounted for in the heat transfer coefficients.

Analysis This problem involves conduction through the wall and convection at its surfaces and can best be handled by making use of the thermal resistance concept and drawing the thermal resistance network, as shown in Fig. 17–51. The heat transfer area of the walls is

$$A = \text{Circumference} \times \text{Height} = (2 \times 30 \text{ ft} + 2 \times 40 \text{ ft})(9 \text{ ft}) = 1260 \text{ ft}^2$$

Then the individual resistances are evaluated from their definitions to be

$$R_i = R_{\text{conv},\,i} = \frac{1}{h_i A} = \frac{1}{(1.46 \text{ Btu/h} \cdot \text{ft}^2 \cdot {}^\circ\text{F})(1260 \text{ ft}^2)} = 0.00054 \text{ h} \cdot {}^\circ\text{F/Btu}$$

$$R_{\text{wall}} = \frac{L}{kA} = \frac{R\text{-value}}{A} = \frac{13 \text{ h} \cdot \text{ft}^2 \cdot {}^\circ\text{F/Btu}}{1260 \text{ ft}^2} = 0.01032 \text{ h} \cdot {}^\circ\text{F/Btu}$$

$$R_o = R_{\text{conv},\,o} = \frac{1}{h_o A} = \frac{1}{(4.0 \text{ Btu/h} \cdot \text{ft}^2 \cdot {}^\circ\text{F})(1260 \text{ ft}^2)} = 0.00020 \text{ h} \cdot {}^\circ\text{F/Btu}$$

Noting that all three resistances are in series, the total resistance is

$$R_{\text{total}} = R_i + R_{\text{wall}} + R_o = 0.00054 + 0.01032 + 0.00020 = 0.01106 \text{ h} \cdot {}^\circ\text{F/Btu}$$

Then the steady rate of heat transfer through the walls of the house becomes

$$\dot{Q} = \frac{T_{\infty 1} - T_{\infty 2}}{R_{\text{total}}} = \frac{(75 - 45){}^\circ\text{F}}{0.01106 \text{ h} \cdot {}^\circ\text{F/Btu}} = 2712 \text{ Btu/h}$$

Finally, the total amount of heat lost through the walls during a 24-h period and its cost to the home owner are

$$Q = \dot{Q}\,\Delta t = (2712 \text{ Btu/h})(24\text{-h/day}) = \mathbf{65{,}100 \text{ Btu/day} = 19.1 \text{ kWh/day}}$$

since 1 kWh = 3412 Btu, and

$$\text{Heating cost} = (\text{Energy lost})(\text{Cost of energy}) = (19.1 \text{ kWh/day})(\$0.075/\text{kWh})$$
$$= \mathbf{\$1.43/day}$$

Discussion The heat losses through the walls of the house that day cost the home owner $1.43 worth of electricity. Most of this loss can be saved by insulation.

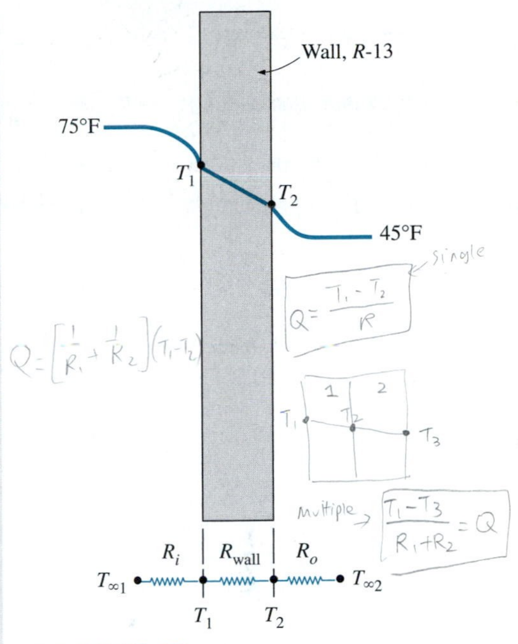

FIGURE 17–51
Schematic for Example 17–15.

SUMMARY

One-dimensional heat transfer through a simple or composite body exposed to convection from both sides to mediums at temperatures $T_{\infty 1}$ and $T_{\infty 2}$ can be expressed as

$$\dot{Q} = \frac{T_{\infty 1} - T_{\infty 2}}{R_{\text{total}}}$$

where R_{total} is the total thermal resistance between the two mediums. For a plane wall exposed to convection on both sides, the total resistance is expressed as

$$R_{\text{total}} = R_{\text{conv},\,1} + R_{\text{wall}} + R_{\text{conv},\,2} = \frac{1}{h_1 A} + \frac{L}{kA} + \frac{1}{h_2 A}$$

This relation can be extended to plane walls that consist of two or more layers by adding an additional resistance for each additional layer. The elementary thermal resistance relations can be expressed as follows:

Conduction resistance (plane wall): $R_{\text{wall}} = \dfrac{L}{kA}$

Conduction resistance (cylinder): $R_{\text{cyl}} = \dfrac{\ln(r_2/r_1)}{2\pi Lk}$

Conduction resistance (sphere): $R_{\text{sph}} = \dfrac{r_2 - r_1}{4\pi r_1 r_2 k}$

Convection resistance: $R_{\text{conv}} = \dfrac{1}{hA}$

Interface resistance: $$R_{\text{interface}} = \frac{1}{h_c A} = \frac{R_c}{A}$$

Radiation resistance: $$R_{\text{rad}} = \frac{1}{h_{\text{rad}} A}$$

where h_c is the thermal contact conductance, R_c is the thermal contact resistance, and the radiation heat transfer coefficient is defined as

$$h_{\text{rad}} = \varepsilon\sigma(T_s^2 + T_{\text{surr}}^2)(T_s + T_{\text{surr}})$$

Once the rate of heat transfer is available, the *temperature drop* across any layer can be determined from

$$\Delta T = \dot{Q}R$$

The thermal resistance concept can also be used to solve steady heat transfer problems involving parallel layers or combined series-parallel arrangements.

Adding insulation to a cylindrical pipe or a spherical shell increases the rate of heat transfer if the outer radius of the insulation is less than the *critical radius of insulation,* defined as

$$r_{\text{cr, cylinder}} = \frac{k_{\text{ins}}}{h}$$

$$r_{\text{cr, sphere}} = \frac{2k_{\text{ins}}}{h}$$

The effectiveness of an insulation is often given in terms of its *R-value*, the thermal resistance of the material per unit surface area, expressed as

$$R\text{-value} = \frac{L}{k} \qquad \text{(flat insulation)}$$

where L is the thickness and k is the thermal conductivity of the material.

Finned surfaces are commonly used in practice to enhance heat transfer. Fins enhance heat transfer from a surface by exposing a larger surface area to convection. The temperature distribution along the fin for very long fins and for fins with negligible heat transfer at the fin tip are given by

Very long fin: $$\frac{T(x) - T_\infty}{T_b - T_\infty} = e^{-x\sqrt{hp/kA_c}}$$

Adiabatic fin tip: $$\frac{T(x) - T_\infty}{T_b - T_\infty} = \frac{\cosh m(L - x)}{\cosh mL}$$

where $m = \sqrt{hp/kA_c}$, p is the perimeter, and A_c is the cross-sectional area of the fin. The rates of heat transfer for both cases are given to be

Very long fin:

$$\dot{Q}_{\text{long fin}} = -kA_c \left.\frac{dT}{dx}\right|_{x=0} = \sqrt{hpkA_c}\,(T_b - T_\infty)$$

Adiabatic fin tip:

$$\dot{Q}_{\text{adiabatic tip}} = -kA_c \left.\frac{dT}{dx}\right|_{x=0} = \sqrt{hpkA_c}\,(T_b - T_\infty) \tanh mL$$

Fins exposed to convection at their tips can be treated as fins with adiabatic tips by using the corrected length $L_c = L + A_c/p$ instead of the actual fin length.

The temperature of a fin drops along the fin, and thus the heat transfer from the fin is less because of the decreasing temperature difference toward the fin tip. To account for the effect of this decrease in temperature on heat transfer, we define *fin efficiency* as

$$\eta_{\text{fin}} = \frac{\dot{Q}_{\text{fin}}}{\dot{Q}_{\text{fin, max}}} = \frac{\text{Actual heat transfer rate from the fin}}{\text{Ideal heat transfer rate from the fin if the entire fin were at base temperature}}$$

When the fin efficiency is available, the rate of heat transfer from a fin can be determined from

$$\dot{Q}_{\text{fin}} = \eta_{\text{fin}} \dot{Q}_{\text{fin, max}} = \eta_{\text{fin}} hA_{\text{fin}} (T_b - T_\infty)$$

The performance of the fins is judged on the basis of the enhancement in heat transfer relative to the no-fin case and is expressed in terms of the *fin effectiveness* ε_{fin}, defined as

$$\varepsilon_{\text{fin}} = \frac{\dot{Q}_{\text{fin}}}{\dot{Q}_{\text{no fin}}} = \frac{\dot{Q}_{\text{fin}}}{hA_b (T_b - T_\infty)} = \frac{\text{Heat transfer rate from the fin of } \textit{base area } A_b}{\text{Heat transfer rate from the surface of } \textit{area } A_b}$$

Here, A_b is the cross-sectional area of the fin at the base and $\dot{Q}_{\text{no fin}}$ represents the rate of heat transfer from this area if no fins are attached to the surface. The *overall effectiveness* for a finned surface is defined as the ratio of the total heat transfer from the finned surface to the heat transfer from the same surface if there were no fins,

$$\varepsilon_{\text{fin, overall}} = \frac{\dot{Q}_{\text{total, fin}}}{\dot{Q}_{\text{total, no fin}}} = \frac{h(A_{\text{unfin}} + \eta_{\text{fin}} A_{\text{fin}})(T_b - T_\infty)}{hA_{\text{no fin}} (T_b - T_\infty)}$$

Fin efficiency and fin effectiveness are related to each other by

$$\varepsilon_{\text{fin}} = \frac{A_{\text{fin}}}{A_b} \eta_{\text{fin}}$$

Certain multidimensional heat transfer problems involve two surfaces maintained at constant temperatures T_1 and T_2. The steady rate of heat transfer between these two surfaces is expressed as

$$\dot{Q} = Sk(T_1 - T_2)$$

where S is the *conduction shape factor* that has the dimension of *length* and k is the thermal conductivity of the medium between the surfaces.

REFERENCES AND SUGGESTED READINGS

1. American Society of Heating, Refrigeration, and Air Conditioning Engineers. *Handbook of Fundamentals*. Atlanta: ASHRAE, 1993.
2. R. V. Andrews. "Solving Conductive Heat Transfer Problems with Electrical-Analogue Shape Factors." *Chemical Engineering Progress* 5 (1955), p. 67.
3. R. Barron. *Cryogenic Systems*. New York: McGraw-Hill, 1967.
4. L. S. Fletcher. "Recent Developments in Contact Conductance Heat Transfer." *Journal of Heat Transfer* 110, no. 4B (1988), pp. 1059–79.
5. E. Fried. "Thermal Conduction Contribution to Heat Transfer at Contacts." *Thermal Conductivity*, vol. 2, ed. R. P. Tye. London: Academic Press, 1969.
6. K. A. Gardner. "Efficiency of Extended Surfaces." *Trans. ASME* 67 (1945), pp. 621–31. Reprinted by permission of ASME International.
7. D. Q. Kern and A. D. Kraus. *Extended Surface Heat Transfer*. New York: McGraw-Hill, 1972.
8. G. P. Peterson. "Thermal Contact Resistance in Waste Heat Recovery Systems." *Proceedings of the 18th ASME/ETCE Hydrocarbon Processing Symposium*. Dallas, TX, 1987, pp. 45–51. Reprinted by permission of ASME International.
9. S. Song, M. M. Yovanovich, and F. O. Goodman. "Thermal Gap Conductance of Conforming Surfaces in Contact." *Journal of Heat Transfer* 115 (1993), p. 533.
10. J. E. Sunderland and K. R. Johnson. "Shape Factors for Heat Conduction through Bodies with Isothermal or Convective Boundary Conditions." *Trans. ASME* 10 (1964), pp. 2317–41.
11. W. M. Edmunds. "Residential Insulation." *ASTM Standardization News* (Jan. 1989), pp. 36–39.

PROBLEMS*

Steady Heat Conduction in Plane Walls

17–1C Consider one-dimensional heat conduction through a cylindrical rod of diameter D and length L. What is the heat transfer area of the rod if (*a*) the lateral surfaces of the rod are insulated and (*b*) the top and bottom surfaces of the rod are insulated?

17–2C Consider heat conduction through a plane wall. Does the energy content of the wall change during steady heat conduction? How about during transient conduction? Explain.

17–3C Consider heat conduction through a wall of thickness L and area A. Under what conditions will the temperature distributions in the wall be a straight line?

17–4C What does the thermal resistance of a medium represent?

17–5C How is the combined heat transfer coefficient defined? What convenience does it offer in heat transfer calculations?

*Problems designated by a "C" are concept questions, and students are encouraged to answer them all. Problems designated by an "E" are in English units, and the SI users can ignore them. Problems with the icon are solved using EES, and complete solutions together with parametric studies are included on the enclosed DVD. Problems with the icon are comprehensive in nature, and are intended to be solved with a computer, preferably using the EES software that accompanies this text.

17–6C Can we define the convection resistance per unit surface area as the inverse of the convection heat transfer coefficient?

17–7C Why are the convection and the radiation resistances at a surface in parallel instead of being in series?

17–8C Consider a surface of area A at which the convection and radiation heat transfer coefficients are h_{conv} and h_{rad}, respectively. Explain how you would determine (*a*) the single equivalent heat transfer coefficient, and (*b*) the equivalent thermal resistance. Assume the medium and the surrounding surfaces are at the same temperature.

17–9C How does the thermal resistance network associated with a single-layer plane wall differ from the one associated with a five-layer composite wall?

17–10C Consider steady one-dimensional heat transfer through a multilayer medium. If the rate of heat transfer $\dot{Q}$ is known, explain how you would determine the temperature drop across each layer.

17–11C Consider steady one-dimensional heat transfer through a plane wall exposed to convection from both sides to environments at known temperatures $T_{\infty 1}$ and $T_{\infty 2}$ with known heat transfer coefficients h_1 and h_2. Once the rate of heat transfer $\dot{Q}$ has been evaluated, explain how you would determine the temperature of each surface.

17–12C Someone comments that a microwave oven can be viewed as a conventional oven with zero convection resistance at the surface of the food. Is this an accurate statement?

17–13C Consider a window glass consisting of two 4-mm-thick glass sheets pressed tightly against each other. Compare the heat transfer rate through this window with that of one consisting of a single 8-mm-thick glass sheet under identical conditions.

17–14C Consider steady heat transfer through the wall of a room in winter. The convection heat transfer coefficient at the outer surface of the wall is three times that of the inner surface as a result of the winds. On which surface of the wall do you think the temperature will be closer to the surrounding air temperature? Explain.

17–15C The bottom of a pan is made of a 4-mm-thick aluminum layer. In order to increase the rate of heat transfer through the bottom of the pan, someone proposes a design for the bottom that consists of a 3-mm-thick copper layer sandwiched between two 2-mm-thick aluminum layers. Will the new design conduct heat better? Explain. Assume perfect contact between the layers.

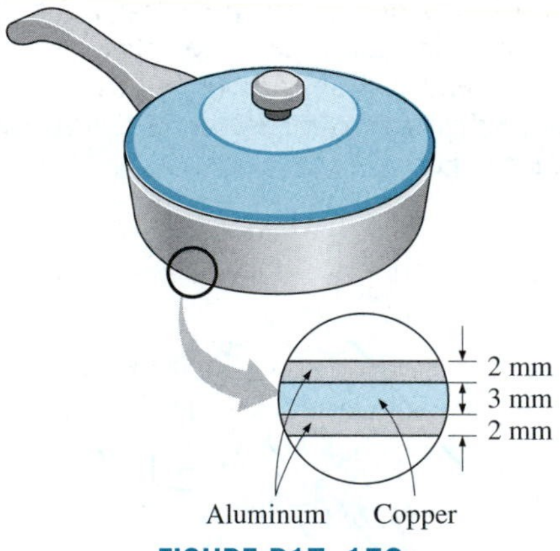

FIGURE P17–15C

17–16C Consider two cold canned drinks, one wrapped in a blanket and the other placed on a table in the same room. Which drink will warm up faster?

17–17 Consider a 3-m-high, 6-m-wide, and 0.3-m-thick brick wall whose thermal conductivity is $k = 0.8$ W/m · °C. On a certain day, the temperatures of the inner and the outer surfaces of the wall are measured to be 14°C and 2°C, respectively. Determine the rate of heat loss through the wall on that day.

17–18 A 1.0 m × 1.5 m double-pane window consists of two 4-mm thick layers of glass ($k = 0.78$ W/m · K) that are the separated by a 5-mm air gap ($k_{air} = 0.025$ W/m · K). The heat flow through the air gap is assumed to be by condition. The inside and outside air temperatures are 20°C and −20°C, respectively, and the inside and outside heat transfer coefficients are 40 and 20 W/m^2 · K. Determine (*a*) the daily rate of heat loss through the window in steady operation and (*b*) the temperature difference across the largest thermal resistence.

17–19 Consider a 1.2-m-high and 2-m-wide glass window whose thickness is 6 mm and thermal conductivity is $k =$ 0.78 W/m · °C. Determine the steady rate of heat transfer through this glass window and the temperature of its inner surface for a day during which the room is maintained at 24°C while the temperature of the outdoors is −5°C. Take the convection heat transfer coefficients on the inner and outer surfaces of the window to be $h_1 = 10$ W/m^2 · °C and $h_2 = 25$ W/m^2 · °C, and disregard any heat transfer by radiation.

17–20 Consider a 1.2-m-high and 2-m-wide double-pane window consisting of two 3-mm-thick layers of glass ($k =$ 0.78 W/m · °C) separated by a 12-mm-wide stagnant air space ($k = 0.026$ W/m · °C). Determine the steady rate of heat transfer through this double-pane window and the temperature of its inner surface for a day during which the room is maintained at 24°C while the temperature of the outdoors is −5°C. Take the convection heat transfer coefficients on the inner and outer surfaces of the window to be $h_1 = 10$ W/m^2 · °C and $h_2 = 25$ W/m^2 · °C, and disregard any heat transfer by radiation. *Answers:* 114 W, 19.2°C

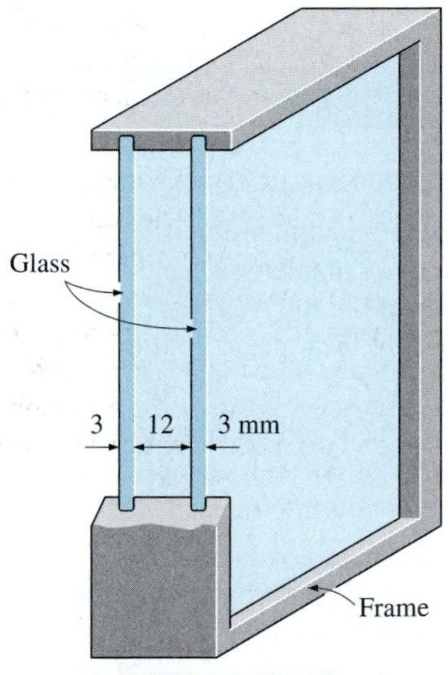

FIGURE P17–20

17–21 Repeat Prob. 17–20, assuming the space between the two glass layers is evacuated.

17–22 Reconsider Prob. 17–20. Using EES (or other) software, plot the rate of heat transfer through the window as a function of the width of air space in the range of 2 mm to 20 mm, assuming pure conduction through the air. Discuss the results.

17–23E Consider an electrically heated brick house ($k =$ 0.40 Btu/h · ft · °F) whose walls are 9 ft high and 1 ft thick. Two of the walls of the house are 50 ft long and the others are 35 ft long. The house is maintained at 70°F at all times while the temperature of the outdoors varies. On a certain day, the temperature of the inner surface of the walls is measured to be at 55°F while the average temperature of the outer surface is observed to remain at 45°F during the day for 10 h and at 35°F at night for 14 h. Determine the amount of heat lost from the house that day. Also determine the cost of that heat loss to the home owner for an electricity price of \$0.09/kWh.

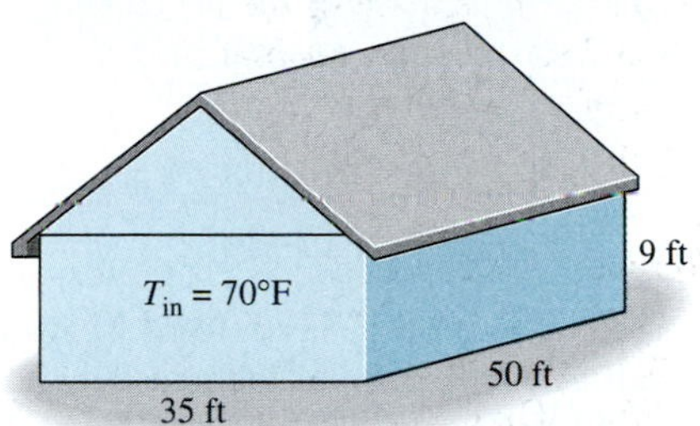

FIGURE P17–23E

17–24 A cylindrical resistor element on a circuit board dissipates 0.15 W of power in an environment at 40°C. The resistor is 1.2 cm long, and has a diameter of 0.3 cm. Assuming heat to be transferred uniformly from all surfaces, determine (*a*) the amount of heat this resistor dissipates during a 24-h period; (*b*) the heat flux on the surface of the resistor, in W/m^2; and (*c*) the surface temperature of the resistor for a combined convection and radiation heat transfer coefficient of 9 W/m^2 · °C.

17–25 Consider a power transistor that dissipates 0.2 W of power in an environment at 30°C. The transistor is 0.4 cm long and has a diameter of 0.5 cm. Assuming heat to be transferred uniformly from all surfaces, determine (*a*) the amount of heat this resistor dissipates during a 24-h period, in kWh; (*b*) the heat flux on the surface of the transistor, in W/m^2; and (*c*) the surface temperature of the resistor for a combined convection and radiation heat transfer coefficient of 18 W/m^2 · °C.

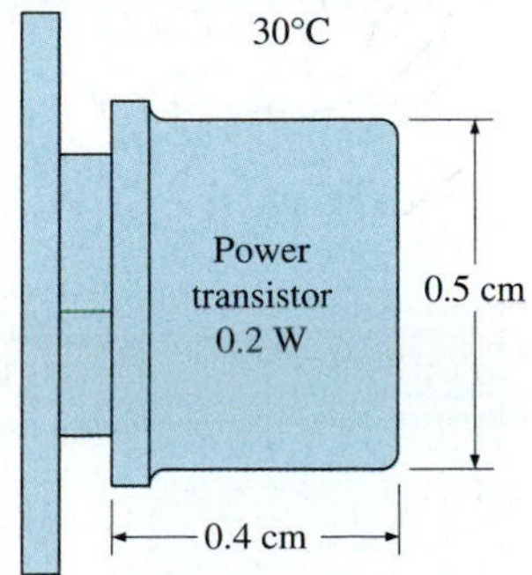

FIGURE P17–25

17–26 A 12-cm × 18-cm circuit board houses on its surface 100 closely spaced logic chips, each dissipating 0.06 W in an environment at 40°C. The heat transfer from the back surface of the board is negligible. If the heat transfer coefficient on the surface of the board is 10 W/m^2 · °C, determine (*a*) the heat flux on the surface of the circuit board, in W/m^2; (*b*) the surface temperature of the chips; and (*c*) the thermal resistance between the surface of the circuit board and the cooling medium, in °C/W.

17–27 Consider a person standing in a room at 20°C with an exposed surface area of 1.7 m^2. The deep body temperature of the human body is 37°C, and the thermal conductivity of the human tissue near the skin is about 0.3 W/m · °C. The body is losing heat at a rate of 150 W by natural convection and radiation to the surroundings. Taking the body temperature 0.5 cm beneath the skin to be 37°C, determine the skin temperature of the person. *Answer*: 35.5°C

17–28 Water is boiling in a 25-cm-diameter aluminum pan (k = 237 W/m · °C) at 95°C. Heat is transferred steadily to the boiling water in the pan through its 0.5-cm-thick flat bottom at a rate of 800 W. If the inner surface temperature of the bottom of the pan is 108°C, determine (*a*) the boiling heat transfer coefficient on the inner surface of the pan and (*b*) the outer surface temperature of the bottom of the pan.

17–29E A wall is constructed of two layers of 0.7-in-thick sheetrock (k = 0.10 Btu/h · ft · °F), which is a plasterboard made of two layers of heavy paper separated by a layer of gypsum, placed 7 in apart. The space between the sheetrocks is filled with fiberglass insulation (k = 0.020 Btu/h · ft · °F). Determine (*a*) the thermal resistance of the wall and (*b*) its *R*-value of insulation in English units.

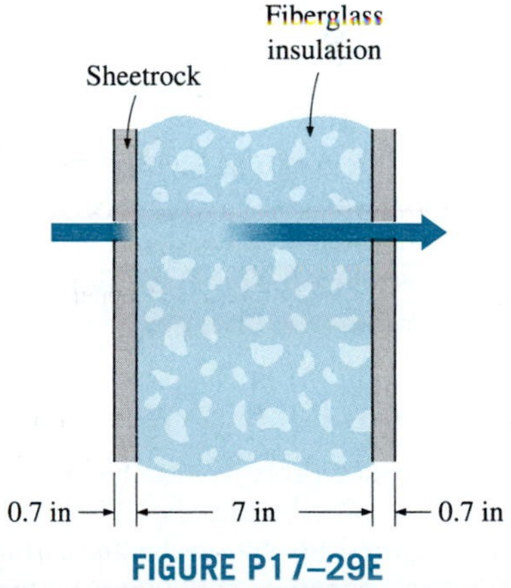

FIGURE P17–29E

17–30 The roof of a house consists of a 15-cm-thick concrete slab (k = 2 W/m · °C) that is 15 m wide and 20 m long. The convection heat transfer coefficients on the inner and outer surfaces of the roof are 5 and 12 W/m^2 · °C, respectively. On a clear winter night, the ambient air is reported to be at 10°C, while the night sky temperature is 100 K. The house and the interior surfaces of the wall are maintained at a constant temperature of 20°C. The emissivity of both surfaces of the concrete roof is 0.9. Considering both radiation and convection heat transfers, determine the rate of heat transfer through the roof, and the inner surface temperature of the roof.

If the house is heated by a furnace burning natural gas with an efficiency of 80 percent, and the price of natural gas is \$1.20/therm (1 therm = 105,500 kJ of energy content), determine the money lost through the roof that night during a 14-h period.

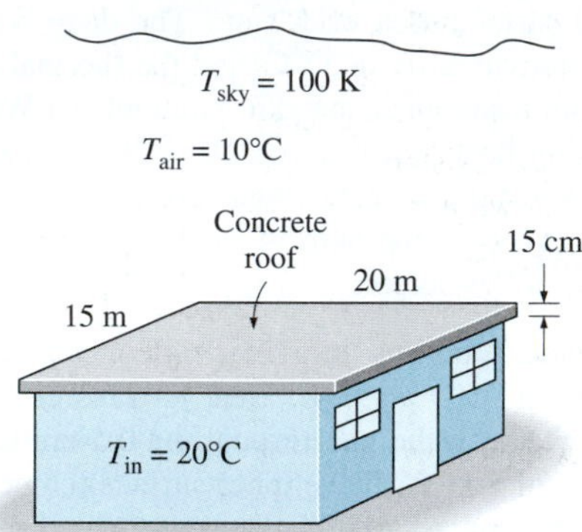

FIGURE P17–30

17–31 A 2-m × 1.5-m section of wall of an industrial furnace burning natural gas is not insulated, and the temperature at the outer surface of this section is measured to be 80°C. The temperature of the furnace room is 30°C, and the combined convection and radiation heat transfer coefficient at the surface of the outer furnace is 10 W/m² · °C. It is proposed to insulate this section of the furnace wall with glass wool insulation (k = 0.038 W/m · °C) in order to reduce the heat loss by 90 percent. Assuming the outer surface temperature of the metal section still remains at about 80°C, determine the thickness of the insulation that needs to be used.

The furnace operates continuously and has an efficiency of 78 percent. The price of the natural gas is \$1.10/therm (1 therm = 105,500 kJ of energy content). If the installation of the insulation will cost \$250 for materials and labor, determine how long it will take for the insulation to pay for itself from the energy it saves.

17–32 Repeat Prob. 17–31 for expanded perlite insulation assuming conductivity is k = 0.052 W/m · °C.

17–33 Reconsider Prob. 17–31. Using EES (or other) software, investigate the effect of thermal conductivity on the required insulation thickness. Plot the thickness of insulation as a function of the thermal conductivity of the insulation in the range of 0.02 W/m · °C to 0.08 W/m · °C, and discuss the results.

17–34E Consider a house whose walls are 12 ft high and 40 ft long. Two of the walls of the house have no windows, while each of the other two walls has four windows made of 0.25-in-thick glass (k = 0.45 Btu/h · ft · °F), 3 ft × 5 ft in size. The walls are certified to have an R-value of 19 (i.e., an L/k value of 19 h · ft² · °F/Btu). Disregarding any direct radiation gain or loss through the windows and taking the heat transfer coefficients at the inner and outer surfaces of the house to be 2 and 4 Btu/h · ft² · °F, respectively, determine the ratio of the heat transfer through the walls with and without windows.

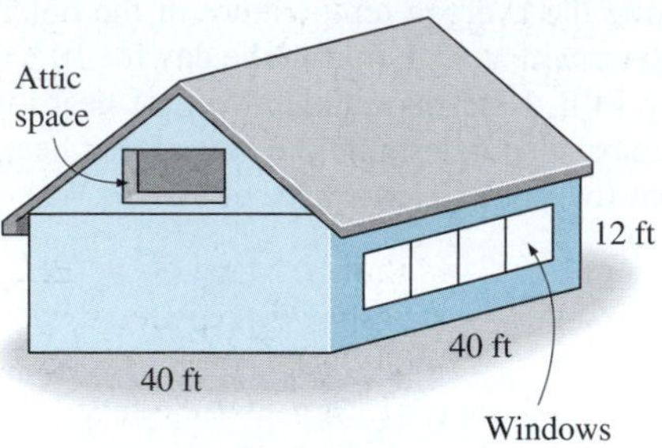

FIGURE P17–34E

17–35 Consider a house that has a 10-m × 20-m base and a 4-m-high wall. All four walls of the house have an R-value of 2.31 m² · °C/W. The two 10-m × 4-m walls have no windows. The third wall has five windows made of 0.5-cm-thick glass (k = 0.78 W/m · °C), 1.2 m × 1.8 m in size. The fourth wall has the same size and number of windows, but they are double-paned with a 1.5-cm-thick stagnant air space (k = 0.026 W/m · °C) enclosed between two 0.5-cm-thick glass layers. The thermostat in the house is set at 24°C and the average temperature outside at that location is 8°C during the seven-month-long heating season. Disregarding any direct radiation gain or loss through the windows and taking the heat transfer coefficients at the inner and outer surfaces of the house to be 7 and 18 W/m² · °C, respectively, determine the average rate of heat transfer through each wall.

If the house is electrically heated and the price of electricity is \$0.08/kWh, determine the amount of money this household will save per heating season by converting the single-pane windows to double-pane windows.

17–36 The wall of a refrigerator is constructed of fiberglass insulation (k = 0.035 W/m · °C) sandwiched between two layers of 1-mm-thick sheet metal (k = 15.1 W/m · °C). The refrigerated space is maintained at 3°C, and the average heat transfer coefficients at the inner and outer surfaces of the wall

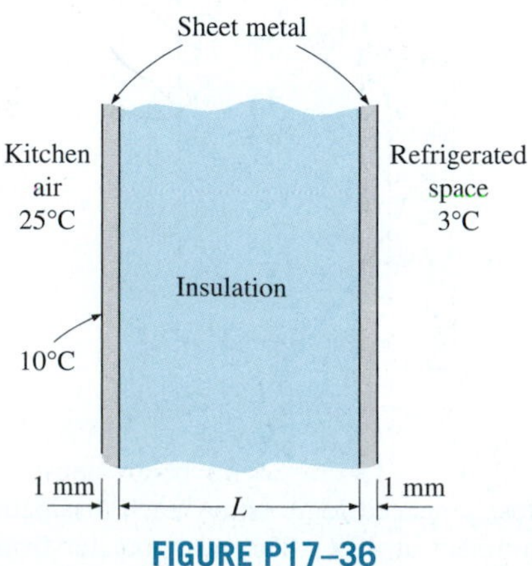

FIGURE P17–36

are 4 W/m^2 · °C and 9 W/m^2 · °C, respectively. The kitchen temperature averages 25°C. It is observed that condensation occurs on the outer surfaces of the refrigerator when the temperature of the outer surface drops to 20°C. Determine the minimum thickness of fiberglass insulation that needs to be used in the wall in order to avoid condensation on the outer surfaces.

17–37 Reconsider Prob. 17–36. Using EES (or other) software, investigate the effects of the thermal conductivities of the insulation material and the sheet metal on the thickness of the insulation. Let the thermal conductivity vary from 0.02 W/m · °C to 0.08 W/m · °C for insulation and 10 W/m · °C to 400 W/m · °C for sheet metal. Plot the thickness of the insulation as the functions of the thermal conductivities of the insulation and the sheet metal, and discuss the results.

17–38 Heat is to be conducted along a circuit board that has a copper layer on one side. The circuit board is 15 cm long and 15 cm wide, and the thicknesses of the copper and epoxy layers are 0.1 mm and 1.2 mm, respectively. Disregarding heat transfer from side surfaces, determine the percentages of heat conduction along the copper (k = 386 W/m · °C) and epoxy (k = 0.26 W/m · °C) layers. Also determine the effective thermal conductivity of the board. *Answers:* 0.8 percent, 99.2 percent, and 29.9 W/m · °C

17–39E A 0.03-in-thick copper plate (k = 223 Btu/h · ft · °F) is sandwiched between two 0.15-in-thick epoxy boards (k = 0.15 Btu/h · ft · °F) that are 7 in × 9 in in size. Determine the effective thermal conductivity of the board along its 9-in-long side. What fraction of the heat conducted along that side is conducted through copper?

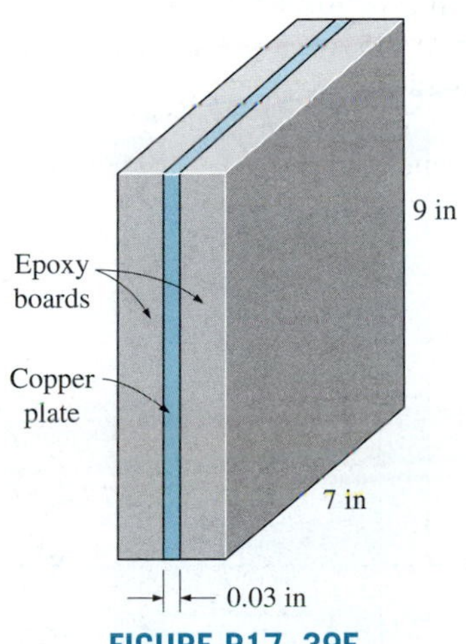

FIGURE P17–39E

Thermal Contact Resistance

17–40C What is thermal contact resistance? How is it related to thermal contact conductance?

17–41C Will the thermal contact resistance be greater for smooth or rough plain surfaces?

17–42C A wall consists of two layers of insulation pressed against each other. Do we need to be concerned about the thermal contact resistance at the interface in a heat transfer analysis or can we just ignore it?

17–43C A plate consists of two thin metal layers pressed against each other. Do we need to be concerned about the thermal contact resistance at the interface in a heat transfer analysis or can we just ignore it?

17–44C Consider two surfaces pressed against each other. Now the air at the interface is evacuated. Will the thermal contact resistance at the interface increase or decrease as a result?

17–45C Explain how the thermal contact resistance can be minimized.

17–46 The thermal contact conductance at the interface of two 1-cm-thick copper plates is measured to be 18,000 W/m^2 · °C. Determine the thickness of the copper plate whose thermal resistance is equal to the thermal resistance of the interface between the plates.

17–47 Six identical power transistors with aluminum casing are attached on one side of a 1.2-cm-thick 20-cm × 30-cm copper plate (k = 386 W/m · °C) by screws that exert an average pressure of 10 MPa. The base area of each transistor is 9 cm^2, and each transistor is placed at the center of a 10-cm × 10-cm section of the plate. The interface roughness is estimated to be about 1.4 μm. All transistors are covered by a thick Plexiglas layer, which is a poor conductor of heat, and thus all the heat generated at the junction of the transistor must be dissipated to the ambient at 23°C through the back surface of the

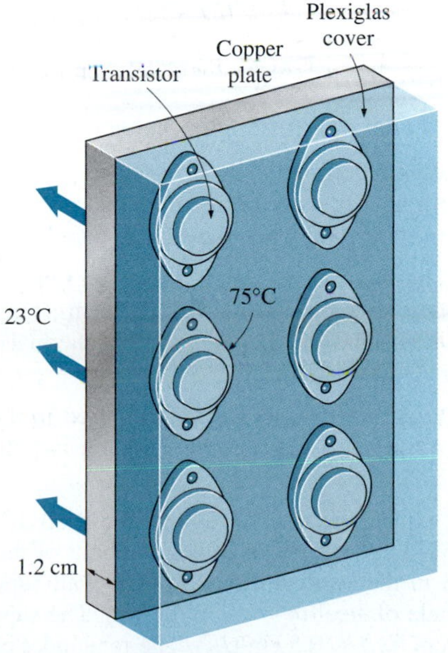

FIGURE P17–47

copper plate. The combined convection/radiation heat transfer coefficient at the back surface can be taken to be 30 W/m^2 · °C. If the case temperature of the transistor is not to exceed 75°C, determine the maximum power each transistor can dissipate safely, and the temperature jump at the case-plate interface.

17–48 Two 5-cm-diameter, 15-cm-long aluminum bars (k = 176 W/m · °C) with ground surfaces are pressed against each other with a pressure of 20 atm. The bars are enclosed in an insulation sleeve and, thus, heat transfer from the lateral surfaces is negligible. If the top and bottom surfaces of the two-bar system are maintained at temperatures of 150°C and 20°C, respectively, determine (*a*) the rate of heat transfer along the cylinders under steady conditions and (*b*) the temperature drop at the interface. *Answers: (a) 142.4 W, (b) 6.4°C*

17–49 A 1-mm-thick copper plate (k = 386 W/m · °C) is sandwiched between two 5-mm-thick epoxy boards (k = 0.26 W/m · °C) that are 15 cm × 20 cm in size. If the thermal contact conductance on both sides of the copper plate is estimated to be 6000 W/m · °C, determine the error involved in the total thermal resistance of the plate if the thermal contact conductances are ignored.

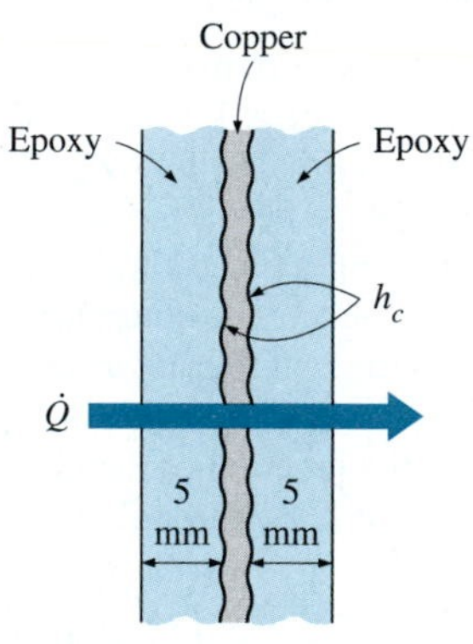

FIGURE P17–49

Generalized Thermal Resistance Networks

17–50C When plotting the thermal resistance network associated with a heat transfer problem, explain when two resistances are in series and when they are in parallel.

17–51C The thermal resistance networks can also be used approximately for multidimensional problems. For what kind of multidimensional problems will the thermal resistance approach give adequate results?

17–52C What are the two approaches used in the development of the thermal resistance network for two-dimensional problems?

17–53 A typical section of a building wall is shown in Fig. P17–53. This section extends in and out of the page and is repeated in the vertical direction. The wall support members are made of steel (k = 50 W/m · K). The support members are 8 cm (t_{23}) × 0.5 cm (L_B). The remainder of the inner wall space is filled with insulation (k = 0.03 W/m · K) and measures 8 cm (t_{23}) × 60 cm (L_B). The inner wall is made of gypsum board (k = 0.5 W/m · K) that is 1 cm thick (t_{12}) and the outer wall is made of brick (k = 1.0 W/m · K) that is 10 cm thick (t_{34}). What is the average heat flux through this wall when T_1 = 20°C and T_4 = 35°C?

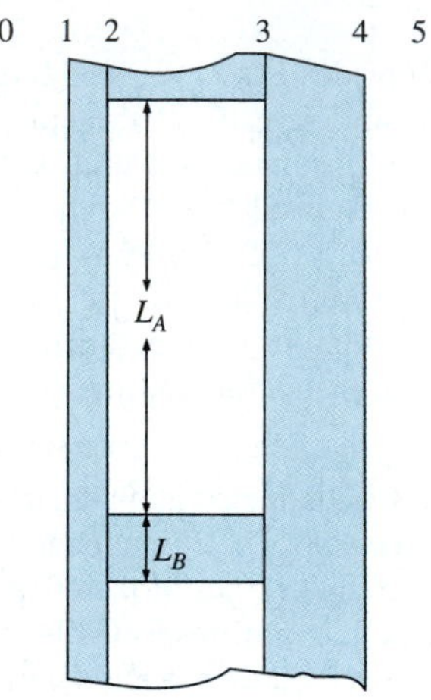

FIGURE P17–53

17–54 A 4-m-high and 6-m-wide wall consists of a long 18-cm × 30-cm cross section of horizontal bricks (k = 0.72 W/m · °C) separated by 3-cm-thick plaster layers (k = 0.22 W/m · °C). There are also 2-cm-thick plaster layers on each side of the wall, and a 2-cm-thick rigid foam (k = 0.026 W/m · °C) on the inner side of the wall. The indoor and the outdoor temperatures are 22°C and −4°C, and the convection heat transfer coefficients on the inner and the outer sides are h_1 = 10 W/m^2 · °C and h_2 = 20 W/m^2 · °C, respectively. Assuming one-dimensional heat

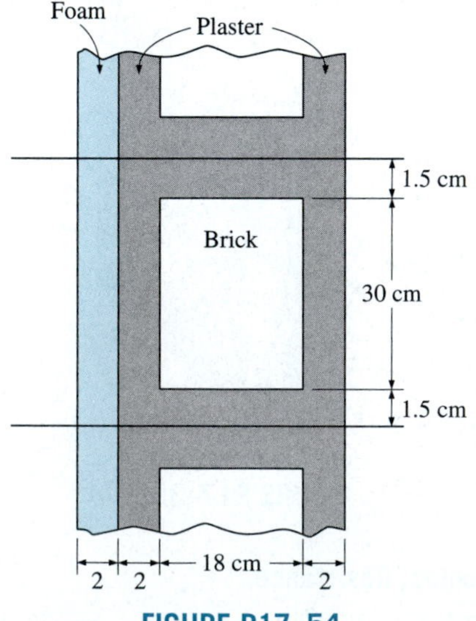

FIGURE P17–54

transfer and disregarding radiation, determine the rate of heat transfer through the wall.

17–55 Reconsider Prob. 17–54. Using EES (or other) software, plot the rate of heat transfer through the wall as a function of the thickness of the rigid foam in the range of 1 cm to 10 cm. Discuss the results.

17–56 A 10-cm-thick wall is to be constructed with 2.5-m-long wood studs ($k = 0.11$ W/m · °C) that have a cross section of 10 cm × 10 cm. At some point the builder ran out of those studs and started using pairs of 2.5-m-long wood studs that have a cross section of 5 cm × 10 cm nailed to eachother instead. The manganese steel nails ($k = 50$ W/m · °C) are 10 cm long and have a diameter of 0.4 cm. A total of 50 nails are used to connect the two studs, which are mounted to the wall such that the nails cross the wall. The temperature difference between the inner and outer surfaces of the wall is 8°C. Assuming the thermal contact resistance between the two layers to be negligible, determine the rate of heat transfer (*a*) through a solid stud and (*b*) through a stud pair of equal length and width nailed to each other. (*c*) Also determine the effective conductivity of the nailed stud pair.

17–57 A 12-m-long and 5-m-high wall is constructed of two layers of 1-cm-thick sheetrock ($k = 0.17$ W/m · °C) spaced 16 cm by wood studs ($k = 0.11$ W/m · °C) whose cross section is 12 cm × 5 cm. The studs are placed vertically 60 cm apart, and the space between them is filled with fiberglass insulation ($k = 0.034$ W/m · °C). The house is maintained at 20°C and the ambient temperature outside is −9°C. Taking the heat transfer coefficients at the inner and outer surfaces of the house to be 8.3 and 34 W/m² · °C, respectively, determine (*a*) the thermal resistance of the wall considering a representative section of it and (*b*) the rate of heat transfer through the wall.

17–58E A 10-in-thick, 30-ft-long, and 10-ft-high wall is to be constructed using 9-in-long solid bricks (k = 0.40 Btu/h · ft · °F) of cross section 7 in × 7 in, or identical size bricks with nine square air holes ($k = 0.015$ Btu/h · ft · °F) that are 9 in long and have a cross section of 1.5 in × 1.5 in. There is a 0.5-in-thick plaster layer ($k = 0.10$ Btu/h · ft · °F) between two adjacent bricks on all four sides and on both sides of the wall. The house is maintained at 80°F and the ambient temperature outside is 30°F. Taking the heat transfer coefficients at the inner and outer surfaces of the wall to be 1.5 and 4 Btu/h · ft² · °F, respectively, determine the rate of heat transfer through the wall constructed of (*a*) solid bricks and (*b*) bricks with air holes.

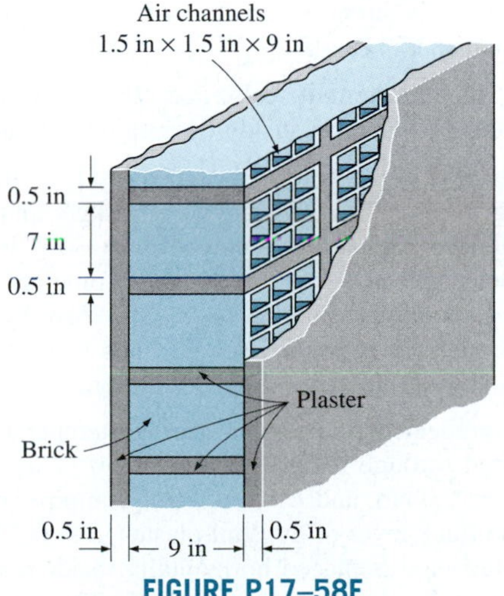

FIGURE P17–58E

17–59 Consider a 5-m-high, 8-m-long, and 0.22-m-thick wall whose representative cross section is as given in the figure. The thermal conductivities of various materials used, in W/m · °C, are $k_A = k_F = 2$, $k_B = 8$, $k_C = 20$, $k_D = 15$, and $k_E = 35$. The left and right surfaces of the wall are maintained at uniform temperatures of 300°C and 100°C, respectively. Assuming heat transfer through the wall to be one-dimensional, determine (*a*) the rate of heat transfer through the wall; (*b*) the temperature at the point where the sections *B*, *D*, and *E* meet; and (*c*) the temperature drop across the section *F*. Disregard any contact resistances at the interfaces.

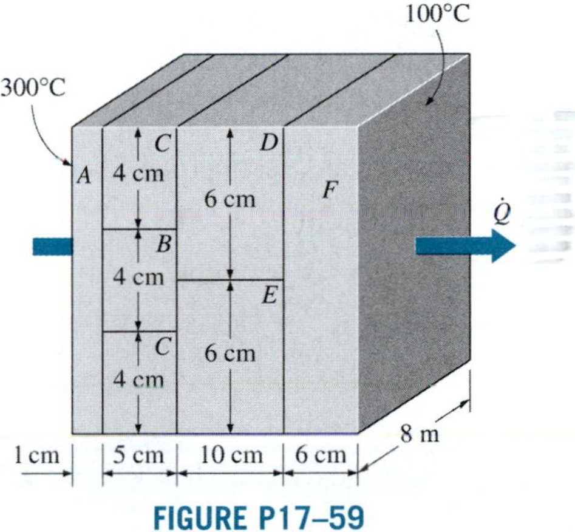

FIGURE P17–59

17–60 Repeat Prob. 17–59 assuming that the thermal contact resistance at the interfaces *D-F* and *E-F* is 0.00012 m² · °C/W.

17–61 Clothing made of several thin layers of fabric with trapped air in between, often called ski clothing, is commonly used in cold climates because it is light, fashionable, and a very effective thermal insulator. So it is no surprise that such clothing has largely replaced thick and heavy old-fashioned coats.

Consider a jacket made of five layers of 0.1-mm-thick synthetic fabric ($k = 0.13$ W/m · °C) with 1.5-mm-thick air space ($k = 0.026$ W/m · °C) between the layers. Assuming the inner surface temperature of the jacket to be 28°C and the surface area to be 1.25 m², determine the rate of heat loss through the jacket when the temperature of the outdoors is 0°C and the heat transfer coefficient at the outer surface is 25 W/m² · °C.

What would your response be if the jacket is made of a single layer of 0.5-mm-thick synthetic fabric? What should be the thickness of a wool fabric (k = 0.035 W/m · °C) if the person is to achieve the same level of thermal comfort wearing a thick wool coat instead of a five-layer ski jacket?

FIGURE P17–61

17–62 Repeat Prob. 17–61 assuming the layers of the jacket are made of cotton fabric (k = 0.06 W/m · °C).

17–63 A 5-m-wide, 4-m-high, and 40-m-long kiln used to cure concrete pipes is made of 20-cm-thick concrete walls and ceiling (k = 0.9 W/m · °C). The kiln is maintained at 40°C by injecting hot steam into it. The two ends of the kiln, 4 m × 5 m in size, are made of a 3-mm-thick sheet metal covered with 2-cm-thick Styrofoam (k = 0.033 W/m · °C). The convection heat transfer coefficients on the inner and the outer surfaces of the kiln are 3000 W/m^2 · °C and 25 W/m^2 · °C, respectively. Disregarding any heat loss through the floor, determine the rate of heat loss from the kiln when the ambient air is at −4°C.

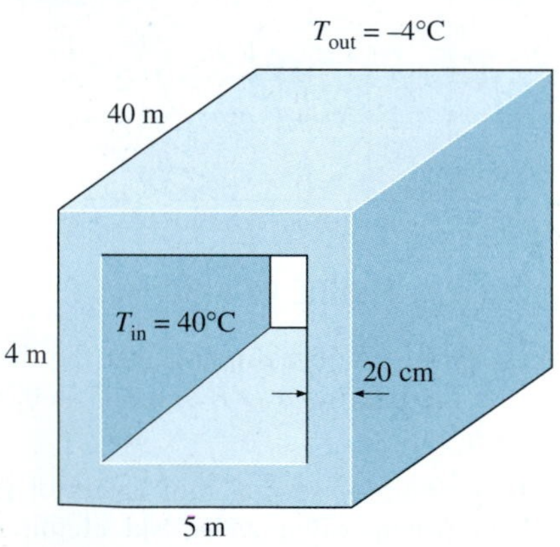

FIGURE P17–63

17–64 Reconsider Prob. 17–63. Using EES (or other) software, investigate the effects of the thickness of the wall and the convection heat transfer coefficient on the outer surface of the rate of heat loss from the kiln. Let the thickness vary from 10 cm to 30 cm and the convection heat transfer coefficient from 5 W/m^2 · °C to 50 W/m^2 · °C. Plot the rate of heat transfer as functions of wall thickness and the convection heat transfer coefficient, and discuss the results.

17–65E Consider a 6-in × 8-in epoxy glass laminate (k = 0.10 Btu/h · ft · °F) whose thickness is 0.05 in. In order to reduce the thermal resistance across its thickness, cylindrical copper fillings (k = 223 Btu/h · ft · °F) of 0.02 in diameter are to be planted throughout the board, with a center-to-center distance of 0.06 in. Determine the new value of the thermal resistance of the epoxy board for heat conduction across its thickness as a result of this modification. *Answer:* 0.00064 h · °F/Btu

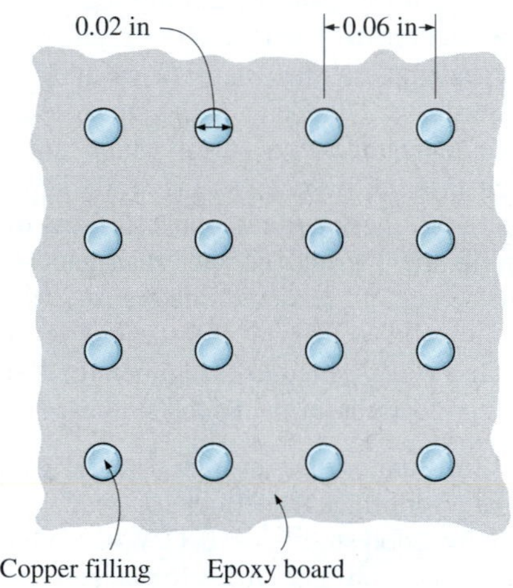

FIGURE P17–65E

Heat Conduction in Cylinders and Spheres

17–66C What is an infinitely long cylinder? When is it proper to treat an actual cylinder as being infinitely long, and when is it not?

17–67C Consider a short cylinder whose top and bottom surfaces are insulated. The cylinder is initially at a uniform temperature T_i and is subjected to convection from its side surface to a medium at temperature T_∞, with a heat transfer coefficient of h. Is the heat transfer in this short cylinder one- or two-dimensional? Explain.

17–68C Can the thermal resistance concept be used for a solid cylinder or sphere in steady operation? Explain.

17–69 Chilled water enters a thin-shelled 5-cm-diameter, 150-m-long pipe at 7°C at a rate of 0.98 kg/s and leaves at 8°C. The pipe is exposed to ambient air at 30°C with a heat transfer coefficient of 9 W/m^2 · °C. If the pipe is to be insulated with glass wool insulation (k = 0.05 W/m · °C) in order to decrease the temperature rise of water to 0.25°C, determine the required thickness of the insulation.

17–70 Superheated steam at an average temperature 200°C is transported through a steel pipe (k = 50 W/m · K, D_o = 8.0 cm, D_i = 6.0 cm, and L = 20.0 m). The pipe is insulated with a 4-cm thick layer of gypsum plaster (k = 0.5 W/m · K). The insulated pipe is placed horizontally inside a warehouse where the average air temperature is 10°C. The steam and the air heat transfer coefficients are estimated to be 800 and 200 W/m² · K, respectively. Calculate (*a*) the daily rate of heat transfer from the superheated steam, and (*b*) the temperature on the outside surface of the gypsum plaster insulation.

17–71 An 8-m-internal-diameter spherical tank made of 1.5-cm-thick stainless steel (k = 15 W/m · °C) is used to store iced water at 0°C. The tank is located in a room whose temperature is 25°C. The walls of the room are also at 25°C. The outer surface of the tank is black (emissivity ε = 1), and heat transfer between the outer surface of the tank and the surroundings is by natural convection and radiation. The convection heat transfer coefficients at the inner and the outer surfaces of the tank are 80 W/m² · °C and 10 W/m² · °C, respectively. Determine (*a*) the rate of heat transfer to the iced water in the tank and (*b*) the amount of ice at 0°C that melts during a 24-h period. The heat of fusion of water at atmospheric pressure is h_{if} = 333.7 kJ/kg.

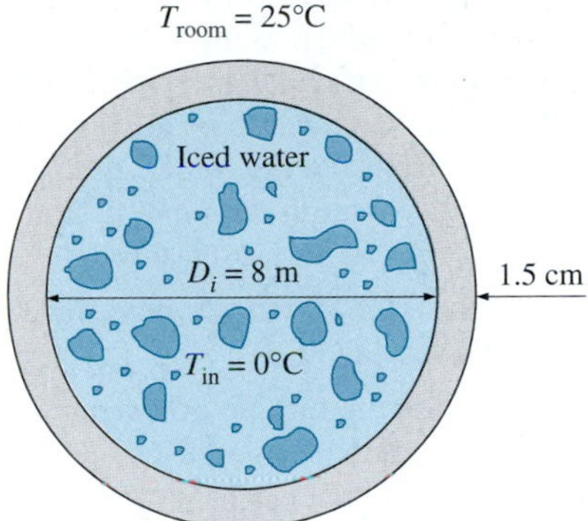

FIGURE P17–71

17–72 Steam at 320°C flows in a stainless steel pipe (k = 15 W/m · °C) whose inner and outer diameters are 5 cm and 5.5 cm, respectively. The pipe is covered with 3-cm-thick glass wool insulation (k = 0.038 W/m · °C). Heat is lost to the surroundings at 5°C by natural convection and radiation, with a combined natural convection and radiation heat transfer coefficient of 15 W/m² · °C. Taking the heat transfer coefficient inside the pipe to be 80 W/m² · °C, determine the rate of heat loss from the steam per unit length of the pipe. Also determine the temperature drops across the pipe shell and the insulation.

17–73 Reconsider Prob. 17–72. Using EES (or other) software, investigate the effect of the thickness of the insulation on the rate of heat loss from the steam and the temperature drop across the insulation layer. Let the insulation thickness vary from 1 cm to 10 cm. Plot the rate of heat loss and the temperature drop as a function of insulation thickness, and discuss the results.

17–74 A 50-m-long section of a steam pipe whose outer diameter is 10 cm passes through an open space at 15°C. The average temperature of the outer surface of the pipe is measured to be 150°C. If the combined heat transfer coefficient on the outer surface of the pipe is 20 W/m² · °C, determine (*a*) the rate of heat loss from the steam pipe; (*b*) the annual cost of this energy lost if steam is generated in a natural gas furnace that has an efficiency of 75 percent and the price of natural gas is \$0.52/therm (1 therm = 105,500 kJ); and (*c*) the thickness of fiberglass insulation (k = 0.035 W/m · °C) needed in order to save 90 percent of the heat lost. Assume the pipe temperature to remain constant at 150°C.

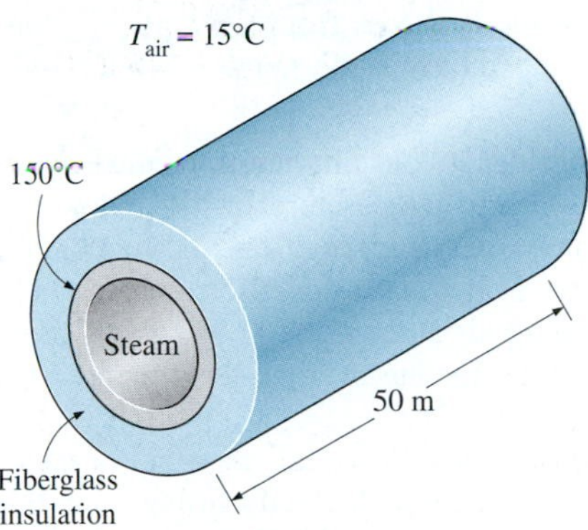

FIGURE P17–74

17–75 Consider a 2-m-high electric hot-water heater that has a diameter of 40 cm and maintains the hot water at 55°C. The tank is located in a small room whose average temperature is 27°C, and the heat transfer coefficients on the inner and outer surfaces of the heater are 50 and 12 W/m² · °C, respectively. The tank is placed in another 46-cm-diameter sheet metal tank of negligible thickness, and the space between the two tanks is filled with foam insulation (k = 0.03 W/m · °C). The thermal resistances of the water tank and the outer thin sheet metal shell are very small and can be neglected. The price of electricity is \$0.08/kWh, and the home owner pays \$280 a year for water heating. Determine the fraction of the hot-water energy cost of this household that is due to the heat loss from the tank.

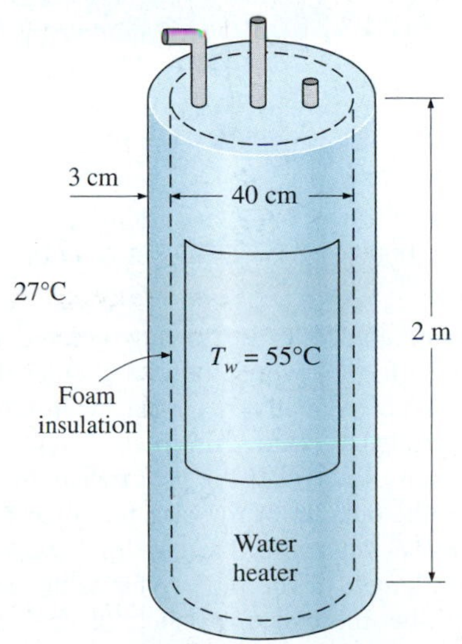

FIGURE P17–75

Hot-water tank insulation kits consisting of 3-cm-thick fiberglass insulation (k = 0.035 W/m · °C) large enough to wrap the entire tank are available in the market for about $30. If such an insulation is installed on this water tank by the home owner himself, how long will it take for this additional insulation to pay for itself? *Answers:* 17.5 percent, 1.5 years

17–76 Reconsider Prob. 17–75. Using EES (or other) software, plot the fraction of energy cost of hot water due to the heat loss from the tank as a function of the hot-water temperature in the range of 40°C to 90°C. Discuss the results.

17–77 Consider a cold aluminum canned drink that is initially at a uniform temperature of 4°C. The can is 12.5 cm high and has a diameter of 6 cm. If the combined convection/radiation heat transfer coefficient between the can and the surrounding air at 25°C is 10 W/m^2 · °C, determine how long it will take for the average temperature of the drink to rise to 15°C.

In an effort to slow down the warming of the cold drink, a person puts the can in a perfectly fitting 1-cm-thick cylindrical rubber insulator (k = 0.13 W/m · °C). Now how long will it take for the average temperature of the drink to rise to 15°C? Assume the top of the can is not covered.

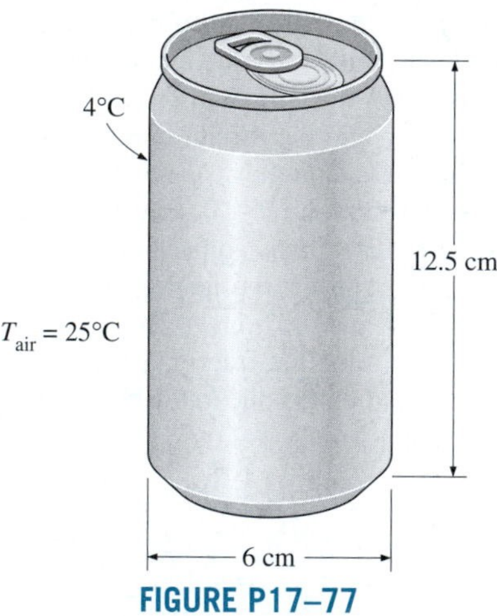

FIGURE P17–77

17–78 Repeat Prob. 17–77, assuming a thermal contact resistance of 0.00008 m^2 · °C/W between the can and the insulation.

17–79E Steam at 450°F is flowing through a steel pipe (k = 8.7 Btu/h · ft · °F) whose inner and outer diameters are 3.5 in and 4.0 in, respectively, in an environment at 55°F. The pipe is insulated with 2-in-thick fiberglass insulation (k = 0.020 Btu/h · ft · °F). If the heat transfer coefficients on the inside and the outside of the pipe are 30 and 5 Btu/h · ft^2 · °F, respectively, determine the rate of heat loss from the steam per foot length of the pipe. What is the error involved in neglecting the thermal resistance of the steel pipe in calculations?

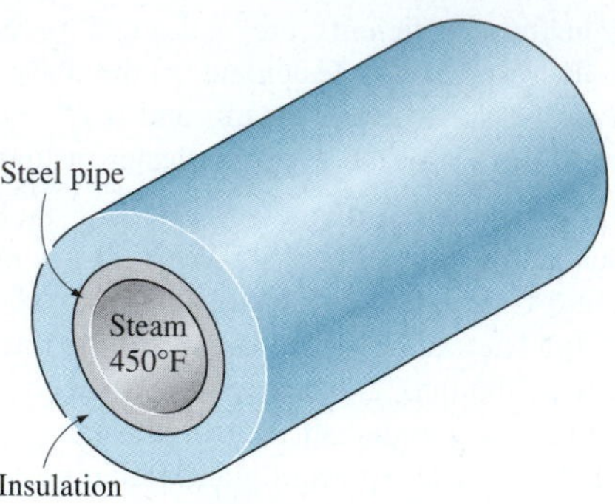

FIGURE P17–79E

17–80 Hot water at an average temperature of 70°C is flowing through a 15-m section of a cast iron pipe (k = 52 W/m · °C) whose inner and outer diameters are 4 cm and 4.6 cm, respectively. The outer surface of the pipe, whose emissivity is 0.7, is exposed to the cold air at 10°C in the basement, with a heat transfer coefficient of 15 W/m^2 · °C. The heat transfer coefficient at the inner surface of the pipe is 120 W/m^2 · °C. Taking the walls of the basement to be at 10°C also, determine the rate of heat loss from the hot water. Also, determine the average velocity of the water in the pipe if the temperature of the water drops by 3°C as it passes through the basement.

17–81 Repeat Prob. 17–80 for a pipe made of copper (k = 386 W/m · °C) instead of cast iron.

17–82E Steam exiting the turbine of a steam power plant at 100°F is to be condensed in a large condenser by cooling water flowing through copper pipes (k = 223 Btu/h · ft · °F) of inner diameter 0.4 in and outer diameter 0.6 in at an average temperature of 70°F. The heat of vaporization of water at 100°F is 1037 Btu/lbm. The heat transfer coefficients are 1500 Btu/h · ft^2 · °F on the steam side and 35 Btu/h · ft^2 · °F on the water side. Determine the length of the tube required to condense steam at a rate of 120 lbm/h. *Answer:* 1148 ft

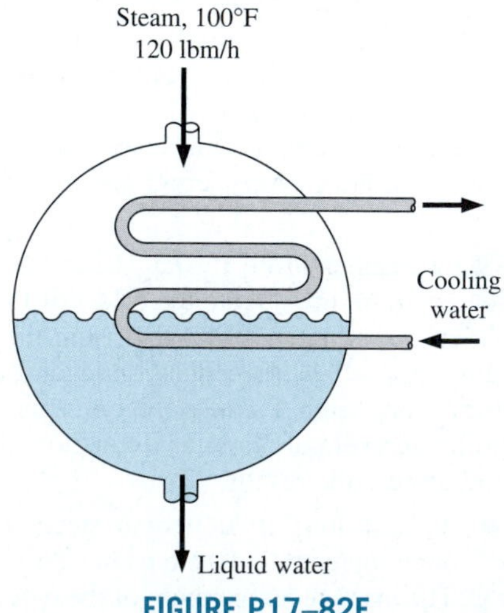

FIGURE P17–82E

17–83E Repeat Prob. 17–82E, assuming that a 0.01-in-thick layer of mineral deposit (k = 0.5 Btu/h · ft · °F) has formed on the inner surface of the pipe.

17–84E Reconsider Prob. 17–82E. Using EES (or other) software, investigate the effects of the thermal conductivity of the pipe material and the outer diameter of the pipe on the length of the tube required. Let the thermal conductivity vary from 10 Btu/h · ft · °F to 400 Btu/h · ft · °F and the outer diameter from 0.5 in to 1.0 in. Plot the length of the tube as functions of pipe conductivity and the outer pipe diameter, and discuss the results.

17–85 The boiling temperature of nitrogen at atmospheric pressure at sea level (1 atm pressure) is −196°C. Therefore, nitrogen is commonly used in low-temperature scientific studies since the temperature of liquid nitrogen in a tank open to the atmosphere will remain constant at −196°C until it is depleted. Any heat transfer to the tank will result in the evaporation of some liquid nitrogen, which has a heat of vaporization of 198 kJ/kg and a density of 810 kg/m^3 at 1 atm.

Consider a 3-m-diameter spherical tank that is initially filled with liquid nitrogen at 1 atm and −196°C. The tank is exposed to ambient air at 15°C, with a combined convection and radiation heat transfer coefficient of 35 W/m^2 · °C. The temperature of the thin-shelled spherical tank is observed to be almost the same as the temperature of the nitrogen inside. Determine the rate of evaporation of the liquid nitrogen in the tank as a result of the heat transfer from the ambient air if the tank is (*a*) not insulated, (*b*) insulated with 5-cm-thick fiberglass insulation (k = 0.035 W/m · °C), and (*c*) insulated with 2-cm-thick superinsulation which has an effective thermal conductivity of 0.00005 W/m · °C.

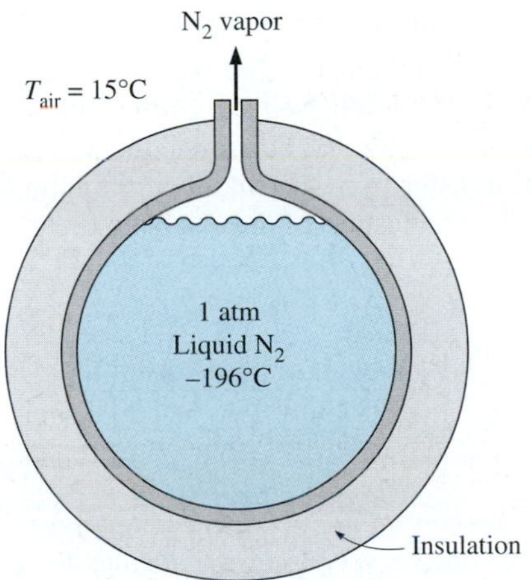

FIGURE P17–85

17–86 Repeat Prob. 17–85 for liquid oxygen, which has a boiling temperature of −183°C, a heat of vaporization of 213 kJ/kg, and a density of 1140 kg/m^3 at 1 atm pressure.

Critical Radius of Insulation

17–87C What is the critical radius of insulation? How is it defined for a cylindrical layer?

17–88C A pipe is insulated such that the outer radius of the insulation is less than the critical radius. Now the insulation is taken off. Will the rate of heat transfer from the pipe increase or decrease for the same pipe surface temperature?

17–89C A pipe is insulated to reduce the heat loss from it. However, measurements indicate that the rate of heat loss has increased instead of decreasing. Can the measurements be right?

17–90C Consider a pipe at a constant temperature whose radius is greater than the critical radius of insulation. Someone claims that the rate of heat loss from the pipe has increased when some insulation is added to the pipe. Is this claim valid?

17–91C Consider an insulated pipe exposed to the atmosphere. Will the critical radius of insulation be greater on calm days or on windy days? Why?

17–92 A 2.2-mm-diameter and 10-m-long electric wire is tightly wrapped with a 1-mm-thick plastic cover whose thermal conductivity is k = 0.15 W/m · °C. Electrical measurements indicate that a current of 13 A passes through the wire and there is a voltage drop of 8 V along the wire. If the insulated wire is exposed to a medium at T_∞ = 30°C with a heat transfer coefficient of h = 24 W/m^2 · °C, determine the temperature at the interface of the wire and the plastic cover in steady operation. Also determine if doubling the thickness of the plastic cover will increase or decrease this interface temperature.

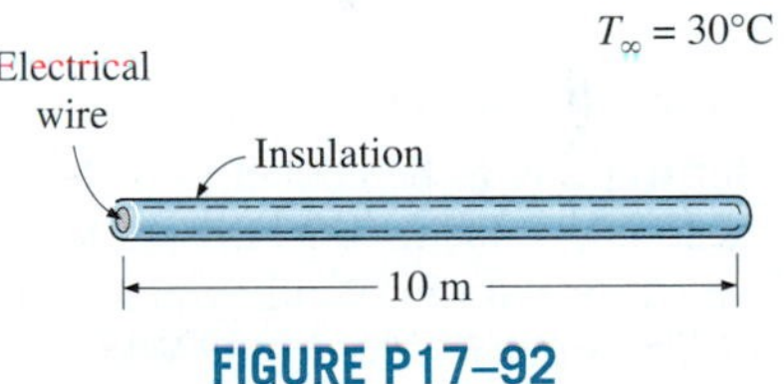

FIGURE P17–92

17–93E A 0.083-in-diameter electrical wire at 90°F is covered by 0.02-in-thick plastic insulation (k = 0.075 Btu/h · ft · °F). The wire is exposed to a medium at 50°F, with a combined convection and radiation heat transfer coefficient of 2.5 Btu/h · ft^2 · °F. Determine if the plastic insulation on the wire will increase or decrease heat transfer from the wire. *Answer:* It helps

17–94E Repeat Prob. 17–93E, assuming a thermal contact resistance of 0.001 h · ft^2 · °F/Btu at the interface of the wire and the insulation.

17–95 A 5-mm-diameter spherical ball at 50°C is covered by a 1-mm-thick plastic insulation (k = 0.13 W/m · °C). The ball is exposed to a medium at 15°C, with a combined convection and radiation heat transfer coefficient of 20 W/m^2 · °C.

Determine if the plastic insulation on the ball will help or hurt heat transfer from the ball.

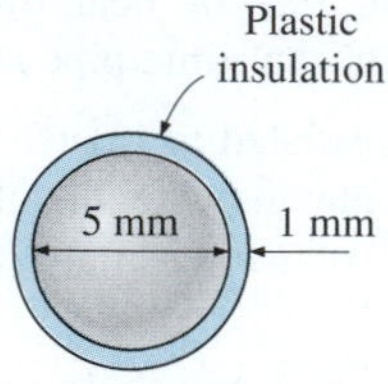

FIGURE P17–95

17–96 Reconsider Prob. 17–95. Using EES (or other) software, plot the rate of heat transfer from the ball as a function of the plastic insulation thickness in the range of 0.5 mm to 20 mm. Discuss the results.

Heat Transfer from Finned Surfaces

17–97C What is the reason for the widespread use of fins on surfaces?

17–98C What is the difference between the fin effectiveness and the fin efficiency?

17–99C The fins attached to a surface are determined to have an effectiveness of 0.9. Do you think the rate of heat transfer from the surface has increased or decreased as a result of the addition of these fins?

17–100C Explain how the fins enhance heat transfer from a surface. Also, explain how the addition of fins may actually decrease heat transfer from a surface.

17–101C How does the overall effectiveness of a finned surface differ from the effectiveness of a single fin?

17–102C Hot water is to be cooled as it flows through the tubes exposed to atmospheric air. Fins are to be attached in order to enhance heat transfer. Would you recommend attaching the fins inside or outside the tubes? Why?

17–103C Hot air is to be cooled as it is forced to flow through the tubes exposed to atmospheric air. Fins are to be added in order to enhance heat transfer. Would you recommend attaching the fins inside or outside the tubes? Why? When would you recommend attaching fins both inside and outside the tubes?

17–104C Consider two finned surfaces that are identical except that the fins on the first surface are formed by casting or extrusion, whereas they are attached to the second surface afterwards by welding or tight fitting. For which case do you think the fins will provide greater enhancement in heat transfer? Explain.

17–105C The heat transfer surface area of a fin is equal to the sum of all surfaces of the fin exposed to the surrounding medium, including the surface area of the fin tip. Under what conditions can we neglect heat transfer from the fin tip?

17–106C Does the (*a*) efficiency and (*b*) effectiveness of a fin increase or decrease as the fin length is increased?

17–107C Two pin fins are identical, except that the diameter of one of them is twice the diameter of the other. For which fin is the (*a*) fin effectiveness and (*b*) fin efficiency higher? Explain.

17–108C Two plate fins of constant rectangular cross section are identical, except that the thickness of one of them is twice the thickness of the other. For which fin is the (*a*) fin effectiveness and (*b*) fin efficiency higher? Explain.

17–109C Two finned surfaces are identical, except that the convection heat transfer coefficient of one of them is twice that of the other. For which finned surface is the (*a*) fin effectiveness and (*b*) fin efficiency higher? Explain.

17–110 Obtain a relation for the fin efficiency for a fin of constant cross-sectional area A_c, perimeter p, length L, and thermal conductivity k exposed to convection to a medium at T_∞ with a heat transfer coefficient h. Assume the fins are sufficiently long so that the temperature of the fin at the tip is nearly T_∞. Take the temperature of the fin at the base to be T_b and neglect heat transfer from the fin tips. Simplify the relation for (*a*) a circular fin of diameter D and (*b*) rectangular fins of thickness t.

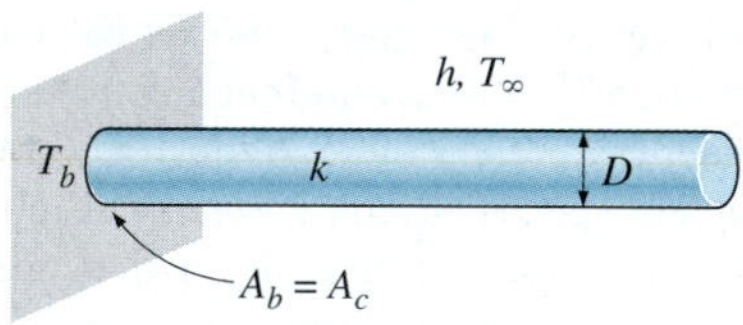

FIGURE P17–110

17–111 The case-to-ambient thermal resistance of a power transistor that has a maximum power rating of 15 W is given to be 25°C/W. If the case temperature of the transistor is not to exceed 80°C, determine the power at which this transistor can be operated safely in an environment at 40°C.

17–112 A 4-mm-diameter and 10-cm-long aluminum fin (k = 237 W/m · °C) is attached to a surface. If the heat transfer coefficient is 12 W/m^2 · °C, determine the percent error in the rate of heat transfer from the fin when the infinitely long fin assumption is used instead of the adiabatic fin tip assumption.

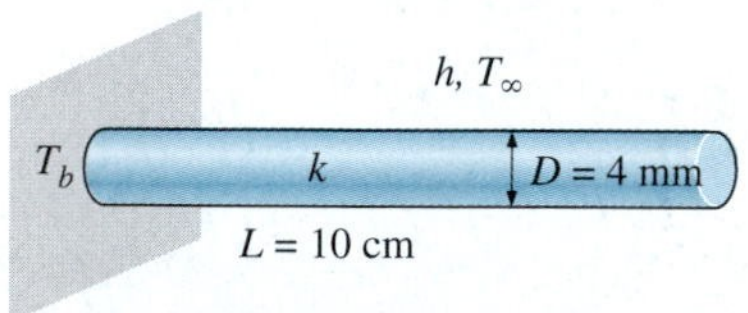

FIGURE P17–112

17–113 Consider a very long rectangular fin attached to a flat surface such that the temperature at the end of the fin is essentially that of the surrounding air, i.e. 20°C. Its width is 5.0 cm; thickness is 1.0 mm; thermal conductivity is 200 W/m · K; and base temperature is 40°C. The heat transfer coefficient is 20 W/m^2 · K. Estimate the fin temperature at a distance of 5.0 cm from the base and the rate of heat loss from the entire fin.

17–114 Circular cooling fins of diameter D = 1 mm and length L = 25.4 mm, made of copper (k = 400 W/m · K), are used to enhance heat transfer from a surface that is maintained at temperature $T_{s1} = 132°C$. Each rod has one end attached to this surface ($x = 0$), while the opposite end ($x = L$) is joined to a second surface, which is maintained at $T_{s2} = 0°C$. The air flowing between the surfaces and the rods is also at $T_\infty = 0°C$, and the convection coefficient is $h = 100\ W/m^2 \cdot K$. For fin with prescribed tip temperature, the temperature distribution and the rate of heat transfer are given by

$$\frac{\theta}{\theta_b} = \frac{\theta_L/\theta_b \sinh(mx) + \sinh[m(L - x)]}{\sinh(mL)} \quad \text{and}$$

$$\dot{Q} = \theta_b \sqrt{hpkA_c}\,\frac{\cosh(mL)}{\sinh(mL)}$$

(*a*) Express the function $\theta(x) = T(x) - T_\infty$ along a fin, and calculate the temperature at $x = L/2$.

(*b*) Determine the rate of heat transferred from the hot surface through each fin and the fin effectiveness. Is the use of fins justified? Why?

(*c*) What is the total rate of heat transfer from a 10-cm by 10-cm section of the wall, which has 625 uniformly distributed fins? Assume the same convection coefficient for the fin and for the unfinned wall surface.

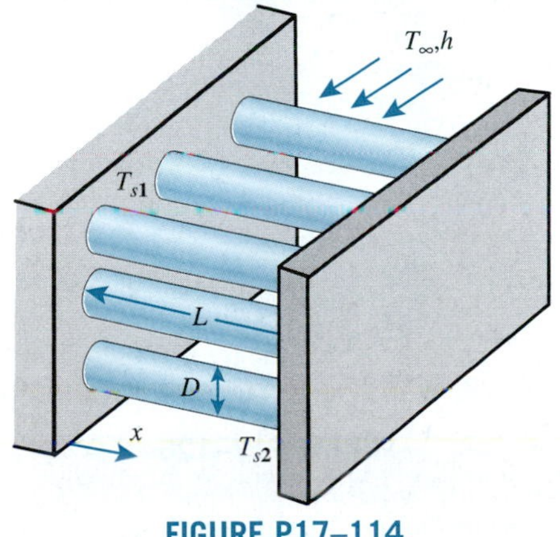

FIGURE P17–114

17–115 A 40-W power transistor is to be cooled by attaching it to one of the commercially available heat sinks shown in Table 17–6. Select a heat sink that will allow the case temperature of the transistor not to exceed 90°C in the ambient air at 20°C.

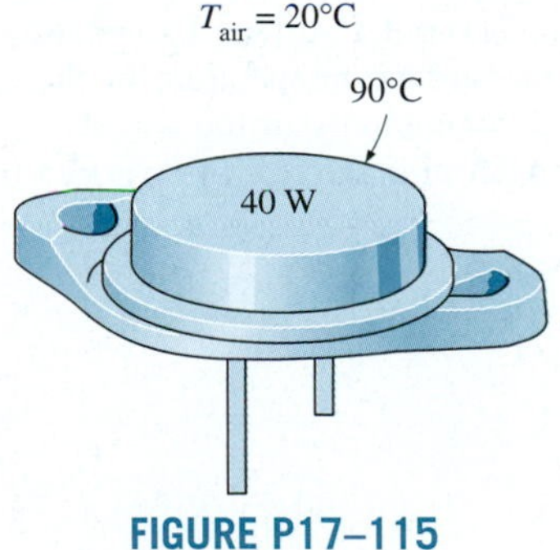

FIGURE P17–115

17–116 A 25-W power transistor is to be cooled by attaching it to one of the commercially available heat sinks shown in Table 17–6. Select a heat sink that will allow the case temperature of the transistor not to exceed 55°C in the ambient air at 18°C.

17–117 Steam in a heating system flows through tubes whose outer diameter is 5 cm and whose walls are maintained at a temperature of 180°C. Circular aluminum alloy 2024-T6 fins ($k = 186\ W/m \cdot °C$) of outer diameter 6 cm and constant thickness 1 mm are attached to the tube. The space between the fins is 3 mm, and thus there are 250 fins per meter length of the tube. Heat is transferred to the surrounding air at $T_\infty = 25°C$, with a heat transfer coefficient of $40\ W/m^2 \cdot °C$. Determine the increase in heat transfer from the tube per meter of its length as a result of adding fins. *Answer:* 2639 W

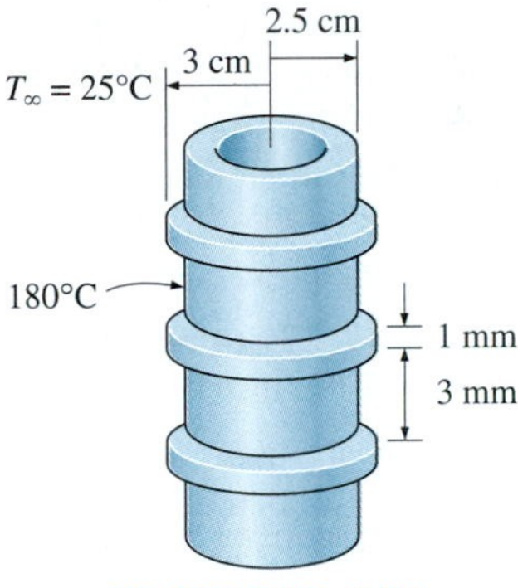

FIGURE P17–117

17–118E Consider a stainless steel spoon ($k = 8.7\ Btu/h \cdot ft \cdot °F$) partially immersed in boiling water at 200°F in a kitchen at 75°F. The handle of the spoon has a cross section of 0.08 in × 0.5 in, and extends 7 in in the air from the free surface of the water. If the heat transfer coefficient at the exposed surfaces of the spoon handle is $3\ Btu/h \cdot ft^2 \cdot °F$, determine the temperature difference across the exposed surface of the spoon handle. State your assumptions. *Answer:* 124.6°F

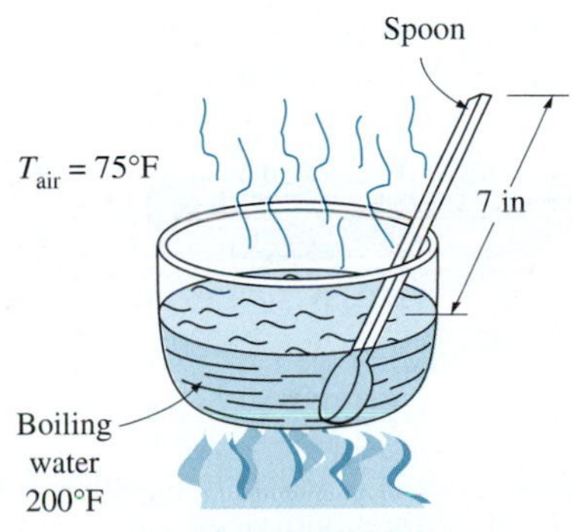

FIGURE P17–118E

17–119E Repeat Prob. 17–118E for a silver spoon ($k = 247\ Btu/h \cdot ft \cdot °F$).

17–120E Reconsider Prob. 17–118E. Using EES (or other) software, investigate the effects of the thermal conductivity of the spoon material and the length of its extension in the air on the temperature difference across the exposed surface of the spoon handle. Let the thermal conductivity vary from 5 Btu/h · ft · °F to 225 Btu/h · ft · °F and the length from 5 in to 12 in. Plot the temperature difference as the functions of thermal conductivity and length, and discuss the results.

17–121 A 0.3-cm-thick, 12-cm-high, and 18-cm-long circuit board houses 80 closely spaced logic chips on one side, each dissipating 0.04 W. The board is impregnated with copper fillings and has an effective thermal conductivity of 30 W/m · °C. All the heat generated in the chips is conducted across the circuit board and is dissipated from the back side of the board to a medium at 40°C, with a heat transfer coefficient of 40 W/m^2 · °C. (*a*) Determine the temperatures on the two sides of the circuit board. (*b*) Now a 0.2-cm-thick, 12-cm-high, and 18-cm-long aluminum plate (k = 237 W/m · °C) with 864 2-cm-long aluminum pin fins of diameter 0.25 cm is attached to the back side of the circuit board with a 0.02-cm-thick epoxy adhesive (k = 1.8 W/m · °C). Determine the new temperatures on the two sides of the circuit board.

17–122 Repeat Prob. 17–121 using a copper plate with copper fins (k = 386 W/m · °C) instead of aluminum ones.

17–123 A hot surface at 100°C is to be cooled by attaching 3-cm-long, 0.25-cm-diameter aluminum pin fins (k = 237 W/m · °C) to it, with a center-to-center distance of 0.6 cm. The temperature of the surrounding medium is 30°C, and the heat transfer coefficient on the surfaces is 35 W/m^2 · °C. Determine the rate of heat transfer from the surface for a 1-m × 1-m section of the plate. Also determine the overall effectiveness of the fins.

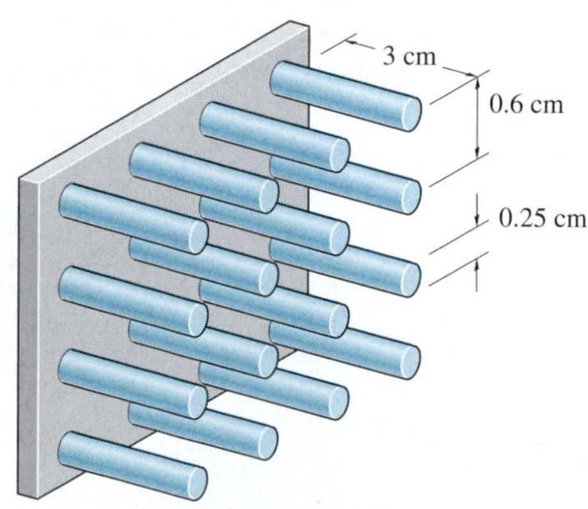

FIGURE P17–123

17–124 Repeat Prob. 17–123 using copper fins (k = 386 W/m · °C) instead of aluminum ones.

17–125 Reconsider Prob. 17–123. Using EES (or other) software, investigate the effect of the center-to-center distance of the fins on the rate of heat transfer from the surface and the overall effectiveness of the fins. Let the center-to-center distance vary from 0.4 cm to 2.0 cm. Plot the rate of heat transfer and the overall effectiveness as a function of the center-to-center distance, and discuss the results.

17–126 Two 3-m-long and 0.4-cm-thick cast iron (k = 52 W/m · °C) steam pipes of outer diameter 10 cm are connected to each other through two 1-cm-thick flanges of outer diameter 20 cm. The steam flows inside the pipe at an average temperature of 200°C with a heat transfer coefficient of 180 W/m^2 · °C. The outer surface of the pipe is exposed to an ambient at 12°C, with a heat transfer coefficient of 25 W/m^2 · °C. (*a*) Disregarding the flanges, determine the average outer surface temperature of the pipe. (*b*) Using this temperature for the base of the flange and treating the flanges as the fins, determine the fin efficiency and the rate of heat transfer from the flanges. (*c*) What length of pipe is the flange section equivalent to for heat transfer purposes?

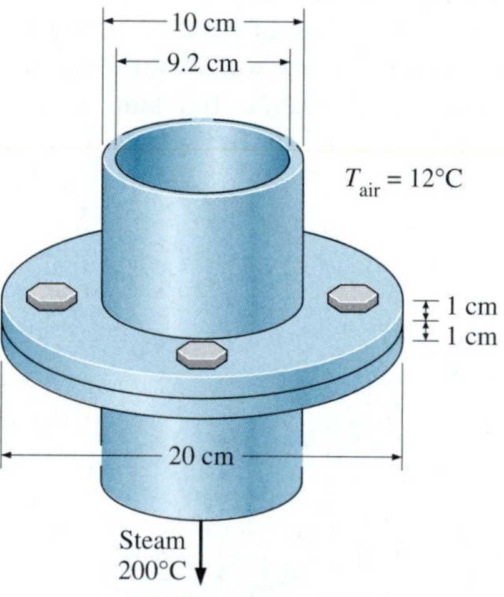

FIGURE P17–126

Heat Transfer in Common Configurations

17–127C What is a conduction shape factor? How is it related to the thermal resistance?

17–128C What is the value of conduction shape factors in engineering?

17–129 A 20-m-long and 8-cm-diameter hot-water pipe of a district heating system is buried in the soil 80 cm below the ground surface. The outer surface temperature of the pipe is 60°C. Taking the surface temperature of the earth to be 5°C and the thermal conductivity of the soil at that location to be 0.9 W/m · °C, determine the rate of heat loss from the pipe.

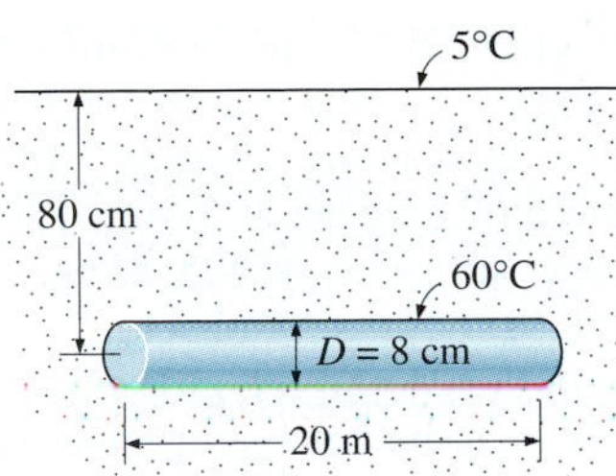

FIGURE P17–129

17–130 Reconsider Prob. 17–129. Using EES (or other) software, plot the rate of heat loss from the pipe as a function of the burial depth in the range of 20 cm to 2.0 m. Discuss the results.

17–131 Hot- and cold-water pipes 8 m long run parallel to each other in a thick concrete layer. The diameters of both pipes are 5 cm, and the distance between the centerlines of the pipes is 40 cm. The surface temperatures of the hot and cold pipes are 60°C and 15°C, respectively. Taking the thermal conductivity of the concrete to be $k = 0.75$ W/m · °C, determine the rate of heat transfer between the pipes. *Answer:* 306 W

17–132 Reconsider Prob. 17–131. Using EES (or other) software, plot the rate of heat transfer between the pipes as a function of the distance between the centerlines of the pipes in the range of 10 cm to 1.0 m. Discuss the results.

17–133E A row of 3-ft-long and 1-in-diameter used uranium fuel rods that are still radioactive are buried in the ground parallel to each other with a center-to-center distance of 8 in at a depth 15 ft from the ground surface at a location where the thermal conductivity of the soil is 0.6 Btu/h · ft · °F. If the surface temperature of the rods and the ground are 350°F and 60°F, respectively, determine the rate of heat transfer from the fuel rods to the atmosphere through the soil.

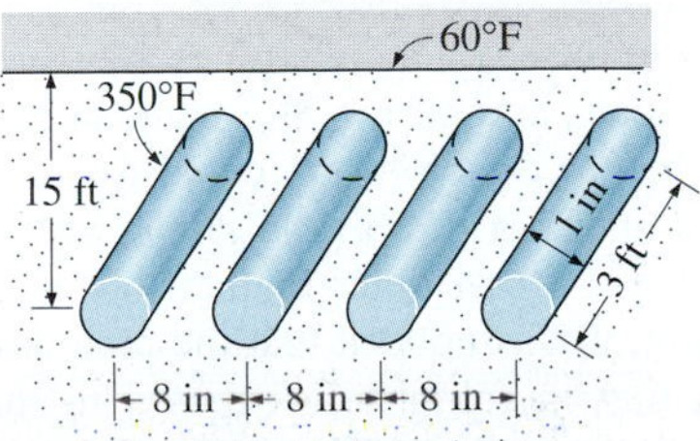

FIGURE P17–133E

17–134 Hot water at an average temperature of 53°C and an average velocity of 0.4 m/s is flowing through a 5-m section of a thin-walled hot-water pipe that has an outer diameter of 2.5 cm. The pipe passes through the center of a 14-cm-thick wall filled with fiberglass insulation ($k = 0.035$ W/m · °C). If the surfaces of the wall are at 18°C, determine (*a*) the rate of heat transfer from the pipe to the air in the rooms and (*b*) the temperature drop of the hot water as it flows through this 5-m-long section of the wall. *Answers:* 19.6 W, 0.024°C

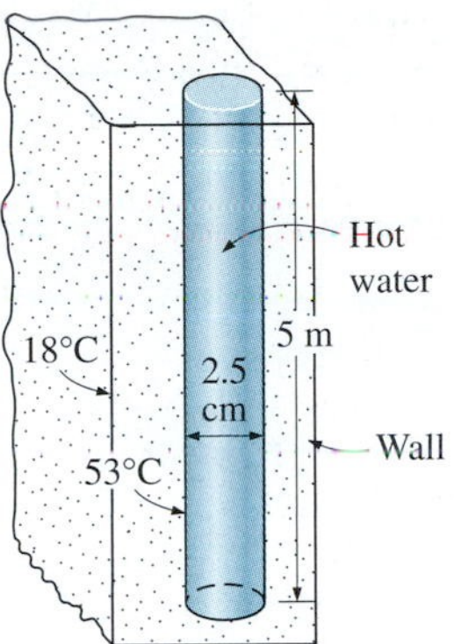

FIGURE P17–134

17–135 Hot water at an average temperature of 80°C and an average velocity of 1.5 m/s is flowing through a 25-m section of a pipe that has an outer diameter of 5 cm. The pipe extends 2 m in the ambient air above the ground, dips into the ground ($k = 1.5$ W/m · °C) vertically for 3 m, and continues horizontally at this depth for 20 m more before it enters the next building. The first section of the pipe is exposed to the ambient air at 8°C, with a heat transfer coefficient of 22 W/m^2 · °C. If the surface of the ground is covered with snow at 0°C, determine (*a*) the total rate of heat loss from the hot water and (*b*) the temperature drop of the hot water as it flows through this 25-m-long section of the pipe.

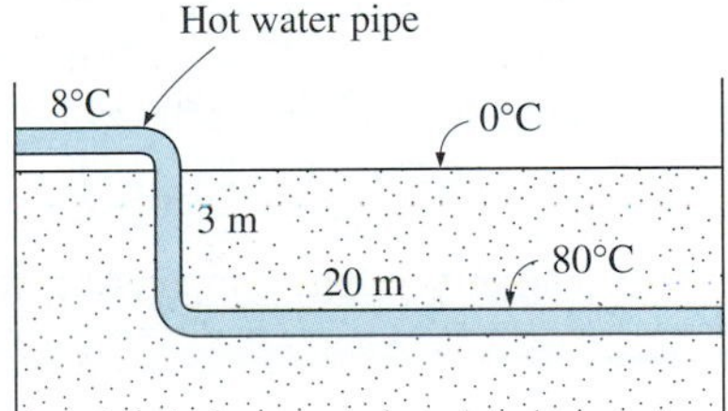

FIGURE P17–135

17–136 Consider a house with a flat roof whose outer dimensions are 12 m × 12 m. The outer walls of the house are 6 m high. The walls and the roof of the house are made of 20-cm-thick concrete ($k = 0.75$ W/m · °C). The temperatures of the inner and outer surfaces of the house are 15°C and 3°C, respectively. Accounting for the effects of the edges of adjoining surfaces, determine the rate of heat loss from the house through its walls and the roof. What is the error involved in ignoring the effects of the edges and corners and treating the roof as a 12 m × 12 m surface and the walls as 6 m × 12 m surfaces for simplicity?

17–137 Consider a 25-m-long thick-walled concrete duct ($k = 0.75$ W/m · °C) of square cross section. The outer dimensions of the duct are 20 cm × 20 cm, and the thickness of the duct wall is 2 cm. If the inner and outer surfaces of the duct are at 100°C and 30°C, respectively, determine the rate of heat transfer through the walls of the duct. *Answer*: 47.1 kW

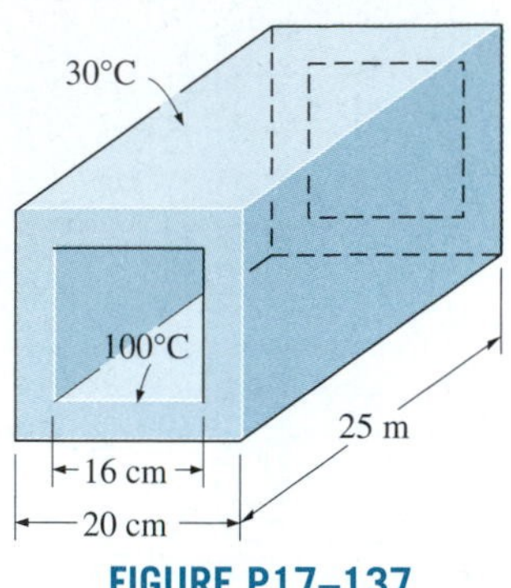

FIGURE P17–137

17–138 A 3-m-diameter spherical tank containing some radioactive material is buried in the ground (k = 1.4 W/m · °C). The distance between the top surface of the tank and the ground surface is 4 m. If the surface temperatures of the tank and the ground are 140°C and 15°C, respectively, determine the rate of heat transfer from the tank.

17–139 Reconsider Prob. 17–138. Using EES (or other) software, plot the rate of heat transfer from the tank as a function of the tank diameter in the range of 0.5 m to 5.0 m. Discuss the results.

17–140 Hot water at an average temperature of 85°C passes through a row of eight parallel pipes that are 4 m long and have an outer diameter of 3 cm, located vertically in the middle of a concrete wall (k = 0.75 W/m · °C) that is 4 m high, 8 m long, and 15 cm thick. If the surfaces of the concrete walls are exposed to a medium at 32°C, with a heat transfer coefficient of 12 W/m^2 · °C, determine the rate of heat loss from the hot water and the surface temperature of the wall.

Review Problems

17–141E Steam is produced in the copper tubes (k = 223 Btu/h · ft · °F) of a heat exchanger at a temperature of 250°F by another fluid condensing on the outside surfaces of the tubes at 350°F. The inner and outer diameters of the tube are 1 in and 1.3 in, respectively. When the heat exchanger was new, the rate of heat transfer per foot length of the tube was 2 × 10^4 Btu/h. Determine the rate of heat transfer per foot length of the tube when a 0.01-in-thick layer of limestone (k = 1.7 Btu/h · ft · °F) has formed on the inner surface of the tube after extended use.

17–142E Repeat Prob. 17–141E, assuming that a 0.01-in-thick limestone layer has formed on both the inner and outer surfaces of the tube.

17–143 A 1.2-m-diameter and 6-m-long cylindrical propane tank is initially filled with liquid propane whose density is 581 kg/m^3. The tank is exposed to the ambient air at 30°C, with a heat transfer coefficient of 25 W/m^2 · °C. Now a crack develops at the top of the tank and the pressure inside drops to 1 atm while the temperature drops to −42°C, which is the boiling temperature of propane at 1 atm. The heat of vaporization of propane at 1 atm is 425 kJ/kg. The propane is slowly vaporized as a result of the heat transfer from the ambient air into the tank, and the propane vapor escapes the tank at −42°C through the crack. Assuming the propane tank to be at about the same temperature as the propane inside at all times, determine how long it will take for the propane tank to empty if the tank is (*a*) not insulated and (*b*) insulated with 5-cm-thick glass wool insulation (k = 0.038 W/m · °C).

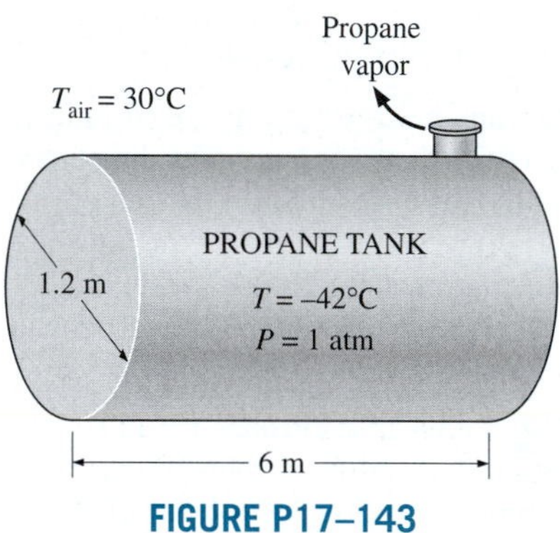

FIGURE P17–143

17–144 Hot water is flowing at an average velocity of 1.5 m/s through a cast iron pipe (k = 52 W/m · °C) whose inner and outer diameters are 3 cm and 3.5 cm, respectively. The pipe passes through a 15-m-long section of a basement whose temperature is 15°C. If the temperature of the water drops from 70°C to 67°C as it passes through the basement and the heat transfer coefficient on the inner surface of the pipe is 400 W/m^2 · °C, determine the combined convection and radiation heat transfer coefficient at the outer surface of the pipe. *Answer:* 272.5 W/m^2 · °C

17–145 Newly formed concrete pipes are usually cured first overnight by steam in a curing kiln maintained at a temperature of 45°C before the pipes are cured for several days outside. The heat and moisture to the kiln is provided by steam flowing in a pipe whose outer diameter is 12 cm. During a plant inspection, it was noticed that the pipe passes through a 10-m section that is completely exposed to the ambient air before it reaches the kiln. The temperature measurements indicate that the average temperature of the outer surface of the steam pipe is 82°C when the ambient temperature is 8°C. The combined convection and radiation heat transfer coefficient at the outer surface of the pipe is estimated to be 35 W/m^2 · °C. Determine the amount of heat lost from the steam during a 10-h curing process that night.

Steam is supplied by a gas-fired steam generator that has an efficiency of 85 percent, and the plant pays $1.20/therm of natural gas (1 therm = 105,500 kJ). If the pipe is insulated and 90 percent of the heat loss is saved as a result, determine the amount of money this facility will save a year as a result of insulating the steam pipes. Assume that the concrete pipes are cured 110 nights a year. State your assumptions.

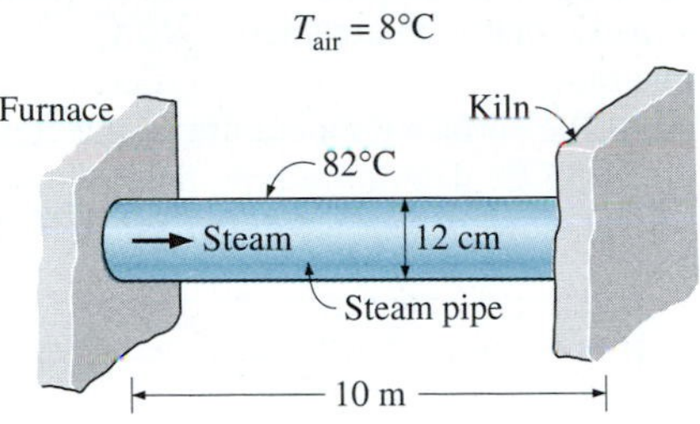

FIGURE P17–145

17–146 Consider an 18-cm × 18-cm multilayer circuit board dissipating 27 W of heat. The board consists of four layers of 0.2-mm-thick copper (k = 386 W/m · °C) and three layers of 1.5-mm-thick epoxy glass (k = 0.26 W/m · °C) sandwiched together, as shown in the figure. The circuit board is attached to a heat sink from both ends, and the temperature of the board at those ends is 35°C. Heat is considered to be uniformly generated in the epoxy layers of the board at a rate of 0.5 W per 1-cm × 18-cm epoxy laminate strip (or 1.5 W per 1-cm × 18-cm strip of the board). Considering only a portion of the board because of symmetry, determine the magnitude and location of the maximum temperature that occurs in the board. Assume heat transfer from the top and bottom faces of the board to be negligible.

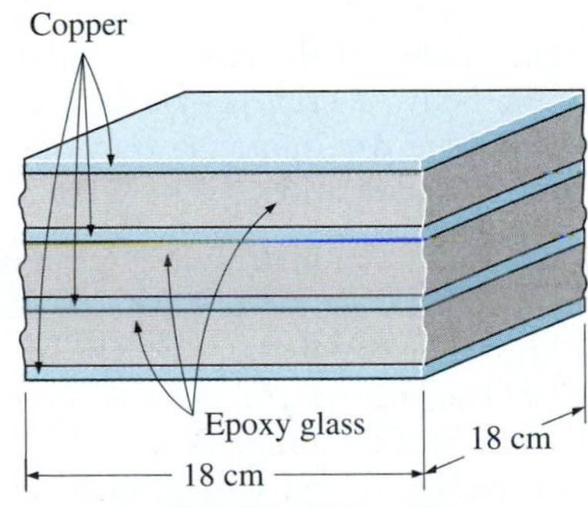

FIGURE P17–146

17–147 The plumbing system of a house involves a 0.5-m section of a plastic pipe (k = 0.16 W/m · °C) of inner diameter 2 cm and outer diameter 2.4 cm exposed to the ambient air. During a cold and windy night, the ambient air temperature remains at about −5°C for a period of 14 h. The combined convection and radiation heat transfer coefficient on the outer surface of the pipe is estimated to be 40 W/m² · °C, and the heat of fusion of water is 333.7 kJ/kg. Assuming the pipe to contain stationary water initially at 0°C, determine if the water in that section of the pipe will completely freeze that night.

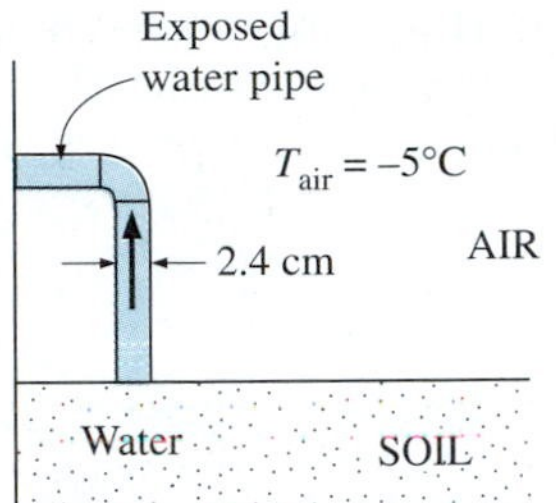

FIGURE P17–147

17–148 Repeat Prob. 17–147 for the case of a heat transfer coefficient of 10 W/m² · °C on the outer surface as a result of putting a fence around the pipe that blocks the wind.

17–149E The surface temperature of a 3-in-diameter baked potato is observed to drop from 300°F to 200°F in 5 min in an environment at 70°F. Determine the average heat transfer coefficient between the potato and its surroundings. Using this heat transfer coefficient and the same surface temperature, determine how long it will take for the potato to experience the same temperature drop if it is wrapped completely in a 0.12-in-thick towel (k = 0.035 Btu/h · ft · °F). You may use the properties of water for potato.

17–150E Repeat Prob. 17–149E assuming there is a 0.02-in-thick air space (k = 0.015 Btu/h · ft · °F) between the potato and the towel.

17–151 An ice chest whose outer dimensions are 30 cm × 40 cm × 50 cm is made of 3-cm-thick Styrofoam (k = 0.033 W/m · °C). Initially, the chest is filled with 50 kg of ice at 0°C, and the inner surface temperature of the ice chest can be taken to be 0°C at all times. The heat of fusion of ice at 0°C is 333.7 kJ/kg, and the heat transfer coefficient between the outer surface of the ice chest and surrounding air at 28°C is 18 W/m² · °C. Disregarding any heat transfer from the 40-cm × 50-cm base of the ice chest, determine how long it will take for the ice in the chest to melt completely.

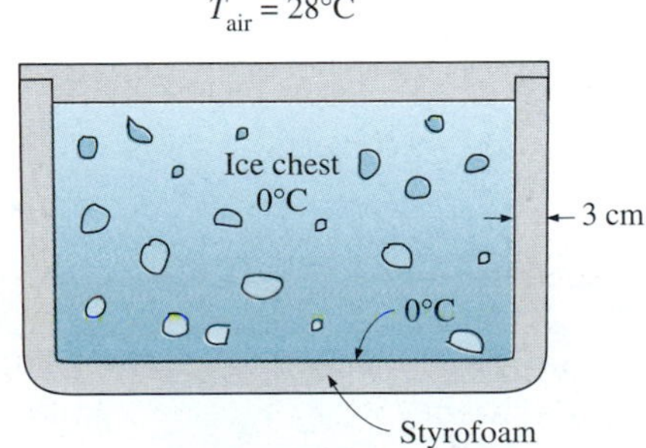

FIGURE P17–151

17–152 A 4-m-high and 6-m-long wall is constructed of two large 2-cm-thick steel plates (k = 15 W/m · °C) separated by 1-cm-thick and 20-cm-wide steel bars placed 99 cm apart. The remaining space between the steel plates is filled with fiberglass insulation (k = 0.035 W/m · °C). If the temperature difference between the inner and the outer surfaces of the walls is

22°C, determine the rate of heat transfer through the wall. Can we ignore the steel bars between the plates in heat transfer analysis since they occupy only 1 percent of the heat transfer surface area?

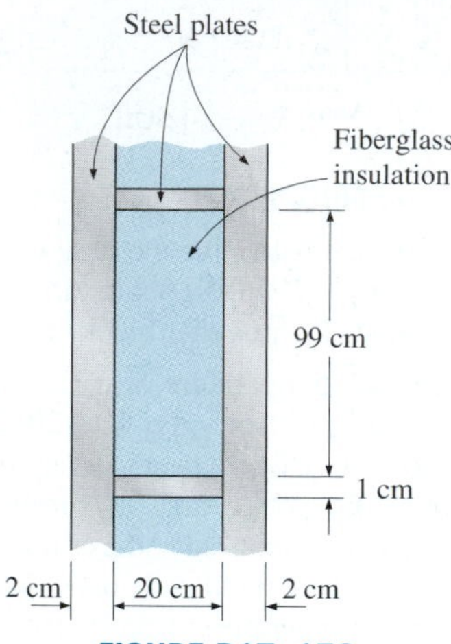

FIGURE P17–152

17–153 A 0.2-cm-thick, 10-cm-high, and 15-cm-long circuit board houses electronic components on one side that dissipate a total of 15 W of heat uniformly. The board is impregnated with conducting metal fillings and has an effective thermal conductivity of 12 W/m · °C. All the heat generated in the components is conducted across the circuit board and is dissipated from the back side of the board to a medium at 37°C, with a heat transfer coefficient of 45 W/m^2 · °C. (*a*) Determine the surface temperatures on the two sides of the circuit board. (*b*) Now a 0.1-cm-thick, 10-cm-high, and 15-cm-long aluminum plate (k = 237 W/m · °C) with 20 0.2-cm-thick, 2-cm-long, and 15-cm-wide aluminum fins of rectangular profile are attached to the back side of the circuit board with a 0.03-cm-thick epoxy adhesive (k = 1.8 W/m · °C). Determine the new temperatures on the two sides of the circuit board.

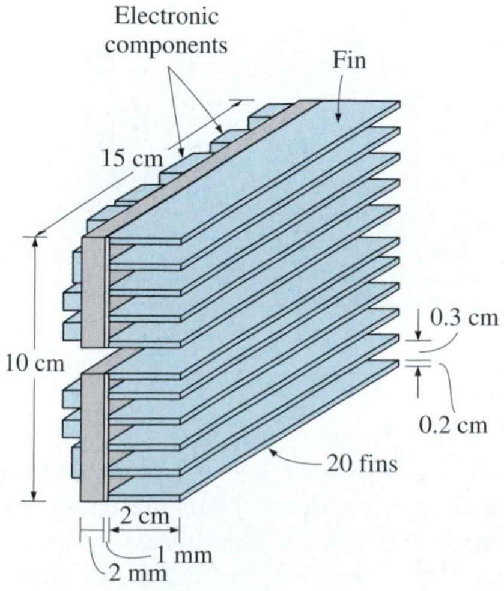

FIGURE P17–153

17–154 Repeat Prob. 17–153 using a copper plate with copper fins (k = 386 W/m · °C) instead of aluminum ones.

17–155 A row of 10 parallel pipes that are 5 m long and have an outer diameter of 6 cm are used to transport steam at 145°C through the concrete floor (k = 0.75 W/m · °C) of a 10-m × 5-m room that is maintained at 20°C. The combined convection and radiation heat transfer coefficient at the floor is 12 W/m^2 · °C. If the surface temperature of the concrete floor is not to exceed 35°C, determine how deep the steam pipes should be buried below the surface of the concrete floor.

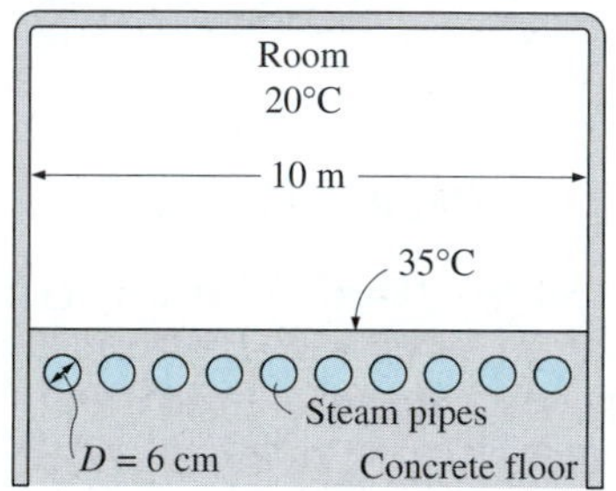

FIGURE P17–155

17–156 Consider two identical people each generating 60 W of metabolic heat steadily while doing sedentary work, and dissipating it by convection and perspiration. The first person is wearing clothes made of 1-mm-thick leather (k = 0.159 W/m · °C) that covers half of the body while the second one is wearing clothes made of 1-mm-thick synthetic fabric (k = 0.13 W/m · °C) that covers the body completely. The ambient air is at 30°C, the heat transfer coefficient at the outer surface is 15 W/m^2 · °C, and the inner surface temperature of the clothes can be taken to be 32°C. Treating the body of each person as a 25-cm-diameter, 1.7-m-long cylinder, determine the fractions of heat lost from each person by perspiration.

17–157 A 6-m-wide, 2.8-m-high wall is constructed of one layer of common brick (k = 0.72 W/m · °C) of thickness 20 cm, one inside layer of light-weight plaster (k = 0.36 W/m · °C) of thickness 1 cm, and one outside layer of cement based covering (k = 1.40 W/m · °C) of thickness 2 cm. The inner surface of the wall is maintained at 23°C while the outer surface is exposed to outdoors at 8°C with a combined convection and radiation heat transfer coefficient of 17 W/m^2 · °C. Determine the rate of heat transfer through the wall and temperature drops across the plaster, brick, covering, and surface-ambient air.

17–158 Reconsider Prob. 17–157. It is desired to insulate the wall in order to decrease the heat loss by 85 percent. For the same inner surface temperature, determine the thickness of insulation and the outer surface temperature if the wall is insulated with (*a*) polyurethane foam (k = 0.025 W/m · °C) and (*b*) glass fiber (k = 0.036 W/m · °C).

17–159 Cold conditioned air at 12°C is flowing inside a 1.5-cm-thick square aluminum (k = 237 W/m · °C) duct of inner cross section 22 cm × 22 cm at a mass flow rate of 0.8 kg/s. The duct is exposed to air at 33°C with a combined convection-radiation heat transfer coefficient of 13 W/m^2 · °C. The convection heat transfer coefficient at the inner surface is 75 W/m^2 · °C. If the air temperature in the duct should not increase by more than 1°C determine the maximum length of the duct.

17–160 When analyzing heat transfer through windows, it is important to consider the frame as well as the glass area. Consider a 2-m-wide, 1.5-m-high wood-framed window with 85 percent of the area covered by 3-mm-thick single-pane glass (k = 0.7 W/m · °C). The frame is 5 cm thick, and is made of pine wood (k = 0.12 W/m · °C). The heat transfer coefficient is 7 W/m^2 · °C inside and 13 W/m^2 · °C outside. The room is maintained at 24°C, and the outdoor temperature is 40°C. Determine the percent error involved in heat transfer when the window is assumed to consist of glass only.

17–161 Steam at 235°C is flowing inside a steel pipe (k = 61 W/m · °C) whose inner and outer diameters are 10 cm and 12 cm, respectively, in an environment at 20°C. The heat transfer coefficients inside and outside the pipe are 105 W/m^2 · °C and 14 W/m^2 · °C, respectively. Determine (*a*) the thickness of the insulation (k = 0.038 W/m · °C) needed to reduce the heat loss by 95 percent and (*b*) the thickness of the insulation needed to reduce the exposed surface temperature of insulated pipe to 40°C for safety reasons.

17–162 When the transportation of natural gas in a pipeline is not feasible for economic or other reasons, it is first liquefied at about −160°C, and then transported in specially insulated tanks placed in marine ships. Consider a 4-m-diameter spherical tank that is filled with liquefied natural gas (LNG) at −160°C. The tank is exposed to ambient air at 24°C with a heat transfer coefficient of 22 W/m^2 · °C. The tank is thin shelled and its temperature can be taken to be the same as the LNG temperature. The tank is insulated with 5-cm-thick super insulation that has an effective thermal conductivity of 0.00008 W/m · °C. Taking the density and the specific heat of LNG to be 425 kg/m^3 and 3.475 kJ/kg · °C, respectively, estimate how long it will take for the LNG temperature to rise to −150°C.

17–163 A 15-cm × 20-cm hot surface at 85°C is to be cooled by attaching 4-cm-long aluminum (k = 237 W/m · °C) fins of 2-mm × 2-mm square cross section. The temperature of surrounding medium is 25°C and the heat transfer coefficient on the surfaces can be taken to be 20 W/m^2 · °C. If it is desired to triple the rate of heat transfer from the bare hot surface, determine the number of fins that needs to be attached.

17–164 EES Reconsider Prob. 17–163. Using EES (or other) software, plot the number of fins as a function of the increase in the heat loss by fins relative to no fin case (i.e., overall effectiveness of the fins) in the range of 1.5 to 5. Discuss the results. Is it realistic to assume the heat transfer coefficient to remain constant?

17–165 A 1.4-m-diameter spherical steel tank filled with iced water at 0°C is buried underground at a location where the thermal conductivity of the soil is k = 0.55 W/m · °C. The distance between the tank center and the ground surface is 2.4 m. For ground surface temperature of 18°C, determine the rate of heat transfer to the iced water in the tank. What would your answer be if the soil temperature were 18°C and the ground surface were insulated?

17–166 A 0.6-m-diameter, 1.9-m-long cylindrical tank containing liquefied natural gas (LNG) at −160°C is placed at the center of a 1.9-m-long 1.4-m × 1.4-m square solid bar made of an insulating material with k = 0.0002 W/m · °C. If the outer surface temperature of the bar is 12°C, determine the rate of heat transfer to the tank. Also, determine the LNG temperature after one month. Take the density and the specific heat of LNG to be 425 kg/m^3 and 3.475 kJ/kg · °C, respectively.

17–167 A typical section of a building wall is shown in Fig. P17–167. This section extends in and out of the page and is repeated in the vertical direction. The wall support members are made of steel (k = 50 W/m · K). The support members are 8 cm (t_{23}) × 0.5 cm (L_B). The remainder of the inner wall space is filled with insulation (k = 0.03 W/m · K) and measures 8 cm (t_{23}) × 60 cm (L_B). The inner wall is made of gypsum board (k = 0.5 W/m · K) that is 1 cm thick (t_{12}) and the outer wall is made of brick (k = 1.0 W/m · K) that is 10 cm thick (t_{34}). What is the temperature on the interior brick surface, 3, when T_1 = 20 °C and T_4 = 35°C?

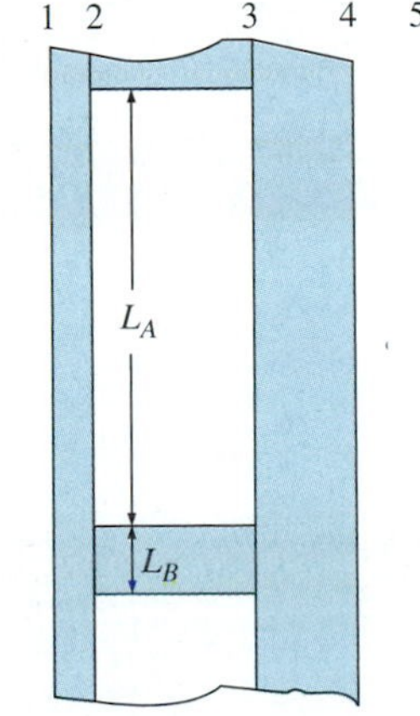

FIGURE P17–167

17–168 A total of 10 rectangular aluminum fins (k = 203 W/m · K) are placed on the outside flat surface of an electronic device. Each fin is 100 mm wide, 20 mm high and 4 mm thick. The fins are located parallel to each other at a center-to-center distance of 8 mm. The temperature at the outside surface of the electronic device is 60°C. The air is at 20°C, and the heat transfer coefficient is 100 W/m^2 · K. Determine (*a*) the rate of heat loss from the electronic device to the surrounding air and (*b*) the fin effectiveness.

17–169 One wall of a refrigerated warehouse is 10.0-m-high and 5.0-m-wide. The wall is made of three layers: 1.0-cm-thick aluminum (k = 200 W/m · K), 8.0-cm-thick fibreglass (k = 0.038 W/m · K), and 3.0-cm thick gypsum board (k = 0.48 W/m · K). The warehouse inside and outside tem-peratures are −10°C and 20°C, respectively, and the average value of both inside and outside heat transfer coefficients is 40 W/m^2 · K.

(*a*) Calculate the rate of heat transfer across the warehouse wall in steady operation.

(*b*) Suppose that 400 metal bolts (k = 43 W/m · K), each 2.0 cm in diameter and 12.0 cm long, are used to fasten (i.e. hold together) the three wall layers. Calculate the rate of heat transfer for the "bolted" wall.

(*c*) What is the percent change in the rate of heat transfer across the wall due to metal bolts?

Design and Essay Problems

17–170 The temperature in deep space is close to absolute zero, which presents thermal challenges for the astronauts who do space walks. Propose a design for the clothing of the astronauts that will be most suitable for the thermal environment in space. Defend the selections in your design.

17–171 In the design of electronic components, it is very desirable to attach the electronic circuitry to a substrate material that is a very good thermal conductor but also a very effective electrical insulator. If the high cost is not a major concern, what material would you propose for the substrate?

17–172 A house with 200-m^2 floor space is to be heated with geothermal water flowing through pipes laid in the ground under the floor. The walls of the house are 4 m high, and there are 10 single-paned windows in the house that are 1.2 m wide and 1.8 m high. The house has R-19 (in h · ft^2 · °F/Btu) insulation in the walls and R-30 on the ceiling. The floor temperature is not to exceed 40°C. Hot geothermal water is available at 90°C, and the inner and outer diameter of the pipes to be used are 2.4 cm and 3.0 cm. Design such a heating system for this house in your area.

17–173 Using a timer (or watch) and a thermometer, conduct this experiment to determine the rate of heat gain of your refrigerator. First, make sure that the door of the refrigerator is not opened for at least a few hours to make sure that steady operating conditions are established. Start the timer when the refrigerator stops running and measure the time Δt_1 it stays off before it kicks in. Then measure the time Δt_2 it stays on. Noting that the heat removed during Δt_2 is equal to the heat gain of the refrigerator during $\Delta t_1 + \Delta t_2$ and using the power consumed by the refrigerator when it is running, determine the average rate of heat gain for your refrigerator, in watts. Take the COP (coefficient of performance) of your refrigerator to be 1.3 if it is not available.

Now, clean the condenser coils of the refrigerator and remove any obstacles on the way of airflow through the coils. By replacing these measurements, determine the improvement in the COP of the refrigerator.

Chapter 18

TRANSIENT HEAT CONDUCTION

The temperature of a body, in general, varies with time as well as position. In rectangular coordinates, this variation is expressed as $T(x, y, z, t)$, where (x, y, z) indicate variation in the x-, y-, and z-directions, and t indicates variation with time. In the preceding chapter, we considered heat conduction under *steady* conditions, for which the temperature of a body at any point does not change with time. This certainly simplified the analysis, especially when the temperature varied in one direction only, and we were able to obtain analytical solutions. In this chapter, we consider the variation of temperature with *time* as well as *position* in one- and multidimensional systems.

We start this chapter with the analysis of *lumped systems* in which the temperature of a body varies with time but remains uniform throughout at any time. Then we consider the variation of temperature with time as well as position for one-dimensional heat conduction problems such as those associated with a large plane wall, a long cylinder, a sphere, and a semi-infinite medium using *transient temperature charts* and analytical solutions. Finally, we consider transient heat conduction in multidimensional systems by utilizing the *product solution*.

Objectives

The objectives of this chapter are to:

- Assess when the spatial variation of temperature is negligible, and temperature varies nearly uniformly with time, making the simplified lumped system analysis applicable,
- Obtain analytical solutions for transient one-dimensional conduction problems in rectangular, cylindrical, and spherical geometries using the method of separation of variables, and understand why a one-term solution is usually a reasonable approximation,
- Solve the transient conduction problem in large mediums using the similarity variable, and predict the variation of temperature with time and distance from the exposed surface, and
- Construct solutions for multi-dimensional transient conduction problems using the product solution approach.

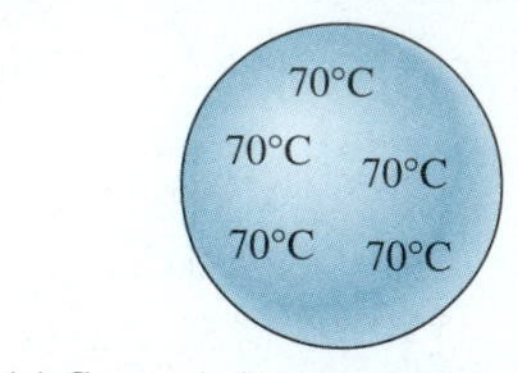

(a) Copper ball

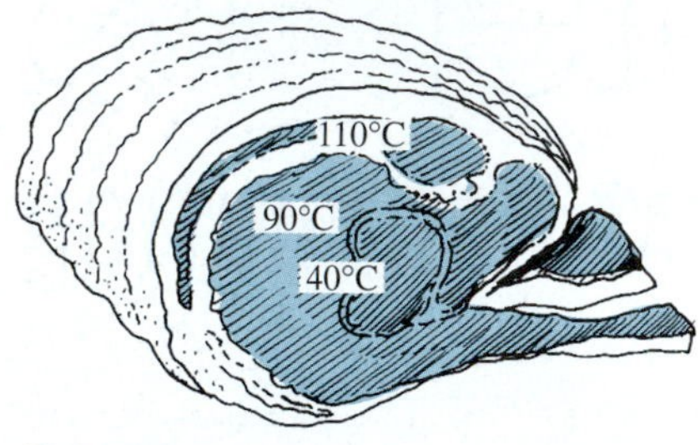

(b) Roast beef

FIGURE 18–1

A small copper ball can be modeled as a lumped system, but a roast beef cannot.

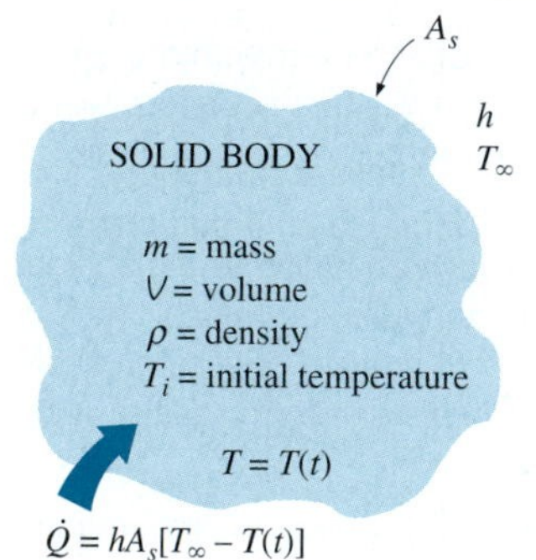

FIGURE 18–2

The geometry and parameters involved in the lumped system analysis.

18–1 ▪ LUMPED SYSTEM ANALYSIS

In heat transfer analysis, some bodies are observed to behave like a "lump" whose interior temperature remains essentially uniform at all times during a heat transfer process. The temperature of such bodies can be taken to be a function of time only, $T(t)$. Heat transfer analysis that utilizes this idealization is known as **lumped system analysis**, which provides great simplification in certain classes of heat transfer problems without much sacrifice from accuracy.

Consider a small hot copper ball coming out of an oven (Fig. 18–1). Measurements indicate that the temperature of the copper ball changes with time, but it does not change much with position at any given time. Thus the temperature of the ball remains nearly uniform at all times, and we can talk about the temperature of the ball with no reference to a specific location.

Now let us go to the other extreme and consider a large roast in an oven. If you have done any roasting, you must have noticed that the temperature distribution within the roast is not even close to being uniform. You can easily verify this by taking the roast out before it is completely done and cutting it in half. You will see that the outer parts of the roast are well done while the center part is barely warm. Thus, lumped system analysis is not applicable in this case. Before presenting a criterion about applicability of lumped system analysis, we develop the formulation associated with it.

Consider a body of arbitrary shape of mass m, volume V, surface area A_s, density ρ, and specific heat c_p initially at a uniform temperature T_i (Fig. 18–2). At time $t = 0$, the body is placed into a medium at temperature T_∞, and heat transfer takes place between the body and its environment, with a heat transfer coefficient h. For the sake of discussion, we assume that $T_\infty > T_i$, but the analysis is equally valid for the opposite case. We assume lumped system analysis to be applicable, so that the temperature remains uniform within the body at all times and changes with time only, $T = T(t)$.

During a differential time interval dt, the temperature of the body rises by a differential amount dT. An energy balance of the solid for the time interval dt can be expressed as

$$\begin{pmatrix}\text{Heat transfer into the body}\\ \text{during } dt\end{pmatrix} = \begin{pmatrix}\text{The increase in the}\\ \text{energy of the body}\\ \text{during } dt\end{pmatrix}$$

or

$$hA_s(T_\infty - T)\,dt = mc_p\,dT \tag{18–1}$$

Noting that $m = \rho V$ and $dT = d(T - T_\infty)$ since T_∞ = constant, Eq. 18–1 can be rearranged as

$$\frac{d(T - T_\infty)}{T - T_\infty} = -\frac{hA_s}{\rho V c_p}\,dt \tag{18–2}$$

Integrating from $t = 0$, at which $T = T_i$, to any time t, at which $T = T(t)$, gives

$$\ln\frac{T(t) - T_\infty}{T_i - T_\infty} = -\frac{hA_s}{\rho V c_p}\,t \tag{18–3}$$

Taking the exponential of both sides and rearranging, we obtain

$$\frac{T(t) - T_\infty}{T_i - T_\infty} = e^{-bt} \qquad \textbf{(18–4)}$$

where

$$b = \frac{hA_s}{\rho V c_p} \qquad (1/\text{s}) \qquad \textbf{(18–5)}$$

is a positive quantity whose dimension is (time)$^{-1}$. The reciprocal of b has time unit (usually s), and is called the **time constant**. Equation 18–4 is plotted in Fig. 18–3 for different values of b. There are two observations that can be made from this figure and the relation above:

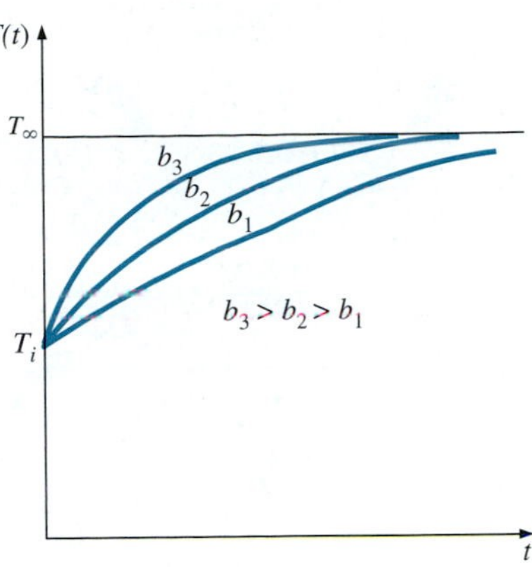

FIGURE 18–3

The temperature of a lumped system approaches the environment temperature as time gets larger.

1. Equation 18–4 enables us to determine the temperature $T(t)$ of a body at time t, or alternatively, the time t required for the temperature to reach a specified value $T(t)$.
2. The temperature of a body approaches the ambient temperature T_∞ exponentially. The temperature of the body changes rapidly at the beginning, but rather slowly later on. A large value of b indicates that the body approaches the environment temperature in a short time. The larger the value of the exponent b, the higher the rate of decay in temperature. Note that b is proportional to the surface area, but inversely proportional to the mass and the specific heat of the body. This is not surprising since it takes longer to heat or cool a larger mass, especially when it has a large specific heat.

Once the temperature $T(t)$ at time t is available from Eq. 18–4, the *rate* of convection heat transfer between the body and its environment at that time can be determined from Newton's law of cooling as

$$\dot{Q}(t) = hA_s[T(t) - T_\infty] \qquad (\text{W}) \qquad \textbf{(18–6)}$$

The *total amount* of heat transfer between the body and the surrounding medium over the time interval $t = 0$ to t is simply the change in the energy content of the body:

$$Q = mc_p[T(t) - T_i] \qquad (\text{kJ}) \qquad \textbf{(18–7)}$$

The amount of heat transfer reaches its upper limit when the body reaches the surrounding temperature T_∞. Therefore, the *maximum* heat transfer between the body and its surroundings is (Fig. 18–4)

$$Q_{\max} = mc_p(T_\infty - T_i) \qquad (\text{kJ}) \qquad \textbf{(18–8)}$$

We could also obtain this equation by substituting the $T(t)$ relation from Eq. 18–4 into the $\dot{Q}(t)$ relation in Eq. 18–6 and integrating it from $t = 0$ to $t \to \infty$.

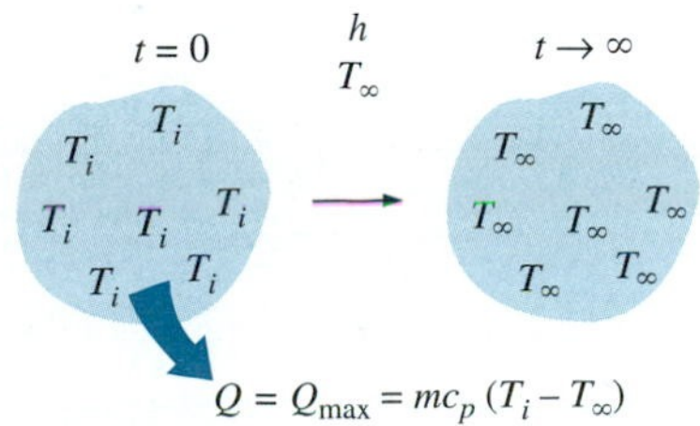

FIGURE 18–4

Heat transfer to or from a body reaches its maximum value when the body reaches the environment temperature.

Criteria for Lumped System Analysis

The lumped system analysis certainly provides great convenience in heat transfer analysis, and naturally we would like to know when it is appropriate to use it. The first step in establishing a criterion for the applicability of the lumped system analysis is to define a **characteristic length** as

$$L_c = \frac{V}{A_s}$$

and a **Biot number** Bi as

$$\text{Bi} = \frac{hL_c}{k} \tag{18–9}$$

It can also be expressed as (Fig. 18–5)

$$\text{Bi} = \frac{h}{k/L_c}\frac{\Delta T}{\Delta T} = \frac{\text{Convection at the surface of the body}}{\text{Conduction within the body}}$$

or

$$\text{Bi} = \frac{L_c/k}{1/h} = \frac{\text{Conduction resistance within the body}}{\text{Convection resistance at the surface of the body}}$$

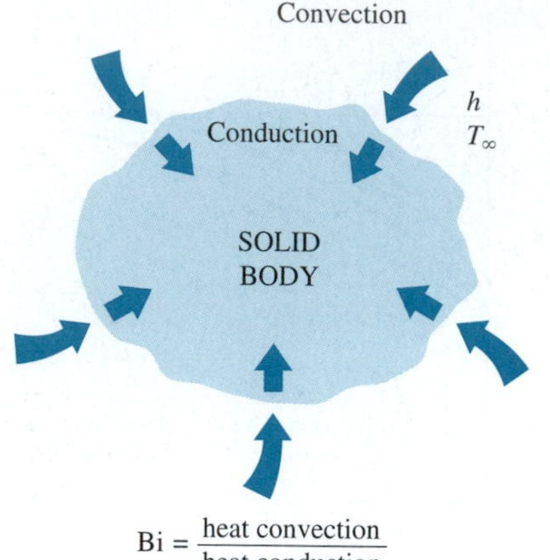

FIGURE 18–5

The Biot number can be viewed as the ratio of the convection at the surface to conduction within the body.

When a solid body is being heated by the hotter fluid surrounding it (such as a potato being baked in an oven), heat is first *convected* to the body and subsequently *conducted* within the body. The Biot number is the *ratio* of the internal resistance of a body to *heat conduction* to its external resistance to *heat convection.* Therefore, a small Biot number represents small resistance to heat conduction, and thus small temperature gradients within the body.

Lumped system analysis assumes a *uniform* temperature distribution throughout the body, which is the case only when the thermal resistance of the body to heat conduction (the *conduction resistance*) is zero. Thus, lumped system analysis is *exact* when Bi $= 0$ and *approximate* when Bi > 0. Of course, the smaller the Bi number, the more accurate the lumped system analysis. Then the question we must answer is, How much accuracy are we willing to sacrifice for the convenience of the lumped system analysis?

Before answering this question, we should mention that a 15 percent uncertainty in the convection heat transfer coefficient h in most cases is considered "normal" and "expected." Assuming h to be *constant* and *uniform* is also an approximation of questionable validity, especially for irregular geometries. Therefore, in the absence of sufficient experimental data for the specific geometry under consideration, we cannot claim our results to be better than ± 15 percent, even when Bi $= 0$. This being the case, introducing another source of uncertainty in the problem will not have much effect on the overall uncertainty, provided that it is minor. It is generally accepted that lumped system analysis is *applicable* if

$$\text{Bi} \leq 0.1$$

When this criterion is satisfied, the temperatures within the body relative to the surroundings (i.e., $T - T_\infty$) remain within 5 percent of each other even for well-rounded geometries such as a spherical ball. Thus, when Bi < 0.1, the variation of temperature with location within the body is slight and can reasonably be approximated as being uniform.

The first step in the application of lumped system analysis is the calculation of the *Biot number*, and the assessment of the applicability of this approach. One may still wish to use lumped system analysis even when the criterion Bi < 0.1 is not satisfied, if high accuracy is not a major concern.

Note that the Biot number is the ratio of the *convection* at the surface to *conduction* within the body, and this number should be as small as possible for lumped system analysis to be applicable. Therefore, *small bodies* with *high thermal conductivity* are good candidates for lumped system analysis, especially when they are in a medium that is a poor conductor of heat (such as air or another gas) and motionless. Thus, the hot small copper ball placed in quiescent air, discussed earlier, is most likely to satisfy the criterion for lumped system analysis (Fig. 18–6).

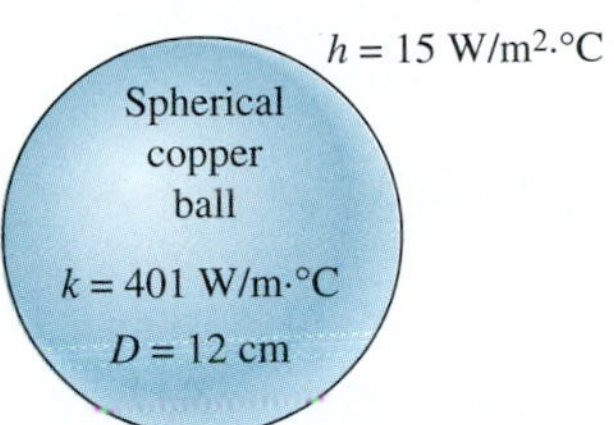

$$L_c = \frac{V}{A_s} = \frac{\frac{1}{6}\pi D^3}{\pi D^2} = \frac{1}{6}D = 0.02 \text{ m}$$

$$\text{Bi} = \frac{hL_c}{k} = \frac{15 \times 0.02}{401} = 0.00075 < 0.1$$

FIGURE 18–6

Small bodies with high thermal conductivities and low convection coefficients are most likely to satisfy the criterion for lumped system analysis.

Some Remarks on Heat Transfer in Lumped Systems

To understand the heat transfer mechanism during the heating or cooling of a solid by the fluid surrounding it, and the criterion for lumped system analysis, consider this analogy (Fig. 18–7). People from the mainland are to go *by boat* to an island whose entire shore is a harbor, and from the harbor to their destinations on the island *by bus*. The overcrowding of people at the harbor depends on the boat traffic to the island and the ground transportation system on the island. If there is an excellent ground transportation system with plenty of buses, there will be no overcrowding at the harbor, especially when the boat traffic is light. But when the opposite is true, there will be a huge overcrowding at the harbor, creating a large difference between the populations at the harbor and inland. The chance of overcrowding is much lower in a small island with plenty of fast buses.

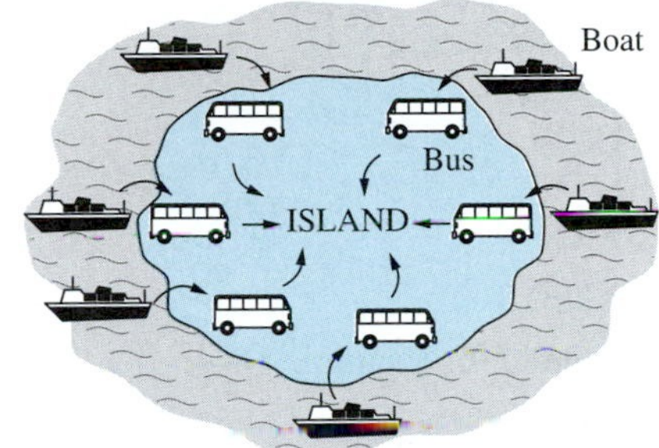

FIGURE 18–7

Analogy between heat transfer to a solid and passenger traffic to an island.

In heat transfer, a poor ground transportation system corresponds to poor heat conduction in a body, and overcrowding at the harbor to the accumulation of thermal energy and the subsequent rise in temperature near the surface of the body relative to its inner parts. Lumped system analysis is obviously not applicable when there is overcrowding at the surface. Of course, we have disregarded radiation in this analogy and thus the air traffic to the island. Like passengers at the harbor, heat changes *vehicles* at the surface from *convection* to *conduction*. Noting that a surface has zero thickness and thus cannot store any energy, heat reaching the surface of a body by convection must continue its journey within the body by conduction.

Consider heat transfer from a hot body to its cooler surroundings. Heat is transferred from the body to the surrounding fluid as a result of a temperature difference. But this energy comes from the region near the surface, and thus the temperature of the body near the surface will drop. This creates a *temperature gradient* between the inner and outer regions of the body and initiates heat transfer by conduction from the interior of the body toward the outer surface.

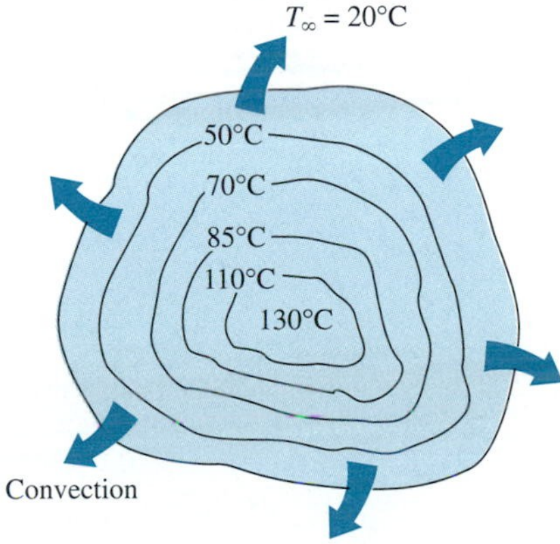

FIGURE 18–8

When the convection coefficient h is high and k is low, large temperature differences occur between the inner and outer regions of a large solid.

When the convection heat transfer coefficient h and thus the rate of convection from the body are high, the temperature of the body near the surface drops quickly (Fig. 18–8). This creates a larger temperature difference between the inner and outer regions unless the body is able to transfer heat from the inner to the outer regions just as fast. Thus, the magnitude of the maximum temperature difference within the body depends strongly on the ability of a body to conduct heat toward its surface relative to the ability of the surrounding medium to convect heat away from the surface. The Biot number is a measure of the relative magnitudes of these two competing effects.

Recall that heat conduction in a specified direction n per unit surface area is expressed as $\dot{q} = -k\,\partial T/\partial n$, where $\partial T/\partial n$ is the temperature gradient and k is the thermal conductivity of the solid. Thus, the temperature distribution in the body will be *uniform* only when its thermal conductivity is *infinite,* and no such material is known to exist. Therefore, temperature gradients and thus temperature differences must exist within the body, no matter how small, in order for heat conduction to take place. Of course, the temperature gradient and the thermal conductivity are inversely proportional for a given heat flux. Therefore, the larger the thermal conductivity, the smaller the temperature gradient.

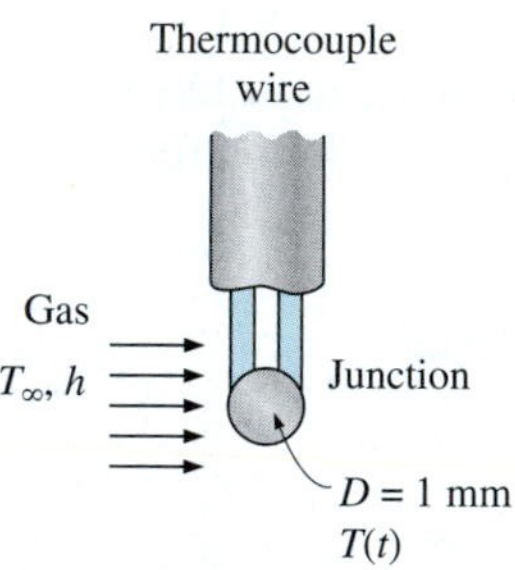

FIGURE 18–9

Schematic for Example 18–1.

EXAMPLE 18–1 Temperature Measurement by Thermocouples

The temperature of a gas stream is to be measured by a thermocouple whose junction can be approximated as a 1-mm-diameter sphere, as shown in Fig. 18–9. The properties of the junction are $k = 35$ W/m · °C, $\rho = 8500$ kg/m^3, and $c_p = 320$ J/kg · °C, and the convection heat transfer coefficient between the junction and the gas is $h = 210$ W/m^2 · °C. Determine how long it will take for the thermocouple to read 99 percent of the initial temperature difference.

Solution The temperature of a gas stream is to be measured by a thermocouple. The time it takes to register 99 percent of the initial ΔT is to be determined.

Assumptions **1** The junction is spherical in shape with a diameter of $D = 0.001$ m. **2** The thermal properties of the junction and the heat transfer coefficient are constant. **3** Radiation effects are negligible.

Properties The properties of the junction are given in the problem statement.

Analysis The characteristic length of the junction is

$$L_c = \frac{V}{A_s} = \frac{\frac{1}{6}\pi D^3}{\pi D^2} = \frac{1}{6}D = \frac{1}{6}(0.001\text{ m}) = 1.67 \times 10^{-4}\text{ m}$$

Then the Biot number becomes

$$\text{Bi} = \frac{hL_c}{k} = \frac{(210\text{ W/m}^2\cdot{}^\circ\text{C})(1.67\times 10^{-4}\text{ m})}{35\text{ W/m}\cdot{}^\circ\text{C}} = 0.001 < 0.1$$

Therefore, lumped system analysis is applicable, and the error involved in this approximation is negligible.

In order to read 99 percent of the initial temperature difference $T_i - T_\infty$ between the junction and the gas, we must have

$$\frac{T(t) - T_\infty}{T_i - T_\infty} = 0.01$$

For example, when $T_i = 0$°C and $T_\infty = 100$°C, a thermocouple is considered to have read 99 percent of this applied temperature difference when its reading indicates $T(t) = 99$°C.

The value of the exponent b is

$$b = \frac{hA_s}{\rho c_p V} = \frac{h}{\rho c_p L_c} = \frac{210 \text{ W/m}^2 \cdot °\text{C}}{(8500 \text{ kg/m}^3)(320 \text{ J/kg} \cdot °\text{C})(1.67 \times 10^{-4} \text{ m})} = 0.462 \text{ s}^{-1}$$

We now substitute these values into Eq. 18–4 and obtain

$$\frac{T(t) - T_\infty}{T_i - T_\infty} = e^{-bt} \longrightarrow 0.01 = e^{-(0.462 \text{ s}^{-1})t}$$

which yields

$$t = \mathbf{10 \text{ s}}$$

Therefore, we must wait at least 10 s for the temperature of the thermocouple junction to approach within 99 percent of the initial junction-gas temperature difference.

Discussion Note that conduction through the wires and radiation exchange with the surrounding surfaces affect the result, and should be considered in a more refined analysis.

EXAMPLE 18–2 Predicting the Time of Death

A person is found dead at 5 PM in a room whose temperature is 20°C. The temperature of the body is measured to be 25°C when found, and the heat transfer coefficient is estimated to be h = 8 W/m² · °C. Modeling the body as a 30-cm-diameter, 1.70-m-long cylinder, estimate the time of death of that person (Fig. 18–10).

FIGURE 18–10

Schematic for Example 18–2.

Solution A body is found while still warm. The time of death is to be estimated.

Assumptions **1** The body can be modeled as a 30-cm-diameter, 1.70-m-long cylinder. **2** The thermal properties of the body and the heat transfer coefficient are constant. **3** The radiation effects are negligible. **4** The person was healthy(!) when he or she died with a body temperature of 37°C.

Properties The average human body is 72 percent water by mass, and thus we can assume the body to have the properties of water at the average temperature of (37 + 25)/2 = 31°C; k = 0.617 W/m · °C, ρ = 996 kg/m³, and c_p = 4178 J/kg · °C (Table A–15).

Analysis The characteristic length of the body is

$$L_c = \frac{V}{A_s} = \frac{\pi r_o^2 L}{2\pi r_o L + 2\pi r_o^2} = \frac{\pi(0.15 \text{ m})^2(1.7 \text{ m})}{2\pi(0.15 \text{ m})(1.7 \text{ m}) + 2\pi(0.15 \text{ m})^2} = 0.0689 \text{ m}$$

Then the Biot number becomes

$$\text{Bi} = \frac{hL_c}{k} = \frac{(8 \text{ W/m}^2 \cdot °\text{C})(0.0689 \text{ m})}{0.617 \text{ W/m} \cdot °\text{C}} = 0.89 > 0.1$$

Therefore, lumped system analysis is *not* applicable. However, we can still use it to get a "rough" estimate of the time of death. The exponent b in this case is

$$b = \frac{hA_s}{\rho c_p V} = \frac{h}{\rho c_p L_c} = \frac{8 \text{ W/m}^2 \cdot °\text{C}}{(996 \text{ kg/m}^3)(4178 \text{ J/kg} \cdot °\text{C})(0.0689 \text{ m})}$$
$$= 2.79 \times 10^{-5} \text{ s}^{-1}$$

We now substitute these values into Eq. 18–4,

$$\frac{T(t) - T_\infty}{T_i - T_\infty} = e^{-bt} \longrightarrow \frac{25 - 20}{37 - 20} = e^{-(2.79 \times 10^{-5} \text{ s}^{-1})t}$$

which yields

$$t = 43{,}860 \text{ s} = \mathbf{12.2 \text{ h}}$$

Therefore, as a rough estimate, the person died about 12 h before the body was found, and thus the time of death is 5 AM.

Discussion This example demonstrates how to obtain "ball park" values using a simple analysis. A similar analysis is used in practice by incorporating constants to account for deviation from lumped system analysis.

18–2 ▪ TRANSIENT HEAT CONDUCTION IN LARGE PLANE WALLS, LONG CYLINDERS, AND SPHERES WITH SPATIAL EFFECTS

In Section 18–1, we considered bodies in which the variation of temperature within the body is negligible; that is, bodies that remain nearly *isothermal* during a process. Relatively small bodies of highly conductive materials approximate this behavior. In general, however, the temperature within a body changes from point to point as well as with time. In this section, we consider the variation of temperature with *time* and *position* in one-dimensional problems such as those associated with a large plane wall, a long cylinder, and a sphere.

Consider a plane wall of thickness $2L$, a long cylinder of radius r_o, and a sphere of radius r_o initially at a *uniform temperature* T_i, as shown in Fig. 18–11. At time $t = 0$, each geometry is placed in a large medium that is at a constant temperature T_∞ and kept in that medium for $t > 0$. Heat transfer takes place between these bodies and their environments by convection with a *uniform* and *constant* heat transfer coefficient h. Note that all three cases possess geometric and thermal symmetry: the plane wall is symmetric about its *center plane* ($x = 0$), the cylinder is symmetric about its *centerline* ($r = 0$), and the sphere is symmetric about its *center point* ($r = 0$). We neglect *radiation* heat transfer between these bodies and their surrounding surfaces, or incorporate the radiation effect into the convection heat transfer coefficient h.

The variation of the temperature profile with *time* in the plane wall is illustrated in Fig. 18–12. When the wall is first exposed to the surrounding medium at $T_\infty < T_i$ at $t = 0$, the entire wall is at its initial temperature T_i. But the wall temperature at and near the surfaces starts to drop as a result of heat transfer from the wall to the surrounding medium. This creates a *temperature gradient* in the wall and initiates heat conduction from the inner

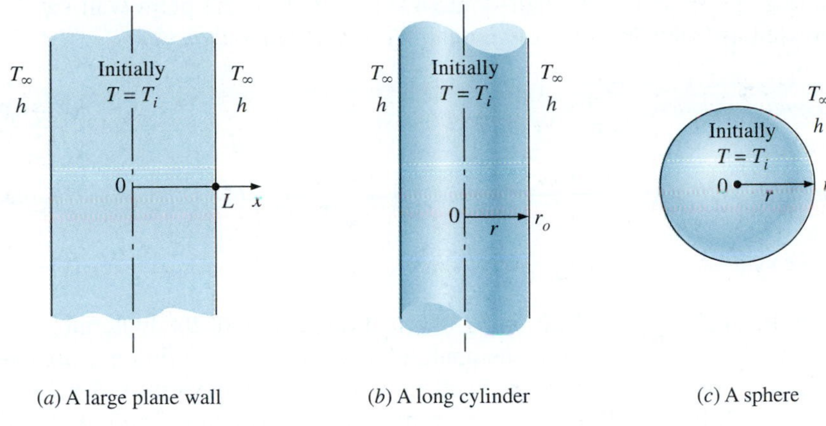

FIGURE 18–11

Schematic of the simple geometries in which heat transfer is one-dimensional.

parts of the wall toward its outer surfaces. Note that the temperature at the center of the wall remains at T_i until $t = t_2$, and that the temperature profile within the wall remains symmetric at all times about the center plane. The temperature profile gets flatter and flatter as time passes as a result of heat transfer, and eventually becomes uniform at $T = T_\infty$. That is, the wall reaches *thermal equilibrium* with its surroundings. At that point, heat transfer stops since there is no longer a temperature difference. Similar discussions can be given for the long cylinder or sphere.

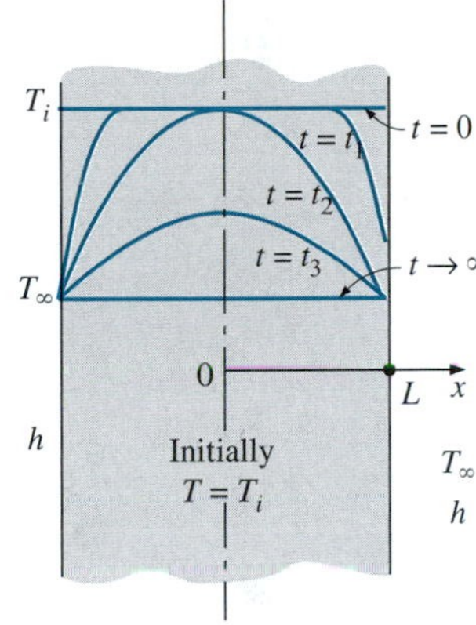

FIGURE 18–12

Transient temperature profiles in a plane wall exposed to convection from its surfaces for $T_i > T_\infty$.

Nondimensionalized One-Dimensional Transient Conduction Problem

The formulation of heat conduction problems for the determination of the one-dimensional transient temperature distribution in a plane wall, a cylinder, or a sphere results in a partial differential equation whose solution typically involves infinite series and transcendental equations, which are inconvenient to use. But the analytical solution provides valuable insight to the physical problem, and thus it is important to go through the steps involved. Below we demonstrate the solution procedure for the case of plane wall.

Consider a plane wall of thickness $2L$ initially at a uniform temperature of T_i, as shown in Fig. 18–11*a*. At time $t = 0$, the wall is immersed in a fluid at temperature T_∞ and is subjected to convection heat transfer from both sides with a convection coefficient of h. The height and the width of the wall are large relative to its thickness, and thus heat conduction in the wall can be approximated to be one-dimensional. Also, there is thermal symmetry about the midplane passing through $x = 0$, and thus the temperature distribution must be symmetrical about the midplane. Therefore, the value of temperature at any $-x$ value in $-L \leq x \leq 0$ at any time t must be equal to the value at $+x$ in $0 \leq x \leq L$ at the same time. This means we can formulate and solve the heat conduction problem in the positive half domain $0 \leq x \leq L$, and then apply the solution to the other half.

Under the conditions of constant thermophysical properties, no heat generation, thermal symmetry about the midplane, uniform initial temperature, and constant convection coefficient, the one-dimensional transient heat conduction problem in the half-domain $0 \leq x \leq L$ of the plain wall can be expressed as (see Chapter 2 of Çengal, *Heat and Mass Transfer*)

Differential equation: $$\frac{\partial^2 T}{\partial x^2} = \frac{1}{\alpha}\frac{\partial T}{\partial t} \quad \text{(18–10a)}$$

Boundary conditions: $$\frac{\partial T(0, t)}{\partial x} = 0 \quad \text{and} \quad -k\frac{\partial T(L, t)}{\partial x} = h[T(L, t) - T_\infty] \quad \text{(18–10b)}$$

Initial condition: $$T(x, 0) = T_i \quad \text{(18–10c)}$$

where the property $\alpha = k/\rho c_p$ is the thermal diffusivity of the material.

We now attempt to nondimensionalize the problem by defining a dimensionless space variable $X = x/L$ and dimensionless temperature $\theta(x, t) = [T(x, t) - T_\infty]/[T_i - T_\infty]$. These are convenient choices since both X and θ vary between 0 and 1. However, there is no clear guidance for the proper form of the dimensionless time variable and the h/k ratio, so we will let the analysis indicate them. We note that

$$\frac{\partial \theta}{\partial X} = \frac{\partial \theta}{\partial (x/L)} = \frac{L}{T_i - T_\infty}\frac{\partial T}{\partial x}, \quad \frac{\partial^2 \theta}{\partial X^2} = \frac{L^2}{T_i - T_\infty}\frac{\partial T}{\partial x} \quad \text{and} \quad \frac{\partial \theta}{\partial t} = \frac{1}{T_i - T_\infty}\frac{\partial T}{\partial t}$$

Substituting into Eqs. 18–10*a* and 18–10*b* and rearranging give

$$\frac{\partial^2 \theta}{\partial X^2} = \frac{L^2}{\alpha}\frac{\partial \theta}{\partial t} \quad \text{and} \quad \frac{\partial \theta(1, t)}{\partial X} = \frac{hL}{k}\theta(1, t) \quad \text{(18–11)}$$

Therefore, the proper form of the dimensionless time is $\tau = \alpha t/L^2$, which is called the **Fourier number** Fo, and we recognize $\text{Bi} = k/hL$ as the Biot number defined in Section 18–1. Then the formulation of the one-dimensional transient heat conduction problem in a plane wall can be expressed in nondimensional form as

Dimensionless differential equation: $$\frac{\partial^2 \theta}{\partial X^2} = \frac{\partial \theta}{\partial \tau} \quad \text{(18–12a)}$$

Dimensionless BC's: $$\frac{\partial \theta(0, \tau)}{\partial X} = 0 \quad \text{and} \quad \frac{\partial \theta(1, \tau)}{\partial X} = -\text{Bi}\theta(1, \tau) \quad \text{(18–12b)}$$

Dimensionless initial condition: $$\theta(X, 0) = 1 \quad \text{(18–12c)}$$

where

$$\theta(X, \tau) = \frac{T(x, t) - T_i}{T_\infty - T_i}$$ *Dimensionless temperature*

$$X = \frac{x}{L}$$ *Dimensionless distance from the center*

$$\text{Bi} = \frac{hL}{k}$$ *Dimensionless heat transfer coefficient* (Biot number)

$$\tau = \frac{\alpha t}{L^2} = \text{Fo}$$ *Dimensionless time* (Fourier number)

The heat conduction equation in cylindrical or spherical coordinates can be nondimensionalized in a similar way. Note that nondimensionalization

(*a*) Original heat conduction problem:

$$\frac{\partial^2 T}{\partial x^2} = \frac{1}{\alpha}\frac{\partial T}{\partial t}, \quad T(x, 0) = T_i$$

$$\frac{\partial T(0, t)}{\partial x} = 0, \quad -k\frac{\partial T(L, t)}{\partial x} = h[T(L, t) - T_\infty]$$

$$T = F(x, L, t, k, \alpha, h, T_i)$$

(*b*) Nondimensionalized problem:

$$\frac{\partial^2 \theta}{\partial X^2} = \frac{\partial \theta}{\partial \tau}, \quad \theta(X, 0) = 1$$

$$\frac{\partial \theta(0, \tau)}{\partial X} = 0, \quad \frac{\partial \theta(1, \tau)}{\partial X} = -\text{Bi}\theta(1, \tau)$$

$$\theta = f(X, \text{Bi}, \tau)$$

FIGURE 18–13

Nondimensionalization reduces the number of independent variables in one-dimensional transient conduction problems from 8 to 3, offering great convenience in the presentation of results.

reduces the number of independent variables and parameters from 8 to 3—from x, L, t, k, α, h, T_i, and T_∞ to X, Bi, and Fo (Fig. 18–13). That is,

$$\theta = f(X, \text{Bi}, \text{Fo}) \tag{18–13}$$

This makes it very practical to conduct parametric studies and to present results in graphical form. Recall that in the case of lumped system analysis, we had $\theta = f(\text{Bi}, \text{Fo})$ with no space variable.

Exact Solution of One-Dimensional Transient Conduction Problem*

The non-dimensionalized partial differential equation given in Eqs. 18–12 together with its boundary and initial conditions can be solved using several analytical and numerical techniques, including the Laplace or other transform methods, the method of separation of variables, the finite difference method, and the finite-element method. Here we use the method of **separation of variables** developed by J. Fourier in 1820s and is based on expanding an arbitrary function (including a constant) in terms of Fourier series. The method is applied by assuming the dependent variable to be a product of a number of functions, each being a function of a single independent variable. This reduces the partial differential equation to a system of ordinary differential equations, each being a function of a single independent variable. In the case of transient conduction in a plane wall, for example, the dependent variable is the solution function $\theta(X, \tau)$, which is expressed as $\theta(X, \tau) = F(X)G(\tau)$, and the application of the method results in two ordinary differential equation, one in X and the other in τ.

The method is applicable if (1) the geometry is simple and finite (such as a rectangular block, a cylinder, or a sphere) so that the boundary surfaces can be described by simple mathematical functions, and (2) the differential equation and the boundary and initial conditions in their most simplified form are linear (no terms that involve products of the dependent variable or its derivatives) and involve only one nonhomogeneous term (a term without the dependent variable or its derivatives). If the formulation involves a number of nonhomogeneous terms, the problem can be split up into an equal number of simpler problems each involving only one nonhomogeneous term, and then combining the solutions by superposition.

Now we demonstrate the use of the method of separation of variables by applying it to the one-dimensional transient heat conduction problem given in Eqs. 18–12. First, we express the dimensionless temperature function $\theta(X, \tau)$ as a product of a function of X only and a function of τ only as

$$\theta(X, \tau) = F(X)G(\tau) \tag{18–14}$$

Substituting Eq. 18–14 into Eq. 18–12a and dividing by the product FG gives

$$\frac{1}{F}\frac{d^2F}{dX^2} = \frac{1}{G}\frac{dG}{d\tau} \tag{18–15}$$

Observe that all the terms that depend on X are on the left-hand side of the equation and all the terms that depend on τ are on the right-hand side. That

*This section can be skipped if desired without a loss of continuity.

is, the terms that are function of different variables are *separated* (and thus the name *separation of variables*). The left-hand side of this equation is a function of X only and the right-hand side is a function of only τ. Considering that both X and τ can be varied independently, the equality in Eq. 18–15 can hold for any value of X and τ only if Eq. 18–15 is equal to a constant. Further, it must be a *negative* constant that we will indicate by $-\lambda^2$ since a positive constant will cause the function $G(\tau)$ to increase indefinitely with time (to be infinite), which is unphysical, and a value of zero for the constant means no time dependence, which is again inconsistent with the physical problem. Setting Eq. 18–15 equal to $-\lambda^2$ gives

$$\frac{d^2F}{dX^2} + \lambda^2 F = 0 \quad \text{and} \quad \frac{dG}{d\tau} + \lambda^2 G = 0 \tag{18–16}$$

whose general solutions are

$$F = C_1\cos(\lambda X) + C_2\sin(\lambda X) \quad \text{and} \quad G = C_3 e^{-\lambda^2\tau} \tag{18–17}$$

and

$$\theta = FG = C_3 e^{-\lambda^2\tau}[C_1\cos(\lambda X) + C_2\sin(\lambda X)] = e^{-\lambda^2\tau}[A\cos(\lambda X) + B\sin(\lambda X)] \tag{18–18}$$

where $A = C_1C_3$ and $B = C_2C_3$ are arbitrary constants. Note that we need to determine only A and B to obtain the solution of the problem.

Applying the boundary conditions in Eq. 18–12*b* gives

$$\frac{\partial\theta(0, \tau)}{\partial X} = 0 \rightarrow -e^{-\lambda^2\tau}(A\lambda\sin 0 + B\lambda\cos 0) = 0 \rightarrow B = 0 \rightarrow \theta = Ae^{-\lambda^2\tau}\cos(\lambda X)$$

$$\frac{\partial\theta(1, \tau)}{\partial X} = -\text{Bi}\theta(1,\tau) \rightarrow -Ae^{-\lambda^2\tau}\lambda\sin\lambda = -\text{Bi}Ae^{-\lambda^2\tau}\cos\lambda \rightarrow \lambda\tan\lambda = \text{Bi}$$

But tangent is a periodic function with a period of π, and the equation $\lambda\tan\lambda = \text{Bi}$ has the root λ_1 between 0 and π, the root λ_2 between π and 2π, the root λ_n between $(n-1)\pi$ and $n\pi$, etc. To recognize that the transcendental equation $\lambda \tan \lambda = \text{Bi}$ has an infinite number of roots, it is expressed as

$$\lambda_n\tan\lambda_n = \text{Bi} \tag{18–19}$$

Eq. 18–19 is called the **characteristic equation** or **eigenfunction**, and its roots are called the **characteristic values** or **eigenvalues**. The characteristic equation is implicit in this case, and thus the characteristic values need to be determined numerically. Then it follows that there are an infinite number of solutions of the form $Ae^{-\lambda^2\tau}\cos(\lambda X)$, and the solution of this linear heat conduction problem is a linear combination of them,

$$\theta = \sum_{n=1}^{\infty} A_n e^{-\lambda_n^2\tau}\cos(\lambda_n X) \tag{18–20}$$

The constants A_n are determined from the initial condition, Eq. 18–12*c*,

$$\theta(X, 0) = 1 \rightarrow 1 = \sum_{n=1}^{\infty} A_n\cos(\lambda_n X) \tag{18–21}$$

This is a Fourier series expansion that expresses a constant in terms of an infinite series of cosine functions. Now we multiply both sides of Eq. 18–21 by $\cos(\lambda_m X)$, and integrate from $X = 0$ to $X = 1$. The right-hand side involves an infinite number of integrals of the form $\int_0^1 \cos(\lambda_m X)\cos(\lambda_n X)dx$. It can be shown that all of these integrals vanish except when $n = m$, and the coefficient A_n becomes

$$\int_0^1 \cos(\lambda_n X)dX = A_n \int_0^1 \cos^2(\lambda_n X)dx \rightarrow A_n = \frac{4\sin\lambda_n}{2\lambda_n + \sin(2\lambda_n)} \tag{18–22}$$

This completes the analysis for the solution of one-dimensional transient heat conduction problem in a plane wall. Solutions in other geometries such as a long cylinder and a sphere can be determined using the same approach. The results for all three geometries are summarized in Table 18–1. The solution for the plane wall is also applicable for a plane wall of thickness L whose left surface at $x = 0$ is insulated and the right surface at $x = L$ is subjected to convection since this is precisely the mathematical problem we solved.

The analytical solutions of transient conduction problems typically involve infinite series, and thus the evaluation of an infinite number of terms to determine the temperature at a specified location and time. This may look intimidating at first, but there is no need to worry. As demonstrated in Fig. 18–14, the terms in the summation decline rapidly as n and thus λ_n increases because of the exponential decay function $e^{-\lambda_n^2\tau}$. This is especially the case when the dimensionless time τ is large. Therefore, the evaluation of the first few terms of the infinite series (in this case just the first term) is usually adequate to determine the dimensionless temperature θ.

$$\theta_n = A_n e^{-\lambda_n^2\tau}\cos(\lambda_n X)$$

$$A_n = \frac{4\sin\lambda_n}{2\lambda_n + \sin(2\lambda_n)}$$

$$\lambda_n \tan\lambda_n = \text{Bi}$$

For Bi = 5, $X = 1$, and $t = 0.2$:

n	λ_n	A_n	θ_n
1	1.3138	1.2402	0.22321
2	4.0336	−0.3442	0.00835
3	6.9096	0.1588	0.00001
4	9.8928	−0.876	0.00000

FIGURE 18–14

The term in the series solution of transient conduction problems decline rapidly as n and thus λ_n increases because of the exponential decay function with the exponent $-\lambda_n\tau$.

Approximate Analytical and Graphical Solutions

The analytical solution obtained above for one-dimensional transient heat conduction in a plane wall involves infinite series and implicit equations, which are difficult to evaluate. Therefore, there is clear motivation to sim-

TABLE 18–1

Summary of the solutions for one-dimensional transient conduction in a plane wall of thickness $2L$, a cylinder of radius r_o and a sphere of radius r_o subjected to convention from all surfaces.*

Geometry	Solution	λ_n's are the roots of
Plane wall	$\theta = \sum_{n=1}^{\infty} \frac{4\sin\lambda_n}{2\lambda_n + \sin(2\lambda_n)} e^{-\lambda_n^2\tau}\cos(\lambda_n x/L)$	$\lambda_n \tan\lambda_n = \text{Bi}$
Cylinder	$\theta = \sum_{n=1}^{\infty} \frac{2}{\lambda_n} \frac{J_1(\lambda_n)}{J_0^2(\lambda_n) + J_1^2(\lambda_n)} e^{-\lambda_n^2\tau} J_0(\lambda_n r/r_o)$	$\lambda_n \frac{J_1(\lambda_n)}{J_0(\lambda_n)} = \text{Bi}$
Sphere	$\theta = \sum_{n=1}^{\infty} \frac{4(\sin\lambda_n - \lambda_n\cos\lambda_n)}{2\lambda_n - \sin(2\lambda_n)} e^{-\lambda_n^2\tau} \frac{\sin(\lambda_n x/L)}{\lambda_n x/L}$	$1 - \lambda_n \cot\lambda_n = \text{Bi}$

*Here $\theta = (T - T_i)/(T_\infty - T_i)$ is the dimensionless temperature, Bi $= hL/k$ or hr_o/k is the Biot number, Fo $= \tau = \alpha t/L^2$ or $\alpha\tau/r_o^2$ is the Fourier number, and J_0 and J_1 are the Bessel functions of the first kind whose values are given in Table 18–3.

plify the analytical solutions and to present the solutions in *tabular* or *graphical* form using simple relations.

The dimensionless quantities defined above for a plane wall can also be used for a *cylinder* or *sphere* by replacing the space variable x by r and the half-thickness L by the outer radius r_o. Note that the characteristic length in the definition of the Biot number is taken to be the *half-thickness* L for the plane wall, and the *radius* r_o for the long cylinder and sphere instead of V/A used in lumped system analysis.

We mentioned earlier that the terms in the series solutions in Table 18–1 converge rapidly with increasing time, and for $\tau > 0.2$, keeping the first term and neglecting all the remaining terms in the series results in an error under 2 percent. We are usually interested in the solution for times with $\tau > 0.2$, and thus it is very convenient to express the solution using this **one-term approximation**, given as

Plane wall: $$\theta_{\text{wall}} = \frac{T(x, t) - T_\infty}{T_i - T_\infty} = A_1 e^{-\lambda_1^2 \tau} \cos(\lambda_1 x/L), \quad \tau > 0.2 \tag{18–23}$$

Cylinder: $$\theta_{\text{cyl}} = \frac{T(r, t) - T_\infty}{T_i - T_\infty} = A_1 e^{-\lambda_1^2 \tau} J_0(\lambda_1 r/r_o), \quad \tau > 0.2 \tag{18–24}$$

Sphere: $$\theta_{\text{sph}} = \frac{T(r, t) - T_\infty}{T_i - T_\infty} = A_1 e^{-\lambda_1^2 \tau} \frac{\sin(\lambda_1 r/r_o)}{\lambda_1 r/r_o}, \quad \tau > 0.2 \tag{18–25}$$

where the constants A_1 and λ_1 are functions of the Bi number only, and their values are listed in Table 18–2 against the Bi number for all three geometries. The function J_0 is the zeroth-order Bessel function of the first kind, whose value can be determined from Table 18–3. Noting that $\cos(0) = J_0(0) = 1$ and the limit of $(\sin x)/x$ is also 1, these relations simplify to the next ones at the center of a plane wall, cylinder, or sphere:

Center of plane wall ($x = 0$): $$\theta_{0,\text{ wall}} = \frac{T_0 - T_\infty}{T_i - T_\infty} = A_1 e^{-\lambda_1^2 \tau} \tag{18–26}$$

Center of cylinder ($r = 0$): $$\theta_{0,\text{ cyl}} = \frac{T_0 - T_\infty}{T_i - T_\infty} = A_1 e^{-\lambda_1^2 \tau} \tag{18–27}$$

Center of sphere ($r = 0$): $$\theta_{0,\text{ sph}} = \frac{T_0 - T_\infty}{T_i - T_\infty} = A_1 e^{-\lambda_1^2 \tau} \tag{18–28}$$

Comparing the two sets of equations above, we notice that the dimensionless temperatures anywhere in a plane wall, cylinder, and sphere are related to the center temperature by

$$\frac{\theta_{\text{wall}}}{\theta_{0,\text{ wall}}} = \cos\left(\frac{\lambda_1 x}{L}\right), \quad \frac{\theta_{\text{cyl}}}{\theta_{0,\text{ cyl}}} = J_0\left(\frac{\lambda_1 r}{r_o}\right), \quad \text{and} \quad \frac{\theta_{\text{sph}}}{\theta_{0,\text{ sph}}} = \frac{\sin(\lambda_1 r/r_o)}{\lambda_1 r/r_o} \tag{18-29}$$

which shows that time dependence of dimensionless temperature within a given geometry is the same throughout. That is, if the dimensionless center temperature θ_0 drops by 20 percent at a specified time, so does the dimensionless temperature θ_0 anywhere else in the medium at the same time.

Once the Bi number is known, these relations can be used to determine the temperature anywhere in the medium. The determination of the constants A_1

TABLE 18–2

Coefficients used in the one-term approximate solution of transient one-dimensional heat conduction in plane walls, cylinders, and spheres (Bi = hL/k for a plane wall of thickness $2L$, and Bi = hr_o/k for a cylinder or sphere of radius r_o)

	Plane Wall		Cylinder		Sphere	
Bi	λ_1	A_1	λ_1	A_1	λ_1	A_1
0.01	0.0998	1.0017	0.1412	1.0025	0.1730	1.0030
0.02	0.1410	1.0033	0.1995	1.0050	0.2445	1.0060
0.04	0.1987	1.0066	0.2814	1.0099	0.3450	1.0120
0.06	0.2425	1.0098	0.3438	1.0148	0.4217	1.0179
0.08	0.2791	1.0130	0.3960	1.0197	0.4860	1.0239
0.1	0.3111	1.0161	0.4417	1.0246	0.5423	1.0298
0.2	0.4328	1.0311	0.6170	1.0483	0.7593	1.0592
0.3	0.5218	1.0450	0.7465	1.0712	0.9208	1.0880
0.4	0.5932	1.0580	0.8516	1.0931	1.0528	1.1164
0.5	0.6533	1.0701	0.9408	1.1143	1.1656	1.1441
0.6	0.7051	1.0814	1.0184	1.1345	1.2644	1.1713
0.7	0.7506	1.0918	1.0873	1.1539	1.3525	1.1978
0.8	0.7910	1.1016	1.1490	1.1724	1.4320	1.2236
0.9	0.8274	1.1107	1.2048	1.1902	1.5044	1.2488
1.0	0.8603	1.1191	1.2558	1.2071	1.5708	1.2732
2.0	1.0769	1.1785	1.5995	1.3384	2.0288	1.4793
3.0	1.1925	1.2102	1.7887	1.4191	2.2889	1.6227
4.0	1.2646	1.2287	1.9081	1.4698	2.4556	1.7202
5.0	1.3138	1.2403	1.9898	1.5029	2.5704	1.7870
6.0	1.3496	1.2479	2.0490	1.5253	2.6537	1.8338
7.0	1.3766	1.2532	2.0937	1.5411	2.7165	1.8673
8.0	1.3978	1.2570	2.1286	1.5526	2.7654	1.8920
9.0	1.4149	1.2598	2.1566	1.5611	2.8044	1.9106
10.0	1.4289	1.2620	2.1795	1.5677	2.8363	1.9249
20.0	1.4961	1.2699	2.2880	1.5919	2.9857	1.9781
30.0	1.5202	1.2717	2.3261	1.5973	3.0372	1.9898
40.0	1.5325	1.2723	2.3455	1.5993	3.0632	1.9942
50.0	1.5400	1.2727	2.3572	1.6002	3.0788	1.9962
100.0	1.5552	1.2731	2.3809	1.6015	3.1102	1.9990
∞	1.5708	1.2732	2.4048	1.6021	3.1416	2.0000

TABLE 18–3

The zeroth- and first-order Bessel functions of the first kind

η	$J_0(\eta)$	$J_1(\eta)$
0.0	1.0000	0.0000
0.1	0.9975	0.0499
0.2	0.9900	0.0995
0.3	0.9776	0.1483
0.4	0.9604	0.1960
0.5	0.9385	0.2423
0.6	0.9120	0.2867
0.7	0.8812	0.3290
0.8	0.8463	0.3688
0.9	0.8075	0.4059
1.0	0.7652	0.4400
1.1	0.7196	0.4709
1.2	0.6711	0.4983
1.3	0.6201	0.5220
1.4	0.5669	0.5419
1.5	0.5118	0.5579
1.6	0.4554	0.5699
1.7	0.3980	0.5778
1.8	0.3400	0.5815
1.9	0.2818	0.5812
2.0	0.2239	0.5767
2.1	0.1666	0.5683
2.2	0.1104	0.5560
2.3	0.0555	0.5399
2.4	0.0025	0.5202
2.6	−0.0968	−0.4708
2.8	−0.1850	−0.4097
3.0	−0.2601	−0.3391
3.2	−0.3202	−0.2613

and λ_1 usually requires interpolation. For those who prefer reading charts to interpolating, these relations are plotted and the one-term approximation solutions are presented in graphical form, known as the *transient temperature charts.* Note that the charts are sometimes difficult to read, and they are subject to reading errors. Therefore, the relations above should be preferred to the charts.

The transient temperature charts in Figs. 18–15, 18–16, and 18–17 for a large plane wall, long cylinder, and sphere were presented by M. P. Heisler in 1947 and are called **Heisler charts**. They were supplemented in 1961 with transient heat transfer charts by H. Gröber. There are *three* charts associated with each geometry: the first chart is to determine the temperature T_0

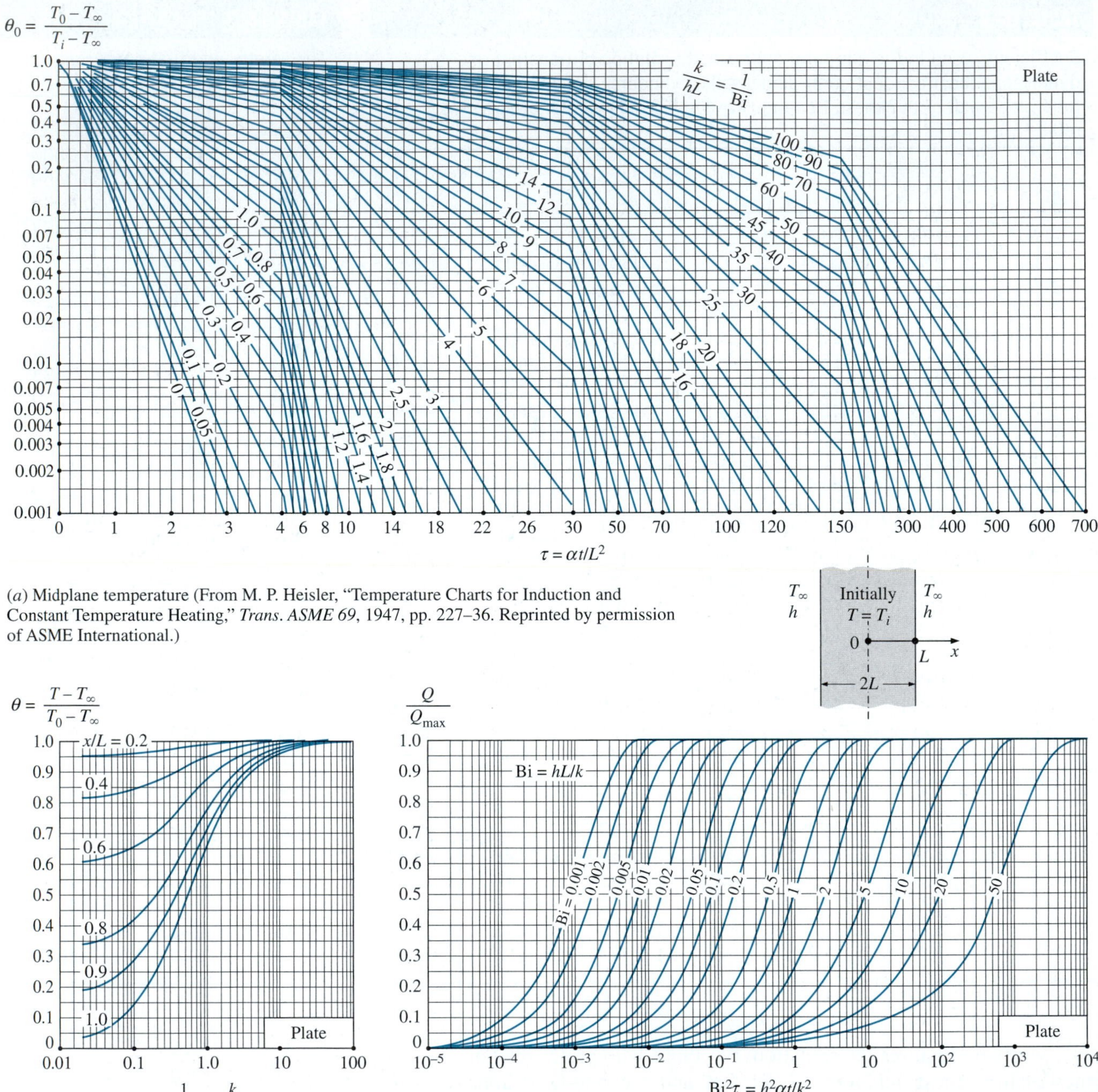

(*a*) Midplane temperature (From M. P. Heisler, "Temperature Charts for Induction and Constant Temperature Heating," *Trans. ASME 69*, 1947, pp. 227–36. Reprinted by permission of ASME International.)

(*b*) Temperature distribution (From M. P. Heisler, "Temperature Charts for Induction and Constant Temperature Heating," *Trans. ASME 69*, 1947, pp. 227–36. Reprinted by permission of ASME International.)

(*c*) Heat transfer (*From H. Gröber et al.*)

FIGURE 18–15

Transient temperature and heat transfer charts for a plane wall of thickness $2L$ initially at a uniform temperature T_i subjected to convection from both sides to an environment at temperature T_∞ with a convection coefficient of h.

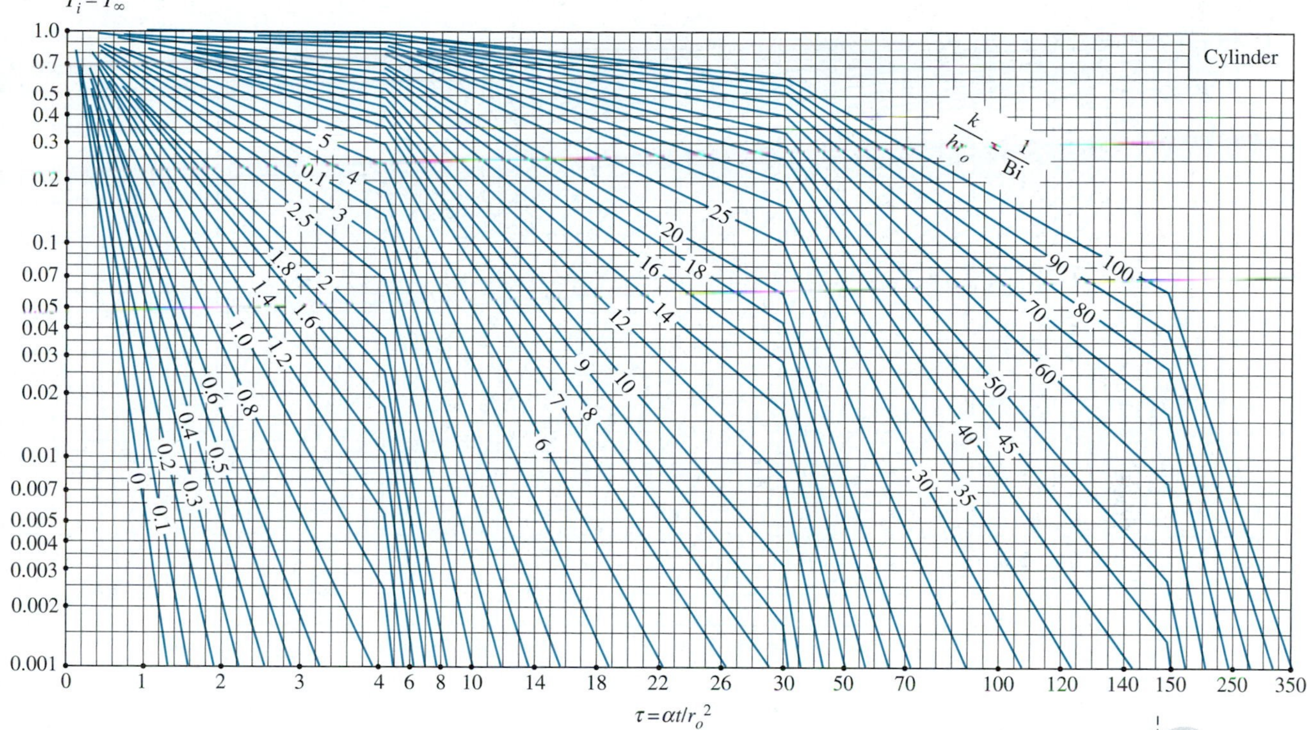

(*a*) Centerline temperature (From M. P. Heisler, Temperat ure Charts for Induction and Constant Temperature Heating," *Trans. ASME 69*, 1947, pp. 227–36. Reprinted by permission of ASME International.)

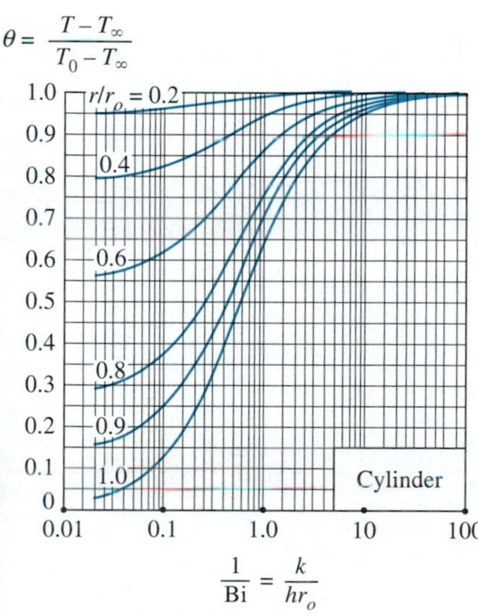

(*b*) Temperature distribution (From M. P. Heisler, "Temperature Charts for Induction and Constant Temperature Heating," *Trans. ASME 69*, 1947, pp. 227–36. Reprinted by permission of ASME International.)

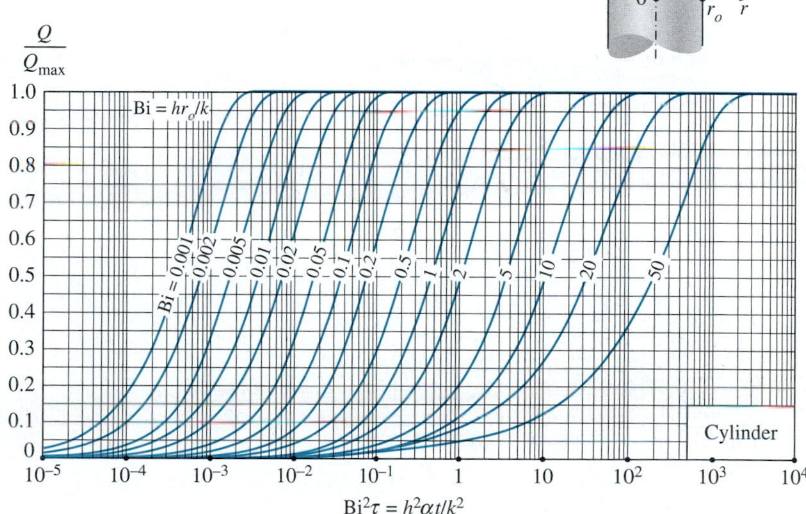

(*c*) Heat transfer (*From H. Gröber et al.*)

FIGURE 18–16

Transient temperature and heat transfer charts for a long cylinder of radius r_o initially at a uniform temperature T_i subjected to convection from all sides to an environment at temperature T_∞ with a convection coefficient of h.

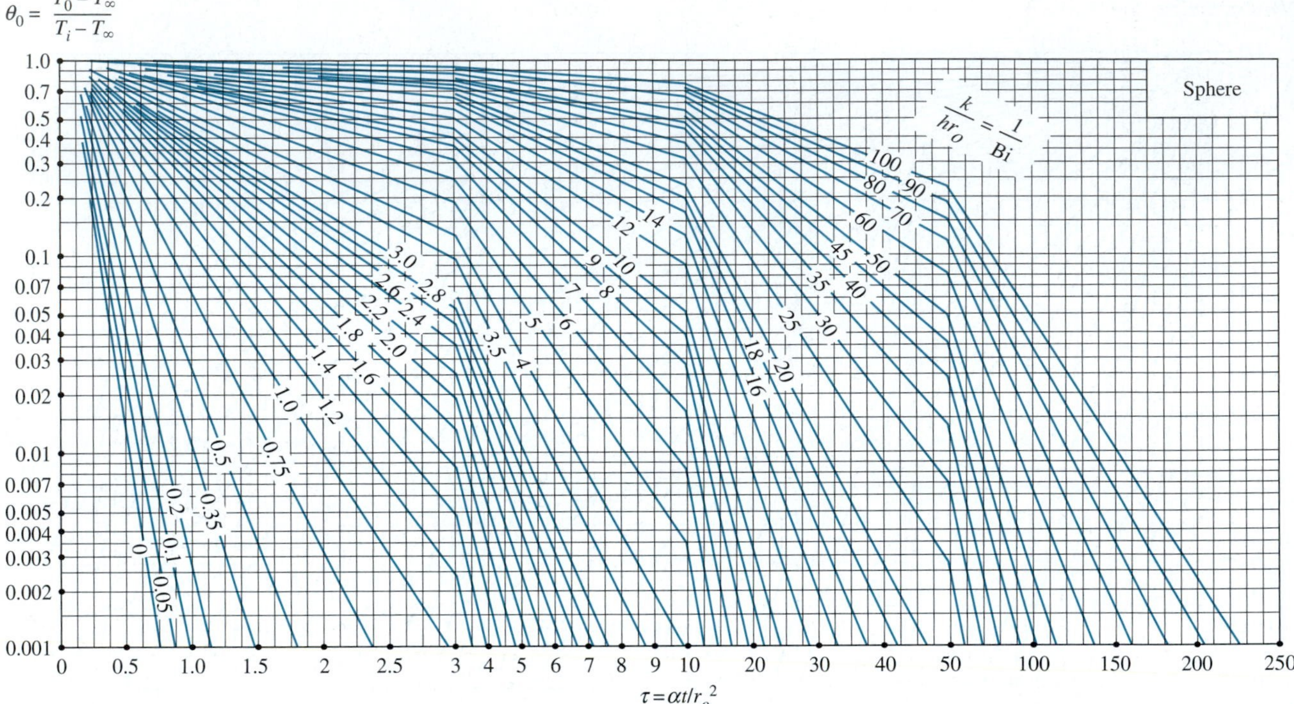

(*a*) Midpoint temperature (From M. P. Heisler, "Temperature Charts for Induction and Constant Temperature Heating," *Trans. ASME 69*, 1947, pp. 227–36. Reprinted by permission of ASME International.)

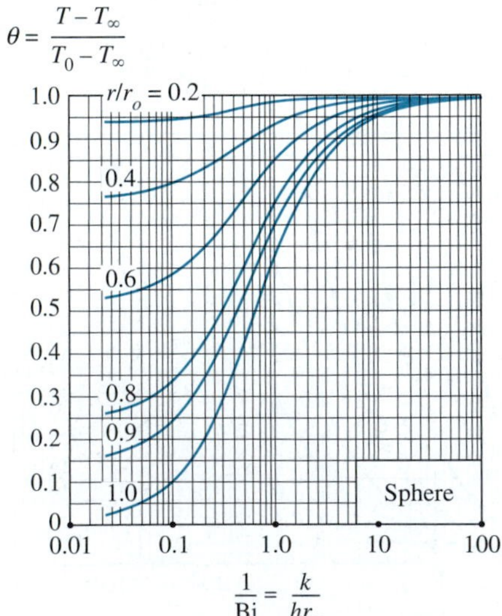

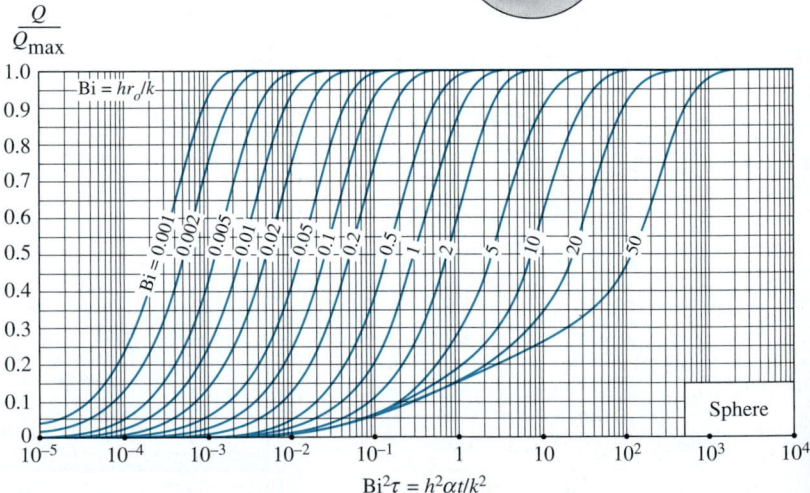

(*b*) Temperature distribution (From M. P. Heisler, "Temperature Charts for Induction and Constant Temperature Heating," *Trans. ASME 69*, 1947, pp. 227–36. Reprinted by permission of ASME International.)

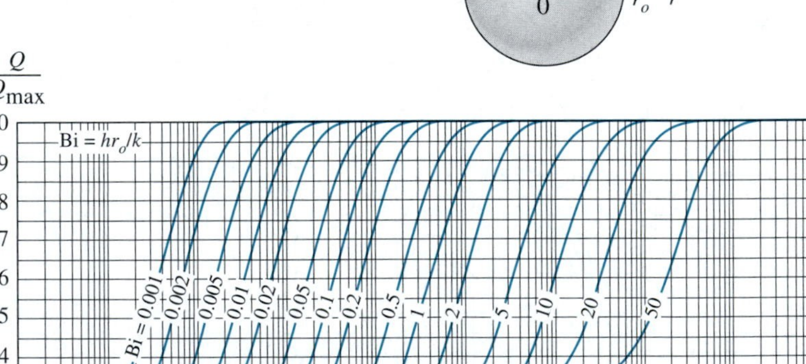

(*c*) Heat transfer (*From H. Gröber et al.*)

FIGURE 18–17

Transient temperature and heat transfer charts for a sphere of radius r_o initially at a uniform temperature T_i subjected to convection from all sides to an environment at temperature T_∞ with a convection coefficient of h.

at the *center* of the geometry at a given time t. The second chart is to determine the temperature at *other locations* at the same time in terms of T_0. The third chart is to determine the total amount of *heat transfer* up to the time t. These plots are valid for $\tau > 0.2$.

Note that the case $1/\text{Bi} = k/hL = 0$ corresponds to $h \to \infty$, which corresponds to the case of *specified surface temperature* T_∞. That is, the case in which the surfaces of the body are suddenly brought to the temperature T_∞ at $t = 0$ and kept at T_∞ at all times can be handled by setting h to infinity (Fig. 18–18).

The temperature of the body changes from the initial temperature T_i to the temperature of the surroundings T_∞ at the end of the transient heat conduction process. Thus, the *maximum* amount of heat that a body can gain (or lose if $T_i > T_\infty$) is simply the *change* in the *energy content* of the body. That is,

$$Q_{\max} = mc_p(T_\infty - T_i) = \rho V c_p(T_\infty - T_i) \qquad \text{(kJ)} \tag{18–30}$$

where m is the mass, V is the volume, ρ is the density, and c_p is the specific heat of the body. Thus, $Q_{\max}$ represents the amount of heat transfer for $t \to \infty$. The amount of heat transfer Q at a finite time t is obviously less than this maximum, and it can be expressed as the sum of the internal energy changes throughout the entire geometry as

$$Q = \int_V \rho c_p[T(x,t) - T_i]dV \tag{18–31}$$

where $T(x, t)$ is the temperature distribution in the medium at time t. Assuming constant properties, the ratio of $Q/Q_{\max}$ becomes

$$\frac{Q}{Q_{\max}} = \frac{\int_V \rho c_p[T(x, t) - T_i]dV}{\rho c_p(T_\infty - T_i)V} = \frac{1}{V}\int_V (1 - \theta)dV \tag{18–32}$$

Using the appropriate nondimensional temperature relations based on the one-term approximation for the plane wall, cylinder, and sphere, and performing the indicated integrations, we obtain the following relations for the fraction of heat transfer in those geometries:

Plane wall: $$\left(\frac{Q}{Q_{\max}}\right)_{\text{wall}} = 1 - \theta_{0,\,\text{wall}}\frac{\sin\lambda_1}{\lambda_1} \tag{18–33}$$

Cylinder: $$\left(\frac{Q}{Q_{\max}}\right)_{\text{cyl}} = 1 - 2\theta_{0,\,\text{cyl}}\frac{J_1(\lambda_1)}{\lambda_1} \tag{18–34}$$

Sphere: $$\left(\frac{Q}{Q_{\max}}\right)_{\text{sph}} = 1 - 3\theta_{0,\,\text{sph}}\frac{\sin\lambda_1 - \lambda_1\cos\lambda_1}{\lambda_1^3} \tag{18–35}$$

These $Q/Q_{\max}$ ratio relations based on the one-term approximation are also plotted in Figures 18–15*c*, 18–16*c*, and 18–17*c*, against the variables Bi and $h^2\alpha t/k^2$ for the large plane wall, long cylinder, and sphere, respectively. Note that once the *fraction* of heat transfer $Q/Q_{\max}$ has been determined from these charts or equations for the given t, the actual amount of heat transfer by that time can be evaluated by multiplying this fraction by $Q_{\max}$. A *negative* sign for $Q_{\max}$ indicates that the body is *rejecting* heat (Fig. 18–19).

The use of the Heisler/Gröber charts and the one-term solutions already discussed is limited to the conditions specified at the beginning of this sec-

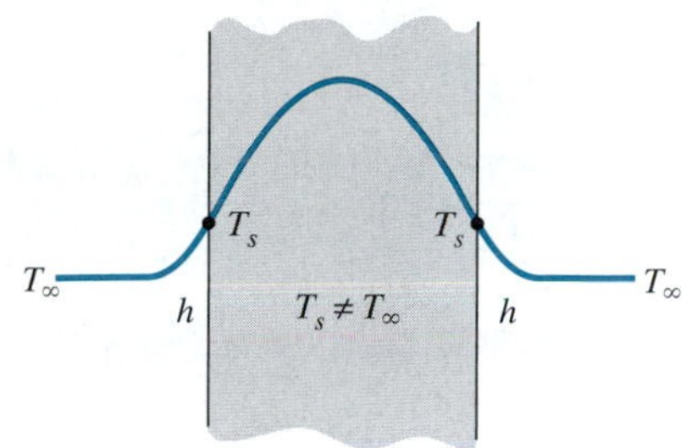

(*a*) Finite convection coefficient

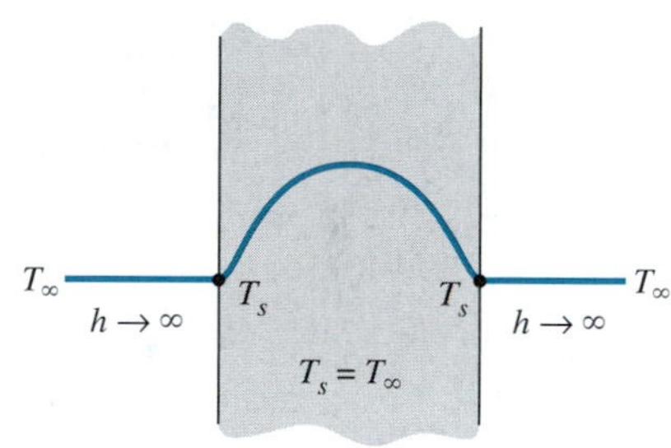

(*b*) Infinite convection coefficient

FIGURE 18–18

The specified surface temperature corresponds to the case of convection to an environment at T_∞ with a convection coefficient h that is *infinite*.

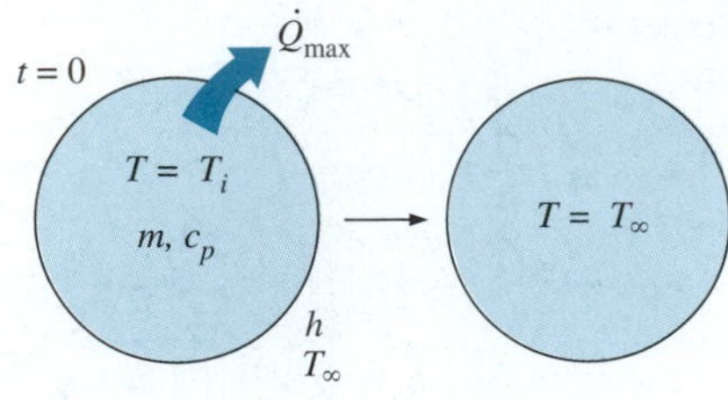

(a) Maximum heat transfer ($t \to \infty$)

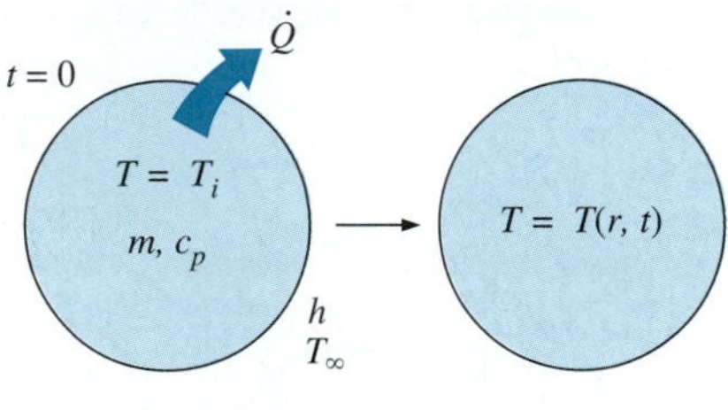

$\text{Bi} = \ldots$

$\frac{h^2\alpha t}{k^2} = \text{Bi}^2\tau = \cdots$ $\Bigg\}$ $\frac{Q}{Q_{max}} = \cdots$

(Gröber chart)

(b) Actual heat transfer for time t

FIGURE 18–19

The fraction of total heat transfer Q/Q_{max} up to a specified time t is determined using the Gröber charts.

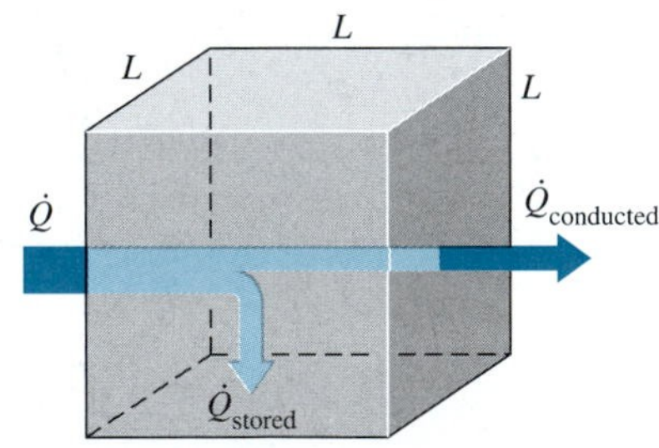

Fourier number: $\tau = \frac{\alpha t}{L^2} = \frac{\dot{Q}_{conducted}}{\dot{Q}_{stored}}$

FIGURE 18–20

Fourier number at time t can be viewed as the ratio of the rate of heat conducted to the rate of heat stored at that time.

tion: the body is initially at a *uniform* temperature, the temperature of the medium surrounding the body and the convection heat transfer coefficient are *constant* and *uniform*, and there is no *heat generation* in the body.

We discussed the physical significance of the *Biot number* earlier and indicated that it is a measure of the relative magnitudes of the two heat transfer mechanisms: *convection* at the surface and *conduction* through the solid. A *small* value of Bi indicates that the inner resistance of the body to heat conduction is *small* relative to the resistance to convection between the surface and the fluid. As a result, the temperature distribution within the solid becomes fairly uniform, and lumped system analysis becomes applicable. Recall that when $\text{Bi} < 0.1$, the error in assuming the temperature within the body to be *uniform* is negligible.

To understand the physical significance of the *Fourier number* τ, we express it as (Fig. 18–20)

$$\tau = \frac{\alpha t}{L^2} = \frac{kL^2\,(1/L)\;\Delta T}{\rho c_p L^3/t\;\Delta T} = \frac{\text{The rate at which heat is } conducted \text{ across } L \text{ of a body of volume } L^3}{\text{The rate at which heat is } stored \text{ in a body of volume } L^3} \tag{18–36}$$

Therefore, the Fourier number is a measure of *heat conducted* through a body relative to *heat stored.* Thus, a large value of the Fourier number indicates faster propagation of heat through a body.

Perhaps you are wondering about what constitutes an infinitely large plate or an infinitely long cylinder. After all, nothing in this world is infinite. A plate whose thickness is small relative to the other dimensions can be modeled as an infinitely large plate, except very near the outer edges. But the edge effects on large bodies are usually negligible, and thus a large plane wall such as the wall of a house can be modeled as an infinitely large wall for heat transfer purposes. Similarly, a long cylinder whose diameter is small relative to its length can be analyzed as an infinitely long cylinder. The use of the transient temperature charts and the one-term solutions is illustrated in Examples 18–3, 18–4, and 18–5.

EXAMPLE 18–3 Boiling Eggs

An ordinary egg can be approximated as a 5-cm-diameter sphere (Fig. 18–21). The egg is initially at a uniform temperature of 5°C and is dropped into boiling water at 95°C. Taking the convection heat transfer coefficient to be $h = 1200\ \text{W/m}^2 \cdot °\text{C}$, determine how long it will take for the center of the egg to reach 70°C.

Solution An egg is cooked in boiling water. The cooking time of the egg is to be determined.

Assumptions **1** The egg is spherical in shape with a radius of $r_o = 2.5$ cm. **2** Heat conduction in the egg is one-dimensional because of thermal symmetry about the midpoint. **3** The thermal properties of the egg and the heat transfer coefficient are constant. **4** The Fourier number is $\tau > 0.2$ so that the one-term approximate solutions are applicable. **5** The internal eggwhite is a solid and admits no convection.

Properties The water content of eggs is about 74 percent, and thus the thermal conductivity and diffusivity of eggs can be approximated by those of water at the average temperature of (5 + 70)/2 = 37.5°C; k = 0.627 W/m · °C and $\alpha = k/\rho c_p = 0.151 \times 10^{-6}$ m²/s (Table A–15).

Analysis The temperature within the egg varies with radial distance as well as time, and the temperature at a specified location at a given time can be determined from the Heisler charts or the one-term solutions. Here we use the latter to demonstrate their use. The Biot number for this problem is

$$\text{Bi} = \frac{hr_o}{k} = \frac{(1200 \text{ W/m}^2 \cdot {}^\circ\text{C})(0.025 \text{ m})}{0.627 \text{ W/m} \cdot {}^\circ\text{C}} = 47.8$$

which is much greater than 0.1, and thus the lumped system analysis is not applicable. The coefficients λ_1 and A_1 for a sphere corresponding to this Bi are, from Table 18–2,

$$\lambda_1 = 3.0754, \qquad A_1 = 1.9958$$

Substituting these and other values into Eq. 18–28 and solving for τ gives

$$\frac{T_0 - T_\infty}{T_i - T_\infty} = A_1 e^{-\lambda_1^2 \tau} \longrightarrow \frac{70 - 95}{5 - 95} = 1.9958 e^{-(3.0754)^2 \tau} \longrightarrow \tau = 0.209$$

which is greater than 0.2, and thus the one-term solution is applicable with an error of less than 2 percent. Then the cooking time is determined from the definition of the Fourier number to be

$$t = \frac{\tau r_o^2}{\alpha} = \frac{(0.209)(0.025 \text{ m})^2}{0.151 \times 10^{-6} \text{ m}^2\text{/s}} = 865 \text{ s} \approx \mathbf{14.4 \text{ min}}$$

Therefore, it will take about 15 min for the center of the egg to be heated from 5°C to 70°C.

Discussion Note that the Biot number in lumped system analysis was defined differently as Bi = $hL_c/k = h(r_o/3)/k$. However, either definition can be used in determining the applicability of the lumped system analysis unless Bi ≈ 0.1.

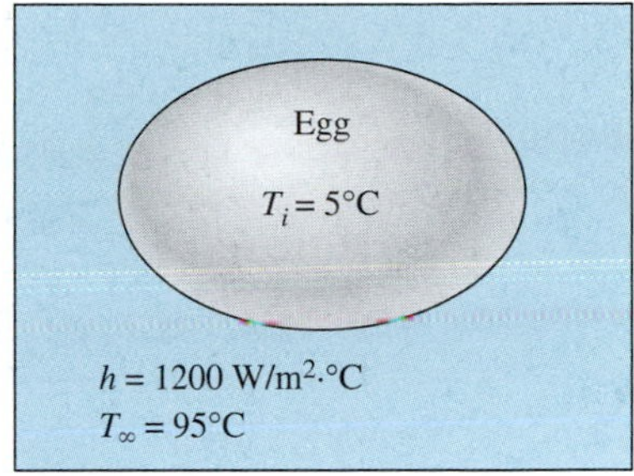

FIGURE 18–21

Schematic for Example 18–3.

EXAMPLE 18–4 Heating of Brass Plates in an Oven

In a production facility, large brass plates of 4-cm thickness that are initially at a uniform temperature of 20°C are heated by passing them through an oven that is maintained at 500°C (Fig. 18–22). The plates remain in the oven for a period of 7 min. Taking the combined convection and radiation heat transfer coefficient to be h = 120 W/m² · °C, determine the surface temperature of the plates when they come out of the oven.

Solution Large brass plates are heated in an oven. The surface temperature of the plates leaving the oven is to be determined.

Assumptions **1** Heat conduction in the plate is one-dimensional since the plate is large relative to its thickness and there is thermal symmetry about the center plane. **2** The thermal properties of the plate and the heat transfer coefficient are constant. **3** The Fourier number is $\tau > 0.2$ so that the one-term approximate solutions are applicable.

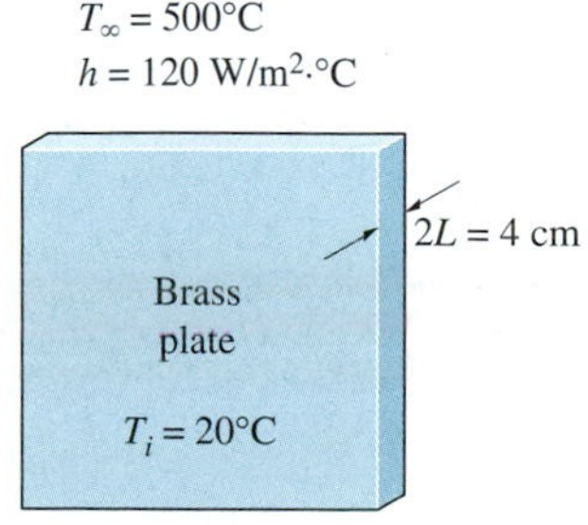

FIGURE 18–22

Schematic for Example 18–4.

Properties The properties of brass at room temperature are $k = 110$ W/m · °C, $\rho = 8530$ kg/m³, $c_p = 380$ J/kg · °C, and $\alpha = 33.9 \times 10^{-6}$ m²/s (Table A–24). More accurate results are obtained by using properties at average temperature.

Analysis The temperature at a specified location at a given time can be determined from the Heisler charts or one-term solutions. Here we use the charts to demonstrate their use. Noting that the half-thickness of the plate is $L = 0.02$ m, from Fig. 18–15 we have

$$\left.\begin{aligned} \frac{1}{\text{Bi}} &= \frac{k}{hL} = \frac{110 \text{ W/m} \cdot {}^\circ\text{C}}{(120 \text{ W/m}^2 \cdot {}^\circ\text{C})(0.02 \text{ m})} = 45.8 \\ \tau &= \frac{\alpha t}{L^2} = \frac{(33.9 \times 10^{-6} \text{ m}^2\text{/s})(7 \times 60 \text{ s})}{(0.02 \text{ m})^2} = 35.6 \end{aligned}\right\} \frac{T_0 - T_\infty}{T_i - T_\infty} = 0.46$$

Also,

$$\left.\begin{aligned} \frac{1}{\text{Bi}} &= \frac{k}{hL} = 45.8 \\ \frac{x}{L} &= \frac{L}{L} = 1 \end{aligned}\right\} \frac{T - T_\infty}{T_0 - T_\infty} = 0.99$$

Therefore,

$$\frac{T - T_\infty}{T_i - T_\infty} = \frac{T - T_\infty}{T_0 - T_\infty}\frac{T_0 - T_\infty}{T_i - T_\infty} = 0.46 \times 0.99 = 0.455$$

and

$$T = T_\infty + 0.455(T_i - T_\infty) = 500 + 0.455(20 - 500) = \mathbf{282^\circ C}$$

Therefore, the surface temperature of the plates will be 282°C when they leave the oven.

Discussion We notice that the Biot number in this case is Bi = 1/45.8 = 0.022, which is much less than 0.1. Therefore, we expect the lumped system analysis to be applicable. This is also evident from $(T - T_\infty)/(T_0 - T_\infty) = 0.99$, which indicates that the temperatures at the center and the surface of the plate relative to the surrounding temperature are within 1 percent of each other. Noting that the error involved in reading the Heisler charts is typically a few percent, the lumped system analysis in this case may yield just as accurate results with less effort.

The heat transfer surface area of the plate is $2A$, where A is the face area of the plate (the plate transfers heat through both of its surfaces), and the volume of the plate is $V = (2L)A$, where L is the half-thickness of the plate. The exponent b used in the lumped system analysis is

$$b = \frac{hA_s}{\rho c_p V} = \frac{h(2A)}{\rho c_p (2LA)} = \frac{h}{\rho c_p L}$$

$$= \frac{120 \text{ W/m}^2 \cdot {}^\circ\text{C}}{(8530 \text{ kg/m}^3)(380 \text{ J/kg} \cdot {}^\circ\text{C})(0.02 \text{ m})} = 0.00185 \text{ s}^{-1}$$

Then the temperature of the plate at $t = 7$ min $= 420$ s is determined from

$$\frac{T(t) - T_\infty}{T_i - T_\infty} = e^{-bt} \longrightarrow \frac{T(t) - 500}{20 - 500} = e^{-(0.00185 \text{ s}^{-1})(420 \text{ s})}$$

It yields

$$T(t) = 279°\text{C}$$

which is practically identical to the result obtained above using the Heisler charts. Therefore, we can use lumped system analysis with confidence when the Biot number is sufficiently small.

EXAMPLE 18–5 Cooling of a Long Stainless Steel Cylindrical Shaft

A long 20-cm-diameter cylindrical shaft made of stainless steel 304 comes out of an oven at a uniform temperature of 600°C (Fig. 18–23). The shaft is then allowed to cool slowly in an environment chamber at 200°C with an average heat transfer coefficient of h = 80 W/m^2 · °C. Determine the temperature at the center of the shaft 45 min after the start of the cooling process. Also, determine the heat transfer per unit length of the shaft during this time period.

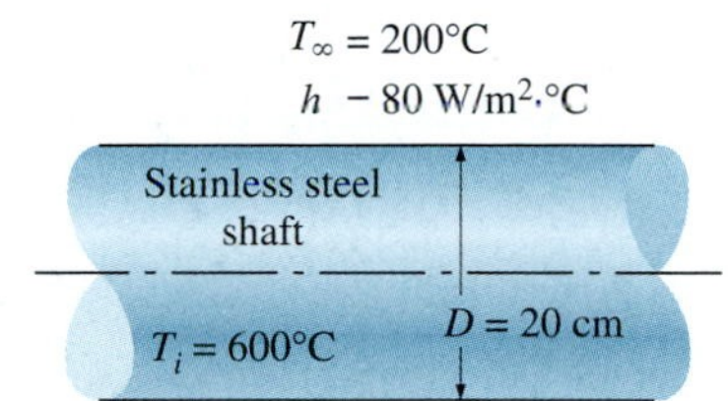

FIGURE 18–23
Schematic for Example 18–5.

Solution A long cylindrical shaft is allowed to cool slowly. The center temperature and the heat transfer per unit length are to be determined.

Assumptions **1** Heat conduction in the shaft is one-dimensional since it is long and it has thermal symmetry about the centerline. **2** The thermal properties of the shaft and the heat transfer coefficient are constant. **3** The Fourier number is $\tau > 0.2$ so that the one-term approximate solutions are applicable.

Properties The properties of stainless steel 304 at room temperature are k = 14.9 W/m · °C, ρ = 7900 kg/m^3, c_p = 477 J/kg · °C, and $\alpha = 3.95 \times 10^{-6}$ m^2/s (Table A–24). More accurate results can be obtained by using properties at average temperature.

Analysis The temperature within the shaft may vary with the radial distance r as well as time, and the temperature at a specified location at a given time can be determined from the Heisler charts. Noting that the radius of the shaft is r_o = 0.1 m, from Fig. 18–16 we have

$$\left.\begin{aligned} \frac{1}{\text{Bi}} &= \frac{k}{hr_o} = \frac{14.9 \text{ W/m} \cdot °\text{C}}{(80 \text{ W/m}^2 \cdot °\text{C})(0.1 \text{ m})} = 1.86 \\ \tau &= \frac{\alpha t}{r_o^2} = \frac{(3.95 \times 10^{-6} \text{ m}^2\text{/s})(45 \times 60 \text{ s})}{(0.1 \text{ m})^2} = 1.07 \end{aligned}\right\} \frac{T_0 - T_\infty}{T_i - T_\infty} = 0.40$$

and

$$T_0 = T_\infty + 0.4(T_i - T_\infty) = 200 + 0.4(600 - 200) = \mathbf{360°C}$$

Therefore, the center temperature of the shaft drops from 600°C to 360°C in 45 min.

To determine the actual heat transfer, we first need to calculate the maximum heat that can be transferred from the cylinder, which is the sensible energy of the cylinder relative to its environment. Taking L = 1 m,

$$\begin{aligned} m &= \rho V = \rho \pi r_o^2 L = (7900 \text{ kg/m}^3)\pi(0.1 \text{ m})^2(1 \text{ m}) = 248.2 \text{ kg} \\ Q_{\text{max}} &= mc_p(T_\infty - T_i) = (248.2 \text{ kg})(0.477 \text{ kJ/kg} \cdot °\text{C})(600 - 200)°\text{C} \\ &= 47{,}350 \text{ kJ} \end{aligned}$$

The dimensionless heat transfer ratio is determined from Fig. 18–16*c* for a long cylinder to be

$$\left.\begin{aligned} \text{Bi} &= \frac{1}{1/\text{Bi}} = \frac{1}{1.86} = 0.537 \\ \frac{h^2\alpha t}{k^2} &= \text{Bi}^2\tau = (0.537)^2(1.07) = 0.309 \end{aligned}\right\} \frac{Q}{Q_{\max}} = 0.62$$

Therefore,

$$Q = 0.62Q_{\max} = 0.62 \times (47{,}350 \text{ kJ}) = \mathbf{29{,}360\ kJ}$$

which is the total heat transfer from the shaft during the first 45 min of the cooling.

Alternative solution We could also solve this problem using the one-term solution relation instead of the transient charts. First we find the Biot number

$$\text{Bi} = \frac{hr_o}{k} = \frac{(80 \text{ W/m}^2 \cdot {}^\circ\text{C})(0.1 \text{ m})}{14.9 \text{ W/m} \cdot {}^\circ\text{C}} = 0.537$$

The coefficients λ_1 and A_1 for a cylinder corresponding to this Bi are determined from Table 18–2 to be

$$\lambda_1 = 0.970, \qquad A_1 = 1.122$$

Substituting these values into Eq. 18–27 gives

$$\theta_0 = \frac{T_0 - T_\infty}{T_i - T_\infty} = A_1 e^{-\lambda_1^2 \tau} = 1.122e^{-(0.970)^2(1.07)} = 0.41$$

and thus

$$T_0 = T_\infty + 0.41(T_i - T_\infty) = 200 + 0.41(600 - 200) = \mathbf{364^\circ C}$$

The value of $J_1(\lambda_1)$ for $\lambda_1 = 0.970$ is determined from Table 18–3 to be 0.430. Then the fractional heat transfer is determined from Eq. 18–34 to be

$$\frac{Q}{Q_{\max}} = 1 - 2\theta_0 \frac{J_1(\lambda_1)}{\lambda_1} = 1 - 2 \times 0.41 \frac{0.430}{0.970} = 0.636$$

and thus

$$Q = 0.636Q_{\max} = 0.636 \times (47{,}350 \text{ kJ}) = \mathbf{30{,}120\ kJ}$$

Discussion The slight difference between the two results is due to the reading error of the charts.

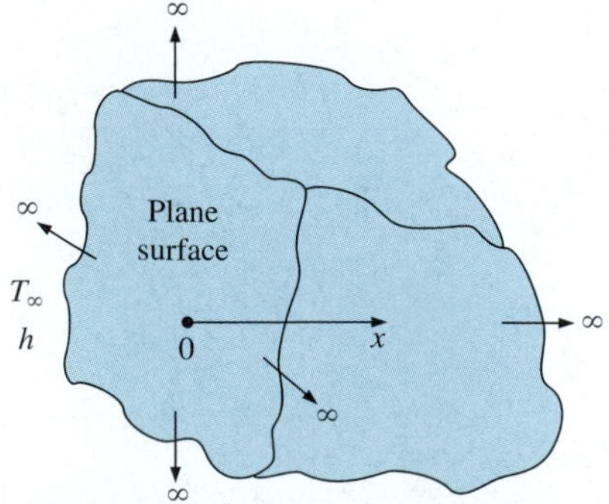

FIGURE 18–24

Schematic of a semi-infinite body.

18–3 ▪ TRANSIENT HEAT CONDUCTION IN SEMI-INFINITE SOLIDS

A semi-infinite solid is an idealized body that has a *single plane surface* and extends to infinity in all directions, as shown in Figure 18–24. This idealized body is used to indicate that the temperature change in the part of the body in which we are interested (the region close to the surface) is due to the thermal conditions on a single surface. The earth, for example, can be considered to be a semi-infinite medium in determining the variation of temperature near its surface. Also, a thick wall can be modeled as a semi-infinite medium if all we are interested in is the variation of temperature in the region near one of the surfaces, and the other surface is too far to have any impact on the region of interest during the time of observation. The temperature in the core region of the wall remains unchanged in this case.

For short periods of time, most bodies can be modeled as semi-infinite solids since heat does not have sufficient time to penetrate deep into the body, and the thickness of the body does not enter into the heat transfer analysis. A steel piece of any shape, for example, can be treated as a semi-infinite solid when it is quenched rapidly to harden its surface. A body whose surface is heated by a laser pulse can be treated the same way.

Consider a semi-infinite solid with constant thermophysical properties, no internal heat generation, uniform thermal conditions on its exposed surface, and initially a uniform temperature of T_i throughout. Heat transfer in this case occurs only in the direction normal to the surface (the x direction), and thus it is one-dimensional. Differential equations are independent of the boundary or initial conditions, and thus Eq. 18–10*a* for one-dimensional transient conduction in Cartesian coordinates applies. The depth of the solid is large ($x \to \infty$) compared to the depth that heat can penetrate, and these phenomena can be expressed mathematically as a boundary condition as $T(x \to \infty, t) = T_i$.

Heat conduction in a semi-infinite solid is governed by the thermal conditions imposed on the exposed surface, and thus the solution depends strongly on the boundary condition at $x = 0$. Below we present a detailed analytical solution for the case of constant temperature T_s on the surface, and give the results for other more complicated boundary conditions. When the surface temperature is changed to T_s at $t = 0$ and held constant at that value at all times, the formulation of the problem can be expressed as

Differential equation:
$$\frac{\partial^2 T}{\partial x^2} = \frac{1}{\alpha}\frac{\partial T}{\partial t} \qquad \textbf{(18–37a)}$$

Boundary conditions:
$$T(0, t) = T_s \quad \text{and} \quad T(x \to \infty, t) = T_i \qquad \textbf{(18–37b)}$$

Initial condition:
$$T(x, 0) = T_i \qquad \textbf{(18–37c)}$$

The separation of variables technique does not work in this case since the medium is infinite. But another clever approach that converts the partial differential equation into an ordinary differential equation by combining the two independent variables x and t into a single variable η, called the **similarity variable**, works well. For transient conduction in a semi-infinite medium, it is defined as

Similarity variable:
$$\eta = \frac{x}{\sqrt{4\alpha t}} \qquad \textbf{(18–38)}$$

Assuming $T = T(\eta)$ (to be verified) and using the chain rule, all derivatives in the heat conduction equation can be transformed into the new variable, as shown in Fig. 18–25. Noting that $\eta = 0$ at $x = 0$ and $\eta \to \infty$ as $x \to \infty$ (and also at $t = 0$) and substituting into Eqs. 18–37 give, after simplification,

$$\frac{\partial^2 T}{\partial x^2} = \frac{1}{\alpha}\frac{\partial T}{\partial t} \quad \text{and} \quad \eta = \frac{x}{\sqrt{4\alpha t}}$$
$$\frac{\partial T}{\partial t} = \frac{dT}{d\eta}\frac{\partial \eta}{\partial t} = \frac{x}{2t\sqrt{4\alpha t}}\frac{dT}{d\eta}$$
$$\frac{\partial T}{\partial x} = \frac{dT}{d\eta}\frac{\partial \eta}{\partial x} = \frac{1}{\sqrt{4\alpha t}}\frac{dT}{d\eta}$$
$$\frac{\partial^2 T}{\partial x^2} = \frac{d}{d\eta}\left(\frac{\partial T}{\partial x}\right)\frac{\partial \eta}{\partial x} = \frac{1}{4\alpha t}\frac{d^2 T}{d\eta^2}$$

FIGURE 18–25
Transformation of variables in the derivatives of the heat conduction equation by the use of chain rule.

$$\frac{d^2T}{d\eta^2} = -2\eta\frac{dT}{d\eta} \quad \textbf{(18–39a)}$$

$$T(0) = T_s \quad \text{and} \quad T(\eta \to \infty) = T_i \quad \textbf{(18–39b)}$$

Note that the second boundary condition and the initial condition result in the same boundary condition. Both the transformed equation and the boundary conditions depend on η only and are independent of x and t. Therefore, transformation is successful, and η is indeed a similarity variable.

To solve the 2nd order ordinary differential equation in Eqs. 18–39, we define a new variable w as $w = dT/d\eta$. This reduces Eq. 18–39*a* into a first order differential equation than can be solved by separating variables,

$$\frac{dw}{d\eta} = -2\eta w \;\rightarrow\; \frac{dw}{w} = -2\eta d\eta \;\rightarrow\; \ln w = -\eta^2 + C_0 \;\rightarrow\; w = C_1 e^{-\eta^2}$$

where $C_1 = \ln C_0$. Back substituting $w = dT/d\eta$ and integrating again,

$$T = C_1 \int_0^\eta e^{-u^2} du + C_2 \quad \textbf{(18–40)}$$

where u is a dummy integration variable. The boundary condition at $\eta = 0$ gives $C_2 = T_s$, and the one for $\eta \to \infty$ gives

$$T_i = C_1 \int_0^\infty e^{-u^2} du + C_2 = C_1\frac{\sqrt{\pi}}{2} + T_s \;\rightarrow\; C_1 = \frac{2(T_i - T_s)}{\sqrt{\pi}} \quad \textbf{(18–41)}$$

Substituting the C_1 and C_2 expressions into Eq. 18–40 and rearranging, the variation of temperature becomes

$$\frac{T - T_s}{T_i - T_s} = \frac{2}{\sqrt{\pi}} \int_0^\eta e^{-u^2} du = \text{erf}(\eta) = 1 - \text{erfc}(\eta) \quad \textbf{(18–42)}$$

where the mathematical functions

$$\text{erf}(\eta) = \frac{2}{\sqrt{\pi}} \int_0^\eta e^{-u^2} du \quad \text{and} \quad \text{erfc}(\eta) = 1 - \frac{2}{\sqrt{\pi}} \int_0^\eta \varepsilon^{-u^2} du \quad \textbf{(18–43)}$$

are called the **error function** and the **complementary error function**, respectively, of argument η (Fig. 18–26). Despite its simple appearance, the integral in the definition of the error function cannot be performed analytically. Therefore, the function erfc(η) is evaluated numerically for different values of η, and the results are listed in Table 18–4.

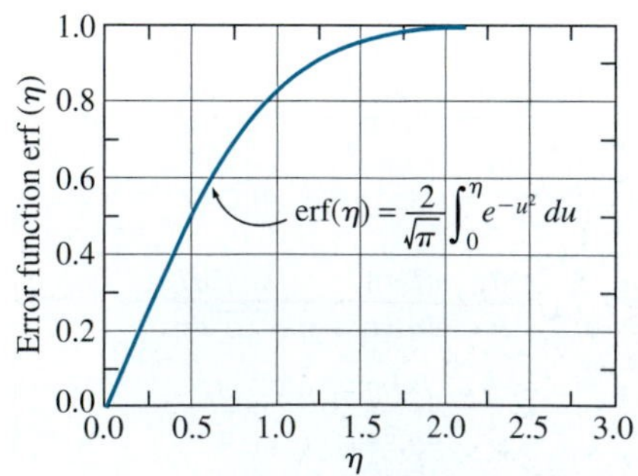

FIGURE 18–26

Error function is a standard mathematical function, just like the sine and cosine functions, whose value varies between 0 and 1.

Knowing the temperature distribution, the heat flux at the surface can be determined from the Fourier's law to be

$$\dot{q}_s = -k\frac{\partial T}{\partial x}\bigg|_{x=0} = -k\frac{dT}{d\eta}\frac{\partial\eta}{\partial x}\bigg|_{\eta=0} = -kC_1 e^{-\eta^2}\frac{1}{\sqrt{4\alpha t}}\bigg|_{\eta=0} = \frac{k(T_s - T_i)}{\sqrt{\pi\alpha t}} \quad \textbf{(18–44)}$$

TABLE 18–4

The complementary error function*

η	erfc (η)	η	erfc (η)	η	erfc (η)	η	erfc (η)	η	erfc (η)	η	erfc (η)
0.00	1.00000	0.38	0.5910	0.76	0.2825	1.14	0.1069	1.52	0.03159	1.90	0.00721
0.02	0.9774	0.40	0.5716	0.78	0.2700	1.16	0.10090	1.54	0.02941	1.92	0.00662
0.04	0.9549	0.42	0.5525	0.80	0.2579	1.18	0.09516	1.56	0.02737	1.94	0.00608
0.06	0.9324	0.44	0.5338	0.82	0.2462	1.20	0.08969	1.58	0.02545	1.96	0.00557
0.08	0.9099	0.46	0.5153	0.84	0.2349	1.22	0.08447	1.60	0.02365	1.98	0.00511
0.10	0.8875	0.48	0.4973	0.86	0.2239	1.24	0.07950	1.62	0.02196	2.00	0.00468
0.12	0.8652	0.50	0.4795	0.88	0.2133	1.26	0.07476	1.64	0.02038	2.10	0.00298
0.14	0.8431	0.52	0.4621	0.90	0.2031	1.28	0.07027	1.66	0.01890	2.20	0.00186
0.16	0.8210	0.54	0.4451	0.92	0.1932	1.30	0.06599	1.68	0.01751	2.30	0.00114
0.18	0.7991	0.56	0.4284	0.94	0.1837	1.32	0.06194	1.70	0.01612	2.40	0.00069
0.20	0.7773	0.58	0.4121	0.96	0.1746	1.34	0.05809	1.72	0.01500	2.50	0.00041
0.22	0.7557	0.60	0.3961	0.98	0.1658	1.36	0.05444	1.74	0.01387	2.60	0.00024
0.24	0.7343	0.62	0.3806	1.00	0.1573	1.38	0.05098	1.76	0.01281	2.70	0.00013
0.26	0.7131	0.64	0.3654	1.02	0.1492	1.40	0.04772	1.78	0.01183	2.80	0.00008
0.28	0.6921	0.66	0.3506	1.04	0.1413	1.42	0.04462	1.80	0.01091	2.90	0.00004
0.30	0.6714	0.68	0.3362	1.06	0.1339	1.44	0.04170	1.82	0.01006	3.00	0.00002
0.32	0.6509	0.70	0.3222	1.08	0.1267	1.46	0.03895	1.84	0.00926	3.20	0.00001
0.34	0.6306	0.72	0.3086	1.10	0.1198	1.48	0.03635	1.86	0.00853	3.40	0.00000
0.36	0.6107	0.74	0.2953	1.12	0.1132	1.50	0.03390	1.88	0.00784	3.60	0.00000

* *Note:* erf = 1 erfc

The solutions in Eqs. 18–42 and 18–44 correspond to the case when the temperature of the exposed surface of the medium is suddenly raised (or lowered) to T_s at $t = 0$ and is maintained at that value at all times. The specified surface temperature case is closely approximated in practice when condensation or boiling takes place on the surface. Using a similar approach or the Laplace transform technique, analytical solutions can be obtained for other boundary conditions on the surface, with the following results.

Case 1: Specified Surface Temperature, T_s = constant (Fig. 18–27).

$$\frac{T(x, t) - T_i}{T_s - T_i} = \text{erfc}\left(\frac{x}{2\sqrt{\alpha t}}\right) \quad \text{and} \quad \dot{q}_s(t) = \frac{k(T_s - T_i)}{\sqrt{\pi \alpha t}} \tag{18–45}$$

Case 2: Specified Surface Heat Flux, $\dot{q}_s$ = constant.

$$T(x, t) - T_i = \frac{\dot{q}_s}{k}\left[\sqrt{\frac{4\alpha t}{\pi}}\exp\left(-\frac{x^2}{4\alpha t}\right) - x\,\text{erfc}\left(\frac{x}{2\sqrt{\alpha t}}\right)\right] \tag{18–46}$$

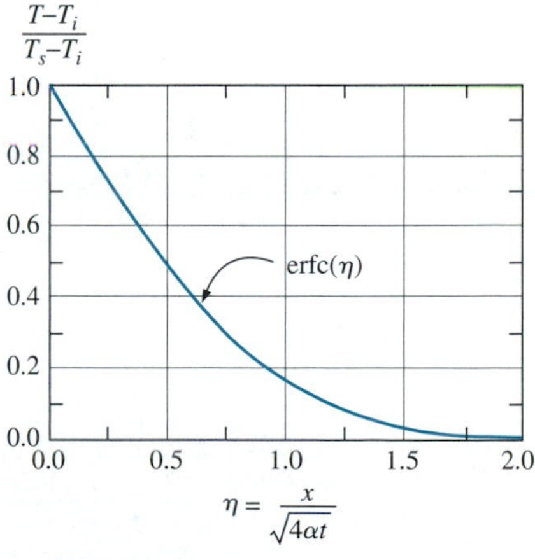

FIGURE 18–27
Dimensionless temperature distribution for transient conduction in a semi-infinite solid whose surface is maintained at a constant temperature T_s.

Case 3: Convection on the Surface, $\dot{q}_s(t) = h[T_\infty - T(0, t)]$.

$$\frac{T(x,t) - T_i}{T_\infty - T_i} = \text{erfc}\left(\frac{x}{2\sqrt{\alpha t}}\right) - \exp\left(\frac{hx}{k} + \frac{h^2 \alpha t}{k^2}\right)\text{erfc}\left(\frac{x}{2\sqrt{\alpha t}} + \frac{h\sqrt{\alpha t}}{k}\right) \tag{18–47}$$

Case 4: Energy Pulse at Surface, e_s = constant.
Energy in the amount of e_s per unit surface area (in J/m^2) is supplied to the semi-infinite body instantaneously at time $t = 0$ (by a laser pulse, for example), and the entire energy is assumed to enter the body, with no heat loss from the surface.

$$T(x, t) - T_i = \frac{e_s}{k\sqrt{\pi t/\alpha}}\exp\left(-\frac{x^2}{4\alpha t}\right) \tag{18–48}$$

Note that Cases 1 and 3 are closely related. In Case 1, the surface $x = 0$ is brought to a temperature T_s at time $t = 0$, and kept at that value at all times. In Case 3, the surface is exposed to convection by a fluid at a constant temperature T_∞ with a heat transfer coefficient h.

The solutions for all four cases are plotted in Fig. 18–28 for a representative case using a large cast iron block initially at 0°C throughout. In Case 1, the surface temperature remains constant at the specified value of T_s, and temperature increases gradually within the medium as heat penetrates deeper into the solid. Note that during initial periods only a thin slice near the surface is affected by heat transfer. Also, the temperature gradient at the surface and thus the rate of heat transfer into the solid decreases with time. In Case 2, heat is continually supplied to the solid, and thus the temperature within the solid, including the surface, increases with time. This is also the case with convection (Case 3), except that the surrounding fluid temperature T_∞ is the highest temperature that the solid body can rise to. In Case 4, the surface is subjected to an instant burst of heat supply at time $t = 0$, such as heating by a laser pulse, and then the surface is covered with insulation. The result is an instant rise in surface temperature, followed by a temperature drop as heat is conducted deeper into the solid. Note that the temperature profile is always normal to the surface at all times. (Why?)

The variation of temperature with position and time in a semi-infinite solid subjected to convection heat transfer is plotted in Fig. 18–29 for the nondimensionalized temperature against the dimensionless similarity variable $\eta = x/\sqrt{4\alpha t}$ for various values of the parameter $h\sqrt{\alpha t}/k$. Although the graphical solution given in Fig. 18–29 is simply a plot of the exact analytical solution, it is subject to reading errors, and thus is of limited accuracy compared to the analytical solution. Also, the values on the vertical axis of Fig. 18–29 correspond to $x = 0$, and thus represent the surface temperature. The curve $h\sqrt{\alpha t}/k = \infty$ corresponds to $h \to \infty$, which corresponds to the case of *specified temperature* T_∞ at the surface at $x = 0$. That is, the case in which the surface of the semi-infinite body is suddenly brought to temperature T_∞ at $t = 0$ and kept at T_∞ at all times can be handled by setting h to infinity. For a *finite* heat transfer coefficient h, the surface temperature approaches the fluid temperature T_∞ as the time t approaches infinity.

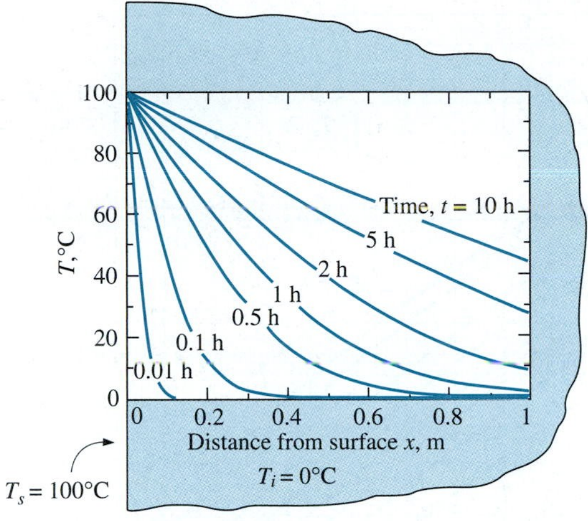

(*a*) Specified surface temperature, T_s = constant.

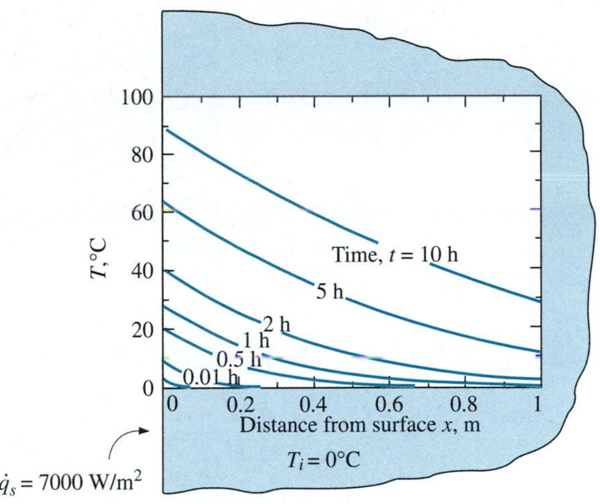

(*b*) Specified surface heat flux, $\dot{q}_s$ = constant.

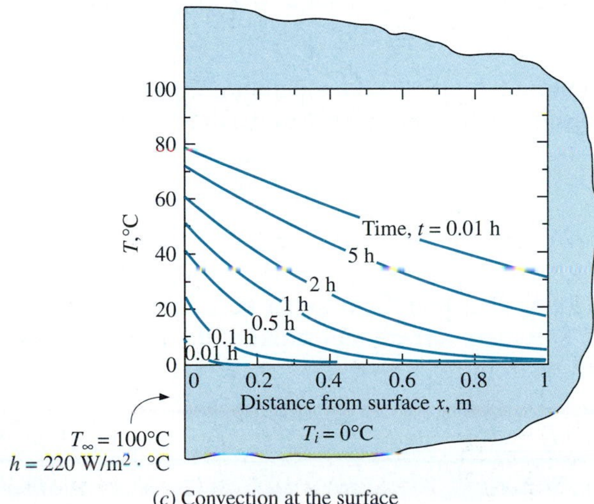

(*c*) Convection at the surface

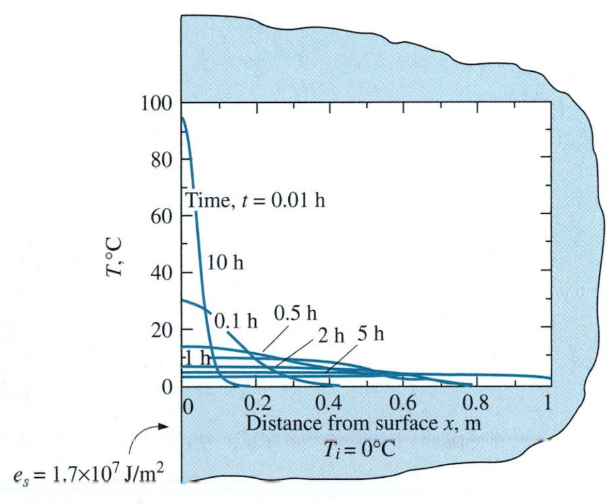

(*d*) Energy pulse at the surface, e_s = constant

FIGURE 18–28

Variations of temperature with position and time in a large cast iron block ($\alpha = 2.31 \times 10^{-5}$ m²/s, $k = 80.2$ W/m · °C) initially at 0 °C under different thermal conditions on the surface.

Contact of Two Semi-Infinite Solids

When two large bodies A and B, initially at uniform temperatures $T_{A,i}$ and $T_{B,i}$ are brought into contact, they instantly achieve temperature equality at the contact surface (temperature equality is achieved over the entire surface if the contact resistance is negligible). If the two bodies are of the same material with constant properties, thermal symmetry requires the contact surface temperature to be the arithmetic average, $T_s = (T_{A,i} + T_{B,i})/2$ and to remain constant at that value at all times.

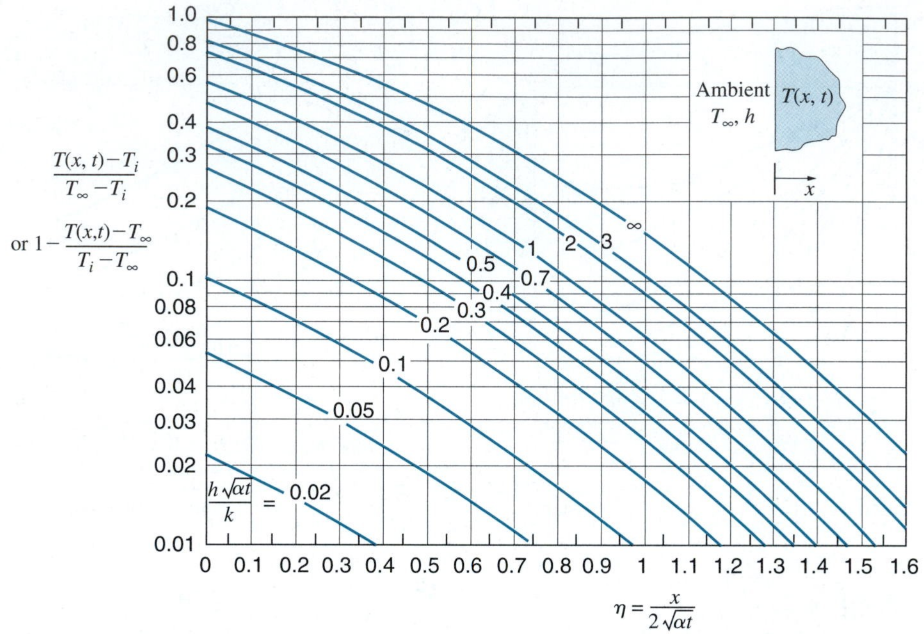

FIGURE 18–29

Variation of temperature with position and time in a semi-infinite solid initially at temperature T_i subjected to convection to an environment at T_∞ with a convection heat transfer coefficient of h (plotted using EES).

If the bodies are of different materials, they still achieve a temperature equality, but the surface temperature T_s in this case will be different than the arithmetic average. Noting that both bodies can be treated as semi-infinite solids with the same specified surface temperature, the energy balance on the contact surface gives, from Eq. 18–45,

$$\dot{q}_{s,A} = \dot{q}_{s,B} \rightarrow -\frac{k_A(T_s - T_{A,i})}{\sqrt{\pi \alpha_A t}} = \frac{k_B(T_s - T_{B,i})}{\sqrt{\pi \alpha_B t}} \rightarrow \frac{T_{A,i} - T_s}{T_s - T_{B,i}} = \sqrt{\frac{(k\rho c_p)_B}{(k\rho c_p)_A}}$$

Then T_s is determined to be (Fig. 18–30)

$$T_s = \frac{\sqrt{(k\rho c_p)_A}T_{A,i} + \sqrt{(k\rho c_p)_B}T_{B,i}}{\sqrt{(k\rho c_p)_A} + \sqrt{(k\rho c_p)_B}} \tag{18–49}$$

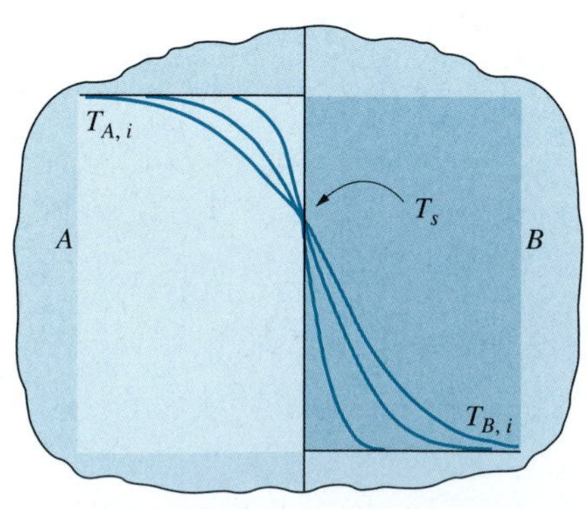

FIGURE 18–30

Contact of two semi-infinite solids of different initial temperatures.

Therefore, the interface temperature of two bodies brought into contact is dominated by the body with the larger $k\rho c_p$. This also explains why a metal at room temperature feels colder than wood at the same temperature. At room temperature, the $\sqrt{k\rho c_p}$ value is 24 kJ/m$^2\cdot$°C for aluminum, 0.38 kJ/m$^2\cdot$°C for wood, and 1.1 kJ/m$^2\cdot$°C for the human flesh. Using Eq. 18–49, if can be shown that when a person with a skin temperature of 35°C touches an aluminum block and then a wood block both at 15°C, the contact surface temperature will be 15.9°C in the case of aluminum and 30°C in the case of wood.

EXAMPLE 18–6 Minimum Burial Depth of Water Pipes to Avoid Freezing

In areas where the air temperature remains below 0°C for prolonged periods of time, the freezing of water in underground pipes is a major concern. Fortunately, the soil remains relatively warm during those periods, and it takes weeks for the subfreezing temperatures to reach the water mains in the ground. Thus, the soil effectively serves as an insulation to protect the water from subfreezing temperatures in winter.

The ground at a particular location is covered with snow pack at −10°C for a continuous period of three months, and the average soil properties at that location are $k = 0.4$ W/m · °C and $\alpha = 0.15 \times 10^{-6}$ m²/s (Fig. 18–31). Assuming an initial uniform temperature of 15°C for the ground, determine the minimum burial depth to prevent the water pipes from freezing.

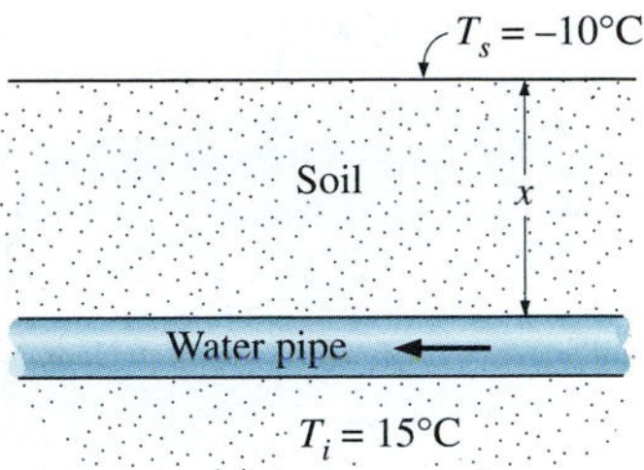

FIGURE 18–31
Schematic for Example 18–6.

Solution The water pipes are buried in the ground to prevent freezing. The minimum burial depth at a particular location is to be determined.

Assumptions **1** The temperature in the soil is affected by the thermal conditions at one surface only, and thus the soil can be considered to be a semi-infinite medium. **2** The thermal properties of the soil are constant.

Properties The properties of the soil are as given in the problem statement.

Analysis The temperature of the soil surrounding the pipes will be 0°C after three months in the case of minimum burial depth. Therefore, from Fig. 18–29, we have

$$\left.\begin{aligned} &\frac{h\sqrt{\alpha t}}{k} = \infty \qquad (\text{since } h \to \infty) \\ &\frac{T(x, t) - T_i}{T_\infty - T_i} = \frac{0 - 15}{-10 - 15} = 0.6 \end{aligned}\right\} \eta = \frac{x}{2\sqrt{\alpha t}} = 0.36$$

We note that

$$t = (90 \text{ days})(24 \text{ h/day})(3600 \text{ s/h}) = 7.78 \times 10^6 \text{ s}$$

and thus

$$x = 2\eta\sqrt{\alpha t} = 2 \times 0.36\sqrt{(0.15 \times 10^{-6} \text{ m}^2/\text{s})(7.78 \times 10^6 \text{ s})} = \mathbf{0.78 \text{ m}}$$

Therefore, the water pipes must be buried to a depth of at least 78 cm to avoid freezing under the specified harsh winter conditions.

ALTERNATIVE SOLUTION The solution of this problem could also be determined from Eq. 18–45:

$$\frac{T(x, t) - T_i}{T_s - T_i} = \text{erfc}\left(\frac{x}{2\sqrt{\alpha t}}\right) \longrightarrow \frac{0 - 15}{-10 - 15} = \text{erfc}\left(\frac{x}{2\sqrt{\alpha t}}\right) = 0.60$$

The argument that corresponds to this value of the complementary error function is determined from Table 18–4 to be $\eta = 0.37$. Therefore,

$$x = 2\eta\sqrt{\alpha t} = 2 \times 0.37\sqrt{(0.15 \times 10^{-6} \text{ m}^2/\text{s})(7.78 \times 10^6 \text{ s})} = \mathbf{0.80 \text{ m}}$$

Again, the slight difference is due to the reading error of the chart.

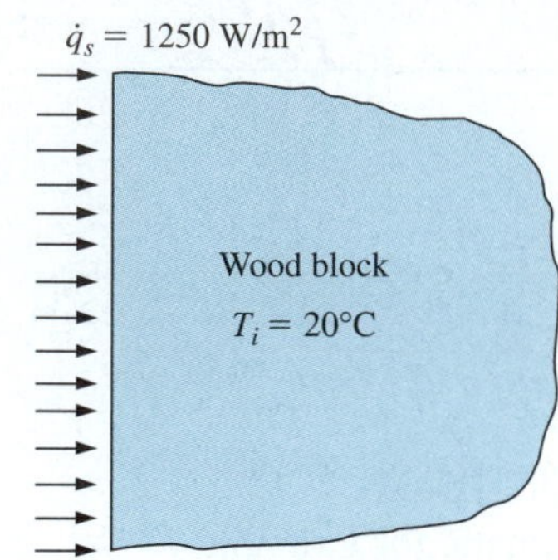

FIGURE 18–32
Schematic for Example 18–7.

EXAMPLE 18–7 Surface Temperature Rise of Heated Blocks

A thick black-painted wood block at 20°C is subjected to constant solar heat flux of 1250 W/m^2 (Fig. 18–32). Determine the exposed surface temperature of the block after 20 minutes. What would your answer be if the block were made of aluminum?

Solution A wood block is subjected to solar heat flux. The surface temperature of the block is to be determined, and to be compared to the value for an aluminum block.

Assumptions **1** All incident solar radiation is absorbed by the block. **2** Heat loss from the block is disregarded (and thus the result obtained is the maximum temperature). **3** The block is sufficiently thick to be treated as a semi-infinite solid, and the properties of the block are constant.

Properties Thermal conductivity and diffusivity values at room temperature are $k = 1.26$ W/m · K and $\alpha = 1.1 \times 10^{-5}$ m^2/s for wood, and $k = 237$ W/m · K and $\alpha = 9.71 \times 10^{-5}$ m^2/s for aluminum.

Analysis This is a transient conduction problem in a semi-infinite medium subjected to constant surface heat flux, and the surface temperature can be expressed from Eq. 18–46 as

$$T_s = T(0, t) = T_i + \frac{\dot{q}_s}{k}\sqrt{\frac{4\alpha t}{\pi}}$$

Substituting the given values, the surface temperatures for both the wood and aluminum blocks are determined to be

$$T_{s,\text{ wood}} = 20°\text{C} + \frac{1250 \text{ W/m}^2}{1.26 \text{ W/m}\cdot°\text{C}}\sqrt{\frac{4(1.1 \times 10^{-5} \text{ m}^2\text{/s})(20 \times 60 \text{ s})}{\pi}} = \mathbf{149°C}$$

$$T_{s,\text{ Al}} = 20°\text{C} + \frac{1250 \text{ W/m}^2}{237 \text{ W/m}\cdot°\text{C}}\sqrt{\frac{4(9.71 \times 10^{-5} \text{ m}^2\text{/s})(20 \times 60 \text{ s})}{\pi}} = \mathbf{22.0°C}$$

Note that thermal energy supplied to the wood accumulates near the surface because of the low conductivity and diffusivity of wood, causing the surface temperature to rise to high values. Metals, on the other hand, conduct the heat they receive to inner parts of the block because of their high conductivity and diffusivity, resulting in minimal temperature rise at the surface. In reality, both temperatures will be lower because of heat losses.

Discussion The temperature profiles for both the wood and aluminum blocks at $t = 20$ min are evaluated and plotted in Fig. 18–33 using EES. At a depth of $x = 0.41$ m, the temperature in both blocks is 20.6°C. At a depth of 0.5 m, the temperatures become 20.1°C for wood and 20.4°C for aluminum block, which confirms that heat penetrates faster and further in metals compared to nonmetals.

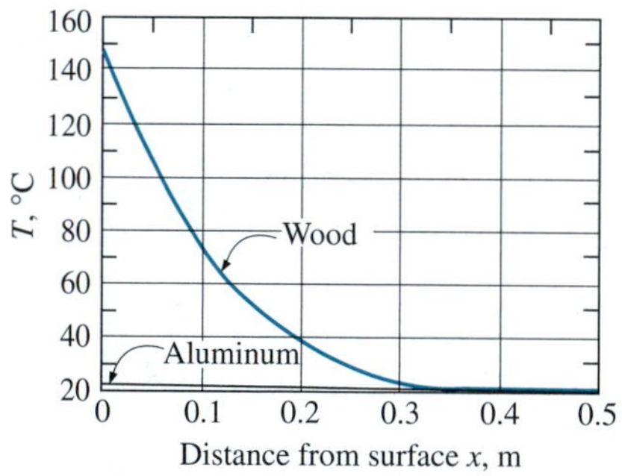

FIGURE 18–33
Variation of temperature within the wood and aluminum blocks at $t = 20$ min.

18–4 ▪ TRANSIENT HEAT CONDUCTION IN MULTIDIMENSIONAL SYSTEMS

The transient temperature charts and analytical solutions presented earlier can be used to determine the temperature distribution and heat transfer in *one-dimensional* heat conduction problems associated with a large plane wall, a long cylinder, a sphere, and a semi-infinite medium. Using a superposition approach called the **product solution**, these charts and solutions can also be used to construct solutions for the *two-dimensional* transient heat conduction problems encountered in geometries such as a short cylinder, a long rectangular bar, or a semi-infinite cylinder or plate, and even *three-dimensional* problems associated with geometries such as a rectangular prism or a semi-infinite rectangular bar, provided that *all* surfaces of the solid are subjected to convection to the *same* fluid at temperature T_∞, with the *same* heat transfer coefficient h, and the body involves no heat generation (Fig. 18–34). The solution in such multidimensional geometries can be expressed as the *product* of the solutions for the one-dimensional geometries whose intersection is the multidimensional geometry.

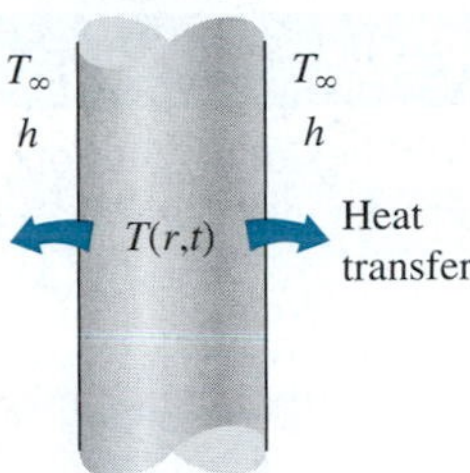

(*a*) Long cylinder

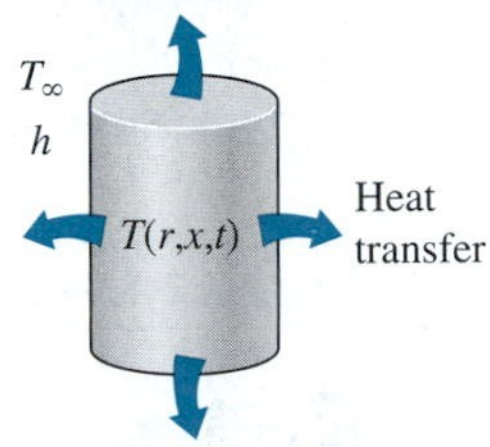

(*b*) Short cylinder (two-dimensional)

FIGURE 18–34
The temperature in a short cylinder exposed to convection from all surfaces varies in both the radial and axial directions, and thus heat is transferred in both directions.

Consider a *short cylinder* of height a and radius r_o initially at a uniform temperature T_i. There is no heat generation in the cylinder. At time $t = 0$, the cylinder is subjected to convection from all surfaces to a medium at temperature T_∞ with a heat transfer coefficient h. The temperature within the cylinder will change with x as well as r and time t since heat transfer occurs from the top and bottom of the cylinder as well as its side surfaces. That is, $T = T(r, x, t)$ and thus this is a two-dimensional transient heat conduction problem. When the properties are assumed to be constant, it can be shown that the solution of this two-dimensional problem can be expressed as

$$\left(\frac{T(r, x, t) - T_\infty}{T_i - T_\infty}\right)_{\substack{\text{short}\\ \text{cylinder}}} = \left(\frac{T(x, t) - T_\infty}{T_i - T_\infty}\right)_{\substack{\text{plane}\\ \text{wall}}} \left(\frac{T(r, t) - T_\infty}{T_i - T_\infty}\right)_{\substack{\text{infinite}\\ \text{cylinder}}} \tag{18–50}$$

That is, the solution for the two-dimensional short cylinder of height a and radius r_o is equal to the *product* of the nondimensionalized solutions for the one-dimensional plane wall of thickness a and the long cylinder of radius r_o, which are the two geometries whose intersection is the short cylinder, as shown in Fig. 18–35. We generalize this as follows: *the solution for a multidimensional geometry is the product of the solutions of the one-dimensional geometries whose intersection is the multidimensional body.*

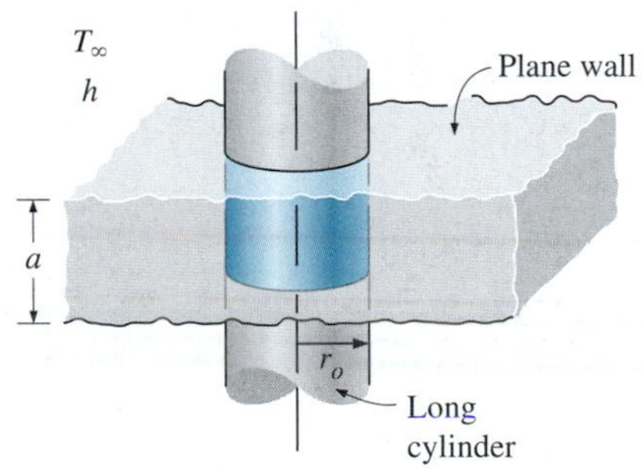

FIGURE 18–35
A short cylinder of radius r_o and height a is the *intersection* of a long cylinder of radius r_o and a plane wall of thickness a.

For convenience, the one-dimensional solutions are denoted by

$$\theta_{\text{wall}}(x, t) = \left(\frac{T(x, t) - T_\infty}{T_i - T_\infty}\right)_{\substack{\text{plane}\\ \text{wall}}}$$

$$\theta_{\text{cyl}}(r, t) = \left(\frac{T(r, t) - T_\infty}{T_i - T_\infty}\right)_{\substack{\text{infinite}\\ \text{cylinder}}}$$

$$\theta_{\text{semi-inf}}(x, t) = \left(\frac{T(x, t) - T_\infty}{T_i - T_\infty}\right)_{\substack{\text{semi-infinite}\\ \text{solid}}} \tag{18–51}$$

For example, the solution for a long solid bar whose cross section is an $a \times b$ rectangle is the intersection of the two infinite plane walls of thicknesses a and b, as shown in Fig. 18–36, and thus the transient temperature distribution for this rectangular bar can be expressed as

$$\left(\frac{T(x, y, t) - T_\infty}{T_i - T_\infty}\right)_{\substack{\text{rectangular}\\ \text{bar}}} = \theta_{\text{wall}}(x, t)\theta_{\text{wall}}(y, t) \tag{18–52}$$

The proper forms of the product solutions for some other geometries are given in Table 18–5. It is important to note that the x-coordinate is measured

TABLE 18–5

Multidimensional solutions expressed as products of one-dimensional solutions for bodies that are initially at a uniform temperature T_i and exposed to convection from all surfaces to a medium at T_∞

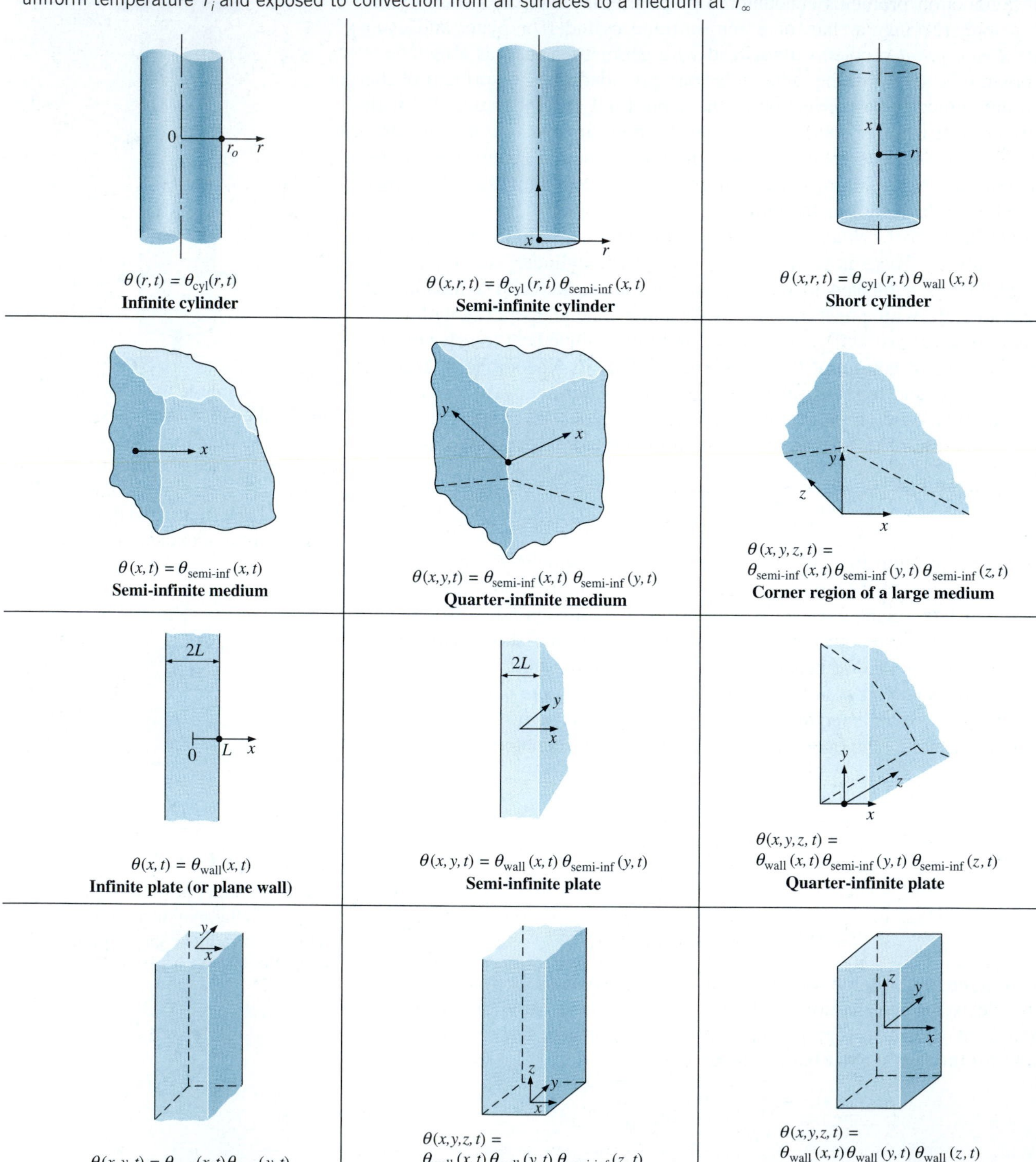

from the *surface* in a semi-infinite solid, and from the *midplane* in a plane wall. The radial distance r is always measured from the centerline.

Note that the solution of a *two-dimensional* problem involves the product of *two* one-dimensional solutions, whereas the solution of a *three-dimensional* problem involves the product of *three* one-dimensional solutions.

A modified form of the product solution can also be used to determine the total transient heat transfer to or from a multidimensional geometry by using the one-dimensional values, as shown by L. S. Langston in 1982. The transient heat transfer for a two-dimensional geometry formed by the intersection of two one-dimensional geometries 1 and 2 is

$$\left(\frac{Q}{Q_{max}}\right)_{total,\ 2D} = \left(\frac{Q}{Q_{max}}\right)_1 + \left(\frac{Q}{Q_{max}}\right)_2\left[1 - \left(\frac{Q}{Q_{max}}\right)_1\right] \quad (18\text{–}53)$$

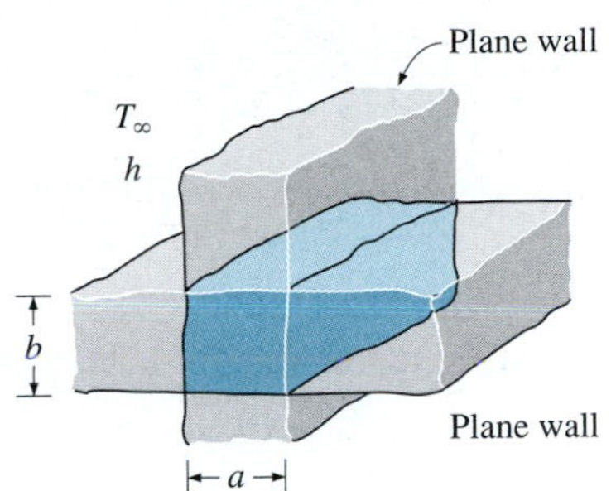

FIGURE 18–36
A long solid bar of rectangular profile $a \times b$ is the *intersection* of two plane walls of thicknesses a and b.

Transient heat transfer for a three-dimensional body formed by the intersection of three one-dimensional bodies 1, 2, and 3 is given by

$$\left(\frac{Q}{Q_{max}}\right)_{total,\ 3D} = \left(\frac{Q}{Q_{max}}\right)_1 + \left(\frac{Q}{Q_{max}}\right)_2\left[1 - \left(\frac{Q}{Q_{max}}\right)_1\right] + \left(\frac{Q}{Q_{max}}\right)_3\left[1 - \left(\frac{Q}{Q_{max}}\right)_1\right]\left[1 - \left(\frac{Q}{Q_{max}}\right)_2\right] \quad (18\text{–}54)$$

The use of the product solution in transient two- and three-dimensional heat conduction problems is illustrated in the following examples.

EXAMPLE 18–8 Cooling of a Short Brass Cylinder

A short brass cylinder of diameter $D = 10$ cm and height $H = 12$ cm is initially at a uniform temperature $T_i = 120°C$. The cylinder is now placed in atmospheric air at 25°C, where heat transfer takes place by convection, with a heat transfer coefficient of $h = 60\ W/m^2 \cdot °C$. Calculate the temperature at (*a*) the center of the cylinder and (*b*) the center of the top surface of the cylinder 15 min after the start of the cooling.

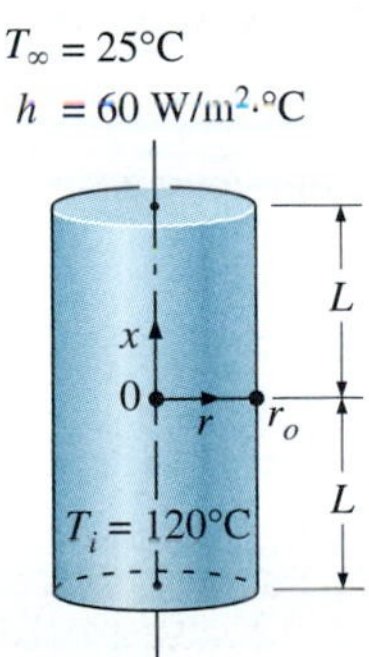

FIGURE 18–37
Schematic for Example 18–8.

Solution A short cylinder is allowed to cool in atmospheric air. The temperatures at the centers of the cylinder and the top surface are to be determined.

Assumptions **1** Heat conduction in the short cylinder is two-dimensional, and thus the temperature varies in both the axial x- and the radial r-directions. **2** The thermal properties of the cylinder and the heat transfer coefficient are constant. **3** The Fourier number is $\tau > 0.2$ so that the one-term approximate solutions are applicable.

Properties The properties of brass at room temperature are $k = 110\ W/m \cdot °C$ and $\alpha = 33.9 \times 10^{-6}\ m^2/s$ (Table A–24). More accurate results can be obtained by using properties at average temperature.

Analysis (*a*) This short cylinder can physically be formed by the intersection of a long cylinder of radius $r_o = 5$ cm and a plane wall of thickness $2L =$

12 cm, as shown in Fig. 18–37. The dimensionless temperature at the center of the plane wall is determined from Fig. 18–15*a* to be

$$\left.\begin{aligned} \tau &= \frac{\alpha t}{L^2} = \frac{(3.39 \times 10^{-5}\ \text{m}^2/\text{s})(900\ \text{s})}{(0.06\ \text{m})^2} = 8.48 \\ \frac{1}{\text{Bi}} &= \frac{k}{hL} = \frac{110\ \text{W/m}\cdot{}^\circ\text{C}}{(60\ \text{W/m}^2\cdot{}^\circ\text{C})(0.06\ \text{m})} = 30.6 \end{aligned}\right\}\ \theta_{\text{wall}}(0, t) = \frac{T(0, t) - T_\infty}{T_i - T_\infty} = 0.8$$

Similarly, at the center of the cylinder, we have

$$\left.\begin{aligned} \tau &= \frac{\alpha t}{r_o^2} = \frac{(3.39 \times 10^{-5}\ \text{m}^2/\text{s})(900\ \text{s})}{(0.05\ \text{m})^2} = 12.2 \\ \frac{1}{\text{Bi}} &= \frac{k}{hr_o} = \frac{110\ \text{W/m}\cdot{}^\circ\text{C}}{(60\ \text{W/m}^2\cdot{}^\circ\text{C})(0.05\ \text{m})} = 36.7 \end{aligned}\right\}\ \theta_{\text{cyl}}(0, t) = \frac{T(0, t) - T_\infty}{T_i - T_\infty} = 0.5$$

Therefore,

$$\left(\frac{T(0, 0, t) - T_\infty}{T_i - T_\infty}\right)_{\substack{\text{short}\\ \text{cylinder}}} = \theta_{\text{wall}}(0, t) \times \theta_{\text{cyl}}(0, t) = 0.8 \times 0.5 = 0.4$$

and

$$T(0, 0, t) = T_\infty + 0.4(T_i - T_\infty) = 25 + 0.4(120 - 25) = \mathbf{63^\circ C}$$

This is the temperature at the center of the short cylinder, which is also the center of both the long cylinder and the plate.

(*b*) The center of the top surface of the cylinder is still at the center of the long cylinder ($r = 0$), but at the outer surface of the plane wall ($x = L$). Therefore, we first need to find the surface temperature of the wall. Noting that $x = L = 0.06$ m,

$$\left.\begin{aligned} \frac{x}{L} &= \frac{0.06\ \text{m}}{0.06\ \text{m}} = 1 \\ \frac{1}{\text{Bi}} &= \frac{k}{hL} = \frac{110\ \text{W/m}\cdot{}^\circ\text{C}}{(60\ \text{W/m}^2\cdot{}^\circ\text{C})(0.06\ \text{m})} = 30.6 \end{aligned}\right\}\ \frac{T(L, t) - T_\infty}{T_0 - T_\infty} = 0.98$$

Then

$$\theta_{\text{wall}}(L, t) = \frac{T(L, t) - T_\infty}{T_i - T_\infty} = \left(\frac{T(L, t) - T_\infty}{T_0 - T_\infty}\right)\left(\frac{T_0 - T_\infty}{T_i - T_\infty}\right) = 0.98 \times 0.8 = 0.784$$

Therefore,

$$\left(\frac{T(L, 0, t) - T_\infty}{T_i - T_\infty}\right)_{\substack{\text{short}\\ \text{cylinder}}} = \theta_{\text{wall}}(L, t)\theta_{\text{cyl}}(0, t) = 0.784 \times 0.5 = 0.392$$

and

$$T(L, 0, t) = T_\infty + 0.392(T_i - T_\infty) = 25 + 0.392(120 - 25) = \mathbf{62.2^\circ C}$$

which is the temperature at the center of the top surface of the cylinder.

EXAMPLE 18–9 Heat Transfer from a Short Cylinder

Determine the total heat transfer from the short brass cylinder (ρ = 8530 kg/m^3, c_p = 0.380 kJ/kg · °C) discussed in Example 18–8.

Solution We first determine the maximum heat that can be transferred from the cylinder, which is the sensible energy content of the cylinder relative to its environment:

$$m = \rho V = \rho\pi r_o^2 H = (8530 \text{ kg/m}^3)\pi(0.05 \text{ m})^2(0.12 \text{ m}) = 8.04 \text{ kg}$$
$$Q_{\max} = mc_p(T_i - T_\infty) = (8.04 \text{ kg})(0.380 \text{ kJ/kg} \cdot °\text{C})(120 - 25)°\text{C} = 290.2 \text{ kJ}$$

Then we determine the dimensionless heat transfer ratios for both geometries. For the plane wall, it is determined from Fig. 18–15*c* to be

$$\left.\begin{aligned} \text{Bi} &= \frac{1}{1/\text{Bi}} = \frac{1}{30.6} = 0.0327 \\ \frac{h^2\alpha t}{k^2} &= \text{Bi}^2\tau = (0.0327)^2(8.48) = 0.0091 \end{aligned}\right\} \left(\frac{Q}{Q_{\max}}\right)_{\substack{\text{plane}\\ \text{wall}}} = 0.23$$

Similarly, for the cylinder, we have

$$\left.\begin{aligned} \text{Bi} &= \frac{1}{1/\text{Bi}} = \frac{1}{36.7} = 0.0272 \\ \frac{h^2\alpha t}{k^2} &= \text{Bi}^2\tau = (0.0272)^2(12.2) = 0.0090 \end{aligned}\right\} \left(\frac{Q}{Q_{\max}}\right)_{\substack{\text{infinite}\\ \text{cylinder}}} = 0.47$$

Then the heat transfer ratio for the short cylinder is, from Eq. 18–53,

$$\begin{aligned} \left(\frac{Q}{Q_{\max}}\right)_{\text{short cyl}} &= \left(\frac{Q}{Q_{\max}}\right)_1 + \left(\frac{Q}{Q_{\max}}\right)_2\left[1 - \left(\frac{Q}{Q_{\max}}\right)_1\right] \\ &= 0.23 + 0.47(1 - 0.23) = 0.592 \end{aligned}$$

Therefore, the total heat transfer from the cylinder during the first 15 min of cooling is

$$Q = 0.592Q_{\max} = 0.592 \times (290.2 \text{ kJ}) = \mathbf{172 \text{ kJ}}$$

EXAMPLE 18–10 Cooling of a Long Cylinder by Water

A semi-infinite aluminum cylinder of diameter D = 20 cm is initially at a uniform temperature T_i = 200°C. The cylinder is now placed in water at 15°C where heat transfer takes place by convection, with a heat transfer coefficient of h = 120 W/m^2 · °C. Determine the temperature at the center of the cylinder 15 cm from the end surface 5 min after the start of the cooling.

Solution A semi-infinite aluminum cylinder is cooled by water. The temperature at the center of the cylinder 15 cm from the end surface is to be determined.

Assumptions **1** Heat conduction in the semi-infinite cylinder is two-dimensional, and thus the temperature varies in both the axial x- and the

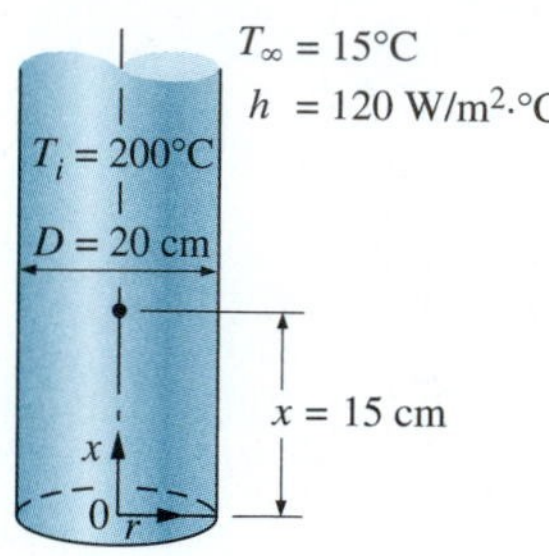

FIGURE 18–38
Schematic for Example 18–10.

radial r-directions. **2** The thermal properties of the cylinder and the heat transfer coefficient are constant. **3** The Fourier number is $\tau > 0.2$ so that the one-term approximate solutions are applicable.

Properties The properties of aluminum at room temperature are k = 237 W/m · °C and α = 9.71 × 10^{-6} m²/s (Table A–24). More accurate results can be obtained by using properties at average temperature.

Analysis This semi-infinite cylinder can physically be formed by the intersection of an infinite cylinder of radius r_o = 10 cm and a semi-infinite medium, as shown in Fig. 18–38.

We solve this problem using the one-term solution relation for the cylinder and the analytic solution for the semi-infinite medium. First we consider the infinitely long cylinder and evaluate the Biot number:

$$\text{Bi} = \frac{hr_o}{k} = \frac{(120 \text{ W/m}^2 \cdot °\text{C})(0.1 \text{ m})}{237 \text{ W/m} \cdot °\text{C}} = 0.05$$

The coefficients λ_1 and A_1 for a cylinder corresponding to this Bi are determined from Table 18–2 to be λ_1 = 0.3126 and A_1 = 1.0124. The Fourier number in this case is

$$\tau = \frac{\alpha t}{r_o^2} = \frac{(9.71 \times 10^{-5} \text{ m}^2\text{/s})(5 \times 60 \text{ s})}{(0.1 \text{ m})^2} = 2.91 > 0.2$$

and thus the one-term approximation is applicable. Substituting these values into Eq. 18–27 gives

$$\theta_0 = \theta_{\text{cyl}}(0, t) = A_1 e^{-\lambda_1^2 \tau} = 1.0124 e^{-(0.3126)^2(2.91)} = 0.762$$

The solution for the semi-infinite solid can be determined from

$$1 - \theta_{\text{semi-inf}}(x, t) = \text{erfc}\left(\frac{x}{2\sqrt{\alpha t}}\right) - \exp\left(\frac{hx}{k} + \frac{h^2 \alpha t}{k^2}\right)\left[\text{erfc}\left(\frac{x}{2\sqrt{\alpha t}} + \frac{h\sqrt{\alpha t}}{k}\right)\right]$$

First we determine the various quantities in parentheses:

$$\eta = \frac{x}{2\sqrt{\alpha t}} = \frac{0.15 \text{ m}}{2\sqrt{(9.71 \times 10^{-5} \text{ m}^2\text{/s})(5 \times 60 \text{ s})}} = 0.44$$

$$\frac{h\sqrt{\alpha t}}{k} = \frac{(120 \text{ W/m}^2 \cdot °\text{C})\sqrt{(9.71 \times 10^{-5} \text{ m}^2\text{/s})(300 \text{ s})}}{237 \text{ W/m} \cdot °\text{C}} = 0.086$$

$$\frac{hx}{k} = \frac{(120 \text{ W/m}^2 \cdot °\text{C})(0.15 \text{ m})}{237 \text{ W/m} \cdot °\text{C}} = 0.0759$$

$$\frac{h^2 \alpha t}{k^2} = \left(\frac{h\sqrt{\alpha t}}{k}\right)^2 = (0.086)^2 = 0.0074$$

Substituting and evaluating the complementary error functions from Table 18–4,

$$\begin{aligned}\theta_{\text{semi-inf}}(x, t) &= 1 - \text{erfc}\,(0.44) + \exp\,(0.0759 + 0.0074)\,\text{erfc}\,(0.44 + 0.086) \\ &= 1 - 0.5338 + \exp\,(0.0833) \times 0.457 \\ &= 0.963\end{aligned}$$

Now we apply the product solution to get

$$\left(\frac{T(x, 0, t) - T_\infty}{T_i - T_\infty}\right)_{\substack{\text{semi-infinite}\\\text{cylinder}}} = \theta_{\text{semi-inf}}(x, t)\theta_{\text{cyl}}(0, t) = 0.963 \times 0.762 = 0.734$$

and

$$T(x, 0, t) = T_\infty + 0.734(T_i - T_\infty) = 15 + 0.734(200 - 15) = \mathbf{151°C}$$

which is the temperature at the center of the cylinder 15 cm from the exposed bottom surface.

EXAMPLE 18–11 Refrigerating Steaks while Avoiding Frostbite

In a meat processing plant, 1-in-thick steaks initially at 75°F are to be cooled in the racks of a large refrigerator that is maintained at 5°F (Fig. 18–39). The steaks are placed close to each other, so that heat transfer from the 1-in-thick edges is negligible. The entire steak is to be cooled below 45°F, but its temperature is not to drop below 35°F at any point during refrigeration to avoid "frostbite." The convection heat transfer coefficient and thus the rate of heat transfer from the steak can be controlled by varying the speed of a circulating fan inside. Determine the heat transfer coefficient *h* that will enable us to meet both temperature constraints while keeping the refrigeration time to a minimum. The steak can be treated as a homogeneous layer having the properties $\rho = 74.9$ lbm/ft^3, $c_p = 0.98$ Btu/lbm · °F, $k = 0.26$ Btu/h · ft · °F, and $\alpha = 0.0035$ ft^2/h.

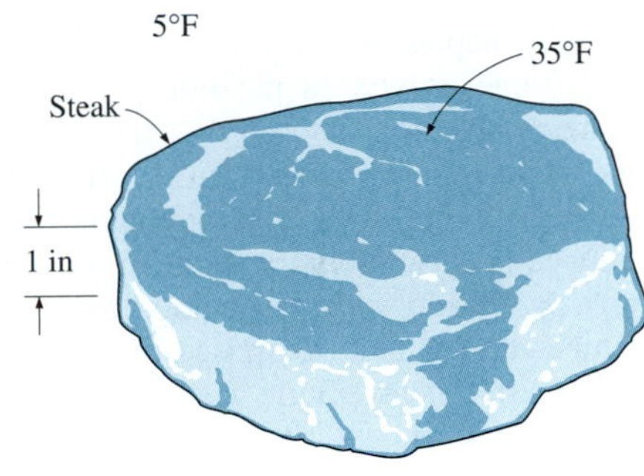

FIGURE 18–39
Schematic for Example 18–11.

Solution Steaks are to be cooled in a refrigerator maintained at 5°F. The heat transfer coefficient that allows cooling the steaks below 45°F while avoiding frostbite is to be determined.

Assumptions **1** Heat conduction through the steaks is one-dimensional since the steaks form a large layer relative to their thickness and there is thermal symmetry about the center plane. **2** The thermal properties of the steaks and the heat transfer coefficient are constant. **3** The Fourier number is $\tau > 0.2$ so that the one-term approximate solutions are applicable.

Properties The properties of the steaks are as given in the problem statement.

Analysis The lowest temperature in the steak occurs at the surfaces and the highest temperature at the center at a given time, since the inner part is the last place to be cooled. In the limiting case, the surface temperature at $x = L = 0.5$ in from the center will be 35°F, while the midplane temperature is 45°F in an environment at 5°F. Then, from Fig. 18–15*b*, we obtain

$$\left.\begin{aligned} \frac{x}{L} &= \frac{0.5\text{ in}}{0.5\text{ in}} = 1 \\ \frac{T(L, t) - T_\infty}{T_0 - T_\infty} &= \frac{35 - 5}{45 - 5} = 0.75 \end{aligned}\right\} \quad \frac{1}{\text{Bi}} = \frac{k}{hL} = 1.5$$

which gives

$$h = \frac{1}{1.5}\frac{k}{L} = \frac{0.26\text{ Btu/h} \cdot \text{ft} \cdot °\text{F}}{1.5(0.5/12\text{ ft})} = 4.16\text{ Btu/h} \cdot \text{ft}^2 \cdot °\text{F}$$

Discussion The convection heat transfer coefficient should be kept below this value to satisfy the constraints on the temperature of the steak during refrigeration. We can also meet the constraints by using a lower heat transfer coefficient, but doing so would extend the refrigeration time unnecessarily.

The restrictions that are inherent in the use of Heisler charts and the one-term solutions (or any other analytical solutions) can be lifted by using the numerical methods.

SUMMARY

In this chapter, we considered the variation of temperature with time as well as position in one- or multidimensional systems. We first considered the *lumped systems* in which the temperature varies with time but remains uniform throughout the system at any time. The temperature of a lumped body of arbitrary shape of mass m, volume V, surface area A_s, density ρ, and specific heat c_p initially at a uniform temperature T_i that is exposed to convection at time $t = 0$ in a medium at temperature T_∞ with a heat transfer coefficient h is expressed as

$$\frac{T(t) - T_\infty}{T_i - T_\infty} = e^{-bt}$$

where

$$b = \frac{hA_s}{\rho c_p V} = \frac{h}{\rho c_p L_c}$$

is a positive quantity whose dimension is $(\text{time})^{-1}$. This relation can be used to determine the temperature $T(t)$ of a body at time t or, alternatively, the time t required for the temperature to reach a specified value $T(t)$. Once the temperature $T(t)$ at time t is available, the *rate* of convection heat transfer between the body and its environment at that time can be determined from Newton's law of cooling as

$$\dot{Q}(t) = hA_s[T(t) - T_\infty]$$

The *total amount* of heat transfer between the body and the surrounding medium over the time interval $t = 0$ to t is simply the change in the energy content of the body,

$$Q = mc_p[T(t) - T_i]$$

The *maximum* heat transfer between the body and its surroundings is

$$Q_{\max} = mc_p(T_\infty - T_i)$$

The error involved in lumped system analysis is negligible when

$$\text{Bi} = \frac{hL_c}{k} < 0.1$$

where Bi is the *Biot number* and $L_c = V/A_s$ is the *characteristic length.*

When the lumped system analysis is not applicable, the variation of temperature with position as well as time can be determined using the *transient temperature charts* given in Figs. 18–15, 18–16, 18–17, and 18–29 for a large plane wall, a long cylinder, a sphere, and a semi-infinite medium, respectively. These charts are applicable for one-dimensional heat transfer in those geometries. Therefore, their use is limited to situations in which the body is initially at a uniform temperature, all surfaces are subjected to the same thermal conditions, and the body does not involve any heat generation. These charts can also be used to determine the total heat transfer from the body up to a specified time t.

Using the *one-term approximation,* the solutions of one-dimensional transient heat conduction problems are expressed analytically as

Plane wall: $$\theta_{\text{wall}} = \frac{T(x, t) - T_\infty}{T_i - T_\infty} = A_1 e^{-\lambda_1^2 \tau} \cos(\lambda_1 x/L)$$

Cylinder: $$\theta_{\text{cyl}} = \frac{T(r, t) - T_\infty}{T_i - T_\infty} = A_1 e^{-\lambda_1^2 \tau} J_0(\lambda_1 r/r_o)$$

Sphere: $$\theta_{\text{sph}} = \frac{T(r, t) - T_\infty}{T_i - T_\infty} = A_1 e^{-\lambda_1^2 \tau} \frac{\sin(\lambda_1 r/r_o)}{\lambda_1 r/r_o}$$

where the constants A_1 and λ_1 are functions of the Bi number only, and their values are listed in Table 18–2 against the Bi number for all three geometries. The error involved in one-term solutions is less than 2 percent when $\tau > 0.2$.

Using the one-term solutions, the fractional heat transfers in different geometries are expressed as

Plane wall: $$\left(\frac{Q}{Q_{\max}}\right)_{\text{wall}} = 1 - \theta_{0,\,\text{wall}} \frac{\sin \lambda_1}{\lambda_1}$$

Cylinder: $$\left(\frac{Q}{Q_{\max}}\right)_{\text{cyl}} = 1 - 2\theta_{0,\,\text{cyl}} \frac{J_1(\lambda_1)}{\lambda_1}$$

Sphere: $$\left(\frac{Q}{Q_{max}}\right)_{sph} = 1 - 3\theta_{0,\,sph}\frac{\sin\lambda_1 - \lambda_1\cos\lambda_1}{\lambda_1^3}$$

The solutions of transient heat conduction in a semi-infinite solid with constant properties under various boundary conditions at the surface are given as follows:

Specified Surface Temperature, T_s = constant:

$$\frac{T(x, t) - T_i}{T_s - T_i} = \text{erfc}\left(\frac{x}{2\sqrt{\alpha t}}\right) \quad \text{and} \quad \dot{q}_s(t) = \frac{k(T_s - T_i)}{\sqrt{\pi\alpha t}}$$

Specified Surface Heat Flux, $\dot{q}_s$ = constant:

$$T(x, t) - T_i = \frac{\dot{q}_s}{k}\left[\sqrt{\frac{4\alpha t}{\pi}}\exp\left(-\frac{x^2}{4\alpha t}\right) - x\,\text{erfc}\left(\frac{x}{2\sqrt{\alpha t}}\right)\right]$$

Convection on the Surface, $\dot{q}_s(t) = h[T_\infty - T(0, t)]$:

$$\frac{T(x, t) - T_i}{T_\infty - T_i} = \text{erfc}\left(\frac{x}{2\sqrt{\alpha t}}\right) - \exp\left(\frac{hx}{k} + \frac{h^2\alpha t}{k^2}\right) \times \text{erfc}\left(\frac{x}{2\sqrt{\alpha t}} + \frac{h\sqrt{\alpha t}}{k}\right)$$

Energy Pulse at Surface, e_s = constant:

$$T(x, t) - T_i = \frac{e_s}{k\sqrt{\pi t/\alpha}}\exp\left(-\frac{x^2}{4\alpha t}\right)$$

where erfc(η) is the *complementary error function* of argument η.

Using a superposition principle called the *product solution* these charts can also be used to construct solutions for the *two-dimensional* transient heat conduction problems encountered in geometries such as a short cylinder, a long rectangular bar, or a semi-infinite cylinder or plate, and even *three-dimensional* problems associated with geometries such as a rectangular prism or a semi-infinite rectangular bar, provided that all surfaces of the solid are subjected to convection to the same fluid at temperature T_∞, with the same convection heat transfer coefficient h, and the body involves no heat generation. The solution in such multidimensional geometries can be expressed as the product of the solutions for the one-dimensional geometries whose intersection is the multidimensional geometry.

The total heat transfer to or from a multidimensional geometry can also be determined by using the one-dimensional values. The transient heat transfer for a two-dimensional geometry formed by the intersection of two one-dimensional geometries 1 and 2 is

$$\left(\frac{Q}{Q_{max}}\right)_{total,\,2D} = \left(\frac{Q}{Q_{max}}\right)_1 + \left(\frac{Q}{Q_{max}}\right)_2\left[1 - \left(\frac{Q}{Q_{max}}\right)_1\right]$$

Transient heat transfer for a three-dimensional body formed by the intersection of three one-dimensional bodies 1, 2, and 3 is given by

$$\left(\frac{Q}{Q_{max}}\right)_{total,\,3D} = \left(\frac{Q}{Q_{max}}\right)_1 + \left(\frac{Q}{Q_{max}}\right)_2\left[1 - \left(\frac{Q}{Q_{max}}\right)_1\right] + \left(\frac{Q}{Q_{max}}\right)_3\left[1 - \left(\frac{Q}{Q_{max}}\right)_1\right]\left[1 - \left(\frac{Q}{Q_{max}}\right)_2\right]$$

REFERENCES AND SUGGESTED READINGS

1. ASHRAE. *Handbook of Fundamentals.* SI version. Atlanta, GA: American Society of Heating, Refrigerating, and Air-Conditioning Engineers, Inc., 1993.
2. ASHRAE. *Handbook of Fundamentals.* SI version. Atlanta, GA: American Society of Heating, Refrigerating, and Air-Conditioning Engineers, Inc., 1994.
3. H. S. Carslaw and J. C. Jaeger. *Conduction of Heat in Solids.* 2nd ed. London: Oxford University Press, 1959.
4. Y. A. Çengel. *Heat and Mass Transfer: A Practical Approach.* 3rd ed. New York: McGraw-Hill, 2007.
5. H. Gröber, S. Erk, and U. Grigull. *Fundamentals of Heat Transfer.* New York: McGraw-Hill, 1961.
6. M. P. Heisler. "Temperature Charts for Induction and Constant Temperature Heating." *ASME Transactions* 69 (1947), pp. 227–36.
7. H. Hillman. *Kitchen Science.* Mount Vernon, NY: Consumers Union, 1981.
8. S. Kakaç and Y. Yener, *Heat Conduction*, New York: Hemisphere Publishing Co., 1985.
9. L. S. Langston. "Heat Transfer from Multidimensional Objects Using One-Dimensional Solutions for Heat Loss." *International Journal of Heat and Mass Transfer* 25 (1982), pp. 149–50.
10. P. J. Schneider. *Conduction Heat Transfer.* Reading, MA: Addison-Wesley, 1955.
11. L. van der Berg and C. P. Lentz. "Factors Affecting Freezing Rate and Appearance of Eviscerated Poultry Frozen in Air." *Food Technology* 12 (1958).

PROBLEMS*

Lumped System Analysis

18–1C What is lumped system analysis? When is it applicable?

18–2C Consider heat transfer between two identical hot solid bodies and the air surrounding them. The first solid is being cooled by a fan while the second one is allowed to cool naturally. For which solid is the lumped system analysis more likely to be applicable? Why?

18–3C Consider heat transfer between two identical hot solid bodies and their environments. The first solid is dropped in a large container filled with water, while the second one is allowed to cool naturally in the air. For which solid is the lumped system analysis more likely to be applicable? Why?

18–4C Consider a hot baked potato on a plate. The temperature of the potato is observed to drop by 4°C during the first minute. Will the temperature drop during the second minute be less than, equal to, or more than 4°C? Why?

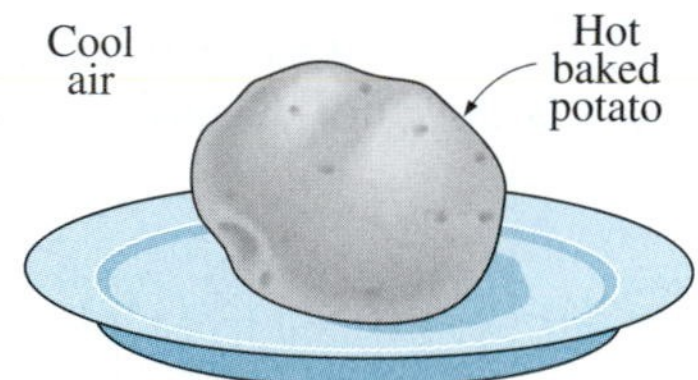

FIGURE P18–4C

18–5C Consider a potato being baked in an oven that is maintained at a constant temperature. The temperature of the potato is observed to rise by 5°C during the first minute. Will the temperature rise during the second minute be less than, equal to, or more than 5°C? Why?

18–6C What is the physical significance of the Biot number? Is the Biot number more likely to be larger for highly conducting solids or poorly conducting ones?

18–7C Consider two identical 4-kg pieces of roast beef. The first piece is baked as a whole, while the second is baked after being cut into two equal pieces in the same oven. Will there be any difference between the cooking times of the whole and cut roasts? Why?

18–8C Consider a sphere and a cylinder of equal volume made of copper. Both the sphere and the cylinder are initially at the same temperature and are exposed to convection in the same environment. Which do you think will cool faster, the cylinder or the sphere? Why?

18–9C In what medium is the lumped system analysis more likely to be applicable: in water or in air? Why?

18–10C For which solid is the lumped system analysis more likely to be applicable: an actual apple or a golden apple of the same size? Why?

18–11C For which kind of bodies made of the same material is the lumped system analysis more likely to be applicable: slender ones or well-rounded ones of the same volume? Why?

18–12 Obtain relations for the characteristic lengths of a large plane wall of thickness $2L$, a very long cylinder of radius r_o, and a sphere of radius r_o.

18–13 Obtain a relation for the time required for a lumped system to reach the average temperature $\frac{1}{2}(T_i + T_\infty)$, where T_i is the initial temperature and T_∞ is the temperature of the environment.

18–14 The temperature of a gas stream is to be measured by a thermocouple whose junction can be approximated as a 1.2-mm-diameter sphere. The properties of the junction are $k = 35$ W/m · °C, $\rho = 8500$ kg/m³, and $c_p = 320$ J/kg · °C, and the heat transfer coefficient between the junction and the gas is $h = 90$ W/m² · °C. Determine how long it will take for the thermocouple to read 99 percent of the initial temperature difference. *Answer: 27.8 s*

18–15E In a manufacturing facility, 2-in-diameter brass balls ($k = 64.1$ Btu/h · ft · °F, $\rho = 532$ lbm/ft³, and $c_p = 0.092$ Btu/lbm · °F) initially at 250°F are quenched in a water bath at 120°F for a period of 2 min at a rate of 120 balls per minute. If the convection heat transfer coefficient is 42 Btu/h · ft² · °F, determine (*a*) the temperature of the balls after quenching and (*b*) the rate at which heat needs to be removed from the water in order to keep its temperature constant at 120°F.

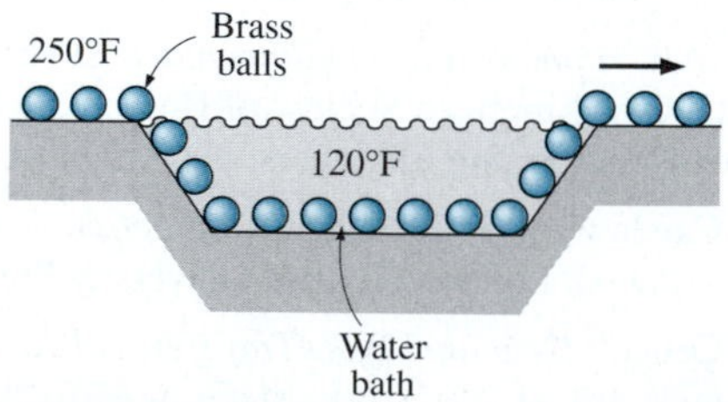

FIGURE P18–15E

18–16E Repeat Prob. 18–15E for aluminum balls.

18–17 To warm up some milk for a baby, a mother pours milk into a thin-walled glass whose diameter is 6 cm. The height of the milk in the glass is 7 cm. She then places the

*Problems designated by a "C" are concept questions, and students are encouraged to answer them all. Problems designated by an "E" are in English units, and the SI users can ignore them. Problems with the icon are solved using EES, and complete solutions together with parametric studies are included on the enclosed DVD. Problems with the icon are comprehensive in nature, and are intended to be solved with a computer, preferably using the EES software that accompanies this text.

glass into a large pan filled with hot water at 60°C. The milk is stirred constantly, so that its temperature is uniform at all times. If the heat transfer coefficient between the water and the glass is 120 W/m^2 · °C, determine how long it will take for the milk to warm up from 3°C to 38°C. Take the properties of the milk to be the same as those of water. Can the milk in this case be treated as a lumped system? Why? *Answer:* 5.8 min

18–18 Repeat Prob. 18–17 for the case of water also being stirred, so that the heat transfer coefficient is doubled to 240 W/m^2 · °C.

18–19 A long copper rod of diameter 2.0 cm is initially at a uniform temperature of 100°C. It is now exposed to an air stream at 20°C with a heat transfer coefficient of 200 W/m^2 · K. How long would it take for the copper road to cool to an average temperature of 25°C?

18–20 Consider a sphere of diameter 5 cm, a cube of side length 5 cm, and a rectangular prism of dimension 4 cm × 5 cm × 6 cm, all initially at 0°C and all made of silver (k = 429 W/m · °C, ρ = 10,500 kg/m^3, c_p = 0.235 kJ/kg · °C). Now all three of these geometries are exposed to ambient air at 33°C on all of their surfaces with a heat transfer coefficient of 12 W/m^2 · °C. Determine how long it will take for the temperature of each geometry to rise to 25°C.

18–21E During a picnic on a hot summer day, all the cold drinks disappeared quickly, and the only available drinks were those at the ambient temperature of 90°F. In an effort to cool a 12-fluid-oz drink in a can, which is 5 in high and has a diameter of 2.5 in, a person grabs the can and starts shaking it in the iced water of the chest at 32°F. The temperature of the drink can be assumed to be uniform at all times, and the heat transfer coefficient between the iced water and the aluminum can is 30 Btu/h · ft^2 · °F. Using the properties of water for the drink, estimate how long it will take for the canned drink to cool to 40°F.

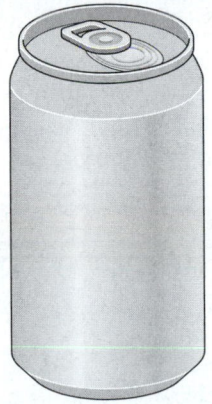

FIGURE P18–21E

18–22 Consider a 1000-W iron whose base plate is made of 0.5-cm-thick aluminum alloy 20218–T6 (ρ = 2770 kg/m^3, c_p = 875 J/kg · °C, α = 7.3 × 10^{-5} m^2/s). The base plate has a surface area of 0.03 m^2. Initially, the iron is in thermal equilibrium with the ambient air at 22°C. Taking the heat transfer coefficient at the surface of the base plate to be 12 W/m^2 · °C and assuming 85 percent of the heat generated in the resistance wires is transferred to the plate, determine how long it will take for the plate temperature to reach 140°C. Is it realistic to assume the plate temperature to be uniform at all times?

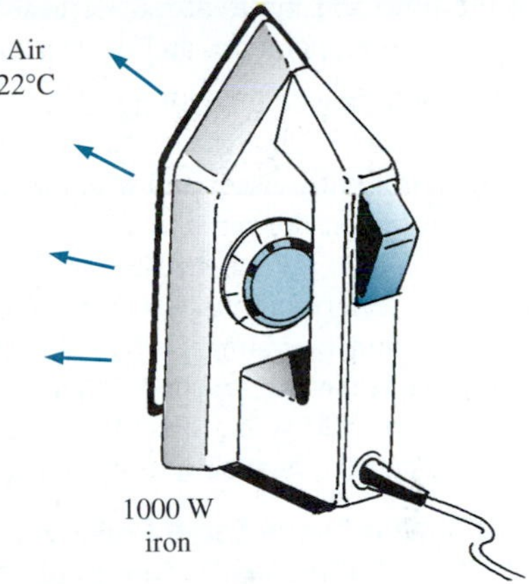

FIGURE P18–22

18–23 Reconsider Prob. 18–22. Using EES (or other) software, investigate the effects of the heat transfer coefficient and the final plate temperature on the time it will take for the plate to reach this temperature. Let the heat transfer coefficient vary from 5 W/m^2 · °C to 25 W/m^2 · °C and the temperature from 30°C to 200°C. Plot the time as functions of the heat transfer coefficient and the temperature, and discuss the results.

18–24 Stainless steel ball bearings (ρ = 8085 kg/m^3, k = 15.1 W/m · °C, c_p = 0.480 kJ/kg · °C, and α = 3.91 × 10^{-6} m^2/s) having a diameter of 1.2 cm are to be quenched in water. The balls leave the oven at a uniform temperature of 900°C and are exposed to air at 30°C for a while before they are dropped into the water. If the temperature of the balls is not to fall below 850°C prior to quenching and the heat transfer coefficient in the air is 125 W/m^2 · °C, determine how long they can stand in the air before being dropped into the water. *Answer:* 3.7 s

18–25 Carbon steel balls (ρ = 7833 kg/m^3, k = 54 W/m · °C, c_p = 0.465 kJ/kg · °C, and α = 1.474 × 10^{-6} m^2/s) 8 mm in diameter are annealed by heating them first to 900°C in a furnace and then allowing them to cool slowly to 100°C in ambient air at 35°C. If the average heat transfer coeffi-cient is 75 W/m^2 · °C, determine how long the annealing process will

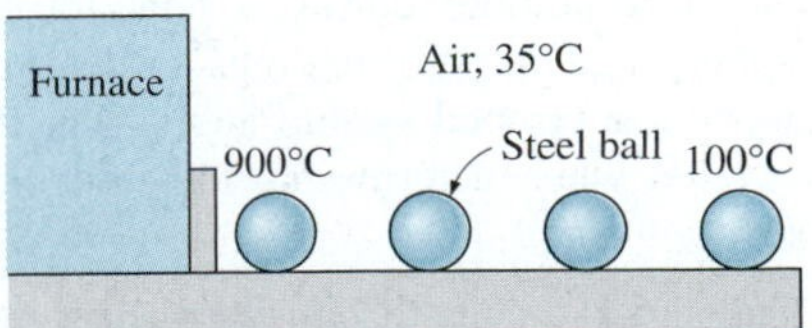

FIGURE P18–25

take. If 2500 balls are to be annealed per hour, determine the total rate of heat transfer from the balls to the ambient air.

18–26 Reconsider Prob. 18–25. Using EES (or other) software, investigate the effect of the initial temperature of the balls on the annealing time and the total rate of heat transfer. Let the temperature vary from 500°C to 1000°C. Plot the time and the total rate of heat transfer as a function of the initial temperature, and discuss the results.

18–27 An electronic device dissipating 20 W has a mass of 20 g, a specific heat of 850 J/kg · °C, and a surface area of 4 cm^2. The device is lightly used, and it is on for 5 min and then off for several hours, during which it cools to the ambient temperature of 25°C. Taking the heat transfer coefficient to be 12 W/m^2 · °C, determine the temperature of the device at the end of the 5-min operating period. What would your answer be if the device were attached to an aluminum heat sink having a mass of 200 g and a surface area of 80 cm^2? Assume the device and the heat sink to be nearly isothermal.

Transient Heat Conduction in Large Plane Walls, Long Cylinders, and Spheres with Spatial Effects

18–28C What is an infinitely long cylinder? When is it proper to treat an actual cylinder as being infinitely long, and when is it not? For example, is it proper to use this model when finding the temperatures near the bottom or top surfaces of a cylinder? Explain.

18–29C Can the transient temperature charts in Fig. 18–15 for a plane wall exposed to convection on both sides be used for a plane wall with one side exposed to convection while the other side is insulated? Explain.

18–30C Why are the transient temperature charts prepared using nondimensionalized quantities such as the Biot and Fourier numbers instead of the actual variables such as thermal conductivity and time?

18–31C What is the physical significance of the Fourier number? Will the Fourier number for a specified heat transfer problem double when the time is doubled?

18–32C How can we use the transient temperature charts when the surface temperature of the geometry is specified instead of the temperature of the surrounding medium and the convection heat transfer coefficient?

18–33C A body at an initial temperature of T_i is brought into a medium at a constant temperature of T_∞. How can you determine the maximum possible amount of heat transfer between the body and the surrounding medium?

18–34C The Biot number during a heat transfer process between a sphere and its surroundings is determined to be 0.02. Would you use lumped system analysis or the transient temperature charts when determining the midpoint temperature of the sphere? Why?

18–35 A student calculates that the total heat transfer from a spherical copper ball of diameter 18 cm initially at 200°C and its environment at a constant temperature of 25°C during the first 20 min of cooling is 3150 kJ. Is this result reasonable? Why?

18–36 An experiment is to be conducted to determine heat transfer coefficient on the surfaces of tomatoes that are placed in cold water at 7°C. The tomatoes (k = 0.59 W/m · °C, $\alpha = 0.141 \times 10^{-6}$ m^2/s, ρ = 999 kg/m^3, c_p = 3.99 kJ/kg · °C) with an initial uniform temperature of 30°C are spherical in shape with a diameter of 8 cm. After a period of 2 hours, the temperatures at the center and the surface of the tomatoes are measured to be 10.0°C and 7.1°C, respectively. Using analytical one-term approximation method (not the Heisler charts), determine the heat transfer coefficient and the amount of heat transfer during this period if there are eight such tomatoes in water.

18–37 An ordinary egg can be approximated as a 5.5 cm-diameter sphere whose properties are roughly k = 0.6 W/m · °C and $\alpha = 0.14 \times 10^{-6}$ m^2/s. The egg is initially at a uniform temperature of 8°C and is dropped into boiling water at 97°C. Taking the convection heat transfer coefficient to be h = 1400 W/m^2 · °C, determine how long it will take for the center of the egg to reach 70°C.

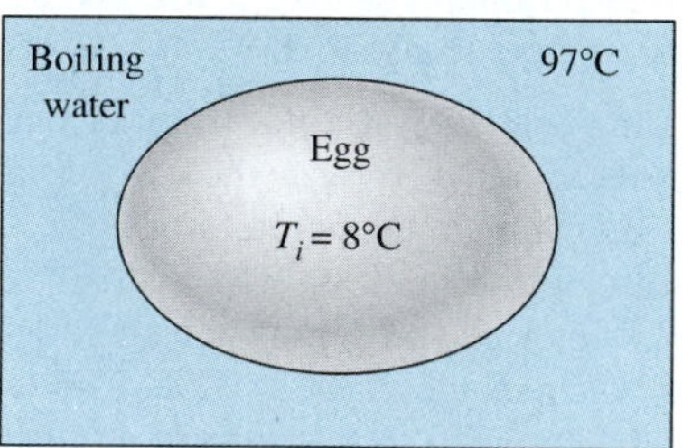

FIGURE P18–37

18–38 Reconsider Prob. 18–37. Using EES (or other) software, investigate the effect of the final center temperature of the egg on the time it will take for the center to reach this temperature. Let the temperature vary from 50°C to 95°C. Plot the time versus the temperature, and discuss the results.

18–39 In a production facility, 3-cm-thick large brass plates (k = 110 W/m · °C, ρ = 8530 kg/m^3, c_p = 380 J/kg · °C, and

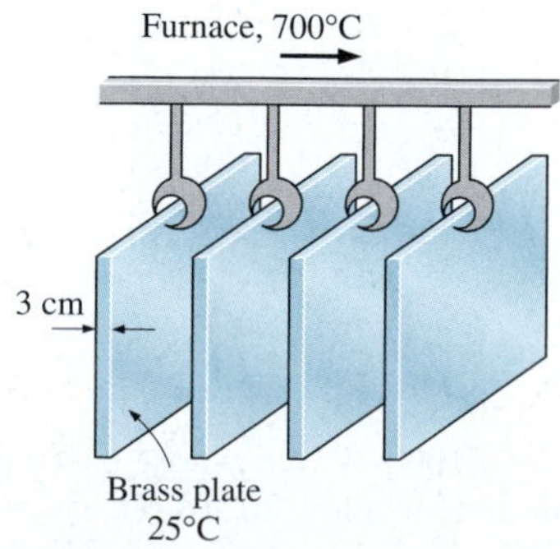

FIGURE P18–39

$\alpha = 33.9 \times 10^{-6}$ m²/s) that are initially at a uniform temperature of 25°C are heated by passing them through an oven maintained at 700°C. The plates remain in the oven for a period of 10 min. Taking the convection heat transfer coefficient to be $h = 80$ W/m² · °C, determine the surface temperature of the plates when they come out of the oven.

18–40 Reconsider Prob. 18–39. Using EES (or other) software, investigate the effects of the temperature of the oven and the heating time on the final surface temperature of the plates. Let the oven temperature vary from 500°C to 900°C and the time from 2 min to 30 min. Plot the surface temperature as the functions of the oven temperature and the time, and discuss the results.

18–41 A long 35-cm-diameter cylindrical shaft made of stainless steel 304 ($k = 14.9$ W/m · °C, $\rho = 7900$ kg/m³, $c_p = 477$ J/kg · °C, and $\alpha = 3.95 \times 10^{-6}$ m²/s) comes out of an oven at a uniform temperature of 400°C. The shaft is then allowed to cool slowly in a chamber at 150°C with an average convection heat transfer coefficient of $h = 60$ W/m² · °C. Determine the temperature at the center of the shaft 20 min after the start of the cooling process. Also, determine the heat transfer per unit length of the shaft during this time period. *Answers:* 390°C, 16,015 kJ/m

18–42 Reconsider Prob. 18–41. Using EES (or other) software, investigate the effect of the cooling time on the final center temperature of the shaft and the amount of heat transfer. Let the time vary from 5 min to 60 min. Plot the center temperature and the heat transfer as a function of the time, and discuss the results.

18–43E Long cylindrical AISI stainless steel rods ($k =$ 7.74 Btu/h · ft · °F and $\alpha = 0.135$ ft²/h) of 4-in-diameter are heat treated by drawing them at a velocity of 7 ft/min through a 21-ft-long oven maintained at 1700°F. The heat transfer coefficient in the oven is 20 Btu/h · ft² · °F. If the rods enter the oven at 70°F, determine their centerline temperature when they leave.

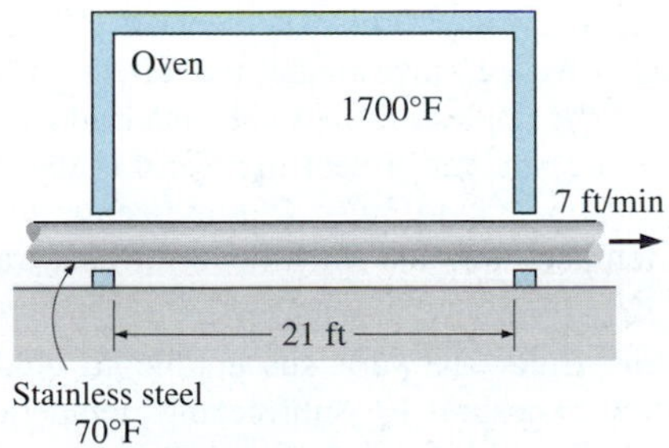

FIGURE P18–43E

18–44 In a meat processing plant, 2-cm-thick steaks ($k =$ 0.45 W/m · °C and $\alpha = 0.91 \times 10^{-7}$ m²/s) that are initially at 25°C are to be cooled by passing them through a refrigeration room at −11°C. The heat transfer coefficient on both sides of the steaks is 9 W/m² · °C. If both surfaces of the steaks are to be cooled to 2°C, determine how long the steaks should be kept in the refrigeration room.

18–45 A long cylindrical wood log ($k = 0.17$ W/m · °C and $\alpha = 1.28 \times 10^{-7}$ m²/s) is 10 cm in diameter and is initially at a uniform temperature of 15°C. It is exposed to hot gases at 550°C in a fireplace with a heat transfer coefficient of 13.6 W/m² · °C on the surface. If the ignition temperature of the wood is 420°C, determine how long it will be before the log ignites.

18–46 In *Betty Crocker's Cookbook,* it is stated that it takes 2 h 45 min to roast a 3.2-kg rib initially at 4.5°C "rare" in an oven maintained at 163°C. It is recommended that a meat thermometer be used to monitor the cooking, and the rib is considered rare done when the thermometer inserted into the center of the thickest part of the meat registers 60°C. The rib can be treated as a homogeneous spherical object with the properties $\rho = 1200$ kg/m³, $c_p = 4.1$ kJ/kg · °C, $k = 0.45$ W/m · °C, and $\alpha = 0.91 \times 10^{-7}$ m²/s. Determine (*a*) the heat transfer coefficient at the surface of the rib; (*b*) the temperature of the outer surface of the rib when it is done; and (*c*) the amount of heat transferred to the rib. (*d*) Using the values obtained, predict how long it will take to roast this rib to "medium" level, which occurs when the innermost temperature of the rib reaches 71°C. Compare your result to the listed value of 3 h 20 min.

If the roast rib is to be set on the counter for about 15 min before it is sliced, it is recommended that the rib be taken out of the oven when the thermometer registers about 4°C below the indicated value because the rib will continue cooking even after it is taken out of the oven. Do you agree with this recommendation? *Answers:* (a) 156.9 W/m² · 8C, (b) 159.58C, (c) 1629 kJ, (d) 3.0 h

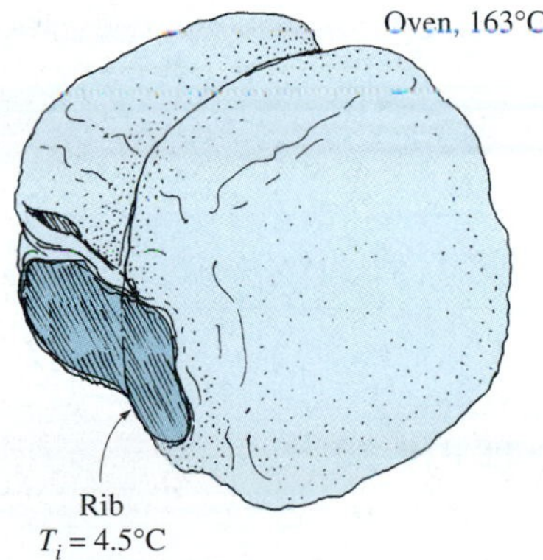

FIGURE P18–46

18–47 Repeat Prob. 18–46 for a roast rib that is to be "well-done" instead of "rare." A rib is considered to be well-done when its center temperature reaches 77°C, and the roasting in this case takes about 4 h 15 min.

18–48 For heat transfer purposes, an egg can be considered to be a 5.5-cm-diameter sphere having the properties of water.

An egg that is initially at 8°C is dropped into the boiling water at 100°C. The heat transfer coefficient at the surface of the egg is estimated to be 800 W/m² · °C. If the egg is considered cooked when its center temperature reaches 60°C, determine how long the egg should be kept in the boiling water.

18–49 Repeat Prob. 18–48 for a location at 1610-m elevation such as Denver, Colorado, where the boiling temperature of water is 94.4°C.

18–50 The author and his then 6-year-old son have conducted the following experiment to determine the thermal conductivity of a hot dog. They first boiled water in a large pan and measured the temperature of the boiling water to be 94°C, which is not surprising, since they live at an elevation of about 1650 m in Reno, Nevada. They then took a hot dog that is 12.5 cm long and 2.2 cm in diameter and inserted a thermocouple into the midpoint of the hot dog and another thermocouple just under the skin. They waited until both thermocouples read 20°C, which is the ambient temperature. They then dropped the hot dog into boiling water and observed the changes in both temperatures. Exactly 2 min after the hot dog was dropped into the boiling water, they recorded the center and the surface temperatures to be 59°C and 88°C, respectively. The density of the hot dog can be taken to be 980 kg/m³, which is slightly less than the density of water, since the hot dog was observed to be floating in water while being almost completely immersed. The specific heat of a hot dog can be taken to be 3900 J/kg · °C, which is slightly less than that of water, since a hot dog is mostly water. Using transient temperature charts, determine (*a*) the thermal diffusivity of the hot dog; (*b*) the thermal conductivity of the hot dog; and (*c*) the convection heat transfer coefficient. *Answers:* (*a*) 2.02×10^{-7} m²/s, (*b*) 0.771 W/m · °C, (*c*) 467 W/m² · °C.

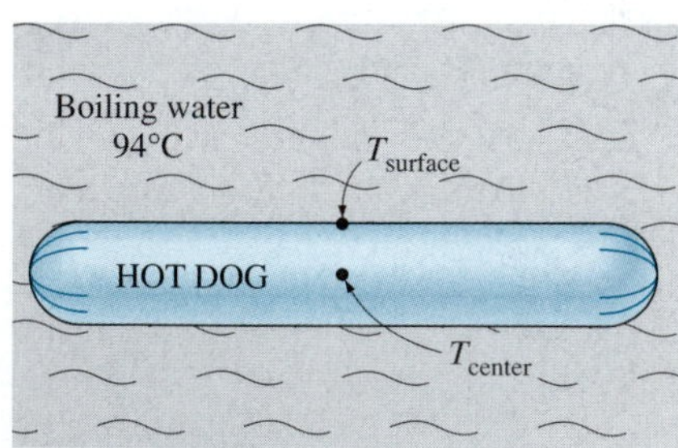

FIGURE P18–50

18–51 Using the data and the answers given in Prob. 18–50, determine the center and the surface temperatures of the hot dog 4 min after the start of the cooking. Also determine the amount of heat transferred to the hot dog.

18–52E In a chicken processing plant, whole chickens averaging 5 lbm each and initially at 65°F are to be cooled in the racks of a large refrigerator that is maintained at 5°F. The entire chicken is to be cooled below 45°F, but the temperature of the chicken is not to drop below 35°F at any point during refrigeration. The convection heat transfer coefficient and thus the rate of heat transfer from the chicken can be controlled by varying the speed of a circulating fan inside. Determine the heat transfer-coefficient that will enable us to meet both temperature constraints while keeping the refrigeration time to a minimum. The chicken can be treated as a homogeneous spherical object having the properties ρ = 74.9 lbm/ft³, c_p = 0.98 Btu/lbm · °F, k = 0.26 Btu/h · ft · °F, and α = 0.0035 ft²/h.

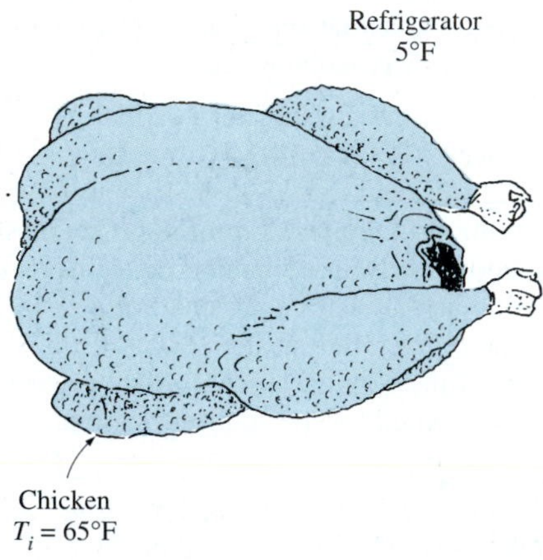

FIGURE P18–52E

18–53 A person puts a few apples into the freezer at −15°C to cool them quickly for guests who are about to arrive. Initially, the apples are at a uniform temperature of 20°C, and the heat transfer coefficient on the surfaces is 8 W/m² · °C. Treating the apples as 9-cm-diameter spheres and taking their properties to be ρ = 840 kg/m³, c_p = 3.81 kJ/kg · °C, k = 0.418 W/m · °C, and $\alpha = 1.3 \times 10^{-7}$ m²/s, determine the center and surface temperatures of the apples in 1 h. Also, determine the amount of heat transfer from each apple.

18–54 EES Reconsider Prob. 18–53. Using EES (or other) software, investigate the effect of the initial temperature of the apples on the final center and surface temperatures and the amount of heat transfer. Let the initial temperature vary from 2°C to 30°C. Plot the center temperature, the surface temperature, and the amount of heat transfer as a function of the initial temperature, and discuss the results.

18–55 Citrus fruits are very susceptible to cold weather, and extended exposure to subfreezing temperatures can destroy them. Consider an 8-cm-diameter orange that is initially at 15°C. A cold front moves in one night, and the ambient temperature suddenly drops to −6°C, with a heat transfer coefficient of 15 W/m² · °C. Using the properties of water for the orange and assuming the ambient conditions to remain constant for 4 h before the cold front moves out, determine if any part of the orange will freeze that night.

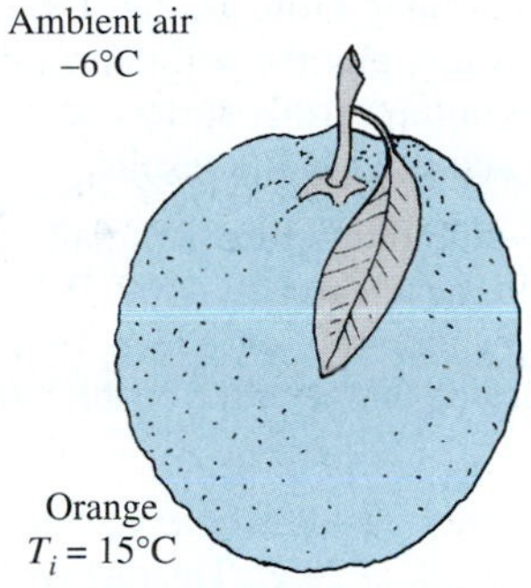

FIGURE P18–55

18–56 A 9-cm-diameter potato (ρ = 1100 kg/m^3, c_p = 3900 J/kg · °C, k = 0.6 W/m · °C, and α = 1.4 × 10^{-7} m^2/s) that is initially at a uniform temperature of 25°C is baked in an oven at 170°C until a temperature sensor inserted to the center of the potato indicates a reading of 70°C. The potato is then taken out of the oven and wrapped in thick towels so that almost no heat is lost from the baked potato. Assuming the heat transfer coefficient in the oven to be 40 W/m^2 · °C, determine (*a*) how long the potato is baked in the oven and (*b*) the final equilibrium temperature of the potato after it is wrapped.

18–57 White potatoes (k = 0.50 W/m · °C and α = 0.13 × 10^{-6} m^2/s) that are initially at a uniform temperature of 25°C and have an average diameter of 6 cm are to be cooled by refrigerated air at 2°C flowing at a velocity of 4 m/s. The average heat transfer coefficient between the potatoes and the air is experimentally determined to be 19 W/m^2 · °C. Determine how long it will take for the center temperature of the potatoes to drop to 6°C. Also, determine if any part of the potatoes will experience chilling injury during this process.

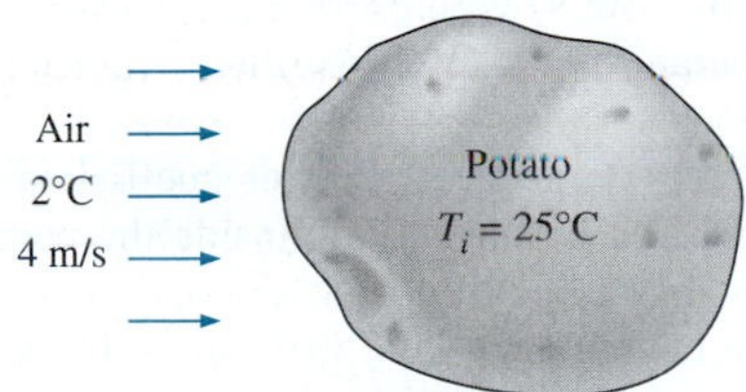

FIGURE P18–57

18–58E Oranges of 2.5-in-diameter (k = 0.26 Btu/h · ft · °F and α = 1.4 × 10^{-6} ft^2/s) initially at a uniform temperature of 78°F are to be cooled by refrigerated air at 25°F flowing at a velocity of 1 ft/s. The average heat transfer coefficient between the oranges and the air is experimentally determined to be 4.6 Btu/h · ft^2 · °F. Determine how long it will take for the center temperature of the oranges to drop to 40°F. Also, determine if any part of the oranges will freeze during this process.

18–59 A 65-kg beef carcass (k = 0.47 W/m · °C and α = 0.13 × 10^{-6} m^2/s) initially at a uniform temperature of 37°C is to be cooled by refrigerated air at −10°C flowing at a velocity of 1.2 m/s. The average heat transfer coefficient between the carcass and the air is 22 W/m^2 · °C. Treating the carcass as a cylinder of diameter 24 cm and height 1.4 m and disregarding heat transfer from the base and top surfaces, determine how long it will take for the center temperature of the carcass to drop to 4°C. Also, determine if any part of the carcass will freeze during this process. *Answer:* 12.2 h

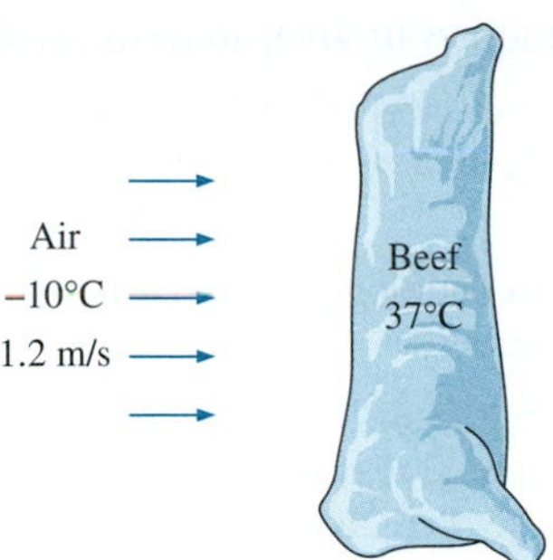

FIGURE P18–59

18–60 Layers of 23-cm-thick meat slabs (k = 0.47 W/m · °C and α = 0.13 × 10^{-6} m^2/s) initially at a uniform temperature of 7°C are to be frozen by refrigerated air at −30°C flowing at a velocity of 1.4 m/s. The average heat transfer coefficient between the meat and the air is 20 W/m^2 · °C. Assuming the size of the meat slabs to be large relative to their thickness, determine how long it will take for the center temperature of the slabs to drop to −18°C. Also, determine the surface temperature of the meat slab at that time.

18–61E Layers of 6-in-thick meat slabs (k = 0.26 Btu/h · ft · °F and α = 1.4 × 10^{-6} ft^2/s) initially at a uniform temperature of 50°F are cooled by refrigerated air at 23°F to a temperature of 36°F at their center in 12 h. Estimate the average heat transfer coefficient during this cooling process. *Answer:* 1.5 Btu/h · ft^2 · °F

18–62 Chickens with an average mass of 1.7 kg (k = 0.45 W/m · °C and α = 0.13 × 10^{-6} m^2/s) initially at a uniform temperature of 15°C are to be chilled in agitated brine at −7°C. The average heat transfer coefficient between the chicken and the brine is determined experimentally to be

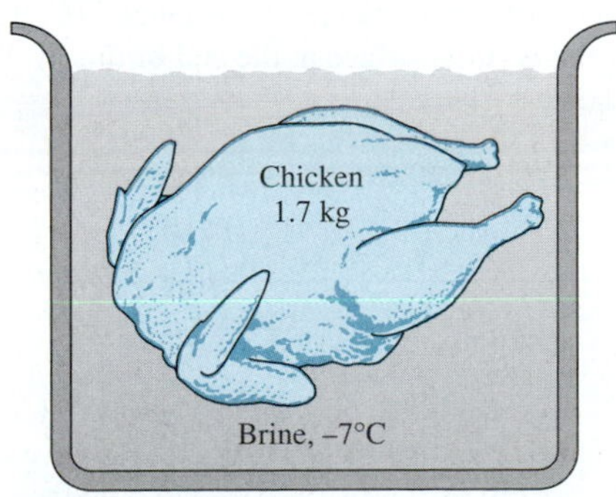

FIGURE P18–62

440 W/m^2 · °C. Taking the average density of the chicken to be 0.95 g/cm^3 and treating the chicken as a spherical lump, determine the center and the surface temperatures of the chicken in 2 h and 45 min. Also, determine if any part of the chicken will freeze during this process.

Transient Heat Conduction in Semi-Infinite Solids

18–63C What is a semi-infinite medium? Give examples of solid bodies that can be treated as semi-infinite mediums for heat transfer purposes.

18–64C Under what conditions can a plane wall be treated as a semi-infinite medium?

18–65C Consider a hot semi-infinite solid at an initial temperature of T_i that is exposed to convection to a cooler medium at a constant temperature of T_∞, with a heat transfer coefficient of h. Explain how you can determine the total amount of heat transfer from the solid up to a specified time t_o.

18–66 In areas where the air temperature remains below 0°C for prolonged periods of time, the freezing of water in underground pipes is a major concern. Fortunately, the soil remains relatively warm during those periods, and it takes weeks for the subfreezing temperatures to reach the water mains in the ground. Thus, the soil effectively serves as an insulation to protect the water from the freezing atmospheric temperatures in winter.

The ground at a particular location is covered with snow pack at −8°C for a continuous period of 60 days, and the average soil properties at that location are k = 0.35 W/m · °C and α = 0.15 × 80^{-6} m^2/s. Assuming an initial uniform temperature of 8°C for the ground, determine the minimum burial depth to prevent the water pipes from freezing.

18–67 The soil temperature in the upper layers of the earth varies with the variations in the atmospheric conditions. Before a cold front moves in, the earth at a location is initially at a uniform temperature of 10°C. Then the area is subjected to a temperature of −10°C and high winds that resulted in a convection heat transfer coefficient of 40 W/m^2 · °C on the earth's surface for a period of 10 h. Taking the properties of the soil at that location to be k = 0.9 W/m · °C and α = 1.6 × 10^{-5} m^2/s, determine the soil temperature at distances 0, 10, 20, and 50 cm from the earth's surface at the end of this 10-h period.

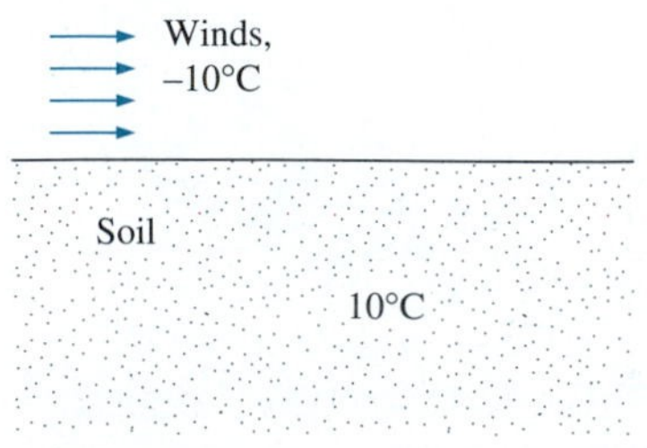

FIGURE P18–67

18–68 Reconsider Prob. 18–67. Using EES (or other) software, plot the soil temperature as a function of the distance from the earth's surface as the distance varies from 0 m to 1m, and discuss the results.

18–69 A thick aluminum block initially at 20°C is subjected to constant heat flux of 4000 W/m^2 by an electric resistance heater whose top surface is insulated. Determine how much the surface temperature of the block will rise after 30 minutes.

18–70 A bare-footed person whose feet are at 32°C steps on a large aluminum block at 20°C. Treating both the feet and the aluminum block as semi-infinite solids, determine the contact surface temperature. What would your answer be if the person stepped on a wood block instead? At room temperature, the $\sqrt{k\rho c_p}$ value is 24 kJ/m^2 · °C for aluminum, 0.38 kJ/m^2 · °C for wood, and 1.1 kJ/m^2 · °C for human flesh.

18–71E The walls of a furnace are made of 1.2-ft-thick concrete (k = 0.64 Btu/h · ft · °F and α = 0.023 ft^2/h). Initially, the furnace and the surrounding air are in thermal equilibrium at 70°F. The furnace is then fired, and the inner surfaces of the furnace are subjected to hot gases at 1800°F with a very large heat transfer coefficient. Determine how long it will take for the temperature of the outer surface of the furnace walls to rise to 70.1°F. *Answer: 116 min*

18–72 A thick wood slab (k = 0.17 W/m · °C and α = 1.28 × 10^{-7} m^2/s) that is initially at a uniform temperature of 25°C is exposed to hot gases at 550°C for a period of 5 min. The heat transfer coefficient between the gases and the wood slab is 35 W/m^2 · °C. If the ignition temperature of the wood is 450°C, determine if the wood will ignite.

18–73 A large cast iron container (k = 52 W/m · °C and α = 1.70 × 10^{-5} m^2/s) with 5-cm-thick walls is initially at a uniform temperature of 0°C and is filled with ice at 0°C. Now the outer surfaces of the container are exposed to hot water at 60°C with a very large heat transfer coefficient. Determine how long it will be before the ice inside the container starts melting. Also, taking the heat transfer coefficient on the inner surface of the container to be 250 W/m^2 · °C, determine the rate of heat transfer to the ice through a 1.2-m-wide and 2-m-high section of the wall when steady operating conditions are reached. Assume the ice starts melting when its inner surface temperature rises to 0.1°C.

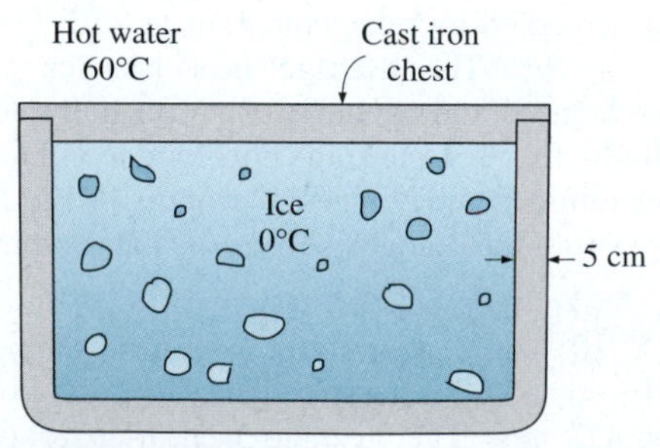

FIGURE P18–73

Transient Heat Conduction in Multidimensional Systems

18–74C What is the product solution method? How is it used to determine the transient temperature distribution in a two-dimensional system?

18–75C How is the product solution used to determine the variation of temperature with time and position in three-dimensional systems?

18–76C A short cylinder initially at a uniform temperature T_i is subjected to convection from all of its surfaces to a medium at temperature T_∞. Explain how you can determine the temperature of the midpoint of the cylinder at a specified time *t*.

18–77C Consider a short cylinder whose top and bottom surfaces are insulated. The cylinder is initially at a uniform temperature T_i and is subjected to convection from its side surface to a medium at temperature T_∞ with a heat transfer coefficient of *h*. Is the heat transfer in this short cylinder one- or two-dimensional? Explain.

18–78 A short brass cylinder (ρ = 8530 kg/m^3, c_p = 0.389 kJ/kg · °C, k = 110 W/m · °C, and α = 3.39 × 10^{-5} m^2/s) of diameter D = 8 cm and height H = 15 cm is initially at a uniform temperature of T_i = 150°C. The cylinder is now placed in atmospheric air at 20°C, where heat transfer takes place by convection with a heat transfer coefficient of h = 40 W/m^2 · °C. Calculate (*a*) the center temperature of the cylinder; (*b*) the center temperature of the top surface of the cylinder; and (*c*) the total heat transfer from the cylinder 15 min after the start of the cooling.

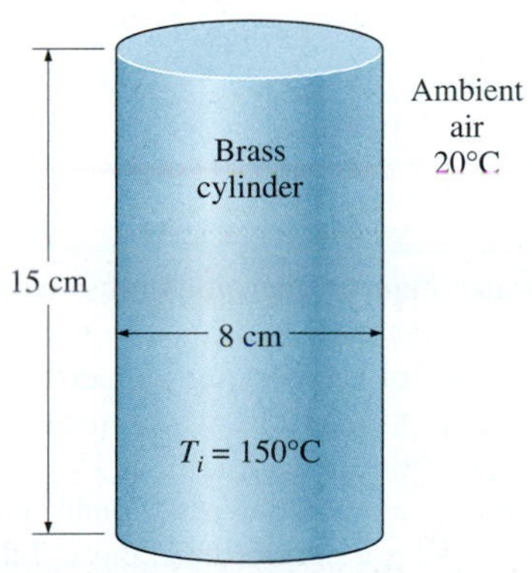

FIGURE P18–78

18–79 Reconsider Prob. 18–78. Using EES (or other) software, investigate the effect of the cooling time on the center temperature of the cylinder, the center temperature of the top surface of the cylinder, and the total heat transfer. Let the time vary from 5 min to 60 min. Plot the center temperature of the cylinder, the center temperature of the top surface, and the total heat transfer as a function of the time, and discuss the results.

18–80 A semi-infinite aluminum cylinder (k = 237 W/m · °C, α = 9.71 × 10^{-5} m^2/s) of diameter D = 15 cm is initially at a uniform temperature of T_i = 115°C. The cylinder is now placed in water at 10°C, where heat transfer takes place by convection with a heat transfer coefficient of h = 140 W/m^2 · °C. Determine the temperature at the center of the cylinder 5 cm from the end surface 8 min after the start of cooling.

18–81E A hot dog can be considered to be a cylinder 5 in long and 0.8 in in diameter whose properties are ρ = 61.2 lbm/ft^3, c_p = 0.93 Btu/lbm · °F, k = 0.44 Btu/h · ft · °F, and α = 0.0077 ft^2/h. A hot dog initially at 40°F is dropped into boiling water at 212°F. If the heat transfer coefficient at the surface of the hot dog is estimated to be 120 Btu/h · ft^2 · °F, determine the center temperature of the hot dog after 5, 10, and 15 min by treating the hot dog as (*a*) a finite cylinder and (*b*) an infinitely long cylinder.

18–82E Repeat Prob. 18–81E for a location at 5300-ft elevation such as Denver, Colorado, where the boiling temperature of water is 202°F.

18–83 A 5-cm-high rectangular ice block (k = 2.22 W/m · °C and α = 0.124 × 10^{-7} m^2/s) initially at −20°C is placed on a table on its square base 4 cm × 4 cm in size in a room at 18°C. The heat transfer coefficient on the exposed surfaces of the ice block is 12 W/m^2 · °C. Disregarding any heat transfer from the base to the table, determine how long it will be before the ice block starts melting. Where on the ice block will the first liquid droplets appear?

FIGURE P18–83

18–84 Reconsider Prob. 18–83. Using EES (or other) software, investigate the effect of the initial temperature of the ice block on the time period before the ice block starts melting. Let the initial temperature vary from −26°C to −4°C. Plot the time versus the initial temperature, and discuss the results.

18–85 A 2-cm-high cylindrical ice block (k = 2.22 W/m · °C and α = 0.124 × 10^{-7} m^2/s) is placed on a table on its base of diameter 2 cm in a room at 24°C. The heat transfer coefficient on the exposed surfaces of the ice block is 13 W/m^2 · °C, and heat transfer from the base of the ice block to the table is negligible. If the ice block is not to start melting at any point for at least 3 h, determine what the initial temperature of the ice block should be.

18–86 Consider a cubic block whose sides are 5 cm long and a cylindrical block whose height and diameter are also 5 cm. Both blocks are initially at 20°C and are made of granite (k = 2.5 W/m · °C and α = 1.15 × 10^{-6} m^2/s). Now both blocks are exposed to hot gases at 500°C in a furnace on all of their surfaces with a heat transfer coefficient of 40 W/m^2 · °C. Determine the center temperature of each geometry after 10, 20, and 60 min.

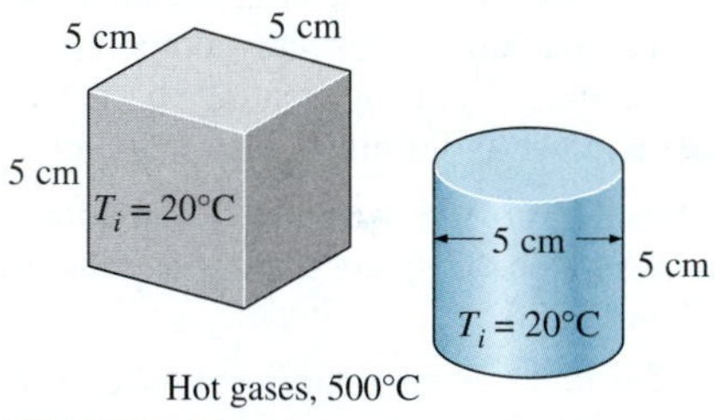

FIGURE P18–86

18–87 Repeat Prob. 18–86 with the heat transfer coefficient at the top and the bottom surfaces of each block being doubled to 80 W/m^2 · °C.

18–88 A 20-cm-long cylindrical aluminum block (ρ = 2702 kg/m^3, c_p = 0.896 kJ/kg · °C, k = 236 W/m · °C, and α = 9.75 × 10^{-5} m^2/s), 15 cm in diameter, is initially at a uniform temperature of 20°C. The block is to be heated in a furnace at 1200°C until its center temperature rises to 300°C. If the heat transfer coefficient on all surfaces of the block is 80 W/m^2 · °C, determine how long the block should be kept in the furnace. Also, determine the amount of heat transfer from the aluminum block if it is allowed to cool in the room until its temperature drops to 20°C throughout.

18–89 Repeat Prob. 18–88 for the case where the aluminum block is inserted into the furnace on a low-conductivity material so that the heat transfer to or from the bottom surface of the block is negligible.

18–90 EES Reconsider Prob. 18–88. Using EES (or other) software, investigate the effect of the final center temperature of the block on the heating time and the amount of heat transfer. Let the final center temperature vary from 50°C to 1000°C. Plot the time and the heat transfer as a function of the final center temperature, and discuss the results.

Review Problems

18–91 Consider two 2-cm-thick large steel plates (k = 43 W/m · °C and α = 1.17 × 10^{-5} m^2/s) that were put on top of each other while wet and left outside during a cold winter night at −15°C. The next day, a worker needs one of the plates, but the plates are stuck together because the freezing of the water between the two plates has bonded them together. In an effort to melt the ice between the plates and separate them, the worker takes a large hair dryer and blows hot air at 50°C all over the exposed surface of the plate on the top. The convection heat transfer coefficient at the top surface is estimated to be 40 W/m^2 · °C. Determine how long the worker must keep blowing hot air before the two plates separate. *Answer: 482 s*

18–92 Consider a curing kiln whose walls are made of 30-cm-thick concrete whose properties are k = 0.9 W/m · °C and α = 0.23 × 10^{-5} m^2/s. Initially, the kiln and its walls are in equilibrium with the surroundings at 6°C. Then all the doors are closed and the kiln is heated by steam so that the temperature of the inner surface of the walls is raised to 42°C and is maintained at that level for 2.5 h. The curing kiln is then opened and exposed to the atmospheric air after the stream flow is turned off. If the outer surfaces of the walls of the kiln were insulated, would it save any energy that day during the period the kiln was used for curing for 2.5 h only, or would it make no difference? Base your answer on calculations.

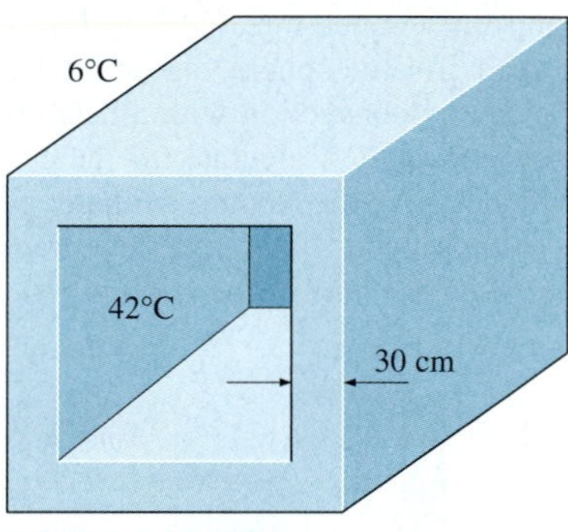

FIGURE P18–92

18–93 The water main in the cities must be placed at sufficient depth below the earth's surface to avoid freezing during extended periods of subfreezing temperatures. Determine the minimum depth at which the water main must be placed at a location where the soil is initially at 15°C and the earth's surface temperature under the worst conditions is expected to remain at −10°C for a period of 75 days. Take the properties of soil at that location to be k = 0.7 W/m · °C and α = 1.4 × 10^{-5} m^2/s. *Answer: 7.05 m*

18–94 A hot dog can be considered to be a 12-cm-long cylinder whose diameter is 2 cm and whose properties are ρ = 980 kg/m^3, c_p = 3.9 kJ/kg · °C, k = 0.76 W/m · °C, and α = 2 × 10^{-7} m^2/s. A hot dog initially at 5°C is dropped into boiling water at 100°C. The heat transfer coefficient at the surface of the hot dog is estimated to be 600 W/m^2 · °C. If the hot dog is considered cooked when its center temperature reaches 80°C, determine how long it will take to cook it in the boiling water.

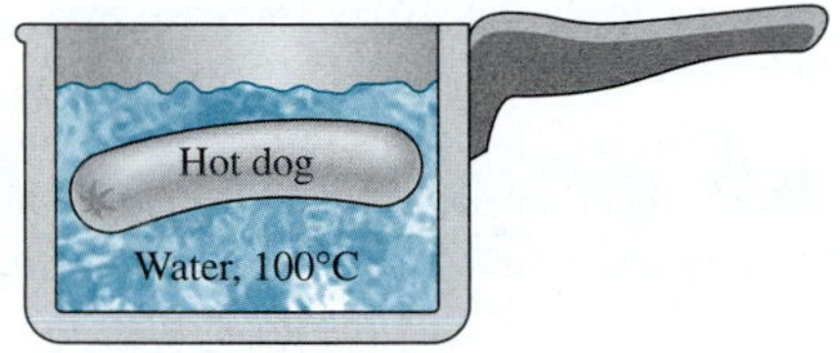

FIGURE P18–94

18–95 A long roll of 2-m-wide and 0.5-cm-thick 1-Mn manganese steel plate coming off a furnace at 820°C is to be quenched in an oil bath (c_p = 2.0 kJ/kg · °C) at 45°C. The metal sheet is moving at a steady velocity of 15 m/min, and the oil bath is 9 m long. Taking the convection heat transfer coefficient on both sides of the plate to be 860 W/m^2 · °C, determine the temperature of the sheet metal when it leaves the oil bath. Also, determine the required rate of heat removal from the oil to keep its temperature constant at 45°C.

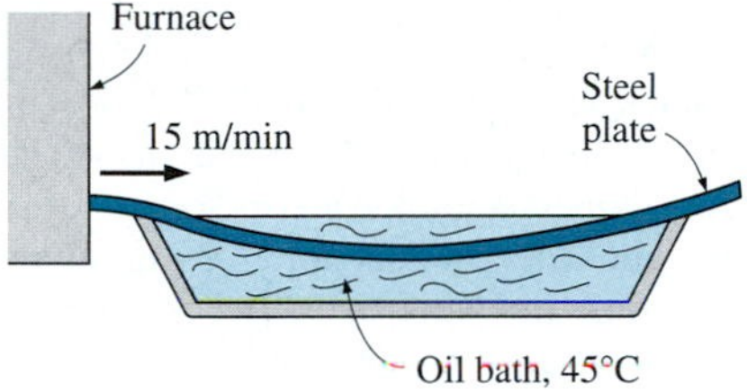

FIGURE P18–95

18–96E In *Betty Crocker's Cookbook*, it is stated that it takes 5 h to roast an 14-lb stuffed turkey initially at 40°F in an oven maintained at 325°F. It is recommended that a meat thermometer be used to monitor the cooking, and the turkey is considered done when the thermometer inserted deep into the thickest part of the breast or thigh without touching the bone registers 185°F. The turkey can be treated as a homogeneous spherical object with the properties ρ = 75 lbm/ft^3, c_p = 0.98 Btu/lbm · °F, k = 0.26 Btu/h · ft · °F, and α = 0.0035 ft^2/h. Assuming the tip of the thermometer is at one-third radial distance from the center of the turkey, determine (*a*) the average heat transfer coefficient at the surface of the turkey; (*b*) the temperature of the skin of the turkey when it is done; and (*c*) the total amount of heat transferred to the turkey in the oven. Will the reading of the thermometer be more or less than 185°F 5 min after the turkey is taken out of the oven?

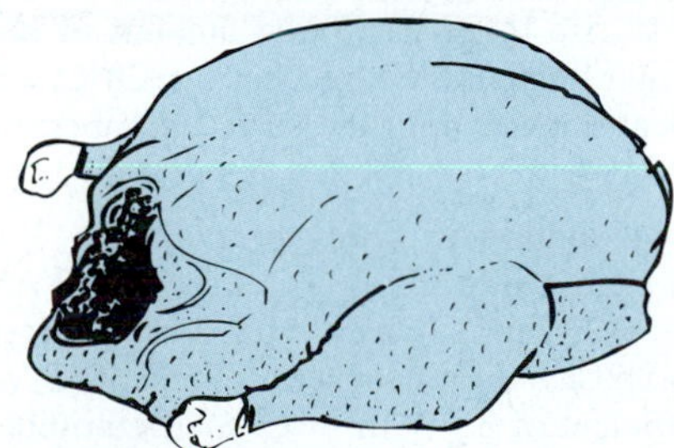

FIGURE P18–96E

18–97 During a fire, the trunks of some dry oak trees (k = 0.17 W/m · °C and α = 1.28 × 10^{-7} m^2/s) that are initially at a uniform temperature of 30°C are exposed to hot gases at 520°C for a period of 5 h, with a heat transfer coefficient of 65 W/m^2 · °C on the surface. The ignition temperature of the trees is 410°C. Treating the trunks of the trees as long cylindrical rods of diameter 20 cm, determine if these dry trees will ignite as the fire sweeps through them.

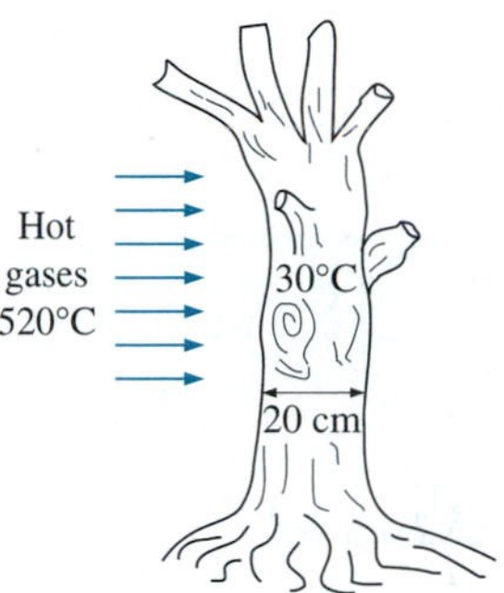

FIGURE P18–97

18–98 We often cut a watermelon in half and put it into the freezer to cool it quickly. But usually we forget to check on it and end up having a watermelon with a frozen layer on the top. To avoid this potential problem a person wants to set the timer such that it will go off when the temperature of the exposed surface of the watermelon drops to 3°C.

Consider a 25-cm-diameter spherical watermelon that is cut into two equal parts and put into a freezer at −12°C. Initially, the entire watermelon is at a uniform temperature of 25°C, and the heat transfer coefficient on the surfaces is 22 W/m^2 · °C. Assuming the watermelon to have the properties of water, determine how long it will take for the center of the exposed cut surfaces of the watermelon to drop to 3°C.

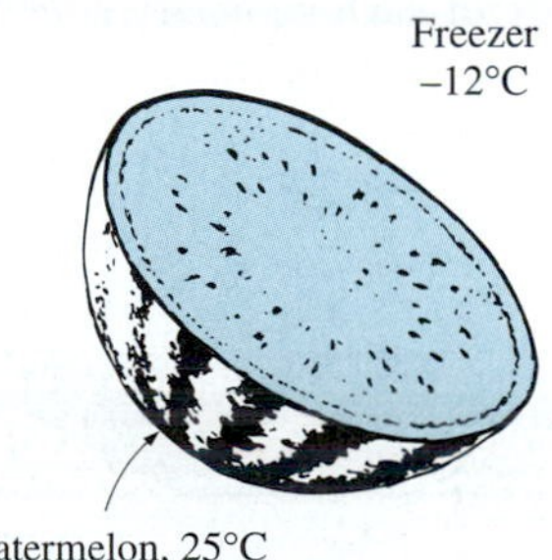

FIGURE P18–98

18–99 The thermal conductivity of a solid whose density and specific heat are known can be determined from the relation $k = \alpha/\rho c_p$ after evaluating the thermal diffusivity α.

Consider a 2-cm-diameter cylindrical rod made of a sample material whose density and specific heat are 3700 kg/m^3

and 920 J/kg · °C, respectively. The sample is initially at a uniform temperature of 25°C. In order to measure the temperatures of the sample at its surface and its center, a thermocouple is inserted to the center of the sample along the centerline, and another thermocouple is welded into a small hole drilled on the surface. The sample is dropped into boiling water at 100°C. After 3 min, the surface and the center temperatures are recorded to be 93°C and 75°C, respectively. Determine the thermal diffusivity and the thermal conductivity of the material.

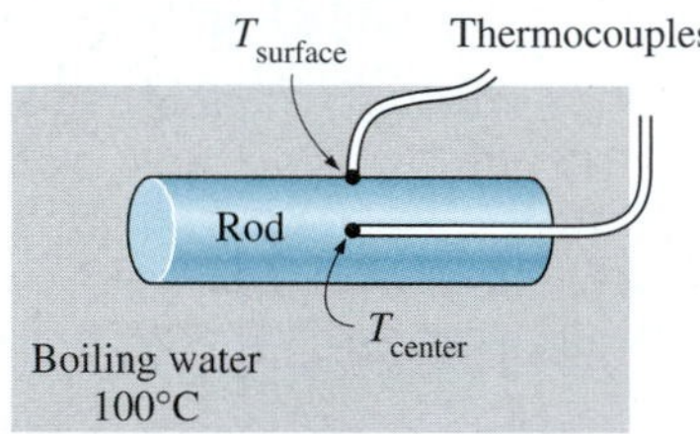

FIGURE P18–99

18–100 In desert climates, rainfall is not a common occurrence since the rain droplets formed in the upper layer of the atmosphere often evaporate before they reach the ground. Consider a raindrop that is initially at a temperature of 5°C and has a diameter of 5 mm. Determine how long it will take for the diameter of the raindrop to reduce to 3 mm as it falls through ambient air at 18°C with a heat transfer coefficient of 400 W/m^2 · °C. The water temperature can be assumed to remain constant and uniform at 5°C at all times.

18–101E Consider a plate of thickness 1 in, a long cylinder of diameter 1 in, and a sphere of diameter 1 in, all initially at 400°F and all made of bronze (k = 15.0 Btu/h · ft · °F and α = 0.333 ft^2/h). Now all three of these geometries are exposed to cool air at 75°F on all of their surfaces, with a heat transfer coefficient of 7 Btu/h · ft^2 · °F. Determine the center temperature of each geometry after 5, 10, and 30 min. Explain why the center temperature of the sphere is always the lowest.

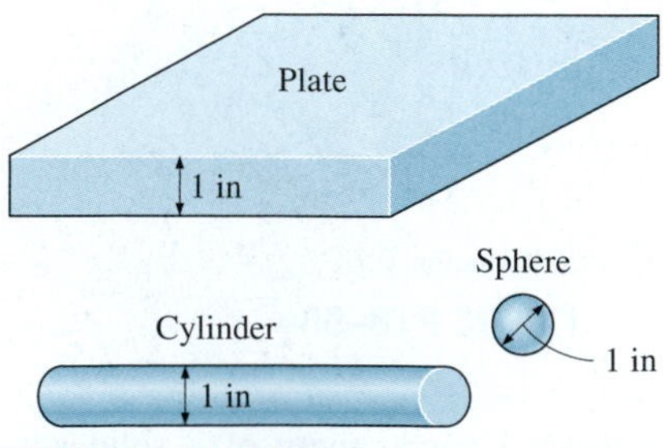

FIGURE P18–101E

18–102E Repeat Prob. 18–101E for cast iron geometries (k = 29 Btu/h · ft · °F and α = 0.61 ft^2/h).

18–103E Reconsider Prob. 18–101E. Using EES (or other) software, plot the center temperature of each geometry as a function of the cooling time as the time varies from 5 min to 60 min, and discuss the results.

18–104 Engine valves (k = 48 W/m · °C, c_p = 440 J/kg · °C, and ρ = 7840 kg/m^3) are heated to 800°C in the heat treatment section of a valve manufacturing facility. The valves are then quenched in a large oil bath at an average temperature of 50°C. The heat transfer coefficient in the oil bath is 800 W/m^2 · °C. The valves have a cylindrical stem with a diameter of 8 mm and a length of 10 cm. The valve head and the stem may be assumed to be of equal surface area, and the volume of the valve head can be taken to be 80 percent of the volume of steam. Determine how long it will take for the valve temperature to drop to (*a*) 400°C, (*b*) 200°C, and (*c*) 51°C, and (*d*) the maximum heat transfer from a single valve.

18–105 A watermelon initially at 35°C is to be cooled by dropping it into a lake at 15°C. After 4 h and 40 min of cooling, the center temperature of watermelon is measured to be 20°C. Treating the watermelon as a 20-cm-diameter sphere and using the properties k = 0.618 W/m · °C, α = 0.15 × 10^{-6} m^2/s, ρ = 995 kg/m^3, and c_p = 4.18 kJ/kg · °C, determine the average heat transfer coefficient and the surface temperature of watermelon at the end of the cooling period.

18–106 10-cm-thick large food slabs tightly wrapped by thin paper are to be cooled in a refrigeration room maintained at 0°C. The heat transfer coefficient on the box surfaces is 25 W/m^2 · °C and the boxes are to be kept in the refrigeration room for a period of 6 h. If the initial temperature of the boxes is 30°C determine the center temperature of the boxes if the boxes contain (*a*) margarine (k = 0.233 W/m · °C and α = 0.11 × 10^{-6} m^2/s), (*b*) white cake (k = 0.082 W/m · °C and α = 0.10 × 10^{-6} m^2/s), and (*c*) chocolate cake (k = 0.106 W/m · °C and α = 0.12 × 10^{-6} m^2/s).

18–107 A 30-cm-diameter, 4-m-high cylindrical column of a house made of concrete (k = 0.79 W/m · °C, α = 5.94 × 10^{-7} m^2/s, ρ = 1600 kg/m^3, and c_p = 0.84 kJ/kg · °C) cooled to 14°C during a cold night is heated again during the day by being exposed to ambient air at an average temperature of 28°C with an average heat transfer coefficient of 14 W/m^2 · °C. Determine (*a*) how long it will take for the column surface temperature to rise to 27°C, (*b*) the amount of heat transfer until the center temperature reaches to 28°C, and (*c*) the amount of heat transfer until the surface temperature reaches to 27°C.

18–108 Long aluminum wires of diameter 3 mm (ρ = 2702 kg/m^3, c_p = 0.896 kJ/kg · °C, k = 236 W/m · °C, and α = 9.75 × 10^{-5} m^2/s) are extruded at a temperature of 350°C and exposed to atmospheric air at 30°C with a heat transfer coefficient of 35 W/m^2 · °C. (*a*) Determine how long

it will take for the wire temperature to drop to 50°C. (*b*) If the wire is extruded at a velocity of 10 m/min, determine how far the wire travels after extrusion by the time its temperature drops to 50°C. What change in the cooling process would you propose to shorten this distance? (*c*) Assuming the aluminum wire leaves the extrusion room at 50°C, determine the rate of heat transfer from the wire to the extrusion room.
Answers: (*a*) 144 s, (*b*) 24 m, (*c*) 856 W

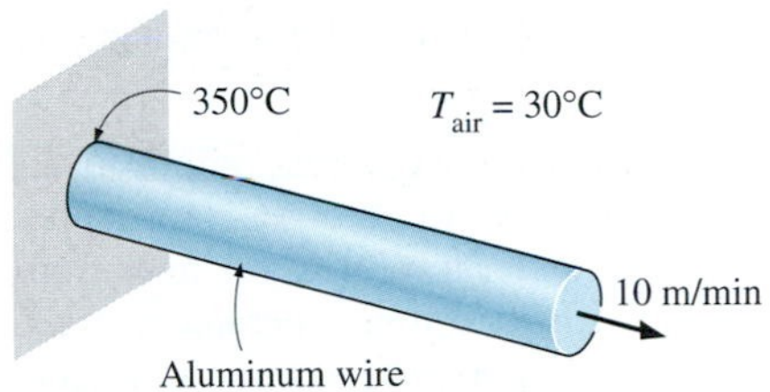

FIGURE P18–108

18–109 Repeat Prob. 18–108 for a copper wire (ρ = 8950 kg/m^3, c_p = 0.383 kJ/kg · °C, k = 386 W/m · °C, and $\alpha = 1.13 \times 10^{-4}$ m^2/s).

18–110 Consider a brick house (k = 0.72 W/m · °C and $\alpha = 0.45 \times 10^{-6}$ m^2/s) whose walls are 10 m long, 3 m high, and 0.3 m thick. The heater of the house broke down one night, and the entire house, including its walls, was observed to be 5°C throughout in the morning. The outdoors warmed up as the day progressed, but no change was felt in the house, which was tightly sealed. Assuming the outer surface temperature of the house to remain constant at 15°C, determine how long it would take for the temperature of the inner surfaces of the walls to rise to 5.1°C.

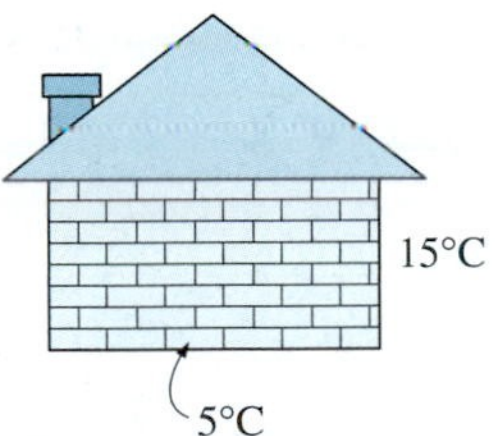

FIGURE P18–110

18–111 A 40-cm-thick brick wall (k = 0.72 W/m · °C, and $\alpha = 1.6 \times 10^{-7}$ m^2/s) is heated to an average temperature of 18°C by the heating system and the solar radiation incident on it during the day. During the night, the outer surface of the wall is exposed to cold air at −3°C with an average heat transfer coefficient of 20 W/m^2 · °C, determine the wall temperatures at distances 15, 30, and 40 cm from the outer surface for a period of 2 h.

18–112 Consider the engine block of a car made of cast iron (k = 52 W/m · °C and $\alpha = 1.7 \times 10^{-5}$ m^2/s). The engine can be considered to be a rectangular block whose sides are 80 cm, 40 cm, and 40 cm. The engine is at a temperature of 150°C when it is turned off. The engine is then exposed to atmospheric air at 17°C with a heat transfer coefficient of 6 W/m^2 · °C. Determine (*a*) the center temperature of the top surface whose sides are 80 cm and 40 cm and (*b*) the corner temperature after 45 min of cooling.

18–113 A man is found dead in a room at 16°C. The surface temperature on his waist is measured to be 23°C and the heat transfer coefficient is estimated to be 9 W/m^2 · °C. Modeling the body as 28-cm diameter, 1.80-m-long cylinder, estimate how long it has been since he died. Take the properties of the body to be k = 0.62 W/m · °C and $\alpha = 0.15 \times 10^{-6}$ m^2/s, and assume the initial temperature of the body to be 36°C.

18–114 An exothermic process occurs uniformly throughout a 10-cm-diameter sphere (k = 300 W/m · K, c_p = 400 J/kg · K, ρ = 7500 kg/m^3), and it generates heat at a constant rate of 1.2 MW/m^3. The sphere initially is at a uniform temperature of 20°C, and the exothermic process is commenced at time t = 0. To keep the sphere temperature under control, it is submerged in a liquid bath maintained at 20°C. The heat transfer coefficient at the sphere surface is 250 W/m^2 · K.

Due to the high thermal conductivity of sphere, the conductive resistance within the sphere can be neglected in comparison to the convective resistance at its surface. Accordingly, this unsteady state heat transfer situation could be analyzed as a lumped system.

(*a*) Show that the variation of sphere temperature T with time t can be expressed as $dT/dt = 0.5 - 0.005T$.
(*b*) Predict the steady-state temperature of the sphere.
(*c*) Calculate the time needed for the sphere temperature to reach the average of its initial and final (steady) temperatures.

18–115 Large steel plates 1.0-cm in thickness are quenched from 600°C to 100°C by submerging them in an oil reservoir held at 30°C. The average heat transfer coefficient for both faces of steel plates is 400 W/m^2 · K. Average steel properties are k = 45 W/m · K, ρ = 7800 kg/m^3, and c_p = 470 J/kg · K. Calculate the quench time for steel plates.

18–116 Aluminium wires, 3 mm in diameter, are produced by extrusion. The wires leave the extruder at an average temperature of 350°C and at a linear rate of 10 m/min. Before leaving the extrusion room, the wires are cooled to an average temperature of 50°C by transferring heat to the surrounding air at 25°C with a heat transfer coefficient of 50 W/m^2 · K. Calculate the necessary length of the wire cooling section in the extrusion room.

Design and Essay Problems

18–117 Conduct the following experiment at home to determine the combined convection and radiation heat transfer coefficient at the surface of an apple exposed to the room air. You will need two thermometers and a clock.

First, weigh the apple and measure its diameter. You can measure its volume by placing it in a large measuring cup halfway filled with water, and measuring the change in volume when it is completely immersed in the water. Refrigerate the apple overnight so that it is at a uniform temperature in the morning and measure the air temperature in the kitchen. Then take the apple out and stick one of the thermometers to its middle and the other just under the skin. Record both temperatures every 5 min for an hour. Using these two temperatures, calculate the heat transfer coefficient for each interval and take their average. The result is the combined convection and radiation heat transfer coefficient for this heat transfer process. Using your experimental data, also calculate the thermal conductivity and thermal diffusivity of the apple and compare them to the values given above.

18–118 Repeat Prob. 18–117 using a banana instead of an apple. The thermal properties of bananas are practically the same as those of apples.

18–119 Conduct the following experiment to determine the time constant for a can of soda and then predict the temperature of the soda at different times. Leave the soda in the refrigerator overnight. Measure the air temperature in the kitchen and the temperature of the soda while it is still in the refrigerator by taping the sensor of the thermometer to the outer surface of the can. Then take the soda out and measure its temperature again in 5 min. Using these values, calculate the exponent b. Using this b-value, predict the temperatures of the soda in 10, 15, 20, 30, and 60 min and compare the results with the actual temperature measurements. Do you think the lumped system analysis is valid in this case?

18–120 Citrus trees are very susceptible to cold weather, and extended exposure to subfreezing temperatures can destroy the crop. In order to protect the trees from occasional cold fronts with subfreezing temperatures, tree growers in Florida usually install water sprinklers on the trees. When the temperature drops below a certain level, the sprinklers spray water on the trees and their fruits to protect them against the damage the subfreezing temperatures can cause. Explain the basic mechanism behind this protection measure and write an essay on how the system works in practice.

Chapter 19

FORCED CONVECTION

So far, we have considered *conduction*, which is the mechanism of heat transfer through a solid or a quiescent fluid. We now consider *convection*, which is the mechanism of heat transfer through a fluid in the presence of bulk fluid motion.

Convection is classified as *natural* (or *free*) and *forced convection*, depending on how the fluid motion is initiated. In forced convection, the fluid is forced to flow over a surface or in a pipe by external means such as a pump or a fan. In natural convection, any fluid motion is caused by natural means such as the buoyancy effect, which manifests itself as the rise of warmer fluid and the fall of the cooler fluid. Convection is also classified as *external* and *internal*, depending on whether the fluid is forced to flow over a surface or in a pipe.

We start this chapter with a general physical description of the *convection* mechanism and *thermal boundary* layer. We continue with the discussion of the dimensionless *Prandtl* and *Nusselt numbers*, and their physical significance. We then present empirical relations for the *heat transfer coefficients* for flow over various geometries such as a flat plate, cylinder, and sphere, for both laminar and turbulent flow conditions. Finally, we discuss the characteristics of flow inside tubes and present the heat transfer correlations associated with it. The relevant concepts from Chaps. 14 and 15 should be reviewed before this chapter is studied.

Objectives

The objectives of this chapter are to:

- Understand the physical mechanism of convection and its classification,
- Visualize the development of thermal boundary layer during flow over surfaces,
- Gain a working knowledge of the dimensionless Prandtl and Nusselt numbers,
- Develop an understanding of the mechanism of heat transfer in turbulent flow,
- Evaluate the heat transfer associated with flow over a flat plate for both laminar and turbulent flow, and flow over cylinders and spheres,
- Have a visual understanding of different flow regions in internal flow, and calculate hydrodynamic and thermal entry lengths,
- Analyze heating and cooling of a fluid flowing in a tube under constant surface temperature and constant surface heat flux conditions, and work with the logarithmic mean temperature difference, and
- Determine the Nusselt number in fully developed turbulent flow using empirical relations, and calculate the heat transfer rate.

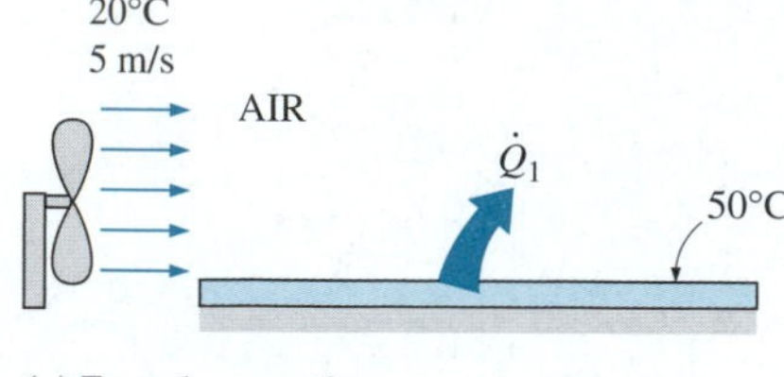

(*a*) Forced convection

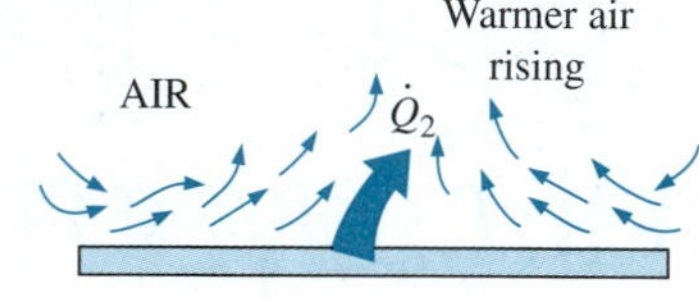

(*b*) Free convection

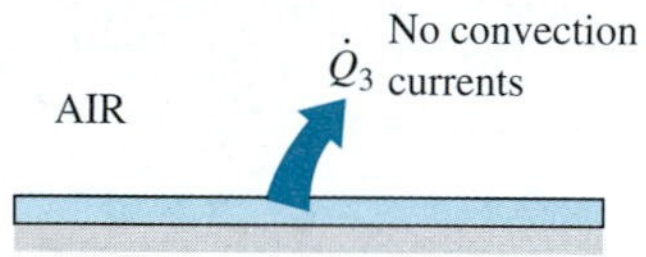

(*c*) Conduction

FIGURE 19–1

Heat transfer from a hot surface to the surrounding fluid by convection and conduction.

19–1 ▪ PHYSICAL MECHANISM OF CONVECTION

We mentioned in Chapter 16 that there are three basic mechanisms of heat transfer: conduction, convection, and radiation. Conduction and convection are similar in that both mechanisms require the presence of a material medium. But they are different in that convection requires the presence of fluid motion.

Heat transfer through a solid is always by conduction, since the molecules of a solid remain at relatively fixed positions. Heat transfer through a liquid or gas, however, can be by conduction or convection, depending on the presence of any bulk fluid motion. Heat transfer through a fluid is by convection in the presence of bulk fluid motion and by conduction in the absence of it. Therefore, conduction in a fluid can be viewed as the limiting case of convection, corresponding to the case of quiescent fluid (Fig. 19–1).

Convection heat transfer is complicated by the fact that it involves fluid motion as well as heat conduction. The fluid motion enhances heat transfer, since it brings warmer and cooler chunks of fluid into contact, initiating higher rates of conduction at a greater number of sites in a fluid. Therefore, the rate of heat transfer through a fluid is much higher by convection than it is by conduction. In fact, the higher the fluid velocity, the higher the rate of heat transfer.

To clarify this point further, consider steady heat transfer through a fluid contained between two parallel plates maintained at different temperatures, as shown in Figure 19–2. The temperatures of the fluid and the plate are the same at the points of contact because of the continuity of temperature. Assuming no fluid motion, the energy of the hotter fluid molecules near the hot plate is transferred to the adjacent cooler fluid molecules. This energy is then transferred to the next layer of the cooler fluid molecules. This energy is then transferred to the next layer of the cooler fluid, and so on, until it is finally transferred to the other plate. This is what happens during conduction through a fluid. Now let us use a syringe to draw some fluid near the hot plate and inject it next to the cold plate repeatedly. You can imagine that this will speed up the heat transfer process considerably, since some energy is carried to the other side as a result of fluid motion.

Consider the cooling of a hot block with a fan blowing air over its top surface. We know that heat is transferred from the hot block to the surrounding cooler air, and the block eventually cools. We also know that the block cools faster if the fan is switched to a higher speed. Replacing air by water enhances the convection heat transfer even more.

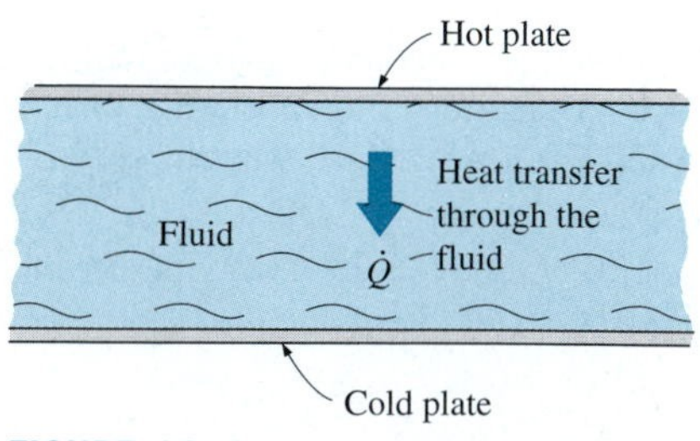

FIGURE 19–2

Heat transfer through a fluid sandwiched between two parallel plates.

Experience shows that convection heat transfer strongly depends on the fluid properties *dynamic viscosity* μ, *thermal conductivity* k, *density* ρ, and *specific heat* c_p, as well as the *fluid velocity* V. It also depends on the *geometry* and the *roughness* of the solid surface, in addition to the *type of fluid flow* (such as being streamlined or turbulent). Thus, we expect the convection heat transfer relations to be rather complex because of the dependence of convection on so many variables. This is not surprising, since convection is the most complex mechanism of heat transfer.

Despite the complexity of convection, the rate of convection heat transfer is observed to be proportional to the temperature difference and is conveniently expressed by **Newton's law of cooling** as

$$\dot{q}_{\text{conv}} = h(T_s - T_\infty) \qquad (\text{W/m}^2) \tag{19–1}$$

or

$$\dot{Q}_{\text{conv}} = hA_s(T_s - T_\infty) \qquad \text{(W)} \tag{19–2}$$

where

h = convection heat transfer coefficient, W/m$^2 \cdot$ °C
A_s = heat transfer surface area, m^2
T_s = temperature of the surface, °C
T_∞ = temperature of the fluid sufficiently far from the surface, °C

Judging from its units, the **convection heat transfer coefficient** h can be defined as *the rate of heat transfer between a solid surface and a fluid per unit surface area per unit temperature difference.*

You should not be deceived by the simple appearance of this relation, because the convection heat transfer coefficient h depends on the several of the mentioned variables, and thus is difficult to determine.

Fluid flow is often confined by solid surfaces, and it is important to understand how the presence of solid surfaces affects fluid flow. Consider the flow of a fluid in a stationary pipe or over a solid surface that is nonporous (i.e., impermeable to the fluid). All experimental observations indicate that a fluid in motion comes to a complete stop at the surface and assumes a zero velocity relative to the surface. That is, a fluid in direct contact with a solid "sticks" to the surface due to viscous effects, and there is no slip. This is known as the **no-slip condition**.

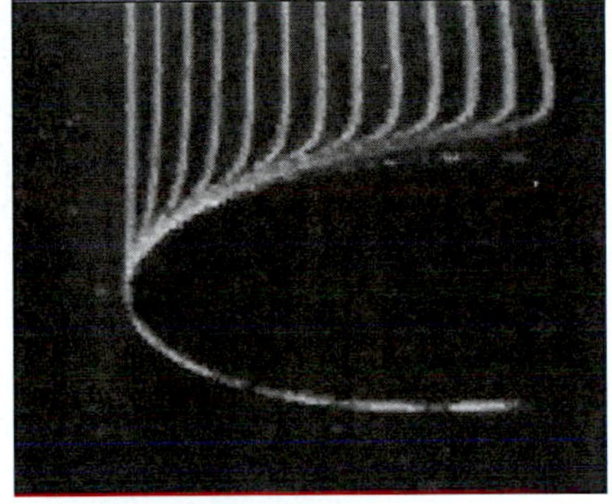

FIGURE 19–3
The development of a velocity profile due to the no-slip condition as a fluid flows over a blunt nose.
"Hunter Rouse: Laminar and Turbulent Flow Film." Copyright IIHR-Hydroscience & Engineering, The University of Iowa. Used by permission.

The photo in Fig. 19–3 obtained from a video clip clearly shows the evolution of a velocity gradient as a result of the fluid sticking to the surface of a blunt nose. The layer that sticks to the surface slows the adjacent fluid layer because of viscous forces between the fluid layers, which slows the next layer, and so on. Therefore, the no-slip condition is responsible for the development of the velocity profile. The flow region adjacent to the wall in which the viscous effects (and thus the velocity gradients) are significant is called the **boundary layer**. The fluid property responsible for the no-slip condition and the development of the boundary layer is *viscosity.*

A fluid layer adjacent to a moving surface has the same velocity as the surface. A consequence of the no-slip condition is that all velocity profiles must have zero values with respect to the surface at the points of contact between a fluid and a solid surface (Fig. 19–4). Another consequence of the no-slip condition is the *surface drag*, which is the force a fluid exerts on a surface in the flow direction.

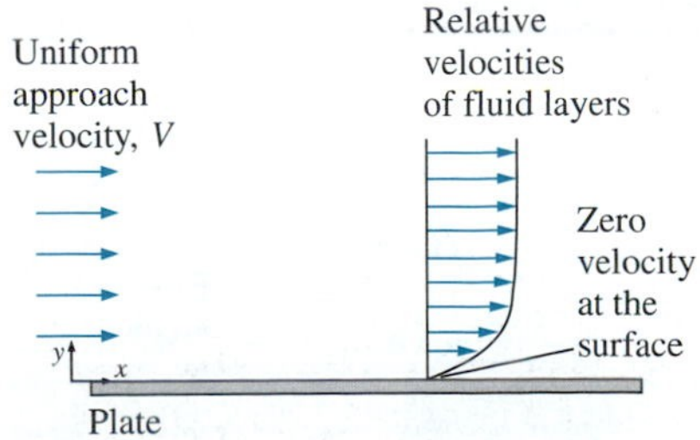

FIGURE 19–4
A fluid flowing over a stationary surface comes to a complete stop at the surface because of the no-slip condition.

An implication of the no-slip condition is that heat transfer from the solid surface to the fluid layer adjacent to the surface is by *pure conduction,* since the fluid layer is motionless, and can be expressed as

$$\dot{q}_{\text{conv}} = \dot{q}_{\text{cond}} = -k_{\text{fluid}} \left.\frac{\partial T}{\partial y}\right|_{y=0} \qquad \text{(W/m}^2\text{)} \tag{19–3}$$

where T represents the temperature distribution in the fluid and $(\partial T/\partial y)_{y=0}$ is the *temperature gradient* at the surface. Heat is then *convected away* from the surface as a result of fluid motion. Note that convection heat transfer from a solid surface to a fluid is merely the conduction heat transfer from the solid surface to the fluid layer adjacent to the surface. Therefore, we can equate Eqs. 19–1 and 19–3 for the heat flux to obtain

$$h = \frac{-k_{\text{fluid}}(\partial T/\partial y)_{y=0}}{T_s - T_\infty} \qquad (\text{W/m}^2 \cdot {}^\circ\text{C}) \tag{19–4}$$

for the determination of the *convection heat transfer coefficient* when the temperature distribution within the fluid is known.

The convection heat transfer coefficient, in general, varies along the flow (or x-) direction. The *average* or *mean* convection heat transfer coefficient for a surface in such cases is determined by properly averaging the *local* convection heat transfer coefficients over the entire surface area A_s or length L as:

$$h = \frac{1}{A_s}\int_{A_s} h_{\text{local}} dA_s \quad \text{and} \quad h = \frac{1}{L}\int_0^L h_x dx \tag{19–5}$$

Nusselt Number

In convection studies, it is common practice to nondimensionalize the governing equations and combine the variables, which group together into *dimensionless numbers* in order to reduce the number of total variables. It is also common practice to nondimensionalize the heat transfer coefficient h with the Nusselt number, defined as

$$\text{Nu} = \frac{hL_c}{k} \tag{19–6}$$

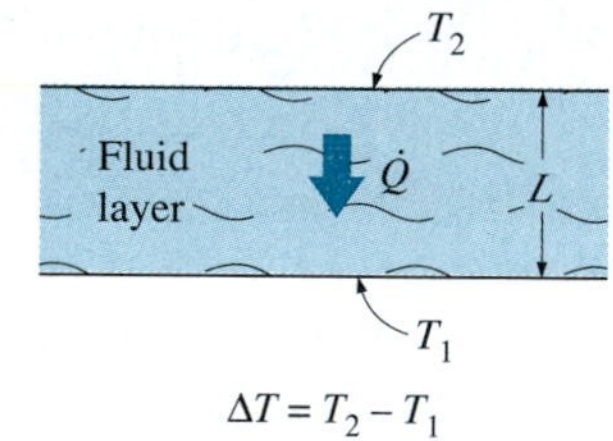

FIGURE 19–5
Heat transfer through a fluid layer of thickness L and temperature difference ΔT.

where k is the thermal conductivity of the fluid and L_c is the *characteristic length.* The Nusselt number is named after Wilhelm Nusselt, who made significant contributions to convective heat transfer in the first half of the twentieth century, and it is viewed as the *dimensionless convection heat transfer coefficient.*

To understand the physical significance of the Nusselt number, consider a fluid layer of thickness L and temperature difference $\Delta T = T_2 - T_1$, as shown in Fig. 19–5. Heat transfer through the fluid layer is by *convection* when the fluid involves some motion and by *conduction* when the fluid layer is motionless. Heat flux (the rate of heat transfer per unit surface area) in either case is

$$\dot{q}_{\text{conv}} = h\Delta T \tag{19–7}$$

and

$$\dot{q}_{\text{cond}} = k\frac{\Delta T}{L} \tag{19–8}$$

Taking their ratio gives

$$\frac{\dot{q}_{\text{conv}}}{\dot{q}_{\text{cond}}} = \frac{h\Delta T}{k\Delta T/L} = \frac{hL}{k} = \text{Nu} \tag{19–9}$$

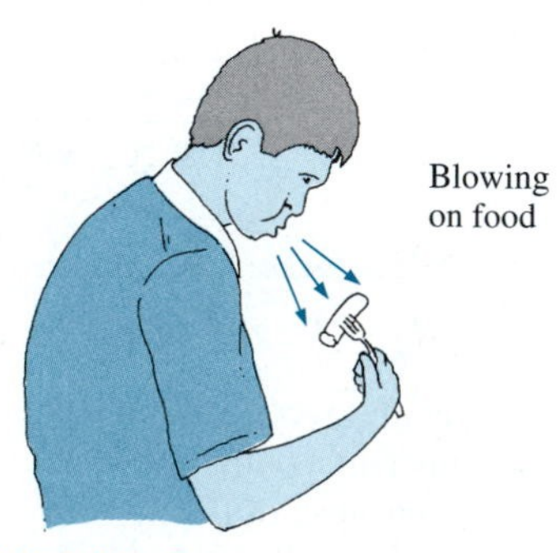

FIGURE 19–6
We resort to forced convection whenever we need to increase the rate of heat transfer.

which is the Nusselt number. Therefore, the Nusselt number represents the enhancement of heat transfer through a fluid layer as a result of convection relative to conduction across the same fluid layer. The larger the Nusselt number, the more effective the convection. A Nusselt number of Nu = 1 for a fluid layer represents heat transfer across the layer by pure conduction.

We use forced convection in daily life more often than you might think (Fig. 19–6). We resort to forced convection whenever we want to increase the rate of heat transfer from a hot object. For example, we turn on the fan on hot summer days to help our body cool more effectively. The higher the fan speed, the better we feel. We *stir* our soup and *blow* on a hot slice of pizza to make them cool faster. The air on windy winter days feels much colder than it actually is. The simplest solution to heating problems in electronics packaging is to use a large enough fan.

19–2 ▪ THERMAL BOUNDARY LAYER

We have seen in Chapter 15 that a velocity boundary layer develops when a fluid flows over a surface as a result of the fluid layer adjacent to the surface assuming the surface velocity (i.e., zero velocity relative to the surface). Also, we defined the velocity boundary layer as the region in which the fluid velocity varies from zero to $0.99V$. Likewise, a *thermal boundary layer* develops when a fluid at a specified temperature flows over a surface that is at a different temperature, as shown in Fig. 19–7.

Consider the flow of a fluid at a uniform temperature of T_∞ over an isothermal flat plate at temperature T_s. The fluid particles in the layer adjacent to the surface reach thermal equilibrium with the plate and assume the surface temperature T_s. These fluid particles then exchange energy with the particles in the adjoining-fluid layer, and so on. As a result, a temperature profile develops in the flow field that ranges from T_s at the surface to T_∞ sufficiently far from the surface. The flow region over the surface in which the temperature variation in the direction normal to the surface is significant is the **thermal boundary layer**. The *thickness* of the thermal boundary layer δ_t at any location along the surface is defined as *the distance from the surface at which the temperature difference* $T - T_s$ *equals* $0.99(T_\infty - T_s)$. Note that for the special case of $T_s = 0$, we have $T = 0.99T_\infty$ at the outer edge of the thermal boundary layer, which is analogous to $u = 0.99V$ for the velocity boundary layer.

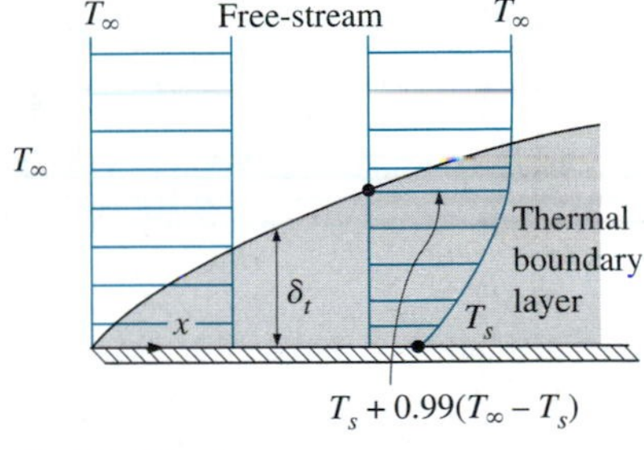

FIGURE 19–7
Thermal boundary layer on a flat plate (the fluid is hotter than the plate surface).

The thickness of the thermal boundary layer increases in the flow direction, since the effects of heat transfer are felt at greater distances from the surface further down stream.

The convection heat transfer rate anywhere along the surface is directly related to the temperature gradient at that location. Therefore, the shape of the temperature profile in the thermal boundary layer dictates the convection heat transfer between a solid surface and the fluid flowing over it. In flow over a heated (or cooled) surface, both velocity and thermal boundary layers develop simultaneously. Noting that the fluid velocity has a strong influence on the temperature profile, the development of the velocity boundary layer relative to the thermal boundary layer will have a strong effect on the convection heat transfer.

Prandtl Number

The relative thickness of the velocity and the thermal boundary layers is best described by the *dimensionless* parameter **Prandtl number**, defined as

$$\text{Pr} = \frac{\text{Molecular diffusivity of momentum}}{\text{Molecular diffusivity of heat}} = \frac{\nu}{\alpha} = \frac{\mu c_p}{k} \tag{19–10}$$

TABLE 19–1

Typical ranges of Prandtl numbers for common fluids

Fluid	Pr
Liquid metals	0.004–0.030
Gases	0.7–1.0
Water	1.7–13.7
Light organic fluids	5–50
Oils	50–100,000
Glycerin	2000–100,000

It is named after Ludwig Prandtl, who introduced the concept of boundary layer in 1904 and made significant contributions to boundary layer theory. The Prandtl numbers of fluids range from less than 0.01 for liquid metals to more than 100,000 for heavy oils (Table 19–1). Note that the Prandtl number is in the order of 10 for water.

The Prandtl numbers of gases are about 1, which indicates that both momentum and heat dissipate through the fluid at about the same rate. Heat diffuses very quickly in liquid metals ($\text{Pr} \ll 1$) and very slowly in oils ($\text{Pr} \gg 1$) relative to momentum. Consequently the thermal boundary layer is much thicker for liquid metals and much thinner for oils relative to the velocity boundary layer.

19–3 ▪ PARALLEL FLOW OVER FLAT PLATES

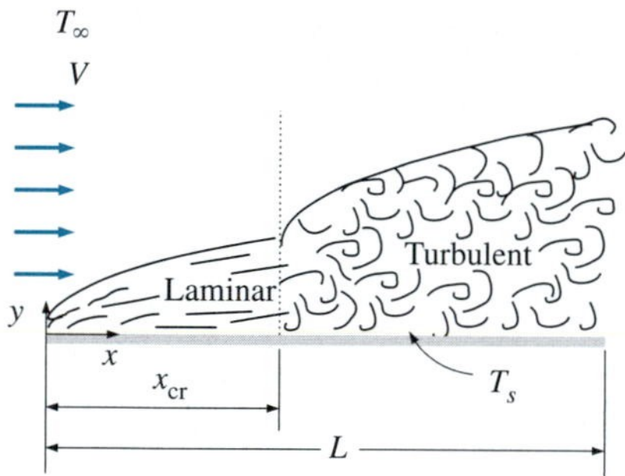

FIGURE 19–8
Laminar and turbulent regions of the boundary layer during flow over a flat plate.

Consider the parallel flow of a fluid over a flat plate of length L in the flow direction, as shown in Fig. 19–8. The x-coordinate is measured along the plate surface from the leading edge in the direction of the flow. The fluid approaches the plate in the x-direction with a uniform velocity V and temperature T_∞. The flow in the velocity boundary layers starts out as laminar, but if the plate is sufficiently long, the flow becomes turbulent at a distance x_{cr} from the leading edge where the Reynolds number reaches its critical value for transition.

The transition from laminar to turbulent flow depends on the *surface geometry, surface roughness, upstream velocity, surface temperature,* and the *type of fluid,* among other things, and is best characterized by the Reynolds number. The Reynolds number at a distance x from the leading edge of a flat plate is expressed as

$$\text{Re}_x = \frac{\rho V x}{\mu} = \frac{Vx}{\nu} \tag{19–11}$$

Note that the value of the Reynolds number varies for a flat plate along the flow, reaching $\text{Re}_L = VL/\nu$ at the end of the plate.

For flow over a flat plate, transition from laminar to turbulent begins at about $\text{Re} \cong 1 \times 10^5$, but does not become fully turbulent before the Reynolds number reaches much higher values, typically around 3×10^6. In engineering analysis, a generally accepted value for the critical Reynold number is

$$\text{Re}_{cr} = \frac{\rho V x_{cr}}{\mu} = 5 \times 10^5 \tag{19–12}$$

The actual value of the engineering critical Reynolds number for a flat plate may vary somewhat from 10^5 to 3×10^6, depending on the surface roughness, the turbulence level, and the variation of pressure along the surface.

The local Nusselt number at a location x for laminar flow over a flat plate may be obtained by solving the differential energy equation to be

$$\textit{Laminar:} \qquad \text{Nu}_x = \frac{h_x x}{k} = 0.332\, \text{Re}_x^{0.5}\, \text{Pr}^{1/3} \qquad \text{Pr} > 0.6 \tag{19–13}$$

The corresponding relation for turbulent flow is

$$\textit{Turbulent:} \quad \text{Nu}_x = \frac{h_x x}{k} = 0.0296\ \text{Re}_x^{0.8}\ \text{Pr}^{1/3} \qquad \begin{array}{c} 0.6 \le \text{Pr} \le 60 \\ 5 \times 10^5 \le \text{Re}_x \le 10^7 \end{array} \tag{19–14}$$

Note that h_x is proportional to $\text{Re}_x^{0.5}$ and thus to $x^{-0.5}$ for laminar flow. Therefore, h_x is *infinite* at the leading edge ($x = 0$) and decreases by a factor of $x^{-0.5}$ in the flow direction. The variation of the boundary layer thickness δ and the friction and heat transfer coefficients along an isothermal flat plate are shown in Fig. 19–9. The local friction and heat transfer coefficients are higher in turbulent flow than they are in laminar flow. Also, h_x reaches its highest values when the flow becomes fully turbulent, and then decreases by a factor of $x^{-0.2}$ in the flow direction, as shown in the figure.

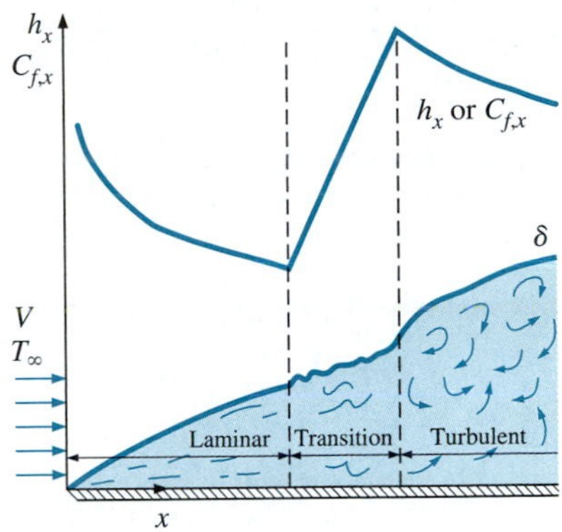

FIGURE 19–9
The variation of the local friction and heat transfer coefficients for flow over a flat plate.

The *average* Nusselt number over the entire plate is determined by substituting the relations above into Eq. 19–5 and performing the integrations. We get

$$\textit{Laminar:} \quad \text{Nu} = \frac{hL}{k} = 0.664\ \text{Re}_L^{0.5}\ \text{Pr}^{1/3} \qquad \text{Re}_L < 5 \times 10^5 \tag{19–15}$$

$$\textit{Turbulent:} \quad \text{Nu} = \frac{hL}{k} = 0.037\ \text{Re}_L^{0.8}\ \text{Pr}^{1/3} \qquad \begin{array}{c} 0.6 \le \text{Pr} \le 60 \\ 5 \times 10^5 \le \text{Re}_L \le 10^7 \end{array} \tag{19–16}$$

The first relation gives the average heat transfer coefficient for the entire plate when the flow is *laminar* over the *entire* plate. The second relation gives the average heat transfer coefficient for the entire plate only when the flow is *turbulent* over the *entire* plate, or when the laminar flow region of the plate is too small relative to the turbulent flow region.

In some cases, a flat plate is sufficiently long for the flow to become turbulent, but not long enough to disregard the laminar flow region. In such cases, the *average* heat transfer coefficient over the entire plate is determined by performing the integration in Eq. 19–5 over two parts as

$$h = \frac{1}{L}\left(\int_0^{x_{\text{cr}}} h_{x,\text{ laminar}}\, dx + \int_{x_{\text{cr}}}^{L} h_{x,\text{ turbulent}}\, dx\right) \tag{19–17}$$

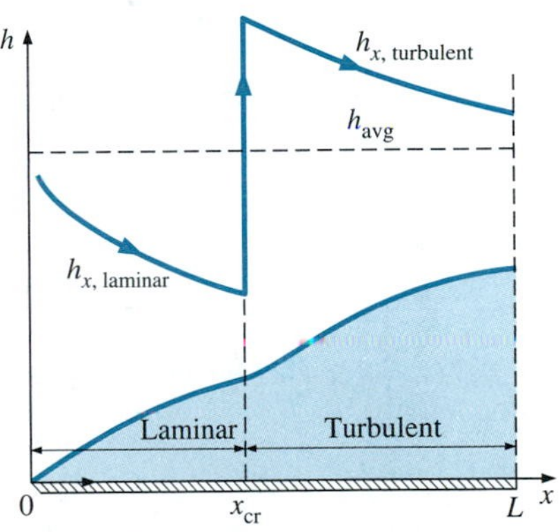

FIGURE 19–10
Graphical representation of the average heat transfer coefficient for a flat plate with combined laminar and turbulent flow.

Again taking the critical Reynolds number to be $\text{Re}_{\text{cr}} = 5 \times 10^5$ and performing the integrations in Eq. 19–17 after substituting the indicated expressions, the *average* Nusselt number over the *entire* plate is determined to be (Fig. 19–10)

$$\text{Nu} = \frac{hL}{k} = (0.037\ \text{Re}_L^{0.8} - 871)\text{Pr}^{1/3} \qquad \begin{array}{c} 0.6 \le \text{Pr} \le 60 \\ 5 \times 10^5 \le \text{Re}_L \le 10^7 \end{array} \tag{19–18}$$

The constants in this relation will be different for different critical Reynolds numbers.

Liquid metals such as mercury have high thermal conductivities, and are commonly used in applications that require high heat transfer rates. However, they have very small Prandtl numbers, and thus the thermal boundary layer develops much faster than the velocity boundary layer. Then we can assume the velocity in the thermal boundary layer to be constant at the free stream value and solve the energy equation. It gives

$$\text{Nu}_x = 0.565(\text{Re}_x\,\text{Pr})^{1/2} \qquad \text{Pr} < 0.05 \tag{19–19}$$

It is desirable to have a single correlation that applies to *all fluids*, including liquid metals. By curve-fitting existing data, Churchill and Ozoe (1973) proposed the following relation which is applicable for *all Prandtl numbers* and is claimed to be accurate to $\pm 1\%$,

$$\text{Nu}_x = \frac{h_x x}{k} = \frac{0.3387\,\text{Pr}^{1/3}\,\text{Re}_x^{1/2}}{[1 + (0.0468/\text{Pr})^{2/3}]^{1/4}} \tag{19–20}$$

These relations have been obtained for the case of *isothermal* surfaces but could also be used approximately for the case of nonisothermal surfaces by assuming the surface temperature to be constant at some average value. Also, the surfaces are assumed to be *smooth,* and the free stream to be *turbulent free*. The effect of variable properties can be accounted for by evaluating all properties at the film temperature.

Flat Plate with Unheated Starting Length

T_∞
V
Thermal boundary layer
Velocity boundary layer
T_s
ξ
x

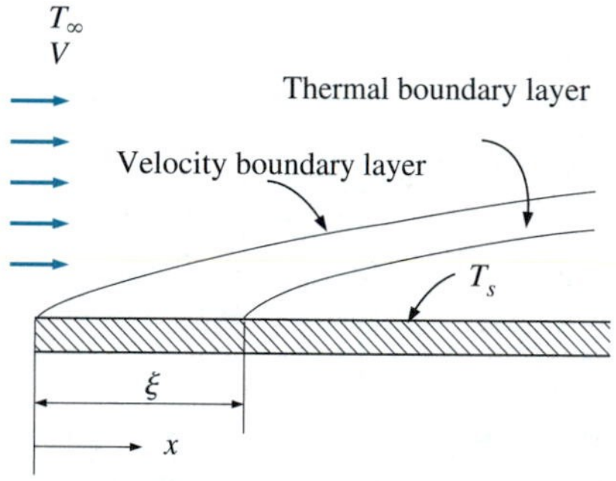

FIGURE 19–11
Flow over a flat plate with an unheated starting length.

So far we have limited our consideration to situations for which the entire plate is heated from the leading edge. But many practical applications involve surfaces with an unheated starting section of length ξ, shown in Fig. 19–11, and thus there is no heat transfer for $0 < x < \xi$. In such cases, the velocity boundary layer starts to develop at the leading edge ($x = 0$), but the thermal boundary layer starts to develop where heating starts ($x = \xi$).

Consider a flat plate whose heated section is maintained at a constant temperature ($T = T_s$ constant for $x > \xi$). Using integral solution methods (see Kays and Crawford, 1994), the local Nusselt numbers for both laminar and turbulent flows are determined to be

Laminar: $$\text{Nu}_x = \frac{\text{Nu}_{x\,(\text{for }\xi=0)}}{[1 - (\xi/x)^{3/4}]^{1/3}} = \frac{0.332\,\text{Re}_x^{0.5}\,\text{Pr}^{1/3}}{[1 - (\xi/x)^{3/4}]^{1/3}} \tag{19–21}$$

Turbulent: $$\text{Nu}_x = \frac{\text{Nu}_{x\,(\text{for }\xi=0)}}{[1 - (\xi/x)^{9/10}]^{1/9}} = \frac{0.0296\,\text{Re}_x^{0.8}\,\text{Pr}^{1/3}}{[1 - (\xi/x)^{9/10}]^{1/9}} \tag{19–22}$$

for $x > \xi$. Note that for $\xi = 0$, these Nu_x relations reduce to $\text{Nu}_{x\,(\text{for }\xi = 0)}$, which is the Nusselt number relation for a flat plate without an unheated starting length. Therefore, the terms in brackets in the denominator serve as correction factors for plates with unheated starting lengths.

The determination of the average Nusselt number for the heated section of a plate requires the integration of the local Nusselt number relations above, which cannot be done analytically. Therefore, integrations must be done numerically. The results of numerical integrations have been correlated for the average convection coefficients (Thomas, 1977) as

Laminar: $$h = \frac{2[1 - (\xi/x)^{3/4}]}{1 - \xi/L}\,h_{x=L} \tag{19–23}$$

Turbulent: $$h = \frac{5[1 - (\xi/x)^{9/10}]}{4(1 - \xi/L)}\,h_{x=L} \tag{19–24}$$

The first relation gives the average convection coefficient for the entire heated section of the plate when the flow is laminar over the entire plate.

Note that for $\xi = 0$ it reduces to $h_L = 2h_{x=L}$, as expected. The second relation gives the average convection coefficient for the case of turbulent flow over the entire plate or when the laminar flow region is small relative to the turbulent region.

Uniform Heat Flux

When a flat plate is subjected to *uniform heat flux* instead of uniform temperature, the local Nusselt number is given by

Laminar: $$\text{Nu}_x = 0.453\,\text{Re}_x^{0.5}\,\text{Pr}^{1/3} \quad (19\text{–}25)$$

Turbulent: $$\text{Nu}_x = 0.0308\,\text{Re}_x^{0.8}\,\text{Pr}^{1/3} \quad (19\text{–}26)$$

These relations give values that are 36 percent higher for laminar flow and 4 percent higher for turbulent flow relative to the isothermal plate case. When the plate involves an unheated starting length, the relations developed for the uniform surface temperature case can still be used provided that Eqs. 19–25 and 19–26 are used for $\text{Nu}_{x(\text{for } \xi = 0)}$ in Eqs. 19–21 and 19–22, respectively.

When heat flux $\dot{q}_s$ is prescribed, the rate of heat transfer to or from the plate and the surface temperature at a distance x are determined from

$$\dot{Q} = \dot{q}_s A_s \quad (19\text{–}27)$$

and

$$\dot{q}_s = h_x[T_s(x) - T_\infty] \quad \rightarrow \quad T_s(x) = T_\infty + \frac{\dot{q}_s}{h_x} \quad (19\text{–}28)$$

where A_s is the heat transfer surface area.

EXAMPLE 19–1 Flow of Hot Oil over a Flat Plate

Engine oil at 60°C flows over the upper surface of a 5-m-long flat plate whose temperature is 20°C with a velocity of 2 m/s (Fig. 19–12). Determine the rate of heat transfer per unit width of the entire plate.

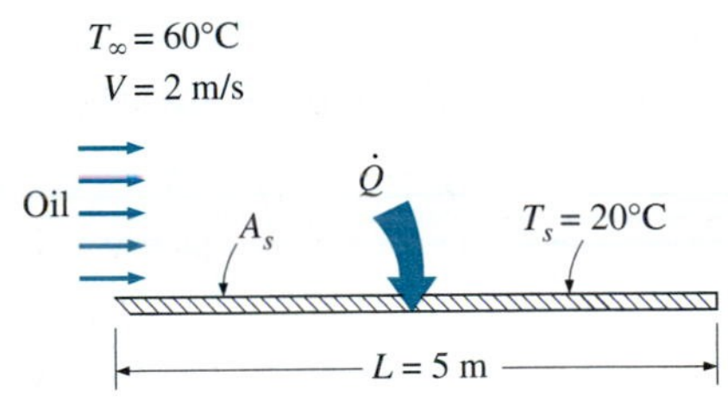

FIGURE 19–12
Schematic for Example 19–1.

Solution Engine oil flows over a flat plate. Rate of heat transfer per unit width of the plate are to be determined.

Assumptions **1** The flow is steady and incompressible. **2** The critical Reynolds number is $\text{Re}_{cr} = 5 \times 10^5$.

Properties The properties of engine oil at the film temperature of $T_f = (T_s + T_\infty)/2 = (20 + 60)/2 = 40°C$ are (Table A–19)

$$\rho = 876 \text{ kg/m}^3 \qquad \text{Pr} = 2962$$
$$k = 0.1444 \text{ W/m}\cdot°\text{C} \qquad \nu = 2.485 \times 10^{-4} \text{ m}^2/\text{s}$$

Analysis Noting that $L = 5$ m, the Reynolds number at the end of the plate is

$$\text{Re}_L = \frac{VL}{\nu} = \frac{(2 \text{ m/s})(5 \text{ m})}{2.485 \times 10^{-4} \text{ m}^2/\text{s}} = 4.024 \times 10^4$$

which is less than the critical Reynolds number. Thus we have *laminar flow* over the entire plate. The Nusselt number is

$$\text{Nu} = \frac{hL}{k} = 0.664\ \text{Re}_L^{0.5}\ \text{Pr}^{1/3} = 0.664 \times (4.024 \times 10^4)^{0.5} \times 2962^{1/3} = 1913$$

Then,

$$h = \frac{k}{L}\text{Nu} = \frac{0.1444\ \text{W/m}\cdot{}^\circ\text{C}}{5\ \text{m}}(1913) = 55.25\ \text{W/m}^2\cdot{}^\circ\text{C}$$

and

$$\dot{Q} = hA_s(T_\infty - T_s) = (55.25\ \text{W/m}^2\cdot{}^\circ\text{C})(5 \times 1\ \text{m}^2)(60 - 20)^\circ\text{C} = \mathbf{11{,}050\ W}$$

Discussion Note that heat transfer is always from the higher-temperature medium to the lower-temperature one. In this case, it is from the oil to the plate. The heat transfer rate is per m width of the plate. The heat transfer for the entire plate can be obtained by multiplying the value obtained by the actual width of the plate.

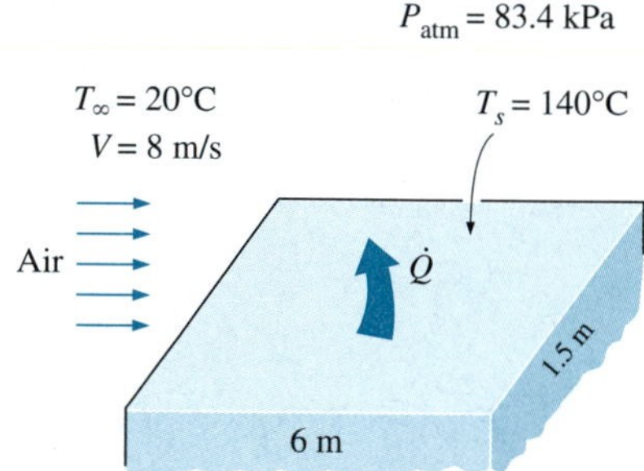

FIGURE 19–13
Schematic for Example 19–2.

EXAMPLE 19–2 Cooling of a Hot Block by Forced Air at High Elevation

The local atmospheric pressure in Denver, Colorado (elevation 1610 m), is 83.4 kPa. Air at this pressure and 20°C flows with a velocity of 8 m/s over a 1.5 m × 6 m flat plate whose temperature is 140°C (Fig. 19–13). Determine the rate of heat transfer from the plate if the air flows parallel to the (*a*) 6-m-long side and (*b*) the 1.5-m side.

Solution The top surface of a hot block is to be cooled by forced air. The rate of heat transfer is to be determined for two cases.

Assumptions **1** Steady operating conditions exist. **2** The critical Reynolds number is $\text{Re}_{cr} = 5 \times 10^5$. **3** Radiation effects are negligible. **4** Air is an ideal gas.

Properties The properties k, μ, c_p, and Pr of ideal gases are independent of pressure, while the properties ν and α are inversely proportional to density and thus pressure. The properties of air at the film temperature of $T_f = (T_s + T_\infty)/2 = (140 + 20)/2 = 80°\text{C}$ and 1 atm pressure are (Table A–22)

$$k = 0.02953\ \text{W/m}\cdot{}^\circ\text{C} \qquad \text{Pr} = 0.7154$$
$$\nu_{@\ 1\ \text{atm}} = 2.097 \times 10^{-5}\ \text{m}^2/\text{s}$$

The atmospheric pressure in Denver is P = (83.4 kPa)/(101.325 kPa/atm) = 0.823 atm. Then the kinematic viscosity of air in Denver becomes

$$\nu = \nu_{@\ 1\ \text{atm}}/P = (2.097 \times 10^{-5}\ \text{m}^2/\text{s})/0.823 = 2.548 \times 10^{-5}\ \text{m}^2/\text{s}$$

Analysis (*a*) When air flow is parallel to the long side, we have L = 6 m, and the Reynolds number at the end of the plate becomes

$$\text{Re}_L = \frac{VL}{\nu} = \frac{(8\ \text{m/s})(6\ \text{m})}{2.548 \times 10^{-5}\ \text{m}^2/\text{s}} = 1.884 \times 10^6$$

which is greater than the critical Reynolds number. Thus, we have combined laminar and turbulent flow, and the average Nusselt number for the entire plate is determined to be

$$\begin{aligned} \text{Nu} &= \frac{hL}{k} = (0.037\ \text{Re}_L^{0.8} - 871)\text{Pr}^{1/3} \\ &= [0.037(1.884 \times 10^6)^{0.8} - 871]0.7154^{1/3} \\ &= 2687 \end{aligned}$$

Then

$$h = \frac{k}{L}\text{Nu} = \frac{0.02953\ \text{W/m}\cdot{}^\circ\text{C}}{6\ \text{m}}(2687) = 13.2\ \text{W/m}^2\cdot{}^\circ\text{C}$$

$$A_s = wL = (1.5\ \text{m})(6\ \text{m}) = 9\ \text{m}^2$$

and

$$\dot{Q} = hA_s(T_s - T_\infty) = (13.2\ \text{W/m}^2\cdot{}^\circ\text{C})(9\ \text{m}^2)(140 - 20)^\circ\text{C} = \mathbf{1.43 \times 10^4\ W}$$

Note that if we disregarded the laminar region and assumed turbulent flow over the entire plate, we would get Nu = 3466 from Eq. 19–16, which is 29 percent higher than the value calculated above.

(*b*) When air flow is along the short side, we have L = 1.5 m, and the Reynolds number at the end of the plate becomes

$$\text{Re}_L = \frac{VL}{\nu} = \frac{(8\ \text{m/s})(1.5\ \text{m})}{2.548 \times 10^{-5}\ \text{m}^2/\text{s}} = 4.71 \times 10^5$$

which is less than the critical Reynolds number. Thus we have laminar flow over the entire plate, and the average Nusselt number is

$$\text{Nu} = \frac{hL}{k} = 0.664\ \text{Re}_L^{0.5}\ \text{Pr}^{1/3} = 0.664 \times (4.71 \times 10^5)^{0.5} \times 0.7154^{1/3} = 408$$

Then

$$h = \frac{k}{L}\text{Nu} = \frac{0.02953\ \text{W/m}\cdot{}^\circ\text{C}}{1.5\ \text{m}}(408) = 8.03\ \text{W/m}^2\cdot{}^\circ\text{C}$$

and

$$\dot{Q} = hA_s(T_s - T_\infty) = (8.03\ \text{W/m}^2\cdot{}^\circ\text{C})(9\ \text{m}^2)(140 - 20)^\circ\text{C} = \mathbf{8670\ W}$$

which is considerably less than the heat transfer rate determined in case (*a*).

Discussion Note that the *direction* of fluid flow can have a significant effect on convection heat transfer to or from a surface (Fig. 19–14). In this case, we can increase the heat transfer rate by 65 percent by simply blowing the air along the long side of the rectangular plate instead of the short side.

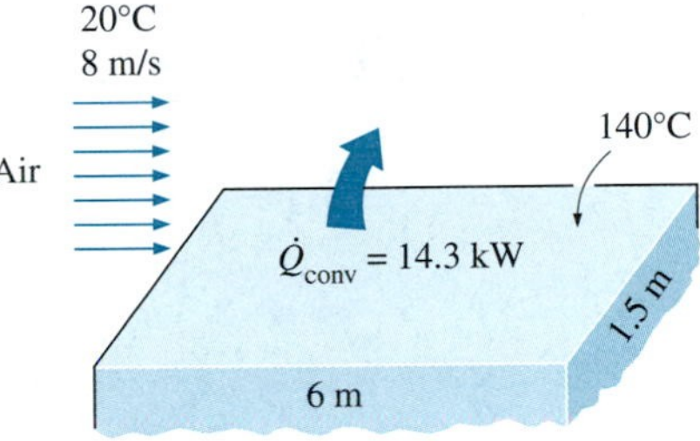

(*a*) Flow along the long side

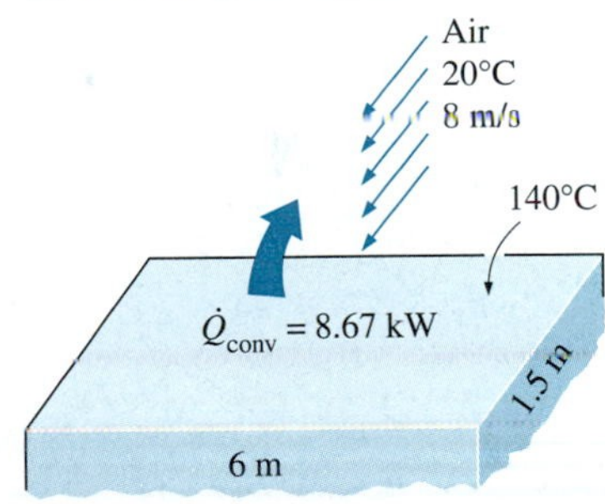

(*b*) Flow along the short side

FIGURE 19–14
The direction of fluid flow can have a significant effect on convection heat transfer.

EXAMPLE 19–3 Cooling of Plastic Sheets by Forced Air

The forming section of a plastics plant puts out a continuous sheet of plastic that is 4 ft wide and 0.04 in thick at a velocity of 30 ft/min. The temperature of the plastic sheet is 200°F when it is exposed to the surrounding air, and a 2-ft-long section of the plastic sheet is subjected to air flow at 80°F at a velocity of 10 ft/s on both sides along its surfaces normal to the direction of motion of the sheet, as shown in Fig. 19–15. Determine (*a*) the rate of heat

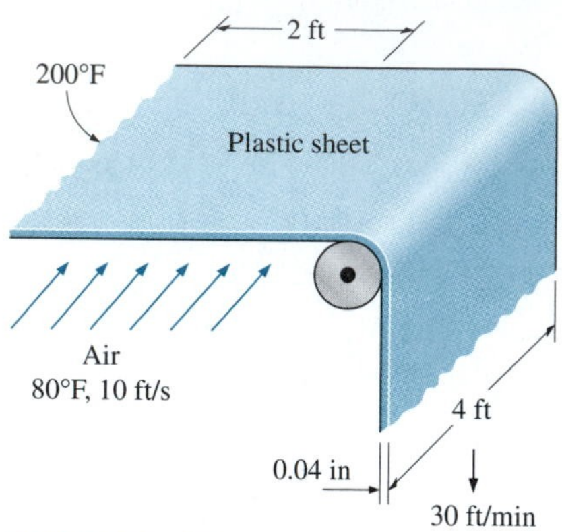

FIGURE 19–15
Schematic for Example 19–3.

transfer from the plastic sheet to air by forced convection and radiation and (*b*) the temperature of the plastic sheet at the end of the cooling section. Take the density, specific heat, and emissivity of the plastic sheet to be ρ = 75 lbm/ft^3, c_p = 0.4 Btu/lbm · °F, and ε = 0.9.

Solution Plastic sheets are cooled as they leave the forming section of a plastics plant. The rate of heat loss from the plastic sheet by convection and radiation and the exit temperature of the plastic sheet are to be determined.

Assumptions **1** Steady operating conditions exist. **2** The critical Reynolds number is $\text{Re}_{cr} = 5 \times 10^5$. **3** Air is an ideal gas. **4** The local atmospheric pressure is 1 atm. **5** The surrounding surfaces are at the temperature of the room air.

Properties The properties of the plastic sheet are given in the problem statement. The properties of air at the film temperature of $T_f = (T_s + T_\infty)/2 = (200 + 80)/2 = 140$°F and 1 atm pressure are (Table A–22E)

$$k = 0.01623 \text{ Btu/h} \cdot \text{ft} \cdot \text{°F} \qquad \text{Pr} = 0.7202$$
$$\nu = 0.204 \times 10^{-3} \text{ ft}^2\text{/s}$$

Analysis (*a*) We expect the temperature of the plastic sheet to drop somewhat as it flows through the 2-ft-long cooling section, but at this point we do not know the magnitude of that drop. Therefore, we assume the plastic sheet to be isothermal at 200°F to get started. We will repeat the calculations if necessary to account for the temperature drop of the plastic sheet.

Noting that L = 4 ft, the Reynolds number at the end of the air flow across the plastic sheet is

$$\text{Re}_L = \frac{VL}{\nu} = \frac{(10 \text{ ft/s})(4 \text{ ft})}{0.204 \times 10^{-3} \text{ ft}^2\text{/s}} = 1.961 \times 10^5$$

which is less than the critical Reynolds number. Thus, we have *laminar flow* over the entire sheet, and the Nusselt number is determined from the laminar flow relations for a flat plate to be

$$\text{Nu} = \frac{hL}{k} = 0.664 \text{ Re}_L^{0.5} \text{ Pr}^{1/3} = 0.664 \times (1.961 \times 10^5)^{0.5} \times (0.7202)^{1/3} = 263.6$$

Then,

$$h = \frac{k}{L}\text{Nu} = \frac{0.01623 \text{ Btu/h} \cdot \text{ft} \cdot \text{°F}}{4 \text{ ft}}(263.6) = 1.07 \text{ Btu/h} \cdot \text{ft}^2 \cdot \text{°F}$$
$$A_s = (2 \text{ ft})(4 \text{ ft})(2 \text{ sides}) = 16 \text{ ft}^2$$

and

$$\begin{aligned}\dot{Q}_{\text{conv}} &= hA_s(T_s - T_\infty)\\ &= (1.07 \text{ Btu/h} \cdot \text{ft}^2 \cdot \text{°F})(16 \text{ ft}^2)(200 - 80)\text{°F}\\ &= 2054 \text{ Btu/h}\end{aligned}$$

$$\begin{aligned}\dot{Q}_{\text{rad}} &= \varepsilon\sigma A_s(T_s^4 - T_{\text{surr}}^4)\\ &= (0.9)(0.1714 \times 10^{-8} \text{ Btu/h} \cdot \text{ft}^2 \cdot \text{R}^4)(16 \text{ ft}^2)[(660 \text{ R})^4 - (540 \text{ R})^4]\\ &= 2585 \text{ Btu/h}\end{aligned}$$

Therefore, the rate of cooling of the plastic sheet by combined convection and radiation is

$$\dot{Q}_{\text{total}} = \dot{Q}_{\text{conv}} + \dot{Q}_{\text{rad}} = 2054 + 2585 = \mathbf{4639\ Btu/h}$$

(*b*) To find the temperature of the plastic sheet at the end of the cooling section, we need to know the mass of the plastic rolling out per unit time (or the mass flow rate), which is determined from

$$\dot{m} = \rho A_c V_{\text{plastic}} = (75\ \text{lbm/ft}^3)\left(\frac{4 \times 0.04}{12}\ \text{ft}^2\right)\left(\frac{30}{60}\ \text{ft/s}\right) = 0.5\ \text{lbm/s}$$

Then, an energy balance on the cooled section of the plastic sheet yields

$$\dot{Q} = \dot{m}c_p(T_2 - T_1) \quad \rightarrow \quad T_2 = T_1 + \frac{\dot{Q}}{\dot{m}c_p}$$

Noting that $\dot{Q}$ is a negative quantity (heat loss) for the plastic sheet and substituting, the temperature of the plastic sheet as it leaves the cooling section is determined to be

$$T_2 = 200°\text{F} + \frac{-4639\ \text{Btu/h}}{(0.5\ \text{lbm/s})(0.4\ \text{Btu/lbm}\cdot°\text{F})}\left(\frac{1\ \text{h}}{3600\ \text{s}}\right) = \mathbf{193.6°F}$$

Discussion The average temperature of the plastic sheet drops by about 6.4°F as it passes through the cooling section. The calculations now can be repeated by taking the average temperature of the plastic sheet to be 196.8°F instead of 200°F for better accuracy, but the change in the results will be insignificant because of the small change in temperature.

19–4 ▪ FLOW ACROSS CYLINDERS AND SPHERES

Flows across cylinders and spheres, in general, involve *flow separation,* which is difficult to handle analytically. Therefore, such flows must be studied experimentally or numerically. Indeed, flow across cylinders and spheres has been studied experimentally by numerous investigators, and several empirical correlations have been developed for the heat transfer coefficient.

The complicated flow pattern across a cylinder greatly influences heat transfer. The variation of the local Nusselt number Nu_θ around the periphery of a cylinder subjected to cross flow of air is given in Fig. 19–16. Note that, for all cases, the value of Nu_θ starts out relatively high at the stagnation point ($\theta = 0°$) but decreases with increasing θ as a result of the thickening of the laminar boundary layer. On the two curves at the bottom corresponding to Re = 70,800 and 101,300, Nu_θ reaches a minimum at $\theta \approx 80°$, which is the separation point in laminar flow. Then Nu_θ increases with increasing θ as a result of the intense mixing in the separated flow region (the wake). The curves at the top corresponding to Re = 140,000 to 219,000 differ from the first two curves in that they have *two* minima for Nu_θ. The sharp increase in Nu_θ at about $\theta \approx 90°$ is due to the transition from laminar to turbulent flow. The later decrease in Nu_θ is again due to the thickening of the boundary layer. Nu_θ reaches its second minimum at about $\theta \approx 140°$, which

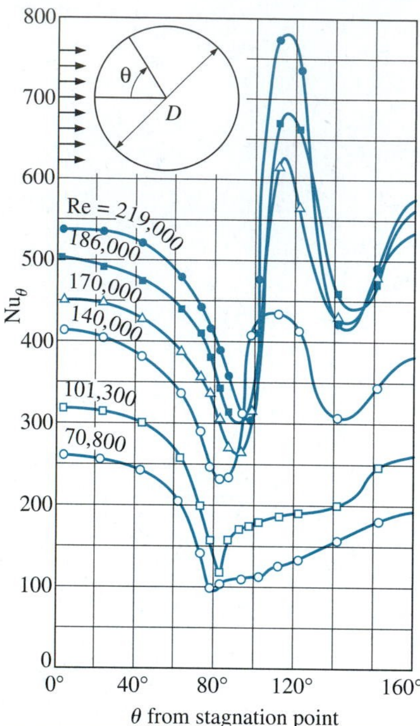

FIGURE 19–16
Variation of the local heat transfer coefficient along the circumference of a circular cylinder in cross flow of air (from Giedt, 1949).

is the flow separation point in turbulent flow, and increases with θ as a result of the intense mixing in the turbulent wake region.

The discussions above on the local heat transfer coefficients are insightful; however, they are of limited value in heat transfer calculations since the calculation of heat transfer requires the *average* heat transfer coefficient over the entire surface. Of the several such relations available in the literature for the average Nusselt number for cross flow over a cylinder, we present the one proposed by Churchill and Bernstein (1977):

$$\text{Nu}_{\text{cyl}} = \frac{hD}{k} = 0.3 + \frac{0.62\ \text{Re}^{1/2}\ \text{Pr}^{1/3}}{[1 + (0.4/\text{Pr})^{2/3}]^{1/4}}\left[1 + \left(\frac{\text{Re}}{282{,}000}\right)^{5/8}\right]^{4/5} \quad \text{(19–29)}$$

This relation is quite comprehensive in that it correlates available data well for RePr > 0.2. The fluid properties are evaluated at the *film temperature* $T_f = \frac{1}{2}(T_\infty + T_s)$, which is the average of the free-stream and surface temperatures.

For flow over a *sphere,* Whitaker (1972) recommends the following comprehensive correlation:

$$\text{Nu}_{\text{sph}} = \frac{hD}{k} = 2 + [0.4\ \text{Re}^{1/2} + 0.06\ \text{Re}^{2/3}]\ \text{Pr}^{0.4}\left(\frac{\mu_\infty}{\mu_s}\right)^{1/4} \quad \text{(19–30)}$$

which is valid for $3.5 \leq \text{Re} \leq 80{,}000$ and $0.7 \leq \text{Pr} \leq 380$. The fluid properties in this case are evaluated at the free-stream temperature T_∞, except for μ_s, which is evaluated at the surface temperature T_s. Although the two relations above are considered to be quite accurate, the results obtained from them can be off by as much as 30 percent.

The average Nusselt number for flow across cylinders can be expressed compactly as

$$\text{Nu}_{\text{cyl}} = \frac{hD}{k} = C\ \text{Re}^m\ \text{Pr}^n \quad \text{(19–31)}$$

where $n = \frac{1}{3}$ and the experimentally determined constants C and m are given in Table 19–2 for circular as well as various noncircular cylinders. The characteristic length D for use in the calculation of the Reynolds and the Nusselt numbers for different geometries is as indicated on the figure. All fluid properties are evaluated at the film temperature.

The relations for cylinders above are for *single* cylinders or cylinders oriented such that the flow over them is not affected by the presence of others. Also, they are applicable to *smooth* surfaces. *Surface roughness* and the *free-stream turbulence* may affect the drag and heat transfer coefficients significantly. Eq. 19–31 provides a simpler alternative to Eq. 19–29 for flow over cylinders. However, Eq. 19–29 is more accurate, and thus should be preferred in calculations whenever possible.

TABLE 19–2

Empirical correlations for the average Nusselt number for forced convection over circular and noncircular cylinders in cross flow (from Zukauskas, 1972 and Jakob, 1949)

Cross-section of the cylinder	Fluid	Range of Re	Nusselt number
Circle	Gas or liquid	0.4–4 4–40 40–4000 4000–40,000 40,000–400,000	$\text{Nu} = 0.989\text{Re}^{0.330}\ \text{Pr}^{1/3}$ $\text{Nu} = 0.911\text{Re}^{0.385}\ \text{Pr}^{1/3}$ $\text{Nu} = 0.683\text{Re}^{0.466}\ \text{Pr}^{1/3}$ $\text{Nu} = 0.193\text{Re}^{0.618}\ \text{Pr}^{1/3}$ $\text{Nu} = 0.027\text{Re}^{0.805}\ \text{Pr}^{1/3}$
Square	Gas	5000–100,000	$\text{Nu} = 0.102\text{Re}^{0.675}\ \text{Pr}^{1/3}$
Square (tilted 45°)	Gas	5000–100,000	$\text{Nu} = 0.246\text{Re}^{0.588}\ \text{Pr}^{1/3}$
Hexagon	Gas	5000–100,000	$\text{Nu} = 0.153\text{Re}^{0.638}\ \text{Pr}^{1/3}$
Hexagon (tilted 45°)	Gas	5000–19,500 19,500–100,000	$\text{Nu} = 0.160\text{Re}^{0.638}\ \text{Pr}^{1/3}$ $\text{Nu} = 0.0385\text{Re}^{0.782}\ \text{Pr}^{1/3}$
Vertical plate	Gas	4000–15,000	$\text{Nu} = 0.228\text{Re}^{0.731}\ \text{Pr}^{1/3}$
Ellipse	Gas	2500–15,000	$\text{Nu} = 0.248\text{Re}^{0.612}\ \text{Pr}^{1/3}$

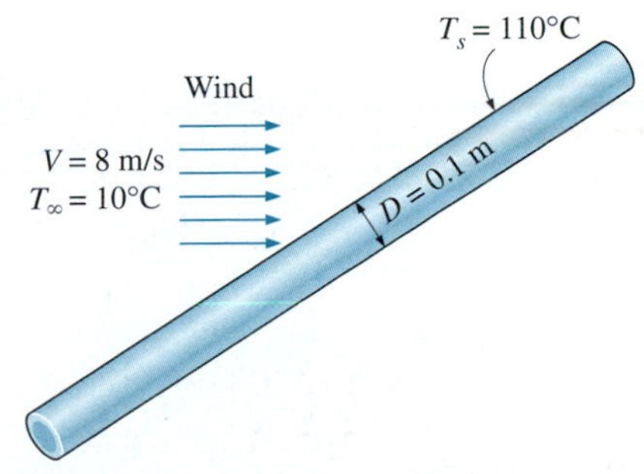

FIGURE 19–17
Schematic for Example 19–4.

EXAMPLE 19–4 Heat Loss from a Steam Pipe in Windy Air

A long 10-cm-diameter steam pipe whose external surface temperature is 110°C passes through some open area that is not protected against the winds (Fig. 19–17). Determine the rate of heat loss from the pipe per unit of its length when the air is at 1 atm pressure and 10°C and the wind is blowing across the pipe at a velocity of 8 m/s.

Solution A steam pipe is exposed to windy air. The rate of heat loss from the steam is to be determined.

Assumptions **1** Steady operating conditions exist. **2** Radiation effects are negligible. **3** Air is an ideal gas.

Properties The properties of air at the average film temperature of $T_f = (T_s + T_\infty)/2 = (110 + 10)/2 = 60°C$ and 1 atm pressure are (Table A–22)

$$k = 0.02808 \text{ W/m} \cdot °\text{C} \qquad \text{Pr} = 0.7202$$
$$\nu = 1.896 \times 10^{-5} \text{ m}^2/\text{s}$$

Analysis The Reynolds number is

$$\text{Re} = \frac{VD}{\nu} = \frac{(8 \text{ m/s})(0.1 \text{ m})}{1.896 \times 10^{-5} \text{ m}^2/\text{s}} = 4.219 \times 10^4$$

The Nusselt number can be determined from

$$\begin{aligned}\text{Nu} = \frac{hD}{k} &= 0.3 + \frac{0.62\,\text{Re}^{1/2}\,\text{Pr}^{1/3}}{[1 + (0.4/\text{Pr})^{2/3}]^{1/4}}\left[1 + \left(\frac{\text{Re}}{282{,}000}\right)^{5/8}\right]^{4/5} \\ &= 0.3 + \frac{0.62(4.219 \times 10^4)^{1/2}(0.7202)^{1/3}}{[1 + (0.4/0.7202)^{2/3}]^{1/4}}\left[1 + \left(\frac{4.219 \times 10^4}{282{,}000}\right)^{5/8}\right]^{4/5} \\ &= 124\end{aligned}$$

and

$$h = \frac{k}{D}\text{Nu} = \frac{0.02808 \text{ W/m} \cdot °\text{C}}{0.1 \text{ m}}(124) = 34.8 \text{ W/m}^2 \cdot °\text{C}$$

Then the rate of heat transfer from the pipe per unit of its length becomes

$$A_s = pL = \pi DL = \pi(0.1 \text{ m})(1 \text{ m}) = 0.314 \text{ m}^2$$
$$\dot{Q} = hA_s(T_s - T_\infty) = (34.8 \text{ W/m}^2 \cdot \text{C})(0.314 \text{ m}^2)(110 - 10)°\text{C} = \mathbf{1093 \text{ W}}$$

The rate of heat loss from the entire pipe can be obtained by multiplying the value above by the length of the pipe in m.

Discussion The simpler Nusselt number relation in Table 19–2 in this case would give Nu = 128, which is 3 percent higher than the value obtained above using Eq. 19–29.

EXAMPLE 19–5 Cooling of a Steel Ball by Forced Air

A 25-cm-diameter stainless steel ball (ρ = 8055 kg/m³, c_p = 480 J/kg · °C) is removed from the oven at a uniform temperature of 300°C (Fig. 19–18). The ball is then subjected to the flow of air at 1 atm pressure and 25°C with a velocity of 3 m/s. The surface temperature of the ball eventually drops to 200°C. Determine the average convection heat transfer coefficient during this cooling process and estimate how long the process will take.

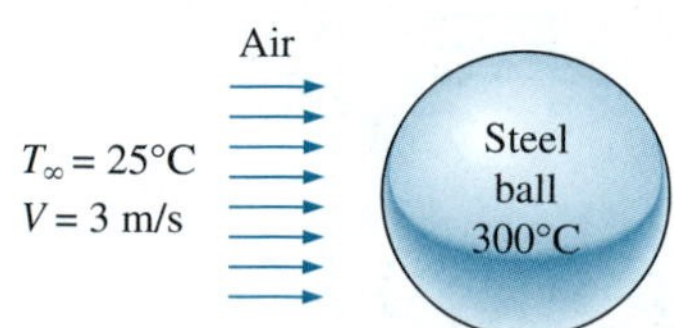

FIGURE 19–18
Schematic for Example 19–5.

Solution A hot stainless steel ball is cooled by forced air. The average convection heat transfer coefficient and the cooling time are to be determined.

Assumptions **1** Steady operating conditions exist. **2** Radiation effects are negligible. **3** Air is an ideal gas. **4** The outer surface temperature of the ball is uniform at all times. **5** The surface temperature of the ball during cooling is changing. Therefore, the convection heat transfer coefficient between the ball and the air will also change. To avoid this complexity, we take the surface temperature of the ball to be constant at the average temperature of (300 + 200)/2 = 250°C in the evaluation of the heat transfer coefficient and use the value obtained for the entire cooling process.

Properties The dynamic viscosity of air at the average surface temperature is $\mu_s = \mu_{@\ 250°C} = 2.76 \times 10^{-5}$ kg/m · s. The properties of air at the free-stream temperature of 25°C and 1 atm are (Table A–22)

$$k = 0.02551 \text{ W/m} \cdot °\text{C} \qquad \nu = 1.562 \times 10^{-5} \text{ m}^2/\text{s}$$
$$\mu = 1.849 \times 10^{-5} \text{ kg/m} \cdot \text{s} \qquad \text{Pr} = 0.7296$$

Analysis The Reynolds number is determined from

$$\text{Re} = \frac{VD}{\nu} = \frac{(3 \text{ m/s})(0.25 \text{ m})}{1.562 \times 10^{\pm 5} \text{ m}^2/\text{s}} = 4.802 \times 10^4$$

The Nusselt number is

$$\text{Nu} = \frac{hD}{k} = 2 + [0.4 \text{ Re}^{1/2} + 0.06 \text{ Re}^{2/3}] \text{ Pr}^{0.4} \left(\frac{\mu_\infty}{\mu_s}\right)^{1/4}$$
$$= 2 + [0.4(4.802 \times 10^4)^{1/2} + 0.06(4.802 \times 10^4)^{2/3}](0.7296)^{0.4}$$
$$\times \left(\frac{1.849 \times 10^{-5}}{2.76 \times 10^{-5}}\right)^{1/4}$$
$$= 135$$

Then the average convection heat transfer coefficient becomes

$$h = \frac{k}{D}\text{Nu} = \frac{0.02551 \text{ W/m} \cdot °\text{C}}{0.25 \text{ m}}(135) = \mathbf{13.8 \text{ W/m}^2 \cdot °\text{C}}$$

In order to estimate the time of cooling of the ball from 300°C to 200°C, we determine the *average* rate of heat transfer from Newton's law of cooling by using the *average* surface temperature. That is,

$$A_s = \pi D^2 = \pi(0.25 \text{ m})^2 = 0.1963 \text{ m}^2$$
$$\dot{Q}_{\text{avg}} = hA_s(T_{s,\text{ avg}} - T_\infty) = (13.8 \text{ W/m}^2 \cdot °\text{C})(0.1963 \text{ m}^2)(250 - 25)°\text{C} = 610 \text{ W}$$

Next we determine the *total* heat transferred from the ball, which is simply the change in the energy of the ball as it cools from 300°C to 200°C:

$$m = \rho V = \rho \tfrac{1}{6}\pi D^3 = (8055 \text{ kg/m}^3)\ \tfrac{1}{6}\pi(0.25 \text{ m})^3 = 65.9 \text{ kg}$$
$$Q_{\text{total}} = mc_p(T_2 - T_1) = (65.9 \text{ kg})(480 \text{ J/kg} \cdot °\text{C})(300 - 200)°\text{C} = 3{,}163{,}000 \text{ J}$$

In this calculation, we assumed that the entire ball is at 200°C, which is not necessarily true. The inner region of the ball will probably be at a higher temperature than its surface. With this assumption, the time of cooling is determined to be

$$\Delta t \approx \frac{Q}{\dot{Q}_{\text{avg}}} = \frac{3{,}163{,}000 \text{ J}}{610 \text{ J/s}} = 5185 \text{ s} = \mathbf{1 \text{ h } 26 \text{ min}}$$

Discussion The time of cooling could also be determined more accurately using the transient temperature charts or relations introduced in Chapter 18. But the simplifying assumptions we made above can be justified if all we need is a ballpark value. It will be naive to expect the time of cooling to be exactly 1 h 26 min, but, using our engineering judgment, it is realistic to expect the time of cooling to be somewhere between one and two hours.

19–5 ■ GENERAL CONSIDERATIONS FOR PIPE FLOW

Liquid or gas flow through pipes or ducts is commonly used in practice in heating and cooling applications. The fluid in such applications is forced to flow by a fan or pump through a conduit that is sufficiently long to accomplish the desired heat transfer.

The general aspects of flow in pipes were considered in Chapter 14. The Reynolds number for flow through a pipe of inside diameter D was defined as

$$\text{Re} = \frac{\rho V_{\text{avg}} D}{\mu} = \frac{V_{\text{avg}} D}{\nu} \tag{19–32}$$

where V_{avg} is the average velocity and $\nu = \mu/\rho$ is the kinematic viscosity of the fluid. Transition from laminar to turbulent flow depends on the Reynolds number as well as the degree of disturbance of the flow by *surface roughness*, *pipe vibrations*, and the *fluctuations in the flow*. Under most practical conditions, the flow in a pipe is laminar for Re $<$ 2300, fully turbulent for Re $>$ 10,000, and transitional in between. But it should be kept in mind that in many cases the flow becomes fully turbulent for Re $>$ 4000. When designing piping networks and determining pumping power, a conservative approach is taken and flows with Re $>$ 4000 are assumed to be turbulent.

When a fluid is heated or cooled as it flows through a tube, the temperature of the fluid at any cross section changes from T_s at the surface of the wall to some maximum (or minimum in the case of heating) at the tube center. In fluid flow it is convenient to work with an **average** or **mean temperature** T_m, which remains constant at a cross section. Unlike the mean velocity, the mean temperature T_m *changes* in the flow direction whenever the fluid is heated or cooled.

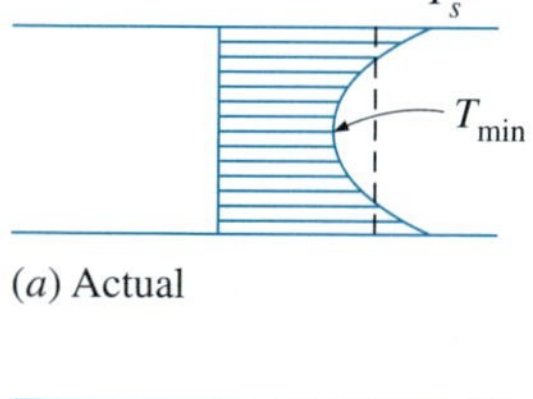

(*a*) Actual

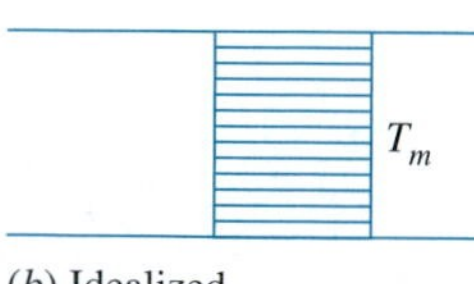

(*b*) Idealized

FIGURE 19–19
Actual and idealized temperature profiles for flow in a tube (the rate at which energy is transported with the fluid is the same for both cases).

The value of the mean temperature T_m is determined from the requirement that the *conservation of energy* principle be satisfied. That is, the energy transported by the fluid through a cross section in actual flow must be equal to the energy that would be transported through the same cross section if the fluid were at a constant temperature T_m. This can be expressed mathematically as (Fig. 19–19)

$$\dot{E}_{\text{fluid}} = \dot{m} c_p T_m = \int_{\dot{m}} c_p T(r) \delta \dot{m} = \int_{A_c} \rho c_p T(r) u(r) V dA_c \tag{19–33}$$

where c_p is the specific heat of the fluid. Note that the product $\dot{m} c_p T_m$ at any cross section along the tube represents the *energy flow* with the fluid at that cross section. Then the mean temperature of a fluid with constant density and specific heat flowing in a circular pipe of radius R can be expressed as

$$T_m = \frac{\int_{\dot{m}} c_p T(r) \delta \dot{m}}{\dot{m} c_p} = \frac{\int_0^R c_p T(r) \rho u(r) 2\pi r dr}{\rho V_{\text{avg}} (\pi R^2) c_p} = \frac{2}{V_{\text{avg}} R^2} \int_0^R T(r) u(r)\, r dr \tag{19–34}$$

Note that the mean temperature T_m of a fluid changes during heating or cooling. Also, the fluid properties in internal flow are usually evaluated at the *bulk mean fluid temperature*, which is the arithmetic average of the mean temperatures at the inlet and the exit. That is, $T_b = (T_{m,\,i} + T_{m,\,e})/2$

Thermal Entrance Region

The development of the velocity boundary layer was discussed in Chapter 14 (Fig. 19–20). Now consider a fluid at a uniform temperature entering a circular tube whose surface is maintained at a different temperature. This time, the fluid particles in the layer in contact with the surface of the tube assume the surface temperature. This initiates convection heat transfer in the tube and the development of a *thermal boundary layer* along the tube. The thickness of this boundary layer also increases in the flow direction until the boundary layer reaches the tube center and thus fills the entire tube, as shown in Fig. 19–21.

The region of flow over which the thermal boundary layer develops and reaches the tube center is called the **thermal entrance region**, and the length of this region is called the **thermal entry length** L_t. Flow in the thermal entrance region is called *thermally developing flow* since this is the region where the temperature profile develops. The region beyond the thermal entrance region in which the dimensionless temperature profile expressed as $(T_s - T)/(T_s - T_m)$ remains unchanged is called the **thermally fully developed region**. The region in which the flow is both hydrodynamically and thermally developed and thus both the velocity and dimensionless temperature profiles remain unchanged is called *fully developed flow*. That is,

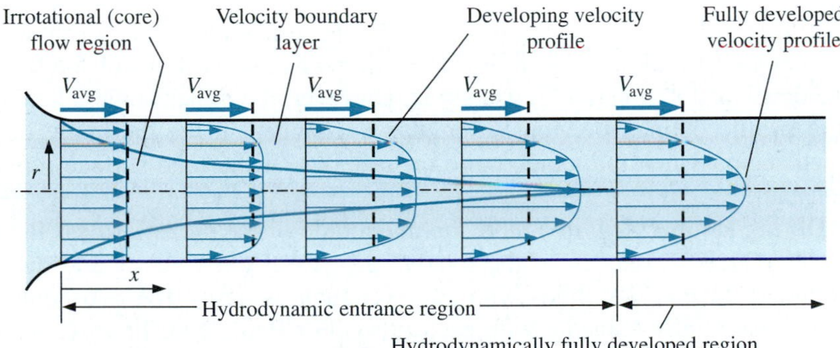

FIGURE 19–20
The development of the velocity boundary layer in a tube. (The developed average velocity profile is parabolic in laminar flow, as shown, but somewhat flatter or fuller in turbulent flow.)

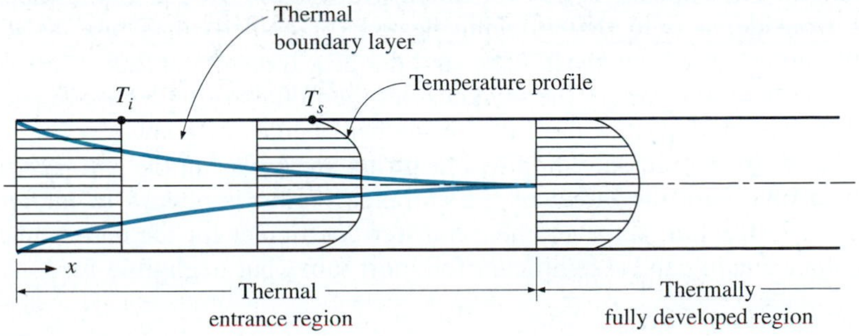

FIGURE 19–21
The development of the thermal boundary layer in a tube. (The fluid in the tube is being cooled.)

Hydrodynamically fully developed: $$\frac{\partial u(r, x)}{\partial x} = 0 \quad \longrightarrow \quad u = u(r) \qquad \text{(19–35)}$$

Thermally fully developed: $$\frac{\partial}{\partial x}\left[\frac{T_s(x) - T(r, x)}{T_s(x) - T_m(x)}\right] = 0 \qquad \text{(19–36)}$$

The shear stress at the tube wall τ_w is related to the slope of the velocity profile at the surface. Noting that the velocity profile remains unchanged in the hydrodynamically fully developed region, the wall shear stress also remains constant in that region. A similar argument can be given for the heat transfer coefficient in the thermally fully developed region.

In a thermally fully developed region, the derivative of $(T_s - T)/(T_s - T_m)$ with respect to x is zero by definition, and thus $(T_s - T)/(T_s - T_m)$ is independent of x. Then the derivative of $(T_s - T)/(T_s - T_m)$ with respect r must also be independent of x. That is,

$$\frac{\partial}{\partial r}\left(\frac{T_s - T}{T_s - T_m}\right)\bigg|_{r=R} = \frac{-(\partial T/\partial r)|_{r=R}}{T_s - T_m} \neq f(x) \qquad \text{(19–37)}$$

Surface heat flux can be expressed as

$$\dot{q}_s = h_x(T_s - T_m) = k\frac{\partial T}{\partial r}\bigg|_{r=R} \quad \longrightarrow \quad h_x = \frac{k(\partial T/\partial r)|_{r=R}}{T_s - T_m} \qquad \text{(19–38)}$$

which, from Eq. 19–37, is independent of x. Thus we conclude that *in the thermally fully developed region of a tube, the local convection coefficient is constant* (does not vary with x). Therefore, both *the friction (which is related to wall shear stress) and convection coefficients remain constant in the fully developed region of a tube.*

Note that the *temperature profile* in the thermally fully developed region may vary with x in the flow direction. That is, unlike the velocity profile, the temperature profile can be different at different cross sections of the tube in the developed region, and it usually is. However, the dimensionless temperature profile defined previously remains unchanged in the thermally developed region when the temperature or heat flux at the tube surface remains constant.

During laminar flow in a tube, the magnitude of the dimensionless Prandtl number Pr is a measure of the relative growth of the velocity and thermal boundary layers. For fluids with Pr $\approx$ 1, such as gases, the two boundary layers essentially coincide with each other. For fluids with Pr $\gg$ 1, such as oils, the velocity boundary layer outgrows the thermal boundary layer. As a result, the hydrodynamic entry length is smaller than the thermal entry length. The opposite is true for fluids with Pr $\ll$ 1 such as liquid metals.

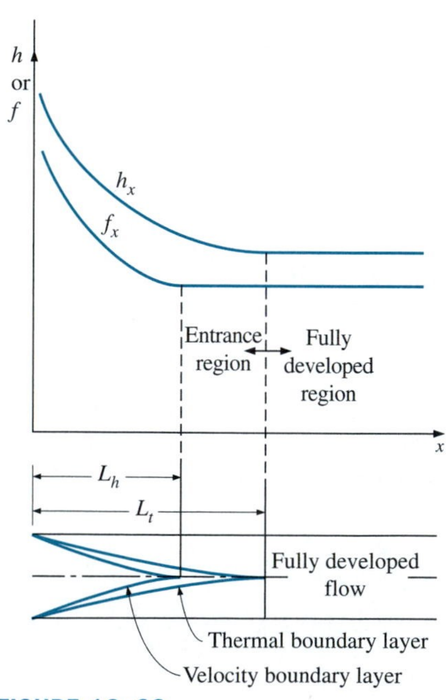

FIGURE 19–22
Variation of the friction factor and the convection heat transfer coefficient in the flow direction for flow in a tube (Pr > 1).

Consider a fluid that is being heated (or cooled) in a tube as it flows through it. The wall shear stress and the heat transfer coefficient are *highest* at the tube inlet where the thickness of the boundary layers is smallest, and decrease gradually to the fully developed values, as shown in Fig. 19–22. Therefore, the pressure drop and heat flux are *higher* in the entrance regions of a tube, and the effect of the entrance region is always to *increase* the average friction factor and heat transfer coefficient for the entire tube. This enhancement can be significant for short tubes but negligible for long ones.

Entry Lengths

The hydrodynamic entry length is usually taken to be the distance from the tube entrance where the wall shear stress (and thus the friction factor) reaches within about 2 percent of the fully developed value. In *laminar flow,* the hydrodynamic and thermal entry lengths are given approximately as (see Kays and Crawford, 1993 and Shah and Bhatti, 1987)

$$L_{h,\ \text{laminar}} \approx 0.05\ \text{Re}\ D \tag{19–39}$$

$$L_{t,\ \text{laminar}} \approx 0.05\ \text{Re}\ \text{Pr}\ D = \text{Pr}\ L_{h,\ \text{laminar}} \tag{19–40}$$

For Re = 20, the hydrodynamic entry length is about the size of the diameter, but increases linearly with velocity. In the limiting case of Re = 2300, the hydrodynamic entry length is $115D$.

In *turbulent flow,* the intense mixing during random fluctuations usually overshadows the effects of molecular diffusion, and therefore the hydrodynamic and thermal entry lengths are of about the same size and independent of the Prandtl number. The hydrodynamic entry length for turbulent flow can be determined from (see Bhatti and Shah, 1987 and Zhi-qing, 1982)

$$L_{h,\ \text{turbulent}} = 1.359D\ \text{Re}^{1/4} \tag{19–41}$$

The entry length is much shorter in turbulent flow, as expected, and its dependence on the Reynolds number is weaker. In many tube flows of practical interest, the entrance effects become insignificant beyond a tube length of 10 diameters, and the hydrodynamic and thermal entry lengths are approximately taken to be

$$L_{h,\ \text{turbulent}} \approx L_{t,\ \text{turbulent}} \approx 10D \tag{19–42}$$

The variation of local Nusselt number along a tube in turbulent flow for both uniform surface temperature and uniform surface heat flux is given in Fig. 19–23 for the range of Reynolds numbers encountered in heat transfer equipment. We make these important observations from this figure:

- The Nusselt numbers and thus the convection heat transfer coefficients are much higher in the entrance region.
- The Nusselt number reaches a constant value at a distance of less than 10 diameters, and thus the flow can be assumed to be fully developed for $x > 10D$.
- The Nusselt numbers for the uniform surface temperature and uniform surface heat flux conditions are identical in the fully developed regions, and nearly identical in the entrance regions. Therefore, Nusselt number is insensitive to the type of thermal boundary condition, and the turbulent flow correlations can be used for either type of boundary condition.

Precise correlations for the friction and heat transfer coefficients for the entrance regions are available in the literature. However, the tubes used in practice in forced convection are usually several times the length of either entrance region, and thus the flow through the tubes is often assumed to be fully developed for the entire length of the tube. This simplistic approach

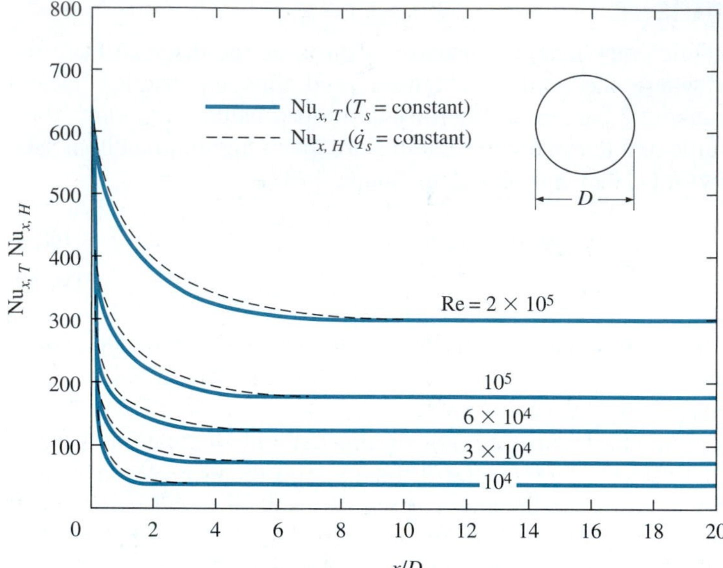

FIGURE 19–23
Variation of local Nusselt number along a tube in turbulent flow for both uniform surface temperature and uniform surface heat flux [Deissler, 1953].

gives *reasonable* results for the rate of heat transfer for long tubes and *conservative* results for short ones.

19–6 ▪ GENERAL THERMAL ANALYSIS

In the absence of any work interactions (such as electric resistance heating), the conservation of energy equation for the steady flow of a fluid in a tube can be expressed as (Fig. 19–24)

$$\dot{Q} = \dot{m}c_p(T_e - T_i) \qquad \text{(W)} \qquad \textbf{(19–43)}$$

where T_i and T_e are the mean fluid temperatures at the inlet and exit of the tube, respectively, and $\dot{Q}$ is the rate of heat transfer to or from the fluid. Note that the temperature of a fluid flowing in a tube remains constant in the absence of any energy interactions through the wall of the tube.

The thermal conditions at the surface can usually be approximated with reasonable accuracy to be *constant surface temperature* (T_s = constant) or *constant surface heat flux* ($\dot{q}_s$ = constant). For example, the constant surface temperature condition is realized when a phase change process such as boiling or condensation occurs at the outer surface of a tube. The constant surface heat flux condition is realized when the tube is subjected to radiation or electric resistance heating uniformly from all directions.

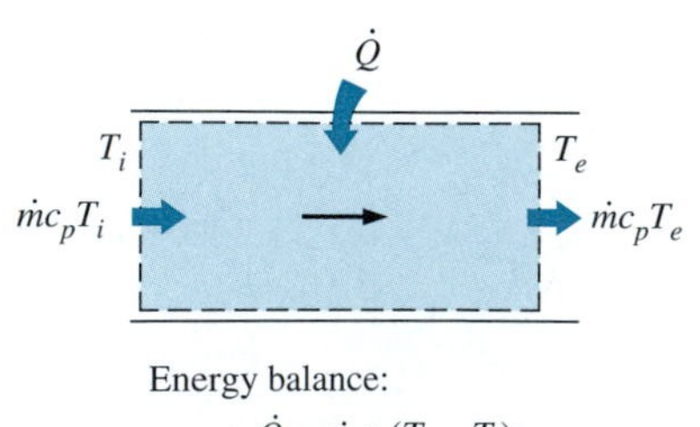

FIGURE 19–24
The heat transfer to a fluid flowing in a tube is equal to the increase in the energy of the fluid.

Surface heat flux is expressed as

$$\dot{q}_s = h_x(T_s - T_m) \qquad \text{(W/m}^2\text{)} \qquad \textbf{(19–44)}$$

where h_x is the *local* heat transfer coefficient and T_s and T_m are the surface and the mean fluid temperatures at that location. Note that the mean fluid temperature T_m of a fluid flowing in a tube must change during heating or cooling. Therefore, when $h_x = h$ = constant, the surface temperature T_s

must change when $\dot{q}_s$ = constant, and the surface heat flux $\dot{q}_s$ must change when T_s = constant. Thus we may have either T_s = constant or $\dot{q}_s$ = constant at the surface of a tube, but not both. Next we consider convection heat transfer for these two common cases.

Constant Surface Heat Flux ($\dot{q}_s$ = constant)

In the case of $\dot{q}_s$ = constant, the rate of heat transfer can also be expressed as

$$\dot{Q} = \dot{q}_s A_s = \dot{m}c_p(T_e - T_i) \qquad \text{(W)} \tag{19–45}$$

Then the mean fluid temperature at the tube exit becomes

$$T_e = T_i + \frac{\dot{q}_s A_s}{\dot{m}c_p} \tag{19–46}$$

Note that the mean fluid temperature increases *linearly* in the flow direction in the case of constant surface heat flux, since the surface area increases linearly in the flow direction (A_s is equal to the perimeter, which is constant, times the tube length).

The surface temperature in the case of constant surface heat flux $\dot{q}_s$ can be determined from

$$\dot{q}_s = h(T_s - T_m) \longrightarrow T_s = T_m + \frac{\dot{q}_s}{h} \tag{19–47}$$

In the fully developed region, the surface temperature T_s will also increase linearly in the flow direction since h is constant and thus $T_s - T_m$ = constant (Fig. 19–25). Of course this is true when the fluid properties remain constant during flow.

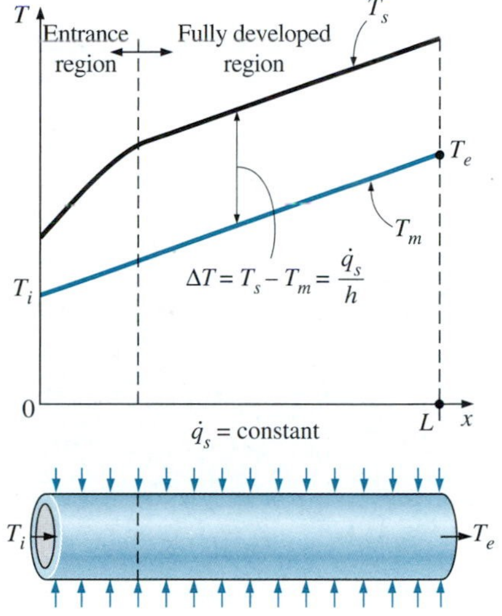

FIGURE 19–25
Variation of the *tube surface* and the *mean fluid* temperatures along the tube for the case of constant surface heat flux.

The slope of the mean fluid temperature T_m on a T-x diagram can be determined by applying the steady-flow energy balance to a tube slice of thickness dx shown in Fig. 19–26. It gives

$$\dot{m}c_p\,dT_m = \dot{q}_s(p\,dx) \longrightarrow \frac{dT_m}{dx} = \frac{\dot{q}_s p}{\dot{m}c_p} = \text{constant} \tag{19–48}$$

where p is the perimeter of the tube.

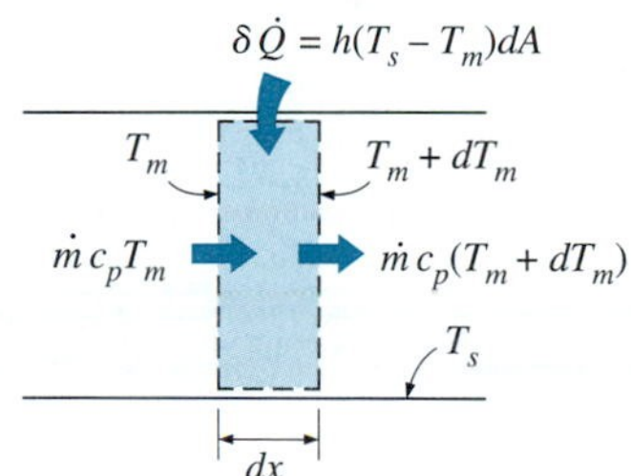

FIGURE 19–26
Energy interactions for a differential control volume in a tube.

Noting that both $\dot{q}_s$ and h are constants, the differentiation of Eq. 19–47 with respect to x gives

$$\frac{dT_m}{dx} = \frac{dT_s}{dx} \tag{19–49}$$

Also, the requirement that the dimensionless temperature profile remains unchanged in the fully developed region gives

$$\frac{\partial}{\partial x}\left(\frac{T_s - T}{T_s - T_m}\right) = 0 \longrightarrow \frac{1}{T_s - T_m}\left(\frac{\partial T_s}{\partial x} - \frac{\partial T}{\partial x}\right) = 0 \longrightarrow \frac{\partial T}{\partial x} = \frac{dT_s}{dx} \tag{19–50}$$

since $T_s - T_m$ = constant. Combining Eqs. 19–48, 19–49, and 19–50 gives

$$\frac{\partial T}{\partial x} = \frac{dT_s}{dx} = \frac{dT_m}{dx} = \frac{\dot{q}_s p}{\dot{m}c_p} = \text{constant} \tag{19–51}$$

Then we conclude that *in fully developed flow in a tube subjected to constant surface heat flux, the temperature gradient is independent of x and thus the shape of the temperature profile does not change along the tube* (Fig. 19–27).

For a circular tube, $p = 2\pi R$ and $\dot{m} = \rho V_{avg} A_c = \rho V_{avg}(\pi R^2)$, and Eq. 19–51 becomes

Circular tube: $$\frac{\partial T}{\partial x} = \frac{dT_s}{dx} = \frac{dT_m}{dx} = \frac{2\dot{q}_s}{\rho V_{avg} c_p R} = \text{constant} \qquad \textbf{(19–52)}$$

where V_{avg} is the mean velocity of the fluid.

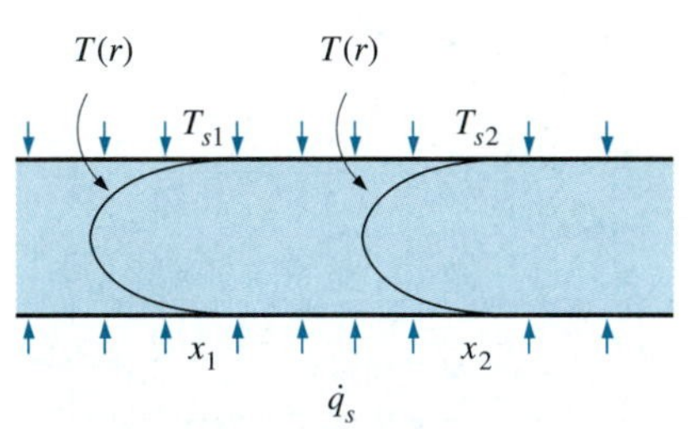

FIGURE 19–27
The shape of the temperature profile remains unchanged in the fully developed region of a tube subjected to constant surface heat flux.

Constant Surface Temperature (T_s = constant)

From Newton's law of cooling, the rate of heat transfer to or from a fluid flowing in a tube can be expressed as

$$\dot{Q} = hA_s\Delta T_{avg} = hA_s(T_s - T_m)_{avg} \qquad \text{(W)} \qquad \textbf{(19–53)}$$

where h is the average convection heat transfer coefficient, A_s is the heat transfer surface area (it is equal to πDL for a circular pipe of length L), and ΔT_{avg} is some appropriate *average* temperature difference between the fluid and the surface. Below we discuss two suitable ways of expressing ΔT_{avg}.

In the constant surface temperature (T_s = constant) case, ΔT_{avg} can be expressed *approximately* by the **arithmetic mean temperature difference** ΔT_{am} as

$$\Delta T_{avg} \approx \Delta T_{am} = \frac{\Delta T_i + \Delta T_e}{2} = \frac{(T_s - T_i) + (T_s - T_e)}{2} = T_s - \frac{T_i + T_e}{2}$$
$$= T_s - T_b \qquad \textbf{(19–54)}$$

where $T_b = (T_i + T_e)/2$ is the *bulk mean fluid temperature,* which is the *arithmetic average* of the mean fluid temperatures at the inlet and the exit of the tube.

Note that the *arithmetic mean temperature difference* ΔT_{am} is simply the *average* of the *temperature differences* between the surface and the fluid at the inlet and the exit of the tube. Inherent in this definition is the assumption that the mean fluid temperature varies linearly along the tube, which is hardly ever the case when T_s = constant. This simple approximation often gives acceptable results, but not always. Therefore, we need a better way to evaluate ΔT_{avg}.

Consider the heating of a fluid in a tube of constant cross section whose inner surface is maintained at a constant temperature of T_s. We know that the mean temperature of the fluid T_m increases in the flow direction as a result of heat transfer. The energy balance on a differential control volume shown in Fig. 19–26 gives

$$\dot{m}c_p dT_m = h(T_s - T_m)dA_s \qquad \textbf{(19–55)}$$

That is, the increase in the energy of the fluid (represented by an increase in its mean temperature by dT_m) is equal to the heat transferred to the fluid from the tube surface by convection. Noting that the differential surface area is

$dA_s = pdx$, where p is the perimeter of the tube, and that $dT_m = -d(T_s - T_m)$, since T_s is constant, the relation above can be rearranged as

$$\frac{d(T_s - T_m)}{T_s - T_m} = -\frac{hp}{\dot{m}c_p}dx \tag{19–56}$$

Integrating from $x = 0$ (tube inlet where $T_m = T_i$) to $x = L$ (tube exit where $T_m = T_e$) gives

$$\ln\frac{T_s - T_e}{T_s - T_i} = -\frac{hA_s}{\dot{m}c_p} \tag{19–57}$$

where $A_s = pL$ is the surface area of the tube and h is the constant *average* convection heat transfer coefficient. Taking the exponential of both sides and solving for T_e gives the following relation which is very useful for the determination of the *mean fluid temperature at the tube exit:*

$$T_e = T_s - (T_s - T_i)\exp(-hA_s/\dot{m}c_p) \tag{19–58}$$

This relation can also be used to determine the mean fluid temperature $T_m(x)$ at any x by replacing $A_s = pL$ by px.

Note that the temperature difference between the fluid and the surface *decays exponentially* in the flow direction, and the rate of decay depends on the magnitude of the exponent $hA_s/\dot{m}c_p$, as shown in Fig. 19–28. This dimensionless parameter is called the *number of transfer units,* denoted by NTU, and is a measure of the effectiveness of the heat transfer systems. For NTU > 5, the exit temperature of the fluid becomes almost equal to the surface temperature, $T_e \approx T_s$ (Fig. 19–29). Noting that the fluid temperature can approach the surface temperature but cannot cross it, an NTU of about 5 indicates that the limit is reached for heat transfer, and the heat transfer does not increase no matter how much we extend the length of the tube. A small value of NTU, on the other hand, indicates more opportunities for heat transfer, and the heat transfer continues to increase as the tube length is increased. A large NTU and thus a large heat transfer surface area (which means a large tube) may be desirable from a heat transfer point of view, but it may be unacceptable from an economic point of view. The selection of heat transfer equipment usually reflects a compromise between heat transfer performance and cost.

Solving Eq. 19–57 for $\dot{m}c_p$ gives

$$\dot{m}c_p = -\frac{hA_s}{\ln[(T_s - T_e)/(T_s - T_i)]} \tag{19–59}$$

Substituting this into Eq. 19–43, we obtain

$$\dot{Q} = hA_s\Delta T_{\ln} \tag{19–60}$$

where

$$\Delta T_{\ln} = \frac{T_i - T_e}{\ln[(T_s - T_e)/(T_s - T_i)]} = \frac{\Delta T_e - \Delta T_i}{\ln(\Delta T_e/\Delta T_i)} \tag{19–61}$$

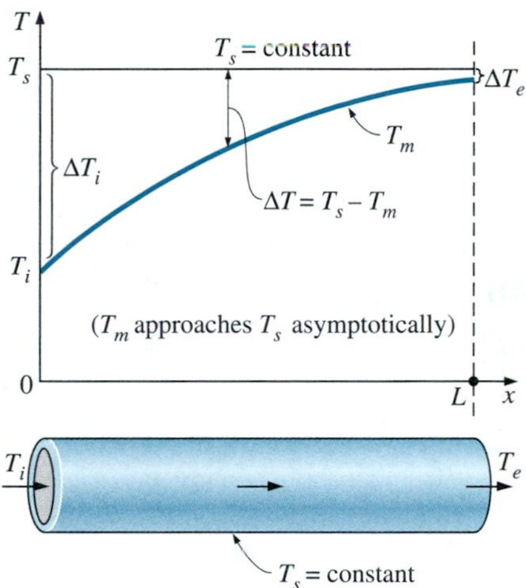

FIGURE 19–28
The variation of the *mean fluid temperature* along the tube for the case of constant temperature.

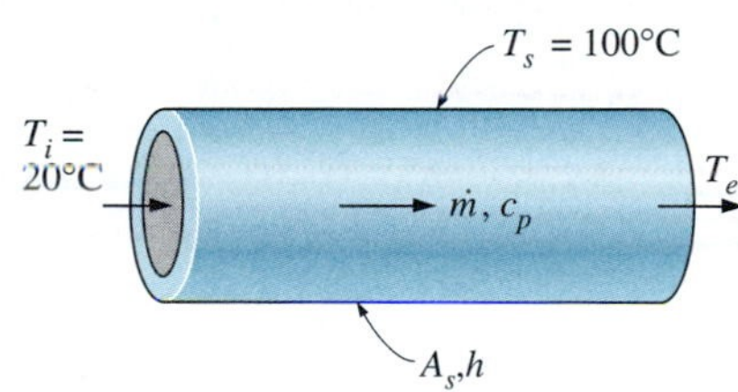

NTU = $hA_s/\dot{m}c_p$	T_e, °C
0.01	20.8
0.05	23.9
0.10	27.6
0.50	51.5
1.00	70.6
5.00	99.5
10.00	100.0

FIGURE 19–29
An NTU greater than 5 indicates that the fluid flowing in a tube will reach the surface temperature at the exit regardless of the inlet temperature.

is the **logarithmic mean temperature difference**. Note that $\Delta T_i = T_s - T_i$ and $\Delta T_e = T_s - T_e$ are the temperature differences between the surface and the fluid at the inlet and the exit of the tube, respectively. This $\Delta T_{\ln}$ relation appears to be prone to misuse, but it is practically fail-safe, since using T_i in place of T_e and vice versa in the numerator and/or the denominator will, at most, affect the sign, not the magnitude. Also, it can be used for both heating ($T_s > T_i$ and T_e) and cooling ($T_s < T_i$ and T_e) of a fluid in a tube.

The logarithmic mean temperature difference $\Delta T_{\ln}$ is obtained by tracing the actual temperature profile of the fluid along the tube, and is an *exact* representation of the *average temperature difference* between the fluid and the surface. It truly reflects the exponential decay of the local temperature difference. When ΔT_e differs from ΔT_i by no more than 40 percent, the error in using the arithmetic mean temperature difference is less than 1 percent. But the error increases to undesirable levels when ΔT_e differs from ΔT_i by greater amounts. Therefore, we should always use the logarithmic mean temperature difference when determining the convection heat transfer in a tube whose surface is maintained at a constant temperature T_s.

EXAMPLE 19–6 Heating of Water in a Tube by Steam

Water enters a 2.5-cm-internal-diameter thin copper tube of a heat exchanger at 15°C at a rate of 0.3 kg/s, and is heated by steam condensing outside at 120°C. If the average heat transfer coefficient is 800 W/m^2 · °C, determine the length of the tube required in order to heat the water to 115°C (Fig. 19–30).

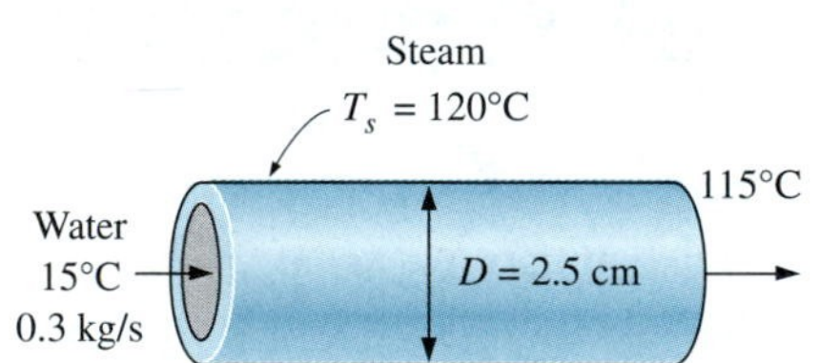

FIGURE 19–30
Schematic for Example 19–6.

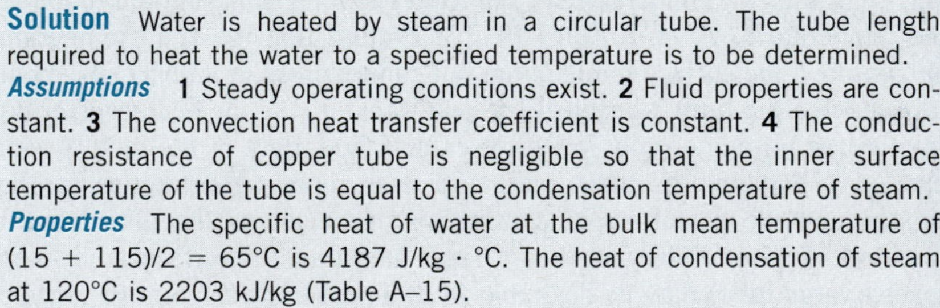

Solution Water is heated by steam in a circular tube. The tube length required to heat the water to a specified temperature is to be determined.

Assumptions **1** Steady operating conditions exist. **2** Fluid properties are constant. **3** The convection heat transfer coefficient is constant. **4** The conduction resistance of copper tube is negligible so that the inner surface temperature of the tube is equal to the condensation temperature of steam.

Properties The specific heat of water at the bulk mean temperature of (15 + 115)/2 = 65°C is 4187 J/kg · °C. The heat of condensation of steam at 120°C is 2203 kJ/kg (Table A–15).

Analysis Knowing the inlet and exit temperatures of water, the rate of heat transfer is determined to be

$$\dot{Q} = \dot{m}c_p(T_e - T_i) = (0.3 \text{ kg/s})(4.187 \text{ kJ/kg} \cdot \text{°C})(115\text{°C} - 15\text{°C})$$
$$= 125.6 \text{ kW}$$

The logarithmic mean temperature difference is

$$\Delta T_e = T_s - T_e = 120\text{°C} - 115\text{°C} = 5\text{°C}$$
$$\Delta T_i = T_s - T_i = 120\text{°C} - 15\text{°C} = 105\text{°C}$$
$$\Delta T_{\ln} = \frac{\Delta T_e - \Delta T_i}{\ln(\Delta T_e/\Delta T_i)} = \frac{5 - 105}{\ln(5/105)} = 32.85\text{°C}$$

The heat transfer surface area is

$$\dot{Q} = hA_s\Delta T_{\ln} \longrightarrow A_s = \frac{\dot{Q}}{h\Delta T_{\ln}} = \frac{125.6 \text{ kW}}{(0.8 \text{ kW/m}^2 \cdot \text{°C})(32.85\text{°C})} = 4.78 \text{ m}^2$$

Then the required tube length becomes

$$A_s = \pi DL \longrightarrow L = \frac{A_s}{\pi D} = \frac{4.78 \text{ m}^2}{\pi(0.025 \text{ m})} = \mathbf{61 \text{ m}}$$

Discussion The bulk mean temperature of water during this heating process is 65°C, and thus the *arithmetic* mean temperature difference is $\Delta T_{am} = 120 - 65 = 55°\text{C}$. Using ΔT_{am} instead of ΔT_{ln} would give $L = 36$ m, which is grossly in error. This shows the importance of using the logarithmic mean temperature in calculations.

19–7 ▪ LAMINAR FLOW IN TUBES

Consider steady laminar flow of a fluid in a circular tube of radius R. The fluid properties ρ, k, and c_p are constant, and the work done by viscous forces is negligible. The fluid flows along the x-axis with velocity u. The flow is fully developed so that u is independent of x and thus $u = u(r)$. Noting that energy is transferred by mass in the x-direction, and by conduction in the r-direction (heat conduction in the x-direction is assumed to be negligible), the steady-flow energy balance for a cylindrical shell element of thickness dr and length dx can be expressed as (Fig. 19–31)

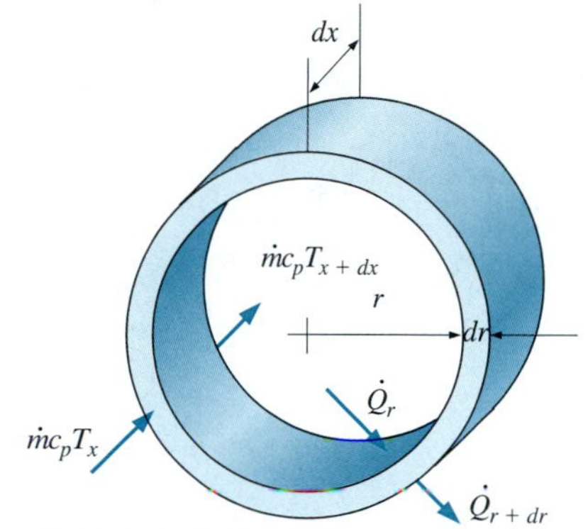

FIGURE 19–31
The differential volume element used in the derivation of energy balance relation.

$$\dot{m}c_pT_x - \dot{m}c_pT_{x+dx} + \dot{Q}_r - \dot{Q}_{r+dr} = 0 \quad \text{(19–62)}$$

where $\dot{m} = \rho u A_c = \rho u(2\pi r dr)$. Substituting and dividing by $2\pi r dr dx$ gives, after rearranging,

$$\rho c_p u \frac{T_{x+dx} - T_x}{dx} = -\frac{1}{2\pi r dx}\frac{\dot{Q}_{r+dr} - \dot{Q}_r}{dr} \quad \text{(19–63)}$$

or

$$u\frac{\partial T}{\partial x} = -\frac{1}{2\rho c_p \pi r dx}\frac{\partial \dot{Q}}{\partial r} \quad \text{(19–64)}$$

But

$$\frac{\partial \dot{Q}}{\partial r} = \frac{\partial}{\partial r}\left(-k2\pi r dx\frac{\partial T}{\partial r}\right) = -2\pi k dx\frac{\partial}{\partial r}\left(r\frac{\partial T}{\partial r}\right) \quad \text{(19–65)}$$

Substituting and using $\alpha = k/\rho c_p$ gives

$$u\frac{\partial T}{\partial x} = \frac{\alpha}{r}\frac{\partial}{dr}\left(r\frac{\partial T}{\partial r}\right) \quad \text{(19–66)}$$

which states that *the rate of net energy transfer to the control volume by mass flow is equal to the net rate of heat conduction in the radial direction.*

Constant Surface Heat Flux

For fully developed flow in a circular tube subjected to constant surface heat flux, we have, from Eq. 19–52,

$$\frac{\partial T}{\partial x} = \frac{dT_s}{dx} = \frac{dT_m}{dx} = \frac{2\dot{q}_s}{\rho V_{avg} c_p R} = \text{constant} \tag{19–67}$$

If heat conduction in the x-direction were considered in the derivation of Eq. 19–66, it would give an additional term $\alpha \partial^2 T/\partial x^2$, which would be equal to zero since $\partial T/\partial x$ = constant and thus $T = T(r)$. Therefore, the assumption that there is no axial heat conduction is satisfied exactly in this case.

Substituting Eq. 19–67 and the relation for velocity profile into Eq. 19–66 gives

$$\frac{4\dot{q}_s}{kR}\left(1 - \frac{r^2}{R^2}\right) = \frac{1}{r}\frac{d}{dr}\left(r\frac{dT}{dr}\right) \tag{19–68}$$

which is a second-order ordinary differential equation. Its general solution is obtained by separating the variables and integrating twice to be

$$T = \frac{\dot{q}_s}{kR}\left(r^2 - \frac{r^4}{4R^2}\right) + C_1 r + C_2 \tag{19–69}$$

The desired solution to the problem is obtained by applying the boundary conditions $\partial T/\partial x = 0$ at $r = 0$ (because of symmetry) and $T = T_s$ at $r = R$. We get

$$T = T_s - \frac{\dot{q}_s R}{k}\left(\frac{3}{4} - \frac{r^2}{R^2} + \frac{r^4}{4R^4}\right) \tag{19–70}$$

The bulk mean temperature T_m is determined by substituting the velocity and temperature profile relation 19–70 into Eq. 19–34 and performing the integration. It gives

$$T_m = T_s - \frac{11}{24}\frac{\dot{q}_s R}{k} \tag{19–71}$$

Combining this relation with $\dot{q}_s = h(T_s - T_m)$ gives

$$h = \frac{24}{11}\frac{k}{R} = \frac{48}{11}\frac{k}{D} = 4.36\frac{k}{D} \tag{19–72}$$

or

Circular tube, laminar ($\dot{q}_s$ = constant): $$\text{Nu} = \frac{hD}{k} = 4.36 \tag{19–73}$$

Therefore, for fully developed laminar flow in a circular tube subjected to constant surface heat flux, the Nusselt number is a constant. There is no dependence on the Reynolds or the Prandtl numbers.

Constant Surface Temperature

A similar analysis can be performed for fully developed laminar flow in a circular tube for the case of constant surface temperature T_s. The solution procedure in this case is more complex as it requires iterations, but the Nusselt number relation obtained is equally simple (Fig. 19–32):

Circular tube, laminar (T_s = constant): $$\text{Nu} = \frac{hD}{k} = 3.66 \qquad \textbf{(19–74)}$$

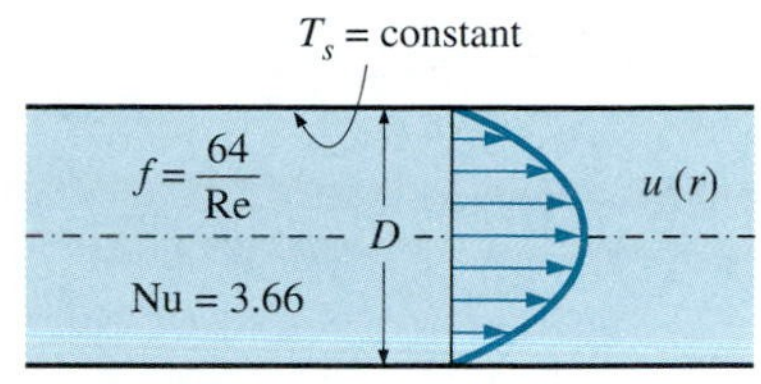

Fully developed laminar flow

FIGURE 19–32
In laminar flow in a tube with constant surface temperature, both the *friction factor* and the *heat transfer coefficient* remain constant in the fully developed region.

The thermal conductivity k for use in the Nu relations above should be evaluated at the bulk mean fluid temperature, which is the arithmetic average of the mean fluid temperatures at the inlet and the exit of the tube. For laminar flow, the effect of *surface roughness* on the friction factor and the heat transfer coefficient is negligible.

Laminar Flow in Noncircular Tubes

Nusselt number relations are given in Table 19–3 for *fully developed laminar flow* in tubes of various cross sections. The Reynolds and Nusselt num-

TABLE 19–3

Nusselt number and friction factor for fully developed laminar flow in tubes of various cross sections ($D_h = 4A_c/p$, $\text{Re} = V_{\text{avg}}D_h/\nu$, and $\text{Nu} = hD_h/k$)

Tube Geometry	a/b or $\theta°$	Nusselt Number T_s = Const.	Nusselt Number $\dot{q}_s$ = Const.
Circle	—	3.66	4.36
Rectangle	a/b		
	1	2.98	3.61
	2	3.39	4.12
	3	3.96	4.79
	4	4.44	5.33
	6	5.14	6.05
	8	5.60	6.49
	∞	7.54	8.24
Ellipse	a/b		
	1	3.66	4.36
	2	3.74	4.56
	4	3.79	4.88
	8	3.72	5.09
	16	3.65	5.18
Isosceles Triangle	θ		
	10°	1.61	2.45
	30°	2.26	2.91
	60°	2.47	3.11
	90°	2.34	2.98
	120°	2.00	2.68

bers for flow in these tubes are based on the hydraulic diameter $D_h = 4A_c/p$, where A_c is the cross sectional area of the tube and p is its perimeter. Once the Nusselt number is available, the convection heat transfer coefficient is determined from $h = k\text{Nu}/D_h$.

Developing Laminar Flow in the Entrance Region

For a circular tube of length L subjected to constant surface temperature, the average Nusselt number for the *thermal entrance region* can be determined from (Edwards et al., 1979)

Entry region, laminar:
$$\text{Nu} = 3.66 + \frac{0.065\,(D/L)\,\text{Re}\,\text{Pr}}{1 + 0.04[(D/L)\,\text{Re}\,\text{Pr}]^{2/3}} \qquad \textbf{(19–75)}$$

Note that the average Nusselt number is larger at the entrance region, as expected, and it approaches asymptotically to the fully developed value of 3.66 as $L \rightarrow \infty$. This relation assumes that the flow is hydrodynamically developed when the fluid enters the heating section, but it can also be used approximately for flow developing hydrodynamically.

When the difference between the surface and the fluid temperatures is large, it may be necessary to account for the variation of viscosity with temperature.

The average Nusselt number for developing laminar flow in a circular tube in that case can be determined from [Sieder and Tate, 1936]

$$\text{Nu} = 1.86\left(\frac{\text{Re}\,\text{Pr}\,D}{L}\right)^{1/3}\left(\frac{\mu_b}{\mu_s}\right)^{0.14} \qquad \textbf{(19–76)}$$

All properties are evaluated at the bulk mean fluid temperature, except for μ_s, which is evaluated at the surface temperature.

The average Nusselt number for the thermal entrance region of flow between *isothermal parallel plates* of length L is expressed as (Edwards et al., 1979)

Entry region, laminar:
$$\text{Nu} = 7.54 + \frac{0.03\,(D_h/L)\,\text{Re}\,\text{Pr}}{1 + 0.016[(D_h/L)\,\text{Re}\,\text{Pr}]^{2/3}} \qquad \textbf{(19–77)}$$

where D_h is the hydraulic diameter, which is twice the spacing of the plates. This relation can be used for $\text{Re} \leq 2800$.

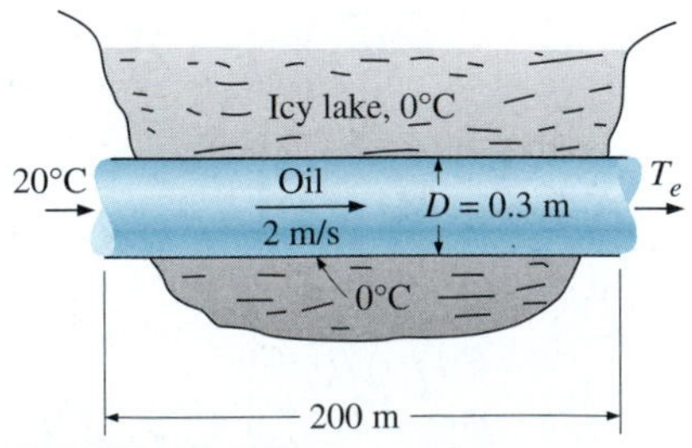

FIGURE 19–33
Schematic for Example 19–7.

EXAMPLE 19–7 Flow of Oil in a Pipeline through a Lake

Consider the flow of oil at 20°C in a 30-cm-diameter pipeline at an average velocity of 2 m/s (Fig. 19–33). A 200-m-long section of the horizontal pipeline passes through icy waters of a lake at 0°C. Measurements indicate that the surface temperature of the pipe is very nearly 0°C. Disregarding the thermal resistance of the pipe material, determine (*a*) the temperature of the oil when the pipe leaves the lake, (*b*) the rate of heat transfer from the oil, and (*c*) the pumping power required to overcome the pressure losses and to maintain the flow of the oil in the pipe.

Solution Oil flows in a pipeline that passes through icy waters of a lake at 0°C. The exit temperature of the oil, the rate of heat loss, and the pumping power needed to overcome pressure losses are to be determined.

Assumptions **1** Steady operating conditions exist. **2** The surface temperature of the pipe is very nearly 0°C. **3** The thermal resistance of the pipe is negligible. **4** The inner surfaces of the pipeline are smooth. **5** The flow is hydrodynamically developed when the pipeline reaches the lake.

Properties We do not know the exit temperature of the oil, and thus we cannot determine the bulk mean temperature, which is the temperature at which the properties of oil are to be evaluated. The mean temperature of the oil at the inlet is 20°C, and we expect this temperature to drop somewhat as a result of heat loss to the icy waters of the lake. We evaluate the properties of the oil at the inlet temperature, but we will repeat the calculations, if necessary, using properties at the evaluated bulk mean temperature. At 20°C we read (Table A-19)

$$\rho = 888.1 \text{ kg/m}^3 \qquad \nu = 9.429 \times 10^{-4} \text{ m}^2\text{/s}$$

$$k = 0.145 \text{ W/m} \cdot {}^\circ\text{C} \qquad c_p = 1880 \text{ J/kg} \cdot {}^\circ\text{C} \qquad \text{Pr} = 10{,}863$$

Analysis (*a*) The Reynolds number is

$$\text{Re} = \frac{V_{\text{avg}} D}{\nu} = \frac{(2 \text{ m/s})(0.3 \text{ m})}{9.429 \times 10^{-4} \text{ m}^2\text{/s}} = 636$$

which is less than the critical Reynolds number of 2300. Therefore, the flow is laminar, and the thermal entry length in this case is roughly

$$L_t \approx 0.05 \text{ Re Pr } D = 0.05 \times 636 \times 10{,}863 \times (0.3 \text{ m}) \approx 103{,}600 \text{ m}$$

which is much greater than the total length of the pipe. This is typical of fluids with high Prandtl numbers. Therefore, we assume thermally developing flow and determine the Nusselt number from

$$\begin{aligned} \text{Nu} = \frac{hD}{k} &= 3.66 + \frac{0.065\,(D/L)\text{ Re Pr}}{1 + 0.04\,[(D/L)\text{ Re Pr}]^{2/3}} \\ &= 3.66 + \frac{0.065(0.3/200) \times 636 \times 10{,}863}{1 + 0.04[(0.3/200) \times 636 \times 10{,}863]^{2/3}} \\ &= 33.7 \end{aligned}$$

Note that this Nusselt number is considerably higher than the fully developed value of 3.66. Then,

$$h = \frac{k}{D}\text{Nu} = \frac{0.145 \text{ W/m} \cdot {}^\circ\text{C}}{0.3 \text{ m}}(33.7) = 16.3 \text{ W/m}^2 \cdot {}^\circ\text{C}$$

Also,

$$A_s = \pi DL = \pi(0.3 \text{ m})(200 \text{ m}) = 188.5 \text{ m}^2$$

$$\dot{m} = \rho A_c V_{\text{avg}} = (888.1 \text{ kg/m}^3)[\tfrac{1}{4}\pi(0.3 \text{ m})^2](2 \text{ m/s}) = 125.6 \text{ kg/s}$$

Next we determine the exit temperature of oil,

$$T_e = T_s - (T_s - T_i)\exp(-hA_s/\dot{m}c_p)$$

$$= 0°\text{C} - [(0 - 20)°\text{C}]\exp\left[-\frac{(16.3\ \text{W/m}^2\cdot°\text{C})(188.5\ \text{m}^2)}{(125.6\ \text{kg/s})(1881\ \text{J/kg}\cdot°\text{C})}\right]$$

$$= \mathbf{19.74°C}$$

Thus, the mean temperature of oil drops by a mere 0.26°C as it crosses the lake. This makes the bulk mean oil temperature 19.87°C, which is practically identical to the inlet temperature of 20°C. Therefore, we do not need to re-evaluate the properties.

(*b*) The logarithmic mean temperature difference and the rate of heat loss from the oil are

$$\Delta T_{\text{ln}} = \frac{T_i - T_e}{\ln\dfrac{T_s - T_e}{T_s - T_i}} = \frac{20 - 19.74}{\ln\dfrac{0 - 19.74}{0 - 20}} = -19.87°\text{C}$$

$$\dot{Q} = hA_s\Delta T_{\text{ln}} = (16.3\ \text{W/m}^2\cdot°\text{C})(188.5\ \text{m}^2)(-19.87°\text{C}) = \mathbf{-6.11\times 10^4\ W}$$

Therefore, the oil will lose heat at a rate of 61.1 kW as it flows through the pipe in the icy waters of the lake. Note that ΔT_{ln} is identical to the arithmetic mean temperature in this case, since $\Delta T_i \approx \Delta T_e$.

(*c*) The laminar flow of oil is hydrodynamically developed. Therefore, the friction factor can be determined from (see Chapter 14)

$$f = \frac{64}{\text{Re}} = \frac{64}{636} = 0.1006$$

Then the pressure drop in the pipe and the required pumping power become

$$\Delta P = f\frac{L}{D}\frac{\rho V_{\text{avg}}^2}{2} = 0.1006\frac{200\ \text{m}}{0.3\ \text{m}}\frac{(888.1\ \text{kg/m}^3)(2\ \text{m/s})^2}{2} = 1.19\times 10^5\ \text{N/m}^2$$

$$\dot{W}_{\text{pump}} = \frac{\dot{m}\Delta P}{\rho} = \frac{(125.6\ \text{kg/s})(1.19\times 10^5\ \text{N/m}^2)}{888.1\ \text{kg/m}^3} = \mathbf{16.8\ kW}$$

Discussion We need a 16.8-kW pump just to overcome the friction in the pipe as the oil flows in the 200-m-long pipe through the lake.

19–8 ▪ TURBULENT FLOW IN TUBES

We mentioned in Chapter 14 that flow in smooth tubes is usually fully turbulent for Re > 10,000. Turbulent flow is commonly utilized in practice because of the higher heat transfer coefficients associated with it. Most correlations for the friction and heat transfer coefficients in turbulent flow are based on experimental studies because of the difficulty in dealing with turbulent flow theoretically.

For *smooth* tubes, the friction factor in turbulent flow can be determined from the explicit *first Petukhov equation* [Petukhov, 1970] given as

Smooth tubes: $\quad f = (0.790\ln\text{Re} - 1.64)^{-2} \qquad 3000 < \text{Re} < 5\times 10^6$ **(19–78)**

The Nusselt number in turbulent flow is related to the friction factor through the *Chilton–Colburn analogy* expressed as

$$\text{Nu} = 0.125 f \text{RePr}^{1/3} \quad \textbf{(19–79)}$$

Once the friction factor is available, this equation can be used conveniently to evaluate the Nusselt number for both smooth and rough tubes.

For fully developed turbulent flow in *smooth tubes,* a simple relation for the Nusselt number can be obtained by substituting the simple power law relation $f = 0.184\ \text{Re}^{-0.2}$ for the friction factor into Eq. 19–79. It gives

$$\text{Nu} = 0.023\ \text{Re}^{0.8}\ \text{Pr}^{1/3} \qquad \begin{pmatrix} 0.7 \leq \text{Pr} \leq 160 \\ \text{Re} > 10{,}000 \end{pmatrix} \quad \textbf{(19–80)}$$

which is known as the *Colburn equation.* The accuracy of this equation can be improved by modifying it as

$$\text{Nu} = 0.023\ \text{Re}^{0.8}\ \text{Pr}^{n} \quad \textbf{(19–81)}$$

where $n = 0.4$ for *heating* and 0.3 for *cooling* of the fluid flowing through the tube. This equation is known as the *Dittus–Boelter equation* (Dittus and Boelter, 1930) and it is preferred to the Colburn equation.

The proceeding equations can be used when the temperature difference between the fluid and wall surface is not large by evaluating all fluid properties at the *bulk mean fluid temperature* $T_b = (T_i + T_e)/2$. When the variation in properties is large due to a large temperature difference, the following equation due to Sieder and Tate (1936) can be used:

$$\text{Nu} = 0.027\ \text{Re}^{0.8}\text{Pr}^{1/3}\left(\frac{\mu}{\mu_s}\right)^{0.14} \qquad \begin{pmatrix} 0.7 \leq \text{Pr} \leq 17{,}600 \\ \text{Re} \geq 10{,}000 \end{pmatrix} \quad \textbf{(19–82)}$$

Here all properties are evaluated at T_b except μ_s, which is evaluated at T_s.

The Nusselt number relations above are fairly simple, but they may give errors as large as 25 percent. This error can be reduced considerably to less than 10 percent by using more complex but accurate relations such as the *second Petukhov equation* expressed as

$$\text{Nu} = \frac{(f/8)\ \text{Re Pr}}{1.07 + 12.7(f/8)^{0.5}\ (\text{Pr}^{2/3} - 1)} \qquad \begin{pmatrix} 0.5 \leq \text{Pr} \leq 2000 \\ 10^4 < \text{Re} < 5 \times 10^6 \end{pmatrix} \quad \textbf{(19–83)}$$

The accuracy of this relation at lower Reynolds numbers is improved by modifying it as (Gnielinski, 1976)

$$\text{Nu} = \frac{(f/8)(\text{Re} - 1000)\ \text{Pr}}{1 + 12.7(f/8)^{0.5}\ (\text{Pr}^{2/3} - 1)} \qquad \begin{pmatrix} 0.5 \leq \text{Pr} \leq 2000 \\ 3 \times 10^3 < \text{Re} < 5 \times 10^6 \end{pmatrix} \quad \textbf{(19–84)}$$

where the friction factor f can be determined from an appropriate relation such as the first Petukhov equation. Gnielinski's equation should be

TABLE 19–4

Equivalent roughness values for new commercial pipes*

Material	Roughness, ε ft	Roughness, ε mm
Glass, plastic	0 (smooth)	
Concrete	0.003–0.03	0.9–9
Wood stave	0.0016	0.5
Rubber, smoothed	0.000033	0.01
Copper or brass tubing	0.000005	0.0015
Cast iron	0.00085	0.26
Galvanized iron	0.0005	0.15
Wrought iron	0.00015	0.046
Stainless steel	0.000007	0.002
Commercial steel	0.00015	0.045

*The uncertainty in these values can be as much as ±60 percent.

preferred in calculations. Again properties should be evaluated at the bulk mean fluid temperature.

The relations above are not very sensitive to the *thermal conditions* at the tube surfaces and can be used for both T_s = constant and $\dot{q}_s$ = constant cases. Despite their simplicity, the correlations already presented give sufficiently accurate results for most engineering purposes. They can also be used to obtain rough estimates of the friction factor and the heat transfer coefficients in the transition region.

The relations given so far do not apply to liquid metals because of their very low Prandtl numbers. For liquid metals ($0.004 < \text{Pr} < 0.01$), the following relations are recommended by Sleicher and Rouse (1975) for $10^4 < \text{Re} < 10^6$:

Liquid metals, T_s = constant:
$$\text{Nu} = 4.8 + 0.0156\,\text{Re}^{0.85}\,\text{Pr}_s^{0.93} \quad \textbf{(19–85)}$$

Liquid metals, $\dot{q}_s$ = constant:
$$\text{Nu} = 6.3 + 0.0167\,\text{Re}^{0.85}\,\text{Pr}_s^{0.93} \quad \textbf{(19–86)}$$

where the subscript s indicates that the Prandtl number is to be evaluated at the surface temperature.

In turbulent flow, wall roughness increases the heat transfer coefficient h by a factor of 2 or more (Dipprey and Sabersky, 1963). The convection heat transfer coefficient for rough tubes can be calculated approximately from the Nusselt number relations such as Eq. 19–84 by using the friction factor determined from the Moody chart or the Colebrook equation. However, this approach is not very accurate since there is no further increase in h with f for $f > 4f_{\text{smooth}}$ (Norris, 1970) and correlations developed specifically for rough tubes should be used when more accuracy is desired.

Developing Turbulent Flow in the Entrance Region

The entry lengths for turbulent flow are typically short, often just 10 tube diameters long, and thus the Nusselt number determined for fully developed turbulent flow can be used approximately for the entire tube. This simple approach gives reasonable results for pressure drop and heat transfer for long tubes and conservative results for short ones. Correlations for the friction and heat transfer coefficients for the entrance regions are available in the literature for better accuracy.

Turbulent Flow in Noncircular Tubes

The velocity and temperature profiles in turbulent flow are nearly straight lines in the core region, and any significant velocity and temperature gradients occur in the viscous sublayer (Fig. 19–34). Despite the small thickness of viscous sublayer (usually much less than 1 percent of the pipe diameter), the characteristics of the flow in this layer are very important since they set the stage for flow in the rest of the pipe. Therefore, pressure drop and heat transfer characteristics of turbulent flow in tubes are dominated by the very thin viscous sublayer next to the wall surface, and the shape of the core region is not of much significance. Consequently, the turbulent flow relations given above for circular tubes can also be used for noncircular tubes with reasonable accuracy by replacing the diameter D in the evaluation of the Reynolds number by the hydraulic diameter $D_h = 4A_c/p$.

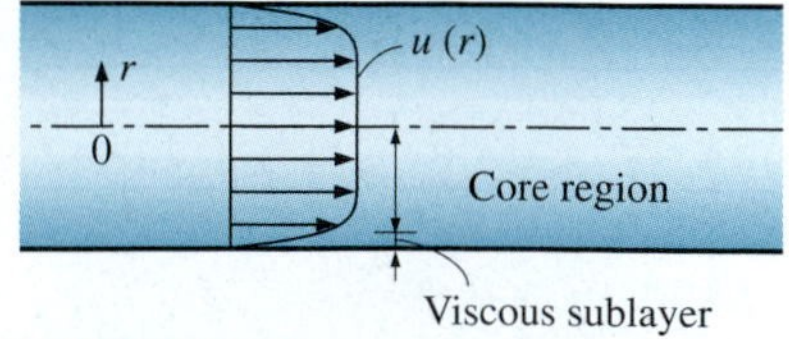

FIGURE 19–34
In turbulent flow, the velocity profile is nearly a straight line in the core region, and any significant velocity gradients occur in the viscous sublayer.

Flow through Tube Annulus

Some simple heat transfer equipments consist of two concentric tubes, and are properly called *double-tube heat exchangers* (Fig. 19–35). In such devices, one fluid flows through the tube while the other flows through the annular space. The governing differential equations for both flows are identical. Therefore, steady laminar flow through an annulus can be studied analytically by using suitable boundary conditions.

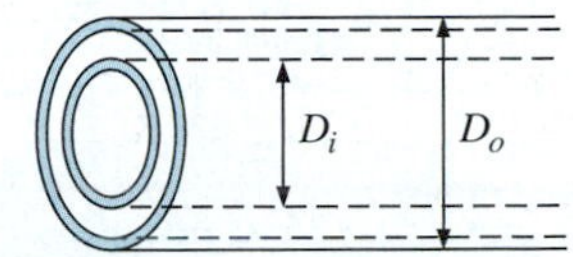

FIGURE 19–35
A double-tube heat exchanger that consists of two concentric tubes.

Consider a concentric annulus of inner diameter D_i and outer diameter D_o. The hydraulic diameter of annulus is

$$D_h = \frac{4A_c}{p} = \frac{4\pi(D_o^2 - D_i^2)/4}{\pi(D_o + D_i)} = D_o - D_i$$

Annular flow is associated with two Nusselt numbers—Nu_i on the inner tube surface and Nu_o on the outer tube surface—since it may involve heat transfer on both surfaces. The Nusselt numbers for fully developed laminar flow with one surface isothermal and the other adiabatic are given in Table 19–5. When Nusselt numbers are known, the convection coefficients for the inner and the outer surfaces are determined from

$$\text{Nu}_i = \frac{h_i D_h}{k} \quad \text{and} \quad \text{Nu}_o = \frac{h_o D_h}{k} \tag{19–87}$$

TABLE 19–5

Nusselt number for fully developed laminar flow in an annulus with one surface isothermal and the other adiabatic (Kays and Perkins, 1972)

D_i/D_o	Nu_i	Nu_o
0	—	3.66
0.05	17.46	4.06
0.10	11.56	4.11
0.25	7.37	4.23
0.50	5.74	4.43
1.00	4.86	4.86

For fully developed turbulent flow, the inner and outer convection coefficients are approximately equal to each other, and the tube annulus can be treated as a noncircular duct with a hydraulic diameter of $D_h = D_o - D_i$. The Nusselt number in this case can be determined from a suitable turbulent flow relation such as the Gnielinski equation. To improve the accuracy of Nusselt numbers obtained from these relations for annular flow, Petukhov and Roizen (1964) recommend multiplying them by the following correction factors when one of the tube walls is adiabatic and heat transfer is through the other wall:

$$F_i = 0.86\left(\frac{D_i}{D_o}\right)^{-0.16} \quad \text{(outer wall adiabatic)} \tag{19–88}$$

$$F_o = 0.86\left(\frac{D_i}{D_o}\right)^{-0.16} \quad \text{(inner wall adiabatic)} \tag{19–89}$$

Heat Transfer Enhancement

Tubes with rough surfaces have much higher heat transfer coefficients than tubes with smooth surfaces. Therefore, tube surfaces are often intentionally *roughened, corrugated,* or *finned* in order to *enhance* the convection heat transfer coefficient and thus the convection heat transfer rate (Fig. 19–36). Heat transfer in turbulent flow in a tube has been increased by as much as 400 percent by roughening the surface. Roughening the surface, of course, also increases the friction factor and thus the power requirement for the pump or the fan.

The convection heat transfer coefficient can also be increased by inducing pulsating flow by pulse generators, by inducing swirl by inserting a twisted tape into the tube, or by inducing secondary flows by coiling the tube.

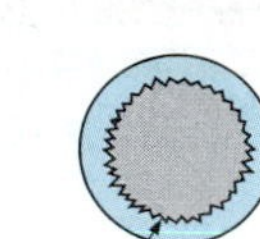

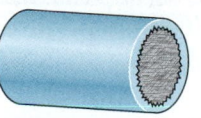

FIGURE 19–36
Tube surfaces are often *roughened, corrugated,* or *finned* in order to *enhance* convection heat transfer.

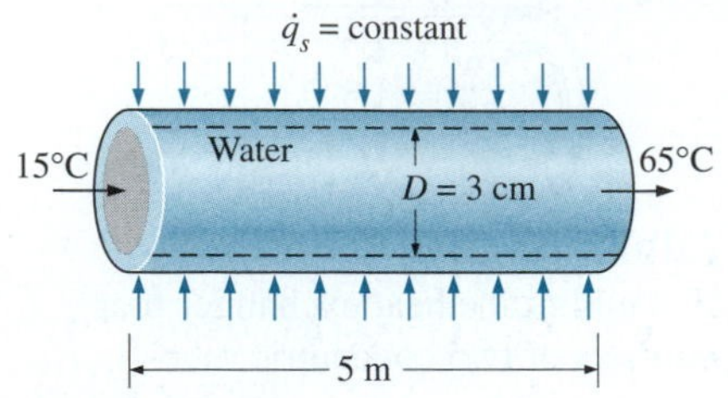

FIGURE 19–37
Schematic for Example 19–8.

EXAMPLE 19–8 Heating of Water by Resistance Heaters in a Tube

Water is to be heated from 15°C to 65°C as it flows through a 3-cm-internal-diameter 5-m-long tube (Fig. 19–37). The tube is equipped with an electric resistance heater that provides uniform heating throughout the surface of the tube. The outer surface of the heater is well insulated, so that in steady operation all the heat generated in the heater is transferred to the water in the tube. If the system is to provide hot water at a rate of 10 L/min, determine the power rating of the resistance heater. Also, estimate the inner surface temperature of the tube at the exit.

Solution Water is to be heated in a tube equipped with an electric resistance heater on its surface. The power rating of the heater and the inner surface temperature at the exit are to be determined.
Assumptions **1** Steady flow conditions exist. **2** The surface heat flux is uniform. **3** The inner surfaces of the tube are smooth.
Properties The properties of water at the bulk mean temperature of $T_b = (T_i + T_e)/2 = (15 + 65)/2 = 40°C$ are (Table A–15)

$$\rho = 992.1 \text{ kg/m}^3 \qquad c_p = 4179 \text{ J/kg} \cdot °\text{C}$$
$$k = 0.631 \text{ W/m} \cdot °\text{C} \qquad \text{Pr} = 4.32$$
$$\nu = \mu/\rho = 0.658 \times 10^{-6} \text{ m}^2/\text{s}$$

Analysis The cross sectional and heat transfer surface areas are

$$A_c = \tfrac{1}{4}\pi D^2 = \tfrac{1}{4}\pi(0.03 \text{ m})^2 = 7.069 \times 10^{-4} \text{ m}^2$$
$$A_s = \pi DL = \pi(0.03 \text{ m})(5 \text{ m}) = 0.471 \text{ m}^2$$

The volume flow rate of water is given as $\dot{V} = 10$ L/min $= 0.01$ m^3/min. Then the mass flow rate becomes

$$\dot{m} = \rho\dot{V} = (992.1 \text{ kg/m}^3)(0.01 \text{ m}^3/\text{min}) = 9.921 \text{ kg/min} = 0.1654 \text{ kg/s}$$

To heat the water at this mass flow rate from 15°C to 65°C, heat must be supplied to the water at a rate of

$$\begin{aligned}\dot{Q} &= \dot{m}c_p(T_e - T_i)\\ &= (0.1654 \text{ kg/s})(4.179 \text{ kJ/kg} \cdot °\text{C})(65 - 15)°\text{C}\\ &= 34.6 \text{ kJ/s} = 34.6 \text{ kW}\end{aligned}$$

All of this energy must come from the resistance heater. Therefore, the power rating of the heater must be **34.6 kW**.

The surface temperature T_s of the tube at any location can be determined from

$$\dot{q}_s = h(T_s - T_m) \quad \rightarrow \quad T_s = T_m + \frac{\dot{q}_s}{h}$$

where h is the heat transfer coefficient and T_m is the mean temperature of the fluid at that location. The surface heat flux is constant in this case, and its value can be determined from

$$\dot{q}_s = \frac{\dot{Q}}{A_s} = \frac{34.6 \text{ kW}}{0.471 \text{ m}^2} = 73.46 \text{ kW/m}^2$$

To determine the heat transfer coefficient, we first need to find the mean velocity of water and the Reynolds number:

$$V_{\text{avg}} = \frac{\dot{V}}{A_c} = \frac{0.010 \text{ m}^3/\text{min}}{7.069 \times 10^{-4} \text{ m}^2} = 14.15 \text{ m/min} = 0.236 \text{ m/s}$$

$$\text{Re} = \frac{V_{\text{avg}} D}{\nu} = \frac{(0.236 \text{ m/s})(0.03 \text{ m})}{0.658 \times 10^{-6} \text{ m}^2/\text{s}} = 10{,}760$$

which is greater than 10,000. Therefore, the flow is turbulent and the entry length is roughly

$$L_h \approx L_t \approx 10D = 10 \times 0.03 = 0.3 \text{ m}$$

which is much shorter than the total length of the tube. Therefore, we can assume fully developed turbulent flow in the entire tube and determine the Nusselt number from

$$\text{Nu} = \frac{hD}{k} = 0.023 \text{ Re}^{0.8} \text{ Pr}^{0.4} = 0.023(10{,}760)^{0.8} (4.32)^{0.4} = 69.4$$

Then,

$$h = \frac{k}{D} \text{Nu} = \frac{0.631 \text{ W/m} \cdot {}^\circ\text{C}}{0.03 \text{ m}} (69.4) = 1460 \text{ W/m}^2 \cdot {}^\circ\text{C}$$

and the surface temperature of the pipe at the exit becomes

$$T_s = T_m + \frac{\dot{q}_s}{h} = 65^\circ\text{C} + \frac{73{,}460 \text{ W/m}^2}{1460 \text{ W/m}^2 \cdot {}^\circ\text{C}} = \mathbf{115^\circ C}$$

Discussion Note that the inner surface temperature of the tube will be 50°C higher than the mean water temperature at the tube exit. This temperature difference of 50°C between the water and the surface will remain constant throughout the fully developed flow region.

EXAMPLE 19–9 Heat Loss from the Ducts of a Heating System

Hot air at atmospheric pressure and 80°C enters an 8-m-long uninsulated square duct of cross section 0.2 m × 0.2 m that passes through the attic of a house at a rate of 0.15 m^3/s (Fig. 19–38). The duct is observed to be nearly isothermal at 60°C. Determine the exit temperature of the air and the rate of heat loss from the duct to the attic space.

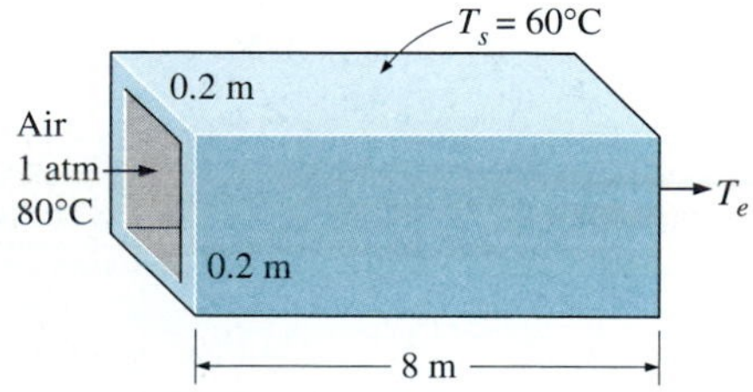

FIGURE 19–38
Schematic for Example 19–9.

Solution Heat loss from uninsulated square ducts of a heating system in the attic is considered. The exit temperature and the rate of heat loss are to be determined.

Assumptions **1** Steady operating conditions exist. **2** The inner surfaces of the duct are smooth. **3** Air is an ideal gas.

Properties We do not know the exit temperature of the air in the duct, and thus we cannot determine the bulk mean temperature of air, which is the temperature at which the properties are to be determined. The temperature of air at the inlet is 80°C and we expect this temperature to drop somewhat as a result of heat loss through the duct whose surface is at 60°C. At 80°C and 1 atm we read (Table A–22)

$$\rho = 0.9994 \text{ kg/m}^3 \qquad c_p = 1008 \text{ J/kg}\cdot{}^\circ\text{C}$$
$$k = 0.02953 \text{ W/m}\cdot{}^\circ\text{C} \qquad \text{Pr} = 0.7154$$
$$\nu = 2.097 \times 10^{-5} \text{ m}^2\text{/s}$$

Analysis The characteristic length (which is the hydraulic diameter), the mean velocity, and the Reynolds number in this case are

$$D_h = \frac{4A_c}{p} = \frac{4a^2}{4a} = a = 0.2 \text{ m}$$

$$V_{\text{avg}} = \frac{\dot{V}}{A_c} = \frac{0.15 \text{ m}^3\text{/s}}{(0.2 \text{ m})^2} = 3.75 \text{ m/s}$$

$$\text{Re} = \frac{V_{\text{avg}} D_h}{\nu} = \frac{(3.75 \text{ m/s})(0.2 \text{ m})}{2.097 \times 10^{-5} \text{ m}^2\text{/s}} = 35{,}765$$

which is greater than 10,000. Therefore, the flow is turbulent and the entry lengths in this case are roughly

$$L_h \approx L_t \approx 10D = 10 \times 0.2 \text{ m} = 2 \text{ m}$$

which is much shorter than the total length of the duct. Therefore, we can assume fully developed turbulent flow in the entire duct and determine the Nusselt number from

$$\text{Nu} = \frac{hD_h}{k} = 0.023 \text{ Re}^{0.8} \text{ Pr}^{0.3} = 0.023(35{,}765)^{0.8}(0.7154)^{0.3} = 91.4$$

Then,

$$h = \frac{k}{D_h}\text{Nu} = \frac{0.02953 \text{ W/m}\cdot{}^\circ\text{C}}{0.2 \text{ m}}(91.4) = 13.5 \text{ W/m}^2\cdot{}^\circ\text{C}$$

$$A_s = 4aL = 4 \times (0.2 \text{ m})(8 \text{ m}) = 6.4 \text{ m}^2$$

$$\dot{m} = \rho\dot{V} = (0.9994 \text{ kg/m}^3)(0.15 \text{ m}^3\text{/s}) = 0.150 \text{ kg/s}$$

Next, we determine the exit temperature of air from

$$\begin{aligned} T_e &= T_s - (T_s - T_i)\exp(-hA_s/\dot{m}c_p) \\ &= 60^\circ\text{C} - [(60 - 80)^\circ\text{C}]\exp\left[-\frac{(13.5 \text{ W/m}^2\cdot{}^\circ\text{C})(6.4 \text{ m}^2)}{(0.150 \text{ kg/s})(1008 \text{ J/kg}\cdot{}^\circ\text{C})}\right] \\ &= \mathbf{71.3^\circ C} \end{aligned}$$

Then the logarithmic mean temperature difference and the rate of heat loss from the air become

$$\Delta T_{\ln} = \frac{T_i - T_e}{\ln \dfrac{T_s - T_e}{T_s - T_i}} = \frac{80 - 71.3}{\ln \dfrac{60 - 71.3}{60 - 80}} = -15.2°\text{C}$$

$$\dot{Q} = hA_s \, \Delta T_{\ln} = (13.5 \text{ W/m}^2 \cdot °\text{C})(6.4 \text{ m}^2)(-15.2°\text{C}) = \mathbf{-1313 \text{ W}}$$

Therefore, air will lose heat at a rate of 1313 W as it flows through the duct in the attic.

Discussion The average fluid temperature is (80 + 71.3)/2 = 75.7°C, which is sufficiently close to 80°C at which we evaluated the properties of air. Therefore, it is not necessary to re-evaluate the properties at this temperature and to repeat the calculations.

SUMMARY

Convection is the mode of heat transfer that involves conduction as well as bulk fluid motion. The rate of convection heat transfer in external flow is expressed by *Newton's law of cooling* as

$$\dot{Q} = hA_s(T_s - T_\infty)$$

where T_s is the surface temperature and T_∞ is the free-stream temperature. The heat transfer coefficient h is usually expressed in the dimensionless form as the *Nusselt number* as $\text{Nu} = hL_c/k$ where L_c is the *characteristic length.* The characteristic length for noncircular tubes is the *hydraulic diameter* D_h defined as $D_h = 4A_c/p$ where A_c is the cross-sectional area of the tube and p is its perimeter. The value of the critical Reynolds number is about 5×10^5 for flow over a flat plate, 2×10^5 for flow over cylinders and spheres, and 2300 for flow inside tubes.

The average Nusselt number relations for flow over a flat plate are:

Laminar: $\text{Nu} = \dfrac{hL}{k} = 0.664 \, \text{Re}_L^{0.5} \, \text{Pr}^{1/3}, \qquad \text{Re}_L < 5 \times 10^5$

Turbulent:

$$\text{Nu} = \frac{hL}{k} = 0.037 \, \text{Re}_L^{0.8} \, \text{Pr}^{1/3}, \qquad \begin{matrix} 0.6 \le \text{Pr} \le 60 \\ 5 \times 10^5 \le \text{Re}_L \le 10^7 \end{matrix}$$

Combined:

$$\text{Nu} = \frac{hL}{k} = (0.037 \, \text{Re}_L^{0.8} - 871) \, \text{Pr}^{1/3}, \qquad \begin{matrix} 0.6 \le \text{Pr} \le 60 \\ 5 \times 10^5 \le \text{Re}_L \le 10^7 \end{matrix}$$

For isothermal surfaces with an unheated starting section of length ξ, the local Nusselt number and the average convection coefficient relations are

Laminar: $$\text{Nu}_x = \frac{\text{Nu}_{x\,(\text{for } \xi=0)}}{[1 - (\xi/x)^{3/4}]^{1/3}} = \frac{0.332 \, \text{Re}_x^{0.5} \, \text{Pr}^{1/3}}{[1 - (\xi/x)^{3/4}]^{1/3}}$$

Turbulent: $$\text{Nu}_x = \frac{\text{Nu}_{x\,(\text{for } \xi=0)}}{[1 - (\xi/x)^{9/10}]^{1/9}} = \frac{0.0296 \, \text{Re}_x^{0.8} \, \text{Pr}^{1/3}}{[1 - (\xi/x)^{9/10}]^{1/9}}$$

Laminar: $$h = \frac{2[1 - (\xi/x)^{3/4}]}{1 - \xi/L} h_{x=L}$$

Turbulent: $$h = \frac{5[1 - (\xi/x)^{9/10}]}{(1 - \xi/L)} h_{x=L}$$

These relations are for the case of *isothermal* surfaces. When a flat plate is subjected to *uniform heat flux,* the local Nusselt number is given by

Laminar: $\text{Nu}_x = 0.453 \, \text{Re}_x^{0.5} \, \text{Pr}^{1/3}$

Turbulent: $\text{Nu}_x = 0.0308 \, \text{Re}_x^{0.8} \, \text{Pr}^{1/3}$

The average Nusselt numbers for cross flow over a *cylinder* and *sphere* are

$$\text{Nu}_{\text{cyl}} = \frac{hD}{k} = 0.3 + \frac{0.62 \, \text{Re}^{1/2} \, \text{Pr}^{1/3}}{[1 + (0.4/\text{Pr})^{2/3}]^{1/4}} \left[1 + \left(\frac{\text{Re}}{282{,}000}\right)^{5/8}\right]^{4/5}$$

which is valid for Re Pr > 0.2, and

$$\text{Nu}_{\text{sph}} = \frac{hD}{k} = 2 + [0.4 \, \text{Re}^{1/2} + 0.06 \, \text{Re}^{2/3}] \text{Pr}^{0.4} \left(\frac{\mu_\infty}{\mu_s}\right)^{1/4}$$

which is valid for $3.5 \le \text{Re} \le 80{,}000$ and $0.7 \le \text{Pr} \le 380$. The fluid properties are evaluated at the film temperature $T_f = (T_\infty + T_s)/2$ in the case of a cylinder, and at the free-stream temperature T_∞ (except for μ_s, which is evaluated at the surface temperature T_s) in the case of a sphere.

The Reynolds number for internal flow and the hydraulic diameter are defined as

$$\text{Re} = \frac{\rho V_{\text{avg}} D}{\mu} = \frac{V_{\text{avg}} D}{\nu} \quad \text{and} \quad D_h = \frac{4A_c}{p}$$

The flow in a tube is laminar for Re < 2300, turbulent for about Re > 10,000, and transitional in between.

The length of the region from the tube inlet to the point at which the boundary layer merges at the centerline is the *hydrodynamic entry length* L_h. The region beyond the entrance region in which the velocity profile is fully developed is the *hydrodynamically fully developed region*. The length of the region of flow over which the thermal boundary layer develops and reaches the tube center is the *thermal entry length* L_t. The region in which the flow is both hydrodynamically and thermally developed is the *fully developed flow region*. The entry lengths are given by

$$L_{h,\text{ laminar}} \approx 0.05 \text{ Re } D$$

$$L_{t,\text{ laminar}} \approx 0.05 \text{ Re Pr } D = \text{Pr } L_{h,\text{ laminar}}$$

$$L_{h,\text{ turbulent}} \approx L_{t,\text{ turbulent}} = 10D$$

For $\dot{q}_s$ = constant, the rate of heat transfer is expressed as

$$\dot{Q} = \dot{q}_s A_s = \dot{m} c_p (T_e - T_i)$$

For T_s = constant, we have

$$\dot{Q} = hA_s \Delta T_{\text{ln}} = \dot{m} c_p (T_e - T_i)$$

$$T_e = T_s - (T_s - T_i)\exp(-hA_s/\dot{m}c_p)$$

$$\Delta T_{\text{ln}} = \frac{T_i - T_e}{\ln[(T_s - T_e)/(T_s - T_i)]} = \frac{\Delta T_e - \Delta T_i}{\ln(\Delta T_e/\Delta T_i)}$$

For fully developed laminar flow in a circular pipe, we have

Circular tube, laminar ($\dot{q}_s$ = constant): $\text{Nu} = \dfrac{hD}{k} = 4.36$

Circular tube, laminar (T_s = constant): $\text{Nu} = \dfrac{hD}{k} = 3.66$

For *developing laminar flow* in the entrance region with constant surface temperature, we have

Circular tube: $\text{Nu} = 3.66 + \dfrac{0.065(D/L)\text{ Re Pr}}{1 + 0.04[(D/L)\text{ Re Pr}]^{2/3}}$

Circular tube: $\text{Nu} = 1.86\left(\dfrac{\text{Re Pr } D}{L}\right)^{1/3}\left(\dfrac{\mu_b}{\mu_s}\right)^{0.14}$

Parallel plates: $\text{Nu} = 7.54 + \dfrac{0.03(D_h/L)\text{ Re Pr}}{1 + 0.016[(D_h/L)\text{ Re Pr}]^{2/3}}$

For *fully developed turbulent flow with smooth surfaces*, we have

$$f = (0.790 \ln \text{Re} - 1.64)^{-2} \qquad 10^4 < \text{Re} < 10^6$$

$$\text{Nu} = 0.125 f \text{ Re Pr}^{1/3}$$

$$\text{Nu} = 0.023 \text{ Re}^{0.8} \text{ Pr}^{1/3} \qquad \begin{pmatrix} 0.7 \le \text{Pr} \le 160 \\ \text{Re} > 10{,}000 \end{pmatrix}$$

$\text{Nu} = 0.023 \text{ Re}^{0.8} \text{ Pr}^n$ with $n = 0.4$ for *heating* and 0.3 for *cooling* of fluid

$$\text{Nu} = \frac{(f/8)(\text{Re} - 1000)\text{ Pr}}{1 + 12.7(f/8)^{0.5}(\text{Pr}^{2/3} - 1)} \qquad \begin{pmatrix} 0.5 \le \text{Pr} \le 2000 \\ 3 \times 10^3 < \text{Re} < 5 \times 10^6 \end{pmatrix}$$

The fluid properties are evaluated at the *bulk mean fluid temperature* $T_b = (T_i + T_e)/2$. For liquid metal flow in the range of $10^4 < \text{Re} < 10^6$ we have:

T_s = constant: $\text{Nu} = 4.8 + 0.0156 \text{ Re}^{0.85} \text{ Pr}_s^{0.93}$

$\dot{q}_s$ = constant: $\text{Nu} = 6.3 + 0.0167 \text{ Re}^{0.85} \text{ Pr}_s^{0.93}$

REFERENCES AND SUGGESTED READING

1. M. S. Bhatti and R. K. Shah. "Turbulent and Transition Flow Convective Heat Transfer in Ducts." In *Handbook of Single-Phase Convective Heat Transfer,* ed. S. Kakaç, R. K. Shah, and W. Aung. New York: Wiley Interscience, 1987.
2. Y. A. Çengel and J. M. Cimbala. *Fluid Mechanics: Fundamentals and Applications.* New York: McGraw-Hill, 2005.
3. S. W. Churchill and M. Bernstein. "A Correlating Equation for Forced Convection from Gases and Liquids to a Circular Cylinder in Cross Flow." *Journal of Heat Transfer* 99 (1977), pp. 300–306.
4. S. W. Churchill and H. Ozoe. "Correlations for Laminar Forced Convection in Flow over an Isothermal Flat Plate and in Developing and Fully Developed Flow in an Isothermal Tube." *Journal of Heat Transfer* 95 (Feb. 1973), pp. 78–84.
5. A. P. Colburn. *Transactions of the AIChE* 26 (1933), p. 174.
6. C. F. Colebrook. "Turbulent flow in Pipes, with Particular Reference to the Transition between the Smooth and Rough Pipe Laws." *Journal of the Institute of Civil Engineers London.* 11 (1939), pp. 133–156.
7. R. G. Deissler. "Analysis of Turbulent Heat Transfer and Flow in the Entrance Regions of Smooth Passages." 1953. Referred to in *Handbook of Single-Phase Convective Heat Transfer,* ed. S. Kakaç, R. K. Shah, and W. Aung. New York: Wiley Interscience, 1987.
8. D. F. Dipprey and D. H. Sabersky. "Heat and Momentum Transfer in Smooth and Rough Tubes at Various Prandtl Numbers." *International Journal of Heat Mass Transfer* 6 (1963), pp. 329–353.

9. F. W. Dittus and L. M. K. Boelter. *University of California Publications on Engineering* 2 (1930), p. 433.

10. D. K. Edwards, V. E. Denny, and A. F. Mills. *Transfer Processes.* 2nd ed. Washington, D.C.: Hemisphere, 1979.

11. W. H. Giedt. "Investigation of Variation of Point Unit-Heat Transfer Coefficient around a Cylinder Normal to an Air Stream." *Transactions of the ASME* 71 (1949), pp. 375–381.

12. V. Gnielinski. "New Equations for Heat and Mass Transfer in Turbulent Pipe and Channel Flow." *International Chemical Engineering* 16 (1976), pp. 359–368.

13. S. E. Haaland. "Simple and Explicit Formulas for the Friction Factor in Turbulent Pipe Flow." *Journal of Fluids Engineering* (March 1983), pp. 89–90.

14. M. Jakob. *Heat Transfer.* Vol. 1. New York: John Wiley & Sons, 1949.

15. S. Kakaç, R. K. Shah, and W. Aung, eds. *Handbook of Single-Phase Convective Heat Transfer.* New York: Wiley Interscience, 1987.

16. W. M. Kays and H. C. Perkins. Chapter 7. In *Handbook of Heat Transfer,* ed. W. M. Rohsenow and J. P. Hartnett. New York: McGraw-Hill, 1972.

17. W. M. Kays and M. E. Crawford. *Convective Heat and Mass Transfer.* 3rd ed. New York: McGraw-Hill, 1993.

18. W. M. Kays and M. E. Crawford. *Convective Heat and Mass Transfer.* 3rd ed. New York: McGraw-Hill, 1993.

19. M. Molki and E. M. Sparrow. "An Empirical Correlation for the Average Heat Transfer Coefficient in Circular Tubes." *Journal of Heat Transfer* 108 (1986), pp. 482–484.

20. L. F. Moody. "Friction Factors for Pipe Flows." *Transactions of the ASME* 66 (1944), pp. 671–684.

21. R. H. Norris. "Some Simple Approximate Heat Transfer Correlations for Turbulent Flow in Ducts with Rough Surfaces." In *Augmentation of Convective Heat Transfer,* ed. A. E. Bergles and R. L. Webb. New York: ASME, 1970.

22. B. S. Petukhov and L. I. Roizen. "Generalized Relationships for Heat Transfer in a Turbulent Flow of a Gas in Tubes of Annular Section." *High Temperature* (USSR) 2 (1964), pp. 65–68.

23. B. S. Petukhov. "Heat Transfer and Friction in Turbulent Pipe Flow with Variable Physical Properties." In *Advances in Heat Transfer,* ed. T. F. Irvine and J. P. Hartnett, Vol. 6. New York: Academic Press, 1970.

24. H. Schlichting. *Boundary Layer Theory,* 7th ed. New York, McGraw-Hill, 1979.

25. R. K. Shah and M. S. Bhatti. "Laminar Convective Heat Transfer in Ducts." In *Handbook of Single-Phase Convective Heat Transfer,* ed. S. Kakaç, R. K. Shah, and W. Aung. New York: Wiley Interscience, 1987.

26. E. N. Sieder and G. E. Tate. "Heat Transfer and Pressure Drop of Liquids in Tubes." *Industrial Engineering Chemistry* 28 (1936), pp. 1429–1435.

27. C. A. Sleicher and M. W. Rouse. "A Convenient Correlation for Heat Transfer to Constant and Variable Property Fluids in Turbulent Pipe Flow." *International Journal of Heat Mass Transfer* 18 (1975), pp. 1429–1435.

28. W. C. Thomas. "Note on the Heat Transfer Equation for Forced Convection Flow over a Flat Plate with an Unheated Starting Length." *Mechanical Engineering News,* 9, no.1 (1977), p. 361.

29. S. Whitaker. "Forced Convection Heat Transfer Correlations for Flow in Pipes, Past Flat Plates, Single Cylinders, and for Flow in Packed Beds and Tube Bundles." *AIChE Journal* 18 (1972), pp. 361–371.

30. W. Zhi-qing. "Study on Correction Coefficients of Laminar and Turbulent Entrance Region Effects in Round Pipes." *Applied Mathematical Mechanics* 3 (1982), p. 433.

PROBLEMS*

Physical Mechanism of Convection

19–1C What is forced convection? How does it differ from natural convection? Is convection caused by winds forced or natural convection?

19–2C What is external forced convection? How does it differ from internal forced convection? Can a heat transfer system involve both internal and external convection at the same time? Give an example.

19–3C In which mode of heat transfer is the convection heat transfer coefficient usually higher, natural convection or forced convection? Why?

19–4C Consider a hot baked potato. Will the potato cool faster or slower when we blow the warm air coming from our lungs on it instead of letting it cool naturally in the cooler air in the room? Explain.

19–5C What is the physical significance of the Nusselt number? How is it defined?

*Problems designated by a "C" are concept questions, and students are encouraged to answer them all. Problems designated by an "E" are in English units, and the SI users can ignore them. Problems with the EES icon are solved using EES, and complete solutions together with parametric studies are included on the enclosed DVD. Problems with the computer icon are comprehensive in nature and are intended to be solved with a computer, preferably using the EES software that accompanies this text.

19–6C When is heat transfer through a fluid conduction and when is it convection? For what case is the rate of heat transfer higher? How does the convection heat transfer coefficient differ from the thermal conductivity of a fluid?

19–7C Define incompressible flow and incompressible fluid. Must the flow of a compressible fluid necessarily be treated as compressible?

19–8 During air cooling of potatoes, the heat transfer coefficient for combined convection, radiation, and evaporation is determined experimentally to be as shown:

Air Velocity, m/s	Heat Transfer Coefficient, W/m² · °C
0.66	14.0
1.00	19.1
1.36	20.2
1.73	24.4

Consider an 8-cm-diameter potato initially at 20°C with a thermal conductivity of 0.49 W/m · °C. Potatoes are cooled by refrigerated air at 5°C at a velocity of 1 m/s. Determine the initial rate of heat transfer from a potato, and the initial value of the temperature gradient in the potato at the surface. *Answers: 5.8 W, −585°C/m*

19–9 An average man has a body surface area of 1.8 m² and a skin temperature of 33°C. The convection heat transfer coefficient for a clothed person walking in still air is expressed as $h = 8.6V^{0.53}$ for $0.5 < V < 2$ m/s, where V is the walking velocity in m/s. Assuming the average surface temperature of the clothed person to be 30°C, determine the rate of heat loss from an average man walking in still air at 10°C by convection at a walking velocity of (*a*) 0.5 m/s, (*b*) 1.0 m/s, (*c*) 1.5 m/s, and (*d*) 2.0 m/s.

19–10 The convection heat transfer coefficient for a clothed person standing in moving air is expressed as $h = 14.8V^{0.69}$ for $0.15 < V < 1.5$ m/s, where V is the air velocity. For a person with a body surface area of 1.7 m² and an average surface temperature of 29°C, determine the rate of heat loss from the person in windy air at 10°C by convection for air velocities of (*a*) 0.5 m/s, (*b*) 1.0 m/s, and (*c*) 1.5 m/s.

19–11 During air cooling of oranges, grapefruit, and tangelos, the heat transfer coefficient for combined convection, radiation, and evaporation for air velocities of $0.11 < V < 0.33$ m/s is determined experimentally and is expressed as $h = 5.05\, k_{air} \text{Re}^{1/3}/D$, where the diameter D is the characteristic length. Oranges are cooled by refrigerated air at 5°C and 1 atm at a velocity of 0.3 m/s. Determine (*a*) the initial rate of heat transfer from a 7-cm-diameter orange initially at 15°C with a thermal conductivity of 0.50 W/m · °C, (*b*) the value of the initial temperature gradient inside the orange at the surface, and (*c*) the value of the Nusselt number.

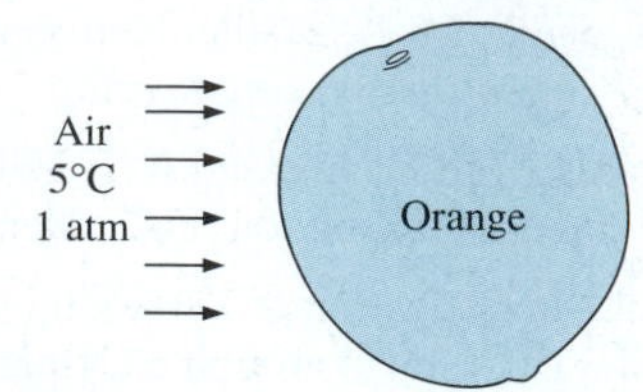

FIGURE P19–11

Flow over Flat Plates

19–12C Consider laminar flow over a flat plate. Will the friction coefficient change with distance from the leading edge? How about the heat transfer coefficient?

19–13C How is the heat transfer coefficient determined in flow over a flat plate?

19–14 Engine oil at 80°C flows over a 10-m-long flat plate whose temperature is 30°C with a velocity of 2.5 m/s. Determine the rate of heat transfer over the entire plate per unit width.

19–15 The local atmospheric pressure in Denver, Colorado (elevation 1610 m), is 83.4 kPa. Air at this pressure and at 30°C flows with a velocity of 6 m/s over a 2.5-m × 8-m flat plate whose temperature is 120°C. Determine the rate of heat transfer from the plate if the air flows parallel to the (*a*) 8-m-long side and (*b*) the 2.5-m side.

19–16 During a cold winter day, wind at 55 km/h is blowing parallel to a 4-m-high and 10-m-long wall of a house. If the air outside is at 5°C and the surface temperature of the wall is 12°C, determine the rate of heat loss from that wall by convection. What would your answer be if the wind velocity was doubled? *Answers: 9081 W, 16,200 W*

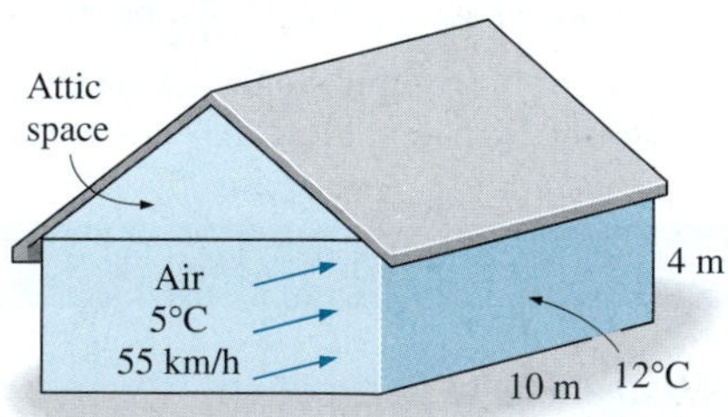

FIGURE P19–16

19–17 Reconsider Prob. 19–16. Using EES (or other) software, investigate the effects of wind velocity and outside air temperature on the rate of heat loss from the wall by convection. Let the wind velocity vary from 10 km/h to 80 km/h and the outside air temperature from 0°C to 10°C. Plot the rate of heat loss as a function of the wind velocity and of the outside temperature, and discuss the results.

19–18E Air at 60°F flows over a 10-ft-long flat plate at 7 ft/s. Determine the local heat transfer coefficients at intervals of 1 ft, and plot the results against the distance from the leading edge.

19–19E Reconsider Prob. 19–18E. Using EES (or other) software, evaluate the local heat transfer coefficient along the plate at intervals of 0.1 ft, and plot them against the distance from the leading edge.

19–20 Water at 43.3°C flows over a large plate at a velocity of 30.0 cm/s. The plate is 1.0 m long (in the flow direction), and its surface is maintained at a uniform temperature of 10.0°C. Calculate the steady rate of heat transfer per unit width of the plate.

19–21 Mercury at 25°C flows over a 3-m-long and 2-m-wide flat plate maintained at 75°C with a velocity of 0.8 m/s. Determine the rate of heat transfer from the entire plate.

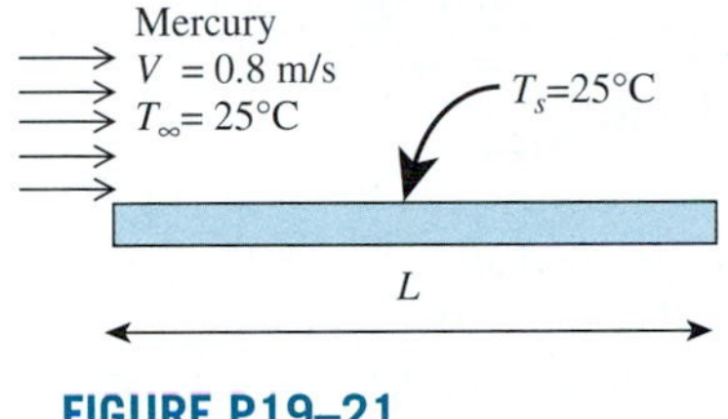

FIGURE P19–21

19–22 Parallel plates form a solar collector that covers a roof, as shown in the figure. The plates are maintained at 15°C, while ambient air at 10°C flows over the roof with V = 2 m/s. Determine the rate of convective heat loss from (*a*) the first plate and (*b*) the third plate.

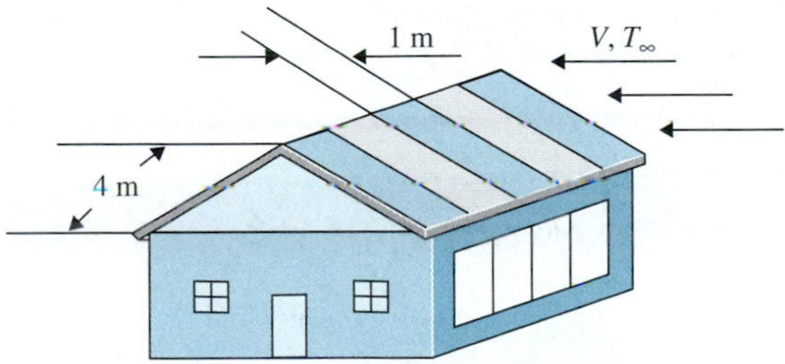

FIGURE P19–22

19–23 Consider a hot automotive engine, which can be approximated as a 0.5-m-high, 0.40-m-wide, and 0.8-m-long rectangular block. The bottom surface of the block is at a temperature of 100°C and has an emissivity of 0.95. The ambient air is at 20°C, and the road surface is at 25°C. Determine the rate of heat transfer from the bottom surface of the engine block by convection and radiation as the car travels at a velocity of 80 km/h. Assume the flow to be turbulent over the entire surface because of the constant agitation of the engine block.

19–24 The forming section of a plastics plant puts out a continuous sheet of plastic that is 1.2 m wide and 2 mm thick at a rate of 15 m/min. The temperature of the plastic sheet is 90°C when it is exposed to the surrounding air, and the sheet is subjected to air flow at 30°C at a velocity of 3 m/s on both sides along its surfaces normal to the direction of motion of the sheet. The width of the air cooling section is such that a fixed point on the plastic sheet passes through that section in 2 s. Determine the rate of heat transfer from the plastic sheet to the air.

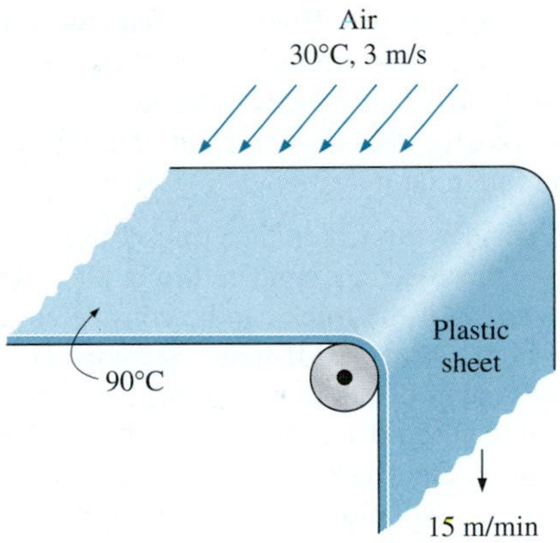

FIGURE P19–24

19–25 The top surface of the passenger car of a train moving at a velocity of 70 km/h is 2.8 m wide and 8 m long. The top surface is absorbing solar radiation at a rate of 200 W/m^2, and the temperature of the ambient air is 30°C. Assuming the roof of the car to be perfectly insulated and the radiation heat exchange with the surroundings to be small relative to convection, determine the equilibrium temperature of the top surface of the car. *Answer:* 35.1°C

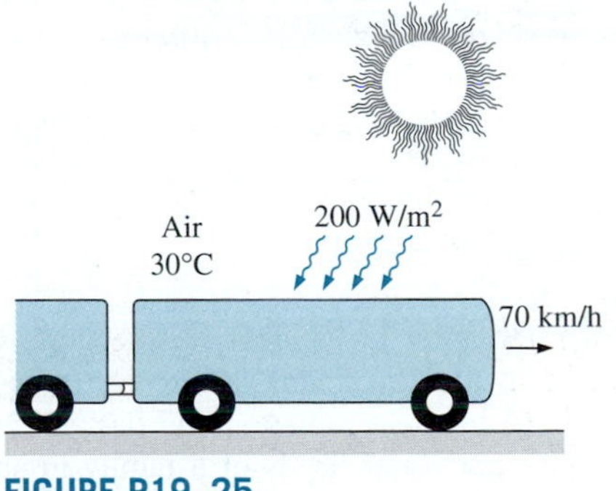

FIGURE P19–25

19–26 Reconsider Prob. 19–25. Using EES (or other) software, investigate the effects of the train velocity and the rate of absorption of solar radiation on the equilibrium temperature of the top surface of the car. Let the train velocity vary from 10 km/h to 120 km/h and the rate of solar absorption from 100 W/m^2 to 500 W/m^2. Plot the equilibrium temperature as functions of train velocity and solar radiation absorption rate, and discuss the results.

19–27 A 15-cm × 15-cm circuit board dissipating 20 W of power uniformly is cooled by air, which approaches the circuit board at 20°C with a velocity of 6 m/s. Disregarding any heat transfer from the back surface of the board, determine the surface temperature of the electronic components (*a*) at the leading edge and (*b*) at the end of the board. Assume the flow to be turbulent since the electronic components are expected to act as turbulators.

19–28 Consider laminar flow of a fluid over a flat plate maintained at a constant temperature. Now the free-stream velocity of the fluid is doubled. Determine the change in the rate of heat transfer between the fluid and the plate. Assume the flow to remain laminar.

19–29E Consider a refrigeration truck traveling at 55 mph at a location where the air temperature is 80°F. The refrigerated compartment of the truck can be considered to be a 9-ft-wide, 8-ft-high, and 20-ft-long rectangular box. The refrigeration system of the truck can provide 3 tons of refrigeration (i.e., it can remove heat at a rate of 600 Btu/min). The outer surface of the truck is coated with a low-emissivity material, and thus radiation heat transfer is very small. Determine the average temperature of the outer surface of the refrigeration compartment of the truck if the refrigeration system is observed to be operating at half the capacity. Assume the air flow over the entire outer surface to be turbulent and the heat transfer coefficient at the front and rear surfaces to be equal to that on side surfaces.

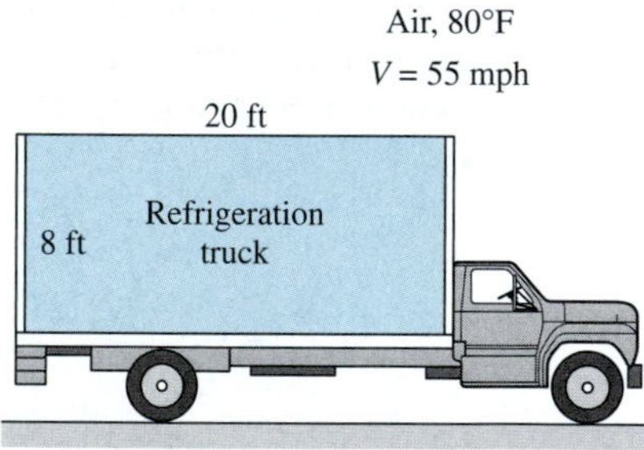

FIGURE P19–29E

19–30 Solar radiation is incident on the glass cover of a solar collector at a rate of 700 W/m^2. The glass transmits 88 percent of the incident radiation and has an emissivity of 0.90. The entire hot water needs of a family in summer can be met by two collectors 1.2 m high and 1 m wide. The two collectors are attached to each other on one side so that they appear like a single collector 1.2 m × 2 m in size. The temperature of the glass cover is measured to be 35°C on a day when the surrounding air temperature is 25°C and the wind is blowing at 30 km/h. The effective sky temperature for radiation exchange between the glass cover and the open sky is −40°C. Water enters the tubes attached to the absorber plate at a rate of 1 kg/min. Assuming the back surface of the absorber plate to be heavily insulated and the only heat loss to occur through the glass cover, determine (*a*) the total rate of heat loss from the collector, (*b*) the collector efficiency, which is the ratio of the amount of heat transferred to the water to the solar energy incident on the collector, and (*c*) the temperature rise of water as it flows through the collector.

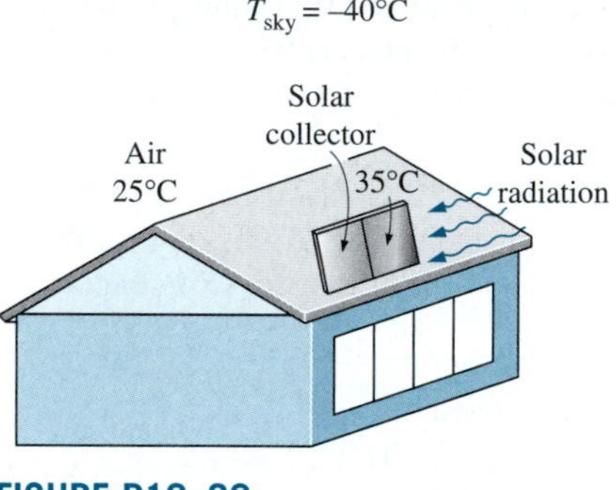

FIGURE P19–30

19–31 A transformer that is 10 cm long, 6.2 cm wide, and 5 cm high is to be cooled by attaching a 10-cm × 6.2-cm wide polished aluminum heat sink (emissivity = 0.03) to its top surface. The heat sink has seven fins, which are 5 mm high, 2 mm thick, and 10 cm long. A fan blows air at 25°C parallel to the passages between the fins. The heat sink is to dissipate 12 W of heat and the base temperature of the heat sink is not to exceed 60°C. Assuming the fins and the base plate to be nearly isothermal and the radiation heat transfer to be negligible, determine the minimum free-stream velocity the fan needs to supply to avoid overheating.

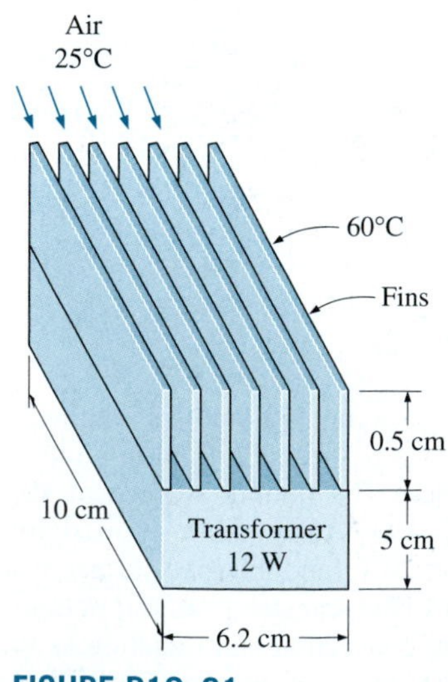

FIGURE P19–31

19–32 Repeat Prob. 19–31 assuming the heat sink to be black-anodized and thus to have an effective emissivity of

0.90. Note that in radiation calculations the base area (10 cm × 6.2 cm) is to be used, not the total surface area.

19–33 An array of power transistors, dissipating 6 W of power each, are to be cooled by mounting them on a 25-cm × 25-cm square aluminum plate and blowing air at 35°C over the plate with a fan at a velocity of 4 m/s. The average temperature of the plate is not to exceed 65°C. Assuming the heat transfer from the back side of the plate to be negligible and disregarding radiation, determine the number of transistors that can be placed on this plate.

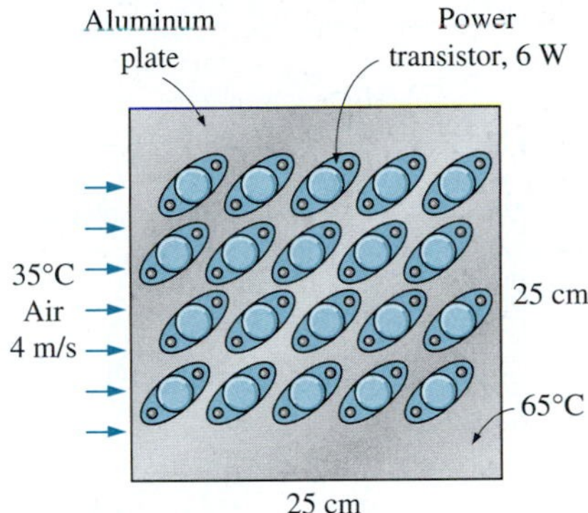

FIGURE P19–33

19–34 Repeat Prob. 19–33 for a location at an elevation of 1610 m where the atmospheric pressure is 83.4 kPa. *Answer:* 4

Flow across Cylinders and Spheres

19–35C Consider laminar flow of air across a hot circular cylinder. At what point on the cylinder will the heat transfer be highest? What would your answer be if the flow were turbulent?

19–36 A long 8-cm-diameter steam pipe whose external surface temperature is 90°C passes through some open area that is not protected against the winds. Determine the rate of heat loss from the pipe per unit of its length when the air is at 1 atm pressure and 7°C and the wind is blowing across the pipe at a velocity of 50 km/h.

19–37 In a geothermal power plant, the used geothermal water at 80°C enters a 15-cm-diameter and 400-m-long uninsulated pipe at a rate of 8.5 kg/s and leaves at 70°C before being reinjected back to the ground. Windy air at 15°C flows normal to the pipe. Disregarding radiation, determine the average wind velocity in km/h.

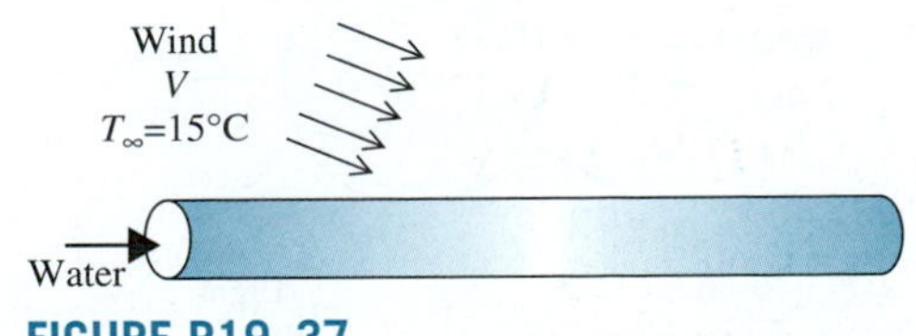

FIGURE P19–37

19–38 A stainless steel ball (ρ = 8055 kg/m^3, c_p = 480 J/kg · °C) of diameter D = 15 cm is removed from the oven at a uniform temperature of 350°C. The ball is then subjected to the flow of air at 1 atm pressure and 30°C with a velocity of 6 m/s. The surface temperature of the ball eventually drops to 250°C. Determine the average convection heat transfer coefficient during this cooling process and estimate how long this process has taken.

19–39 EES Reconsider Prob. 19–38. Using EES (or other) software, investigate the effect of air velocity on the average convection heat transfer coefficient and the cooling time. Let the air velocity vary from 1 m/s to 10 m/s. Plot the heat transfer coefficient and the cooling time as a function of air velocity, and discuss the results.

19–40E A person extends his uncovered arms into the windy air outside at 54°F and 20 mph in order to feel nature closely. Initially, the skin temperature of the arm is 86°F. Treating the arm as a 2-ft-long and 3-in-diameter cylinder, determine the rate of heat loss from the arm.

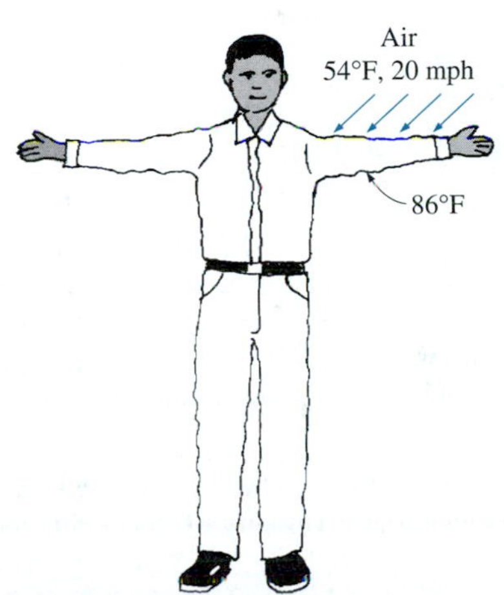

FIGURE P19–40E

19–41E EES Reconsider Prob. 19–40E. Using EES (or other) software, investigate the effects of air temperature and wind velocity on the rate of heat loss from the arm. Let the air temperature vary from 20°F to 80°F and the wind velocity from 10 mph to 40 mph. Plot the rate of heat loss as a function of air temperature and of wind velocity, and discuss the results.

19–42 An average person generates heat at a rate of 84 W while resting. Assuming one-quarter of this heat is lost from the head and disregarding radiation, determine the average surface temperature of the head when it is not covered and is subjected to winds at 10°C and 25 km/h. The head can be approximated as a 30-cm-diameter sphere. *Answer:* 13.2°C

19–43 Consider the flow of a fluid across a cylinder maintained at a constant temperature. Now the free-stream velocity of the fluid is doubled. Determine the change the rate of heat transfer between the fluid and the cylinder.

19–44 A 6-mm-diameter electrical transmission line carries an electric current of 50 A and has a resistance of 0.002 ohm per meter length. Determine the surface temperature of the wire during a windy day when the air temperature is 10°C and the wind is blowing across the transmission line at 40 km/h.

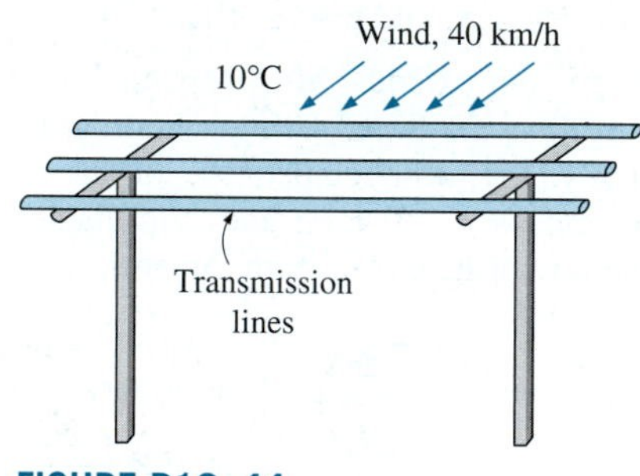

FIGURE P19–44

19–45 Reconsider Prob. 19–44. Using EES (or other) software, investigate the effect of the wind velocity on the surface temperature of the wire. Let the wind velocity vary from 10 km/h to 80 km/h. Plot the surface temperature as a function of wind velocity, and discuss the results.

19–46 A heating system is to be designed to keep the wings of an aircraft cruising at a velocity of 900 km/h above freezing temperatures during flight at 12,200-m altitude where the standard atmospheric conditions are −55.4°C and 18.8 kPa. Approximating the wing as a cylinder of elliptical cross section whose minor axis is 50 cm and disregarding radiation, determine the average convection heat transfer coefficient on the wing surface and the average rate of heat transfer per unit surface area.

19–47 A long aluminum wire of diameter 3 mm is extruded at a temperature of 370°C. The wire is subjected to cross air flow at 30°C at a velocity of 6 m/s. Determine the rate of heat transfer from the wire to the air per meter length when it is first exposed to the air.

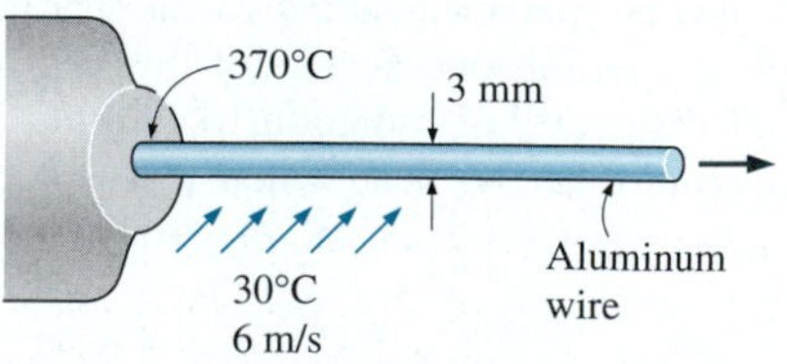

FIGURE P19–47

19–48E Consider a person who is trying to keep cool on a hot summer day by turning a fan on and exposing his entire body to air flow. The air temperature is 85°F and the fan is blowing air at a velocity of 6 ft/s. If the person is doing light work and generating sensible heat at a rate of 300 Btu/h, determine the average temperature of the outer surface (skin or clothing) of the person. The average human body can be treated as a 1-ft-diameter cylinder with an exposed surface area of 18 ft^2. Disregard any heat transfer by radiation. What would your answer be if the air velocity were doubled? *Answers:* 95.1°F, 91.6°F

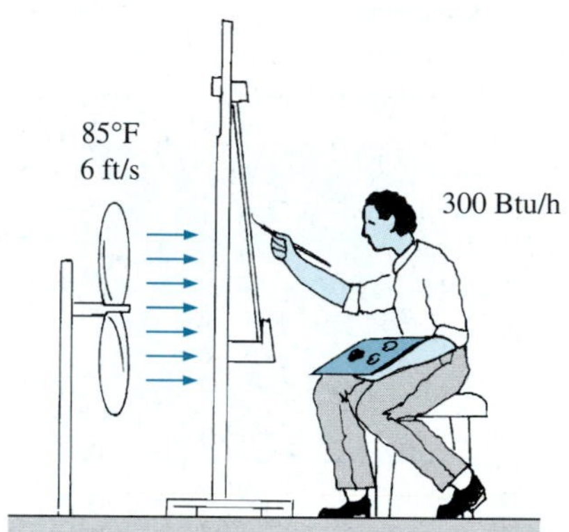

FIGURE P19–48E

19–49 An incandescent lightbulb is an inexpensive but highly inefficient device that converts electrical energy into light. It converts about 10 percent of the electrical energy it consumes into light while converting the remaining 90 percent into heat. (A fluorescent lightbulb will give the same amount of light while consuming only one-fourth of the electrical energy, and it will last 10 times longer than an incandescent lightbulb.) The glass bulb of the lamp heats up very quickly as a result of absorbing all that heat and dissipating it to the surroundings by convection and radiation.

Consider a 10-cm-diameter 100-W lightbulb cooled by a fan that blows air at 30°C to the bulb at a velocity of 2 m/s. The surrounding surfaces are also at 30°C, and the emissivity of the glass is 0.9. Assuming 10 percent of the energy passes through the glass bulb as light with negligible absorption and the rest of the energy is absorbed and dissipated by the bulb itself, determine the equilibrium temperature of the glass bulb.

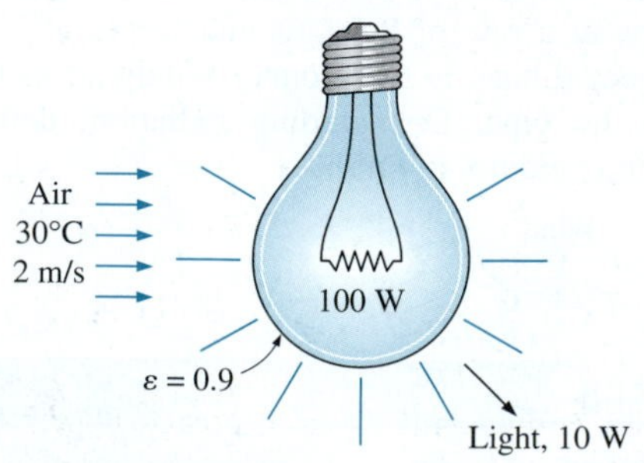

FIGURE P19–49

19–50 During a plant visit, it was noticed that a 12-m-long section of a 10-cm-diameter steam pipe is completely exposed to the ambient air. The temperature measurements indicate that the average temperature of the outer surface of the steam pipe is 75°C when the ambient temperature is 5°C. There are also light winds in the area at 10 km/h. The emissivity of the outer surface of the pipe is 0.8, and the average temperature of the surfaces surrounding the pipe, including the sky, is estimated to be 0°C. Determine the amount of heat lost from the steam during a 10-h-long work day.

Steam is supplied by a gas-fired steam generator that has an efficiency of 80 percent, and the plant pays $1.05/therm of natural gas. If the pipe is insulated and 90 percent of the heat loss is saved, determine the amount of money this facility will save a year as a result of insulating the steam pipes. Assume the plant operates every day of the year for 10 h. State your assumptions.

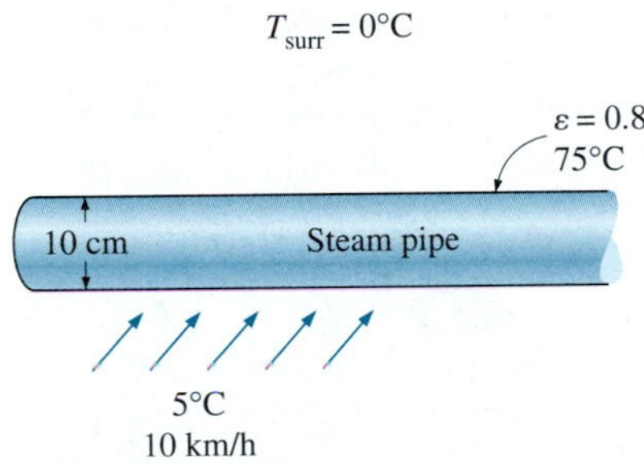

FIGURE P19–50

19–51 Reconsider Prob. 19–50. There seems to be some uncertainty about the average temperature of the surfaces surrounding the pipe used in radiation calculations, and you are asked to determine if it makes any significant difference in overall heat transfer. Repeat the calculations for average surrounding and surface temperatures of −20°C and 25°C, respectively, and determine the change in the values obtained.

19–52E A 12-ft-long, 1.5-kW electrical resistance wire is made of 0.1-in-diameter stainless steel (k = 8.7 Btu/h · ft · °F). The resistance wire operates in an environment at 85°F. Determine the surface temperature of the wire if it is cooled by a fan blowing air at a velocity of 20 ft/s.

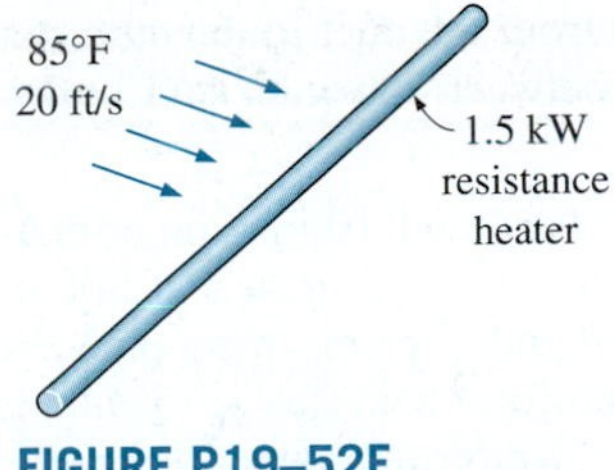

FIGURE P19–52E

19–53 The components of an electronic system are located in a 1.5-m-long horizontal duct whose cross section is 20 cm × 20 cm. The components in the duct are not allowed to come into direct contact with cooling air, and thus are cooled by air at 30°C flowing over the duct with a velocity of 200 m/min. If the surface temperature of the duct is not to exceed 65°C, determine the total power rating of the electronic devices that can be mounted into the duct. *Answer:* 640 W

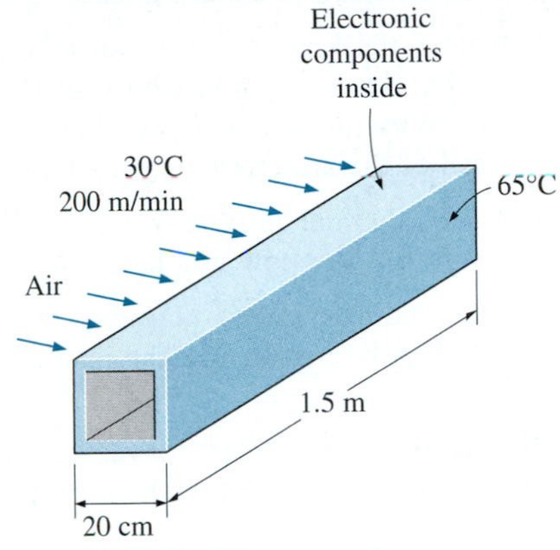

FIGURE P19–53

19–54 Repeat Prob. 19–53 for a location at 4000-m altitude where the atmospheric pressure is 61.66 kPa.

19–55 A 0.4-W cylindrical electronic component with diameter 0.3 cm and length 1.8 cm and mounted on a circuit board is cooled by air flowing across it at a velocity of 240 m/min. If the air temperature is 35°C, determine the surface temperature of the component.

19–56 Consider a 50-cm-diameter and 95-cm-long hot water tank. The tank is placed on the roof of a house. The water inside the tank is heated to 80°C by a flat-plate solar collector during the day. The tank is then exposed to windy air at 18°C with an average velocity of 40 km/h during the night. Estimate the temperature of the tank after a 45-min period. Assume the tank surface to be at the same temperature as the water inside, and the heat transfer coefficient on the top and bottom surfaces to be the same as that on the side surface.

19–57 EES Reconsider Prob. 19–56. Using EES (or other) software, plot the temperature of the tank as a function of the cooling time as the time varies from 30 min to 5 h, and discuss the results.

19–58 A 1.8-m-diameter spherical tank of negligible thickness contains iced water at 0°C. Air at 25°C flows over the tank with a velocity of 7 m/s. Determine the rate of heat transfer to the tank and the rate at which ice melts. The heat of fusion of water at 0°C is 333.7 kJ/kg.

19–59 A 10-cm-diameter, 30-cm-high cylindrical bottle contains cold water at 3°C. The bottle is placed in windy air at 27°C. The water temperature is measured to be 11°C after 45 min of cooling. Disregarding radiation effects and heat transfer from the top and bottom surfaces, estimate the average wind velocity.

Flow in Tubes

19–60C What is the physical significance of the number of transfer units NTU $= hA_s/\dot{m}c_p$? What do small and large NTU values tell about a heat transfer system?

19–61C What does the logarithmic mean temperature difference represent for flow in a tube whose surface temperature is constant? Why do we use the logarithmic mean temperature instead of the arithmetic mean temperature?

19–62C How is the thermal entry length defined for flow in a tube? In what region is the flow in a tube fully developed?

19–63C Consider laminar forced convection in a circular tube. Will the heat flux be higher near the inlet of the tube or near the exit? Why?

19–64C Consider turbulent forced convection in a circular tube. Will the heat flux be higher near the inlet of the tube or near the exit? Why?

19–65C In the fully developed region of flow in a circular tube, will the velocity profile change in the flow direction? How about the temperature profile?

19–66C Consider the flow of oil in a tube. How will the hydrodynamic and thermal entry lengths compare if the flow is laminar? How would they compare if the flow were turbulent?

19–67C Consider the flow of mercury (a liquid metal) in a tube. How will the hydrodynamic and thermal entry lengths compare if the flow is laminar? How would they compare if the flow were turbulent?

19–68C What do the average velocity V_{avg} and the mean temperature T_m represent in flow through circular tubes of constant diameter?

19–69C Consider fluid flow in a tube whose surface temperature remains constant. What is the appropriate temperature difference for use in Newton's law of cooling with an average heat transfer coefficient?

19–70 Air enters a 25-cm-diameter 12-m-long underwater duct at 50°C and 1 atm at a mean velocity of 7 m/s, and is cooled by the water outside. If the average heat transfer coefficient is 85 W/m^2 · °C and the tube temperature is nearly equal to the water temperature of 10°C, determine the exit temperature of air and the rate of heat transfer.

19–71 Cooling water available at 10°C is used to condense steam at 30°C in the condenser of a power plant at a rate of 0.15 kg/s by circulating the cooling water through a bank of 5-m-long 1.2-cm-internal-diameter thin copper tubes. Water enters the tubes at a mean velocity of 4 m/s, and leaves at a temperature of 24°C. The tubes are nearly isothermal at 30°C. Determine the average heat transfer coefficient between the water and the tubes, and the number of tubes needed to achieve the indicated heat transfer rate in the condenser.

19–72 Repeat Prob. 19–71 for steam condensing at a rate of 0.60 kg/s.

19–73 Combustion gases passing through a 3-cm-internal-diameter circular tube are used to vaporize waste water at atmospheric pressure. Hot gases enter the tube at 115 kPa and 250°C at a mean velocity of 5 m/s, and leave at 150°C. If the average heat transfer coefficient is 120 W/m^2 · °C and the inner surface temperature of the tube is 110°C, determine (*a*) the tube length and (*b*) the rate of evaporation of water.

19–74 Repeat Prob. 19–73 for a heat transfer coefficient of 40 W/m^2 · °C.

Laminar and Turbulent Flow in Tubes

19–75 Determine the convection heat transfer coefficient for the flow of (*a*) air and (*b*) water at a velocity of 2 m/s in an 8-cm-diameter and 7-m-long tube when the tube is subjected to uniform heat flux from all surfaces. Use fluid properties at 25°C.

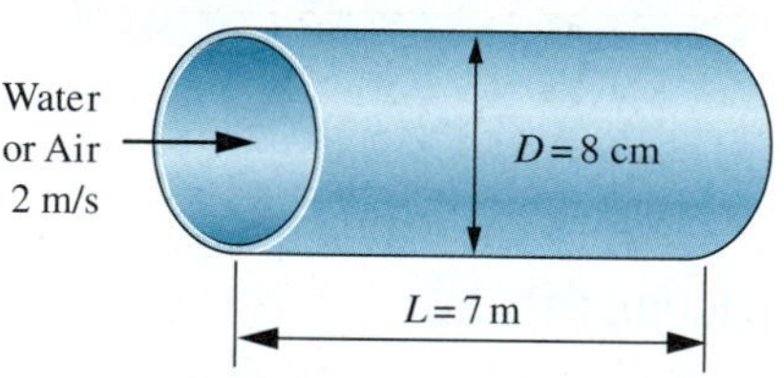

FIGURE P19–75

19–76 Air at 10°C enters a 12-cm-diameter and 5-m-long pipe at a rate of 0.065 kg/s. The inner surface of the pipe has a roughness of 0.22 mm and the pipe is nearly isothermal at 50°C. Determine the rate of heat transfer to air using the Nusselt number relation given by (*a*) Eq. 19–79 and (*b*) Eq. 19–84.

19–77 An 8-m long, uninsulated square duct of cross section 0.2 m × 0.2 m and relative roughness 10^{-8} passes through the attic space of a house. Hot air enters the duct at 1 atm and 80°C at a volume flow rate of 0.15 m^3/s. The duct surface is nearly isothermal at 60°C. Determine the rate of heat loss from the duct to the attic space and the pressure difference between the inlet and outlet sections of the duct.

19–78 A 10-m long and 10-mm inner-diameter pipe made of commercial steel is used to heat a liquid in an industrial process. The liquid enters the pipe with $T_i = 25°C$, $V = 0.8$ m/s. A uniform heat flux is maintained by an electric resistance heater wrapped around the outer surface of the pipe, so that the fluid exits at 75°C. Assuming fully developed flow and taking the average fluid properties to be $\rho = 1000$ kg/m^3, $c_p = 4000$ J/kg · K, $\mu = 2 \times 10^{-3}$ kg/m · s,

$k = 0.48$ W/m · K, and Pr = 10, determine (*a*) the required surface heat flux $\dot{q}_s$, produced by the heater, and (*b*) the surface temperature at the exit, T_s.

19–79 Water is to be heated from 10°C to 80°C as it flows through a 2-cm-internal-diameter, 13-m-long tube. The tube is equipped with an electric resistance heater, which provides uniform heating throughout the surface of the tube. The outer surface of the heater is well insulated, so that in steady operation all the heat generated in the heater is transferred to the water in the tube. If the system is to provide hot water at a rate of 5 L/min, determine the power rating of the resistance heater. Also, estimate the inner surface temperature of the pipe at the exit.

19–80 Hot air at atmospheric pressure and 85°C enters a 10-m-long uninsulated square duct of cross section 0.15 m × 0.15 m that passes through the attic of a house at a rate of 0.10 m³/s. The duct is observed to be nearly isothermal at 70°C. Determine the exit temperature of the air and the rate of heat loss from the duct to the air space in the attic. *Answers:* 75.7°C, 941 W

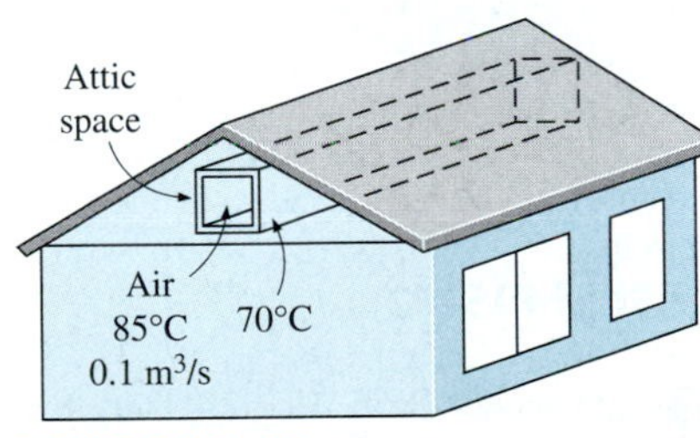

FIGURE P19–80

19–81 Reconsider Prob. 19–80. Using EES (or other) software, investigate the effect of the volume flow rate of air on the exit temperature of air and the rate of heat loss. Let the flow rate vary from 0.05 m³/s to 0.15 m³/s. Plot the exit temperature and the rate of heat loss as a function of flow rate, and discuss the results.

19–82 Consider an air solar collector that is 1 m wide and 5 m long and has a constant spacing of 3 cm between the glass cover and the collector plate. Air enters the collector at 30°C at a rate of 0.15 m³/s through the 1-m-wide edge and flows along the 5-m-long passage way. If the average temperatures of the glass cover and the collector plate are 20°C and 60°C, respectively, determine (*a*) the net rate of heat transfer to the air in the collector and (*b*) the temperature rise of air as it flows through the collector.

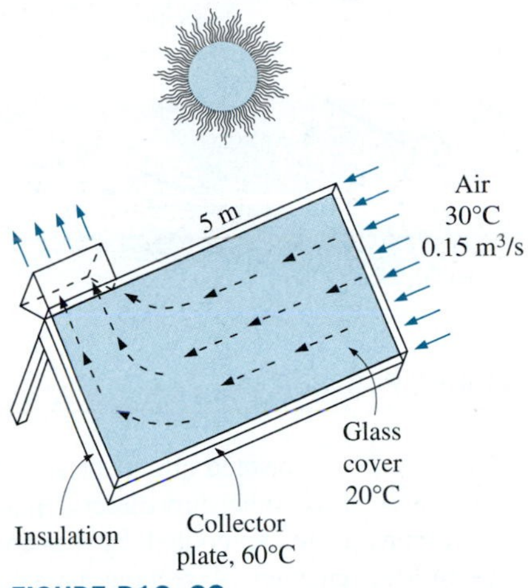

FIGURE P19–82

19–83 Consider the flow of oil at 10°C in a 40-cm-diameter pipeline at an average velocity of 0.5 m/s. A 1500-m-long section of the pipeline passes through icy waters of a lake at 0°C. Measurements indicate that the surface temperature of the pipe is very nearly 0°C. Disregarding the thermal resistance of the pipe material, determine (*a*) the temperature of the oil when the pipe leaves the lake, (*b*) the rate of heat transfer from the oil, and (*c*) the pumping power required to overcome the pressure losses and to maintain the flow oil in the pipe.

19–84 Consider laminar flow of a fluid through a square channel maintained at a constant temperature. Now the mean velocity of the fluid is doubled. Determine the change in the pressure drop and the change in the rate of heat transfer between the fluid and the walls of the channel. Assume the flow regime remains unchanged. Assume fully developed flow and disregard any changes in ΔT_{ln}.

19–85 Repeat Prob. 19–84 for turbulent flow.

19–86E The hot water needs of a household are to be met by heating water at 55°F to 200°F by a parabolic solar collector at a rate of 4 lbm/s. Water flows through a 1.25-in-diameter thin aluminum tube whose outer surface is blackanodized in order to maximize its solar absorption ability. The centerline of the tube coincides with the focal line of the collector, and a glass sleeve is placed outside the tube to minimize the heat losses. If solar energy is transferred to water at a net rate of 350 Btu/h per ft length of the tube, determine the required length of the parabolic collector to meet the hot water requirements of this house. Also, determine the surface temperature of the tube at the exit.

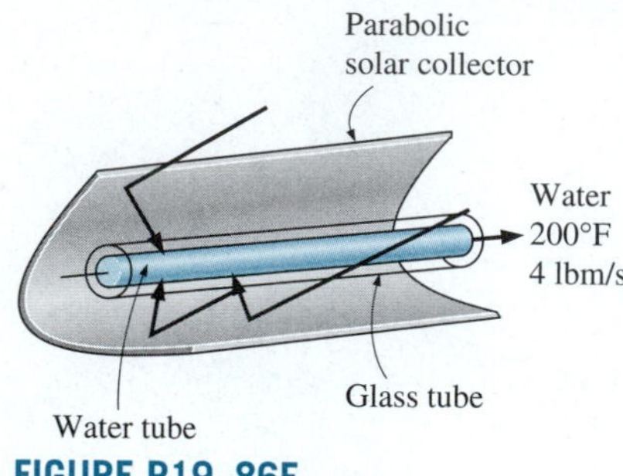

FIGURE P19–86E

19–87 A 15-cm × 20-cm printed circuit board whose components are not allowed to come into direct contact with air for reliability reasons is to be cooled by passing cool air through a 20-cm-long channel of rectangular cross section 0.2 cm × 14 cm drilled into the board. The heat generated by the electronic components is conducted across the thin layer of the board to the channel, where it is removed by air that enters the channel at 15°C. The heat flux at the top surface of the channel can be considered to be uniform, and heat transfer through other surfaces is negligible. If the velocity of the air at the inlet of the channel is not to exceed 4 m/s and the surface temperature of the channel is to remain under 50°C, determine the maximum total power of the electronic components that can safely be mounted on this circuit board.

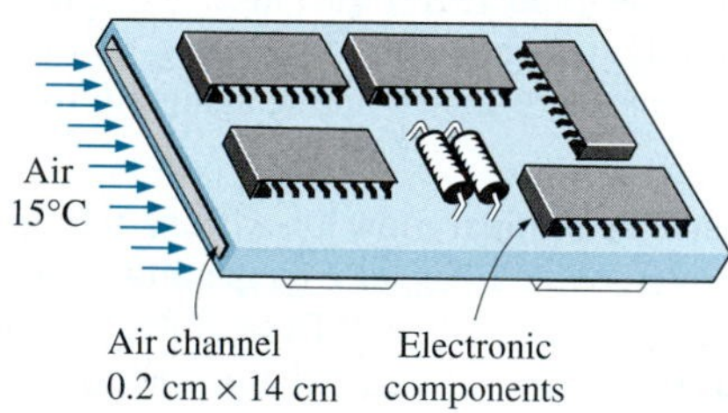

FIGURE P19–87

19–88 Repeat Prob. 19–87 by replacing air with helium, which has six times the thermal conductivity of air.

19–89 Reconsider Prob. 19–87. Using EES (or other) software, investigate the effects of air velocity at the inlet of the channel and the maximum surface temperature on the maximum total power dissipation of electronic components. Let the air velocity vary from 1 m/s to 10 m/s and the surface temperature from 30°C to 90°C. Plot the power dissipation as functions of air velocity and surface temperature, and discuss the results.

19–90 Air enters a 7-m-long section of a rectangular duct of cross section 15 cm × 20 cm at 50°C at an average velocity of 7 m/s. If the walls of the duct are maintained at 10°C, determine (*a*) the outlet temperature of the air, (*b*) the rate of heat transfer from the air, and (*c*) the fan power needed to overcome the pressure losses in this section of the duct. *Answers:* (*a*) 34.2°C, (*b*) 3775 W, (*c*) 4.7 W

19–91 Reconsider Prob. 19–90. Using EES (or other) software, investigate the effect of air velocity on the exit temperature of air, the rate of heat transfer, and the fan power. Let the air velocity vary from 1 m/s to 10 m/s. Plot the exit temperature, the rate of heat transfer, and the fan power as a function of the air velocity, and discuss the results.

19–92 Hot air at 60°C leaving the furnace of a house enters a 12-m-long section of a sheet metal duct of rectangular cross section 20 cm × 20 cm at an average velocity of 4 m/s. The thermal resistance of the duct is negligible, and the outer surface of the duct, whose emissivity is 0.3, is exposed to the cold air at 10°C in the basement, with a convection heat transfer coefficient of 10 W/m² · °C. Taking the walls of the basement to be at 10°C also, determine (*a*) the temperature at which the hot air will leave the basement and (*b*) the rate of heat loss from the hot air in the duct to the basement.

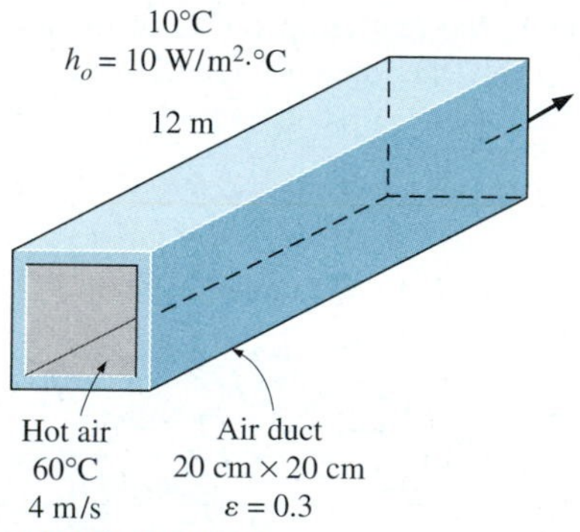

FIGURE P19–92

19–93 Reconsider Prob. 19–92. Using EES (or other) software, investigate the effects of air velocity and the surface emissivity on the exit temperature of air and the rate of heat loss. Let the air velocity vary from 1 m/s to 10 m/s and the emissivity from 0.1 to 1.0. Plot the exit temperature and the rate of heat loss as functions of air velocity and emissivity, and discuss the results.

19–94 The components of an electronic system dissipating 180 W are located in a 1-m-long horizontal duct whose cross section is 16 cm × 16 cm. The components in the duct are cooled by forced air, which enters at 27°C at a rate of 0.65 m³/min. Assuming 85 percent of the heat generated inside is transferred to air flowing through the duct and the remaining 15 percent is lost through the outer surfaces of the duct, determine (*a*) the exit temperature of air and (*b*) the highest component surface temperature in the duct.

19–95 Repeat Prob. 19–94 for a circular horizontal duct of 15-cm diameter.

19–96 Consider a hollow-core printed circuit board 12 cm high and 18 cm long, dissipating a total of 20 W. The width of the air gap in the middle of the PCB is 0.25 cm. The cooling air enters the 12-cm-wide core at 32°C at a rate of 0.8 L/s. Assuming the heat generated to be uniformly distributed over the two side surfaces of the PCB, determine (*a*) the temperature at which the air leaves the hollow core and (*b*) the highest temperature on the inner surface of the core. *Answers:* (*a*) 54.0°C, (*b*) 72.2°C

19–97 Repeat Prob. 19–96 for a hollow-core PCB dissipating 35 W.

19–98 A computer cooled by a fan contains eight PCBs, each dissipating 10 W of power. The height of the PCBs is 12 cm and the length is 18 cm. The clearance between the tips of the components on the PCB and the back surface of the adjacent PCB is 0.3 cm. The cooling air is supplied by a 10-W fan mounted at the inlet. If the temperature rise of air as it flows through the case of the computer is not to exceed 10°C, determine (*a*) the flow rate of the air that the fan needs to deliver, (*b*) the fraction of the temperature rise of air that is due to the heat generated by the fan and its motor, and (*c*) the highest allowable inlet air temperature if the surface temperature of the components is not to exceed 70°C anywhere in the system. Use air properties at 25°C.

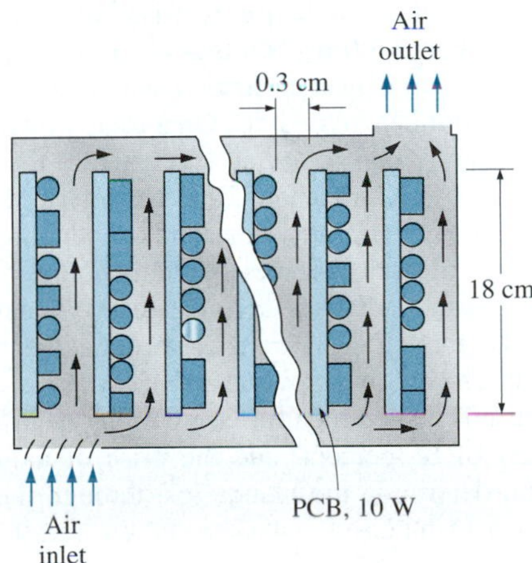

FIGURE P19–98

Review Problems

19–99 Consider a house that is maintained at 22°C at all times. The walls of the house have *R*-3.38 insulation in SI units (i.e., an *L*/*k* value or a thermal resistance of 3.38 m² · °C/W). During a cold winter night, the outside air temperature is 6°C and wind at 50 km/h is blowing parallel to a 4-m-high and 8-m-long wall of the house. If the heat transfer coefficient on the interior surface of the wall is 8 W/m² · °C, determine the rate of heat loss from that wall of the house. Draw the thermal resistance network and disregard radiation heat transfer. *Answer:* 145 W

19–100 An automotive engine can be approximated as a 0.4-m-high, 0.60-m-wide, and 0.7-m-long rectangular block. The bottom surface of the block is at a temperature of 75°C and has an emissivity of 0.92. The ambient air is at 5°C, and the road surface is at 10°C. Determine the rate of heat transfer from the bottom surface of the engine block by convection and radiation as the car travels at a velocity of 60 km/h. Assume the flow to be turbulent over the entire surface because of the constant agitation of the engine block. How will the heat transfer be affected when a 2-mm-thick gunk (k = 3 W/m · °C) has formed at the bottom surface as a result of the dirt and oil collected at that surface over time? Assume the metal temperature under the gunk still to be 75°C.

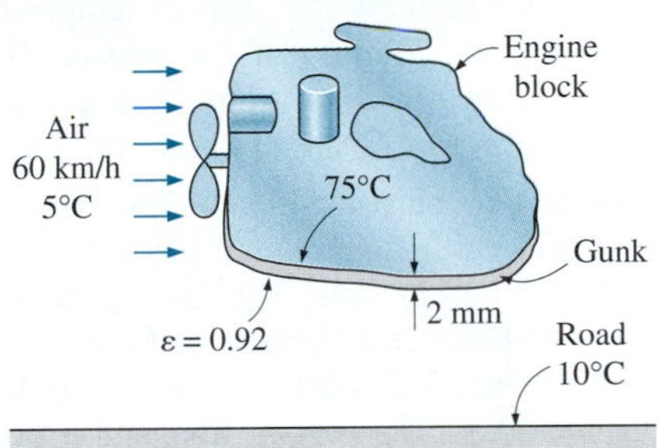

FIGURE P19–100

19–101E The passenger compartment of a minivan traveling at 60 mph can be modeled as a 3.2-ft-high, 6-ft-wide, and 11-ft-long rectangular box whose walls have an insulating value of *R*-3 (i.e., a wall thickness-to-thermal conductivity ratio of 3 h · ft² · °F/Btu). The interior of a minivan is maintained at an average temperature of 70°F during a trip at night while the outside air temperature is 90°F. The average heat transfer coefficient on the interior surfaces of the van is 1.2 Btu/h · ft² · °F. The air flow over the exterior surfaces can be assumed to be turbulent because of the intense vibrations involved, and the heat transfer coefficient on the front and back surfaces can be taken to be equal to that on the top surface. Disregarding any heat gain or loss by radiation, determine the rate of heat transfer from the ambient air to the van.

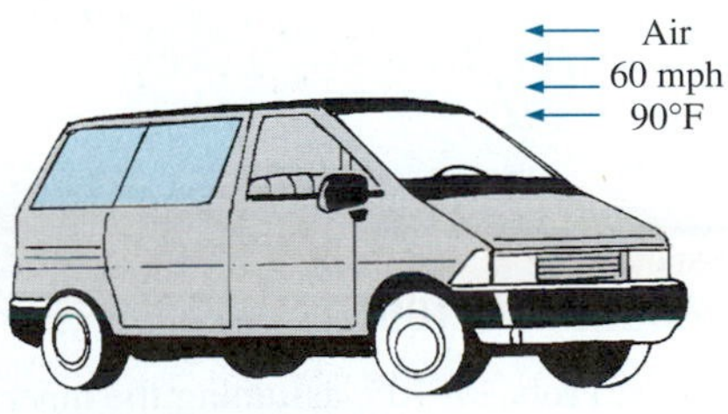

FIGURE P19–101E

19–102 Consider a house that is maintained at a constant temperature of 22°C. One of the walls of the house has three single-pane glass windows that are 1.5 m high and 1.8 m long. The glass (k = 0.78 W/m · °C) is 0.5 cm thick, and the heat transfer coefficient on the inner surface of the glass is

8 W/m² · C. Now winds at 35 km/h start to blow parallel to the surface of this wall. If the air temperature outside is −2°C, determine the rate of heat loss through the windows of this wall. Assume radiation heat transfer to be negligible.

19–103 Consider a person who is trying to keep cool on a hot summer day by turning a fan on and exposing his body to air flow. The air temperature is 32°C, and the fan is blowing air at a velocity of 5 m/s. The surrounding surfaces are at 40°C, and the emissivity of the person can be taken to be 0.9. If the person is doing light work and generating sensible heat at a rate of 90 W, determine the average temperature of the outer surface (skin or clothing) of the person. The average human body can be treated as a 30-cm-diameter cylinder with an exposed surface area of 1.7 m². *Answer:* 36.2°C

19–104 Four power transistors, each dissipating 12 W, are mounted on a thin vertical aluminum plate (k = 237 W/m · °C) 22 cm × 22 cm in size. The heat generated by the transistors is to be dissipated by both surfaces of the plate to the surrounding air at 20°C, which is blown over the plate by a fan at a velocity of 250 m/min. The entire plate can be assumed to be nearly isothermal, and the exposed surface area of the transistor can be taken to be equal to its base area. Determine the temperature of the aluminum plate.

19–105 A 3-m-internal-diameter spherical tank made of 1-cm-thick stainless steel (k = 15 W/m · °C) is used to store iced water at 0°C. The tank is located outdoors at 30°C and is subjected to winds at 25 km/h. Assuming the entire steel tank to be at 0°C and thus its thermal resistance to be negligible, determine (*a*) the rate of heat transfer to the iced water in the tank and (*b*) the amount of ice at 0°C that melts during a 24-h period. The heat of fusion of water at atmospheric pressure is h_{if} = 333.7 kJ/kg. Disregard any heat transfer by radiation.

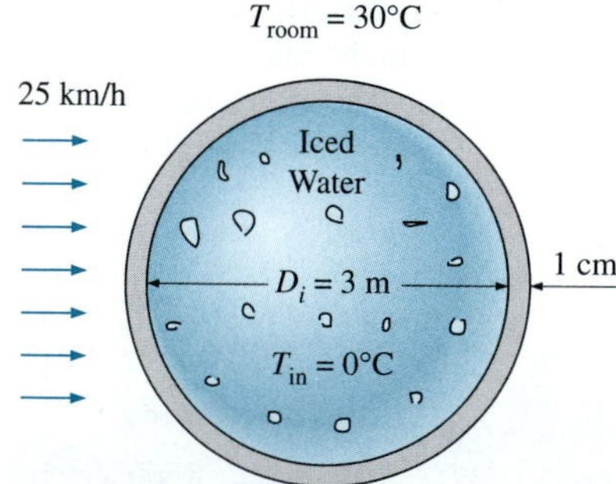

FIGURE P19–105

19–106 Repeat Prob. 19–105, assuming the inner surface of the tank to be at 0°C but by taking the thermal resistance of the tank and heat transfer by radiation into consideration. Assume the average surrounding surface temperature for radiation exchange to be 25°C and the outer surface of the tank to have an emissivity of 0.75. *Answers:* (*a*) 10,530 W, (*b*) 2727 kg

19–107E A transistor with a height of 0.25 in and a diameter of 0.22 in is mounted on a circuit board. The transistor is cooled by air flowing over it at a velocity of 500 ft/min. If the air temperature is 120°F and the transistor case temperature is not to exceed 180°F, determine the amount of power this transistor can dissipate safely.

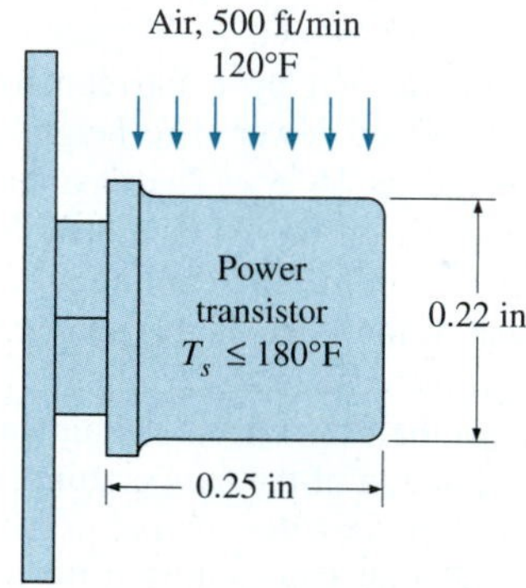

FIGURE P19–107E

19–108 The roof of a house consists of a 15-cm-thick concrete slab (k = 2 W/m · °C) that is 15 m wide and 20 m long. The convection heat transfer coefficient on the inner surface of the roof is 5 W/m² · °C. On a clear winter night, the ambient air is reported to be at 10°C, while the night sky temperature is 100 K. The house and the interior surfaces of the wall are maintained at a constant temperature of 20°C. The emissivity of both surfaces of the concrete roof is 0.9. Considering both radiation and convection heat transfer, determine the rate of heat transfer through the roof when wind at 60 km/h is blowing over the roof.

If the house is heated by a furnace burning natural gas with an efficiency of 85 percent, and the price of natural gas is $1.20/therm, determine the money lost through the roof that night during a 14-h period. *Answers:* 28 kW, $18.9

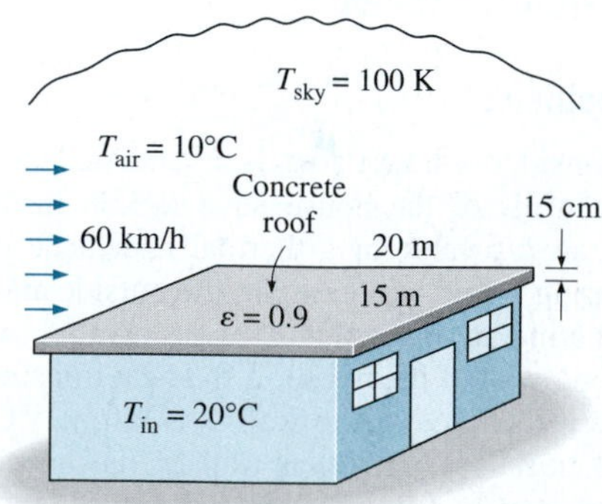

FIGURE P19–108

19–109 Steam at 250°C flows in a stainless steel pipe (k = 15 W/m · °C) whose inner and outer diameters are 4 cm and

4.6 cm, respectively. The pipe is covered with 3.5-cm-thick glass wool insulation (k = 0.038 W/m · °C) whose outer surface has an emissivity of 0.3. Heat is lost to the surrounding air and surfaces at 3°C by convection and radiation. Taking the heat transfer coefficient inside the pipe to be 80 W/m² · °C, determine the rate of heat loss from the steam per unit length of the pipe when air is flowing across the pipe at 4 m/s.

19–110 The boiling temperature of nitrogen at atmospheric pressure at sea level (1 atm pressure) is −196°C. Therefore, nitrogen is commonly used in low-temperature scientific studies, since the temperature of liquid nitrogen in a tank open to the atmosphere will remain constant at −196°C until it is depleted. Any heat transfer to the tank will result in the evaporation of some liquid nitrogen, which has a heat of vaporization of 198 kJ/kg and a density of 810 kg/m³ at 1 atm.

Consider a 4-m-diameter spherical tank that is initially filled with liquid nitrogen at 1 atm and −196°C. The tank is exposed to 20°C ambient air and 40 km/h winds. The temperature of the thin-shelled spherical tank is observed to be almost the same as the temperature of the nitrogen inside. Disregarding any radiation heat exchange, determine the rate of evaporation of the liquid nitrogen in the tank as a result of heat transfer from the ambient air if the tank is (*a*) not insulated, (*b*) insulated with 5-cm-thick fiberglass insulation (k = 0.035 W/m · °C), and (*c*) insulated with 2-cm-thick superinsulation that has an effective thermal conductivity of 0.00005 W/m · °C.

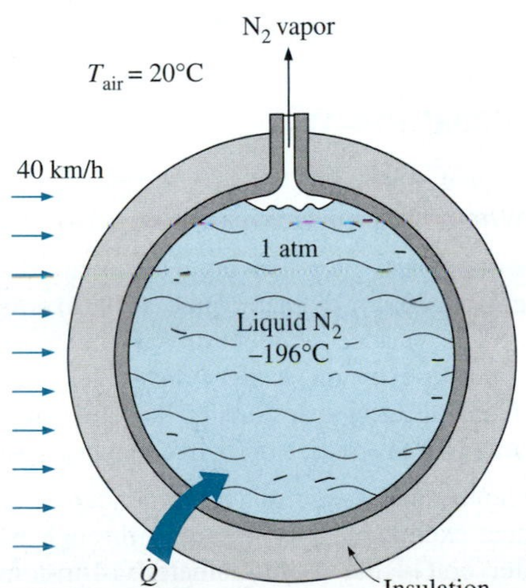

FIGURE P19–110

19–111 Repeat Prob. 19–110 for liquid oxygen, which has a boiling temperature of −183°C, a heat of vaporization of 213 kJ/kg, and a density of 1140 kg/m³ at 1 atm pressure.

19–112 A 0.5-cm-thick, 12-cm-high, and 18-cm-long circuit board houses 80 closely spaced logic chips on one side, each dissipating 0.06 W. The board is impregnated with copper fillings and has an effective thermal conductivity of 16 W/m · °C. All the heat generated in the chips is conducted across the circuit board and is dissipated from the back side of the board to the ambient air at 30°C, which is forced to flow over the surface by a fan at a free-stream velocity of 300 m/min. Determine the temperatures on the two sides of the circuit board.

19–113 A silicon chip is cooled by passing water through microchannel etched in the back of the chip, as shown in Fig. P19–113. The channels are covered with a silicon cap. Consider a 10-mm × 10-mm square chip in which N = 50 rectangular microchannels, each of width W = 50 μm and height H = 200 μm have been etched. Water enters the microchannels at a temperature T_i = 290 K, and a total flow rate of 0.005 kg/s. The chip and cap are maintained at a uniform temperature of 350 K. Assuming that the flow in the channels is fully developed, all the heat generated by the circuits on the top surface of the chip is transferred to the water, and using circular tube correlations, determine:

(*a*) The water outlet temperature, T_e

(*b*) The chip power dissipation, $\dot{W}_e$

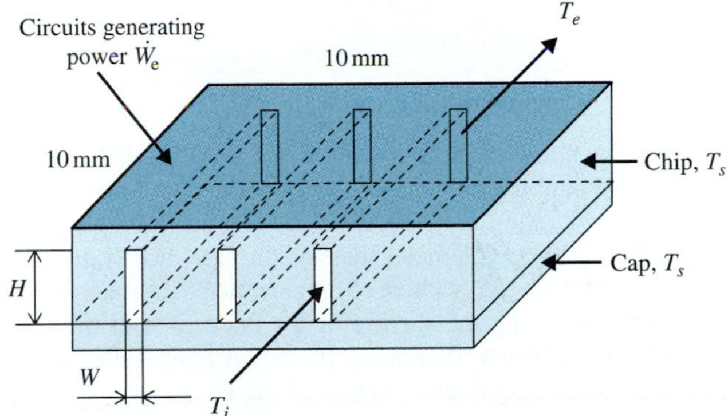

FIGURE P19–113

19–114 Water is heated at a rate of 10 kg/s from a temperature of 15°C to 35°C by passing it through five identical tubes, each 5.0 cm in diameter, whose surface temperature is 60.0°C. Estimate (*a*) the steady rate of heat transfer and (*b*) the length of tubes necessary to accomplish this task.

19–115 Repeat Prob. 19–114 for a flow rate of 20 kg/s.

19–116 Water at 1500 kg/h and 10°C enters a 10-mm diameter smooth tube whose wall temperature is maintained at 49°C. Calculate (*a*) the tube length necessary to heat the water to 40°C, and (*b*) the water outlet temperature if the tube length is doubled. Assume average water properties to be the same as in (*a*).

19–117 The compressed air requirements of a manufacturing facility are met by a 150-hp compressor located in a room that is maintained at 20°C. In order to minimize the compressor work, the intake port of the compressor is connected to the outside through an 11-m-long, 20-cm-diameter duct made of thin aluminum sheet. The compressor takes in air at a rate of 0.27 m³/s at the outdoor conditions of 10°C and 95 kPa. Disregarding the thermal resistance of the duct and taking the heat transfer coefficient on the outer surface of the duct to be 10 W/m² · °C, determine (*a*) the power used by the compressor to overcome the pressure drop in this duct, (*b*) the rate of heat transfer to the incoming cooler air, and (*c*) the temperature rise of air as it flows through the duct.

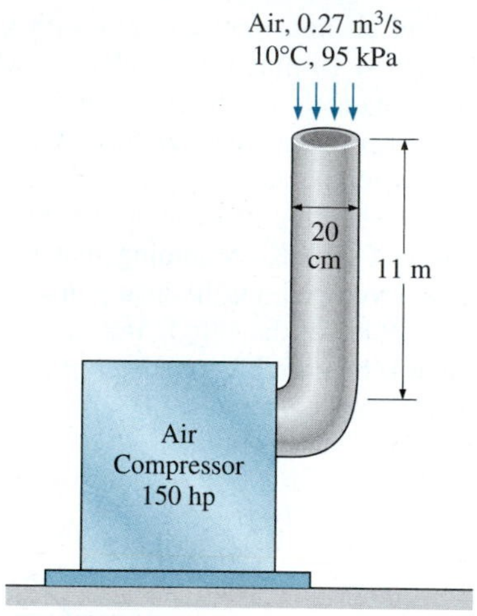

FIGURE P19–117

19–118 A house built on a riverside is to be cooled in summer by utilizing the cool water of the river, which flows at an average temperature of 15°C. A 15-m-long section of a circular duct of 20-cm diameter passes through the water. Air enters the underwater section of the duct at 25°C at a velocity of 3 m/s. Assuming the surface of the duct to be at the temperature of the water, determine the outlet temperature of air as it leaves the underwater portion of the duct.

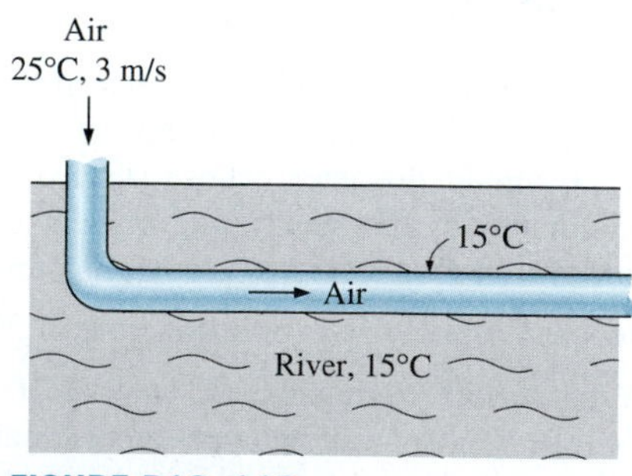

FIGURE P19–118

19–119 Repeat Prob. 19–118 assuming that a 0.25-mm-thick layer of mineral deposit (k = 3 W/m · °C) formed on the inner surface of the pipe.

19–120E The exhaust gases of an automotive engine leave the combustion chamber and enter a 8-ft-long and 3.5-in-diameter thin-walled steel exhaust pipe at 800°F and 15.5 psia at a rate of 0.2 lbm/s. The surrounding ambient air is at a temperature of 80°F, and the heat transfer coefficient on the outer surface of the exhaust pipe is 3 Btu/h · ft² · °F. Assuming the exhaust gases to have the properties of air, determine (*a*) the velocity of the exhaust gases at the inlet of the exhaust pipe and (*b*) the temperature at which the exhaust gases will leave the pipe and enter the air.

19–121 Hot water at 90°C enters a 15-m section of a cast iron pipe (k = 52 W/m · °C) whose inner and outer diameters are 4 and 4.6 cm, respectively, at an average velocity of 1.2 m/s. The outer surface of the pipe, whose emissivity is 0.7, is exposed to the cold air at 10°C in a basement, with a convection heat transfer coefficient of 12 W/m² · °C. Taking the walls of the basement to be at 10°C also, determine (*a*) the rate of heat loss from the water and (*b*) the temperature at which the water leaves the basement.

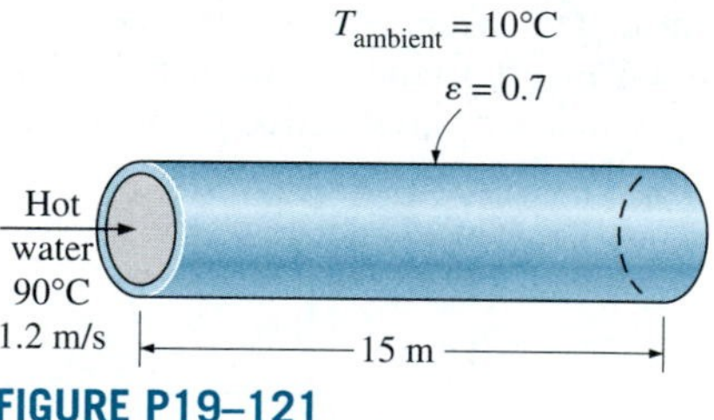

FIGURE P19–121

19–122 Repeat Prob. 19–121 for a pipe made of copper (k = 386 W/m · °C) instead of cast iron.

19–123 Hot exhaust gases leaving a stationary diesel engine at 450°C enter a 15-cm-diameter pipe at an average velocity of 4.5 m/s. The surface temperature of the pipe is 180°C. Determine the pipe length if the exhaust gases are to leave the pipe at 250°C after transferring heat to water in a heat recovery unit. Use properties of air for exhaust gases.

19–124 Geothermal steam at 165°C condenses in the shell side of a heat exchanger over the tubes through which water flows. Water enters the 4-cm-diameter, 14-m-long tubes at 20°C at a rate of 0.8 kg/s. Determine the exit temperature of water and the rate of condensation of geothermal steam.

19–125 Cold air at 5°C enters a 12-cm-diameter 20-m-long isothermal pipe at a velocity of 2.5 m/s and leaves at 19°C. Estimate the surface temperature of the pipe.

19–126 Oil at 15°C is to be heated by saturated steam at 1 atm in a double-pipe heat exchanger to a temperature of 25°C. The inner and outer diameters of the annular space are 3 cm and 5 cm, respectively, and oil enters at with a mean velocity of 0.8 m/s. The inner tube may be assumed to be isothermal at 100°C, and the outer tube is well insulated. Assuming fully developed flow for oil, determine the tube length required to heat the oil to the indicated temperature. In reality, will you need a shorter or longer tube? Explain.

Design and Essay Problems

19–127 On average, superinsulated homes use just 15 percent of the fuel required to heat the same size conventional home built before the energy crisis in the 1970s. Write an essay on superinsulated homes, and identify the features that make them so energy efficient as well as the problems associated with them. Do you think superinsulated homes will be economically attractive in your area?

19–128 Conduct this experiment to determine the heat loss coefficient of your house or apartment in W/°C or Btu/h · °F. First make sure that the conditions in the house are steady and the house is at the set temperature of the thermostat. Use an outdoor thermometer to monitor outdoor temperature. One evening, using a watch or timer, determine how long the heater was on during a 3-h period and the average outdoor temperature during that period. Then using the heat output rating of your heater, determine the amount of heat supplied. Also, estimate the amount of heat generation in the house during that period by noting the number of people, the total wattage of lights that were on, and the heat generated by the appliances and equipment. Using that information, calculate the average rate of heat loss from the house and the heat loss coefficient.

19–129 Write an article on forced convection cooling with air, helium, water, and a dielectric liquid. Discuss the advantages and disadvantages of each fluid in heat transfer. Explain the circumstances under which a certain fluid will be most suitable for the cooling job.

19–130 Electronic boxes such as computers are commonly cooled by a fan. Write an essay on forced air cooling of electronic boxes and on the selection of the fan for electronic devices.

19–131 Design a heat exchanger to pasteurize milk by steam in a dairy plant. Milk is to flow through a bank of 1.2-cm internal diameter tubes while steam condenses outside the tubes at 1 atm. Milk is to enter the tubes at 4°C, and it is to be heated to 72°C at a rate of 15 L/s. Making reasonable assumptions, you are to specify the tube length and the number of tubes, and the pump for the heat exchanger.

19–132 A desktop computer is to be cooled by a fan. The electronic components of the computer consume 80 W of power under full-load conditions. The computer is to operate in environments at temperatures up to 50°C and at elevations up to 3000 m where the atmospheric pressure is 70.12 kPa. The exit temperature of air is not to exceed 60°C to meet the reliability requirements. Also, the average velocity of air is not to exceed 120 m/min at the exit of the computer case, where the fan is installed to keep the noise level down. Specify the flow rate of the fan that needs to be installed and the diameter of the casing of the fan.

Chapter 20

NATURAL CONVECTION

In Chapter 19, we considered heat transfer by *forced convection*, where a fluid was *forced* to move over a surface or in a tube by external means such as a pump or a fan. In this chapter, we consider *natural convection*, where any fluid motion occurs by natural means such as buoyancy. The fluid motion in forced convection is quite *noticeable*, since a fan or a pump can transfer enough momentum to the fluid to move it in a certain direction. The fluid motion in natural convection, however, is often not noticeable because of the low velocities involved.

The convection heat transfer coefficient is a strong function of *velocity*: the higher the velocity, the higher the convection heat transfer coefficient. The fluid velocities associated with natural convection are low, typically less than 1 m/s. Therefore, the heat transfer coefficients encountered in natural convection are usually much lower than those encountered in forced convection. Yet several types of heat transfer equipment are designed to operate under natural convection conditions instead of forced convection, because natural convection does not require the use of a fluid mover.

We start this chapter with a discussion of the physical mechanism of *natural convection* and the *Grashof number*. We then present the correlations to evaluate heat transfer by natural convection for various geometries, including finned surfaces and enclosures.

Objectives

The objectives of this chapter are to:

- Understand the physical mechanism of natural convection,
- Derive the governing equations of natural convection, and obtain the dimensionless Grashof number by nondimensionalizing them,
- Evaluate the Nusselt number for natural convection associated with vertical, horizontal, and inclined plates as well as cylinders and spheres,
- Examine natural convection from finned surfaces, and determine the optimum fin spacing,
- Analyze natural convection inside enclosures such as double-pane windows.

20–1 ■ PHYSICAL MECHANISM OF NATURAL CONVECTION

Many familiar heat transfer applications involve natural convection as the primary mechanism of heat transfer. Some examples are cooling of electronic equipment such as power transistors, TVs, and DVDs; heat transfer from electric baseboard heaters or steam radiators; heat transfer from the refrigeration coils and power transmission lines; and heat transfer from the bodies of animals and human beings. Natural convection in gases is usually accompanied by radiation of comparable magnitude except for low-emissivity surfaces.

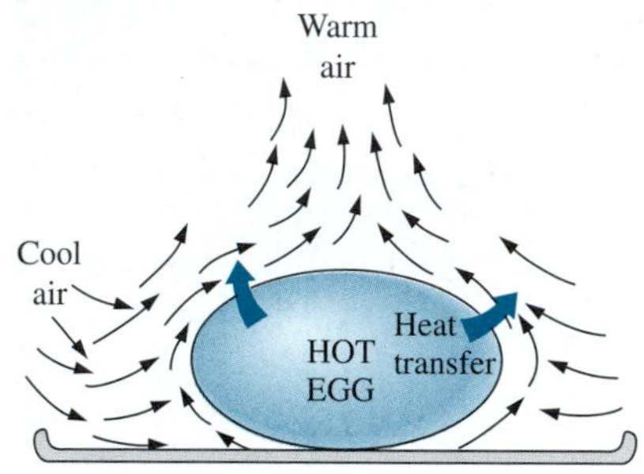

We know that a hot boiled egg (or a hot baked potato) on a plate eventually cools to the surrounding air temperature (Fig. 20–1). The egg is cooled by transferring heat by convection to the air and by radiation to the surrounding surfaces. Disregarding heat transfer by radiation, the physical mechanism of cooling a hot egg (or any hot object) in a cooler environment can be explained as follows:

FIGURE 20–1
The cooling of a boiled egg in a cooler environment by natural convection.

As soon as the hot egg is exposed to cooler air, the temperature of the outer surface of the egg shell drops somewhat, and the temperature of the air adjacent to the shell rises as a result of heat conduction from the shell to the air. Consequently, the egg is surrounded by a thin layer of warmer air, and heat is then transferred from this warmer layer to the outer layers of air. The cooling process in this case is rather slow since the egg would always be blanketed by warm air, and it has no direct contact with the cooler air farther away. We may not notice any air motion in the vicinity of the egg, but careful measurements would indicate otherwise.

The temperature of the air adjacent to the egg is higher and thus its density is lower, since at constant pressure the density of a gas is inversely proportional to its temperature. Thus, we have a situation in which some low-density or "light" gas is surrounded by a high-density or "heavy" gas, and the natural laws dictate that the *light gas rise.* This is no different than the oil in a vinegar-and-oil salad dressing rising to the top (since $\rho_{\text{oil}} < \rho_{\text{vinegar}}$). This phenomenon is characterized incorrectly by the phrase "heat rises," which is understood to mean *heated air rises.* The space vacated by the warmer air in the vicinity of the egg is replaced by the cooler air nearby, and the presence of cooler air in the vicinity of the egg speeds up the cooling process. The rise of warmer air and the flow of cooler air into its place continues until the egg is cooled to the temperature of the surrounding air. The motion that results from the continual replacement of the heated air in the vicinity of the egg by the cooler air nearby is called a **natural convection current**, and the heat transfer that is enhanced as a result of this natural convection current is called **natural convection heat transfer**. Note that in the absence of natural convection currents, heat transfer from the egg to the air surrounding it would be by conduction only, and the rate of heat transfer from the egg would be much lower.

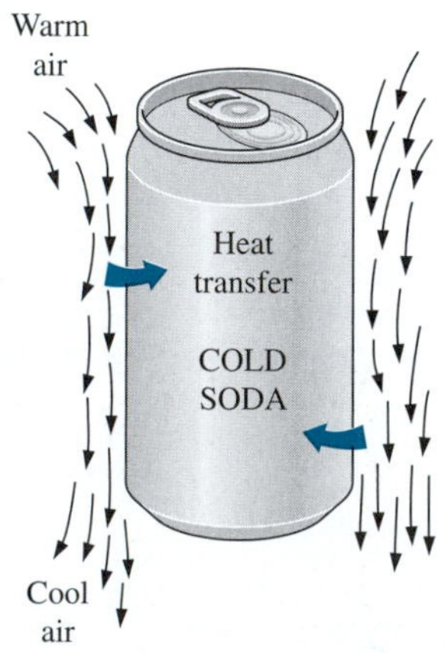

FIGURE 20–2
The warming up of a cold drink in a warmer environment by natural convection.

Natural convection is just as effective in the heating of cold surfaces in a warmer environment as it is in the cooling of hot surfaces in a cooler environment, as shown in Fig. 20–2. Note that the direction of fluid motion is reversed in this case.

In a gravitational field, there is a net force that pushes upward a light fluid placed in a heavier fluid. The upward force exerted by a fluid on a body completely or partially immersed in it is called the **buoyancy force**. The magnitude of the buoyancy force is equal to the weight of the *fluid displaced* by the body. That is,

$$F_{\text{buoyancy}} = \rho_{\text{fluid}}\, g V_{\text{body}} \tag{20–1}$$

where ρ_{fluid} is the average density of the *fluid* (not the body), g is the gravitational acceleration, and V_{body} is the volume of the portion of the body immersed in the fluid (for bodies completely immersed in the fluid, it is the total volume of the body). In the absence of other forces, the net vertical force acting on a body is the difference between the weight of the body and the buoyancy force. That is,

$$\begin{aligned} F_{\text{net}} &= W - F_{\text{buoyancy}} \\ &= \rho_{\text{body}}\, g V_{\text{body}} - \rho_{\text{fluid}}\, g V_{\text{body}} \\ &= (\rho_{\text{body}} - \rho_{\text{fluid}})\, g V_{\text{body}} \end{aligned} \tag{20–2}$$

Note that this force is proportional to the difference in the densities of the fluid and the body immersed in it. Thus, a body immersed in a fluid will experience a "weight loss" in an amount equal to the weight of the fluid it displaces. This is known as *Archimedes' principle.*

To have a better understanding of the buoyancy effect, consider an egg dropped into water. If the average density of the egg is greater than the density of water (a sign of freshness), the egg settles at the bottom of the container. Otherwise, it rises to the top. When the density of the egg equals the density of water, the egg settles somewhere in the water while remaining completely immersed, acting like a "weightless object" in space. This occurs when the upward buoyancy force acting on the egg equals the weight of the egg, which acts downward.

The *buoyancy effect* has far-reaching implications in life. For one thing, without buoyancy, heat transfer between a hot (or cold) surface and the fluid surrounding it would be by *conduction* instead of by *natural convection.* The natural convection currents encountered in the oceans, lakes, and the atmosphere owe their existence to buoyancy. Also, light boats as well as heavy warships made of steel float on water because of buoyancy (Fig. 20–3). Ships are designed on the basis of the principle that the entire weight of a ship and its contents is equal to the weight of the water that the submerged volume of the ship can contain. The "chimney effect" that induces the upward flow of hot combustion gases through a chimney is also due to the buoyancy effect, and the upward force acting on the gases in the chimney is proportional to the difference between the densities of the hot gases in the chimney and the cooler air outside. Note that there is *no noticable gravity* in space, and thus there can be no natural convection heat transfer in a spacecraft, even if the spacecraft is filled with atmospheric air.

In heat transfer studies, the primary variable is *temperature,* and it is desirable to express the net buoyancy force (Eq. 20–2) in terms of temperature differences. But this requires expressing the density difference in terms of a temperature difference, which requires a knowledge of a property that represents the *variation of the density of a fluid with temperature at constant pressure. The property that provides that information is the* **volume expansion coefficient** β, defined as (Fig. 20–4)

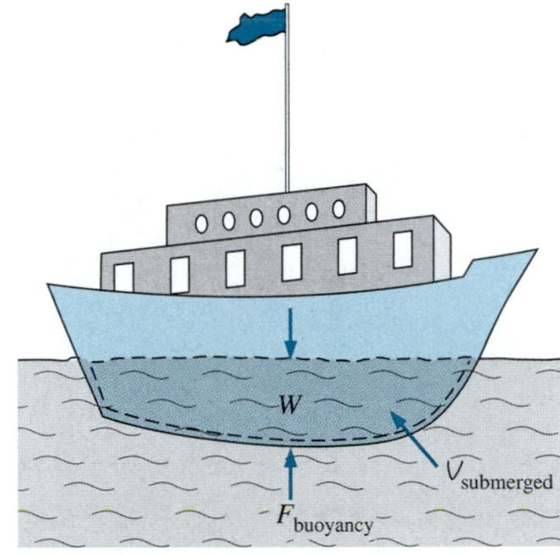

FIGURE 20–3
It is the buoyancy force that keeps the ships afloat in water ($W = F_{\text{buoyancy}}$ for floating objects).

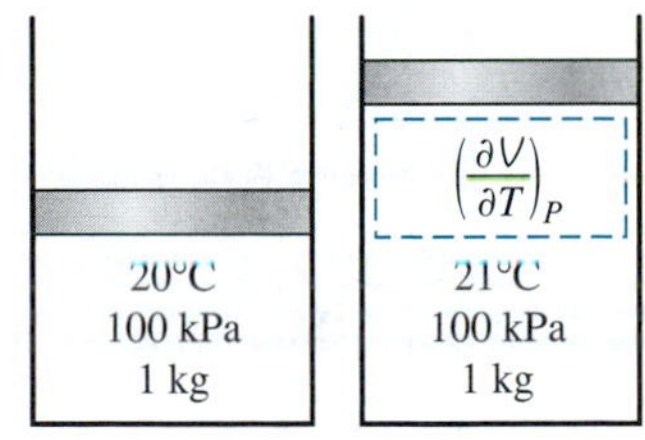

(*a*) A substance with a large β

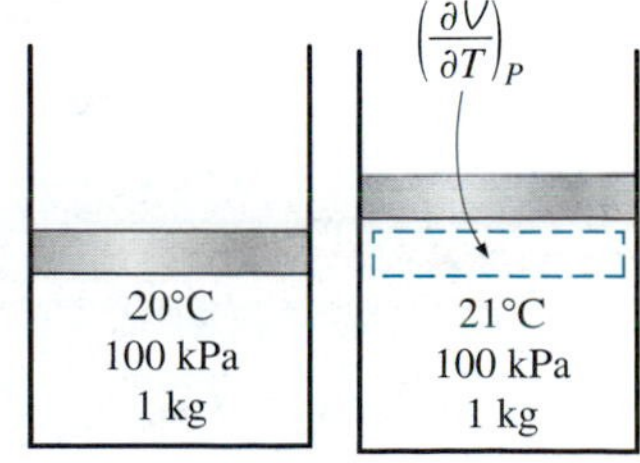

(*b*) A substance with a small β

FIGURE 20–4
The coefficient of volume expansion is a measure of the change in volume of a substance with temperature at constant pressure.

$$\beta = \frac{1}{v}\left(\frac{\partial v}{\partial T}\right)_P = -\frac{1}{\rho}\left(\frac{\partial \rho}{\partial T}\right)_P \qquad \text{(1/K)} \tag{20–3}$$

In natural convection studies, the condition of the fluid sufficiently far from the hot or cold surface is indicated by the subscript "infinity" to serve as a reminder that this is the value at a distance where the presence of the surface is not felt. In such cases, the volume expansion coefficient can be expressed approximately by replacing differential quantities by differences as

$$\beta \approx -\frac{1}{\rho}\frac{\Delta \rho}{\Delta T} = -\frac{1}{\rho}\frac{\rho_\infty - \rho}{T_\infty - T} \qquad \text{(at constant } P\text{)} \tag{20–4}$$

or

$$\rho_\infty - \rho = \rho\beta(T - T_\infty) \qquad \text{(at constant } P\text{)} \tag{20–5}$$

where ρ_∞ is the density and T_∞ is the temperature of the quiescent fluid away from the surface.

We can show easily that the volume expansion coefficient β of an *ideal gas* ($P = \rho RT$) at a temperature T is equivalent to the inverse of the temperature:

$$\beta_{\text{ideal gas}} = \frac{1}{T} \qquad \text{(1/K)} \tag{20–6}$$

where T is the *thermodynamic* temperature. Note that a large value of β for a fluid means a large change in density with temperature, and that the product $\beta\Delta T$ represents the fraction of volume change of a fluid that corresponds to a temperature change ΔT at constant pressure. Also note that the buoyancy force is proportional to the *density difference,* which is proportional to the *temperature difference* at constant pressure. Therefore, the larger the temperature difference between the fluid adjacent to a hot (or cold) surface and the fluid away from it, the *larger* the buoyancy force and the *stronger* the natural convection currents, and thus the *higher* the heat transfer rate.

The magnitude of the natural convection heat transfer between a surface and a fluid is directly related to the *flow rate* of the fluid. The higher the flow rate, the higher the heat transfer rate. In fact, it is the very high flow rates that increase the heat transfer coefficient by orders of magnitude when forced convection is used. In natural convection, no blowers are used, and therefore the flow rate cannot be controlled externally. The flow rate in this case is established by the dynamic balance of *buoyancy* and *friction.*

As we have discussed earlier, the buoyancy force is caused by the density difference between the heated (or cooled) fluid adjacent to the surface and the fluid surrounding it, and is proportional to this density difference and the volume occupied by the warmer fluid. It is also well known that whenever two bodies in contact (solid–solid, solid–fluid, or fluid–fluid) move relative to each other, a *friction force* develops at the contact surface in the direction opposite to that of the motion. This opposing force slows down the fluid and thus reduces the flow rate of the fluid. Under steady conditions, the airflow rate driven by buoyancy is established at the point where these two effects *balance* each other. The friction force increases as more and more solid surfaces are introduced, seriously disrupting the fluid flow and heat transfer. For that reason, heat sinks with closely spaced fins are not suitable for natural convection cooling.

Most heat transfer correlations in natural convection are based on experimental measurements. The instrument often used in natural convection experiments is the *Mach–Zehnder interferometer,* which gives a plot of isotherms in the fluid in the vicinity of a surface. The operation principle of interferometers is based on the fact that at low pressure, the lines of constant temperature for a gas correspond to the lines of constant density, and that the index of refraction of a gas is a function of its density. Therefore, the degree of refraction of light at some point in a gas is a measure of the temperature gradient at that point. An interferometer produces a map of interference fringes, which can be interpreted as lines of *constant temperature* as shown in Fig. 20–5. The smooth and parallel lines in (*a*) indicate that the flow is *laminar,* whereas the eddies and irregularities in (*b*) indicate that the flow is *turbulent.* Note that the lines are closest near the surface, indicating a *higher temperature gradient.*

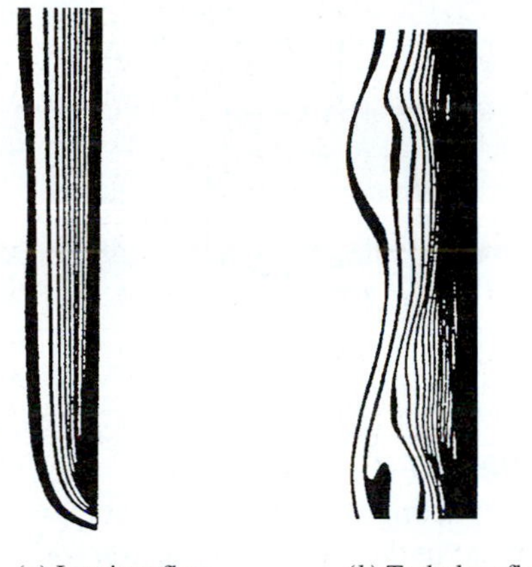

FIGURE 20–5
Isotherms in natural convection over a hot plate in air.

20–2 ▪ EQUATION OF MOTION AND THE GRASHOF NUMBER

In this section we derive the equation of motion that governs the natural convection flow in laminar boundary layer. The conservation of mass and energy equations derived for forced convection are also applicable for natural convection, but the momentum equation needs to be modified to incorporate buoyancy.

Consider a vertical hot flat plate immersed in a quiescent fluid body. We assume the natural convection flow to be steady, laminar, and two-dimensional, and the fluid to be Newtonian with constant properties, including density, with one exception: the density difference $\rho - \rho_\infty$ is to be considered since it is this density difference between the inside and the outside of the boundary layer that gives rise to buoyancy force and sustains flow. (This is known as the *Boussinesq approximation.*) We take the upward direction along the plate to be x, and the direction normal to surface to be y, as shown in Fig. 20–6. Therefore, gravity acts in the $-x$-direction. Noting that the flow is steady and two-dimensional, the x- and y-components of velocity within boundary layer are $u = u(x, y)$ and $v = v(x, y)$, respectively.

The velocity and temperature profiles for natural convection over a vertical hot plate are also shown in Fig. 20–6. Note that as in forced convection, the thickness of the boundary layer increases in the flow direction. Unlike forced convection, however, the fluid velocity is *zero* at the outer edge of the velocity boundary layer as well as at the surface of the plate. This is expected since the fluid beyond the boundary layer is motionless. Thus, the fluid velocity increases with distance from the surface, reaches a maximum, and gradually decreases to zero at a distance sufficiently far from the surface. At the surface, the fluid temperature is equal to the plate temperature, and gradually decreases to the temperature of the surrounding fluid at a distance sufficiently far from the surface, as shown in the figure. In the case of *cold surfaces,* the shape of the velocity and temperature profiles remains the same but their direction is reversed.

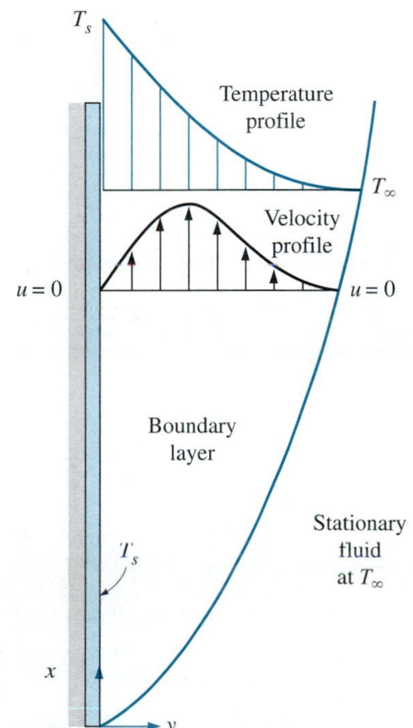

FIGURE 20–6
Typical velocity and temperature profiles for natural convection flow over a hot vertical plate at temperature T_s inserted in a fluid at temperature T_∞.

Consider a differential volume element of height dx, length dy, and unit depth in the z-direction (normal to the paper) for analysis. The forces acting on this volume element are shown in Fig. 20–7. Newton's second law of

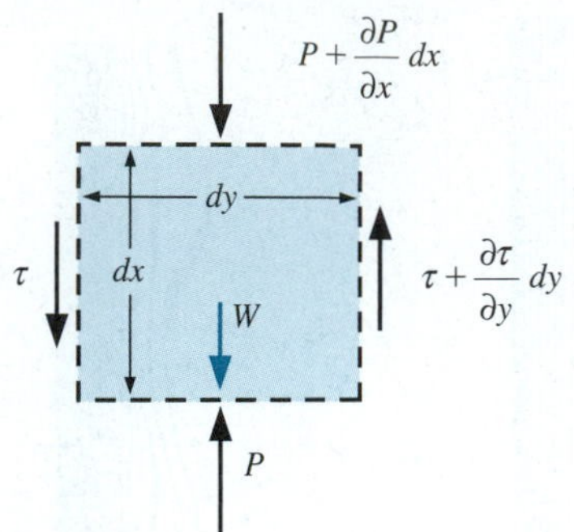

FIGURE 20–7
Forces acting on a differential volume element in the natural convection boundary layer over a vertical flat plate.

motion for this volume element can be expressed as

$$\delta m \cdot a_x = F_x \tag{20–7}$$

where $\delta m = \rho(dx \cdot dy \cdot 1)$ is the mass of the fluid within the volume element. The acceleration in the x-direction is obtained by taking the total differential of $u(x, y)$, which is $du = (\partial u/\partial x)dx + (\partial u/\partial y)dy$, and dividing it by dt. We get

$$a_x = \frac{du}{dt} = \frac{\partial u}{\partial x}\frac{dx}{dt} + \frac{\partial u}{\partial y}\frac{dy}{dt} = u\frac{\partial u}{\partial x} + v\frac{\partial u}{\partial y} \tag{20–8}$$

The forces acting on the differential volume element in the vertical direction are the pressure forces acting on the top and bottom surfaces, the shear stresses acting on the side surfaces (the normal stresses acting on the top and bottom surfaces are small and are disregarded), and the force of gravity acting on the entire volume element. Then the net force acting in the x-direction becomes

$$\begin{aligned} F_x &= \left(\frac{\partial \tau}{\partial y} dy\right)(dx \cdot 1) - \left(\frac{\partial P}{\partial x} dx\right)(dy \cdot 1) - \rho g(dx \cdot dy \cdot 1) \\ &= \left(\mu \frac{\partial^2 u}{\partial y^2} - \frac{\partial P}{\partial x} - \rho g\right)(dx \cdot dy \cdot 1) \end{aligned} \tag{20–9}$$

since $\tau = \mu(\partial u/\partial y)$. Substituting Eqs. 20–8 and 20–9 into Eq. 20–7 and dividing by $\rho \cdot dx \cdot dy \cdot 1$ gives Newton's second-law of motion in the x-direction (the x-momentum equation) as

$$\rho\left(u\frac{\partial u}{\partial x} + v\frac{\partial u}{\partial y}\right) = \mu\frac{\partial^2 u}{\partial y^2} - \frac{\partial P}{\partial x} - \rho g \tag{20–10}$$

The x-momentum equation in the quiescent fluid outside the boundary layer can be obtained from the relation above as a special case by setting $u = 0$. It gives

$$\frac{\partial P_\infty}{\partial x} = -\rho_\infty g \tag{20–11}$$

which is simply the relation for the variation of hydrostatic pressure in a quiescent fluid with height, as expected. Also, noting that $v \ll u$ in the boundary layer and thus $\partial v/\partial x \approx \partial v/\partial y \approx 0$, and that there are no body forces (including gravity) in the y-direction, the force balance in that direction gives $\partial P/\partial y = 0$. That is, the variation of pressure in the direction normal to the surface is negligible, and for a given x the pressure in the boundary layer is equal to the pressure in the quiescent fluid. Therefore, $P = P(x) = P_\infty(x)$ and $\partial P/\partial x = \partial P_\infty/\partial x = -\rho_\infty g$. Substituting into Eq. 20–10,

$$\rho\left(u\frac{\partial u}{\partial x} + v\frac{\partial u}{\partial y}\right) = \mu\frac{\partial^2 u}{\partial y^2} + (\rho_\infty - \rho)g \tag{20–12}$$

The last term represents the net upward force per unit volume of the fluid (the difference between the buoyant force and the fluid weight). This is the force that initiates and sustains convection currents.

From Eq. 20–5, we have $\rho_\infty - \rho = \rho\beta(T - T_\infty)$. Substituting it into the last equation and dividing both sides by ρ gives the desired form of the x-momentum equation,

$$u\frac{\partial u}{\partial x} + v\frac{\partial u}{\partial y} = \nu\frac{\partial^2 u}{\partial y^2} + g\beta(T - T_\infty) \tag{20–13}$$

This is the equation that governs the fluid motion in the boundary layer due to the effect of buoyancy. Note that the momentum equation involves the temperature, and thus the momentum and energy equations must be solved simultaneously.

The Grashof Number

The governing equations of natural convection and the boundary conditions can be nondimensionalized by dividing all dependent and independent variables by suitable constant quantities: all lengths by a characteristic length L_c, all velocities by an arbitrary reference velocity V (which, from the definition of Reynolds number, is taken to be $V = \text{Re}_L\, \nu/L_c$), and temperature by a suitable temperature difference (which is taken to be $T_s - T_\infty$) as

$$x^* = \frac{x}{L_c} \qquad y^* = \frac{y}{L_c} \qquad u^* = \frac{u}{V} \qquad v^* = \frac{v}{V} \qquad \text{and} \qquad T^* = \frac{T - T_\infty}{T_s - T_\infty}$$

where asterisks are used to denote nondimensional variables. Substituting them into the momentum equation and simplifying give

$$u^* \frac{\partial u^*}{\partial x^*} + v^* \frac{\partial u^*}{\partial y^*} = \left[\frac{g\beta(T_s - T_\infty)L_c^3}{\nu^2}\right] \frac{T^*}{\text{Re}_L^2} + \frac{1}{\text{Re}_L} \frac{\partial^2 u^*}{\partial y^{*2}} \tag{20–14}$$

The dimensionless parameter in the brackets represents the natural convection effects, and is called the **Grashof number** Gr_L,

$$\text{Gr}_L = \frac{g\beta(T_s - T_\infty)L_c^3}{\nu^2} \tag{20–15}$$

where

g = gravitational acceleration, m/s^2
β = coefficient of volume expansion, 1/K ($\beta = 1/T$ for ideal gases)
T_s = temperature of the surface, °C
T_∞ = temperature of the fluid sufficiently far from the surface, °C
L_c = characteristic length of the geometry, m
ν = kinematic viscosity of the fluid, m^2/s

We mentioned in the preceding chapters that the flow regime in forced convection is governed by the dimensionless *Reynolds number,* which represents the ratio of inertial forces to viscous forces acting on the fluid. The flow regime in natural convection is governed by the dimensionless *Grashof number,* which represents the ratio of the *buoyancy force* to the *viscous force* acting on the fluid (Fig. 20–8).

The role played by the Reynolds number in forced convection is played by the Grashof number in natural convection. As such, the Grashof number provides the main criterion in determining whether the fluid flow is laminar or turbulent in natural convection. For vertical plates, for example, the critical Grashof number is observed to be about 10^9. Therefore, the flow regime on a vertical plate becomes *turbulent* at Grashof numbers greater than 10^9.

When a surface is subjected to external flow, the problem involves both natural and forced convection. The relative importance of each mode of heat

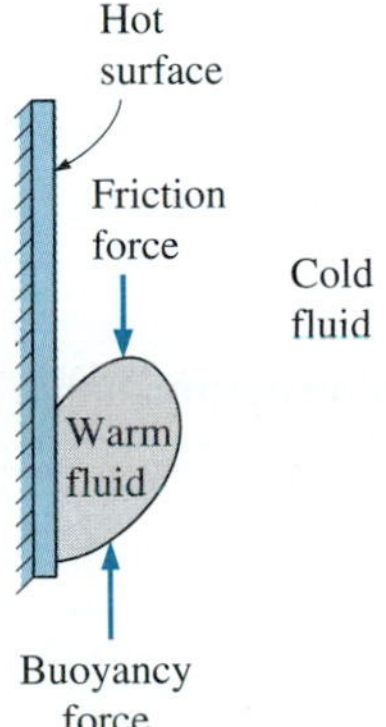

FIGURE 20–8
The Grashof number Gr is a measure of the relative magnitudes of the *buoyancy force* and the opposing *viscous force* acting on the fluid.

transfer is determined by the value of the coefficient Gr_L/Re_L^2: Natural convection effects are negligible if $Gr_L/Re_L^2 \ll 1$, free convection dominates and the forced convection effects are negligible if $Gr_L/Re_L^2 \gg 1$, and both effects are significant and must be considered if $Gr_L/Re_L^2 \approx 1$.

20–3 ▪ NATURAL CONVECTION OVER SURFACES

Natural convection heat transfer on a surface depends on the geometry of the surface as well as its orientation. It also depends on the variation of temperature on the surface and the thermophysical properties of the fluid involved.

Although we understand the mechanism of natural convection well, the complexities of fluid motion make it very difficult to obtain simple analytical relations for heat transfer by solving the governing equations of motion and energy. Some analytical solutions exist for natural convection, but such solutions lack generality since they are obtained for simple geometries under some simplifying assumptions. Therefore, with the exception of some simple cases, heat transfer relations in natural convection are based on experimental studies. Of the numerous such correlations of varying complexity and claimed accuracy available in the literature for any given geometry, we present here the ones that are best known and widely used.

The simple empirical correlations for the average *Nusselt number* Nu in natural convection are of the form (Fig. 20–9)

$$\text{Nu} = \frac{hL_c}{k} = C(\text{Gr}_L\,\text{Pr})^n = C\,\text{Ra}_L^n \qquad \textbf{(20–16)}$$

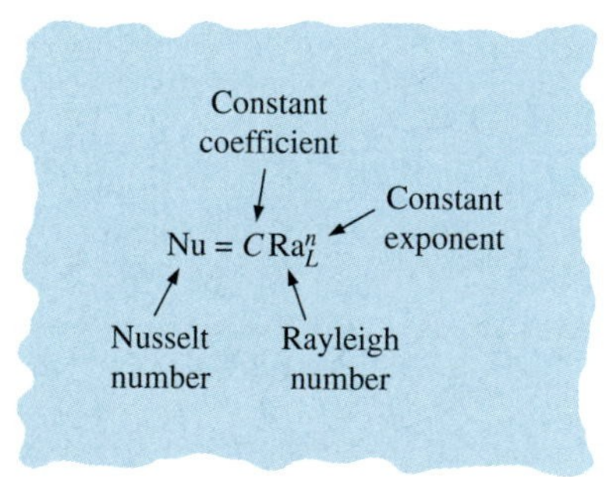

FIGURE 20–9
Natural convection heat transfer correlations are usually expressed in terms of the Rayleigh number raised to a constant n multiplied by another constant C, both of which are determined experimentally.

where Ra_L is the **Rayleigh number**, which is the product of the Grashof and Prandtl numbers:

$$\text{Ra}_L = \text{Gr}_L\,\text{Pr} = \frac{g\beta(T_s - T_\infty)L_c^3}{\nu^2}\,\text{Pr} \qquad \textbf{(20–17)}$$

The values of the constants C and n depend on the *geometry* of the surface and the *flow regime,* which is characterized by the range of the Rayleigh number. The value of n is usually $\frac{1}{4}$ for laminar flow and $\frac{1}{3}$ for turbulent flow. The value of the constant C is normally less than 1.

Simple relations for the average Nusselt number for various geometries are given in Table 20–1, together with sketches of the geometries. Also given in this table are the characteristic lengths of the geometries and the ranges of Rayleigh number in which the relation is applicable. All fluid properties are to be evaluated at the film temperature $T_f = \frac{1}{2}(T_s + T_\infty)$.

When the average Nusselt number and thus the average convection coefficient is known, the rate of heat transfer by natural convection from a solid surface at a uniform temperature T_s to the surrounding fluid is expressed by Newton's law of cooling as

$$\dot{Q}_{\text{conv}} = hA_s(T_s - T_\infty) \qquad (\text{W}) \qquad \textbf{(20–18)}$$

where A_s is the heat transfer surface area and h is the average heat transfer coefficient on the surface.

TABLE 20–1

Empirical correlations for the average Nusselt number for natural convection over surfaces

Geometry	Characteristic length L_c	Range of Ra	Nu
Vertical plate (L, T_s)	L	10^4–10^9 10^9–10^{13} Entire range	$\text{Nu} = 0.59\text{Ra}_L^{1/4}$ **(20–19)** $\text{Nu} = 0.1\text{Ra}_L^{1/3}$ **(20–20)** $\text{Nu} = \left\{0.825 + \dfrac{0.387\text{Ra}_L^{1/6}}{[1 + (0.492/\text{Pr})^{9/16}]^{8/27}}\right\}^2$ **(20–21)** (complex but more accurate)
Inclined plate (L, θ)	L		Use vertical plate equations for the upper surface of a cold plate and the lower surface of a hot plate Replace g by $g \cos\theta$ for $\text{Ra} < 10^9$
Horizontal plate (Surface area A and perimeter p) (*a*) Upper surface of a hot plate (or lower surface of a cold plate) Hot surface, T_s (*b*) Lower surface of a hot plate (or upper surface of a cold plate) T_s, Hot surface	A_s/p	10^4–10^7 10^7–10^{11} 10^5–10^{11}	$\text{Nu} = 0.54\text{Ra}_L^{1/4}$ **(20–22)** $\text{Nu} = 0.15\text{Ra}_L^{1/3}$ **(20–23)** $\text{Nu} = 0.27\text{Ra}_L^{1/4}$ **(20–24)**
Vertical cylinder (L, T_s)	L		A vertical cylinder can be treated as a vertical plate when $D \geq \dfrac{35L}{\text{Gr}_L^{1/4}}$
Horizontal cylinder (T_s, D)	D	$\text{Ra}_D \leq 10^{12}$	$\text{Nu} = \left\{0.6 + \dfrac{0.387\text{Ra}_D^{1/6}}{[1 + (0.559/\text{Pr})^{9/16}]^{8/27}}\right\}^2$ **(20–25)**
Sphere (D)	D	$\text{Ra}_D \leq 10^{11}$ ($\text{Pr} \geq 0.7$)	$\text{Nu} = 2 + \dfrac{0.589\text{Ra}_D^{1/4}}{[1 + (0.469/\text{Pr})^{9/16}]^{4/9}}$ **(20–26)**

Vertical Plates (T_s = constant)

For a vertical flat plate, the characteristic length is the plate height L. In Table 20–1 we give three relations for the average Nusselt number for an isothermal vertical plate. The first two relations are very simple. Despite its complexity, we suggest using the third one (Eq. 20–21) recommended by Churchill and Chu (1975) since it is applicable over the entire range of Rayleigh number. This relation is most accurate in the range of $10^{-1} < \text{Ra}_L < 10^9$.

Vertical Plates ($\dot{q}_s$ = constant)

In the case of constant surface heat flux, the rate of heat transfer is known (it is simply $\dot{Q} = \dot{q}_s A_s$), but the surface temperature T_s is not. In fact, T_s increases with height along the plate. It turns out that the Nusselt number relations for the constant surface temperature and constant surface heat flux cases are nearly identical (Churchill and Chu, 1975). Therefore, the relations for isothermal plates can also be used for plates subjected to uniform heat flux, provided that the plate midpoint temperature $T_{L/2}$ is used for T_s in the evaluation of the film temperature, Rayleigh number, and the Nusselt number. Noting that $h = \dot{q}_s/(T_{L/2} - T_\infty)$, the average Nusselt number in this case can be expressed as

$$\text{Nu} = \frac{hL}{k} = \frac{\dot{q}_s L}{k(T_{L/2} - T_\infty)} \tag{20–27}$$

The midpoint temperature $T_{L/2}$ is determined by iteration so that the Nusselt numbers determined from Eqs. 20–21 and 20–27 match.

Vertical Cylinders

An outer surface of a vertical cylinder can be treated as a vertical plate when the diameter of the cylinder is sufficiently large so that the curvature effects are negligible. This condition is satisfied if

$$D \geq \frac{35L}{\text{Gr}_L^{1/4}} \tag{20–28}$$

When this criteria is met, the relations for vertical plates can also be used for vertical cylinders. Nusselt number relations for slender cylinders that do not meet this criteria are available in the literature (e.g., Cebeci, 1974).

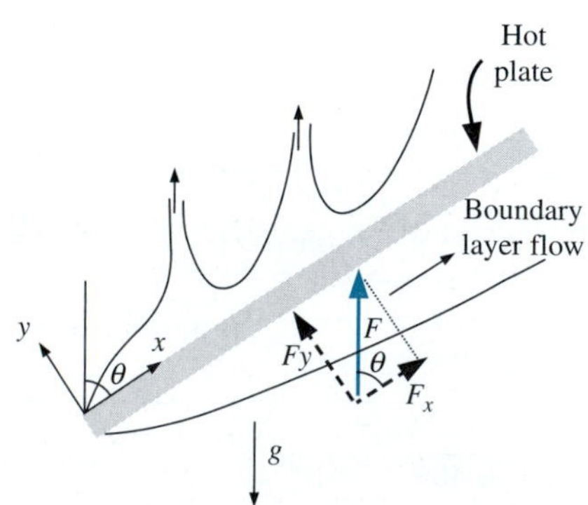

FIGURE 20–10
Natural convection flows on the upper and lower surfaces of an inclined hot plate.

Inclined Plates

Consider an inclined hot plate that makes an angle θ from the vertical, as shown in Fig. 20–10, in a cooler environment. The net force $F = g(\rho_\infty - \rho)$ (the difference between the buoyancy and gravity) acting on a unit volume of the fluid in the boundary layer is always in the vertical direction. In the case of inclined plate, this force can be resolved into two components: $F_y = F \cos \theta$ parallel to the plate that drives the flow along the plate, and $F_y = F \sin \theta$ normal to the plate. Noting that the force that drives the motion is reduced, we expect the convection currents to be weaker, and the rate of heat transfer to be lower relative to the vertical plate case.

The experiments confirm what we suspect for the lower surface of a hot plate, but the opposite is observed on the upper surface. The reason for this curious behavior for the upper surface is that the force component F_y initiates upward motion in addition to the parallel motion along the plate, and thus the boundary layer breaks up and forms plumes, as shown in the figure. As a result, the thickness of the boundary layer and thus the resistance to heat transfer decreases, and the rate of heat transfer increases relative to the vertical orientation.

In the case of a cold plate in a warmer environment, the opposite occurs as expected: The boundary layer on the upper surface remains intact with weaker boundary layer flow and thus lower rate of heat transfer, and the boundary layer on the lower surface breaks apart (the colder fluid falls down) and thus enhances heat transfer.

When the boundary layer remains intact (the lower surface of a hot plate or the upper surface of a cold plate), the Nusselt number can be determined from the vertical plate relations provided that g in the Rayleigh number relation is replaced by $g \cos\theta$ for $\theta < 60°$. Nusselt number relations for the other two surfaces (the upper surface of a hot plate or the lower surface of a cold plate) are available in the literature (e.g., Fuji and Imura, 1972).

Horizontal Plates

The rate of heat transfer to or from a horizontal surface depends on whether the surface is facing upward or downward. For a hot surface in a cooler environment, the net force acts upward, forcing the heated fluid to rise. If the hot surface is facing upward, the heated fluid rises freely, inducing strong natural convection currents and thus effective heat transfer, as shown in Fig. 20–11. But if the hot surface is facing downward, the plate blocks the heated fluid that tends to rise (except near the edges), impeding heat transfer. The opposite is true for a cold plate in a warmer environment since the net force (weight minus buoyancy force) in this case acts downward, and the cooled fluid near the plate tends to descend.

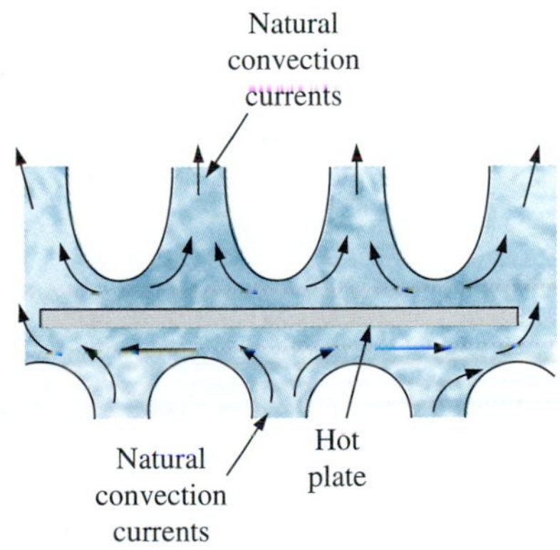

FIGURE 20–11
Natural convection flows on the upper and lower surfaces of a horizontal hot plate.

The average Nusselt number for horizontal surfaces can be determined from the simple power-law relations given in Table 20–1. The characteristic length for horizontal surfaces is calculated from

$$L_c = \frac{A_s}{p} \tag{20–29}$$

where A_s is the surface area and p is the perimeter. Note that $L_c = a/4$ for a horizontal square surface of length a, and $D/4$ for a horizontal circular surface of diameter D.

Horizontal Cylinders and Spheres

The boundary layer over a hot horizontal cylinder starts to develop at the bottom, increasing in thickness along the circumference, and forming a rising plume at the top, as shown in Fig. 20–12. Therefore, the local Nusselt number

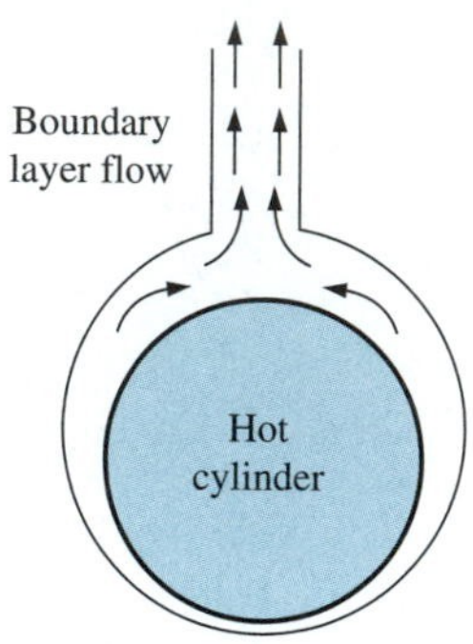

FIGURE 20–12
Natural convection flow over a horizontal hot cylinder.

is highest at the bottom, and lowest at the top of the cylinder when the boundary layer flow remains laminar. The opposite is true in the case of a cold horizontal cylinder in a warmer medium, and the boundary layer in this case starts to develop at the top of the cylinder and ending with a descending plume at the bottom.

The average Nusselt number over the entire surface can be determined from Eq. 20–25 (Churchill and Chu, 1975) for an isothermal horizontal cylinder, and from Eq. 20–26 for an isothermal sphere (Churchill, 1983) both given in Table 20–1.

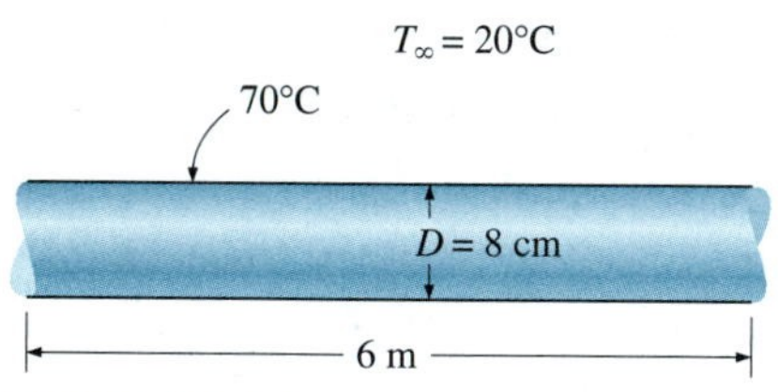

FIGURE 20–13
Schematic for Example 20–1.

EXAMPLE 20–1 Heat Loss from Hot-Water Pipes

A 6-m-long section of an 8-cm-diameter horizontal hot-water pipe shown in Fig. 20–13 passes through a large room whose temperature is 20°C. If the outer surface temperature of the pipe is 70°C, determine the rate of heat loss from the pipe by natural convection.

Solution A horizontal hot-water pipe passes through a large room. The rate of heat loss from the pipe by natural convection is to be determined.

Assumptions **1** Steady operating conditions exist. **2** Air is an ideal gas. **3** The local atmospheric pressure is 1 atm.

Properties The properties of air at the film temperature of $T_f = (T_s + T_\infty)/2 = (70 + 20)/2 = 45°C$ and 1 atm are (Table A–22)

$$k = 0.02699 \text{ W/m} \cdot °\text{C} \qquad \text{Pr} = 0.7241$$

$$\nu = 1.750 \times 10^{-5} \text{ m}^2/\text{s} \qquad \beta = \frac{1}{T_f} = \frac{1}{318 \text{ K}}$$

Analysis The characteristic length in this case is the outer diameter of the pipe, $L_c = D = 0.08$ m. Then the Rayleigh number becomes

$$\text{Ra}_D = \frac{g\beta(T_s - T_\infty)D^3}{\nu^2}\text{Pr}$$

$$= \frac{(9.81 \text{ m/s}^2)[1/(318 \text{ K})](70 - 20 \text{ K})(0.08 \text{ m})^3}{(1.750 \times 10^{-5} \text{ m}^2/\text{s})^2}(0.7241) = 1.867 \times 10^6$$

The natural convection Nusselt number in this case can be determined from Eq. 20–25 to be

$$\text{Nu} = \left\{0.6 + \frac{0.387\text{Ra}_D^{1/6}}{[1 + (0.559/\text{Pr})^{9/16}]^{8/27}}\right\}^2 = \left\{0.6 + \frac{0.387(1.867 \times 10^6)^{1/6}}{[1 + (0.559/0.7241)^{9/16}]^{8/27}}\right\}^2$$

$$= 17.39$$

Then,

$$h = \frac{k}{D}\text{Nu} = \frac{0.02699 \text{ W/m} \cdot °\text{C}}{0.08 \text{ m}}(17.39) = 5.867 \text{ W/m} \cdot °\text{C}$$

$$A_s = \pi DL = \pi(0.08 \text{ m})(6 \text{ m}) = 1.508 \text{ m}^2$$

and

$$\dot{Q} = hA_s(T_s - T_\infty) = (5.867 \text{ W/m}^2 \cdot °\text{C})(1.508 \text{ m}^2)(70 - 20)°\text{C} = \mathbf{442 \text{ W}}$$

Therefore, the pipe loses heat to the air in the room at a rate of 442 W by natural convection.

Discussion The pipe loses heat to the surroundings by radiation as well as by natural convection. Assuming the outer surface of the pipe to be black (emissivity $\varepsilon = 1$) and the inner surfaces of the walls of the room to be at room temperature, the radiation heat transfer is determined to be (Fig. 20–14)

$$\begin{aligned}\dot{Q}_{\text{rad}} &= \varepsilon A_s \sigma (T_s^4 - T_{\text{surr}}^4) \\ &= (1)(1.508\ \text{m}^2)(5.67 \times 10^{-8}\ \text{W/m}^2 \cdot \text{K}^4)[(70 + 273\ \text{K})^4 - (20 + 273\ \text{K})^4] \\ &= 553\ \text{W}\end{aligned}$$

which is larger than natural convection. The emissivity of a real surface is less than 1, and thus the radiation heat transfer for a real surface will be less. But radiation will still be significant for most systems cooled by natural convection. Therefore, a radiation analysis should normally accompany a natural convection analysis unless the emissivity of the surface is low.

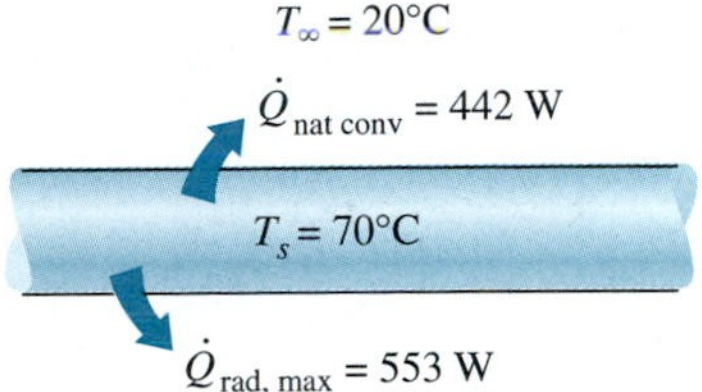

FIGURE 20–14
Radiation heat transfer is usually comparable to natural convection in magnitude and should be considered in heat transfer analysis.

EXAMPLE 20–2 Cooling of a Plate in Different Orientations

Consider a 0.6-m × 0.6-m thin square plate in a room at 30°C. One side of the plate is maintained at a temperature of 90°C, while the other side is insulated, as shown in Fig. 20–15. Determine the rate of heat transfer from the plate by natural convection if the plate is (*a*) vertical, (*b*) horizontal with hot surface facing up, and (*c*) horizontal with hot surface facing down.

Solution A hot plate with an insulated back is considered. The rate of heat loss by natural convection is to be determined for different orientations.

Assumptions **1** Steady operating conditions exist. **2** Air is an ideal gas. **3** The local atmospheric pressure is 1 atm.

Properties The properties of air at the film temperature of $T_f = (T_s + T_\infty)/2 = (90 + 30)/2 = 60°\text{C}$ and 1 atm are (Table A–22)

$$k = 0.02808\ \text{W/m} \cdot °\text{C} \qquad \text{Pr} = 0.7202$$

$$\nu = 1.896 \times 10^{-5}\ \text{m}^2/\text{s} \qquad \beta = \frac{1}{T_f} = \frac{1}{333\ \text{K}}$$

Analysis (*a*) *Vertical.* The characteristic length in this case is the height of the plate, which is $L = 0.6$ m. The Rayleigh number is

$$\begin{aligned}\text{Ra}_L &= \frac{g\beta(T_s - T_\infty)L^3}{\nu^2}\text{Pr} \\ &= \frac{(9.81\ \text{m/s}^2)[1/(333\ \text{K})](90 - 30\ \text{K})(0.6\ \text{m})^3}{(1.896 \times 10^{-5}\ \text{m}^2/\text{s})^2}(0.7202) = 7.649 \times 10^8\end{aligned}$$

Then the natural convection Nusselt number can be determined from Eq. 20–21 to be

$$\begin{aligned}\text{Nu} &= \left\{0.825 + \frac{0.387\text{Ra}_L^{1/6}}{[1 + (0.492/\text{Pr})^{9/16}]^{8/27}}\right\}^2 \\ &= \left\{0.825 + \frac{0.387(7.649 \times 10^8)^{1/6}}{[1 + (0.492/0.7202)^{9/16}]^{8/27}}\right\}^2 = 113.3\end{aligned}$$

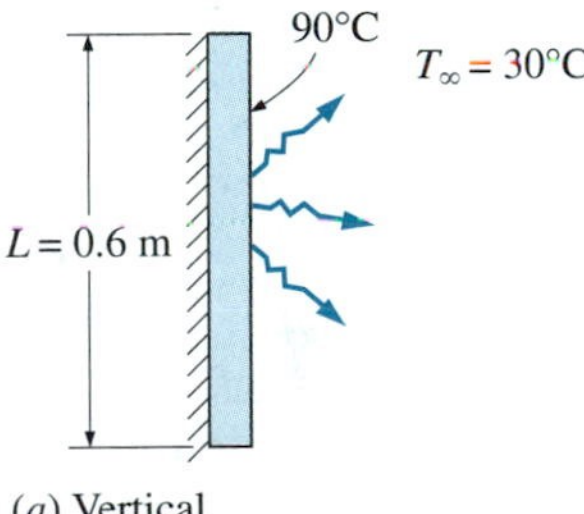

(*a*) Vertical

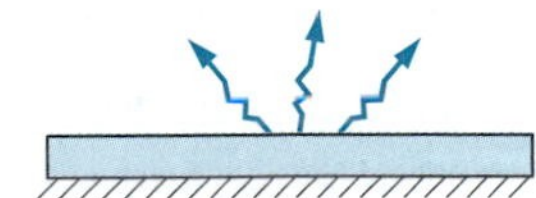

(*b*) Hot surface facing up

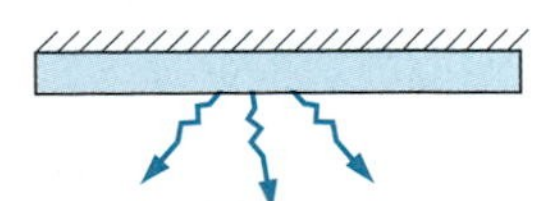

(*c*) Hot surface facing down

FIGURE 20–15
Schematic for Example 20–2.

Note that the simpler relation Eq. 20–19 would give $\text{Nu} = 0.59\ \text{Ra}_L^{1/4} = 98.12$, which is 13 percent lower. Then,

$$h = \frac{k}{L}\text{Nu} = \frac{0.02808\ \text{W/m}\cdot{}^\circ\text{C}}{0.6\ \text{m}}(113.3) = 5.302\ \text{W/m}^2\cdot{}^\circ\text{C}$$

$$A_s = L^2 = (0.6\ \text{m})^2 = 0.36\ \text{m}^2$$

and

$$\dot{Q} = hA_s(T_s - T_\infty) = (5.302\ \text{W/m}^2\cdot{}^\circ\text{C})(0.36\ \text{m}^2)(90 - 30)^\circ\text{C} = \mathbf{115\ W}$$

(*b*) *Horizontal with hot surface facing up.* The characteristic length and the Rayleigh number in this case are

$$L_c = \frac{A_s}{p} = \frac{L^2}{4L} = \frac{L}{4} = \frac{0.6\ \text{m}}{4} = 0.15\ \text{m}$$

$$\text{Ra}_L = \frac{g\beta(T_s - T_\infty)L_c^3}{\nu^2}\text{Pr}$$

$$= \frac{(9.81\ \text{m/s}^2)[1/(333\ \text{K})](90 - 30\ \text{K})(0.15\ \text{m})^3}{(1.896\times10^{-5}\ \text{m}^2/\text{s})^2}(0.7202) = 1.195\times10^7$$

The natural convection Nusselt number can be determined from Eq. 20–22 to be

$$\text{Nu} = 0.54\text{Ra}_L^{1/4} = 0.54(1.195\times10^7)^{1/4} = 31.75$$

Then,

$$h = \frac{k}{L_c}\text{Nu} = \frac{0.02808\ \text{W/m}\cdot{}^\circ\text{C}}{0.15\ \text{m}}(31.75) = 5.944\ \text{W/m}^2\cdot{}^\circ\text{C}$$

and

$$\dot{Q} = hA_s(T_s - T_\infty) = (5.944\ \text{W/m}^2\cdot{}^\circ\text{C})(0.36\ \text{m}^2)(90 - 30)^\circ\text{C} = \mathbf{128\ W}$$

(*c*) *Horizontal with hot surface facing down.* The characteristic length and the Rayleigh number in this case are the same as those determined in (*b*). But the natural convection Nusselt number is to be determined from Eq. 20–24,

$$\text{Nu} = 0.27\text{Ra}_L^{1/4} = 0.27(1.195\times10^7)^{1/4} = 15.87$$

Then,

$$h = \frac{k}{L_c}\text{Nu} = \frac{0.02808\ \text{W/m}\cdot{}^\circ\text{C}}{0.15\ \text{m}}(15.87) = 2.971\ \text{W/m}^2\cdot{}^\circ\text{C}$$

and

$$\dot{Q} = hA_s(T_s - T_\infty) = (2.971\ \text{W/m}^2\cdot{}^\circ\text{C})(0.36\ \text{m}^2)(90 - 30)^\circ\text{C} = \mathbf{64.2\ W}$$

Note that the natural convection heat transfer is the lowest in the case of the hot surface facing down. This is not surprising, since the hot air is "trapped" under the plate in this case and cannot get away from the plate easily. As a result, the cooler air in the vicinity of the plate will have difficulty reaching the plate, which results in a reduced rate of heat transfer.

Discussion The plate will lose heat to the surroundings by radiation as well as by natural convection. Assuming the surface of the plate to be black (emissivity $\varepsilon = 1$) and the inner surfaces of the walls of the room to be at room temperature, the radiation heat transfer in this case is determined to be

$$\dot{Q}_{rad} = \varepsilon A_s \sigma (T_s^4 - T_{surr}^4)$$
$$= (1)(0.36 \text{ m}^2)(5.67 \times 10^{-8} \text{ W/m}^2 \cdot \text{K}^4)[(90 + 273 \text{ K})^4 - (30 + 273 \text{ K})^4]$$
$$= 182 \text{ W}$$

which is larger than that for natural convection heat transfer for each case. Therefore, radiation can be significant and needs to be considered in surfaces cooled by natural convection.

Note the characteristic length is different for the vertical and horizontal plates, which is important because Gr and Ra vary as L^3.

20–4 ▪ NATURAL CONVECTION FROM FINNED SURFACES AND PCBs

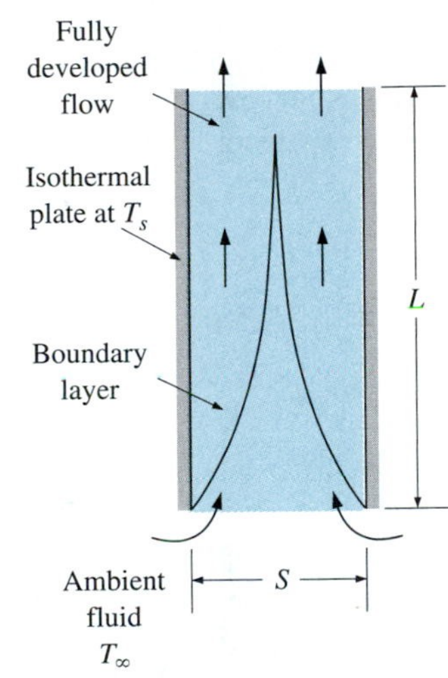

FIGURE 20–16
Natural convection flow through a channel between two isothermal vertical plates.

Natural convection flow through a channel formed by two parallel plates as shown in Fig. 20–16 is commonly encountered in practice. When the plates are hot ($T_s > T_\infty$), the ambient fluid at T_∞ enters the channel from the lower end, rises as it is heated under the effect of buoyancy, and the heated fluid leaves the channel from the upper end. The plates could be the fins of a finned heat sink, or the PCBs (printed circuit boards) of an electronic device. The plates can be approximated as being isothermal (T_s = constant) in the first case, and isoflux ($\dot{q}_s$ = constant) in the second case.

Boundary layers start to develop at the lower ends of opposing surfaces, and eventually merge at the midplane if the plates are vertical and sufficiently long. In this case, we will have fully developed channel flow after the merger of the boundary layers, and the natural convection flow is analyzed as channel flow. But when the plates are short or the spacing is large, the boundary layers of opposing surfaces never reach each other, and the natural convection flow on a surface is not affected by the presence of the opposing surface. In that case, the problem should be analyzed as natural convection from two independent plates in a quiescent medium, using the relations given for surfaces, rather than natural convection flow through a channel.

Natural Convection Cooling of Finned Surfaces (T_s = constant)

Finned surfaces of various shapes, called *heat sinks,* are frequently used in the cooling of electronic devices. Energy dissipated by these devices is transferred to the heat sinks by conduction and from the heat sinks to the ambient air by natural or forced convection, depending on the power dissipation requirements. Natural convection is the preferred mode of heat transfer since it involves no moving parts, like the electronic components themselves. However, in the natural convection mode, the components are more likely to run at a higher temperature and thus undermine reliability. A properly selected heat sink may considerably lower the operation temperature of the components and thus reduce the risk of failure.

Natural convection from vertical finned surfaces of rectangular shape has been the subject of numerous studies, mostly experimental. Bar-Cohen and

Rohsenow (1984) have compiled the available data under various boundary conditions, and developed correlations for the Nusselt number and optimum spacing. The characteristic length for vertical parallel plates used as fins is usually taken to be the spacing between adjacent fins S, although the fin height L could also be used. The Rayleigh number is expressed as

$$\text{Ra}_S = \frac{g\beta(T_s - T_\infty)S^3}{\nu^2}\text{Pr} \quad \text{and} \quad \text{Ra}_L = \frac{g\beta(T_s - T_\infty)L^3}{\nu^2}\text{Pr} = \text{Ra}_S\frac{L^3}{S^3} \tag{20–30}$$

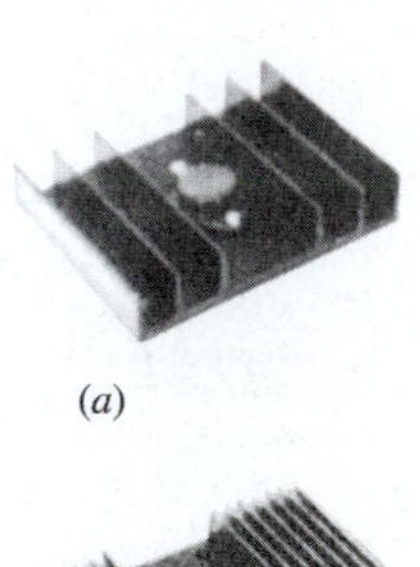

(*a*)

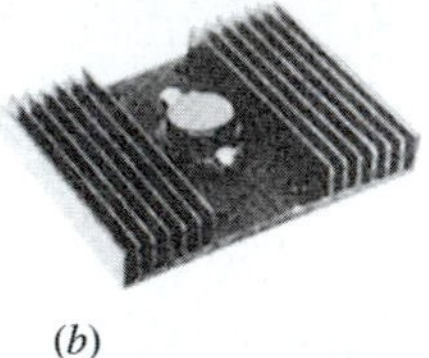

(*b*)

FIGURE 20–17
Heat sinks with (*a*) widely spaced and (*b*) closely packed fins.
Courtesy of Yunus Çengel

The recommended relation for the average Nusselt number for vertical isothermal parallel plates is

$$T_s = \text{constant:} \qquad \text{Nu} = \frac{hS}{k} = \left[\frac{576}{(\text{Ra}_S S/L)^2} + \frac{2.873}{(\text{Ra}_S S/L)^{0.5}}\right]^{-0.5} \tag{20–31}$$

A question that often arises in the selection of a heat sink is whether to select one with *closely packed* fins or *widely spaced* fins for a given base area (Fig. 20–17). A heat sink with closely packed fins will have greater surface area for heat transfer but a smaller heat transfer coefficient because of the extra resistance the additional fins introduce to fluid flow through the interfin passages. A heat sink with widely spaced fins, on the other hand, will have a higher heat transfer coefficient but a smaller surface area. Therefore, there must be an *optimum spacing* that maximizes the natural convection heat transfer from the heat sink for a given base area WL, where W and L are the width and height of the base of the heat sink, respectively, as shown in Fig. 20–18. When the fins are essentially isothermal and the fin thickness t is small relative to the fin spacing S, the optimum fin spacing for a vertical heat sink is determined by Bar-Cohen and Rohsenow to be

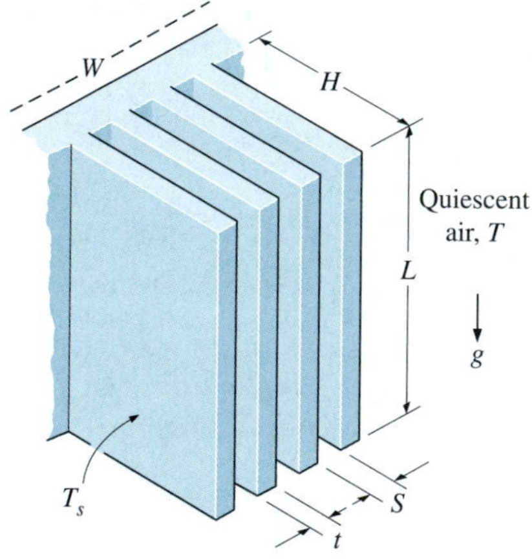

FIGURE 20–18
Various dimensions of a finned surface oriented vertically.

$$T_s = \text{constant:} \qquad S_{\text{opt}} = 2.714\left(\frac{S^3L}{\text{Ra}_S}\right)^{0.25} = 2.714\frac{L}{\text{Ra}_L^{0.25}} \tag{20–32}$$

It can be shown by combining the three equations above that when $S = S_{\text{opt}}$, the Nusselt number is a constant and its value is 1.307,

$$S = S_{\text{opt}}: \qquad \text{Nu} = \frac{hS_{\text{opt}}}{k} = 1.307 \tag{20–33}$$

The rate of heat transfer by natural convection from the fins can be determined from

$$\dot{Q} = h(2nLH)(T_s - T_\infty) \tag{20–34}$$

where $n = W/(S + t) \approx W/S$ is the number of fins on the heat sink and T_s is the surface temperature of the fins. All fluid properties are to be evaluated at the average temperature $T_{\text{avg}} = (T_s + T_\infty)/2$.

Natural Convection Cooling of Vertical PCBs ($\dot{q}_s$ = constant)

Arrays of printed circuit boards used in electronic systems can often be modeled as parallel plates subjected to uniform heat flux $\dot{q}_s$ (Fig. 20–19). The plate temperature in this case increases with height, reaching a maximum at the upper edge of the board. The modified Rayleigh number for uniform heat flux on both plates is expressed as

$$\text{Ra}_S^* = \frac{g\beta\dot{q}_s S^4}{k\upsilon^2}\text{Pr} \tag{20–35}$$

The Nusselt number at the upper edge of the plate where maximum temperature occurs is determined from (Bar-Cohen and Rohsenow, 1984)

$$\text{Nu}_L = \frac{h_L S}{k} = \left[\frac{48}{\text{Ra}_S^* S/L} + \frac{2.51}{(\text{Ra}_L^* S/L)^{0.4}}\right]^{-0.5} \tag{20–36}$$

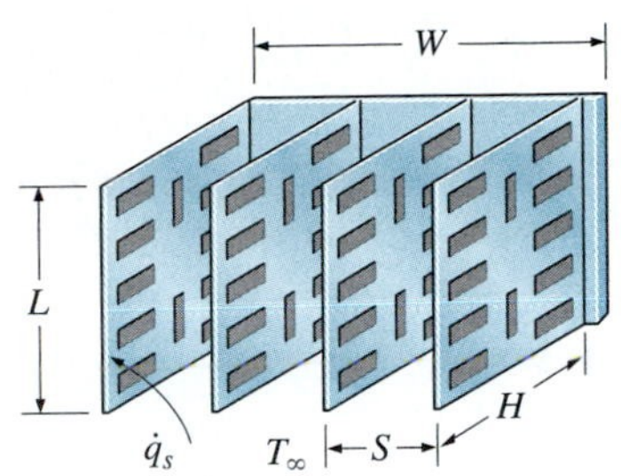

FIGURE 20–19
Arrays of vertical printed circuit boards (PCBs) cooled by natural convection.

The optimum fin spacing for the case of uniform heat flux on both plates is given as

$$\dot{q}_s = \text{constant:} \qquad S_{opt} = 2.12\left(\frac{S^4 L}{\text{Ra}_S^*}\right)^{0.2} \tag{20–37}$$

The total rate of heat transfer from the plates is

$$\dot{Q} = \dot{q}_s A_s = \dot{q}_s(2nLH) \tag{20–38}$$

where $n = W/(S + t) \approx W/S$ is the number of plates. The critical surface temperature T_L occurs at the upper edge of the plates, and it can be determined from

$$\dot{q}_s = h_L(T_L - T_\infty) \tag{20–39}$$

All fluid properties are to be evaluated at the average temperature $T_{avg} = (T_L + T_\infty)/2$.

Mass Flow Rate through the Space between Plates

As we mentioned earlier, the magnitude of the natural convection heat transfer is directly related to the mass flow rate of the fluid, which is established by the dynamic balance of two opposing effects: *buoyancy* and *friction.*

The fins of a heat sink introduce both effects: *inducing extra buoyancy* as a result of the elevated temperature of the fin surfaces and *slowing down the fluid* by acting as an added obstacle on the flow path. As a result, increasing the number of fins on a heat sink can either enhance or reduce natural convection, depending on which effect is dominant. The buoyancy-driven fluid flow rate is established at the point where these two effects balance each other. The friction force increases as more and more solid surfaces are introduced, seriously disrupting fluid flow and heat transfer. Under some conditions, the increase in friction may more than offset the increase in buoyancy. This in turn will tend to reduce the flow rate and thus the heat transfer. For that reason, heat sinks with closely spaced fins are not suitable for natural convection cooling.

When the heat sink involves closely spaced fins, the narrow channels formed tend to block or "suffocate" the fluid, especially when the heat sink is long. As a result, the blocking action produced overwhelms the extra buoyancy and downgrades the heat transfer characteristics of the heat sink. Then, at a fixed power setting, the heat sink runs at a higher temperature relative to the no-shroud case. When the heat sink involves widely spaced fins, the shroud does not introduce a significant increase in resistance to flow, and the buoyancy effects dominate. As a result, heat transfer by natural convection may improve, and at a fixed power level the heat sink may run at a lower temperature.

When extended surfaces such as fins are used to enhance natural convection heat transfer between a solid and a fluid, the flow rate of the fluid in the vicinity of the solid adjusts itself to incorporate the changes in buoyancy and friction. It is obvious that this enhancement technique will work to advantage only when the increase in buoyancy is greater than the additional friction introduced. One does not need to be concerned with pressure drop or pumping power when studying natural convection since no pumps or blowers are used in this case. Therefore, an enhancement technique in natural convection is evaluated on heat transfer performance alone.

The failure rate of an electronic component increases almost exponentially with operating temperature. The cooler the electronic device operates, the more reliable it is. A rule of thumb is that the semiconductor failure rate is halved for each 10°C reduction in junction operating temperature. The desire to lower the operating temperature without having to resort to forced convection has motivated researchers to investigate enhancement techniques for natural convection. Sparrow and Prakash (1987) have demonstrated that, under certain conditions, the use of discrete plates in lieu of continuous plates of the same surface area increases heat transfer considerably. In other experimental work, using transistors as the heat source, Çengel and Zing (1987) have demonstrated that temperature recorded on the transistor case dropped by as much as 30°C when a shroud was used, as opposed to the corresponding no-shroud case.

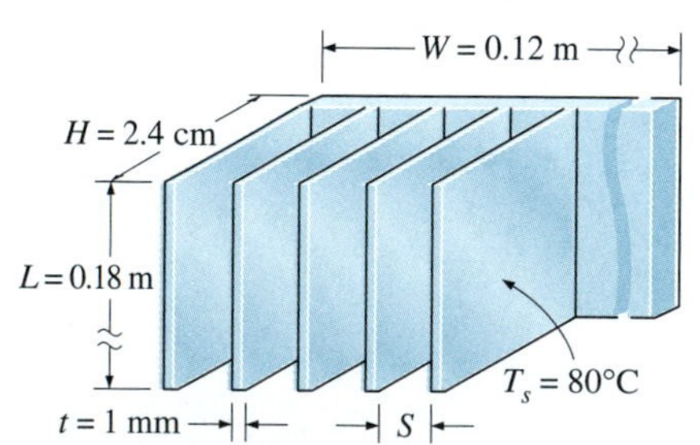

FIGURE 20–20
Schematic for Example 20–3.

EXAMPLE 20–3 Optimum Fin Spacing of a Heat Sink

A 12-cm-wide and 18-cm-high vertical hot surface in 30°C air is to be cooled by a heat sink with equally spaced fins of rectangular profile (Fig. 20–20). The fins are 0.1 cm thick and 18 cm long in the vertical direction and have a height of 2.4 cm from the base. Determine the optimum fin spacing and the rate of heat transfer by natural convection from the heat sink if the base temperature is 80°C.

Solution A heat sink with equally spaced rectangular fins is to be used to cool a hot surface. The optimum fin spacing and the rate of heat transfer are to be determined.

Assumptions **1** Steady operating conditions exist. **2** Air is an ideal gas. **3** The atmospheric pressure at that location is 1 atm. **4** The thickness t of the fins is very small relative to the fin spacing S so that Eq. 20–32 for optimum fin spacing is applicable. **5** All fin surfaces are isothermal at base temperature.

Properties The properties of air at the film temperature of $T_f = (T_s + T_\infty)/2 = (80 + 30)/2 = 55°\text{C}$ and 1 atm pressure are (Table A–22)

$$k = 0.02772 \text{ W/m} \cdot °\text{C} \qquad \text{Pr} = 0.7215$$
$$\nu = 1.847 \times 10^{-5} \text{ m}^2\text{/s} \qquad \beta = 1/T_f = 1/328 \text{ K}$$

Analysis We take the characteristic length to be the length of the fins in the vertical direction (since we do not know the fin spacing). Then the Rayleigh number becomes

$$\mathrm{Ra}_L = \frac{g\beta(T_s - T_\infty)L^3}{\nu^2}\,\mathrm{Pr}$$

$$= \frac{(9.81\ \mathrm{m/s^2})[1/(328\ \mathrm{K})](80 - 30\ \mathrm{K})(0.18\ \mathrm{m})^3}{(1.847 \times 10^{-5}\ \mathrm{m^2/s})^2}(0.7215) = 1.845 \times 10^7$$

The optimum fin spacing is determined from Eq. 20–32 to be

$$S_{\mathrm{opt}} = 2.714\frac{L}{\mathrm{Ra}_L^{0.25}} = 2.714\frac{0.18\ \mathrm{m}}{(1.845 \times 10^7)^{0.25}} = 7.45 \times 10^{-3}\ \mathrm{m} = \mathbf{7.45\ mm}$$

which is about seven times the thickness of the fins. Therefore, the assumption of negligible fin thickness in this case is acceptable. The number of fins and the heat transfer coefficient for this optimum fin spacing case are

$$n = \frac{W}{S + t} = \frac{0.12\ \mathrm{m}}{(0.00745 + 0.001)\ \mathrm{m}} \approx 14\ \mathrm{fins}$$

The convection coefficient for this optimum fin spacing case is, from Eq. 20–33,

$$h = \mathrm{Nu}_{\mathrm{opt}}\frac{k}{S_{\mathrm{opt}}} = 1.307\frac{0.02772\ \mathrm{W/m\cdot °C}}{0.00745\ \mathrm{m}} = 4.863\ \mathrm{W/m^2\cdot °C}$$

Then the rate of natural convection heat transfer becomes

$$\dot{Q} = hA_s(T_s - T_\infty) = h(2nLH)(T_s - T_\infty)$$
$$= (4.863\ \mathrm{W/m^2\cdot °C})[2 \times 14(0.18\ \mathrm{m})(0.024\ \mathrm{m})](80 - 30)°\mathrm{C} = \mathbf{29.4\ W}$$

Therefore, this heat sink can dissipate heat by natural convection at a rate of 29.4 W.

20–5 ▪ NATURAL CONVECTION INSIDE ENCLOSURES

A considerable portion of heat loss from a typical residence occurs through the windows. We certainly would insulate the windows, if we could, in order to conserve energy. The problem is finding an insulating material that is transparent. An examination of the thermal conductivities of the insulating materials reveals that *air* is a *better insulator* than most common insulating materials. Besides, it is transparent. Therefore, it makes sense to insulate the windows with a layer of air. Of course, we need to use another sheet of glass to trap the air. The result is an *enclosure,* which is known as a *double-pane window* in this case. Other examples of enclosures include wall cavities, solar collectors, and cryogenic chambers involving concentric cylinders or spheres.

Enclosures are frequently encountered in practice, and heat transfer through them is of practical interest. Heat transfer in enclosed spaces is complicated by the fact that the fluid in the enclosure, in general, does not remain stationary. In a vertical enclosure, the fluid adjacent to the hotter surface rises and the fluid adjacent to the cooler one falls, setting off a rotationary motion within the enclosure that enhances heat transfer through the enclosure. Typical flow patterns in vertical and horizontal rectangular enclosures are shown in Figs. 20–21 and 20–22.

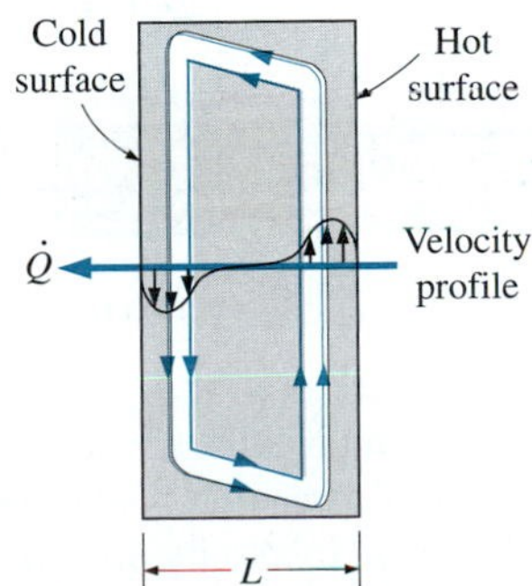

FIGURE 20–21
Convective currents in a vertical rectangular enclosure.

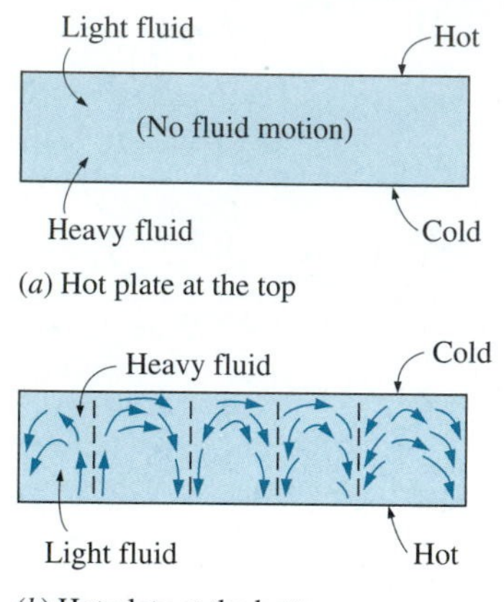

FIGURE 20–22
Convective currents in a horizontal enclosure with (*a*) hot plate at the top and (*b*) hot plate at the bottom.

The characteristics of heat transfer through a horizontal enclosure depend on whether the hotter plate is at the top or at the bottom, as shown in Fig. 20–22. When the hotter plate is at the *top,* no convection currents develop in the enclosure, since the lighter fluid is always on top of the heavier fluid. Heat transfer in this case is by *pure conduction,* and we have Nu = 1. When the hotter plate is at the *bottom,* the heavier fluid will be on top of the lighter fluid, and there will be a tendency for the lighter fluid to topple the heavier fluid and rise to the top, where it comes in contact with the cooler plate and cools down. Until that happens, however, heat transfer is still by *pure conduction* and Nu = 1. When Ra > 1708, the buoyant force overcomes the fluid resistance and initiates natural convection currents, which are observed to be in the form of hexagonal cells called *Bénard cells.* For Ra $> 3 \times 10^5$, the cells break down and the fluid motion becomes turbulent.

The Rayleigh number for an enclosure is determined from

$$\text{Ra}_L = \frac{g\beta(T_1 - T_2)L_c^3}{\nu^2}\,\text{Pr} \tag{20–40}$$

where the characteristic length L_c is the distance between the hot and cold surfaces, and T_1 and T_2 are the temperatures of the hot and cold surfaces, respectively. All fluid properties are to be evaluated at the average fluid temperature $T_{avg} = (T_1 + T_2)/2$.

Effective Thermal Conductivity

When the Nusselt number is known, the rate of heat transfer through the enclosure can be determined from

$$\dot{Q} = hA_s(T_1 - T_2) = k\text{Nu}A_s\frac{T_1 - T_2}{L_c} \tag{20–41}$$

since $h = k\text{Nu}/L$. The rate of steady heat conduction across a layer of thickness L_c, area A_s, and thermal conductivity k is expressed as

$$\dot{Q}_{cond} = kA_s\frac{T_1 - T_2}{L_c} \tag{20–42}$$

where T_1 and T_2 are the temperatures on the two sides of the layer. A comparison of this relation with Eq. 20–41 reveals that the convection heat transfer in an enclosure is analogous to heat conduction across the fluid layer in the enclosure provided that the thermal conductivity k is replaced by kNu. That is, *the fluid in an enclosure behaves like a fluid whose thermal conductivity is kNu as a result of convection currents.* Therefore, the quantity kNu is called the **effective thermal conductivity** of the enclosure. That is,

$$k_{eff} = k\text{Nu} \tag{20–43}$$

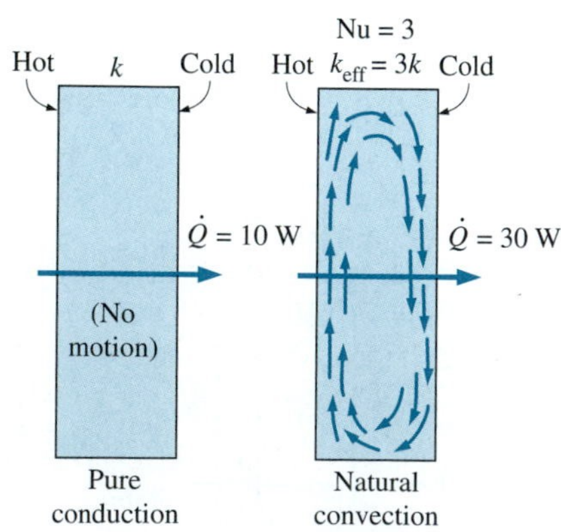

FIGURE 20–23
A Nusselt number of 3 for an enclosure indicates that heat transfer through the enclosure by *natural convection* is three times that by *pure conduction.*

Note that for the special case of Nu = 1, the effective thermal conductivity of the enclosure becomes equal to the conductivity of the fluid. This is expected since this case corresponds to pure conduction (Fig. 20–23).

Natural convection heat transfer in enclosed spaces has been the subject of many experimental and numerical studies, and numerous correlations for the Nusselt number exist. Simple power-law type relations in the form of

$Nu = CRa_L^n$, where C and n are constants, are sufficiently accurate, but they are usually applicable to a narrow range of Prandtl and Rayleigh numbers and aspect ratios. The relations that are more comprehensive are naturally more complex. Next we present some widely used relations for various types of enclosures.

Horizontal Rectangular Enclosures

We need no Nusselt number relations for the case of the hotter plate being at the top, since there are no convection currents in this case and heat transfer is downward by conduction ($Nu = 1$). When the hotter plate is at the bottom, however, significant convection currents set in for $Ra_L > 1708$, and the rate of heat transfer increases (Fig. 20–24).

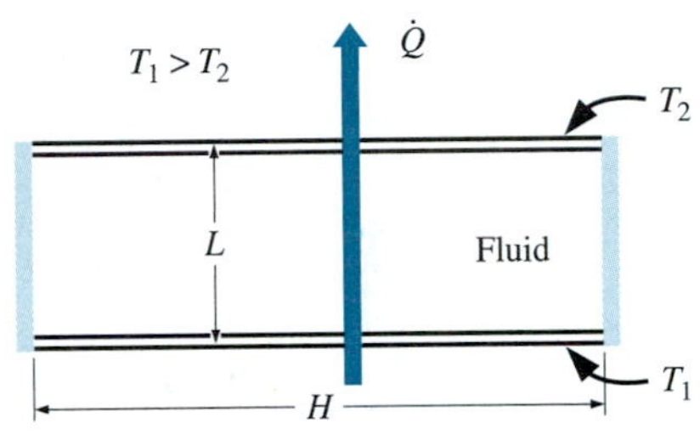

FIGURE 20–24
A horizontal rectangular enclosure with isothermal surfaces.

For horizontal enclosures that contain air, Jakob (1949) recommends the following simple correlations

$$Nu = 0.195Ra_L^{1/4} \qquad 10^4 < Ra_L < 4 \times 10^5 \qquad \textbf{(20–44)}$$

$$Nu = 0.068Ra_L^{1/3} \qquad 4 \times 10^5 < Ra_L < 10^7 \qquad \textbf{(20–45)}$$

These relations can also be used for other gases with $0.5 < Pr < 2$. Using water, silicone oil, and mercury in their experiments, Globe and Dropkin (1959) obtained this correlation for horizontal enclosures heated from below,

$$Nu = 0.069Ra_L^{1/3}\,Pr^{0.074} \qquad 3 \times 10^5 < Ra_L < 7 \times 10^9 \qquad \textbf{(20–46)}$$

Based on experiments with air, Hollands et al. (1976) recommend this correlation for horizontal enclosures,

$$Nu = 1 + 1.44\left[1 - \frac{1708}{Ra_L}\right]^+ + \left[\frac{Ra_L^{1/3}}{18} - 1\right]^+ \qquad Ra_L < 10^8 \qquad \textbf{(20–47)}$$

The notation $[\ ]^+$ indicates that if the quantity in the bracket is negative, it should be set equal to zero. This relation also correlates data well for liquids with moderate Prandtl numbers for $Ra_L < 10^5$, and thus it can also be used for water.

Inclined Rectangular Enclosures

Air spaces between two inclined parallel plates are commonly encountered in flat-plate solar collectors (between the glass cover and the absorber plate) and the double-pane skylights on inclined roofs. Heat transfer through an inclined enclosure depends on the **aspect ratio** H/L as well as the tilt angle θ from the horizontal (Fig. 20–25).

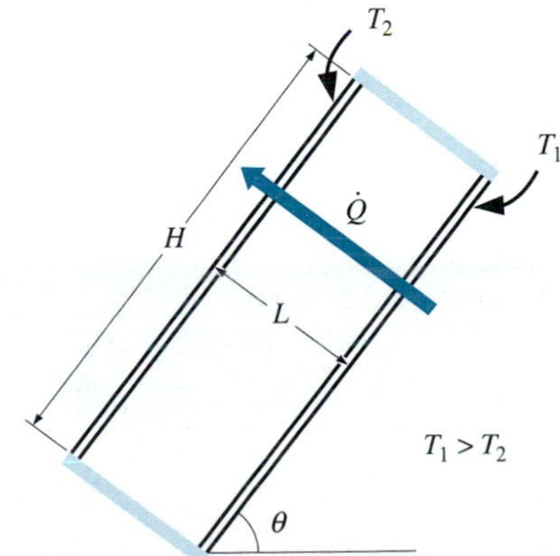

FIGURE 20–25
An inclined rectangular enclosure with isothermal surfaces.

For large aspect ratios ($H/L \geq 12$), this equation (Hollands et al., 1976) correlates experimental data extremely well for tilt angles up to 70°,

$$Nu = 1 + 1.44\left[1 - \frac{1708}{Ra_L \cos\theta}\right]^+\left(1 - \frac{1708(\sin 1.8\theta)^{1.6}}{Ra_L \cos\theta}\right) + \left[\frac{(Ra_L \cos\theta)^{1/3}}{18} - 1\right]^+ \qquad \textbf{(20–48)}$$

for $Ra_L < 10^5$, $0 < \theta < 70°$, and $H/L \geq 12$. Again any quantity in $[\ ]^+$ should be set equal to zero if it is negative. This is to ensure that $Nu = 1$ for $Ra_L \cos\theta < 1708$. Note that this relation reduces to Eq. 20–47 for horizontal enclosures for $\theta = 0°$, as expected.

For enclosures with smaller aspect ratios ($H/L < 12$), the next correlation can be used provided that the tilt angle is less than the critical value θ_{cr} listed in Table 20–2 (Catton, 1978)

$$\text{Nu} = \text{Nu}_{\theta=0^\circ}\left(\frac{\text{Nu}_{\theta=90^\circ}}{\text{Nu}_{\theta=0^\circ}}\right)^{\theta/\theta_{cr}}(\sin\theta_{cr})^{\theta/(4\theta_{cr})} \qquad 0^\circ < \theta < \theta_{cr} \tag{20–49}$$

For tilt angles greater than the critical value ($\theta_{cr} < \theta < 90°$), the Nusselt number can be obtained by multiplying the Nusselt number for a vertical enclosure by $(\sin\theta)^{1/4}$ (Ayyaswamy and Catton, 1973),

$$\text{Nu} = \text{Nu}_{\theta=90^\circ}(\sin\theta)^{1/4} \qquad \theta_{cr} < \theta < 90^\circ,\ \text{any } H/L \tag{20–50}$$

For enclosures tilted more than 90°, the recommended relation is [Arnold et al. (1974)]

$$\text{Nu} = 1 + (\text{Nu}_{\theta=90^\circ} - 1)\sin\theta \qquad 90^\circ < \theta < 180^\circ,\ \text{any } H/L \tag{20–51}$$

More recent but more complex correlations are also available in the literature (e.g., ElSherbiny et al., 1982).

TABLE 20–2

Critical angles for inclined rectangular enclosures

Aspect ratio, H/L	Critical angle, θ_{cr}
1	25°
3	53°
6	60°
12	67°
>12	70°

Vertical Rectangular Enclosures

For vertical enclosures (Fig. 20–26), Catton (1978) recommends these two correlations due to Berkovsky and Polevikov (1977),

$$\text{Nu} = 0.18\left(\frac{\text{Pr}}{0.2+\text{Pr}}\text{Ra}_L\right)^{0.29} \qquad \begin{array}{c} 1 < H/L < 2 \\ \text{any Prandtl number} \\ \text{Ra}_L\,\text{Pr}/(0.2+\text{Pr}) > 10^3 \end{array} \tag{20–52}$$

$$\text{Nu} = 0.22\left(\frac{\text{Pr}}{0.2+\text{Pr}}\text{Ra}_L\right)^{0.28}\left(\frac{H}{L}\right)^{-1/4} \qquad \begin{array}{c} 2 < H/L < 10 \\ \text{any Prandtl number} \\ \text{Ra}_L < 10^{10} \end{array} \tag{20–53}$$

For vertical enclosures with larger aspect ratios, the following correlations can be used (MacGregor and Emery, 1969)

$$\text{Nu} = 0.42\text{Ra}_L^{1/4}\,\text{Pr}^{0.012}\left(\frac{H}{L}\right)^{-0.3} \qquad \begin{array}{c} 10 < H/L < 40 \\ 1 < \text{Pr} < 2\times10^4 \\ 10^4 < \text{Ra}_L < 10^7 \end{array} \tag{20–54}$$

$$\text{Nu} = 0.46\text{Ra}_L^{1/3} \qquad \begin{array}{c} 1 < H/L < 40 \\ 1 < \text{Pr} < 20 \\ 10^6 < \text{Ra}_L < 10^9 \end{array} \tag{20–55}$$

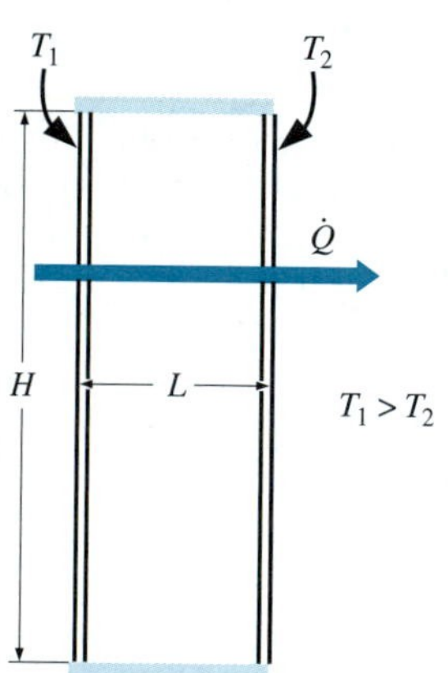

FIGURE 20–26
A vertical rectangular enclosure with isothermal surfaces.

Again all fluid properties are to be evaluated at the average temperature $(T_1 + T_2)/2$.

Concentric Cylinders

Consider two long concentric horizontal cylinders maintained at uniform but different temperatures of T_i and T_o, as shown in Fig. 20–27. The diameters of the inner and outer cylinders are D_i and D_o, respectively, and the characteristic length is the spacing between the cylinders, $L_c = (D_o - D_i)/2$. The rate of heat transfer through the annular space between the cylinders by natural

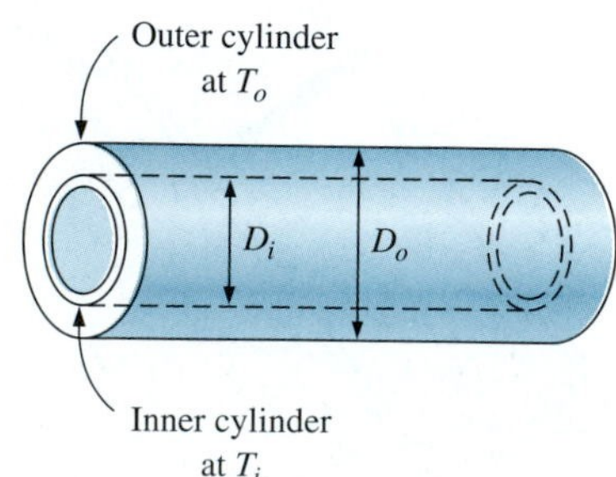

FIGURE 20–27
Two concentric horizontal isothermal cylinders.

convection per unit length is expressed as

$$\dot{Q} = \frac{2\pi k_{\text{eff}}}{\ln(D_o/D_i)}(T_i - T_o) \qquad \text{(W/m)} \tag{20–56}$$

The recommended relation for effective thermal conductivity is (Raithby and Hollands, 1975)

$$\frac{k_{\text{eff}}}{k} = 0.386\left(\frac{\text{Pr}}{0.861 + \text{Pr}}\right)^{1/4}(F_{\text{cyl}}\text{Ra}_L)^{1/4} \tag{20–57}$$

where the geometric factor for concentric cylinders F_{cyl} is

$$F_{\text{cyl}} = \frac{[\ln(D_o/D_i)]^4}{L_c^3(D_i^{-3/5} + D_o^{-3/5})^5} \tag{20–58}$$

The k_{eff} relation in Eq. 20–57 is applicable for $0.70 \leq \text{Pr} \leq 6000$ and $10^2 \leq F_{\text{cyl}}\text{Ra}_L \leq 10^7$. For $F_{\text{cyl}}\text{Ra}_L < 100$, natural convection currents are negligible and thus $k_{\text{eff}} = k$. Note that k_{eff} cannot be less than k, and thus we should set $k_{\text{eff}} = k$ if $k_{\text{eff}}/k < 1$. The fluid properties are evaluated at the average temperature of $(T_i + T_o)/2$.

Concentric Spheres

For concentric isothermal spheres, the rate of heat transfer through the gap between the spheres by natural convection is expressed as (Fig. 20–28)

$$\dot{Q} = k_{\text{eff}}\frac{\pi D_i D_o}{L_c}(T_i - T_o) \qquad \text{(W)} \tag{20–59}$$

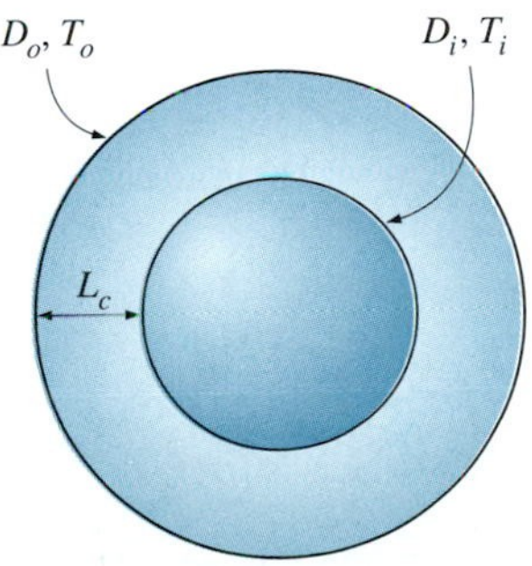

FIGURE 20–28
Two concentric isothermal spheres.

where $L_c = (D_o - D_i)/2$ is the characteristic length. The recommended relation for effective thermal conductivity is (Raithby and Hollands, 1975)

$$\frac{k_{\text{eff}}}{k} = 0.74\left(\frac{\text{Pr}}{0.861 + \text{Pr}}\right)^{1/4}(F_{\text{sph}}\text{Ra}_L)^{1/4} \tag{20–60}$$

where the geometric factor for concentric spheres F_{sph} is

$$F_{\text{sph}} = \frac{L_c}{(D_i D_o)^4(D_i^{-7/5} + D_o^{-7/5})^5} \tag{20–61}$$

The k_{eff} relation in Eq. 20–60 is applicable for $0.70 \leq \text{Pr} \leq 4200$ and $10^2 \leq F_{\text{sph}}\text{Ra}_L \leq 10^4$. If $k_{\text{eff}}/k < 1$, we should set $k_{\text{eff}} = k$.

Combined Natural Convection and Radiation

Gases are nearly transparent to radiation, and thus heat transfer through a gas layer is by simultaneous convection (or conduction, if the gas is quiescent) and radiation. Natural convection heat transfer coefficients are typically very low compared to those for forced convection. Therefore, radiation is usually disregarded in forced convection problems, but it must be considered in natural convection problems that involve a gas. This is especially the case for surfaces with high emissivities. For example, about half of the heat transfer through the air space of a double-pane window is by radiation. The total rate of heat transfer is determined by adding the convection and radiation components,

$$\dot{Q}_{\text{total}} = \dot{Q}_{\text{conv}} + \dot{Q}_{\text{rad}} \tag{20–62}$$

Radiation heat transfer from a surface at temperature T_s surrounded by surfaces at a temperature T_{surr} (both in K) is determined from

$$\dot{Q}_{rad} = \varepsilon\sigma A_s(T_s^4 - T_{surr}^4) \qquad \text{(W)} \tag{20–63}$$

where ε is the emissivity of the surface, A_s is the surface area, and σ = 5.67×10^{-8} W/m² · K⁴ is the Stefan–Boltzmann constant.

When the end effects are negligible, radiation heat transfer between two large parallel plates at temperatures T_1 and T_2 is expressed as (see Chapter 21 for details)

$$\dot{Q}_{rad} = \frac{\sigma A_s(T_1^4 - T_2^4)}{1/\varepsilon_1 + 1/\varepsilon_2 - 1} = \varepsilon_{effective}\,\sigma A_s(T_1^4 - T_2^4) \qquad \text{(W)} \tag{20–64}$$

where ε_1 and ε_2 are the emissivities of the plates, and $\varepsilon_{effective}$ is the *effective emissivity* defined as

$$\varepsilon_{effective} = \frac{1}{1/\varepsilon_1 + 1/\varepsilon_2 - 1} \tag{20–65}$$

The emissivity of an ordinary glass surface, for example, is 0.84. Therefore, the effective emissivity of two parallel glass surfaces facing each other is 0.72. Radiation heat transfer between concentric cylinders and spheres is discussed in Chapter 21.

Note that in some cases the temperature of the surrounding medium may be below the surface temperature ($T_\infty < T_s$), while the temperature of the surrounding surfaces is above the surface temperature ($T_{surr} > T_s$). In such cases, convection and radiation heat transfers are subtracted from each other instead of being added since they are in opposite directions. Also, for a metal surface, the radiation effect can be reduced to negligible levels by polishing the surface and thus lowering the surface emissivity to a value near zero.

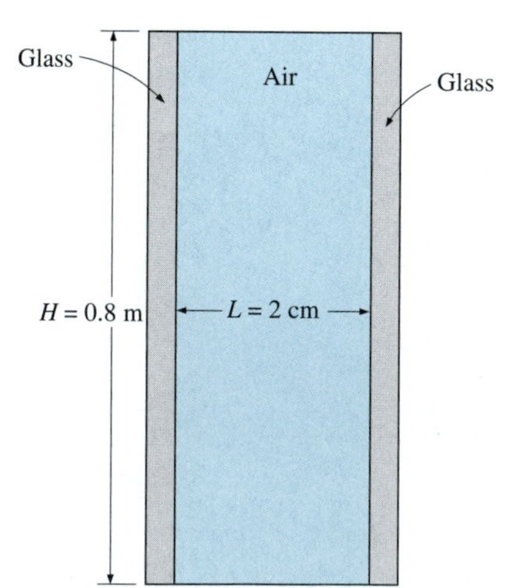

FIGURE 20–29
Schematic for Example 20–4.

EXAMPLE 20–4 Heat Loss through a Double-Pane Window

The vertical 0.8-m-high, 2-m-wide double-pane window shown in Fig. 20–29 consists of two sheets of glass separated by a 2-cm air gap at atmospheric pressure. If the glass surface temperatures across the air gap are measured to be 12°C and 2°C, determine the rate of heat transfer through the window.

Solution Two glasses of a double-pane window are maintained at specified temperatures. The rate of heat transfer through the window is to be determined.

Assumptions **1** Steady operating conditions exist. **2** Air is an ideal gas. **3** Radiation heat transfer is not considered.

Properties The properties of air at the average temperature of $T_{avg} = (T_1 + T_2)/2 = (12 + 2)/2 = 7$°C and 1 atm pressure are (Table A–22)

$$k = 0.02416 \text{ W/m} \cdot \text{°C} \qquad \text{Pr} = 0.7344$$

$$\nu = 1.400 \times 10^{-5} \text{ m}^2\text{/s} \qquad \beta = \frac{1}{T_{avg}} = \frac{1}{280 \text{ K}}$$

Analysis We have a rectangular enclosure filled with air. The characteristic length in this case is the distance between the two glasses, $L_c = L = 0.02$ m. Then the Rayleigh number becomes

$$\text{Ra}_L = \frac{g\beta(T_1 - T_2)L_c^3}{\nu^2}\text{Pr}$$

$$= \frac{(9.81\ \text{m/s}^2)[1/(280\ \text{K})](12 - 2\ \text{K})(0.02\ \text{m})^3}{(1.400 \times 10^{-5}\ \text{m}^2/\text{s})^2}(0.7344) = 1.050 \times 10^4$$

The aspect ratio of the geometry is $H/L = 0.8/0.02 = 40$. Then the Nusselt number in this case can be determined from Eq. 20–54 to be

$$\text{Nu} = 0.42\text{Ra}_L^{1/4}\,\text{Pr}^{0.012}\left(\frac{H}{L}\right)^{-0.3}$$

$$= 0.42(1.050 \times 10^4)^{1/4}(0.7344)^{0.012}\left(\frac{0.8}{0.02}\right)^{-0.3} = 1.40$$

Then,

$$A_s = H \times W = (0.8\ \text{m})(2\ \text{m}) = 1.6\ \text{m}^2$$

and

$$\dot{Q} = hA_s(T_1 - T_2) = k\text{Nu}A_s\frac{T_1 - T_2}{L}$$

$$= (0.02416\ \text{W/m}\cdot{}^\circ\text{C})(1.40)(1.6\ \text{m}^2)\frac{(12 - 2)^\circ\text{C}}{0.02\ \text{m}} = \mathbf{27.1\ W}$$

Therefore, heat is lost through the window at a rate of 27.1 W.

Discussion Recall that a Nusselt number of Nu = 1 for an enclosure corresponds to pure conduction heat transfer through the enclosure. The air in the enclosure in this case remains still, and no natural convection currents occur in the enclosure. The Nusselt number in our case is 1.40, which indicates that heat transfer through the enclosure is 1.40 times that by pure conduction. The increase in heat transfer is due to the natural convection currents that develop in the enclosure.

EXAMPLE 20–5 Heat Transfer through a Spherical Enclosure

The two concentric spheres of diameters $D_i = 20$ cm and $D_o = 30$ cm shown in Fig. 20–30 are separated by air at 1 atm pressure. The surface temperatures of the two spheres enclosing the air are $T_i = 320$ K and $T_o = 280$ K, respectively. Determine the rate of heat transfer from the inner sphere to the outer sphere by natural convection.

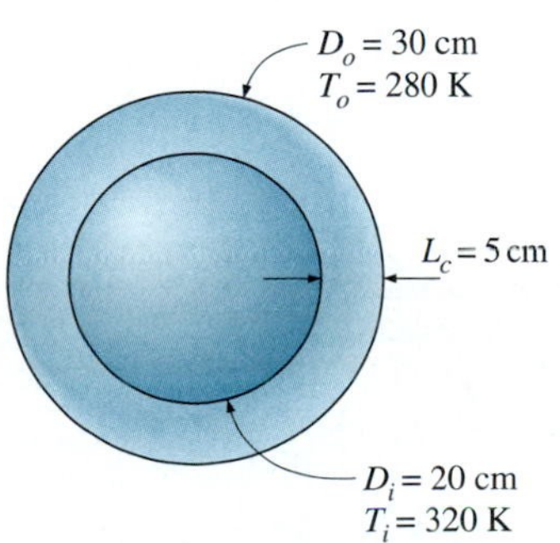

FIGURE 20–30
Schematic for Example 20–5.

Solution Two surfaces of a spherical enclosure are maintained at specified temperatures. The rate of heat transfer through the enclosure is to be determined.

Assumptions **1** Steady operating conditions exist. **2** Air is an ideal gas. **3** Radiation heat transfer is not considered.

Properties The properties of air at the average temperature of $T_{avg} = (T_i + T_o)/2 = (320 + 280)/2 = 300\ \text{K} = 27^\circ\text{C}$ and 1 atm pressure are (Table A–22)

$$k = 0.02566\ \text{W/m}\cdot{}^\circ\text{C} \qquad \text{Pr} = 0.7290$$

$$\nu = 1.580 \times 10^{-5}\ \text{m}^2/\text{s} \qquad \beta = \frac{1}{T_{avg}} = \frac{1}{300\ \text{K}}$$

Analysis We have a spherical enclosure filled with air. The characteristic length in this case is the distance between the two spheres,

$$L_c = (D_o - D_i)/2 = (0.3 - 0.2)/2 = 0.05 \text{ m}$$

The Rayleigh number is

$$\begin{aligned} \text{Ra}_L &= \frac{g\beta(T_i - T_o)L_c^3}{\nu^2}\text{Pr} \\ &= \frac{(9.81 \text{ m/s}^2)[1/(300 \text{ K})](320 - 280 \text{ K})(0.05 \text{ m})^3}{(1.58 \times 10^{-5} \text{ m}^2/\text{s})^2}(0.729) = 4.775 \times 10^5 \end{aligned}$$

The effective thermal conductivity is

$$\begin{aligned} F_{\text{sph}} &= \frac{L_c}{(D_iD_o)^4(D_i^{-7/5} + D_o^{-7/5})^5} \\ &= \frac{0.05 \text{ m}}{[(0.2 \text{ m})(0.3 \text{ m})]^4[(0.2 \text{ m})^{-7/5} + (0.3 \text{ m})^{-7/5}]^5} = 0.005229 \end{aligned}$$

$$\begin{aligned} k_{\text{eff}} &= 0.74k\left(\frac{\text{Pr}}{0.861 + \text{Pr}}\right)^{1/4}(F_{\text{sph}}\text{Ra}_L)^{1/4} \\ &= 0.74(0.02566 \text{ W/m} \cdot {}^\circ\text{C})\left(\frac{0.729}{0.861 + 0.729}\right)^{1/4}(0.005229 \times 4.775 \times 10^5)^{1/4} \\ &= 0.1105 \text{ W/m} \cdot {}^\circ\text{C} \end{aligned}$$

Then the rate of heat transfer between the spheres becomes

$$\begin{aligned} \dot{Q} &= k_{\text{eff}}\frac{\pi D_iD_o}{L_c}(T_i - T_o) \\ &= (0.1105 \text{ W/m} \cdot {}^\circ\text{C})\frac{\pi(0.2 \text{ m})(0.3 \text{ m})}{0.05 \text{ m}}(320 - 280) \text{ K} = \mathbf{16.7 \text{ W}} \end{aligned}$$

Therefore, heat will be lost from the inner sphere to the outer one at a rate of 16.7 W.

Discussion Note that the air in the spherical enclosure acts like a stationary fluid whose thermal conductivity is $k_{\text{eff}}/k = 0.1105/0.02566 = 4.3$ times that of air as a result of natural convection currents. Also, radiation heat transfer between spheres is usually significant, and should be considered in a complete analysis.

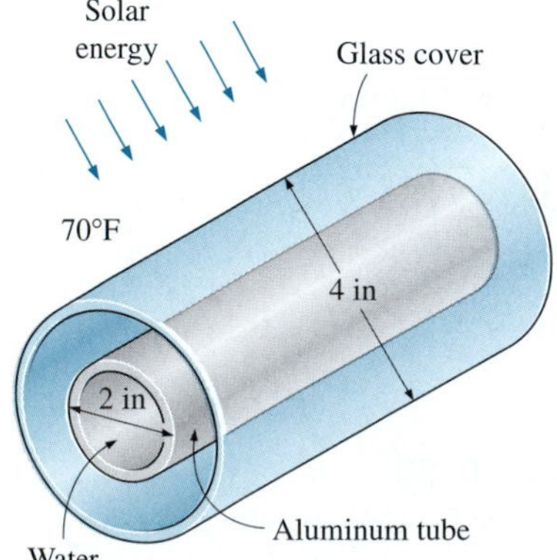

FIGURE 20–31
Schematic for Example 20–6.

EXAMPLE 20–6 Heating Water in a Tube by Solar Energy

A solar collector consists of a horizontal aluminum tube having an outer diameter of 2 in enclosed in a concentric thin glass tube of 4-in-diameter (Fig. 20–31). Water is heated as it flows through the tube, and the annular space between the aluminum and the glass tubes is filled with air at 1 atm pressure. The pump circulating the water fails during a clear day, and the water temperature in the tube starts rising. The aluminum tube absorbs solar radiation at a rate of 30 Btu/h per foot length, and the temperature of the ambient air outside is 70°F. Disregarding any heat loss by radiation, determine the temperature of the aluminum tube when steady operation is established (i.e., when the rate of heat loss from the tube equals the amount of solar energy gained by the tube).

Solution The circulating pump of a solar collector that consists of a horizontal tube and its glass cover fails. The equilibrium temperature of the tube is to be determined.

Assumptions **1** Steady operating conditions exist. **2** The tube and its cover are isothermal. **3** Air is an ideal gas. **4** Heat loss by radiation is negligible.

Properties The properties of air should be evaluated at the average temperature. But we do not know the exit temperature of the air in the duct, and thus we cannot determine the bulk fluid and glass cover temperatures at this point, and we cannot evaluate the average temperatures. Therefore, we assume the glass temperature to be 110°F, and use properties at an anticipated average temperature of (70 + 110)/2 = 90°F (Table A–22E),

$$k = 0.01505 \text{ Btu/h}\cdot\text{ft}\cdot{}^\circ\text{F} \qquad \text{Pr} = 0.7275$$

$$\nu = 1.753 \times 10^{-4} \text{ ft}^2/\text{s} \qquad \beta = \frac{1}{T_{\text{avg}}} = \frac{1}{550 \text{ R}}$$

Analysis We have a horizontal cylindrical enclosure filled with air at 1 atm pressure. The problem involves heat transfer from the aluminum tube to the glass cover and from the outer surface of the glass cover to the surrounding ambient air. When steady operation is reached, these two heat transfer rates must equal the rate of heat gain. That is,

$$\dot{Q}_{\text{tube-glass}} = \dot{Q}_{\text{glass-ambient}} = \dot{Q}_{\text{solar gain}} = 30 \text{ Btu/h} \qquad \text{(per foot of tube)}$$

The heat transfer surface area of the glass cover is

$$A_o = A_{\text{glass}} = (\pi D_o L) = \pi(4/12 \text{ ft})(1 \text{ ft}) = 1.047 \text{ ft}^2 \qquad \text{(per foot of tube)}$$

To determine the Rayleigh number, we need to know the surface temperature of the glass, which is not available. Therefore, it is clear that the solution will require a trial-and-error approach. Assuming the glass cover temperature to be 110°F, the Rayleigh number, the Nusselt number, the convection heat transfer coefficient, and the rate of natural convection heat transfer from the glass cover to the ambient air are determined to be

$$\begin{aligned}\text{Ra}_{D_o} &= \frac{g\beta(T_s - T_\infty)D_o^3}{\nu^2}\text{Pr}\\ &= \frac{(32.2 \text{ ft/s}^2)[1/(550 \text{ R})](110 - 70 \text{ R})(4/12 \text{ ft})^3}{(1.753 \times 10^{-4} \text{ ft}^2/\text{s})^2}(0.7275) = 2.054 \times 10^6\end{aligned}$$

$$\begin{aligned}\text{Nu} &= \left\{0.6 + \frac{0.387\,\text{Ra}_D^{1/6}}{[1 + (0.559/\text{Pr})^{9/16}]^{8/27}}\right\}^2 = \left\{0.6 + \frac{0.387(2.054 \times 10^6)^{1/6}}{[1 + (0.559/0.7275)^{9/16}]^{8/27}}\right\}^2\\ &= 17.89\end{aligned}$$

$$h_o = \frac{k}{D_o}\text{Nu} = \frac{0.01505 \text{ Btu/h}\cdot\text{ft}\cdot{}^\circ\text{F}}{4/12 \text{ ft}}(17.89) = 0.8077 \text{ Btu/h}\cdot\text{ft}^2\cdot{}^\circ\text{F}$$

$$\begin{aligned}\dot{Q}_o &= h_o A_o(T_o - T_\infty) = (0.8077 \text{ Btu/h}\cdot\text{ft}^2\cdot{}^\circ\text{F})(1.047 \text{ ft}^2)(110 - 70)^\circ\text{F}\\ &= 33.8 \text{ Btu/h}\end{aligned}$$

which is more than 30 Btu/h. Therefore, the assumed temperature of 110°F for the glass cover is high. Repeating the calculations with lower temperatures, the glass cover temperature corresponding to 30 Btu/h is determined to be 106°F.

The temperature of the aluminum tube is determined in a similar manner using the natural convection relations for two horizontal concentric cylinders. The characteristic length in this case is the distance between the two cylinders, which is

$$L_c = (D_o - D_i)/2 = (4 - 2)/2 = 1 \text{ in} = 1/12 \text{ ft}$$

We start the calculations by assuming the tube temperature to be 200°F, and thus an average temperature of (106 + 200)/2 = 153°F = 613 R. Using air properties at this temperature gives

$$\text{Ra}_L = \frac{g\beta(T_i - T_o)L_c^3}{\nu^2}\text{Pr}$$

$$= \frac{(32.2 \text{ ft/s}^2)[1/613 \text{ R})](200 - 106 \text{ R})(1/12 \text{ ft})^3}{(2.117 \times 10^{-4} \text{ ft}^2\text{/s})^2}(0.7184) = 4.580 \times 10^4$$

The effective thermal conductivity is

$$F_{\text{cyl}} = \frac{[\ln(D_o/D_i)]^4}{L_c^3(D_i^{-3/5} + D_o^{-3/5})^5}$$

$$= \frac{[\ln(4/2)]^4}{(1/12 \text{ ft})^3[(2/12 \text{ ft})^{-3/5} + (4/12 \text{ ft})^{-3/5}]^5} = 0.1466$$

$$k_{\text{eff}} = 0.386k\left(\frac{\text{Pr}}{0.861 + \text{Pr}}\right)^{1/4}(F_{\text{cyl}}\text{Ra}_L)^{1/4}$$

$$= 0.386(0.01653 \text{ Btu/h} \cdot \text{ft} \cdot \text{°F})\left(\frac{0.7184}{0.861 + 0.7184}\right)^{1/4}$$

$$\times (0.1466 \times 4.580 \times 10^4)^{1/4}$$

$$= 0.04743 \text{ Btu/h} \cdot \text{ft} \cdot \text{°F}$$

Then the rate of heat transfer between the cylinders becomes

$$\dot{Q} = \frac{2\pi k_{\text{eff}}}{\ln(D_o/D_i)}(T_i - T_o)$$

$$= \frac{2\pi(0.04743 \text{ Btu/h} \cdot \text{ft} \cdot \text{°F})}{\ln(4/2)}(200 - 106)\text{°F} = 40.4 \text{ Btu/h}$$

which is more than 30 Btu/h. Therefore, the assumed temperature of 200°F for the tube is high. By trying other values, the tube temperature corresponding to 30 Btu/h is determined to be **180°F**. Therefore, the tube will reach an equilibrium temperature of 180°F when the pump fails.

Discussion Note that we have not considered heat loss by radiation in the calculations, and thus the tube temperature determined is probably too high. This problem is considered in Chapter 21 by accounting for the effect of radiation heat transfer.

SUMMARY

In this chapter, we have considered natural convection heat transfer where any fluid motion occurs by natural means such as buoyancy. The volume expansion coefficient of a substance represents the variation of the density of that substance with temperature at constant pressure, and for an ideal gas, it is expressed as $\beta = 1/T$, where T is the absolute temperature in K or R.

The flow regime in natural convection is governed by a dimensionless number called the *Grashof number,* which represents the ratio of the buoyancy force to the viscous force acting on the fluid and is expressed as

$$\text{Gr}_L = \frac{g\beta(T_s - T_\infty)L_c^3}{\nu^2}$$

where L_c is the *characteristic length,* which is the height L for a vertical plate and the diameter D for a horizontal cylinder. The correlations for the Nusselt number $\text{Nu} = hL_c/k$ in natural convection are expressed in terms of the *Rayleigh number* defined as

$$\text{Ra}_L = \text{Gr}_L \text{Pr} = \frac{g\beta(T_s - T_\infty)L_c^3}{\nu^2}\text{Pr}$$

Nusselt number relations for various surfaces are given in Table 20–1. All fluid properties are evaluated at the film temperature of $T_f = \frac{1}{2}(T_s + T_\infty)$. The outer surface of a vertical cylinder can be treated as a vertical plate when the curvature effects are negligible. The characteristic length for a horizontal surface is $L_c = A_s/p$, where A_s is the surface area and p is the perimeter.

The average Nusselt number for vertical isothermal *parallel plates* of spacing S and height L is given as

$$\text{Nu} = \frac{hS}{k} = \left[\frac{576}{(\text{Ra}_S S/L)^2} + \frac{2.873}{(\text{Ra}_S S/L)^{0.5}}\right]^{-0.5}$$

The optimum fin spacing for a vertical heat sink and the Nusselt number for optimally spaced fins is

$$S_{\text{opt}} = 2.714\left(\frac{S^3L}{\text{Ra}_S}\right)^{0.25} = 2.714\frac{L}{\text{Ra}_L^{0.25}} \text{ and Nu} = \frac{hS_{\text{opt}}}{k} = 1.307$$

In a *horizontal rectangular enclosure* with the hotter plate at the top, heat transfer is by pure conduction and Nu = 1. When the hotter plate is at the bottom, the Nusselt number is

$$\text{Nu} = 1 + 1.44\left[1 - \frac{1708}{\text{Ra}_L}\right]^+ + \left[\frac{\text{Ra}_L^{1/3}}{18} - 1\right]^+ \qquad \text{Ra}_L < 10^8$$

The notation $[\ ]^+$ indicates that if the quantity in the bracket is negative, it should be set equal to zero. For *vertical horizontal enclosures,* the Nusselt number can be determined from

$$\text{Nu} = 0.18\left(\frac{\text{Pr}}{0.2 + \text{Pr}}\text{Ra}_L\right)^{0.29} \qquad \begin{array}{c} 1 < H/L < 2 \\ \text{any Prandtl number} \\ \text{Ra}_L\text{Pr}/(0.2 + \text{Pr}) > 10^3 \end{array}$$

$$\text{Nu} = 0.22\left(\frac{\text{Pr}}{0.2 + \text{Pr}}\text{Ra}_L\right)^{0.28}\left(\frac{H}{L}\right)^{-1/4} \qquad \begin{array}{c} 2 < H/L < 10 \\ \text{any Prandtl number} \\ \text{Ra}_L < 10^{10} \end{array}$$

For aspect ratios greater than 10, Eqs. 20–54 and 20–55 should be used. For inclined enclosures, Eqs. 20–48 through 20–51 should be used.

For *concentric horizontal cylinders,* the rate of heat transfer through the annular space between the cylinders by natural convection per unit length is

$$\dot{Q} = \frac{2\pi k_{\text{eff}}}{\ln(D_o/D_i)}(T_i - T_o)$$

where

$$\frac{k_{\text{eff}}}{k} = 0.386\left(\frac{\text{Pr}}{0.861 + \text{Pr}}\right)^{1/4}(F_{\text{cyl}}\text{Ra}_L)^{1/4}$$

and

$$F_{\text{cyl}} = \frac{[\ln(D_o/D_i)]^4}{L_c^3(D_i^{-3/5} + D_o^{-3/5})^5}$$

For a *spherical enclosure,* the rate of heat transfer through the space between the spheres by natural convection is expressed as

$$\dot{Q} = k_{\text{eff}}\frac{\pi D_i D_o}{L_c}(T_i - T_o)$$

where

$$\frac{k_{\text{eff}}}{k} = 0.74\left(\frac{\text{Pr}}{0.861 + \text{Pr}}\right)^{1/4}(F_{\text{sph}}\text{Ra}_L)^{1/4}$$

$$L_c = (D_o - D_i)/2$$

$$F_{\text{sph}} = \frac{L_c}{(D_i D_o)^4(D_i^{-7/5} + D_o^{-7/5})^5}$$

The quantity kNu is called the *effective thermal conductivity* of the enclosure, since a fluid in an enclosure behaves like a quiescent fluid whose thermal conductivity is kNu as a result of convection currents. The fluid properties are evaluated at the average temperature of $(T_i + T_o)/2$.

REFERENCES AND SUGGESTED READINGS

1. J. N Arnold, I. Catton, and D. K. Edwards. "Experimental Investigation of Natural Convection in Inclined Rectangular Region of Differing Aspects Ratios." ASME Paper No. 75-HT-62, 1975.
2. P. S. Ayyaswamy and I. Catton. "The Boundary-Layer Regime for Natural Convection in a Differently Heated Tilted Rectangular Cavity." *Journal of Heat Transfer* 95 (1973), p. 543.
3. A. Bar-Cohen. "Fin Thickness for an Optimized Natural Convection Array of Rectangular Fins." *Journal of Heat Transfer* 101 (1979), pp. 564–566.
4. A. Bar-Cohen and W. M. Rohsenow. "Thermally Optimum Spacing of Vertical Natural Convection Cooled Parallel Plates." *Journal of Heat Transfer* 106 (1984), p. 116.
5. B. M. Berkovsky and V. K. Polevikov. "Numerical Study of Problems on High-Intensive Free Convection." In *Heat Transfer and Turbulent Buoyant Convection,* eds. D. B. Spalding and N. Afgan, pp. 443–445. Washington, DC: Hemisphere, 1977.
6. I. Catton. "Natural Convection in Enclosures." *Proceedings of Sixth International Heat Transfer Conference.* Toronto: Canada, 1978, Vol. 6, pp. 13–31.
7. T. Cebeci. "Laminar Free Convection Heat Transfer from the Outer Surface of a Vertical Slender Circular Cylinder." *Proceedings of Fifth International Heat Transfer Conference* paper NCI.4, 1974 pp. 15–19.
8. Y. A. Çengel and P. T. L. Zing. "Enhancement of Natural Convection Heat Transfer from Heat Sinks by Shrouding." *Proceedings of ASME/JSME Thermal Engineering Conference.* Honolulu: HA, 1987, Vol. 3, pp. 451–475.
9. S. W. Churchill. "Free Convection around Immersed Bodies." In *Heat Exchanger Design Handbook,* ed. E. U. Schlünder, Section 2.5.7. New York: Hemisphere, 1983.
10. S. W. Churchill. "Combined Free and Forced Convection around Immersed Bodies." In *Heat Exchanger Design Handbook,* Section 2.5.9. New York: Hemisphere Publishing, 1986.
11. S. W. Churchill and H. H. S. Chu. "Correlating Equations for Laminar and Turbulent Free Convection from a Horizontal Cylinder." *International Journal of Heat Mass Transfer* 18 (1975), p. 1049.
12. S. W. Churchill and H. H. S. Chu. "Correlating Equations for Laminar and Turbulent Free Convection from a Vertical Plate." *International Journal of Heat Mass Transfer* 18 (1975), p. 1323.
13. E. R. G. Eckert and E. Soehngen. "Studies on Heat Transfer in Laminar Free Convection with Zehnder–Mach Interferometer." USAF Technical Report 5747, December 1948.
14. E. R. G. Eckert and E. Soehngen. "Interferometric Studies on the Stability and Transition to Turbulence of a Free Convection Boundary Layer." *Proceedings of General Discussion, Heat Transfer ASME-IME,* London, 1951.
15. S. M. ElSherbiny, G. D. Raithby, and K. G. T. Hollands. "Heat Transfer by Natural Convection Across Vertical and Inclined Air Layers. *Journal of Heat Transfer* 104 (1982), pp. 96–102.
16. T. Fuji and H. Imura. "Natural Convection Heat Transfer from a Plate with Arbitrary Inclination." *International Journal of Heat Mass Transfer* 15 (1972), p. 755.
17. K. G. T. Hollands, T. E. Unny, G. D. Raithby, and L. Konicek. "Free Convective Heat Transfer Across Inclined Air Layers." *Journal of Heat Transfer* 98 (1976), pp. 182–193.
18. M. Jakob. *Heat Transfer.* New York: Wiley, 1949.
19. W. M. Kays and M. E. Crawford. *Convective Heat and Mass Transfer.* 3rd ed. New York: McGraw-Hill, 1993.
20. R. K. MacGregor and A. P. Emery. "Free Convection Through Vertical Plane Layers: Moderate and High Prandtl Number Fluids." *Journal of Heat Transfer* 91 (1969), p. 391.
21. S. Ostrach. "An Analysis of Laminar Free Convection Flow and Heat Transfer About a Flat Plate Parallel to the Direction of the Generating Body Force." National Advisory Committee for Aeronautics, Report 1111, 1953.
22. G. D. Raithby and K. G. T. Hollands. "A General Method of Obtaining Approximate Solutions to Laminar and Turbulent Free Convection Problems." In *Advances in Heat Transfer,* ed. F. Irvine and J. P. Hartnett, Vol. II, pp. 265–315. New York: Academic Press, 1975.
23. E. M. Sparrow and J. L. Gregg. "Laminar Free Convection from a Vertical Flat Plate." *Transactions of the ASME* 78 (1956), p. 438.
24. E. M. Sparrow and J. L. Gregg. "Laminar Free Convection Heat Transfer from the Outer Surface of a Vertical Circular Cylinder." *ASME* 78 (1956), p. 1823.
25. E. M. Sparrow and C. Prakash. "Enhancement of Natural Convection Heat Transfer by a Staggered Array of Vertical Plates." *Journal of Heat Transfer* 102 (1980), pp. 215–220.
26. E. M. Sparrow and S. B. Vemuri. "Natural Convection/Radiation Heat Transfer from Highly Populated Pin Fin Arrays." *Journal of Heat Transfer* 107 (1985), pp. 190–197.

PROBLEMS*

Physical Mechanism of Natural Convection

20–1C What is natural convection? How does it differ from forced convection? What force causes natural convection currents?

20–2C In which mode of heat transfer is the convection heat transfer coefficient usually higher, natural convection or forced convection? Why?

20–3C Consider a hot boiled egg in a spacecraft that is filled with air at atmospheric pressure and temperature at all times. Will the egg cool faster or slower when the spacecraft is in space instead of on the ground? Explain.

20–4C What is buoyancy force? Compare the relative magnitudes of the buoyancy force acting on a body immersed in these mediums: (*a*) air, (*b*) water, (*c*) mercury, and (*d*) an evacuated chamber.

20–5C When will the hull of a ship sink in water deeper: when the ship is sailing in fresh water or in seawater? Why?

20–6C A person weighs himself on a waterproof spring scale placed at the bottom of a 1-m-deep swimming pool. Will the person weigh more or less in water? Why?

20–7C Consider two fluids, one with a large coefficient of volume expansion and the other with a small one. In what fluid will a hot surface initiate stronger natural convection currents? Why? Assume the viscosity of the fluids to be the same.

20–8C Consider a fluid whose volume does not change with temperature at constant pressure. What can you say about natural convection heat transfer in this medium?

20–9C What do the lines on an interferometer photograph represent? What do closely packed lines on the same photograph represent?

20–10C Physically, what does the Grashof number represent? How does the Grashof number differ from the Reynolds number?

20–11 Show that the volume expansion coefficient of an ideal gas is $\beta = 1/T$, where T is the absolute temperature.

Natural Convection over Surfaces

20–12C How does the Rayleigh number differ from the Grashof number?

20–13C Under what conditions can the outer surface of a vertical cylinder be treated as a vertical plate in natural convection calculations?

20–14C Will a hot horizontal plate whose back side is insulated cool faster or slower when its hot surface is facing down instead of up?

20–15C Consider laminar natural convection from a vertical hot-plate. Will the heat flux be higher at the top or at the bottom of the plate? Why?

20–16 Consider a thin 16-cm-long and 20-cm-wide horizontal plate suspended in air at 20°C. The plate is equipped with electric resistance heating elements with a rating of 20 W. Now the heater is turned on and the plate temperature rises. Determine the temperature of the plate when steady operating conditions are reached. The plate has an emissivity of 0.90 and the surrounding surfaces are at 17°C.

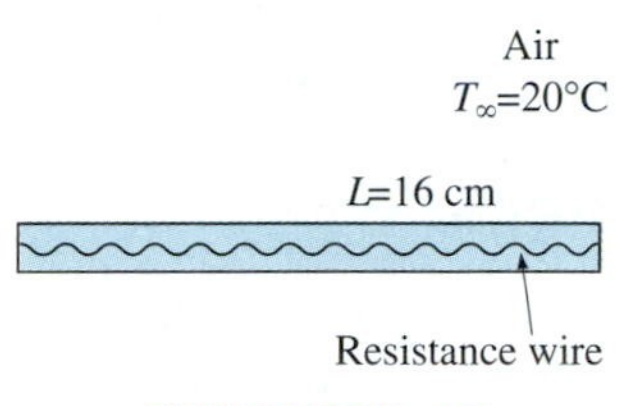

FIGURE P20–16

20–17 Flue gases from an incinerator are released to atmosphere using a stack that is 0.6 m in diameter and 10.0 m high. The outer surface of the stack is at 40°C and the surrounding air is at 10°C. Determine the rate of heat transfer from the stack assuming (*a*) there is no wind and (*b*) the stack is exposed to 20 km/h winds.

20–18 Thermal energy generated by the electrical resistance of a 5-mm-diameter and 4-m-long bare cable is dissipated to the surrounding air at 20°C. The voltage drop and the electric current across the cable in steady operation are measured to be 60 V and 1.5 A, respectively. Disregarding radiation, estimate the surface temperature of the cable.

20–19 A 10-m-long section of a 6-cm-diameter horizontal hot-water pipe passes through a large room whose temperature is 27°C. If the temperature and the emissivity of the outer surface of the pipe are 73°C and 0.8, respectively, determine the rate of heat loss from the pipe by (*a*) natural convection and (*b*) radiation.

20–20 Consider a wall-mounted power transistor that dissipates 0.18 W of power in an environment at 35°C. The transistor is 0.45 cm long and has a diameter of 0.4 cm. The emissivity of the outer surface of the transistor is 0.1, and the average temperature of the surrounding surfaces is 25°C. Disregarding any heat transfer from the base surface, determine

*Problems designated by a "C" are concept questions, and students are encouraged to answer them all. Problems designated by an "E" are in English units, and the SI users can ignore them. Problems with the EES icon are solved using EES, and complete solutions together with parametric studies are included on the enclosed DVD. Problems with the computer icon are comprehensive in nature, and are intended to be solved with a computer, preferably using the EES software that accompanies this text.

the surface temperature of the transistor. Use air properties at 100°C. *Answer:* 183°C

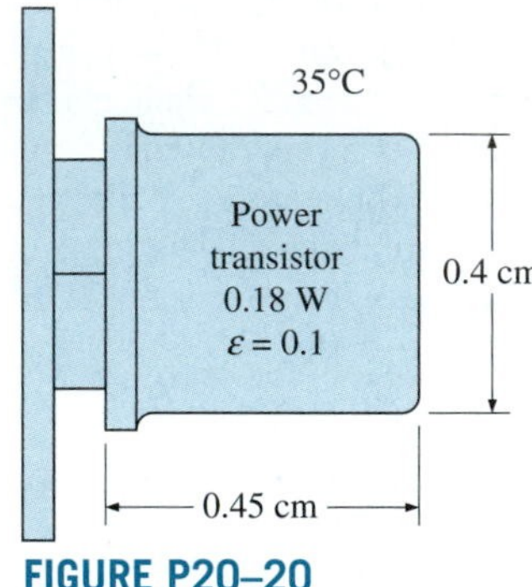

FIGURE P20–20

20–21 Reconsider Prob. 20–20. Using EES (or other) software, investigate the effect of ambient temperature on the surface temperature of the transistor. Let the environment temperature vary from 10°C to 40°C and assume that the surrounding surfaces are 10°C colder than the environment temperature. Plot the surface temperature of the transistor versus the environment temperature, and discuss the results.

20–22E Consider a 2-ft × 2-ft thin square plate in a room at 75°F. One side of the plate is maintained at a temperature of 130°F, while the other side is insulated. Determine the rate of heat transfer from the plate by natural convection if the plate is (*a*) vertical; (*b*) horizontal with hot surface facing up; and (*c*) horizontal with hot surface facing down.

20–23E Reconsider Prob. 20–22E. Using EES (or other) software, plot the rate of natural convection heat transfer for different orientations of the plate as a function of the plate temperature as the temperature varies from 80°F to 180°F, and discuss the results.

20–24 A 300-W cylindrical resistance heater is 0.75 m long and 0.5 cm in diameter. The resistance wire is placed horizontally in a fluid at 20°C. Determine the outer surface temperature of the resistance wire in steady operation if the fluid is (*a*) air and (*b*) water. Ignore any heat transfer by radiation. Use properties at 500°C for air and 40°C for water.

20–25 Water is boiling in a 12-cm-deep pan with an outer diameter of 25 cm that is placed on top of a stove. The ambient air and the surrounding surfaces are at a temperature of 25°C, and the emissivity of the outer surface of the pan is 0.80. Assuming the entire pan to be at an average temperature of 98°C, determine the rate of heat loss from the cylindrical side surface of the pan to the surroundings by (*a*) natural convection and (*b*) radiation. (*c*) If water is boiling at a rate of 1.5 kg/h at 100°C, determine the ratio of the heat lost from the side surfaces of the pan to that by the evaporation of water. The enthalpy of vaporization of water at 100°C is 2257 kJ/kg. *Answers:* 46.2 W, 47.3 W, 0.099

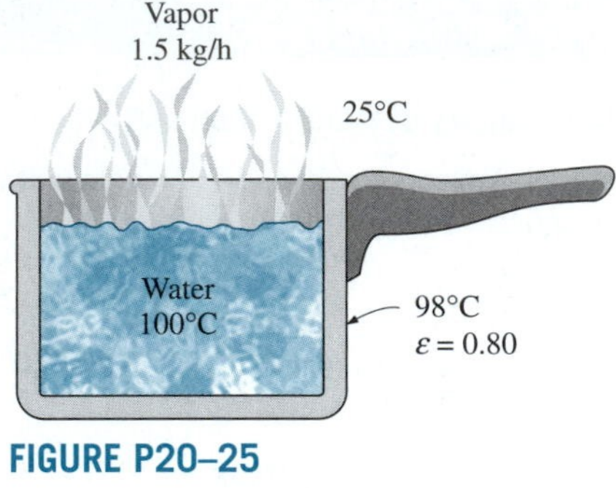

FIGURE P20–25

20–26 Repeat Prob. 20–25 for a pan whose outer surface is polished and has an emissivity of 0.1.

20–27 In a plant that manufactures canned aerosol paints, the cans are temperature-tested in water baths at 55°C before they are shipped to ensure that they withstand temperatures up to 55°C during transportation and shelving. The cans, moving on a conveyor, enter the open hot water bath, which is 0.5 m deep, 1 m wide, and 3.5 m long, and move slowly in the hot water toward the other end. Some of the cans fail the test and explode in the water bath. The water container is made of sheet metal, and the entire container is at about the same temperature as the hot water. The emissivity of the outer surface of the container is 0.7. If the temperature of the surrounding air and surfaces is 20°C, determine the rate of heat loss from the four side surfaces of the container (disregard the top surface, which is open).

The water is heated electrically by resistance heaters, and the cost of electricity is $0.085/kWh. If the plant operates 24 h a day 365 days a year and thus 8760 h a year, determine the annual cost of the heat losses from the container for this facility.

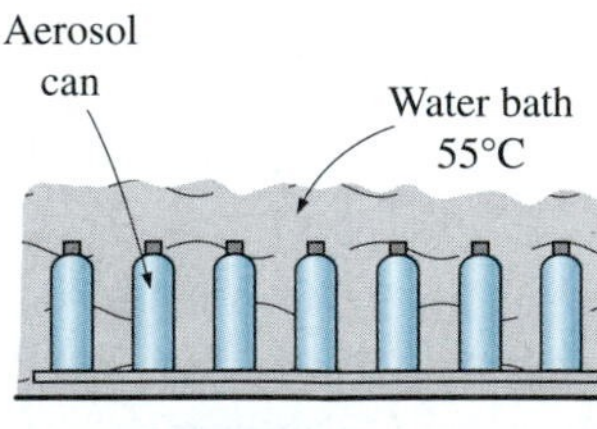

FIGURE P20–27

20–28 Reconsider Prob. 20–27. In order to reduce the heating cost of the hot water, it is proposed to insulate the side and bottom surfaces of the container with 5-cm-thick fiberglass insulation (k = 0.035 W/m · °C) and to wrap the insulation with aluminum foil (ε = 0.1) in order to minimize the heat loss by radiation. An estimate is obtained from a local insulation contractor, who proposes to do the insulation job for $350, including materials and labor. Would you support this proposal? How long will it take for the insulation to pay for itself from the energy it saves?

20–29 Consider a 15-cm × 20-cm printed circuit board (PCB) that has electronic components on one side. The board is placed in a room at 20°C. The heat loss from the back surface of the board is negligible. If the circuit board is dissipating 8 W of power in steady operation, determine the average temperature of the hot surface of the board, assuming the board is (*a*) vertical; (*b*) horizontal with hot surface facing up; and (*c*) horizontal with hot surface facing down. Take the emissivity of the surface of the board to be 0.8 and assume the surrounding surfaces to be at the same temperature as the air in the room. *Answers:* (*a*) 46.6°C, (*b*) 42.6°C, (*c*) 50.7°C

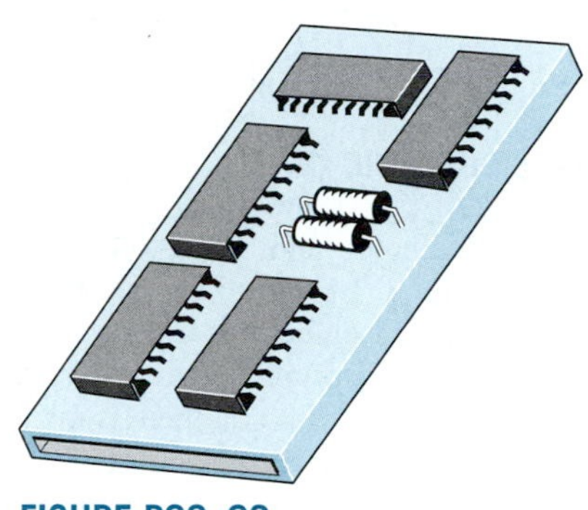

FIGURE P20–29

20–30 Reconsider Prob. 20–29. Using EES (or other) software, investigate the effects of the room temperature and the emissivity of the board on the temperature of the hot surface of the board for different orientations of the board. Let the room temperature vary from 5°C to 35°C and the emissivity from 0.1 to 1.0. Plot the hot surface temperature for different orientations of the board as the functions of the room temperature and the emissivity, and discuss the results.

20–31 A manufacturer makes absorber plates that are 1.2 m × 0.8 m in size for use in solar collectors. The back side of the plate is heavily insulated, while its front surface is coated with black chrome, which has an absorptivity of 0.87 for solar radiation and an emissivity of 0.09. Consider such a plate placed horizontally outdoors in calm air at 25°C. Solar radiation is incident on the plate at a rate of 700 W/m^2. Taking the effective sky temperature to be 10°C, determine the equilibrium temperature of the absorber plate. What would your answer be if the absorber plate is made of ordinary aluminum plate that has a solar absorptivity of 0.28 and an emissivity of 0.07?

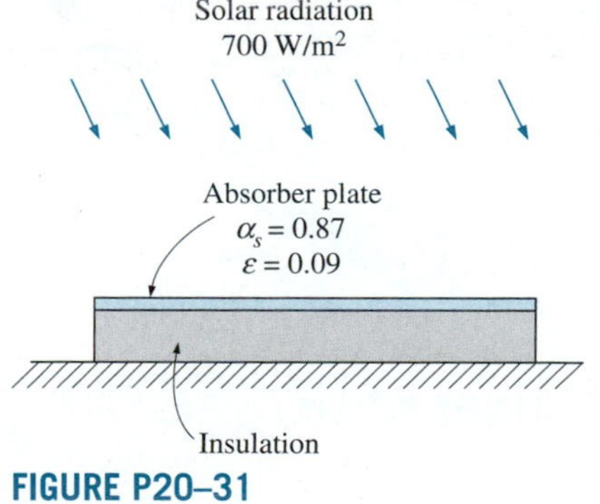

FIGURE P20–31

20–32 Repeat Prob. 20–31 for an aluminum plate painted flat black (solar absorptivity 0.98 and emissivity 0.98) and also for a plate painted white (solar absorptivity 0.26 and emissivity 0.90).

20–33 The following experiment is conducted to determine the natural convection heat transfer coefficient for a horizontal cylinder that is 80 cm long and 2 cm in diameter. A 80-cm-long resistance heater is placed along the centerline of the cylinder, and the surfaces of the cylinder are polished to minimize the radiation effect. The two circular side surfaces of the cylinder are well insulated. The resistance heater is turned on, and the power dissipation is maintained constant at 60 W. If the average surface temperature of the cylinder is measured to be 120°C in the 20°C room air when steady operation is reached, determine the natural convection heat transfer coefficient. If the emissivity of the outer surface of the cylinder is 0.1 and a 5 percent error is acceptable, do you think we need to do any correction for the radiation effect? Assume the surrounding surfaces to be at 20°C also.

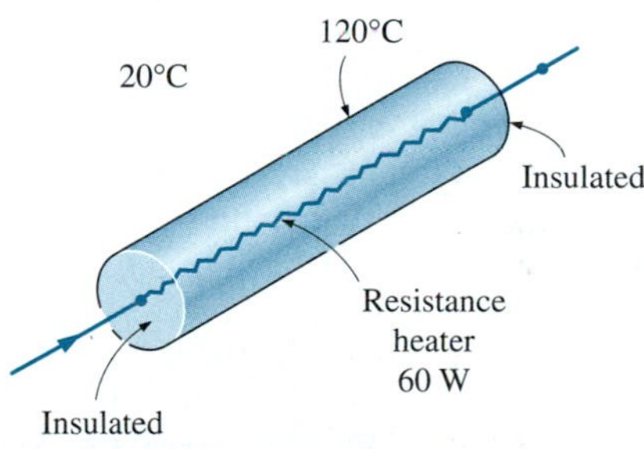

FIGURE P20–33

20–34 Thick fluids such as asphalt and waxes and the pipes in which they flow are often heated in order to reduce the viscosity of the fluids and thus to reduce the pumping costs. Consider the flow of such a fluid through a 100-m-long pipe of outer diameter 30 cm in calm ambient air at 0°C. The pipe is heated electrically, and a thermostat keeps the outer surface temperature of the pipe constant at 25°C. The emissivity of the

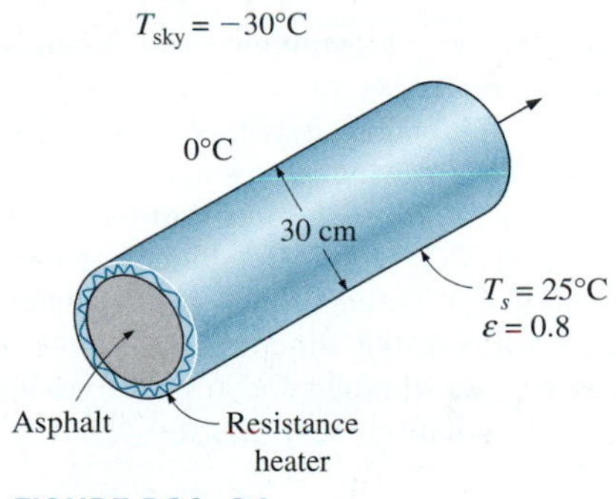

FIGURE P20–34

outer surface of the pipe is 0.8, and the effective sky temperature is −30°C. Determine the power rating of the electric resistance heater, in kW, that needs to be used. Also, determine the cost of electricity associated with heating the pipe during a 10-h period under the above conditions if the price of electricity is \$0.09/kWh. *Answers:* 29.1 kW, \$26.2

20–35 Reconsider Prob. 20–34. To reduce the heating cost of the pipe, it is proposed to insulate it with sufficiently thick fiberglass insulation ($k = 0.035$ W/m · °C) wrapped with aluminum foil ($\varepsilon = 0.1$) to cut down the heat losses by 85 percent. Assuming the pipe temperature to remain constant at 25°C, determine the thickness of the insulation that needs to be used. How much money will the insulation save during this 10-h period? *Answers:* 1.3 cm, \$22.3

20–36E Consider an industrial furnace that resembles a 13-ft-long horizontal cylindrical enclosure 8 ft in diameter whose end surfaces are well insulated. The furnace burns natural gas at a rate of 48 therms/h. The combustion efficiency of the furnace is 82 percent (i.e., 18 percent of the chemical energy of the fuel is lost through the flue gases as a result of incomplete combustion and the flue gases leaving the furnace at high temperature). If the heat loss from the outer surfaces of the furnace by natural convection and radiation is not to exceed 1 percent of the heat generated inside, determine the highest allowable surface temperature of the furnace. Assume the air and wall surface temperature of the room to be 75°F, and take the emissivity of the outer surface of the furnace to be 0.85. If the cost of natural gas is \$1.15/therm and the furnace operates 2800 h per year, determine the annual cost of this heat loss to the plant.

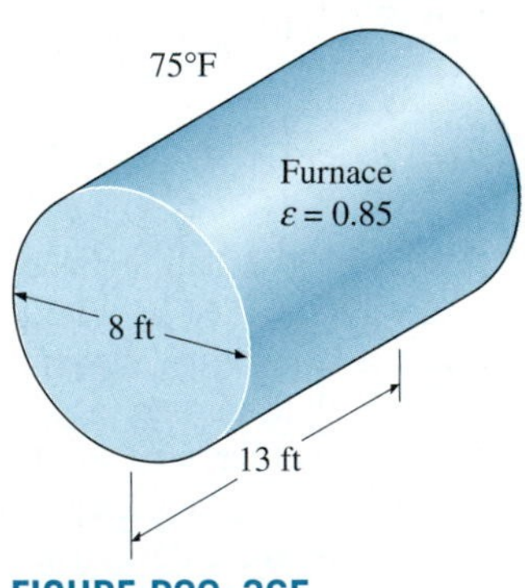

FIGURE P20–36E

20–37 Consider a 1.2-m-high and 2-m-wide glass window with a thickness of 6 mm, thermal conductivity $k = 0.78$ W/m · °C, and emissivity $\varepsilon = 0.9$. The room and the walls that face the window are maintained at 25°C, and the average temperature of the inner surface of the window is measured to be 5°C. If the temperature of the outdoors is −5°C, determine (*a*) the convection heat transfer coefficient on the inner surface of the window, (*b*) the rate of total heat transfer through the window, and (*c*) the combined natural convection and radiation heat transfer coefficient on the outer surface of the window. Is it reasonable to neglect the thermal resistance of the glass in this case?

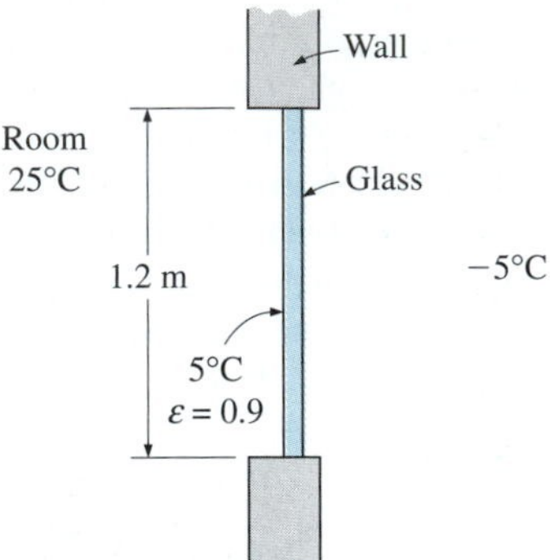

FIGURE P20–37

20–38 A 3-mm-diameter and 12-m-long electric wire is tightly wrapped with a 1.5-mm-thick plastic cover whose thermal conductivity and emissivity are $k = 0.20$ W/m · °C and $\varepsilon = 0.9$. Electrical measurements indicate that a current of 10 A passes through the wire and there is a voltage drop of 7 V along the wire. If the insulated wire is exposed to calm atmospheric air at $T_\infty = 30$°C, determine the temperature at the interface of the wire and the plastic cover in steady operation. Take the surrounding surfaces to be at about the same temperature as the air.

20–39 During a visit to a plastic sheeting plant, it was observed that a 60-m-long section of a 2-in nominal (6.03-cm-outer-diameter) steam pipe extended from one end of the plant to the other with no insulation on it. The temperature measurements at several locations revealed that the average temperature of the exposed surfaces of the steam pipe was 170°C, while the temperature of the surrounding air was 20°C. The outer surface of the pipe appeared to be oxidized, and its emissivity can be taken to be 0.7. Taking the temperature of the surrounding surfaces to be 20°C also, determine the rate of heat loss from the steam pipe.

Steam is generated in a gas furnace that has an efficiency of 78 percent, and the plant pays \$1.10 per therm (1 therm = 105,500 kJ) of natural gas. The plant operates 24 h a day 365 days a year, and thus 8760 h a year. Determine the annual cost of the heat losses from the steam pipe for this facility.

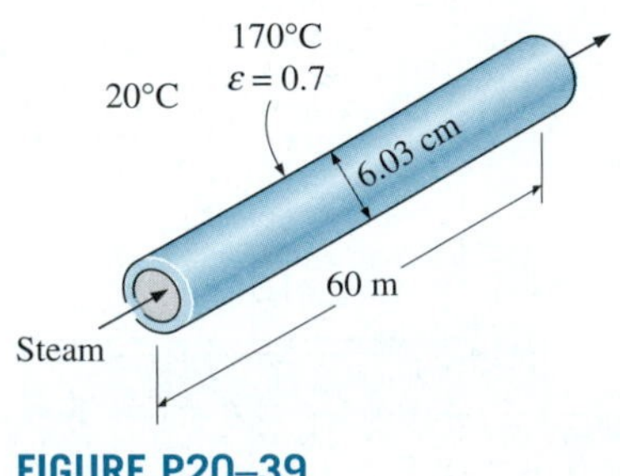

FIGURE P20–39

20–40 Reconsider Prob. 20–39. Using EES (or other) software, investigate the effect of the surface temperature of the steam pipe on the rate of heat loss from the pipe and the annual cost of this heat loss. Let the surface temperature vary from 100°C to 200°C. Plot the rate of heat loss and the annual cost as a function of the surface temperature, and discuss the results.

20–41 Reconsider Prob. 20–39. In order to reduce heat losses, it is proposed to insulate the steam pipe with 5-cm-thick fiberglass insulation (k = 0.038 W/m · °C) and to wrap it with aluminum foil (ε = 0.1) in order to minimize the radiation losses. Also, an estimate is obtained from a local insulation contractor, who proposed to do the insulation job for $750, including materials and labor. Would you support this proposal? How long will it take for the insulation to pay for itself from the energy it saves? Assume the temperature of the steam pipe to remain constant at 170°C.

20–42 A 50-cm × 50-cm circuit board that contains 121 square chips on one side is to be cooled by combined natural convection and radiation by mounting it on a vertical surface in a room at 25°C. Each chip dissipates 0.18 W of power, and the emissivity of the chip surfaces is 0.7. Assuming the heat transfer from the back side of the circuit board to be negligible, and the temperature of the surrounding surfaces to be the same as the air temperature of the room, determine the surface temperature of the chips. *Answer:* 36.2°C

20–43 Repeat Prob. 20–42 assuming the circuit board to be positioned horizontally with (*a*) chips facing up and (*b*) chips facing down.

20–44 The side surfaces of a 2-m-high cubic industrial furnace burning natural gas are not insulated, and the temperature at the outer surface of this section is measured to be 110°C. The temperature of the furnace room, including its surfaces, is 30°C, and the emissivity of the outer surface of the furnace is 0.7. It is proposed that this section of the furnace wall be insulated with glass wool insulation (k = 0.038 W/m · °C) wrapped by a reflective sheet (ε = 0.2) in order to reduce the heat loss by 90 percent. Assuming the outer surface temperature of the metal section still remains at about 110°C, determine the thickness of the insulation that needs to be used.

The furnace operates continuously throughout the year and has an efficiency of 78 percent. The price of the natural gas is $0.55/therm (1 therm = 105,500 kJ of energy content). If the installation of the insulation will cost $550 for materials and labor, determine how long it will take for the insulation to pay for itself from the energy it saves.

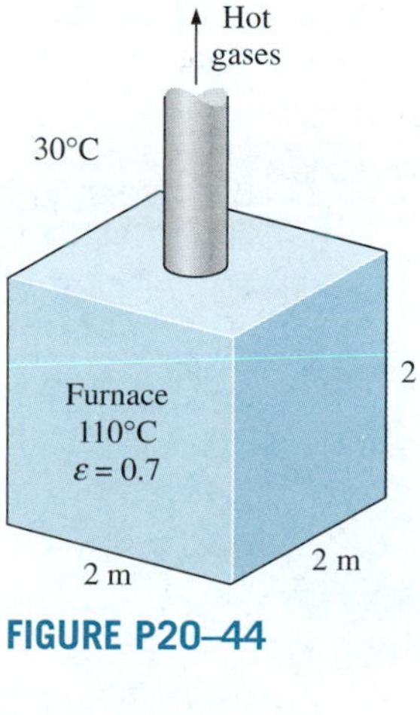

FIGURE P20–44

20–45 A 1.5-m-diameter, 4-m-long cylindrical propane tank is initially filled with liquid propane, whose density is 581 kg/m^3. The tank is exposed to the ambient air at 25°C in calm weather. The outer surface of the tank is polished so that the radiation heat transfer is negligible. Now a crack develops at the top of the tank, and the pressure inside drops to 1 atm while the temperature drops to −42°C, which is the boiling temperature of propane at 1 atm. The heat of vaporization of propane at 1 atm is 425 kJ/kg. The propane is slowly vaporized as a result of the heat transfer from the ambient air into the tank, and the propane vapor escapes the tank at −42°C through the crack. Assuming the propane tank to be at about the same temperature as the propane inside at all times, determine how long it will take for the tank to empty if it is not insulated.

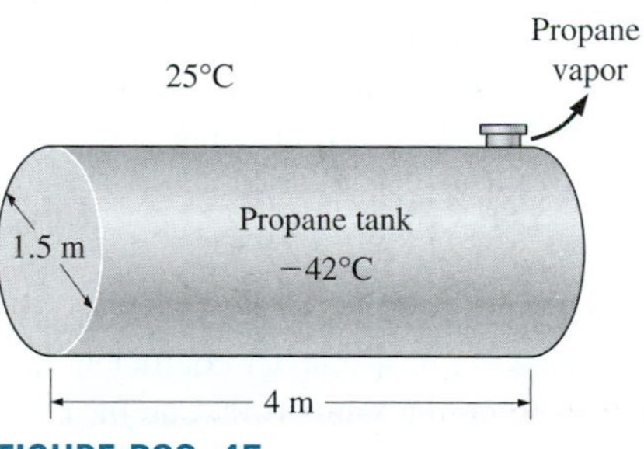

FIGURE P20–45

20–46E An average person generates heat at a rate of 240 Btu/h while resting in a room at 70°F. Assuming one-quarter of this heat is lost from the head and taking the emissivity of the skin to be 0.9, determine the average surface temperature of the head when it is not covered. The head can be approximated as a 12-in-diameter sphere, and the interior surfaces of the room can be assumed to be at the room temperature.

20–47 An incandescent lightbulb is an inexpensive but highly inefficient device that converts electrical energy into light. It converts about 10 percent of the electrical energy it consumes into light while converting the remaining 90 percent into heat. The glass bulb of the lamp heats up very quickly as a result of absorbing all that heat and dissipating it to the surroundings by convection and radiation. Consider an 8-cm-diameter 60-W lightbulb in a room at 25°C. The emissivity of the glass is 0.9. Assuming that 10 percent of the energy passes through the glass bulb as light with negligible absorption and the rest of the energy is absorbed and dissipated by the bulb itself by natural convection and radiation, determine the equilibrium temperature of the glass bulb. Assume the interior surfaces of the room to be at room temperature. *Answer:* 169°C

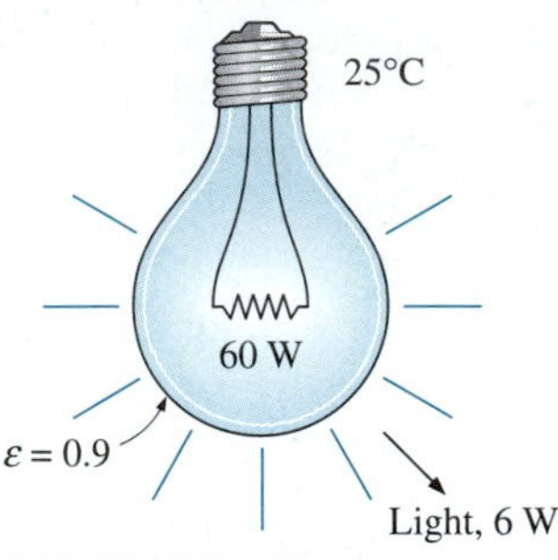

FIGURE P20–47

20–48 A 40-cm-diameter, 110-cm-high cylindrical hot-water tank is located in the bathroom of a house maintained at 20°C. The surface temperature of the tank is measured to be 44°C and its emissivity is 0.4. Taking the surrounding surface temperature to be also 20°C, determine the rate of heat loss from all surfaces of the tank by natural convection and radiation.

20–49 A 28-cm-high, 18-cm-long, and 18-cm-wide rectangular container suspended in a room at 24°C is initially filled with cold water at 2°C. The surface temperature of the container is observed to be nearly the same as the water temperature inside. The emissivity of the container surface is 0.6, and the temperature of the surrounding surfaces is about the same as the air temperature. Determine the water temperature in the container after 3 h, and the average rate of heat transfer to the water. Assume the heat transfer coefficient on the top and bottom surfaces to be the same as that on the side surfaces.

20–50 Reconsider Prob. 20–49. Using EES (or other) software, plot the water temperature in the container as a function of the heating time as the time varies from 30 min to 10 h, and discuss the results.

20–51 A room is to be heated by a coal-burning stove, which is a cylindrical cavity with an outer diameter of 32 cm and a height of 70 cm. The rate of heat loss from the room is estimated to be 1.5 kW when the air temperature in the room is maintained constant at 24°C. The emissivity of the stove surface is 0.85 and the average temperature of the surrounding wall surfaces is 14°C. Determine the surface temperature of the stove. Neglect the transfer from the bottom surface and take the heat transfer coefficient at the top surface to be the same as that on the side surface.

The heating value of the coal is 30,000 kJ/kg, and the combustion efficiency is 65 percent. Determine the amount of coal burned a day if the stove operates 14 h a day.

20–52 The water in a 40-L tank is to be heated from 15°C to 45°C by a 6-cm-diameter spherical heater whose surface temperature is maintained at 85°C. Determine how long the heater should be kept on.

Natural Convection from Finned Surfaces and PCBs

20–53C Why are finned surfaces frequently used in practice? Why are the finned surfaces referred to as heat sinks in the electronics industry?

20–54C Why are heat sinks with closely packed fins not suitable for natural convection heat transfer, although they increase the heat transfer surface area more?

20–55C Consider a heat sink with optimum fin spacing. Explain how heat transfer from this heat sink will be affected by (*a*) removing some of the fins on the heat sink and (*b*) doubling the number of fins on the heat sink by reducing the fin spacing. The base area of the heat sink remains unchanged at all times.

20–56 Aluminum heat sinks of rectangular profile are commonly used to cool electronic components. Consider a 7.62-cm-long and 9.68-cm-wide commercially available heat sink whose cross section and dimensions are as shown in Fig. P20–56. The heat sink is oriented vertically and is used to cool a power transistor that can dissipate up to 125 W of power. The back surface of the heat sink is insulated. The surfaces of the heat sink are untreated, and thus they have a low emissivity (under 0.1). Therefore, radiation heat transfer from the heat sink can be neglected. During an experiment conducted in room air at 22°C, the base temperature of the heat sink was measured to be 120°C when the power dissipation of the transistor was 15 W. Assuming the entire heat sink to be at the base temperature, determine the average natural convection heat transfer coefficient for this case. *Answer:* 7.13 W/m² · °C

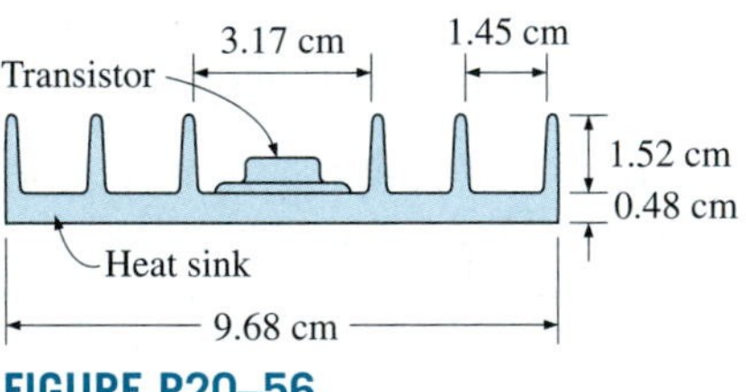

FIGURE P20–56

20–57 Reconsider the heat sink in Prob. 20–56. In order to enhance heat transfer, a shroud (a thin rectangular metal plate) whose surface area is equal to the base area of the heat

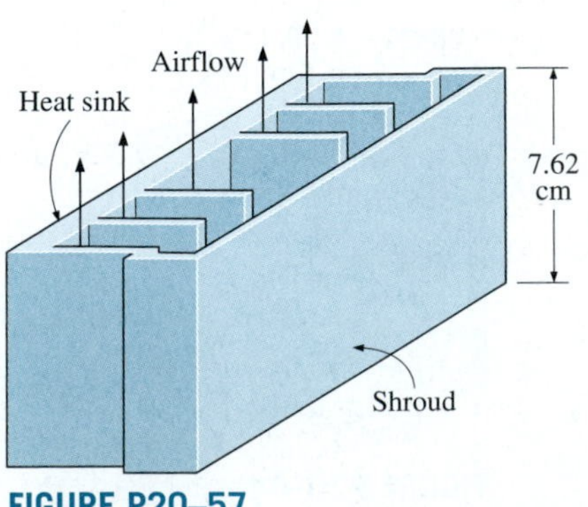

FIGURE P20–57

sink is placed very close to the tips of the fins such that the interfin spaces are converted into rectangular channels. The base temperature of the heat sink in this case was measured to be 108°C. Noting that the shroud loses heat to the ambient air from both sides, determine the average natural convection heat transfer coefficient in this shrouded case. (For complete details, see Çengel and Zing, 1987.)

20–58E A 6-in-wide and 8-in-high vertical hot surface in 78°F air is to be cooled by a heat sink with equally spaced fins of rectangular profile. The fins are 0.08 in thick and 8 in long in the vertical direction and have a height of 1.2 in from the base. Determine the optimum fin spacing and the rate of heat transfer by natural convection from the heat sink if the base temperature is 180°F.

20–59E EES Reconsider Prob. 20–58E. Using EES (or other) software, investigate the effect of the length of the fins in the vertical direction on the optimum fin spacing and the rate of heat transfer by natural convection. Let the fin length vary from 2 in to 10 in. Plot the optimum fin spacing and the rate of convection heat transfer as a function of the fin length, and discuss the results.

20–60 A 15-cm-wide and 18-cm-high vertical hot surface in 25°C air is to be cooled by a heat sink with equally spaced fins of rectangular profile. The fins are 0.1 cm thick and 18 cm long in the vertical direction. Determine the optimum fin height and the rate of heat transfer by natural convection from the heat sink if the base temperature is 85°C.

The criteria for optimum fin height H in the literature is given by $H = \sqrt{hA_c/pk}$. Take the thermal conductivity of fin material to be 177 W/m · °C

Natural Convection inside Enclosures

20–61C The upper and lower compartments of a well-insulated container are separated by two parallel sheets of glass with an air space between them. One of the compartments is to be filled with a hot fluid and the other with a cold fluid. If it is desired that heat transfer between the two compartments be minimal, would you recommend putting the hot fluid into the upper or the lower compartment of the container? Why?

20–62C Someone claims that the air space in a double-pane window enhances the heat transfer from a house because of the natural convection currents that occur in the air space and recommends that the double-pane window be replaced by a single sheet of glass whose thickness is equal to the sum of the thicknesses of the two glasses of the double-pane window to save energy. Do you agree with this claim?

20–63C Consider a double-pane window consisting of two glass sheets separated by a 1-cm-wide air space. Someone suggests inserting a thin vinyl sheet in the middle of the two glasses to form two 0.5-cm-wide compartments in the window in order to reduce natural convection heat transfer through the window. From a heat transfer point of view, would you be in favor of this idea to reduce heat losses through the window?

20–64C What does the effective conductivity of an enclosure represent? How is the ratio of the effective conductivity to thermal conductivity related to the Nusselt number?

20–65 Show that the thermal resistance of a rectangular enclosure can be expressed as $R = L_c/(Ak\ \text{Nu})$, where k is the thermal conductivity of the fluid in the enclosure.

20–66 Determine the U-factors for the center-of-glass section of a double-pane window and a triple-pane window. The heat transfer coefficients on the inside and outside surfaces are 6 and 25 W/m^2 · °C, respectively. The thickness of the air layer is 1.5 cm and there are two such air layers in triple-pane window. The Nusselt number across an air layer is estimated to be 1.2. Take the thermal conductivity of air to be 0.025 W/m · °C and neglect the thermal resistance of glass sheets. Also, assume that the effect of radiation through the air space is of the same magnitude as the convection.

Considering that about 70 percent of total heat transfer through a window is due to center-of-glass section, estimate the percentage decrease in total heat transfer when triple-pane window is used in place of double-pane window.

20–67 A vertical 1.5-m-high and 3.0-m-wide enclosure consists of two surfaces separated by a 0.4-m air gap at atmospheric pressure. If the surface temperatures across the air gap are measured to be 280 K and 336 K and the surface emissivities to be 0.15 and 0.90, determine the fraction of heat transferred through the enclosure by radiation. *Answer:* 0.30

20–68E A vertical 4-ft-high and 6-ft-wide double-pane window consists of two sheets of glass separated by a 1-in air gap at atmospheric pressure. If the glass surface temperatures across the air gap are measured to be 65°F and 40°F, determine the rate of heat transfer through the window by (*a*) natural convection and (*b*) radiation. Also, determine the R-value of insulation of this window such that multiplying the inverse of the R-value by the surface area and the temperature difference gives the total rate of heat transfer through the window. The effective emissivity for use in radiation calculations between two large parallel glass plates can be taken to be 0.82.

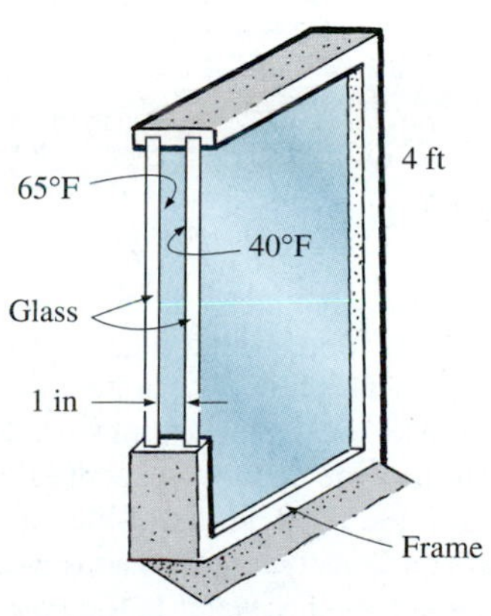

FIGURE P20–68E

20–69E Reconsider Prob. 20–68E. Using EES (or other) software, investigate the effect of the air gap thickness on the rates of heat transfer by natural convection and radiation, and the *R*-value of insulation. Let the air gap thickness vary from 0.2 in to 2.0 in. Plot the rates of heat transfer by natural convection and radiation, and the *R*-value of insulation as a function of the air gap thickness, and discuss the results.

20–70 Two concentric spheres of diameters 15 cm and 25 cm are separated by air at 1 atm pressure. The surface temperatures of the two spheres enclosing the air are $T_1 = 350$ K and $T_2 = 275$ K, respectively. Determine the rate of heat transfer from the inner sphere to the outer sphere by natural convection.

20–71 Reconsider Prob. 20–70. Using EES (or other) software, plot the rate of natural convection heat transfer as a function of the hot surface temperature of the sphere as the temperature varies from 300 K to 500 K, and discuss the results.

20–72 Flat-plate solar collectors are often tilted up toward the sun in order to intercept a greater amount of direct solar radiation. The tilt angle from the horizontal also affects the rate of heat loss from the collector. Consider a 1.5-m-high and 3-m-wide solar collector that is tilted at an angle θ from the horizontal. The back side of the absorber is heavily insulated. The absorber plate and the glass cover, which are spaced 2.5 cm from each other, are maintained at temperatures of 80°C and 40°C, respectively. Determine the rate of heat loss from the absorber plate by natural convection for $\theta = 0°, 30°$, and 90°.

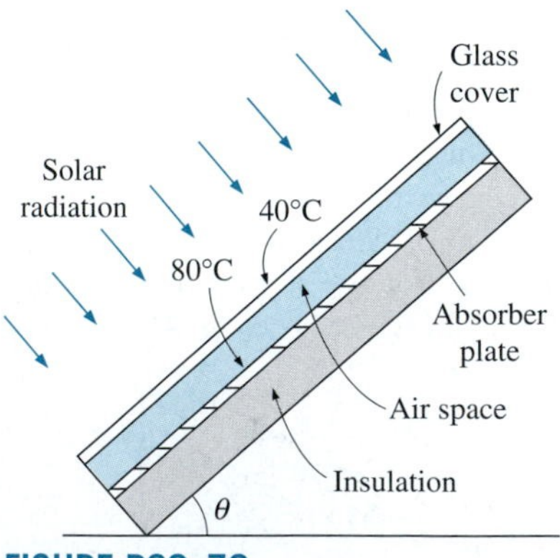

FIGURE P20–72

20–73 A simple solar collector is built by placing a 5-cm-diameter clear plastic tube around a garden hose whose outer diameter is 1.6 cm. The hose is painted black to maximize solar absorption, and some plastic rings are used to keep the spacing between the hose and the clear plastic cover constant. During a clear day, the temperature of the hose is measured to be 65°C, while the ambient air temperature is 26°C. Determine the rate of heat loss from the water in the hose per meter of its length by natural convection. Also, discuss how the performance of this solar collector can be improved. *Answer:* 8.2 W

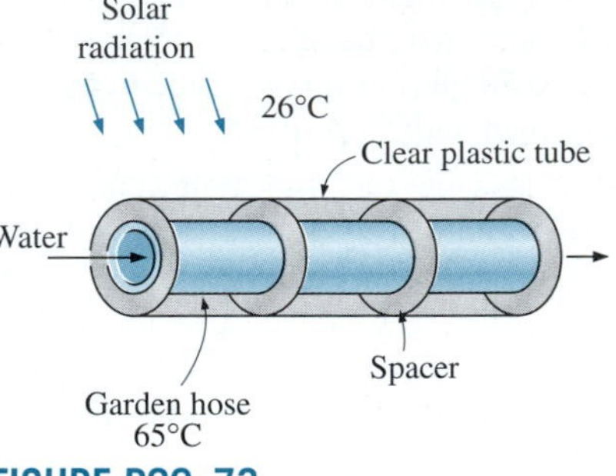

FIGURE P20–73

20–74 Reconsider Prob. 20–73. Using EES (or other) software, plot the rate of heat loss from the water by natural convection as a function of the ambient air temperature as the temperature varies from 4°C to 40°C, and discuss the results.

20–75 A vertical 1.3-m-high, 2.8-m-wide double-pane window consists of two layers of glass separated by a 2.2-cm air gap at atmospheric pressure. The room temperature is 26°C while the inner glass temperature is 18°C. Disregarding radiation heat transfer, determine the temperature of the outer glass layer and the rate of heat loss through the window by natural convection.

20–76 Consider two concentric horizontal cylinders of diameters 55 cm and 65 cm, and length 125 cm. The surfaces of the inner and outer cylinders are maintained at 54°C and 106°C, respectively. Determine the rate of heat transfer between the cylinders by natural convection if the annular space is filled with (*a*) water and (*b*) air.

Review Problems

20–77 A 10-cm-diameter and 10-m-long cylinder with a surface temperature of 10°C is placed horizontally in air at 40°C. Calculate the steady rate of heat transfer for the cases of (*a*) free-stream air velocity of 10 m/s due to normal winds and (*b*) no winds and thus a free stream velocity of zero.

20–78 A spherical vessel, with 30.0-cm outside diameter, is used as a reactor for a slow endothermic reaction. The vessel is completely submerged in a large water-filled tank, held at a constant temperature of 30°C. The outside surface temperature of the vessel is 20°C. Calculate the rate of heat transfer in steady operation for the following cases: (*a*) the water in the tank is still, (*b*) the water in the tank is still (as in a part a), however, the buoyancy force caused by the difference in water density is assumed to be negligible, and (*c*) the water in the tank is circulated at an average velocity of 20 cm/s.

20–79 A vertical cylindrical pressure vessel is 1.0 m in diameter and 3.0 m in height. Its outside average wall temperature is 60°C, while the surrounding air is at 0°C. Calculate the rate of heat loss from the vessel's cylindrical surface when there is (*a*) no wind and (*b*) a crosswind of 20 km/h.

20–80 Consider a solid sphere, 50 cm in diameter embedded with electrical heating elements such that its surface temperature is always maintained constant at 60°C. The sphere is placed in a large pool of oil held at a constant temperature of 20°C. Using the oil properties tabulated below, calculate the rate of heat transfer in steady operation for each of the following scenarios.

(*a*) Heat flow in the oil is assumed to occur only by conduction.

(*b*) The oil is circulated across the sphere at an average velocity of 1.50 m/s.

(*c*) The pump causing the oil circulation in part (*b*) has broken down.

T, °C	k, W/m · K	ρ, kg/m³	c_p, J/kg · K	μ, mPa · s	β, K⁻¹
20.0	0.22	888.0	1880	10.0	0.00070
40.0	0.21	876.0	1965	7.0	0.00070
60.0	0.20	864.0	2050	4.0	0.00070

20–81E A 0.1-W small cylindrical resistor mounted on a lower part of a vertical circuit board is 0.3 in long and has a diameter of 0.2 in. The view of the resistor is largely blocked by another circuit board facing it, and the heat transfer through the connecting wires is negligible. The air is free to flow through the large parallel flow passages between the boards as a result of natural convection currents. If the air temperature at the vicinity of the resistor is 120°F, determine the approximate surface temperature of the resistor. *Answer:* 212°F

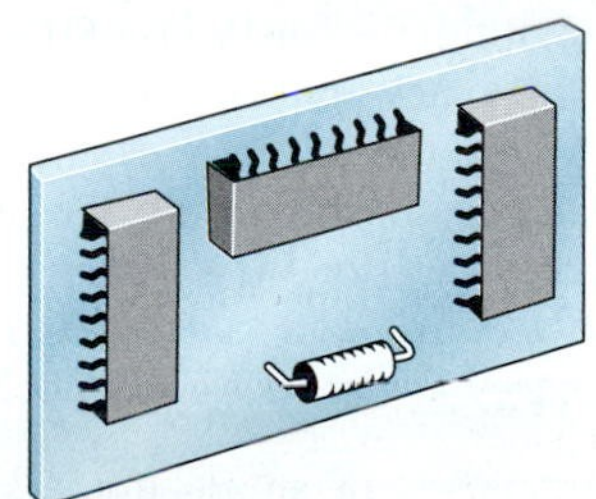

FIGURE P20–81E

20–82 An ice chest whose outer dimensions are 30 cm × 40 cm × 40 cm is made of 3-cm-thick Styrofoam (k = 0.033 W/m · °C). Initially, the chest is filled with 30 kg of ice at 0°C, and the inner surface temperature of the ice chest can be taken to be 0°C at all times. The heat of fusion of water at 0°C is 333.7 kJ/kg, and the surrounding ambient air is at 20°C. Disregarding any heat transfer from the 40 cm × 40 cm base of the ice chest, determine how long it will take for the ice in the chest to melt completely if the ice chest is subjected to (*a*) calm air and (*b*) winds at 50 km/h. Assume the heat transfer coefficient on the front, back, and top surfaces to be the same as that on the side surfaces.

20–83 An electronic box that consumes 200 W of power is cooled by a fan blowing air into the box enclosure. The dimensions of the electronic box are 15 cm × 50 cm × 50 cm, and all surfaces of the box are exposed to the ambient except the base surface. Temperature measurements indicate that the box is at an average temperature of 32°C when the ambient temperature and the temperature of the surrounding walls are 25°C. If the emissivity of the outer surface of the box is 0.75, determine the fraction of the heat lost from the outer surfaces of the electronic box.

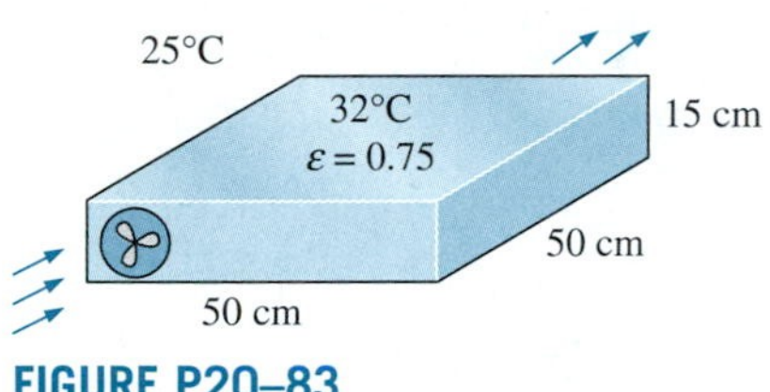

FIGURE P20–83

20–84 A 6-m-internal-diameter spherical tank made of 1.5-cm-thick stainless steel (k = 15 W/m · °C) is used to store iced water at 0°C in a room at 20°C. The walls of the room are also at 20°C. The outer surface of the tank is black (emissivity ε = 1), and heat transfer between the outer surface of the tank and the surroundings is by natural convection and radiation. Assuming the entire steel tank to be at 0°C and thus the thermal resistance of the tank to be negligible, determine (*a*) the rate of heat transfer to the iced water in the tank and (*b*) the amount of ice at 0°C that melts during a 24-h period. The heat of fusion of water is 333.7 kJ/kg *Answers:* (*a*) 15.4 kW, (*b*) 3988 kg

20–85 Consider a 1.2-m-high and 2-m-wide double-pane window consisting of two 3-mm-thick layers of glass (k = 0.78 W/m · °C) separated by a 3-cm-wide air space. Determine the steady rate of heat transfer through this window and the temperature of its inner surface for a day during which the room is maintained at 20°C while the temperature of the outdoors is 0°C. Take the heat transfer coefficients on the inner and outer surfaces of the window to be h_1 = 10 W/m² · °C and h_2 = 25 W/m² · °C and disregard any heat transfer by radiation.

20–86 An electric resistance space heater is designed such that it resembles a rectangular box 50 cm high, 80 cm long, and 15 cm wide filled with 45 kg of oil. The heater is to be placed against a wall, and thus heat transfer from its back surface is negligible. The surface temperature of the heater is not to exceed 75°C in a room at 25°C for safety considerations. Disregarding heat transfer from the bottom and top surfaces of the heater in anticipation that the top surface will be used as a shelf, determine the power rating of the heater in W. Take the emissivity of the outer surface of the heater to be 0.8 and the average temperature of the ceiling and wall surfaces to be the same as the room air temperature.

Also, determine how long it will take for the heater to reach steady operation when it is first turned on (i.e., for the oil temperature to rise from 25°C to 75°C). State your assumptions in the calculations.

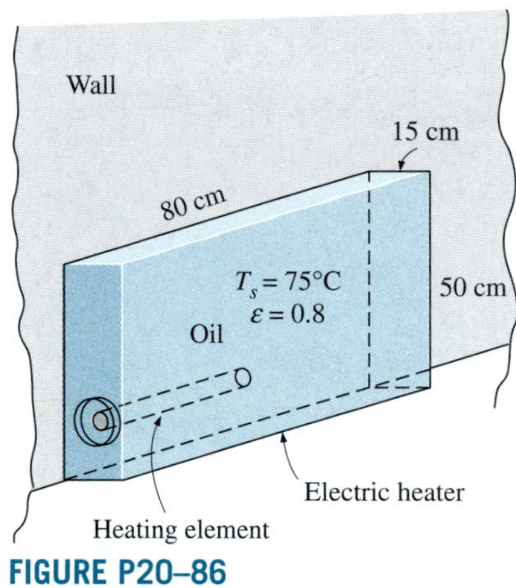

FIGURE P20–86

20–87 Skylights or "roof windows" are commonly used in homes and manufacturing facilities since they let natural light in during day time and thus reduce the lighting costs. However, they offer little resistance to heat transfer, and large amounts of energy are lost through them in winter unless they are equipped with a motorized insulating cover that can be used in cold weather and at nights to reduce heat losses. Consider a 1-m-wide and 2.5-m-long horizontal skylight on the roof of a house that is kept at 20°C. The glazing of the skylight is made of a single layer of 0.5-cm-thick glass (k = 0.78 W/m · °C and ε = 0.9). Determine the rate of heat loss through the skylight when the air temperature outside is −10°C and the effective sky temperature is −30°C. Compare your result with the rate of heat loss through an equivalent surface area of the roof that has a common R-5.34 construction in SI units (i.e., a thickness-to-effective-thermal-conductivity ratio of 5.34 m² · °C/W).

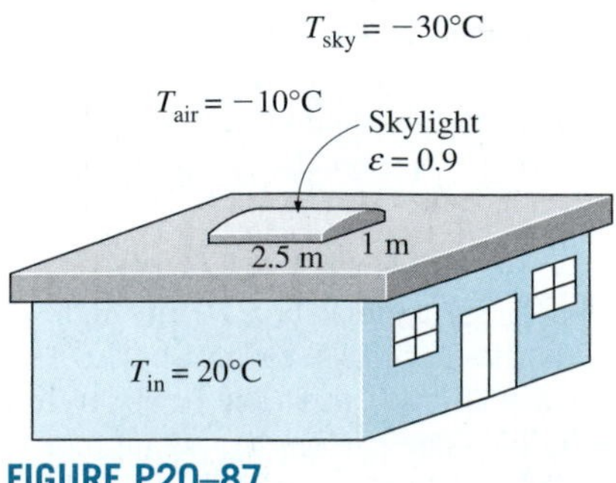

FIGURE P20–87

20–88 A solar collector consists of a horizontal copper tube of outer diameter 5 cm enclosed in a concentric thin glass tube of 9 cm diameter. Water is heated as it flows through the tube, and the annular space between the copper and glass tube is filled with air at 1 atm pressure. During a clear day, the temperatures of the tube surface and the glass cover are measured to be 60°C and 32°C, respectively. Determine the rate of heat loss from the collector by natural convection per meter length of the tube. *Answer: 17.4 W*

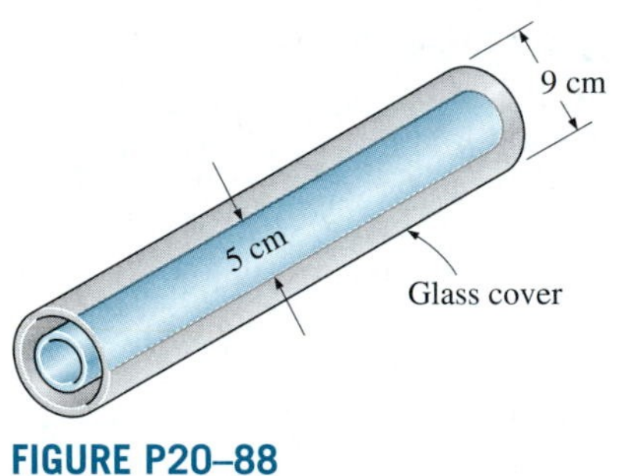

FIGURE P20–88

20–89 A solar collector consists of a horizontal aluminum tube of outer diameter 5 cm enclosed in a concentric thin glass tube of 7 cm diameter. Water is heated as it flows through the aluminum tube, and the annular space between the aluminum and glass tubes is filled with air at 1 atm pressure. The pump circulating the water fails during a clear day, and the water temperature in the tube starts rising. The aluminum tube absorbs solar radiation at a rate of 20 W per meter length, and the temperature of the ambient air outside is 30°C. Approximating the surfaces of the tube and the glass cover as being black (emissivity ε = 1) in radiation calculations and taking the effective sky temperature to be 20°C, determine the temperature of the aluminum tube when equilibrium is established (i.e., when the net heat loss from the tube by convection and radiation equals the amount of solar energy absorbed by the tube).

20–90E The components of an electronic system dissipating 180 W are located in a 4-ft-long horizontal duct whose cross section is 6 in × 6 in. The components in the duct are cooled by forced air, which enters at 85°F at a rate of 22 cfm and leaves at 100°F. The surfaces of the sheet metal duct are not painted, and thus radiation heat transfer from the outer surfaces is negligible. If the ambient air temperature is 80°F, determine (*a*) the heat transfer from the outer surfaces of the duct to the ambient air by natural convection and (*b*) the average temperature of the duct.

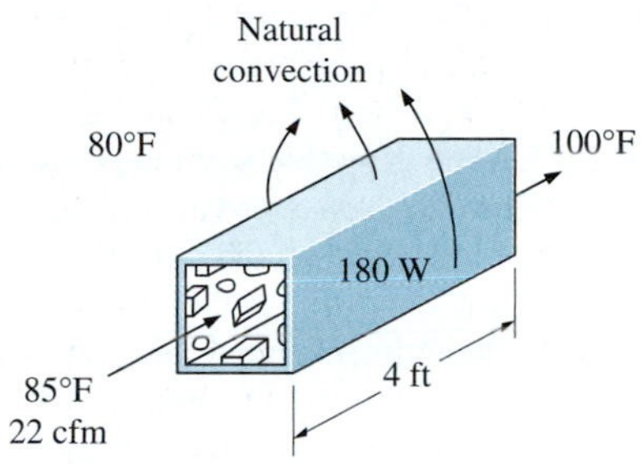

FIGURE P20–90E

20–91E Repeat Prob. 20–90E for a circular horizontal duct of diameter 4 in.

20–92E Repeat Prob. 20–90E assuming the fan fails and thus the entire heat generated inside the duct must be rejected to the ambient air by natural convection through the outer surfaces of the duct.

20–93 Consider a cold aluminum canned drink that is initially at a uniform temperature of 5°C. The can is 12.5 cm high and has a diameter of 6 cm. The emissivity of the outer surface of the can is 0.6. Disregarding any heat transfer from the bottom surface of the can, determine how long it will take for the average temperature of the drink to rise to 7°C if the surrounding air and surfaces are at 25°C. *Answer:* 12.1 min

20–94 Consider a 2-m-high electric hot-water heater that has a diameter of 40 cm and maintains the hot water at 60°C. The tank is located in a small room at 20°C whose walls and the ceiling are at about the same temperature. The tank is placed in a 44-cm-diameter sheet metal shell of negligible thickness, and the space between the tank and the shell is filled with foam insulation. The average temperature and emissivity of the outer surface of the shell are 40°C and 0.7, respectively. The price of electricity is \$0.08/kWh. Hot-water tank insulation kits large enough to wrap the entire tank are available on the market for about \$60. If such an insulation is installed on this water tank by the home owner himself, how long will it take for this additional insulation to pay for itself? Disregard any heat loss from the top and bottom surfaces, and assume the insulation to reduce the heat losses by 80 percent.

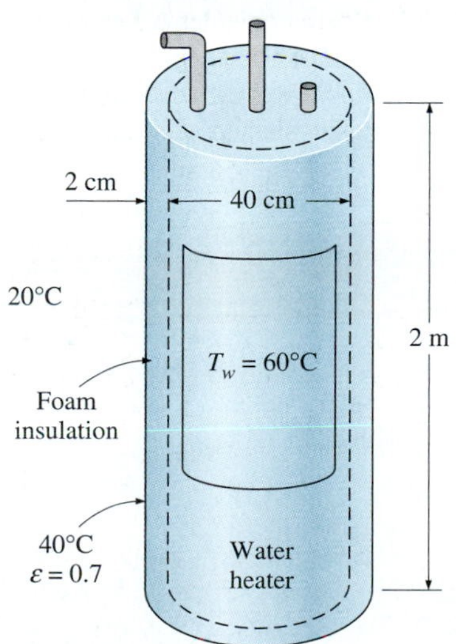

FIGURE P20–94

20–95 During a plant visit, it was observed that a 1.5-m-high and 1-m-wide section of the vertical front section of a natural gas furnace wall was too hot to touch. The temperature measurements on the surface revealed that the average temperature of the exposed hot surface was 110°C, while the temperature of the surrounding air was 25°C. The surface appeared to be oxidized, and its emissivity can be taken to be 0.7. Taking the temperature of the surrounding surfaces to be 25°C also, determine the rate of heat loss from this furnace.

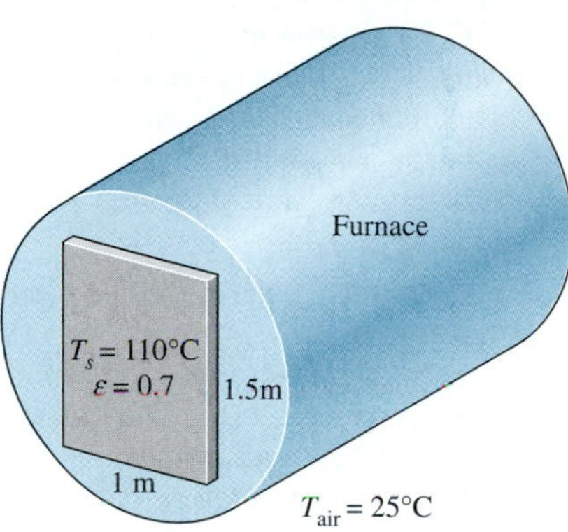

FIGURE P20–95

The furnace has an efficiency of 79 percent, and the plant pays \$1.20 per therm of natural gas. If the plant operates 10 h a day, 310 days a year, and thus 3100 h a year, determine the annual cost of the heat loss from this vertical hot surface on the front section of the furnace wall.

20–96 A group of 25 power transistors, dissipating 1.5 W each, are to be cooled by attaching them to a black-anodized

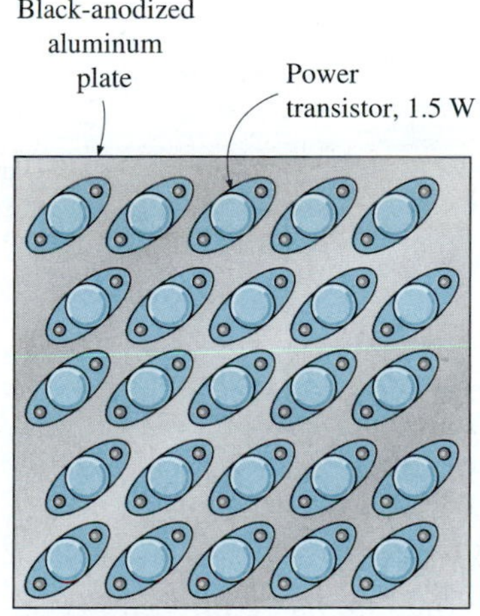

FIGURE P20–96

square aluminum plate and mounting the plate on the wall of a room at 30°C. The emissivity of the transistor and the plate surfaces is 0.9. Assuming the heat transfer from the back side of the plate to be negligible and the temperature of the surrounding surfaces to be the same as the air temperature of the room, determine the size of the plate if the average surface temperature of the plate is not to exceed 50°C. *Answer:* 43 cm × 43 cm

20–97 Repeat Prob. 20–96 assuming the plate to be positioned horizontally with (*a*) transistors facing up and (*b*) transistors facing down.

20–98E Hot water is flowing at an average velocity of 4 ft/s through a cast iron pipe (k = 30 Btu/h · ft · °F) whose inner and outer diameters are 1.0 in and 1.2 in, respectively. The pipe passes through a 50-ft-long section of a basement whose temperature is 60°F. The emissivity of the outer surface of the pipe is 0.5, and the walls of the basement are also at about 60°F. If the inlet temperature of the water is 150°F and the heat transfer coefficient on the inner surface of the pipe is 30 Btu/h · ft^2 · °F, determine the temperature drop of water as it passes through the basement.

20–99 Consider a flat-plate solar collector placed horizontally on the flat roof of a house. The collector is 1.5 m wide and 6 m long, and the average temperature of the exposed surface of the collector is 42°C. Determine the rate of heat loss from the collector by natural convection during a calm day when the ambient air temperature is 8°C. Also, determine the heat loss by radiation by taking the emissivity of the collector surface to be 0.9 and the effective sky temperature to be −15°C. *Answers:* 1750 W, 2490 W

20–100 Solar radiation is incident on the glass cover of a solar collector at a rate of 650 W/m^2. The glass transmits 88 percent of the incident radiation and has an emissivity of 0.90. The hot water needs of a family in summer can be met completely by a collector 1.5 m high and 2 m wide, and tilted 40° from the horizontal. The temperature of the glass cover is measured to be 40°C on a calm day when the surrounding air temperature is 20°C. The effective sky temperature for radiation exchange between the glass cover and the open sky is −40°C. Water enters the tubes attached to the absorber plate at a rate of 1 kg/min. Assuming the back surface of the absorber plate to be heavily insulated and the only heat loss occurs through the glass cover, determine (*a*) the total rate of heat loss from the collector; (*b*) the collector efficiency, which is the ratio of the amount of heat transferred to the water to the solar energy incident on the collector; and (*c*) the temperature rise of water as it flows through the collector.

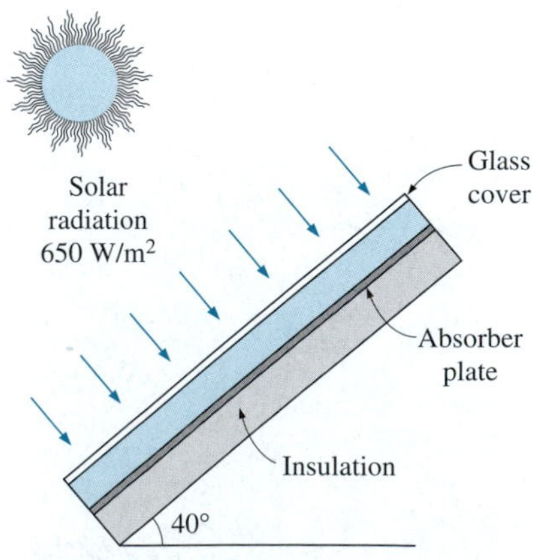

FIGURE P20–100

Design and Essay Problems

20–101 Write a computer program to optimize the spacing between the two glasses of a double-pane window. Assume the spacing is filled with dry air at atmospheric pressure. The program should evaluate the recommended practical value of the spacing to minimize the heat losses and list it when the size of the window (the height and the width) and the temperatures of the two glasses are specified.

20–102 Contact a manufacturer of aluminum heat sinks and obtain their product catalog for cooling electronic components by natural convection and radiation. Write an essay on how to select a suitable heat sink for an electronic component when its maximum power dissipation and maximum allowable surface temperature are specified.

20–103 The top surfaces of practically all flat-plate solar collectors are covered with glass in order to reduce the heat losses from the absorber plate underneath. Although the glass cover reflects or absorbs about 15 percent of the incident solar radiation, it saves much more from the potential heat losses from the absorber plate, and thus it is considered to be an essential part of a well-designed solar collector. Inspired by the energy efficiency of double-pane windows, someone proposes to use double glazing on solar collectors instead of a single glass. Investigate if this is a good idea for the town in which you live. Use local weather data and base your conclusion on heat transfer analysis and economic considerations.

Chapter 21

RADIATION HEAT TRANSFER

So far, we have considered the conduction and convection modes of heat transfer, which are related to the nature of the media involved and the presence of fluid motion, among other things. We now turn our attention to the third mechanism of heat transfer: *radiation,* which is characteristically different from the other two.

We start this chapter with a discussion of *electromagnetic waves* and the *electromagnetic spectrum,* with particular emphasis on *thermal radiation.* Then we introduce the idealized *blackbody, blackbody radiation, and blackbody radiation function,* together with the *Stefan–Boltzmann law, Planck's law,* and *Wien's displacement law.* Various radiation fluxes such as *emissive power, irradiation,* and *radiosity* are defined. This is followed by a discussion of radiative properties of materials such as *emissivity, absorptivity, reflectivity,* and *transmissivity* and their dependence on wavelength and temperature. The *greenhouse effect* is presented as an example of the consequences of the wavelength dependence of radiation properties.

We continue with a discussion of view factors and the rules associated with them. View factor expressions and charts for some common configurations are given, and the crossed-strings method is presented. Finally, we discuss radiation heat transfer, first between black surfaces and then between nonblack surfaces using the radiation network approach.

Objectives

The objectives of this chapter are to:

- Classify electromagnetic radiation, and identify thermal radiation,
- Understand the idealized blackbody, and calculate the total and spectral blackbody emissive power,
- Calculate the fraction of radiation emitted in a specified wavelength band using the blackbody radiation functions,
- Develop a clear understanding of the properties emissivity, absorptivity, reflectivity, and transmissivity on spectral and total basis,
- Apply Kirchhoff law's to determine the absorptivity of a surface when its emissivity is known,
- Define view factor, and understand its importance in radiation heat transfer calculations,
- Calculate radiation heat transfer between black surfaces,
- Obtain relations for net rate of radiation heat transfer between the surfaces of a two-zone enclosure, including two large parallel plates, two long concentric cylinders, and two concentric spheres,
- Understand radiation heat transfer in three-surface enclosures.

21–1 ▪ INTRODUCTION

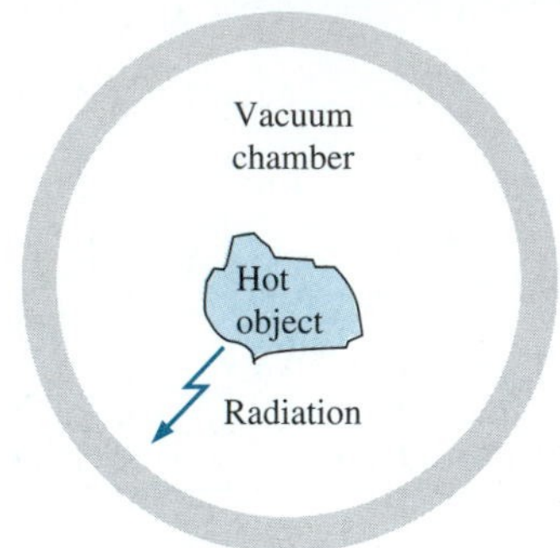

FIGURE 21–1
A hot object in a vacuum chamber loses heat by radiation only.

Consider a hot object that is suspended in an evacuated chamber whose walls are at room temperature (Fig. 21–1). The hot object will eventually cool down and reach thermal equilibrium with its surroundings. That is, it will lose heat until its temperature reaches the temperature of the walls of the chamber. Heat transfer between the object and the chamber could not have taken place by conduction or convection, because these two mechanisms cannot occur in a vacuum. Therefore, heat transfer must have occurred through another mechanism that involves the emission of the internal energy of the object. This mechanism is *radiation*.

Radiation differs from the other two heat transfer mechanisms in that it does not require the presence of a material medium to take place. In fact, energy transfer by radiation is fastest (at the speed of light) and it suffers no attenuation in a *vacuum*. Also, radiation transfer occurs in solids as well as liquids and gases. In most practical applications, all three modes of heat transfer occur concurrently at varying degrees. But heat transfer through an evacuated space can occur only by radiation. For example, the energy of the sun reaches the earth by radiation.

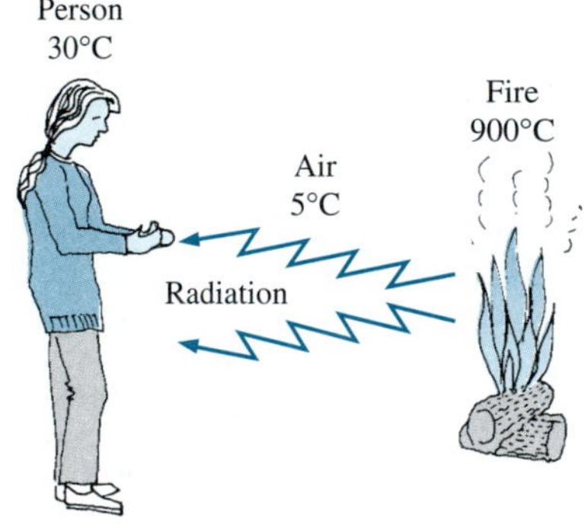

FIGURE 21–2
Unlike conduction and convection, heat transfer by radiation can occur between two bodies, even when they are separated by a medium colder than both.

You will recall that heat transfer by conduction or convection takes place in the direction of decreasing temperature; that is, from a high-temperature medium to a lower-temperature one. It is interesting that radiation heat transfer can occur between two bodies separated by a medium colder than both bodies (Fig. 21–2). For example, solar radiation reaches the surface of the earth after passing through cold air layers at high altitudes. Also, the radiation-absorbing surfaces inside a greenhouse reach high temperatures even when its plastic or glass cover remains relatively cool.

The theoretical foundation of radiation was established in 1864 by physicist James Clerk Maxwell, who postulated that accelerated charges or changing electric currents give rise to electric and magnetic fields. These rapidly moving fields are called **electromagnetic waves** or **electromagnetic radiation**, and they represent the energy emitted by matter as a result of the changes in the electronic configurations of the atoms or molecules. In 1887, Heinrich Hertz experimentally demonstrated the existence of such waves. Electromagnetic waves transport energy just like other waves, and all electromagnetic waves travel at the *speed of light* in a vacuum, which is $c_0 = 2.9979 \times 10^8$ m/s. Electromagnetic waves are characterized by their *frequency* ν or *wavelength* λ. These two properties in a medium are related by

$$\lambda = \frac{c}{\nu} \tag{21–1}$$

where c is the speed of propagation of a wave in that medium. The speed of propagation in a medium is related to the speed of light in a vacuum by $c = c_0/n$, where n is the *index of refraction* of that medium. The refractive index is essentially unity for air and most gases, about 1.5 for glass, and 1.33 for water. The commonly used unit of wavelength is the *micrometer* (μm) or micron, where 1 μm $= 10^{-6}$ m. Unlike the wavelength and the speed of propagation, the frequency of an electromagnetic wave depends only on the source and is independent of the medium through which the wave travels. The *frequency* (the number of oscillations per second) of an electromagnetic wave can range from less than a million Hz to a septillion Hz or higher, depending on the source. Note from Eq. 21–1 that the wavelength and the frequency of electromagnetic radiation are inversely proportional.

It has proven useful to view electromagnetic radiation as the propagation of a collection of discrete packets of energy called **photons** or **quanta**, as proposed by Max Planck in 1900 in conjunction with his *quantum theory*. In this view, each photon of frequency ν is considered to have an energy of

$$e = h\nu = \frac{hc}{\lambda} \tag{21–2}$$

where $h = 6.626069 \times 10^{-34}$ J · s is *Planck's constant*. Note from the second part of Eq. 21–2 that the energy of a photon is inversely proportional to its wavelength. Therefore, shorter-wavelength radiation possesses larger photon energies. It is no wonder that we try to avoid very-short-wavelength radiation such as gamma rays and X-rays since they are highly destructive.

21–2 ▪ THERMAL RADIATION

Although all electromagnetic waves have the same general features, waves of different wavelength differ significantly in their behavior. The electromagnetic radiation encountered in practice covers a wide range of wavelengths, varying from less than 10^{-10} μm for cosmic rays to more than 10^{10} μm for electrical power waves. The **electromagnetic spectrum** also includes gamma rays, X-rays, ultraviolet radiation, visible light, infrared radiation, thermal radiation, microwaves, and radio waves, as shown in Fig. 21–3.

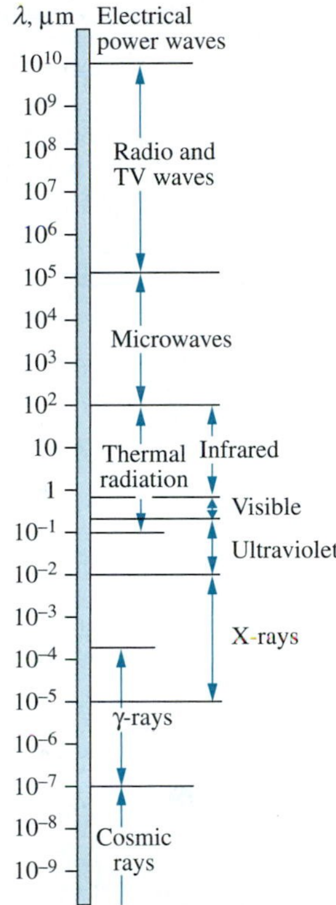

FIGURE 21–3
The electromagnetic wave spectrum.

Different types of electromagnetic radiation are produced through various mechanisms. For example, *gamma rays* are produced by nuclear reactions, *X-rays* by the bombardment of metals with high-energy electrons, *microwaves* by special types of electron tubes such as klystrons and magnetrons, and *radio waves* by the excitation of some crystals or by the flow of alternating current through electric conductors.

The short-wavelength gamma rays and X-rays are primarily of concern to nuclear engineers, while the long-wavelength microwaves and radio waves are of concern to electrical engineers. The type of electromagnetic radiation that is pertinent to heat transfer is the **thermal radiation** emitted as a result of energy transitions of molecules, atoms, and electrons of a substance. Temperature is a measure of the strength of these activities at the microscopic level, and the rate of thermal radiation emission increases with increasing temperature. Thermal radiation is continuously emitted by all matter whose temperature is above absolute zero. That is, everything around us such as walls, furniture, and our friends constantly emits (and absorbs) radiation (Fig. 21–4). Thermal radiation is also defined as the portion of the electromagnetic spectrum that extends from about 0.1 to 100 μm, since the radiation emitted by bodies due to their temperature falls almost entirely into this wavelength range. Thus, thermal radiation includes the entire visible and infrared (IR) radiation as well as a portion of the ultraviolet (UV) radiation.

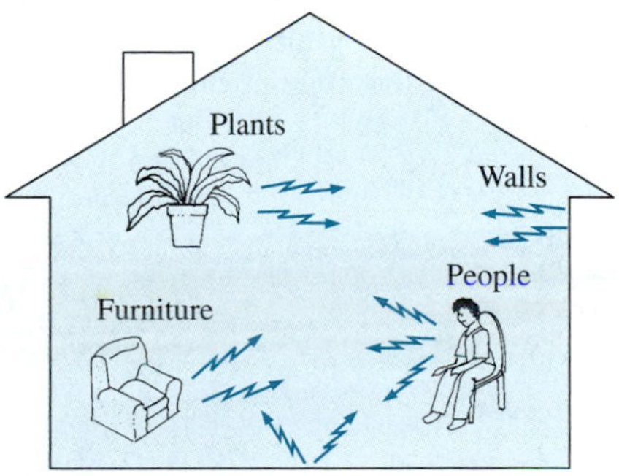

FIGURE 21–4
Everything around us constantly emits thermal radiation.

What we call **light** is simply the *visible* portion of the electromagnetic spectrum that lies between 0.40 and 0.76 μm. Light is characteristically no different than other electromagnetic radiation, except that it happens to trig-

TABLE 21–1

The wavelength ranges of different colors

Color	Wavelength band
Violet	0.40–0.44 μm
Blue	0.44–0.49 μm
Green	0.49–0.54 μm
Yellow	0.54–0.60 μm
Orange	0.60–0.67 μm
Red	0.63–0.76 μm

ger the sensation of seeing in the human eye. Light, or the visible spectrum, consists of narrow bands of color from violet (0.40–0.44 μm) to red (0.63–0.76 μm), as shown in Table 21–1.

A body that emits some radiation in the visible range is called a light source. The sun is obviously our primary light source. The electromagnetic radiation emitted by the sun is known as **solar radiation**, and nearly all of it falls into the wavelength band 0.3–3 μm. Almost *half* of solar radiation is light (i.e., it falls into the visible range), with the remaining being ultraviolet and infrared.

The radiation emitted by bodies at room temperature falls into the **infrared** region of the spectrum, which extends from 0.76 to 100 μm. Bodies start emitting noticeable visible radiation at temperatures above 800 K. The tungsten filament of a lightbulb must be heated to temperatures above 2000 K before it can emit any significant amount of radiation in the visible range.

The **ultraviolet** radiation includes the low-wavelength end of the thermal radiation spectrum and lies between the wavelengths 0.01 and 0.40 μm. Ultraviolet rays are to be avoided since they can kill microorganisms and cause serious damage to humans and other living beings. About 12 percent of solar radiation is in the ultraviolet range, and it would be devastating if it were to reach the surface of the earth. Fortunately, the ozone (O_3) layer in the atmosphere acts as a protective blanket and absorbs most of this ultraviolet radiation. The ultraviolet rays that remain in sunlight are still sufficient to cause serious sunburns to sun worshippers, and prolonged exposure to direct sunlight is the leading cause of skin cancer, which can be lethal. Recent discoveries of "holes" in the ozone layer have prompted the international community to ban the use of ozone-destroying chemicals such as the refrigerant Freon-12 in order to save the earth. Ultraviolet radiation is also produced artificially in fluorescent lamps for use in medicine as a bacteria killer and in tanning parlors as an artificial tanner.

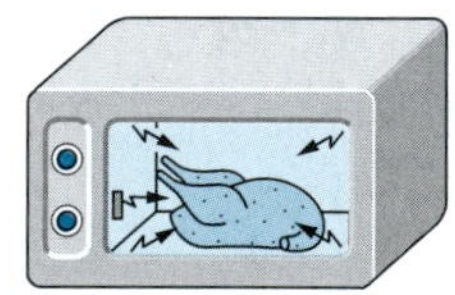

FIGURE 21–5
Food is heated or cooked in a microwave oven by absorbing the electromagnetic radiation energy generated by the magnetron of the oven.

Microwave ovens utilize electromagnetic radiation in the **microwave** region of the spectrum generated by microwave tubes called *magnetrons*. Microwaves in the range of 10^2–10^5 μm are very suitable for use in cooking since they are *reflected* by metals, *transmitted* by glass and plastics, and *absorbed* by food (especially water) molecules. Thus, the electric energy converted to radiation in a microwave oven eventually becomes part of the internal energy of the food. The fast and efficient cooking of microwave ovens has made them one of the essential appliances in modern kitchens (Fig. 21–5).

Radars and cordless telephones also use electromagnetic radiation in the microwave region. The wavelength of the electromagnetic waves used in radio and TV broadcasting usually ranges between 1 and 1000 m in the **radio wave** region of the spectrum.

In heat transfer studies, we are interested in the energy emitted by bodies because of their temperature only. Therefore, we limit our consideration to *thermal radiation*, which we simply call *radiation*. The relations developed below are restricted to thermal radiation only and may not be applicable to other forms of electromagnetic radiation.

The electrons, atoms, and molecules of all solids, liquids, and gases above absolute zero temperature are constantly in motion, and thus radiation is constantly emitted, as well as being absorbed or transmitted throughout the entire volume of matter. That is, radiation is a **volumetric phenomenon**.

However, for opaque (nontransparent) solids such as metals, wood, and rocks, radiation is considered to be a **surface phenomenon**, since the radiation emitted by the interior regions can never reach the surface, and the radiation incident on such bodies is usually absorbed within a few microns from the surface (Fig. 21–6). Note that the radiation characteristics of surfaces can be changed completely by applying thin layers of coatings on them.

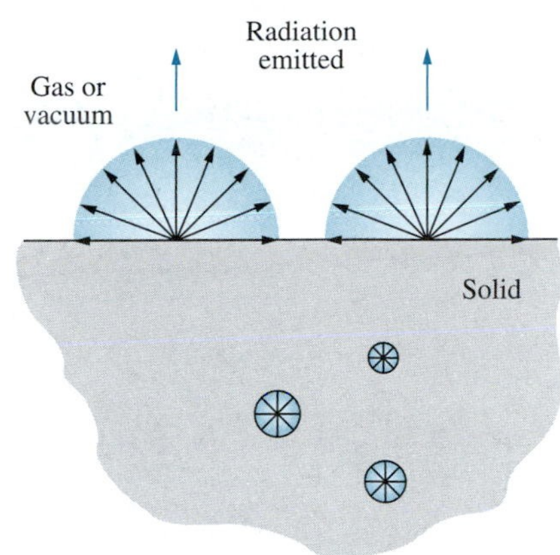

FIGURE 21–6
Radiation in opaque solids is considered a surface phenomenon since the radiation emitted only by the molecules at the surface can escape the solid.

21–3 ▪ BLACKBODY RADIATION

A body at a thermodynamic (or absolute) temperature above zero emits radiation in all directions over a wide range of wavelengths. The amount of radiation energy emitted from a surface at a given wavelength depends on the material of the body and the condition of its surface as well as the surface temperature. Therefore, different bodies may emit different amounts of radiation per unit surface area, even when they are at the same temperature. Thus, it is natural to be curious about the *maximum* amount of radiation that can be emitted by a surface at a given temperature. Satisfying this curiosity requires the definition of an idealized body, called a *blackbody*, to serve as a standard against which the radiative properties of real surfaces may be compared.

A **blackbody** is defined as *a perfect emitter and absorber of radiation.* At a specified temperature and wavelength, no surface can emit more energy than a blackbody. A blackbody absorbs *all* incident radiation, regardless of wavelength and direction. Also, a blackbody emits radiation energy uniformly in all directions per unit area normal to direction of emission (Fig. 21–7). That is, a blackbody is a *diffuse* emitter. The term *diffuse* means "independent of direction."

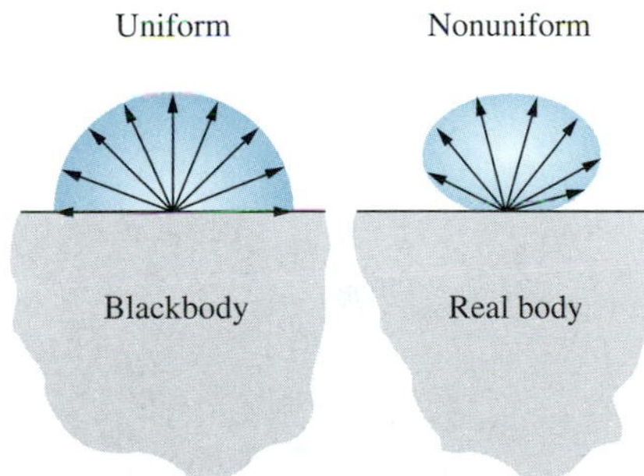

FIGURE 21–7
A blackbody is said to be a *diffuse* emitter since it emits radiation energy uniformly in all directions.

The radiation energy emitted by a blackbody per unit time and per unit surface area was determined experimentally by Joseph Stefan in 1879 and expressed as

$$E_b(T) = \sigma T^4 \qquad (\text{W/m}^2) \qquad \textbf{(21–3)}$$

where $\sigma = 5.670 \times 10^{-8}$ W/m$^2 \cdot$ K^4 is the *Stefan–Boltzmann constant* and T is the absolute temperature of the surface in K. This relation was theoretically verified in 1884 by Ludwig Boltzmann. Equation 21–3 is known as the **Stefan–Boltzmann law** and E_b is called the **blackbody emissive power**. Note that the emission of thermal radiation is proportional to the *fourth power* of the absolute temperature.

Although a blackbody would appear *black* to the eye, a distinction should be made between the idealized blackbody and an ordinary black surface. Any surface that absorbs light (the visible portion of radiation) would appear black to the eye, and a surface that reflects it completely would appear white. Considering that visible radiation occupies a very narrow band of the spectrum from 0.4 to 0.76 μm, we cannot make any judgments about the blackness of a surface on the basis of visual observations. For example, snow and white paint reflect light and thus appear white. But they are essentially black for infrared radiation since they strongly absorb long-wavelength radiation. Surfaces coated with lampblack paint approach idealized blackbody behavior.

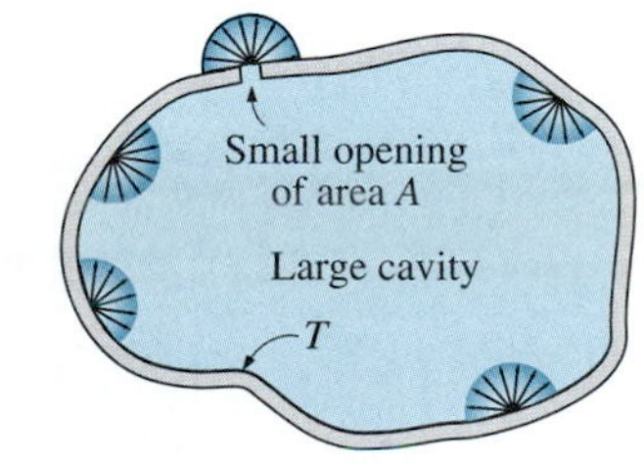

FIGURE 21–8
A large isothermal cavity at temperature T with a small opening of area A closely resembles a blackbody of surface area A at the same temperature.

Another type of body that closely resembles a blackbody is a *large cavity with a small opening*, as shown in Fig. 21–8. Radiation coming in through the opening of area A undergoes multiple reflections, and thus it has several chances to be absorbed by the interior surfaces of the cavity before any part of it can possibly escape. Also, if the surface of the cavity is isothermal at temperature T, the radiation emitted by the interior surfaces streams through the opening after undergoing multiple reflections, and thus it has a diffuse nature. Therefore, the cavity acts as a perfect absorber and perfect emitter, and the opening will resembles a blackbody of surface area A at temperature T, regardless of the actual radiative properties of the cavity.

The Stefan–Boltzmann law in Eq. 21–3 gives the *total* blackbody emissive power E_b, which is the sum of the radiation emitted over all wavelengths. Sometimes we need to know the **spectral blackbody emissive power**, which is *the amount of radiation energy emitted by a blackbody at a thermodynamic temperature T per unit time, per unit surface area, and per unit wavelength about the wavelength* λ. For example, we are more interested in the amount of radiation an incandescent lightbulb emits in the visible wavelength spectrum than we are in the total amount emitted.

The relation for the spectral blackbody emissive power $E_{b\lambda}$ was developed by Max Planck in 1901 in conjunction with his famous quantum theory. This relation is known as **Planck's law** and is expressed as

$$E_{b\lambda}(\lambda, T) = \frac{C_1}{\lambda^5[\exp(C_2/\lambda T) - 1]} \qquad (\text{W/m}^2 \cdot \mu\text{m}) \tag{21–4}$$

where

$$C_1 = 2\pi h c_0^2 = 3.74177 \times 10^8 \text{ W} \cdot \mu\text{m}^4/\text{m}^2$$
$$C_2 = hc_0/k = 1.43878 \times 10^4 \ \mu\text{m} \cdot \text{K}$$

Also, T is the absolute temperature of the surface, λ is the wavelength of the radiation emitted, and $k = 1.38065 \times 10^{-23}$ J/K is *Boltzmann's constant*. This relation is valid for a surface in a *vacuum* or a *gas*. For other mediums, it needs to be modified by replacing C_1 by C_1/n^2, where n is the index of refraction of the medium. Note that the term *spectral* indicates dependence on wavelength.

The variation of the spectral blackbody emissive power with wavelength is plotted in Fig. 21–9 for selected temperatures. Several observations can be made from this figure:

1. The emitted radiation is a continuous function of *wavelength*. At any specified temperature, it increases with wavelength, reaches a peak, and then decreases with increasing wavelength.
2. At any wavelength, the amount of emitted radiation *increases* with increasing temperature.
3. As temperature increases, the curves shift to the left to the shorter-wavelength region. Consequently, a larger fraction of the radiation is emitted at *shorter wavelengths* at higher temperatures.
4. The radiation emitted by the *sun*, which is considered to be a blackbody at 5780 K (or roughly at 5800 K), reaches its peak in the visible region of the spectrum. Therefore, the sun is in tune with our eyes. On the other hand, surfaces at $T \leq 800$ K emit almost entirely in the infrared region and thus are not visible to the eye unless they reflect light coming from other sources.

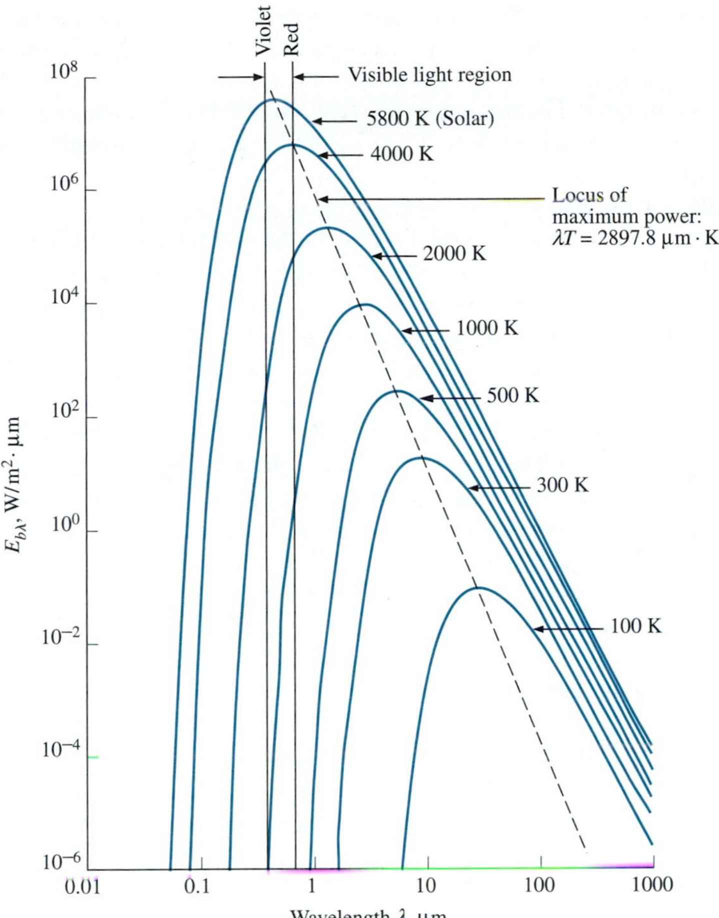

FIGURE 21–9
The variation of the blackbody emissive power with wavelength for several temperatures.

As the temperature increases, the peak of the curve in Fig. 21–9 shifts toward shorter wavelengths. The wavelength at which the peak occurs for a specified temperature is given by **Wien's displacement law** as

$$(\lambda T)_{\text{max power}} = 2897.8\ \mu\text{m}\cdot\text{K} \tag{21–5}$$

This relation was originally developed by Willy Wien in 1894 using classical thermodynamics, but it can also be obtained by differentiating Eq. 21–4 with respect to λ while holding T constant and setting the result equal to zero. A plot of Wien's displacement law, which is the locus of the peaks of the radiation emission curves, is also given in Fig. 21–9.

The peak of the solar radiation, for example, occurs at $\lambda = 2897.8/5780 = 0.50\ \mu\text{m}$, which is near the middle of the visible range. The peak of the radiation emitted by a surface at room temperature ($T = 298$ K) occurs at 9.72 μm, which is well into the infrared region of the spectrum.

An electrical resistance heater starts radiating heat soon after it is plugged in, and we can feel the emitted radiation energy by holding our hands against the heater. But this radiation is entirely in the infrared region and thus cannot be sensed by our eyes. The heater would appear dull red when its temperature reaches about 1000 K, since it starts emitting a detectable amount (about 1 $W/m^2 \cdot \mu m$) of visible red radiation at that temperature. As the temperature rises even more, the heater appears bright red and is said to be *red hot.* When the temperature reaches about 1500 K, the heater emits enough radiation in the entire visible range of the spectrum to appear almost *white* to the eye, and it is called *white hot*.

Although it cannot be sensed directly by the human eye, infrared radiation can be detected by infrared cameras, which transmit the information to microprocessors to display visual images of objects at night. *Rattlesnakes* can sense the infrared radiation or the "body heat" coming off warm-blooded animals, and thus they can see at night without using any instruments. Similarly, honeybees are sensitive to ultraviolet radiation. A surface that reflects all of the light appears *white*, while a surface that absorbs all of the light incident on it appears black. (Then how do we see a black surface?)

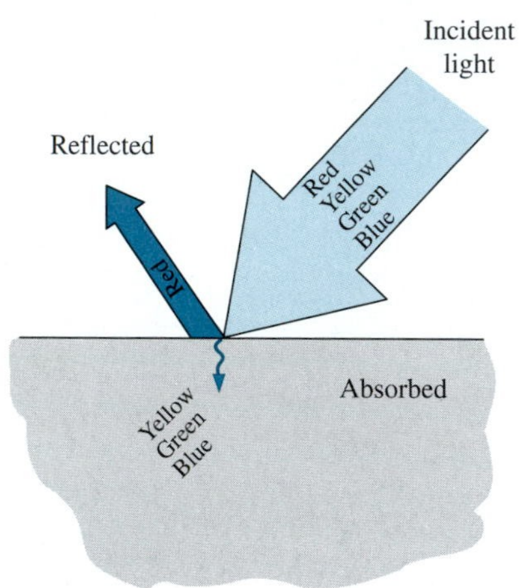

FIGURE 21–10
A surface that reflects red while absorbing the remaining parts of the incident light appears red to the eye.

It should be clear from this discussion that the color of an object is not due to emission, which is primarily in the infrared region, unless the surface temperature of the object exceeds about 1000 K. Instead, the color of a surface depends on the absorption and reflection characteristics of the surface and is due to selective absorption and reflection of the incident visible radiation coming from a light source such as the sun or an incandescent lightbulb. A piece of clothing containing a pigment that reflects red while absorbing the remaining parts of the incident light appears "red" to the eye (Fig. 21–10). Leaves appear "green" because their cells contain the pigment chlorophyll, which strongly reflects green while absorbing other colors.

It is left as an exercise to show that integration of the *spectral* blackbody emissive power $E_{b\lambda}$ over the entire wavelength spectrum gives the *total* blackbody emissive power E_b:

$$E_b(T) = \int_0^\infty E_{b\lambda}(\lambda, T)\, d\lambda = \sigma T^4 \qquad \text{(W/m}^2\text{)} \tag{21-6}$$

Thus, we obtained the Stefan–Boltzmann law (Eq. 21–3) by integrating Planck's law (Eq. 21–4) over all wavelengths. Note that on an $E_{b\lambda}$–λ chart, $E_{b\lambda}$ corresponds to any value on the curve, whereas E_b corresponds to the area under the entire curve for a specified temperature (Fig. 21–11). Also, the term *total* means "integrated over all wavelengths."

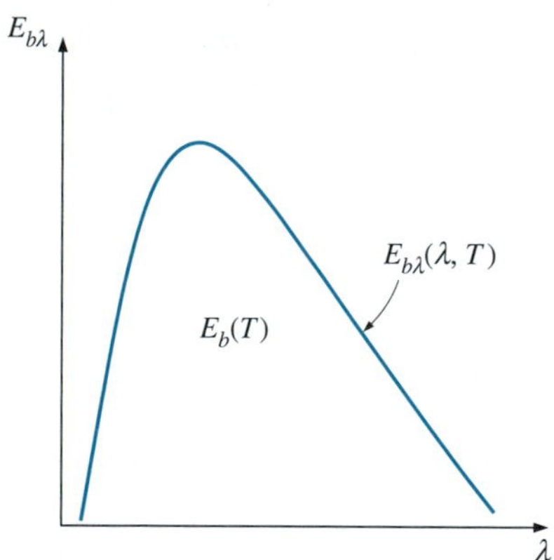

FIGURE 21–11
On an $E_{b\lambda}$–λ chart, the area under a curve for a given temperature represents the total radiation energy emitted by a blackbody at that temperature.

EXAMPLE 21–1 Radiation Emission from a Black Ball

Consider a 20-cm-diameter spherical ball at 800 K suspended in air as shown in Fig. 21–12. Assuming the ball closely approximates a blackbody, determine (*a*) the total blackbody emissive power, (*b*) the total amount of radiation emitted by the ball in 5 min, and (*c*) the spectral blackbody emissive power at a wavelength of 3 μm.

Solution An isothermal sphere is suspended in air. The total blackbody emissive power, the total radiation emitted in 5 min, and the spectral blackbody emissive power at 3 μm are to be determined.

Assumptions The ball behaves as a blackbody.

Analysis (*a*) The total blackbody emissive power is determined from the Stefan–Boltzmann law to be

$$E_b = \sigma T^4 = (5.67 \times 10^{-8}\ \text{W/m}^2 \cdot \text{K}^4)(800\ \text{K})^4 = \mathbf{23.2\ kW/m^2}$$

That is, the ball emits 23.2 kJ of energy in the form of electromagnetic radiation per second per m^2 of the surface area of the ball.

(*b*) The total amount of radiation energy emitted from the entire ball in 5 min is determined by multiplying the blackbody emissive power obtained above by the total surface area of the ball and the given time interval:

$$A_s = \pi D^2 = \pi(0.2\ \text{m})^2 = 0.1257\ \text{m}^2$$

$$\Delta t = (5\ \text{min})\left(\frac{60\ \text{s}}{1\ \text{min}}\right) = 300\ \text{s}$$

$$Q_{\text{rad}} = E_b A_s\, \Delta t = (23.2\ \text{kW/m}^2)(0.1257\ \text{m}^2)(300\ \text{s})\left(\frac{1\ \text{kJ}}{1\ \text{kW} \cdot \text{s}}\right) = \mathbf{875\ kJ}$$

That is, the ball loses 875 kJ of its internal energy in the form of electromagnetic waves to the surroundings in 5 min, which is enough energy to heat 20 kg of water from 0°C to 100°C. Note that the surface temperature of the ball cannot remain constant at 800 K unless there is an equal amount of energy flow to the surface from the surroundings or from the interior regions of the ball through some mechanisms such as chemical or nuclear reactions.

(*c*) The spectral blackbody emissive power at a wavelength of 3 μm is determined from Planck's distribution law to be

$$E_{b\lambda} = \frac{C_1}{\lambda^5\left[\exp\left(\dfrac{C_2}{\lambda T}\right) - 1\right]} = \frac{3.74177 \times 10^8\ \text{W} \cdot \mu\text{m}^4/\text{m}^2}{(3\ \mu\text{m})^5\left[\exp\left(\dfrac{1.43878 \times 10^4\ \mu\text{m} \cdot \text{K}}{(3\ \mu\text{m})(800\ \text{K})}\right) - 1\right]} = \mathbf{3846\ W/m^2 \cdot \mu m}$$

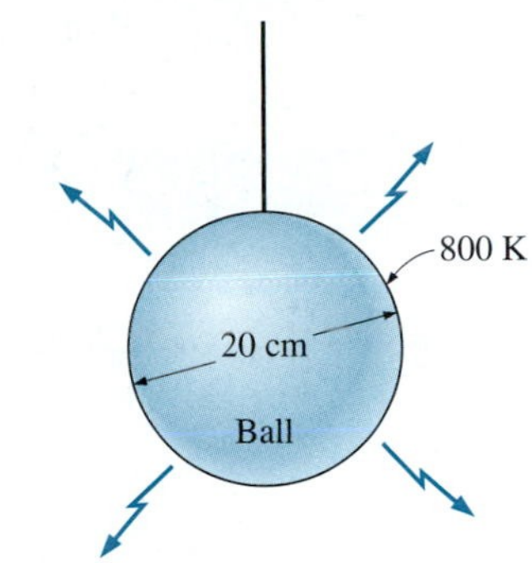

FIGURE 21–12
The spherical ball considered in Example 21–1.

The Stefan–Boltzmann law $E_b(T) = \sigma T^4$ gives the *total* radiation emitted by a blackbody at all wavelengths from $\lambda = 0$ to $\lambda = \infty$. But we are often interested in the amount of radiation emitted over *some wavelength band*. For example, an incandescent lightbulb is judged on the basis of the radiation it emits in the visible range rather than the radiation it emits at all wavelengths.

The radiation energy emitted by a blackbody per unit area over a wavelength band from $\lambda = 0$ to λ is determined from (Fig. 21–13)

$$E_{b,\,0\text{–}\lambda}(T) = \int_0^{\lambda} E_{b\lambda}(\lambda, T)\, d\lambda \qquad (\text{W/m}^2) \tag{21-7}$$

It looks like we can determine $E_{b,\,0\text{–}\lambda}$ by substituting the $E_{b\lambda}$ relation from Eq. 21–4 and performing this integration. But it turns out that this integration does not have a simple closed-form solution, and performing a numerical integration each time we need a value of $E_{b,0\text{–}\lambda}$ is not practical. Therefore, we define a dimensionless quantity f_λ called the **blackbody radiation function** as

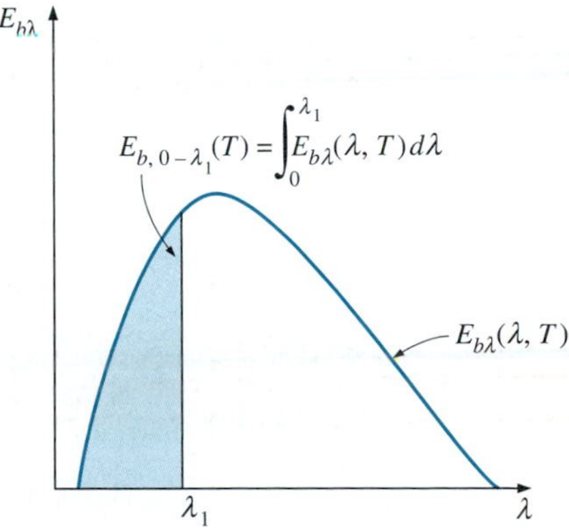

FIGURE 21–13
On an $E_{b\lambda}$–λ chart, the area under the curve to the left of the $\lambda = \lambda_1$ line represents the radiation energy emitted by a blackbody in the wavelength range 0–λ_1 for the given temperature.

TABLE 21–2

Blackbody radiation functions f_λ

λT, μm · K	f_λ	λT, μm · K	f_λ
200	0.000000	6200	0.754140
400	0.000000	6400	0.769234
600	0.000000	6600	0.783199
800	0.000016	6800	0.796129
1000	0.000321	7000	0.808109
1200	0.002134	7200	0.819217
1400	0.007790	7400	0.829527
1600	0.019718	7600	0.839102
1800	0.039341	7800	0.848005
2000	0.066728	8000	0.856288
2200	0.100888	8500	0.874608
2400	0.140256	9000	0.890029
2600	0.183120	9500	0.903085
2800	0.227897	10,000	0.914199
3000	0.273232	10,500	0.923710
3200	0.318102	11,000	0.931890
3400	0.361735	11,500	0.939959
3600	0.403607	12,000	0.945098
3800	0.443382	13,000	0.955139
4000	0.480877	14,000	0.962898
4200	0.516014	15,000	0.969981
4400	0.548796	16,000	0.973814
4600	0.579280	18,000	0.980860
4800	0.607559	20,000	0.985602
5000	0.633747	25,000	0.992215
5200	0.658970	30,000	0.995340
5400	0.680360	40,000	0.997967
5600	0.701046	50,000	0.998953
5800	0.720158	75,000	0.999713
6000	0.737818	100,000	0.999905

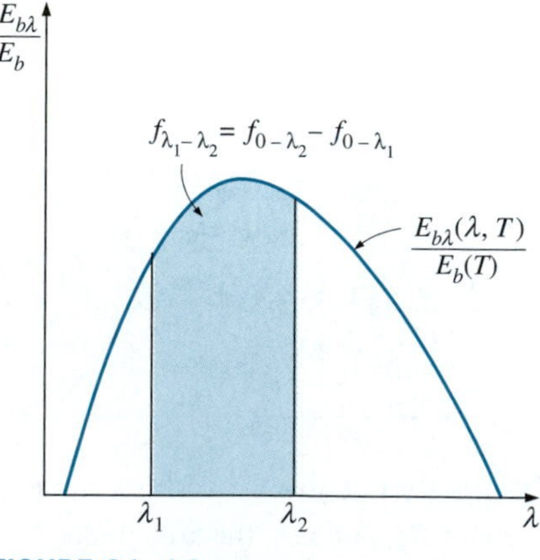

FIGURE 21–14
Graphical representation of the fraction of radiation emitted in the wavelength band from λ_1 to λ_2.

$$f_\lambda(T) = \frac{\int_0^\lambda E_{b\lambda}(\lambda, T)\, d\lambda}{\sigma T^4} \qquad \textbf{(21–8)}$$

The function f_λ represents *the fraction of radiation emitted from a blackbody at temperature T in the wavelength band from $\lambda = 0$ to λ*. The values of f_λ are listed in Table 21–2 as a function of λT, where λ is in μm and T is in K.

The fraction of radiation energy emitted by a blackbody at temperature T over a finite wavelength band from $\lambda = \lambda_1$ to $\lambda = \lambda_2$ is determined from (Fig. 21–14)

$$f_{\lambda_1-\lambda_2}(T) = f_{\lambda_2}(T) - f_{\lambda_1}(T) \qquad \textbf{(21–9)}$$

where $f_{\lambda_1}(T)$ and $f_{\lambda_2}(T)$ are blackbody radiation functions corresponding to $\lambda_1 T$ and $\lambda_2 T$, respectively.

EXAMPLE 21–2 Emission of Radiation from a Lightbulb

The temperature of the filament of an incandescent lightbulb is 2500 K. Assuming the filament to be a blackbody, determine the fraction of the radiant energy emitted by the filament that falls in the visible range. Also, determine the wavelength at which the emission of radiation from the filament peaks.

Solution The temperature of the filament of an incandescent lightbulb is given. The fraction of visible radiation emitted by the filament and the wavelength at which the emission peaks are to be determined.

Assumptions The filament behaves as a blackbody.

Analysis The visible range of the electromagnetic spectrum extends from $\lambda_1 = 0.4$ μm to $\lambda_2 = 0.76$ μm. Noting that $T = 2500$ K, the blackbody radiation functions corresponding to $\lambda_1 T$ and $\lambda_2 T$ are determined from Table 21–2 to be

$$\lambda_1 T = (0.40\ \mu\text{m})(2500\ \text{K}) = 1000\ \mu\text{m}\cdot\text{K} \longrightarrow f_{\lambda_1} = 0.000321$$

$$\lambda_2 T = (0.76\ \mu\text{m})(2500\ \text{K}) = 1900\ \mu\text{m}\cdot\text{K} \longrightarrow f_{\lambda_2} = 0.053035$$

That is, 0.03 percent of the radiation is emitted at wavelengths less than 0.4 μm and 5.3 percent at wavelengths less than 0.76 μm. Then the fraction of radiation emitted between these two wavelengths is (Fig. 21–15)

$$f_{\lambda_1-\lambda_2} = f_{\lambda_2} - f_{\lambda_1} = 0.053035 - 0.000321 = \mathbf{0.052714}$$

Therefore, only about 5 percent of the radiation emitted by the filament of the lightbulb falls in the visible range. The remaining 95 percent of the radiation appears in the infrared region in the form of radiant heat or "invisible light," as it used to be called. This is certainly not a very efficient way of converting electrical energy to light and explains why fluorescent tubes are a wiser choice for lighting.

The wavelength at which the emission of radiation from the filament peaks is easily determined from Wien's displacement law to be

$$(\lambda T)_{\text{max power}} = 2897.8\ \mu\text{m}\cdot\text{K} \rightarrow \lambda_{\text{max power}} = \frac{2897.8\ \mu\text{m}\cdot\text{K}}{2500\ \text{K}} = \mathbf{1.16\ \mu m}$$

Discussion Note that the radiation emitted from the filament peaks in the infrared region.

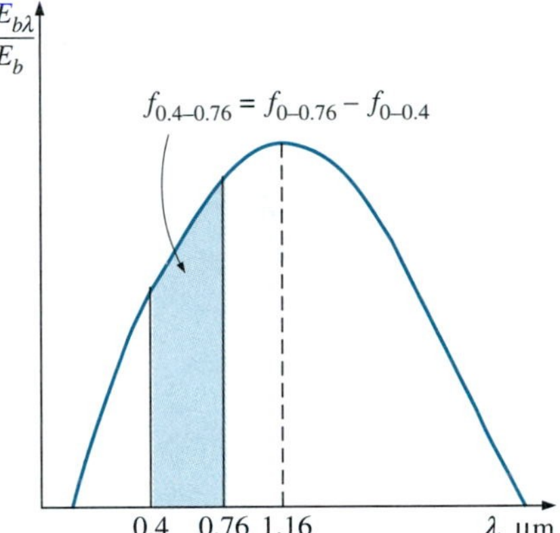

FIGURE 21–15
Graphical representation of the fraction of radiation emitted in the visible range in Example 21–2.

21–4 ▪ RADIATIVE PROPERTIES

Most materials encountered in practice, such as metals, wood, and bricks, are *opaque* to thermal radiation, and radiation is considered to be a *surface phenomenon* for such materials. That is, thermal radiation is emitted or absorbed within the first few microns of the surface, and thus we speak of radiative properties of *surfaces* for opaque materials.

Some other materials, such as glass and water, allow visible radiation to penetrate to considerable depths before any significant absorption takes place. Radiation through such *semitransparent* materials obviously cannot be considered to be a surface phenomenon since the entire volume of the material interacts with radiation. On the other hand, both glass and water are practically opaque to infrared radiation. Therefore, materials can exhibit different behavior at different wavelengths, and the dependence on wavelength is an important consideration in the study of radiative properties such as emissivity, absorptivity, reflectivity, and transmissivity of materials.

In the preceding section, we defined a *blackbody* as a perfect emitter and absorber of radiation and said that no body can emit more radiation than a blackbody at the same temperature. Therefore, a blackbody can serve as a convenient *reference* in describing the emission and absorption characteristics of real surfaces.

Emissivity

The **emissivity** of a surface represents *the ratio of the radiation emitted by the surface at a given temperature to the radiation emitted by a blackbody at the same temperature*. The emissivity of a surface is denoted by ε, and it varies between zero and one, $0 \leq \varepsilon \leq 1$. Emissivity is a measure of how closely a surface approximates a blackbody, for which $\varepsilon = 1$.

The emissivity of a real surface is not a constant. Rather, it varies with the *temperature* of the surface as well as the *wavelength* and the *direction* of the emitted radiation. Therefore, different emissivities can be defined for a surface, depending on the effects considered. For example, the emissivity of a surface at a specified wavelength is called *spectral emissivity* and is denoted by ε_λ. Likewise, the emissivity in a specified direction is called *directional emissivity*, denoted by ε_θ, where θ is the angle between the direction of radiation and the normal of the surface. The most elemental emissivity of a surface at a given temperature is the *spectral directional emissivity* $\varepsilon_{\lambda,\theta}$, which is the emissivity at a specified wavelength in a specified direction. Note that blackbody radiation is independent of direction.

In practice, it is usually more convenient to work with radiation properties averaged over all directions, called *hemispherical properties*. Noting that the integral of the rate of radiation energy emitted at a specified wavelength per unit surface area over the entire hemisphere is *spectral emissive power*, the **spectral hemispherical emissivity** can be expressed as

$$\varepsilon_\lambda(\lambda, T) = \frac{E_\lambda(\lambda, T)}{E_{b\lambda}(\lambda, T)} \tag{21-10}$$

Note that the emissivity of a surface at a given wavelength can be different at different temperatures since the spectral distribution of emitted radiation (and thus the amount of radiation emitted at a given wavelength) changes with temperature.

The **total hemispherical emissivity** is defined in terms of the radiation energy emitted over all wavelengths in all directions as

$$\varepsilon(T) = \frac{E(T)}{E_b(T)} \tag{21-11}$$

Therefore, the total hemispherical emissivity (or simply the "average emissivity") of a surface at a given temperature represents the ratio of the total radiation energy emitted by the surface to the radiation emitted by a blackbody of the same surface area at the same temperature.

Noting that $E = \int_0^\infty E_\lambda d\lambda$ and $E_\lambda(\lambda, T) = \varepsilon_\lambda(\lambda, T)E_{b\lambda}(\lambda, T)$, the total hemispherical emissivity can also be expressed as

$$\varepsilon(T) = \frac{E(T)}{E_b(T)} = \frac{\int_0^\infty \varepsilon_\lambda(\lambda, T)E_{b\lambda}(\lambda, T)d\lambda}{\sigma T^4} \quad \text{(21–12)}$$

since $E_b(T) = \sigma T^4$. To perform this integration, we need to know the variation of spectral emissivity with wavelength at the specified temperature. The integrand is usually a complicated function, and the integration has to be performed numerically. However, the integration can be performed quite easily by dividing the spectrum into a sufficient number of *wavelength bands* and assuming the emissivity to remain constant over each band; that is, by expressing the function $\varepsilon_\lambda(\lambda, T)$ as a step function. This simplification offers great convenience for little sacrifice of accuracy, since it allows us to transform the integration into a summation in terms of blackbody emission functions.

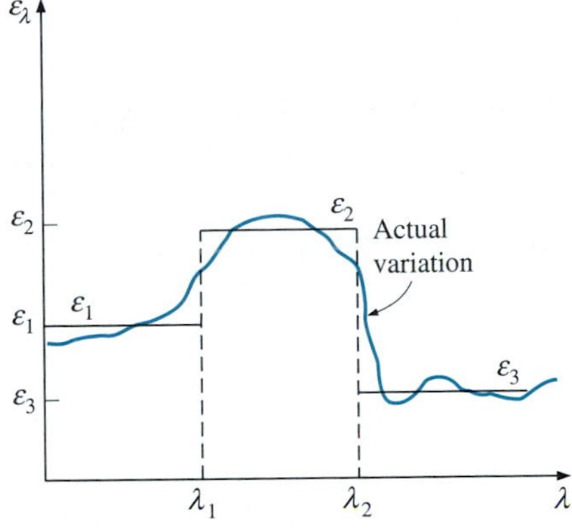

FIGURE 21–16
Approximating the actual variation of emissivity with wavelength by a step function.

As an example, consider the emissivity function plotted in Fig. 21–16. It seems like this function can be approximated reasonably well by a step function of the form

$$\varepsilon_\lambda = \begin{cases} \varepsilon_1 = \text{constant}, & 0 \le \lambda < \lambda_1 \\ \varepsilon_2 = \text{constant}, & \lambda_1 \le \lambda < \lambda_2 \\ \varepsilon_3 = \text{constant}, & \lambda_2 \le \lambda < \infty \end{cases} \quad \text{(21–13)}$$

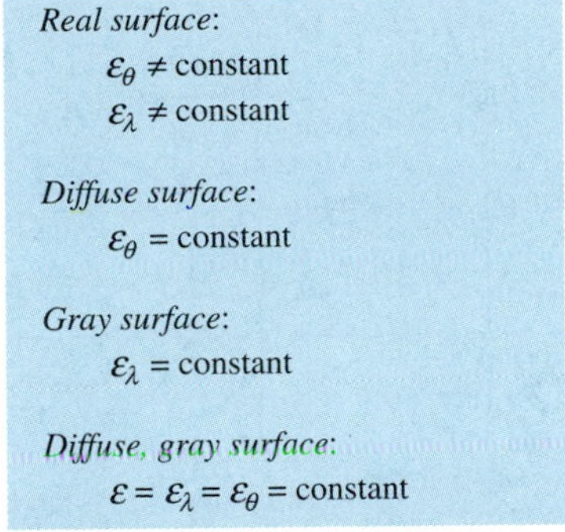

Real surface:
$\varepsilon_\theta \neq$ constant
$\varepsilon_\lambda \neq$ constant
Diffuse surface:
ε_θ = constant
Gray surface:
ε_λ = constant
Diffuse, gray surface:
$\varepsilon = \varepsilon_\lambda = \varepsilon_\theta$ = constant

FIGURE 21–17
The effect of diffuse and gray approximations on the emissivity of a surface.

Then the average emissivity can be determined from Eq. 21–12 by breaking the integral into three parts and utilizing the definition of the blackbody radiation function as

$$\varepsilon(T) = \frac{\varepsilon_1 \int_0^{\lambda_1} E_{b\lambda}\, d\lambda}{E_b} + \frac{\varepsilon_2 \int_{\lambda_1}^{\lambda_2} E_{b\lambda}\, d\lambda}{E_b} + \frac{\varepsilon_3 \int_{\lambda_2}^{\infty} E_{b\lambda}\, d\lambda}{E_b} \quad \text{(21–14)}$$

$$= \varepsilon_1 f_{0-\lambda_1}(T) + \varepsilon_2 f_{\lambda_1-\lambda_2}(T) + \varepsilon_3 f_{\lambda_2-\infty}(T)$$

Radiation is a complex phenomenon as it is, and the consideration of the wavelength and direction dependence of properties, assuming sufficient data exist, makes it even more complicated. Therefore, the *gray* and *diffuse* approximations are often utilized in radiation calculations. A surface is said to be *diffuse* if its properties are *independent of direction*, and *gray* if its properties are *independent of wavelength*. Therefore, the emissivity of a gray, diffuse surface is simply the total hemispherical emissivity of that surface because of independence of direction and wavelength (Fig. 21–17).

A few comments about the validity of the diffuse approximation are in order. Although real surfaces do not emit radiation in a perfectly diffuse manner as a blackbody does, they often come close. The variation of emissivity with direction for both electrical conductors and nonconductors is given in Fig. 21–18. Here θ is the angle measured from the normal of the surface, and thus $\theta = 0$ for radiation emitted in a direction normal to the surface. Note that ε_θ remains nearly constant for about $\theta < 40°$ for conductors such as metals and for $\theta < 70°$ for nonconductors such as plastics. Therefore, the directional

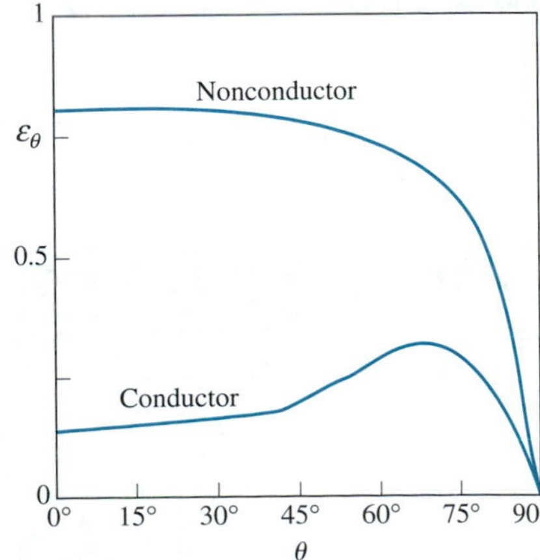

FIGURE 21–18
Typical variations of emissivity with direction for electrical conductors and nonconductors.

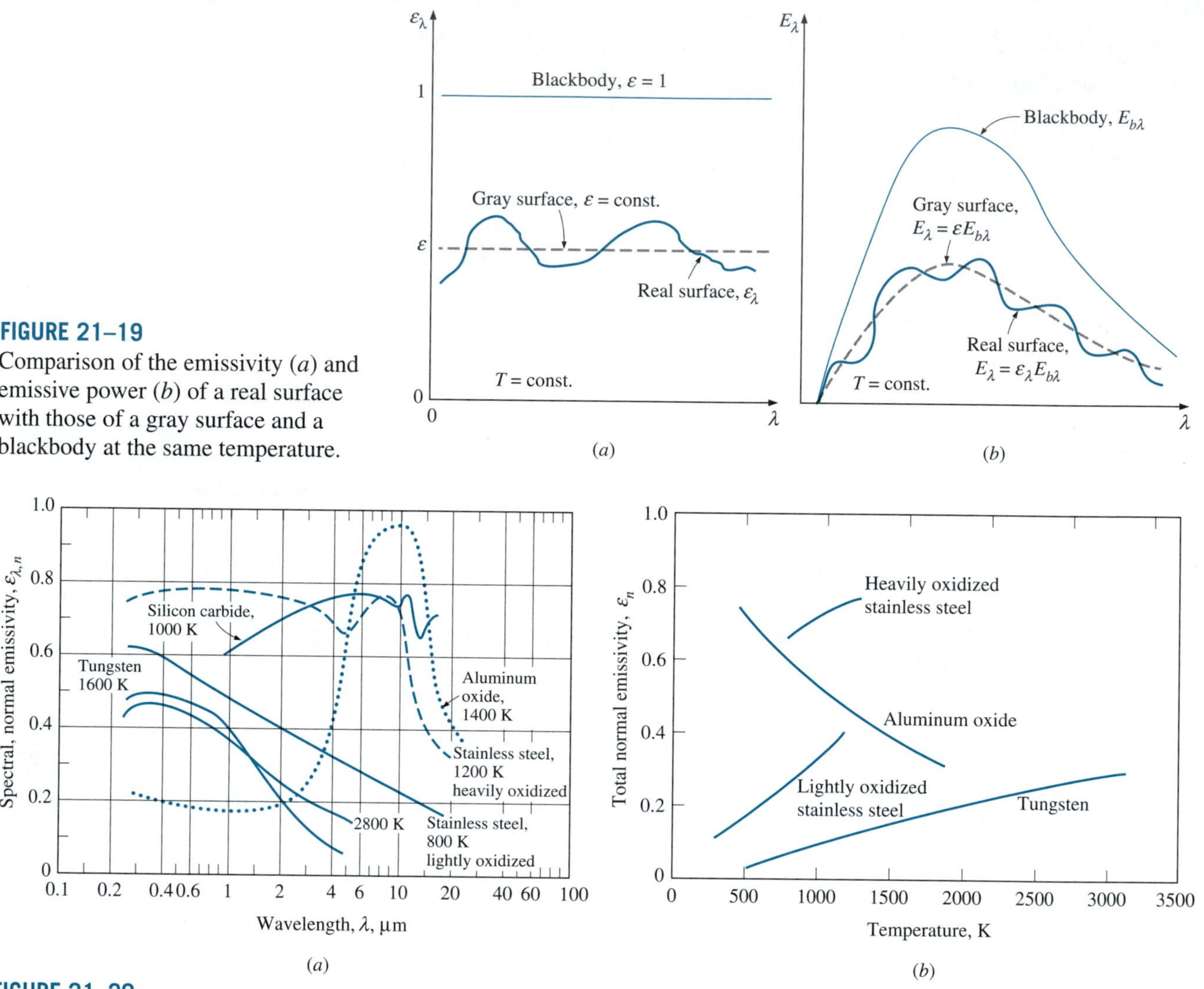

FIGURE 21–19
Comparison of the emissivity (a) and emissive power (b) of a real surface with those of a gray surface and a blackbody at the same temperature.

FIGURE 21–20
The variation of normal emissivity with (a) wavelength and (b) temperature for various materials.

emissivity of a surface in the normal direction is representative of the hemispherical emissivity of the surface. In radiation analysis, it is common practice to assume the surfaces to be diffuse emitters with an emissivity equal to the value in the normal ($\theta = 0$) direction.

The effect of the gray approximation on emissivity and emissive power of a real surface is illustrated in Fig. 21–19. Note that the radiation emission from a real surface, in general, differs from the Planck distribution, and the emission curve may have several peaks and valleys. A gray surface should emit as much radiation as the real surface it represents at the same temperature. Therefore, the areas under the emission curves of the real and gray surfaces must be equal.

The emissivities of common materials are listed in Table A–26 in the appendix, and the variation of emissivity with wavelength and temperature is illustrated in Fig. 21–20. Typical ranges of emissivity of various materials are given in Fig. 21–21. Note that metals generally have low emissivities, as low as 0.02 for polished surfaces, and nonmetals such as ceramics and organic materials have high ones. The emissivity of metals increases with temperature. Also, oxidation causes significant increases in the emissivity of metals. Heavily oxidized metals can have emissivities comparable to those of nonmetals.

Care should be exercised in the use and interpretation of radiation property data reported in the literature, since the properties strongly depend on the surface conditions such as oxidation, roughness, type of finish, and cleanliness. Consequently, there is considerable discrepancy and uncertainty in the reported values. This uncertainty is largely due to the difficulty in characterizing and describing the surface conditions precisely.

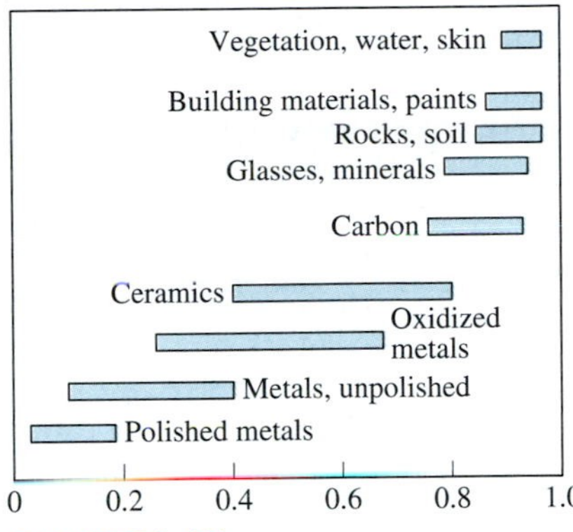

FIGURE 21–21
Typical ranges of emissivity for various materials.

EXAMPLE 21–3 Emissivity of a Surface and Emissive Power

The spectral emissivity function of an opaque surface at 800 K is approximated as (Fig. 21–22)

$$\varepsilon_\lambda = \begin{cases} \varepsilon_1 = 0.3, & 0 \leq \lambda < 3\ \mu\text{m} \\ \varepsilon_2 = 0.8, & 3\ \mu\text{m} \leq \lambda < 7\ \mu\text{m} \\ \varepsilon_3 = 0.1, & 7\ \mu\text{m} \leq \lambda < \infty \end{cases}$$

Determine the average emissivity of the surface and its emissive power.

SOLUTION The variation of emissivity of a surface at a specified temperature with wavelength is given. The average emissivity of the surface and its emissive power are to be determined.

Analysis The variation of the emissivity with wavelength is given as a step function. Therefore, the average emissivity of the surface can be determined from Eq. 21–12 by breaking the integral into three parts,

$$\varepsilon(T) = \frac{\varepsilon_1 \int_0^{\lambda_1} E_{b\lambda}\, d\lambda}{\sigma T^4} + \frac{\varepsilon_2 \int_{\lambda_1}^{\lambda_2} E_{b\lambda}\, d\lambda}{\sigma T^4} + \frac{\varepsilon_3 \int_{\lambda_2}^{\infty} E_{b\lambda}\, d\lambda}{\sigma T^4}$$

$$= \varepsilon_1 f_{0-\lambda_1}(T) + \varepsilon_2 f_{\lambda_1-\lambda_2}(T) + \varepsilon_3 f_{\lambda_2-\infty}(T)$$

$$= \varepsilon_1 f_{\lambda_1} + \varepsilon_2(f_{\lambda_2} - f_{\lambda_1}) + \varepsilon_3(1 - f_{\lambda_2})$$

where f_{λ_1} and f_{λ_2} are blackbody radiation functions and are determined from Table 21–2 to be

$$\lambda_1 T = (3\ \mu\text{m})(800\ \text{K}) = 2400\ \mu\text{m}\cdot\text{K} \;\rightarrow\; f_{\lambda_1} = 0.140256$$

$$\lambda_2 T = (7\ \mu\text{m})(800\ \text{K}) = 5600\ \mu\text{m}\cdot\text{K} \;\rightarrow\; f_{\lambda_2} = 0.701046$$

Note that $f_{0-\lambda_1} = f_{\lambda_1} - f_0 = f_{\lambda_1}$ since $f_0 = 0$, and $f_{\lambda_2-\infty} = f_\infty - f_{\lambda_2} = 1 - f_{\lambda_2}$ since $f_\infty = 1$. Substituting,

$$\varepsilon = 0.3 \times 0.140256 + 0.8(0.701046 - 0.140256) + 0.1(1 - 0.701046)$$

$$= \mathbf{0.521}$$

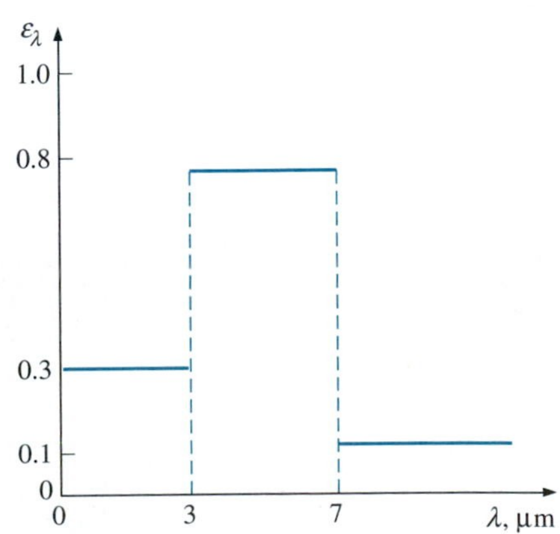

FIGURE 21–22
The spectral emissivity of the surface considered in Example 21–3.

That is, the surface will emit as much radiation energy at 800 K as a gray surface having a constant emissivity of $\varepsilon = 0.521$. The emissive power of the surface is

$$E = \varepsilon\sigma T^4 = 0.521(5.67 \times 10^{-8}\ \text{W/m}^2 \cdot \text{K}^4)(800\ \text{K})^4 = \mathbf{12{,}100\ W/m^2}$$

Discussion Note that the surface emits 12.1 kJ of radiation energy per second per m^2 area of the surface.

Absorptivity, Reflectivity, and Transmissivity

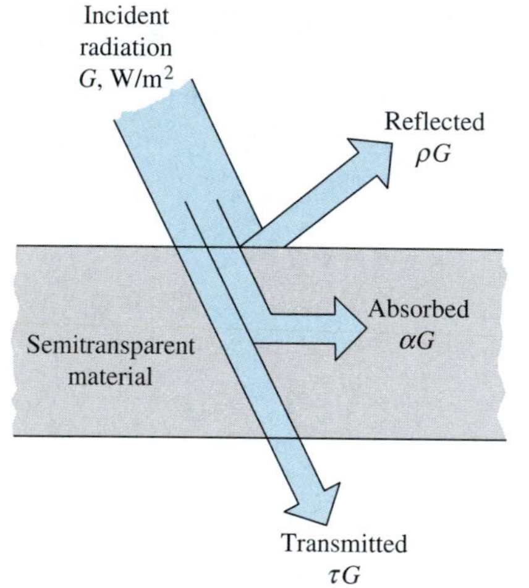

FIGURE 21–23
The absorption, reflection, and transmission of incident radiation by a semitransparent material.

Everything around us constantly emits radiation, and the emissivity represents the emission characteristics of those bodies. This means that every body, including our own, is constantly bombarded by radiation coming from all directions over a range of wavelengths. Recall that radiation flux *incident on a surface* is called **irradiation** and is denoted by G.

When radiation strikes a surface, part of it is absorbed, part of it is reflected, and the remaining part, if any, is transmitted, as illustrated in Fig. 21–23. *The fraction of irradiation absorbed by the surface* is called the **absorptivity** α, *the fraction reflected by the surface* is called the **reflectivity** ρ, and *the fraction transmitted* is called the **transmissivity** τ. That is,

Absorptivity:
$$\alpha = \frac{\text{Absorbed radiation}}{\text{Incident radiation}} = \frac{G_{abs}}{G}, \qquad 0 \le \alpha \le 1 \tag{21–15}$$

Reflectivity:
$$\rho = \frac{\text{Reflected radiation}}{\text{Incident radiation}} = \frac{G_{ref}}{G}, \qquad 0 \le \rho \le 1 \tag{21–16}$$

Transmissivity:
$$\tau = \frac{\text{Transmitted radiation}}{\text{Incident radiation}} = \frac{G_{tr}}{G}, \qquad 0 \le \tau \le 1 \tag{21–17}$$

where G is the radiation flux incident on the surface, and G_{abs}, G_{ref}, and G_{tr} are the absorbed, reflected, and transmitted portions of it, respectively. The first law of thermodynamics requires that the sum of the absorbed, reflected, and transmitted radiation be equal to the incident radiation. That is,

$$G_{abs} + G_{ref} + G_{tr} = G \tag{21–18}$$

Dividing each term of this relation by G yields

$$\alpha + \rho + \tau = 1 \tag{21–19}$$

For opaque surfaces, $\tau = 0$, and thus

$$\alpha + \rho = 1 \tag{21–20}$$

This is an important property relation since it enables us to determine both the absorptivity and reflectivity of an opaque surface by measuring either of these properties.

These definitions are for *total hemispherical* properties, since G represents the radiation flux incident on the surface from all directions over the hemispherical space and over all wavelengths. Thus, α, ρ, and τ are the *average* properties of a medium for all directions and all wavelengths. However, like emissivity, these properties can also be defined for a specific wavelength and/or direction. For example, the **spectral hemispherical absorptivity** and **spectral hemispherical reflectivity** of a surface are defined as

$$\alpha_\lambda(\lambda) = \frac{G_{\lambda,\,\text{abs}}(\lambda)}{G_\lambda(\lambda)} \quad \text{and} \quad \rho_\lambda(\lambda) = \frac{G_{\lambda,\,\text{ref}}(\lambda)}{G_\lambda(\lambda)} \tag{21-21}$$

where G_λ is the spectral irradiation (in W/m^2 · μm) incident on the surface, and $G_{\lambda,\,\text{abs}}$ and $G_{\lambda,\,\text{ref}}$ are the reflected and absorbed portions of it, respectively.

Similar quantities can be defined for the transmissivity of semitransparent materials. For example, the **spectral hemispherical transmissivity** of a medium can be expressed as

$$\tau_\lambda(\lambda) = \frac{G_{\lambda,\,\text{tr}}(\lambda)}{G_\lambda(\lambda)} \tag{21-22}$$

The average absorptivity, reflectivity, and transmissivity of a surface can also be defined in terms of their spectral counterparts as

$$\alpha = \frac{\int_0^\infty \alpha_\lambda G_\lambda\, d\lambda}{\int_0^\infty G_\lambda\, d\lambda}, \quad \rho = \frac{\int_0^\infty \rho_\lambda G_\lambda\, d\lambda}{\int_0^\infty G_\lambda\, d\lambda}, \quad \tau = \frac{\int_0^\infty \tau_\lambda G_\lambda\, d\lambda}{\int_0^\infty G_\lambda\, d\lambda} \tag{21-23}$$

The reflectivity differs somewhat from the other properties in that it is *bidirectional* in nature. That is, the value of the reflectivity of a surface depends not only on the direction of the incident radiation but also the direction of reflection. Therefore, the reflected rays of a radiation beam incident on a real surface in a specified direction forms an irregular shape, as shown in Fig. 21–24. Such detailed reflectivity data do not exist for most surfaces, and even if they did, they would be of little value in radiation calculations since this would usually add more complication to the analysis.

In practice, for simplicity, surfaces are assumed to reflect in a perfectly *specular* or *diffuse* manner. In **specular** (or *mirrorlike*) **reflection**, *the angle of reflection equals the angle of incidence of the radiation beam.* In **diffuse reflection**, *radiation is reflected equally in all directions*, as shown in Fig. 21–24. Reflection from smooth and polished surfaces approximates specular reflection, whereas reflection from rough surfaces approximates diffuse reflection. In radiation analysis, smoothness is defined relative to wavelength. A surface is said to be *smooth* if the height of the surface roughness is much smaller than the wavelength of the incident radiation.

Unlike emissivity, the absorptivity of a material is practically independent of surface temperature. However, the absorptivity depends strongly on the temperature of the source at which the incident radiation is originating. This is also evident from Fig. 21–25, which shows the absorptivities of various materials at room temperature as functions of the temperature of the radiation source. For example, the absorptivity of the concrete roof of a house is about 0.6 for solar radiation (source temperature: 5780 K) and 0.9 for radiation originating from the surrounding trees and buildings (source temperature: 300 K), as illustrated in Fig. 21–26.

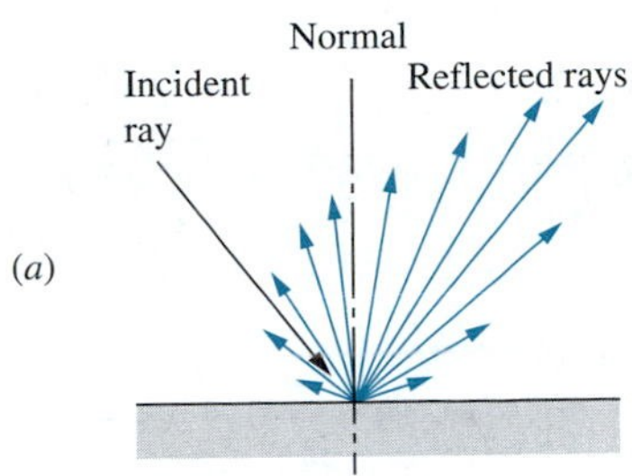

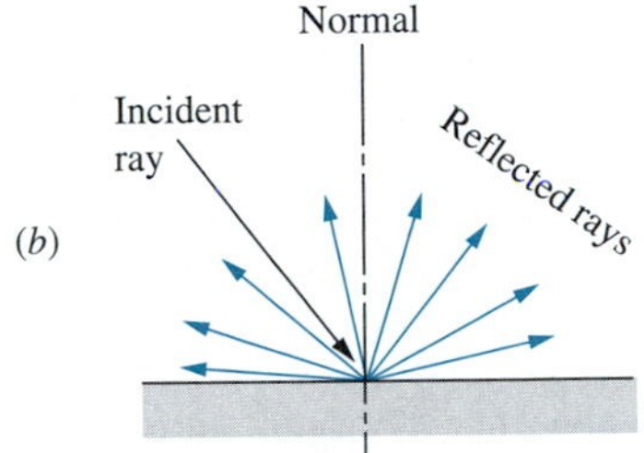

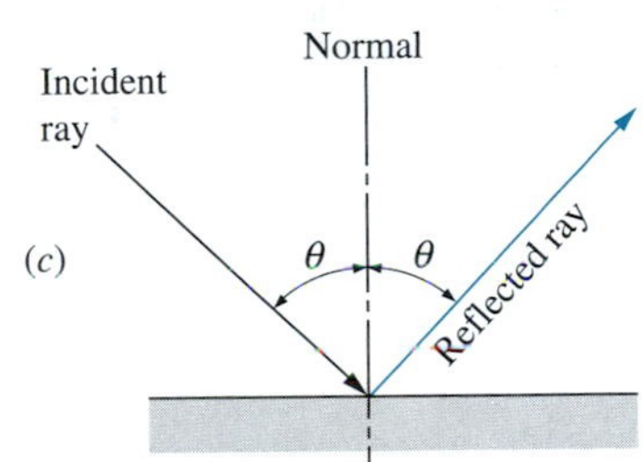

FIGURE 21–24
Different types of reflection from a surface: (*a*) actual or irregular, (*b*) diffuse, and (*c*) specular or mirrorlike.

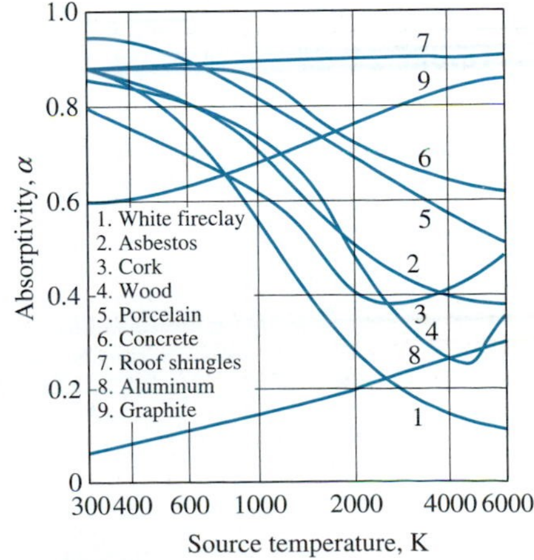

FIGURE 21–25
Variation of absorptivity with the temperature of the source of irradiation for various common materials at room temperature.

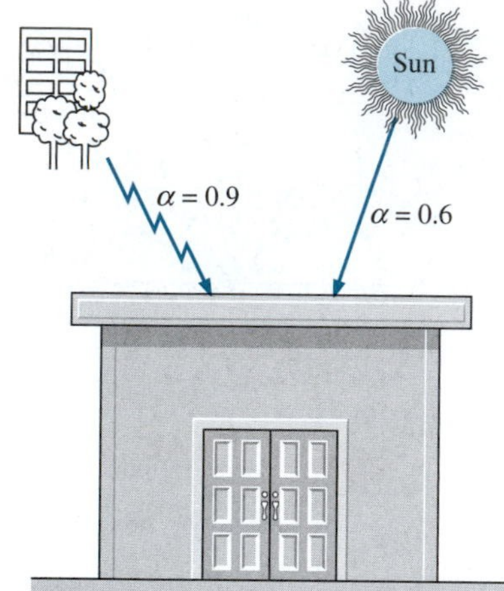

FIGURE 21–26
The absorptivity of a material may be quite different for radiation originating from sources at different temperatures.

Notice that the absorptivity of aluminum increases with the source temperature, a characteristic for metals, and the absorptivity of electric nonconductors, in general, decreases with temperature. This decrease is most pronounced for surfaces that appear white to the eye. For example, the absorptivity of a white painted surface is low for solar radiation, but it is rather high for infrared radiation.

Kirchhoff's Law

Consider a small body of surface area A_s, emissivity ε, and absorptivity α at temperature T contained in a large isothermal enclosure at the same temperature, as shown in Fig. 21–27. Recall that a large isothermal enclosure forms a blackbody cavity regardless of the radiative properties of the enclosure surface, and the body in the enclosure is too small to interfere with the blackbody nature of the cavity. Therefore, the radiation incident on any part of the surface of the small body is equal to the radiation emitted by a blackbody at temperature T. That is, $G = E_b(T) = \sigma T^4$, and the radiation absorbed by the small body per unit of its surface area is

$$G_{\text{abs}} = \alpha G = \alpha \sigma T^4$$

The radiation emitted by the small body is

$$E_{\text{emit}} = \varepsilon \sigma T^4$$

Considering that the small body is in thermal equilibrium with the enclosure, the net rate of heat transfer to the body must be zero. Therefore, the radiation emitted by the body must be equal to the radiation absorbed by it:

$$A_s \varepsilon \sigma T^4 = A_s \alpha \sigma T^4 \tag{21–24}$$

Thus, we conclude that

$$\varepsilon(T) = \alpha(T) \tag{21–25}$$

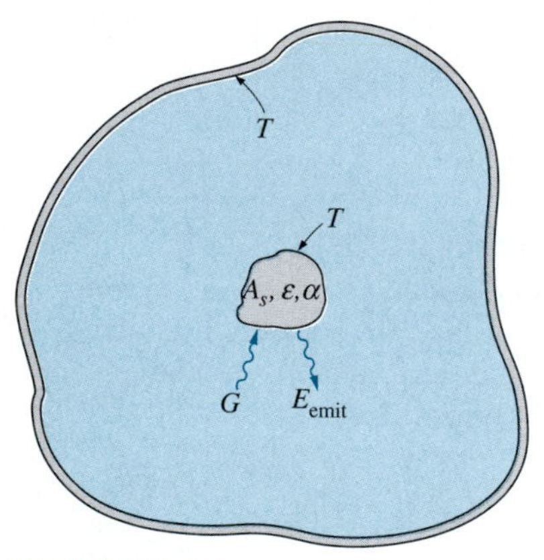

FIGURE 21–27
The small body contained in a large isothermal enclosure used in the development of Kirchhoff's law.

That is, *the total hemispherical emissivity of a surface at temperature T is equal to its total hemispherical absorptivity for radiation coming from a blackbody at the same temperature*. This relation, which greatly simplifies the radiation analysis, was first developed by Gustav Kirchhoff in 1860 and is now called **Kirchhoff's law**. Note that this relation is derived under the condition that the surface temperature is equal to the temperature of the source of irradiation, and the reader is cautioned against using it when considerable difference (more than a few hundred degrees) exists between the surface temperature and the temperature of the source of irradiation.

The derivation above can also be repeated for radiation at a specified wavelength to obtain the *spectral* form of Kirchhoff's law:

$$\varepsilon_\lambda(T) = \alpha_\lambda(T) \tag{21–26}$$

This relation is valid when the irradiation or the emitted radiation is independent of direction. The form of Kirchhoff's law that involves no restrictions is the *spectral directional* form expressed as $\varepsilon_{\lambda,\theta}(T) = \alpha_{\lambda,\theta}(T)$. That is, the emissivity of a surface at a specified wavelength, direction, and temperature is always equal to its absorptivity at the same wavelength, direction, and temperature.

It is very tempting to use Kirchhoff's law in radiation analysis since the relation $\varepsilon = \alpha$ together with $\rho = 1 - \alpha$ enables us to determine all three properties of an opaque surface from a knowledge of only *one* property. Although Eq. 21–25 gives acceptable results in most cases, in practice, care should be exercised when there is considerable difference between the surface temperature and the temperature of the source of incident radiation.

The Greenhouse Effect

You have probably noticed that when you leave your car under direct sunlight on a sunny day, the interior of the car gets much warmer than the air outside, and you may have wondered why the car acts like a *heat trap*. The answer lies in the spectral transmissivity curve of the *glass*, which resembles an inverted U, as shown in Fig. 21–28. We observe from this figure that glass at thicknesses encountered in practice transmits over 90 percent of radiation in the visible range and is practically opaque (nontransparent) to radiation in the longer-wavelength infrared regions of the electromagnetic spectrum (roughly $\lambda > 3$ μm). Therefore, glass has a transparent window in the wavelength range 0.3 μm $< \lambda <$ 3 μm in which over 90 percent of solar radiation is emitted. On the other hand, the entire radiation emitted by surfaces at room temperature falls in the infrared region. Consequently, glass allows the solar radiation to enter but does not allow the infrared radiation from the interior surfaces to escape. This causes a rise in the interior temperature as a result of the energy buildup in the car. This heating effect, which is due to the nongray characteristic of glass (or clear plastics), is known as the **greenhouse effect**, since it is utilized extensively in greenhouses (Fig. 21–29).

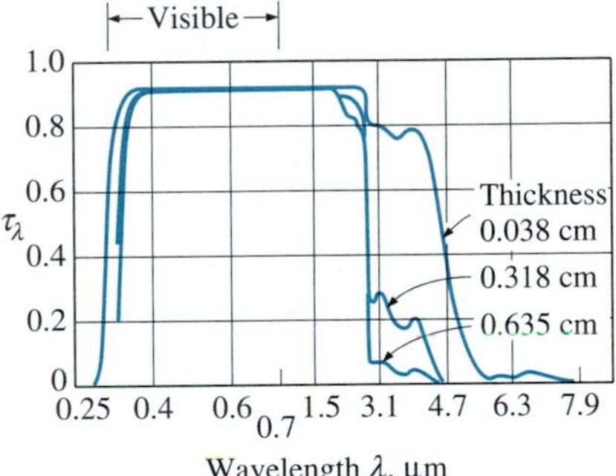

FIGURE 21–28
The spectral transmissivity of low-iron glass at room temperature for different thicknesses.

The greenhouse effect is also experienced on a larger scale on earth. The surface of the earth, which warms up during the day as a result of the absorption of solar energy, cools down at night by radiating its energy into deep space as infrared radiation. The combustion gases such as CO_2 and water vapor in the atmosphere transmit the bulk of the solar radiation but absorb the infrared radiation emitted by the surface of the earth. Thus, there is concern that the energy trapped on earth will eventually cause global warming and thus drastic changes in weather patterns.

In *humid* places such as coastal areas, there is not a large change between the daytime and nighttime temperatures, because the humidity acts as a barrier on the path of the infrared radiation coming from the earth, and thus slows down the cooling process at night. In areas with clear skies such as deserts, there is a large swing between the daytime and nighttime temperatures because of the absence of such barriers for infrared radiation.

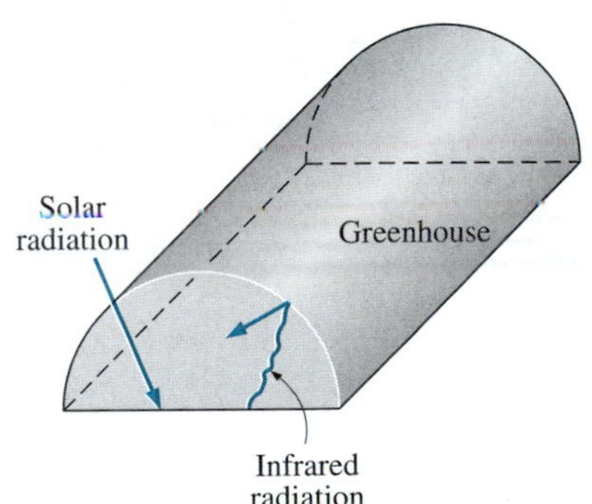

FIGURE 21–29
A greenhouse traps energy by allowing the solar radiation to come in but not allowing the infrared radiation to go out.

21–5 ▪ THE VIEW FACTOR

Radiation heat transfer between surfaces depends on the *orientation* of the surfaces relative to each other as well as their radiation properties and

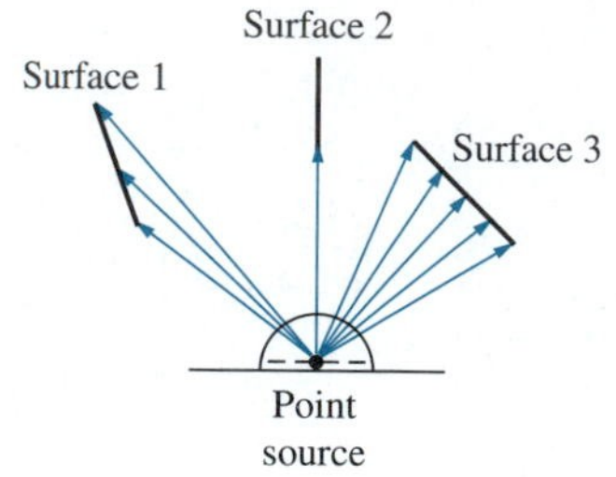

FIGURE 21–30

Radiation heat exchange between surfaces depends on the *orientation* of the surfaces relative to each other, and this dependence on orientation is accounted for by the *view factor.*

temperatures, as illustrated in Fig. 21–30. For example, a camper can make the most use of a campfire on a cold night by standing as close to the fire as possible and by blocking as much of the radiation coming from the fire by turning his or her front to the fire instead of the side. Likewise, a person can maximize the amount of solar radiation incident on him or her and take a sunbath by lying down on his or her back instead of standing.

To account for the effects of orientation on radiation heat transfer between two surfaces, we define a new parameter called the *view factor*, which is a purely geometric quantity and is independent of the surface properties and temperature. It is also called the *shape factor*, *configuration factor*, and *angle factor*. The view factor based on the assumption that the surfaces are diffuse emitters and diffuse reflectors is called the *diffuse view factor*, and the view factor based on the assumption that the surfaces are diffuse emitters but specular reflectors is called the *specular view factor*. In this book, we consider radiation exchange between diffuse surfaces only, and thus the term *view factor* simply means *diffuse view factor*.

The view factor from a surface i to a surface j is denoted by $F_{i \to j}$ or just F_{ij}, and is defined as

$$F_{ij} = \textit{the fraction of the radiation leaving surface i that strikes surface j directly}$$

The notation $F_{i \to j}$ is instructive for beginners, since it emphasizes that the view factor is for radiation that travels from surface i to surface j. However, this notation becomes rather awkward when it has to be used many times in a problem. In such cases, it is convenient to replace it by its *shorthand* version F_{ij}.

The view factor F_{12} represents the fraction of radiation leaving surface 1 that strikes surface 2 directly, and F_{21} represents the fraction of radiation leaving surface 2 that strikes surface 1 directly. Note that the radiation that strikes a surface does not need to be absorbed by that surface. Also, radiation that strikes a surface after being reflected by other surfaces is not considered in the evaluation of view factors.

For the special case of $j = i$, we have

$$F_{i \to i} = \textit{the fraction of radiation leaving surface i that strikes itself directly}$$

Noting that in the absence of strong electromagnetic fields radiation beams travel in straight paths, the view factor from a surface to itself is zero unless the surface "sees" itself. Therefore, $F_{i \to i} = 0$ for *plane* or *convex* surfaces and $F_{i \to i} \neq 0$ for concave surfaces, as illustrated in Fig. 21–31.

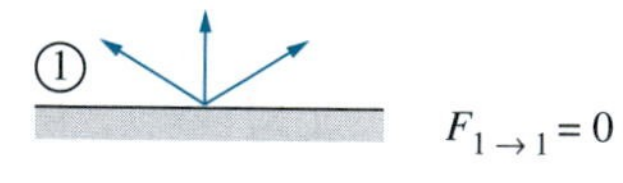

(*a*) Plane surface

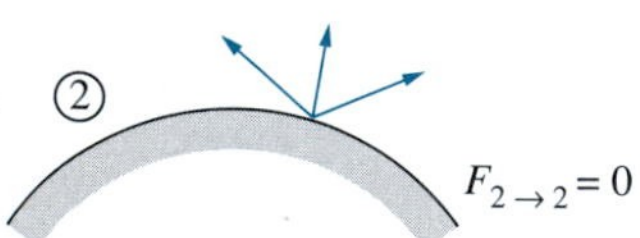

(*b*) Convex surface

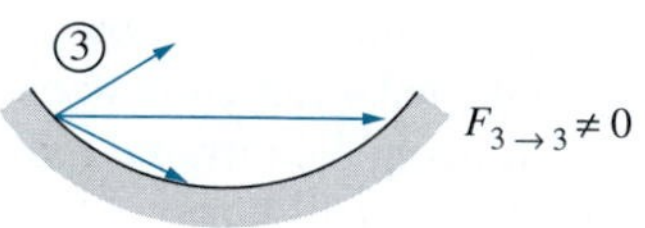

(*c*) Concave surface

FIGURE 21–31

The view factor from a surface to itself is *zero* for *plane* or *convex* surfaces and *nonzero* for *concave* surfaces.

The value of the view factor ranges between *zero* and *one*. The limiting case $F_{i \to j} = 0$ indicates that the two surfaces do not have a direct view of each other, and thus radiation leaving surface i cannot strike surface j directly. The other limiting case $F_{i \to j} = 1$ indicates that surface j completely surrounds surface i, so that the entire radiation leaving surface i is intercepted by surface j. For example, in a geometry consisting of two concentric spheres, the entire radiation leaving the surface of the smaller sphere (surface 1) strikes the larger sphere (surface 2), and thus $F_{1 \to 2} = 1$, as illustrated in Fig. 21–32.

The view factor has proven to be very useful in radiation analysis because it allows us to express the *fraction of radiation* leaving a surface that strikes another surface in terms of the orientation of these two surfaces relative to each other. The underlying assumption in this process is that the radiation a surface receives from a source is directly proportional to the angle the surface subtends when viewed from the source. This would be the case only if the radiation coming off the source is *uniform* in all directions throughout its surface and the medium between the surfaces does not *absorb*, *emit*, or *scatter* radiation. That is, it is the case when the surfaces are *isothermal* and *diffuse* emitters and reflectors and the surfaces are separated by a *nonparticipating* medium such as a vacuum or air.

The view factor $F_{1 \to 2}$ between two surfaces A_1 and A_2 can be determined in a systematic manner first by expressing the view factor between two differential areas dA_1 and dA_2 in terms of the spatial variables and then by performing the necessary integrations. However, this approach is not practical, since, even for simple geometries, the resulting integrations are usually very complex and difficult to perform.

View factors for hundreds of common geometries are evaluated and the results are given in analytical, graphical, and tabular form in several publications. View factors for selected geometries are given in Tables 21–3 and 21–4 in *analytical* form and in Figs. 21–33 to 21–36 in *graphical* form. The view factors in Table 21–3 are for three-dimensional geometries. The view factors in Table 21–4, on the other hand, are for geometries that are *infinitely long* in the direction perpendicular to the plane of the paper and are therefore two-dimensional.

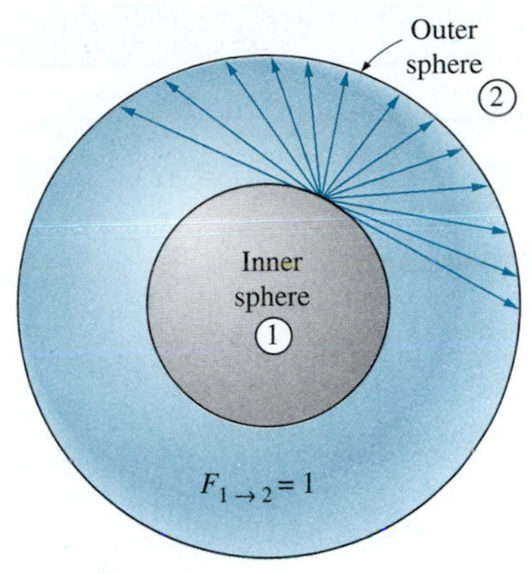

FIGURE 21–32

In a geometry that consists of two concentric spheres, the view factor $F_{1 \to 2} = 1$ since the entire radiation leaving the surface of the smaller sphere is intercepted by the larger sphere.

View Factor Relations

Radiation analysis on an enclosure consisting of N surfaces requires the evaluation of N^2 view factors, and this evaluation process is probably the most time-consuming part of a radiation analysis. However, it is neither practical nor necessary to evaluate all of the view factors directly. Once a sufficient number of view factors are available, the rest of them can be determined by utilizing some fundamental relations for view factors, as discussed next.

1 The Reciprocity Relation

The view factors $F_{i \to j}$ and $F_{j \to i}$ are *not* equal to each other unless the areas of the two surfaces are. That is,

$$F_{j \to i} = F_{i \to j} \quad \text{when} \quad A_i = A_j$$
$$F_{j \to i} \neq F_{i \to j} \quad \text{when} \quad A_i \neq A_j$$

We have shown earlier that the pair of view factors $F_{i \to j}$ and $F_{j \to i}$ are related to each other by

$$A_i F_{i \to j} = A_j F_{j \to i} \tag{21-27}$$

TABLE 21–3

View factor expressions for some common geometries of finite size (3-D)

Geometry	Relation
Aligned parallel rectangles L, j, i, Y, X	$\bar{X} = X/L$ and $\bar{Y} = Y/L$ $F_{i\to j} = \frac{2}{\pi \bar{X}\bar{Y}}\left\{\ln\left[\frac{(1+\bar{X}^2)(1+\bar{Y}^2)}{1+\bar{X}^2+\bar{Y}^2}\right]^{1/2} + \bar{X}(1+\bar{Y}^2)^{1/2}\tan^{-1}\frac{\bar{X}}{(1+\bar{Y}^2)^{1/2}} + \bar{Y}(1+\bar{X}^2)^{1/2}\tan^{-1}\frac{\bar{Y}}{(1+\bar{X}^2)^{1/2}} - \bar{X}\tan^{-1}\bar{X} - \bar{Y}\tan^{-1}\bar{Y}\right\}$
Coaxial parallel disks r_j, j, r_i, L, i	$R_i = r_i/L$ and $R_j = r_j/L$ $S = 1 + \frac{1+R_j^2}{R_i^2}$ $F_{i\to j} = \frac{1}{2}\left\{S - \left[S^2 - 4\left(\frac{r_j}{r_i}\right)^2\right]^{1/2}\right\}$
Perpendicular rectangles with a common edge Z, j, i, Y, X	$H = Z/X$ and $W = Y/X$ $F_{i\to j} = \frac{1}{\pi W}\left(W\tan^{-1}\frac{1}{W} + H\tan^{-1}\frac{1}{H} - (H^2+W^2)^{1/2}\tan^{-1}\frac{1}{(H^2+W^2)^{1/2}} + \frac{1}{4}\ln\left\{\frac{(1+W^2)(1+H^2)}{1+W^2+H^2}\left[\frac{W^2(1+W^2+H^2)}{(1+W^2)(W^2+H^2)}\right]^{W^2} \times \left[\frac{H^2(1+H^2+W^2)}{(1+H^2)(H^2+W^2)}\right]^{H^2}\right\}\right)$

This relation is referred to as the **reciprocity relation** or the **reciprocity rule**, and it enables us to determine the counterpart of a view factor from a knowledge of the view factor itself and the areas of the two surfaces. When determining the pair of view factors $F_{i\to j}$ and $F_{j\to i}$, it makes sense to evaluate first the easier one directly and then the more difficult one by applying the reciprocity relation.

2 The Summation Rule

The radiation analysis of a surface normally requires the consideration of the radiation coming in or going out in all directions. Therefore, most radiation problems encountered in practice involve enclosed spaces. When for-

TABLE 21–4

View factor expressions for some infinitely long (2-D) geometries

Geometry	Relation
Parallel plates with midlines connected by perpendicular line 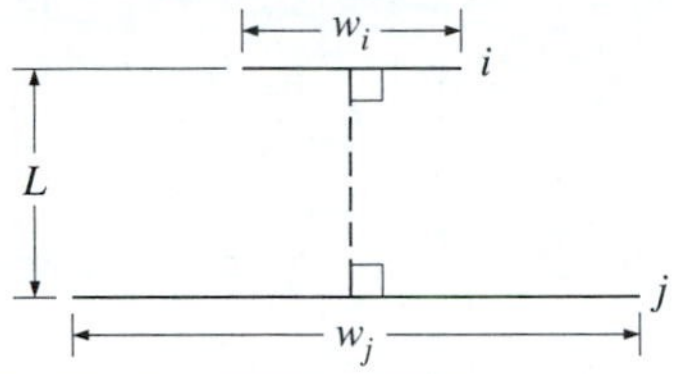	$W_i = w_i/L$ and $W_j = w_j/L$ $F_{i \to j} = \dfrac{[(W_i + W_j)^2 + 4]^{1/2} - (W_j - W_i)^2 + 4]^{1/2}}{2W_i}$
Inclined plates of equal width and with a common edge 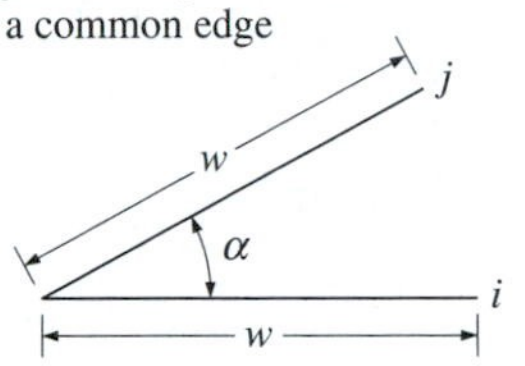	$F_{i \to j} = 1 - \sin \frac{1}{2}\alpha$
Perpendicular plates with a common edge 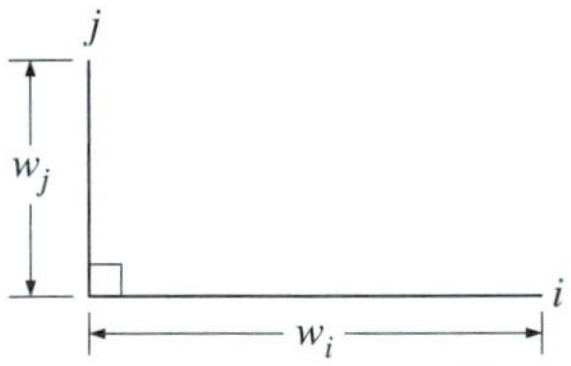	$F_{i \to j} = \frac{1}{2}\left\{1 + \frac{w_j}{w_i} - \left[1 + \left(\frac{w_j}{w_i}\right)^2\right]^{1/2}\right\}$
Three-sided enclosure 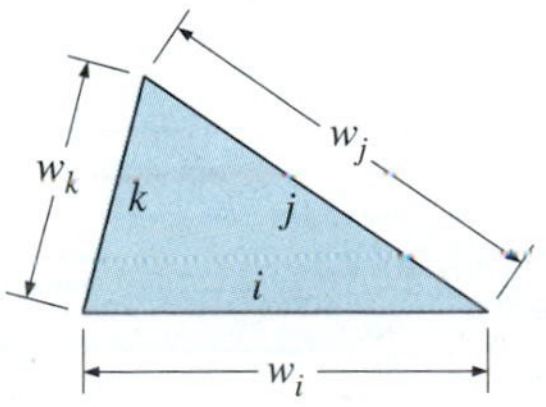	$F_{i \to j} = \dfrac{w_i + w_j - w_k}{2w_i}$
Infinite plane and row of cylinders 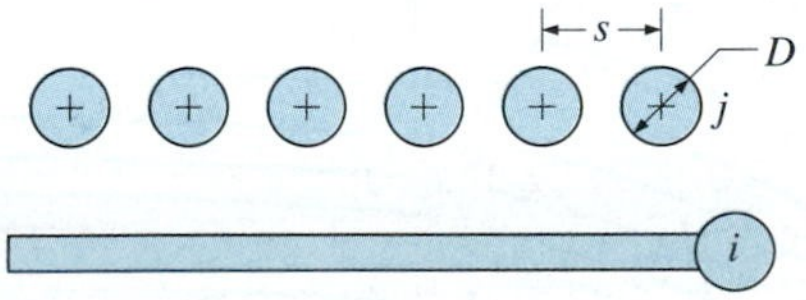	$F_{i \to j} = 1 - \left[1 - \left(\frac{D}{s}\right)^2\right]^{1/2} + \frac{D}{s}\tan^{-1}\left(\frac{s^2 - D^2}{D^2}\right)^{1/2}$

mulating a radiation problem, we usually form an *enclosure* consisting of the surfaces interacting radiatively. Even openings are treated as imaginary surfaces with radiation properties equivalent to those of the opening.

The conservation of energy principle requires that the entire radiation leaving any surface i of an enclosure be intercepted by the surfaces of

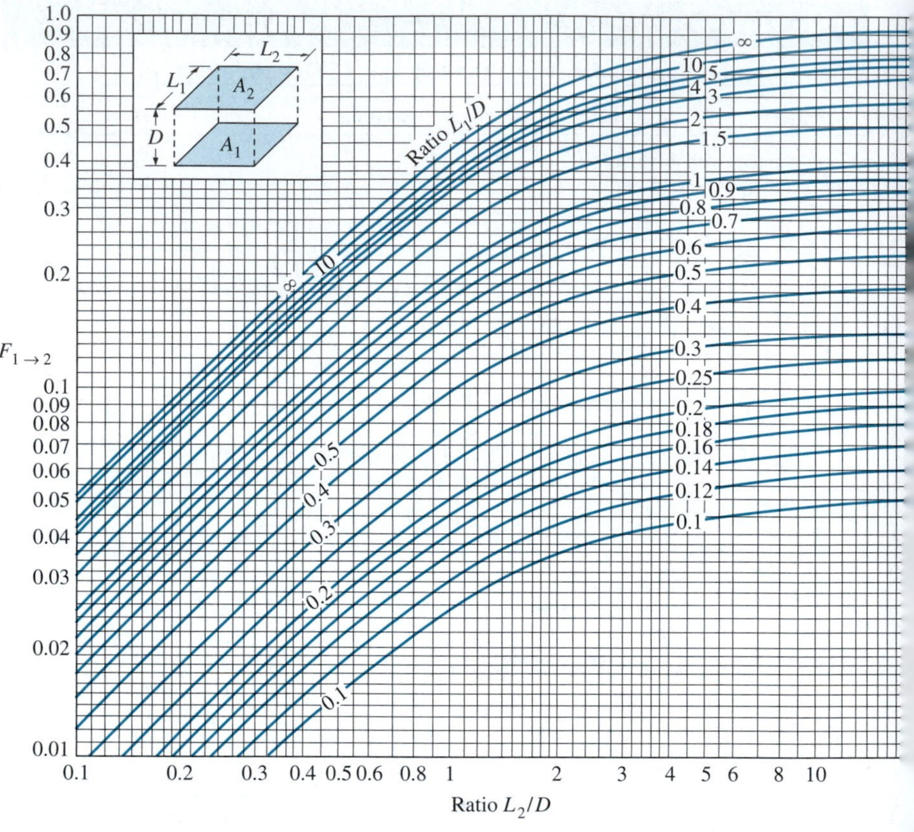

FIGURE 21–33

View factor between two aligned parallel rectangles of equal size.

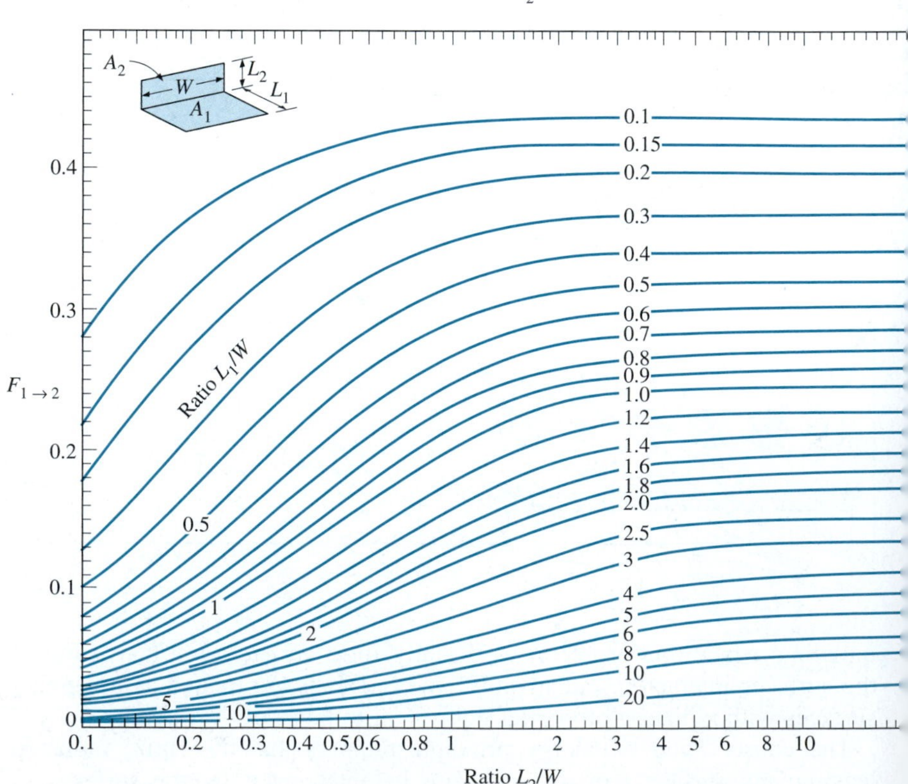

FIGURE 21–34

View factor between two perpendicular rectangles with a common edge.

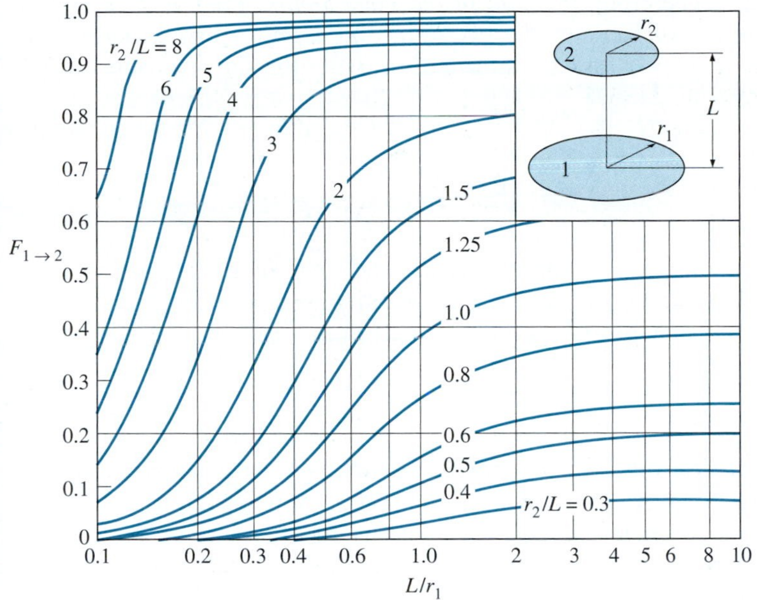

FIGURE 21–35

View factor between two coaxial parallel disks.

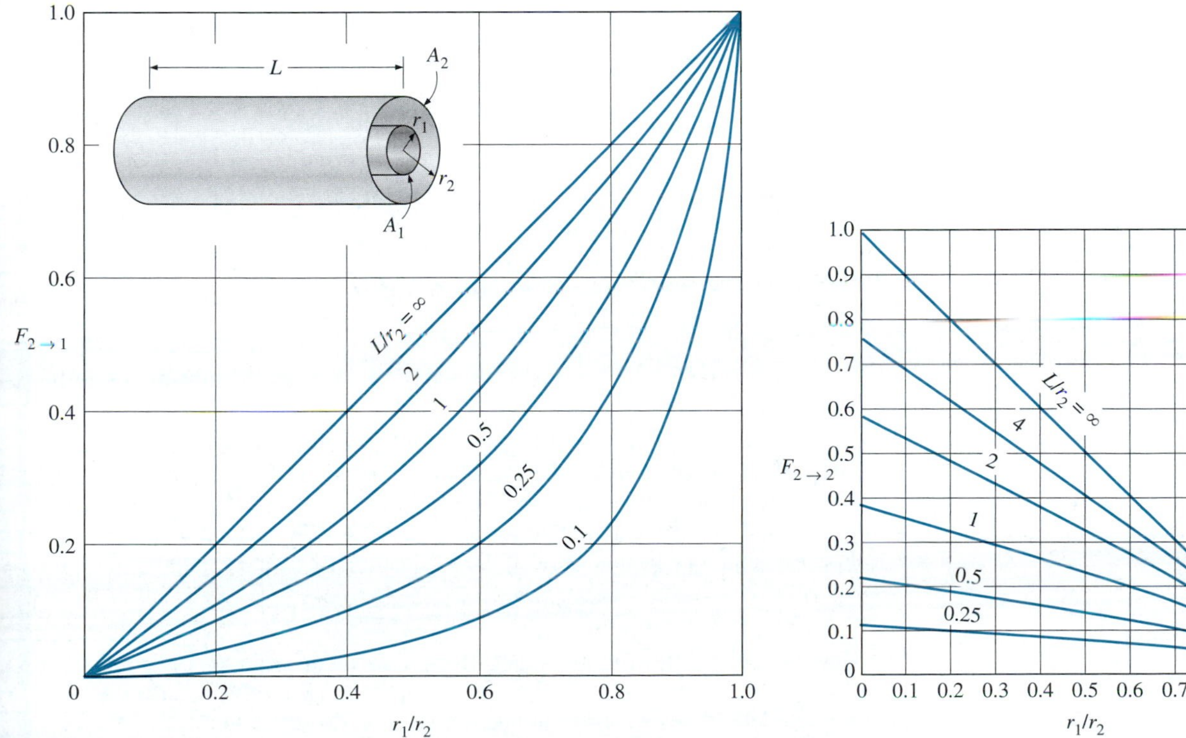

FIGURE 21–36

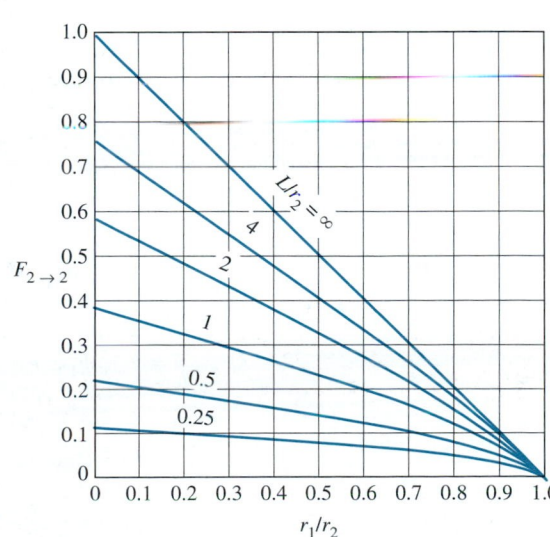

View factors for two concentric cylinders of finite length: (*a*) outer cylinder to inner cylinder; (*b*) outer cylinder to itself.

FIGURE 21–37

Radiation leaving any surface i of an enclosure must be intercepted completely by the surfaces of the enclosure. Therefore, the sum of the view factors from surface i to each one of the surfaces of the enclosure must be unity.

the enclosure. Therefore, *the sum of the view factors from surface i of an enclosure to all surfaces of the enclosure, including to itself, must equal unity*. This is known as the **summation rule** for an enclosure and is expressed as (Fig. 21–37)

$$\sum_{j=1}^{N} F_{i \to j} = 1 \tag{21–28}$$

where N is the number of surfaces of the enclosure. For example, applying the summation rule to surface 1 of a three-surface enclosure yields

$$\sum_{j=1}^{3} F_{1 \to j} = F_{1 \to 1} + F_{1 \to 2} + F_{1 \to 3} = 1$$

The summation rule can be applied to each surface of an enclosure by varying i from 1 to N. Therefore, the summation rule applied to each of the N surfaces of an enclosure gives N relations for the determination of the view factors. Also, the reciprocity rule gives $\frac{1}{2}N(N-1)$ additional relations. Then the total number of view factors that need to be evaluated directly for an N-surface enclosure becomes

$$N^2 - [N + \tfrac{1}{2}N(N-1)] = \tfrac{1}{2}N(N-1)$$

For example, for a six-surface enclosure, we need to determine only $\frac{1}{2} \times 6(6-1) = 15$ of the $6^2 = 36$ view factors directly. The remaining 21 view factors can be determined from the 21 equations that are obtained by applying the reciprocity and the summation rules.

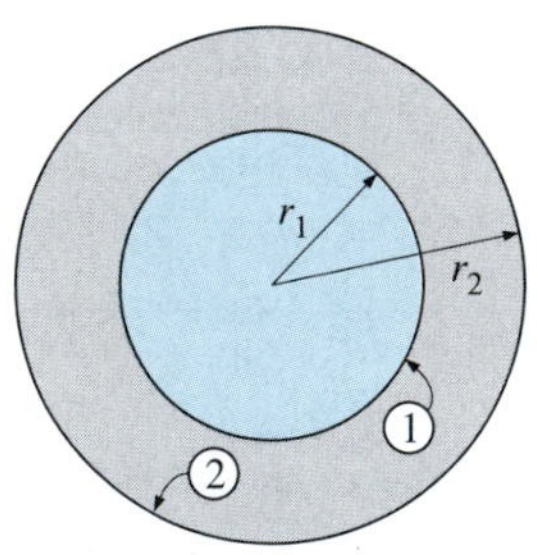

FIGURE 21–38

The geometry considered in Example 21–4.

EXAMPLE 21–4 View Factors Associated with Two Concentric Spheres

Determine the view factors associated with an enclosure formed by two concentric spheres, shown in Fig. 21–38.

Solution The view factors associated with two concentric spheres are to be determined.

Assumptions The surfaces are diffuse emitters and reflectors.

Analysis The outer surface of the smaller sphere (surface 1) and inner surface of the larger sphere (surface 2) form a two-surface enclosure. Therefore, $N = 2$ and this enclosure involves $N^2 = 2^2 = 4$ view factors, which are F_{11}, F_{12}, F_{21}, and F_{22}. In this two-surface enclosure, we need to determine only

$$\tfrac{1}{2}N(N-1) = \tfrac{1}{2} \times 2(2-1) = 1$$

view factor directly. The remaining three view factors can be determined by the application of the summation and reciprocity rules. But it turns out that we can determine not only one but *two* view factors directly in this case by a simple *inspection:*

$F_{11} = 0$, since no radiation leaving surface 1 strikes itself

$F_{12} = 1$, since all radiation leaving surface 1 strikes surface 2

Actually it would be sufficient to determine only one of these view factors by inspection, since we could always determine the other one from the summation rule applied to surface 1 as $F_{11} + F_{12} = 1$.

The view factor F_{21} is determined by applying the reciprocity relation to surfaces 1 and 2:

$$A_1F_{12} = A_2F_{21}$$

which yields

$$F_{21} = \frac{A_1}{A_2}F_{12} = \frac{4\pi r_1^2}{4\pi r_2^2} \times 1 = \left(\frac{r_1}{r_2}\right)^2$$

Finally, the view factor F_{22} is determined by applying the summation rule to surface 2:

$$F_{21} + F_{22} = 1$$

and thus

$$F_{22} = 1 - F_{21} = 1 - \left(\frac{r_1}{r_2}\right)^2$$

Discussion Note that when the outer sphere is much larger than the inner sphere ($r_2 \gg r_1$), F_{22} approaches one. This is expected, since the fraction of radiation leaving the outer sphere that is intercepted by the inner sphere will be negligible in that case. Also note that the two spheres considered above do not need to be concentric. However, the radiation analysis will be most accurate for the case of concentric spheres, since the radiation is most likely to be uniform on the surfaces in that case.

3 The Superposition Rule

Sometimes the view factor associated with a given geometry is not available in standard tables and charts. In such cases, it is desirable to express the given geometry as the sum or difference of some geometries with known view factors, and then to apply the **superposition rule**, which can be expressed as *the view factor from a surface i to a surface j is equal to the sum of the view factors from surface i to the parts of surface j*. Note that the reverse of this is not true. That is, the view factor from a surface j to a surface i is *not* equal to the sum of the view factors from the parts of surface j to surface i.

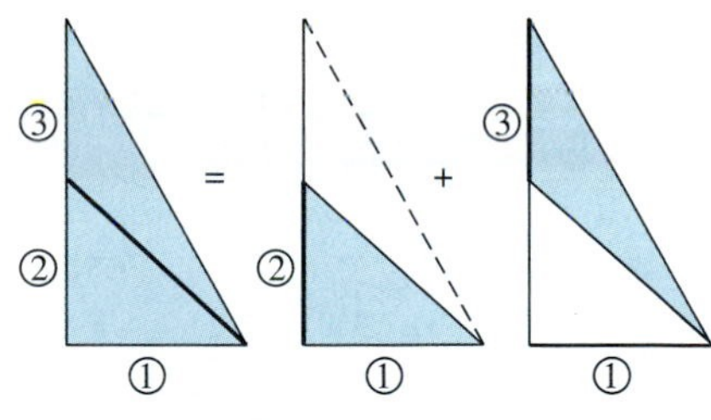

FIGURE 21–39

The view factor from a surface to a composite surface is equal to the sum of the view factors from the surface to the parts of the composite surface.

Consider the geometry in Fig. 21–39, which is infinitely long in the direction perpendicular to the plane of the paper. The radiation that leaves surface 1 and strikes the combined surfaces 2 and 3 is equal to the sum of the radiation that strikes surfaces 2 and 3. Therefore, the view factor from surface 1 to the combined surfaces of 2 and 3 is

$$F_{1\to(2,3)} = F_{1\to2} + F_{1\to3} \qquad \textbf{(21–29)}$$

Suppose we need to find the view factor $F_{1\to3}$. A quick check of the view factor expressions and charts in this section reveals that such a view factor cannot be evaluated directly. However, the view factor $F_{1 \to 3}$ can be determined from Eq. 21–29 after determining both $F_{1 \to 2}$ and $F_{1 \to (2, 3)}$ from the chart in Table 21–4. Therefore, it may be possible to determine some difficult view factors with relative ease by expressing one or both of the areas as the sum or differences of areas and then applying the superposition rule.

To obtain a relation for the view factor $F_{(2, 3) \to 1}$, we multiply Eq. 21–29 by A_1,

$$A_1F_{1 \to (2, 3)} = A_1F_{1 \to 2} + A_1F_{1 \to 3}$$

and apply the reciprocity relation to each term to get

$$(A_2 + A_3)F_{(2, 3) \to 1} = A_2F_{2 \to 1} + A_3F_{3 \to 1}$$

or

$$F_{(2, 3) \to 1} = \frac{A_2 F_{2 \to 1} + A_3 F_{3 \to 1}}{A_2 + A_3} \tag{21–30}$$

Areas that are expressed as the sum of more than two parts can be handled in a similar manner.

EXAMPLE 21–5 Fraction of Radiation Leaving through an Opening

Determine the fraction of the radiation leaving the base of the cylindrical enclosure shown in Fig. 21–40 that escapes through a coaxial ring opening at its top surface. The radius and the length of the enclosure are $r_1 = 10$ cm and $L = 10$ cm, while the inner and outer radii of the ring are $r_2 = 5$ cm and $r_3 = 8$ cm, respectively.

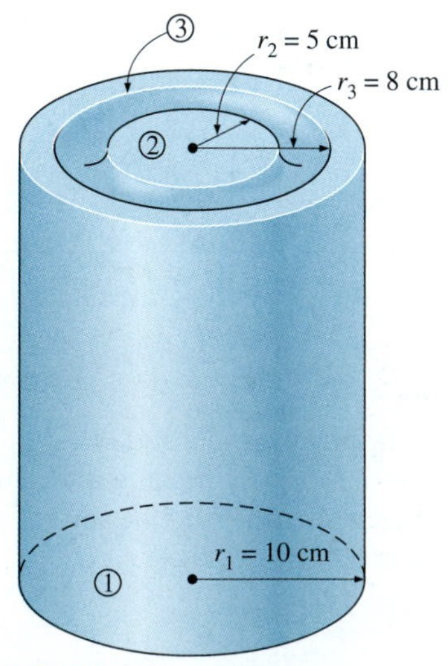

FIGURE 21–40
The cylindrical enclosure considered in Example 21–5.

Solution The fraction of radiation leaving the base of a cylindrical enclosure through a coaxial ring opening at its top surface is to be determined.

Assumptions The base surface is a diffuse emitter and reflector.

Analysis We are asked to determine the fraction of the radiation leaving the base of the enclosure that escapes through an opening at the top surface. Actually, what we are asked to determine is simply the *view factor* $F_{1 \to \text{ring}}$ from the base of the enclosure to the ring-shaped surface at the top.

We do not have an analytical expression or chart for view factors between a circular area and a coaxial ring, and so we cannot determine $F_{1 \to \text{ring}}$ directly. However, we do have a chart for view factors between two coaxial parallel disks, and we can always express a ring in terms of disks.

Let the base surface of radius $r_1 = 10$ cm be surface 1, the circular area of $r_2 = 5$ cm at the top be surface 2, and the circular area of $r_3 = 8$ cm be surface 3. Using the superposition rule, the view factor from surface 1 to surface 3 can be expressed as

$$F_{1 \to 3} = F_{1 \to 2} + F_{1 \to \text{ring}}$$

since surface 3 is the sum of surface 2 and the ring area. The view factors $F_{1\to 2}$ and $F_{1\to 3}$ are determined from the chart in Fig. 21–35.

$$\frac{L}{r_1} = \frac{10\text{ cm}}{10\text{ cm}} = 1 \quad\text{and}\quad \frac{r_2}{L} = \frac{5\text{ cm}}{10\text{ cm}} = 0.5 \xrightarrow{\text{(Fig. 21–35)}} F_{1\to 2} = 0.11$$

$$\frac{L}{r_1} = \frac{10\text{ cm}}{10\text{ cm}} = 1 \quad\text{and}\quad \frac{r_3}{L} = \frac{8\text{ cm}}{10\text{ cm}} = 0.8 \xrightarrow{\text{(Fig. 21–35)}} F_{1\to 3} = 0.28$$

Therefore,

$$F_{1\to \text{ring}} = F_{1\to 3} - F_{1\to 2} = 0.28 - 0.11 = \mathbf{0.17}$$

which is the desired result. Note that $F_{1\to 2}$ and $F_{1\to 3}$ represent the fractions of radiation leaving the base that strike the circular surfaces 2 and 3, respectively, and their difference gives the fraction that strikes the ring area.

4 The Symmetry Rule

The determination of the view factors in a problem can be simplified further if the geometry involved possesses some sort of symmetry. Therefore, it is good practice to check for the presence of any *symmetry* in a problem before attempting to determine the view factors directly. The presence of symmetry can be determined *by inspection*, keeping the definition of the view factor in mind. Identical surfaces that are oriented in an identical manner with respect to another surface will intercept identical amounts of radiation leaving that surface. Therefore, the **symmetry rule** can be expressed as *two (or more) surfaces that possess symmetry about a third surface will have identical view factors from that surface* (Fig. 21–41).

The symmetry rule can also be expressed as *if the surfaces j and k are symmetric about the surface* i *then* $F_{i\to j} = F_{i\to k}$. Using the reciprocity rule, we can show that the relation $F_{j\to i} = F_{k\to i}$ is also true in this case.

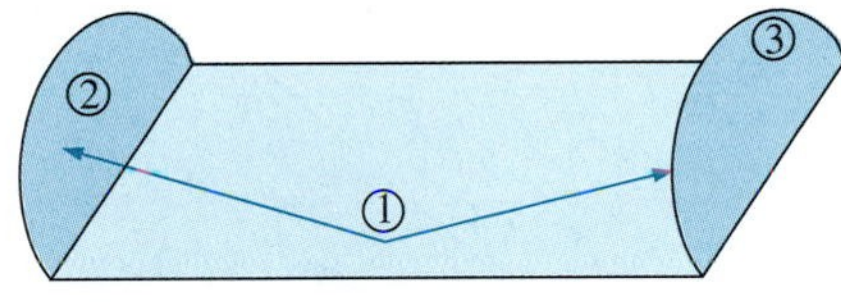

$F_{1\to 2} = F_{1\to 3}$
(Also, $F_{2\to 1} = F_{3\to 1}$)

FIGURE 21–41

Two surfaces that are symmetric about a third surface will have the same view factor from the third surface.

EXAMPLE 21–6 View Factors Associated with a Tetragon

Determine the view factors from the base of the pyramid shown in Fig. 21–42 to each of its four side surfaces. The base of the pyramid is a square, and its side surfaces are isosceles triangles.

Solution The view factors from the base of a pyramid to each of its four side surfaces for the case of a square base are to be determined.

Assumptions The surfaces are diffuse emitters and reflectors.

Analysis The base of the pyramid (surface 1) and its four side surfaces (surfaces 2, 3, 4, and 5) form a five-surface enclosure. The first thing we notice about this enclosure is its symmetry. The four side surfaces are symmetric about the base surface. Then, from the *symmetry rule*, we have

$$F_{12} = F_{13} = F_{14} = F_{15}$$

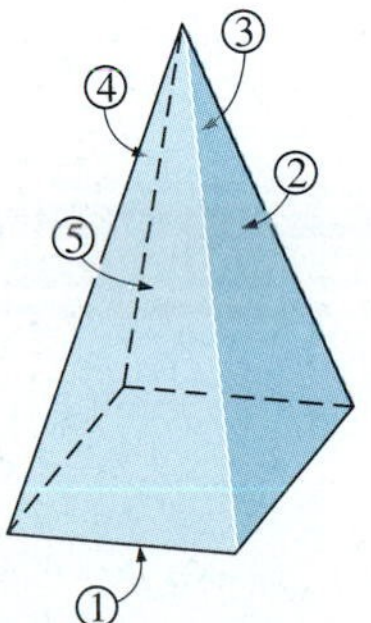

FIGURE 21–42

The pyramid considered in Example 21–6.

Also, the *summation rule* applied to surface 1 yields

$$\sum_{j=1}^{5} F_{1j} = F_{11} + F_{12} + F_{13} + F_{14} + F_{15} = 1$$

However, $F_{11} = 0$, since the base is a *flat* surface. Then the two relations above yield

$$F_{12} = F_{13} = F_{14} = F_{15} = \mathbf{0.25}$$

Discussion Note that each of the four side surfaces of the pyramid receive one-fourth of the entire radiation leaving the base surface, as expected. Also note that the presence of symmetry greatly simplified the determination of the view factors.

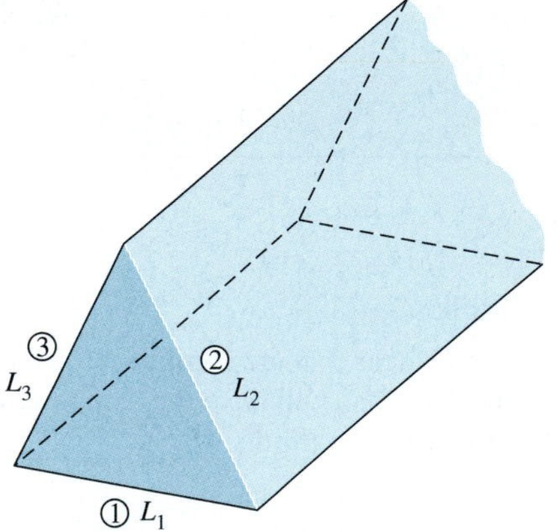

FIGURE 21–43

The infinitely long triangular duct considered in Example 21–7.

EXAMPLE 21–7 View Factors Associated with a Triangular Duct

Determine the view factor from any one side to any other side of the infinitely long triangular duct whose cross section is given in Fig. 21–43.

Solution The view factors associated with an infinitely long triangular duct are to be determined.

Assumptions The surfaces are diffuse emitters and reflectors.

Analysis The widths of the sides of the triangular cross section of the duct are L_1, L_2, and L_3, and the surface areas corresponding to them are A_1, A_2, and A_3, respectively. Since the duct is infinitely long, the fraction of radiation leaving any surface that escapes through the ends of the duct is negligible. Therefore, the infinitely long duct can be considered to be a three-surface enclosure, $N = 3$.

This enclosure involves $N^2 = 3^2 = 9$ view factors, and we need to determine

$$\tfrac{1}{2}N(N - 1) = \tfrac{1}{2} \times 3(3 - 1) = 3$$

of these view factors directly. Fortunately, we can determine all three of them by inspection to be

$$F_{11} = F_{22} = F_{33} = 0$$

since all three surfaces are flat. The remaining six view factors can be determined by the application of the summation and reciprocity rules.

Applying the summation rule to each of the three surfaces gives

$$F_{11} + F_{12} + F_{13} = 1$$

$$F_{21} + F_{22} + F_{23} = 1$$

$$F_{31} + F_{32} + F_{33} = 1$$

Noting that $F_{11} = F_{22} = F_{33} = 0$ and multiplying the first equation by A_1, the second by A_2, and the third by A_3 gives

$$A_1F_{12} + A_1F_{13} = A_1$$

$$A_2F_{21} + A_2F_{23} = A_2$$

$$A_3F_{31} + A_3F_{32} = A_3$$

Finally, applying the three reciprocity relations $A_1F_{12} = A_2F_{21}$, $A_1F_{13} = A_3F_{31}$, and $A_2F_{23} = A_3F_{32}$ gives

$$A_1F_{12} + A_1F_{13} = A_1$$

$$A_1F_{12} + A_2F_{23} = A_2$$

$$A_1F_{13} + A_2F_{23} = A_3$$

This is a set of three algebraic equations with three unknowns, which can be solved to obtain

$$F_{12} = \frac{A_1 + A_2 - A_3}{2A_1} = \frac{L_1 + L_2 - L_3}{2L_1}$$

$$F_{13} = \frac{A_1 + A_3 - A_2}{2A_1} = \frac{L_1 + L_3 - L_2}{2L_1}$$

$$F_{23} = \frac{A_2 + A_3 - A_1}{2A_2} = \frac{L_2 + L_3 - L_1}{2L_2} \quad \text{(21–31)}$$

Discussion Note that we have replaced the areas of the side surfaces by their corresponding widths for simplicity, since $A = Ls$ and the length s can be factored out and canceled. We can generalize this result as *the view factor from a surface of a very long triangular duct to another surface is equal to the sum of the widths of these two surfaces minus the width of the third surface, divided by twice the width of the first surface.*

View Factors between Infinitely Long Surfaces: The Crossed-Strings Method

Many problems encountered in practice involve geometries of constant cross section such as channels and ducts that are *very long* in one direction relative to the other directions. Such geometries can conveniently be considered to be *two-dimensional*, since any radiation interaction through their end surfaces is negligible. These geometries can subsequently be modeled as being *infinitely long*, and the view factor between their surfaces can be determined by the amazingly simple *crossed-strings method* developed by H. C. Hottel in the 1950s. The surfaces of the geometry do not need to be flat; they can be convex, concave, or any irregular shape.

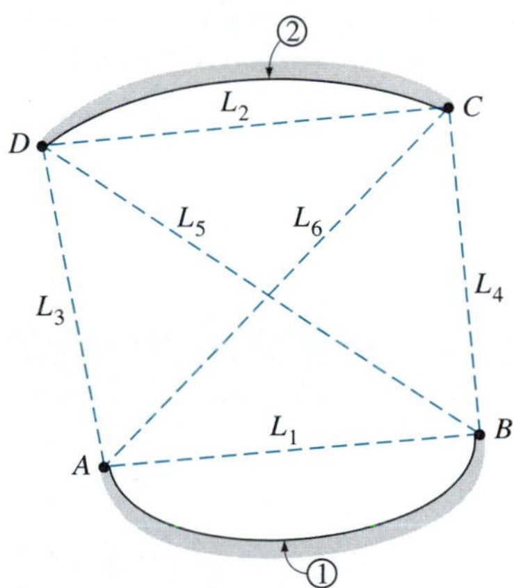

FIGURE 21–44

Determination of the view factor $F_{1 \to 2}$ by the application of the crossed-strings method.

To demonstrate this method, consider the geometry shown in Fig. 21–44, and let us try to find the view factor $F_{1 \to 2}$ between surfaces 1 and 2. The first thing we do is identify the endpoints of the surfaces (the points A, B, C, and D) and connect them to each other with tightly stretched strings, which are indicated by dashed lines. Hottel has shown that the view factor $F_{1 \to 2}$

can be expressed in terms of the lengths of these stretched strings, which are straight lines, as

$$F_{1\to 2} = \frac{(L_5 + L_6) - (L_3 + L_4)}{2L_1} \tag{21-32}$$

Note that $L_5 + L_6$ is the sum of the lengths of the *crossed strings*, and $L_3 + L_4$ is the sum of the lengths of the *uncrossed strings* attached to the endpoints. Therefore, Hottel's crossed-strings method can be expressed verbally as

$$F_{i\to j} = \frac{\Sigma\,(\text{Crossed strings}) - \Sigma\,(\text{Uncrossed strings})}{2 \times (\text{String on surface } i)} \tag{21-33}$$

The crossed-strings method is applicable even when the two surfaces considered share a common edge, as in a triangle. In such cases, the common edge can be treated as an imaginary string of zero length. The method can also be applied to surfaces that are partially blocked by other surfaces by allowing the strings to bend around the blocking surfaces.

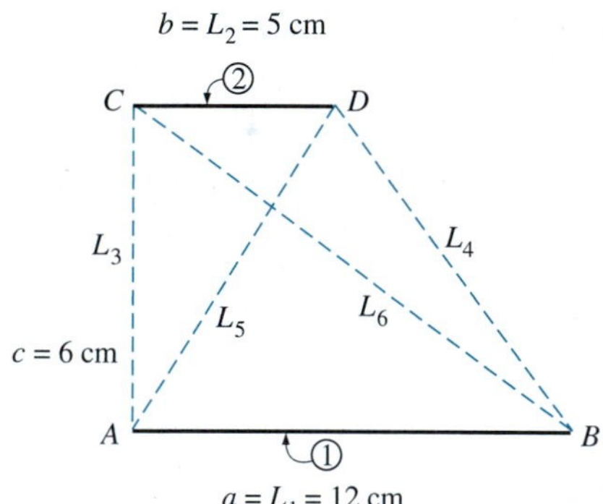

FIGURE 21–45

The two infinitely long parallel plates considered in Example 21–8.

EXAMPLE 21–8 The Crossed-Strings Method for View Factors

Two infinitely long parallel plates of widths $a = 12$ cm and $b = 5$ cm are located a distance $c = 6$ cm apart, as shown in Fig. 21–45. (*a*) Determine the view factor $F_{1\to 2}$ from surface 1 to surface 2 by using the crossed-strings method. (*b*) Derive the crossed-strings formula by forming triangles on the given geometry and using Eq. 21–31 for view factors between the sides of triangles.

Solution The view factors between two infinitely long parallel plates are to be determined using the crossed-strings method, and the formula for the view factor is to be derived.

Assumptions The surfaces are diffuse emitters and reflectors.

Analysis (*a*) First we label the endpoints of both surfaces and draw straight dashed lines between the endpoints, as shown in Fig. 21–45. Then we identify the crossed and uncrossed strings and apply the crossed-strings method (Eq. 21–33) to determine the view factor $F_{1\to 2}$:

$$F_{1\to 2} = \frac{\Sigma\,(\text{Crossed strings}) - \Sigma\,(\text{Uncrossed strings})}{2 \times (\text{String on surface 1})} = \frac{(L_5 + L_6) - (L_3 + L_4)}{2L_1}$$

where

$$L_1 = a = 12 \text{ cm} \qquad L_4 = \sqrt{7^2 + 6^2} = 9.22 \text{ cm}$$
$$L_2 = b = 5 \text{ cm} \qquad L_5 = \sqrt{5^2 + 6^2} = 7.81 \text{ cm}$$
$$L_3 = c = 6 \text{ cm} \qquad L_6 = \sqrt{12^2 + 6^2} = 13.42 \text{ cm}$$

Substituting,

$$F_{1\to 2} = \frac{[(7.81 + 13.42) - (6 + 9.22)] \text{ cm}}{2 \times 12 \text{ cm}} = \mathbf{0.250}$$

(*b*) The geometry is infinitely long in the direction perpendicular to the plane of the paper, and thus the two plates (surfaces 1 and 2) and the two openings (imaginary surfaces 3 and 4) form a four-surface enclosure. Then applying the summation rule to surface 1 yields

$$F_{11} + F_{12} + F_{13} + F_{14} = 1$$

But $F_{11} = 0$ since it is a flat surface. Therefore,

$$F_{12} = 1 - F_{13} - F_{14}$$

where the view factors F_{13} and F_{14} can be determined by considering the triangles *ABC* and *ABD*, respectively, and applying Eq. 21–31 for view factors between the sides of triangles. We obtain

$$F_{13} = \frac{L_1 + L_3 - L_6}{2L_1}, \qquad F_{14} = \frac{L_1 + L_4 - L_5}{2L_1}$$

Substituting,

$$F_{12} = 1 - \frac{L_1 + L_3 - L_6}{2L_1} - \frac{L_1 + L_4 - L_5}{2L_1}$$

$$= \frac{(L_5 + L_6) - (L_3 + L_4)}{2L_1}$$

which is the desired result. This is also a miniproof of the crossed-strings method for the case of two infinitely long plain parallel surfaces.

21–6 ▪ RADIATION HEAT TRANSFER: BLACK SURFACES

So far, we have considered the nature of radiation, the radiation properties of materials, and the view factors, and we are now in a position to consider the rate of heat transfer between surfaces by radiation. The analysis of radiation exchange between surfaces, in general, is complicated because of reflection: a radiation beam leaving a surface may be reflected several times, with partial reflection occurring at each surface, before it is completely absorbed. The analysis is simplified greatly when the surfaces involved can be approximated as blackbodies because of the absence of reflection. In this section, we consider radiation exchange between *black surfaces* only; we extend the analysis to reflecting surfaces in the next section.

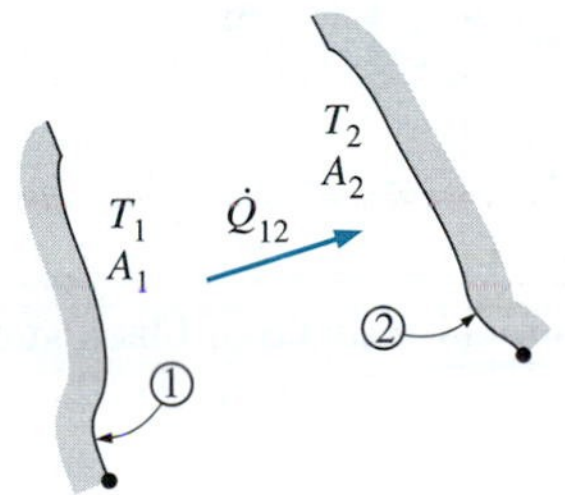

FIGURE 21–46

Two general black surfaces maintained at uniform temperatures T_1 and T_2.

Consider two black surfaces of arbitrary shape maintained at uniform temperatures T_1 and T_2, as shown in Fig. 21–46. Recognizing that radiation leaves a black surface at a rate of $E_b = \sigma T^4$ per unit surface area and that the view factor $F_{1\rightarrow 2}$ represents the fraction of radiation leaving surface 1 that strikes surface 2, the *net* rate of radiation heat transfer from surface 1 to surface 2 can be expressed as

$$\dot{Q}_{1\rightarrow 2} = \begin{pmatrix} \text{Radiation leaving} \\ \text{the entire surface 1} \\ \text{that strikes surface 2} \end{pmatrix} - \begin{pmatrix} \text{Radiation leaving} \\ \text{the entire surface 2} \\ \text{that strikes surface 1} \end{pmatrix}$$

$$= A_1E_{b1}F_{1\rightarrow 2} - A_2E_{b2}F_{2\rightarrow 1} \qquad \text{(W)} \tag{21-34}$$

Applying the reciprocity relation $A_1F_{1\rightarrow 2} = A_2F_{2\rightarrow 1}$ yields

$$\dot{Q}_{1\rightarrow 2} = A_1F_{1\rightarrow 2}\,\sigma(T_1^4 - T_2^4) \qquad \text{(W)} \tag{21-35}$$

which is the desired relation. A negative value for $\dot{Q}_{1\rightarrow 2}$ indicates that net radiation heat transfer is from surface 2 to surface 1.

Now consider an *enclosure* consisting of *N black* surfaces maintained at specified temperatures. The *net* radiation heat transfer *from* any surface *i* of this enclosure is determined by adding up the net radiation heat transfers from surface *i* to each of the surfaces of the enclosure:

$$\dot{Q}_i = \sum_{j=1}^{N} \dot{Q}_{i\rightarrow j} = \sum_{j=1}^{N} A_iF_{i\rightarrow j}\,\sigma(T_i^4 - T_j^4) \qquad \text{(W)} \tag{21-36}$$

Again a negative value for $\dot{Q}$ indicates that net radiation heat transfer is *to* surface *i* (i.e., surface *i gains* radiation energy instead of losing). Also, the net heat transfer from a surface to itself is zero, regardless of the shape of the surface.

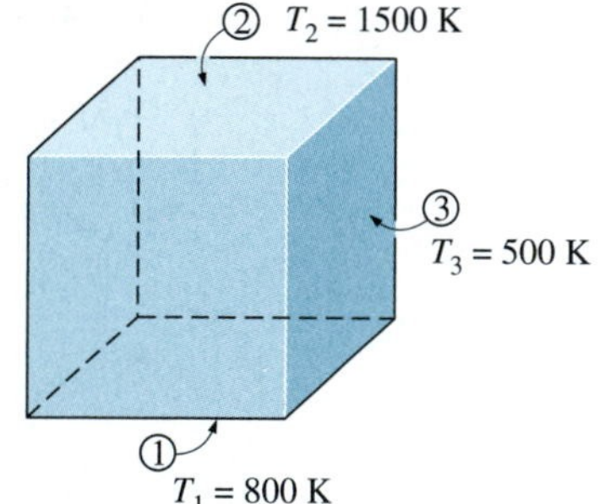

FIGURE 21–47

The cubical furnace of black surfaces considered in Example 21–9.

EXAMPLE 21–9 Radiation Heat Transfer in a Black Furnace

Consider the 5-m × 5-m × 5-m cubical furnace shown in Fig. 21–47, whose surfaces closely approximate black surfaces. The base, top, and side surfaces of the furnace are maintained at uniform temperatures of 800 K, 1500 K, and 500 K, respectively. Determine (*a*) the net rate of radiation heat transfer between the base and the side surfaces, (*b*) the net rate of radiation heat transfer between the base and the top surface, and (*c*) the net radiation heat transfer from the base surface.

Solution The surfaces of a cubical furnace are black and are maintained at uniform temperatures. The net rate of radiation heat transfer between the base and side surfaces, between the base and the top surface, and from the base surface are to be determined.

Assumptions The surfaces are black and isothermal.

Analysis (*a*) The geometry involves six surfaces, and thus we may be tempted at first to treat the furnace as a six-surface enclosure. However, the four side surfaces possess the same properties, and thus we can treat them as a single side surface in radiation analysis. We consider the base surface to be surface 1, the top surface to be surface 2, and the side surfaces to be surface 3. Then the problem reduces to determining $\dot{Q}_{1\rightarrow 3}$, $\dot{Q}_{1\rightarrow 2}$, and $\dot{Q}_1$.

The net rate of radiation heat transfer $\dot{Q}_{1\rightarrow 3}$ from surface 1 to surface 3 can be determined from Eq. 21–35, since both surfaces involved are black, by replacing the subscript 2 by 3:

$$\dot{Q}_{1\rightarrow 3} = A_1F_{1\rightarrow 3}\,\sigma(T_1^4 - T_3^4)$$

But first we need to evaluate the view factor $F_{1\rightarrow 3}$. After checking the view factor charts and tables, we realize that we cannot determine this view factor directly. However, we can determine the view factor $F_{1\rightarrow 2}$ from Fig. 21–33 to be $F_{1\to 2} = 0.2$, and we know that $F_{1\to 1} = 0$ since surface 1 is a plane. Then applying the summation rule to surface 1 yields

$$F_{1\to 1} + F_{1\to 2} + F_{1\to 3} = 1$$

or

$$F_{1\to 3} = 1 - F_{1\to 1} - F_{1\to 2} = 1 - 0 - 0.2 = 0.8$$

Substituting,

$$\begin{aligned}\dot{Q}_{1\to 3} &= (25\ \text{m}^2)(0.8)(5.67 \times 10^{-8}\ \text{W/m}^2\cdot\text{K}^4)[(800\ \text{K})^4 - (500\ \text{K})^4]\\ &= \mathbf{394\ kW}\end{aligned}$$

(*b*) The net rate of radiation heat transfer $\dot{Q}_{1\to 2}$ from surface 1 to surface 2 is determined in a similar manner from Eq. 21–35 to be

$$\begin{aligned}\dot{Q}_{1\to 2} &= A_1F_{1\to 2}\sigma(T_1^4 - T_2^4)\\ &= (25\ \text{m}^2)(0.2)(5.67 \times 10^{-8}\ \text{W/m}^2\cdot\text{K}^4)[(800\ \text{K})^4 - (1500\ \text{K})^4]\\ &= \mathbf{-1319\ kW}\end{aligned}$$

The negative sign indicates that net radiation heat transfer is from surface 2 to surface 1.

(*c*) The net radiation heat transfer from the base surface $\dot{Q}_1$ is determined from Eq. 21–36 by replacing the subscript i by 1 and taking $N = 3$:

$$\begin{aligned}\dot{Q}_1 &= \sum_{j=1}^{3}\dot{Q}_{1\to j} = \dot{Q}_{1\to 1} + \dot{Q}_{1\to 2} + \dot{Q}_{1\to 3}\\ &= 0 + (-1319\ \text{kW}) + (394\ \text{kW})\\ &= \mathbf{-925\ kW}\end{aligned}$$

Again the negative sign indicates that net radiation heat transfer is *to* surface 1. That is, the base of the furnace is gaining net radiation at a rate of 925 kW.

21–7 ▪ RADIATION HEAT TRANSFER: DIFFUSE, GRAY SURFACES

The analysis of radiation transfer in enclosures consisting of black surfaces is relatively easy, as we have seen, but most enclosures encountered in practice involve nonblack surfaces, which allow multiple reflections to occur. Radiation analysis of such enclosures becomes very complicated unless some simplifying assumptions are made.

To make a simple radiation analysis possible, it is common to assume the surfaces of an enclosure to be *opaque, diffuse,* and *gray*. That is, the surfaces are nontransparent, they are diffuse emitters and diffuse reflectors, and their radiation properties are independent of wavelength. Also, each surface of the

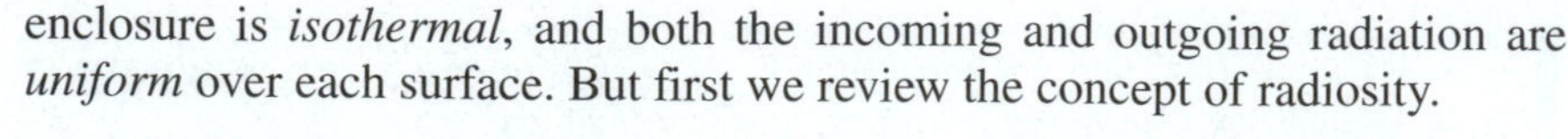

enclosure is *isothermal*, and both the incoming and outgoing radiation are *uniform* over each surface. But first we review the concept of radiosity.

Radiosity

Surfaces emit radiation as well as reflect it, and thus the radiation leaving a surface consists of emitted and reflected parts. The calculation of radiation heat transfer between surfaces involves the *total* radiation energy streaming away from a surface, with no regard for its origin. The *total radiation energy leaving a surface per unit time and per unit area* is the **radiosity** and is denoted by J (Fig. 21–48).

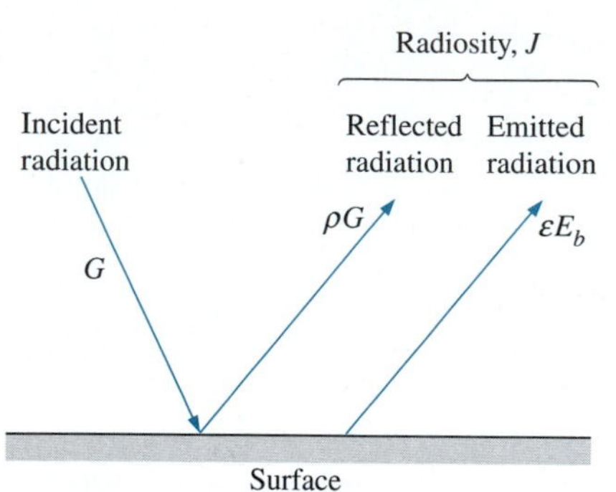

FIGURE 21–48

Radiosity represents the sum of the radiation energy emitted and reflected by a surface.

For a surface i that is *gray* and *opaque* ($\varepsilon_i = \alpha_i$ and $\alpha_i + \rho_i = 1$), the radiosity can be expressed as

$$\begin{aligned} J_i &= \begin{pmatrix}\text{Radiation emitted}\\ \text{by surface } i\end{pmatrix} + \begin{pmatrix}\text{Radiation reflected}\\ \text{by surface } i\end{pmatrix} \\ &= \varepsilon_i E_{bi} + \rho_i G_i \\ &= \varepsilon_i E_{bi} + (1 - \varepsilon_i) G_i \qquad (\text{W/m}^2) \end{aligned} \tag{21-37}$$

where $E_{bi} = \sigma T_i^4$ is the blackbody emissive power of surface i and G_i is irradiation (i.e., the radiation energy incident on surface i per unit time per unit area).

For a surface that can be approximated as a *blackbody* ($\varepsilon_i = 1$), the radiosity relation reduces to

$$J_i = E_{bi} = \sigma T_i^4 \qquad \text{(blackbody)} \tag{21-38}$$

That is, *the radiosity of a blackbody is equal to its emissive power*. This is expected, since a blackbody does not reflect any radiation, and thus radiation coming from a blackbody is due to emission only.

Net Radiation Heat Transfer to or from a Surface

During a radiation interaction, a surface *loses* energy by emitting radiation and *gains* energy by absorbing radiation emitted by other surfaces. A surface experiences a net gain or a net loss of energy, depending on which quantity is larger. The *net* rate of radiation heat transfer from a surface i of surface area A_i is denoted by $\dot{Q}_i$ and is expressed as

$$\begin{aligned} \dot{Q}_i &= \begin{pmatrix}\text{Radiation leaving}\\ \text{entire surface } i\end{pmatrix} - \begin{pmatrix}\text{Radiation incident}\\ \text{on entire surface } i\end{pmatrix} \\ &= A_i(J_i - G_i) \qquad (\text{W}) \end{aligned} \tag{21-39}$$

Solving for G_i from Eq. 21–37 and substituting into Eq. 21–39 yields

$$\dot{Q}_i = A_i\left(J_i - \frac{J_i - \varepsilon_i E_{bi}}{1 - \varepsilon_i}\right) = \frac{A_i \varepsilon_i}{1 - \varepsilon_i}(E_{bi} - J_i) \qquad (\text{W}) \tag{21-40}$$

In an electrical analogy to Ohm's law, this equation can be rearranged as

$$\dot{Q}_i = \frac{E_{bi} - J_i}{R_i} \qquad (\text{W}) \tag{21-41}$$

where

$$R_i = \frac{1 - \varepsilon_i}{A_i \varepsilon_i} \tag{21–42}$$

is the **surface resistance** to radiation. The quantity $E_{bi} - J_i$ corresponds to a *potential difference* and the net rate of radiation heat transfer corresponds to *current* in the electrical analogy, as illustrated in Fig. 21–49.

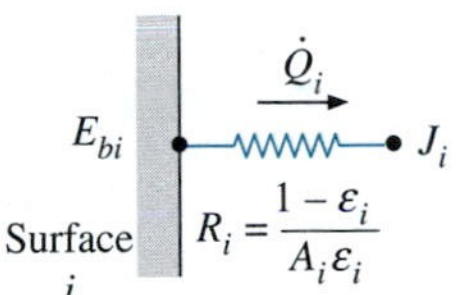

FIGURE 21–49
Electrical analogy of surface resistance to radiation.

The direction of the net radiation heat transfer depends on the relative magnitudes of J_i (the radiosity) and E_{bi} (the emissive power of a blackbody at the temperature of the surface). It is *from* the surface if $E_{bi} > J_i$ and *to* the surface if $J_i > E_{bi}$. A negative value for $\dot{Q}_i$ indicates that heat transfer is *to* the surface. All of this radiation energy gained must be removed from the other side of the surface through some mechanism if the surface temperature is to remain constant.

The surface resistance to radiation for a *blackbody* is *zero* since $\varepsilon_i = 1$ and $J_i = E_{bi}$. The net rate of radiation heat transfer in this case is determined directly from Eq. 21–39.

Some surfaces encountered in numerous practical heat transfer applications are modeled as being *adiabatic* since their back sides are well insulated and the net heat transfer through them is zero. When the convection effects on the front (heat transfer) side of such a surface is negligible and steady-state conditions are reached, the surface must lose as much radiation energy as it gains, and thus $\dot{Q}_i = 0$. In such cases, the surface is said to *reradiate* all the radiation energy it receives, and such a surface is called a **reradiating surface**. Setting $\dot{Q}_i = 0$ in Eq. 21–41 yields

$$J_i = E_{bi} = \sigma T_i^4 \qquad \text{(W/m}^2\text{)} \tag{21–43}$$

Therefore, the *temperature* of a reradiating surface under steady conditions can easily be determined from the equation above once its radiosity is known. Note that the temperature of a reradiating surface is *independent of its emissivity*. In radiation analysis, the surface resistance of a reradiating surface is disregarded since there is no net heat transfer through it. (This is like the fact that there is no need to consider a resistance in an electrical network if no current is flowing through it.)

Net Radiation Heat Transfer between Any Two Surfaces

Consider two diffuse, gray, and opaque surfaces of arbitrary shape maintained at uniform temperatures, as shown in Fig. 21–50. Recognizing that the radiosity J represents the rate of radiation leaving a surface per unit surface area and that the view factor $F_{i \to j}$ represents the fraction of radiation leaving surface i that strikes surface j, the *net* rate of radiation heat transfer from surface i to surface j can be expressed as

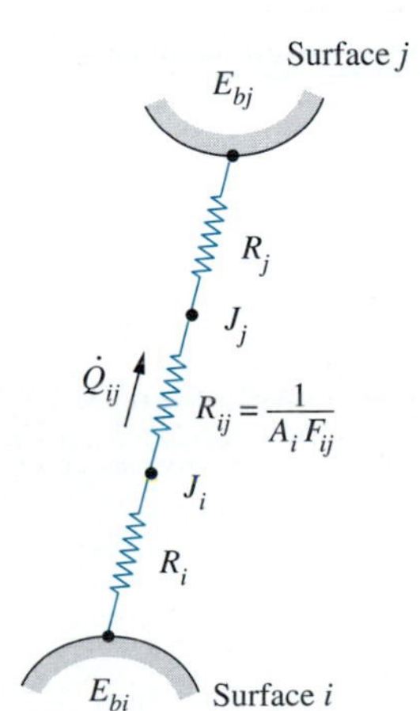

FIGURE 21–50
Electrical analogy of space resistance to radiation.

$$\dot{Q}_{i \to j} = \begin{pmatrix} \text{Radiation leaving} \\ \text{the entire surface } i \\ \text{that strikes surface } j \end{pmatrix} - \begin{pmatrix} \text{Radiation leaving} \\ \text{the entire surface } j \\ \text{that strikes surface } i \end{pmatrix} \tag{21–44}$$

$$= A_i J_i F_{i \to j} - A_j J_j F_{j \to i} \qquad \text{(W)}$$

Applying the reciprocity relation $A_i F_{i \to j} = A_j F_{j \to i}$ yields

$$\dot{Q}_{i \to j} = A_i F_{i \to j} (J_i - J_j) \qquad \text{(W)} \tag{21–45}$$

Again in analogy to Ohm's law, this equation can be rearranged as

$$\dot{Q}_{i \to j} = \frac{J_i - J_j}{R_{i \to j}} \qquad \text{(W)} \tag{21–46}$$

where

$$R_{i \to j} = \frac{1}{A_i F_{i \to j}} \tag{21–47}$$

is the **space resistance** to radiation. Again the quantity $J_i - J_j$ corresponds to a *potential difference,* and the net rate of heat transfer between two surfaces corresponds to *current* in the electrical analogy, as illustrated in Fig. 21–50.

The direction of the net radiation heat transfer between two surfaces depends on the relative magnitudes of J_i and J_j. A positive value for $\dot{Q}_{i \to j}$ indicates that net heat transfer is *from* surface *i to* surface *j*. A negative value indicates the opposite.

In an N-surface enclosure, the conservation of energy principle requires that the net heat transfer from surface i be equal to the sum of the net heat transfers from surface i to each of the N surfaces of the enclosure. That is,

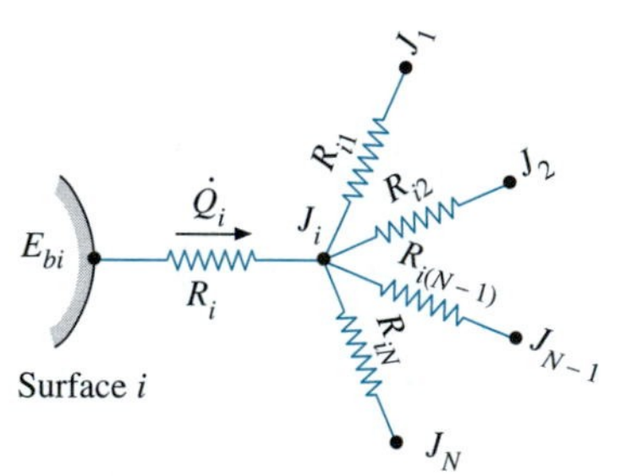

FIGURE 21–51
Network representation of net radiation heat transfer from surface i to the remaining surfaces of an N-surface enclosure.

$$\dot{Q}_i = \sum_{j=1}^{N} \dot{Q}_{i \to j} = \sum_{j=1}^{N} A_i F_{i \to j} (J_i - J_j) = \sum_{j=1}^{N} \frac{J_i - J_j}{R_{i \to j}} \qquad \text{(W)} \tag{21–48}$$

The network representation of net radiation heat transfer from surface i to the remaining surfaces of an N-surface enclosure is given in Fig. 21–51. Note that $\dot{Q}_{i \to i}$ (the net rate of heat transfer from a surface to itself) is zero regardless of the shape of the surface. Combining Eqs. 21–41 and 21–48 gives

$$\frac{E_{bi} - J_i}{R_i} = \sum_{j=1}^{N} \frac{J_i - J_j}{R_{i \to j}} \qquad \text{(W)} \tag{21–49}$$

which has the electrical analogy interpretation that *the net radiation flow from a surface through its surface resistance is equal to the sum of the radiation flows from that surface to all other surfaces through the corresponding space resistances.*

Methods of Solving Radiation Problems

In the radiation analysis of an enclosure, either the temperature or the net rate of heat transfer must be given for each of the surfaces to obtain a unique solution for the unknown surface temperatures and heat transfer rates. There are two methods commonly used to solve radiation problems.

In the first method, Eqs. 21–48 (for surfaces with specified heat transfer rates) and 21–49 (for surfaces with specified temperatures) are simplified and rearranged as

Surfaces with specified net heat transfer rate Q

$$\dot{Q}_i = A_i \sum_{j=1}^{N} F_{i \to j}(J_i - J_j) \quad \text{(21–50)}$$

Surfaces with specified temperature T_i

$$\sigma T_i^4 = J_i + \frac{1 - \varepsilon_i}{\varepsilon_i} \sum_{j=1}^{N} F_{i \to j}(J_i - J_j) \quad \text{(21–51)}$$

Note that $\dot{Q}_i = 0$ for insulated (or reradiating) surfaces, and $\sigma T_i^4 = J_i$ for black surfaces since $\varepsilon_i = 1$ in that case. Also, the term corresponding to $j = i$ drops out from either relation since $J_i - J_j = J_i - J_i = 0$ in that case.

The equations above give N linear algebraic equations for the determination of the N unknown radiosities for an N-surface enclosure. Once the radiosities $J_1, J_2, \ldots, J_N$ are available, the unknown heat transfer rates can be determined from Eq. 21–50 while the unknown surface temperatures can be determined from Eq. 21–51. The temperatures of insulated or reradiating surfaces can be determined from $\sigma T_i^4 = J_i$. A positive value for $\dot{Q}_i$ indicates net radiation heat transfer *from* surface i to other surfaces in the enclosure while a negative value indicates net radiation heat transfer *to* the surface.

The systematic approach described above for solving radiation heat transfer problems is very suitable for use with today's popular equation solvers such as EES, Mathcad, and Matlab, especially when there are a large number of surfaces, and is known as the **direct method** (formerly, the *matrix method,* since it resulted in matrices and the solution required a knowledge of linear algebra). The second method described below, called the **network method**, is based on the electrical network analogy.

The network method was first introduced by A. K. Oppenheim in the 1950s and found widespread acceptance because of its simplicity and emphasis on the physics of the problem. The application of the method is straightforward: draw a surface resistance associated with each surface of an enclosure and connect them with space resistances. Then solve the radiation problem by treating it as an electrical network problem where the radiation heat transfer replaces the current and radiosity replaces the potential.

The network method is not practical for enclosures with more than three or four surfaces, however, because of the increased complexity of the network. Next we apply the method to solve radiation problems in two- and three-surface enclosures.

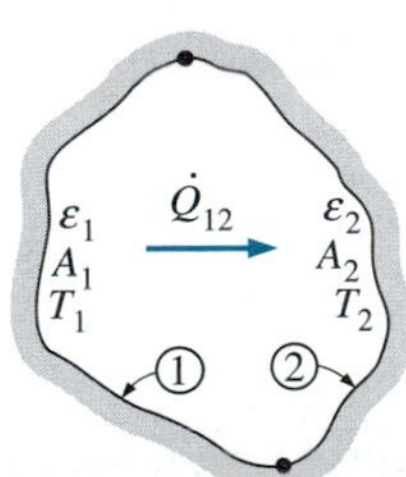

FIGURE 21–52
Schematic of a two-surface enclosure and the radiation network associated with it.

Radiation Heat Transfer in Two-Surface Enclosures

Consider an enclosure consisting of two opaque surfaces at specified temperatures T_1 and T_2, as shown in Fig. 21–52, and try to determine the net rate of radiation heat transfer between the two surfaces with the network method. Surfaces 1 and 2 have emissivities ε_1 and ε_2 and surface areas A_1 and A_2 and are maintained at uniform temperatures T_1 and T_2, respectively. There are only two surfaces in the enclosure, and thus we can write

$$\dot{Q}_{12} = \dot{Q}_1 = -\dot{Q}_2$$

That is, the net rate of radiation heat transfer from surface 1 to surface 2 must equal the net rate of radiation heat transfer *from* surface 1 and the net rate of radiation heat transfer *to* surface 2.

The radiation network of this two-surface enclosure consists of two surface resistances and one space resistance, as shown in Fig. 21–52. In an electrical network, the electric current flowing through these resistances connected in series would be determined by dividing the potential difference between points A and B by the total resistance between the same two points. The net rate of radiation transfer is determined in the same manner and is expressed as

$$\dot{Q}_{12} = \frac{E_{b1} - E_{b2}}{R_1 + R_{12} + R_2} = \dot{Q}_1 = -\dot{Q}_2$$

or

$$\dot{Q}_{12} = \frac{\sigma(T_1^4 - T_2^4)}{\dfrac{1-\varepsilon_1}{A_1\varepsilon_1} + \dfrac{1}{A_1F_{12}} + \dfrac{1-\varepsilon_2}{A_2\varepsilon_2}} \quad \text{(W)} \qquad \textbf{(21–52)}$$

This important result is applicable to any two gray, diffuse, and opaque surfaces that form an enclosure. The view factor F_{12} depends on the geometry and must be determined first. Simplified forms of Eq. 21–52 for some familiar arrangements that form a two-surface enclosure are given in Table 21–5. Note that $F_{12} = 1$ for all of these special cases.

EXAMPLE 21–10 Radiation Heat Transfer between Parallel Plates

Two very large parallel plates are maintained at uniform temperatures $T_1 = 800$ K and $T_2 = 500$ K and have emissivities $\varepsilon_1 = 0.2$ and $\varepsilon_2 = 0.7$, respectively, as shown in Fig. 21–53. Determine the net rate of radiation heat transfer between the two surfaces per unit surface area of the plates.

Solution Two large parallel plates are maintained at uniform temperatures. The net rate of radiation heat transfer between the plates is to be determined.

Assumptions Both surfaces are opaque, diffuse, and gray.

Analysis The net rate of radiation heat transfer between the two plates per unit area is readily determined from Eq. 21–54 to be

$$\dot{q}_{12} = \frac{\dot{Q}_{12}}{A} = \frac{\sigma(T_1^4 - T_2^4)}{\dfrac{1}{\varepsilon_1} + \dfrac{1}{\varepsilon_2} - 1} = \frac{(5.67 \times 10^{-8}\ \text{W/m}^2 \cdot \text{K}^4)[(800\ \text{K})^4 - (500\ \text{K})^4]}{\dfrac{1}{0.2} + \dfrac{1}{0.7} - 1}$$

$$= \mathbf{3625\ W/m^2}$$

Discussion Note that heat at a net rate of 3625 W is transferred from plate 1 to plate 2 by radiation per unit surface area of either plate.

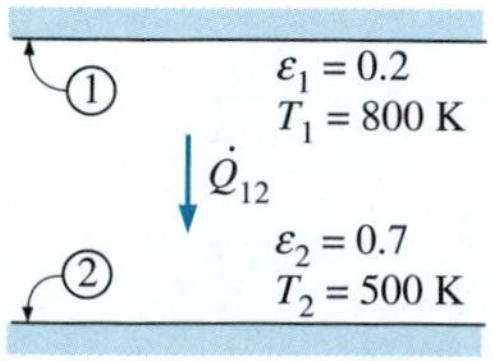

$\varepsilon_1 = 0.2$
$T_1 = 800$ K
(1)
$\dot{Q}_{12}$
$\varepsilon_2 = 0.7$
$T_2 = 500$ K
(2)

FIGURE 21–53
The two parallel plates considered in Example 21–10.

TABLE 21–5

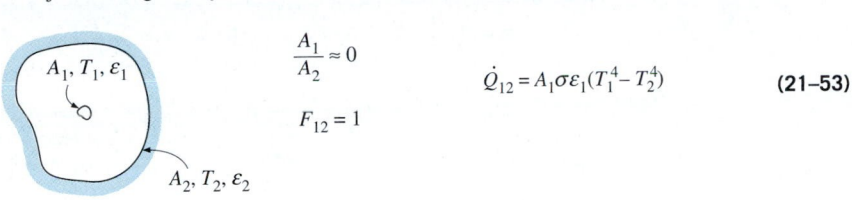

Geometry	Relations	Net radiation transfer	Eq.
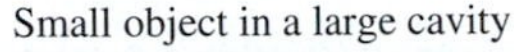 Small object in a large cavity A_1, T_1, ε_1 A_2, T_2, ε_2	$\frac{A_1}{A_2} \approx 0$ $F_{12} = 1$	$\dot{Q}_{12} = A_1\sigma\varepsilon_1(T_1^4 - T_2^4)$	(21–53)
Infinitely large parallel plates A_1, T_1, ε_1 A_2, T_2, ε_2	$A_1 = A_2 = A$ $F_{12} = 1$	$\dot{Q}_{12} = \dfrac{A\sigma(T_1^4 - T_2^4)}{\dfrac{1}{\varepsilon_1} + \dfrac{1}{\varepsilon_2} - 1}$	(21–54)
Infinitely long concentric cylinders 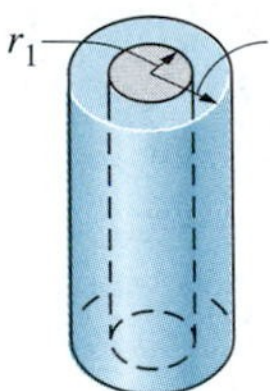	$\frac{A_1}{A_2} = \frac{r_1}{r_2}$ $F_{12} = 1$	$\dot{Q}_{12} = \dfrac{A_1\sigma(T_1^4 - T_2^4)}{\dfrac{1}{\varepsilon_1} + \dfrac{1 - \varepsilon_2}{\varepsilon_2}\left(\dfrac{r_1}{r_2}\right)}$	(21–55)
Concentric spheres 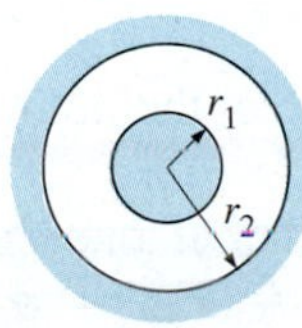	$\frac{A_1}{A_2} = \left(\frac{r_1}{r_2}\right)^2$ $F_{12} = 1$	$\dot{Q}_{12} = \dfrac{A_1\sigma(T_1^4 - T_2^4)}{\dfrac{1}{\varepsilon_1} + \dfrac{1 - \varepsilon_2}{\varepsilon_2}\left(\dfrac{r_1}{r_2}\right)^2}$	(21–56)

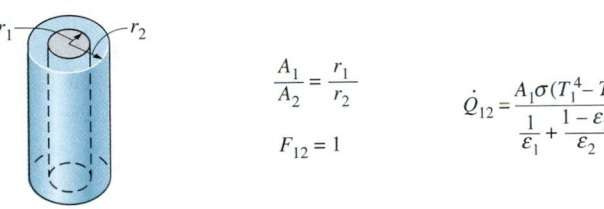

Radiation Heat Transfer in Three-Surface Enclosures

We now consider an enclosure consisting of three opaque, diffuse, and gray surfaces, as shown in Fig. 21–54. Surfaces 1, 2, and 3 have surface areas A_1, A_2, and A_3; emissivities ε_1, ε_2, and ε_3; and uniform temperatures T_1, T_2, and T_3, respectively. The radiation network of this geometry is constructed by following the standard procedure: draw a surface resistance associated with each of the three surfaces and connect these surface resistances with space resistances, as shown in the figure. Relations for the surface and space resistances are given by Eqs. 21–42 and 21–47. The three endpoint potentials E_{b1}, E_{b2}, and E_{b3} are considered known, since the surface temperatures

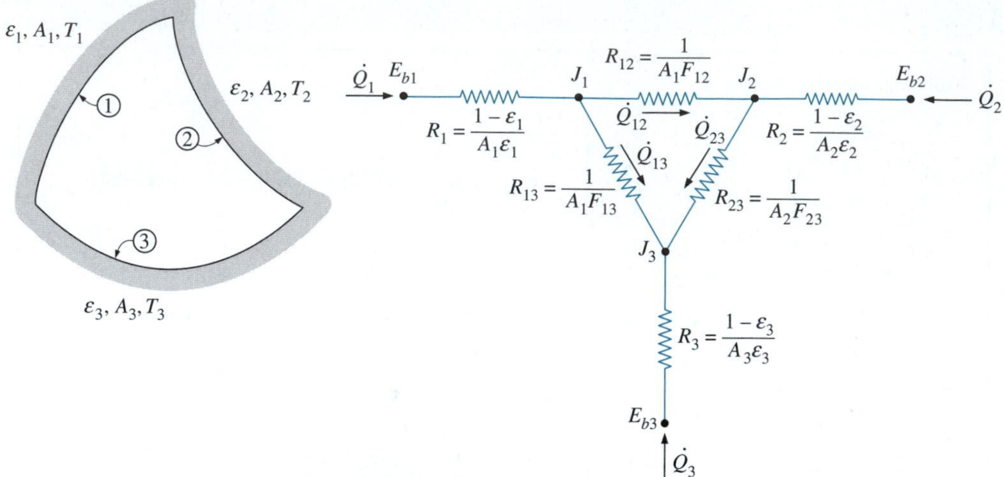

FIGURE 21–54
Schematic of a three-surface enclosure and the radiation network associated with it.

are specified. Then all we need to find are the radiosities J_1, J_2, and J_3. The three equations for the determination of these three unknowns are obtained from the requirement that *the algebraic sum of the currents (net radiation heat transfer) at each node must equal zero.* That is,

$$\frac{E_{b1} - J_1}{R_1} + \frac{J_2 - J_1}{R_{12}} + \frac{J_3 - J_1}{R_{13}} = 0$$
$$\frac{J_1 - J_2}{R_{12}} + \frac{E_{b2} - J_2}{R_2} + \frac{J_3 - J_2}{R_{23}} = 0$$
$$\frac{J_1 - J_3}{R_{13}} + \frac{J_2 - J_3}{R_{23}} + \frac{E_{b3} - J_3}{R_3} = 0 \quad \textbf{(21–57)}$$

Once the radiosities J_1, J_2, and J_3 are available, the net rate of radiation heat transfers at each surface can be determined from Eq. 21–48.

The set of equations above simplify further if one or more surfaces are "special" in some way. For example, $J_i = E_{bi} = \sigma T_i^4$ for a *black* or *reradiating* surface. Also, $\dot{Q}_i = 0$ for a reradiating surface. Finally, when the net rate of radiation heat transfer $\dot{Q}_i$ is specified at surface i instead of the temperature, the term $(E_{bi} - J_i)/R_i$ should be replaced by the specified $\dot{Q}_i$.

EXAMPLE 21–11 Radiation Heat Transfer in a Cylindrical Furnace

Consider a cylindrical furnace with $r_o = H = 1$ m, as shown in Fig. 21–55. The top (surface 1) and the base (surface 2) of the furnace have emissivities $\varepsilon_1 = 0.8$ and $\varepsilon_2 = 0.4$, respectively, and are maintained at uniform temperatures $T_1 = 700$ K and $T_2 = 500$ K. The side surface closely approximates a blackbody and is maintained at a temperature of $T_3 = 400$ K. Determine the net rate of radiation heat transfer at each surface during steady operation and explain how these surfaces can be maintained at specified temperatures.

Solution The surfaces of a cylindrical furnace are maintained at uniform temperatures. The net rate of radiation heat transfer at each surface during steady operation is to be determined.

Assumptions **1** Steady operating conditions exist. **2** The surfaces are opaque, diffuse, and gray. **3** Convection heat transfer is not considered.

Analysis We will solve this problem systematically using the direct method to demonstrate its use. The cylindrical furnace can be considered to be a three-surface enclosure with surface areas of

$$A_1 = A_2 = \pi r_o^2 = \pi(1 \text{ m})^2 = 3.14 \text{ m}^2$$
$$A_3 = 2\pi r_o H = 2\pi(1 \text{ m})(1 \text{ m}) = 6.28 \text{ m}^2$$

The view factor from the base to the top surface is, from Fig. 21–35, $F_{12} = 0.38$. Then the view factor from the base to the side surface is determined by applying the summation rule to be

$$F_{11} + F_{12} + F_{13} = 1 \quad \rightarrow \quad F_{13} = 1 - F_{11} - F_{12} = 1 - 0 - 0.38 = 0.62$$

since the base surface is flat and thus $F_{11} = 0$. Noting that the top and bottom surfaces are symmetric about the side surface, $F_{21} = F_{12} = 0.38$ and $F_{23} = F_{13} = 0.62$. The view factor F_{31} is determined from the reciprocity relation,

$$A_1F_{13} = A_3F_{31} \quad \rightarrow \quad F_{31} = F_{13}(A_1/A_3) = (0.62)(0.314/0.628) = 0.31$$

Also, $F_{32} = F_{31} = 0.31$ because of symmetry. Now that all the view factors are available, we apply Eq. 21–51 to each surface to determine the radiosities:

Top surface $(i = 1)$: $\sigma T_1^4 = J_1 + \dfrac{1 - \varepsilon_1}{\varepsilon_1}[F_{12}(J_1 - J_2) + F_{13}(J_1 - J_3)]$

Bottom surface $(i = 2)$: $\sigma T_2^4 = J_2 + \dfrac{1 - \varepsilon_2}{\varepsilon_2}[F_{21}(J_2 - J_1) + F_{23}(J_2 - J_3)]$

Side surface $(i = 3)$: $\sigma T_3^4 = J_3 + 0$ (since surface 3 is black and thus $\varepsilon_3 = 1$)

Substituting the known quantities,

$$(5.67 \times 10^{-8} \text{ W/m}^2 \cdot \text{K}^4)(700 \text{ K})^4 = J_1 + \frac{1 - 0.8}{0.8}[0.38(J_1 - J_2) + 0.62(J_1 - J_3)]$$

$$(5.67 \times 10^{-8} \text{ W/m}^2 \cdot \text{K}^4)(500 \text{ K})^4 = J_2 + \frac{1 - 0.4}{0.4}[0.38(J_2 - J_1) + 0.62(J_2 - J_3)]$$

$$(5.67 \times 10^{-8} \text{ W/m}^2 \cdot \text{K}^4)(400 \text{ K})^4 = J_3$$

Solving these equations for J_1, J_2, and J_3 gives

$$J_1 = 11{,}418 \text{ W/m}^2, J_2 = 4562 \text{ W/m}^2, \quad \text{and} \quad J_3 = 1452 \text{ W/m}^2$$

Then the net rates of radiation heat transfer at the three surfaces are determined from Eq. 21–50 to be

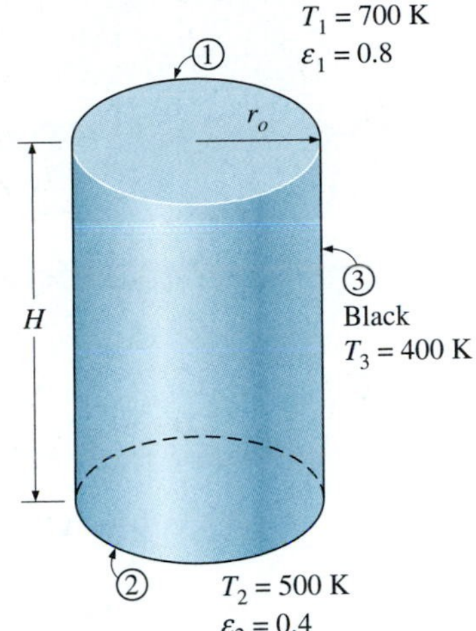

FIGURE 21–55
The cylindrical furnace considered in Example 21–11.

$$\dot{Q}_1 = A_1[F_{1\rightarrow 2}(J_1 - J_2) + F_{1\rightarrow 3}(J_1 - J_3)]$$
$$= (3.14\ \text{m}^2)[0.38(11{,}418 - 4562) + 0.62(11{,}418 - 1452)]\ \text{W/m}^2$$
$$= \mathbf{27.6\ kW}$$
$$\dot{Q}_2 = A_2[F_{2\rightarrow 1}(J_2 - J_1) + F_{2\rightarrow 3}(J_2 - J_3)]$$
$$= (3.14\ \text{m}^2)[0.38(4562 - 11{,}418) + 0.62(4562 - 1452)]\ \text{W/m}^2$$
$$= \mathbf{-2.13\ kW}$$
$$\dot{Q}_3 = A_3[F_{3\rightarrow 1}(J_3 - J_1) + F_{3\rightarrow 2}(J_3 - J_2)]$$
$$= (6.28\ \text{m}^2)[0.31(1452 - 11{,}418) + 0.31(1452 - 4562)]\ \text{W/m}^2$$
$$= \mathbf{-25.5\ kW}$$

Note that the direction of net radiation heat transfer is *from* the top surface *to* the base and side surfaces, and the algebraic sum of these three quantities must be equal to zero. That is,

$$\dot{Q}_1 + \dot{Q}_2 + \dot{Q}_3 = 27.6 + (-2.13) + (-25.5) \cong 0$$

Discussion To maintain the surfaces at the specified temperatures, we must supply heat to the top surface continuously at a rate of 27.6 kW while removing 2.13 kW from the base and 25.5 kW from the side surfaces.

The direct method presented here is straightforward, and it does not require the evaluation of radiation resistances. Also, it can be applied to enclosures with any number of surfaces in the same manner.

EXAMPLE 21–12 Radiation Heat Transfer in a Triangular Furnace

A furnace is shaped like a long equilateral triangular duct, as shown in Fig. 21–56. The width of each side is 1 m. The base surface has an emissivity of 0.7 and is maintained at a uniform temperature of 600 K. The heated left-side surface closely approximates a blackbody at 1000 K. The right-side surface is well insulated. Determine the rate at which heat must be supplied to the heated side externally per unit length of the duct in order to maintain these operating conditions.

Solution Two of the surfaces of a long equilateral triangular furnace are maintained at uniform temperatures while the third surface is insulated. The external rate of heat transfer to the heated side per unit length of the duct during steady operation is to be determined.

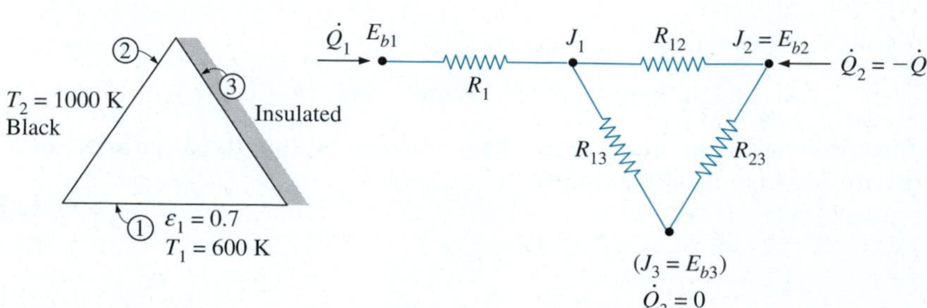

FIGURE 21–56
The triangular furnace considered in Example 21–12.

Assumptions **1** Steady operating conditions exist. **2** The surfaces are opaque, diffuse, and gray. **3** Convection heat transfer is not considered.

Analysis The furnace can be considered to be a three-surface enclosure with a radiation network as shown in the figure, since the duct is very long and thus the end effects are negligible. We observe that the view factor from any surface to any other surface in the enclosure is 0.5 because of symmetry. Surface 3 is a reradiating surface since the net rate of heat transfer at that surface is zero. Then we must have $\dot{Q}_1 = -\dot{Q}_2$, since the entire heat lost by surface 1 must be gained by surface 2. The radiation network in this case is a simple series–parallel connection, and we can determine $\dot{Q}_1$ directly from

$$\dot{Q}_1 = \frac{E_{b1} - E_{b2}}{R_1 + \left(\dfrac{1}{R_{12}} + \dfrac{1}{R_{13} + R_{23}}\right)^{-1}} = \frac{E_{b1} - E_{b2}}{\dfrac{1 - \varepsilon_1}{A_1\varepsilon_1} + \left(A_1F_{12} + \dfrac{1}{1/A_1F_{13} + 1/A_2F_{23}}\right)^{-1}}$$

where

$$A_1 = A_2 = A_3 = wL = 1 \text{ m} \times 1 \text{ m} = 1 \text{ m}^2 \qquad \text{(per unit length of the duct)}$$

$$F_{12} = F_{13} = F_{23} = 0.5 \qquad \text{(symmetry)}$$

$$E_{b1} = \sigma T_1^4 = (5.67 \times 10^{-8} \text{ W/m}^2 \cdot \text{K}^4)(600 \text{ K})^4 = 7348 \text{ W/m}^2$$

$$E_{b2} = \sigma T_2^4 = (5.67 \times 10^{-8} \text{ W/m}^2 \cdot \text{K}^4)(1000 \text{ K})^4 = 56{,}700 \text{ W/m}^2$$

Substituting,

$$\dot{Q}_1 = \frac{(56{,}700 - 7348) \text{ W/m}^2}{\dfrac{1 - 0.7}{0.7 \times 1 \text{ m}^2} + \left[(0.5 \times 1 \text{ m}^2) + \dfrac{1}{1/(0.5 \times 1 \text{ m}^2) + 1/(0.5 \times 1 \text{ m}^2)}\right]^{-1}}$$

$$= \mathbf{28.0 \text{ kW}}$$

Therefore, heat at a rate of 28 kW must be supplied to the heated surface per unit length of the duct to maintain steady operation in the furnace.

EXAMPLE 21–13 Heat Transfer through a Tubular Solar Collector

A solar collector consists of a horizontal aluminum tube having an outer diameter of 2 in enclosed in a concentric thin glass tube of 4-in diameter, as shown in Fig. 21–57. Water is heated as it flows through the tube, and the space between the aluminum and the glass tubes is filled with air at 1 atm pressure. The pump circulating the water fails during a clear day, and the water temperature in the tube starts rising. The aluminum tube absorbs solar radiation at a rate of 30 Btu/h per foot length, and the temperature of the ambient air outside is 70°F. The emissivities of the tube and the glass cover are 0.95 and 0.9, respectively. Taking the effective sky temperature to be 50°F, determine the temperature of the aluminum tube when steady operating conditions are established (i.e., when the rate of heat loss from the tube equals the amount of solar energy gained by the tube).

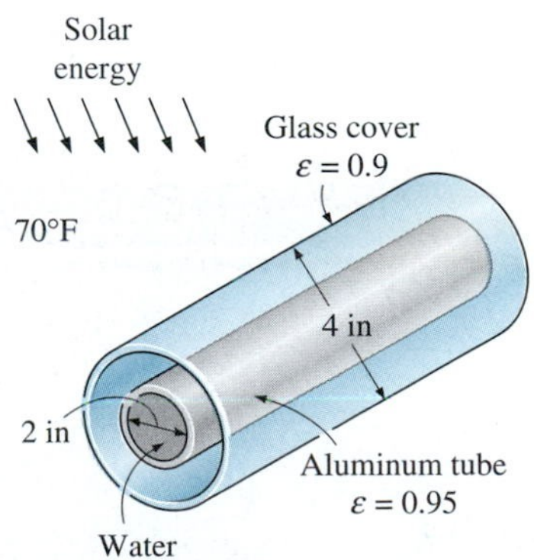

FIGURE 21–57
Schematic for Example 21–13.

Solution The circulating pump of a solar collector that consists of a horizontal tube and its glass cover fails. The equilibrium temperature of the tube is to be determined.

Assumptions **1** Steady operating conditions exist. **2** The tube and its cover are isothermal. **3** Air is an ideal gas. **4** The surfaces are opaque, diffuse, and gray for infrared radiation. **5** The glass cover is transparent to solar radiation.

Properties The properties of air should be evaluated at the average temperature. But we do not know the temperature of the air in the duct, and thus we cannot determine the bulk fluid and glass cover temperatures at this point, and thus we cannot evaluate the average temperatures. Therefore, we assume the glass temperature to be 110°F, and use properties at an anticipated average temperature of (70 + 110)/2 = 90°F (Table A–22E),

$$k = 0.01505 \text{ Btu/h}\cdot\text{ft}\cdot{}^\circ\text{F} \qquad \text{Pr} = 0.7275$$

$$\nu = 1.753 \times 10^{-4} \text{ ft}^2\text{/s} \qquad \beta = \frac{1}{T_{\text{ave}}} = \frac{1}{550 \text{ R}}$$

Analysis This problem was solved in Chapter 20 by disregarding radiation heat transfer. Now we repeat the solution by considering natural convection and radiation occurring simultaneously.

We have a horizontal cylindrical enclosure filled with air at 1 atm pressure. The problem involves heat transfer from the aluminum tube to the glass cover and from the outer surface of the glass cover to the surrounding ambient air. When steady operation is reached, these two heat transfer rates must equal the rate of heat gain. That is,

$$\dot{Q}_{\text{tube-glass}} = \dot{Q}_{\text{glass-ambient}} = \dot{Q}_{\text{solar gain}} = 30 \text{ Btu/h} \qquad \text{(per foot of tube)}$$

The heat transfer surface area of the glass cover is

$$A_o = A_{\text{glass}} = (\pi D_o L) = \pi(4/12 \text{ ft})(1 \text{ ft}) = 1.047 \text{ ft}^2 \qquad \text{(per foot of tube)}$$

To determine the Rayleigh number, we need to know the surface temperature of the glass, which is not available. Therefore, it is clear that the solution requires a trial-and-error approach unless we use an equation solver such as EES. Assuming the glass cover temperature to be 110°F, the Rayleigh number, the Nusselt number, the convection heat transfer coefficient, and the rate of natural convection heat transfer from the glass cover to the ambient air are determined to be

$$\text{Ra}_{D_o} = \frac{g\beta(T_o - T_\infty)D_o^3}{\nu^2}\text{Pr}$$

$$= \frac{(32.2 \text{ ft/s}^2)[1/(550 \text{ R})](110 - 70 \text{ R})(4/12 \text{ ft})^3}{(1.753 \times 10^{-4} \text{ ft}^2\text{/s})^2}(0.7275) = 2.053 \times 10^6$$

$$\text{Nu} = \left\{0.6 + \frac{0.387\,\text{Ra}_{D_o}^{1/6}}{[1 + (0.559/\text{Pr})^{9/16}]^{8/27}}\right\}^2 = \left\{0.6 + \frac{0.387(2.053 \times 10^6)^{1/6}}{[1 + (0.559/0.7275)^{9/16}]^{8/27}}\right\}^2$$

$$= 17.88$$

$$h_o = \frac{k}{D_o}\text{Nu} = \frac{0.01505 \text{ Btu/h}\cdot\text{ft}\cdot{}^\circ\text{F}}{4/12 \text{ ft}}(17.88) = 0.8073 \text{ Btu/h}\cdot\text{ft}^2\cdot{}^\circ\text{F}$$

$$\dot{Q}_{o,\text{conv}} = h_o A_o(T_o - T_\infty) = (0.8073 \text{ Btu/h}\cdot\text{ft}^2\cdot{}^\circ\text{F})(1.047 \text{ ft}^2)(110 - 70)^\circ\text{F}$$

$$= 33.8 \text{ Btu/h}$$

Also,

$$\dot{Q}_{o,\text{ rad}} = \varepsilon_o \sigma A_o(T_o^4 - T_{\text{sky}}^4)$$
$$= (0.9)(0.1714 \times 10^{-8} \text{ Btu/h} \cdot \text{ft}^2 \cdot \text{R}^4)(1.047 \text{ ft}^2)[(570 \text{ R})^4 - (510 \text{ R})^4]$$
$$= 61.2 \text{ Btu/h}$$

Then the total rate of heat loss from the glass cover becomes

$$\dot{Q}_{o,\text{ total}} = \dot{Q}_{o,\text{ conv}} + \dot{Q}_{o,\text{ rad}} = 33.8 + 61.2 = 95.0 \text{ Btu/h}$$

which is much larger than 30 Btu/h. Therefore, the assumed temperature of 110°F for the glass cover is high. Repeating the calculations with lower temperatures (including the evaluation of properties), the glass cover temperature corresponding to 30 Btu/h is determined to be 78°F (it would be 106°F if radiation were ignored).

The temperature of the aluminum tube is determined in a similar manner using the natural convection and radiation relations for two horizontal concentric cylinders. The characteristic length in this case is the distance between the two cylinders, which is

$$L_c = (D_o - D_i)/2 = (4 - 2)/2 = 1 \text{ in} = 1/12 \text{ ft}$$

Also,

$$A_i = A_{\text{tube}} = (\pi D_i L) = \pi(2/12 \text{ ft})(1 \text{ ft}) = 0.5236 \text{ ft}^2 \qquad \text{(per foot of tube)}$$

We start the calculations by assuming the tube temperature to be 122°F, and thus an average temperature of (78 + 122)/2 = 100°F = 560 R. Using properties at 100°F,

$$\text{Ra}_L = \frac{g\beta(T_i - T_o)L_c^3}{\nu^2}\text{Pr}$$
$$= \frac{(32.2 \text{ ft/s}^2)[1/(560 \text{ R})](122 - 78 \text{ R})(1/12 \text{ ft})^3}{(1.809 \times 10^{\pm 4} \text{ ft}^2/\text{s})^2}(0.726) = 3.246 \times 10^4$$

The effective thermal conductivity is

$$F_{\text{cyl}} = \frac{[\ln(D_o/D_i)]^4}{L_c^3(D_i^{-3/5} + D_o^{-3/5})^5} = \frac{[\ln(4/2)]^4}{(1/12 \text{ ft})^3[(2/12 \text{ ft})^{-3/5} + (4/12 \text{ ft})^{-3/5}]^5} = 0.1466$$

$$k_{\text{eff}} = 0.386k\left(\frac{\text{Pr}}{0.861 + \text{Pr}}\right)^{1/4}(F_{\text{cyl}}\text{Ra}_L)^{1/4}$$
$$= 0.386(0.01529 \text{ Btu/h} \cdot \text{ft} \cdot °\text{F})\left(\frac{0.726}{0.861 + 0.726}\right)^{1/4}(0.1466 \times 3.248 \times 10^4)^{1/4}$$
$$= 0.04032 \text{ Btu/h} \cdot \text{ft} \cdot °\text{F}$$

Then the rate of heat transfer between the cylinders by convection becomes

$$\dot{Q}_{i,\text{ conv}} = \frac{2\pi k_{\text{eff}}}{\ln(D_o/D_i)}(T_i - T_o)$$
$$= \frac{2\pi(0.04032 \text{ Btu/h} \cdot \text{ft °F})}{\ln(4/2)}(122 - 78)°\text{F} = 16.1 \text{ Btu/h}$$

Also,

$$\dot{Q}_{i,\text{ rad}} = \frac{\sigma A_i\,(T_i^4 - T_o^4)}{\dfrac{1}{\varepsilon_i} + \dfrac{1-\varepsilon_o}{\varepsilon_o}\left(\dfrac{D_i}{D_o}\right)}$$

$$= \frac{(0.1714 \times 10^{-8}\ \text{Btu/h}\cdot\text{ft}^2\cdot\text{R}^4)(0.5236\ \text{ft}^2)[(582\ \text{R})^4 - (538\ \text{R})^4]}{\dfrac{1}{0.95} + \dfrac{1-0.9}{0.9}\left(\dfrac{2\ \text{in}}{4\ \text{in}}\right)}$$

$$= 25.1\ \text{Btu/h}$$

Then the total rate of heat loss from the glass cover becomes

$$\dot{Q}_{i,\text{ total}} = \dot{Q}_{i,\text{ conv}} + \dot{Q}_{i,\text{ rad}} = 16.1 + 25.1 = 41.2\ \text{Btu/h}$$

which is larger than 30 Btu/h. Therefore, the assumed temperature of 122°F for the tube is high. By trying other values, the tube temperature corresponding to 30 Btu/h is determined to be **112°F** (it would be 180°F if radiation were ignored). Therefore, the tube will reach an equilibrium temperature of 112°F when the pump fails.

Discussion It is clear from the results obtained that radiation should always be considered in systems that are heated or cooled by natural convection, unless the surfaces involved are polished and thus have very low emissivities.

SUMMARY

Radiation propagates in the form of electromagnetic waves. The *frequency* ν and *wavelength* λ of electromagnetic waves in a medium are related by $\lambda = c/\nu$, where c is the speed of propagation in that medium. All matter continuously emits *thermal radiation* as a result of vibrational and rotational motions of molecules, atoms, and electrons of a substance.

A *blackbody* is defined as a perfect emitter and absorber of radiation. At a specified temperature and wavelength, no surface can emit more energy than a blackbody. A blackbody absorbs *all* incident radiation, regardless of wavelength and direction. The radiation energy emitted by a blackbody per unit time and per unit surface area is called the *blackbody emissive power* E_b and is expressed by the *Stefan–Boltzmann law* as

$$E_b(T) = \sigma T^4$$

where $\sigma = 5.670 \times 10^{-8}\ \text{W/m}^2\cdot\text{K}^4$ is the *Stefan–Boltzmann constant* and T is the absolute temperature of the surface in K. At any specified temperature, the spectral blackbody emissive power $E_{b\lambda}$ increases with wavelength, reaches a peak, and then decreases with increasing wavelength. The wavelength at which the peak occurs for a specified temperature is given by *Wien's displacement law* as

$$(\lambda T)_{\text{max power}} = 2897.8\ \mu\text{m}\cdot\text{K}$$

The *blackbody radiation function* f_λ represents the fraction of radiation emitted by a blackbody at temperature T in the wavelength band from $\lambda = 0$ to λ. The fraction of radiation energy emitted by a blackbody at temperature T over a finite wavelength band from $\lambda = \lambda_1$ to $\lambda = \lambda_2$ is determined from

$$f_{\lambda_1-\lambda_2}(T) = f_{\lambda_2}(T) - f_{\lambda_1}(T)$$

where $f_{\lambda_1}(T)$ and $f_{\lambda_2}(T)$ are the blackbody radiation functions corresponding to $\lambda_1 T$ and $\lambda_2 T$, respectively.

The *emissivity* of a surface represents the ratio of the radiation emitted by the surface at a given temperature to the radiation emitted by a blackbody at the same temperature. Different emissivities are defined as

Spectral hemispherical emissivity:

$$\varepsilon_\lambda(\lambda, T) = \frac{E_\lambda(\lambda, T)}{E_{b\lambda}(\lambda, T)}$$

Total hemispherical emissivity:

$$\varepsilon(T) = \frac{E(T)}{E_b(T)} = \frac{\displaystyle\int_0^\infty \varepsilon_\lambda(\lambda, T)E_{b\lambda}(\lambda, T)\,d\lambda}{\sigma T^4}$$

Emissivity can also be expressed as a step function by dividing the spectrum into a sufficient number of *wavelength bands* of constant emissivity as, for example,

$$\varepsilon(T) = \varepsilon_1 f_{0-\lambda_1}(T) + \varepsilon_2 f_{\lambda_1-\lambda_2}(T) + \varepsilon_3 f_{\lambda_2-\infty}(T)$$

The *total hemispherical emissivity* ε of a surface is the average emissivity over all directions and wavelengths.

When radiation strikes a surface, part of it is absorbed, part of it is reflected, and the remaining part, if any, is transmitted. The fraction of incident radiation (irradiation G) absorbed by the surface is called the *absorptivity*, the fraction reflected by the surface is called the *reflectivity*, and the fraction transmitted is called the *transmissivity*. Various absorptivities, reflectivities, and transmissivities for a medium are expressed as

$$\alpha_\lambda(\lambda) = \frac{G_{\lambda,\,\text{abs}}(\lambda)}{G_\lambda(\lambda)}, \quad \rho_\lambda(\lambda) = \frac{G_{\lambda,\,\text{ref}}(\lambda)}{G_\lambda(\lambda)}, \quad \text{and} \quad \tau_\lambda(\lambda) = \frac{G_{\lambda,\,\text{tr}}(\lambda)}{G_\lambda(\lambda)}$$

$$\alpha = \frac{G_{\text{abs}}}{G}, \quad \rho = \frac{G_{\text{ref}}}{G}, \quad \text{and} \quad \tau = \frac{G_{\text{tr}}}{G}$$

The consideration of wavelength and direction dependence of properties makes radiation calculations very complicated. Therefore, the *gray* and *diffuse* approximations are commonly utilized in radiation calculations. A surface is said to be *diffuse* if its properties are independent of direction and *gray* if its properties are independent of wavelength.

The sum of the absorbed, reflected, and transmitted fractions of radiation energy must be equal to unity,

$$\alpha + \rho + \tau = 1$$

For *opaque* surfaces, $\tau = 0$, and thus

$$\alpha + \rho = 1$$

Surfaces are usually assumed to reflect in a perfectly *specular* or *diffuse* manner for simplicity. In *specular* (or *mirror-like*) *reflection*, the angle of reflection equals the angle of incidence of the radiation beam. In *diffuse reflection*, radiation is reflected equally in all directions. Reflection from smooth and polished surfaces approximates specular reflection, whereas reflection from rough surfaces approximates diffuse reflection. *Kirchhoff's law* of radiation is expressed as

$$\varepsilon_{\lambda,\,\theta}(T) = \alpha_{\lambda,\,\theta}(T), \qquad \varepsilon_\lambda(T) = \alpha_\lambda(T), \qquad \text{and} \qquad \varepsilon(T) = \alpha(T)$$

Radiaton heat transfer between surfaces depends on the orientation of the surfaces relative to each other. In a radiation analysis, this effect is accounted for by the geometric parameter *view factor*. The *view factor* from a surface i to a surface j is denoted by $F_{i\to j}$ or F_{ij}, and is defined as the fraction of the radiation leaving surface i that strikes surface j directly. The view factor $F_{i\to i}$ represents the fraction of the radiation leaving a surface i that strikes itself directly; $F_{i\to i} = 0$ for *plane* or *convex surfaces* and $F_{i\to i} \neq 0$ for *concave surfaces*.

For view factors, the *reciprocity rule* is expressed as

$$A_i F_{i\to j} = A_j F_{j\to i}$$

The sum of the view factors from surface i of an enclosure to all surfaces of the enclosure, including to itself, must equal unity. This is known as the *summation rule* for an enclosure. The *superposition rule* is expressed as the view factor from a surface i to a surface j is equal to the sum of the view factors from surface i to the parts of surface j. The symmetry rule is expressed as if the surfaces j and k are symmetric about the surface i then $F_{i\to j} = F_{i\to k}$.

The rate of net radiation heat transfer between two *black* surfaces is determined from

$$\dot{Q}_{1\to 2} = A_1 F_{1\to 2}\sigma(T_1^4 - T_2^4)$$

The *net* radiation heat transfer from any surface i of a *black* enclosure is determined by adding up the net radiation heat transfers from surface i to each of the surfaces of the enclosure:

$$\dot{Q}_i = \sum_{j=1}^{N} \dot{Q}_{i\to j} = \sum_{j=1}^{N} A_i F_{i\to j}\sigma(T_i^4 - T_j^4)$$

The total radiation energy leaving a surface per unit time and per unit area is called the *radiosity* and is denoted by J. The *net* rate of radiation heat transfer from a surface i of surface area A_i is expressed as

$$\dot{Q}_i = \frac{E_{bi} - J_i}{R_i} \quad \text{where} \quad R_i = \frac{1 - \varepsilon_i}{A_i\,\varepsilon_i}$$

is the *surface resistance* to radiation. The *net* rate of radiation heat transfer from surface i to surface j can be expressed as

$$\dot{Q}_{i\to j} = \frac{J_i - J_j}{R_{i\to j}} \quad \text{where} \quad R_{i\to j} = \frac{1}{A_i F_{i\to j}}$$

is the *space resistance* to radiation. The *network method* is applied to radiation enclosure problems by drawing a surface resistance associated with each surface of an enclosure and connecting them with space resistances. Then the problem is solved by treating it as an electrical network problem where the radiation heat transfer replaces the current and the radiosity replaces the potential. The *direct method* is based on the following two equations:

Surfaces with specified net heat transfer rate $\dot{Q}_i$ $\qquad \dot{Q}_i = A_i \sum_{j=1}^{N} F_{i\to j}(J_i - J_j)$

Surfaces with specified temperature T_i

$$\sigma T_i^4 = J_i + \frac{1 - \varepsilon_i}{\varepsilon_i} \sum_{j=1}^{N} F_{i \to j}(J_i - J_j)$$

The first and the second groups of equations give N linear algebraic equations for the determination of the N unknown radiosities for an N-surface enclosure. Once the radiosities J_1, J_2, . . . , J_N are available, the unknown surface temperatures and heat transfer rates can be determined from the equations just shown.

The net rate of radiation transfer between any two gray, diffuse, opaque surfaces that form an enclosure is given by

$$\dot{Q}_{12} = \frac{\sigma(T_1^4 - T_2^4)}{\dfrac{1 - \varepsilon_1}{A_1 \varepsilon_1} + \dfrac{1}{A_1 F_{12}} + \dfrac{1 - \varepsilon_2}{A_2 \varepsilon_2}}$$

REFERENCES AND SUGGESTED READINGS

1. American Society of Heating, Refrigeration, and Air Conditioning Engineers, *Handbook of Fundamentals*, Atlanta, ASHRAE, 1993.
2. A. G. H. Dietz. "Diathermanous Materials and Properties of Surfaces." In *Space Heating with Solar Energy,* ed. R. W. Hamilton. Cambridge, MA: MIT Press, 1954.
3. D. C. Hamilton and W. R. Morgan. "Radiation Interchange Configuration Factors." National Advisory Committee for Aeronautics, Technical Note 2836, 1952.
4. H. C. Hottel. "Radiant Heat Transmission." In *Heat Transmission*. 3rd ed., ed. W. H. McAdams. New York: McGraw-Hill, 1954.
5. J. R. Howell. *A Catalog of Radiation Configuration Factors*. New York: McGraw-Hill, 1982.
6. M. F. Modest. *Radiative Heat Transfer*. New York: McGraw-Hill, 1993.
7. A. K. Oppenheim. "Radiation Analysis by the Network Method." *Transactions of the ASME* 78 (1956), pp. 725–735.
8. M. Planck. *The Theory of Heat Radiation*. New York: Dover, 1959.
9. W. Sieber. *Zeitschrift für Technische Physics* 22 (1941), pp. 130–135.
10. R. Siegel and J. R. Howell. *Thermal Radiation Heat Transfer*. 3rd ed. Washington, DC: Hemisphere, 1992.
11. Y. S. Touloukain and D. P. DeWitt. "Nonmetallic Solids." In *Thermal Radiative Properties*. Vol. 8. New York: IFI/Plenum, 1970.
12. Y. S. Touloukian and D. P. DeWitt. "Metallic Elements and Alloys." In *Thermal Radiative Properties,* Vol. 7. New York: IFI/Plenum, 1970.

PROBLEMS*

Electromagnetic and Thermal Radiation

21–1C What is an electromagnetic wave? How does it differ from a sound wave?

21–2C By what properties is an electromagnetic wave characterized? How are these properties related to each other?

21–3C What is visible light? How does it differ from the other forms of electromagnetic radiation?

21–4C How do ultraviolet and infrared radiation differ? Do you think your body emits any radiation in the ultraviolet range? Explain.

21–5C What is thermal radiation? How does it differ from the other forms of electromagnetic radiation?

21–6C What is the cause of color? Why do some objects appear blue to the eye while others appear red? Is the color of a surface at room temperature related to the radiation it emits?

21–7C Why is radiation usually treated as a surface phenomenon?

21–8C Why do skiers get sunburned so easily?

*Problems designated by a "C" are concept questions, and students are encouraged to answer them all. Problems designated by an "E" are in English units, and the SI users can ignore them. Problems with the icon are solved using EES, and complete solutions together with parametric studies are included on the enclosed DVD. Problems with the icon are comprehensive in nature, and are intended to be solved with a computer, preferably using the EES software that accompanies this text.

21–9C How does microwave cooking differ from conventional cooking?

21–10 The speed of light in vacuum is given to be 3.0×10^8 m/s. Determine the speed of light in air ($n = 1$), in water ($n = 1.33$), and in glass ($n = 1.5$).

21–11 Electricity is generated and transmitted in power lines at a frequency of 60 Hz (1 Hz = 1 cycle per second). Determine the wavelength of the electromagnetic waves generated by the passage of electricity in power lines.

21–12 A microwave oven is designed to operate at a frequency of 2.2×10^9 Hz. Determine the wavelength of these microwaves and the energy of each microwave.

21–13 A radio station is broadcasting radio waves at a wavelength of 200 m. Determine the frequency of these waves.
Answer: 1.5×10^6 Hz

21–14 A cordless telephone is designed to operate at a frequency of 8.5×10^8 Hz. Determine the wavelength of these telephone waves.

Blackbody Radiation

21–15C What is a blackbody? Does a blackbody actually exist?

21–16C Define the total and spectral blackbody emissive powers. How are they related to each other? How do they differ?

21–17C Why did we define the blackbody radiation function? What does it represent? For what is it used?

21–18C Consider two identical bodies, one at 1000 K and the other at 1500 K. Which body emits more radiation in the shorter-wavelength region? Which body emits more radiation at a wavelength of 20 μm?

21–19 Consider a surface at a uniform temperature of 800 K. Determine the maximum rate of thermal radiation that can be emitted by this surface, in W/m^2.

21–20 Consider a 20-cm × 20-cm × 20-cm cubical body at 750 K suspended in the air. Assuming the body closely approximates a blackbody, determine (*a*) the rate at which the cube emits radiation energy, in W and (*b*) the spectral blackbody emissive power at a wavelength of 4 μm.

21–21E The sun can be treated as a blackbody at an effective surface temperature of 10,400 R. Determine the rate at which infrared radiation energy ($\lambda = 0.76$–100 μm) is emitted by the sun, in Btu/h · ft^2.

21–22 [EES] The sun can be treated as a blackbody at 5780 K. Using EES (or other) software, calculate and plot the spectral blackbody emissive power $E_{b\lambda}$ of the sun versus wavelength in the range of 0.01 μm to 1000 μm. Discuss the results.

21–23 The temperature of the filament of an incandescent lightbulb is 3200 K. Treating the filament as a blackbody, determine the fraction of the radiant energy emitted by the filament that falls in the visible range. Also, determine the wavelength at which the emission of radiation from the filament peaks.

21–24 [EES] Reconsider Prob. 21–23. Using EES (or other) software, investigate the effect of temperature on the fraction of radiation emitted in the visible range. Let the surface temperature vary from 1000 K to 4000 K, and plot fraction of radiation emitted in the visible range versus the surface temperature.

21–25 An incandescent lightbulb is desired to emit at least 15 percent of its energy at wavelengths shorter than 0.8 μm. Determine the minimum temperature to which the filament of the lightbulb must be heated.

21–26 It is desired that the radiation energy emitted by a light source reach a maximum in the blue range ($\lambda = 0.47$ μm). Determine the temperature of this light source and the fraction of radiation it emits in the visible range ($\lambda = 0.40$–0.76 μm).

21–27 A 3-mm-thick glass window transmits 90 percent of the radiation between $\lambda = 0.3$ and 3.0 μm and is essentially opaque for radiation at other wavelengths. Determine the rate of radiation transmitted through a 2-m × 2-m glass window from blackbody sources at (*a*) 5800 K and (*b*) 1000 K.
Answers: (*a*) 218,400 kW, (*b*) 55.8 kW

Radiation Properties

21–28C Define the properties emissivity and absorptivity. When are these two properties equal to each other?

21–29C Define the properties reflectivity and transmissivity and discuss the different forms of reflection.

21–30C What is a graybody? How does it differ from a blackbody? What is a diffuse gray surface?

21–31C What is the greenhouse effect? Why is it a matter of great concern among atmospheric scientists?

21–32C We can see the inside of a microwave oven during operation through its glass door, which indicates that visible radiation is escaping the oven. Do you think that the harmful microwave radiation might also be escaping?

21–33 The spectral emissivity function of an opaque surface at 1000 K is approximated as

$$\varepsilon_\lambda = \begin{cases} \varepsilon_1 = 0.4, & 0 \leq \lambda < 2\ \mu\text{m} \\ \varepsilon_2 = 0.7, & 2\ \mu\text{m} \leq \lambda < 6\ \mu\text{m} \\ \varepsilon_3 = 0.3, & 6\ \mu\text{m} \leq \lambda < \infty \end{cases}$$

Determine the average emissivity of the surface and the rate of radiation emission from the surface, in W/m^2.
Answers: 0.575, 32.6 kW/m^2

21–34 The reflectivity of aluminum coated with lead sulfate is 0.35 for radiation at wavelengths less than 3 μm and 0.95 for radiation greater than 3 μm. Determine the average reflectivity of this surface for solar radiation ($T \approx 5800$ K) and radiation coming from surfaces at room temperature ($T \approx 300$ K). Also, determine the emissivity and absorptivity of this surface at both temperatures. Do you think this material is suitable for use in solar collectors?

21–35 A furnace that has a 40-cm × 40-cm glass window can be considered to be a blackbody at 1200 K. If the transmissivity of the glass is 0.7 for radiation at wavelengths less than 3 μm and zero for radiation at wavelengths greater than 3 μm, determine the fraction and the rate of radiation coming from the furnace and transmitted through the window.

21–36 The emissivity of a tungsten filament can be approximated to be 0.5 for radiation at wavelengths less than 1 μm and 0.15 for radiation at greater than 1 μm. Determine the average emissivity of the filament at (*a*) 2000 K and (*b*) 3000 K. Also, determine the absorptivity and reflectivity of the filament at both temperatures.

21–37 The variations of the spectral emissivity of two surfaces are as given in Fig. P21–37. Determine the average emissivity of each surface at $T = 3000$ K. Also, determine the average absorptivity and reflectivity of each surface for radiation coming from a source at 3000 K. Which surface is more suitable to serve as a solar absorber?

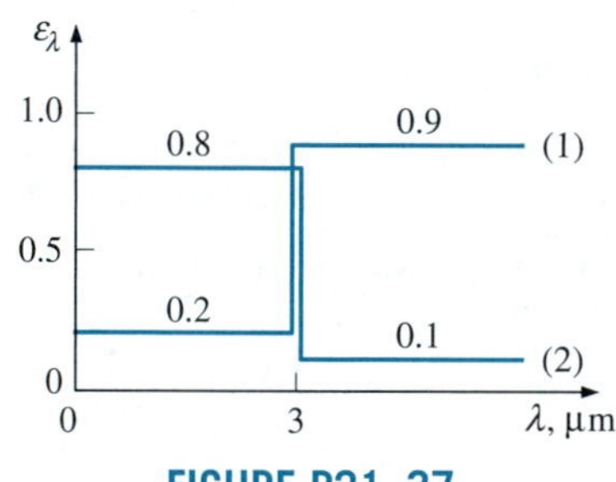

FIGURE P21–37

21–38 The emissivity of a surface coated with aluminum oxide can be approximated to be 0.15 for radiation at wavelengths less than 5 μm and 0.9 for radiation at wavelengths greater than 5 μm. Determine the average emissivity of this surface at (*a*) 5800 K and (*b*) 300 K. What can you say about the absorptivity of this surface for radiation coming from sources at 5800 K and 300 K? *Answers:* (*a*) 0.153, (*b*) 0.89

21–39 The variation of the spectral absorptivity of a surface is as given in Fig. P21–39. Determine the average absorptivity and reflectivity of the surface for radiation that originates from a source at $T = 2500$ K. Also, determine the average emissivity of this surface at 3000 K.

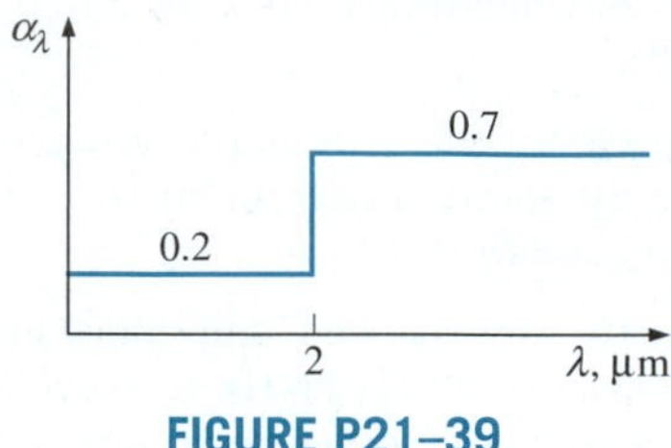

FIGURE P21–39

21–40E A 5-in-diameter spherical ball is known to emit radiation at a rate of 550 Btu/h when its surface temperature is 950 R. Determine the average emissivity of the ball at this temperature.

21–41 The variation of the spectral transmissivity of a 0.6-cm-thick glass window is as given in Fig. P21–41. Determine the average transmissivity of this window for solar radiation ($T \approx 5800$ K) and radiation coming from surfaces at room temperature ($T \approx 300$ K). Also, determine the amount of solar radiation transmitted through the window for incident solar radiation of 650 W/m^2. *Answers:* 0.848, 0.00015, 551.1 W/m^2

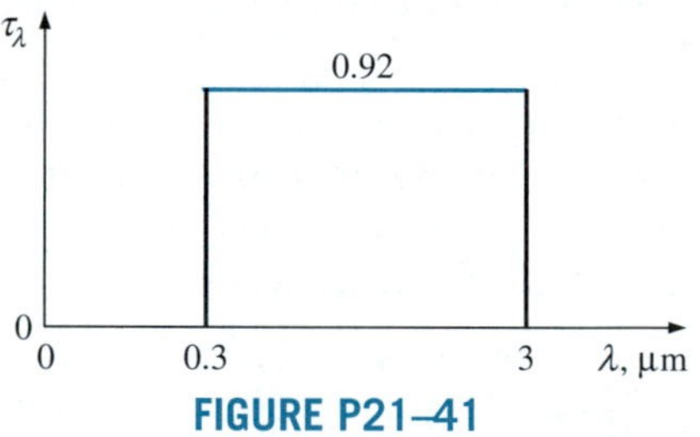

FIGURE P21–41

The View Factor

21–42C What does the view factor represent? When is the view factor from a surface to itself not zero?

21–43C How can you determine the view factor F_{12} when the view factor F_{21} and the surface areas are available?

21–44C What are the summation rule and the superposition rule for view factors?

21–45C What is the crossed-strings method? For what kind of geometries is the crossed-strings method applicable?

21–46 Consider an enclosure consisting of eight surfaces. How many view factors does this geometry involve? How many of these view factors can be determined by the application of the reciprocity and the summation rules?

21–47 Consider an enclosure consisting of five surfaces. How many view factors does this geometry involve? How many of these view factors can be determined by the application of the reciprocity and summation rules?

21–48 Consider an enclosure consisting of 12 surfaces. How many view factors does this geometry involve? How many of these view factors can be determined by the application of the reciprocity and the summation rules? *Answers:* 144, 78

21–49 Determine the view factors F_{13} and F_{23} between the rectangular surfaces shown in Fig. P21–49.

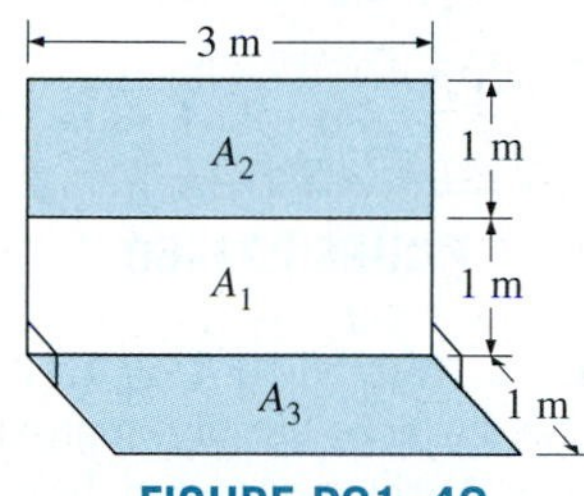

FIGURE P21–49

21–50 Consider a cylindrical enclosure whose height is twice the diameter of its base. Determine the view factor from the side surface of this cylindrical enclosure to its base surface.

21–51 Consider a hemispherical furnace with a flat circular base of diameter D. Determine the view factor from the dome of this furnace to its base. *Answer:* 0.5

21–52 Determine the view factors F_{12} and F_{21} for the very long ducts shown in Fig. P21–52 without using any view factor tables or charts. Neglect end effects.

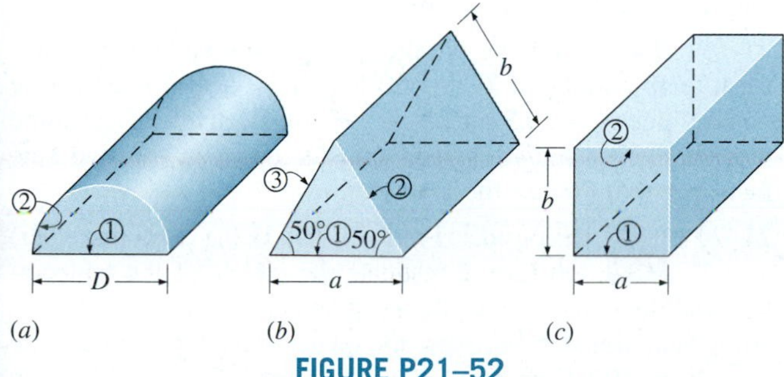

FIGURE P21–52

21–53 Determine the view factors from the very long grooves shown in Fig. P21–53 to the surroundings without using any view factor tables or charts. Neglect end effects.

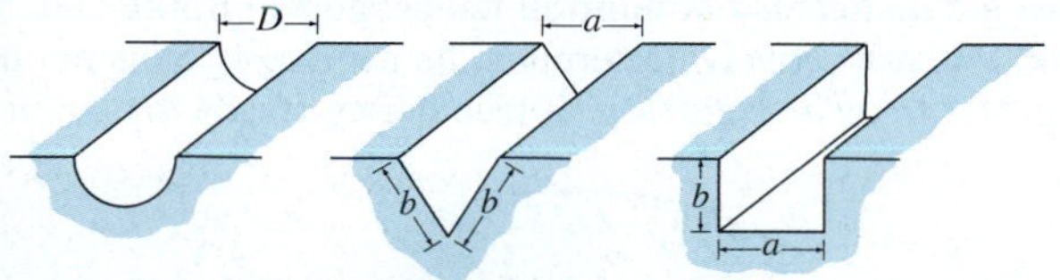

(*a*) Semicylindrical. (*b*) Triangular groove. (*c*) Rectangular groove.

FIGURE P21–53

21–54 Determine the view factors from the base of a cube to each of the other five surfaces.

21–55 Consider a conical enclosure of height h and base diameter D. Determine the view factor from the conical side surface to a hole of diameter d located at the center of the base.

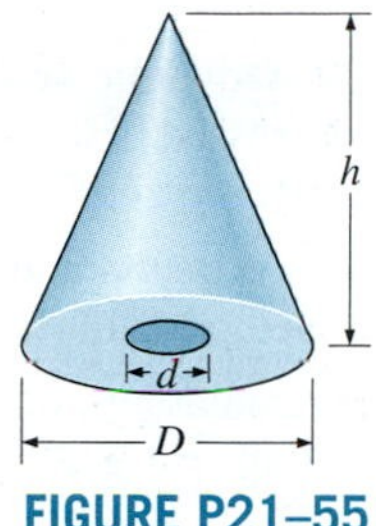

FIGURE P21–55

21–56 Determine the four view factors associated with an enclosure formed by two very long concentric cylinders of radii r_1 and r_2. Neglect the end effects.

21–57 Determine the view factor F_{12} between the rectangular surfaces shown in Fig. P21–57.

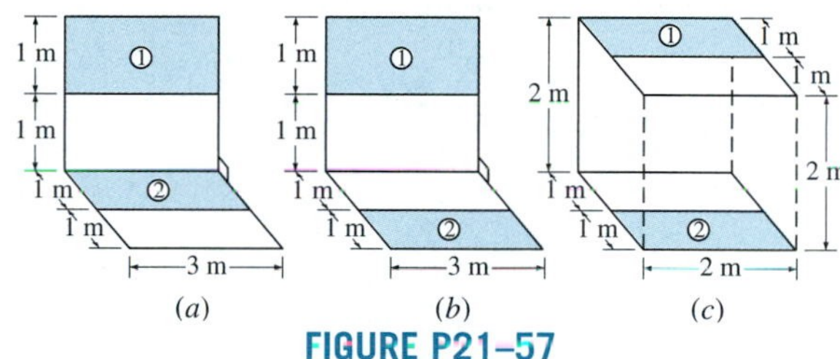

FIGURE P21–57

21–58 Two infinitely long parallel cylinders of diameter D are located a distance s apart from each other. Determine the view factor F_{12} between these two cylinders.

21–59 Three infinitely long parallel cylinders of diameter D are located a distance s apart from each other. Determine the view factor between the cylinder in the middle and the surroundings.

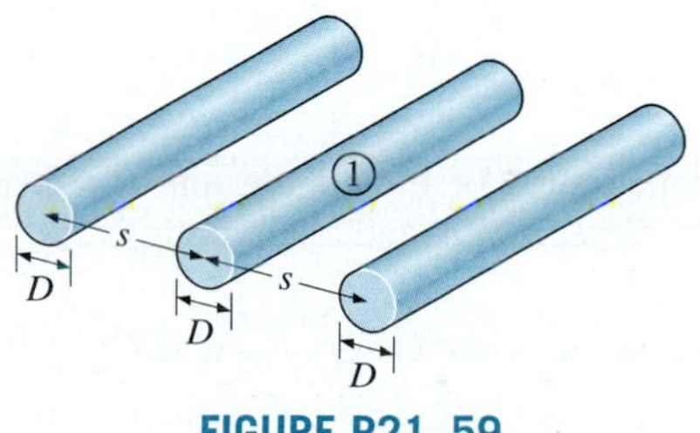

FIGURE P21–59

Radiation Heat Transfer between Surfaces

21–60C Why is the radiation analysis of enclosures that consist of black surfaces relatively easy? How is the rate of radiation heat transfer between two surfaces expressed in this case?

21–61C How does radiosity for a surface differ from the emitted energy? For what kind of surfaces are these two quantities identical?

21–62C What are the radiation surface and space resistances? How are they expressed? For what kind of surfaces is the radiation surface resistance zero?

21–63C What are the two methods used in radiation analysis? How do these two methods differ?

21–64C What is a reradiating surface? What simplifications does a reradiating surface offer in the radiation analysis?

21–65 A solid sphere of 1 m diameter at 500 K is kept in an evacuated, long, equilateral triangular enclosure whose sides are 2 m long. The emissivity of the sphere is 0.45 and the temperature of the enclosure is 380 K. If heat is generated uniformly within the sphere at a rate of 3100 W, determine (*a*) the view factor from the enclosure to the sphere and (*b*) the emissivity of the enclosure.

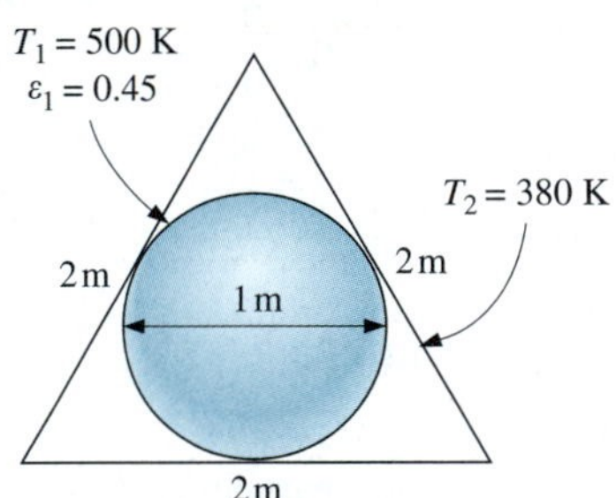

FIGURE P21–65

21–66 This question deals with steady–state radiation heat transfer between a sphere (r_1 = 30 cm) and a circular disk (r_2 = 120 cm), which are separated by a center-to-center distance h = 60 cm. When the normal to the center of disk passes through the center of sphere, the radiation view factor is given by

$$F_{12} = 0.5\left\{1 - \left[1 + \left(\frac{r_2}{h}\right)^2\right]^{-0.5}\right\}$$

Surface temperatures of the sphere and the disk are 600°C and 200°C, respectively; and their emissivities are 0.9 and 0.5, respectively.

(*a*) Calculate the view factors F_{12} and F_{21}.

(*b*) Calculate the net rate of radiation heat exchange between the sphere and the disk.

(*c*) For the given radii and temperatures of the sphere and the disk, the following four possible modifications could increase the net rate of radiation heat exchange: paint each of the two surfaces to alter their emissivities, adjust the distance between them, and provide an (refractory) enclosure. Calculate the net rate of radiation heat exchange between the two bodies if the *best values* are selected for each of the above modifications.

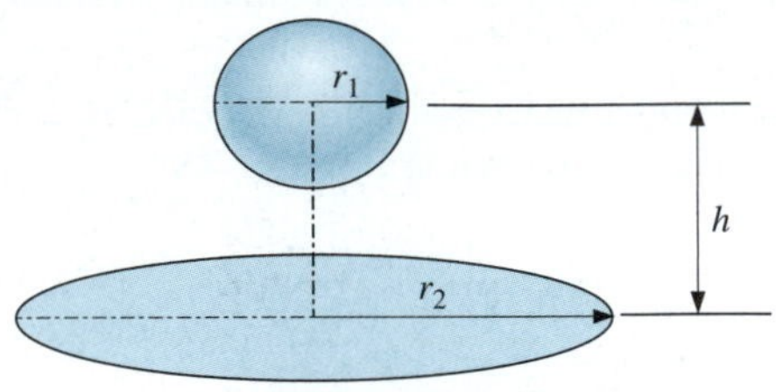

FIGURE P21–66

21–67E Consider a 10-ft × 10-ft × 10-ft cubical furnace whose top and side surfaces closely approximate black surfaces and whose base surface has an emissivity ε = 0.7. The base, top, and side surfaces of the furnace are maintained at uniform temperatures of 800 R, 1600 R, and 2400 R, respectively. Determine the net rate of radiation heat transfer between (*a*) the base and the side surfaces and (*b*) the base and the top surfaces. Also, determine the net rate of radiation heat transfer to the base surface.

21–68E Reconsider Prob. 21–67E. Using EES (or other) software, investigate the effect of base surface emissivity on the net rates of radiation heat transfer between the base and the side surfaces, between the base and top surfaces, and to the base surface. Let the emissivity vary from 0.1 to 0.9. Plot the rates of heat transfer as a function of emissivity, and discuss the results.

21–69 Two very large parallel plates are maintained at uniform temperatures of T_1 = 600 K and T_2 = 400 K and have emissivities ε_1 = 0.5 and ε_2 = 0.9, respectively. Determine the net rate of radiation heat transfer between the two surfaces per unit area of the plates.

21–70 Reconsider Prob. 21–69. Using EES (or other) software, investigate the effects of the temperature and the emissivity of the hot plate on the net rate of radiation heat transfer between the plates. Let the temperature vary from 500 K to 1000 K and the emissivity from 0.1 to 0.9. Plot the net rate of radiation heat transfer as functions of temperature and emissivity, and discuss the results.

21–71 A furnace is of cylindrical shape with $R = H = 2$ m. The base, top, and side surfaces of the furnace are all black and are maintained at uniform temperatures of 500, 700, and 1400 K, respectively. Determine the net rate of radiation heat transfer to or from the top surface during steady operation.

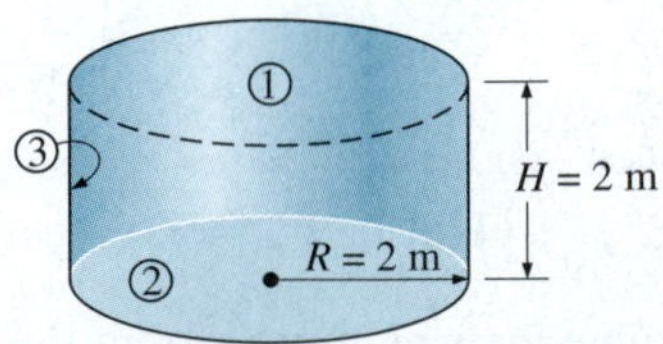

FIGURE P21–71

21–72 Consider a hemispherical furnace of diameter $D = 5$ m with a flat base. The dome of the furnace is black, and the base has an emissivity of 0.7. The base and the dome of the furnace are maintained at uniform temperatures of 400 and 1000 K, respectively. Determine the net rate of radiation heat transfer from the dome to the base surface during steady operation. *Answer:* 759 kW

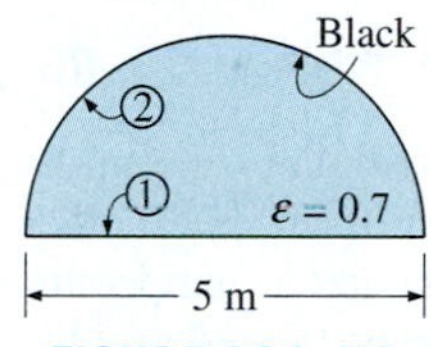

FIGURE P21–72

21–73 Two very long concentric cylinders of diameters $D_1 = 0.35$ m and $D_2 = 0.5$ m are maintained at uniform temperatures of $T_1 = 950$ K and $T_2 = 500$ K and have emissivities $\varepsilon_1 = 1$ and $\varepsilon_2 = 0.55$, respectively. Determine the net rate of radiation heat transfer between the two cylinders per unit length of the cylinders.

21–74 This experiment is conducted to determine the emissivity of a certain material. A long cylindrical rod of diameter $D_1 = 0.01$ m is coated with this new material and is placed in an evacuated long cylindrical enclosure of diameter $D_2 = 0.1$ m and emissivity $\varepsilon_2 = 0.95$, which is cooled externally and maintained at a temperature of 200 K at all times. The rod is heated by passing electric current through it. When steady operating conditions are reached, it is observed that the rod is dissipating electric power at a rate of 8 W per unit of its length and its surface temperature is 500 K. Based on these measurements, determine the emissivity of the coating on the rod.

21–75E A furnace is shaped like a long semicylindrical duct of diameter $D = 15$ ft. The base and the dome of the furnace have emissivities of 0.5 and 0.9 and are maintained at uniform temperatures of 550 and 1800 R, respectively. Determine the net rate of radiation heat transfer from the dome to the base surface per unit length during steady operation.

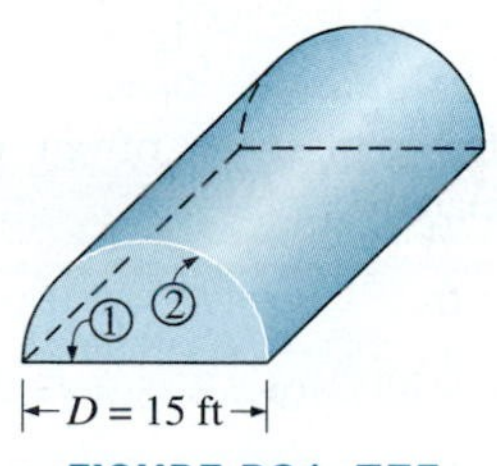

FIGURE P21–75E

21–76 Two parallel disks of diameter $D = 0.6$ m separated by $L = 0.4$ m are located directly on top of each other. Both disks are black and are maintained at a temperature of 450 K. The back sides of the disks are insulated, and the environment that the disks are in can be considered to be a blackbody at 300 K. Determine the net rate of radiation heat transfer from the disks to the environment. *Answer:* 781 W

21–77 A furnace is shaped like a long equilateral-triangular duct where the width of each side is 2 m. Heat is supplied from the base surface, whose emissivity is $\varepsilon_1 = 0.8$, at a rate of 800 W/m^2 while the side surfaces, whose emissivities are 0.5, are maintained at 500 K. Neglecting the end effects, determine the temperature of the base surface. Can you treat this geometry as a two-surface enclosure?

21–78 Reconsider Prob. 21–77. Using EES (or other) software, investigate the effects of the rate of the heat transfer at the base surface and the temperature of the side surfaces on the temperature of the base surface. Let the rate of heat transfer vary from 500 W/m^2 to 1000 W/m^2 and the temperature from 300 K to 700 K. Plot the temperature of the base surface as functions of the rate of heat transfer and the temperature of the side surfaces, and discuss the results.

21–79 Consider a 4-m × 4-m × 4-m cubical furnace whose floor and ceiling are black and whose side surfaces are reradiating. The floor and the ceiling of the furnace are maintained at temperatures of 550 K and 1100 K, respectively. Determine the net rate of radiation heat transfer between the floor and the ceiling of the furnace.

21–80 Two concentric spheres of diameters $D_1 = 0.3$ m and $D_2 = 0.4$ m are maintained at uniform temperatures $T_1 = 700$ K and $T_2 = 500$ K and have emissivities $\varepsilon_1 = 0.5$ and $\varepsilon_2 = 0.7$, respectively. Determine the net rate of radiation heat transfer between the two spheres. Also, determine the convection heat transfer coefficient at the outer surface if both the surrounding medium and the surrounding surfaces are at 30°C. Assume the emissivity of the outer surface is 0.35.

21–81 A spherical tank of diameter $D = 2$ m that is filled with liquid nitrogen at 100 K is kept in an evacuated cubic enclosure whose sides are 3 m long. The emissivities of the spherical tank and the enclosure are $\varepsilon_1 = 0.1$ and $\varepsilon_2 = 0.8$, respectively. If the temperature of the cubic enclosure is measured to be 240 K, determine the net rate of radiation heat transfer to the liquid nitrogen. *Answer:* 228 W

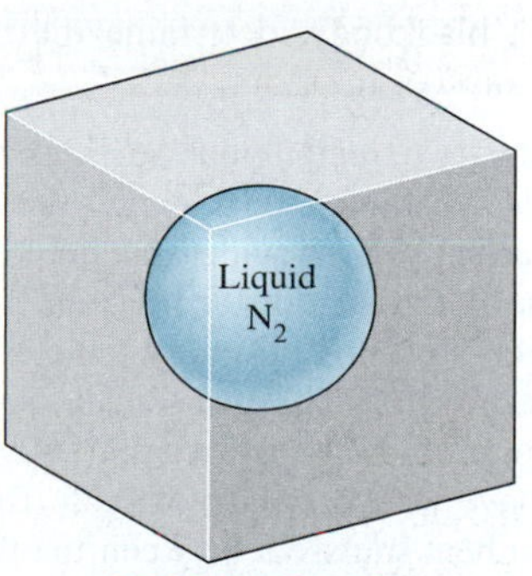

FIGURE P21–81

21–82 Repeat Prob. 21–81 by replacing the cubic enclosure by a spherical enclosure whose diameter is 3 m.

21–83 Reconsider Prob. 21–81. Using EES (or other) software, investigate the effects of the side length and the emissivity of the cubic enclosure, and the emissivity of the spherical tank on the net rate of radiation heat transfer. Let the side length vary from 2.5 m to 5.0 m and both emissivities from 0.1 to 0.9. Plot the net rate of radiation heat transfer as functions of side length and emissivities, and discuss the results.

21–84 Consider a circular grill whose diameter is 0.3 m. The bottom of the grill is covered with hot coal bricks at 950 K, while the wire mesh on top of the grill is covered with steaks initially at 5°C. The distance between the coal bricks and the steaks is 0.20 m. Treating both the steaks and the coal bricks as blackbodies, determine the initial rate of radiation heat transfer from the coal bricks to the steaks. Also, determine the initial rate of radiation heat transfer to the steaks if the side opening of the grill is covered by aluminum foil, which can be approximated as a reradiating surface. *Answers:* 928 W, 2085 W

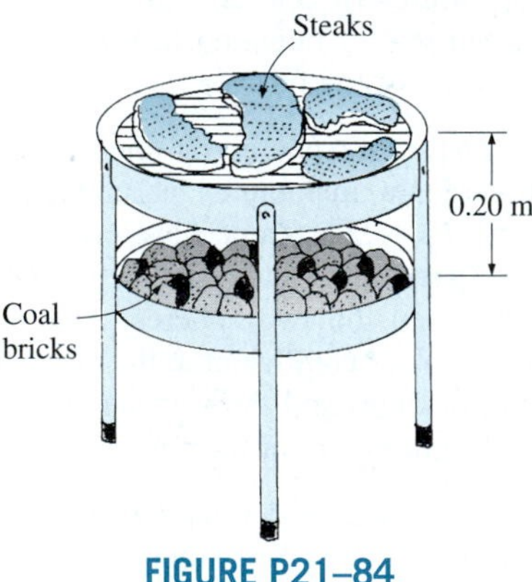

FIGURE P21–84

21–85E A 19-ft-high room with a base area of 12 ft × 12 ft is to be heated by electric resistance heaters placed on the ceiling, which is maintained at a uniform temperature of 90°F at all times. The floor of the room is at 65°F and has an emissivity of 0.8. The side surfaces are well insulated. Treating the ceiling as a blackbody, determine the rate of heat loss from the room through the floor.

21–86 Consider two rectangular surfaces perpendicular to each other with a common edge which is 1.6 m long. The horizontal surface is 0.8 m wide and the vertical surface is 1.2 m high. The horizontal surface has an emissivity of 0.75 and is maintained at 400 K. The vertical surface is black and is maintained at 550 K. The back sides of the surfaces are insulated. The surrounding surfaces are at 290 K, and can be considered to have an emissivity of 0.85. Determine the net rate of radiation heat transfers between the two surfaces, and between the horizontal surface and the surroundings.

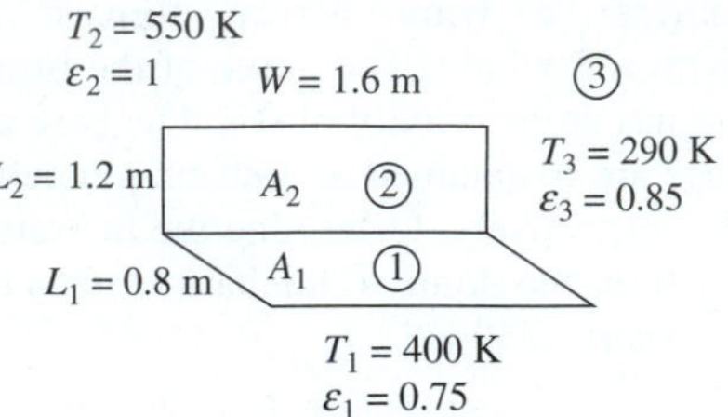

FIGURE P21–86

21–87 Two long parallel 20-cm-diameter cylinders are located 30 cm apart from each other. Both cylinders are black, and are maintained at temperatures 425 K and 275 K. The surroundings can be treated as a blackbody at 300 K. For a 1-m-long section of the cylinders, determine the rates of radiation heat transfer between the cylinders and between the hot cylinder and the surroundings.

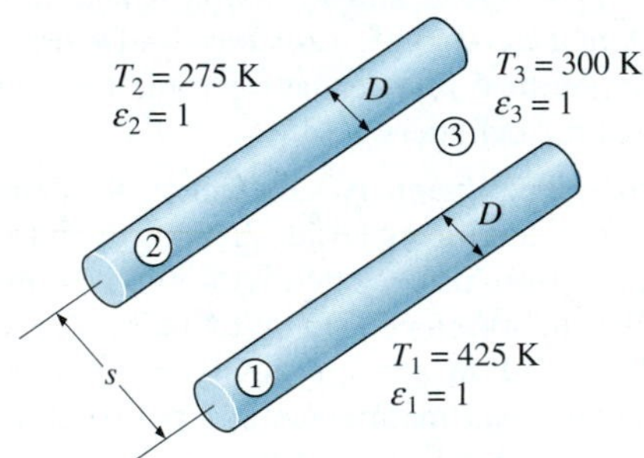

FIGURE P21–87

21–88 Consider a long semicylindrical duct of diameter 1.0 m. Heat is supplied from the base surface, which is black, at a rate of 1200 W/m^2, while the side surface with an emissivity of 0.4 are is maintained at 650 K. Neglecting the end effects, determine the temperature of the base surface.

21–89 Consider a 20-cm-diameter hemispherical enclosure. The dome is maintained at 600 K and heat is supplied from the dome at a rate of 50 W while the base surface with an emissivity is 0.55 is maintained at 400 K. Determine the emissivity of the dome.

Review Problems

21–90 The spectral emissivity of an opaque surface at 1500 K is approximated as

$$\varepsilon_1 = 0 \quad \text{for} \quad \lambda < 2 \ \mu\text{m}$$
$$\varepsilon_2 = 0.85 \quad \text{for} \quad 2 \leq \lambda \leq 6 \ \mu\text{m}$$
$$\varepsilon_3 = 0 \quad \text{for} \quad \lambda > 6 \ \mu\text{m}$$

Determine the total emissivity and the emissive flux of the surface.

21–91 The spectral transmissivity of a 3-mm-thick regular glass can be expressed as

$$\tau_1 = 0 \quad \text{for} \quad \lambda < 0.35\ \mu\text{m}$$
$$\tau_2 = 0.85 \quad \text{for} \quad 0.35 < \lambda < 2.5\ \mu\text{m}$$
$$\tau_3 = 0 \quad \text{for} \quad \lambda > 2.5\ \mu\text{m}$$

Determine the transmissivity of this glass for solar radiation. What is the transmissivity of this glass for light?

21–92 A 1-m-diameter spherical cavity is maintained at a uniform temperature of 600 K. Now a 5-mm-diameter hole is drilled. Determine the maximum rate of radiation energy streaming through the hole. What would your answer be if the diameter of the cavity were 3 m?

21–93 The spectral absorptivity of an opaque surface is as shown on the graph. Determine the absorptivity of the surface for radiation emitted by a source at (*a*) 1000 K and (*b*) 3000 K.

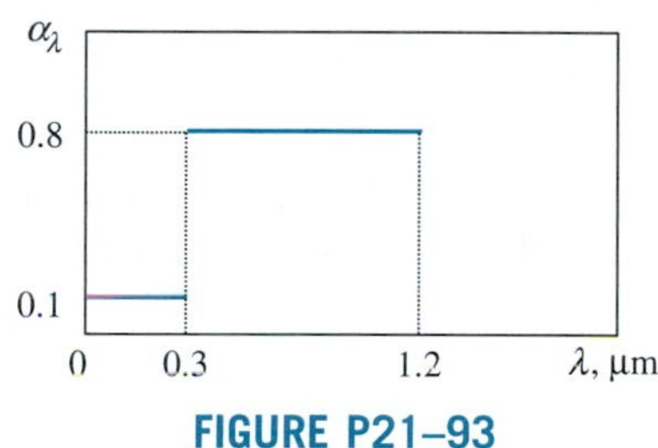

FIGURE P21–93

21–94 The surface in Prob. 21–93 receives solar radiation at a rate of 470 W/m². Determine the solar absorptivity of the surface and the rate of absorption of solar radiation.

21–95 The spectral transmissivity of a glass cover used in a solar collector is given as

$$\tau_1 = 0 \quad \text{for} \quad \lambda < 0.3\ \mu\text{m}$$
$$\tau_2 = 0.9 \quad \text{for} \quad 0.3 < \lambda < 3\ \mu\text{m}$$
$$\tau_3 = 0 \quad \text{for} \quad \lambda > 3\ \mu\text{m}$$

Solar radiation is incident at a rate of 950 W/m², and the absorber plate, which can be considered to be black, is maintained at 340 K by the cooling water. Determine (*a*) the solar flux incident on the absorber plate, (*b*) the transmissivity of the glass cover for radiation emitted by the absorber plate, and (*c*) the rate of heat transfer to the cooling water if the glass cover temperature is also 340 K.

21–96 A thermocouple used to measure the temperature of hot air flowing in a duct whose walls are maintained at T_w = 500 K shows a temperature reading of T_{th} = 850 K. Assuming the emissivity of the thermocouple junction to be ε = 0.6 and the convection heat transfer coefficient to be h = 60 W/m² · °C, determine the actual temperature of air. *Answer:* 1111 K

21–97 A large number of long tubes, each of diameter D, are placed parallel to each other and at a center-to-center distance of s. Since all of the tubes are geometrically similar and at the same temperature, these could be treated collectively as one surface (A_j) for radiation heat transfer calculations. As shown in Fig. P21–97, the tube-bank (A_j) is placed opposite a large flat wall (A_i) such that the tube-bank is parallel to the wall. The radiation view factor, F_{ij}, for this arrangement is given by

$$F_{ij} = 1 - \left[1 - \left(\frac{D}{s}\right)^2\right]^{0.5} + \frac{D}{s}\left\{\tan^{-1}\left[\left(\frac{s}{D}\right)^2 - 1\right]^{0.5}\right\}$$

(*a*) Calculate the view factors F_{ij} and F_{ji} for s = 3.0 cm and D = 1.5 cm.

(*b*) Calculate the net rate of radiation heat transfer between the wall and the tube-bank per unit area of the wall when T_i = 900°C, T_j = 60°C, ε_i = 0.8, and ε_j = 0.9.

(*c*) A fluid flows through the tubes at an average temperature of 40°C, resulting in a heat transfer coefficient of 2.0 kW/m² · K. Assuming T_i = 900°C, ε_i = 0.8 and ε_j = 0.9 (as above) and neglecting the tube wall thickness and convection from the outer surface, calculate the temperature of the tube surface in steady operation.

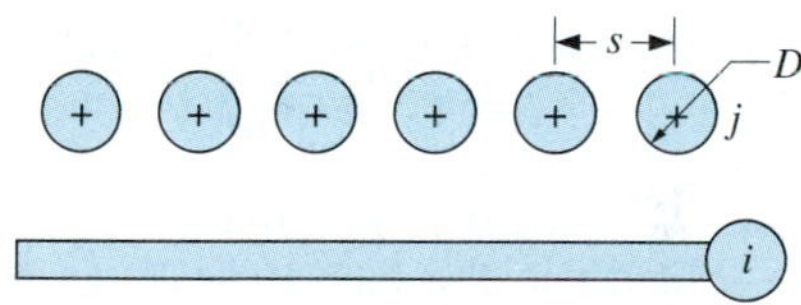

FIGURE P21–97

21–98E Consider a sealed 8-in-high electronic box whose base dimensions are 12 in × 12 in placed in a vacuum chamber. The emissivity of the outer surface of the box is 0.95. If the electronic components in the box dissipate a total of 90 W of power and the outer surface temperature of the box is not to exceed 130°F, determine the highest temperature at which the surrounding surfaces must be kept if this box is to be cooled by radiation alone. Assume the heat transfer from the bottom surface of the box to the stand to be negligible. *Answer:* 54°F

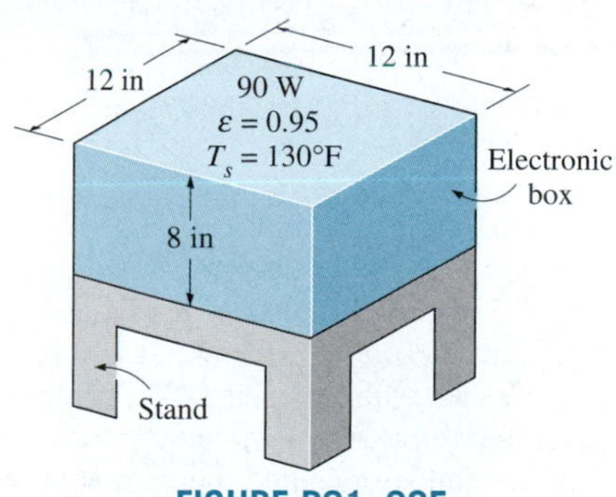

FIGURE P21–98E

21–99 A 2-m-internal-diameter double-walled spherical tank is used to store iced water at 0°C. Each wall is 0.5 cm thick, and the 1.5-cm-thick air space between the two walls of the tank is evacuated in order to minimize heat transfer. The surfaces surrounding the evacuated space are polished so that each surface has an emissivity of 0.15. The temperature of the outer wall of the tank is measured to be 20°C. Assuming the inner wall of the steel tank to be at 0°C, determine (*a*) the rate of heat transfer to the iced water in the tank and (*b*) the amount of ice at 0°C that melts during a 24-h period.

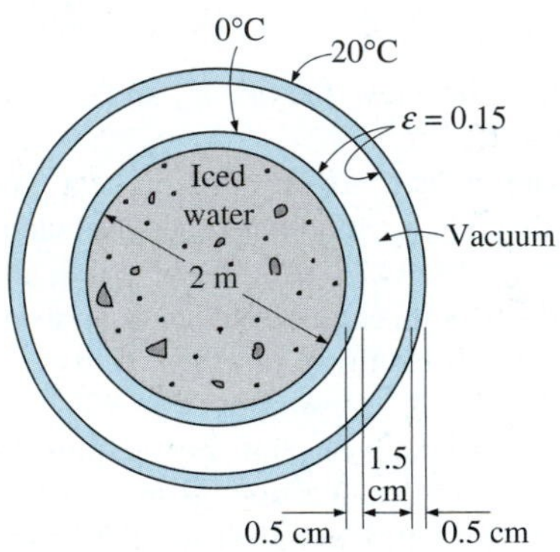

FIGURE P21–99

21–100 Two concentric spheres of diameters $D_1 = 15$ cm and $D_2 = 25$ cm are separated by air at 1 atm pressure. The surface temperatures of the two spheres enclosing the air are $T_1 = 350$ K and $T_2 = 275$ K, respectively, and their emissivities are 0.75. Determine the rate of heat transfer from the inner sphere to the outer sphere by (*a*) natural convection and (*b*) radiation.

21–101 Consider a 1.5-m-high and 3-m-wide solar collector that is tilted at an angle 20° from the horizontal. The distance between the glass cover and the absorber plate is 3 cm, and the back side of the absorber is heavily insulated. The absorber plate and the glass cover are maintained at temperatures of 80°C and 32°C, respectively. The emissivity of the glass surface is 0.9 and that of the absorber plate is 0.8. Determine the rate of heat loss from the absorber plate by natural convection and radiation. *Answers:* 750 W, 1289 W

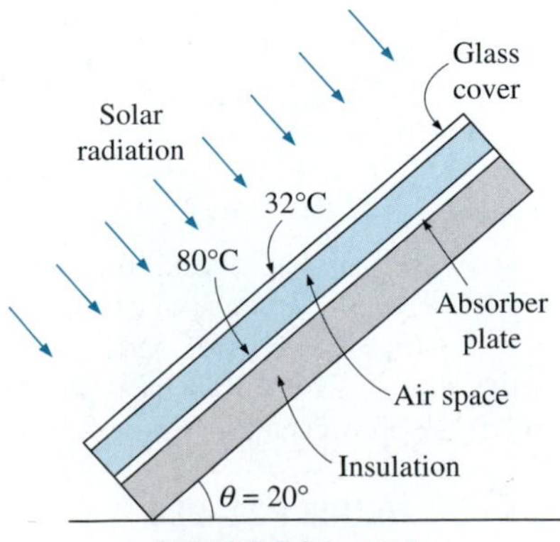

FIGURE P21–101

21–102E A solar collector consists of a horizontal aluminum tube having an outer diameter of 2.5 in enclosed in a concentric thin glass tube of diameter 5 in. Water is heated as it flows through the tube, and the annular space between the aluminum and the glass tube is filled with air at 0.5 atm pressure. The pump circulating the water fails during a clear day, and the water temperature in the tube starts rising. The aluminum tube absorbs solar radiation at a rate of 30 Btu/h per foot length, and the temperature of the ambient air outside is 75°F. The emissivities of the tube and the glass cover are 0.9. Taking the effective sky temperature to be 60°F, determine the temperature of the aluminum tube when thermal equilibrium is established (i.e., when the rate of heat loss from the tube equals the amount of solar energy gained by the tube).

21–103 A vertical 2-m-high and 5-m-wide double-pane window consists of two sheets of glass separated by a 3-cm-thick air gap. In order to reduce heat transfer through the window, the air space between the two glasses is partially evacuated to 0.3 atm pressure. The emissivities of the glass surfaces are 0.9. Taking the glass surface temperatures across the air gap to be 15°C and 5°C, determine the rate of heat transfer through the window by natural convection and radiation.

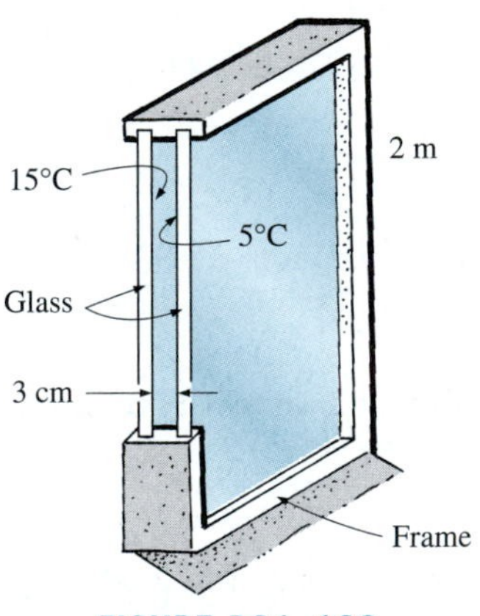

FIGURE P21–103

21–104 A simple solar collector is built by placing a 6-cm-diameter clear plastic tube around a garden hose whose outer diameter is 2 cm. The hose is painted black to maximize solar absorption, and some plastic rings are used to keep the spacing between the hose and the clear plastic cover constant. The emissivities of the hose surface and the glass cover are 0.9, and the effective sky temperature is estimated to be 15°C. The temperature of the plastic tube is measured to be 40°C, while the ambient air temperature is 25°C. Determine the rate of heat loss from the water in the hose by natural convection and radiation per meter of its length under steady conditions. *Answers:* 5.2 W, 26.2 W

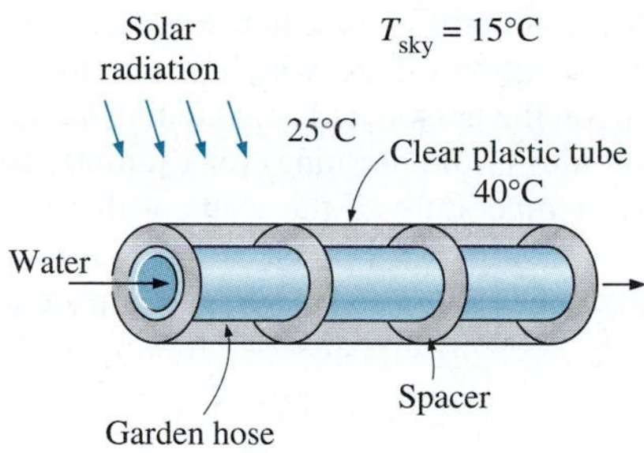

FIGURE P21–104

21–105 A solar collector consists of a horizontal copper tube of outer diameter 5 cm enclosed in a concentric thin glass tube of diameter 12 cm. Water is heated as it flows through the tube, and the annular space between the copper and the glass tubes is filled with air at 1 atm pressure. The emissivities of the tube surface and the glass cover are 0.85 and 0.9, respectively. During a clear day, the temperatures of the tube surface and the glass cover are measured to be 60°C and 40°C, respectively. Determine the rate of heat loss from the collector by natural convection and radiation per meter length of the tube.

21–106 A furnace is of cylindrical shape with a diameter of 1.2 m and a length of 1.2 m. The top surface has an emissivity of 0.70 and is maintained at 500 K. The bottom surface has an emissivity of 0.50 and is maintained at 650 K. The side surface has an emissivity of 0.40. Heat is supplied from the base surface at a net rate of 1400 W. Determine the temperature of the side surface and the net rates of heat transfer between the top and the bottom surfaces, and between the bottom and side surfaces.

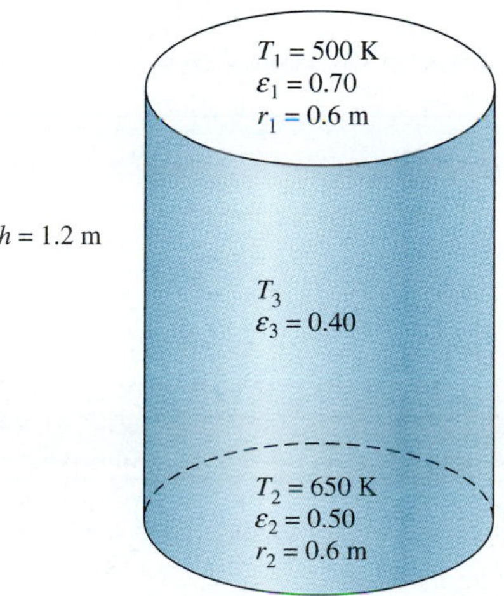

FIGURE P21–106

21–107 Consider a cubical furnace with a side length of 3 m. The top surface is maintained at 700 K. The base surface has an emissivity of 0.90 and is maintained at 950 K. The side surface is black and is maintained at 450 K. Heat is supplied from the base surface at a rate of 340 kW. Determine the emissivity of the top surface and the net rates of heat transfer between the top and the bottom surfaces, and between the bottom and side surfaces.

21–108 Two square plates, with the sides a and b (and $b > a$), are coaxial and parallel to each other, as shown in Fig. P21–108, and they are separated by a center-to-center distance of L. The radiation view factor from the smaller to the larger plate, F_{ab}, is given by

$$F_{ab} = \frac{1}{2A}\left\{[(B + A)^2 + 4]^{0.5} - [(B - A)^2 + 4]^{0.5}\right\}$$

where, $A = a/L$ and $B = b/L$.

(*a*) Calculate the view factors F_{ab} and F_{ba} for $a = 20$ cm, $b = 60$ cm, and $L = 40$ cm.

(*b*) Calculate the net rate of radiation heat exchange between the two plates described above if $T_a = 800°C$, $T_b = 200°C$, $\varepsilon_a = 0.8$, and $\varepsilon_b = 0.4$.

(*c*) A large square plate (with the side $c = 2.0$ m, $\varepsilon_c = 0.1$, and negligible thickness) is inserted symmetrically between the two plates such that it is parallel to and equidistant from them. For the data given above, calculate the temperature of this third plate when steady operating conditions are established.

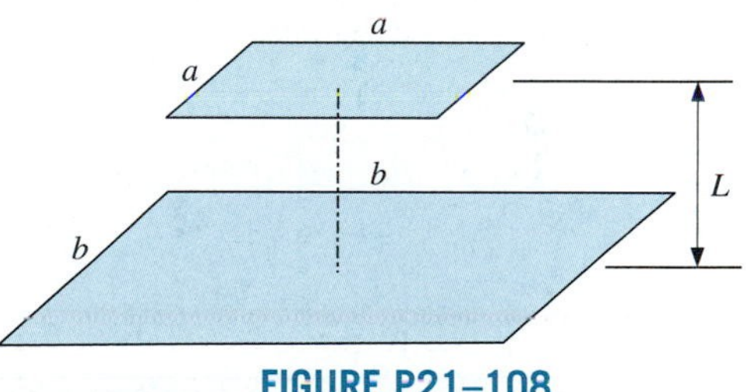

FIGURE P21–108

21–109 Two parallel concentric disks, 20 cm and 40 cm in diameter, are separated by a distance of 10 cm. The smaller disk ($\varepsilon = 0.80$) is at a temperature of 300°C. The larger disk ($\varepsilon = 0.60$) is at a temperature of 800°C.

(*a*) Calculate the radiation view factors.

(*b*) Determine the rate of radiation heat exchange between the two disks.

(*c*) Suppose that the space between the two disks is completely surrounded by a reflective surface. Estimate the rate of radiation heat exchange between the two disks.

Design and Essay Problems

21–110 Write an essay on the radiation properties of selective surfaces used on the absorber plates of solar collectors. Find out about the various kinds of such surfaces, and discuss the performance and cost of each type. Recommend a selective surface that optimizes cost and performance.

21–111 According to an Atomic Energy Commission report, a hydrogen bomb can be approximated as a large fireball at a temperature of 7200 K. You are to assess the impact if such a bomb exploded 5 km above a city. Assume the diameter of the fireball to be 1 km, and the blast to last 15 s. Investigate the level of radiation energy people, plants, and houses will be exposed to, and how adversely they will be affected by the blast.

21–112 Consider an enclosure consisting of *N* diffuse and gray surfaces. The emissivity and temperature of each surface as well as the view factors between the surfaces are specified. Write a program to determine the net rate of radiation heat transfer for each surface.

21–113 Thermal comfort in a house is strongly affected by the so-called radiation effect, which is due to radiation heat transfer between the person and surrounding surface. A person feels much colder in the morning, for example, becaus of the lower surface temperature of the walls at that time, although the thermostat setting of the house is fixed. Write an essay on the radiation effect, how it affect human comfort, and how it is accounted for in heating and air-conditioning applications.

Chapter 22

HEAT EXCHANGERS

Heat exchangers are devices that facilitate the *exchange of heat* between two fluids that are at different temperatures while keeping them from mixing with each other. Heat exchangers are commonly used in practice in a wide range of applications, from heating and air-conditioning systems in a household, to chemical processing and power production in large plants. Heat exchangers differ from mixing chambers in that they do not allow the two fluids involved to mix.

Heat transfer in a heat exchanger usually involves *convection* in each fluid and *conduction* through the wall separating the two fluids. In the analysis of heat exchangers, it is convenient to work with an *overall heat transfer coefficient U* that accounts for the contribution of all these effects on heat transfer. The rate of heat transfer between the two fluids at a location in a heat exchanger depends on the magnitude of the temperature difference at that location, which varies along the heat exchanger.

Heat exchangers are manufactured in a variety of types, and thus we start this chapter with the *classification* of heat exchangers. We then discuss the determination of the overall heat transfer coefficient in heat exchangers, and the *logarithmic mean temperature difference* (LMTD) for some configurations. We then introduce the *correction factor F* to account for the deviation of the mean temperature difference from the LMTD in complex configurations. Next we discuss the effectiveness–NTU method, which enables us to analyze heat exchangers when the outlet temperatures of the fluids are not known. Finally, we discuss the selection of heat exchangers.

Objectives

The objectives of this chapter are to:

- Recognize numerous types of heat exchangers, and classify them,
- Develop an awareness of fouling on surfaces, and determine the overall heat transfer coefficient for a heat exchanger,
- Perform a general energy analysis on heat exchangers,
- Obtain a relation for the logarithmic mean temperature difference for use in the LMTD method, and modify it for different types of heat exchangers using the correction factor,
- Develop relations for effectiveness, and analyze heat exchangers when outlet temperatures are not known using the effectiveness-NTU method,
- Know the primary considerations in the selection of heat exchangers.

22–1 ▪ TYPES OF HEAT EXCHANGERS

Different heat transfer applications require different types of hardware and different configurations of heat transfer equipment. The attempt to match the heat transfer hardware to the heat transfer requirements within the specified constraints has resulted in numerous types of innovative heat exchanger designs.

The simplest type of heat exchanger consists of two concentric pipes of different diameters, as shown in Fig. 22–1, called the **double-pipe** heat exchanger. One fluid in a double-pipe heat exchanger flows through the smaller pipe while the other fluid flows through the annular space between the two pipes. Two types of flow arrangement are possible in a double-pipe heat exchanger: in **parallel flow**, both the hot and cold fluids enter the heat exchanger at the same end and move in the *same* direction. In **counter flow**, on the other hand, the hot and cold fluids enter the heat exchanger at opposite ends and flow in *opposite* directions.

Another type of heat exchanger, which is specifically designed to realize a large heat transfer surface area per unit volume, is the **compact** heat exchanger. The ratio of the heat transfer surface area of a heat exchanger to its volume is called the *area density* β. A heat exchanger with $\beta > 700$ m^2/m^3 (or 200 ft^2/ft^3) is classified as being compact. Examples of compact heat exchangers are car radiators ($\beta \approx 1000$ m^2/m^3), glass-ceramic gas turbine heat exchangers ($\beta \approx 6000$ m^2/m^3), the regenerator of a Stirling engine ($\beta \approx 15{,}000$ m^2/m^3), and the human lung ($\beta \approx 20{,}000$ m^2/m^3). Compact

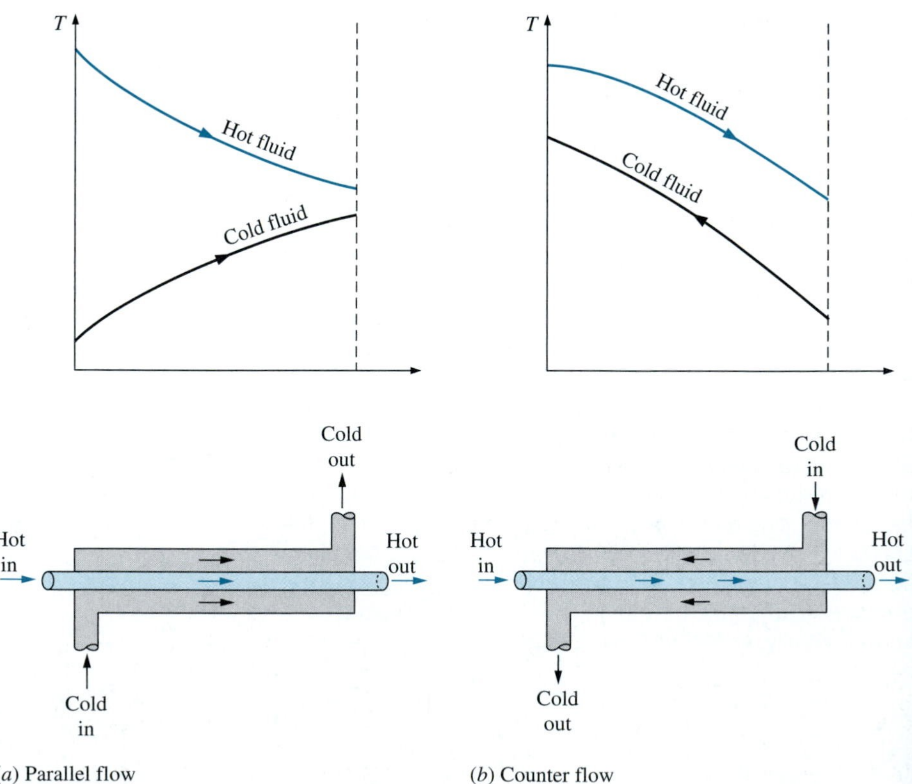

FIGURE 22–1
Different flow regimes and associated temperature profiles in a double-pipe heat exchanger.

heat exchangers enable us to achieve high heat transfer rates between two fluids in a small volume, and they are commonly used in applications with strict limitations on the weight and volume of heat exchangers (Fig. 22–2).

The large surface area in compact heat exchangers is obtained by attaching closely spaced *thin plate* or *corrugated fins* to the walls separating the two fluids. Compact heat exchangers are commonly used in gas-to-gas and gas-to-liquid (or liquid-to-gas) heat exchangers to counteract the low heat transfer coefficient associated with gas flow with increased surface area. In a car radiator, which is a water-to-air compact heat exchanger, for example, it is no surprise that fins are attached to the air side of the tube surface.

FIGURE 22–2
A gas-to-liquid compact heat exchanger for a residential air-conditioning system.
(© Yunus Çengel)

In compact heat exchangers, the two fluids usually move *perpendicular* to each other, and such flow configuration is called **cross-flow**. The cross-flow is further classified as *unmixed* and *mixed flow,* depending on the flow configuration, as shown in Fig. 22–3. In (*a*) the cross-flow is said to be *unmixed* since the plate fins force the fluid to flow through a particular interfin spacing and prevent it from moving in the transverse direction (i.e., parallel to the tubes). The cross-flow in (*b*) is said to be *mixed* since the fluid now is free to move in the transverse direction. Both fluids are unmixed in a car radiator. The presence of mixing in the fluid can have a significant effect on the heat transfer characteristics of the heat exchanger.

Perhaps the most common type of heat exchanger in industrial applications is the **shell-and-tube** heat exchanger, shown in Fig. 22–4. Shell-and-tube heat exchangers contain a large number of tubes (sometimes several hundred) packed in a shell with their axes parallel to that of the shell. Heat transfer takes place as one fluid flows inside the tubes while the other fluid flows outside the tubes through the shell. *Baffles* are commonly placed in the shell to force the shell-side fluid to flow across the shell to enhance heat transfer and to maintain uniform spacing between the tubes. Despite their widespread use, shell-and-tube heat exchangers are not suitable for use in automotive and aircraft applications because of their relatively large size and weight. Note that the tubes in a shell-and-tube heat exchanger open to some large flow areas called *headers* at both ends of the shell, where the tube-side fluid accumulates before entering the tubes and after leaving them.

Shell-and-tube heat exchangers are further classified according to the number of shell and tube passes involved. Heat exchangers in which all the tubes make one U-turn in the shell, for example, are called *one-shell-pass and two-tube-passes* heat exchangers. Likewise, a heat exchanger that

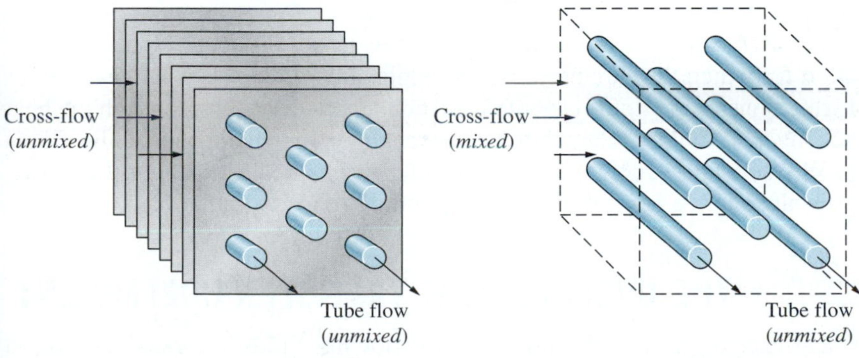

FIGURE 22–3
Different flow configurations in cross-flow heat exchangers.

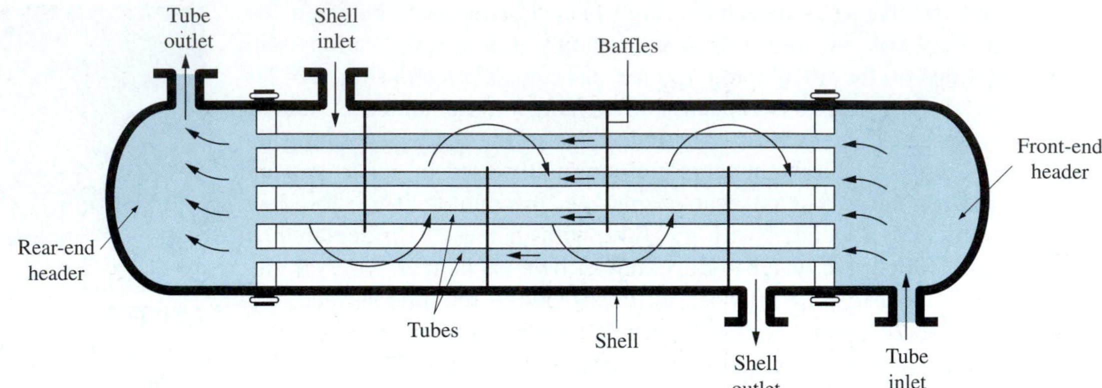

FIGURE 22–4
The schematic of a shell-and-tube heat exchanger (one-shell pass and one-tube pass).

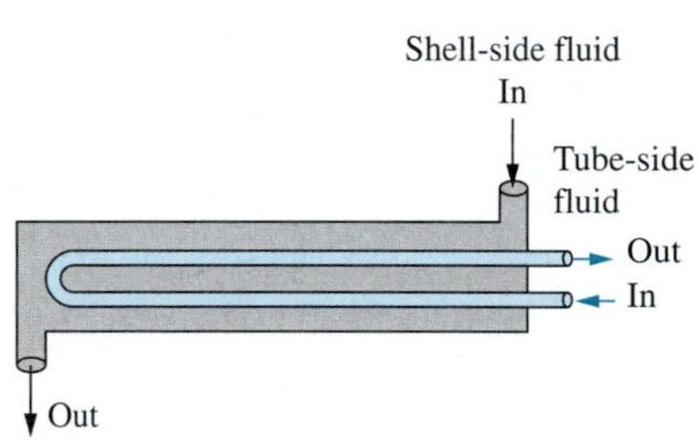

(*a*) One-shell pass and two-tube passes

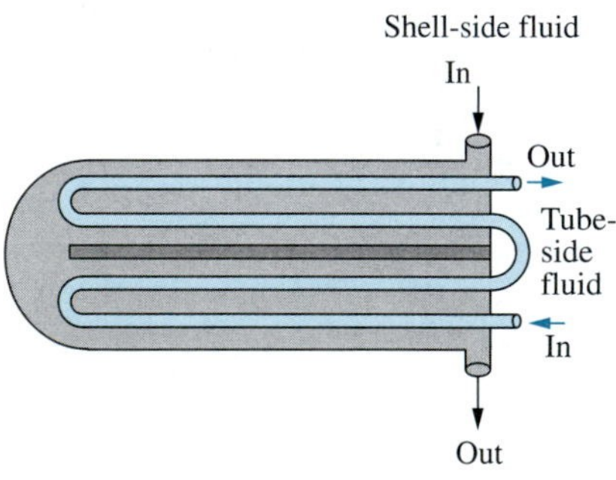

(*b*) Two-shell passes and four-tube passes

FIGURE 22–5
Multipass flow arrangements in shell-and-tube heat exchangers.

involves two passes in the shell and four passes in the tubes is called a *two-shell-passes and four-tube-passes* heat exchanger (Fig. 22–5).

An innovative type of heat exchanger that has found widespread use is the **plate and frame** (or just plate) heat exchanger, which consists of a series of plates with corrugated flat flow passages (Fig. 22–6). The hot and cold fluids flow in alternate passages, and thus each cold fluid stream is surrounded by two hot fluid streams, resulting in very effective heat transfer. Also, plate heat exchangers can grow with increasing demand for heat transfer by simply mounting more plates. They are well suited for liquid-to-liquid heat exchange applications, provided that the hot and cold fluid streams are at about the same pressure.

Another type of heat exchanger that involves the alternate passage of the hot and cold fluid streams through the same flow area is the **regenerative** heat exchanger. The *static*-type regenerative heat exchanger is basically a porous mass that has a large heat storage capacity, such as a ceramic wire mesh. Hot and cold fluids flow through this porous mass alternatively. Heat is transferred from the hot fluid to the matrix of the regenerator during the flow of the hot fluid, and from the matrix to the cold fluid during the flow of the cold fluid. Thus, the matrix serves as a temporary heat storage medium.

The *dynamic*-type regenerator involves a rotating drum and continuous flow of the hot and cold fluid through different portions of the drum so that any portion of the drum passes periodically through the hot stream, storing heat, and then through the cold stream, rejecting this stored heat. Again the drum serves as the medium to transport the heat from the hot to the cold fluid stream.

Heat exchangers are often given specific names to reflect the specific application for which they are used. For example, a *condenser* is a heat exchanger in which one of the fluids is cooled and condenses as it flows through the heat exchanger. A *boiler* is another heat exchanger in which one of the fluids absorbs heat and vaporizes. A *space radiator* is a heat exchanger that transfers heat from the hot fluid to the surrounding space by radiation.

22–2 ▪ THE OVERALL HEAT TRANSFER COEFFICIENT

A heat exchanger typically involves two flowing fluids separated by a solid wall. Heat is first transferred from the hot fluid to the wall by *convection,*

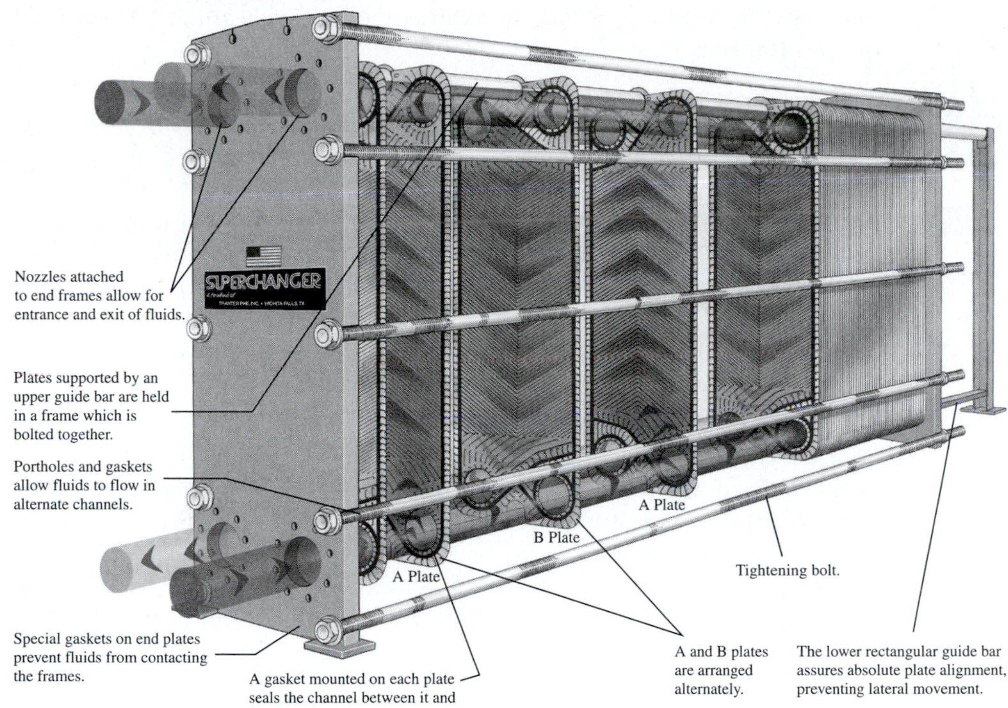

FIGURE 22–6
A plate-and-frame liquid-to-liquid heat exchanger.
(Courtesy of Tranter PHE, Inc.)

through the wall by *conduction,* and from the wall to the cold fluid again by *convection.* Any radiation effects are usually included in the convection heat transfer coefficients.

The thermal resistance network associated with this heat transfer process involves two convection and one conduction resistances, as shown in Fig. 22–7. Here the subscripts i and o represent the inner and outer surfaces of the inner tube. For a double-pipe heat exchanger, the *thermal resistance* of the tube wall is

$$R_{\text{wall}} = \frac{\ln (D_o/D_i)}{2\pi kL} \qquad \textbf{(22–1)}$$

where k is the thermal conductivity of the wall material and L is the length of the tube. Then the *total thermal resistance* becomes

$$R = R_{\text{total}} = R_i + R_{\text{wall}} + R_o = \frac{1}{h_iA_i} + \frac{\ln (D_o/D_i)}{2\pi kL} + \frac{1}{h_oA_o} \qquad \textbf{(22–2)}$$

The A_i is the area of the inner surface of the wall that separates the two fluids, and A_o is the area of the outer surface of the wall. In other words, A_i and A_o are surface areas of the separating wall wetted by the inner and the outer fluids, respectively. When one fluid flows inside a circular tube and the other outside of it, we have $A_i = \pi D_iL$ and $A_o = \pi D_oL$ (Fig. 22–8).

In the analysis of heat exchangers, it is convenient to combine all the thermal resistances in the path of heat flow from the hot fluid to the cold one

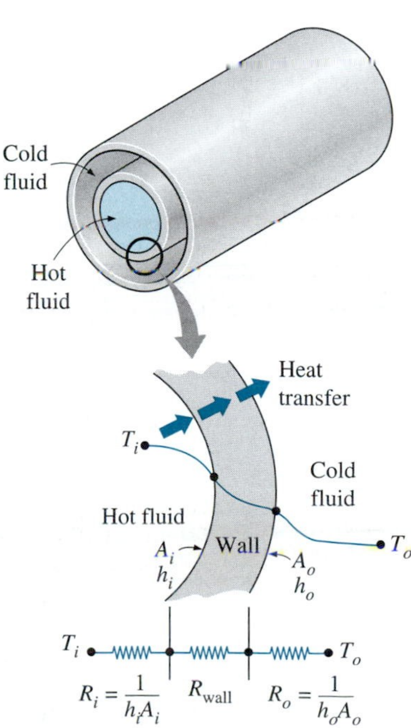

FIGURE 22–7
Thermal resistance network associated with heat transfer in a double-pipe heat exchanger.

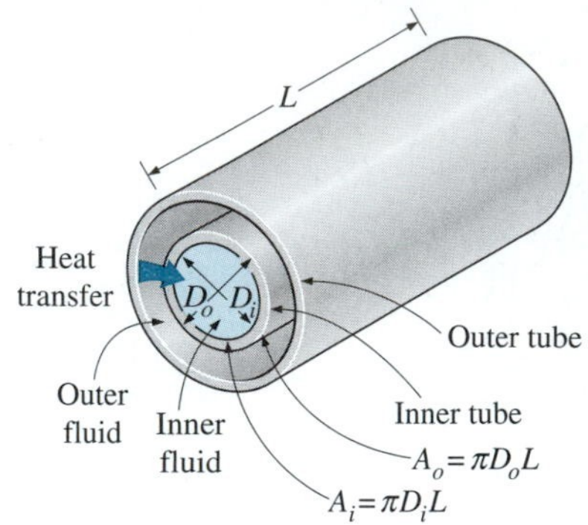

FIGURE 22–8

The two heat transfer surface areas associated with a double-pipe heat exchanger (for thin tubes, $D_i \approx D_o$ and thus $A_i \approx A_o$).

into a single resistance R, and to express the rate of heat transfer between the two fluids as

$$\dot{Q} = \frac{\Delta T}{R} = UA\Delta T = U_i A_i \Delta T = U_o A_o \Delta T \tag{22–3}$$

where U is the **overall heat transfer coefficient**, whose unit is W/m$^2 \cdot$ °C, which is identical to the unit of the ordinary convection coefficient h. Canceling ΔT, Eq. 22–3 reduces to

$$\frac{1}{UA_s} = \frac{1}{U_i A_i} = \frac{1}{U_o A_o} = R = \frac{1}{h_i A_i} + R_{\text{wall}} + \frac{1}{h_o A_o} \tag{22–4}$$

Perhaps you are wondering why we have two overall heat transfer coefficients U_i and U_o for a heat exchanger. The reason is that every heat exchanger has two heat transfer surface areas A_i and A_o, which, in general, are not equal to each other.

Note that $U_i A_i = U_o A_o$, but $U_i \neq U_o$ unless $A_i = A_o$. Therefore, the overall heat transfer coefficient U of a heat exchanger is meaningless unless the area on which it is based is specified. This is especially the case when one side of the tube wall is finned and the other side is not, since the surface area of the finned side is several times that of the unfinned side.

When the wall thickness of the tube is small and the thermal conductivity of the tube material is high, as is usually the case, the thermal resistance of the tube is negligible ($R_{\text{wall}} \approx 0$) and the inner and outer surfaces of the tube are almost identical ($A_i \approx A_o \approx A_s$). Then Eq. 22–4 for the overall heat transfer coefficient simplifies to

$$\frac{1}{U} \approx \frac{1}{h_i} + \frac{1}{h_o} \tag{22–5}$$

where $U \approx U_i \approx U_o$. The individual convection heat transfer coefficients inside and outside the tube, h_i and h_o, are determined using the convection relations discussed in earlier chapters.

The overall heat transfer coefficient U in Eq. 22–5 is dominated by the *smaller* convection coefficient, since the inverse of a large number is small. When one of the convection coefficients is *much smaller* than the other (say, $h_i \ll h_o$), we have $1/h_i \gg 1/h_o$, and thus $U \approx h_i$. Therefore, the smaller heat transfer coefficient creates a *bottleneck* on the path of heat transfer and seriously impedes heat transfer. This situation arises frequently when one of the fluids is a gas and the other is a liquid. In such cases, fins are commonly used on the gas side to enhance the product UA and thus the heat transfer on that side.

Representative values of the overall heat transfer coefficient U are given in Table 22–1. Note that the overall heat transfer coefficient ranges from about 10 W/m$^2 \cdot$ °C for gas-to-gas heat exchangers to about 10,000 W/m$^2 \cdot$ °C for heat exchangers that involve phase changes. This is not surprising, since gases have very low thermal conductivities, and phase-change processes involve very high heat transfer coefficients.

TABLE 22–1

Representative values of the overall heat transfer coefficients in heat exchangers

Type of heat exchanger	U, W/m$^2\cdot$°C*
Water-to-water	850–1700
Water-to-oil	100–350
Water-to-gasoline or kerosene	300–1000
Feedwater heaters	1000–8500
Steam-to-light fuel oil	200–400
Steam-to-heavy fuel oil	50–200
Steam condenser	1000–6000
Freon condenser (water cooled)	300–1000
Ammonia condenser (water cooled)	800–1400
Alcohol condensers (water cooled)	250–700
Gas-to-gas	10–40
Water-to-air in finned tubes (water in tubes)	30–60†
	400–850†
Steam-to-air in finned tubes (steam in tubes)	30–300†
	400–4000‡

*Multiply the listed values by 0.176 to convert them to Btu/h · ft^2 · °F.

†Based on air-side surface area.

‡Based on water- or steam-side surface area.

When the tube is *finned* on one side to enhance heat transfer, the total heat transfer surface area on the finned side becomes

$$A_s = A_{\text{total}} = A_{\text{fin}} + A_{\text{unfinned}} \tag{22-6}$$

where A_{fin} is the surface area of the fins and A_{unfinned} is the area of the unfinned portion of the tube surface. For short fins of high thermal conductivity, we can use this total area in the convection resistance relation $R_{\text{conv}} = 1/hA_s$ since the fins in this case will be very nearly isothermal. Otherwise, we should determine the effective surface area A from

$$A_s = A_{\text{unfinned}} + \eta_{\text{fin}} A_{\text{fin}} \tag{22-7}$$

where η_{fin} is the fin efficiency. This way, the temperature drop along the fins is accounted for. Note that $\eta_{\text{fin}} = 1$ for isothermal fins, and thus Eq. 22–7 reduces to Eq. 22–6 in that case.

Fouling Factor

The performance of heat exchangers usually deteriorates with time as a result of accumulation of *deposits* on heat transfer surfaces. The layer of deposits represents *additional resistance* to heat transfer and causes the rate of heat transfer in a heat exchanger to decrease. The net effect of these accumulations on heat transfer is represented by a **fouling factor** R_f, which is a measure of the *thermal resistance* introduced by fouling.

The most common type of fouling is the *precipitation* of solid deposits in a fluid on the heat transfer surfaces. You can observe this type of fouling even in your house. If you check the inner surfaces of your teapot after prolonged use, you will probably notice a layer of calcium-based deposits on the surfaces at which boiling occurs. This is especially the case in areas where the water is hard. The scales of such deposits come off by scratching, and the surfaces can be cleaned of such deposits by chemical treatment. Now imagine those mineral deposits forming on the inner surfaces of fine tubes in a heat exchanger (Fig. 22–9) and the detrimental effect it may have on the flow passage area and the heat transfer. To avoid this potential problem, water in power and process plants is extensively treated and its solid contents are removed before it is allowed to circulate through the system. The solid ash particles in the flue gases accumulating on the surfaces of air preheaters create similar problems.

Another form of fouling, which is common in the chemical process industry, is *corrosion* and other *chemical fouling.* In this case, the surfaces are fouled by the accumulation of the products of chemical reactions on the surfaces. This form of fouling can be avoided by coating metal pipes with glass or using plastic pipes instead of metal ones. Heat exchangers may also be fouled by the growth of algae in warm fluids. This type of fouling is called *biological fouling* and can be prevented by chemical treatment.

In applications where it is likely to occur, fouling should be considered in the design and selection of heat exchangers. In such applications, it may be necessary to select a larger and thus more expensive heat exchanger to ensure that it meets the design heat transfer requirements even after fouling occurs. The periodic cleaning of heat exchangers and the resulting down time are additional penalties associated with fouling.

The fouling factor is obviously zero for a new heat exchanger and increases with time as the solid deposits build up on the heat exchanger surface. The fouling factor depends on the *operating temperature* and the *velocity* of the fluids, as well as the length of service. Fouling increases with *increasing temperature* and *decreasing velocity.*

FIGURE 22–9
Precipitation fouling of ash particles on superheater tubes.
(From Steam: Its Generation, and Use, Babcock and Wilcox Co., 1978. Reprinted by permission.)

The overall heat transfer coefficient relation given above is valid for clean surfaces and needs to be modified to account for the effects of fouling on both the inner and the outer surfaces of the tube. For an unfinned shell-and-tube heat exchanger, it can be expressed as

$$\frac{1}{UA_s} = \frac{1}{U_iA_i} = \frac{1}{U_oA_o} = R = \frac{1}{h_iA_i} + \frac{R_{f,i}}{A_i} + \frac{\ln(D_o/D_i)}{2\pi kL} + \frac{R_{f,o}}{A_o} + \frac{1}{h_oA_o} \quad \text{(22–8)}$$

where $R_{f,i}$ and $R_{f,o}$ are the fouling factors at those surfaces.

Representative values of fouling factors are given in Table 22–2. More comprehensive tables of fouling factors are available in handbooks. As you would expect, considerable uncertainty exists in these values, and they should be used as a guide in the selection and evaluation of heat exchangers to account for the effects of anticipated fouling on heat transfer. Note that most fouling factors in the table are of the order of 10^{-4} m² · °C/W, which is equivalent to the thermal resistance of a 0.2-mm-thick limestone layer (k = 2.9 W/m · °C) per unit surface area. Therefore, in the absence of specific data, we can assume the surfaces to be coated with 0.2 mm of limestone as a starting point to account for the effects of fouling.

TABLE 22–2

Representative fouling factors (thermal resistance due to fouling for a unit surface area)

Fluid	R_f, m² · °C/W
Distilled water, sea-water, river water, boiler feedwater:	
Below 50°C	0.0001
Above 50°C	0.0002
Fuel oil	0.0009
Steam (oil-free)	0.0001
Refrigerants (liquid)	0.0002
Refrigerants (vapor)	0.0004
Alcohol vapors	0.0001
Air	0.0004

(*Source:* Tubular Exchange Manufacturers Association.)

EXAMPLE 22–1 Overall Heat Transfer Coefficient of a Heat Exchanger

Hot oil is to be cooled in a double-tube counter-flow heat exchanger. The copper inner tubes have a diameter of 2 cm and negligible thickness. The inner diameter of the outer tube (the shell) is 3 cm. Water flows through the tube at a rate of 0.5 kg/s, and the oil through the shell at a rate of 0.8 kg/s. Taking the average temperatures of the water and the oil to be 45°C and 80°C, respectively, determine the overall heat transfer coefficient of this heat exchanger.

Solution Hot oil is cooled by water in a double-tube counter-flow heat exchanger. The overall heat transfer coefficient is to be determined.

Assumptions **1** The thermal resistance of the inner tube is negligible since the tube material is highly conductive and its thickness is negligible. **2** Both the oil and water flow are fully developed. **3** Properties of the oil and water are constant.

Properties The properties of water at 45°C are (Table A–15)

$$\rho = 990.1 \text{ kg/m}^3 \qquad \text{Pr} = 3.91$$
$$k = 0.637 \text{ W/m} \cdot {}^\circ\text{C} \qquad \nu = \mu/\rho = 0.602 \times 10^{-6} \text{ m}^2/\text{s}$$

The properties of oil at 80°C are (Table A–19)

$$\rho = 852 \text{ kg/m}^3 \qquad \text{Pr} = 499.3$$
$$k = 0.138 \text{ W/m} \cdot {}^\circ\text{C} \qquad \nu = 3.794 \times 10^{-5} \text{ m}^2/\text{s}$$

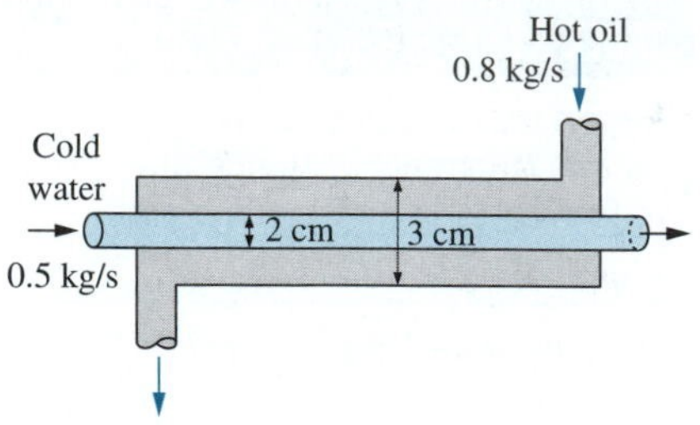

FIGURE 22–10
Schematic for Example 22–1.

Analysis The schematic of the heat exchanger is given in Fig. 22–10. The overall heat transfer coefficient U can be determined from Eq. 22–5:

$$\frac{1}{U} \approx \frac{1}{h_i} + \frac{1}{h_o}$$

where h_i and h_o are the convection heat transfer coefficients inside and outside the tube, respectively, which are to be determined using the forced convection relations.

The hydraulic diameter for a circular tube is the diameter of the tube itself, $D_h = D = 0.02$ m. The average velocity of water in the tube and the Reynolds number are

$$V = \frac{\dot{m}}{\rho A_c} = \frac{\dot{m}}{\rho(\frac{1}{4}\pi D^2)} = \frac{0.5 \text{ kg/s}}{(990.1 \text{ kg/m}^3)[\frac{1}{4}\pi(0.02 \text{ m})^2]} = 1.61 \text{ m/s}$$

and

$$\text{Re} = \frac{VD}{\nu} = \frac{(1.61 \text{ m/s})(0.02 \text{ m})}{0.602 \times 10^{-6} \text{ m}^2/\text{s}} = 53{,}490$$

which is greater than 10,000. Therefore, the flow of water is turbulent. Assuming the flow to be fully developed, the Nusselt number can be determined from

$$\text{Nu} = \frac{hD}{k} = 0.023 \text{ Re}^{0.8}\text{Pr}^{0.4} = 0.023(53{,}490)^{0.8}(3.91)^{0.4} = 240.6$$

Then,

$$h = \frac{k}{D}\text{Nu} = \frac{0.637 \text{ W/m}\cdot{}^\circ\text{C}}{0.02 \text{ m}}(240.6) = 7663 \text{ W/m}^2\cdot{}^\circ\text{C}$$

Now we repeat the analysis above for oil. The properties of oil at 80°C are

$$\rho = 852 \text{ kg/m}^3 \qquad \nu = 37.5 \times 10^{-6} \text{ m}^2/\text{s}$$
$$k = 0.138 \text{ W/m}\cdot{}^\circ\text{C} \qquad \text{Pr} = 490$$

The hydraulic diameter for the annular space is

$$D_h = D_o - D_i = 0.03 - 0.02 = 0.01 \text{ m}$$

The average velocity and the Reynolds number in this case are

$$V = \frac{\dot{m}}{\rho A_c} = \frac{\dot{m}}{\rho[\frac{1}{4}\pi(D_o^2 - D_i^2)]} = \frac{0.8 \text{ kg/s}}{(852 \text{ kg/m}^3)[\frac{1}{4}\pi(0.03^2 - 0.02^2)] \text{ m}^2} = 2.39 \text{ m/s}$$

and

$$\text{Re} = \frac{VD}{\nu} = \frac{(2.39 \text{ m/s})(0.01 \text{ m})}{3.794 \times 10^{-5} \text{ m}^2/\text{s}} = 630$$

which is less than 2300. Therefore, the flow of oil is laminar. Assuming fully developed flow, the Nusselt number on the tube side of the annular space Nu_i corresponding to $D_i/D_o = 0.02/0.03 = 0.667$ can be determined from Table 22–3 by interpolation to be

$$\text{Nu} = 5.45$$

TABLE 22–3

Nusselt number for fully developed laminar flow in a circular annulus with one surface insulated and the other isothermal (Kays and Perkins, 1972)

D_i/D_o	Nu_i	Nu_o
0.00	—	3.66
0.05	17.46	4.06
0.10	11.56	4.11
0.25	7.37	4.23
0.50	5.74	4.43
1.00	4.86	4.86

and

$$h_o = \frac{k}{D_h}\text{Nu} = \frac{0.138 \text{ W/m}\cdot°\text{C}}{0.01 \text{ m}}(5.45) = 75.2 \text{ W/m}^2\cdot°\text{C}$$

Then the overall heat transfer coefficient for this heat exchanger becomes

$$U = \frac{1}{\frac{1}{h_i}+\frac{1}{h_o}} = \frac{1}{\frac{1}{7663 \text{ W/m}^2\cdot°\text{C}}+\frac{1}{75.2 \text{ W/m}^2\cdot°\text{C}}} = \mathbf{74.5 \text{ W/m}^2\cdot°\text{C}}$$

Discussion Note that $U \approx h_o$ in this case, since $h_i \gg h_o$. This confirms our earlier statement that the overall heat transfer coefficient in a heat exchanger is dominated by the smaller heat transfer coefficient when the difference between the two values is large.

To improve the overall heat transfer coefficient and thus the heat transfer in this heat exchanger, we must use some enhancement techniques on the oil side, such as a finned surface.

EXAMPLE 22–2 Effect of Fouling on the Overall Heat Transfer Coefficient

A double-pipe (shell-and-tube) heat exchanger is constructed of a stainless steel (k = 15.1 W/m · °C) inner tube of inner diameter D_i = 1.5 cm and outer diameter D_o = 1.9 cm and an outer shell of inner diameter 3.2 cm. The convection heat transfer coefficient is given to be h_i = 800 W/m^2 · °C on the inner surface of the tube and h_o = 1200 W/m^2 · °C on the outer surface. For a fouling factor of $R_{f,i}$ = 0.0004 m^2 · °C/W on the tube side and $R_{f,o}$ = 0.0001 m^2 · °C/W on the shell side, determine (*a*) the thermal resistance of the heat exchanger per unit length and (*b*) the overall heat transfer coefficients, U_i and U_o based on the inner and outer surface areas of the tube, respectively.

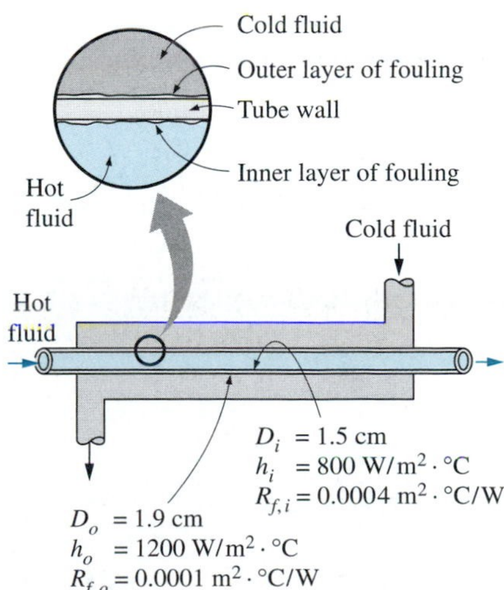

FIGURE 22–11
Schematic for Example 22–2.

SOLUTION The heat transfer coefficients and the fouling factors on the tube and shell sides of a heat exchanger are given. The thermal resistance and the overall heat transfer coefficients based on the inner and outer areas are to be determined.

Assumptions The heat transfer coefficients and the fouling factors are constant and uniform.

Analysis (*a*) The schematic of the heat exchanger is given in Fig. 22–11. The thermal resistance for an unfinned shell-and-tube heat exchanger with fouling on both heat transfer surfaces is given by Eq. 22–8 as

$$R = \frac{1}{UA_s} = \frac{1}{U_iA_i} = \frac{1}{U_oA_o} = \frac{1}{h_iA_i} + \frac{R_{f,i}}{A_i} + \frac{\ln(D_o/D_i)}{2\pi kL} + \frac{R_{f,o}}{A_o} + \frac{1}{h_oA_o}$$

where

$$A_i = \pi D_i L = \pi(0.015 \text{ m})(1 \text{ m}) = 0.0471 \text{ m}^2$$
$$A_o = \pi D_o L = \pi(0.019 \text{ m})(1 \text{ m}) = 0.0597 \text{ m}^2$$

Substituting, the total thermal resistance is determined to be

$$R = \frac{1}{(800 \text{ W/m}^2 \cdot °\text{C})(0.0471 \text{ m}^2)} + \frac{0.0004 \text{ m}^2 \cdot °\text{C/W}}{0.0471 \text{ m}^2}$$

$$+ \frac{\ln (0.019/0.015)}{2\pi(15.1 \text{ W/m} \cdot °\text{C})(1 \text{ m})}$$

$$+ \frac{0.0001 \text{ m}^2 \cdot °\text{C/W}}{0.0597 \text{ m}^2} + \frac{1}{(1200 \text{ W/m}^2 \cdot °\text{C})(0.0597 \text{ m}^2)}$$

$$= (0.02654 + 0.00849 + 0.0025 + 0.00168 + 0.01396)°\text{C/W}$$

$$= \mathbf{0.0532°C/W}$$

Note that about 19 percent of the total thermal resistance in this case is due to fouling and about 5 percent of it is due to the steel tube separating the two fluids. The rest (76 percent) is due to the convection resistances.

(*b*) Knowing the total thermal resistance and the heat transfer surface areas, the overall heat transfer coefficients based on the inner and outer surfaces of the tube are

$$U_i = \frac{1}{RA_i} = \frac{1}{(0.0532 \text{ °C/W})(0.0471 \text{ m}^2)} = \mathbf{399 \text{ W/m}^2 \cdot °C}$$

and

$$U_o = \frac{1}{RA_o} = \frac{1}{(0.0532 \text{ °C/W})(0.0597 \text{ m}^2)} = \mathbf{315 \text{ W/m}^2 \cdot °C}$$

Discussion Note that the two overall heat transfer coefficients differ significantly (by 27 percent) in this case because of the considerable difference between the heat transfer surface areas on the inner and the outer sides of the tube. For tubes of negligible thickness, the difference between the two overall heat transfer coefficients would be negligible.

22–3 ▪ ANALYSIS OF HEAT EXCHANGERS

Heat exchangers are commonly used in practice, and an engineer often finds himself or herself in a position to *select a heat exchanger* that will achieve a *specified temperature change* in a fluid stream of known mass flow rate, or to *predict the outlet temperatures* of the hot and cold fluid streams in a *specified heat exchanger.*

In upcoming sections, we discuss the two methods used in the analysis of heat exchangers. Of these, the *log mean temperature difference* (or LMTD) method is best suited for the first task and the *effectiveness–NTU* method for the second task. But first we present some general considerations.

Heat exchangers usually operate for long periods of time with no change in their operating conditions. Therefore, they can be modeled as *steady-flow* devices. As such, the mass flow rate of each fluid remains constant, and the fluid properties such as temperature and velocity at any inlet or outlet remain the same. Also, the fluid streams experience little or no change in their velocities and elevations, and thus the kinetic and potential energy changes are negligible. The specific heat of a fluid, in general, changes with temperature. But, in a specified temperature range, it can be treated as a constant at some average value with little loss in accuracy. Axial heat conduction along the tube is usually insignificant and can be considered negligible. Finally, the outer surface of the heat exchanger is assumed to be *perfectly insulated,* so that there is no heat loss to the surrounding medium, and any heat transfer occurs between the two fluids only.

The idealizations stated above are closely approximated in practice, and they greatly simplify the analysis of a heat exchanger with little sacrifice from accuracy. Therefore, they are commonly used. Under these assumptions, the *first law of thermodynamics* requires that the rate of heat transfer from the hot fluid be equal to the rate of heat transfer to the cold one. That is,

$$\dot{Q} = \dot{m}_c c_{pc}(T_{c,\text{ out}} - T_{c,\text{ in}}) \tag{22–9}$$

and

$$\dot{Q} = \dot{m}_h c_{ph}(T_{h,\text{ in}} - T_{h,\text{ out}}) \tag{22–10}$$

where the subscripts c and h stand for *cold* and *hot* fluids, respectively, and

$$\dot{m}_c, \dot{m}_h = \text{mass flow rates}$$
$$c_{pc}, c_{ph} = \text{specific heats}$$
$$T_{c,\text{ out}}, T_{h,\text{ out}} = \text{outlet temperatures}$$
$$T_{c,\text{ in}}, T_{h,\text{ in}} = \text{inlet temperatures}$$

Note that the heat transfer rate $\dot{Q}$ is taken to be a positive quantity, and its direction is understood to be from the hot fluid to the cold one in accordance with the second law of thermodynamics.

In heat exchanger analysis, it is often convenient to combine the product of the mass flow rate and the specific heat of a fluid into a single quantity. This quantity is called the **heat capacity rate** and is defined for the hot and cold fluid streams as

$$C_h = \dot{m}_h c_{ph} \quad \text{and} \quad C_c = \dot{m}_c c_{pc} \tag{22–11}$$

The heat capacity rate of a fluid stream represents the rate of heat transfer needed to change the temperature of the fluid stream by 1°C as it flows through a heat exchanger. Note that in a heat exchanger, the fluid with a *large* heat capacity rate experiences a *small* temperature change, and the fluid with a *small* heat capacity rate experiences a *large* temperature change. Therefore, *doubling* the mass flow rate of a fluid while leaving everything else unchanged will *halve* the temperature change of that fluid.

With the definition of the heat capacity rate above, Eqs. 22–9 and 22–10 can also be expressed as

$$\dot{Q} = C_c(T_{c,\text{ out}} - T_{c,\text{ in}}) \tag{22–12}$$

and

$$\dot{Q} = C_h(T_{h,\text{ in}} - T_{h,\text{ out}}) \tag{22–13}$$

That is, the heat transfer rate in a heat exchanger is equal to the heat capacity rate of either fluid multiplied by the temperature change of that fluid. Note that *the only time the temperature rise of a cold fluid is equal to the temperature drop of the hot fluid is when the heat capacity rates of the two fluids are equal to each other* (Fig. 22–12).

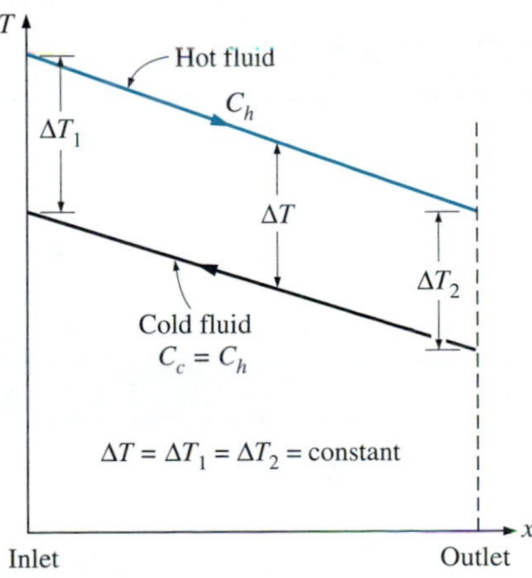

FIGURE 22–12
Two fluid streams that have the same capacity rates experience the same temperature change in a well-insulated heat exchanger.

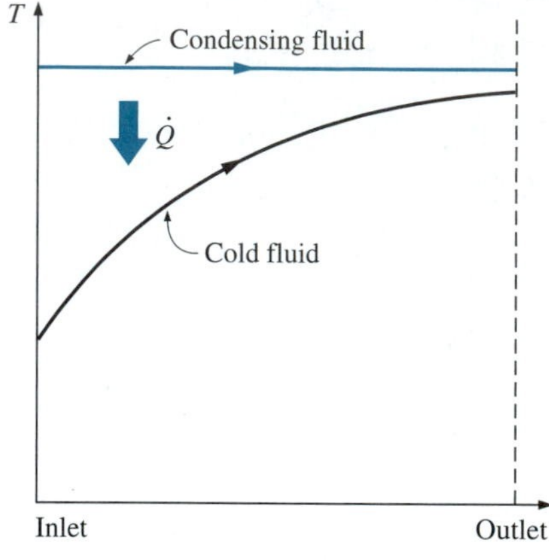

(*a*) Condenser ($C_h \to \infty$)

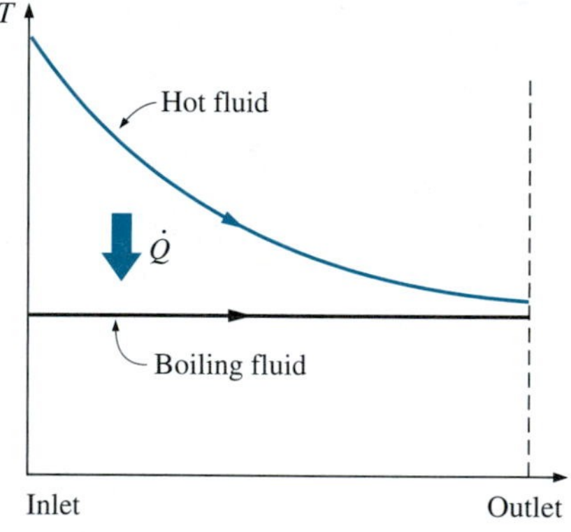

(*b*) Boiler ($C_c \to \infty$)

FIGURE 22–13
Variation of fluid temperatures in a heat exchanger when one of the fluids condenses or boils.

Two special types of heat exchangers commonly used in practice are *condensers* and *boilers.* One of the fluids in a condenser or a boiler undergoes a phase-change process, and the rate of heat transfer is expressed as

$$\dot{Q} = \dot{m}h_{fg} \tag{22–14}$$

where $\dot{m}$ is the rate of evaporation or condensation of the fluid and h_{fg} is the enthalpy of vaporization of the fluid at the specified temperature or pressure.

An ordinary fluid absorbs or releases a large amount of heat essentially at constant temperature during a phase-change process, as shown in Fig. 22–13. The heat capacity rate of a fluid during a phase-change process must approach infinity since the temperature change is practically zero. That is, $C = \dot{m}c_p \to \infty$ when $\Delta T \to 0$, so that the heat transfer rate $\dot{Q} = \dot{m}c_p\,\Delta T$ is a finite quantity. Therefore, in heat exchanger analysis, a condensing or boiling fluid is conveniently modeled as a fluid whose heat capacity rate is *infinity.*

The rate of heat transfer in a heat exchanger can also be expressed in an analogous manner to Newton's law of cooling as

$$\dot{Q} = UA_s\,\Delta T_m \tag{22–15}$$

where U is the overall heat transfer coefficient, A_s is the heat transfer area, and ΔT_m is an appropriate average temperature difference between the two fluids. Here the surface area A_s can be determined precisely using the dimensions of the heat exchanger. However, the overall heat transfer coefficient U and the temperature difference ΔT between the hot and cold fluids, in general, may vary along the heat exchanger.

The average value of the overall heat transfer coefficient can be determined as described in the preceding section by using the average convection coefficients for each fluid. It turns out that the appropriate form of the average temperature difference between the two fluids is *logarithmic* in nature, and its determination is presented in Section 22–4.

22–4 ▪ THE LOG MEAN TEMPERATURE DIFFERENCE METHOD

Earlier, we mentioned that the temperature difference between the hot and cold fluids varies along the heat exchanger, and it is convenient to have a *mean temperature difference* ΔT_m for use in the relation $\dot{Q} = UA_s\,\Delta T_m$.

In order to develop a relation for the equivalent average temperature difference between the two fluids, consider the *parallel-flow double-pipe* heat exchanger shown in Fig. 22–14. Note that the temperature difference ΔT between the hot and cold fluids is large at the inlet of the heat exchanger but decreases exponentially toward the outlet. As you would expect, the temperature of the hot fluid decreases and the temperature of the cold fluid increases along the heat exchanger, but the temperature of the cold fluid can never exceed that of the hot fluid no matter how long the heat exchanger is.

Assuming the outer surface of the heat exchanger to be well insulated so that any heat transfer occurs between the two fluids, and disregarding any changes in kinetic and potential energy, an energy balance on each fluid in a differential section of the heat exchanger can be expressed as

$$\delta \dot{Q} = -\dot{m}_h c_{ph}\, dT_h \tag{22-16}$$

and

$$\delta \dot{Q} = \dot{m}_c c_{pc}\, dT_c \tag{22-17}$$

That is, the rate of heat loss from the hot fluid at any section of a heat exchanger is equal to the rate of heat gain by the cold fluid in that section. The temperature change of the hot fluid is a *negative* quantity, and so a *negative sign* is added to Eq. 22–16 to make the heat transfer rate $\dot{Q}$ a positive quantity. Solving the equations above for dT_h and dT_c gives

$$dT_h = -\frac{\delta \dot{Q}}{\dot{m}_h c_{ph}} \tag{22-18}$$

and

$$dT_c = \frac{\delta \dot{Q}}{\dot{m}_c c_{pc}} \tag{22-19}$$

Taking their difference, we get

$$dT_h - dT_c = d(T_h - T_c) = -\delta \dot{Q}\left(\frac{1}{\dot{m}_h c_{ph}} + \frac{1}{\dot{m}_c c_{pc}}\right) \tag{22-20}$$

The rate of heat transfer in the differential section of the heat exchanger can also be expressed as

$$\delta \dot{Q} = U(T_h - T_c)\, dA_s \tag{22-21}$$

Substituting this equation into Eq. 22–20 and rearranging give

$$\frac{d(T_h - T_c)}{T_h - T_c} = -U\, dA_s\left(\frac{1}{\dot{m}_h c_{ph}} + \frac{1}{\dot{m}_c c_{pc}}\right) \tag{22-22}$$

Integrating from the inlet of the heat exchanger to its outlet, we obtain

$$\ln \frac{T_{h,\,\text{out}} - T_{c,\,\text{out}}}{T_{h,\,\text{in}} - T_{c,\,\text{in}}} = -UA_s\left(\frac{1}{\dot{m}_h c_{ph}} + \frac{1}{\dot{m}_c c_{pc}}\right) \tag{22-23}$$

Finally, solving Eqs. 22–9 and 22–10 for $\dot{m}_c c_{pc}$ and $\dot{m}_h c_{ph}$ and substituting into Eq. 22–23 give, after some rearrangement,

$$\dot{Q} = UA_s\, \Delta T_{\text{lm}} \tag{22-24}$$

where

$$\Delta T_{\text{lm}} = \frac{\Delta T_1 - \Delta T_2}{\ln (\Delta T_1/\Delta T_2)} \tag{22-25}$$

is the **log mean temperature difference**, which is the suitable form of the average temperature difference for use in the analysis of heat exchangers. Here ΔT_1 and ΔT_2 represent the temperature difference between the two fluids at the two ends (inlet and outlet) of the heat exchanger. It makes no dif-

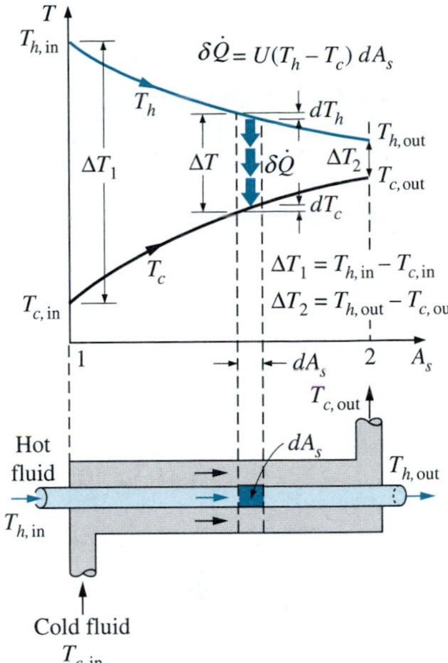

FIGURE 22–14
Variation of the fluid temperatures in a parallel-flow double-pipe heat exchanger.

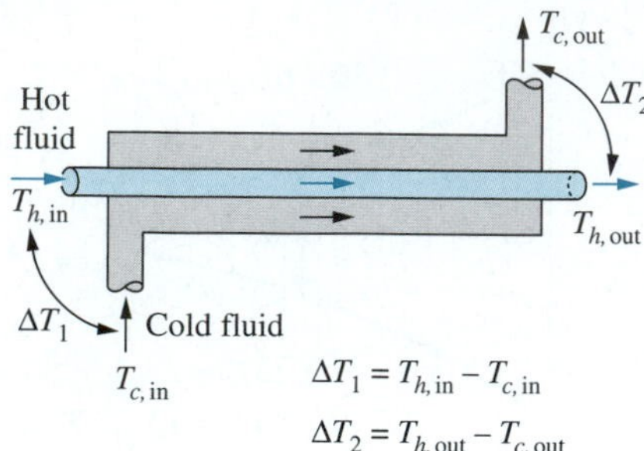

(*a*) Parallel-flow heat exchangers

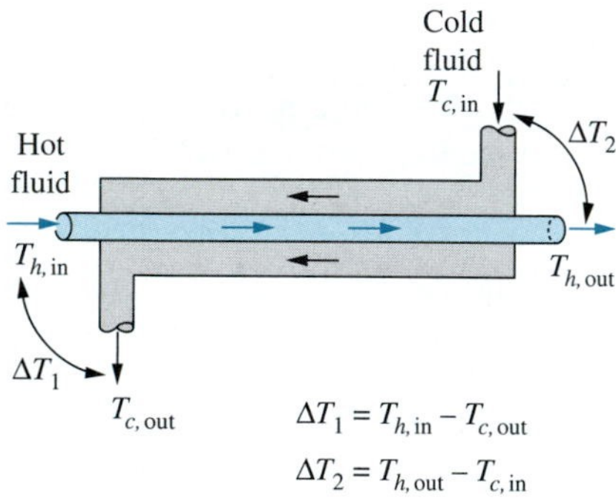

(*b*) Counter-flow heat exchangers

FIGURE 22–15
The ΔT_1 and ΔT_2 expressions in parallel-flow and counter-flow heat exchangers.

ference which end of the heat exchanger is designated as the inlet or the outlet (Fig. 22–15).

The temperature difference between the two fluids decreases from ΔT_1 at the inlet to ΔT_2 at the outlet. Thus, it is tempting to use the arithmetic mean temperature $\Delta T_{am} = \frac{1}{2}(\Delta T_1 + \Delta T_2)$ as the average temperature difference. The logarithmic mean temperature difference ΔT_{lm} is obtained by tracing the actual temperature profile of the fluids along the heat exchanger and is an *exact* representation of the *average temperature difference* between the hot and cold fluids. It truly reflects the exponential decay of the local temperature difference.

Note that ΔT_{lm} is always less than ΔT_{am}. Therefore, using ΔT_{am} in calculations instead of ΔT_{lm} will overestimate the rate of heat transfer in a heat exchanger between the two fluids. When ΔT_1 differs from ΔT_2 by no more than 40 percent, the error in using the arithmetic mean temperature difference is less than 1 percent. But the error increases to undesirable levels when ΔT_1 differs from ΔT_2 by greater amounts. Therefore, we should always use the *logarithmic mean temperature difference* when determining the rate of heat transfer in a heat exchanger.

Counter-Flow Heat Exchangers

The variation of temperatures of hot and cold fluids in a counter-flow heat exchanger is given in Fig. 22–16. Note that the hot and cold fluids enter the heat exchanger from opposite ends, and the outlet temperature of the *cold fluid* in this case may exceed the outlet temperature of the *hot fluid.* In the limiting case, the cold fluid will be heated to the inlet temperature of the hot fluid. However, the outlet temperature of the cold fluid can *never* exceed the inlet temperature of the hot fluid, since this would be a violation of the second law of thermodynamics.

The relation already given for the log mean temperature difference is developed using a parallel-flow heat exchanger, but we can show by repeating the analysis for a counter-flow heat exchanger that is also applicable to counter-flow heat exchangers. But this time, ΔT_1 and ΔT_2 are expressed as shown in Fig. 22–15.

For specified inlet and outlet temperatures, the log mean temperature difference for a counter-flow heat exchanger is always greater than that for a parallel-flow heat exchanger. That is, $\Delta T_{lm,\,CF} > \Delta T_{lm,\,PF}$, and thus a smaller surface area (and thus a smaller heat exchanger) is needed to achieve a specified heat transfer rate in a counter-flow heat exchanger. Therefore, it is common practice to use counter-flow arrangements in heat exchangers.

In a counter-flow heat exchanger, the temperature difference between the hot and the cold fluids remains constant along the heat exchanger when the *heat capacity rates* of the two fluids are *equal* (that is, ΔT = constant when $C_h = C_c$ or $\dot{m}_h c_{ph} = \dot{m}_c c_{pc}$). Then we have $\Delta T_1 = \Delta T_2$, and the log mean temperature difference relation gives $\Delta T_{lm} = \frac{0}{0}$, which is indeterminate. It can be shown by the application of l'Hôpital's rule that in this case we have $\Delta T_{lm} = \Delta T_1 = \Delta T_2$, as expected.

A *condenser* or a *boiler* can be considered to be either a parallel- or counterflow heat exchanger since both approaches give the same result.

Multipass and Cross-Flow Heat Exchangers: Use of a Correction Factor

The log mean temperature difference ΔT_{lm} relation developed earlier is limited to parallel-flow and counter-flow heat exchangers only. Similar relations are also developed for *cross-flow* and *multipass shell-and-tube* heat exchangers, but the resulting expressions are too complicated because of the complex flow conditions.

In such cases, it is convenient to relate the equivalent temperature difference to the log mean temperature difference relation for the counter-flow case as

$$\Delta T_{lm} = F\,\Delta T_{lm,\,CF} \tag{22–26}$$

where F is the **correction factor**, which depends on the *geometry* of the heat exchanger and the inlet and outlet temperatures of the hot and cold fluid streams. The $\Delta T_{lm,\,CF}$ is the log mean temperature difference for the case of a *counter-flow* heat exchanger with the same inlet and outlet temperatures and is determined from Eq. 22–25 by taking $\Delta T_1 = T_{h,\,in} - T_{c,\,out}$ and $\Delta T_2 = T_{h,\,out} - T_{c,\,in}$ (Fig. 22–17).

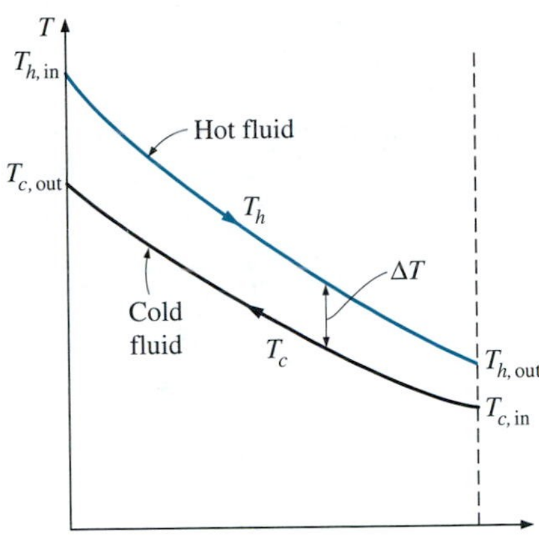

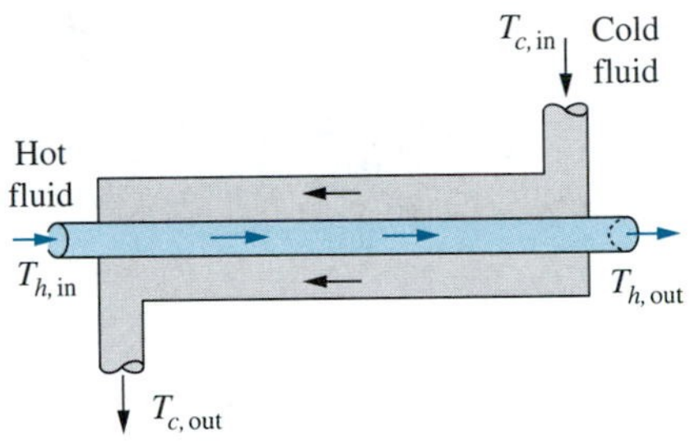

FIGURE 22–16
The variation of the fluid temperatures in a counter-flow double-pipe heat exchanger.

The correction factor is less than unity for a cross-flow and multipass shell-and-tube heat exchanger. That is, $F \leq 1$. The limiting value of $F = 1$ corresponds to the counter-flow heat exchanger. Thus, the correction factor F for a heat exchanger is *a measure of deviation of the* ΔT_{lm} *from the corresponding values for the counter-flow case.*

The correction factor F for common cross-flow and shell-and-tube heat exchanger configurations is given in Fig. 22–18 versus two temperature ratios P and R defined as

$$P = \frac{t_2 - t_1}{T_1 - t_1} \tag{22–27}$$

and

$$R = \frac{T_1 - T_2}{t_2 - t_1} = \frac{(\dot{m}c_p)_{\text{tube side}}}{(\dot{m}c_p)_{\text{shell side}}} \tag{22–28}$$

where the subscripts 1 and 2 represent the *inlet* and *outlet,* respectively. Note that for a shell-and-tube heat exchanger, T and t represent the *shell-* and *tube-side* temperatures, respectively, as shown in the correction factor charts. It makes no difference whether the hot or the cold fluid flows through the shell or the tube. The determination of the correction factor F requires the availability of the inlet and the outlet temperatures for both the cold and hot fluids.

Note that the value of P ranges from 0 to 1. The value of R, on the other hand, ranges from 0 to infinity, with $R = 0$ corresponding to the phase-change (condensation or boiling) on the shell-side and $R \to \infty$ to phase-change on the tube side. The correction factor is $F = 1$ for both of these limiting cases. Therefore, the correction factor for a condenser or boiler is $F = 1$, regardless of the configuration of the heat exchanger.

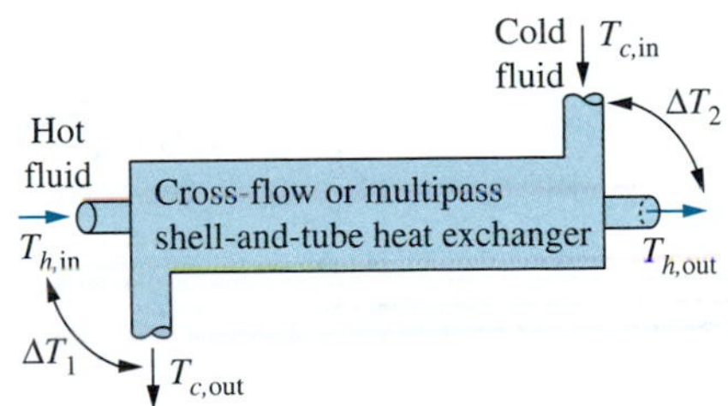

Heat transfer rate:

$$\dot{Q} = UA_s F \Delta T_{lm,CF}$$

where $\Delta T_{lm,CF} = \dfrac{\Delta T_1 - \Delta T_2}{\ln(\Delta T_1/\Delta T_2)}$

$\Delta T_1 = T_{h,in} - T_{c,out}$

$\Delta T_2 = T_{h,out} - T_{c,in}$

and $F = \ldots$ (Fig. 22–18)

FIGURE 22–17
The determination of the heat transfer rate for cross-flow and multipass shell-and-tube heat exchangers using the correction factor.

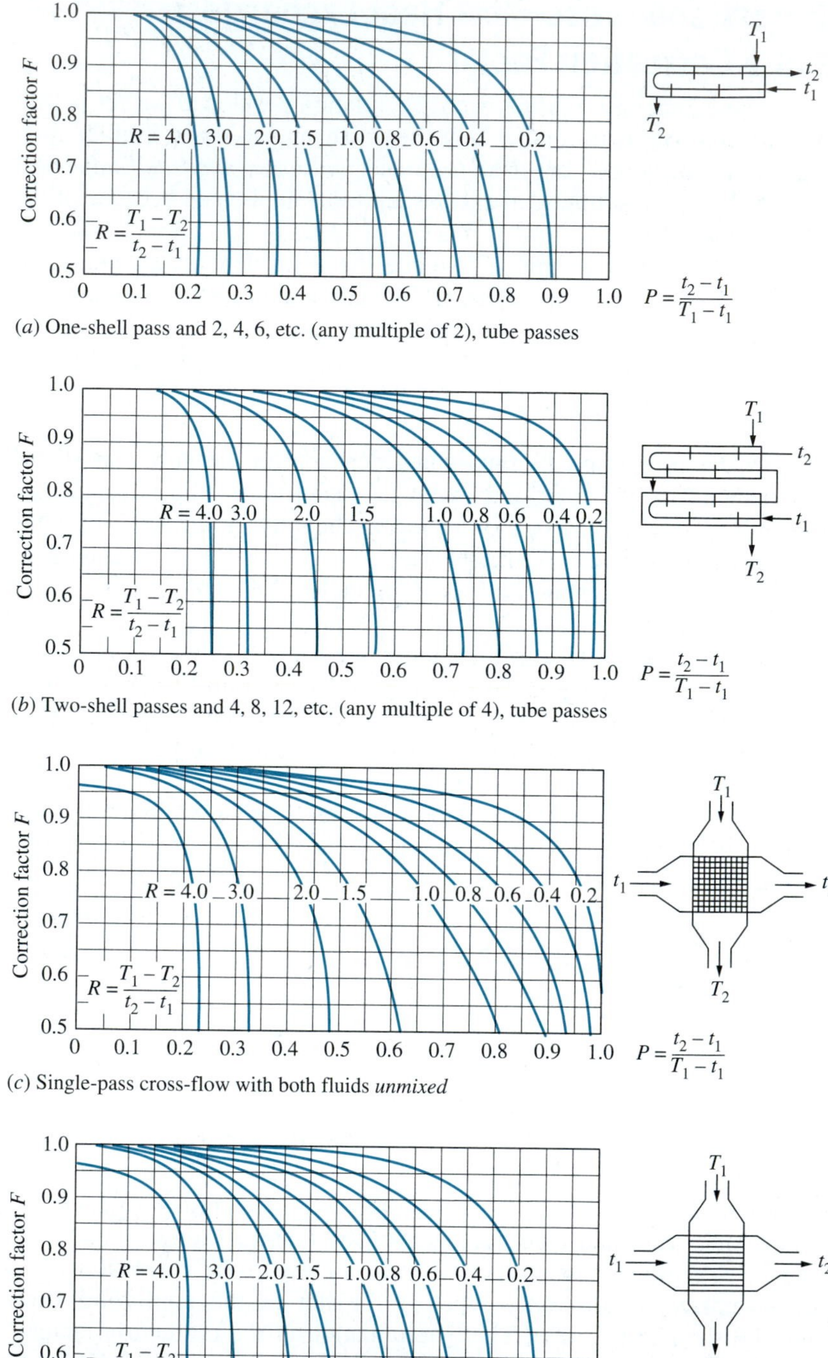

FIGURE 22–18
Correction factor F charts for common shell-and-tube and cross-flow heat exchangers.
(From Bowman, Mueller, and Nagle, 1940.)

EXAMPLE 22–3 The Condensation of Steam in a Condenser

Steam in the condenser of a power plant is to be condensed at a temperature of 30°C with cooling water from a nearby lake, which enters the tubes of the condenser at 14°C and leaves at 22°C. The surface area of the tubes is 45 m^2, and the overall heat transfer coefficient is 2100 W/m^2 · °C. Determine the mass flow rate of the cooling water needed and the rate of condensation of the steam in the condenser.

Solution Steam is condensed by cooling water in the condenser of a power plant. The mass flow rate of the cooling water and the rate of condensation are to be determined.

Assumptions **1** Steady operating conditions exist. **2** The heat exchanger is well insulated so that heat loss to the surroundings is negligible. **3** Changes in the kinetic and potential energies of fluid streams are negligible. **4** There is no fouling. **5** Fluid properties are constant.

Properties The heat of vaporization of water at 30°C is h_{fg} = 2431 kJ/kg and the specific heat of cold water at the average temperature of 18°C is c_p = 4184 J/kg · °C (Table A–15).

Analysis The schematic of the condenser is given in Fig. 22–19. The condenser can be treated as a counter-flow heat exchanger since the temperature of one of the fluids (the steam) remains constant.

The temperature difference between the steam and the cooling water at the two ends of the condenser is

$$\Delta T_1 = T_{h,\,\text{in}} - T_{c,\,\text{out}} = (30 - 22)°\text{C} = 8°\text{C}$$

$$\Delta T_2 = T_{h,\,\text{out}} - T_{c,\,\text{in}} = (30 - 14)°\text{C} = 16°\text{C}$$

That is, the temperature difference between the two fluids varies from 8°C at one end to 16°C at the other. The proper average temperature difference between the two fluids is the *logarithmic mean temperature difference* (not the arithmetic), which is determined from

$$\Delta T_{\text{lm}} = \frac{\Delta T_1 - \Delta T_2}{\ln(\Delta T_1/\Delta T_2)} = \frac{8 - 16}{\ln(8/16)} = 11.5°\text{C}$$

This is a little less than the arithmetic mean temperature difference of $\frac{1}{2}$ (8 + 16) = 12°C. Then the heat transfer rate in the condenser is determined from

$$\dot{Q} = UA_s\,\Delta T_{\text{lm}} = (2100\ \text{W/m}^2\cdot°\text{C})(45\ \text{m}^2)(11.5°\text{C}) = 1.087 \times 10^6\ \text{W} = 1087\ \text{kW}$$

Therefore, steam will lose heat at a rate of 1087 kW as it flows through the condenser, and the cooling water will gain practically all of it, since the condenser is well insulated.

The mass flow rate of the cooling water and the rate of the condensation of the steam are determined from $\dot{Q} = [\dot{m}c_p(T_{\text{out}} - T_{\text{in}})]_{\text{cooling water}} = (\dot{m}h_{fg})_{\text{steam}}$ to be

$$\dot{m}_{\text{cooling water}} = \frac{\dot{Q}}{c_p(T_{\text{out}} - T_{\text{in}})} = \frac{1087\ \text{kJ/s}}{(4.184\ \text{kJ/kg}\cdot°\text{C})(22 - 14)°\text{C}} = \mathbf{32.5\ kg/s}$$

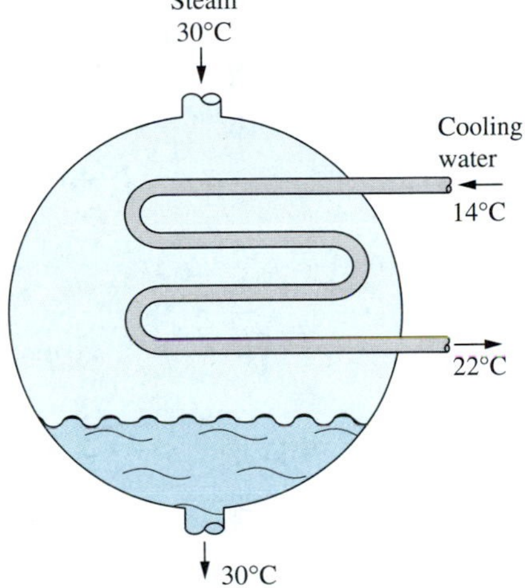

FIGURE 22–19
Schematic for Example 22–3.

and

$$\dot{m}_{\text{steam}} = \frac{\dot{Q}}{h_{fg}} = \frac{1087 \text{ kJ/s}}{2431 \text{ kJ/kg}} = \mathbf{0.45 \text{ kg/s}}$$

Therefore, we need to circulate about 72 kg of cooling water for each 1 kg of steam condensing to remove the heat released during the condensation process.

EXAMPLE 22–4 Heating Water in a Counter-Flow Heat Exchanger

A counter-flow double-pipe heat exchanger is to heat water from 20°C to 80°C at a rate of 1.2 kg/s. The heating is to be accomplished by geothermal water available at 160°C at a mass flow rate of 2 kg/s. The inner tube is thin-walled and has a diameter of 1.5 cm. If the overall heat transfer coefficient of the heat exchanger is 640 W/m² · °C, determine the length of the heat exchanger required to achieve the desired heating.

Solution Water is heated in a counter-flow double-pipe heat exchanger by geothermal water. The required length of the heat exchanger is to be determined.

Assumptions **1** Steady operating conditions exist. **2** The heat exchanger is well insulated so that heat loss to the surroundings is negligible. **3** Changes in the kinetic and potential energies of fluid streams are negligible. **4** There is no fouling. **5** Fluid properties are constant.

Properties We take the specific heats of water and geothermal fluid to be 4.18 and 4.31 kJ/kg · °C, respectively.

Analysis The schematic of the heat exchanger is given in Fig. 22–20. The rate of heat transfer in the heat exchanger can be determined from

$$\dot{Q} = [\dot{m}c_p(T_{\text{out}} - T_{\text{in}})]_{\text{water}} = (1.2 \text{ kg/s})(4.18 \text{ kJ/kg} \cdot °\text{C})(80 - 20)°\text{C} = 301 \text{ kW}$$

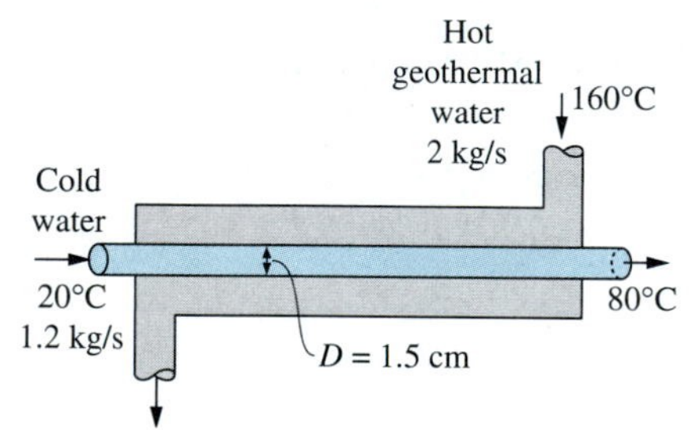

FIGURE 22–20
Schematic for Example 22–4.

Noting that all of this heat is supplied by the geothermal water, the outlet temperature of the geothermal water is determined to be

$$\dot{Q} = [\dot{m}c_p(T_{\text{in}} - T_{\text{out}})]_{\text{geothermal}} \longrightarrow T_{\text{out}} = T_{\text{in}} - \frac{\dot{Q}}{\dot{m}c_p}$$

$$= 160°\text{C} - \frac{301 \text{ kW}}{(2 \text{ kg/s})(4.31 \text{ kJ/kg} \cdot °\text{C})}$$

$$= 125°\text{C}$$

Knowing the inlet and outlet temperatures of both fluids, the logarithmic mean temperature difference for this counter-flow heat exchanger becomes

$$\Delta T_1 = T_{h,\text{ in}} - T_{c,\text{ out}} = (160 - 80)°\text{C} = 80°\text{C}$$
$$\Delta T_2 = T_{h,\text{ out}} - T_{c,\text{ in}} = (125 - 20)°\text{C} = 105°\text{C}$$

and

$$\Delta T_{\text{lm}} = \frac{\Delta T_1 - \Delta T_2}{\ln (\Delta T_1/\Delta T_2)} = \frac{80 - 105}{\ln (80/105)} = 91.9°\text{C}$$

Then the surface area of the heat exchanger is determined to be

$$\dot{Q} = UA_s\,\Delta T_{\text{lm}} \longrightarrow A_s = \frac{\dot{Q}}{U\,\Delta T_{\text{lm}}} = \frac{301{,}000 \text{ W}}{(640 \text{ W/m}^2 \cdot °\text{C})(91.9°\text{C})} = 5.12 \text{ m}^2$$

To provide this much heat transfer surface area, the length of the tube must be

$$A_s = \pi DL \longrightarrow L = \frac{A_s}{\pi D} = \frac{5.12 \text{ m}^2}{\pi(0.015 \text{ m})} = \mathbf{109 \text{ m}}$$

Discussion The inner tube of this counter-flow heat exchanger (and thus the heat exchanger itself) needs to be over 100 m long to achieve the desired heat transfer, which is impractical. In cases like this, we need to use a plate heat exchanger or a multipass shell-and-tube heat exchanger with multiple passes of tube bundles.

EXAMPLE 22–5 Heating of Glycerin in a Multipass Heat Exchanger

A 2-shell passes and 4-tube passes heat exchanger is used to heat glycerin from 20°C to 50°C by hot water, which enters the thin-walled 2-cm-diameter tubes at 80°C and leaves at 40°C (Fig. 22–21). The total length of the tubes in the heat exchanger is 60 m. The convection heat transfer coefficient is 25 W/m^2 · °C on the glycerin (shell) side and 160 W/m^2 · °C on the water (tube) side. Determine the rate of heat transfer in the heat exchanger (*a*) before any fouling and (*b*) after fouling with a fouling factor of 0.0006 m^2 · °C/W occurs on the outer surfaces of the tubes.

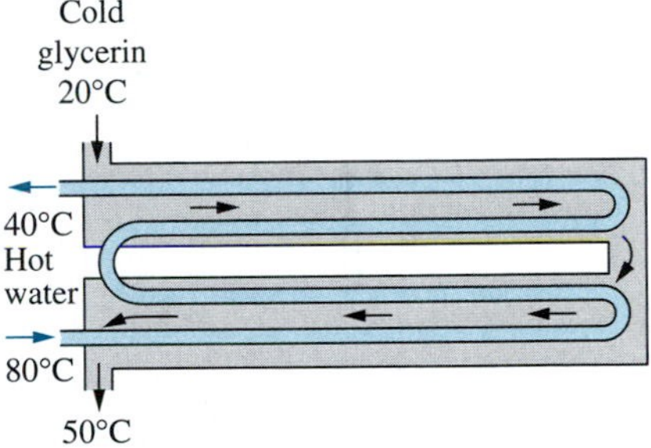

FIGURE 22–21
Schematic for Example 22–5.

Solution Glycerin is heated in a 2-shell passes and 4-tube passes heat exchanger by hot water. The rate of heat transfer for the cases of fouling and no fouling are to be determined.

Assumptions **1** Steady operating conditions exist. **2** The heat exchanger is well insulated so that heat loss to the surroundings is negligible. **3** Changes in the kinetic and potential energies of fluid streams are negligible. **4** Heat transfer coefficients and fouling factors are constant and uniform. **5** The thermal resistance of the inner tube is negligible since the tube is thin-walled and highly conductive.

Analysis The tubes are said to be thin-walled, and thus it is reasonable to assume the inner and outer surface areas of the tubes to be equal. Then the heat transfer surface area becomes

$$A_s = \pi DL = \pi(0.02 \text{ m})(60 \text{ m}) = 3.77 \text{ m}^2$$

The rate of heat transfer in this heat exchanger can be determined from

$$\dot{Q} = UA_s F \, \Delta T_{\text{lm, CF}}$$

where F is the correction factor and $\Delta T_{\text{lm, CF}}$ is the log mean temperature difference for the counter-flow arrangement. These two quantities are determined from

$$\Delta T_1 = T_{h,\text{ in}} - T_{c,\text{ out}} = (80 - 50)°\text{C} = 30°\text{C}$$

$$\Delta T_2 = T_{h,\text{ out}} - T_{c,\text{ in}} = (40 - 20)°\text{C} = 20°\text{C}$$

$$\Delta T_{\text{lm, CF}} = \frac{\Delta T_1 - \Delta T_2}{\ln(\Delta T_1/\Delta T_2)} = \frac{30 - 20}{\ln(30/20)} = 24.7°\text{C}$$

and

$$\left.\begin{aligned} P &= \frac{t_2 - t_1}{T_1 - t_1} = \frac{40 - 80}{20 - 80} = 0.67 \\ R &= \frac{T_1 - T_2}{t_2 - t_1} = \frac{20 - 50}{40 - 80} = 0.75 \end{aligned}\right\} F = 0.91 \qquad \text{(Fig. 22–18}b\text{)}$$

(*a*) In the case of no fouling, the overall heat transfer coefficient U is

$$U = \frac{1}{\dfrac{1}{h_i} + \dfrac{1}{h_o}} = \frac{1}{\dfrac{1}{160 \text{ W/m}^2 \cdot {}^\circ\text{C}} + \dfrac{1}{25 \text{ W/m}^2 \cdot {}^\circ\text{C}}} = 21.6 \text{ W/m}^2 \cdot {}^\circ\text{C}$$

Then the rate of heat transfer becomes

$$\dot{Q} = UA_s F\,\Delta T_{\text{lm, CF}} = (21.6 \text{ W/m}^2 \cdot {}^\circ\text{C})(3.77 \text{ m}^2)(0.91)(24.7^\circ\text{C}) = \mathbf{1830 \text{ W}}$$

(*b*) When there is fouling on one of the surfaces, we have

$$U = \frac{1}{\dfrac{1}{h_i} + \dfrac{1}{h_o} + R_f} = \frac{1}{\dfrac{1}{160 \text{ W/m}^2 \cdot {}^\circ\text{C}} + \dfrac{1}{25 \text{ W/m}^2 \cdot {}^\circ\text{C}} + 0.0006 \text{ m}^2 \cdot {}^\circ\text{C/W}}$$

$$= 21.3 \text{ W/m}^2 \cdot {}^\circ\text{C}$$

and

$$\dot{Q} = UA_s F\,\Delta T_{\text{lm, CF}} = (21.3 \text{ W/m}^2 \cdot {}^\circ\text{C})(3.77 \text{ m}^2)(0.91)(24.7^\circ\text{C}) = \mathbf{1805 \text{ W}}$$

Discussion Note that the rate of heat transfer decreases as a result of fouling, as expected. The decrease is not dramatic, however, because of the relatively low convection heat transfer coefficients involved.

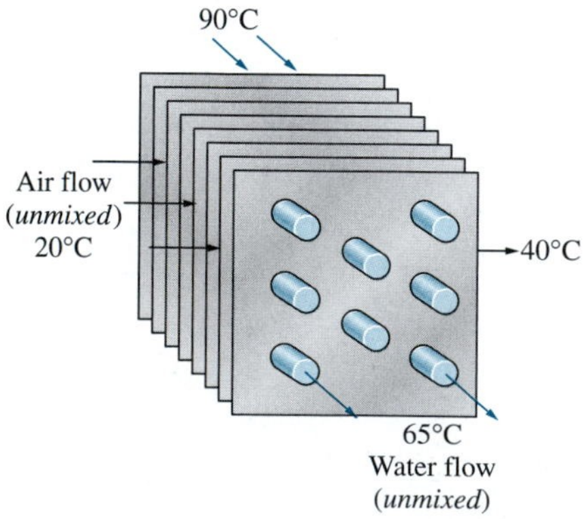

90°C

Air flow (*unmixed*) 20°C

40°C

65°C Water flow (*unmixed*)

FIGURE 22–22
Schematic for Example 22–6.

EXAMPLE 22–6 Cooling of Water in an Automotive Radiator

A test is conducted to determine the overall heat transfer coefficient in an automotive radiator that is a compact cross-flow water-to-air heat exchanger with both fluids (air and water) unmixed (Fig. 22–22). The radiator has 40 tubes of internal diameter 0.5 cm and length 65 cm in a closely spaced plate-finned matrix. Hot water enters the tubes at 90°C at a rate of 0.6 kg/s and leaves at 65°C. Air flows across the radiator through the interfin spaces and is heated from 20°C to 40°C. Determine the overall heat transfer coefficient U_i of this radiator based on the inner surface area of the tubes.

Solution During an experiment involving an automotive radiator, the inlet and exit temperatures of water and air and the mass flow rate of water are measured. The overall heat transfer coefficient based on the inner surface area is to be determined.

Assumptions **1** Steady operating conditions exist. **2** Changes in the kinetic and potential energies of fluid streams are negligible. **3** Fluid properties are constant.

Properties The specific heat of water at the average temperature of (90 + 65)/2 = 77.5°C is 4.195 kJ/kg · °C (Table A–15).

Analysis The rate of heat transfer in this radiator from the hot water to the air is determined from an energy balance on water flow,

$$\begin{aligned}\dot{Q} &= [\dot{m}c_p(T_{in} - T_{out})]_{water} = (0.6 \text{ kg/s})(4.195 \text{ kJ/kg} \cdot {}^\circ\text{C})(90 - 65)^\circ\text{C} \\ &= 62.93 \text{ kW}\end{aligned}$$

The tube-side heat transfer area is the total surface area of the tubes, and is determined from

$$A_i = n\pi D_i L = (40)\pi(0.005 \text{ m})(0.65 \text{ m}) = 0.408 \text{ m}^2$$

Knowing the rate of heat transfer and the surface area, the overall heat transfer coefficient can be determined from

$$\dot{Q} = U_i A_i F \Delta T_{lm, CF} \longrightarrow U_i = \frac{\dot{Q}}{A_i F \Delta T_{lm, CF}}$$

where F is the correction factor and $\Delta T_{lm, CF}$ is the log mean temperature difference for the counter-flow arrangement. These two quantities are found to be

$$\begin{aligned}\Delta T_1 &= T_{h, in} - T_{c, out} = (90 - 40)^\circ\text{C} = 50^\circ\text{C} \\ \Delta T_2 &= T_{h, out} - T_{c, in} = (65 - 20)^\circ\text{C} = 45^\circ\text{C} \\ \Delta T_{lm, CF} &= \frac{\Delta T_1 - \Delta T_2}{\ln(\Delta T_1/\Delta T_2)} = \frac{50 - 45}{\ln(50/45)} = 47.5^\circ\text{C}\end{aligned}$$

and

$$\left.\begin{aligned}P &= \frac{t_2 - t_1}{T_1 - t_1} = \frac{65 - 90}{20 - 90} = 0.36 \\ R &= \frac{T_1 - T_2}{t_2 - t_1} = \frac{20 - 40}{65 - 90} = 0.80\end{aligned}\right\} F = 0.97 \qquad \text{(Fig. 22–18}c\text{)}$$

Substituting, the overall heat transfer coefficient U_i is determined to be

$$U_i = \frac{\dot{Q}}{A_i F \Delta T_{lm, CF}} = \frac{62{,}930 \text{ W}}{(0.408 \text{ m}^2)(0.97)(47.5^\circ\text{C})} = \mathbf{3347 \text{ W/m}^2 \cdot {}^\circ\text{C}}$$

Discussion Note that the overall heat transfer coefficient on the air side will be much lower because of the large surface area involved on that side.

22–5 ▪ THE EFFECTIVENESS–NTU METHOD

The log mean temperature difference (LMTD) method discussed in Section 22–4 is easy to use in heat exchanger analysis when the inlet and the outlet temperatures of the hot and cold fluids are known or can be determined from an energy balance. Once ΔT_{lm}, the mass flow rates, and the overall heat transfer coefficient are available, the heat transfer surface area of the heat exchanger can be determined from

$$\dot{Q} = UA_s \, \Delta T_{\text{lm}}$$

Therefore, the LMTD method is very suitable for determining the *size* of a heat exchanger to realize prescribed outlet temperatures when the mass flow rates and the inlet and outlet temperatures of the hot and cold fluids are specified.

With the LMTD method, the task is to *select* a heat exchanger that will meet the prescribed heat transfer requirements. The procedure to be followed by the selection process is:

1. Select the type of heat exchanger suitable for the application.
2. Determine any unknown inlet or outlet temperature and the heat transfer rate using an energy balance.
3. Calculate the log mean temperature difference ΔT_{lm} and the correction factor F, if necessary.
4. Obtain (select or calculate) the value of the overall heat transfer coefficient U.
5. Calculate the heat transfer surface area A_s.

The task is completed by selecting a heat exchanger that has a heat transfer surface area equal to or larger than A_s.

A second kind of problem encountered in heat exchanger analysis is the determination of the *heat transfer rate* and the *outlet temperatures* of the hot and cold fluids for prescribed fluid mass flow rates and inlet temperatures when the *type* and *size* of the heat exchanger are specified. The heat transfer surface area of the heat exchanger in this case is known, but the *outlet temperatures* are not. Here the task is to determine the heat transfer performance of a specified heat exchanger or to determine if a heat exchanger available in storage will do the job.

The LMTD method could still be used for this alternative problem, but the procedure would require tedious iterations, and thus it is not practical. In an attempt to eliminate the iterations from the solution of such problems, Kays and London came up with a method in 1955 called the **effectiveness–NTU method**, which greatly simplified heat exchanger analysis.

This method is based on a dimensionless parameter called the **heat transfer effectiveness** ε, defined as

$$\varepsilon = \frac{\dot{Q}}{\dot{Q}_{\max}} = \frac{\text{Actual heat transfer rate}}{\text{Maximum possible heat transfer rate}} \tag{22–29}$$

The *actual* heat transfer rate in a heat exchanger can be determined from an energy balance on the hot or cold fluids and can be expressed as

$$\dot{Q} = C_c(T_{c,\text{ out}} - T_{c,\text{ in}}) = C_h(T_{h,\text{ in}} - T_{h,\text{ out}}) \tag{22–30}$$

where $C_c = \dot{m}_c c_{pc}$ and $C_h = \dot{m}_c c_{ph}$ are the heat capacity rates of the cold and hot fluids, respectively.

To determine the maximum possible heat transfer rate in a heat exchanger, we first recognize that the *maximum temperature difference* in a heat exchanger is the difference between the *inlet* temperatures of the hot and cold fluids. That is,

$$\Delta T_{max} = T_{h,\, in} - T_{c,\, in} \quad \text{(22–31)}$$

The heat transfer in a heat exchanger will reach its maximum value when (1) the cold fluid is heated to the inlet temperature of the hot fluid or (2) the hot fluid is cooled to the inlet temperature of the cold fluid. These two limiting conditions will not be reached simultaneously unless the heat capacity rates of the hot and cold fluids are identical (i.e., $C_c = C_h$). When $C_c \neq C_h$, which is usually the case, the fluid with the *smaller* heat capacity rate will experience a larger temperature change, and thus it will be the first to experience the maximum temperature, at which point the heat transfer will come to a halt. Therefore, the maximum possible heat transfer rate in a heat exchanger is (Fig. 22–23)

$$\dot{Q}_{max} = C_{min}(T_{h,\, in} - T_{c,\, in}) \quad \text{(22–32)}$$

where C_{min} is the smaller of C_h and C_c. This is further clarified by Example 22–7.

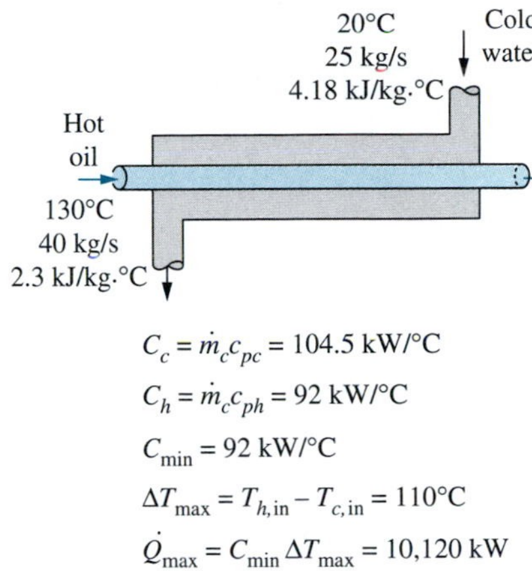

FIGURE 22–23
The determination of the maximum rate of heat transfer in a heat exchanger.

EXAMPLE 22–7 Upper Limit for Heat Transfer in a Heat Exchanger

Cold water enters a counter-flow heat exchanger at 10°C at a rate of 8 kg/s, where it is heated by a hot-water stream that enters the heat exchanger at 70°C at a rate of 2 kg/s. Assuming the specific heat of water to remain constant at $c_p = 4.18$ kJ/kg · °C, determine the maximum heat transfer rate and the outlet temperatures of the cold- and the hot-water streams for this limiting case.

Solution Cold- and hot-water streams enter a heat exchanger at specified temperatures and flow rates. The maximum rate of heat transfer in the heat exchanger and the outlet temperatures are to be determined.

Assumptions **1** Steady operating conditions exist. **2** The heat exchanger is well insulated so that heat loss to the surroundings is negligible. **3** Changes in the kinetic and potential energies of fluid streams are negligible. **4** Fluid properties are constant.

Properties The specific heat of water is given to be $c_p = 4.18$ kJ/kg · °C.

Analysis A schematic of the heat exchanger is given in Fig. 22–24. The heat capacity rates of the hot and cold fluids are

$$C_h = \dot{m}_h c_{ph} = (2 \text{ kg/s})(4.18 \text{ kJ/kg} \cdot \text{°C}) = 8.36 \text{ kW/°C}$$

and

$$C_c = \dot{m}_c c_{pc} = (8 \text{ kg/s})(4.18 \text{ kJ/kg} \cdot \text{°C}) = 33.4 \text{ kW/°C}$$

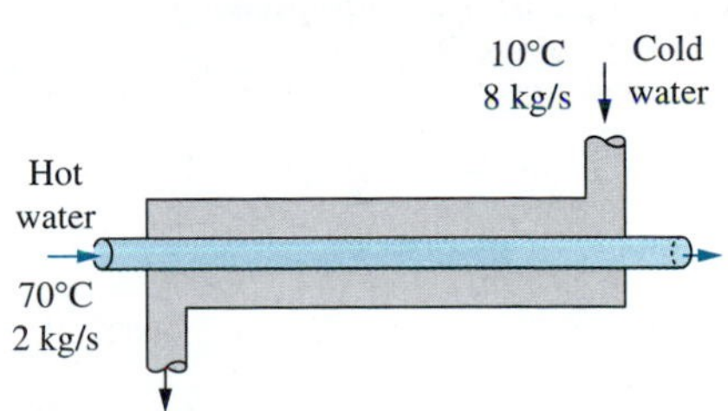

FIGURE 22–24
Schematic for Example 22–7.

Therefore,

$$C_{\min} = C_h = 8.36 \text{ kW/°C}$$

which is the smaller of the two heat capacity rates. Then the maximum heat transfer rate is determined from Eq. 22–32 to be

$$\begin{aligned}\dot{Q}_{\max} &= C_{\min}(T_{h,\text{ in}} - T_{c,\text{ in}})\\ &= (8.36 \text{ kW/°C})(70 - 10)\text{°C}\\ &= \mathbf{502 \text{ kW}}\end{aligned}$$

That is, the maximum possible heat transfer rate in this heat exchanger is 502 kW. This value would be approached in a counter-flow heat exchanger with a *very large* heat transfer surface area.

The maximum temperature difference in this heat exchanger is $\Delta T_{\max} = T_{h,\text{ in}} - T_{c,\text{ in}} = (70 - 10)\text{°C} = 60\text{°C}$. Therefore, the hot water cannot be cooled by more than 60°C (to 10°C) in this heat exchanger, and the cold water cannot be heated by more than 60°C (to 70°C), no matter what we do. The outlet temperatures of the cold and the hot streams in this limiting case are determined to be

$$\dot{Q} = C_c(T_{c,\text{ out}} - T_{c,\text{ in}}) \longrightarrow T_{c,\text{ out}} = T_{c,\text{ in}} + \frac{\dot{Q}}{C_c} = 10\text{°C} + \frac{502 \text{ kW}}{33.4 \text{ kW/°C}} = \mathbf{25°C}$$

$$\dot{Q} = C_h(T_{h,\text{ in}} - T_{h,\text{ out}}) \longrightarrow T_{h,\text{ out}} = T_{h,\text{ in}} - \frac{\dot{Q}}{C_h} = 70\text{°C} - \frac{502 \text{ kW}}{8.38 \text{ kW/°C}} = \mathbf{10°C}$$

Discussion Note that the hot water is cooled to the limit of 10°C (the inlet temperature of the cold-water stream), but the cold water is heated to 25°C only when maximum heat transfer occurs in the heat exchanger. This is not surprising, since the mass flow rate of the hot water is only one-fourth that of the cold water, and, as a result, the temperature of the cold water increases by 0.25°C for each 1°C drop in the temperature of the hot water.

You may be tempted to think that the cold water should be heated to 70°C in the limiting case of maximum heat transfer. But this will require the temperature of the hot water to drop to −170°C (below 10°C), which is impossible. Therefore, heat transfer in a heat exchanger reaches its maximum value when the fluid with the smaller heat capacity rate (or the smaller mass flow rate when both fluids have the same specific heat value) experiences the maximum temperature change. This example explains why we use $C_{\min}$ in the evaluation of $\dot{Q}_{\max}$ instead of $C_{\max}$.

We can show that the hot water will leave at the inlet temperature of the cold water and vice versa in the limiting case of maximum heat transfer when the mass flow rates of the hot- and cold-water streams are identical (Fig. 22–25). We can also show that the outlet temperature of the cold water will reach the 70°C limit when the mass flow rate of the hot water is greater than that of the cold water.

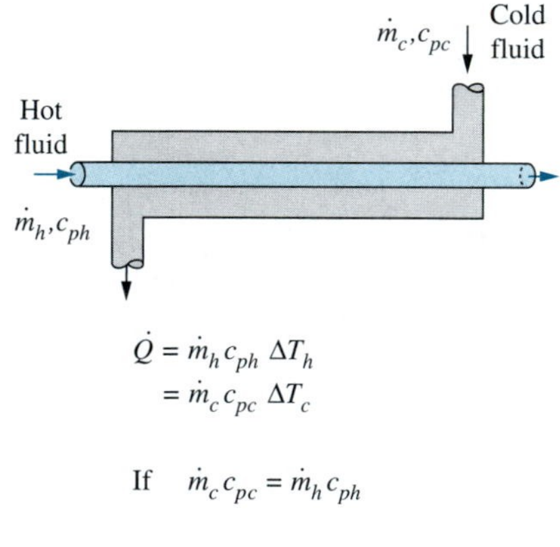

$$\begin{aligned}\dot{Q} &= \dot{m}_h c_{ph} \Delta T_h\\ &= \dot{m}_c c_{pc} \Delta T_c\end{aligned}$$

If $\dot{m}_c c_{pc} = \dot{m}_h c_{ph}$

then $\Delta T_h = \Delta T_c$

FIGURE 22–25
The temperature rise of the cold fluid in a heat exchanger will be equal to the temperature drop of the hot fluid when the heat capcity rates of the hot and cold fluids are identical.

The determination of $\dot{Q}_{\max}$ requires the availability of the *inlet temperature* of the hot and cold fluids and their *mass flow rates,* which are usually specified. Then, once the effectiveness of the heat exchanger is known, the actual heat transfer rate $\dot{Q}$ can be determined from

$$\dot{Q} = \varepsilon \dot{Q}_{\max} = \varepsilon C_{\min}(T_{h,\text{ in}} - T_{c,\text{ in}}) \qquad \textbf{(22–33)}$$

Therefore, the effectiveness of a heat exchanger enables us to determine the heat transfer rate without knowing the *outlet temperatures* of the fluids.

The effectiveness of a heat exchanger depends on the *geometry* of the heat exchanger as well as the *flow arrangement.* Therefore, different types of heat exchangers have different effectiveness relations. Below we illustrate the development of the effectiveness ε relation for the double-pipe *parallel-flow* heat exchanger.

Equation 22–23 developed in Section 22–4 for a parallel-flow heat exchanger can be rearranged as

$$\ln \frac{T_{h,\text{ out}} - T_{c,\text{ out}}}{T_{h,\text{ in}} - T_{c,\text{ in}}} = -\frac{UA_s}{C_c}\left(1 + \frac{C_c}{C_h}\right) \tag{22–34}$$

Also, solving Eq. 22–30 for $T_{h,\text{ out}}$ gives

$$T_{h,\text{ out}} = T_{h,\text{ in}} - \frac{C_c}{C_h}(T_{c,\text{ out}} - T_{c,\text{ in}}) \tag{22–35}$$

Substituting this relation into Eq. 22–34 after adding and subtracting $T_{c,\text{ in}}$ gives

$$\ln \frac{T_{h,\text{ in}} - T_{c,\text{ in}} + T_{c,\text{ in}} - T_{c,\text{ out}} - \dfrac{C_c}{C_h}(T_{c,\text{ out}} - T_{c,\text{ in}})}{T_{h,\text{ in}} - T_{c,\text{ in}}} = -\frac{UA_s}{C_c}\left(1 + \frac{C_c}{C_h}\right)$$

which simplifies to

$$\ln \left[1 - \left(1 + \frac{C_c}{C_h}\right)\frac{T_{c,\text{ out}} - T_{c,\text{ in}}}{T_{h,\text{ in}} - T_{c,\text{ in}}}\right] = -\frac{UA_s}{C_c}\left(1 + \frac{C_c}{C_h}\right) \tag{22–36}$$

We now manipulate the definition of effectiveness to obtain

$$\varepsilon = \frac{\dot{Q}}{\dot{Q}_{\max}} = \frac{C_c(T_{c,\text{ out}} - T_{c,\text{ in}})}{C_{\min}(T_{h,\text{ in}} - T_{c,\text{ in}})} \longrightarrow \frac{T_{c,\text{ out}} - T_{c,\text{ in}}}{T_{h,\text{ in}} - T_{c,\text{ in}}} = \varepsilon\frac{C_{\min}}{C_c}$$

Substituting this result into Eq. 22–36 and solving for ε gives the following relation for the effectiveness of a *parallel-flow* heat exchanger:

$$\varepsilon_{\text{parallel flow}} = \frac{1 - \exp\left[-\dfrac{UA_s}{C_c}\left(1 + \dfrac{C_c}{C_h}\right)\right]}{\left(1 + \dfrac{C_c}{C_h}\right)\dfrac{C_{\min}}{C_c}} \tag{22–37}$$

Taking either C_c or C_h to be $C_{\min}$ (both approaches give the same result), the relation above can be expressed more conveniently as

$$\varepsilon_{\text{parallel flow}} = \frac{1 - \exp\left[-\dfrac{UA_s}{C_{\min}}\left(1 + \dfrac{C_{\min}}{C_{\max}}\right)\right]}{1 + \dfrac{C_{\min}}{C_{\max}}} \tag{22–38}$$

Again $C_{\min}$ is the *smaller* heat capacity ratio and $C_{\max}$ is the larger one, and it makes no difference whether $C_{\min}$ belongs to the hot or cold fluid.

Effectiveness relations of the heat exchangers typically involve the *dimensionless* group UA_s/C_{min}. This quantity is called the **number of transfer units NTU** and is expressed as

$$\text{NTU} = \frac{UA_s}{C_{min}} = \frac{UA_s}{(\dot{m}c_p)_{min}} \quad \text{(22–39)}$$

where U is the overall heat transfer coefficient and A_s is the heat transfer surface area of the heat exchanger. Note that NTU is proportional to A_s. Therefore, for specified values of U and C_{min}, the value of NTU *is a measure of the heat transfer surface area* A_s. Thus, the larger the NTU, the larger the heat exchanger.

In heat exchanger analysis, it is also convenient to define another dimensionless quantity called the **capacity ratio *c*** as

$$c = \frac{C_{min}}{C_{max}} \quad \text{(22–40)}$$

It can be shown that the effectiveness of a heat exchanger is a function of the number of transfer units NTU and the capacity ratio c. That is,

$$\varepsilon = \text{function}\,(UA_s/C_{min}, C_{min}/C_{max}) = \text{function}\,(\text{NTU}, c)$$

Effectiveness relations have been developed for a large number of heat exchangers, and the results are given in Table 22–4. The effectivenesses of some common types of heat exchangers are also plotted in Fig. 22–26.

TABLE 22–4

Effectiveness relations for heat exchangers: NTU $= UA_s/C_{min}$ and $c = C_{min}/C_{max} = (\dot{m}c_p)_{min}/(\dot{m}c_p)_{max}$

Heat exchanger type	Effectiveness relation
1 *Double pipe:*	
Parallel-flow	$\varepsilon = \dfrac{1 - \exp\,[-\text{NTU}(1 + c)]}{1 + c}$
Counter-flow	$\varepsilon = \dfrac{1 - \exp\,[-\text{NTU}(1 - c)]}{1 - c\exp\,[-\text{NTU}(1 - c)]}$
2 *Shell-and-tube:* One-shell pass 2, 4, . . . tube passes	$\varepsilon = 2\left\{1 + c + \sqrt{1 + c^2}\,\dfrac{1 + \exp\,[-\text{NTU}\sqrt{1 + c^2}]}{1 - \exp\,[-\text{NTU}\sqrt{1 + c^2}]}\right\}^{-1}$
3 *Cross-flow (single-pass)*	
Both fluids unmixed	$\varepsilon = 1 - \exp\left\{\dfrac{\text{NTU}^{0.22}}{c}\,[\exp\,(-c\,\text{NTU}^{0.78}) - 1]\right\}$
C_{max} mixed, C_{min} unmixed	$\varepsilon = \dfrac{1}{c}(1 - \exp\,\{-c[1 - \exp\,(-\text{NTU})]\})$
C_{min} mixed, C_{max} unmixed	$\varepsilon = 1 - \exp\left\{-\dfrac{1}{c}[1 - \exp\,(-c\,\text{NTU})]\right\}$
4 *All heat exchangers with c = 0*	$\varepsilon = 1 - \exp(-\text{NTU})$

From W. M. Kays and A. L. London. *Compact Heat Exchangers, 3/e*. McGraw-Hill, 1984. Reprinted by permission of William M. Kays.

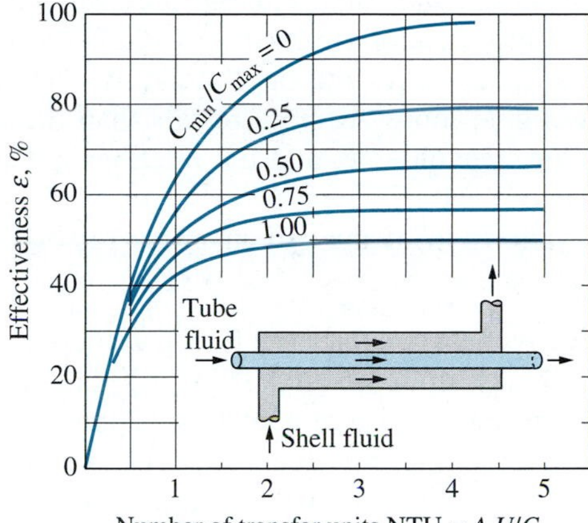

(*a*) Parallel-flow

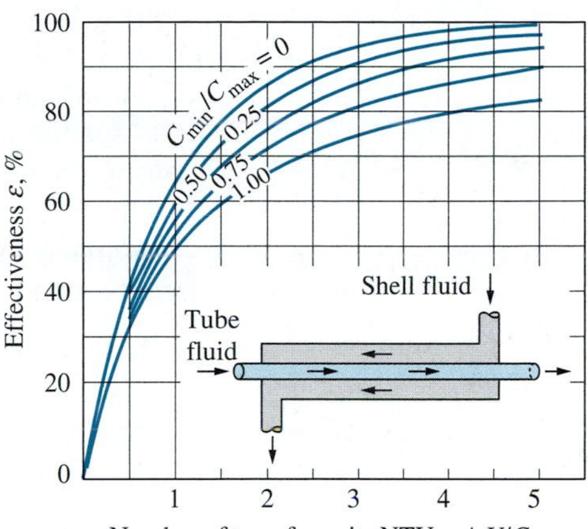

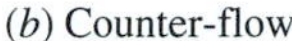

(*b*) Counter-flow

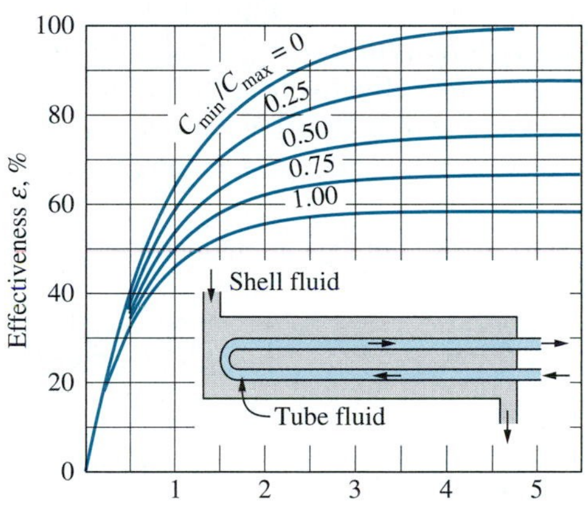

(*c*) One-shell pass and 2, 4, 6, … tube passes

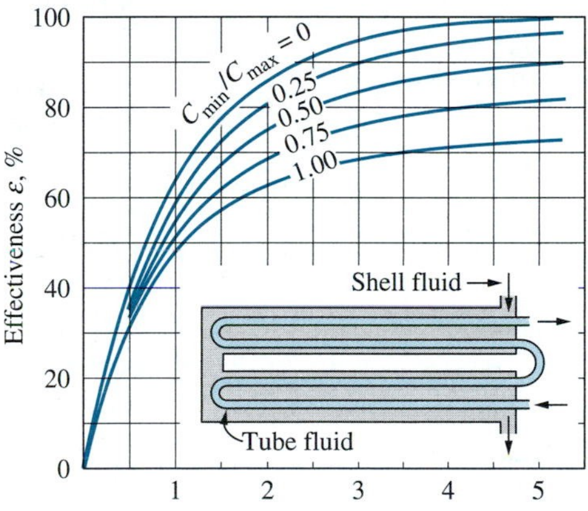

(*d*) Two-shell passes and 4, 8, 12, … tube passes

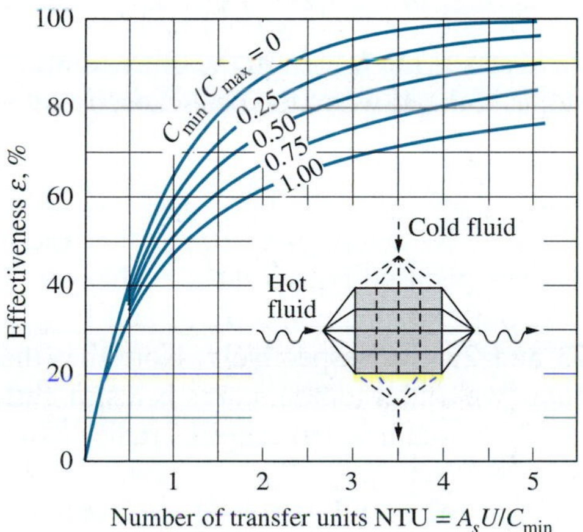

(*e*) Cross-flow with both fluids unmixed

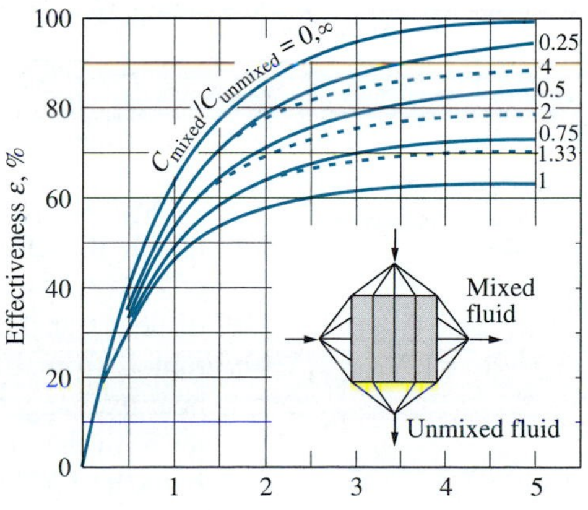

(*f*) Cross-flow with one fluid mixed and the other unmixed

FIGURE 22–26

Effectiveness for heat exchangers.

(From Kays and London, 1984.)

More extensive effectiveness charts and relations are available in the literature. The dashed lines in Fig. 22–26*f* are for the case of C_{min} unmixed and C_{max} mixed and the solid lines are for the opposite case. The analytic relations for the effectiveness give more accurate results than the charts, since reading errors in charts are unavoidable, and the relations are very suitable for computerized analysis of heat exchangers.

We make these observations from the effectiveness relations and charts already given:

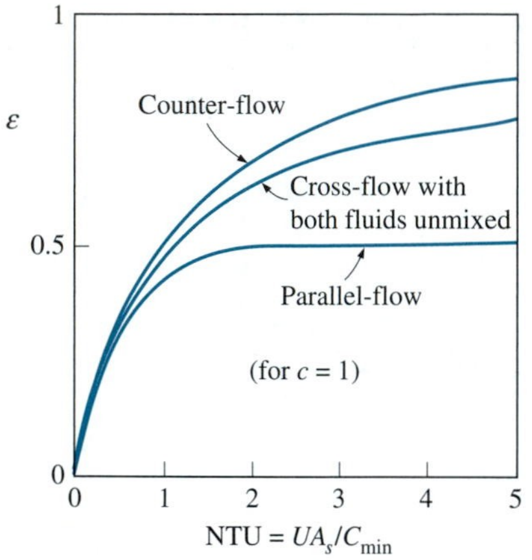

1. The value of the effectiveness ranges from 0 to 1. It increases rapidly with NTU for small values (up to about NTU $= 1.5$) but rather slowly for larger values. Therefore, the use of a heat exchanger with a large NTU (usually larger than 3) and thus a large size cannot be justified economically, since a large increase in NTU in this case corresponds to a small increase in effectiveness. Thus, a heat exchanger with a very high effectiveness may be desirable from a heat transfer point of view but undesirable from an economical point of view.
2. For a given NTU and capacity ratio $c = C_{min}/C_{max}$, the *counter-flow* heat exchanger has the *highest* effectiveness, followed closely by the cross-flow heat exchangers with both fluids unmixed. As you might expect, the lowest effectiveness values are encountered in parallel-flow heat exchangers (Fig. 22–27).
3. The effectiveness of a heat exchanger is independent of the capacity ratio c for NTU values of less than about 0.3.
4. The value of the capacity ratio c ranges between 0 and 1. For a given NTU, the effectiveness becomes a *maximum* for $c = 0$ and a *minimum* for $c = 1$. The case $c = C_{min}/C_{max} \to 0$ corresponds to $C_{max} \to \infty$, which is realized during a phase-change process in a *condenser* or *boiler*. All effectiveness relations in this case reduce to

$$\varepsilon = \varepsilon_{max} = 1 - \exp(-\text{NTU}) \qquad \textbf{(22–41)}$$

regardless of the type of heat exchanger (Fig. 22–28). Note that the temperature of the condensing or boiling fluid remains constant in this case. The effectiveness is the *lowest* in the other limiting case of $c = C_{min}/C_{max} = 1$, which is realized when the heat capacity rates of the two fluids are equal.

FIGURE 22–27
For a specified NTU and capacity ratio c, the counter-flow heat exchanger has the highest effectiveness and the parallel-flow the lowest.

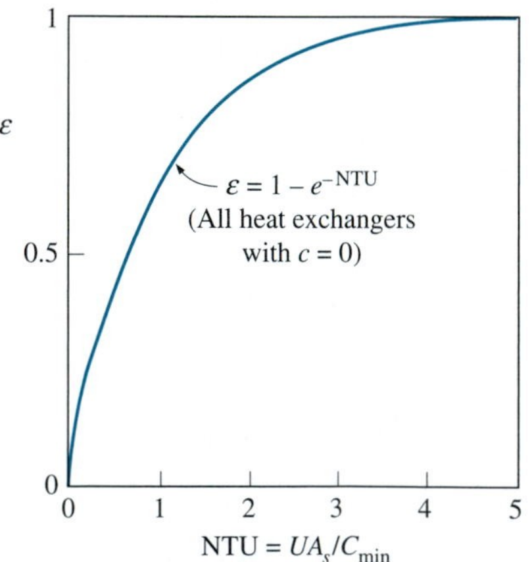

FIGURE 22–28
The effectiveness relation reduces to $\varepsilon = \varepsilon_{max} = 1 - \exp(-\text{NTU})$ for all heat exchangers when the capacity ratio $c = 0$.

Once the quantities $c = C_{min}/C_{max}$ and NTU $= UA_s/C_{min}$ have been evaluated, the effectiveness ε can be determined from either the charts or the effectiveness relation for the specified type of heat exchanger. Then the rate of heat transfer $\dot{Q}$ and the outlet temperatures $T_{h,\text{ out}}$ and $T_{c,\text{ out}}$ can be determined from Eqs. 22–33 and 22–30, respectively. Note that the analysis of heat exchangers with unknown outlet temperatures is a straightforward matter with the effectiveness–NTU method but requires rather tedious iterations with the LMTD method.

We mentioned earlier that when all the inlet and outlet temperatures are specified, the *size* of the heat exchanger can easily be determined using the LMTD method. Alternatively, it can also be determined from the effectiveness–NTU method by first evaluating the effectiveness ε from its definition (Eq. 22–29) and then the NTU from the appropriate NTU relation in Table 22–5.

TABLE 22–5

NTU relations for heat exchangers: NTU $= UA_s/C_{min}$ and $c = C_{min}/C_{max} = (\dot{m}c_p)_{min}/(\dot{m}c_p)_{max}$

Heat exchanger type	NTU relation
1 *Double-pipe:* Parallel-flow	$\text{NTU} = -\dfrac{\ln[1-\varepsilon(1+c)]}{1+c}$
Counter-flow	$\text{NTU} = \dfrac{1}{c-1}\ln\left(\dfrac{\varepsilon-1}{\varepsilon c-1}\right)$
2 *Shell and tube:* One-shell pass 2, 4, . . . tube passes	$\text{NTU} = -\dfrac{1}{\sqrt{1+c^2}}\ln\left(\dfrac{2/\varepsilon-1-c-\sqrt{1+c^2}}{2/\varepsilon-1-c+\sqrt{1+c^2}}\right)$
3 *Cross-flow (single-pass):* C_{max} mixed, C_{min} unmixed	$\text{NTU} = -\ln\left[1+\dfrac{\ln(1-\varepsilon c)}{c}\right]$
C_{min} mixed, C_{max} unmixed	$\text{NTU} = -\dfrac{\ln[c\ln(1-\varepsilon)+1]}{c}$
4 *All heat exchangers with $c = 0$*	$\text{NTU} = -\ln(1-\varepsilon)$

From W. M. Kays and A. L. London. *Compact Heat Exchangers, 3/e.* McGraw-Hill, 1984. Reprinted by permission of William M. Kays.

Note that the relations in Table 22–5 are equivalent to those in Table 22–4. Both sets of relations are given for convenience. The relations in Table 22–4 give the effectiveness directly when NTU is known, and the relations in Table 22–5 give the NTU directly when the effectiveness ε is known.

EXAMPLE 22–8 Using the Effectiveness–NTU Method

Repeat Example 22–4, which was solved with the LMTD method, using the effectiveness–NTU method.

Solution The schematic of the heat exchanger is redrawn in Fig. 22–29, and the same assumptions are utilized.

Analysis In the effectiveness–NTU method, we first determine the heat capacity rates of the hot and cold fluids and identify the smaller one:

$$C_h = \dot{m}_h c_{ph} = (2\text{ kg/s})(4.31\text{ kJ/kg}\cdot{}^\circ\text{C}) = 8.62\text{ kW/}^\circ\text{C}$$
$$C_c = \dot{m}_c c_{pc} = (1.2\text{ kg/s})(4.18\text{ kJ/kg}\cdot{}^\circ\text{C}) = 5.02\text{ kW/}^\circ\text{C}$$

Therefore,

$$C_{min} = C_c = 5.02\text{ kW/}^\circ\text{C}$$

and

$$c = C_{min}/C_{max} = 5.02/8.62 = 0.582$$

Then the maximum heat transfer rate is determined from Eq. 22–32 to be

$$\begin{aligned}\dot{Q}_{max} &= C_{min}(T_{h,\text{ in}} - T_{c,\text{ in}})\\ &= (5.02\text{ kW/}^\circ\text{C})(160-20)^\circ\text{C}\\ &= 702.8\text{ kW}\end{aligned}$$

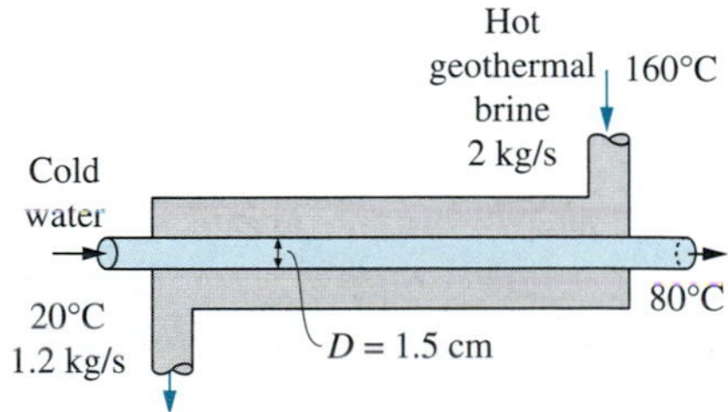

FIGURE 22–29
Schematic for Example 22–8.

That is, the maximum possible heat transfer rate in this heat exchanger is 702.8 kW. The actual rate of heat transfer is

$$\dot{Q} = [\dot{m}c_p(T_{\text{out}} - T_{\text{in}})]_{\text{water}} = (1.2\text{ kg/s})(4.18\text{ kJ/kg}\cdot°\text{C})(80 - 20)°\text{C} = 301.0\text{ kW}$$

Thus, the effectiveness of the heat exchanger is

$$\varepsilon = \frac{\dot{Q}}{\dot{Q}_{\max}} = \frac{301.0\text{kW}}{702.8\text{ kW}} = 0.428$$

Knowing the effectiveness, the NTU of this counter-flow heat exchanger can be determined from Fig. 22–26*b* or the appropriate relation from Table 22–5. We choose the latter approach for greater accuracy:

$$\text{NTU} = \frac{1}{c-1}\ln\left(\frac{\varepsilon - 1}{\varepsilon c - 1}\right) = \frac{1}{0.582 - 1}\ln\left(\frac{0.428 - 1}{0.428 \times 0.582 - 1}\right) = 0.651$$

Then the heat transfer surface area becomes

$$\text{NTU} = \frac{UA_s}{C_{\min}} \longrightarrow A_s = \frac{\text{NTU}\,C_{\min}}{U} = \frac{(0.651)(5020\text{ W/°C})}{640\text{ W/m}^2\cdot°\text{C}} = 5.11\text{ m}^2$$

To provide this much heat transfer surface area, the length of the tube must be

$$A_s = \pi DL \longrightarrow L = \frac{A_s}{\pi D} = \frac{5.11\text{ m}^2}{\pi(0.015\text{ m})} = \mathbf{108\text{ m}}$$

Discussion Note that we obtained practically the same result with the effectiveness–NTU method in a systematic and straightforward manner.

EXAMPLE 22–9 Cooling Hot Oil by Water in a Multipass Heat Exchanger

Hot oil is to be cooled by water in a 1-shell-pass and 8-tube-passes heat exchanger. The tubes are thin-walled and are made of copper with an internal diameter of 1.4 cm. The length of each tube pass in the heat exchanger is 5 m, and the overall heat transfer coefficient is 310 W/m^2 · °C. Water flows through the tubes at a rate of 0.2 kg/s, and the oil through the shell at a rate of 0.3 kg/s. The water and the oil enter at temperatures of 20°C and 150°C, respectively. Determine the rate of heat transfer in the heat exchanger and the outlet temperatures of the water and the oil.

Solution Hot oil is to be cooled by water in a heat exchanger. The mass flow rates and the inlet temperatures are given. The rate of heat transfer and the outlet temperatures are to be determined.

Assumptions **1** Steady operating conditions exist. **2** The heat exchanger is well insulated so that heat loss to the surroundings is negligible. **3** The thickness of the tube is negligible since it is thin-walled. **4** Changes in the kinetic and potential energies of fluid streams are negligible. **5** The overall heat transfer coefficient is constant and uniform.

Properties We take the specific heats of water and oil to be 4.18 and 2.13 kJ/kg · °C, respectively.

Analysis The schematic of the heat exchanger is given in Fig. 22–30. The outlet temperatures are not specified, and they cannot be determined from an energy balance. The use of the LMTD method in this case will involve tedious iterations, and thus the ε–NTU method is indicated. The first step in the ε–NTU method is to determine the heat capacity rates of the hot and cold fluids and identify the smaller one:

$$C_h = \dot{m}_h c_{ph} = (0.3 \text{ kg/s})(2.13 \text{ kJ/kg} \cdot °\text{C}) = 0.639 \text{ kW/°C}$$

$$C_c = \dot{m}_c c_{pc} = (0.2 \text{ kg/s})(4.18 \text{ kJ/kg} \cdot °\text{C}) = 0.836 \text{ kW/°C}$$

Therefore,

$$C_{\min} = C_h = 0.639 \text{ kW/°C} \quad \text{and} \quad c = \frac{C_{\min}}{C_{\max}} = \frac{0.639}{0.836} = 0.764$$

Then the maximum heat transfer rate is determined from Eq. 22–32 to be

$$\dot{Q}_{\max} = C_{\min}(T_{h,\text{ in}} - T_{c,\text{ in}}) = (0.639 \text{ kW/°C})(150 - 20)°\text{C} = 83.1 \text{ kW}$$

That is, the maximum possible heat transfer rate in this heat exchanger is 83.1 kW. The heat transfer surface area is

$$A_s = n(\pi DL) = 8\pi(0.014 \text{ m})(5 \text{ m}) = 1.76 \text{ m}^2$$

Then the NTU of this heat exchanger becomes

$$\text{NTU} = \frac{UA_s}{C_{\min}} = \frac{(310 \text{ W/m}^2 \cdot °\text{C})(1.76 \text{ m}^2)}{639 \text{ W/°C}} = 0.854$$

The effectiveness of this heat exchanger corresponding to $c = 0.764$ and NTU = 0.854 is determined from Fig. 22–26*c* to be

$$\varepsilon = 0.47$$

We could also determine the effectiveness from the third relation in Table 22–4 more accurately but with more labor. Then the actual rate of heat transfer becomes

$$\dot{Q} = \varepsilon \dot{Q}_{\max} = (0.47)(83.1 \text{ kW}) = \mathbf{39.1 \text{ kW}}$$

Finally, the outlet temperatures of the cold and the hot fluid streams are determined to be

$$\dot{Q} = C_c(T_{c,\text{ out}} - T_{c,\text{ in}}) \longrightarrow T_{c,\text{ out}} = T_{c,\text{ in}} + \frac{\dot{Q}}{C_c}$$

$$= 20°\text{C} + \frac{39.1 \text{ kW}}{0.836 \text{ kW/°C}} = \mathbf{66.8°C}$$

$$\dot{Q} = C_h(T_{h,\text{ in}} - T_{h,\text{ out}}) \longrightarrow T_{h,\text{ out}} = T_{h,\text{ in}} - \frac{\dot{Q}}{C_h}$$

$$= 150°\text{C} - \frac{39.1 \text{ kW}}{0.639 \text{ kW/°C}} = \mathbf{88.8°C}$$

Therefore, the temperature of the cooling water will rise from 20°C to 66.8°C as it cools the hot oil from 150°C to 88.8°C in this heat exchanger.

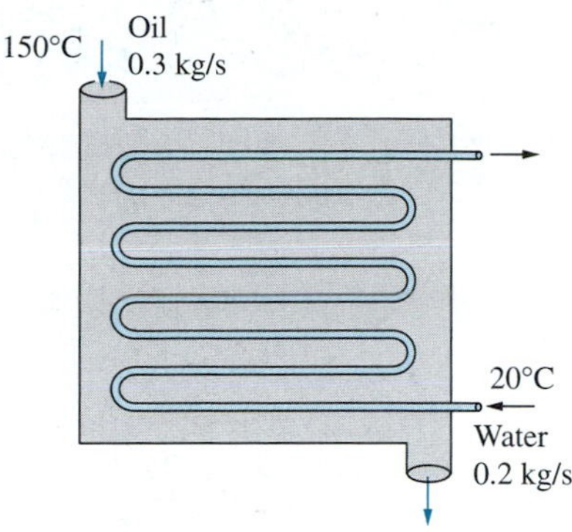

FIGURE 22–30
Schematic for Example 22–9.

22–6 ▪ SELECTION OF HEAT EXCHANGERS

Heat exchangers are complicated devices, and the results obtained with the simplified approaches presented above should be used with care. For example, we assumed that the overall heat transfer coefficient U is constant throughout the heat exchanger and that the convection heat transfer coefficients can be predicted using the convection correlations. However, it should be kept in mind that the uncertainty in the predicted value of U can exceed 30 percent. Thus, it is natural to tend to overdesign the heat exchangers in order to avoid unpleasant surprises.

Heat transfer enhancement in heat exchangers is usually accompanied by *increased pressure drop,* and thus *higher pumping power.* Therefore, any gain from the enhancement in heat transfer should be weighed against the cost of the accompanying pressure drop. Also, some thought should be given to which fluid should pass through the tube side and which through the shell side. Usually, the *more viscous fluid is more suitable for the shell side* (larger passage area and thus lower pressure drop) and *the fluid with the higher pressure for the tube side.*

Engineers in industry often find themselves in a position to select heat exchangers to accomplish certain heat transfer tasks. Usually, the goal is to heat or cool a certain fluid at a known mass flow rate and temperature to a desired temperature. Thus, the *rate of heat transfer* in the prospective heat exchanger is

$$\dot{Q} = \dot{m}c_p(T_{\text{in}} - T_{\text{out}})$$

which gives the heat transfer requirement of the heat exchanger before having any idea about the heat exchanger itself.

An engineer going through catalogs of heat exchanger manufacturers will be overwhelmed by the type and number of readily available off-the-shelf heat exchangers. The proper selection depends on several factors.

Heat Transfer Rate

This is the most important quantity in the selection of a heat exchanger. A heat exchanger should be capable of transferring heat at the specified rate in order to achieve the desired temperature change of the fluid at the specified mass flow rate.

Cost

Budgetary limitations usually play an important role in the selection of heat exchangers, except for some specialized cases where "money is no object." An off-the-shelf heat exchanger has a definite cost advantage over those made to order. However, in some cases, none of the existing heat exchangers will do, and it may be necessary to undertake the expensive and time-consuming task of designing and manufacturing a heat exchanger from scratch to suit the needs. This is often the case when the heat exchanger is an integral part of the overall device to be manufactured.

The operation and maintenance costs of the heat exchanger are also important considerations in assessing the overall cost.

Pumping Power

In a heat exchanger, both fluids are usually forced to flow by pumps or fans that consume electrical power. The annual cost of electricity associated with the operation of the pumps and fans can be determined from

$$\text{Operating cost} = (\text{Pumping power, kW}) \times (\text{Hours of operation, h}) \times (\text{Unit cost of electricity, \$/kWh})$$

where the pumping power is the total electrical power consumed by the motors of the pumps and fans. For example, a heat exchanger that involves a 1-hp pump and a $\frac{1}{3}$-hp fan (1 hp = 0.746 kW) operating at full load 8 h a day and 5 days a week will consume 2069 kWh of electricity per year, which will cost $166 at an electricity cost of 8 cents/kWh.

Minimizing the pressure drop and the mass flow rate of the fluids will minimize the operating cost of the heat exchanger, but it will maximize the size of the heat exchanger and thus the initial cost. As a rule of thumb, doubling the mass flow rate will reduce the initial cost by *half* but will increase the pumping power requirements by a factor of roughly *eight.*

Typically, fluid velocities encountered in heat exchangers range between 0.7 and 7 m/s for liquids and between 3 and 30 m/s for gases. Low velocities are helpful in avoiding erosion, tube vibrations, and noise as well as pressure drop.

Size and Weight

Normally, the *smaller* and the *lighter* the heat exchanger, the better it is. This is especially the case in the *automotive* and *aerospace* industries, where size and weight requirements are most stringent. Also, a larger heat exchanger normally carries a higher price tag. The space available for the heat exchanger in some cases limits the length of the tubes that can be used.

Type

The type of heat exchanger to be selected depends primarily on the type of *fluids* involved, the *size* and *weight* limitations, and the presence of any *phase-change* processes. For example, a heat exchanger is suitable to cool a liquid by a gas if the surface area on the gas side is many times that on the liquid side. On the other hand, a plate or shell-and-tube heat exchanger is very suitable for cooling a liquid by another liquid.

Materials

The materials used in the construction of the heat exchanger may be an important consideration in the selection of heat exchangers. For example, the thermal and structural *stress effects* need not be considered at pressures below 15 atm or temperatures below 150°C. But these effects are major considerations above 70 atm or 550°C and seriously limit the acceptable materials of the heat exchanger.

A temperature difference of 50°C or more between the tubes and the shell will probably pose *differential thermal expansion* problems and needs to be considered. In the case of corrosive fluids, we may have to select expensive *corrosion-resistant* materials such as stainless steel or even titanium if we are not willing to replace low-cost heat exchangers frequently.

Other Considerations

There are other considerations in the selection of heat exchangers that may or may not be important, depending on the application. For example, being *leak-tight* is an important consideration when *toxic* or *expensive* fluids are involved. Ease of servicing, low maintenance cost, and safety and reliability are some other important considerations in the selection process. Quietness is one of the primary considerations in the selection of liquid-to-air heat exchangers used in heating and air-conditioning applications.

EXAMPLE 22–10 Installing a Heat Exchanger to Save Energy and Money

In a dairy plant, milk is pasteurized by hot water supplied by a natural gas furnace. The hot water is then discharged to an open floor drain at 80°C at a rate of 15 kg/min. The plant operates 24 h a day and 365 days a year. The furnace has an efficiency of 80 percent, and the cost of the natural gas is \$1.10 per therm (1 therm = 105,500 kJ). The average temperature of the cold water entering the furnace throughout the year is 15°C. The drained hot water cannot be returned to the furnace and recirculated, because it is contaminated during the process.

In order to save energy, installation of a water-to-water heat exchanger to preheat the incoming cold water by the drained hot water is proposed. Assuming that the heat exchanger will recover 75 percent of the available heat in the hot water, determine the heat transfer rating of the heat exchanger that needs to be purchased and suggest a suitable type. Also, determine the amount of money this heat exchanger will save the company per year from natural gas savings.

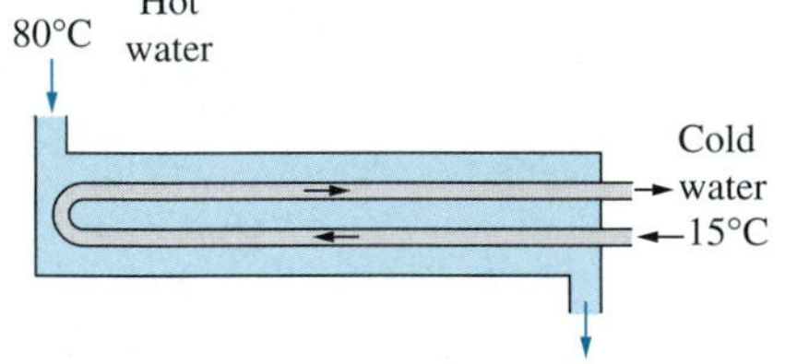

FIGURE 22–31
Schematic for Example 22–10.

Solution A water-to-water heat exchanger is to be installed to transfer energy from drained hot water to the incoming cold water to preheat it. The rate of heat transfer in the heat exchanger and the amounts of energy and money saved per year are to be determined.

Assumptions **1** Steady operating conditions exist. **2** The effectiveness of the heat exchanger remains constant.

Properties We use the specific heat of water at room temperature, c_p = 4.18 kJ/kg · °C, and treat it as a constant.

Analysis A schematic of the prospective heat exchanger is given in Fig. 22–31. The heat recovery from the hot water will be a maximum when it leaves the heat exchanger at the inlet temperature of the cold water. Therefore,

$$\begin{aligned}\dot{Q}_{\max} &= \dot{m}_h c_p (T_{h,\,\text{in}} - T_{c,\,\text{in}})\\ &= \left(\frac{15}{60}\ \text{kg/s}\right)(4.18\ \text{kJ/kg}\cdot{}^\circ\text{C})(80 - 15)^\circ\text{C}\\ &= 67.9\ \text{kJ/s}\end{aligned}$$

That is, the existing hot-water stream has the potential to supply heat at a rate of 67.9 kJ/s to the incoming cold water. This value would be approached in a counter-flow heat exchanger with a *very large* heat transfer surface area. A heat exchanger of reasonable size and cost can capture 75 percent of this heat transfer potential. Thus, the heat transfer rating of the prospective heat exchanger must be

$$\dot{Q} = \varepsilon \dot{Q}_{max} = (0.75)(67.9 \text{ kJ/s}) = \mathbf{50.9 \text{ kJ/s}}$$

That is, the heat exchanger should be able to deliver heat at a rate of 50.9 kJ/s from the hot to the cold water. An ordinary plate or *shell-and-tube* heat exchanger should be adequate for this purpose, since both sides of the heat exchanger involve the same fluid at comparable flow rates and thus comparable heat transfer coefficients. (Note that if we were heating air with hot water, we would have to specify a heat exchanger that has a large surface area on the air side.)

The heat exchanger will operate 24 h a day and 365 days a year. Therefore, the annual operating hours are

$$\text{Operating hours} = (24 \text{ h/day})(365 \text{ days/year}) = 8760 \text{ h/year}$$

Noting that this heat exchanger saves 50.9 kJ of energy per second, the energy saved during an entire year will be

$$\begin{aligned}\text{Energy saved} &= (\text{Heat transfer rate})(\text{Operation time})\\ &= (50.9 \text{ kJ/s})(8760 \text{ h/year})(3600 \text{ s/h})\\ &= 1.605 \times 10^9 \text{ kJ/year}\end{aligned}$$

The furnace is said to be 80 percent efficient. That is, for each 80 units of heat supplied by the furnace, natural gas with an energy content of 100 units must be supplied to the furnace. Therefore, the energy savings determined above result in fuel savings in the amount of

$$\begin{aligned}\text{Fuel saved} &= \frac{\text{Energy saved}}{\text{Furnace efficiency}} = \frac{1.605 \times 10^9 \text{ kJ/year}}{0.80}\left(\frac{1 \text{ therm}}{105{,}500 \text{ kJ}}\right)\\ &= 19{,}020 \text{ therms/year}\end{aligned}$$

Noting that the price of natural gas is $1.10 per therm, the amount of money saved becomes

$$\begin{aligned}\text{Money saved} &= (\text{Fuel saved}) \times (\text{Price of fuel})\\ &= (19{,}020 \text{ therms/year})(\$1.10/\text{therm})\\ &= \mathbf{\$20{,}920/\text{year}}\end{aligned}$$

Therefore, the installation of the proposed heat exchanger will save the company $20,920 a year, and the installation cost of the heat exchanger will probably be paid from the fuel savings in a short time.

SUMMARY

Heat exchangers are devices that allow the exchange of heat between two fluids without allowing them to mix with each other. Heat exchangers are manufactured in a variety of types, the simplest being the *double-pipe* heat exchanger. In a *parallel-flow* type, both the hot and cold fluids enter the heat exchanger at the same end and move in the same direction, whereas in a *counter-flow* type, the hot and cold fluids enter the heat exchanger at opposite ends and flow in opposite directions. In *compact* heat exchangers, the two fluids move perpendicular to each other, and such a flow configuration is called *cross-flow.* Other common types of heat exchangers in industrial applications are the *plate* and the *shell-and-tube* heat exchangers.

Heat transfer in a heat exchanger usually involves convection in each fluid and conduction through the wall separating the two fluids. In the analysis of heat exchangers, it is convenient to work with an *overall heat transfer coefficient U* or a *total thermal resistance R*, expressed as

$$\frac{1}{UA_s} = \frac{1}{U_iA_i} = \frac{1}{U_oA_o} = R = \frac{1}{h_iA_i} + R_{\text{wall}} + \frac{1}{h_oA_o}$$

where the subscripts i and o stand for the inner and outer surfaces of the wall that separates the two fluids, respectively. When the wall thickness of the tube is small and the thermal conductivity of the tube material is high, the relation simplifies to

$$\frac{1}{U} \approx \frac{1}{h_i} + \frac{1}{h_o}$$

where $U \approx U_i \approx U_o$. The effects of fouling on both the inner and the outer surfaces of the tubes of a heat exchanger can be accounted for by

$$\frac{1}{UA_s} = \frac{1}{U_iA_i} = \frac{1}{U_oA_o} = R$$
$$= \frac{1}{h_iA_i} + \frac{R_{f,i}}{A_i} + \frac{\ln(D_o/D_i)}{2\pi kL} + \frac{R_{f,o}}{A_o} + \frac{1}{h_oA_o}$$

where $A_i = \pi D_i L$ and $A_o = \pi D_o L$ are the areas of the inner and outer surfaces and $R_{f,i}$ and $R_{f,o}$ are the fouling factors at those surfaces.

In a well-insulated heat exchanger, the rate of heat transfer from the hot fluid is equal to the rate of heat transfer to the cold one. That is,

$$\dot{Q} = \dot{m}_c c_{pc}(T_{c,\text{out}} - T_{c,\text{in}}) = C_c(T_{c,\text{out}} - T_{c,\text{in}})$$

and

$$\dot{Q} = \dot{m}_h c_{ph}(T_{h,\text{in}} - T_{h,\text{out}}) = C_h(T_{h,\text{in}} - T_{h,\text{out}})$$

where the subscripts c and h stand for the cold and hot fluids, respectively, and the product of the mass flow rate and the specific heat of a fluid $\dot{m}c_p$ is called the *heat capacity rate.*

Of the two methods used in the analysis of heat exchangers, the *log mean temperature difference* (or LMTD) method is best suited for determining the size of a heat exchanger when all the inlet and the outlet temperatures are known. The *effectiveness–NTU* method is best suited to predict the outlet temperatures of the hot and cold fluid streams in a specified heat exchanger. In the LMTD method, the rate of heat transfer is determined from

$$\dot{Q} = UA_s\,\Delta T_{\text{lm}}$$

where

$$\Delta T_{\text{lm}} = \frac{\Delta T_1 - \Delta T_2}{\ln(\Delta T_1/\Delta T_2)}$$

is the *log mean temperature difference,* which is the suitable form of the average temperature difference for use in the analysis of heat exchangers. Here ΔT_1 and ΔT_2 represent the temperature differences between the two fluids at the two ends (inlet and outlet) of the heat exchanger. For cross-flow and multipass shell-and-tube heat exchangers, the logarithmic mean temperature difference is related to the counter-flow one $\Delta T_{\text{lm, CF}}$ as

$$\Delta T_{\text{lm}} = F\,\Delta T_{\text{lm, CF}}$$

where F is the *correction factor,* which depends on the geometry of the heat exchanger and the inlet and outlet temperatures of the hot and cold fluid streams.

The *effectiveness* of a heat exchanger is defined as

$$\varepsilon = \frac{\dot{Q}}{\dot{Q}_{\text{max}}} = \frac{\text{Actual heat transfer rate}}{\text{Maximum possible heat transfer rate}}$$

where

$$\dot{Q}_{\text{max}} = C_{\text{min}}(T_{h,\text{in}} - T_{c,\text{in}})$$

and C_{min} is the smaller of $C_h = \dot{m}_h c_{ph}$ and $C_c = \dot{m}_c c_{pc}$. The effectiveness of heat exchangers can be determined from effectiveness relations or charts.

The selection or design of a heat exchanger depends on several factors such as the heat transfer rate, cost, pressure drop, size, weight, construction type, materials, and operating environment.

REFERENCES AND SUGGESTED READINGS

1. N. Afgan and E. U. Schlunder. *Heat Exchanger: Design and Theory Sourcebook.* Washington, DC: McGraw-Hill/Scripta, 1974.
2. R. A. Bowman, A. C. Mueller, and W. M. Nagle. "Mean Temperature Difference in Design." *Trans. ASME* 62, 1940, p. 283. Reprined with permission of ASME International.
3. A. P. Fraas. *Heat Exchanger Design.* 2d ed. New York: John Wiley & Sons, 1989.
4. K. A. Gardner. "Variable Heat Transfer Rate Correction in Multipass Exchangers, Shell Side Film Controlling." *Transactions of the ASME* 67 (1945), pp. 31–38.
5. W. M. Kays and A. L. London. *Compact Heat Exchangers.* 3rd ed. New York: McGraw-Hill, 1984.

6. W. M. Kays and H. C. Perkins. In *Handbook of Heat Transfer*, ed. W. M. Rohsenow and J. P. Hartnett. New York: McGraw-Hill, 1972, Chap. 7.

7. A. C. Mueller. "Heat Exchangers." In *Handbook of Heat Transfer*, ed. W. M. Rohsenow and J. P. Hartnett. New York: McGraw-Hill, 1972, Chap. 18.

8. M. N. Özişik. *Heat Transfer—A Basic Approach.* New York: McGraw-Hill, 1985.

9. E. U. Schlunder. *Heat Exchanger Design Handbook.* Washington, DC: Hemisphere, 1982.

10. *Standards of Tubular Exchanger Manufacturers Association.* New York: Tubular Exchanger Manufacturers Association, latest ed.

11. R. A. Stevens, J. Fernandes, and J. R. Woolf. "Mean Temperature Difference in One, Two, and Three Pass Crossflow Heat Exchangers." *Transactions of the ASME* 79 (1957), pp. 287–297.

12. J. Taborek, G. F. Hewitt, and N. Afgan. *Heat Exchangers: Theory and Practice.* New York: Hemisphere, 1983.

13. G. Walker. *Industrial Heat Exchangers.* Washington, DC: Hemisphere, 1982.

PROBLEMS*

Types of Heat Exchangers

22–1C Classify heat exchangers according to flow type and explain the characteristics of each type.

22–2C Classify heat exchangers according to construction type and explain the characteristics of each type.

22–3C When is a heat exchanger classified as being compact? Do you think a double-pipe heat exchanger can be classified as a compact heat exchanger?

22–4C How does a cross-flow heat exchanger differ from a counter-flow one? What is the difference between mixed and unmixed fluids in cross-flow?

22–5C What is the role of the baffles in a shell-and-tube heat exchanger? How does the presence of baffles affect the heat transfer and the pumping power requirements? Explain.

22–6C Draw a 1-shell-pass and 6-tube-passes shell-and-tube heat exchanger. What are the advantages and disadvantages of using 6 tube passes instead of just 2 of the same diameter?

22–7C Draw a 2-shell-passes and 8-tube-passes shell-and-tube heat exchanger. What is the primary reason for using so many tube passes?

22–8C What is a regenerative heat exchanger? How does a static type of regenerative heat exchanger differ from a dynamic type?

* Problems designated by a "C" are concept questions, and students are encouraged to answer them all. Problems designated by an "E" are in English units, and the SI users can ignore them. Problems with the EES icon are solved using EES, and complete solutions together with parametric studies are included on the enclosed DVD. Problems with the computer icon are comprehensive in nature and are intended to be solved with a computer, preferably using the EES software that accompanies this text.

The Overall Heat Transfer Coefficient

22–9C What are the heat transfer mechanisms involved during heat transfer from the hot to the cold fluid?

22–10C Under what conditions is the thermal resistance of the tube in a heat exchanger negligible?

22–11C Consider a double-pipe parallel-flow heat exchanger of length L. The inner and outer diameters of the inner tube are D_1 and D_2, respectively, and the inner diameter of the outer tube is D_3. Explain how you would determine the two heat transfer surface areas A_i and A_o. When is it reasonable to assume $A_i \approx A_o \approx A_s$?

22–12C Is the approximation $h_i \approx h_o \approx h$ for the convection heat transfer coefficient in a heat exchanger a reasonable one when the thickness of the tube wall is negligible?

22–13C Under what conditions can the overall heat transfer coefficient of a heat exchanger be determined from $U = (1/h_i + 1/h_o)^{-1}$?

22–14C What are the restrictions on the relation $UA_s = U_iA_i = U_oA_o$ for a heat exchanger? Here A_s is the heat transfer surface area and U is the overall heat transfer coefficient.

22–15C In a thin-walled double-pipe heat exchanger, when is the approximation $U = h_i$ a reasonable one? Here U is the overall heat transfer coefficient and h_i is the convection heat transfer coefficient inside the tube.

22–16C What are the common causes of fouling in a heat exchanger? How does fouling affect heat transfer and pressure drop?

22–17C How is the thermal resistance due to fouling in a heat exchanger accounted for? How do the fluid velocity and temperature affect fouling?

22–18 A double-pipe heat exchanger is constructed of a copper ($k = 380$ W/m · °C) inner tube of internal diameter $D_i = 1.2$ cm and external diameter $D_o = 1.6$ cm and an outer tube of diameter 3.0 cm. The convection heat transfer coefficient is reported to be $h_i = 700\ \text{W/m}^2 \cdot °\text{C}$ on the inner surface of the tube and $h_o = 1400\ \text{W/m}^2 \cdot °\text{C}$ on its outer surface. For a fouling factor $R_{f,i} = 0.0005\ \text{m}^2 \cdot °\text{C/W}$ on the tube side and $R_{f,o} = 0.0002\ \text{m}^2 \cdot °\text{C/W}$ on the shell side, determine (*a*) the thermal resistance of the heat exchanger per unit length and (*b*) the overall heat transfer coefficients U_i and U_o based on the inner and outer surface areas of the tube, respectively.

22–19 Reconsider Prob. 22–18. Using EES (or other) software, investigate the effects of pipe conductivity and heat transfer coefficients on the thermal resistance of the heat exchanger. Let the thermal conductivity vary from 10 W/m · °C to 400 W/m · °C, the convection heat transfer coefficient from 500 W/m² · °C to 1500 W/m² · °C on the inner surface, and from 1000 W/m² · °C to 2000 W/m² · °C on the outer surface. Plot the thermal resistance of the heat exchanger as functions of thermal conductivity and heat transfer coefficients, and discuss the results.

22–20 A jacketted-agitated vessel, fitted with a turbine agitator, is used for heating a water stream from 10°C to 54°C. The average heat transfer coefficient for water at the vessel's inner-wall can be estimated from $\text{Nu} = 0.76\text{Re}^{2/3}\text{Pr}^{1/3}$. Saturated steam at 100°C condenses in the jacket, for which the average heat transfer coefficient in kW/m² · K is: $h_o = 13.1(T_g - T_w)^{-0.25}$. The vessel dimensions are: $D_t = 0.6$ m, $H = 0.6$ m and $D_a = 0.2$ m. The agitator speed is 60 rpm. Calculate the mass rate of water that can be heated in this agitated vessel steadily.

22–21 Water at an average temperature of 110°C and an average velocity of 3.5 m/s flows through a 5-m-long stainless steel tube ($k = 14.2$ W/m · °C) in a boiler. The inner and outer diameters of the tube are $D_i = 1.0$ cm and $D_o = 1.4$ cm, respectively. If the convection heat transfer coefficient at the outer surface of the tube where boiling is taking place is $h_o = 8400\ \text{W/m}^2 \cdot °\text{C}$, determine the overall heat transfer coefficient U_i of this boiler based on the inner surface area of the tube.

22–22 Repeat Prob. 22–21, assuming a fouling factor $R_{f,i} = 0.0005\ \text{m}^2 \cdot °\text{C/W}$ on the inner surface of the tube.

22–23 Reconsider Prob. 22–21. Using EES (or other) software, plot the overall heat transfer coefficient based on the inner surface as a function of fouling factor as it varies from 0.0001 m² · °C/W to 0.0008 m² · °C/W, and discuss the results.

22–24 A long thin-walled double-pipe heat exchanger with tube and shell diameters of 1.0 cm and 2.5 cm, respectively, is used to condense refrigerant-134a by water at 20°C. The refrigerant flows through the tube, with a convection heat transfer coefficient of $h_i = 5000\ \text{W/m}^2 \cdot °\text{C}$. Water flows through the shell at a rate of 0.3 kg/s. Determine the overall heat transfer coefficient of this heat exchanger. *Answer:* 2020 W/m² · °C

22–25 Repeat Prob. 22–24 by assuming a 2-mm-thick layer of limestone ($k = 1.3$ W/m · °C) forms on the outer surface of the inner tube.

22–26 Reconsider Prob. 22–25. Using EES (or other) software, plot the overall heat transfer coefficient as a function of the limestone thickness as it varies from 1 mm to 3 mm, and discuss the results.

22–27E Water at an average temperature of 180°F and an average velocity of 4 ft/s flows through a thin-walled $\frac{3}{4}$-in-diameter tube. The water is cooled by air that flows across the tube with a velocity of 12 ft/s at an average temperature of 80°F. Determine the overall heat transfer coefficient.

Analysis of Heat Exchangers

22–28C What are the common approximations made in the analysis of heat exchangers?

22–29C Under what conditions is the heat transfer relation

$$\dot{Q} = \dot{m}_c c_{pc}(T_{c,\text{ out}} - T_{c,\text{ in}}) = \dot{m}_h c_{ph}\,(T_{h,\text{ in}} - T_{h,\text{ out}})$$

valid for a heat exchanger?

22–30C What is the heat capacity rate? What can you say about the temperature changes of the hot and cold fluids in a heat exchanger if both fluids have the same capacity rate? What does a heat capacity of infinity for a fluid in a heat exchanger mean?

22–31C Consider a condenser in which steam at a specified temperature is condensed by rejecting heat to the cooling water. If the heat transfer rate in the condenser and the temperature rise of the cooling water is known, explain how the rate of condensation of the steam and the mass flow rate of the cooling water can be determined. Also, explain how the total thermal resistance R of this condenser can be evaluated in this case.

22–32C Under what conditions will the temperature rise of the cold fluid in a heat exchanger be equal to the temperature drop of the hot fluid?

The Log Mean Temperature Difference Method

22–33C In the heat transfer relation $\dot{Q} = UA_s\,\Delta T_{\text{lm}}$ for a heat exchanger, what is ΔT_{lm} called? How is it calculated for a parallel-flow and counter-flow heat exchanger?

22–34C How does the log mean temperature difference for a heat exchanger differ from the arithmetic mean temperature difference? For specified inlet and outlet temperatures, which one of these two quantities is larger?

22–35C The temperature difference between the hot and cold fluids in a heat exchanger is given to be ΔT_1 at one end and ΔT_2 at the other end. Can the logarithmic temperature difference ΔT_{lm} of this heat exchanger be greater than both ΔT_1 and ΔT_2? Explain.

22–36C Can the logarithmic mean temperature difference ΔT_{lm} of a heat exchanger be a negative quantity? Explain.

22–37C Can the outlet temperature of the cold fluid in a heat exchanger be higher than the outlet temperature of the hot fluid in a parallel-flow heat exchanger? How about in a counter-flow heat exchanger? Explain.

22–38C For specified inlet and outlet temperatures, for what kind of heat exchanger will the ΔT_{lm} be greatest: double-pipe parallel-flow, double-pipe counter-flow, cross-flow, or multipass shell-and-tube heat exchanger?

22–39C In the heat transfer relation $\dot{Q} = UA_s F\Delta T_{\text{lm}}$ for a heat exchanger, what is the quantity F called? What does it represent? Can F be greater than one?

22–40C When the outlet temperatures of the fluids in a heat exchanger are not known, is it still practical to use the LMTD method? Explain.

22–41C Explain how the LMTD method can be used to determine the heat transfer surface area of a multipass shell-and-tube heat exchanger when all the necessary information, including the outlet temperatures, is given.

22–42 Ethylene glycol is heated from 20°C to 40°C at a rate of 1.0 kg/s in a horizontal copper tube (k = 386 W/m · K) with an inner diameter of 2.0 cm and an outer diameter of 2.5 cm. A saturated vapor (T_g = 110°C) condenses on the outside-tube surface with the heat transfer coefficient (in kW/m^2 · K) given by $9.2/(T_g - T_w)^{0.25}$, where T_w is the average outside-tube wall temperature. What tube length must be used? Take the properties of ethylene glycol to be ρ = 1109 kg/m^3, c_p = 2428 kj/kg · K, k = 0.253 W/m · °C, μ = 0.01545 kg/m · s, and Pr = 148.5.

22–43 A double-pipe parallel-flow heat exchanger is used to heat cold tap water with hot water. Hot water (c_p = 4.25 kJ/kg · °C) enters the tube at 85°C at a rate of 1.4 kg/s and leaves at 50°C. The heat exchanger is not well insulated, and it is estimated that 3 percent of the heat given up by the hot fluid is lost from the heat exchanger. If the overall heat transfer coefficient and the surface area of the heat exchanger are 1150 W/m^2 · °C and 4 m^2, respectively, determine the rate of heat transfer to the cold water and the log mean temperature difference for this heat exchanger.

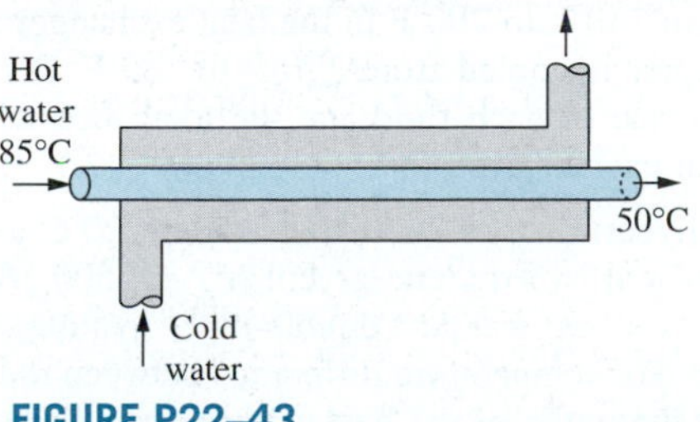

FIGURE P22–43

22–44 A stream of hydrocarbon (c_p = 2.2 kJ/kg · K) is cooled at a rate of 720 kg/h from 150°C to 40°C in the tube side of a double-pipe counter-flow heat exchanger. Water (c_p = 4.18 kJ/kg · K) enters the heat exchanger at 10°C at a rate of 540 kg/h. The outside diameter of the inner tube is 2.5 cm, and its length is 6.0 m. Calculate the overall heat transfer coefficient.

22–45 A shell-and-tube heat exchanger is used for heating 10 kg/s of oil (c_p = 2.0 kJ/kg · K) from 25°C to 46°C. The heat exchanger has 1-shell pass and 6-tube passes. Water enters the shell side at 80°C and leaves at 60°C. The overall heat transfer coefficient is estimated to be 1000 W/m^2 · K. Calculate the rate of heat transfer and the heat transfer area.

22–46 Steam in the condenser of a steam power plant is to be condensed at a temperature of 50°C (h_{fg} = 2383 kJ/kg) with cooling water (c_p = 4180 J/kg · °C) from a nearby lake, which enters the tubes of the condenser at 18°C and leaves at 27°C. The surface area of the tubes is 42 m^2, and the overall heat transfer coefficient is 2400 W/m^2 · °C. Determine the mass flow rate of the cooling water needed and the rate of condensation of the steam in the condenser. *Answers: 73.1 kg/s, 1.15 kg/s*

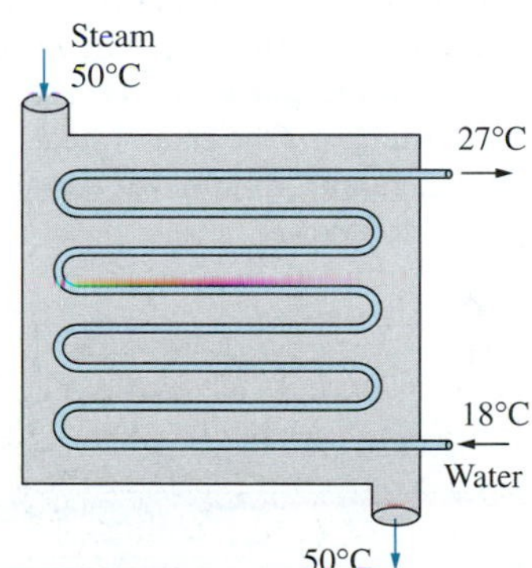

FIGURE P22–46

22–47 A double-pipe parallel-flow heat exchanger is to heat water (c_p = 4180 J/kg · °C) from 25°C to 60°C at a rate of 0.2 kg/s. The heating is to be accomplished by geothermal water (c_p = 4310 J/kg · °C) available at 140°C at a mass flow rate of 0.3 kg/s. The inner tube is thin-walled and has a diameter of 0.8 cm. If the overall heat transfer coefficient of the heat exchanger is 550 W/m^2 · °C, determine the length of the tube required to achieve the desired heating.

22–48 Reconsider Prob. 22–47. Using EES (or other) software, investigate the effects of temperature and mass flow rate of geothermal water on the length of the tube. Let the temperature vary from 100°C to 200°C, and the mass flow rate from 0.1 kg/s to 0.5 kg/s. Plot the length of the tube as functions of temperature and mass flow rate, and discuss the results.

22–49E A 1-shell-pass and 8-tube-passes heat exchanger is used to heat glycerin (c_p = 0.60 Btu/lbm · °F) from 65°F to 140°F by hot water (c_p = 1.0 Btu/lbm · °F) that enters the thin-walled 0.5-in-diameter tubes at 175°F and leaves at 120°F. The total length of the tubes in the heat exchanger is 500 ft. The convection heat transfer coefficient is 4 Btu/h · ft^2 · °F on the glycerin (shell) side and 50 Btu/h · ft^2 · °F on the water (tube) side. Determine the rate of heat transfer in the heat exchanger (*a*) before any fouling occurs and (*b*) after fouling with a fouling factor of 0.002 h · ft^2 · °F/Btu on the outer surfaces of the tubes.

22–50 A test is conducted to determine the overall heat transfer coefficient in a shell-and-tube oil-to-water heat exchanger that has 24 tubes of internal diameter 1.2 cm and length 2 m in a single shell. Cold water (c_p = 4180 J/kg · °C) enters the tubes at 20°C at a rate of 3 kg/s and leaves at 55°C. Oil (c_p = 2150 J/kg · °C) flows through the shell and is cooled from 120°C to 45°C. Determine the overall heat transfer coefficient U_i of this heat exchanger based on the inner surface area of the tubes. *Answer:* 8.31 kW/m^2 · °C

22–51 A double-pipe counter-flow heat exchanger is to cool ethylene glycol (c_p = 2560 J/kg · °C) flowing at a rate of 3.5 kg/s from 80°C to 40°C by water (c_p = 4180 J/kg · °C) that enters at 20°C and leaves at 55°C. The overall heat transfer coefficient based on the inner surface area of the tube is 250 W/m^2 · °C. Determine (*a*) the rate of heat transfer, (*b*) the mass flow rate of water, and (*c*) the heat transfer surface area on the inner side of the tube.

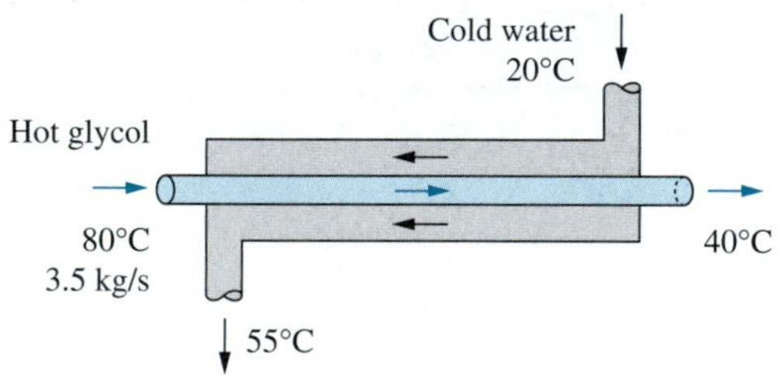

FIGURE P22–51

22–52 Water (c_p = 4180 J/kg · °C) enters the 2.5-cm-internal-diameter tube of a double-pipe counter-flow heat exchanger at 17°C at a rate of 3 kg/s. It is heated by steam condensing at 120°C (h_{fg} = 2203 kJ/kg) in the shell. If the overall heat transfer coefficient of the heat exchanger is 1500 W/m^2 · °C, determine the length of the tube required in order to heat the water to 80°C.

22–53 A thin-walled double-pipe counter-flow heat exchanger is to be used to cool oil (c_p = 2200 J/kg · °C) from 150°C to 40°C at a rate of 2 kg/s by water (c_p = 4180 J/kg · °C) that enters at 22°C at a rate of 1.5 kg/s. The diameter of the tube is 2.5 cm, and its length is 6 m. Determine the overall heat transfer coefficient of this heat exchanger.

22–54 EES Reconsider Prob. 22–53. Using EES (or other) software, investigate the effects of oil exit temperature and water inlet temperature on the overall heat transfer coefficient of the heat exchanger. Let the oil exit temperature vary from 30°C to 70°C and the water inlet temperature from 5°C to 25°C. Plot the overall heat transfer coefficient as functions of the two temperatures, and discuss the results.

22–55 Consider a water-to-water double-pipe heat exchanger whose flow arrangement is not known. The temperature measurements indicate that the cold water enters at 20°C and leaves at 50°C, while the hot water enters at 80°C and leaves at 45°C. Do you think this is a parallel-flow or counter-flow heat exchanger? Explain.

22–56 Cold water (c_p = 4180 J/kg · °C) leading to a shower enters a thin-walled double-pipe counter-flow heat exchanger at 15°C at a rate of 1.25 kg/s and is heated to 45°C by hot water (c_p = 4190 J/kg · °C) that enters at 100°C at a rate of 3 kg/s. If the overall heat transfer coefficient is 880 W/m^2 · °C, determine the rate of heat transfer and the heat transfer surface area of the heat exchanger.

22–57 Engine oil (c_p = 2100 J/kg · °C) is to be heated from 20°C to 60°C at a rate of 0.3 kg/s in a 2-cm-diameter thin-walled copper tube by condensing steam outside at a temperature of 130°C (h_{fg} = 2174 kJ/kg). For an overall heat transfer coefficient of 650 W/m^2 · °C, determine the rate of heat transfer and the length of the tube required to achieve it. *Answers:* 25.2 kW, 7.0 m

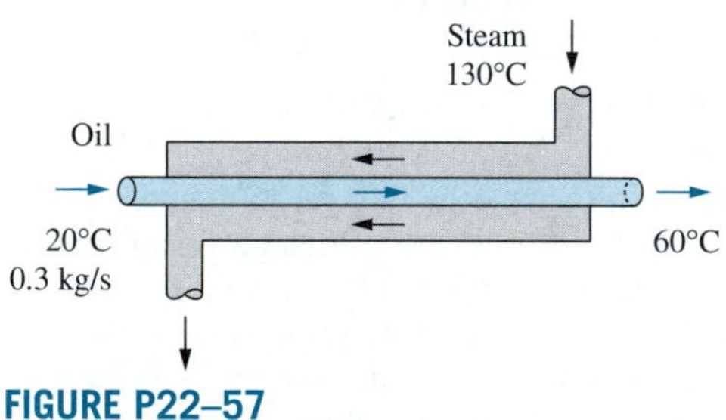

FIGURE P22–57

22–58E Geothermal water (c_p = 1.03 Btu/lbm · °F) is to be used as the heat source to supply heat to the hydronic heating system of a house at a rate of 40 Btu/s in a double-pipe counter-flow heat exchanger. Water (c_p = 1.0 Btu/lbm · °F) is heated from 140°F to 200°F in the heat exchanger as the geothermal water is cooled from 270°F to 180°F. Determine the mass flow rate of each fluid and the total thermal resistance of this heat exchanger.

22–59 Glycerin (c_p = 2400 J/kg · °C) at 20°C and 0.3 kg/s is to be heated by ethylene glycol (c_p = 2500 J/kg · °C) at 60°C in a thin-walled double-pipe parallel-flow heat exchanger. The temperature difference between the two fluids is 15°C at the outlet of the heat exchanger. If the overall heat transfer coefficient is 240 W/m^2 · °C and the heat transfer surface area is 3.2 m^2, determine (*a*) the rate of heat transfer, (*b*) the outlet temperature of the glycerin, and (*c*) the mass flow rate of the ethylene glycol.

22–60 Air (c_p = 1005 J/kg · °C) is to be preheated by hot exhaust gases in a cross-flow heat exchanger before it enters the furnace. Air enters the heat exchanger at 95 kPa and 20°C at a rate of 0.8 m^3/s. The combustion gases (c_p = 1100 J/kg · °C) enter at 180°C at a rate of 1.1 kg/s and leave at 95°C. The product of the overall heat transfer coefficient and the heat transfer surface area is UA_s = 1200 W/°C. Assuming both fluids to be unmixed, determine the rate of heat transfer and the outlet temperature of the air.

22–61 A shell-and-tube heat exchanger with 2-shell passes and 12-tube passes is used to heat water (c_p = 4180 J/kg · °C) in the tubes from 20°C to 70°C at a rate of 4.5 kg/s. Heat is supplied by hot oil (c_p = 2300 J/kg · °C) that enters the shell side at 170°C at a rate of 10 kg/s. For a tube-side overall heat transfer coefficient of 350 W/m^2 · °C, determine the heat transfer surface area on the tube side. *Answer:* 25.6 m^2

22–62 Repeat Prob. 22–61 for a mass flow rate of 2 kg/s for water.

22–63 A shell-and-tube heat exchanger with 2-shell passes and 8-tube passes is used to heat ethyl alcohol (c_p = 2670 J/kg · °C) in the tubes from 25°C to 70°C at a rate of 2.1 kg/s. The heating is to be done by water (c_p = 4190 J/kg · °C) that enters the shell side at 95°C and leaves at 45°C. If the overall heat transfer coefficient is 950 W/m^2 · °C, determine the heat transfer surface area of the heat exchanger.

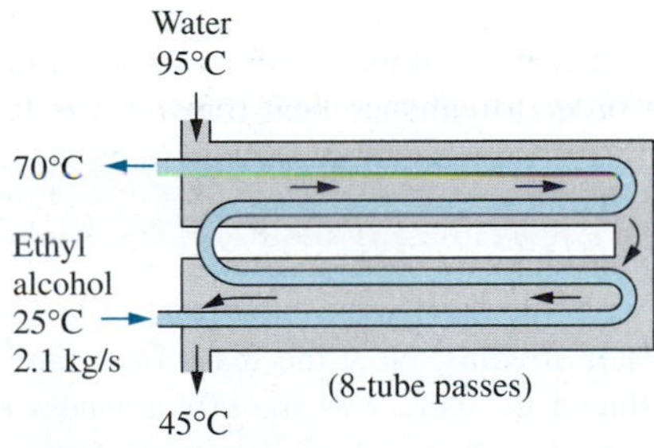

FIGURE P22–63

22–64 A shell-and-tube heat exchanger with 2-shell passes and 12-tube passes is used to heat water (c_p = 4180 J/kg · °C) with ethylene glycol (c_p = 2680 J/kg · °C). Water enters the tubes at 22°C at a rate of 0.8 kg/s and leaves at 70°C. Ethylene glycol enters the shell at 110°C and leaves at 60°C. If the overall heat transfer coefficient based on the tube side is 280 W/m^2 · °C, determine the rate of heat transfer and the heat transfer surface area on the tube side.

22–65 Reconsider Prob. 22–64. Using EES (or other) software, investigate the effect of the mass flow rate of water on the rate of heat transfer and the tube-side surface area. Let the mass flow rate vary from 0.4 kg/s to 2.2 kg/s. Plot the rate of heat transfer and the surface area as a function of the mass flow rate, and discuss the results.

22–66E Steam is to be condensed on the shell side of a 1-shell-pass and 8-tube-passes condenser, with 50 tubes in each pass at 90°F (h_{fg} = 1043 Btu/lbm). Cooling water (c_p = 1.0 Btu/lbm · °F) enters the tubes at 60°F and leaves at 73°F. The tubes are thin walled and have a diameter of 3/4 in and length of 5 ft per pass. If the overall heat transfer coefficient is 600 Btu/h · ft^2 · °F, determine (*a*) the rate of heat transfer, (*b*) the rate of condensation of steam, and (*c*) the mass flow rate of cold water.

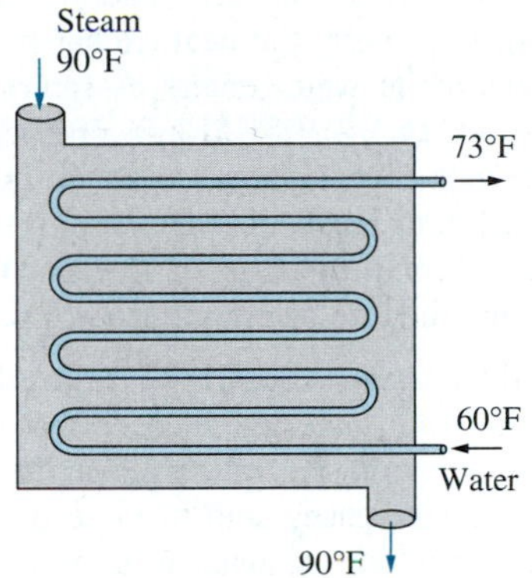

FIGURE P22–66E

22–67E Reconsider Prob. 22–66E. Using EES (or other) software, investigate the effect of the condensing steam temperature on the rate of heat transfer, the rate of condensation of steam, and the mass flow rate of cold water. Let the steam temperature vary from 80°F to 120°F. Plot the rate of heat transfer, the condensation rate of steam, and the mass flow rate of cold water as a function of steam temperature, and discuss the results.

22–68 A shell-and-tube heat exchanger with 1-shell pass and 20–tube passes is used to heat glycerin (c_p = 2480 J/kg · °C) in the shell, with hot water in the tubes. The tubes are thin-walled and have a diameter of 4 cm and length of 2 m per pass. The water enters the tubes at 100°C at a rate of 0.5 kg/s and leaves at 55°C. The glycerin enters the shell at 15°C and leaves at 55°C. Determine the mass flow rate of the glycerin and the overall heat transfer coefficient of the heat exchanger.

22–69 In a binary geothermal power plant, the working fluid isobutane is to be condensed by air in a condenser at 75°C (h_{fg} = 255.7 kJ/kg) at a rate of 2.7 kg/s. Air enters the condenser at 21°C and leaves at 28°C (see Fig. P22–69 on the next page). The heat transfer surface area based on the isobutane side is 24 m^2. Determine the mass flow rate of air and the overall heat transfer coefficient.

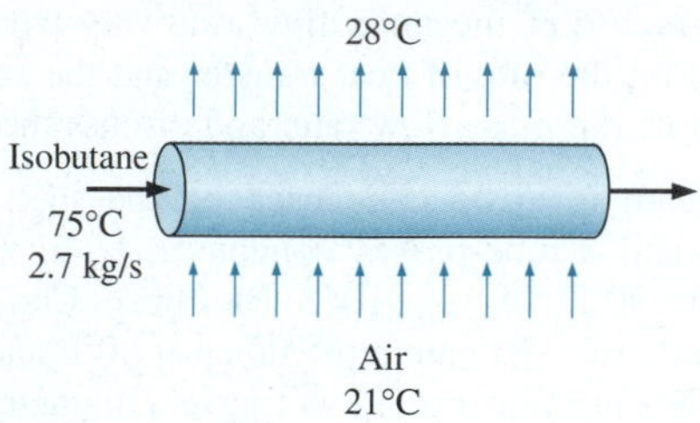

FIGURE P22–69

22–70 Hot exhaust gases of a stationary diesel engine are to be used to generate steam in an evaporator. Exhaust gases (c_p = 1051 J/kg · °C) enter the heat exchanger at 550°C at a rate of 0.25 kg/s while water enters as saturated liquid and evaporates at 200°C (h_{fg} = 1941 kJ/kg). The heat transfer surface area of the heat exchanger based on water side is 0.5 m^2 and overall heat transfer coefficient is 1780 W/m^2 · °C. Determine the rate of heat transfer, the exit temperature of exhaust gases, and the rate of evaporation of water.

22–71 Reconsider Prob. 22–71. Using EES (or other) software, investigate the effect of the exhaust gas inlet temperature on the rate of heat transfer, the exit temperature of exhaust gases, and the rate of evaporation of water. Let the temperature of exhaust gases vary from 300°C to 600°C. Plot the rate of heat transfer, the exit temperature of exhaust gases, and the rate of evaporation of water as a function of the temperature of the exhaust gases, and discuss the results.

22–72 In a textile manufacturing plant, the waste dyeing water (c_p = 4295 J/kg · °C) at 75°C is to be used to preheat fresh water (c_p = 4180 J/kg · °C) at 15°C at the same flow rate in a double-pipe counter-flow heat exchanger. The heat transfer surface area of the heat exchanger is 1.65 m^2 and the overall heat transfer coefficient is 625 W/m^2 · °C. If the rate of heat transfer in the heat exchanger is 35 kW, determine the outlet temperature and the mass flow rate of each fluid stream.

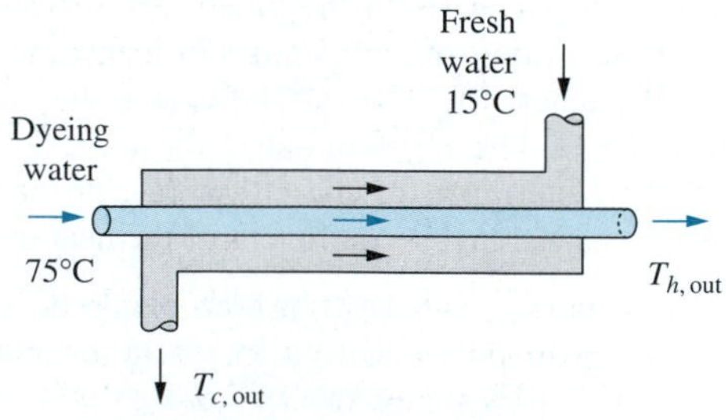

FIGURE P22–72

The Effectiveness–NTU Method

22–73C Under what conditions is the effectiveness–NTU method definitely preferred over the LMTD method in heat exchanger analysis?

22–74C What does the effectiveness of a heat exchanger represent? Can effectiveness be greater than one? On what factors does the effectiveness of a heat exchanger depend?

22–75C For a specified fluid pair, inlet temperatures, and mass flow rates, what kind of heat exchanger will have the highest effectiveness: double-pipe parallel-flow, double-pipe counter-flow, cross-flow, or multipass shell-and-tube heat exchanger?

22–76C Explain how you can evaluate the outlet temperatures of the cold and hot fluids in a heat exchanger after its effectiveness is determined.

22–77C Can the temperature of the hot fluid drop below the inlet temperature of the cold fluid at any location in a heat exchanger? Explain.

22–78C Can the temperature of the cold fluid rise above the inlet temperature of the hot fluid at any location in a heat exchanger? Explain.

22–79C Consider a heat exchanger in which both fluids have the same specific heats but different mass flow rates. Which fluid will experience a larger temperature change: the one with the lower or higher mass flow rate?

22–80C Explain how the maximum possible heat transfer rate $\dot{Q}_{max}$ in a heat exchanger can be determined when the mass flow rates, specific heats, and the inlet temperatures of the two fluids are specified. Does the value of $\dot{Q}_{max}$ depend on the type of the heat exchanger?

22–81C Consider two double-pipe counter-flow heat exchangers that are identical except that one is twice as long as the other one. Which heat exchanger is more likely to have a higher effectiveness?

22–82C Consider a double-pipe counter-flow heat exchanger. In order to enhance heat transfer, the length of the heat exchanger is now doubled. Do you think its effectiveness will also double?

22–83C Consider a shell-and-tube water-to-water heat exchanger with identical mass flow rates for both the hot- and cold-water streams. Now the mass flow rate of the cold water is reduced by half. Will the effectiveness of this heat exchanger increase, decrease, or remain the same as a result of this modification? Explain. Assume the overall heat transfer coefficient and the inlet temperatures remain the same.

22–84C Under what conditions can a counter-flow heat exchanger have an effectiveness of one? What would your answer be for a parallel-flow heat exchanger?

22–85C How is the NTU of a heat exchanger defined? What does it represent? Is a heat exchanger with a very large NTU (say, 10) necessarily a good one to buy?

22–86C Consider a heat exchanger that has an NTU of 4. Someone proposes to double the size of the heat exchanger and thus double the NTU to 8 in order to increase the effectiveness of the heat exchanger and thus save energy. Would you support this proposal?

22–87C Consider a heat exchanger that has an NTU of 0.1. Someone proposes to triple the size of the heat exchanger and thus triple the NTU to 0.3 in order to increase the effectiveness of the heat exchanger and thus save energy. Would you support this proposal?

22–88 The radiator in an automobile is a cross-flow heat exchanger (UA_s = 10 kW/K) that uses air (c_p = 1.00 kJ/kg · K) to cool the engine-coolant fluid (c_p = 4.00 kJ/kg · K). The engine fan draws 30°C air through this radiator at a rate of 10 kg/s while the coolant pump circulates the engine coolant at a rate of 5 kg/s. The coolant enters this radiator at 80°C. Under these conditions, the effectiveness of the radiator is 0.4. Determine (*a*) the outlet temperature of the air and (*b*) the rate of heat transfer between the two fluids.

22–89 During an experiment, a shell-and-tube heat exchanger that is used to transfer heat from a hot-water stream to a cold-water stream is tested, and the following measurements are taken:

	Hot-Water Stream	Cold-Water Stream
Inlet temperature, °C	71.5	19.7
Outlet temperature, °C	58.2	27.8
Volume flow rate, L/min	1.05	1.55

The heat transfer area is calculated to be 0.0200 m^2.

(*a*) Calculate the rate of heat transfer to the cold water.

(*b*) Calculate the overall heat transfer coefficient.

(*c*) Determine if the heat exchanger is truly adiabatic. If not, determine the fraction of heat loss and calculate the heat transfer efficiency.

(*d*) Determine the effectiveness and the NTU values of the heat exchanger.

Also, discuss if the measured values are reasonable.

22–90 Cold water (c_p = 4.18 kJ/kg · °C) enters a cross-flow heat exchanger at 14°C at a rate of 0.35 kg/s where it is heated by hot air (c_p = 1.0 kJ/kg · °C) that enters the heat exchanger at 65°C at a rate of 0.8 kg/s and leaves at 25°C. Determine the maximum outlet temperature of the cold water and the effectiveness of this heat exchanger.

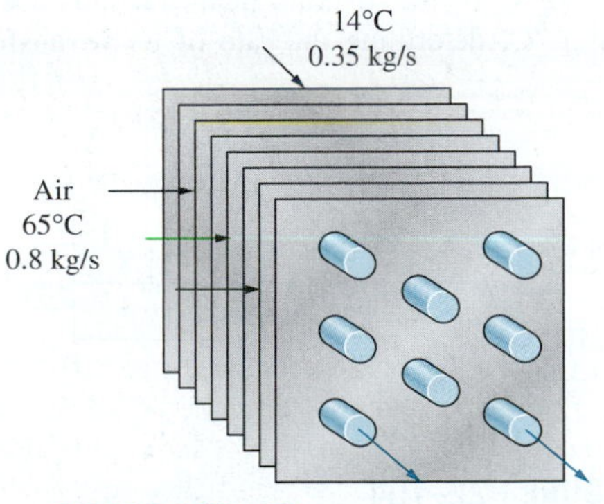

FIGURE P22–90

22–91 Water from a lake is used as the cooling agent in a power plant. To achieve condensation of 2.5 kg/s of steam exiting the turbine, a shell-and-tube heat exchanger is used, which has a single shell and 300 thin-walled, 25-mm-diameter tubes, each tube making two passes. Steam flows through the shell, while cooling water flows through the tubes. Steam enters as saturated vapor at 60°C and leaves as saturated liquid. Cooling water at 20°C is available at a rate of 200 kg/s. The convection coefficient at the outer surface of the tubes is 8500 W/m^2 · K. Determine (*a*) the temperature of the cooling water leaving the condenser and (*b*) the required tube length per pass. (Use the following average properties for water: c_p = 4180 J/kg · K, $\mu = 8 \times 10^{-4}$ N · s/m^2, k = 0.6 W/m · K, Pr = 6).

22–92 Air (c_p = 1005 J/kg · °C) enters a cross-flow heat exchanger at 20°C at a rate of 3 kg/s, where it is heated by a hot water stream (c_p = 4190 J/kg · °C) that enters the heat exchanger at 70°C at a rate of 1 kg/s. Determine the maximum heat transfer rate and the outlet temperatures of both fluids for that case.

22–93 Hot oil (c_p = 2200 J/kg · °C) is to be cooled by water (c_p = 4180 J/kg · °C) in a 2-shell-passes and 12-tube-passes heat exchanger. The tubes are thin-walled and are made of copper with a diameter of 1.8 cm. The length of each tube pass in the heat exchanger is 3 m, and the overall heat transfer coefficient is 340 W/m^2 · °C. Water flows through the tubes at a total rate of 0.1 kg/s, and the oil through the shell at a rate of 0.2 kg/s. The water and the oil enter at temperatures 18°C and 160°C, respectively. Determine the rate of heat transfer in the heat exchanger and the outlet temperatures of the water and the oil. *Answers:* 36.2 kW, 104.6°C, 77.7°C

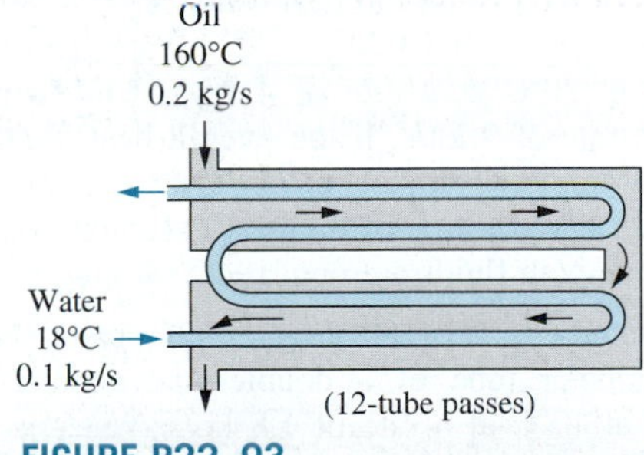

FIGURE P22–93

22–94 Consider an oil-to-oil double-pipe heat exchanger whose flow arrangement is not known. The temperature measurements indicate that the cold oil enters at 20°C and leaves at 55°C, while the hot oil enters at 80°C and leaves at 45°C. Do you think this is a parallel-flow or counter-flow heat exchanger? Why? Assuming the mass flow rates of both fluids to be identical, determine the effectiveness of this heat exchanger.

22–95E Hot water enters a double-pipe counter-flow water-to-oil heat exchanger at 190°F and leaves at 100°F. Oil enters at 70°F and leaves at 130°F. Determine which fluid has the smaller heat capacity rate and calculate the effectiveness of this heat exchanger.

22–96 A thin-walled double-pipe parallel-flow heat exchanger is used to heat a chemical whose specific heat is 1800 J/kg · °C with hot water (c_p = 4180 J/kg · °C). The chemical enters at 20°C at a rate of 3 kg/s, while the water enters at 110°C at a rate of 2 kg/s. The heat transfer surface area of the heat exchanger is 7 m^2 and the overall heat transfer coefficient is 1200 W/m^2 · °C. Determine the outlet temperatures of the chemical and the water.

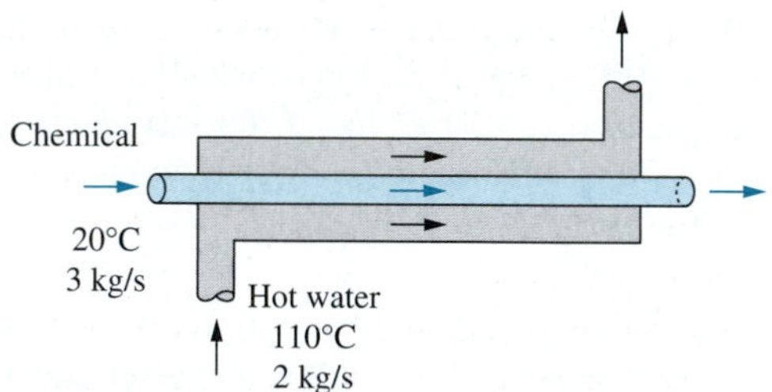

FIGURE P22–96

22–97 Reconsider Prob. 22–96. Using EES (or other) software, investigate the effects of the inlet temperatures of the chemical and the water on their outlet temperatures. Let the inlet temperature vary from 10°C to 50°C for the chemical and from 80°C to 150°C for water. Plot the outlet temperature of each fluid as a function of the inlet temperature of that fluid, and discuss the results.

22–98 A cross-flow air-to-water heat exchanger with an effectiveness of 0.65 is used to heat water (c_p = 4180 J/kg · °C) with hot air (c_p = 1010 J/kg · °C). Water enters the heat exchanger at 20°C at a rate of 4 kg/s, while air enters at 100°C at a rate of 9 kg/s. If the overall heat transfer coefficient based on the water side is 260 W/m^2 · °C, determine the heat transfer surface area of the heat exchanger on the water side. Assume both fluids are unmixed. *Answer:* 52.4 m^2

22–99 Water (c_p = 4180 J/kg · °C) enters the 2.5-cm-internal-diameter tube of a double-pipe counter-flow heat exchanger at 17°C at a rate of 1.8 kg/s. Water is heated by steam condensing at 120°C (h_{fg} = 2203 kJ/kg) in the shell. If the overall heat transfer coefficient of the heat exchanger is 700 W/m^2 · °C, determine the length of the tube required in order to heat the water to 80°C using (*a*) the LMTD method and (*b*) the ε–NTU method.

22–100 Ethanol is vaporized at 78°C (h_{fg} = 846 kJ/kg) in a double-pipe parallel-flow heat exchanger at a rate of 0.03 kg/s by hot oil (c_p = 2200 J/kg · °C) that enters at 120°C. If the heat transfer surface area and the overall heat transfer coefficients are 6.2 m^2 and 320 W/m^2 · °C, respectively, determine the outlet temperature and the mass flow rate of oil using (*a*) the LMTD method and (*b*) the ε–NTU method.

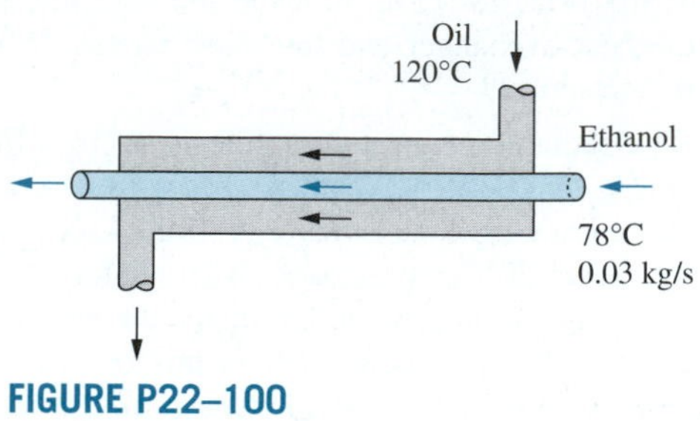

FIGURE P22–100

22–101 Water (c_p = 4180 J/kg · °C) is to be heated by solar-heated hot air (c_p = 1010 J/kg · °C) in a double-pipe counter-flow heat exchanger. Air enters the heat exchanger at 90°C at a rate of 0.3 kg/s, while water enters at 22°C at a rate of 0.1 kg/s. The overall heat transfer coefficient based on the inner side of the tube is given to be 80 W/m^2 · °C. The length of the tube is 12 m and the internal diameter of the tube is 1.2 cm. Determine the outlet temperatures of the water and the air.

22–102 Reconsider Prob. 22–101. Using EES (or other) software, investigate the effects of the mass flow rate of water and the tube length on the outlet temperatures of water and air. Let the mass flow rate vary from 0.05 kg/s to 1.0 kg/s and the tube length from 5 m to 25 m. Plot the outlet temperatures of the water and the air as functions of the mass flow rate and the tube length, and discuss the results.

22–103E A thin-walled double-pipe heat exchanger is to be used to cool oil (c_p = 0.525 Btu/lbm · °F) from 300°F to 105°F at a rate of 5 lbm/s by water (c_p = 1.0 Btu/lbm · °F) that enters at 70°F at a rate of 3 lbm/s. The diameter of the tube is 5 in and its length is 200 ft. Determine the overall heat transfer coefficient of this heat exchanger using (*a*) the LMTD method and (*b*) the ε–NTU method.

22–104 Cold water (c_p = 4180 J/kg · °C) leading to a shower enters a thin-walled double-pipe counter-flow heat exchanger at 15°C at a rate of 0.25 kg/s and is heated to 45°C by hot water (c_p = 4190 J/kg · °C) that enters at 100°C at a rate of 3 kg/s. If the overall heat transfer coefficient is 950 W/m^2 · °C, determine the rate of heat transfer and the

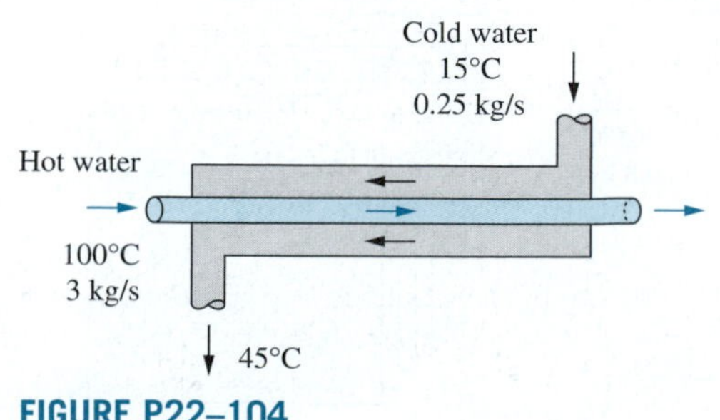

FIGURE P22–104

heat transfer surface area of the heat exchanger using the ε–NTU method. *Answers:* 31.35 kW, 0.482 m^2

22–105 Reconsider Prob. 22–104. Using EES (or other) software, investigate the effects of the inlet temperature of hot water and the heat transfer coefficient on the rate of heat transfer and the surface area. Let the inlet temperature vary from 60°C to 120°C and the overall heat transfer coefficient from 750 W/m^2 · °C to 1250 W/m^2 · °C. Plot the rate of heat transfer and surface area as functions of the inlet temperature and the heat transfer coefficient, and discuss the results.

22–106 Glycerin (c_p = 2400 J/kg · °C) at 20°C and 0.3 kg/s is to be heated by ethylene glycol (c_p = 2500 J/kg · °C) at 60°C and the same mass flow rate in a thin-walled double-pipe parallel-flow heat exchanger. If the overall heat transfer coefficient is 380 W/m^2 · °C and the heat transfer surface area is 5.3 m^2, determine (*a*) the rate of heat transfer and (*b*) the outlet temperatures of the glycerin and the glycol.

22–107 A cross-flow heat exchanger consists of 80 thin-walled tubes of 3-cm diameter located in a duct of 1 m × 1 m cross section. There are no fins attached to the tubes. Cold water (c_p = 4180 J/kg · °C) enters the tubes at 18°C with an average velocity of 3 m/s, while hot air (c_p = 1010 J/kg · °C) enters the channel at 130°C and 105 kPa at an average velocity of 12 m/s. If the overall heat transfer coefficient is 130 W/m^2 · °C, determine the outlet temperatures of both fluids and the rate of heat transfer.

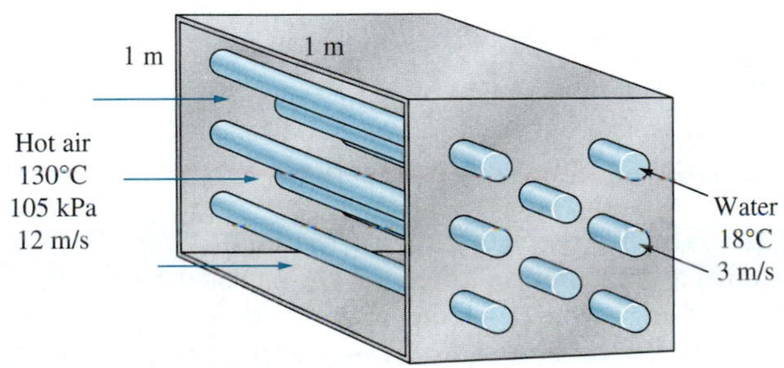

FIGURE P22–107

22–108 A shell-and-tube heat exchanger with 2-shell passes and 8-tube passes is used to heat ethyl alcohol (c_p = 2670 J/kg · °C) in the tubes from 25°C to 70°C at a rate of 2.1 kg/s. The heating is to be done by water (c_p = 4190 J/kg · °C) that enters the shell at 95°C and leaves at 60°C. If the overall heat transfer coefficient is 800 W/m^2 · °C, determine the heat transfer surface area of the heat exchanger using (*a*) the LMTD method and (*b*) the ε–NTU method. *Answer:* (*a*) 11.4 m^2

22–109 Steam is to be condensed on the shell side of a 1-shell-pass and 8-tube-passes condenser, with 50 tubes in each pass, at 30°C (h_{fg} = 2431 kJ/kg). Cooling water (c_p = 4180 J/kg · °C) enters the tubes at 15°C at a rate of 1800 kg/h. The tubes are thin-walled, and have a diameter of 1.5 cm and length of 2 m per pass. If the overall heat transfer coefficient is 3000 W/m^2 · °C, determine (*a*) the rate of heat transfer and (*b*) the rate of condensation of steam.

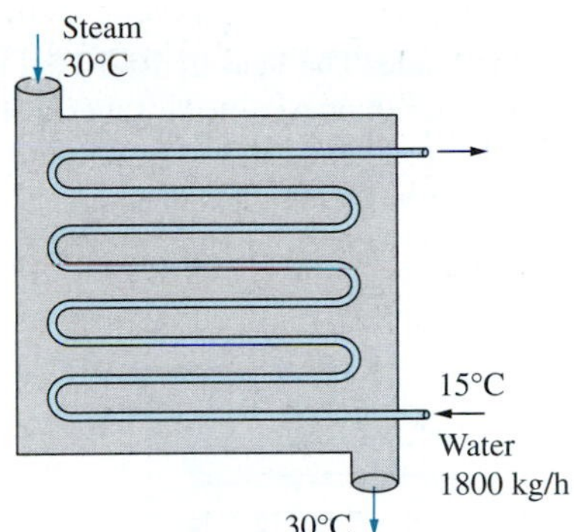

FIGURE P22–109

22–110 Reconsider Prob. 22–109. Using EES (or other) software, investigate the effects of the condensing steam temperature and the tube diameter on the rate of heat transfer and the rate of condensation of steam. Let the steam temperature vary from 20°C to 70°C and the tube diameter from 1.0 cm to 2.0 cm. Plot the rate of heat transfer and the rate of condensation as functions of steam temperature and tube diameter, and discuss the results.

22–111 Cold water (c_p = 4180 J/kg · °C) enters the tubes of a heat exchanger with 2-shell passes and 23-tube passes at 14°C at a rate of 3 kg/s, while hot oil (c_p = 2200 J/kg · °C) enters the shell at 200°C at the same mass flow rate. The overall heat transfer coefficient based on the outer surface of the tube is 300 W/m^2 · °C and the heat transfer surface area on that side is 20 m^2. Determine the rate of heat transfer using (*a*) the LMTD method and (*b*) the ε–NTU method.

Selection of Heat Exchangers

22–112C A heat exchanger is to be selected to cool a hot liquid chemical at a specified rate to a specified temperature. Explain the steps involved in the selection process.

22–113C There are two heat exchangers that can meet the heat transfer requirements of a facility. One is smaller and cheaper but requires a larger pump, while the other is larger and more expensive but has a smaller pressure drop and thus requires a smaller pump. Both heat exchangers have the same life expectancy and meet all other requirements. Explain which heat exchanger you would choose under what conditions.

22–114C There are two heat exchangers that can meet the heat transfer requirements of a facility. Both have the same pumping power requirements, the same useful life, and the same price tag. But one is heavier and larger in size. Under what conditions would you choose the smaller one?

22–115 A heat exchanger is to cool oil (c_p = 2200 J/kg · °C) at a rate of 13 kg/s from 120°C to 50°C by air. Determine the heat transfer rating of the heat exchanger and propose a suitable type.

22–116 A shell-and-tube process heater is to be selected to heat water (c_p = 4190 J/kg · °C) from 20°C to 90°C by steam flowing on the shell side. The heat transfer load of the heater is 600 kW. If the inner diameter of the tubes is 1 cm and the velocity of water is not to exceed 3 m/s, determine how many tubes need to be used in the heat exchanger.

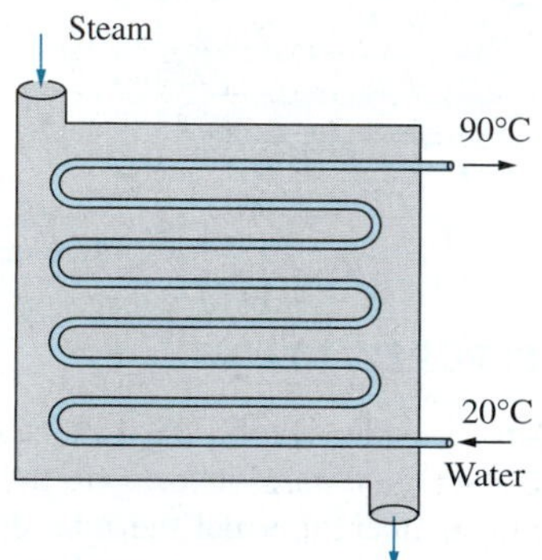

FIGURE P22–116

22–117 Reconsider Prob. 22–116. Using EES (or other) software, plot the number of tube passes as a function of water velocity as it varies from 1 m/s to 8 m/s, and discuss the results.

22–118 The condenser of a large power plant is to remove 500 MW of heat from steam condensing at 30°C (h_{fg} = 2431 kJ/kg). The cooling is to be accomplished by cooling water (c_p = 4180 J/kg · °C) from a nearby river, which enters the tubes at 18°C and leaves at 26°C. The tubes of the heat exchanger have an internal diameter of 2 cm, and the overall heat transfer coefficient is 3500 W/m^2 · °C. Determine the total length of the tubes required in the condenser. What type of heat exchanger is suitable for this task? *Answer:* 312.3 km

22–119 Repeat Prob. 22–118 for a heat transfer load of 50 MW.

Review Problems

22–120 The mass flow rate, specific heat, and inlet temperature of the tube-side stream in a double-pipe, parallel-flow heat exchanger are 2700 kg/h, 2.0 kJ/kg · K, and 120°C, respectively. The mass flow rate, specific heat, and inlet temperature of the other stream are 1800 kg/h, 4.2 kJ/kg · K, and 20°C, respectively. The heat transfer area and overall heat transfer coefficient are 0.50 m^2 and 2.0 kW/m^2 · K, respectively. Find the outlet temperatures of both streams in steady operation using (*a*) the LMTD method and (*b*) the effectiveness–NTU method.

22–121 A shell-and-tube heat exchanger is used for cooling 47 kg/s of a process stream flowing through the tubes from 160°C to 100°C. This heat exchanger has a total of 100 identical tubes, each with an inside diameter of 2.5 cm and negligible wall thickness. The average properties of the process stream are: ρ = 950 kg/m^3, k = 0.50 W/m · K, c_p = 3.5 kJ/kg · K and μ = 2.0 mPa · s. The coolant stream is water (c_p = 4.18 kJ/kg · K) at a flow rate of 66 kg/s and an inlet temperature of 10°C, which yields an average shell-side heat transfer coefficient of 4.0 kW/m^2 · K. Calculate the tube length if the heat exchanger has (*a*) a 1-shell pass and a 1-tube pass and (*b*) a 1-shell pass and 4-tube passes.

22–122 A 2-shell passes and 4-tube passes heat exchanger is used for heating a hydrocarbon stream (c_p = 2.0 kJ/kg · K) steadily from 20°C to 50°C. A water stream enters the shell-side at 80°C and leaves at 40°C. There are 160 thin-walled tubes, each with a diameter of 2.0 cm and length of 1.5 m. The tube-side and shell-side heat transfer coefficients are 1.6 and 2.5 kW/m^2 · K, respectively. (*a*) Calculate the rate of heat transfer and the mass rates of water and hydrocarbon streams. (*b*) With usage, the outlet hydrocarbon-stream temperature was found to decrease by 5°C due to the deposition of solids on the tube surface. Estimate the magnitude of fouling factor.

22–123 Hot water at 60°C is cooled to 36°C through the tube side of a 1–shell pass and 2-tube passes heat exchanger. The coolant is also a water stream, for which the inlet and outlet temperatures are 7°C and 31°C, respectively. The overall heat transfer coefficient and the heat transfer area are 950 W/m^2 · K and 15 m^2, respectively. Calculate the mass flow rates of hot and cold water streams in steady operation.

22–124 Hot oil is to be cooled in a multipass shell-and-tube heat exchanger by water. The oil flows through the shell, with a heat transfer coefficient of h_o = 35 W/m^2 · °C, and the water flows through the tube with an average velocity of 3 m/s. The tube is made of brass (k = 110 W/m · °C) with internal and external diameters of 1.3 cm and 1.5 cm, respectively. Using water properties at 25°C, determine the overall heat transfer coefficient of this heat exchanger based on the inner surface.

22–125 Repeat Prob. 22–124 by assuming a fouling factor $R_{f,o}$ = 0.0004 m^2 · °C/W on the outer surface of the tube.

22–126 Cold water (c_p = 4180 J/kg · °C) enters the tubes of a heat exchanger with 2-shell passes and 20-tube passes at 20°C at a rate of 3 kg/s, while hot oil (c_p = 2200 J/kg · °C) enters the shell at 130°C at the same mass flow rate and leaves at 60°C. If the overall heat transfer coefficient based on the outer surface of the tube is 220 W/m^2 · °C, determine (*a*) the rate of heat transfer and (*b*) the heat transfer surface area on the outer side of the tube. *Answers:* (*a*) 462 kW, (*b*) 39.8 m^2

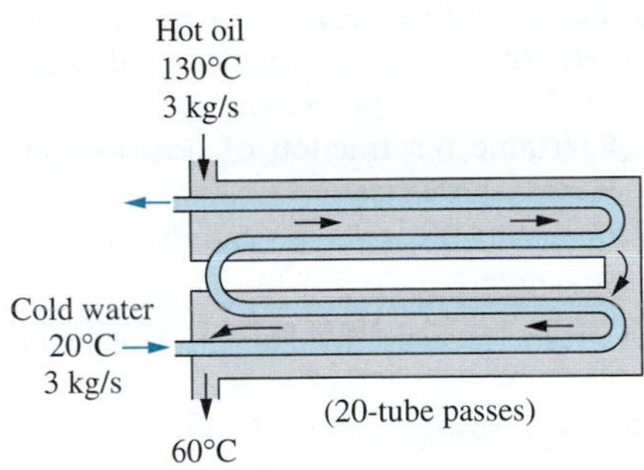

FIGURE P22–126

22–127E Water (c_p = 1.0 Btu/lbm · °F) is to be heated by solar-heated hot air (c_p = 0.24 Btu/lbm · °F) in a double-pipe counter-flow heat exchanger. Air enters the heat exchanger at 190°F at a rate of 0.7 lbm/s and leaves at 135°F. Water enters at 70°F at a rate of 0.35 lbm/s. The overall heat transfer coefficient based on the inner side of the tube is given to be 20 Btu/h · ft² · °F. Determine the length of the tube required for a tube internal diameter of 0.5 in.

22–128 By taking the limit as $\Delta T_2 \to \Delta T_1$, show that when $\Delta T_1 = \Delta T_2$ for a heat exchanger, the ΔT_{lm} relation reduces to $\Delta T_{lm} = \Delta T_1 = \Delta T_2$.

22–129 The condenser of a room air conditioner is designed to reject heat at a rate of 15,000 kJ/h from refrigerant-134a as the refrigerant is condensed at a temperature of 40°C. Air (c_p = 1005 J/kg · °C) flows across the finned condenser coils, entering at 25°C and leaving at 35°C. If the overall heat transfer coefficient based on the refrigerant side is 150 W/m² · °C, determine the heat transfer area on the refrigerant side. *Answer:* 3.05 m²

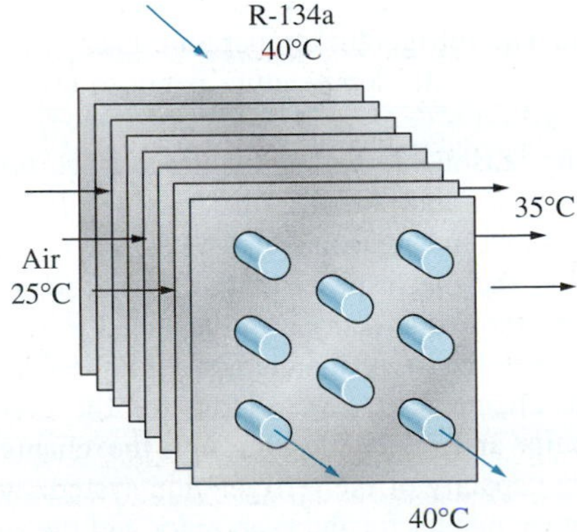

FIGURE P22–129

22–130 Air (c_p = 1005 J/kg · °C) is to be preheated by hot exhaust gases in a cross-flow heat exchanger before it enters the furnace. Air enters the heat exchanger at 95 kPa and 20°C at a rate of 0.4 m³/s. The combustion gases (c_p = 1100 J/kg · °C) enter at 180°C at a rate of 0.65 kg/s and leave at 95°C. The product of the overall heat transfer coefficient and the heat transfer surface area is UA_s = 1620 W/°C. Assuming both fluids to be unmixed, determine the rate of heat transfer.

22–131 In a chemical plant, a certain chemical is heated by hot water supplied by a natural gas furnace. The hot water (c_p = 4180 J/kg · °C) is then discharged at 60°C at a rate of 8 kg/min. The plant operates 8 h a day, 5 days a week, 52 weeks a year. The furnace has an efficiency of 78 percent, and the cost of the natural gas is $1.00 per therm (1 therm = 105,500 kJ). The average temperature of the cold water entering the furnace throughout the year is 14°C. In order to save energy, it is proposed to install a water-to-water heat exchanger to preheat the incoming cold water by the drained hot water. Assuming that the heat exchanger will recover 72 percent of the available heat in the hot water, determine the heat transfer rating of the heat exchanger that needs to be purchased and suggest a suitable type. Also, determine the amount of money this heat exchanger will save the company per year from natural gas savings.

22–132 A shell-and-tube heat exchanger with 1-shell pass and 14-tube passes is used to heat water in the tubes with geothermal steam condensing at 120°C (h_{fg} = 2203 kJ/kg) on the shell side. The tubes are thin-walled and have a diameter of 2.4 cm and length of 3.2 m per pass. Water (c_p = 4180 J/kg · °C) enters the tubes at 22°C at a rate of 3.9 kg/s. If the temperature difference between the two fluids at the exit is 46°C, determine (*a*) the rate of heat transfer, (*b*) the rate of condensation of steam, and (*c*) the overall heat transfer coefficient.

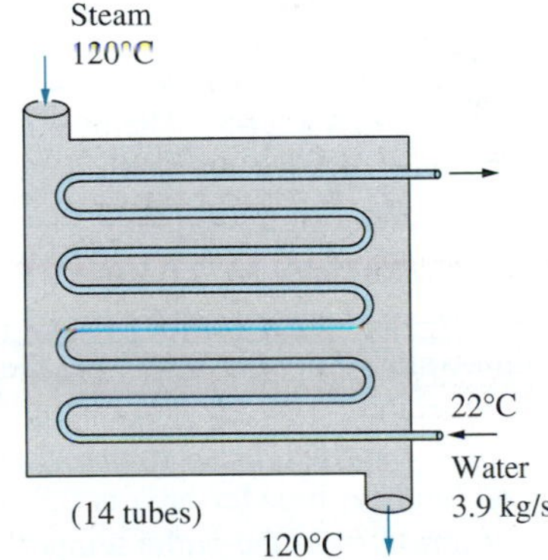

FIGURE P22–132

22–133 Geothermal water (c_p = 4250 J/kg · °C) at 75°C is to be used to heat fresh water (c_p = 4180 J/kg · °C) at 17°C at a rate of 1.2 kg/s in a double-pipe counter-flow heat exchanger. The heat transfer surface area is 25 m², the overall heat transfer coefficient is 480 W/m² · °C, and the mass flow rate of geothermal water is larger than that of fresh water. If the effectiveness of the heat exchanger is desired to be 0.823, determine the mass flow rate of geothermal water and the outlet temperatures of both fluids.

22–134 Air at 18°C (c_p = 1006 J/kg · °C) is to be heated to 70°C by hot oil at 80°C (c_p = 2150 J/kg · °C) in a cross-flow heat exchanger with air mixed and oil unmixed. The product of heat transfer surface area and the overall heat transfer coefficient is 750 W/°C and the mass flow rate of air is twice that of oil. Determine (*a*) the effectiveness of the heat exchanger, (*b*) the mass flow rate of air, and (*c*) the rate of heat transfer.

22–135 Consider a water-to-water counter-flow heat exchanger with these specifications. Hot water enters at 95°C while cold water enters at 20°C. The exit temperature of hot water is 15°C greater than that of cold water, and the mass flow rate of hot water is 50 percent greater than that of cold water. The product of heat transfer surface area and the overall heat transfer coefficient is 1400 W/°C. Taking the specific heat of both cold and hot water to be c_p = 4180 J/kg · °C, determine (*a*) the outlet temperature of the cold water, (*b*) the effectiveness of the heat exchanger, (*c*) the mass flow rate of the cold water, and (*d*) the heat transfer rate.

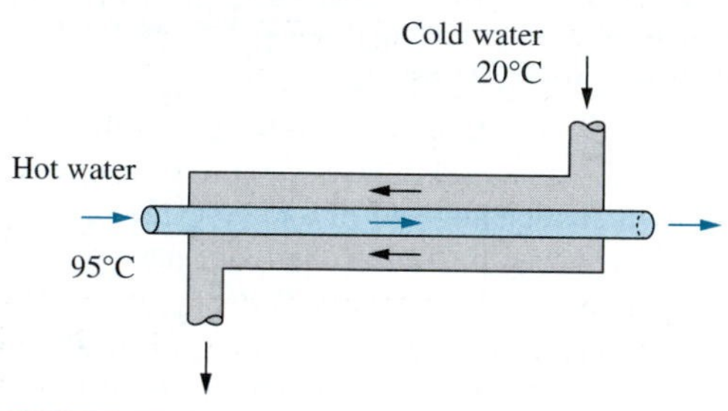

FIGURE P22–135

22–136 A shell-and-tube heat exchanger with 2-shell passes and 4-tube passes is used for cooling oil (c_p = 2.0 kJ/kg · K) from 125°C to 55°C. The coolant is water, which enters the shell side at 25°C and leaves at 46°C. The overall heat transfer coefficient is 900 W/m^2 · K. For an oil flow rate of 10 kg/s, calculate the cooling water flow rate and the heat transfer area.

22–137 A polymer solution (c_p = 2.0 kJ/kg · K) at 20°C and 0.3 kg/s is heated by ethylene glycol (c_p = 2.5 kJ/kg · K) at 60°C in a thin-walled double-pipe parallel-flow heat exchanger. The temperature difference between the two outlet fluids is 15°C. The overall heat transfer coefficient is 240 W/m^2 · K and the heat transfer area is 0.8 m^2. Calculate (*a*) the rate of heat transfer, (*b*) the outlet temperature of polymer solution, and (*c*) the mass flow rate of ethylene glycol.

22–138 During an experiment, a plate heat exchanger that is used to transfer heat from a hot-water stream to a cold-water stream is tested, and the following measurements are taken:

	Hot Water Stream	Cold Water Stream
Inlet temperature, °C	38.9	14.3
Outlet temperature, °C	27.0	19.8
Volume flow rate, L/min	2.5	4.5

The heat transfer area is calculated to be 0.0400 m^2.

(*a*) Calculate the rate of heat transfer to the cold water.
(*b*) Calculate the overall heat transfer coefficient.
(*c*) Determine if the heat exchanger is truly adiabatic. If not, determine the fraction of heat loss and calculate the heat transfer efficiency.
(*d*) Determine the effectiveness and the NTU values of the heat exchanger.

Also, discuss if the measured values are reasonable.

Design and Essay Problems

22–139 Water flows through a shower head steadily at a rate of 8 kg/min. The water is heated in an electric water heater from 15°C to 45°C. In an attempt to conserve energy, it is proposed to pass the drained warm water at a temperature of 38°C through a heat exchanger to preheat the incoming cold water. Design a heat exchanger that is suitable for this task, and discuss the potential savings in energy and money for your area.

22–140 Open the engine compartment of your car and search for heat exchangers. How many do you have? What type are they? Why do you think those specific types are selected? If you were redesigning the car, would you use different kinds? Explain.

22–141 Design a hydrocooling unit that can cool fruits and vegetables from 30°C to 5°C at a rate of 20,000 kg/h under the following conditions:

The unit will be of flood type that will cool the products as they are conveyed into the channel filled with water. The products will be dropped into the channel filled with water at one end and picked up at the other end. The channel can be as wide as 3 m and as high as 90 cm. The water is to be circulated and cooled by the evaporator section of a refrigeration system. The refrigerant temperature inside the coils is to be –2°C, and the water temperature is not to drop below 1°C and not to exceed 6°C.

Assuming reasonable values for the average product density, specific heat, and porosity (the fraction of air volume in a box), recommend reasonable values for the quantities related to the thermal aspects of the hydrocooler, including (*a*) how long the fruits and vegetables need to remain in the channel, (*b*) the length of the channel, (*c*) the water velocity through the channel, (*d*) the velocity of the conveyor and thus the fruits and vegetables through the channel, (*e*) the refrigeration capacity of the refrigeration system, and (*f*) the type of heat exchanger for the evaporator and the surface area on the water side.

22–142 A company owns a refrigeration system whose refrigeration capacity is 200 tons (1 ton of refrigeration = 211 kJ/min), and you are to design a forced-air cooling system for fruits whose diameters do not exceed 7 cm under the following conditions:

The fruits are to be cooled from 28°C to an average temperature of 8°C. The air temperature is to remain above −2°C and below 10°C at all times, and the velocity of air approaching the fruits must remain under 2 m/s. The cooling section can be as wide as 3.5 m and as high as 2 m.

Assuming reasonable values for the average fruit density, specific heat, and porosity (the fraction of air volume in a box), recommend reasonable values for the quantities related to the thermal aspects of the forced-air cooling, including (*a*) how long the fruits need to remain in the cooling section, (*b*) the length of the cooling section, (*c*) the air velocity approaching the cooling section, (*d*) the product cooling capacity of the system, in kg · fruit/h, (*e*) the volume flow rate of air, and (*f*) the type of heat exchanger for the evaporator and the surface area on the air side.

22–143 A counterflow double-pipe heat exchanger with $A_s = 9.0\ \text{m}^2$ is used for cooling a liquid stream (c_p = 3.15 kJ/kg · K) at a rate of 10.0 kg/s with an inlet temperature of 90°C. The coolant (c_p = 4.2 kJ/kg · K) enters the heat exchanger at a rate of 8.0 kg/s with an inlet temperature of 10°C. The plant data gave the following equation for the overall heat transfer coefficient in W/m² · K: $U = 600/(1/\dot{m}_c^{0.8} + 2/\dot{m}_h^{0.8})$, where $\dot{m}_c$ and $\dot{m}_h$ are the cold-and hot-stream flow rates in kg/s, respectively. (*a*) Calculate the rate of heat transfer and the outlet stream temperatures for this unit. (*b*) The existing unit is to be replaced. A vendor is offering a very attractive discount on two identical heat exchangers that are presently stocked in its warehouse, each with $A_s = 5\ \text{m}^2$. Because the tube diameters in the existing and new units are the same, the above heat transfer coefficient equation is expected to be valid for the new units as well. The vendor is proposing that the two new units could be operated in parallel, such that each unit would process exactly one-half the flow rate of each of the hot and cold streams in a counterflow manner; hence, they together would meet (or exceed) the present plant heat duty. Give your recommendation, with supporting calculations, on this replacement proposal.

Appendix 1

PROPERTY TABLES AND CHARTS (SI UNITS)

TABLE A – 1

Molar mass, gas constant, and critical-point properties

			Gas	*Critical-point properties*		
Substance	Formula	Molar mass, *M* kg/kmol	constant, *R* kJ/kg · K*	Temperature, K	Pressure, MPa	Volume, m^3/kmol
Air	—	28.97	0.2870	132.5	3.77	0.0883
Ammonia	NH_3	17.03	0.4882	405.5	11.28	0.0724
Argon	Ar	39.948	0.2081	151	4.86	0.0749
Benzene	C_6H_6	78.115	0.1064	562	4.92	0.2603
Bromine	Br_2	159.808	0.0520	584	10.34	0.1355
n-Butane	C_4H_{10}	58.124	0.1430	425.2	3.80	0.2547
Carbon dioxide	CO_2	44.01	0.1889	304.2	7.39	0.0943
Carbon monoxide	CO	28.011	0.2968	133	3.50	0.0930
Carbon tetrachloride	CCl_4	153.82	0.05405	556.4	4.56	0.2759
Chlorine	Cl_2	70.906	0.1173	417	7.71	0.1242
Chloroform	$CHCl_3$	119.38	0.06964	536.6	5.47	0.2403
Dichlorodifluoromethane (R-12)	CCl_2F_2	120.91	0.06876	384.7	4.01	0.2179
Dichlorofluoromethane (R-21)	$CHCl_2F$	102.92	0.08078	451.7	5.17	0.1973
Ethane	C_2H_6	30.070	0.2765	305.5	4.48	0.1480
Ethyl alcohol	C_2H_5OH	46.07	0.1805	516	6.38	0.1673
Ethylene	C_2H_4	28.054	0.2964	282.4	5.12	0.1242
Helium	He	4.003	2.0769	5.3	0.23	0.0578
n-Hexane	C_6H_{14}	86.179	0.09647	507.9	3.03	0.3677
Hydrogen (normal)	H_2	2.016	4.1240	33.3	1.30	0.0649
Krypton	Kr	83.80	0.09921	209.4	5.50	0.0924
Methane	CH_4	16.043	0.5182	191.1	4.64	0.0993
Methyl alcohol	CH_3OH	32.042	0.2595	513.2	7.95	0.1180
Methyl chloride	CH_3Cl	50.488	0.1647	416.3	6.68	0.1430
Neon	Ne	20.183	0.4119	44.5	2.73	0.0417
Nitrogen	N_2	28.013	0.2968	126.2	3.39	0.0899
Nitrous oxide	N_2O	44.013	0.1889	309.7	7.27	0.0961
Oxygen	O_2	31.999	0.2598	154.8	5.08	0.0780
Propane	C_3H_8	44.097	0.1885	370	4.26	0.1998
Propylene	C_3H_6	42.081	0.1976	365	4.62	0.1810
Sulfur dioxide	SO_2	64.063	0.1298	430.7	7.88	0.1217
Tetrafluoroethane (R-134a)	CF_3CH_2F	102.03	0.08149	374.2	4.059	0.1993
Trichlorofluoromethane (R-11)	CCl_3F	137.37	0.06052	471.2	4.38	0.2478
Water	H_2O	18.015	0.4615	647.1	22.06	0.0560
Xenon	Xe	131.30	0.06332	289.8	5.88	0.1186

*The unit kJ/kg · K is equivalent to kPa · m^3/kg · K. The gas constant is calculated from $R = R_u/M$, where R_u = 8.31447 kJ/kmol · K and *M* is the molar mass.

Source: K. A. Kobe and R. E. Lynn, Jr., *Chemical Review* 52 (1953), pp. 117–236; and ASHRAE, *Handbook of Fundamentals* (Atlanta, GA: American Society of Heating, Refrigerating and Air-Conditioning Engineers, Inc., 1993), pp. 16.4 and 36.1.

TABLE A–2

Ideal-gas specific heats of various common gases

(*a*) At 300 K

Gas	Formula	Gas constant, R kJ/kg · K	c_p kJ/kg · K	c_v kJ/kg · K	k
Air	—	0.2870	1.005	0.718	1.400
Argon	Ar	0.2081	0.5203	0.3122	1.667
Butane	C_4H_{10}	0.1433	1.7164	1.5734	1.091
Carbon dioxide	CO_2	0.1889	0.846	0.657	1.289
Carbon monoxide	CO	0.2968	1.040	0.744	1.400
Ethane	C_2H_6	0.2765	1.7662	1.4897	1.186
Ethylene	C_2H_4	0.2964	1.5482	1.2518	1.237
Helium	He	2.0769	5.1926	3.1156	1.667
Hydrogen	H_2	4.1240	14.307	10.183	1.405
Methane	CH_4	0.5182	2.2537	1.7354	1.299
Neon	Ne	0.4119	1.0299	0.6179	1.667
Nitrogen	N_2	0.2968	1.039	0.743	1.400
Octane	C_8H_{18}	0.0729	1.7113	1.6385	1.044
Oxygen	O_2	0.2598	0.918	0.658	1.395
Propane	C_3H_8	0.1885	1.6794	1.4909	1.126
Steam	H_2O	0.4615	1.8723	1.4108	1.327

Note: The unit kJ/kg · K is equivalent to kJ/kg · °C.

Source: Chemical and Process Thermodynamics 3/E by Kyle, B. G., © 2000. Adapted by permission of Pearson Education, Inc., Upper Saddle River, NJ.

TABLE A–2

Ideal-gas specific heats of various common gases (*Continued*)

(*b*) At various temperatures

Temperature, K	c_p kJ/kg · K	c_v kJ/kg · K	k	c_p kJ/kg · K	c_v kJ/kg · K	k	c_p kJ/kg · K	c_v kJ/kg · K	k
	Air			*Carbon dioxide, CO_2*			*Carbon monoxide, CO*		
250	1.003	0.716	1.401	0.791	0.602	1.314	1.039	0.743	1.400
300	1.005	0.718	1.400	0.846	0.657	1.288	1.040	0.744	1.399
350	1.008	0.721	1.398	0.895	0.706	1.268	1.043	0.746	1.398
400	1.013	0.726	1.395	0.939	0.750	1.252	1.047	0.751	1.395
450	1.020	0.733	1.391	0.978	0.790	1.239	1.054	0.757	1.392
500	1.029	0.742	1.387	1.014	0.825	1.229	1.063	0.767	1.387
550	1.040	0.753	1.381	1.046	0.857	1.220	1.075	0.778	1.382
600	1.051	0.764	1.376	1.075	0.886	1.213	1.087	0.790	1.376
650	1.063	0.776	1.370	1.102	0.913	1.207	1.100	0.803	1.370
700	1.075	0.788	1.364	1.126	0.937	1.202	1.113	0.816	1.364
750	1.087	0.800	1.359	1.148	0.959	1.197	1.126	0.829	1.358
800	1.099	0.812	1.354	1.169	0.980	1.193	1.139	0.842	1.353
900	1.121	0.834	1.344	1.204	1.015	1.186	1.163	0.866	1.343
1000	1.142	0.855	1.336	1.234	1.045	1.181	1.185	0.888	1.335
	Hydrogen, H_2			*Nitrogen, N_2*			*Oxygen, O_2*		
250	14.051	9.927	1.416	1.039	0.742	1.400	0.913	0.653	1.398
300	14.307	10.183	1.405	1.039	0.743	1.400	0.918	0.658	1.395
350	14.427	10.302	1.400	1.041	0.744	1.399	0.928	0.668	1.389
400	14.476	10.352	1.398	1.044	0.747	1.397	0.941	0.681	1.382
450	14.501	10.377	1.398	1.049	0.752	1.395	0.956	0.696	1.373
500	14.513	10.389	1.397	1.056	0.759	1.391	0.972	0.712	1.365
550	14.530	10.405	1.396	1.065	0.768	1.387	0.988	0.728	1.358
600	14.546	10.422	1.396	1.075	0.778	1.382	1.003	0.743	1.350
650	14.571	10.447	1.395	1.086	0.789	1.376	1.017	0.758	1.343
700	14.604	10.480	1.394	1.098	0.801	1.371	1.031	0.771	1.337
750	14.645	10.521	1.392	1.110	0.813	1.365	1.043	0.783	1.332
800	14.695	10.570	1.390	1.121	0.825	1.360	1.054	0.794	1.327
900	14.822	10.698	1.385	1.145	0.849	1.349	1.074	0.814	1.319
1000	14.983	10.859	1.380	1.167	0.870	1.341	1.090	0.830	1.313

Source: Kenneth Wark, *Thermodynamics,* 4th ed. (New York: McGraw-Hill, 1983), p. 783, Table A–4M. Originally published in *Tables of Thermal Properties of Gases,* NBS Circular 564, 1955.

TABLE A–2

Ideal-gas specific heats of various common gases (*Concluded*)

(*c*) As a function of temperature

$$\bar{c}_p = a + bT + cT^2 + dT^3$$

$$(T \text{ in K, } c_p \text{ in kJ/kmol} \cdot \text{K})$$

Substance	Formula	*a*	*b*	*c*	*d*	Temperature range, K	% error Max.	% error Avg.
Nitrogen	N_2	28.90	-0.1571×10^{-2}	0.8081×10^{-5}	-2.873×10^{-9}	273–1800	0.59	0.34
Oxygen	O_2	25.48	1.520×10^{-2}	-0.7155×10^{-5}	1.312×10^{-9}	273–1800	1.19	0.28
Air	—	28.11	0.1967×10^{-2}	0.4802×10^{-5}	-1.966×10^{-9}	273–1800	0.72	0.33
Hydrogen	H_2	29.11	-0.1916×10^{-2}	0.4003×10^{-5}	-0.8704×10^{-9}	273–1800	1.01	0.26
Carbon monoxide	CO	28.16	0.1675×10^{-2}	0.5372×10^{-5}	-2.222×10^{-9}	273–1800	0.89	0.37
Carbon dioxide	CO_2	22.26	5.981×10^{-2}	-3.501×10^{-5}	7.469×10^{-9}	273–1800	0.67	0.22
Water vapor	H_2O	32.24	0.1923×10^{-2}	1.055×10^{-5}	-3.595×10^{-9}	273–1800	0.53	0.24
Nitric oxide	NO	29.34	-0.09395×10^{-2}	0.9747×10^{-5}	-4.187×10^{-9}	273–1500	0.97	0.36
Nitrous oxide	N_2O	24.11	5.8632×10^{-2}	-3.562×10^{-5}	10.58×10^{-9}	273–1500	0.59	0.26
Nitrogen dioxide	NO_2	22.9	5.715×10^{-2}	-3.52×10^{-5}	7.87×10^{-9}	273–1500	0.46	0.18
Ammonia	NH_3	27.568	2.5630×10^{-2}	0.99072×10^{-5}	-6.6909×10^{-9}	273–1500	0.91	0.36
Sulfur	S_2	27.21	2.218×10^{-2}	-1.628×10^{-5}	3.986×10^{-9}	273–1800	0.99	0.38
Sulfur dioxide	SO_2	25.78	5.795×10^{-2}	-3.812×10^{-5}	8.612×10^{-9}	273–1800	0.45	0.24
Sulfur trioxide	SO_3	16.40	14.58×10^{-2}	-11.20×10^{-5}	32.42×10^{-9}	273–1300	0.29	0.13
Acetylene	C_2H_2	21.8	9.2143×10^{-2}	-6.527×10^{-5}	18.21×10^{-9}	273–1500	1.46	0.59
Benzene	C_6H_6	−36.22	48.475×10^{-2}	-31.57×10^{-5}	77.62×10^{-9}	273–1500	0.34	0.20
Methanol	CH_4O	19.0	9.152×10^{-2}	-1.22×10^{-5}	-8.039×10^{-9}	273–1000	0.18	0.08
Ethanol	C_2H_6O	19.9	20.96×10^{-2}	-10.38×10^{-5}	20.05×10^{-9}	273–1500	0.40	0.22
Hydrogen chloride	HCl	30.33	-0.7620×10^{-2}	1.327×10^{-5}	-4.338×10^{-9}	273–1500	0.22	0.08
Methane	CH_4	19.89	5.024×10^{-2}	1.269×10^{-5}	-11.01×10^{-9}	273–1500	1.33	0.57
Ethane	C_2H_6	6.900	17.27×10^{-2}	-6.406×10^{-5}	7.285×10^{-9}	273–1500	0.83	0.28
Propane	C_3H_8	−4.04	30.48×10^{-2}	-15.72×10^{-5}	31.74×10^{-9}	273–1500	0.40	0.12
n-Butane	C_4H_{10}	3.96	37.15×10^{-2}	-18.34×10^{-5}	35.00×10^{-9}	273–1500	0.54	0.24
i-Butane	C_4H_{10}	−7.913	41.60×10^{-2}	-23.01×10^{-5}	49.91×10^{-9}	273–1500	0.25	0.13
n-Pentane	C_5H_{12}	6.774	45.43×10^{-2}	-22.46×10^{-5}	42.29×10^{-9}	273–1500	0.56	0.21
n-Hexane	C_6H_{14}	6.938	55.22×10^{-2}	-28.65×10^{-5}	57.69×10^{-9}	273–1500	0.72	0.20
Ethylene	C_2H_4	3.95	15.64×10^{-2}	-8.344×10^{-5}	17.67×10^{-9}	273–1500	0.54	0.13
Propylene	C_3H_6	3.15	23.83×10^{-2}	-12.18×10^{-5}	24.62×10^{-9}	273–1500	0.73	0.17

Source: Chemical and Process Thermodynamics 3/E by Kyle, B. G., © 2000. Adapted by permission of Pearson Education, Inc., Upper Saddle River, NJ.

TABLE A–3

Properties of common liquids, solids, and foods

(*a*) Liquids

	Boiling data at 1 atm		Freezing data		Liquid properties		
Substance	Normal boiling point, °C	Latent heat of vaporization h_{fg}, kJ/kg	Freezing point, °C	Latent heat of fusion h_{if}, kJ/kg	Temperature, °C	Density ρ, kg/m^3	Specific heat c_p, kJ/kg · K
Ammonia	−33.3	1357	−77.7	322.4	−33.3	682	4.43
					−20	665	4.52
					0	639	4.60
					25	602	4.80
Argon	−185.9	161.6	−189.3	28	−185.6	1394	1.14
Benzene	80.2	394	5.5	126	20	879	1.72
Brine (20% sodium chloride by mass)	103.9	—	−17.4	—	20	1150	3.11
n-Butane	−0.5	385.2	−138.5	80.3	−0.5	601	2.31
Carbon dioxide	−78.4*	230.5 (at 0°C)	−56.6		0	298	0.59
Ethanol	78.2	838.3	−114.2	109	25	783	2.46
Ethyl alcohol	78.6	855	−156	108	20	789	2.84
Ethylene glycol	198.1	800.1	−10.8	181.1	20	1109	2.84
Glycerine	179.9	974	18.9	200.6	20	1261	2.32
Helium	−268.9	22.8	—	—	−268.9	146.2	22.8
Hydrogen	−252.8	445.7	−259.2	59.5	−252.8	70.7	10.0
Isobutane	−11.7	367.1	−160	105.7	−11.7	593.8	2.28
Kerosene	204–293	251	−24.9	—	20	820	2.00
Mercury	356.7	294.7	−38.9	11.4	25	13,560	0.139
Methane	−161.5	510.4	−182.2	58.4	−161.5	423	3.49
					−100	301	5.79
Methanol	64.5	1100	−97.7	99.2	25	787	2.55
Nitrogen	−195.8	198.6	−210	25.3	−195.8	809	2.06
					−160	596	2.97
Octane	124.8	306.3	−57.5	180.7	20	703	2.10
Oil (light)					25	910	1.80
Oxygen	−183	212.7	−218.8	13.7	−183	1141	1.71
Petroleum	—	230–384			20	640	2.0
Propane	−42.1	427.8	−187.7	80.0	−42.1	581	2.25
					0	529	2.53
					50	449	3.13
Refrigerant-134a	−26.1	217.0	−96.6	—	−50	1443	1.23
					−26.1	1374	1.27
					0	1295	1.34
					25	1207	1.43
Water	100	2257	0.0	333.7	0	1000	4.22
					25	997	4.18
					50	988	4.18
					75	975	4.19
					100	958	4.22

* Sublimation temperature. (At pressures below the triple-point pressure of 518 kPa, carbon dioxide exists as a solid or gas. Also, the freezing-point temperature of carbon dioxide is the triple-point temperature of −56.5°C.)

TABLE A–3

Properties of common liquids, solids, and foods (*Concluded*)

(*b*) Solids (values are for room temperature unless indicated otherwise)

Substance	Density, ρ kg/m^3	Specific heat, c_p kJ/kg · K	Substance	Density, ρ kg/m^3	Specific heat, c_p kJ/kg · K
Metals			Nonmetals		
Aluminum			Asphalt	2110	0.920
200 K		0.797	Brick, common	1922	0.79
250 K		0.859	Brick, fireclay (500°C)	2300	0.960
300 K	2,700	0.902	Concrete	2300	0.653
350 K		0.929	Clay	1000	0.920
400 K		0.949	Diamond	2420	0.616
450 K		0.973	Glass, window	2700	0.800
500 K		0.997	Glass, pyrex	2230	0.840
Bronze (76% Cu, 2% Zn, 2% Al)	8,280	0.400	Graphite	2500	0.711
			Granite	2700	1.017
Brass, yellow (65% Cu, 35% Zn)	8,310	0.400	Gypsum or plaster board	800	1.09
			Ice		
Copper			200 K		1.56
−173°C		0.254	220 K		1.71
−100°C		0.342	240 K		1.86
−50°C		0.367	260 K		2.01
0°C		0.381	273 K	921	2.11
27°C	8,900	0.386	Limestone	1650	0.909
100°C		0.393	Marble	2600	0.880
200°C		0.403	Plywood (Douglas Fir)	545	1.21
Iron	7,840	0.45	Rubber (soft)	1100	1.840
Lead	11,310	0.128	Rubber (hard)	1150	2.009
Magnesium	1,730	1.000	Sand	1520	0.800
Nickel	8,890	0.440	Stone	1500	0.800
Silver	10,470	0.235	Woods, hard (maple, oak, etc.)	721	1.26
Steel, mild	7,830	0.500	Woods, soft (fir, pine, etc.)	513	1.38
Tungsten	19,400	0.130			

(*c*) Foods

Food	Water content, % (mass)	Freezing point, °C	Specific heat, kJ/kg · K: Above freezing	Specific heat, kJ/kg · K: Below freezing	Latent heat of fusion, kJ/kg	Food	Water content, % (mass)	Freezing point, °C	Specific heat, kJ/kg · K: Above freezing	Specific heat, kJ/kg · K: Below freezing	Latent heat of fusion, kJ/kg
Apples	84	−1.1	3.65	1.90	281	Lettuce	95	−0.2	4.02	2.04	317
Bananas	75	−0.8	3.35	1.78	251	Milk, whole	88	−0.6	3.79	1.95	294
Beef round	67	—	3.08	1.68	224	Oranges	87	−0.8	3.75	1.94	291
Broccoli	90	−0.6	3.86	1.97	301	Potatoes	78	−0.6	3.45	1.82	261
Butter	16	—	—	1.04	53	Salmon fish	64	−2.2	2.98	1.65	214
Cheese, swiss	39	−10.0	2.15	1.33	130	Shrimp	83	−2.2	3.62	1.89	277
Cherries	80	−1.8	3.52	1.85	267	Spinach	93	−0.3	3.96	2.01	311
Chicken	74	−2.8	3.32	1.77	247	Strawberries	90	−0.8	3.86	1.97	301
Corn, sweet	74	−0.6	3.32	1.77	247	Tomatoes, ripe	94	−0.5	3.99	2.02	314
Eggs, whole	74	−0.6	3.32	1.77	247	Turkey	64	—	2.98	1.65	214
Ice cream	63	−5.6	2.95	1.63	210	Watermelon	93	−0.4	3.96	2.01	311

Source: Values are obtained from various handbooks and other sources or are calculated. Water content and freezing-point data of foods are from *ASHRAE, Handbook of Fundamentals,* SI version (Atlanta, GA: American Society of Heating, Refrigerating and Air-Conditioning Engineers, Inc., 1993), Chapter 30, Table 1. Freezing point is the temperature at which freezing starts for fruits and vegetables, and the average freezing temperature for other foods.

TABLE A–4

Saturated water—Temperature table

Temp., T °C	Sat. press., P_{sat} kPa	Specific volume, m^3/kg: Sat. liquid, v_f	Sat. vapor, v_g	Internal energy, kJ/kg: Sat. liquid, u_f	Evap., u_{fg}	Sat. vapor, u_g	Enthalpy, kJ/kg: Sat. liquid, h_f	Evap., h_{fg}	Sat. vapor, h_g	Entropy, kJ/kg · K: Sat. liquid, s_f	Evap., s_{fg}	Sat. vapor, s_g
0.01	0.6117	0.001000	206.00	0.000	2374.9	2374.9	0.001	2500.9	2500.9	0.0000	9.1556	9.1556
5	0.8725	0.001000	147.03	21.019	2360.8	2381.8	21.020	2489.1	2510.1	0.0763	8.9487	9.0249
10	1.2281	0.001000	106.32	42.020	2346.6	2388.7	42.022	2477.2	2519.2	0.1511	8.7488	8.8999
15	1.7057	0.001001	77.885	62.980	2332.5	2395.5	62.982	2465.4	2528.3	0.2245	8.5559	8.7803
20	2.3392	0.001002	57.762	83.913	2318.4	2402.3	83.915	2453.5	2537.4	0.2965	8.3696	8.6661
25	3.1698	0.001003	43.340	104.83	2304.3	2409.1	104.83	2441.7	2546.5	0.3672	8.1895	8.5567
30	4.2469	0.001004	32.879	125.73	2290.2	2415.9	125.74	2429.8	2555.6	0.4368	8.0152	8.4520
35	5.6291	0.001006	25.205	146.63	2276.0	2422.7	146.64	2417.9	2564.6	0.5051	7.8466	8.3517
40	7.3851	0.001008	19.515	167.53	2261.9	2429.4	167.53	2406.0	2573.5	0.5724	7.6832	8.2556
45	9.5953	0.001010	15.251	188.43	2247.7	2436.1	188.44	2394.0	2582.4	0.6386	7.5247	8.1633
50	12.352	0.001012	12.026	209.33	2233.4	2442.7	209.34	2382.0	2591.3	0.7038	7.3710	8.0748
55	15.763	0.001015	9.5639	230.24	2219.1	2449.3	230.26	2369.8	2600.1	0.7680	7.2218	7.9898
60	19.947	0.001017	7.6670	251.16	2204.7	2455.9	251.18	2357.7	2608.8	0.8313	7.0769	7.9082
65	25.043	0.001020	6.1935	272.09	2190.3	2462.4	272.12	2345.4	2617.5	0.8937	6.9360	7.8296
70	31.202	0.001023	5.0396	293.04	2175.8	2468.9	293.07	2333.0	2626.1	0.9551	6.7989	7.7540
75	38.597	0.001026	4.1291	313.99	2161.3	2475.3	314.03	2320.6	2634.6	1.0158	6.6655	7.6812
80	47.416	0.001029	3.4053	334.97	2146.6	2481.6	335.02	2308.0	2643.0	1.0756	6.5355	7.6111
85	57.868	0.001032	2.8261	355.96	2131.9	2487.8	356.02	2295.3	2651.4	1.1346	6.4089	7.5435
90	70.183	0.001036	2.3593	376.97	2117.0	2494.0	377.04	2282.5	2659.6	1.1929	6.2853	7.4782
95	84.609	0.001040	1.9808	398.00	2102.0	2500.1	398.09	2269.6	2667.6	1.2504	6.1647	7.4151
100	101.42	0.001043	1.6720	419.06	2087.0	2506.0	419.17	2256.4	2675.6	1.3072	6.0470	7.3542
105	120.90	0.001047	1.4186	440.15	2071.8	2511.9	440.28	2243.1	2683.4	1.3634	5.9319	7.2952
110	143.38	0.001052	1.2094	461.27	2056.4	2517.7	461.42	2229.7	2691.1	1.4188	5.8193	7.2382
115	169.18	0.001056	1.0360	482.42	2040.9	2523.3	482.59	2216.0	2698.6	1.4737	5.7092	7.1829
120	198.67	0.001060	0.89133	503.60	2025.3	2528.9	503.81	2202.1	2706.0	1.5279	5.6013	7.1292
125	232.23	0.001065	0.77012	524.83	2009.5	2534.3	525.07	2188.1	2713.1	1.5816	5.4956	7.0771
130	270.28	0.001070	0.66808	546.10	1993.4	2539.5	546.38	2173.7	2720.1	1.6346	5.3919	7.0265
135	313.22	0.001075	0.58179	567.41	1977.3	2544.7	567.75	2159.1	2726.9	1.6872	5.2901	6.9773
140	361.53	0.001080	0.50850	588.77	1960.9	2549.6	589.16	2144.3	2733.5	1.7392	5.1901	6.9294
145	415.68	0.001085	0.44600	610.19	1944.2	2554.4	610.64	2129.2	2739.8	1.7908	5.0919	6.8827
150	476.16	0.001091	0.39248	631.66	1927.4	2559.1	632.18	2113.8	2745.9	1.8418	4.9953	6.8371
155	543.49	0.001096	0.34648	653.19	1910.3	2563.5	653.79	2098.0	2751.8	1.8924	4.9002	6.7927
160	618.23	0.001102	0.30680	674.79	1893.0	2567.8	675.47	2082.0	2757.5	1.9426	4.8066	6.7492
165	700.93	0.001108	0.27244	696.46	1875.4	2571.9	697.24	2065.6	2762.8	1.9923	4.7143	6.7067
170	792.18	0.001114	0.24260	718.20	1857.5	2575.7	719.08	2048.8	2767.9	2.0417	4.6233	6.6650
175	892.60	0.001121	0.21659	740.02	1839.4	2579.4	741.02	2031.7	2772.7	2.0906	4.5335	6.6242
180	1002.8	0.001127	0.19384	761.92	1820.9	2582.8	763.05	2014.2	2777.2	2.1392	4.4448	6.5841
185	1123.5	0.001134	0.17390	783.91	1802.1	2586.0	785.19	1996.2	2781.4	2.1875	4.3572	6.5447
190	1255.2	0.001141	0.15636	806.00	1783.0	2589.0	807.43	1977.9	2785.3	2.2355	4.2705	6.5059
195	1398.8	0.001149	0.14089	828.18	1763.6	2591.7	829.78	1959.0	2788.8	2.2831	4.1847	6.4678
200	1554.9	0.001157	0.12721	850.46	1743.7	2594.2	852.26	1939.8	2792.0	2.3305	4.0997	6.4302

(Continued)

TABLE A–4

Saturated water—Temperature table (*Continued*)

		Specific volume, m³/kg		*Internal energy,* kJ/kg			*Enthalpy,* kJ/kg			*Entropy,* kJ/kg · K		
Temp., T °C	Sat. press., P_{sat} kPa	Sat. liquid, v_f	Sat. vapor, v_g	Sat. liquid, u_f	Evap., u_{fg}	Sat. vapor, u_g	Sat. liquid, h_f	Evap., h_{fg}	Sat. vapor, h_g	Sat. liquid, s_f	Evap., s_{fg}	Sat. vapor, s_g
205	1724.3	0.001164	0.11508	872.86	1723.5	2596.4	874.87	1920.0	2794.8	2.3776	4.0154	6.3930
210	1907.7	0.001173	0.10429	895.38	1702.9	2598.3	897.61	1899.7	2797.3	2.4245	3.9318	6.3563
215	2105.9	0.001181	0.094680	918.02	1681.9	2599.9	920.50	1878.8	2799.3	2.4712	3.8489	6.3200
220	2319.6	0.001190	0.086094	940.79	1660.5	2601.3	943.55	1857.4	2801.0	2.5176	3.7664	6.2840
225	2549.7	0.001199	0.078405	963.70	1638.6	2602.3	966.76	1835.4	2802.2	2.5639	3.6844	6.2483
230	2797.1	0.001209	0.071505	986.76	1616.1	2602.9	990.14	1812.8	2802.9	2.6100	3.6028	6.2128
235	3062.6	0.001219	0.065300	1010.0	1593.2	2603.2	1013.7	1789.5	2803.2	2.6560	3.5216	6.1775
240	3347.0	0.001229	0.059707	1033.4	1569.8	2603.1	1037.5	1765.5	2803.0	2.7018	3.4405	6.1424
245	3651.2	0.001240	0.054656	1056.9	1545.7	2602.7	1061.5	1740.8	2802.2	2.7476	3.3596	6.1072
250	3976.2	0.001252	0.050085	1080.7	1521.1	2601.8	1085.7	1715.3	2801.0	2.7933	3.2788	6.0721
255	4322.9	0.001263	0.045941	1104.7	1495.8	2600.5	1110.1	1689.0	2799.1	2.8390	3.1979	6.0369
260	4692.3	0.001276	0.042175	1128.8	1469.9	2598.7	1134.8	1661.8	2796.6	2.8847	3.1169	6.0017
265	5085.3	0.001289	0.038748	1153.3	1443.2	2596.5	1159.8	1633.7	2793.5	2.9304	3.0358	5.9662
270	5503.0	0.001303	0.035622	1177.9	1415.7	2593.7	1185.1	1604.6	2789.7	2.9762	2.9542	5.9305
275	5946.4	0.001317	0.032767	1202.9	1387.4	2590.3	1210.7	1574.5	2785.2	3.0221	2.8723	5.8944
280	6416.6	0.001333	0.030153	1228.2	1358.2	2586.4	1236.7	1543.2	2779.9	3.0681	2.7898	5.8579
285	6914.6	0.001349	0.027756	1253.7	1328.1	2581.8	1263.1	1510.7	2773.7	3.1144	2.7066	5.8210
290	7441.8	0.001366	0.025554	1279.7	1296.9	2576.5	1289.8	1476.9	2766.7	3.1608	2.6225	5.7834
295	7999.0	0.001384	0.023528	1306.0	1264.5	2570.5	1317.1	1441.6	2758.7	3.2076	2.5374	5.7450
300	8587.9	0.001404	0.021659	1332.7	1230.9	2563.6	1344.8	1404.8	2749.6	3.2548	2.4511	5.7059
305	9209.4	0.001425	0.019932	1360.0	1195.9	2555.8	1373.1	1366.3	2739.4	3.3024	2.3633	5.6657
310	9865.0	0.001447	0.018333	1387.7	1159.3	2547.1	1402.0	1325.9	2727.9	3.3506	2.2737	5.6243
315	10,556	0.001472	0.016849	1416.1	1121.1	2537.2	1431.6	1283.4	2715.0	3.3994	2.1821	5.5816
320	11,284	0.001499	0.015470	1445.1	1080.9	2526.0	1462.0	1238.5	2700.6	3.4491	2.0881	5.5372
325	12,051	0.001528	0.014183	1475.0	1038.5	2513.4	1493.4	1191.0	2684.3	3.4998	1.9911	5.4908
330	12,858	0.001560	0.012979	1505.7	993.5	2499.2	1525.8	1140.3	2666.0	3.5516	1.8906	5.4422
335	13,707	0.001597	0.011848	1537.5	945.5	2483.0	1559.4	1086.0	2645.4	3.6050	1.7857	5.3907
340	14,601	0.001638	0.010783	1570.7	893.8	2464.5	1594.6	1027.4	2622.0	3.6602	1.6756	5.3358
345	15,541	0.001685	0.009772	1605.5	837.7	2443.2	1631.7	963.4	2595.1	3.7179	1.5585	5.2765
350	16,529	0.001741	0.008806	1642.4	775.9	2418.3	1671.2	892.7	2563.9	3.7788	1.4326	5.2114
355	17,570	0.001808	0.007872	1682.2	706.4	2388.6	1714.0	812.9	2526.9	3.8442	1.2942	5.1384
360	18,666	0.001895	0.006950	1726.2	625.7	2351.9	1761.5	720.1	2481.6	3.9165	1.1373	5.0537
365	19,822	0.002015	0.006009	1777.2	526.4	2303.6	1817.2	605.5	2422.7	4.0004	0.9489	4.9493
370	21,044	0.002217	0.004953	1844.5	385.6	2230.1	1891.2	443.1	2334.3	4.1119	0.6890	4.8009
373.95	22,064	0.003106	0.003106	2015.7	0	2015.7	2084.3	0	2084.3	4.4070	0	4.4070

Source: Tables A–4 through A–8 are generated using the Engineering Equation Solver (EES) software developed by S. A. Klein and F. L. Alvarado. The routine used in calculations is the highly accurate Steam_IAPWS, which incorporates the 1995 Formulation for the Thermodynamic Properties of Ordinary Water Substance for General and Scientific Use, issued by The International Association for the Properties of Water and Steam (IAPWS). This formulation replaces the 1984 formulation of Haar, Gallagher, and Kell (NBS/NRC Steam Tables, Hemisphere Publishing Co., 1984), which is also available in EES as the routine STEAM. The new formulation is based on the correlations of Saul and Wagner (J. Phys. Chem. Ref. Data, 16, 893, 1987) with modifications to adjust to the International Temperature Scale of 1990. The modifications are described by Wagner and Pruss (J. Phys. Chem. Ref. Data, 22, 783, 1993). The properties of ice are based on Hyland and Wexler, "Formulations for the Thermodynamic Properties of the Saturated Phases of H_2O from 173.15 K to 473.15 K," *ASHRAE Trans.*, Part 2A, Paper 2793, 1983.

TABLE A–5

Saturated water—Pressure table

		Specific volume, m³/kg		*Internal energy,* kJ/kg			*Enthalpy,* kJ/kg			*Entropy,* kJ/kg · K		
Press., P kPa	Sat. temp., T_{sat} °C	Sat. liquid, v_f	Sat. vapor, v_g	Sat. liquid, u_f	Evap., u_{fg}	Sat. vapor, u_g	Sat. liquid, h_f	Evap., h_{fg}	Sat. vapor, h_g	Sat. liquid, s_f	Evap., s_{fg}	Sat. vapor, s_g
1.0	6.97	0.001000	129.19	29.302	2355.2	2384.5	29.303	2484.4	2513.7	0.1059	8.8690	8.9749
1.5	13.02	0.001001	87.964	54.686	2338.1	2392.8	54.688	2470.1	2524.7	0.1956	8.6314	8.8270
2.0	17.50	0.001001	66.990	73.431	2325.5	2398.9	73.433	2459.5	2532.9	0.2606	8.4621	8.7227
2.5	21.08	0.001002	54.242	88.422	2315.4	2403.8	88.424	2451.0	2539.4	0.3118	8.3302	8.6421
3.0	24.08	0.001003	45.654	100.98	2306.9	2407.9	100.98	2443.9	2544.8	0.3543	8.2222	8.5765
4.0	28.96	0.001004	34.791	121.39	2293.1	2414.5	121.39	2432.3	2553.7	0.4224	8.0510	8.4734
5.0	32.87	0.001005	28.185	137.75	2282.1	2419.8	137.75	2423.0	2560.7	0.4762	7.9176	8.3938
7.5	40.29	0.001008	19.233	168.74	2261.1	2429.8	168.75	2405.3	2574.0	0.5763	7.6738	8.2501
10	45.81	0.001010	14.670	191.79	2245.4	2437.2	191.81	2392.1	2583.9	0.6492	7.4996	8.1488
15	53.97	0.001014	10.020	225.93	2222.1	2448.0	225.94	2372.3	2598.3	0.7549	7.2522	8.0071
20	60.06	0.001017	7.6481	251.40	2204.6	2456.0	251.42	2357.5	2608.9	0.8320	7.0752	7.9073
25	64.96	0.001020	6.2034	271.93	2190.4	2462.4	271.96	2345.5	2617.5	0.8932	6.9370	7.8302
30	69.09	0.001022	5.2287	289.24	2178.5	2467.7	289.27	2335.3	2624.6	0.9441	6.8234	7.7675
40	75.86	0.001026	3.9933	317.58	2158.8	2476.3	317.62	2318.4	2636.1	1.0261	6.6430	7.6691
50	81.32	0.001030	3.2403	340.49	2142.7	2483.2	340.54	2304.7	2645.2	1.0912	6.5019	7.5931
75	91.76	0.001037	2.2172	384.36	2111.8	2496.1	384.44	2278.0	2662.4	1.2132	6.2426	7.4558
100	99.61	0.001043	1.6941	417.40	2088.2	2505.6	417.51	2257.5	2675.0	1.3028	6.0562	7.3589
101.325	99.97	0.001043	1.6734	418.95	2087.0	2506.0	419.06	2256.5	2675.6	1.3069	6.0476	7.3545
125	105.97	0.001048	1.3750	444.23	2068.8	2513.0	444.36	2240.6	2684.9	1.3741	5.9100	7.2841
150	111.35	0.001053	1.1594	466.97	2052.3	2519.2	467.13	2226.0	2693.1	1.4337	5.7894	7.2231
175	116.04	0.001057	1.0037	486.82	2037.7	2524.5	487.01	2213.1	2700.2	1.4850	5.6865	7.1716
200	120.21	0.001061	0.88578	504.50	2024.6	2529.1	504.71	2201.6	2706.3	1.5302	5.5968	7.1270
225	123.97	0.001064	0.79329	520.47	2012.7	2533.2	520.71	2191.0	2711.7	1.5706	5.5171	7.0877
250	127.41	0.001067	0.71873	535.08	2001.8	2536.8	535.35	2181.2	2716.5	1.6072	5.4453	7.0525
275	130.58	0.001070	0.65732	548.57	1991.6	2540.1	548.86	2172.0	2720.9	1.6408	5.3800	7.0207
300	133.52	0.001073	0.60582	561.11	1982.1	2543.2	561.43	2163.5	2724.9	1.6717	5.3200	6.9917
325	136.27	0.001076	0.56199	572.84	1973.1	2545.9	573.19	2155.4	2728.6	1.7005	5.2645	6.9650
350	138.86	0.001079	0.52422	583.89	1964.6	2548.5	584.26	2147.7	2732.0	1.7274	5.2128	6.9402
375	141.30	0.001081	0.49133	594.32	1956.6	2550.9	594.73	2140.4	2735.1	1.7526	5.1645	6.9171
400	143.61	0.001084	0.46242	604.22	1948.9	2553.1	604.66	2133.4	2738.1	1.7765	5.1191	6.8955
450	147.90	0.001088	0.41392	622.65	1934.5	2557.1	623.14	2120.3	2743.4	1.8205	5.0356	6.8561
500	151.83	0.001093	0.37483	639.54	1921.2	2560.7	640.09	2108.0	2748.1	1.8604	4.9603	6.8207
550	155.46	0.001097	0.34261	655.16	1908.8	2563.9	655.77	2096.6	2752.4	1.8970	4.8916	6.7886
600	158.83	0.001101	0.31560	669.72	1897.1	2566.8	670.38	2085.8	2756.2	1.9308	4.8285	6.7593
650	161.98	0.001104	0.29260	683.37	1886.1	2569.4	684.08	2075.5	2759.6	1.9623	4.7699	6.7322
700	164.95	0.001108	0.27278	696.23	1875.6	2571.8	697.00	2065.8	2762.8	1.9918	4.7153	6.7071
750	167.75	0.001111	0.25552	708.40	1865.6	2574.0	709.24	2056.4	2765.7	2.0195	4.6642	6.6837

(Continued)

TABLE A–5

Saturated water—Pressure table (*Continued*)

		Specific volume, m^3/kg		*Internal energy, kJ/kg*			*Enthalpy, kJ/kg*			*Entropy, kJ/kg · K*		
Press., P kPa	Sat. temp., T_{sat} °C	Sat. liquid, v_f	Sat. vapor, v_g	Sat. liquid, u_f	Evap., u_{fg}	Sat. vapor, u_g	Sat. liquid, h_f	Evap., h_{fg}	Sat. vapor, h_g	Sat. liquid, s_f	Evap., s_{fg}	Sat. vapor, s_g
800	170.41	0.001115	0.24035	719.97	1856.1	2576.0	720.87	2047.5	2768.3	2.0457	4.6160	6.6616
850	172.94	0.001118	0.22690	731.00	1846.9	2577.9	731.95	2038.8	2770.8	2.0705	4.5705	6.6409
900	175.35	0.001121	0.21489	741.55	1838.1	2579.6	742.56	2030.5	2773.0	2.0941	4.5273	6.6213
950	177.66	0.001124	0.20411	751.67	1829.6	2581.3	752.74	2022.4	2775.2	2.1166	4.4862	6.6027
1000	179.88	0.001127	0.19436	761.39	1821.4	2582.8	762.51	2014.6	2777.1	2.1381	4.4470	6.5850
1100	184.06	0.001133	0.17745	779.78	1805.7	2585.5	781.03	1999.6	2780.7	2.1785	4.3735	6.5520
1200	187.96	0.001138	0.16326	796.96	1790.9	2587.8	798.33	1985.4	2783.8	2.2159	4.3058	6.5217
1300	191.60	0.001144	0.15119	813.10	1776.8	2589.9	814.59	1971.9	2786.5	2.2508	4.2428	6.4936
1400	195.04	0.001149	0.14078	828.35	1763.4	2591.8	829.96	1958.9	2788.9	2.2835	4.1840	6.4675
1500	198.29	0.001154	0.13171	842.82	1750.6	2593.4	844.55	1946.4	2791.0	2.3143	4.1287	6.4430
1750	205.72	0.001166	0.11344	876.12	1720.6	2596.7	878.16	1917.1	2795.2	2.3844	4.0033	6.3877
2000	212.38	0.001177	0.099587	906.12	1693.0	2599.1	908.47	1889.8	2798.3	2.4467	3.8923	6.3390
2250	218.41	0.001187	0.088717	933.54	1667.3	2600.9	936.21	1864.3	2800.5	2.5029	3.7926	6.2954
2500	223.95	0.001197	0.079952	958.87	1643.2	2602.1	961.87	1840.1	2801.9	2.5542	3.7016	6.2558
3000	233.85	0.001217	0.066667	1004.6	1598.5	2603.2	1008.3	1794.9	2803.2	2.6454	3.5402	6.1856
3500	242.56	0.001235	0.057061	1045.4	1557.6	2603.0	1049.7	1753.0	2802.7	2.7253	3.3991	6.1244
4000	250.35	0.001252	0.049779	1082.4	1519.3	2601.7	1087.4	1713.5	2800.8	2.7966	3.2731	6.0696
5000	263.94	0.001286	0.039448	1148.1	1448.9	2597.0	1154.5	1639.7	2794.2	2.9207	3.0530	5.9737
6000	275.59	0.001319	0.032449	1205.8	1384.1	2589.9	1213.8	1570.9	2784.6	3.0275	2.8627	5.8902
7000	285.83	0.001352	0.027378	1258.0	1323.0	2581.0	1267.5	1505.2	2772.6	3.1220	2.6927	5.8148
8000	295.01	0.001384	0.023525	1306.0	1264.5	2570.5	1317.1	1441.6	2758.7	3.2077	2.5373	5.7450
9000	303.35	0.001418	0.020489	1350.9	1207.6	2558.5	1363.7	1379.3	2742.9	3.2866	2.3925	5.6791
10,000	311.00	0.001452	0.018028	1393.3	1151.8	2545.2	1407.8	1317.6	2725.5	3.3603	2.2556	5.6159
11,000	318.08	0.001488	0.015988	1433.9	1096.6	2530.4	1450.2	1256.1	2706.3	3.4299	2.1245	5.5544
12,000	324.68	0.001526	0.014264	1473.0	1041.3	2514.3	1491.3	1194.1	2685.4	3.4964	1.9975	5.4939
13,000	330.85	0.001566	0.012781	1511.0	985.5	2496.6	1531.4	1131.3	2662.7	3.5606	1.8730	5.4336
14,000	336.67	0.001610	0.011487	1548.4	928.7	2477.1	1571.0	1067.0	2637.9	3.6232	1.7497	5.3728
15,000	342.16	0.001657	0.010341	1585.5	870.3	2455.7	1610.3	1000.5	2610.8	3.6848	1.6261	5.3108
16,000	347.36	0.001710	0.009312	1622.6	809.4	2432.0	1649.9	931.1	2581.0	3.7461	1.5005	5.2466
17,000	352.29	0.001770	0.008374	1660.2	745.1	2405.4	1690.3	857.4	2547.7	3.8082	1.3709	5.1791
18,000	356.99	0.001840	0.007504	1699.1	675.9	2375.0	1732.2	777.8	2510.0	3.8720	1.2343	5.1064
19,000	361.47	0.001926	0.006677	1740.3	598.9	2339.2	1776.8	689.2	2466.0	3.9396	1.0860	5.0256
20,000	365.75	0.002038	0.005862	1785.8	509.0	2294.8	1826.6	585.5	2412.1	4.0146	0.9164	4.9310
21,000	369.83	0.002207	0.004994	1841.6	391.9	2233.5	1888.0	450.4	2338.4	4.1071	0.7005	4.8076
22,000	373.71	0.002703	0.003644	1951.7	140.8	2092.4	2011.1	161.5	2172.6	4.2942	0.2496	4.5439
22,064	373.95	0.003106	0.003106	2015.7	0	2015.7	2084.3	0	2084.3	4.4070	0	4.4070

- Measure T_{DP} (with chilled mirror hygrometer)
- Find P_v at T_{DP}
- Measure T_{DB} (dry bulb)
- Find P_g at T_{DB}

$$\phi = \frac{P_v \text{ at } T_{DP}}{P_g \text{ at } T_{DB}}$$

2) Adiabatic saturation apparatus → T_{DB}

$w_1 = c_p a(T_2 - T_1) + w_2(h_{fg2})$; $h_{g1} - h_{f2}$

wet bulb thermometer

- unsaturated air passes over wet wick; water in wick evaporates
- Energy for latent heat of vaporization comes from water
- Temp of water drops
- Eventually ... = gained by air
- water temp stabilizes & thermometer reads $T_{wb} < T_{DB}$

sling psychrometer

TABLE A–6

Superheated water

T °C	v m³/kg	u kJ/kg	h kJ/kg	s kJ/kg · K	v m³/kg	u kJ/kg	h kJ/kg	s kJ/kg · K	v m³/kg	u kJ/kg	h kJ/kg	s kJ/kg · K
	P = 0.01 MPa (45.81°C)*				P = 0.05 MPa (81.32°C)				P = 0.10 MPa (99.61°C)			
Sat.†	14.670	2437.2	2583.9	8.1488	3.2403	2483.2	2645.2	7.5931	1.6941	2505.6	2675.0	7.3589
50	14.867	2443.3	2592.0	8.1741								
100	17.196	2515.5	2687.5	8.4489	3.4187	2511.5	2682.4	7.6953	1.6959	2506.2	2675.8	7.3611
150	19.513	2587.9	2783.0	8.6893	3.8897	2585.7	2780.2	7.9413	1.9367	2582.9	2776.6	7.6148
200	21.826	2661.4	2879.6	8.9049	4.3562	2660.0	2877.8	8.1592	2.1724	2658.2	2875.5	7.8356
250	24.136	2736.1	2977.5	9.1015	4.8206	2735.1	2976.2	8.3568	2.4062	2733.9	2974.5	8.0346
300	26.446	2812.3	3076.7	9.2827	5.2841	2811.6	3075.8	8.5387	2.6389	2810.7	3074.5	8.2172
400	31.063	2969.3	3280.0	9.6094	6.2094	2968.9	3279.3	8.8659	3.1027	2968.3	3278.6	8.5452
500	35.680	3132.9	3489.7	9.8998	7.1338	3132.6	3489.3	9.1566	3.5655	3132.2	3488.7	8.8362
600	40.296	3303.3	3706.3	10.1631	8.0577	3303.1	3706.0	9.4201	4.0279	3302.8	3705.6	9.0999
700	44.911	3480.8	3929.9	10.4056	8.9813	3480.6	3929.7	9.6626	4.4900	3480.4	3929.4	9.3424
800	49.527	3665.4	4160.6	10.6312	9.9047	3665.2	4160.4	9.8883	4.9519	3665.0	4160.2	9.5682
900	54.143	3856.9	4398.3	10.8429	10.8280	3856.8	4398.2	10.1000	5.4137	3856.7	4398.0	9.7800
1000	58.758	4055.3	4642.8	11.0429	11.7513	4055.2	4642.7	10.3000	5.8755	4055.0	4642.6	9.9800
1100	63.373	4260.0	4893.8	11.2326	12.6745	4259.9	4893.7	10.4897	6.3372	4259.8	4893.6	10.1698
1200	67.989	4470.9	5150.8	11.4132	13.5977	4470.8	5150.7	10.6704	6.7988	4470.7	5150.6	10.3504
1300	72.604	4687.4	5413.4	11.5857	14.5209	4687.3	5413.3	10.8429	7.2605	4687.2	5413.3	10.5229
	P = 0.20 MPa (120.21°C)				P = 0.30 MPa (133.52°C)				P = 0.40 MPa (143.61°C)			
Sat.	0.88578	2529.1	2706.3	7.1270	0.60582	2543.2	2724.9	6.9917	0.46242	2553.1	2738.1	6.8955
150	0.95986	2577.1	2769.1	7.2810	0.63402	2571.0	2761.2	7.0792	0.47088	2564.4	2752.8	6.9306
200	1.08049	2654.6	2870.7	7.5081	0.71643	2651.0	2865.9	7.3132	0.53434	2647.2	2860.9	7.1723
250	1.19890	2731.4	2971.2	7.7100	0.79645	2728.9	2967.9	7.5180	0.59520	2726.4	2964.5	7.3804
300	1.31623	2808.8	3072.1	7.8941	0.87535	2807.0	3069.6	7.7037	0.65489	2805.1	3067.1	7.5677
400	1.54934	2967.2	3277.0	8.2236	1.03155	2966.0	3275.5	8.0347	0.77265	2964.9	3273.9	7.9003
500	1.78142	3131.4	3487.7	8.5153	1.18672	3130.6	3486.6	8.3271	0.88936	3129.8	3485.5	8.1933
600	2.01302	3302.2	3704.8	8.7793	1.34139	3301.6	3704.0	8.5915	1.00558	3301.0	3703.3	8.4580
700	2.24434	3479.9	3928.8	9.0221	1.49580	3479.5	3928.2	8.8345	1.12152	3479.0	3927.6	8.7012
800	2.47550	3664.7	4159.8	9.2479	1.65004	3664.3	4159.3	9.0605	1.23730	3663.9	4158.9	8.9274
900	2.70656	3856.3	4397.7	9.4598	1.80417	3856.0	4397.3	9.2725	1.35298	3855.7	4396.9	9.1394
1000	2.93755	4054.8	4642.3	9.6599	1.95824	4054.5	4642.0	9.4726	1.46859	4054.3	4641.7	9.3396
1100	3.16848	4259.6	4893.3	9.8497	2.11226	4259.4	4893.1	9.6624	1.58414	4259.2	4892.9	9.5295
1200	3.39938	4470.5	5150.4	10.0304	2.26624	4470.3	5150.2	9.8431	1.69966	4470.2	5150.0	9.7102
1300	3.63026	4687.1	5413.1	10.2029	2.42019	4686.9	5413.0	10.0157	1.81516	4686.7	5412.8	9.8828
	P = 0.50 MPa (151.83°C)				P = 0.60 MPa (158.83°C)				P = 0.80 MPa (170.41°C)			
Sat.	0.37483	2560.7	2748.1	6.8207	0.31560	2566.8	2756.2	6.7593	0.24035	2576.0	2768.3	6.6616
200	0.42503	2643.3	2855.8	7.0610	0.35212	2639.4	2850.6	6.9683	0.26088	2631.1	2839.8	6.8177
250	0.47443	2723.8	2961.0	7.2725	0.39390	2721.2	2957.6	7.1833	0.29321	2715.9	2950.4	7.0402
300	0.52261	2803.3	3064.6	7.4614	0.43442	2801.4	3062.0	7.3740	0.32416	2797.5	3056.9	7.2345
350	0.57015	2883.0	3168.1	7.6346	0.47428	2881.6	3166.1	7.5481	0.35442	2878.6	3162.2	7.4107
400	0.61731	2963.7	3272.4	7.7956	0.51374	2962.5	3270.8	7.7097	0.38429	2960.2	3267.7	7.5735
500	0.71095	3129.0	3484.5	8.0893	0.59200	3128.2	3483.4	8.0041	0.44332	3126.6	3481.3	7.8692
600	0.80409	3300.4	3702.5	8.3544	0.66976	3299.8	3701.7	8.2695	0.50186	3298.7	3700.1	8.1354
700	0.89696	3478.6	3927.0	8.5978	0.74725	3478.1	3926.4	8.5132	0.56011	3477.2	3925.3	8.3794
800	0.98966	3663.6	4158.4	8.8240	0.82457	3663.2	4157.9	8.7395	0.61820	3662.5	4157.0	8.6061
900	1.08227	3855.4	4396.6	9.0362	0.90179	3855.1	4396.2	8.9518	0.67619	3854.5	4395.5	8.8185
1000	1.17480	4054.0	4641.4	9.2364	0.97893	4053.8	4641.1	9.1521	0.73411	4053.3	4640.5	9.0189
1100	1.26728	4259.0	4892.6	9.4263	1.05603	4258.8	4892.4	9.3420	0.79197	4258.3	4891.9	9.2090
1200	1.35972	4470.0	5149.8	9.6071	1.13309	4469.8	5149.6	9.5229	0.84980	4469.4	5149.3	9.3898
1300	1.45214	4686.6	5412.6	9.7797	1.21012	4686.4	5412.5	9.6955	0.90761	4686.1	5412.2	9.5625

*The temperature in parentheses is the saturation temperature at the specified pressure.

† Properties of saturated vapor at the specified pressure.

TABLE A–6

Superheated water (*Continued*)

T °C	*v* m³/kg	*u* kJ/kg	*h* kJ/kg	*s* kJ/kg · K	*v* m³/kg	*u* kJ/kg	*h* kJ/kg	*s* kJ/kg · K	*v* m³/kg	*u* kJ/kg	*h* kJ/kg	*s* kJ/kg · K
	P = 1.00 MPa (179.88°C)				*P* = 1.20 MPa (187.96°C)				*P* = 1.40 MPa (195.04°C)			
Sat.	0.19437	2582.8	2777.1	6.5850	0.16326	2587.8	2783.8	6.5217	0.14078	2591.8	2788.9	6.4675
200	0.20602	2622.3	2828.3	6.6956	0.16934	2612.9	2816.1	6.5909	0.14303	2602.7	2803.0	6.4975
250	0.23275	2710.4	2943.1	6.9265	0.19241	2704.7	2935.6	6.8313	0.16356	2698.9	2927.9	6.7488
300	0.25799	2793.7	3051.6	7.1246	0.21386	2789.7	3046.3	7.0335	0.18233	2785.7	3040.9	6.9553
350	0.28250	2875.7	3158.2	7.3029	0.23455	2872.7	3154.2	7.2139	0.20029	2869.7	3150.1	7.1379
400	0.30661	2957.9	3264.5	7.4670	0.25482	2955.5	3261.3	7.3793	0.21782	2953.1	3258.1	7.3046
500	0.35411	3125.0	3479.1	7.7642	0.29464	3123.4	3477.0	7.6779	0.25216	3121.8	3474.8	7.6047
600	0.40111	3297.5	3698.6	8.0311	0.33395	3296.3	3697.0	7.9456	0.28597	3295.1	3695.5	7.8730
700	0.44783	3476.3	3924.1	8.2755	0.37297	3475.3	3922.9	8.1904	0.31951	3474.4	3921.7	8.1183
800	0.49438	3661.7	4156.1	8.5024	0.41184	3661.0	4155.2	8.4176	0.35288	3660.3	4154.3	8.3458
900	0.54083	3853.9	4394.8	8.7150	0.45059	3853.3	4394.0	8.6303	0.38614	3852.7	4393.3	8.5587
1000	0.58721	4052.7	4640.0	8.9155	0.48928	4052.2	4639.4	8.8310	0.41933	4051.7	4638.8	8.7595
1100	0.63354	4257.9	4891.4	9.1057	0.52792	4257.5	4891.0	9.0212	0.45247	4257.0	4890.5	8.9497
1200	0.67983	4469.0	5148.9	9.2866	0.56652	4468.7	5148.5	9.2022	0.48558	4468.3	5148.1	9.1308
1300	0.72610	4685.8	5411.9	9.4593	0.60509	4685.5	5411.6	9.3750	0.51866	4685.1	5411.3	9.3036
	P = 1.60 MPa (201.37°C)				*P* = 1.80 MPa (207.11°C)				*P* = 2.00 MPa (212.38°C)			
Sat.	0.12374	2594.8	2792.8	6.4200	0.11037	2597.3	2795.9	6.3775	0.09959	2599.1	2798.3	6.3390
225	0.13293	2645.1	2857.8	6.5537	0.11678	2637.0	2847.2	6.4825	0.10381	2628.5	2836.1	6.4160
250	0.14190	2692.9	2919.9	6.6753	0.12502	2686.7	2911.7	6.6088	0.11150	2680.3	2903.3	6.5475
300	0.15866	2781.6	3035.4	6.8864	0.14025	2777.4	3029.9	6.8246	0.12551	2773.2	3024.2	6.7684
350	0.17459	2866.6	3146.0	7.0713	0.15460	2863.6	3141.9	7.0120	0.13860	2860.5	3137.7	6.9583
400	0.19007	2950.8	3254.9	7.2394	0.16849	2948.3	3251.6	7.1814	0.15122	2945.9	3248.4	7.1292
500	0.22029	3120.1	3472.6	7.5410	0.19551	3118.5	3470.4	7.4845	0.17568	3116.9	3468.3	7.4337
600	0.24999	3293.9	3693.9	7.8101	0.22200	3292.7	3692.3	7.7543	0.19962	3291.5	3690.7	7.7043
700	0.27941	3473.5	3920.5	8.0558	0.24822	3472.6	3919.4	8.0005	0.22326	3471.7	3918.2	7.9509
800	0.30865	3659.5	4153.4	8.2834	0.27426	3658.8	4152.4	8.2284	0.24674	3658.0	4151.5	8.1791
900	0.33780	3852.1	4392.6	8.4965	0.30020	3851.5	4391.9	8.4417	0.27012	3850.9	4391.1	8.3925
1000	0.36687	4051.2	4638.2	8.6974	0.32606	4050.7	4637.6	8.6427	0.29342	4050.2	4637.1	8.5936
1100	0.39589	4256.6	4890.0	8.8878	0.35188	4256.2	4889.6	8.8331	0.31667	4255.7	4889.1	8.7842
1200	0.42488	4467.9	5147.7	9.0689	0.37766	4467.6	5147.3	9.0143	0.33989	4467.2	5147.0	8.9654
1300	0.45383	4684.8	5410.9	9.2418	0.40341	4684.5	5410.6	9.1872	0.36308	4684.2	5410.3	9.1384
	P = 2.50 MPa (223.95°C)				*P* = 3.00 MPa (233.85°C)				*P* = 3.50 MPa (242.56°C)			
Sat.	0.07995	2602.1	2801.9	6.2558	0.06667	2603.2	2803.2	6.1856	0.05706	2603.0	2802.7	6.1244
225	0.08026	2604.8	2805.5	6.2629								
250	0.08705	2663.3	2880.9	6.4107	0.07063	2644.7	2856.5	6.2893	0.05876	2624.0	2829.7	6.1764
300	0.09894	2762.2	3009.6	6.6459	0.08118	2750.8	2994.3	6.5412	0.06845	2738.8	2978.4	6.4484
350	0.10979	2852.5	3127.0	6.8424	0.09056	2844.4	3116.1	6.7450	0.07680	2836.0	3104.9	6.6601
400	0.12012	2939.8	3240.1	7.0170	0.09938	2933.6	3231.7	6.9235	0.08456	2927.2	3223.2	6.8428
450	0.13015	3026.2	3351.6	7.1768	0.10789	3021.2	3344.9	7.0856	0.09198	3016.1	3338.1	7.0074
500	0.13999	3112.8	3462.8	7.3254	0.11620	3108.6	3457.2	7.2359	0.09919	3104.5	3451.7	7.1593
600	0.15931	3288.5	3686.8	7.5979	0.13245	3285.5	3682.8	7.5103	0.11325	3282.5	3678.9	7.4357
700	0.17835	3469.3	3915.2	7.8455	0.14841	3467.0	3912.2	7.7590	0.12702	3464.7	3909.3	7.6855
800	0.19722	3656.2	4149.2	8.0744	0.16420	3654.3	4146.9	7.9885	0.14061	3652.5	4144.6	7.9156
900	0.21597	3849.4	4389.3	8.2882	0.17988	3847.9	4387.5	8.2028	0.15410	3846.4	4385.7	8.1304
1000	0.23466	4049.0	4635.6	8.4897	0.19549	4047.7	4634.2	8.4045	0.16751	4046.4	4632.7	8.3324
1100	0.25330	4254.7	4887.9	8.6804	0.21105	4253.6	4886.7	8.5955	0.18087	4252.5	4885.6	8.5236
1200	0.27190	4466.3	5146.0	8.8618	0.22658	4465.3	5145.1	8.7771	0.19420	4464.4	5144.1	8.7053
1300	0.29048	4683.4	5409.5	9.0349	0.24207	4682.6	5408.8	8.9502	0.20750	4681.8	5408.0	8.8786

TABLE A–6

Superheated water (*Continued*)

T °C	*v* m³/kg	*u* kJ/kg	*h* kJ/kg	*s* kJ/kg · K	*v* m³/kg	*u* kJ/kg	*h* kJ/kg	*s* kJ/kg · K	*v* m³/kg	*u* kJ/kg	*h* kJ/kg	*s* kJ/kg · K
	P = 4.0 MPa (250.35°C)				*P* = 4.5 MPa (257.44°C)				*P* = 5.0 MPa (263.94°C)			
Sat.	0.04978	2601.7	2800.8	6.0696	0.04406	2599.7	2798.0	6.0198	0.03945	2597.0	2794.2	5.9737
275	0.05461	2668.9	2887.3	6.2312	0.04733	2651.4	2864.4	6.1429	0.04144	2632.3	2839.5	6.0571
300	0.05887	2726.2	2961.7	6.3639	0.05138	2713.0	2944.2	6.2854	0.04535	2699.0	2925.7	6.2111
350	0.06647	2827.4	3093.3	6.5843	0.05842	2818.6	3081.5	6.5153	0.05197	2809.5	3069.3	6.4516
400	0.07343	2920.8	3214.5	6.7714	0.06477	2914.2	3205.7	6.7071	0.05784	2907.5	3196.7	6.6483
450	0.08004	3011.0	3331.2	6.9386	0.07076	3005.8	3324.2	6.8770	0.06332	3000.6	3317.2	6.8210
500	0.08644	3100.3	3446.0	7.0922	0.07652	3096.0	3440.4	7.0323	0.06858	3091.8	3434.7	6.9781
600	0.09886	3279.4	3674.9	7.3706	0.08766	3276.4	3670.9	7.3127	0.07870	3273.3	3666.9	7.2605
700	0.11098	3462.4	3906.3	7.6214	0.09850	3460.0	3903.3	7.5647	0.08852	3457.7	3900.3	7.5136
800	0.12292	3650.6	4142.3	7.8523	0.10916	3648.8	4140.0	7.7962	0.09816	3646.9	4137.7	7.7458
900	0.13476	3844.8	4383.9	8.0675	0.11972	3843.3	4382.1	8.0118	0.10769	3841.8	4380.2	7.9619
1000	0.14653	4045.1	4631.2	8.2698	0.13020	4043.9	4629.8	8.2144	0.11715	4042.6	4628.3	8.1648
1100	0.15824	4251.4	4884.4	8.4612	0.14064	4250.4	4883.2	8.4060	0.12655	4249.3	4882.1	8.3566
1200	0.16992	4463.5	5143.2	8.6430	0.15103	4462.6	5142.2	8.5880	0.13592	4461.6	5141.3	8.5388
1300	0.18157	4680.9	5407.2	8.8164	0.16140	4680.1	5406.5	8.7616	0.14527	4679.3	5405.7	8.7124
	P = 6.0 MPa (275.59°C)				*P* = 7.0 MPa (285.83°C)				*P* = 8.0 MPa (295.01°C)			
Sat.	0.03245	2589.9	2784.6	5.8902	0.027378	2581.0	2772.6	5.8148	0.023525	2570.5	2758.7	5.7450
300	0.03619	2668.4	2885.6	6.0703	0.029492	2633.5	2839.9	5.9337	0.024279	2592.3	2786.5	5.7937
350	0.04225	2790.4	3043.9	6.3357	0.035262	2770.1	3016.9	6.2305	0.029975	2748.3	2988.1	6.1321
400	0.04742	2893.7	3178.3	6.5432	0.039958	2879.5	3159.2	6.4502	0.034344	2864.6	3139.4	6.3658
450	0.05217	2989.9	3302.9	6.7219	0.044187	2979.0	3288.3	6.6353	0.038194	2967.8	3273.3	6.5579
500	0.05667	3083.1	3423.1	6.8826	0.048157	3074.3	3411.4	6.8000	0.041767	3065.4	3399.5	6.7266
550	0.06102	3175.2	3541.3	7.0308	0.051966	3167.9	3531.6	6.9507	0.045172	3160.5	3521.8	6.8800
600	0.06527	3267.2	3658.8	7.1693	0.055665	3261.0	3650.6	7.0910	0.048463	3254.7	3642.4	7.0221
700	0.07355	3453.0	3894.3	7.4247	0.062850	3448.3	3888.3	7.3487	0.054829	3443.6	3882.2	7.2822
800	0.08165	3643.2	4133.1	7.6582	0.069856	3639.5	4128.5	7.5836	0.061011	3635.7	4123.8	7.5185
900	0.08964	3838.8	4376.6	7.8751	0.076750	3835.7	4373.0	7.8014	0.067082	3832.7	4369.3	7.7372
1000	0.09756	4040.1	4625.4	8.0786	0.083571	4037.5	4622.5	8.0055	0.073079	4035.0	4619.6	7.9419
1100	0.10543	4247.1	4879.7	8.2709	0.090341	4245.0	4877.4	8.1982	0.079025	4242.8	4875.0	8.1350
1200	0.11326	4459.8	5139.4	8.4534	0.097075	4457.9	5137.4	8.3810	0.084934	4456.1	5135.5	8.3181
1300	0.12107	4677.7	5404.1	8.6273	0.103781	4676.1	5402.6	8.5551	0.090817	4674.5	5401.0	8.4925
	P = 9.0 MPa (303.35°C)				*P* = 10.0 MPa (311.00°C)				*P* = 12.5 MPa (327.81°C)			
Sat.	0.020489	2558.5	2742.9	5.6791	0.018028	2545.2	2725.5	5.6159	0.013496	2505.6	2674.3	5.4638
325	0.023284	2647.6	2857.1	5.8738	0.019877	2611.6	2810.3	5.7596				
350	0.025816	2725.0	2957.3	6.0380	0.022440	2699.6	2924.0	5.9460	0.016138	2624.9	2826.6	5.7130
400	0.029960	2849.2	3118.8	6.2876	0.026436	2833.1	3097.5	6.2141	0.020030	2789.6	3040.0	6.0433
450	0.033524	2956.3	3258.0	6.4872	0.029782	2944.5	3242.4	6.4219	0.023019	2913.7	3201.5	6.2749
500	0.036793	3056.3	3387.4	6.6603	0.032811	3047.0	3375.1	6.5995	0.025630	3023.2	3343.6	6.4651
550	0.039885	3153.0	3512.0	6.8164	0.035655	3145.4	3502.0	6.7585	0.028033	3126.1	3476.5	6.6317
600	0.042861	3248.4	3634.1	6.9605	0.038378	3242.0	3625.8	6.9045	0.030306	3225.8	3604.6	6.7828
650	0.045755	3343.4	3755.2	7.0954	0.041018	3338.0	3748.1	7.0408	0.032491	3324.1	3730.2	6.9227
700	0.048589	3438.8	3876.1	7.2229	0.043597	3434.0	3870.0	7.1693	0.034612	3422.0	3854.6	7.0540
800	0.054132	3632.0	4119.2	7.4606	0.048629	3628.2	4114.5	7.4085	0.038724	3618.8	4102.8	7.2967
900	0.059562	3829.6	4365.7	7.6802	0.053547	3826.5	4362.0	7.6290	0.042720	3818.9	4352.9	7.5195
1000	0.064919	4032.4	4616.7	7.8855	0.058391	4029.9	4613.8	7.8349	0.046641	4023.5	4606.5	7.7269
1100	0.070224	4240.7	4872.7	8.0791	0.063183	4238.5	4870.3	8.0289	0.050510	4233.1	4864.5	7.9220
1200	0.075492	4454.2	5133.6	8.2625	0.067938	4452.4	5131.7	8.2126	0.054342	4447.7	5127.0	8.1065
1300	0.080733	4672.9	5399.5	8.4371	0.072667	4671.3	5398.0	8.3874	0.058147	4667.3	5394.1	8.2819

TABLE A–6

Superheated water (*Concluded*)

T °C	*v* m³/kg	*u* kJ/kg	*h* kJ/kg	*s* kJ/kg · K	*v* m³/kg	*u* kJ/kg	*h* kJ/kg	*s* kJ/kg · K	*v* m³/kg	*u* kJ/kg	*h* kJ/kg	*s* kJ/kg · K
	P = 15.0 MPa (342.16°C)				*P* = 17.5 MPa (354.67°C)				*P* = 20.0 MPa (365.75°C)			
Sat.	0.010341	2455.7	2610.8	5.3108	0.007932	2390.7	2529.5	5.1435	0.005862	2294.8	2412.1	4.9310
350	0.011481	2520.9	2693.1	5.4438								
400	0.015671	2740.6	2975.7	5.8819	0.012463	2684.3	2902.4	5.7211	0.009950	2617.9	2816.9	5.5526
450	0.018477	2880.8	3157.9	6.1434	0.015204	2845.4	3111.4	6.0212	0.012721	2807.3	3061.7	5.9043
500	0.020828	2998.4	3310.8	6.3480	0.017385	2972.4	3276.7	6.2424	0.014793	2945.3	3241.2	6.1446
550	0.022945	3106.2	3450.4	6.5230	0.019305	3085.8	3423.6	6.4266	0.016571	3064.7	3396.2	6.3390
600	0.024921	3209.3	3583.1	6.6796	0.021073	3192.5	3561.3	6.5890	0.018185	3175.3	3539.0	6.5075
650	0.026804	3310.1	3712.1	6.8233	0.022742	3295.8	3693.8	6.7366	0.019695	3281.4	3675.3	6.6593
700	0.028621	3409.8	3839.1	6.9573	0.024342	3397.5	3823.5	6.8735	0.021134	3385.1	3807.8	6.7991
800	0.032121	3609.3	4091.1	7.2037	0.027405	3599.7	4079.3	7.1237	0.023870	3590.1	4067.5	7.0531
900	0.035503	3811.2	4343.7	7.4288	0.030348	3803.5	4334.6	7.3511	0.026484	3795.7	4325.4	7.2829
1000	0.038808	4017.1	4599.2	7.6378	0.033215	4010.7	4592.0	7.5616	0.029020	4004.3	4584.7	7.4950
1100	0.042062	4227.7	4858.6	7.8339	0.036029	4222.3	4852.8	7.7588	0.031504	4216.9	4847.0	7.6933
1200	0.045279	4443.1	5122.3	8.0192	0.038806	4438.5	5117.6	7.9449	0.033952	4433.8	5112.9	7.8802
1300	0.048469	4663.3	5390.3	8.1952	0.041556	4659.2	5386.5	8.1215	0.036371	4655.2	5382.7	8.0574
	P = 25.0 MPa				*P* = 30.0 MPa				*P* = 35.0 MPa			
375	0.001978	1799.9	1849.4	4.0345	0.001792	1738.1	1791.9	3.9313	0.001701	1702.8	1762.4	3.8724
400	0.006005	2428.5	2578.7	5.1400	0.002798	2068.9	2152.8	4.4758	0.002105	1914.9	1988.6	4.2144
425	0.007886	2607.8	2805.0	5.4708	0.005299	2452.9	2611.8	5.1473	0.003434	2253.3	2373.5	4.7751
450	0.009176	2721.2	2950.6	5.6759	0.006737	2618.9	2821.0	5.4422	0.004957	2497.5	2671.0	5.1946
500	0.011143	2887.3	3165.9	5.9643	0.008691	2824.0	3084.8	5.7956	0.006933	2755.3	2997.9	5.6331
550	0.012736	3020.8	3339.2	6.1816	0.010175	2974.5	3279.7	6.0403	0.008348	2925.8	3218.0	5.9093
600	0.014140	3140.0	3493.5	6.3637	0.011445	3103.4	3446.8	6.2373	0.009523	3065.6	3399.0	6.1229
650	0.015430	3251.9	3637.7	6.5243	0.012590	3221.7	3599.4	6.4074	0.010565	3190.9	3560.7	6.3030
700	0.016643	3359.9	3776.0	6.6702	0.013654	3334.3	3743.9	6.5599	0.011523	3308.3	3711.6	6.4623
800	0.018922	3570.7	4043.8	6.9322	0.015628	3551.2	4020.0	6.8301	0.013278	3531.6	3996.3	6.7409
900	0.021075	3780.2	4307.1	7.1668	0.017473	3764.6	4288.8	7.0695	0.014904	3749.0	4270.6	6.9853
1000	0.023150	3991.5	4570.2	7.3821	0.019240	3978.6	4555.8	7.2880	0.016450	3965.8	4541.5	7.2069
1100	0.025172	4206.1	4835.4	7.5825	0.020954	4195.2	4823.9	7.4906	0.017942	4184.4	4812.4	7.4118
1200	0.027157	4424.6	5103.5	7.7710	0.022630	4415.3	5094.2	7.6807	0.019398	4406.1	5085.0	7.6034
1300	0.029115	4647.2	5375.1	7.9494	0.024279	4639.2	5367.6	7.8602	0.020827	4631.2	5360.2	7.7841
	P = 40.0 MPa				*P* = 50.0 MPa				*P* = 60.0 MPa			
375	0.001641	1677.0	1742.6	3.8290	0.001560	1638.6	1716.6	3.7642	0.001503	1609.7	1699.9	3.7149
400	0.001911	1855.0	1931.4	4.1145	0.001731	1787.8	1874.4	4.0029	0.001633	1745.2	1843.2	3.9317
425	0.002538	2097.5	2199.0	4.5044	0.002009	1960.3	2060.7	4.2746	0.001816	1892.9	2001.8	4.1630
450	0.003692	2364.2	2511.8	4.9449	0.002487	2160.3	2284.7	4.5896	0.002086	2055.1	2180.2	4.4140
500	0.005623	2681.6	2906.5	5.4744	0.003890	2528.1	2722.6	5.1762	0.002952	2393.2	2570.3	4.9356
550	0.006985	2875.1	3154.4	5.7857	0.005118	2769.5	3025.4	5.5563	0.003955	2664.6	2901.9	5.3517
600	0.008089	3026.8	3350.4	6.0170	0.006108	2947.1	3252.6	5.8245	0.004833	2866.8	3156.8	5.6527
650	0.009053	3159.5	3521.6	6.2078	0.006957	3095.6	3443.5	6.0373	0.005591	3031.3	3366.8	5.8867
700	0.009930	3282.0	3679.2	6.3740	0.007717	3228.7	3614.6	6.2179	0.006265	3175.4	3551.3	6.0814
800	0.011521	3511.8	3972.6	6.6613	0.009073	3472.2	3925.8	6.5225	0.007456	3432.6	3880.0	6.4033
900	0.012980	3733.3	4252.5	6.9107	0.010296	3702.0	4216.8	6.7819	0.008519	3670.9	4182.1	6.6725
1000	0.014360	3952.9	4527.3	7.1355	0.011441	3927.4	4499.4	7.0131	0.009504	3902.0	4472.2	6.9099
1100	0.015686	4173.7	4801.1	7.3425	0.012534	4152.2	4778.9	7.2244	0.010439	4130.9	4757.3	7.1255
1200	0.016976	4396.9	5075.9	7.5357	0.013590	4378.6	5058.1	7.4207	0.011339	4360.5	5040.8	7.3248
1300	0.018239	4623.3	5352.8	7.7175	0.014620	4607.5	5338.5	7.6048	0.012213	4591.8	5324.5	7.5111

TABLE A–7

Compressed liquid water

T °C	v m³/kg	u kJ/kg	h kJ/kg	s kJ/kg · K	v m³/kg	u kJ/kg	h kJ/kg	s kJ/kg · K	v m³/kg	u kJ/kg	h kJ/kg	s kJ/kg · K
	P = 5 MPa (263.94°C)				P = 10 MPa (311.00°C)				P = 15 MPa (342.16°C)			
Sat.	0.0012862	1148.1	1154.5	2.9207	0.0014522	1393.3	1407.9	3.3603	0.0016572	1585.5	1610.3	3.6848
0	0.0009977	0.04	5.03	0.0001	0.0009952	0.12	10.07	0.0003	0.0009928	0.18	15.07	0.0004
20	0.0009996	83.61	88.61	0.2954	0.0009973	83.31	93.28	0.2943	0.0009951	83.01	97.93	0.2932
40	0.0010057	166.92	171.95	0.5705	0.0010035	166.33	176.37	0.5685	0.0010013	165.75	180.77	0.5666
60	0.0010149	250.29	255.36	0.8287	0.0010127	249.43	259.55	0.8260	0.0010105	248.58	263.74	0.8234
80	0.0010267	333.82	338.96	1.0723	0.0010244	332.69	342.94	1.0691	0.0010221	331.59	346.92	1.0659
100	0.0010410	417.65	422.85	1.3034	0.0010385	416.23	426.62	1.2996	0.0010361	414.85	430.39	1.2958
120	0.0010576	501.91	507.19	1.5236	0.0010549	500.18	510.73	1.5191	0.0010522	498.50	514.28	1.5148
140	0.0010769	586.80	592.18	1.7344	0.0010738	584.72	595.45	1.7293	0.0010708	582.69	598.75	1.7243
160	0.0010988	672.55	678.04	1.9374	0.0010954	670.06	681.01	1.9316	0.0010920	667.63	684.01	1.9259
180	0.0011240	759.47	765.09	2.1338	0.0011200	756.48	767.68	2.1271	0.0011160	753.58	770.32	2.1206
200	0.0011531	847.92	853.68	2.3251	0.0011482	844.32	855.80	2.3174	0.0011435	840.84	858.00	2.3100
220	0.0011868	938.39	944.32	2.5127	0.0011809	934.01	945.82	2.5037	0.0011752	929.81	947.43	2.4951
240	0.0012268	1031.6	1037.7	2.6983	0.0012192	1026.2	1038.3	2.6876	0.0012121	1021.0	1039.2	2.6774
260	0.0012755	1128.5	1134.9	2.8841	0.0012653	1121.6	1134.3	2.8710	0.0012560	1115.1	1134.0	2.8586
280					0.0013226	1221.8	1235.0	3.0565	0.0013096	1213.4	1233.0	3.0410
300					0.0013980	1329.4	1343.3	3.2488	0.0013783	1317.6	1338.3	3.2279
320									0.0014733	1431.9	1454.0	3.4263
340									0.0016311	1567.9	1592.4	3.6555

T °C	v m³/kg	u kJ/kg	h kJ/kg	s kJ/kg · K	v m³/kg	u kJ/kg	h kJ/kg	s kJ/kg · K	v m³/kg	u kJ/kg	h kJ/kg	s kJ/kg · K
	P = 20 MPa (365.75°C)				P = 30 MPa				P = 50 MPa			
Sat.	0.0020378	1785.8	1826.6	4.0146								
0	0.0009904	0.23	20.03	0.0005	0.0009857	0.29	29.86	0.0003	0.0009767	0.29	49.13	−0.0010
20	0.0009929	82.71	102.57	0.2921	0.0009886	82.11	111.77	0.2897	0.0009805	80.93	129.95	0.2845
40	0.0009992	165.17	185.16	0.5646	0.0009951	164.05	193.90	0.5607	0.0009872	161.90	211.25	0.5528
60	0.0010084	247.75	267.92	0.8208	0.0010042	246.14	276.26	0.8156	0.0009962	243.08	292.88	0.8055
80	0.0010199	330.50	350.90	1.0627	0.0010155	328.40	358.86	1.0564	0.0010072	324.42	374.78	1.0442
100	0.0010337	413.50	434.17	1.2920	0.0010290	410.87	441.74	1.2847	0.0010201	405.94	456.94	1.2705
120	0.0010496	496.85	517.84	1.5105	0.0010445	493.66	525.00	1.5020	0.0010349	487.69	539.43	1.4859
140	0.0010679	580.71	602.07	1.7194	0.0010623	576.90	608.76	1.7098	0.0010517	569.77	622.36	1.6916
160	0.0010886	665.28	687.05	1.9203	0.0010823	660.74	693.21	1.9094	0.0010704	652.33	705.85	1.8889
180	0.0011122	750.78	773.02	2.1143	0.0011049	745.40	778.55	2.1020	0.0010914	735.49	790.06	2.0790
200	0.0011390	837.49	860.27	2.3027	0.0011304	831.11	865.02	2.2888	0.0011149	819.45	875.19	2.2628
220	0.0011697	925.77	949.16	2.4867	0.0011595	918.15	952.93	2.4707	0.0011412	904.39	961.45	2.4414
240	0.0012053	1016.1	1040.2	2.6676	0.0011927	1006.9	1042.7	2.6491	0.0011708	990.55	1049.1	2.6156
260	0.0012472	1109.0	1134.0	2.8469	0.0012314	1097.8	1134.7	2.8250	0.0012044	1078.2	1138.4	2.7864
280	0.0012978	1205.6	1231.5	3.0265	0.0012770	1191.5	1229.8	3.0001	0.0012430	1167.7	1229.9	2.9547
300	0.0013611	1307.2	1334.4	3.2091	0.0013322	1288.9	1328.9	3.1761	0.0012879	1259.6	1324.0	3.1218
320	0.0014450	1416.6	1445.5	3.3996	0.0014014	1391.7	1433.7	3.3558	0.0013409	1354.3	1421.4	3.2888
340	0.0015693	1540.2	1571.6	3.6086	0.0014932	1502.4	1547.1	3.5438	0.0014049	1452.9	1523.1	3.4575
360	0.0018248	1703.6	1740.1	3.8787	0.0016276	1626.8	1675.6	3.7499	0.0014848	1556.5	1630.7	3.6301
380					0.0018729	1782.0	1838.2	4.0026	0.0015884	1667.1	1746.5	3.8102

TABLE A–8

Saturated ice–water vapor

		Specific volume, m³/kg		Internal energy, kJ/kg			Enthalpy, kJ/kg			Entropy, kJ/kg · K		
Temp., T °C	Sat. press., P_{sat} kPa	Sat. ice, v_i	Sat. vapor, v_g	Sat. ice, u_i	Subl., u_{ig}	Sat. vapor, u_g	Sat. ice, h_i	Subl., h_{ig}	Sat. vapor, h_g	Sat. ice, s_i	Subl., s_{ig}	Sat. vapor, s_g
0.01	0.61169	0.001091	205.99	−333.40	2707.9	2374.5	−333.40	2833.9	2500.5	−1.2202	10.374	9.154
0	0.61115	0.001091	206.17	−333.43	2707.9	2374.5	−333.43	2833.9	2500.5	−1.2204	10.375	9.154
−2	0.51772	0.001091	241.62	−337.63	2709.4	2371.8	−337.63	2834.5	2496.8	−1.2358	10.453	9.218
−4	0.43748	0.001090	283.84	−341.80	2710.8	2369.0	−341.80	2835.0	2493.2	−1.2513	10.533	9.282
−6	0.36873	0.001090	334.27	−345.94	2712.2	2366.2	−345.93	2835.4	2489.5	−1.2667	10.613	9.347
−8	0.30998	0.001090	394.66	−350.04	2713.5	2363.5	−350.04	2835.8	2485.8	−1.2821	10.695	9.413
−10	0.25990	0.001089	467.17	−354.12	2714.8	2360.7	−354.12	2836.2	2482.1	−1.2976	10.778	9.480
−12	0.21732	0.001089	554.47	−358.17	2716.1	2357.9	−358.17	2836.6	2478.4	−1.3130	10.862	9.549
−14	0.18121	0.001088	659.88	−362.18	2717.3	2355.2	−362.18	2836.9	2474.7	−1.3284	10.947	9.618
−16	0.15068	0.001088	787.51	−366.17	2718.6	2352.4	−366.17	2837.2	2471.0	−1.3439	11.033	9.689
−18	0.12492	0.001088	942.51	−370.13	2719.7	2349.6	−370.13	2837.5	2467.3	−1.3593	11.121	9.761
−20	0.10326	0.001087	1131.3	−374.06	2720.9	2346.8	−374.06	2837.7	2463.6	−1.3748	11.209	9.835
−22	0.08510	0.001087	1362.0	−377.95	2722.0	2344.1	−377.95	2837.9	2459.9	−1.3903	11.300	9.909
−24	0.06991	0.001087	1644.7	−381.82	2723.1	2341.3	−381.82	2838.1	2456.2	−1.4057	11.391	9.985
−26	0.05725	0.001087	1992.2	−385.66	2724.2	2338.5	−385.66	2838.2	2452.5	−1.4212	11.484	10.063
−28	0.04673	0.001086	2421.0	−389.47	2725.2	2335.7	−389.47	2838.3	2448.8	−1.4367	11.578	10.141
−30	0.03802	0.001086	2951.7	−393.25	2726.2	2332.9	−393.25	2838.4	2445.1	−1.4521	11.673	10.221
−32	0.03082	0.001086	3610.9	−397.00	2727.2	2330.2	−397.00	2838.4	2441.4	−1.4676	11.770	10.303
−34	0.02490	0.001085	4432.4	−400.72	2728.1	2327.4	−400.72	2838.5	2437.7	−1.4831	11.869	10.386
−36	0.02004	0.001085	5460.1	−404.40	2729.0	2324.6	−404.40	2838.4	2434.0	−1.4986	11.969	10.470
−38	0.01608	0.001085	6750.5	−408.07	2729.9	2321.8	−408.07	2838.4	2430.3	−1.5141	12.071	10.557
−40	0.01285	0.001084	8376.7	−411.70	2730.7	2319.0	−411.70	2838.3	2426.6	−1.5296	12.174	10.644

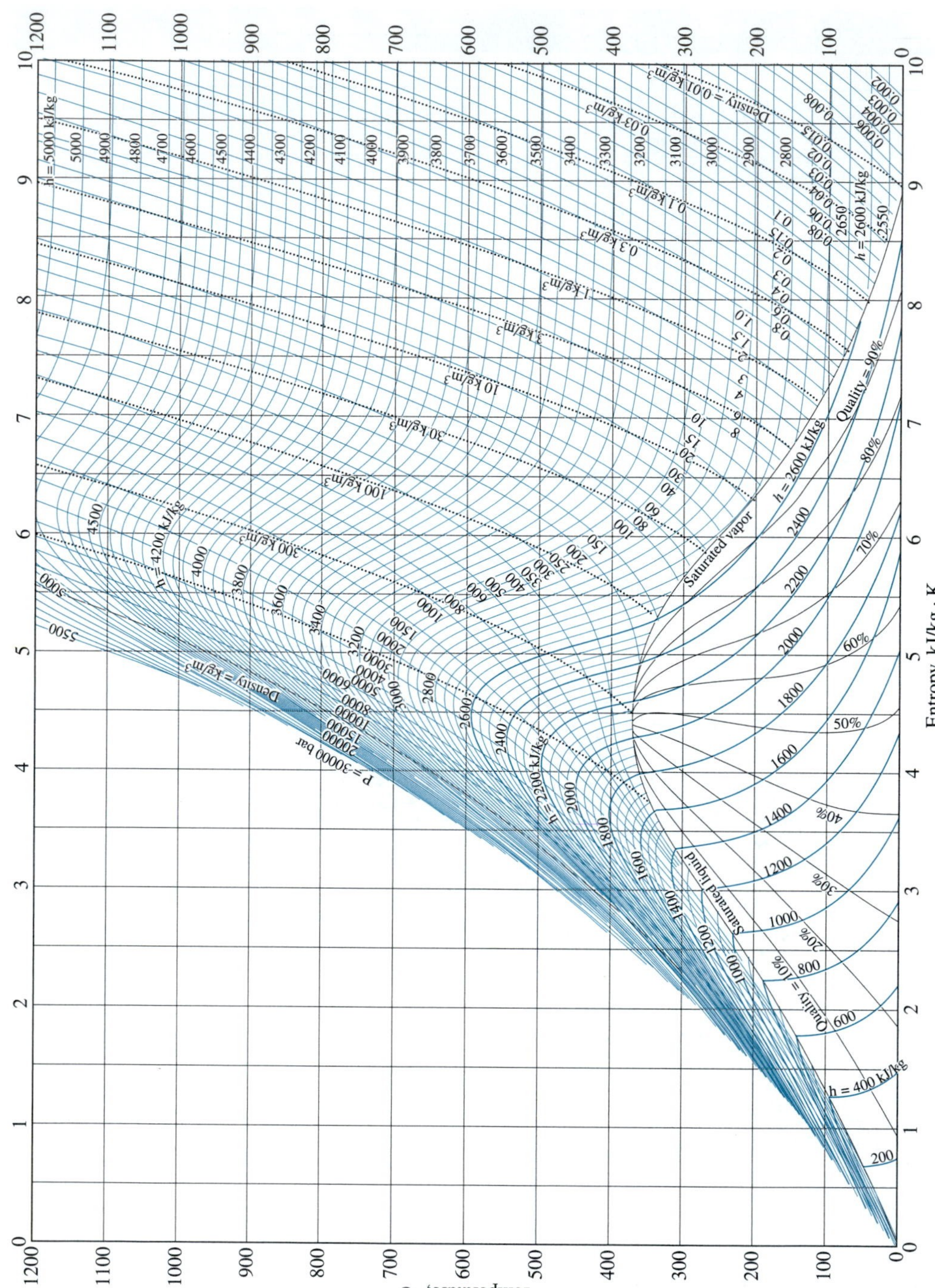

FIGURE A–9

T-s diagram for water.

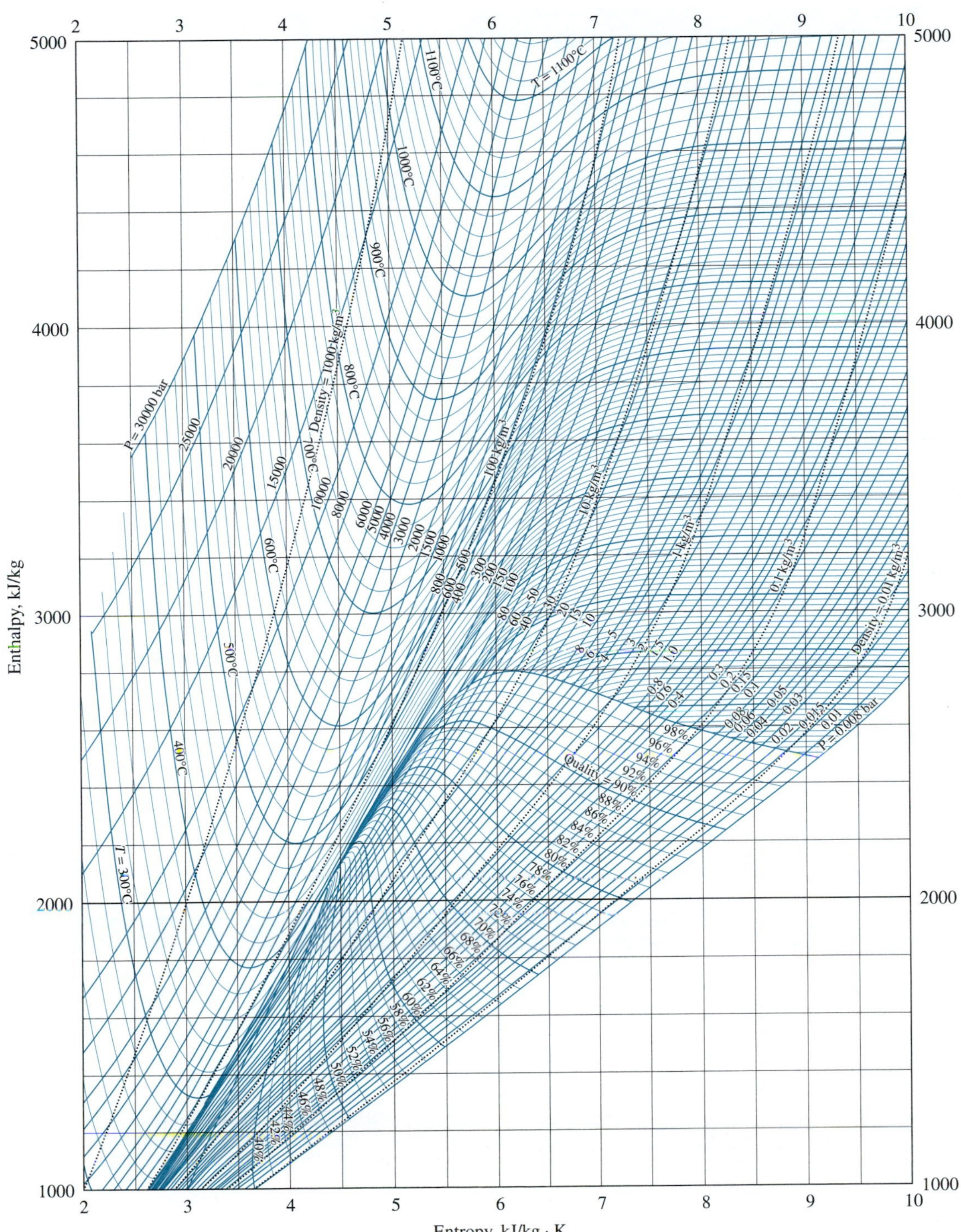

FIGURE A–10

Mollier diagram for water.

TABLE A–11

Saturated refrigerant-134a—Temperature table

		Specific volume, m^3/kg		*Internal energy,* kJ/kg			*Enthalpy,* kJ/kg			*Entropy,* kJ/kg · K		
Temp., T °C	Sat. press., P_{sat} kPa	Sat. liquid, v_f	Sat. vapor, v_g	Sat. liquid, u_f	Evap., u_{fg}	Sat. vapor, u_g	Sat. liquid, h_f	Evap., h_{fg}	Sat. vapor, h_g	Sat. liquid, s_f	Evap., s_{fg}	Sat. vapor, s_g
−40	51.25	0.0007054	0.36081	−0.036	207.40	207.37	0.000	225.86	225.86	0.00000	0.96866	0.96866
−38	56.86	0.0007083	0.32732	2.475	206.04	208.51	2.515	224.61	227.12	0.01072	0.95511	0.96584
−36	62.95	0.0007112	0.29751	4.992	204.67	209.66	5.037	223.35	228.39	0.02138	0.94176	0.96315
−34	69.56	0.0007142	0.27090	7.517	203.29	210.81	7.566	222.09	229.65	0.03199	0.92859	0.96058
−32	76.71	0.0007172	0.24711	10.05	201.91	211.96	10.10	220.81	230.91	0.04253	0.91560	0.95813
−30	84.43	0.0007203	0.22580	12.59	200.52	213.11	12.65	219.52	232.17	0.05301	0.90278	0.95579
−28	92.76	0.0007234	0.20666	15.13	199.12	214.25	15.20	218.22	233.43	0.06344	0.89012	0.95356
−26	101.73	0.0007265	0.18946	17.69	197.72	215.40	17.76	216.92	234.68	0.07382	0.87762	0.95144
−24	111.37	0.0007297	0.17395	20.25	196.30	216.55	20.33	215.59	235.92	0.08414	0.86527	0.94941
−22	121.72	0.0007329	0.15995	22.82	194.88	217.70	22.91	214.26	237.17	0.09441	0.85307	0.94748
−20	132.82	0.0007362	0.14729	25.39	193.45	218.84	25.49	212.91	238.41	0.10463	0.84101	0.94564
−18	144.69	0.0007396	0.13583	27.98	192.01	219.98	28.09	211.55	239.64	0.11481	0.82908	0.94389
−16	157.38	0.0007430	0.12542	30.57	190.56	221.13	30.69	210.18	240.87	0.12493	0.81729	0.94222
−14	170.93	0.0007464	0.11597	33.17	189.09	222.27	33.30	208.79	242.09	0.13501	0.80561	0.94063
−12	185.37	0.0007499	0.10736	35.78	187.62	223.40	35.92	207.38	243.30	0.14504	0.79406	0.93911
−10	200.74	0.0007535	0.099516	38.40	186.14	224.54	38.55	205.96	244.51	0.15504	0.78263	0.93766
−8	217.08	0.0007571	0.092352	41.03	184.64	225.67	41.19	204.52	245.72	0.16498	0.77130	0.93629
−6	234.44	0.0007608	0.085802	43.66	183.13	226.80	43.84	203.07	246.91	0.17489	0.76008	0.93497
−4	252.85	0.0007646	0.079804	46.31	181.61	227.92	46.50	201.60	248.10	0.18476	0.74896	0.93372
−2	272.36	0.0007684	0.074304	48.96	180.08	229.04	49.17	200.11	249.28	0.19459	0.73794	0.93253
0	293.01	0.0007723	0.069255	51.63	178.53	230.16	51.86	198.60	250.45	0.20439	0.72701	0.93139
2	314.84	0.0007763	0.064612	54.30	176.97	231.27	54.55	197.07	251.61	0.21415	0.71616	0.93031
4	337.90	0.0007804	0.060338	56.99	175.39	232.38	57.25	195.51	252.77	0.22387	0.70540	0.92927
6	362.23	0.0007845	0.056398	59.68	173.80	233.48	59.97	193.94	253.91	0.23356	0.69471	0.92828
8	387.88	0.0007887	0.052762	62.39	172.19	234.58	62.69	192.35	255.04	0.24323	0.68410	0.92733
10	414.89	0.0007930	0.049403	65.10	170.56	235.67	65.43	190.73	256.16	0.25286	0.67356	0.92641
12	443.31	0.0007975	0.046295	67.83	168.92	236.75	68.18	189.09	257.27	0.26246	0.66308	0.92554
14	473.19	0.0008020	0.043417	70.57	167.26	237.83	70.95	187.42	258.37	0.27204	0.65266	0.92470
16	504.58	0.0008066	0.040748	73.32	165.58	238.90	73.73	185.73	259.46	0.28159	0.64230	0.92389
18	537.52	0.0008113	0.038271	76.08	163.88	239.96	76.52	184.01	260.53	0.29112	0.63198	0.92310

(Continued)

TABLE A–11

Saturated refrigerant-134a—Temperature table *(Continued)*

		Specific volume, m³/kg		Internal energy, kJ/kg			Enthalpy, kJ/kg			Entropy, kJ/kg · K		
Temp., T °C	Sat. press., P_{sat} kPa	Sat. liquid, v_f	Sat. vapor, v_g	Sat. liquid, u_f	Evap., u_{fg}	Sat. vapor, u_g	Sat. liquid, h_f	Evap., h_{fg}	Sat. vapor, h_g	Sat. liquid, s_f	Evap., s_{fg}	Sat. vapor, s_g
20	572.07	0.0008161	0.035969	78.86	162.16	241.02	79.32	182.27	261.59	0.30063	0.62172	0.92234
22	608.27	0.0008210	0.033828	81.64	160.42	242.06	82.14	180.49	262.64	0.31011	0.61149	0.92160
24	646.18	0.0008261	0.031834	84.44	158.65	243.10	84.98	178.69	263.67	0.31958	0.60130	0.92088
26	685.84	0.0008313	0.029976	87.26	156.87	244.12	87.83	176.85	264.68	0.32903	0.59115	0.92018
28	727.31	0.0008366	0.028242	90.09	155.05	245.14	90.69	174.99	265.68	0.33846	0.58102	0.91948
30	770.64	0.0008421	0.026622	92.93	153.22	246.14	93.58	173.08	266.66	0.34789	0.57091	0.91879
32	815.89	0.0008478	0.025108	95.79	151.35	247.14	96.48	171.14	267.62	0.35730	0.56082	0.91811
34	863.11	0.0008536	0.023691	98.66	149.46	248.12	99.40	169.17	268.57	0.36670	0.55074	0.91743
36	912.35	0.0008595	0.022364	101.55	147.54	249.08	102.33	167.16	269.49	0.37609	0.54066	0.91675
38	963.68	0.0008657	0.021119	104.45	145.58	250.04	105.29	165.10	270.39	0.38548	0.53058	0.91606
40	1017.1	0.0008720	0.019952	107.38	143.60	250.97	108.26	163.00	271.27	0.39486	0.52049	0.91536
42	1072.8	0.0008786	0.018855	110.32	141.58	251.89	111.26	160.86	272.12	0.40425	0.51039	0.91464
44	1130.7	0.0008854	0.017824	113.28	139.52	252.80	114.28	158.67	272.95	0.41363	0.50027	0.91391
46	1191.0	0.0008924	0.016853	116.26	137.42	253.68	117.32	156.43	273.75	0.42302	0.49012	0.91315
48	1253.6	0.0008996	0.015939	119.26	135.29	254.55	120.39	154.14	274.53	0.43242	0.47993	0.91236
52	1386.2	0.0009150	0.014265	125.33	130.88	256.21	126.59	149.39	275.98	0.45126	0.45941	0.91067
56	1529.1	0.0009317	0.012771	131.49	126.28	257.77	132.91	144.38	277.30	0.47018	0.43863	0.90880
60	1682.8	0.0009498	0.011434	137.76	121.46	259.22	139.36	139.10	278.46	0.48920	0.41749	0.90669
65	1891.0	0.0009750	0.009950	145.77	115.05	260.82	147.62	132.02	279.64	0.51320	0.39039	0.90359
70	2118.2	0.0010037	0.008642	154.01	108.14	262.15	156.13	124.32	280.46	0.53755	0.36227	0.89982
75	2365.8	0.0010372	0.007480	162.53	100.60	263.13	164.98	115.85	280.82	0.56241	0.33272	0.89512
80	2635.3	0.0010772	0.006436	171.40	92.23	263.63	174.24	106.35	280.59	0.58800	0.30111	0.88912
85	2928.2	0.0011270	0.005486	180.77	82.67	263.44	184.07	95.44	279.51	0.61473	0.26644	0.88117
90	3246.9	0.0011932	0.004599	190.89	71.29	262.18	194.76	82.35	277.11	0.64336	0.22674	0.87010
95	3594.1	0.0012933	0.003726	202.40	56.47	258.87	207.05	65.21	272.26	0.67578	0.17711	0.85289
100	3975.1	0.0015269	0.002630	218.72	29.19	247.91	224.79	33.58	258.37	0.72217	0.08999	0.81215

Source: Tables A–11 through A–13 are generated using the Engineering Equation Solver (EES) software developed by S. A. Klein and F. L. Alvarado. The routine used in calculations is the R134a, which is based on the fundamental equation of state developed by R. Tillner-Roth and H.D. Baehr, "An International Standard Formulation for the Thermodynamic Properties of 1,1,1,2-Tetrafluoroethane (HFC-134a) for temperatures from 170 K to 455 K and Pressures up to 70 MPa," *J. Phys. Chem, Ref. Data*, Vol. 23, No. 5, 1994. The enthalpy and entropy values of saturated liquid are set to zero at −40°C (and −40°F).

TABLE A–12

Saturated refrigerant-134a—Pressure table

		Specific volume, m^3/kg		Internal energy, kJ/kg			Enthalpy, kJ/kg			Entropy, kJ/kg · K		
Press., P kPa	Sat. temp., T_{sat} °C	Sat. liquid, v_f	Sat. vapor, v_g	Sat. liquid, u_f	Evap., u_{fg}	Sat. vapor, u_g	Sat. liquid, h_f	Evap., h_{fg}	Sat. vapor, h_g	Sat. liquid, s_f	Evap., s_{fg}	Sat. vapor, s_g
60	−36.95	0.0007098	0.31121	3.798	205.32	209.12	3.841	223.95	227.79	0.01634	0.94807	0.96441
70	−33.87	0.0007144	0.26929	7.680	203.20	210.88	7.730	222.00	229.73	0.03267	0.92775	0.96042
80	−31.13	0.0007185	0.23753	11.15	201.30	212.46	11.21	220.25	231.46	0.04711	0.90999	0.95710
90	−28.65	0.0007223	0.21263	14.31	199.57	213.88	14.37	218.65	233.02	0.06008	0.89419	0.95427
100	−26.37	0.0007259	0.19254	17.21	197.98	215.19	17.28	217.16	234.44	0.07188	0.87995	0.95183
120	−22.32	0.0007324	0.16212	22.40	195.11	217.51	22.49	214.48	236.97	0.09275	0.85503	0.94779
140	−18.77	0.0007383	0.14014	26.98	192.57	219.54	27.08	212.08	239.16	0.11087	0.83368	0.94456
160	−15.60	0.0007437	0.12348	31.09	190.27	221.35	31.21	209.90	241.11	0.12693	0.81496	0.94190
180	−12.73	0.0007487	0.11041	34.83	188.16	222.99	34.97	207.90	242.86	0.14139	0.79826	0.93965
200	−10.09	0.0007533	0.099867	38.28	186.21	224.48	38.43	206.03	244.46	0.15457	0.78316	0.93773
240	−5.38	0.0007620	0.083897	44.48	182.67	227.14	44.66	202.62	247.28	0.17794	0.75664	0.93458
280	−1.25	0.0007699	0.072352	49.97	179.50	229.46	50.18	199.54	249.72	0.19829	0.73381	0.93210
320	2.46	0.0007772	0.063604	54.92	176.61	231.52	55.16	196.71	251.88	0.21637	0.71369	0.93006
360	5.82	0.0007841	0.056738	59.44	173.94	233.38	59.72	194.08	253.81	0.23270	0.69566	0.92836
400	8.91	0.0007907	0.051201	63.62	171.45	235.07	63.94	191.62	255.55	0.24761	0.67929	0.92691
450	12.46	0.0007985	0.045619	68.45	168.54	237.00	68.81	188.71	257.53	0.26465	0.66069	0.92535
500	15.71	0.0008059	0.041118	72.93	165.82	238.75	73.33	185.98	259.30	0.28023	0.64377	0.92400
550	18.73	0.0008130	0.037408	77.10	163.25	240.35	77.54	183.38	260.92	0.29461	0.62821	0.92282
600	21.55	0.0008199	0.034295	81.02	160.81	241.83	81.51	180.90	262.40	0.30799	0.61378	0.92177
650	24.20	0.0008266	0.031646	84.72	158.48	243.20	85.26	178.51	263.77	0.32051	0.60030	0.92081
700	26.69	0.0008331	0.029361	88.24	156.24	244.48	88.82	176.21	265.03	0.33230	0.58763	0.91994
750	29.06	0.0008395	0.027371	91.59	154.08	245.67	92.22	173.98	266.20	0.34345	0.57567	0.91912
800	31.31	0.0008458	0.025621	94.79	152.00	246.79	95.47	171.82	267.29	0.35404	0.56431	0.91835
850	33.45	0.0008520	0.024069	97.87	149.98	247.85	98.60	169.71	268.31	0.36413	0.55349	0.91762
900	35.51	0.0008580	0.022683	100.83	148.01	248.85	101.61	167.66	269.26	0.37377	0.54315	0.91692
950	37.48	0.0008641	0.021438	103.69	146.10	249.79	104.51	165.64	270.15	0.38301	0.53323	0.91624
1000	39.37	0.0008700	0.020313	106.45	144.23	250.68	107.32	163.67	270.99	0.39189	0.52368	0.91558
1200	46.29	0.0008934	0.016715	116.70	137.11	253.81	117.77	156.10	273.87	0.42441	0.48863	0.91303
1400	52.40	0.0009166	0.014107	125.94	130.43	256.37	127.22	148.90	276.12	0.45315	0.45734	0.91050
1600	57.88	0.0009400	0.012123	134.43	124.04	258.47	135.93	141.93	277.86	0.47911	0.42873	0.90784
1800	62.87	0.0009639	0.010559	142.33	117.83	260.17	144.07	135.11	279.17	0.50294	0.40204	0.90498
2000	67.45	0.0009886	0.009288	149.78	111.73	261.51	151.76	128.33	280.09	0.52509	0.37675	0.90184
2500	77.54	0.0010566	0.006936	166.99	96.47	263.45	169.63	111.16	280.79	0.57531	0.31695	0.89226
3000	86.16	0.0011406	0.005275	183.04	80.22	263.26	186.46	92.63	279.09	0.62118	0.25776	0.87894

TABLE A–13

Superheated refrigerant-134a

T °C	v m³/kg	u kJ/kg	h kJ/kg	s kJ/kg · K	v m³/kg	u kJ/kg	h kJ/kg	s kJ/kg · K	v m³/kg	u kJ/kg	h kJ/kg	s kJ/kg · K
	P = 0.06 MPa (T_{sat} = −36.95°C)				P = 0.10 MPa (T_{sat} = −26.37°C)				P = 0.14 MPa (T_{sat} = −18.77°C)			
Sat.	0.31121	209.12	227.79	0.9644	0.19254	215.19	234.44	0.9518	0.14014	219.54	239.16	0.9446
−20	0.33608	220.60	240.76	1.0174	0.19841	219.66	239.50	0.9721				
−10	0.35048	227.55	248.58	1.0477	0.20743	226.75	247.49	1.0030	0.14605	225.91	246.36	0.9724
0	0.36476	234.66	256.54	1.0774	0.21630	233.95	255.58	1.0332	0.15263	233.23	254.60	1.0031
10	0.37893	241.92	264.66	1.1066	0.22506	241.30	263.81	1.0628	0.15908	240.66	262.93	1.0331
20	0.39302	249.35	272.94	1.1353	0.23373	248.79	272.17	1.0918	0.16544	248.22	271.38	1.0624
30	0.40705	256.95	281.37	1.1636	0.24233	256.44	280.68	1.1203	0.17172	255.93	279.97	1.0912
40	0.42102	264.71	289.97	1.1915	0.25088	264.25	289.34	1.1484	0.17794	263.79	288.70	1.1195
50	0.43495	272.64	298.74	1.2191	0.25937	272.22	298.16	1.1762	0.18412	271.79	297.57	1.1474
60	0.44883	280.73	307.66	1.2463	0.26783	280.35	307.13	1.2035	0.19025	279.96	306.59	1.1749
70	0.46269	288.99	316.75	1.2732	0.27626	288.64	316.26	1.2305	0.19635	288.28	315.77	1.2020
80	0.47651	297.41	326.00	1.2997	0.28465	297.08	325.55	1.2572	0.20242	296.75	325.09	1.2288
90	0.49032	306.00	335.42	1.3260	0.29303	305.69	334.99	1.2836	0.20847	305.38	334.57	1.2553
100	0.50410	314.74	344.99	1.3520	0.30138	314.46	344.60	1.3096	0.21449	314.17	344.20	1.2814
	P = 0.18 MPa (T_{sat} = −12.73°C)				P = 0.20 MPa (T_{sat} = −10.09°C)				P = 0.24 MPa (T_{sat} = −5.38°C)			
Sat.	0.11041	222.99	242.86	0.9397	0.09987	224.48	244.46	0.9377	0.08390	227.14	247.28	0.9346
−10	0.11189	225.02	245.16	0.9484	0.09991	224.55	244.54	0.9380				
0	0.11722	232.48	253.58	0.9798	0.10481	232.09	253.05	0.9698	0.08617	231.29	251.97	0.9519
10	0.12240	240.00	262.04	1.0102	0.10955	239.67	261.58	1.0004	0.09026	238.98	260.65	0.9831
20	0.12748	247.64	270.59	1.0399	0.11418	247.35	270.18	1.0303	0.09423	246.74	269.36	1.0134
30	0.13248	255.41	279.25	1.0690	0.11874	255.14	278.89	1.0595	0.09812	254.61	278.16	1.0429
40	0.13741	263.31	288.05	1.0975	0.12322	263.08	287.72	1.0882	0.10193	262.59	287.06	1.0718
50	0.14230	271.36	296.98	1.1256	0.12766	271.15	296.68	1.1163	0.10570	270.71	296.08	1.1001
60	0.14715	279.56	306.05	1.1532	0.13206	279.37	305.78	1.1441	0.10942	278.97	305.23	1.1280
70	0.15196	287.91	315.27	1.1805	0.13641	287.73	315.01	1.1714	0.11310	287.36	314.51	1.1554
80	0.15673	296.42	324.63	1.2074	0.14074	296.25	324.40	1.1983	0.11675	295.91	323.93	1.1825
90	0.16149	305.07	334.14	1.2339	0.14504	304.92	333.93	1.2249	0.12038	304.60	333.49	1.2092
100	0.16622	313.88	343.80	1.2602	0.14933	313.74	343.60	1.2512	0.12398	313.44	343.20	1.2356
	P = 0.28 MPa (T_{sat} = −1.25°C)				P = 0.32 MPa (T_{sat} = 2.46°C)				P = 0.40 MPa (T_{sat} = 8.91°C)			
Sat.	0.07235	229.46	249.72	0.9321	0.06360	231.52	251.88	0.9301	0.051201	235.07	255.55	0.9269
0	0.07282	230.44	250.83	0.9362								
10	0.07646	238.27	259.68	0.9680	0.06609	237.54	258.69	0.9544	0.051506	235.97	256.58	0.9305
20	0.07997	246.13	268.52	0.9987	0.06925	245.50	267.66	0.9856	0.054213	244.18	265.86	0.9628
30	0.08338	254.06	277.41	1.0285	0.07231	253.50	276.65	1.0157	0.056796	252.36	275.07	0.9937
40	0.08672	262.10	286.38	1.0576	0.07530	261.60	285.70	1.0451	0.059292	260.58	284.30	1.0236
50	0.09000	270.27	295.47	1.0862	0.07823	269.82	294.85	1.0739	0.061724	268.90	293.59	1.0528
60	0.09324	278.56	304.67	1.1142	0.08111	278.15	304.11	1.1021	0.064104	277.32	302.96	1.0814
70	0.09644	286.99	314.00	1.1418	0.08395	286.62	313.48	1.1298	0.066443	285.86	312.44	1.1094
80	0.09961	295.57	323.46	1.1690	0.08675	295.22	322.98	1.1571	0.068747	294.53	322.02	1.1369
90	0.10275	304.29	333.06	1.1958	0.08953	303.97	332.62	1.1840	0.071023	303.32	331.73	1.1640
100	0.10587	313.15	342.80	1.2222	0.09229	312.86	342.39	1.2105	0.073274	312.26	341.57	1.1907
110	0.10897	322.16	352.68	1.2483	0.09503	321.89	352.30	1.2367	0.075504	321.33	351.53	1.2171
120	0.11205	331.32	362.70	1.2742	0.09775	331.07	362.35	1.2626	0.077717	330.55	361.63	1.2431
130	0.11512	340.63	372.87	1.2997	0.10045	340.39	372.54	1.2882	0.079913	339.90	371.87	1.2688
140	0.11818	350.09	383.18	1.3250	0.10314	349.86	382.87	1.3135	0.082096	349.41	382.24	1.2942

TABLE A–13

Superheated refrigerant-134a (*Continued*)

T °C	v m³/kg	u kJ/kg	h kJ/kg	s kJ/kg · K	v m³/kg	u kJ/kg	h kJ/kg	s kJ/kg · K	v m³/kg	u kJ/kg	h kJ/kg	s kJ/kg · K
	P = 0.50 MPa (T_{sat} = 15.71°C)				P = 0.60 MPa (T_{sat} = 21.55°C)				P = 0.70 MPa (T_{sat} = 26.69°C)			
Sat.	0.041118	238.75	259.30	0.9240	0.034295	241.83	262.40	0.9218	0.029361	244.48	265.03	0.9199
20	0.042115	242.40	263.46	0.9383								
30	0.044338	250.84	273.01	0.9703	0.035984	249.22	270.81	0.9499	0.029966	247.48	268.45	0.9313
40	0.046456	259.26	282.48	1.0011	0.037865	257.86	280.58	0.9816	0.031696	256.39	278.57	0.9641
50	0.048499	267.72	291.96	1.0309	0.039659	266.48	290.28	1.0121	0.033322	265.20	288.53	0.9954
60	0.050485	276.25	301.50	1.0599	0.041389	275.15	299.98	1.0417	0.034875	274.01	298.42	1.0256
70	0.052427	284.89	311.10	1.0883	0.043069	283.89	309.73	1.0705	0.036373	282.87	308.33	1.0549
80	0.054331	293.64	320.80	1.1162	0.044710	292.73	319.55	1.0987	0.037829	291.80	318.28	1.0835
90	0.056205	302.51	330.61	1.1436	0.046318	301.67	329.46	1.1264	0.039250	300.82	328.29	1.1114
100	0.058053	311.50	340.53	1.1705	0.047900	310.73	339.47	1.1536	0.040642	309.95	338.40	1.1389
110	0.059880	320.63	350.57	1.1971	0.049458	319.91	349.59	1.1803	0.042010	319.19	348.60	1.1658
120	0.061687	329.89	360.73	1.2233	0.050997	329.23	359.82	1.2067	0.043358	328.55	358.90	1.1924
130	0.063479	339.29	371.03	1.2491	0.052519	338.67	370.18	1.2327	0.044688	338.04	369.32	1.2186
140	0.065256	348.83	381.46	1.2747	0.054027	348.25	380.66	1.2584	0.046004	347.66	379.86	1.2444
150	0.067021	358.51	392.02	1.2999	0.055522	357.96	391.27	1.2838	0.047306	357.41	390.52	1.2699
160	0.068775	368.33	402.72	1.3249	0.057006	367.81	402.01	1.3088	0.048597	367.29	401.31	1.2951
	P = 0.80 MPa (T_{sat} = 31.31°C)				P = 0.90 MPa (T_{sat} = 35.51°C)				P = 1.00 MPa (T_{sat} = 39.37°C)			
Sat.	0.025621	246.79	267.29	0.9183	0.022683	248.85	269.26	0.9169	0.020313	250.68	270.99	0.9156
40	0.027035	254.82	276.45	0.9480	0.023375	253.13	274.17	0.9327	0.020406	251.30	271.71	0.9179
50	0.028547	263.86	286.69	0.9802	0.024809	262.44	284.77	0.9660	0.021796	260.94	282.74	0.9525
60	0.029973	272.83	296.81	1.0110	0.026146	271.60	295.13	0.9976	0.023068	270.32	293.38	0.9850
70	0.031340	281.81	306.88	1.0408	0.027413	280.72	305.39	1.0280	0.024261	279.59	303.85	1.0160
80	0.032659	290.84	316.97	1.0698	0.028630	289.86	315.63	1.0574	0.025398	288.86	314.25	1.0458
90	0.033941	299.95	327.10	1.0981	0.029806	299.06	325.89	1.0860	0.026492	298.15	324.64	1.0748
100	0.035193	309.15	337.30	1.1258	0.030951	308.34	336.19	1.1140	0.027552	307.51	335.06	1.1031
110	0.036420	318.45	347.59	1.1530	0.032068	317.70	346.56	1.1414	0.028584	316.94	345.53	1.1308
120	0.037625	327.87	357.97	1.1798	0.033164	327.18	357.02	1.1684	0.029592	326.47	356.06	1.1580
130	0.038813	337.40	368.45	1.2061	0.034241	336.76	367.58	1.1949	0.030581	336.11	366.69	1.1846
140	0.039985	347.06	379.05	1.2321	0.035302	346.46	378.23	1.2210	0.031554	345.85	377.40	1.2109
150	0.041143	356.85	389.76	1.2577	0.036349	356.28	389.00	1.2467	0.032512	355.71	388.22	1.2368
160	0.042290	366.76	400.59	1.2830	0.037384	366.23	399.88	1.2721	0.033457	365.70	399.15	1.2623
170	0.043427	376.81	411.55	1.3080	0.038408	376.31	410.88	1.2972	0.034392	375.81	410.20	1.2875
180	0.044554	386.99	422.64	1.3327	0.039423	386.52	422.00	1.3221	0.035317	386.04	421.36	1.3124
	P = 1.20 MPa (T_{sat} = 46.29°C)				P = 1.40 MPa (T_{sat} = 52.40°C)				P = 1.60 MPa (T_{sat} = 57.88°C)			
Sat.	0.016715	253.81	273.87	0.9130	0.014107	256.37	276.12	0.9105	0.012123	258.47	277.86	0.9078
50	0.017201	257.63	278.27	0.9267								
60	0.018404	267.56	289.64	0.9614	0.015005	264.46	285.47	0.9389	0.012372	260.89	280.69	0.9163
70	0.019502	277.21	300.61	0.9938	0.016060	274.62	297.10	0.9733	0.013430	271.76	293.25	0.9535
80	0.020529	286.75	311.39	1.0248	0.017023	284.51	308.34	1.0056	0.014362	282.09	305.07	0.9875
90	0.021506	296.26	322.07	1.0546	0.017923	294.28	319.37	1.0364	0.015215	292.17	316.52	1.0194
100	0.022442	305.80	332.73	1.0836	0.018778	304.01	330.30	1.0661	0.016014	302.14	327.76	1.0500
110	0.023348	315.38	343.40	1.1118	0.019597	313.76	341.19	1.0949	0.016773	312.07	338.91	1.0795
120	0.024228	325.03	354.11	1.1394	0.020388	323.55	352.09	1.1230	0.017500	322.02	350.02	1.1081
130	0.025086	334.77	364.88	1.1664	0.021155	333.41	363.02	1.1504	0.018201	332.00	361.12	1.1360
140	0.025927	344.61	375.72	1.1930	0.021904	343.34	374.01	1.1773	0.018882	342.05	372.26	1.1632
150	0.026753	354.56	386.66	1.2192	0.022636	353.37	385.07	1.2038	0.019545	352.17	383.44	1.1900
160	0.027566	364.61	397.69	1.2449	0.023355	363.51	396.20	1.2298	0.020194	362.38	394.69	1.2163
170	0.028367	374.78	408.82	1.2703	0.024061	373.75	407.43	1.2554	0.020830	372.69	406.02	1.2421
180	0.029158	385.08	420.07	1.2954	0.024757	384.10	418.76	1.2807	0.021456	383.11	417.44	1.2676

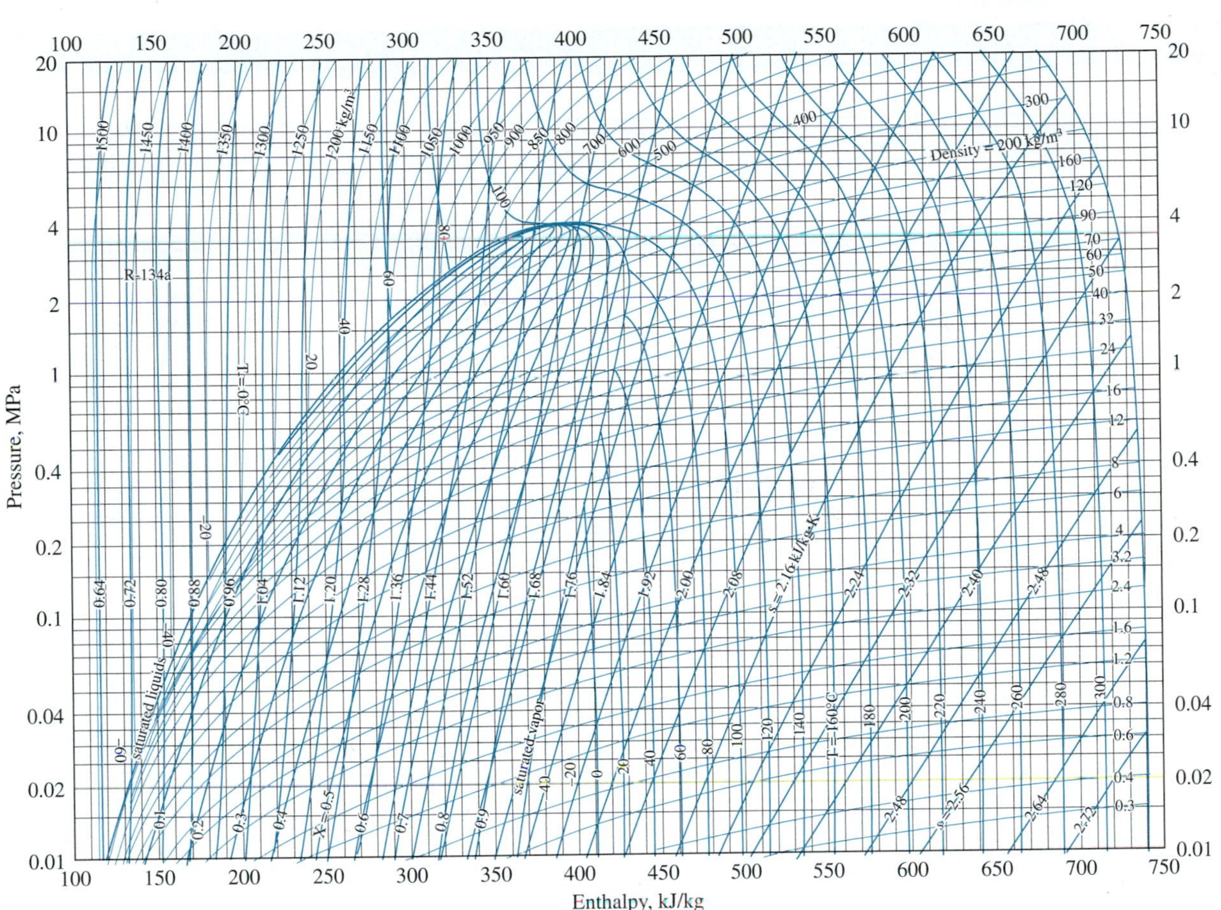

FIGURE A–14

P-h diagram for refrigerant-134a.

Note: The reference point used for the chart is different than that used in the R-134a tables. Therefore, problems should be solved using all property data either from the tables or from the chart, but not from both.

Reprinted by permission of American Society of Heating, Refrigerating, and Air-Conditioning Engineers, Inc., Atlanta, GA.

TABLE A–15

Properties of saturated water

Temp. T, °C	Saturation Pressure P_{sat}, kPa	Density ρ, kg/m^3 Liquid	Density Vapor	Enthalpy of Vaporization h_{fg}, kJ/kg	Specific Heat c_p, J/kg · K Liquid	Specific Heat Vapor	Thermal Conductivity k, W/m · k Liquid	Thermal Conductivity Vapor	Dynamic Viscosity μ, kg/m · s Liquid	Dynamic Viscosity Vapor	Prandtl Number Pr Liquid	Prandtl Number Vapor	Volume Expansion Coefficient β, 1/K Liquid
0.01	0.6113	999.8	0.0048	2501	4217	1854	0.561	0.0171	1.792×10^{-3}	0.922×10^{-5}	13.5	1.00	-0.068×10^{-3}
5	0.8721	999.9	0.0068	2490	4205	1857	0.571	0.0173	1.519×10^{-3}	0.934×10^{-5}	11.2	1.00	0.015×10^{-3}
10	1.2276	999.7	0.0094	2478	4194	1862	0.580	0.0176	1.307×10^{-3}	0.946×10^{-5}	9.45	1.00	0.733×10^{-3}
15	1.7051	999.1	0.0128	2466	4185	1863	0.589	0.0179	1.138×10^{-3}	0.959×10^{-5}	8.09	1.00	0.138×10^{-3}
20	2.339	998.0	0.0173	2454	4182	1867	0.598	0.0182	1.002×10^{-3}	0.973×10^{-5}	7.01	1.00	0.195×10^{-3}
25	3.169	997.0	0.0231	2442	4180	1870	0.607	0.0186	0.891×10^{-3}	0.987×10^{-5}	6.14	1.00	0.247×10^{-3}
30	4.246	996.0	0.0304	2431	4178	1875	0.615	0.0189	0.798×10^{-3}	1.001×10^{-5}	5.42	1.00	0.294×10^{-3}
35	5.628	994.0	0.0397	2419	4178	1880	0.623	0.0192	0.720×10^{-3}	1.016×10^{-5}	4.83	1.00	0.337×10^{-3}
40	7.384	992.1	0.0512	2407	4179	1885	0.631	0.0196	0.653×10^{-3}	1.031×10^{-5}	4.32	1.00	0.377×10^{-3}
45	9.593	990.1	0.0655	2395	4180	1892	0.637	0.0200	0.596×10^{-3}	1.046×10^{-5}	3.91	1.00	0.415×10^{-3}
50	12.35	988.1	0.0831	2383	4181	1900	0.644	0.0204	0.547×10^{-3}	1.062×10^{-5}	3.55	1.00	0.451×10^{-3}
55	15.76	985.2	0.1045	2371	4183	1908	0.649	0.0208	0.504×10^{-3}	1.077×10^{-5}	3.25	1.00	0.484×10^{-3}
60	19.94	983.3	0.1304	2359	4185	1916	0.654	0.0212	0.467×10^{-3}	1.093×10^{-5}	2.99	1.00	0.517×10^{-3}
65	25.03	980.4	0.1614	2346	4187	1926	0.659	0.0216	0.433×10^{-3}	1.110×10^{-5}	2.75	1.00	0.548×10^{-3}
70	31.19	977.5	0.1983	2334	4190	1936	0.663	0.0221	0.404×10^{-3}	1.126×10^{-5}	2.55	1.00	0.578×10^{-3}
75	38.58	974.7	0.2421	2321	4193	1948	0.667	0.0225	0.378×10^{-3}	1.142×10^{-5}	2.38	1.00	0.607×10^{-3}
80	47.39	971.8	0.2935	2309	4197	1962	0.670	0.0230	0.355×10^{-3}	1.159×10^{-5}	2.22	1.00	0.653×10^{-3}
85	57.83	968.1	0.3536	2296	4201	1977	0.673	0.0235	0.333×10^{-3}	1.176×10^{-5}	2.08	1.00	0.670×10^{-3}
90	70.14	965.3	0.4235	2283	4206	1993	0.675	0.0240	0.315×10^{-3}	1.193×10^{-5}	1.96	1.00	0.702×10^{-3}
95	84.55	961.5	0.5045	2270	4212	2010	0.677	0.0246	0.297×10^{-3}	1.210×10^{-5}	1.85	1.00	0.716×10^{-3}
100	101.33	957.9	0.5978	2257	4217	2029	0.679	0.0251	0.282×10^{-3}	1.227×10^{-5}	1.75	1.00	0.750×10^{-3}
110	143.27	950.6	0.8263	2230	4229	2071	0.682	0.0262	0.255×10^{-3}	1.261×10^{-5}	1.58	1.00	0.798×10^{-3}
120	198.53	943.4	1.121	2203	4244	2120	0.683	0.0275	0.232×10^{-3}	1.296×10^{-5}	1.44	1.00	0.858×10^{-3}
130	270.1	934.6	1.496	2174	4263	2177	0.684	0.0288	0.213×10^{-3}	1.330×10^{-5}	1.33	1.01	0.913×10^{-3}
140	361.3	921.7	1.965	2145	4286	2244	0.683	0.0301	0.197×10^{-3}	1.365×10^{-5}	1.24	1.02	0.970×10^{-3}
150	475.8	916.6	2.546	2114	4311	2314	0.682	0.0316	0.183×10^{-3}	1.399×10^{-5}	1.16	1.02	1.025×10^{-3}
160	617.8	907.4	3.256	2083	4340	2420	0.680	0.0331	0.170×10^{-3}	1.434×10^{-5}	1.09	1.05	1.145×10^{-3}
170	791.7	897.7	4.119	2050	4370	2490	0.677	0.0347	0.160×10^{-3}	1.468×10^{-5}	1.03	1.05	1.178×10^{-3}
180	1,002.1	887.3	5.153	2015	4410	2590	0.673	0.0364	0.150×10^{-3}	1.502×10^{-5}	0.983	1.07	1.210×10^{-3}
190	1,254.4	876.4	6.388	1979	4460	2710	0.669	0.0382	0.142×10^{-3}	1.537×10^{-5}	0.947	1.09	1.280×10^{-3}
200	1,553.8	864.3	7.852	1941	4500	2840	0.663	0.0401	0.134×10^{-3}	1.571×10^{-5}	0.910	1.11	1.350×10^{-3}
220	2,318	840.3	11.60	1859	4610	3110	0.650	0.0442	0.122×10^{-3}	1.641×10^{-5}	0.865	1.15	1.520×10^{-3}
240	3,344	813.7	16.73	1767	4760	3520	0.632	0.0487	0.111×10^{-3}	1.712×10^{-5}	0.836	1.24	1.720×10^{-3}
260	4,688	783.7	23.69	1663	4970	4070	0.609	0.0540	0.102×10^{-3}	1.788×10^{-5}	0.832	1.35	2.000×10^{-3}
280	6,412	750.8	33.15	1544	5280	4835	0.581	0.0605	0.094×10^{-3}	1.870×10^{-5}	0.854	1.49	2.380×10^{-3}
300	8,581	713.8	46.15	1405	5750	5980	0.548	0.0695	0.086×10^{-3}	1.965×10^{-5}	0.902	1.69	2.950×10^{-3}
320	11,274	667.1	64.57	1239	6540	7900	0.509	0.0836	0.078×10^{-3}	2.084×10^{-5}	1.00	1.97	
340	14,586	610.5	92.62	1028	8240	11,870	0.469	0.110	0.070×10^{-3}	2.255×10^{-5}	1.23	2.43	
360	18,651	528.3	144.0	720	14,690	25,800	0.427	0.178	0.060×10^{-3}	2.571×10^{-5}	2.06	3.73	
374.14	22,090	317.0	317.0	0	—	—	—	—	0.043×10^{-3}	4.313×10^{-5}			

Note 1: Kinematic viscosity ν and thermal diffusivity α can be calculated from their definitions, $\nu = \mu/\rho$ and $\alpha = k/\rho c_p = \nu/\text{Pr}$. The temperatures 0.01°C, 100°C, and 374.14°C are the triple-, boiling-, and critical-point temperatures of water, respectively. The properties listed above (except the vapor density) can be used at any pressure with negligible error except at temperatures near the critical-point value.

Note 2: The unit kJ/kg · °C for specific heat is equivalent to kJ/kg · K, and the unit W/m · °C for thermal conductivity is equivalent to W/m · K.

Source: Viscosity and thermal conductivity data are from J. V. Sengers and J. T. R. Watson, *Journal of Physical and Chemical Reference Data* 15 (1986), pp. 1291–1322. Other data are obtained from various sources or calculated.

TABLE A–16

Properties of saturated refrigerant-134a

Temp. T, °C	Saturation Pressure P, kPa	Density ρ, kg/m^3 Liquid	Density Vapor	Enthalpy of Vaporization h_{fg}, kJ/kg	Specific Heat c_p, J/kg · K Liquid	Specific Heat Vapor	Thermal Conductivity k, W/m · K Liquid	Thermal Conductivity Vapor	Dynamic Viscosity μ, kg/m · s Liquid	Dynamic Viscosity Vapor	Prandtl Number Pr Liquid	Prandtl Number Vapor	Volume Expansion Coefficient β, 1/K Liquid	Surface Tension, N/m
−40	51.2	1418	2.773	225.9	1254	748.6	0.1101	0.00811	4.878×10^{-4}	2.550×10^{-6}	5.558	0.235	0.00205	0.01760
−35	66.2	1403	3.524	222.7	1264	764.1	0.1084	0.00862	4.509×10^{-4}	3.003×10^{-6}	5.257	0.266	0.00209	0.01682
−30	84.4	1389	4.429	219.5	1273	780.2	0.1066	0.00913	4.178×10^{-4}	3.504×10^{-6}	4.992	0.299	0.00215	0.01604
−25	106.5	1374	5.509	216.3	1283	797.2	0.1047	0.00963	3.882×10^{-4}	4.054×10^{-6}	4.757	0.335	0.00220	0.01527
−20	132.8	1359	6.787	213.0	1294	814.9	0.1028	0.01013	3.614×10^{-4}	4.651×10^{-6}	4.548	0.374	0.00227	0.01451
−15	164.0	1343	8.288	209.5	1306	833.5	0.1009	0.01063	3.371×10^{-4}	5.295×10^{-6}	4.363	0.415	0.00233	0.01376
−10	200.7	1327	10.04	206.0	1318	853.1	0.0989	0.01112	3.150×10^{-4}	5.982×10^{-6}	4.198	0.459	0.00241	0.01302
−5	243.5	1311	12.07	202.4	1330	873.8	0.0968	0.01161	2.947×10^{-4}	6.709×10^{-6}	4.051	0.505	0.00249	0.01229
0	293.0	1295	14.42	198.7	1344	895.6	0.0947	0.01210	2.761×10^{-4}	7.471×10^{-6}	3.919	0.553	0.00258	0.01156
5	349.9	1278	17.12	194.8	1358	918.7	0.0925	0.01259	2.589×10^{-4}	8.264×10^{-6}	3.802	0.603	0.00269	0.01084
10	414.9	1261	20.22	190.8	1374	943.2	0.0903	0.01308	2.430×10^{-4}	9.081×10^{-6}	3.697	0.655	0.00280	0.01014
15	488.7	1244	23.75	186.6	1390	969.4	0.0880	0.01357	2.281×10^{-4}	9.915×10^{-6}	3.604	0.708	0.00293	0.00944
20	572.1	1226	27.77	182.3	1408	997.6	0.0856	0.01406	2.142×10^{-4}	1.075×10^{-5}	3.521	0.763	0.00307	0.00876
25	665.8	1207	32.34	177.8	1427	1028	0.0833	0.01456	2.012×10^{-4}	1.160×10^{-5}	3.448	0.819	0.00324	0.00808
30	770.6	1188	37.53	173.1	1448	1061	0.0808	0.01507	1.888×10^{-4}	1.244×10^{-5}	3.383	0.877	0.00342	0.00742
35	887.5	1168	43.41	168.2	1471	1098	0.0783	0.01558	1.772×10^{-4}	1.327×10^{-5}	3.328	0.935	0.00364	0.00677
40	1017.1	1147	50.08	163.0	1498	1138	0.0757	0.01610	1.660×10^{-4}	1.408×10^{-5}	3.285	0.995	0.00390	0.00613
45	1160.5	1125	57.66	157.6	1529	1184	0.0731	0.01664	1.554×10^{-4}	1.486×10^{-5}	3.253	1.058	0.00420	0.00550
50	1318.6	1102	66.27	151.8	1566	1237	0.0704	0.01720	1.453×10^{-4}	1.562×10^{-5}	3.231	1.123	0.00455	0.00489
55	1492.3	1078	76.11	145.7	1608	1298	0.0676	0.01777	1.355×10^{-4}	1.634×10^{-5}	3.223	1.193	0.00500	0.00429
60	1682.8	1053	87.38	139.1	1659	1372	0.0647	0.01838	1.260×10^{-4}	1.704×10^{-5}	3.229	1.272	0.00554	0.00372
65	1891.0	1026	100.4	132.1	1722	1462	0.0618	0.01902	1.167×10^{-4}	1.771×10^{-5}	3.255	1.362	0.00624	0.00315
70	2118.2	996.2	115.6	124.4	1801	1577	0.0587	0.01972	1.077×10^{-4}	1.839×10^{-5}	3.307	1.471	0.00716	0.00261
75	2365.8	964	133.6	115.9	1907	1731	0.0555	0.02048	9.891×10^{-5}	1.908×10^{-5}	3.400	1.612	0.00843	0.00209
80	2635.2	928.2	155.3	106.4	2056	1948	0.0521	0.02133	9.011×10^{-5}	1.982×10^{-5}	3.558	1.810	0.01031	0.00160
85	2928.2	887.1	182.3	95.4	2287	2281	0.0484	0.02233	8.124×10^{-5}	2.071×10^{-5}	3.837	2.116	0.01336	0.00114
90	3246.9	837.7	217.8	82.2	2701	2865	0.0444	0.02357	7.203×10^{-5}	2.187×10^{-5}	4.385	2.658	0.01911	0.00071
95	3594.1	772.5	269.3	64.9	3675	4144	0.0396	0.02544	6.190×10^{-5}	2.370×10^{-5}	5.746	3.862	0.03343	0.00033
100	3975.1	651.7	376.3	33.9	7959	8785	0.0322	0.02989	4.765×10^{-5}	2.833×10^{-5}	11.77	8.326	0.10047	0.00004

Note 1: Kinematic viscosity ν and thermal diffusivity α can be calculated from their definitions, $\nu = \mu/\rho$ and $\alpha = k/\rho c_p = \nu/\text{Pr}$. The properties listed here (except the vapor density) can be used at any pressures with negligible error except at temperatures near the critical-point value.

Note 2: The unit kJ/kg · °C for specific heat is equivalent to kJ/kg · K, and the unit W/m · °C for thermal conductivity is equivalent to W/m · K.

Source: Data generated from the EES software developed by S. A. Klein and F. L. Alvarado. Original sources: R. Tillner-Roth and H. D. Baehr, "An International Standard Formulation for the Thermodynamic Properties of 1,1,1,2-Tetrafluoroethane (HFC-134a) for Temperatures from 170 K to 455 K and Pressures up to 70 MPa," *J. Phys. Chem, Ref. Data*, Vol. 23, No. 5, 1994; M.J. Assael, N. K. Dalaouti, A. A. Griva, and J. H. Dymond, "Viscosity and Thermal Conductivity of Halogenated Methane and Ethane Refrigerants," *IJR*, Vol. 22, pp. 525–535, 1999; NIST REFPROP 6 program (M. O. McLinden, S. A. Klein, E. W. Lemmon, and A. P. Peskin, Physical and Chemical Properties Division, National Institute of Standards and Technology, Boulder, CO 80303, 1995).

TABLE A–17

Properties of saturated ammonia

Temp. T, °C	Saturation Pressure P, kPa	Density ρ, kg/m^3 Liquid	Vapor	Enthalpy of Vaporization h_{fg}, kJ/kg	Specific Heat c_p, J/kg · K Liquid	Vapor	Thermal Conductivity k, W/m · K Liquid	Vapor	Dynamic Viscosity μ, kg/m · s Liquid	Vapor	Prandtl Number Pr Liquid	Vapor	Volume Expansion Coefficient β, 1/K Liquid	Surface Tension, N/m
−40	71.66	690.2	0.6435	1389	4414	2242	—	0.01792	2.926×10^{-4}	7.957×10^{-6}	—	0.9955	0.00176	0.03565
−30	119.4	677.8	1.037	1360	4465	2322	—	0.01898	2.630×10^{-4}	8.311×10^{-6}	—	1.017	0.00185	0.03341
−25	151.5	671.5	1.296	1345	4489	2369	0.5968	0.01957	2.492×10^{-4}	8.490×10^{-6}	1.875	1.028	0.00190	0.03229
−20	190.1	665.1	1.603	1329	4514	2420	0.5853	0.02015	2.361×10^{-4}	8.669×10^{-6}	1.821	1.041	0.00194	0.03118
−15	236.2	658.6	1.966	1313	4538	2476	0.5737	0.02075	2.236×10^{-4}	8.851×10^{-6}	1.769	1.056	0.00199	0.03007
−10	290.8	652.1	2.391	1297	4564	2536	0.5621	0.02138	2.117×10^{-4}	9.034×10^{-6}	1.718	1.072	0.00205	0.02896
−5	354.9	645.4	2.886	1280	4589	2601	0.5505	0.02203	2.003×10^{-4}	9.218×10^{-6}	1.670	1.089	0.00210	0.02786
0	429.6	638.6	3.458	1262	4617	2672	0.5390	0.02270	1.896×10^{-4}	9.405×10^{-6}	1.624	1.107	0.00216	0.02676
5	516	631.7	4.116	1244	4645	2749	0.5274	0.02341	1.794×10^{-4}	9.593×10^{-6}	1.580	1.126	0.00223	0.02566
10	615.3	624.6	4.870	1226	4676	2831	0.5158	0.02415	1.697×10^{-4}	9.784×10^{-6}	1.539	1.147	0.00230	0.02457
15	728.8	617.5	5.729	1206	4709	2920	0.5042	0.02492	1.606×10^{-4}	9.978×10^{-6}	1.500	1.169	0.00237	0.02348
20	857.8	610.2	6.705	1186	4745	3016	0.4927	0.02573	1.519×10^{-4}	1.017×10^{-5}	1.463	1.193	0.00245	0.02240
25	1003	602.8	7.809	1166	4784	3120	0.4811	0.02658	1.438×10^{-4}	1.037×10^{-5}	1.430	1.218	0.00254	0.02132
30	1167	595.2	9.055	1144	4828	3232	0.4695	0.02748	1.361×10^{-4}	1.057×10^{-5}	1.399	1.244	0.00264	0.02024
35	1351	587.4	10.46	1122	4877	3354	0.4579	0.02843	1.288×10^{-4}	1.078×10^{-5}	1.372	1.272	0.00275	0.01917
40	1555	579.4	12.03	1099	4932	3486	0.4464	0.02943	1.219×10^{-4}	1.099×10^{-5}	1.347	1.303	0.00287	0.01810
45	1782	571.3	13.8	1075	4993	3631	0.4348	0.03049	1.155×10^{-4}	1.121×10^{-5}	1.327	1.335	0.00301	0.01704
50	2033	562.9	15.78	1051	5063	3790	0.4232	0.03162	1.094×10^{-4}	1.143×10^{-5}	1.310	1.371	0.00316	0.01598
55	2310	554.2	18.00	1025	5143	3967	0.4116	0.03283	1.037×10^{-4}	1.166×10^{-5}	1.297	1.409	0.00334	0.01493
60	2614	545.2	20.48	997.4	5234	4163	0.4001	0.03412	9.846×10^{-5}	1.189×10^{-5}	1.288	1.452	0.00354	0.01389
65	2948	536.0	23.26	968.9	5340	4384	0.3885	0.03550	9.347×10^{-5}	1.213×10^{-5}	1.285	1.499	0.00377	0.01285
70	3312	526.3	26.39	939.0	5463	4634	0.3769	0.03700	8.879×10^{-5}	1.238×10^{-5}	1.287	1.551	0.00404	0.01181
75	3709	516.2	29.90	907.5	5608	4923	0.3653	0.03862	8.440×10^{-5}	1.264×10^{-5}	1.296	1.612	0.00436	0.01079
80	4141	505.7	33.87	874.1	5780	5260	0.3538	0.04038	8.030×10^{-5}	1.292×10^{-5}	1.312	1.683	0.00474	0.00977
85	4609	494.5	38.36	838.6	5988	5659	0.3422	0.04232	7.646×10^{-5}	1.322×10^{-5}	1.338	1.768	0.00521	0.00876
90	5116	482.8	43.48	800.6	6242	6142	0.3306	0.04447	7.284×10^{-5}	1.354×10^{-5}	1.375	1.871	0.00579	0.00776
95	5665	470.2	49.35	759.8	6561	6740	0.3190	0.04687	6.946×10^{-5}	1.389×10^{-5}	1.429	1.999	0.00652	0.00677
100	6257	456.6	56.15	715.5	6972	7503	0.3075	0.04958	6.628×10^{-5}	1.429×10^{-5}	1.503	2.163	0.00749	0.00579

Note 1: Kinematic viscosity ν and thermal diffusivity α can be calculated from their definitions, $\nu = \mu/\rho$ and $\alpha = k/\rho c_p = \nu/\text{Pr}$. The properties listed here (except the vapor density) can be used at any pressures with negligible error except at temperatures near the critical-point value.

Note 2: The unit kJ/kg · °C for specific heat is equivalent to kJ/kg · K, and the unit W/m · °C for thermal conductivity is equivalent to W/m · K.

Source: Data generated from the EES software developed by S. A. Klein and F. L. Alvarado. Original sources: Tillner-Roth, Harms-Watzenberg, and Baehr, "Eine neue Fundamentalgleichung fur Ammoniak," DKV-Tagungsbericht 20:167–181, 1993; Liley and Desai, "Thermophysical Properties of Refrigerants," *ASHRAE*, 1993, ISBN 1-1883413-10-9.

TABLE A–18

Properties of saturated propane

Temp. T, °C	Saturation Pressure P, kPa	Density ρ, kg/m^3 Liquid	Density ρ, kg/m^3 Vapor	Enthalpy of Vaporization h_{fg}, kJ/kg	Specific Heat c_p, J/kg · K Liquid	Specific Heat c_p, J/kg · K Vapor	Thermal Conductivity k, W/m · K Liquid	Thermal Conductivity k, W/m · K Vapor	Dynamic Viscosity μ, kg/m · s Liquid	Dynamic Viscosity μ, kg/m · s Vapor	Prandtl Number Pr Liquid	Prandtl Number Pr Vapor	Volume Expansion Coefficient β, 1/K Liquid	Surface Tension, N/m
−120	0.4053	664.7	0.01408	498.3	2003	1115	0.1802	0.00589	6.136×10^{-4}	4.372×10^{-6}	6.820	0.827	0.00153	0.02630
−110	1.157	654.5	0.03776	489.3	2021	1148	0.1738	0.00645	5.054×10^{-4}	4.625×10^{-6}	5.878	0.822	0.00157	0.02486
−100	2.881	644.2	0.08872	480.4	2044	1183	0.1672	0.00705	4.252×10^{-4}	4.881×10^{-6}	5.195	0.819	0.00161	0.02344
−90	6.406	633.8	0.1870	471.5	2070	1221	0.1606	0.00769	3.635×10^{-4}	5.143×10^{-6}	4.686	0.817	0.00166	0.02202
−80	12.97	623.2	0.3602	462.4	2100	1263	0.1539	0.00836	3.149×10^{-4}	5.409×10^{-6}	4.297	0.817	0.00171	0.02062
−70	24.26	612.5	0.6439	453.1	2134	1308	0.1472	0.00908	2.755×10^{-4}	5.680×10^{-6}	3.994	0.818	0.00177	0.01923
−60	42.46	601.5	1.081	443.5	2173	1358	0.1407	0.00985	2.430×10^{-4}	5.956×10^{-6}	3.755	0.821	0.00184	0.01785
−50	70.24	590.3	1.724	433.6	2217	1412	0.1343	0.01067	2.158×10^{-4}	6.239×10^{-6}	3.563	0.825	0.00192	0.01649
−40	110.7	578.8	2.629	423.1	2258	1471	0.1281	0.01155	1.926×10^{-4}	6.529×10^{-6}	3.395	0.831	0.00201	0.01515
−30	167.3	567.0	3.864	412.1	2310	1535	0.1221	0.01250	1.726×10^{-4}	6.827×10^{-6}	3.266	0.839	0.00213	0.01382
−20	243.8	554.7	5.503	400.3	2368	1605	0.1163	0.01351	1.551×10^{-4}	7.136×10^{-6}	3.158	0.848	0.00226	0.01251
−10	344.4	542.0	7.635	387.8	2433	1682	0.1107	0.01459	1.397×10^{-4}	7.457×10^{-6}	3.069	0.860	0.00242	0.01122
0	473.3	528.7	10.36	374.2	2507	1768	0.1054	0.01576	1.259×10^{-4}	7.794×10^{-6}	2.996	0.875	0.00262	0.00996
5	549.8	521.8	11.99	367.0	2547	1814	0.1028	0.01637	1.195×10^{-4}	7.970×10^{-6}	2.964	0.883	0.00273	0.00934
10	635.1	514.7	13.81	359.5	2590	1864	0.1002	0.01701	1.135×10^{-4}	8.151×10^{-6}	2.935	0.893	0.00286	0.00872
15	729.8	507.5	15.85	351.7	2637	1917	0.0977	0.01767	1.077×10^{-4}	8.339×10^{-6}	2.909	0.905	0.00301	0.00811
20	834.4	500.0	18.13	343.4	2688	1974	0.0952	0.01836	1.022×10^{-4}	8.534×10^{-6}	2.886	0.918	0.00318	0.00751
25	949.7	492.2	20.68	334.8	2742	2036	0.0928	0.01908	9.702×10^{-5}	8.738×10^{-6}	2.866	0.933	0.00337	0.00691
30	1076	484.2	23.53	325.8	2802	2104	0.0904	0.01982	9.197×10^{-5}	8.952×10^{-6}	2.850	0.950	0.00358	0.00633
35	1215	475.8	26.72	316.2	2869	2179	0.0881	0.02061	8.710×10^{-5}	9.178×10^{-6}	2.837	0.971	0.00384	0.00575
40	1366	467.1	30.29	306.1	2943	2264	0.0857	0.02142	8.240×10^{-5}	9.417×10^{-6}	2.828	0.995	0.00413	0.00518
45	1530	458.0	34.29	295.3	3026	2361	0.0834	0.02228	7.785×10^{-5}	9.674×10^{-6}	2.824	1.025	0.00448	0.00463
50	1708	448.5	38.79	283.9	3122	2473	0.0811	0.02319	7.343×10^{-5}	9.950×10^{-5}	2.826	1.061	0.00491	0.00408
60	2110	427.5	49.66	258.4	3283	2769	0.0765	0.02517	6.487×10^{-5}	1.058×10^{-5}	2.784	1.164	0.00609	0.00303
70	2580	403.2	64.02	228.0	3595	3241	0.0717	0.02746	5.649×10^{-5}	1.138×10^{-5}	2.834	1.343	0.00811	0.00204
80	3127	373.0	84.28	189.7	4501	4173	0.0663	0.03029	4.790×10^{-5}	1.249×10^{-5}	3.251	1.722	0.01248	0.00114
90	3769	329.1	118.6	133.2	6977	7239	0.0595	0.03441	3.807×10^{-5}	1.448×10^{-5}	4.465	3.047	0.02847	0.00037

Note 1: Kinematic viscosity ν and thermal diffusivity α can be calculated from their definitions, $\nu = \mu/\rho$ and $\alpha = k/\rho c_p = \nu/\text{Pr}$. The properties listed here (except the vapor density) can be used at any pressures with negligible error except at temperatures near the critical-point value.

Note 2: The unit kJ/kg · °C for specific heat is equivalent to kJ/kg · K, and the unit W/m · °C for thermal conductivity is equivalent to W/m · K.

Source: Data generated from the EES software developed by S. A. Klein and F. L. Alvarado. Original sources: Reiner Tillner-Roth, "Fundamental Equations of State," Shaker, Verlag, Aachan, 1998; B. A. Younglove and J. F. Ely, "Thermophysical Properties of Fluids. II Methane, Ethane, Propane, Isobutane, and Normal Butane," *J. Phys. Chem. Ref. Data*, Vol. 16, No. 4, 1987; G.R. Somayajulu, "A Generalized Equation for Surface Tension from the Triple-Point to the Critical-Point," *International Journal of Thermophysics*, Vol. 9, No. 4, 1988.

TABLE A–19

Properties of liquids

Temp. T, °C	Density ρ, kg/m^3	Specific Heat c_p, J/kg·K	Thermal Conductivity k, W/m·K	Thermal Diffusivity α, m^2/s	Dynamic Viscosity μ, kg/m·s	Kinematic Viscosity ν, m^2/s	Prandtl Number Pr	Volume Expansion Coeff. β, 1/K
				Methane [CH_4]				
−160	420.2	3492	0.1863	1.270×10^{-7}	1.133×10^{-4}	2.699×10^{-7}	2.126	0.00352
−150	405.0	3580	0.1703	1.174×10^{-7}	9.169×10^{-5}	2.264×10^{-7}	1.927	0.00391
−140	388.8	3700	0.1550	1.077×10^{-7}	7.551×10^{-5}	1.942×10^{-7}	1.803	0.00444
−130	371.1	3875	0.1402	9.749×10^{-8}	6.288×10^{-5}	1.694×10^{-7}	1.738	0.00520
−120	351.4	4146	0.1258	8.634×10^{-8}	5.257×10^{-5}	1.496×10^{-7}	1.732	0.00637
−110	328.8	4611	0.1115	7.356×10^{-8}	4.377×10^{-5}	1.331×10^{-7}	1.810	0.00841
−100	301.0	5578	0.0967	5.761×10^{-8}	3.577×10^{-5}	1.188×10^{-7}	2.063	0.01282
−90	261.7	8902	0.0797	3.423×10^{-8}	2.761×10^{-5}	1.055×10^{-7}	3.082	0.02922
				Methanol [$CH_3(OH)$]				
20	788.4	2515	0.1987	1.002×10^{-7}	5.857×10^{-4}	7.429×10^{-7}	7.414	0.00118
30	779.1	2577	0.1980	9.862×10^{-8}	5.088×10^{-4}	6.531×10^{-7}	6.622	0.00120
40	769.6	2644	0.1972	9.690×10^{-8}	4.460×10^{-4}	5.795×10^{-7}	5.980	0.00123
50	760.1	2718	0.1965	9.509×10^{-8}	3.942×10^{-4}	5.185×10^{-7}	5.453	0.00127
60	750.4	2798	0.1957	9.320×10^{-8}	3.510×10^{-4}	4.677×10^{-7}	5.018	0.00132
70	740.4	2885	0.1950	9.128×10^{-8}	3.146×10^{-4}	4.250×10^{-7}	4.655	0.00137
				Isobutane (R600a)				
−100	683.8	1881	0.1383	1.075×10^{-7}	9.305×10^{-4}	1.360×10^{-6}	12.65	0.00142
−75	659.3	1970	0.1357	1.044×10^{-7}	5.624×10^{-4}	8.531×10^{-7}	8.167	0.00150
−50	634.3	2069	0.1283	9.773×10^{-8}	3.769×10^{-4}	5.942×10^{-7}	6.079	0.00161
−25	608.2	2180	0.1181	8.906×10^{-8}	2.688×10^{-4}	4.420×10^{-7}	4.963	0.00177
0	580.6	2306	0.1068	7.974×10^{-8}	1.993×10^{-4}	3.432×10^{-7}	4.304	0.00199
25	550.7	2455	0.0956	7.069×10^{-8}	1.510×10^{-4}	2.743×10^{-7}	3.880	0.00232
50	517.3	2640	0.0851	6.233×10^{-8}	1.155×10^{-4}	2.233×10^{-7}	3.582	0.00286
75	478.5	2896	0.0757	5.460×10^{-8}	8.785×10^{-5}	1.836×10^{-7}	3.363	0.00385
100	429.6	3361	0.0669	4.634×10^{-8}	6.483×10^{-5}	1.509×10^{-7}	3.256	0.00628
				Glycerin				
0	1276	2262	0.2820	9.773×10^{-8}	10.49	8.219×10^{-3}	84,101	
5	1273	2288	0.2835	9.732×10^{-8}	6.730	5.287×10^{-3}	54,327	
10	1270	2320	0.2846	9.662×10^{-8}	4.241	3.339×10^{-3}	34,561	
15	1267	2354	0.2856	9.576×10^{-8}	2.496	1.970×10^{-3}	20,570	
20	1264	2386	0.2860	9.484×10^{-8}	1.519	1.201×10^{-3}	12,671	
25	1261	2416	0.2860	9.388×10^{-8}	0.9934	7.878×10^{-4}	8,392	
30	1258	2447	0.2860	9.291×10^{-8}	0.6582	5.232×10^{-4}	5,631	
35	1255	2478	0.2860	9.195×10^{-8}	0.4347	3.464×10^{-4}	3,767	
40	1252	2513	0.2863	9.101×10^{-8}	0.3073	2.455×10^{-4}	2,697	
				Engine Oil (unused)				
0	899.0	1797	0.1469	9.097×10^{-8}	3.814	4.242×10^{-3}	46,636	0.00070
20	888.1	1881	0.1450	8.680×10^{-8}	0.8374	9.429×10^{-4}	10,863	0.00070
40	876.0	1964	0.1444	8.391×10^{-8}	0.2177	2.485×10^{-4}	2,962	0.00070
60	863.9	2048	0.1404	7.934×10^{-8}	0.07399	8.565×10^{-5}	1,080	0.00070
80	852.0	2132	0.1380	7.599×10^{-8}	0.03232	3.794×10^{-5}	499.3	0.00070
100	840.0	2220	0.1367	7.330×10^{-8}	0.01718	2.046×10^{-5}	279.1	0.00070
120	828.9	2308	0.1347	7.042×10^{-8}	0.01029	1.241×10^{-5}	176.3	0.00070
140	816.8	2395	0.1330	6.798×10^{-8}	0.006558	8.029×10^{-6}	118.1	0.00070
150	810.3	2441	0.1327	6.708×10^{-8}	0.005344	6.595×10^{-6}	98.31	0.00070

Source: Data generated from the EES software developed by S. A. Klein and F. L. Alvarado. Originally based on various sources.

TABLE A–20

Properties of liquid metals

Temp. T, °C	Density ρ, kg/m³	Specific Heat c_p, J/kg · K	Thermal Conductivity k, W/m · K	Thermal Diffusivity α, m²/s	Dynamic Viscosity μ, kg/m · s	Kinematic Viscosity ν, m²/s	Prandtl Number Pr	Volume Expansion Coeff. β, 1/K
				Mercury (Hg) Melting Point: −39°C				
0	13595	140.4	8.18200	4.287×10^{-6}	1.687×10^{-3}	1.241×10^{-7}	0.0289	1.810×10^{-4}
25	13534	139.4	8.51533	4.514×10^{-6}	1.534×10^{-3}	1.133×10^{-7}	0.0251	1.810×10^{-4}
50	13473	138.6	8.83632	4.734×10^{-6}	1.423×10^{-3}	1.056×10^{-7}	0.0223	1.810×10^{-4}
75	13412	137.8	9.15632	4.956×10^{-6}	1.316×10^{-3}	9.819×10^{-8}	0.0198	1.810×10^{-4}
100	13351	137.1	9.46706	5.170×10^{-6}	1.245×10^{-3}	9.326×10^{-8}	0.0180	1.810×10^{-4}
150	13231	136.1	10.07780	5.595×10^{-6}	1.126×10^{-3}	8.514×10^{-8}	0.0152	1.810×10^{-4}
200	13112	135.5	10.65465	5.996×10^{-6}	1.043×10^{-3}	7.959×10^{-8}	0.0133	1.815×10^{-4}
250	12993	135.3	11.18150	6.363×10^{-6}	9.820×10^{-4}	7.558×10^{-8}	0.0119	1.829×10^{-4}
300	12873	135.3	11.68150	6.705×10^{-6}	9.336×10^{-4}	7.252×10^{-8}	0.0108	1.854×10^{-4}
				Bismuth (Bi) Melting Point: 271°C				
350	9969	146.0	16.28	1.118×10^{-5}	1.540×10^{-3}	1.545×10^{-7}	0.01381	
400	9908	148.2	16.10	1.096×10^{-5}	1.422×10^{-3}	1.436×10^{-7}	0.01310	
500	9785	152.8	15.74	1.052×10^{-5}	1.188×10^{-3}	1.215×10^{-7}	0.01154	
600	9663	157.3	15.60	1.026×10^{-5}	1.013×10^{-3}	1.048×10^{-7}	0.01022	
700	9540	161.8	15.60	1.010×10^{-5}	8.736×10^{-4}	9.157×10^{-8}	0.00906	
				Lead (Pb) Melting Point: 327°C				
400	10506	158	15.97	9.623×10^{-6}	2.277×10^{-3}	2.167×10^{-7}	0.02252	
450	10449	156	15.74	9.649×10^{-6}	2.065×10^{-3}	1.976×10^{-7}	0.02048	
500	10390	155	15.54	9.651×10^{-6}	1.884×10^{-3}	1.814×10^{-7}	0.01879	
550	10329	155	15.39	9.610×10^{-6}	1.758×10^{-3}	1.702×10^{-7}	0.01771	
600	10267	155	15.23	9.568×10^{-6}	1.632×10^{-3}	1.589×10^{-7}	0.01661	
650	10206	155	15.07	9.526×10^{-6}	1.505×10^{-3}	1.475×10^{-7}	0.01549	
700	10145	155	14.91	9.483×10^{-6}	1.379×10^{-3}	1.360×10^{-7}	0.01434	
				Sodium (Na) Melting Point: 98°C				
100	927.3	1378	85.84	6.718×10^{-5}	6.892×10^{-4}	7.432×10^{-7}	0.01106	
200	902.5	1349	80.84	6.639×10^{-5}	5.385×10^{-4}	5.967×10^{-7}	0.008987	
300	877.8	1320	75.84	6.544×10^{-5}	3.878×10^{-4}	4.418×10^{-7}	0.006751	
400	853.0	1296	71.20	6.437×10^{-5}	2.720×10^{-4}	3.188×10^{-7}	0.004953	
500	828.5	1284	67.41	6.335×10^{-5}	2.411×10^{-4}	2.909×10^{-7}	0.004593	
600	804.0	1272	63.63	6.220×10^{-5}	2.101×10^{-4}	2.614×10^{-7}	0.004202	
				Potassium (K) Melting Point: 64°C				
200	795.2	790.8	43.99	6.995×10^{-5}	3.350×10^{-4}	4.213×10^{-7}	0.006023	
300	771.6	772.8	42.01	7.045×10^{-5}	2.667×10^{-4}	3.456×10^{-7}	0.004906	
400	748.0	754.8	40.03	7.090×10^{-5}	1.984×10^{-4}	2.652×10^{-7}	0.00374	
500	723.9	750.0	37.81	6.964×10^{-5}	1.668×10^{-4}	2.304×10^{-7}	0.003309	
600	699.6	750.0	35.50	6.765×10^{-5}	1.487×10^{-4}	2.126×10^{-7}	0.003143	
				Sodium-Potassium (%22Na-%78K) Melting Point: −11°C				
100	847.3	944.4	25.64	3.205×10^{-5}	5.707×10^{-4}	6.736×10^{-7}	0.02102	
200	823.2	922.5	26.27	3.459×10^{-5}	4.587×10^{-4}	5.572×10^{-7}	0.01611	
300	799.1	900.6	26.89	3.736×10^{-5}	3.467×10^{-4}	4.339×10^{-7}	0.01161	
400	775.0	879.0	27.50	4.037×10^{-5}	2.357×10^{-4}	3.041×10^{-7}	0.00753	
500	751.5	880.1	27.89	4.217×10^{-5}	2.108×10^{-4}	2.805×10^{-7}	0.00665	
600	728.0	881.2	28.28	4.408×10^{-5}	1.859×10^{-4}	2.553×10^{-7}	0.00579	

Source: Data generated from the EES software developed by S. A. Klein and F. L. Alvarado. Originally based on various sources.

TABLE A–21

Ideal-gas properties of air

T K	h kJ/kg	P_r	u kJ/kg	v_r	$s°$ kJ/kg · K	T K	h kJ/kg	P_r	u kJ/kg	v_r	$s°$ kJ/kg · K
200	199.97	0.3363	142.56	1707.0	1.29559	580	586.04	14.38	419.55	115.7	2.37348
210	209.97	0.3987	149.69	1512.0	1.34444	590	596.52	15.31	427.15	110.6	2.39140
220	219.97	0.4690	156.82	1346.0	1.39105	600	607.02	16.28	434.78	105.8	2.40902
230	230.02	0.5477	164.00	1205.0	1.43557	610	617.53	17.30	442.42	101.2	2.42644
240	240.02	0.6355	171.13	1084.0	1.47824	620	628.07	18.36	450.09	96.92	2.44356
250	250.05	0.7329	178.28	979.0	1.51917	630	638.63	19.84	457.78	92.84	2.46048
260	260.09	0.8405	185.45	887.8	1.55848	640	649.22	20.64	465.50	88.99	2.47716
270	270.11	0.9590	192.60	808.0	1.59634	650	659.84	21.86	473.25	85.34	2.49364
280	280.13	1.0889	199.75	738.0	1.63279	660	670.47	23.13	481.01	81.89	2.50985
285	285.14	1.1584	203.33	706.1	1.65055	670	681.14	24.46	488.81	78.61	2.52589
290	290.16	1.2311	206.91	676.1	1.66802	680	691.82	25.85	496.62	75.50	2.54175
295	295.17	1.3068	210.49	647.9	1.68515	690	702.52	27.29	504.45	72.56	2.55731
298	298.18	1.3543	212.64	631.9	1.69528	700	713.27	28.80	512.33	69.76	2.57277
300	300.19	1.3860	214.07	621.2	1.70203	710	724.04	30.38	520.23	67.07	2.58810
305	305.22	1.4686	217.67	596.0	1.71865	720	734.82	32.02	528.14	64.53	2.60319
310	310.24	1.5546	221.25	572.3	1.73498	730	745.62	33.72	536.07	62.13	2.61803
315	315.27	1.6442	224.85	549.8	1.75106	740	756.44	35.50	544.02	59.82	2.63280
320	320.29	1.7375	228.42	528.6	1.76690	750	767.29	37.35	551.99	57.63	2.64737
325	325.31	1.8345	232.02	508.4	1.78249	760	778.18	39.27	560.01	55.54	2.66176
330	330.34	1.9352	235.61	489.4	1.79783	780	800.03	43.35	576.12	51.64	2.69013
340	340.42	2.149	242.82	454.1	1.82790	800	821.95	47.75	592.30	48.08	2.71787
350	350.49	2.379	250.02	422.2	1.85708	820	843.98	52.59	608.59	44.84	2.74504
360	360.58	2.626	257.24	393.4	1.88543	840	866.08	57.60	624.95	41.85	2.77170
370	370.67	2.892	264.46	367.2	1.91313	860	888.27	63.09	641.40	39.12	2.79783
380	380.77	3.176	271.69	343.4	1.94001	880	910.56	68.98	657.95	36.61	2.82344
390	390.88	3.481	278.93	321.5	1.96633	900	932.93	75.29	674.58	34.31	2.84856
400	400.98	3.806	286.16	301.6	1.99194	920	955.38	82.05	691.28	32.18	2.87324
410	411.12	4.153	293.43	283.3	2.01699	940	977.92	89.28	708.08	30.22	2.89748
420	421.26	4.522	300.69	266.6	2.04142	960	1000.55	97.00	725.02	28.40	2.92128
430	431.43	4.915	307.99	251.1	2.06533	980	1023.25	105.2	741.98	26.73	2.94468
440	441.61	5.332	315.30	236.8	2.08870	1000	1046.04	114.0	758.94	25.17	2.96770
450	451.80	5.775	322.62	223.6	2.11161	1020	1068.89	123.4	776.10	23.72	2.99034
460	462.02	6.245	329.97	211.4	2.13407	1040	1091.85	133.3	793.36	23.29	3.01260
470	472.24	6.742	337.32	200.1	2.15604	1060	1114.86	143.9	810.62	21.14	3.03449
480	482.49	7.268	344.70	189.5	2.17760	1080	1137.89	155.2	827.88	19.98	3.05608
490	492.74	7.824	352.08	179.7	2.19876	1100	1161.07	167.1	845.33	18.896	3.07732
500	503.02	8.411	359.49	170.6	2.21952	1120	1184.28	179.7	862.79	17.886	3.09825
510	513.32	9.031	366.92	162.1	2.23993	1140	1207.57	193.1	880.35	16.946	3.11883
520	523.63	9.684	374.36	154.1	2.25997	1160	1230.92	207.2	897.91	16.064	3.13916
530	533.98	10.37	381.84	146.7	2.27967	1180	1254.34	222.2	915.57	15.241	3.15916
540	544.35	11.10	389.34	139.7	2.29906	1200	1277.79	238.0	933.33	14.470	3.17888
550	555.74	11.86	396.86	133.1	2.31809	1220	1301.31	254.7	951.09	13.747	3.19834
560	565.17	12.66	404.42	127.0	2.33685	1240	1324.93	272.3	968.95	13.069	3.21751
570	575.59	13.50	411.97	121.2	2.35531						

(Continued)

TABLE A–21

Ideal-gas properties of air (*Concluded*)

T K	h kJ/kg	P_r	u kJ/kg	v_r	$s°$ kJ/kg · K	T K	h kJ/kg	P_r	u kJ/kg	v_r	$s°$ kJ/kg · K
1260	1348.55	290.8	986.90	12.435	3.23638	1600	1757.57	791.2	1298.30	5.804	3.52364
1280	1372.24	310.4	1004.76	11.835	3.25510	1620	1782.00	834.1	1316.96	5.574	3.53879
1300	1395.97	330.9	1022.82	11.275	3.27345	1640	1806.46	878.9	1335.72	5.355	3.55381
1320	1419.76	352.5	1040.88	10.747	3.29160	1660	1830.96	925.6	1354.48	5.147	3.56867
1340	1443.60	375.3	1058.94	10.247	3.30959	1680	1855.50	974.2	1373.24	4.949	3.58335
1360	1467.49	399.1	1077.10	9.780	3.32724	1700	1880.1	1025	1392.7	4.761	3.5979
1380	1491.44	424.2	1095.26	9.337	3.34474	1750	1941.6	1161	1439.8	4.328	3.6336
1400	1515.42	450.5	1113.52	8.919	3.36200	1800	2003.3	1310	1487.2	3.994	3.6684
1420	1539.44	478.0	1131.77	8.526	3.37901	1850	2065.3	1475	1534.9	3.601	3.7023
1440	1563.51	506.9	1150.13	8.153	3.39586	1900	2127.4	1655	1582.6	3.295	3.7354
1460	1587.63	537.1	1168.49	7.801	3.41247	1950	2189.7	1852	1630.6	3.022	3.7677
1480	1611.79	568.8	1186.95	7.468	3.42892	2000	2252.1	2068	1678.7	2.776	3.7994
1500	1635.97	601.9	1205.41	7.152	3.44516	2050	2314.6	2303	1726.8	2.555	3.8303
1520	1660.23	636.5	1223.87	6.854	3.46120	2100	2377.7	2559	1775.3	2.356	3.8605
1540	1684.51	672.8	1242.43	6.569	3.47712	2150	2440.3	2837	1823.8	2.175	3.8901
1560	1708.82	710.5	1260.99	6.301	3.49276	2200	2503.2	3138	1872.4	2.012	3.9191
1580	1733.17	750.0	1279.65	6.046	3.50829	2250	2566.4	3464	1921.3	1.864	3.9474

Note: The properties P_r (relative pressure) and v_r (relative specific volume) are dimensionless quantities used in the analysis of isentropic processes, and should not be confused with the properties pressure and specific volume.

Source: Kenneth Wark, *Thermodynamics,* 4th ed. (New York: McGraw-Hill, 1983), pp. 785–86, table A–5. Originally published in J. H. Keenan and J. Kaye, *Gas Tables* (New York: John Wiley & Sons, 1948).

TABLE A–22

Properties of air at 1 atm pressure

Temp. T, °C	Density ρ, kg/m^3	Specific Heat c_p, J/kg · K	Thermal Conductivity k, W/m · K	Thermal Diffusivity α, m^2/s	Dynamic Viscosity μ, kg/m · s	Kinematic Viscosity ν, m^2/s	Prandtl Number Pr
−150	2.866	983	0.01171	4.158×10^{-6}	8.636×10^{-6}	3.013×10^{-6}	0.7246
−100	2.038	966	0.01582	8.036×10^{-6}	1.189×10^{-6}	5.837×10^{-6}	0.7263
−50	1.582	999	0.01979	1.252×10^{-5}	1.474×10^{-5}	9.319×10^{-6}	0.7440
−40	1.514	1002	0.02057	1.356×10^{-5}	1.527×10^{-5}	1.008×10^{-5}	0.7436
−30	1.451	1004	0.02134	1.465×10^{-5}	1.579×10^{-5}	1.087×10^{-5}	0.7425
−20	1.394	1005	0.02211	1.578×10^{-5}	1.630×10^{-5}	1.169×10^{-5}	0.7408
−10	1.341	1006	0.02288	1.696×10^{-5}	1.680×10^{-5}	1.252×10^{-5}	0.7387
0	1.292	1006	0.02364	1.818×10^{-5}	1.729×10^{-5}	1.338×10^{-5}	0.7362
5	1.269	1006	0.02401	1.880×10^{-5}	1.754×10^{-5}	1.382×10^{-5}	0.7350
10	1.246	1006	0.02439	1.944×10^{-5}	1.778×10^{-5}	1.426×10^{-5}	0.7336
15	1.225	1007	0.02476	2.009×10^{-5}	1.802×10^{-5}	1.470×10^{-5}	0.7323
20	1.204	1007	0.02514	2.074×10^{-5}	1.825×10^{-5}	1.516×10^{-5}	0.7309
25	1.184	1007	0.02551	2.141×10^{-5}	1.849×10^{-5}	1.562×10^{-5}	0.7296
30	1.164	1007	0.02588	2.208×10^{-5}	1.872×10^{-5}	1.608×10^{-5}	0.7282
35	1.145	1007	0.02625	2.277×10^{-5}	1.895×10^{-5}	1.655×10^{-5}	0.7268
40	1.127	1007	0.02662	2.346×10^{-5}	1.918×10^{-5}	1.702×10^{-5}	0.7255
45	1.109	1007	0.02699	2.416×10^{-5}	1.941×10^{-5}	1.750×10^{-5}	0.7241
50	1.092	1007	0.02735	2.487×10^{-5}	1.963×10^{-5}	1.798×10^{-5}	0.7228
60	1.059	1007	0.02808	2.632×10^{-5}	2.008×10^{-5}	1.896×10^{-5}	0.7202
70	1.028	1007	0.02881	2.780×10^{-5}	2.052×10^{-5}	1.995×10^{-5}	0.7177
80	0.9994	1008	0.02953	2.931×10^{-5}	2.096×10^{-5}	2.097×10^{-5}	0.7154
90	0.9718	1008	0.03024	3.086×10^{-5}	2.139×10^{-5}	2.201×10^{-5}	0.7132
100	0.9458	1009	0.03095	3.243×10^{-5}	2.181×10^{-5}	2.306×10^{-5}	0.7111
120	0.8977	1011	0.03235	3.565×10^{-5}	2.264×10^{-5}	2.522×10^{-5}	0.7073
140	0.8542	1013	0.03374	3.898×10^{-5}	2.345×10^{-5}	2.745×10^{-5}	0.7041
160	0.8148	1016	0.03511	4.241×10^{-5}	2.420×10^{-5}	2.975×10^{-5}	0.7014
180	0.7788	1019	0.03646	4.593×10^{-5}	2.504×10^{-5}	3.212×10^{-5}	0.6992
200	0.7459	1023	0.03779	4.954×10^{-5}	2.577×10^{-5}	3.455×10^{-5}	0.6974
250	0.6746	1033	0.04104	5.890×10^{-5}	2.760×10^{-5}	4.091×10^{-5}	0.6946
300	0.6158	1044	0.04418	6.871×10^{-5}	2.934×10^{-5}	4.765×10^{-5}	0.6935
350	0.5664	1056	0.04721	7.892×10^{-5}	3.101×10^{-5}	5.475×10^{-5}	0.6937
400	0.5243	1069	0.05015	8.951×10^{-5}	3.261×10^{-5}	6.219×10^{-5}	0.6948
450	0.4880	1081	0.05298	1.004×10^{-4}	3.415×10^{-5}	6.997×10^{-5}	0.6965
500	0.4565	1093	0.05572	1.117×10^{-4}	3.563×10^{-5}	7.806×10^{-5}	0.6986
600	0.4042	1115	0.06093	1.352×10^{-4}	3.846×10^{-5}	9.515×10^{-5}	0.7037
700	0.3627	1135	0.06581	1.598×10^{-4}	4.111×10^{-5}	1.133×10^{-4}	0.7092
800	0.3289	1153	0.07037	1.855×10^{-4}	4.362×10^{-5}	1.326×10^{-4}	0.7149
900	0.3008	1169	0.07465	2.122×10^{-4}	4.600×10^{-5}	1.529×10^{-4}	0.7206
1000	0.2772	1184	0.07868	2.398×10^{-4}	4.826×10^{-5}	1.741×10^{-4}	0.7260
1500	0.1990	1234	0.09599	3.908×10^{-4}	5.817×10^{-5}	2.922×10^{-4}	0.7478
2000	0.1553	1264	0.11113	5.664×10^{-4}	6.630×10^{-5}	4.270×10^{-4}	0.7539

Note: For ideal gases, the properties c_p, k, μ, and Pr are independent of pressure. The properties ρ, ν, and α at a pressure P (in atm) other than 1 atm are determined by multiplying the values of ρ at the given temperature by P and by dividing ν and α by P.

Source: Data generated from the EES software developed by S. A. Klein and F. L. Alvarado. Original sources: Keenan, Chao, Keyes, Gas Tables, Wiley, 1984; and Thermophysical Properties of Matter. Vol. 3: Thermal Conductivity, Y. S. Touloukian, P. E. Liley, S. C. Saxena, Vol. 11: Viscosity, Y. S. Touloukian, S. C. Saxena, and P. Hestermans, IFI/Plenun, NY, 1970, ISBN 0-306067020-8.

TABLE A–23

Properties of gases at 1 atm pressure

Temp. T, °C	Density ρ, kg/m^3	Specific Heat c_p, J/kg · K	Thermal Conductivity k, W/m · K	Thermal Diffusivity α, m^2/s	Dynamic Viscosity μ, kg/m · s	Kinematic Viscosity ν, m^2/s	Prandtl Number Pr
Carbon Dioxide, CO_2							
−50	2.4035	746	0.01051	5.860×10^{-6}	1.129×10^{-5}	4.699×10^{-6}	0.8019
0	1.9635	811	0.01456	9.141×10^{-6}	1.375×10^{-5}	7.003×10^{-6}	0.7661
50	1.6597	866.6	0.01858	1.291×10^{-5}	1.612×10^{-5}	9.714×10^{-6}	0.7520
100	1.4373	914.8	0.02257	1.716×10^{-5}	1.841×10^{-5}	1.281×10^{-5}	0.7464
150	1.2675	957.4	0.02652	2.186×10^{-5}	2.063×10^{-5}	1.627×10^{-5}	0.7445
200	1.1336	995.2	0.03044	2.698×10^{-5}	2.276×10^{-5}	2.008×10^{-5}	0.7442
300	0.9358	1060	0.03814	3.847×10^{-5}	2.682×10^{-5}	2.866×10^{-5}	0.7450
400	0.7968	1112	0.04565	5.151×10^{-5}	3.061×10^{-5}	3.842×10^{-5}	0.7458
500	0.6937	1156	0.05293	6.600×10^{-5}	3.416×10^{-5}	4.924×10^{-5}	0.7460
1000	0.4213	1292	0.08491	1.560×10^{-4}	4.898×10^{-5}	1.162×10^{-4}	0.7455
1500	0.3025	1356	0.10688	2.606×10^{-4}	6.106×10^{-5}	2.019×10^{-4}	0.7745
2000	0.2359	1387	0.11522	3.521×10^{-4}	7.322×10^{-5}	3.103×10^{-4}	0.8815
Carbon Monoxide, CO							
−50	1.5297	1081	0.01901	1.149×10^{-5}	1.378×10^{-5}	9.012×10^{-6}	0.7840
0	1.2497	1048	0.02278	1.739×10^{-5}	1.629×10^{-5}	1.303×10^{-5}	0.7499
50	1.0563	1039	0.02641	2.407×10^{-5}	1.863×10^{-5}	1.764×10^{-5}	0.7328
100	0.9148	1041	0.02992	3.142×10^{-5}	2.080×10^{-5}	2.274×10^{-5}	0.7239
150	0.8067	1049	0.03330	3.936×10^{-5}	2.283×10^{-5}	2.830×10^{-5}	0.7191
200	0.7214	1060	0.03656	4.782×10^{-5}	2.472×10^{-5}	3.426×10^{-5}	0.7164
300	0.5956	1085	0.04277	6.619×10^{-5}	2.812×10^{-5}	4.722×10^{-5}	0.7134
400	0.5071	1111	0.04860	8.628×10^{-5}	3.111×10^{-5}	6.136×10^{-5}	0.7111
500	0.4415	1135	0.05412	1.079×10^{-4}	3.379×10^{-5}	7.653×10^{-5}	0.7087
1000	0.2681	1226	0.07894	2.401×10^{-4}	4.557×10^{-5}	1.700×10^{-4}	0.7080
1500	0.1925	1279	0.10458	4.246×10^{-4}	6.321×10^{-5}	3.284×10^{-4}	0.7733
2000	0.1502	1309	0.13833	7.034×10^{-4}	9.826×10^{-5}	6.543×10^{-4}	0.9302
Methane, CH_4							
−50	0.8761	2243	0.02367	1.204×10^{-5}	8.564×10^{-6}	9.774×10^{-6}	0.8116
0	0.7158	2217	0.03042	1.917×10^{-5}	1.028×10^{-5}	1.436×10^{-5}	0.7494
50	0.6050	2302	0.03766	2.704×10^{-5}	1.191×10^{-5}	1.969×10^{-5}	0.7282
100	0.5240	2443	0.04534	3.543×10^{-5}	1.345×10^{-5}	2.567×10^{-5}	0.7247
150	0.4620	2611	0.05344	4.431×10^{-5}	1.491×10^{-5}	3.227×10^{-5}	0.7284
200	0.4132	2791	0.06194	5.370×10^{-5}	1.630×10^{-5}	3.944×10^{-5}	0.7344
300	0.3411	3158	0.07996	7.422×10^{-5}	1.886×10^{-5}	5.529×10^{-5}	0.7450
400	0.2904	3510	0.09918	9.727×10^{-5}	2.119×10^{-5}	7.297×10^{-5}	0.7501
500	0.2529	3836	0.11933	1.230×10^{-4}	2.334×10^{-5}	9.228×10^{-5}	0.7502
1000	0.1536	5042	0.22562	2.914×10^{-4}	3.281×10^{-5}	2.136×10^{-4}	0.7331
1500	0.1103	5701	0.31857	5.068×10^{-4}	4.434×10^{-5}	4.022×10^{-4}	0.7936
2000	0.0860	6001	0.36750	7.120×10^{-4}	6.360×10^{-5}	7.395×10^{-4}	1.0386
Hydrogen, H_2							
−50	0.11010	12635	0.1404	1.009×10^{-4}	7.293×10^{-6}	6.624×10^{-5}	0.6562
0	0.08995	13920	0.1652	1.319×10^{-4}	8.391×10^{-6}	9.329×10^{-5}	0.7071
50	0.07603	14349	0.1881	1.724×10^{-4}	9.427×10^{-6}	1.240×10^{-4}	0.7191
100	0.06584	14473	0.2095	2.199×10^{-4}	1.041×10^{-5}	1.582×10^{-4}	0.7196
150	0.05806	14492	0.2296	2.729×10^{-4}	1.136×10^{-5}	1.957×10^{-4}	0.7174
200	0.05193	14482	0.2486	3.306×10^{-4}	1.228×10^{-5}	2.365×10^{-4}	0.7155

(Continued)

TABLE A–23

Properties of gases at 1 atm pressure *(Continued)*

Temp. T, °C	Density ρ, kg/m^3	Specific Heat c_p, J/kg · K	Thermal Conductivity k, W/m · K	Thermal Diffusivity α, m^2/s	Dynamic Viscosity μ, kg/m · s	Kinematic Viscosity ν, m^2/s	Prandtl Number Pr
300	0.04287	14481	0.2843	4.580×10^{-4}	1.403×10^{-5}	3.274×10^{-4}	0.7149
400	0.03650	14540	0.3180	5.992×10^{-4}	1.570×10^{-5}	4.302×10^{-4}	0.7179
500	0.03178	14653	0.3509	7.535×10^{-4}	1.730×10^{-5}	5.443×10^{-4}	0.7224
1000	0.01930	15577	0.5206	1.732×10^{-3}	2.455×10^{-5}	1.272×10^{-3}	0.7345
1500	0.01386	16553	0.6581	2.869×10^{-3}	3.099×10^{-5}	2.237×10^{-3}	0.7795
2000	0.01081	17400	0.5480	2.914×10^{-3}	3.690×10^{-5}	3.414×10^{-3}	1.1717
				Nitrogen, N_2			
−50	1.5299	957.3	0.02001	1.366×10^{-5}	1.390×10^{-5}	9.091×10^{-6}	0.6655
0	1.2498	1035	0.02384	1.843×10^{-5}	1.640×10^{-5}	1.312×10^{-5}	0.7121
50	1.0564	1042	0.02746	2.494×10^{-5}	1.874×10^{-5}	1.774×10^{-5}	0.7114
100	0.9149	1041	0.03090	3.244×10^{-5}	2.094×10^{-5}	2.289×10^{-5}	0.7056
150	0.8068	1043	0.03416	4.058×10^{-5}	2.300×10^{-5}	2.851×10^{-5}	0.7025
200	0.7215	1050	0.03727	4.921×10^{-5}	2.494×10^{-5}	3.457×10^{-5}	0.7025
300	0.5956	1070	0.04309	6.758×10^{-5}	2.849×10^{-5}	4.783×10^{-5}	0.7078
400	0.5072	1095	0.04848	8.727×10^{-5}	3.166×10^{-5}	6.242×10^{-5}	0.7153
500	0.4416	1120	0.05358	1.083×10^{-4}	3.451×10^{-5}	7.816×10^{-5}	0.7215
1000	0.2681	1213	0.07938	2.440×10^{-4}	4.594×10^{-5}	1.713×10^{-4}	0.7022
1500	0.1925	1266	0.11793	4.839×10^{-4}	5.562×10^{-5}	2.889×10^{-4}	0.5969
2000	0.1502	1297	0.18590	9.543×10^{-4}	6.426×10^{-5}	4.278×10^{-4}	0.4483
				Oxygen, O_2			
−50	1.7475	984.4	0.02067	1.201×10^{-5}	1.616×10^{-5}	9.246×10^{-6}	0.7694
0	1.4277	928.7	0.02472	1.865×10^{-5}	1.916×10^{-5}	1.342×10^{-5}	0.7198
50	1.2068	921.7	0.02867	2.577×10^{-5}	2.194×10^{-5}	1.818×10^{-5}	0.7053
100	1.0451	931.8	0.03254	3.342×10^{-5}	2.451×10^{-5}	2.346×10^{-5}	0.7019
150	0.9216	947.6	0.03637	4.164×10^{-5}	2.694×10^{-5}	2.923×10^{-5}	0.7019
200	0.8242	964.7	0.04014	5.048×10^{-5}	2.923×10^{-5}	3.546×10^{-5}	0.7025
300	0.6804	997.1	0.04751	7.003×10^{-5}	3.350×10^{-5}	4.923×10^{-5}	0.7030
400	0.5793	1025	0.05463	9.204×10^{-5}	3.744×10^{-5}	6.463×10^{-5}	0.7023
500	0.5044	1048	0.06148	1.163×10^{-4}	4.114×10^{-5}	8.156×10^{-5}	0.7010
1000	0.3063	1121	0.09198	2.678×10^{-4}	5.732×10^{-5}	1.871×10^{-4}	0.6986
1500	0.2199	1165	0.11901	4.643×10^{-4}	7.133×10^{-5}	3.243×10^{-4}	0.6985
2000	0.1716	1201	0.14705	7.139×10^{-4}	8.417×10^{-5}	4.907×10^{-4}	0.6873
				Water Vapor, H_2O			
−50	0.9839	1892	0.01353	7.271×10^{-6}	7.187×10^{-6}	7.305×10^{-6}	1.0047
0	0.8038	1874	0.01673	1.110×10^{-5}	8.956×10^{-6}	1.114×10^{-5}	1.0033
50	0.6794	1874	0.02032	1.596×10^{-5}	1.078×10^{-5}	1.587×10^{-5}	0.9944
100	0.5884	1887	0.02429	2.187×10^{-5}	1.265×10^{-5}	2.150×10^{-5}	0.9830
150	0.5189	1908	0.02861	2.890×10^{-5}	1.456×10^{-5}	2.806×10^{-5}	0.9712
200	0.4640	1935	0.03326	3.705×10^{-5}	1.650×10^{-5}	3.556×10^{-5}	0.9599
300	0.3831	1997	0.04345	5.680×10^{-5}	2.045×10^{-5}	5.340×10^{-5}	0.9401
400	0.3262	2066	0.05467	8.114×10^{-5}	2.446×10^{-5}	7.498×10^{-5}	0.9240
500	0.2840	2137	0.06677	1.100×10^{-4}	2.847×10^{-5}	1.002×10^{-4}	0.9108
1000	0.1725	2471	0.13623	3.196×10^{-4}	4.762×10^{-5}	2.761×10^{-4}	0.8639
1500	0.1238	2736	0.21301	6.288×10^{-4}	6.411×10^{-5}	5.177×10^{-4}	0.8233
2000	0.0966	2928	0.29183	1.032×10^{-3}	7.808×10^{-5}	8.084×10^{-4}	0.7833

Note: For ideal gases, the properties c_p, k, μ, and Pr are independent of pressure. The properties ρ, ν, and α at a pressure P (in atm) other than 1 atm are determined by multiplying the values of ρ at the given temperature by ρ and by dividing ν and α by P.

Source: Data generated from the EES software developed by S. A. Klein and F. L. Alvarado. Originally based on various sources.

TABLE A–24

Properties of solid metals

Composition	Melting Point, K	ρ kg/m³	c_p J/kg · K	k W/m · K	$\alpha \times 10^6$ m²/s	100	200	400	600	800	1000
		Properties at 300 K				Properties at Various Temperatures (K), k(W/m · K)/c_p(J/kg · K)					
Aluminum:											
Pure	933	2702	903	237	97.1	302	237	240	231	218	
						482	798	949	1033	1146	
Alloy 2024-T6 (4.5% Cu, 1.5% Mg, 0.6% Mn)	775	2770	875	177	73.0	65	163	186	186		
						473	787	925	1042		
Alloy 195, Cast (4.5% Cu)		2790	883	168	68.2			174	185		
Beryllium	1550	1850	1825	200	59.2	990	301	161	126	106	90.8
						203	1114	2191	2604	2823	3018
Bismuth	545	9780	122	7.86	6.59	16.5	9.69	7.04			
						112	120	127			
Boron	2573	2500	1107	27.0	9.76	190	55.5	16.8	10.6	9.60	9.85
						128	600	1463	1892	2160	2338
Cadmium	594	8650	231	96.8	48.4	203	99.3	94.7			
						198	222	242			
Chromium	2118	7160	449	93.7	29.1	159	111	90.9	80.7	71.3	65.4
						192	384	484	542	581	616
Cobalt	1769	8862	421	99.2	26.6	167	122	85.4	67.4	58.2	52.1
						236	379	450	503	550	628
Copper:											
Pure	1358	8933	385	401	117	482	413	393	379	366	352
						252	356	397	417	433	451
Commercial bronze (90% Cu, 10% Al)	1293	8800	420	52	14		42	52	59		
							785	160	545		
Phosphor gear bronze (89% Cu, 11% Sn)	1104	8780	355	54	17		41	65	74		
							—	—	—		
Cartridge brass (70% Cu, 30% Zn)	1188	8530	380	110	33.9	75	95	137	149		
							360	395	425		
Constantan (55% Cu, 45% Ni)	1493	8920	384	23	6.71	17	19				
						237	362				
Germanium	1211	5360	322	59.9	34.7	232	96.8	43.2	27.3	19.8	17.4
						190	290	337	348	357	375
Gold	1336	19,300	129	317	127	327	323	311	298	284	270
						109	124	131	135	140	145
Iridium	2720	22,500	130	147	50.3	172	153	144	138	132	126
						90	122	133	138	144	153
Iron:											
Pure	1810	7870	447	80.2	23.1	134	94.0	69.5	54.7	43.3	32.8
						216	384	490	574	680	975
Armco (99.75% pure)		7870	447	72.7	20.7	95.6	80.6	65.7	53.1	42.2	32.3
						215	384	490	574	680	975
Carbon steels:											
Plain carbon (Mn ≤ 1% Si ≤ 0.1%)		7854	434	60.5	17.7			56.7	48.0	39.2	30.0
								487	559	685	1169
AISI 1010		7832	434	63.9	18.8			58.7	48.8	39.2	31.3
							487	559	685	1168	
Carbon–silicon (Mn ≤ 1% 0.1% < Si ≤ 0.6%)		7817	446	51.9	14.9			49.8	44.0	37.4	29.3
								501	582	699	971

(Continued)

TABLE A–24

Properties of solid metals *(Continued)*

Composition	Melting Point, K	Properties at 300 K: ρ kg/m^3	c_p J/kg · K	k W/m · K	$\alpha \times 10^6$ m^2/s	Properties at Various Temperatures (K), k(W/m · K)/c_p(J/kg · K): 100	200	400	600	800	1000
Carbon–manganese–silicon (1% < Mn < 1.65% 0.1% < Si < 0.6%)		8131	434	41.0	11.6			42.2	39.7	35.0	27.6
								487	559	685	1090
Chromium (low) steels:											
$\frac{1}{2}$ Cr–$\frac{1}{4}$ Mo–Si (0.18% C, 0.65% Cr, 0.23% Mo, 0.6% Si)		7822	444	37.7	10.9			38.2	36.7	33.3	26.9
								492	575	688	969
1 Cr–$\frac{1}{2}$ Mo (0.16% C, 1% Cr, 0.54% Mo, 0.39% Si)		7858	442	42.3	12.2			42.0	39.1	34.5	27.4
								492	575	688	969
1 Cr–V (0.2% C, 1.02% Cr, 0.15% V)		7836	443	48.9	14.1			46.8	42.1	36.3	28.2
								492	575	688	969
Stainless steels:											
AISI 302		8055	480	15.1	3.91			17.3	20.0	22.8	25.4
								512	559	585	606
AISI 304	1670	7900	477	14.9	3.95	9.2	12.6	16.6	19.8	22.6	25.4
						272	402	515	557	582	611
AISI 316		8238	468	13.4	3.48			15.2	18.3	21.3	24.2
								504	550	576	602
AISI 347		7978	480	14.2	3.71			15.8	18.9	21.9	24.7
								513	559	585	606
Lead	601	11,340	129	35.3	24.1	39.7	36.7	34.0	31.4		
						118	125	132	142		
Magnesium	923	1740	1024	156	87.6	169	159	153	149	146	
						649	934	1074	1170	1267	
Molybdenum	2894	10,240	251	138	53.7	179	143	134	126	118	112
						141	224	261	275	285	295
Nickel:											
Pure	1728	8900	444	90.7	23.0	164	107	80.2	65.6	67.6	71.8
					232	383	485	592	530	562	
Nichrome (80% Ni, 20% Cr)	1672	8400	420	12	3.4			14	16	21	
								480	525	545	
Inconel X-750 (73% Ni, 15% Cr, 6.7% Fe)	1665	8510	439	11.7	3.1	8.7	10.3	13.5	17.0	20.5	24.0
						—	372	473	510	546	626
Niobium	2741	8570	265	53.7	23.6	55.2	52.6	55.2	58.2	61.3	64.4
						188	249	274	283	292	301
Palladium	1827	12,020	244	71.8	24.5	76.5	71.6	73.6	79.7	86.9	94.2
						168	227	251	261	271	281
Platinum:											
Pure	2045	21,450	133	71.6	25.1	77.5	72.6	71.8	73.2	75.6	78.7
						100	125	136	141	146	152
Alloy 60Pt–40Rh (60% Pt, 40% Rh)	1800	16,630	162	47	17.4			52	59	65	69
								—	—	—	—
Rhenium	3453	21,100	136	47.9	16.7	58.9	51.0	46.1	44.2	44.1	44.6
						97	127	139	145	151	156
Rhodium	2236	12,450	243	150	49.6	186	154	146	136	127	121
						147	220	253	274	293	311

(Continued)

TABLE A–24

Properties of solid metals *(Concluded)*

Composition	Melting Point, K	*Properties at 300 K* ρ kg/m^3	c_p J/kg · K	k W/m · K	$\alpha \times 10^6$ m^2/s	*Properties at Various Temperatures (K),* k(W/m · K)/c_p(J/kg · K) 100	200	400	600	800	1000
Silicon	1685	2330	712	148	89.2	884	264	98.9	61.9	42.4	31.2
						259	556	790	867	913	946
Silver	1235	10,500	235	429	174	444	430	425	412	396	379
						187	225	239	250	262	277
Tantalum	3269	16,600	140	57.5	24.7	59.2	57.5	57.8	58.6	59.4	60.2
						110	133	144	146	149	152
Thorium	2023	11,700	118	54.0	39.1	59.8	54.6	54.5	55.8	56.9	56.9
						99	112	124	134	145	156
Tin	505	7310	227	66.6	40.1	85.2	73.3	62.2			
						188	215	243			
Titanium	1953	4500	522	21.9	9.32	30.5	24.5	20.4	19.4	19.7	20.7
						300	465	551	591	633	675
Tungsten	3660	19,300	132	174	68.3	208	186	159	137	125	118
						87	122	137	142	146	148
Uranium	1406	19,070	116	27.6	12.5	21.7	25.1	29.6	34.0	38.8	43.9
						94	108	125	146	176	180
Vanadium	2192	6100	489	30.7	10.3	35.8	31.3	31.3	33.3	35.7	38.2
						258	430	515	540	563	597
Zinc	693	7140	389	116	41.8	117	118	111	103		
						297	367	402	436		
Zirconium	2125	6570	278	22.7	12.4	33.2	25.2	21.6	20.7	21.6	23.7
						205	264	300	332	342	362

From Frank P. Incropera and David P. DeWitt, *Fundamentals of Heat and Mass Transfer, 3rd ed.*, 1990. This material is used by permission of John Wiley & Sons, Inc.

TABLE A–25

Properties of solid non-metals

Composition	Melting Point, K	Properties at 300 K				Properties at Various Temperatures (K), k (W/m · K)/c_p(J/kg · K)					
		ρ kg/m³	c_p J/kg · K	k W/m · K	$\alpha \times 10^6$ m²/s	100	200	400	600	800	1000
Aluminum oxide, sapphire	2323	3970	765	46	15.1	450	82	32.4	18.9	13.0	10.5
						—	—	940	1110	1180	1225
Aluminum oxide, polycrystalline	2323	3970	765	36.0	11.9	133	55	26.4	15.8	10.4	7.85
						—	—	940	1110	1180	1225
Beryllium oxide	2725	3000	1030	272	88.0			196	111	70	47
								1350	1690	1865	1975
Boron	2573	2500	1105	27.6	9.99	190	52.5	18.7	11.3	8.1	6.3
						—	—	1490	1880	2135	2350
Boron fiber epoxy (30% vol.) composite	590	2080									
k, ∥ to fibers				2.29		2.10	2.23	2.28			
k, ⊥ to fibers				0.59		0.37	0.49	0.60			
c_p			1122			364	757	1431			
Carbon											
Amorphous	1500	1950	—	1.60	—	0.67	1.18	1.89	21.9	2.37	2.53
						—	—	—	—	—	—
Diamond, type IIa insulator	—	3500	509	2300		10,000	4000	1540			
						21	194	853			
Graphite, pyrolytic	2273	2210									
k, ∥ to layers				1950		4970	3230	1390	892	667	534
k, ⊥ to layers				5.70		16.8	9.23	4.09	2.68	2.01	1.60
c_p			709			136	411	992	1406	1650	1793
Graphite fiber epoxy (25% vol.) composite	450	1400									
k, heat flow ∥ to fibers				11.1		5.7	8.7	13.0			
k, heat flow ⊥ to fibers				0.87	0.46	0.68	1.1				
c_p			935			337	642	1216			
Pyroceram, Corning 9606	1623	2600	808	3.98	1.89	5.25	4.78	3.64	3.28	3.08	2.96
						—	—	908	1038	1122	1197
Silicon carbide	3100	3160	675	490	230			—	—	—	87
								880	1050	1135	1195
Silicon dioxide, crystalline (quartz)	1883	2650									
k, ∥ to c-axis				10.4		39	16.4	7.6	5.0	4.2	
k, ⊥ to c-axis				6.21		20.8	9.5	4.70	3.4	3.1	
c_p			745			—	—	885	1075	1250	
Silicon dioxide, polycrystalline (fused silica)	1883	2220	745	1.38	0.834	0.69	1.14	1.51	1.75	2.17	2.87
						—	—	905	1040	1105	1155
Silicon nitride	2173	2400	691	16.0	9.65	—	—	13.9	11.3	9.88	8.76
						—	578	778	937	1063	1155
Sulfur	392	2070	708	0.206	0.141	0.165	0.185				
						403	606				
Thorium dioxide	3573	9110	235	13	6.1			10.2	6.6	4.7	3.68
								255	274	285	295
Titanium dioxide, polycrystalline	2133	4157	710	8.4	2.8			7.01	5.02	8.94	3.46
								805	880	910	930

TABLE A–26

Emissivities of surfaces
(*a*) Metals

Material	Temperature, K	Emissivity, ε
Aluminum		
Polished	300–900	0.04–0.06
Commercial sheet	400	0.09
Heavily oxidized	400–800	0.20–0.33
Anodized	300	0.8
Bismuth, bright	350	0.34
Brass		
Highly polished	500–650	0.03–0.04
Polished	350	0.09
Dull plate	300–600	0.22
Oxidized	450–800	0.6
Chromium, polished	300–1400	0.08–0.40
Copper		
Highly polished	300	0.02
Polished	300–500	0.04–0.05
Commercial sheet	300	0.15
Oxidized	600–1000	0.5–0.8
Black oxidized	300	0.78
Gold		
Highly polished	300–1000	0.03–0.06
Bright foil	300	0.07
Iron		
Highly polished	300–500	0.05–0.07
Case iron	300	0.44
Wrought iron	300–500	0.28
Rusted	300	0.61
Oxidized	500–900	0.64–0.78
Lead		
Polished	300–500	0.06–0.08
Unoxidized, rough	300	0.43
Oxidized	300	0.63
Magnesium, polished	300–500	0.07–0.13
Mercury	300–400	0.09–0.12
Molybdenum		
Polished	300–2000	0.05–0.21
Oxidized	600–800	0.80–0.82
Nickel		
Polished	500–1200	0.07–0.17
Oxidized	450–1000	0.37–0.57
Platinum, polished	500–1500	0.06–0.18
Silver, polished	300–1000	0.02–0.07
Stainless steel		
Polished	300–1000	0.17–0.30
Lightly oxidized	600–1000	0.30–0.40
Highly oxidized	600–1000	0.70–0.80
Steel		
Polished sheet	300–500	0.08–0.14
Commercial sheet	500–1200	0.20–0.32
Heavily oxidized	300	0.81
Tin, polished	300	0.05
Tungsten		
Polished	300–2500	0.03–0.29
Filament	3500	0.39
Zinc		
Polished	300–800	0.02–0.05
Oxidized	300	0.25

TABLE A–26

Emissivities of surfaces *(Concluded)*
(*b*) Nonmetals

Material	Temperature, K	Emissivity, ε	Material	Temperature, K	Emissivity, ε
Alumina	800–1400	0.65–0.45	Paper, white	300	0.90
Aluminum oxide	600–1500	0.69–0.41	Plaster, white	300	0.93
Asbestos	300	0.96	Porcelain, glazed	300	0.92
Asphalt pavement	300	0.85–0.93	Quartz, rough, fused	300	0.93
Brick			Rubber		
Common	300	0.93–0.96	Hard	300	0.93
Fireclay	1200	0.75	Soft	300	0.86
Carbon filament	2000	0.53	Sand	300	0.90
Cloth	300	0.75–0.90	Silicon carbide	600–1500	0.87–0.85
Concrete	300	0.88–0.94	Skin, human	300	0.95
Glass			Snow	273	0.80–0.90
Window	300	0.90–0.95	Soil, earth	300	0.93–0.96
Pyrex	300–1200	0.82–0.62	Soot	300–500	0.95
Pyroceram	300–1500	0.85–0.57	Teflon	300–500	0.85–0.92
Ice	273	0.95–0.99	Water, deep	273–373	0.95–0.96
Magnesium oxide	400–800	0.69–0.55	Wood		
Masonry	300	0.80	Beech	300	0.94
Paints			Oak	300	0.90
Aluminum	300	0.40–0.50			
Black, lacquer, shiny	300	0.88			
Oils, all colors	300	0.92–0.96			
Red primer	300	0.93			
White acrylic	300	0.90			
White enamel	300	0.90			

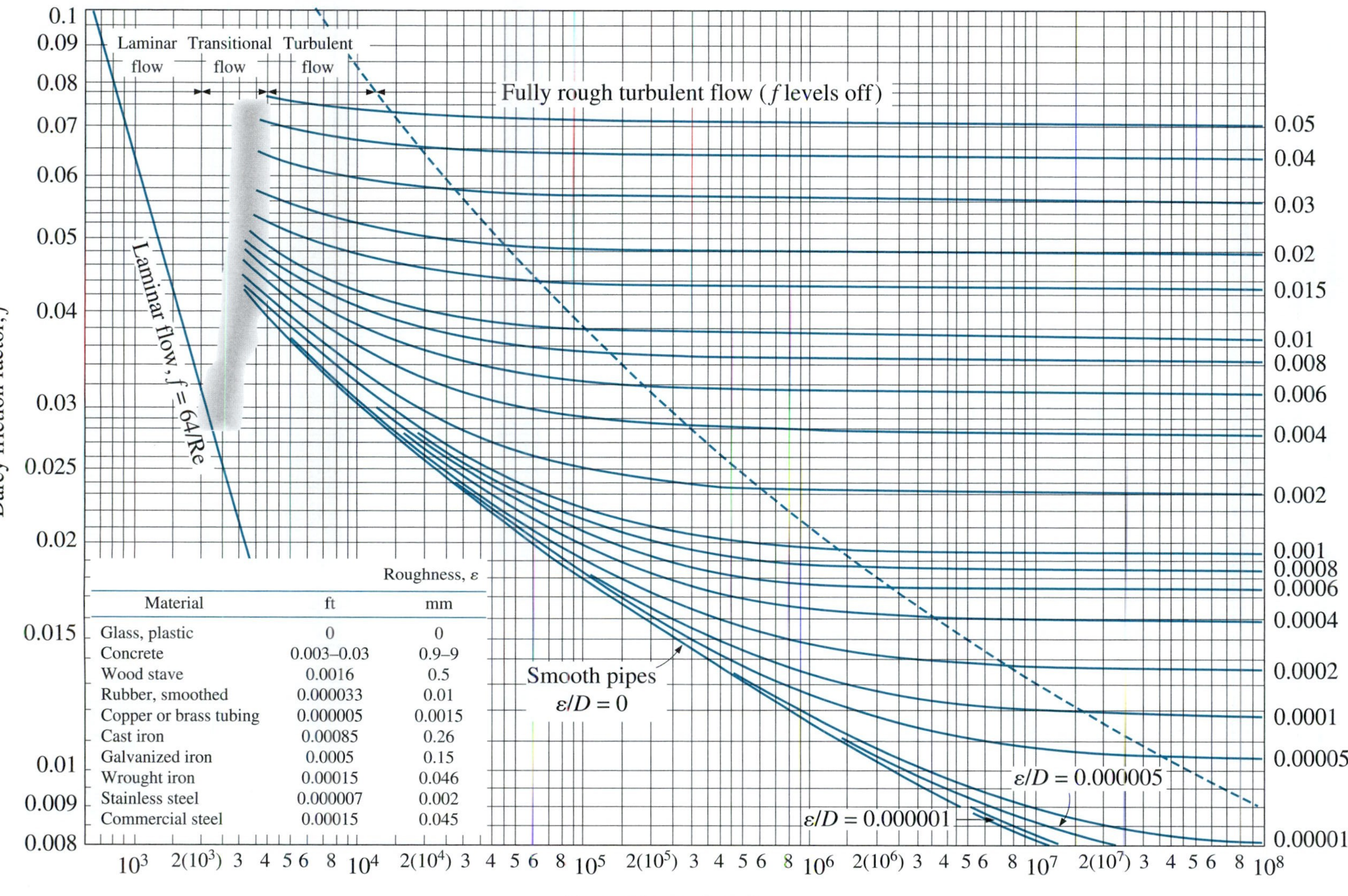

Material	Roughness, ε ft	Roughness, ε mm
Glass, plastic	0	0
Concrete	0.003–0.03	0.9–9
Wood stave	0.0016	0.5
Rubber, smoothed	0.000033	0.01
Copper or brass tubing	0.000005	0.0015
Cast iron	0.00085	0.26
Galvanized iron	0.0005	0.15
Wrought iron	0.00015	0.046
Stainless steel	0.000007	0.002
Commercial steel	0.00015	0.045

FIGURE A–27

The Moody chart for the friction factor for fully developed flow in circular pipes for use in the head loss relation $h_L = f\frac{L}{D}\frac{V^2}{2g}$. Friction factors in the turbulent flow are evaluated from the Colebrook equation $\frac{1}{\sqrt{f}} = -2\log_{10}\left(\frac{\varepsilon/D}{3.7} + \frac{2.51}{\mathrm{Re}\sqrt{f}}\right)$.

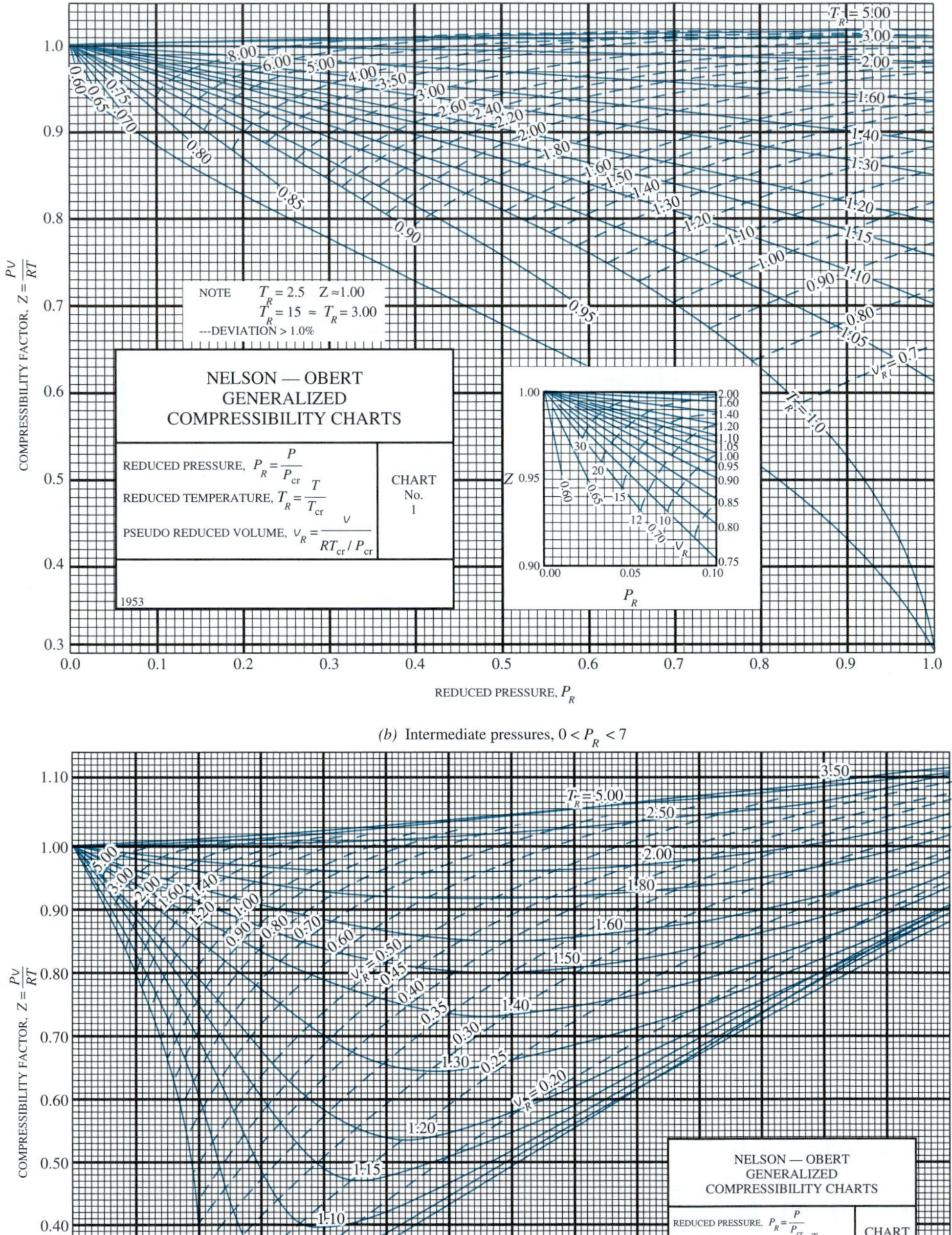

FIGURE A–28

Nelson–Obert generalized compressibility chart.

Used with permission of Dr. Edward E. Obert, University of Wisconsin.

Appendix 2

PROPERTY TABLES AND CHARTS (ENGLISH UNITS)

TABLE A–1E

Molar mass, gas constant, and critical-point properties

			Gas constant, R		Critical-point properties		
Substance	Formula	Molar mass, M lbm/lbmol	Btu/lbm · R*	psia · ft^3/lbm · R*	Temperature, R	Pressure, psia	Volume, ft^3/lbmol
Air	—	28.97	0.06855	0.3704	238.5	547	1.41
Ammonia	NH_3	17.03	0.1166	0.6301	729.8	1636	1.16
Argon	Ar	39.948	0.04971	0.2686	272	705	1.20
Benzene	C_6H_6	78.115	0.02542	0.1374	1012	714	4.17
Bromine	Br_2	159.808	0.01243	0.06714	1052	1500	2.17
n-Butane	C_4H_{10}	58.124	0.03417	0.1846	765.2	551	4.08
Carbon dioxide	CO_2	44.01	0.04513	0.2438	547.5	1071	1.51
Carbon monoxide	CO	28.011	0.07090	0.3831	240	507	1.49
Carbon tetrachloride	CCl_4	153.82	0.01291	0.06976	1001.5	661	4.42
Chlorine	Cl_2	70.906	0.02801	0.1517	751	1120	1.99
Chloroform	$CHCl_3$	119.38	0.01664	0.08988	965.8	794	3.85
Dichlorodifluoromethane (R-12)	CCl_2F_2	120.91	0.01643	0.08874	692.4	582	3.49
Dichlorofluoromethane (R-21)	$CHCl_2F$	102.92	0.01930	0.1043	813.0	749	3.16
Ethane	C_2H_6	30.020	0.06616	0.3574	549.8	708	2.37
Ethyl alcohol	C_2H_5OH	46.07	0.04311	0.2329	929.0	926	2.68
Ethylene	C_2H_4	28.054	0.07079	0.3825	508.3	742	1.99
Helium	He	4.003	0.4961	2.6809	9.5	33.2	0.926
n-Hexane	C_6H_{14}	86.178	0.02305	0.1245	914.2	439	5.89
Hydrogen (normal)	H_2	2.016	0.9851	5.3224	59.9	188.1	1.04
Krypton	Kr	83.80	0.02370	0.1280	376.9	798	1.48
Methane	CH_4	16.043	0.1238	0.6688	343.9	673	1.59
Methyl alcohol	CH_3OH	32.042	0.06198	0.3349	923.7	1154	1.89
Methyl chloride	CH_3Cl	50.488	0.03934	0.2125	749.3	968	2.29
Neon	Ne	20.183	0.09840	0.5316	80.1	395	0.668
Nitrogen	N_2	28.013	0.07090	0.3830	227.1	492	1.44
Nitrous oxide	N_2O	44.013	0.04512	0.2438	557.4	1054	1.54
Oxygen	O_2	31.999	0.06206	0.3353	278.6	736	1.25
Propane	C_3H_8	44.097	0.04504	0.2433	665.9	617	3.20
Propylene	C_3H_6	42.081	0.04719	0.2550	656.9	670	2.90
Sulfur dioxide	SO_2	64.063	0.03100	1.1675	775.2	1143	1.95
Tetrafluoroethane (R-134a)	CF_3CH_2F	102.03	0.01946	0.1052	673.6	588.7	3.19
Trichlorofluoromethane (R-11)	CCl_3F	137.37	0.01446	0.07811	848.1	635	3.97
Water	H_2O	18.015	0.1102	0.5956	1164.8	3200	0.90
Xenon	Xe	131.30	0.01513	0.08172	521.55	852	1.90

*Calculated from $R = R_u/M$, where R_u = 1.98588 Btu/lbmol · R = 10.7316 psia · ft^3/lbmol · R and M is the molar mass.

Source: K. A. Kobe and R. E. Lynn, Jr., *Chemical Review* 52 (1953), pp. 117–236, and ASHRAE, *Handbook of Fundamentals* (Atlanta, GA: American Society of Heating, Refrigerating, and Air-Conditioning Engineers, Inc., 1993), pp. 16.4 and 36.1.

TABLE A–2E

Ideal-gas specific heats of various common gases
(*a*) At 80°F

Gas	Formula	Gas constant, *R* Btu/lbm · R	c_p Btu/lbm · R	c_v Btu/lbm · R	*k*
Air	—	0.06855	0.240	0.171	1.400
Argon	Ar	0.04971	0.1253	0.0756	1.667
Butane	C_4H_{10}	0.03424	0.415	0.381	1.09
Carbon dioxide	CO_2	0.04513	0.203	0.158	1.285
Carbon monoxide	CO	0.07090	0.249	0.178	1.399
Ethane	C_2H_6	0.06616	0.427	0.361	1.183
Ethylene	C_2H_4	0.07079	0.411	0.340	1.208
Helium	He	0.4961	1.25	0.753	1.667
Hydrogen	H_2	0.9851	3.43	2.44	1.404
Methane	CH_4	0.1238	0.532	0.403	1.32
Neon	Ne	0.09840	0.246	0.1477	1.667
Nitrogen	N_2	0.07090	0.248	0.177	1.400
Octane	C_8H_{18}	0.01742	0.409	0.392	1.044
Oxygen	O_2	0.06206	0.219	0.157	1.395
Propane	C_3H_8	0.04504	0.407	0.362	1.124
Steam	H_2O	0.1102	0.445	0.335	1.329

Source: Gordon J. Van Wylen and Richard E. Sonntag, *Fundamentals of Classical Thermodynamics*, English/SI Version, 3rd ed. (New York: John Wiley & Sons, 1986), p. 687, Table A–8E.

TABLE A–2E

Ideal-gas specific heats of various common gases (*Continued*)
(*b*) At various temperatures

Temp., °F	c_p Btu/lbm · R	c_v Btu/lbm · R	k	c_p Btu/lbm · R	c_v Btu/lbm · R	k	c_p Btu/lbm · R	c_v Btu/lbm · R	k
	Air			*Carbon dioxide, CO_2*			*Carbon monoxide, CO*		
40	0.240	0.171	1.401	0.195	0.150	1.300	0.248	0.177	1.400
100	0.240	0.172	1.400	0.205	0.160	1.283	0.249	0.178	1.399
200	0.241	0.173	1.397	0.217	0.172	1.262	0.249	0.179	1.397
300	0.243	0.174	1.394	0.229	0.184	1.246	0.251	0.180	1.394
400	0.245	0.176	1.389	0.239	0.193	1.233	0.253	0.182	1.389
500	0.248	0.179	1.383	0.247	0.202	1.223	0.256	0.185	1.384
600	0.250	0.182	1.377	0.255	0.210	1.215	0.259	0.188	1.377
700	0.254	0.185	1.371	0.262	0.217	1.208	0.262	0.191	1.371
800	0.257	0.188	1.365	0.269	0.224	1.202	0.266	0.195	1.364
900	0.259	0.191	1.358	0.275	0.230	1.197	0.269	0.198	1.357
1000	0.263	0.195	1.353	0.280	0.235	1.192	0.273	0.202	1.351
1500	0.276	0.208	1.330	0.298	0.253	1.178	0.287	0.216	1.328
2000	0.286	0.217	1.312	0.312	0.267	1.169	0.297	0.226	1.314
	Hydrogen, H_2			*Nitrogen, N_2*			*Oxygen, O_2*		
40	3.397	2.412	1.409	0.248	0.177	1.400	0.219	0.156	1.397
100	3.426	2.441	1.404	0.248	0.178	1.399	0.220	0.158	1.394
200	3.451	2.466	1.399	0.249	0.178	1.398	0.223	0.161	1.387
300	3.461	2.476	1.398	0.250	0.179	1.396	0.226	0.164	1.378
400	3.466	2.480	1.397	0.251	0.180	1.393	0.230	0.168	1.368
500	3.469	2.484	1.397	0.254	0.183	1.388	0.235	0.173	1.360
600	3.473	2.488	1.396	0.256	0.185	1.383	0.239	0.177	1.352
700	3.477	2.492	1.395	0.260	0.189	1.377	0.242	0.181	1.344
800	3.494	2.509	1.393	0.262	0.191	1.371	0.246	0.184	1.337
900	3.502	2.519	1.392	0.265	0.194	1.364	0.249	0.187	1.331
1000	3.513	2.528	1.390	0.269	0.198	1.359	0.252	0.190	1.326
1500	3.618	2.633	1.374	0.283	0.212	1.334	0.263	0.201	1.309
2000	3.758	2.773	1.355	0.293	0.222	1.319	0.270	0.208	1.298

Note: The unit Btu/lbm · R is equivalent to Btu/lbm · °F.

Source: Kenneth Wark, *Thermodynamics*, 4th ed. (New York: McGraw-Hill, 1983), p. 830, Table A–4. Originally published in *Tables of Properties of Gases*, NBS Circular 564, 1955.

TABLE A–2E

Ideal-gas specific heats of various common gases (*Concluded*)
(*c*) As a function of temperature

$$\bar{c}_p = a + bT + cT^2 + dT^3$$

(T in R, c_p in Btu/lbmol · R)

Substance	Formula	a	b	c	d	Temperature range, R	% error Max.	% error Avg.
Nitrogen	N_2	6.903	-0.02085×10^{-2}	0.05957×10^{-5}	-0.1176×10^{-9}	491-3240	0.59	0.34
Oxygen	O_2	6.085	0.2017×10^{-2}	-0.05275×10^{-5}	0.05372×10^{-9}	491-3240	1.19	0.28
Air	—	6.713	0.02609×10^{-2}	0.03540×10^{-5}	-0.08052×10^{-9}	491-3240	0.72	0.33
Hydrogen	H_2	6.952	-0.02542×10^{-2}	0.02952×10^{-5}	-0.03565×10^{-9}	491-3240	1.02	0.26
Carbon monoxide	CO	6.726	0.02222×10^{-2}	0.03960×10^{-5}	-0.09100×10^{-9}	491-3240	0.89	0.37
Carbon dioxide	CO_2	5.316	0.79361×10^{-2}	-0.2581×10^{-5}	0.3059×10^{-9}	491-3240	0.67	0.22
Water vapor	H_2O	7.700	0.02552×10^{-2}	0.07781×10^{-5}	-0.1472×10^{-9}	491-3240	0.53	0.24
Nitric oxide	NO	7.008	-0.01247×10^{-2}	0.07185×10^{-5}	-0.1715×10^{-9}	491-2700	0.97	0.36
Nitrous oxide	N_2O	5.758	0.7780×10^{-2}	-0.2596×10^{-5}	0.4331×10^{-9}	491-2700	0.59	0.26
Nitrogen dioxide	NO_2	5.48	0.7583×10^{-2}	-0.260×10^{-5}	0.322×10^{-9}	491-2700	0.46	0.18
Ammonia	NH_3	6.5846	0.34028×10^{-2}	0.073034×10^{-5}	-0.27402×10^{-9}	491-2700	0.91	0.36
Sulfur	S_2	6.499	0.2943×10^{-2}	-0.1200×10^{-5}	0.1632×10^{-9}	491-3240	0.99	0.38
Sulfur dioxide	SO_2	6.157	0.7689×10^{-2}	-0.2810×10^{-5}	0.3527×10^{-9}	491-3240	0.45	0.24
Sulfur trioxide	SO_3	3.918	1.935×10^{-2}	-0.8256×10^{-5}	1.328×10^{-9}	491-2340	0.29	0.13
Acetylene	C_2H_2	5.21	1.2227×10^{-2}	-0.4812×10^{-5}	0.7457×10^{-9}	491-2700	1.46	0.59
Benzene	C_6H_6	−8.650	6.4322×10^{-2}	-2.327×10^{-5}	3.179×10^{-9}	491-2700	0.34	0.20
Methanol	CH_4O	4.55	1.214×10^{-2}	-0.0898×10^{-5}	-0.329×10^{-9}	491-1800	0.18	0.08
Ethanol	C_2H_6O	4.75	2.781×10^{-2}	-0.7651×10^{-5}	0.821×10^{-9}	491-2700	0.40	0.22
Hydrogen chloride	HCl	7.244	-0.1011×10^{-2}	0.09783×10^{-5}	-0.1776×10^{-9}	491-2740	0.22	0.08
Methane	CH_4	4.750	0.6666×10^{-2}	0.09352×10^{-5}	-0.4510×10^{-9}	491-2740	1.33	0.57
Ethane	C_2H_6	1.648	2.291×10^{-2}	-0.4722×10^{-5}	0.2984×10^{-9}	491-2740	0.83	0.28
Propane	C_3H_8	−0.966	4.044×10^{-2}	-1.159×10^{-5}	1.300×10^{-9}	491-2740	0.40	0.12
n-Butane	C_4H_{10}	0.945	4.929×10^{-2}	-1.352×10^{-5}	1.433×10^{-9}	491-2740	0.54	0.24
i-Butane	C_4H_{10}	−1.890	5.520×10^{-2}	-1.696×10^{-5}	2.044×10^{-9}	491-2740	0.25	0.13
n-Pentane	C_5H_{12}	1.618	6.028×10^{-2}	-1.656×10^{-5}	1.732×10^{-9}	491-2740	0.56	0.21
n-Hexane	C_6H_{14}	1.657	7.328×10^{-2}	-2.112×10^{-5}	2.363×10^{-9}	491-2740	0.72	0.20
Ethylene	C_2H_4	0.944	2.075×10^{-2}	-0.6151×10^{-5}	0.7326×10^{-9}	491-2740	0.54	0.13
Propylene	C_3H_6	0.753	3.162×10^{-2}	-0.8981×10^{-5}	1.008×10^{-9}	491-2740	0.73	0.17

Source: Chemical and Process Thermodynamics 3/E by Kyle, B. G., © 2000. Adapted by permission of Pearson Education, Inc., Upper Saddle River, NJ.

TABLE A–3E

Properties of common liquids, solids, and foods
(*a*) Liquids

	Boiling data at 1 atm		*Freezing data*		*Liquid properties*		
Substance	Normal boiling point, °F	Latent heat of vaporization, h_{fg} Btu/lbm	Freezing point, °F	Latent heat of fusion, h_{if} Btu/lbm	Temperature, °F	Density, ρ lbm/ft^3	Specific heat, c_p Btu/lbm · R
Ammonia	−27.9	24.54	−107.9	138.6	−27.9	42.6	1.06
					0	41.3	1.083
					40	39.5	1.103
					80	37.5	1.135
Argon	−302.6	69.5	−308.7	12.0	−302.6	87.0	0.272
Benzene	176.4	169.4	41.9	54.2	68	54.9	0.411
Brine (20% sodium chloride by mass)	219.0	—	0.7	—	68	71.8	0.743
n-Butane	31.1	165.6	−217.3	34.5	31.1	37.5	0.552
Carbon dioxide	−109.2*	99.6 (at 32°F)	−69.8	—	32	57.8	0.583
Ethanol	172.8	360.5	−173.6	46.9	77	48.9	0.588
Ethyl alcohol	173.5	368	−248.8	46.4	68	49.3	0.678
Ethylene glycol	388.6	344.0	12.6	77.9	68	69.2	0.678
Glycerine	355.8	419	66.0	86.3	68	78.7	0.554
Helium	−452.1	9.80	—	—	−452.1	9.13	5.45
Hydrogen	−423.0	191.7	−434.5	25.6	−423.0	4.41	2.39
Isobutane	10.9	157.8	−255.5	45.5	10.9	37.1	0.545
Kerosene	399–559	108	−12.8	—	68	51.2	0.478
Mercury	674.1	126.7	−38.0	4.90	77	847	0.033
Methane	−258.7	219.6	296.0	25.1	−258.7	26.4	0.834
					−160	20.0	1.074
Methanol	148.1	473	−143.9	42.7	77	49.1	0.609
Nitrogen	−320.4	85.4	−346.0	10.9	−320.4	50.5	0.492
					−260	38.2	0.643
Octane	256.6	131.7	−71.5	77.9	68	43.9	0.502
Oil (light)	—	—			77	56.8	0.430
Oxygen	−297.3	91.5	−361.8	5.9	−297.3	71.2	0.408
Petroleum	—	99–165			68	40.0	0.478
Propane	−43.7	184.0	−305.8	34.4	−43.7	36.3	0.538
					32	33.0	0.604
					100	29.4	0.673
Refrigerant-134a	−15.0	93.3	−141.9	—	−40	88.5	0.283
					−15	86.0	0.294
					32	80.9	0.318
					90	73.6	0.348
Water	212	970.1	32	143.5	32	62.4	1.01
					90	62.1	1.00
					150	61.2	1.00
					212	59.8	1.01

*Sublimation temperature. (At pressures below the triple-point pressure of 75.1 psia, carbon dioxide exists as a solid or gas. Also, the freezing-point temperature of carbon dioxide is the triple-point temperature of −69.8°F.)

TABLE A–3E

Properties of common liquids, solids, and foods (*Continued*)

(*b*) Solids (values are for room temperature unless indicated otherwise)

Substance	Density, ρ lbm/ft^3	Specific heat, c_p Btu/lbm · R
Metals		
Aluminum		
−100°F		0.192
32°F		0.212
100°F	170	0.218
200°F		0.224
300°F		0.229
400°F		0.235
500°F		0.240
Bronze (76% Cu, 2% Zn, 2% Al)	517	0.0955
Brass, yellow (65% Cu, 35% Zn)	519	0.0955
Copper		
−60°F		0.0862
0°F		0.0893
100°F	555	0.0925
200°F		0.0938
390°F		0.0963
Iron	490	0.107
Lead	705	0.030
Magnesium	108	0.239
Nickel	555	0.105
Silver	655	0.056
Steel, mild	489	0.119
Tungsten	1211	0.031

Substance	Density, ρ lbm/ft^3	Specific heat, c_p Btu/lbm · R
Nonmetals		
Asphalt	132	0.220
Brick, common	120	0.189
Brick, fireclay (500°C)	144	0.229
Concrete	144	0.156
Clay	62.4	0.220
Diamond	151	0.147
Glass, window	169	0.191
Glass, pyrex	139	0.200
Graphite	156	0.170
Granite	169	0.243
Gypsum or plaster board	50	0.260
Ice		
−50°F		0.424
0°F		0.471
20°F		0.491
32°F	57.5	0.502
Limestone	103	0.217
Marble	162	0.210
Plywood (Douglas fir)	34.0	
Rubber (soft)	68.7	
Rubber (hard)	71.8	
Sand	94.9	
Stone	93.6	
Woods, hard (maple, oak, etc.)	45.0	
Woods, soft (fir, pine, etc.)	32.0	

(*c*) Foods

Food	Water content, % (mass)	Freezing point, °F	Specific heat, Btu/lbm · R: Above freezing	Specific heat, Btu/lbm · R: Below freezing	Latent heat of fusion, Btu/lbm
Apples	84	30	0.873	0.453	121
Bananas	75	31	0.801	0.426	108
Beef round	67	—	0.737	0.402	96
Broccoli	90	31	0.921	0.471	129
Butter	16	—	—	0.249	23
Cheese, Swiss	39	14	0.513	0.318	56
Cherries	80	29	0.841	0.441	115
Chicken	74	27	0.793	0.423	106
Corn, sweet	74	31	0.793	0.423	106
Eggs, whole	74	31	0.793	0.423	106
Ice cream	63	22	0.705	0.390	90
Lettuce	95	32	0.961	0.487	136
Milk, whole	88	31	0.905	0.465	126
Oranges	87	31	0.897	0.462	125
Potatoes	78	31	0.825	0.435	112
Salmon fish	64	28	0.713	0.393	92
Shrimp	83	28	0.865	0.450	119
Spinach	93	31	0.945	0.481	134
Strawberries	90	31	0.921	0.471	129
Tomatoes, ripe	94	31	0.953	0.484	135
Turkey	64	—	0.713	0.393	92
Watermelon	93	31	0.945	0.481	134

Source: Values are obtained from various handbooks and other sources or are calculated. Water content and freezing-point data of foods are from ASHRAE, *Handbook of Fundamentals*, I-P version (Atlanta, GA: American Society of Heating, Refrigerating, and Air-Conditioning Engineers, Inc., 1993), Chap. 30, Table 1. Freezing point is the temperature at which freezing starts for fruits and vegetables, and the average freezing temperature for other foods.

TABLE A–4E

Saturated water—Temperature table

		Specific volume, ft³/lbm		Internal energy, Btu/lbm			Enthalpy, Btu/lbm			Entropy, Btu/lbm · R		
Temp., T °F	Sat. press., P_{sat} psia	Sat. liquid, v_f	Sat. vapor, v_g	Sat. liquid, u_f	Evap., u_{fg}	Sat. vapor, u_g	Sat. liquid, h_f	Evap., h_{fg}	Sat. vapor, h_g	Sat. liquid, s_f	Evap., s_{fg}	Sat. vapor, s_g
32.018	0.08871	0.01602	3299.9	0.000	1021.0	1021.0	0.000	1075.2	1075.2	0.00000	2.18672	2.1867
35	0.09998	0.01602	2945.7	3.004	1019.0	1022.0	3.004	1073.5	1076.5	0.00609	2.17011	2.1762
40	0.12173	0.01602	2443.6	8.032	1015.6	1023.7	8.032	1070.7	1078.7	0.01620	2.14271	2.1589
45	0.14756	0.01602	2035.8	13.05	1012.2	1025.3	13.05	1067.8	1080.9	0.02620	2.11587	2.1421
50	0.17812	0.01602	1703.1	18.07	1008.9	1026.9	18.07	1065.0	1083.1	0.03609	2.08956	2.1256
55	0.21413	0.01603	1430.4	23.07	1005.5	1028.6	23.07	1062.2	1085.3	0.04586	2.06377	2.1096
60	0.25638	0.01604	1206.1	28.08	1002.1	1030.2	28.08	1059.4	1087.4	0.05554	2.03847	2.0940
65	0.30578	0.01604	1020.8	33.08	998.76	1031.8	33.08	1056.5	1089.6	0.06511	2.01366	2.0788
70	0.36334	0.01605	867.18	38.08	995.39	1033.5	38.08	1053.7	1091.8	0.07459	1.98931	2.0639
75	0.43016	0.01606	739.27	43.07	992.02	1035.1	43.07	1050.9	1093.9	0.08398	1.96541	2.0494
80	0.50745	0.01607	632.41	48.06	988.65	1036.7	48.07	1048.0	1096.1	0.09328	1.94196	2.0352
85	0.59659	0.01609	542.80	53.06	985.28	1038.3	53.06	1045.2	1098.3	0.10248	1.91892	2.0214
90	0.69904	0.01610	467.40	58.05	981.90	1040.0	58.05	1042.4	1100.4	0.11161	1.89630	2.0079
95	0.81643	0.01612	403.74	63.04	978.52	1041.6	63.04	1039.5	1102.6	0.12065	1.87408	1.9947
100	0.95052	0.01613	349.83	68.03	975.14	1043.2	68.03	1036.7	1104.7	0.12961	1.85225	1.9819
110	1.2767	0.01617	264.96	78.01	968.36	1046.4	78.02	1031.0	1109.0	0.14728	1.80970	1.9570
120	1.6951	0.01620	202.94	88.00	961.56	1049.6	88.00	1025.2	1113.2	0.16466	1.76856	1.9332
130	2.2260	0.01625	157.09	97.99	954.73	1052.7	97.99	1019.4	1117.4	0.18174	1.72877	1.9105
140	2.8931	0.01629	122.81	107.98	947.87	1055.9	107.99	1013.6	1121.6	0.19855	1.69024	1.8888
150	3.7234	0.01634	96.929	117.98	940.98	1059.0	117.99	1007.8	1125.7	0.21508	1.65291	1.8680
160	4.7474	0.01639	77.185	127.98	934.05	1062.0	128.00	1001.8	1129.8	0.23136	1.61670	1.8481
170	5.9999	0.01645	61.982	138.00	927.08	1065.1	138.02	995.88	1133.9	0.24739	1.58155	1.8289
180	7.5197	0.01651	50.172	148.02	920.06	1068.1	148.04	989.85	1137.9	0.26318	1.54741	1.8106
190	9.3497	0.01657	40.920	158.05	912.99	1071.0	158.08	983.76	1141.8	0.27874	1.51421	1.7930
200	11.538	0.01663	33.613	168.10	905.87	1074.0	168.13	977.60	1145.7	0.29409	1.48191	1.7760
210	14.136	0.01670	27.798	178.15	898.68	1076.8	178.20	971.35	1149.5	0.30922	1.45046	1.7597
212	14.709	0.01671	26.782	180.16	897.24	1077.4	180.21	970.09	1150.3	0.31222	1.44427	1.7565
220	17.201	0.01677	23.136	188.22	891.43	1079.6	188.28	965.02	1153.3	0.32414	1.41980	1.7439
230	20.795	0.01684	19.374	198.31	884.10	1082.4	198.37	958.59	1157.0	0.33887	1.38989	1.7288
240	24.985	0.01692	16.316	208.41	876.70	1085.1	208.49	952.06	1160.5	0.35342	1.36069	1.7141
250	29.844	0.01700	13.816	218.54	869.21	1087.7	218.63	945.41	1164.0	0.36779	1.33216	1.6999
260	35.447	0.01708	11.760	228.68	861.62	1090.3	228.79	938.65	1167.4	0.38198	1.30425	1.6862
270	41.877	0.01717	10.059	238.85	853.94	1092.8	238.98	931.76	1170.7	0.39601	1.27694	1.6730
280	49.222	0.01726	8.6439	249.04	846.16	1095.2	249.20	924.74	1173.9	0.40989	1.25018	1.6601
290	57.573	0.01735	7.4607	259.26	838.27	1097.5	259.45	917.57	1177.0	0.42361	1.22393	1.6475
300	67.028	0.01745	6.4663	269.51	830.25	1099.8	269.73	910.24	1180.0	0.43720	1.19818	1.6354
310	77.691	0.01755	5.6266	279.79	822.11	1101.9	280.05	902.75	1182.8	0.45065	1.17289	1.6235
320	89.667	0.01765	4.9144	290.11	813.84	1104.0	290.40	895.09	1185.5	0.46396	1.14802	1.6120
330	103.07	0.01776	4.3076	300.46	805.43	1105.9	300.80	887.25	1188.1	0.47716	1.12355	1.6007
340	118.02	0.01787	3.7885	310.85	796.87	1107.7	311.24	879.22	1190.5	0.49024	1.09945	1.5897
350	134.63	0.01799	3.3425	321.29	788.16	1109.4	321.73	870.98	1192.7	0.50321	1.07570	1.5789
360	153.03	0.01811	2.9580	331.76	779.28	1111.0	332.28	862.53	1194.8	0.51607	1.05227	1.5683
370	173.36	0.01823	2.6252	342.29	770.23	1112.5	342.88	853.86	1196.7	0.52884	1.02914	1.5580
380	195.74	0.01836	2.3361	352.87	761.00	1113.9	353.53	844.96	1198.5	0.54152	1.00628	1.5478
390	220.33	0.01850	2.0842	363.50	751.58	1115.1	364.25	835.81	1200.1	0.55411	0.98366	1.5378

(Continued)

TABLE A–4E

Saturated water—Temperature table (*Concluded*)

		Specific volume, ft³/lbm		*Internal energy,* Btu/lbm			*Enthalpy,* Btu/lbm			*Entropy,* Btu/lbm · R		
Temp., T °F	Sat. press., P_{sat} psia	Sat. liquid, v_f	Sat. vapor, v_g	Sat. liquid, u_f	Evap., u_{fg}	Sat. vapor, u_g	Sat. liquid, h_f	Evap., h_{fg}	Sat. vapor, h_g	Sat. liquid, s_f	Evap., s_{fg}	Sat. vapor, s_g
400	247.26	0.01864	1.8639	374.19	741.97	1116.2	375.04	826.39	1201.4	0.56663	0.96127	1.5279
410	276.69	0.01878	1.6706	384.94	732.14	1117.1	385.90	816.71	1202.6	0.57907	0.93908	1.5182
420	308.76	0.01894	1.5006	395.76	722.08	1117.8	396.84	806.74	1203.6	0.59145	0.91707	1.5085
430	343.64	0.01910	1.3505	406.65	711.80	1118.4	407.86	796.46	1204.3	0.60377	0.89522	1.4990
440	381.49	0.01926	1.2178	417.61	701.26	1118.9	418.97	785.87	1204.8	0.61603	0.87349	1.4895
450	422.47	0.01944	1.0999	428.66	690.47	1119.1	430.18	774.94	1205.1	0.62826	0.85187	1.4801
460	466.75	0.01962	0.99510	439.79	679.39	1119.2	441.48	763.65	1205.1	0.64044	0.83033	1.4708
470	514.52	0.01981	0.90158	451.01	668.02	1119.0	452.90	751.98	1204.9	0.65260	0.80885	1.4615
480	565.96	0.02001	0.81794	462.34	656.34	1118.7	464.43	739.91	1204.3	0.66474	0.78739	1.4521
490	621.24	0.02022	0.74296	473.77	644.32	1118.1	476.09	727.40	1203.5	0.67686	0.76594	1.4428
500	680.56	0.02044	0.67558	485.32	631.94	1117.3	487.89	714.44	1202.3	0.68899	0.74445	1.4334
510	744.11	0.02067	0.61489	496.99	619.17	1116.2	499.84	700.99	1200.8	0.70112	0.72290	1.4240
520	812.11	0.02092	0.56009	508.80	605.99	1114.8	511.94	687.01	1199.0	0.71327	0.70126	1.4145
530	884.74	0.02118	0.51051	520.76	592.35	1113.1	524.23	672.47	1196.7	0.72546	0.67947	1.4049
540	962.24	0.02146	0.46553	532.88	578.23	1111.1	536.70	657.31	1194.0	0.73770	0.65751	1.3952
550	1044.8	0.02176	0.42465	545.18	563.58	1108.8	549.39	641.47	1190.9	0.75000	0.63532	1.3853
560	1132.7	0.02207	0.38740	557.68	548.33	1106.0	562.31	624.91	1187.2	0.76238	0.61284	1.3752
570	1226.2	0.02242	0.35339	570.40	532.45	1102.8	575.49	607.55	1183.0	0.77486	0.59003	1.3649
580	1325.5	0.02279	0.32225	583.37	515.84	1099.2	588.95	589.29	1178.2	0.78748	0.56679	1.3543
590	1430.8	0.02319	0.29367	596.61	498.43	1095.0	602.75	570.04	1172.8	0.80026	0.54306	1.3433
600	1542.5	0.02362	0.26737	610.18	480.10	1090.3	616.92	549.67	1166.6	0.81323	0.51871	1.3319
610	1660.9	0.02411	0.24309	624.11	460.73	1084.8	631.52	528.03	1159.5	0.82645	0.49363	1.3201
620	1786.2	0.02464	0.22061	638.47	440.14	1078.6	646.62	504.92	1151.5	0.83998	0.46765	1.3076
630	1918.9	0.02524	0.19972	653.35	418.12	1071.5	662.32	480.07	1142.4	0.85389	0.44056	1.2944
640	2059.3	0.02593	0.18019	668.86	394.36	1063.2	678.74	453.14	1131.9	0.86828	0.41206	1.2803
650	2207.8	0.02673	0.16184	685.16	368.44	1053.6	696.08	423.65	1119.7	0.88332	0.38177	1.2651
660	2364.9	0.02767	0.14444	702.48	339.74	1042.2	714.59	390.84	1105.4	0.89922	0.34906	1.2483
670	2531.2	0.02884	0.12774	721.23	307.22	1028.5	734.74	353.54	1088.3	0.91636	0.31296	1.2293
680	2707.3	0.03035	0.11134	742.11	269.00	1011.1	757.32	309.57	1066.9	0.93541	0.27163	1.2070
690	2894.1	0.03255	0.09451	766.81	220.77	987.6	784.24	253.96	1038.2	0.95797	0.22089	1.1789
700	3093.0	0.03670	0.07482	801.75	146.50	948.3	822.76	168.32	991.1	0.99023	0.14514	1.1354
705.10	3200.1	0.04975	0.04975	866.61	0	866.6	896.07	0	896.1	1.05257	0	1.0526

Source: Tables A-4E through A-8E are generated using the Engineering Equation Solver (EES) software developed by S. A. Klein and F. L. Alvarado. The routine used in calculations is the highly accurate Steam_IAPWS, which incorporates the 1995 Formulation for the Thermodynamic Properties of Ordinary Water Substance for General and Scientific Use, issued by The International Association for the Properties of Water and Steam (IAPWS). This formulation replaces the 1984 formulation of Haar, Gallagher, and Kell (NBS/NRC Steam Tables, Hemisphere Publishing Co., 1984), which is also available in EES as the routine STEAM. The new formulation is based on the correlations of Saul and Wagner (J. Phys. Chem. Ref. Data, 16, 893, 1987) with modifications to adjust to the International Temperature Scale of 1990. The modifications are described by Wagner and Pruss (J. Phys. Chem. Ref. Data, 22, 783, 1993). The properties of ice are based on Hyland and Wexler, "Formulations for the Thermodynamic Properties of the Saturated Phases of H_2O from 173.15 K to 473.15 K," *ASHRAE Trans.*, Part 2A, Paper 2793, 1983.

TABLE A–5E

Saturated water—Pressure table

		Specific volume, ft³/lbm		Internal energy, Btu/lbm			Enthalpy, Btu/lbm			Entropy, Btu/lbm · R		
Press., P psia	Sat. temp., T_{sat} °F	Sat. liquid, v_f	Sat. vapor, v_g	Sat. liquid, u_f	Evap., u_{fg}	Sat. vapor, u_g	Sat. liquid, h_f	Evap., h_{fg}	Sat. vapor, h_g	Sat. liquid, s_f	Evap., s_{fg}	Sat. vapor, s_g
1	101.69	0.01614	333.49	69.72	973.99	1043.7	69.72	1035.7	1105.4	0.13262	1.84495	1.9776
2	126.02	0.01623	173.71	94.02	957.45	1051.5	94.02	1021.7	1115.8	0.17499	1.74444	1.9194
3	141.41	0.01630	118.70	109.39	946.90	1056.3	109.40	1012.8	1122.2	0.20090	1.68489	1.8858
4	152.91	0.01636	90.629	120.89	938.97	1059.9	120.90	1006.0	1126.9	0.21985	1.64225	1.8621
5	162.18	0.01641	73.525	130.17	932.53	1062.7	130.18	1000.5	1130.7	0.23488	1.60894	1.8438
6	170.00	0.01645	61.982	138.00	927.08	1065.1	138.02	995.88	1133.9	0.24739	1.58155	1.8289
8	182.81	0.01652	47.347	150.83	918.08	1068.9	150.86	988.15	1139.0	0.26757	1.53800	1.8056
10	193.16	0.01659	38.425	161.22	910.75	1072.0	161.25	981.82	1143.1	0.28362	1.50391	1.7875
14.696	211.95	0.01671	26.805	180.12	897.27	1077.4	180.16	970.12	1150.3	0.31215	1.44441	1.7566
15	212.99	0.01672	26.297	181.16	896.52	1077.7	181.21	969.47	1150.7	0.31370	1.44441	1.7549
20	227.92	0.01683	20.093	196.21	885.63	1081.8	196.27	959.93	1156.2	0.33582	1.39606	1.7319
25	240.03	0.01692	16.307	208.45	876.67	1085.1	208.52	952.03	1160.6	0.35347	1.36060	1.7141
30	250.30	0.01700	13.749	218.84	868.98	1087.8	218.93	945.21	1164.1	0.36821	1.33132	1.6995
35	259.25	0.01708	11.901	227.92	862.19	1090.1	228.03	939.16	1167.2	0.38093	1.30632	1.6872
40	267.22	0.01715	10.501	236.02	856.09	1092.1	236.14	933.69	1169.8	0.39213	1.28448	1.6766
45	274.41	0.01721	9.4028	243.34	850.52	1093.9	243.49	928.68	1172.2	0.40216	1.26506	1.6672
50	280.99	0.01727	8.5175	250.05	845.39	1095.4	250.21	924.03	1174.2	0.41125	1.24756	1.6588
55	287.05	0.01732	7.7882	256.25	840.61	1096.9	256.42	919.70	1176.1	0.41958	1.23162	1.6512
60	292.69	0.01738	7.1766	262.01	836.13	1098.1	262.20	915.61	1177.8	0.42728	1.21697	1.6442
65	297.95	0.01743	6.6560	267.41	831.90	1099.3	267.62	911.75	1179.4	0.43443	1.20341	1.6378
70	302.91	0.01748	6.2075	272.50	827.90	1100.4	272.72	908.08	1180.8	0.44112	1.19078	1.6319
75	307.59	0.01752	5.8167	277.31	824.09	1101.4	277.55	904.58	1182.1	0.44741	1.17895	1.6264
80	312.02	0.01757	5.4733	281.87	820.45	1102.3	282.13	901.22	1183.4	0.45335	1.16783	1.6212
85	316.24	0.01761	5.1689	286.22	816.97	1103.2	286.50	898.00	1184.5	0.45897	1.15732	1.6163
90	320.26	0.01765	4.8972	290.38	813.62	1104.0	290.67	894.89	1185.6	0.46431	1.14737	1.6117
95	324.11	0.01770	4.6532	294.36	810.40	1104.8	294.67	891.89	1186.6	0.46941	1.13791	1.6073
100	327.81	0.01774	4.4327	298.19	807.29	1105.5	298.51	888.99	1187.5	0.47427	1.12888	1.6032
110	334.77	0.01781	4.0410	305.41	801.37	1106.8	305.78	883.44	1189.2	0.48341	1.11201	1.5954
120	341.25	0.01789	3.7289	312.16	795.79	1107.9	312.55	878.20	1190.8	0.49187	1.09646	1.5883
130	347.32	0.01796	3.4557	318.48	790.51	1109.0	318.92	873.21	1192.1	0.49974	1.08204	1.5818
140	353.03	0.01802	3.2202	324.45	785.49	1109.9	324.92	868.45	1193.4	0.50711	1.06858	1.5757
150	358.42	0.01809	3.0150	330.11	780.69	1110.8	330.61	863.88	1194.5	0.51405	1.05595	1.5700
160	363.54	0.01815	2.8347	335.49	776.10	1111.6	336.02	859.49	1195.5	0.52061	1.04405	1.5647
170	368.41	0.01821	2.6749	340.62	771.68	1112.3	341.19	855.25	1196.4	0.52682	1.03279	1.5596
180	373.07	0.01827	2.5322	345.53	767.42	1113.0	346.14	851.16	1197.3	0.53274	1.02210	1.5548
190	377.52	0.01833	2.4040	350.24	763.31	1113.6	350.89	847.19	1198.1	0.53839	1.01191	1.5503
200	381.80	0.01839	2.2882	354.78	759.32	1114.1	355.46	843.33	1198.8	0.54379	1.00219	1.5460
250	400.97	0.01865	1.8440	375.23	741.02	1116.3	376.09	825.47	1201.6	0.56784	0.95912	1.5270
300	417.35	0.01890	1.5435	392.89	724.77	1117.7	393.94	809.41	1203.3	0.58818	0.92289	1.5111
350	431.74	0.01912	1.3263	408.55	709.98	1118.5	409.79	794.65	1204.4	0.60590	0.89143	1.4973
400	444.62	0.01934	1.1617	422.70	696.31	1119.0	424.13	780.87	1205.0	0.62168	0.86350	1.4852
450	456.31	0.01955	1.0324	435.67	683.52	1119.2	437.30	767.86	1205.2	0.63595	0.83828	1.4742
500	467.04	0.01975	0.92819	447.68	671.42	1119.1	449.51	755.48	1205.0	0.64900	0.81521	1.4642
550	476.97	0.01995	0.84228	458.90	659.91	1118.8	460.93	743.60	1204.5	0.66107	0.79388	1.4550
600	486.24	0.02014	0.77020	469.46	648.88	1118.3	471.70	732.15	1203.9	0.67231	0.77400	1.4463

(Continued)

TABLE A–5E

Saturated water—Pressure table (*Concluded*)

		Specific volume, ft³/lbm		*Internal energy,* Btu/lbm			*Enthalpy,* Btu/lbm			*Entropy,* Btu/lbm · R		
Press., P psia	Sat. temp., T_{sat} °F	Sat. liquid, v_f	Sat. vapor, v_g	Sat. liquid, u_f	Evap., u_{fg}	Sat. vapor, u_g	Sat. liquid, h_f	Evap., h_{fg}	Sat. vapor, h_g	Sat. liquid, s_f	Evap., s_{fg}	Sat. vapor, s_g
700	503.13	0.02051	0.65589	488.96	627.98	1116.9	491.62	710.29	1201.9	0.69279	0.73771	1.4305
800	518.27	0.02087	0.56920	506.74	608.30	1115.0	509.83	689.48	1199.3	0.71117	0.70502	1.4162
900	532.02	0.02124	0.50107	523.19	589.54	1112.7	526.73	669.46	1196.2	0.72793	0.67505	1.4030
1000	544.65	0.02159	0.44604	538.58	571.49	1110.1	542.57	650.03	1192.6	0.74341	0.64722	1.3906
1200	567.26	0.02232	0.36241	566.89	536.87	1103.8	571.85	612.39	1184.2	0.77143	0.59632	1.3677
1400	587.14	0.02307	0.30161	592.79	503.50	1096.3	598.76	575.66	1174.4	0.79658	0.54991	1.3465
1600	604.93	0.02386	0.25516	616.99	470.69	1087.7	624.06	539.18	1163.2	0.81972	0.50645	1.3262
1800	621.07	0.02470	0.21831	640.03	437.86	1077.9	648.26	502.35	1150.6	0.84144	0.46482	1.3063
2000	635.85	0.02563	0.18815	662.33	404.46	1066.8	671.82	464.60	1136.4	0.86224	0.42409	1.2863
2500	668.17	0.02860	0.13076	717.67	313.53	1031.2	730.90	360.79	1091.7	0.91311	0.31988	1.2330
3000	695.41	0.03433	0.08460	783.39	186.41	969.8	802.45	214.32	1016.8	0.97321	0.18554	1.1587
3200.1	705.10	0.04975	0.04975	866.61	0	866.6	896.07	0	896.1	1.05257	0	1.0526

TABLE A–6E

Superheated water

T °F	*v* ft^3/lbm	*u* Btu/lbm	*h* Btu/lbm	*s* Btu/lbm · R	*v* ft^3/lbm	*u* Btu/lbm	*h* Btu/lbm	*s* Btu/lbm · R	*v* ft^3/lbm	*u* Btu/lbm	*h* Btu/lbm	*s* Btu/lbm · R
	P = 1.0 psia (101.69°F)*				*P* = 5.0 psia (162.18°F)				*P* = 10 psia (193.16°F)			
Sat.†	333.49	1043.7	1105.4	1.9776	73.525	1062.7	1130.7	1.8438	38.425	1072.0	1143.1	1.7875
200	392.53	1077.5	1150.1	2.0509	78.153	1076.2	1148.5	1.8716	38.849	1074.5	1146.4	1.7926
240	416.44	1091.2	1168.3	2.0777	83.009	1090.3	1167.1	1.8989	41.326	1089.1	1165.5	1.8207
280	440.33	1105.0	1186.5	2.1030	87.838	1104.3	1185.6	1.9246	43.774	1103.4	1184.4	1.8469
320	464.20	1118.9	1204.8	2.1271	92.650	1118.4	1204.1	1.9490	46.205	1117.6	1203.1	1.8716
360	488.07	1132.9	1223.3	2.1502	97.452	1132.5	1222.6	1.9722	48.624	1131.9	1221.8	1.8950
400	511.92	1147.1	1241.8	2.1722	102.25	1146.7	1241.3	1.9944	51.035	1146.2	1240.6	1.9174
440	535.77	1161.3	1260.4	2.1934	107.03	1160.9	1260.0	2.0156	53.441	1160.5	1259.4	1.9388
500	571.54	1182.8	1288.6	2.2237	114.21	1182.6	1288.2	2.0461	57.041	1182.2	1287.8	1.9693
600	631.14	1219.4	1336.2	2.2709	126.15	1219.2	1335.9	2.0933	63.029	1219.0	1335.6	2.0167
700	690.73	1256.8	1384.6	2.3146	138.09	1256.7	1384.4	2.1371	69.007	1256.5	1384.2	2.0605
800	750.31	1295.1	1433.9	2.3553	150.02	1294.9	1433.7	2.1778	74.980	1294.8	1433.5	2.1013
1000	869.47	1374.2	1535.1	2.4299	173.86	1374.2	1535.0	2.2524	86.913	1374.1	1534.9	2.1760
1200	988.62	1457.1	1640.0	2.4972	197.70	1457.0	1640.0	2.3198	98.840	1457.0	1639.9	2.2433
1400	1107.8	1543.7	1748.7	2.5590	221.54	1543.7	1748.7	2.3816	110.762	1543.6	1748.6	2.3052
	P = 15 psia (212.99°F)				*P* = 20 psia (227.92°F)				*P* = 40 psia (267.22°F)			
Sat.	26.297	1077.7	1150.7	1.7549	20.093	1081.8	1156.2	1.7319	10.501	1092.1	1169.8	1.6766
240	27.429	1087.8	1163.9	1.7742	20.478	1086.5	1162.3	1.7406				
280	29.085	1102.4	1183.2	1.8010	21.739	1101.4	1181.9	1.7679	10.713	1097.3	1176.6	1.6858
320	30.722	1116.9	1202.2	1.8260	22.980	1116.1	1201.2	1.7933	11.363	1112.9	1197.1	1.7128
360	32.348	1131.3	1221.1	1.8496	24.209	1130.7	1220.2	1.8171	11.999	1128.1	1216.9	1.7376
400	33.965	1145.7	1239.9	1.8721	25.429	1145.1	1239.3	1.8398	12.625	1143.1	1236.5	1.7610
440	35.576	1160.1	1258.8	1.8936	26.644	1159.7	1258.3	1.8614	13.244	1157.9	1256.0	1.7831
500	37.986	1181.9	1287.3	1.9243	28.458	1181.6	1286.9	1.8922	14.165	1180.2	1285.0	1.8143
600	41.988	1218.7	1335.3	1.9718	31.467	1218.5	1334.9	1.9398	15.686	1217.5	1333.6	1.8625
700	45.981	1256.3	1383.9	2.0156	34.467	1256.1	1383.7	1.9837	17.197	1255.3	1382.6	1.9067
800	49.967	1294.6	1433.3	2.0565	37.461	1294.5	1433.1	2.0247	18.702	1293.9	1432.3	1.9478
1000	57.930	1374.0	1534.8	2.1312	43.438	1373.8	1534.6	2.0994	21.700	1373.4	1534.1	2.0227
1200	65.885	1456.9	1639.8	2.1986	49.407	1456.8	1639.7	2.1668	24.691	1456.5	1639.3	2.0902
1400	73.836	1543.6	1748.5	2.2604	55.373	1543.5	1748.4	2.2287	27.678	1543.3	1748.1	2.1522
1600	81.784	1634.0	1861.0	2.3178	61.335	1633.9	1860.9	2.2861	30.662	1633.7	1860.7	2.2096
	P = 60 psia (292.69°F)				*P* = 80 psia (312.02°F)				*P* = 100 psia (327.81°F)			
Sat.	7.1766	1098.1	1177.8	1.6442	5.4733	1102.3	1183.4	1.6212	4.4327	1105.5	1187.5	1.6032
320	7.4863	1109.6	1192.7	1.6636	5.5440	1105.9	1187.9	1.6271				
360	7.9259	1125.5	1213.5	1.6897	5.8876	1122.7	1209.9	1.6545	4.6628	1119.8	1206.1	1.6263
400	8.3548	1140.9	1233.7	1.7138	6.2187	1138.7	1230.8	1.6794	4.9359	1136.4	1227.8	1.6521
440	8.7766	1156.1	1253.6	1.7364	6.5420	1154.3	1251.2	1.7026	5.2006	1152.4	1248.7	1.6759
500	9.4005	1178.8	1283.1	1.7682	7.0177	1177.3	1281.2	1.7350	5.5876	1175.9	1279.3	1.7088
600	10.4256	1216.5	1332.2	1.8168	7.7951	1215.4	1330.8	1.7841	6.2167	1214.4	1329.4	1.7586
700	11.4401	1254.5	1381.6	1.8613	8.5616	1253.8	1380.5	1.8289	6.8344	1253.0	1379.5	1.8037
800	12.4484	1293.3	1431.5	1.9026	9.3218	1292.6	1430.6	1.8704	7.4457	1292.0	1429.8	1.8453
1000	14.4543	1373.0	1533.5	1.9777	10.8313	1372.6	1532.9	1.9457	8.6575	1372.2	1532.4	1.9208
1200	16.4525	1456.2	1638.9	2.0454	12.3331	1455.9	1638.5	2.0135	9.8615	1455.6	1638.1	1.9887
1400	18.4464	1543.0	1747.8	2.1073	13.8306	1542.8	1747.5	2.0755	11.0612	1542.6	1747.2	2.0508
1600	20.438	1633.5	1860.5	2.1648	15.3257	1633.3	1860.2	2.1330	12.2584	1633.2	1860.0	2.1083
1800	22.428	1727.6	1976.6	2.2187	16.8192	1727.5	1976.5	2.1869	13.4541	1727.3	1976.3	2.1622
2000	24.417	1825.2	2096.3	2.2694	18.3117	1825.0	2096.1	2.2376	14.6487	1824.9	2096.0	2.2130

*The temperature in parentheses is the saturation temperature at the specified pressure.

† Properties of saturated vapor at the specified pressure.

TABLE A–6E

Superheated water (*Continued*)

T °F	*v* ft^3/lbm	*u* Btu/lbm	*h* Btu/lbm	*s* Btu/lbm · R	*v* ft^3/lbm	*u* Btu/lbm	*h* Btu/lbm	*s* Btu/lbm · R	*v* ft^3/lbm	*u* Btu/lbm	*h* Btu/lbm	*s* Btu/lbm · R
	P = 120 psia (341.25°F)				*P* = 140 psia (353.03°F)				*P* = 160 psia (363.54°F)			
Sat.	3.7289	1107.9	1190.8	1.5883	3.2202	1109.9	1193.4	1.5757	2.8347	1111.6	1195.5	1.5647
360	3.8446	1116.7	1202.1	1.6023	3.2584	1113.4	1197.8	1.5811				
400	4.0799	1134.0	1224.6	1.6292	3.4676	1131.5	1221.4	1.6092	3.0076	1129.0	1218.0	1.5914
450	4.3613	1154.5	1251.4	1.6594	3.7147	1152.6	1248.9	1.6403	3.2293	1150.7	1246.3	1.6234
500	4.6340	1174.4	1277.3	1.6872	3.9525	1172.9	1275.3	1.6686	3.4412	1171.4	1273.2	1.6522
550	4.9010	1193.9	1302.8	1.7131	4.1845	1192.7	1301.1	1.6948	3.6469	1191.4	1299.4	1.6788
600	5.1642	1213.4	1328.0	1.7375	4.4124	1212.3	1326.6	1.7195	3.8484	1211.3	1325.2	1.7037
700	5.6829	1252.2	1378.4	1.7829	4.8604	1251.4	1377.3	1.7652	4.2434	1250.6	1376.3	1.7498
800	6.1950	1291.4	1429.0	1.8247	5.3017	1290.8	1428.1	1.8072	4.6316	1290.2	1427.3	1.7920
1000	7.2083	1371.7	1531.8	1.9005	6.1732	1371.3	1531.3	1.8832	5.3968	1370.9	1530.7	1.8682
1200	8.2137	1455.3	1637.7	1.9684	7.0367	1455.0	1637.3	1.9512	6.1540	1454.7	1636.9	1.9363
1400	9.2149	1542.3	1746.9	2.0305	7.8961	1542.1	1746.6	2.0134	6.9070	1541.8	1746.3	1.9986
1600	10.2135	1633.0	1859.8	2.0881	8.7529	1632.8	1859.5	2.0711	7.6574	1632.6	1859.3	2.0563
1800	11.2106	1727.2	1976.1	2.1420	9.6082	1727.0	1975.9	2.1250	8.4063	1726.9	1975.7	2.1102
2000	12.2067	1824.8	2095.8	2.1928	10.4624	1824.6	2095.7	2.1758	9.1542	1824.5	2095.5	2.1610
	P = 180 psia (373.07°F)				*P* = 200 psia (381.80°F)				*P* = 225 psia (391.80°F)			
Sat.	2.5322	1113.0	1197.3	1.5548	2.2882	1114.1	1198.8	1.5460	2.0423	1115.3	1200.3	1.5360
400	2.6490	1126.3	1214.5	1.5752	2.3615	1123.5	1210.9	1.5602	2.0728	1119.7	1206.0	1.5427
450	2.8514	1148.7	1243.7	1.6082	2.5488	1146.7	1241.0	1.5943	2.2457	1144.1	1237.6	1.5783
500	3.0433	1169.8	1271.2	1.6376	2.7247	1168.2	1269.0	1.6243	2.4059	1166.2	1266.3	1.6091
550	3.2286	1190.2	1297.7	1.6646	2.8939	1188.9	1296.0	1.6516	2.5590	1187.2	1293.8	1.6370
600	3.4097	1210.2	1323.8	1.6897	3.0586	1209.1	1322.3	1.6771	2.7075	1207.7	1320.5	1.6628
700	3.7635	1249.8	1375.2	1.7361	3.3796	1249.0	1374.1	1.7238	2.9956	1248.0	1372.7	1.7099
800	4.1104	1289.5	1426.5	1.7785	3.6934	1288.9	1425.6	1.7664	3.2765	1288.1	1424.5	1.7528
900	4.4531	1329.7	1478.0	1.8179	4.0031	1329.2	1477.3	1.8059	3.5530	1328.5	1476.5	1.7925
1000	4.7929	1370.5	1530.1	1.8549	4.3099	1370.1	1529.6	1.8430	3.8268	1369.5	1528.9	1.8296
1200	5.4674	1454.3	1636.5	1.9231	4.9182	1454.0	1636.1	1.9113	4.3689	1453.6	1635.6	1.8981
1400	6.1377	1541.6	1746.0	1.9855	5.5222	1541.4	1745.7	1.9737	4.9068	1541.1	1745.4	1.9606
1600	6.8054	1632.4	1859.1	2.0432	6.1238	1632.2	1858.8	2.0315	5.4422	1632.0	1858.6	2.0184
1800	7.4716	1726.7	1975.6	2.0971	6.7238	1726.5	1975.4	2.0855	5.9760	1726.4	1975.2	2.0724
2000	8.1367	1824.4	2095.4	2.1479	7.3227	1824.3	2095.3	2.1363	6.5087	1824.1	2095.1	2.1232
	P = 250 psia (400.97°F)				*P* = 275 psia (409.45°F)				*P* = 300 psia (417.35°F)			
Sat.	1.8440	1116.3	1201.6	1.5270	1.6806	1117.0	1202.6	1.5187	1.5435	1117.7	1203.3	1.5111
450	2.0027	1141.3	1234.0	1.5636	1.8034	1138.5	1230.3	1.5499	1.6369	1135.6	1226.4	1.5369
500	2.1506	1164.1	1263.6	1.5953	1.9415	1162.0	1260.8	1.5825	1.7670	1159.8	1257.9	1.5706
550	2.2910	1185.6	1291.5	1.6237	2.0715	1183.9	1289.3	1.6115	1.8885	1182.1	1287.0	1.6001
600	2.4264	1206.3	1318.6	1.6499	2.1964	1204.9	1316.7	1.6380	2.0046	1203.5	1314.8	1.6270
650	2.5586	1226.8	1345.1	1.6743	2.3179	1225.6	1343.5	1.6627	2.1172	1224.4	1341.9	1.6520
700	2.6883	1247.0	1371.4	1.6974	2.4369	1246.0	1370.0	1.6860	2.2273	1244.9	1368.6	1.6755
800	2.9429	1287.3	1423.5	1.7406	2.6699	1286.5	1422.4	1.7294	2.4424	1285.7	1421.3	1.7192
900	3.1930	1327.9	1475.6	1.7804	2.8984	1327.3	1474.8	1.7694	2.6529	1326.6	1473.9	1.7593
1000	3.4403	1369.0	1528.2	1.8177	3.1241	1368.5	1527.4	1.8068	2.8605	1367.9	1526.7	1.7968
1200	3.9295	1453.3	1635.0	1.8863	3.5700	1452.9	1634.5	1.8755	3.2704	1452.5	1634.0	1.8657
1400	4.4144	1540.8	1745.0	1.9488	4.0116	1540.5	1744.6	1.9381	3.6759	1540.2	1744.2	1.9284
1600	4.8969	1631.7	1858.3	2.0066	4.4507	1631.5	1858.0	1.9960	4.0789	1631.3	1857.7	1.9863
1800	5.3777	1726.2	1974.9	2.0607	4.8882	1726.0	1974.7	2.0501	4.4803	1725.8	1974.5	2.0404
2000	5.8575	1823.9	2094.9	2.1116	5.3247	1823.8	2094.7	2.1010	4.8807	1823.6	2094.6	2.0913

TABLE A–6E

Superheated water (*Continued*)

T °F	*v* ft³/lbm	*u* Btu/lbm	*h* Btu/lbm	*s* Btu/lbm · R	*v* ft³/lbm	*u* Btu/lbm	*h* Btu/lbm	*s* Btu/lbm · R	*v* ft³/lbm	*u* Btu/lbm	*h* Btu/lbm	*s* Btu/lbm · R
	P = 350 psia (431.74°F)				*P* = 400 psia (444.62°F)				*P* = 450 psia (456.31°F)			
Sat.	1.3263	1118.5	1204.4	1.4973	1.1617	1119.0	1205.0	1.4852	1.0324	1119.2	1205.2	1.4742
450	1.3739	1129.3	1218.3	1.5128	1.1747	1122.5	1209.4	1.4901				
500	1.4921	1155.2	1251.9	1.5487	1.2851	1150.4	1245.6	1.5288	1.1233	1145.4	1238.9	1.5103
550	1.6004	1178.6	1282.2	1.5795	1.3840	1174.9	1277.3	1.5610	1.2152	1171.1	1272.3	1.5441
600	1.7030	1200.6	1310.9	1.6073	1.4765	1197.6	1306.9	1.5897	1.3001	1194.6	1302.8	1.5737
650	1.8018	1221.9	1338.6	1.6328	1.5650	1219.4	1335.3	1.6158	1.3807	1216.9	1331.9	1.6005
700	1.8979	1242.8	1365.8	1.6567	1.6507	1240.7	1362.9	1.6401	1.4584	1238.5	1360.0	1.6253
800	2.0848	1284.1	1419.1	1.7009	1.8166	1282.5	1417.0	1.6849	1.6080	1280.8	1414.7	1.6706
900	2.2671	1325.3	1472.2	1.7414	1.9777	1324.0	1470.4	1.7257	1.7526	1322.7	1468.6	1.7117
1000	2.4464	1366.9	1525.3	1.7791	2.1358	1365.8	1523.9	1.7636	1.8942	1364.7	1522.4	1.7499
1200	2.7996	1451.7	1633.0	1.8483	2.4465	1450.9	1632.0	1.8331	2.1718	1450.1	1631.0	1.8196
1400	3.1484	1539.6	1743.5	1.9111	2.7527	1539.0	1742.7	1.8960	2.4450	1538.4	1742.0	1.8827
1600	3.4947	1630.8	1857.1	1.9691	3.0565	1630.3	1856.5	1.9541	2.7157	1629.8	1856.0	1.9409
1800	3.8394	1725.4	1974.0	2.0233	3.3586	1725.0	1973.6	2.0084	2.9847	1724.6	1973.2	1.9952
2000	4.1830	1823.3	2094.2	2.0742	3.6597	1823.0	2093.9	2.0594	3.2527	1822.6	2093.5	2.0462
	P = 500 psia (467.04°F)				*P* = 600 psia (486.24°F)				*P* = 700 psia (503.13°F)			
Sat.	0.92815	1119.1	1205.0	1.4642	0.77020	1118.3	1203.9	1.4463	0.65589	1116.9	1201.9	1.4305
500	0.99304	1140.1	1231.9	1.4928	0.79526	1128.2	1216.5	1.4596				
550	1.07974	1167.1	1267.0	1.5284	0.87542	1158.7	1255.9	1.4996	0.72799	1149.5	1243.8	1.4730
600	1.15876	1191.4	1298.6	1.5590	0.94605	1184.9	1289.9	1.5325	0.79332	1177.9	1280.7	1.5087
650	1.23312	1214.3	1328.4	1.5865	1.01133	1209.0	1321.3	1.5614	0.85242	1203.4	1313.8	1.5393
700	1.30440	1236.4	1357.0	1.6117	1.07316	1231.9	1351.0	1.5877	0.90769	1227.2	1344.8	1.5666
800	1.44097	1279.2	1412.5	1.6576	1.19038	1275.8	1408.0	1.6348	1.01125	1272.4	1403.4	1.6150
900	1.57252	1321.4	1466.9	1.6992	1.30230	1318.7	1463.3	1.6771	1.10921	1316.0	1459.7	1.6581
1000	1.70094	1363.6	1521.0	1.7376	1.41097	1361.4	1518.1	1.7160	1.20381	1359.2	1515.2	1.6974
1100	1.82726	1406.2	1575.3	1.7735	1.51749	1404.4	1572.9	1.7522	1.29621	1402.5	1570.4	1.7341
1200	1.95211	1449.4	1630.0	1.8075	1.62252	1447.8	1627.9	1.7865	1.38709	1446.2	1625.9	1.7685
1400	2.1988	1537.8	1741.2	1.8708	1.82957	1536.6	1739.7	1.8501	1.56580	1535.4	1738.2	1.8324
1600	2.4430	1629.4	1855.4	1.9291	2.0340	1628.4	1854.2	1.9085	1.74192	1627.5	1853.1	1.8911
1800	2.6856	1724.2	1972.7	1.9834	2.2369	1723.4	1971.8	1.9630	1.91643	1722.7	1970.9	1.9457
2000	2.9271	1822.3	2093.1	2.0345	2.4387	1821.7	2092.4	2.0141	2.08987	1821.0	2091.7	1.9969
	P = 800 psia (518.27°F)				*P* = 1000 psia (544.65°F)				*P* = 1250 psia (572.45°F)			
Sat.	0.56920	1115.0	1199.3	1.4162	0.44604	1110.1	1192.6	1.3906	0.34549	1102.0	1181.9	1.3623
550	0.61586	1139.4	1230.5	1.4476	0.45375	1115.2	1199.2	1.3972				
600	0.67799	1170.5	1270.9	1.4866	0.51431	1154.1	1249.3	1.4457	0.37894	1129.5	1217.2	1.3961
650	0.73279	1197.6	1306.0	1.5191	0.56411	1185.1	1289.5	1.4827	0.42703	1167.5	1266.3	1.4414
700	0.78330	1222.4	1338.4	1.5476	0.60844	1212.4	1325.0	1.5140	0.46735	1198.7	1306.8	1.4771
750	0.83102	1246.0	1369.1	1.5735	0.64944	1237.6	1357.8	1.5418	0.50344	1226.4	1342.9	1.5076
800	0.87678	1268.9	1398.7	1.5975	0.68821	1261.7	1389.0	1.5670	0.53687	1252.2	1376.4	1.5347
900	0.96434	1313.3	1456.0	1.6413	0.76136	1307.7	1448.6	1.6126	0.59876	1300.5	1439.0	1.5826
1000	1.04841	1357.0	1512.2	1.6812	0.83078	1352.5	1506.2	1.6535	0.65656	1346.7	1498.6	1.6249
1100	1.13024	1400.7	1568.0	1.7181	0.89783	1396.9	1563.1	1.6911	0.71184	1392.2	1556.8	1.6635
1200	1.21051	1444.6	1623.8	1.7528	0.96327	1441.4	1619.7	1.7263	0.76545	1437.4	1614.5	1.6993
1400	1.36797	1534.2	1736.7	1.8170	1.09101	1531.8	1733.7	1.7911	0.86944	1528.7	1729.8	1.7649
1600	1.52283	1626.5	1851.9	1.8759	1.21610	1624.6	1849.6	1.8504	0.97072	1622.2	1846.7	1.8246
1800	1.67606	1721.9	1970.0	1.9306	1.33956	1720.3	1968.2	1.9053	1.07036	1718.4	1966.0	1.8799
2000	1.82823	1820.4	2091.0	1.9819	1.46194	1819.1	2089.6	1.9568	1.16892	1817.5	2087.9	1.9315

TABLE A–6E

Superheated water (*Concluded*)

T °F	*v* ft³/lbm	*u* Btu/lbm	*h* Btu/lbm	*s* Btu/lbm·R	*v* ft³/lbm	*u* Btu/lbm	*h* Btu/lbm	*s* Btu/lbm·R	*v* ft³/lbm	*u* Btu/lbm	*h* Btu/lbm	*s* Btu/lbm·R
	P = 1500 psia (596.26°F)				*P* = 1750 psia (617.17°F)				*P* = 2000 psia (635.85°F)			
Sat.	0.27695	1092.1	1169.0	1.3362	0.22681	1080.5	1153.9	1.3112	0.18815	1066.8	1136.4	1.2863
600	0.28189	1097.2	1175.4	1.3423								
650	0.33310	1147.2	1239.7	1.4016	0.26292	1122.8	1207.9	1.3607	0.20586	1091.4	1167.6	1.3146
700	0.37198	1183.6	1286.9	1.4433	0.30252	1166.8	1264.7	1.4108	0.24894	1147.6	1239.8	1.3783
750	0.40535	1214.4	1326.9	1.4771	0.33455	1201.5	1309.8	1.4489	0.28074	1187.4	1291.3	1.4218
800	0.43550	1242.2	1363.1	1.5064	0.36266	1231.7	1349.1	1.4807	0.30763	1220.5	1334.3	1.4567
850	0.46356	1268.2	1396.9	1.5328	0.38835	1259.3	1385.1	1.5088	0.33169	1250.0	1372.8	1.4867
900	0.49015	1293.1	1429.2	1.5569	0.41238	1285.4	1419.0	1.5341	0.35390	1277.5	1408.5	1.5134
1000	0.54031	1340.9	1490.8	1.6007	0.45719	1334.9	1482.9	1.5796	0.39479	1328.7	1474.9	1.5606
1100	0.58781	1387.3	1550.5	1.6402	0.49917	1382.4	1544.1	1.6201	0.43266	1377.5	1537.6	1.6021
1200	0.63355	1433.3	1609.2	1.6767	0.53932	1429.2	1603.9	1.6572	0.46864	1425.1	1598.5	1.6400
1400	0.72172	1525.7	1726.0	1.7432	0.61621	1522.6	1722.1	1.7245	0.53708	1519.5	1718.3	1.7081
1600	0.80714	1619.8	1843.8	1.8033	0.69031	1617.4	1840.9	1.7852	0.60269	1615.0	1838.0	1.7693
1800	0.89090	1716.4	1963.7	1.8589	0.76273	1714.5	1961.5	1.8410	0.66660	1712.5	1959.2	1.8255
2000	0.97358	1815.9	2086.1	1.9108	0.83406	1814.2	2084.3	1.8931	0.72942	1812.6	2082.6	1.8778
	P = 2500 psia (668.17°F)				*P* = 3000 psia (695.41°F)				*P* = 3500 psia			
Sat.	0.13076	1031.2	1091.7	1.2330	0.08460	969.8	1016.8	1.1587				
650									0.02492	663.7	679.9	0.8632
700	0.16849	1098.4	1176.3	1.3072	0.09838	1005.3	1059.9	1.1960	0.03065	760.0	779.9	0.9511
750	0.20327	1154.9	1249.0	1.3686	0.14840	1114.1	1196.5	1.3118	0.10460	1057.6	1125.4	1.2434
800	0.22949	1195.9	1302.0	1.4116	0.17601	1167.5	1265.3	1.3676	0.13639	1134.3	1222.6	1.3224
850	0.25174	1230.1	1346.6	1.4463	0.19771	1208.2	1317.9	1.4086	0.15847	1183.8	1286.5	1.3721
900	0.27165	1260.7	1386.4	1.4761	0.21640	1242.8	1362.9	1.4423	0.17659	1223.4	1337.8	1.4106
950	0.29001	1289.1	1423.3	1.5028	0.23321	1273.9	1403.3	1.4716	0.19245	1257.8	1382.4	1.4428
1000	0.30726	1316.1	1458.2	1.5271	0.24876	1302.8	1440.9	1.4978	0.20687	1289.0	1423.0	1.4711
1100	0.33949	1367.3	1524.4	1.5710	0.27732	1356.8	1510.8	1.5441	0.23289	1346.1	1496.9	1.5201
1200	0.36966	1416.6	1587.6	1.6103	0.30367	1408.0	1576.6	1.5850	0.25654	1399.3	1565.4	1.5627
1400	0.42631	1513.3	1710.5	1.6802	0.35249	1507.0	1702.7	1.6567	0.29978	1500.7	1694.8	1.6364
1600	0.48004	1610.1	1832.2	1.7424	0.39830	1605.3	1826.4	1.7199	0.33994	1600.4	1820.5	1.7006
1800	0.53205	1708.6	1954.8	1.7991	0.44237	1704.7	1950.3	1.7773	0.37833	1700.8	1945.8	1.7586
2000	0.58295	1809.4	2079.1	1.8518	0.48532	1806.1	2075.6	1.8304	0.41561	1802.9	2072.1	1.8121
	P = 4000 psia				*P* = 5000 psia				*P* = 6000 psia			
650	0.02448	657.9	676.1	0.8577	0.02379	648.3	670.3	0.8485	0.02325	640.3	666.1	0.8408
700	0.02871	742.3	763.6	0.9347	0.02678	721.8	746.6	0.9156	0.02564	708.1	736.5	0.9028
750	0.06370	962.1	1009.2	1.1410	0.03373	821.8	853.0	1.0054	0.02981	788.7	821.8	0.9747
800	0.10520	1094.2	1172.1	1.2734	0.05937	986.9	1041.8	1.1581	0.03949	897.1	941.0	1.0711
850	0.12848	1156.7	1251.8	1.3355	0.08551	1092.4	1171.5	1.2593	0.05815	1018.6	1083.1	1.1819
900	0.14647	1202.5	1310.9	1.3799	0.10390	1155.9	1252.1	1.3198	0.07584	1103.5	1187.7	1.2603
950	0.16176	1240.7	1360.5	1.4157	0.11863	1203.9	1313.6	1.3643	0.09010	1163.7	1263.7	1.3153
1000	0.17538	1274.6	1404.4	1.4463	0.13128	1244.0	1365.5	1.4004	0.10208	1211.4	1324.7	1.3578
1100	0.19957	1335.1	1482.8	1.4983	0.15298	1312.2	1453.8	1.4590	0.12211	1288.4	1424.0	1.4237
1200	0.22121	1390.3	1554.1	1.5426	0.17185	1372.1	1531.1	1.5070	0.13911	1353.4	1507.8	1.4758
1300	0.24128	1443.0	1621.6	1.5821	0.18902	1427.8	1602.7	1.5490	0.15434	1412.5	1583.8	1.5203
1400	0.26028	1494.3	1687.0	1.6182	0.20508	1481.4	1671.1	1.5868	0.16841	1468.4	1655.4	1.5598
1600	0.29620	1595.5	1814.7	1.6835	0.23505	1585.6	1803.1	1.6542	0.19438	1575.7	1791.5	1.6294
1800	0.33033	1696.8	1941.4	1.7422	0.26320	1689.0	1932.5	1.7142	0.21853	1681.1	1923.7	1.6907
2000	0.36335	1799.7	2068.6	1.7961	0.29023	1793.2	2061.7	1.7689	0.24155	1786.7	2054.9	1.7463

TABLE A–7E

Compressed liquid water

T °F	*v* ft³/lbm	*u* Btu/lbm	*h* Btu/lbm	*s* Btu/lbm · R	*v* ft³/lbm	*u* Btu/lbm	*h* Btu/lbm	*s* Btu/lbm · R	*v* ft³/lbm	*u* Btu/lbm	*h* Btu/lbm	*s* Btu/lbm · R
	P = 500 psia (467.04°F)				*P* = 1000 psia (544.65°F)				*P* = 1500 psia (596.26°F)			
Sat.	0.019750	447.68	449.51	0.64900	0.021595	538.58	542.57	0.74341	0.023456	605.07	611.58	0.80836
32	0.015994	0.01	1.49	0.00001	0.015966	0.03	2.99	0.00005	0.015939	0.05	4.48	0.00008
50	0.015998	18.03	19.51	0.03601	0.015972	17.99	20.95	0.03593	0.015946	17.95	22.38	0.03584
100	0.016107	67.86	69.35	0.12930	0.016083	67.69	70.67	0.12899	0.016059	67.53	71.98	0.12869
150	0.016317	117.70	119.21	0.21462	0.016292	117.42	120.43	0.21416	0.016267	117.14	121.66	0.21369
200	0.016607	167.70	169.24	0.29349	0.016580	167.31	170.38	0.29289	0.016553	166.92	171.52	0.29229
250	0.016972	218.04	219.61	0.36708	0.016941	217.51	220.65	0.36634	0.016911	217.00	221.69	0.36560
300	0.017417	268.92	270.53	0.43641	0.017380	268.24	271.46	0.43551	0.017345	267.57	272.39	0.43463
350	0.017954	320.64	322.30	0.50240	0.017910	319.77	323.08	0.50132	0.017866	318.91	323.87	0.50025
400	0.018609	373.61	375.33	0.56595	0.018552	372.48	375.91	0.56463	0.018496	371.37	376.51	0.56333
450	0.019425	428.44	430.24	0.62802	0.019347	426.93	430.51	0.62635	0.019271	425.47	430.82	0.62472
500					0.020368	484.03	487.80	0.68764	0.020258	482.01	487.63	0.68550
550									0.021595	542.50	548.50	0.74731
	P = 2000 psia (635.85°F)				*P* = 3000 psia (695.41°F)				*P* = 5000 psia			
Sat.	0.025634	662.33	671.82	0.86224	0.034335	783.39	802.45	0.97321				
32	0.015912	0.07	5.96	0.00010	0.015859	0.10	8.90	0.00011	0.015756	0.13	14.71	0.00002
50	0.015921	17.91	23.80	0.03574	0.015870	17.83	26.64	0.03554	0.015773	17.65	32.25	0.03505
100	0.016035	67.36	73.30	0.12838	0.015988	67.04	75.91	0.12776	0.015897	66.41	81.12	0.12652
200	0.016527	166.54	172.66	0.29170	0.016475	165.79	174.94	0.29053	0.016375	164.36	179.51	0.28824
300	0.017310	266.92	273.33	0.43376	0.017242	265.65	275.22	0.43204	0.017112	263.24	279.07	0.42874
400	0.018442	370.30	377.12	0.56205	0.018338	368.22	378.41	0.55959	0.018145	364.35	381.14	0.55492
450	0.019199	424.06	431.16	0.62314	0.019062	421.36	431.94	0.62010	0.018812	416.40	433.80	0.61445
500	0.020154	480.08	487.54	0.68346	0.019960	476.45	487.53	0.67958	0.019620	469.94	488.10	0.67254
560	0.021739	552.21	560.26	0.75692	0.021405	546.59	558.47	0.75126	0.020862	537.08	556.38	0.74154
600	0.023317	605.77	614.40	0.80898	0.022759	597.42	610.06	0.80086	0.021943	584.42	604.72	0.78803
640					0.024765	654.52	668.27	0.85476	0.023358	634.95	656.56	0.83603
680					0.028821	728.63	744.64	0.92288	0.025366	690.67	714.14	0.88745
700									0.026777	721.78	746.56	0.91564

TABLE A–8E

Saturated ice—water vapor

Temp., T °F	Sat. press., P_{sat} psia	Specific volume, ft³/lbm: Sat. ice, v_i	Sat. vapor, v_g	Internal energy, Btu/lbm: Sat. ice, u_i	Subl., u_{ig}	Sat. vapor, u_g	Enthalpy, Btu/lbm: Sat. ice, h_i	Subl., h_{ig}	Sat. vapor, h_g	Entropy, Btu/lbm · R: Sat. ice, s_i	Subl., s_{ig}	Sat. vapor, s_g
32.018	0.08871	0.01747	3299.6	−143.34	1164.2	1020.9	−143.34	1218.3	1075.0	−0.29146	2.4779	2.1864
32	0.08864	0.01747	3302.6	−143.35	1164.2	1020.9	−143.35	1218.4	1075.0	−0.29148	2.4779	2.1865
30	0.08086	0.01747	3605.8	144.35	1164.6	1020.2	−144.35	1218.5	1074.2	−0.29353	2.4883	2.1948
25	0.06405	0.01746	4505.8	−146.85	1165.4	1018.6	−146.85	1218.8	1072.0	−0.29865	2.5146	2.2160
20	0.05049	0.01746	5657.6	−149.32	1166.2	1016.9	−149.32	1219.1	1069.8	−0.30377	2.5414	2.2376
15	0.03960	0.01745	7138.9	−151.76	1167.0	1015.2	−151.76	1219.3	1067.6	−0.30889	2.5687	2.2598
10	0.03089	0.01744	9054.0	−154.18	1167.8	1013.6	−154.18	1219.5	1065.4	−0.31401	2.5965	2.2825
5	0.02397	0.01743	11,543	−156.57	1168.5	1011.9	−156.57	1219.7	1063.1	−0.31913	2.6248	2.3057
0	0.01850	0.01743	14,797	−158.94	1169.2	1010.3	−158.94	1219.9	1060.9	−0.32426	2.6537	2.3295
−5	0.01420	0.01742	19,075	−161.28	1169.9	1008.6	−161.28	1220.0	1058.7	−0.32938	2.6832	2.3538
−10	0.01083	0.01741	24,731	−163.60	1170.6	1007.0	−163.60	1220.1	1056.5	−0.33451	2.7133	2.3788
−15	0.00821	0.01740	32,257	−165.90	1171.2	1005.3	−165.90	1220.2	1054.3	−0.33964	2.7440	2.4044
−20	0.00619	0.01740	42,335	−168.16	1171.8	1003.6	−168.16	1220.3	1052.1	−0.34478	2.7754	2.4306
−25	0.00463	0.01739	55,917	−170.41	1172.4	1002.0	−170.41	1220.3	1049.9	−0.34991	2.8074	2.4575
−30	0.00344	0.01738	74,345	−172.63	1173.0	1000.3	−172.63	1220.3	1047.7	−0.35505	2.8401	2.4850
−35	0.00254	0.01738	99,526	−174.83	1173.5	998.7	−174.83	1220.3	1045.5	−0.36019	2.8735	2.5133
−40	0.00186	0.01737	134,182	−177.00	1174.0	997.0	−177.00	1220.3	1043.3	−0.36534	2.9076	2.5423

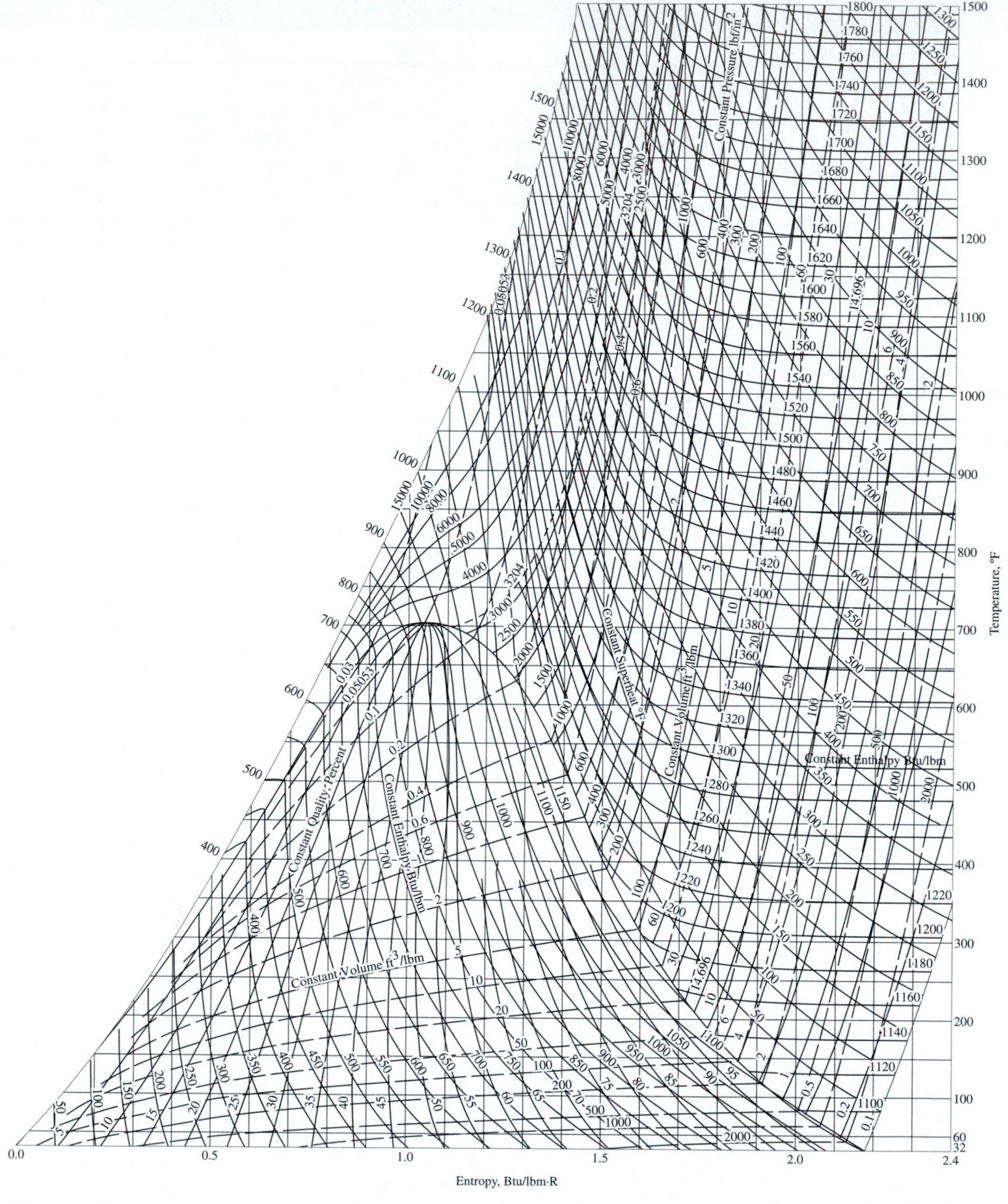

FIGURE A–9E

T-s diagram for water.

Source: Joseph H. Keenan, Frederick G. Keyes, Philip G. Hill, and Joan G. Moore, Steam Tables *(New York: John Wiley & Sons, 1969).*

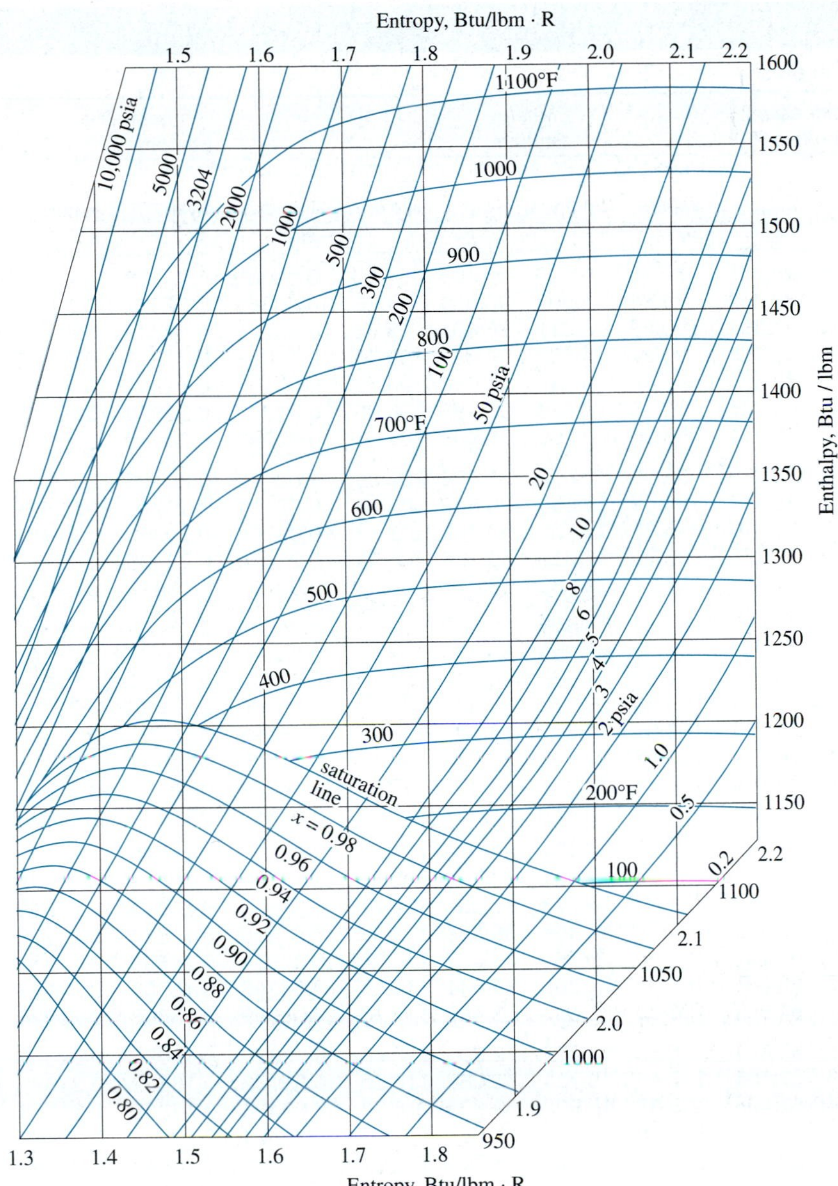

FIGURE A–10E

Mollier diagram for water.

Source: Joseph H. Keenan, Frederick G. Keyes, Philip G. Hill, and Joan G. Moore, Steam Tables *(New York: John Wiley & Sons, 1969).*

TABLE A–11E

Saturated refrigerant-134a—Temperature table

		Specific volume, ft³/lbm		Internal energy, Btu/lbm			Enthalpy, Btu/lbm			Entropy, Btu/lbm · R		
Temp., T °F	Sat. press., P_{sat} psia	Sat. liquid, v_f	Sat. vapor, v_g	Sat. liquid, u_f	Evap., u_{fg}	Sat. vapor, u_g	Sat. liquid, h_f	Evap., h_{fg}	Sat. vapor, h_g	Sat. liquid, s_f	Evap., s_{fg}	Sat. vapor, s_g
−40	7.432	0.01130	5.7796	−0.016	89.167	89.15	0.000	97.100	97.10	0.00000	0.23135	0.23135
−35	8.581	0.01136	5.0509	1.484	88.352	89.84	1.502	96.354	97.86	0.00355	0.22687	0.23043
−30	9.869	0.01143	4.4300	2.990	87.532	90.52	3.011	95.601	98.61	0.00708	0.22248	0.22956
−25	11.306	0.01150	3.8988	4.502	86.706	91.21	4.526	94.839	99.36	0.01058	0.21817	0.22875
−20	12.906	0.01156	3.4426	6.019	85.874	91.89	6.047	94.068	100.12	0.01405	0.21394	0.22798
−15	14.680	0.01163	3.0494	7.543	85.036	92.58	7.574	93.288	100.86	0.01749	0.20978	0.22727
−10	16.642	0.01171	2.7091	9.073	84.191	93.26	9.109	92.498	101.61	0.02092	0.20569	0.22660
−5	18.806	0.01178	2.4137	10.609	83.339	93.95	10.650	91.698	102.35	0.02431	0.20166	0.22598
0	21.185	0.01185	2.1564	12.152	82.479	94.63	12.199	90.886	103.08	0.02769	0.19770	0.22539
5	23.793	0.01193	1.9316	13.702	81.610	95.31	13.755	90.062	103.82	0.03104	0.19380	0.22485
10	26.646	0.01201	1.7345	15.259	80.733	95.99	15.318	89.226	104.54	0.03438	0.18996	0.22434
15	29.759	0.01209	1.5612	16.823	79.846	96.67	16.889	88.377	105.27	0.03769	0.18617	0.22386
20	33.147	0.01217	1.4084	18.394	78.950	97.34	18.469	87.514	105.98	0.04098	0.18243	0.22341
25	36.826	0.01225	1.2732	19.973	78.043	98.02	20.056	86.636	106.69	0.04426	0.17874	0.22300
30	40.813	0.01234	1.1534	21.560	77.124	98.68	21.653	85.742	107.40	0.04752	0.17509	0.22260
35	45.124	0.01242	1.0470	23.154	76.195	99.35	23.258	84.833	108.09	0.05076	0.17148	0.22224
40	49.776	0.01251	0.95205	24.757	75.253	100.01	24.873	83.907	108.78	0.05398	0.16791	0.22189
45	54.787	0.01261	0.86727	26.369	74.298	100.67	26.497	82.963	109.46	0.05720	0.16437	0.22157
50	60.175	0.01270	0.79136	27.990	73.329	101.32	28.131	82.000	110.13	0.06039	0.16087	0.22127
55	65.957	0.01280	0.72323	29.619	72.346	101.97	29.775	81.017	110.79	0.06358	0.15740	0.22098
60	72.152	0.01290	0.66195	31.258	71.347	102.61	31.431	80.013	111.44	0.06675	0.15396	0.22070
65	78.780	0.01301	0.60671	32.908	70.333	103.24	33.097	78.988	112.09	0.06991	0.15053	0.22044
70	85.858	0.01312	0.55681	34.567	69.301	103.87	34.776	77.939	112.71	0.07306	0.14713	0.22019
75	93.408	0.01323	0.51165	36.237	68.251	104.49	36.466	76.866	113.33	0.07620	0.14375	0.21995
80	101.45	0.01334	0.47069	37.919	67.181	105.10	38.169	75.767	113.94	0.07934	0.14038	0.21972
85	110.00	0.01347	0.43348	39.612	66.091	105.70	39.886	74.641	114.53	0.08246	0.13703	0.21949
90	119.08	0.01359	0.39959	41.317	64.979	106.30	41.617	73.485	115.10	0.08559	0.13368	0.21926
95	128.72	0.01372	0.36869	43.036	63.844	106.88	43.363	72.299	115.66	0.08870	0.13033	0.21904
100	138.93	0.01386	0.34045	44.768	62.683	107.45	45.124	71.080	116.20	0.09182	0.12699	0.21881
105	149.73	0.01400	0.31460	46.514	61.496	108.01	46.902	69.825	116.73	0.09493	0.12365	0.21858
110	161.16	0.01415	0.29090	48.276	60.279	108.56	48.698	68.533	117.23	0.09804	0.12029	0.21834
115	173.23	0.01430	0.26913	50.054	59.031	109.08	50.512	67.200	117.71	0.10116	0.11693	0.21809
120	185.96	0.01446	0.24909	51.849	57.749	109.60	52.346	65.823	118.17	0.10428	0.11354	0.21782
130	213.53	0.01482	0.21356	55.495	55.071	110.57	56.080	62.924	119.00	0.11054	0.10670	0.21724
140	244.06	0.01521	0.18315	59.226	52.216	111.44	59.913	59.801	119.71	0.11684	0.09971	0.21655
150	277.79	0.01567	0.15692	63.059	49.144	112.20	63.864	56.405	120.27	0.12321	0.09251	0.21572
160	314.94	0.01619	0.13410	67.014	45.799	112.81	67.958	52.671	120.63	0.12970	0.08499	0.21469
170	355.80	0.01681	0.11405	71.126	42.097	113.22	72.233	48.499	120.73	0.13634	0.07701	0.21335
180	400.66	0.01759	0.09618	75.448	37.899	113.35	76.752	43.726	120.48	0.14323	0.06835	0.21158
190	449.90	0.01860	0.07990	80.082	32.950	113.03	81.631	38.053	119.68	0.15055	0.05857	0.20911
200	504.00	0.02009	0.06441	85.267	26.651	111.92	87.140	30.785	117.93	0.15867	0.04666	0.20533
210	563.76	0.02309	0.04722	91.986	16.498	108.48	94.395	19.015	113.41	0.16922	0.02839	0.19761

Source: Tables A–11E through A–13E are generated using the Engineering Equation Solver (EES) software developed by S. A. Klein and F. L. Alvarado. The routine used in calculations is the R134a, which is based on the fundamental equation of state developed by R. Tillner-Roth and H.D. Baehr, "An International Standard Formulation for the Thermodynamic Properties of 1,1,1,2-Tetrafluoroethane (HFC-134a) for Temperatures from 170 K to 455 K and Pressures up to 70 MPa," *J. Phys. Chem, Ref. Data,* Vol. 23, No. 5, 1994. The enthalpy and entropy values of saturated liquid are set to zero at −40°C (and −40°F).

TABLE A–12E

Saturated refrigerant-134a—Pressure table

		Specific volume, ft³/lbm		Internal energy, Btu/lbm			Enthalpy, Btu/lbm			Entropy, Btu/lbm · R		
Press., P psia	Sat. temp., T °F	Sat. liquid, v_f	Sat. vapor, v_g	Sat. liquid, u_f	Evap., u_{fg}	Sat. vapor, u_g	Sat. liquid, h_f	Evap., h_{fg}	Sat. vapor, h_g	Sat. liquid, s_f	Evap., s_{fg}	Sat. vapor, s_g
5	−53.09	0.01113	8.3785	−3.918	91.280	87.36	−3.907	99.022	95.11	−0.00945	0.24353	0.23408
10	−29.52	0.01144	4.3753	3.135	87.453	90.59	3.156	95.528	98.68	0.00742	0.22206	0.22948
15	−14.15	0.01165	2.9880	7.803	84.893	92.70	7.835	93.155	100.99	0.01808	0.20908	0.22715
20	−2.43	0.01182	2.2772	11.401	82.898	94.30	11.445	91.282	102.73	0.02605	0.19962	0.22567
25	7.17	0.01196	1.8429	14.377	81.231	95.61	14.432	89.701	104.13	0.03249	0.19213	0.22462
30	15.37	0.01209	1.5492	16.939	79.780	96.72	17.006	88.313	105.32	0.03793	0.18589	0.22383
35	22.57	0.01221	1.3369	19.205	78.485	97.69	19.284	87.064	106.35	0.04267	0.18053	0.22319
40	29.01	0.01232	1.1760	21.246	77.307	98.55	21.337	85.920	107.26	0.04688	0.17580	0.22268
45	34.86	0.01242	1.0497	23.110	76.221	99.33	23.214	84.858	108.07	0.05067	0.17158	0.22225
50	40.23	0.01252	0.94791	24.832	75.209	100.04	24.948	83.863	108.81	0.05413	0.16774	0.22188
55	45.20	0.01261	0.86400	26.435	74.258	100.69	26.564	82.924	109.49	0.05733	0.16423	0.22156
60	49.84	0.01270	0.79361	27.939	73.360	101.30	28.080	82.030	110.11	0.06029	0.16098	0.22127
65	54.20	0.01279	0.73370	29.357	72.505	101.86	29.510	81.176	110.69	0.06307	0.15796	0.22102
70	58.30	0.01287	0.68205	30.700	71.688	102.39	30.867	80.357	111.22	0.06567	0.15512	0.22080
75	62.19	0.01295	0.63706	31.979	70.905	102.88	32.159	79.567	111.73	0.06813	0.15245	0.22059
80	65.89	0.01303	0.59750	33.201	70.151	103.35	33.394	78.804	112.20	0.07047	0.14993	0.22040
85	69.41	0.01310	0.56244	34.371	69.424	103.79	34.577	78.064	112.64	0.07269	0.14753	0.22022
90	72.78	0.01318	0.53113	35.495	68.719	104.21	35.715	77.345	113.06	0.07481	0.14525	0.22006
95	76.02	0.01325	0.50301	36.578	68.035	104.61	36.811	76.645	113.46	0.07684	0.14307	0.21991
100	79.12	0.01332	0.47760	37.623	67.371	104.99	37.869	75.962	113.83	0.07879	0.14097	0.21976
110	85.00	0.01347	0.43347	39.612	66.091	105.70	39.886	74.641	114.53	0.08246	0.13703	0.21949
120	90.49	0.01360	0.39644	41.485	64.869	106.35	41.787	73.371	115.16	0.08589	0.13335	0.21924
130	95.64	0.01374	0.36491	43.258	63.696	106.95	43.589	72.144	115.73	0.08911	0.12990	0.21901
140	100.51	0.01387	0.33771	44.945	62.564	107.51	45.304	70.954	116.26	0.09214	0.12665	0.21879
150	105.12	0.01400	0.31401	46.556	61.467	108.02	46.945	69.795	116.74	0.09501	0.12357	0.21857
160	109.50	0.01413	0.29316	48.101	60.401	108.50	48.519	68.662	117.18	0.09774	0.12062	0.21836
170	113.69	0.01426	0.27466	49.586	59.362	108.95	50.035	67.553	117.59	0.10034	0.11781	0.21815
180	117.69	0.01439	0.25813	51.018	58.345	109.36	51.497	66.464	117.96	0.10284	0.11511	0.21795
190	121.53	0.01452	0.24327	52.402	57.349	109.75	52.912	65.392	118.30	0.10524	0.11250	0.21774
200	125.22	0.01464	0.22983	53.743	56.371	110.11	54.285	64.335	118.62	0.10754	0.10998	0.21753
220	132.21	0.01490	0.20645	56.310	54.458	110.77	56.917	62.256	119.17	0.11192	0.10517	0.21710
240	138.73	0.01516	0.18677	58.746	52.591	111.34	59.419	60.213	119.63	0.11603	0.10061	0.21665
260	144.85	0.01543	0.16996	61.071	50.757	111.83	61.813	58.192	120.00	0.11992	0.09625	0.21617
280	150.62	0.01570	0.15541	63.301	48.945	112.25	64.115	56.184	120.30	0.12362	0.09205	0.21567
300	156.09	0.01598	0.14266	65.452	47.143	112.60	66.339	54.176	120.52	0.12715	0.08797	0.21512
350	168.64	0.01672	0.11664	70.554	42.627	113.18	71.638	49.099	120.74	0.13542	0.07814	0.21356
400	179.86	0.01757	0.09642	75.385	37.963	113.35	76.686	43.798	120.48	0.14314	0.06848	0.21161
450	190.02	0.01860	0.07987	80.092	32.939	113.03	81.641	38.041	119.68	0.15056	0.05854	0.20911
500	199.29	0.01995	0.06551	84.871	27.168	112.04	86.718	31.382	118.10	0.15805	0.04762	0.20566

TABLE A–13E

Superheated refrigerant-134a

T °F	v ft³/lbm	u Btu/lbm	h Btu/lbm	s Btu/lbm · R	v ft³/lbm	u Btu/lbm	h Btu/lbm	s Btu/lbm · R	v ft³/lbm	u Btu/lbm	h Btu/lbm	s Btu/lbm · R
	P = 10 psia (T_{sat} = −29.52°F)				P = 15 psia (T_{sat} = −14.15°F)				P = 20 psia (T_{sat} = −2.43°F)			
Sat.	4.3753	90.59	98.68	0.22948	2.9880	92.70	100.99	0.22715	2.2772	94.30	102.73	0.22567
−20	4.4856	92.13	100.43	0.23350								
0	4.7135	95.41	104.14	0.24174	3.1001	95.08	103.68	0.23310	2.2922	94.72	103.20	0.22671
20	4.9380	98.77	107.91	0.24976	3.2551	98.48	107.52	0.24127	2.4130	98.19	107.12	0.23504
40	5.1600	102.20	111.75	0.25761	3.4074	101.95	111.41	0.24922	2.5306	101.70	111.07	0.24311
60	5.3802	105.72	115.67	0.26531	3.5577	105.50	115.38	0.25700	2.6461	105.28	115.07	0.25097
80	5.5989	109.32	119.68	0.27288	3.7064	109.13	119.42	0.26463	2.7600	108.93	119.15	0.25866
100	5.8165	113.01	123.78	0.28033	3.8540	112.84	123.54	0.27212	2.8726	112.66	123.29	0.26621
120	6.0331	116.79	127.96	0.28767	4.0006	116.63	127.74	0.27950	2.9842	116.47	127.52	0.27363
140	6.2490	120.66	132.22	0.29490	4.1464	120.51	132.02	0.28677	3.0950	120.37	131.82	0.28093
160	6.4642	124.61	136.57	0.30203	4.2915	124.48	136.39	0.29393	3.2051	124.35	136.21	0.28812
180	6.6789	128.65	141.01	0.30908	4.4361	128.53	140.84	0.30100	3.3146	128.41	140.67	0.29521
200	6.8930	132.77	145.53	0.31604	4.5802	132.66	145.37	0.30798	3.4237	132.55	145.22	0.30221
220	7.1068	136.98	150.13	0.32292	4.7239	136.88	149.99	0.31487	3.5324	136.78	149.85	0.30912
	P = 30 psia (T_{sat} = 15.37°F)				P = 40 psia (T_{sat} = 29.01°F)				P = 50 psia (T_{sat} = 40.23°F)			
Sat.	1.5492	96.72	105.32	0.22383	1.1760	98.55	107.26	0.22268	0.9479	100.04	108.81	0.22188
20	1.5691	97.56	106.27	0.22581								
40	1.6528	101.17	110.35	0.23414	1.2126	100.61	109.58	0.22738				
60	1.7338	104.82	114.45	0.24219	1.2768	104.34	113.79	0.23565	1.0019	103.84	113.11	0.23031
80	1.8130	108.53	118.59	0.25002	1.3389	108.11	118.02	0.24363	1.0540	107.68	117.43	0.23847
100	1.8908	112.30	122.80	0.25767	1.3995	111.93	122.29	0.25140	1.1043	111.55	121.77	0.24637
120	1.9675	116.15	127.07	0.26517	1.4588	115.82	126.62	0.25900	1.1534	115.48	126.16	0.25406
140	2.0434	120.08	131.42	0.27254	1.5173	119.78	131.01	0.26644	1.2015	119.47	130.59	0.26159
160	2.1185	124.08	135.84	0.27979	1.5750	123.81	135.47	0.27375	1.2488	123.53	135.09	0.26896
180	2.1931	128.16	140.34	0.28693	1.6321	127.91	140.00	0.28095	1.2955	127.66	139.65	0.27621
200	2.2671	132.32	144.91	0.29398	1.6887	132.10	144.60	0.28803	1.3416	131.87	144.28	0.28333
220	2.3408	136.57	149.56	0.30092	1.7449	136.36	149.27	0.29501	1.3873	136.15	148.98	0.29036
240	2.4141	140.89	154.29	0.30778	1.8007	140.70	154.03	0.30190	1.4326	140.50	153.76	0.29728
260	2.4871	145.30	159.10	0.31456	1.8562	145.12	158.86	0.30871	1.4776	144.93	158.60	0.30411
280	2.5598	149.78	163.99	0.32126	1.9114	149.61	163.76	0.31543	1.5223	149.44	163.53	0.31086
	P = 60 psia (T_{sat} = 49.84°F)				P = 70 psia (T_{sat} = 58.30°F)				P = 80 psia (T_{sat} = 65.89°F)			
Sat.	0.7936	101.30	110.11	0.22127	0.6821	102.39	111.22	0.22080	0.59750	103.35	112.20	0.22040
60	0.8179	103.31	112.39	0.22570	0.6857	102.73	111.62	0.22155				
80	0.8636	107.23	116.82	0.23407	0.7271	106.76	116.18	0.23016	0.62430	106.26	115.51	0.22661
100	0.9072	111.16	121.24	0.24211	0.7662	110.76	120.68	0.23836	0.66009	110.34	120.11	0.23499
120	0.9495	115.14	125.68	0.24991	0.8037	114.78	125.19	0.24628	0.69415	114.42	124.69	0.24304
140	0.9908	119.16	130.16	0.25751	0.8401	118.85	129.73	0.25398	0.72698	118.52	129.29	0.25083
160	1.0312	123.25	134.70	0.26496	0.8756	122.97	134.31	0.26149	0.75888	122.68	133.91	0.25841
180	1.0709	127.41	139.30	0.27226	0.9105	127.15	138.94	0.26885	0.79003	126.89	138.58	0.26583
200	1.1101	131.63	143.96	0.27943	0.9447	131.40	143.63	0.27607	0.82059	131.16	143.31	0.27310
220	1.1489	135.93	148.69	0.28649	0.9785	135.71	148.39	0.28317	0.85065	135.49	148.09	0.28024
240	1.1872	140.30	153.48	0.29344	1.0118	140.10	153.21	0.29015	0.88030	139.90	152.93	0.28726
260	1.2252	144.75	158.35	0.30030	1.0449	144.56	158.10	0.29704	0.90961	144.37	157.84	0.29418
280	1.2629	149.27	163.29	0.30707	1.0776	149.10	163.06	0.30384	0.93861	148.92	162.82	0.30100
300	1.3004	153.87	168.31	0.31376	1.1101	153.71	168.09	0.31055	0.96737	153.54	167.86	0.30773
320	1.3377	158.54	173.39	0.32037	1.1424	158.39	173.19	0.31718	0.99590	158.24	172.98	0.31438

(Continued)

TABLE A–13E

Superheated refrigerant-134a (*Concluded*)

T °F	*v* ft³/lbm	*u* Btu/lbm	*h* Btu/lbm	*s* Btu/lbm · R	*v* ft³/lbm	*u* Btu/lbm	*h* Btu/lbm	*s* Btu/lbm · R	*v* ft³/lbm	*u* Btu/lbm	*h* Btu/lbm	*s* Btu/lbm · R
	P = 90 psia (T_{sat} = 72.78°F)				P = 100 psia (T_{sat} = 79.12°F)				P = 120 psia (T_{sat} = 90.49°F)			
Sat.	0.53113	104.21	113.06	0.22006	0.47760	104.99	113.83	0.21976	0.39644	106.35	115.16	0.21924
80	0.54388	105.74	114.80	0.22330	0.47906	105.18	114.05	0.22016				
100	0.57729	109.91	119.52	0.23189	0.51076	109.45	118.90	0.22900	0.41013	108.48	117.59	0.22362
120	0.60874	114.04	124.18	0.24008	0.54022	113.66	123.65	0.23733	0.43692	112.84	122.54	0.23232
140	0.63885	118.19	128.83	0.24797	0.56821	117.86	128.37	0.24534	0.46190	117.15	127.41	0.24058
160	0.66796	122.38	133.51	0.25563	0.59513	122.08	133.09	0.25309	0.48563	121.46	132.25	0.24851
180	0.69629	126.62	138.22	0.26311	0.62122	126.35	137.85	0.26063	0.50844	125.79	137.09	0.25619
200	0.72399	130.92	142.97	0.27043	0.64667	130.67	142.64	0.26801	0.53054	130.17	141.95	0.26368
220	0.75119	135.27	147.78	0.27762	0.67158	135.05	147.47	0.27523	0.55206	134.59	146.85	0.27100
240	0.77796	139.69	152.65	0.28468	0.69605	139.49	152.37	0.28233	0.57312	139.07	151.80	0.27817
260	0.80437	144.19	157.58	0.29162	0.72016	143.99	157.32	0.28931	0.59379	143.61	156.79	0.28521
280	0.83048	148.75	162.58	0.29847	0.74396	148.57	162.34	0.29618	0.61413	148.21	161.85	0.29214
300	0.85633	153.38	167.64	0.30522	0.76749	153.21	167.42	0.30296	0.63420	152.88	166.96	0.29896
320	0.88195	158.08	172.77	0.31189	0.79079	157.93	172.56	0.30964	0.65402	157.62	172.14	0.30569
	P = 140 psia (T_{sat} = 100.50°F)				P = 160 psia (T_{sat} = 109.50°F)				P = 180 psia (T_{sat} = 117.69°F)			
Sat.	0.33771	107.51	116.26	0.21879	0.29316	108.50	117.18	0.21836	0.25813	109.36	117.96	0.21795
120	0.36243	111.96	121.35	0.22773	0.30578	111.01	120.06	0.22337	0.26083	109.94	118.63	0.21910
140	0.38551	116.41	126.40	0.23628	0.32774	115.62	125.32	0.23230	0.28231	114.77	124.17	0.22850
160	0.40711	120.81	131.36	0.24443	0.34790	120.13	130.43	0.24069	0.30154	119.42	129.46	0.23718
180	0.42766	125.22	136.30	0.25227	0.36686	124.62	135.49	0.24871	0.31936	124.00	134.64	0.24540
200	0.44743	129.65	141.24	0.25988	0.38494	129.12	140.52	0.25645	0.33619	128.57	139.77	0.25330
220	0.46657	134.12	146.21	0.26730	0.40234	133.64	145.55	0.26397	0.35228	133.15	144.88	0.26094
240	0.48522	138.64	151.21	0.27455	0.41921	138.20	150.62	0.27131	0.36779	137.76	150.01	0.26837
260	0.50345	143.21	156.26	0.28166	0.43564	142.81	155.71	0.27849	0.38284	142.40	155.16	0.27562
280	0.52134	147.85	161.35	0.28864	0.45171	147.48	160.85	0.28554	0.39751	147.10	160.34	0.28273
300	0.53895	152.54	166.50	0.29551	0.46748	152.20	166.04	0.29246	0.41186	151.85	165.57	0.28970
320	0.55630	157.30	171.71	0.30228	0.48299	156.98	171.28	0.29927	0.42594	156.66	170.85	0.29656
340	0.57345	162.13	176.98	0.30896	0.49828	161.83	176.58	0.30598	0.43980	161.53	176.18	0.30331
360	0.59041	167.02	182.32	0.31555	0.51338	166.74	181.94	0.31260	0.45347	166.46	181.56	0.30996
	P = 200 psia (T_{sat} = 125.22°F)				P = 300 psia (T_{sat} = 156.09°F)				P = 400 psia (T_{sat} = 179.86°F)			
Sat.	0.22983	110.11	118.62	0.21753	0.14266	112.60	120.52	0.21512	0.09642	113.35	120.48	0.21161
140	0.24541	113.85	122.93	0.22481								
160	0.26412	118.66	128.44	0.23384	0.14656	113.82	121.95	0.21745				
180	0.28115	123.35	133.76	0.24229	0.16355	119.52	128.60	0.22802	0.09658	113.41	120.56	0.21173
200	0.29704	128.00	138.99	0.25035	0.17776	124.78	134.65	0.23733	0.11440	120.52	128.99	0.22471
220	0.31212	132.64	144.19	0.25812	0.19044	129.85	140.42	0.24594	0.12746	126.44	135.88	0.23500
240	0.32658	137.30	149.38	0.26565	0.20211	134.83	146.05	0.25410	0.13853	131.95	142.20	0.24418
260	0.34054	141.99	154.59	0.27298	0.21306	139.77	151.59	0.26192	0.14844	137.26	148.25	0.25270
280	0.35410	146.72	159.82	0.28015	0.22347	144.70	157.11	0.26947	0.15756	142.48	154.14	0.26077
300	0.36733	151.50	165.09	0.28718	0.23346	149.65	162.61	0.27681	0.16611	147.65	159.94	0.26851
320	0.38029	156.33	170.40	0.29408	0.24310	154.63	168.12	0.28398	0.17423	152.80	165.70	0.27599
340	0.39300	161.22	175.77	0.30087	0.25246	159.64	173.66	0.29098	0.18201	157.97	171.44	0.28326
360	0.40552	166.17	181.18	0.30756	0.26159	164.70	179.22	0.29786	0.18951	163.15	177.18	0.29035

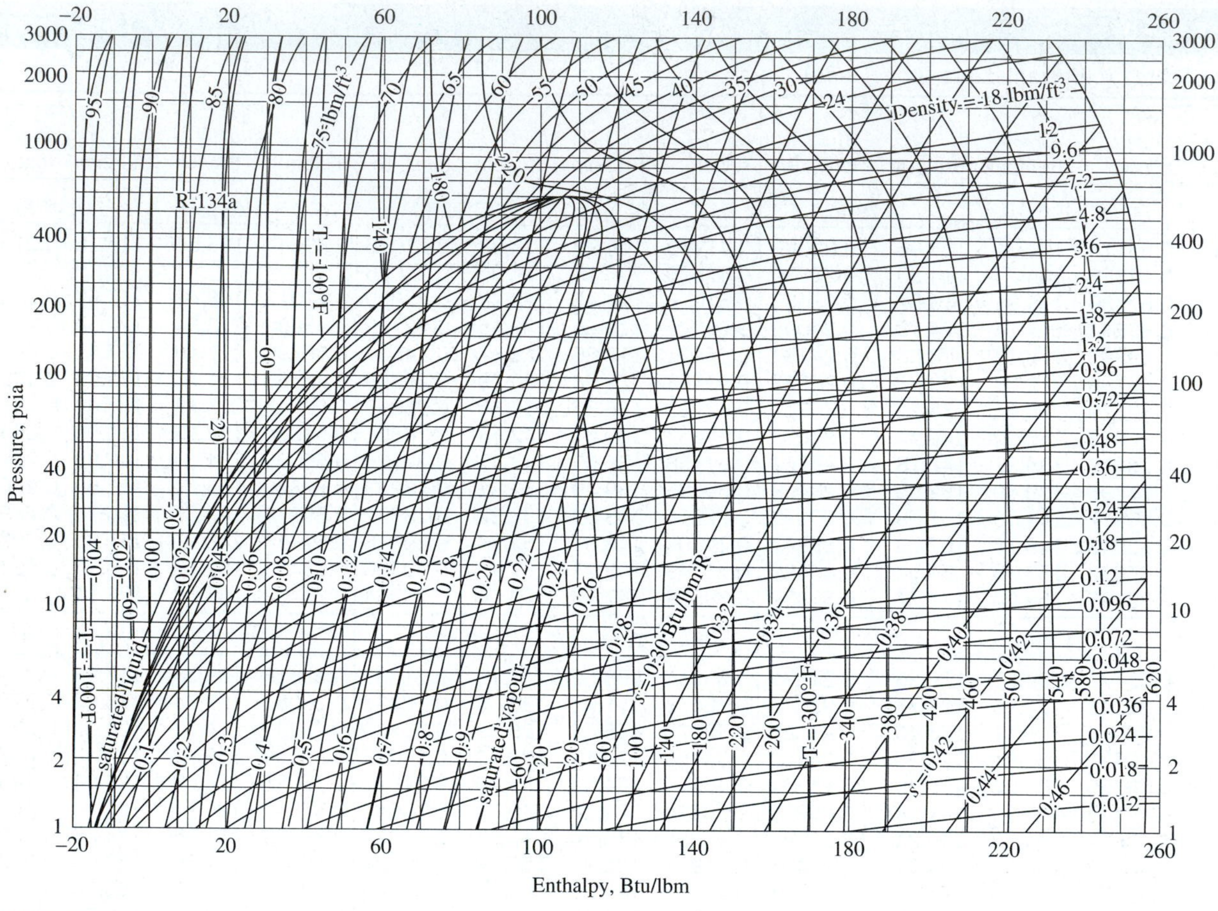

FIGURE A–14E

P-h diagram for refrigerant-134a.

Reprinted by permission of American Society of Heating, Refrigerating, and Air-Conditioning Engineers, Inc., Atlanta, GA.

TABLE A–15E

Properties of saturated water

Temp. T, °F	Saturation Pressure P_{sat}, psia	Density ρ, lbm/ft^3 Liquid	Vapor	Enthalpy of Vaporization h_{fg}, Btu/lbm	Specific Heat c_p, Btu/lbm · R Liquid	Vapor	Thermal Conductivity k, Btu/h · ft · R Liquid	Vapor	Dynamic Viscosity μ, lbm/ft · s Liquid	Vapor	Prandtl Number Pr Liquid	Vapor	Volume Expansion Coefficient β, 1/R Liquid
32.02	0.0887	62.41	0.00030	1075	1.010	0.446	0.324	0.0099	1.204×10^{-3}	6.194×10^{-6}	13.5	1.00	-0.038×10^{-3}
40	0.1217	62.42	0.00034	1071	1.004	0.447	0.329	0.0100	1.308×10^{-3}	6.278×10^{-6}	11.4	1.01	A0.003 $\times 10^{-3}$
50	0.1780	62.41	0.00059	1065	1.000	0.448	0.335	0.0102	8.781×10^{-4}	6.361×10^{-6}	9.44	1.01	0.047×10^{-3}
60	0.2563	62.36	0.00083	1060	0.999	0.449	0.341	0.0104	7.536×10^{-4}	6.444×10^{-6}	7.95	1.00	0.080×10^{-3}
70	0.3632	62.30	0.00115	1054	0.999	0.450	0.347	0.0106	6.556×10^{-4}	6.556×10^{-6}	6.79	1.00	0.115×10^{-3}
80	0.5073	62.22	0.00158	1048	0.999	0.451	0.352	0.0108	5.764×10^{-4}	6.667×10^{-6}	5.89	1.00	0.145×10^{-3}
90	0.6988	62.12	0.00214	1043	0.999	0.453	0.358	0.0110	5.117×10^{-4}	6.778×10^{-6}	5.14	1.00	0.174×10^{-3}
100	0.9503	62.00	0.00286	1037	0.999	0.454	0.363	0.0112	4.578×10^{-4}	6.889×10^{-6}	4.54	1.01	0.200×10^{-3}
110	1.2763	61.86	0.00377	1031	0.999	0.456	0.367	0.0115	4.128×10^{-4}	7.000×10^{-6}	4.05	1.00	0.224×10^{-3}
120	1.6945	61.71	0.00493	1026	0.999	0.458	0.371	0.0117	3.744×10^{-4}	7.111×10^{-6}	3.63	1.00	0246 $\times 10^{-3}$
130	2.225	61.55	0.00636	1020	0.999	0.460	0.375	0.0120	3.417×10^{-4}	7.222×10^{-6}	3.28	1.00	0.267×10^{-3}
140	2.892	61.38	0.00814	1014	0.999	0.463	0.378	0.0122	3.136×10^{-4}	7.333×10^{-6}	2.98	1.00	0.287×10^{-3}
150	3.722	61.19	0.0103	1008	1.000	0.465	0.381	0.0125	2.889×10^{-4}	7.472×10^{-6}	2.73	1.00	0.306×10^{-3}
160	4.745	60.99	0.0129	1002	1.000	0.468	0.384	0.0128	2.675×10^{-4}	7.583×10^{-6}	2.51	1.00	0.325×10^{-3}
170	5.996	60.79	0.0161	996	1.001	0.472	0.386	0.0131	2.483×10^{-4}	7.722×10^{-6}	2.90	1.00	0.346×10^{-3}
180	7.515	60.57	0.0199	990	1.002	0.475	0.388	0.0134	2.317×10^{-4}	7.833×10^{-6}	2.15	1.00	0.367×10^{-3}
190	9.343	60.35	0.0244	984	1.004	0.479	0.390	0.0137	2.169×10^{-4}	7.972×10^{-6}	2.01	1.00	0.382×10^{-3}
200	11.53	60.12	0.0297	978	1.005	0.483	0.391	0.0141	2.036×10^{-4}	8.083×10^{-6}	1.88	1.00	0.395×10^{-3}
210	14.125	59.87	0.0359	972	1.007	0.487	0.392	0.0144	1.917×10^{-4}	8.222×10^{-6}	1.77	1.00	0.412×10^{-3}
212	14.698	59.82	0.0373	970	1.007	0.488	0.392	0.0145	1.894×10^{-4}	8.250×10^{-6}	1.75	1.00	0.417×10^{-3}
220	17.19	59.62	0.0432	965	1.009	0.492	0.393	0.0148	1808×10^{-4}	8.333×10^{-6}	1.67	1.00	0.429×10^{-3}
230	20.78	59.36	0.0516	959	1.011	0.497	0.394	0.0152	1.711×10^{-4}	8.472×10^{-6}	1.58	1.00	0.443×10^{-3}
240	24.97	59.09	0.0612	952	1.013	0.503	0.394	0.0156	1.625×10^{-4}	8.611×10^{-6}	1.50	1.00	0.462×10^{-3}
250	29.82	58.82	0.0723	946	1.015	0.509	0.395	0.0160	1.544×10^{-4}	8.611×10^{-6}	1.43	1.00	0.480×10^{-3}
260	35.42	58.53	0.0850	939	1.018	0.516	0.395	0.0164	1.472×10^{-4}	8.861×10^{-6}	1.37	1.00	0.497×10^{-3}
270	41.85	58.24	0.0993	932	1.020	0.523	0.395	0.0168	1.406×10^{-4}	9.000×10^{-6}	1.31	1.01	0.514×10^{-3}
280	49.18	57.94	0.1156	926	1.023	0.530	0.395	0.0172	1.344×10^{-4}	9.111×10^{-6}	1.25	1.01	0.532×10^{-3}
290	57.53	57.63	0.1360	918	1.026	0.538	0.395	0.0177	1.289×10^{-4}	9.250×10^{-6}	1.21	1.01	0.549×10^{-3}
300	66.98	57.31	0.1545	910	1.029	0.547	0.394	0.0182	1.236×10^{-4}	9.389×10^{-6}	1.16	1.02	0.566×10^{-3}
320	89.60	56.65	0.2033	895	1.036	0.567	0.393	0.0191	1.144×10^{-4}	9.639×10^{-6}	1.09	1.03	0.636×10^{-3}
340	117.93	55.95	0.2637	880	1.044	0.590	0.391	0.0202	1.063×10^{-4}	9.889×10^{-6}	1.02	1.04	0.656×10^{-3}
360	152.92	56.22	0.3377	863	1.054	0.617	0.389	0.0213	9.972×10^{-5}	1.013×10^{-5}	0.973	1.06	0.681×10^{-3}
380	195.60	54.46	0.4275	845	1.065	0.647	0.385	0.0224	9.361×10^{-5}	1.041×10^{-5}	0.932	1.08	0.720×10^{-3}
400	241.1	53.65	0.5359	827	1.078	0.683	0.382	0.0237	8.833×10^{-5}	1.066×10^{-5}	0.893	1.11	0.771×10^{-3}
450	422.1	51.46	0.9082	775	1.121	0.799	0.370	0.0271	7.722×10^{-5}	1.130×10^{-5}	0.842	1.20	0.912×10^{-3}
500	680.0	48.95	1.479	715	1.188	0.972	0.352	0.0312	6.833×10^{-5}	1.200×10^{-5}	0.830	1.35	1.111×10^{-3}
550	1046.7	45.96	4.268	641	1.298	1.247	0.329	0.0368	6.083×10^{-5}	1.280×10^{-5}	0.864	1.56	1.445×10^{-3}
600	1541	42.32	3.736	550	1.509	1.759	0.299	0.0461	5.389×10^{-5}	1.380×10^{-5}	0.979	1.90	1.883×10^{-3}
650	2210	37.31	6.152	422	2.086	3.103	0.267	0.0677	4.639×10^{-5}	1.542×10^{-5}	1.30	2.54	
700	3090	27.28	13.44	168	13.80	25.90	0.254	0.1964	3.417×10^{-5}	2.044×10^{-5}	6.68	9.71	
705.44	3204	19.79	19.79	0	∞	∞	∞	∞	2.897×10^{-5}	2.897×10^{-5}			

Note 1: Kinematic viscosity ν and thermal diffusivity α can be calculated from their definitions, $\nu = \mu/\rho$ and $\alpha = k/\rho c_p = \nu/\text{Pr}$. The temperatures 32.02°F, 212°F, and 705.44°F are the triple-, boiling-, and critical-point temperatures of water, respectively. All properties listed above (except the vapor density) can be used at any pressures with negligible error except at temperatures near the critical-point value.

Note 2: The unit Btu/lbm·°F for specific heat is equivalent to Btu/lbm·R, and the unit Btu/h·ft·°F for thermal conductivity is equivalent to Btu/h·ft·R.

Source: Viscosity and thermal conductivity data are from J. V. Sengers and J. T. T. Watson, *Journal of Physical and Chemical Reference Data* 15 (1986), pp. 1291–1322. Other data are obtained from various sources or calculated.

TABLE A–16E

Properties of saturated refrigerant-134a

Temp.	Saturation Pressure	Density ρ, lbm/ft^3		Enthalpy of Vaporization	Specific Heat c_p, Btu/lbm·R		Thermal Conductivity k, Btu/h·ft·R		Dynamic Viscosity μ, lbm/ft·s		Prandtl Number Pr		Volume Expansion Coefficient β, 1/R	Surface Tension
T, °F	P_{sat}, psia	Liquid	Vapor	h_{fg}, Btu/lbm	Liquid	Vapor	Liquid	Vapor	Liquid	Vapor	Liquid	Vapor	Liquid	lbf/ft
−40	7.4	88.51	0.1731	97.1	0.2996	0.1788	0.0636	0.00466	3.278×10^{-4}	1.714×10^{-6}	5.558	0.237	0.00114	0.001206
−30	9.9	87.5	0.2258	95.6	0.3021	0.1829	0.0626	0.00497	3.004×10^{-4}	2.053×10^{-6}	5.226	0.272	0.00117	0.001146
−20	12.9	86.48	0.2905	94.1	0.3046	0.1872	0.0613	0.00529	2.762×10^{-4}	2.433×10^{-6}	4.937	0.310	0.00120	0.001087
−10	16.6	85.44	0.3691	92.5	0.3074	0.1918	0.0602	0.00559	2.546×10^{-4}	2.856×10^{-6}	4.684	0.352	0.00124	0.001029
0	21.2	84.38	0.4635	90.9	0.3103	0.1966	0.0589	0.00589	2.345×10^{-4}	3.314×10^{-6}	4.463	0.398	0.00128	0.000972
10	26.6	83.31	0.5761	89.3	0.3134	0.2017	0.0576	0.00619	2.181×10^{-4}	3.811×10^{-4}	4.269	0.447	0.00132	0.000915
20	33.1	82.2	0.7094	87.5	0.3167	0.2070	0.0563	0.00648	2.024×10^{-4}	4.342×10^{-6}	4.098	0.500	0.00132	0.000859
30	40.8	81.08	0.866	85.8	0.3203	0.2127	0.0550	0.00676	1.883×10^{-4}	4.906×10^{-6}	3.947	0.555	0.00142	0.000803
40	49.8	79.92	1.049	83.9	0.3240	0.2188	0.0536	0.00704	1.752×10^{-4}	5.494×10^{-6}	3.814	0.614	0.00149	0.000749
50	60.2	78.73	1.262	82.0	0.3281	0.2253	0.0522	0.00732	1.633×10^{-4}	6.103×10^{-6}	3.697	0.677	0.00156	0.000695
60	72.2	77.51	1.509	80.0	0.3325	0.2323	0.0507	0.00758	1.522×10^{-4}	6.725×10^{-6}	3.594	0.742	0.00163	0.000642
70	85.9	76.25	1.794	78.0	0.3372	0.2398	0.0492	0.00785	1.420×10^{-4}	7.356×10^{-6}	3.504	0.810	0.00173	0.000590
80	101.4	74.94	2.122	75.8	0.3424	0.2481	0.0476	0.00810	1.324×10^{-4}	7.986×10^{-6}	3.425	0.880	0.00183	0.000538
90	119.1	73.59	2.5	73.5	0.3481	0.2572	0.0460	0.00835	1.234×10^{-4}	8.611×10^{-6}	3.357	0.955	0.00195	0.000488
100	138.9	72.17	2.935	71.1	0.3548	0.2674	0.0444	0.00860	1.149×10^{-4}	9.222×10^{-6}	3.303	1.032	0.00210	0.000439
110	161.2	70.69	3.435	68.5	0.3627	0.2790	0.0427	0.00884	1.068×10^{-4}	9.814×10^{-6}	3.262	1.115	0.00227	0.000391
120	186.0	69.13	4.012	65.8	0.3719	0.2925	0.0410	0.00908	9.911×10^{-5}	1.038×10^{-5}	3.235	1.204	0.00248	0.000344
130	213.5	67.48	4.679	62.9	0.3829	0.3083	0.0392	0.00931	9.175×10^{-5}	1.092×10^{-5}	3.223	1.303	0.00275	0.000299
140	244.1	65.72	5.455	59.8	0.3963	0.3276	0.0374	0.00954	8.464×10^{-5}	1.144×10^{-5}	3.229	1.416	0.00308	0.000255
150	277.8	63.83	6.367	56.4	0.4131	0.3520	0.0355	0.00976	7.778×10^{-5}	1.195×10^{-5}	3.259	1.551	0.00351	0.000212
160	314.9	61.76	7.45	52.7	0.4352	0.3839	0.0335	0.00998	7.108×10^{-5}	1.245×10^{-5}	3.324	1.725	0.00411	0.000171
170	355.8	59.47	8.762	48.5	0.4659	0.4286	0.0314	0.01020	6.450×10^{-5}	1.298×10^{-5}	3.443	1.963	0.00498	0.000132
180	400.7	56.85	10.4	43.7	0.5123	0.4960	0.0292	0.01041	5.792×10^{-5}	1.366×10^{-5}	3.661	2.327	0.00637	0.000095
190	449.9	53.75	12.53	38.0	0.5929	0.6112	0.0267	0.01063	5.119×10^{-5}	1.431×10^{-5}	4.090	2.964	0.00891	0.000061
200	504.0	49.75	15.57	30.7	0.7717	0.8544	0.0239	0.01085	4.397×10^{-5}	1.544×10^{-5}	5.119	4.376	0.01490	0.000031
210	563.8	43.19	21.18	18.9	1.4786	1.6683	0.0199	0.01110	3.483×10^{-5}	1.787×10^{-5}	9.311	9.669	0.04021	0.000006

Note 1: KInematic viscosity ν and thermal diffusivity α can be calculated from their definitions, $\nu = \mu/\rho$ and $\alpha = k/\rho c_p = \nu/\text{Pr}$. The properties listed here (except the vapor density) can be used at any pressures with negligible error except at temperatures near the critical-point value.

Note 2: The unit Btu/lbm . °F for specific heat is equivalent to Btu/lbm·R, and the unit Btu/h·ft·°F for thermal conductivity is equivalent to Btu/h·ft·R.

Source: Data generated from the EES software developed by S. A. Klein and F. L. Alvarado. Original sources: R. Tilner-Roth and H. D. Baehr, "An International Standard Formulation for the Thermodynamic Properties of 1,1,1,2-Tetrafluorethane (HFC-134a) for Temperatures from 170 K to 455 K and Pressures up to 70 Mpa," *J. Phys. Chem. Ref. Data*, Vol. 23, No.5, 1994: M. J. Assael, N. K. Dalaouti, A. A. Griva, and J. H. Dymond, "Viscosity and Thermal Conductivity of Halogenated Methane and Ethane Refrigerants," *IJR*, Vol. 22, pp. 525–535, 1999: NIST REPROP 6 program (M. O. McLinden, S. A. Klein, E. W. Lemmon, and A. P. Peskin, Physicial and Chemical Properties Division, National Institute of Standards and Technology, Boulder, CO 80303. 1995).

TABLE A–17E

Properties of saturated ammonia

Temp. T, °F	Saturation Pressure P_{sat}, psia	Density ρ, lbm/ft^3 Liquid	Vapor	Enthalpy of Vaporization h_{fg}, Btu/lbm	Specific Heat c_p, Btu/lbm·R Liquid	Vapor	Thermal Conductivity k, Btu/h·ft·R Liquid	Vapor	Dynamic Viscosity μ, lbm/ft·s Liquid	Vapor	Prandtl Number Pr Liquid	Vapor	Volume Expansion Coefficient β, 1/R Liquid	Surface Tension lbf/ft
−40	10.4	43.08	0.0402	597.0	1.0542	0.5354	–	0.01026	1.966×10^{-4}	5.342×10^{-6}	–	1.003	0.00098	0.002443
−30	13.9	42.66	0.0527	590.2	1.0610	0.5457	–	0.01057	1.853×10^{-4}	5.472×10^{-6}	–	1.017	0.00101	0.002357
−20	18.3	42.33	0.0681	583.2	1.0677	0.5571	0.3501	0.01089	1.746×10^{-4}	5.600×10^{-6}	1.917	1.031	0.00103	0.002272
−10	23.7	41.79	0.0869	575.9	1.0742	0.5698	0.3426	0.01121	1.645×10^{-4}	5.731×10^{-6}	1.856	1.048	0.00106	0.002187
0	30.4	41.34	0.1097	568.4	1.0807	0.5838	0.3352	0.01154	1.549×10^{-4}	5.861×10^{-6}	1.797	1.068	0.00109	0.002103
10	38.5	40.89	0.1370	560.7	1.0873	0.5992	0.3278	0.01187	1.458×10^{-4}	5.994×10^{-6}	1.740	1.089	0.00112	0.002018
20	48.2	40.43	0.1694	552.6	1.0941	0.6160	0.3203	0.01220	1.371×10^{-4}	6.125×10^{-6}	1.686	1.113	0.00116	0.001934
30	59.8	39.96	0.2075	544.4	1.1012	0.6344	0.3129	0.01254	1290×10^{-4}	6.256×10^{-6}	1.634	1.140	0.00119	0.001850
40	73.4	39.48	0.2521	535.8	1.1087	0.6544	0.3055	0.01288	1.213×10^{-4}	6.389×10^{-6}	1.585	1.168	0.00123	0.001767
50	89.2	38.99	0.3040	526.9	1.1168	0.6762	0.2980	0.01323	1.140×10^{-4}	6.522×10^{-6}	1.539	1.200	0.00128	0.001684
60	107.7	38.50	0.3641	517.7	1.1256	0.6999	0.2906	0.01358	1.072×10^{-4}	6.656×10^{-6}	1.495	1.234	0.00132	0.001601
70	128.9	37.99	0.4332	508.1	1.1353	0.7257	0.2832	0.01394	1.008×10^{-4}	6.786×10^{-6}	1.456	1.272	0.00137	0.001518
80	153.2	37.47	0.5124	498.2	1.1461	0.7539	0.2757	0.01431	9.486×10^{-5}	6.922×10^{-6}	1.419	1.313	0.00143	0.001436
90	180.8	36.94	0.6029	487.8	1.1582	0.7846	0.2683	0.01468	8.922×10^{-5}	7.056×10^{-6}	1.387	1.358	0.00149	0.001354
100	212.0	36.40	0.7060	477.0	1.1719	0.8183	0.2609	0.01505	8.397×10^{-5}	7.189×10^{-6}	1.358	1.407	0.00156	0.001273
110	247.2	35.83	0.8233	465.8	1.1875	0.8554	0.2535	0.01543	7.903×10^{-5}	7.325×10^{-6}	1.333	1.461	0.00164	0.001192
120	286.5	35.26	0.9564	454.1	1.2054	0.8965	0.2460	0.01582	7.444×10^{-5}	7.458×10^{-6}	1.313	1.522	0.00174	0.001111
130	330.4	34.66	1.1074	441.7	1.2261	0.9425	0.2386	0.01621	7.017×10^{-5}	7.594×10^{-6}	1.298	1.589	0.00184	0.001031
140	379.4	34.04	1.2786	428.8	1.2502	0.9943	0.2312	0.01661	6.617×10^{-5}	7.731×10^{-6}	1.288	1.666	0.00196	0.000951
150	433.2	33.39	1.4730	415.2	1.2785	1.0533	0.2237	0.01702	6.244×10^{-5}	7.867×10^{-4}	1.285	1.753	0.00211	0.000872
160	492.7	32.72	1.6940	400.8	1.3120	1.1214	0.2163	0.01744	5.900×10^{-5}	8.006×10^{-6}	1.288	1.853	0.00228	0.000794
170	558.2	32.01	1.9460	385.4	1.3523	1.2012	0.2089	0.01786	5.578×10^{-5}	8.142×10^{-6}	1.300	1.971	0.00249	0.000716
180	630.1	31.26	2.2346	369.1	1.4015	1.2965	0.2014	0.01829	5.278×10^{-5}	8.281×10^{-6}	1.322	2.113	0.00274	0.000638
190	708.5	30.47	2.5670	351.6	1.4624	1.4128	0.1940	0.01874	5.000×10^{-5}	8.419×10^{-6}	1.357	2.286	0.00306	0.000562
200	794.4	29.62	2.9527	332.7	1.5397	1.5586	0.1866	0.01919	4.742×10^{-5}	8.561×10^{-6}	1.409	2.503	0.00348	0.000486
210	887.9	28.70	3.4053	312.0	1.6411	1.7473	0.1791	0.01966	4500×10^{-5}	8.703×10^{-6}	1.484	2.784	0.00403	0.000411
220	989.5	27.69	3.9440	289.2	1.7798	2.0022	0.1717	0.02015	4.275×10^{-5}	8.844×10^{-6}	1.595	3.164	0.00480	0.000338
230	1099.0	25.57	4.5987	263.5	1.9824	2.3659	0.1643	0.02065	4.064×10^{-5}	8.989×10^{-6}	1.765	3.707	0.00594	0.000265
240	1219.4	25.28	5.4197	234.0	2.3100	2.9264	0.1568	0.02119	3.864×10^{-5}	9.136×10^{-6}	2.049	4.542	0.00784	0.000194

Note 1: Kinematic viscosity ν and thermal diffusivity α can be calculated from their definitions, $\nu = \mu/\rho$ and $\alpha = k/\rho c_p = \nu/\text{Pr}$. The properties listed here (except the vapor density) can be used at any pressures with negligible error except at temperatures near the critical-point value.

Note 2: The unit Btu/lbm·°F for specific heat is equivalent to Btu/lbm·R, and the unit Btu/h·ft·°F for thermal conductivity is equivalent to Btu/h·ft·R.

Source: Data generated from the EES software developed by S. A. Klein and F. L. Alvarado. Orginal sources: Tillner-Roth, Harms-Watzenterg, and Baehr, "Eine neue Fundamentalgleichung fur Ammoniak," *DKV-Tagungsbericht* 20: 167–181, 1993; Liley and Desai, "Thermophysical Properties of Refrigerants," *ASHRAE,* 1993, ISBN 1-1883413-10-9.

TABLE A–18E

Properties of saturated propane

Temp. T, °F	Saturation Pressure P_{sat}, psia	Density ρ, lbm/ft³ Liquid	Density ρ, lbm/ft³ Vapor	Enthalpy of Vaporization h_{fg}, Btu/lbm	Specific Heat c_p, Btu/lbm·R Liquid	Specific Heat c_p, Btu/lbm·R Vapor	Thermal Conductivity k, Btu/h·ft·R Liquid	Thermal Conductivity k, Btu/h·ft·R Vapor	Dynamic Viscosity μ, lbm/ft·s Liquid	Dynamic Viscosity μ, lbm/ft·s Vapor	Prandtl Number Pr Liquid	Prandtl Number Pr Vapor	Volume Expansion Coefficient β, 1/R Liquid	Surface Tension lbf/ft
−200	0.0201	42.06	0.0003	217.7	0.4750	0.2595	0.1073	0.00313	5.012×10^{-4}	2.789×10^{-6}	7.991	0.833	0.00083	0.001890
−180	0.0752	41.36	0.0011	213.4	0.4793	0.2680	0.1033	0.00347	3.941×10^{-4}	2.975×10^{-6}	6.582	0.826	0.00086	0.001780
−160	0.2307	40.65	0.0032	209.1	0.4845	0.2769	0.0992	0.00384	3.199×10^{-4}	3.164×10^{-6}	5.626	0.821	0.00088	0.001671
−140	0.6037	39.93	0.0078	204.8	0.4907	0.2866	0.0949	0.00423	2.660×10^{-4}	3.358×10^{-6}	4.951	0.818	0.00091	0.001563
−120	1.389	39.20	0.0170	200.5	0.4982	0.2971	0.0906	0.00465	2.252×10^{-4}	3.556×10^{-6}	4.457	0.817	0.00094	0.001455
−100	2.878	38.46	0.0334	196.1	0.5069	0.3087	0.0863	0.00511	1.934×10^{-4}	3.756×10^{-6}	4.087	0.817	0.00097	0.001349
−90	4.006	38.08	0.0453	193.9	0.5117	0.3150	0.0842	0.00534	1.799×10^{-4}	3.858×10^{-6}	3.936	0.819	0.00099	0.001297
−80	5.467	37.70	0.0605	191.6	0.5169	0.3215	0.0821	0.00559	1.678×10^{-4}	3.961×10^{-6}	3.803	0.820	0.00101	0.001244
−70	7.327	37.32	0.0793	189.3	0.5224	0.3284	0.0800	0.00585	1.569×10^{-4}	4.067×10^{-6}	3.686	0.822	0.00104	0.001192
−60	9.657	36.93	0.1024	186.9	0.5283	0.3357	0.0780	0.00611	1.469×10^{-4}	4.172×10^{-6}	3.582	0.825	0.00106	0.001140
−50	12.54	36.54	0.1305	184.4	0.5345	0.3433	0.0760	0.00639	1.378×10^{-4}	4.278×10^{-6}	3.490	0.828	0.00109	0.001089
−40	16.05	36.13	0.1641	181.9	0.5392	0.3513	0.0740	0.00568	1.294×10^{-4}	4.386×10^{-6}	3.395	0.831	0.00112	0.001038
−30	20.29	35.73	0.2041	179.3	0.5460	0.3596	0.0721	0.00697	1.217×10^{-4}	4.497×10^{-6}	3.320	0.835	0.00115	0.000987
−20	25.34	35.31	0.2512	176.6	0.5531	0.3684	0.0702	0.00728	1.146×10^{-4}	4.611×10^{-6}	3.253	0.840	0.00119	0.000937
−10	31.3	34.89	0.3063	173.8	0.5607	0.3776	0.0683	0.00761	1.079×10^{-4}	4.725×10^{-6}	3.192	0.845	0.00123	0.000887
0	38.28	34.46	0.3703	170.9	0.5689	0.3874	0.0665	0.00794	1.018×10^{-4}	4.842×10^{-6}	3.137	0.850	0.00127	0.000838
10	46.38	34.02	0.4441	167.9	0.5775	0.3976	0.0647	0.00829	9.606×10^{-5}	4.961×10^{-6}	3.088	0.857	0.00132	0.000789
20	55.7	33.56	0.5289	164.8	0.5867	0.4084	0.0629	0.00865	9.067×10^{-5}	5.086×10^{-6}	3.043	0.864	0.00138	0.000740
30	66.35	33.10	0.6259	161.6	0.5966	0.4199	0.0512	0.00903	8.561×10^{-5}	5.211×10^{-6}	3.003	0.873	0.00144	0.000692
40	78.45	32.62	0.7365	158.1	0.6072	0.4321	0.0595	0.00942	8.081×10^{-5}	5.342×10^{-6}	2.967	0.882	0.00151	0.000644
50	92.12	32.13	0.8621	154.6	0.6187	0.4452	0.0579	0.00983	7.631×10^{-5}	5.478×10^{-6}	2.935	0.893	0.00159	0.000597
60	107.5	31.63	1.0046	150.8	0.6311	0.4593	0.0563	0.01025	7.200×10^{-5}	5.617×10^{-6}	2.906	0.906	0.00168	0.000551
70	124.6	31.11	1.1659	146.8	0.6447	0.4746	0.0547	0.01070	6.794×10^{-5}	5.764×10^{-6}	2.881	0.921	0.00179	0.000505
80	143.7	30.56	1.3484	142.7	0.6596	0.4915	0.0532	0.01116	6.406×10^{-5}	5.919×10^{-6}	2.860	0.938	0.00191	0.000460
90	164.8	30.00	1.5549	138.2	0.6762	0.5103	0.0517	0.01165	6.033×10^{-5}	6.081×10^{-6}	2.843	0.959	0.00205	0.000416
100	188.1	29.41	1.7887	133.6	0.6947	0.5315	0.0501	0.01217	5.675×10^{-5}	6.256×10^{-6}	2.831	0.984	0.00222	0.000372
120	241.8	28.13	2.3562	123.2	0.7403	0.5844	0.0472	0.01328	5.000×10^{-6}	6.644×10^{-6}	2.825	1.052	0.00267	0.000288
140	306.1	26.69	3.1003	111.1	0.7841	0.6613	0.0442	0.01454	4.358×10^{-5}	7.111×10^{-6}	2.784	1.164	0.00338	0.000208
160	382.4	24.98	4.1145	96.4	0.8696	0.7911	0.0411	0.01603	3.733×10^{-5}	7.719×10^{-6}	2.845	1.371	0.00459	0.000133
180	472.9	22.79	5.6265	77.1	1.1436	1.0813	0.0376	0.01793	3.083×10^{-5}	8.617×10^{-6}	3.380	1.870	0.00791	0.000065

Note 1: Kinematic viscosity ν and thermal diffusivity α can be calculated from their definitions, $\nu = \mu/\rho$ and $\alpha = k/\rho c_p = \nu/\text{Pr}$. The properties listed here (except the vapor density) can be used at any pressures with negligible error at temperatures near the critical-point value.

Note 2: The unit Btu/lbm·°F for specific heat is equivalent to Btu/lbm·R, and the unit Btu/h·ft·°F for thermal conductivity is equivalent to Btu/h·ft·R.

Source: Data generated from the EES software developed by S. A. Klein and F. L. Alvarado. Original sources: Reiner Tillner-Roth, "Fundamental Equations of State," Shaker, Verlag, Aachan, 1998; B. A. Younglove and J. F. Ely. "Thermophysical Properties of Fluids. II Methane, Ethane, Propane, Isobutane, and Normal Butane," *J. Phys. Chem. Ref. Data*, Vol. 16, No. 4, 1987; G. R. Somayajulu, "A Generalized Equation for Surface Tension from the Triple-Point to the Critical-Point," *International Journal of Thermophysics*, Vol. 9, No. 4, 1988.

TABLE A–19E

Properties of liquids

Temp. T, °F	Density ρ, lbm/ft^3	Specific Heat c_p, Btu/lbm·R	Thermal Conductivity k, Btu/h·ft·R	Thermal Diffusivity α, ft^2/s	Dynamic Viscosity μ, lbm/ft·s	Kinematic Viscosity ν, ft^2/s	Prandtl Number Pr	Volume Expansion Coeff. β, 1/R
				Methane (CH_4)				
−280	27.41	0.8152	0.1205	1.497×10^{-6}	1.057×10^{-4}	3.857×10^{-6}	2.575	0.00175
−260	26.43	0.8301	0.1097	1.389×10^{-6}	8.014×10^{-5}	3.032×10^{-6}	2.183	0.00192
−240	25.39	0.8523	0.0994	1.276×10^{-6}	6.303×10^{-5}	2.482×10^{-6}	1.945	0.00215
−220	24.27	0.8838	0.0896	1.159×10^{-6}	5.075×10^{-5}	2.091×10^{-6}	1.803	0.00247
−200	23.04	0.9314	0.0801	1.036×10^{-6}	4.142×10^{-5}	1.798×10^{-6}	1.734	0.00295
−180	21.64	1.010	0.0709	9.008×10^{-7}	3.394×10^{-5}	1.568×10^{-6}	1.741	0.00374
−160	19.99	1.158	0.0616	7.397×10^{-7}	2.758×10^{-5}	1.379×10^{-6}	1.865	0.00526
−140	17.84	1.542	0.0518	5.234×10^{-7}	2.168×10^{-5}	1.215×10^{-6}	2.322	0.00943
				Methanol [$CH_3(OH)$]				
70	49.15	0.6024	0.1148	1.076×10^{-6}	3.872×10^{-4}	7.879×10^{-6}	7.317	0.000656
90	48.50	0.6189	0.1143	1.057×10^{-6}	3.317×10^{-4}	6.840×10^{-6}	6.468	0.000671
110	47.85	0.6373	0.1138	1.036×10^{-6}	2.872×10^{-4}	6.005×10^{-6}	5.793	0.000691
130	47.18	0.6576	0.1133	1.014×10^{-6}	2.513×10^{-4}	5.326×10^{-6}	5.250	0.000716
150	46.50	0.6796	0.1128	9.918×10^{-7}	2.218×10^{-4}	4.769×10^{-6}	4.808	0.000749
170	45.80	0.7035	0.1124	9.687×10^{-7}	1.973×10^{-4}	4.308×10^{-6}	4.447	0.000789
				Isobutane (R600a)				
−150	42.75	0.4483	0.0799	1.157×10^{-6}	6.417×10^{-4}	1.500×10^{-5}	12.96	0.000785
−100	41.06	0.4721	0.0782	1.120×10^{-6}	3.669×10^{-4}	8.939×10^{-6}	7.977	0.000836
−50	39.31	0.4986	0.0731	1.036×10^{-6}	2.376×10^{-4}	6.043×10^{-6}	5.830	0.000908
0	37.48	0.5289	0.0664	9.299×10^{-7}	1.651×10^{-4}	4.406×10^{-6}	4.738	0.001012
50	35.52	0.5643	0.0591	8.187×10^{-7}	1.196×10^{-4}	3.368×10^{-6}	4.114	0.001169
100	33.35	0.6075	0.0521	7.139×10^{-7}	8.847×10^{-5}	2.653×10^{-6}	3.716	0.001421
150	30.84	0.6656	0.0457	6.188×10^{-7}	6.558×10^{-5}	2.127×10^{-6}	3.437	0.001883
200	27.73	0.7635	0.0400	5.249×10^{-7}	4.750×10^{-5}	1.713×10^{-6}	3.264	0.002970
				Glycerin				
32	79.65	0.5402	0.163	1.052×10^{-6}	7.047	0.08847	84101	
40	79.49	0.5458	0.1637	1.048×10^{-6}	4.803	0.06042	57655	
50	79.28	0.5541	0.1645	1.040×10^{-6}	2.850	0.03594	34561	
60	79.07	0.5632	0.1651	1.029×10^{-6}	1.547	0.01956	18995	
70	78.86	0.5715	0.1652	1.018×10^{-6}	0.9422	0.01195	11730	
80	78.66	0.5794	0.1652	1.007×10^{-6}	0.5497	0.00699	6941	
90	78.45	0.5878	0.1652	9.955×10^{-7}	0.3756	0.004787	4809	
100	78.24	0.5964	0.1653	9.841×10^{-7}	0.2277	0.00291	2957	
				Engine Oil (unused)				
32	56.12	0.4291	0.0849	9.792×10^{-7}	2.563	4.566×10^{-2}	46636	0.000389
50	55.79	0.4395	0.08338	9.448×10^{-7}	1.210	2.169×10^{-2}	22963	0.000389
75	55.3	0.4531	0.08378	9.288×10^{-7}	0.4286	7.751×10^{-3}	8345	0.000389
100	54.77	0.4669	0.08367	9.089×10^{-7}	0.1630	2.977×10^{-3}	3275	0.000389
125	54.24	0.4809	0.08207	8.740×10^{-7}	7.617×10^{-2}	1.404×10^{-3}	1607	0.000389
150	53.73	0.4946	0.08046	8.411×10^{-7}	3.833×10^{-2}	7.135×10^{-4}	848.3	0.000389
200	52.68	0.5231	0.07936	7.999×10^{-7}	1.405×10^{-2}	2.668×10^{-4}	333.6	0.000389
250	51.71	0.5523	0.07776	7.563×10^{-7}	6.744×10^{-3}	1.304×10^{-4}	172.5	0.000389
300	50.63	0.5818	0.07673	7.236×10^{-7}	3.661×10^{-3}	7.232×10^{-5}	99.94	0.000389

Source: Data generated from the EES software developed by S. A. Klein and F. L. Alvarado. Originally based on various sources.

TABLE A–20E

Properties of liquid metals

Temp. T, °F	Density ρ, lbm/ft^3	Specific Heat c_p, Btu/lbm·R	Thermal Conductivity k, Btu/h·ft·R	Thermal Diffusivity α, ft^2/s	Dynamic Viscosity μ, lbm/ft·s	Kinematic Viscosity ν, ft^2/s	Prandtl Number Pr	Volume Expansion Coeff. β, 1/R
				Mercury (Hg) Melting Point: −38°F				
32	848.7	0.03353	4.727	4.614×10^{-5}	1.133×10^{-3}	1.335×10^{-6}	0.02895	1.005×10^{-4}
50	847.2	0.03344	4.805	4.712×10^{-5}	1.092×10^{-3}	1.289×10^{-6}	0.02737	1.005×10^{-4}
100	842.9	0.03319	5.015	4.980×10^{-5}	9.919×10^{-4}	1.176×10^{-6}	0.02363	1.005×10^{-4}
150	838.7	0.03298	5.221	5.244×10^{-5}	9.122×10^{-4}	1.087×10^{-6}	0.02074	1.005×10^{-4}
200	834.5	0.03279	5.422	5.504×10^{-5}	8.492×10^{-4}	1.017×10^{-6}	0.01849	1.005×10^{-4}
300	826.2	0.03252	5.815	6.013×10^{-5}	7.583×10^{-4}	9.180×10^{-7}	0.01527	1.005×10^{-4}
400	817.9	0.03236	6.184	6.491×10^{-5}	6.972×10^{-4}	8.524×10^{-7}	0.01313	1.008×10^{-4}
500	809.6	0.03230	6.518	6.924×10^{-5}	6.525×10^{-4}	8.061×10^{-7}	0.01164	1.018×10^{-4}
600	801.3	0.03235	6.839	7.329×10^{-5}	6.186×10^{-4}	7.719×10^{-7}	0.01053	1.035×10^{-4}
				Bismuth (Bi) Melting Point: 520°F				
700	620.7	0.03509	9.361	1.193×10^{-4}	1.001×10^{-3}	1.614×10^{-6}	0.01352	
800	616.5	0.03569	9.245	1.167×10^{-4}	9.142×10^{-4}	1.482×10^{-6}	0.01271	
900	612.2	0.0363	9.129	1.141×10^{-4}	8.267×10^{-4}	1.350×10^{-6}	0.01183	
1000	608.0	0.0369	9.014	1.116×10^{-4}	7.392×10^{-4}	1.215×10^{-6}	0.0109	
1100	603.7	0.0375	9.014	1.105×10^{-4}	6.872×10^{-4}	1.138×10^{-6}	0.01029	
				Lead (Pb) Melting Point: 621°F				
700	658	0.03797	9.302	1.034×10^{-4}	1.612×10^{-3}	2.450×10^{-6}	0.02369	
800	654	0.03750	9.157	1.037×10^{-4}	1.453×10^{-3}	2.223×10^{-6}	0.02143	
900	650	0.03702	9.013	1.040×10^{-4}	1.296×10^{-3}	1.994×10^{-6}	0.01917	
1000	645.7	0.03702	8.912	1.035×10^{-4}	1.202×10^{-3}	1.862×10^{-6}	0.01798	
1100	641.5	0.03702	8.810	1.030×10^{-4}	1.108×10^{-3}	1.727×10^{-6}	0.01676	
1200	637.2	0.03702	8.709	1.025×10^{-4}	1.013×10^{-3}	1.590×10^{-6}	0.01551	
				Sodium (Na) Melting Point: 208°F				
300	57.13	0.3258	48.19	7.192×10^{-4}	4.136×10^{-4}	7.239×10^{-6}	0.01007	
400	56.28	0.3219	46.58	7.142×10^{-4}	3.572×10^{-4}	6.350×10^{-6}	0.008891	
500	55.42	0.3181	44.98	7.087×10^{-4}	3.011×10^{-4}	5.433×10^{-6}	0.007667	
600	54.56	0.3143	43.37	7.026×10^{-4}	2.448×10^{-4}	4.488×10^{-6}	0.006387	
800	52.85	0.3089	40.55	6.901×10^{-4}	1.772×10^{-4}	3.354×10^{-6}	0.004860	
1000	51.14	0.3057	38.12	6.773×10^{-4}	1.541×10^{-4}	3.014×10^{-6}	0.004449	
				Potassium (K) Melting Point: 147°F				
300	50.40	0.1911	26.00	7.500×10^{-4}	2.486×10^{-4}	4.933×10^{-6}	0.006577	
400	49.58	0.1887	25.37	7.532×10^{-4}	2.231×10^{-4}	4.500×10^{-6}	0.005975	
500	48.76	0.1863	24.73	7.562×10^{-4}	1.976×10^{-4}	4.052×10^{-6}	0.005359	
600	47.94	0.1839	24.09	7.591×10^{-4}	1.721×10^{-4}	3.589×10^{-6}	0.004728	
800	46.31	0.1791	22.82	7.643×10^{-4}	1.210×10^{-4}	2.614×10^{-6}	0.003420	
1000	44.62	0.1791	21.34	7.417×10^{-4}	1.075×10^{-4}	2.409×10^{-6}	0.003248	
				Sodium-Potassium (%22Na-%78K) Melting Point: 12°F				
200	52.99	0.2259	14.79	3.432×10^{-4}	3.886×10^{-4}	7.331×10^{-6}	0.02136	
300	52.16	0.2230	14.99	3.580×10^{-4}	3.467×10^{-4}	6.647×10^{-6}	0.01857	
400	51.32	0.2201	15.19	3.735×10^{-4}	3.050×10^{-4}	5.940×10^{-6}	0.0159	
600	49.65	0.2143	15.59	4.070×10^{-4}	2.213×10^{-4}	4.456×10^{-6}	0.01095	
800	47.99	0.2100	15.95	4.396×10^{-4}	1.539×10^{-4}	3.207×10^{-6}	0.007296	
1000	46.36	0.2103	16.20	4.615×10^{-4}	1.353×10^{-4}	2.919×10^{-6}	0.006324	

Source: Data generated from the EES software developed by S. A. Klein and F. L. Alvarado. Originally based on various sources.

TABLE A–21E

Ideal-gas properties of air

T R	h Btu/lbm	P_r	u Btu/lbm	v_r	s° Btu/lbm · R	T R	h Btu/lbm	P_r	u Btu/lbm	v_r	s° Btu/lbm · R
360	85.97	0.3363	61.29	396.6	0.50369	1600	395.74	71.13	286.06	8.263	0.87130
380	90.75	0.4061	64.70	346.6	0.51663	1650	409.13	80.89	296.03	7.556	0.87954
400	95.53	0.4858	68.11	305.0	0.52890	1700	422.59	90.95	306.06	6.924	0.88758
420	100.32	0.5760	71.52	270.1	0.54058	1750	436.12	101.98	316.16	6.357	0.89542
440	105.11	0.6776	74.93	240.6	0.55172	1800	449.71	114.0	326.32	5.847	0.90308
460	109.90	0.7913	78.36	215.33	0.56235	1850	463.37	127.2	336.55	5.388	0.91056
480	114.69	0.9182	81.77	193.65	0.57255	1900	477.09	141.5	346.85	4.974	0.91788
500	119.48	1.0590	85.20	174.90	0.58233	1950	490.88	157.1	357.20	4.598	0.92504
520	124.27	1.2147	88.62	158.58	0.59173	2000	504.71	174.0	367.61	4.258	0.93205
537	128.10	1.3593	91.53	146.34	0.59945	2050	518.71	192.3	378.08	3.949	0.93891
540	129.06	1.3860	92.04	144.32	0.60078	2100	532.55	212.1	388.60	3.667	0.94564
560	133.86	1.5742	95.47	131.78	0.60950	2150	546.54	223.5	399.17	3.410	0.95222
580	138.66	1.7800	98.90	120.70	0.61793	2200	560.59	256.6	409.78	3.176	0.95919
600	143.47	2.005	102.34	110.88	0.62607	2250	574.69	281.4	420.46	2.961	0.96501
620	148.28	2.249	105.78	102.12	0.63395	2300	588.82	308.1	431.16	2.765	0.97123
640	153.09	2.514	109.21	94.30	0.64159	2350	603.00	336.8	441.91	2.585	0.97732
660	157.92	2.801	112.67	87.27	0.64902	2400	617.22	367.6	452.70	2.419	0.98331
680	162.73	3.111	116.12	80.96	0.65621	2450	631.48	400.5	463.54	2.266	0.98919
700	167.56	3.446	119.58	75.25	0.66321	2500	645.78	435.7	474.40	2.125	0.99497
720	172.39	3.806	123.04	70.07	0.67002	2550	660.12	473.3	485.31	1.996	1.00064
740	177.23	4.193	126.51	65.38	0.67665	2600	674.49	513.5	496.26	1.876	1.00623
760	182.08	4.607	129.99	61.10	0.68312	2650	688.90	556.3	507.25	1.765	1.01172
780	186.94	5.051	133.47	57.20	0.68942	2700	703.35	601.9	518.26	1.662	1.01712
800	191.81	5.526	136.97	53.63	0.69558	2750	717.83	650.4	529.31	1.566	1.02244
820	196.69	6.033	140.47	50.35	0.70160	2800	732.33	702.0	540.40	1.478	1.02767
840	201.56	6.573	143.98	47.34	0.70747	2850	746.88	756.7	551.52	1.395	1.03282
860	206.46	7.149	147.50	44.57	0.71323	2900	761.45	814.8	562.66	1.318	1.03788
880	211.35	7.761	151.02	42.01	0.71886	2950	776.05	876.4	573.84	1.247	1.04288
900	216.26	8.411	154.57	39.64	0.72438	3000	790.68	941.4	585.04	1.180	1.04779
920	221.18	9.102	158.12	37.44	0.72979	3050	805.34	1011	596.28	1.118	1.05264
940	226.11	9.834	161.68	35.41	0.73509	3100	820.03	1083	607.53	1.060	1.05741
960	231.06	10.61	165.26	33.52	0.74030	3150	834.75	1161	618.82	1.006	1.06212
980	236.02	11.43	168.83	31.76	0.74540	3200	849.48	1242	630.12	0.955	1.06676
1000	240.98	12.30	172.43	30.12	0.75042	3250	864.24	1328	641.46	0.907	1.07134
1040	250.95	14.18	179.66	27.17	0.76019	3300	879.02	1418	652.81	0.8621	1.07585
1080	260.97	16.28	186.93	24.58	0.76964	3350	893.83	1513	664.20	0.8202	1.08031
1120	271.03	18.60	194.25	22.30	0.77880	3400	908.66	1613	675.60	0.7807	1.08470
1160	281.14	21.18	201.63	20.29	0.78767	3450	923.52	1719	687.04	0.7436	1.08904
1200	291.30	24.01	209.05	18.51	0.79628	3500	938.40	1829	698.48	0.7087	1.09332
1240	301.52	27.13	216.53	16.93	0.80466	3550	953.30	1946	709.95	0.6759	1.09755
1280	311.79	30.55	224.05	15.52	0.81280	3600	968.21	2068	721.44	0.6449	1.10172
1320	322.11	34.31	231.63	14.25	0.82075	3650	983.15	2196	732.95	0.6157	1.10584
1360	332.48	38.41	239.25	13.12	0.82848	3700	998.11	2330	744.48	0.5882	1.10991
1400	342.90	42.88	246.93	12.10	0.83604	3750	1013.1	2471	756.04	0.5621	1.11393
1440	353.37	47.75	254.66	11.17	0.84341	3800	1028.1	2618	767.60	0.5376	1.11791
1480	363.89	53.04	262.44	10.34	0.85062	3850	1043.1	2773	779.19	0.5143	1.12183
1520	374.47	58.78	270.26	9.578	0.85767	3900	1058.1	2934	790.80	0.4923	1.12571
1560	385.08	65.00	278.13	8.890	0.86456	3950	1073.2	3103	802.43	0.4715	1.12955

(Continued)

TABLE A–21E

Ideal-gas properties of air (*Concluded*)

T R	h Btu/lbm	P_r	u Btu/lbm	v_r	s° Btu/lbm · R	T R	h Btu/lbm	P_r	u Btu/lbm	v_r	s° Btu/lbm · R
4000	1088.3	3280	814.06	0.4518	1.13334	4600	1270.4	6089	955.04	0.2799	1.17575
4050	1103.4	3464	825.72	0.4331	1.13709	4700	1300.9	6701	978.73	0.2598	1.18232
4100	1118.5	3656	837.40	0.4154	1.14079	4800	1331.5	7362	1002.5	0.2415	1.18876
4150	1133.6	3858	849.09	0.3985	1.14446	4900	1362.2	8073	1026.3	0.2248	1.19508
4200	1148.7	4067	860.81	0.3826	1.14809	5000	1392.9	8837	1050.1	0.2096	1.20129
4300	1179.0	4513	884.28	0.3529	1.15522	5100	1423.6	9658	1074.0	0.1956	1.20738
4400	1209.4	4997	907.81	0.3262	1.16221	5200	1454.4	10,539	1098.0	0.1828	1.21336
4500	1239.9	5521	931.39	0.3019	1.16905	5300	1485.3	11,481	1122.0	0.1710	1.21923

Note: The properties P_r (relative pressure) and v_r (relative specific volume) are dimensionless quantities used in the analysis of isentropic processes, and should not be confused with the properties pressure and specific volume.

Source: Kenneth Wark, *Thermodynamics*, 4th ed. (New York: McGraw-Hill, 1983), pp. 832–33, Table A–5. Originally published in J. H. Keenan and J. Kaye, *Gas Tables* (New York: John Wiley & Sons, 1948).

TABLE A–22E

Properties of air at 1 atm pressure

Temp. T, °F	Density ρ, lbm/ft^3	Specific Heat c_p, Btu/lbm·R	Thermal Conductivity k, Btu/h·ft·R	Thermal Diffusivity α, ft^2/s	Dynamic Viscosity μ, lbm/ft·s	Kinematic Viscosity ν, ft^2/s	Prandtl Number Pr
−300	0.24844	0.5072	0.00508	1.119×10^{-5}	4.039×10^{-6}	1.625×10^{-5}	1.4501
−200	0.15276	0.2247	0.00778	6.294×10^{-5}	6.772×10^{-6}	4.433×10^{-5}	0.7042
−100	0.11029	0.2360	0.01037	1.106×10^{-4}	9.042×10^{-6}	8.197×10^{-5}	0.7404
−50	0.09683	0.2389	0.01164	1.397×10^{-4}	1.006×10^{-5}	1.039×10^{-4}	0.7439
0	0.08630	0.2401	0.01288	1.726×10^{-4}	1.102×10^{-5}	1.278×10^{-4}	0.7403
10	0.08446	0.2402	0.01312	1.797×10^{-4}	1.121×10^{-5}	1.328×10^{-4}	0.7391
20	0.08270	0.2403	0.01336	1.868×10^{-4}	1.140×10^{-5}	1.379×10^{-4}	0.7378
30	0.08101	0.2403	0.01361	1.942×10^{-4}	1.158×10^{-5}	1.430×10^{-4}	0.7365
40	0.07939	0.2404	0.01385	2.016×10^{-4}	1.176×10^{-5}	1.482×10^{-4}	0.7350
50	0.07783	0.2404	0.01409	2.092×10^{-4}	1.194×10^{-5}	1.535×10^{-4}	0.7336
60	0.07633	0.2404	0.01433	2.169×10^{-4}	1.212×10^{-5}	1.588×10^{-4}	0.7321
70	0.07489	0.2404	0.01457	2.248×10^{-4}	1.230×10^{-5}	1.643×10^{-4}	0.7306
80	0.07350	0.2404	0.01481	2.328×10^{-4}	1.247×10^{-5}	1.697×10^{-4}	0.7290
90	0.07217	0.2404	0.01505	2.409×10^{-4}	1.265×10^{-5}	1.753×10^{-4}	0.7275
100	0.07088	0.2405	0.01529	2.491×10^{-4}	1.281×10^{-5}	1.809×10^{-4}	0.7260
110	0.06963	0.2405	0.01552	2.575×10^{-4}	1.299×10^{-5}	1.866×10^{-4}	0.7245
120	0.06843	0.2405	0.01576	2.660×10^{-4}	1.316×10^{-5}	1.923×10^{-4}	0.7230
130	0.06727	0.2405	0.01599	2.746×10^{-4}	1.332×10^{-5}	1.981×10^{-4}	0.7216
140	0.06615	0.2406	0.01623	2.833×10^{-4}	1.349×10^{-5}	2.040×10^{-4}	0.7202
150	0.06507	0.2406	0.01646	2.921×10^{-4}	1.365×10^{-5}	2.099×10^{-4}	0.7188
160	0.06402	0.2406	0.01669	3.010×10^{-4}	1.382×10^{-5}	2.159×10^{-4}	0.7174
170	0.06300	0.2407	0.01692	3.100×10^{-4}	1.398×10^{-5}	2.220×10^{-4}	0.7161
180	0.06201	0.2408	0.01715	3.191×10^{-4}	1.414×10^{-5}	2.281×10^{-4}	0.7140
190	0.06106	0.2408	0.01738	3.284×10^{-4}	1.430×10^{-5}	2.343×10^{-4}	0.7136
200	0.06013	0.2409	0.01761	3.377×10^{-4}	1.446×10^{-5}	2.406×10^{-4}	0.7124
250	0.05590	0.2415	0.01874	3.857×10^{-4}	1.524×10^{-5}	2.727×10^{-4}	0.7071
300	0.05222	0.2423	0.01985	4.358×10^{-4}	1.599×10^{-5}	3.063×10^{-4}	0.7028
350	0.04899	0.2433	0.02094	4.879×10^{-4}	1.672×10^{-5}	3.413×10^{-4}	0.6995
400	0.04614	0.2445	0.02200	5.419×10^{-4}	1.743×10^{-5}	3.777×10^{-4}	0.6971
450	0.04361	0.2458	0.02305	5.974×10^{-4}	1.812×10^{-5}	4.154×10^{-4}	0.6953
500	0.04134	0.2472	0.02408	6.546×10^{-4}	1.878×10^{-5}	4.544×10^{-4}	0.6942
600	0.03743	0.2503	0.02608	7.732×10^{-4}	2.007×10^{-5}	5.361×10^{-4}	0.6934
700	0.03421	0.2535	0.02800	8.970×10^{-4}	2.129×10^{-5}	6.225×10^{-4}	0.6940
800	0.03149	0.2568	0.02986	1.025×10^{-3}	2.247×10^{-5}	7.134×10^{-4}	0.6956
900	0.02917	0.2599	0.03164	1.158×10^{-3}	2.359×10^{-5}	8.087×10^{-4}	0.6978
1000	0.02718	0.2630	0.03336	1.296×10^{-3}	2.467×10^{-5}	9.080×10^{-4}	0.7004
1500	0.02024	0.2761	0.04106	2.041×10^{-3}	2.957×10^{-5}	1.460×10^{-3}	0.7158
2000	0.01613	0.2855	0.04752	2.867×10^{-3}	3.379×10^{-5}	2.095×10^{-3}	0.7308
2500	0.01340	0.2922	0.05309	3.765×10^{-3}	3.750×10^{-5}	2.798×10^{-3}	0.7432
3000	0.01147	0.2972	0.05811	4.737×10^{-3}	4.082×10^{-5}	3.560×10^{-3}	0.7516
3500	0.01002	0.3010	0.06293	5.797×10^{-3}	4.381×10^{-5}	4.373×10^{-3}	0.7543
4000	0.00889	0.3040	0.06789	6.975×10^{-3}	4.651×10^{-5}	5.229×10^{-3}	0.7497

Note: For ideal gases, the properties c_p, k, μ, and Pr are independent of pressure. The properties ρ, ν, and α at a pressure P (in atm) other than 1 atm are determined by multiplying the values of ρ at the given temperature by P and by dividing ν and α by P.

Source: Data generated from the EES software developed by S. A. Klein and F. L. Alvarado. Original sources: Keenan, Chao, Keyes, Gas Tables, Wiley, 1984; and *Thermophysical Properties of Matter*, Vol. 3: *Thermal Conductivity*, Y. S. Touloukian, P. E. Liley, S. C. Saxena, Vol. 11: *Viscosity*, Y. S. Touloukian, S. C. Saxena, and P. Hestermans, IFI/Plenun, NY, 1970, ISBN 0-306067020-8.

TABLE A–23E

Properties of gases at 1 atm pressure

Temp. T, °F	Density ρ, lbm/ft³	Specific Heat c_p, Btu/lbm·R	Thermal Conductivity k, Btu/h·ft·R	Thermal Diffusivity α, ft²/s	Dynamic Viscosity μ, lbm/ft·s	Kinematic Viscosity ν, ft²/s	Prandtl Number Pr
			Carbon Dioxide, CO_2				
−50	0.14712	0.1797	0.00628	6.600×10^{-5}	7.739×10^{-6}	5.261×10^{-5}	0.7970
0	0.13111	0.1885	0.00758	8.522×10^{-5}	8.661×10^{-6}	6.606×10^{-5}	0.7751
50	0.11825	0.1965	0.00888	1.061×10^{-4}	9.564×10^{-6}	8.086×10^{-5}	0.7621
100	0.10769	0.2039	0.01017	1.286×10^{-4}	1.045×10^{-5}	9.703×10^{-5}	0.7543
200	0.09136	0.2171	0.01273	1.784×10^{-4}	1.217×10^{-5}	1.332×10^{-4}	0.7469
300	0.07934	0.2284	0.01528	2.341×10^{-4}	1.382×10^{-5}	1.743×10^{-4}	0.7445
500	0.06280	0.2473	0.02027	3.626×10^{-4}	1.696×10^{-5}	2.700×10^{-4}	0.7446
1000	0.04129	0.2796	0.03213	7.733×10^{-4}	2.381×10^{-5}	5.767×10^{-4}	0.7458
1500	0.03075	0.2995	0.04281	1.290×10^{-3}	2.956×10^{-5}	9.610×10^{-4}	0.7445
2000	0.02450	0.3124	0.05193	1.885×10^{-3}	3.451×10^{-5}	1.408×10^{-3}	0.7474
			Carbon Monoxide, CO				
−50	0.09363	0.2571	0.01118	1.290×10^{-4}	9.419×10^{-6}	1.005×10^{-4}	0.7798
0	0.08345	0.2523	0.01240	1.636×10^{-4}	1.036×10^{-5}	1.242×10^{-4}	0.7593
50	0.07526	0.2496	0.01359	2.009×10^{-4}	1.127×10^{-5}	1.498×10^{-4}	0.7454
100	0.06854	0.2484	0.01476	2.408×10^{-4}	1.214×10^{-5}	1.772×10^{-4}	0.7359
200	0.05815	0.2485	0.01702	3.273×10^{-4}	1.379×10^{-5}	2.372×10^{-4}	0.7247
300	0.05049	0.2505	0.01920	4.217×10^{-4}	1.531×10^{-5}	3.032×10^{-4}	0.7191
500	0.03997	0.2567	0.02331	6.311×10^{-4}	1.802×10^{-5}	4.508×10^{-4}	0.7143
1000	0.02628	0.2732	0.03243	1.254×10^{-3}	2.334×10^{-5}	8.881×10^{-4}	0.7078
1500	0.01957	0.2862	0.04049	2.008×10^{-3}	2.766×10^{-5}	1.413×10^{-3}	0.7038
2000	0.01559	0.2958	0.04822	2.903×10^{-3}	3.231×10^{-5}	2.072×10^{-3}	0.7136
			Methane, CH_4				
−50	0.05363	0.5335	0.01401	1.360×10^{-4}	5.861×10^{-6}	1.092×10^{-4}	0.8033
0	0.04779	0.5277	0.01616	1.780×10^{-4}	6.506×10^{-6}	1.361×10^{-4}	0.7649
50	0.04311	0.5320	0.01839	2.228×10^{-4}	7.133×10^{-6}	1.655×10^{-4}	0.7428
100	0.03925	0.5433	0.02071	2.698×10^{-4}	7.742×10^{-6}	1.972×10^{-4}	0.7311
200	0.03330	0.5784	0.02559	3.690×10^{-4}	8.906×10^{-6}	2.674×10^{-4}	0.7245
300	0.02892	0.6226	0.03077	4.748×10^{-4}	1.000×10^{-5}	3.457×10^{-4}	0.7283
500	0.02289	0.7194	0.04195	7.075×10^{-4}	1.200×10^{-5}	5.244×10^{-4}	0.7412
1000	0.01505	0.9438	0.07346	1.436×10^{-3}	1.620×10^{-5}	1.076×10^{-3}	0.7491
1500	0.01121	1.1162	0.10766	2.390×10^{-3}	1.974×10^{-5}	1.760×10^{-3}	0.7366
2000	0.00893	1.2419	0.14151	3.544×10^{-3}	2.327×10^{-5}	2.605×10^{-3}	0.7353
			Hydrogen, H_2				
−50	0.00674	3.0603	0.08246	1.110×10^{-3}	4.969×10^{-6}	7.373×10^{-4}	0.6638
0	0.00601	3.2508	0.09049	1.287×10^{-3}	5.381×10^{-6}	8.960×10^{-4}	0.6960
50	0.00542	3.3553	0.09818	1.500×10^{-3}	5.781×10^{-6}	1.067×10^{-3}	0.7112
100	0.00493	3.4118	0.10555	1.742×10^{-3}	6.167×10^{-6}	1.250×10^{-3}	0.7177
200	0.00419	3.4549	0.11946	2.295×10^{-3}	6.911×10^{-6}	1.652×10^{-3}	0.7197
300	0.00363	3.4613	0.13241	2.924×10^{-3}	7.622×10^{-6}	2.098×10^{-3}	0.7174
500	0.00288	3.4572	0.15620	4.363×10^{-3}	8.967×10^{-6}	3.117×10^{-3}	0.7146
1000	0.00189	3.5127	0.20989	8.776×10^{-3}	1.201×10^{-5}	6.354×10^{-3}	0.7241
1500	0.00141	3.6317	0.26381	1.432×10^{-2}	1.477×10^{-5}	1.048×10^{-2}	0.7323
2000	0.00112	3.7656	0.31923	2.098×10^{-2}	1.734×10^{-5}	1.544×10^{-2}	0.7362

(Continued)

TABLE A–23E

Properties of gases at 1 atm pressure *(Continued)*

Temp. T, °F	Density ρ, lbm/ft^3	Specific Heat c_p, Btu/lbm·R	Thermal Conductivity k, Btu/h·ft·R	Thermal Diffusivity α, ft^2/s	Dynamic Viscosity μ, lbm/ft·s	Kinematic Viscosity ν, ft^2/s	Prandtl Number Pr
			Nitrogen, N_2				
−50	0.09364	0.2320	0.01176	1.504×10^{-4}	9.500×10^{-6}	1.014×10^{-4}	0.6746
0	0.08346	0.2441	0.01300	1.773×10^{-4}	1.043×10^{-5}	1.251×10^{-4}	0.7056
50	0.07527	0.2480	0.01420	2.113×10^{-4}	1.134×10^{-5}	1.507×10^{-4}	0.7133
100	0.06854	0.2489	0.01537	2.502×10^{-4}	1.221×10^{-5}	1.783×10^{-4}	0.7126
200	0.05815	0.2487	0.01760	3.379×10^{-4}	1.388×10^{-5}	2.387×10^{-4}	0.7062
300	0.05050	0.2492	0.01970	4.349×10^{-4}	1.543×10^{-5}	3.055×10^{-4}	0.7025
500	0.03997	0.2535	0.02359	6.466×10^{-4}	1.823×10^{-5}	4.559×10^{-4}	0.7051
1000	0.02628	0.2697	0.03204	1.255×10^{-3}	2.387×10^{-5}	9.083×10^{-4}	0.7232
1500	0.01958	0.2831	0.04002	2.006×10^{-3}	2.829×10^{-5}	1.445×10^{-3}	0.7202
2000	0.01560	0.2927	0.04918	2.992×10^{-3}	3.212×10^{-5}	2.059×10^{-3}	0.6882
			Oxygen, O_2				
−50	0.10697	0.2331	0.01216	1.355×10^{-4}	1.104×10^{-5}	1.032×10^{-4}	0.7622
0	0.09533	0.2245	0.01346	1.747×10^{-4}	1.218×10^{-5}	1.277×10^{-4}	0.7312
50	0.08598	0.2209	0.01475	2.157×10^{-4}	1.326×10^{-5}	1.543×10^{-4}	0.7152
100	0.07830	0.2200	0.01601	2.582×10^{-4}	1.429×10^{-5}	1.826×10^{-4}	0.7072
200	0.06643	0.2221	0.01851	3.484×10^{-4}	1.625×10^{-5}	2.446×10^{-4}	0.7020
300	0.05768	0.2262	0.02096	4.463×10^{-4}	1.806×10^{-5}	3.132×10^{-4}	0.7018
500	0.04566	0.2352	0.02577	6.665×10^{-4}	2.139×10^{-5}	4.685×10^{-4}	0.7029
1000	0.03002	0.2520	0.03698	1.357×10^{-3}	2.855×10^{-5}	9.509×10^{-4}	0.7005
1500	0.02236	0.2626	0.04701	2.224×10^{-3}	3.474×10^{-5}	1.553×10^{-3}	0.6985
2000	0.01782	0.2701	0.05614	3.241×10^{-3}	4.035×10^{-5}	2.265×10^{-3}	0.6988
			Water Vapor, H_2O				
−50	0.06022	0.4512	0.00797	8.153×10^{-5}	4.933×10^{-6}	8.192×10^{-5}	1.0050
0	0.05367	0.4484	0.00898	1.036×10^{-4}	5.592×10^{-6}	1.041×10^{-4}	1.0049
50	0.04841	0.4472	0.01006	1.291×10^{-4}	6.261×10^{-6}	1.293×10^{-4}	1.0018
100	0.04408	0.4473	0.01121	1.579×10^{-4}	6.942×10^{-6}	1.574×10^{-4}	0.9969
200	0.03740	0.4503	0.01372	2.263×10^{-4}	8.333×10^{-6}	2.228×10^{-4}	0.9845
300	0.03248	0.4557	0.01648	3.093×10^{-4}	9.756×10^{-6}	3.004×10^{-4}	0.9713
500	0.02571	0.4707	0.02267	5.204×10^{-4}	1.267×10^{-5}	4.931×10^{-4}	0.9475
1000	0.01690	0.5167	0.04134	1.314×10^{-3}	2.014×10^{-5}	1.191×10^{-3}	0.9063
1500	0.01259	0.5625	0.06315	2.477×10^{-3}	2.742×10^{-5}	2.178×10^{-3}	0.8793
2000	0.01003	0.6034	0.08681	3.984×10^{-3}	3.422×10^{-5}	3.411×10^{-3}	0.8563

Note: For ideal gases, the properties c_p, k, μ, and Pr are independent of pressure. The properties ρ, ν, and α at a pressure P (in atm) other than 1 atm are determined by multiplying the values of ρ at the given temperature by P and by dividing ν and α by P.

Source: Data generated from the EES software developed by S. A. Klein and F. L. Alvarado. Originally based on various sources.

TABLE A–24E

Properties of solid metals

		Properties at 540 R				Properties at Various Temperatures (R), k (Btu/h · ft · R)/c_p(Btu/lbm · R)					
Composition	Melting Point, R	ρ lbm/ft^3	c_p (Btu/lbm · R)	k (Btu/h · ft · R)	$\alpha \times 10^6$ ft^2/s	180	360	720	1080	1440	1800
Aluminum	1679	168	0.216	137	1045	174.5	137	138.6	133.4	126	
Pure						0.115	0.191	0.226	0.246	0.273	
Allov 2024-T6	1395	173	0.209	102.3	785.8	37.6	94.2	107.5	107.5		
(4.5% Cu, 1.5% Mg,											
0.6% Mn)						0.113	0.188	0.22	0.249		
Alloy 195, cast (4.5% Cu)		174.2	0.211	97	734		100.5	106.9			
Beryllium	2790	115.5	0.436	115.6	637.2	572	174	93	72.8	61.3	52.5
						0.048	0.266	0.523	0.621	0.624	0.72
Bismuth	981	610.5	0.029	4.6	71	9.5	5.6	4.06			
						0.026	0.028	0.03			
Boron	4631	156	0.264	15.6	105	109.7	32.06	9.7	6.1	5.5	5.7
						0.03	0.143	0.349	0.451	0.515	0.558
Cadmium	1069	540	0.055	55.6	521	117.3	57.4	54.7			
						0.047	0.053	0.057			
Chromium	3812	447	0.107	54.1	313.2	91.9	64.1	52.5	46.6	41.2	37.8
						0.045	0.091	0.115	0.129	0.138	0.147
Cobalt	3184	553.2	0.101	57.3	286.3	96.5	70.5	49.3	39	33.6	80.1
						0.056	0.09	0.107	0.12	0.131	0.145
Copper	2445	559	0.092	231.7	1259.3	278.5	238.6	227.07	219	212	203.4
Pure						0.06	0.085	0.094	0.01	0.103	0.107
Commercial bronze	2328	550	0.1	30	150.7	24.3	30	34			
(90% Cu, 10% Al)							0.187	0.109	0.130		
Phosphor gear bronze	1987	548.1	0.084	31.2	183	23.7	37.6	42.8			
(89% Cu, 11% Sn)							—	—	—		
Cartridge brass	2139	532.5	0.09	63.6	364.9	43.3	54.9	79.2	86.0		
(70% Cu, 30% Zn)							0.09	0.09	0.101		
Constantan	2687	557	0.092	13.3	72.3	9.8	1.1				
(55% Cu, 45% Ni)						0.06	0.09				
Germanium	2180	334.6	0.08	34.6	373.5	134	56	25	15.7	11.4	10.05
						0.045	0.069	0.08	0.083	0.085	0.089
Gold	2405	1205	0.03	183.2	1367	189	186.6	179.7	172.2	164.09	156
						0.026	0.029	0.031	0.032	0.033	0.034
Iridium	4896	1404.6	0.031	85	541.4	99.4	88.4	83.2	79.7	76.3	72.8
						0.021	0.029	0.031	0.032	0.034	0.036
Iron:	3258	491.3	0.106	46.4	248.6	77.4	54.3	40.2	31.6	25.01	19
Pure						0.051	0.091	0.117	0.137	0.162	0.232
Armco		491.3	0.106	42	222.8	55.2	46.6	38	30.7	24.4	18.7
(99.75% pure)						0.051	0.091	0.117	0.137	0.162	0.233
Carbon steels		490.3	0.103	35	190.6			32.8	27.7	22.7	17.4
Plain carbon (Mn ≤ 1%.											
Si ≤ 0.1%)								0.116	0.113	0.163	0.279
AISI 1010		489	0.103	37	202.4			33.9	28.2	22.7	18
								0.116	0.133	0.163	0.278
Carbon-silicon		488	0.106	30	160.4			28.8	25.4	21.6	17
(Mn ≤ 1%,											
0.1% < Si ≤ 0.6%)								0.119	0.139	0.166	0.231
Carbon-manganese-silicon		508	0.104	23.7	125			24.4	23	20.2	16
(1% < Mn ≤ 1.65%,								0.116	0.133	0.163	0.260
0.1% < Si ≤ 0.6%)											
Chromium (low) steels:		488.3	0.106	21.8	117.4			22	21.2	19.3	15.6
$\frac{1}{2}$ Cr–$\frac{1}{4}$ Mo–Si								0.117	0.137	0.164	0.23
(0.18% C. 0.65% Cr.											
0.23% Mo. 0.6% Si)1											
1 Cr–$\frac{1}{2}$ Mo (0.16% C,		490.6	0.106	24.5	131.3			24.3	22.6	20	15.8
1% Cr. 0.54% Mo. 0.39% Si)								0.117	0.137	0.164	0.231
1 Cr–V		489.2	0.106	28.3	151.8			27.0	24.3	21	16.3
(0.2% C, 1.02% Cr, 0.15% V)								0.117	0.137	0.164	0.231

(Continued)

TABLE A–24E

Properties of solid metals

Composition	Melting Point, R	ρ lbm/ft^3	c_p(Btu/lbm · R)	k (Btu/h · ft · R)	$\alpha \times 10^6$ ft^2/s	180	360	720	1080	1440	1800
		Properties at 540 R				Properties at Various Temperatures (R), k (Btu/h · ft · R)/c_p(Btu/lbm · R)					
Stainless steels:		503	0.114	8.7	42			10	11.6	13.2	14.7
AISI 302								0.122	0.133	0.140	0.144
AISI 304	3006	493.2	0.114	8.6	42.5	5.31	7.3	9.6	11.5	13	14.7
						0.064	0.096	0.123	0.133	0.139	0.145
AISI 316		514.3	0.111	7.8	37.5			8.8	10.6	12.3	14
						0.12	0.131	0.137	0.143		
AISI 347		498	0.114	8.2	40			9.1	1.1	12.7	14.3
								0.122	0.133	0.14	0.144
Lead	1082	708	0.03	20.4	259.4	23	21.2	19.7	18.1		
						0.028	0.029	0.031	0.034		
Magnesium	1661	109	0.245	90.2	943	87.9	91.9	88.4	86.0	84.4	
						0.155	0.223	0.256	0.279	0.302	
Molybdenum	5209	639.3	0.06	79.7	578	1034	82.6	77.4	72.8	68.2	64.7
						0.038	0.053	0.062	0.065	0.068	0.070
Nickel:	3110	555.6	0.106	52.4	247.6	94.8	61.8	46.3	37.9	39	41.4
Pure						0.055	0.091	0.115	0.141	0.126	0.134
Nichrome	3010	524.4	0.1	6.9	36.6			8.0	9.3	12.2	
(80% Ni, 20% Cr)							0.114	0.125	0.130		
Inconel X-750	2997	531.3	0.104	6.8	33.4	5	5.9	7.8	9.8	11.8	13.9
(73% Ni, 15% Cr, 6.7% Fe)						—	0.088	0.112	0.121	0.13	0.149
Niobium	4934	535	0.063	31	254	31.9	30.4	32	33.6	35.4	32.2
						0.044	0.059	0.065	0.067	0.069	0.071
Palladium	3289	750.4	0.058	41.5	263.7	44.2	41.4	42.5	46	50	54.4
						0.04	0.054	0.059	0.062	0.064	0.067
Platinum:	3681	1339	0.031	41.4	270	44.7	42	41.5	42.3	43.7	45.5
Pure						0.024	0.03	0.032	0.034	0.035	0.036
Alloy 60Pt-40Rh	3240	1038.2	0.038	27.2	187.3			30	34	37.5	40
(60% Pt, 40% Rh)								—	—	—	
Rhenium	6215	1317.2	0.032	27.7	180	34	30	26.6	25.5	25.4	25.8
						0.023	0.03	0.033	0.034	0.036	0.037
Rhodium	4025	777.2	0.058	86.7	534	107.5	89	84.3	78.5	73.4	70
						0.035	0.052	0.06	0.065	0.069	0.074
Silicon	3033	145.5	0.17	85.5	960.2	510.8	152.5	57.2	35.8	24.4	18.0
						0.061	0.132	0.189	0.207	0.218	0.226
Silver	2223	656	0.056	248	1873	257	248.4	245.5	238	228.8	219
						0.044	0.053	0.057	0.059	0.062	0.066
Tantalum	5884	1036.3	0.033	33.2	266	34.2	33.2	33.4	34	34.3	34.8
						0.026	0.031	0.034	0.035	0.036	0.036
Thorium	3641	730.4	0.028	31.2	420.9	34.6	31.5	31.4	32.2	32.9	32.9
						0.024	0.027	0.029	0.032	0.035	0.037
Tin	909	456.3	0.054	38.5	431.6	49.2	42.4	35.9			
						0.044	0.051	0.058			
Titanium	3515	281	0.013	12.7	100.3	17.6	14.2	11.8	11.2	11.4	12
						0.071	0.111	0.131	0.141	0.151	0.161
Tungsten	6588	1204.9	0.031	100.5	735.2	120.2	107.5	92	79.2	72.2	68.2
						0.020	0.029	0.032	0.033	0.034	0.035
Uranium	2531	1190.5	0.027	16	134.5	12.5	14.5	17.1	19.6	22.4	25.4
						0.022	0.026	0.029	0.035	0.042	0.043
Vanadium	3946	381	0.117	17.7	110.9	20.7	18	18	19.3	20.6	22.0
						0.061	0.102	0.123	0.128	0.134	0.142
Zinc	1247	445.7	0.093	67	450	67.6	68.2	64.1	59.5		
						0.07	0.087	0.096	0.104		
Zirconium	3825	410.2	0.067	13.1	133.5	19.2	14.6	12.5	12	12.5	13.7
						0.049	0.063	0.072	0.77	0.082	0.087

Source: Tables A–24E and A–25E are obtained from the respective tables in SI units in Appendix 1 using proper conversion factors.

TABLE A–25E

Properties of solid nonmentals

Composition	Melting Point, R	Properties at 540 R: ρ lbm/ft^3	c_p(Btu/ lbm · R)	k (Btu/ h · ft · R)	$\alpha \times 10^6$ ft^2/s	Properties at Various Temperatures (R), k (Btu/h · ft · R)/c_p(Btu/lbm · R): 180	360	720	1080	1440	1800
Aluminum oxide,	4181	247.8	0.182	26.6	162.5	260	47.4	18.7	11	7.5	6
sapphire						—	—	0.224	0.265	0.281	0.293
Aluminum oxide	4181	247.8	0.182	20.8	128	76.8	31.7	15.3	9.3	6	4.5
polycrystalline								0.244	0.265	0.281	0.293
Beryilium oxide	4905	187.3	0.246	157.2	947.3			113.2	64.2	40.4	27.2
								0.322	0.40	0.44	0.459
Boron	4631	156	0.264	16	107.5	109.8	30.3	10.8	6.5	4.6	3.6
								0.355	0.445	0.509	0.561
Boron fiber epoxy	1062	130									
(30% vol.) composite											
k, ∥ to fibers				1.3		1.2	1.3	1.31			
k, ⊥ to fibers				0.34		0.21	0.28	0.34			
c_p			0.268			0.086	0.18	0.34			
Carbon	2700	121.7	—	0.92	—	0.38	0.68	1.09	1.26	1.36	1.46
Amorphous Diamond,											
type IIa	—	219	0.121	1329	—	5778	2311.2	889.8			
insulator							0.005	0.046	0.203		
Graphite, pyrolytic	4091	138									
k, ∥ to layers				1126.7		2871.6	1866.3	803.2	515.4	385.4	308.5
k, ⊥ to layers				3.3		9.7	5.3	2.4	1.5	1.16	0.92
c_p			0.169			0.32	0.098	0.236	0.335	0.394	0.428
Graphite fiber epoxy	810	87.4									
(25% vol.) composite											
k, heat flow ∥ to fibers				6.4		3.3	5.0	7.5			
k, heat flow ⊥ to fibers				0.5	5	0.4	0.63				
c_p			0.223			0.08	0.153	0.29			
Pyroceram, Corning 9606	2921	162.3	0.193	2.3	20.3	3.0	2.3	2.1	1.9	1.7	1.7
Silicon carbide,	5580	197.3	0.161	283.1	2475.7			—	—	—	50.3
								0.210	0.25	0.27	0.285
Silicon dioxide,											
crystalline (quartz)	3389	165.4									
k, ∥ to c-axis				6		22.5	9.5	4.4	2.9	2.4	
k, ⊥ to c-axis				3.6		12.0	5.9	2.7	2	1.8	
c_p			0.177					0.211	0.256	0.298	
Silicon dioxide,	3389	138.6	0.177	0.79	9	0.4	0.65	0.87	1.01	1.25	1.65
polycrystalline (fused silica)						—	—	0.216	0.248	0.264	0.276
Silicon nitride	3911	150	0.165	9.2	104	—	—	8.0	6.5	5.7	5.0
						—	0.138	0.185	0.223	0.253	0.275
Sulfur	706	130	0.169	0.1	1.51	0.095	0.1				
						0.962	0.144				
Thorium dioxide	6431	568.7	0.561	7.5	65.7			5.9	3.8	2.7	2.12
								0.609	0.654	0.680	0.704
Titanium dioxide,	3840	259.5	0.170	4.9	30.1			4.0	2.9	2.3	2
polycrystalline								0.192	0.210	0.217	0.222

INDEX

C

D

E

F

G

H

I

J

K

L

M

N

Q

R

S

T

U

V

W

Z